PROCEEDINGS

MICROSCOPY AND MICROANALYSIS 1995

Microscopy Society of America
53rd Annual Meeting

The Histochemical Society
46th Annual Meeting

Kansas City, Missouri
August 13-17, 1995

Edited by

G. W. Bailey
M. H. Ellisman
R. A. Hennigar
N. J. Zaluzec

Jones & Begell Publishing
New York • Boston

PROCEEDINGS Microscopy and Microanalysis 1995

Direct all inquiries to Jones and Begell Publishing, 79 Madison Avenue, New York, NY 10016

International Standard Book Number: 1-56700-032-0
International Standard Serial Number: 1083-0375

MICROSCOPY AND MICROANALYSIS 1995

PROGRAM COMMITTEE

Mark H. Ellisman, *Chair*
Nestor J. Zaluzec, *Vice Chair*
Randolph A. Hennigar, *Vice Chair*

Channing Ahn
Charles Allen
Susan Babcock
Andrew Belmont
Steven Bradley
Mary Buckett
Wah Chiu
Scott Chumbley
Gary Fan
Louis Germinario
Barbara Hamkalo
Brian Herman
Jan Hoh
Gwen Jacobs

Jay Jerome
Louis Kerr
Earl Kirkland
Allen Koretsky
Abraham Koster
Fred Lanni
Giles L'Esperance
Yang-Pi Lin
John Mansfield
Maryann Martone
Stuart McKernan
Terence Mitchell
Janet Oliver
Margaret Olsen

Klaus-Ruediger Peters
Brian Robertson
John Robinson
Michael Schmid
Sheldon Schultz
Sandra Silvers
David Spector
Phoebe Stewart
Elizabeth Unger
Karen Winey
Nan Yao
Robert Youngman
Jian Min Zuo

MSA SUSTAINING MEMBERS

3M Computer Tape Technology Division
4pi Analysis, Inc.
Advanced Microbeam, Inc.
Advanced Microscopy Techniques
Amray, Inc.
Asteco, Inc.
Bal-Tec Products, Inc.
Bio-Rad
Charles Evans & Associates
Degroot Industries International, Inc.
Delaware Diamond Knives, Inc.
Denton Vacuum, Inc.
Diatome U.S.
Digital Instruments, Inc.
E. A. Fischione Instruments, Inc.
Eastman Kodak Co.
Edax International
Edgecraft Corporation
Edwards High Vacuum International

Egoltronics Corporation
Electron Microscopy Sciences
Electroscan Corp.
Ernest F. Fullam, Inc.
ETP-USA Electron Detectors
FEI Company
Fisons Instruments
G. W. Electronics, Inc.
Gatan, Inc.
Harris Diamond Corp.
Hitachi Instruments, Inc.
Jeol-USA, Inc.
Mager Scientific, Inc.
Materials Analytical Services
Micron, Inc.
Microscopy, Marketing and Education
Molecular Dynamics
NSA/Hitachi Scientific Instruments
Noran Instruments
Olympus America, Inc.
Optronics Engineering

Oxford Instruments
Park Scientific Instruments
Philips Electronic Instruments Co.
Princeton Gamma-Tech, Inc.
Raith USA, Inc.
RJ Lee Group, Inc.
RMC, Inc.
Semicaps, Inc.
Scanalytics
SCIEX
SPI Supplies
Technical Instrument Co.
Ted Pella, Inc.
Topcon Technologies, Inc.
Topometrix
Universal Imaging Corp.
Virtual Laboratories
Vital Image Technology
Carl Zeiss, Inc.

In Memoriam

Morton D Maser
1934-1995

This book is dedicated to the
memory of our beloved friend,
Morton D Maser.

1995 MICROSCOPY SOCIETY OF AMERICA AWARDS

Distinguished Scientist Awards

Shinya Inoué
Biological Sciences

David B. Wittry
Physical Sciences

Dr. Shinya Inoué's 50-year career has spanned the fields of cell biology and biophysics. Among his accomplishments, in 1951 Dr. Inoué settled a long-standing debate about whether the spindle fibers that were thought to be responsible for moving chromosomes within the cell during cell division (mitosis) were real or artifacts. Dr. Inoué settled this debate by documenting the presence of the spindle fibers with a polarizing light microscope of his own design.

Dr. Inoué has spent nearly four decades designing, building, and fine-tuning his six-foot tall rectified polarizing microscope, one of the most powerful light microscopes in the world. In 1980, he attached a video camera, a procedure that turned out to be profoundly significant for cell biology and light microscopy. The camera could see what the human eye could not. Dr. Inoué is expanding the imaging ability of the light microscope to four dimensions—three dimensions and time—using video and sophisticated computer programs. The study of living organisms and their cellular processes has changed dramatically thanks to video technology and the efforts of Dr. Inoué whose monograph <u>Video Microscopy</u> was published in 1986.

Dr. Inoué continues to teach courses on the subject and continues his work in cell biology and light microscopy as Director of the Marine Biological Laboratory's Living Cells Program.

Professor Wittry received the Ph.D. in physics from Caltech in 1957. He has been at the University of Southern California since 1959 and has joint appointments as Professor in the Materials Science Department and the Electrical Engineering Department. He has published more than 140 papers and has been granted 14 U.S. Patents.

Professor Wittry has received a Guggenheim Fellowship, the Japan Society for the Promotion of Science Visiting Scientist Award, the Presidential Award from the Microbeam Analysis for outstanding scientific contributions, honorary membership in MAS, two certificates of recognition from NASA, and twice received the Birks Award for the best paper presented at the MAS annual meeting. He has served as president of EMSA and MAS.

Professor Wittry's research interests include charged particle and x-ray optics, instrumentation for materials characterization using electron, ion or x-ray beams, and studies of the electronic properties of semiconductors.

Burton Medal

Joanna L. Batstone

Dr. Batstone is a Research Staff Member in the Computer Sciences Department at IBM's T.J. Watson Research Center in Yorktown Heights, NY. She received a B.Sc. in Chemical Physics in 1982 and a Ph.D. in Physics in 1985 from the University of Bristol, UK, which were followed by postdoctoral work at AT&T Bell Laboratories in New Jersey, and a Lectureship in the Department of Materials Science and Engineering at the University of Liverpool, UK, before joining IBM in 1989.

Dr. Batstone's research has centered around the structural and electronic properties of defects in semiconductors, utilizing the techniques of cathodoluminescence and transmission electron microscopy. She organized Symposia at the 1990, 1991 and 1992 MSA/MAS Annual Meetings and the 1992 American Physical Society March Meeting. She is a MAS Director (since 1993) and was the 1992 MAS Tour Speaker. She received the 1989 MAS Cosslett Award and 1991 Robert Lansing Hardy Gold Medal from the Minerals, Metals and Materials Society. She served on the Editorial Boards of the MSA Bulletin and the Journal of Microscopy.

The Morton D Maser Distinguished Service Award

Raymond K. Hart

Raymond K. Hart graduated from Sydney Technical College in 1949; received a DIC from Imperial College, London, in 1952; the Ph.D. in Metallurgy from Cambridge University in 1955; and a Juris Doctorate (Law) from Kennedy-Western University in 1991. He was Research Scientist at the Aeronautical Research Laboratories in Melbourne, Senior Scientist and Manager of the High Voltage Microscope Program at the Argonne National Laboratory and, in 1974, formed a metallurgy consultancy in Atlanta, practicing as a litigation consultant in civil and criminal cases.

Dr. Hart joined MSA (EMSA) in 1959, was Chairman of the 1967 MSA Chicago meeting and the 1976 MSA Miami Beach meeting. He was MSA Physical Sciences Director, 1969-72, and the first MSA Bulletin Editor, 1971-1976. Dr. Hart was a Charter Member of the Midwest Society for the Electron Microscopy, serving as Director in 1962, President in 1964, and received their President's Award in 1986. For the past 25 years Dr. Hart has been active in the Southeastern Electron Microscopy Society, serving as President in 1985 and receiving their Distinguished Scientist Award in 1993.

MSA Outstanding Technologist Award

Kai Chien

Kai Chien was born in China and raised in Hong Kong. His majors in college were Mechanical Engineering, Education and Chinese Literature and History. He came to the United States in 1969 and studied biology specializing in electron microscopy. In 1973, Kai Chien became associated with the Department of Pathology, Cedars-Sinai Medical Center, Los Angeles, where he has remained.

Electron microscopy is a major hobby for him. He likes to solve problems. He studies the current methods and tries to develop some new techniques. He has had over thirty publications and several inventions. In 1989, Kai Chen was selected to the "Technical Exchange" Program and gave lectures and workshops in China. He was invited to teach in China for four consecutive years and was promoted to "Guest Professor" by The Second Military Medical University in Shanghai in 1992. Currently he is serving as an officer of the MSA Technologist Certification Board.

MSA DISTINGUISHED SCIENTIST AWARD

Biological Sciences

1975	Keith Porter
1976	L. L. Marton
1977	Robley Williams
1978	Thomas Anderson
1979	Daniel Pease
1980	George Palade
1981	Sanford Palay
1982	Richard Eakin
1983	Hans Ris
1984	Cecil Hall
1985	Gaston Dupouy
1986	F. O. Schmitt
1987	Marilyn Farquhar
1988	Morris Karnovsky
1989	Don W. Fawcett
1990	Audrey M. Glauert
1991	Hugh E. Huxley
1992	Fritiof Sjöstrand
1993	Jean-Paul Revel
1994	Andrew Somlyo

Physical Sciences

1975	Robert Heidenreich
1976	Albert Crewe
1977	James Hillier
1978	V. E. Cosslet
1979	John Cowley
1980	Gareth Thomas
1981	Vladimir Zworykin
1982	Benjamin M. Siegel
1983	Otto Scherzer
1984	Sir Charles Oatley
1985	Ernst Ruska
1986	Peter Hirsch
1987	Jan LePoole
1988	Hatsujiro Hashimoto
1989	Elmar Zeitler
1990	Gertrude F. Rempfer
1991	Archie Howie
1992	Oliver Wells
1993	Ken Smith
1994	Dennis McMullan

MSA BURTON MEDALIST

1975	James Lake
1976	Michael Isaacson
1977	Robert Sinclair
1978	David Joy
1979	Norton B. Gilula
1980	John Spence
1981	Barbara Panessa-Warren
1982	Nestor Zaluzec
1983	Ronald Gronsky
1984	David B. Williams
1985	Richard Leapman
1986	J. Murray Gibson
1987	Ronald Milligan
1988	A. D. Romig, Jr.
1989	Laurence D. Marks
1990	W. Mason Skiff
1991	Joseph R. Michael
1992	Kannan Krishnan
1993	Joseph A. N. Zasadzinski
1994	Jan M. Chabala

MSA DISTINGUISHED SERVICE AWARD

1992	Ronald Anderson
	G. W. "Bill" Bailey
	Frances Ball
	Blair Bowers
	Deborah Clayton
	Joseph Harb
	Kenneth Lawless
	Morton Maser
	Caroline Schooley
	John H L Watson
1993	E. Laurence Thurston
1994	Richard F. E. Crang

MSA OUTSTANDING TECHNOLOGIST AWARD

1993	Ben O. Spurlock
1994	Bernard J. Kestel

MSA PAST PRESIDENTS

1942	G. L. Clark[1]		1969	W. C. Bigelow
1943	R. Bowling Barnes[2]		1970	Russell Steere
1944	R. Bowling Barnes		1971	Robert M. Fisher
1945	James Hillier		1972	Daniel C. Pease
1946	David Harker		1973	Benjamin Siegel
1947	William G. Kinsinger		1974	Russell J. Barnett
1948	Perry C. Smith		1975	Gareth Thomas
1949	F. O. Schmitt		1976	Etienne de Harven
1950	Ralph W. G. Wyckoff		1977	T. E. Everhart
1951	Robley C. Williams		1978	Myron Ledbetter
1952	R. D. Heidenreich		1979	John Silcox
1953	Cecil E. Hall		1980	Michael Beer
1954	Robert G. Picard		1981	John Hren
1955	Thomas F. Anderson		1982	Lee Peachey
1956	William L. Grube		1983	David Wittry
1957	John H. L. Watson		1984	J. David Robertson
1958	Max Swerdlow		1985	Dale Johnson
1959	John H. Reisner		1986	Robert Glaeser
1960	D. Gordon Sharp		1987	Linn W. Hobbs
1961	D. Maxwell Teague		1988	John-Paul Revel
1962	Keith R. Porter		1989	Ray Carpenter
1963	Charles Schwartz		1990	Keith R. Porter
1964	Sidney S. Breese		1991	Charles Lyman
1965	Virgil G. Peck		1992	Patricia Calarco
1966	Walter Frajola		1993	Michael S. Isaacson
1967	Jospeh J. Comer		1994	Robert R. Cardell
1968	John H. Luft			

[1]Chair of committee to arrange first meeting
[2]Temporary (pre-constitution)

MICROSCOPY AND MICROANALYSIS 1995

Terence E. Mitchell
MSA President

John M. Robinson
HSC President

Al L. Chapman
Local Arrangements Chair

G. Michael Veith
Local Arrangements Treasurer

Mark H. Ellisman
Program Chair

1995 MSA Presidential Scholars

E.C. Dickey
Northwestern University
Materials Science and Engineering
Evanston, IL 60208

W. F. Marshall
University of California-San Francisco
Biochemistry and Biophysics
San Francisco, CA 94143

R. Grimm
Max Planck Institut für Biochemie
Molekulare Strukturbiologie
82152 Martinsried, GERMANY

Kai Tang
Stanford University
Materials Science and Engineering
Stanford, CA 94305

P.G. Kotula
University of Minnesota
Chemical Engineering
   and Materials Science
Minneapolis, MN 55455-0132

S.A. Walker
University of California-Santa Barbara
Chemical Engineering
Santa Barbara, CA 93106

Tan-Chen Lee
Cornell University
Materials Science and Engineering
Ithaca, NY 14853

Zara Weng-Sieh
University of California-Berkeley
Materials Science and
   Mineral Engineering
Berkeley, CA 94720-1760

FOREWORD

This volume records the Proceedings of the Fifty-Third Annual Meeting of the Microscopy Society of America and the Forty-Sixth Annual Meeting of the Histochemical Society jointly held in Kansas City. It also represents a watershed for the Microscopy Society in as much as it is the first Proceedings to be a part of our new Journal (the *Journal of the Microscopy Society of America*). It is my privilege to be able to write the first foreword under these circumstances. However, I must immediately add that all of the hard work in bringing this about was performed by my predecessors, past-presidents Charlie Lyman, Pat Calarco, Mike Isaacson and Bob Cardell. We should also acknowledge with thanks the many years of dedicated high-quality service provided by Charles Susskind and San Francisco Press who published the Proceedings through 1994.

The Proceedings include papers from twenty different symposia. In accordance with the change of our name from EMSA to MSA two years ago, the topics include non-electron microscopies such as light microscopy, scanned probe microscopy and MRI imaging. Computing and digital imaging continue to be important themes in microscopy. And applications from macromolecules and the nervous system to nano-materials and ceramics continue to command great interest. The symposia with their world-class invited speakers tend to attract the most interest, but the quality of the contributed sessions is extraordinarily high and considerable pleasure will be gained by browsing through the abstracts. I find that many of these abstracts contain as much information crammed into two pages as a full-length paper!

The high quality of the scientific program is due not only to the contributing scientists but also to the organizational and inspirational talents of the hard-working program committee led by Mark Ellisman as Chair and Nestor Zaluzec as Vicechair. Excellent co-ordination was provided by the HCS Program Chair, Randolph Hennigar.

The most significant and saddest event for MSA this year was the untimely death of Mort Maser, who ran the Woods Hole office for MSA, was our Annual Meeting Manager and was the Executive Secretary to MSA Council. Mort was a devoted friend of MSA and will be sorely missed. In memory of his lasting contributions to MSA, Council has decided that the Distinguished Service Award henceforth be known as the Morton D Maser Distinguished Service Award. I am pleased to announce that the first recipient will be Ray Hart. He is to be congratulated as are the other award recipients: Shinya Inoué, Distinguished Scientist in the biological sciences; David Wittry, Distinguished Scientist in the physical sciences; Joanna Batstone, Burton Medal; and Kai Chien, Outstanding Technologist. In addition, Presidential Student Awards are being given to eight young scientists from various fields of microscopy, out of thirty-four nominees. I congratulate them and look forward to their growing involvement in MSA as their careers progress.

For those who have never been involved in the Local Arrangements Committee, it is hard to imagine how much dedication and hard work is required for a successful meeting. For those who have, the magnificent job that the Kansas City folks have done is even more greatly appreciated. Their efforts under the leadership of Al Chapman, Mike Vieth, Barb Fegley, John Robinson (for HSC) and others are to be congratulated.

Bill Bailey, as ever, has done a fine job putting together all the disparate components that make up these Proceedings into an elegant package. Regrettably, Mort Maser was not around this year to help; but he, and we, can be proud of the valiant efforts of Larry Maser for his part in organizing the Proceedings.

Terence E. Mitchell
President, Microscopy Society of America
August 1995

THRUST AREAS IN HIGH-RESOLUTION AND INTERMEDIATE- AND HIGH-VOLTAGE EM IN BIOLOGY AND MATERIALS SCIENCE

MAGNETIC RESONANCE IMAGING

EMERGING AND RE-EMERGING TECHNOLOGIES
FOR MICROSCOPIC CHARACTERIZATION OF MATERIALS

DEVELOPMENTS AND APPLICATIONS
IN QUANTITATIVE ELECTRON DIFFRACTION

MICROSCOPY OF NANO-PHASE AND NANO-COMPOSITE MATERIALS

IN SITU MICROSCOPY STUDIES IN MATERIALS RESEARCH

HIGH SPATIAL RESOLUTION MICROANALYSIS AND MICROSCOPY

ENERGY-LOSS AND ENERGY-FILTERED SPECTROSCOPY

ADVANCES IN ANALYTICAL MICROSCOPY OF CERAMICS

MAGNETIC MATERIALS

POLYMERS

SPECIMEN PREPARATION

METALS AND ALLOYS

GENERAL OPTICS AND IMAGING

SPECTROSCOPY AND HOLOGRAPHY

SCANNING ELECTRON MICROSCOPY

IMAGE PROCESSING, ANALYSIS AND MODELING

DEVELOPMENTS IN COMPUTER CONTROL

TUTORIALS

TECHNOLOGISTS' FORUM SYMPOSIUM

TECHNOLOGISTS' FORUM: CORRELATIVE MICROSCOPIES

SCANNED PROBE MICROSCOPY

EM TOMOGRAPHY IN BIOLOGY

HISTOCHEMICAL ASPECTS OF NUCLEIC ACID DETECTION

PROBING THE FUNCTIONAL ORGANIZATION OF THE CELL NUCLEUS

MEMBRANE MICRODOMAINS: LESSONS FROM MICROSCOPY

APPLIED BIOMEDICAL MICROSCOPY

ADVANCES IN FUNCTIONAL IMAGING OF THE NERVOUS SYSTEM

HISTOCHEMISTRY/CYTOCHEMISTRY

MACROMOLECULAR MICROSCOPY

PATHOLOGY I

PLANTS

PATHOLOGY II

IMMUNOCHEMISTRY/CYTOCHEMISTRY/HISTOCHEMISTRY

SPECIMEN PREPARATION

THREE-DIMENSIONAL VISUALIZATION

FINDING OUT MORE AND MORE ABOUT LESS AND LESS
-- FIFTY YEARS OF MILESTONES IN MICROANALYSIS

D.B. Wittry, Departments of Materials Science and Engineering, and Electrical Engineering
University of Southern California, Los Angeles, CA 90089-0241

During the past half century, a wide variety of techniques of microanalysis have been developed to a practical stage and many of these are presently undergoing refinement and improvement. In fact, practically every means of bombarding a specimen with radiation and observing signals from the interaction of this radiation with the specimen has been explored, as shown in Fig. 1.[1] Many of the techniques used in microanalysis today are based on physical principles that were known for many years. But they became practicable only with improved technology, e.g. improved vacuum systems, sources, detectors, lenses, spectrometers, signal amplifiers and computers.

A "road map" of the history of microanalysis would contain some of the milestones along the way, similar to the list of significant events in the history of microscopy published in the EMSA Bulletin in Fall of 1983.[2] My own (and highly subjective) list is given in the Table. It is not surprising that many of the key figures contributors on this list have been previous recipients of the EMSA Distinguished Scientist Award (names printed in bold face). Clearly, there are many others who should be considered for this award, including some that I have inadvertently omitted. To those who have made significant contributions and are not listed, I offer my apologies and the hope that there will be an opportunity to make corrections in the future.

In compiling the list of milestones, I am grateful for input from E. Etz, R. Pinizzotto, J. Reffner, and O. Wells.

References
1. D. B. Wittry, "Spectroscopy in Microscopy and Microanalysis: The Search for the Ultimate Analytical Technique," *Electron Microscopy 3(*1980)14-20. Modified version of the figures was reprinted on the cover of the inaugural issue of *Microbeam Analysis 1*(1992).
2. R.F. Pinizzotto, "The History of Microscopy," EMSA Bulletin 13(1983)108.

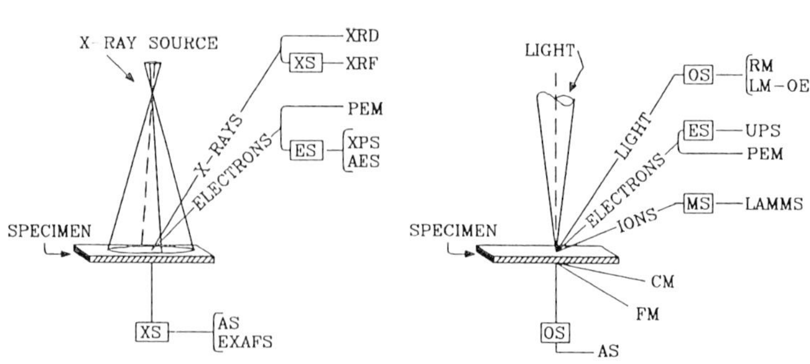

Fig. 1. Techniques of microanalysis using various exciting beams and spectrometers:

    XS = x-ray spectrometer
    OS = optical spectrometer
    MS = mass spectrometer
    ES = electron energy spectrometer
    IS = ion energy spectrometer

Proc. Microscopy and Microanalysis 1995, edited by G.W. Bailey, M.H. Ellisman, R.A. Hennigar, and N.J. Zaluzec

MILESTONES IN MICROANALYSIS

Year	Researchers(s)	Contributions
1942	**Zworykin**, Hillier & Snyder	SEM with detection by fluorescent screen & photomultiplier
1944	**LePoole**	selected area electron diffraction in TEM
1944	**Hillier** & Baker	electron microanalyzer based on EELS
1949	Castaing & Guinier	electron probe x-ray microanalyzer
1951	Steinhardt & Sternglass	x-ray photoelecton and Auger electron energy measurements
1953	Lander	first study of surfaces with Auger electron spectroscopy
1953	V. J. Coates	micro infrared spectroscopy
1955	**McMullen, Oatley**	scanning electron microscopy of surfaces
1955	**Smith** & Oatley	backscattered electron detection in SEM
1956	Everhart & Thornley	improved secondary electron detection in SEM
1956	**Cosslett** & Duncumb	scanning electron probe x-ray microscopy
1957	Siegbahn	chemical shifts in x-ray photoelectron spectroscopy
1957	Long & Cosslett	micro x-ray fluorescence
1960	Castaing & Slodzian	secondary ion microscope
1962	Castaing & Henry	imaging energy filter on TEM
1962	Riecke	focussed electron probe diffraction
1963	**Wells,** Everhart	EBIC & voltage contrast of microcircuits
1963	Henke	Langmuir-Blodgett films as x-ray diffractors
1964	Wittry & Kyser	cathodoluminescence microscopy of semiconductors
1967	D.C. Coates	electron beam channeling patterns
1967	Liebl	ion microprobe mass analyzer
1967	D.P. Smith	ion scattering spectrometry as an analytical tool
1968	Fitzgerald et al.	energy-dispersive spectrometry on SEM
1968	Duncumb	wavelength-dispersive spectrometry on TEM
1968	Wittry, Ferrier & Cosslett	inner-shell EELS with magnetic prism on TEM
1968	Harris	study of materials with Auger electron spectra
1968	Muller, Panitz & McLane	atom probe field ion microscope
1969	Palmberg, Bohn & Tracy	Auger electron spectroscopy with cylindrical mirror analyzer
1969	Broers	electron gun with lanthanum hexaboride cathode
1970	**Crewe**, Wall & Langmore	STEM with field emission gun, single atom imaging, Z contrast
1970	Sayers, Lytle & Stern	extended x-ray absorption fine structure
1971	Curtis & Silcox	high resolution EELS with Wein filter on TEM
1971	Wells	low loss imaging in SEM
1972	Young, Ward & Scire	micro topographiner
1974	Rosasco, Etz & Cassatt	micro Raman spectroscopy
1974	Delhaye & Dhamelincourt	micro Raman spectroscopy
1974	Egerton & Whelan	atomic bonding effects by EELS in TEM
1975	Clampitt, Aitken & Jefferies	needle-type liquid metal ion source
1975	Hurst, Payne & Wagner	resonance ionization for analytical spectrometry
1975	Hillenkamp et al.	high sensitivity laser microprobe mass spectrometer
1976	Barbee & Keith	sputter-deposited multilayer x-ray diffractors
1977	Petroff & Lang	scanning deep level transient spectroscopy
1980	Kellogg & Tsong	pulsed laser atom probe
1981	Johnson, Csillage & Stern	extended energy loss fine structure, parallel EELS detection
1982	Bennig & Rohrer	scanning tunneling microscopy
1982	Spence & Tafto	atom localization by channeling enhanced microanalysis
1983	Levi-Setti et al.	scanning ion microscopy
1986	Bennig, Quate & Gerber	atomic force microscopy

ial semiconductor thin films for the semiconductor device industry. The III-V and II-VI semi-

STRUCTURAL AND ELECTRONIC PROPERTIES OF DEFECTS IN SEMICONDUCTORS

J.L. Batstone

IBM T.J.Watson Research Center, P.O.Box 704, Yorktown Heights, NY 10598

The development of growth techniques such as metal organic chemical vapor deposition (MOCVD) and molecular beam epitaxy during the last fifteen years has resulted in the growth of high quality epitaxial semiconductor thin films for the semiconductor device industry. The III-V and II-VI semi-conductors exhibit a wide range of fundamental band gap energies, enabling the fabrication of sophis-ticated optoelectronic devices such as lasers and electroluminescent displays. However, the radiative efficiency of such devices is strongly affected by the presence of optically and electrically active defects within the epitaxial layer; thus an understanding of factors influencing the defect densities is required.

Extended defects such as dislocations, twins, stacking faults and grain boundaries can occur during epitaxial growth to relieve the misfit strain that builds up. Such defects can nucleate either at surfaces or thin film/substrate interfaces and the growth and nucleation events can be determined by *in situ* transmission electron microscopy (TEM). In order to improve radiative efficiencies of luminescent thin films, a correlation between the structure of defects and their electronic properties must be performed. Techniques such as cathodoluminescence (CL) and electron beam induced conductivity (in both the scanning electron microscope (SEM) and TEM) rely on electron beam excitation of carriers across the fundamental band gap in semiconductors and insulators. The subsequent relaxation of carriers via radiative recombination results in photon emission or CL. Detailed spectroscopic information can be obtained at high spatial resolution from defect states related to both impurity species and crystallographic defects. At room temperature, energy levels due to donors, acceptors, dislocations, and excitons are thermalized and cannot be resolved from the conduction or valence bands. However cooling the material of interest to 5-10K within the microscope enables individual energy levels to be resolved. Cl wavelength-imaging allows the correlation with defects observed using the complementary techniques of secondary electron or backscattered electron imaging within the SEM or by TEM diffraction contrast techniques where individual defects in a thin film can be identified.

Examples of TEM-CL, SEM-CL and high resolution TEM analysis of interfacial defects will be pre-sented from thin films of ZnSe/GaAs[1] and Porous Silicon.[2] *In situ* growth experiments will also show the propagation of dislocations and steps in epitaxial metal silicide films on silicon.[3] An example of defect analysis utilizing CL in the SEM is shown in Figure 1 for a $2\mu m$ film of epitaxial MOCVD ZnSe/GaAs(001).[1] The sample was cooled to <10K by an Oxford Instruments liquid He cold stage in a JEOL JSM6400 SEM in order to resolve spectral features associated with dislocations. CL was collected by an Oxford Instruments CL system and dispersed by a monochromator to allow both spectra and wavelength-selective images to be collected.[2] Figures 1a and 1b show monochromatic CL images recorded at 475nm (2.6eV) and Figure 1c shows two spectra obtained from two different areas on the sample, marked by the positions of the crossed lines. The bright area corresponds to a dense tangle of dislocations in the ZnSe film, giving rise to dislocation luminescence. CL was excited by a 20kV, 0.5nA electron beam and collected by a cooled GaAs photomultiplier. The spectrum obtained $\approx 20\mu m$ from the defect (a) shows a strong Al donor bound exciton transition at $\approx 445nm$ (2.78eV) which decreases in intensity at the defect (b). At the defect, the dislocation energy level, known as Y,

Proc. Microscopy and Microanalysis 1995, edited by G.W. Bailey, M.H. Ellisman, R.A. Hennigar, and N.J. Zaluzec
Copyright © 1995 MSA. Published by Jones and Begell Publishing, 79 Madison Ave., New York, NY 10016

appears at 475nm.[1] CL imaging at 445nm shows a complementary reduction in bound exciton emission in the vicinity of the defect.[4]

References

1. J.L. Batstone J.W. and Steeds, *Inst. Phys. Conf. Ser. No.* 76 (1985) 383.
2. J.L. Batstone et al., *Appl. Phys. Lett.* 62 (1992) 2667.
3. J.L. Batstone and C. Hayzelden, *Solid State Phenomena* 37-38 (1994) 257
4. Technical discussions with J.W. Steeds, J.M. Gibson, J.M. Phillips, C.J. Humphrey, D.A. Smith and C. Hayzelden are gratefully acknowledged.

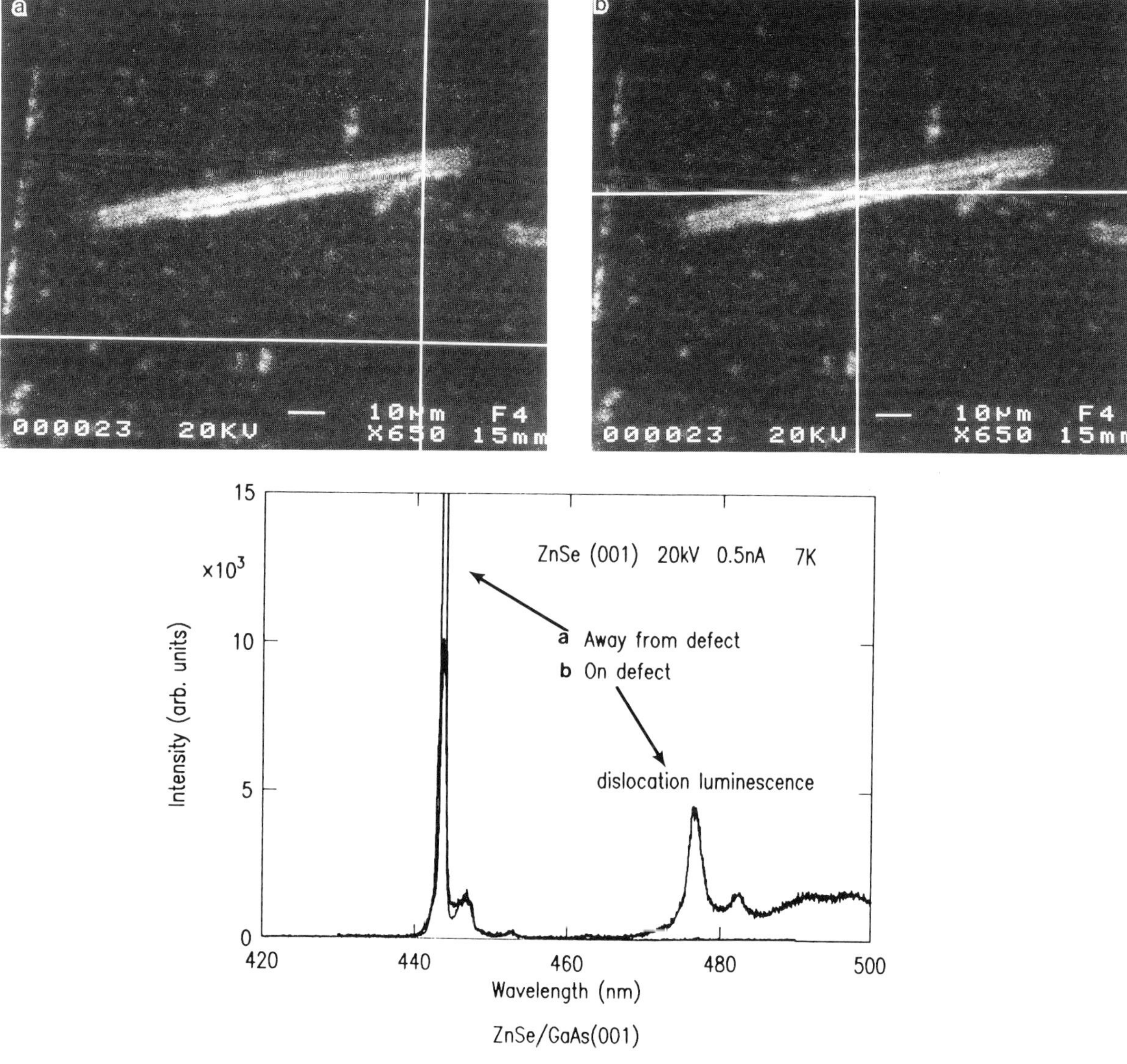

Fig.1 -- (a) and (b) Monochromatic CL images from defect in ZnSe/GaAs(001) film. CL spectra were obtained from the two positions of the cross wires, away from (a) and on (b) the defect. (c) CL spectra to show the enhanced luminescence at 2.6eV associated with the defect.

PROGRESS IN DESIGN AND APPLICATIONS OF CCD CAMERAS FOR ELECTRON MICROSCOPY

Kenneth H. Downing

Life Science Division, Lawrence Berkeley Lab., Berkeley CA 94720

Over the last several years the long-awaited revolution in direct-digital readout systems has begun, with the introduction of efficient slow-scan CCD cameras. Earlier, the introduction of video cameras to electron microscopes had brought a quantum leap in the speed and efficiency of carrying out a host of operations. The high sensitivity of the video cameras provided the ability to see the image in much more detail and at a lower beam intensity than had been possible previously by viewing the fluorescent screen. The ability to assess, on line, characteristics such as specimen quality and image focus, even qualitatively, gave feedback to the operator that previously took hours to obtain. Due to the low resolution of these video systems, however, they were rarely useful for data recording.

The current generation of CCD cameras goes a long way toward meeting the needs of data recording in many, though certainly not all, applications. The immediate access to the data recorded in digital form has provided another quantum leap in efficiency. With their high dynamic range, linearity, and increased resolution, CCDs are capable of replacing photographic data recording in a number of applications. Avoiding the tedious processes of developing and printing or digitizing negatives, as well as on-line, quantitative analysis of the image and imaging conditions, allow a tremendous improvement in data collection. Digital image information can be fed back to a control computer for automating a number of complex microscope operations, such as the long repetitive sequences of steps involved in tomographic data collection.

For some applications, such as electron crystallography of proteins, the limited number of pixels in the CCD still poses a serious problem. We discuss here some of these limitations, along with our approach to circumventing them by use of a larger-format CCD. Most present CCDs consist of an array of 1024 pixels on edge. CCD cameras coupled to electron microscopes also suffer from a rather poor modulation transfer function (MTF) that limits the useful array size to something closer to 512 pixels. In contrast, arrays of 6000 or even 10000 pixels on edge are now frequently used when micrographs on film are digitized. Thus the transition from the use of film to CCD requires careful consideration of our requirements and the camera system limitations.

A recent series of experiments (Perkins, Downing and Glaeser, in press) was aimed at determining the minimum array size that could be efficiently used for the crystallographic extraction of high resolution data from crystal images. We found that data from a set of 1000x1000-pixel images of purple membrane could be combined to provide the same quality as would be obtained from larger images of the same total number of unit cells. With smaller images, however, parameters such as lattice vectors could not be determined with sufficient accuracy to yield data of the same quality at high resolution. This experiment sets a lower limit on the actual image array size that is required for extraction of high resolution data from images of protein crystals.

In an effort to overcome the limitations of current CCDs, we did a careful study of ways to increase the number of pixels in the image, and still have a system that would perform at least as well as film in terms of both MTF and detective quantum efficiency (DQE). As the system is intended primarily for low-dose work, we need to see how the DQE at low exposure will compare to that of film. We assume for the present that images are to be digitized at around one angstrom per pixel, either on film or on the CCD At an exposure of 10 e/Å^2, the exposure is then about 10 electrons per pixel. The DQE of the image on film is limited at low exposure by the fog level of the film. The fog has a typical optical density of around 0.1, which corresponds to an effective exposure of ~5 e/Å^2 when 10 μm pixels are used. The system

Proc. Microscopy and Microanalysis 1995, edited by G.W. Bailey, M.H. Ellisman, R.A. Hennigar, and N.J. Zaluzec
Copyright © 1995 MSA. Published by Jones and Begell Publishing, 79 Madison Ave., New York, NY 10016

noise for a slow scan CCD is on the order of one primary electron per pixel, so it is expected that the CCD will outperform film more and more as the exposure level is decreased.

The MTF is principally limited by the point spread function (PSF) of the scintillator. One needs a reasonably thick scintillator to produce enough photons from each electron to give an adequate signal. On the other hand, the width of the PSF increases with the scintillator thickness. Most present systems employ scintillators that have a PSF on the order of 50 μm in diameter. CCDs, on the other hand, have pixel sizes only up to around 25 μm. One approach to improving the resolution is to demagnify the image formed on the scintillator onto the CCD using either lens coupling or a reducing fiber optic. Either of these, however, has a light coupling efficiency roughly proportional to one over the square of the demagnification. To compensate for a two-fold demagnification, which would decrease the light incident on the CCD by a factor of four, the scintillator could be made four times as thick, but the resultant degradation of its PSF would more than offset the gain from demagnifying the image. The most efficient approach, thus, seems to be to use direct coupling (e.g., 1:1 fiber optics) of a thin scintillator to the CCD.

We are led to the conclusion that the best approach to matching the scintillator PSF and the CCD is to bin adjacent pixels during readout, doubling the effective pixel size. This is a straightforward way to meet our requirements, although it is expensive, since it requires us to use a larger CCD in order to obtain the necessary field of view. The camera uses a Tektronix 2048x2048 CCD that has 24 μm pixels. There are a number of less expensive CCDs available with even larger pixel arrays, but they have smaller pixels that make them unsuitable for this application. As shown in fig. 1, the MTF of this camera at the Nyquist limit with 2x2 binned pixels is nearly the same as that of film with 10μm sampling, as is frequently used for digitizing micrographs.

There are a number of additional benefits of using the larger format. A further enhancement of the DQE results from reducing the cross-talk between pixels that is caused by the PSF.[2] For electron diffraction studies, magnifying the pattern by a factor of two decreases the effect of light diffusion within the fiber optic that arises from the high intensity of the direct beam, and increases by a factor of four the exposure required to saturate the CCD, which reduces blooming from the direct beam.

REFERENCES
1. K. H. Downing and D. A. Grano Ultramicroscopy **7**, 381 (1982)
2. K. Ishizuka Ultramicroscopy **52**, 1 (1993)
3. O. L. Krivanek and P. E. Mooney Ultramicroscopy **49**, 95 (1993)

Figure 1. Modulation transfer functions for the Gatan Slow Scan CCD camera with gadolinium oxysulfide scintillator[3] and for Kodak SO-163 film.[1] Frequencies are indicated that correspond to the 24μm pixel size of the Tektronix 2048 CCD and to the effective pixel size with 2x2 binning. With 48 μm pixels, the MTF is nearly the same as that of film with 10 mm pixels.

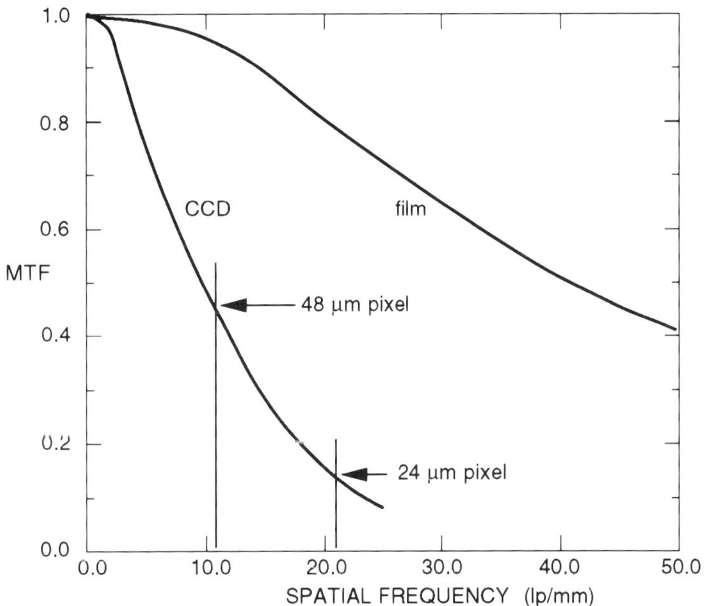

APPLICATION OF A SLOW–SCAN CCD CAMERA IN PROTEIN ELECTRON CRYSTALLOGRAPHY AT 400 KV

Jaap Brink, Michael B. Sherman and Wah Chiu

W.M. Keck Center for Computational Biology and Verna and Marrs McLean Department of Biochemistry, Baylor College of Medicine, One Baylor Plaza, Houston, TX 77030, U.S.A.

In protein electron crystallography, high resolution amplitudes and phases are required for a 3D reconstruction of a protein in which one can trace the C_α–backbone. Recording of these amplitudes and phases has so far been done on photographic film. In principle they can be obtained with higher fidelity using an electronic device, such as a slow–scan charge–coupled device (CCD) camera. We use a 1024 x 1024 Gatan CCD camera (model 679) attached to a JEOL 4000EX intermediate voltage electron cryo–microscope. The camera has a 20–30 μm thin P43 phosphor scintillator and an extra set of fiber optics, that reduces the number of x–ray pixels in acquired frames. In this abstract, we will focus on several issues related to the CCD camera in obtaining high resolution diffraction patterns and images from thin protein crystals.

To obtain electron diffraction patterns with sufficient statistical definition, one needs to use an adequate electron dose. However, at exposures well above 0.25 e/$Å^2$, the pattern will suffer from blooming due to the central beam[1]. This can be avoided by summing multiple patterns each acquired at a lower exposure, for instance 0.04 e/$Å^2$. Figure 1 shows a sum of 4 patterns from a glucose–embedded crotoxin complex crystal with reflections out to 2.7 Å and an R_{sym} for Friedel–related reflections of 0.04. Individual patterns, each acquired at 0.04 e/$Å^2$, have an R_{sym} of 0.07. Summing, however, will result in a higher standard deviation of the background for each reflection as compared to a single pattern acquired at a higher dose.

Due to the CCD camera's low background and high linearity, multiple patterns can be acquired at low exposure from a single crystal tilted to different angles[2]. This approach is practically useful because of the reduction in the amount of work involved in merging large data sets from suitable crystals, that may differ in thickness, or that may be difficult to obtain due to a lack of flatness. We have recorded tilt series of electron diffraction patterns of crotoxin complex crystals 128 Å thick, with up to 120 patterns per series at a total dose of ~2–3 e/$Å^2$. A new merging scheme has been developed to scale and merge the intensities using a high tilt pattern as reference to which the low tilts ones are merged. A 3D data set was obtained with 55,000 reflections out to 3.3 Å corresponding to 75% completeness. The overall R_{merg} is 0.26 with an R_{sym} of 0.15. Some lattice lines show a good fit as jugded from its R_{merg} of 0.16 (Fig. 2a); others fit poorly at 0.45 R_{merg} (Fig. 2b). The poor fit may be related to anomalies in first, the angular refinement of the low–tilt patterns, and second, in the diffraction intensities themselves due to radiation damage.

The major concerns in recording images of protein crystals using the CCD camera are the resolution and the MTF. Based on the camera's pixel size (24 μm), an image acquired at 132,000x could potentially yield 3.6 Å–resolution phases. Figure 3 shows a spot–scan image of an ice–embedded catalase crystal acquired at such a magnification. The IQ plot of its Fourier transform shows data extending to beyond 4 Å (Fig. 4). Even though high resolution phases can be obtained, the accompanying amplitudes of the structure factors will be dampened significantly by the MTF of the camera. This was determined in separate experiments from images acquired with the CCD camera of the electron beam without specimen in the beam path, and from carbon film. These data strongly suggest the importance of a correction of high resolution amplitudes retrieved from images acquired with a CCD camera in the absence of electron diffraction amplitudes.

References

1. J. Brink and W. Chiu, *J. Struct. Biol.,* 113 (1994) 23.
2. J. Brink and W. Chiu, Proc. 52nd MSA (1994) 104.
3. W. Chiu *et al.,* Proc. 52nd MSA (1994) 98.
4. Research supported by grants from the NIH (RR02250) and the W.M. Keck Foundation.

Proc. Microscopy and Microanalysis 1995, edited by G.W. Bailey, M.H. Ellisman, R.A. Hennigar, and N.J. Zaluzec
Copyright © 1995 MSA. Published by Jones and Begell Publishing, 79 Madison Ave., New York, NY 10016

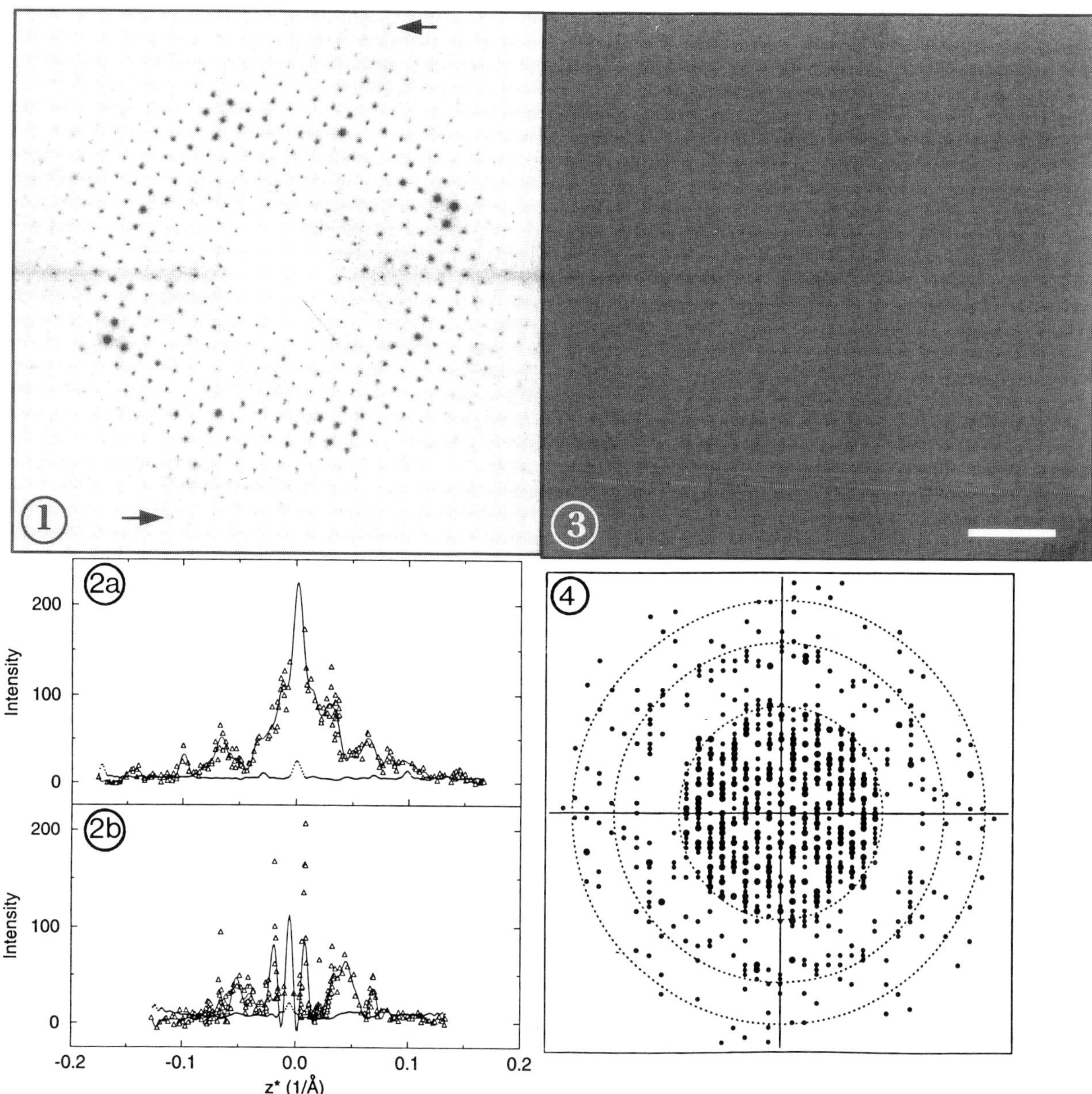

FIG. 1—Sum of 4 electron diffraction patterns of glucose–embedded crotoxin complex crystal acquired on the CCD camera at 0.04 e/Å^2 each. The pattern is background–corrected and has the central spike computationally removed. Bragg reflections close to the edge of the CCD detector are at 2.7 Å resolution as indicated by the arrow.

FIG. 2—(a) Lattice rod (6,–7) at 4.2 Å resolution after merging 353 patterns in $P1$ spacegroup with an R$_{merg}$ of 0.16. Shown are the observed intensity (Δ), fitted curve through the observed intensity (—), and the fitting error ($\cdot$) as function of z*; (b) as (a), but now for lattice rod (7,7) at 3.9 Å resolution with an R$_{merg}$ of 0.45.

FIG. 3—Spot–scan image of vitreous ice–embedded catalase crystal acquired on the CCD camera at an effective magnification of 132,000x at an electron exposure of ~14 e/Å^2. The scale bar is 500 Å.

FIG. 4—Plot of quality index of reflections in the Fourier transform of crystal shown in figure 3. Highest quality reflections, whose amplitude is more than 6x the local background, are shown with the largest dots. Resolution circles at 8, 5 and 4 Å are indicated.

CCD Quantum Efficiency at the Theoretical Limit with Molecular Beam Epitaxy

Shouleh Nikzad, M.E. Hoenk, P.J. Grunthaner, and F.J. Grunthaner

Center for Space Microelectronics Technology, Microdevices Laboratory, Jet Propulsion Laboratory, California Institute of Technology, Pasadena, CA 91109

A variety of biological, medical, astronomical and materials science measurements require real-time imaging of phenomena which produce low light levels. Charge-coupled devices (CCDs), which have high resolution and linear response over a large dynamic range, are often used as imaging sensors in conjunction with image intensifiers or phosphors to improve the signal-to-noise at low light levels, particularly in the blue and UV regions of the spectrum. Conventional CCDs typically have quantum efficiency (QE) as low as ~10% in the range of 300-500 nm.[1] This region of the spectrum is very important in biological experiments which require detecting the fluorescence of dyes, in astronomical observations of spectral lines in the blue and near UV, and in a variety of applications which use image intensifiers to amplify the signal. Improvement of CCD quantum efficiency with respect to the conventional technology is critical for most of these experiments. The detection of blue and UV photons in a silicon CCD is challenging due to the short absorption length of these photons in silicon (e.g., 4 nm absorption length at 270 nm) and the existence of a backside potential well (caused by positive charge at the interface of Si and SiO_2). By growing a thin, highly doped layer of silicon on a commercial CCD in conjunction with a single layer anti-reflection coating, we have obtained 100% internal quantum efficiency and total quantum efficiency greater than 80% in this critical region of the spectrum. This quantum efficiency is intrinsically stable, and highly uniform over the imaging area.

Commercial backside-illuminated CCDs were modified by growing a delta-doped silicon layer on the back surface using molecular beam epitaxy.[2,3] During the silicon growth on the CCD, 30% of a monolayer of boron atoms are deposited on the surface, followed by a 15-20 Å silicon layer for surface passivation. The boron is nominally incorporated within a single atomic layer at the back surface of the device, resulting in the effective elimination of the backside potential well. Delta-doped CCDs exhibit stable and uniform, 100% internal quantum efficiency, so that the overall quantum efficiency of the device is determined by reflection from the silicon surface (Fig. 1). Reduction of the reflection by deposition of HfO_2 on the CCD back surface further increased the total QE to over 80% in 250-400 nm region of the spectrum (Fig. 1).[2] Long term stability was tested showing no degradation of the quantum efficiency over sixteen months. Measurements of the spectral and lateral response of delta-doped CCDs exhibited excellent uniformity of the device (Fig. 2). Measurements in the region below 200 nm have shown that the device performs as expected with higher than unity quantum yield. We will discuss these results as well as the delta-doped CCD concept and process, biological and astronomy applications.

References

1. J. Janesick, T. Elliot, G. Fraschetti, S. Collins, M. Blouke, and B. Corrie, SPIE Proceedings, **1071**, pp. 153-169 (1989).

2. M.E. Hoenk, P.J. Grunthaner, F.J. Grunthaner, R.W. Terhune, M. Fattahi, and H-F. Tseng, Appl. Phys. Lett., **61**, pp. 1084-1086 (1992).

3. S. Nikzad, M.E. Hoenk, P.J. Grunthaner, F.J. Grunthaner, R.W. Terhune, R. Winzenread, M. Fattahi, H-F. Tseng, and M. Lesser, SPIE Proceedings, **2198**, pp. 907-915 (1994).

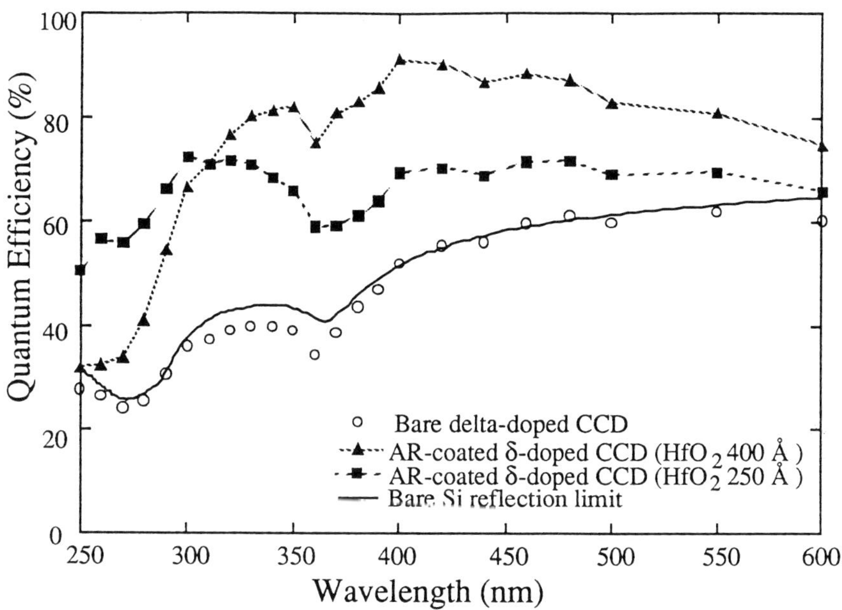

Fig. 1

Fig. 2

<u>Figure 1</u>. Quantum efficiency data from bare and antireflection-coated regions of a delta-doped CCD. The solid line is the silicon transmittance which shows that the bare delta-doped CCD performs at the theoretical reflection limit. The two antireflection coatings of HfO_2 layers 250 Å thick and 400 Å thick were designed for the 270 nm region and the 300-400 nm region respectively.

<u>Figure 2.</u> Flat field response of a delta-doped Reticon 512 x 512 CCD taken at 350 nm shows excellent uniformity of response at this wavelength.

DEVELOPMENT OF THE 25 MICRON PIXEL IMAGING PLATE SYSTEM FOR TEM

N. Ogura,* T. Yamada,**

Fuji Photo Film Co., Ltd.
*Miyanodai, Kaisei-machi Ashigarakami-gun, Kanagawa, 258 Japan
** Equipment Products Div., Nishiazabu, Minato-ku, Tokyo, 106 Japan

In 1986, we reported that the Imaging Plate (IP) could be applied to the recording material for TEM.[1] In 1990, we reported the first commercial IP system for TEM, which was co-developed with JEOL Ltd.[2] This time, we developed a new IP system that has the following remarkable concepts.

1. Enormous amount of information and high image quality
The IP is read by the IP Reader at a pixel size of 25 microns and at a gray level of 16384 (14 bit). One image consists of 3760×3000 pixels. This means the image satisfies two essential requirements, i.e., high resolution and large area, at the same time. As the new sophisticated scanning system and the 400 dpi high-quality color printer (Fujix Pictrography 3000) have been combined, the final hard copies are almost identical to photographic films.

2. Applicable to most TEMs
The size of the IP is 99.6 x 80.9 mm, which is the same as that of photographic films mainly used in the USA, so the IP can be applied to most TEMs of any manufacturer in the world. Once adequate cassettes are obtained, IPs can be used in place of photographic films.

The image data in the DDS cartridge can be transferred to the host computer (Macintosh) and displayed and processed there. As the host computer is not connected to the IP Reader physically, image capturing, processing or analysis can be done separately at the same time. We have developed software on the Macintosh suited for TEM images that consists of processing, such as FFT and flexible filtering; analysis, such as quantitating electron diffraction intensities; and image management, such as file archiving. The software can handle extraordinarily large images (22.56M bytes) without a decrease in handling speed.

We developed a new type of IP. To improve system resolution, we adopted fine-grain phosphors and a thin protective layer of fluoropolymer. Easy IP insertion to TEM cassettes is achieved by decreasing the total IP thickness to about 300 microns.

This system has many advantages, as mentioned above, and we have succeeded in reducing system and running costs. We believe that this IP system will replace the conventional film systems in the near future.

References

1. N. Mori,T. Katoh,T. Oikawa. et al (1986) Proc.XIth Int.Cong. on Electron Microscopy, 29-32
2. N. Mori,T. Oikawa,Y. Harada et al (1990) J Electron Microsc 39:433-436

Proc. Microscopy and Microanalysis 1995, edited by G.W. Bailey, M.H. Ellisman, R.A. Hennigar, and N.J. Zaluzec
Copyright © 1995 MSA. Published by Jones and Begell Publishing, 79 Madison Ave., New York, NY 10016

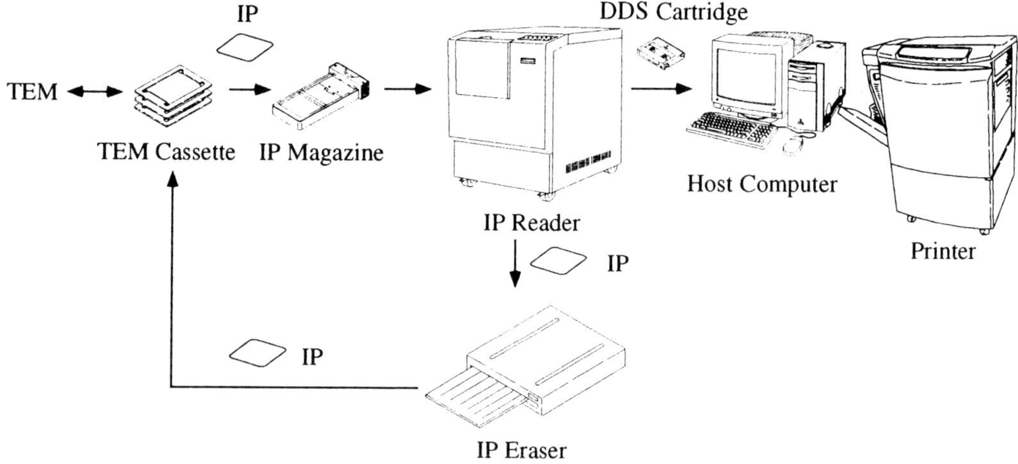

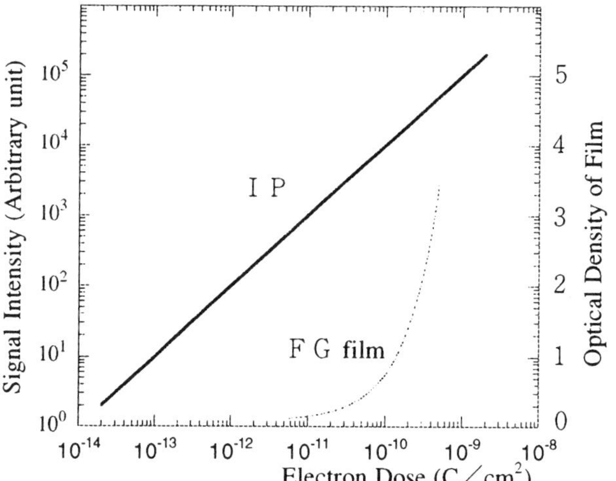

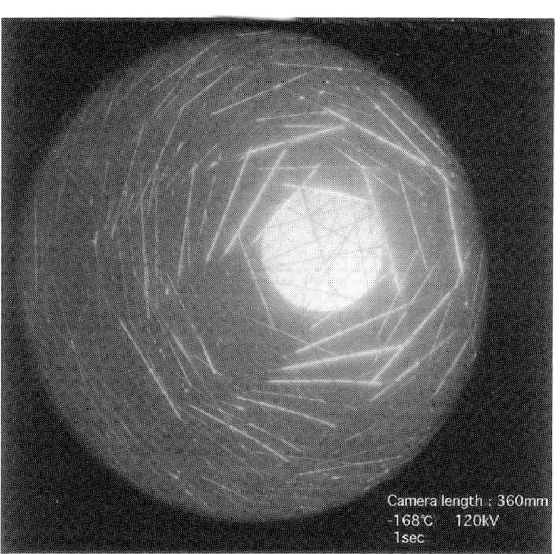

Fig. 1—Schematic diagram of the new IP system
Fig. 1 explains the usage and the basic components of the system. IPs should be exposed in the TEM just like photographic films. The exposed IPs are transferred into the IP Magazine, which can store up to 32 IPs. Miscellaneous information, such as TEM operating conditions, file name, comments and reading parameters, can be recorded prior to reading. After setting, the IP Reader reads all the IPs automatically, one by one. The total reading range of the electron dose is 2×10^{-14} to 2×10^{-9} c/cm² (1 to 100,000 electrons/pixel), and dynamic range of a single frame is 4 orders of magnitude. The IP Reader stores the image data onto Digital Data Storage (DDS) simultaneously while reading the IPs If a printer is connected to the IP Reader, image hard copies will also be available during the reading procedure. Another advantage of this IP system is that high-quality hard copies can be obtained without a dark room.

Fig.2—Sensitivity characteristics
The maximum sensitivity for 100KeV electron beam is 2×20^{-14}C/mm², which is equivalent to a detection capacity of 0.8 electrons for each 25μm pixel (Fig. 2). This implies that the sensitivity is as high as 1,000 times that of film.

Fig.3—α-AI_2O_3 large-angle convergent electron diffraction image (data provided by Prof. Tomokiyo, Kyushu University)

Tele-Presence Microscopy: An Interactive Multi-User Environment For Collaborative Research using High Speed Networks and The Internet

Nestor J. Zaluzec

Materials Science Division, Kinetics & Irradiation Effects Group
Argonne National Laboratory, Argonne, Illinois 60439 , USA

Tele-Presence Microscopy (*TPM*) is an advanced concept in the integration of computers and high speed networks with scientific instruments for operation, control, communication and research which makes use of ANL's Advanced Analytical Electron Microscope (AAEM) and Analytical Scanning Electron Microscope (ASEM) as development/testbed sites[1]. The implementation of a Tele–Presence Microscopy Facility allows a user from a remote location to either observe and/or control state–of–the–art instrumentation in a real time interactive mode. Using TPM, a user will be able to *actively* participate in scientific investigations at unique resources such as user facilities without being physically present at those locations. Manufacturers would be able to configure demonstration equipment which are accessible via the TPM system and thus allow prospective customers to remotely evaluate instrumentation before purchase. After acquisition, implementation by the manufacturer of a TPM system would allow remote service/diagnostics by a systems engineer who resides at the manufacturing site. Finally in an educational environment, students can initiate tele-presence operation of instruments which may not be available at their host institution, allowing widest possible access to unique facilities, Alternatively, should students have access to local equipment, they will have the opportunity of consulting an advisor or non-local expert in the field in an on-line mode during their actual experimental session, thus freeing valuable time which would be otherwise wasted during unproductive experiments. Several remote operation concepts have been suggested and partially implemented[2-4], *TPM* when linked with the *LabSpace* [1] represents a superset of these ideas, in that it is applicable to not only all microscopies but all scientific instrumentation in general.

The generic TPM/LabSpace system is composed of both software and hardware, which operate in a client/server relationship (figure 1). The local operator would be able to control access to the instrument, via a secure login system granting either observation mode or control mode to a remote access request. The remote TPM user would be present at a workstation displaying system configured windows showing relevant experimental details (microscope room & inhabitants, images, spectra, instrument &/or control parameters. see figure 2.), which would be either active or passive, depending upon the granted operational mode. Hot links would be provided so that it will be possible to integrate special resources (such as massively parallel computers) needed for data processing and analysis, when they are available either locally or remotely. The TPM workstation client software is being designed to be as platform independent as possible (i.e. capable of running on most high end workstations Mac's, PC's, Sun's....) and will therefore allow the user community widest access at minimum cost. The TPM/LabSpace proposal is a multi-year program to develop a permanent Tele-Presence Facility providing a continuous presence on the Internet for the microscopy community. Generic TPM client software for this instrumentation will be distributed free of charge to all interested organizations. The current status of the TPM project can be checked by accessing the Microscopy and Microanalysis WWW Site (URL= *http://www.amc.anl.gov*). This site provides a demonstration of some of the capabilities of a passive link using conventional WWW browsing tools .

Proc. Microscopy and Microanalysis 1995, edited by G.W. Bailey, M.H. Ellisman, R.A. Hennigar, and N.J. Zaluzec
Copyright © 1995 MSA. Published by Jones and Begell Publishing, 79 Madison Ave., New York, NY 10016

TPM/LabSpace Architecture Overview

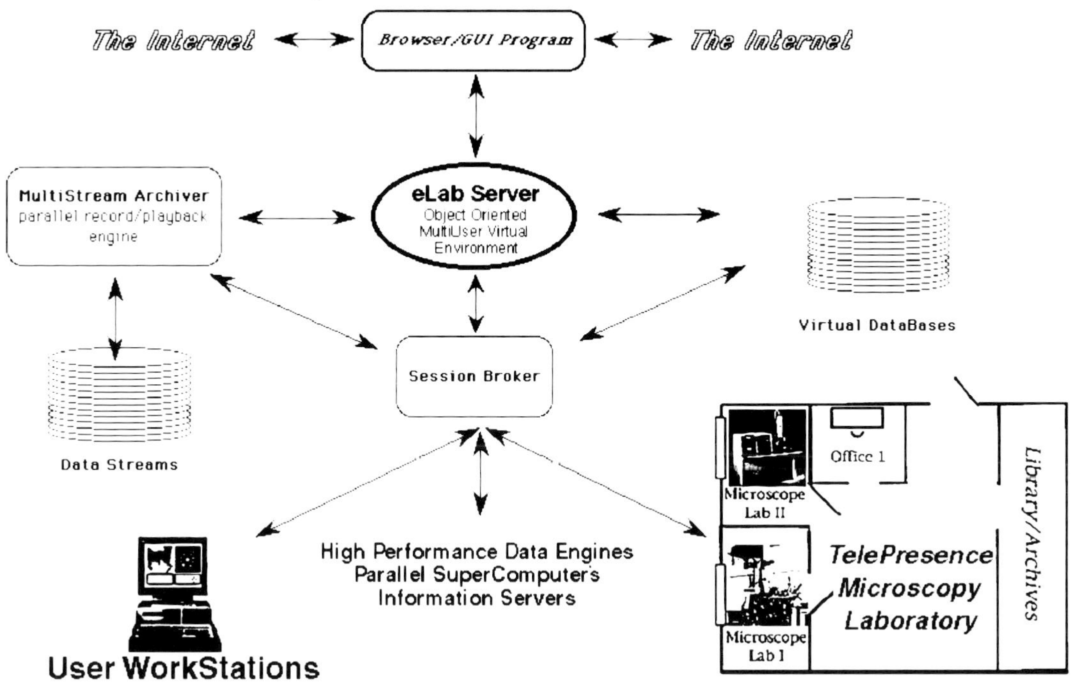

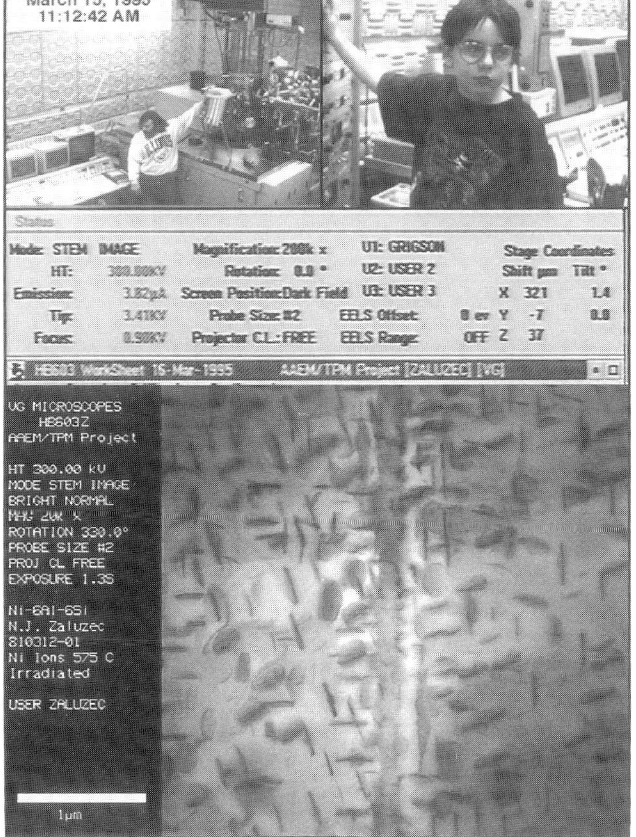

References:
1.) Lab Space and The National Electronic Laboratory Infrastructure : Documentation available upon request P.I.'s R. Stevens, R. Evard, T. Disz, E. May and N.J. Zaluzec ANL
2.) R. Weinstein etal, Am. Journal of Clinical Pathology, Vol 91, pg 39 1989.
3.) S. Tremblay etal, Proc. of the Microscopical Society of Canada, Vol. XXI, 1994 page 72
4.) M.Ellisman etal , Proc. of the Microscopy Society of America, 52nd Ann. Meeting, 1994 pg 10.
4.) This work was supported in part by US. DoE under contracts BES-MS W-31-109-Eng-38, and DE-AC03-76SF00098.

Figure 1: System Architecture for TPM/LabSpace Project.

Figure 2: WWW Screen Images from site URL= http://www.amc.anl.gov showing information available to a passive TPM/WWW user. Microscope Room Views (top), Instrument Status Screen(middle), Microscope Image & Data(bottom).

CREATING A STANDARD METHOD OF CONTROLLING DIGITAL MICROSCOPES:
THE MICROSCOPE ACCESS PROTOCOL (MAP)

T. A. Dodson, E. Völkl, L. F. Allard, and T. A. Nolan

High Temperature Materials Laboratory, Oak Ridge National Laboratory (ORNL), Oak Ridge, TN 37831-6064

The process of moving to a fully digital microscopy laboratory requires changes in instrumentation, computing hardware, computing software, data storage systems, and data networks, as well as in the operating procedures of each facility. Moving from analog to digital systems in the microscopy laboratory is similar to the instrumentation projects being undertaken in many scientific labs. A central problem of any of these projects is to create the best combination of hardware and software to effectively control the parameters of data collection and then to actually acquire data from the instrument. This problem is particularly acute for the microscopist who wishes to "digitize" the operation of a transmission or scanning electron microscope. Although the basic physics of each type of instrument and the type of data (images & spectra) generated by each are very similar, each manufacturer approaches automation differently. The communications interfaces vary as well as the command language used to control the instrument.

As manufacturers of microscopes become more sophisticated in their use of instrument control computers, one hopes that standard higher speed interfaces will emerge on the instruments of tomorrow. These interfaces might include HIPPI for high performance local connection or FDDI / ATM for high performance network attachment. Yet even as high performance interfaces emerge, the problem remains that different command languages pose a significant problem that greatly impedes progress in developing innovative, automated control systems for electron microscopes. A standard language that is not specific to the features of a single microscope, microscope manufacturer, or even a hardware interface would alleviate this severe obstacle.

Some might argue that a standard language without these specificities would be of no use. The best counter argument for this point is an analogy in the computer printing world. Before the advent of PostScript (Adobe Systems), the command language used to control printers was specific to single printers, the printers from particular vendors, and in some case, even to the hardware interface used for connection. Output from personal computer depended on elaborate drivers for each of the myriad printers. When Adobe introduced the PostScript language for printers, this problem rapidly disappeared. PostScript is a page description language (used to control printers and some computer displays) that is hardware and interface independent. Today, desktop publishing works because output created in PostScript for a desktop laser printer remains the same for any quality of PostScript typesetter with little to no modification.

We believe that a standard command language will generate great returns for both microscopists and microscope manufacturers. For the scientist/software developer, it would enable control algorithms to be rapidly applied to any instrument that has the hardware capability to perform the steps in the algorithm. It would eliminate the need to write instrument drivers for each microscope found in a facility. For the microscope manufacturer, it would eliminate developing and debugging different command interpreters for each instrument. The net effect of these benefits would be to increase speed of automation and control of microscopes, thereby increasing their usability and their usefulness to the end-user.

To this end, we offer the Microscope Access Protocol (MAP) as a first draft of the features might be included in a standardized microscope command language.

The data types that a device must handle play a major role in structuring automated systems and command languages to control the device. Table 1 lists the data types that must be addressed for electron microscopes.

Based upon our experience in developing control software for the JEOL 4000EX transmission, Hitachi HF-2000 transmission, and Hitachi S-4500 scanning electron microscopes, we have listed some of the commands in table 2 that we have identified as essential to the language. In the table, each command is listed along with its

Proc. Microscopy and Microanalysis 1995, edited by G.W. Bailey, M.H. Ellisman, R.A. Hennigar, and N.J. Zaluzec
Copyright © 1995 MSA. Published by Jones and Begell Publishing, 79 Madison Ave., New York, NY 10016

parameters. Brief examples are included after the list of paramters. A careful examination of the sample commands show that instrument-specific features are accessed with instrument-specific arguments but without instrument-specific commands. It is essential to the development and acceptance of MAP that arguments referring to the same instrument functions are identical from microscope to microscope.

We hope that this draft will spark discussion and development that will ultimately lead to an industry standard command set for controlling electron microscopes.

<table>
<tr><td colspan="3">1</td></tr>
<tr><td>Data type</td><td>Variable Types</td><td>Comments</td></tr>
<tr><td>Single value</td><td colspan="2">integer, floating point, or character</td></tr>
<tr><td></td><td colspan="2">used to set and check operating parameters, analogous to the front panel knob position</td></tr>
<tr><td>Spectra-like</td><td colspan="2">1 dimensional array of integer, floating point, or character</td></tr>
<tr><td></td><td colspan="2">used by functions requiring lists of values and generated by the instrument from acquisition of spectral data</td></tr>
<tr><td>Image-like</td><td colspan="2">2 dimensional array of integer, floating point, or character</td></tr>
<tr><td></td><td colspan="2">generated by the instrument from a single image acquisition</td></tr>
<tr><td>Stream</td><td colspan="2">compound type consisiting of a series of integer time stamps + any one of previous data types</td></tr>
<tr><td></td><td colspan="2">generated as a status check through time or as data is acquired through time</td></tr>
</table>

<table>
<tr><td colspan="2">2</td></tr>
<tr><td colspan="2">Static Commands — supplies and retrieves data immediately for a single point in time</td></tr>
<tr><td>set   (parameter_name, value)</td><td>error_code</td></tr>
<tr><td>get   (parameter_name)</td><td>error_code, requested data</td></tr>
<tr><td>error   (error_code)</td><td>text translation of error_code</td></tr>
</table>

Examples — set and check miscellaneous microscope parameters.

```
get("magnification")      returns
          0               <=      error_code, indicates successful execution
          25,000x         <=      present magnifification setting
set("accelerating_voltage",200.0)
          0               <=      error_code, indicates successful execution
error(0)                  returns
          Command successfully executed
```

Stream Commands — create & manage time-dependent data flows from microscope

```
initialize   (stream_ID, parameter_name, update interval, duration)   error_code
begin        (stream_ID)                                              error_code
end          (stream_ID)                                              error_code
clear        (stream_ID)                                              error_code
```

The initialize command creates an I/O channel for returning streams of data, e.g. the images display during a rapid-scan on an SEM. The begin command starts the transmission of stream data specified in the initialization. The end command stops the transmission of stream data before its specified duration. The clear command removes the stream specification from the instrument memory.

Example — sets small raster image area and requests 5 seconds of images taken every 0.1 seconds.

```
set("raster_size",128)
initialize(0, "image_raster", 0.1, 5.0)
begin(0)
clear(0)
```

TABLE 1.—Data types commonly encountered in electron microscope control and data acquisition.
TABLE 2.—Draft command set for Microscope Access Protocol (MAP) with sample commands.

LENS-FIELD AXIS ALIGNMENT: A NEW OBJECTIVE LENS ALIGNMENT FOR HIGH-RESOLUTION ELECTRON MICROSCOPY

K. Ishizuka* and K. Shirota**

* 14-48 Matsukazedai, Higashimatsuyama, Saitama, 355 Japan
** Topcon, Electron Beam Eng. Dept., Hasunuma, Itabashi, Tokyo, 174 Japan

1. Introduction

Accurate alignment of an electron microscope is important when the highest achievable resolution is demanded. In a conventional alignment for high-resolution electron microscopy, the incident beam direction is aligned for the object point imaged at a viewing-screen center. It should be noted however that the object point corresponding to the viewing-screen center does not usually located on the magnetic-field axis of an objective lens due to lens misorientation and/or beam deflection caused by a leakage field around the objective lens.[1] Then, the voltage-center and coma-free conditions cannot be satisfied simultaneously, even if a magnetic-field is rotationally symmetric. We implemented last year a proposed scheme[1] to satisfy both coma-free and voltage-center conditions. This alignment scheme however requires a suitable deflector at the objective image plane.[2]

In this report, we summarize ordinary alignment procedures for high resolution electron microscopy and propose a new alignment procedure, which we call a lens-field axis alignment. Using this alignment we can satisfy the coma-free and voltage-center conditions simultaneously.

2. Voltage-center and coma-free alignments for an off-axis object

We consider here an aplanatic imaging condition and ignore the beam rotation due to the magnetic field by using a rotating coordinate. Then, the wave aberration can be expressed as a function of the lens coordinate which directly relates with the scattering angle. Since the object point under consideration is close to the lens-field axis, each scattering cone originated from the object point suffers with the same aberration that is applied to the object point on the lens-field axis, when the cone axis passes through the lens center. Therefore, the direction connecting the object and the lens center can be considered as an optic axis. This direction is thus free from coma for a rotation symmetric lens-field. When the lens-field introduces an intrinsic coma, this coma can be canceled out with coma generated from the spherical aberration by tilting the incident beam direction off the optic axis. The coma-free direction thus depends on the object point and the inherent coma.

An incident beam direction passing through the lens center is free from magnification chromatic aberration. In the case of a magnetic lens, however, we have to consider rotation chromatic aberration for the object point off the lens-field axis. Thus, the optic axis is not free from chromatic dispersion except for the object point on the lens-field axis. In the voltage-center alignment the screen center is made dispersion-free against a voltage fluctuation by adjusting the incident beam direction. When there is a deflecting field within or close to the objective lens-field, deflection chromatic aberration can be canceled by adjusting the incident beam direction.

Therefore, any incident beam direction will not satisfy both voltage-center and coma-free conditions for the object point located off the lens-field axis. A procedure to find an object point on the lens-field axis is thus required in order to simultaneously satisfy these conditions.

3. Lens-field axis alignment

We assume here that objective lens misorientation and inherent coma are small, while beam deflection caused by a leakage field is significant as shown in the Fig. 1. This deflection can be compensated out by using a beam deflector just below the objective. Most of the microscopes have a deflector (usually called the image shift coils) which can be used for this purpose. The optic axis is found by coma-free alignment for any object point under consideration. In this case, the object point suffers from rotation chromatic aberration caused by the lens-field and deflection chromatic aberration caused by the leakage field plus the image shift field. We can investigate different object points using the image shift. At a certain image shift setting we can cancel a displacement caused by

Proc. Microscopy and Microanalysis 1995, edited by G.W. Bailey, M.H. Ellisman, R.A. Hennigar, and N.J. Zaluzec
Copyright © 1995 MSA. Published by Jones and Begell Publishing, 79 Madison Ave., New York, NY 10016

the leakage field. Both rotation and deflection chromatic aberrations will then disappear, and both voltage-center and coma-free conditions can be simultaneously satisfied.

We can cancel the chromatic dispersion by voltage-center alignment by introducing magnification chromatic aberration. The required amount of beam tilt for the voltage-center alignment from the coma-free condition is proportional to the chromatic dispersion or the image displacement. The object point on the lens-field axis is thus determined from either the image displacement caused by a voltage variation or the required beam tilt for the voltage-center alignment. Note that the required amount of beam tilt can also be estimated by the coma-free alignment followed by the voltage-center alignment.

4. Results and discussion

The lens-field axis alignment explained above was carried out for a Topcon EM-002A microscope. The images were acquired using a TV camera through DigitalMicrograph.[3] Coma-free alignment was carried out using a routine[4] based on the defocus determined by diffractogram analysis.[5] The voltage center was easily estimated by looking a real-time TV image. Fig. 2 shows required beam tilts between coma-free and voltage-center directions as a function of the image shift. The diagram shows that there is a coma-free and voltage-center point which corresponds to the object point on the lens-field axis. This diagram thus demonstrates that the lens misorientation and the inherent coma are negligibly small for this microscope.

Recently, coma-free alignment becomes a routine task with an advent of simple algorithms based on the diffractogram analysis.[4,5] In this report it is demonstrated that using the coma-free alignment we can find an object point on the lens-field axis. Then, both coma-free and voltage-center conditions can be simultaneously satisfied.

References

1. T. Yanaka et al., Proc. 8th Int'l Cong. Electron Microscopy, Canberra, 1 (1974) 128.
2. K. Ishizuka and K. Shirota, Proc. 52nd MSA meet., New Orleans (1994) 410.
3. DigitalMicrograph (a software), Gatan R&D, 6678 Owens Dr., Pleasanton, CA 94588, USA.
4. K. Ishizuka, Ultramicroscopy 55 (1994) 407.
5. O.L. Krivanek and G.Y. Fan, Proc. 10th Pfefferkorn Conf., Cambridge (1992) 105.

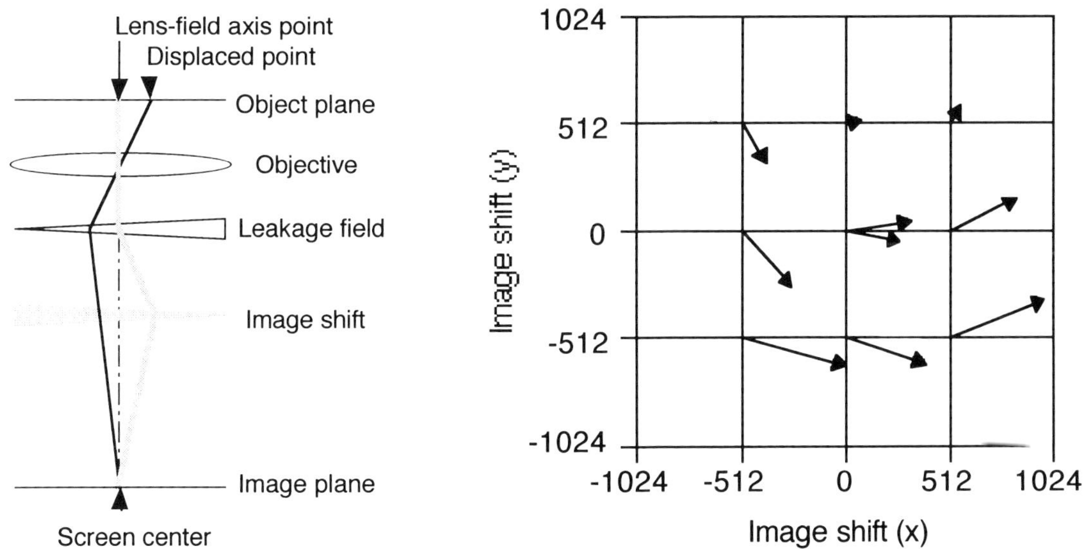

(Left) Fig. 1. Beam deflection caused by a leakage field around the objective lens. Image shift can cancel this beam deflection (gray line).
(Right) Fig. 2. Differences of beam tilts between coma-free and voltage-center directions.

P20 PHOSPHOR ON POLYIMIDE AS A LARGE AREA HIGH RESOLUTION TRANSMISSION SCREEN FOR SLOW SCAN CCD SYSTEMS

C. C. Ahn, S. Karnes*, M. Lvovsky*, C. M. Garland, H. A. Atwater and B. Fultz
Division of Engineering and Applied Science, Caltech, Pasadena, California 91125
**CRT Scientific, Van Nuys, CA 91405*

The bane of CCD imaging systems for transmission electron microscopy at intermediate and high voltages has been their relatively poor modulation transfer function (MTF), or line pair resolution. The problem originates primarily with the phosphor screen. On the one hand, screens should be thick so that as many incident electrons as possible are converted to photons, yielding a high detective quantum efficiency(DQE). The MTF diminishes as a function of scintillator thickness however, and to some extent as a function of fluorescence within the scintillator substrates. Fan[1] has noted that the use of a thin layer of phosphor beneath a self supporting 2μ thick Al substrate might provide the most appropriate compromise for high DQE and MTF in transmission electron microcscopes which operate at higher voltages. Monte Carlo simulations of high energy electron trajectories reveal that only little beam broadening occurs within this thickness of Al film. Consequently, the MTF is limited predominantly by broadening within the thin phosphor underlayer. There are difficulties however, in the practical implementation of this design, associated mostly with the mechanical stability of the Al support film.

We have adopted a similar approach in our design of an electron transparent support. We use a 7.5μm thick polyimide film, available commercially from DuPont Electronics as Kapton 30HN. In spite of a low density of 1.42 g/cc, polyimide is a robust material for a screen support and can be handled physically without difficulty. It has the added advantage of low x-ray fluorescence production. Concentric rings of PVC are used to clamp the polyimide into a flat film. A 50 nm layer of Al is evaporated onto each side of the polyimide to minimize charge buildup within the film. A uniform layer of 30μm of P20 phosphor is then deposited onto the Al facing away from the incident beam. A simulation of electron trajectories[2] through the support material is shown in Fig. 1. Because polyimide is composed primarily of C, the trajectory broadening for 300 keV electrons falls within 2μm through the 7.5μ thickness.

A schematic of our CCD system is shown in Fig. 2. It is mounted on a Philips EM430 and uses a Photometrics PXL camera which digitizes to 14 bits at an 800 khz readout rate, and uses a SITe anti-reflection coated, backside thinned TK1024 for enhanced QE response, typically $\sim$80% over most of the visible range. While the peak wavelength for our P20 is $\sim$500nm, the CCD has over 60% QE at 400nm, a consideration for the new generation of high efficiency phosphors based on Lu oxyorthosilicates[3]. The scintillator screen is contained wholly within the microscope camera vacuum chamber and imaged through a Schott SF6 leaded glass viewport. Optical coupling of the scintillator screen is accomplished using a Nikon 60mm micro lens which provides from 2:1 to 1:1 coupling of object to image at a screen to lens distance of 3.5 inches. A test of this system imaging the 1951 USAF test pattern through the leaded window resolves 20 lp/mm when the pattern is aligned with the CCD array, the resolution limit of the CCD.

Fig. 3 shows a "knife-edge" comparison using an opaque sample and shows the extent of broadening which occurs in our transmission phosphor screen when compared with a phosphor which has been deposited onto and imaged through a fused silica substrate. Slight 'tailing' of the beam occurs with the thick fused silica substrate although the broadening is comparable for each setup.

References

1. G. Y. Fan, D. G. Dunkelberger and M. H. Ellisman, Ultramicroscopy **55** (1), 7 (1994).
2. Calculated using MCSim by Z. J. Radzimski, based on an original code by R. Myklebust and D. C. Joy.
3. C. L. Melcher and J. S. Schweitzer, Nucl. Instr and Meth A. **314**, 212 (1992).

Proc. Microscopy and Microanalysis 1995, edited by G.W. Bailey, M.H. Ellisman, R.A. Hennigar, and N.J. Zaluzec
Copyright © 1995 MSA. Published by Jones and Begell Publishing, 79 Madison Ave., New York, NY 10016

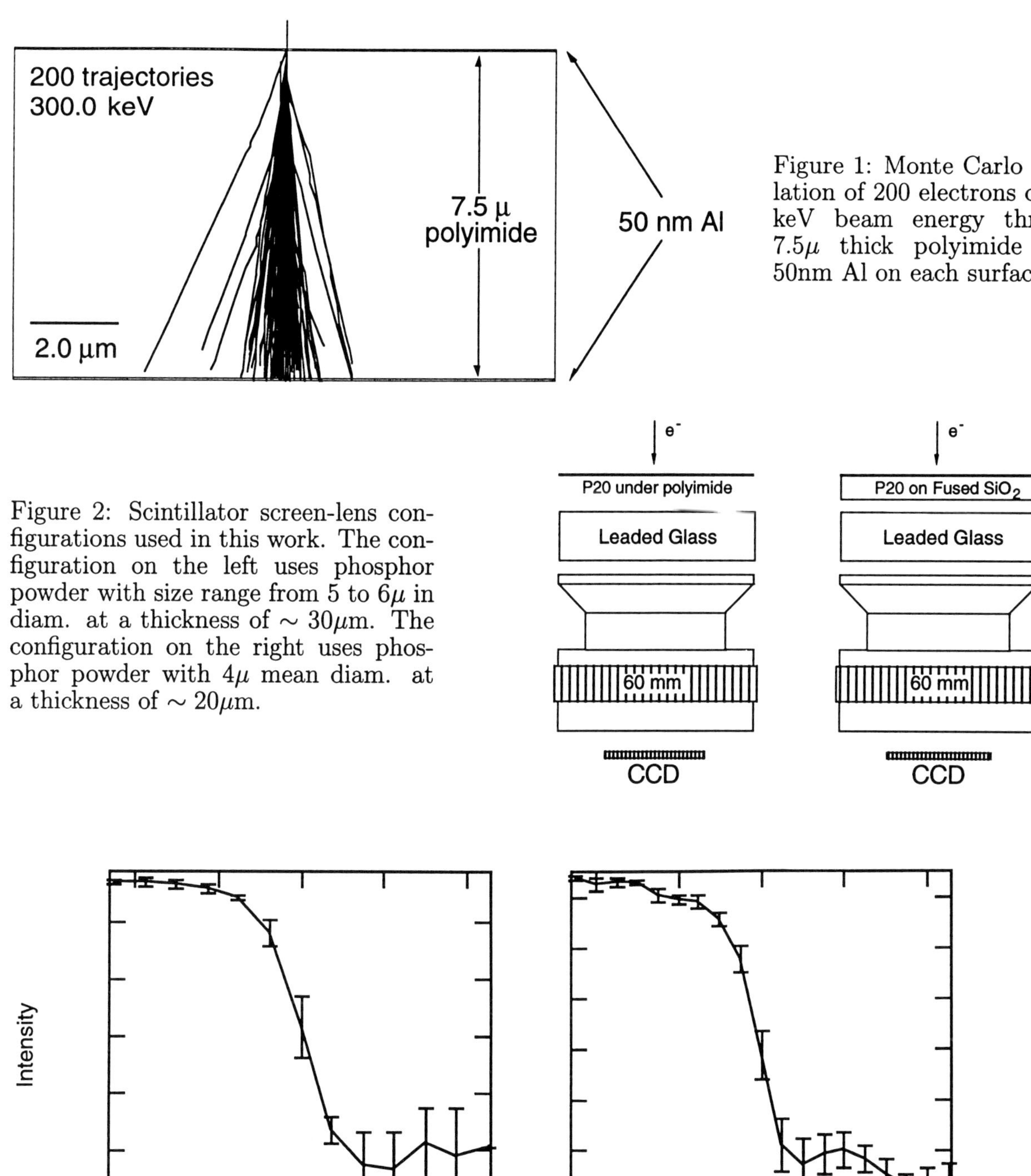

Figure 1: Monte Carlo simulation of 200 electrons of 300 keV beam energy through 7.5μ thick polyimide with 50nm Al on each surface.

Figure 2: Scintillator screen-lens configurations used in this work. The configuration on the left uses phosphor powder with size range from 5 to 6μ in diam. at a thickness of $\sim 30\mu$m. The configuration on the right uses phosphor powder with 4μ mean diam. at a thickness of $\sim 20\mu$m.

Figure 3: Intensity profiles for an opaque 'knife-edge' across the polyimide screen (left) and fused silica substrate (right). The broadening is comparable in each case, even though the thickness is 50% greater in the left screen. The larger error bars on the right of each plot result from granularity in the phosphor screen which has not been normalized out in this data.

COMPUTER CONTROL OF TRANSMISSION ELECTRON MICROSCOPES POSSIBILITIES, CONCEPTS AND PRESENT LIMITATIONS

E. Völkl, L.F. Allard, T.A. Dodson and T.A. Nolan

High Temperature Materials Laboratory, Oak Ridge National Laboratory, Oak Ridge, TN 37831-6064

The electron microscope laboratory at the High Temperature Materials Laboratory in Oak Ridge National Laboratory runs essentially film free. This is possible due to the use of TV-rate and slow-scan CCD cameras together with fast desktop computers connected through Ethernet to network servers for image storage, and a variety of hard copy output devices. In our experience, the functionality of a film-free laboratory goes beyond the simple replacement of film material with some other storage media. For example, the time and effort to produce a final hardcopy image has been reduced effectively from several hours to several minutes. At the same time, data accuracy has increased due to the high linearity of the CCD cameras and data safety is improved due to automated nightly backups.

The present all-digital setup of our TEMs has already changed the routines on how an TEM is run. Two situations occur regularly: 1) one microscopist moves back and forth between the TEM and the computer and 2) two microscopists run the computer and/or TEM as a team. Unfortunately, the tasks of recording an image and running the microscope are still very much separated. Procedures such as stage movements, microscope alignment and many other routines are run from the microscope. Images are recorded and immediately processed at the computer to refine data. Our intention is to eliminate this separation and to optimize the usage of electron microscopes in general.

At our Hitachi HF2000 TEM, a Macintosh Quadra 950 with a 80 MFlop array processor is connected through an RS232 interface with the internal computer of the microscope as sketched in figure 1. A partial list of accessible functions is displayed in figure 3. This approach is, of course, not new.[1] Some routines are already controlled directly from the computer. For example, the software package DigitalMicrograph allows an automated fine-tuning of the microscope on a suitable specimen area using beam deflection controls.[2] We have gone one step farther than partial microscope control and implemented all accessible microscope functions into DigitalMicrograph, i.e. the software that runs the CCD camera (and/or TV-camera) also runs the electron microsope.

The actual use of complex microscope control functions reveals major hardware problems which are present to some degree in almost every modern microscope. The primary obstacle is the response time of the microscopes to an external command. On the HF2000, the response time ranges between 0.2 and 0.3 seconds for a single command. For complex microscope control functions – other than setting or receiving a single parameter – we find that, on the average, about 10-20 SET- and GET- commands are necessary. To the human operator, those control functions should work without noticeable delay. Assuming a time delay of < 0.2 seconds is not noticeable, we conclude that the average response time of the microscope to a single SET-GET command should be < 10 ms.

The response-time of the beam shutter is another major obstacle. The shortest practical exposure time our present mechanical TEM beam shutter can handle is > 0.2 sec. We found that in using complex microscope control functions for analytical applications (such as EDS elemental profile mapping) it is *essential* to locate the position of the fully convergent electron beam. With our present beam shutter, the position can be located only for very small beam currents. For normal or high beam currents as used for analytical work, the CCD chip becomes oversaturated, i.e. the position cannot be evaluated exactly and artefacts are created in successive images. We consider a beam shutter that allows exposure time of < 20 ms sufficient, and we are considering substituting a fast electromechanical device for the present mechanical shutter.

For the future we hope that manufacturers of electron microscopes will drastically improve the capability for external computer control of their microscopes. Practice has shown that present microscopes are far

Proc. Microscopy and Microanalysis 1995, edited by G.W. Bailey, M.H. Ellisman, R.A. Hennigar, and N.J. Zaluzec
Copyright © 1995 MSA. Published by Jones and Begell Publishing, 79 Madison Ave., New York, NY 10016

from being satisfactory for external computer control for the following reasons: 1) not all microscope functions are accessible (some functions are still analog only), 2) the response time of the microscopes is too slow (by about a factor of 1000), 3) the beam shutter in a TEM maybe too slow, and 4) the SET- GET commands vary for different microscopes. Standard command sets would improve the potential to integrate every manufacturers instruments into the digitally controlled microscopy environment of the future.[3]

References

1. O. Krivanek et al., Proc. MSA (1993) 546
2. Gatan Inc., 6678 Owens Drive, Pleasanton, CA 94588
3. Research sponsored in part by the Laboratory Directed R&D Program of Oak Ridge National Laboratory, managed for the DOE by Martin Marietta Energy Systems, Inc. under contract DE-AC05-84OR21400 and supported by an appointment (E.V.) to the Oak Ridge National Laboratory Postdoctoral Research Program administered by the Oak Ridge Institute for Science and Education.

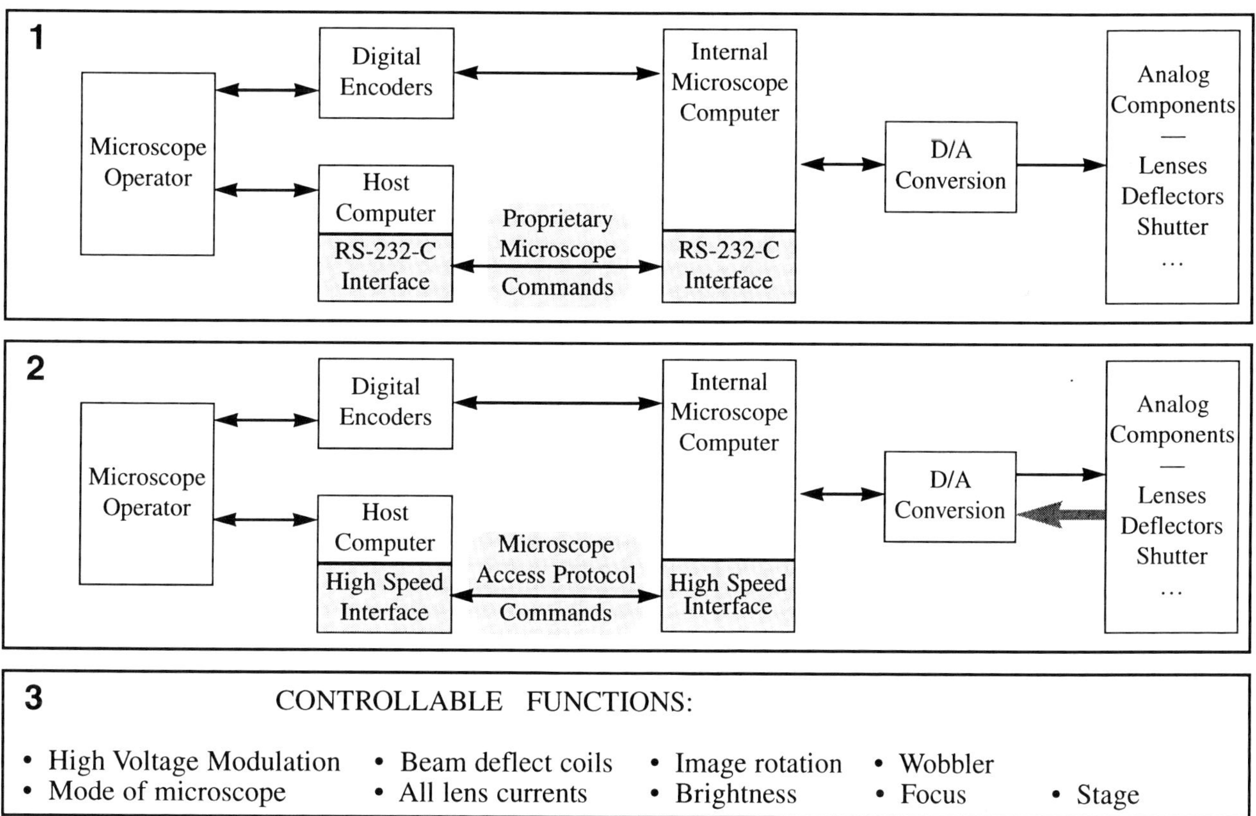

FIG. 1. – External control of the Hitachi HF2000 TEM relies on the exchange of proprietary commands over a RS-232-C serial line. This imposes great transmission delays in addition to interpretation delays.

FIG. 2. – External control of the TEM should rely on the exchange of standardized commands over a high speed interface such as HIPPI, ATM, or FDDI. In addition, the interpreter should be designed for high-speed command execution. Finally, actual analog status changes should be returned to the controlling computer.

FIG 3. – Partial list of microscope functions implemented in DigitalMicrograph to run camera and HF2000 microscope. These microscope functions are building blocks for complex programs that add functionality to microscope and simplify TEM microscopy.

HREM AUTOTUNING ON CRYSTALLINE MATERIALS

M. Pan and O.L. Krivanek

Gatan R&D, 6678 Owens Drive, Pleasanton, CA 94588

Complete autotuning of a high resolution electron microscope has been well established[1,2]. It performs the following tasks: align the electron beam along the true electron-optical axis of objective lens (autoalignment), correct the astigmatism (autostigmation), and set the defocus to a user defined value (autofocus). It can also characterize the coefficient of 3-fold astigmatism while performing the autoalignment[3,4]. Based on diffractogram analysis current HREM autotuning algorithm only works on amorphous materials. In reality, however, most of the HREM practice is performed on crystalline materials. Therefore it is highly desirable to extend the current HREM autotuning algorithm to crystalline specimens. In this abstract we report preliminary studies on attempting to analyze diffractograms from a mix of crystalline and amorphous materials.

For crystalline specimens observed in most high resolution electron microscopes, except under UHV conditions, there is typically a thin layer of amorphous contamination due to either sample preparation or poor vacuum conditions. This amorphous layer can be easily seen at the edge of a crystalline sample in the microscope. Diffractograms from HREM images of such crystalline samples often show sharp and strong Bragg reflections. However, if these strong reflections are removed, the diffractogram may show the typical amorphous rings pattern depending upon the amount of amorphous materials present on the crystalline surface. This is the basic idea of our attempt to perform HREM autotuning using crystalline specimens.

In our preliminary studies, a diffractogram from an amorphous sample (figure 1) was recorded on a Philips CM12ST electron microscope operated at 120 kV and analyzed by the existing HREM autotuning routine[5], yielding the defocus value and the amount of astigmatism present in the image as shown at the bottom of figure 1. Then strong Bragg reflections were artificially created and added to the amorphous diffractogram, resulting in a mixture of rings and sharp spots as shown in figure 2. It is noted that the strong Bragg reflections have made the amorphous rings less visible. These Bragg reflections can be successfully removed with a script that compares the peak intensity to the average intensity from typically 6×6 neighboring pixels around a particular peak (figure 3). The remaining diffractogram was analyzed again by the HREM autotuning routine and almost identical results were obtained as shown at the bottom of figure 4.

Our preliminary investigations have shown that it is very likely that the current HREM autotuning routines can be extended to crystalline materials by removing strong Bragg peaks in a diffractogram using simple and fast algorithm.

References

1. O.L. Krivanek and G.Y. Fan, Proc. 50th . Meeting of MSA (1992) 96.
2. O.L. Krivanek and G.Y. Fan, Scan. Micros. Suppl. 6 (1992) 105.
3. O.L. Krivanek and M.L. Leber, Proc. 51th . Meeting of MSA (1993) 972.
4. O.L. Krivanek, Ultramicroscopy 55 (1994) 419.
5. G.Y. Fan and O.L. Krivanek, Proc. 12th Int. Congr. on Elec. Micros., Seattle, vol.1 (1990) 532.

Proc. Microscopy and Microanalysis 1995, edited by G.W. Bailey, M.H. Ellisman, R.A. Hennigar, and N.J. Zaluzec
Copyright © 1995 MSA. Published by Jones and Begell Publishing, 79 Madison Ave., New York, NY 10016

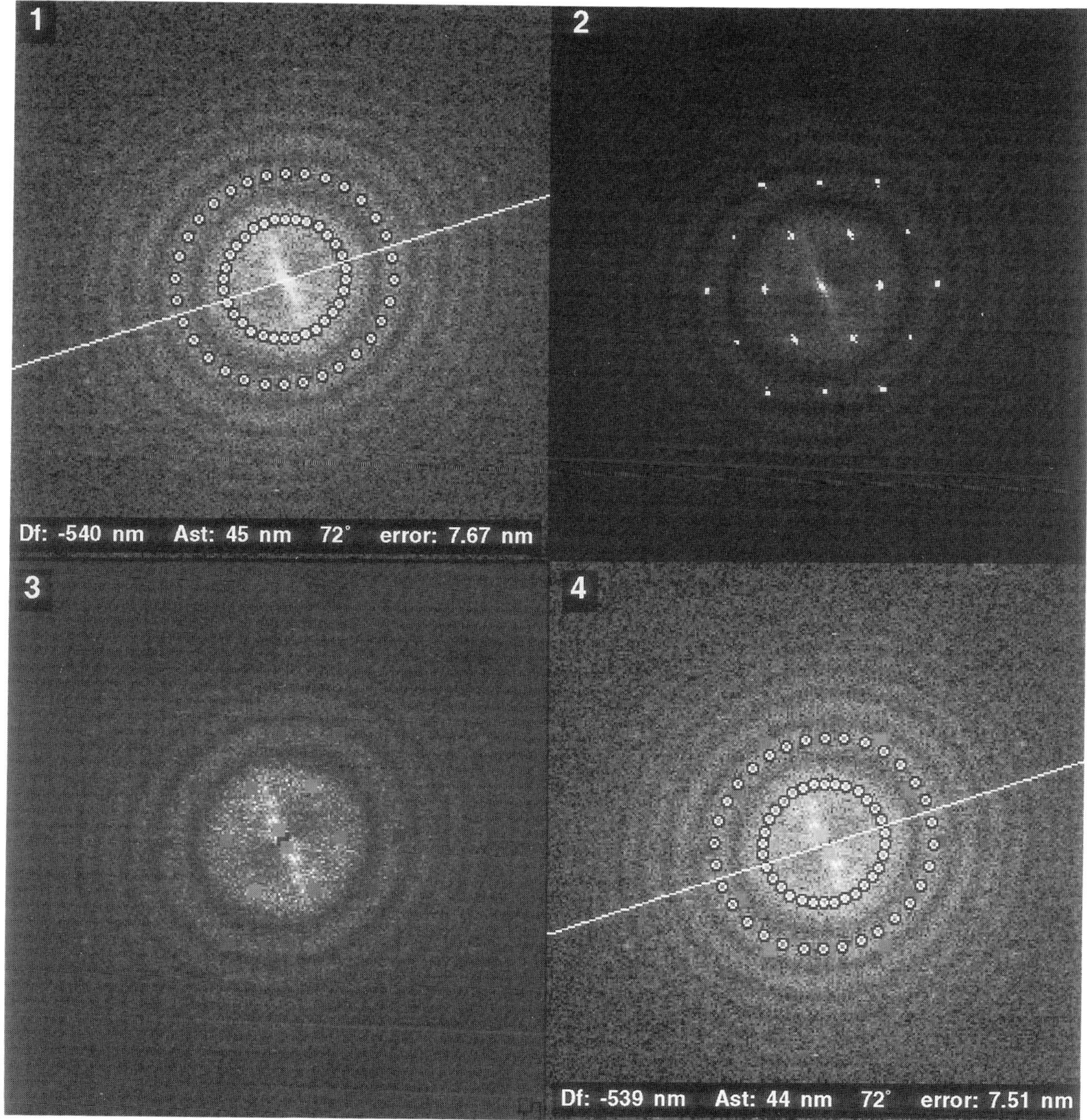

Fig.1 Diffractogram from an amorphous image recorded on Philips CM12ST TEM with a slow-scan CCD camera. Results of diffractogram analysis are shown at the bottom. The line indicates the direction of astigmatism.

Fig.2 Diffractogram in fig.1 mixed with artificially created sharp Bragg reflections. The amorphous rings are less visible due to the presence of Bragg reflections.

Fig.3 Diffractogram after the Bragg reflections are removed by comparing the peak intensity to the average intensity from the neighboring 6×6 pixels in fig.2.

Fig.4 Diffractogram in fig.3 is analyzed with results shown at the bottom.

AUTOMATED ELECTRON TOMOGRAPHY: FROM DATA COLLECTION TO IMAGE PROCESSING

Weiping Liu, Jennifer Fung, W.J. de Ruijter, Hans Chen, John W. Sedat and David A. Agard

Department of Biochemistry and Biophysics and the Howard Hughes Medical Institute, University of California at San Francisco, San Francisco, CA 94143-0448

Electron tomography is a technique where many projections of an object are collected from the transmission electron microscope (TEM), and are then used to reconstruct the object in its entirety, allowing internal structure to be viewed. As vital as is the 3-D structural information and with no other 3-D imaging technique to compete in its resolution range, electron tomography of amorphous structures has been exercised only sporadically over the last ten years. Its general lack of popularity can be attributed to the tediousness of the entire process starting from the data collection, image processing for reconstruction, and extending to the 3-D image analysis. We have been investing effort to automate all aspects of electron tomography. Our systems of data collection and tomographic image processing will be briefly described.

To date, we have developed a second generation automated data collection system based on an SGI workstation (Fig. 1) (The previous version used a micro VAX[1]). The computer takes full control of the microscope operations with its graphical menu driven environment. This is made possible by the direct digital recording of images using the CCD camera. Commands are issued via screen buttons, accelerator keys, command line input, graphic set-up menu (Fig. 2), or batch files. Automated tomography requires that many functions be precisely calibrated (focus, image shift, beam shift, CCD correction files, angle of goniometer tilt axis, and stage motors, etc.), this complex procedure now only requires a few mouse clicks. In its application to tomography, a data set of over 100 projections can be collected within a few hours, with the additional advantage of a superior quality by reducing the total electron dose >100 fold compared with the manual operation. Each projectional view typically requires four steps: 1) tilt specimen; 2) make image and beam shifts; 3) make focus change; 4) take and save image. Steps 2-3 are needed to track the 3-D movement of specimen during tilt. A necessary transitional step is back-tracking: i.e., after the collection of views between tilt 0° to 75°, the imaging condition of 0° view needs to be recovered to start 0° to -75°. Applying the coil current values of the initial 0° might not work due to potentially large accumulated specimen drift. We correct such drift incrementally by back-tracking via intermediate tilts using saved images as references. Once under computer control, beside tomography, the system can readily be used for many other applications such as stereo imaging, restoration from defocus series, serial reconstructions, and montage imaging, etc.

For the tomographic image processing stage, we have developed the software system EMCAT (Electron Microscopic Computerized Axial Tomography)[2-4]. EMCAT is designed to maximize the degree of automation while maintaining the flexibility to solve any specific problems. It also attempts to devise optimal approaches in its every step so that superior reconstructions are achieved in a fast and easy fashion. There are currently three sub-systems in EMCAT: BALIGN and SALIGN for the projection alignment step[3,4]; and EMRECON for the mass normalization and reconstruction steps[2]. Alignment is the most involved and a resolution-limiting step. The approach with BALIGN is based on fiducial markers (i.e., gold beads)[3]. The modules of bead-like feature finding, bead feature indexing, and least-square fitting can be dynamically combined to solve any problem during bead alignment. For isolated structures (e.g., with little background), the approach with SALIGN can be adopted[4], which avoids the tedious task of having to spread beads onto the specimen for bead alignment. The computation involves the iterative comparisons of re-projections and projections, a procedure that explores the information of self-consistency between projections of different views.

With our automation of data collection and image processing in electron tomography, getting a reconstruction from a prepared specimen is now achieved within one day. Since the essence of telemicroscopy (e.g., to operate the microscope from remote sites) is automation, our systems represents a major step in fulfilling the promises of telemicroscopy.

Proc. Microscopy and Microanalysis 1995, edited by G.W. Bailey, M.H. Ellisman, R.A. Hennigar, and N.J. Zaluzec

References:

1. A.J. Koster et al., Mic. Soc. of Am. Bul. 23(2)(1993)176.
2. W. Liu et al., Proc. Ann MSA Meeting of MAS 52(1994)418.
3. W. Liu et al., Proc. Ann MSA Meeting of MAS 52(1994)932.
4. W. Liu et al., Proc. Ann MSA Meeting of MAS 51(1993)554.

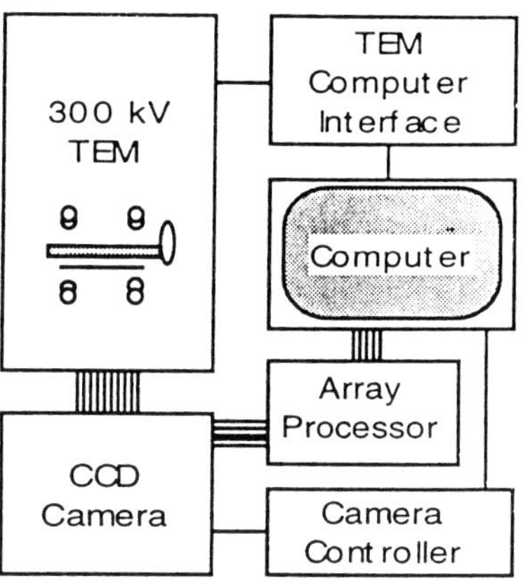

FIG. 1: Instrumental setup at UCSF: The TEM is a Philips EM430 operating at 300 kV. The cooled slow scan CCD camera has 1024^2 pixels of $19*19$ um^2 at a digital precision of 12 bits and a readout rate of 200,000 pixels/s. The SGI Personel Iris 4D35 workstation is equipped with four 80 Mflops Sky array processors for some repetitive image processing tasks. The computer controls all functions of the TEM (via a Philips C400 interface unit) and the CCD camera (via a Photometrics controller).

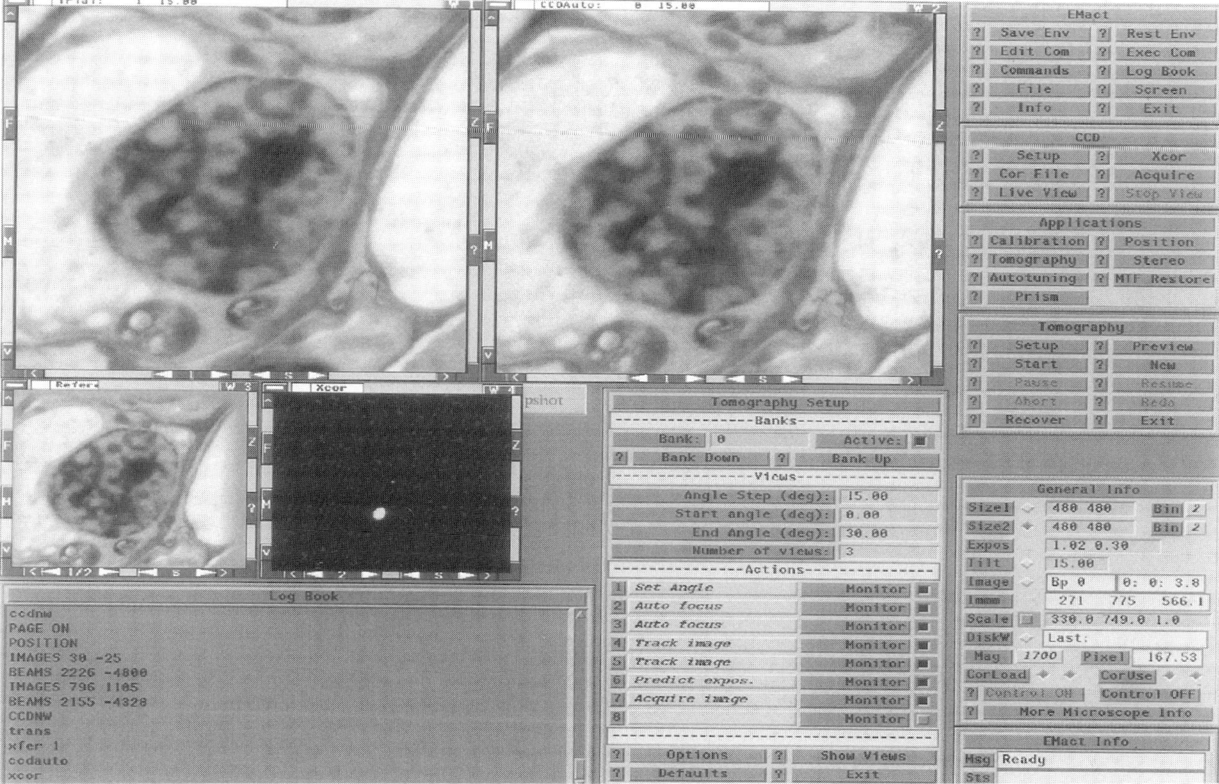

FIG. 2: User interface of the computer control system --- a snapshot of the computer screen that displays some windows relevant to tomographic data collection: There are four graphic windows for image displays: images to be saved in W1, images used for tracking and focus in W2, reference images for cross-correlations in W3, and cross-correlation maps in W4. The buttons of saving/ restoring environment (in upper-right window) take care of the system status while the log-book (lower-left window) records the system history. Also, the general information window (lower-right) indicates the system status. The data collection protocol is dictated by the tomography setup menu (upper-middle) that is composed of banks. Each bank specifies the views to be collected (as defined by the tilt range and step) and the operations needed for each view. The data collection process can be paused, resumed and aborted. There is also the capability of re-collecting selected views and recovering from a system crash. Additionally, the progress can be remotely monitored (by displaying the log-book content) and commands can be remotely issued.

MICROSTRUCTURAL CHARACTERIZATION OF AL-8900 SUPERPLASTIC MATERIALS USING ORIENTATION IMAGING MICROSCOPY

H. Garmestani*, P. Kalu*, K. Mehr* and D. Dingley**
*FAMU/FSU College of Engineering, Tallahassee, Florida
** Bristol University, On leave to TSL company, Provo, Utah

Electron Backscattered Pattern (EBSP) technique has been shown to be an effective tool for measuring lattice orientation of bulk polycrystalline materials in an SEM[1,2]. Automation of this technique through Orientation Imaging Microscopy (OIM) opened a new era of materials characterization which had only existed using Transmission Electron Microscope (TEM) for thin foil specimens[3]. Investigation of grain orientation and grain to grain misorientation are among a number of parameters which are crucial in studying the mechanisms of deformation in materials, especially superplastic materials. The microstructure and texture changes corresponding to different stages of Superplastic Forming (SPF) has been traditionally characterized using conventional optical, TEM and X-ray Techniques. TEM analysis of grain structure and subgrain formation could only provide information for individual grains or at best a small region of the microstructure, hence do not provide a global picture. On the other hand, the data obtained by OIM technique and the regeneration of the microstructure based on different definitions of grain boundary can provide a better understanding of Superplasticity.

This paper presents the results of the study on Al-8090 superplastic material using the OIM technique. Previous work had shown that the as-received material is unrecrystallized and consists of elongated grains[4]. Tensile specimens of this material were deformed to different strains at 520° C and 5×10^{-4} strain rate. By assigning different values for the misorientation angles (0-5, 0-10) different geometries were created and the analysis of which revealed two distinct regions of high and low angles. Detailed texture study clearly identified the presence of a recrystallization texture at the early stage and texture randomization at the later stages of deformation.[5]

Figure 1 shows the microstructure of a 15% deformed specimen (early stage of superplastic forming). The micrograph is a regeneration of the quality index image map resembling that of an optical micrograph. This data became available at points 1 micron apart (a six sided hexagon has been chosen for each block). The contrast inside each grain indicates the amount of deformation or substructure within the microstructure. Figure 2 is the regeneration of the same data based on the misorientation angle between adjacent points. Grain boundaries (sub-grain boundaries) were created using different line widths for different low ($1-4^\circ$, $4-10^\circ$) and high (10° and above) misorientation angles. Both micrographs show the presence of equiaxed structure. However, if the classical definition of 10° is chosen to represent the cut-off for high angle boundary a new microstructure is obtained which is shown in figure 3. This picture gives us a completely new insight about the actual microstructure of the material: a microstructure consisting of clusters of fine grains sandwiched by two large grains of the same texture (Fig. 4).

References:

1.D. J. Dingley in J.S. Kallend and G. Gottstein, Eds., Proc. ICOTOM 8, Warrendale:TMS. (1988)189.
2.K. Kunze et al., Textures Microstruct. 20(1993)41.
3.S. I. Wright and B.L. Adams, Metall, Tran. A 23(1992)759.
4. A.W. Bowen, *Textures and Microstructures* 8,9(1988)233.
5. H. Garmestani et al., Proc. Light Element Alloys of TMS, Feb 10-15, 1995.

6. The authors gratefully acknowledge the excellent technical assistance of D. Field of TSL company. This research was supported by NASA Grant and the Center for Materials Research and Technology.

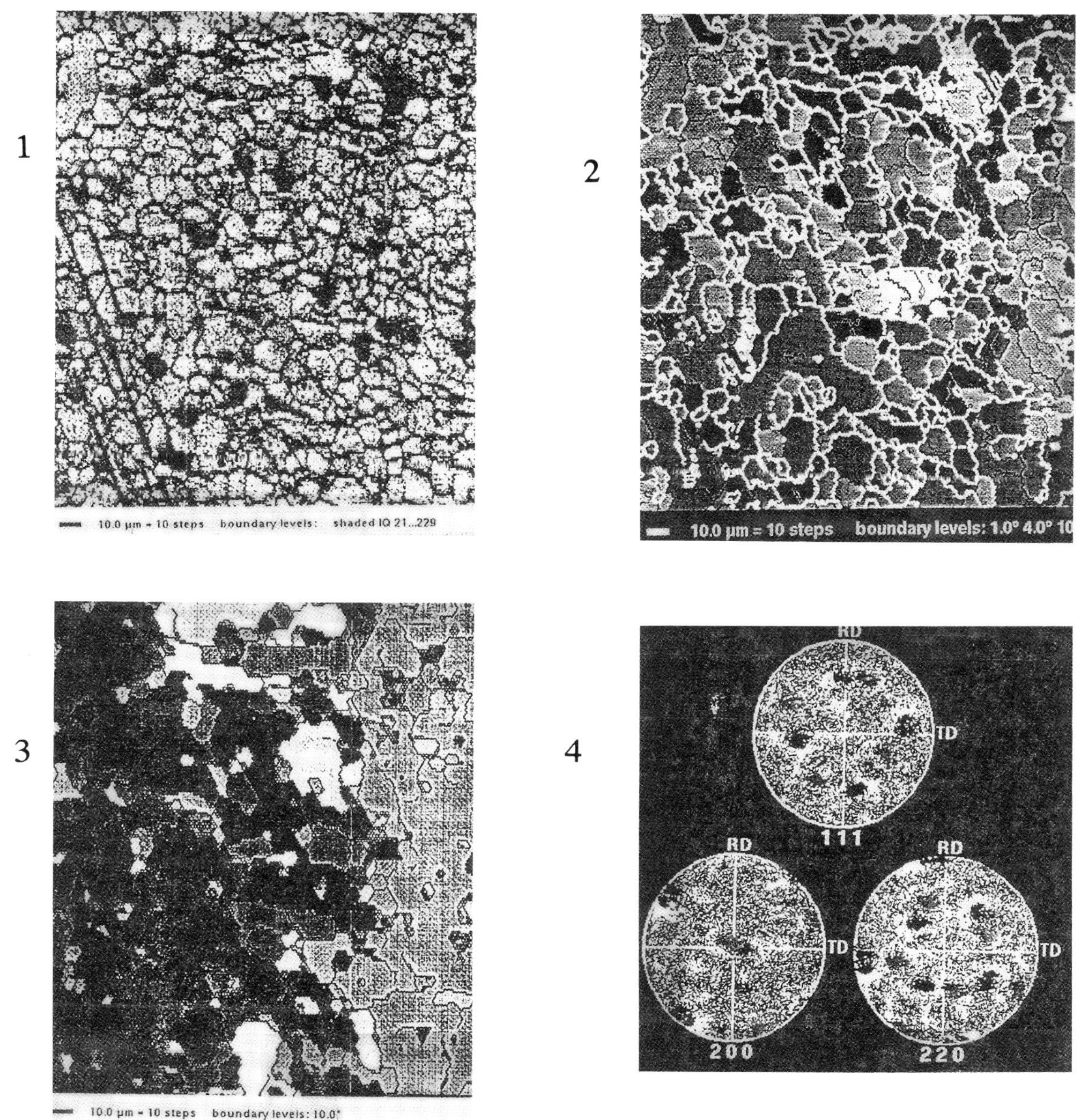

FIG 1.- A micrograph for a 15% deformed SPF sample of Al-8900 material obtained from the Orientation Imaging Microscope. This resembles an optical micrograph representing an equiaxed grain structure of a superplastic material.

FIG. 2- OIM Micrograph with white representing high angle grain boundaries ($>10°$) and thinner black lines depict low angle grain boundaries.

FIG. 3- Same as the previous figure with black lines depicting high angle boundaries.

FIG. 4- Pole figure representation of the previous figures. Black and gray high intensity poles represent the left and right regions of the microstructure.

THIN-FOIL BASED TEM SCREENS WITH ELECTROPHORETIC DEPOSITION OF PHOSPHORS

G.Y. Fan, Ben Russ,* Jan B. Talbot,* Jeff Price** and Mark H. Ellisman

San Diego Microscopy and Imaging Resource, Department of Neurosciences,
*Department of AMES/Chemical Engineering,
**Department of Bioengineering,
University of California at San Diego, La Jolla, CA 92093, USA

The availability of 2kx2k or larger format CCD chips has made direct digital imaging more practical in electron microscopy. But the suboptimal performance of scintillating screens, particularly their inferior resolution as compared to film, is still an obstacle to a broader adaptation of digital imaging technology to electron microscopy. Thin-foil substrate screens[1] have improved brightness and resolution over the more commonly used glass substrate screens, particularly at higher operating voltages, but further improvements are required for the optimum performance of the CCD imaging systems. Self-supporting single crystal YAG screens, approximately 30 μm in thickness, provide better resolution and detection-quantum-efficiency as compared to the phosphor screens, but their lower brightness, about 4 - 8 times lower, makes them less than ideal at least for low dose applications. Until brighter single crystal scintillators become available, powder phosphor screens may offer performance advantages for applications where radiation damage to the specimen is a concern.

Powder phosphor screens are conventionally made by the sedimentation method. In sedimentation, the substrate is immersed in a solution consisting of a mixture of phosphor and a binder material, typically in 1:1 ratio. The solution is then slowly drained away and a layer of phosphor sediments onto the upper surface of the substrate. This is a slow process, and the quality of the screen can be difficult to predict. More importantly, the thickness of the phosphor layer can be hard to control and to reproduce. Electrophoretic deposition of phosphors[2,3] appears to overcome these difficulties. And the thin Al foil based screens are particularly suited for this type of process, since Al is an excellent conductor. An apparatus for this electrophoretic process is schematically illustrated in Fig. 1. The main advantages of this method are:
    (1) Speed. A screen can be coated typically in a few minutes.
    (2) Reproducibility. Since the rate of deposition can be controlled precisely, the phosphor layer to be coated can be consistently reproduced.
    (3) Uniformity. Screens are coated more evenly over a large area.

With the parameters shown in the Fig. 1, a 10 μm layer of phosphor can be deposited on a thin Al foil in 3.5 minutes. Quantitative comparisons of brightness and resolution of screens made using electrophoresis and sedimentation methods are under way and will be reported at the meeting.[4]

References:

1. G.Y. Fan, D. G. Dunkelberger and M. H. Ellisman, Ultramicroscopy 55 (1994) 7.
2. M.J. Shane, J.B. Talbot, R.D. Schreiber, C.L. Ross, E. Sluzky and K.R. Hesse, J. Colloid and Interface Science 165 (1994) 325.
3. M.J. Shane, J.B. Talbot, B.G. Kinney, E. Sluzky and K.R. Hesse, J. Colloid and Interface Science 165 (1994) 334.
4. This work was supported by NIH grants: NS14718, NS26739, RR04050.

Proc. Microscopy and Microanalysis 1995, edited by G.W. Bailey, M.H. Ellisman, R.A. Hennigar, and N.J. Zaluzec
Copyright © 1995 MSA. Published by Jones and Begell Publishing, 79 Madison Ave., New York, NY 10016

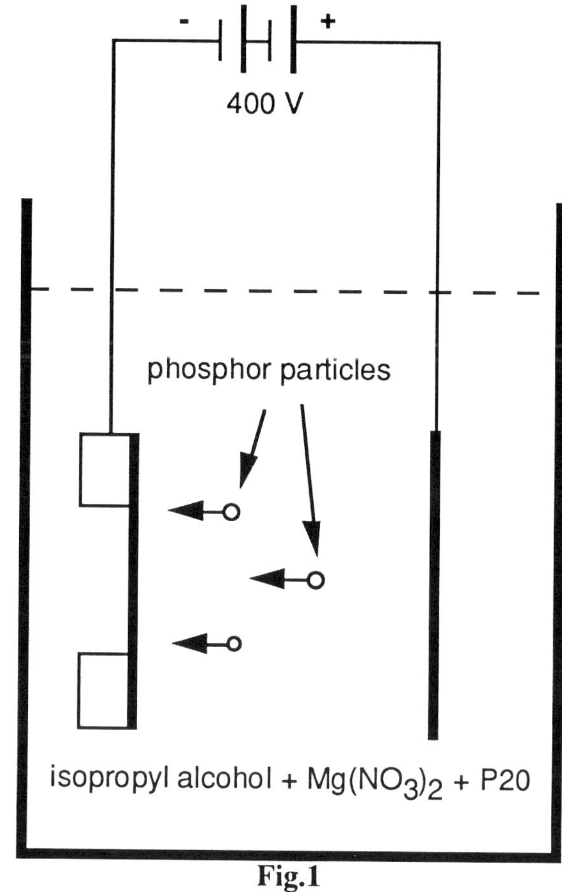

Fig.1

FIG.1--Electrophoretic deposition of phosphors for making TEM scintillating screens. A 10 μm layer of phosphor can be coated onto the Al foil mounted on a brass ring in 3.5 minutes.

BUILDING DATA STORAGE SYSTEMS AND DATA NETWORKS
TO SUPPORT DIGITAL MICROSCOPY

T. A. Dodson, T. A. Nolan, L. F. Allard, and E. Völkl

High Temperature Materials Laboratory, Oak Ridge National Laboratory, Oak Ridge, TN 37831-6064

As laboratories like the Materials Analysis User Center (MAUC) in the HTML at ORNL move from analog to digital imaging systems, the process for acquiring, processing, analyzing, and publishing research results is changing. In this case since original scientific data exist only in digital form, analog systems for gathering, storing, and transmitting data are being set aside in favor of digital systems. In order to adequately protect original scientific data and to ensure that digital laboratories have the same (or greater) functionality as analog laboratories, scientists must focus on building information systems that make data highly available, highly reliable, and quickly accessible. Meeting these three objectives imposes special requirements for both data storage systems and data networks.

A data storage system for a digital microscopy laboratory must have a very large capacity (at MAUC, 30 GB). In addition, the data stored within the system must be *highly available* and *highly reliable*. Highly available means that the data must be available both locally, and across the internet, to any of the many different computer platforms that might be needed to process or analyze digital data. In our laboratory as in others, this includes Macintoshes, DOS/Windows PC's, and UNIX workstations. Highly reliable means that the data must be safeguarded against tampering and secured for long periods of time. We have developed a simple architecture representing the data flows in our facility (Figure 1). Its shared storage system was developed to manage the data flows shown in the architecture and to answer the requirements of high reliability and availability.

The shared storage system incorporates components for storing data on-line and for archiving data off-line in a client/server arrangement that allows access via NFS from Macintoshes, DOS/Windows PC's, Unix Workstations, and via FTP from authorized users on the internet. To configure the on-line portion of this system, one must decide the amount of time that research data must be immediately accessible and work backwards to determine how much on-line storage is required. Figure 2 illustrates a sample calculation for our laboratory. CD-ROM's serve as the off-line archive in the MAUC system. Writable CD-ROM's are used because of their long media life (30 years), relatively low cost ($10/blank), and wide readability (on Mac, PC, and Unix). Figure 3 shows how the pieces of our system fit together on the network. Operating this system requires great attention to detail. Regular backups are required to assure that scientific data is never lost. Timely archiving is required to prevent the data storage pool from becoming full and stopping further data acquisition. Security review and server access management are also required to avoid unauthorized access.

Since all the components of the information system used in the digital microscopy laboratory are interconnected via a network, the *reliability, availability*, and *speed* of the network become limiting factors on the performance of the entire system. For networks, reliability refers to the quality of the data connection, availability refers to the amount of time that the network is up and useable, and the speed refers to the amount of information that the network is capable of transmitting per unit time. A range of network technologies exist that deliver different levels of reliability and speed. Most laboratories begin with Ethernet (10 Mbps, shared) and move to faster systems as they see performance bottlenecks. Bottlenecks arise soon after the conversion to digital begins. Users transmitting 2 MB images soon appreciate speeds (100 Mbps and greater) that were previously reserved for connecting large campus networks.

When evaluating proposed network upgrades, consider how fast each network operates and whether this bandwidth is guaranteed to each port (switched) or shared between all the users on the network. In general, the higher the bandwidth guaranteed to each port, the better. We currently use a combination of ethernet (10 Mbps, shared), switched ethernet (10 Mbps, dedicated), and FDDI (100 Mbps, shared) interconnected as shown in Figure 4. As our bandwidth requirements grow, it is anticipated that switched FDDI and ultimately Asynchronous Transfer Mode (ATM, 155 Mbps, dedicated) will be installed to the desktop.

Proc. Microscopy and Microanalysis 1995, edited by G.W. Bailey, M.H. Ellisman, R.A. Hennigar, and N.J. Zaluzec
Copyright © 1995 MSA. Published by Jones and Begell Publishing, 79 Madison Ave., New York, NY 10016

Although the systems described above are adequate for the data types handled today (images, spectra, and reports), digital microscopy is rapidly evolving into remote microscopy. As systems for TelePresence Microscopy constructed as part of an Electronic Laboratory (E-lab) mature, bi-directional streams of image data and audio will further increase the requirements for both networks and storage systems.[1] It is impossible to accurately estimate performance requirements for the Electronic Laboratory. Certainly laboratories that have gradually improved their data storage systems and data networks will be closest to operating an Electronic Laboratory. However, improving the speed of data network facilities and the capacity/structure of data storage systems can require significant funds and long lead times. Microscopists should begin communicating the vision of these new systems to managers and support staff while planning to meet new requirements today.

References

1. N.J. Zaluzec et al, *LabSpace/TPM–A National Electronic Laboratory Infrastructure Initiative,* Argonne National Laboratory (1994).

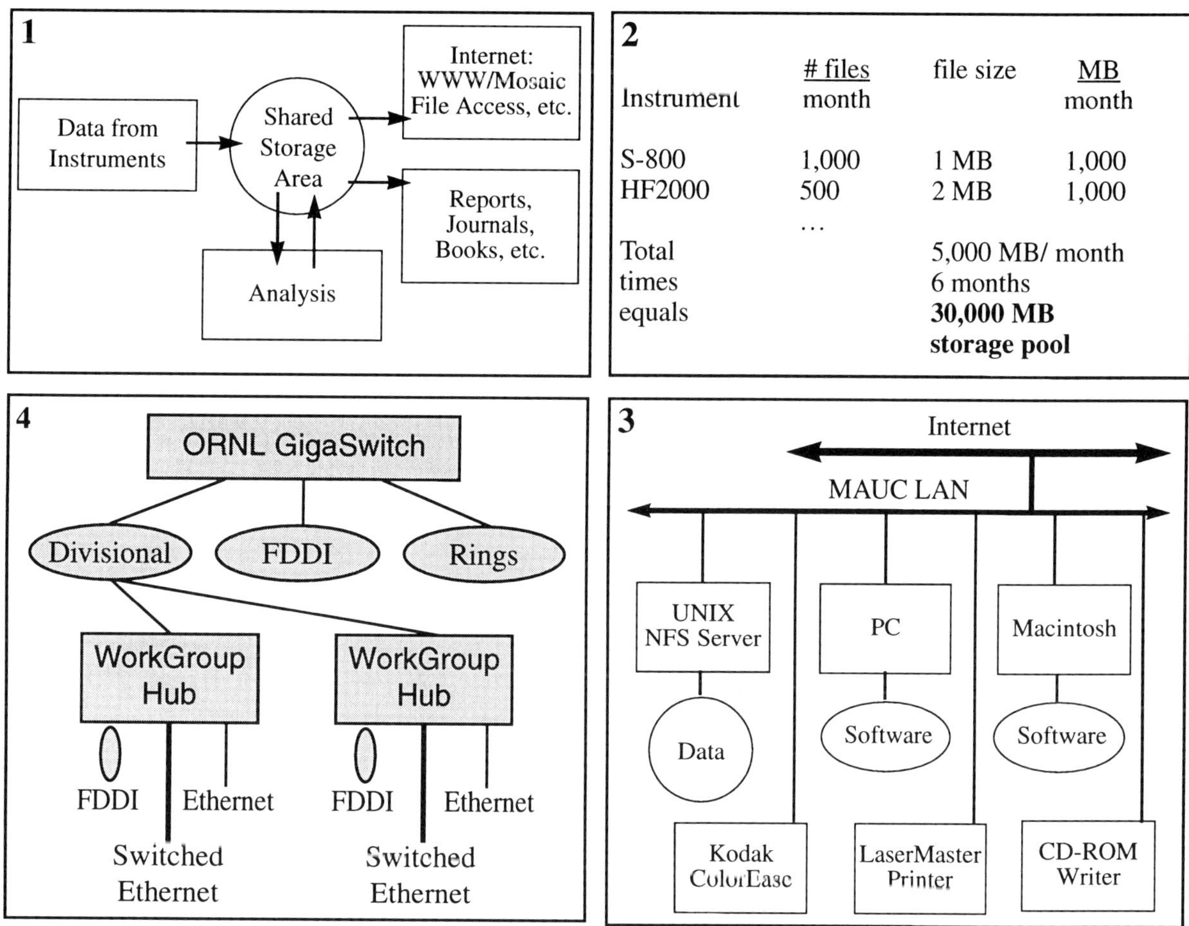

FIGURE 1.—Overview of information flows handled by the MAUC storage architecture.
FIGURE 2.—Sample calculation for determining the size of on-line storage requirements for the MAUC system.
FIGURE 3.—Diagram of the implementation of the MAUC storage architecture for digital microscopy data.
FIGURE 4.—Diagram showing the current structure of MAUC data network.

QUANTITATIVE DETERMINATION OF MODULATION TRANSFER FUNCTION AND DETECTION QUANTUM EFFICIENCY FOR STANDARD AND ANTI-REFLECTION YAG SCINTILLATOR SLOW-SCAN CCD-CAMERA

A.L. Weickenmeier, W. Nüchter, and J. Mayer

Max-Planck-Institut für Metallforschung, Institut für Werkstoffwissenschaft, Seestraße 92, D-70174 Stuttgart, Germany

Introduction

The modulation transfer function (MTF) of a slow-scan CCD camera (Gatan modell 679 attached to our Zeiss EM 912 Omega) has been determined with high precision for the standard and the new anti-reflection YAG scintillator. It is shown that deconvolution of experimental patterns allows the reconstruction of image details down to pixel size. From the analysis of deconvoluted noise patterns we found that for this camera the detection quantum efficiency (DQE) is 0.6 in the typical working range of 300 to 3000 electrons per pixel and does not depend on the scintillator. A full discussion of this work can be found in [1].

Experiment

To determine the point spread function (PSF) we punched a hole in a sheet film holder which then was partially covered by one or two knife edges (slit). These masks were projected onto the scintillator. About 600 line scans perpendicular to the edge or slit were extracted and averaged to reduce noise (Fig. 1). The resulting profiles represent an ideal step function (knife edge) or top-hat function (slit) convoluted with the PSF projected onto one co-ordinate axis[1]. To extract the PSF we convoluted the step or top-hat function with a model function depending on a set of fit parameters and minimized the difference between the experimental and theoretical profiles with respect to the parameters. From the PSF we finally obtained the MTF by Fourier transformation, results are depicted in Fig. 2. To check the quality of the PSF we deconvoluted our experimental profiles (Fig. 3). As can be seen we are able to perfectly restore the step or top-hat function from the blurred experimental data by deconvolution. It also evident that the anti-reflection YAG shows much less point spread and is clearly a big improvement compared to the standard YAG. The DQE has been determined by recording images with homogeneous illumination of the CCD camera. Gain normalization and dark current subtraction were applied. From the unprocessed (i.e. not deconvoluted) patterns we obtained DQE values of 10 and 3 for the standard and anti-reflection scintillator, respectively, whereas the DQE by definition cannot exceed one. These high values result from the high spatial frequency filter behavior of the camera according to the MTF. The true DQE of 0.6 for electron doses in the range of 300 to 3000 electrons per pixel was revealed by deconvoluting the patterns before analyzing the noise statistics and did not depend on the scintillator.

Discussion

The MTF for the same camera type with standard scintillator has also been measured by Zuo[2] and de Ruijter and Weiss[3]. Our results are in good agreement with those of de Ruijter and Weiss while Zuo found a much worse high spatial frequency behavior, probably due to the low quality of the

Proc. Microscopy and Microanalysis 1995, edited by G.W. Bailey, M.H. Ellisman, R.A. Hennigar, and N.J. Zaluzec

aperture he used to generate a step function. The anti-reflection YAG offers superior image quality in terms of MTF. However, for quantitative evaluation of images recorded with the CCD camera a proper deconvolution still seems to be indispensable. The DQE value of 0.6 is significantly lower than the value of 0.99 found by de Ruijter and Weiss[3] for the standard YAG. This value was obtained without deconvolution, however, and is thus to be compared to our value of 10.

References

1. A.L. Weickenmeier et. al., *Optik*, accepted for publication.
2. J. M. Zuo, *Proc. 13th Int. Congr. on Electr. Microsc.* Vol. 1 (1994)215.
3. W.J. de Ruijter and J.K. Weiss, *Rev. Sci. Instrum.* 63(1992)4314.

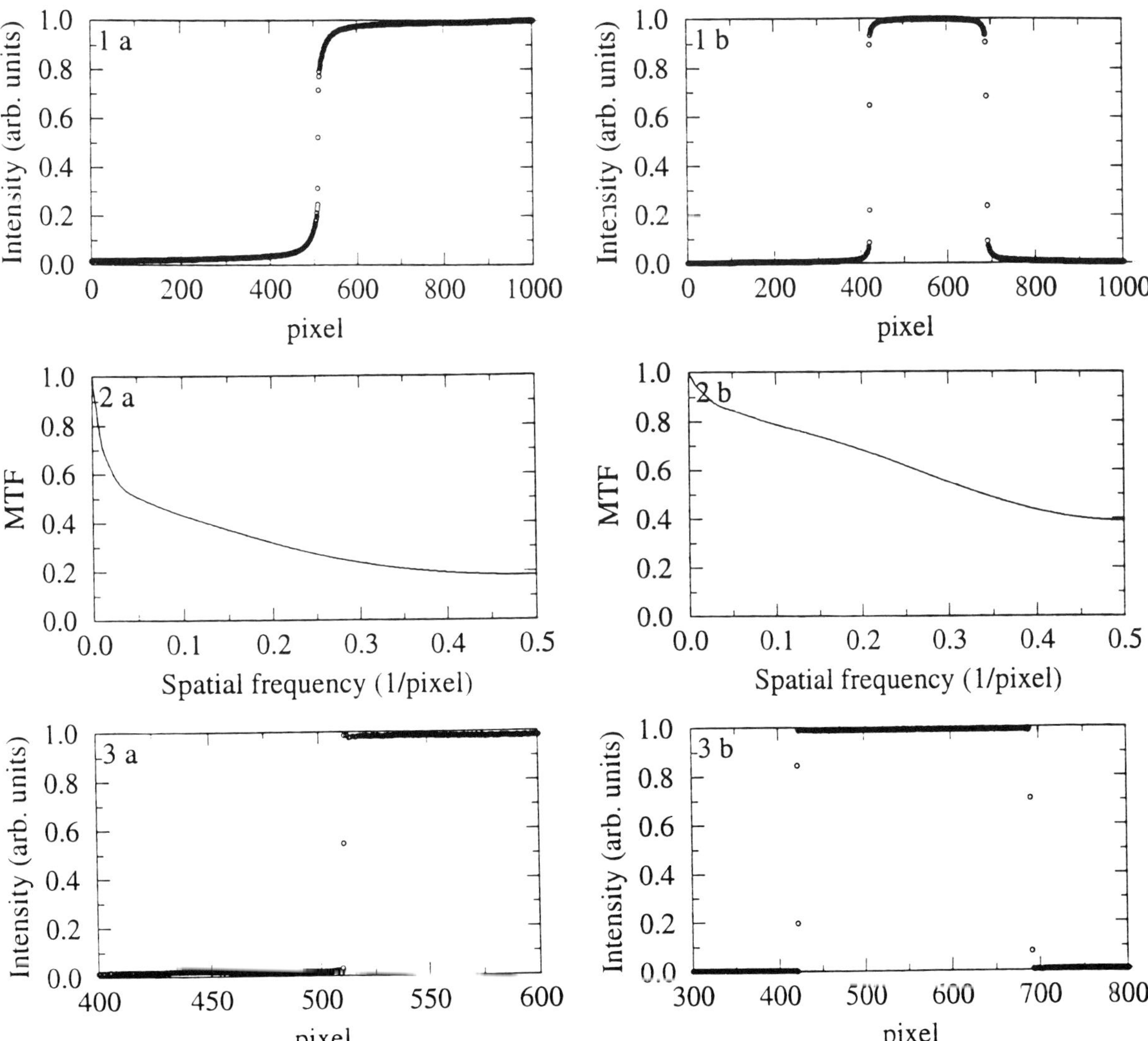

FIG. 1 (a) Experimental profiles for knife edge (standard YAG) and (b) slit (anti-reflection YAG)
FIG. 2 (a) MTF for the standard YAG and (b) anti-reflection YAG
FIG. 3 (a) Deconvoluted knife edge profile and (b) deconvoluted slit profile

ADVANCED COMPUTING FOR ULTRA-HIGH RESOLUTION IN TEM

A. Thust and W. M. J. Coene

Philips Research Laboratories, Prof. Holstlaan 4, 5656 AA Eindhoven, The Netherlands

1. Introduction

Attempts to push the resolution of electron microscopes towards 1 Å follow presently two different strategies. One approach takes advantage of the short electron wavelength provided by high-voltage instruments ($E > 1$ MeV) resulting in an "interpretable" point resolution close to the information limit. An alternative strategy is based on the idea to extend the information limit of conventional medium-voltage instruments ($E \approx 200 - 300$ keV) by taking advantage of the excellent coherence properties of field-emission guns (FEG). In the latter approach, however, the gain of extra information beyond the "interpretable" point resolution is of no direct use for structure interpretation. Generally, the interpretability of single high-resolution images suffers from a loss of phase information and from contrast-delocalization effects, the latter being caused by the spherical aberration and the defocussing of the objective lens. These delocalization effects become drastically apparent when aiming at the ultra-high resolution regime ($d < 1.5$ Å) which is routinely accessible with medium-voltage FEG-TEMs.[1]

2. Focal series reconstruction

In order to make use of the full information up to the information limit of a microscope, a series of approximately 20 images is recorded from the object using equidistantly spaced values of the objective lens defocus. Thereby it is guaranteed that both amplitude *and* phase information originating from the object plane are stored in the recorded data set, albeit in a highly encoded form. The goal of the algorithms described in the following is to decode the recorded data and to retrieve the wavefunction at the exit plane of the object (EPW) which is free from all imaging artifacts caused by the observation instrument.

3. Algorithms and implementation

The Philips/BRITE-EURAM reconstruction software consists of two different algorithms which are used sequentially in a sort of tandem package.[2] A quite reliable initial guess for the EPW is obtained by employing the *Paraboloid Method* (PAM) which is originally a linear reconstruction method and which has been extended towards a perturbational treatment of non-linear contributions to the image contrast.[3,4] The solution obtained by the application of the PAM algorithm serves then as input for the so called *Maximum Likelihood* (MAL) algorithm, of which the basic concept has been developed by Kirkland.[5] The MAL algorithm follows a recursive least-squares scheme which takes into account the full formalism of partially coherent non-linear contrast theory. By making advantage of (a) a factorized approach for modelling the spatial coherence of a FEG, of (b) a newly designed explicit focal averaging technique for modelling temporal coherence effects and of (c) a new algorithmic approach treating time-consuming calculations logically in parallel, the computation speed is enhanced by 3-4 orders of magnitude with respect to Kirkland's initial formulation.[6] On a fast workstation (≈ 100 MFlops) reconstructions of frames comprising typically 768^2 pixels can be obtained now within a few minutes of CPU time.

4. Correction for residual aberrations

As any other phase-retrieval technique, focal series reconstruction allows one to remove residual aberrations from the EPW which may be due to beam misalignment, 2- and 3-fold astigmatism. The parameters describing these aberrations (6 in total) can be determined from an extra beam-tilt series taken from a neighbouring amorphous area yielding a unique "fingerprint" of the optical recording conditions. An alternative way for the parameter determination is to make use of prior knowledge about the object itself, e.g. about special symmetry properties of crystalline regions.[7] A phase plate compensating for the residual aberrations can be directly incorporated into the reconstruction or – much more effectively in the case of a knowledge-based determination – applied a-posteriori to the reconstructed EPW.

Proc. Microscopy and Microanalysis 1995, edited by G.W. Bailey, M.H. Ellisman, R.A. Hennigar, and N.J. Zaluzec

5. Experimental example

Fig. 1 shows a small area of the first image belonging to a focal series taken with a Philips CM300 SuperTWIN FEG microscope. This area contains in its center the core of a 60° dislocation located at a GaAs/InGaAs interface. In Fig. 2 the correcting phase plate applied to the reconstructed EPW is displayed (Fourier space). Fig. 3 shows the uncorrected and Fig. 4 the corrected phase of the reconstructed EPW. Note that the typical dumbbell contrast ($d = 1.4$ Å) is prominently visible after the correction.

References

1. W. Coene and A.J.E.M. Jansen, *Scanning Microsc. Suppl.* 6 (1992) 379.
2. A. Thust et al,. submitted to *Ultramicroscopy*.
3. W.O. Saxton, in: *Advances in Electronics and Electron Physics Suppl.* 10 (1978) 236.
4. D. Van Dyck et al., *Optik* 93 No. 3 (1993) 103.
5. E.J. Kirkland, *Ultramicroscopy* 15 (1984) 151.
6. W.M.J. Coene et al., submitted to *Ultramicroscopy*.
7. M. Lehmann and H. Lichte, *Proc. 13th Int. Congr. for Electron Microscopy*, Paris, Vol. 1 (1994) 293.
8. The experimental cooperation with M. Lentzen, KFA Jülich, is gratefully acknowledged.

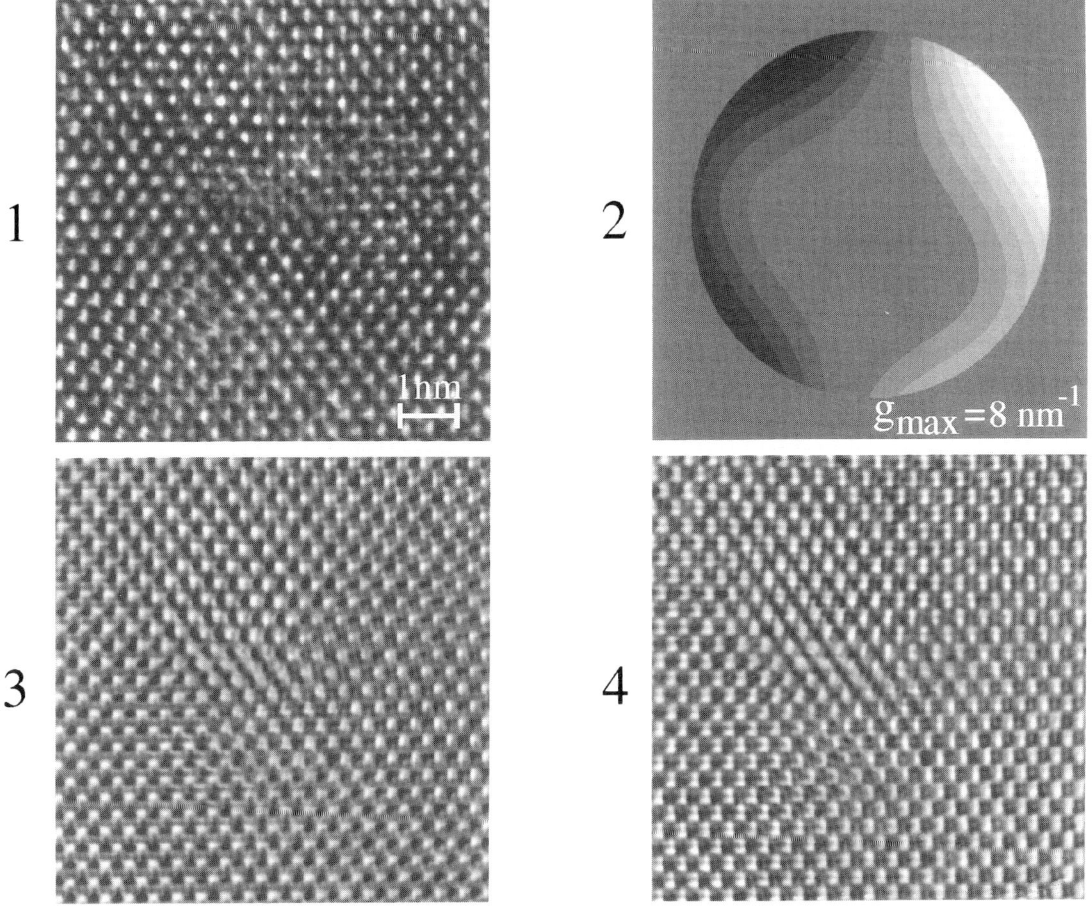

Fig. 1 - Small area of first image of a focal series recorded with a Philips CM300 FEG microscope. The core region of a 60° dislocation located at a GaAs/InGaAs interface is imaged along the [110] zone axis.
Fig. 2 - Phase plate used for residual aberration correction. One graylevel represents a phase shift of $\pi/2$.
Fig. 3 - Phase of reconstructed EPW without correction for residual aberrations.
Fig. 4 - Phase of reconstructed EPW after correction with the compensating phase plate shown in Fig. 2.

ADVANCES IN IMAGE SIMULATION FOR HIGH RESOLUTION TEM

Michael A. O'Keefe

National Center for Electron Microscopy, University of California, LBL B72, Berkeley, CA 94720

The original high-resolution transmission electron microscope (HRTEM) image simulation program was written as a tool to confirm interpretation of HRTEM images of niobium oxides.[1,2] Thorough testing on known structures showed that image simulation could reliably duplicate the imaging process occurring in the HRTEM, and could thus be confidently used to interpret images of unknown structures.[3] Mainstream application of image simulation to routine structure determination by HRTEM was ushered in by the establishment of the wide applicability of the SHRLI (simulated high-resolution lattice image) programs.[4,5] Structure determination of the mineral takéuchiite by HRTEM and image simulation was the first such determination accepted by the IUCr without x-ray data.[6] Of course, once the reliability of image simulation had been established, it was realized that the technique could be put to work for applications other than structure determination. Early on, simulations were used to explore various HRTEM imaging parameters, including specimen ionicity, validity of the projection approximation, and the resolution-limiting effects of incident-beam convergence.[7-10] Since the inception of HRTEM image simulation, its range of uses has continued to expand, and so has the number of programs available; distribution of the SHRLI code spawned improved versions as well as some new programs.[11-16]

Advances in various image simulation programs have occurred on several fronts. Importantly, there has been a steady improvement in modeling the process of HRTEM image formation. It was soon realized that image resolution was limited, not solely by the physical objective aperture, but by partial coherence effects due to incident beam convergence and spread of focus. HRTEM incident-beam convergence was first included in simulation programs in the form of a real-space average,[10] then as a transmission cross-coefficient with accompanying spread of focus.[17] Specimen vibration was added to account for observed damping of the experimental power spectrum,[17] and higher-order aberrations such as three-fold astigmatism continue to be added to the simulation process as microscope resolutions improve sufficiently to warrant their addition.[18] Improved modeling accuracy of the HRTEM imaging process enables simulations to be used to characterize HRTEMs,[19] and to explore the effects of various imaging parameters on the image.[20] HRTEM image simulation has also become a valuable tool for testing the accuracy of techniques like measurement of grain-boundary volume expansion,[21] and two- and three-dimensional HRTEM image reconstruction schemes.[22,23]

The most important recent advance in HRTEM image simulation is the result of quantitative matching of experimental and simulated images. Instead of relying on a visual comparison to judge the simulated image that fits best to the experimental one, the matching process is handled objectively by the computer. Once the experimental image has been digitized, it is compared with simulations and a "goodness-of-fit" parameter computed.[24] For periodic specimens, this comparison can be carried out in real space, by comparing the intensities of the images pixel by pixel,[25] or in reciprocal space by comparing components of the image intensity spectrum.[26] The matching process can be automated by allowing the computer to vary its values of the experimental imaging parameters within designated ranges of uncertainty while searching for the best simulated image to optimize the fitting parameter. In this manner the fit of the simulation can be improved to yield the best possible matching image that can be produced from the model crystal structure.[27] Alternatively, by using a specimen with a known structure to produce the experimental image, it is possible to obtain very accurate measures of the experimental imaging parameters.[27,28]

A two-step process can be used for refinement of an unknown structure in the case where an experimental image is available containing the unknown structure close to a known one. First an image match for the

Proc. Microscopy and Microanalysis 1995, edited by G.W. Bailey, M.H. Ellisman, R.A. Hennigar, and N.J. Zaluzec
Copyright © 1995 MSA. Published by Jones and Begell Publishing, 79 Madison Ave., New York, NY 10016

known structure is used to establish the imaging parameters, then these parameters are fixed and images of the unknown structure are simulated and matched with experiment while modifying the model of the unknown structure. For example, to refine the structure of a Σ99 grain boundary in aluminum, a defocus-thickness image map of perfect [110] aluminum images was first simulated. The map was used to choose the best match with the experimental aluminum image, and thus to establish the experimental defocus and specimen thickness. Images of the grain boundary were then simulated at these values of defocus and specimen thickness, and the model of the grain boundary was modified to achieve the best match of the simulation with the experimental grain boundary image.[29]

Automated quantitative image matching can be used to automate the two-step structure refinement process. King and Campbell have used an automated non-linear least-squares image matching process to establish image parameters from an image of niobium (all the parameters, not just defocus and thickness), and then refined the structure of the Σ5 grain boundary visible in the image, starting from an initial model of the boundary and allowing the program to vary atom positions to achieve the best possible match and thus produce a refined structure model.[27,30,31] A similar procedure, using a Gauss-Seidel bivariant search method was used by Möbus and Rühle to refine the structure of a niobium-sapphire interface.[32] Although automated image matching, and especially automated structure refinement, require powerful computing hardware, they will become more accessible as improvements in hardware continue to appear.[33]

1. J.G. Allpress, E.A. Hewat, A.F. Moodie and J.V. Sanders, *Acta Cryst.* A**28** (1972) 528-536.
2. D.F. Lynch and M.A. O'Keefe, *Acta Cryst.* A**28** (1972) 536-548.
3. J.G. Allpress and J.V. Sanders, *J. Appl. Cryst.* **6** (1973) 165-190.
4. M.A. O'Keefe, P.R. Buseck and S. Iijima, *Nature* **274** (1978) 322-324.
5. M.A. O'Keefe and P. R. Buseck, *Transactions of the ACA* **15** (1979) 27-46.
6. Jan-Olov Bovin, M. O'Keeffe and M.A. O'Keefe, *Acta Cryst.* A **35** (1981) 32-46.
7. G.R. Anstis, D.F. Lynch, A.F. Moodie and M.A. O'Keefe, *Acta Cryst.* A**29** (1973) 138-147.
8. M.A. O'Keefe, *Acta Cryst.* A**29** (1973) 389-401.
9. D. F. Lynch, A. F. Moodie and M.A. O'Keefe, *Acta Cryst.* A**31** (1975) 300-307.
10. M.A. O'Keefe and J.V. Sanders, *Acta Cryst.* A**31** (1975) 307-310.
11. K. Ishizuka and N. Uyeda, *Acta Cryst* A**38** (1982) 740-749.
12. R. Kilaas, *45th Ann. Proc. EMSA* (1985) 66-69.
13. P.A. Stadelmann, *Ultramicroscopy* **21** (1987) 131-146.
14. Michael A. O'Keefe and Roar Kilaas, *Scanning Microscopy suppl. 2* (1988) 225-244.
15. P.G. Self, M.A. O'Keefe, P.R. Buseck and A.E.C. Spargo, *Ultramicroscopy* **11** (1983) 35-52.
16. M.A. O'Keefe, *52nd Ann. Proc. MSA*, New Orleans, Louisiana (1994) 394-395.
17. M.A. O'Keefe, *37th Ann. Proc. EMSA*, San Antonio, Texas (1979) 556-557.
18. W.O. Saxton, *52nd Ann. Proc. MSA*, New Orleans, Louisiana (1994) 908-909.
19. R.M. Fisher and T. Imura, *Ultramicroscopy* **3** (1978) 3-18.
20. D.J. Smith et al, *Ultramicroscopy* **11** (1983) 263-282.
21. M.I. Buckett and K.L. Merkle, *Ultramicroscopy* **56** (1994) 71-78.
22. W. Dong et al, *Nature* **355** (1992) 605-609.
23. K.H. Downing, Hu Meisheng, H.-R. Wenk and M.A. O'Keefe, *Nature* **348** (1990) 525-528.
24. Adam Rae Smith and LeRoy Eyring, *Ultramicroscopy* **8** (1982) 65-78.
25. W.E. King and B.S. Lawver, in *Microbeam Analysis,* Ed. D.G. Howitt (SF Press, 1991) 217.
26. A. Thust and K. Urban, *Ultramicroscopy* **45** (1992) 23.
27. W.E. King and G.H. Campbell, *Ultramicroscopy* **51** (1993) 128.
28. R. Bierwolf and M. Hohenstein, *Ultramicroscopy* **56** (1994) 32-45.
29. M.A. O'Keefe et al, *Mat. Res. Soc. Symp. Proc.* **159** (1989) 453-458.
30. W.E. King, M.A. O'Keefe & G.H. Campbell, *52nd Ann. Proc. MSA*, New Orleans (1994) 716-717.
31. W.E. King and G.H. Campbell, *Ultramicroscopy* **56** (1994) 46-53.
32. G. Möbus and M. Rühle, *Ultramicroscopy* **56** (1994) 54-70.
33. Work supported by the Director, Office of Energy Research, Office of Basic Energy Sciences, Material Sciences Division of the U.S. Department of Energy, under contract No. DE-AC03-76SF00098.

SIMULATION OF HIGH RESOLUTION ANNULAR DARK FIELD STEM IMAGES

E. J. Kirkland

School of Applied and Engineering Physics, Cornell University, Ithaca, NY 14853

In a STEM an electron beam is focused into a small probe on the specimen. This probe is raster scanned across the specimen to form an image from the electrons transmitted through the specimen. The objective lens is positioned *before* the specimen instead of *after* the specimen as in a CTEM. Because the probe is focused and scanned before the specimen, accurate annular dark field (ADF) STEM image simulation is more difficult than CTEM simulation. Instead of an incident uniform plane wave, ADF-STEM simulation starts with a probe wavefunction focused at a specified position on the specimen. The wavefunction is then propagated through the specimen one atomic layer (or slice) at a time with Fresnel diffraction between slices using the multislice method.[1] After passing through the specimen the wavefunction is diffracted onto the detector. The ADF signal for one position of the probe is formed by integrating all electrons scattered outside of an inner angle large compared with the objective aperture. The bright field (BF) signal is formed by the electrons that have been scattered into angles small compared with the objective aperture. (By reciprocity BF-STEM with a small collector angle is equivalent to BF-CTEM with a small illumination angle.) The multislice calculation is repeated for each position of the incident focused probe and must also be accurate at large scattering angles for the ADF signal.[2] This leads to a large amount of computer time when compared to CTEM. A smaller number of pixels can be used in the probe than the specimen slices for improved speed.

The YBCO superconductor has orthorhombic symmetry (a=0.38231, b=0.38864, c=1.16807nm).[3] The b axis of the crystal will be along the optic axis of the microscope. The specimen naturally divides into two rectangular layers A and B, each 0.19nm thick (Fig.1) with an ABAB stacking sequence. Micrographs were simulated for 100, 200, 300 and 400keV (Fig.2). Scherzer conditions were used for the objective aperture and defocus. Spherical aberration was fixed at 1.3mm. There are several different sampling requirements. The pixel size must be small enough in both real space and reciprocal space and the largest scattering angle must include the FOLZ line for the ADF signal. Each layer was sampled at 1024x1024 pixels with a supercell size of 9a x 3c. The probe wavefunction was 512x512 pixels. The projected atomic potentials were calculated by cubic spline interpolation of the electron scattering factor tables.[4,5,6] The specimen was 25 unit cells (10nm) thick. Each image was calculated for 12x40 pixels in one unit cell and expanded symmetrically to display an image area of 6a x 2c. This figure took about 38 hr. of time on an IBM RS6000/560 computer. Note that several detector geometries can be calculated at the same time. Heavy atoms should appear white in ADF with some variation in intensity due to atomic number and heavy atoms should appear black in BF.

The ADF-STEM images have a noticeably better resolution than the CTEM images, particularly at low voltage. ADF-STEM is primarily amplitude contrast and well suited for medium to high atomic numbers . It also shows much less variation with thickness and defocus than CTEM although it is not totally immune from these effects. For this specimen 100keV ADF-STEM is comparable to 400keV CTEM in resolution. Even at 400keV CTEM shows contrast reversals but 100keV ADF-STEM does not. 100keV has much less displacement radiation damage than 400keV and is more economical. CTEM may have some advantages for low atomic number specimens however ADF-STEM is an interesting alternative to CTEM for materials with medium to high atomic numbers.[7]

[1] P.Goodman, A.F. Moody, Acta. Cryst. A30 (1974) 280.
[2] E.J.Kirkland et al, Ultramicroscopy 23 (1987) 77.
[3] M.A Beno et al, Appl. Phys. Lett. 51(1987) 57.
[4] P.A.Doyle, P.S.Turner, Acta. Cryst. A24 (1968) 390.
[5] P.A.Doyle J.M.Cowley, in: Int'l Tables for Cryst. IV (Kynoch Press, 1974) Table 2.4.6A.
[6] A.G.Fox et al, Acta Cryst. A45 (1989) 786.
[7] This work was supported by DOE Grant No. DE-FG02-87ER45322.

Proc. Microscopy and Microanalysis 1995, edited by G.W. Bailey, M.H. Ellisman, R.A. Hennigar, and N.J. Zaluzec

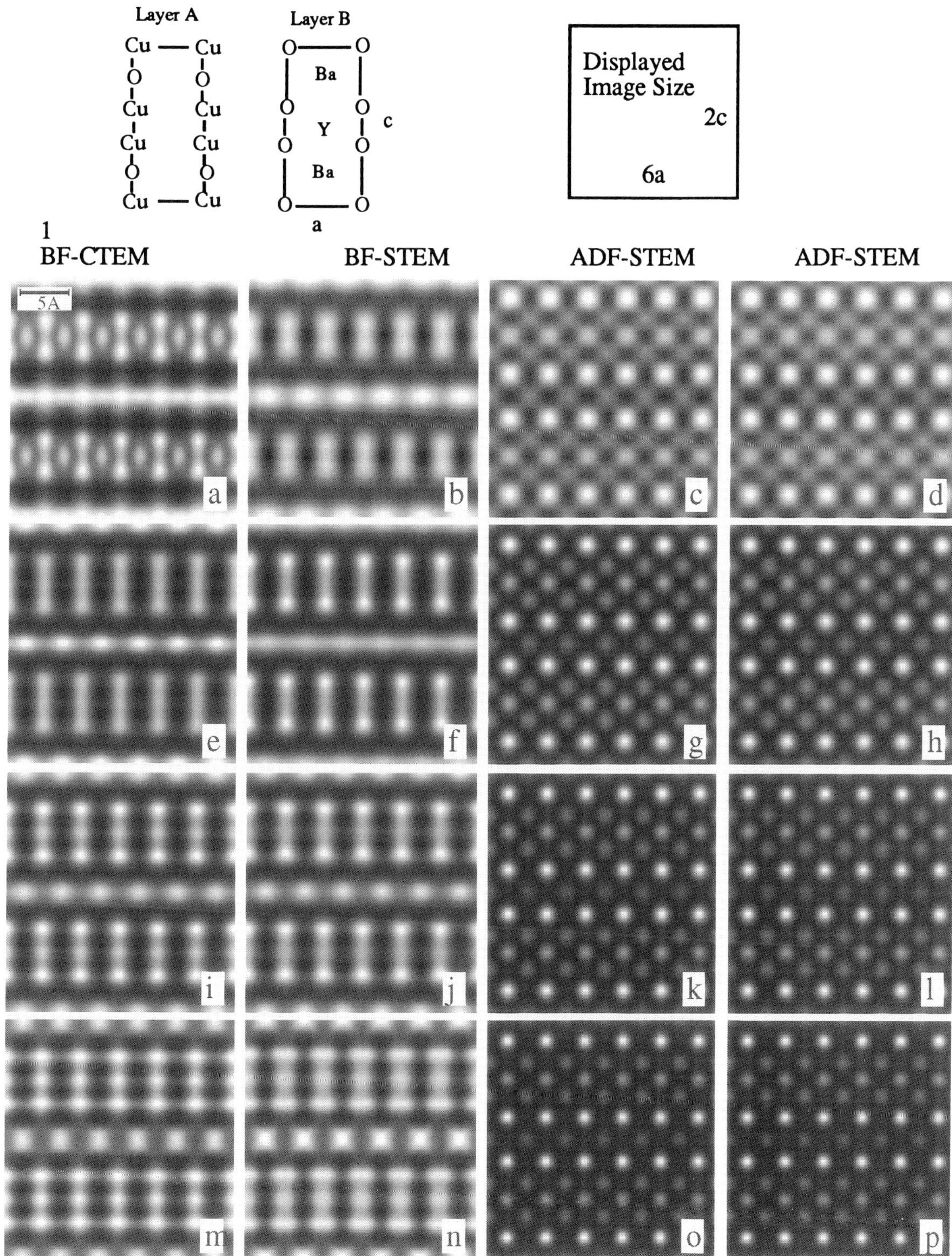

Figure 1. Schematic of the structure of the YBCO superconductor following Beno et al[3]

Figure 2. Simulated images for a YBCO thickness of 10nm. a-d), 100kev; e-h) 200keV; i-l) 300keV; m-p) 400keV; a,e,i,m), BF-CTEM; b,f,j,n) BF-STEM detector < (obj.apert)/3; c,g,k,o) ADF-STEM detector > 4 (obj. apert.); d,h,l,p) ADF-STEM detector > 10(obj.apert).

SEMI-AUTOMATED METHODS FOR THREE-DIMENSIONAL RECONSTRUCTION OF
MACROMOLECULAR COMPLEXES.

B. Carragher* and M. Whittaker**

*Beckman Institute and Department of Cell and Structural Biology, University of Illinois at
Urbana-Champaign, Urbana, IL 61821.
**Department of Cell Biology, The Scripps Research Institute, La Jolla, CA 92093.

Techniques for three-dimensional reconstruction of macromolecular complexes from electron
micrographs have been successfully used for many years[1]. These include methods which take
advantage of the natural symmetry properties of the structure (for example helical or icosahedral) as
well as those that use single axis or other tilting geometries to reconstruct from a set of projection
images. These techniques have traditionally relied on a very experienced operator to manually
perform the often numerous and time consuming steps required to obtain the final reconstruction.
While the guidance and oversight of an experienced and critical operator will always be an essential
component of these techniques, recent advances in computer technology, microprocessor
controlled microscopes and the availability of high quality CCD cameras have provided the means
to automate many of the individual steps.

During the acquisition of data automation provides benefits not only in terms of convenience and
time saving but also in circumstances where manual procedures limit the quality of the final
reconstruction. This includes the collection of a set of projection images using single tilt axis
geometry, where the total dose required during manual focusing, recentering and collecting the data
set has, until recently, made this technique inapplicable to beam sensitive specimens. By
automating these steps and using a CCD camera for data collection, the cumulative dose can now
be kept low enough to apply this technique to ice-embedded specimens[2]. Automatic data collection
may also be applicable to other reconstruction geometries, for instance in the case of single
particles where individual images may be automatically identified and selected from an acquired
image. It is less clear how to acquire digital images of long filamentous structures at sufficiently
high resolution. The limiting factor at present is the size of the CCD chip which is typically 1024^2.
Although 2048^2 chips are starting to be available it is clear that it will be some time before the
resolution of CCD's matches that of film. While it may be possible to use the microprocessor
controlled microscope stage to montage individual arrays into a long digital image this has not yet
been demonstrated in practice and it is not clear whether this can be achieved under the required
low dose conditions.

Following data acquisition there are a series of steps required to align the data and determine the
reconstruction parameters before a three-dimensional volume can be calculated. The exact steps
depend on the particular reconstruction geometry. For instance in the case of a single axis
reconstruction these steps would include the alignment of the individual images in the tilt series and
the determination of the tilt axis and tilt angles. For a helical reconstruction the particle axis has to
be determined and layer lines must be identified from the Fourier transform of the straightened
filament (see Fig. 1). These procedures can be both time consuming and tedious if purely manual
methods are used. The judicious application of computational tools to automate many of these
steps effectively increases the number of reconstructions that can be performed in practice. In the
case of reconstructions of structures with a fixed geometry this increases the number of individual
particles that contribute to the final reconstructed volume and leads to an improved resolution in the
three-dimensional map (see Fig. 2). Finally, the implementation of automatic procedures facilitates
the application of iterative alignment steps in the processing, for example using an average 3-D
map to better define the orientation of the individual particles used to calculate the map.

Proc. Microscopy and Microanalysis 1995, edited by G.W. Bailey, M.H. Ellisman, R.A. Hennigar, and N.J. Zaluzec
Copyright © 1995 MSA. Published by Jones and Begell Publishing, 79 Madison Ave., New York, NY 10016

References

1.   J. Frank, in J. Frank, Ed., *Electron Tomography: Three-Dimensional Imaging with the Transmission Microscope*, Plenum Press, N.Y. (1987)1.
2. K. Dierkson et al., Ultramicroscopy 49(1993)109.

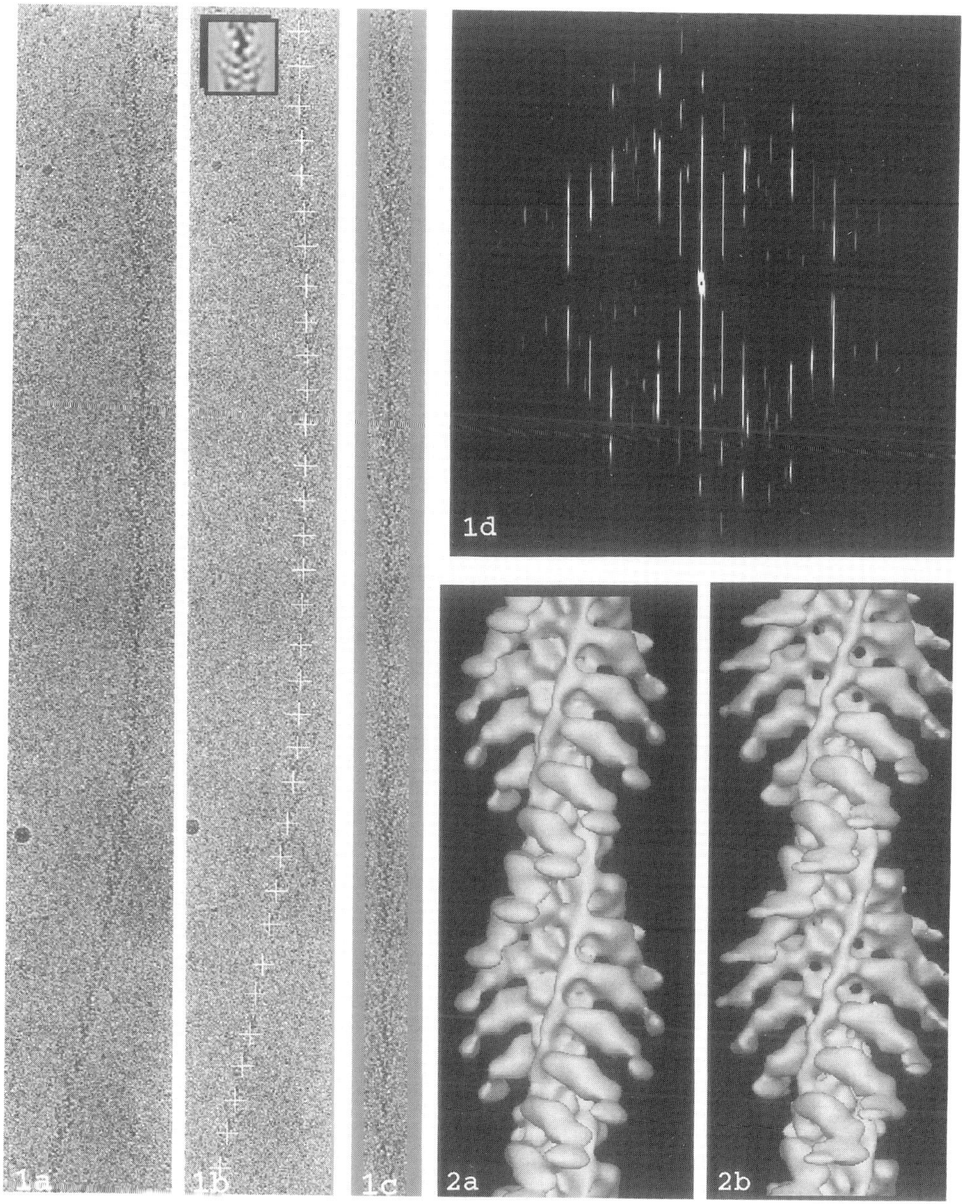

FIG. 1. --Steps in the processing of helical filaments. (a) Densitometered filament image. (b) The axis of the filament is identified by cross correlation to a template (shown as inset). (c) The filament is then straightened by mapping the cubic spline fitted to the axis points to a straight line and background corrected.   (d) Power spectrum of the straightened filament shown in (c).
FIG. 2. -- Surface views of a density map of a decorated thin filament calculated using (a) manual methods and (b) automated procedures.

COMBINING STRUCTURES FROM CRYO-ELECTRON MICROSCOPY AND X-RAY CRYSTALLOGRAPHY

P.L. Stewart* and G.R. Nemerow**

* Department of Molecular and Medical Pharmacology, UCLA School of Medicine, Los Angeles, CA 90095
** Department of Immunology, The Scripps Research Institute, La Jolla, CA 92037

Cryo-electron microscopy (cryo-EM) combined with three-dimensional image reconstruction techniques has produced structures of large macromolecular assemblies. Interpretation of low resolution (25-35 Å) cryo-EM density can be greatly enhanced by mapping in crystallographic structures of component molecules. Difference imaging between the cryo-EM structure of the human adenovirus particle and a capsid calculated from the crystal structure of the major structural protein, hexon, revealed numerous minor structural components in the viral capsid.[1] In addition, the atomic binding sites of the minor protein components were visualized on the crystallographic structure of hexon.

Current studies are focused on examining the structural events during adenovirus cell entry. The receptor that triggers adenovirus internalization has been shown to be αv integrin.[2] Cryo-EM is being used to solve the structure of adenovirus with bound Fab fragments from a monoclonal antibody that blocks binding of the virus to the receptor (Fig. 1). The antibody recognizes a linear epitope of nine amino acids that includes an Arg-Gly-Asp (RGD) integrin binding motif.[3] A preliminary, 70Å resolution, image reconstruction of the virus/Fab complex is shown in Fig. 2. After determination of a high resolution (25 Å) cryo-EM structure of the complex and deconvolution of the density, to correct for the effects of the contrast transfer function of the electron microscope, a crystallographic Fab structure may be mapped within the cryo-EM density. In this manner, the RGD virus/receptor interaction site will be visualized.[4]

References

1. P.L. Stewart et al., *EMBO J* 12 (1993) 2589.
2. T.J. Wickham et al., *Cell* 73 (1993) 309.
3. G.R. Nemerow et al., in preparation.
4. The authors gratefully acknowledge Dr. R. Milligan for use of his electron microscope during the initial stages of this project.

Proc. Microscopy and Microanalysis 1995, edited by G.W. Bailey, M.H. Ellisman, R.A. Hennigar, and N.J. Zaluzec
Copyright © 1995 MSA. Published by Jones and Begell Publishing, 79 Madison Ave., New York, NY 10016

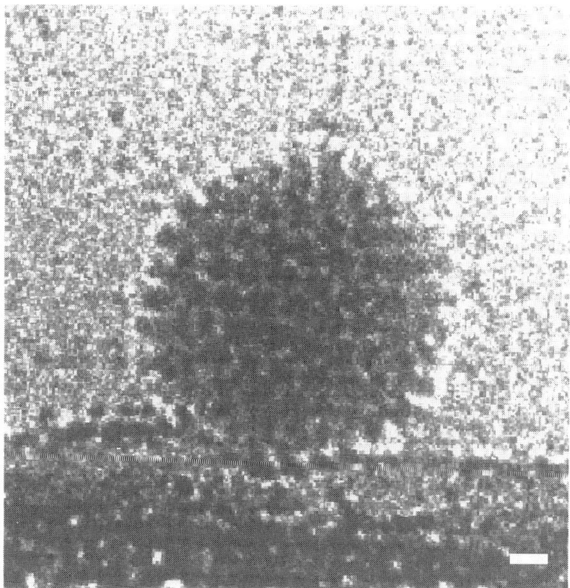

FIG. 1. Cryo-electron micrograph of human adenovirus with bound Fab fragments from a function-blocking monoclonal antibody. Fab density is observed at the base of the long protruding fibers. Bar = 100 Å.

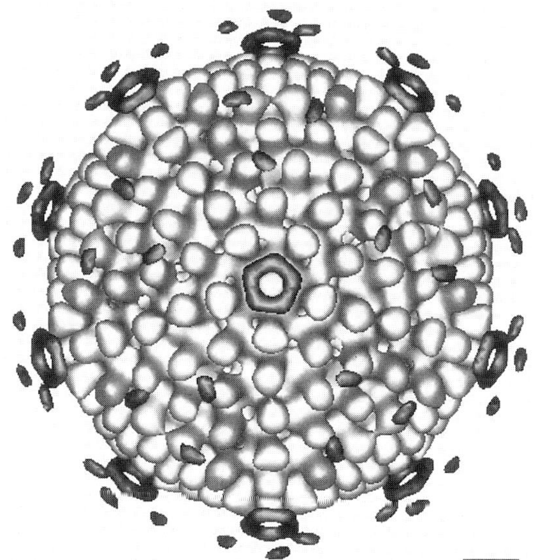

FIG. 2. Preliminary low resolution (70 Å) image reconstruction of adenovirus with bound Fab fragments based on images similar to that shown in Fig.1. Fab density (dark gray) is observed as rings floating above the vertices of the viral capsid. A higher resolution reconstruction will enable modeling of the Fab structure within the cryo-EM density and thus identification of the Arg-Gly-Asp (RGD) binding site of the antibody on the virus. Bar = 100 Å.

RECENT DEVELOPMENTS WITH BLIND DECONVOLUTION FOR 3D LIGHT MICROSCOPY

* **T.J. Holmes**, *S. Bhattacharyya, * **N. O'Connor, *H. Ancin, * **J. Gregorio, * **B. Roysam, ***D. Szarowski and * ****J.N. Turner

* Biomedical Engineering Department, Rensselaer Polytechnic Institute, Troy NY 12180
** AutoQuant Imaging Inc., Troy NY, 12180
*** Wadsworth Center for Laboratories and Research, NY State Dept. Health, Albany NY 12201

A novel deblurring method is under development for 3D light microscopic image deblurring[1], including the widefield fluorescence, confocal fluorescence and transmitted light brightfield modalities.

The main practical advantage of the approach is that it does not require the point spread function (PSF) to be measured. It produces a reconstruction of the PSF along with the deblurred image, which is possible because we mathematically constrain the PSF to have physically justified and reasonable properties, such as bandlimitedness, a missing frequency cone (for widefield) and nonnegative values.

Recent developments have included robustizing the reconstruction for confocal data, so that it will repeatedly perform well under various adverse conditions and for a wide variety of sample types. Among other refinements, additional PSF mathematical constraints are being integrated.

Evaluation testing is underway with known fabricated test objects. An example is shown in Fig. 1. A specially drawn glass micropipette tip was filled with 0.2% eosin, mounted in Refractive Index Liquid (n=1.460; Cargille Labs, NJ) and scanned with a BioRad MRC 600 confocal microscope having a 40X, 1.0 N.A. oil objective lens. The field size shown is 15.1x15.1x12.8 μm (x,y,z; 64x64x64 sampling). Fig. 2 shows a segmented version of the same data set, using a totally automatic segmentation scheme, which is designed for the automatic measurement of dendrite volumes in neurons[2]. Both Fig's. 1 and 2 show noise and axial smear. This smear will cause erroneously large volume measurements which are important for any quantitative study and critical to studies in neuroscience where accurate determination of dendritic and axonal geometric parameters are required[3]. A true representation of the micropipette, needed for exact volume measurements, would show the inner dimension of the micropipette along z as being the same as along x and y. Fig's. 3 and 4 show the reconstructed image and its volume segmentation, respectively. These are a more exact representation of the micropipette tip, which will result in more exact volume measurements.

1. T.J. Holmes et al. "Light Microscopic Images Reconstructed by Maximum Likelihood Deconvolution," <u>Handbook of Biological Confocal Microscopy</u> 2nd Ed., J. Pawley, Plenum, in press, 1995.
2. A.R. Cohen, B. Roysam and J.N. Turner, "Automated Tracing and Volume Measurements of Neurons from 3-D Confocal Fluorescence Microscopy Data," J. of Micr., 173(2), Feb. 1994.
3. J.N. Turner et al. "Three-Dimensional Imaging and Image Analysis of Hippocampal Neurons: Confocal and Digitally Enhanced Wide Field Microscopy," Micr. Res. and Tech., 29: 269-278, 1994.
4. The following funding sources are gratefully acknowledged: NSF DIR-9108492 (Wadsworth); NIH NCRR RR01219 (Wadsworth); NSF BES-94130371 (Rensselaer); NIH 1-R43-MH53691-01 (AutoQuant); NIH 1-R43-MH53692-01 (AutoQuant).

Proc. Microscopy and Microanalysis 1995, edited by G.W. Bailey, M.H. Ellisman, R.A. Hennigar, and N.J. Zaluzec
Copyright © 1995 MSA. Published by Jones and Begell Publishing, 79 Madison Ave., New York, NY 10016

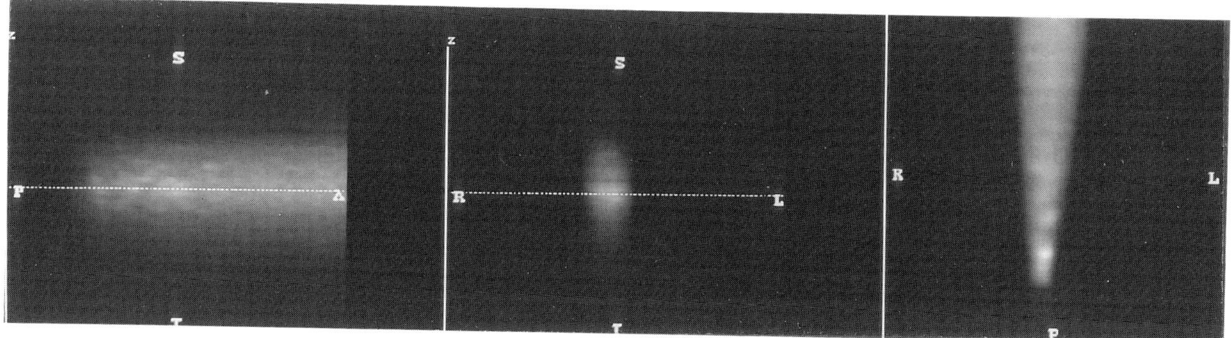

Fig. 1. Sagittal, coronal and transverse central cross sections, respectively: raw confocal data set.

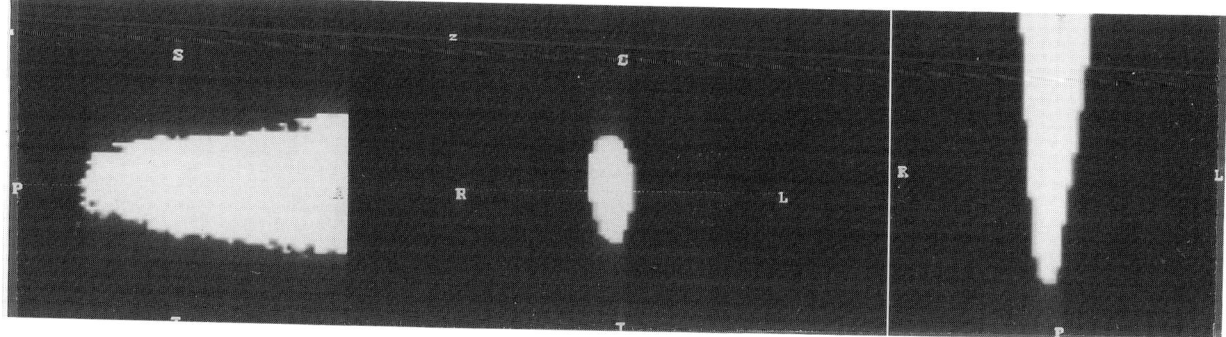

Fig. 2 Automatically segmented respective cross sections, corresponding to the raw data of Fig. 1.

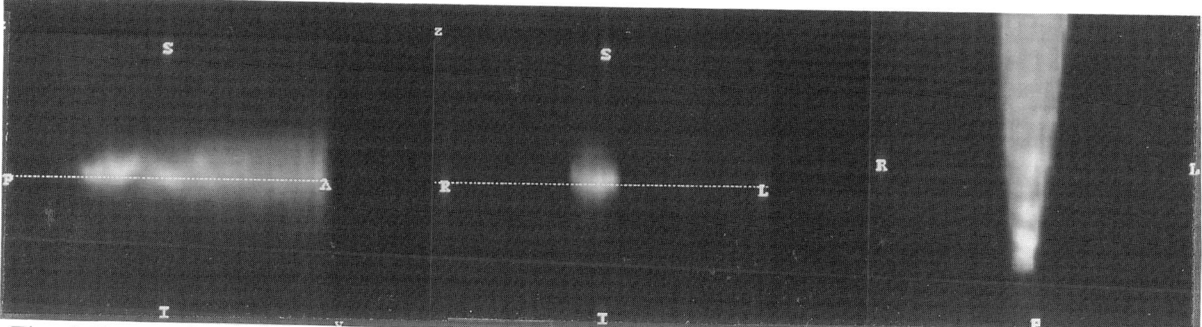

Fig. 3 Respective cross sections after 100 iterations of blind deconvolution.

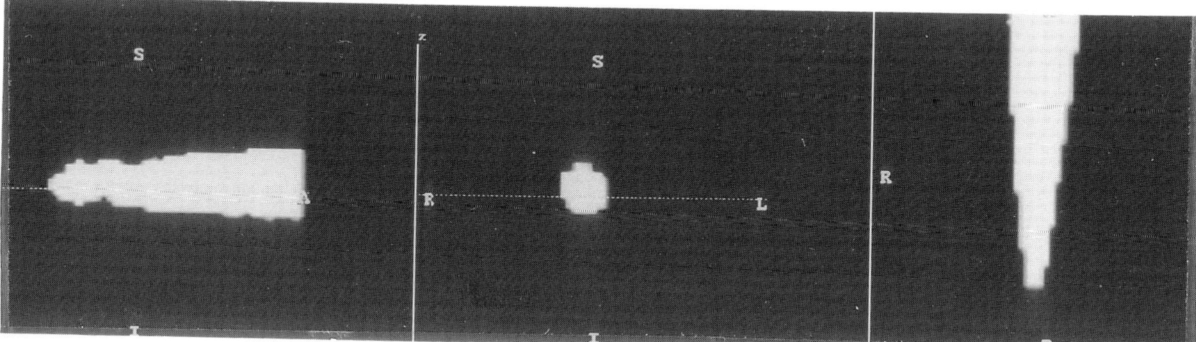

Fig. 4. Automatically segmented respective cross sections from the data of Fig. 3.

NEAR-FIELD PLASMON-RESONANCE SCANNING MICROSCOPY

Sheldon Schultz

Department of Physics, and Center for Magnetic Recording Research,
University of California, San Diego, 9500 Gilman Drive, La Jolla, CA
92093-0319

In the past few years the field of near-field scanning optical
microscopy (NSOM) has developed rapidly with applications spanning all
the physical sciences.[1] A key goal of this form of microscopy is to
obtain resolution at levels well beyond those possible with the usual
far-field optics. In contrast to far-field optics, which is bounded by
the well known limits imposed by diffraction, near-field optics has no
"in principle" fundamental lower limit in lateral size, at least down to
atomic dimensions, although in practice, signal-to-noise considerations
may restrict the application of NSOM to a few nanometers.

The simplest form of NSOM to visualize is based on the principle of a
sub-wavelength aperture (with $D/\lambda <<1$) in an opaque plane. Light
impinging on this aperture may only be transmitted through the diameter
D, and, indeed, were it observed in the far-field, would be spread out
over the entire half space due to diffraction. However, if the sample
to be studied is placed in the near-field of the aperture, say within a
distance D away, the region illuminated will also be restricted to a
lateral dimension very close to D. The light impinging on the sample
may result in all the usual consequences of transmission, reflection,
excitation, absorption etc., with subsequent optical observations of
elastic scanning, fluorescence, etc., all made in the far-field. When
the sample is then scanned past the aperture, resulting data images
constitute the NSOM technique. The practical realization of such an
aperture was developed by E. Betzig, et al., who showed that heating and
pulling a commercial optical fiber down to ~50nm diameter, followed by
evaporative coating of the tapered walls with aluminum, produced tips
with enough throughput for effective NSOM imaging and spectroscopy.[2]

The essence of NSOM is to have an optical element which produces
extremely spatially limited optical fields, i.e., via the rapid
attenuation of the near-fields. One also wishes to have a bright
source, and the ability to perform the scanning while holding the
dimensional variations to nm tolerances. We have developed an
alternative form of NSOM based on the defining optical element being an
individual Ag spherical particle, ~ 40 nm in diameter, that is excited
at the peak of its surface plasmon resonance. Techniques for making,
mounting, characterizing, and scanning such a particle past a reflective
thin film sample have been developed. The primary excitation of such a
plasmon resonance by an incident plane wave is the dipole mode, whose
fields decay as $1/r^3$, providing the requisite rapid attenuation. We
find that such a configuration also permits retention of plane
polarization of the light, and have reported on the application of the
Ag plasmon ball resonator to observe the magnetic Kerr rotation in sub-
micron written bits in perpendicular magneto-optic thin films.[3]

Proc. Microscopy and Microanalysis 1995, edited by G.W. Bailey, M.H. Ellisman, R.A. Hennigar, and N.J. Zaluzec
Copyright © 1995 MSA. Published by Jones and Begell Publishing, 79 Madison Ave., New York, NY 10016

Having demonstrated the practical application of plasmon based NSOM for Kerr imaging, we have undertaken a theoretical and experimental investigation of the ultimate resolution of such a system for diverse applications, including fluorescence, Raman, as well as Kerr microscopy. Our presentation will present a discussion on the results of these investigations.[4]

References

1. D.W. Pohl and D Courjon, Eds. *Near-field Optics*, NATO ASI series : *Applied Sciences*,242,(1993) Kluwer Academic Publishers. Also note that the most recent near-field conference was held in Brno Czech Republic, May, 1995.

2. E. Betzig, et al., 'Breaking the diffraction barrier:optical microscopy on a nanometric scale', *Science*, 251, (1991) 1468.

3. T. J. Silva,et al., 'Scanning near-field optical microscope for the imaging of magnetic domains in optically opaque materials', *Appl. Phys. Lett.*, 65 (1994) 658.

4. The author is pleased to acknowledge the collaboration of Dr. T. J. Silva, whose thesis project was the development of the Kerr plasmon based NSOM microscope. Support for the studies to be reported is by National Science Foundation grants DMR-93-02913 and DMR-94-00439.

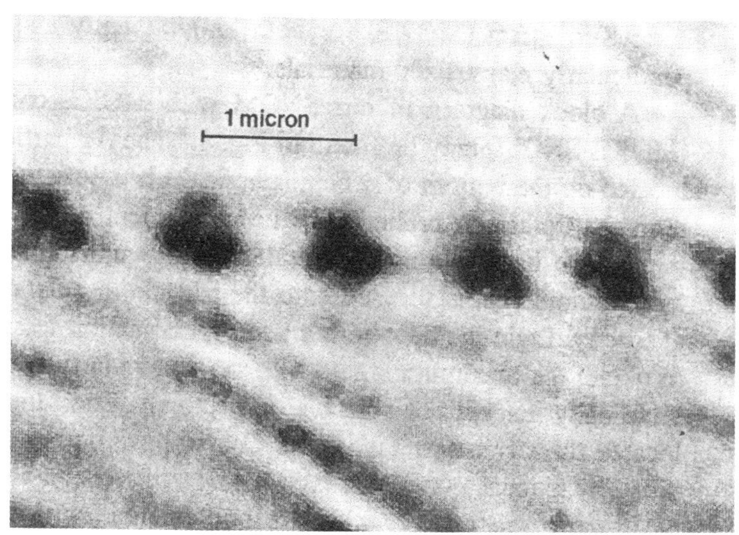

FIG. 1. - A magneto-optic NSOM image of five individual 0.5 μm-diameter magnetic domains in the Co/Pt multilayer film. The diagonal banding is due to the surface texturing of the disk. The irregular features for any given image are reproducible. The width of the domain walls in the image suggests a current resolution in excess of 100 nm.

Molecular Microscopy of DNA for Genome Analysis: Optical Mapping of Restriction Digests.

Edward J. Huff,[*] Weiwen Cai,[*] Xinghua Hu,[*] John Huang,[*] Junping Jing,[*] Xun Meng,[*] Eugene Perlov,[*] Jason Reed,[*] and David C. Schwartz[*]

[*]W. M. Keck Laboratory for Biomolecular Imaging, Department of Chemistry, New York University, New York, NY 10003

Optical microscopy of individual DNA molecules has been an interesting technique for the past 15 years, but until recently has not been useful for genome analysis.[1] We have developed Optical Mapping, an emerging single molecule approach for the rapid generation of ordered restriction maps.[2,3,4,5] Many identical individual DNA molecules from a single clone are elongated and fixed onto derivatized glass surfaces, digested with a restriction enzyme which cuts the DNA wherever a specific sequence pattern is found, stained with YOYO, and imaged with a cooled CCD camera attached to an automated epifluorescence microscope. Images are automatically processed to correct for non-uniform illumination, remove background, locate the DNA fragments, reject objects which do not look like single DNA molecules, recognize which fragments originate from an original uncut molecule, and calculate the relative sizes of the fragments by apparent length and fluorescence intensity. Results from many molecules are combined by clustering to recognize a consistent cutting pattern. Molecules which match the pattern are averaged to improve the sizing accuracy. This system permits rapid generation of detailed restriction maps from large numbers of clones and for several different enzymes per clone. Usually all of the clones derive from a much longer parental DNA strand, for instance part of one chromosome. The maps are searched for identical patterns, which indicate clones deriving from the same region of the parental molecule. These results are used to assemble the clones into contigs.

Individual DNA molecules elongated and fixed to a surface are essentially one dimensional objects. They are invisible to all forms of light microscopy except fluorescence. In practice, the DNA must be stained to be detectable. We have found that the dye YOYO gives a very low background. The dye is essentially non-fluorescent except when bound to DNA.[6] It also has a high binding constant, so that nearly all of the dye molecules remain closely associated with the DNA. Photobleaching is minimized by using 15% β-mercapto ethanol. After focusing using a SIT video camera and an autofocus unit, the DNA molecules are all in a single plane and in focus throughout their length. The resulting CCD images permit accurate calculation of relative fragment sizes by fluorescence intensity. Also, since the fixation techniques produce reasonably uniform elongation, the relative apparent length is a useful measure of fragment size. The size of restriction fragments down to 800 base pairs in size have been reliably measured using fluorescence intensity and length calculations averaged over a number of consistent images. An 800 b.p. fragment of B-DNA is under 0.3 µm. Since the DNA comes from clones in which both ends of each molecule have been engineered to produce fragments of widely different lengths, it is always possible to tell one end of the molecule from the other, and hence, to average results from many identical molecules.[7]

References

1. S. Matsumoto et. al. *J. Molec. Biol.* **152** (1981) 501.
2. D. C. Schwartz et. al., *Science* **262** (1993) 110.
3. Y.-K. Wang et. al., *Proc. Natl. Acad. Sci. U.S.A.* **92**(1995)165.
4. X. Meng et. al., *Nature Genetics*, in the press.
5. W. Cai et. al., *Proc. Natl. Acad. Sci. U.S.A.*, in the press.
6. Haugland, R., *Handbook of Fluorescent Probes* , Molecular Probes, Eugene OR (1992) 225.
7. We thank Ilya Lisanskiy, Jung-Shih Lo, Brett Porter, Thomas Anatharaman, Bhubaneswar Mishra, and Davi Geiger of the NYU Computer Science Dept. for assistance with software development. This work was supported by grants from the NIH (HG00225-02), the NSF, the W.M. Keck Foundation and the Lucille P. Markey Charitable Trust. D.C.S. is a Lucille P. Markey Scholar.

Proc. Microscopy and Microanalysis 1995, edited by G.W. Bailey, M.H. Ellisman, R.A. Hennigar, and N.J. Zaluzec
Copyright © 1995 MSA. Published by Jones and Begell Publishing, 79 Madison Ave., New York, NY 10016

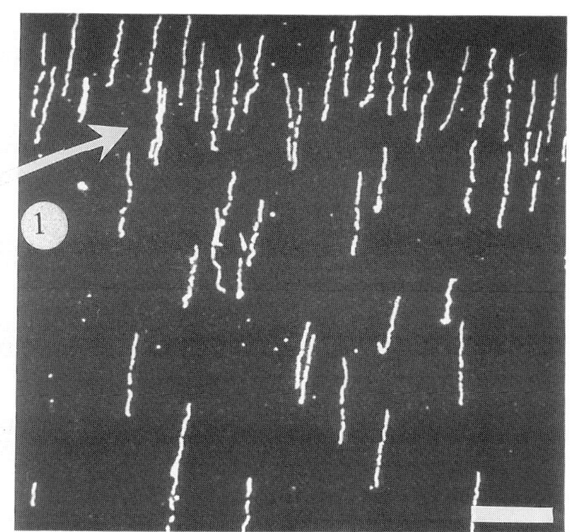

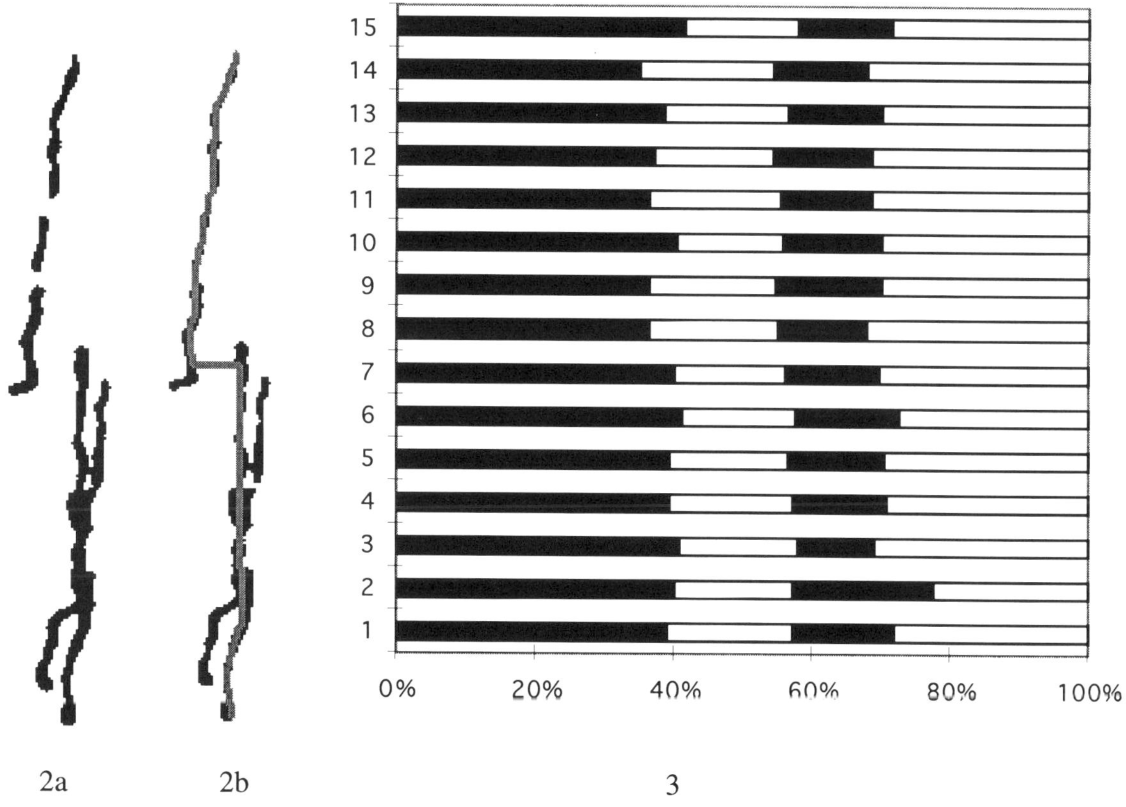

FIG. 1.—Typical image of restriction digested DNA molecules from a single clone. Molecules are 35kb, bar = 10μm. Arrow, molecules shown in Fig. 2.

FIG. 2.—Automatic rejection of unsuitable data. (a) Original thresholded image. (b) Dilated least cost path backbone, gray, residual pixels, black. Additional techniques are also used to reject objects.

FIG. 3.—Some clustered maps from one clone before averaging.

ORIENTATION OF ACTIN FILAMENTS DURING MOTION IN MOTILITY ASSAY

J. Borejdo & S. Burlacu.
Baylor Research Institute
Baylor University Medical Center
3812 Elm St., Dallas, TX 75226

Polarization of fluorescence is a classical method to assess orientation or mobility of macromolecules[1]. It has been a common practice to measure polarization of fluorescence through a microscope to characterize orientation or mobility of intracellular organelles, for example anisotropic bands in striated muscle[2]. Recently, we have extended this technique to characterize single protein molecules[3]. The scientific question concerned the current problem in muscle motility: whether myosin heads or actin filaments change orientation during contraction. The classical view is that the force-generating step in muscle is caused by change in orientation of myosin head (subfragment-1 or S1) relative to the axis of thin filament[4]. The molecular impeller which causes this change resides at the interface between actin and S1[5], but it is not clear whether only the myosin head or both S1 and actin change orientation during contraction. Most studies assume that observed orientational change in myosin head is a reflection of the fact that myosin is an active entity and actin serves merely as a passive "rail" on which myosin moves[6]. However changes in attitude of S1 can also result from the changes in flexibility of F-actin or from the fact that the series elasticity may reside in an actin protomer[7]. Recently, it has been explicitly suggested that actin plays an active role in contraction of muscle[8]. The objective of this work was to test whether, during muscle contraction, orientational changes occurred in actin. To this end we measured polarization of fluorescence of rhodamine-phalloidin labeled actin filaments moving in *in vitro* motility assay. In this assay, actin filaments labeled with phalloidin move in the presence of ATP on a substrate covered with myosin. This movement is thought to be an analog of filament sliding occurring during muscle contraction. We measured polarization of a single actin molecule using a novel video technique. The results showed that, within the accuracy of the experiments (18%), the stationary and moving filaments were polarized to the same extent. Simple modelling led us to conclude that during activity actin filaments did not become more flexible and that monomers did not undergo major rotational motion.

METHODS: The video microscope used to measure the polarization of fluorescence from individual actin filaments is shown in **Fig. 1**. The light source was an argon laser operating at 514.5 nm. The laser beam was passed through the stabilization accessory (Liconix Model 50SA, Sunnyvale, CA) which reduced the noise in the light output to less than 0.01% rms. The beam was attenuated by neutral density filters (ND) and its direction of polarization was defined by a polarization rotator PR (Spectra Physics, Model 310A). The light was directed by a dichroic mirror (DM) on a sample (S). The emitted light was split by a Wollaston prism (WP) which created two images of a filament. The left image was created by emitted light which was polarized horizontally, and the right one was created by light which was polarized vertically. Similar arrangement was originally used by Kinosita et al.[9] using unpolarized excitation light and an arrangement of mirrors to split fluorescent light. The image was formed on the faceplate of a SIT camera (Model SIT-68, Dage-MTI, Michigan City, IN) and recorded on a VCR (Sharp VC6610U, Mahwah, NJ). Images were transferred from a VCR tape to a computer by a frame grabber (Data Translation 2861, Marlboro, MA) connected to a DT 2868 auxiliary frame processor. The relative intensities of the two images carry the information about polarization properties of sample. The intensities were analyzed by Image Pro Plus image analysis

Proc. Microscopy and Microanalysis 1995, edited by G.W. Bailey, M.H. Ellisman, R.A. Hennigar, and N.J. Zaluzec
Copyright © 1995 MSA. Published by Jones and Begell Publishing, 79 Madison Ave., New York, NY 10016

program (Media Cybernetics, Silver Spring, MD) with an accuracy better than 0.5%. Only those filaments which happened to be oriented horizontally were analyzed. **RESULTS:** We first looked at filaments immobilized by binding to myosin-coated coverslip. **Fig. 2** (bar is 10 µm) shows a typical image of actin filaments in the absence of ATP. In A the excitation light was polarized horizontally. The emitted light which was polarized horizontally created an image at the left of the frame, and the emitted light which was polarized vertically created an image at the right of the frame. At the center the two images overlap, creating background with higher intensity. Care was taken not to analyze filaments present in this area. Only the filaments with long axis oriented horizontally and outside of overlap area were considered. One such filament is pointed to by arrows. The intensity of a filament and its horizontal (P_h) and vertical (P_v) polarizations were measured by Image Pro. In the absence of ATP the average P_h was 0.196±0.0180 and average P_v was 0.011±0.0210. To induce motion samples contained in addition 2 mM ATP. Filaments moved with an average velocity of 6 µm/sec (SEM=2 µm/sec). **Fig. 3** shows the image of moving filaments. Now different images were obtained when filaments were illuminated with horizontally polarized light (**Fig. 3A**) and with vertically polarized light (**Fig. 3B**) because about 20 sec elapsed between recordings. Now average P_h was 0.225±0.110 and average P_v was 0.039±0.022. A t-test revealed no statistically significant difference between both polarizations. **CONCLUSION:** Results suggest that mechanochemical interactions between myosin heads and actin do not deform the polymer structure of actin filaments. S1 reorientations observed during contraction[10] were most likely confined to the myosin heads.

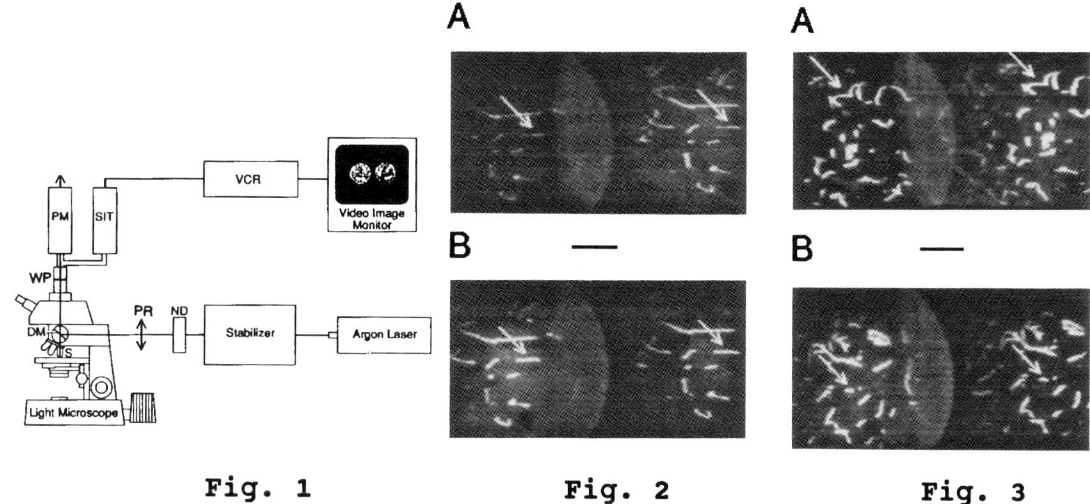

Fig. 1   Fig. 2   Fig. 3

REFERENCES:
1. G. Weber (1966) In:"Fluorescence and Phosphorescence Analysis" D. Hercules, ed. J. Wiley, NY.
2. J. Aronson, Detert, A.M. & Morales, M.F. (1968) J. Royal Mic. Soc. 88, 3.
3. J. Borejdo & Burlacu, S. (1994) Biophys. J. 66, 1319-1327.
4. H.E. Huxley (1969) Science 164, 1356-1366.
5. T. Nihei et al. (1974) Biophys. J. 14, 236-242.
6. J. Borejdo et al. (1982) J. Mol. Biol. 158, 391-414.
7. A.F. Huxley & Simmons, R.M. (1971) Nature 233, 533-538.
8. C.G. Schutt & Lindberg, U. (1990). In: "Molecular Mechanisms of Muscular Contraction" (J.M. Squire ed.) CRC Press, Boca Raton, 49-60.
9. K. Kinosita et al. (1991) J. Cell Biol. 115, 67-73.
10. J. Borejdo et al. (1979) Proc. Natl. Acad. Sci. 76, 6345-6350.

HIGH RESOLUTION MEASUREMENT OF BIREFRINGENT FINE STRUCTURE IN LIVING CELLS USING A NEW POLARIZED LIGHT MICROSCOPE

Rudolf Oldenbourg

Marine Biological Laboratory, Woods Hole MA 02543, USA

The recent renaissance of the light microsope is fueled in part by technological advances in components on the periphery of the microscope, such as the laser as illumination source, electronic image recording (video), computer assisted image analysis and the biochemistry of fluorescent dyes for labeling specimens. After great progress in these peripheral parts, it seems timely to examine the optics itself and ask how progress in the periphery facilitates the use of new optical components and of new optical designs inside the microscope. Some results of this fruitful reflection are presented in this symposium.

We have considered the polarized light microscope, and developed a design that replaces the traditional compensator, typically a birefringent crystal plate, with a precision universal compensator made of two liquid crystal variable retarders. A video camera and digital image processing system provide fast measurements of specimen anisotropy (retardance magnitude and azimuth) at ALL POINTS of the image forming the field of view. The images document fine structural and molecular organization within a thin optical section of the specimen. The sensitivity of the instrument is 0.1 nm of specimen retardance, measured with data gathered in 0.43 seconds at all 640x480 image points.[1,2]

To illustrate the workings of the new polarized light microscope, we show images of an aster (next page, top row) formed in lysate prepared from eggs of the surf clam.[3] The camera images are recorded using circularly and elliptically polarized light. The computed images are the result of algorithms developed by us and applied as image arithmetic functions to the camera images. The retardance image shows the magnitude of retardance, irrespective of slow axis orientation, which is represented separately in the azimuth image. In the azimuth image, black corresponds to horizontal orientation, medium gray to vertical, and white to horizontal orientation again.

Additional retardance images on the next page highlight specific fine structural details inside living cells. The images were obtained without the requirement of labeling, staining or other chemical modification and disturbance of live cell architecture and function.

Future development goals include increase of time resolution by reducing the elapsed time for collecting four camera images to less than 30 ms and increase of spatial resolution to 0.2 μm in all three dimensions by applying deconvolution procedures.[4]

References

1. Mei, G. & Oldenbourg, R. "Fast imaging polarimetry with precision universal compensator" *SPIE-Proceedings*.2265 (1994) 29.
2. Oldenbourg, R. & Mei, G. "New polarized light microscope with precision universal compensator" *Journal of Microscopy* (submitted for publication).
3. Oldenbourg, R., Mei, G. & Palazzo, R.E. *Biological Bulletin* 185 (1993) 288 .
4. I greatfully acknowledge the inspiration and support from Shinya Inoué. I thank my collaborators Robert E. Palazzo, University of Kansas (aster), Edward D. Salmon, University of North Carolina (Newt lung epithelial cell), and Yoshio Fukui, North Western University (Dictyostelium ameba). The instrument development is supported by the National Institutes of Health grant R01 GM49210.

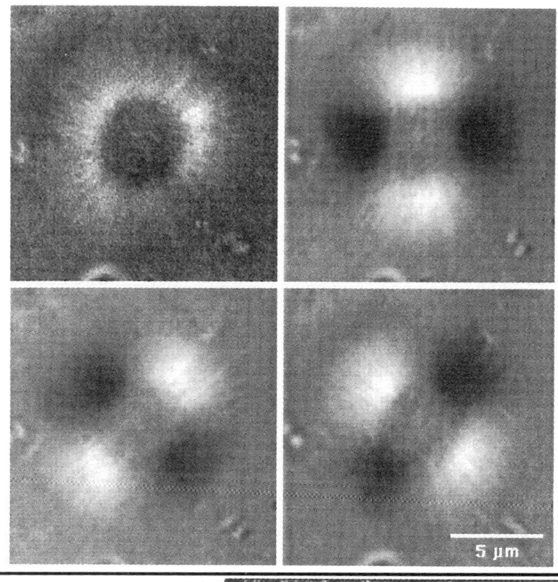

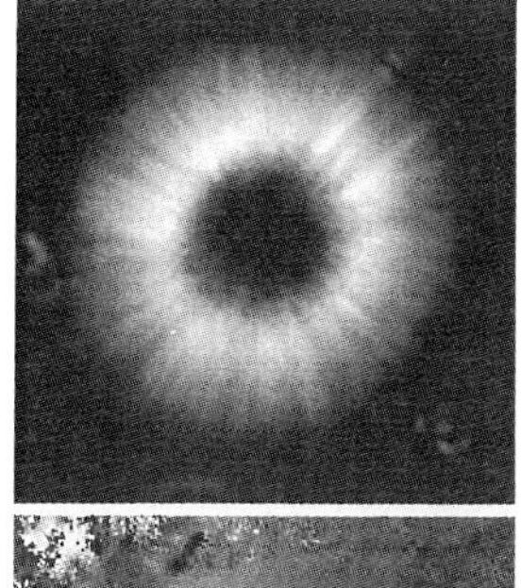

Left: Four camera images of single aster consisting of microtubules radiating from the centrosome. Images were recorded using the new polarized light microscope.
Right: Two images computed from raw image data shown to the left. The top represents retardance magnitude, the bottom represents azimuth orientation of slow axis.

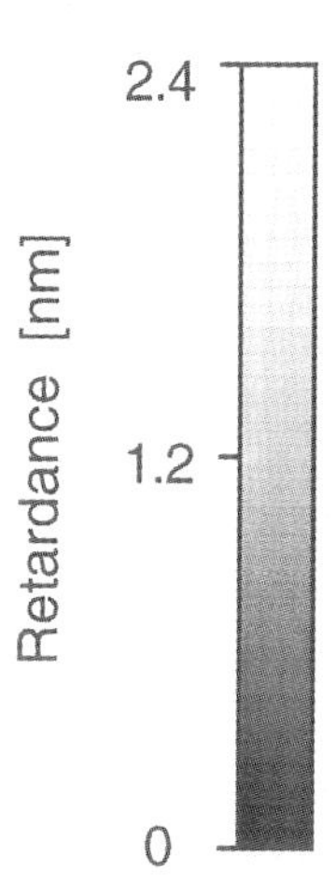

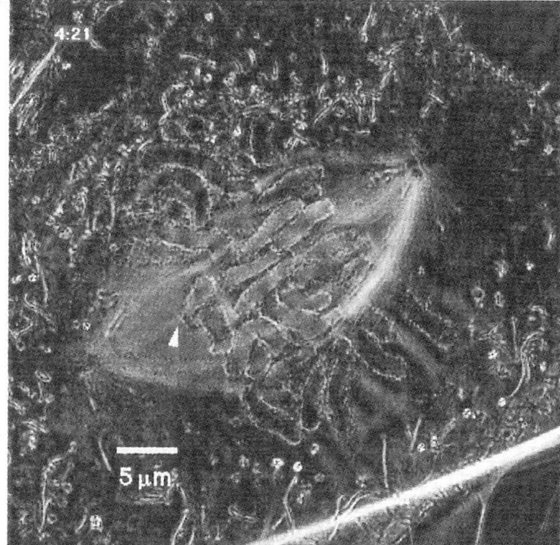

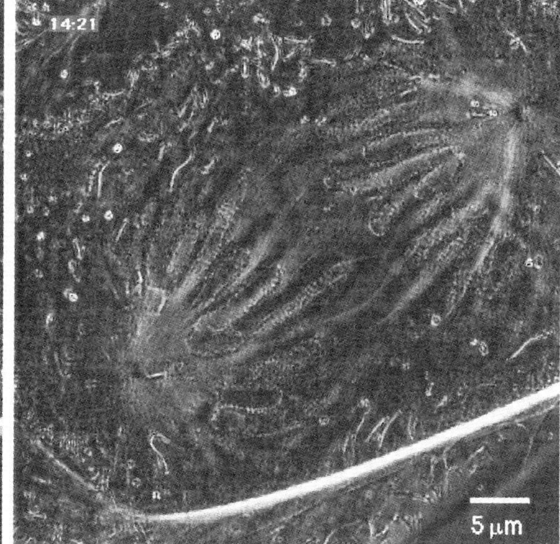

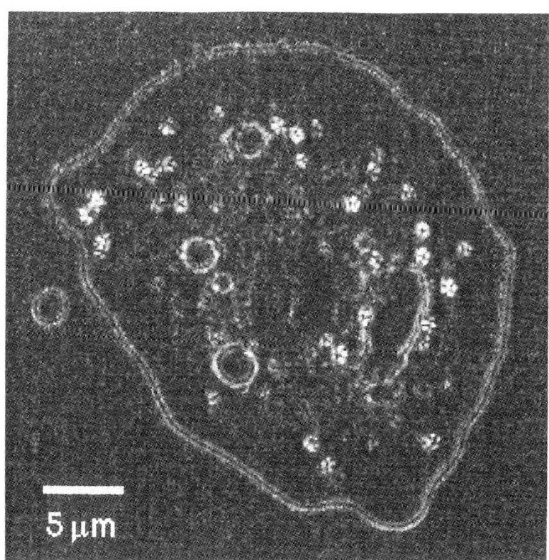

Top left: Retardance image of newt lung epithelial cell in mitosis showing birefringence of spindle fibers. White arrow head marks kinetochore microtubules of moving chromosome (beginning anaphase). Chromosomes appear to be outlined by edge birefringence. Other components are stress fiber with strong off scale retardance, mitochondria (thin filaments within cell) and golgi complexes (small, bright spherical objects). (Image brightness proportional to retardance; see scale.)
Top right: Anaphase of same cell as on left.
Bottom left: Retardance image of dictyostelium ameba depicting cell surface layers and highly birefringent organelles, some with distinct fivefold symmetry. All images Plan Apo 60x/1.4 NA oil.

RESOLUTION BEYOND THE DIFFRACTION LIMIT: 4PI-CONFOCAL-, STED-, AND GSD- FLUORESCENCE MICROSCOPY

Stefan W. Hell [ab], Pekka E. Hänninen [a], Martin Schrader [ab], Tony Wilson [b], Erkki Soini [a]

a)Department of Medical Physics & Chemistry, University of Turku, PO Box 123, Turku, Finland
b)Department of Engineering Science, University of Oxford, Parks Road, Oxford OX1 3PJ, England

In far-field light microscopy resolution is determined by diffraction. In a far-field light microscope such as the confocal scanning light microscope, the resolution is governed by the extent of the squared intensity distribution in the focal region. Precise measurements of the confocal PSF have shown that the axial and lateral resolution of a confocal microscope (NA=1.4 oil, l= 633 nm) is 520nm and 200nm (FWHM), respectively. At a wavelength of 375nm, this amounts to a resolution of 300 nm (axial) and 120 nm (lateral), obtainable with a standard confocal microscope of high aperture[1-2].

A 3-7 fold increase in axial resolution is achieved with a 4Pi-confocal microscope[2-4]. The 4Pi-confocal microscope uses two high numerical aperture objective lenses that are used coherently for illuminating or detecting the same point in the object space. The present paper deals with the latest developments in the field of 4Pi-confocal microscopy. The Optical Transfer Functions (OTF) of 4Pi-confocal microscopies with 4Pi-illumination (type A), 4Pi-detection (type B), and 4Pi illumination and detection (type C) are measured and compared with their standard confocal counterpart. The application of a three point deconvolution[7] demonstrates an axial resolution of 140 nm (4Pi type A,B and 2-photon excitation type A) of 75nm (4Pi type C) at a (comparatively long!) wavelength of 633 nm. We discuss the conceptual pros and cons when comparing 4Pi-confocal microscopy with Standing-wave-microscopy[5-6]. We show that the increased axial bandwidth of the 4Pi-confocal OTF[7-8] can be used to determine the axial distance between two point objects with a precision[9] of 15 nm.

Furthermore we discuss STED-fluorescence microscopy which is a novel concept we proposed to overcome the diffraction resolution limit in lateral direction[10]. The resolution increase is based on the reduction of the lateral extent of the effective PSF in a scanning fluorescence microscope by preventing exited fluorescent molecules from undergoing spontaneous emission. This is achieved by depleting the excited molecules at the edge of the focus by stimulated emission. The physics of the lateral resolution increase as well as the depletion process is discussed and demonstrated. An alternative but related method of depleting the ground state of the fluorophore (Ground State Depletion Microscopy) is introduced[11].

Proc. Microscopy and Microanalysis 1995, edited by G.W. Bailey, M.H. Ellisman, R.A. Hennigar, and N.J. Zaluzec
Copyright © 1995 MSA. Published by Jones and Begell Publishing, 79 Madison Ave., New York, NY 10016

References:

1. G. J. Brakenhoff et al, *J Micr*, 117, (1979).
2. S. W. Hell, et al, Appl Phys Lett, 64, (1994) 1335.
3. M. Schrader, S.W. Hell, T. Wilson, SPIE, in press(1995).
4. S. Hell & E. Stelzer, JOSA A, (1992) 2159.
5. F. Lanni, *Application of Fluorescence in Biol.Sciences*, D.L. Taylor, Ed., New York:Liss (1986)
6. Bailey et al, *Nature*, 366, (1993) 44.
7. P. E. Hänninen et al, *Appl Phys Lett*, 66, 13 (1995).
8. M. Gu, C.J. R. Sheppard, *JOSA A*, (1994).
9. Hell et al, submitted (1995).
10. S.W. Hell & Wichmann, *Opt Lett*, 19, (1994) 780.
11. S. W. Hell & M. Kroug, *Appl. Phys. B*, in press (1995).

1 Measured Optical Transfer Functions (NA=1.4*, oil, λ = 633 nm)

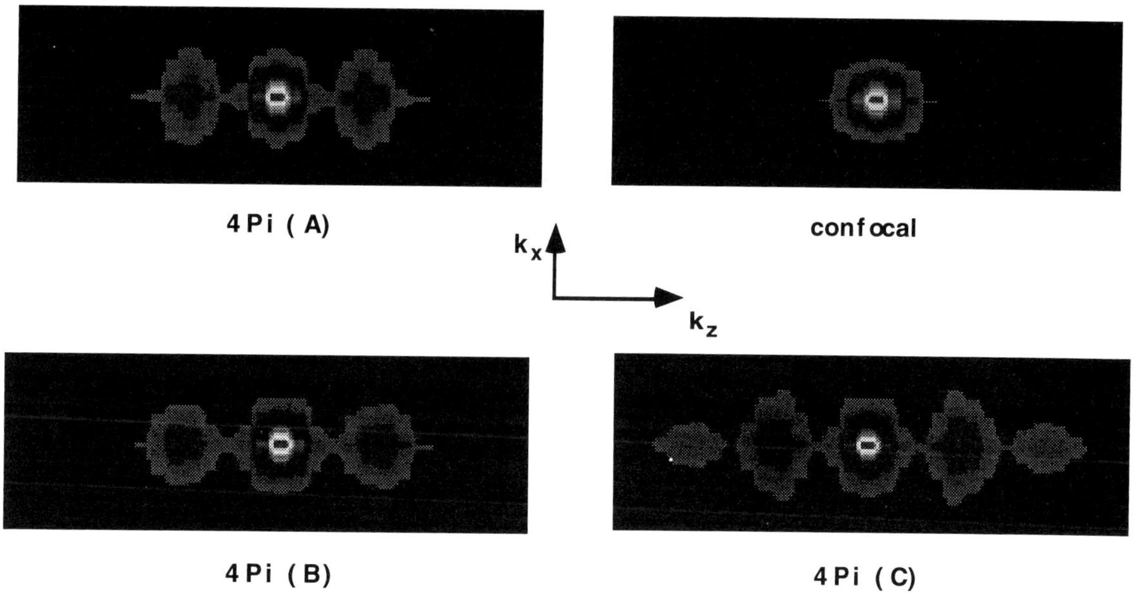

Fig. 1.—The measured Optical Transfer Functions (OTF) for the different types of 4Pi-confocal microscopy. The lateral axis shows the spatial frequencies along the optical axis whereas the vertical axis displays the spatial ferquencies in lateral direction. Both axes are in arbitrary units. The 4Pi-confocal microscope with coherent illumination and detection (type C) exhibits a considerably improved axial bandwith over the standard confocal microscope.

NEW APPLICATIONS OF NEAR-FIELD OPTICAL STORAGE

G. S. Kino*, S. Hayashi*, and I. Ichimura**

*Ginzton Laboratory, Stanford University, Stanford CA
**Opto-Electronics Research Dept., Sony Co., Tokyo, Japan. Presently at Stanford University.

Some time ago, we described a new type of near-field imaging system using the *Solid Immersion Lens*.[1,2] A confocal microscope using this principle has demonstrated sub-100 nm resolution with visible light. In this paper, we will describe how this device is being applied to increasing the density of optical storage in CD ROMs and magneto-optical storage systems.

The SIL is placed between the objective and the disk. As illustrated in Fig. 1(a), if the SIL is in the form of a hemisphere, rays enter it along its radii and converge at its center. If the refractive index of the SIL is n, the wavelength in this material is reduced by $1/n$ and consequently increases the effective NA of the objective lens by the refractive index n and decreases the spot size by $1/n$. With this configuration, we were able to obtain good quality images of a grating with 100 nm gaps and strip widths. The system is a near-field system because ray

s at a large angle to the axis are totally internally reflected at the lower surface of the SIL and the fields fall off exponentially with distance from this surface. Consequently, the air gap has to be very small.

Recently, working with the Sony Corporation and IBM, we have become interested in applying this near-field technique to increasing the density of optical storage in both CD optical ROMs and magneto-optical storage disks. One configuration, which Sony is pursuing, is illustrated in Fig. 1(b). In this case an effective numerical aperture of less than 1 is used with an SIL with a refractive index equal to that of the transparent material covering the disk. Consequently there is no total internal reflection, and rays passing from the SIL into the disk are not aberrated, provided that the air gap is sufficiently thin, and the thickness of the disk is decreased to compensate for the thickness of the transparent layer. If the spherical aberrations due to the air gap are taken into account, an aspheric objective can be made and the air gap may be as large as 100 μm. A major advantage of this configuration is that it is not too sensitive to dust, and can be operated in an unsealed environment. Its disadvantage is that a high numerical aperture is not used so the increase in area density is limited.

By using a solid immersion lens with an effective numerical aperture 0.83 and green light from a doubled semiconductor laser at 532 nm, the theoretical density should be ~ 5 Gbs/inch^2, about ten times the density presently used. So far, densities of the order of 3-4 Gbs/inch^2 have been observed with this configuration at Stanford and at Sony.

A further improvement is to use a *stigmatic focusing* SIL which we call a *supersphere*. In this case, the beam emerging from the objective is focused to a point distant a/n from the center of the lens, and the effective NA of the objective is increased by a factor n^2, where a is the radius of the SIL. Just as with the hemispherical SIL, the beam can be focused through a transparent layer of the disk if it is chosen to be of the same refractive index as the lens. In both cases, this is a near-field system, for the air gap must be small. However, to avoid an exponential fall off of the fields in the air gap, it is usually wise to avoid total internal reflection at the surface of the SIL. Thus, in principle, the maximum effective NA can be equal to the refractive index n. This makes it possible to obtain a spot size of the order of 120 nm, with a wavelength $\lambda = 450$ nm and a refractive index of 2. This could lead to a storage density greater than 30 Gbs/inch^2.

In this system, the lens must be placed very close to the disk (typically less than 100 nm), and it is obviously advantageous to work with materials with as high a refractive index as possible. We have carried out calculations of the air gap which can be used. It is apparent that if the air gap is kept to less than 100 nm, the aberrations would not be severe, and would not deteriorate the spot size. We conclude that the best way to make the system is to use an SIL in the form of a floating head, much as

Proc. Microscopy and Microanalysis 1995, edited by G.W. Bailey, M.H. Ellisman, R.A. Hennigar, and N.J. Zaluzec
Copyright © 1995 MSA. Published by Jones and Begell Publishing, 79 Madison Ave., New York, NY 10016

is used in standard magnetic disk storage. Following this line of reasoning, it is possible to obtain suitable lenses with a refractive index of 1.96 of the order of 0.5-1 mm diameter which could be used in such a system.

Another difficulty which could arise in the use of this system with a magneto-optical disk is that the polarization will be changed in the lens. We have carried out theoretical analyses which show that this problem should not cause serious difficulties. In this analysis, we have been able to show that the operation of a magneto-optical microscopy system, because it takes the product of the rotated polarization signal with the directly reflected signal, is equivalent to that of the correlation microscope, or of a confocal microscope. Thus, similar definions to that of the confocal microscope should be obtained.

Experiments carried out by Terris et al. at IBM Almaden and at Stanford have also shown that loss of polarization information is not a major problem.[3] Terris et al have been able to read spots in a magneto-optical disks made for this purpose without difficulty. Furthermore, with the stigmatic lens, they were able to obtain a measured gaussian beam profile with a half width of 317 nm, in good agreement with the calculated value of 304 nm, using a 780 nm laser source with an objective with NA = 0.55 and an n = 1.9 index glass for a stigmatic SIL.   The potential storage density was, therefore, increased from that of the objective alone by an order of magnitude with the use of the SIL.

A floating air bearing can be constructed for use with the SIL. The control system for the objective can be very similar to that presently used, since the critical focus distance are magnified by n^2 at the objective. Furthermore, since this system can provide images over a finite field of view of the order of 100 spot widths in diameter, it should be possible to use standard tracking methods and thus make a viable and practical high density optical storage system.

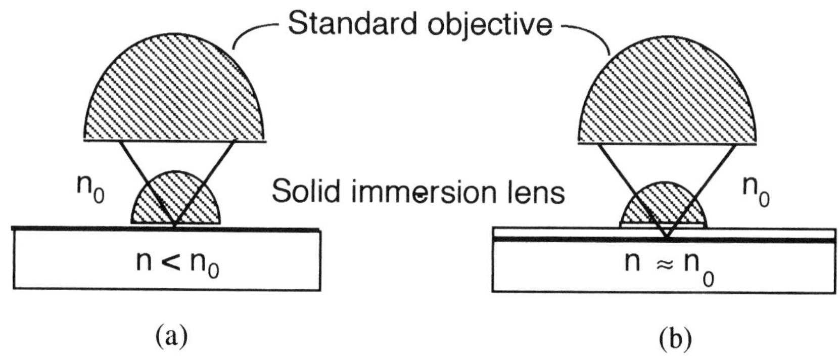

Fig. 1 Solid immersion lens: (a) hemispherical lens; with focusing at the bottom surface; (b) lens of same refractive index as the CD disk.

References

1.   S. M. Mansfield and G. S. Kino, Appl. Phys. Lett. 57(1990)2615.
2.   S. M. Mansfield et al., Opt. Lett. 18(1993)305.
3.   B. D Terris et al., Appl. Phys. Lett. 65(1994)388.
4.   G. S. Kino and S. S. C. Chim, Appl. Opt. 29(1990)3775.

REAL-TIME OBSERVATION OF SINGLE MOLECULES BY CONFOCAL FLUORESCENCE MICROSCOPY

Shuming Nie,[+] Daniel T. Chiu, and Richard N. Zare

Department of Chemistry, Stanford University, Stanford, California 94305.
+Present Address: Department of Chemistry, Indiana University, Bloomington, Indiana 47405

The ability to detect, identify, and manipulate individual molecules offer exciting possibilities in many fields, including chemical analysis, materials research, and the biological sciences. A particularly powerful approach is to combine the exquisite sensitivity of laser-induced fluorescence and the spatial localization and imaging capabilities of diffraction-limited or near-field optical microscopes.[1,2] Unlike scanning tunneling microscopy (STM)[3] and atomic force microscopy (AFM),[4] which lack molecular specificity, optical spectroscopy and microscopy techniques can be used for real-time monitoring and molecular identification at nanometer dimensions or in ultrasmall volumes.

We report the use of confocal fluorescence microscopy coupled with a diffraction-limit laser beam and a high-efficiency photodiode for real-time detection of single fluorescent molecules in solution at room temperature.[5] Rigler and Eigen have also demonstrated single-molecule detection with a confocal microscope and fluorescence correlation spectroscopy.[6] The probe (or sampling) volume is effectively an elongated cylinder, with its radius being determined by optical diffraction and length by spherical aberration.[7] For a propagating gaussian laser beam focused by a diffraction-limited objective, the $1/e^2$ radius at the focal plane is $r_0 = \lambda f / n\pi d_0$, where λ is the laser wavelength in vacuum, f is the focal length of the objective, n is the refractive index of the media (immersion oil n = 1.52), and d_0 is the $1/e^2$ radius of the input laser beam.[8] The diffraction-limited radius was calculated to be 250-260 nm under our experimental conditions (TEM$_{00}$ laser beam radius 0.65 mm, objective focal length 1.6 mm, laser wavelength 488.0 nm or 514.5 nm). In theory, the $1/e^2$ probe depth (z_0) in confocal microscopy can be estimated by using the point spread function (PSF) and collection efficiency function (CEF);[9] in practice, however, it is primarily determined by spherical aberration of the objective and has been experimentally measured to be ca. 1.0 μm. The cylindrical probe volume is thus estimated to be ~ 5.0×10^{-16} L or 0.5 fL. This tiny probe volume is expected to contain an average of one dye molecule in a 3.3×10^{-9} M solution, but the actual number of molecules in the probe volume will fluctuate between 0 and 1, 1 and 2, etc. In more dilute solutions the detection events are increasingly dominated by single molecules because the probabilities for two or more molecules in the probe volume become vanishingly small. Figure 1 shows fluctuating fluorescence signals detected in a 1×10^{-10} M rhodamine 6G (R6G) solution, indicating that the probe volume contains 0 or 1 dye molecule and rarely 2 or more molecules. Concentration studies provide evidence that the observed fluorescence signals are truly single-molecule events and are not caused by molecular aggregates or artifacts such as dust scattering. Additional criteria we have used to ascertain single-molecule events include signal intensity dependence on the nature of solvent and pH and fluorescence saturation with increasing laser intensity.

With an unlimited excitation throughput and a low background level, this technique allows fluorescence detection of single rhodamine molecules with a signal-to-noise ratio of ~10 in 1 millisecond, which approaches the theoretical limit set by fluorescence saturation. Real-time measurements at a speed of 500,000 data points per second yield single-molecule fluorescence records that show not only the actual transit time of a particular molecule but also contain characteristically long (~50 μs) and short (~4 μs) dark gaps (Figure 2). Random-walk simulations of single fluorescent molecules provide evidence that these long and short dark periods are mainly caused by boundary recrossing motions of a single molecule at the probe volume periphery and by intersystem crossing into and out of the dark triplet state. We have also extended the use of confocal

Proc. Microscopy and Microanalysis 1995, edited by G.W. Bailey, M.H. Ellisman, R.A. Hennigar, and N.J. Zaluzec
Copyright © 1995 MSA. Published by Jones and Begell Publishing, 79 Madison Ave., New York, NY 10016

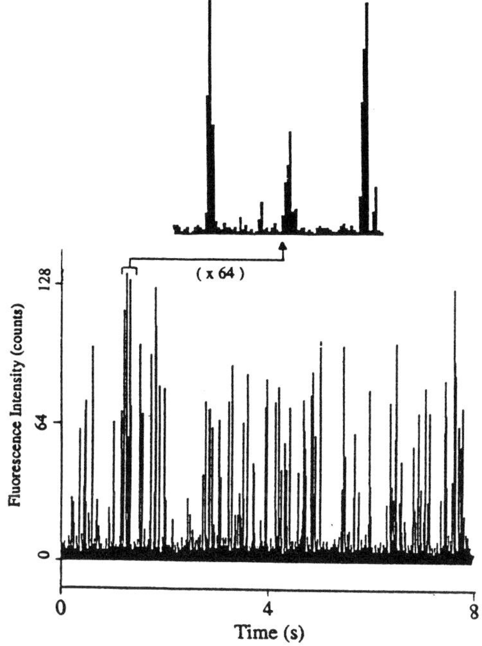

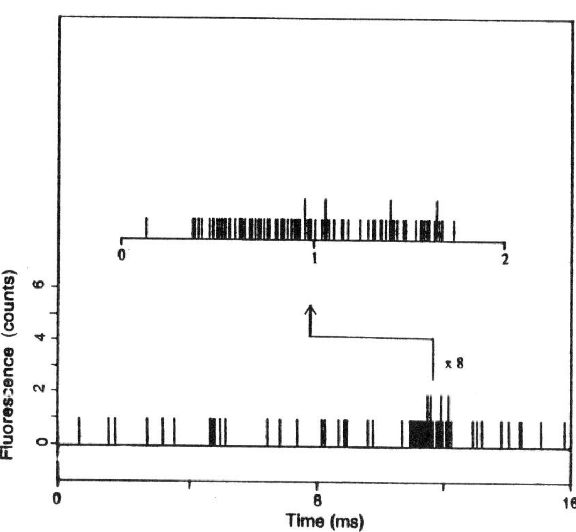

Figure 1. Fluorescence photon bursts detected from single rhodamine 6G molecules in ethanol solution (1 x 10^{-10} M). The data were acquired at 1000 points per second (1-ms integration) with 1 mW CW laser excitation (514.5 nm) and an emission bandpass filter (560 nm /40-nm FWHM).

Figure 2. Fluorescence record of a single rhodamine 6G molecule in ethanol obtained at 500,000 data points per second (2-μs integration) and a CW laser power of 1 mW (514.5 nm).

fluorescence microscopy to study individual, fluorescently tagged biomolecules including deoxynucleotides, single-stranded primers, and double-stranded DNA. The achieved sensitivity permits dynamic structural studies of individual λ-phage DNA molecules labeled with intercalating fluorescent dyes, and the results reveal large-amplitude DNA structural fluctuations that occur on the millisecond time scale.

Because all of the equipment we used is inexpensive and readily available, we believe that we can look forward to the routine application of real-time measurements of single molecules in solution. Following a single molecule and chemical or biochemical reactions such a molecule may undergo can reveal new structural and dynamical features hidden in conventional measurements. Single-molecule techniques may also be used for ultrasensitive DNA analysis, real-time monitoring of intracellular protein transport, and for rapid screening of rare molecules and nanostructures in large chemical libraries.

References

1. W. E. Moerner, *Science* 265 (1994) 46-53; and references therein.
2. E. Betzig and R. J. Chichester, *Science* 262 (1993) 1422-1425.
3. P. K. Hansma,V. B. Elings, D. Marti, and C. E. Bracker,*Science* 242 (1988) 209.
4. M. Radmacher, R. W. Tillman, M. Fritz, and H. E. Gaub, *Science* 257 (1992) 1900.
5. S. Nie, D. T. Chiu, and Zare, *Science* 266 (1994) 1018-1021.
6. M. Eigen and R. Rigler, *Proc. Natl. Acad. Sci. USA* 91 (1994) 5740-5747.
7. M. B. Schneider and W. W. Webb, *Appl. Optics* 20 (1981) 1382-1388.
8. L. D. Dickson, *Appl. Optics* 9 (1970) 1854-1861.
9. H. Qian and E. L. Elson, *Appl. Optics* 30 (1991) 1185-1185.

TWO-PHOTON EXCITATION MICROSCOPY IN CELLULAR BIOPHYSICS

David W. Piston*, Brian D. Bennett*, and Robert G. Summers**

*Department of Molecular Physiology and Biophysics, Vanderbilt University, Nashville, TN 37232.
*Department of Anatomical Sciences, SUNY-Buffalo, Buffalo, NY 14214

Two-photon excitation microscopy (TPEM) provides attractive advantages over confocal microscopy for three-dimensionally resolved fluorescence imaging and photochemistry.[1,2] Two-photon excitation arises from the simultaneous absorption of two photons in a single quantitized event whose probability is proportional to the square of the instantaneous intensity.[3] For example, two red photons can cause the transition to an excited electronic state normally reached by absorption in the ultraviolet. In practice, two-photon excitation is made possible by the very high local instantaneous intensity provided by a combination of diffraction-limited focusing of a single laser beam in the microscope and the temporal concentration of 100 femtosecond pulses generated by a mode-locked laser. Resultant peak excitation intensities are 10^6 times greater than the CW intensities used in confocal microscopy, but the pulse duty cycle of 10^{-5} maintains the average input power on the order of 10 mW, only slightly greater than the power normally used in confocal microscopy.

Three properties TPEM give this method a tremendous advantage over conventional optical sectioning microscopies for the study of thick samples: **1)** The excitation is limited to the focal volume because of the intensity-squared dependence of the two-photon absorption. This inherent localization provides three-dimensional resolution and eliminates background *equivalent to an ideal confocal microscope* without requiring a confocal spatial filter, whose absence enhances fluorescence collection efficiency. Confinement of excitation to the focal volume also *minimizes photobleaching and photodamage* - the ultimate limiting factors in fluorescence microscopy of living cells and tissues. **2)** The two-photon technique allows imaging of UV fluorophores with conventional visible light optics in both the scanning and imaging systems, because both the red excitation light (~700 nm) and the blue fluorescence (>400 nm) are within the visible spectrum. **3)** Red or infrared light is far less damaging to most living cells and tissues than bluer light because fewer biological molecules absorb at the higher wavelengths. Longer wavelength excitation also reduces scattering of the incident light by the specimen, thus allowing more of the input power to reach the focal plane. This relative transparency of biological specimens to 700 nm light *permits deeper sectioning*, since both absorbance and scattering are reduced.

Fluorescence from the natural reduced pyridine nucleotides (NAD(P)H) in cells is a potential indicator of cellular respiration. However, NAD(P)H is not a "good" fluorophore (it has a small absorption cross-section and a low quantum yield), and previous examinations of cellular dynamics by NADH autofluorescence have been limited by severe photobleaching and photodamage problems. Two-photon excitation of NADH yields minimal photodamage and allows 3-D metabolic mapping of cellular redox state. Figure 1 shows the result of a typical optical sectioning experiment of islet NADH autofluorescence through the middle of a pancreatic islet (approximately 40 μm into the islet from the cover slip). The outlines of single cells are clearly visible as are the nuclei, which appear dark. We have examined the glucose response of the autofluorescence intensity and find that it follows a dose response pattern similar to the islets insulin response.[4] Useful NAD(P)H images have also been obtained by TPEM from cornea[5], cardiac myocytes, and neural cells.

Another powerful application of TPEM is its use in photoactivation of caged compounds.[6] Caged compounds have a chemical cage attached to them so that they are basically invisible. However, when certain light (such as the two-photon excitation laser) is shined upon the compound, the caging group falls off, and the compound becomes active. Cell lineage patterns of the sea urchin embryo have been visualized by two-photon photorelease of caged fluorophore. The pinpoint control of excitation in TPEM can be utilized to photorelease specifically within a single cell. For example, a

Proc. Microscopy and Microanalysis 1995, edited by G.W. Bailey, M.H. Ellisman, R.A. Hennigar, and N.J. Zaluzec.
Copyright © 1995 MSA. Published by Jones and Begell Publishing, 79 Madison Ave., New York, NY 10016

single cell out of the 8 cells after third division is irradiated by the two-photon excitation laser beam. The caged fluorophore becomes active and "lights up" this single cell. The dye is coupled to 10,000 MW dextran, so it does not travel between cells. Thus, only descendants of the original marked cell will be marked with visible dye. The exact locations of the marked descendent cells can be tracked and correlated with the morphology of the developing larva. A 3-D reconstruction of an embryo in which a single cell was marked at the 8 cell stage is shown in figure 2.

Intracellular calcium activity ($[Ca^{2+}]_i$) can be measured using the ratiometric calcium indicator dye Indo-1 in TPEM.[7] In this case, use of TPEM permits visualization of both standing calcium gradients and transient heterogeneities in $[Ca^{2+}]_i$ with sufficient background rejection to allow quantitation in thick and strongly scattering samples. One still largely unexplored application of TPEM is imaging of visible absorbing fluorophores with ~1 µm excitation, which may become a powerful method to perform 3-D reconstructions of easily photobleached samples.[8]

[1] Denk W., J.H. Strickler and W.W. Webb *Science* **248**:73-76 (1990).

[2] Williams, R.M., D.W. Piston, W.W. Webb. *FASEB J.* **8**:804-813 (1994).

[3] Goppert-Mayer, M. *Ann. Phys.* **9**:273-294 (1931).

[4] Piston, D.W., B.D. Bennett, G. Ying. *J. Micro. Soc. Amer.* **1**:In Press (1995).

[5] Piston, D.W., B.R. Masters, W.W. Webb. *J. Microscopy* **177**:In Press (1995).

[6] Denk, W. *Proc. Nat. Acad. Sci. USA* **91**:6629-6633 (1994).

[7] Piston, D.W., M.S. Kirby, H. Cheng, W.J. Lederer, W.W. Webb. *App. Optics.* **33**:662-669 (1994).

[8] The author thanks Watt W. Webb, Winfried Denk, and Dave Sandison for many helpful discussions about two-photon excitation microscopy. Experiments described were aided by technical assistance from Valerie Blackwell and Guangtao Ying. DWP is a Beckman Young Investigator of the Arnold and Mabel Beckman Foundation, Irvine, CA. Support for this work also comes from the Whitaker Foundation, the Vanderbilt Diabetes Research and Training Center, and the Vanderbilt Cancer Center.

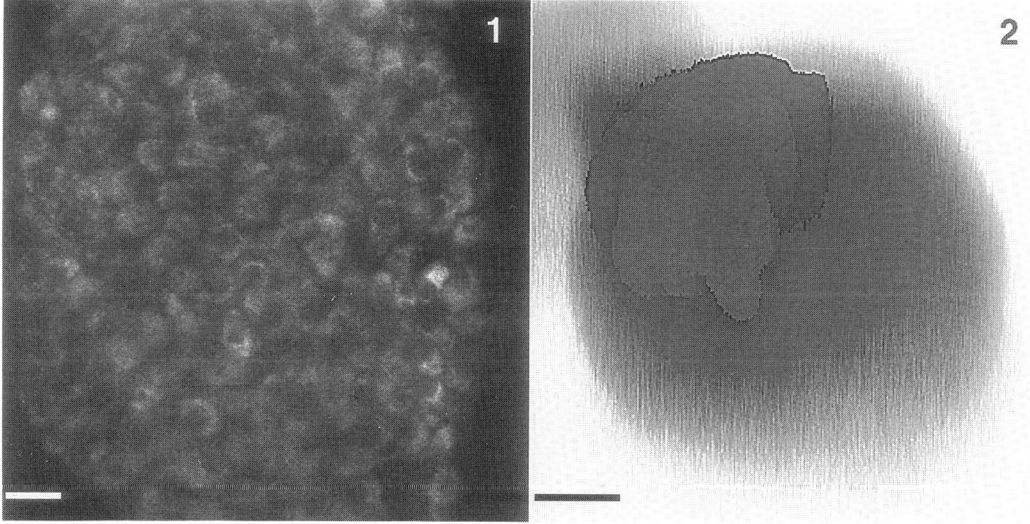

FIG 1. NAD(P)H Autofluorescence from an isolated islet (8 sec exposure, bar = 10 µm)
FIG 2. Three-dimensional reconstruction of a series of confocal micrographs of a sea urchin embryo. Data was taken six hours after uncaging in 1 of 8 cells. The now contains ~160 cells and ~1/8 of the cells contain uncaged marker (bar = 20 µm).

3D OPTICAL TRANSFER IN EXCITATION FIELD SYNTHESIS MICROSCOPES

Vijay Krishnamurthi*, Brent Bailey**, and Frederick Lanni***

Center for Light Microscope Imaging and Biotechnology, *Biomedical Engineering
Program, **Department of Physics, and ***Department of Biological Sciences,
Carnegie Mellon University, 4400 Fifth Avenue, Pittsburgh, Pennsylvania 15213 USA

Excitation field synthesis (EFS) refers to the use of an interference optical system in a direct-imaging microscope to improve 3D resolution by axially-selective excitation of fluorescence within a specimen [1,2]. The excitation field can be thought of as a weighting factor for the point-spread function (PSF) of the microscope, so that the optical transfer function (OTF) gets expanded by convolution with the Fourier transform of the field intensity. The simplest EFS system is the standing-wave fluorescence microscope, in which an axially-periodic excitation field is set up through the specimen by interference of a pair of collimated, coherent, s-polarized beams that enter the specimen from opposite sides at matching angles. In this case, spatial information about the object is recovered in the central OTF passband, plus two symmetric, axially-shifted sidebands. Gaps between these bands represent "lost" information about the 3D structure of the object. Because the sideband shift is equal to the spatial frequency of the standing-wave (SW) field, more complete recovery of information is possible by superposition of fields having different periods. When all of the fields have an antinode at a common plane (set to be coincident with the in-focus plane), the "synthesized" field is peaked in a narrow in-focus zone [1]. Away from the in-focus plane, the field is no longer periodic, but decays to an average value equal to 50% of the peak. Here, we describe a fully-interferometric EFS microscope in which an extended light source is used to generate a continuum of superposed SW fields.

A single pair of beams (for example, from a and a' in Fig. 1a) produces the interference pattern
$$I_{ex}(z)/2I_0 \; = \; 1 \; + \; \cos[(4\pi n/\lambda_{ex})(\cos\theta)\, z \; + \; \phi] \; = \; 1 \; + \; \cos[Kz + \phi],$$
where θ is the beam angle relative to the microscope axis. The parameter ϕ depends upon the relative phase of the interfering beams and sets the position of the standing-wave antinode and node planes relative to the in-focus plane of the microscope. The node plane spacing of the field is $\lambda_{ex}/2n \cos\theta$, ranging from a minimum of $\lambda_{ex}/2n$ for beams counter-propagating on-axis, to a maximum set by the numerical aperture of the objective lenses. Typically, node spacing ranges from 0.18-0.36 μm. When multiple beam pairs are interfered, cross-interference must be minimized to prevent the appearance of transverse nodes in the field of illumination. In laser-based systems, this problem can be solved by time-multiplexing a series of SW fields by sweeping the beam crossing angle during acquisition of each image. With an extended, "incoherent" source, such as an arc lamp, mutually-incoherent SW fields can be generated simultaneously.

The essential features of an interferometric EFS microscope are shown in Figure 1. A beam splitter and lens system are used to form matching images of the light source in the rear focal planes of two identical, aligned objectives positioned on opposite sides of the specimen and focused to a common plane. The characteristics of the light source and intermediate optical system are critical to the performance of the instrument. Fig. 1a shows that matched source image points (a and a', b and b', ...) produce the required beam pairs. Each pair of source points will be coherent, provided the two paths from the splitter to the lenses are balanced to well within the longitudinal coherence length (L‖) of the light. L‖ is given by $\lambda^2/\Delta\lambda$, where $\Delta\lambda$ is the bandwidth. In fluorescence microscopy, excitation bandwidths of 10-30 nm are typical, corresponding to L‖ in the range 8-25 μm. However, greater precision is required to guarantee that all of the superposed SW fields have an antinode plane at the common in-focus plane of the objectives. For this, the optical path from corresponding points on the source images to the in-focus plane must be matched to well within the smallest node-plane spacing. Nominally, the source image points a, b, and c are mutually incoherent because they correspond to independent sources in the lamp. The finite f/# of the intermediate optical system introduces some mutual coherence. This can be minimized by forming the source images at the lowest possible f/#,

Proc. Microscopy and Microanalysis 1995, edited by G.W. Bailey, M.H. Ellisman, R.A. Hennigar, and N.J. Zaluzec
Copyright © 1995 MSA. Published by Jones and Begell Publishing, 79 Madison Ave., New York, NY 10016

which is equivalent to maximizing the illuminated field-of-view in the specimen. The size and irradiance distribution of the source images sets the relative contribution of different beam crossing angles to the total field. A wide range is essential to average out interference fringes that persist away from the in-focus plane but within L‖. The complete layout of the EFS system is diagrammed in Fig. 1b. A light source (L) is monochromated by bandpass filtering (BPF), further modified by masking and polarization (AM,P), then imaged through a 45° beam splitter (BS) onto a moveable beam splitter (MBS). MBS is symmetrically-positioned relative to the specimen (Sp) in a rectangular interferometer loop defined by BS, mirror M, and two dichroic reflectors (DR). This optical arrangement (one extra reflection in one beam path) forms the two source images with matching orientation and opposite handedness. Fluorescence images are recorded conventionally, with an electronic camera (C) and emission filter (Em). Piezoelectric axial movement of MBS provides a means to adjust the two optical paths to the specimen. When MBS is at the balance point, the excitation field is peaked at the in-focus plane. As described in [1], when MBS is then shifted by $\lambda/4$, the field will be nulled at the in-focus plane, provided that corresponding pairs of source image points (i.e., a and a' in Fig. 1a) produce s-polarized beams in the specimen. The difference image will be weighted to show fluorescent features in a zone that is approximately five-fold more sharp than the diffraction-limited depth-of-field.[3]

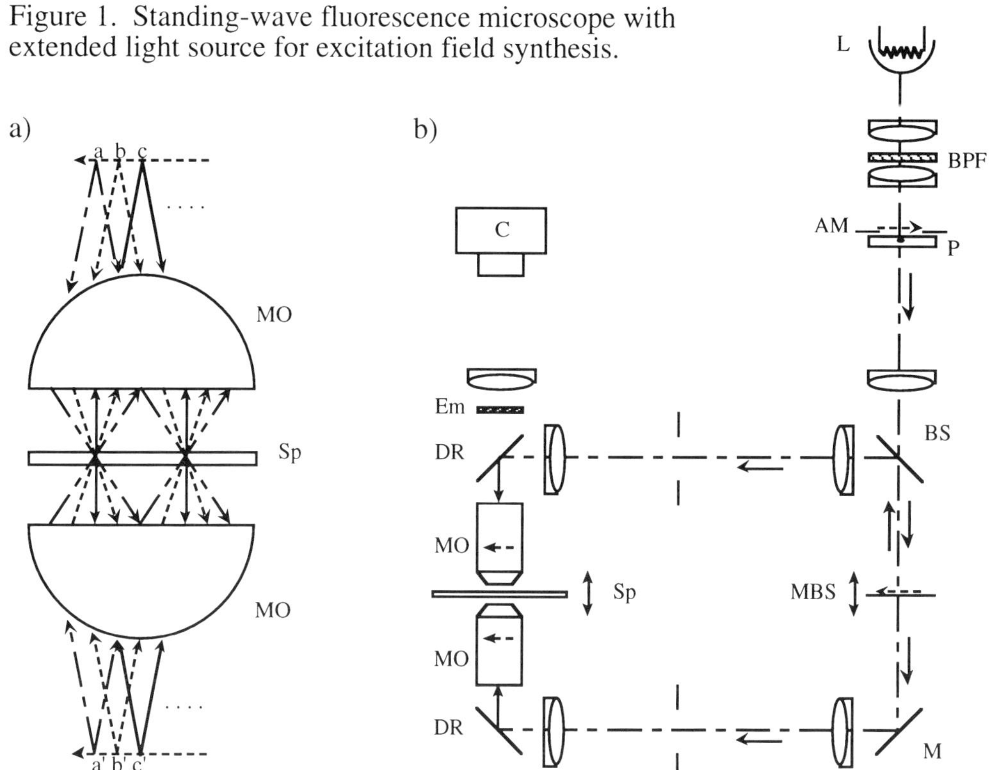

Figure 1. Standing-wave fluorescence microscope with extended light source for excitation field synthesis.

References
1 Lanni, F., B. Bailey, D.L. Farkas, and D.L. Taylor (1993). Excitation field synthesis as a means for obtaining enhanced axial resolution in fluorescence microscopes. Bioimaging 1: 187-196.
2. Lanni, F., D.L. Taylor, and B. Bailey (1995). Field synthesis and optical subsectioning for standing-wave microscopy. U.S. Patent 5,394,268.

3. Acknowledgement: This research is supported by the National Science Foundation Instrument Development Program (BIR-9217217), and through the Science and Technology Centers Program (MCB-8920118).

DESIGN OF AN INTERMEDIATE HIGH VOLTAGE EM FOR 3-D STUDIES OF BIOLOGICAL MATERIAL AND IT'S INTEGRATION WITH A SYSTEM FOR REMOTE ACCESS

Mark H. Ellisman

San Diego Microscopy and Imaging Resource, University of California San Diego, La Jolla, CA 92093-0608

The increased availability of High Performance Computing and Communications (HPCC) offers scientists and students the potential for effective remote interactive use of centralized, specialized, and expensive instrumentation and computers. Examples of instruments capable of remote operation that may be usefully controlled from a distance are increasing. Some in current use include telescopes, networks of remote geophysical sensing devices and more recently, the intermediate high voltage electron microscope developed at the San Diego Microscopy and Imaging Resource (SDMIR) in La Jolla. In this presentation the imaging capabilities of a specially designed JEOL 4000EX IVEM will be described. This instrument was developed mainly to facilitate the extraction of 3-dimensional information from thick sections. In addition, progress will be described on a project now underway to develop a more advanced version of the Telemicroscopy software we previously demonstrated as a tool to for providing remote access to this IVEM (Mercurio et al., 1992; Fan et al., 1992).

The goal of this new project is to design and implement a collaborative computational environment, providing a researcher at a remote site distributed interaction with a unique instrument, such as the IVEM, for the acquisition and manipulation of images of biological structure. The initial focus of this project is on the development of an integrated system for remote interactive acquisition and analysis of 2- and 3-dimensional electron microscopic data. To accomplish this task, a multidisciplinary team representing several national laboratories has been assembled. Participants include computer scientists specializing in HPCC, volume visualization, and visual image management, as well as biologists with expertise in computer-aided light and electron microscopy.

Although the software being developed is designed to facilitate use of many types of instruments, the main test-bed instrument in the first phases of the project will be the SDMIR IVEM which has been specially designed for digital acquisition of three dimensional data from biological specimens. The remote user's system will provide interactive control of image acquisition from the IVEM from any laboratory on the Internet.

A sophisticated image analysis, visualization, and data management environment will be described that is intended to provide enhanced capabilities for the scientists using the microscope from remote sites. The system design will provide transparent distribution of tasks that require extensive computation to high performance computers accessible on the network. These tasks include, the derivation of three dimensional structure using electron microscope tomography and procedures for the automatic extraction of image features. Examples will be given of biological projects that are expected to benefit from the development of a stable remote access system such as that summarized above.

References:

Mercurio, P.J., Elvins, T.T., Young, S.J., Cohen, P.S., Fall, K.R. and Ellisman, M.H. (1992) The distributed laboratory: An interactive visualization environment for electron microscopy and three- dimensional imaging. *Comm.Assoc.Comp.M ach*.35(6): 54-63.

Fan, G.Y., Mercurio, P.J., Young, S.J. and Ellisman, M.H. (1993) Telemicroscopy. *Ultramicroscopy* 52:499-503.

Proc. Microscopy and Microanalysis 1995, edited by G.W. Bailey, M.H. Ellisman, R.A. Hennigar, and N.J. Zaluzec
Copyright © 1995 MSA. Published by Jones and Begell Publishing, 79 Madison Ave., New York, NY 10016

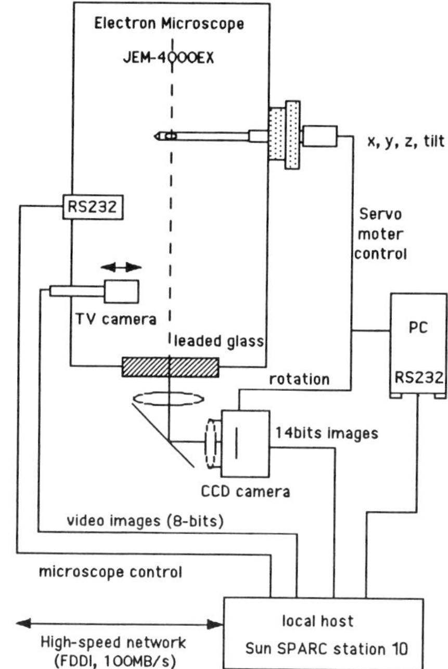

Figure 1. Telemicroscopy setup at the microscope. The host computer is on an FDDI ring, connecting the microscope to the Supercomputer Center where it links to the Internet.

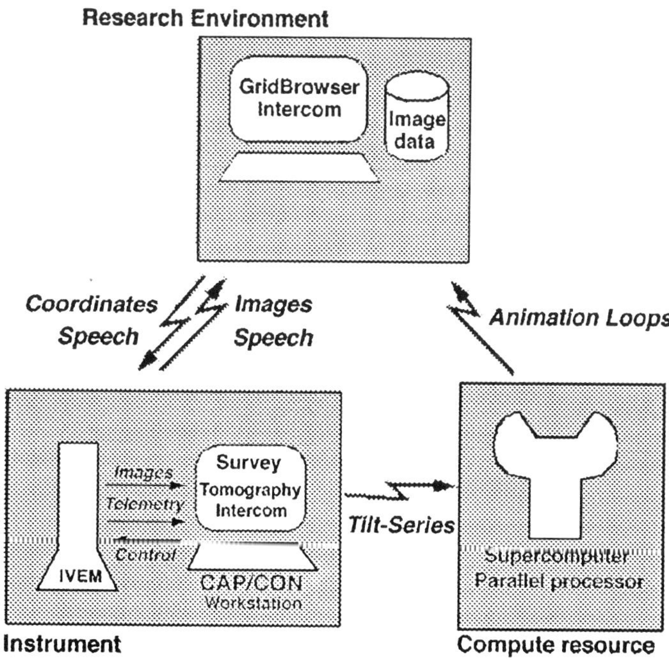

Figure 2. The triumvirate of interconnected sites: The remote user's "visible" third of the triumvirate is the workstation. The computer controlled microscope and super computing resources are the other two components.

Visualization of β-sheets and some amino acid side chains in 400 kV electron images of streptavidin crystallized on phospholipid monolayers.

Agustín J. Avila-Sakar* and Wah Chiu*†

*Dept. of Molecular Physiology and Biophysics, Baylor College of Medicine, Houston TX 77030;
†Verna and Marrs McLean Dept. of Biochemistry, Baylor College of Medicine, Houston TX 77030.

Two-dimensional (2-D) protein crystallization on lipid monolayers is an emerging technique with potential application to structure elucidation of biological macromolecules (Uzgiris and Kornberg, 1983). Among more than a dozen proteins or macromolecular assemblies which have been crystallized by this technique, streptavidin 2-D crystals diffract electrons beyond 3 Å (Kubalek, *et al.*, 1991), indicating that structural information from them is retrievable to near-atomic resolution. This 15 kD protein, synthesized by *Streptomyces avidinii,* is probably the single most widely used protein in modern biochemical assays. It forms nearly irreversible complexes (Kd ≈ 10^{-15} M) with the vitamin biotin and biotinylated molecules, which can be detected by a variety of methods.

2-D streptavidin crystals were grown on biotinylated lipid monolayers by a method similar to that of Kubalek, *et al.*, 1991. They were preserved in vitreous ice for electron cryomicroscopy. Electron images were recorded at 400 kV under flood beam illumination at a dose of 8 e-/Å^2. Phases from four micrographs were merged. The merging phase residual for all reflections with signal/noise greater than 1.6 up to 3 Å resolution was 18°. The phases had apparent *cmm* symmetry. The phase residuals for two in plane perpendicular mirror lines were 18° and 15°, respectively, for reflections with signal/noise greater than 1.6. The phase residual for a 2-fold rotation about an axis perpendicular to the projection plane at the intersection of those mirror lines was 24°.

The merged phases were combined with amplitudes from a 100 kV electron diffraction pattern. That pattern had a Friedel R-factor of 9% up to 2.5 Å resolution. The *cmm* symmetry R-factors were 20%, probably indicating slight tilt of the sample. A projection map was generated with the combined amplitudes from electron diffraction intensities and phases from images including data to 3 Å resolution with a *cmm* symmetry enforcement. To evaluate the merit of the reconstruction, we used the atomic coordinates of streptavidin, retrieved from the Brookhaven Protein Data Bank (Weber, *et al.*, 1989). Streptavidin is an eight-stranded β-barrel which forms tetramers with 222 point group symmetry. Therefore, the view could only be along one of the three mutually perpendicular 2-fold axes of the molecule. By projecting the atomic model along each of those axes, we could easily decide which was the correct one by its resemblance to the projection map. The other two views did not match the projection either in size or details.

In Figure 1, the model backbone from the x-ray crystal structure is overlaid on the projection map computed from electron microscopy data. One unit cell (**a=b**= 82.3 Å) is shown with the protein density in white contrast. For clarity, only two subunits of the tetramer of the model structure are shown. The labels **8,7,6** and **5** refer to the respective β-strands of the barrel. Notice that along the horizontal axis of the map densities add up from symmetry-related components of two subunits. In a region with densities from only one subunit, a pair of overlapping aromatic side chains generates a strong feature in the map (label **a**).

The experimental structure factors were compared to the model amplitudes and phases. 80% of the experimental phases up to 3 Å resolution are identical to the model. Most of those phases correspond to strong reflections. On the other hand, the R-factor was 0.39. This high value could be due in part to the inaccuracy introduced in the *cmm* averaging of the amplitudes. However, discrepancies in amplitudes and phases could also be a consequence of possible conformational changes of streptavidin on lipid monolayers.

In conclusion, we provide the first near-atomic view of a protein crystallized on lipid monolayers. Densities from β-strands and some amino acid side chains are identified. The methodology of 2-

Proc. Microscopy and Microanalysis 1995, edited by G.W. Bailey, M.H. Ellisman, R.A. Hennigar, and N.J. Zaluzec
Copyright © 1995 MSA. Published by Jones and Begell Publishing, 79 Madison Ave., New York, NY 10016

dimensional crystallization on phospholipid monolayer may be applicable to study a broad spectrum
of macromolecular complexes to high resolution by electron crystallography.

References:

Kubalek, E.W., Kornberg, R.D. & Darst, S.A. *Ultramicroscopy* **35**, 295-304 (1991).
Uzgiris, E.E. & Kornberg, R.D. *Nature* **301**, 125-129 (1983).
Weber, P.C., Ohlendorf, D.H., Wendoloski, J.J. & Salemme, F.R. *Science* **243**, 85-88 (1989).
Acknowledgments: This work was supported by grants from NCRR of NIH (R002250), R.A.
Welch Foundation and W.M. Keck Foundation.

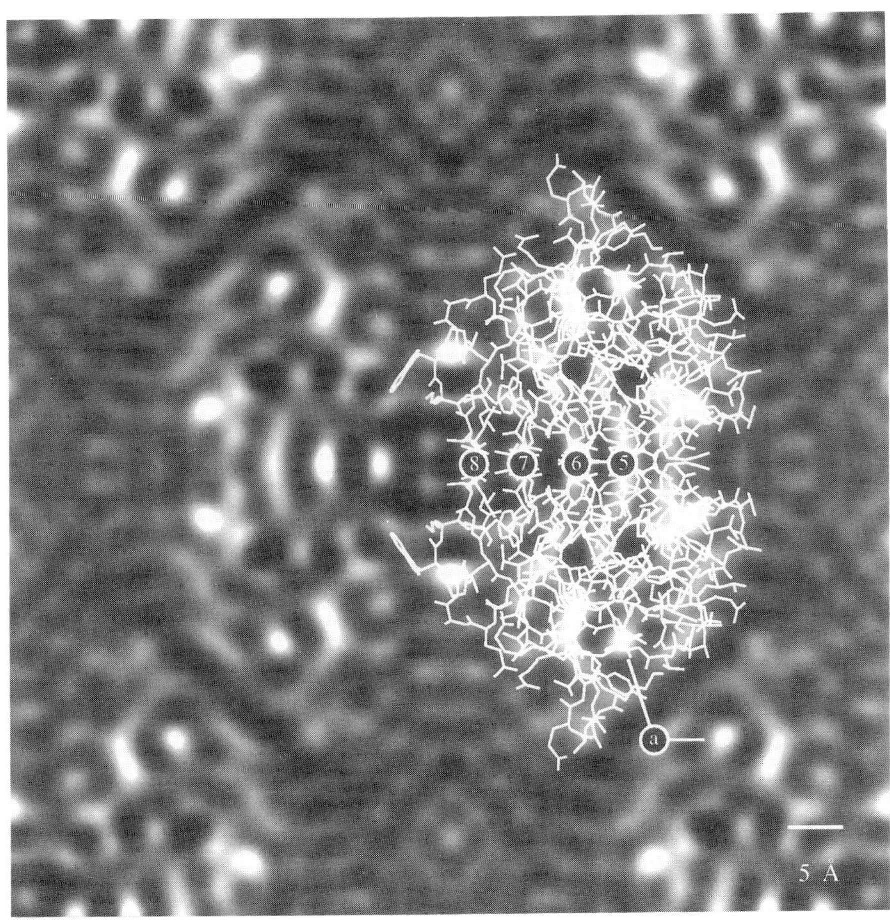

Figure 1. Projection reconstruction of streptavidin by electron crystallography of 2-D crystals
grown on biotinylated phospholipid monolayers. One unit cell is represented (**a=b**=82.3 Å). The
reconstruction includes data to 3 Å resolution. The atomic model of the molecule (retrieved from
Brookhaven Protein Data Bank; Weber, *et al.*, 1989) is overlaid on the projection map. For clarity,
only two subunits of the tetramer are shown. The labels **8**,**7**,**6** and **5** refer to β-strands of the
streptavidin β-barrel. Subunit overlap enhances the projected densities at the level of the horizontal 2-
fold symmetry axis (where the labels are placed). The label **a** points at a density where a pair of
overlapping aromatic side chains generates a strong feature in the map.

A MICROSCOPIC SYSTEM FOR HIGH RESOLUTION ELECTRON CRYSTALLOGRAPHY

Y. Fujiyoshi,* K. Murata,* K. Mitsuoka,* T. Hirai,* A. Miyazawa,** and Y. Kimura**

*International Institute for Advanced Research, Matsushita Electric Industrial, 3-4, Hikaridai, Seika, Kyoto, 619-02, JAPAN
**Protein Engineering Research Institute, 6-2-3, Furuedai, Suita, Osaka, 565, JAPAN

High-resolution electron cryo-microscopy is one of good candidate for structure analysis of membrane-protein, and also actually analyzed the structure of membrane-proteins such as bacteriorhodopsin (bR)[1] and plant light-harvesting complex (LHC)[2]. By developing an expeditious method for structure analysis up to atomic or near atomic resolution, we would like to interpret a function of protein from the structural point of view. However, there are some difficulties in electron microscopy for structure analysis of protein. Especially, the most serious problems are the specimen damage caused by electron irradiation, the denaturation of biomolecules caused by dehydration and missing high-resolution data on electron micrographs at high-tilted angle.

The irradiation damage at 8K has been found to be reduced to 1/20 compared with that at room temperature. We have, therefore, developed a high-resolution electron cryo-microscope and improved it by which images can be recorded with higher resolution than 3 Å at a specimen-stage temperature of 4.2 K, even when the specimen is highly tilted. The highly tilted data are essential for reduction of the missing corn effect.

The new cryo-microscope, our 3rd version, is equipped with FEG and Slow scan CCD camera (SSCCD)[3]. We have also developed an easily operated cryo-transfer device for our electron cryo-microscope (Fig. 1). This instrument and ice embedding technique enable us to take both high-resolution images and diffraction patterns of bacteriorhodopsin at various pH conditions between pH 4.0 and 10.0. The SSCCD increased efficiency of the data collection of electron diffraction and also improved the resolution limit as shown in Fig. 2 a,b. A typical example of high-resolution images is shown in Fig. 3 which is Fourier transform of an image of a two-dimensional crystal of bR. Numerical grades (IQ value) for each spot are shown as the size of the symbols in this figure. In accordance with a paper of Henderson[4], we calculated the contrast versus resolution for our image of bR as shown in Fig. 4 which includes data of both images taken by new cryo-microscope with FEG and the 2nd version of our microscope with LaB6 filament. The image taken by new microscope is apparently better than that by 2nd version.

Our microscope is very easy for operation as well as maintenance compare with the other microscope. Recent experiments indicate that it is possible to collect a complete phase data from two dimensional crystals with in weeks instead of years which has been the last a big obstacle except crystallization in electron crystallography, and in case of membrane proteins, we will have a chance to compete with X-ray crystallographers. Though apparently we are behind from x-ray crystallographers, we did surely take a step forward.

References

1. R. Henderson et al., *J. Mol. Biol.* 213 (1990) 899.
2. W. Kühlbrandt et al., *Nature* 367 (1994) 614.
3. O. L. Krivanek and P. E. Mooney, *Ultramicroscopy* 49 (1993) 95.
4. R. Henderson, *Ultramicroscopy* 46 (1992) 1.

Proc. Microscopy and Microanalysis 1995, edited by G.W. Bailey, M.H. Ellisman, R.A. Hennigar, and N.J. Zaluzec

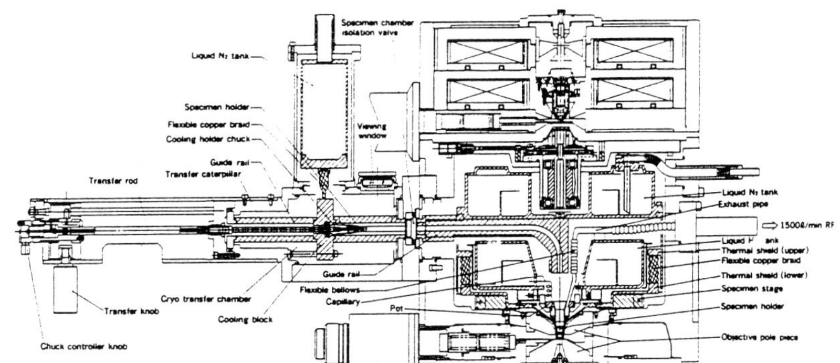

FIG. 1. Cross sectional drawing parallel to electron beam path of superfluid helium stage and cryo-transfer device.

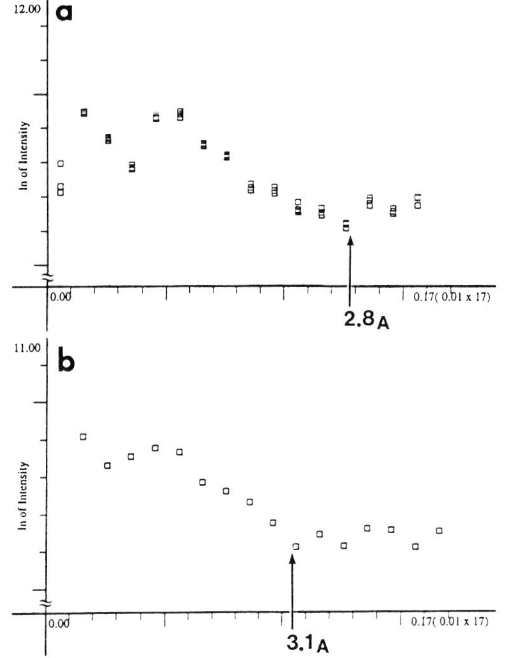

FIG. 2. Wilson plots of electron diffraction patterns at 60 degree tilt taken by Slow Scan CCD camera (a) and film (b).

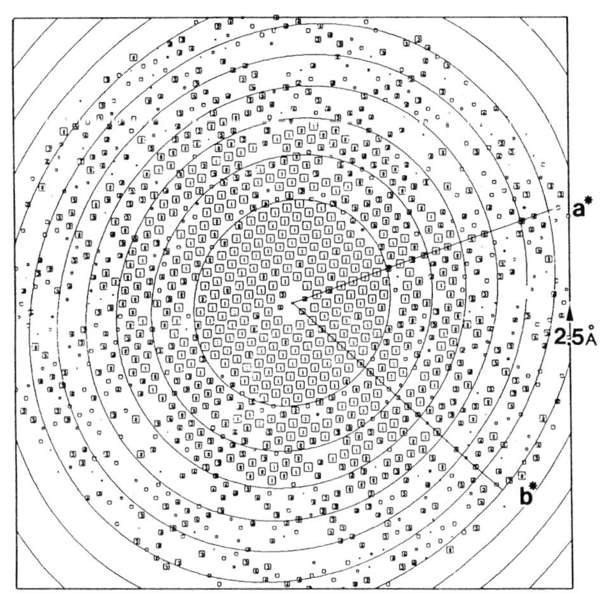

FIG. 3. Fourier transform of an image of bacteriorhodopsin. The size of the squares is proportional to the quality of the phase information. a* and b* are the unit cell vectors of the reciprocal lattice.

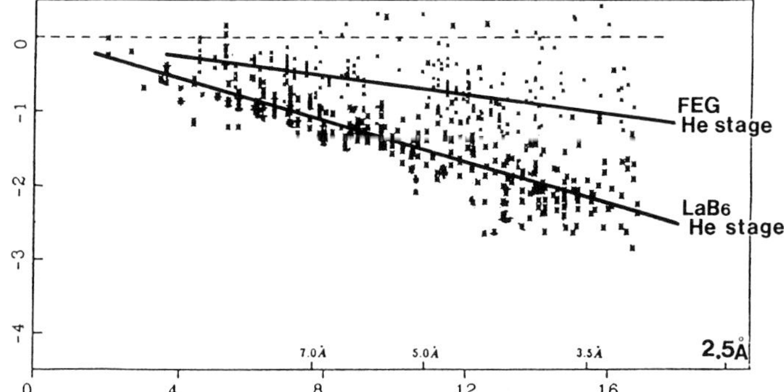

FIG. 4. Fall-off in amplitudes of images taken by cryo-microscope with FEG and without FEG veresus resolution.

SOPHIE, A HELIUM COOLED SUPERCONDUCTING ELECTRON MICROSCOPE WITH A SCHOTTKY FIELD EMITTER

H. Stark,* E. Beckmann,* R. Henderson** and F. Zemlin*

* Fritz-Haber-Institut der Max-Planck-Gesellschaft, 14195 Berlin, Germany
** Medical Research Council, Cambridge, UK

High-resolution cryo-electron microscopy yielded remarkable results in biological molecular structure research [1,2]. These successes encouraged to proceed in developing even better instruments and procedures. In our laboratory we built the cryo-electron microscope "SOPHIE", a modified Philips CM 20 FEG. SOPHIE stands as an acronym for Superconducting Objective in a PHIlips Electronmicroscope. The main features of this microscope are the following:
1. 200kV acceleration voltage
2. Schottky field emitter
3. Helium cooled superconducting objective
The Helium-cooled cryo-objective from the former microscope SULEIKA [3] was adapted into the column of the original Philips CM20 FEG [Fig.1]. This new SOPHIE microscope, thus, combines two important features: a modern computer controlled electron microscope with a high coherent beam source and the proven advantages of the superconducting cryo-objective. The advantage of the field emission gun at 200 kV over a thermionic cathode at 100 kV can be best visualized while comparing the respective phase-contrast-transfer functions at Scherzer focus [Fig.2]. Moreover, SOPHIE is equipped with the small spot scanning procedure [4]. Using a FEG in the small-spot scanning mode the coherence of the illumination remains even if spots smaller than 1000 Å in diameter are employed.

The advantage of low specimen temperature is the reduction of electron radiation damage. Electron micrographs of crystal sheets of purple membrane obtained at 4K enabled us to solve the structure of the protein bacteriorhodopsin at nearly atomic resolution (2.8 Å).

To be able to image ice embedded specimens, the microscope is equipped with a cryotransfer airlock. The cryotransfer procedure is carried out under liquid nitrogen temperature and during the transfer the specimen is under an additional shield, assuring a "contamination-free" transfer. A very helpful facility associated with the microscope is the on-line digital Fourier transformation and its display. The continuous monitoring of the power spectra assists in adjusting focus conditions and in correcting astigmatism. With the use of the on-line FFT we could recognize any disturbing magnetic stray field and corrected it with a dynamical Helmholtz field compensation.

The work utilizing the advantages of SOPHIE is under progress and results shall be presented at the conference.

[1] R. Henderson et al., *J. Mol. Biol.* (1990) **213**, 899-929
[2] W. Kühlbrandt et al., *Nature* (1994) **367**, 614-621
[3] I. Dietrich et al., *Ultramicroscopy* (1977) **2**, 241-249
[4] I. Tews, Diplomarbeit (1993)
[5] Acknowledgement: The authors are grateful to Elmar Zeitler for his continuous support. This work was financially supported by the comission of the European Communities, Science Plan ERBSC1*CT000734.

Proc. Microscopy and Microanalysis 1995, edited by G.W. Bailey, M.H. Ellisman, R.A. Hennigar, and N.J. Zaluzec
Copyright © 1995 MSA. Published by Jones and Begell Publishing, 79 Madison Ave., New York, NY 10016

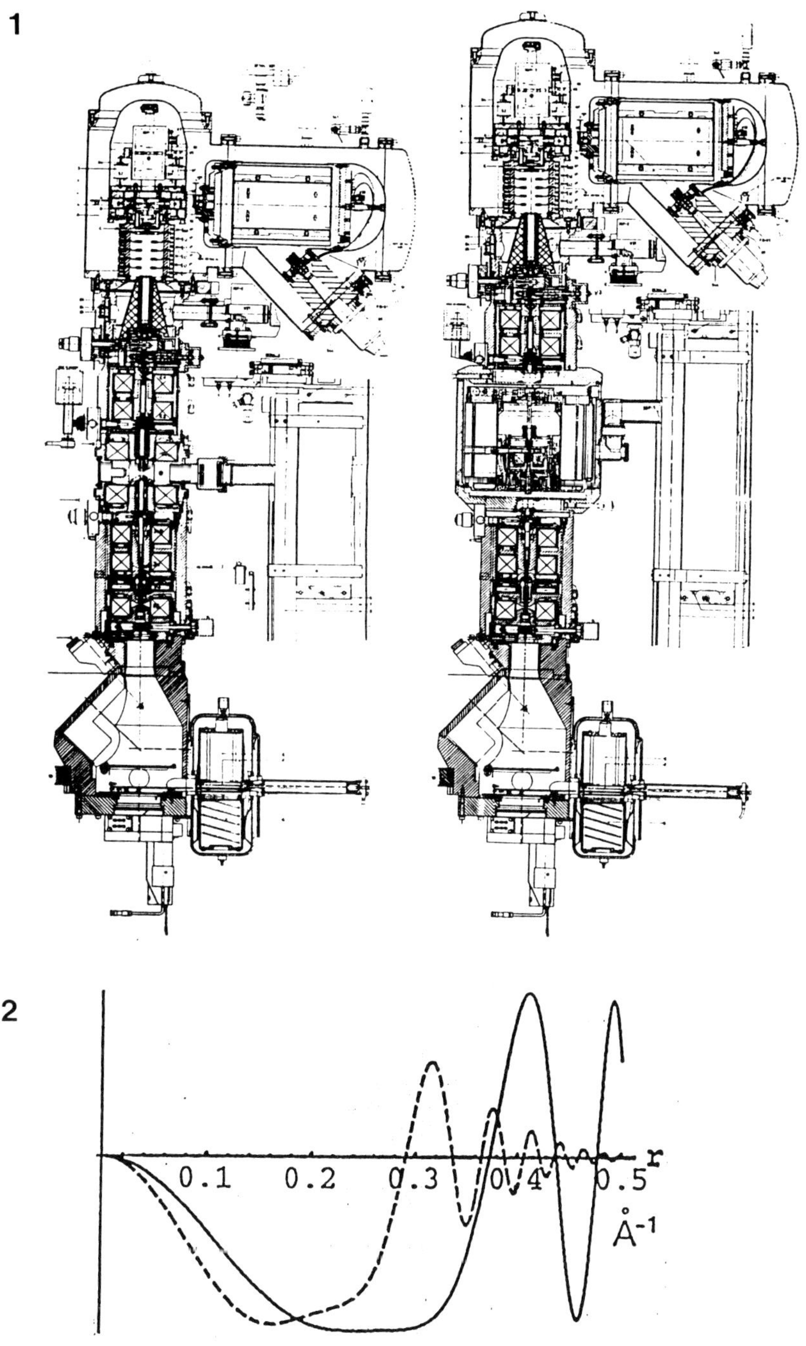

Fig.1: Cross-section of the conventionel microscope CM20 FEG (left) and of SOPHIE (right)

Fig.2: Phase-contrast-transfer functions at Scherzer focus of the cryo-objective
dashed: thermionic beam source at 100 kV (SULEIKA)
line: FEG at 200 kV (SOPHIE)

NEW APPROACHES TO CHARACTERIZING ADVANCED MATERIALS

David J. Smith

Center for Solid State Science and Department of Physics and Astronomy
Arizona State University, Tempe AZ 85287

Motivations for using the electron microscope are obviously many and varied. For example, engineers in the semiconductor industry might be primarily interested in establishing reasons for device failure. Chemists in the petrochemical industry could be concerned with analyzing the composition and structure of novel catalytic materials. Many researchers seek to characterize microstructure and establish definitive connections with preparation conditions and/or some pertinent macroscopic behavior. Instrumentation for the high-resolution electron microscope (HREM) has continued to evolve to the extent that imaging on the atomic scale and microanalysis on the sub-nanometer scale are oftentimes available from the same microscope. Such instruments are thus highly attractive to all those people interested in characterizing advanced materials. Our purpose here is to provide a brief overview of some recent developments in instrumentation and techniques and to highlight their relevance for materials science applications.

It is just over a decade since custom-designed HREMs first successfully exceeded the 0.2nm interpretable resolution limit on a routine basis.[1,2] Intermediate- and high-voltage electron microscopes soon became available commercially that could also regularly attain this performance level (for further details, see Reference 3). Further improvements have recently been reported. The highly coherent illumination available with 300-kiloVolt field-emission-gun HREMs has facilitated the achievement of instrumental resolution approaching 0.1nm aided either by through-focal series reconstruction[4] or electron holography.[5] The latest generation of high-voltage HREMs have even succeeded in closely approaching the long-sought-after goal of 0.1nm without reversals in the contrast transfer function,[6-8] thereby eliminating the need for any off-line image processing and enabling direct image interpretation to that level. Unexpectedly, structural damage due to electron irradiation is not proving to be a serious limitation in these latter instruments during extended observation of many materials[9]: it appears that recombination processes are effective in healing, at least partially, any damage resulting from knock-on displacements. These two types of instruments with ~0.1nm resolution limits should play a major role in future analysis of structural defects, in particular since imaging at improved resolution or in higher-order projections allows more reliable three-dimensional structural models of defects and interfaces to be developed. An excellent example worthy of note is the recent study of Cu twin boundaries[10]: much improved structural sensitivity was achieved by imaging at 0.1nm resolution and 1.25MeV rather than 0.17nm at 400keV.

Further developments in instrumentation of much potential benefit to materials characterization warrant being mentioned. Energy-filtered imaging over respectable fields of view (1024x1024 pixels) has been realized utilizing either in-column,[11] or post-column[12] filter systems. Comparable performance levels have been achieved with both systems: either type allows enhanced recording and interpretation of convergent beam electron diffraction patterns, as well as compositional analysis with chemical imaging and elemental mapping at resolution levels primarily determined by the de-localized nature of the inelastic scattering processes or by limited signal-to-noise ratios for higher energy losses. The addition of the field emission gun to a high-resolution TEM already equipped with facilities for energy-dispersive Xray spectroscopy (EDS) and electron-energy-loss spectroscopy (EELS) likewise represents a powerful new tool for materials science.[13] As well as the convenience of standard, high-resolution imaging, chemical analysis on the sub-nanometer scale from readily identifiable regions is achievable. Finally, mention should be made of the capability for *in situ* experimentation with negligible compromise of microscope performance, as recently demonstrated by the addition of ion-beam interfaces to study irradiation phenomena,[14] and the utilization of specimen heating holders to follow surface reactions[15] and thin-film transformation mechanisms.[16]

Proc. Microscopy and Microanalysis 1995, edited by G.W. Bailey, M.H. Ellisman, R.A. Hennigar, and N.J. Zaluzec
Copyright © 1995 MSA. Published by Jones and Begell Publishing, 79 Madison Ave., New York, NY 10016

The great attraction of the TEM is its ability to provide images of structural irregularities on a highly localized scale in a way that is unattainable with other bulk, diffraction or microscopy techniques. Over the years, the electron microscope has played a major role in furthering our knowledge and understanding of many different types of advanced materials, far too many to be mentioned individually here. For example, the high-Tc superconductors have received much attention from the electron microscopy community.[17] In recent years, novel types have been discovered and investigated by TEM, such as the oxycarbonates[18] which must be prepared under very high pressures of several GPa. Following closely on the discovery of fullerenes based on the Carbon-60 cluster, electron microscopists have been highly active in this new materials area, for example in characterizing structural defects[19] and studying carbon nanotubes.[20] Ultrathin films based on magnetic superlattices have intriguing magnetoresistive behavior resulting from anti-ferromagnetic coupling between closely spaced magnetic layers: we have undertaken a comprehensive TEM study aimed at elucidating the relationship between microstructure, deposition parameters and magnetic properties.[21] The demonstrated ability of high-voltage HREMs to distinguish individual atomic columns in compound semiconductors[7,8] should prove to be of immense value in studying semiconductor heterostructures, particularly for establishing the nature of interfacial structure and bonding and its relationship to crystal polarity.

Steady improvements in microscope performance and auxiliary instrumentation have consistently in the past led to an increase in the range of materials problems that can be usefully tackled by microscopy methods. If success is achieved in the ongoing efforts to correct spherical aberration,[22] then further expansion will undoubtedly occur. Meanwhile, through-focal series reconstruction or electron holography as methods for determining exit-surface wavefunctions remain time-consuming and highly specialized: development of routine, user-friendly image processing software would allow considerable progress to be made towards truly quantitative structure refinement.[23] Electron holography has traditionally been regarded as a method for extracting phase information: more attention to the information contained in the amplitude signal could be especially valuable in some materials applications.[24] Overall, however, it can be concluded that an impressive array of instrumentation and techniques is already available that should be gainfully employed in characterizing advanced materials.

References

1. D.J. Smith et al., J. Microscopy 130 (1983) 127.
2. K. Yagi et al., Proc. 7th. Int. Conf. HVEM (Berkeley, 1983) pp. 11-14.
3. D.J. Smith, J. Electr. Micr. Techn., 12 (1989) 11.
4. W. Coene et al., MSA Bull., 24 (1994) 472.
5. A. Orchowski et al. Phys. Rev. Letts., 74 (1995) 399
6. Y. Matsui et al., Ultramicroscopy 39 (1991) 8.
7. F. Phillipp et al., Ultramicroscopy 56 (1994) 1.
8. H. Ichinose, reported at Int. Workshop on HV & HREM, Stuttgart, 21-24 Feb., 1994,
9. M. Ruhle, reported at 2nd. NIRIM Int. Symp. on Advanced Materials, Tsukuba, 6-10 March, 1995.
10. D. Hofmann and F. Ernst, Ultramicroscopy 53 (1994) 205.
11. A. Berger et al., Ultramicroscopy 55 (1994) 101.
12. O.L. Krivanek et al., Microscopy, Microanal., Microstr., 2 (1991) 315.
13. Y. Bando et al., Electron Microscopy 1994, Vol. 1, pp.591-592.
14. N. Ishikawa and K. Furuya, Ultramicroscopy 56 (1994) 211.
15. D.J. Smith et al., Ultramicroscopy 49 (1993) 26.
16. R. Sinclair, Mater. Res. Soc. Bull., 19(6) (1994) 26.
17. H. Zandbergen, Physica C 156 (1988) 325.
18. J. Akimitsu et al. Physica C 211 (1992) 320.
19. G. van Tendeloo et al., Europhys. Lett. 15 (1991) 295.
20. S. Iijima, Nature 354 (1991) 56.
21. D.J. Smith et al. Scripta Met. et Mater., 30 (1994) 689.
22. H. Rose, Ultramicroscopy 56 (1994) 11.
23. D.J. Smith et al. Ultramicroscopy 52 (1993) 591.
24. M.R. McCartney and M. Gajdardziska-Josifovska, Ultramicroscopy 53 (1994) 283.

HIGH RESOLUTION TRANSMISSION ELECTRON MICROSCOPY OF DISLOCATIONS IN INTERMETALLIC COMPOUNDS

M. J. Mills

Department of Materials Science and Engineering, The Ohio State University, Columbus, OH 43210

The fine structure of dislocations plays a critical role in determining the macroscopic mechanical behavior of intermetallic compounds. Many of the technologically important characteristics of these compounds, for example their strength at high temperatures, appear to be determined by intricate details of dislocation structure at the atomic level[1]. High resolution transmission electron microscopy (HREM) offers the potential to obtain structural information at this level by observing these line defects in an "end-on" configuration.

Examples of HREM images of several important dislocation types in Ni3Al and TiAl are shown in Figures 1 through 3. Each of these particular dislocation types (i.e. Burgers vectors and line directions) tend to be strongly favored in these compounds, indicating that along these line directions the dislocations are likely to have either low mobility or low energy. As a result, HREM of these dislocation types is particularly relevant in terms of the deformation mechanisms. In addition, the alignment of dislocations along relatively low index crystal directions makes the HREM observations feasible.

Analysis of such core images requires extensive comparison with computer simulated images in order to account for several imaging artifacts which can arise. One is the discrepancy between image intensities and atomic column positions (or tunnels between columns), particularly close to the "core" of the dislocation. A second issue is the possible rearrangement of dislocations in the thin-foils required for HREM observation. These artifacts have been addressed by performing image simulations based on atomistic calculations. These calculations which have been based on the Embedded Atom Method (EAM) not only provide reasonable model structures of the dislocation cores with which to compare to experiment, but also enable a more rigorous accounting of the effects of the thin-foil geometry used in HREM.[2]

The images of screw superdislocations in Ni3Al (L1$_2$ structure) and TiAl (L1$_0$ structure) shown in Figures 1a and 2a show the dissociation of perfect a[$\bar{1}$01] into non-planar configurations in which the Burgers vector content of the dislocation is actually lying on more than one crystallographic plane. The specific configurations deduced from comparison with simulations are shown in the corresponding diagrams shown in Figures 1b and 2b.[3] In both cases, an anti-phase boundary (APB) is geometrically necessary, although it is not visible in the viewing direction shown. Bounding this APB are a/2[$\bar{1}$01] superpartials which dissociate *on different {111} planes*, resulting in a non-planar, overall configuration. These non-planar configurations can arise through a thermally-activated cross-slip-pinning process[4], which causes the anomalous increase in the flow strength in both Ni3Al and TiAl with increasing temperature. While the dissociation of these superdislocations into superpartials can be detected using weak-beam imaging, the subsequent dissociation of the superpartials is much more difficult, if not impossible, to observe using diffraction contrast approaches. An additional advantage of the HREM technique is that the non-planar configuration of the dislocations can be determined directly without the need to perform multiple observations in several viewing directions as is necessary using weak-beam imaging. Nevertheless, the diffraction contrast imaging of dislocations lying parallel to the foil plane provides information concerning the variation of dislocation structure along the line direction – essential information which is complementary to the HREM results.

A second type of dislocation observed in TiAl is the a/2[110] 60° dislocation shown in Figure 3. In contrast to the other dislocation configurations described here, this core is remarkably compact with no apparent dissociation out of its glide plane. The fact that this dislocation is also present under conditions for which the strength of TiAl increases anomalously with temperature implies that a very different mechanism from the cross-slip pinning mechanism described above must also be operative in TiAl. Detailed comparison between observed images and atomistic calculations have provided the first experimental estimate of the complex stacking fault energy in TiAl.[5]

Proc. Microscopy and Microanalysis 1995, edited by G.W. Bailey, M.H. Ellisman, R.A. Hennigar, and N.J. Zaluzec
Copyright © 1995 MSA. Published by Jones and Begell Publishing, 79 Madison Ave., New York, NY 10016

These HREM observations have been conducted using a JEOL 4000EX with a LaB$_6$ filament.[6] Similar studies are currently being pursued using a Philips CM200 field emission gun microscope equipped with an energy filter. The latter instrument offers the potential of: (a) higher spatial resolution; (b) imaging using only elastically-scattered electrons and (c) local chemical information at the dislocation cores using electron energy loss spectroscopy. Progress in the application of this technique to the study of dislocation core structure and chemistry will also be reported.

References

1. M. S. Duesbery, in F. R. N. Nabarro, Ed., *Dislocations in Solids*, Amsterdam: Elsevier (1989) 67.
2. M. J. Mills, M. S. Daw and S. M. Foiles, *Ultramicroscopy*, **56** (1994) 79.
3. K. J. Hemker, B. Viguier and M. J. Mills, *Mater. Sci.and Eng. A*, **164** (1993) 391.
4. V. Paidar, D. P. Pope and V. Vitek, *Acta metall. mater.*, **40** (1984) 435.
5. J. P. Simmons, S. Rao and M. J. Mills, in J. H. Horton, et al., Eds., *High Temperature Ordered Intermetallic Compounds VI*, Pittsburgh: Materials Research Society, in press.
6. The partial support of the U.S. Department of Energy, Office of Basic Energy Sciences under Contract No. DE-AC04-94AL85000 is gratefully acknowledged.

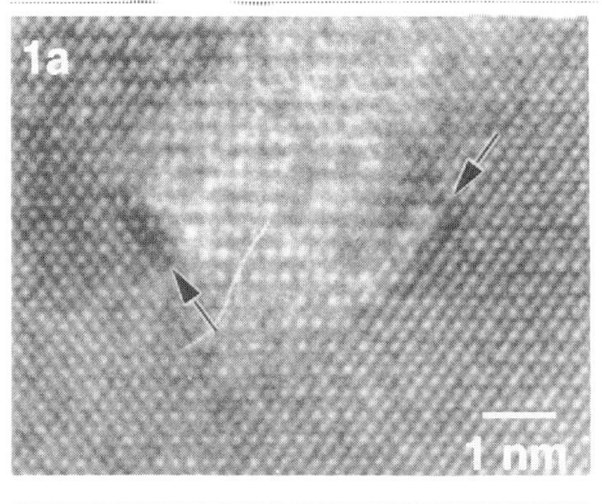

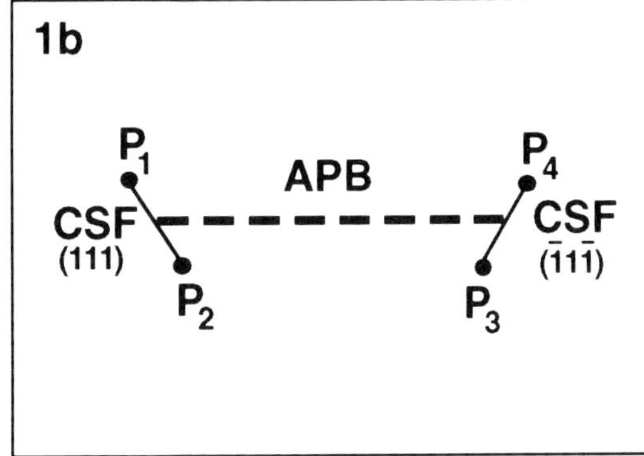

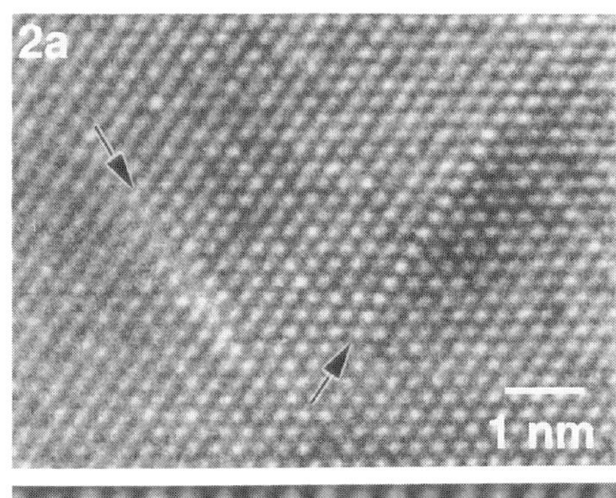

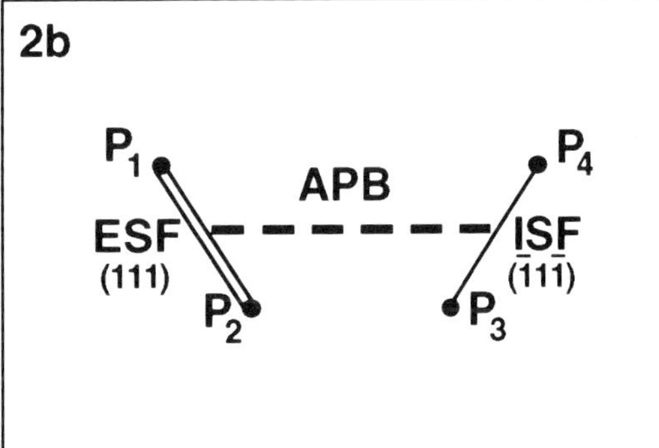

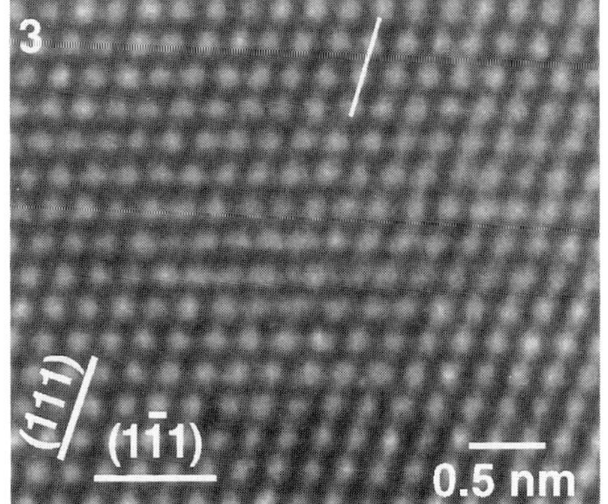

FIG. 1.–(a) HREM image of an a[$\bar{1}$01] dislocation in Ni$_3$Al; (b) proposed dissociation into a/6<112> partials (P) creating complex stacking faults (CSFs).

FIG. 2.–(a) HREM image of an a[$\bar{1}$01] in TiAl; (b) proposed dissociation into a/6<112> partials (P) creating an extrinsic and an intrinsic stacking fault (ESF and ISF).

FIG.3.– HREM image of an a/2[110] dislocation in TiAl exhibiting an extremely compact core structure. Line indicates fringe associated with extra half-plane.

CORRELATING IMAGING AND SPECTROSCOPY AT ATOMIC RESOLUTION IN THE STEM

N. D. Browning*, A. J. McGibbon**, D. E. Jesson** and S. J. Pennycook**

*Dept. Physics (M/C 273), U. Illinois at Chicago, 845 W. Taylor St., Chicago, IL 60607-7059. USA
**Solid State Division, Oak Ridge National Laboratory, P. O. Box 2008, Oak Ridge, TN 37831. USA

The study of materials by electron microscopy has traditionally been divided into two sub-fields, imaging and microanalysis. While both of these have advanced to the level of atomic sensitivity, a complete characterization of a materials structure, composition and bonding requires a correlation between the two. This correlation has been achieved through the combination of Z-contrast imaging and electron energy loss spectrocopy (EELS) in the scanning transmission electron microscope (STEM)[1-3].

For the highly coherent and convergent probe in the STEM, the scattering to a high-angle detector (40-150 mrad) shows an intensity which is proportional to the Z^2-dependence of the Rutherford scattering cross-section. The resultant "Z-contrast" image, which is generated point-by-point as the probe is scanned over the surface of the specimen, can be described as a simple convolution of the probe intensity profile, $P^2_{eff}(\mathbf{R})$, and a specimen object function $O(\mathbf{R},t)$[4],

$$I(\mathbf{R},t) = O(\mathbf{R},t) * P^2_{eff}(\mathbf{R}) \tag{1}$$

This incoherent description of image intensity holds for crystalline materials in zone-axis orientations where dynamical diffraction manifests itself as a columnar channeling effect. Therefore, in cases where the atomic spacing is larger than the incident probe size, a direct, intuitive map of the atomic column locations in the specimen is generated. This simple description of the image intensity permits the specimen object function to be retrieved through the maximum entropy processing technique[5]. Applied to Z-contrast images[6], maximum entropy retains the intensities from the image and can quantify atomic column locations to ~0.2Å. An example of this is shown in figure 1, where the maximum entropy processing has been applied to the Z-contrast image of a 25° [001] symmetric tilt boundary in $SrTiO_3$ (Note that the maximum entropy routine takes the column locations down to individual pixels, but to aid in visualization of the result a blurring operator has been applied).

For EELS, provided the signal is sufficiently localized, the same description of the the resultant intensity in terms of a convolution of a probe intensity profile with a specimen object function can be applied[7] (achieved in practice from the core-loss region of the spectrum). Additionally, since the Z-contrast image uses only the high-angle scattering, energy loss spectroscopy can be performed simultaneously with the image. Spectra can be taken from atomically precise locations and correlated directly with the structure observed in the image. For the 25° [001] tilt boundary in $SrTiO_3$, both oxygen K-edge and titanium L-edge profiles were acquired across the boundary. Shown in figure 2 are the spectra from the bulk and the boundary plane for comparison. For the titanium L-edge, the onset, total edge intensity and L_2/L_3 branching ratio all remained constant, indicating that the average valence remains constant across the boundary. For the oxygen K-edge, the only variation is in the relative intensity of the π^* and σ^* peaks, consistent with a slight distortion from the linear Ti-O-Ti coordination at the boundary.

In figure 1, it is noticeable that there are positions where two columns are seperated by distances much shorter than in the bulk material. Since the Z-contrast technique is incoherent in nature, both of these columns must exist. To avoid problems of like-ion repulsion, a simple model for partial occupancy which assumes the columns are alternately occupied through the thickness of the material is employed. Rationalization of the spectroscopic information with the atomic structure is achieved through bond-valence sum analysis[8]. In these calculations, the amount that a particular bond contributes to the overall valence of the elements involved in the bond is dependent on the bond-length. For the boundary here, the metal column positions were taken directly from the maximum entropy object function and oxygen atoms were positioned so as to maintain the bulk valences and preserve linear titanium oxygen

Proc. Microscopy and Microanalysis 1995, edited by G.W. Bailey, M.H. Ellisman, R.A. Hennigar, and N.J. Zaluzec
Copyright © 1995 MSA. Published by Jones and Begell Publishing, 79 Madison Ave., New York, NY 10016

coordination (figure 3). From this analysis, the boundary plane is stoichiometric and the average valences are all within 0.1 of their bulk values with no individual valence varying by more than 0.5 from the average. Additionally, a dilatation of (0.6 ± 0.2)Å normal to the boundary plane and a rigid body translation parallel to the boundary of (0 ± 0.2)Å have been determined directly from the image. It should be noted that this determination of the boundary structure comes directly from the experimental results and can therefore serve as a starting model or independent comparison for theoretical simulations[9].

References

1. N. D. Browning et al, *Nature* 366(1993)146
2. P. E. Batson, *Nature* 366(1993)727
3. M. M. McGibbon et al, *Science* 266(1994)102
4. S. J. Pennycook and D. E. Jesson, *Phys.Rev.Lett* 64(1990)938
5. S. F. Gull and G. J. Daniell, *Nature* 272(1978)686
6. A. J. McGibbon et al, submitted *J. Microscopy*
7. S. J. Pennycook et al, in press *Nucl. Inst & Methods*
8. N. D. Browning et al, in press *Interface Science*
9. Research sponsored by DOE-DMS, under contract DE-AC05-84OR21400 with Martin Marietta Energy Systems Inc, and by the ORNL Postdoctoral Research Program administered by ORISE.

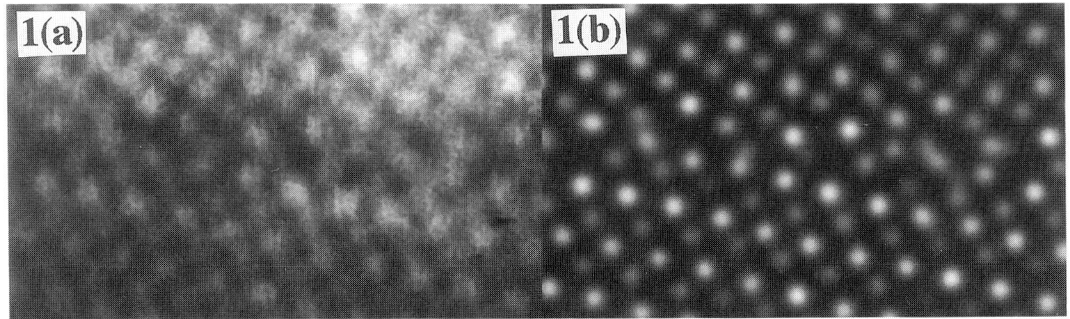

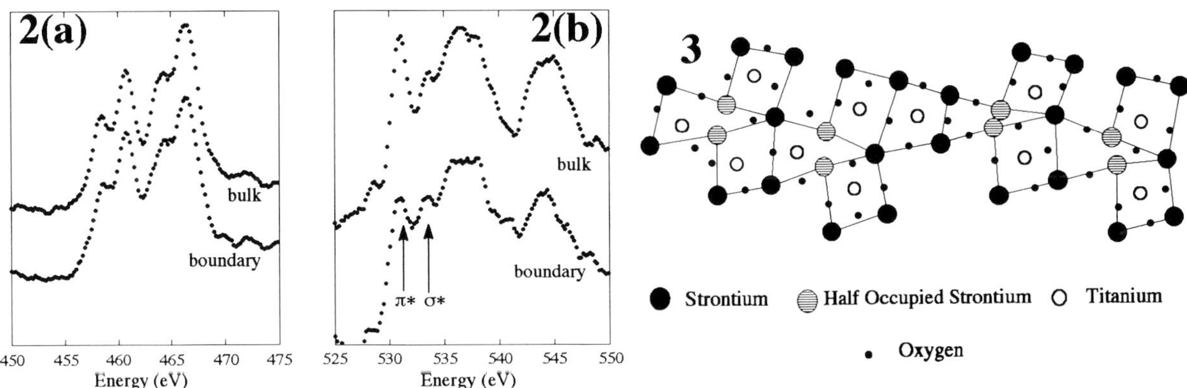

Figure 1: (a) Z-contrast image and (b) maximum entropy object for a 25° [001] tilt boundary in $SrTiO_3$.
Figure 2: (a) Titanium L-edge and (b) Oxygen K-edge spectra from the bulk and boundary plane.
Figure 3: Boundary structure determined from the experimental data through bond valence sum analysis.

HIGH VOLTAGE ELECTRON MICROSCOPY: WHERE IT IS AND WHERE IT MAY BE GOING

Charles W. Allen

Electron Microscopy Center for Materials Research—HVEM-Tandem Facility[1], Materials Science Division, Argonne National Laboratory, Argonne, IL 60439 USA

High voltage TEMs were introduced commercially thirty years ago, with the installations of 500 kV Hitachi instruments at the Universities of Nogoya and Tokyo. Since that time a total of 51 commercial instruments, having maximum accelerating potentials of 0.5–3.5 MV, have been delivered. Prices have gone from about a dollar per volt for the early instruments to roughly twenty dollars per volt today, which is not so unreasonable considering inflation and vastly improved electronics and other improvements. The most expensive HVEM (the 3.5 MV instrument at Osaka University) cost about 5 percent of the construction cost of the USA's latest synchrotron.

Table 1 briefly traces the development of HVEM in this country for the materials sciences. There are now only three available instruments at two sites: the 1.2 MeV HVEM at Argonne National Lab, and 1.0 and 1.5 MeV instruments at Lawrence Berkeley National Lab. Fortunately, both sites are user facilities funded by DOE for the materials research community. The U.S. life science community has had three HVEMs available since the 1970's, with AEI instruments at the University of Wisconsin (Madison) and the Wadsworth Center for Laboratories and Research (Albany) and a JEOL instrument at the University of Colorado (Boulder), all of which having been funded by NIH.

HVEMs are the instrument of choice for several types of studies reflecting certain characteristics of these instruments. Higher energy electrons (1) penetrate thicker specimens (more bulk-like behavior), (2) produce Frenkel pairs in crystalline materials (irradiation effects studies), (3) cause less ionization damage in organic materials and (4) allow higher spatial resolution for imaging (ultra-high resolution TEM). The more generous space associated with the objective lenses of many HVEMs accommodates more robust and complex specimen holders and other apparatus for an experiment such as an environmental cell or even in situ material processing. The wider range of electron energies associated with HVEMs allows more applications involving the critical voltage effect and irradiation effects studies involving a broad range of specimen damage rates, including no damage in many material systems.

Table 2 briefly summarizes the status of the world's living HVEMs, including their principal use in three broad categories: high resolution imaging analysis, in situ experimental studies and life science imaging analysis. The classification 'In Situ (I)' means that one or more ion accelerators have been interfaced to a particular HVEM for conducting in situ ion irradiation effects studies. The newest instrument in this category is a side entry JEOL 1000 in the National Research Institute for Metals (Tsukuba, Japan) which is interfaced to two ion sources; its point-to-point resolution for imaging is 0.13 nm. The most recent instrument dedicated to ultra-high image resolution is the JEOL 1250 in the Max-Planck-Institute (Stuttgart, Germany). the point-to-point resolution of which is 0.1 nm or better; this instrument may be converted to side entry for later in situ experiments with the installation of a second objective lens section. The Hitachi 3500 top entry HVEM, installation of which is to be completed in 1995, is an instrument for in situ experiments primarily and is unique in several ways including remote controlled operation and specialty top entry holders including a 2200 K heating holder. All of these facilities welcome users from other institutuons.

Reference

1. Electron Microscopy Center for Materials Research is supported by U. S. Department of Energy, BES-Materials Sciences, under Contract W-31-109-Eng-38.

Proc. Microscopy and Microanalysis 1995, edited by G.W. Bailey, M.H. Ellisman, R.A. Hennigar, and N.J. Zaluzec
Copyright © 1995 MSA. Published by Jones and Begell Publishing, 79 Madison Ave., New York, NY 10016

Table 1.—Development of HVEM for Materials Research in USA. *Italics = decommissioned.*

Installed	Location	Manufacturer	Maximum Voltage (MV)
1967	*U. S. Steel Corp. Bain Res. Lab.*	*RCA*	*1.0*
1968	*U. of Virginia*	*RCA*	*0.5*
1968	*U. of California, Berkeley*	*Hitachi*	*0.65*
1970	*Oak Ridge Natl. Lab.*	*Hitachi*	*0.65 (1.0)*
1970	*Case-Western Reserve U.*	*Hitachi*	*0.65*
1972	*Westinghouse Hanford Engr. Dev. Center*	*JEOL*	*1.0*
1979	Argonne Natl. Lab.	AEI	1.2
1981	Lawrence Berkeley Lab	AEI	1.5
1983	Lawrence Berkeley Lab.	JEOL	1.0

Table 2.—Operational HVEMs throughout the world in 1995 and their primary use.

Installed	Location	Manufacturer	Maximum Voltage (MV)	Use
1969	Toulouse	(In House)	3.1	In Situ
1971	U. of Wisconsin—Madison	AEI	1.0	Life Sci
1971	Max-Planck-Inst.—Halle	JEOL	1.0	In Situ
1972	Faculty of Engr., Nagoya U.	Hitachi	1.0	In Situ
1972	U, of Colorado—Boulder	JEOL	1.0	Life Sci
1974	Japan Atomic Energy Res. Inst.—Tokai	JEOL	1.0	In Situ
1974	U. of Antwerp	JEOL	1.25	HREM
1974	Shin Nippon Steel Corp.	Hitachi	1.0	In Situ
1975	Baikov Inst. of Metallurgy, Moscow	JEOL	1.0	In Situ
1976	U. of Kyushu	JEOL	1.25	In Situ (I)*
1977	Centr. Res. Inst. of Non-Fer. Metals, Beijing	JEOL	1.0	In Situ
1978	Wadsworth Centr for Labs and Res., Albany	AEI	1.0	Life Sci
1978	Max-Planck-Inst.—Stuttgart	AEI	1.0	In Situ
1979	Argonne Nat'l. Lab	AEI	1.2	In Situ (I)
1980	Faculty of Engr., Hokkaido U.	Hitachi	1.3	In Situ (I)
1980	Saclay	AEI	1.2	In Situ
1980	Tokyo Inst. of Tech.	Hitachi	1.25	HREM
1981	Lawrence Berkeley Nat'l. Lab	AEI	1.5	In Situ
1981	Nat'l Inst. for Physiological Sci. (Japan)	Hitachi	1.25	Life Sci
1982	Faculty of Engr., Nagoya U.	Hitachi	1.25(STEM)	In Situ
1982	Lawrence Berkeley Nat'l. Lab	JEOL	1.0	HREM
1989	Inst. for Chem. Research, Kyoto U.	JEOL	1.0	Org. Matls
1990	Natl Inst. for Res. in Inorg. Mat.—Tsukuba	Hitachi	1.3	HREM
1991	Tohoku U.	JEOL	1.25	HREM
1992	U. of Tokyo	JEOL	1.25	HREM
1993	Max-Planck-Inst.—Stuttgart	JEOL	1.25	HREM
1994	Nat'l Res. Inst. for Metals—Tsukuba	JEOL	1.25	In Situ (I)
1995	Res. Center for UHVEM, Osaka U.	Hitachi	3.5	In Situ

*   Upgraded electronics in 1990. In Situ (I) means in situ ion beam capability.

A PROJECT FOR ON-LINE REMOTE CONTROL OF A HIGH-VOLTAGE TEM

B. Parvin,⋄ D. Agarwal,⋄ D. Owen,☆ M.A. O'Keefe,☆ K.H. Westmacott,☆ U. Dahmen,☆ and R. Gronsky*

⋄Information and Computing Sciences, University of California, LBL B50B Berkeley, CA 94720
☆National Center for Electron Microscopy, University of California, LBL B72, Berkeley, CA 94720
*Materials Science & Mineral Engineering, University of California, Berkeley, CA 94720

On-line microscopy commenced with demonstrations for biological and SEM applications.[1-2] For materials science, a project has recently been established to provide users of the NCEM with remote on-line access to a 1.5MeV Kratos EM-1500 high-voltage transmission electron microscope via existing wide area networks. Within this project we are developing and implementing a set of tools, protocols, and interfaces to bring transmission electron microscopy on-line for collaborative research. Initially applied to the Kratos, the project will provide increased utilization of this unique instrument with its heretofore restricted access due to its sensitive components and demand for sophisticated operator skills. Additionally, the project will provide computer tools for capturing and manipulating real-time audio and video signals. These tools will be integrated into a standardized user interface that may be used for remote access to any transmission electron microscope equipped with a suitable control computer.

Dynamic study of a specimen in an electron microscope requires continuous adjustment of specimen position and orientation under the beam, illumination conditions, focus adjustments, and corrections for aberrations, all based upon the video signal coming from the imaging system. Requirements for dynamic studies cannot be met over wide area networks due to bandwidth limitation and the inability to offer real-time delivery. Our approach is to integrate sufficient capability into the system, based on advanced computer vision algorithms, so that the real-time delivery requirement over the wide area network is eliminated for a large class of applications. By partitioning the operating tasks in this manner, we ensure that only those servo loop functionalities that require man-machine interfaces are performed over the wide area network, while those functionalities that can be fully automated (for example, correction of specimen drift) are performed using the local computers acting over a local area network (fig.1).

The operator area of the 1.5MeV Kratos (fig.1a) includes mechanical knobs and switches, which provide the necessary currents and voltages for driving the microscope and thus controlling its functionalities (magnification, focus, specimen tilt, etc.). Only the specimen translation controls are mechanical, but have recently been fitted with stepper motors that can be driven from a PC DAC-server. Experiments carried out on the Kratos EM-1500 include characterization of the microstructures and dynamic behavior of a wide range of materials during heating, cooling and tensile testing. Current protocol for conducting these experiments requires the user to adjust for thermal drift, as well as to control the usual microscope parameters, dynamically at the microscope console. This is often a tedious task, and is nearly impossible for remote collaborators due to several constraints in network technologies.

To bring any electron microscope on-line for collaborative research, we need to address the following set of issues in the context of accessibility over the wide area network. We need to determine the appropriate system architecture, appropriate microscope instrumentation for safe remote control, the type of video analysis tools required, the properties of the video information that are best for servo-loop control, video-conferencing tool integration for remote collaboration, arbitration mechanisms for microscope operation, and required authentication services. The computational environment must be able to handle the large volume of video data, process these data, and provide means for manipulating the microscope under the computer control. These requirements are met in a local area network using three computers (fig.1b). The PC contains all the data acquisition hardware needed for control of the microscope. The Sun workstation is currently designed as a client processor that acquires the video, dispenses the video to the DEC to extract desirable features from the video stream, and directs the content of the processed video to the PC for manipulating the microscope. The multiprocessor DEC (2100 running OSF) is connected to the Sun workstation via an FDDI ring, and is primarily used for CPU intensive operations. The Sun workstation is currently used as a client for local in-situ studies. For collaborative operation, it will be used to provide both reliable and unreliable multicasts of messages to the remote collaborators.[3]

Proc. Microscopy and Microanalysis 1995, edited by G.W. Bailey, M.H. Ellisman, R.A. Hennigar, and N.J. Zaluzec
Copyright © 1995 MSA. Published by Jones and Begell Publishing, 79 Madison Ave., New York, NY 10016

Not all of the microscope functionalities are required for remote collaboration, and some are not suitable for safe remote operation of the instrument. For example, control of the filament current is a feature that will not be provided to the remote users; even though numerous safety features are built into the microscope, a novice operator (or an intruder) may accidentally damage the filament under certain conditions. In our system, critical functions can be adjusted only by the local operator. The remaining set of functionalities have limit switches to inhibit remote users from going beyond specified boundaries. Within these limits, remote collaborators will have access to x and y translation, sample tilt on two axes, temperature control, magnification, focus, beam size and shift, aperture control, and camera control.

Dynamic video analysis will act locally to compensate for artifacts such as thermal drift. We have defined two sets of tools for execution on the video-server, and the implementation uses the OSF threads to achieve parallel processing within the server. One set of tools is aimed at the motion estimation problem based on a flow field technique. A suitable algorithm has been implemented on a Maspar computer and demonstrated over a high speed wide area network between Washington DC and Lawrence Berkeley Laboratory.[4] Our current implementation runs at 2Hz on the DEC multiprocessor, sufficient for maintaining thermal stability for in-situ electron microscopy. The other set of tools aims at quantifying shape features based on deformable models. It makes a strong use of high level constraints, and is based on dynamic programming.[5] This set of tools has been partially implemented.

This work is a first step towards a new kind of availability for unique microscopes located at central facilities. It will allow collaborations where multiple people at remote locations can monitor the experiment and discuss results while it is in progress. It will provide much greater accessability to the microscope, and the new methods of controlling the microscope will greatly ease the operators task, leaving them free to concentrate on the science rather than the mechanics of running the microscope.[6]

1. Mark H. Ellisman, Gabriel E. Soto & Maryann E. Martone, *52nd Ann. Proc. EMSA*, New Orleans, Louisiana (1994) 64-65.
2. Nestor Zaluzec, TelePresence Microscopy Project, http://www.amc.anl.gov/, ANL (1995).
3. D. A. Agarwal, "A Reliable Ordered Delivery Protocol for Interconnected Local-Area Networks", Ph.D. thesis #94-29, University of California, Santa Barbara (1994).
4. B. Parvin, D. Robertson, B. Crowley, and W. Johnston, "Visual servoing over a High Speed Wide-Area Network", *Demonstration at SuperComputing 1994*, Washington (1994).
5. B. Parvin, C. Peng, M. Meastre, and W. Johnston, "Tracking of Tubular Molecules for Scientific Applications", *IEEE Trans. on Pattern Analysis and Machine Intelligence* (1995, submitted).
6. Work supported by the Director, Office of Energy Research, Office of Basic Energy Sciences, Material Sciences Division of the U.S. Department of Energy, contract No. DE-AC03-76SF00098.

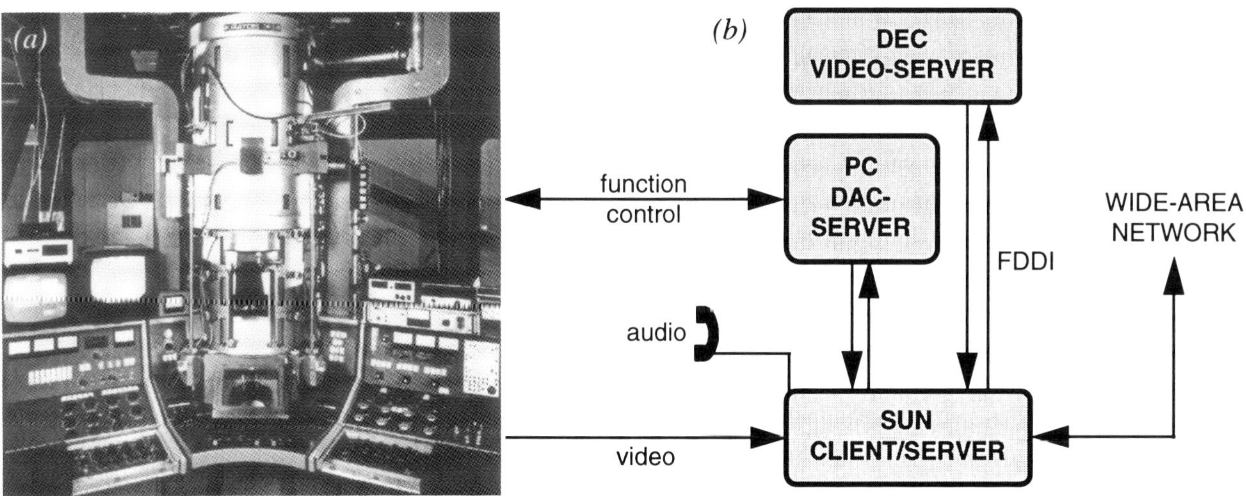

Fig.1. (a) View of operator area of the 1.5MeV Kratos high-voltage transmission electron microscope. (b) System architecture for handling the video and servo loop.

Functional Information from Magnetic Resonance Imaging

Alan P. Koretsky, Afonso Costa e Silva, and Yi-Jen Lin

Department of Biological Sciences, Pittsburgh NMR Center for Biomedical Research and Center for the Neural Basis of Cognition, Carnegie Mellon University, Pittsburgh, PA 15213

Magnetic resonance imaging (MRI) has become established as an important imaging modality for the clinical management of disease. This is primarily due to the great tissue contrast inherent in magnetic resonance images of normal and diseased organs. Due to the wide availability of high field magnets and the ability to generate large and rapidly switched magnetic field gradients there is growing interest in applying high resolution MRI to obtain microscopic information. This symposium on MRI microscopy highlights new developments that are leading to increased resolution. The application of high resolution MRI to significant problems in developmental biology and cancer biology will illustrate the potential of these techniques.

In combination with a growing interest in obtaining high resolution MRI there is also a growing interest in obtaining functional information from MRI. The great success of MRI in clinical applications is due to the inherent contrast obtained from different tissues leading to anatomical information. By manipulating the way images are obtained or by adding agents that alter contrast, functional information can be obtained. For example, MRI can be acquired in a way that is sensitive to the amount of paramagnetic, deoxyhemoglobin leading to maps of hemodynamics that allow spatial mapping of brain activity (1, 2). Monitoring the time course of changes in image contrast due to administration of paramagnetic substances allows determination of a variety of tissue functions (3). We have been developing novel ways of obtaining functional information from MRI that have the potential to allow high resolution mapping of biological function. Three of these techniques are discussed below.

Most MRI is based on detection of water hydrogen. The fact that images can be generated means that water in different regions of an animal can be specifically labeled for MRI detection. By magnetically labeling endogenous water in arteries and detecting the effects of that label on signal in the tissue of interest, quantitative maps of regional tissue blood flow can be obtained. This technique has been applied to measuring regional tissue perfusion in brain, heart, and kidney (4). The figure shows a series of MRI slices through the rat brain. The top row are the anatomical images and the second row are the perfusion images obtained from the same slices. This technique has been applied to mapping perfusion in the human brain and kidney and can be used to obtain high spatial resolution images of perfusion in the mouse brain.

In addition to blood flow, we are developing techniques that will allow high resolution images that are sensitive to calcium influx. Manganese ion is known to enter cells on voltage dependent calcium channels. Manganese is paramagnetic and can thus have large effects on MRI contrast. By administering appropriate levels of manganese to a rat we have demonstrated that the influx of manganese into regions of the brain can be imaged and correlates with areas that are active or that have been activated via pharmacological manipulation. These initial results open the possibility of obtaining high resolution images of neuronal activation using MRI.

Proc. Microscopy and Microanalysis 1995, edited by G.W. Bailey, M.H. Ellisman, R.A. Hennigar, and N.J. Zaluzec
Copyright © 1995 MSA. Published by Jones and Begell Publishing, 79 Madison Ave., New York, NY 10016

Finally we are developing techniques that will allow us to manipulate tissue contrast under genetic control. The idea is to genetically manipulate cells so that they express proteins that lead to detectable changes in magnetic resonance signals. In this way MRI can be used to monitor gene expression. Being able to monitor gene expression should have applications to developmental biology and gene therapy. Initial results demonstrating the feasibility of the approaches taken will be presented.

References:

1) K. K. Kwong et al., Proc. Natl. Acad. Sci. (USA) 89(1992)5675.
2) S. Ogawa et al., Proc. Natl. Acad. Sci. (USA) 89(1992)5951.
3) B. R. Rosen et al., Magn. Resonan. Med. 14(1990)249.
4) J. A. Detre et al., NMR in Biomed. 7(1994)75.

Figure Caption:

Top Row (A): Conventional, multi-slice proton magnetic resonance images of the rat brain showing anatomical information.

Bottom Row (B): Perfusion magnetic resonance images obtained from the same slices shown in the top row using arterial spin labeling techniques.

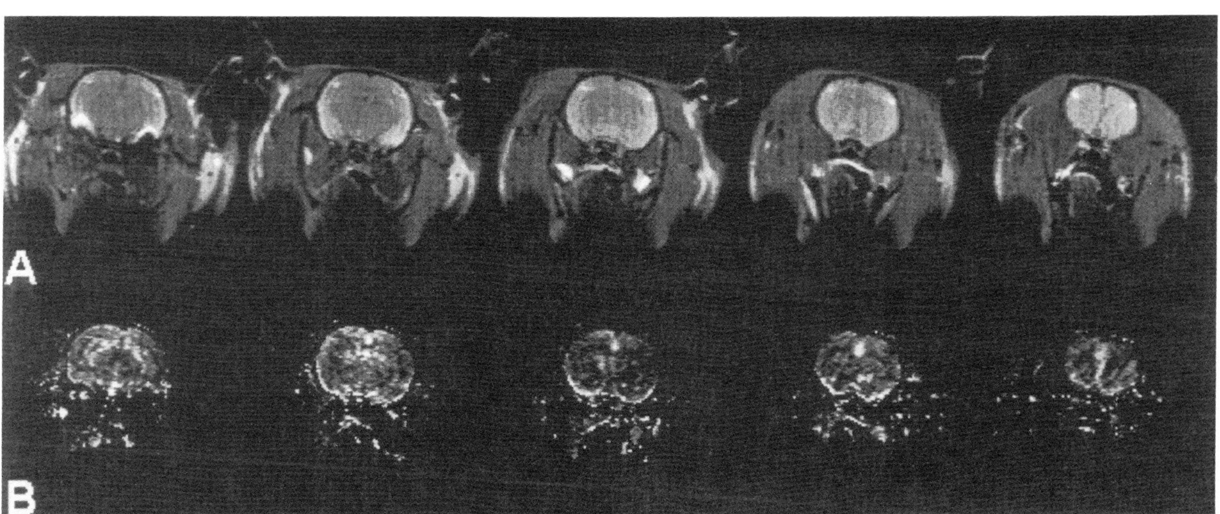

SNR Gains with Superconducting Coils and Hyperpolarized Inert Gas

R. D. Black

Depts. of Radiology and Biomedical Engineering, Duke U., Durham, NC, 27708-3302

Magnetic resonance microscopy (MRM) has advanced from a technical challenge to a practical tool in a wide range of basic sciences (1). The nondestructive nature of the technique allows for repeated studies of the same sample, retrospective studies through any arbitrary plane, registered studies using different contrast mechanisms, and examination of valuable specimens. "Proton contrast" is provided by making use of T_1, T_2, $T_{1\rho}$, diffusion, and magnetization contrast weightings, thus permitting direct examination of the state of water in tissues; something not possible with other microscopic techniques. The technical hurdles imposed by a limited signal-to-noise ratio (SNR) have been partially overcome by improved RF coil design and by three dimensional Fourier transform encoding for large arrays. We have introduced two additional technologies that help to further redress this SNR deficit: superconducting RF coils and inert gas hyperpolarization.

It has been argued elsewhere (2,3) that the dominant source of noise in the receiving chain for small objects in high fields is the pickup coil and the preamplifier. Specifically, for a field of 7 Tesla, the sample and coil noise contributions become roughly equal at a linear dimension of around 5 cm for "typical" biological tissue. For sample sizes that are well below this point of equivalence, it makes sense to take steps to reduce coil noise. Superconductors are natural candidates for coil materials given that they exhibit a low AC resistance at low temperatures. Thus Johnson/Nyquist noise power, which depends on the product of resistance and temperature, is lowered through the action of two parameters.

Two working microscopy probes have been constructed (2,4) using the high-temperature superconductor $YBa_2Cu_3O_7$ (YBCO) in thin-film form. The YBCO coil is 1.8 cm in diameter and operates at a temperature of 10-15 K in a 7 or 9.4 Tesla field. The quality factor of the coil is on the order of 30,000 - 50,000 in this environment. A sample region, approximately 1 cm in diameter, is maintained at room temperature in close proximity (3 mm) to the coil.

We shall present the results of several imaging studies that show the comparative improvement of the YBCO coils relative to those constructed of copper. For example, a 3-D, spin warp experiment was run on a fixed rat liver specimen with a T_R = 400 ms and a T_E = 10 ms. Lesions were created in the rat liver by controlled exposure to bromobenzene. The model is histologically well characterized, producing regions of necrosis distributed throughout the liver. An image with a resolution of 30x30x55 µm per voxel (128 x 128 x 64 points) was acquired. An image was acquired with a geometrically identical copper probe,

Proc. Microscopy and Microanalysis 1995, edited by G.W. Bailey, M.H. Ellisman, R.A. Hennigar, and N.J. Zaluzec
Copyright © 1995 MSA. Published by Jones and Begell Publishing, 79 Madison Ave., New York, NY 10016

has recently been demonstrated and the instrument should ultimately be capable of a factor of 50 improvement.

Another very exciting advance in MR microscopy involves the use of optically polarized inert gas as the NMR signal source (5,6). The key feature of the hyperpolarized gases is the extremely large, non-equilibrium polarization in the nuclei. Even accounting for the different gyromagnetic ratios, the MR signal from the gas at atmospheric pressure can be more than an order of magnitude larger than that from an equal volume of water.

For example, if we assume a 30% ^{3}He polarization, a value that has been achieved in large volumes (170 cc), then a given volume of ^{3}He at one atmosphere will produce 10.3 times more signal than an equivalent volume of water in a 2 Tesla field. If the ^{3}He signal is acquired in a noise background produced by biological fluids, then the sample noise will be dominated by the Johnson noise from the fluids at the ^{3}He Larmor frequency. That is, the linear dependence of sample noise on frequency (2) means less sample noise will be seen at 64 MHz (the Larmor frequency of ^{3}He at 2 T) than at 85 MHz. Thus, SNR per voxel in a ^{3}He image under the conditions stated above will be 13.5 times greater than that which would be realized in a proton scan (assuming no substantial noise from the coil and electronics).

Magnetic resonance images of the lungs of a guinea pig have been produced using hyperpolarized helium as the source of the MR signal. The resulting images are not yet sufficiently optimized to reveal fine structural detail within the lung, but the spectacular signal from this normally signal-deficient organ system offers great promise for eventual *in vivo* imaging experiments. Fast 2-D and 3-D GRASS sequences with very small flip angles were employed to conserve the non-renewable longitudinal magnetization. A fast projection reconstruction routine has recently been used successfully with gas phantoms. We shall discuss various unique features associated with performing MRI with hyperpolarized gases, such as the selection of the noble gas species, polarization technique, and constraints on the MR pulse sequence.

1) G.A. Johnson, H. Benveniste, R.D. Black, L.W. Hedlund, R.R. Maronpot, and B.R. Smith,*Mag. Res. Quarterly* **9**, 1 (1993).
2) R.D. Black, T.A. Early, P.B. Roemer, O.M. Mueller, A. Mogro-Campero, L.G. Turner, and G.A. Johnson,*Science* **259**, 793 (1993).
3) W.A. Edelstein, G.H. Glover, C.J. Hardy, R.W. Redington, *Magn. Reson. Med.* **3**, 604 (1986).
4) R.D. Black, T.A. Early, and G.A. Johnson, *J. Mag. Res.*, to appear 3/95.
5) M. Albert, G. Cates, B. Driehuys, W. Happer, B. Saam, C. J. Springer, A. Wishnia,*Nature* **370**, 199-201 (1994).
6) H. Middleton, R.D. Black, B. Saam, G.D. Cates, G.P. Cofer, R. Guenther, W. Happer, L.W. Hedlund, G.A. Johnson, K. Juvan, J. Swartz, *Magn. Res. Med.* **33**, 271 (1995).
7) This was made possible through collaboration with the coworkers cited in the references. Support at Duke was obtained through: NIH P41-RR05959, NSF CDR-8622201, and NIH R43-RR09244 and at Princeton through: AFOSR F49620-92-J-0211, ARPA DAMD17-94-J-4469, and ARO DAAH04-94-0204.

NMR MICROSCOPY STUDIES OF TUMOR MICROENVIRONMENTS

Michal Neeman*, Yael S. Schiffenbauer,* Rinat Abramovitch,* and Gila Meir*

*Department of Hormone Research, Weizmann Institute of Science, Rehovot 76100 Israel

The microenvironmental heterogeneity, characteristic of solid tumors 500 μm in diameter or larger, arises from insufficient nutrient supply due to tumor growth exceeding the rate of vascularization. Multicellular spheroids were proposed as an in vitro experimental model system for simulating these stress environments.[1] As in hypoxic regions of tumors, inner cell layers in spheroids show resistance to radiation and conventional chemotherapy,[1] as well as enhanced expression of the angiogenic vascular endothelial growth factor (VEGF).[2] We have applied NMR microscopy for the study of the effects of these microenvironments on tumor physiology.

NMR microscopy experiments were developed for mapping gradients in water diffusion and metabolite content across intact, perfused spheroids (1 mm in diameter).[3] These methods were tested in a study of the cytotoxic effects of cyclocreatine, a synthetic analogue of creatine. In contrast with most of the common anti cancer drugs, accumulation of cyclocreatine was observed across the entire viable rim of C6 glioma spheroids. The profiles were obtained by one dimensional chemical shift imaging, using a 9.4 T narrow bore NMR spectrometer equipped with actively shielded magnetic field gradients (in plane pixel resolution of 23 μm). The spherical symmetry of the spheroids enabled us to deduce the metabolic profiles from one dimensional projections. Diffusion NMR microscopy measurements showed that cyclocreatine treatment led to massive water imbibition and cellular swelling.

The primary role of angiogenesis in growth of solid tumors has been postulated previously and provides an attractive mode of possible therapy.[4] NMR microimaging has been used here to follow tumor growth and neovascularization in vivo in a model system of multicellular C6 glioma spheroids implanted in nude mice.[5] Experiments were conducted on a 4.7 T horizontal bore spectrometer using a 2 cm surface coil (in plane pixel resolution of 100 μm). By positioning the spheroid at a distance of approximately 1 cm from the site of incision both vascularization of the tumor and wound healing were spatially separated and could be simultaneously followed (Fig 1). Enhancement of T_2^* relaxation due to deoxyhaemoglobin provided an intrinsic NMR marker for blood containing vessels. The implanted spheroid induced directed vessel growth within 4 days of implantation, that was geometrically oriented towards the spheroid and distinct from the wound healing at the site of incision. NMR imaging was also used to measure the full time course of tumor growth from each animal. A lag of 4 days followed by Gompertz kinetics of tumor growth was observed. These results represent a direct non invasive measurement of the correlation between tumor growth and its neovascularization. The methodology proposed here for the study of primary tumor angiogenesis provides well defined initial conditions with respect to three dimensional tumor geometry and cell proliferative status. The new vessels proximal to the tumor were highly permeable to the NMR contrast reagent Gd-DTPA, in accord with the enhanced expression of VEGF, a potent vessel permeability factor.[6]

References

1. R.M. Sutherland, *Science* 240 (1988)177.
2. D. Shweiki et al. *Proc. Natl. Acad. Sci. USA* (1995) in press.
3. Y.S. Schiffenbauer et al. *Cancer Res.* 55 (1994)153.
4. J. Folkman *Nature Medicine* 1(1995)27.
5. R. Abramovitch, G. Meir and M. Neeman, *Cancer Research* (1995) in press.
6. The authors gratefully acknowledge valuable discussions with Professors Mildred Cohn and Eli Keshet. This work was supported by The United States - Israel Binational Science Foundation 93-00073. M.N. is incumbent of the Dr. Phil Gold career development chair in cancer research.

Proc. Microscopy and Microanalysis 1995, edited by G.W. Bailey, M.H. Ellisman, R.A. Hennigar, and N.J. Zaluzec
Copyright © 1995 MSA. Published by Jones and Begell Publishing, 79 Madison Ave., New York, NY 10016

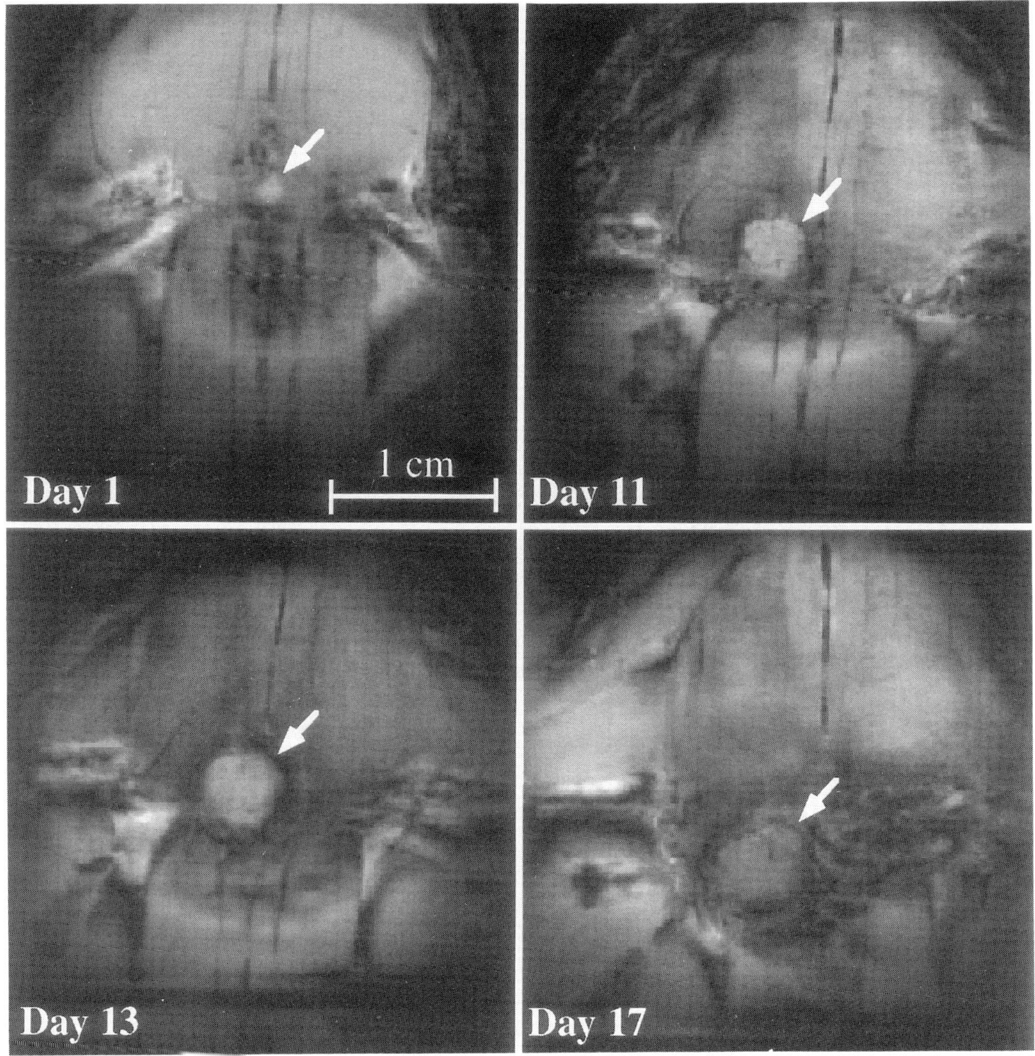

Figure 1: In vivo NMR microimaging of growth and neovascularization of an implanted C6 glioma spheroid. Gradient echo images shown here were obtained from a single mouse at the indicated days after implantation. The mouse was placed supine with the spheroid, implanted subcutaneous in the lower back, centered on a 2 cm surface coil. The tail direction is at the top of the images, and the tumor is indicated by the white arrow. TE=10.5 ms, TR=100 ms, 110 m in plane resolution. Note the darkening of the tumor periphery due to the increased density of blood vessels, that accompany tumor growth.

USING MICRO-MRI TO WATCH DEVELOPMENT LIVE AND IN THREE DIMENSIONS

R. E. Jacobs*, J. Allman*, A. Barr**, S. E. Fraser*, and T. Meade*

*Beckman Institute, Divisions of Biology and **Enginering & Applied Science,
California Institute of Technology, Pasadena, CA 91125

The ability of MRI to provide three dimensional images of thick opaque samples in a noninvasive manner has made it an extremely important clinical tool. In addition, the large number of types of contrast mechanisms in a MR experiment offer the clinician and research scientist the possibility of adapting the image contrast to fit the problem of interest. While typical resolutions employed clinically are on the order of a millimeter, the notion of using MRI at microscopic resolutions arose early in the development of this technique. Spatial resolution in biological samples is typically limited by a number of physical effects as well as signal-to-noise ratio (S/N) considerations. Estimates of the *theoretical* limits of resolution in the MR image arising from these phenomena range from 2 to 0.5 micron. The *practical* spatial resolution is currently determined by the S/N which is often limited by the amount of time available to actually acquire the image (*i.e.* the temporal resolution). The challenge in MRI microscopy is to optimize the experimental setup (hardware and software) to overcome the poor intrinsic S/N, optimize contrast, etc. in order to obtain images incorporating the type of information desired by the investigators. Toward this end, we are applying teleological modeling methods newly developed in the computer graphics realm. We are currently implementing these methodologies on our MR microscopic system where the combination of high magnetic field strength (11.7 Tesla) and restricted sample volume present special opportunities.

Magnetic Resonance Imaging studies on a number of biological systems are being carried out at the Biological Imaging Center of the Beckman Institute. These studies rely on both the intrinsic contrast available within the sample and the use of cell-autonomous MR contrast agents. To be able to compare and contrast information across species we are using three model systems: amphibian (frog), rodent (mouse), and primate (lemur). High resolution (10-100 micron) three dimensional images of developing systems obtained at 11.7T with a variety of imaging schemes will be discussed.

We note the support of the Beckman Institute, the National Institute of Child Health and Human Development (HD25390), and the Human Brain Project funded jointly by the National Institute of Drug Abuse, the National Institute of Mental Health, and the National Science Foundation. We thank David Haring of the Duke Primate Center for the generous gifts of the mouse lemur specimens used in this work.

Proc. Microscopy and Microanalysis 1995, edited by G.W. Bailey, M.H. Ellisman, R.A. Hennigar, and N.J. Zaluzec
Copyright © 1995 MSA. Published by Jones and Begell Publishing, 79 Madison Ave., New York, NY 10016

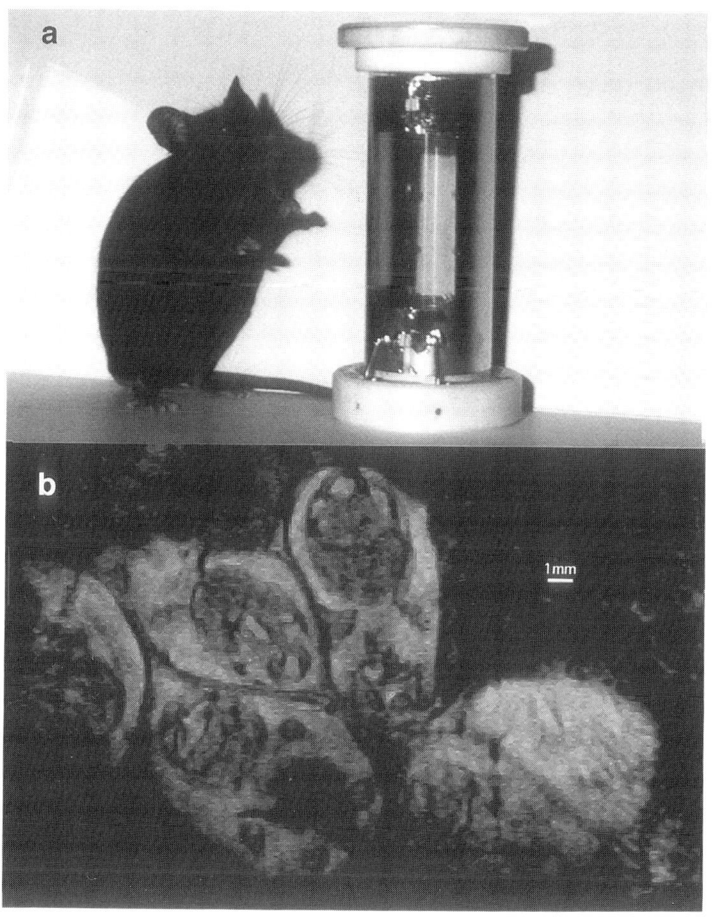

FIG 1. - (a) A twelve day pregnant female mouse is shown in along side the RF coil used to obtain the MR images shown in the bottom two frames. (b) A portion of one MRI slice taken parallel to the long axis of the female. Four mouse embryos are visible, three in almost coronal sections. Ventricles in the developing brain appear at relatively high intensity.

ELEMENTAL, CHEMICAL AND ORIENTATIONAL CHARACTERIZATION OF EASILY DAMAGEABLE MATERIALS WITH X-RAY MICROSCOPY

H. Ade

Dept. of Physics, North Carolina State University, Raleigh, NC 27695-8202

Many of the transmission x-ray microscopy developments since the late seventies, including the ones at the National Synchrotron Light Source (NSLS)[1], were driven primarily by the potential of x-ray microscopy to image wet and unstained biological samples with low radiation dose. High image contrast between an aqueous environment and carbon based materials can be achieved with photon energies between the carbon and oxygen K absorption edges (290-540 eV). This contrast is based on differences in cross section of the elements oxygen and carbon in this energy range. Going beyond this "elemental" sensitivity, Ade et al. recently demonstrated how chemical (valence) sensitivity can be achieved via the X-ray Absorption Near Edge Structure (XANES) at the carbon K edge.[2] In addition, linear dichroism microscopy can exploit the dependence of x-ray absorption resonances on the bond orientation relative to the linearly polarized x rays.[3] For an application of the latter technique see A.P. Smith et al. [4]

The Scanning Transmission X-ray Microscope (STXM) located at the National Synchrotron Light Source at Brookhaven National Laboratory has a spatial resolution of 50 nm, and can acquire XANES spectra from 0.1 μm^2 areas. As an example of XANES imaging to polymers, which are generally radiation sensitive materials, we present here the investigation of the phase morphology of a liquid crystalline polyester based on several aromatic monomers. Electron microscopy (EM) and differential scanning calorimetry indicated the existence of phase separation in materials based on this polyester, but these techniques could not determine whether the phases are chemically distinct, and if so how many phases there are, or whether the observed phase separation is solely due to differences in crystallinity. We compared melt-screened and un-screened samples. We found that the un-screened material had four chemically distinct phases. As an illustration we show micrographs acquired at (Fig. 1A) 285.0 eV, (Fig. 1B) 286.8 eV, (Fig. 1C) 296.2 eV, and (Fig. 1D) 281.8 eV. Micrograph 1A emphasizes aromatic content and a discontinuous phase with domains smaller then 100 nm, as well as continuous phases with dimensions of a few microns are clearly discernible. Fig. 1B reverses the contrast between the continuous phases, while there is virtually no contrast between the small features. Fig. 1C is acquired at a photon energy that has only residual chemical sensitivity and is predominantly a "density map" that most closely resembles the EM micrographs of this material. Micrograph 1D is acquired below the carbon edge, and emphasizes elements other than carbon. Given the elemental constituents and the low level of metal contamination in this polymer, the dark features in Fig. 1D are regions rich in oxygen. This complex morphology observed in the un-screened material was completely absent in the screened material, which illuminates and illustrates the structural dependence of the materials on the processing route.

References

1. J. Kirz, R. Burge, and H. Rarback, Ann. NY. Acad. Sci **342,** 135-147 (1980), J. Kirz *et al.*, Rev. Sci. Instrum. **63,** 557-563 (1992).
2. H. Ade, X. Zhang, S. Cameron, C. Costello, J. Kirz, and S. Williams, Science **258,** 972 (1992).
3. H. Ade and B. Hsiao Science 262, (1993) 1427.
4. A. P. Smith, et al. (these proceedings).
5. We would like to thank Profs. J. Kirz and C. Jacobsen and their groups for the maintenance and construction of the NSLS STXM, A.P. Smith for help with the data acquisition and B. Wood and I. Plotzker for providing the samples.

Proc. Microscopy and Microanalysis 1995, edited by G.W. Bailey, M.H. Ellisman, R.A. Hennigar, and N.J. Zaluzec
Copyright © 1995 MSA. Published by Jones and Begell Publishing, 79 Madison Ave., New York, NY 10016

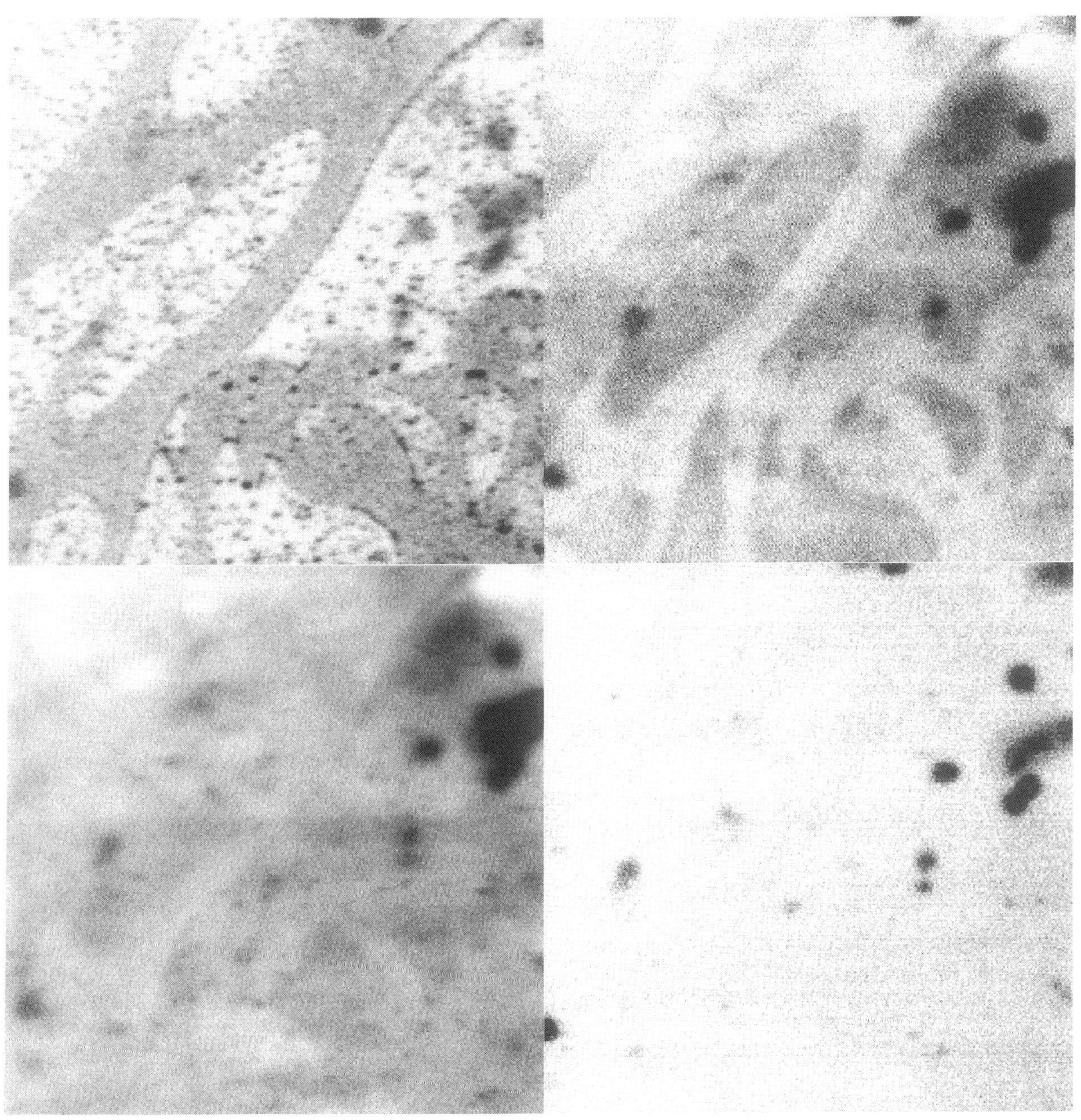

Fig. 1. --- Micrographs of the same region of a thin section of an aromatic liquid crystalline polyester imaged at a photon energy of (A, top left) 285.0 eV, (B, top right) 286.8 eV, (C, bottom left) 296.2 eV, and (D, bottom right) 281.8 eV. Field of view is 9 x 9 μm.

X-RAY STANDING WAVE MEASUREMENTS OF ADSORBATES ON METAL SURFACES

D. Heskett*, X. Shi*, C. Su*, P. Xu*, D. Corridon*, L. E. Berman**, C.-C. Kao**, and M.J. Bedzyk***

*Department of Physics, University of Rhode Island, Kingston, RI 02881
**National Synchrotron Light Source, Brookhaven National Laboratory, Upton, NY 11973
***Department of Materials Science and Engineering, Northwestern University, Evanston, IL 60208, and Argonne National Laboratory, Materials Science Division

One of the main goals in surface science is the determination of the geometric structure of the surface, including the local registry of adsorbed atoms and molecules on surfaces. The X-ray Standing Wave (XSW) technique offers a powerful method of locating the positions of atoms and molecules in and on single crystals[1]. This technique is now being employed with increasing frequency with the availability of x-ray radiation synchrotron sources, but is still underutilized with regards to structural determinations of adsorbates on metal surfaces.

When x rays are Bragg-reflected from a crystal, the incident and diffracted waves interfere to set up a standing wave field parallel to and having the same spatial periodicity as the reflecting planes. The exact location of the peaks of this standing wave field shift relative to the atomic scattering planes as one scans through the region of total reflectivity associated with the Bragg condition. By measuring a yield characteristic of a bulk atom or an adsorbate excited by the standing wave field, such as Auger electron emission, core level photoemission, or x-ray fluorescence, the atom's position relative to the diffraction planes can be determined. By combining results of standing wave measurements using sets of diffraction planes which are not parallel, it is in addition possible to triangulate the atom's position in three dimensions, thus determining an adsorbate's bonding site unambiguously.

Compared with other surface structural methods, the XSW technique offers definite advantages over some and is complementary to others. First, the technique is *element- and chemical state-specific*, in contrast to many other structural probes. Second, it can provide a *very high degree of positional accuracy* (~0.01 Å) which is as good as or better than all other available techniques. Third, *data analysis is uncomplicated and straightforward*; elaborate model-dependent-calculations are not required to extract atomic position information from the data. Fourth, along with accurate geometric information of adsorbate positions, the *degree of static and dynamic order or disorder* is also derived from XSW data. Fifth, as mentioned above, *bond sites* can also be directly obtained through triangulation measurements.

As an example of the XSW technique, we present some results from an investigation of the Rb/Cu(111) system[2,3]. In Figure 1 are presented a series of electron energy distribution curves for the (2x2) Rb/Cu(111) overlayer system obtained at different incident x-ray photon energies. The results and our analysis of the Rb/Cu(111) data are presented in Figure 2. From this data set we obtained for Rb/Cu(111) a coherent position of $0.39_{\pm}.005$ and a coherent fraction of $0.71_{\pm}.02$. This coherent position corresponds to an Rb-Cu(111) bulk lattice plane spacing of $2.89_{\pm}.01$Å. The coherent fraction indicates a relatively high degree of ordering of the Rb atoms in the direction perpendicular to the

Proc. Microscopy and Microanalysis 1995, edited by G.W. Bailey, M.H. Ellisman, R.A. Hennigar, and N.J. Zaluzec
Copyright © 1995 MSA. Published by Jones and Begell Publishing, 79 Madison Ave., New York, NY 10016

surface. To triangulate to the Rb bond position, we repeated the XSW measurement in a second geometry. The combination of these measurements allowed us to determine a top site as the Rb bonding position for the (2x2) layer.

References
1. J. Zegenhagen, *Surf. Sci. Rpts.* 18, 199 (1993).
2. X. Shi, C. Su, D. Heskett, L. Berman, C.-C. Kao, and M.J. Bedzyk, *Phys. Rev.* B49, 14638 (1994).
3. P. Xu, D. Heskett, L. Berman, C.-C. Kao, and M.J. Bedzyk, submitted to *Surface Science.*

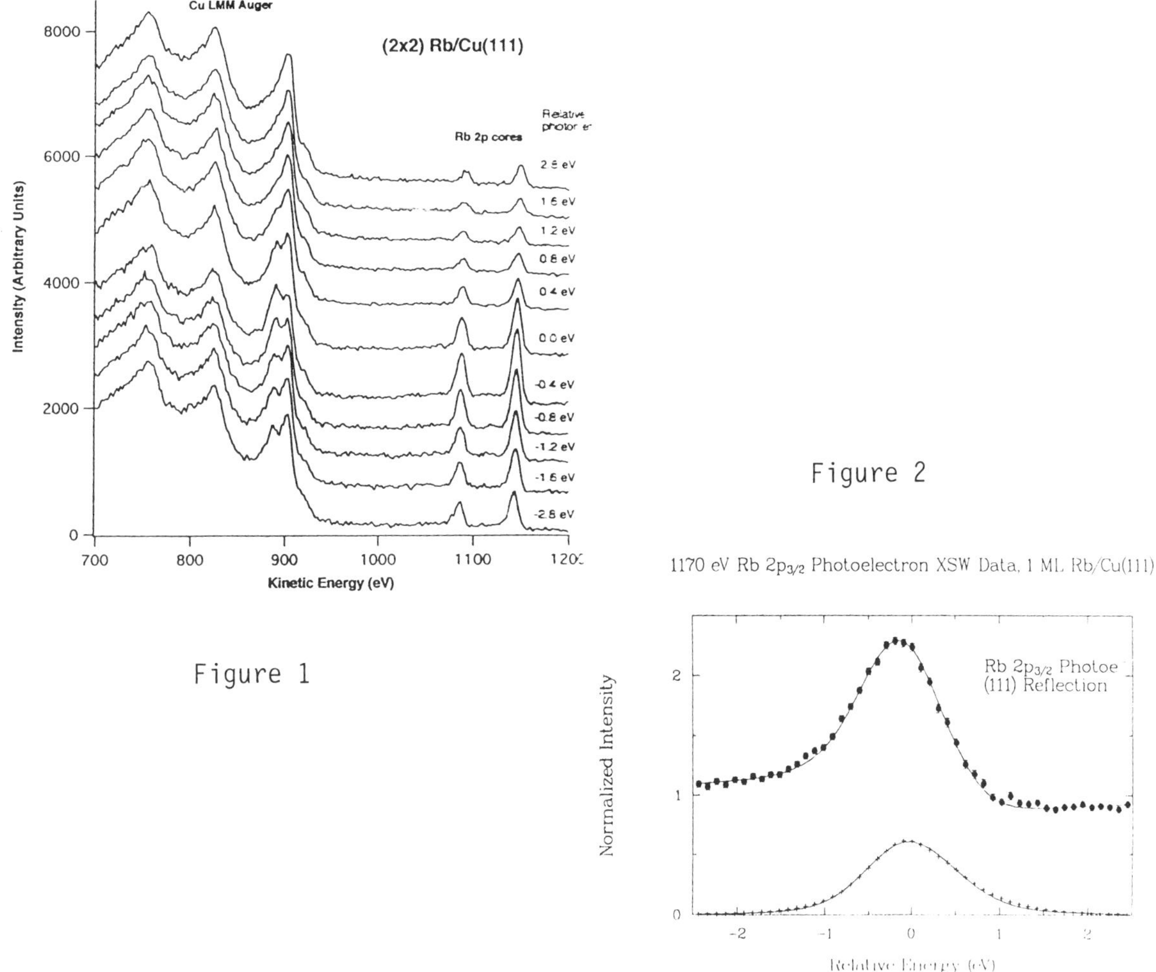

Figure 2

Figure 1

FIG. 1. - Electron Energy Distribution spectra for one monolayer of Rb on Cu(111) as a function of photon energy relative to the rocking curve maximum (at 2974.5 eV). The Cu LMM Auger peaks and the Rb 2p core level photoemission peaks are indicated. The x-ray photon beam was incident normal to the crystal in the (111) direction.

FIG. 2. - Photon energy dependence of the Rb $2p_{3/2}$ photoelectron standing-wave yield for the (111) reflection plane (upper data set). The lower data set is the measured reflectivity, and the solid lines are fits using dynamical diffraction theory.

Photoelectron Diffraction and Holography:
New Directions in Imaging Atoms Near Surfaces[†]

Charles S. Fadley

Department of Physics, University of California-Davis, Davis, CA 95616, and
Materials Sciences Division, Lawrence Berkeley Laboratory, Berkeley, CA 94720

Photoelectron spectroscopy is by now a widely applied characterization technique in physics, chemistry, and materials science in which the energy, direction, and perhaps also spin of electrons ejected from a solid surface are measured, thereby providing a broad range of information on atomic composition and electronic, atomic, and magnetic structure. However, further measuring the directional dependence of electrons excited from core levels yields what has been termed photoelectron diffraction, a versatile probe of near-surface atomic positions and magnetic order [1-4]. Photoelectron diffraction has been used to study a wide variety of surface structure problems, including adsorption, epitaxial growth, and surface structural and magnetic phase transitions. This technique is also unique in permitting the separate study of chemically, structurally, or magnetically distinct atoms of a given type, through high-resolution measurements of core binding energy shifts and multiplet splittings in open-shell systems. Such diffraction data can also be analyzed with a holographic fourier-transform methodology, thus yielding the exciting prospect of direct three-dimensional images of both atomic positions and magnetic order. As one example of this kind of holographic imaging, Fig. 1 shows two perpendicular cross sections through a single-energy experimental image of a sulfur overlayer on a (001) surface of nickel. The nearest-neighbor sulfur atoms to the emitter are clearly localized in three dimensions, with positions that are within a few tenths of an Angstrom of the known structure. Multienergy holograms should significantly improve image quality. The sensitivity of such measurements to magnetism can also be enhanced by studying exchange-induced multiplet splittings, by using circularly-polarized synchrotron radiation for excitation, and by directly measuring the photoelectron spin. The fundamental physics of these techniques will be reviewed, including discussion of several new directions of development. The use of these photoelectron-based methods in conjunction with other surface structure probes such as scanning tunneling microscopy and low energy electron diffraction is beginning to provide a powerful set of complementary information, as will be illustrated for the epitaxial growth of magnetic metals and oxides on non-magnetic substrates. Third-generation synchrotron radiation sources (with the first of these being the Advanced Light Source in Berkeley), are also now providing ultraviolet and soft x-ray radiation of unprecedented brightness that can be used to significantly expand the capabilities of photoelectron diffraction and holography for near-surface atomic imaging. A newly-installed system for performing such measurements at the ALS will be discussed, and some of the first results obtained with it presented.

References

[†]Work supported by the Department of Energy (Contract No. DE-AC03-76SF0098) and the Office of Naval Research (Contract No. N00014-94-1-0162).

[1] C.S. Fadley, in Synchrotron Radiation Research: Advances in Surface and Interface Science, Vol. 1-Techniques, R.Z. Bachrach, Ed. (Plenum Press, 1992).

[2] C.S. Fadley, Surf. Sci. Repts. 19 (1993) 231.

[3] C.S. Fadley, in the Proceedings of the Fourth International Conference on the Structure of Surfaces: Structure of Surfaces IV, X. Xie, M. Van Hove, and S.Y. Tong, Eds. (World Publishing Co., 1994).

[4] C.S. Fadley, S. Thevuthasan, A.P. Kaduwela, C. Westphal, Y.J. Kim, R. Ynzunza, P. Len, E. Tober, F. Zhang, Z. Wang, S. Ruebush, A. Budge and M.A. Van Hove, in the Proceedings of the Fifth International Conference on Electron Spectroscopy, J. Electron Spectrosc. <u>68</u> (1994) 19.

[5] S. Thevuthasan, R.X. Ynzunza, E.D. Tober, C.S. Fadley, A.P. Kaduwela, and M.A. Van Hove, Phys. Rev. Letters <u>70</u> (1993) 595.

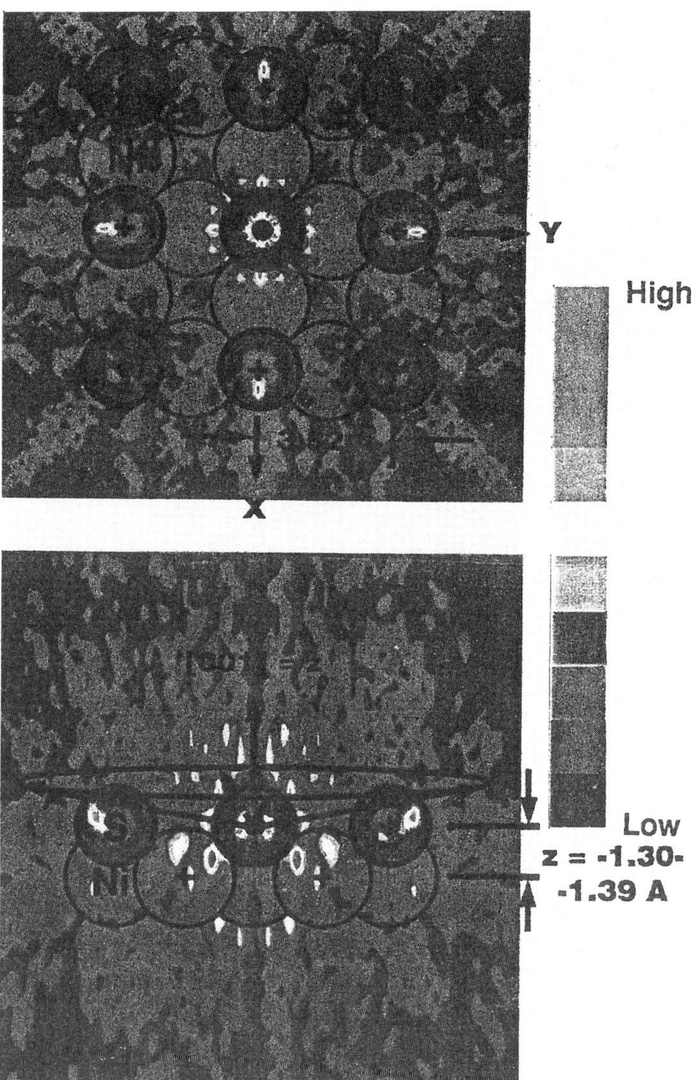

Figure 1: Experimental photoelectron holographic images of the sulfur atoms in an ordered overlayer on a (001) surface of nickel. The images were obtained from a sulfur 2p photoelectron hologram at 1327 eV kinetic energy, using a scattered-wave correction in taking the fourier transforms. They are in two perpendicular cross sections: top = xy plane containing the sulfur atoms, bottom = yz plane perpendicular to this, with the known atomic geometry overlaid in each case.

FURTHER ADVANCES IN ORIENTATION IMAGING MICROSCOPY

D.J.Dingley*

TexSEM Laboratories, Provo Utah 84604
*On leave from Bristol University, Bristol BS8 1TL United Kingdom

Orientation Imaging Microscopy, OIM, is a relatively new technique which provides an image of the surface of polycrystalline material in which the grains are distinguished by their orientation differences, by the strain within them and the type of grain boundaries that separate them.[1] The technique evolved from the work of Dingley [2] and Venables [3] on application of electron backscatter diffraction EBSD in the scanning electron microscope. In OIM, electron backscatter diffraction patterns are obtained successively at regularly spaced points on a sample surface. At each point, the diffraction pattern is captured, transferred to a computer and automatically indexed. Crystal orientation and diffraction line width are measured. Recent advances have been concerned with post data collection image processing.

In the following illustration orientation imaging microscopy was used to investigate the microstructure of submicron aluminium, vapour deposited onto single crystal silicon coated with silicon dioxide. The experimental procedure described in reference 2 was adapted using a Philips XL 30 SEM fitted with a tungsten electron gun. Microscope operating conditions were, accelerating voltage 20kV, electron probe current 0.4nA, and specimen operating position, 10mm below the objective lens. OIMs were obtained over areas measuring 5 μm x 5 μm with a step distance between data collection points of 0.1 μm. Error checking routines involved determination of a confidence index for the correctness of each measurement and a comparison of each measurement with that of its immediate neighbours.

Figure 1 shows pole figures of the as-recorded orientations and figure 1b the corresponding pole figures after processing to remove suspect measurements. The checking procedure has strengthened the observed [111] fibre. The OIM formed from the processed data, figure 2a, has been shaded to reveal the residual strain, dark denotes high strain. Low angle boundaries (2°-5°) have been drawn as white lines, and intermediate (5°-15°) and high angle boundaries, (15°-62°) drawn as grey and black lines respectively. Figure 2b is a similar OIM but drawn to reveal $\Sigma3$ (white), $\Sigma7$ (grey) and $\Sigma13b$ (dark grey) coincident site lattice boundaries. Five percent of the boundary length was of $\Sigma3$ type, two percent $\Sigma7$ and two percent $\Sigma13b$. Grain size distribution is shown in figure 3 and the misorientation angle distribution in figure 4. Dashed and dotted lines in figure 4 show respectively the form of the distribution if the structure had been completely random or had a perfect [111] fibre texture.

References

1) B. L. Adams, S.I. Wright and K Kunze, *Metall. Trans.*, 24A,(1993)819.
2) D.J.Dingley and G Burns, *Scanning Electron Microscopy 11* (1986)383.
3) J.A Venables and C. J Harland, *Philos. Mag.* 27(1973)1193

Proc. Microscopy and Microanalysis 1995, edited by G.W. Bailey, M.H. Ellisman, R.A. Hennigar, and N.J. Zaluzec
Copyright © 1995 MSA. Published by Jones and Begell Publishing, 79 Madison Ave., New York, NY 10016

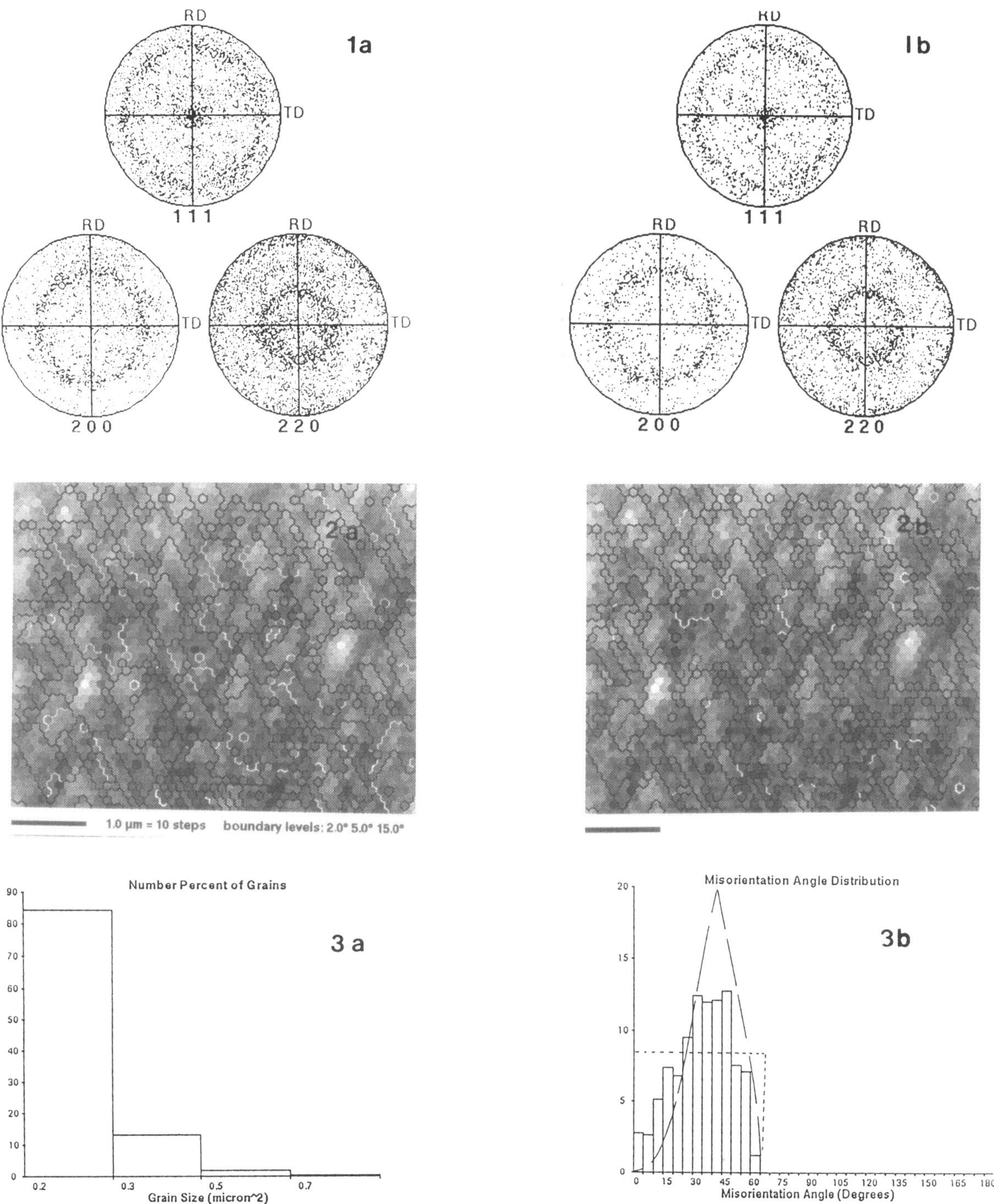

FIG.1.-(a) Pole figures representing aluminium grain orientations from as collected data; (b) corresponding pole figures for corrected data.

FIG.2.-(a) OIM from aluminium showing low and high angle boundaries; (b) corresponding OIM showing in addition low order CSL boundaries.

FIG.3.-Grain size histogram for grain data shown in FIG.2.

FIG.4.-Misorientation distribution for grain data shown in FIG.2.

STUDY OF CRYSTALLINE INTERFACES USING ΔG APPROACH

W.-Z. Zhang

C-CORE, Memorial University of Newfoundland, St. John's, NF, Canada, A1B 3X5

The study of interfacial structure is a fundamental step to achieve the beneficial structure/property relationship in materials. A knowledge of phase boundaries may provide useful insights for a better understanding of the phase transformation. The O-lattice theory,[1] containing the coincidence site lattice (CSL) model as a subset, has been widely applied and shown to account for boundaries of preferred orientation. Examples include low-Σ CSL grain boundaries with high-Γ (Σ=the ratio between the volumes of unit cells of CSL and the crystal lattice; Γ=the planar density of CSL points),[2] and facets or habit plane of interfaces of precipitates.[3-6] The indices of the crystals adjoining the boundary are commonly used to define the boundary plane.[2,3,5] These indices do not reveal a direct connection between the O-lattice planes and the boundaries.

This paper emphasizes the application of a simple parameter, $\Delta\mathbf{g}$, for the study of interfaces and its relationship with the O-lattice planes. A $\Delta\mathbf{g}$ is defined as $\Delta\mathbf{g} = \mathbf{g}_1 - \mathbf{g}_2$, where $\mathbf{g}_1$ and $\mathbf{g}_2$ are the reciprocal vectors from the adjacent crystals. It can be measured from the superimposed diffraction pattern taken from the two crystals adjoining at the boundary, for the diffraction spots define the $\mathbf{g}$ vectors. If the two $\mathbf{g}$s are related by the transformation strain, the associated $\Delta\mathbf{g}$ has been proved to be identical to a reciprocal vector of the O-lattice and termed as $\Delta\mathbf{g}^o$ to distinguish from the others.[7] Semicoherent boundaries of preferred orientations, with a minimum net strain at the forced coherent regions, are possibly parallel to the primary O-lattice planes. When either $\mathbf{g}_1$ or $\mathbf{g}_2$ represents a set of closest packed planes of a crystal, the associated $\Delta\mathbf{g}^o$ defines the primary O-lattice planes.[7] The possible dislocation network in the boundary parallel to this primary O-lattice plane can be described by the Burgers vectors lying in the $\mathbf{g}_1$ plane. One can denote this type of boundaries with the corresponding $\Delta\mathbf{g}^o_{g1}$, such as $\Delta\mathbf{g}^o_{200}$ in Fig. 1. This parameter, identified with a low index $\mathbf{g}_1$, displays the explicit connection between the boundary plane and the O-lattice plane in terms of the displacement field in the boundary. Moreover, using the simple $\Delta\mathbf{g}^o_{g1}$, rather than the high index of crystal planes, one obtains more rational parallels between different precipitation systems. In special cases, such as an invariant line or an invariant plane strain, $\Delta\mathbf{g}^o$s are related by some known relationships (e.g., many $\Delta\mathbf{g}^o$s can be parallel to each other).[8,9] The analysis of $\Delta\mathbf{g}^o$s guided by these relationships can provide useful information about the displacement field in real space as well as a direct test of the assumption of the models. The advantage of using the $\Delta\mathbf{g}$ approach may also be applied to the study of the interfaces of martensite plates, especially the irrational habit plane. It can be proved that a habit plane is usually perpendicular to a group of $\Delta\mathbf{g}^o$s, if the habit plane can be interpreted by the phenomenological theory of martensitic crystallography.[10]

Besides defining a boundary normal with the direction of the vector, a $\Delta\mathbf{g}^o$, being a reciprocal vector, specifies the spacing of the corresponding O-lattice plane by its inverse magnitude. This value implies a possible unit height of equilibrium steps at the boundary plane normal to the $\Delta\mathbf{g}^o$. The ratio of $|\mathbf{g}_1|/|\Delta\mathbf{g}^o_{g1}|$ provides an indication of the net Burgers vector content in the boundary plane normal to the $\Delta\mathbf{g}^o_{g1}$.[7] The restriction of the transformation strain does not apply to the study of planes of CSL. The relationship between $\Delta\mathbf{g}$ and the CSL planes is clearly manifested using the reciprocity relation.[11] The reciprocal lattice of CSL is defined by DSCL*, which is the DSC (Displacement Shift Complete) lattice in reciprocal space. By the definition, all $\Delta\mathbf{g}$s belong to the vectors of DSCL*. For a fixed Σ, the density of CSL points at a boundary normal to a $\Delta\mathbf{g}$ is inversely proportional to $|\Delta\mathbf{g}|$, since Σ and Γ are simply related by $\Gamma\Sigma = \rho_{cr}/|\Delta\mathbf{g}|$, where ρ_{cr} is the density of the crystal lattice. Because the CSL is independent of a transformation strain, the association of $\Delta\mathbf{g}$ with corresponding $\mathbf{g}_1$ and $\mathbf{g}_2$ is not as significant as in the case of the primary O-lattice. At a CSL orientation, a $\Delta\mathbf{g}$ can often be expressed as a fraction of a rational g. A $\Delta\mathbf{g}^o$ can be parallel to low index $\mathbf{g}_1$ and $\mathbf{g}_2$, when two phases are related by a well defined rational orientation, such as a cubic-cubic orientation. Though $\mathbf{g}_1$ and/or $\mathbf{g}_2$ is often

Proc. Microscopy and Microanalysis 1995, edited by G.W. Bailey, M.H. Ellisman, R.A. Hennigar, and N.J. Zaluzec

used to describe the boundary plane, the $\Delta\mathbf{g}$ approach is still beneficial for providing more information of the boundary, e.g., $|\Delta\mathbf{g}^o|$.

Measurement of boundary planes in terms of $\Delta\mathbf{g}^o_{g1}$ can be made by searching the intersection of low index Kikuchi bands (e.g., $\mathbf{g}_1$ vs. $\mathbf{g}_2$) in the convergent beam electron diffraction mode while keeping the interface at an edge-on orientation. One can also follow the Kikuchi bands of the known $\mathbf{g}_1$ and $\mathbf{g}_2$ according to the O-lattice calculations to arrive at the right orientation. The $\Delta\mathbf{g}$s, that relate the nearest neighbours of diffraction spots in a superimposed (selected area) diffraction pattern, are of most interest. Only those in the vicinity of the origin may belong to $\Delta\mathbf{g}^o$. Furthermore, an analysis of the O-lattice in terms of $\Delta\mathbf{g}^o$s can be carried out when the boundary is not at the edge-on orientation. In special cases, e.g., when $\mathbf{g}_1$ and $\mathbf{g}_2$ are both parallel to the electron beam and normal to a boundary, it is possible to construct a two-dimensional O-lattice in the boundary according to the primary $\Delta\mathbf{g}^o$s in the corresponding diffraction pattern.[6] The relationship of a $\Delta\mathbf{g}$ and the boundary plane result in two useful features in the TEM image mode. One feature is the moiré fringes which must be normal to the associated $\Delta\mathbf{g}$, helpful for identifying the relationship of the $\Delta\mathbf{g}$ with the boundary. In Fig. 1, the edge-on habit plane is seen normal to $\Delta\mathbf{g}^o_{200}$ and is parallel to the moiré fringes on the right side of the interface, where a portion of projected boundary deviates from the ideal orientation of the habit plane. Another important feature of an interface normal to a $\Delta\mathbf{g}^o$ is contained in a high resolution TEM image. As can be demonstrated using the geometry of moiré fringes, the correlated planes from the two sides of the boundary must be in exact registry at the plane normal to a $\Delta\mathbf{g}^o$.[7] This type of relationship has been observed in the habit plane of an α precipitate in a Zr-Nb alloy.[12] It may be interpreted as a cause of a fully coherent martensite interface in an Fe-8Cr-1C alloy.[13]

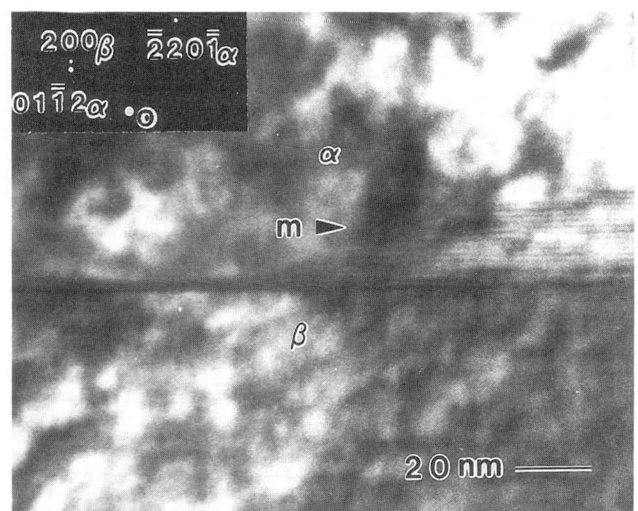

Fig. 1. Determination of habit plane using $\Delta\mathbf{g}^o_{200}$.[4]

References

1. W. Bollmann, *Crystal Defects and Crystalline Interfaces* Berlin:Springer (1970).
2. H. Jang et al., *J. Mater. Res.*, 7(1992)1707.
3. M.G. Hall and H. I. Aaronson, *Acta Metall.* 34(1986)1409.
4. W.-Z. Zhang and G.R. Purdy, *Acta Metall. Mater.* 41(1993)543.
5. C.P. Luo et al., *Acta Metall. Mater.* 42(1994)1923.
6. D. Duly et al. *Phil. Mag.* 71A(1995)187.
7. W.-Z. Zhang and G.R. Purdy, *Phil. Mag.* 68A(1993)279.
8. W.-Z. Zhang and G.R. Purdy, *Phil. Mag.* 68A(1993)291.
9. A.G. Khachaturyan, *The Theory of Structural Transformations in Solids* New York:John Wiley & Sons (1983).
10. W.-Z. Zhang, to be submitted to *Acta Metall. Mater.*
11. H. Grimmer, *Scripta Metall.* 8(1974)1221.
12. W.-Z. Zhang et al., to be submitted to *Phil. Mag. Let.*
13. G.J. Mahon ct al, *Phil. Mag. Let.* 59(1989)273.
14. The author is indebted to Drs. G.R. Purdy and G.C. Weatherly for valuable discussions, to Dr. Y.-P. Lin and the Organizing Committee for the invitation to develop and present this paper, and to Ms. E. Nesbitt for proofreading the manuscript. This paper grew mainly from the author's research at McMaster University.

THE QUANTITATIVE COMPARISON OF EXPERIMENTAL AND SIMULATED DIFFRACTION CONTRAST IMAGES

Stuart McKernan

University of Minnesota, Center for Interfacial Engineering Microscopy Facility, 100 Union St. S. E., Minneapolis, Minnesota 55455

For many years the concept of quantitative diffraction contrast experiments might have consisted of the determination of dislocation Burgers vectors using a $\mathbf{g.b} = 0$ criterion from several different 2-beam images. Since the advent of the personal computer revolution, the available computing power for performing image-processing and image-simulation calculations is enormous and ubiquitous. Several programs now exist to perform simulations of diffraction contrast images using various approximations. The most common approximations are the use of only 2-beams or a single systematic row to calculate the image contrast, or calculating the image using a column approximation. The increasing amount of literature showing comparisons of experimental and simulated images shows that it is possible to obtain very close agreement between the two images; although the choice of parameters used, and the assumptions made, in performing the calculation must be properly dealt with. The simulation of the images of defects in materials has, in many cases, therefore become a tractable problem. A variety of different diffraction contrast images may now be simulated, including isolated line and planar defects as well as arrays of such defects, and the Fresnel fringes at discontinuities in the crystal mean inner potential. Applications covering these different defect morphologies, as well as the effects of using some of these approximations, will be briefly reviewed.

To perform a quantitative comparison between experimental and simulated images the experimental image must be in a digital form; either acquired directly using a CCD camera on the microscope, or digitized from the negative or print. Since the quantitative comparisons rely on the measurement of the absolute intensity of the images, the calibration the intensity scale must allow for the non-linear response of the photographic process. With the recent advances in microscopy techniques it is now possible to obtain energy-filtered images from the TEM - either with an energy filter in the column of the microscope or immediately prior to the CCD. By forming images with only the elastically scattered electrons, the effects of absorption in thicker regions of the specimen can be greatly reduced. These energy-filtered images show much more contrast over a greater thickness range than unfiltered images, which enables a more detailed comparison with calculated images.

The dependence of the simulated images on the input parameters of the calculation will be illustrated with reference to specific defect structures. For a planar boundary across which a displacement occurs, fringe contrast is observed under appropriate imaging conditions. The form of this contrast may vary from a simple sinusoid to much more complex profiles. One of the least complex such defects to analyze is the anti-phase boundary (APB) in III-V semiconductors, since bonding between like atoms occurs across this boundary. Even in this simple case we find that the materials parameters required to calculate the fringe profile are not adequately known in the literature, and several refinements are required to produce a "best fit" to the experimental data. A knowledge of the effects of the input parameters on the image allows us to not only select which parameters need to be refined, but also to determine figures of merit for the match between the experimental and calculated image which are less sensitive to those parameters which are less well known such as the local deviation parameter, the effective extinction distance, and the anomalous absorption coefficient.

Figure 1 shows the experimental and calculated fringes arising from the rigid-body translation at a faceted antiphase boundary in GaAs. The theoretical images were calculated using the MaCOMIS program [3], which uses a many-beam, non-column-approximation formalism based on the Howie-Whelan equation. The deviation parameter for the fringes determined from a diffraction pattern represents an average value over the whole region defined by the selected area aperture. The actual value used in the simulations was determined by comparison of the ratio of the contrast of the outermost fringes at the top and bottom

Proc. Microscopy and Microanalysis 1995, edited by G.W. Bailey, M.H. Ellisman, R.A. Hennigar, and N.J. Zaluzec
Copyright © 1995 MSA. Published by Jones and Begell Publishing, 79 Madison Ave., New York, NY 10016

surfaces of the APB. The value of the translation thus determined was refined by comparison of the experimental and simulated intensity profiles by matching the contrast of the outermost fringes. This procedure was determined as being a sensitive probe of the interface structure whilst being relatively insensitive to the absorption parameter [2]. To further reduce the effects of absorption on the contrast of the outer fringes, relatively thick specimens were used and the contrast of the outer fringes was normalized with the sum of the background intensity of the bright-field and dark-field images. By combining data from two different images recorded under well controlled conditions, the total information content of the data is increased, and it is possible to reduce the dependency of the matching to particular parameters. The fringes at the top and bottom surfaces were treated independently, and good agreement was obtained, as is shown in figure 2, and give a value for a rigid-body translation of 0.019 ±0.04 nm. [3]

References

1. D. R. Rasmussen, S. McKernan and C. B. Carter, Phil Mag A, **63**, 1299, (1991)
2. D. R. Rasmussen and C. B. Carter, J. Elec. Micr. Tech., **18**, 429-436 (1991)
3. This research is supported, in part, by the Microscopy Facility at the University of Minnesota Center for Interfacial Engineering, an NSF Engineering Research Center.

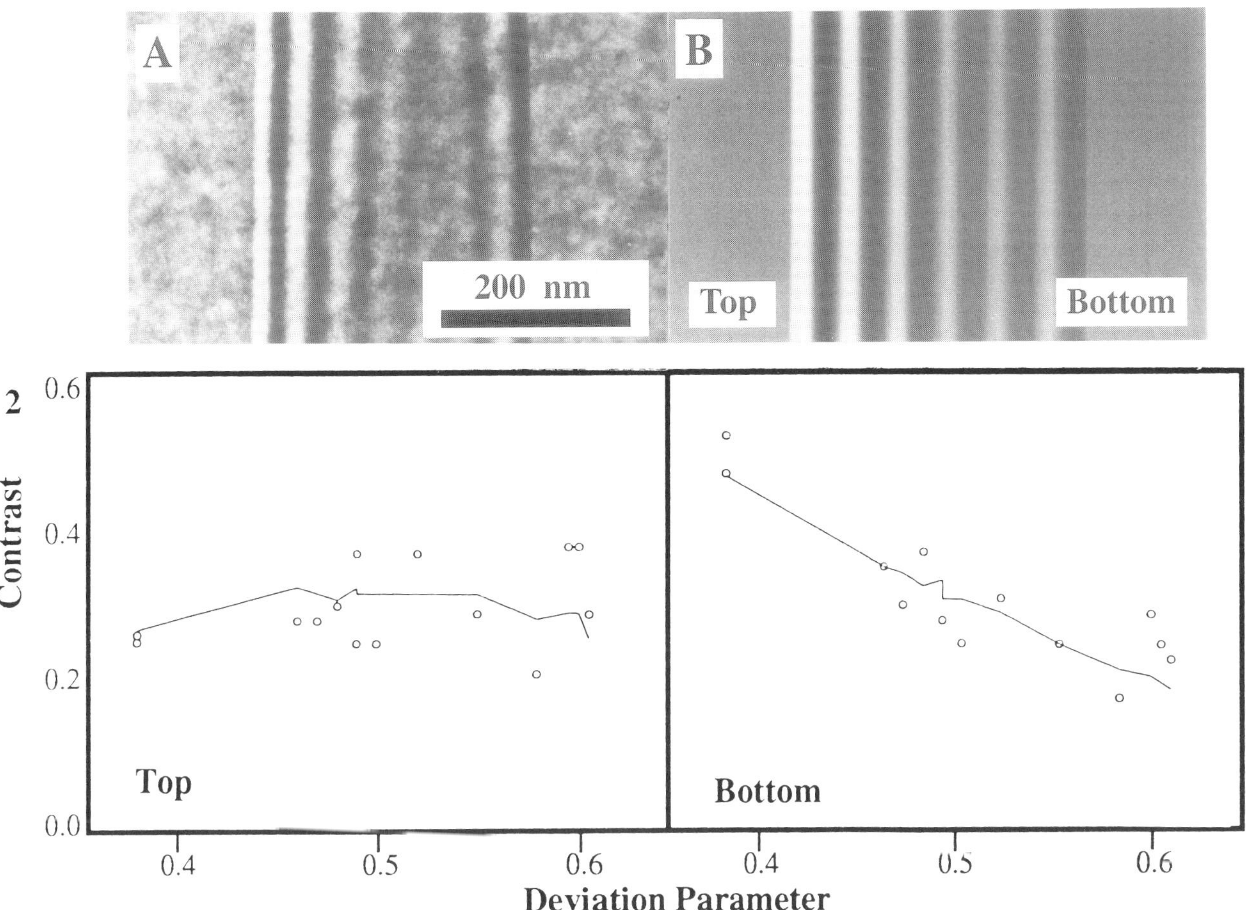

FIG 1. A) Experimental and B) Simulated (2$\bar{2}$0) dark-field images from a GaAs APB recorded under strong two-beam conditions. The top and bottom surfaces of the APB are marked.
FIG 2. Comparison of experimental and theoretical contrast curves from the top and bottom surfaces of the defect shown in figure 1.

PREPARATION AND CHARACTERIZATION OF THIN FILMS OF CARBON NITRIDE

L. Wan and R. F. Egerton

Department of Physics, University of Alberta, Edmonton, Canada T6G 2J1

INTRODUCTION

Recently, a new compound carbon nitride (CN_x) has captured the attention of materials scientists, resulting from the prediction of a metastable crystal structure [1] β-C_3N_4. Calculations showed that the mechanical properties of β-C_3N_4 are close to those of diamond. Various methods, including high pressure synthesis, ion beam deposition, chemical vapor deposition, plasma enhanced evaporation, and reactive sputtering, have been used in an attempt to make this compound. In this paper, we present the results of electron energy loss spectroscopy (EELS) analysis of composition and bonding structure of CN_x films deposited by two different methods.

SPECIMEN PREPARATION

Specimens were prepared by arc-discharge evaporation and reactive sputtering. The apparatus for evaporation is similar to the traditional setup of vacuum arc-discharge evaporation, but working in a 0.05 torr ambient of nitrogen or ammonia. A bias was applied between the carbon source and the substrate in order to generate more ions and electrons and change their energy. During deposition, this bias causes a secondary discharge between the source and the substrate. RF and DC magnetron sputtering of a graphite target were performed in nitrogen, ammonia, and a mixture of argon and nitrogen. The substrate holder was maintained at a constant temperature between 20°C and 600°C. The substrates were rocksalt, mica and {100} silicon.

ANALYSIS METHODS

Specimens were studied using a JEOL-2010 transmission electron microscope (TEM) equipped with a Gatan 666 electron energy-loss spectrometer. The spectral data were processed in a Macintosh IIfx computer by Gatan EL/P software. From the low loss spectrum, we can find the plasmon energy E_p and the thickness of a film t in units of the inelastic scattering mean free path λ; from K-shell edges, we can get the atomic ratio x of nitrogen and carbon [2]. Quantifying the fine structure of the carbon K-edge by fitting the π^* peak with a Lorentzian (fig.1), we can find the fraction of π electrons (proportional to the π^* intensity). If only sp^2 and sp^3 hybridizations exist in the film, this fraction is a measure of the percentage of sp^2 bonding [3]. We need a standard (a 100% sp^2 bonded material, eg. graphite) to obtain the absolute value R. Plural scattering has the effect of reducing the intensity of the π^* peak but can be accommodated, either by a correction factor $\exp(t/\lambda)$ or by removing the plural-scattering intensity by Fourier deconvolution.

RESULTS AND DISCUSSION

The diffuse rings in the electron diffraction patterns of CN_x films revealed that they were amorphous. There was no evidence of a crystalline β-C_3N_4 phase. The electrical conductivity was about $10^8 \Omega$cm. There was no nitrogen in films deposited by evaporation in nitrogen ambient, but about 5% in films deposited in ammonia. With a substrate bias, the carbon/nitrogen atomic ratio increased to 0.1-0.2. For sputtered films, x ranged from 0.3 to 0.7 depending on the substrate temperature and sputtering power. Annealing the

Proc. Microscopy and Microanalysis 1995, edited by G.W. Bailey, M.H. Ellisman, R.A. Hennigar, and N.J. Zaluzec
Copyright © 1995 MSA. Published by Jones and Begell Publishing, 79 Madison Ave., New York, NY 10016

specimens at 300°C, there was little change of x; but at 800°C, up to 75% of nitrogen was lost, suggesting that most of the nitrogen in the films is weakly bonded with carbon. From the fact that x for a film deposited at 600°C was much higher than for a film deposited at room temperature but annealed at 600°C, we conclude that higher substrate temperature is helpful in forming strong C-N bonds.

The R-value increased as nitrogen content increased (fig.2). and in most cases this value exceeded 1. One possible reason is the existence of sp hybridization. Since our graphite standard is entirely sp^2 bonded (π electron fraction = 1/4), the R-value of a entirely sp bonded material (π electron fraction = 2/4) would be 2. This explanation accounts for all our experiment results (0.54 to 1.95) and agrees with the results of Fourier transform infrared spectroscopy [4]. Another possible explanation is charge transfer between carbon and nitrogen atoms: π electrons of carbon gravitate towards a neighboring nitrogen atom because the electronegativity of nitrogen is higher. Emptying the carbon π band will lead to extra intensity in the π^* peak, since the π and π^* bands overlap [5], again resulting in an abnormally high R-value. Further measurements are needed to distinguish between these two possibilities.

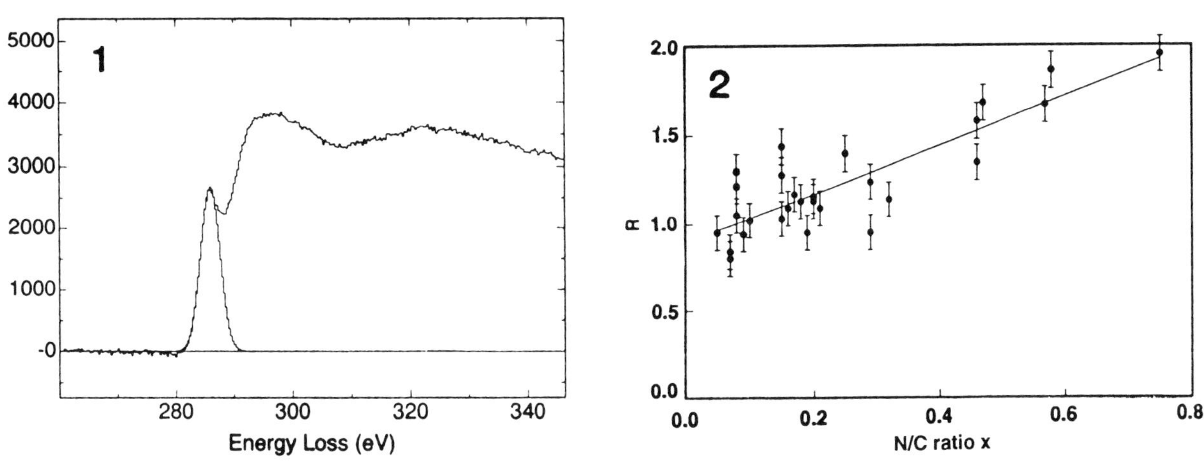

Fig. 1 Carbon K-edge from CN_x, showing the π^* peak fitted to a Lorentzian function

Fig. 2 Dependence of R-value on nitrogen- to-carbon atomic ratio

REFERENCES

1. A. Y. Liu and M. L. Cohen, *Science* 245(1989)841
2. R. F. Egerton, *Electron Energy-Loss Spectroscopy,* New York: Plenum Publishing Corp. (1986) 293,316
3. S. D. Berger, D. R. McKenzie and P. J. Martin, *Philos. Mag. Lett.* 57(1988)285
4. P. A.. Cox, *The Electronic Structure and Chemistry of Solids*, Oxford University Press(1987)116
5. A. Mansour and D. Ugolini, *Phys. Rev.* B47(1993)10201

HIGH RESOLUTION AND DIFFRACTION CONTRAST IMAGING OF LINEAR ISLANDS FORMED DURING MOLECULAR BEAM EPITAXY

D. Loretto*, F. M. Ross** and C. A. Lucas*

*Materials Sciences Division, Lawrence Berkeley Laboratory, 1 Cyclotron Road, Berkeley, CA 94720
**National Center for Electron Microscopy, Lawrence Berkeley Laboratory, 1 Cyclotron Road, Berkeley, CA 94720

There is currently a great deal of interest in the growth and properties of 'one-dimensional' structures. This is motivated by scientific curiosity and by the practical urge to fabricate faster, smaller and more efficient electronic and optical devices. It has recently been reported that under certain conditions linear islands can form spontaneously during growth.[1,2] In this work we investigate linear CaF_2 islands which form on Si during molecular beam epitaxy. By combining low resolution diffraction contrast images from plan-view specimens with high resolution images from cross-section specimens we determine that these islands are extremely uniform in width, can be as narrow as 5 nm, and extend over many 10s of µm. We postulate a mechanism for the formation of these novel features based on surface energy minimization.

The experimental conditions have been reported in detail elsewhere.[3] Briefly, CaF_2 is deposited from an effusion cell onto a Si substrate at 700°C in a vacuum of 10^{-10} Torr. Samples are capped with a 5 nm protective layer of amorphous Si at room temperature before removal from the vacuum chamber.

Figure 1(a) is a diffraction pattern taken from a plan-view specimen with a total CaF_2 coverage equivalent to around one monolayer. The strong reflections are from the Si substrate. The weaker reflections originate from two symmetry-equivalent epitaxial orientations of CaF_2 in which CaF_2<110> is parallel with the Si[001] surface normal. The two orientations are related by a 90° rotation around Si[001]. For each of the two orientations there is a good match at the interface between CaF_2{220} and Si{220} planes and a very poor match at the interface between CaF_2{220} planes and Si{200} planes. This asymmetry is reflected in the morphology of the CaF_2 islands. Figure 1(b), a dark-field image taken using reflections specific to CaF_2, shows two sets of linear islands each elongated along the direction of good match with the substrate.

Figure 2 is a high resolution image taken along a Si[011] zone axis showing a section through a CaF_2 island. The epitaxial orientation of the CaF_2 is apparent from the rotation of the {111} planes between the Si and the CaF_2. It can also be seen that the CaF_2 island has a triangular cross section defined by the inclined {111} facets which intersect at the island apex.

The strain exerted by the linear islands on the substrate gives striking contrast in two-beam images formed with Si reflections. Figure 3 shows two-beam, dark-field images taken from the same region of a CaF_2/Si(001) sample using orthogonal Si<220> reflections. The two sets of islands appear either as broad parallel black and white bands or as narrow black or white lines. The broad black and white bands are caused by strain in the substrate along the direction of high mismatch. The change in contrast from black to white in the narrow lines may be related to steps on the Si surface.[4]

References

1. J. Tersoff and R. M. Tromp, Phys. Rev. Lett. 70, 2782 (1993).
2. E. Hahn, E. Kampshoff, A. Fricke, J.P. Bucher and K. Kern, Surf. Sci. 319, 277 (1994).
3. C A. Lucas, D. Loretto and G. C. L. Wong, Phys. Rev. B 50, 14 340 (1994).
4 . Supported by the Director, Office of Basic Energy Sciences, Materials Sciences Division of the U. S. Department of Energy under contract No. ACO3-76SF00098.

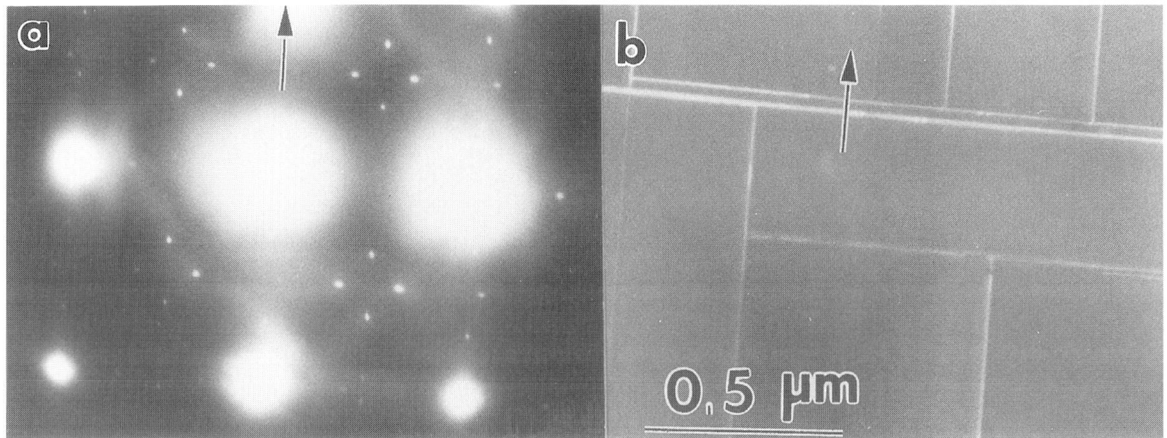

FIG. 1. (a) Selected area TED pattern from CaF_2 /Si(001). (b) Dark-field image formed using CaF_2 reflections only. The beam direction is close to Si[001]. The arrow lies along Si[220].

FIG. 2 HREM image taken along the Si[011] zone axis showing an [011] section through a CaF_2 island. The $CaF_2\{111\}$ facets are clearly visible. Image recorded at 800 keV using a JEOL 1000 ARM. The interface has been partly amorphized because of radiation damage.

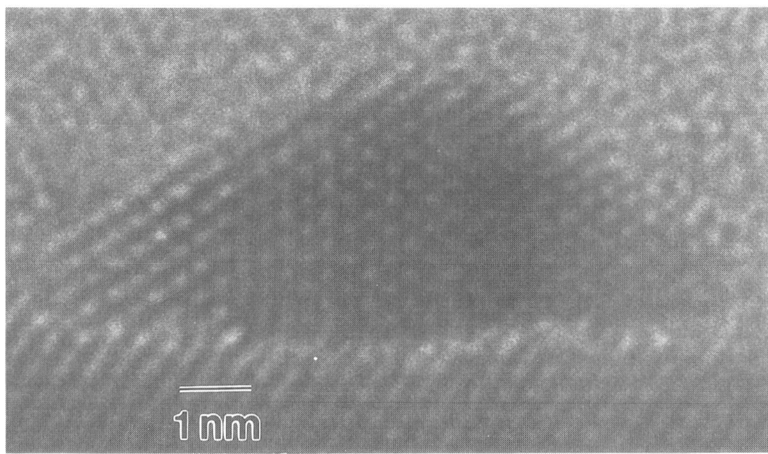

FIG. 3. Two-beam, dark-field images taken from the same region of a CaF_2 /Si(001) sample using orthogonal Si<220> reflections. The diffraction vectors are marked as arrows and the beam direction is close to Si[001] in both images

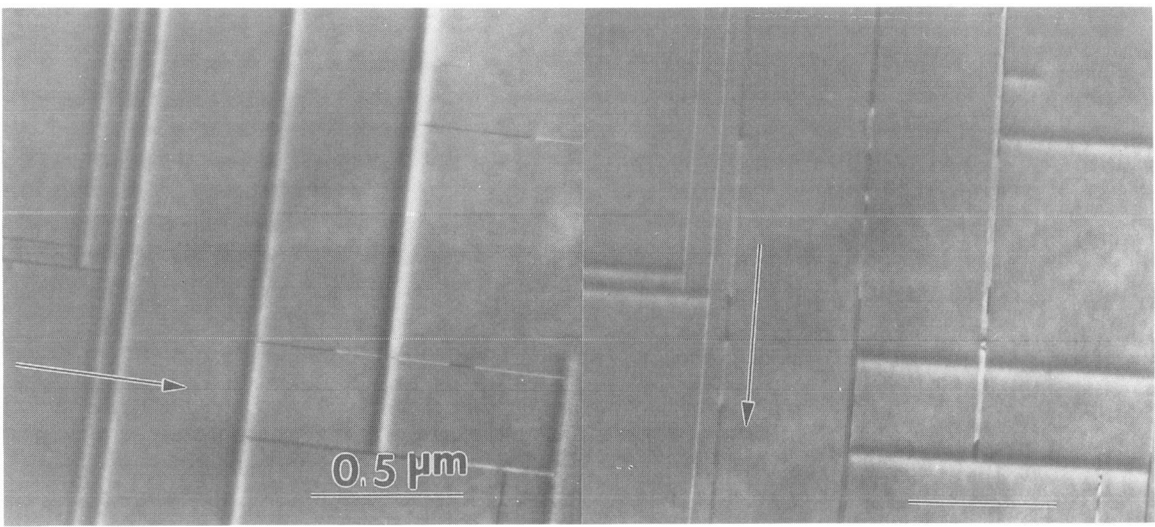

CORRELATION OF MICROMAGNETIC STRUCTURE AND MICROSTRUCTURE IN THIN FILM LONGITUDINAL MAGNETIC RECORDING MEDIA

Kai Tang

Department of Materials Science and Engineering, Stanford University, Stanford, CA 94305

Lorentz transmission electron microscopy (LTEM) has long been used as a tool for imaging the magnetic domain structure in materials and has occasionally been applied to thin film longitudinal magnetic recording media.[1,2] Conventional/high-resolution electron microscopy (TEM) has shown its power in revealing the detailed microstructure of these materials.[3] The combination of these two techniques can result in a more complete understanding of the interdependence of micromagnetic structure and microstructure and provide insight into how to improve the magnetic performance.

LTEM requires specimens with large uniformly thin areas so that the deflection angle of the incident electron beam is proportional to the magnetic field within the specimen plane. Conventional specimen preparation methods, which normally generate a wedge, are therefore no longer suitable. Previous work has used specially designed substrates to facilitate preparation of LTEM specimens of Co alloy/Cr magnetic media.[2] In the present study a chemical etching technique was introduced to successfully produce LTEM specimens directly from the C/Co alloy/Cr/NiP/Al(substrate) structures, typical of those currently used in the magnetic disk industry. A flowchart of this technique is shown in Figure 1. The key point is the removal of the NiP layer by concentrated nitric acid. Taking use of the fact that both Cr and Al are insoluble in this acid, the Cr underlayer can now act as a protective layer, isolating the Co alloy magnetic layer from the etchant while the Al substrate remains as mechanical support for the resulting thin metallic film. Removal of the Cr underlayer and C overcoat can be accomplished using low-angle ion-milling. This technique has generated LTEM specimens with uniformly thick electron transparent areas $\geq 2000 \ \mu m^2$ for films on both smooth and mechanically textured substrates.

Figure 2a is part of an underfocused Fresnel LTEM image of $Co_{84}Cr_{10}Ta_6$/Cr film (H_c=1590 Oe) on a smooth substrate in a bits-written magnetic state, which confirms that a large thin area is successfully obtained using this chemical etching technique. The bits are written in alternating direction of magnetization along tracks and the regions between tracks are in the as-deposited magnetic state, with magnetization direction perpendicular to that within the bits, as indicated in the picture. Alternating dark and light domain walls along the track direction are observed between the bits and the regions between tracks, consistent with alternating direction of magnetization of the bits. Longitudinal magnetic ripple structure is found within the bits and the regions between tracks. Figure 2b shows a transition region at a higher magnification. It consists of alternating dark and light dots, consistent with a previous observation on $Co_{86}Cr_{12}Ta_2$/Cr film deposited on a carbon-precoated Si(substrate), which can be interpreted as vortices with alternating sense of rotation.[2] Vortices observed in ac-demagnetized samples are an order of magnitude larger.

While it is believed that the micromagnetic structure in bit transition regions plays an important role in limiting recording density, a full understanding of this structure is not possible without detailed knowledge of the film microstructure. Numerous parameters, including film composition and thickness, deposition conditions, and substrate character, strongly affect the resulting microstructure. Figures 3a and 4a are bright field images of $Co_{86}Cr_8Ta_6$ film on Cr/NiP/Al(substrate) with thicknesses of 40 nm and 60 nm, respectively. Figures 3b and 4b are high-resolution images of these films. BF pictures show strong grain boundary contrast in the 60nm, but not the 40nm film. HREM pictures indicate sharp crystalline grain boundaries in the 40nm film and separation of grains in the 60nm film. This difference is thought to arise from growth morphology, that is, as the polycrystalline film grows thicker, the center part of a crystal grain tends to "outrun" its surroundings. HREM images also demonstrate clusters of Co alloy grains with perpendicular c-axis orientations as observed by previous

Proc. Microscopy and Microanalysis 1995, edited by G.W. Bailey, M.H. Ellisman, R.A. Hennigar, and N.J. Zaluzec
Copyright © 1995 MSA. Published by Jones and Begell Publishing, 79 Madison Ave., New York, NY 10016

authors.[3] These microstructural features are the basis of micromagnetic structure and will ultimately determine the magnetic performance of the media.

The complexity of thin film magnetic media requires knowledge of both micromagnetic structure and microstructure in order to fully understand their performance. A recently acquired Philips CM 20 FEG TEM equipped with both a Twin2 Lorentz imaging lens (3nm resolution) and a SuperTwin objective lens (0.24nm resolution) allows direct correlation between micromagnetic structural features, e.g. vortices, and microstructural features, e.g. crystal grains, of the same sample area, and when combined with special specimen preparation methods, will provide much new information in this important scientific and technological area.[4]

References
1. T. Chen, *IEEE Trans. Magn.*, MAG-17(1981)1181
2. G.P. Cameron and J.H. Judy, *IEEE Trans. Magn.*, 29(1993)4177
3. T.P. Nolan et al., *Ultramicroscopy*, 47(1992)437
4. Funding from Komag Inc. and contributions from C.A. Ross, R. Ranjan, and T. Yamashita of Komag, and M.R. Visokay and R. Sinclair of Stanford are greatly appreciated.

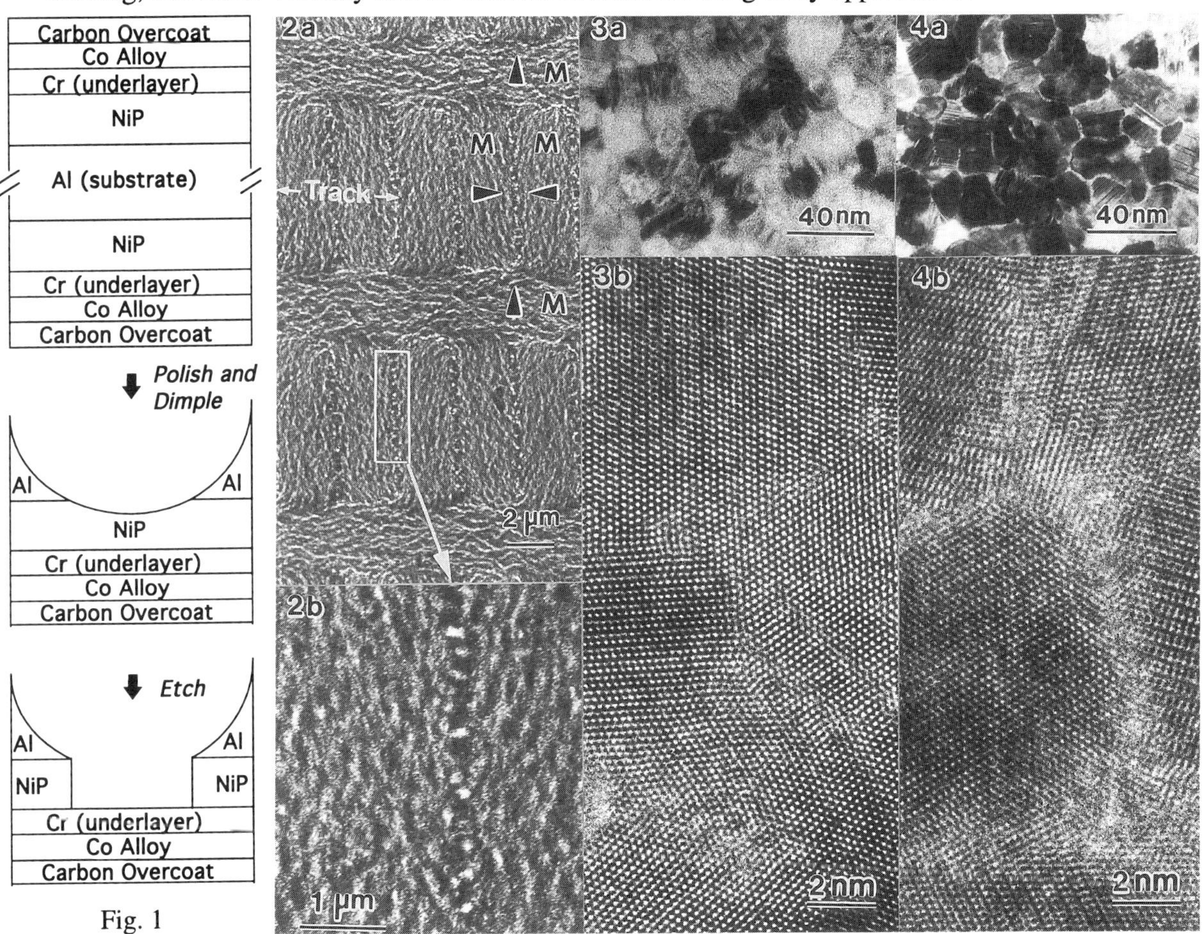

Fig. 1

Fig.1: Flowchart of the etching technique. The etchant is concentrated nitric acid.
Fig.2: a) LTEM image of $Co_{84}Cr_{10}Ta_6$/Cr film on a smooth substrate in bits-written magnetic state; b) Bit transition region at a higher magnification.
Fig.3: a)/b) BF/HREM images of a 40 nm thick $Co_{86}Cr_8Ta_6$ film on Cr/NiP/Al(substrate).
Fig.4: a)/b) BF/HREM images of a 60 nm thick $Co_{86}Cr_8Ta_6$ film on Cr/NiP/Al(substrate).

DYNAMICAL LORENTZ TEM OBSERVATION OF EXCHANGE COUPLING IN MR $NI_{0.80}FE_{0.20}/NIO$ FILMS

Wei Cao

Department of Materials Science and Mineral Engineering, University of California, Berkeley, CA 94720

In our research program involved with materials showing magnetoresistive (MR) phenomena, attention has been given to detailed microstructural analyses in order to understand physical properties and other characteristics. One of the biggest problems in the MR sensor application is the noise associated with Barkhausen jumps due to magnetic domain wall motion in the MR sensor, usually permalloy films $(Ni_{0.80}Fe_{0.20})$. It has been found that when a permalloy film was deposited together with an antiferromagnetic film such as FeMn, the noise could be suppressed.[1] It is known that there is an exchange coupling field (He) in the $Ni_{0.80}Fe_{0.20}/FeMn$ bilayers through the interface, and it is this exchange coupling field that makes the MR sensor a single domain which suppresses the noise. However, FeMn antiferromagnetic films have poor corrosion resistance, and this provides a motivation to find a substitute material for FeMn.

Antiferromagnetic NiO and CoO have received some attention as alternative materials because of their corrosion resistance. Recent research shows that comparable magnetic properties have been obtained in the sputtered $Ni_{0.80}Fe_{0.20}/Ni_xCo_{1-x}O$ bilayers. [2] In our previous research, the results of the epitaxial oxide films on several different substrates, the dependence of the growth orientation of the $Ni_{0.80}Fe_{0.20}$ layer on the oxide underlayer and the interface of $Ni_{0.80}Fe_{0.20}/Ni_xCo_{1-x}O$ layers have been reported. [3] In this paper, we will report the research results of Lorentz microscopy observing magnetic domain and domain wall motion in the permalloy which exchange couples with an antiferromagnetic NiO film.

The materials studied were polycrystalline $Ni_{0.80}Fe_{0.20}$ films coupled to a NiO (500Å) film on a Si substrate. Lorentz magnetic imaging was done in a Phillips EM 301 electron microscope. During the experiment, the objective lens is turned off so as to obtain Fresnel and Foucault modes to image the domain walls and the domain structures as shown in Figure 1. Fig. 1a and 1b are the magnetic domain wall images (Fresnel mode) at overfocus and defocus condition, Fig.1c and 1d are the domain images (Foucault mode) with shifted objective aperture. The average domain size is measured directly from the image to be about 2μm, and the domain wall width is about 1000Å. Figure 2 shows the domain wall motion in a weak magnetic field of objective lens as the sample is tilted from its horizontal position. Compared to domain wall motion in the pure permalloy, domain wall motion in the permalloy coupled to the NiO film is very smooth and slow. This indicates that the domain walls in the permalloy/NiO experience a strong force from the exchange coupling at the interface. Lorentz imaging is a useful method for understanding ferro- and antiferromagnet coupling in complex multilayers.[4]

References

1. C. Tsang, N. Heiman and Kenneth Lee, *J. Appl. Phys.* 50, 247(1981).
2. M.J. Carey and A.E. Berkowitz, *Appl. Phys. Lett.* 60, 3060(1992).
3. W. Cao, G. Thomas, M.J. Carey and A. E. Berkowitz, *J. Electr. Mater.* 23, 1043(1994).
4. The author would like to thank Prof. G. Thomas for his encouragement, discussions, and comments on this work. This research was supported by NSF Grant No. DMR 90-10908. Research facilities at the NCEM, LBL are gratefully acknowledged. The specimens were obtained from Prof. A.E. Berkowitz and Mr. K. Tanaka of UCSD.

Proc. Microscopy and Microanalysis 1995, edited by G.W. Bailey, M.H. Ellisman, R.A. Hennigar, and N.J. Zaluzec
Copyright © 1995 MSA. Published by Jones and Begell Publishing, 79 Madison Ave., New York, NY 10016

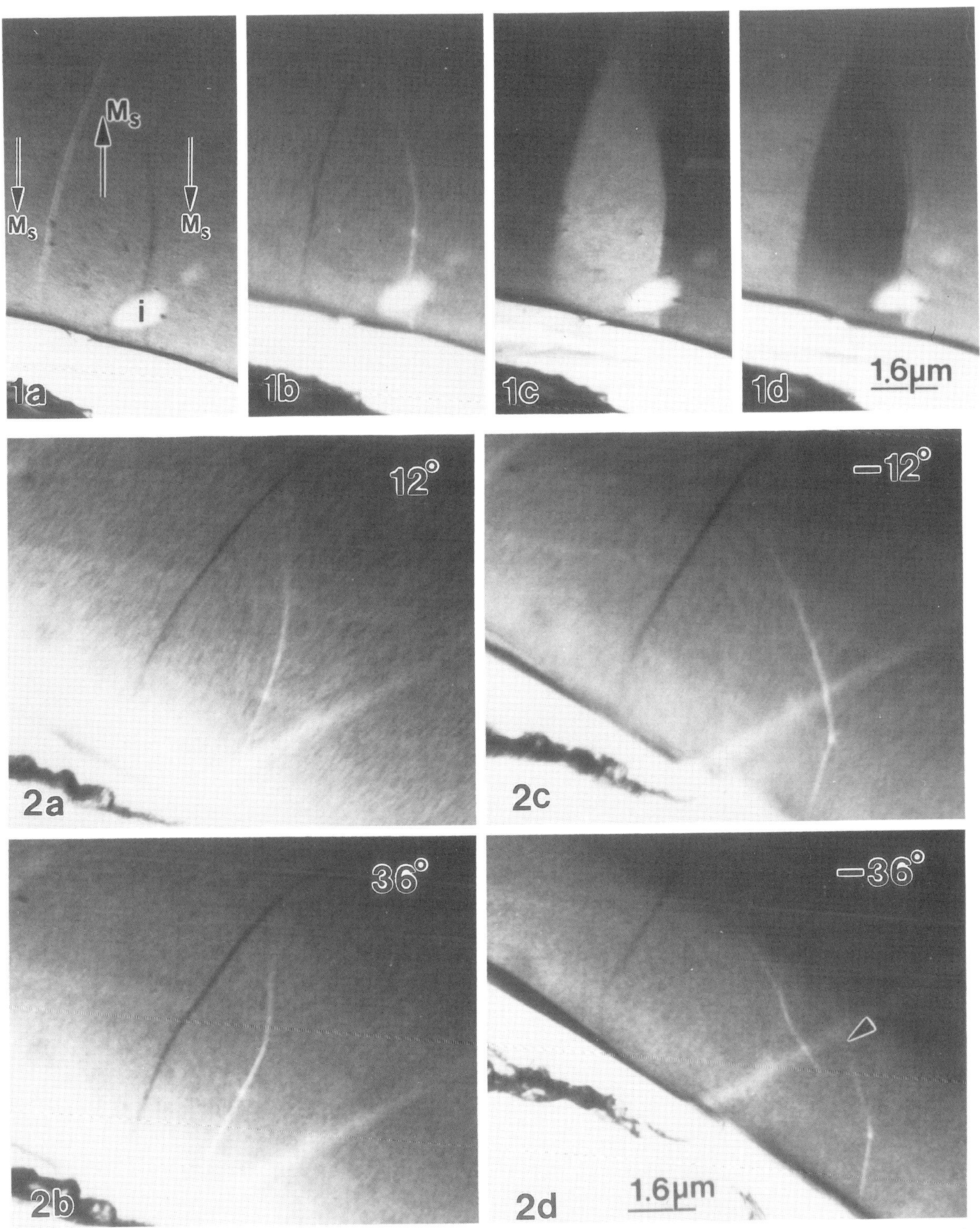

FIG.1.- Domain structure images, a) Fresnel mode at overfocus, b) Fresnel mode at defocus. c) and d) Foucault mode.

FIG.2.- Lorentz microscopy images of the domain wall motion in the Ni-Fe permalloy film at different tilting angles.

ADVANCED MICROSCOPY -- THE NEW HIGH RESOLUTION ZONE-PLATE MICROSCOPE AT THE ADVANCED LIGHT SOURCE IN BERKELEY

W. Meyer-Ilse, H. Medecki, C. Magowan, R. Balhorn*, M. Moronne, D. Attwood

Lawrence Berkeley Laboratory, Center for X-ray Optics, 1 Cyclotron Rd., Berkeley, CA 94720
*Lawrence Livermore National Laboratory, 7000 East Ave., Livermore, CA 94550

A new x-ray microscope (XM-1) has been installed at the Advanced Light Source in Berkeley. This transmission microscope uses zone-plates for a resolution exceeding visible light microscopies. Samples can be as thick as 10 microns, for wet or dry specimens. These features make x-ray microscopy a valuable complement to other advanced techniques.

There are two types of x-ray microscopes, scanning and conventional (imaging) microscopes. The scanning type minimizes radiation dose to the sample and is convenient for high resolution use of fluorescent labels; however, it requires a spatially coherent x-ray source and as a result involves long exposure times.[1] The conventional type provides a higher potential for ultimate resolution as there is no scanning stage needed, and it can operate with an incoherent light source. It therefore has a shorter exposure time, but does require a higher radiation dose due to lens inefficiencies. The new XM-1 is of the second type. Its optical layout is very similar to the Göttingen x-ray microscope operated at the BESSY facility in Berlin, Germany.[2]

The microscope design uses novel methods for precise motion controls and fully integrates state-of-the-art visible light microscopes that permit examination of the sample before and after x-ray imaging. To minimize radiation dose to the sample, zone plates with high diffraction efficiency and a high efficiency x-ray CCD camera are used for image formation and detection.

The mechanical design incorporates two visible-light microscopes. The first visible light microscope is used to view the sample before mounting in the x-ray microscope (external microscope). It is a Zeiss Axioplan employing several contrast methods including phase contrast and fluorescence. Using the external microscope, the sample is precisely mapped and focused for significant features. Their locations are stored digitally and later used in the x-ray microscope for positioning and focusing. The sample carrier uses kinematic mounts to assure precise positioning of the specimen in either microscope. The second visible light microscope is mechanically built into the x-ray microscope (alignment microscope) and is used to view the sample once it is mounted in the x-ray microscope stage. With this system we were able to study samples that, in order to be located, require phase contrast or fluorescence and thus could not be studied with other x-ray microscopes.

First experiments with XM-1 have been performed and are reported here. Sample images are shown in the figures of the next page. Figure 1 shows an image of a malaria infected human red blood cell. The parasite *Plasmodium falciparum* is in the ring stage of its development. The cells were fixed with 2% glutaraldehyde and imaged in aqueous environment. In the ring stage the parasite can not be seen with a transmission light microscope without staining. The intracellular parasite was labeled with Hoescht nuclear stain, so that infected cells could be recognized using fluorescence in the visible light microscope. Once identified, the infected cells were imaged with x-rays. The advantages of using the XM-1 for these investigations are that intact cells can be analyzed. Compared to electron microscopy studies, where sectioning is required to investigate whole cells, a higher throughput of the x-ray microscope allows analysis of a significantly larger number of samples. The fact that infected cells can be identified with the

Proc. Microscopy and Microanalysis 1995, edited by G.W. Bailey, M.H. Ellisman, R.A. Hennigar, and N.J. Zaluzec
Copyright © 1995 MSA. Published by Jones and Begell Publishing, 79 Madison Ave., New York, NY 10016

visible light microscope and that they are precisely positioned on the XM-1 stage before the first x-ray exposure makes these experiments with XM-1 much more effective compared to other x-ray microscopes.

Figure 2 shows an image of the sperm head from a marsupial mouse *(Sminthopsis)*. The sample was fixed in glutaraldehyde and embedded in a physiological buffer solution. Electron microscope images from stained sections of these samples showed two morphologically different areas of the chromatin distribution. The tip area appears to exhibits large variations in chromatin density, whereas in the inner area of the sperm head the chromatin appears homogeneous. This agrees well with electron microscope studies[3] of stained sections from the same type of specimen.

Over the next year a low temperature (cryo) sample holder will be introduced and increased access provided to visiting scientists.[4]

1. J. Kirz, et al.: "Soft X-ray Microscopes and Their Biological Applications", to appear in *Quarterly Review of Biophysics* (1995)

2. G. Schmahl, et al.: "X-ray microscopy studies", *Optik* **93**, No. 3 (1993) 95-102

3. W.G. Breed, et al.: "Unusual nuclear structure of the spermatozoon in a marsupial, *Sminthopsis crassicaudata*", *Molecular Reproduction and Development* 37(1), 78-86 (1994).

4. We gratefully acknowledge the collaboration with our colleagues G. Schmahl, D. Rudolph, and their co-workers from the University of Göttingen, Germany. Without their continuing contributions in zone plate and microscope development, this instrument would not have been possible. Special thanks also go to E. Anderson, who fabricated the objective zone plates in a collaboration with the IBM research center in Yorktown Heights, NY., and W. Breed and L. Soon from the University of Adelaide, who prepared the sample in figure 2. This work is supported by the US. Department of Energy, Office of Basic Energy Sciences, and the Office of Health and Environmental Research under contract number DE-AC03-76SF00098. CM is supported by NIH grant DK 32094.

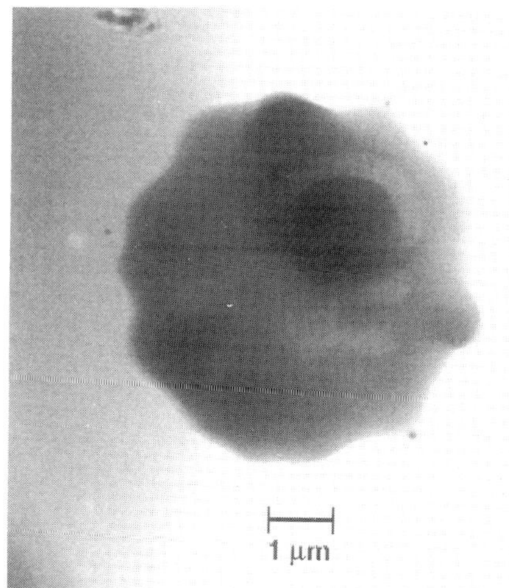

Figure 1: Human red blood cell infected with malaria parasite *Plasmodium falciparum*, shown in the ring structured stage of the parasite's development

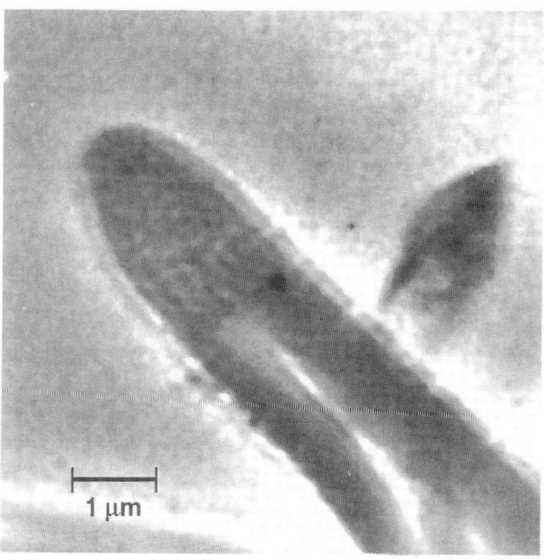

Figure 2: Head of a sperm cell from a marsupial mouse *(Sminthopsis)*. Structured details of the chromatin distribution can be seen in the tip of the sperm head.

TEM CONTRIBUTION TO SOLVE THE CRYSTAL STRUCTURE OF $Sr_5Ir_3O_{11}$

Z. G. Li, R.L. Harlow, W.J. Marshall, M.K. Crawford, M.A.Subramanian

Du Pont Central Research and Development, P. O. Box 80228, Wilmington, DE 19880-0228

X-ray diffraction plays a dominant role in the crystal structure determinations. Typically, diffraction data of a selected "single" crystal is first collected by a diffractometer, and then the crystal structure can be solved and refined. The quality of the structure determination is measured by the crystallographic residual value R. If the R value is reasonably small, the structural model is assumed to be correct. Usually, models found by x-ray diffraction do describe real structures and these can be confirmed by other techniques. We report here an unusual case, a plausible structural model which yielded a low R value from data collected on a crystal that was NOT "single": the structure derived from the "single" crystal x-ray diffraction data is completely bogus. TEM played an important role in solving this problem.

In the context of high-temperature superconductivity, we have recently studied Ir^{+4} containing layered oxides in the $SrO\text{-}IrO_2$ system [1,2]. Sr_2IrO_4 has a K_2NiF_4-type structure with a rotational distortion (see fig. 1a). The same type of distortion was also observed in the second member of this series, $Sr_3Ir_2O_7$, which contains double perovsite-type layers (see fig. 1b). Further investigation of compositions with various Sr/Ir ratios led to the discovery of crystals with an approximate stoichiometry of $Sr_5Ir_3O_{11}$ [3]. A plate-shaped, presumably "single" crystal of $Sr_5Ir_3O_{11}$ was studied on an Enraf-Nonius CAD4 x-ray diffractometer. The structural solution yielded a perfectly plausible model which contained alternating single and double layers of IrO_2. The "refined" model of $Sr_5Ir_3O_{11}$, which converged at R = 0.050, is shown in figure 1c.

Although the solution of the structure proceeded in a straightforward fashion, several questions were raised regarding its validity. First, the c-axis of the refined unit cell had a very large estimated standard deviation. Second, several of the atoms had non-positive definite thermal parameters. In an attempt to sort out these problems, it was decided to examine the crystal by transmission electron microscopy. Both bright field and selected area electron diffraction images were obtained. A typical image is showed in figure 2; the a-b plane is parallel to the electron beam direction. Clearly the crystals are simple intergrowths of the "single layer" (Sr_2IrO_4) and "double layer" ($Sr_3Ir_2O_7$) phases, and the structure derived from the "single" crystal x-ray diffraction data is incorrect. HREM images show district "single-layer" regions with a spacing of 6.1Å and "double layer" regions with a spacing of 10.0 Å. Such stacking faults are common materials where there is an epitaxial match of lattice constants.

It is surprising that a crystal which is composed of random layers of two phases would produce a reasonable structure model. This example serves as a cautionary tale for those who rely solely on x-ray diffractometer data for their structural analysis; there is a real danger of making a serious mistake unless they pay close attention to all aspects of the "automated" data collection and warning signs that the refinement produces, particularly in the form of unusual thermal parameters. In these cases, electron microscopy can help X-ray diffraction to CORRECTLY determine the crystal structure.

References

1. M.K.Crawford et al., Phys. Rev. B49, 9198 (1994).
2. M.A.Subramanian, R.L. Harlow and M.K.Crawford, Mat. Res. Bull. 29, 645 (1994).
3. R.L. Harlow, Z. G. Li, W.J. Marshall, M.K. Crawford, and M.A.Subramanian, Mat. Res. Bull. 30, 217 (1995).

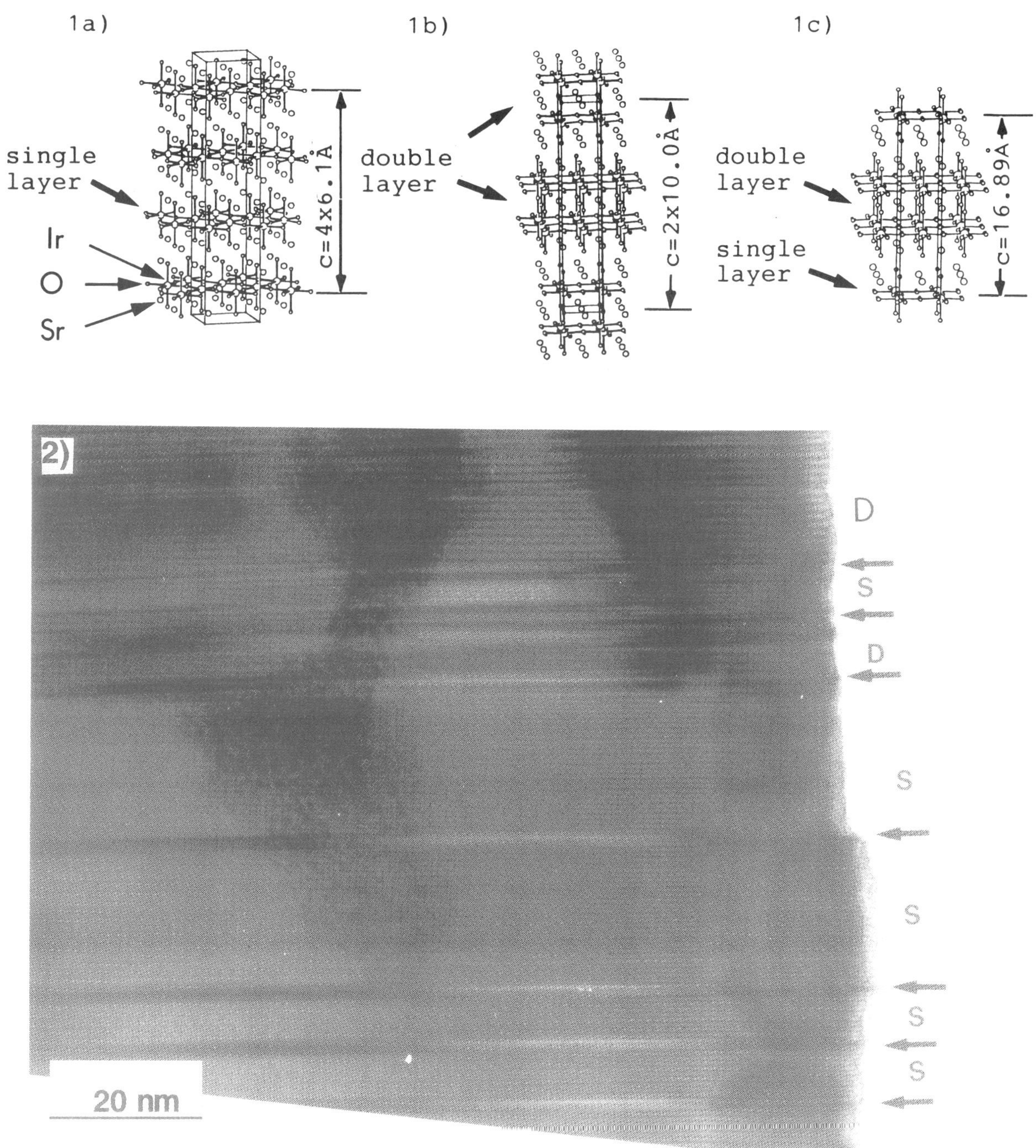

Figure 1. -- Stereo drawing showing the structures of (a) "single layer" Sr_2IrO_4, (b) "double layer" $Sr_3Ir_2O_7$, and (c) new "single" crystal $Sr_5Ir_3O_{11}$ found by "automated" x-ray diffraction.
Figure 2. -- HREM image of a crystal of $Sr_5Ir_3O_{11}$ (electron beam is parallel to the a-b plane). "Single-layer" and "double-layer" regions are indicated by the letters "S" and "D", respectively.

REFLECTION DIFFRACTED BEAM INTERFEROMETRY, RDBI, APPLIED TO THE STUDY OF SURFACES

Rodney A. Herring

604 Aldershot Rd., Baltimore MD 21229 USA

Diffracted beam interferometry, DBI, [1,2,3] (previously referred to as Convergent Beam Electron Diffraction + Electron Biprism Interferometry, CBED+EBI) which uses an electron biprism to deflect diffracted beams (convergent or parallel) can produce an interferogram between any two beams within the information envelope of the microscope such that the beam's amplitude and phase can be measured and studied. As well, the electron source need not be highly coherent. So far, DBI has been applied only to transmission electron diffraction, although there is no reason why it shouldn't be applicable to all electron diffraction methods including reflection high (low) energy electron diffraction, RH(L)EED and, possibly, back-scattered electron diffraction, BSED, in the SEM for the study of surfaces. DBI has already shown that substantial phase information, such as strain at interfaces and dislocations, compositional gradients and small defect clusters which are on the size scale of the unit cell, are retrievable from its holograms [4,5,6]. Already, reflection off-axis electron holography by Osakabe et. al. [7] has shown that phase information, in the form of a deformed surface produced by a single-atomic shear dislocation, is retrievable by phase-amplification of the reconstructed hologram. As well, when imaging is performed using RHEED-produced diffracted beams, good compositional information can be seen [8]. The reflection diffracted beam interferometry, RDBI, method is quite straight-forward and illustrated in Fig. 1 where a convergent beam which impinges a surface at a low angle is diffracted into high ordered beams which are then made to interfere with each other by their deflection induced by an electron biprism placed in a post-specimen position. The high order diffracted beams are elongated for a clean flat surface or oddly shaped such as oblong for rough surfaces. The shape of the diffracted beams does not limit their ability to interfere as illustrated in Fig. 1b and Fig. 1c for the interference of two highly stigmated, oddly shaped beams. The amplitude and phase of the interferogram will depend on the individual set-up of the RHEED microscope but should be essentially the same as that given for DBI [1,3], especially since RHEED is easily performed in a TEM [8]. Other than the phase additions due to the biprism and transfer functions of the imaging system, the phase of reflection DBI should consist of atomic configurations, surface roughness, compositions and contamination, electrostatic and magnetic potentials. If the acceleration energy of the beams becomes low, reflection DBI would become very sensitive to the phase of surface electrostatic and magnetic potentials and would provide a RLEED method which is currently unable to study these potentials. Structure factor equations would be able to model the phases due to atomic configurations and compositions and reflection DBI would be very useful for determining whether the surface is buckled or not [9]. The DBI method as applied to BSED in the SEM would also be sensitive to surface and sub-surface structures but the role of the interaction volume, which may provide a large source for the diffracted electrons, would need to be investigated as it may require BSED-DBI to use a more highly-coherent primary beam.

The support given by the Tonomura Electron Wavefront Project, ERATO, JRDC, its members, and Professor Giulio Pozzi are gratefully acknowledged.

1. R.A. Herring, these proceedings.
2. R.A. Herring, G. Pozzi, T. Tanji, and A. Tonomura, Ultramicroscopy 50 (1993) 94.
3. R.A. Herring, G. Pozzi, T. Tanji, and A. Tonomura, Ultramicroscopy, in press.
4. R.A. Herring and T. Tanji, ICEM 13-Paris (1994) 325.
5. R.A. Herring and T. Tanji, 51st MSA meeting, (1993) 1086.
6. R.A. Herring, J.E. Bonevich, T. Tanji, and A. Tonomura, Mat. Res. Soc. Proc. 332 (1994) 231.
7. N. Osakabe, J. Endo, T. Matsuda, A. Tonomura and A. Fukuhara, Phys. Rev. Lett. 62 25 (1989) 2969.
8. M. Gajdardziska-Josifovska, 51st MSA meeting, (1993) 828.
9. Y. Ma, S. Lordi and J.A. Eades, ICEM 13 (Paris, 1994) V1 953.

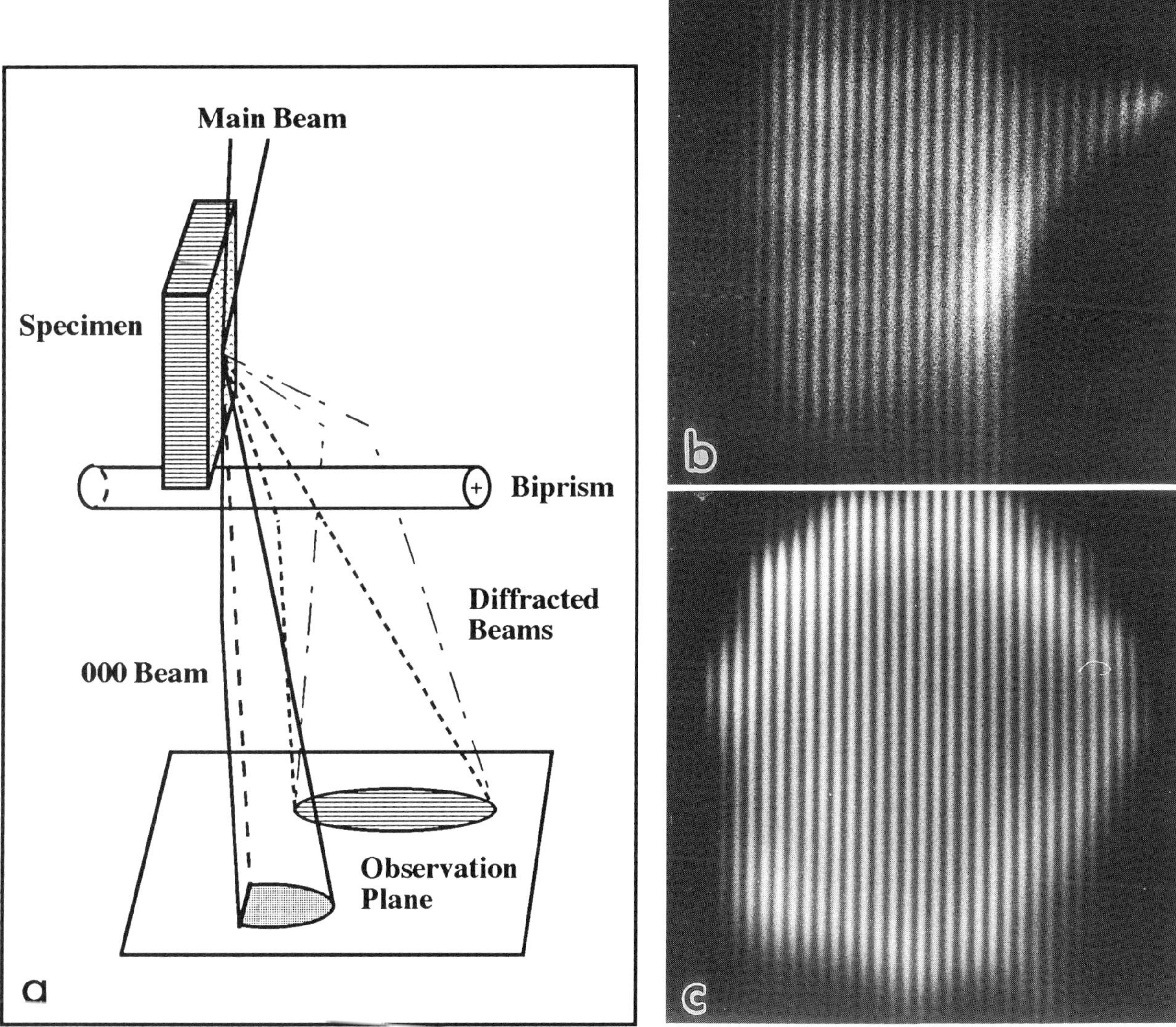

Fig. 1 In a) the illustration is an electron ray diagram of the set-up needed for Reflection Diffracted Beam Interferometry, RDBI, where a positive potential applied to an electron biprism, which is placed below the specimen, is able to deflect the diffracted beam sufficiently for their overlap and formation of an interferogram which possesses amplitude and phase information in the fringe contrast and fringe spacing, respectively. In b) and c) are two oddly shaped interferograms produced by the DBI method which implies that the interference and study of highly-stigmated surface-diffracted beams is surely possible.

IMAGE PROCESSING OF CONVENTIONAL TEM IMAGES: A CATALYST CASE STUDY

J. M. Briceño-Valero[*], L. Martínez[*], K.Moore[**] and V. Grassian[***]

[*] Applied Physics Group , Department of Physics, Faculty of Science. Universidad de los Andes. Mérida. Venezuela.
[**] Central Electron Microscopy Research Facility. University of Iowa. Iowa. 52242. USA.
[***]Chemistry Department, University of Iowa, Iowa City, Iowa 52242. USA.

Particle characterization with the Transmission Electron Microscope (TEM) has the inherent problem of interpreting three dimensional objects from two-dimensional projections. Relevant questions when measuring small particles include how projection of the image will distort measurement of crystallite size,[1] and how will the image distort the morphology of the particle Another difficulty is the effect of overlapping images from two or more particles. In the case of supported catalysts, an additional problem may be termed "differential contrast" which interferes with distinguishing between the supported metal crystallites and the support itself. While these situations may appear to be quite different, the is a shared concern of how the general appearance of the particle image related to its real shape. Image processing technology can be used to extract and display hidden information from conventional TEM micrographs. In this study, platinum particles supported on silica were examines.

TEM images were taken at 80,000X magnification with a Hitachi H-600 at 100 kV, digitally acquired using an Imapro QCS 3200 flatbed scanner at 3200 dots per inch. The scanner was controlled by an Adobe Photoshop Plug In module run on a Macintosh Centris 650, stored in a tagged Image File Format (TIFF) and processed with the aid of a Mtrace software written by the University of Iowa Image Analysis Facility. Analysis was restricted to a selected region of interest (ROI) associated to a cluster of Pt-particles.

Figure 1-a is a magnified image of a highly contrasted feature in a TEM picture after digitization. Following contrast enhancement, distinctive features are observed within the ROI such as those depicted in figure 1-B and labeled 1, 2 and 3. The black round spot in figure 1-a was carefully observed prior to contrast enhancement and did not reveal specific features such as those in figure 1-c. Pseudo-coloring (not shown here) of the gray level values from figure 1-c for maximum color contrast provided a much better indication of the presence of details within the ROI. This result indicates that without image analysis, information will be hidden within a conventional TEM micrograph. With the aid of image analysis, which is an easy and affordable method, interpretation of TEM micrographs can be improved. The technique can be used to count and measure particle size in the platinum supported on silica catalyst (Pt/SiO$_2$) and significally improve statistics, as seen in the increasing population of platinum particles from 129 to 388 and the standard deviation changing from 2.37 to 0.45. Several considerations, such as morphology of the particles, clustering, and overlapping of image projection were taken into account during this analysis[2]

Proc. Microscopy and Microanalysis 1995, edited by G.W. Bailey, M.H. Ellisman, R.A. Hennigar, and N.J. Zaluzec
Copyright © 1995 MSA. Published by Jones and Begell Publishing, 79 Madison Ave., New York, NY 10016

References:

[1] R.J. Maty, L. H. Schartz, and J.B. Butt. *Cata. Rev.-Sci. Eng.,* **29**(1), 41 (1987).
[2] J. M. Briceño-Valero, *et. al.,* To be published (1995).

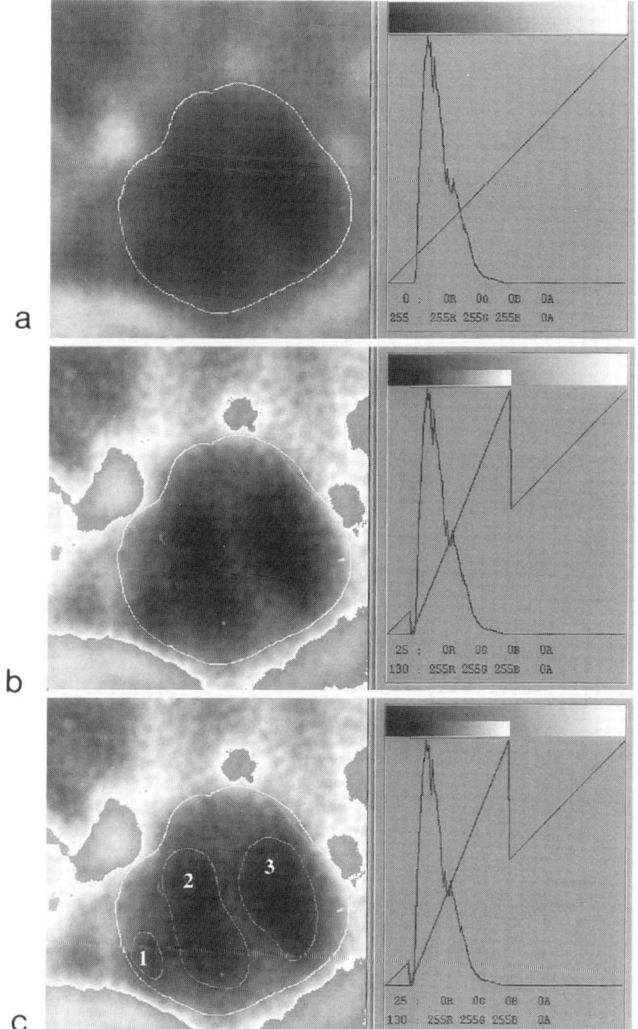

Fig. 1 (a) Digitized Image of a region of interest (ROI) for Pt/SiO_2.
   (b) Digitized Image with contrast enhancement.
   (c) Digitized Image showing Pt particles within the ROI.
   Histograms of gray level values for the ROI are shown at the right of each image.

INTERPRETATION OF SECONDARY ELECTRON CONTRAST FROM NEGATIVE ELECTRON AFFINITY DIAMOND SURFACES

D.P. Malta, J.B. Posthill, T.P. Humphreys, and R.J. Markunas
Research Triangle Institute, Research Triangle Park, North Carolina 27709

Diamond is a wide band-gap semiconductor with many unique physical properties that make it an attractive technological material. One such property is the negative electron affinity (NEA) behavior of the surface when properly terminated with hydrogen[1,2] or a thin metal layer[3]. The NEA diamond surface exhibits an unusually large secondary electron (SE) yield[4] which is desirable for applications in cold cathode electron emitters of flat panel displays. Examination of NEA diamond surfaces by scanning electron microscopy (SEM)[5] has indicated that a unique mechanism appears to be responsible for the SE contrast in which sub-surface microstructural information is contained. This paper provides a brief interpretation of the origin of SE contrast from the NEA diamond surface.

The electron affinity of a semiconductor surface, χ, is defined by the position of the vacuum energy level, E_o, relative to the conduction band minimum, E_c (Fig. 1a). If $\chi>0$, excited conduction band electrons must migrate to the surface and arrive with sufficient kinetic energy to overcome the surface energy barrier in order to escape into vacuum. The probability, p(x), that an electron generated at a depth x will migrate to the surface with sufficient escape energy is:

$$p(x) = C \exp(-x / L_{eff})$$

where C is a constant <1 and L_{eff} is the effective mean escape depth[6]. For χ small and positive, $L_{eff} \approx L_T$, the thermalization length defined as the average distance an excited electron travels before losing excess energy with respect to the conduction band edge. L_{eff} is typically about 0.5-1.5nm in metals and 10-20nm in insulators[7]. For a short L_{eff}, SE contrast originates very near the surface, the effective emission area, d_{eff}, is very small allowing high spatial resolution, and surface topography usually provides the principle source of contrast. Fig. 2a is a SE image of a positive electron affinity "clean" natural diamond surface. Only topographical contrast is observed. For $\chi<0$ (Fig. 1b), the escape depth is no longer limited by the thermalization length since no excess kinetic energy is required to escape the surface. If $\chi<0$, then $L_{eff} \approx L_D$, the diffusion length defined by the average distance a minority free electron can travel before being annihilated by recombination. Since typically, $L_D \gg L_T$, the escape depth from the NEA surface is much larger. This results in a higher SE yield, a larger effective escape area, d_{eff}, and poorer spatial resolution. A large component of the total SE signal, I_{SE}, from the NEA surface carries sub-surface information which easily dominates the topographical signal component. If a free electron is generated in the vicinity of a crystalline defect, L_D is considerably shortened, as these defects typically act as free electron traps. Since the intensity of SE emission is proportional to p(x), emission near defects will be suppressed creating defect contrast in SE images of NEA diamond surfaces (Fig. 2b).

The unique SE contrast observed from NEA diamond surfaces is, therefore, attributed to an unusually large effective escape depth for SE electrons resulting from a barrier-free surface condition. The SE signal is dominated by sub-surface information and contrast is provided by microcrystalline defects acting as free electron traps[8].

Proc. Microscopy and Microanalysis 1995, edited by G.W. Bailey, M.H. Ellisman, R.A. Hennigar, and N.J. Zaluzec
Copyright © 1995 MSA. Published by Jones and Begell Publishing, 79 Madison Ave., New York, NY 10016

References

1. F.J. Himpsel, J.A. Knapp, J.A. van Vechten and D.E. Eastman, Phys.Rev. B **20**, 624 (1979).

2. B. B. Pate, P.M. Stefan, C. Binns, P.J. Jupiter, M.L. Shek, I. Lindau and W.E. Spicer, J. Vac. Sci. Technology. **19**, 349 (1981).

3. J. van der Weide and R.J. Nemanich, J. Vac. Sci. Technol. B., Vol. 10, 1940 (1992)

4. D.P. Malta, J.B. Posthill, T.P. Humphreys, R.E. Thomas, G.G. Fountain, R.A. Rudder, G.C. Hudson, M.J. Mantini, and R.J. Markunas, Appl. Phys. Lett. **64**, 1929 (1994).

5. D.P. Malta, J.B. Posthill, T.P. Humphreys, R.E. Thomas, R.A. Rudder, G.C. Hudson, M.J. Mantini and R.J. Markunas, Proceedings of the 28th Annual Microbeam Analysis Society Meeting, (Ed. J.J. Friel), July 31-August 5, (1994), 205

6. J. van Laar, Acta Electronica, **16**,3,(1973) 215

7. H. Seiler, Z. Angew Phys. **22** (1967) 249

8. The authors gratefully acknowledge support of this work by BMDO/IST through ONR (Contract No. N00014-92-C-0081) and the helpful assistance of Janice Stephens for manuscript preparation.

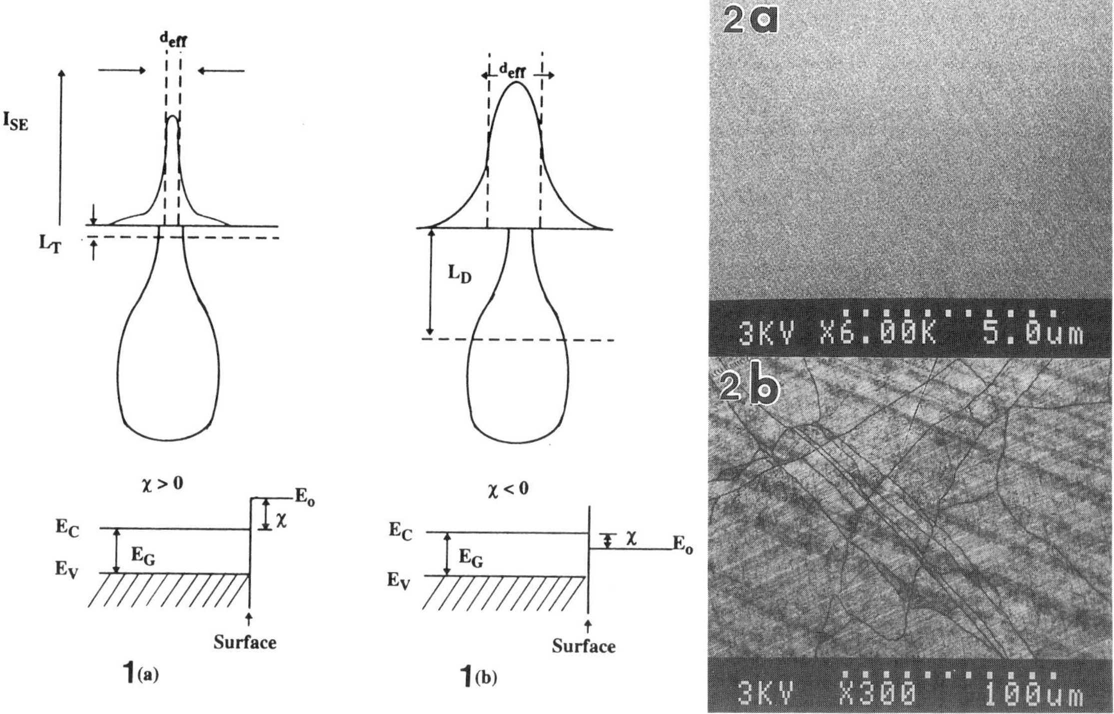

Fig. 1 Schematic diagrams of energy band configurations and origins of secondary electron signal from diamond (a) positive electron affinity surface, (b) negative electron affinity surface.

Fig. 2 Secondary electron images of a polished natural type IIa diamond surface: (a) "clean" surface; χ >0; topographical polishing scratches observed only; (b) H-terminated surface; χ<0; defect contrast is observed; surface orientation is (110); defects show directional clustering (diagonal lines) and cellular clustering (dark random lines).

MICROSTRUCTURE AND MICROTEXTURE OF A Nb-16%Si (DS) ALLOY

J. A. Sutliff and B. P. Bewlay

GE Corporate R&D, P.O. Box 8, Schenectady, NY 12301

In-situ composite Nb-Si alloys have been studied by several investigators as potential high temperature structural materials.[1,2,3,4] The two major processing routes used to fabricate these composites are directional solidification and extrusion of arc-cast solidified ingots. In both cases a stable microstructure of primary Nb dendrites in a eutectoid of Nb and Nb_5Si_3 phases is developed after heat treatment. The Nb_5Si_3 phase is stable at room temperature and forms as a decomposition product of the high temperature Nb_3Si phase. The anisotropic microstructures developed by both directional solidification and extrusion require evaluation of the texture to fully interpret the fracture and other orientation dependent mechanical behavior of these composites.

In this paper we report on the microstructural characterization of a directionally solidified (DS) and heat treated Nb-16 at.%Si alloy. The microtexture of each of the phases (Nb, Nb_5Si_3) was determined using the Electron BackScattering Pattern (EBSP) technique for electron diffraction in the scanning electron microscope.[5] A system employing automatic diffraction pattern recognition, crystallographic analysis, and sample or beam scanning was used to acquire the microtexture data.[6] The experimental conditions used were 30 kV, ~2 nA, 70° tilt, and ~35mm screen to specimen distance. The samples were metallographically prepared with a final vibratory polish using colloidal silica. Generally, there are difficulties in performing multiphase microtexture analysis using automated EBSP[7], but in this study the samples produced good patterns from each phase and the analysis was not subject to any crystallographic ambiguity between the phases. For each pattern six bands were detected and indexed to obtain the orientation. The data presented are from a 2 mm^2 scan containing 5200 patterns, 82% of which accurately and unambiguously index as Nb or Nb_5Si_3; the remaining were not indexed due to poor pattern quality.

Figure 1 is a backscattered electron image (BEI) of the DS Nb-16at.%Si alloy after a 1500C/100hr heat treatment. In this image the Nb phase appears in both dendrites and eutectoid and is lighter than the Nb_5Si_3 phase. Figure 2 contains typical EBSP diffraction patterns obtained from a) the Nb and b) the Nb_5Si_3 phases. $\{002\}_{Nb}$ and $\{001\}_{Nb5Si3}$ pole figures as well as inverse pole figures indicating the crystallographic direction parallel to the growth direction for the Nb and Nb_5Si_3 phases, respectively, are presented in Figures 3a through 3d.

No subgrain structure or evidence of recrystallization was detected in the Nb dendrites of this DS alloy. Nor were precipitates observed in the dendrites, although very small precipitates may go undetected by scanning electron microscopy. These findings are unlike those reported for extruded Nb-Si alloys.[1] The crystallographic texture is complex as indicated by the pole figures. The primary Nb dendrites alone showed little preferred orientation, but in the eutectic/eutectoid a distinct texture is found with a preferred growth direction in the vicinity of $[113]_{Nb}$. The Nb_5Si_3 phase exhibited a limited texture with [001] predominantly perpendicular to the growth direction. Eutectic grains were visible in the microstructure and inspection of the orientation data confirms clustering of both Nb and Nb_5Si_3 orientations in regions corresponding to a eutectic grain. As the eutectic grain structure results from the solidification of Nb and the high temperature Nb_3Si phase, and not the Nb_5Si_3 phase, larger

Proc. Microscopy and Microanalysis 1995, edited by G.W. Bailey, M.H. Ellisman, R.A. Hennigar, and N.J. Zaluzec
Copyright © 1995 MSA. Published by Jones and Begell Publishing, 79 Madison Ave., New York, NY 10016

variation in orientations of Nb_5Si_3 might result from variants in the Nb_3Si decomposition. Texture measurements in extruded materials as well as the fracture behavior of both DS and extruded materials will be discussed in a forthcoming paper.[8]

References

1. I. Weiss, M. Thirukkonda, and R. Srinivasan, *Mat. Res Soc. Proc.* **332** (1994) 377
2. M. Saqib, M. Thirukkonda, B. Cockeram, R. Srinivasan, and I. Weiss, *PRICM-I*,
Changxu Shi, Hengde Li, and A. Scott, eds. (TMS, Inc., Warrendale, PA 1992) p. 711
3. M.G. Mendiratta, J.L. Lewandowski and D.M. Dimiduk, *Met. Trans*, **22A** (1991) 1573
4. B.P. Bewlay, H.A. Lipsitt, W.J. Reeder, M.R. Jackson and J.A. Sutliff, in *Processing and Fabrication of Advanced Materials for High TemperatureApplications III*, Ed. V.A. Ravi, T.S. Srivatsan and J.J. Moore (TMS, Warrendale, PA, 1994) p. 547
5. J.A. Venables and C.J. Harland, *Philosophical Magazine*, **27** (1973) 1193
6. N.H. Schmidt, *CHANNEL++* v.3.0, (DSS, Inc., Randers, Denmark, 1995)
7. J. A. Sutliff, *Scanning 95* (FAMS, Inc., Mawah, NJ, 1995) in press
8. Funding from the Air Force Office of Scientific Research is gratefully acknowledged.

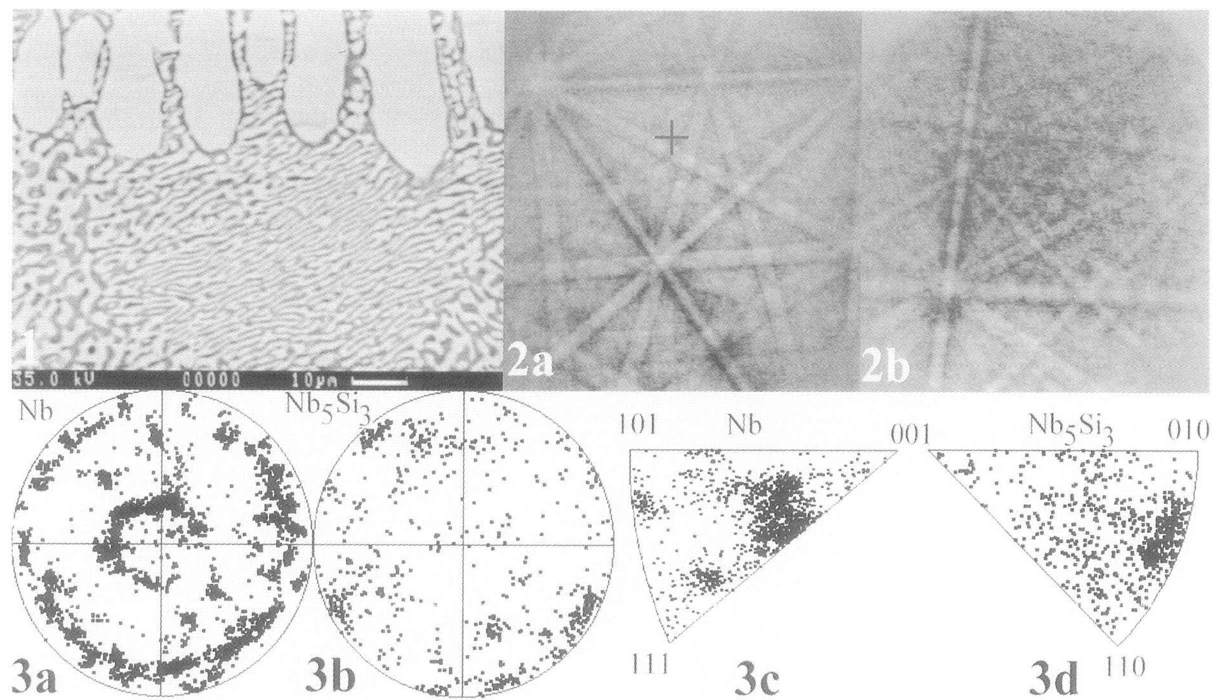

FIG. 1 - Backscattered electron image of DS and heat treated Nb-16at.%Si alloy.
FIG. 2 - EBSP diffraction patterns of a) Nb and b) Nb_5Si_3 phases.
FIG. 3 - a) {002}Nb and b) (001)Nb_5Si_3 pole figures; inverse pole figures for growth direction of c) Nb and d) Nb_5Si_3 phases

IONIZATION DELOCALIZATION AND ALCHEMI OF B2-ORDERED ALLOYS

I. M. Anderson and J. Bentley

Metals & Ceramics Division, Oak Ridge National Laboratory, P.O. Box 2008, Oak Ridge, TN 37831

Recent formulations of the ALCHEMI technique have explicitly accounted for ionization delocalization of an element i with a single linear parameter L_i that represents the extent of localization.[1,2] The x-ray emission of a spectral line characteristic of the element is assumed to be proportional to a linear combination of the thickness-averaged electron intensity at (1) the site of the atom (fraction: L_i) and (2) averaged over the unit cell (fraction: $(1 - L_i)$). The parameter L_i is assumed to be characteristic of the elemental shell and independent of the diffracting conditions. In consequence of these assumptions, it has been shown that ionization delocalization correction can be performed in conjunction with "statistical" ALCHEMI by correcting only the extracted correlation coefficients.[2]

The parameter $r_i(\xi) = I_i(\xi) / I_i(0)$ is defined to be the ratio of the characteristic intensity (I) of element i at a crystal orientation (ξ) to its characteristic intensity at a reference orientation (0), which is ideally free of electron channeling effects. Coefficients α_{kj} correlating the r-values of each alloying element k with those of the n host elements j are extracted using multivariate analysis following Rossouw et al.,[3]

$$r_k = \sum_{j=0}^{n} \alpha_{kj} r_j , \qquad (1)$$

except with an additional coefficient α_{k0} to account for the delocalized fractions of each element, with $r_0 \equiv 1$.[1] The correlation coefficients can be corrected for ionization delocalization with the relation:[2]

$$\chi_{kj} = \frac{L_j}{L_k} \alpha_{kj} \equiv L_{jk} \alpha_{kj} . \qquad (2)$$

These corrected coefficients equal the fractional occupancies p_{kj} of alloying element k on site j if the host elements are perfect markers for their respective sites. Whether the host elements are perfect markers or not, the sum of χ_{kj} over all sites j should equal unity, just as the site occupancies sum to unity. The purpose of the present paper is to demonstrate that the major assumption underlying this formulation is justified: that the degree of ionization localization of an elemental shell can be accounted for by a linear coefficient; and to introduce a potential method, which would be applicable to B2-ordered alloys, of independently extracting the ratio of coefficients L_{jk} necessary for delocalization correction.

For B2-ordered alloys, the major low-index planes are {110} and {200}. The {110} planes are of equal composition: for a stoichiometric alloy AB, the composition of each {110} plane is AB (Fig. 1). Conversely, the two crystallographic sites are completely separated on alternating {100} planes and have alternating compositions A and B in a fully ordered alloy. When dynamical diffracting conditions are set up at these two systematics orientations, {200} should provide good discrimination between the 'A' and 'B' sites of the crystal and {110} none at all, as long as contributions from nonsystematic reflections are insignificant. Ionization delocalization should influence the variation of the characteristic x-ray intensities at both planar channeling conditions. However, at the {110} systematics orientation, the effect of ionization delocalization should be isolated.

A Cr-doped FeAl alloy of nominal composition $Fe_{60}Al_{35}Cr_5$ and a series of Fe-doped NiAl alloys with Fe-concentrations between 0.25 and 12 at.% were analyzed for this study. Energy-dispersive x-ray (EDX) spectra were acquired with a Philips CM12 operated at 120 kV equipped with an EDAX 9900 spectrometer and superUTW Si(Li) detector. Fig. 2 shows the variation of r-values of the host elements Fe and Al relative to that of the alloying element Cr for spectra acquired near <117> beam direction at a range of {110} excitations between symmetry and the {220} Bragg position. The slopes of these lines, which are the ratios L_{jk} for the {110} systematics orientation, are shown in Fig. 2. These values, which indicate that Fe is slightly more localized and Al substantially less localized than Cr, are plausible. The excellent linearity of the data substantiates the use of linear coefficients to model ionization localization.

Proc. Microscopy and Microanalysis 1995, edited by G.W. Bailey, M.H. Ellisman, R.A. Hennigar, and N.J. Zaluzec

Furthermore, we have investigated whether the L_{jk} acquired at a $\{110\}$ systematics orientation could be accurately applied to ALCHEMI data acquired at $\{200\}$. Other studies have indicated that the degree of localization varies with interplanar spacing, and that this variation may be strong for lower energy ionizations such as Al K.[4] Correlation coefficients α_{kj} acquired at $\{200\}$ were corrected with L_{jk} ratios determined at $\{110\}$, as indicated by equation (2). This method yields independent χ_{kj}, and one can check to make sure that their sum, $\Sigma \equiv \chi_{CrFe} + \chi_{CrAl}$, is unity. It would be expected that a significant difference in L_{Al} between $\{110\}$ and $\{200\}$ would effect a significant deviation of Σ from unity, particulary for alloys in which the alloying element resides predominantly on the 'Al'-site. This test was performed by applying the L_{jk} ratios in Fig. 2 to ALCHEMI data acquired from the Cr-doped FeAl specimen near <014> beam direction at a range of $\{200\}$ excitations between symmetry and the $\{400\}$ Bragg position. The result is $\Sigma = 1.04 \pm 0.03$. Similarly, localization coefficients L_{jk} were extracted from an alloy of stoichiometry $Ni_{50}Al_{40}Fe_{10}$ and applied to the correlation coefficients of a series of sixteen Fe-doped NiAl alloys. The result of this more extensive study is $<\Sigma> = 1.01 \pm 0.01$. Also, there was no measurable variation in the Σ values with site-occupancy. These results seem to be in better agreement with similar L_{jk} for the $\{110\}$ and $\{200\}$ systematics orientations than with L_{jk} that vary significantly with interplanar spacing. The outlined method for delocalization correction would allow accurate ALCHEMI analyses that are free from reliance on adjustable parameters.[5]

References

1.   M.G. Walls, Microsc. Microanal. Microstruct. 3(1992)443.
2.   I.M. Anderson and J. Bentley, Proc. 13th ICEM: Electron Microscopy 1994 1(1994)609.
3.   C.J. Rossouw et al., Phil. Mag. Lett. 60(1989)225.
4.   S.J. Pennycook, Ultramicroscopy 26(1988)239.
5.   This research was supported by the Division of Materials Sciences, U.S. Department of Energy, under contract DE-AC05-84OR21400 with Martin Marietta Energy Systems, Inc. This work was also partially supported by an appointment (IMA) to the Oak Ridge National Laboratory Postdoctoral Research Associates Program, which is administered jointly by the Oak Ridge Institute of Science and Education and Oak Ridge National Laboratory.

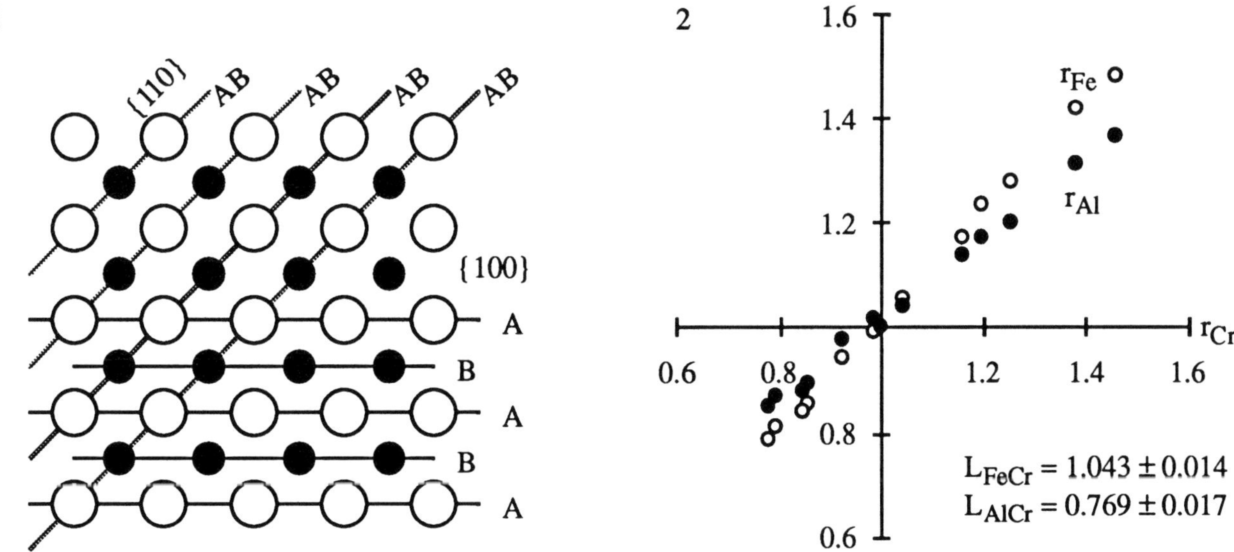

FIG. 1.   Schematic representation of a B2-ordered alloy "AB" viewed along <001>.
FIG. 2.   Variation of the intensity ratios of the host and alloying elements in Cr-doped FeAl for $\{110\}$ systematics orientation. The values L_{jk} are the slopes of the linear regression fits to the data.

A SYSTEMATIC ALCHEMI STUDY OF Fe-DOPED NiAl ALLOYS

I.M. Anderson,* A.J. Duncan,† and J. Bentley*

*Metals & Ceramics Division, Oak Ridge National Laboratory, P.O. Box 2008, Oak Ridge, TN 37831
†Department of Materials Science & Engineering, University of Florida, Gainesville, FL 32611

ALCHEMI site-occupation studies of alloying additions to ordered aluminide intermetallic alloys have been performed with varying degrees of success, depending on the ionization delocalization correction procedure applied to the data.[1,2] Most published studies have been applied to just one or two alloys in order to ascertain qualitatively the site-preference of the alloying element. However, quantitative site-occupancies can be extracted from carefully performed ALCHEMI experiments as long as the data are properly treated for ionization delocalization and anti-site occupancies.[3,4] Systematic studies can therefore yield the dependence of site-occupancies in alloys on independent thermodynamic variables. The purpose of the present study is to examine the variation in the site-occupancy of Fe in B2-ordered NiAl as a function of solute concentration and alloy stoichiometry.

Twelve different alloys were prepared for this study with stoichiometries $Ni_{50-x}Al_{50}Fe_x$ (Ni-deficient), $Ni_{50-x/2}Al_{50-x/2}Fe_x$ (intermediate), and $Ni_{50}Al_{50-x}Fe_x$ (Al-deficient), with x = 0.25, 2, 5, and 10. All alloys were homogenized at 1300°C for 5 h, step-cooled to room temperature, then annealed at 800°C for 72 h and water quenched. Three-millimeter discs were spark-cut from slices of each alloy and thinned by electrojet polishing. Energy-dispersive x-ray (EDX) spectra were acquired with a Philips CM12 operated at 120 kV and equipped with a LaB_6 filament, an EDAX 9900 spectrometer and superUTW Si(Li) detector. Experiments were performed with a beam divergence semi-angle of 3 milliradians, which is approximately one-quarter of the Bragg angle of a {200} reflection. Spectra were acquired near <014> beam direction at ~10 distinct {200} excitations between symmetry and the {600} Bragg position. Diffracting conditions were chosen to minimize the excitation of nonsystematic reflections and the beam current was monitored and held at a constant value.

Data analysis was performed by correlating the characteristic x-ray intensities of the Fe alloying element with those of the Ni and Al host elements with linear regression analysis, as discussed elsewhere.[4] Ionization delocalization correction was performed with coefficients extracted from data acquired at a {110} systematics orientation.[5] Because of the low counting statistics of the 0.25%-Fe-containing alloys, the delocalization-corrected site distributions χ_{kj} of these alloys were normalized so that $\Sigma = \chi_{FeNi} + \chi_{FeAl} = 1$. This procedure was deemed valid because the average (before normalization) of the Σ values among the alloys was unity, so that systematic error should not be introduced. This procedure effectively utilizes the counts from all spectra, rather than just one spectrum, to determine the k-factor for the dilute Fe-impurity. Two independent ALCHEMI analyses were performed for each of the 0.25%-Fe-containing alloys because of the large statistical error of these analyses.

The results of these analyses are shown in Fig. 1. Here, the fraction of Fe on the 'Ni'-site, $p_{Fe'Ni'}$, is plotted as a function of Fe-concentration, x. Different symbols are used for the three different Ni-Al stoichiometries, as indicated. The results of the two analyses from each 0.25%-Fe-containing alloy are plotted separated by the width of the symbol on either side of x = 0.25, for clarity. The site-occupancies extracted from the two analyses of each 0.25%-Fe-containing alloy agree to well within one standard deviation. This agreement shows that the normalization procedure enhances the precision of the measurement, as expected. The good separation among the data from alloys of the three stoichiometries clearly shows that the site occupancy of iron depends upon the relative concentrations of the Ni and Al host elements. However, a preference for the 'Ni'-site is clearly indicated: more than half of the iron occupies the 'Ni'-site in all of the intermediate alloys, and the iron occupies a higher fraction of 'Ni'-sites in the Ni-deficient alloys than it does 'Al'-sites in the Al-deficient alloys. Definite trends in the dependence of site-occupancy on solute concentration are apparent for each alloy stoichiometry. The site-occupancies in alloys of intermediate stoichiometry are relatively independent of Fe-concentration. Conversely, in the Al-deficient alloys the fraction of Fe occupying 'Ni'-sites decreases with increasing

Proc. Microscopy and Microanalysis 1995, edited by G.W. Bailey, M.H. Ellisman, R.A. Hennigar, and N.J. Zaluzec
Copyright © 1995 MSA. Published by Jones and Begell Publishing, 79 Madison Ave., New York, NY 10016

solute concentration. In the Ni-deficient alloys, the 'Ni'-site-occupancy of Fe at solute concentrations ≥5% is enhanced (~95%) relative to that (~90%) for concentrations ≤2%.

The similarity in the site-occupancies of the dilute 0.25%-Fe-containing alloys of intermediate and 'Al'-deficient stoichiometries is noteworthy. Because of the difficulty in assuring alloy stoichiometry at such low solute levels, the increase of $p_{Fe'Ni'}$ in the Al-deficient alloy relative to the members of this family having higher solute levels is suspicious in the absence of intermediate data points. However, the three 0.25% alloys were arc-melted from the same ingot of stoichiometric NiAl, with appropriate additions of the solute and host elements. The site-occupancies of the dilute alloys of Ni-deficient and intermediate stoichiometries are consistent with those of higher solute levels in their respective families, hence it is unlikely that the change in site-occupancy of the Al-deficient alloy results entirely from a systematic error in stoichiometry. Moreover, the constitutional point-defect concentration needed to compensate the relatively high 'Ni'-site occupancy at the 0.25% solute level is smaller than that in any of the three 2%-Fe-containing alloys. A tendency of the site-occupancy to approach a constant low-concentration value, as indicated by the site-occupancies of the alloys of intermediate stoichiometry, is plausible.[6]

References

1.   J. Bentley, Proc. Ann. EMSA Meeting 44(1986)704.
2.   P.R. Monroe and I. Baker, J. Mat. Res. 7(1992)2119.
3.   M.G. Walls, Microsc. Microanal. Microstruct. 3(1992)443.
4.   I.M. Anderson and J. Bentley, Proc. 13th ICEM: Electron Microscopy 1994 1(1994)609.
5.   I.M. Anderson and J. Bentley, these proceedings (1995).
6.   This research is supported by the Division of Materials Sciences and by the Assistant Secretary for Energy Efficiency and Renewable Energy, Office of Industrial Technologies, Advanced Industrial Materials Program, U.S. Department of Energy, under contract DE-AC05-84OR21400 with Martin Marietta Energy Systems, Inc. This work is also partially supported by an appointment (IMA) to the Oak Ridge National Laboratory (ORNL) Postdoctoral Research Associates Program, which is administered jointly by the Oak Ridge Institute for Science and Education and ORNL.

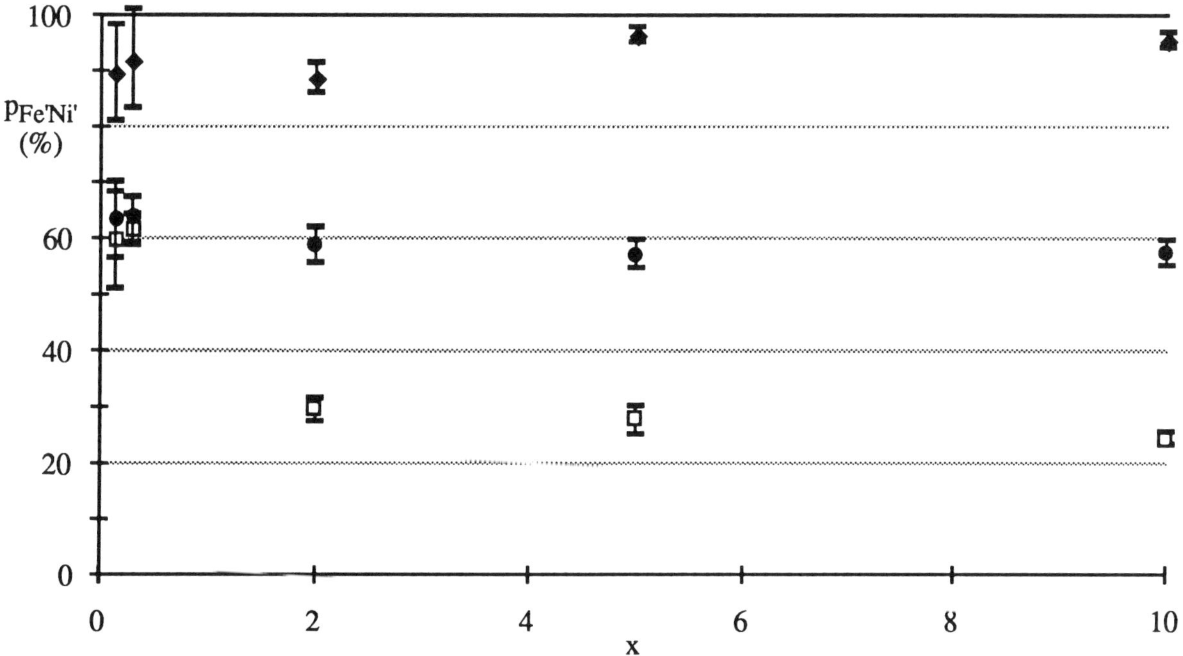

FIG. 1.   Site-occupancies of Fe in B2-ordered NiAl alloys as a function of at.% Fe, x, in Ni-deficient (diamonds), Al-deficient (squares), and intermediate (circles) Ni-Al stoichiometries.

A NEW APPROACH TO ALCHEMI : THE ORDERING TIE LINE METHOD

D.H. Hou and H.L. Fraser

Dept. of Materials Science and Engineering, The Ohio State University, OH43210

The ALCHEMI (Atom Location by CHanneling Enhanced Microanalysis) technique has been first developed by Spence and Taftø[1] to determine site occupancies of alloying elements in ordered crystals. Based on their work, a number of formulations have been developed[2-8] to improve the accuracy by using multivariant analysis[5], to extend the analysis to a more general case[2,3], and to incorporated anti-site defects and delocalization effect[6-8]. However, all these formulations can be replaced by a new formulation based on the Ordering Tie Line (OTL) concept.

The OTL concept can be illustrated by using a ternary compound ($C^o = A_x B_y C_z$), with a B2 crystal structure, as an example. In the B2 crystal structure (Fig.1), the sublattices of α and β can be treated as two 'phases' with compositions C^α and C^β (for B2 ordering $\Leftrightarrow C^\alpha \neq C^\beta$). The OTL is then defined as the tie-line connecting C^α and C^β(Fig.1). Similar to a regular tie-line, the OTL describes mass conservation in a B2 crystal and its mid point will locate at C^o. The ordering state (i.e. atom configuration) of the B2 crystal is then given by C^α and C^β. As seen in Fig.1, the advantage of using OTL to describe an ordered crystal is that it provides a graphical method to represent the ordering state, which is the parameter/ property measured in an ALCHEMI experiment.

Recalling that the Alchemi experiment is basically a microanalysis experiment, for each channeling experiment (e.g. $100 \pm s$ channeling in B2 crystals) an "apparent composition ($C^{app.}$)" can be deduced from the measured x-ray spectrum and k-factors. It can be proven[9] that, assuming no delocalization effect, each measured apparent composition is simply a linear combination of the two sublattice compositions C^α and C^β. It is noted that the apparent compositions can never be equal to the sublattice compositions since the channeling effect will not be so strong that all electrons channel through one sublattice. Therefore, the sublattice compositions (i.e. site occupancies) cannot be determined in the ALCHEMI experiment. However, the trace of the OTL can be determined by fitting a straight line to the apparent compositions obtained from several channeling experiments. This fitting process is equivalent to the multivariant ALCHEMI analysis[5-8]. If results of one channeling experiment (C^{app}) and one kinematical experiment (C^o) are used, the OTL slope can also be determined analytically. This is similar to the algorithm used in the conventional ALCHEMI analysis[1-4].

Even though only the slope of the OTL is determined by the ALCHEMI experiment, site occupancies can still be deduced by making appropriate assumptions. For example, the apparent compositions of a L1$_0$ crystal, TiAl+Ga with $C^o = Ti_{49.9}Al_{45.2}Nb_{4.9}$, are measured to be $Ti_{44.9}Al_{49.5}Nb_{5.6}$ and $Ti_{63.3}Al_{33.1}Nb_{3.6}$ at 110_{+s} and 110_{-s} channeling conditions, respectively[10]. By fitting these compositions, an OTL slope of $dC_{Ti}{}^{\alpha,\beta}/dC_{Al}{}^{\alpha,\beta} = -1.115$ is obtained. The site occupancies of Ti and Al atoms can be assumed to be unity since no significant redistribution of host atoms will occur with the addition of small amount of Ga atoms. This assumption has been made implicitly in the original ALCHEMI formulation, and is equivalent to stretch the OTL outward until one end intersects the composition diagram boundary, as shown in Fig.2. The final sublattice compositions are found to be $C^\alpha = Ti_{97.2}Al_{2.8}Nb_{0.0}$ and $C^\beta = Ti_{2.7}Al_{87.6}Nb_{9.7}$, and the site occupancy of Ga atom can then be deduced as $f_{Ga} = 1.00$ on the Al-rich site, which is the same as that deduced by the conventional formulation[10]. It is noted that a different site occupancy could be obtained if there are significant amounts of anti-site defects, i.e. $f_{Ga} = 0.71$ if 30% of Ti atoms are on the Al site. Therefore, the importance of host atom distribution is demonstrated.

The same analysis can also be applied to crystals exhibiting no apparent host atoms and stoichiometry, such as ternary B2 alloys of $Ti_{61.5}Al_{24.5}Nb_{14}$[11] and $Ti_{40}Al_{15}Nb_{45}$[12]. The slope of the OTL of the former crystal, as deduced from the original data, are found to be -1.225 which yields the site occupancies of $f_{Nb} = 0.70$ and $f_{Ti} = 0.26$ on Al-rich site by assuming $f_{Al} = 1.0$. Apparent compositions and OTL trace of the latter alloy are shown in Fig.3. Site occupancies of $f_{Nb} = 0.52$ and $f_{Ti} = 0.29$ on Al-rich site are obtained by

Proc. Microscopy and Microanalysis 1995, edited by G.W. Bailey, M.H. Ellisman, R.A. Hennigar, and N.J. Zaluzec
Copyright © 1995 MSA. Published by Jones and Begell Publishing, 79 Madison Ave., New York, NY 10016

assuming again f_{Al}=1.0. It is noted that the assumption of f_{Al}=1.0 can be justified by the "orientation" of the OTL, i.e. the crystal will be in its "perfectly ordered" state if all the Al atoms occupy only one site. Should the measured OTL trace be parallel to the Nb-Ti boundary, the appropriate assumption will then be f_{Nb}=1.0 or f_{Ti}=1.0. Nevertheless, the exact site occupancies in these ternary B2 alloys should not be over-emphasized since they can not be determined without knowing the distribution of at least one element. However, the slope of OTL is a parameter that can be compared between B2 crystals of different compositions.

References

1. J.C.H. Spence and J. Taftø, *J. Microscopy* , **130**(1983)147.
2. K.M. Krishnan and G. Thomas, *J. Microscopy* , **136**(1984)97.
3. S. Matsumura, T. Morimura and K. Oki, *Mat. Trans., JIM* , **32**(1991)905.
4. P.R. Munroe and I. Baker, *J. Mater. Res.* , **6**,(1991)943.
5. C.J. Rossouw, P.S. Turner, T.J. White and A.J. O'Connor, *Phil. Mag. Lett.* , **60**(1989)25.
6. M.G. Walls, *Microsc. Microanal. Microstruct.* , **3**(1992)443.
7. I.M. Anderson and J. Bentley, in *Electron Microscopy 1994*, Eds. B. Jouffrey and C. Colliex, 13th International Congress on Electron Microscopy (Paris, France, 1994), vol. 1, pp. 609.
8. M.G. Walls, *J. De Physique IV* , **3**(1993)2155.
9. D.H. Hou, I.P. Jones and H.L. Fraser, (1995), in preparation.
10. Y. Ren, G. Chen and B.F. Oliver, *Scripta Metall.* , **25**(1991)249.
11. D. Banerjee, T.K. Nandy and A.K. Gogia, *Scripta Metall.* , **21**(1987)597.
12. D.H. Hou, J. Shyue, S.-S. Yang and H.L. Fraser, in *Alloy Modeling and Design*, Eds. Stock and Turchi, Symposium on "Alloy Modeling and Design" (Pittsburgh, PA, 1993), pp. 276.

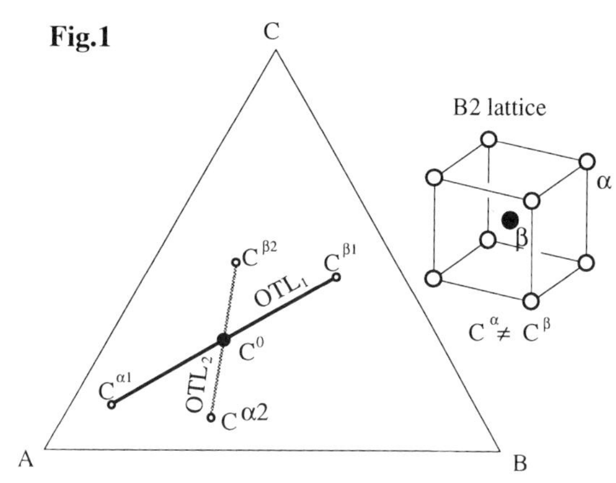

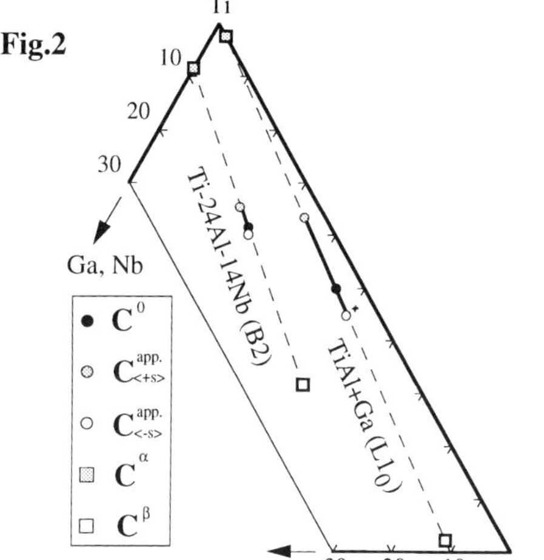

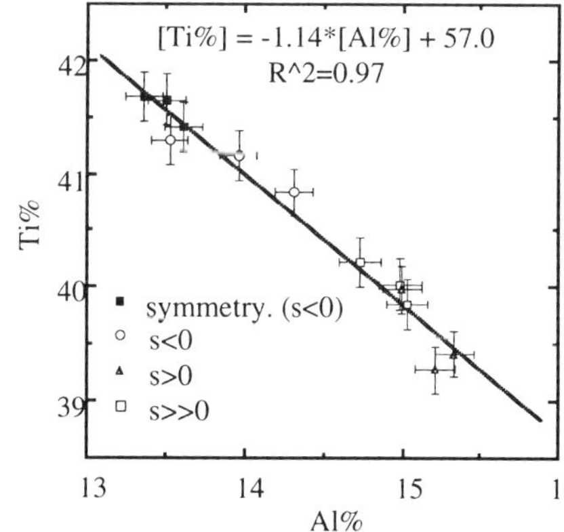

Fig.1 Schematic diagram illustrating Ordering Tie Line (OTL) concept. Two possible ordering states of ternary B2 crystal are represented by two different OTL's.

Fig.2 Apparent compositions and corresponding OTL's of L10 ordered alloy (TiAl+Ga[10]) and ternary B2 alloy (Ti$_{62}$Al$_{24}$Nb$_{14}$[11]).

Fig.3 Apparent compositions and corresponding OTL of ternary B2 alloy Nb$_{45}$Al$_{15}$Ti$_{40}$.

QUANTITATIVE CBED: AN OVERVIEW

D.M. Bird

School of Physics, University of Bath, Bath BA2 7AY, U.K.

In this abstract I will focus on one particular topic which has widespread relevance in the quantitative analysis of CBED; namely the role of computer simulations of CBED patterns. The basic point is that with modern, high-powered workstations we can perform highly accurate and detailed simulations of all types of CBED patterns from crystals in a relatively short amount of time. Most CBED simulations are based on diagonalisation of the many-beam equations,[1] and, for example, on a DEC Alpha workstation one can perform this calculation on a 150×150 matrix (sufficient for a highly accurate simulation in many systems) in around 1.8s. A full, two-dimensional CBED pattern with, say, 50 orientations across each disc (which is a high-resolution simulation) can then be calculated in less than 1 hour. How can such simulations be used in a quantitative analysis of CBED patterns? I will address this by looking at some specific examples covering various aspects of quantitative CBED.

Measurement of lattice parameters   The matching of experimental and simulated HOLZ line maps has become a very widely used technique for the accurate measurement of lattice parameters.[1] The simulations used here are generally very simple and use kinematic theory to generate the positions of the HOLZ lines. However, more detailed simulations, based on dynamical theory, also have a role. A particularly striking example of the use of theoretical simulations is provided by the work of Mansfield et al.[2] who showed that HOLZ line positions appear to oscillate with thickness when strong, zero-layer dynamical diffraction effects are present. This effect would not have been found without high quality simulations. The work clearly shows that crystal orientations where dynamical diffraction effects are minimised must be used for accurate lattice parameter measurements.

Symmetry determination   Determination of point and space group symmetry is one of the main uses of CBED.[1,3,4] Computer simulations are very valuable in addressing the question of the sensitivity of CBED patterns to small structural changes which alter symmetry. For example, Tanaka et al. have investigated the sensitivity of CBED symmetry to atomic displacements in FeS_2.[3] The sensitivity to compositional changes can also be analysed. Consider the simple case of a fictitious Si crystal whose two fcc sublattices have site occupancies of $1 + \epsilon$ and $1 - \epsilon$. (Note that computer experiments can easily and usefully be performed on systems not available to experimentalists.) This reduces the usual $2mm$ symmetry of the zero-layer Si[110] CBED pattern to m, as, for example, seen in GaAs[110]. For small ϵ the intensity change is purely symmetry breaking and we can quantitatively compare $2mm$ and m patterns using the χ^2 measure[1]

$$\chi^2 = \frac{1}{N} \sum_{i=1}^{N} \frac{(I_\epsilon^i - I_0^i)^2}{I_0^i}$$

where I_0 and I_ϵ are the calculated intensities for real and fictitious Si respectively, and the sum is over the N intensities included in the comparison (in this case 8160 points, covering the first 7 discs of the [110] pattern). The many-beam calculations include 67 beams, the accelerating voltage is 200kV and the intensities are scaled to a maximum of 4096 to simulate the effects of a CCD detector. Plots of χ^2 as a function of ϵ are shown in Fig. 1. A value of χ^2 greater than 1 implies a symmetry breaking above the noise level expected in a 12 bit counting system. It can be seen that compositional changes of around 1% should be measurable in this case, with the surprising feature that the pattern from a thinner specimen is more sensitive to change than that from a thicker one.

Refinement of atomic coordinates   HOLZ reflections are particularly sensitive to small changes in atomic coordinates, and this can be used for accurate structure refinement. For example, Tanaka et al. have

Proc. Microscopy and Microanalysis 1995, edited by G.W. Bailey, M.H. Ellisman, R.A. Hennigar, and N.J. Zaluzec.

used large-scale dynamical simulations of HOLZ reflections in SrTiO$_3$ to fit experimental patterns and determine the rotation of the oxygen octahedron.[4]

Accurate structure factor measurement   There has been much recent interest in the matching of experimental and simulated CBED patterns to determine highly accurate values for the low-order structure factors which are sensitive to bonding effects in crystals.[1,5] Very accurate simulations are vital for this, and one important issue is to ensure that the theoretical intensities are fully converged with respect to the number of beams included in the many-beam equations. The use of Bethe potentials can dramatically reduce the size of the many-beam matrix required for full convergence.[1] This is demonstrated by the results shown in Figure 2, which refer to calculations on the Si[110] zone axis. Tests show that the exact many-beam equations are well converged with 400 beams (we use independent atom structure factors for both real and absorptive potentials, an accelerating voltage of 200kV, a crystal thickness of 2500Å and a temperature factor of 0.26Å^2, which corresponds to liquid N$_2$ temperature). We then define a χ^2 measure similar to that above and calculate the χ^2 difference between an "exact" simulation with 444×444 matrices and simulations with smaller numbers of beams, both with and without the use of Bethe potentials. When Bethe potentials are used, extra beams are included out to a total of 444. As above, the intensities are scaled to a maximum of 4096 and a total of 1183 data points are used which cover the first 7 discs in the [110] CBED pattern. Figure 2 shows that when Bethe potentials are used, χ^2 falls below 1 for matrices larger than about 80×80 and reaches the effectively negligible value of 0.1 for about 115 beams. Conversely, the unperturbed many-beam equations are clearly not converged even for matrices as large as 200×200.

1. J.C.H. Spence and J.M. Zuo, *Electron Microdiffraction*, New York: Plenum Press (1992).
2. J.F. Mansfield, D.M. Bird and M. Saunders, *Ultramicroscopy*, **48** (1993) 1.
3. M. Tanaka et al., *Convergent Beam Electron Diffraction II*, Tokyo: JEOL Ltd (1988).
4. M. Tanaka et al., *Convergent Beam Electron Diffraction III*, Tokyo: JEOL Ltd (1994).
5. M. Saunders et al., in B. Jouffrey and C. Colliex, Eds., *Electron Microscopy 1994, Proc. 13th ICEM, Vol. 1*, Les Editions de Physique (1994) 847.

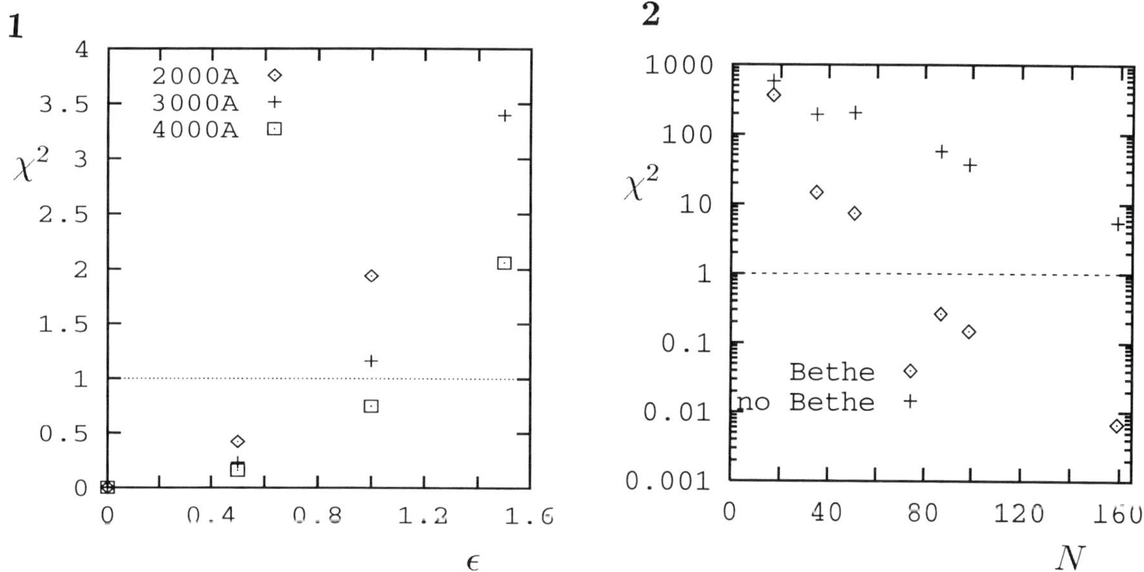

FIG. 1. Plot of χ^2 against occupancy parameter ϵ (in %), for crystal thicknesses of 2000Å, 3000Å and 4000Å, demonstrating sensitivity of CBED patterns to symmetry breaking. See text for details.
FIG. 2. Plot of χ^2 against number of beams included in dynamical calculations, showing convergence of simulations with and without Bethe potentials. See text for details.

MULTIPLE PLASMON SCATTERING CONTRIBUTIONS TO CBED CONTRAST

Knut Marthinsen[*], Ragnvald Høier[**] and Randi Holmestad[**]

[*]SINTEF Applied Physics, N-7034 Trondheim, Norway.
[**]Dept. of Physics and Mathematics, Univ. of Trondheim-NTH, N-7034 Trondheim, Norway.

One of the most successful applications of quantitative convergent beam electron diffraction (CBED) which have developed over the last years are methods for accurate determination of low-order structure factors.[1] Obvious applications are determination of electron charge densities and bonding effects.[2] However, the use of these methods is still not straight forward, and they are demanding and expensive both experimenatlly and computationally. The experiments have in general to be based on digital intensity recording and energy filtering for partly removing the background of inelastically scattered electrons. Computationally the methods are based on multi-parameter least squares fitting between experiment and theory, and since the complexity of the problem is $O(n^3)$, where n is the number of beams included in the calculations, one of the central problems here is to limit n without loss of accuracy.[3] The present work is focused on a theoretical description of the multiple inelastic scattering contributions, in particular plasmon scattering contributions, to CBED contrast.

The theory combines multiple inelastic scattering and Bragg scattering. It is based on a slice model where many-beam Bragg scattering takes place in the upper part, z, of the crystal. Inelastic scattering takes place in a slice dz at dept z, followed by Bragg scattering of the inelastically scattered electrons, in the lower part, t-z, of the crystal. Assuming that the inelastic scattering is dominated by low-angle scattering, the total intensity at thickness t in the direction $\mathbf{k_o}+\mathbf{s}+\mathbf{h}$ within the h-disk is given by

$$I(\mathbf{k_o}+\mathbf{s}+\mathbf{h}, t) = \int \left| \Sigma_i C_0^{i*} C_h^i \exp(2\pi i \gamma^i t) \exp(-2\pi\mu^i) \right|^2$$

$$\times \left[\delta(\mathbf{s}-\mathbf{s}') + \Sigma_{n>0} \frac{(2\pi\mu't)^n}{n!} F_n(\mathbf{s}-\mathbf{s}') \delta(E-E_n) \right] ds' \tag{1}$$

i.e. the total intensity depend on the standard elastic scattering intensity, the number of inelastic scattering events, n, and the cross-section for n times scattering, F_n, with energy change E_n.[4,5] Considering only plasmon scattering, μ' in equation 1 refers to the intensity absorption coefficient for plasmons, and the convolution is performed over angles typical for plasmon scattering. Theoretical profiles for the individual plasmon scattering contributions, according to equation 1, have been calculated across the (220) CBED disk in Si for the systematic (2n,2n,0)-case and they are shown in Fig. 1a. The total intensity, adding up the individual contributions, is shown in Fig. 1b.

The theoretical calculations have been compared with corresponding experimental intensity profiles. The experiments have been performed with a Philips CM30 at 300 keV and at -156°C using a modified PEELS analyzer for one-dimensional energy filtering. The observed profiles are shown in Fig. 1c. The crystal thickness is 1750 Å. Plasmon profiles across the (220) CBED disk for the order 1 to 4 have been collected with an energy window of 6 eV centered around the characteristic plasmon peaks. The dark current of the diode array has been subtracted, the intensities are corrected for unlinearities in the detector response, and the profiles are deconvoluted with the point spread function (PSF) of the PEELS analyzer.[6] Fig. 1d shows the corresponding total (unfiltered) experimental intensity. Comparing the experimental and theoretical profiles it is observed that on a semi-quantitative level the agreement is surprisingly good. However, two effects are not fully taken care of in the present description. The first is the large experimental background which has a maximum at the Bragg position. A possible explanation is thermal diffuse scattering (TDS) which is known to be both strongly peaked at the Bragg positions and generally anisotropic. The second effect relates to the relative intensities for different n. This latter effect may be due to the fact that TDS is not properly accounted for, but another explanation

Proc. Microscopy and Microanalysis 1995, edited by G.W. Bailey, M.H. Ellisman, R.A. Hennigar, and N.J. Zaluzec
Copyright © 1995 MSA. Published by Jones and Begell Publishing, 79 Madison Ave., New York, NY 10016

is single electron excitations which are not included in the present treatment. Inelastic scattering by single electron excitations scatter to higher angles, and may to first order be included by increasing the characteristic angle of integration. The inclusion of single electron excitations will affect the relative heights directly through μ', but also through a further smearing out of the calculated profiles which also will change the relative heights of the different plasmon contributions. If the inelastic scattering contributions can be calculated analytically with sufficient accuracy, it may make numerical filtering of CBED patterns possible and thus be an important alternative to experimental energy filtering.

References

1. J.C.H. Spence, and J.M. Zuo, *Electron Microdiffraction*, New York: Plenum Press (1992).
2. R. Holmestad et al., *Phil. Mag.*, in press (1995).
3. C. Birkeland et al., *This proceedings.*
4. R. Høier, *Acta Cryst.* A29(1973)663-672; *Acta Cryst.* A30(1973)457.
5. K. Marthinsen, R. Holmestad, and R. Høier, *Ultramicroscopy* 55(1994)268.
6. The aid of Dr. J.M. Zuo is gratefully acknowledged.

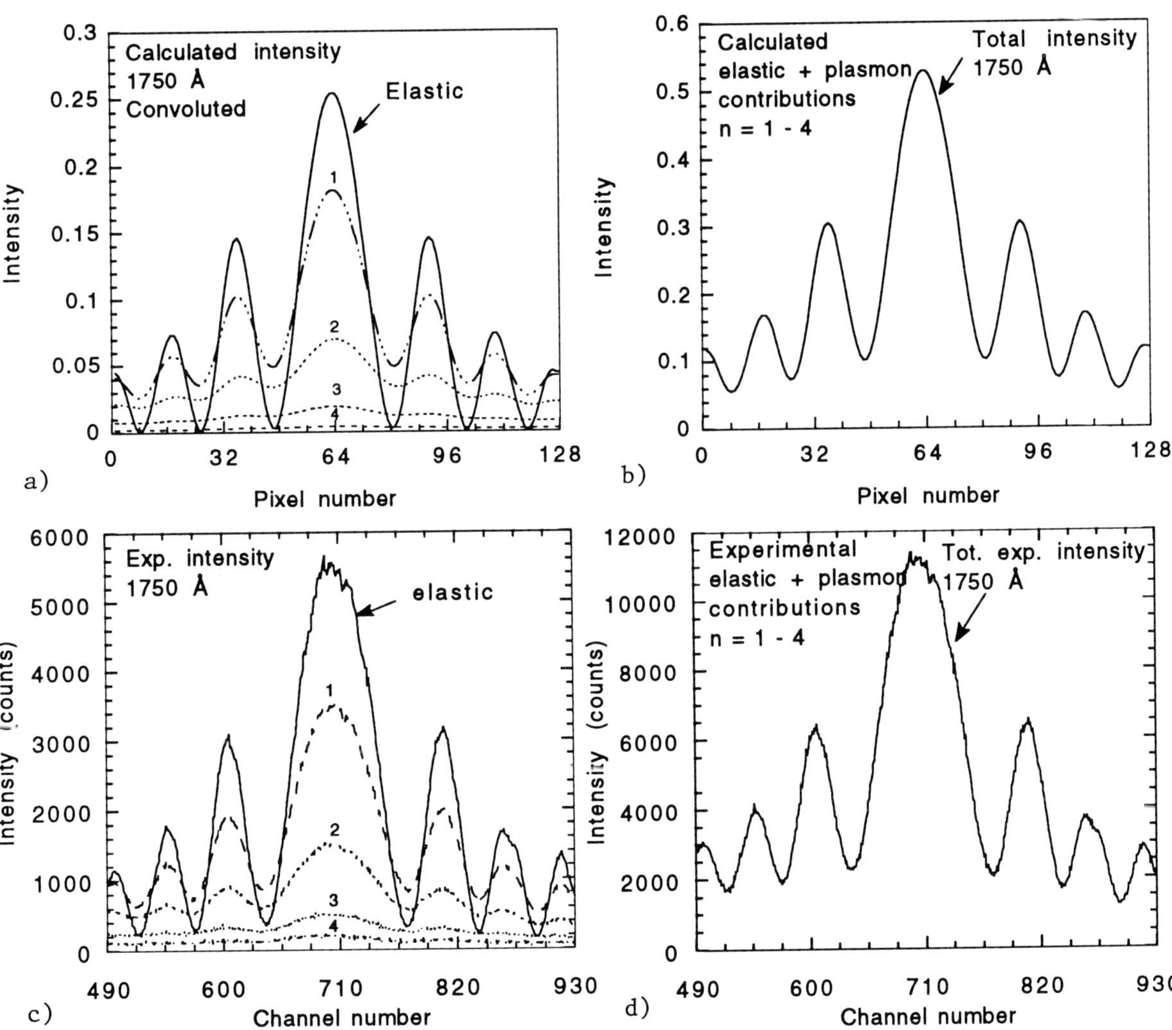

FIG. 1. Si 220-disk, 300 keV, 1750 Å. a) Calculated elastic and plasmon profiles. b) Calculated total intensity. c) Experimental profiles corresponding to (a). d) Total (unfiltered) experimental intensity.

EVALUATION OF PHONON SCATTERING IN ELECTRON DIFFRACTION USING IMAGE PLATES AND ELECTRON ENERGY FILTERING

R. Høier[*,**], M.Y. Kim[*], J.M. Zuo[*], J.C.H. Spence[*] and D. Shindo[***]

[*]Department of Physics, Arizona State University, Tempe, AZ 85287-1504, USA.
[**]Perm. addr.: Department of Physics, University of Trondheim-NTH, N7034 Trondheim, Norway.
[***]Institute for Advanced Materials Processing, Tohoku University, Katahira, Sendai, 980-77 Japan

There have been few studies of thermal diffuse scattering (TDS) by electron diffraction[1,2,3] although this scattering is easily observed. The TDS intensity distribution is as a rule strongly anisotropic which can be ascribed to scattering from individual phonon modes. Results reported so far in the literature are all based on qualitative or semi-quantitative use of the observed intensity distributions. However, at present several new methods (energy filters, imaging plate, CCD, digitized film) are available to extract data for detailed quantitative interpretation. In the present work the Zeiss 912 Omega energy filtering microscope has been used, combined with the Fuji Imaging Plate and a low temperature specimen holder. This opens new possibilities both for the study of soft modes and other TDS phenomena related to phase transitions, and also static disorder and order-disorder effects. By comparison with synchrotron work on TDS[4], crystals may be very much smaller and count rates are much higher. However, multiple scattering and Kikuchi lines may complicate the interpretation unless thin samples are used.

Fig. 1 shows the diffuse phonon scattering lines recorded on film from the Si [001] zone at 105°K and 120 kV. The thickness is estimated to be about 40 nm. The streaks observed through the Bragg spots in the <110> type directions, result from transvers, acoustic phonon modes whose wave vector $\mathbf{q}$ are along [110] and [-1,1,0] with polarization vectors $\mathbf{e}$ along [-1,1,0] and [110], respectively. The intensity along the diffuse streaks which are directly related to the phonon dispersion relation, depend on factors of type $\mathbf{e \cdot g}$ and falls off as q^{-2} (ω^{-2}) with distance $\mathbf{q}$ from the nearest Bragg spot $\mathbf{g}$ .Three-dimensional maps indicate both lines and planes of diffuse scattering[1]. Fig. 2 shows a contour map on a logarithmic scale obtained from an Imaging Plate (IP) recording from the same region as Fig.1 while Fig. 3 shows absolute intensities for a line scan taken from the IP, demonstrating the large dynamic range. Fig. 4 shows the intensity difference between a scan through the [-6,6,0] spot along the TDS streak, line 1 in Fig.2, and one along an orthogonal line (2) through the same Bragg spot. The elongation of the spot intensity in the diffuse streak directions in Figs. 1 and 2, being particularly clear in the $\mathbf{g}$=<220> type systematics, should be noticed. This effect is ascribed to an orthogonal transverse mode. Along the <400> systematics two orthogonal phonon modes contribute equally much and a more square like intensity distribution obtains. The exposure time was 15 s for the IP, Fig.2, as compared to 40 s for film, Fig.1. The camera length was 360 mm and emission current 5 μA.

The IP is erased by light, has 3000 X 3760 X 25 micron pixels, excellent linearity, and about twice the speed of film. When read-out by optical scanning on the FDL5000 reader a 14 bit dynamic range is obtained, with more expected in future. The MTF (Modulated Transfer Function) reflects a narrow impulse response without tails[5,7]. Plates with larger pixels and lower dynamic range have been used for many years for X-rays. They have recently been evaluated for electron microscopy[6,7]. Our plates were mailed cooled by dry ice to Japan for processing, since no 14 bit readers exist in USA at present. At room temperature there is a 20% latent image loss per day. The 20-30 Mbyte data files per IP image (14 bits) are demanding for computer storage. File transfer and image read-out times are large. Plates may only be read out a few times with loss of signal after each reading and so cannot be used for data storage, but they can be reused. For the study of the 2-D distribution of the weak diffuse scattering(where high sensitivity and high dynamic range are needed), for electron holography (where large numbers of pixels are needed) and possibly for high resolution biological work (where high sensitivity, DQE and many small pixels are needed), the image plate offers valuable advantages, particularly in the form of the MTF (compared to CCD), linearity, dynamic range and large numbers of pixels. (An earlier evaluation of the IP for biology[8]used larger pixels than those used here[9]).

Proc. Microscopy and Microanalysis 1995, edited by G.W. Bailey, M.H. Ellisman, R.A. Hennigar, and N.J. Zaluzec
Copyright © 1995 MSA. Published by Jones and Begell Publishing, 79 Madison Ave., New York, NY 10016

References

1. G.Honjo et al., *J. Phys. Soc. Japan* 19(1964)351.
2. J.M.Zuo and P.Rez, *Proc. Ann. MSA Meeting* 48(1990)420.
3. G.L.Hua et al., *J. Phys.: Condens. Matter* 1(1989)2305.
4. N.Takesue et al., *Phys. Rev. Abstr.* 26(1995)No 3, 28.
5. N. Mori et al., *Ultramicr.* 25(1988)195.
6. D. Shindo et al., *J. Electr. Micros.* 39(1990)449.
7. See J.M. Zuo et al., These proceedings for applications to CBED.
8. S. Isoda et al., *Ultramicr.* 35(1991)329.
9. Supported by NSF award DMR 9015867 and NFR. Many thanks also to Drs. Tomokiyo and Yase.

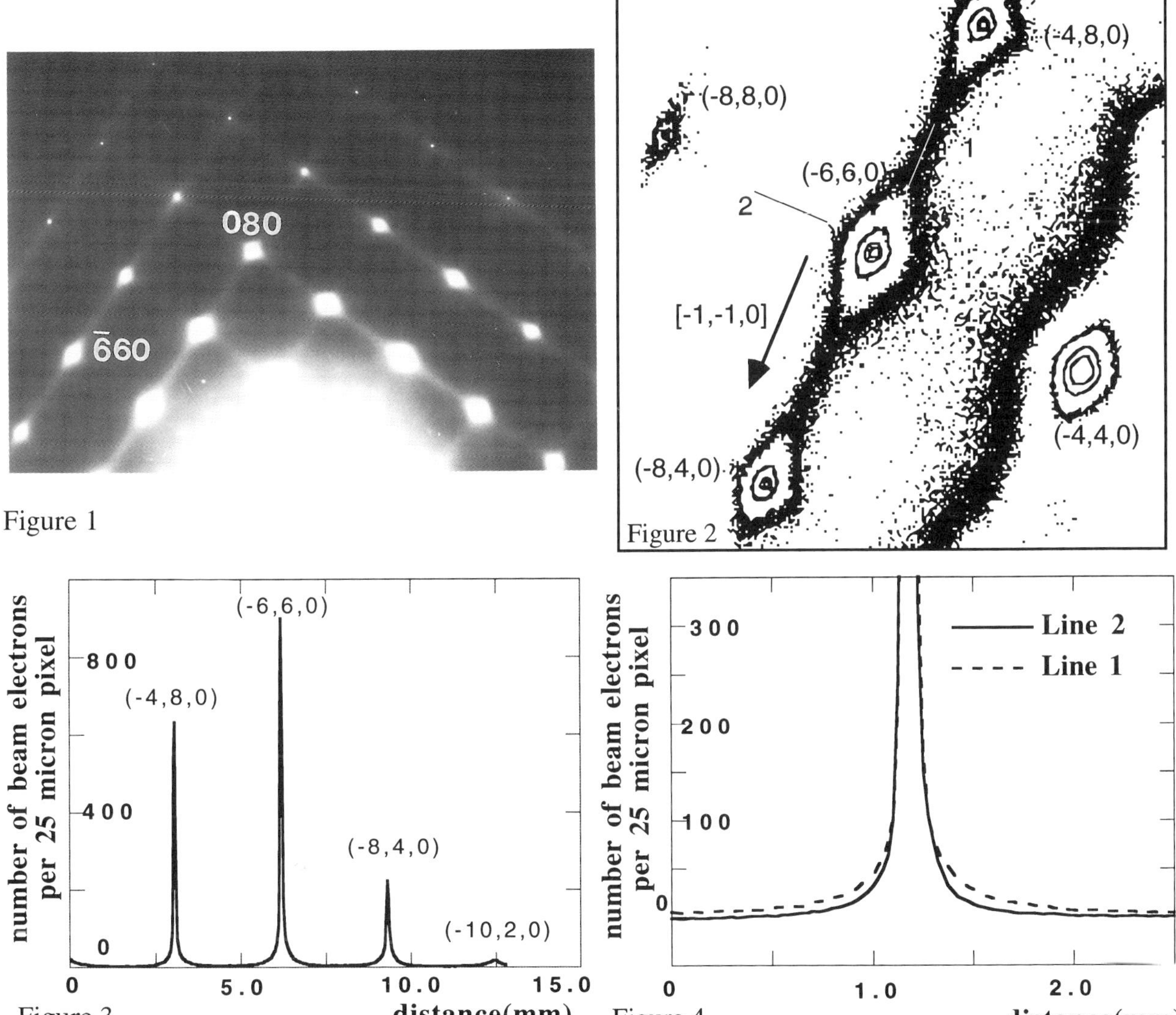

FIG 1 —Elastic filtered [001] TED pattern from Si at 120 KV and 105°K showing TDS from phonon modes recorded on Kodak SO163 film.

FIG 2.—Contour map (log. scale) around spot (-6,6,0) from IP. Filtered, 120 kV, 105°K.

FIG 3.—Line trace from IP along [-1,-1,0], through (-6,6,0)

FIG 4.—[-1,-1,0] scan through (-6,6,0), line 1 in FIG.2, and [-1,1,0], line 2. Background subtracted.

NEW BEAM-SELECTION CRITERIA FOR QUANTITATIVE CONVERGENT BEAM ELECTRON DIFFRACTION

C. Birkeland*, R. Holmestad*, K. Marthinsen** and R. Høier*

*Dept. of Physics, Univ. of Trondheim, NTH Norway.
**SINTEF Applied Physics, Trondheim Norway.

A successful technique of quantitative CBED is the simulation and refinement of intensity line-scans taken across energy-filtered CBED-disks recorded with a CCD-camera.[1] Using many-beam dynamical calculations, line-scans simulated theoretically are fitted to the experiment using a least-square fitting procedure. In the present studies, the systematic row method and modified standard software are used. A description of this technique and its use in intermetallic alloys are explained elsewhere.[2,3] The Bloch-wave method requires diagonalisation of a non-Hermitian matrix and a large number of beams is needed to give accurate results. In practice, the problem is solved using a limited number of beams. The selection criteria used to select these beams are therefore crucial. Too many diagonalised beams in the calculations result in numerical instabilities and very time consuming computations. Very little has been published on this beam-selection problem. Recently suggested criteria are:[3] 1) the proximity to the Ewald sphere ($|2KS_g|_{max}$), 2) the kinematic cut-off criteria used to cut off high order reflections (g_{max}), and 3) the perturbation strength ($|U_h/2KS_h|_{min}|$). These three criteria are adjusted in order to obtain convergence in the Bloch-wave calculations. Weak beams are included using the effective Bethe-potential. Instead of diagonalising a very large dynamical matrix, the diagonalisation is performed on a smaller matrix with modified matrix elements.

As alternative to use the proximity to the Ewald sphere as sole criterion for distinguishing beams to be fully diagonalised from beams included by Bethe perturbation, an additional new criterion is now presented. It has proven useful and effective when the calculations are done with few beams, and the convergence in the Bloch wave method seems to be improved dramatically. This new criterion which is closely related to the Bethe correction is called B_{max}. The basic idea is to include in the diagonalisation weak beams that give large contribution to the Bethe perturbation. The following term in this perturbation is used to study beam h's effect on the dynamical matrix element U_g:

$$\Delta U_h = \frac{U_{g-h}\, U_h}{2KS_h\, U_g}$$

Here g is the reciprocal lattice vector of a diagonalised "strong" beam, S_h the excitation error of a non-diagonalised "weak" beam, K the wave vector and U_g the electron diffraction structure factor. ΔU_h is calculated for all so-called weak beams according to the proximity to the Ewald sphere criteria. Of these, the ones satisfying $\Delta U_h \geq B_{max}$ are then given strong beam status. This process is repeated until convergence. This procedure guarantees that no single beam contributes (in the Bethe perturbation) by more than B_{max}.

The CBED-pattern studied is a systematic **200**-row in **TiAl**. The experimental data is taken from a line-scan made up of 354 pixels. By defining a fit index χ^2 and minimising this function with respect to five geometric parameters and two complex structure factors, small changes in the structure factors are found.[2] The three criteria described above are used to generate a large set of beams (445 in this example) that will be analysed subsequently according to their proximity

Proc. Microscopy and Microanalysis 1995, edited by G.W. Bailey, M.H. Ellisman, R.A. Hennigar, and N.J. Zaluzec
Copyright © 1995 MSA. Published by Jones and Begell Publishing, 79 Madison Ave., New York, NY 10016

to the Ewald sphere and using the new B_{max} criteria. $B_{max} = 0.10$ (corresponding to a relative change of 10%) is found to be a reasonable value. The χ^2-value after refinement and the final refined values for U_{200} and U_{400} may be used as a reference to analyse the best selection criteria. The ultimate goal is to have the best possible fit diagonalising as few beams as possible. When calculating the average number of strong beams (taken over the line-scan), the cubic average will be used since the complexity of the diagonalisation procedure is $O(n^3)$.

$2KS_g$ [Å^{-1}]	B_{max}	$(\sum_i x_i^3)^{\frac{1}{3}}$	χ^2	U_{200} [Å^{-2}]	U_{400} [Å^{-2}]
0.1	—	6.5	3.00	0.05630	0.02203
0.2	—	9.6	2.64	0.05636	0.02228
0.3	—	12.7	2.60	0.05658	0.02246
0.5	—	18.9	3.10	0.05670	0.02245
0.7	—	24.7	2.32	0.05700	0.02208
1.0	—	32.8	2.35	0.05684	0.02218
1.5	—	48.9	2.30	0.05684	0.02219
2.0	—	69.9	2.29	0.05680	0.02235
3.0	—	120.6	2.25	0.05685	0.02215
0.1	0.1	13.3	2.63	0.05680	0.02212
0.2	0.1	16.0	2.33	0.05688	0.02241
0.3	0.1	19.1	2.37	0.05682	0.02248
0.5	0.1	23.8	2.34	0.05685	0.02227
0.7	0.1	28.3	2.35	0.05685	0.02215
1.0	0.1	33.7	2.34	0.05683	0.02224
1.5	0.1	48.9	2.31	0.05683	0.02225

Table 1: Fit parameter, refined structure factors and average number of beams as function of beam selection parameters.

Structure factors for neutral atom:
$$U_{200} = 0.05424 \text{ Å}^{-2}$$
$$U_{400} = 0.02184 \text{ Å}^{-2}$$

If the selection criteria are reasonable, the fit (χ^2) should get better for increasing number of reflections included in the dynamical calculations. As seen from the table, this is not the case when the excitation criteria alone is used. The great increase in χ^2 when going from $S_g = 0.3$ to $S_g = 0.5$ is a major problem. In general, unstabilities may be found when $S_g < 1.0$, indicating that $S_g = 1.0$ should be a minimum value when doing refinements. The results found when the additional criterion $B_{max}(= 0.1)$ is used are more stable, the irregularities around $S_g = 0.5$ have disappeared. It is therefore reasonable to have more confidence in these refinements.

We conclude from these preliminary results that the use of smart beam selection criteria can give a high level of accuracy only including a relatively small number of beams by diagonalisation. This assures better control on the numerical instabilities and much faster calculations. From the table, we can now compare for example the refinements done using the beam selection criteria *1. Proximity to Ewald sphere only ($S_g = 1.0$)* and *2. Proximity to Ewald sphere ($S_g = 0.3$)* **and** Maximum Bethe-perturbation ($B_{max} = 0.1$). These two give the same accuracy in the results, and because the complexity of the problem is $O(n^3)$, the new beam selection criteria gives an **80%** reduction of the computing time.[5]

References

1. J. C. H. Spence and J. M. Zuo, *Electron microdiffraction,* Plenum (1992).
2. R. Holmestad et al., *Phil.Mag.,* in press (1995).
3. J. M. Zuo and A. L. Weickenmeier, *Ultramicroscopy,* in press (1995).
4. R. Holmestad et al., *This proceedings.*
5. We are greatful to Drs. J.M. Zuo and A.L. Weickenmeier for showing us their results before publication. Supported by NFR.

EXPERIMENTAL MEASUREMENT OF DEBYE-WALLER FACTORS

J. M. Zuo*[1], R. Holmestad[2], Y. Tomokiyo[3] and K. Yase[4]
[1]Physcis Dept., Arizona State University, Tempe, AZ 85287; [2]Univ. of Trondheim-NTH, Trondheim, Norway; [3]HVEM Lab, Kyushu Univ., Fukuoka 812, Japan; [4]Dept. of Polymer Physics, NIMC,Japan

Effects of atomic thermal vibration in a crystal on the diffracted intensities is described by the Debye-Waller factor (DWF)[1]. In general, DWF is a 3x3 symmetric matrix with six independent elements. For isotropic atomic vibration or atoms with cubic site symmetry, a single DWF B suffices:

$$F_g = \sum_i f_i \exp(-B_i s^2) \exp(2\pi i \mathbf{g} \cdot \mathbf{r}_i)$$

here s=sinθ_b/λ. DWFs are among the refined parameters in structure analysis in x-ray or neutron diffraction. The accuracy of DWF in such measurements often depends on the quality of the crystal. In quantitative electron diffraction, accurate DWF is needed for 1) converting Fourier coefficients of potential to that of charge density, 2) comparison with theoretical calculations of static crystals, 3) estimation of high order structure factor and absorption coefficients.

DWF can be determined from measured amplitude of high order structure factors where the effects of bonding is minimal and DWF largest. Fig. 1 shows a plot of logrithmitic of ratios of measured MgO electron structure factors[2] and calculated ones without temperature factor. The slope of this plot gives the averaged DWF of MgO as 0.32 Å^2, which is in good agreement with the best x-ray and neutron diffraction measurements[3]. Individual DWFs can be found by refinement method with χ^2 or R-factor. Fig. 2 shows a R-factor map of DWFs for Mg and O using the published x-ray data[4]. Refinement with electron structure factors was unsecessful because of bonding effects in the low order reflections used.

A means of measuring DWF by electron diffraction is desired for two reasons: 1) electron diffraction is from a smaller volume of crystals where crystal imperfection and their effects on DWFs can be carefully avoided and 2) electron diffraction is often done at low temperature where existing measurements are few. There are a number of attempts at direct measurement of DWFs by electron diffraction recently[5-7]. Direct measurement from intensities of high order CBED reflections in γ-TiAl using the kinematic approximation was not successful[5]. Measurement based on two-beam theory and Bethe potential was better, but failed when individual DWFs were refined. χ^2 refinement based on many-beam dynamic theory gives isotropic B$_{Ti}$ = 0.17 and B$_{Al}$ = 0.38[5]. Fully automated refinement of DWF with many high order reflection rocking curves included is under development, which may be the ultimate solution[6].

To obtain DWFs with accuracy comparable to the best x-ray measurements, electron structure factor need to be measured at accuracy of about 1% for high order reflections. To achieve this experimentally, energy-filering of about 5° angular range and large dynamic range and linear detection of the whole CBED pattern is desired. This can be done with the combination of Ω-energy filter in Zeiss 912 electron microscope and new Fuji Imaging Plate with 25 μm pixel. Fig. 4a and b show a large angle energy-filered Si CBED pattern recorded on the imaging plate, and the absolute intensity profile showing large dynamic range. The tail typical in slow-scan CCD camera response is absent in point spread function of imaging plate as shown in fig. 3. Imaging plate is also linear, about twice as sensitive as film and dynamic range of 10^5 [8]. Quantitative characterization of this detector is under way. Numerical convergence becomes more important in calculating high order intensities[9]. Convergence in Bloch wave is discussed in reference 10. In conclusion, measurement of DWF presents a new challenge to quantitative electron diffraction which may ultimately lead to better understanding of bonding in crystals.

Reference

1. T. M. Willis and A. W. Pryor, Thermal vibrations in crystallography, Cambridge, 1975
2. J. M. Zuo and J.C.H. Spence, ICEM 13-Paris, vol **1**, 851, 1994
3. T. H. K. Barron, Acta Cryst., A**33**, 602, 1977; 4. J. L. Lawrence, Acta Cryst., A**29**, 94, 1973
5. R. Holmestad et. al., Inst. Phys. Conf. Ser. **138**, 141; 6. A. Weickenmeier, this proceedings
7. W. Burgess et. al. ICEM 13-Paris, vol **1**, 849, 1994

8. N. Mori et al, Ultramicrocopy, 25, 195, 1988
9. J. M. Zuo, Acta Cryst, A**49**, 429, 1993
10. J. M. Zuo and A. Weickenmeier, Ultramicroscopy, in press, 1995
*This work is supported by NSF grant 9015867

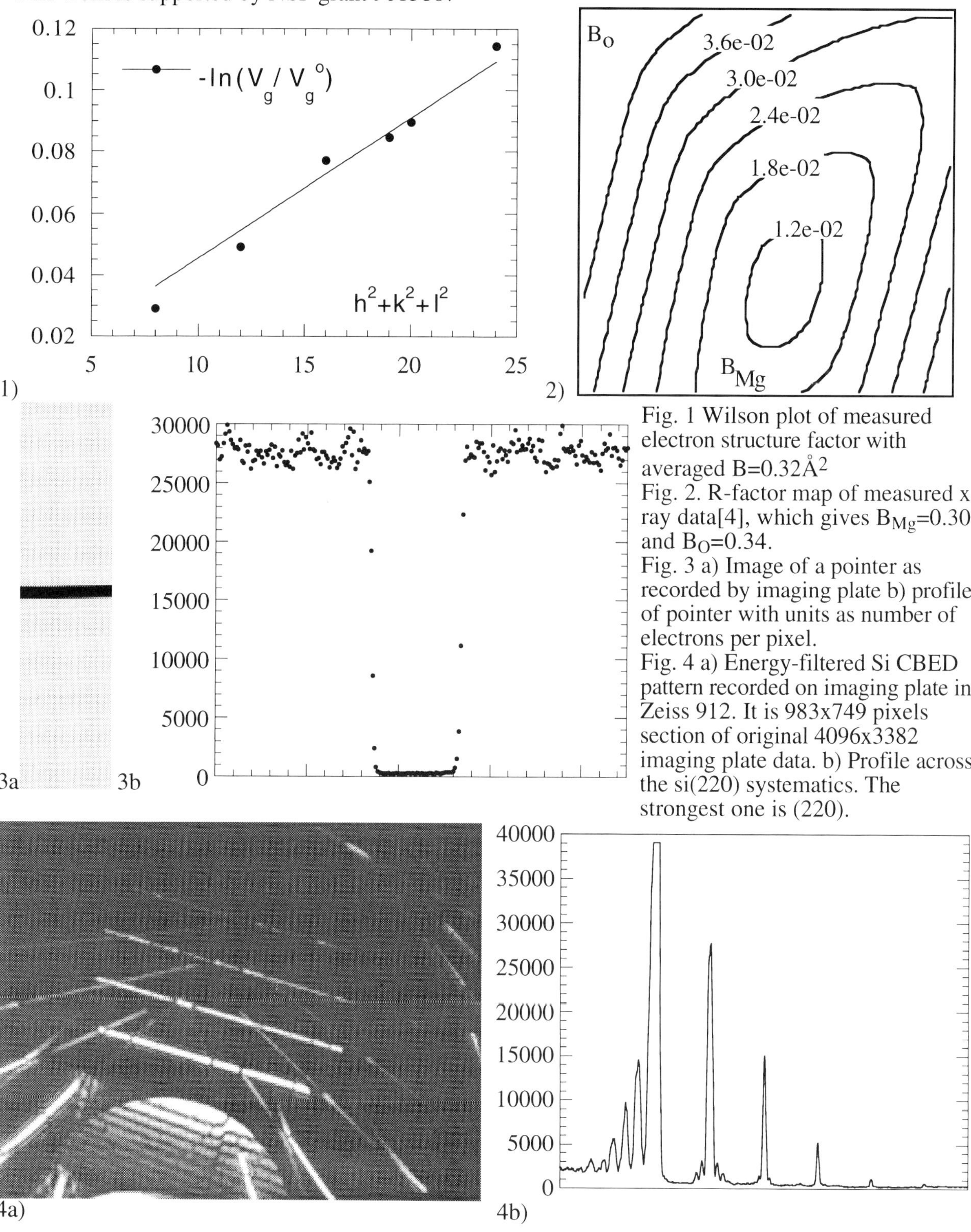

Fig. 1 Wilson plot of measured electron structure factor with averaged B=0.32Å^2

Fig. 2. R-factor map of measured x-ray data[4], which gives B_{Mg}=0.30 and B_O=0.34.

Fig. 3 a) Image of a pointer as recorded by imaging plate b) profile of pointer with units as number of electrons per pixel.

Fig. 4 a) Energy-filtered Si CBED pattern recorded on imaging plate in Zeiss 912. It is 983x749 pixels section of original 4096x3382 imaging plate data. b) Profile across the si(220) systematics. The strongest one is (220).

HIGH PRECISION MEASUREMENT OF DEBYE–WALLER FACTORS FOR NiAl

W. Nüchter, A.L. Weickenmeier, and J. Mayer

Max-Planck-Institut für Metallforschung, Institut für Werkstoffwissenschaft, Seestraße 92, D-70174 Stuttgart, Germany

Motivation

Quantitative convergent beam electron diffraction (CBED) is increasingly appreciated as a tool to determine bonding charge densities of crystalline materials. Simulated CBED patterns are fitted to experimental ones to derive the structure factors. These are converted by means of the Mott formula to yield the total charge density. Finally a neutral atom total charge density is subtracted and the difference is interpreted as the bonding charge density. Accounting for the temperature by a Debye-Waller factor (DWF) the g-th Fourier coefficient of the bonding charge density is then given by

$$\rho_{\vec{g}}^{\text{bond}} = \rho_{\vec{g}}^{\text{Exp}} - \rho_{\vec{g}}^{\text{Atom}} \exp[-\frac{1}{2}u^2 g^2] \, , \tag{1}$$

where u denotes the thermal root mean square atomic displacement. Here we have assumed the simplest case of identical isotropic atomic vibrations for all atoms in the crystal. In order to estimate the error in $\rho_{\vec{g}}^{\text{bond}}$ due to the uncertainty in u we insert the results obtained by Fox and Tabbernor[1] for NiAl at room temperature. They found differences between atomic and measured values of $\rho_{\vec{g}}$ in the order of 2 percent. In comparison, an error of 20 percent for the u value (which roughly is the spread of X-ray results found in literature[1]) causes a change of 3 percent in the atomic reference. Thus a new high precision Debye-Waller factor determination is indispensable, in particular for liquid Nitrogen temperatures, where no literature data could be found.

Experiment

With our Zeiss EM912 Omega equipped with a Gatan model 679 slow-scan CCD camera we have recorded high order reflections of NiAl in systematic row orientation (Fig. 1). Liquid Nitrogen cooling was used and the CBED patterns were filtered at zero energy loss. Excitation of low-indexed reflections was carefully avoided, its influence, however, included in simulations. A rather large convergence angle (larger than Bragg angle) was used to fulfill the Bragg conditon for at least 3 reflections. After removing the point spread function of the camera by deconvolution[2] we extracted line profiles from these patterns (plus the zero disc) to which we fitted simulated profiles (Fig. 2).

Results

The χ^2-function as a measure for the difference between experiment and simulation was minimized with respect to the individual DWFs of Ni and Al, respectively. We evaluated several patterns showing the (100) and (110) sytematic rows for different thicknesses. χ^2-values in the range of 6 to 10 were typically achieved. The best fit is shown in Fig. 2, where χ^2 is 5.97. The corresponding u-values are (Ni) 0.050±0.001 Ångstroem and (Al) 0.051±0.003 Ångstroem, respectively.

Acknowledgement
Financial support by the Stiftung Volkswagenwerk is gratefully acknowledged.

Proc. Microscopy and Microanalysis 1995, edited by G.W. Bailey, M.H. Ellisman, R.A. Hennigar, and N.J. Zaluzec
Copyright © 1995 MSA. Published by Jones and Begell Publishing, 79 Madison Ave., New York, NY 10016

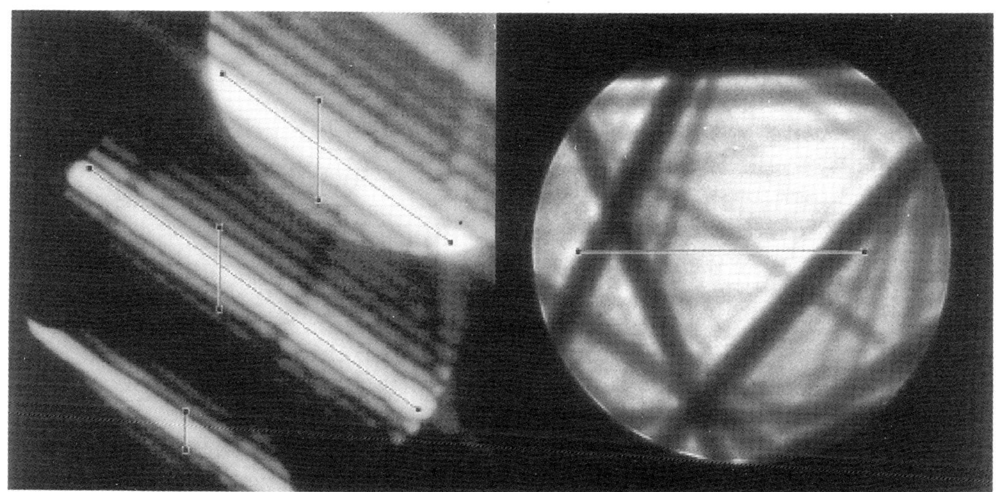

FIG. 1 Left: The systematic row reflections (660) (bottom left), (550) and (440) reflections. Right: The corresponding (000) disc. Indicated are the scan locations. The contrast is processed for better visibility of details.

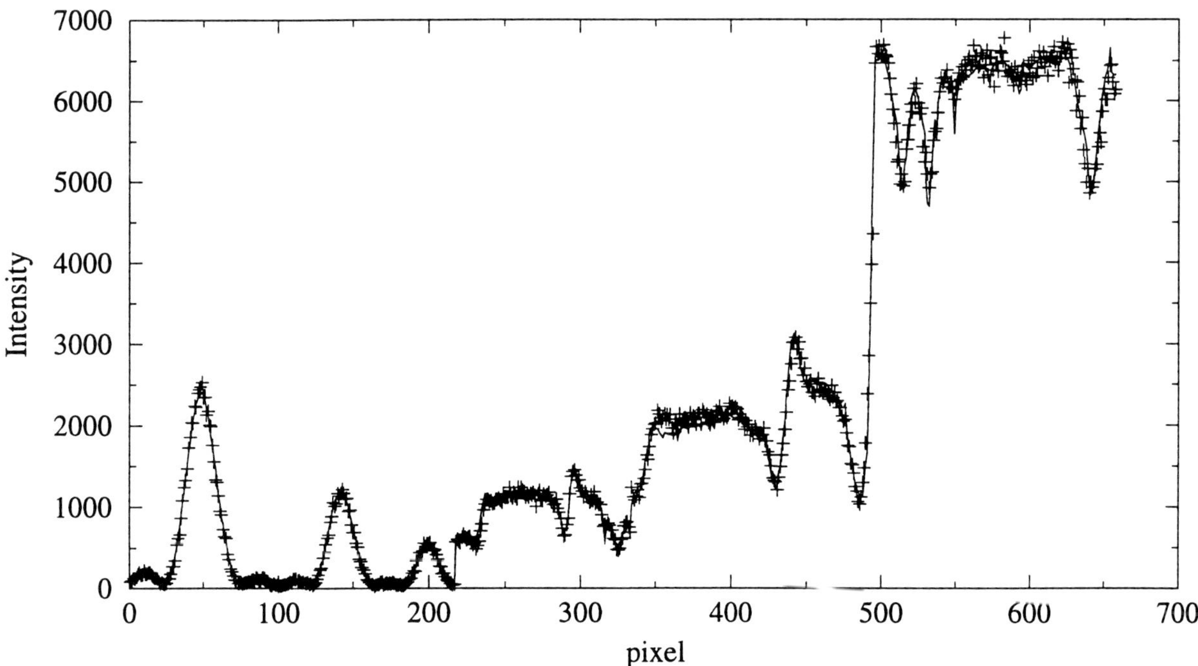

FIG. 2 Experimental data extracted from the experimental pattern as indicated in Fig. 1 (+) and best fit (—).

References
1. A.G. Fox and M.A. Tabbernor, *Acta metall. mater.* 39(1991)669.
2. A.L. Weickenmeier et. al., *Optik*, accepted for publication.

STRUCTURE DETERMINATION OF A NEW METASTABLE Au-Sn PHASE USING CONVERGENT BEAM ELECTRON DIFFRACTION.

M.E. Sleight, P.A. Midgley, and R. Vincent

H.H. Wills Physics Laboratory, University of Bristol, Tyndall Avenue, Bristol, BS8 1TL, UK.

Reports of a new metastable Au-Sn phase [1] led to the investigation of this system by Convergent Beam Electron Diffraction (CBED). A novel growth route was used, in which a single metallic layer of Sn was evaporated onto rock salt and allowed to form a thin oxide layer, before a film of gold was evaporated directly on top. Subsequent in-situ heating in a TEM supplied enough energy to allow solid state diffusion of Sn through the oxide to react with the Au, forming a number of intermetallic phases.

A new tetragonal $AuSn_4$ phase was found [2], space group P4/nbm and lattice parameters a=6.63Å, c=5.92Å, and it was proposed that this new phase was isostructural with $PtPb_4$[3]. Further work described here confirms this structure type and shows how the atomic positions were refined using First Order Laue Zone (FOLZ) reflections within a quasi-kinematic approximation [4].

Fig. 1 shows a section of the FOLZ at the [001] zone axis of this new Au-Sn phase. Two excess lines or 'branches' are visible, corresponding to diffraction out of two excited Bloch states. Dynamical calculations using the $PtPb_4$ structure indicate that the inner branch corresponds to a Bloch state localised on the Au atom strings, whilst the outer branch is related to a hybridised state on both the Au and Sn strings. Fig. 2 shows these states. Within the $PtPb_4$ structure, only the Sn atoms have free atomic parameters at the 8(m) positions. Of the two branches, only the outer branch, therefore, will be sensitive to these parameters. The intensity of the outer branch was measured for FOLZ reflections from [001] zone axis patterns for a sequence of voltages, ranging from 150kV to 270kV in 10kV steps. There were sufficient reflections present in consecutive patterns to allow cross calibration. A conditional Patterson transform was constructed using these reflections in which peaks indicated Sn-Sn interatomic vectors, as shown in Fig. 3. The limited data set available gives rise to a highly truncated Fourier series from which to construct the Patterson transform. The high background 'ripple' is a reflection of this truncation. By measuring the peak positions, the x co-ordinate of the Sn was estimated as x=0.17(1). This value was then used as a starting point in a least-squares refinement of the intensity data [4] giving x=0.167(1) and z=0.25(4). A [001] projection of this new phase is shown in Fig. 4. It is closely linked to the equilibrium orthorhombic phase [5] which can be recovered by a small displacement of alternate Sn planes, and a concurrent doubling of the c-axis lattice parameter [6].

References

1. D.C. Dufner and L. Eyring, *J. Solid State Chemistry* 62 (1986) 112
2. P.A. Midgley et al., in *Proceedings ICEM-13 Paris* (1994) 873.
3. U. Rösler and K. Schubert, *Naturwissenschaften* 38 (1951) 331.
4. R. Vincent et al., *Phil. Mag. A* 50 (1984) 745.
5. R. Kubiak, *J. Less Common Metals* 80 (1981) P53
6. The authors acknowledge support from a Royal Society University Research Fellowship (PAM), ICI (MES) and the EPSRC.

Proc. Microscopy and Microanalysis 1995, edited by G.W. Bailey, M.H. Ellisman, R.A. Hennigar, and N.J. Zaluzec

Copyright © 1995 MSA. Published by Jones and Begell Publishing, 79 Madison Ave., New York, NY 10016

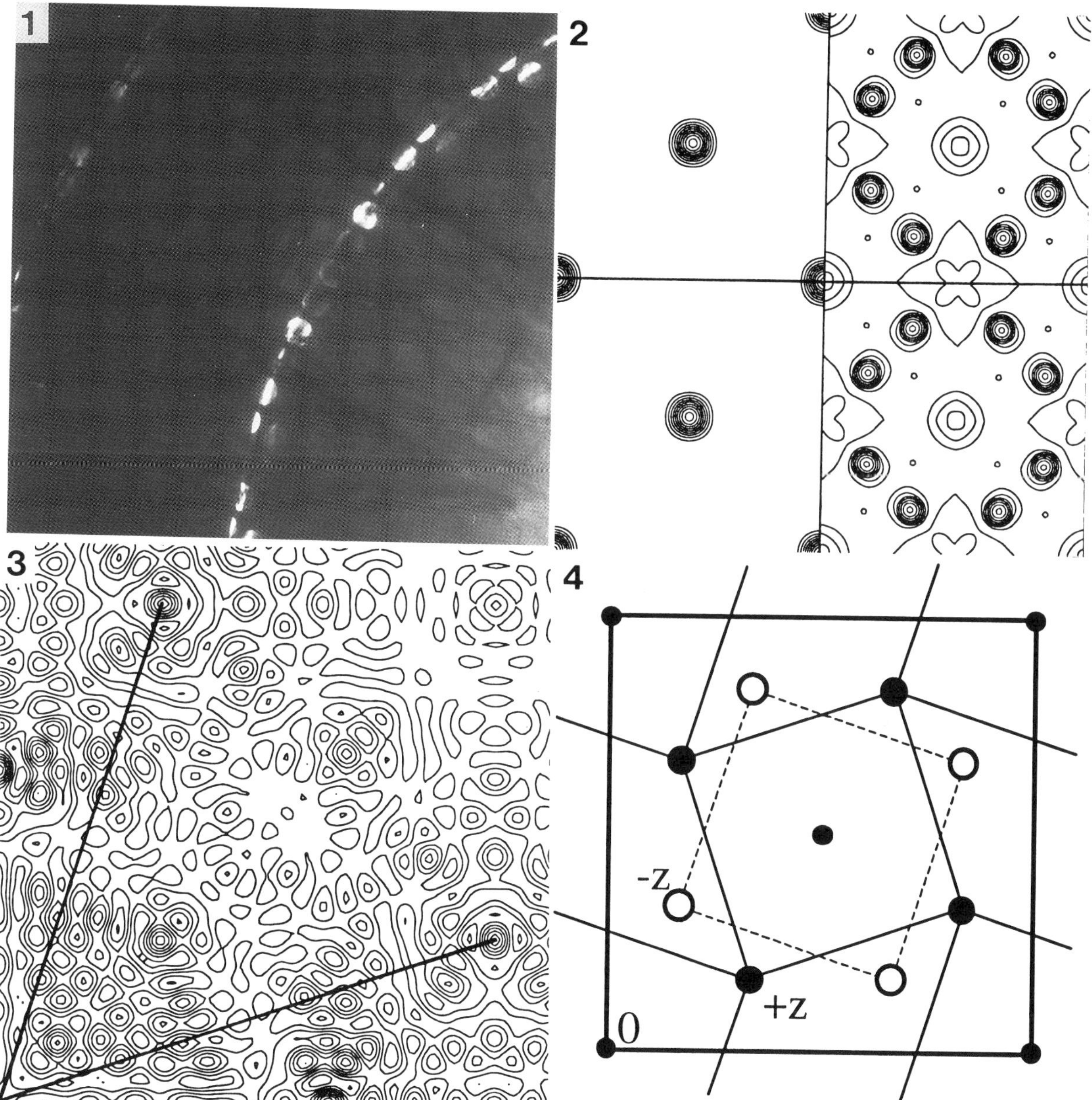

FIG. 1.- Section of the HOLZ from the [001] zone axis of the new AuSn$_4$ phase clearly showing two excess lines, or 'branches' in the first and second layer. The pattern was taken at 90kV and at liquid nitrogen temperatures to reduce thermal diffuse scattering.

FIG. 2.- The Bloch states associated with the two excess lines. The Bloch state in the left half of the figure lies on gold strings only, and is related to the inner branch; the right half shows a hybridised Bloch state lying on both gold and tin strings, related to the outer branch. Four unit cells are shown.

FIG. 3.- Quadrant of the conditional Patterson transform constructed from FOLZ reflections. The indicated peaks define the tin-tin interatomic vectors, and a measurement of the peak position is directly related to the x co-ordinate. The origin peak has been removed for clarity.

FIG. 4.- The [001] projected structure of the AuSn$_4$ tetragonal phase. Small filled circles represent Au atoms at z=0, and larger circles represent Sn atoms at ±z.

WHAT IS THE ULTIMATE PRECISION OF CBED STRUCTURE FACTOR MEASUREMENT? APPLICATION TO GROUP IV AND III-V SEMICONDUCTORS

M. Saunders, P.A. Midgley and R. Vincent

H.H. Wills Physics Laboratory, University of Bristol, Tyndall Avenue, Bristol, BS8 1TL, UK.

We have previously demonstrated the feasibility of fitting calculated zone-axis Convergent Beam Electron Diffraction (CBED) patterns to zero-loss energy-filtered experimental data.[1] The aim of this pattern matching procedure is to refine accurate low-order structure factors that can be used to reconstruct the bonding charge density. Initial tests on silicon [110] zone-axis patterns have shown that this new electron diffraction technique is capable of accuracies comparable with existing X-ray methods and that the results are in good agreement with solid state theory calculations.[2,3] The general applicability of the CBED pattern matching technique is now investigated by studying the bond charge of two other Group IV materials, i.e. germanium and diamond, and the III-V semiconductor GaAs.

The (110) plane of the reconstructed bond charge density for each of the Group IV materials is shown in Figure 1. The agreement between these results and those predicted by solid state theory[3] is excellent despite the use of only five independent Fourier components in the reconstruction. These measurements show that the technique is not restricted to measuring the charge density in silicon alone! Work on GaAs is still ongoing but initial results indicate that similar accuracies are achievable.

Now that experience has been gained in applying this technique some comments can be made regarding the factors which limit the accuracy of the fitted structure factors. First, a clean, unstrained sample is a necessity. Specimen preparation is crucial to the success of the technique. The presence of an amorphous surface layer degrades the quality of the experimental data and local strain makes it impossible to acquire patterns with the correct symmetry. Second, the point spread function (psf) of the detector must be accurately determined.[4] Errors can easily be introduced by deconvoluting the recorded data with an inaccurate psf. Third, the specimen thickness must be chosen carefully to produce sufficient dynamical interaction to make the patterns sensitive to the required structure factors. Due to the use of an energy-filter, thicknesses ~5000Å can routinely be studied from the Group IV materials under investigation. This also reduces the effect of the surface layer as it then only forms a small percentage of the total thickness. Care must however be taken that the sharp features present in patterns from thick regions of the sample can be accurately recorded using the limited resolution of the detector (1024×1024 for the GIF). Fourth, the composition of the sample must be known accurately. Finally, Debye-Waller factors must be available.[1,2] This point will be addressed in detail elsewhere in this session. If these considerations are taken into account, accurate charge density measurements should be possible from a wide range of materials.[5]

References

1. D.M. Bird and M. Saunders, *Ultramicroscopy* 45 (1992) 241.
2. M. Saunders et al., in *Proceedings ICEM-13 Paris* (1994) 847.
3. Z.W. Lu et al., *Phys. Rev. B* 47:15 (1993) 9385.
4. M. Saunders et al., in *Proceedings ICEM-13 Paris* (1994) 739.
5. This work is supported by the EPSRC.

Proc. Microscopy and Microanalysis 1995, edited by G.W. Bailey, M.H. Ellisman, R.A. Hennigar, and N.J. Zaluzec
Copyright © 1995 MSA. Published by Jones and Begell Publishing, 79 Madison Ave., New York, NY 10016

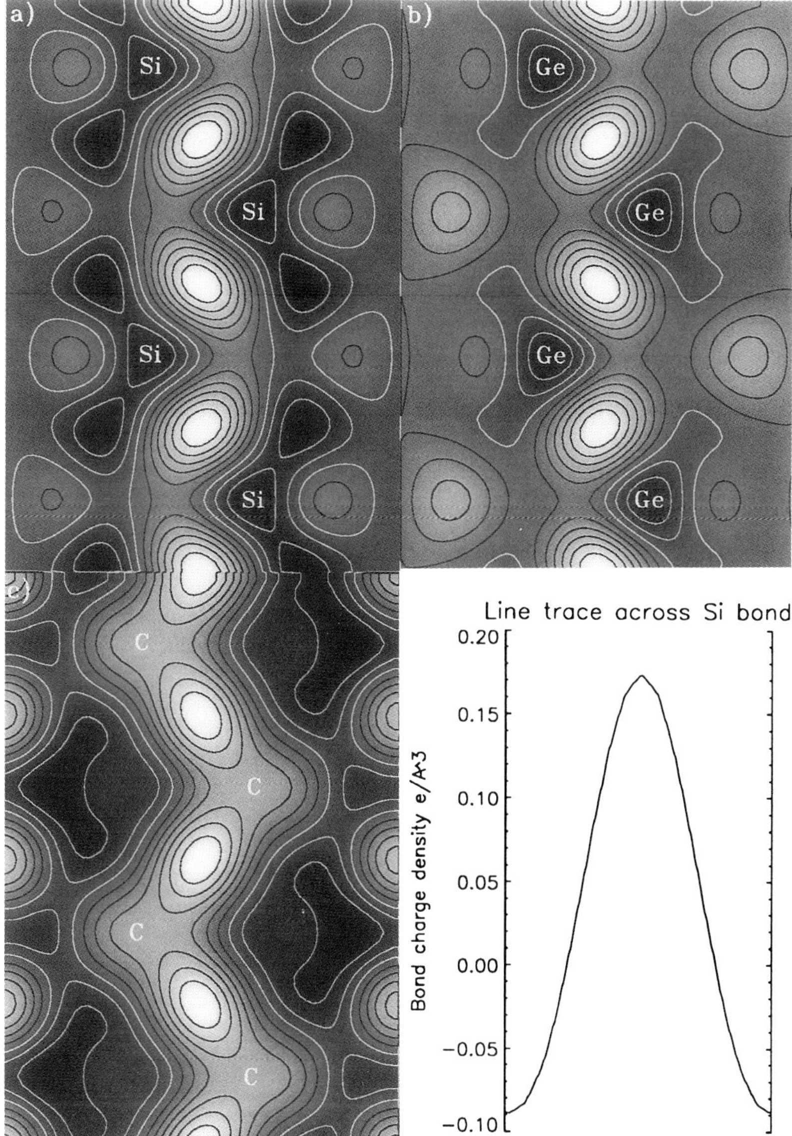

Figure 1: (110) plane of the measured bonding charge density in (a) Si, (b) Ge and (c) diamond with arbitrary contours. A line trace between two atoms across the Si bond is shown in (d). The charge densities are reconstructed using the refined values of the {111}, {220}, {113}, {222} and {004} structure factors. The structure factors are refined by minimising the sum-of-squares difference between a theoretical calculation and a set of intensities extracted from a zero-loss energy-filtered zone-axis CBED pattern. The minimisation parameters include the real and imaginary parts of the six lowest-order structure factors, i.e. the earlier list plus {331}, the sample thickness and a set of background levels. The CBED patterns have been acquired using a Gatan Imaging Filter (GIF) attached to a Hitachi HF2000 TEM. The accelerating voltage has been measured as 198.8±0.2kV, the probe size is ~3nm and the samples are cooled to liquid nitrogen temperatures to reduce thermal diffuse scattering. In order to remove the effects of the detector, the data is first deconvoluted with the measured psf of the GIF.

CHARGE DENSITY IN TiAl INTERMETALLICS STUDIED BY QUANTITATIVE CBED

R. Holmestad*, J.M. Zuo**, J.C.H. Spence**, R. Høier*' **** and Z. Horita***

* Dept. of Physics, University of Trondheim-NTH, 7034 Trondheim, NORWAY
** Dept. of Physics and Astronomy, Arizona State University, Tempe, AZ85287, USA
*** Dept. of Material Science and Engineering, Kyushu University, Fukuora, 812 JAPAN
**** SINTEF Applied Physics, 7034 Trondheim, NORWAY

The relationship between mechanical properties and charge density topology in TiAl intermetallics has attracted considerable interest during the last few years.[1-4] TiAl is a candidate for use in the aerospace industry, but a problem is the undesired brittleness at low temperatures. Doping (with i.e. Mn) is known to increase the ductility, and the question is if this doping is followed by detectable changes in the bonding charge density. It has been suggested for example that the largely covalent Ti-Ti bond is diluted by Mn, reducing brittleness, and that the depth of the Peierls valley encountered by dislocations may also be controlled by doping. In this paper we use the systematic row convergent beam electron diffraction (CBED) technique[5] to determine low order structure factors in TiAl intermetallics with and without manganese doping. Subsequently, from the structure factors found, electronic bonding charge density maps are made. Also TiAl with other doping elements are now under examination.

Polycrystalline ingots of equiatomic γ-TiAl (50-50) and TiAl doped with Mn (47.5-47.5-5) were made by arc melting in argon atmosphere and homogenised at 1473 K for one day. Materials now under investigation are doped with V and Cr. TEM specimens were prepared by electropolishing in a solution of CH_3OH (300 ml) and C_4H_9OH (175) ml plus $HClO_4$ (30 ml). After doping, two phases (gamma and alpha) exist; however, most of the material still consists of the tetragonal γ-phase, which is further studied. Lattice parameters after doping have been determined by HOLZ line matching[6], and the site occupancy of dopant atoms is found using the ALCHEMI method.[7]

An overview of the quantitative CBED procedure is shown in figure 1. A Zeiss Omega energy filtering microscope (120 kV) is used, equipped with a CCD camera and a cold stage double tilt holder. The beam corresponding to the structure factor to be refined is set at the Bragg position, and the CCD pattern acquired. The point spread function of the CCD camera is removed by deconvolution. A line scan is extracted along the systematic row, and a corresponding theoretical line scan calculated using the Bloch wave method. These two line scans are quantitatively compared, and the difference between them minimised. The input parameters are changed until the theory fits the experiment. From the refined structure factors, x-ray structure factors are determined and charge density maps made. Figure 2 shows the electron bonding charge deformation map in the $(1\bar{1}1)$ slip plane for (a) pure and (b) Mn-doped TiAl created from the nine lowest structure factors. We see large change in the charge density due to doping. The general effect of Mn doping is to draw charge away from the strong nearest neighbour covalent bonds and redistribute it more uniformly between second nearest neighbours.[8]

One problem limiting the accuracy of structure factors and the interpretation of experimental charge density maps in the TiAl intermetallic system, is the stochiometry.[1] TiAl exists in a large stochiometric range, and when two phases exist, the local composition may alter. It seems like the qualitative form of the charge density deformation maps is not changed significantly by small composition changes, but better analysis, further understanding of the local effects of doping and the relation to microstructure seem to be essential. Work is now in progress, calculating corresponding charge density maps using the LAPW method.[9] These theoretical maps will be compared with the experimental maps. In this way, the local effects of dopant atoms and stochiometry can be studied in more details.[10]

References

1. S. Subramanian, Ph.D. thesis, Ohio State University, USA (1994).
2. Y. Song et al., *Phil. Mag.* B70 (1994) 987.
3. Z. W. Lu et al., *Acta Met et Mat.*, 42 (1994) 3929.

Proc. Microscopy and Microanalysis 1995, edited by G.W. Bailey, M.H. Ellisman, R.A. Hennigar, and N.J. Zaluzec
Copyright © 1995 MSA. Published by Jones and Begell Publishing, 79 Madison Ave., New York, NY 10016

4. R. Holmestad, Ph.D. thesis, University of Trondheim-NTH, Norway (1994).
5. J.C.H. Spence and J.M. Zuo *Electron Microdiffraction* New York: Plenum (1992).
6. J.M. Zuo *Ultramicr.*, 41 (1992) 211.
7. J.C.H. Spence and J. Taftø, *Journ. of Micr.*, 130 (1982) 147.
8. R. Holmestad et al. Subm. to *Phil Mag.* (1995).
9. P. Blaha et al. *Comp. Phys. Commun.* 59 (1990) 399.
10. Thanks to Drs. A. Weickenmeier, K. Marthinsen, J. Henderson and Mr. C Birkeland for discussions and contributions. Supported by NSF award DMR-9015867 and the Norwegian Research Council.

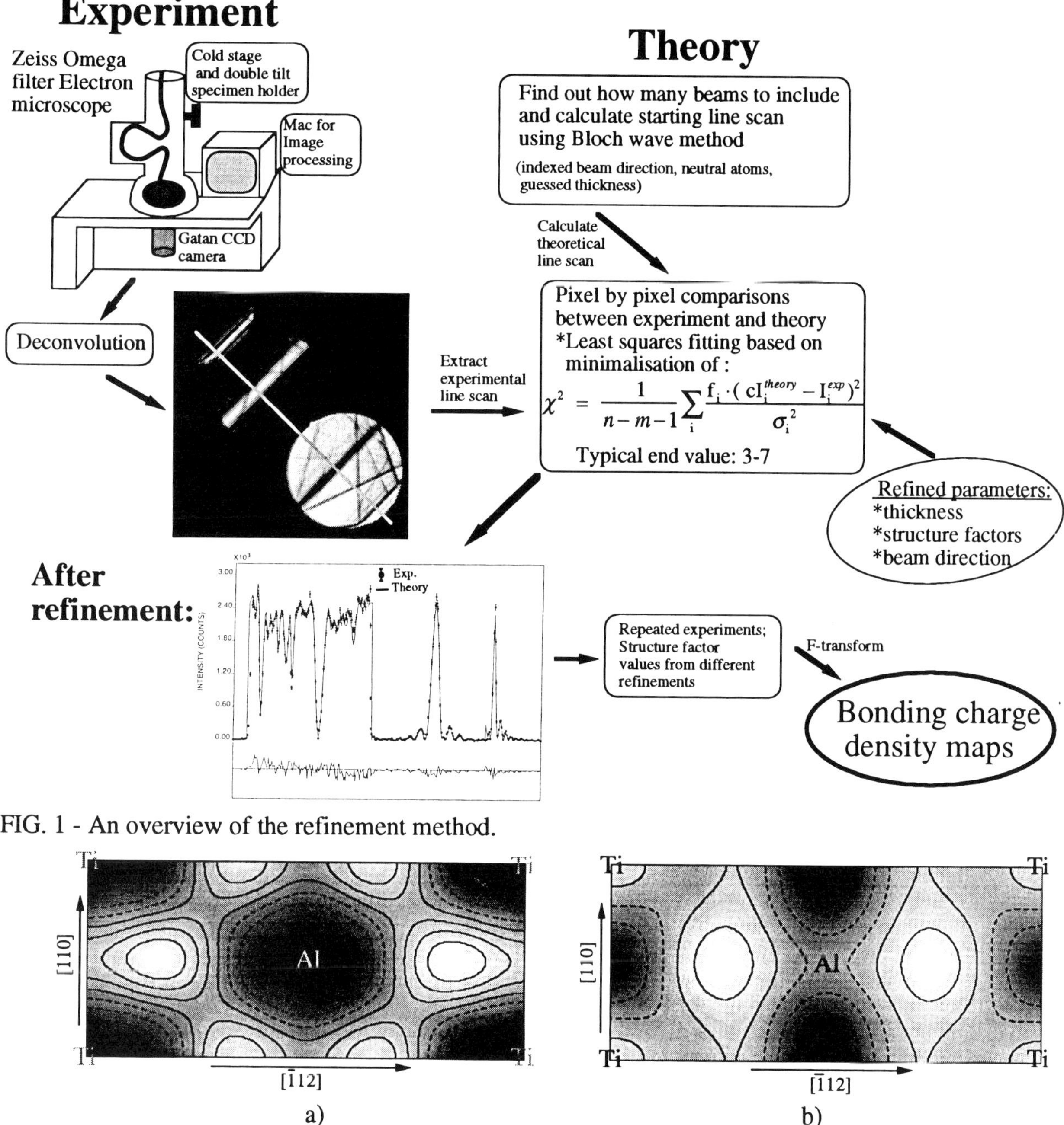

$$\chi^2 = \frac{1}{n-m-1}\sum_i \frac{f_i \cdot (\, cI_i^{theory} - I_i^{exp}\,)^2}{\sigma_i^2}$$

FIG. 1 - An overview of the refinement method.

FIG. 2. - Bonding charge density maps for (a) TiAl (contour level 0.5 e/cell volume≈0.008 e/Å^3) and (b) TiAl doped with Mn (contour level 1 e/cell volume ≈0.016 e/Å^3) in the (1 Ī 1) plane.

CHEMICAL SENSITIVITY OF CONVERGENT BEAM ELECTRON DIFFRACTION ON THE STOICHIOMETRY OF GaN

Zuzanna Liliental-Weber, Christian Kisielowski, and Jack Washburn

Materials Science Division, Lawrence Berkeley Laboratory, Berkeley CA 94720

III-V nitride thin film growth has attracted considerable attention because it now seems feasible to engineer semiconductor band gaps between 2.1 and 6.2 eV. One of the challenges coming with this development is related to the fact that structural perfection seems not to correlate directly with optical properties such as the emission of blue-green or UV light in GaN. In order to better understand this material High Resolution Transmission Electron Microscopy (HREM) and Convergent Beam Electron Diffraction (CBED) experiments were used to study structural defects in GaN thin films. Experiments were performed with a Topcon 002B and ARM operating at 200 and 800 KeV, respectively, and were guided by image simulations. Results of parallel luminescence studies will be published elsewhere.

Plan-view micrographs of GaN grown on the (0001) basal plane of Al_2O_3 with a lattice mismatch of 14% show small angle grain boundaries which divide the layer into large subgrains of about 800 nm diameter. Other defects visible in the plan-view micrographs are threading dislocations and planar defects lying parallel to the $\{10\bar{1}0\}$ planes of the GaN.

Two other planar defects were also found in cross-sectioned samples: one along $(\bar{1}100)$ GaN planes perpendicular to the growth surface and the second parallel to the (0001) GaN growth surface. The contrast of the defects parallel to the growth surface disappears for the diffraction vector parallel to $[10\bar{1}0]$ GaN. These defects were found by high resolution microscopy to be stacking faults parallel to the (0001) growth surface and can be described as localized changes of the stacking sequence also commonly observed in SiC and its polytypes. The density of these stacking faults was observed to decrease with increasing distance from the substrate.

Most intriguing were the defects perpendicular to the interface (Fig. 1a). Their contrast disappears for the (0001) diffraction vector. They start at the interface and extend through the entire layer thickness. They are of 8-10 nm wide and spaced at the distance of about 45 nm. The apparent width of these defects does not change significantly through the entire layer. Two significant features characteristic of this defect can be observed on the high resolution micrograph: the interface between the defect and the surrounding matrix appears as a straight line and the lattice planes across the defect bend in opposite directions depending on local sample thickness as indicated by arrows on (Fig. 1a) (the thickness of the sample in this micrograph is increasing from the top to the bottom). In addition, the same image pattern can be found in different areas of the micrograph (compare the areas marked by "A" outside the defect with "a" inside the defect as well as "B" and "b".Two reasons could be responsible for this pattern shift: the thickness difference or different composition of the sample within the defect area. Since the pattern of the matrix on both sides of the defect is identical (see "A" and "A" on Fig. 1a), a change of thickness within the defect area seems unlikely. Therefore, these vertical defects appear to be microtwins with a composition slightly different than in the surrounding matrix.

In order to monitor compositional changes within the layer convergent beam electron diffraction was applied to plan view and cross-sectional samples. The CBED method is known to be very sensitive to the lattice parameter, space group or structural defects. We were interested in determining how sensitive the intensity distribution within the CBED discs is to compositional fluctuations. For a [0001] zone axis we obtained good agreement of the experimental CBED pattern with a simulation (Fig. 1(b)). However, we also find from the image simulation that this particular zone axis exhibits little sensitivity to compositional change within the unit cell.

In contrast simulated CBED pattern for the $[1\bar{1}00]$ zone axis predict significant changes of the intensity distribution in CBED discs due to compositional changes. The chemical sensitivity is found to dominate over reasonable changes of the lattice parameter, which might occur, as well. Compositional changes down to 10% can be observed by visual image inspection, only, and they are characteristic for different defects that cause non-stoichiometry such as: V_{Ga}, V_N, Ga_N, N_{Ga}, and Ga_I (Fig. 1(c)). The origin of this chemical sensitivity of the CBED pattern is explored in order to establish alternatives to existing methods for local composition determination and to present experimental data on the stoichiometry of GaN.

Proc. Microscopy and Microanalysis 1995, edited by G.W. Bailey, M.H. Ellisman, R.A. Hennigar, and N.J. Zaluzec.

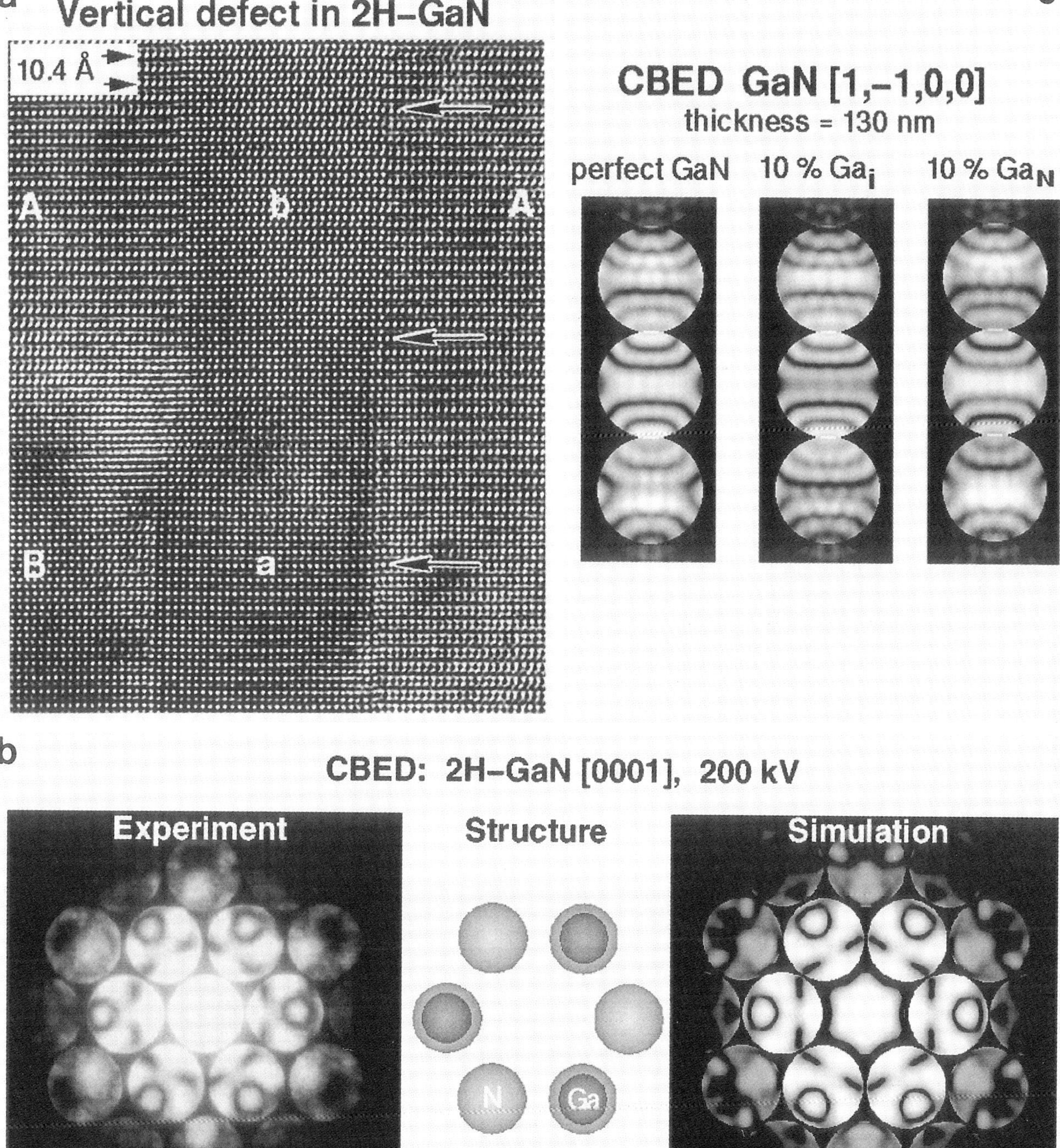

Fig. 1. (a) High resolution image of the vertical defect present in GaN grown by MBE on (0001) basal plane of sapphire. Capital and small letters show positions having the same image pattern inside and outside the defect. The arrows indicate bending of the planes inside the defect compared to the same plane in the matrix; (b) an experimental CBED pattern along [0001] zone axis, the structure of GaN along the same zone axis and a simulated CBED pattern; (c) simulated CBED pattern for [1 100] zone axis assuming perfect GaN crystal and 10% Ga$_I$, and Ga$_N$, respectively.

STRUCTURE FACTOR MEASUREMENT IN TiAl AND SILICON

S. Swaminathan, J. M. Wiezorek, I. P. Jones*, N. J. Zaluzec**, D. M. Maher*** and H. L. Fraser.

Dept. of Materials Science and Engineering, The Ohio State University, Columbus, OH
*School of Metallurgy and Materials, University of Birmingham, Birmingham, UK
**Center for Electron Microscopy, Argonne National Laboratory, Argonne, IL
***Dept. of Materials Science and Engineering, North Carolina State University, Raleigh, NC

The accurate measurement of low order structure factors is required for the determination of the electron charge density distribution in crystals. In this work the energy-filtered convergent beam electron diffraction (CBED) rocking curve method has been used for accurate structure factor measurements. This CBED method for structure factor refinement involves matching of the experimental CBED intensities to those calculated using dynamical electron diffraction theory. The CBED experiments were conducted with a Philips EM420 Transmission Electron Microscope coupled with a custom built energy-filtering attachment enabling single electron counting. The theoretical pattern matching was performed using FORTRAN programs which were developed by Swaminathan.[1] Initially the experimental plan involved an attempt to refine structure factors of TiAl by two dimensional Bloch wave calculations. The results of this project have been reported elsewhere.[2] Subsequently it proved impossible to obtain results with sufficient precision for TiAl reproducibly, i.e. less than 0.1%, from samples of different thicknesses. It is imperative that such consistency of results be obtained. Thus, in order to ascertain the origin of this inconsistency the present study concentrates on the development of the experimental and theoretical tools required for such refinement and measurements for Si, which has been well characterized by other techniques, have been performed.

Experimental rocking curves for **g**=*220* obtained from three different sample thicknesses were matched to the theoretically calculated curves. An IMSL library routine incorporating the quasi-Newton method with a finite difference gradient has been used for structure factor refinement.[3] Five variables including the coefficient of the crystal potential (V_g), the absorption coefficient (V'_g), the sample thickness (t) and the constant background intensity under the bright field and the dark field rocking curves were fitted by minimizing the quantity χ^2 defined by

$$\chi^2 = \frac{1}{n}\sum_{i=1}^{n}\frac{\left[c_*I_{calc} - I_{expt}\right]^2}{I_{expt}} \tag{1}$$

where n is the total number of points in the rocking curves. The fitted values of the V_g and structure factors (F_g) are given in table 1. These values were tested for local minima by repeating the calculations using different starting points and arriving consistently at the same minimum for the χ^2. Recently Deininger et al.[4] have suggested that the above definition of the χ^2 overestimates the significance of the low intensity values and have proposed the use of a χ^2 defined as

$$\chi^2 = \frac{1}{n}\sum_{i=1}^{n}\left[c_*I_{calc} - I_{expt}\right]^2 \tag{2}$$

The results of calculations using the χ^2 defined in (2) are also shown in table 1 and the fitted values of

Proc. Microscopy and Microanalysis 1995, edited by G.W. Bailey, M.H. Ellisman, R.A. Hennigar, and N.J. Zaluzec

V_g agree within the estimated error irrespective of the definition of χ^2.

Treatment of anomalous absorption : Without considering anomalous absorption the structure matrix of a centrosymmetric crystal is real symmetric because the imaginary part of V_g is zero. In the Bloch wave treatment anomalous absorption can be treated either by perturbation theory[5] or by adding a constant to the imaginary part of V_g. The former does not change the structure matrix. However, the latter treatment produces a complex symmetric structure matrix. Consequently matrix diagonalization requires longer computer time in the latter case. The complex matrix formulation has been recommended for high precision structure factor measurements.[5] However, the present calculations show no marked differences between the complex matrix formulation and the perturbation theory treatment (table 1). Hence, the computationally more efficient perturbation treatment may be utilized to treat anomalous absorption for centrosymmetric crystals.

Table 1 : Fitted values of the V_g and F_g. The fitted F_g values strongly agree with the x-ray *Pendellösung* measurements[6] (F_g = 67.11) and CBED rocking curve measurement[4] (F_g = 66.53).

Thickness	Best fit : χ^2 by (1)		χ^2 by (2)		Complex Struct. Matrix	
(t) nm	V_g (volts)	F_g	V_g (volts)	F_g	V_g (volts)	F_g
136.9	4.387	66.951	4.383	66.988	4.390	66.920
144.6	4.385	66.975	4.387	66.929	4.387	66.987
193.4	4.375	67.065	4.378	67.018	4.378	67.035
Average	4.382±.004	67.00±.04	4.384±.003	66.96±.02	4.385±.004	66.97±0.03

Structure Factor Measurements in TiAl : A recent study[7] of the site occupancy in single phase TiAl suggests that the accuracy of structure factor measurements by electron diffraction methods would critically depend on the composition of the material volume sampled by the electron beam and on the accuracy of the Debye-Waller factors used to determine the structure factors. Thus, the validity of recently reported structure factor measurements in TiAl appears to be affected.[2, 8] Work is in progress to study the compositional and Debye-Waller factor dependency of structure factors in TiAl. The results of these studies will be presented at the meeting.

References

1. S. Swaminathan, Ph. D thesis, The Ohio State University (1994), Columbus, OH.
2. S. Swaminathan et al., *Materials Science and Engineering*, (1993), **A170**, 227.
3. J.E. Dennis and R.B. Schnabel, Numerical Methods for Unconstrained Optimization and Nonlinear Equations, Prentice-Hall, (1983) Englewood Cliffs, NJ.
4. C. Deininger et al., *Ultramicroscopy*, (1994), **54**, 15.
5. C. J. Humphreys and E. G. Bithell, *Electron Diffraction Techniques*, Ed. J. M. Cowley, Oxford Univ. Press, NY.
6. S. Cummings and M. Hart, *Australian Journal of Physics*, (1988), **41**, 423.
7. S. Swaminathan et al., *MRS Proceedings : Fall meeting* , (1994), in press.
8. Z. W. Lu et al., *Acta Metallurgica*, (1994), **42**, 3929.
9. This project is sponsored by National Science Foundation.

PLATE- AND NEEDLE-LIKE CRYSTAL STRUCTURE DETERMINATION BY COMBINING ELECTRON AND SYNCHROTRON DIFFRACTION

Z. G. Li[1], R. L. Harlow[1], K. H. Gardner[1], L. Liang[2], D. L. Dorset[3], J. M. Zou[4]

[1]DuPont Central Research and Development, P. O. Box 80228, Wilmington, DE 19880-0228
[2]DuPont-Merck, Pharmaceutical Division, P. O. Box 80402, Wilmington, DE 19880-0402
[3]Electron Diffraction Department, Hauptman-Woodward Medical Research Institute, Buffalo, NY
[4]Arizona State University, Dept. of Physics, Tempe, AZ 85287

In recent years, significant progress in determining light-element crystal structures using electron diffraction has been made to meet the increasing needs from the material science community. The set of techniques for carrying out electron crystallography, including sample preparation, data collection and recording, diffraction micrograph digitization, and confirmation by direct phasing methods and structure refinements, have being developed [1]. However, only a very limited number of structures have been determinated by electron crystallography because of a number of severe problems and difficulties. Meanwhile, the progress in determining crystal structures using x-ray diffraction has been rapidly increasing, particularly because of the more extensive use of powerful synchrotron diffraction techniques. In this case, the minimum crystal size required for a single-crystal study has decreased from 100 microns to about 10 microns. Even so, many new materials can only be obtained in microcrystalline form with crystallite sizes well below one micron. In this submicron regime, ab initio structure solutions from synchrotron powder-diffraction patterns have proven to be quite powerful. Our experience, however, suggests that the ab initio powder method fails as often as it succeeds. Problems such as the presence of multiple phases, Laue group ambiguities, and considerable peak overlap put serious limitations on powder determinations. In this abstract, we report our preliminary results of crystal structure determinations by combining electron single-crystal diffraction and synchrotron powder-diffraction techniques [2].

The plate- and needle-like crystals (with dimensions typically less than 0.1 micron) we have chosen to study represent a class of industrially important materials. We are, for example, presently studying a recently synthesized phase of HAlF4 [3] which contains only light elements and only exists as very thin crystals where the kinematic approximation (with some correction) can be assumed. A JEM 2000EX (accelerating voltage: 200 kv) and 1200EX electron microscopes (accelerating voltage: 80 kv) have been used with very low beam densities to avoid radiation damage. Electron microscopy has revealed that HAlF4 has a needle-like habit, with average sizes in two short dimensions of about 15 nm and 200 nm. Figure 1a is an image of the nanocrystals viewed perpendicular to the shortest dimension. Figure 1b is an electron diffraction pattern of the crystal shown in figure 1a with the electron beam perpendicular to the a*-b* plane. Trial unit cells were derived from a combination of symmetry and d-spacings obtained from the EM patterns and from in-house (making use of the natural preferred orientation in this sample) and synchrotron (Brookhaven National Laboratory) x-ray powder diffraction patterns. We have used these trial unit cells along with IdealMicroscopc software [4] in guiding us to obtain and index EM diffraction patterns in different zones. The measured d-spacings of the crystal in the different electron diffraction patterns were systematically compared and adjusted using the much more accurate values provided by synchrotron data.

Electron and synchrotron diffraction are complementary techniques for determining the crystal structures of the plate- and needle-like crystals. Electron diffraction data can be used to get initial unit cell dimensions and symmetries and, often, partial crystal structure information. Refinement of the structure using synchrotron powder data provides accurate cell dimensions as well as more accurate atomic parameters. Two examples, one in which synchrotron data was used to verify a crystal

Proc. Microscopy and Microanalysis 1995, edited by G.W. Bailey, M.H. Ellisman, R.A. Hennigar, and N.J. Zaluzec
Copyright © 1995 MSA. Published by Jones and Begell Publishing, 79 Madison Ave., New York, NY 10016

structure originally determinated by electron crystallography and one in which a completely new crystal structure was determination by combining electron and x-ray diffraction techniques, will be reported.

References

1. D. L. Dorset, Advances in Electronics and Electron Physics, Vol. 88, 111 (1994).
2. Z. G. Li, et al., (unpublished results).
3. N. Herron et al., Chem. of Materials, **7**, 75-83 (1995).
4. "IdealMicros." software is developed by EMLab Inc., 16203 S. 26th Place, Phoenix, AZ 85048.

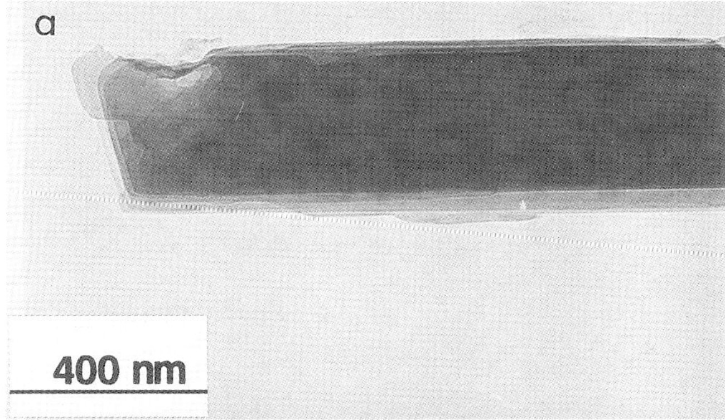

Figure 1. -- a) Electron microscope image, and b) diffraction pattern of needle-like HAlF4 crystal. Electron beam is perpendicular to a*-b* plane of the crystal.

USE OF DIRECT PHASING IN DETERMINATION OF ATOMIC CO-ORDINATES OF ANTHANTHRENE, $C_{22}H_{12}$.

C. E. Leslie, J. R. Fryer, C. J. Gilmore and W. Nicolson

Chemistry Department, University of Glasgow, Glasgow, U. K.

Imaging of organic crystals in the electron microscope is limited by their sensitivity to the electron beam, with even the most stable molecules only giving a resolution of ~3Å.[1] By contrast, electron diffraction requires a lower beam intensity, and so can give much better resolution, down to ~1Å under favourable conditions. Intensities may be measured directly from the diffraction pattern, and the phase determination problem can be surmounted by applying the maximum entropy method, which chooses phases by likelihood estimation.[2,3]

It was decided to apply this method to the organic molecule anthanthrene, $C_{22}H_{12}$ (fig. 1), since x-ray studies found that the crystals were monoclinic with a = 12.10Å, b = 10.34Å, c = 10.74Å, β = 92.2°, space group $P2_1/a$ and Z = 4, but no details of the atomic co-ordinates were given in the literature.[4] A previous study of the molecule, using KCl as the substrate, produced epitaxial layers showing the ac-projection being most common at room temperature.[5] It was decided therefore to use a different substrate, NaCl, in an attempt to produce another orientation, since the lattice constraints would, in theory, make it difficult for the crystals to form an even layer in the same orientation.

The specimens were formed by evaporation of anthanthrene onto the freshly cleaved [100] surface of NaCl at 25°C. The organic film was then strengthened with carbon, floated off the substrate and picked up on grids. Study of the films was carried out on a JEOL 1200 microscope operating at 120kV, using normal electron microscope film. Figure 2(a) shows the organic layer of small, needlelike, crystals showing little alignment relative to each other, or to the substrate. Figure 2(b) shows the diffraction pattern obtained from these crystals. Primary reflections give d-spacings of 10.3Å and 10.7Å, with the corresponding angle 90°. This the bc-projection, showing that choice of substrate can alter the orientation of the layer produced.

This diffraction pattern was analysed using the electron diffraction pattern processing function in CRISP, which found 85 reflections and their associated intensities.[6] These were analysed and normalised, then first run over 128 nodes in the maximum entropy program MICE.[7] This gave rise to 8 nodes which had reasonable likelihoods. These 8 nodes were then used as the basis for a second level analysis, this time run over 512 nodes. This in turn gave 11 nodes with good likelihood values. The larger the likelihood value, the more likely the phase choices made are to be correct. Once this analysis had been carried out, the maximum entropy distributions associated with the nodes giving the highest likelihood values were converted to centroid maps, which are comparable to electron density maps, and may be examined in the normal way.

Figure 3 shows a two dimensional bc-projection centroid map with a likelihood value of 0.89. It shows the herringbone packing exhibited by the molecules within the unit cell, and the molecular overlap. The y axis corresponds to 1.034nm and the z axis to 1.072nm.

1. J. R. Fryer, *MSA Bulletin*, 23, 1 (1993) 44-56

Proc. Microscopy and Microanalysis 1995, edited by G.W. Bailey, M.H. Ellisman, R.A. Hennigar, and N.J. Zaluzec
Copyright © 1995 MSA. Published by Jones and Begell Publishing, 79 Madison Ave., New York, NY 10016

2. G. Bricogne, *Acta Cryst.,* A44 (1988) 517
3. C. J. Gilmore, K. Shankland and G. Bricogne, *Proc. R. Soc. Lond.* A442 (1993) 97-111
4. G. W. Smith and J. C. Stalley, *Acta Cryst.,* A31 (1975) S122
5. J. R. Fryer, *Acta Cryst.,* A34 (1978) 603-607
6. X. Zou, Y. Sukharev and S. Hovmöller, *Ultramicroscopy,* 49 (1993) 147-158
7. G. Bricogne and C. J. Gilmore, *Acta Cryst.,* A46 (1990) 284-297

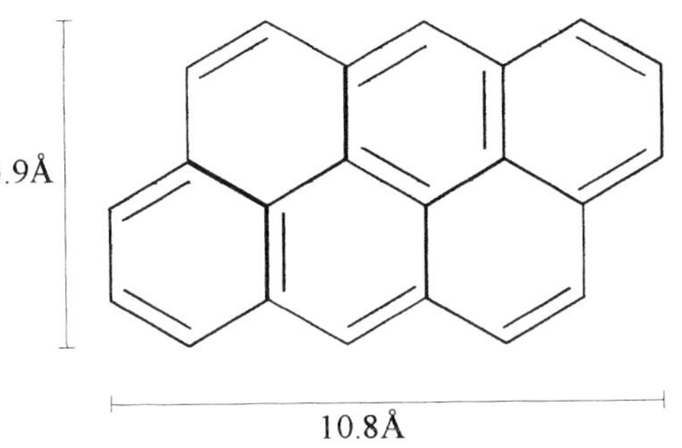

Figure 1. Structure of anthanthrene

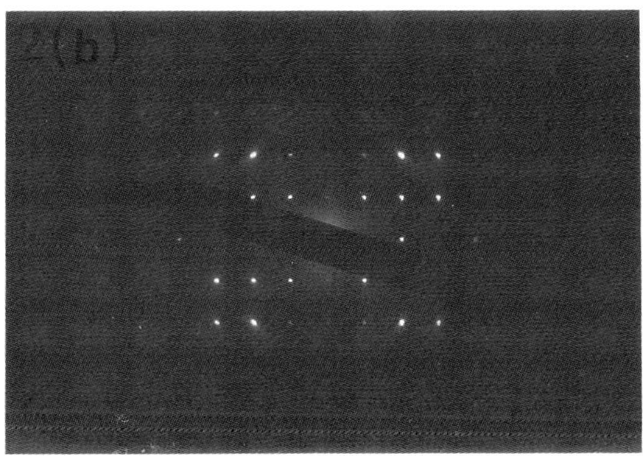

(b) Electron diffraction pattern given by above crystals

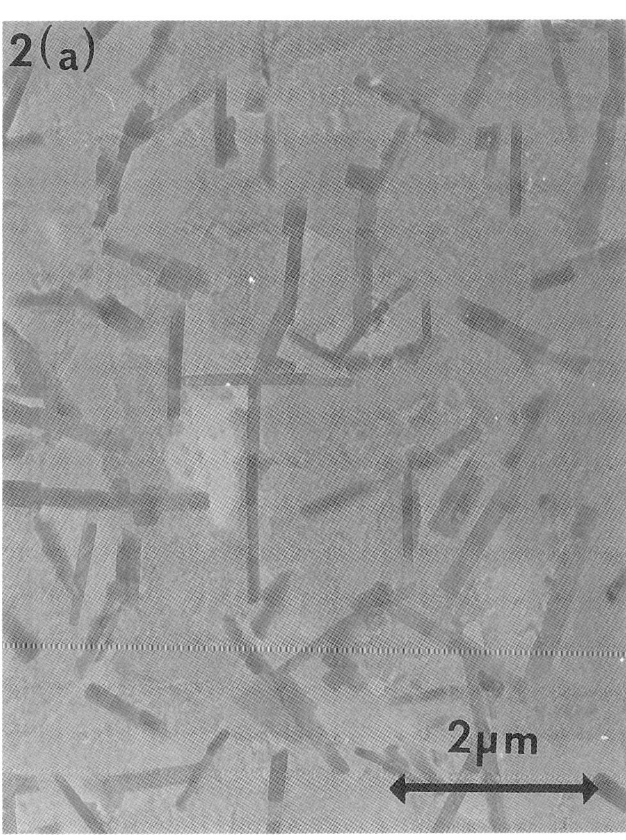

Figure 2 (a). Crystals of anthanthrene grown on 25°C NaCl

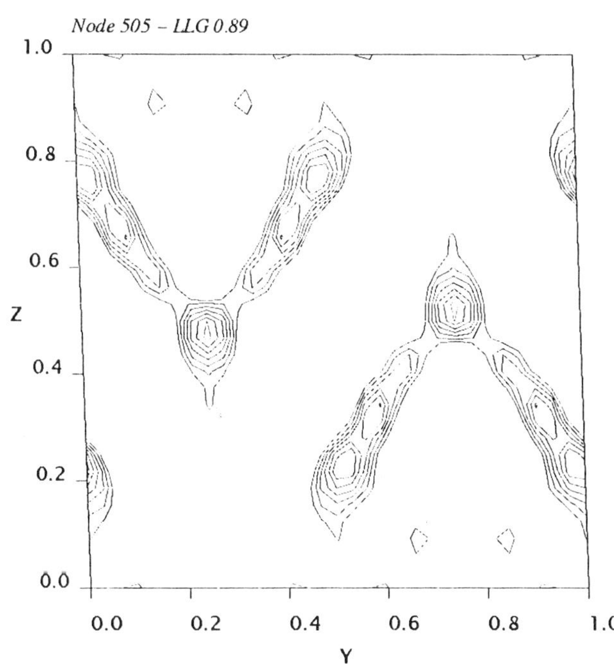

Figure 3. Centroid map given by node with highest
likelihood value after analysis via MICE

UNITARITY AND THE REAL WORLD: APPLICATIONS OF DIRECT PHASING TO ELECTRON DIFFRACTION DATA

W. F. Tivol[*]

[*]Wadsworth Center and the School of Public Health, the University at Albany, Empire State Plaza, PO Box 509, Albany NY 12201-0509

The Sayre equation can be derived as a special case of unitarity, and recently it was shown that the derivation applies to dynamical scattering and electron diffraction.[1,2] Since both the triplet formula and the tangent formula can be derived from the Sayre equation, use of these direct methods to find the phases for electron diffraction amplitudes is fully justified.

The most transparent form of the unitary equation is

$$\langle f|\hat{A}|i\rangle - \langle f|\hat{A}^{\dagger}|i\rangle = \frac{ik}{2\pi}\Sigma\langle f|\hat{A}^{\dagger}|n\rangle\langle n|\hat{A}|i\rangle, \tag{1}$$

where $\hat{A}$ is the scattering amplitude operator, f is the final state, i the initial state, and n is a complete set of intermediate states; the sum over these states becomes an integral where the intermediate states form a continuum.

If there is Friedel symmetry, the form of the Sayre equation derived from equation (1) is

$$A(\boldsymbol{H}) = \frac{A(\boldsymbol{0})}{\Sigma'|A(\boldsymbol{K})|^2}\Sigma'A(\boldsymbol{H}-\boldsymbol{K})A(\boldsymbol{K}). \tag{2}$$

The prime on the sum means $\boldsymbol{K} \neq \boldsymbol{0}$. If Friedel symmetry is violated, $\Sigma'|A(\boldsymbol{K})|^2$ becomes

$$\Sigma'A(\boldsymbol{K})A(-\boldsymbol{K})\left[1+\frac{2\pi i}{k}\left(\frac{A(\boldsymbol{H})-A^*(-\boldsymbol{H})}{A(\boldsymbol{H})A(\boldsymbol{0})}\right)\right].$$

This equation is a quantitative relationship which not only has an essentially unique solution for the phases, but which also constrains the magnitudes of the intensities.

There are some assumptions in the derivation of equation (2) which affect how that equation applies to real electron diffraction data. First, the equation was derived for zero excitation error. Although it is possible to obtain the appropriate scattered intensities to put into equation (1) (*e.g.*, from convergent beam data), often only the diffraction pattern where the incident beam is parallel to a zone axis is obtained. Second, the set of intermediate states must be complete--*i.e.*, backscattering and inelastic scattering must be included.

Proc. Microscopy and Microanalysis 1995, edited by G.W. Bailey, M.H. Ellisman, R.A. Hennigar, and N.J. Zaluzec
Copyright © 1995 MSA. Published by Jones and Begell Publishing, 79 Madison Ave., New York, NY 10016

The excitation error consideration enters equation (1) in the $<f \mid \hat{A}^\dagger \mid n>$ term, which is equal to the A(**H-K**) term in equation (2) for zero excitation error. In principle, the difference between the two can be measured by tilting the crystal to the angle (about 0.01 radian) of the beam **K**, or by using convergent beam diffraction; however, imperfections in the crystal and spread of the incident beam momentum have effects of about this order. For a resolution of 0.05 nm and an accelerating voltage of 1.2 MeV, the excitation error is about 0.04 nm^{-1}, which is comparable to the reflection half-width of a 10 nm thick crystal.[3] Thus, for the use of high voltage to obtain electron diffraction patterns from thin crystals, there is no appreciable error in using the measured intensities from a single zone axis pattern in equation (2).

Closure error--the neglect of some intermediate states--is not always easy to analyze, but is also reasonably small for an accelerating voltage of 1.2 MV and a crystal thickness of about 10 nm. Elastic intermediate states which are usually not included in the sums in equations (1) and (2) include backscattering. For a transition to a forward-directed final state, the terms including a backscattered intermediate state also include the transition from the backward-directed intermediate to the final state and, thus, are proportional to the square of the backscattering cross-section. For high voltage and thin crystals, calculations of backscattering yields are on the order of 10^{-7} for typical cases.[4] Inelastic intermediate states include those with both small and large energy losses. For small losses, *e.g.*, plasmon production, unless energy filtering is used, the diffracted electron is very near the location of an elastically scattered diffraction spot, and thus such intermediate states will be included in the integrated intensities measured. For larger losses, the electron optics of the microscope will not focus these electrons, and they will not be included. The transition from such an intermediate state to a final state with energy equal (or nearly so) to that of the elastically scattered electrons is, however, zero, so neglect of these states results only in the need to renormalize the observed intensities.

The final test of the applicability of direct methods to real electron diffraction data is the ability to obtain correct structures from these data. So far *ab initio* determinations from small model compounds and phase extensions from both small molecules and macromolecules[5-9] have been successful.[10]

References

1. A.F. Mishnev, in H. Schenk, Ed., *Direct Methods of Solving Crystal Structures*, New York, London: Plenum Press (1991)399.
2. W.F. Tivol, *Acta Cryst.* (1995) in press.
3. P.B. Hirsch et al., *Electron Microscopy of Thin Crystals*, 1st ed. New York: Plenum Press, London: Butterworths (1965)97.
4. D.C. Joy, *Monte Carlo Simulation for Microscopy and Microanalysis*, New York: Oxford University Press (1995).
5. D.L. Dorset, *MSA Bulletin* 23(1993)99.
6. D.L. Dorset et al., *Ultramicroscopy* 38(1991)41.
7. D.L. Dorset et al., *Acta Cryst.* A48(1992)562.
8. D.L. Dorset et al., *J. Appl. Cryst.* 26(1993)778.
9. D.L. Dorset et al., *Ultramicroscopy* 57(1995)59.
10. Supported by Biotechnological Resource grant RR01219, awarded by the National Center for Research Resources, Department of Health and Human Services/Public Health Service, to support the Wadsworth Center's Biological Microscopy and Image Reconstruction Facility as a National Biotechnological Resource.

ELECTRON DIFFRACTION OF COPPER PERFLUOROPHTHALOCYANINE ON THE HVEM

W. F. Tivol[*] and J. R. Fryer[**]

[*] Wadsworth Center and the School of Public Health, the University at Albany, Empire State Plaza, PO Box 509, Albany NY 12201-0509
[**] Chemistry Department, University of Glasgow, Glasgow G12 8QQ, Scotland, U.K.

The use of high-energy (1200 kV) electrons has been shown to be advantageous in the *ab initio* structure analysis from electron diffraction of organic compounds.[1,2,3] Previous studies showed that *ab initio* analysis of copper perchlorophthalocyanine could be accomplished at accelerating voltages at or above 1000 kV, but not at 400 kV for crystals which are about 10 nm thick.[1,2] Copper perbromophthalocyanine could also be analyzed *ab initio* at 1200 kV, but the presence of severe dynamical scattering precluded such analysis at lower voltages.[3]

Copper perfluorophthalocyanine (Cu FPC) was grown epitaxially from the vapor phase onto a clean KCl crystal face. Electron diffraction patterns were obtained from crystals tilted at 20° and oriented so that the electron beam was parallel to the c-axis.

Electron doses were kept to a minimum by the use of a 100 μm condenser aperture and a highly excited first condenser lens. A video system allows scanning the specimen to find a crystal suitable for the correct orientation of the grid by rotating it within its own plane and focusing the objective lens, all at an illumination level below that necessary to distinguish anything by eye on the HVEM's phosphor screen.

The AEI EM7 high-voltage electron microscope was used at a voltage of 1200 kV in diffraction mode with a 10 μm selected area aperture. The data were recorded on DuPont Lo-dose Mammography film. Diffraction spots are seen out to a resolution of 0.07 nm. Crystals of Cu FPC have been found in two morphologies (Fig. 1) which give different diffraction patterns (Fig. 2). One of the forms is very similar to the chlorinated analog (Figs. 1a and 2a), while the other form is similar to the unhalogenated analog (Figs. 1b and 2b). Since different parts of the grid depart from the nominal value of the tilt by a few degrees, so no significance can be attached to the fact that each form showed a zone axis at a tilt of 20°. Analysis of the intensities to determine the structure of Cu FPC is in progress.

References

1. Dorset, D.L., et al., *Ultramicroscopy* 38 (1991) 41.
2. Tivol, W.F., et al., *MSA Bulletin* 23 (1993) 91.
3. Dorset, D.L., et al., *Acta Cryst.* A48 (1992) 562.
4. Supported by Biotechnological Resource grant RR01219, awarded by the National Center for Research Resources, Department of Health and Human Services/Public Health Service, to support the Wadsworth Center's Biological Microscopy and Image Reconstruction Facility as a National Biotechnological Resource.

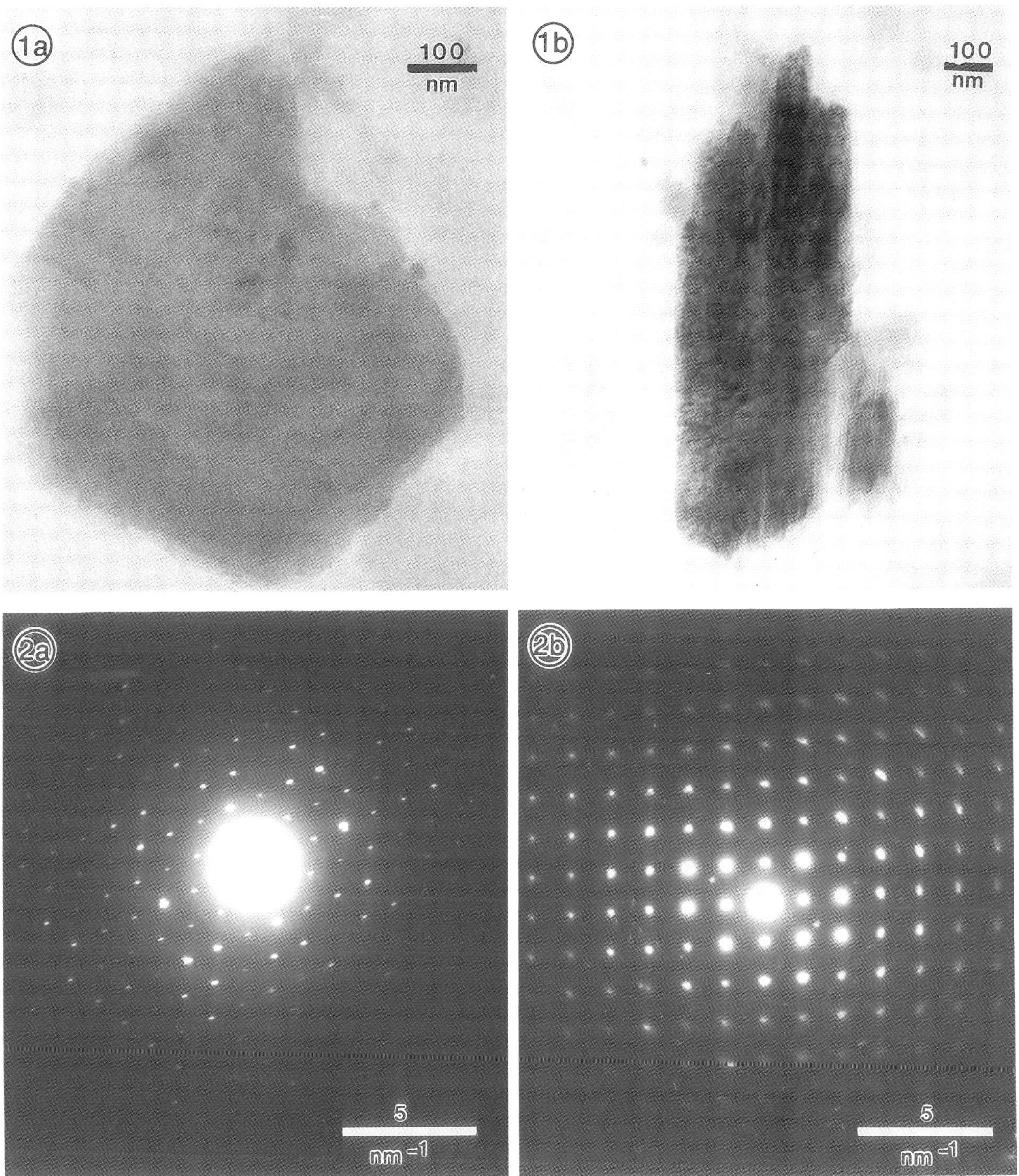

FIG. 1.—Images of Cu FPC crystals. a) form similar to chlorinated analog, b) form similar to unhalogenated analog.

FIG. 2.—Electron diffraction patterns of Cu FPC obtained at 1200 kV accelerating voltage. a) form similar to chlorinated analog, b) form similar to unhalogenated analog.

ELECTRON DIFFRACTION ANALYSIS OF A METASTABLE STATE IN ION-IRRADIATED MgAl$_2$O$_4$ SPINEL

Ram Devanathan, Ning Yu, Kurt E. Sickafus and Michael Nastasi

Materials Science and Technology Division, Los Alamos National Laboratory, Los Alamos, NM 87545

Magnesium aluminate spinel is considered to be a remarkably radiation resistant material. Single crystal spinel exhibits very little damage when irradiated with fast neutrons or energetic ions.[1-2] Recently, amorphization of single-crystal MgAl$_2$O$_4$ was observed, for the first time, following irradiation with 400 keV Xe^{2+} at about 100 K to a peak dose of 25 dpa.[3] Electron diffraction patterns from spinel subjected to different doses show the extinction of first-order fundamental reflections prior to the occurence of amorphization. This suggests that, enroute to becoming amorphous, ion-irradiated spinel transforms to a metastable crystalline state with a lattice constant that is half that of spinel. In an effort to understand this phase transformation, we have analyzed the structure of the metastable state using electron diffraction.

Spinel belongs to the spacegroup Fd3m (fcc Bravais lattice). The unit cell has a lattice constant of about 0.8 nm and is made of two sets of four octants. One set has Al^{3+} in octahedral coordination with O^{2-}, while the other has Mg^{2+} in tetrahedral coordination with O^{2-}. The oxygen ions form a fcc sublattice with a periodicity of about 0.4 nm. This is an interesting fact, because the experimental electron diffraction patterns indicate that the metastable state is fcc with a lattice constant of about 0.4 nm. In addition, we found that the intensities of the diffraction spots showed the following trends:
a) The (111), (200), and (220) reflections had comparable intensities.
b) The (311) reflection was stronger than the (420) and (422) reflections.
Keeping these experimental facts in mind, we have performed a diffraction analysis of more than thirty model structures. For each structure, the square of the structure factor, F^2, was calculated for several reflections. By comparing the variation of F^2 as a function of d-spacing with the experimental diffraction intensity distribution, we have determined the structure of the metastable state in spinel.

We started with a 0.4 nm fcc oxygen lattice and varied the cation distribution to generate different model structures. Space considerations permit us to discuss only a few of the structures examined. The first choice was to place all cations in the octahedral interstices at random. The resulting NaCl structure was unsuitable, because it gave rise to extremely weak (111) peaks. The next approach was to distribute the cations in either half (ZnS) or all (CaF$_2$) tetrahedral sites. In both cases, the (200) reflections were very weak. Combinations of octahedrally and tetrahedrally distributed cations were also unsuitable because the (220) spot was four times as intense as the (111) and (200). In addition, the (311) was weaker than (422). The model that provides the best fit to experiment is obtained by placing the cations in the 16d interstice of the Fd3m space group (three-fold co-ordination) in addition to the octahedral and tetrahedral positions. The 16d interstice can be stabilized if the anion lattice is distorted, which we believe is true. The intensity distributions for some of the model structures are shown in Figure 1.

In conclusion, low-temperature ion-irradiated MgAl$_2$O$_4$ transforms to a metastable crystalline state prior to amorphization. This structure consists of a 0.4 nm fcc oxygen lattice with cation interstitials.

References

1. F. W. Clinard Jr., G. F. Hurley, and L. W. Hobbs, *J. Nucl. Mater.* 108-109 (1982) 655.
2. S. J. Zinkle, *J. Am. Ceram. Soc.* 72 (1989) 1343.
3. N. Yu, K. E. Sickafus, and M. Nastasi, *Phil. Mag. Lett.*. 70 (1994) 235.

This research is sponsored by the US Department of Energy, Office of Basic Energy Sciences.

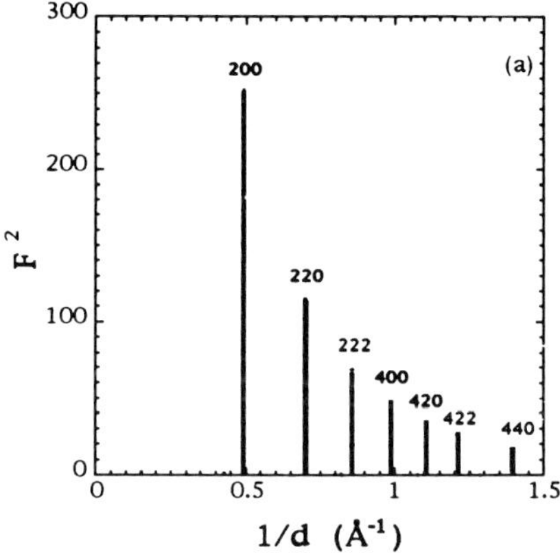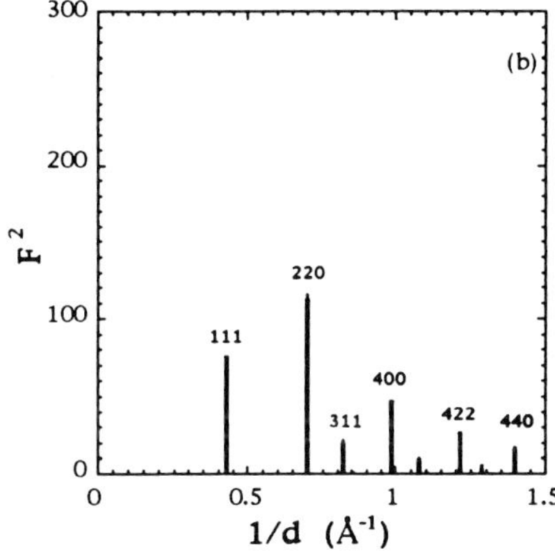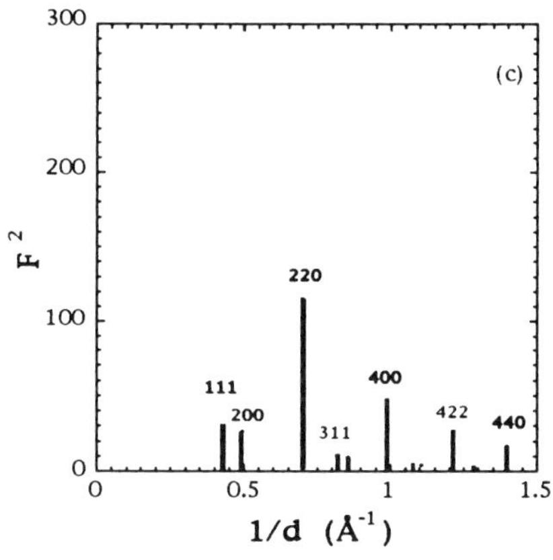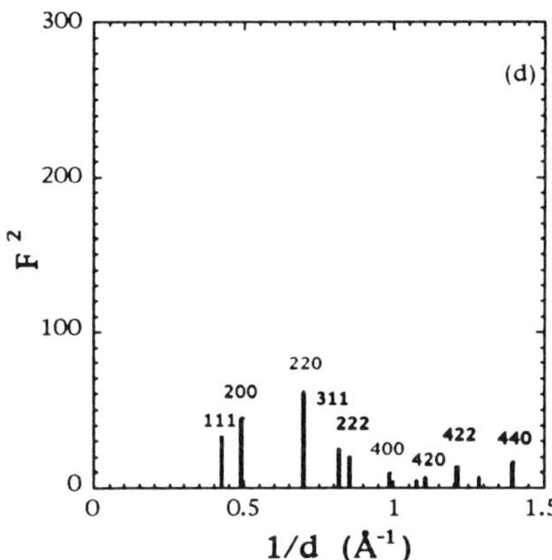

FIG. 1. The square of the structure factor as a function of the reciprocal of d-spacing for the following models: (a) octahedral cation interstitials (NaCl-type); (b) tetrahedral cation interstitials (CaF_2-type); (c) octahedral and tetrahedral cation interstitials; (d) octahedral, tetrahedral, and three-fold cation interstitials.

HIGH-ANGLE ELECTRON SCATTERING FROM AMORPHOUS SILICON

M. Libera,* J.A. Ott,* K. Siangchaew,* and L. Tsung **
*   Stevens Institute of Technology, Hoboken, New Jersey
** Philips Electronic Instruments, Inc. Mahwah, New Jersey

Channeling occurs when fast electrons follow atomic strings in a crystal where there is a minimum in the potential energy (1). Channeling has a strong effect on high-angle scattering. Deviations in atomic position along a channel due to structural defects or thermal vibrations increase the probability of scattering (2-5). Since there are no extended channels in an amorphous material the question arises: for a given material with constant thickness, will the high-angle scattering be higher from a crystal or a glass?

Figure 1a shows a HAADF STEM image collected using a Philips CM20 FEG TEM/STEM with inner and outer collection angles of 35mrad and 100mrad. The specimen (6) was a cross section of single-crystal Si containing: amorphous Si (region A), defective Si containing many stacking faults (B), two coherent Ge layers (C1; C2), and a contamination layer (D). CBED patterns (fig. 1b), PEELS spectra, and HAADF signals (fig. 1c) were collected at 106K and 300K along the indicated line. The CBED patterns show that the specimen is aligned along the [110] zone axis. This orientation is constant. Thickness (fig. 1d) was measured using a t/λ analysis of PEELS data collected at T=106K. The thickness is roughly constant. The HAADF intensity (fig. 1c) was recorded using a storage oscilloscope externally triggered by the STEM scan generator (7). Fig. 2 summarizes the results. Crystalline Si scatters greater high-angle intensity than amorphous Si at both 106K and 300K. Without composition and thickness variations, these differences can be attributed to channeling. The probability of high-angle scattering in crystalline Si will be high, since the fast electrons are channeled and are in greater proximity to the atom cores. With no long-range order, electrons incident upon a glassy specimen are less localized on the atom cores, and the phonon-scattering probability is lower. Fig. 2 also shows that the crystal and glass HAADF intensities both increase with temperature. The rate of increase is higher for the glass. This can be attributed to the fact that amorphous Si has a higher density of unsaturated bonds leading to a weaker average atomic restoring force and larger vibrational amplitudes. Finally, fig. 2 indicates that the HAADF scattering from the defective Si shows little temperature dependence. Apparently, a physical interruption to a channel such as a stacking fault is sufficiently significant that any additional effects due to a temperature change of 200K are secondary.

Differences in local atomic number give rise to significant HAADF signal via Rutherford scattering, hence the technique is often referred to as Z-contrast imaging (8). We have shown elsewhere (7) that atomic number differences are also manifested in the thermal-diffuse scattering. Nevertheless, the present results indicate that other scattering mechanisms that do not rely on Z differences can give rise to meaningful contrast in HAADF STEM.

References

1. J. Spence, in *Electron Diffraction Techniques*, ed. J.M. Cowley (International Union of Crystallography, Oxford Univ., 1992) 360.
2. J.M. Cowley and Y. Huang, Ultramicroscopy 40 (1992) 171-180.

Proc. Microscopy and Microanalysis 1995, edited by G.W. Bailey, M.H. Ellisman, R.A. Hennigar, and N.J. Zaluzec
Copyright © 1995 MSA. Published by Jones and Begell Publishing, 79 Madison Ave., New York, NY 10016

3. J. Liu and J.M. Cowley, Proc. 50th Annual EMSA Meeting, ed. G.W. Bailey et al., (San Francisco Press, San Francisco, 1992) 1224-1225.

4. R. Loane, P. Xu, and J. Silcox, Ultramicroscopy 40 (1992) 121-38.

5. Z.L. Wang and J.M. Cowley, Ultramicroscopy 31 (1989) 437-454.

6. M. Libera et al., Ultramicroscopy 52, (1993) 564-569

7. S.J. Pennycook, Ultramicroscopy 30 (1989) 58-69.

8. M. Libera et al., submitted to Ultramicroscopy.

9. Funding was provided by the NJ Commission on Science & Tech.

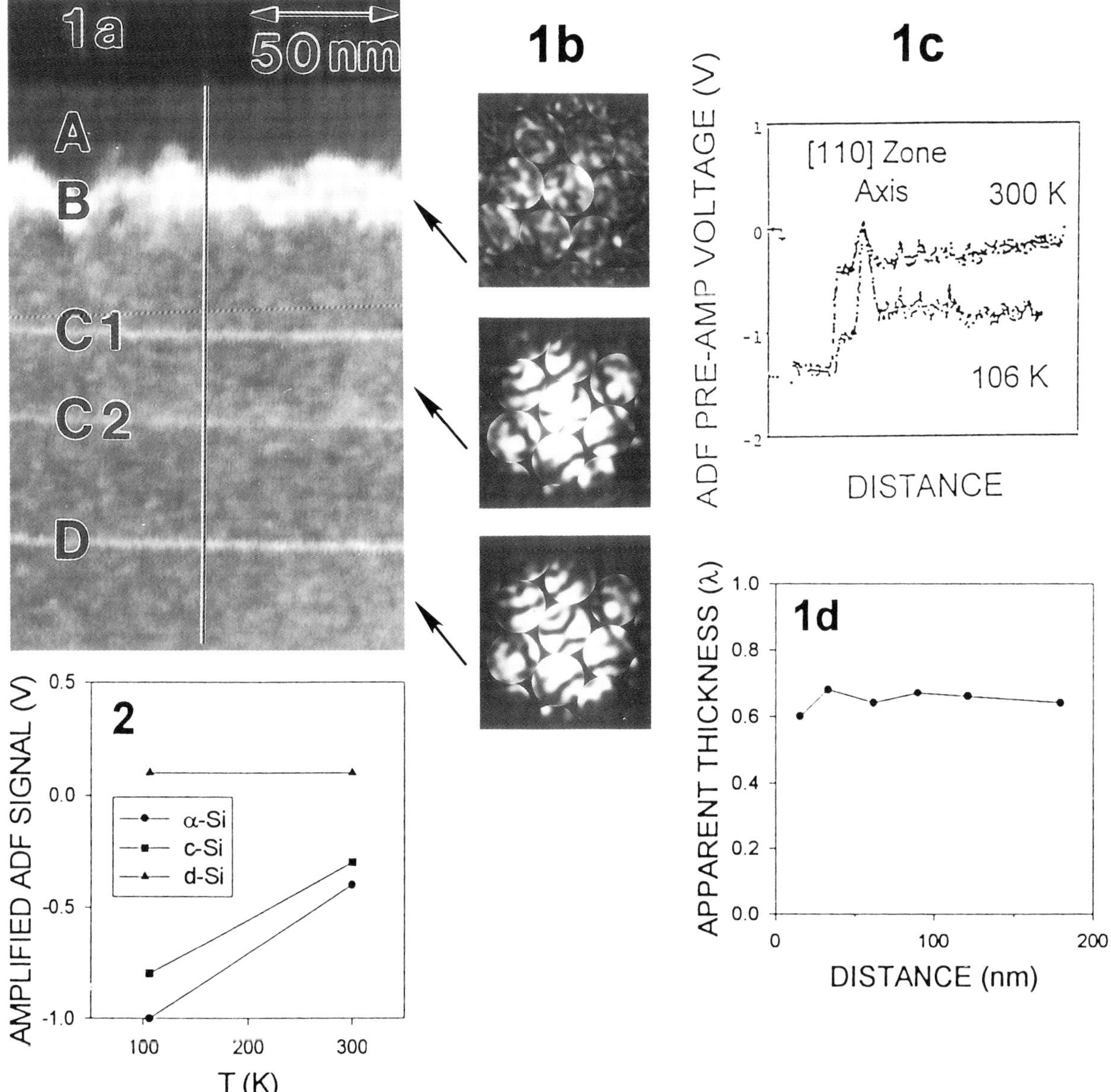

Fig. 1 (a) HAADF STEM image of Si cross section with: (b) CBED data; (c) HAADF line scans; and (d) t/λ PEELS data.

Fig. 2 HAADF signals for amorphous, crystalline, and defective Si.

ALTERNATIVE INTERPRETATION OF ELECTRON DIFFRACTION PATTERNS FROM CHRYSOTILE ASBESTOS FIBERS

P. Ling[†], H. Ling[†], H. Kawayoshi[†], F. Tsai[†], R. Gronsky[*] and J. Washburn[*]
[†]Advanced Materials Engineering Research, Inc., 250 Santa Ana Court, Sunnyvale, CA 94086
[*]Department of Materials Science & Mineral Engineering, University of California, Berkeley, CA 94720-1760

Chrysotile asbestos fibers have been the object of intensive study for many years, and their structure and growth mechanisms have perplexed investigators ever since the original 1930 X-ray diffraction work of Warren and Bragg[1] confused superlattice for fundamental reflections along two of the three unit cell directions. Even after the lattice parameters were accurately found[2], the origins of the characteristic and plentiful diffuse scattering seen in diffraction patterns challenged the most careful analyses. In 1956, Whittaker[3] performed very detailed calculations of the scattering from fibers that were assumed to adopt the structure of concentric circular cylinders, yet could not obtain agreement with experimental results, especially the intensity distribution of subsidiary maxima(diffuse scattering) bordering Bragg reflections on a single side. It took the beautiful high-resolution electron microscopy of Yada[4] to show that the fibers were actually spiral cylinders rather than circular cylinders, an observation that removed some of the minor discrepancies in Whittaker's model.

However, Yada[5] later reported a difference in separation of h0l and h0-l reflections on different sides of the a-axis, and the appearance of unusual streaks oriented away from the fiber axis when the incident beam was normal to the fiber axis. These differences were the subject of a theoretical analysis by Chisholm[6], but remained mysterious, and they were confirmed to persist through large-angle tilting in the experimental work of Ling et al[7]. Another interesting but unexplained feature in this early work was the presence of both sharp and diffuse rings in diffraction patterns recorded when the incident beam was parallel to the fiber axis. All of these findings assume significance in light of the Asbestos Hazard Emergency Response Act (AHERA) regulations requiring transmission electron microscopy and diffraction in the identification of asbestos fibers.

The present paper introduces a new concept to assist in the interpretation of electron diffraction patterns from chrysotile asbestos fibers. The concept is the decomposition of the structure of the fiber into very thin slabs, each of which passes through the center of the fiber axis and possesses a near-perfect orthohombic structure ($\alpha = \beta = \gamma = 90°$; no bond-bending associated with a macroscopic spiral). As shown in Fig.1, when any one of these slabs is fanned through $360°$, it sweeps out the cylindrical structure of the fiber. Only within the slab is the crystal properly oriented for Bragg scattering, but outside of the slab the lattice planes are bent away from the Bragg scattering angle. This simplification is further shown in Fig.2a, wherein the slab is only two unit cells thick, causing relrods in the reciprocal space as indicated in Fig.2b. If the electron beam is incident upon the fiber along [100], the fiber axis, the intersection between the relrods in Fig.2b and Ewald Sphere generate the observed diffraction effects. the 010 reciprocal circle is severely broadened due to streaking along [010] directions, but the 001 reciprocal circle is only moderately broadened, explaining the dual occurrence of sharp and diffused rings. If the electron beam is incident upon the fiber along [001], diffraction spots will be sharp. But along [010], diffraction spots will be streaked, with the directions of the flares indicated by the two slabs shown in Fig.3.

Fig.4 shows the SAED pattern of a chrysotile asbestos fiber at 100 kV when the electron beam is incident perpendicularly on the fiber. It can be noticed that the SAED pattern consist of sharp and streaked diffraction spots. The streaking at {110} and {310} can be interpreted as being predicated in the analysis of Fig.2a and b. The slight difference of of h0l and h0-l can also be understood. From Fig.3a and b, it can be interpreted as the slight shifts of positions of each diffraction spots caused by irregularity of the fiber.

[1] B. E. Warren and W. L Bragg, *Z. Kristallogr.* **76**, 201 (1930).
[2] B. E. Warren and K. W. Herring, *Phys. Rev.* **59**, 925 (1941).
[3] E. J. W. Whittaker, *Acta Cryst.* **10**, 149 (1957).
[4] K. Yada, *Acta Cryst.* **23**, 704 (1967).
[5] K. Yada, *Can. Mineral.* **62**, (679 (1979).
[6] J. E. Chisholm, *Acta Cryst.* **A44**, 70 (1988).
[7] P. Ling, H. Kawayoshi, T. Tien and L Martinez, in <u>Proc. 49th Ann Mtg. Micros. Soc. Amer.</u>, G. W. Bailey, Ed., San Francisco Press, p.976, (1991).

Proc. Microscopy and Microanalysis 1995, edited by G.W. Bailey, M.H. Ellisman, R.A. Hennigar, and N.J. Zaluzec
Copyright © 1995 MSA. Published by Jones and Begell Publishing, 79 Madison Ave., New York, NY 10016

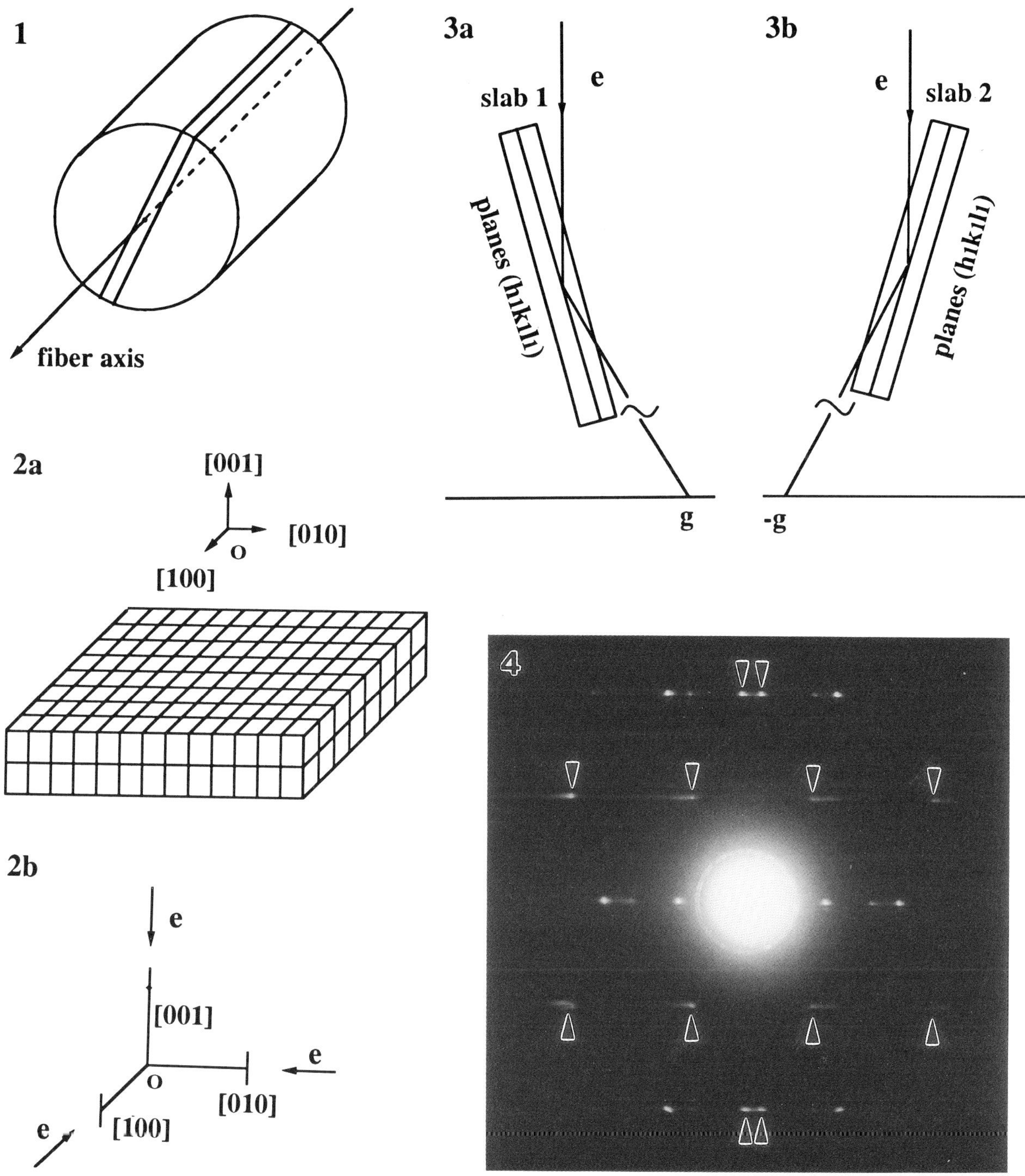

Fig.1 Schematic showing "slab" structural element relative to a full fiber.

Fig.2 Real space structure (a) and reciprocal lattice structure (b) of a component slab. The shapes of diffracted intensities are determined by intersections between these relrods and Ewald's Sphere.

Fig.3 Two opposing slabs oriented for exact Bragg scattering into +g and -g directions.

Fig.4 A SAED pattern of a non-tilted asbestos fiber showing the splittings and streakings, as indicated by arrows.

ENERGY FILTERED DIFFRACTION CONTRAST DEFECT IMAGING USING A FIELD EMISSION TEM

J.M.K. Wiezorek, S. Swaminathan and H.L. Fraser

Department of Materials Science and Engineering, The Ohio State University, Columbus OH 43210

The dynamical theory of electron diffraction describes most of the effects observed in diffraction contrast defect images very well. However, quantitative discrepancies between appropriately simulated and experimental defect image contrast have been reported and have generally been attributed to the inadequate theoretical treatment of absorption.[1] Thus, it is expected that the comparison of energy filtered defect images with many-beam dynamical simulations will facilitate a quantitative study of the contributions from inelastically scattered electrons to the image. Presently, two commercially available instruments allow the acquisition of energy-filtered imaging data in TEM: (1) the Zeiss EM912 Omega equipped with an in-column filter and (2) a post-column system, the Gatan Imaging PEELS (GIF) which, in principle, can be added to any TEM.[2,3] At the Ohio State University (OSU) the GIF has been attached to a Philips CM200 FEG TEM and this paper will report on some preliminary experimental results of diffraction contrast imaging of stacking fault (SF) pyramids in <001>-Si, which have been chosen because they have been well characterized in the past.[4]

Unfiltered, conventionally collected (using the TEM camera) bright and dark field (BF and DF) images have been used to characterize the SF pyramids. The fringe contrast behaviour in the BF-DF pair shown in fig. 1 clearly indicates that the base of the SF pyramid lies at the bottom of the foil and that the faults (A, B) are extrinsic and (C, D) are intrinsic.[5] The GIF imaging was performed with the FEG TEM operating at 198.1 kV (determined from CBED) and an energy resolution of 1.6 eV using a 10 eV slid and acquisition times between 1 to 18 s. The GIF magnifies any entrance object by a constant factor which for the equipment at OSU has been determined as 19.94. Furthermore, the maximum size of the entrance object for GIF imaging is limited to 3 mm in diameter. Thus, in order to collect the image of an entire SF pyramid (about 2.5 μm^2) with the slow scan camera (SSC) a FEG TEM magnification setting of 2.3kx was chosen and a further magnification reduction by a factor of about four was introduced using the GIF software to control the intermediate and projector lenses of the microscope. The contrast improvement in defect images due to the removal of the majority of inelastically scattered electrons (zero-loss imaging) has been reported previously for a TEM equipped with a LaB_6 gun[6] and is demonstrated for the FEG TEM in fig. 2 for weak beam (WB) images. For both, BF and WB, quantitatively similar contrast improvements were obtained in the zero-loss (ZL) image relative to the unfiltered image. WB images which have been acquired unfiltered and energy filtered for various parts of the electron energy loss spectrum (EELS) are compared in fig. 3. Reasonably strong fault contrast is evident in the images formed with the first (-21 eV) second (-42 eV) and third (-63 eV) plasmon in descending order of magnitude, whereas the image formed with the diffuse background at -8 eV energy loss shows no significant fault contrast. The graph in fig. 4 depicts more quantitatively the variation of the ratio of average intensity in the SF region to the perfect crystal region, (I_{SF}/I_{back}), for the WB images shown in fig. 4. These preliminary experimental results indicate that electrons which have suffered plasmon losses carry diffraction information whereas the electrons of the diffuse background do not to any significant extent. Hence, the dramatic contrast improvement in ZL images can be attributed mainly to the removal of plasmon losses, whereas the contributions of the thermally diffuse background to image contrast is more uniform and of relatively small intensity.

The comparison of high quality defect imaging data acquired experimentally with the GIF equipped FEG TEM with dynamical many-beam simulations, which is planned as the next step in this project, facilitates the quantitative study of the diffraction contrast image formation and promises to allow a more convincing separation of the effects of beam convergence, thickness variations and absorption on defect images than previously possible.[7]

References
1. E.J. Bithell, P.E. Donavan and W.M. Stobbs, *Phil.Mag. A*59(1987)63.
2. J. Bihr et al., *Proc. 49th Ann.Meeting EMSA* , San Francisco: San Francisco Press (1991).

3. A.J. Gubbens and O.L. Krivanek, *Ultramicroscopy* 51(1993)146.
4. G.R. Booker and W.J. Tunstall, *Phil.Mag.* 13(1966)71.
5. P. Hirsch et al, Electron Microscopy of Thin Crystals, Malabar: Krieger (1977).
6. J. Mayer, *Proc. Electron Microscopy Vol 1, EUREM 92*, Granada: Spain (1992) 269.
7. Provision of the <001>-Si foil by I.P. Jones and financial supprot by NSF are acknowledged.

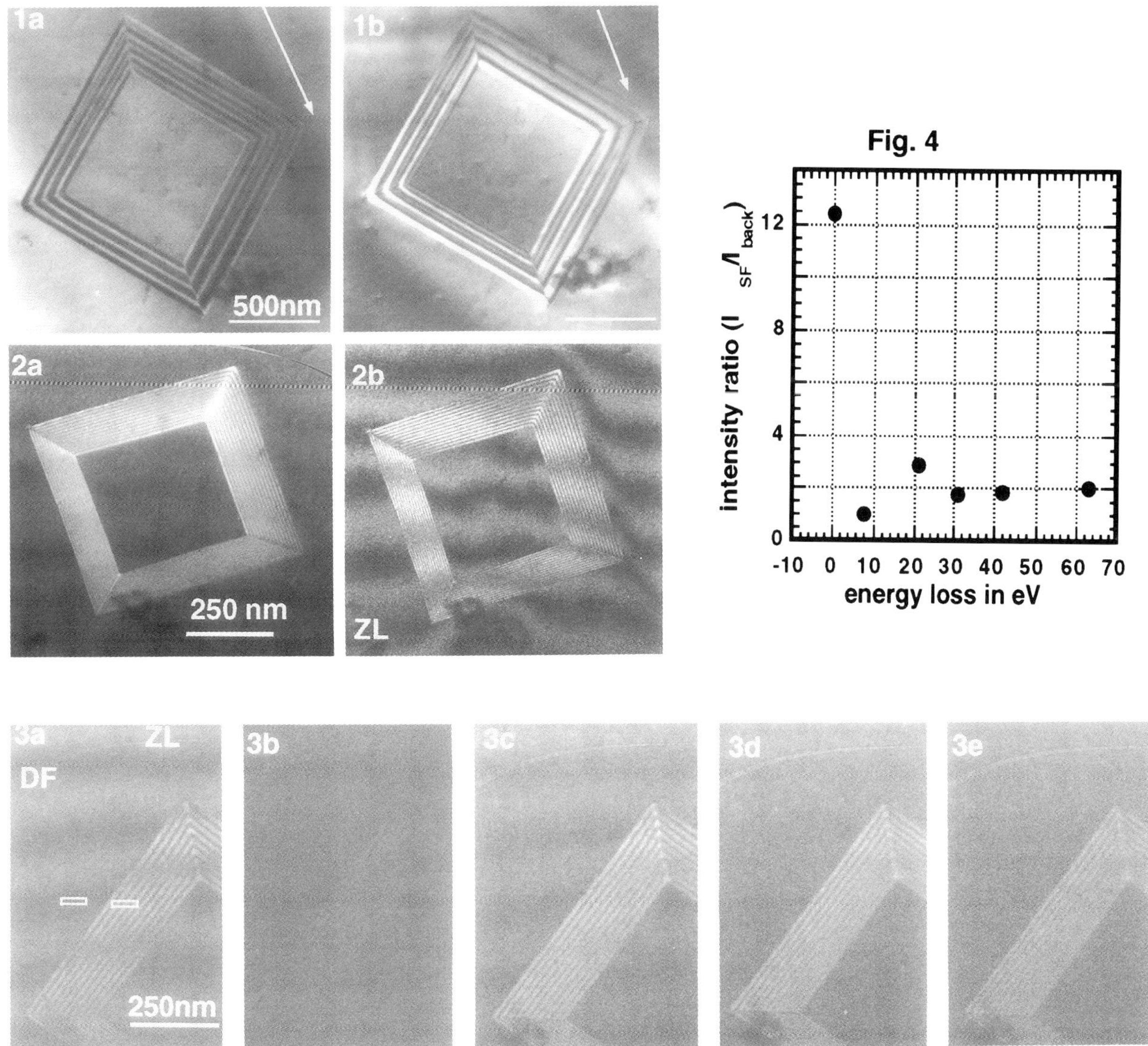

FIG. 1.-(a) BF and (b) DF TEM images; the arrow indicates the direction $\mathbf{g}{=}040$, near [001].

FIG. 2.- WB images of SF pyramid as fig. 1, (-$\mathbf{g}$,$\mathbf{g}$), $\mathbf{g}{=}040$ at [001] (a) unfiltered and (b) zero-loss (ZL); note the 90° rotation of images due to flipped over foil.

FIG. 3.- WB images as in fig. 2, (-$\mathbf{g}$,$\mathbf{g}$), $\mathbf{g}{=}0\bar{4}0$ at [001] for various slit positions (a) ZL, 0 eV, (b) diffuse Background, -8 eV, (c) first plasmon, -21 eV, second plasmon, -42 eV, (e) third plasmon, -63 eV; areas marked in (a) have been used to determine the averaged intensities compared for each energy in fig. 4.

FIG. 4.-Comparison of ratio of relative intensity in SF region to perfect crystal region as function of position of the energy selecting slid in WB images of fig. 3.

STRAIN ANALYSIS WITH NANOMETER RESOLUTION USING A CONVENTIONAL TRANSMISSION ELECTRON MICROSCOPY TECHNIQUE: ELECTRON DIFFRACTION CONTRAST IMAGING REVISITED

Koenraad G F Janssens*, Omer Van der Biest*, Jan Vanhellemont**, Herman E Maes** and Robert Hull***

*Leuven University (KUL), Department of Metallurgy and Materials Engineering (MTM), de Croylaan 2, B-3001 Leuven, Belgium.
**IMEC, Kapeldreef 75, B-3001 Leuven, Belgium.
***University of Virginia, Department of Materials Science, Charlottesville, VA 22903-2442.

There is a growing need for elastic strain characterization techniques with submicrometer resolution in several engineering technologies. In advanced material science and engineering the quantitative knowledge of elastic strain, e.g. at small particles or fibers in reinforced composite materials, can lead to a better understanding of the underlying physical mechanisms and thus to an optimization of material production processes. In advanced semiconductor processing and technology, the current size of micro-electronic devices requires an increasing effort in the analysis and characterization of localized strain. More than 30 years have passed since electron diffraction contrast imaging (EDCI) was used for the first time to analyse the local strain field in and around small coherent precipitates[1]. In later stages the same technique was used to identify straight dislocations by simulating the EDCI contrast resulting from the strain field of a dislocation and comparing it with experimental observations.[2] Since then the technique was developed further by a small number of researchers, most of whom programmed their own dedicated algorithms to solve the problem of EDCI image simulation for the particular problem they were studying at the time. Programming dedicated algorithms was necessary at that time because the computation power was still limited.

In view of the fact that computers have become much faster during the past decade, it was decided to develop an algorithm which can solve the EDCI image simulation problem for local strain fields of arbitrary geometry.[3] In the present contribution we will present the main results of 4 years of (Ph.D. thesis oriented) research efforts on this matter. Apart from describing the capabilities of the SIMCON software comprizing the image simulation algorithms, some examples of application will be given. In addition the critical factors limiting the use of EDCI for the *quantitative* characterization of local strain fields will be discussed. Limiting factors which are most critical are the spatial resolution of EDCI and the surface relaxation of the TEM specimen. Other essential parameters are Bragg deviation and specimen thickness which should be determined with great care. The effect of energy filtering of the images will also be discussed.

An overview of the EDCI procedure can be found in figure 1. On the left the experiment is depicted. Standard but state of the art TEM is used to record a *picture* of the strain field present in the thin film specimen. The information in the picture which relates to the strain in the specimen is the *contrast* in the EDCI image. The intensity of a particular pixel in the EDCI image depends on the orientation of the crystal lattice planes through which the electrons pass. If a localized strain field exists in the specimen the lattice plane orientation will locally change, and thus a contrast will be generated in the image. The right half of figure 1 depicts the simulation part of EDCI. Because of the dynamic aspects of high energy electron diffraction, the correlation between EDCI image contrast and the strain field in the specimen is not straightforward. The only way to overcome this is to simulate the EDCI process and to compare the experimentally taken image with the simulated image. Based on this comparison the strain field in the specimen can be characterized. The procedure of image simulation comprises two stages. In the first stage the strain field is modelled. For this purpose analytically formulated models as well as finite element formulated ones can be implemented. In the second stage the strain field model is used as input for EDCI simulation. Iteration of both stages is necessary to obtain a fitted image from which statements can be made regarding the characteristics of the strain field.

Proc. Microscopy and Microanalysis 1995, edited by G.W. Bailey, M.H. Ellisman, R.A. Hennigar, and N.J. Zaluzec.

An application example is shown in figure 2. On the left an experimental EDCI observation of a semiconductor laser structure is shown. The image in the middle is an image simulated assuming plane stress conditions, while the image on the right is a simulated image based on a full 3 dimensional finite element calculation of the strain in the TEM specimen. This particular example shows that is essential to include specimen surface relaxation when interpreting the EDCI images in a quantitative manner.

1. M F Ashby and L M Brown, Phil. Mag 8(1963)1083.
2. A K Head et al., Computed Electron Micrographs and Defect Identification, North Holland (1973).
3. K G F Janssens et al., Ultramicroscopy 45(1992)323.
4. Koenraad G F Janssens is indebted to the Flemish IWT Foundation for his Fellowship. AT&T is acknowledged for provision of the samples.

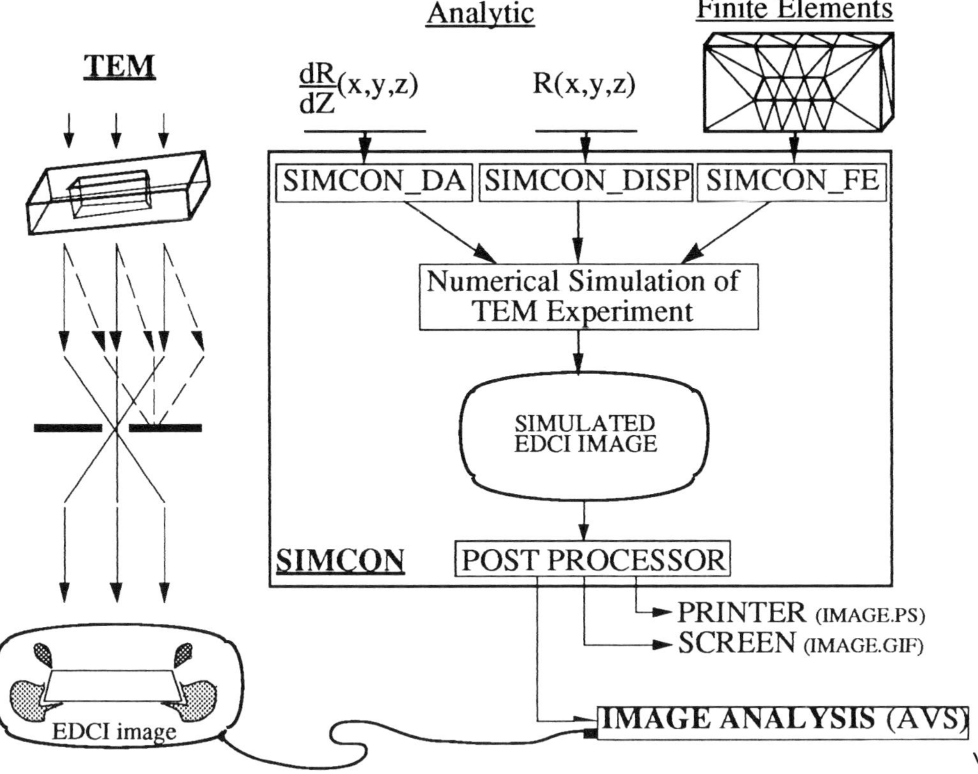

Figure 1: Electron Diffraction Contrast Imaging (EDCI): experiment (left) and interpretation (right).

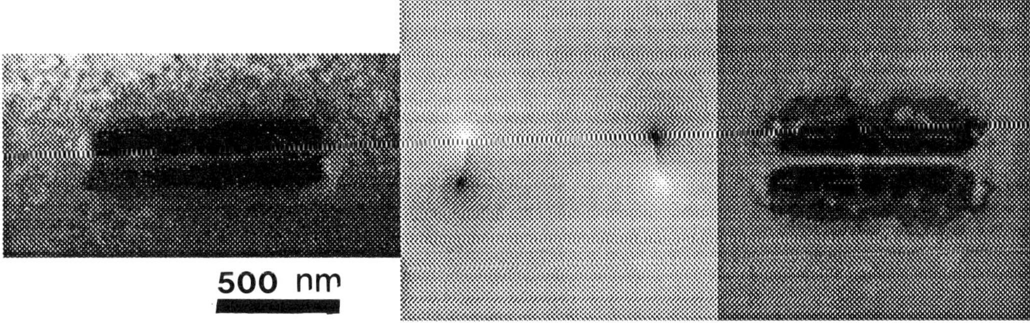

Figure 2: An application example of EDCI, see text for more details.

AUTOMATION IN ELECTRON DIFFRACTION ANALYSIS

J. M. Zuo
Physics Department, Arizona State University, Tempe, AZ 85287

Automated lattice parameter measurement and orientation analysis are often needed in the characterization of microstructures using electron diffraction, and is made possible with increasingly popular use of slow scan CCD camera and imaging plates. Both of these two detectors are largely linear and digital. Typical electron diffraction analysis has three steps: 1) diffraction pattern measurement, 2) diffraction pattern indexing and 3) solution. Full automation in all these three steps is desired, however, may be hard to achieve especially for complex crystal structures. The importance of automation in each of these steps depends on the type of analysis and the number of analysis needed. Full automation is necessary in the type of applications where the same analysis is repeated many times, such as in texture analysis. In table 1, various applications of electron diffraction and automation needed are listed.

Table 1. Type of electron diffraction analysis and automation needed

Type of application	Measurement	Indexing	Solution
Lattice parameter determination			√*
Strain measurement	√		√
Phase identification		√	√
Beam direction measurement			
Texture analysis	√	√	√

* √ automation needed

There are two types of approach to the automatic indexing. The commonly used method is to matching a calculated list of g's with the measured ones. This may give multiple number of choices, which has to be narrowed in the final solution stage. A potentially less ambiguous approach is to use Niggli cell by measuring the primitive cell of the zone axis[1]. In practice, both method are limited by the accuracy of reciprocal lattice vector g measurements, especially in high order zone axes and/or complex crystal structures. Problem of phase identification can be viewed as a extension of automatic indexing problem. Both approaches and problems described above apply.

Lattice parameter and beam direction measurements are the same type of problem as shown below. The difference is that the relative position of zero spot or center of zero disk rather than lattice parameter need to be measured additionally in beam direction determination. The problem is summarized in figure 1, where the HOLZ or Kikuchi lines are recorded on a film coordinates (x,y). In this film coordinate, the center of zone axis is at (x_0,y_0), (x',y') is the zone axis coordinated chosen for the convenience of calculation and a is the angle between x' and x. Relation between the theoretical and film coordinate is given by

$$\begin{pmatrix} x' \\ y' \end{pmatrix} = \frac{1}{c}\begin{pmatrix} \cos\alpha & -\sin\alpha \\ \sin\alpha & \cos\alpha \end{pmatrix}\begin{pmatrix} x \\ y \end{pmatrix} - \frac{1}{c}\begin{pmatrix} x_o \\ y_o \end{pmatrix}$$

c=λL is the camera constant. In theory, a HOLZ or Kikuchi line is given by [2]

$$x\cos(\theta'-\alpha) + y\sin(\theta'-\alpha) = cp' + x_o\cos\theta' + y_o\cos\theta'$$

Here $\theta' = \cos^{-1} g_{x'} / \sqrt{g_{x'}^2 + g_{y'}^2}, p' = (g_z K_z - g^2/2)/\sqrt{g_{x'}^2 + g_{y'}^2}$

Experimentally recorded lines are measured as

$$x\cos\theta^x + y\sin\theta^x = p^x$$

with uncertainties σ_θ and σ_p. The rotation angle can be found by

$$\alpha = \sum_i (\theta_i^x - \theta_i^t)/\sigma_{\theta_i^x}^2 \Big/ \sum_i 1/\sigma_{\theta_i^x}^2$$

Proc. Microscopy and Microanalysis 1995, edited by G.W. Bailey, M.H. Ellisman, R.A. Hennigar, and N.J. Zaluzec

while the remaining parameters can found by minimizing the following chi-square

$$\chi^2 = \sum_i \frac{1}{\sigma_{p_i}^2} \left(p_i^x - c p_i^t - x_o \cos\theta^t - y_o \cos\theta^t \right)^2$$

If only c and (x_o,y_o) are parameters, the minimum χ^2 can be found by solving a linear equation of order of 3, otherwise the minimum can be found using optimization methods as in case of lattice parameter refinement . The above definition of χ^2 is better than the previous one [1] in its clearer meaning. Once the camera constant and origin of zone axis center, the beam direction can be found by measuring the position of the direct spot.

Automatic measurements of diffraction pattern include locating spots or disks and lines in the presence of noise. Sensitivity and accuracy in locating these features in a digitally recorded diffraction patterns are the important factors in evaluating different schemes. Diffraction spots often has very strong signal and can be easily distinguished from the background. Features of lines and disks can be isolated via preprocessing of differentiation or other methods. Existence of lines or disks can be detected by parameter space projection such as Hough transformation. Figure 2, 3 and 4 show simulated line with Possion noise, 2nd order differentiation with the sobel operator and Hough transformation. The strong peak in the hough transformation indicates the existence of a line.

In summary, various stage of diffraction analysis can be automated. The degree of success depends on the accuracy of reciprocal lattice spacing and angles and the complexity of crystals.

References

[1] J. M. Zuo, Ultramicroscopy, **52**, 459, 1993
[2] J. C. H. Spence and J. M. Zuo, Electron Microdiffraction, Plenum, 1992
[3] I. Krivy and B. Gruber, Acta Cryst., A **32** (1976), 297-298
[4] J. M. Zuo, Ultramicroscopy, 41 (1992), 211-223
*Supported by NSF grant DMR 9015867

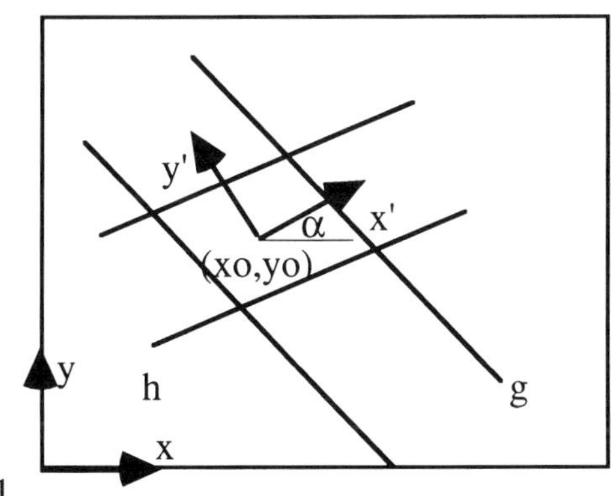

Fig. 1 Schematic relationship between theoretical and experimental coordinates of diffraction pattern.
Fig. 2 A simulated deficiency line of Gaussian profile with 2 pixel width and 50 counts height in a white background of 1000 counts. Poison noise is added to simulate experimental condition.
Fig. 3 Laplacian of figure 2 after thresholding of 80.
Fig. 4 Hough transformation of fig. 3. The range of x coordinates is from -90° to 90°

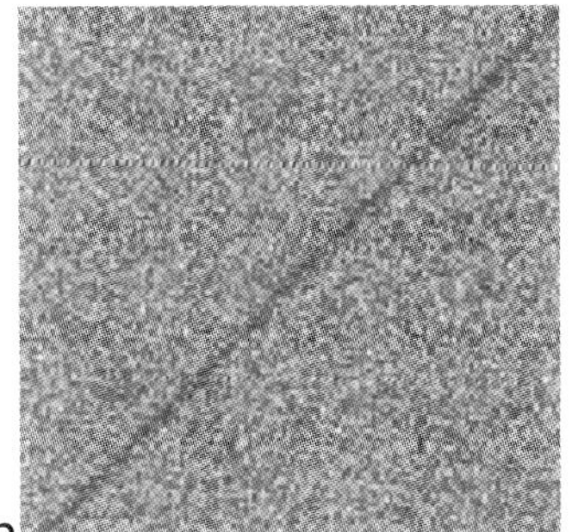

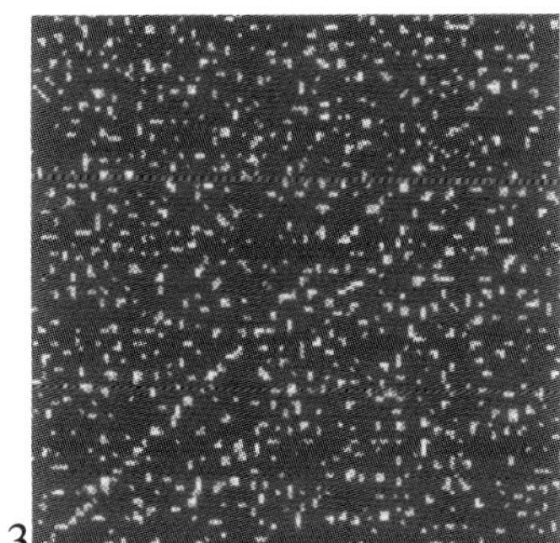

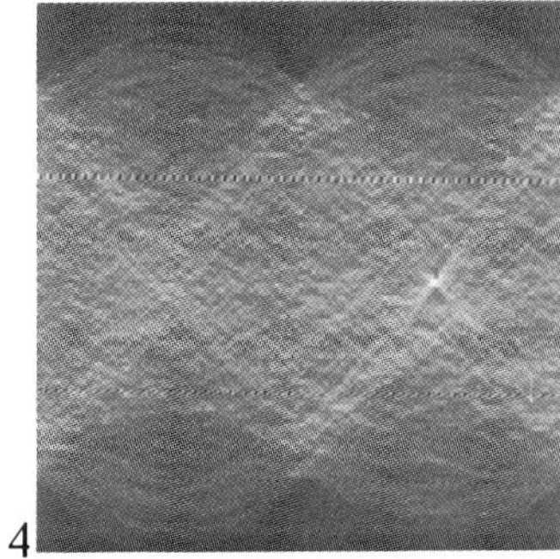

HIGH RESOLUTION ELECTRON MICROSCOPY OF NANOSTRUCTURED MATERIALS.

M. José-Yacamán

Instituto de Física, Universidad Nacional Autónoma de México, Apdo. Postal 20-364, Deleg. Alvaro Obregón, 01000 México, D.F., México

Electron microscopy is a fundamental tool in materials characterization. In the case of nanostructured materials we are looking for features with a size in the nanometer range. Therefore often the conventional TEM techniques are not enough for characterization of nanophases. High Resolution Electron Microscopy (HREM), is a key technique in order to characterize those materials with a resolution of ~ 1.7Å. High resolution studies of metallic nanostructured materials has been also reported in the literature. It is concluded that boundaries in nanophase materials are similar in structure to the regular grain boundaries. That work therefore did not confirm the early hipothesis on the field that grain boundaries in nanostructured materials have a special behavior. We will show in this paper that by a combination of HREM image processing, and image calculations, it is possible to prove that small particles and coalesced grains have a significant surface roughness, as well as large internal strain.

HREM can also be invaluable to characterize coalesced particles. Figure 1 shows an example of a gold coalesced particle and its corresponding fast Fourier transform (FFT). By indexing the FFT we can obtain direct information about the orientation of the sample. Figure 2a) shows the particle at lower magnification and Figure 2b), and c) at higher magnification (two different portions). Several interesting aspects of this image can be pointed out:
- There are large orientation changes between areas across the particle
- In the neck region Figure 2c) a region out of contrast (most likely in a orientation off the zone axis) is seen. This region seems to be the "glue" between crystalline regions
- Large areas of the particle are formed by twins with a five-fold symmetry. The five fold structure is retained after the coalescence process.

Figure 1a) shows another consolidated a gold particle imaged at high resolution and its corresponding image processed high resolution image Figure 1b). There is surprisingly large bending of planes of atoms in some regions such as the one marked with an arrow. This particle is a fine example of a nanocrystalline structure in which the grains are oriented in many ways rotating about an arbitrary axis.
In the described experimental conditions the substrate plays a fundamental role on the sintering process. The fact that large strained regions in the "fan-like shape" can exist in a metastable form indicates that in nanocrystalline materials large strains can be produced. While the unstrained misoriented grains have sizes of ~ 20Å typically, the fan like structures sizes up to 80-100 nm in length. This kind of effect will have to be considered in predicting the properties of nanophase materials.

An interesting feature is that under experimental conditions in which large grains are allow to grow before they coalesce, produces a structure in which individual grains mantain their structure.

Proc. Microscopy and Microanalysis 1995, edited by G.W. Bailey, M.H. Ellisman, R.A. Hennigar, and N.J. Zaluzec
Copyright © 1995 MSA. Published by Jones and Begell Publishing, 79 Madison Ave., New York, NY 10016

This is in sharp contrast with the case of nanoclusters in the size range of ~20Å, where the coalescence leads to a recrystallization in which a single crystalline grain usually results. It should be noted that some of the observed features are consistent with the theoretical simulations where that use the embedded atom method and molecular dynamics to simulate the consolidation process. This kind of calculations appear to be very useful and in the future it might have an impact on the understanding of experimental data. It has been shown that HREM could be used to determine the roughness of small particles at the atomic level. Our results indicate that in the case of noble metals the surface structure of the particle is very rough and contains many kinks. In the case of coalesced particles it has been shown that large strained areas can exist in which lines of atoms have the shape of a fan. On the other hand the individual grains of the coalesced particles seem to retain their crystal structure and orientation.

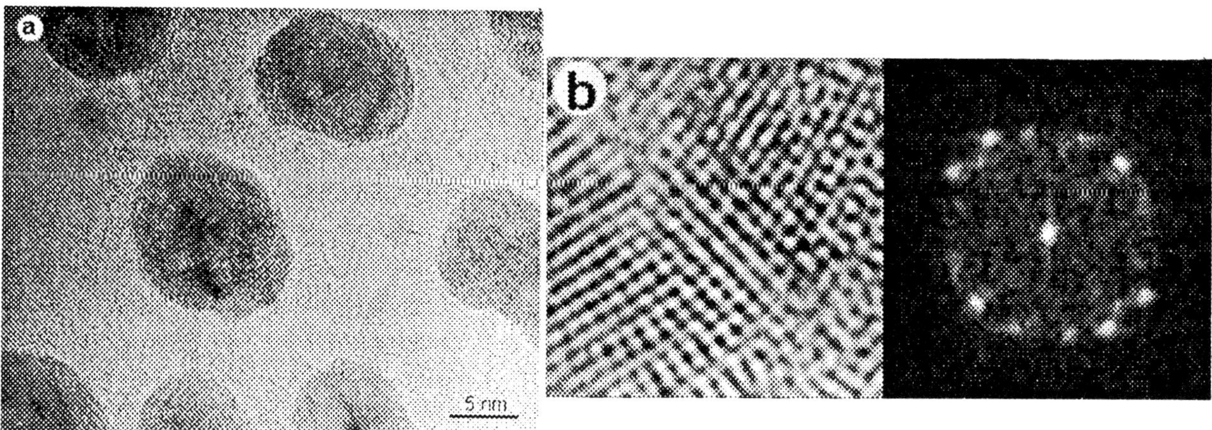

FIG. 1. HREM image of a gold particle: a) low magnification and b) high magnification showing the fan-like distortion and the polycrystalline nature of the particle and the corresponding Fourier transform of that portion of the image.

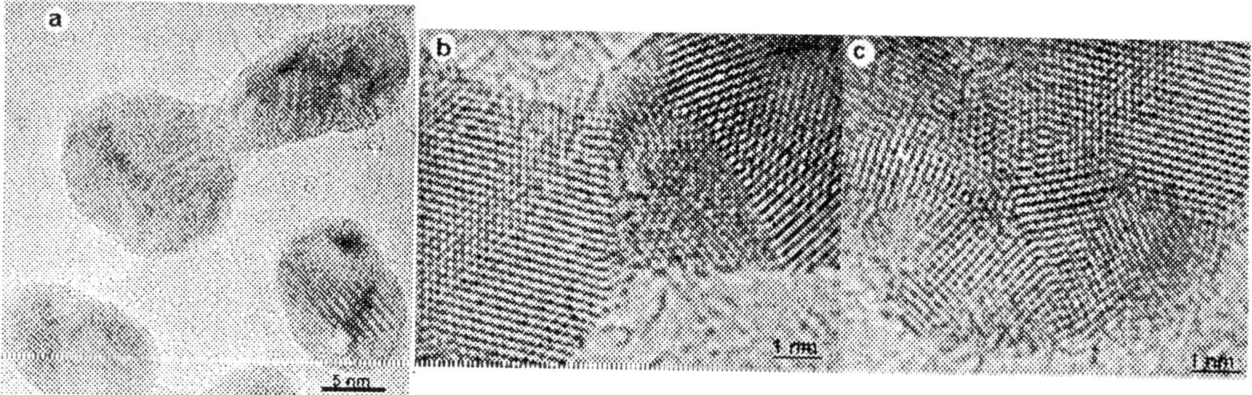

FIG. 2. HREM of a coalesced particle of gold. a) the particle is shown at smaller magnification. b) a portion of the particle at higher magnification, c) another position of the particle at higher magnification.

IN Z-CONTRAST, NOT ALL THAT GLISTENS IS HIGH-Z

M. M. J. Treacy*, M. E. Bisher*, A. J. Jacobson** and J. M. Gibson***

*NEC Research Institute, Inc., 4 Independence Way, Princeton, NJ 08540-6685
**University of Houston, Department of Chemistry, Houston, TX 77204-5641
***University of Illinois, Dept. Physics & Matls. Science, 104 S. Goodwin Ave., Urbana-Champaign, IL 61801

It is by now a given that high resolution Z-contrast is a useful imaging technique for detecting high atomic number (Z) domains in nanophase materials.[1-3] However, image intensity also depends on specimen orientation, electron-optical geometry, thickness and local strain fields, as well as local structural ordering.[4] In this paper we will present examples from our materials studies where contrasts from strain and short-range order in hollow-cone dark-field images provide important structural insights.

Strain fields were found to be responsible for the unusual "blister' contrasts observed in hollow-cone dark field images of delaminated sheets of calcium sodium niobate layered materials.[5] Wherever two sheets overlap, the average intensity increases and dark disks of diameter ~20 nm, with bright central spots, appear (Fig. 1). For large hollow-cone angles, $q_1 > 20$ nm^{-1}, the blisters vanish, although faint fine-scale features persist. Elasticity analysis and image simulations (Fig. 2) reveal that the blisters are pockets of gas or liquid that became trapped between sheets during the drying stage of the specimen preparation process, when individual sheets come into contact at random mutual angles and reseal (Fig. 3). Simulations indicate that a typical blister contains about 2000 molecules of gas or liquid, and is pressurized to about 10^3 atmospheres. The work of adhesion for two calcium niobate layers is found to be ≈ 45 mJ m^{-2}. Radiation damage destroys blisters, presumably by allowing the trapped molecules to leak to the microscope vacuum. High angle hollow-cone "Z-contrast" images do not reveal any high-Z particles inside the blisters.

Short-range order in amorphous materials is known to produce speckle contrast in small angle hollow-cone images (see Fig 4). Speckle arises because of chance alignments (and misalignments) of atom pairs within the scattering coherence volume.[6] The coherence volume generally shrinks as the hollow-cone angle q_1 increases, thereby providing a method of selectively tuning the contribution from pair-pair correlations.[7] At resolutions larger than the typical inter-atomic spacings, perfectly ordered single crystals and perfectly disordered random structures will yield images with little or no speckle due to fluctuations in local ordering – images will be featureless if there are no mass-thickness variations. At high resolution, a perfect crystal will exhibit a periodic contrast, but no speckle. A randomly disordered specimen will exhibit statistical mass thickness fluctuations. However, specimens that are intermediate between crystalline perfection and random disorder will exhibit a speckle whose statistics will depend on the details of the short-range ordering. Speckle analysis of amorphous germanium specimens as a function of coherence volume reveals that the speckle "signature" of an annealed germanium foil (annealed below the crystallization temperature) is distinct from that for a freshly-evaporated amorphous film, even though electron diffraction patterns indicate that both specimens have no long range ordering. Our analysis reveals that the speckle contrast is sensitive to medium range ordering (> 1 nm) and has additional sensitivity over diffraction. Preliminary experimental results will be presented.

References

(1) M. M. J. Treacy, A. Howie and C. J. Wilson, *Philos. Mag.* 38 (1979) 569.

Proc. Microscopy and Microanalysis 1995, edited by G.W. Bailey, M.H. Ellisman, R.A. Hennigar, and N.J. Zaluzec

(2) M. M. J. Treacy and S. B. Rice, *J. Microsc.* 156 (1989) 211.
(3) S. J. Pennycook and L.A. Boatner, *Nature* 336 (1988) 565.
(4) M. M. J. Treacy, J. M. Gibson, K. T. Short and S. B. Rice, *Ultramicroscopy* 26 (1988) 133.
(5) M. M. J. Treacy, M. E. Bisher and A. J. Jacobson, *Philos. Mag.* (1995) accepted for publication.
(6) J. M. Gibson and A. Howie, *Chem. Scripta* 14 (1978) 109.
(7) M. M. J. Treacy and J. M. Gibson, *Ultramicroscopy* 52 (1993) 31.

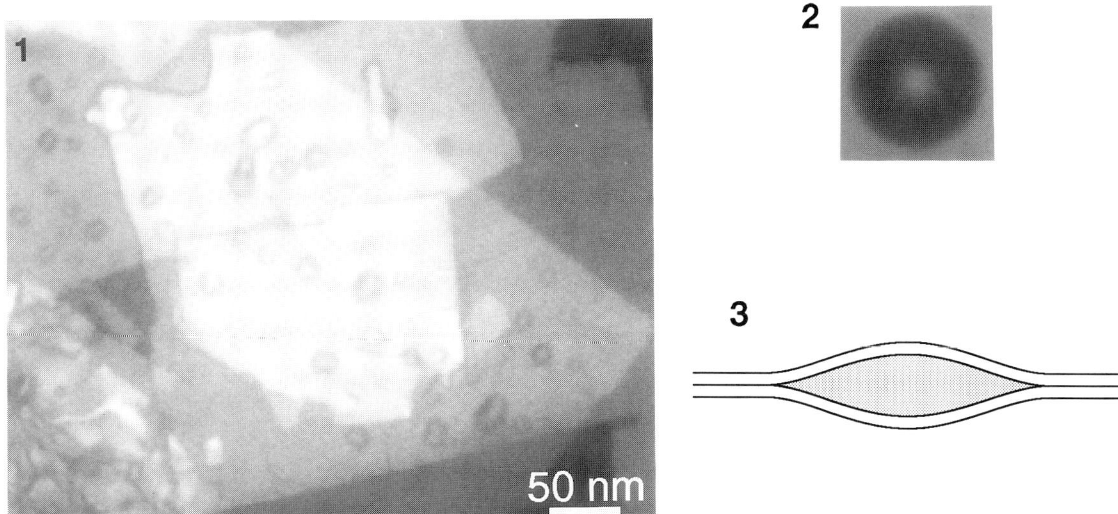

FIG. 1.—Hollow-cone dark image of ~1.4 nm thick delaminated calcium niobate layers that are supported on amorphous carbon. Where two or more layers overlap, a "blister" type of contrast appears.
FIG. 2.—Image simulation of a blister 20 nm in diameter. Note the darkened disk and the central white spot.
FIG. 3.—Cross-sectional view of a blister as computed by elasticity theory. The shaded region represents gas or liquid that has become trapped between calcium niobate layers after layers have come into contact and re-sealed.

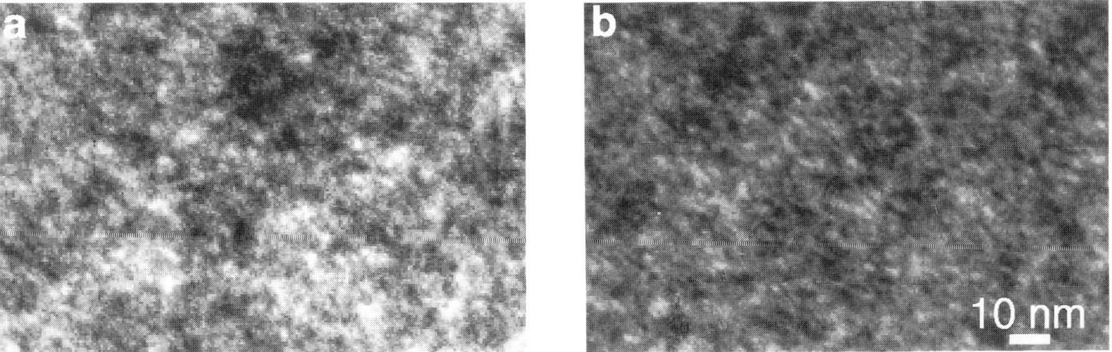

FIG. 4.—Hollow-cone dark-field images of amorphous germanium showing speckle. a) Hollow-cone inner illumination vector $q_1 \approx 3nm^{-1}$ b) $q_1 \approx 7nm^{-1}$. In each case the outer illumination vector is $q_2 \approx q_1 + 1 nm^{-1}$.

175

ENERGY-FILTERED HIGH-RESOLUTION ELECTRON MICROSCOPY OF NANOSTRUCTURED MATERIALS

Z. L. Wang*

Metallurgy Division, National Institute of Standards and Technology, Gaithersburg, MD 20899.
* Currently at: School of Materials Science and Engineering, Georgia Institute of Technology, Atlanta GA 30332-0245.

The interaction between an incident electron and the atoms in condensed matter results in various inelastic scattering processes [1]. Thermal diffuse scattering or phonon scattering is the result of atomic vibrations in crystals. This process does not introduce any significant energy-loss (< 0.1 eV) but produces large momentum transfer. Valence-loss (or plasmon for metals and semiconductors) excitation, which characterizes the transitions of electrons from the valence band to the conduction band, involves an energy-loss in the range of 1-50 eV. Atomic inner-shell ionization is excited by the energy transfer of the incident electron, resulting in an ejected electron from the deep-core states. Continuous energy-loss spectra can also be generated by an electron which penetrates into the specimen and undergoes collisions with the atoms in it, resulting in Bremsstrahlung and leading to emission of x-rays with continuous energy. The electron Compton scattering refers to the collision of the incident electron with an electron belonging to the specimen. In an electron energy-loss spectrum (EELS), the zero-loss peak is composed of elastically and thermal diffusely scattered electrons. The low-loss region is dominated by valence-excitations. The characteristic features observed in the high energy-loss region are usually due to atomic core-shell excitations. The background observed in EELS is produced by scattering processes of multiple valence excitations, Bremsstrahlung and electron Compton scattering. With an energy-filter, it is possible to form images (or diffraction patterns) using electrons with specific energy-losses [2, 3]. Chemical images can be formed using ionization edge energy-filtered electrons, and the image contrast is proportional to the projected number of atoms characterized by the selected ionization edges. Thus, each energy-selected core-loss electron image provides a thickness integrated chemical image of the specimen.

The energy-filtered HREM experiments were performed at 300 kV using the Gatan image filtering system equipped on an JEOL 3010 HREM [4]. Figure 1a is a 0-loss HREM image of an Al/Ti interface, and Figures 1b-d are the chemical images obtained using the electrons corresponding to different ionization edges of Al and Ti, respectively. A dislocation line, which is located at the twin boundary of two Al grains, is seen in Figure 1a. The Ti layer is at the right-hand side. The Al and Ti layers have a twin relationship. The interface as seen in the 0-loss image (Figure 1a) is not sharp. Diffraction contrast due to interface mismatch and stress makes the interface appears broad. Figure 1b is an energy-selected image of the interface using the Al-$L_{2,3}$ edge electrons with energy-losses from 73 to 83 eV, where the Al layer is clearly resolved. The change of contrast in the image reflects the variation of the projected (or thickness integrated) local density of Al. The image formed by the Ti-M edge, located at energy-loss about 40 eV, clearly reveals the existence of the Ti layer (Figure 1c). The image (Figure 1d) recorded using the Ti-$L_{2,3}$ edge, located at energy-loss of 455 eV, also shows the distribution of Ti. The Ti distribution given by the two images is consistent, in that the high Ti concentration seen at the upper-right hand corner of Figure 1d is also shown in the same region in Figure 1c.

The spatial resolution in the energy-filtered chemical image is determined by: the signal-to-noise ratio, the chromatic aberration effect (or the width of the energy-window) and specimen drift. To see the optimum spatial resolution that can be achieved with the energy-filter, the image acquisition time was reduced to 2s. Figure 2 shows an intensity line scan across the Al/Ti interface in the Al-$L_{2,3}$ edge energy-filtered image. The intensity variation from pixel to pixel is due to the statistical noise, but an average intensity drop between the Al and Ti sides is apparent. The Al concentration shows a sharp jump at the interface and a spatial resolution of about 0.4 nm is obtained.

Figure 3 shows a group of images of the 0-loss (Figure 3a), Al-plasmon (Figure 3b) and Ti-plasmon (Figure 3c) energy-selected electrons. It is apparent that the Al layer and Ti layer show bright contrast in Figures 3b and 3c, respectively, which represent high local atomic concentrations. But the spatial resolution of this type of images is limited by the delocalized scattering effect [5, 6]

1. Z.L. Wang, Elastic and Inelastic Scattering in Electron Diffraction and Imaging, Plenum Press (New York, 1995), Chapter 6.
2. H. Shuman, C.F. Chang and A.P. Somlyo, Ultramicroscopy **19** (1986) 121.

Proc. Microscopy and Microanalysis 1995, edited by G.W. Bailey, M.H. Ellisman, R.A. Hennigar, and N.J. Zaluzec
Copyright © 1995 MSA. Published by Jones and Begell Publishing, 79 Madison Ave., New York, NY 10016

3. O.L. Krivanek, A.J. Gubbens and N. Dellby, Microsc. Microanal. Microstruct. **2** (1991) 315.
4. Reference to commercial equipment does not imply NIST endorsement.
5. Z.L. Wang and A.J. Shapiro, in these proceedings.
6. Thanks to Dr. D. van Heerden for kindly providing the Al/Ti multilayer specimen.

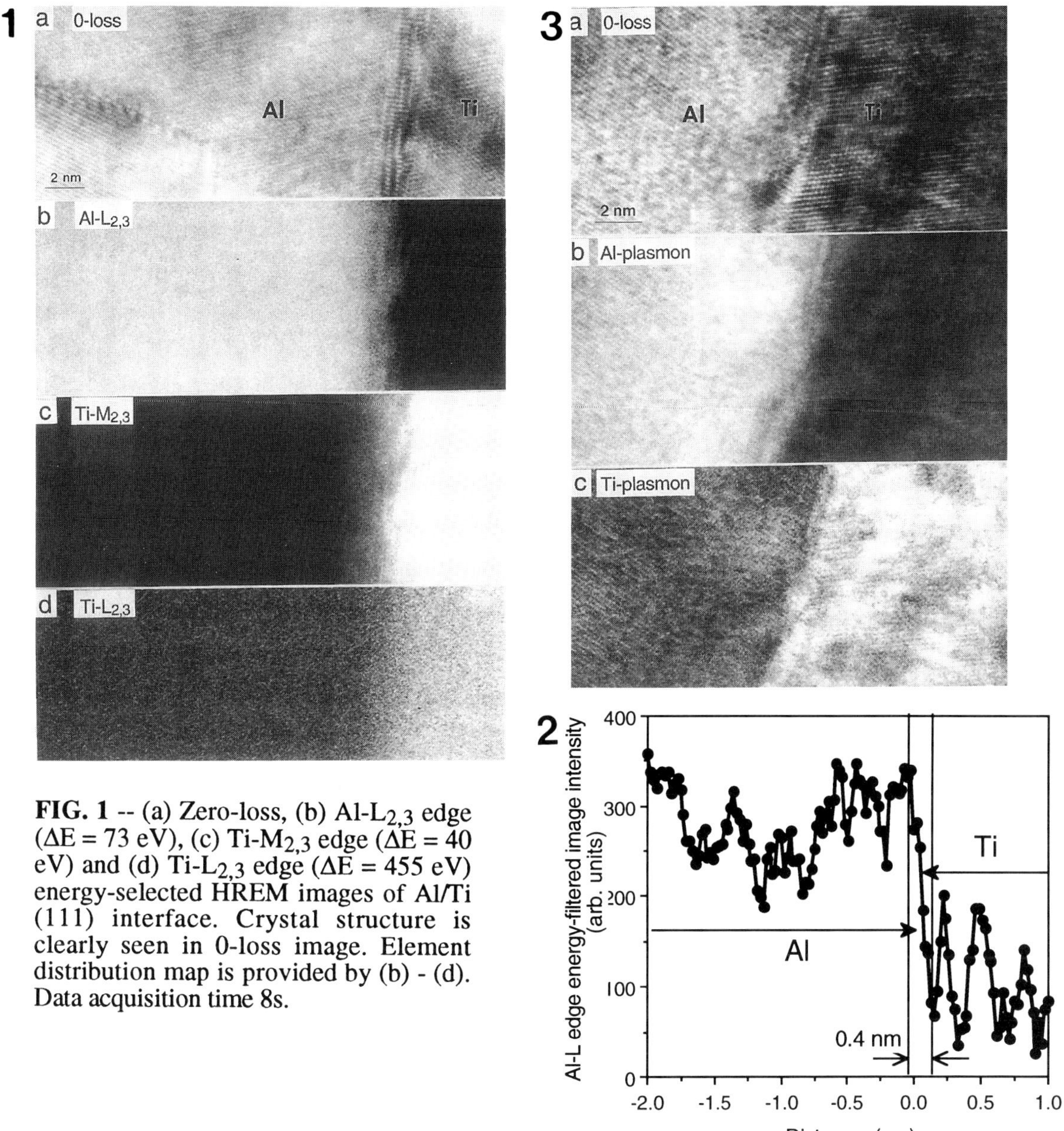

FIG. 1 -- (a) Zero-loss, (b) Al-L$_{2,3}$ edge (ΔE = 73 eV), (c) Ti-M$_{2,3}$ edge (ΔE = 40 eV) and (d) Ti-L$_{2,3}$ edge (ΔE = 455 eV) energy-selected HREM images of Al/Ti (111) interface. Crystal structure is clearly seen in 0-loss image. Element distribution map is provided by (b) - (d). Data acquisition time 8s.

FIG. 2 -- Intensity line scan across Al/Ti interface from image recorded using Al-L$_{2,3}$ edge energy-loss electrons, showing spatial resolution obtained in core-loss electron imaging. Data acquisition time 2s.

FIG. 3 -- (a) Zero-loss, (b) Al-plasmon (ΔE = 15 eV) and (c) Ti-plasmon (ΔE = 22 eV) energy-selected HREM images of Al/Ti (111) interface, showing compositional sensitivity. But image contrast is largely influenced by diffraction contrast effect. Data acquisition time 2s. Energy window 5 eV

MICROSTRUCTURAL CHARACTERIZATION OF GRANULAR Au-Co GMR MATERIAL BY LORENTZ MICROSCOPY

J. Bernardi[*], A. Hütten[**], C. Nelson[***], and G. Thomas

Department of Materials Science and Mineral Engineering, University of California and
Lawrence Berkeley Laboratory, Berkeley, CA 94720 USA,
[*] Permanent address: Inst. of Appl. and Techn. Physics, Techn. Univ. Vienna, A-1040 Vienna, Austria
[**] Present address: Institut für Festkörperforschung, D-01069 Dresden, Germany
[***]National Center for Electron Microscopy, Lawrence Berkeley Laboratory, Berkeley, CA 94720 USA

Giant magnetoresistance (GMR) has recently been achieved in multilayers [1] as well as in granular alloys prepared by sputtering or melt spinning [2]. In these granular alloys magnetic Co particles are usually embedded in a nonferromagnetic metallic matrix (Cu, Ag, Au) and have a statistical magnetic moment distribution. The GMR effect originates mainly from spin dependent scattering of conduction electrons within the ferromagnetic Co particles and at their interfaces and is controlled by the size distribution and the volume fraction of the Co particles as well as the roughness of the interfaces between the Co particles and the matrix. Au-Co was chosen as a model system to investigate systematically the microstructural influence on the transport properties. The large difference of the atomic scattering factors of Au and Co enables the distribution of ferromagnetic Co particles be distinguished by structure factor contrast. The eutectic decomposition in Au-Co (Liq. $\Leftrightarrow$ Au + Co at T_E = 995.6 °C and 28.4 at% Co) allows a multilayered structure in rapidly quenched Au-Co to be synthesized [3].

Melt-spun $Au_{71.6}Co_{28.4}$ was investigated by analytical TEM and its magnetic properties were characterized in a vibration sample magnetometer (VSM). Lorentz microscopy was performed at the JEOL atomic resolution microscope at the NCEM, Berkeley, operating at 800 kV in diffraction mode and using the intermediate lenses for magnification. The high accelerating voltage enables thicker regions of the ribbons to be investigated and to obtain information about the bulk magnetic properties. The magnetic contrast results from the magnetization of the Co particles as well as their magnetic strayfield. Fig. 1 shows the changes of magnetic contrast of Co particles in the as quenched sample at a large defocus of the first intermediate lens {$\Delta f = \pm 1.8$ mm}. Simulations suggest that the symmetrical contrast shown by some Co precipitates can be attributed to a curling like domain state. An asymmetric contrast of the Co particles in the defocused images results most likely from uniformly magnetized particles with the direction of magnetization deviated from the direction of the incident electrons.

Annealing the samples at 480 °C causes a drastic growth of Co precipitates (Fig. 2). The saturation magnetization and the amount of ferromagnetic particles increases due to a decreasing amount of superparamagnetic Co particles. The coercive field decreases as a function of the increasing number of large ferromagnetic particles which show multidomain or curling like behavior (Fig. 3). GMR decreases in the annealed samples due to a larger interparticle distance. Multidomain Co particles were observed by Lorentz microscopy with an average diameter D > 200 nm.

References
1. A. Barthelemy et al., *J. Appl. Phys.* 67 (1990) 5908.
2. A.E. Berkowitz et al., , *Phys. Rev. Lett.* 68 (1992) 3744.
3. J. Bernardi, A.Hütten, S.Friedrichs, C.E.Echer and G.Thomas, *phys.stat.sol.. (a)* <u>147</u>, (1995), 165.

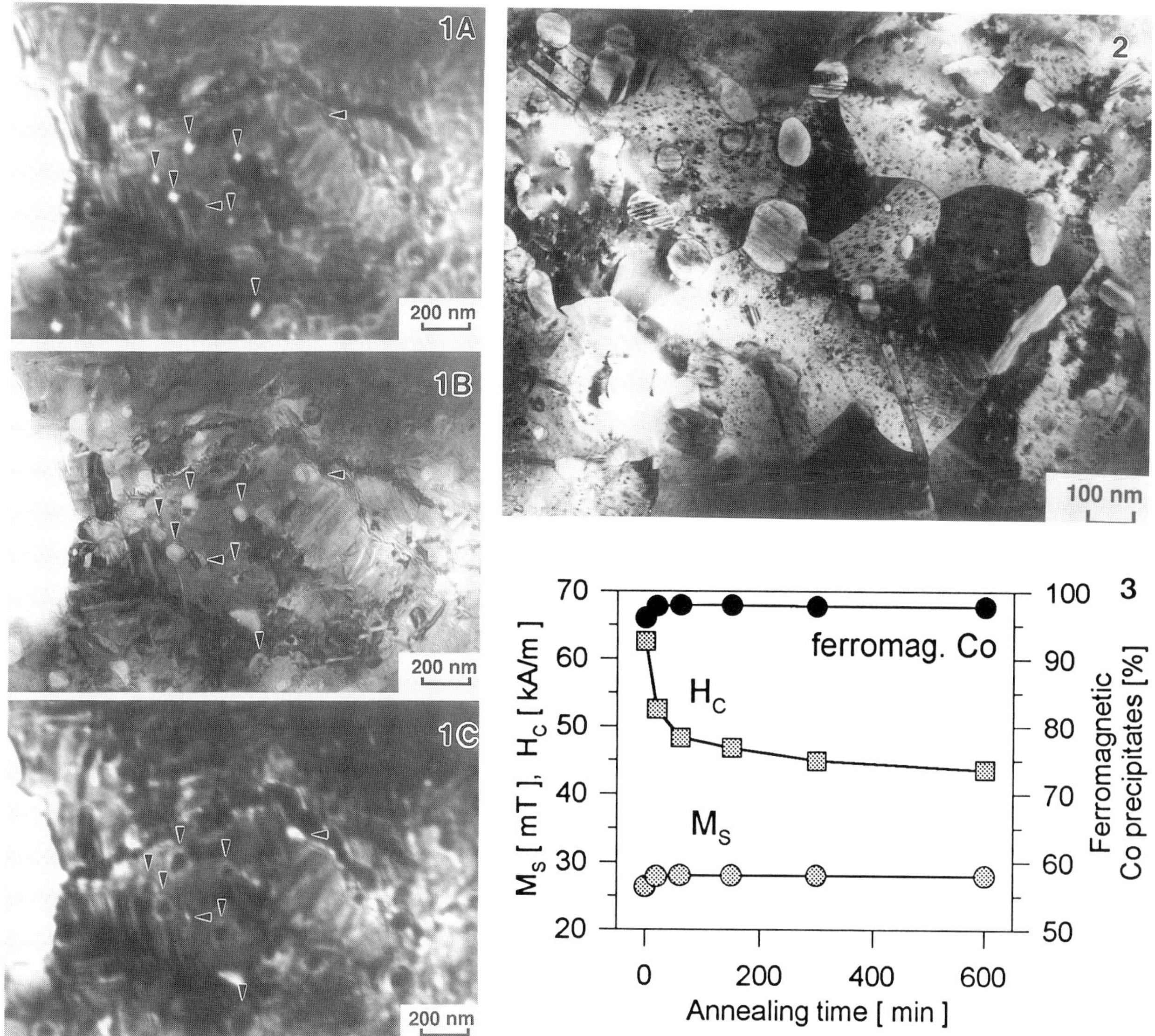

Fig 1.   Lorentz micrograph (Fresnel mode) of as quenched $Au_{71.6}Co_{28.4}$. a) overfocus $\Delta f = 1.8$ mm, b) in focus, c) underfocus $\Delta F = 1.8$ mm. Co particles with a strong magnetic contrast are marked by arrows. The image was taken in diffraction mode and magnified by adjusting the intermediate lenses.

Fig. 2. Microstructure of melt spun $Au_{71.6}Co_{28.4}$ annealed for 20 min.

Fig. 3. Change of saturation magnetization M_S, coercive field H_C and amount of ferromagnetic particles during annealing at 480 °C.

Acknowledgment

This work was supported by the Director, Office of Energy Research, Office of Basic Energy Sciences, Materials Sciences Division of the U.S. Department of Energy under Contract No. DE-AC03-76SF00098. J. Bernardi acknowledges the "Schrödinger Scholarship" from the Austrian National Science Foundation.

SELF-ASSEMBLY OF SURFACTANT-SILICATE NANOPHASES

S. A. Walker and J. A. Zasadzinski

Department of Chemical Engineering, University of California, Santa Barbara, CA 93106

Nanostructured zeolites are ubiquitous throughout chemistry as catalysts and sorption media.[1] A new variety of zeolites, constructed by templating inorganic silica on surfactant liquid crystalline phases, consist of ordered arrays of cylindrical pores with diameters ranging from 20 Å to greater than 500 Å.[2] The nanostructure of these materials can be tailored to a variety of applications, each of which makes use of the zeolite network's large surface area and prescribed microstructure. Recent studies indicate that the initial step in the nanophase formation is a cooperative self-assembly of inorganic silicate species and organic surfactant species, dictated by ionic interaction between silicate anions and cationic surfactants.[3] Freeze-fracture transmission electron microscopy (FF-TEM) is the only technique that can visualize this self-assembly process *in situ* to determine phase boundaries and examine the formation mechanism.

The nanophases form on mixing aqueous cetyl trimethylammonium bromide (CTAB) with silicate (SiO_2; prepared from CAB-O-SIL M-5, tetramethylammonium hydroxide (TMA-OH), trimethyl benzene (TMB), methanol (CH_3OH), and water).[3] The formation of the CTAB nanostructures is directed by the silicate polyanions, which assemble several CTAB cations into a neutral complex that resembles a multi-tailed surfactant. Such multi-tailed surfactants are expected to form higher order aggregates such as lamellar and hexagonal phases. The final CTAB nanostructure can be controlled by changing the initial CTAB concentration or SiO_2-CTAB mole ratio. In contrast, aqueous solutions of CTAB at the concentrations used (without silicate) form isotropic micellar phases.

Fig. 1A shows a hexagonal phase formed by preparing a 2.8:1 SiO_2- CTAB (mole ratio) solution. FF-TEM of this nanophase revealed an assembly of well-ordered rod-like structures (Fig. 1A) consistent with hexagonal packing of indefinitely long cylindrical aggregates. The 7.0±0.5 nm repeat spacing determined by Fourier Transform image analysis (right inset, Fig. 1A) is consistent with a staggered fracture along a minimum energy fracture path (path 1 in Fig. 1A, left inset), which corresponds to a 4.0±0.3 nm rod center-to-center spacing (path 2 in Fig. 1A, left inset), consistent with x-ray diffraction.[3] Unlike the lamellar phase (Fig. 2), no obvious layering is apparent. Fig. 1B shows regions in which hexagonally close-packed rods fractured end-on, roughly perpendicular to the fracture surface, as evidenced by the small bumps and divots. This image indicates that the hexagonal phase is far from perfectly ordered, showing only a few local regions of hexagonally close-packed rods (arrow). The extent of order in the hexagonal phase, which is only observable by FF-TEM, is crucial to growth of defect free zeolites.

Fig. 2 shows a lamellar phase formed by preparing a 5.6:1 SiO_2-CTAB (mole ratio) solution. At low resolution, FF-TEM of this nanophase revealed a typical lamellar phase, with layering seen in fractured cross sections. However, higher resolution images of this nanophase revealed that the lamellae were rippled. Fig. 2A shows these undulations whose directionality is preserved over many lamellar surfaces. The internal lamellar spacing is observed in Fig. 2B. The FT of the image (inset, Fig. 2B) shows two periodicities corresponding to (1) ~4.0 nm wavelength ripple undulations on the lamellar surfaces and (2) ~4.0 nm lamellar spacing as seen within the terraced discontinuities. The ripple wavelength, bilayer spacing, and hexagonal rod spacing, each ~4.0 nm, are likely set by the length of two CTAB molecules.

1. C.T. Kresge et al., *Nature* 359(1992)710; J.S. Beck et al., *J. Am. Chem. Soc.* 114 (1992)10834.
2. IUPAC Manual of Symbols and Terminology, Appendix 2, Part 1, *Colloid and Surface Chemistry*, *Pure Appl. Chem.* 31(1972)578.
3. A. Firouzi et al., *Science* 267(1995)1138.
4. The authors thank A. Firouzi and D. Kumar for sample preparation, and A. Firouzi, B.F. Chmelka, L.L. Madsen, and G.Stucky for useful discussions. The authors acknowledge support National Science Foundation grant CTS-9319447, Office of Naval Research grant N00014-90-J-1551, and National Institute of Health grant GM 47334.

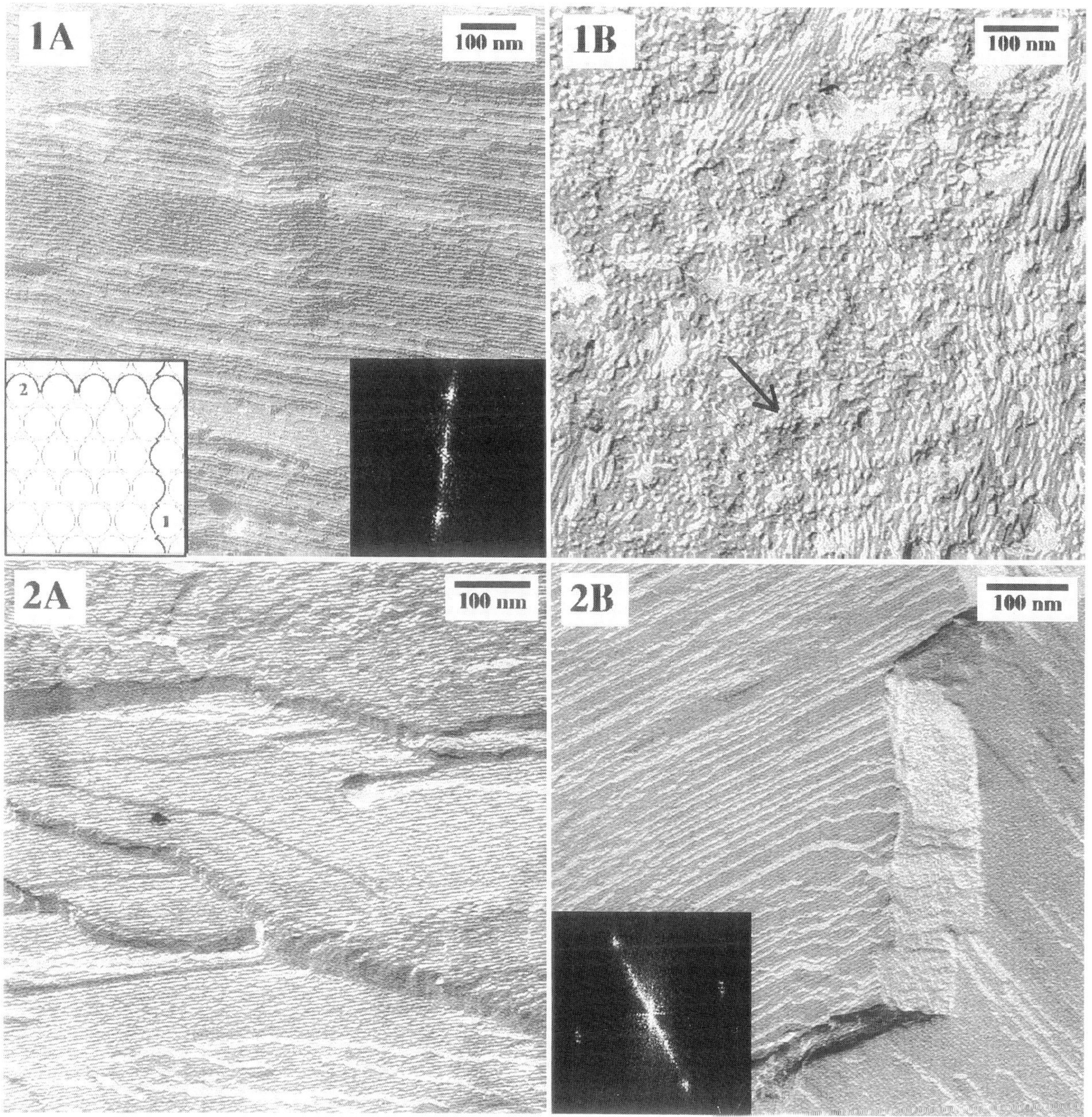

FIG. 1.--Freeze-fracture electron micrographs of surfactant-silicate hexagonal nanophase. (A) Region of well-aligned rod-like aggregates. Right inset: Fourier Transform (FT) of image. Left inset: Possible fracture paths. (B) A region in which the rods have fractured end-on. Molar composition: 1.4 SiO_2 - 196 Water - 0.5 $(TMA)_2O$ - 0.5 $(C_{16}TMA)_2O$ - 18.8 CH_3OH - 1.2 TMB. Final CTAB conc. is 7.5 wt%.
FIG. 2.--Freeze-fracture electron micrographs of surfactant-silicate lamellar nanophase. (A) A region clearly showing the novel ripple undulations existing on the lamellar surfaces. The ripple directionality is preserved over several lamellar surfaces. (B) A region showing two distinct periodicities, as seen in the FT. The lamellae are clearly seen in this region. Molar composition: 1.4 SiO_2 - 196 Water - 0.5 $(TMA)_2O$ - 0.25 $(C_{16}TMA)_2O$ - 18.8 CH_3OH - 1.2 TMB. Final CTAB conc. is 6.8 wt%.

MORPHOLOGY OF POLYMER-POLYMER DEWETTING IN THIN FILMS

K. I. Winey,* A. Faldi,** and R. J. Composto*

* Materials Science and Engineering Department, University of Pennsylvania, Philadelphia, PA 19104-6272
** Exxon Chemical Co., P.O. Box 5200, Baytown, Texas 77522-5200

Two or more thin polymer layers are frequently combined for industrial applications such as protective coatings, lubricants, packaging, and adhesives. One might refer to such multi-layered polymeric structures as planar nanocomposites. The use of multiple polymer thin films requires the control of spreading and dewetting of these films. Our studies have identified a number of important characteristics of the kinetics and morphology of polymer melt / polymer melt dewetting.[1,2] Particularly noteworthy is our ability to distinguish the two polymer melts in a cross-sectional view.

At 190°C a 200 nm film of polycarbonate (PC) was found to dewet a 200 nm film of poly(styrene-*co*-acrylonitrile) (SAN) which had been deposited onto a rigid substrate. An optical micrograph of an intermediate stage of polymer-polymer dewetting shows the characteristic features: a circular hole in the upper layer (light gray), a rim surrounding the hole, and a small dimple in the center of the hole (dark gray), Figure 1. Auger electron spectroscopy measurements indicate that the near surface composition inside the hole is SAN and away from the hole is PC, which confirms that the circular feature is indeed a hole in the PC layer which exposes the lower layer of SAN. With increasing annealing times the rims impinge upon one another and eventually the rims break into isolated droplets of PC. We have monitored the rate of growth of isolated holes as a function of temperature and AN composition; these results are reported separately.[1] Figure 2 shows an AFM image produced using the tapping mode on the surface of the PC/SAN bilayers; the black bands correspond to a fixed height difference. The rim of the hole rises gradually from the unperturbed bilayer and drops more steeply into the pit. The AFM image indicates that the dimple in the center of the pit is a slight depression.

Cross-sectional TEM (Figure 3) and LV-SEM (not shown) were performed on microtomed sections from samples having similar thermal histories as the samples shown in Figures 1 and 2. The shape of the SAN layer is clearly non-planar in the region of the rim. This surprising result illustrates that as the upper layer retracts it pulls the lower layer along at a characteristic contact angle. Furthermore, the SAN layer within the pit is thinner than the original SAN layer thickness. The cross-sectional view provides unique and insightful information which highlight the profound differences between the mechanisms of polymer melt / solid and polymer melt / polymer melt dewetting. We expect the nature of the cross-sectional profile to change during the course of the dewetting process.[3]

References
1. A. Faldi et al., *Materials Research Society Symposium Proc.*, Boston 1994, in press.
2. A. Faldi et al., *Langmiur*, submitted.
3. Support: NSF Young Investigator grants DMR91-58462(RJC) & DMR94-57997(KIW); Petroleum Research Fund(AF, RJC and KIW), administered by ACS. Central facilities of the Laboratory for Research on the Structure of Matter, NSF DMR91-20668.

Proc. Microscopy and Microanalysis 1995, edited by G.W. Bailey, M.H. Ellisman, R.A. Hennigar, and N.J. Zaluzec
Copyright © 1995 MSA. Published by Jones and Begell Publishing, 79 Madison Ave., New York, NY 10016

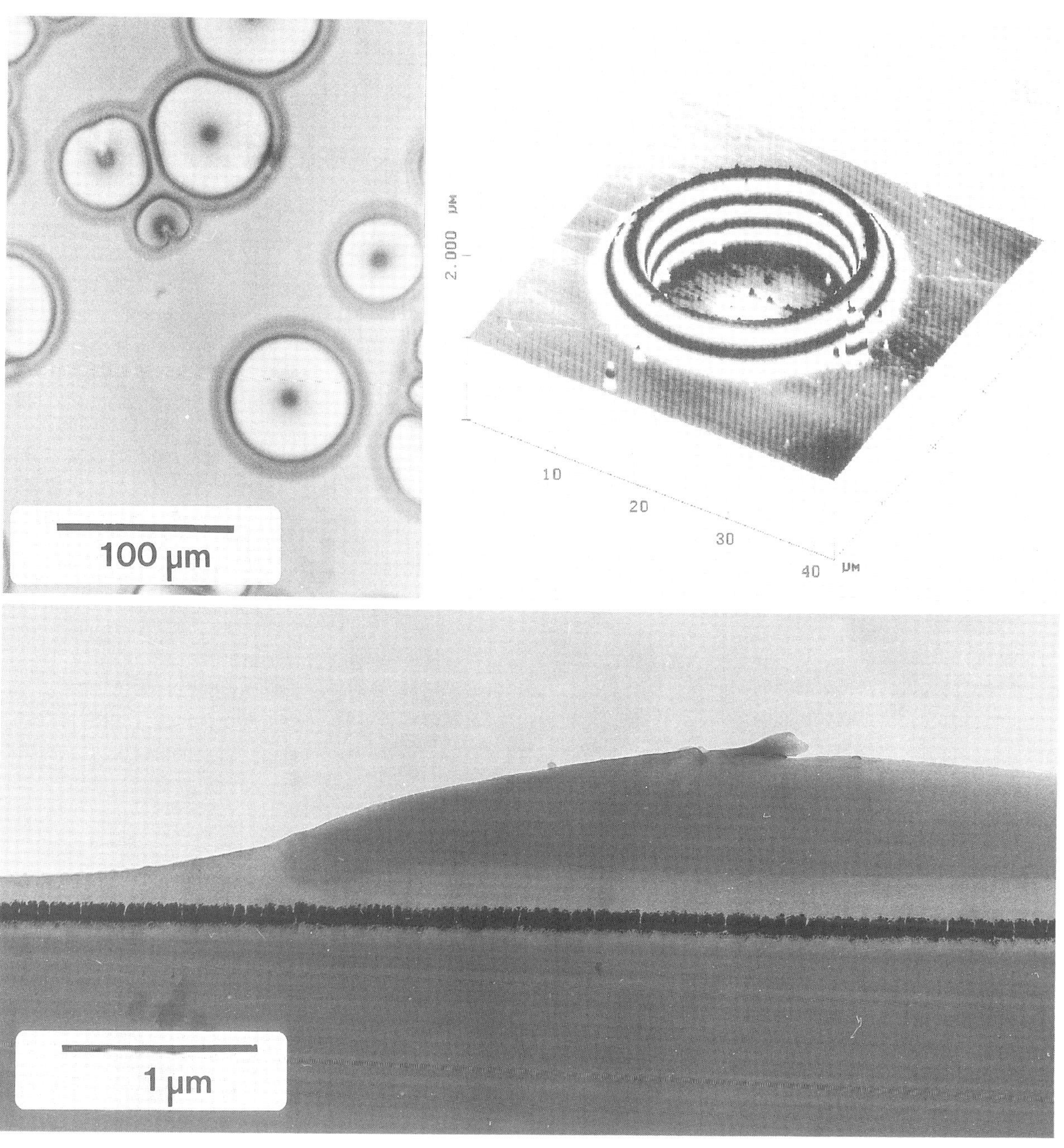

FIG. 1.-Optical micrograph of intermediate stage of PC thin film (200 nm) dewetting from SAN thin film (200 nm) on Si substrate showing holes in the PC layer surround by rims.

FIG. 2.-AFM image of a pit caused by dewetting in a PC/SAN bilayer. Note the slight depression in the center of the hole which we have termed a "dimple."

FIG. 3.-Cross-sectional TEM image of PC/SAN bilayer showing only one rim of a hole. From bottom: polyimide substrate, gold marker layer (black), SAN layer, and PC layer.

MORPHOLOGICAL TRANSFORMATIONS IN SOLUTION-CAST Styrene-Butadiene-Styrene (SBS) TRIBLOCK COPOLYMER THIN FILMS

Ginam Kim, W. Marsillo and M. Libera
Stevens Institute of Technology, Hoboken, New Jersey 07030

The fact that block copolymers can assume a range of morphologies depending upon such variables as relative block length and molecular weight is now well known. [1-5] In the case of poly(styrene)[PS]-poly(butadiene)[PB]-poly(styrene) (SBS) triblock copolymer, the morphologies range from spheres (roughly ~20% minor component), to cylinders (roughly 20%~35% minor component), to lamellae (roughly equal component fractions)[1,2,3] Most recently, there has been increasing interest in transformations between morphologies by thermal annealing.[4,5] This paper describes initial results studying the effect of solvent evaporation rate and post-casting annealing treatment on the morphology of SBS thin films.

TEM specimens were prepared by solution casting electron transparent films. 50 µl of 0.1 wt% SBS (30% styrene, M_w=14,000, Scientific Polymer Products, Inc.) dissolved in toluene was deposited on a polished NaCl single crystal substrate placed in a small dish. After solvent evaporation the film was cut into small squares, floated from the salt in water, and each square was collected on a Cu grid. The solvent evaporation rate could be slowed from minutes to hours by increasing the toluene partial pressure (add extra toluene to the dish) and restricting its removal (cover the dish). Annealing was done either in flowing nitrogen or in vacuum for 18 hrs at 155 ^{0}C. Each film was then stained by exposure to OsO_4 vapor. OsO_4 reacts with the unsaturated carbon bond in PB.[6]

Figure 1 summarizes results for a matrix of specimen preparation procedures: fast (~3hrs) or slow (~10 hrs) solvent evaporation time with/without a post-casting anneal. Fast evaporation (fig.1a) generates a microstructure consisting entirely of hexagonally packed PS cylinders in a PB matrix. This microstructure coarsens but does not undergo an overall morphological transformation(fig. 1b). Slow evaporation (fig. 1c) produces a mixed microstructure. Some regions have a lamellar morphology, and others in the same specimen have the hexagonal-cylinder morphology. Annealing converts this mixed microstructure into a fully lamellar microstructure (fig. 1d). The dark spots in the annealed specimens are apparently due to local oxidation effects. These are absent when vacuum is used in place of nitrogen atmosphere.

Increased temperature and solvent concentration both enhance macromolecular mobility. Slower evaporation time coupled with post-cast annealing would thus provide longer time under conditions of relatively high mobility where this polymer should assume its most stable microstructure. Such treatment produces the fully lamellar microstructure of fig.1d. There is apparently a relatively large barrier to the nucleation of the lamellar phase from the cylinder phase, since the particular anneal used is unable to convert a fully cylindrical microstructure. This same anneal is, however, able to to convert the mixed microstructure where the lamellar phase is already present. More careful experiments involving a broader array of evaporation rates and annealing times are required to better understand this morphological transformation and establish to what extent it may be reversible if at all.

Proc. Microscopy and Microanalysis 1995, edited by G.W. Bailey, M.H. Ellisman, R.A. Hennigar, and N.J. Zaluzec
Copyright © 1995 MSA. Published by Jones and Begell Publishing, 79 Madison Ave., New York, NY 10016

References

1. S.L. Aggarwal, Polymer 17 (1976) 938-956.
2. E. Helfand and Z.R. Wasserman, Macromolecules 13 (1980) 994-998.
3. S.P. Gido, J. Gunther, and E.L. Thomas, Macromolecules 17 (1993) No.17 4506-4520.
4. D.A. Hajduk, S.M. Gruner, et.al., Macromolecules 27 (1994) 490-501.
5. S. Sakurai, T. Momii, K. Taie, et.al, Macromolecules 26 (1993) 485-491.
6. L.C. Sawyer and D.T. Grubb "Polymer Microscopy" Chapman and Hall, New York (1987).
7. This research has been supported by the Army Research Office (Grant #DAAH 04-93-0239).

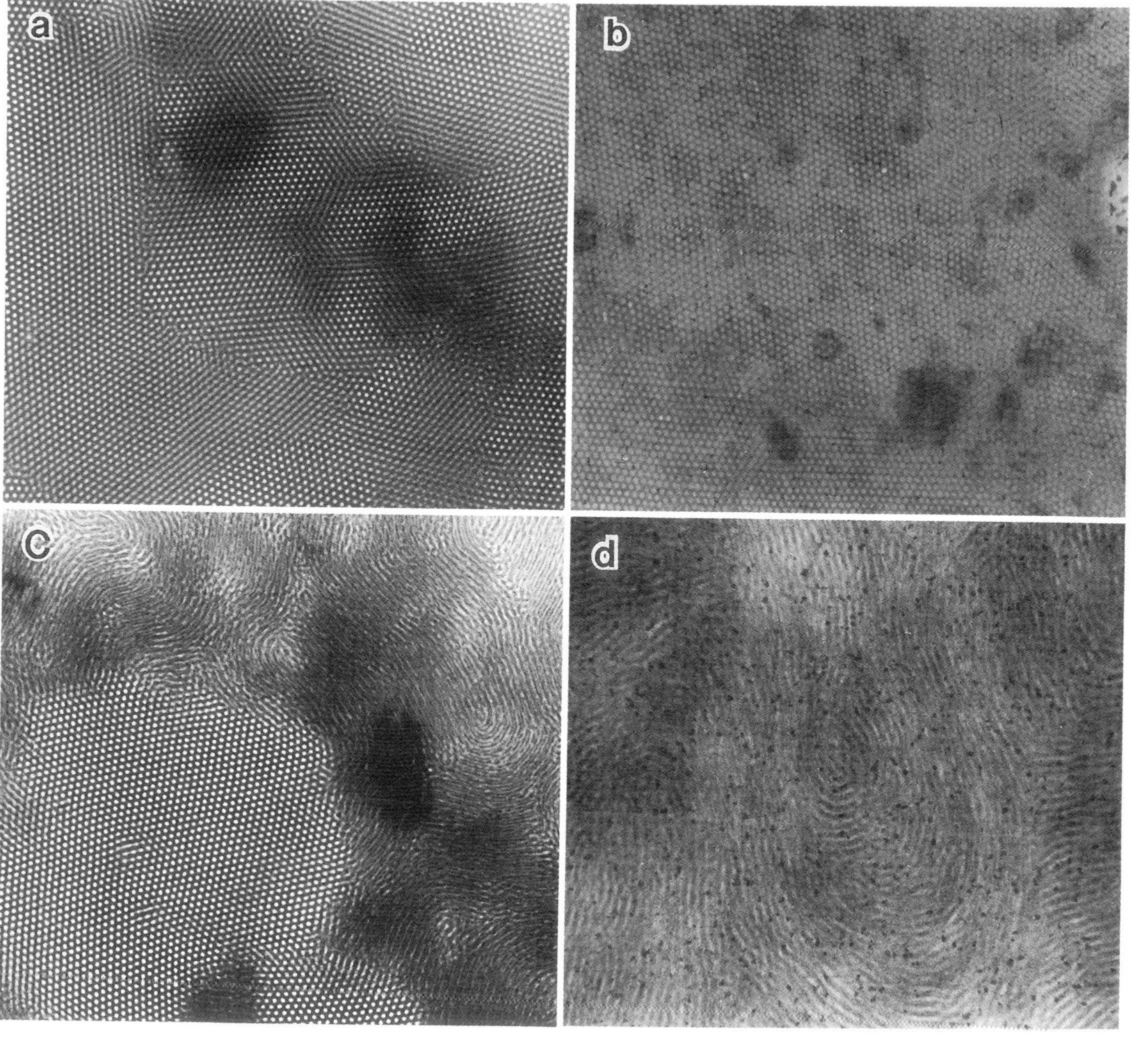

━━━0.5 μm

Fig. 1 Morphologies for SBS triblock copolymer: a: fast evaporation; b: fast evaporation and annealing; c: slow evaporation; d: slow evaporation and annealing

ELECTRON MICROSCOPE INVESTIGATION OF CRACK PATH SELECTION AND CRACK INTERFACE INTERACTIONS IN A W-ZrO$_2$(Y$_2$O$_3$) DIRECTIONALLY SOLIDIFIED EUTECTIC (DSE).

T.C. Isabell,* V.P.Dravid,* and D.N. Hill**

*Department of Materials Science and Engineering, Northwestern University, Evanston, IL 60208
** School of Materials Science and Engineering, Georgia Institute of Technology, Atlanta, GA 30332

Metal-ceramic composites play a vital materials role in many technologically important industries due to their unique electrical, thermal and structural properties. The mechanical properties of metal-oxide materials are largely controlled by the properties of the metal-ceramic interface, specifically its structure and chemistry which govern the bonding between the constituents.[1,2] For a brittle matrix composite, the fracture properties of the interface can determine the failure response of the composite as a whole.

The W-ZrO$_2$(Y$_2$O$_3$) directionally solidified eutectic was grown by a skull melting technique described elsewhere.[3] Composites were grown with 10 mol % yttria and ~ 6 w% tungsten. Figure 1 is a scanning electron micrograph showing that the tungsten grows as parallel fibers approximately 1μm in diameter in a zirconia matrix. Selected area diffraction of the composite (Figure 2) reveals the growth direction to be: <110>$_{ZrO2}$ // <111>$_W$.

Electron microscopy of microindentation cracks, of pre-existing microcracks due to thermal expansion anisotropy and of short cracks inadvertently introduced during specimen preparation shows four different crack-interface interactions: crack deflection along the metal-ceramic interface, crack deflection away from the interface, crack penetration through the interface, and crack-interface bridging.[4] Figure 3 is a TEM micrograph, showing crack deflection around a fiber-matrix interface. In this example and in all examples seen of this phenomena, the angle of crack impingement upon the metal-ceramic interface was less than 60°.   Figure 4 is an example of crack deflection away from the interface in a TEM thin foil. In this case, the crack approached the fiber at an oblique angle of approximately 55°. An example of crack penetration is shown in Figure 5, where the angle of crack incidence is about 90°. Figures 6a and b show TEM micrographs indicating bridging, in transverse and longitudinal sections, respectively. In every occurrence of bridging, the matrix crack approached the interface at an angle between 75° and 90°.

He and Hutchinson have studied the competition between penetration of a crack into a second phase and deflection of it at and along the interface.[5] The tendency for a crack to deflect at an interface or to pass through as a function of angle of crack impingement will be discussed within the framework of this model.[6]

References

1. A.G. Evans and M. Rühle, p. 654-661 in *Materials Interfaces*, D. Wolf and S. Yip, Ed., Chapman and Hall, London (1992).
2. M. Rühle and A.G. Evans, *Mater. Res. Soc. Symp. Proc.*, **120** 293-311 (1988).
3. A.T. Chapman and W. Clark, *J. Am. Ceram. Soc.* **48** [9] 494 (1965).
4. T.C. Isabell, V.P. Dravid and D.N. Hill, *submitted to J. Am. Ceram.Soc.*
5. M.Y. He and J.W. Hutchinson, *Inter. J. Solids. Struct.*, **25** [9] 1053-1067 (1989).
6. Research supported by the National Science Foundation, grant no. DMR-9203722 and NSF Materials Research Center, grant no. MRL-9120521.

Proc. Microscopy and Microanalysis 1995, edited by G.W. Bailey, M.H. Ellisman, R.A. Hennigar, and N.J. Zaluzec
Copyright © 1995 MSA. Published by Jones and Begell Publishing, 79 Madison Ave., New York, NY 10016

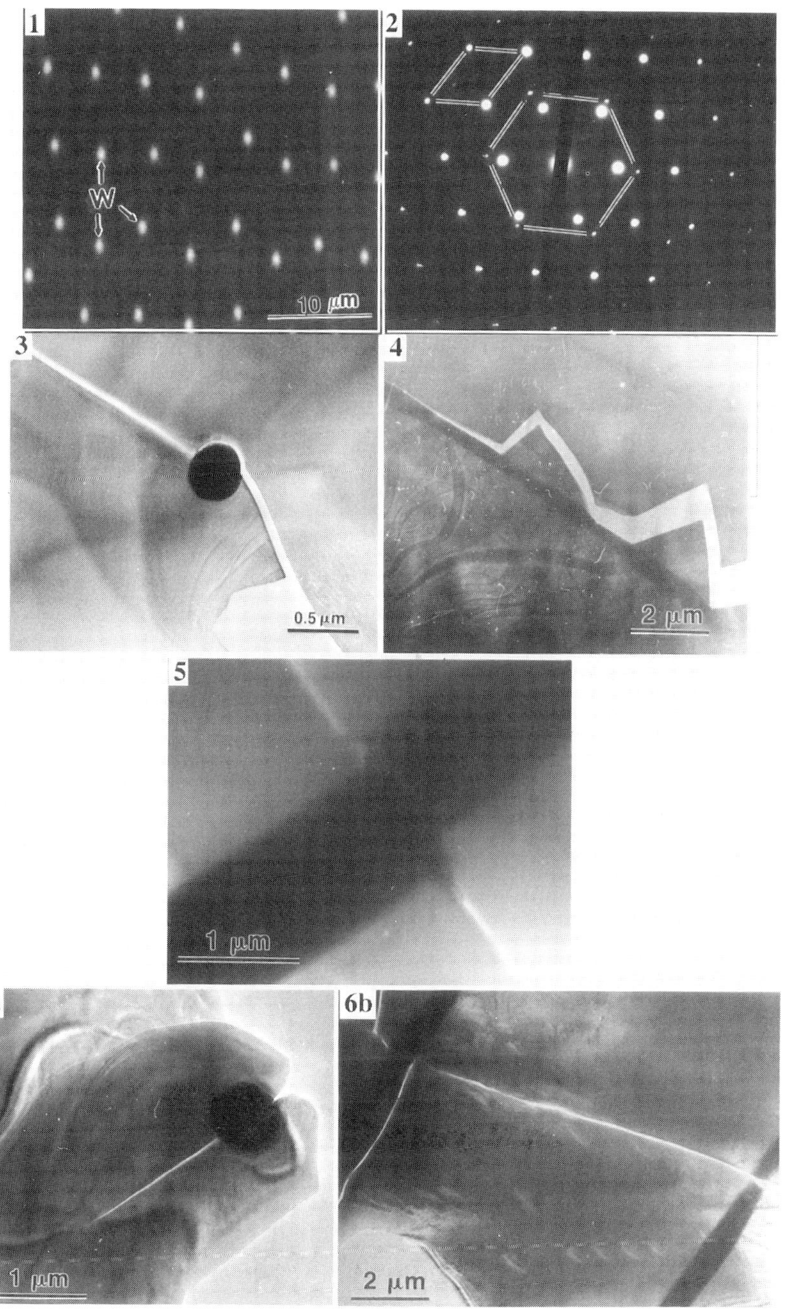

Figure 1. SEM micrograph of W-ZrO2 microstructure.
Figure 2. Selected Area Diffraction showing orientation relationship between fibers and matrix.
Figure 3. TEM micrograph of crack deflection around a fiber.
Figure 4. TEM micrograph of crack deflection away from the interface
Figure 5. TEM micrograph of interface penetration
Figure 6. Crack-interface bridging in a)transverse and b)longitudinal cross section

AGGREGATION GROWTH OF NANOMETER-SIZED BaTiO₃ PARTICLES

C. M. Chun,[†*] A. Navrotsky[†*], and I. A. Aksay[#*]

[†]Department of Geological and Geophysical Sciences, [#]Department of Chemical Engineering, and
*Princeton Materials Institute, Princeton University, Princeton, New Jersey   08544-5211

Nanocrystalline $BaTiO_3$ particles are synthesized under hydrothermal conditions by the reaction of nano-sized TiO_2 colloids with aqueous solutions of $Ba(OH)_2$ at 80°C.[1] A dissolution-precipitation mechanism is proposed for the formation of $BaTiO_3$ by this route, because suspended TiO_2 nano-particles dissolve into $Ba(OH)_2$ solution, supersaturate the solution phase, and subsequently precipitate $BaTiO_3$ particles.[2] The origin of the rough "raspberry-like" $BaTiO_3$ particles during growth stages can be explained by one of two mechanisms: (i) the morphological instability at the growth front during continuous growth, or (ii) aggregation of colloidal particles by multiple clustering. Our TEM studies on samples prepared by controlled seeding experiments show that a multiplicity of low or high angle grain boundaries occur during growth, supporting the aggregation growth.

In order to separate the nucleation step from the growth step more efficiently, seed $BaTiO_3$ particles were prepared by adding TiO_2 colloids (Degussa, P-25) to 2.23 M $Ba(OH)_2$ solution and hydrothermally reacting at 80°C under atmospheric pressure for 48 hours in polyethylene bottles. The resulting seed particles were recovered by centrifugation, rinsed with CO_2-free deionized water and freeze-dried. Final particles were prepared by adding these seeds (50 nm) into a supernatant prepared after a 1 hour reaction of TiO_2 in 1 M $Ba(OH)_2$ solution at 80°C and aged for 30 min and 48 hours at 80°C.

At the early stages of growth of the seeds, primary $BaTiO_3$ nuclei of 5~10 nm in size were observed on the surface of seed particles (Fig. 1(a)). In contrast, at the final stage of growth of seeds, the size of $BaTiO_3$ seeds increased to 120 nm and the morphology was more or less spherical (Fig. 1(b)). HREM images corresponding to the early stages of growth show low angle tilt boundaries (Fig. 2), misfit orientation, and kink in lattice fringes (Fig. 3). Because randomly oriented primary nuclei are attached to the seeds, there is a misfit at the interface equivalent to the insertion of a row of dislocations. The angle of misfit is small and the boundary consists of regions of perfect fit and regions of misfit which result in the formation of dislocations. For a seed with relatively many primary nuclei, crystal lattices are completely matched (Fig. 4) suggesting that initially misaligned primary nuclei are rapidly restructured to form a single crystallite with the same crystallographic orientation. These observations then strongly support the growth mechanism where the primary particles first homogeneously nucleate within the solution and then aggregate with the seed particles. Homogeneous nucleation appears to be favored due to relatively high saturation levels used in our experiments. These observations also illustrate that unless the interfaces are examined at the early stages of clustering, "raspberry-like" morphologies do not provide sufficient information on the exact growth mechanism.

References:

1. W. Hertl, *J. Am. Ceram. Soc.*, **71**(1988)879.
2. R. J. Pugh and L. Bergstrom, Eds., *Surface and colloid chemistry in advanced ceramic processing*, Marcel Dekker (1994)29-69.
3. This work made use of MRSEC Shared Facilities supported by the NSF under Award Number DMR-940032 and was supported by a grant from the U. S. Air Force Office of Scientific Research AFOSR-F49620-93-1-0259.

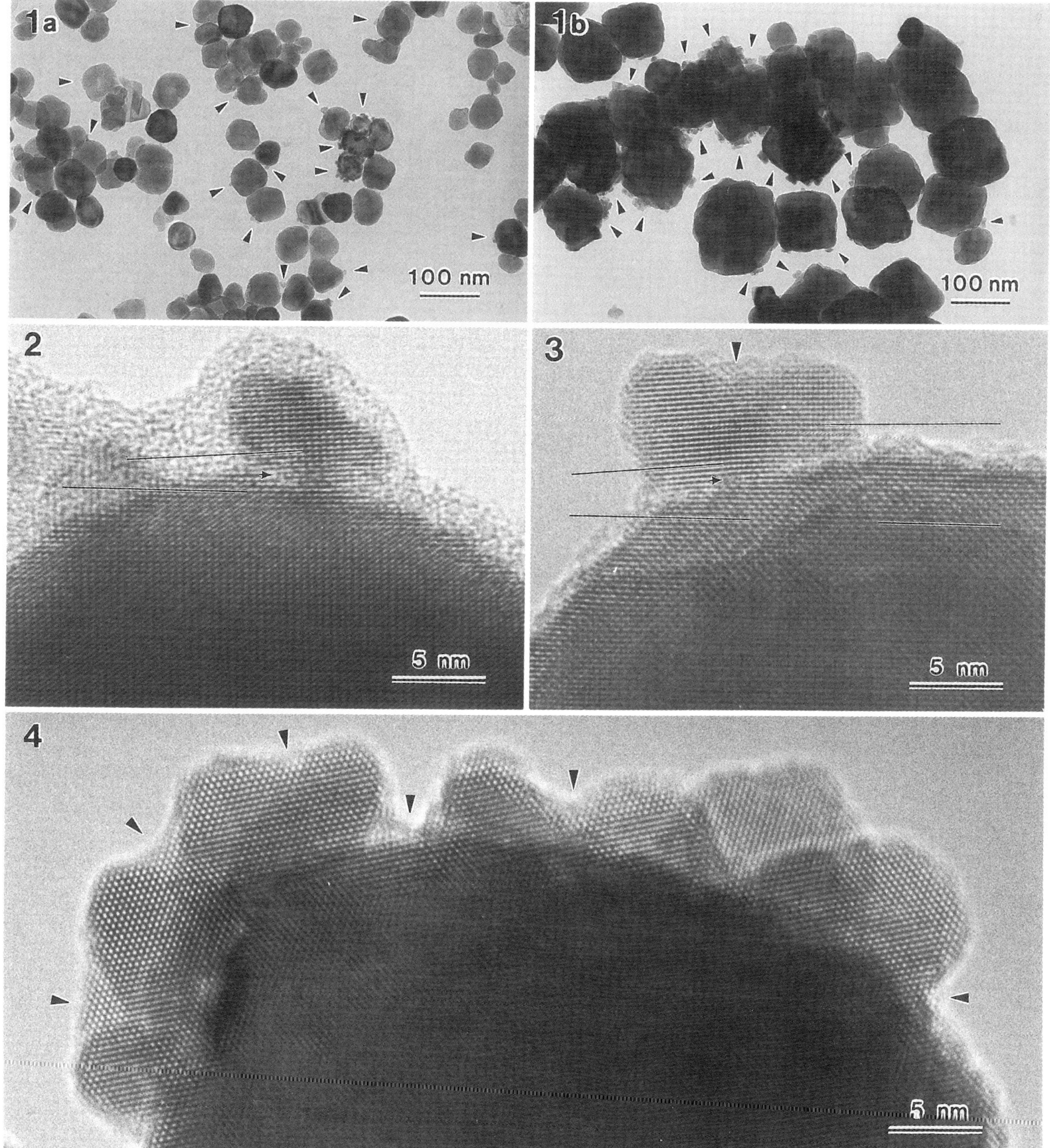

FIG. 1.—TEM images of BaTiO$_3$ particles processed by adding BaTiO$_3$ seeds into a supernatant and aged at 80°C (a) for 30 min, the early stage of growth (b) for 48 hours, the final stage of growth.

FIG. 2.—HREM image of BaTiO$_3$ particles corresponding to the early stage of growth, showing low angle tilt boundary between a primary BaTiO$_3$ nuclei and a seed

FIG. 3.—HREM image of BaTiO$_3$ particles corresponding to the early stage of growth, showing both misfit orientation between primary BaTiO$_3$ nuclei and a seed and kink in lattice fringes between two neighboring nuclei.

FIG. 4.—HREM image of BaTiO$_3$ particles for a seed with relatively many primary nuclei.

THE INFLUENCE OF NITROGEN ON THE STABILITY OF NANOPHASE MOLYBDENUM DISILICIDE

H. Kung, T. R. Jervis, J-P Hirvonen*, T. E. Mitchell and M. Nastasi

Los Alamos National Laboratory, Materials Science and Technology Division, Los Alamos, NM 87545
*Technical Research Centre of Finland, Espoo, Finland

The intermetallic compound, $MoSi_2$, has a combination of interesting properties, ranging from a high melting point to superior high temperature oxidation and corrosion resistance, which makes it a potential candidate for high temperature structural applications [1]. It was shown previously that the addition of nitrogen into $MoSi_2$ forms $MoSi_2N_x$ (x:3-4). The new phase has a very high crystallization temperature (> 1000°C) [2] and is thermally compatible with both Mo and $MoSi_2$ [3]. As suggested by these results, the introduction of nitrogen into $MoSi_2$ may cause significant changes in its properties, yet the role of nitrogen in influencing the structure and phase stability of $MoSi_2$ is still not known. In this study, we have systematically investigated the evolution of structure in $MoSi_2$ in the presence of different nitrogen contents and annealing conditions.

The system investigated is single phase $MoSi_2N_x$. Films were prepared by sputter deposition using a planar magnetron $MoSi_2$ target at a DC power of 100-200 W. The nitrogen concentration, with x varying between 0 and 4.2, was controlled by varying the N_2/Ar partial pressure. The nitrogen concentration was determined by Rutherford Backscattering Spectroscopy. Single crystal silicon {100} substrates were used for all the sputtered films. The nominal thickness of single phase $MoSi_2N_x$ films was 1 μm. The films were then annealed at 900°C in a vacuum of 10^{-8} torr. Cross-sectional transmission electron microscopy (XTEM) was used to examine the structure, phase and grain size distribution of the as-sputtered and annealed films.

The as-sputtered films exhibit an amorphous structure in all cases. Fig. 1 shows the XTEM images of $MoSi_2N_x$ after 900°C annealing, with x= 0 (a), 1.5 (b), 2.95 (c), and 4.2 (d). The corresponding selected area diffraction patterns are shown as insets at the top right hand corner. In pure $MoSi_2$ (x=0), nanocrystalline $MoSi_2$ with the tetragonal $C11_b$ phase forms after the annealing. The addition of 33 at% of nitrogen (x=1.5) results in the formation of metastable $C40-MoSi_2$. In addition, the average grain size decreases from ~30nm to ~8nm with the nitrogen addition. With even higher nitrogen content (50 at% for x=2.95 and 60 at% for x=4.2), no crystalline phases were found after the annealing. The amorphous structure observed in Fig. 1(c) and (d) is confirmed by the corresponding SAD patterns which show diffuse amorphous rings. The amorphous structure exhibits a phase separation (spinodal decomposition) into a uniformly distributed darker phase surrounded by a lighter background. The actual composition of the two phases has not been determined yet. It is suspected that the darker phase has a higher Mo content than the lighter phase.

In summary, the addition of nitrogen changes the microstructure and phase stability of $MoSi_2N_x$ films. After annealing at 900°C, different microstructural changes occur depending on the nitrogen content. The resulting microstructure consists of nanocrystalline $C11_b-MoSi_2$ for x=0, nanocrystalline C40 phase for x= 1.5 and amorphous for x=2.95 and 4.2. The nitrogen addition also results in a decrease in average grain size. It is suspected that the nitrogen addition may have hindered self-diffusion and inhibited extensive grain growth [4].

References
1. S.M. Tuominen and J.M. Dahl, J. Less-Common Met. **81**, 249 (1981).
2. J-P. Hirvonen, I. Suni, H. Kattelus, R. Lappalainen, P. Torri, H. Kung, T.R. Jervis and M. Nastasi, in <u>High Temperature Silicides and Refractory Alloys</u>, MRS. Proc. <u>322</u>, 279 (1993).
3. H. Kung, T.R. Jervis, J-P. Hirvonen, T.E. Mitchell and M. Nastasi, J. Vacuum Science and Technology A, 1995 (in press).
4. This research is supported by a DOE program in Basic Energy Sciences/Advanced Energy Project.

Proc. Microscopy and Microanalysis 1995, edited by G.W. Bailey, M.H. Ellisman, R.A. Hennigar, and N.J. Zaluzec
Copyright © 1995 MSA. Published by Jones and Begell Publishing, 79 Madison Ave., New York, NY 10016

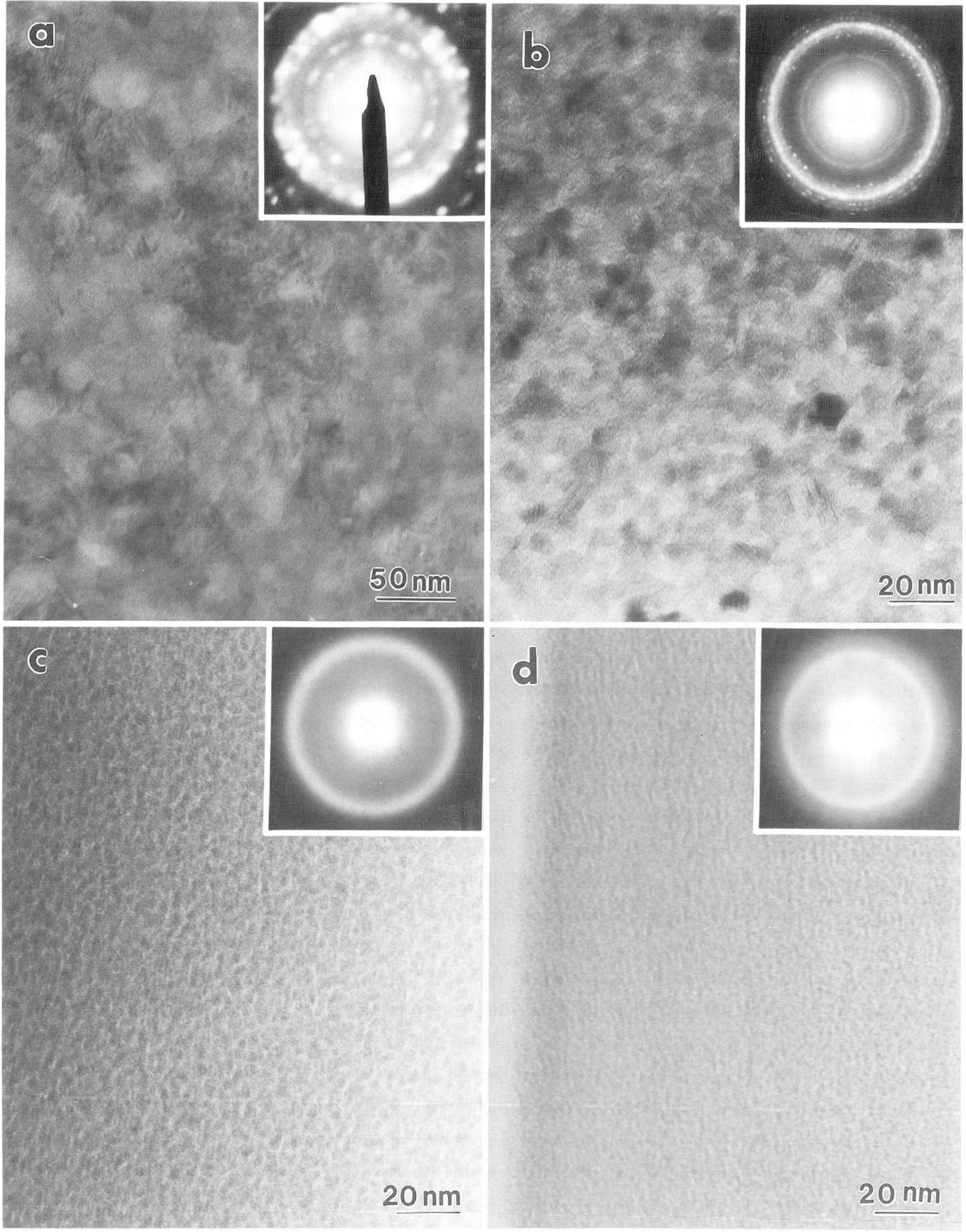

FIG. 1 XTEM micrographs showing 900°C-3h annealed $MoSi_2N_x$ films with different nitrogen content. The corresponding SAD patterns are shown in insets at the upper right hand corners.
(a) x=0, pure $MoSi_2$ with the $C11_b$ structure. The average grain size is ~30nm.
(b) x=1.5, 33 at% addition of nitrogen results in the hexagonal C40 structure. The average grain size is ~ 8nm.
(c) x=2.95, 50 at% addition of nitrogen results in the amorphous structure.
(d) x=4.2, 60 at% addition of nitrogen results in the amorphous structure.

Intergrainular Structure and Deformation Mechanism in Nanocrystalline Materials*

H.Q.YE, P.H.Ping, D.X.Li, J.Y.Huang and Y.K.Wu

Laboratory of Atomic Imaging of Solids, Institute of Metal Research, Chinese Academy of Sciences, Shenyang 110015, P.R.China

It is recognized that boundary structure (GB) characterization is essential in order to understand the structure-properties relationship of nanocrystalline (NC). In most cases, NC materials have to suffer deformation during compacting or ball milling techniques, a deep and systematic study on deformation mechanism is also necessary. In this paper, characterization of microstructure in NC materials synthesized by three methods has been presented.

1. Ordered and Disordered Regions at GBs of NC Pd

The NC Pd samples were synthesized by the inert gas condensation and in situ compacting technique. The results of X-ray diffraction and HREM observations showed that the average grain size of the NC Pd is about 10 nm. It can be seen that most of the GBs have ordered structure and no 'gas-like' feature has been observed. Some disordered GB regions, such as nanovoid formed during compacting process, are also detected as marked by "V" in Fig.1. The structural modification from the disordered state was found during in situ HREM investigation. Fig.1 shows that the disordered regions indicated by arrow between two grains (labeled 1 and 2 respectively) have changed into more ordered area after the electron beam irradiation of 290 mins. Defect structures, such as dislocations, multiple twins and twin junctions, are also seen at GBs and inside grains.

2. More Ordered Structure at GBs of NC $Ti_{70}Ni_{20}Si_{10}$ and $(Fe_{0.99}Mo_{0.01})_{78}Si_9B_{13}$ Made by the Amorphous Crystallization Method

Fig.2 shows the NC FeMoSiB annealed at 560°C for 1h with average grain size about 20 nm. It can be seen that no highly disorder regions and nanovoids can be detected at the interfaces. The width of the strained regions at the interfaces is less than 1 nm. Fig.3 shows the NC TiNiSi annealed at 600°C for 1h with grain size about 35 nm. The atomic structure of GBs is close to that typically observed in coarse-grained materials. However, the core areas of the interfaces showed a little strained appearance and lattice fringes are curved near the interfaces. The experimental results revealed that much less defect structure had been detected inside the crystallites and at the interfaces in this kind of NC samples.

3. Twinning Deformation Process in NC of Ball-milled Cu

The pure Cu powder was employed to study the effects of ball-milling on the development of structure and properties of ductile metals. The average grain size of resultant spheres formed after 20 hours of milling is 10-100 nm. The microhardness increased from 45 Mpa (unmilled) to 220 MPa (milled). HREM investigations showed that the deformation of ball-milled Cu proceed by <112>{111} twinning or high order twinning (Fig.4). In addition to twinning, dislocation slip plays another important role in the deformation process. The motion and interaction of (a/6)[112] twinning partial and 60° dislocations can lead to the formation of subgrains. Both low- and high-angle GBs with local strain and high density of dislocations were observed.

----

*Project supported by NNSFC and NAMCC.

Proc. Microscopy and Microanalysis 1995, edited by G.W. Bailey, M.H. Ellisman, R.A. Hennigar, and N.J. Zaluzec

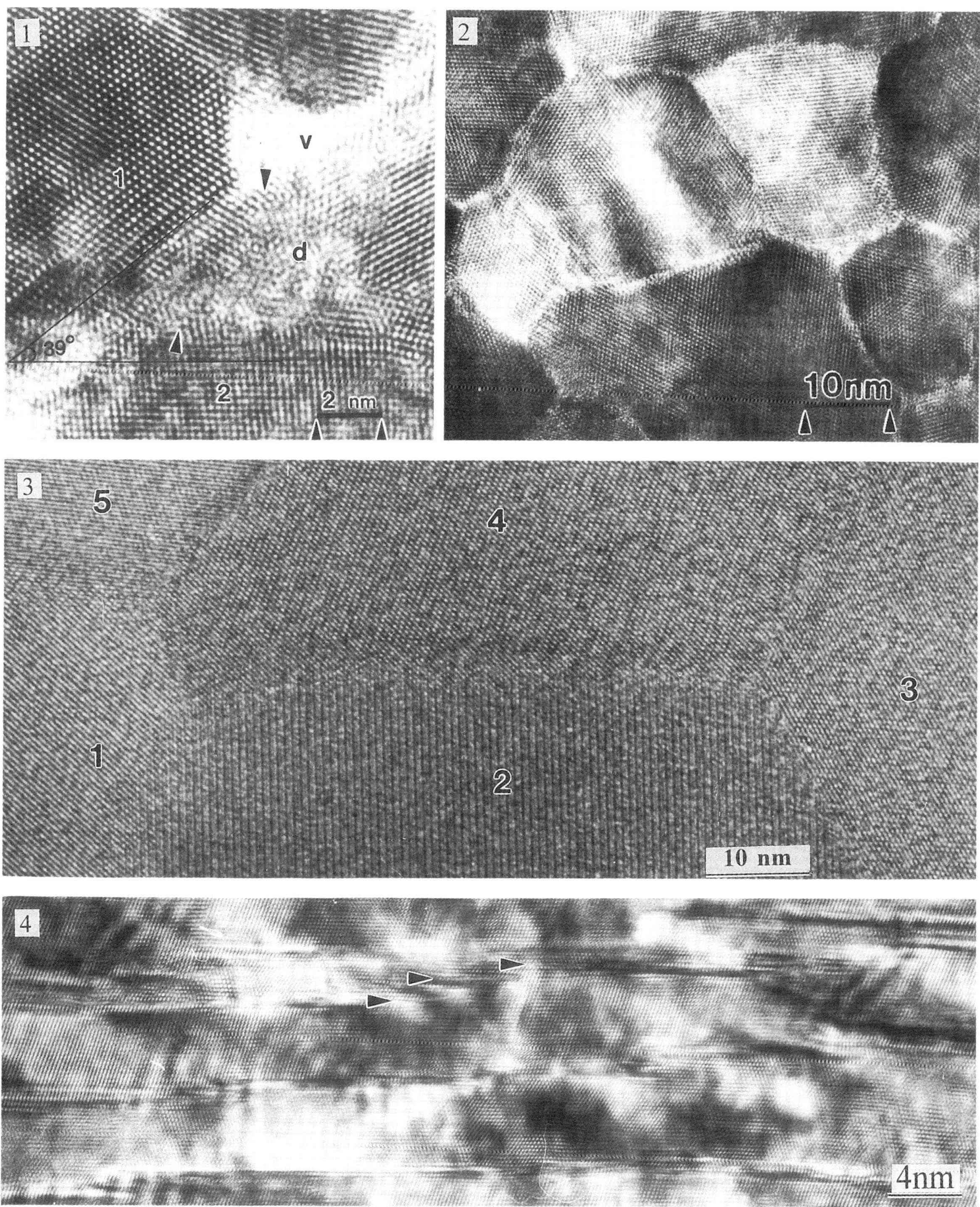

Fig.1 A typical GB area in NC Pd.
Fig.2 Interfaces in NC FeMoSiB alloy annealed at 560°C for 1 h.
Fig.3 Several crystallites (denoted by arabic numerals) and interfaces in NC TiNiSi samples
      annealed at 600°C for 1 h.
Fig.4 Multipletwins in ball-milled NC Cu with interface steps marked by arrows.

CLOSED GRAPHENE NANOSTRUCTURES

Sumio Iijima

R & D Group, NEC Corporation, 34 Miyukigaoka, Tsukuba, Japan 305

Graphene is named a single sheet of graphite, or a 2-D carbon hexagons network. The graphene structure has been observed in partially graphitized carbon which is familiar to electron microscopists and its HRTEM image properties were analyzed previously in detail[1-3]. C_{60} molecules, which has brought a great excitements in interdisciplinary fields of science and technology[4], is basically the same graphene structure with a curvature. The individual C_{60} molecule can be imaged without difficulty by HRTEM. Many of fundamental problems with the molecule however are not solved by the HRTEM technique. On the other hand, carbon nanotubes, a family of the fullerene and discovered serendipitously by the present author[5], are an ideal subject for the HRTEM investigation and in fact their structural details can only be analyzed by this technique. The present talk reviews the carbon nanotubes and related structures with an emphasis of usefulness of the HRTEM.

Search for the multi-shell graphite particles leads to unexpected discovery of carbon nanotubes which grow on a cathode in a carbon-arc chamber for the C_{60} production[5]. HRTEM images of carbon nanotubes disclosed a novel structure such as a coaxial arrangement of several cylindrical graphene sheets with a nanometer dimension. Micro-beam diffraction studies on individual tubules concluded the consistent presence of 2mm symmetry with respect to the tubule axis (Fig. 1), suggesting a helical arrangement of the hexagons. It is of interest to note that a single graphene tube as thin as 1nm in diameter diffracts intensities enough for recording on a photographic plate.

Pentagons in fullerene molecule structures is to introduce a positive gaussian curvature, or disclination, into a flat hexagon network. A complete enclosure of the network needs twelve pentagons according to the Euler theorem. A variety of tubule cap morphologies are explained by this rule. Negative curvatures, which is caused by heptagons or octagons, are also present in our nanotubes, and suggest possible 3-D graphene network. Non-hexagon carbon rings were found to be more reactive in oxidation than hexagon rings[7,8]. This observation implies the importance of these disclinations in the development of the practical activated carbon.

Electronic band structure calculations of nanotubes[9] predict a semiconductive nature when the tubules are single-shelled and less than 2nm in diameter. Such an ideal nanotube has been prepared successfully in a carbon-arc deposition by feeding metal catalyst[10,11]. The thinnest tubule was only 0.7nm in diameter which is the same as that of C_{60}. An electron diffraction pattern recorded from such an individual tubule shows 2mm symmetry, which confirms the presence of the helicity having a different pitch from tubule to tubule. It is needed crucially to grow well-defined tubules in macroscopic quantity for the characterization of their electronic properties since they are dependent on their structures. Tubule morphologies are associated with their growth and thus their analysis enables us to understand the mechanism to some extent[12].

Figure 2 shows three different types of graphitic structures, nanotubes, giant balloons and conventional graphite layer formed on Fe carbide surfaces. The largest balloon (indicated by an arrow) is measured 3 nm in diameter, being close to a fullerene molecules of C_{1500} but not perfectly spherical. We believe these graphene structures originate from certain active growth centers on the carbide surfaces and some of them can grow as tubules. From these observations we proposed that our single-shell nanotubes grow in the root growth mechanism[13].

An alternative approach to elucidate the growth of not only tubules but fullerenes is to examine the carbon soot which is one of co-products in fullerenes or nanotubes generation. For this purpose we prepared the soot by laser ablation at different carrier gas temperatures since all these structures are generated simultaneously so that they will be related structurally each other. One of findings was an agglomeration of incomplete fullerenes with various sizes which forms loosely bound 3-D networks

Proc. Microscopy and Microanalysis 1995, edited by G.W. Bailey, M.H. Ellisman, R.A. Hennigar, and N.J. Zaluzec

(Fig. 3). The soot was synthesized at 1200℃. It is surprising that a glassy carbon-like soot is formed abundantly at the room temperature ablation.
Growth of graphene structures will be discussed.

References

1 S. Iijima, *Chemica Scripta*, 14(1978-9)117.
2 S. Iijima, *J. Microscopy* 119(1980)99.
3 S. Iijima, *J. Cryst. Growth* 50(1980)657.
4 H.W. Kroto, J.R. Heath., S.C. O'Brien, R.F. Curl & R.E. Smalley, *Nature* 318(1985)162.
5 S. Iijima, *Nature* 354(1991)56.
6 S. Iijima, T. Ichihashi & Y. Ando, *Nature* 356(1992)776.
7 P.M. Ajayan, & S. Iijima, *Nature* 361(1993)333
8 P.M. Ajayan, T.W. Ebbssen, T. Ichihashi, S. Iijima, K. Tanigaki, & H. Hiura, *Nature* 362(1993)522.
9 N. Hamada, S. Sawada & A. Oshiyama, *Phys. Rev. Lett.* 68(1992)1579.
10 S. Iijima & T. Ichihashi, *Nature* 361(1993)603.
11 D.S. Bethune, C-H. Kiang, M.S. deVries, G. Gorman, R. Savoy, J. Vazques, & R. Beyers, *Nature* 353(1994)605
12 S. Iijima, *Mat. Sci. & Engin.*, B19(1993)172.
13 S. Iijima, *MRS Bulletin*, XIX(1994)43.

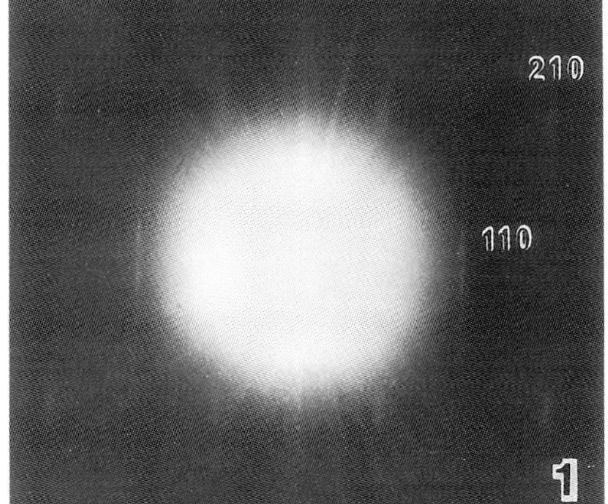

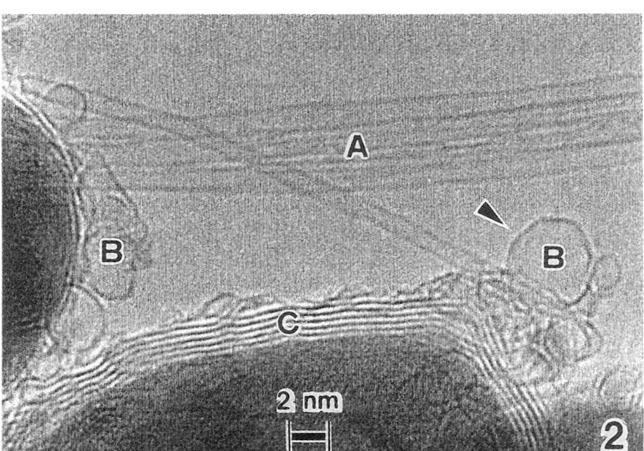

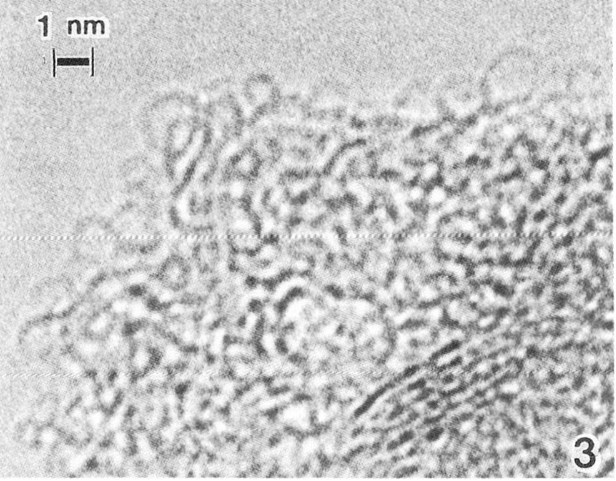

Fig. 1 An electron diffraction pattern from a 0.9nm diameter carbon nanotube. About 200 carbon atoms contribute the pattern.

Fig. 2 Three different graphene structures formed on Fe-carbide particles. Nanotubes(A), giant fullerenes(B), and conventional layered graphite covering the particles(C).

Fig. 3 Laser ablated carbon soot showing 3-D networks comprising loosly bound incomplete fullerenes. The specimen was prepared at 1200℃

THE BREAKUP OF THE INTERMEDIATE GOLD AGGREGATES

Nan Yao,[1] Wan Y. Shih,[1,2] Daniel M. Dabbs,[1,2] and Ilhan A. Aksay[1,2]

[1]Princeton Materials Institute, and the
[2]Department of Chemical Engineering, Princeton University, New Jersey 08540-5211

The synthesis of nanometer scale gold particles is well known historically.[1,2] Stable and highly colored suspensions result from the reduction of gold cations in solution. But the formation of the gold particles is not monotonic, that is, the mechanism is not simply nucleation and growth. Rather, the pronounced color changes in the evolving suspension indicate that the mechanism is nucleation, aggregation, deaggregation, and growth.[3,4] Until recently, the structure of the intermediate aggregate had been assumed to be an aggregation of primary particles, each 4-7 nm in diameter.[4] HRTEM studies have shown that the intermediates are continuous ramified structures (Figure 1) which undergo a desintering process to form the final particles (~24-26 nm diameter).[3,5] In this paper, we discuss a model which explains the role of the surfactant on the gold intermediate in the breakup of the intermediate structure.

Gold colloids were prepared by reducing $Au^{III}Cl_3$ (aq) with sodium citrate as in previous studies:[3-5] $Au^{III}Cl_3$ (aq) + Cit^{3-} (aq) $\rightarrow Au^0$. For the present study, 95mL of an aqueous solution of $AuCl_3$ (50 mg Au/L) was heated to 95°C and 5 mL of 1 wt% sodium citrate (aq) added to yield a solution of $C_{Au} = 2.54 \times 10^{-4}$ M and $C_c = 1.9 \times 10^{-3}$ M. TEM samples were prepared by dipping holey carbon support grids into the suspension followed by drying on filter paper. HREM studies were conducted with a Philips CM-20 ST TEM operated at 200 keV. The electron dose was about 10-20 A/cm^2 under moderate vacuum of 10^{-6}-10^{-7} Torr. Figure 1 shows three TEM micrographs for samples collected at various times: (a) at t = 5 min when the suspension color was purple and D ~ 90 nm, (b) at t = 15 min when the suspension started to redden and D ~ 50 nm, and (c) at t = 35 min when the suspension was ruby red and D stabilized at ~ 20 nm. The aggregates shown in Figure 1(a) are not clusters of well-separated particles. A detailed structure is shown in Figure 2. The aggregates appear continuous with a high population of twin boundaries in the intermediate sausage-like structure. The high density of nanometer-sized twin boundaries in the gold structure is due to the minimization of the total interfacial free energy of the gold particles.[6] Also, due to the small diameter in the sausage structure, the surface area of the intermediate is orders of magnitude larger than in bulk material. It is known that diffusion coefficients can be orders of magnitude higher at grain boundaries than inside grains. The surfaces and twin boundaries may then provide transport paths, allowing much faster redistribution of matter within the aggregates.

Mass transport is first facilitated by the movement of the adsorbed citrate ions on the conductive gold surface under a constant potential. In a corner of angle β between the two faces of the corner (Figure 3), the electrical field normal to the surface, $E_\phi(\rho)$, varies with ρ as $E_\phi(\rho) = -(\alpha_1/\beta)\rho^{(\pi/\beta)-1}$, where ρ is the radial distance from the origin and α_1 a constant related to the surface potential.[7] The repulsive force diminishes near the origin and the area over which the repulsive force diminishes increases with decreasing angle β. The breakup mechanism can be explained as follows: (1) The repulsive electric force on the adsorbed charges diminishes near the corner of a cusp and (2) the surface area with a diminishing electric force increases as the cusp gets sharper. So the aggregate surface is unstable against a cusp-like perturbation. Once a cusp is formed, the cusp will sharpen to allow more adsorbed charges around the cusp, leading eventually to breakup. The formation and sharpening of the cusps are facilitated by the high diffusivity of gold atoms.[8]

Proc. Microscopy and Microanalysis 1995, edited by G.W. Bailey, M.H. Ellisman, R.A. Hennigar, and N.J. Zaluzec.
Copyright © 1995 MSA. Published by Jones and Begell Publishing, 79 Madison Ave., New York, NY 10016

References:

1. M. Faraday, *Phil. Trans.*, **147**(1857)145.
2. For example, see P. C. Hiemenz, *Principles of Colloid and Surface Chemistry*, 2nd edition, Marcel Dekker (1986).
3. J. Liu, M. Sarikaya, and I. A. Aksay, *Proc. Ann. MSA,* **46**(1988)716.
4. M. K. Chow and C. F. Zukoski, *J. Colloid and Interf. Sci.*, **165**(1994) 97.
5. D. M. Dabbs, N. Yao, and I. A. Aksay, *Proc. Ann. MSA,* **52**(1994)446.
6. L. Marks and A. Howie, *Philos. Mag.* **A49**(1984)81.
7. J. D. Jackson, *Classical Electrodynamics* John Wiley & Sons (1975)77.
8. This work made use of MRSEC Shared Facilities supported by the NSF under Award Number DMR-940032 and was supported by a grant from the U. S. Air Force Office of Scientific Research AFOSR-F49620-93-1-0259.

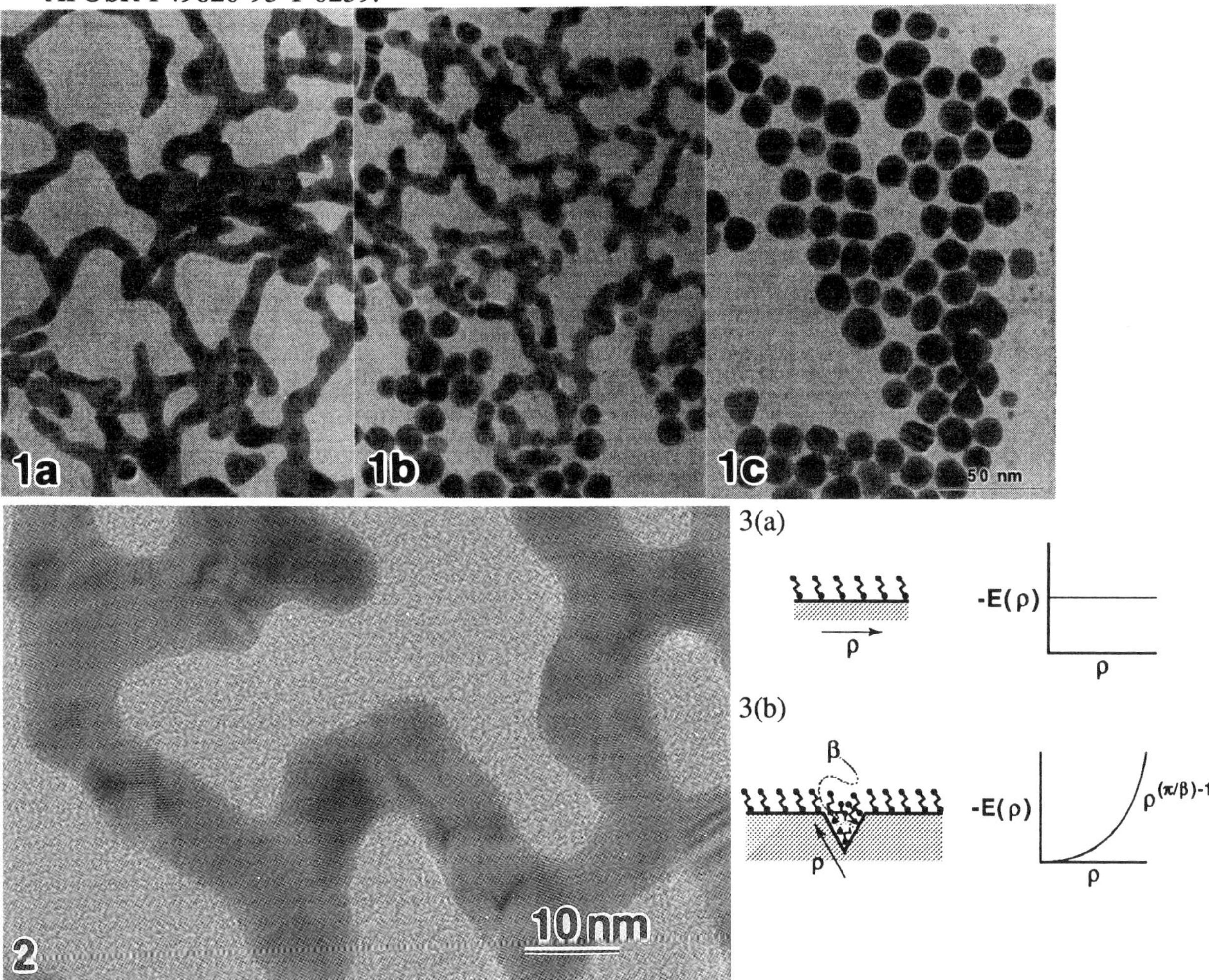

FIG. 1.—Electron microscope images showing the morphology of gold particles collected at various times: (a) t=5 min., (b) t=15 min., and (c) t=30 min. Note the sausage-like structure at t=5 min.

FIG. 2.—TEM image showing the high density of twin boundaries within sausage-like gold structure.

FIG. 3.—Schematic diagram of a gold surface under (a) a constant potential and (b) with the presence of a cusp on the surface of angle β.

MARTENSITIC TRANSFORMATION IN ZIRCONIA-ALUMINA NANOLAMINATES

M. Gajdardziska-Josifovska[*] and C. R. Aita[**]

* Department of Physics and Laboratory for Surface Studies, University of Wisconsin Milwaukee, P. O. Box 413, Milwaukee WI 53201
** Materials Department and Laboratory for Surface Studies, University of Wisconsin Milwaukee, P. O. Box 784, Milwaukee WI 53201

With an eye towards developing transformation-toughening ceramic coatings, we grew multilayers of polycrystalline zirconia and amorphous alumina in which the layer spacing was scaled to insure nanosize zirconia crystallites. [1,2] In this manner, nanolaminates with a high volume fraction of tetragonal zirconia (t-ZrO_2) were produced, independent of the deposition parameters and without the use of dopants.

For a coating to be of practical use, not only must it contain a significant amount of t-ZrO_2, but this phase must also transform *locally* to the monoclinic phase (m-ZrO_2) in response to stress. In bulk zirconia-alumina composites, with dopant stabilized tetragonal zirconia, the martensitic t $\rightarrow$ m transition is auto-catalytic, resulting in widespread transformation of the parent phase. [3] Twinning and slip are the recognized transformation mechanisms. [3] In this work, we study the transformation mechanism in zirconia-alumina nanolaminates using high resolution electron microscopy (HREM).

The multilayers were grown by reactive sputter deposition in a multiple target rf diode system.[1] Si (111) wafers were used as substrates for the multilayers studied by microscopy. Targets of Zr and Al were sputtered using 10^{-2} Torr O_2 discharges operated at -1.4 kV and -1.1 kV (p-p) respectively. The temperature at the nanolaminate growth interface was 564 ($\pm$10) K, due to plasma heating. The zirconia layer growth rate was 1.5 nm/min and the alumina layer growth rate was 0.5 nm/min. Different samples were grown by varying the number and thickness of bilayers. The present study is based on a nanolaminate with 50 zirconia-alumina bilayers. The nominal thickness of each zirconia layer was 4.5 nm, and each alumina layer thickness was 3.7 nm. The total nanolaminate thickness was 0.4 μm.

HREM studies were performed with a JEM 4000EX transmission electron microscope at Arizona State University. The nanolaminates were viewed in cross-section, using specimen prepared by standard dimpling and ion-milling procedures. The HREM images were digitized with an optical CCD camera. Numerical diffractograms were used to measure lattice spacings to 1% accuracy, allowing unique phase identification of individual nano-crystallites. [4] Approximately 75% of the analyzed particles were found to be tetragonal, with the remainder identified as monoclinic. [5]

The correlation between zirconia phases is of particular interest in elucidating the origin of the monoclinic phase. Two possibilities exist: i) m-ZrO_2 and t-ZrO_2 crystallites co-exist since nucleation; ii) m-ZrO_2 crystallites are product of martensitic transformation. Close investigation of all identified m-ZrO_2 grains supported a transformation mechanism. Namely, all m-ZrO_2 were invariably found next to t-ZrO_2 grains, while the reverse was not the case. Furthermore, a distinct crystallographic orientation relationship was found between neighboring m- and t-ZrO_2 grains with a twinned region in the monoclinic phase.

Figure 1 illustrates the crystallographic relationship between the zirconia phases. The HREM image in (a) shows a biphasic area in the first zirconia layer adjacent to the substrate. The boxed area was used to calculate the numerical diffractogram shown in (b). The lattice spacings obtained from the diffractogram indicate that the top part of the area is tetragonal while the bottom part is monoclinic. The lattice image shows that one set of planes remains invariant in spacing and angle with respect to the substrate plane. This lattice invariant is denoted t,m(200). The zone axes for these regions are m[011] and t[01-1]. It can be seen from the geometry of Fig. 1 that these zone axes are parallel to each other. It can be calculated that in each case, the zone axis makes an angle of 44° with the c-axis of the lattice. Therefore, the c-axes of the parent and daughter phases must be parallel to each other. We can write the following plane and direction invariants: m(100)//t(100) and m[001]//t[001].

In conclusion, nano-scale martensitic transformation was found in zirconia-alumina nanolaminates. The transformation was limited to within individual zirconia layers. Furthermore, it was localized in a nano-sized region adjacent to the parent crystallite. We identified the crystallographic correspondence between the parent and daughter phases, and twinning as the mode of transformation.[6]

Proc. Microscopy and Microanalysis 1995, edited by G.W. Bailey, M.H. Ellisman, R.A. Hennigar, and N.J. Zaluzec.

References

1. C.M. Scanlan, M. Gajdardziska-Josifovska, and C.R. Aita, *Appl. Phys. Lett.* 64 (1994) 3548.
2. C.M. Scanlan et al. *MRS Proceedings* 343 (1994) 481.
3. D.J. Green, R.H.J. Hannink, M.V. Swain, *Transformation Toughening of Ceramics* (CRC Press, Boca Raton, FL, 1989).
4. W. J. deRuijter et al. *Scanning Microsc.* 6 (1992) 347.
5. M. Gajdardziska-Josifovska et al. Proc. 52nd Ann. MSA Meeting 52 (1994) 754.
6. The work was supported by US-ARO grant No. DAAH04-93-G-0238. Microscopy was performed at the Center for HREM at Arizona State University supported by NSF grant No. DMR-9115680.

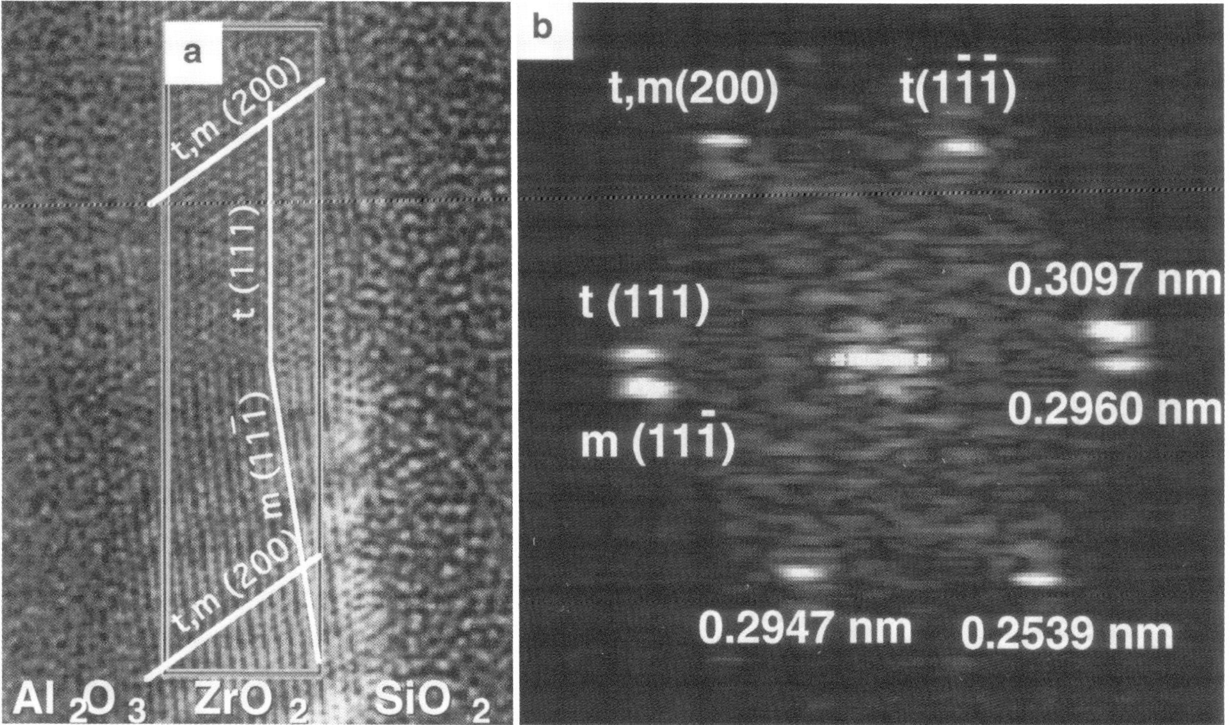

FIG. 1. Orientation relationship between parent tetragonal (t) and daughter monoclinic (m) phases in nano-crystalline zirconia-alumina multilayers.
(a)   Digitized HREM image of a biphasic region with denoted plane invariant m(100)//t(100). Distinct angle between t(111) and m(11-1) yielded direction invariant m[001]//t[001]. Boxed region, 4 nm wide, was used for diffractogram in (b);
(b)   Numerical diffractogram of biphasic region with denoted indexes and measured lattice spacings.

OXIDATION AND PHASE IDENTIFICATION IN SILVER / IRON-OXIDE NANOCOMPOSITES

P. Bandaru*, T. Yamamoto° and F. Cosandey*

* Materials Science and Engineering, Rutgers University, Piscataway, NJ 08855-0909
° Research Institute and Advanced Science and Technology, University of Osaka
  Prefecture, Gakuen, Sakai, Osaka 593, Japan.

In the last few years, there has been great interest in magnetic nanocomposite materials because of their unique properties such as giant magnetoresistance and enhanced magnetocaloric effects.[1,2] We have recently synthesized bulk Ag/FeO_x nanocomposites by inert gas phase condensation (GPC).[3] In this process, Ag and Fe particles were first formed by co-evaporation followed by vapor phase condensation and collection on a LN_2 cooled collector. The particles were then passivated in oxygen, collected and compacted in-situ. Selected samples were then annealed in an inert atmosphere of helium at 250°C for 1h or in an oxidizing atmosphere of $10\%-O_2/He$ at 230°C for 15h. These annealing treatments were found to modify the effective magnetic moment size and saturation value of magnetization. For the O annealed sample, a superparamagnetic nanocomposite has been obtained with a transition temperature Tc of 160 °K. In order to understand these magnetization changes, an effort has been made to characterize the microstructure and to identify the oxide phases by combined Mössbauer, X-ray absorption spectroscopy (XAS) and transmission electron microscopy (TEM). In this paper, phase identification by combined selected area diffraction and direct high resolution imaging are presented.

In the Fe-O system, there are two ferrimagnetic $\gamma-Fe_2O_3$, Fe_3O_4 and two antiferromagnetic FeO, $\alpha-Fe_2O_3$ phases. The ferrimagnetic oxide, $\gamma-Fe_2O_3$ is based on the same inverse spinel structure as Fe_3O_4 but with vacancies in the octahedral sites corresponding to the chemical formula $Fe(Fe_{5/3} [\ ]_{1/3})O_4$. The distribution of the cation vacancies [ ] within the octahedral sites can be either disordered (D), partially ordered (P) or ordered (O), giving rise to unit cells with different crystal symmetries [4-6](c. f. Table I).
A typical TEM image of Ag/FeO_x as prepared powder after passivation in oxygen is shown in Figure 1. The distribution of Ag and FeO_x particles have been determined by EDS and EELS spectroscopy techniques and their sizes are summarized in Table II. A distinction between $\gamma-Fe_2O_3$ (D) and Fe_3O_4 phases is only possible by detecting the presence of Fe^{2+} ions which cannot be obtained from TEM directly, but the presence of Fe_3O_4 has been confirmed by XAS and Mössbauer data. In Figure 2, the kinematical diffraction lines of Fe_3O_4 and $\gamma-Fe_2O_3$ (P) phases are represented and with the observed lines indexed. The presence of the $\gamma-Fe_2O_3$ (P) phase is detected primarily from the presence of the {421} reflection in the diffraction pattern. A lattice image of the oxygen annealed sample is shown in Figure 3. Fourier diffraction patterns taken from various particles confirm the presence of $\gamma-Fe_2O_3$ (P) phase. Evidence for the occurence of antiferromagnetic phases other than Ag that could explain the observed superparamagnetic behavior has not been found. Also, the ordered tetragonal phase, with cell dimension c=3a, has not been detected. One possible explanation is the small oxide particles size of 20 nm which is below the critical size of 30 nm required for the occurence of this ordered structure.[7]

References

1. R. D. Shull and L. H. Bennett, Nanostructured Materials, 1 (1992) 83.
2. G. Xiao, J. Q. Wang and P. Xiong, Appl. Phys. Lett. 62 (1993) 420.
3. T. Yamamoto, R. D. Shull, P. R. Bandaru, F. Cosandey and H. Hahn, Jpn. J. Appl. Phys.
   33 (1994) L1301.
4. C. Greaves, J. Solid State Chem. 32-33 (1893) 325.
5. Pearson's Handbook of Crystallographic Data for Intermetallic Phases, ASM International (1991)
6. M. P. Morales, C. Pecharroman, T. Gonzalez Carreño and C. J. Serna, J. Solid State Chem.
   108 (1994) 158.
7. K. Haneda and A. H. Morish, Solid State Comm. 22 (1977) 106.

Proc. Microscopy and Microanalysis 1995, edited by G.W. Bailey, M.H. Ellisman, R.A. Hennigar, and N.J. Zaluzec

Table I. Space group and lattice parameter of various FeO$_x$ phases (After Ref. 4-6)

Phases	Space Group	Lattice Parameter
FeO	Fm3m	a=0.4331 nm
α-Fe$_2$O$_3$	R-3c	a=0.5014 nm c=1.367 nm
γ-Fe$_2$O$_3$ (D)	Fd3m	a=0.840 nm
γ-Fe$_2$O$_3$ (P)	P4$_1$32	a=0.835 nm
γ-Fe$_2$O$_3$ (O)	P4$_1$2$_1$2	a=0.834 nm c/a=3
Fe$_3$O$_4$	Fd3m	a=0.840 nm

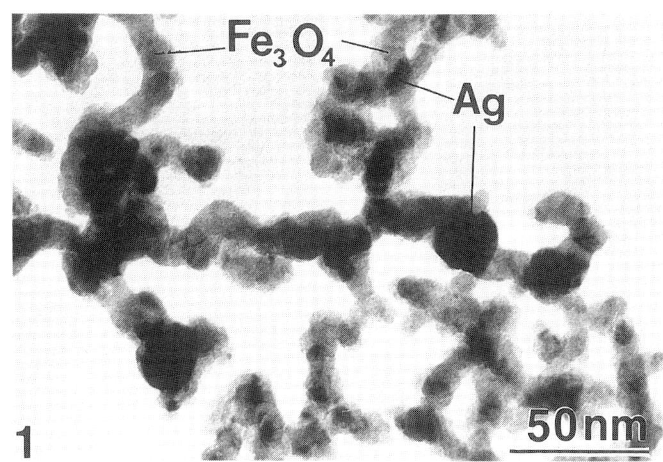

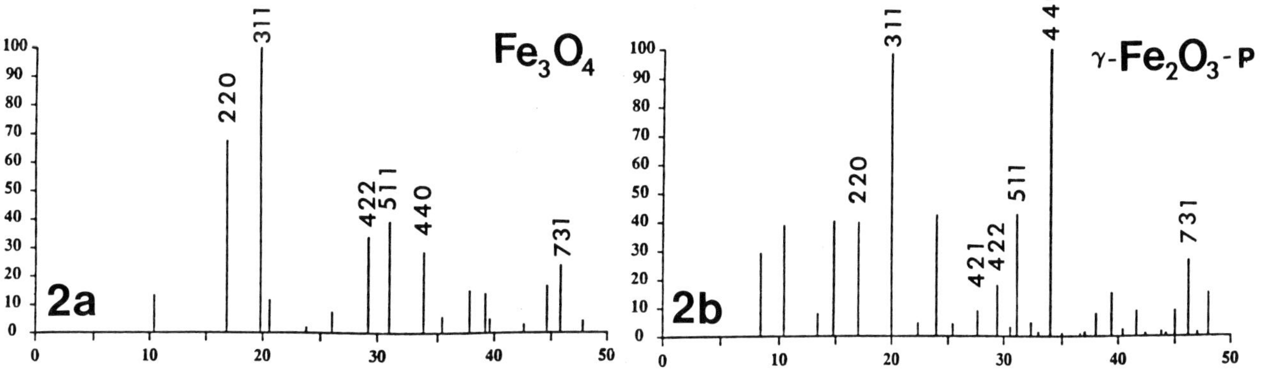

Table II. Particle sizes and oxide phases in Ag/FeO$_x$ nanocomposites .

	As-prepared	As-compacted	He-annealed	O-annealed
Ag Particles	15 nm	25 nm	30 nm	35 nm
FeO$_x$ Particles	15 nm	15 nm	20 nm	20 nm
Oxide Phases	Fe$_3$O$_4$	Fe$_3$O$_4$ + γ-Fe$_2$O$_3$	γ-Fe$_2$O$_3$ (P)	γ-Fe$_2$O$_3$ (P)

FIG. 1.   TEM micrograph of loose Ag/FeO$_x$ powder containing 40% Fe.
FIG. 2.   Kinematical diffraction intensities of (a) Fe$_3$O$_4$ and (b) γ-Fe$_2$O$_3$ (P) powders.
FIG. 3.   Lattice image of an O-annealed Ag/FeO$_x$ compacted sample containing 40% Fe.

RECENT DEVELOPMENTS IN SCANNING ION CONDUCTANCE MICROSCOPY

Roger Proksch, Ratnesh Lal and Paul K. Hansma
Department of Physics, University of California, Santa Barbara 93106

Scanning probe microscopes sensitive to two or more localized physical or chemical interactions are finding increased application in science and technology. These microscopes usually measure surface topography and another property such as electric field[1], magnetic field[2] or optical propeties such as fluorescence.[3] These simultaneously imaged properties can be correlated to provide a multimodal perspective of the sample. In order to measure ion transport through membranes and surface topography, we have developed a combination Scanning Ion Conductance Microscope (SICM) and Atomic Force Microscope (AFM). A schematic diagram of this microscope is shown in Figure 1. The SICM/AFM uses a flexible micropipette which has been pulled to a very small diameter (c.a. 50-100 nm).[4] The end of the pipette is bent at 90° and a small gold covered piece of mica is glued to the top of the bend allowing it to function as a cantilever for AFM. The pipette is filled with an electrolyte and an Ag/AgCl electrode is inserted, allowing it to serve as a conductance probe. The pipette is scanned over the surface of a specimen immersed in an electrolyte (typically 0.1M KCl). We use the deflection of the cantilevered pipette to image the sample topography and provide feedback so the tip is not pressed into the sample surface. Simultaneously, we acquire the associated ionic conductance across the sample.

An example of the pipette deflection and ionic conductance as a function of the base of the pipette-sample separation is shown in Figure 2. We have defined the contact point, where the pipette begins to deflect as 0 nm. As the base of the pipette moves past the contact point, the pipette deflection increases and the conductivity decreases. We believe the decrease in conductivity arises from the rim of the pipette being pressed with increasing force onto the sample surface, thus limiting ionic transport. Past a certain point (~1µm) the conductivity begins to increase again. This can be explained if the edge of the tip tilts up in response to a torsional force from the base of the pipette.

Figure 3 shows an example of the simultaneous SICM and AFM images of a polycabonate membrane (400 nm pore size, Milli-Pore Inc.). It shows the ionic conductance on the left and the topography on the right. There is a one to one correspondence between the ionic conductance (left panel, bright) and the holes visible in the topographic image (right panel, dark). The spatial resolution was about 200 nm for both the AFM and the SICM. This surface was imaged for several hours with no noticeable change in the AFM or SICM images suggesting negligible tip-induced sample damage. Interestingly, the ionic conductivity profile of the pores was rarely the same, even if the topographic image indicated they were the same size and shape, reflecting intrapore inhomogeneities in the polycarbonate membrane.

References

1. C. A. Huber et. al., Science, **263**, 800 (1994); 2. R. Proksch et. al., J. Appl. Phys., **75**, 5776 (1993); 3. R. Lal, et. al., Am. J. Physiol., **268**, in press (1995); 4. Nanonics Ltd. Jerusalem, Israel
This work was supported by the Materials Research Program of the National Science Foundation under grant # NSF-DMR-9123048

Proc. Microscopy and Microanalysis 1995, edited by G.W. Bailey, M.H. Ellisman, R.A. Hennigar, and N.J. Zaluzec
Copyright © 1995 MSA. Published by Jones and Begell Publishing, 79 Madison Ave., New York, NY 10016

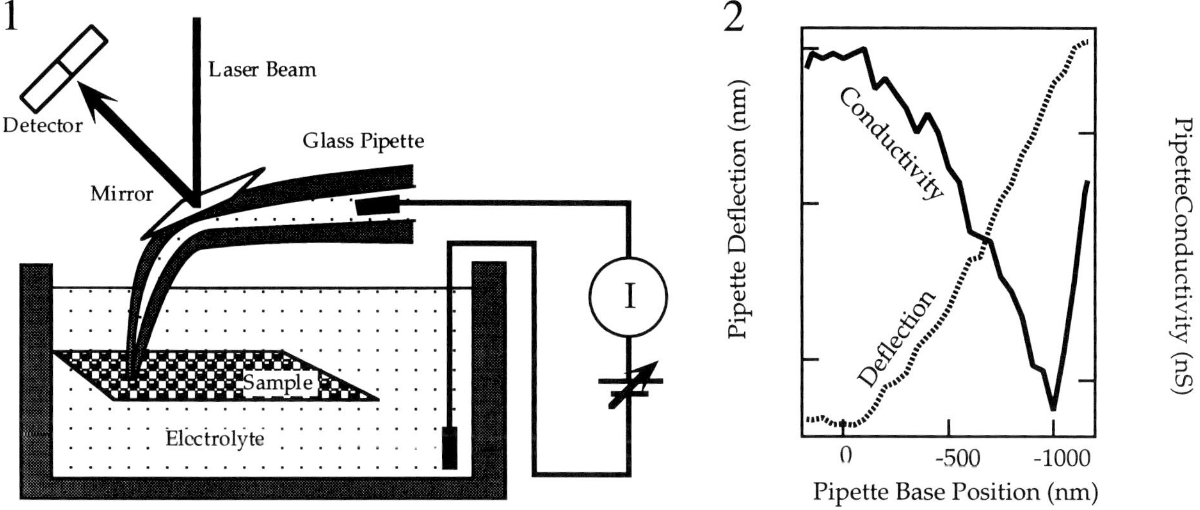

FIG. 1: Schematic diagram of the combined SICM/AFM. A laser beam reflecting off a mirror glued to the back of the pipette provides the deflection signal for the topographic image. Electrodes inside the pipette and in the sample bath provide the ionic conductivity signal.

FIG. 2: Typical deflection and ionic conductivity curves of a pipette as a function of the vertical position of the base of the pipette. The point of contact between the pipette and the membrane surface is designated as 0 nm. As the tip is pressed into the sample the deflection goes up and the conductivity decreases as the seal between the membrane and the tip gets tighter. Finally, at ~-1000nm torsional forces on the pipette cause the seal to break and the conductivity increases.

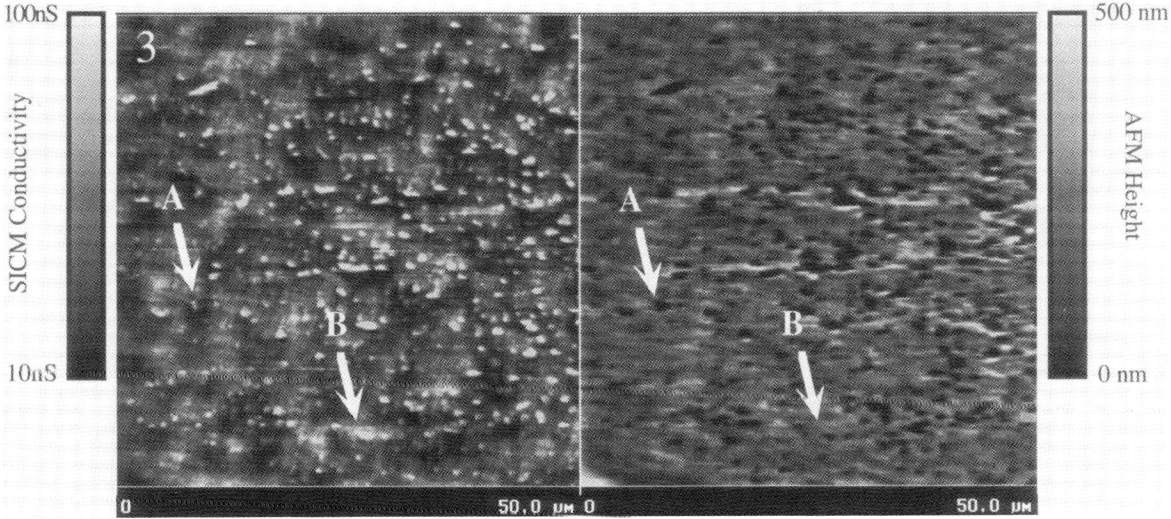

FIG. 3: Large scale image of simultaneously acquired SICM (left panel) and AFM (right panel) data. Note the one to one correspondence between most of the the ionic conductivity high spots and the topographical lows. The spatial resolution of both the SICM and AFM is about 200nm in this figure. Point A labels the location of a large pore in the topographic image. The same spot on the SICM image shows a very small conductivity. Point B labels an area where the reverse is true, there being a large conductivity in an area that is apparently closed in the topographic image.

SCANNING FORCE MICROSCOPE Z-CALIBRATION USING STEPS ON CHEMICALLY ETCHED MICA

L. Fei* and P. Fraundorf*

*Center for Molecular Electronics and Department of Physics & Astronomy, University of Missouri-St. Louis, St. Louis, MO 63121

Obtaining reliable dimensional information in all three directions is very important in scanning force microscopy (SFM). Calibration standards for SFM should be easy to produce and reliable. For example, mica is often used as a lateral calibration standard, because lattice fringes are relatively easy to obtain.[1] However, reliable height information is hard to get. Pits formed by lithography (180 nm in depth) are used for vertical calibration by some SFM manufacturers.[1] These standards have 10variability and are large on the size scale of monolayers. Colloidal gold particles have been proposed as one kind of SFM vertical standard,[2] but the size variability of these particles (e.g. ranging from 5 to 24 nm) makes their use of limited practical value.

One solution to the problem is to use chemically etched mica as height calibration standard. Mica has a layered structure, and its c-axis is well defined and weakly bonded. In this study, we used muscovite mica which has c-axis lattice spacing of 10 Å. A suitably sized piece of muscovite mica was etched with HF (52-55%, Fisher Chemical Co., Fair Lawn, NJ) for about 4 h, and rinsed with distilled water. In general, SFM reveals two kinds distinct structure regions of mica surface. One is "normal etched regions" with irregular monolayer (10 Å in height) and/or bilayer (20 Å in height) steps as shown in Fig. 1a and 1b. These regions result from etching because HF proceeds by lateral motion along the basal plane of steps, probably nucleated at clusters of point defects.[3] The other is "etched track pit regions" as shown in Fig. 2. In these regions, etch pits begin at defects (like nuclear particle tracks from actinide fissions in the mica) and develop by the creation and motion of single steps, which merge into pairs along certain crystallographic directions and remain separate along others.[3]

All of these steps immediately provide information on your microscope's height calibration. In cases where specimen surfaces are almost level, direct measurment of single layer step heights gives a good standard. However, if a specimen is tilted, as happens when line scans orthogonal to the line of steps show a saw tooth shape rather than a stair case shape, vertical height determination is slightly complicated. For images like these to be useful for vertical calibration, one needs to untilt the surface. This operation can be done by adding a planar ramp to the tilted image. We've written macros in the SEMPER image processing language to determine this ramp by first extracting the regions belong to the same step, and second least squire fitting with a plan to these regions to get the surface normal for the step, and then, averaging the surface normal of all identifiable step surfaces. One complication encountered along the way is that one first needs to convert all units in the image to real physical units to aid in decoupling the x- and z- direction sensitivities.

References

Proc. Microscopy and Microanalysis 1995, edited by G.W. Bailey, M.H. Ellisman, R.A. Hennigar, and N.J. Zaluzec
Copyright © 1995 MSA. Published by Jones and Begell Publishing, 79 Madison Ave., New York, NY 10016

1. See, for example, *NanoScope III Instruction Manual*, Digital Instruments, Santa Barbara, CA. (1993)121.

2. J. Vesenka *et al.*, *Biophys. J.* **65**(1993)992.

3. D. Snowden-Ifft *et al.*, *Phys. Rev. Lett.* **70**(1993)2348.

4. This work is supported in part by the U.S. Department of Energy, Grant No. DE-FG02-92CH10499.

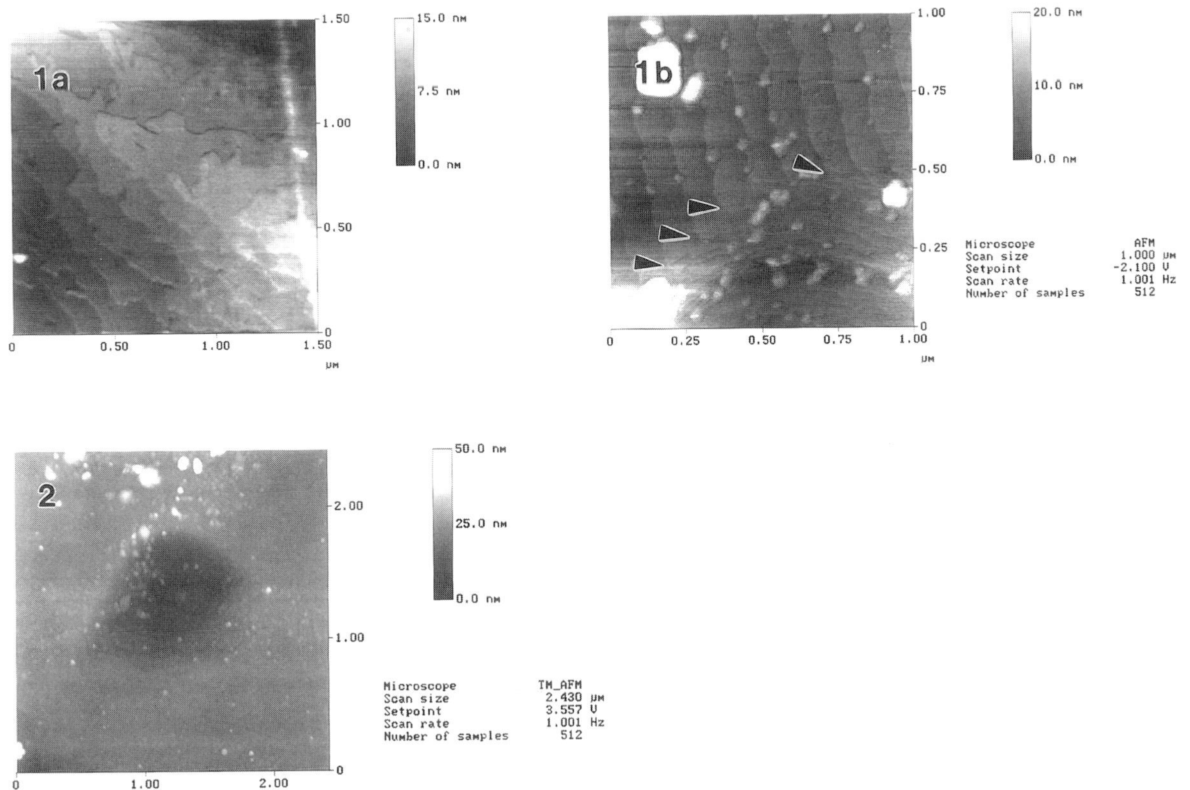

FIG. 1.—(a) Monolayer steps on an etched mica surface. Steps are irregular in shape but uniform in height. (b) bilayer steps on an etched mica surface. Steps are relatively straight. Arrows point to where bilayer steps bificate into monolayer steps.

FIG. 2.—A typical region where an etched pit is seen. The side of this kind of pits has regular shaped steps along certain crystallographic direction.

STRUCTURE OF HIGH HARDNESS CATHODIC-ARC DEPOSITED CARBON

D. L. Callahan[1], S. D. McAdams[1], S. Anders[2], A. Anders[2], I.G. Brown[2], and C. S. Bhatia[3]

[1] Dept. Mech. Engg. and Materials Science, Rice University, Houston, tx 77251-1892
[2] Lawrence Berkeley Laboratory, Berkeley, CA 94720
[3] Almaden Research Center, IBM Corporation, San Jose, CA 95120

Carbon films exhibiting a peak hardness of 59 GPa have been produced using pulse-biased cathodic-arc deposition.[1] This value is much greater than that expected of an amorphous "diamond-like" carbon (dlc) film and is well within the hardness range of chemically vapor-deposited diamond. Furthermore, this peak hardness is observed at an indentation depth of approximately 50 nm and not on the upper surface of the film. A structural analysis by transmission electron microscopy (TEM) has been conducted in order to explain these mechanical phenomena.

Details of the sputtering apparatus are described elsewhere.[2] Films for this study were deposited on silicon using a two-step process in which a thin layer, approximately 10 nm, was first deposited at a relatively high substrate bias (~2 kV) followed by growth of the majority of the film at a lower bias (~200 V). Plan-view examination of carbon films revealed both uniformly amorphous regions and extensive nanocrystalline phase regions. Examination of inclined boundaries indicate that the nanocrystalline phase may exist between two amorphous regions. The crystallites in these regions are on the order of 10 nm in diameter (Fig. 1). The diffraction patterns from this area (Fig. 2) and similar nanocrystalline regions correspond to that of polycrystalline diamond and include a {200} ring resulting from {110} double-diffraction.

Figure 3 is a bright field/diffuse dark field pair showing the film to be approximately 350 nm thick in cross-section. The diffuse dark field image (Fig. 3, right) indicates that there is an amorphous layer at both the surface of the film and at the carbon-silicon interface. The thickness of each of these layers was determined to be at most 50 nm and more probably in the range of 20-40 nm. Analysis of the interior of the thin film in cross-section has been inconclusive to date due to substantial thickness gradients.

These observations are consistent with a layered structure consisting of a hard, predominantly nanocrystalline inner layer bounded on each surface by softer amorphous regions. The amorphous interlayer is the result of high energy implantation / deposition designed to provide an adhesive layer between the silicon substrate and the hard carbon film. The hard inner layer containing nanocrystalline diamond is interpreted as the result of subplantation at energies in excess of the graphite displacement energy and resulting high compressive stresses. The amorphous surface may indicate the minimum effective depth of the subplantation process.

1. G. M. Pharr, et al, submitted to *Applied Physics Letters*.
2. A. Anders, S. Anders, and I. G. Brown, in press, *Plasma Sources Sci. Tech.*
3. This work was supported by the Advanced Research Projects Agency through the National Storage Information Consortium program in Ultra High Density Recording

Proc. Microscopy and Microanalysis 1995, edited by G.W. Bailey, M.H. Ellisman, R.A. Hennigar, and N.J. Zaluzec
Copyright © 1995 MSA. Published by Jones and Begell Publishing, 79 Madison Ave., New York, NY 10016

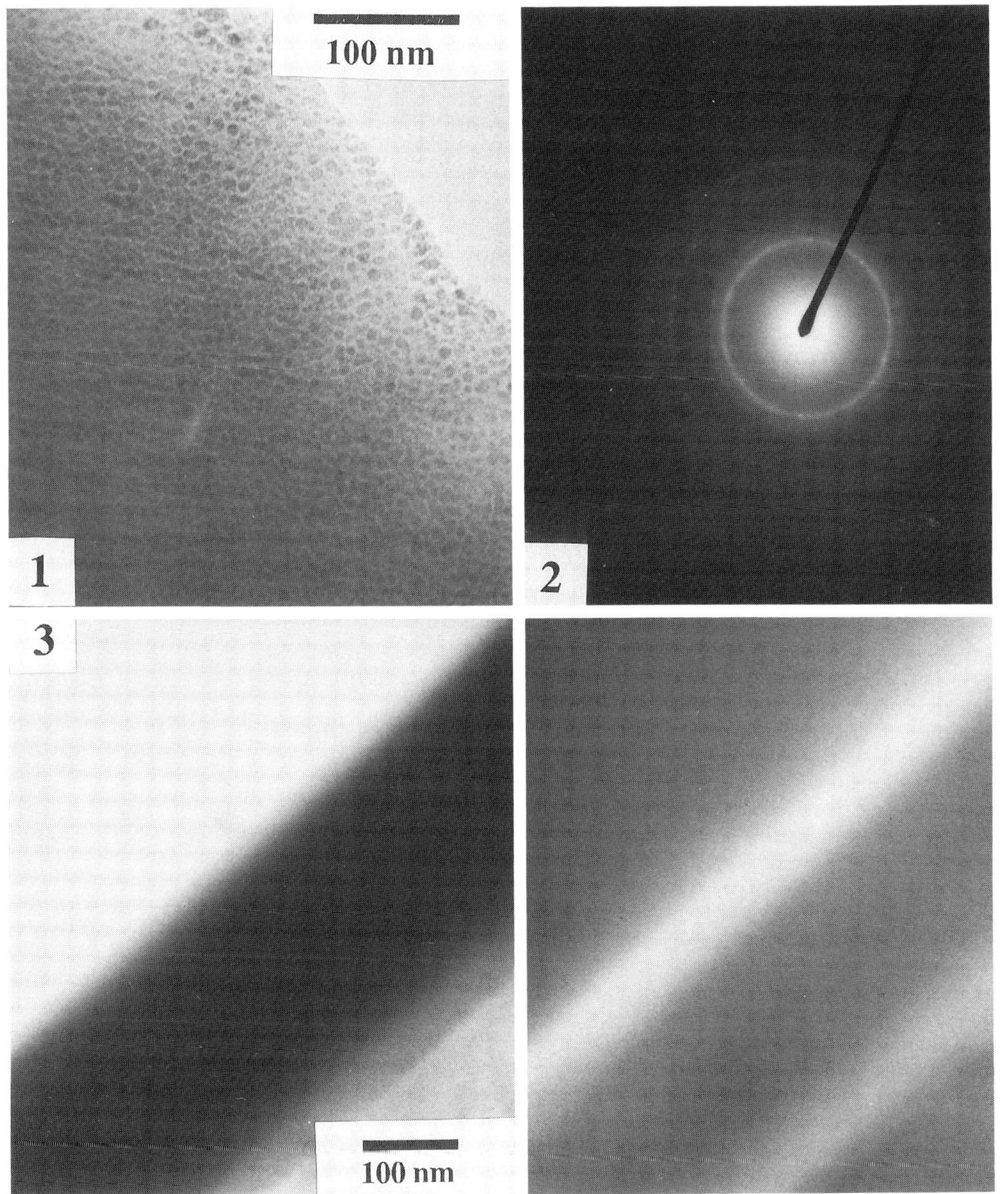

FIG. 1. -- Bright field TEM micrograph of nanocrystalline diamond region from cathodic-arc deposited carbon.

FIG. 2. -- Selected area diffraction pattern of nanocrystalline diamond

FIG. 3 -- left: bright field TEM micrograph of carbon film in cross-section
right: corresponding diffuse dark field TEM micrograph

DIRECT OBSERVATION OF HYMENES IN DIATOMS BY LOW VOLTAGE SEM

Nan Yao[1] and Ilhan A. Aksay[1,2]

[1]Princeton Materials Institute and the [2]Department of Chemical Engineering,
Princeton University, New Jersey 08540-5211

From the early days of Kolkwitz and Marsson (1908)[1] diatoms have proved to be valuable in pollution studies, especially of river systems. Economic applications have developed because the remains of sedimentary diatoms were found to be useful for many industrial processes owing to the fine structure and inert nature of the siliceous materials. Their ecological significance arises from observing that diatoms that grew in oceanic and freshwater sediments contain indications of past environmental conditions; more recently, diatoms have been used to detect pH changes attributed to 'acid rain.' The first record of a diatom can be traced back to the early 1700s when an English gentleman looked at roots of the pond-weed *Lemna* using his simple microscope.[2] Today, extensive studies using TEM and SEM techniques have provided much new information about the fixed and variable details of the diatom, yet research focusing on the morphology of the shell structure remains very active. In this paper, we discuss LVSEM observations of hymenes in the diatom ultrastructure which have been often obscured by investigators using conventional SEM and TEM techniques.

Diatoms are algae, microscopic unicellular plants that range in size from approximately 5 µm to 2 mm and live wherever there is moisture, even in soil. The hallmark of the diatom is its cell wall which is highly differentiated and almost always heavily impregnated with silica ($SiO_2 \cdot nH_2O$). The cell wall is composed of at least two large structural siliceous elements, the valve and the girdle. The valves lie at each end of the cell and the girdle elements surround the regions in between. This structure is so regularly reproduced that for many years it has been the main criterion for the systematic classification of diatoms. Figure 1 is a low magnification SEM image of a diatom from freshwater, recorded with a standard Everhart-Thornley SE detector positioned to one side of an uncoated specimen. The valve is lanceolate of about 30 µm long and 5 µm wide and the valve face is gently curved in transapical section. The valve contains small round pores in linear arrays slightly curved across the valve face. According to Barber and Haworth,[3] the structure of the valve clearly places this genus in the class of *Synedra*.

One of the major factors determining the visible contrast in SEM images is the nature of the interaction between the incident electron beam and solid. The production rate of secondary electrons as a result of such interaction is proportional to the stopping power of the electron as described by Bethe law,

$$D_B = \int_{E_{min}}^{E_{inc}} [-dE / dl]^{-1} \cdot dE,$$ where dE/dl is the stopping power of an incident electron with energy E_{inc}

and E_{min} is a suitable lower energy limit for the interaction. It is known that the electron stopping power of about 1 eV/Å at 10 keV increases dramatically to about 10 eV/Å when energies decrease below 1 keV.[4] In accordance with this, the electron penetration depth falls significantly from the micrometer regime to only tens of nanometers, and the production of secondary electrons rises rapidly in the low energy range. Therefore, the two dimensional information of a specimen surface might be inevitably overlooked due to the deep penetration depth at voltages above 10 keV. Figures 2(a)-(c) are three high magnification SEM images obtained at 3 keV, 2 keV, and 0.8 keV, respectively, showing details of pore structures of about 10 nm in diameter on the diatom valve face as seen in Figure 1. The hymene layer located near the external end of pores is almost invisible even at 3 keV electron beam incidence because

Proc. Microscopy and Microanalysis 1995, edited by G.W. Bailey, M.H. Ellisman, R.A. Hennigar, and N.J. Zaluzec
Copyright © 1995 MSA. Published by Jones and Begell Publishing, 79 Madison Ave., New York, NY 10016

the majority of electrons traverse the siliceous thin layer without producing any interaction. At voltages less than this, the underlying structure around the pore end becomes progressively more visible (Figure 2(b)). At 0.8 keV, all of the electron interaction volume is contained within the thickness of the hymene layer, and the resultant secondary electron emissions make it appear solid and opaque with contrast similar to the valve framework background. Based on the above analysis, the thickness of the hymene layer can be estimated to be less than 20 nm which is compatible with the depth of information for SEM imaging. The fact that the diatom valve pores which appear unoccluded even at 3 keV image are indeed occluded by the hymene layer is a good example indicating that low voltage SEM is the method of choice for resolving detailed surface structures.[5]

References:

1. R. Kolkwitz and M. Marsson, *Ber. Deutsch. Bot. Ges.,* **26**(1908)505.
2. F. E. Round, R. M. Crawford, and D. G. Mann, *The Diatoms,* Cambridge University Press (1990).
3. H. G. Barber and E. Y. Haworth, *A Guide to the Morphology of the Diatom Frustule,* Fresh Water Association, Scientific Publication No. 44(1981).
4. D. C. Joy and C. S. Joy, *Micron,* in press.
5. This work made use of MRSEC Shared Facilities supported by the National Science Foundation under Award Number DMR-940032

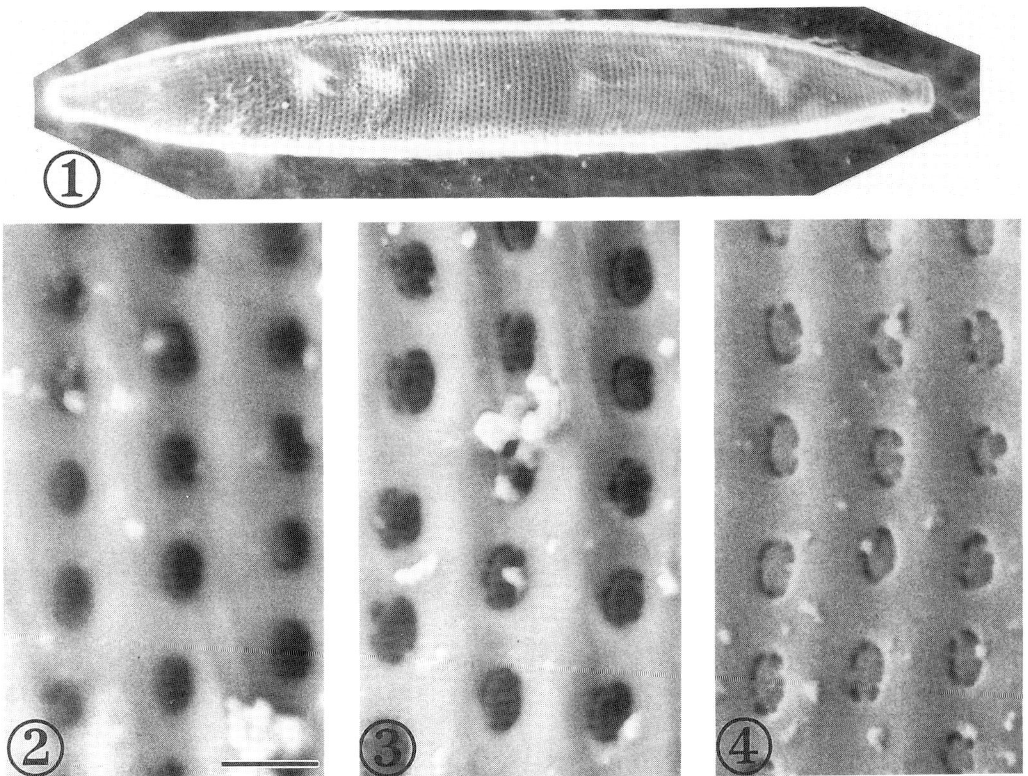

FIG. 1.—Low magnification SE image of the valve view showing the morphology of a freshwater diatom.

FIG. 2.—SEM images of the same diatom obtained at (a) 3 keV, (b) 2 keV and (c) 0.8 keV showing the change in image contrast. It is noteworthy that the hymene layer appears solid and opaque only at 0.8 keV. Images recorded from JEOL 6030 FEG SEM.

AMORPHOUS SILICA COATING ON α-ALUMINA PARTICLES

J. W. Mellowes,[#*] C. M. Chun,[‡*] and I. A. Aksay[#*]

Departments of [#]Chemical Engineering and [‡]Geological and Geophysical Sciences; and
*Princeton Materials Institute, Princeton University, Princeton, New Jersey 08544-5211

Mullite ($3Al_2O_3 \cdot 2SiO_2$) can be fabricated by transient viscous sintering using composite particles which consist of inner cores of α-alumina and outer coatings of amorphous silica.[1,2] Powder compacts prepared with these particles are sintered to almost full density at relatively low temperatures ($\sim 1300°C$) and converted to dense, fine-grained mullite at higher temperatures ($>1500°C$) by reaction between the alumina core and the silica coating. In order to achieve complete mullitization, optimal conditions for coating alumina particles with amorphous silica must be achieved. Formation of amorphous silica can occur in solution (homogeneous nucleation) or on the surface of alumina (heterogeneous nucleation) depending on the degree of supersaturation of the solvent in which the particles are immersed. Successful coating of silica on alumina occurs when heterogeneous nucleation is promoted and homogeneous nucleation is suppressed.[3] Therefore, one key to successful coating is an understanding of the factors such as pH and concentration that control silica nucleation in aqueous solutions. In the current work, we use TEM to determine the optimal conditions of this processing.

Silica coated alumina particles were prepared by dispersing α-alumina (AKP-30, Sumitomo Chemical) in 60 ml of ethanol and adding tetraethoxysilane (TEOS). After mixing with a magnetic stir bar for 1 hour, silica is precipitated onto the alumina particles by adding 30 ml ammoniated water. Finally, the suspension of coated particles is filtered after 24 hours of stirring and the particles are dried. TEM images (Figs. 3 and 4) indicate that α-alumina core particles are coated with amorphous silica. From the thickness of a coating and by knowing the surface area of the powder and density of amorphous silica, the mass of the coating is calculated. When the mass of the coating is compared to the amount of TEOS used initially, it is possible to determine whether homogeneous or heterogeneous nucleation is occurring within the solution.

The nucleation of amorphous silica is dependent on both pH and concentration of TEOS. Keeping the concentration of TEOS constant, the effects of altering the pH of the ammoniated water were investigated (Fig. 1). At pH < 10, no coatings were present. With increasing pH, the thickness and fraction of coating increased. At pH > 11, particles were completely coated without silica precipitation out of solution. Because the solubility of silica in water increases at higher pH, coatings without precipitation were promoted. Holding the pH of the ammoniated water constant at 11.5, the effects of changing the concentrations of TEOS added were investigated (Fig. 2). At low concentrations, all the TEOS added coated the alumina and therefore the coating thickness was directly related to the amount added. At higher concentrations, the thickness of the coating decreased. In these situations, the degree of supersaturation was high, thus forming many nucleation sites within the solution, making homogeneous nucleation the favored mechanism. This study illustrates the crucial role of optimization within the pH and concentration regime in order to promote silica coating without any homogeneous nucleation.[4]

Proc. Microscopy and Microanalysis 1995, edited by G.W. Bailey, M.H. Ellisman, R.A. Hennigar, and N.J. Zaluzec
Copyright © 1995 MSA. Published by Jones and Begell Publishing, 79 Madison Ave., New York, NY 10016

References:

1. M. D. Sacks *et al.*, *J. Am. Ceram. Soc.*, **74**(1991)2428.
2. I. A. Aksay *et al.*, *J. Am. Ceram. Soc.*, **74**(1991)2343.
3. B. C. Bunker *et al.*, *Science*, **264**(1994)48.
4. This work was supported by a grant from the U. S. Air Force Office of Scientific Research AFOSR-F49620-93-1-0259 and made use of MRSEC Shared Facilities supported by the NSF under Award Number DMR-940032.

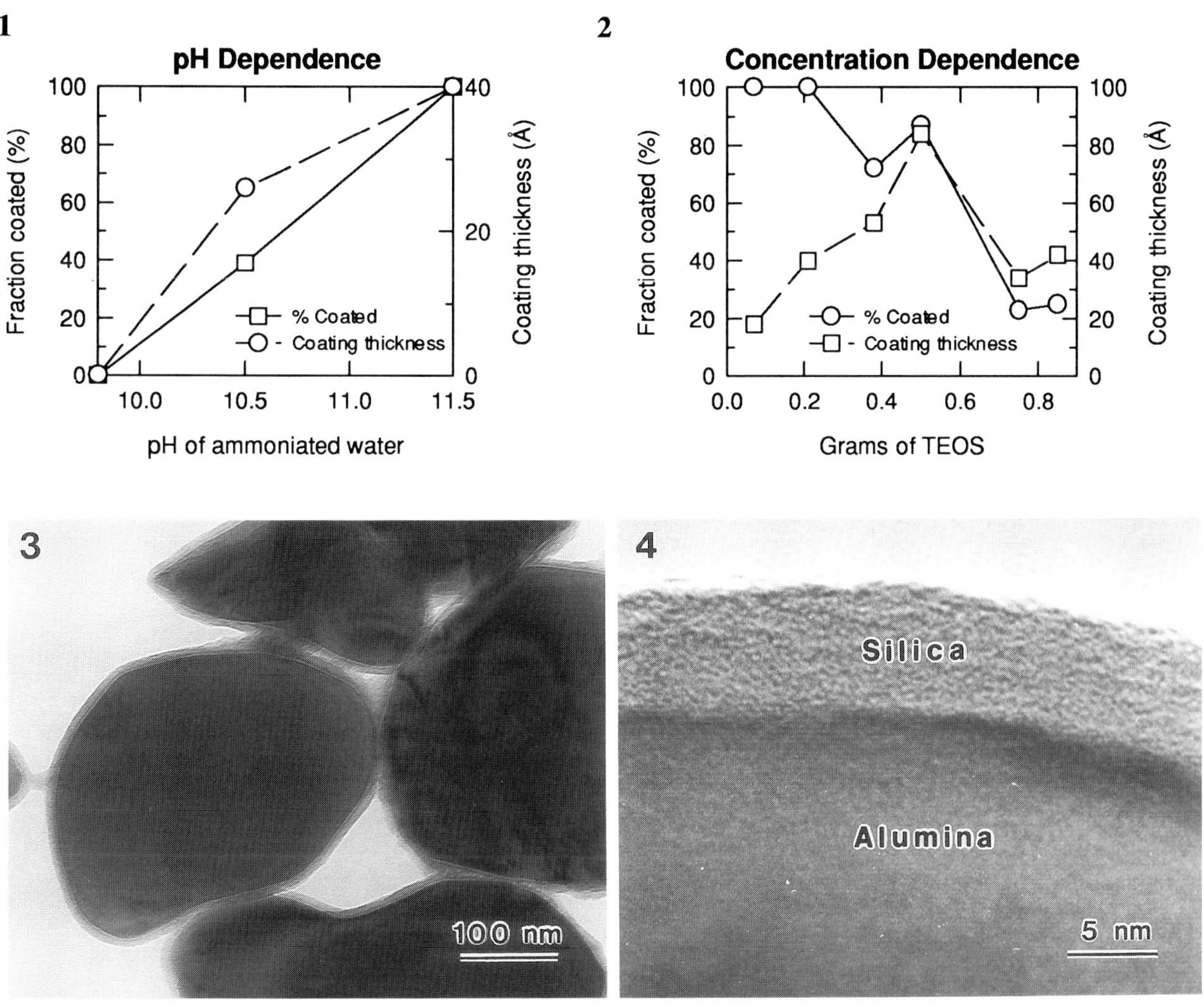

FIG.1.—Coating thickness and coating fraction (mass of silica coating/total silica used in experiment) of composite particles vs. pH of ammoniated water at constant TEOS concentration.

FIG.2.—Coating thickness and coating fraction of composite particles vs. various TEOS concentrations at constant pH of 11.5.

FIG.3.—TEM image of α-alumina core particles coated with amorphous silica.

FIG.4.—HREM image of interface between α-alumina and amorphous silica.

NANOCOMPOSITE PROCESSING VIA INFILTRATION OF MESOPOROUS SILICA

U. Srinivasan,[#*] I. Homma,[#*] C. M. Chun,[‡*] D. M. Dabbs,[#*] D. A. Hajduk,[†*] S. M. Gruner,[†*] and I. A. Aksay[#*]

Departments of [#]Chemical Engineering, [‡]Geological and Geophysical Sciences, and [†]Physics; and *Princeton Materials Institute, Princeton University, Princeton, New Jersey 08544-5211

Synthesis of materials with nanoscale (1~100 nm) organization is important in various applications. Recently, scientists at Mobil described a new family of mesoporous molecular sieves.[1] These materials have regular arrays of uniform pore channels ranging from 1.6~10 nm in diameter in contrast to other mesoporous solids such as amorphous silicas and modified layered clays and silicates.[2] A surfactant-silicate co-assembly model has been proposed to explain the formation of these materials.[3] According to this pathway, the matching of the charge density at the organic-inorganic interface controls the assembly of the mesophases and four distinct silica mesophases have been observed, lamellae, hexagonally packed tubes, and two bicontinuous structures of cubic symmetry. Different phases are constructed by varying the synthesis parameters such as the surfactant/silicate ratio and the acidity. Here, we report the synthesis of a new amorphous mesoporous phase with short range order and with no long range crystallinity. Due to its interpenetrating network structure, this amorphous mesoporous silica can be used as a matrix for nanocomposite processing. Here, we use it as a host to process ruby glass.

In order to synthesize this amorphous mesoporous silica (AMS) phase, 1.11 g of C_{16}TMACl (29% aqueous solution, Akzo Chem.), 11.2 g of 10 N HCl, and 400 g of deionized water are stirred for 10 min. Then, 32 g of TEOS (Alpha Chem.) is added and the solution is stirred for 15 min or until the solution becomes transparent. The mixture is poured into a flat dish and is allowed to evaporate in a fume hood until a transparent solid forms. These (AMS) samples were ground and dispersed in methanol using ultrasonication. TEM (Fig. 1) shows no periodic order. XRD (Fig. 2) shows a single broad maximum consistent with disordered pores with ca. 8 nm spacing. The transparent samples were calcined at 600°C for 4 hours to remove the surfactant from the mesopores. After heat treatment, the AMS samples remain transparent and colorless. BET surface area analysis (Micromeritics Flow Sorb II2300) gives surface areas of 680 m^2/g. Since calcined AMS materials are transparent and have high surface areas, they are prime candidates for infiltration with another material to form a nanocomposite.

In order to dope the AMS samples with nanosized metallic gold, calcined AMS pieces were soaked overnight in a 10 wt% aqueous solution of gold (III) chloride (Aldrich Chem.). Next, the pieces were carefully washed with deionized water to remove the excess gold chloride ions adsorbed onto their surfaces. Then, the pieces were soaked in a 1 wt% aqueous solution of citric acid, trisodium salt dihydrate (Aldrich Chem.) to reduce the adsorbed anions to metallic gold inclusions. At this stage, the thick films became purple while maintaining transparency. This color change can be attributed to the plasmon absorption of the nanosized gold particles encapsulated within the mesochannels of the AMS films. TEM (Fig. 3) shows that following reduction, the particles in the mesopores aggregate with those from neighboring channels to form branched clusters approximately 10 nm in diameter. In comparison to physical infiltration techniques which involve ion implantation, high-vacuum evaporation and plasma processing, the low-temperature route to nanocomposite synthesis described in this paper is economical and can be used to create other nanocomposites for electronic and optical applications.

Proc. Microscopy and Microanalysis 1995, edited by G.W. Bailey, M.H. Ellisman, R.A. Hennigar, and N.J. Zaluzec

References:

1. J. S. Beck *et al.*, *J. Am. Chem. Soc.*, **114**(1992)10834.
2. C. T. Kresge *et al.*, *Nature*, **359**(1992)710.
3. M. D. McGehee *et al.*, *Proc. Ann MSA Meeting*, **52**(1994)448.
4. This work was supported by a grant from the U. S. Air Force Office of Scientific Research AFOSR-F49620-93-1-0259 and made use of MRSEC Shared Facilities supported by the NSF under Award Number DMR-940032.

1   2

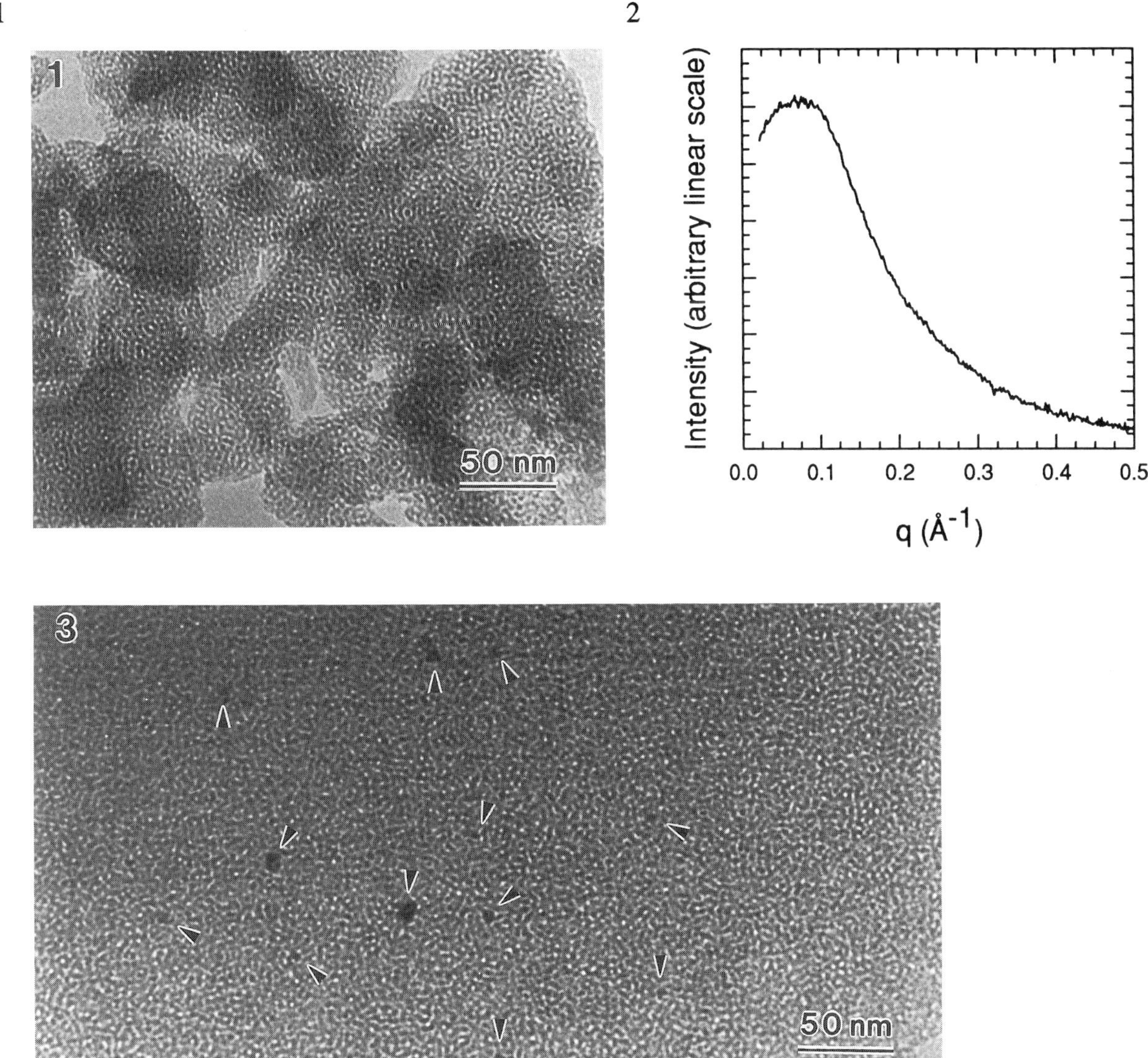

FIG.1.—TEM image of an amorphous mesoporous silicate (AMS) material.
FIG.2.—Low angle XRD spectra for an AMS material.
FIG.3.—TEM image of calcined AMS material infiltrated with nanosized gold inclusions.

CRYSTALLIZATION OF NANO-SIZE THALLOUS OXIDE DURING ION IRRADIATION OF Tl-Ba-Ca-Cu-O HIGH TEMPERATURE SUPERCONDUCTORS

P. P. Newcomer,* L. M. Wang,** M. L. Miller,** R. C. Ewing,**

*Sandia National Lab., Albuquerque, NM, 87185
**Univ. of New Mexico, Albuquerque, NM 87131

The Tl-Ba-Ca-Cu-O class of type-II high temperature superconductors (HTS) have Tc's as high as 125K. Although they have good critical current values, when a field is applied the weak pinning and consequent flow of magnetic vortices are a major impediment to the usefulness of these materials. Ion irradiation has been shown to enhance the pinning.[1] High quality single crystals, as determined with x-ray precession and HRTEM, with sharp HTS Meissner signals, were irradiated with 1.5 MeV Kr^+ and Xe^+ ions using the HVEM-Tandem facility at Argonne National Laboratory. Ion beam microstructural modification was studied in-situ using electron diffraction and after irradiation using HRTEM and nano-beam EDS on Tl-1212 and Tl-2212 (numbers designate the stoichiometry Tl-Ba-Ca-Cu-O) single-crystal HTS. After irradiation, microstructure was studied using the JEOL 2010 in the Earth and Planetary Science Department at the University of New Mexico in order to characterize the resulting irradiation-induced nano-size precipitates.[2]

Electron diffraction, during irradiation at various temperatures, shows an amorphous halo developed quickly, after only 10^{12} ions/cm^2, in all specimens. The dose was increased until complete modification of the original structure was complete, Dc. Unexpectedly, when irradiating at very low temperature, a temperature where quenching of the amorphized state is expected,[3] polycrystalline rings developed. As expected, when irradiating at high temperatures additional diffraction spots are observed and a polycrystalline ring pattern developed, indicating radiation induced polycrystallization occurs. Diffusion resulting in damage recovery and/or nucleation of a second phase can be induced in the displacement cascade during the thermal spike.[4]

After a low dose of 8.5×10^{12} ions/cm^2 at 22K, nano-size precipitates and isolated amorphized domains were observed. HRTEM of the damaged regions revealed circular, misoriented, cascade-size (~5 nm) precipitates and amorphous domains as shown by Moire fringes, lattice fringes, and non-coherent domains. Precipitates show dark contrast in bright field TEM. Digital HRTEM images were processed using the methods of Sattler and O'Keefe, and an intensity profile across the processed image approximates the size of the damaged regions.[1] After high dose Dc, at very low temperature, the precipitates remain crystalline and are well dispersed in an amorphized matrix. Nano-beam EDS, with 1 to 3 nm beams, shows the dark contrast precipitates are thallium rich. Micro selected area-diffration reveals the thallous oxide phase. This suggests the Thallium is reducing from Tl^{3+} to Tl^{1+} as in annealed Tl-type HTS.[5] Perhaps the heat from the thermal spike was not dissipated due to poor thermal conductivity and thallous oxide nucleated (melts at 300 C). Various models are analyzed in order to interpret the ion beam modification mechanism with these materials. Work supported by BES under DOE contract numbers DE-AC04-94AL85000 and DE-FG03-93ER45498.

References

1. J.C. Barbour, E. L. Venturini, D.S. Ginley, J.F. Kwak, Nucl. Instr. and Methods B65 (1992) 531.
2. L.M. Wang, P.P. Newcomer, R.C. Ewing, B. Morosin, ICEM 13-Paris, 17-22-July (1994) 957.
3. W.J. Weber, R.C.Ewing, L.M. Wang, J. Mater. Res. 9(3) (1994).
4. R.S. Averback, Nucl. Instr. and Methods in Phys. Research B15 (1986) 675.
5. B. Morosin, E.L. Venturini, D.S. Ginley, Physica C 183 (1991) 90.

Proc. Microscopy and Microanalysis 1995, edited by G.W. Bailey, M.H. Ellisman, R.A. Hennigar, and N.J. Zaluzec
Copyright © 1995 MSA. Published by Jones and Begell Publishing, 79 Madison Ave., New York, NY 10016

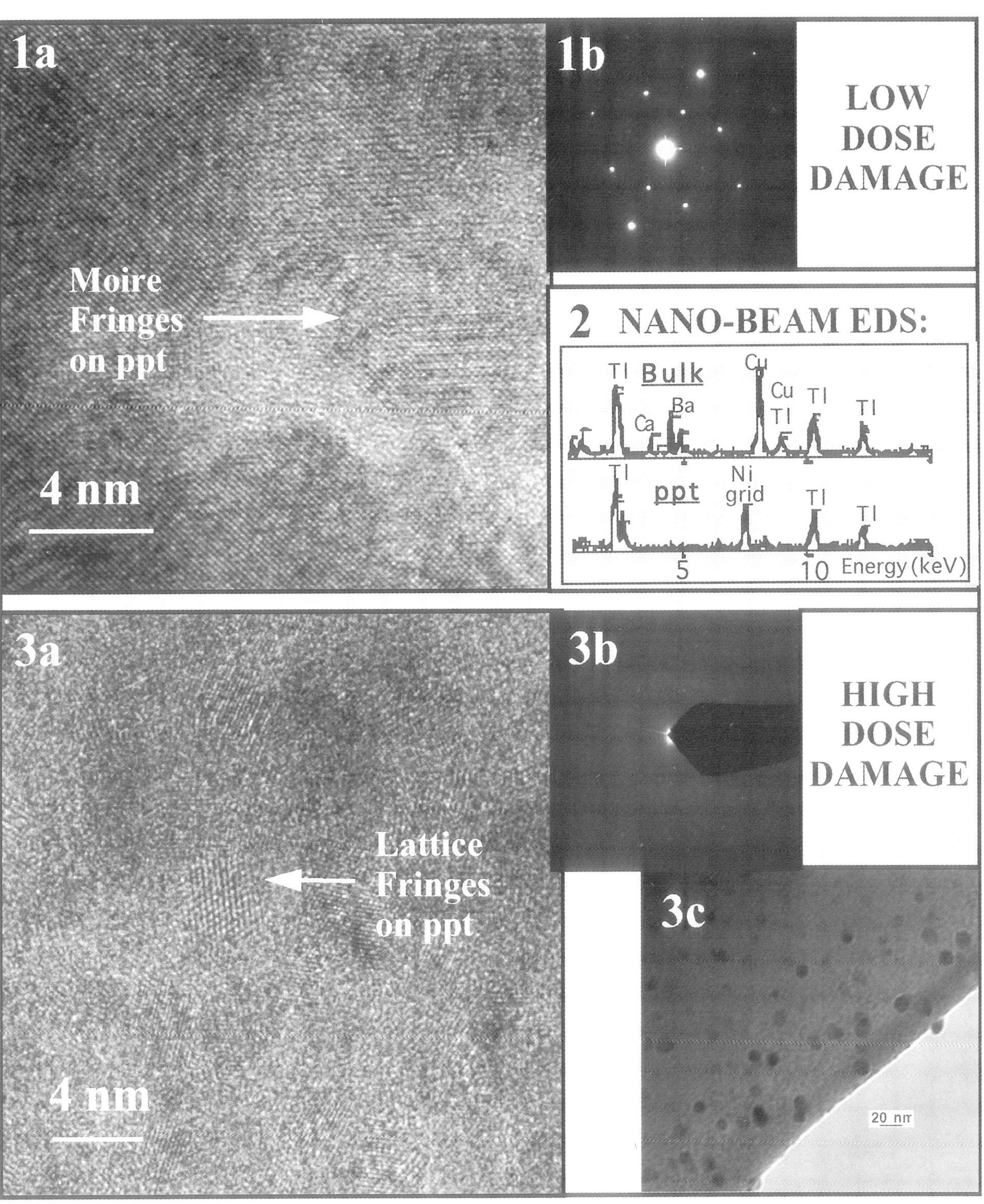

FIG. 1 1a) Moire fringes after low dose (8.5×10^{12} ions/cm^2) show that precipitates are induced in the crystalline matrix during irradiation, 1b) electron diffraction of area. 2) Nano-beam EDS of a precipitate compared to the bulk. 3a) Crystalline precipitates exist in amorphous matrix, after high dose and 3b) electron diffraction of area. 3c) Low magnification image shows the dark contrast precipitates in the amorphous matrix.

AEM CHARACTERIZATION OF MAGNETIC NANOPARTICLES ENCASED IN GRAPHITE SHELLS

J. J. Host, M. H. Teng, J. H. Hwang, B. R. Elliott, J. R. Weertman, T. O. Mason, D. L. Johnson, and V. P. Dravid

Department of Materials Science and Engineering, Northwestern University, Evanston, IL 60208

Since it was discovered that crystalline materials could be encased in graphite shells, protection in this manner has been the focus of significant research. Previously, production of graphite encapsulated crystalline nanoparticles has been reported using an arc discharge between an anode stuffed with a metal oxide and a graphite cathode, resulting in a mixture of carbides, encapsulated carbides and carbonaceous debris[1-4]. More recently, an arc discharge[5] between metal electrodes has been used to produce large quantities of pure metal nanoparticles, with the addition of graphite to the anode resulting in the large scale production[6] of graphite encapsulated metal nanoparticles (GEMN). In earlier studies, these materials were separated from the other products of the arc discharge only by a magnetic gradient, which does not remove the non-encapsulated crystalline particles. Therefore, the immersion of the product mixture in an acid bath was added as a subsequent processing step. The combined arc discharge/acid bath technique reduces both production and separation problems, producing a large quantity of Fe, Co, and Ni GEMN free from any non-encapsulated metal particles. Subsequent characterization of the resulting material has led to a better understanding of the GEMN produced.

AEM characterization revealed the size distribution, chemical composition, and structural details of the encapsulated metals. The methods utilized included conventional TEM using both the Hitachi HF-2000 and the Hitachi H-700H TEMs, high resolution electron microscopy (HF-2000), Energy Dispersive Spectrometry (HF-2000), Electron Energy Loss Spectroscopy (HF-2000), and Selected Area Electron Diffraction (H-700H). Conventional TEM demonstrated that the particles were spherical in shape, and that the particle diameter was 10 to 50 nm. Furthermore, the average particle size can be controlled by adjusting a processing parameter[5]. High resolution electron microscopy of the GEMN (Fig. 1) showed that the graphite layer was usually curved to follow the spherical contour of the particle, and that larger particles were usually coated with a larger number of graphite sheets than the smaller particles. Energy Dispersive Spectrometry (EDS) using the Oxford-Link x-ray detector showed (Fig 2) that each of the encapsulated samples was composed solely of carbon (largely from the protective graphite layers) and either Fe, Co, or Ni (the encapsulated material). EELS was used to investigate the presence of carbon-metal bonds in the bulk of the particles (Fig 3). Lastly, selected area electron diffraction using the Hitachi H-700H TEM was used to determine the phase of the metal particles[7].

References

1.   Y. Saito, et al, *Chem. Phys. Let.* **212** (1993) 379
2.   S. Subramoney, et al., *Carbon,* **32** (1994) 507
3.   R. Ruoff, et al., *Science,* **259** (1993) 346
4.   M. Tomita, et al., Jpn. J. Appl. Phys. **32** (1993) 280
5.   M. H. Teng, et al., *J. Mater. Res.,* **10**, (1995) 233
6.   V. P. Dravid, et al., *Nature* (1995) Accepted for Publication
7.   This research is supported by NSF Grant DMR-9202574

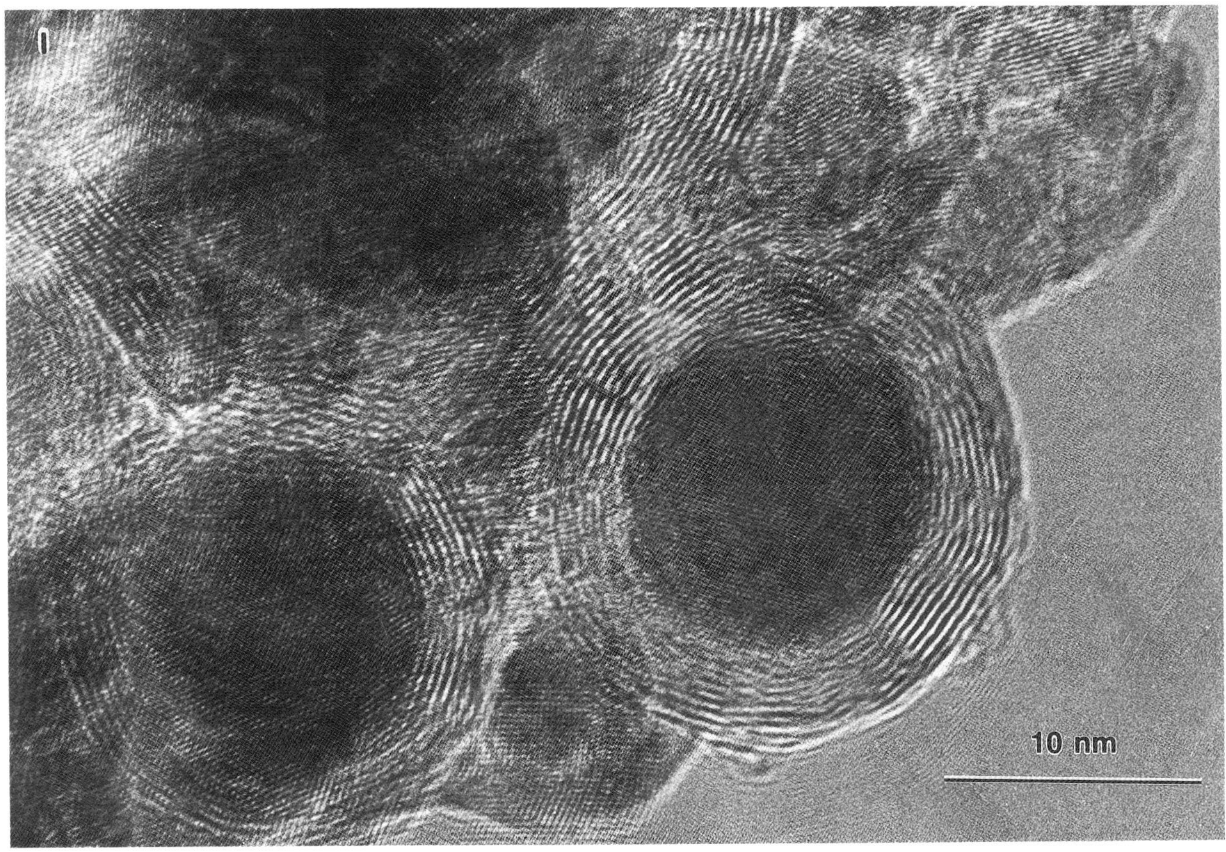

FIG. 1 - HREM image of a graphite encapsulated nickel nanoparticle.   Note curvature of graphite layers, following the contour of the particle.

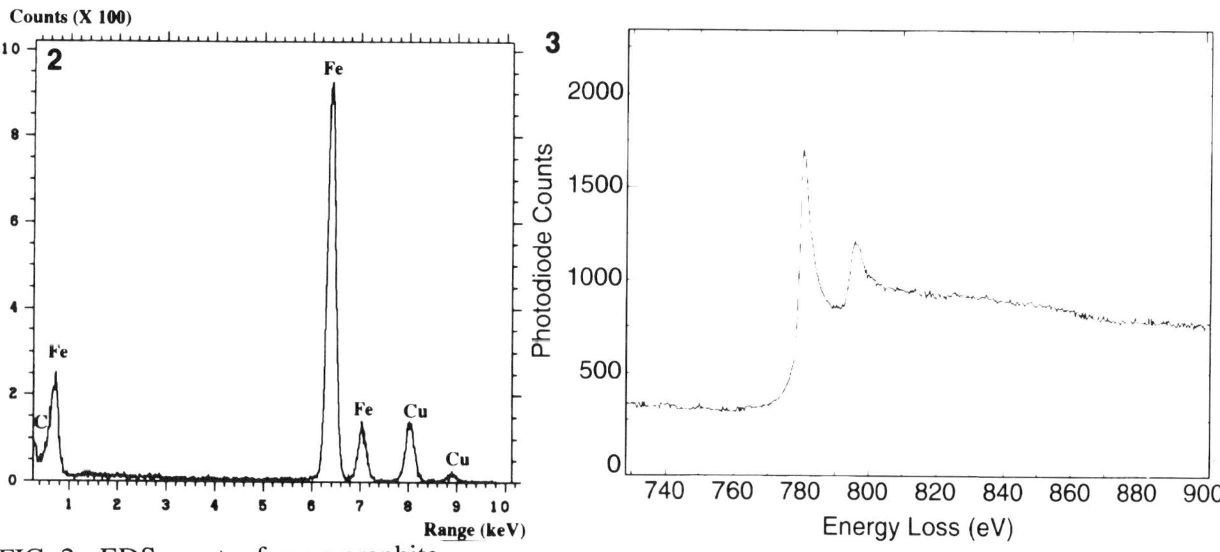

FIG. 2 - EDS spectra from a graphite encapsulated iron sample, showing that the sample is composed only of Fe, C (largely from the encapsulating graphite layers), and Cu from the support grid.

FIG. 3 - Core loss EELS spectra from a graphite encapsulated cobalt particle.

ANALYSIS OF NANOMETER SIZED PYROGENIC PARTICLES BY SCANNING TRANSMISSION ELECTRON MICROSCOPY

DJ Wallis[*][1], ND Browning[*] and CM Megaridis[**]

[*]University of Illinois at Chicago, Dept. Physics (M/C 273), 845 W. Taylor, Chicago, Il 60607 - 7059
[**]University of Illinois at Chicago, Dept. Mech. Eng. (M/C 251), 842 W. Taylor, Chicago, Il 60607

Iron is a ubiquitous element on the earth's surface, and is thus involved in most naturally occurring fires. Iron organometalic compounds have also been known to suppress carbonaceous soot emissions under certain operating conditions of practical combustors. In order to unravel the physical and chemical mechanisms of influence, of iron on the emission of carbonaceous pyrogenic particles, fine-scale characterization techniques need to be implemented.

The combined techniques of Z-contrast imaging and electron energy loss spectroscopy (EELS) in a VG HB-501 dedicated STEM are ideally suited to study such a system. The sensitivity of the Z-contrast imaging technique[2] to high-Z materials makes it ideal for location of the iron particles within the much lower atomic number matrix. As only the high-angle scattering is used in the image formation, EELS can be performed simultaneously from a position defined in the image. This accurate positioning of the probe by the Z-contrast image permits both compositional and bonding information to be obtained with a spatial resolution approaching the atomic scale[2-3]. Additionally, the high-efficiency of modern parallel EELS devices enables the detection of trace components consisting of only a few atoms[4,5].

Figure 1 shows a Z-contrast image of Iron bearing particles produced by the introduction of ferrocene, $(C_5H_5)_2Fe$, into a combustion mixture (ethylene + air). Particles were collected onto a holey carbon support grid by passing it through the flame. In figure 1 particles ranging in size from 20nm to about 1nm in diameter can be identified and give very high contrast compared to the a-C support film. Particles of less than a few nanometers, contained within an a-C matrix, can not normally be imaged using conventional TEM techniques[7]. Fe L_3 and L_2-edges collected from a single particle are shown in figure 2. The upper curve is for a probe positioned in the centre of a particle and the lower curve is for a probe positioned at its edge. L_3 and L_2-edges represent respectively transitions from $2p_{3/2}$ and $2p_{1/2}$ core states to unfilled conduction band states with d- and s-symmetry. The "white lines" seen at the edge onsets are mostly due to transitions to empty d-states.[8]

The intensity ratio of the L_3 and L_2 white lines may be correlated to the charge at the Fe site[9]. For the curves shown, the L_3/L_2 ratio at the centre of the particle is 3.3 and at its edge is 2.9. A value of 3.3 is in good agreement with the value given by reference 9, for an Fe film partially oxidized in air and containing Fe, Fe_2O_3 and Fe_3O_4. This suggests that the bulk of the particle contains Fe in a range of oxidation states. The lower L_3/L_2 ratio measured for the edge of the particle is confirmed to be a real effect since a decrease in intensity is also seen in the first peak of the O K-edges on going from the centre of the particle to its edge (figure 3). This indicates that the surface of the particle is more highly oxidized than the bulk[9].

The current study has demonstrated the potential of Z-contrast imaging and EELS in the STEM to provide new insight on the atomic interaction of metals with the carbonaceous matrix in nanoscale combustion aerosols. Such information is of critical importance for the verification or rejection of theories postulating that the soot suppressing role of Fe is due to catalytically enhanced removal of carbon by molecular oxygen in the presence of Fe compounds[10].

References
1. Contact address, ORNL, Solid State Division, P.O. Box 2008, Oak Ridge, TN 37831 - 6031
2. S.J. Pennycook and D.E. Jesson, *Phys. Rev. Lett.* 64(1990)938.
3. N.D. Browning, M.F. Chisholm and S.J. Pennycook, *Nature* 366(1993)143.

Proc. Microscopy and Microanalysis 1995, edited by G.W. Bailey, M.H. Ellisman, R.A. Hennigar, and N.J. Zaluzec

4. P.E. Batson, *Nature* 366 (1993)727.
5. R.D. Leapman, in *Transmission EELS in Material Science* (1992)47.
6. N.D. Browning and S.J. Pennycook, *Microbeam Analysis* 2(1993)81.
7. J. Zhang, and C.M. Megaridis, *Proc 25th Symp. on Combustion*, Combustion Inst., Pittsburgh (1994)
8. R.F. Egerton, *Electron Energy Loss spectroscopy in the Electron Microscope* (1986)205.
9 L.A. Grunes, *Proc 38th Ann. MSA Meeting* (1980)122.
10. The authors gratefully acknowledge the provision of experimental facilities by S. Pennycook of the Solid State Division at ORNL under DOE contract DE-AC05-84OR21400 with Martin Marietta Energy Systems, and the support of the National Science Foundation under Grant No. CTS-9109166.

FIG. 1.

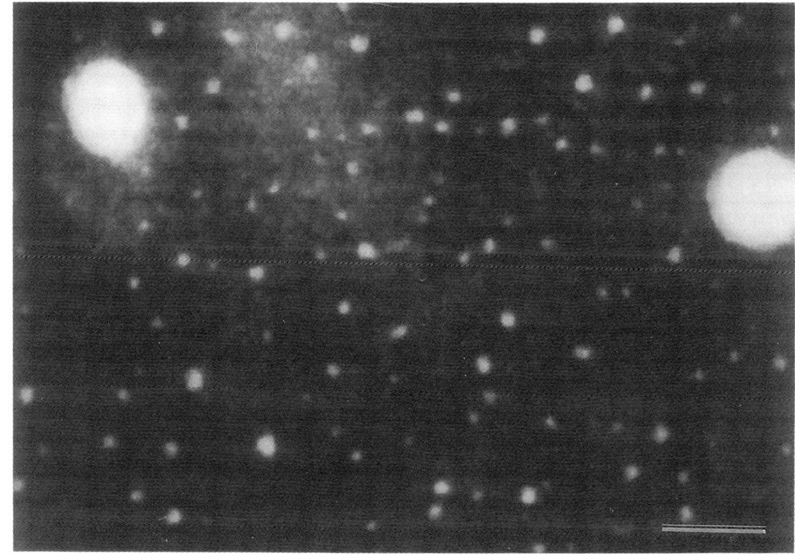

FIG. 2.   FIG. 3.

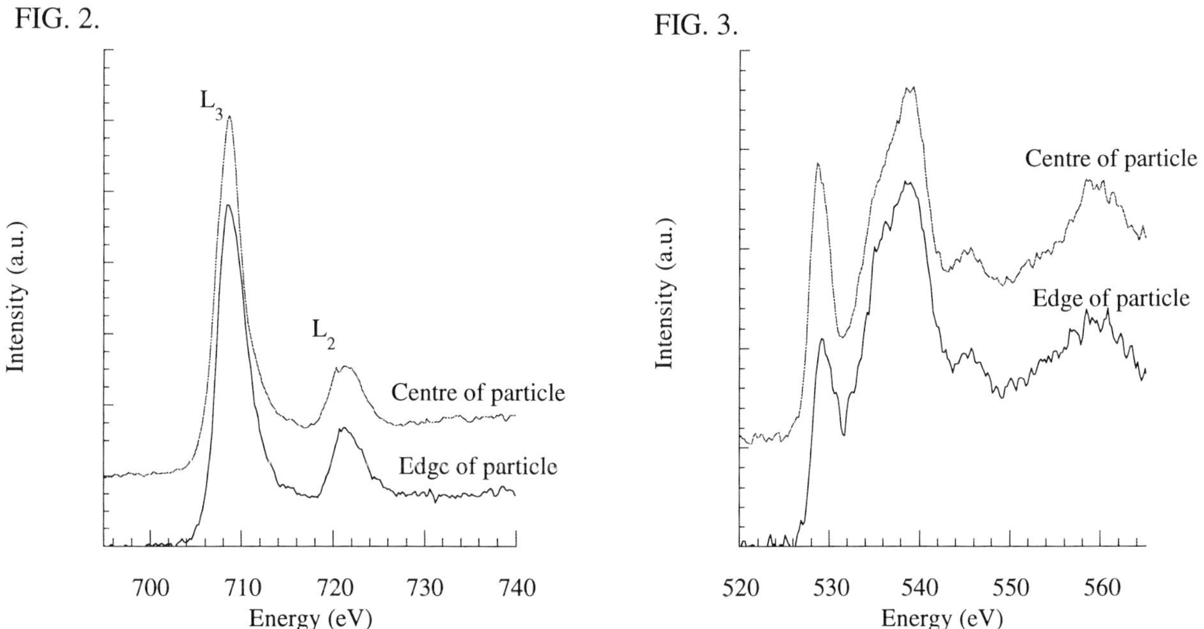

FIG. 1. Z-contrast image of Iron bearing particles on an a-carbon support film. Bar = 20nm.
FIG. 2. Core loss Fe L$_{23}$-edges from the centre of a particle (upper curve) and its edge (lower curve).
FIG. 3. Core loss Oxygen K-edge from the centre of a particle (upper curve) and its edge (lower curve).

STABILIZING AQUEOUS CLAY SUSPENSIONS BY AOT VESICULAR SOLUTIONS: A CRYO-TEM STUDY

Xiangbing Li

Department of Chemical Engineering and Materials Science, and NSF Center for Interfacial Engineering, University of Minnesota, Minneapolis, MN 55455

Production, storage, and performance of suspensions of colloidal solid particles depend critically on suspensions stability. Past studies mostly focused on roles of attractive van der waals force and repulsive electrostatic force between particles[1-2]. In contrast, this paper reports that aqueous suspensions of clay particles of maximum dimension of 5 to 20 nm can be greatly stabilized by adding vesicular solutions of the anionic surfactant Aerosol OT (AOT).

How does the interaction between the clay particles, which bear charge, and multiple bilayers of AOT vesicles lead to more stable suspensions? Cryogenic transmission electron microscopy (cryo-TEM) was brought to bear on this question because it can capture the microstructure in fluids[3] without staining or drying artifacts[4]. Cryo-TEM samples were made by rapidly freezing thin liquid films (200-500 nm) in a controlled environment vitrification system (CEVS)[5] to preserve the original structures. Clay particles (often hexagonal) were flat plates and tended to associate face-to-face (Fig. 1a). Cryo-TEM images show that 3 wt% and 5 wt% AOT formed unilamellar and multilamellar vesicles in solutions (Fig. 1b–c).

Adding the AOT vesicular solutions dramatically lengthened the time needed to flocculate the clay suspensions by centrifugation at a given speed. Two major mechanisms were uncovered that account for the enhanced colloid stability. The first is the encapsulation of platelets by AOT vesicles. This is demonstrated in the cryo-TEM micrograph (Fig. 2a) of 5 wt% AOT and 2 wt% clay. However in a more concentrated suspension (5 wt% AOT and 5 wt% clay), no encapsulated particle was found and yet this suspension was even more stable. What were found were large AOT bilayers and spheroidal vesicles which occupied most of the volume and confined the particles (Fig. 2b). Thus the second contribution is obstruction of sedimentation and aggregation of clay platelets. The greater resistance of the mixture in Fig. 2b to centrifugation indicates that the second mechanism is more effective in stabilizing suspensions. Indeed, lowering the AOT concentration to 3 wt% reduced the resistance to sedimentation and the micrograph (Fig. 2c) shows mostly multilamellar vesicles few of which encapsulate platelets.

The addition of salt to the AOT-containing clay suspensions shortened the time to separation by centrifugation. Micrograph of 5 wt% AOT solutions with 0.8 wt% NaCl indicate two differences in the microstructures (Fig. 3a): (1) the bilayers are far more irregular, and (2) the mean size of the vesicles is smaller, which indicartes shrinking of large vesicles. These differences suggest more flexible and fragile bilayers and less resistance to sedimentation of clay platelets by either encapsulation or obstruction. Micrographs of salt-containing clay-AOT suspensions (Fig. 3b and 3c) show not only the putatively softer and weaker surfactant structures, but also less well dispersed clay platelets.

The study also demonstrated that Cryogenic transmission electron microscopy (cryo-TEM) provides crucial information which is not easily obtainable with other indirect techniques.

References

1. H. van Olphen, *An Introduction to Clay Colloid Chemistry*. New York:Wiley 1977.
2. B. Alince and T. G. M. van de Ven, *J. Colloid Interface Sci.*, 155(1993)465
3. P. K. Vinson et al., *J. Colloid Interface Sci.*, 142(1991)74.
4. P. K. Kilpatrick, W. G. Miller, Y. Talmon, *J. Colloid Interface Sci.*, 107(1985)146.
5. J. R. Bellare et al., *J. Electron Microsc. Tech.* 10(1988) 87.
6. Thanks go to Dr. Y. Guo and Profs. Y. Talmon, L. E. Scriven, H. T. Davis for helpful discussions

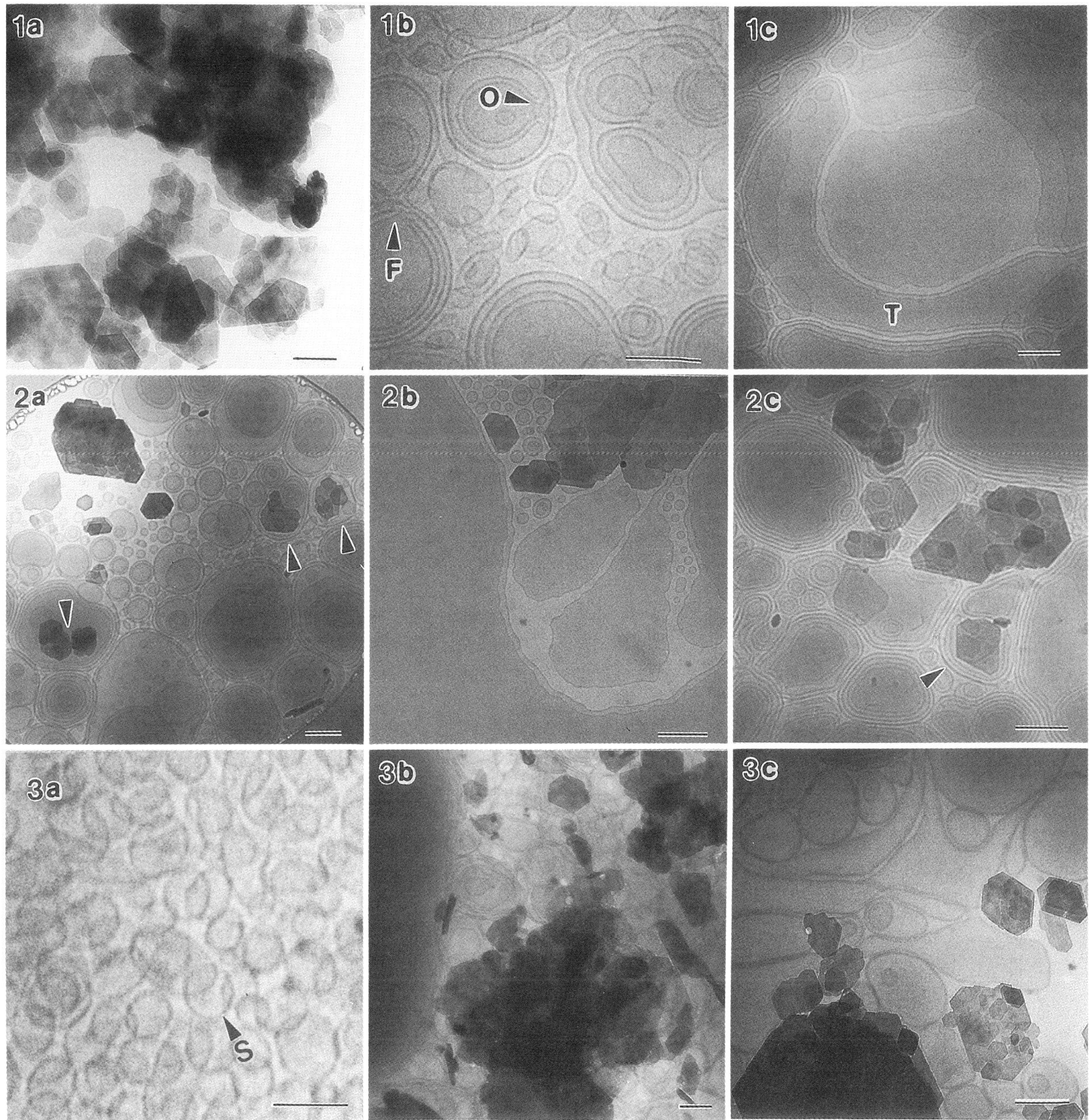

FIG. 1.— Cryo-TEM images of (a) 5 wt% clay in water, particles are mostly face-to-face associated plates; (b) 3 wt % AOT in water, note unilamellar vesicles, multilamellar vesicles. Some vesicles open up (O), and some fuse together (F); (c) 5 wt% AOT in water, there are large unilamellar and multilamellar vesicles. Some vesicles are long tubes (T). Bars = 100 nm.

FIG. 2.— Cryo-TEM images of Clay-AOT-H_2O suspensions. (a) 5 wt% AOT and 2 wt% clay, arrows indicate encapsulated clay platelets in vesicles; (b) 5 wt% AOT and 5 wt% clay, particles were confined by large double-walled vesicles and large and small unilamellar vesicles; (c) 3 wt% AOT and 5 wt% clay, note stacked plates, vesicles, and encapsulated clay platelets (arrow). Bars = 200 nm.

FIG. 3.— Effect of added salt. (a) 5 wt% AOT-0.8 wt% NaCl, there are numerous small and irregular vesicles (S), Bar = 100 nm; (b) 5 wt% AOT, 5 wt% clay and 0.8 wt% NaCl, clay particles aggregated in an environment of small and large vesicles with soft bilayers. Bar = 200 nm; (c) 3 wt% AOT, 5 wt% clay and 0.8% NaCl, there are long tubes and irregularly shaped vesicles. Bar = 200 nm.

IN SITU ELECTRON MICROSCOPY AND ITS APPLICATIONS TO SEMICONDUCTOR REACTIONS AT HIGH RESOLUTION

Robert Sinclair

Department of Materials Science and Engineering, Stanford University, Stanford, California, 94305

In situ electron microscopy experiments can provide the most revealing insights into material behavior. However, in order to take full advantage of the observations, quantitative measurements are required so that the underlying mechanisms are completely interpreted. This approach also ensures that specimen and environmental artifacts do not play a role and that real "bulk" processes are being studied. These points are illustrated in this paper by reference to work on reactions in semiconductor systems, especially at high resolution.

The technique and practice of in situ microscopy are quite exacting. Thus it is often necessary to record changes in the same specimen area for extensive periods of time (e.g., hours), under identical imaging conditions. One can never be sure when a significant event will take place, or sometimes whether it has actually occurred -- accordingly a high degree of acuity on behalf of the observer is essential. A number of procedures is recommended to check that the results are representative and reproducible, including comparing the structural evolution with that from ex situ samples both qualitatively and quantitatively (e.g., [1]). Some contemporary applications are given in a recent publication.[2]

High resolution in situ investigations can show directly the atomic events associated with a reaction, as described in reference [3]. A most striking example concerns the metal-mediated crystallization of amorphous semiconductors.[4] As metallic crystals (e.g., Ag, Al) migrate into amorphous Si or Ge, leaving behind crystalline products, it is clearly seen in the recordings that the metallic lattice is stationary (e.g., [5, 6]). This can only be interpreted by a substitutional diffusion mechanism of the Si/Ge atoms through the metal. Measurements of the crystal growth rate as a function of temperature show classical Arrhenius variation, with the activation energy consistent with this view (e.g., Fig. 1). Thus the quantitative data are in accord with the qualitative observations and with a satisfactory atomic diffusion model.[6] In a similar vein, Fig. 2 shows experimental results for the growth of an amorphous phase during the solid-state amorphization of a Zr-Si interface.[7] The in situ and ex situ data points lie on exactly the same curves giving a strong indication that the in situ experiment is a reliable reproduction of the bulk behavior.

Our studies over the last ten years have shown that the in situ approach works well in a surprisingly large number of situations. Fig. 3 represents a remarkable example. A thin titanium nitride layer is employed in integrated circuits as a "diffusion barrier" to the reaction between the semiconductor silicon and the metallization aluminum alloy. Thermal stability is a key requirement. In situ experiments documenting the barrier breakdown at elevated temperatures, as in Fig. 3, show identical results to those in ex situ, furnace annealed samples.[8] This is a very exacting test of the efficacy of in situ experiments.

In conclusion, it can be stated that in situ observations provide unique knowledge about material reactions, so long as they are carried out with care.[9]

References

1. D. H. Ko and R. Sinclair, *Ultramicroscopy* 54(1994)166.
2. F. M. Ross, *Mats. Res. Soc. Bull.* 19(6)(1994)17 et seq.
3. R. Sinclair, *ibid* p. 26.
4. R. Sinclair and T. J. Konno, *Ultramicroscopy* 56(1994)225.
5. T. J. Konno and R. Sinclair, *Phil. Mag.* B71(1995)163.
6. T. J. Konno and R. Sinclair, *ibid* p. 179.

Proc. Microscopy and Microanalysis 1995, edited by G.W. Bailey, M.H. Ellisman, R.A. Hennigar, and N.J. Zaluzec

7. H. Tanaka et al., submitted.
8. H. J. Lee et al., *Mats. Res. Soc. Symp. Proc.* submitted.
9. Partial financial support for this work was provided by the National Science Foundation (Grant No. DMR 89-022332). Illustrations were kindly provided by H. J. Lee, H. Tanaka and T. J. Konno.

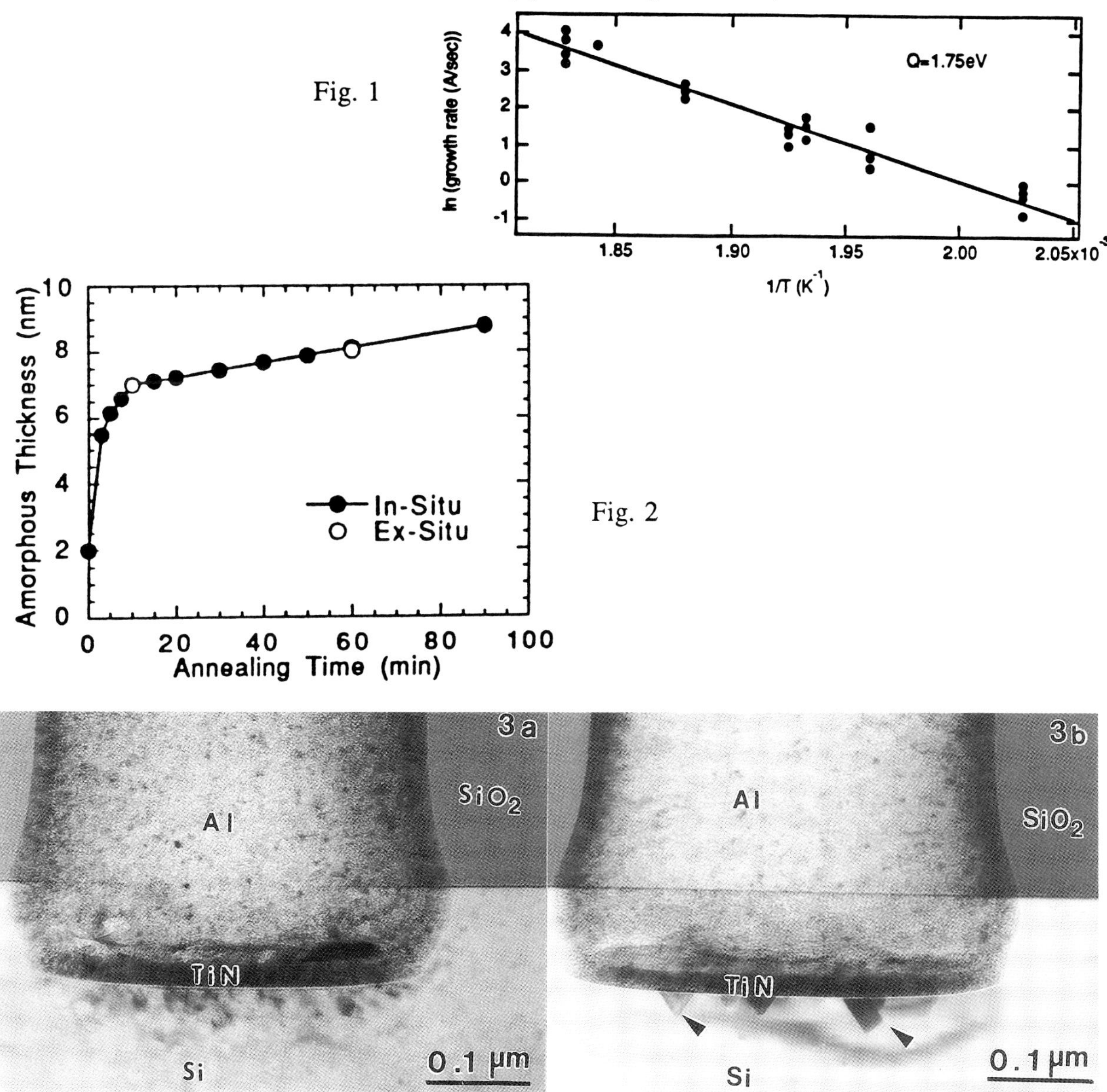

FIG. 1.—Arrhenius plot for the growth of crystalline Ge in the Ag-Ge system as revealed by in situ HREM yielding an activation energy of 1·75eV.[6]

FIG. 2.—Variation of thickness of amorphous ZrSi mixture during annealing of a Zr/Si interface at 400°C.[7]

FIG. 3.—Bright field images of the bottom of a 0·5 μm wide integrated circuit contact (a) before (b) after in situ annealing at temperatures up to 630°C. The breakdown of the TiN diffusion barrier is indicated (arrows), the appearance of which being very similar to that developed by ex situ furnace annealing.[8]

INVESTIGATION OF IRRADIATION-INDUCED NUCLEATION PROCESSES IN SILICON AND GERMANIUM

Harry A. Atwater, C.M. Yang and K.V. Shcheglov,

California Institute of Technology, Pasadena, CA 91125

Studies of the initial stages of nucleation of silicon and germanium have yielded insights that point the way to achievement of engineering control over crystal size evolution at the nanometer scale. In addition to their importance in understanding fundamental issues in nucleation, these studies are relevant to efforts to (i) control the size distributions of silicon and germanium "quantum dots", which will in turn enable control of the optical properties of these materials, (ii) and control the kinetics of crystallization of amorphous silicon and germanium films on amorphous insulating substrates so as to, e.g., produce crystalline grains of essentially arbitrary size.

Ge quantum dot nanocrystals with average sizes between 2 nm and 9 nm were formed by room temperature ion implantation into SiO_2, followed by precipitation during thermal anneals at temperatures between 30°C and 1200°C[1]. Surprisingly, it was found that Ge nanocrystal nucleation occurs at room temperature as shown in Fig. 1, and that subsequent microstructural evolution occurred via coarsening of the initial distribution. Little coarsening was seen at 600°C (Fig. 2), but anneals at 1000°C resulted in significant coarsening (Fig. 3). Nanocrystal size distributions obtained by transmission electron microscopy are shown in Fig. 4. Nanocrystals were in all cases spheroidal and approximately 20% were found to contain twin boundaries or other defects. Notably, this approach to synthesis has produced the narrowest size distributions produced to date for group IV semiconductors. The size distributions obtained from electron microscopy were combined with theoretical estimates for the variation in energy gap and the oscillator strength for recombination of quantum-confined excitons[2] to correlate size with observed and calculated visible photoluminescence spectra. Qualitative agreement between calculated and observed spectra could be obtained when: (i) nonradiative and radiative rates were of the same order of magnitude, (ii) the spatial extent of the nanocrystal confinement potential is assumed to be slightly smaller than the physical size of the nanocrystal, and (iii) both decay rates were saturated by the pump for particles of large radius.

Ion irradiation during crystallization of amorphous silicon can also be used to tailor the population of crystals near the critical size for thermodynamic stability. *In situ* electron microscopy studies permit extremely efficient and detailed investigation of crystallization kinetics, including transient nucleation rates. One can variously enhance or suppress silicon crystallization during ion irradiation. The crystallization rate for silicon is enhanced by 5-7 orders of magnitude relative to the thermal crystallization rates at the same temperature by steady state irradiation at high temperatures (600 °C). The increased crystallization rate is correlated with an increase in the stored defect enthalpy, and is consistent with an increase in the thermodynamic driving force for crystallization under irradiation. A cyclic ion irradiation anneal method allows the silicon crystallization rate to be suppressed so as to be essentially unobservable at anneal temperatures that permit crystal growth of existing grains in the amorphous silicon matrix.

This work is supported by the Department of Energy under grant DE-FG03-89ER45395.

[1] H.A. Atwater, K.V. Shcheglov, S.S. Wong, K.J. Vahala, R.C. Flagan, M.L. Brongersma and A. Polman, Mat. Res. Soc. Symp. Proc. 316, 409 (1994).

[2] T. Takagahara and K. Takeda, Phys. Rev., B46, 15578 (1992).

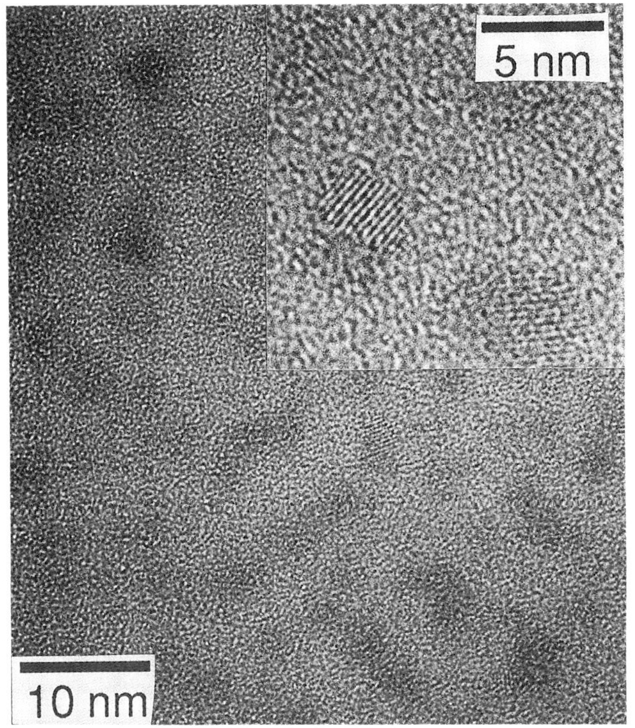

Fig. 1. Ge nanocrystals following ion implantation at room temperature.

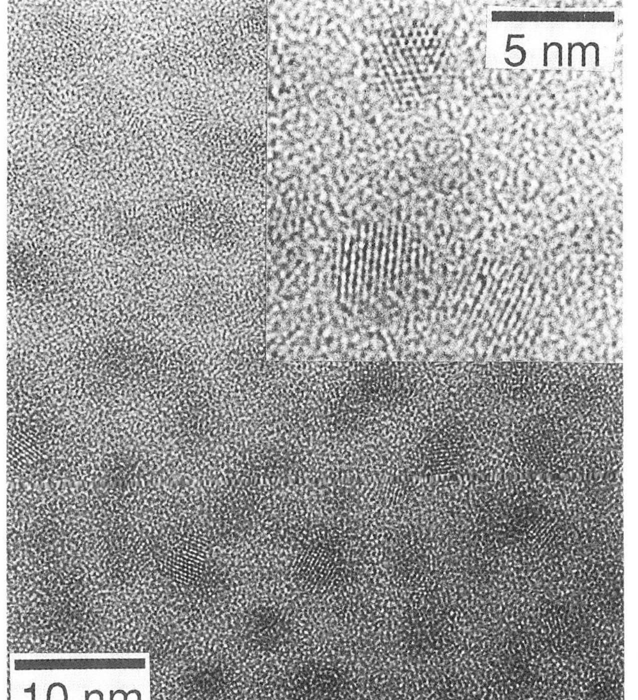

Fig. 2. Ge nanocrystals following 600 °C anneal; little coarsening is observed.

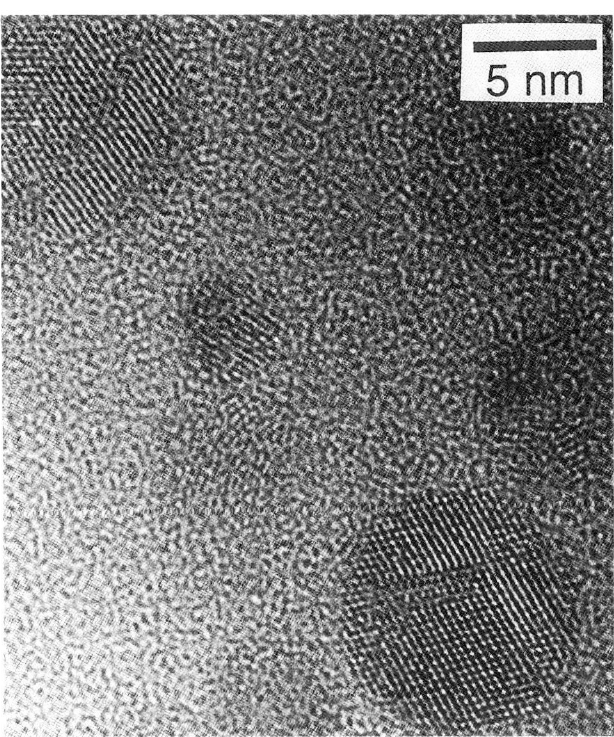

Fig. 3. Ge nanocrystals following 1200 °C anneal; significant coarsening is observed.

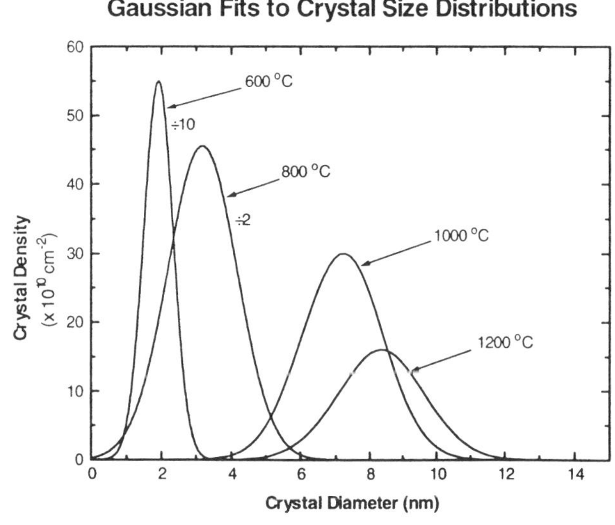

Fig. 4. Nanocrystal size distributions resulting from electron microscopy.

IN-SITU REM STUDY OF INDIUM PARTICLE EVOLUTION ON InP(110) SURFACES

M. Gajdardziska-Josifovska[*], M. H. Malay[*] and David J. Smith[**]

[*] Department of Physics and Laboratory for Surface Studies, University of Wisconsin Milwaukee, P.O. Box 413, Milwaukee WI 53201

[**] Department of Physics and Center for Solid State Science, Arizona State University, Tempe AZ 85287

Reflection electron microscopy (REM) has been used by several groups for *in situ* studies of terrace and step dynamics induced by phase transformations, film growth, and surface electromigration. [1] Real-time viewing and video-recording of these phenomena has become possible with the use of intensified TV cameras under surface resonance diffracting conditions. Recently we captured the dynamics of nucleation and growth of In particles on InP(110) surfaces induced by thermal annealing in an ultrahigh-vacuum transmission electron microscope. [2,3] In addition to the projection problem, also encountered in TEM, the REM image interpretation is further complicated by foreshortening effects. At present, there are no developed methods for measurement of particle sizes from REM images. This work quantifies the In particle shape and size evolution from *in-situ* REM experiments.

Samples were prepared by cleavage of an undoped (001) InP wafer along the (110) crystallographic plane. The cleavage was performed in air, immediately before loading into a Philips 430ST HREM, modified for UHV operation and equipped with a single-tilt specimen heating holder. [4] REM images were formed with 100keV electrons at 1.36° grazing angle of incidence. The resulting foreshortening in the electron beam direction was then ~40x. An azimuthal tilt of 0.57° from the [001] zone was used to achieve a surface resonance condition for the 660 specular reflection. REM images of In particle growth were recorded using a Gatan 622 TV camera. The video sequences were digitized with a Macintosh Centris 660AV computer running Video Monitor. Four consecutive video frames were averaged, using NIH Image.

The dark features visible in Fig. 1 are In particles that have nucleated following the non-congruent evaporation of P from the InP surface. Their formation was not initiated until an annealing temperature of 650°C was reached but, as shown by this time sequence of images, their growth was then comparatively rapid. Two different types of particle morphology and growth behavior are visible, as indicated by the particles labeled A and B which are followed in Fig. 2. Both particles have nucleated on the same terrace, but show different growth dynamics resulting in two distinct but typical shapes at the end of the annealing cycle.

Particle image sizes were measured by calculating a first derivative of each image and fitting ellipses to the traced particle edges. The major and minor axis of each ellipse yielded a particle image size in the non-foreshortened and foreshortened directions respectively. A model was developed for image interpretation of two dimensional (2D) and three dimensional (3D) particles. In both cases, the image size in the non-foreshortened direction gave a direct measure of the length of the particle base perpendicular to the beam direction. For a 2D particle, the image size in the foreshortened direction would give the foreshortened length of the particle parallel to the beam direction. For a 3D particle, half of the foreshortened image size would correspond to the particle height. Knowledge of the REM imaging geometry, with additional bright field imaging after annealing, resulted in use of the 3D model for the particle sizes plotted in Fig. 2. Particle A grew as a hemisphere in the first 100 seconds from its nucleation. The growth rate in the first 30 seconds was fast and almost constant (the particle diameter increased at ~3.5 nm/sec), followed by a saturation growth mode to a final diameter of ~160 nm. The growth resumed abruptly at the base of the particle at the expense of the particle height, which corresponded to decreased contact angle between the particle and the substrate. Two growth rates were detected, with a transition occurring abruptly at particle size of ~260 nm. For comparison, particle B grew rapidly in the direction parallel to the substrate, but had almost constant height and therefore constant dimension in the foreshortened direction. The particle grew in the [1-10] direction with an approximately constant rate of ~3.2 nm/sec during the first 100 sec, after which time the growth effectively stopped, at least for the remaining minute of the recording period.[5]

References
1. K. Yagi, *Surf. Sci. Reports* 17 (1993) 305.
2. M. Gajdardziska-Josifovska and D. J. Smith, *Electron Microscopy, Vol.* 2B (1994) 1067
3. M. Gajdardziska-Josifovska, M. H. Malay and D. J. Smith, *Surf. Sci.* in press.
4. D. J. Smith et al. *Ultramicroscopy* 49 (1993) 26.
5. Microscopy was conducted at CHREM at Arizona State University, supported by NSF grant DMR-9115680.

Proc. Microscopy and Microanalysis 1995, edited by G.W. Bailey, M.H. Ellisman, R.A. Hennigar, and N.J. Zaluzec
Copyright © 1995 MSA. Published by Jones and Begell Publishing, 79 Madison Ave., New York, NY 10016

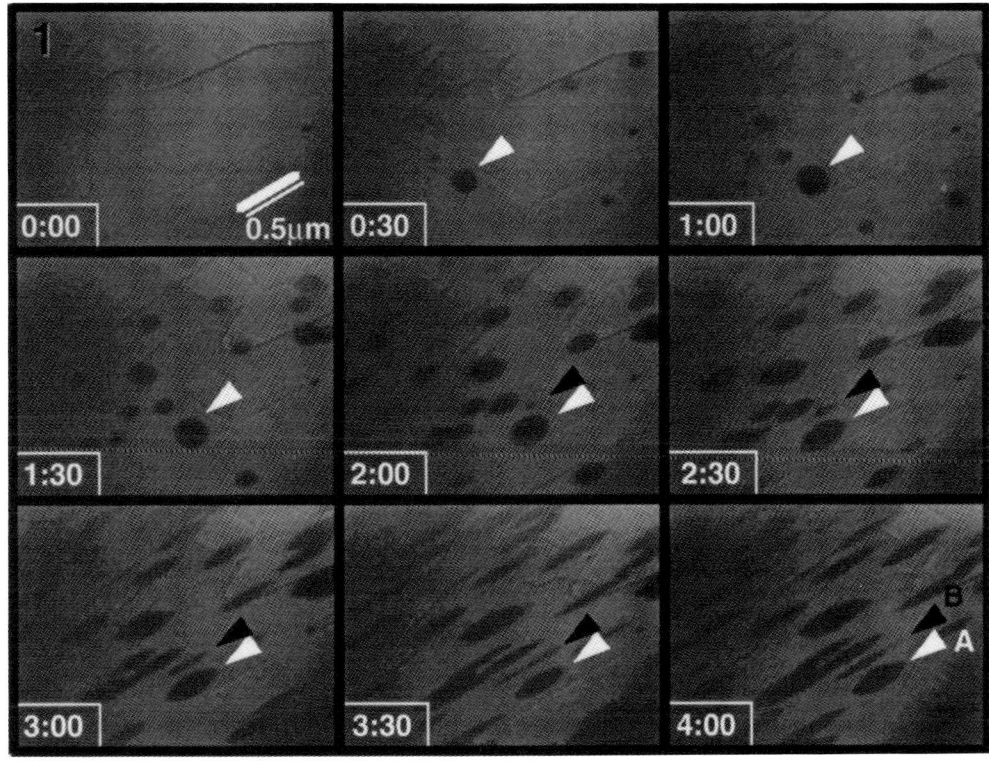

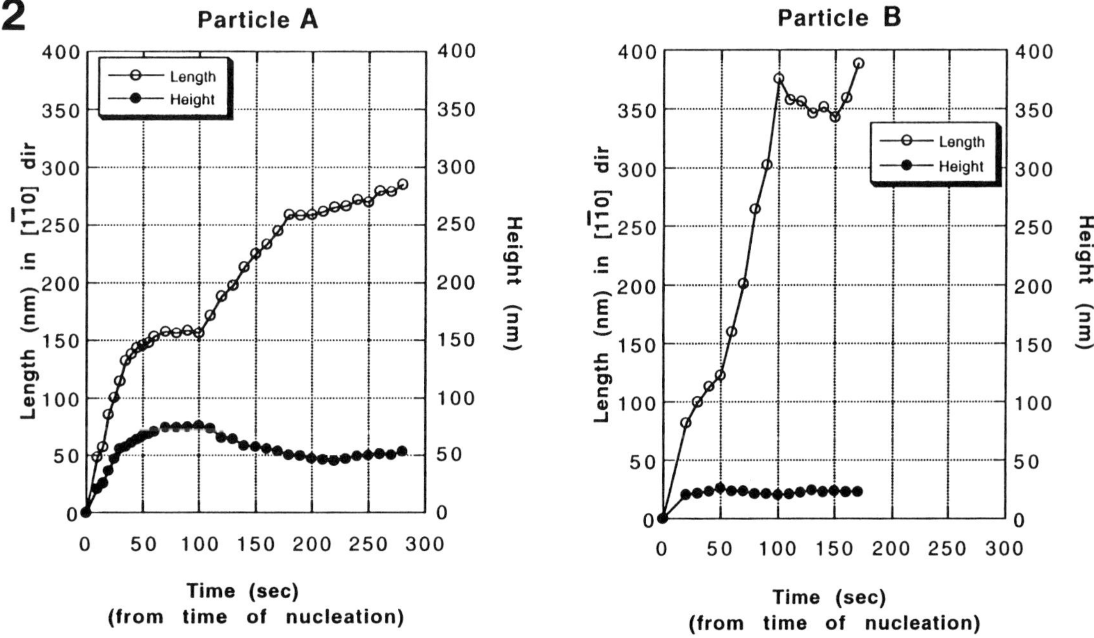

FIG.1.   Series of REM images from video recording showing nucleation and growth of In particles at 650°C. Evolution of particle shape and size differs for In islands of type A and type B, as shown in Fig. 2.

FIG. 2.   Particle length at base measured in [1-10] direction (solid circles) and particle height (open circles) as function of annealing time at 650°C: a) particle A; and b) particle B from Fig. 1.

IN SITU HOT-STAGE HIGH-RESOLUTION TRANSMISSION ELECTRON MICROSCOPE STUDIES OF THE MECHANISMS AND KINETICS OF PRECIPITATION REACTIONS

James M. Howe

Department of Materials Science and Engineering, University of Virginia, Charlottesville, VA 22903

In situ hot-stage high-resolution transmission electron microscopy (HRTEM) provides unique capabilities for quantifying the mechanisms and kinetics of precipitation reactions at the atomic level. Such information is required to understand phase transformations and the behavior of material interfaces. This paper provides a brief summary of the *in situ* hot-stage HRTEM technique and illustrates the use of this technique to obtain information about heterogeneous nucleation processes in precipitation and crystallization reactions. Examples of other types of *in situ* HRTEM studies can be found in previous papers by Sinclair et al.[1] and Howe et al.[2]

The specimen and microscope requirements for *in situ* hot-stage HRTEM are the same as those for static HRTEM except that one must have a hot-stage specimen holder and equipment for recording and analyzing dynamic images. A high-resolution television-rate camera connected to a standard videocassette recorder (VCR) can be used to store hours of data for low cost. The dynamic images can then be analyzed directly during playback or sent to a computer for image processing and analysis. In the present investigation, the *in situ* hot-stage experiments were performed on a JEOL 4000EX HRTEM equipped with a UHP40X hot-stage pole piece and double-tilt specimen holder and DigitalMicrograph^{TM} software[3] was used to capture and analyze the dynamic images during playback of the videocassettes.

Figure 1 shows a HRTEM image of a large growth ledge on the face of a θ plate viewed edge-on along a $[1\bar{2}1]_\alpha||[001]_\theta$ direction during precipitation of the θ phase in an Al-Cu-Mg-Ag alloy.[4,5] The habit plane of the θ plate is $(111)_\alpha$ and the ledge was observed to move across the habit plane in the direction indicated by the arrow labelled v_L in Fig.1. The temperature was about 220°C and the average velocity of the ledge determined from the VCR was 0.08 nm/s. Of particular interest is the mechanism by which the large ledge moved across the precipitate face. The large ledge was found to advance by the nucleation of kinks at the base of the ledge which propagated upward across the face of the ledge. Such a nucleus was captured during playback of the videocasette and is highlighted by a dashed line in Fig. 1. This nucleus is only a few atom spacings wide in the projection of the image but probably extends through the thickness of the foil along the base of the ledge. The dark contrast characteristic of the Cu-rich precipitate was observed to fluctuate over a few atom spacings before becoming stable and moving upward across the ledges in the direction labelled v_k in Fig. 1. The nucleation rate of the kinks also fluctuated but was about 0.2 nuclei/s on average during the observation period. Thus, the *in situ* hot-stage HRTEM directly revealed both the mechanisms and kinetics of ledge motion during precipitate growth. It would be extremely difficult to obtain this information by any other technique.

Additional capabilites of *in situ* hot-stage HRTEM are provided by an example of crystallization in an amorphous Pd-Si alloy. The image in Fig. 2 was taken during growth of a eutectic mixture of Pd_3Si and Pd_9Si_2 lamellae from the amorphous $Pd_{80}Si_{20}$ matrix at about 225°C in the HRTEM.[6,7] In this reaction, the Pd_3Si lamellae are usually faceted and grow by a two-dimensional nucleation and growth process[8] along low-index directions such as [010] while the Pd_9Si_2 lamellae are nonfaceted and appear to grow along relatively high-index directions that are rough on an atomic scale. In the sequence shown in Fig. 2, the Pd_3Si lamella was observed to grow by ledges which nucleated at the three-phase junction labelled A in the figure. The ledges then propagated across the tip of the lamella in the direction labelled v_L. In contrast, the Pd_9Si_2 lamella appeared to grow continuously into the amorphous matrix without evidence of a nucleation barrier to growth.

Proc. Microscopy and Microanalysis 1995, edited by G.W. Bailey, M.H. Ellisman, R.A. Hennigar, and N.J. Zaluzec

It is possible to explore the details of the atomic processes which occur during precipitation reactions at temperature directly from the videocassettes using image processing software. For example, it is possible to obtain intensity traces across terraces and ledges at temperature to examine the degree of order across the interfaces. The intensity trace in Fig. 3 was taken along the line labelled B-B' in Fig. 2. The distinct changes in average intensity and periodicity across this interface indicate that the structural transition from crystalline Pd_3Si to amorphous $Pd_{80}Si_{20}$ matrix occurs fairly abruptly across the terrace with little evidence for conformation of the amorphous matrix to the crystalline solid.[9] The diffraction pattern in Fig. 4 was obtained by fast Fourier transform (FFT) of the the amorphous matrix enclosed by the box in Fig. 2. The pattern displays characteristics typical of an amorphous solid even at the base of the three-phase junction. Many other possibilities for image analysis at temperature exist during *in situ* hot-stage HRTEM studies.[10]

References

1. R. Sinclair et al., *Acta Cryst.* A44 (1988) 965.
2. J. M. Howe et al., in M. Sarikaya, H. K. Wickramasinghe and M. Isaacson, Eds., *Mater. Res. Soc. Symp. Proc. Vol. 332*, Pittsburgh, PA: MRS (1994) 65.
3. *DigitalMicrograph™ Instruction Manual*, Pleasanton, CA: Gatan, Inc. (1992).
4. A. Garg and J. M. Howe, *Acta Metall. Mater.* 39 (1991) 1939.
5. A. Garg, Y. C. Chang and J. M. Howe, *Acta Metall. Mater.* 41 (1993) 235.
6. W. H. Brearley, P.-C. Shieh and J. M. Howe, *Metall. Trans.* A22 (1991) 1287.
7. J. M. Howe, *Mater. Sci. Forum* 126-128 (1993) 467.
8. J. W. Cahn, W. B. Hillig and J. W. Sears, *Acta Metall.* 12 (1964) 1421.
9. J. Q. Broughton, A. Bonissent and F. F. Abraham, *J. Chem. Phys.* 74 (1981) 4029.
10. This research was supported by NSF under grant DMR-9302493. The author gratefully acknowledges the collaboration of Dr. A. Garg during the Al-Cu-Mg-Ag studies.

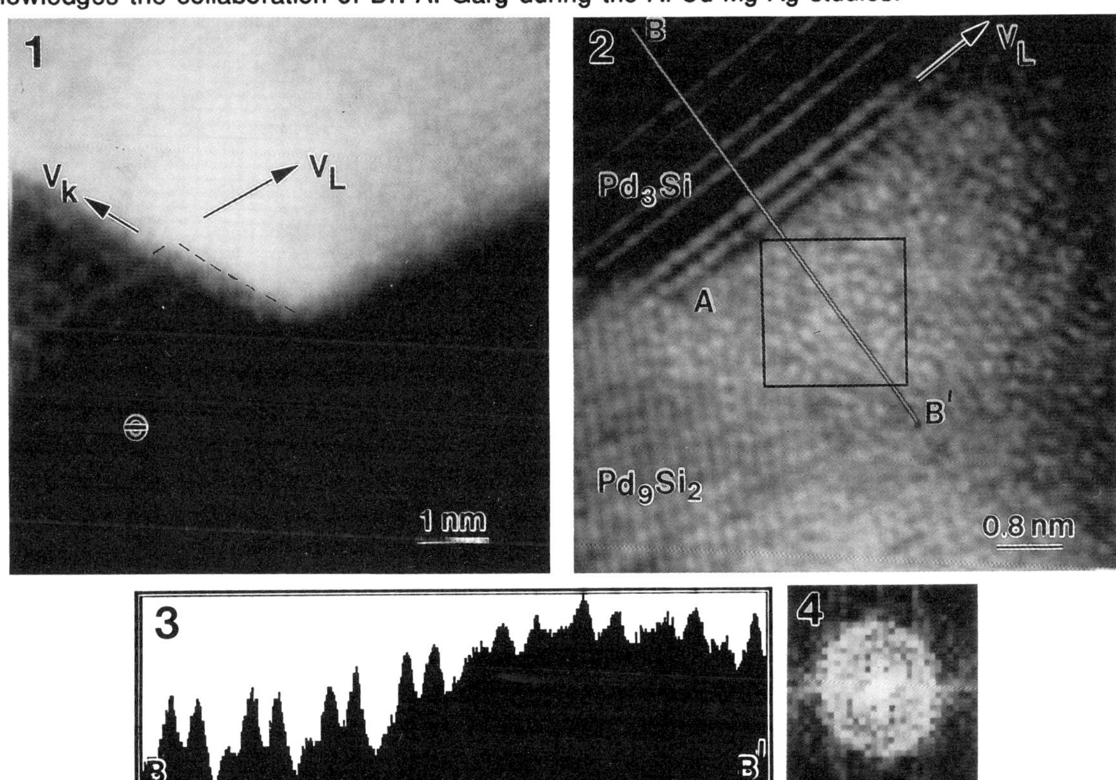

FIG. 1 - HRTEM image of kink nucleating at base of ledge on θ precipitate plate in Al-Cu-Mg-Ag alloy.
FIG. 2 - HRTEM image of ledge on Pd_3Si crystal nucleating at three-phase junction in Pd-Si alloy.
FIG. 3 - Intensity trace across terrace along the line labelled B-B' in Fig. 2.
FIG. 4 - Diffraction pattern (FFT) of enclosed region near three-phase junction in Fig. 2.

IN-SITU STUDY OF TETRAGONAL-ORTHORHOMBIC STRUCTURAL TRANSFORMATION IN $La_{2-x}Ba_xCuO_4$

Yimei Zhu and J. Tafto[1]
Materials Science Division, Brookhaven National Laboratory, Upton, NY 11973

In the concentration range of the local T_c minimum in $La_{2-x}Ba_xCuO_4$ ($0.10 \leq x \leq 0.15$), below 60K, there is a low temperature tetragonal (LTT) structure (space group $P4_2/ncm$).[2] With an increase in temperature, the structure becomes orthorhombic (LTO, space group Bmab), then above 140K becomes tetragonal (HTT) with the space group I4/mmm.

The transition from HTT to LTO during cooling generates twins, and the twin morphology persists even below 60K. This is inconsistent with a pure LTT phase, since twinning is associated with an orthorhombic symmetry. To address the issue of the LTO/LTT phase transition, we performed *in-situ* experiments with a 200 keV TEM equipped with a liquid H_2 specimen holder.

Fig.1(a) and (b) show selected area diffraction patterns taken at 15K in the (001)* and (101)* zone, respectively. During *in- situ* cooling and the LTO/LTT transition, the (101)* diffraction pattern remains the same because h+l=2n reflections are allowed in both LTO and LTT. However, below 60K in the (001)* projection, additional superstructure spots appear for h+k=2n with h and k odd. The presence of (110) type reflections, which are forbidden in LTO phase, are allowed for the LTT phase, and also for the space group Pccn (low temperature less orthorhombic (LTLO) phase) that was proposed to account for the presence of a considerable orthorhombic strain.[2]

A phase mixture of LTO-LTT was observed at temperatures much below 60K. The location of the phase was determined using reflections associated with the LTO or LTT phases. Fig.2(a) is a bright-field image showing an areal morphology with the twins edge-on in the (001) orientation at about 20K. Fig.2(b) is a dark-field image using the (121) reflection in the (101)* projection showing the [101] and the twin-related [011] orientation; only every second twin domains are bright. An intriguing observation is evident, when we used the LTT-related (110) reflection in the (001)* projection, showing narrow bright lines that clearly occurred at the twin boundaries (Fig.2(c)).

The presence of the LTT phase at the LTO twin boundary is consistent with the tipping of the CuO_6 octahedra (Fig.3). In the LTO phase, the CuO_6 octahedra are tilted a few degrees around the <100> axis, while, in the LTT phase, they are tilted around the <110> axis. The tilt of the LTO phase around the [100] and [010] axes results in a conflict at the twin boundary such that the two twin domains have displacements in orthogonal directions, as indicated in Fig.4(a). Thus, an ideally sharp twin boundary, which is likely to have a very high interfacial energy (especially near the LTO/LTT transition temperature, due to the drastically increased orthorhombicity of the LTO phase), is incompatible with the required tilting of the octahedron. The possible configuration is a delocalized boundary with LTT structure at the center of the twin boundary, gradually changing via the orthorhombic Pccn (LTLO) to an orthorhombic LTO in the center of the twin matrix, as shown in Fig.4(b).[3] Further cooling of the specimens causes an increasing twin density, and thus, a larger volume of the LTT phase, consistent with the neutron diffraction study.[4]

REFERENCE 1. Visiting scientist from University of Oslo, Norway. 2. J.D. Axe, et al, *Phys.Rev.Lett.*, **62** 2751 (1989). 3. Yimei Zhu, et al, *Phys.Rev.Lett.*, **73** 3026 (1994). 4. Work supported by US DOE, contract No.DE-AC02-76CH00016.

Proc. Microscopy and Microanalysis 1995, edited by G.W. Bailey, M.H. Ellisman, R.A. Hennigar, and N.J. Zaluzec
Copyright © 1995 MSA. Published by Jones and Begell Publishing, 79 Madison Ave., New York, NY 10016

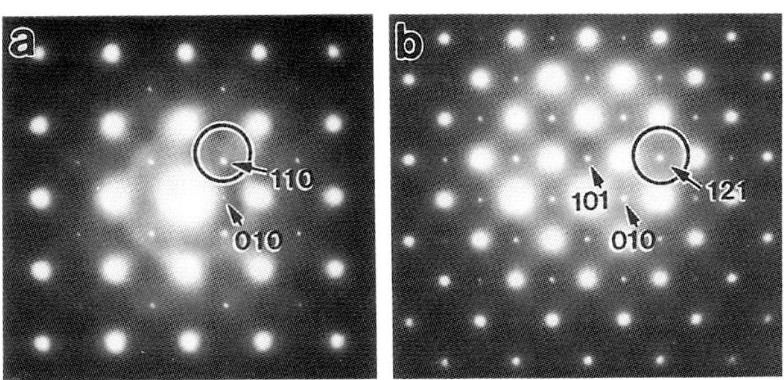

Fig.1 SAD of (001)* (a) and (101)* (b) at 20K.

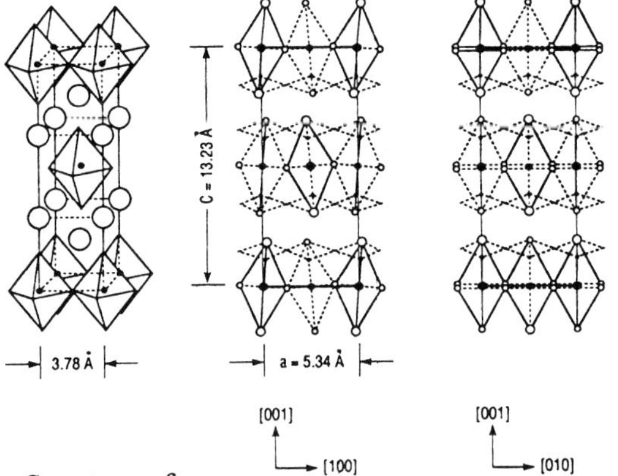

Fig.3 Structure of La$_{2-x}$Ba$_x$CuO$_4$ showing how the CuO$_6$ octrahedra are tilted.

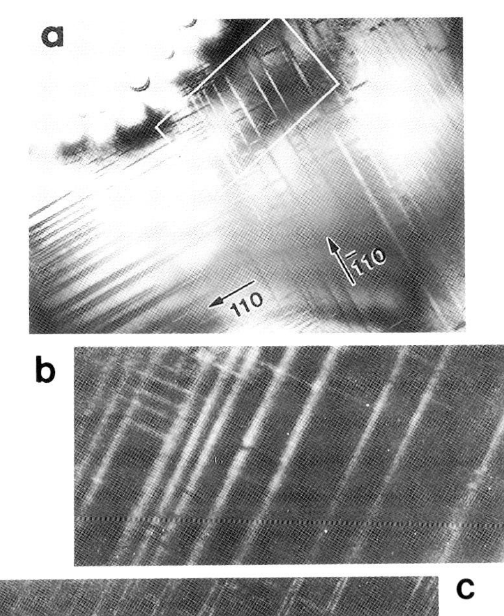

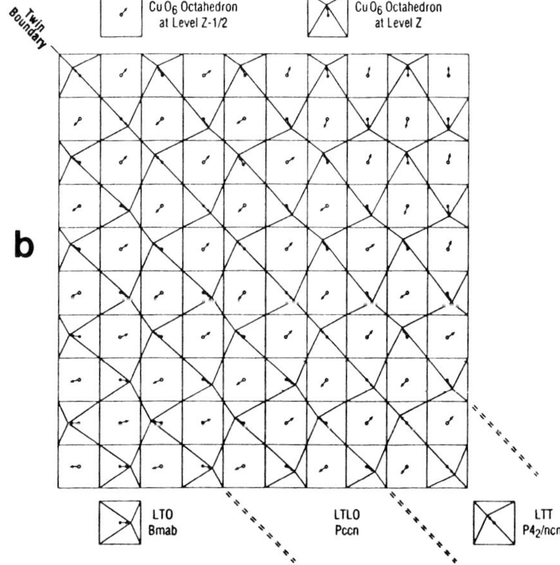

Fig.2 Diffraction contrast observed at 20K. (a) BF, (b) DF, using (121) reflection, and (c) DF, using the (110) LTT-reflection. (b) and (c) are the same area bounded by a rectangular box in (a).

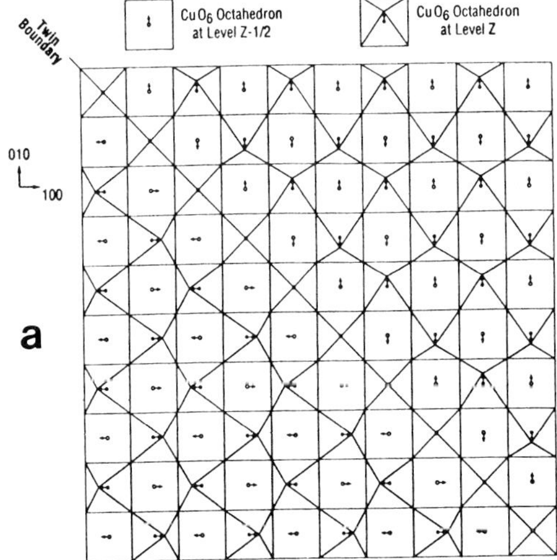

Fig.4 (a) A twin boundary in the LTO structure results in a conflict between tilting of the CuO$_6$ octahedra around the [100] and [010] axes. (b) The proposed model in which the LTT structure (tipping axis [110]) exists at the center of the twin boundaries with a gradual transition, via the LTLO structure, to LTO (tipping axis [100]) in the center of the twin domain.

IN SITU MICROSCOPY OF THE ANODIC ETCHING OF SILICON

Frances M. Ross* and Peter C. Searson**

*National Center for Electron Microscopy, Lawrence Berkeley Laboratory, Berkeley, CA 94720
**Department of Materials Science and Engineering, The Johns Hopkins University, Baltimore, MD 21218

Porous semiconductors represent a relatively new class of materials formed by the selective etching of a single or polycrystalline substrate. Although porous silicon has received considerable attention due to its novel optical properties[1], porous layers can be formed in other semiconductors such as GaAs and GaP [2-4]. These materials are characterised by very high surface area and by electrical, optical and chemical properties that may differ considerably from bulk. The properties depend on the pore morphology, which can be controlled by adjusting the processing conditions and the dopant concentration. A number of novel structures can be fabricated using selective etching. For example, self-supporting membranes can be made by growing pores through a wafer[5], films with modulated pore structure can be fabricated by varying the applied potential during growth[6], composite structures can be prepared by depositing a second phase into the pores and silicon-on-insulator structures can be formed by oxidising a buried porous layer[7]. In all these applications the ability to grow nanostructures controllably is critical.

We are attempting to gain a better understanding of the formation mechanism of porous silicon by allowing pore formation to occur in real time within the electron microscope. We have designed and constructed a specimen holder for a JEOL 200CX featuring a sealed reservoir which can be filled with an electrolyte such as HF (figure 1). Electrical feedthroughs are provided so that a potential can be applied between the specimen and electrolyte while under observation in the microscope. The specimen itself requires a complex design to avoid the release of HF during the etching process (figure 2). The initial substrate is a lightly doped SIMOX wafer. MBE is used to add a sandwich structure consisting of a lightly doped barrier layer, a heavily doped "active" layer in which etching will be confined, and a further barrier layer. The wafer is diced and etched from the back to create an electron transparent membrane. Wax is used to protect the edges of the specimen and the specimen is sealed into the the reservoir using HF-resistant epoxy to leave an area of approximately 1mm^2 in contact with the electrolyte.

During the electrochemical etching of silicon, current densities of 1-100 mAcm^{-2} flow, silicon goes into solution as H_2SiF_6 and a network of unetched silicon is left behind. H_2 gas is a product of the reaction. In our system the application of 2-3V to the n-type specimens of figure 2 results in a total current of the order of 0.01mA and pore growth at 10-100 nm s^{-1}. A porous layer formed *ex situ* (i.e. at 1 atm.) is shown in figure 3. Pores have formed in the heavily doped layer, leaving behind walls less than 50nm thick. The bright and dark bands were created by modulating the voltage. This allows us to mark the pore front at selected times as well as calculate etch rate as a function of voltage. We are presently working on achieving a reliable vacuum seal between the specimen and reservoir so that this result can be duplicated *in situ*.

Our aim in observing the dynamics of the etching process is to to make a detailed study of models proposed to explain the morphology of n-type silicon[8-11].We can characterise the porosity and etch rate as a function of voltage and doping level and observe the process of pore branching. As the voltage increases, the system enters the electropolishing régime where material is removed isotropically. We expect to find changes in morphology, such as an increase in dimensions, near this point. Another area of interest is the behaviour of the etch rate when the voltage is removed, which could indicate the presence of certain species at the Si/HF interface. Finally, the flow of current along dislocations can be studied using partially relaxed Ge_xSi_{1-x}/Si substrates with misfit dislocations at the interfaces. There are analogous systems for which this type of *in situ* experiment is valuable, such as n-type GaAs, where the pores run in <111>a directions and have triangular or hexagonal cross sections and interesting branching geometry[12]. Modifications of the holder and specimen design can extend these experiments further, and we are fabricating specimens in which we can study the dynamic behaviour of ionic conductors under an applied field.

Proc. Microscopy and Microanalysis 1995, edited by G.W. Bailey, M.H. Ellisman, R.A. Hennigar, and N.J. Zaluzec
Copyright © 1995 MSA. Published by Jones and Begell Publishing, 79 Madison Ave., New York, NY 10016

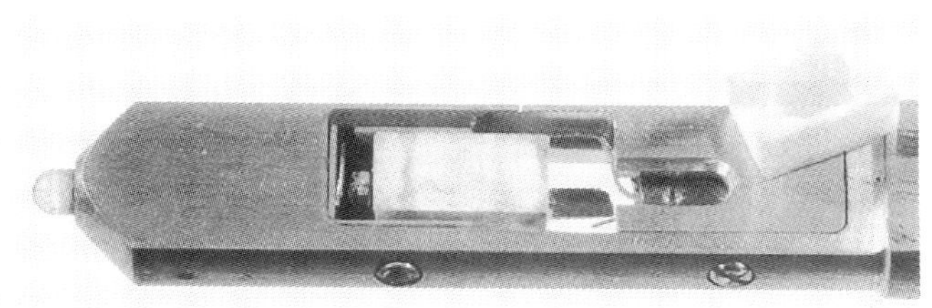

Figure 1. Electrochemical biasing holder. The reservoir is 3x4x6mm and is made of machinable ceramic. A sectioned reservoir is placed on the holder for illustration. Electrical contact to the HF is made by a Pt screw which also acts as a means to fill the reservoir. Contact to the specimen is via the thin wire to the right.

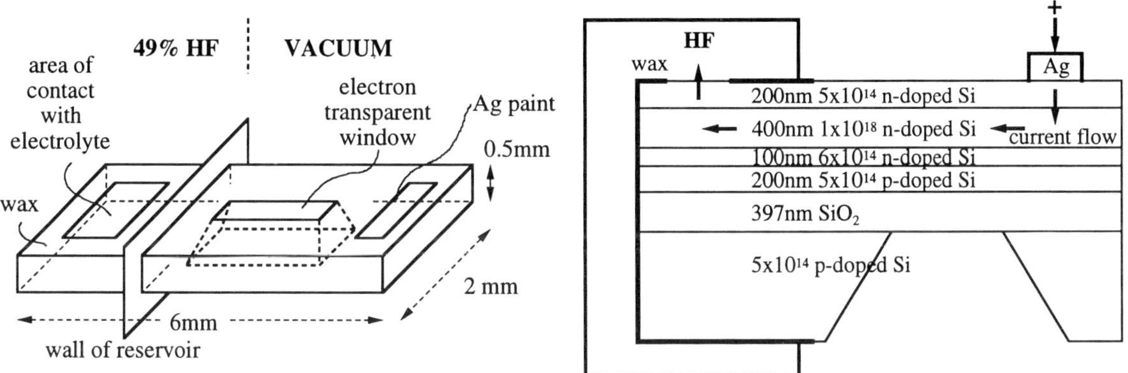

Figure 2. Schematic diagram showing the overall specimen geometry and details of the layers.

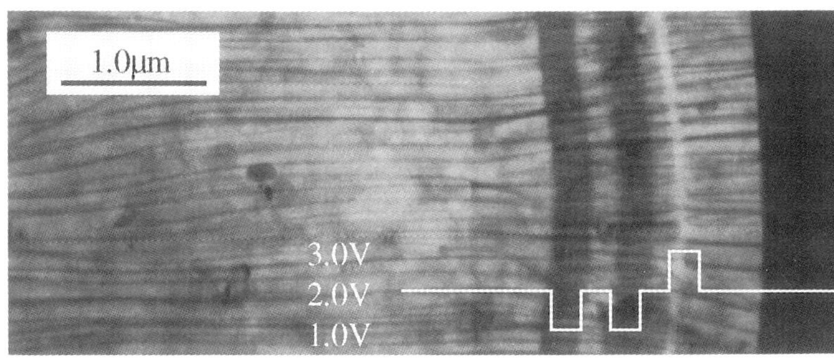

Figure 3. Bright field, plan view image of the pore front after etching *ex situ* for 32 minutes at 2.0V and 0.016mA in the (001) direction. The pores have propagated about 250µm from the initial window to the electron transparent region. XS-SEM confirms that etching occurs in the n[+] layer, so the aspect ratio of the pores is about 600. Inset is the time-varying voltage which was applied to create the bands of varying porosity.

1. L.T. Canham (1990), Appl. Phys. Lett. **57**, 1046
2. J.-P. Krumme and M. E. Straumanis (1967), Trans. Met. Soc. AIME **239**, 396
3. M. M. Faktor, D. G. Fiddyment and M. R. Taylor (1975), J. Electrochem. Soc. **122**, 1566
4. B. D. Chase and D. B. Holt (1972), J. Electrochem. Soc. **119**, 314
5. P. C. Searson (1991), Appl. Phys. Lett. **59**, 832
6. H. Munder, M. Berger, S. Frohnhoff, M. Thonissen and H. Luth (1993), J. Luminescence **57**, 5
7. S. S. Tsao, T. R. Guilinger, M. J. Kelly and P. J. Clews (1989), J. Electrochem. Soc. **136**, 586.
8. R. L. Smith, S. F. Chuang and S. D. Collins, J. Electronic Materials **17** (1988), 533
9. M. I. J. Beale, N. G. Chew, M. J. Uren, A. G. Cullis and J. D. Benjamin (1985a), Appl. Phys. Lett. **46**, 86
10 M. I. J. Beale, J. D. Benjamin, M. J. Uren, N. G. Chew and A. G. Cullis (1985b), J. Cryst. Growth **73**, 622
11. J. Erlebacher, K. Sieradzki and P. C. Searson (1994), J. Appl. Phys. **76**, 182
12. F. M. Ross, G. Oskam, P. C. Searson, J. M. Macaulay and J. A. Liddle (1995b), in preparation

We gratefully acknowledge John C. Bean and J. Alex Liddle of AT&T Bell Laboratories, NJ for MBE growth and SEM, respectively, Andrew Wittkower of SOITEC/USA, MA, for kindly providing the SIMOX wafers, and M. L. Tech, CA, for construction of the specimen holder. This work was supported by the Director, Office of Energy Research, Materials Science Division, U. S. Department of Energy under contract DE AC-03-76SF00098, and by the National Science Foundation under grant DMR-9202645.

IN SITU IRRADIATION EFFECTS STUDIES IN MEDIUM AND HIGH VOLTAGE TEM

Charles W. Allen

Electron Microscopy Center for Materials Research—HVEM-Tandem Facility, Materials Science Division, Argonne National Laboratory, Argonne, IL 60439 USA

With respect to structural consequences within a material, energetic electrons, above a threshold value of energy characteristic of a particular material, produce vacancy-interstial pairs (Frenkel pairs) by displacement of individual atoms, as illustrated for several materials in Table 1.[1] Ion projectiles produce cascades of Frenkel pairs. Such displacement cascades result from high energy primary knock-on atoms which produce many secondary defects. These defects rearrange to form a variety of defect complexes on the time scale of tens of picoseconds following the primary displacement. A convenient measure of the extent of irradiation damage, both for electrons and ions, is the number of displacements per atom (dpa). 1 dpa means, on average, each atom in the irradiated region of material has been displaced once from its original lattice position. Displacement rate (dpa/s) is proportional to particle flux ($cm^{-2}s^{-1}$), the proportionality factor being the "displacement cross-section" σ_D (cm^2). The cross-section σ_D depends mainly on the masses of target and projectile and on the kinetic energy of the projectile particle.

The effects of irradiation for a given material depend on temperature and particle flux as well as on the nature of the radiation and include a variety of phenomena, which may be classed as induced (not thermodynamically favorable) or enhanced (thermodynamically favorable). Thus under some conditions irradiation may produce thermodynamically 'uphill' processes (e.g., formation of metastable phases such as intermetallic glasses or segregation in stable solid solutions) and under other conditions, it may accelerate 'downhill' processes (e.g., precipitation from solid solution at otherwise kinetically too low a temperature). Thus, while many people think of irradiation effects as synonymous with irradiation damage, in fact radiation is an important variable in materials research (just like temperature, stress or magnetic field), which, when appropriately controlled, opens new possibilities for exploration and modification of much of materials behavior, including of course damage behavior *per se*.

Because TEM is a principal tool for evaluating the effects of irradiation on materials, conducting irradiation effects studies in situ is a very efficient process. In this manner the analysis can be performed either simultaneously during or intermittently between increments of the irradiation. For electron irradiations, σ_D for light elements such as Al or Si is adequate with 300–400 kV electrons; however, for a majority of materials, higher energies are required involving an HVEM. In the USA there are now only three available instruments at two sites: the 1.2 MeV HVEM at Argonne National Lab, and 1.0 and 1.5 MeV instruments at Lawrence Berkeley National Lab. Fortunately, both sites are user facilities funded by DOE for the materials research community. Studies involving in situ ion irradiation are complicated by the need to interface one or more ion beamlines to a TEM. A typical configuration of this sort is depicted very schematically in Fig. 1 in which the positions of several Faraday cup assemblies inside the TEM column, which are required for dosimetry of ions and electrons, at the Argonne HVEM and a new medium voltage TEM are emphasized.[2] The ion Faraday cup and the electron Faraday cup 1 determine total incident currents; electron Faraday cup 2 profiles the electron beam. While these in situ capabilities are unique within this hemisphere, nearly a dozen installations of this sort are operational in Japan and one in France.[3,4]

Proc. Microscopy and Microanalysis 1995, edited by G.W. Bailey, M.H. Ellisman, R.A. Hennigar, and N.J. Zaluzec
Copyright © 1995 MSA. Published by Jones and Begell Publishing, 79 Madison Ave., New York, NY 10016

References

1. For much more detailed data and original references, see W. E. King et al., *Ultramicroscopy* 23 (1987) 345.
2. C. W. Allen et al., *Nucl. Instrum. Methods B* 40/41 (1989) 553.
3. S. Ishino, *J. Nucl. Mater.* 256 (1993) 139.
4. C. W. Allen, *Ultramicroscopy* 56 (1994) 200.
5. Electron Microscopy Center for Materials Research is supported by U. S. Department of Energy, BES-Materials Sciences, under Contract W-31-109-Eng-38.

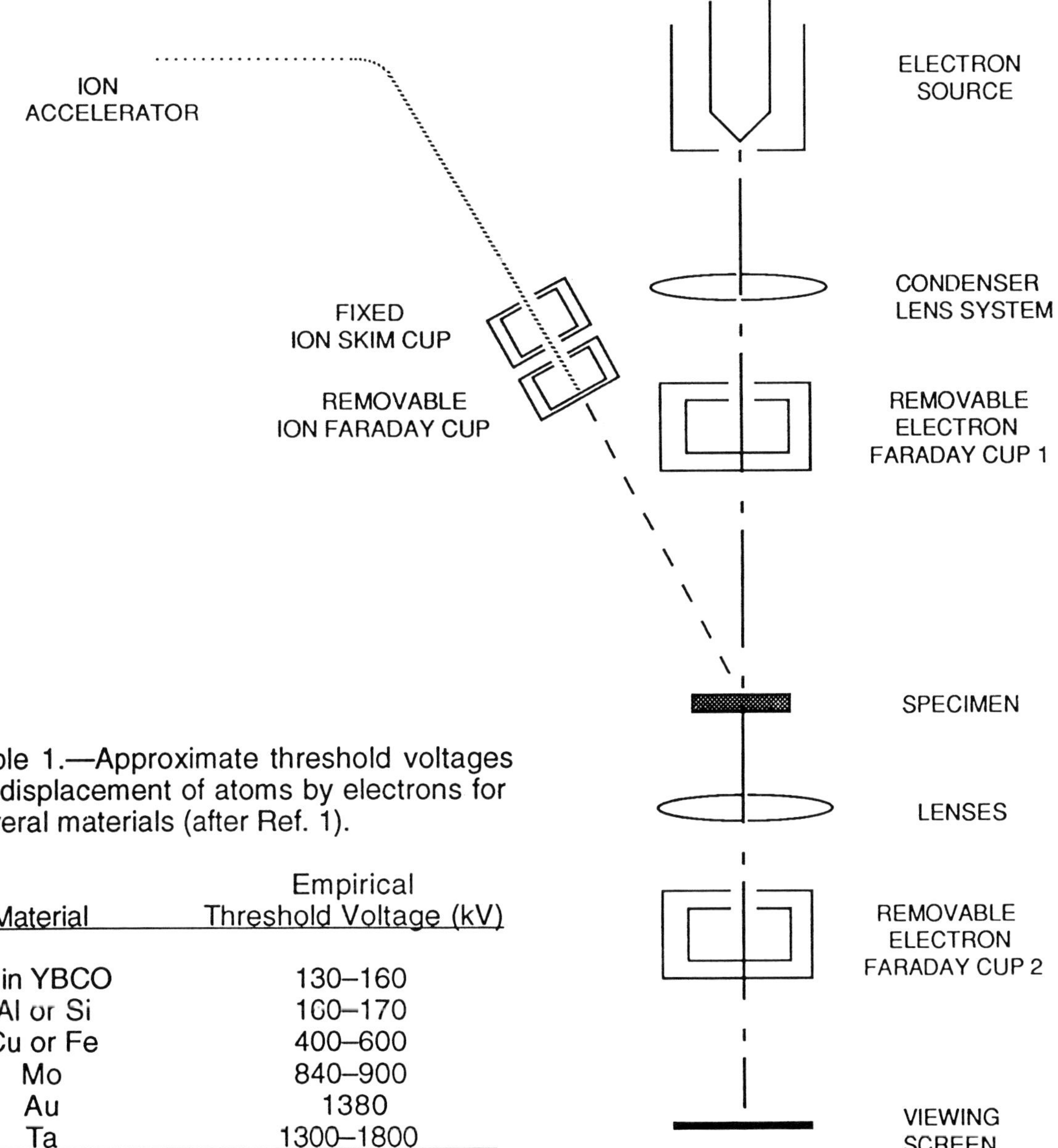

FIG. 1.—Schematic of TEM with in situ ion beam capability, emphasizing positions of Faraday cups for ion and electron dosimetry.

Table 1.—Approximate threshold voltages for displacement of atoms by electrons for several materials (after Ref. 1).

Material	Empirical Threshold Voltage (kV)
O in YBCO	130–160
Al or Si	160–170
Cu or Fe	400–600
Mo	840–900
Au	1380
Ta	1300–1800

235

ANALYTICAL ULTRA HIGH VOLTAGE ELECTRON MICROSCOPE EQUIPPED WITH EDS AND ENERGY FILTERED IMAGING SYSTEM AT NRIM

K. Furuya*, M. Osaki**, S. Hagiwara**, T. Saito*

*National Research Institute for Metals, Sengen, Tsukuba, 305 JAPAN
**JEOL LTD Co., Musashino, Akishima, 196 JAPAN

The ultra high voltage transmission electron microscope (UHV-TEM) having an acceleration voltage of 1000 kV or more has been used for the observation of atomic structure of materials because of its high resolution approaching to 0.1 nm[1-2]. Few effort has been done to develop an analytical UHV-TEM with energy dispersive x-ray spectroscope (EDS) for in-situ experiments such as ion irradiation, because high density of hard x-ray accompanied with primary electrons is supposed to prevent to measure the characteristic x-ray from the specimen. However, the observation of the heterostructures and interfaces requires analytical capability of determining the local chemical composition compatible with high resolution electron microscopy. This paper describes about a new analytical UHV-TEM (AHV-TEM) equipped with EDS and energy filtered imaging system and its application to Ni silicide thin film.

The AHV-TEM in this study is based on JEM-ARM1000 with a side entry goniometer. Figure 1 shows the comparison of calculated CTF^2 and optical diffractogram of amorphous Ge taken near Scherzer defocus with E=1000 keV, Cs=2.6 mm, Cc=3.4 mm and Qi=0.3 mrad. A minimum of intensity at spatial frequency of about 7.7 nm-1 clearly indicate a point resolution of about 0.13 nm which coincides with theoretical value. A typical example of the results of high resolution observation is shown in figure 2 for Si thin film projected along [110] direction. Clearly seen is the structures of black dumbbells of 0.136 nm interatomic spacing.

Figure 3 displays a ultra thin window EDS detector attached with AHV-TEM. Since ion introduction ports for in-situ experiment is designed to occupied the high angle position, the detector was set in a horizontal direction and hard x-ray shield for the primary electrons of 400 keV was carried out inside of condenser lens. EDS measurement was performed on $NiSi_2$/Si thin film at 400 keV and with 30 nm probe size. $NiSi_2$ was grown epitaxially ,but inhomogeneously distributed on Si(100) surface. The HRTEM image of $NiSi_2$ and Si boundary shown in figure 4 indicated the change in the morphology between $NiSi_2$ and Si. Corresponding EDS spectrum of figure 5 taken at left part of figure 4 clearly shows the existence of Ni. although the quantitative information could not be obtained because of the layer structure of $NiSi_2$ on Si. An energy filtered imaging system of GATAN HV-GIF[3] was also attached beneath the camera chamber of the AHV-TEM and maps of Ni were taken with M edge at 1000 keV. The results in figure 6 indicate an existence and local map of Ni, but detailed information at high magnification is not clear so far due to very low signal to noise ratio. The results in this study clearly indicated the usefulness of the AHV-TEM for the characteristics of semiconductors and future work will be done on in-situ ion implantation and composition changes of materials.

References.
1. F. Phillipp, R Höschen, M. Osaki, G. Möbus and M. Rühle, Ultramicroscopy 56(1994)1-10
2. Y. Matsui, S. Horiuchi, Y. Bando, Y. Kitami, M. Yokoyama and S. Suehara, Ultramicroscopy 39(1991)8-20
3. O. L. Krivanek, A. J. Gubbens and N. Dellby, Microscopy, Microanal., Microstr. 2(1992)187

Proc. Microscopy and Microanalysis 1995, edited by G.W. Bailey, M.H. Ellisman, R.A. Hennigar, and N.J. Zaluzec
Copyright © 1995 MSA. Published by Jones and Begell Publishing, 79 Madison Ave., New York, NY 10016

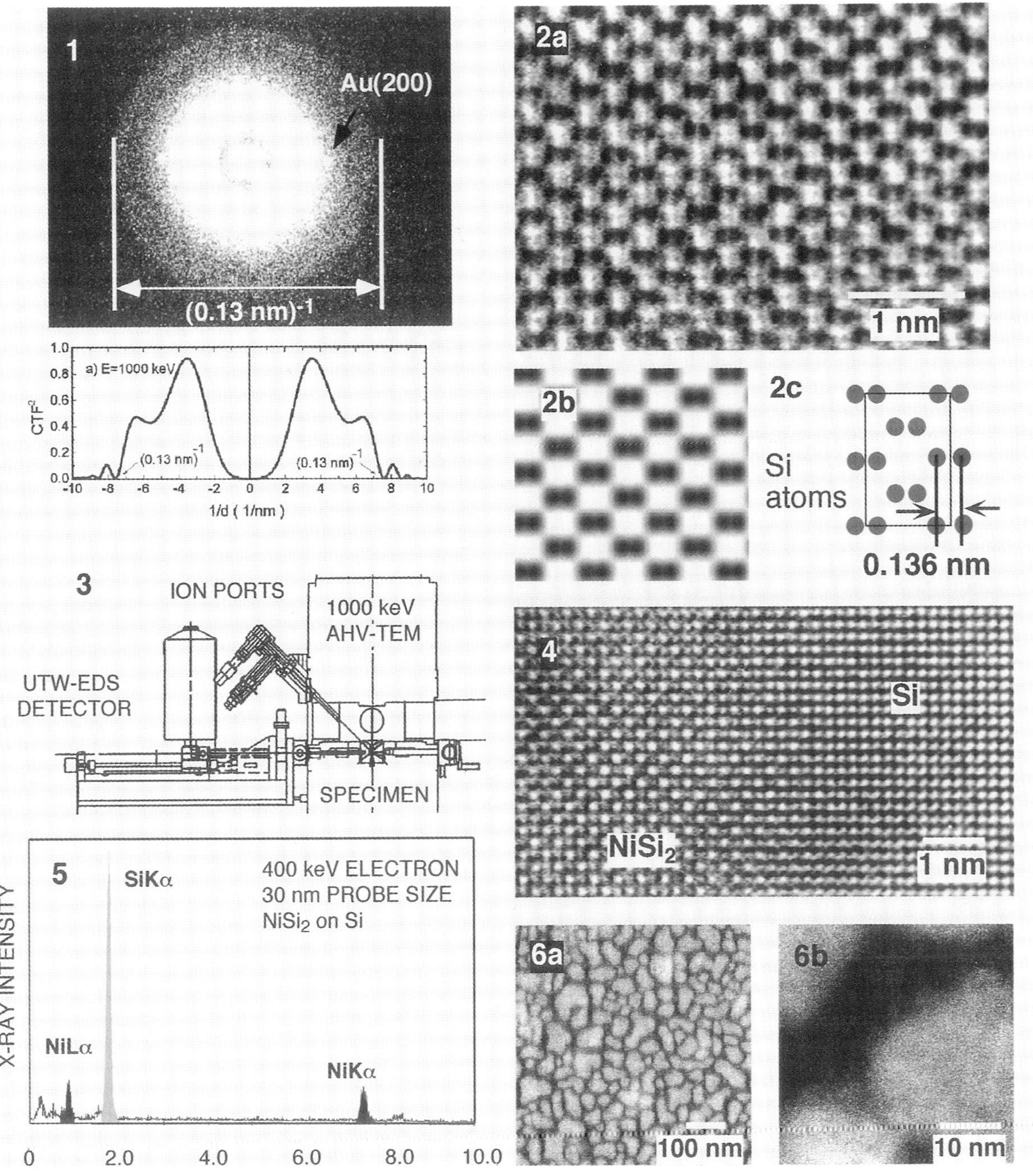

Fig. 1. -- Optical diffractogram of a photograph of a thin amorphous Ge film taken near Scherzer defocus (-57.5 nm) with a curve of CTF^2 of AHV-TEM. E=1000 keV, Cs=2.6 mm, Cc=3.4 mm, Qi=0.3 mrad.

Fig. 2. -- Structure image of Si(110) taken with AHV-TEM at near Scherzer defocus. (a) structure image, (b) the result of computer simulation and (c) structure model of Si crystal projected along |110|.

Fig. 3. -- Schematic drawing of UTW-EDS interface of AHV-TEM.

Fig. 4. -- HRTEM image of $NiSi_2$ on Si(100) thin film.

Fig. 5. -- A typical example of EDS spectrum of $NiSi_2$ region (left part of fig. 4.)

Fig. 6. -- Map of Ni taken with M edge of $NiSi_2$ by GATAN HV-GIF. (a) low mag. and (b) high mag.

IN SITU ELECTRON MICROSCOPY OBSERVATIONS OF STRUCTURAL TRANSFORMATIONS
IN SINGLE CRYSTAL LANTHANUM ALUMINATE

M. Grant Norton*, James Bentley**, and Rand R. Biggers***

*   Mechanical and Materials Engineering, Washington State University, Pullman WA 99164
**   Metals and Ceramics Division, Oak Ridge National Laboratory, Oak Ridge TN 37831
***   US Air Force Wright Laboratory, Wright Patterson Air Force Base OH 45433

Single crystals of lanthanum aluminate ($LaAlO_3$) are of interest as substrates to support the growth of films of the superconducting oxide $YBa_2Cu_3O_7$. In order to facilitate epitaxy the thin films will often be deposited onto heated ($T \geq 700°C$) substrates. $LaAlO_3$ has a cubic structure at elevated temperatures and undergoes a transformation on cooling to the rhombohedral crystal structure — the stable form at room temperature. The phase transformation from the cubic to the rhombohedral structure leads to the formation of twins. The twin boundary is predominantly the (100) plane of the pseudo-cubic cell. Several potential problems associated with the phase transformation in $LaAlO_3$ and the concomitant formation of twins have been postulated. Firstly, defects related to the twin boundaries might act as preferred nucleation sites during film deposition. Secondly, on cooling through the transformation temperature, after film deposition, twinning of the substrate might introduce strains into the thin film deposited on the initially untwinned substrate. It is therefore of interest to examine the behavior of the twin boundaries in $LaAlO_3$ during thermal cycling through temperatures close to those used for film growth.

On heating a $LaAlO_3$ sample in the transmission electron microscope (TEM) from room temperature no change in the position or appearance of the twins was noticed until temperatures of ~ 320°C. Beyond this temperature, the intensity of the contrast in the bright field image due to the twins became gradually weaker until at temperatures of ~ 400 - 420°C the twins were no longer visible. In many cases, the twins were observed to reappear with little or no change in position on cycling through the transformation temperature. This memory effect is illustrated in the sequence of images in Figure 1. Where there was a complex arrangement of twins in a particular region of the specimen it was often observed that there was some similarity between the initial and final images of a sequence but there was not always an exact one-to-one correspondence of the features [1]. Some of the prethinned $LaAlO_3$ samples were annealed at 1400°C in air prior to examination in the TEM. Observations of twins in the annealed thin foils revealed that, in general, there was a much higher incidence of them reappearing in the same morphology after the crystal had passed through the transformation temperature. In the example in Figure 2 a number of twin boundaries are visible. It can be seen in Fig. 2(a) that there is some strain associated with the twin boundaries where they intersect in the crystal. Strain contrast is visible at the two intersections A and B, indicated by arrows in Fig. 2(a). On cooling the sample through the transformation temperature the twin boundaries reappeared almost unchanged. The only region that reappeared in a different morphology was the area indicated by B in Fig. 2(a). After cycling through the transformation temperature this boundary has changed direction and this change appears to have reduced the length of the boundary and also reduced the strain contrast visible in the bright field image. In this presentation the results of these *in situ* TEM studies will be discussed together with recent reflection electron microscopy (REM) observations of the (001) surface of $LaAlO_3$ [2].

References

1. M.G. Norton and R.R. Biggers, Scripta Metall. *et* Mater. <u>32</u>, 481 (1995).
2. This research has been supported by the Wright Laboratory, Aeronautical Systems Center, Air Force Materiel Command, USAF, and the Advanced Research Projects Agency (ARPA) under grant number, F33615-94-1-5802. The work at ORNL was sponsored by the Division of Materials Science, US Department of Energy, under contract DE-AC05-84OR21400 with Martin Marietta Energy Systems, Inc., and through the SHaRE Program under contract DE-AC05-76OR00033 with the Oak Ridge Institute for Science and Education.

Proc. Microscopy and Microanalysis 1995, edited by G.W. Bailey, M.H. Ellisman, R.A. Hennigar, and N.J. Zaluzec
Copyright © 1995 MSA. Published by Jones and Begell Publishing, 79 Madison Ave., New York, NY 10016

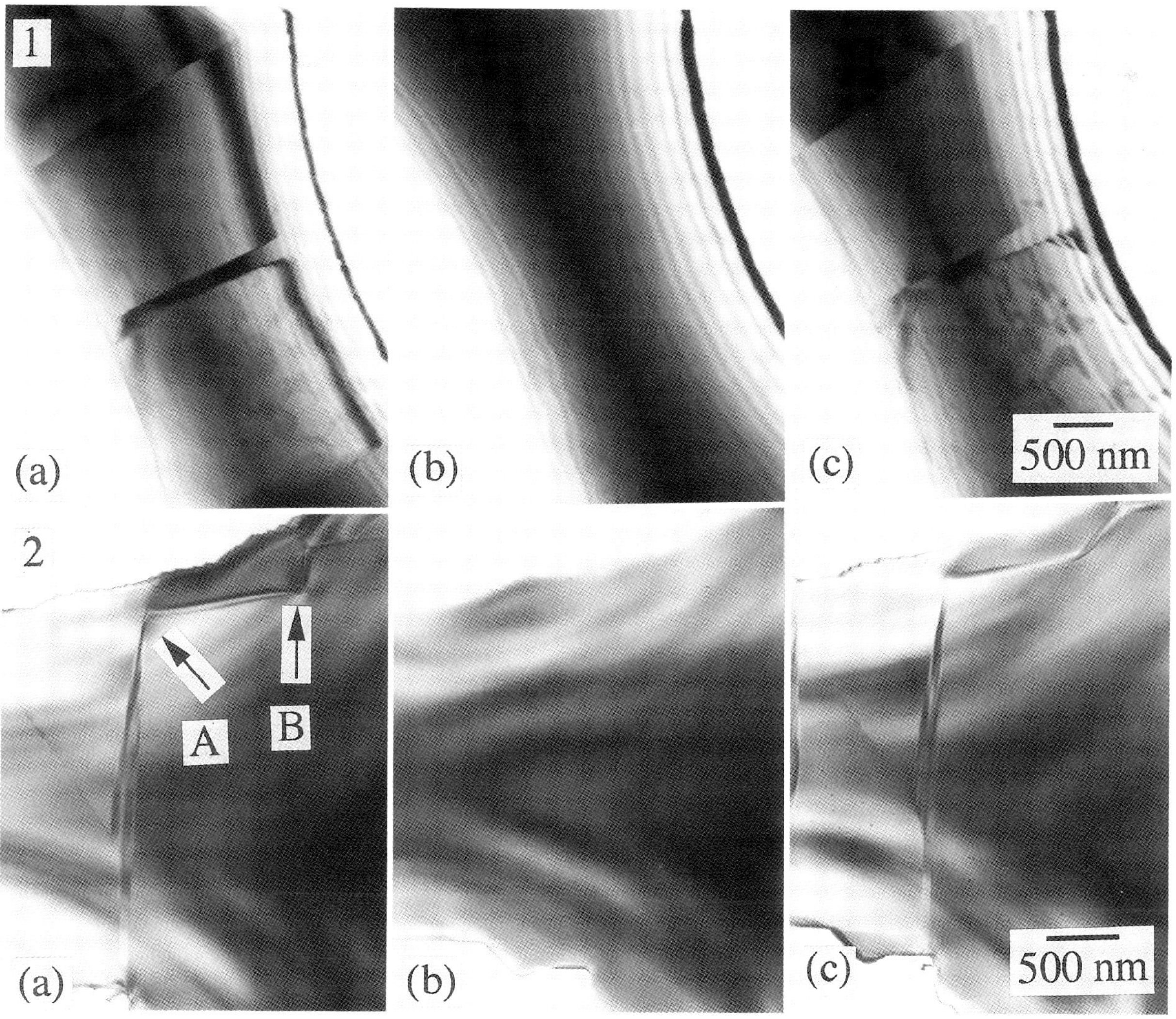

Figure 1.   Sequence of micrographs illustrating changes in the appearance of twin boundaries in LaAlO$_3$ as a function of temperature. Micrograph (a) was recorded at 22°C while heating; (b) at 750°C still heating; (c) at 56°C while cooling.

Figure 2.   Sequence of micrographs illustrating changes in the appearance of twin boundaries in LaAlO$_3$ annealed at high temperature prior to examination in the TEM. Micrograph (a) was recorded at 41°C while heating; (b) at 748°C still heating; (c) at 67°C while cooling.

TECHNIQUES FOR IN SITU HVEM MECHANICAL DEFORMATION OF NANOSTRUCTURED MATERIALS

M.A. Wall*, T.W. Barbee Jr.* and U. Dahmen**

*Lawrence Livermore National Laboratory
**Lawrence Berkeley Laboratory

We have developed two in situ HVEM experimental techniques which allow us to begin fundamental investigations into the mechanisms of deformation and fracture in nanostructured materials. First, a procedure for the observation of tensile deformation and failure of multilayer (ML) materials in cross-section is detailed. Second, the development of an in situ HVEM nanoindentor of surfaces and films on surfaces in cross-section is presented.

Nanostructred ML materials often exhibit enhanced physical properties[1] such as increased hardness and strength. There have been few (if any) direct observations of the mechanisms by which nano-scale structures affect deformation thereby enhancing mechanical performance. Nanoindenting[2] has become a primary technique for measuring the mechanical properties of small volume materials. The mechanical properties of small volumes can vary greatly from bulk values. The elastic and plastic response and microstructure evolution in these small volumes under the indentor tip has never been dynamically observed before. Observation of the behavior of material under these conditions would no doubt further our understanding of the mechanical behavior of nanostructured materials.

Our development of in situ tensile deformation technique of multilayers in cross-section is focused upon ML sample design and specimen preparation. Sample design consisted of the synthesis of a ML sample having materials with different mechanical properties, a small period (<100nm) yet large enough to view at moderate magnification and a large total thickness for handling during specimen preparation. A Cu/Zr ML with a period of 90nm (Cu 80nm/Zr 10nm) and a total thickness of 120 μm was synthesized by magnetron sputter deposition. Specimens were prepared by a modification to a technique detailed by Wall,1994[3]. Free standing ML foils were electroplated to bulk dimensions, sliced, lapped, dimpled and low angle ion milled. The final step was to use of focused ion beam to mill a micro gauge section. This determines the location at which deformation and fracture will be first observed at high magnification. Figure 1 is SEM micrograph of the micro gauge section in the tensile specimen. The as deposited structure in the micro gauge section is shown in figure 2a, point P. After straining and observing dislocation motion within the Cu layers a series of Zr layers failed and Cu layers deformed to form a row of voids and ligaments, (figure 2b). The voids and ligaments form a crack which under continued loading grows by necking of the Cu layers. Crack growth continues by void formation in front of the crack tip and Cu ligament elongation, leading to failure.

In situ nanoindenting of surfaces in cross-section required a new specimen holder design that had to be built and tested. The holder features 3 axis mechanical and piezo position control and a replaceable indentor tip. Cleaved or cross-sectioned samples are mounted on a removable specimen platform. The results of initial testing are shown in figure 3. The sapphire indentor is positioned on the surface of a cleaved Si wedge in cross-section, (fig. 3a). The estimated specimen thickness is 3 μm. When the indentor is depressed into the surface the elastic response of the Si is seen by the formation of strain contours radiating out from the indentor tip, (fig.3b). Further indenting plasitcally deforms the Si, (fig. 3c).

Correspondence of the microstructural scale and the imaging capabilities of the HVEM make it productive to apply in situ characterization techniques to nanostructured materials. These two novel techniques provide two new tools which will generate new insight into the micromechanisims of deformation and failure in nanostructured materials.

Proc. Microscopy and Microanalysis 1995, edited by G.W. Bailey, M.H. Ellisman, R.A. Hennigar, and N.J. Zaluzec
Copyright © 1995 MSA. Published by Jones and Begell Publishing, 79 Madison Ave., New York, NY 10016

All in situ experiments were performed on a Kratos 1.5Mev HVEM at Lawrence Berkeley Laboratory.

References

1. Suryanarayana, C. and Froes, F.H., Met. Trans. A,23A(1992)1071.
2. Oliver, W. C., Properties and Deformation Behavior of Materials Having Ultra Fine Microstructures, Boston:Academic Press,(1993)417-428.
3. Wall, M. A., Microscopy Research and Technique, 27(1994)262-267.
4. The authors wish to thank Tim Weihs and Richard Gross of Lawrence Livermore National Laboratory and Doug Owen of Lawrence Berkeley Laboratory for their scientific and technical support. This work was performed under the auspices of the US Department of Energy by the Lawrence Livermore National Laboratory under contract W-7405-Engr-48.

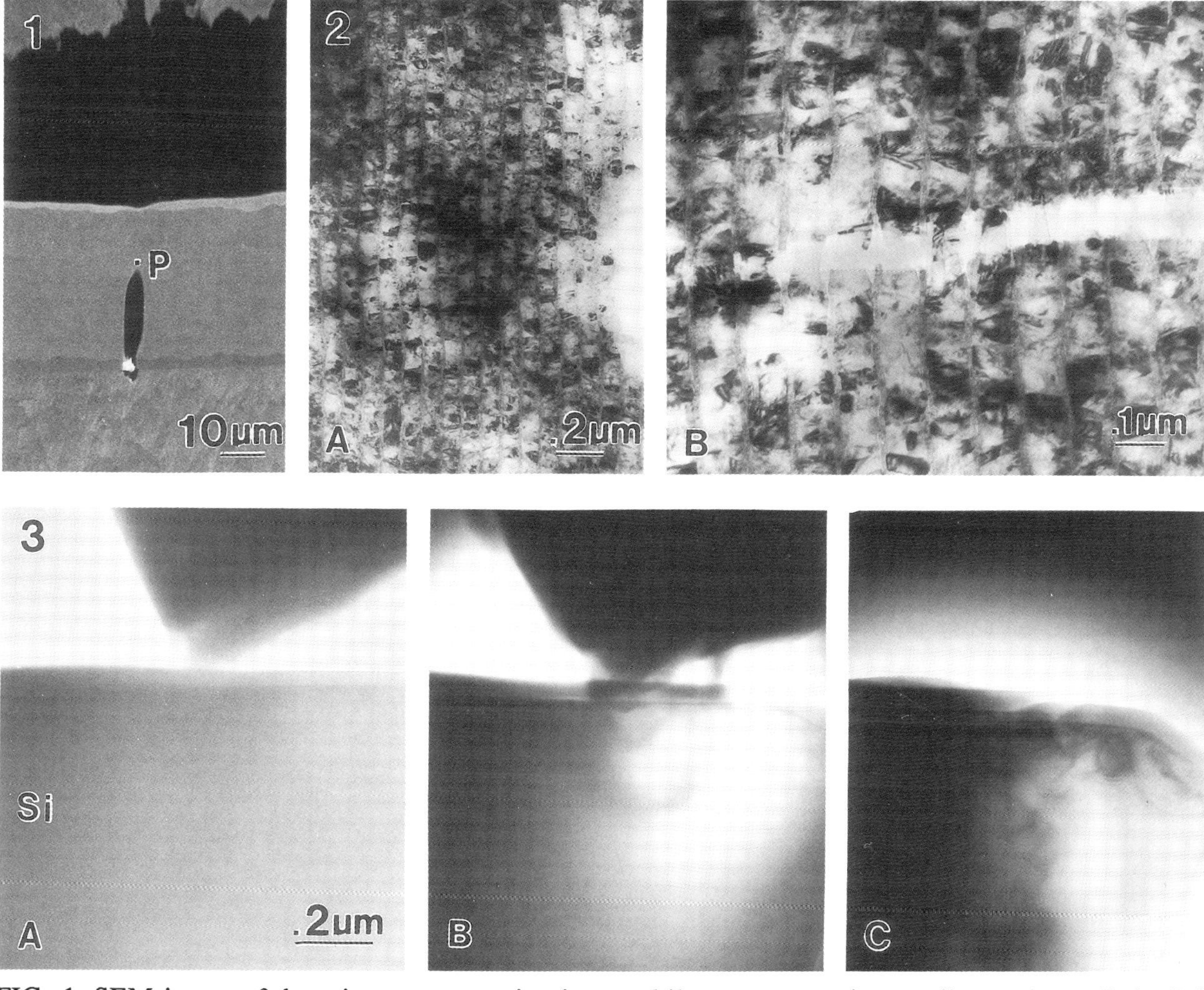

FIG. 1.-SEM image of the micro gauge section in a multilayer cross-section tensile specimen. Point P is the electron transparent viewing region during the tensile deformation experiment.

FIG. 2.-(a) The as deposited Cu/Zr multilayer structure viewed in cross-section at point P in figure 1; (b) Crack formation consisting of voids originating at the Zr layers and elongated Cu ligaments.

FIG. 3.-(a) A sapphire indentor tip on the surface of a cleaved chip of Si oriented in cross-section; (b) Indentor tip elastically deforming bulk Si; (c) Indent and plastic deformation of bulk Si.

IN-SITU CRYSTALLIZATION OF NANO-WIRES AND DOTS BY ELECTRON IRRADIATION IN A FIELD-EMISSION TEM/STEM

M. Libera
Dept. of Materials Science and Engineering
Stevens Institute of Technology, Hoboken, New Jersey

Specimen damage and heating by high-energy electron irradiation can be exploited to affect local specimen composition/temperature to drive phase transformations. Field-emission (FEG) sources are especially prone to inducing damage/heating, because their high brightness provides focused ~1nm FWHM/1.0nA probes. The physics of beam heating and damage are reviewed by Reimer (1,2). Heating is related to energy-loss mechanisms broadly summarized by the Bethe formula for stopping power. Radiation damage can cause atomic displacement, ionization, and bond scission. Displacement increases the point-defect concentration. Composition changes occur when displacements eject atoms from the specimen. This paper outlines first results using focused-probe irradiation in a Philips CM20 FEG TEM/STEM to locally crystallize an amorphous 80nm $Ge_{48}Te_{52}$ film. *Ex-situ* annealing leads to spherulitic crystallites. The crystallization kinetics have been studied by time-resolved reflectivity (3) and *in-situ* hot-stage optical microscopy (4). *In-situ* TEM annealing was done with a single-tilt hot stage. Films were floated from carbon-coated mica onto Cu grids. *In-situ* annealing under a parallel (low dose) beam leads to crystallization much like the *ex-situ* studies except that the crystallite morphology is highly anisotropic due to non-uniform film heating (5).

Local crystallization can be induced by focused-probe irradiation. The hot stage maintains an average temperature far below the bulk crystallization temperature (~180°C). The dose rate is controlled by the second condenser lens (spot size). The probe position is controlled using the x-y controls in STEM spot mode. Local crystallization is suspended by switching to STEM raster mode. Fig. 1 shows a typical result (hot stage T=40°C). The two larger crystal dots were each formed by ~4 sec irradiation under a stationary ~3nA probe (spot size #2). The connecting crystalline wire was formed by CW exposure from a smaller probe (~0.5nA, spot #4) by irradiating existing crystal at each of the two end dots. The width of the connecting wire is less than 35nm. Though unclear why, this wire in places has two lines grown during a single pass of the probe. Fig. 2 shows a time-resolved sequence of BF STEM images illustrating one case of nucleation (stage T=50°C; ~1.0nA probe/spot #3). A light dot appears and grows. Unclear at this point is whether the light contrast is due to mass loss or to Bragg diffraction. After 240 secs a crystal (dark contrast) appears and rapidly grows. Inducing nucleation appears more difficult than driving growth of existing crystal. Focused-probe irradiation induces local changes in specimen temperature, composition, and defect concentration which, when coupled with probe positioning capabilities can be exploited to synthesize patterned structures (fig. 3). One can envision that this process could be used for nano-synthesis applications involving crystallization and precipitation phenomena in a variety of systems.

References

1. L. Reimer, *Scanning Electron Microscopy* (Springer, 1985).
2. L. Reimer, *Transmission Electron Microscopy* (Springer, 1989).
3. M. Libera and M. Chen, J. Applied Phys. 73 (1993) 2272-2282.
4. Q.M. Lu and M. Libera, J. Applied Phys. 77 (1995) 517.
5. M. Libera, MRS Symposium Procs V343, 89-100 (1994).
6. Microscopy resources were provided by the NJCST and the NSF.

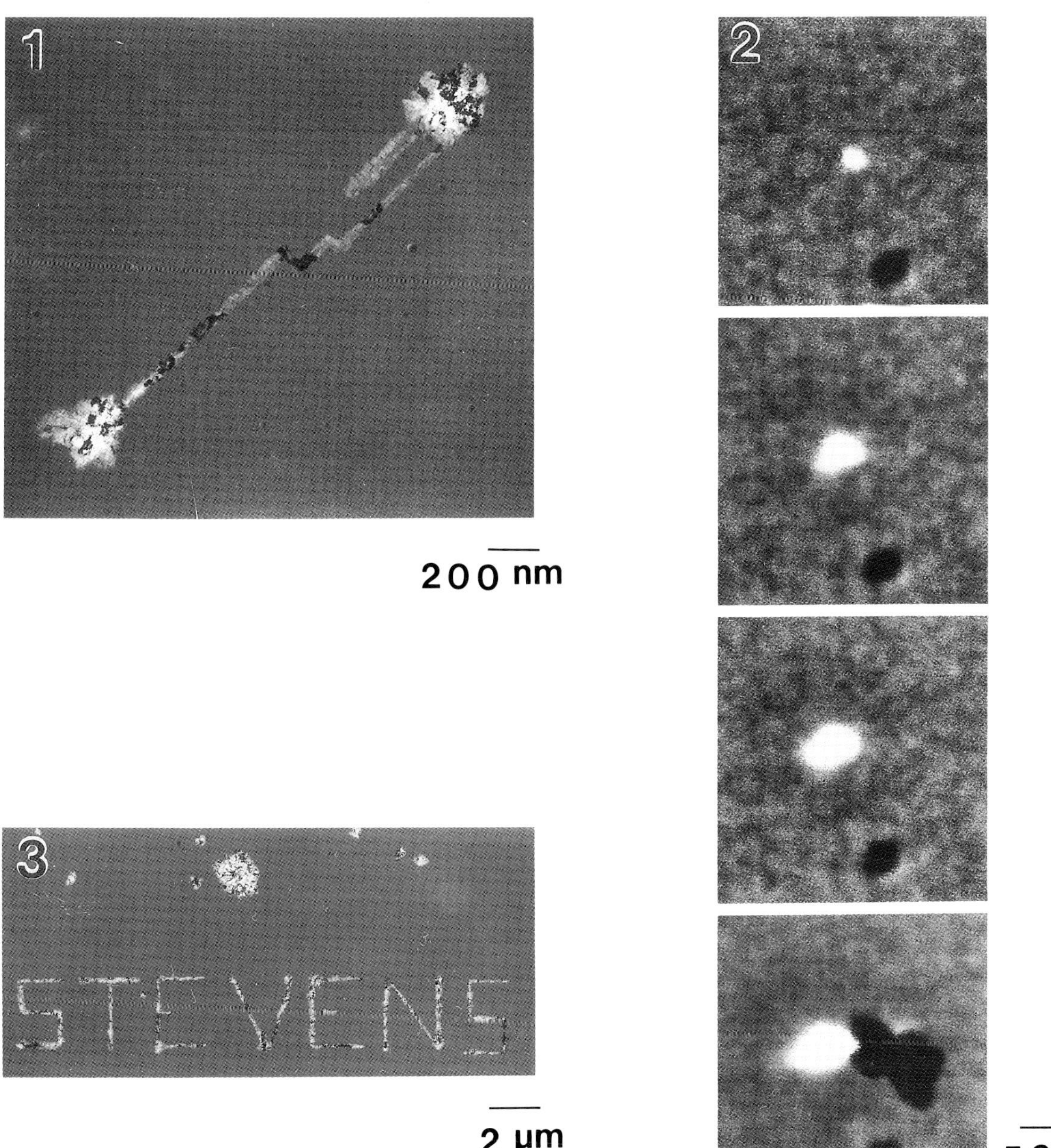

Fig. 1 - BF TEM image of focused-probe induced nanocrystalline wire.
Fig. 2 - Time-resolved BF STEM images of probe-induced nucleation.
Fig. 3 - Patterned wires can be made by selective probe positioning.

IN SITU OBSERVATIONS OF ELECTROMIGRATION INDUCED VOID BEHAVIOR

T. Marieb†, J. C. Bravman*, P. Flinn*†, D. Gardner†, and M. Madden†

*Department of Materials Science and Engineering, Stanford University, Stanford, CA 94305
†Intel Corporation, 3065 Bowers Ave., Santa Clara, CA 95052

Electromigration and stress voiding have been active areas of research in the microelectronics industry for many years. While accelerated testing of these phenomena has been performed for the last 25 years [1-2], only recently has the introduction of high voltage scanning electron microscopy (HVSEM) made possible *in situ* testing of realistic, passivated, full thickness samples at high resolution [3-5] .

With a combination of *in situ* HVSEM and post-testing transmission electron microscopy (TEM) , electromigration void nucleation sites in both normal polycrystalline and near-bamboo pure Al were investigated. The effect of the microstructure of the lines on the void motion was also studied.

The HVSEM used was a slightly modified JEOL 1200 EX II scanning TEM with a backscatter electron detector placed above the sample [3] . To observe electromigration *in situ* the sample was heated and the line had current supplied to it to accelerate the voiding process. After testing lines were prepared for TEM by employing the plan-view wedge technique [6] . The line remained encapsulated within the thinned Si substrate and the nitride passivation. This guaranteed that the preparation technique did not modify the void morphology. The samples were examined with a Philips 430 ST TEM operated at 300 kV. The voids seen in the HVSEM were easily identified, and no new voids were observed, indicating that the resolution of the HVSEM is adequate to observe all stable voids.

Two different types of lines were investigated; one deposited under high purity conditions with a near-bamboo microstructure, and one deposited under conventional conditions with a polycrystalline microstructure with stress voids present. Electromigration voids and stress voids were seen to nucleate at very specific, unique, microstructural sites that require the intersection of a grain boundary with the line sidewall [5], as an example see Figure 1. Void movement after initiation is also dictated by microstructure, with voids only growing into and causing failure in grains oriented with a {111} plane near perpendicular to the line [5], as is shown in Figures 2 and 3. The conditions for void nucleation suggest that heterogeneities at the line sidewall are necessary for void nucleation.

References

.[1] I.A. Blech. and E.S. Meieran, Appl. Phys. Lett. **11**, 263 (1967).
[2] E. Levine and J. Kitcher, Proceedings of the 22nd Annual IEEE International Reliability Physics Symposium, USA (1984), p. 242.
[3] T. Marieb, J.C. Bravman, P. Flinn, D.S. Gardner, and M. Madden, Appl. Phys. Lett. **64**, 2424 (1994).
[4] D.M. Follstaedt, D.M., J.A. Van Den Avyle, A.D. Romig Jr., and J.A. Knapp, Mater. Res. Soc. Symp. Proc. **225**, 225 (1991).
[5] T. Marieb, J. C. Bravman, P. Flinn, D. Gardner, and M. Madden, to be published in J. Appl. Phys.
[6] J. Benedict, R. Anderson, and S.J. Klepeis, Mater. Res. Soc. Symp. Proc. **254**, 121 (1992).

Proc. Microscopy and Microanalysis 1995, edited by G.W. Bailey, M.H. Ellisman, R.A. Hennigar, and N.J. Zaluzec
Copyright © 1995 MSA. Published by Jones and Begell Publishing, 79 Madison Ave., New York, NY 10016

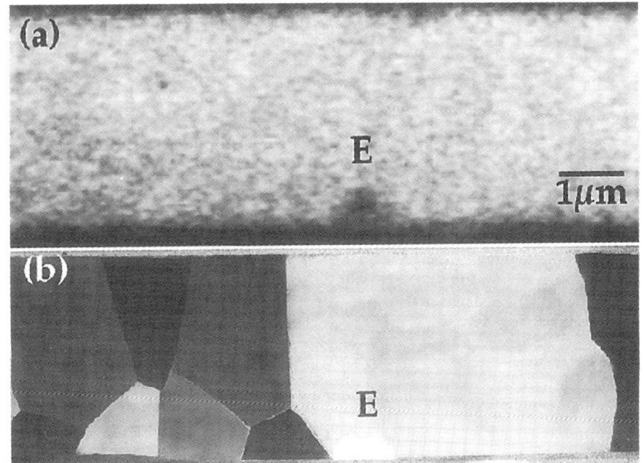

FIG. 1.-(a) HVSEM image of electromigration void in near-bamboo passivated Al line; (b) TEM image of same area taken post-mortem (after Ref. 5).

FIG. 2.-Sequence of HVSEM images showing void nucleation and growth during *in situ* testing. Current density in the line was 6.0 MA/cm², line temperature was 320°C, time elapsed was (a) 40 min.,(b) 1 hr.,(c) 1.75 hrs., (d) 2 hrs., (e) 2.25 hrs., (f) 2.5 hrs., (g) 2.58 hrs. and (h) 2.62 hrs (after Ref. 5).

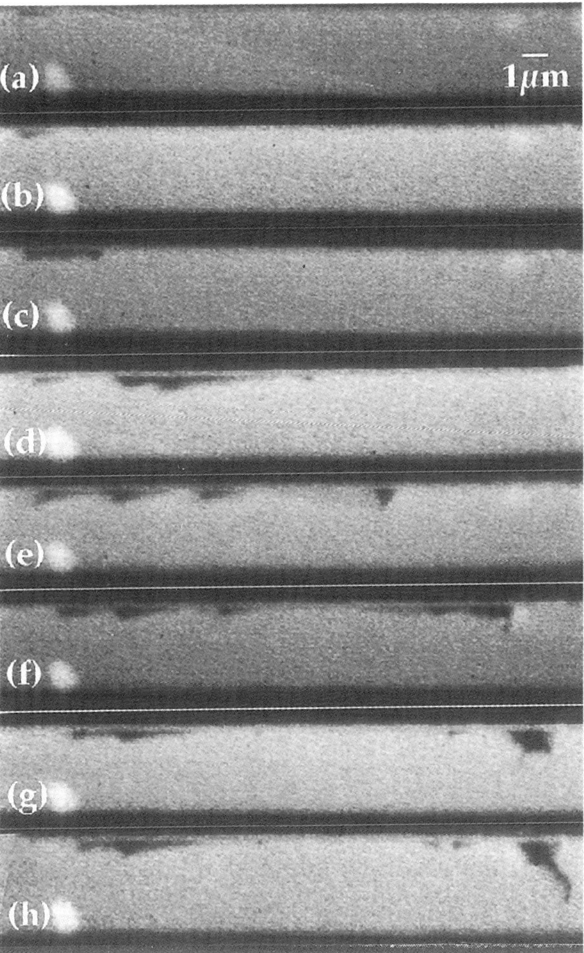

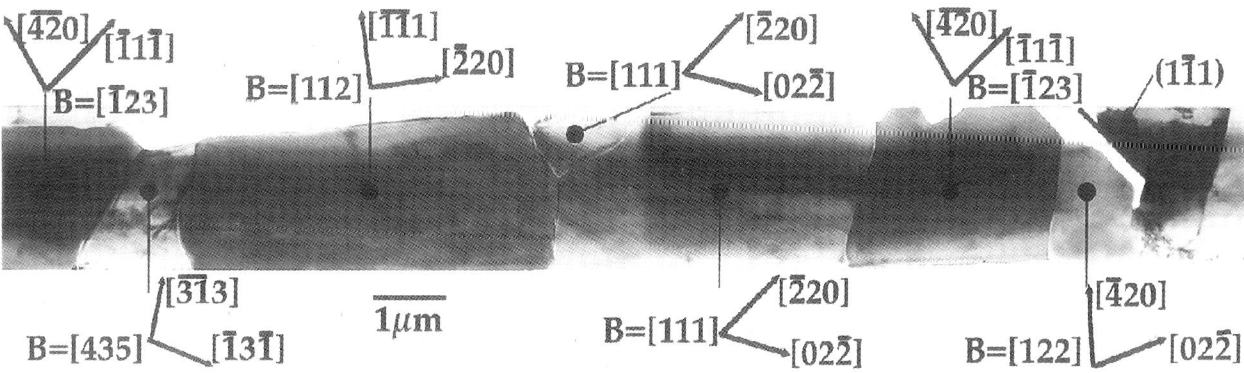

FIG. 3.-Post-mortem TEM image of same area as shown in FIG. 2 (after Ref. 5).

IN-SITU OBSERVATION OF ORIENTATION CHANGES ON METALLIC SURFACES

H. Weiland*, D.P. Field# and B.L. Adams**
*Aluminum Company of America, Alcoa Center, Pa. 15069
#TexSEM Laboratories, Provo, UT 84604
**Carnegie Mellon University, Pittsburgh, Pa. 15213

The characterization of crystalline aggregates by the crystallographic orientations of their grains and subgrains has become a subject of increasing interest. The information obtained is not only used for the characterization of materials, but also more importantly for the determination of properties. To mention only a few, applications have been found in the areas of fracture analysis, recrystallization, and plastic deformation[1-5].

Most commonly, crystallographic orientations are determined from Backscattered Kikuchi Diffraction (BKD) in the SEM and from Kikuchi patterns obtained by microdiffraction in the TEM. Since the development of fully automatic pattern analysis routines for the BKD[6,7], the SEM based techniques currently finds the most applications. In conjunction with computer controlled stage or beam displacements, the technique is known as Orientation Imaging Microscopy (OIM)[8]. In this manner, thousands of diffraction patterns are analyzed automatically within a short time. This leads to a statistical description of the distribution of crystallographic orientations, which sufficiently represent the bulk material. Another advantage of SEM based orientation analysis over the TEM based techniques is the much simpler specimen preparation, which helps to increase the accuracy of tracking macroscopic sample directions in the microscope. However, TEM based OIM is in its development stages as well[9,10] and will find its application where a high spatial resolution for the orientation analysis is required.

The analysis of static conditions have already given much new insight into fundamental processes of physical metallurgy. However, processing of materials always involves the evolution of the microstructure and therefore makes the in-situ observation of orientation changes desirable. For example, it is of interest to study, how grains with different activated slip systems break up into subgrains during deformation and how the dislocation structure evolves during this event. Also for studying superplastic deformation it would be of interest to observe grain rotation and grain sliding by tracking crystallographic orientations of individual grains in-situ. Recrystallization events in deformed materials would benefit from studies of observing growing nuclei within a deformed matrix with regard to their respective crystallographic orientation relationships. This kind of analysis requires the in-situ observation of orientation changes.

We have now developed the capabilities for in-situ orientation analysis. An existing straining-heating stage was modified to fit in our SEM under BKD conditions. Straining and/or heating stages are typically placed in untilted conditions for in-situ observations of microstructural changes. However, in order to obtain BKD patterns, a sample and therefore also the stage had to be tilted to about 70° with respect to the incident electron beam (Figure 1). Because of the relative large size of the stage, the working distance between sample and pole piece is with about 25 mm longer than usually applied for diffraction analysis in the SEM. The stage allows heating up to 500°C. In order to remove excessive heat, the heating block is water cooled. The stage can apply a maximum load of 40 kg to the sample. This set-up is used to study the evolution of orientation changes during deformation of strain-hardening aluminum alloys as well as the formation of recrystallized grains at the surface of hot deformed aluminum samples.

Proc. Microscopy and Microanalysis 1995, edited by G.W. Bailey, M.H. Ellisman, R.A. Hennigar, and N.J. Zaluzec
Copyright © 1995 MSA. Published by Jones and Begell Publishing, 79 Madison Ave., New York, NY 10016

References

1 Y. Pan, T. Olsen and B.L. Adams, *A probabilistic Model of Intergranular Fracture in Polycrystals*, to appear in Proc. 33rd CIM Conference, Toronto, Aug. 1994
2 T.A. Mason and B.L. Adams, *The Application of Orientation Imaging Microscopy*, JOM, Vol. 46, No. 10, pp. 43-45
3 S.I.Wright, A.J. Beaudoin, and G.T. Gray III, *Texture Gradient Effects in Tantulum*, Materials Science Forum, Volumes 157-163, Trans Tech Publications, pp. 1695-1700
4 H. Weiland, T. N. Rouns, and J. Liu, *The Role of Particle Stimulated Nucleation during Recrystallization of an Aluminum-Manganese Alloy*, Z. Metallkunde 85 (1994) pp. 592-597
5 S. Panchanadeeswaran, R. Becker, R.D. Doherty, and K. Kunze, *Direct Observation of Orientation Changes following Channel Die Compression of Polycrystalline Aluminum - Comparison between Experiment and Model*, Materials Science Forum, Volumes 157-163, Trans Tech Publications, pp. 1277-1282
6 S.I. Wright and B.L. Adams, *Automatic Analysis of Electron Backscatter Diffraction Patterns*, Met.Trans A, 23 (1992), pp. 759-767
7 N.C.K. Lassen, K. Conradsen, and D. Juul-Jensen, *Image Processing Procedures for Analysis of Electron Backscattering Patterns*, Scanning Microscopy, 6 (1992), pp. 115-121
8 B.L. Adams, S.I. Wright, and K. Kunze, *Orientation Imaging: The Emergence of a New Microscopy*, Met. Trans. A, 24 (1993), pp.819-831
9 H. Weiland and D.P. Field, *Automated Indexing of Kikuchi Patterns for Orientation Determination*, Proc. 52nd Annual Meeting of the Microscopy Society of America, Eds.: G.W. Bailey and A.J. Garett-Reed, San Francisco Press, 1994, pp. 900-901
10 S. Zaefferer, R.A. Schwarzer, *On-Line Interpretation of Spot and Kikuchi Patterns*, Materials Science Forum, Volumes 157-163, Trans Tech Publications, pp. 247-250

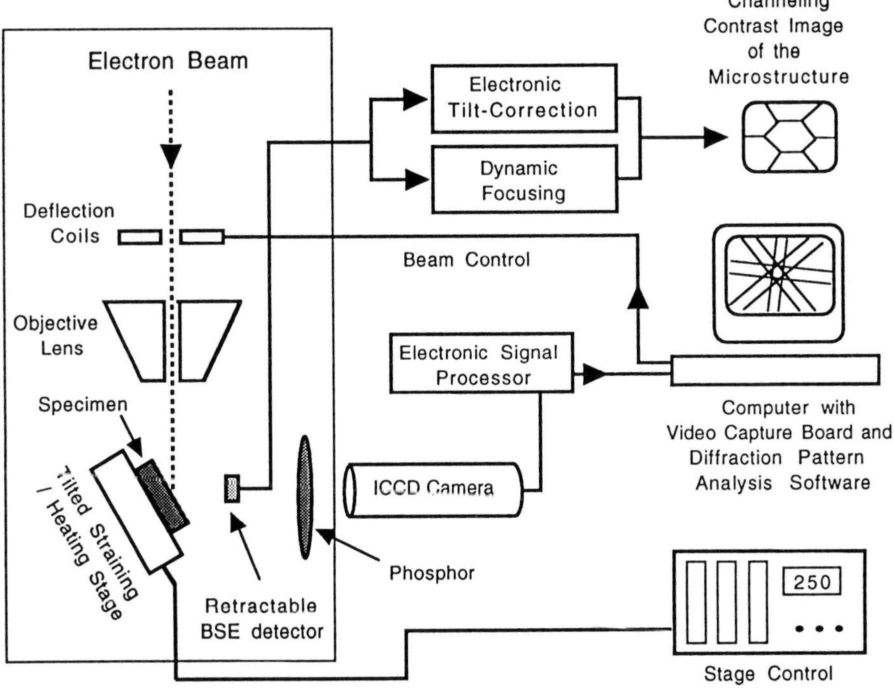

Fig. 1 Set-up for in-situ observation of crystallographic orientation changes in the SEM

IN SITU REVERSED DEFORMATION OF ALUMINUM SINGLE CRYSTALS IN THE HVEM: A NEW X-Y STRAINING STAGE

M. A. Wall*, M. E. Kassner**

*Lawrence Livermore National Laboratory
**Oregon State University

The metal fatigue phenomenon is poorly understood, partially because the dislocation dynamics of reversed deformation have not been well characterized. Little success has been realized with the direct observation of dislocations during in situ cyclic deformation. Problems associated with buckling of the specimen foil occur during applied shear[1], bending[2] or compression. Buckling can preclude adequate imaging conditions and further complicate the thin foil stress-state. Recent experiments have shown that dislocation movement can be reversed by tensile stressing in alternate perpendicular directions (i.e. 90° rotation of the tensile stress to X and Y directions results in the reversal of the shear stress)[3]. Buckling of the specimen foil is also reduced in these experiments.

Results from recent in situ reversed deformation experiments are presented here. The experiments were performed on a new straining stage which is a modification of the Sleeswyk- Kassner X-Y stage[3]. The specimen design and preparation procedures have also been modified to facilitate better and more reproducible in situ experiments. The new X-Y straining stage is illustrated in figure 1. There are three primary components of the stage: three movable platens, a push-pull (P-P) activation rod and threaded specimen grips. Tensile loading is achieved by actuating the P-P rod in the +X direction engaging the X platen. The specimen is stressed between screws in the X platen and the two Y platens, (Fig.1b). Reversing the stress state by 90° is achieved by reversing the P-P rod to the -X direction until it wedges between the two Y platens. The Y platens then move apart, orthogonal to the X axis.

The specimen design consists of a small 1.0mm diameter <u>non</u>perforated electron transparent dished region at point P and six holes for specimen clamping screws, (Fig. 1b). A perforation in the specimen complicates the stress state and the specimens bend and fracture easily during loading. The aluminum specimens are single crystals, preoriented for single slip of dislocations having Burgers vector $[0,1,-1]$ on the (111) plane, (Fig. 2). The (111) plane is also inclined approximately 25° to the foil normal to allow good projection of the (111) plane dislocations. The $[0,1,-1]$ direction is at 45° to the X and Y tensile axes. The specimen rod is tilted 14° to a $g(1,1,-1)$ • $b[0,1,-1] = 1$ imaging condition. The specimen orientation and holder tilt maximize the + and - Schmidt factors necessary for reversal of the stress state and still allow for good projection and imaging conditions of the single slip dislocations. Relatively little specimen foil bending was observed during X or Y loading.

The results of an in situ test are illustrated in figure 3 taken from a video sequence. The in situ experiment was performed using the Kratos HVEM at Argonne National Laboratory operating at 1.0Mev. Two dislocations labeled A and B move up towards a low angle boundary during X loading, (Fig. 3a). With Y loading, the dislocations reverse their motion, (Fig 3b). When the stress is rotated back to the X direction, the dislocations return to the same position as in figure 3a, (Fig. 3c). Trace analysis shows that these dislocations are moving on the (111) plane. We believe that b is $[0,1,-1]$, but this has not been unambiguously confirmed by g•b experiments. The $[0,1,-1]$ Burgers vector lies in the (111) plane, provides the largest + and - schmidt factors and g•b>0.

This new modified X-Y stage has successfully reversed dislocations in a well characterized crystal orientation. This stage operates more smoothly than its predecessor and is accommodated in Kratos HVEMs. The smaller electropolishing dimple appears to reduce bending of the thin specimen foil. Bulk cycled crystals are being prepared to be studied using in situ HVEM tests. We believe that these new techniques and future tests will lead to further insight on the dynamics of dislocation behavior under reversed deformation.

Proc. Microscopy and Microanalysis 1995, edited by G.W. Bailey, M.H. Ellisman, R.A. Hennigar, and N.J. Zaluzec

References

1. J. Lepinoux et al., Philosophical Magazine A. 51(1985)No.5. 675.
2. A.Yamamoto et al., Proc. of Fifth Int. Conf. on HVEM.Kyoto:Nakanishi Print. (1977)395.
3. M.E. Kassner et al., J. Electon Microscopy Technique. 5(1987)189.

We wish to thank the Argonne National Laboratory and the NECM at Lawrence Berkeley Laboratory for the use of the Kratos HVEMs and the support of their technical staffs.

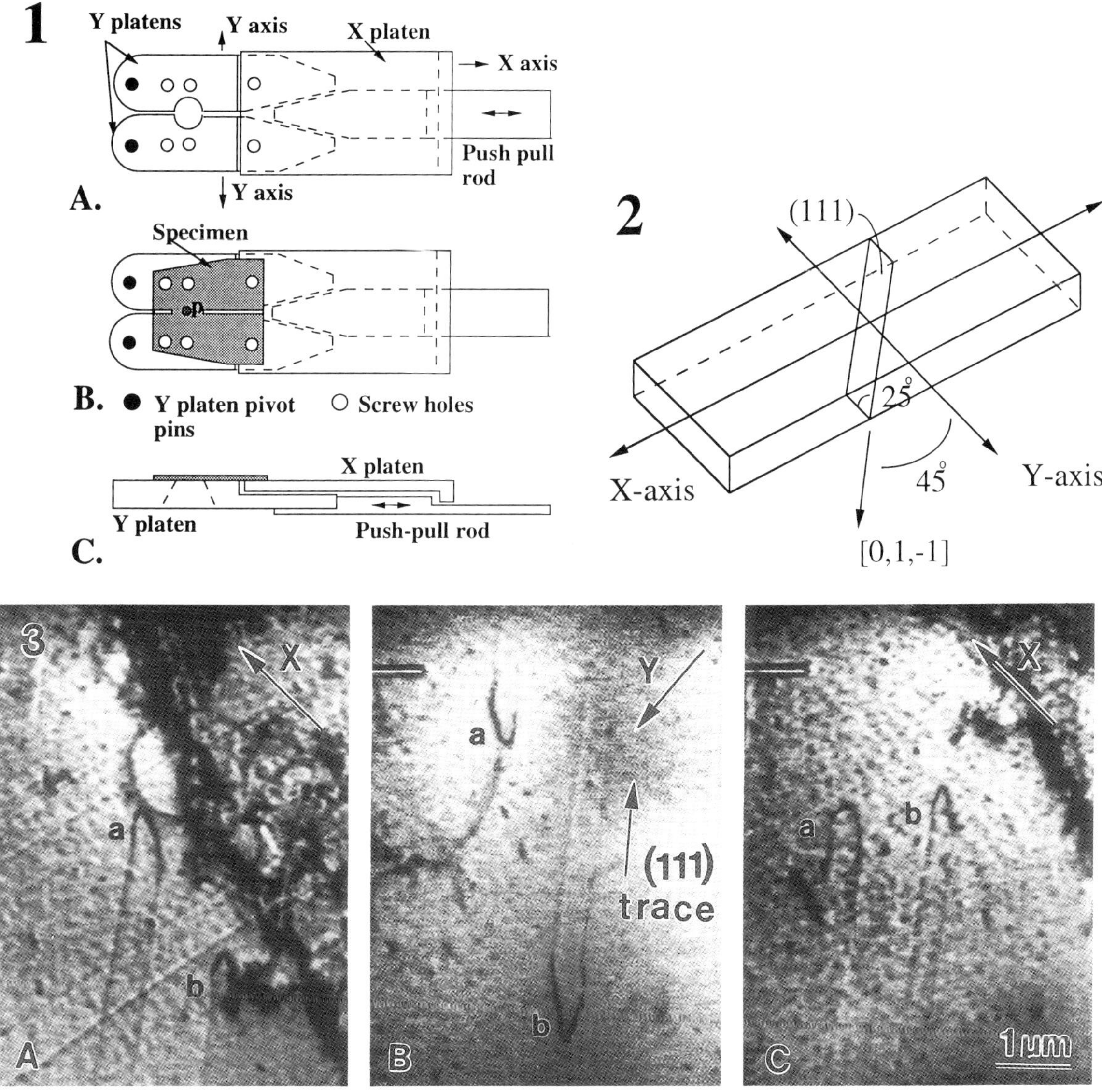

Fig. 1-(a) X-Y deformation stage specimen platens and push-pull rod; (b) Drawing of specimen shape mounted on the platens; (c) Side view of Fig. 1b.
Fig. 2-Aluminum single crystal oriented for single slip reverse deformation experiments.
Fig. 3-Sequence from an in situ HVEM reverse deformation experiment, 1 1/2 cycles showing reversal of dislocation motion on the (111) plane. (a) X loading; (b) Y loading; (c) X loading.

IN-SITU GRAIN GROWTH STUDY IN ELECTRODEPOSITED NANOCRYSTALLINE Ni-1.2WT.%P ALLOY

D. A. Smith*, S. C. Mehta and U. Erb**, Department of Materials Science and Engineering, Stevens Institute of Technology, Hoboken, NJ 07030
* Present address: Department of Materials Science and Engineering, Whitaker Laboratory, Lehigh University, Bethlehem, PA 18015.
** Department of Materials Science and Metallurgy, Queens' University, Kingston, Ontario, Canada K7L 3N6

Nanocrystalline (nc) materials, owing to their ultrafine grain sizes, exhibit several physical and mechanical properties which are significantly different from, and in many cases, superior to those of their polycrystalline counterparts and therefore promise their usage in a number of applications. However, very poor resistance of nc pure metals against thermal grain growth[1-3] severely limits the application of these novel materials. Several attempts are being made to improve the thermal stability of nc materials. The conventional approaches of grain boundary pinning by second phase precipitates and solute drag effects are commonly being employed. However, the kinetics of grain growth and the roles of solute atoms and second phase precipitates in affecting grain growth inhibition in these nanolength scale materials are still not well understood.

A transmission electron microscope study has been made of the microstructural evolution of electrodeposited nanocrystalline Ni-1.2wt.%P alloy and pure nickel during heating experiments performed in-situ. The continous heating rate DSC scans were performed to evaluate the activation energy for grain growth in the Ni-P alloy. The microstructural evolution in Ni-P alloy as a function of annealing temperature is shown in fig. 1. As can be seen from fig. 1, the nanoscale grain structure of Ni-P alloy is retained with only modest grain coarsening up to a temperature of 360°C but the rate of grain growth is rapid in the temperature range of 360-480°C. The electron diffraction patterns of microstructures at various temperatures indicate that the onset of rapid grain growth in Ni-P alloy is concurrent with the Ni_3P pecipitation, suggesting the essential role of phosphorus in supersaturated solid solution in inhibiting the grain growth below a temperature of 360°C in the Ni-P alloy. Normal grain growth was observed in the Ni-1.2wt.%P at higher annealing temperatures. Subsequent microstructural analysis by TEM revealed pinning of grain boundaries by Ni_3P precipitates. A Kissinger plot (fig. 2) from the continuous scan rate DSC experiments gave the value of activation energy for grain growth in Ni-P alloy to be 2.25eV. This relatively high value of activation energy is consistent with the observed grain size stability below 360°C. Microstructural evolution in nc Ni, on the other hand, was characterized by the rapid onset of abnormal grain growth at temperatures as low as 260°C (fig. 3). At 320°C, the microstructure had largely transformed from the nanocrystalline to the microcrystalline state. The conclusion of our study is that nc grain structure of Ni can be retained at higher temperatures by alloying with phosphorus. This finding is practically significant for all applications except those where a low value of electrical resistivity is required.

References:
1. B. Gunther et al., *Scripta Metall. et. Mater.*, 27(1992)833.
2. V. Y. Gertsman et al., *Scripta Metall. et Mater.*, 30(1994)577.
3. K. Ganapathi et al., *Scripta Metall. et Mater.*, 25(1991)2699.

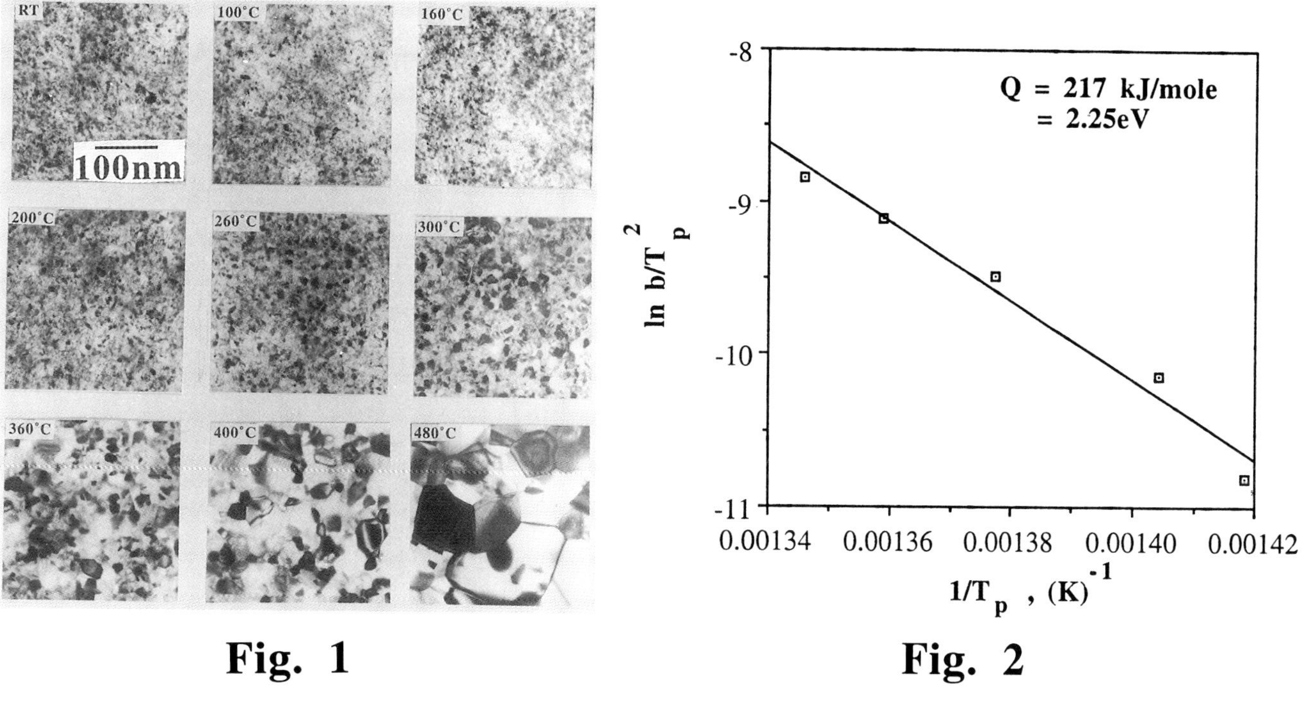

Fig. 1   Fig. 2

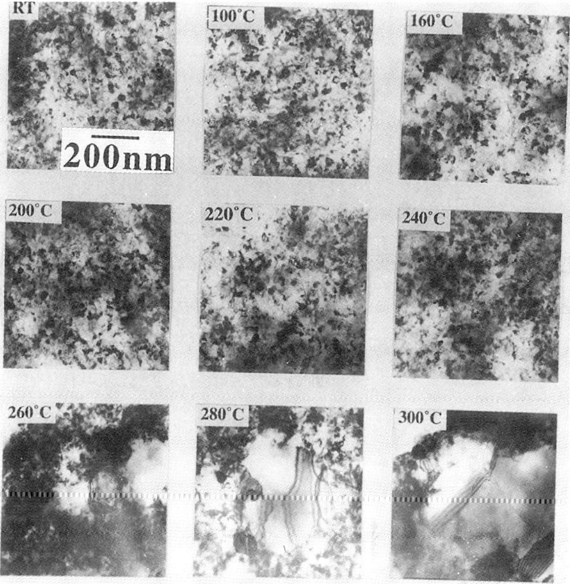

Fig. 3

Fig. 1. Bright field TEM images showing the microstructural evolution at different temperatures during an in-situ annealing of electrodeposited nanocrystalline Ni-1.2wt.%P alloy.

Fig. 2. Kissinger's analysis for grain growth in Ni-P alloy.

Fig. 3. Bright field TEM images showing the microstructural evolution at various temperatures in electrodeposited nanocrystalline Ni during an in-situ annealing experiment.

TEM INVESTIGATION OF THE LOW TEMPERATURE PHASE OF HfV$_2$

F. Chu and T. E. Mitchell

Center for Materials Science, Mail Stop K-765, Los Alamos National Laboratory,

Los Alamos, NM 87545, U. S. A.

C15 Laves phase intermetallic compounds are potential high temperature structural materials, of which C15 HfV$_2$ is particularly attractive[1]. At room temperatures (RT), the C15 Laves phase has a fcc-based structure with a lattice parameter a=7.4 Å, as shown in Fig. 1. Sensitive specific heat measurements indicate that HfV$_2$ undergoes a structural transformation at 115 K, as shown in Fig. 2. The crystal structure of the low temperature (LT) phase of HfV$_2$ is not unambiguously known, although some studies have been done [2-3]. A HfV$_2$ alloy was made by arc-melting. The alloy was homogenized at 1200°C for 120h. In a preliminary study we have concentrated on selected area diffraction patterns (SADs) along different zone axes of C15 structure at 83 K, using a Philips CM 30 microscope with a liquid nitrogen cold stage.

A <100>$_{C15}$ SAD at RT is shown in Fig. 3 (a). At 83 K, SADs along some <100>$_{C15}$ show (100), (200), and (300) superlattice spots, as shown in Fig 3 (b), while SADs along other <100>$_{C15}$ indicate (110) and (200) superlattice spots, as shown in Fig. 3 (c). These SADs indicate that the LT HfV$_2$ phase is not tetragonal based on the original cubic structure, i.e. not as claimed by Ref. 3. At 83 K, the SAD along one particular [111]$_{C15}$ reveals no superlattice spots, i.e., it is identical with <111> SADs at RT. The RT SAD along <110>$_{C15}$ is shown in Fig. 4 (a). At 83 K, two types of SADs along <110>$_{C15}$, as shown in Fig. 4 (b-c), reveal 1/2(111), 1/2(113), and 1/2(331) superlattice spots. The appearance of (111)/2 superlattice spots means that the LT HfV$_2$ phase has larger lattice parameters, which is not consistent with Refs. 2-3. LT SADs along other zone axes, e.g., <112>$_{C15}$, <013>$_{C15}$, <114>$_{C15}$, and <332>$_{C15}$, reveal the same types, i.e., 1/2(111), 1/2(113), and 1/2(331), superlattice spots.

Using a combination of these SADs, a tentative reciprocal lattice of LT HfV$_2$ is constructed, as shown in Fig. 5, It can be seen from Fig. 5 that the LT HfV$_2$ shows no 4-fold symmetry and probably has an orthorhombic structure with a doubling of the unit cell parameters[4]. Further and detailed studies are under way, using a combination of CBED and synchrotron or neutron diffraction to determine unambiguously the crystal symmetry, lattice parameters, and atomic positions of the LT HfV$_2$ phase.

References:
1. F. Chu and D. P. Pope, Mater. Sci. Eng. **A170**, (1993) 39.

2. A. C. Lawson and W. H. Zachariasen, Phys. Lett. **38A**, (1972) 1.

3. V. A. Finkel and E. A. Pushkarev, The Reports of the Academy of Science of the USSR, Physical Chemistry, **228**, (1976) 119.

4. This research was performed under the auspices of DOE-OBES.

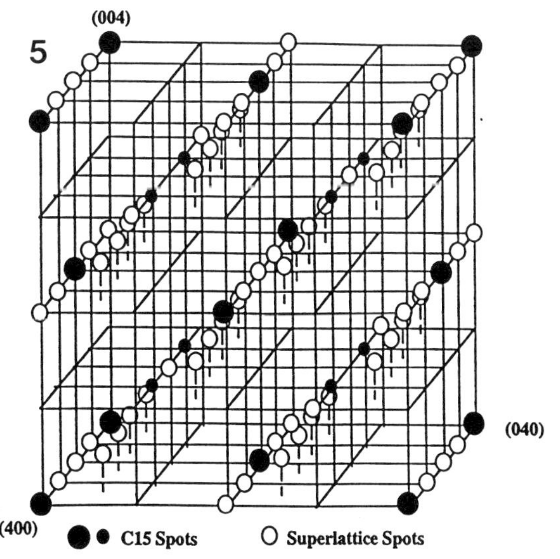

FIG.1 C15 crystal structure.

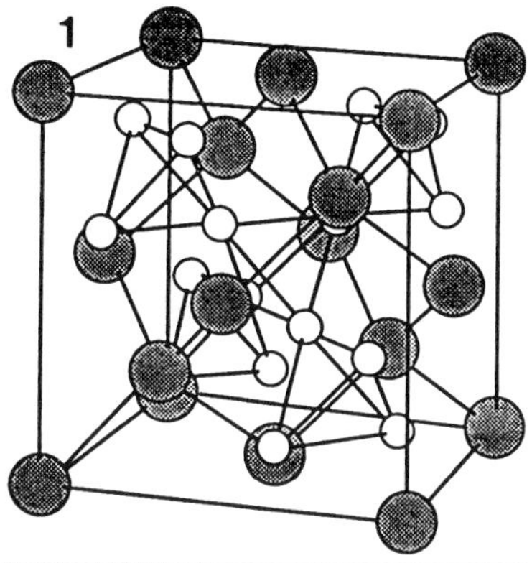

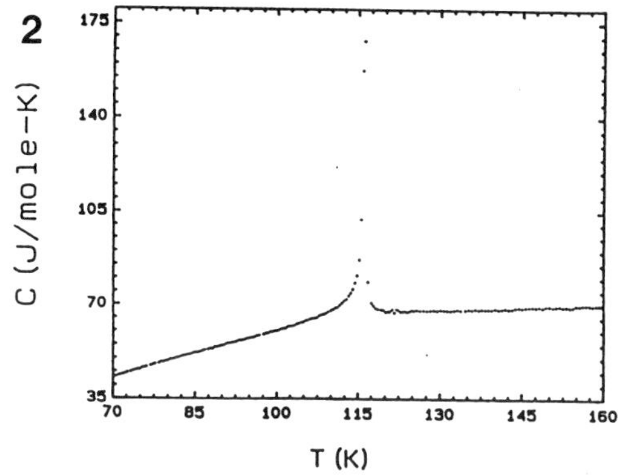

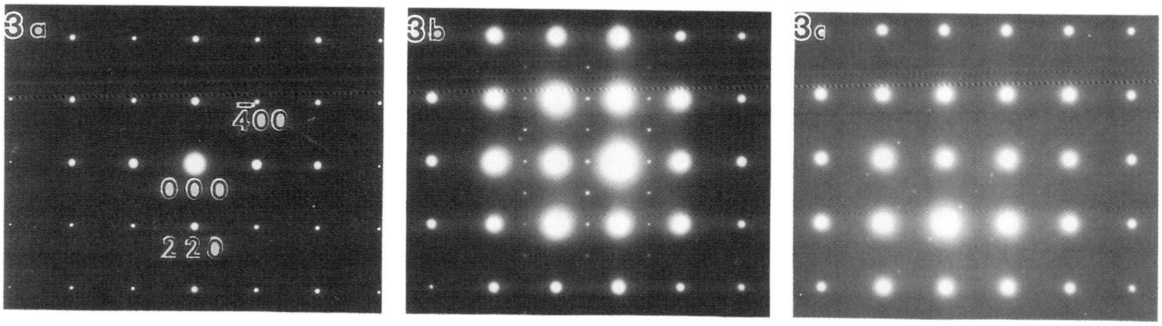

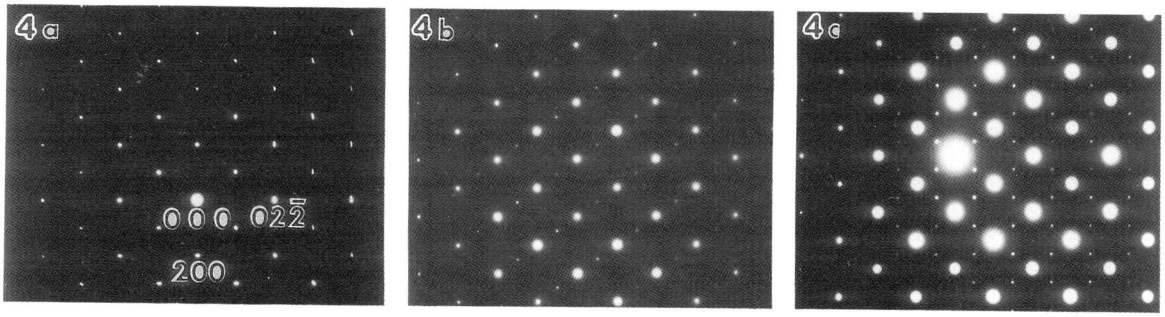

FIG.2 Specific heat vs. temperature of HfV_2, indicating a structural transformation at 115.5 K.

FIG.3 SAD along $<100>_{C15}$: (a) at RT, (b) at 83 K, and (c) at 83 K.

FIG.4 SAD along $<110>_{C15}$: (a) at RT, (b) at 83 K, and (c) at 83 K.

FIG.5 Reciprocal lattice of the LT HfV_2.

253

IN-SITU ATOMIC FORCE MICROSCOPY OF Pt / Ti FILM MORPHOLOGY CHANGES ON A MICROELECTRONIC GAS SENSOR OPERATING AT ELEVATED TEMPERATURES

M. DiBattista*, S. V. Patel*, J. F. Mansfield**, J. L. Gland***, J. W. Schwank*

*Department of Chemical Engineering, University of Michigan, Ann Arbor, MI 48109
**Electron Microbeam Analysis Laboratory, University of Michigan, Ann Arbor, MI 48109
***Department of Chemistry, University of Michigan, Ann Arbor, MI 48109

Thin film electronic devices that employ resistance change responses of Pt / Ti films to detect gas species have been microfabricated at the University of Michigan. Atomic force microscopy (AFM) is used to investigate morphology of the Pt / Ti sensing films deposited on the microfabricated device. These Pt / Ti sensing films are strongly influenced by many factors, making it difficult to determine the exact relationship between film structure, chemical sensitivity, and selectivity. In-situ AFM investigations of Pt / Ti films on this device at elevated temperatures provides the opportunity for real time observation of film morphology changes under controlled conditions, testing sensing film stability during device operation, and correlating film structure to resistance.

Observation of the Pt / Ti film surface and in-situ resistance measurements at elevated temperatures are possible due to the construction of the sensing device. The sensors are based on chemically active thin films deposited on a micromachined silicon window, supported by a 300 μm thick silicon rim. The gas sensor window is a dielectric membrane composed of silicon dioxide (3500Å), silicon nitride (5000 Å), and low pressure chemical vapor deposition (LPCVD) silicon dioxide (4000 Å), and is used to support the sensing film as well as a set of four electrodes for measuring the film conductivity. The four point probe configuration allows accurate measurement of film resistance changes due to changes in the film or its environment. A deep boron doped 5 μm thick silicon heater and two temperature sensing resistors (TSR) are located underneath the window. Contacts to the heater and TSR's are made with an iridium / titanium metallization layer (2000 Å / 250Å). This gas sensor device has been successfully used in detecting parts per million (ppm) levels of oxygen in CF_4.

Results of in-situ investigation with AFM under controlled atmospheres show real time Pt / Ti film morphology changes, the dispersion of the deposited Pt / Ti film at elevated temperatures, and removal of surface carbon in the presence of oxygen. The differences in treating films in air and 5% H_2 / 95% N_2 environments will be discussed. This work demonstrates that resistance measurements can be directly related to the film morphology observed by in-situ AFM. This new ability aids in the understanding and development of thin films for sensing applications.

References

1. C. L. Johnson, J. W. Schwank, and K. D. Wise, Integrated Ultra Thin Film Gas Sensor, *Sensors and Actuators B*, 20 (1994) 55
2. N. Najafi, K.D. Wise, R. Merchant, and J. W. Schwank, An Integrated Multi-Element Ultra Thin Film Gas Analyzer, *IEEE Transactions on Electronic Devices*, 41 (1994) 1770

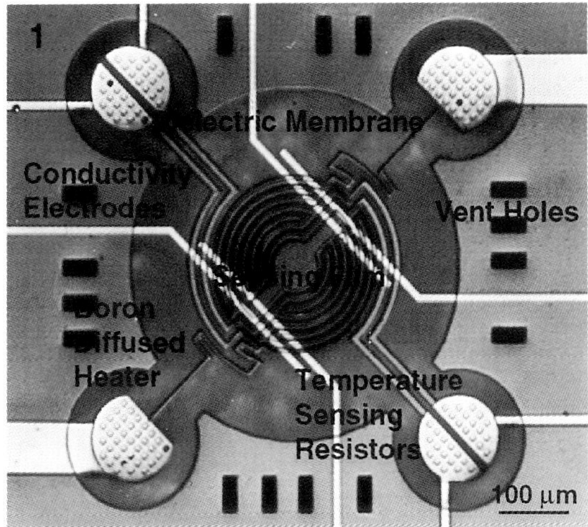

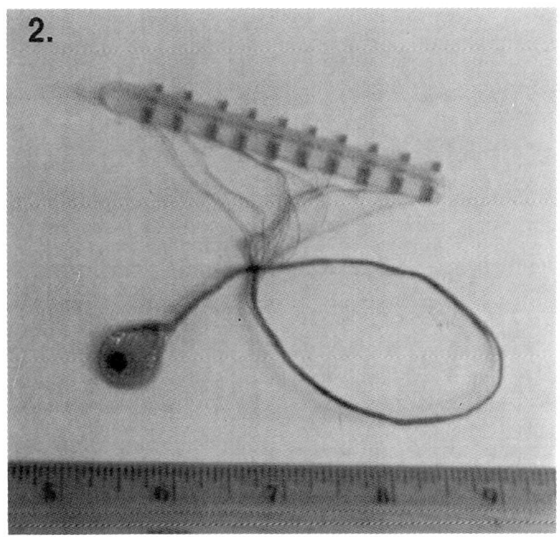

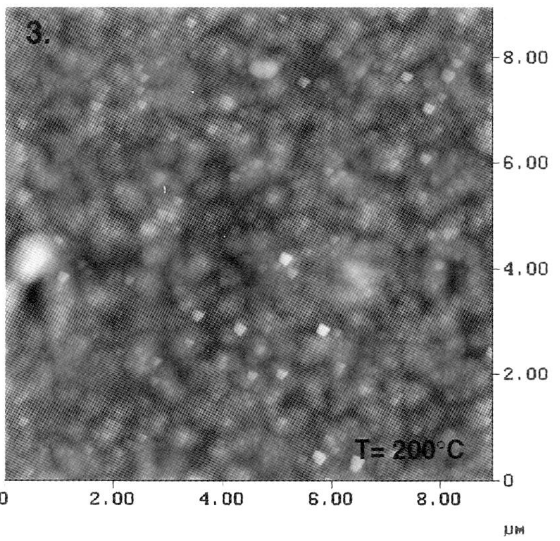

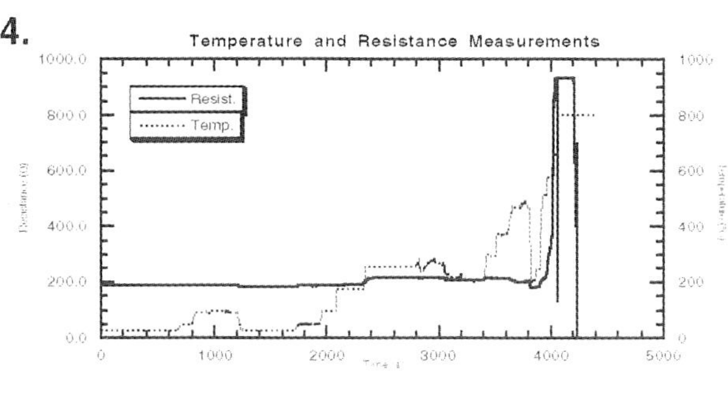

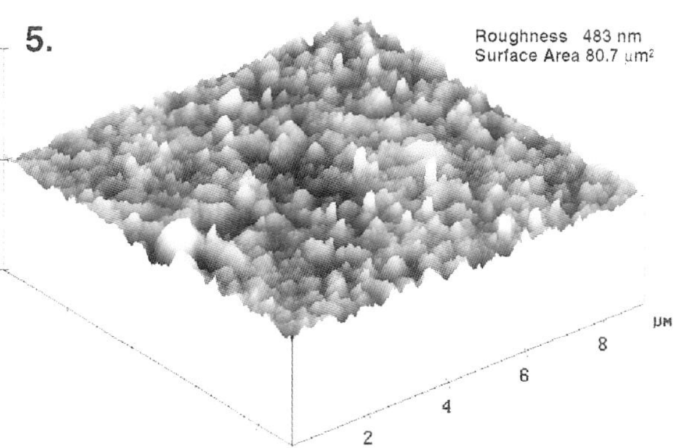

FIG. 1.- Optical microscope image of the important components on the sensor device platform.

FIG. 2.- 35 mm image of the sensor device mounted on a support designed for in-situ operation and analysis of the Pt/Ti film.

FIG. 3.- AFM Image of the Pt/Ti sensing film during operation of the sensor device in air at atmospheric pressure.

FIG. 4.- Graph of the film resistance and temperature control during operation while the film was being examined with AFM.

FIG. 5.- 3-D surface view of the Pt/Ti film with surface area and roughness calculation.

IN SITU STUDY OF ELECTRON BEAM-INDUCED CHEMICAL VAPOR DEPOSITION IN OF Au IN AN ENVIRONMENTAL TEM

J. Drucker,[*]R. Sharma, J. Kouvetakis[+] and K.H.J. Weiss

Center for Solid State Science, Arizona State University, Tempe, AZ 85287-1704
[*]Present address: Dept. of Physics, University of Texas at El Paso, El Paso, TX
[+]Dept. of Chemistry, Arizona State University, Tempe, AZ 85287-1704

Patterning of metals is a key element in the fabrication of integrated microelectronics. For circuit repair and engineering changes constructive lithography, writing techniques, based on electron, ion or photon beam-induced decomposition of precursor molecule and its deposition on top of a structure have gained wide acceptance[1] Recently, scanning probe techniques have been used for line drawing and wire growth of W on a silicon substrate for quantum effect devices.[2] The kinetics of electron beam induced W deposition from WF_6 gas has been studied by adsorbing the gas on SiO_2 surface and measuring the growth in a TEM for various exposure times.[3] Our environmental cell allows us to control not only electron exposure time but also the gas pressure flow and the temperature. We have studied the growth kinetics of Au Chemical vapor deposition (CVD), *in situ*, at different temperatures with/without the electron beam on highly clean Si surfaces in an environmental cell fitted inside a TEM column.

Cross-sectional Si samples were prepared by regular ion beam thinning techniques. These samples were cleaned by etching in 10% HF solution and exposed to ≈10-4 vapor pressure of ethyl (trimethylphosphine)gold(I)[4] in an environmental cell attached to PHILLIPS 400-T microscope operated at 120 KV.[5] The particle size distribution was measured at different temperatures from the images taken after 5, 10, 20 and 40 minutes interval from a region unexposed to the electron beam. For beam induced study, the nucleation and growth of Au particles was recorded real time on a video tape continuously under isothermal conditions. The samples were analyzed by energy dispersive x-ray spectroscopy (EDS) and scanning tunneling (STM) microscopy *ex-situ* for purity of Au and vertical growth of Au islands.

Several Au particles were observed to nucleate at the same time forming a polycrystalline film as expected. The larger size of grains for depositions and nearly continous film under the electron beam than without electron beam confirms that electron beam stimulates the breaking precursor molecule, thus increasing the growth rate by a factor of three (Fig. 1b and Ia). The rings in the selected area electron diffraction pattern taken after 40 minutes of deposition without electron beam at 230°C (Fig. 2) can be indexed as Au while the spot pattern belongs to Si substrate. The STM studies show the growth rate for both beam induced and without beam deposition to be ≈10 times higher perpendicular to the beam than perpendicular substrate surface (Fig. 3) due to different growth rates in different crystallographic directions. The growth rates at various temperatures are used to obtain kinetic information while the growth rates with or without the beam are used to quantify the effects of electron beam on CVD process.

References

1. Anthony D. Dell Art, John and Carl V., J. Vac. Sci. Technol., B11 (1993)2195.
2. K.T. Kohlmann-von Platen, J. Chlebeck, M. Weiss, K. Reimer, H. Oertel and W.H. Brunger, J. Vac. Sci. Technol. B11 (1993)2219.
3. Toshari Ichihashi and Shinji Matsui, J. Vac. Sci. Technol., B6(1988)1869.
4. M.M. Banaszak Holl, P.F. Seidler, S.P. Kowalczyk and F.R. McFeely, Appl. Phys. Lett. 62 (1993)1476.
5. R. Sharma, K. Weiss, M. McKelvy and W. Glaunsinger, Proc. 52nd Ann. Meet. Microscopy Society of America 1994, 494.

Proc. Microscopy and Microanalysis 1995, edited by G.W. Bailey, M.H. Ellisman, R.A. Hennigar, and N.J. Zaluzec
Copyright © 1995 MSA. Published by Jones and Begell Publishing, 79 Madison Ave., New York, NY 10016

6. The research support through AFOSR (DARPA) award # F49620-93-C-0014 and NSF (DMR) grant # 9314326 and Dr. B.L. Ramakrishna's help for STM images is greatfully acknowledged.

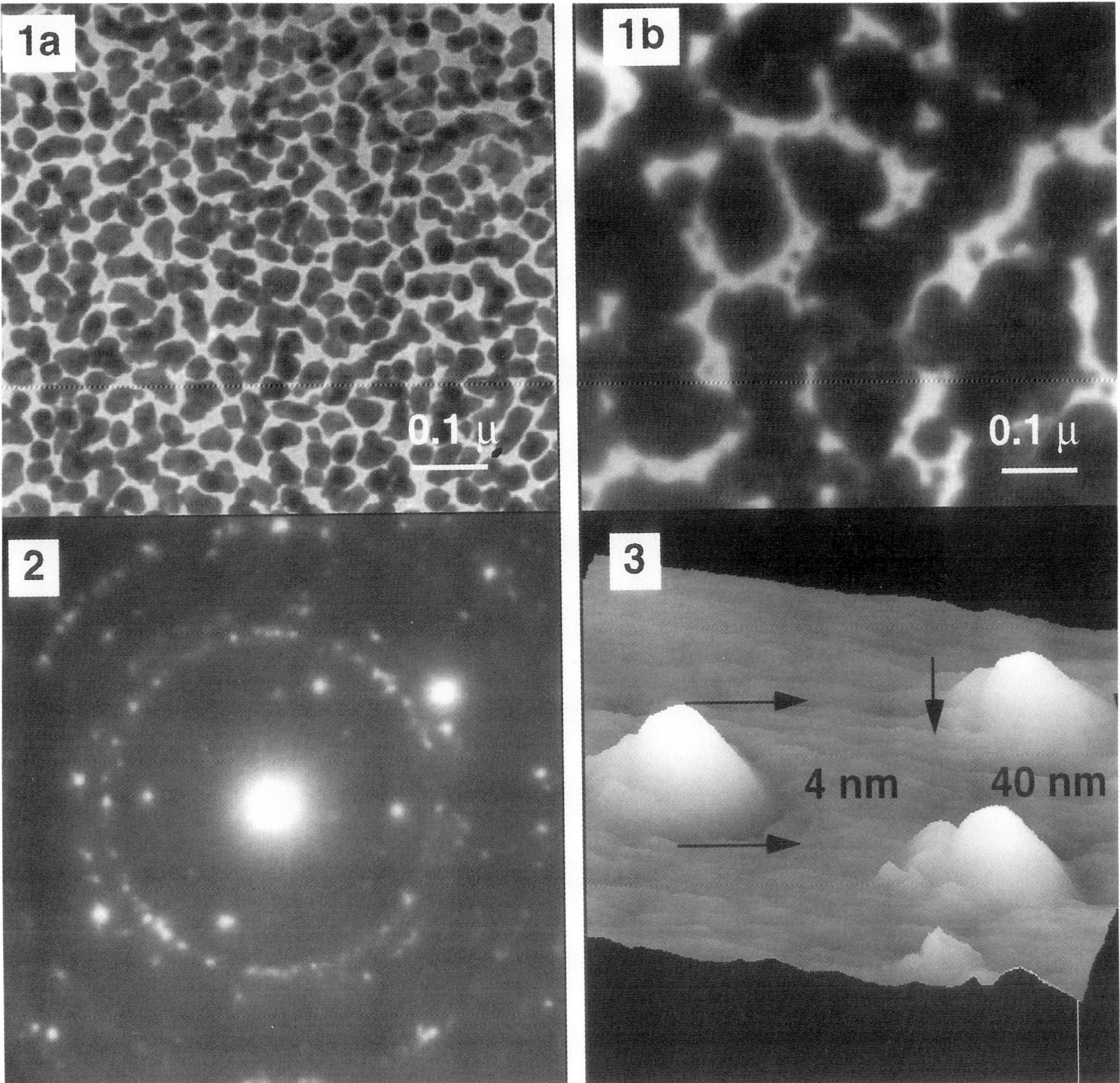

Fig. 1. Low resolution TEM image of gold particles grown by CVD on a clean Si surface at 230°C (a) without and (b) with the electron beam.

Fig. 2. The selected area electron diffraction after 40 minutes showing the polycrystalline ring pattern of Au films. The spots pattern is from Si substrate.

Fig. 3. A scanning tunneling microscopy image showing the different growth rates parallel and perpendicular to the electron beam direction.

CRYSTALLIZATION OF BARIUM TITANATE THIN FILMS SYNTHESIZED VIA SOL-GEL PROCESSING

M. C. Gust*, N. D. Evans**, and M. L. Mecartney*

*Materials Science and Engineering, University of California, Irvine, CA 92717
**Oak Ridge Institute for Science and Education, P.O. Box 117, Oak Ridge, TN 37831-0117

Sol-gel processing of oxide thin films offers the potential to generate different microstructures by varying process parameters such as water of hydrolysis, chemical precursor, or substrate.[1] For this work, barium titanate ($BaTiO_3$) thin films were prepared by the sol-gel method in order to try and generate different microstructures for dielectric measurements. However, for $BaTiO_3$ prepared using alkoxide precursors, the final microstructures of the films were found to be almost process independent. Films using different precursors, different substrates, and different amounts of water of hydrolysis all showed similar TEM microstructures.[2] All the films after final heat treatment were polycrystalline, nanoporous, and randomly oriented $BaTiO_3$; they were also fine-grained, with an average uniform grain size of 25-50 nm (Fig. 1). This fine-grained microstructure has also been observed in other studies of sol-gel derived $BaTiO_3$ thin films.[3,4] In addition to $BaTiO_3$, electron diffraction patterns showed the presence of trace amounts of rutile TiO_2 that were not observed by x-ray diffraction (XRD).

Sol-gel films were also spin coated onto single crystal $BaTiO_3$ substrates grown epitaxially by MBE on (100) Si. However, XRD studies revealed that the gel crystallized as a polycrystalline material, with no preferred orientation. To understand this crystallization behavior, in-situ TEM heating studies were performed. The in-situ studies were conducted using a Philips CM30 microscope and a single tilt Gatan heating holder. The holder was ramped initially to 300°C and held for 5 minutes to drive off water and contaminants. The holder was then ramped to the desired temperature within a few minutes.

A sequence of in-situ temperature holds at 500°, 600°, 650°, and 700°C was used. The gel film remained amorphous at temperatures lower than 600°C. After 25 minutes at 600°C, the gel film began to crystallize (Fig. 2). As shown in the dark field cross-section TEM micrograph (Fig. 2b), finely dispersed particles approximately 5-10 nm in diameter were formed. The corresponding electron diffraction pattern (Fig. 2c) shows a bright ring at 0.32 nm which corresponds to the 100% intensity (110) spacing of the rutile phase of TiO_2. $BaTiO_3$ in the bulk of the gel film began to crystallize at 700°C, and the intensity of the TiO_2 diffraction ring began to decrease. Regions of epitaxial growth appeared along the interface between the gel film and the $BaTiO_3$ substrate. Although epitaxial growth of the $BaTiO_3$ gel film increased with time, even after 5.5 hours at 700°C there was still some polycrystalline $BaTiO_3$ and a trace amount of TiO_2 in the film, as seen in the electron diffraction pattern (Fig. 3).

These results suggest that the formation of finely dispersed TiO_2 prior to crystallization of $BaTiO_3$ acts as a seeding technique for nucleation of $BaTiO_3$, resulting in a uniform microstructure that is independent of process parameters. Experiments are underway to prevent this intermediate formation of TiO_2 by using chemical additives to more tightly bind the Ti and Ba alkoxides in the sol and gel.[5]

References
1. V. Joshi and M. L. Mecartney, *J. Mater. Res.*, **8** [10] (1993) 2668.
2. M. C. Gust et al., to be published in *Proceedings of the International Symposium on Sol-Gel Science and Technology* (1994).
3. M. H. Frey and D. A. Payne, *Appl. Phys. Lett.*, **63** [20] (1993) 2753.
4. T. Hayashi et al., *Jpn. J. Appl. Phys.*, **32** (1993) 4092.
5. This work was supported by Hughes Research Labs, Malibu, California, and the DOE sponsored SHaRE Program under contract DE-AC05-76OR00033 with Oak Ridge Associated Universities.
R. A. McKee of ORNL is acknowledged for providing the epitaxial $BaTiO_3$ substrates.

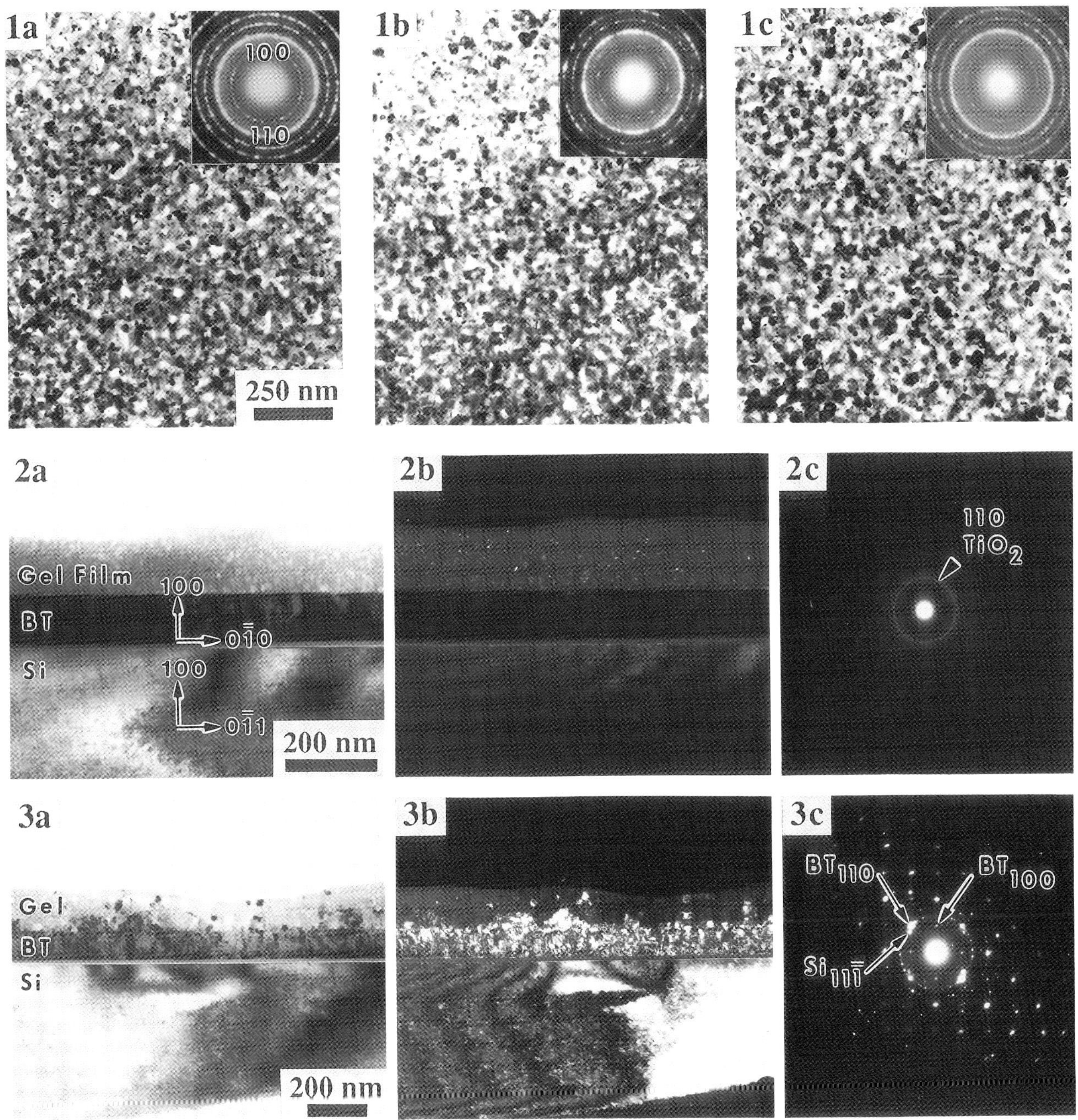

FIG. 1. - TEM micrographs showing the similar microstructures produced for BaTiO₃ thin films processed using: (a) standard process (Si substrate, no added water of hydrolysis, and Ba-Ti methoxypropoxide precursor); (b) Pt substrate; and (c) added water of hydrolysis.

FIG. 2. - Cross-section TEM images of film held at 600°C for 25 minutes shows formation of TiO₂: (a) bright field; (b) dark field; and (c) diffraction pattern of gel film.

FIG. 3. - Cross-section TEM images of film held at 700°C for 5.5 hours showing presence of epitaxial regions of BaTiO₃ grown on (100) BaTiO₃: (a) bright field; (b) dark field; and (c) diffraction pattern.

ENERGY-FILTERED IMAGING OF INTERFACIAL COMPOSITION IN COMPLEX Ni-BASE SUPERALLOYS

E. L. Hall* and J. Bentley[+]

*GE Corporate Research & Development, PO Box 8, Schenectady, NY 12301
[+]Metals and Ceramics Division, Oak Ridge National Laboratory, PO Box 2008, Oak Ridge, TN 37831-6376

In a wide variety of important engineering materials, including structural Ni, Fe, and Al-base alloys, the study of compositional gradients that occur at grain boundaries and other interfaces due to the precipitation of solute-rich second phases is of critical importance. X-ray spectrometry in the analytical electron microscope (AEM) has been used for many years to study solute depletion at interfaces in these alloys. It has been particularly effective at providing accurate, quantitative measurements in cases where the interface precipitates are spaced far apart and where relatively wide compositional gradients result from diffusion-controlled "collector-plate" mechanisms. However, in complex alloys, with heavy grain boundary coverage, multiple precipitate types, and/or multiphase matrices, x-ray microanalysis is more limited due to beam spreading and the associated 5-20nm spatial resolution achieved in these alloys with AEMs equipped with thermionic sources.

Elemental mapping in an AEM equipped with an imaging energy-filter is an alternative method for the study of interface compositional gradients. It has been shown to be effective for this purpose in relatively simple situations in steels.[1] Energy-filtered imaging is a good complement to x-ray spectrometry, providing higher spatial resolution and more efficient data collection. However, the accuracy of quantitation with this approach has not been fully explored, and some elements may not be easily accessible. The purpose of this investigation is to examine the use of an imaging energy-filter to map compositions in complex alloys.

The alloy chosen for study was an advanced Ni-base alloy, Rene' 95, with nominal composition 61.5%Ni, 7.9%Co, 14.5%Cr, 2.1%Mo, 1.1%W, 7.5%Al, 3%Ti, 2.2%Nb, 0.03%Zr, 0.24%C, and 0.05%B in at. %. Grain boundary composition has the potential to strongly affect the properties of this alloy, particularly intergranular fracture and fatigue mechanisms. The microstructure of the alloy consists of a Ni solid solution (γ) containing coherent misfitting γ' ($Ni_3(Al, Ti)$) particles of varying sizes.[2] At the grain boundaries, both large and small γ' particles are present, as well as occasional Cr/Mo/W-rich particles. Elemental mapping was performed at 300 kV with a Gatan Imaging Filter (GIF™) attached to a Philips CM30 AEM. Images were recorded with exposure times of up to 15 s and with 30eV windows before and after ionization edges of interest. Zero- and low-loss images were also recorded with 5, 30, and 60eV windows to aid in data interpretation and quantification. Core-loss and jump-ratio elemental maps were produced either by AE^{-r} background extrapolation from two pre-edge windows, or by simple division of the post-edge image by a pre-edge one, respectively.

Figure 1a shows a bright-field image (low-loss 60eV window) of a segment of grain boundary between two large γ' particles. Figures 1b,c,d are the corresponding Cr, Ni, and Ti maps. Figure 1b shows the expected enrichment of the γ phase and depletion of the γ' in Cr.[2] The small γ', invisible in Fig. 1a due to low contrast, are clearly seen in the elemental map. Further, the image shows that equilibrium has been reached at both the grain boundary and at the particle/matrix interfaces, and that no significant Cr depletion or enrichment occurs. Figure 1c, the Ni core-loss map, reflects the expected Ni partitioning. Figure 1d is a jump-ratio image showing that Ti is preferentially present in the γ'. The jump-ratio image has better signal-to-noise than the corresponding core-loss image for this low-concentration element.

Figure 2 is a Cr core-loss map showing a γ/γ' interface at a grain boundary in this alloy. In this case the Cr intensity can be mapped adjacent to the large γ' particle, which is not possible with x-ray spectrometry due to the beam spread. Figure 3 shows an example of high spatial resolution microanalysis in this alloy. This Cr core-loss image reveals a small strip of Cr-enriched γ material at the grain boundary between two large γ' particles. The width of the γ material is on the order of a few pixels, or approximately 5 nm. Whereas quantification of thin grain boundary regions such as this is complicated by uncertainties about "chemical" resolution, the modulation transfer function of the phosphor-CCD camera system, and interface inclination, this result clearly shows the capability for qualitative high-resolution mapping.

Proc. Microscopy and Microanalysis 1995, edited by G.W. Bailey, M.H. Ellisman, R.A. Hennigar, and N.J. Zaluzec.

These preliminary results show that the GIF can be an important tool for compositional assessment in complex alloys. Further work is in progress on extracting quantitative data from the images and comparing these with the results from previous x-ray microanalysis.[2]

1. J. Bentley and E. A. Kenik, Proc. Ann. MSA Meeting 52(1994)1000.
2. E. L. Hall, C. P. Blankenship, and M. F. Henry, Proc. Ann. MSA Meeting 52(1994)964.
3. Research partially sponsored by the Division of Materials Sciences, U.S. Department of Energy, under contract DE-AC05-84OR21400 with Martin Marietta Energy Systems Inc., and through the SHaRE Program under contract DE-AC05-76OR00033 with Oak Ridge Associated Universities.

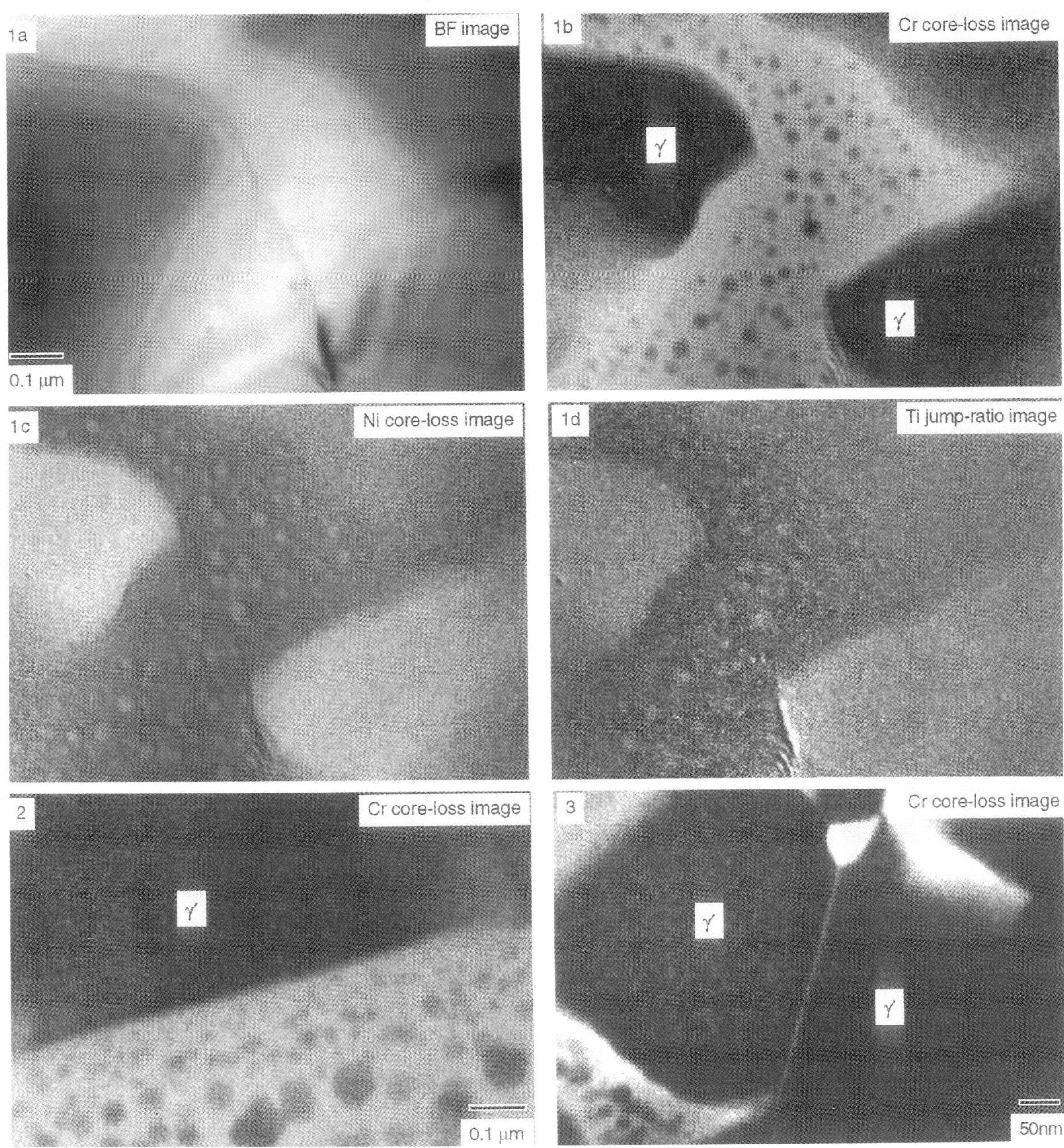

FIG. 1 - (a) Bright-field image , (b,c) Cr, Ni core-loss images , (d) Ti jump-ration image of grain boundary segment in Rene' 95.
FIG. 2 - Cr core-loss image of γ/γ′ interface at grain boundary.
FIG. 3 - Cr core-loss image of thin strip of γ at grain boundary between two large γ′ particles.

ATOMIC RESOLUTION IN ENERGY-LOSS IMAGES, FACT OR ARTEFACT?

Max T. Otten

Philips Electron Optics Applications Laboratory, Building AAE, 5600 MD Eindhoven, The Netherlands

Since the introduction of energy-filtered imaging, the ultimate resolution has been a point of debate. Berger and Kohl [1] concluded that resolution is limited to ~1 nm. In contrast, one group of investigators [2,3] has argued that atomic resolution can be achieved. The conclusions from the latter group have been questioned previously [4]. In order to provide more insight into the origin of the fine detail in some energy-loss images, a number of experiments have been performed to try and resolve this issue.

The main issue of the high-resolution detail in energy-loss images (i.e., images from electrons that have lost an amount of energy) is whether it comes from elastic or to inelastic scattering. In the former case, a combination of elastic scattering and inelastic scattering, the high-resolution energy-loss image is an artefact. In the latter case, inelastic scattering only, it would be real.

On the basis of theory energy-loss images are unlikely to achieve atomic resolution. The imaging conditions used [2,3] are not conducive to producing atomic-resolution detail as the slit width used (5 eV) is too wide, which results in a focus spread worse than 30 nm and loss of resolution detail (in the energy-loss image but not the conventional elastic high-resolution image). In addition, delocalisation of the inelastic scattering is expected to be ~1 nm for 100 eV energy losses, far from the 0.2 nm required for atomic resolution.

From a practical point of view the biggest difficulty is the impossibility of separating elastic and inelastic electrons. Even for the small thicknesses typically used for high-resolution imaging, combined elastic-inelastic scattering is widespread, partly because the channelling conditions concentrate the electrons at the atom positions, thereby increasing the inelastic scattering. One of our experiments consisted of obtaining a high-resolution image of silicon and then recording a series of diffraction patterns at different energies. Even though this area is thin (EELS spectroscopy indicates 0.25 mean free path), Bragg spots are observed in diffraction patterns from a wide range of energies (Fig. 1). The only limitation to observing these spots is the broadening from the additional inelastic-scattering angle, which makes the spots indistinct around 750 eV energy loss. We also looked at the high-resolution image detail. The behaviour of normal high-resolution images with their strong sensitivity to focus setting should differ from that of energy-loss high-resolution images which should have just one suitable focus setting. Away from this focus setting, the contrast should disappear. We have collected through-focus series at plasmon energies and energies just below and above the silicon L edge. All these images show the same behaviour - the image contrast changes as a function of focus (Fig. 2), with the same periodicity as the zero-loss high-resolution image.

One final point is the background correction. Since the inner-shell energy-loss edges reside on a high background (especially for low energy losses), background correction is necessary to extract the real inner-shell edge signal. At small energy losses, background correction is, however, fraught with difficulties because the background does not follow the Power Law and extrapolation underneath the edge is difficult. Assuming that a workable background subtraction for the whole area also works for individual pixels is therefore hazardous.

Proc. Microscopy and Microanalysis 1995, edited by G.W. Bailey, M.H. Ellisman, R.A. Hennigar, and N.J. Zaluzec
Copyright © 1995 MSA. Published by Jones and Begell Publishing, 79 Madison Ave., New York, NY 10016

What is needed to prove the attainability of atomic resolution in energy-loss images is a demonstration that: 1. the focus behaviour of energy-loss images differs from that of zero-loss images; 2. the contrast in the energy-loss image is insensitive to background correction; 3. Bragg spots in energy-filtered diffraction are absent (and not due to the fact that the microscope used is unable to produce such patterns for small crystals as for [2]). In practice, this may be very difficult to achieve, with a possible exception for amorphous material. In conclusion, there is as yet no proof that energy-filtered imaging can resolve detail finer than 1 nm.

References
1. A. Berger and H. Kohl, *Optik* 92(1993)175.
2. Y. Makita et al., *Microsc. Microanal. Microstruct.*, 4(1993)143.
3. H. Endoh et al., Ultramicroscopy, 56(1994)108.
4. O.L. Krivanek et al., *Microsc. Microanal. Microstruct.*, 3(1992)187.

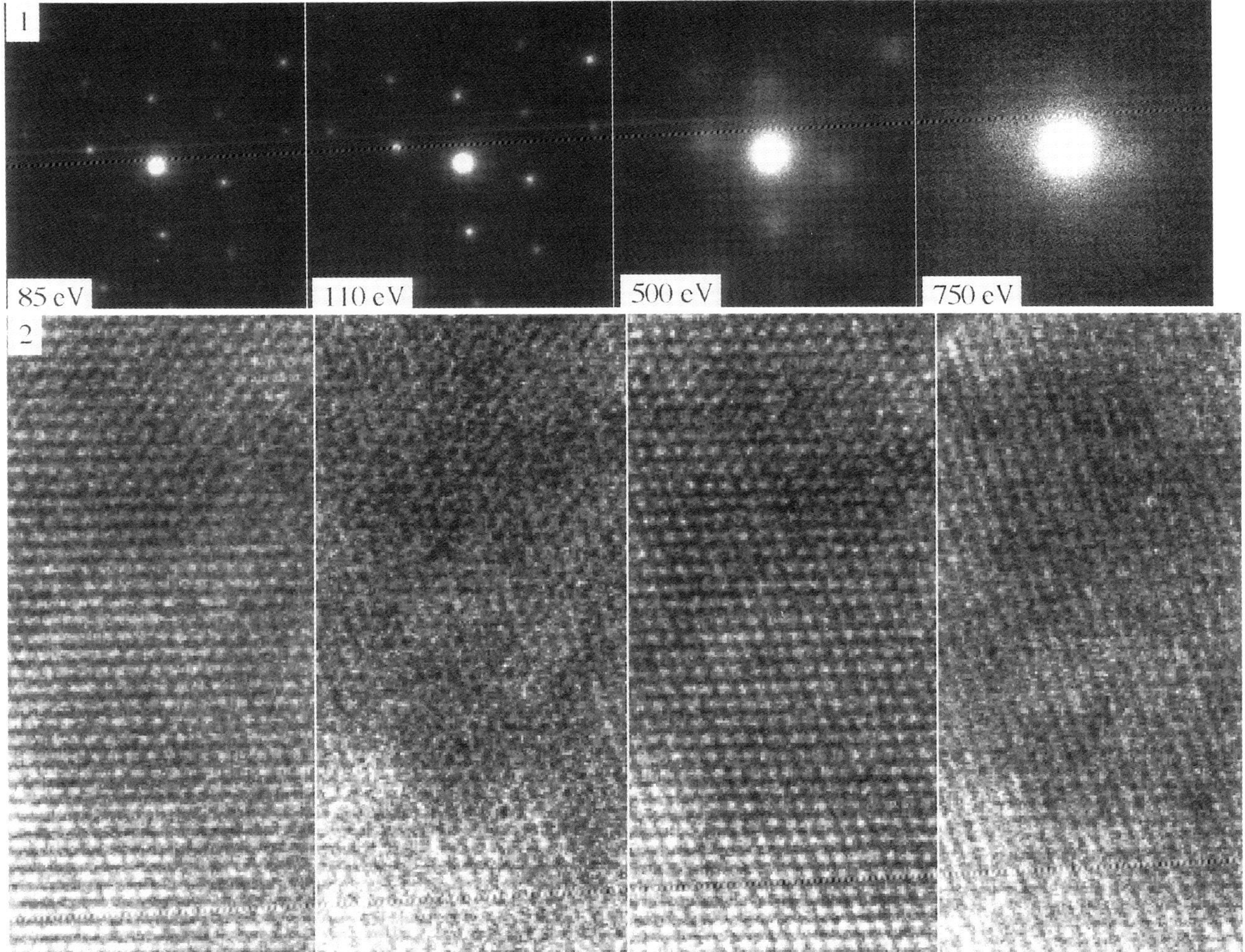

Fig. 1. Energy-loss diffraction patterns recorded at various energies demonstrate combined elastic-inelastic scattering up to at least 750 eV energy loss.
Fig. 2. Images from a through-focus series recorded at plasmon energies reveal the same contrast periodicity as zero-loss high-resolution images.

DELOCALIZATION EFFECT IN ENERGY-FILTERED ELECTRON IMAGES OF SURFACES AND INTERFACES

Z. L. Wang* and A.J. Shapiro

Metallurgy Division, National Institute of Standards and Technology, Gaithersburg, MD 20899.
* Currently at: School of Materials Science and Engineering, Georgia Institute of Technology, Atlanta GA 30332-0245.

Accurate structure determination is feasible using zero energy-loss images, since image simulations can be done accurately based on elastic scattering theory [1]. Two factors can be assessed from the energy-filtered images: one is the contribution of inelastically scattered electrons to HREM images; and two is the effect of absorption by inelastic scattering on images obtained using 0-loss electrons. In this paper, MgO(100) surface and Al/Ti interface are used to demonstrate the delocalization effect of valence-excitation in energy-filtered surface and interface images, respectively.

Figure 1 shows a pair of MgO lattice images recorded using all the transmitted (Figure 1a) and just the elastically transmitted (Figure 1b) electrons, respectively. The MgO cube is oriented along [110], so that the contrast variation as a continuous function of the specimen thickness is revealed by a single image. When the specimen is thinner than about 20 nm, no significant differences are seen between the filtered and unfiltered images. In the unfiltered image, the image contrast is dramatically decreased when the specimen thickness is about 40 nm (as indicated by an arrowhead), but the 0-loss filtered image shows significantly better contrast. The image contrast has been improved from 7% to 11% when the energy-filtering is applied. The images can be used to determine the inelastic mean-free-path of MgO, and the result is $\Lambda \approx 181$ nm at 300 kV. In surface profile imaging, atomic structures of crystal surfaces are imaged when the surface is parallel to the incident beam. The excitation probability of the surface valence states depends sensitively on the impact distance of the incident electron from the surface [2]. Thus, the contrast variation along [100] can be used to study the delocalization effect in surface excitation. The subtraction of the intensity of the 0-loss filtered image (Figure 1b) from that of the unfiltered image (Figure 1a) gives the image formed by all the inelastically scattered electrons, an intensity line scan across the (100) surface from this processed image being shown in Figure 2. Disregarding the intensity variation at the edge of the surface due to Fresnel fringes, the average intensity profile drops slightly when the beam impact distance approaches the surface, indicating less inelastic absorption near the surface.

Composition-sensitive image can be formed using plasmon-loss electrons. Figure 3 shows a pair of images of the Al/Ti interface using the 0-loss and Al-plasmon (15 eV) energy-selected electrons. Clearly, the Al layer is pronounced in Figure 3b. However, the spatial resolution of the image is limited due to the delocalized scattering of valence-loss electrons. A line scan is made across the Al/Ti interface from the image recorded using the Al plasmon (Figure 4). The fine oscillation in the intensity profile is due to the lattice fringes. The intensity profile drops within a distance of 2 nm around the interface. Thus, the spatial resolution is approximately 1 nm. This result is interpreted using the calculated result of dielectric response theory for interface excitations [2].

Using the energy-filter, HREM lattice images of crystals can be formed with electrons that have suffered energy-losses in a well-defined, narrow band due to inelastic interactions with the crystal (Figure 3c) [3, 4]. The lattice image is the interference between electrons elastically scattered to different Bragg beams both before and after inelastic excitations. The coherent interference between these beams, producing lattice images, is possible because electrons inelastically scattered by the same crystal state (and therefore having the same energy-loss and momentum transfer) are still coherent [3]. Since there are many crystal states of different energies and momenta, the final image contrast is reduced due to the incoherent superposition of the electrons scattered by different crystal states [5].

1. Z.L. Wang, in these proceedings.
2. R. Garcia-Molina et al., J. Phys. C: Solid State Phys. **18** (1985) 5335.
3. Z.L. Wang and J. Bentley, Microscopy Microstructure Microanalysis **2** (1992) 569.
4. H. Endoh, H. Hashimoto and Y. Makita, Ultramicroscopy **56** (1994) 108.
5. Thanks to Dr. D. van Heerden for kindly providing the Al/Ti multilayer specimen.

Proc. Microscopy and Microanalysis 1995, edited by G.W. Bailey, M.H. Ellisman, R.A. Hennigar, and N.J. Zaluzec
Copyright © 1995 MSA. Published by Jones and Begell Publishing, 79 Madison Ave., New York, NY 10016

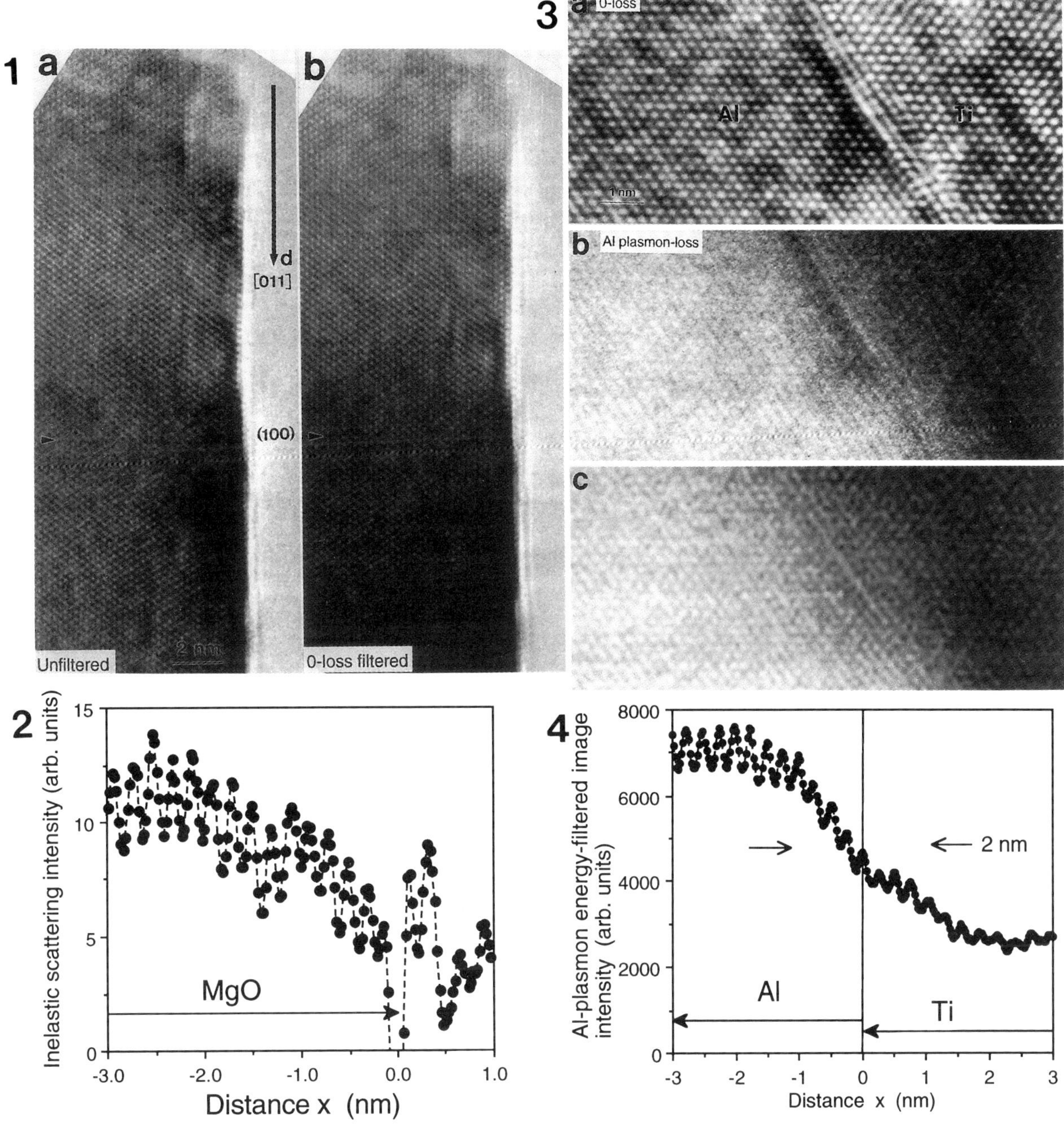

FIG. 1 -- Energy_unfiltered (a) and zero-loss filtered (b) HREM profile images of MgO(100) surface, oriented along [01$\bar{1}$]. Specimen thickness increases linearly from zero along direction indicated due to cubic shape of MgO.

FIG. 2 -- Intensity line scan across MgO(100) surface, showing variation of absorption function across surface. Intensity oscillation at surface ($x = 0$) is due to Fresnel fringes.

FIG. 3 -- (a) Zero-loss and (b) 15 eV Al-plasmon-loss energy-selected HREM images of Al/Ti (111) interface. (c) is processed image of (b) after reducing noise. Energy window $\Delta = 8$ eV.

FIG. 4 -- Intensity line scan across Al/Ti interface from image recorded using 15 eV Al-plasmon-loss electrons, showing dependence of inelastic absorption function on impact distance x of electron from interface. This curve determines spatial resolution of valence-loss electron imaging.

ENERGY-FILTERED PLASMON IMAGES OF $MgAl_2O_4$ IMPLANTED WITH Al^+ AND Mg^+ IONS

N. D. Evans[+], J. Bentley[*], and S. J. Zinkle[*]

[+]Oak Ridge Institute for Science and Education, P.O. Box 117, Oak Ridge, TN 37831-0117
[*]Metals and Ceramics Division, Oak Ridge National Laboratory, P.O. Box 2008, Oak Ridge, TN 37831-6376

Magnesium aluminate spinel ($MgAl_2O_4$) is a candidate material for specialized applications in proposed fusion reactors, and previously, has been irradiated with Al^+ or Mg^+ ions to assess the effects of high-dose irradiation. Electron energy-loss spectrometry (EELS) has been used to confirm the identity of metallic aluminum colloids located in the ion-implanted region of the spinel because electron diffraction experiments were inconclusive for phase identification.[1] In the present study, energy-filtered plasmon images of the ion-implanted region have been obtained to reveal this colloid distribution.

Following implantation with 2 MeV Al^+ ions to a fluence of 3.8×10^{21} ions/m^2 at 923 K, or with 2.4 MeV Mg^+ to a fluence of 2.8×10^{21} ions/m^2 at room temperature, spinel specimens were prepared in cross-section for analytical electron microscopy.[2] Energy-filtered images were obtained using a Philips CM30 microscope with an attached Gatan Imaging Filter. Acquired images were 512×512 pixels in size and gain normalized. Filtered images were recorded with 5-eV-wide windows and 1 s exposure times; windows were centered about 0, 10, 15, 20, and 25 eV losses. Images acquired using 15-eV-loss electrons (I_B) contained contributions from both spinel matrix, I_S, and metallic colloids, I_M, and $I_B \sim I_S + I_M$. For some images, an estimate of I_S was subtracted from I_B to improve the contrast between the colloids and the matrix. This background subtraction of the spinel contribution was accomplished by estimating $I_S \sim I_A + (I_C - I_A)/n$, where I_A and I_C were images with losses centered at 10 and 20 eV, respectively, and n is a suitable weighting factor (typically $2 \leq n \leq 3$).[3]

Energy-filtered bright-field images were acquired from both Al^+ and Mg^+ ion-implanted spinel. These images appear similar to conventional bright-field images, and show little diffraction contrast between the spinel matrix and the colloids. A filtered bright-field image of the Mg^+ implanted spinel is shown in Fig. 1. Figures 2, 3, and 4 are energy-filtered images of the same region shown in Fig. 1. Figure 2 was acquired with loss electrons corresponding to the spinel volume plasmon (~25 eV), so the continuous spinel matrix appears bright. Because the zero-loss electrons have been attenuated by the filter, the hole in the specimen appears dark at the top of the figure. The dark features in the spinel image, which are complementary to bright features in Figs. 3 and 4, are metallic aluminum colloids. Figure 3 was acquired with 15-eV-loss electrons corresponding to the metallic aluminum volume plasmon. In this figure, the colloids in the implanted-ion region have the greatest intensity, although there is significant intensity from the spinel matrix. This is expected as the spinel volume plasmon can be excited over the loss range of ~10-40 eV. However, with the spinel background subtracted from Fig. 3, the contrast between the colloids and the spinel matrix is improved (Fig. 4). Similar features were observed in the spinel implanted with Al^+ ions. Presently, the background subtraction routine assumes the weighting factor n is constant everywhere in the whole image. A possible further improvement in background subtraction could be made by allowing n to vary with thickness (position) in the image. A suitable routine for accomplishing this is being examined. Small colloids with diameters ~2 nm are clearly resolvable near the edge of the Mg^+ (Fig. 5) and the Al^+ implanted-ion regions (Fig. 6), respectively. Because of delocalization effects, the "chemical" resolution in a volume plasmon image has been suggested to be of the order of 10 nm.[4] However, an advantage of energy-filtering is clearly demonstrated here: very small features, similar to the matrix in chemistry but having different volume plasmon spectra, can be imaged in non-specific diffracting conditions.[5]

1. N.D. Evans et al., *Proc. Ann MSA Meeting* **49**(1991)728; *Proc. Ann MSA Meeting* **52**(1994)980.
2. N.D. Evans et al., in *Microstructure of Irradiated Materials, Mat. Res. Soc. Symp. Proc.* **373** (1995) in press.
3. N.D. Evans et al., *Proc. 29th Ann MAS Meeting* (1995), to be published.
4. R. F. Egerton, *Electron Energy-Loss Spectroscopy in the Electron Microscope*, New York: Plenum Press (1986)306.
5. Research sponsored by the Division of Materials Sciences and Office of Fusion Energy, U.S. Department of Energy, under contract DE-AC05-84OR21400 with Martin Marietta Energy Systems, Inc., and through the SHaRE Program under contract DE-AC05-76OR00033 with Oak Ridge Associated Universities.

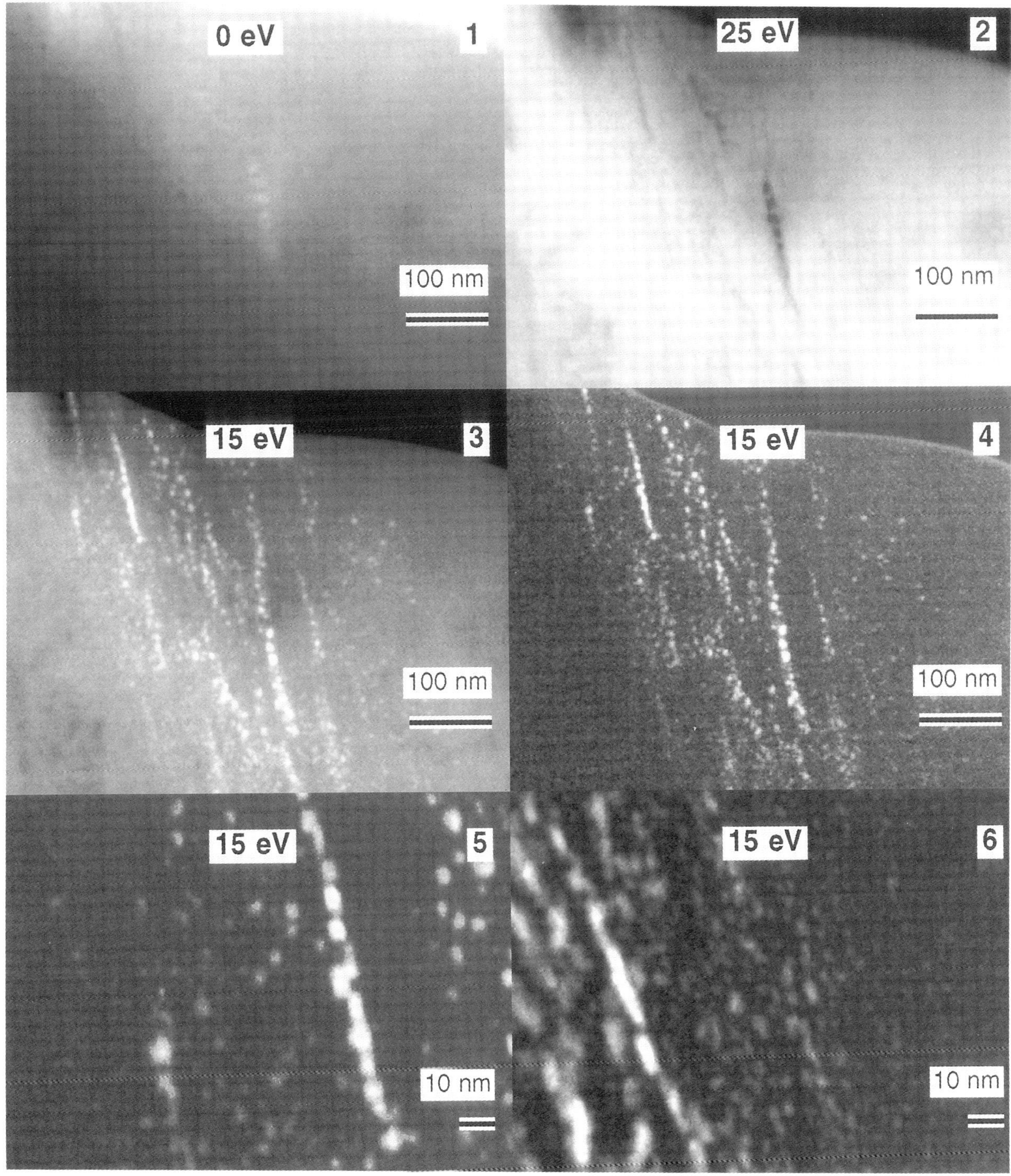

FIG. 1.--Energy-filtered bright-field image of implanted Mg$^+$ region of MgAl$_2$O$_4$ spinel.
FIG. 2.--25 eV loss plasmon image, implanted Mg$^+$ region.
FIG. 3.--15 eV loss plasmon image, implanted Mg$^+$ region.
FIG. 4.--Background subtracted 15 eV loss plasmon image, implanted Mg$^+$ region.
FIG. 5.--nanometer-sized metallic aluminum colloids in Mg$^+$ implanted spinel.
FIG. 6.--nanometer-sized metallic aluminum colloids in Al$^+$ implanted spinel.

QUANTITATIVE ELEMENTAL CONCENTRATIONS BY ENERGY-FILTERED IMAGING

J. Bentley,* E. L. Hall,† and E. A. Kenik*

*Metals & Ceramics Division, Oak Ridge National Laboratory, PO Box 2008, Oak Ridge, TN 37831-6376
†GE Corporate Research & Development, PO Box 8, Schenectady, NY 12301

There is widespread interest in elemental distribution maps produced from energy-filtered core-loss images obtained with commercial imaging energy-filters and slow-scan charge-coupled device (CCD) cameras on transmission electron microscopes (TEMs).[1] Earlier work on the feasibility of mapping solute segregation in stainless steels by energy-filtered imaging confirmed the utility of jump-ratio images (created by division of a post-edge image by a pre-edge one) for rapid assessments of elemental distributions.[2] The effects of diffraction contrast and thickness variations are largely corrected for in such images. However, quantitative compositional information requires the use of net core-loss intensities following subtraction of an extrapolated background. Such core-loss intensities are influenced by diffraction contrast and thickness variations; corrections for these effects may be necessary for a quantitative interpretation. In the present work, energy-filtered images are treated similarly to quantitative electron energy-loss spectrometry (EELS) data. An image showing number of atoms per unit area, n_x, is obtained by dividing the core-loss intensity image, $S_x(\Delta,\beta)$, by the low-loss image, $J_l(\Delta,\beta)$, obtained with identical energy window Δ and collection half-angle β, and by the partial ionization cross-section, $\sigma_x(\Delta,\beta)$.[3] Further normalization by specimen thickness, t, yields an image showing elemental concentration in atoms per unit volume:

$$C_x = n_x \,/\, t = S_x(\Delta,\beta) \,/\, [J_l(\Delta,\beta) \cdot \sigma_x(\Delta,\beta) \cdot t]$$

Experiments were performed with a Gatan Imaging Filter (GIF™)[4] interfaced to a Philips CM30 TEM operated at 300 kV with a LaB$_6$ emitter, and DigitalMicrograph software on Macintosh computers. Gain-normalized images with 512 x 512 pixels (2× binning), 15s exposure times and 30eV windows (for pre- and post-edge images), TEM magnifications of ~2000×, and probe currents of ~100 nA, were typically used. Zero-loss I_0 (5eV window) and "total" intensity I_T (60eV window or no slits) images were used to produce maps of $t/\lambda = \ln(I_T \,/\, I_0)$, where λ is the total inelastic mean free path. Core-loss maps were produced by AE^{-r} background extrapolation from two pre-edge windows.[1,3] All images were recorded with no reflections strongly excited.

Figures 1-6 show an example from elemental mapping of René 95, a complex nickel-base superalloy with coherent misfitting small and large γ' particles in a γ matrix.[5] Electropolishing preferentially thins the matrix so that the large γ' particle at the left of the images appears darker in the bright-field low-loss image (30 eV) of Fig. 1. The t/λ map of Fig. 2 shows the thickness variation quantitatively; t/λ varies from ~0.5 in the large γ' to ~0.2 in the γ. The Ni and Cr L$_{23}$ signals of Figs. 3 and 4 qualitatively confirm the expected Ni enrichment and Cr depletion in the γ'. However, in Fig. 3 the intensity ratio R for the large γ' relative to the γ matrix is ~2, much larger than the expected Ni partitioning. Division of Fig. 3 by Fig. 1 produces Fig. 5, where intensities are proportional to the areal density of Ni, and R~3.7. The final processing step, the division of Fig. 5 by Fig. 2, yields Fig. 6, in which intensities are proportional to Ni concentration (atoms/unit volume) and R~1.2. This result is in excellent agreement with quantitative x-ray microanalysis[6] (R = 1.25) and PHACOMP thermodynamic predictions (R = 1.19). Since there are no apparent local thickness variations associated with the small γ' particles, their image contrast relative to the nearby matrix is not influenced by the low-loss and t/λ normalizations; however, because of possible matrix overlap the measurement of quantitative Ni concentrations is questionable. Also, because of strong Cr partitioning to the γ matrix, the Cr signal is not affected nearly as much by low-loss and t/λ processing.

With the use of appropriate values for $\sigma_x(\Delta,\beta)$ and λ, elemental concentrations can be put on an absolute basis. With $\sigma_x(\Delta,\beta)$ data alone, concentration ratios for multiple elements can be mapped without normalization to low-loss or t/λ. Optimum image processing procedures that minimize noise amplification (apparent in Fig. 6), such as the use of smoothed t/λ and pre-edge (but **not** post-edge) images, are being investigated. Alignment of the images is critical to avoid loss of resolution and the generation of image artifacts; customized cross-correlation procedures are being optimized. Applications to other materials, including steels and ceramics, are in progress.[7]

Proc. Microscopy and Microanalysis 1995, edited by G.W. Bailey, M.H. Ellisman, R.A. Hennigar, and N.J. Zaluzec
Copyright © 1995 MSA. Published by Jones and Begell Publishing, 79 Madison Ave., New York, NY 10016

1. A.J. Gubbens and O.L. Krivanek, *Ultramicroscopy* **51**(1993)146.
2. J. Bentley and E. A. Kenik, *Proc. Ann MSA Meeting* **52**(1994)1000.
3. R.F. Egerton, *Electron-Energy-Loss Spectroscopy in the Electron Microscope*, New York: Plenum (1986).
4. O.L. Krivanek et al. *Microsc. Microanal. Microstruct.* **2**(1991)315; **3**(1992)187.
5. E. L. Hall and J. Bentley, *these proceedings*.
6. E.L. Hall et al. *Proc. Ann MSA Meeting* **52**(1994)964.

7. Research partially sponsored by the Division of Materials Sciences, U.S. Department of Energy, under contract DE-AC05-84OR21400 with Martin Marietta Energy Systems Inc., and through the SHaRE Program under contract DE-AC05-76OR00033 with Oak Ridge Associated Universities.

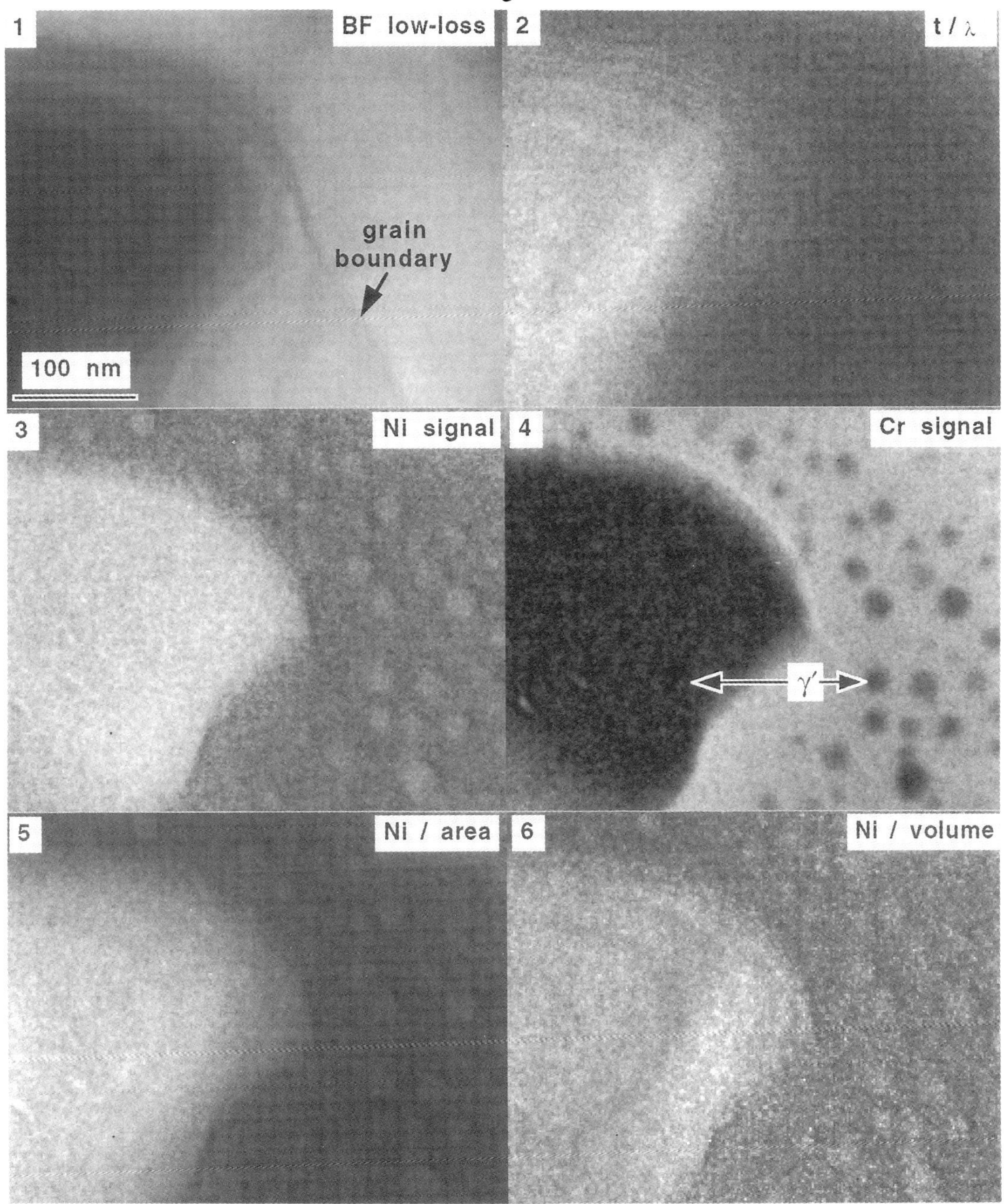

FIG. 1 - Bright-field low-loss (30eV window) of René 95. FIG. 2 - t/λ map. FIG. 3 - Ni L_{23} core-loss map. FIG. 4 - Cr L_{23} core-loss map. FIG. 5 - Ni areal density map. FIG. 6 - Ni concentration map.

COMBINED X-RAY, EELS MICROANALYSIS AND ELECTRON HOLOGRAPHY OF ELECTRICALLY ACTIVE ELECTROCERAMIC INTERFACES

V. Ravikumar, N. Wilcox, R. P. Rodrigues and Vinayak P. Dravid

Dept. of Materials Science & Engineering, Northwestern University, Evanston, IL 60208, USA.

A large number of electroceramics derive their useful properties from the presence of electrically active interfaces.[1] Varistors and GB capacitors are examples of important electroceramics where the electrical activity of grain boundary (GB) region controls the macroscopic response of the bulk devices. Electrical activity of GBs is often induced by doping and thermal processing. As a part of a broad program, we are investigating the inter-relationship among GB variables (structure, chemistry, and electronic structure) and GB induced properties in model electroceramics such as $SrTiO_3$.

GB doping of $SrTiO_3$ polycrystals was achieved by infiltrating Mn into the GBs under appropriate thermal processing cycle.[2] A quick survey of the microstructure by SIMS revealed that virtually all GBs are decorated with Mn. Subsequently, thin foils were examined for AEM analysis using Hitachi HF-2000 cold field emission gun TEM (FEG TEM), equipped with an Oxford Pentafet x-ray detector and a Gatan 666 PEELS system. A probe size of ~ 1 nm was utilized for microanalysis. Several x-ray microanalysis line scans across many GBs consistently showed substantial enrichment of Mn at the Gb core (Fig. 1A). EELS microanalysis also showed enrichment of Mn at the GB core (Fig. 1B) in accord with x-ray microanalysis.[2] Since Mn can be present in various oxidation states (+2, +3 and +4), the Mn L23 edge was analyzed for fine structure and the L2/L3 intensity ratio. Figure 2 shows two sets of EELS spectra, one from GB and the other from the bulk. The background substracted Mn L23 edge is consistent with predominant presence of Mn^{+2} with possibly some contribution from Mn^{+3}. Since Mn is expected to replace Ti (for example due to size similarity), the observed decoration of Mn at $SrTiO_3$ GBs indicates enrichment of negatively charged defects (Mn_{Ti}" and some Mn_{Ti}'). Thus, Mn-doped GBs in $SrTiO_3$ should exhibit negatively charged GBs with compensating positive space-charge.

The inference of negatively charged GBs and positive space-charge for Mn-doped $SrTiO_3$ was directly confirmed using electron holography. Figure 3A is an electron hologram of an Mn-doped GB in $SrTiO_3$. The biprism fringes are clearly shifted towards the vacuum indicating the GB charge is indeed negative. The phase image of this hologram (Fig. 3B) reveals the spatially varying potential (inset) across the GB. The second derivative of the potential provides the charge density distribution which is shown as Fig. 3C. This is, to our knowledge, the first real space evidence for spatially varying charge density distribution across electrically active interfaces in electroceramics. The spatial extent of space-charge region is about 2-5 nm, consistent with the expected Debye decay length.

The presentation will highlight the above results and emphasize the powerful approach of synergistic combination of diverse high spatial resolution electron beam technique to problems of interfaces and defects in solids.[3]

References

1. *Grain Boundary Phenomena in Electroceramics, Advances in Ceramics Vol. 1*, L. M. Levinson and D. C. Hill, Eds. (Am. Ceram. Soc., Columbus, OH, 1986).
2. N. Wilcox et al., *Solid State Ionics*, in press.
3. This research is supported by US DOE (Grant No. DE-FG02-92ER45475.)

Fig. 1.--(A) Mn/Ti XES grain boundary segregation profile in Fe/Mn doped $SrTiO_3$ polycrystal
      (B) Mn/Ti EELS grain boundary segregation profile in Fe/Mn doped $SrTiO_3$ polycrystal.
Fig. 2.--Comparison of EELS core loss at GB and bulk in Fe/Mn doped $SrTiO_3$ polycrystal.

Proc. Microscopy and Microanalysis 1995, edited by G.W. Bailey, M.H. Ellisman, R.A. Hennigar, and N.J. Zaluzec
Copyright © 1995 MSA. Published by Jones and Begell Publishing, 79 Madison Ave., New York, NY 10016

Fig. 3.--(A) Electron hologram of a Mn doped GB in SrTiO3
(B) Phase reconstruction of the hologram, and (C) corresponding charge density profile.

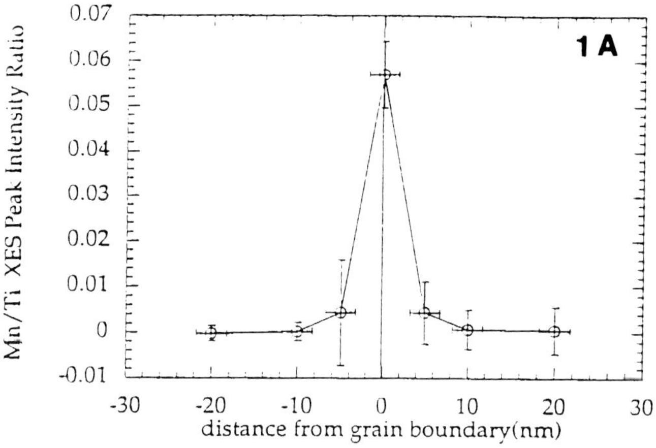

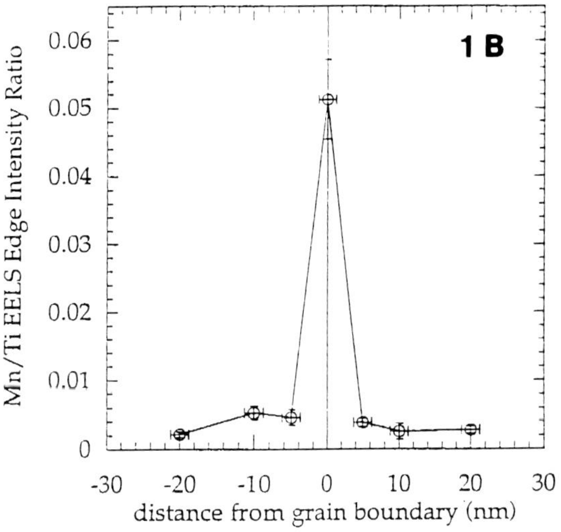

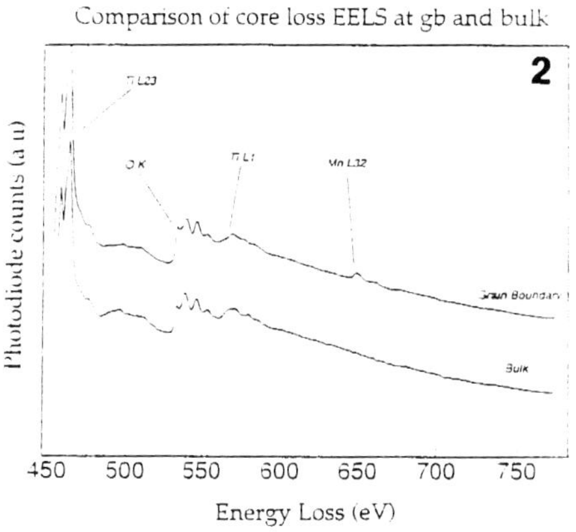

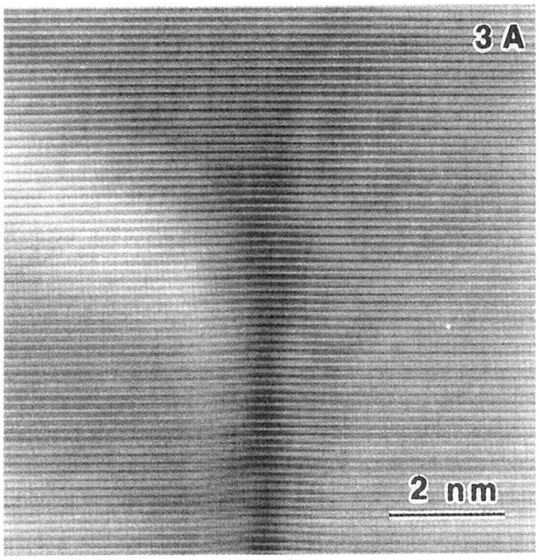

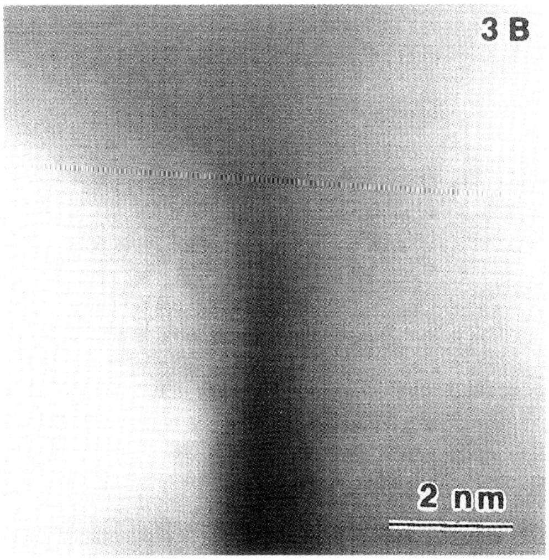

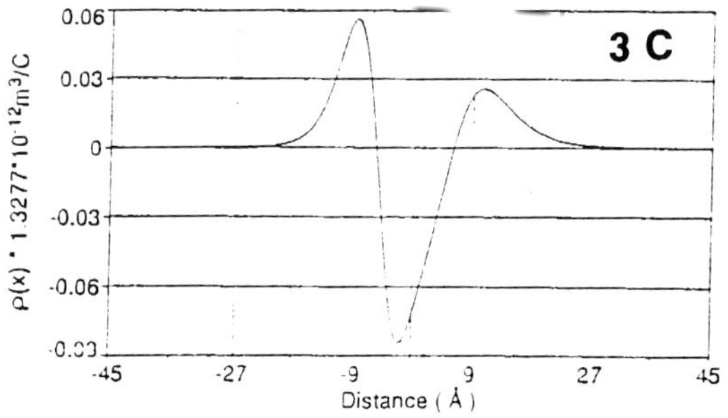

High Resoultion TEM and Energy Filtering TEM Study of Interfacial Structure and Reaction of Diamond/ Si in Chemical Vapor Deposition

Fu-Rong Chen [*], L. Chang[**], C. J. Chen[*] and T. S. Lin[***]

[*]   Materials Science Center, National Tsing Hua University, HsinChu, Taiwan, R. O. C.
[**]   Division of Engineering, National Science Council, Taipei, Taiwan, R. O. C.
[***] Materials Research Laboratories, Industrial Technology Research Institute, Hsinchu, Taiwan, R. O. C.

Diamond film was grown using microwave plasma CVD technique which consists of three steps: carburization, bias and growth . The high resolution TEM (HRTEM) in cross-sectional view has been used to observe the evolution of interfacial structure in each processing step [1]. The chemistry near the interface was characterized with elemental mapping using energy-filtered imaging technique with Gatan imaging Filter (GIF) [2]. At the carburization stage, β-SiC, diamond particles and graphite plates have been observed in an amorphous layer. This amorphous layer was analysized to be carbon by energy filtering technique.

As shown in the Fig. 1, β-SiC can form in epitaxıal orientation with Si in the following stage of biasing. Graphite was not observed after the bias was applied. At the bias stage there is an interlayer of 6 nm thick between diamond and silicon substrate . From the high resoultion image in Fig. 2 (a), most of the regions of the interlayer are of amorphous characteristics which presents a barrier to identify the elemental compositions. Fig. 2 (b) and (c) are the Si and C jump ratio maps of the energy selecting images, respectively. The jump ratio maps were obtained by dividing the post-edge loass image by the pre-edge image. The post-edge image of Si and C were recorded at the 119 and 303ev which corresponds to 20ev beyond $L_{2,3}$ edge of Si and K edge loss of C, respectively. The pre-edge image were recorded at 79 and 263 ev for the Si and C maps, respectively. The width of the energy slit was set to be 10 ev. There are three contrast levels in both of the elemental images in Fig. 2 (b) and (c). The interlayer shows a contrast higher than carbon laryer on the top in the Si map and a higher contrast than the substrate Si in the carbon map. We can confirm from these images that the interfacial layer is SiC. It therefore suggests that the thin interlayer transforms from carbon in the process of carburization to SiC in following bias stage.   The detail relation of the evolution of microstructure and the chemistry near the interface with the processing conditions will be discussed.

References
1. C. J. Chen, L. Chang, T. S. Lin and F.-R. Chen, submitted to *J. Mater. Res.*
2. Gatan Inc., 6678 Owens Dr, Pleasanton, CA
3. The authors gratefully acknowledge the support from the National Science Council.

Proc. Microscopy and Microanalysis 1995, edited by G.W. Bailey, M.H. Ellisman, R.A. Hennigar, and N.J. Zaluzec
Copyright © 1995 MSA. Published by Jones and Begell Publishing, 79 Madison Ave., New York, NY 10016

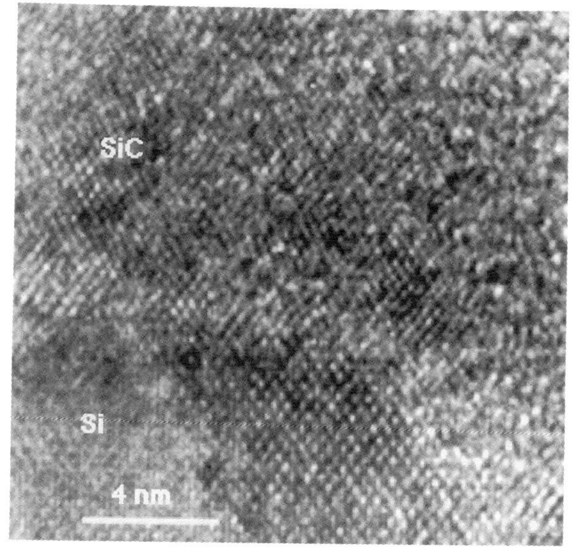

FIG. 1 High resolution image shows β-SiC in epitaxial orientation with respect to the substrate Si in the stage of biasing

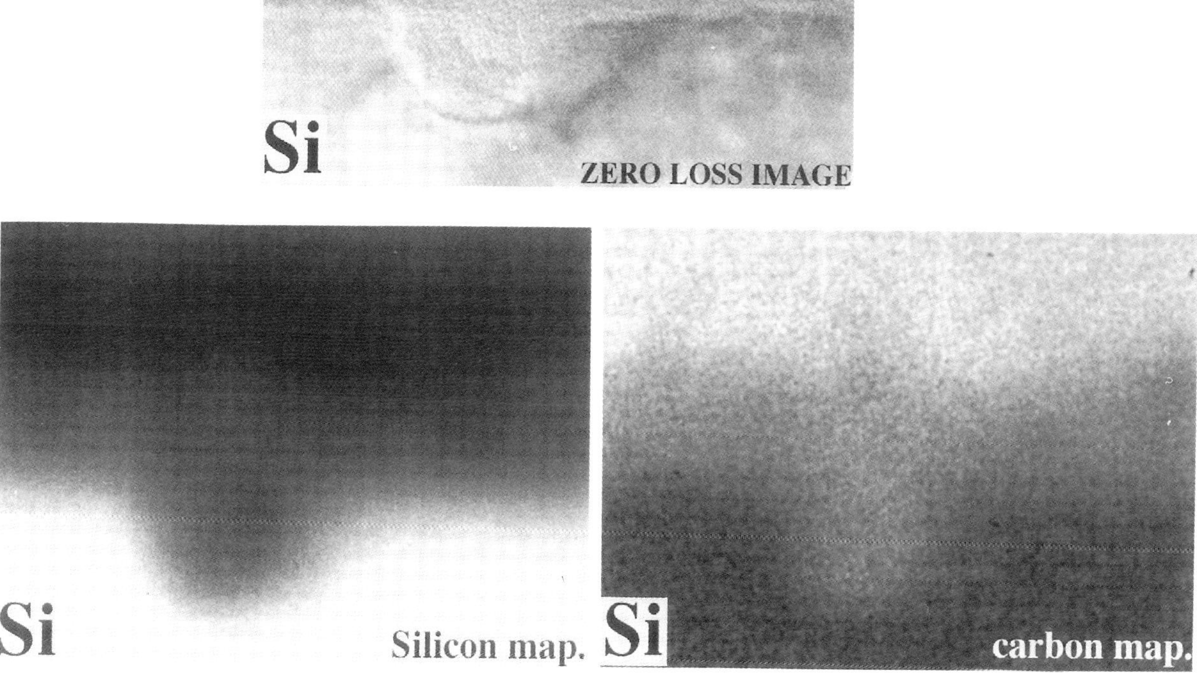

FIG. 2 (a) Zero loss image of diamond/ Si interface in the biasing stage, (b) Si jump ratio map of (a), (c) C jump ratio map of (a), 1 cm is 3 nm in the jump ratio maps

DECOMPOSITION OF ELECTRON ENERGY LOSS SPECTRUM-LINE PROFILES BY TARGET FACTOR ANALYSIS

S.J. Splinter, J. Bruley, and D.A. Smith

Department of Materials Science and Engineering, Lehigh University, Bethlehem, PA 18017

Interfaces play a ubiquitous role in governing the behavior of materials, but the fundamental means by which they exert their influence are surprisingly little understood. EELS offers a path towards elucidating the nature of both the coordination and local chemistry at interfaces; however, successful application requires the development of quantitative techniques for the deconvolution of various contributions to the spectra. The present work describes a first application of Factor Analysis (FA) to this problem. FA is a multiparameter statistical method used to extract subtle information from complicated data sets, the mathematical basis of which has been well documented in the literature.[1,2] In brief, FA seeks to reproduce a given data set within experimental error using the minimum number of statistically non-zero "factors" or "components". In this study, FA has been successfully applied to the decomposition of an Al $L_{2,3}$ spectrum-line profile across a magnesium aluminate spinel/sapphire interface. One of us (JB) recently examined this heterophase structure in detail using multiple least squares decomposition of the two endpoint components (spinel and sapphire).[3] Gaussian fitting of the edge structures was subsequently used in order to highlight subtle bonding changes present near the interface. It is shown below that the same information can be retrieved with a minimum of effort using target factor analysis.

Fig. 1 shows the Al $L_{2,3}$ spectrum-line profile across the spinel/sapphire interface.[3] In order to perform FA, the data in Fig.1 were first assembled into a matrix of size (1024 energy channels x 38 spectra) and then subjected to a standard eigenanalysis. The number of principal components contributing to the changing data set was determined by searching for the number of statistically non-zero eigenvalues for each given eigenvector. Various non-parametric tests were employed[2], all pointing towards the presence of three principal components. This result is significant in that the presence of an independent interfacial component cannot be inferred from the data by any other rigorous means. Once the correct number of principal components was established, the data space was compressed and a solution existing in real factor space was sought. This was achieved by Target Transformation[2] (TT) using the endpoints of the data set as test vectors for the first two components (spinel, sapphire). The unknown interface spectrum was determined via an iterative procedure using subsets of measured spectra[4] and used as the third test vector. Fig. 2(a) shows the shape of the factors predicted by TT and Fig. 2(b) shows the composition percentages of these two components as a function of incident beam position. The small differences existing between the "spinel" and "interfacial" factors can be attributed to an increased degree of Al site inversion near the boundary.[3]

References

1. H.H. Harmon, *Modern Factor Analysis*, Chicago: University of Chicago Press (1976).
2. E.R. Malinowski and D.G. Howery, *Factor Analysis in Chemistry*, New York: Wiley (1980).
3. J. Bruley, M.W. Tseng and D.B. Williams, *Microsc., Microanal., Microstruct.* in press.
4. M. Sarkar et al., *Surf. Interface Anal.* 20(1993)60.

Proc. Microscopy and Microanalysis 1995, edited by G.W. Bailey, M.H. Ellisman, R.A. Hennigar, and N.J. Zaluzec
Copyright © 1995 MSA. Published by Jones and Begell Publishing, 79 Madison Ave., New York, NY 10016

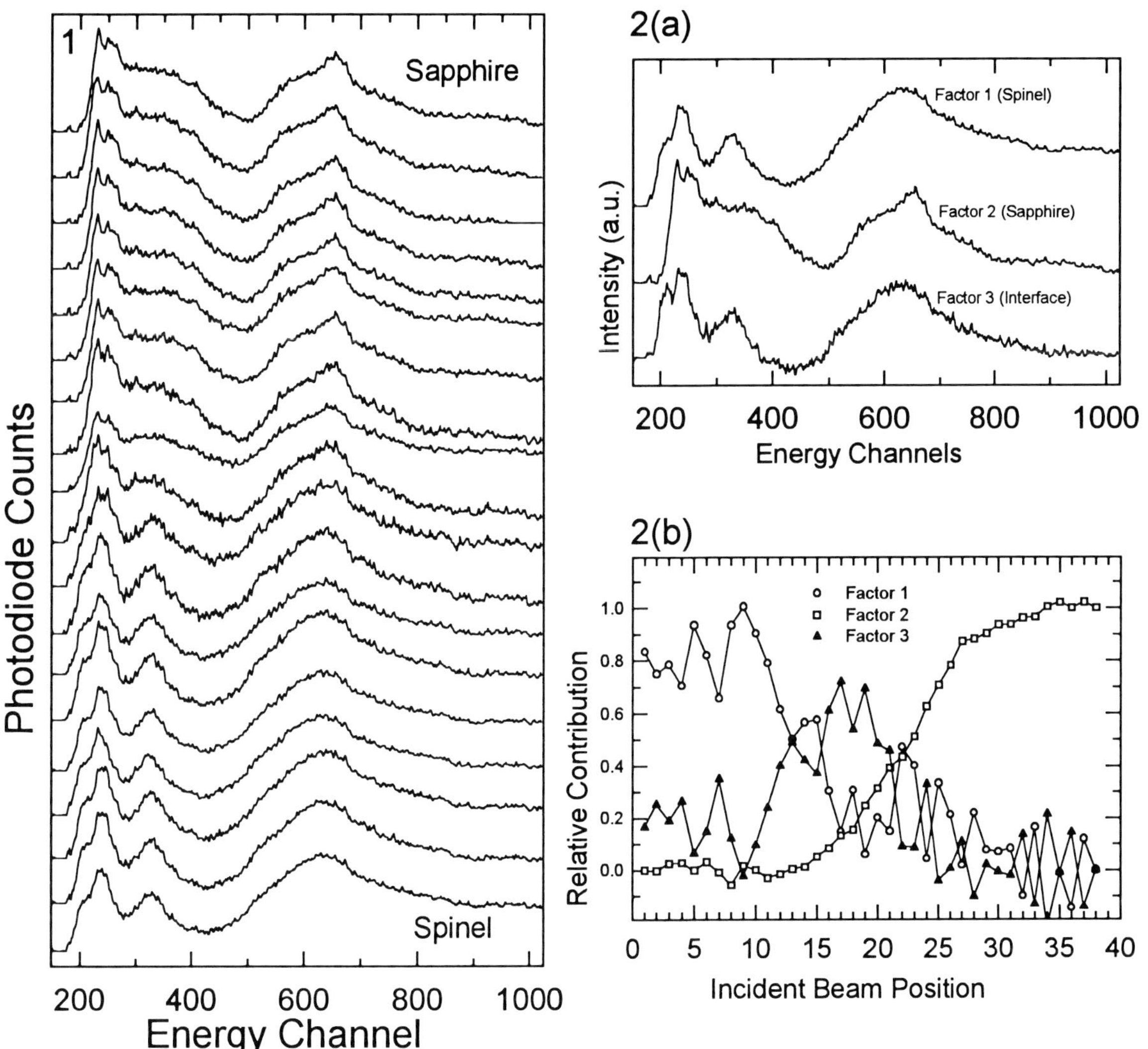

Fig. 1: Al $L_{2,3}$ spectrum-line profile across magnesium aluminate spinel/sapphire interface.
Fig. 2: (a) Predicted shapes of three principal factors calculated by target factor analysis;
(b) Composition percentages (factor loadings) of three principal factors as a function of
incident beam position.

ATOMIC AND ELECTRONIC STRUCTURE OF A 60° MISFIT DISLOCATION

P.E. Batson

IBM Thomas J. Watson Research Center
Yorktown Heights, New York 10598

Strained Si quantum wells provide high electron mobility channels for devices within the standard Si-based technology. These wells consist of 8-12nm thick layers, epitaxially grown over relaxed Si-Ge alloys. The 1.4% lattice mismatch between Si and a $Ge_{40}Si_{60}$ alloy produces a Si layer strained in tension in the plane of the layer and in compression in the direction perpendicular to the layer, splitting the Si conduction band into a higher energy four-fold degenerate band and a lower energy two-fold degenerate band. The latter band is a high mobility channel for electron conduction. If the Si layer is made thicker than the critical thickness for epitaxy, misfit dislocations result to accommodate the strain. For growth of Si on (001) $Ge_{40}Si_{60}$ these are of the 60° type, dissociated into a 30° partial dislocation at the Si/alloy interface, a 90° partial some tens of Angstroms away in the alloy, and a connecting stacking fault. In the presence of this strain relieving structure, the electron mobility is degraded.[1] It is of interest, therefore, to investigate the atomic and electronic structure of this dislocation, so that it can be correlated with the local band offset or electron mobility.

Recent work has shown that Spatially Resolved EELS can obtain the positions of conduction band critical points in this alloy system.[2] Briefly, we measure the position and shape of the Si $2p \rightarrow$ conduction band absorption edge with an energy resolution of about 0.25eV and an accuracy of ± 20 meV. This is accomplished using a Wien filter spectrometer built at IBM and described elsewhere.[3] We next model the local symmetry projected final density of states, and calculate the expected shape of the absorption edge. Comparing this result with the measurement then provides a way of measuring positions of conduction band symmetry positions: the band minimum, Δ_1, the minimum in the [111] directions, L_1, and other points. This has been applied to a range of alloy compositions and to the Si quantum well heterojunction band offset measurement.[4] Using high resolution Annular Dark Field imaging in the STEM.[5,6] these measurements can be combined with structure measurements. At IBM I have modified the HB501 STEM to boost the voltage to 120 kV, and to include the high resolution polepiece ($C_s = 1.3mm$) This produces a probe size of about 1.95Å. Using this instrument, the Si (022) lattice fringes (1.92 Å) have been resolved, with the clear limit being mechanical and electrical instabilities. An initial look at bonding information with this spatial resolution has been accomplished.[7]

I report here dark field images of the atomic structure of the misfit dislocation and show some electronic structure information. Figure 1a shows an ADF image of the object, spatially filtered to exclude periodicities smaller than about 2.0Å. This smooths noise and electrical instabilities. The image is also a digital framestore sum of eight 0.5sec frames consisting of 450 lines with 530 pixels per line. The misfit dislocation is located within the $Ge_{40}Si_{60}$ alloy (bright regions), below the Si well (dark region at top). In Figure 1b, periodicities longer than about 15Å are eliminated to suppress atomic number information while preserving atomic structure. In Figure 1c, the 30° partial dislocation is cut out of the image and enlarged. In this image, the stacking can be deduced, confirming that the fault is intrinsic, and that the partial is of the glide type suggested by Olsen and Spence[8] and confirmed by others.[9] Correspondence of the measured column positions with the glide model prediction is very good at the dislocation core, but some deviation occurs beyond 2-3 atomic spacings away. It should be cautioned that we do not yet know whether this image is a faithful rendering of the structure, particularly when the 1.35Å dumbell spacing less than the point resolution of the instrument, and when the forbidden (002) reflection in the image transform indicates that dynamical scattering conditions may be present.

Figure 2a shows a $2p_{3/2}$ spectrum from the middle of the stacking fault (closed circles), compared with a spectrum from the relaxed alloy (open circles). We can see that the conduction band onset at Δ_1 is unaffected, but the minimum in the [111] direction appears to be shifted downwards and split

Proc. Microscopy and Microanalysis 1995, edited by G.W. Bailey, M.H. Ellisman, R.A. Hennigar, and N.J. Zaluzec

into two components. This is predicted by theory that considers band folding within the stacking fault.[10] Figure 2b shows a result for the core of the 30° partial, with an attempted fit to a likely projected density of states. There is a modification of the onset at Δ_1, suggesting that some electronic states reside within the gap. Calculations suggest that this dislocation core should reconstruct to clear the gap of electronic states.[11] Thus, kinks may be responsible for the observed gap intensity. There is some evidence for these in the elongation of the terminating atom column in the partial core in the [00-1] direction, but confirmation of their presence will have to wait for ADF imaging and EELS spectra using [111] oriented specimens. I would like to thank F.K. LeGoues and K. Ismail for discussion and access to specimens.

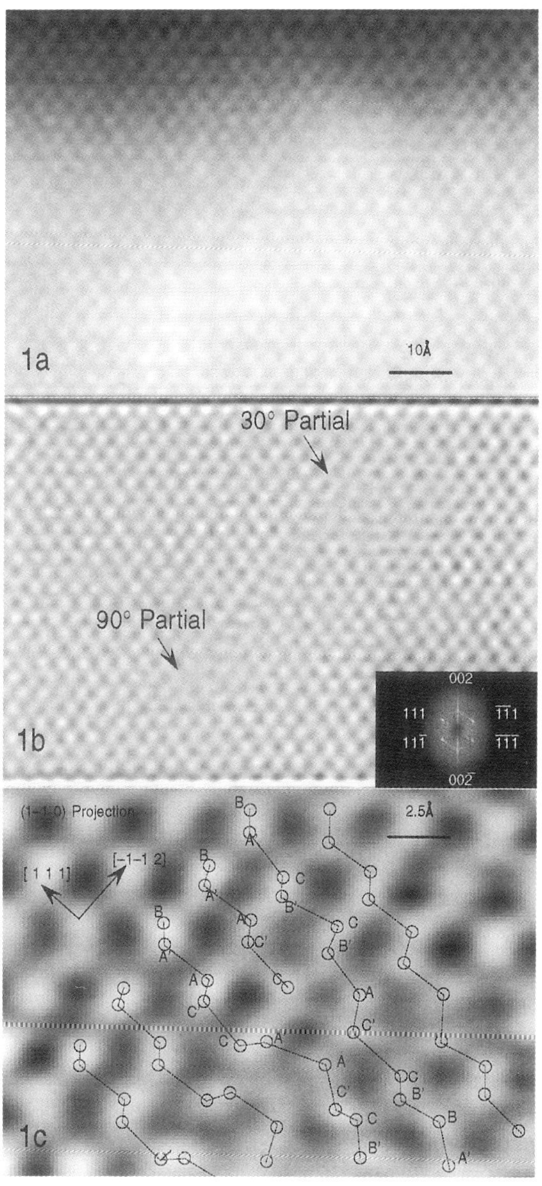

1. K. Ismail, F.K. LeGoues, K.L. Saenger, M. Arafa, J.O. Chu, P.M Mooney, and B.S. Meyerson, Phys. Rev. Lett. <u>73</u> 3447 (1994).
2. P.E. Batson, and J.F. Morar Phys. Rev. Lett. <u>71</u> 609 (1993).
3. P.E. Batson, Rev. Sci. Inst. <u>57</u> 43 (1986), <u>59</u> 1132 (1988).
4. P.E. Batson, Ultramicroscopy, in press.
5. S.J. Pennycook and Boatner Nature, <u>336</u> 565 (1988).
6. P. Xu, E.J. Kirkland, J. Silcox, and R. Keyse, Ultramicroscopy, <u>32</u> 93 (1993).
7. P.E. Batson, Nature, <u>366</u> 727 (1993).
8. A. Olsen and J.C.H. Spence, Phil. Mag. <u>A43</u> 945 (1981).
9. J.C. Barry and H. Alexander, Proc. 45th EMSA (ed. Bailey, San Fransisco Press, 1987), p. 242.
10. L.F. Mattheis and J.R. Patel Phys. Rev. <u>B23</u> 5384 (1981).
11. J.R. Chelikowsky, Phys. Rev. Lett. <u>49</u> 1569 (1982).

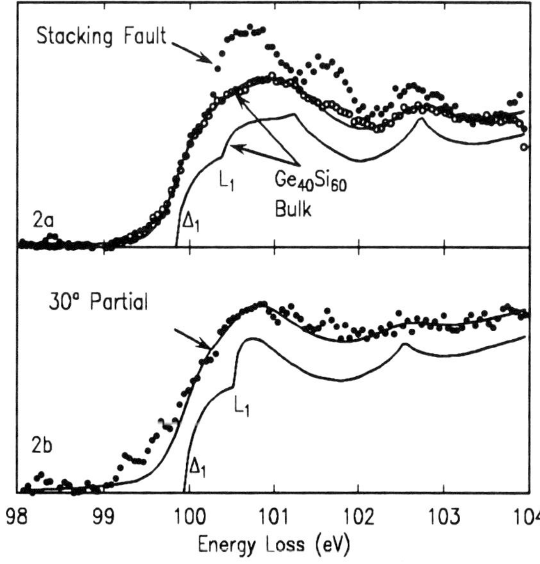

Figure 1. a) ADF image of a 60° misfit dislocation. b) Atomic number information suppressed. c) 30° partial extracted and enlarged. Connectivity of columns is indicated, and the stacking sequence has been lettered.

Figure 2. a) Si $2p_{3/2}$ stacking fault absorption compared with the bulk relaxed alloy. Δ_1 is unchanged, but L_1 is shifted and split into two peaks. b) At the 30° partial dislocation, states within the gap appear to be present.

THEORETICAL MODELLING OF THE SIZE AND SHAPE OF ATOMIC IMAGES FORMED WITH INELASTICALLY SCATTERED ELECTRONS

O.F. Holbrook and D.M. Bird

School of Physics, University of Bath, Bath BA2 7AY, U.K.

In electron diffraction the study of inelastic scattering processes has been revolutionised with the development of imaging PEELS. It is now relatively straightforward to produce images from fast electrons which have lost a specific amount of energy (to within about 5eV).[1] Therefore electrons which have been inelastically scattered via an atomic transition, loosing a characteristic amount of energy, can in principle be used to form an equally characteristic image. It should be noted that such images will not be perfect since other single and multiple inelastic events may result in a similar energy loss which will tend to obscure the ideal picture. An important issue is the ultimate resolution limit of this type of atomic imaging.[2,3] Although practical questions (such as signal-to-noise ratio) are important it is still interesting to ask what theoretical limit might be imposed by the underlying physics of the inelastic scattering and imaging processes. The work presented here is an initial investigation of the typical radial size and shape which can be expected when imaging a single atom and how these depend upon the transition and energy loss being considered.

The system chosen for these calculations is that of an isolated hydrogenic atom where the atomic number is varied to give different energy losses. Analytic expressions are therefore available for the initial and final atomic states. A further simplification is that elastic scattering is ignored, so that both incident and final fast electron states are described by plane waves. Although very simple, these calculations will indicate the type of shape and approximate radial size of an image, allowing a more informed choice of material for further investigation (and experimentation) to be made. The equation for the image is given by,[4]

$$I(\mathbf{R}) = \left| \int d\mathbf{Q} \frac{\exp(-i\mathbf{Q}.\mathbf{R})F(\mathbf{Q}, k\theta_E)}{Q^2 + k^2\theta_E^2} \right|^2.$$

The scattering geometry we consider is shown in Fig. 1, $\mathbf{k}$ and $\mathbf{k}'$ are the incident and final wavevectors respectively, $\mathbf{q}$ is the scattering wavevector with components $\mathbf{Q}$ in the (x, y) plane (ie the imaging $\mathbf{R}$ plane) and q_z in the z direction. For simplicity we assume q_z is fixed at its value for $\mathbf{Q} = 0$, ie $q_z = k\theta_E$, where $\theta_E = \Delta E/2E$. Here, E is the energy of the incident electron (assumed to be 200keV) and ΔE is the energy loss. $F(\mathbf{Q}, k\theta_E)$ is the atomic matrix element,

$$F(\mathbf{q}) = < \psi_f|\exp(i\mathbf{q}.\mathbf{r})|\psi_i >,$$

where ψ_i and ψ_f are the initial and final atomic states respectively. With hydrogenic wavefunctions, analytic expressions can be derived for $F(\mathbf{q})$.[4]

Some calculated images are shown in Fig. 2. Fig. 2a shows images due to transitions from the $1s$ to the $2p$ state, corresponding to an energy loss of 164eV ($Z = 4$). Figs. 2b and 2c result from 2p to 3d transitions at energies 189eV ($Z = 10$) and 758eV ($Z = 20$) respectively. The full set of possible transitions has not been shown here since the dipole forbidden transitions result in very weak images. The figures reveal that the peak in the calculated images is inside 1Å, even for quite moderate energy losses. They imply that the underlying physics of the scattering process does not preclude atomic resolution images using inelastically scattered electrons.

1. e.g. H. Hashimoto et al., in B. Jouffrey and C. Colliex, Eds., *Proc. 13th ICEM*, Les Editions de Physique (1994)727.

Proc. Microscopy and Microanalysis 1995, edited by G.W. Bailey, M.H. Ellisman, R.A. Hennigar, and N.J. Zaluzec

2. J. Mayer et al., in B. Jouffrey and C. Colliex, Eds., *Proc. 13th ICEM*, Les Editions de Physique (1994)615.
3. U. Golla and H. Kohl, in B. Jouffrey and C. Colliex, Eds., *Proc. 13th ICEM*, Les Editions de Physique (1994)743.
4. O.F Holbrook and D.M. Bird, to be published.

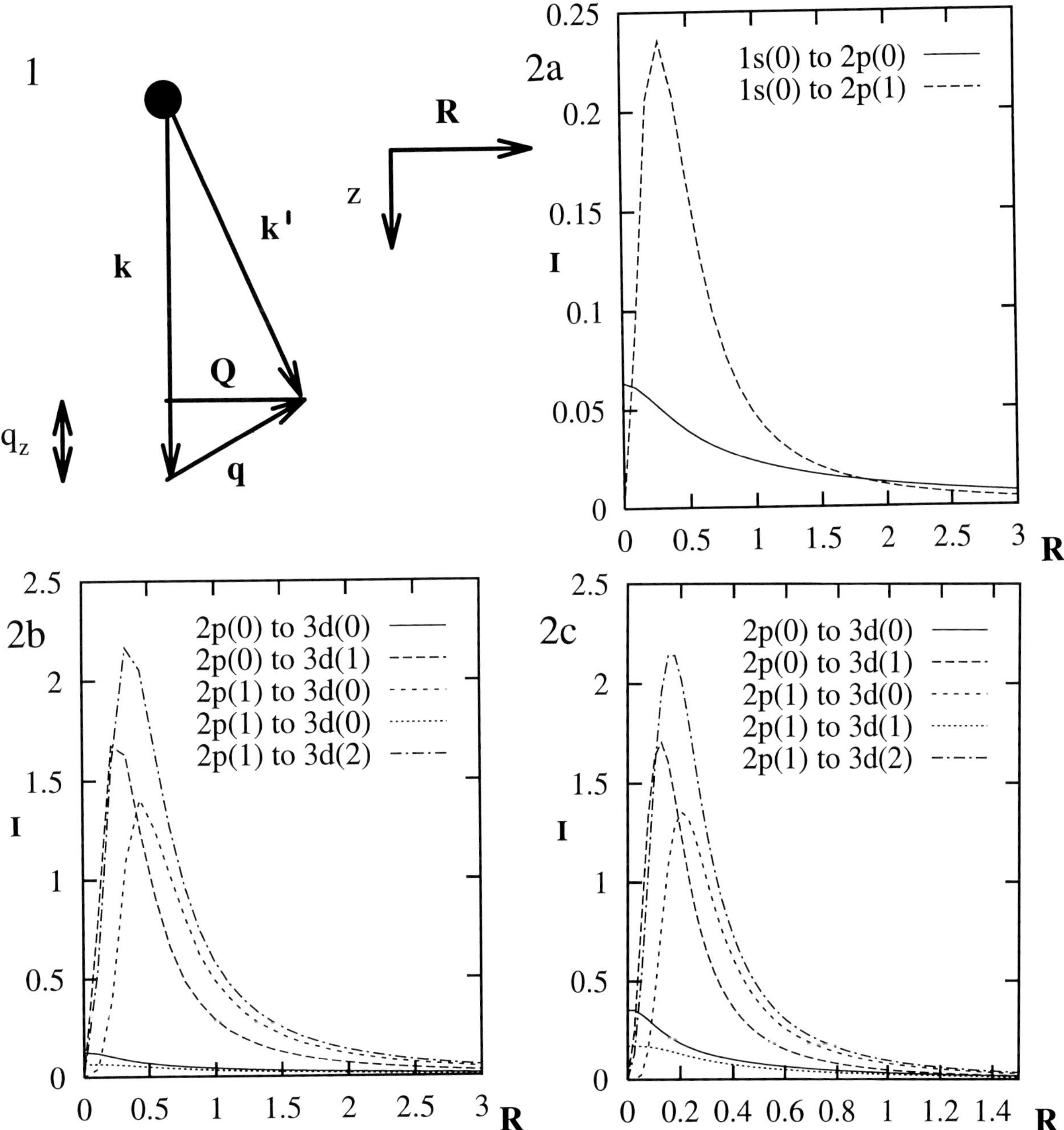

FIG1:- Inelastic scattering geometry.

FIG2:- Plots of image intensity (arbitrary units) as a function of radial distance R in Å. a) 1s to 2p with energy loss 164eV b) 2p to 3d with energy loss 189eV c) 2p to 3d with energy loss 758eV. The number in brackets in the key is the magnetic quantum number of the atomic state.

WHERE ARE THE LIMITS TO SPATIAL RESOLUTION IN THE HRTEM?

Michael A. O'Keefe

National Center for Electron Microscopy, University of California, LBL B72, Berkeley, CA 94720

In discussing resolution limits, it is important to distinguish between the resolution of a microscope and the resolutions of the images produced by the microscope. The spatial resolution of a high-resolution transmission electron microscope (HRTEM) is an upper bound on the resolution of images that can be produced by that microscope, but there is no guarantee that any particular image from the microscope will attain this bound. In addition to instrumental limits, properties of the specimen will act to degrade image resolutions.[1] For specimens that beam-damage rapidly, image resolution depends upon electron energy and electron dose. For small-cell crystalline specimens, Bragg's law mandates that allowable resolutions be quantized, preventing image resolution from reaching microscope resolution. However, improving the resolution of a HRTEM gives it the capability to produce higher resolution images.

Image "resolution" is a measure of the spatial frequencies transferred from the image amplitude spectrum (specimen exit-surface wavefunction) into the image intensity spectrum (Fourier transform of the image intensity). This transfer is affected by several factors -- the scattering angles and phases of the diffracted beams exiting the sample surface, phase changes imposed by the angular-dependent objective lens defocus and spherical aberration, the physical objective aperture, and coherence effects that can be characterized by microscope spread-of-focus and incident beam convergence. These factors combine to impose two limits on the information produced by the HRTEM. The "resolution" has traditionally been defined in terms of its Scherzer resolution limit at optimum defocus.[2] However, even beyond the Scherzer limit, it is possible to transfer spatial frequencies from the specimen to the image, out to the so-called information limit of the electron microscope.[3] Broadly speaking, the information limit measures the highest spatial frequency that can be transferred from the exit-surface wavefunction into an image, whereas the Scherzer resolution measures the highest spatial frequency that can be transferred into the image with the same phase as all the lower frequencies contributing to that image.[3]

Scherzer resolution applies only for specimens thin enough to be approximated as weak phase objects. Under this condition, and at the Scherzer optimum defocus of $-\sqrt{(1.5C_S\lambda)}$, the Scherzer spatial frequency is $u_S=1.49C_S^{-\frac{1}{4}}\lambda^{-\frac{3}{4}}$ leading to a Scherzer resolution of $d_S=0.67C_S^{\frac{1}{4}}\lambda^{\frac{3}{4}}$. A Scherzer image shows the projected potential of the specimen, as limited by the microscope resolution. This resolution is dependent on two instrumental factors -- electron wavelength, and spherical aberration present in the objective lens. Because it is strongly dependent on electron wavelength, d_S can be improved by increasing the voltage of the microscope. Current high-voltage HREMs have resolutions approaching 1Å (table 1).

The information limit of the HRTEM is determined by the amount of focus-spread present in the final image. Although for thicker crystals, the frequency-limiting action of the focus-spread effect is complex and must be described in terms of the transmission cross-coefficient,[4] for a thin crystal the damping action can best be described by quasi-coherent imaging theory[5] in terms of envelope functions[6,7] imposed on the usual phase-contrast transfer function (CTF). The envelope for the spread of focus produced by partial temporal coherence is $E_\Delta(u)=\exp\{-\frac{1}{2}\pi^2\lambda^2\Delta^2u^4\}$ where Δ is the halfwidth of a Gaussian focus-spread. $E_\Delta(u)$ drops to a value of $\exp(-2)$ at $u_\Delta=(\pi\lambda\Delta/2)^{-\frac{1}{2}}$, giving an information limit of $d_\Delta=\sqrt{(\pi\lambda\Delta/2)}$. Focus-spread arises from the effects of several microscope parameters; it includes energy spread in the electron beam used to illuminate the sample (δE), variations in microscope accelerating voltage (δV) and objective lens current (δI), and specimen vibration with respect to the image recording system. Setting aside vibration, we can express the focus-spread as $\Delta=C_C\sqrt{[(\delta V/V)^2+(2\delta I/I)^2+(\delta E/E)^2]}$, where C_C is the chromatic aberration coefficient of the objective lens. Improvements in the information limit can be obtained by decreasing C_C and by improving the ripple in high-voltage and lens power supplies. One factor under user control is energy spread; reducing the beam current lowers δE and thus Δ.[8] A larger improvement in δE can be obtained by using a field-emission source. With such a source, some mid-voltage FEG-TEMs can approach the information limits of high-voltage HRTEMs (table 1).

Proc. Microscopy and Microanalysis 1995, edited by G.W. Bailey, M.H. Ellisman, R.A. Hennigar, and N.J. Zaluzec
Copyright © 1995 MSA. Published by Jones and Begell Publishing, 79 Madison Ave., New York, NY 10016

It is possible to successively tune the HRTEM to pass different spatial frequency bands by choosing suitable defocus settings, then to combine these images into one image with a resolution that approaches the information limit of the microscope. Methods of focus-series reconstruction range from the linear-image Schiske algorithm[9] to the non-linear procedures of Kirkland,[10] the paraboloid method,[11] and maximum likelihood.[12] With these methods, a microscope's resolution will no longer be determined by its Scherzer limit, but will then become identical to its information limit .[13]

1. F.A Ponce and M.A. O'Keefe, *Proc. 44th Ann. Meeting EMSA*, Albuquerque, N.M. (1986) 522-525.
2. O. Scherzer, J. Appl. Phys. **20** (1949) 20.
3. M.A. O'Keefe, *Ultramicroscopy* **47** (1992) 282-297.
4. M.A. O'Keefe, *Proc. 37th Ann. Meeting EMSA*, San Antonio, Texas (1979) 556-557.
5. M.A. O'Keefe and W.O. Saxton, *Proc. 41st Ann. Meeting EMSA*, Phx, Arizona (1983) 288-289.
6. J. Frank, *Optik* **38** (1973) 519-536.
7. R.H. Wade and J. Frank, *Optik* **49** (1977) 81-92.
8. O.L. Krivanek, *Optik* **43** (1975) 361-372.
9. P. Schiske, in Image Proc. and CAD, Ed. P.W. Hawkes, Acad. Press, London, (1973) 82-90.
10. E.J. Kirkland, *Proc. 42nd Ann. Meeting EMSA*, Detroit, Michigan, (1984) 432-433.
11. D. Van Dyck and M. Op de Beeck, *Proc. 12th Int. Congr. Electr. Mic.*, Seattle, Wash. (1990) 26-27.
12. W. Coene *Proc. 50th Ann. Meeting EMSA*, Boston, Massachusetts, pp. 986-987, 1992.
13. Work supported by the Director, Office of Energy Research, Office of Basic Energy Sciences, Material Sciences Division of the U.S. Department of Energy, under contract No. DE-AC03-76SF00098.

Table 1. HRTEMs with point-to-point resolution better than 2.5Å[a]

Microscope	Maximum energy (keV)	C_S (mm)	C_C (mm)	Theoretical[b] resolutions (Å) Scherzer[c]	Information Limit[d]
Hitachi H8100	200	0.7	1.1	2.07	1.54
Hitachi HF2000[e]	200	1.2	1.5	2.37	1.26
Hitachi HF2000-UHR[e]	200	0.7	1.0	2.07	1.03
Hitachi H9000-UHR	300	0.9	1.5	1.84	1.36
Hitachi H9000-NAR	300	0.68	1.4	1.71	1.32
Hitachi 300keV FEG[e]	300	0.9	1.5	1.84	1.07
Hitachi HU1250	1250	2.5	3.5	1.13	0.98
Hitachi H-1500	1300	1.85	3.4	1.03	0.95
ISI/Topcon 002A	120	0.3	0.6	2.08	1.65
ISI/Topcon 002B	200	0.4	0.8	1.80	1.31
JEOL 2000EX	200	0.9	1.2	2.21	1.61
JEOL 2010HT	200	1.0	1.4	2.26	1.74
JEOL 2010UHR	200	0.5	1.0	1.90	1.47
JEOL 2010F[e]	200	0.5	1.0	1.90	1.03
JEOL 3010HT	300	1.4	2.2	2.05	1.65
JEOL 3010UHR	300	0.6	1.4	1.66	1.32
JEOL 4000EX	400	0.9	1.65	1.61	1.20
JEOL HAREM	500	1.0	1.4	1.48	0.97
JEOL ARM-1000	1000	2.3	3.4	1.26	1.07
JEOL Kyoto-1000	1000	1.7	3.6	1.17	1.10
JEOL ARM-1250	1250	1.6	4.0	1.02	1.05
Philips CM20/ST	200	1.2	1.2	2.37	1.61
Philips CM200FEG/ST[e]	200	1.2	1.2	2.37	1.13
Philips CM20/UT	200	0.5	1.0	1.90	1.47
Philips CM200FEG/UT[e]	200	0.5	1.0	1.90	1.03
Philips CM30/T	300	2.0	2.0	2.25	1.58
Philips CM300FEG/T[e]	300	2.0	2.0	2.25	1.23
Philips CM30/ST	300	1.2	1.5	1.98	1.36
Philips CM300FEG/ST[e]	300	1.2	1.5	1.98	1.07
Philips CM300FEG/UT[e]	300	0.7	1.5	1.73	1.07

[a] From ref.2. [b] Theoretical resolution for intercomparison only -- experimental resolution may be worse.
[c] CTF crossover at a defocus of 1.225 Scherzers. [d] $1/|u|_\Delta$, where $/|u|_\Delta$ is spatial frequency at which the $E_\Delta(\mathbf{u})$ envelope drops to exp(-2), assuming 1ppm ripple in lens and high-voltage power supplies, zero vibration, and energy spread of 1eV (LaB$_6$) or 0.3eV (FEG). [e] Equipped with field emission gun (FEG).

APPLICATION OF THROUGH-FOCAL SERIES RESTORATION FOR RESOLUTION ENHANCEMENT

M. R. McCartney

Center for Solid State Science, Arizona State University, Tempe, AZ 85287-1704

Apart from instrumental instabilities, the point resolution for phase contrast imaging is determined by the spherical aberration of the objective lens and the electron wavelength. The 1Å spatial resolution barrier in transmission electron microscopy has recently been broken or very closely approached by several research groups.[1,2] These efforts have concentrated in two directions; reducing the wavelength and post-correction of the aberration function. The advantage of increasing the accelerating voltage is that images of sufficiently thin specimens are immediately interpretable, with the disadvantages being increased radiation damage and enormous cost. The alternate method involves application of a complex phase plate which is the inverse of the objective lens transfer function. One must then have access to the phase and amplitude of the image wave, rather than the intensity which is normally obtained. Two promising methods for this complex wave retrieval are electron holography,[3] and through-focal-series restoration.[4] This presentation will address practical issues in the use of the "paraboloid" through-focal-series method.[5] This method utilizes the functional form of the Fresnel propagator which relates wave functions modified by different defoci. In reciprocal space, the wave function can be written as:

$$\Psi(\mathbf{r},z) = C + \sum_{g \neq 0} \varphi_g \exp\left[-i\pi\lambda z g^2\right] \exp(2\pi i \mathbf{g} \bullet \mathbf{r})$$

where the constant term is separated out and z is the defocus. The Fourier transform of the image intensity $|\Psi(r,z)|^2$ with respect to both space and defocus variables yields:

$$I(\mathbf{g},\xi) = \left|C^2\right|\delta(\mathbf{g}) + C^*\varphi(\mathbf{g})\,\delta\!\left(\xi - \frac{\lambda g^2}{2}\right) + C\,\varphi^*(\mathbf{g})\,\delta\!\left(\xi + \frac{\lambda g^2}{2}\right)$$

$$+ \int_{\substack{g' \neq 0 \\ g+g' \neq 0}} d\mathbf{g}'\; \varphi^*(\mathbf{g}')\,\varphi(\mathbf{g}+\mathbf{g}')\,\delta\!\left\{\xi - \frac{\lambda}{2}\left[(\mathbf{g}+\mathbf{g}')^2 - \mathbf{g}'^2\right]\right\}$$

where δ is the Dirac functional and ξ is the conjugate of z. The first term contributes only at the origin, the second and third represent linear contributions lying on paraboloids in reciprocal space. The last term gives non-linear contributions distributed throughout reciprocal space. The most successful restorations involve use of the paraboloid method to provide a highly linear first guess for a more computationally demanding non-linear solution such as the "maximum-likelihood method".[6]

Fig. 1 shows four members of a focal series of a Si-doped AlN polytypoid in a $[11\overline{2}0]$ projection: the image contrast is severely affected by various transfer functions. Figs. 2a, b show the amplitude and phase, respectively, of the image wave reconstructed by the paraboloid method using parameters appropriate to the JEOL 4000EX and assuming a focal step size of 5.17nm. This complex image is the back projection of the linear parts of the 30 member series onto the initial defocus plane, with an improvement in statistics on the order of $N^{1/2}$, where N is the number of images. Figs. 2c,d show the amplitude and phase, respectively, of the reconstructed exit-surface wave obtained from Figs. 2a,b assuming an initial defocus value of -194nm. Fig. 3 shows the behavior of the intensity on the parabola as a function of defocus stepsize as the calculation was iterated using the estimated images to improve the image alignment.[7] Fig 4 shows the behavior of the contrast and mean value of the amplitude of the reconstructed exit wave for various starting defocus values. Evaluation of error in determination of relevant experimental parameters and image alignment is of utmost importance in any quantitative reconstruction scheme. A simulated focal series of an oxygen-containing, planar inversion domain boundary in AlN will be used to estimate the precision of the technique to reconstruct non-periodic features. The effects of noise, misalignment, thickness and non-constant stepsize will be discussed.

Proc. Microscopy and Microanalysis 1995, edited by G.W. Bailey, M.H. Ellisman, R.A. Hennigar, and N.J. Zaluzec
Copyright © 1995 MSA. Published by Jones and Begell Publishing, 79 Madison Ave., New York, NY 10016

References
1. F. Phillipp, R. Höschen, M. Osaki, G. Möbus and M. Rühle, *Ultramicroscopy* 56(1994)1.
2. W. M. Coene, et al. *Proc. 13th ICEM*, Vol. 1.(1994)461; Y. Ishida and H. Ichinose, ibid, 161; W.-D. Rau and H. Lichte, ibid, 291.
3. H. Lichte, *Adv. Opt. Electron Microsc.* 12(1991)25.
4. D. Van Dyck, M. Op de Beeck, *Proc. of the XIIth ICEM* (1990)26.
5. D. Van Dyck, M. Op de Beeck and W. Coene *Optik* 93(1993)103.
6. W. Coene, et al., *Phys. Rev. Lett.* 69(1992)3743.
7. W.J. de Ruijter, Ph.D. Thesis, Delft University of Technology,The Netherlands, 1992.

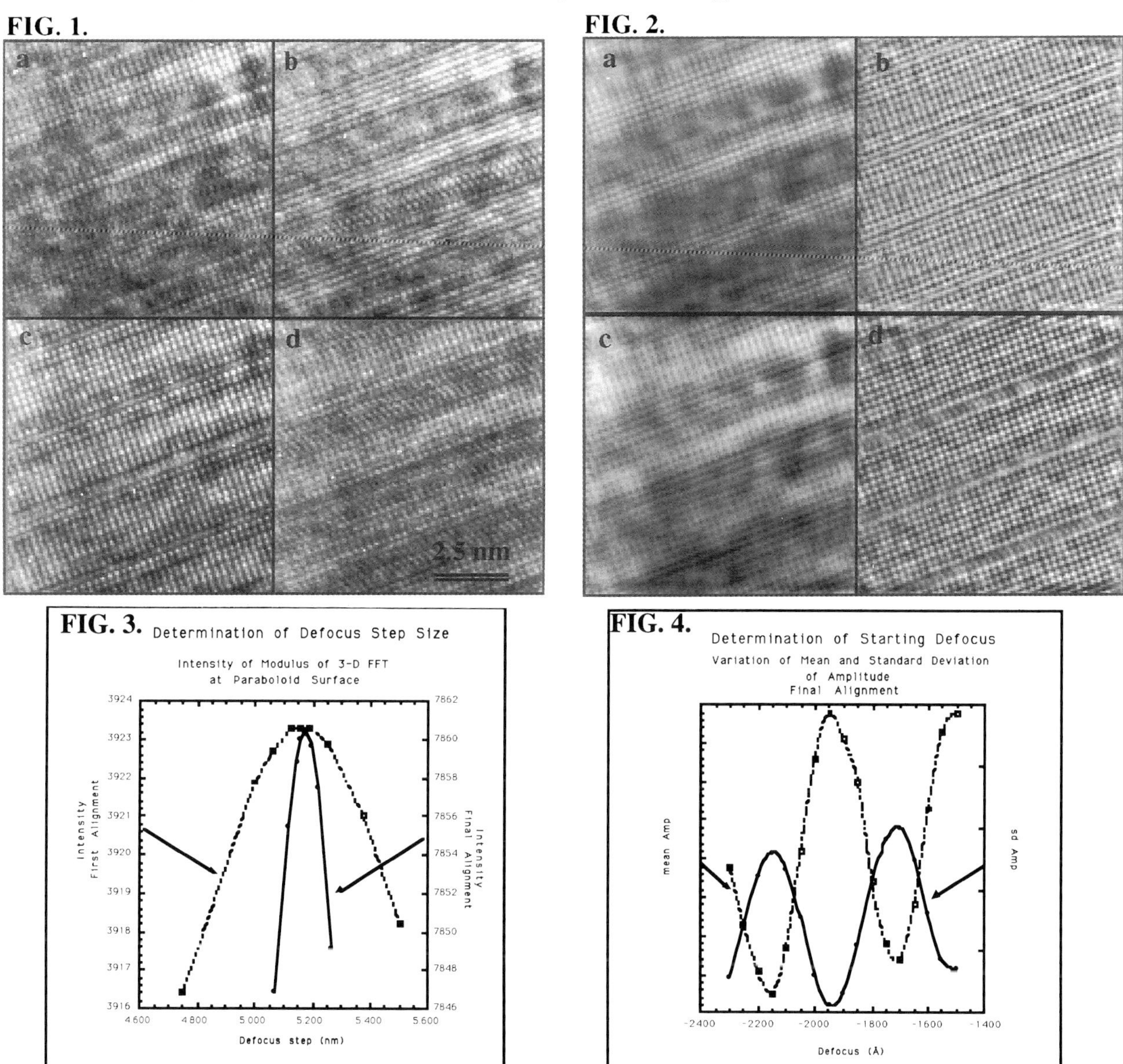

FIG. 1 Members of focal series of AlN polytypoid for defoci (in nm); a)-173; b)-153; c)-132; d)-111.
FIG. 2. a) amplitude and b) phase of reconstructed image wave; c) amplitude and d) phase of exit wave.
FIG. 3. Intensity on parabola increases and becomes more sharply peaked as alignment is improved.
FIG. 4. Maximum mean value and minimum contrast of the corrected exit wave amplitude at initial defocus value of -194nm.

High Spatial Resolution X-ray Microanalysis of Interfaces

J. R. Michael, Materials and Process Sciences Center, Sandia National Laboratories, Albuquerque, NM 87185-0342

X-ray microanalysis in the analytical electron microscope (AEM) refers to a technique by which chemical composition can be determined on spatial scales of less than 10 nm. There are many factors that influence the quality of x-ray microanalysis. The minimum probe size with sufficient current for microanalysis that can be generated determines the ultimate spatial resolution of each individual microanalysis.[1] However, it is also necessary to collect efficiently the x-rays generated. Modern high brightness field emission gun equipped AEMs can now generate probes that are less than 1 nm in diameter with high probe currents.[2] Improving the x-ray collection solid angle of the solid state energy dispersive spectrometer (EDS) results in more efficient collection of x-ray generated by the interaction of the electron probe with the specimen, thus reducing the minimum detectability limit. The combination of decreased interaction volume due to smaller electron probe size and the increased collection efficiency due to larger solid angle of x-ray collection should enhance our ability to study interfacial segregation. This paper will discuss the detection and measurement of interfacial segregation in the AEM and the expected improvements brought about by the use of medium voltage (200-300 kV) instruments with field emission electron sources.

AEM has been used to study interfacial segregation in many different materials. The addition of Cu to Al for integrated circuit interconnects is known to greatly increase the lifetime of the conductor by reducing damage caused by electromigration. The mechanism of how Cu improves the electromigration resistance of Al interconnects is not known. Figure 1a is a micrograph of an Al-4wt%Cu thin film that has been aged to promote segregation of Cu to the grain boundaries. Precipitation of $CuAl_2$ has occurred on the grain boundaries. Compositional profiles across grain boundaries were obtained in a Vacuum Generators HB-501 dedicated scanning transmission electron microscope operated at 100 kV. The electron probe size was determined to be 1 nm (FWHM) in diameter. Figure 1b shows a typical compositional profile obtained across one of the grain boundaries shown in Figure 1a. The compositional profile demonstrates that Cu has segregated to the Al grain boundaries.

The development of medium voltage AEMs equipped with field emission electron sources should result in improvement in the detection of interfacial segregation.[3] The medium voltage microscopes are not expected to produce electron probes significantly smaller than 1.0 nm with sufficient current for microanalysis. The interaction of a 1 nm diameter electron probe with a 1.0 nm wide region of $CuAl_2$ in a 100 nm thick specimen was modeled using Monte Carlo electron trajectory simulations.[4] Figure 2 shows the results of these calculations as raw intensity ratios of Cu to Al for accelerating voltages of 100 and 300 kV. The profiles calculated for 100 and 300 kV are nearly identical in width, but have very different peak concentrations. These calculations demonstrate the somewhat surprising result that given identical electron probe diameters, higher accelerating voltages do not significantly increase the spatial resolution. This result indicates that in a 100 nm thick specimen of Al, the electron probe size is the dominant factor in determining the spatial resolution. Thus, microanalysis conducted with similar electron probe size will result in similar profile widths. The use of higher accelerating voltages does result in a reduced amount of beam broadening as evidenced by the higher measured Cu levels when the electron probe is at the grain boundary position. The increased peak concentration is important as it indicates that higher accelerating voltages should allow the detection of lower levels of segregation than is possible at lower accelerating voltages.

These results show that the use of medium voltage AEM to study interfacial segregation may not provide any further information about the width of the segregated region due to the probe size dominating the spatial resolution. The spatial resolution for x-ray microanalysis is determined by the electron probe size in moderate thickness specimens. However, the decreased electron scattering at higher accelerating voltages should improve the detectability limits for interfacial segregation when compared to lower accelerating voltages.

Proc. Microscopy and Microanalysis 1995, edited by G.W. Bailey, M.H. Ellisman, R.A. Hennigar, and N.J. Zaluzec
Copyright © 1995 MSA. Published by Jones and Begell Publishing, 79 Madison Ave., New York, NY 10016

1. D. B. Williams et al., *Ultramicros* 47 (1992) 121.
2. J. R. Michael et al., *J. Microsc.* 147 (1987) 289.
3. A. J. Garratt-Reed, *Proc. Ann MSA Meeting* 51 (1993) 600.
4. S. J. Plimpton, J. R. Michael and A. D. Romig,jr *J. Supercomputing* 6 (1992) 139.
5. This work was supported by the U. S. department of Energy under contract #DE-AC04-94AL85000.

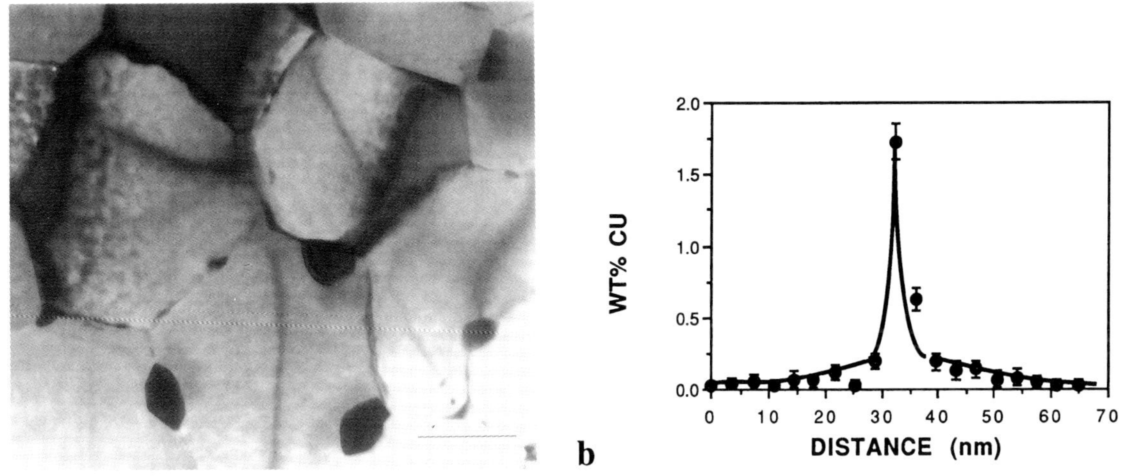

Figure 1. a)Plan view of an Al-2 wt% Cu metallization treated at 200°C for 10 days (scale bar is 1 μm). b) Cu concentration profile obtained across a grain boundary in Figure 1a.

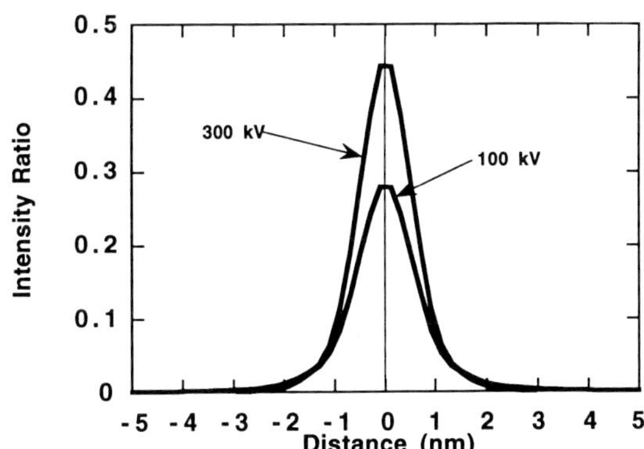

Figure 2. Calculated raw intensity ratios (I_{Cu}/I_{Al}) for a 100 and 300 kV 1.0 nm electron probe scanned across a 1.0nm layer of $CuAl_2$ in a 100 nm thick specimen.

INTERFACE ANALYSIS IN THE VG MICROSCOPES' HB603 300 keV STEM.

John Bruley, David W. Ackland, Jianxin Fang and David B. Williams

Department of Materials Science and Engineering, Lehigh University, Bethlehem, PA 18015-3195.

The optimum operating conditions for microanalysis (either EDX or EELS) would be those which maximize the signal-to-noise ratio, SNR, and signal-to-background ratio, SBR, and at the same time minimize the probe size so as to locate the signal to as small a volume as possible. In practice however these three quantities can not be optimized simultaneously. For many materials problems, the analyst seeks first to determine if a solute has segregated to an interface. We therefore consider the conditions required to operate the analytical microscope to maximize the detection sensitivity of interfacial segregants.

The choice of lens operating conditions for carrying out interface analysis on a STEM is not necessarily intuitive. Apart from the geometrical optics, instrumental artifacts may adversely influence the observed SNR and SBR. We have carried out a series of EDX experiments to compare the working performance of interface analysis with that expected using a simple planar interface segregation model and geometrical optics. The microscope was the VG Microscopes' HB603 STEM operating at 300 keV. The effect of specimen damage at 300 keV with a finely focused probe was minimized by maintaining a low beam current (0.4 nA). A 50 μm virtual objective aperture (VOA) was inserted between the gun lens and the first condenser lens (C1). Throughout these experiments a focus was maintained on the selected area plane by adjustment of the 2nd condenser (C2). EDX spectra were acquired for 20 seconds. Using a VOA the incident beam current on the specimen remains constant, whilst the beam illumination angle onto the specimen is continuously varied. The current was measured using a Faraday cup on the selected area plane and the size of the illumination disk was measured on the microdiffraction plane. Figure 1 shows how incident angle changes as a function of C1 current. The specimen was Al_2O_3 containing small amounts of Y_2O_3. Y is known to segregate to grain boundaries. A flat segment of thin boundary oriented parallel to the beam direction was chosen. To minimize beam spreading effects to less than 0.5 nm, a sample thickness of less than about 20 nm is needed. During EDX acquisition the beam was scanned over an area of about 1 nm^2 which included the interface. The specimen was moved by about 3 nm between measurements so that fresh segments of boundary were observed.

By normalizing the total Al signal by the total integrated x-ray count we observe a constant ratio independent of C1 setting. This is to be expected because both signals are derived from the bulk and therefore not sensitive to the probe size. Normalization corrects for thickness and beam current fluctuations. Any slight variations may reflect some beam damage, or slight change to in the background signal. The SNR and SBR for the Y exhibit two maxima, as seen in Figure 3. The first is located with C1 off and the second with C1 excited by 1.5 A. Assuming the simple model composed of a low coverage of interfacial segregant an estimate of how SNR and SBR varies with C1 settings is shown in Figure 4. Comparison to figure 3 reveals a discrepancy in the relative height of the 2nd maximum with respect to the first maximum. The for most VOA apertures, the smallest probe is, in principle, only achievable with C1 and C2 both excited. The 50 μm aperture results in an illumination semi-angle at the specimen of less than 4 mrad whereas the optimum convergence semi-angle for the objective lens with C_S=4.5 mm is 6.5 mrad. This convergence is achieved with a C1 current set at around 1.6 A. Since the Al signal was not affected by C1 there appears to be little indication that extraneous x-rays were generated at higher C1 settings. A more likely explanation relates to the variation of probe-size with C1 excitation. This might be due to slight misalignment of C1 and C2. Stray magnetic fields or microscope instabilities would also contribute to the overall probe size. Probably the limiting factor was the choice of the scanning area mode during analysis. A probe diameter as low as 0.3 nm should be achievable under optimum conditions. Further experiments are necessary to distinguish these effects. The results show that there are no spurious contributions of x-rays at different excitations of C1 and C2.[1]

Proc. Microscopy and Microanalysis 1995, edited by G.W. Bailey, M.H. Ellisman, R.A. Hennigar, and N.J. Zaluzec
Copyright © 1995 MSA. Published by Jones and Begell Publishing, 79 Madison Ave., New York, NY 10016

Reference

1. Research supported by NSF DMR 930625

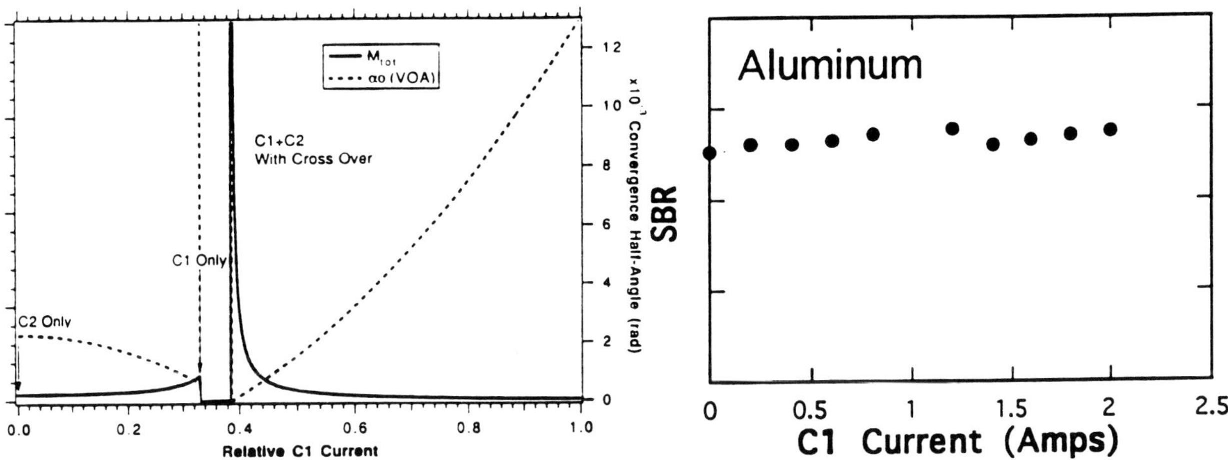

Fig. 1: Variation of beam convergence angle (dashed line) and geometrical magnification (solid) as with C1 excitation

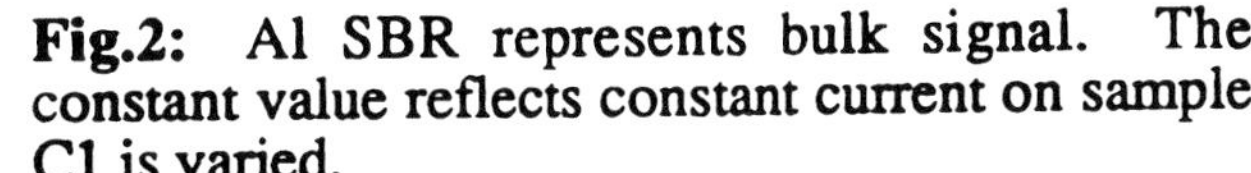

Fig.2: Al SBR represents bulk signal. The constant value reflects constant current on sample C1 is varied.

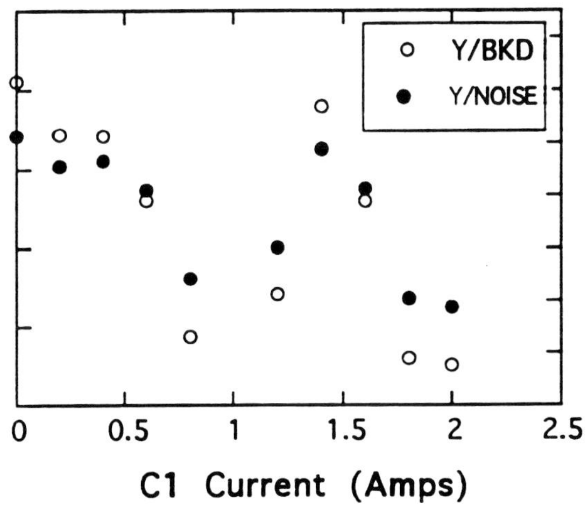

Fig. 3: Variation of SNR and SBR for interface signal with C1 lens current

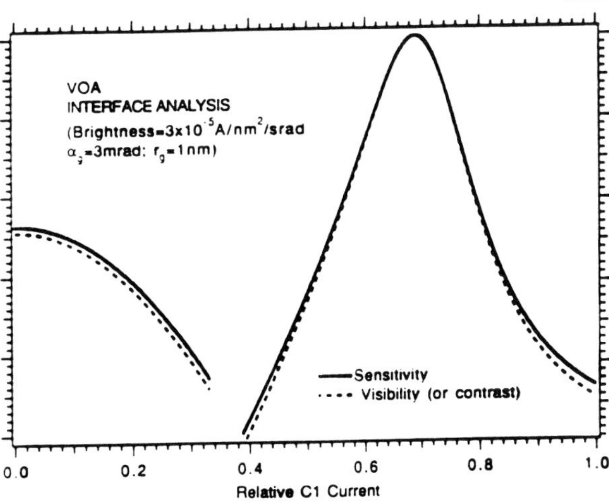

Fig. 4: The relative variation of SNR and SBR with C1 current for an interfacial signal.

QUANTITATIVE HIGH SPATIAL RESOLUTION MICROCHEMICAL ANALYSIS OF Fe-Co-Ni TRANSFORMATION-TOUGHENED STEELS

H.E. Lippard, V.P. Dravid, G.B. Olson

Department of Materials Science & Engineering, Northwestern University, Evanston, IL 60208, USA

Superior strength-toughness combinations can be obtained in Aermet 100 by an austenite→martensite phase transformation of small intralath austenite precipitates. The enhanced toughness in these alloys has been attributed to the presence of "metastable" austenite precipitates, the stability of which is intimately related to the size and composition. The austenite precipitates require high stability to resist transformation under all stress conditions except when in the triaxial stress field of a crack tip. The two primary factors determining austenite stability are size and composition. Thus the microstructure-property correlations and further alloy development require quantitative microchemical analysis of the 50-150Å austenite precipitates.

Of the two obvious analytical electron microscopy techniques, energy dispersive spectrometry (EDS) is superior for quantitative results. Several artifacts must be considered when performing quantitative EDS, for example; the continuum background, absorption/fluorescence effects, k-factor measurement and peak overlap. The continuum background fitting is elementary due to the high peak/background ratio and extensive research in this area. The absorption/fluorescence effects were calculated from theoretical models[1,2] and found to be <3% at 100 nm for the substructure composition range. The k-factors were measured experimentally from appropriately homogenized standard samples. The major difficulty for valid quantitative analysis is the peak overlap of (FeK$_\beta$, CoK$_\alpha$) and (CoK$_\beta$, NiK$_\alpha$). Figure 1 displays a generated spectrum for the alloy composition showing the relative intensities of the overlapping peaks. Figure 2 is an experimental spectrum from a lath martensite region where the four overlapped peaks have merged together into two broad peaks. Initially, extraction of the integrated peak intensities was done using theoretical K$_\alpha$/K$_\beta$ ratios[3] and the unconvoluted FeK$_\alpha$ and NiK$_\beta$ peaks; however, systematic errors were noted when comparing the NiK$_\alpha$ intensities calculated from each unconvoluted peak. For austenite spectra, the estimation of NiK$_\alpha$ from NiK$_\beta$ was consistently higher than the estimation from FeK$_\alpha$ through the CoK$_\alpha$ and CoK$_\beta$ peaks. The opposite case was found for matrix martensite spectra.

A methodology has been developed to deconvolute the overlapped FeK$_\beta$, CoK$_\alpha$, CoK$_\beta$, and NiK$_\alpha$ peaks in energy dispersive spectra based on experimental K$_\alpha$/K$_\beta$ ratios obtained from single element standards and the unconvoluted FeK$_\alpha$ and NiK$_\beta$ peaks. The austenite precipitates were found in a range of nickel enriched compositions from 15-30wt% Ni while the lath martensite matrix was ~10wt% Ni. Figure 3 is a dark field image displaying the three morphological forms of austenite found - large interlath precipitates, thin lath boundary films, and small intralath precipitates. These results and a complete discussion of the methodology will be presented.[4]

References

1. J.I. Goldstein et al., in O. Johari, Ed. *SEM/77*, Chicago, IITRI, 1 (1977) 315.
2. C. Nockolds et al., in T. Mulvey, Ed.,*Electron Microscopy and Analysis 1979* , Bristol and London: Inst. of Physics (1980) 417.
3. T.P. Schreiber and A.M. Wims, *X-Ray Spectrometry* 11 (1982) 42.
4. This work was supported by ARO Grant # DAAH04-93-G-0471/P01 and an NSF Graduate Fellowship (HL).

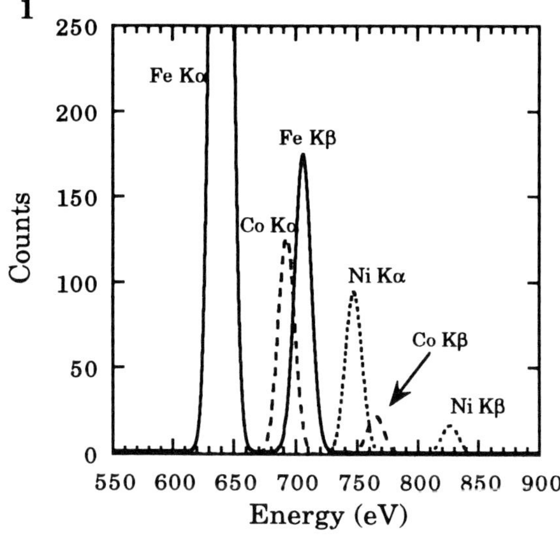

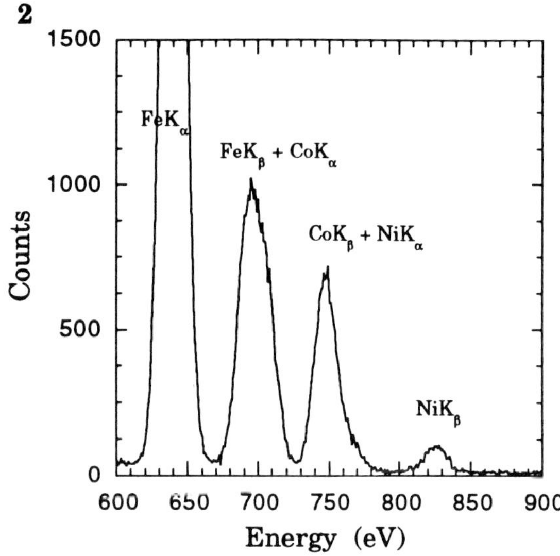

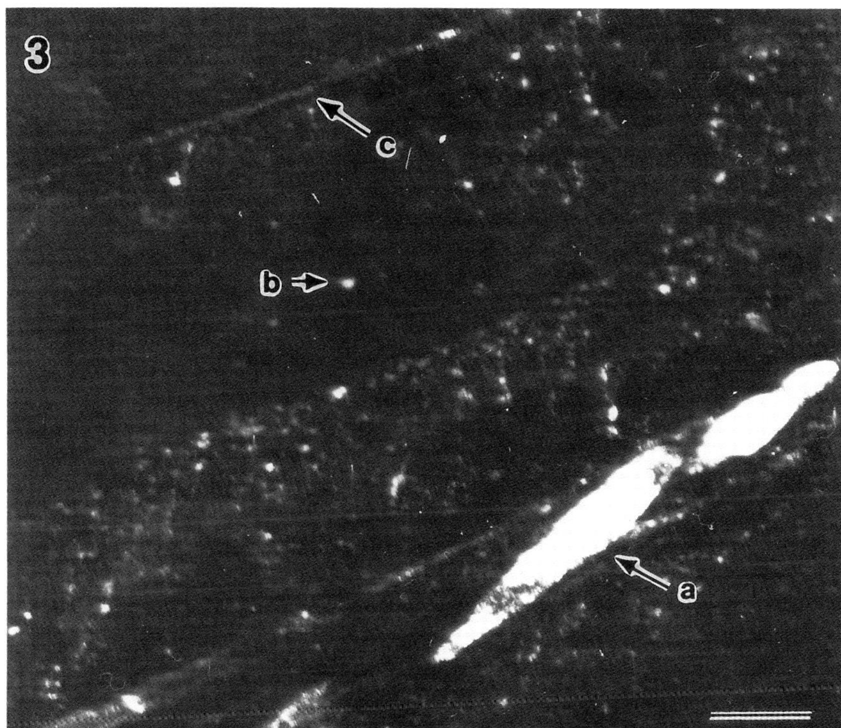

Table 1. - Alloy composition

	Composition (wt%)
Fe	balance
Co	13.42
Ni	11.03
Cr	3.00
Mo	1.18
C	0.238

FIG. 1. - Simulated energy dispersive spectrum for Fe-14Co-11Ni (wt%) composition alloy demonstrating overlap potential for the FeK_β, CoK_α, CoK_β, and NiK_α peaks.

FIG. 2. - Experimental energy dispersive spectrum from a matrix martensite lath region confirming peak overlap.

FIG. 3. - TEM dark field of austenite precipitates showing: (a) large interlath precipitates, (b) small intralath precipitates, and (c) thin lath boundary films.

ATOM PROBES AS ANALYTICAL MICROSCOPES

T. F. Kelly[*,**,***], P. P. Camus[**,***], D. J. Larson[*,***], S. S. Bajikar[*,***], and L. M. Holzman[*,***]

[*]Materials Science Program, [**]Department of Materials Science and Engineering, [***]Applied Superconductivity Center, University of Wisconsin, Madison, WI 53706

Analytical microscopes are used to provide compositional and image information at spatial scales from the atomic (nm) to the macroscopic (mm). There are limitations in these techniques, however, and three in particular are noteworthy. 1) Compositional information at the atomic scale is not readily available. 2) The precision of the data is not high (limited to about a few part per thousand). 3) It is difficult or impossible to determine information about the composition in three dimensions.

Atom probe field ion microscopy (APFIM) has historically been able to complement analytical electron microscopes (AEM) at the atomic scale but has not competed with them otherwise. For example, atomic-scale compositional information is inherent to APFIM but it has been difficult to extend this to large dimensions. Furthermore, APFIMs achieve single atom sensitivity which is difficult for AEMs in all but limited situations. In this paper, we consider how atom probes are evolving with capabilities which allow them to not only overcome some of the shortcomings of AEM, but also to compete directly with AEM against its strengths.

Because APFIMs provide three-dimensional images at the atomic scale, they have an inherent advantage over AEMs for compositional imaging since, in general, compositional features are three-dimensional. This is especially true with the advent of fully three-dimensional atom probes (3DAP)[1,2]. Even for problems which are usually considered to be two-dimensional like segregation to planar interfaces, recent evidence suggests that three-dimensional composition variations are present[3].

AEMs offer clear advantages for structure analysis because of their ability to produce atomic structure information either directly in the image or in diffraction patterns. APFIMs provide structural information in a field ion image, however, these images are not always readily interpreted and they are a glimpse of the surface only. There is hope that structural information may become available in 3DAP images. Thus far, aberrations in 3DAP images makes is difficult or impossible to see the underlying atomic structure directly. Some progress has been made with pattern recognition techniques for finding the spatial frequencies in 3DAP images[4]. Correction of the trajectory aberrations which prevent the direct interpretation of structure, is another possibility for determining the atomic positions. If either approach can be made to work, then 3D structural information will be available on an atomic scale in real space from 3DAP images. This possibility would be even more powerful than AEM currently offers.

APFIMs have been limited by sample geometry in two principal ways: a) it is difficult to locate an analyzed volume at a particular position in a sample especially if the feature of interest is laying on a surface like a thin film; b) it is nearly impossible to survey large areas of a sample. A sample geometry which was recently proposed by Nishikawa and Kimoto[5], overcomes most of these problems. A planar sample would be prepared with a large number of tips normal to the surface and a local electrode would be positioned above any tip to apply the high field (Fig. 1a and b). Samples may be made by a variety of methods like ion milling with masking particles placed randomly[6] (Fig. 2) or positioned. Thousands of tips are created in this process and the operator could obviously choose only the best shapes and locations for analysis (Fig. 3). This geometry is expected to make it possible to use low extraction voltages ($<$1 kV) to achieve very high repetition rates (10^6 atoms per second) and high mass resolution ($<$ 1 part in 1000) in a 3DAP[7] which we call a local electrode atom probe (LEAP). Because of the high data collection rates, images may be recorded quickly and surveying the sample over many millimeters with this approach is a realistic possibility.

What about precision and detection limits for composition analysis? Let's consider an example problem from the semiconductor industry. Shallow dopant implants ($\sim$30 nm deep at less than 10^{15} cm^{-3} concentrations) are being used increasingly in microelectronic device structures. In order to even detect such a low concentration (about 40

Proc. Microscopy and Microanalysis 1995, edited by G.W. Bailey, M.H. Ellisman, R.A. Hennigar, and N.J. Zaluzec
Copyright © 1995 MSA. Published by Jones and Begell Publishing, 79 Madison Ave., New York, NY 10016

part per billion) we will need to collect on the order of 1 billion atoms from this shallow region. Since 30 nm depth is about 150 atomic layers, this means that about 7×10^6 atoms per layer must be recorded. This is an image of about 2500 atoms diameter which would require a tip of about 0.5 micron radius of curvature. This should be possible with the local electrode geometry since the electric field on the tip can be maintained with a modest total applied voltage (~3 kV). There is currently no instrument which can approach this level of sensitivity from a submicron analysis region. Furthermore, 100 part per million precision can be achieved for major constituents.

Thus, in the near future, APFIMs may be able to provide part per billion sensitivity and high precision from specific locations over macroscopic distances, with high speed, convenience, atomic-scale image resolution, and structure determination. This combination of capabilities will make a very powerful analytical microscope.[8]

1.   G. D. W. Smith et al., Proc. 52nd Ann. MSA Mtg., (San Francisco Press, Inc., 1994) p. 828.
2.   B. Deconihot et al., Appl. Surf. Sci., 76/77 (1994) 145.
3.   D. N. Seidman, Mat. Sci. and Eng., A137 (1991) 57.
4.   P. P. Camus, et al., Appl. Surf. Sci., 87/88 (1995) 305 .
5.   O. Nishikawa and M. Kimoto, Appl. Surf. Sci., 76/77 (1994) 424.
6.   D. J. Larson, et al., Appl. Surf. Sci., 87/88 (1995) 446 .
7.   T. F. Kelly, et al., US Patent No. 08/272,204, allowed February 1995.
8.   This work sponsored by the National Science Foundation, grant #DMR-8911332 (Drs. L. Haworth and U. Strom) and Electric Power Research Institute under agreement #RP8009-5 (Drs. T. Schnieder and P. Grant).

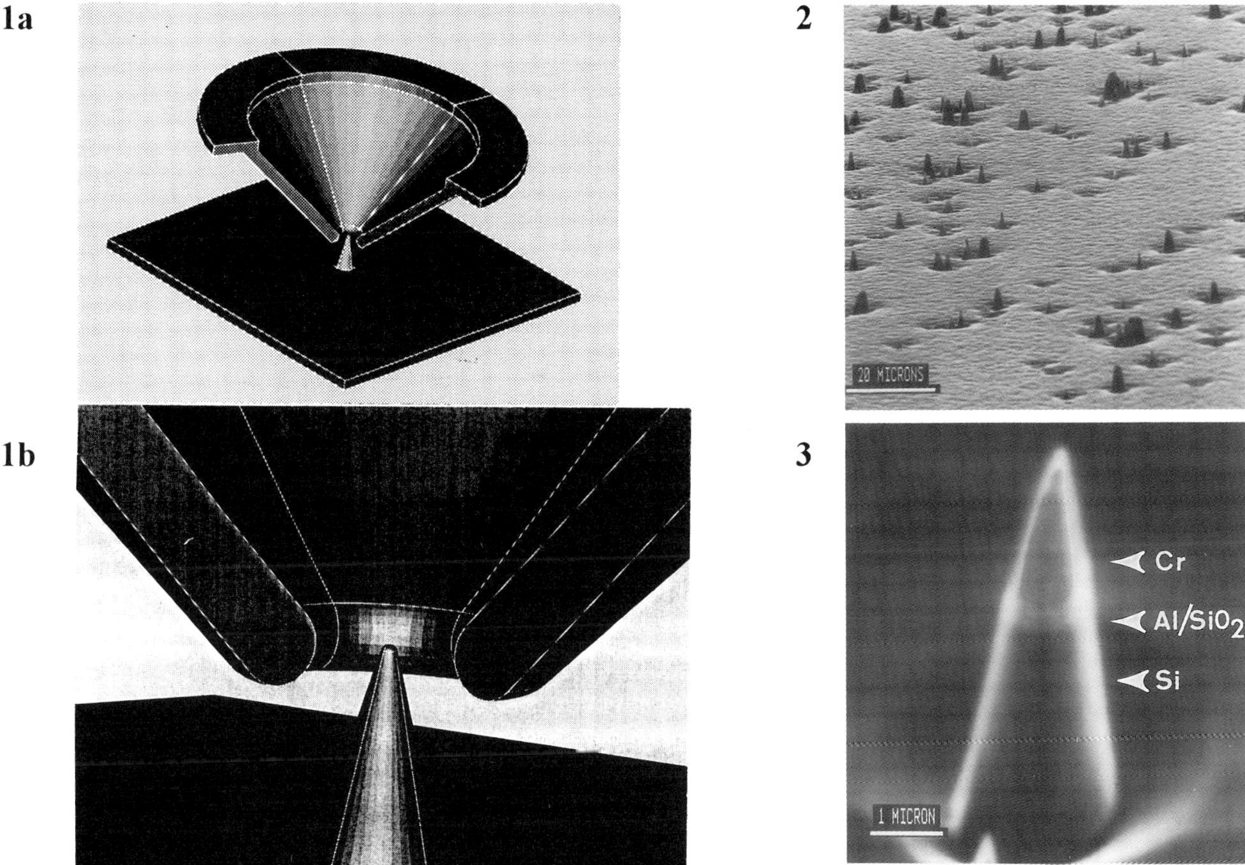

Fig. 1   Schematic diagrams of local extraction electrode for LEAP configuration of APFIM.

Fig. 2   SEM image of many tips from metal-oxide-semiconductor (MOS) device.

Fig. 3   SEM image of single tip from metal-oxide-semiconductor (MOS) device. Cr layer was placed on the surface to make it easier to find the metal-oxide interface.

COMBINING ATOMIC IMAGING, GROWTH, AND SURFACE SPECTROSCOPY

L. D. Marks,[*] C. Collazo,[*] N Doraiswamy,[*] D. Grozea,[*] G. Jayaram,[*] E. Landree,[*] R. Plass,[*] and P. C. Stair[**]

[*]Department of Materials Science and Engineering, Northwestern University, Evanston, IL 60208
[**]Department of Chemistry, Northwestern University, Evanston, IL 60208

UHV transmission electron microscopy brings several advantages to the study of surfaces including direct imaging of surface structures and surface defects with contrast based on atomic scattering and the ability to obtain information from subsurface layers. Over the last several months our research group has implemented several image and diffraction analysis techniques to enhance these advantages and help determine four noble metal induced silicon surface structures. Perhaps the most significant of these techniques are Wiener and Parametric Wiener image filtering[1] to improve signal to noise levels in off axis HREM images of surface structures, two examples of which are shown in Figs. 1 and 2. Figure 1 is a Parametric Wiener filtered, near Schertzer, off-zone HREM image of Si(111)-(5x2) Au which clearly shows the rows of gold and also contains some information on the silicon structure between the gold rows. With data from images and χ^2 diffraction analysis we have found this structure is basically a Si surface dislocation decorated by two rows of gold atoms. Figure 2 is a Wiener filtered off-zone HREM image of roughly 0.3 ML of Ag on Si(100) and illustrates some of the growth studies possible with UHV-TEM. The dark features of this image are mostly single crystal Ag islands plus some silver multiply twinned particles (MTP's). The size and type distributions of these MTP's agree with studies of gold MTP's on amorphous substrates[2] except for a surprising absence of icosohedral MTP's. Another approach to growth studies is to track, in real time, the decay of surface diffraction spot intensities with metal deposition to obtain the surface structure decomposition mechanism. The decay of the strongest diffraction spots of the Si(111)-(5x2) Au structure upon room temperature gold deposition followed zero order kinetics after the first 0.25 ML and was inversely proportional to the gold deposition rate, i. e., faster gold deposition caused a slower linear decline of spot intensity with respect to deposited gold coverage. While more work is needed, this result would suggest a zipper like unravelling of the structure. The 5x2 Au zero order kinetics result contrasts the first order decay kinetics we have observed for gold on Si(111)-($\sqrt{3}$x$\sqrt{3}$) Au which suggests a low gold surface mobility / attack-from- above type disordering mechanism.

In addition to these data analysis enhancements mentioned above a set UHV chambers (shown in Fig. 3) has recently been attached to our Hitachi UHV-H9000 TEM. These chambers (named S.P.E.A.R for Surface Preparation Evaluation Analysis Research) allow for much better control of surface preparation than we had in the past, detailed surface chemical analysis using XPS and AES, and the preparation of III-V semiconductor surfaces and thin films. An example of the system's surface chemical sensitivity is shown in Figs 4a and b which are before and after XPS spectra of a GaAs(100) surface after annealing at 620°C in with a 6.5×10^6 Torr As back pressure, the drop in the C and O peaks is clear. Further details of these and other studies will be presented.

References

[1] H. C. Andrews and B. R. Hunt, *Digital Image Restoration*, Englewood NJ: Prentice-Hall (1977)150.
[2] N. Doraiswamy and L. D. Marks, *Phil. Mag.* 71(1995)291.

Proc. Microscopy and Microanalysis 1995, edited by G.W. Bailey, M.H. Ellisman, R.A. Hennigar, and N.J. Zaluzec
Copyright © 1995 MSA. Published by Jones and Begell Publishing, 79 Madison Ave., New York, NY 10016

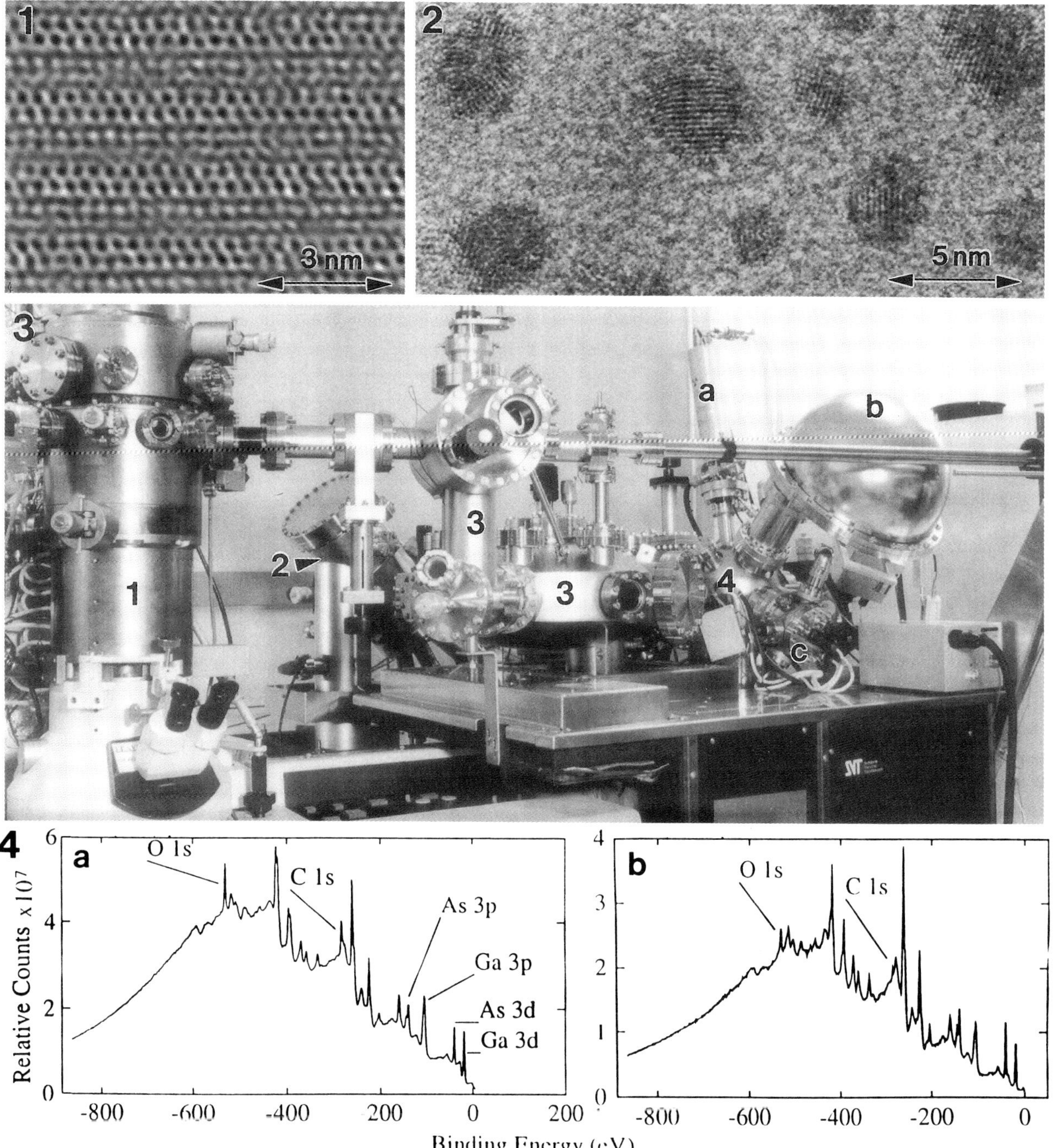

FIG. 1.-Noise filtered off-zone HREM image of Si(111)-(5x2) Au showing rows of gold atoms.
FIG. 2.-Noise filtered off-zone HREM image of silver multiply twinned particles on Si(100).
FIG. 3.-Side view photograph of the UHV-TEM facility at Northwestern Univ. Shown are (1) the UHV-H9000 TEM, (2) MBE deposition chamber, (3) sample transfer chamber, (4) analytical chamber with the (a) X-ray source for XPS, (b) hemispherical electron energy analyzer, and (c) field emission electron gun for AES and SEM. Not shown here are the duoplasmatron ion gun and the load lock.
FIG. 4.-(a) and (b) Before and after XPS spectra of GaAs(100) upon annealing at 620°C for 4 min. in a As atmosphere of 6.5×10^6 Torr.

QUANTITATIVE COMPOSITIONAL MAPPING OF $In_xGa_{1-x}As$/GaAs QUANTUM WELLS BY ANNULAR DARK FIELD IMAGING

S. Hillyard,* Y.-P., Chen,** W.J. Schaff,** L.F. Eastman** and J. Silcox*

* School of Applied and Engineering Physics, Cornell University, Ithaca, NY 14853
** School of Electrical Engineering, Cornell University, Ithaca, NY 14853

Annular dark field imaging in the scanning transmission electron microscope (STEM) exhibits both high resolution and Z-contrast.[1] It is intrinsically quantitative since image data can be recorded directly from linear detectors into digital memory. Annular dark field imaging has been used, along with energy filtered imaging to correct for sample thickness variation, to map out the In concentration in $In_xGa_{1-x}As$ quantum wells with near atomic resolution and sensitivity. This approach is similar to "chemical lattice imaging",[2] which maps out composition variation using a conventional transmission electron microscope image and a vector pattern recognition algorithm.

The quantum wells were grown by molecular-beam epitaxy (MBE). Figure 1 shows a typical high resolution annular dark field image of a 50 Å wide nominal $In_{0.3}Ga_{0.7}As$/GaAs quantum well. The linescan in figure 2 gives the actual numbers making up the image. Barring contaminants and lattice imperfections, the change in intensity with position is caused by two things: variation of In concentration and thickness. Changes of In concentration have been found by both multislice simulations and experiments using wells of differing nominal In value, to be linear in effect. Therefore, provided compensation for thickness variations is possible, In concentration can be directly determined.

In an annular dark field STEM image each pixel consists of the number of electrons scattered to high angle with each position of a small (in this case about 2.9 Å at 100 keV energy) probe. This allows the directly transmitted portion of the beam, consisting of the electrons scattered at smaller angles, to be simultaneously collected in perfect registration. Both a zero loss and the first plasmon loss images were recorded. The ratio of these should be an image with intensity proportional to the thickness of the specimen since the two plasmon energies are the same.[3] With this thickness map, the intensity change due to sample thickness variation in the annular dark field image can be eliminated giving the In concentration.

Figure 3 shows the histogram of pixels in figure 1. Three distinct peaks are seen, two consisting of different thickness GaAs areas, and one of the $In_xGa_{1-x}As$. After breaking down the image into atomic columns and correcting for thickness, a histogram as shown in figure 4 is obtained. As can be seen, the GaAs has converged to a single peak. In addition, the average thickness of the sample was approximately 450 Å, meaning an $In_{0.3}Ga_{0.7}As$ column should have approximately 34 In atoms. Therefore, the bin size of the histogram as chosen corresponds an intensity change of a single In atom, and the width of the GaAs peak implies a sensitivity of ± 1 In atom for each column.[4]

[1] S.J. Pennycook, Ultramicroscopy 30 (1989) 58.

[2] A. Ourmazd et al., Ultramicroscopy 34 (1990) 237.

[3] R.F. Egerton, Electron Energy-Loss Spectroscopy in the Electron Microscope, New York: Plenum Press (1989) 159.

[4] This work was supported by DoE (Grant #DE-FG02-87ER45322). The UHV-STEM was acquired through NSF (Grant #DMR-8314255) and is operated by the Cornell material Science Center (NSF Grant #DMR-912654).

Proc. Microscopy and Microanalysis 1995, edited by G.W. Bailey, M.H. Ellisman, R.A. Hennigar, and N.J. Zaluzec.
Copyright © 1995 MSA. Published by Jones and Begell Publishing, 79 Madison Ave., New York, NY 10016

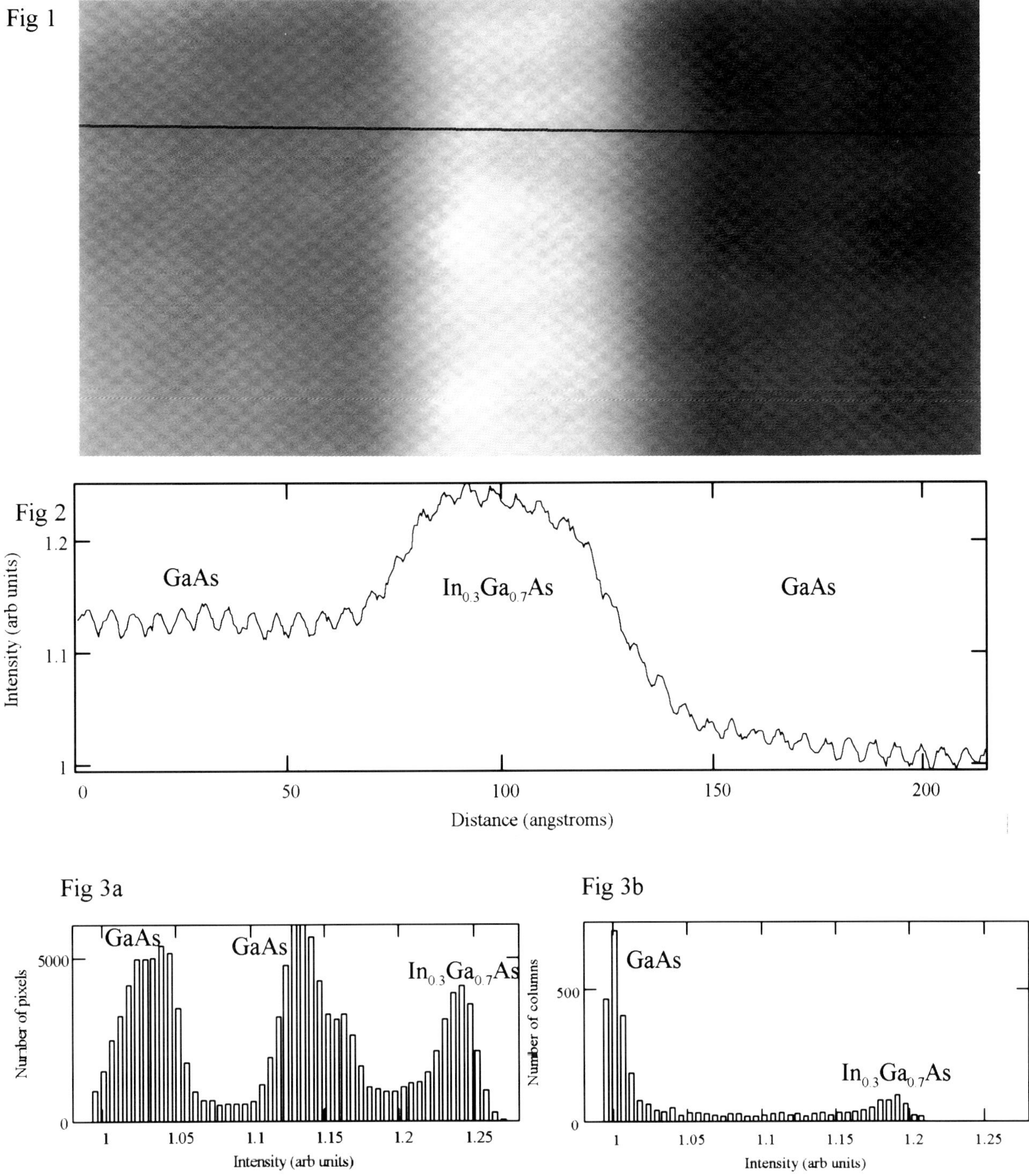

Figure 1. Annular Dark Field Image of GaAs/In$_{0.3}$Ga$_{0.7}$As quantum well.

Figure 2. Linescan along the line shown in figure 1 showing the relative intensities.

Figure 3. a) histogram of pixels in figure one. b) histogram of column intensities after thickness correction.

GRAIN GROWTH STUDY OF Al(Cu) LINES BY FOCUSED ION BEAM MICROSCOPY AND TRANSMISSION ELECTRON MICROSCOPY

M.H. Sohn*, D.A. Smith**, K.P. Rodbell***, and Roy Izai****

* Department of Materials Science and Engineering, Stevens Institute Technology, Hoboken, NJ 07030
** Department of Materials Science and Engineering, Lehigh University, Bethlehem, PA 18015
*** IBM T.J. Watson Research Center, P.O. Box 218, Yorktown Heights, NY 10598
**** Surface Science Division, Seiko Instruments USA Inc., San Jose, CA 95131

Interconnects for integrated circuits now have widths which are of the same order as the grain size. Consequently, their reliability is sensitive to the specific grain boundary properties and configurations rather than being averaged over many grains as in earlier technology when the line width was much greater than the grain diameter. From the point of view of electromigration resistance a bamboo grain structure is desirable. However, the conditions for forming a bamboo structure are not established. Walton et al. developed a model which predicts that the grain structure of aluminum lines becomes a bamboo structure, which is stagnant, upon annealing after particular time at specific temperature when the width to height ratio of the line is lower than a critical ratio (2.1-3.0). The precise value of the critical ratio depends on the ratio of the surface and grain boundary energies. Based on the modeling, we expect that the grain structure of pure aluminum stagnates after annealing at 500 °C for 2.3 hrs.

The grain structure evolution in fine lines has been analyzed in an attempt to verify experimentally the prediction of Walton et al. Al(Cu) alloys were deposited by sputtering onto oxidized silicon wafers. The substrate temperature was less than 200 °C. The films were patterned into parallel lines, 0.8 μm wide and 1 μm high. There was no passivation layer deposited on top of the lines. Grain sizes have been measured in as-deposited, 400 °C, 500 °C, and 550 °C annealed Al(0.5 wt% Cu) lines. Figure 1. shows the Focused Ion Beam (FIB) micrographs of as-deposited and annealed Al(0.5 wt% Cu) lines. On the basis of these measurements, we see that the stagnant grain structure observed experimentally is qualitatively consistent with the grain growth simulation. Since the modeling considered only the grain boundary pinning by surface grooving, it is different from our experiment which also includes grain boundary pinning by solutes and precipitates. In an effort to further study the effect of precipitates on grain structure evolution, Al(2 wt% Cu) (Fig.2) and Al(4 wt% Cu) fine line samples (Fig.3) were also heat treated at the same temperatures. Since precipitates at the grain boundaries inhibit the migration of these boundaries, the grains not pinned by the precipitates may grow larger. Therefore, as was observed in our experiments, the grain size distributions widen, as the amount of copper increases in the line. Finally, fine lines were annealed in the TEM to observe real-time grain growth and to confirm grain growth stagnation. The results of this experiment accord qualitatively with the modeling; however, the activation energy of grain growth plays a critical role in quantitatively fitting the results to the analytical model of Walton et al.

Reference

1. D.T. Walton, H.J. Frost, and C.V. Thompson, MRS Proceedings Series, 225, 1991.

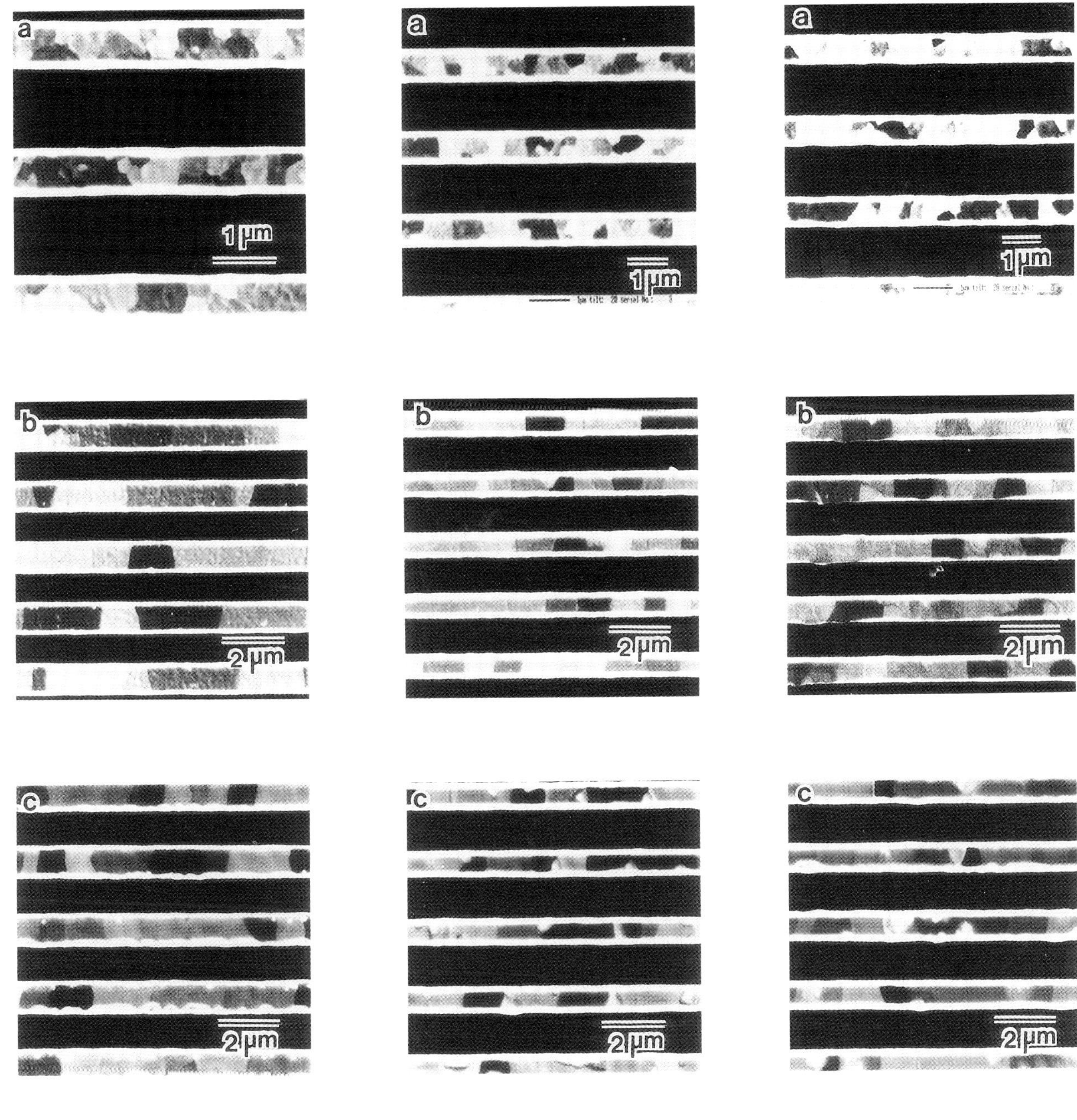

Fig. 1.   Fig. 2.   Fig. 3.

Fig. 1. FIB micrographs of Al(0.5 wt% Cu) parallel lines; (a) as-deposited, (b) annealed at 400 °C for 10 hrs, (c) annealed at 500 °C for 3 hrs.

Fig. 2. FIB micrographs of Al(2 wt % Cu) parallel lines; (a) as-deposited, (b) annealed at 400 °C for 10 hrs, (c) annealed at 500 °C for 3 hrs.

Fig.3. FIB micrographs of Al(4 wt% Cu) parallel lines; (a) as deposited, (b) annealed at 400 °C for 10 hrs, (c) annealed at 500 °C for 3 hrs.

SYMMETRY OF MOLECULAR WAVEFUNCTION IN BaBiO$_3$ DETERMINED BY q-RESOLVED EELS

Y.Y. Wang,* V.P. Dravid,* and F.C. Zhang[#]

*Department of Materials Science and Engineering, Northwestern University, Evanston, IL 60208
[#]Physics Department, University of Cincinnati, Cincinnati, OH 45221

Momentum-transfer (q) resolved electron energy loss spectroscopy is a unique technique to investigate the symmetry of electronic excitations in materials. Under small momentum-transfer conditions, the dielectric function derived from electron energy loss spectrum (EELS) is comparable to the dielectric function obtained by optical measurement. However, relatively large momentum-transfer conditions, which optical spectroscopy cannot achieve, can be met with high energy electrons and such q-resolved EELS data can provide dynamic information of the excitations as well as possible location of optically forbidden transitions.

According to the Born approximation, the differential cross section of inelastic scattering can be written as:

$$\frac{d^2\sigma}{dEd\Omega} \sim q^{-4}\sum_f \left|\langle \Psi_f|\exp(i\boldsymbol{q}\bullet r)|\Psi_0\rangle\right|^2 \delta(E_f - E_0 - E), \tag{1}$$

where Ψ_0 and Ψ_f are the initial- and final-state wave functions with energies as E_0 and E_f, respectively, q is the momentum transfer. If r_c is the effective radius of the excitation, then for $q<1/r_c$, one can write:

$$\exp(i\boldsymbol{q}\bullet r) \cong 1 + (i\boldsymbol{q}\bullet r) + (i\boldsymbol{q}\bullet r)^2/2!+\cdots. \tag{2}$$

For small q, the second term dominates the integral, permitting dipole-allowed transitions. For large $\boldsymbol{q}$, the third term, which contains monopole and quadrupole transitions, increases in strength relative to the dipole transitions. By mapping the oscillator strength of the forbidden transitions along different crystal orientations in solids, one can obtain information about the symmetry of the wave functions.

In the present study, we have employed momentum-transfer (q) resolved electron energy loss spectroscopy (EELS) to probe the forbidden transition along the different orientation in BaBiO$_3$ and we propose a molecular orbital model to interpret the forbidden transition observed in the experiments. Figures 1(a) and 1(b) show the loss functions (after removal of zero loss function) at different q for BaBiO$_3$ along [100] and [110], respectively. A broad excitation appears at about 4.5 eV with increasing q, as can be seen in Fig.1. The oscillator strength of this feature is stronger along [100] than along [110] and it increases as q^2 relative to the dipole transition, which is the indication of an optically forbidden transition according to eq.(1) and eq.(2). This forbidden transition is strongly anisotropy as shown in Fig.1. The ratio of the oscillator strength along [100] and [110] is estimated to be ~ 3.1. This excitation is assigned to be a transition from the non-bonding oxygen $2p\sigma$ orbital of the valence band to the empty Bi $6s$ state. The wavefunctions of $2p\sigma$ can be expressed as:

$$\phi_0 = \frac{1}{\sqrt{6}}(p_x + p_y + p_z - p_{-x} - p_{-y} - p_{-z}),$$

$$\phi_1 = \frac{1}{2}(p_x - p_y - p_{-x} + p_{-y}),$$

$$\phi_2 = \frac{1}{\sqrt{12}}(p_x + p_y - 2p_z - p_{-x} - p_{-y} + 2p_{-z}),$$

(3)

where p_i ($i = \pm$ x, y or z) is the O ($2p\sigma$) wave function at the i atom position. ϕ_0 as a bonding and anti-bonding wavefunction has an s-symmetry (bonding-antibonding) shown in Fig.2(a), ϕ_1(non-bonding) has a $d(x^2\text{-}y^2)$ symmetry as illustrated in Fig.2(b), and ϕ_2(non-bonding) has a $d(3z^2\text{-}r^2)$ symmetry. Because of the symmetry, these wavefunctions are orthogonal to the Bi $6s$ state, and the transitions from these states to Bi $6s$ are optically forbidden. Since the forbidden transition is between s-like and d-like states, it is referred to as a quadrupole transition. Furthermore, the amplitude of ϕ_1 is large along [100], and zero along [110]. This leads to a strong anisotropy of the strength for the quadrupole transition as observed in the experiments. The quantitative calculations indicate that the ratio of the transition strength between [100] and [110] is 4, which is close to the experimental result of ~ 3.1.

This experiment illustrates a unique method for investigating the wave function symmetry of solid state materials by angle resolved electron energy loss spectroscopy.

This work was supported by National Science Foundation with grant no. DMR-91-20000.

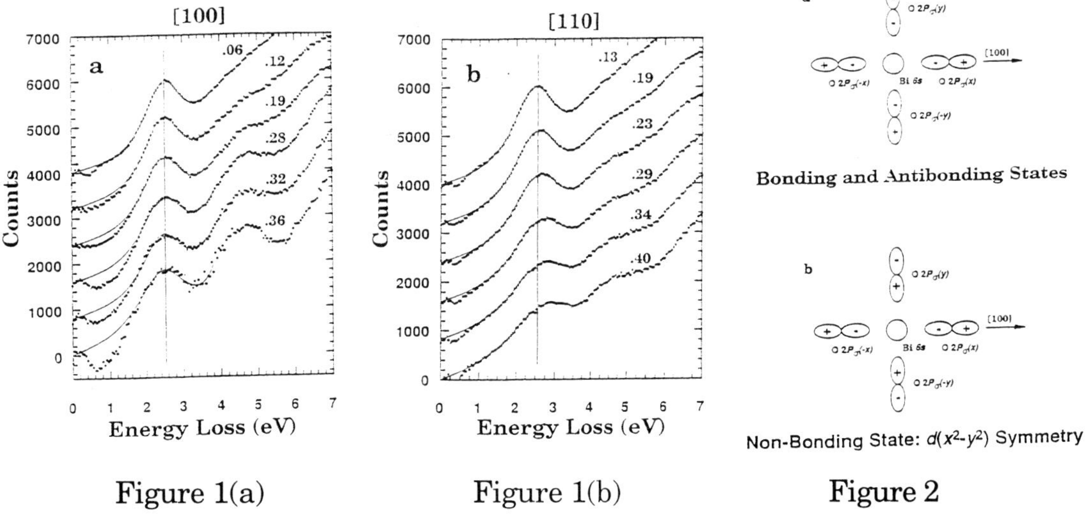

Figure 1(a)　　　　Figure 1(b)　　　　Figure 2

Fig.1, Energy loss spectra of $BaBiO_3$ with different momentum-transfer q in [100] and [110] direction. The unit of the insert numbers is Å^{-1}.

Fig.2, Illustration of bonding-antibonding wave function ϕ_0, which has an s-symmetry and the non-bonding wave function ϕ_1, which has a $d(x^2\text{-}y^2)$ symmetry. The amplitude of the non-bonding wave function is zero along [110] direction, which results in weaker oscillator strength of the forbidden transition along [110] direction.

THE MEASUREMENTS OF THE ELASTIC-INELASTIC MULTIPLE SCATTERING ELECTRON INTENSITY IN EELS

S. C. Cheng, Y. Y. Wang, V. P. Dravid

Department of Materials Science & Engineering, Northwestern University, Evanston, IL. 60208

The Electron energy loss function in the low energy range is determined by collective excitation of valence electrons and charge carriers, i.e. plasmons, as well as interband and intraband excitations. The explicit dependence of the cross-section on the momentum transfer q allows the observation of nonvertical interband transition and a measurement of the dispersion of plasmon excitations.[1] The drawback of the momentum resolved electron spectroscopy is the multiple scattering, which often obscure the single scattering events. Under relatively small scattering angles, both strong elastic-inelastic multiple (E-I-M) scattering and elastic scattering events compared to the inelastic scattering have been reported.[2] In order to find out in what momentum range the E-I-M scattering intensity can be ignored in the momentum resolved electron spectroscopy, we have measured the angular dependency of the intensities of the E-I-M scattering electrons Ie+in. The intensities of the elastic scattering electrons Ie as well as of the inelastic scattering electrons Iin were also measured and are presented in this paper together. A simple relationship between Ie and Ie+in is found.

The experiments were carried out on the Hitachi HF-2000 cold field emission TEM equipped with a Gatan 666 parallel EELS spectrometer. The best momentum resolution was about 0.09 Å^{-1}. The thickness of the Al sample was chosen to be about 380 nm to increase the intensity of the E-I-M scattering electrons. With such specimen thickness the intensity magnitudes of the E-I-M scattering electrons and the elastic scattering electrons are comparable, thereby providing more precise assessment of their intensity relationship.

For spectra obtained at q < 0.9 Å^{-1} the areas under the zero loss peak represents the measured Ie; and that under the first plasmon peak represents the possible combination of Iin and Ie+in. At q >= 0.9 Å^{-1} the E-I-M peak showed up and partially resolved from the inelastic scattering peak as shown in Fig. 1. Its intensity was measured by removing the background first, then applying the Lorentz function by means of the Peakfit data fitting program to fit the two peaks, and finally calculating the areas of resulted two Lorentz peaks. Fig. 2 shows the resolved Ie+in and Iin peaks. The intensities of Ie, Iin, Ie+in are plotted as a function of q in Fig. 3(a) with logarithmic scale, and plotted in the enlarged linear scale for q > 0.9 Å^{-1} as Fig. 3(b). The numbers on the y axis correspond the counts per 0.1 second integration time reading from the photodiode array.

The angular dependency of the measured Iin for Al is in good agreement with the previous experimental work and theoretical calculations.[3,4] The Ie data measured at q < 0.2 Å^{-1} are eliminated, because that the unscattered electrons may have mixed with elastic electrons. Figure 3 (b) also shows that at high q values (from 0.9 Å^{-1} to 1.59 Å^{-1}), the Ie data are within the same magnitude with some fluctuations. The Ie+in data show the similar trend. As a result, the calculated ratios of Ie+in/Ie in this q range are nearly constant. The observed near constant ratio of Ie+in/Ie in the relatively high q region can be explained by means of probability analysis of scattering events. The detail of this analysis will be presented in other paper. The constant ratio of Ie+in/Ie is used to estimate Ie+in in the q region where Ie+in can not be measured directly. The dotted line in Fig. 3 represent the intensity of Ie+in, which is obtained by calculation. Overall, the intensity of E-I-M scattering peak can be either estimated by the intensity of the elastic scattering or directly measured. It is estimated that when an Al sample with a thickness of a few hundreds Å is used, the intensity of Ie+in is only about 1% of the intensity Iin in the range of q<0.9 Å^{-1}.

Proc. Microscopy and Microanalysis 1995, edited by G.W. Bailey, M.H. Ellisman, R.A. Hennigar, and N.J. Zaluzec.

Under these conditions the intensity of Ie+in may be ignored in the measurement of the dispersion relationship of low energy excitation in EELS.

References

1.   H. Raether *Excitation of Plasmons and Interband Transitions by Electrons*, Springer Trancts in Modern Physics, Vol. 88, Springer-Verlag, New York (1980) 74.

2. P. E. Batson and J. Silcox, *Phys. Rev. B* 27 (1983) 5224.

3. P. Schmuser, *Z. Phys*, 180 (1964) 105.

4. Y. Y. Wang et al. *Microbeam Analysis*, 3(1994) 87, Y. Y. Wang et al. *Ultramicroscopy* (1995) in press.

5. this research is supported by the NSF 91-20000 through the Science and Technology Center for Superconductivity

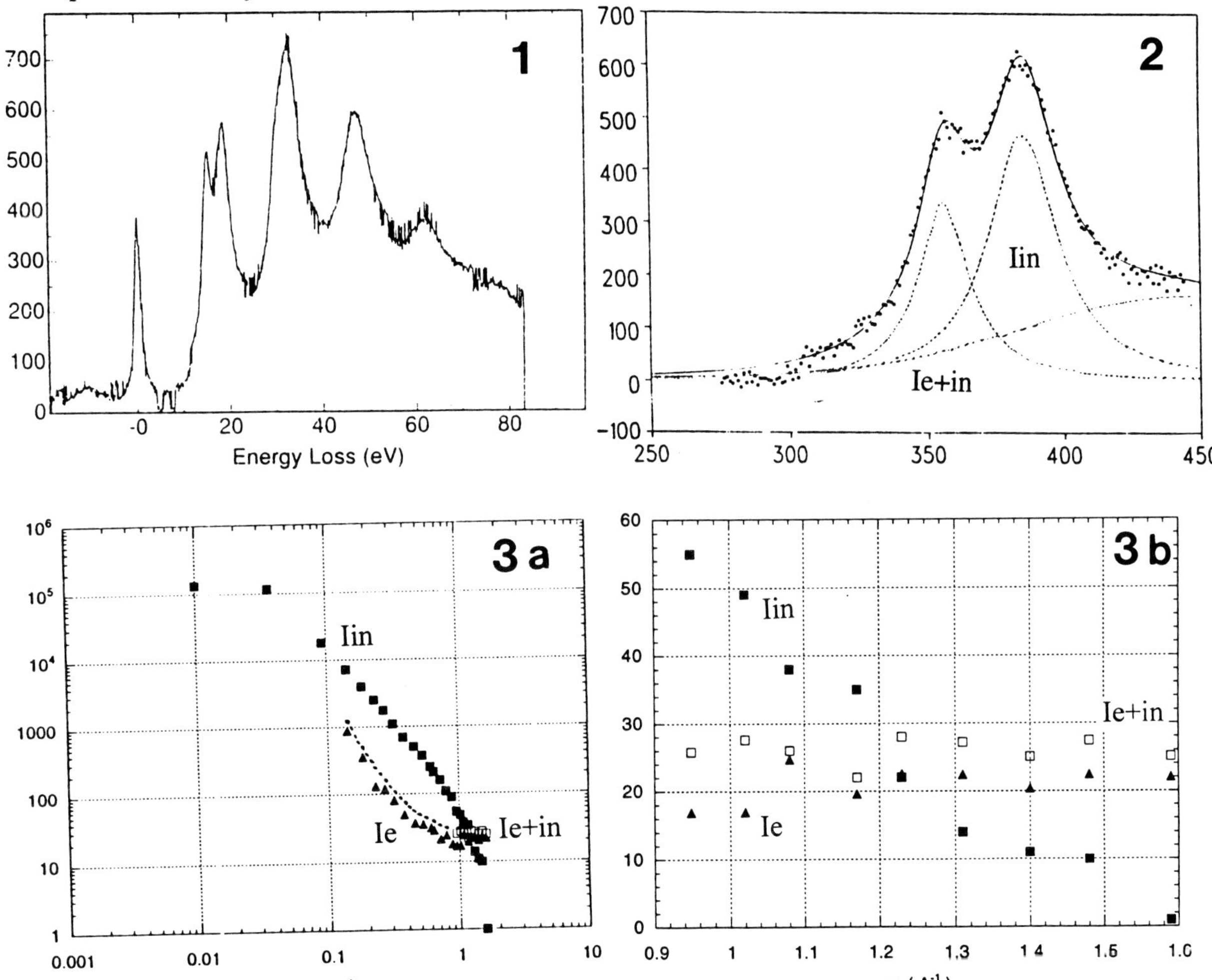

FIG. 1 - EELS spectrum obtained at q = 1.08 Å^{-1}.

FIG. 2 - The resolved E-I-M scattering peak and inelastic peak by Peakfit data fitting program. The numbers on energy axis have been multiplied by a constant.

FIG. 3 - The intensities of Ie, Iin and Ie+in as function of q with logarithmic scale in (a) and with the enlarged linear scale in (b).

REMOVING EDGE OVERLAPPING FOR MAXIMUM USAGE OF EXELFS SIGNAL

Maoxu Qian,* Mehmet Sarikaya,* and Edward A. Stern+

*Materials Science and Engineering and +Physics, University of Washington, Seattle, WA 98195

Extended energy loss fine structure (EXELFS) is an unique tool to determine the local atomic structure of various materials.[1] Compared to X-ray absorption fine structure (XAFS), EXELFS has several advantages, such as having much greater spatial resolution that can be used to investigate phenomena that are spatially inhomogeneous samples, e.g., nanograined materials. In addition, the region being investigated can be characterized by the full capabilities of the TEM such as by imaging and diffraction. Recently, the development of high quality data acquisition[2] and sophisticated data analysis[3] techniques has removed the obstacles for routine use of the EXELFS technique. The applicability of EXELFS is often limited by the edge-overlapping problem due to the close proximity of the edges and the overlap of their fine structures. When this happens, either the desired edge is completely unanalyzable or the usable fine structure range is very narrow and, hence, the uncertainty in the deduced information is too large for reasonable conclusion. In this paper, we describe a method to overcome this problem which makes it possible to deconvolute extended energy loss fine structure information and maximizes the usage of the spectra with overlapping edges.

A good example of edge overlapping is in spectra obtained from carbonitride samples (e.g., cubic C_4N_4).[4] As shown in Fig. 1, the energy difference between C K-edge (284 eV) and N K-edge (401 eV) is only 117 eV which corresponds to 5.5 $Å^{-1}$ in k-space. The jump of the N K-edge and the overlapping of fine structure of C-K edge beyond the threshold of the N K-edge limits the use of the C K-edge fine structure beyond the N K-edge causing fewer independent points and higher uncertainty. To overcome this difficulty, we first fit the pre-edge background of N and subtract the N K-edge from the whole spectrum to obtain a somewhat "pure" C-spectrum with all the structural information after the threshold value of 401 eV missing, as shown in Fig. 2. Next, the post-edge background of the N-edge is determined and the overlapped fine structure is deduced, as shown in Fig. 3, and is added back to the "pure" C-edge spectrum to construct out an extended the C K-edge spectrum (Fig. 4). UW XAFS/EXELFS software[5] were modified to accept this specially treated EXELFS data and perform background removing and normalization correctly. The χ-data (Fig. 5a) contains three regions: that with the pure C K-edge information, that with overlapping mixed C and N

edge information, and a narrow empty gap in between. Fig. 5b shows the Fourier transform of the χ data. The N-edge fine structure information in the new spectrum (Fig. 4) is distorted because its edge energy is shifted to the threshold energy of the C K-edge. The Fourier transforms of such distorted fine structures only give a broad background mainly after the second shell, as illustrated in Fig. 6 using the O K-edge of an MgO EXELFS spectrum. The first (sometimes up to the second) shell r-space fitting can be done without considering the interference by the N K-edge. If necessary, however, distorted N-edge contribution can also be fitted simultaneously with C-edge for a better accuracy.

In principle, this method can be applied to all materials producing EELS spectra with overlapping edges. Considerable future work, however, is required to test various different overlapping cases, considering important parameters such as the separation of the threshold energies and the background determination which are now under consideration for further investigation.

1. R. D. Leapman, and V. E. Cosslett, *J. Phys.,* **D 9 L**, 29-32, (1976).
2. M. Qian, M. Sarikaya, and E. A. Stern, *Ultramicroscopy* (1995).
3. E. A. Stern, M. Qian, and M. Sarikaya, MRS Proc., Vol. 332 (1994) pp. 3-14.
4. M. Qian, M. Sarikaya, J. Martin Gil, F. Martin Gil, and M. J. Yacaman, to appear in Nature (1995).
5. For example, see the copyrighted UWXAFS 2.0 and UWEXELFS 1.0 software.
6. This work is supported by an AFOSR and an URI/ARO Grants.

Proc. Microscopy and Microanalysis 1995, edited by G.W. Bailey, M.H. Ellisman, R.A. Hennigar, and N.J. Zaluzec
Copyright © 1995 MSA. Published by Jones and Begell Publishing, 79 Madison Ave., New York, NY 10016

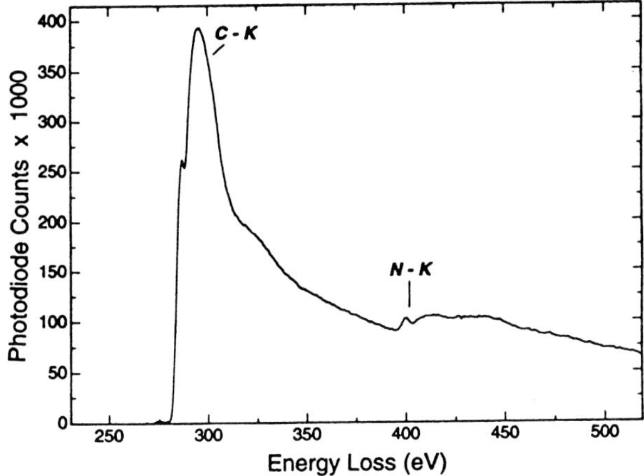

Fig. 1- Crystaline Carbon nitride C_4N_4 EELS spectrum.

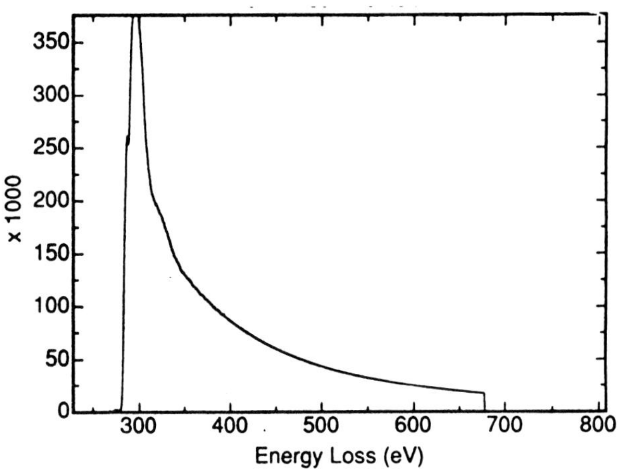

Fig. 2 - The same as Fig. 1, but the N K-edge is removed, only smooth background remains after 401 eV.

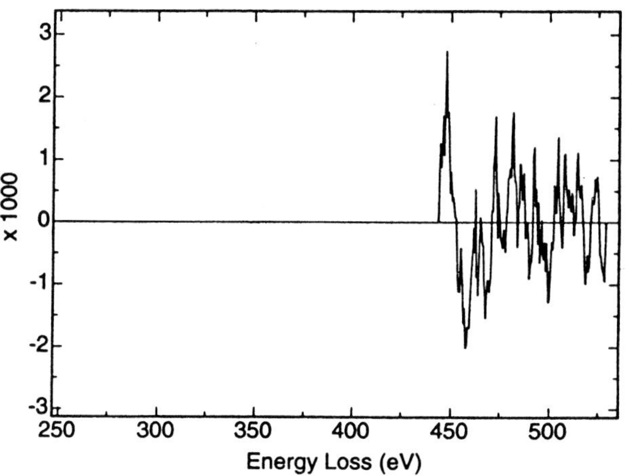

Fig. 3 - Overlapped fine structure of both C and N K-edges at the far end of the spectrum.

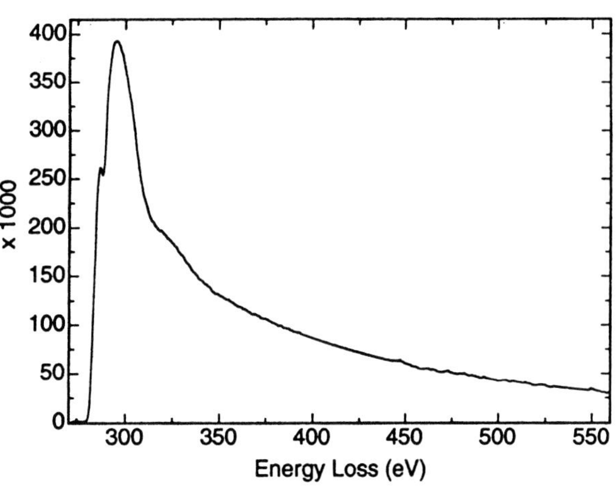

Fig. 4 - The reconstructed C K-edge spectrum with extended energy range.

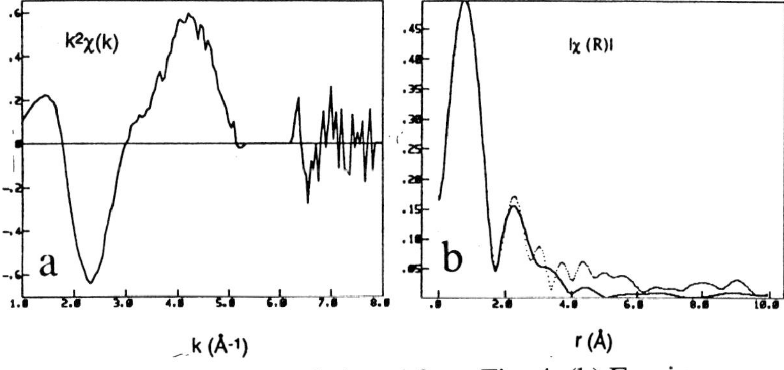

Fig. 5 (a) The χ-data deduced from Fig. 4. (b) Fourier transform of (a). Both pre-overlapping (2-5.2 Å⁻¹, solid line) and overlapping (2-8 Å⁻¹, dotted line) k-ranges wrere used.

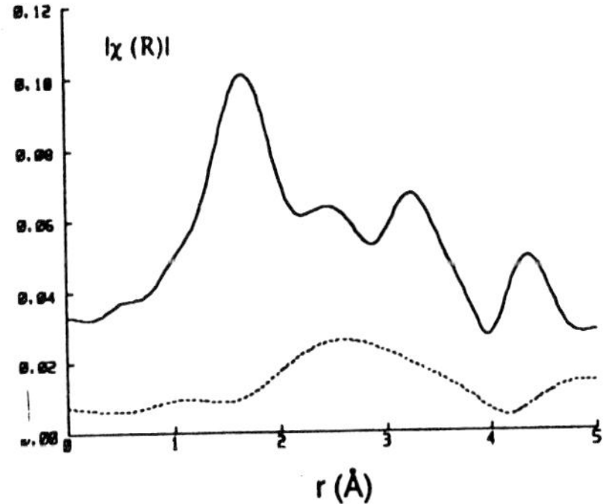

Fig. 6 - Fourier transform of O K-edge of MgO crystal. Conditions are solid line: E_0=532 eV k= 3-9.5 Å⁻¹, k weight =1; dotted line: E_0 shifted by -250 eV k=9.5-12.5 Å⁻¹.

PRACTICAL LIMITS ON THE SPATIAL RESOLUTION IN ENERGY-FILTERED MAPPING

P.A. Crozier

Center for Solid State Science, Arizona State University, Tempe, AZ 85287-1704

One parameter of importance in energy-filtered elemental mapping is the achievable spatial resolution. In practice, the resolution in a particular experiment depends on the energy loss used to form the image. Moreover, the importance of critical factors for defining the resolution, namely quantum localization, chromatic aberration and statistics, change with energy loss.

The fundamental limit on the spatial resolution in any energy-loss experiment is determined by the localization of the inelastic scattering process. Kohl and Rose (1985) have undertaken detailed quantum mechanical calculations of the influence of inelastic scattering on high resolution images. Fig 1 shows the results from these calculations for the change in spatial resolution as a function of energy loss for typical experimental conditions on the Zeiss 912.

Chromatic aberration in the electron optical column causes electrons of different energy to be focused in different planes causing image blurring. The resulting diameter d of the disk of least confusion depends on the energy and angular spread of the beam through the equation:

$$d = C_c \, \Delta \, \beta \, / \, E_0 \qquad (1)$$

where E_0 is the incident energy, Δ is the slit width, β the collection angle and C_c the chromatic aberration coefficient of the objective lens. This equation will give a reasonably accurate estimate of the chromatic blurring when the inelastic angular distribution uniformly fills the collector aperture i.e. when Bragg beams lie close to β or when $\theta_E >> \beta$ where θ_E is the characteristic scattering angle related to the incident energy E_0 and energy loss E through $\theta_E = E/2E_0$. When the angular distribution is much narrower than β, it is more accurate to replace β with an effective scattering angle θ_{eff} containing 58% of the total inelastic intensity collected by the aperture β.

The intensity in an energy-filtered image drops rapidly with increasing energy-loss and this together with the limited brightness of the thermal electron sources cause energy-filtered images to be noisy. For fixed illumination, the sensitivity increases with increasing detector pixel size and it is possible to determine the relationship between the effective resolution d_{eff} and the sensitivity. For a thin sample, d_{eff} can be written in terms of inelastic scattering behavior and instrumental parameters through:

$$d^2_{eff} = K^2 \, / \, J \, t \, N_s \, c_s \, \sigma_s \qquad (2)$$

The effective resolution depends inversely on the acquisition time t, incident electron density J, the atomic density N_s and the inelastic scattering cross section σ and the signal-to-background ratio c_s for the inner-shell edge being used to form the elemental map. The factor K incorporates uncertainties in background determination. The inelastic scattering cross section decreases rapidly with increasing energy-loss and can be represented by putting $\sigma = AE^{-r}$ into eqn(2) to get :

$$d^2_{eff} = K^2 \, E^r \, / \, J \, t \, N_s \, c_s \, A \qquad (3)$$

For K and L shell scattering, $r \approx 3$ is reasonable choice for the exponent and eqn(3) shows that the spatial resolution increases rapidly with increasing energy loss.

The behavior of the main factors affecting the spatial resolution are plotted in fig 1 as a function of energy-loss for the Zeiss 912 (E_0 = 120 kV and C_c = 2.7 mm) for typical experimental conditions (Δ = 25 eV and the collection angle β = 4.3 mrad). Eqn(3) is evaluated with K = 4.2, J= 10^3 e/ Å^2 , t = 10 s,

Proc. Microscopy and Microanalysis 1995, edited by G.W. Bailey, M.H. Ellisman, R.A. Hennigar, and N.J. Zaluzec

$c_s = 0.2$ assuming a sample thickness of approximately 500 Å. Mapping with both K and L shell losses are considered.

For the low loss region, the contribution due to localization is several nanometers dropping to approximately 5 Å for energy losses of about 1000 eV. For the energy range 100-500 eV, the spatial resolution is dominated by the blurring due to chromatic aberration. At energy losses above about 500 eV, statistical noise determines the actual spatial resolution obtained in the elemental map. J and t may be increased to reduce d_{eff} if the sample does not drift and is not beam sensitive. (It is usually necessary to align the energy-filtered images from a series before background subtraction procedures can be employed). It may be necessary to use larger collection angles and slit widths in order to achieve acceptable statistics. Typically the resolution will lie somewhere between 20 - 50 Å range depending on the details of the analysis conditions. Fig 1 indicates that improved resolutions down to 10 Å may be most easily achieved for K and L edges in the range 100 and 500 eV.

References

1. H. Kohl and H. Rose Adv. Electronics and Electron Physics. 65(1985)173
2. This work was supported by the HREM Industrial Associates Program of Arizona State University (ASU) and was performed in the Center for HREM at ASU.

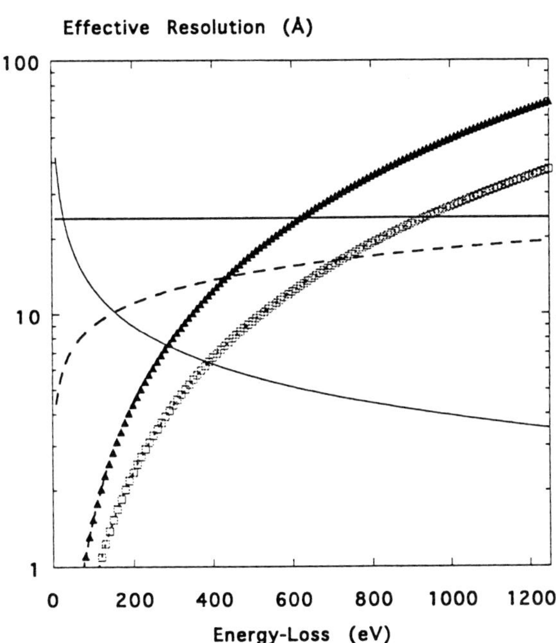

Fig1: The behavior of the main factors affecting the spatial resolution in energy-filtered mapping as a function of energy loss for typical experimental conditions in the Zeiss 912. The falling solid curve represents quantum localization, the solid line at about 25 Å is chromatic broadening determined from eqn(1). The dotted curve is chromatic broadening assuming an effective scattering angle θ_{eff}. The loss in spatial resolution due to signal-to-noise considerations is shown for K (solid triangles) and L shell (open squares) for the conditions described in the text.

ENERGY FILTERED MAGNETIC DOMAIN IMAGING AT 400 kV

N.T. Nuhfer, J. Dooley, and M. De Graef

Department of Materials Science & Engineering, Carnegie Mellon University, Pittsburgh, PA 15213

Lorentz microscopy provides an important technique for the study of advanced magnetic recording media. Recording densities as high as 10 Gbit/in^2 may be within reach in the future. Microstructural characterization of both the crystallographic and magnetic structure of these thin films is needed at the highest spatial resolution. Another area where Lorentz microscopy has proven to be an extremely valuable tool is in magnetic actuator applications, in particular for magnetostrictive alloys, such as Terfenol-D.[1] The standard Lorentz techniques (Fresnel and Foucault imaging) require a low-field environment, so that the sample is not completely saturated during observations. Depending on the microscope type this low-field area can be obtained by either switching off the objective lens and using the weak post-field to focus the image, or by using special low-field pole pieces. The attainable magnification in both cases is usually lower than for conventional TEM. Recently a new low-field lens for the JEOL 4000EX TEM was reported[2] (AMG40, 1 Gauss field strength) with a useful magnification range upto ×200,000. In this paper we report an alternative method to obtain high magnification magnetic domain images at 400 kV.

When a JEOL 4000EX TEM is operated in LOW MAG mode, the objective lens is completely switched off. The specimen is then located in an effectively field free region. The maximum attainable magnification in this mode is only ×3000, which would be insufficient for most observations. However, we have recently installed a Gatan Imaging Filter (GIF) on this microscope. The GIF has an internal magnification of about ×18.75. For Lorentz microscopy we can make efficient use of the internal magnification of the GIF; the total magnification at the CCD camera (with the microscope in LOW MAG mode) is about 3,000×18.75=56,250. The image captured on the CCD can then be digitally enlarged by a factor of 2 to 10, without substantial loss of resolution, leading to a useful magnification of over ×500,000. Simultaneously one can remove most of the inelastically scattered electrons from the image and hence increase the domain contrast.

The ability of the GIF interface control software (IFC) to modify the lens currents for all intermediate lenses independently is crucial for Lorentz microscopy. The following procedure was found to yield high quality images, of both the Fresnel and Foucault types. The microscope is operated in LOW MAG mode and the Objective Minilens (OM) is used to focus the image. The OM back focal plane is located approximately at the level of the Selected Area Aperture (SAA). One can obtain a diffraction pattern by decreasing the current of the IL2 lens to its minimum value; this can be achieved via the *Magnification Reduction* control panel of the GIF-IFC program (an example of a diffraction pattern obtained in this way is shown in Figure 1b of reference 3). The third intermediate lens current (IL3) can then be modified to focus the diffraction pattern onto the viewing screen. The projector lens controls are used to position the central beam onto the entrance aperture of the energy filter. Inserting the SAD aperture around the transmitted beam one can effectively obtain a bright field image of the sample on the GIF-CCD camera (the IL2 current must be slightly adjusted to obtain a focused image). At this point one can obtain Fresnel contrast by defocusing the OM lens and Foucault contrast by moving the Selected Area aperture to exclude a certain component of the (split) central beam. An example of an unfiltered and an energy filtered (slit width 40 eV) Fresnel image of magnetic domains in [$\bar{2}$11] oriented Terfenol-D is shown in Figure 1. Figure 2 shows a filtered underfocus image with a total of 4 different domain orientations. The composite image of Figure 3 shows 4 different energy filtered images of a region around a growth twin in Terfenol-D; the details of this configuration are explained in a companion paper in this proceedings.[3] Figure 3a and b are two Foucault images with different aperture displacements, and 3c and d are underfocus and overfocus Fresnel images of the same region. Note that the thin twin boundary indicated by arrows in 3b is only 8 nm thick.

References

1. M. Al-Jiboory and D.G. Lord, *IEEE Trans. Magn.*, 26(1990)2583

Proc. Microscopy and Microanalysis 1995, edited by G.W. Bailey, M.H. Ellisman, R.A. Hennigar, and N.J. Zaluzec
Copyright © 1995 MSA. Published by Jones and Begell Publishing, 79 Madison Ave., New York, NY 10016

2. R.C. Doole and A.K. Petford-Long, *JEOL News*, 31E(1994)23
3. J. Dooley, N.T. Nuhfer, and M. De Graef, this proceedings (1995).

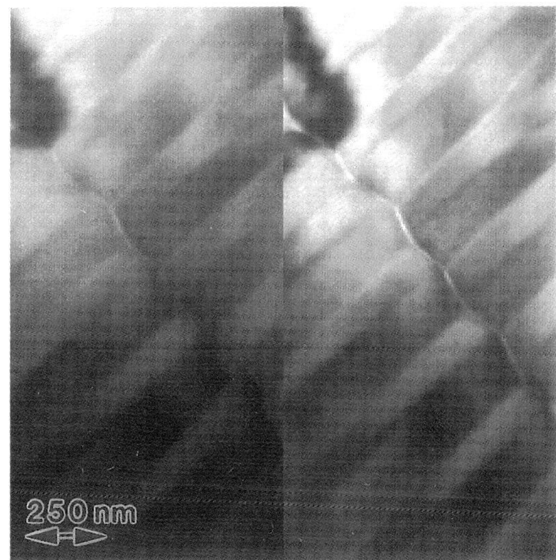

Figure 1: Unfiltered (left) and filtered (right) Foucault images of magnetic domain boundaries in [$\bar{2}$11]-oriented Terfenol-D.

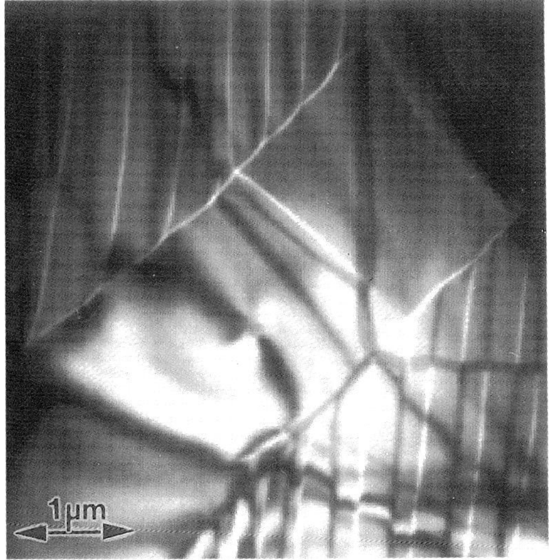

Figure 2: Energy filtered Fresnel image of magnetic domain boundaries in [$\bar{2}$11]-oriented Terfenol-D.

Figure 3: Energy-filtered domain images of [$\bar{1}$10]-oriented Terfenol-D around a growth twin; a) and b) Foucault images, c) and d) underfocus and overfocus Fresnel images. Magnetic domain contrast can be observed in the 8 nm thick growth twin indicated by arrows in b).

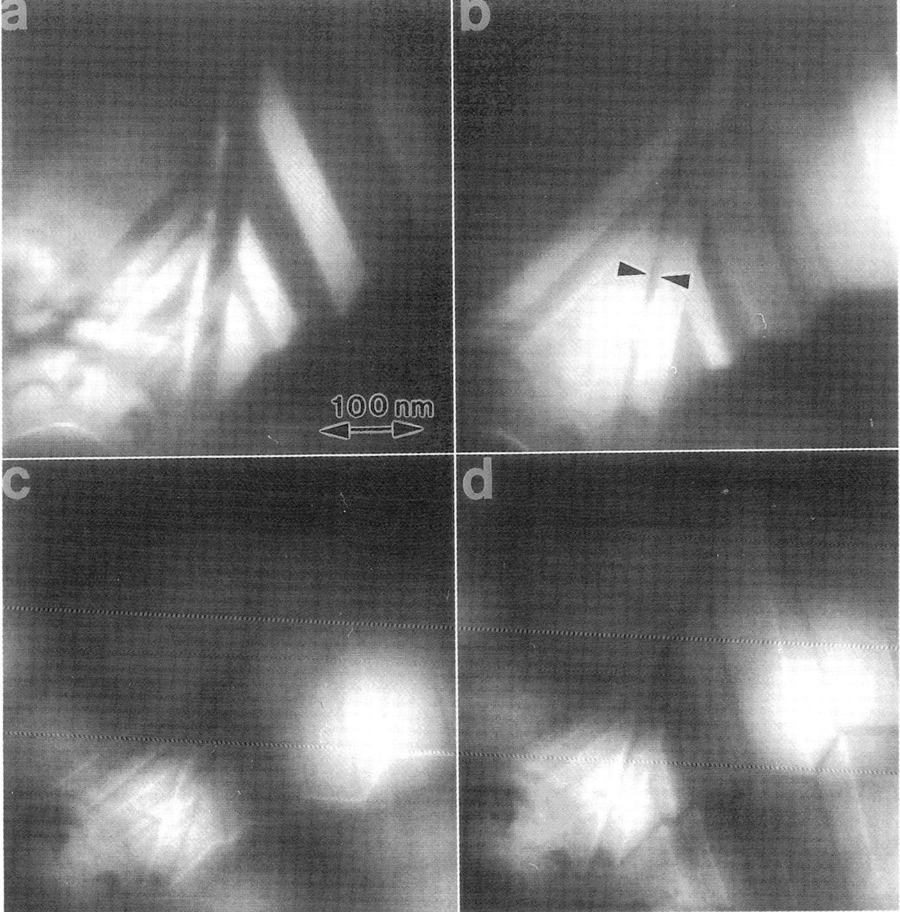

Electron Optical Design Of An EFTEM Concerning Optimum Selection Of Imaging Parameters For Different Sizes Of Detection Systems

G. Benner, W. Probst and R. Rilk

Carl Zeiss, Electron Optics Division, P.O.Box 1380, D-73446 Oberkochen, Germany

The amount of information which can be gained about an object is considerably improved by the incorporation of an imaging energy filter in a TEM. Besides the conventional modes of operation elastic brightfield (EBF) and darkfield (EDF) imaging or diffraction are provided by an Energy Filtering Transmission Electron Microscope (EFTEM) as well as Electron Spectroscopic Imaging (ESI) or Electron Energy Loss Spectroscopy (EELS). Thus, the demands for carefully designed highly flexible electron optics are much more sophisticated as compared to a CTEM.

One of the most important parameters for elastic imaging and ESI is the width of the energy window, which can be selected by a slit aperture of adjustable width according to the particular needs. This energy window defines the size of the transferable specimen area or diffraction pattern, respectively, as well as the acceptance angle of the spectrometer. In order to meet the optimum parameters concerning the energy window and the size of the detection system (e.g. photographic plate, image plate, Slow Scan CCD camera (SSCCD), TV camera, electron detector, etc.) the magnifications of the pre- and of the post-spectrometer projector lens systems has to be independently adjustable.

The EM 912 OMEGA has got such a flexible imaging system. The imaging system consists of an objective lens, a first projector lens system, a spectrometer, and a second projector lens system. For optimum aberration correction, minimum adjustment needs, and stable alignment any spectrometer has to be operated with constant excitation resulting in fixed entrance planes (pupil and image) as well as fixed exit planes (energy dispersive and achromatic image plane). These optical requirements of an EFTEM must be strictly fulfilled by the projector lens systems.

The magnification M of the EM 912 OMEGA is given by the product of the magnifications of the objective lens (M_0), of the first projector lens system (M_I), and the second projector lens system (M_{II}).

$$M = M_0 \times M_I \times M_{II}$$

The magnification of the objective lens is kept constant. Therefore a variation of the image magnification or the camera length, respectively, can be achieved using the first or the second projector lens systems. In order to achieve the maximum acceptance angle the portion of the magnification of the first projector lens system M_I in relation to the total magnification should be as large as possible. The magnification of the second projector lens system (M_{II}) is mainly defined by the size of the detection system ($\varnothing_{DET}$). It can be calculated using the size of the spectrometer achromatic image ($\varnothing_{ACI}$) which is shown in Fig.1 for different energy widths. The algorithm for the calculation is

$$M_{II} \quad >/= \quad \varnothing_{DET}/\varnothing_{ACI}$$

In the EM 912 OMEGA the magnification of the second projector lens system can be varied from 12.5x to 100x which allows optimum adaptation of parameters for any available detection system whether it is large like photographic negative or image plate or small like SSCCD camera. The size of the respective object field ($\varnothing_{OBJ}$) can be calculated using the inversed magnifications of the first projector lens system (M_I^{-1})and the objective lens (M_0^{-1}). Table 1 shows a list of important data concerning optimum integration of the spectrometer in the EFTEM EM 912 OMEGA.

Proc. Microscopy and Microanalysis 1995, edited by G.W. Bailey, M.H. Ellisman, R.A. Hennigar, and N.J. Zaluzec
Copyright © 1995 MSA. Published by Jones and Begell Publishing, 79 Madison Ave., New York, NY 10016

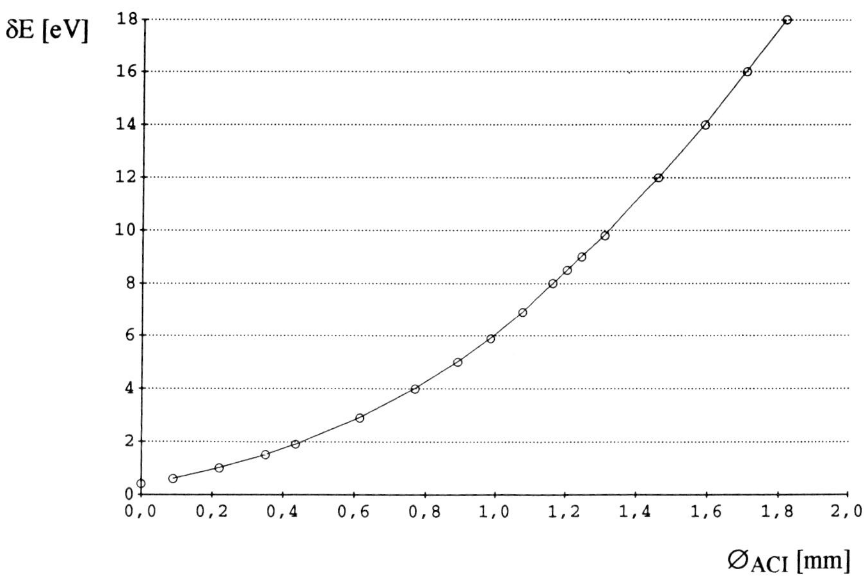

Fig. 1   Measured diameter of the achromatic image ($\varnothing_{ACI}$) as a function of the energy width.

Table 1: Data concerning OMEGA Spectrometer and Imaging Optics	
Parameter	**Specification**
image magnification entrance image to exit image	1x
magnification of entrance object at slit	1x
magnification range (image) of the first projector lens system (pre spectrometer)	1.2x to 200x
magnification range (image) of the second projector lens system (post spectrometer)	12.5x to 100x
possible detectors for image, diffraction pattern, or EELS (all mounted in parallel)	TV-(image intensifier) in 35 mm-camera plane; 35 mm camera microscope's fluorescent screen sheet film (3 1/4" x 4") image plate serial EELS detector SSCCD (image and parallel EELS)
diameter of filtered image for 1 eV energy resolution	18 mm (screen),+ 22 mm (negative, image plate), 24 mm (SSCCD without STEM), 36 mm (SSCCD with STEM)
energy dispersion (120 kV) at energy dispersive plane (spectrum mode)	1.0 μm/eV
energy dispersion in final image plane (spectrum mode)	40 μm/eV to 0.3 mm/eV (continous)
largest acceptance halfangle for 1 eV resolution (at specimen)	30 mrad
largest image area (on the specimen) for 1 eV energy window	0.3 μm
deflection system for entrance crossover	yes
geometric distortions	none
spectrometer alignment	factory aligned
differences concerning microscope lens and deflection systems settings and alignment with and without spectrometer in use	none

ELECTRON SPECTROSCOPIC IMAGING: DETECTION AND RESOLUTION LIMITS, QUANTITATIVE EVALUATION

J. Mayer and J. Plitzko

Max-Planck-Institut für Metallforschung, Seestr. 92, D-70174 Stuttgart, F. R. Germany

One of the main applications of the new imaging energy filters is to obtain elemental distribution images by electron spectroscopic imaging (ESI). Before the technique can successfully be applied to new materials problems it is important to evaluate the detection and resolution limits which have to be expected. Since the intensities in ESI images recorded with inner shell loss electrons are very small, the detection limit is governed by the signal to noise ratio (S/N). In a systematic study we have evaluated the performance of three different electron microscopes, a Zeiss EM 912 Omega, a Philips CM 20, and a Jeol ARM 1250, the latter two being equipped with a Gatan imaging filter (GIF). As experimental systems we chose amorphous oxide grain boundary films in Si_3N_4 ceramics and thin diamond films on silicon substrates. Furthermore, new routines were developed to extract EELS spectra from a whole series of ESI images. These EELS spectra can be used to study the local variation of the near edge fine structure (ELNES) or, after quantitative evaluation, of the chemical concentration of certain elements.

Elemental distribution images were obtained using the three-window technique. As model system we have used Si_3N_4 ceramics. During sintering with oxidic additives, an amorphous grain boundary film forms which possesses an equilibrium thickness. For our investigations, materials with a film thickness of 0.7, 1.0, and 1.5 nm were used. On the Zeiss EM 912 Omega, elemental distribution images for oxygen revealed the presence of the grain boundary films for all three materials (Fig. 1). For comparison, the material with 1.0 nm film thickness was investigated on a Philips CM 20 with GIF and the material with 1.5 nm film thickness on a JEOL ARM 1250, which is equipped with a high voltage GIF. The S/N ratio was determined for a number of grain boundaries in each system and the results are shown in Fig. 2. The S/N ratio was obtained by evaluating line profiles parallel to the boundary. The experimental results are in good agreement with the results of theoretical calculations (Fig. 2).[1] The detection limits for thin films depend on a number of parameters and can be as low as one monolayer.

ESI can not only be used to detect the presence of an element, but also to reveal the presence of different bonding states by mapping the ELNES. Thin diamond layers were grown by CVD on Si substrates.[2] High resolution electron microscopy reveals the presence of an amorphous layer with a thickness of 1-2 nm at the interface. The amorphous layer consists mainly of carbon and can thus not be distinguished from diamond in elemental distribution images. However, the energy loss near edge structure (ELNES) of the carbon K-edge shows marked differences for the two phases. In particular, the π^* peak, which is only present in the amorphous phase, can be used for a distinction of the two phases. In order to map the characteristic differences in the ELNES we have acquired a series of 20 ESI images with an energy window width of 5 eV and a step width of 2 eV around the C-K edge. The ESI image in which the amorphous layer becomes brighter due to π^* excitation (Fig. 3a) can clearly be distinguished from the subsequent images in which the diamond layer also becomes bright. Background subtracted images for the π^* and the σ peak are shown in Fig. 3b and c, respectively. Furthermore, EELS spectra showing the ELNES from the amorphous carbon layer and the diamond film can be extracted from the whole series of ESI images (Fig. 4). EELS spectra which include several inner shell loss edges can be analysed quantitatively, which yields a two-dimensional map of the concentration of these elements.[3]

REFERENCES
1. A. Berger et al., *Ultramicroscopy* **55**, 101 (1994).
2. M. Rösler et al., Proc. 2nd Int. Conf. Appl. Diamond Films and Related Mat., Tokyo, (1993).
3. We thank F. Hofer for letting us use the Philips CM 20 with GIF at the Technical University Graz and the Stiftung Volkswagenwerk for support under contract I/69931.

Proc. Microscopy and Microanalysis 1995, edited by G.W. Bailey, M.H. Ellisman, R.A. Hennigar, and N.J. Zaluzec

Copyright © 1995 MSA. Published by Jones and Begell Publishing, 79 Madison Ave., New York, NY 10016

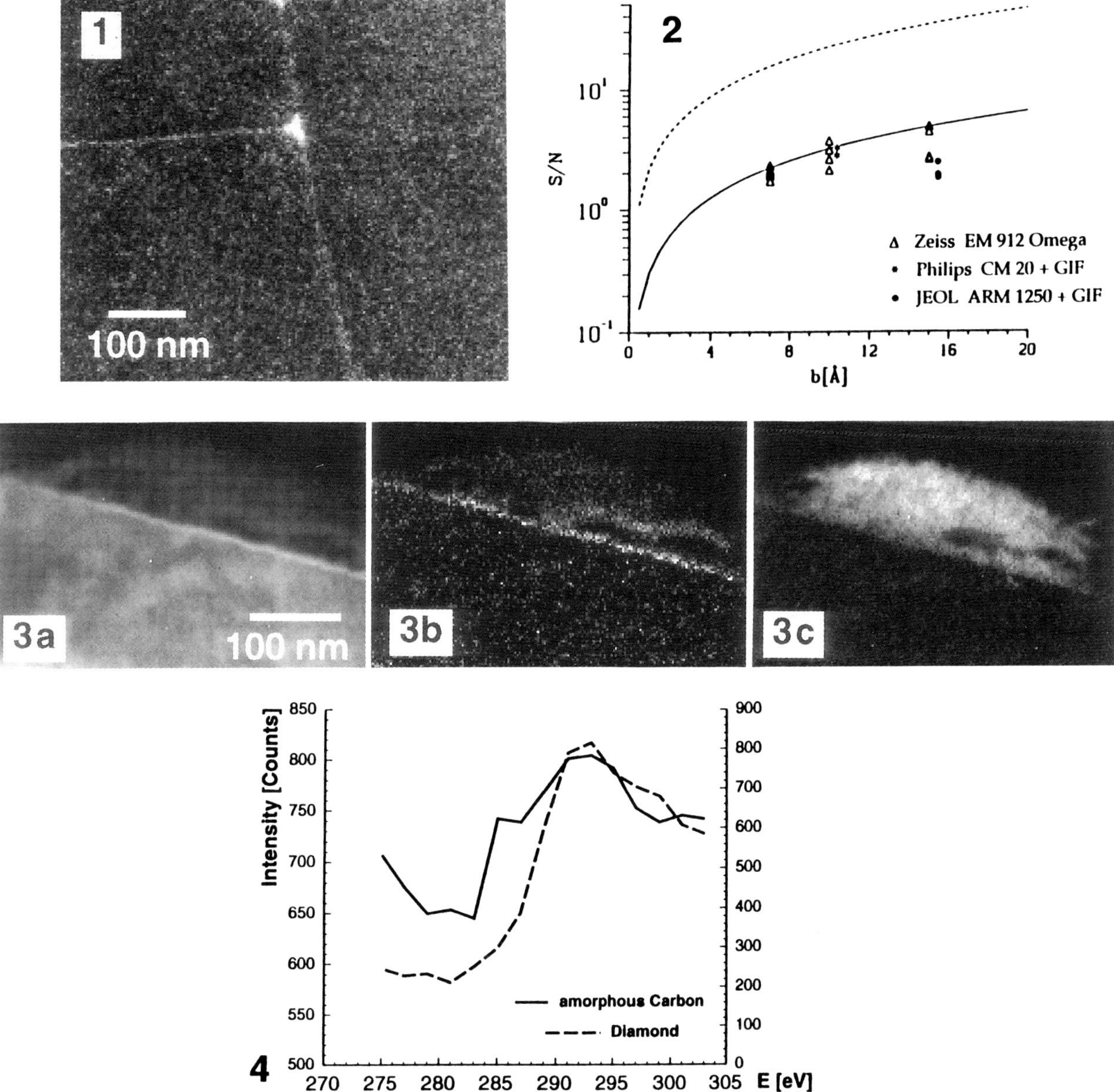

Fig. 1 Oxygen distribution image of a Si_3N_4 sample with 0.7 nm thick grain boundary films, acquired on the Zeiss EM 912 Omega. A triple junction and the amorphous oxide films at the grain boundary can clearly be seen.

Fig. 2 Measured and calculated[1] (solid line) S/N ratio in the oxygen distribution images of the grain boundary films. Integration over 50 pixcls parallel to the boundary increases the S/N ratio to the dashed curve. Experimental results were obtained on the Zeiss EM 912 Omega (0.7, 1.0 and 1.5 nm films), a Philips CM 20 with GIF (1.0 nm films) and the Jeol ARM 1250 with GIF (1.5 nm films).

Fig. 3 Cross-section of a thin diamond film grown by CVD on a Si substrate. a) ESI image obtained at $\Delta E = 283$ eV. b) Background subtracted image for $\Delta E = 283$ eV, showing the amorphous carbon layer, and c) for $\Delta E = 289$ eV, showing the diamond film.

Fig. 4 EELS spectra which were extracted from a series of ESI images in the regions containing the amorphous carbon layer and the diamond film (left and right intensity scale for the amorphous carbon and the diamond film, respectively).

POINT-DEFECT DISTRIBUTIONS IN CERAMICS BY SPECTRUM-LINE PROFILING

John Bruley, David B. Williams and Ming-Wei Tseng

Department of Materials Science and Engineering, Lehigh University, Bethlehem, PA 18015-3195.

Electron energy-loss spectroscopy is well suited to the study of interfaces because the ionization edges can be used to identify the atomic species present and the near edge fine structures on the edges reveal information on the electronic structure surrounding these atoms. This information can be extracted from regions of sample approaching atomic dimensions. A challenge is to be able to map and interpret spectral features with a level of confidence that will allow explanations of physical properties of interest.

The term Spectrum-Line is used here to refer to a linear array of spectra, recorded sequentially along a line traced out by the electron probe and is the 1 dimensional analog of the Spectrum-Image. Processing Spectrum-Lines to generate quantitative profiles can provide a vivid picture of changes in composition, a valence state, or site occupancy across an interface. To aid interpretation and guide further spectrum analysis the near-edge fine structures can be compared to spectra from compounds whose atomic coordinations are well known or modeled using the results of band-structure or molecular orbital calculations of electronic structures. Multiple least squares analysis can be carried out to extract composition or phase variations assuming there exists a relatively small number of suspected components. Very recently a target factorial analysis, which is able to transform and reduce large data sets into a much smaller statistically significant number of independent principal components, has been demonstrated.[1] Such data reduction is important for Spectrum-Imaging because even relatively small images result in extremely large data sets composed of more than 1000 spectra. Viewing and processing each spectrum individually is impractical and so the development of robust spectrum processing and feature recognition tools is a primary requirement.

A study of the interrelationship between the site occupancy of Al cations at a spinel-sapphire interface and the degree of non-stoichiometry provides a suitable demonstration of these points.[2] The ELNES on the Al $L_{2,3}$ edge is sensitive to local changes in the chemical environment and bonding of the atom and is thus an ideal probe for charge transfer, and coordination symmetry. The preliminary assessment is carried out with the Al $L_{2,3}$ absorption edge recorded from a series of spinel standards, selected for their known range of tetrahedral and octahedral occupancies, Figure 1. In bulk materials, the study of chemical and bonding changes are relatively easy because the EELS signal is strong, and there are virtually no spatial resolution limitations. The situation is different at an interface because the changes are localized to a nanometer or less, and the signal changes may be quite small. To maximize the interface signal to background ratio, the probe must sample as much of the interface as possible whilst reducing the overlap with the neighboring bulk. Spatially resolved chemical composition and spectroscopic line profiles have been gathered from the space charge region across a spinel/sapphire interface using a dedicated scanning transmission electron microscope operating at 100 keV. The electron energy-loss spectra reveal an excess of Mg (relative to stoichiometric spinel) along with Cr segregated to the interfacial zone, Figure 2. A least squares decomposition of a spectrum-line profile of the Al $L_{2,3}$ absorption edge fine structures into two standard components clearly highlights the transition from sapphire into spinel; closer analysis by Gaussean peak fitting indicates that the degree of site inversion, which is the fractional occupancy of tetrahedral sites by trivalent Al, increases within 5 nm of the boundary, Figure 3. The overall charge neutrality at the interface is maintained by the increased concentration of negative charge, probably interstitial O anions which are also present in excess quantity relative to stoichiometric spinel and sapphire.[3]

References

1. S.J. Splinter, J. Bruley, D.A. Smith, MAS proceedings, Breckenridge, Co (1995)
2. J. Bruley, M.W. Tseng, D. B. Williams, Microsc. Microstruc. Microanal., **6** no. 1 (1995)
3. Research supported through NSF DMR 930 625

Proc. Microscopy and Microanalysis 1995, edited by G.W. Bailey, M.H. Ellisman, R.A. Hennigar, and N.J. Zaluzec
Copyright © 1995 MSA. Published by Jones and Begell Publishing, 79 Madison Ave., New York, NY 10016

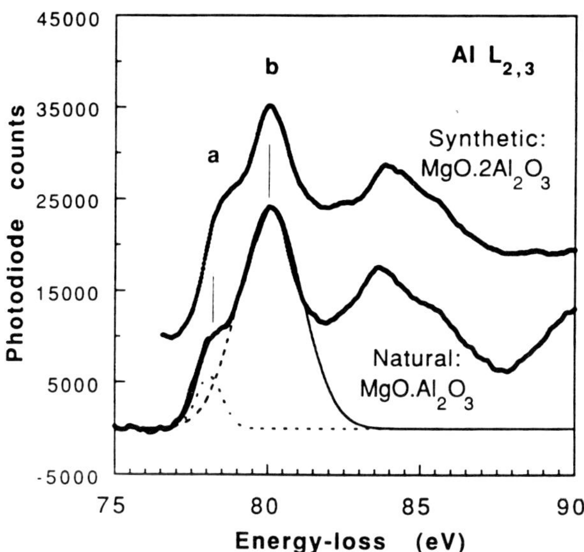

Figure 1 Al $L_{2,3}$ edge for partially inverse and partially normal spinel. Peak-a scales with tetrahedral occupancy.

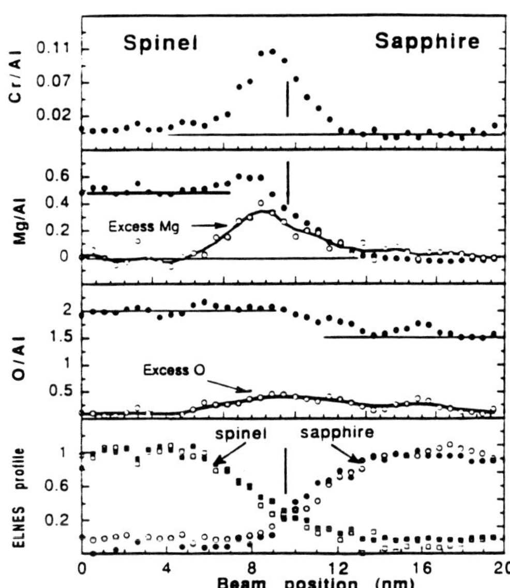

Figure 2 Composition line-profiles of spinel/sapphire interface. Revealing Cr segregation and slight excess of Mg. Lower phase composition curves are multiple least squares fits to standard edge shapes take from end members of a Spectrum-Line

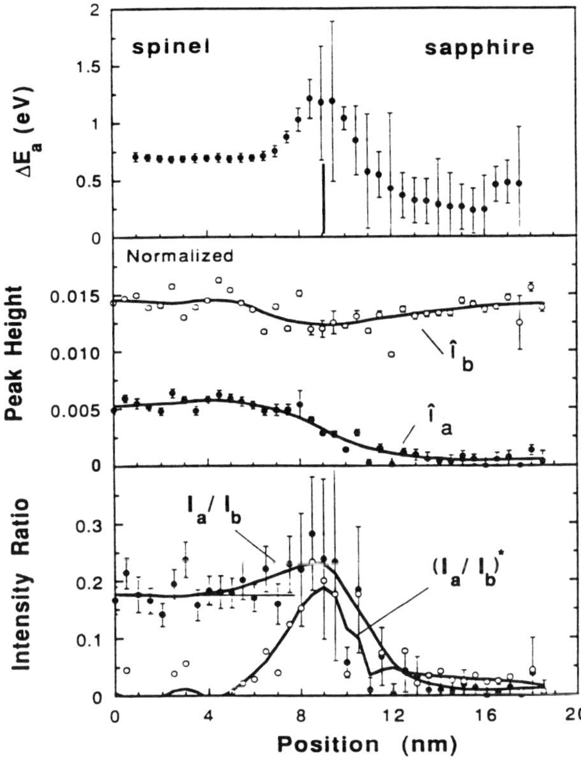

Figure 3 Line-profiles extracted from Spectrum-Lines by gausean peak fitting to near edge fine structures. The increased intensity of the peak a is indicative of increased site inversion.

QUANTITATIVE ANALYSIS OF SPATIALLY RESOLVED VALENCE ELECTRON ENERGY-LOSS SPECTRA FOR ELECTRONIC STRUCTURE STUDIES OF GRAIN BOUNDARIES AND THIN INTERGRANULAR FILMS

Roger H. French

DuPont Co. Central Research, E356-323 Exp. Sta., Wilmington DE 19880

The spatial variation of the electronic structure at interfaces is critical to both interatomic bonding at atomically abrupt interfaces such as grain boundaries[1] and also to the development of van der Waals (vdW) attraction forces at partially wetted interfaces[2,3]. This interfacial electronic structure, as represented by the interband transition strength $[J_{cv} = \frac{\hbar^2 \omega^2}{8\pi^2} i \left(\varepsilon_1(\omega) + i\varepsilon_2(\omega) \right)^*]$[4], can be determined by Kramers Kronig (KK) analysis of either vacuum ultraviolet (VUV) optical reflectance spectra or spatially resolved valence electron energy loss (SR-VEEL) spectra. Quantitative analysis of SR-VEELS requires accurate spectral line shapes coupled with single scattering deconvolution, convergence correction, and KK analysis. Both the energy loss functions (Fig. 1) and the interband transitions (Fig. 2) determined for α-Al_2O_3 using SR-VEELS compare well with the VUV results. In addition the use of the spectral line scan method, whereby typically 200 SR-VEEL spectra are acquired along a scan line of 20 nm, helps overcome many uncertainties in the data acquisition and analysis.

Si_3N_4 exhibits equilibrium thin intergranular films[5] whose thickness is determined by a force balance, where the contribution of the van der Waals (vdW) attractive force is dictated by the interband transition strengths of the electrons in the interatomic bonds. The vdW attraction forces, as given by the Hamaker constant (A)[6] can be calculated directly from the interfacial electronic structure using Liftshitz's quantum field theory results. The London dispersion transform of $J_{cv}(\omega)$ is used to calculate $\varepsilon_2(\xi)$ for the phases present and then spectral differences produce the Hamaker constant.

Hamaker constants (A_{121})of the van der Waals interaction for Si_3N_4 as material 1 and various intergranular materials as material 2. Values are in zepto joules (10^{-21} J)					
Method	Bulk Glass	Intergranular	Silicon	Undoped SiO_2	Vacuum
VUV	8.3		29.4	33.3	174.3
SR-VEEL	4.3	22.5			190.2

Interband transition strengths and full spectral Hamaker constants for Si_3N_4 samples containing a SiYAlON glass can be determined using SR-VEELS from grains and grain boundaries and compare with results from bulk VUV spectroscopy[7]. The VEELS method permits the determination of vdW forces based upon actual local compositions and structure, which may differ noticeably from bulk standards. For these polycrystalline Si_3N_4 samples, the grain boundary glass composition is found to be of lower index of refraction, which is consistent with an increased oxygen content in the intergranular glass, and this leads to an increased value of the full spectral Hamaker constant (24 zJ) determined in situ from the SR-VEELS of the grain boundary film.

[1]   H. Müllejans, J. Bruley, R. H. French, P. A. Morris, MRS Proceedings, ed. by M. Sarikaya, (1994).

[2]   D. R. Clarke, **J. Am. Ceram. Soc., 70**,15-22, (1987).

[3]   Y. M. Chiang, L. E. Silverman, R. H. French, R. M. Cannon, **J. Amer. Ceram. Soc., 77**, 1143-52, (1994).

[4]   S. Loughin, R. H. French, W. Y. Ching, Y. N. Xu, G. A. Slack, **Appl. Phys. Lett., 63**, 9, 1182-4, (1993).

[5]   H.-J. Kleebe, M. K. Cinibulk, R. M. Cannon, M. Rühle, **J. Am. Ceram. Soc., 76** 1969-77 (1993).

[6]   R. H. French, R.. M. Cannon, L. K. DeNoyer, Y.-M. Chiang, **Solid State Ionics**, (1995).

[7]   R. H. French et al., MRS Proceedings, ed. by D. Bonnell, (1995).

Proc. Microscopy and Microanalysis 1995, edited by G.W. Bailey, M.H. Ellisman, R.A. Hennigar, and N.J. Zaluzec

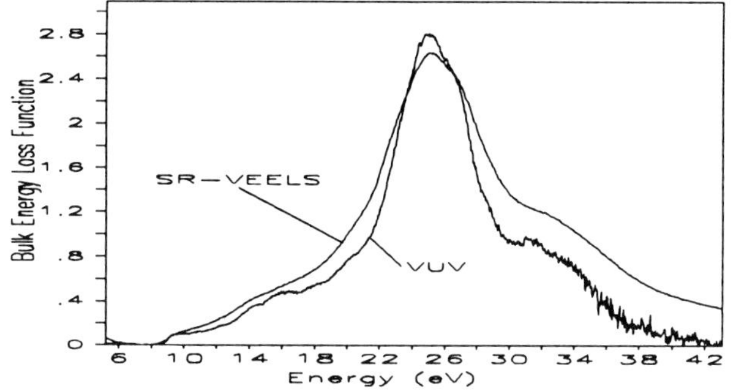

Figure 1. Energy loss function of α-Al$_2$O$_3$ from VUV and SR-VEEL spectroscopy

Figure 2. Interband Transition Strength of Bulk α-Al$_2$O$_3$ from VUV and SR-VEEL spectroscopy.

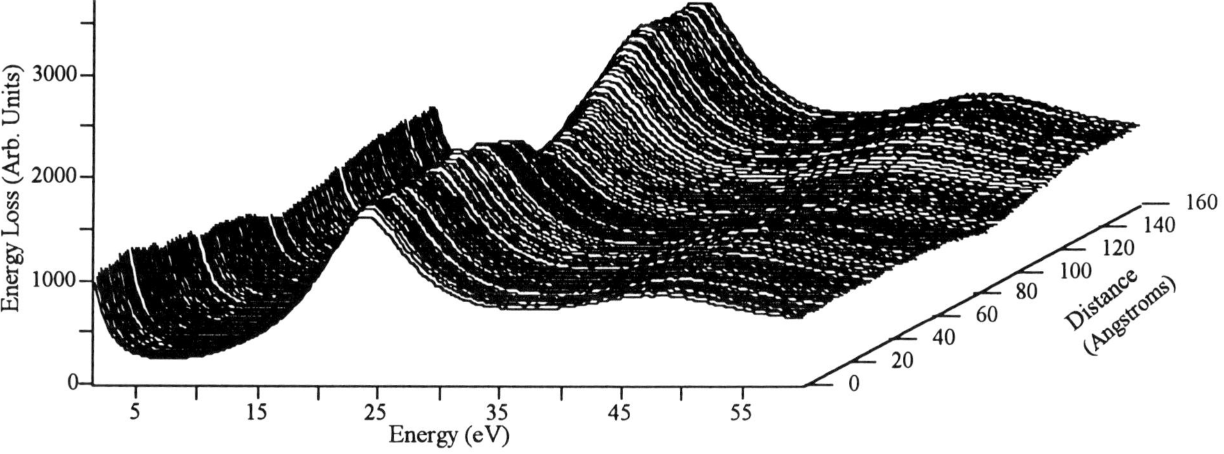

Figure 3 As acquired SR-VEELS spectra across a intergranular film in Si$_3$N$_4$.

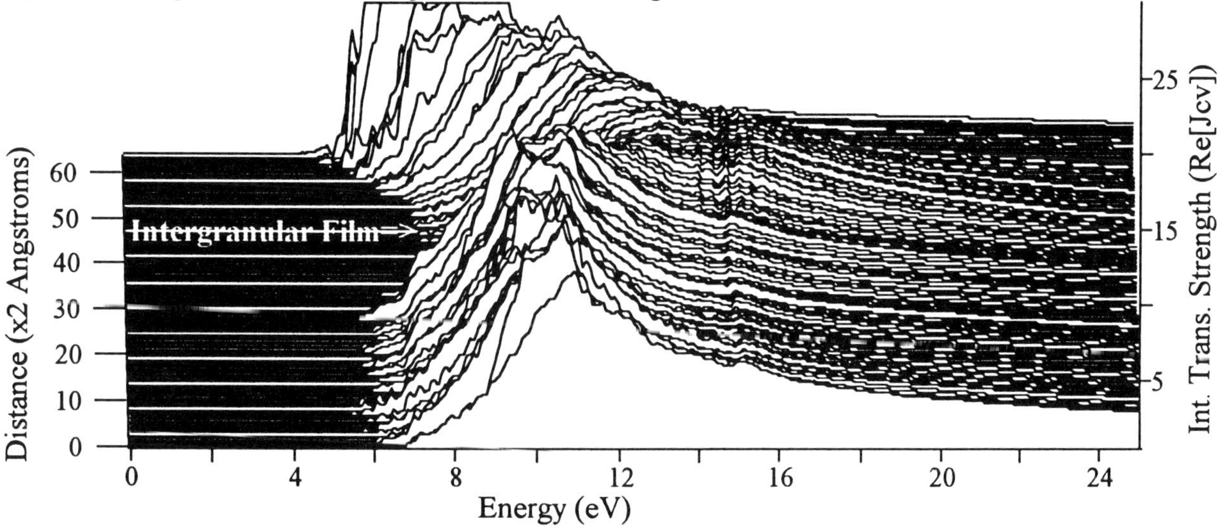

Figure 4. Interband Transition Strengths (J$_{cv}$) of Si$_3$N$_4$ and its intergranular film in the SiYAlON doped system. These results obtained from quantitative analysis of the SR-VEEL spectra of Figure 3. Notice the increase of the band gap energy from 6 eV in Si$_3$N$_4$ to 8 eV in the intergranular film.

Acknowledgments: D. J. Jones, S. Loughin, L. K. DeNoyer, C. Argento, C. Scheu, G. Duscher, H. Müllejans, M. J. Hoffmann,. M. Rühle, R. M. Cannon M. Walls, A. Howie, H. Ackler, Y. M. Chiang,

DIRECT IMAGING OF SPATIALLY VARYING POTENTIAL AND CHARGE ACROSS INTERNAL INTERFACES IN SOLIDS

V. Ravikumar, R. P. Rodrigues and V. P. Dravid

Department of Materials Science and Engineering, Northwestern University, Evanston, IL 60208, USA

The importance of spatially varying potential (and thus charge) across lattice discontinuities in solids has been recognized in many technologically important systems, especially those containing electrically active interfaces, e.g. electroceramics. The presence of spatially varying potential across electroceramic interfaces has been indirectly deduced and analyzed using predominantly bulk measurement techniques like I-V, C-V curves and impedance spectroscopy.[1] Direct imaging of spatially varying electrostatic potential profile and determination of the sign, magnitude and spatial distribution of the associated interface- and space- charge (and therefore defect density) in electroceramics have remained elusive.

We have utilized the technique of transmission high energy electron holography to directly image and quantify the electrostatic potential across grain boundaries (GBs) in $SrTiO_3$, a functional electroceramic. The phase of the exit wave function at the GB region can be altered by : (i) variation in local mean inner potential (related to the change in density of atoms at the GB), (ii) differential diffraction conditions across the interface, (iii) change in the local specimen thickness and, (iv) presence of local electrostatic (electrical charge) and magnetic potential.[2] Under weak diffraction conditions or certainly in the case of tilt bicrystals, the differential phase change across the GB due to diffraction is minimal, and the changes in the phase of electron at and in the vicinity of the GB can be readily interpreted. Fig. 1 is a high resolution electron hologram of an undoped (electrically inactive) $\Sigma = 5$ (310) symmetrical tilt GB in $SrTiO_3$. The reconstructed phase image (Fig. 2) has virtually no contrast at the GB. However, a line scan of the phase profile across the GB averaged along the symmetry axis of the interface (inset, Fig. 2) clearly indicates a small increase in the phase (ranging between 0.1 and 0.125 radians for several boundaries) at the GB core (within 1.8±0.6nm), due to a reduction in the mean inner potential which is a manifestation of the reduction in density of atoms at the GB. Fig. 3 shows an electron hologram of a Mn-doped GB in polycrystalline $SrTiO_3$ taken with both crystals under very weak diffraction conditions. Mn has been found to be ubiquitous at all GBs (using SIMS) and confined to the GB core (EDS and EELS) predominantly with a valence of +2.[3] The shifting of the electron biprism fringes at the GB core towards vacuum is evident, indicating a negative GB charge. The reconstructed phase image (Fig. 4A) and the corresponding line scan of the phase (Fig. 4B) across the GB shows a substantial decrease in phase at the GB core. The double Schottky barrier height at this GB is determined to be about 8V. The spatial extent of the space-charge region is found to be about 2.2 nm on either side of the GB, consistent with the typical Debye length. A sizable variation in the magnitude of the phase decrease (and thus the double Schottky barrier height (from 1 to 8 V) and the net charge) was observed for the various Mn-doped polycrystalline GBs. Fig. 4C is the charge density profile obtained from the phase profile, and it clearly shows the presence of negative charge centered at the GB core and a compensating positive space-charge across the GB. This indicates that Mn^{+2} substitutes Ti^{+4} at the GB core resulting in negatively populated defects at the GB core which are compensated by positively charged oxygen vacancies or holes. The charge density profile yields the net charge accumulation at the GBs, which corresponds to 30% to 70% substitution of Ti with Mn^{+2} at the core of the GB. This is the first direct real space evidence for GB charge and the compensating space-charge in electroceramics. This report demonstrates that electron holography is a viable bulk sensitive tool to quantitatively probe the changes in electrostatic potential, and thus the magnitude, sign and spatial extent of electrical charge at and across internal interfaces in solids.[4]

References

1. *Grain Boundary Phenomena in Electroceramics, Advances in Ceramics Vol. 1*, L. M. Levinson and D. C. Hill, Eds. (Am. Ceram. Soc., Columbus, OH, 1986).

Proc. Microscopy and Microanalysis 1995, edited by G.W. Bailey, M.H. Ellisman, R.A. Hennigar, and N.J. Zaluzec
Copyright © 1995 MSA. Published by Jones and Begell Publishing, 79 Madison Ave., New York, NY 10016

2. A. Tonomura, *Reviews of Modern Physics* **59(3),** 639-668 (1987); M. Gajdardziska-Josifovska et al.,*Ultramicroscopy* **50,** 285-299 (1993).
3. M. M. McGibbon *et al., Science* **266,** 102 (1994); N. Wilcox et al., *Solid State Ionics*, in press.
4. Research supported by US DOE, Grant No. DE-FG02-92ER45475.

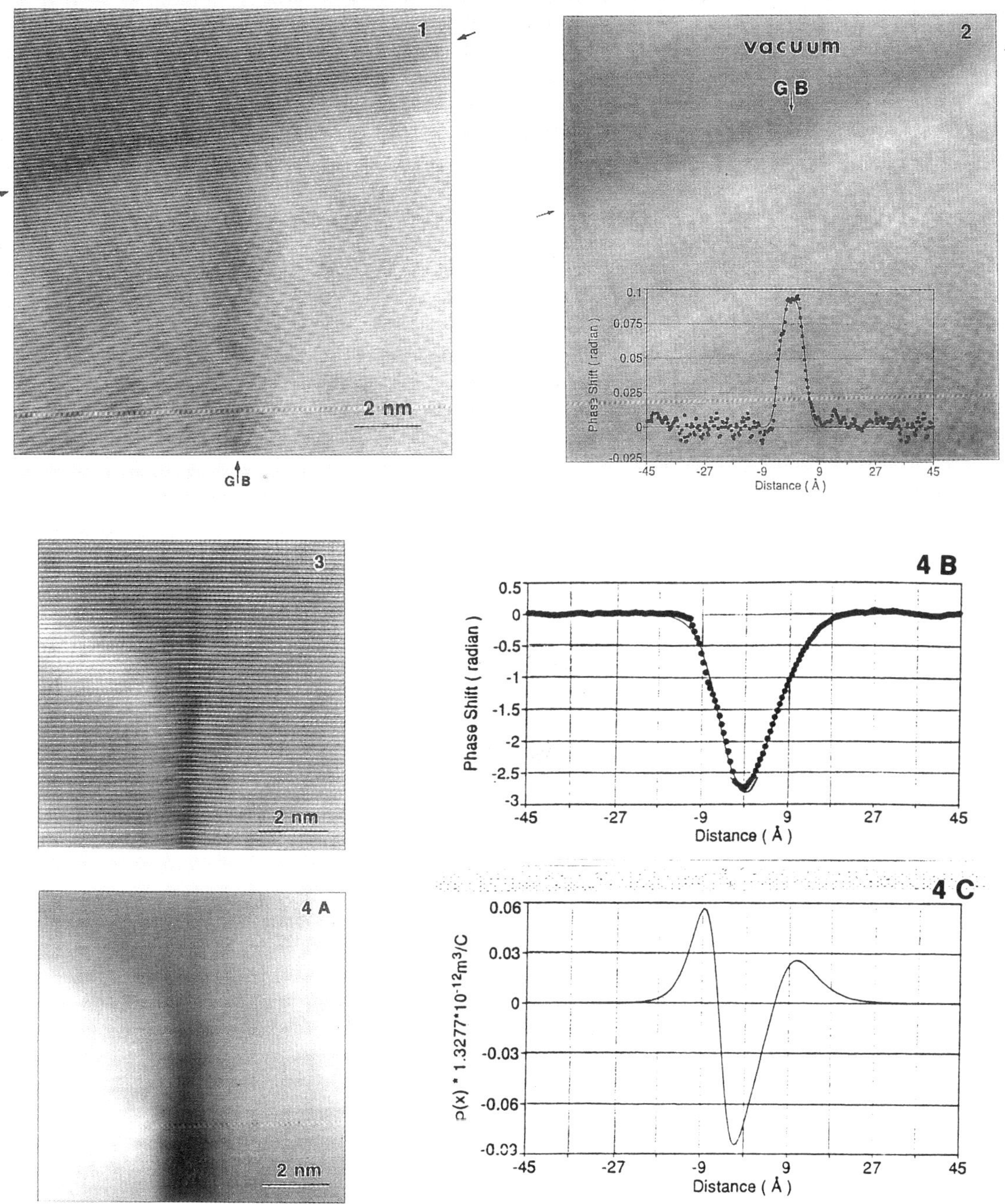

Fig. 1.--Electron hologram of undoped Σ=5 (310) symmetrical tilt GB in SrTiO$_3$
Fig. 2.--Phase reconstruction of the hologram in Fig. 1. Phase profile across GB is shown as inset.
Fig. 3.--Hologram of a Mn doped GB in SrTiO$_3$
Fig. 4.--(A) Phase reconstruction of the hologram in Fig. 3. (B) Phase scan across the Mn doped GB in Fig. 3, and (C) corresponding charge density profile.

APPLICATION OF AEM TO CHEMICAL AND STRUCTURAL CHARACTERIZATION OF THE AlN-Al₂O₃ AND AlN-Al₂O₃-SiO₂ SYSTEMS

A. D. Westwood

IBM, Thomas J. Watson Research Center, Yorktown Heights, NY 10598

The thermal, electronic and mechanical properties of aluminum nitride (AlN) make it an attractive material to a number of technologically important areas; microelectronic packaging, high temperature semiconductors, opto-electronic and piezoelectric devices and structural ceramics. It is well established that the concentration and distribution of impurities can control the macroscopic properties of materials. AlN is a classic example, with oxygen (O) being the dominant controlling impurity. The role of O in point defect formation and its detrimental effect on thermal conductivity was first documented by Slack.[1] Oxygen has subsequently been shown to cause the formation of two-dimensional extended defects of which two variants exist, planar and curved.[2-6] Both defects have been identified as O-rich inversion domain boundaries (IDB's)[3] (Fig.1). Because of the controlling influence that O has on thermal conductivity, it is important to fully understand the defect structures and chemistries that exist as a result of O incorporation. AlN-Al₂O₃ is a prototype system for understanding the role that non-stoichiometry (impurities) plays in IDB formation.

The interfacial structure and chemistry of the planar and curved IDB's have been studied using a large array of analytical techniques, conventional Transmission electron microscopy (CTEM), analytical electron microscopy (AEM), convergent beam electron diffraction (CBED), high-resolution transmission electron microscopy (HRTEM) and atomistic computer simulations. The planar IDB structure consists of a basal plane of Al that is octahedrally coordinated to O and N in a ratio of 3:1, with structural relaxations at the interface in the form of a contraction of the adjacent basal planes due to the presence of Al-O bonds rather than Al-N bonds, and a small rotation about $<0001>$ across the interface.[4] The displacement vector across the interface is $\mathbf{R} = 1/3 <10\bar{1}0> + 0.387 <0001>$. The planar IDB structure described here is similar to the original model proposed by Harris et al.;[2] however, the c-axis displacement is $\simeq 17\%$ larger, the interface is electrically neutral and structural relaxations are incorporated. The curved IDB is far more complex than its planar counterpart, exhibiting three distinct morphologies; curved, facetted $\{10\bar{1}1\}$ and corrugated.[5] The facetted segments possess a $\mathbf{R} = 0.23 <0001>$ displacement. The displacement vector across the curved portions was found to vary between $\mathbf{R} = 0.10 <0001>$ and $0.43 <0001>$. The variation is believed to be due to local variations in chemistry. Measurements of O concentration across and along the curved IDB's support this hypothesis. Consistently high O concentrations were found on the facetted segments with low fluctuating concentrations being present on the curved sections. In addition, atomistic calculations indicate that the curved IDB interface was most stable when both Al vacancies and O ions were present in the form of an atmosphere around the interface. The calculations also reveal that the interface energy was independent of the displacement vector in the range of $\mathbf{R} = 0.05 <0001>$ to $0.35 <0001>$.

With continued oxygen incorporation into AlN, polytypoid phases form, which have commonly been described as arrays of stacking faults.[7] A direct extension of the studies on isolated planar and curved IDB's reveal that the polytypoid phases are not comprised of stacking faults, but rather alternating planar and corrugated IDB's (Fig.2).[8] A corrugated IDB is a curved IDB that has been forced to lie on a basal plane due to the close proximity and constraining influence of adjacent planar IDB's. The influence of small Si additions ($\simeq 1000$ ppm) was also studied and found to have a dramatic effect on the formation and structure of polytypoids in the SiAlON ceramic system. First, it appears that Si causes the creation of new distinct planar IDB different from the planar IDB observed in the AlN-

Proc. Microscopy and Microanalysis 1995, edited by G.W. Bailey, M.H. Ellisman, R.A. Hennigar, and N.J. Zaluzec
Copyright © 1995 MSA. Published by Jones and Begell Publishing, 79 Madison Ave., New York, NY 10016

Al$_2$O$_3$ polytypoids. Second, the addition of Si alters the structure of the corrugated IDB, such that it appears to become planar, and third, the Si-containing polytypoid phases form more readily than the AlN-Al$_2$O$_3$ polytypoid phases.

References

1. G. A. Slack, *J. Phys. Chem. Solids* **34**, (1973) 321.
2. J. H. Harris et al., *J. Mater. Res.*, **5**, (1990), 1763.
3. A. D. Westwood and M. R. Notis, *J. Am. Ceram. Soc.*, **74**, (1991),1226.
4. A. D. Westwood et al., *J. Mater. Res.*, **10[5]**, (1995).
5. A. D. Westwood et al., *J. Mater. Res.*, **10[5]**, (1995).
6. A. D. Westwood et al., *J. Microsc.*, **167**, (1992), 287.
7. K. H. Jack, *J. Mat. Sci.*, **11**, (1976), 1135.
8. R. A. Youngman et al., *Mat. Res. Soc. Symp. Proc.*, **319**, (1994), 45.
9. The author gratefully acknowledges the contributions of R. A. Youngman, J. R. Michael, M. R. McCartney, M. R. Notis, A. N. Cormack, J. Bruley and S. Witek, during the course of this work.

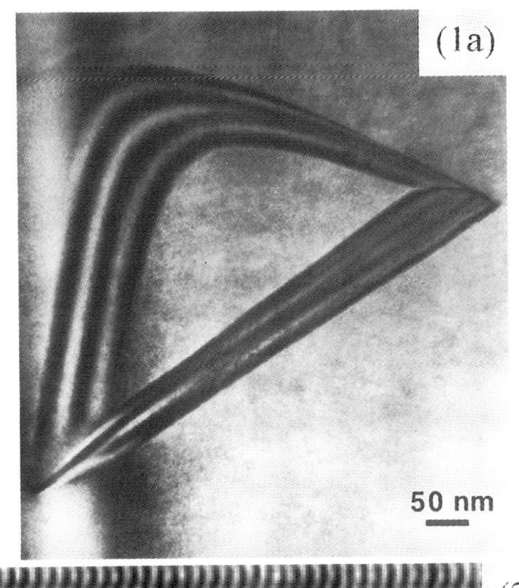

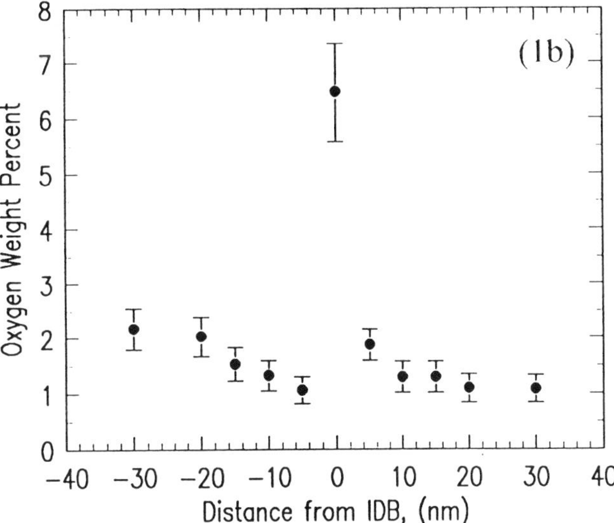

Fig.1.-(a) Conventional TEM bright field image of a planar and curved IDB: (b) Oxygen concentration profile across a planar IDB showing strong segregation to the interface.

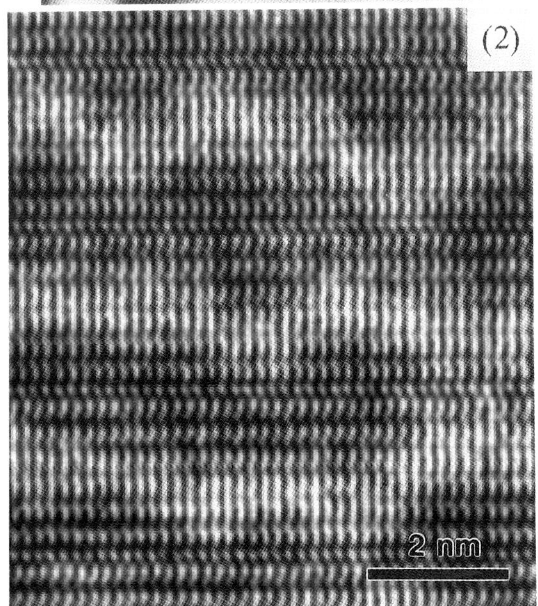

Fig.2.-High-resolution TEM image of a polytypoid structure in AlN-Al$_2$O$_3$ using imaging conditions to observe the corrugated IDB (bright contrast) lying between the planar IDB's.

ATOMIC RESOLUTION DETERMINATION OF THE STRUCTURE AND CHEMISTRY OF CERAMIC GRAIN BOUNDARIES

N. D. Browning*, M. M. McGibbon**, M. F. Chisholm** and S. J. Pennycook**

*Dept. Physics (M/C 273), U. Illinois at Chicago, 845 W. Taylor St., Chicago, IL 60607-7059. USA
**Solid State Division, Oak Ridge National Laboratory, P. O. Box 2008, Oak Ridge, TN 37831. USA

Characterization of grain boundaries in ceramics is complicated by the multicomponent nature of the materials, the presence of secondary phases, and the tendency for the grain boundary plane to "wander" on the length scale of a few nanometers. However, recent developments in the scanning transmission electron microscope (STEM) have now made it possible to correlate directly the structure, composition and bonding at grain boundaries on the atomic scale. This direct experimental characterization of grain boundaries is achieved through the combination of Z-contrast imaging (structure) and electron energy loss spectroscopy (EELS) (composition and bonding). For crystalline materials in zone-axis orientations, where the atomic spacing is larger than the probe size, the Z-contrast technique provides a direct image of the metal (high Z) columns[1]. This image, being formed from only the high-angle scattering, can be used to position the electron probe with atomic precision for simultaneous EELS[2]. Under certain collection conditions, the spectrum can have the same atomic spatial resolution as the image[3], thus permitting the spectra to be correlated with a known atomic location.

Utilizing these combined techniques, a series of [001] tilt boundaries in commercially available $SrTiO_3$ bicrystals were investigated. Maximum entropy processed images[4] of a 25° symmetric and a 45° asymmetric boundary are shown in figures 1(a) and 2(a). The symmetric boundary has a dilatation normal to the boundary plane of $(0.6\pm0.2)Å$ and a rigid body translation parallel to the boundary of $(0\pm0.2)Å$, while the asymmetric boundary shows a contraction of $(0.56\pm0.16)Å$. Titanium L-edge and oxygen K-edge profiles, taken in 4Å steps across both boundaries, indicate that the titanium valence remains bulk-like (+4) in the boundary and the linear titanium-oxygen coordination is preserved. To rationalize the spectroscopic data with the heavy metal atom locations determined directly from the image, bond valence sum analysis was performed. In these calculations, the amount that a particular bond contributes to the overall valence of the elements involved in the bond is dependent on the bond-length. For the two grain boundaries here, oxygen atoms were positioned so as to maintain the bulk valences and preserve linear titanium oxygen coordination (Figures 1(b) and 2(b))[5].

Noticeable in the structures are positions where two columns are seperated by distances much shorter than in the bulk material. Since the Z-contrast technique is incoherent in nature, both of these columns must exist, and one possible explanation is partial occupancy. Assuming the simplest model is that each of the columns is alternately occupied through the thickness of the material avoids this problem of like-ion repulsion. The boundary structures shown in figures 1 and 2 can also be seen to be composed of regular repeating units. From studies of other grain boundaries in $SrTiO_3$, a series of repeating "structural units" have been identified for both symmetric (figure 3a) and asymmetric (figure 3b) boundaries. These structural units can now be applied using the principles of continuity of grain boundary structure[6], to predict the grain boundary structure of all symmetric[7] and asymmetric[8] [001] tilt boundaries in $SrTiO_3$[9].

References

1. S. J. Pennycook and D. E. Jesson, *Phys.Rev.Lett* 64(1990)938
2. N. D. Browning et al, *Nature* 366(1993)146
3. S. J. Pennycook et al, in press *Nucl. Inst & Methods*
4. A. J. McGibbon et al, submitted *J. Microscopy*
5. M. M. McGibbon et al, *Science* 266(1994)102
6. A. P. Sutton & V. Vitek, *Phil. Trans. R. Soc. Lond. A* 309(1983)1
7. N. D. Browning et al, in press *Interface Science*

8. M. M. McGibbon et al, submitted *Phil.Mag*
9. Research sponsored by DOE-DMR, under contract DE-AC05-84OR21400 with Martin Marietta
   Energy systems Inc, and by the ORNL Postdoctoral Research Program administered by ORISE.

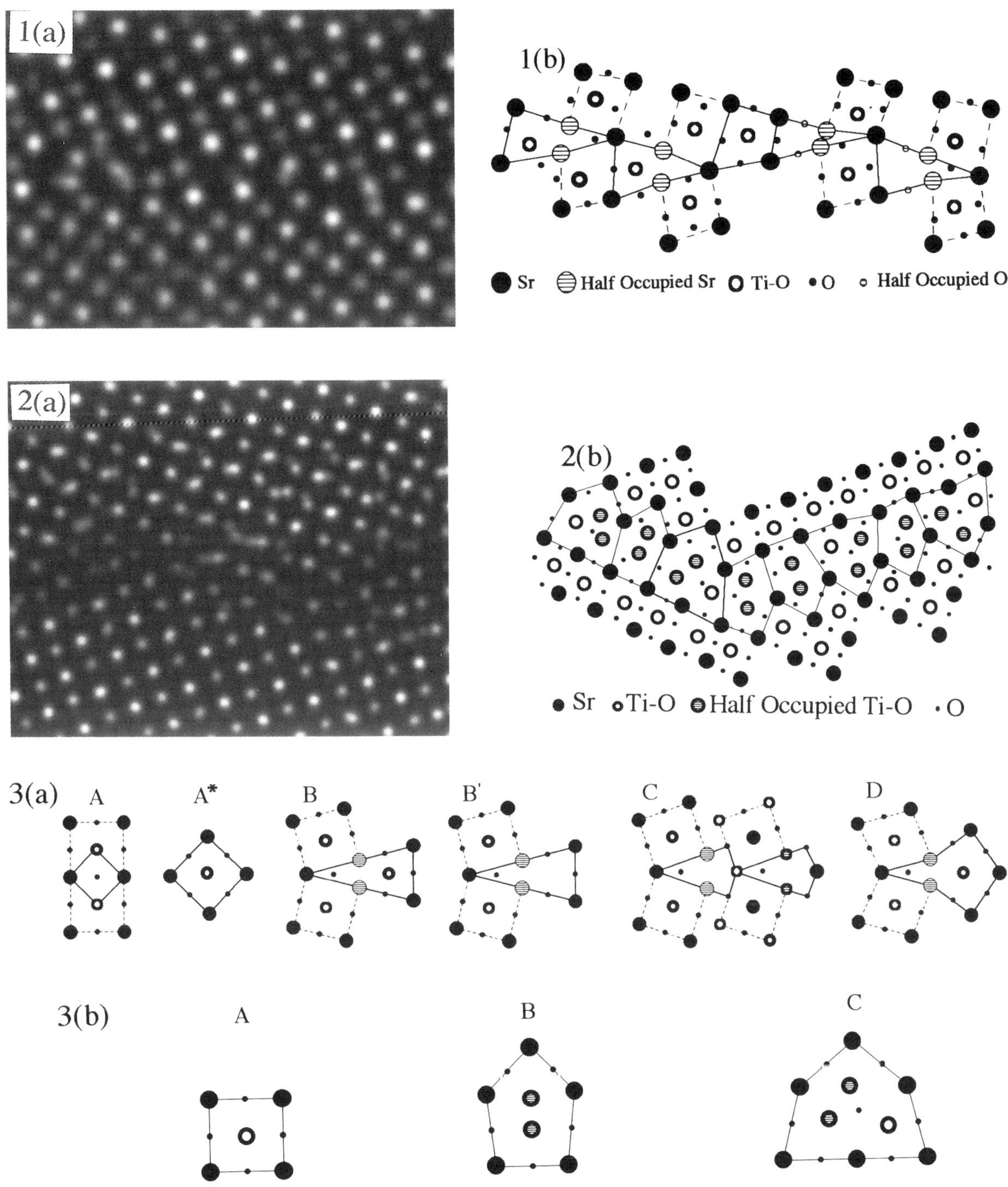

Figure 1:(a) Maximum entropy image and (b) model of a symmetric 25° [001] tilt boundary in SrTiO₃
Figure 2:(a) Maximum entropy image and (b) model of an asymmetric 45° [001] tilt boundary in SrTiO₃
Figure 3: Structural units for (a) symmetric and (b) asymmetric [001] tilt boundaries in SrTiO₃.

QUANTITATIVE ANALYSIS OF HOLE DENSITY AND LATTICE PARAMETER AT GRAIN BOUNDARIES IN $YBa_2Cu_3O_{7-\delta}$

Yimei Zhu and J.M. Zuo*

Department of Applied Science, Brookhaven National Laboratory, Upton, NY 11973
*Physics Department, Arizona State University, Tempe, AZ 85287

One challenge for high T_c superconductor research is to overcome the low intergranular transport currents which hinder practical applications. Evidence shows that lattice mismatch at the boundaries contributes to the depletion in charge carriers (holes), while the strain fields surrounding the cores of boundary dislocations suppress the superconducting order parameter. Using high-spatial resolution EDX, EELS and CBED techniques with a 20 Å field emission probe, we studied cation segregation, variations in oxygen/hole content, and lattice parameters at and across grain boundaries for various Σ and near-Σ boundaries in bulk $YBa_2Cu_3O_{7-\delta}$. The geometry of those arbitrary grain boundaries was modeled using the constrained Coincidence Site Lattice (cCSL) theory, which requires c^2/a^2 being rational for coincidence boundaries in non-cubic crystals.[1,2]

In this presentation, we report our quantitative and simultaneous measurements of oxygen/hole content (Fig.1) and of lattice parameters (Fig.3) at the boundary regions. The oxygen/hole density was quantified by guassian fitting to the observed oxygen K-edge spectra (Fig.2) and comparing them with those from the same batch which had undergone oxygen reduction process (oxygen content was determined by thermal gravimetric analysis and mass-measurement techniques). The local lattice parameters of the same grain boundary were determined by matching the simulated high-order Laue zone (HOLZ) patterns with experimental ones using a least-square algorithm. Dynamic correction due to multiple scattering was also included.[3] Examples of experimental and simulated HOLZ patterns are shown in Fig.3.

To ensure the boundary is edge-on during measurement, we chose boundaries which have the a-b plane as their boundary plane. An example of the change of hole content and lattice parameters at three large-angle grain boundaries ($\Sigma10$, $\Sigma13$, and $\Sigma26$) are shown in Fig.4. In general, we found that large-angle grain boundaries in $YBa_2Cu_3O_{7-\delta}$ often exhibit a volume expansion. The increase of the c-lattice parameter near the boundary can be attributed to lattice distortion and/or to an oxygen deficiency at the boundary. The lattice distortion extends less than 30Å, while the deficiency can extend more than 150Å.[4] Comparison of the lattice parameters (or c/a ratio) in the boundary region and crystal matrix suggests that to reduce boundary lattice mismatch, or to form a coincidence boundary in $YBa_2Cu_3O_{7-\delta}$, the required grain-boundary constraint can be realized by adjusting the oxygen content at the boundary. Our observations convincingly demonstrate the capability of quantitative nano-probe analytical electron microscopy, and the validity of the cCSL model for studying grain boundaries in non-cubic system.[5]

References

1. Y. Zhu et al, *Philo. Mag. A*, **67** (1993) 11; *Interface Science* **1** (1993) 361.
2. Y. Zhu, *Philo. Mag. A*, **69** (1994) 717. · 3. J.M. Zuo, *Ultramicroscopy*, **41** (1992) 211.
4. Y. Zhu, J.M. Zuo, A.R. Moodenbaugh and M. Suenaga, *Philo. Mag. A*, **70** (1994) 969.
5. Work supported by the U.S. DOE DE-AC02-76H00016 and NSF 9015867.

Proc. Microscopy and Microanalysis 1995, edited by G.W. Bailey, M.H. Ellisman, R.A. Hennigar, and N.J. Zaluzec
Copyright © 1995 MSA. Published by Jones and Begell Publishing, 79 Madison Ave., New York, NY 10016

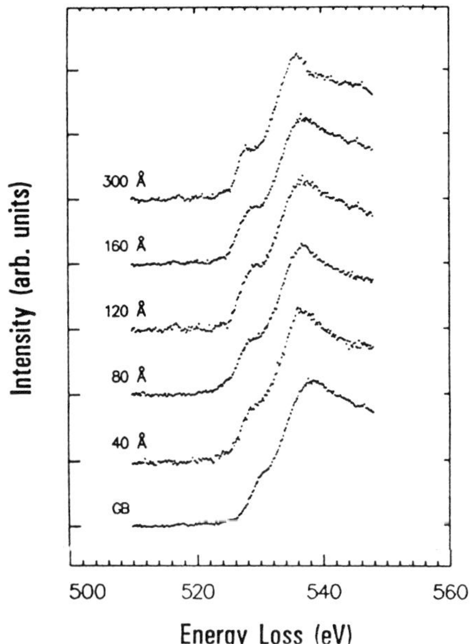

Fig.1 A series of EELS spectra of the oxygen K edge across an oxygen-deficient grain boundary.

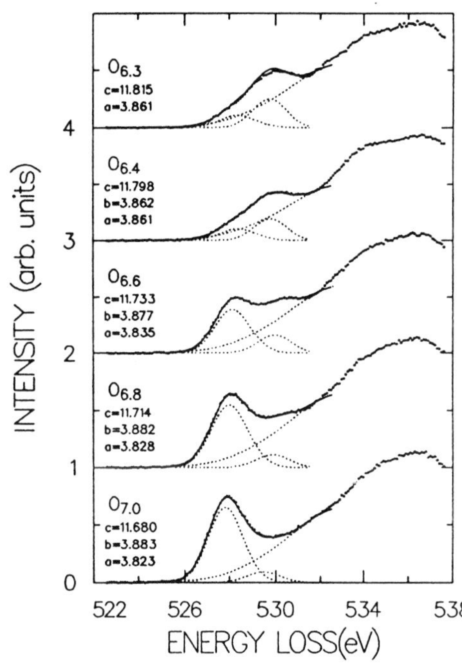

Fig.2 Spectra of oxygen K-edge for $YBa_2Cu_3O_{7-\delta}$ ($\delta\approx0\sim0.7$). The integrated intensity of the K-edge was fit by three gaussian functions.

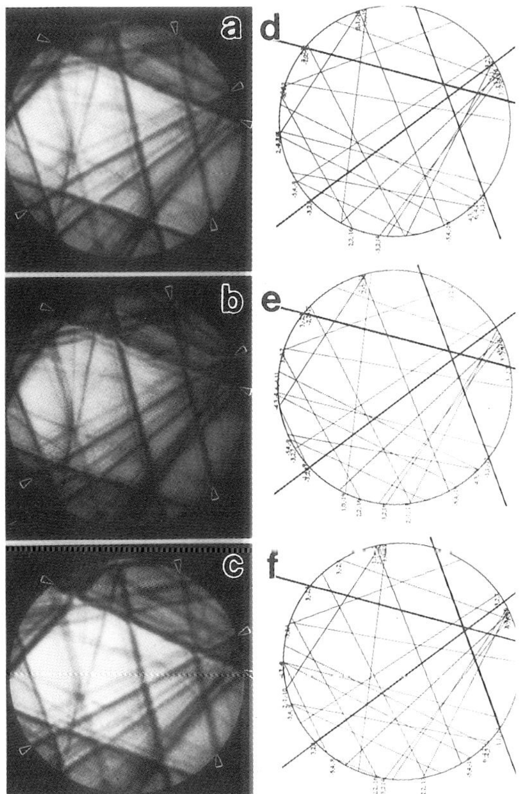

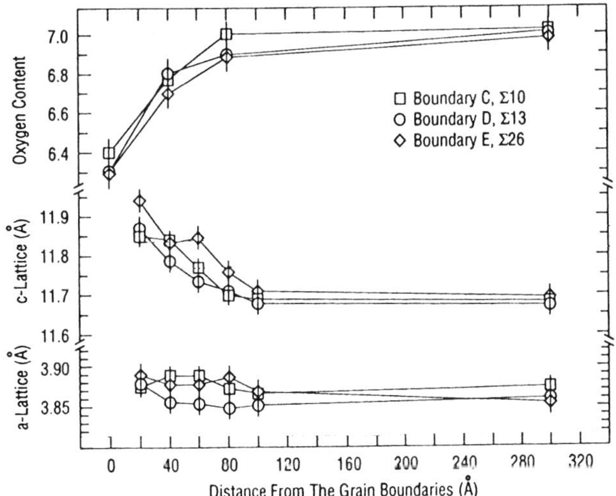

Fig.4 Hole density (described by the equivalent oxygen content) and lattice parameter as a function of the distance from the boundaries. The experimentally determined boundary misorientations are: Boundary A, 64.10° [0.998 0.003 0.006], corresponding system: Σ10, 66.42° [100], $\Delta\Theta=2.55°$; Boundary B, 83.78° [0.620 0.618 0.008], corresponding system: Σ13, 85.59° [110], $\Delta\Theta=2.88°$; and Boundary C, 58.10° [0.843 0.422 0.002], corresponding system: Σ26 52.02°[210], $\Delta\Theta=6.09°$.

Fig.3 An example of consecutive experimental HOLZ patterns recorded near a grain boundary (a)-(c). The lattice parameters retrieved from the best fit to the HOLZ patterns are: (d) $c=11.736$, $a=b=3.8793$, (e) $c=11.641$, $a=b=3.8485$, and (f) $c=11.655$, $a=b=3.8409$.

OBSERVATION OF GRAIN BOUNDARY SEGREGATION IN ADVANCED CERAMICS USING HIGH RESOLUTION IMAGING SIMS

R. Levi-Setti[1], K. K. Soni[1], J. M. Chabala[1], and A. M. Thompson[2]

1. Enrico Fermi Institute and Department of Physics, The University of Chicago, Chicago, IL 60637.
2. Department of Materials Science and Engineering, Lehigh University, Bethlehem, PA 18015.

The significance of grain boundaries in controlling processing and properties of ceramics is widely acknowledged. Through the addition of suitable dopants to ceramics, their processability and properties can be improved. These dopants may segregate to grain boundaries, but the characterization of boundary chemistry is a challenging task. Studies of segregation phenomena require the application of high-lateral-resolution techniques such as STEM/AEM or surface sensitive techniques such as AES, XPS. These techniques require rigorous sample preparation and have their limitations.

The scanning ion microprobe is a powerful tool that has exhibited unprecedented potential in the characterization of grain boundaries in ceramics.[1,2] When interfaced to a mass spectrometer (magnetic sector in our case), this instrument allows mapping of many trace elements at nanometer level in bulk specimens.[2] The combination of excellent sensitivity and high spatial resolution enables direct imaging of grain boundary segregants. The results thus obtained are free from artifacts that typically complicate analysis with broad beam, non-imaging techniques. Also, the nature (isotropic versus anisotropic) and extent of segregation (or depletion) can be evaluated with one map. This technique constitutes a novel method of characterizing the microchemistry of ceramics and complements traditional microstructural characterization.

Recently it has been found that 1000 ppm yttrium addition to polycrystalline Al_2O_3 reduces its tensile creep rate by two orders of magnitude.[3] Figure 1 shows SIMS images of a polished, unetched surface of a bulk alumina specimen doped with 1000 ppm Y/Al. The ISI (ion-induced secondary ion) image displays topographic contrast of the mapped area; occasional polishing scratches and residual pores are present. Two distinct features in the Y^+ SIMS map are: a cobweb-like structure representing the grain boundaries, and small secondary particles (presumably YAG precipitates) in the top left corner. All grain boundaries in this sample area are uniformly enriched in Y with respect to the grain interiors. Segregation of Y to grain boundaries in alumina was also detected by Dupau et $al.$ using STEM/EDS.[4] Comparison of the ISI and the Y^+ maps reveals that the surfaces of the isolated pores are also enriched with yttrium, as indicated by arrows. The Al_2O_3 grain size is reduced by the YAG precipitates: the Al_2O_3 grain size is larger in regions containing fewer precipitates. The analysis of a similar alumina sample hot-pressed with 1000 ppm La/Al is described in Fig. 2: La segregates isotropically to the grain boundaries. Note that the boundaries are completely and homogeneously enriched in La, similar to the yttria dooped sample. In addition, coarse second-phases (La_2O_3) are also present along the grain boundaries and pores are enriched in La, as indicated.

A major drawback of the SIMS method is that quantification of the secondary ion signal is often problematic, complicated by variations in secondary particle sputtering yields and by the emission of neutral species. However, by performing rigorous calibration involving suitable standards, it is possible to obtain reasonable quantitative analyses. Often, it is sufficient to obtain relative concentration data, and in some cases, a feature in the image (e.g., a known second-phase) serves as a "built-in" standard. For example, in the present study, it is possible to calculate segregation ratios by comparing the signal emitted from the boundaries and that from grain interiors.

The ability to examine solute segregation at grain boundaries, pore surfaces, and precipitates within a single map underscores the useful capabilities of this imaging-SIMS technique. This methodology has been applied to the study of segregation phenomena in a variety of ceramic systems such as Mg and Ca in MgO doped alumina, impurities in YBCO[1], dopants in ZnO and $SrTiO_3$ varistors.[2,5]

Proc. Microscopy and Microanalysis 1995, edited by G.W. Bailey, M.H. Ellisman, R.A. Hennigar, and N.J. Zaluzec.

1. K. K. Soni *et al.*, *Surf. Interface Anal.* 21 (1994), 117.
2. R. Levi-Setti *et al.*, *Scanning Microscopy* 7 (1994), 1161.
3. J. D. French, Ph.D. Dissertation, Lehigh University, (1993).
4. F. Dupau *et al.*, ICEM 13-Paris (1994), 885.
5. This research was sponsored by the NSF through the MRSEC at The University of Chicago.

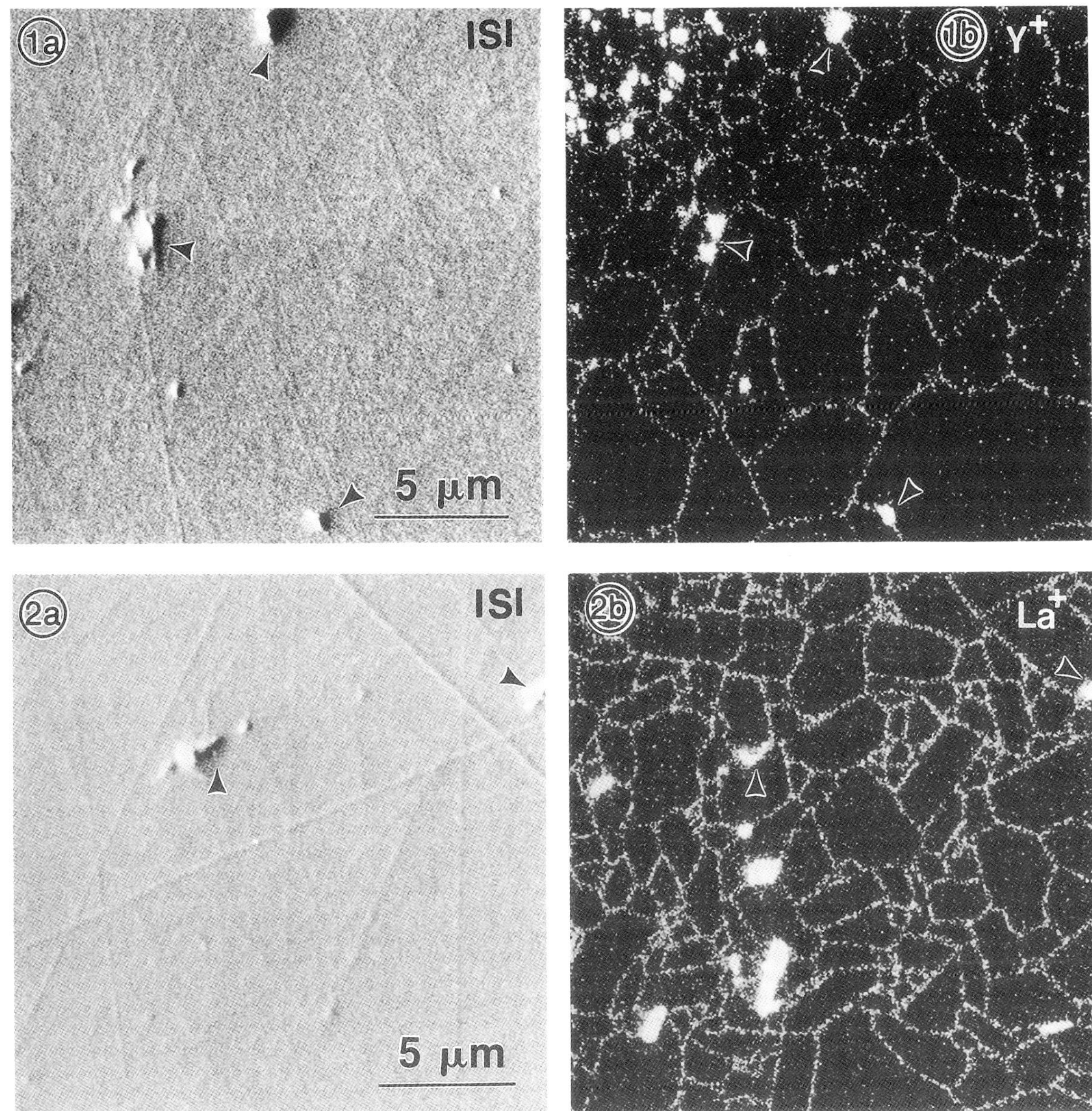

High spatial resolution SIMS images of polycrystalline alumina hot-pressed with 1000 ppm of Y_2O_3 (Fig. 1) and La_2O_3 (Fig. 2). All images were acquired from a 20 µm x 20 µm area of a polished section. ISI (ion-induced secondary ion) images display topographic contrast of the mapped area. Each SIMS map was acquired in 524 sec using a 512 x 512 raster and contains no noise signal. Grain boundary segregation and second-phases can be discerned easily in these maps. The thickness of the segregation/boundary layer is ~0.1 µm in this image. On average, grain boundary layers appear to be ~30X (Fig. 1b) and ~20X (Fig. 2b) brighter than the grains. Assuming the segregation width to be two monolayers, segregation ratios (grain boundary/grain concentrations) are estimated to be approximately 900 and 500 for Y and La, respectively.

INTERFACIAL COMPOSITION ANALYSIS OF SILICON NITRIDE WHISKER REINFORCED OXYNITRIDE GLASS SYSTEMS

E. Y. Sun, K. B. Alexander, P. F. Becher, and S. L. Hwang

Metals and Ceramics Division, Oak Ridge National Laboratory, Oak Ridge, TN 37831-6068

Microstructural and compositional analysis in many structural ceramics requires sophisticated analytical electron microscopy because of the light elements (O, N, C and B) present in the materials and the nano-scale amorphous phases at the grain boundaries. In the present study, scanning transmission electron microscopy (STEM) combined with energy-dispersive x-ray spectroscopy (EDS) and parallel-detection electron energy-loss spectroscopy (PEELS) was used to analyze the composition at interfaces in β-Si_3N_4 whisker reinforced oxynitride glass systems. Composition profiles were acquired by either manually stepping the probe across the interface or by sequential integration during slow-rate line-scanning with a < 2.0 nm probe in the STEM mode of a Philips EM400T/FEG microscope.

Silicon nitride, which has been recognized as one of the most promising candidates for high temperature applications, is typically densified by employing various sintering aids such as yttria and alumina. Reaction of the sintering aids and the oxide surface layer on Si_3N_4 particles forms secondary intergranular phases that are primarily silicon metal oxynitride glasses.[1] To study how the boundary phases influence crack-bridging toughening mechanisms, such as interfacial debonding and pull-out of the elongated β-Si_3N_4 grains,[2] model systems of β-Si_3N_4 whisker-reinforced SiAlYON glasses were fabricated. The nominal starting compositions of the glasses are listed in Table 1. For whiskers embedded in glasses with low Al:Y ratio (composite A and B), no interfacial layers were formed between the glass and the β-Si_3N_4 grains, as shown in the high resolution image in Fig. 1. In composites formed from glasses with high Al:Y ratio, the interfacial structure varied with the nitrogen content. A crystallized layer ~60 nm thick was present at the whisker/glass interface in the high-nitrogen systems (composite E) as shown in Fig. 2, whereas in the low-nitrogen glasses (composites C and D), no interfacial layer was observed. The composition profiles shown in Fig. 3 indicate that the interfacial layer formed in composite E contained Si and Al, with very low Y solubility, which is typical for SiAlON. SiAlON formation at interfaces during β-Si_3N_4 grain growth has also been observed via scanning electron microscopy (SEM) by Kramer $et\ al.$[3] Interfacial debonding experiments performed on these composites indicated that glass chemistry had a strong influence on the debonding behavior.[4] With the formation of a SiAlON solid solution layer at the interface, the energy required to debond the Si_3N_4–glass interface increases significantly. In silicon nitride ceramics, such interfacial debonding processes are critical to the formation of bridging grains to achieve toughening effects. The above results suggest that the fracture toughness of silicon nitride might be optimized by tailoring the intergranular amorphous phases.

References

1. P. Drew and M. H. Lewis, $J.\ Mat.\ Sci.,$ 9 (1974) 261-269.
2. P. F. Becher $et\ al.$, $Ceram.\ Acta.$ 2(6) (1990) 9-25.
3. M. Kramer $et\ al.$, $Acta.\ Metall.\ Mater.$ 41(10) (1993) 2939-47.
4. P. F. Becher et al., $MRS\ Bull.$ 20(2) (1995) 21-27.
5. This research is supported by Office of Basic Energy Sciences, U.S. Department of Energy under contract DE-AC05-84OR21400 with Martin Marietta Energy Systems, Inc. and also by an appointment to the ORNL Postdoctoral Research Associates Program administered jointly by ORNL and by ORISE.

Proc. Microscopy and Microanalysis 1995, edited by G.W. Bailey, M.H. Ellisman, R.A. Hennigar, and N.J. Zaluzec
Copyright © 1995 MSA. Published by Jones and Begell Publishing, 79 Madison Ave., New York, NY 10016

Table 1. Composition of the oxynitride glasses.

Glass	Si eq%	Al eq%	Y eq%	N eq%	O eq%
A	55	10	35	10	90
B	55	10	35	20	80
C	55	25	20	5	95
D	55	25	20	10	90
E	55	25	20	20	80

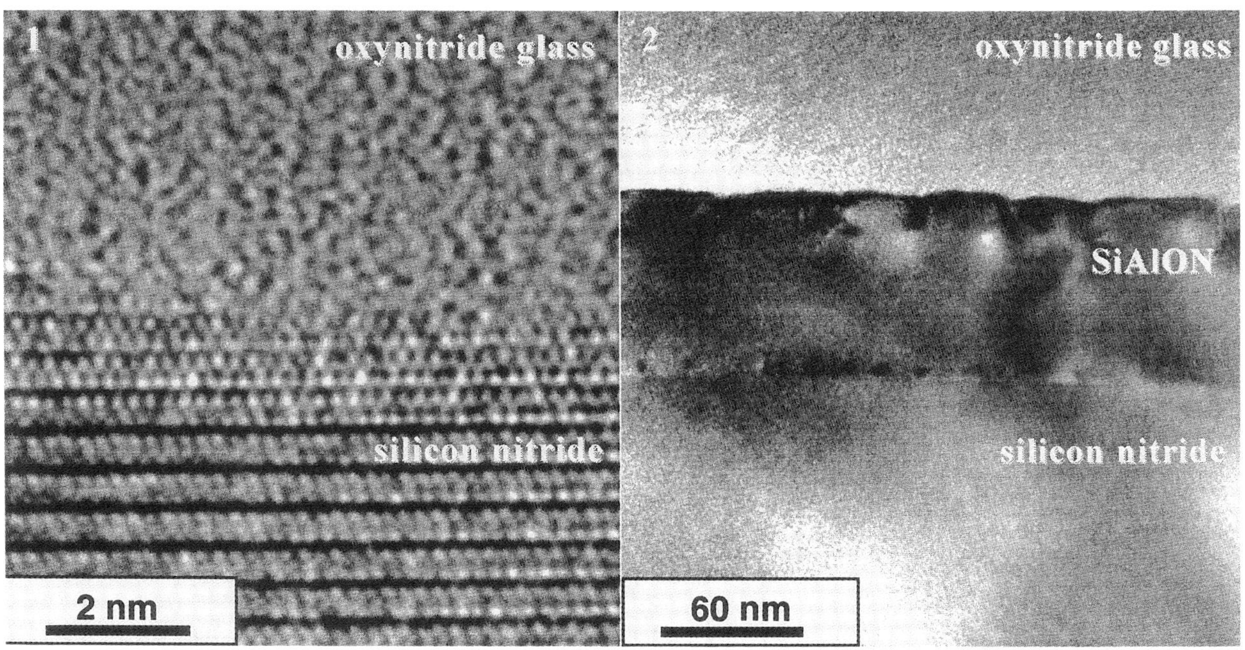

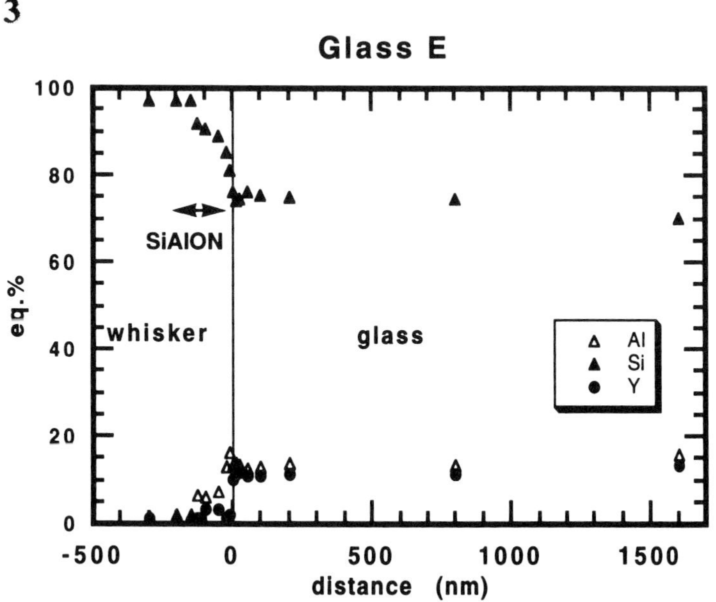

FIG. 1.-High resolution TEM image of silicon nitride/oxynitride glass interface in composite B.

FIG. 2.-TEM image of SiAlON formation at interface between silicon nitride whisker and oxynitride glass in composite E.

FIG. 3.-Composition profiles as determined by EDS analysis of cation species near interface in composite E.

THREE-DIMENSIONAL ANALYSIS OF OXIDE-OXIDE DIRECTIONALLY SOLIDIFIED
EUTECTIC INTERFACES BY HIGH RESOLUTION ELECTRON MICROSCOPY AND
ELECTRON ENERGY LOSS SPECTROSCOPY

E.C. Dickey*, V.P. Dravid, A. Revcolevschi**

*Materials Science and Engineering Department, Northwestern University, Evanston, IL 60208

**Laboratoire de Chimie des Solides, Universite de Paris-Sud, Orsay Cedex (France)

It is well appreciated that the strength of heterophase interfaces in ceramic composites plays an extremely influential role in the macroscopic mechanical behavior of the bulk material.[1-3] It is therefore useful to understand the structure of such interfaces in order to make structure/ property relationships between interface structure and mechanical behavior. High Resolution Electron Microscopy (HREM) is an extremely useful technique for elucidating the atomic structure of heterophase interfaces in ceramic materials, but HREM has typically been used to image lattice planes along only one zone axis. Often, however, it is useful to obtain three-dimensional information about the interface in order to fully appreciate and understand the complete relaxation and structure of the interface. Directionally solidified eutectics (DSEs) offer a unique opportunity to study heterophase interfaces in three dimensions, because they contain numerous, identical interfaces with consistent crystallographic orientation relationships between the two phases. Of particular interest are those eutectics that have lamellar morphologies, because the planar interfaces may be viewed along two orthogonal directions (Fig.1).

The atomic structure of interphase boundaries in several oxide DSEs, including NiO-ZrO_2(Cubic) and NiO-Y_2O_3 have been investigated by HREM. The eutectics were grown at the Universite de Paris-Sud in a floating zone device associated with a biellipsoid image furnace.[4,5] Although only the NiO-ZrO_2 eutectics are presented in this abstract, analogous results have been obtained for the NiO-Y_2O_3 system. Both of these DSEs contain planar interfaces and have identical orientation relationships, for example $(111)NiO//(001)ZrO_2//$interface, $[1\bar{1}0]NiO//[100]ZrO_2//$growth direction. Due to the orientation of the two phases, an anion-cation sequence of planes is maintained across the interface, so it has been hypothesized that the interface is a common oxygen plane[6] (Fig.2). Along the growth direction, the lattice mismatch between the two phases is very small , only 0.6% in the NiO-ZrO_2 system. HREM along the growth direction reveals atomically sharp interfaces with no detectable relaxation of the two phases (Fig. 3). Orthogonal to the growth direction, however, the mismatch between the two phases is quite large, 14.9% in the NiO-ZrO_2 system. Along this direction, HREM reveals lattice distortion in the NiO within 1.2nm of the interface (Fig. 4), indicating that the mismatch is being accommodated by relaxation of the NiO. Because these systems are highly ionic, crystal chemistry principles must be considered when developing a model for the interface structure. Consequently, high spatial resolution Electron Energy Loss Spectroscopy (EELS) is used to obtain chemical information about the interface including Ni valence and O coordination. We use this chemical information, the HREM images, and crystal chemistry principles to predict a plausible relaxation mechanism. Multislice image simulations are used to check the consistency of interface models with experimental HREM images.

Acknowledgments

This research is supported under an NSF Graduate Research Fellowship and NSF-DMR Grant No. DMR-9203722.

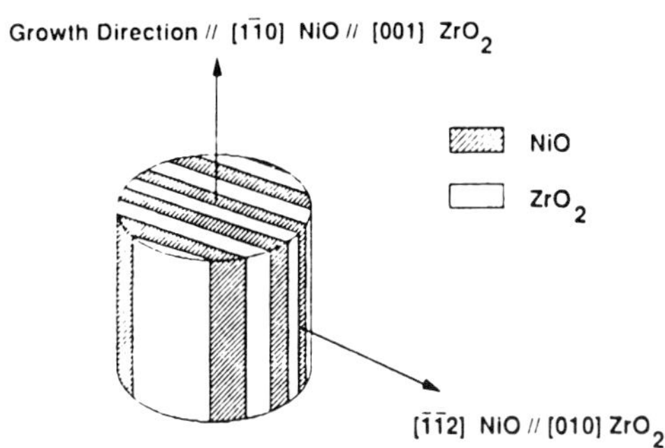

Fig. 1 Schematic representation of bulk eutectic system

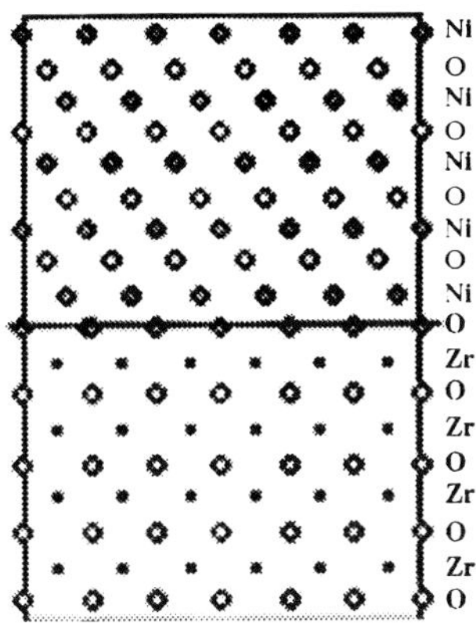

Fig. 2 Model of NiO-ZrO2 interface showing the cation-anion sequence across the interface.

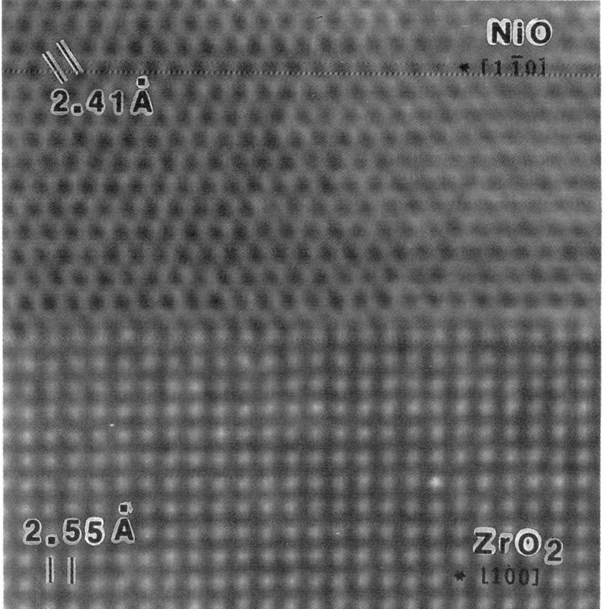

Fig.3 HREM image of NiO-ZrO2 interface perpendicular to the growth direction showing atomically sharp interface and virtually no mismatch between the two lattices.

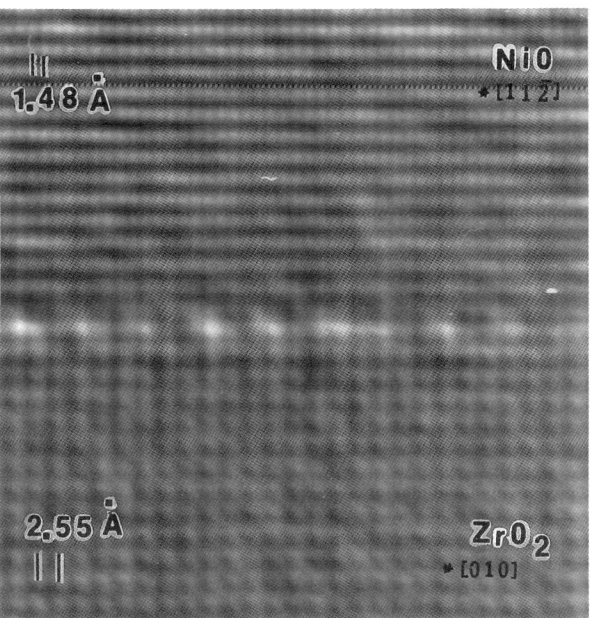

Fig. 4 HREM image of NiO-ZrO2 interface parallel to the growth direction showing distortion of NiO (220) near the interface.

References

1. A.G. Evans, *Mat. Sci. and Engr.* A107(1989)227.
2. M.Y. He and J.W. Hutchinson, *Intl. J. Sol. Struct.*, 25(1989)1053.
3. A.S. Argon and V. Gupta in Material Interfaces eds. D. Wolf and S. Yip, London: Chapman and Hall (1992) 641.
4. M. Fragneau and A. Revcolevschi, *J.Am. Cer. Soc.*, July (1982)C-102.
5. G. Dhalenne and A. Revcolevschi, *J. Crys. Growth*, 69(1984)616.
6. A. Revcolevschi and G. Dhalenne, Advanced Materials, 5(1993)657.

THE MEASUREMENT OF THIN-FILM REACTION LAYERS WITH HIGH-RESOLUTION FESEM

Paul G. Kotula and C. Barry Carter

Department of Chemical Engineering and Materials Science, University of Minnesota,
421 Washington Avenue SE, Minneapolis, MN 55455-0132

Thin-film reactions in ceramic systems are of increasing importance as materials such as oxide superconductors and ferroelectrics are applied in thin-film form. In fact, reactions have been found to occur during the growth of $YBa_2Cu_3O_{6+x}$ on ZrO_2.[1] Additionally, thin-film reactions have also been intentionally initiated for the production of buffer layers for the subsequent growth of high-T_c superconductor thin films.[2] The problem is that the kinetics of ceramic thin-film reactions are not well understood when the reaction layer is very thin; that is, when the rate-limiting step is a phase-boundary reaction as opposed to diffusion of the reactants through the product layer.[3] In this case, the reaction layer is likely to be laterally non-uniform. In the present study, the measurement of thin reaction-product layers is accomplished by first digitally acquiring backscattered-electron images in a high-resolution field-emission scanning electron microscope (FESEM) followed by image analysis. Furthermore, the problem of measuring such small thicknesses (e.g., 20-500nm) over lengths of interfaces longer than 3mm is addressed.

For the study of ceramic thin-film reactions the model oxide system, $NiO/NiAl_2O_4/\alpha$-Al_2O_3, has been chosen. Thin (~20nm) buffer-layers of $NiAl_2O_4$ were grown on single-crystal alumina substrates followed by ~400nm of nickel oxide in order to avoid having to nucleate the spinel reaction product. This nucleation step has been shown to obscure the measurement of the thin film reaction kinetics.[4,5] The starting thin-film reaction geometry (i.e., $NiO/NiAl_2O_4/Al_2O_3$) is shown schematically in Figure 1.

Thin-film reaction couples were glued with an acetone-soluble adhesive to alumina blanks for the purpose of preparing cross-sections of the reaction layers by tripod polishing[6] (i.e., one surface was polished with successively finer grades of diamond lapping-film). This sample preparation method was necessary for producing the 3.5mm lengths of essentially damage-free interface. The samples were then coated with less than 1nm of platinum to prevent charging in the FESEM. The thin-film reaction couples were imaged at an accelerating potential of 5kV in a Hitachi S-900 immersion-objective-lens FESEM equipped with an Autrata backscattered-electron (BSE) detector.[7] This BSE detector is capable of providing the signal-to-noise required to obtain high-resolution images at low voltages. The estimated BSE resolution of this system at 5kV is better than 2nm.

In order to determine the reaction kinetics, individual specimens were imaged in the FESEM and then taken apart by dissolving the glue. The specimens were subsequently heat-treated and prepared as above for further observation in the FESEM. A profile of reaction-layer thickness versus time at reaction temperature is the result of multiple iterations of this procedure. For the purpose of determining the thickness of the spinel layer for a given reaction couple, approximately 20 images were collected over the 3.5mm length of each cross-section at 175 µm intervals. The individual images were acquired digitally at a microscope magnification of 50kX with a Silicon Graphics IRIS (SGI)[8] and then transferred via ethernet to a Macintosh Quadra 900. NIH Image[9] was then used for measuring the thickness of the reaction layer; the image was calibrated, smoothed to remove noise and then a density slice (i.e., range of image gray levels) corresponding to the spinel layer was selected and measured giving the area of the spinel layer in nm^2 in the image. This area was then divided by the measured length of the reaction layer (~2.5 µm) to arrive at the spinel layer thickness for the particular image. Figure 2 is a plot of the reaction-layer thickness versus time at 1100°C for a thin-film reaction. The bars on the data points represent 2 standard deviations of the scatter in the measured data. The scale of this scatter is due not to error in measuring, which is much smaller, but rather to an actual variation in the reaction layer thickness along the length of the specimen. The as-deposited $NiAl_2O_4$ layer was quite uniform as seen by the small scatter in measurements for the t=0 data point. Figure 3 is a digital BSE

Proc. Microscopy and Microanalysis 1995, edited by G.W. Bailey, M.H. Ellisman, R.A. Hennigar, and N.J. Zaluzec

Copyright © 1995 MSA. Published by Jones and Begell Publishing, 79 Madison Ave., New York, NY 10016

image of a reaction couple heated to 1100°C for 8 hours (one of the images used to determine the third data point in figure 2) which shows that the reaction layer is laterally non-uniform (i.e., the interfaces are not planar). The spinel reaction product is intermediate in intensity (gray level) as it is intermediate in average atomic number between NiO and Al_2O_3. The area of the spinel layer has been measured as described above and is outlined by a white line which corresponds to the phase boundaries.[10]

References

1. L. A. Tietz et al., *J. Mat. Res.* 4[5](1989)1072-1081.
2. H. Haefke et al., *Appl. Phys. Lett.* 61[19](1992)2359-2361.
3. H. Schmalzried, *Polish J. Chem.* 67[1993]167-190.
4. P. G. Kotula and C. B. Carter, *J. Am. Ceram. Soc.* 78[1](1995)
5. P. G. Kotula and C. B. Carter, *These Proceedings*.
6. S. J. Klepeis et al., *Mat. Res. Soc. Symp. Proc.* 115[1988]179-183.
7. R. Autrata and J. Hejna, *Scanning* 13[1991]275.
8. S. L. Erlandsen et al., *Proc. 52nd Annual Mtg. MSA* 1994)1034-1035.
9. NIH-Image, W. Rasband, National Institutes of Health, Bethesda, MD

10. The authors acknowledge research support from the Center for Interfacial Engineering (CIE), an NSF Engineering Research Center. PGK acknowledges a graduate student fellowship from IBM and from the Graduate School at the U of MN. The authors also thank Stan Erlandsen for access to the Hitachi S-900 and Chris Frethem for technical assistance.

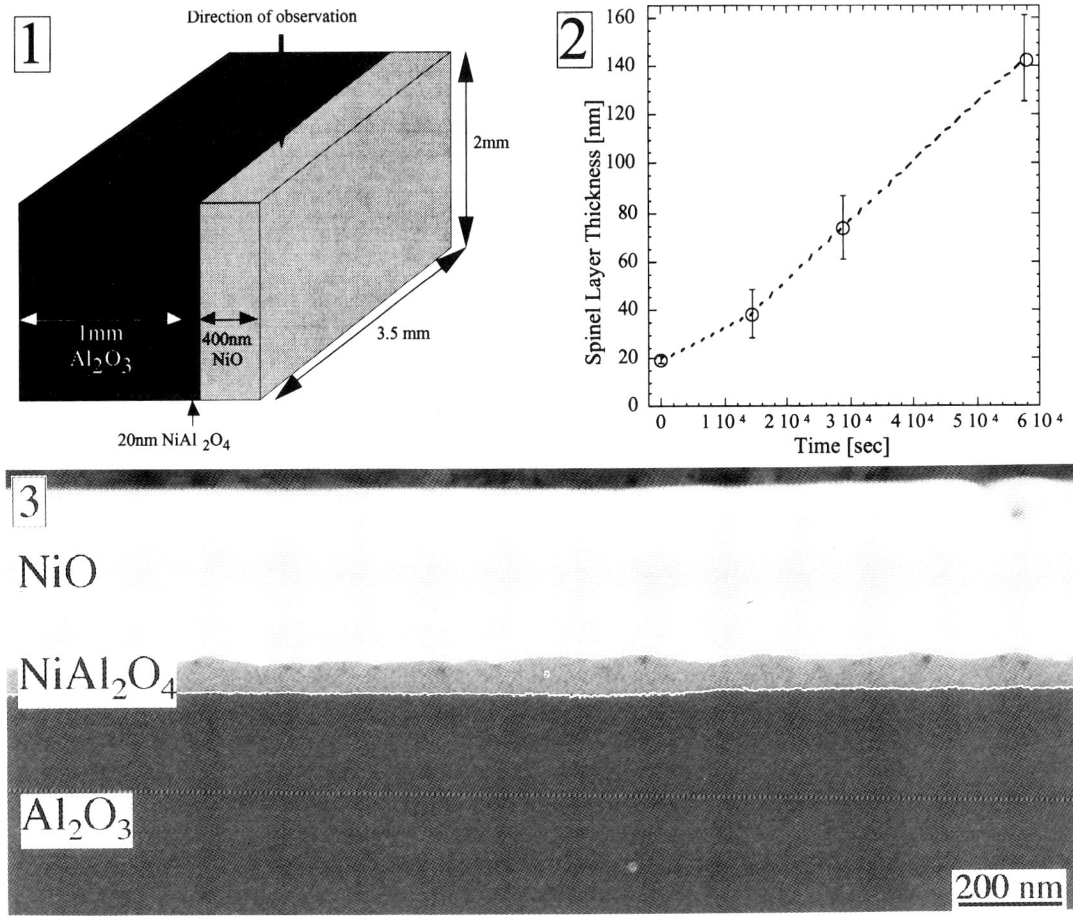

Fig 1. Schematic of the thin-film reaction geometry. Fig 2. Plot of reaction-layer thickness versus time for a thin-film reaction couple. Fig 3. Digital BSE image demonstrating the measurement process.

AFM STUDIES OF ANNEALED SINGLE-CRYSTAL α-ALUMINA SURFACES

Jason R. Heffelfinger, Michael W. Bench and C. Barry Carter

Department of Chemical Engineering and Materials Science, University of Minnesota,
421 Washington Ave. S.E., Minneapolis, MN 55455

Since the invention of atomic-force microscopy (AFM) in 1986,[1] the technique has found an invaluable niche in the imaging of insulating surfaces.[2-4] AFM allows for analysis of topographical details at the atomic level with minimum sample preparation,[5] but the technique is subject to artifacts such as broadening of surface structures and ghost images of the tip due to the finite size and shape of the contacting probe.[6] In the present investigation, transmission electron microscopy (TEM) and scanning electron microscopy (SEM) were used as supporting techniques to AFM in the study of annealed single-crystal α-alumina surfaces. Although discrepancies were found between the EM data and the AFM images which point to artifacts inherent to the AFM, useful information on surface step heights and roughness of terraces was gained using AFM.

AFM studies were performed on a Nanoscope III (Digital Instruments, Santa Barbara, CA) using microfabricated Si_3N_4 cantilevers (Ultralevers, Park Inst., Sunnyvale, CA). All images were recorded in air with a nominal applied force of 10-15 nN. The alumina samples were prepared from polished single crystals supplied by Crystal Systems (Salem, MA). Samples with the (0001), (10$\bar{1}$0), and (11$\bar{2}$0) surface orientations were acid cleaned and annealed at 1400°C for 6 hours in air. Special precautions were taken to minimize the exposure of the surfaces to any contamination from the furnace environment.

For analyzing the **(0001) surface**, a TEM sample was prepared by conventional dimpling and ion milling prior to annealing. The surface of the TEM sample, as seen by the AFM in Figure 1a, shows large terraces separated by surface steps. Step heights varied from larger steps measuring near 4 nm to smaller steps such as the one indicated by the arrow in Figure 1a, which measured 0.9 nm. These steps are observed to be shallowly inclined to the surface and have a horizontal distance of ~40 nm for a 4 nm step height. Figure 1b is a TEM image of an annealed (0001) foil and shows contrast variation due to surface steps on both sides of the sample. Previous research on annealed (0001) TEM foils found step heights to be multiples of 1.3 nm, which is the unit cell height for alumina in the [0001] direction.[7] In comparison to the AFM, the TEM found surface steps of similar height to be an order of magnitude steeper than those observed by the AFM. The difference between the AFM and EM observations points to an AFM artifact and the tendency for the AFM to distort lateral dimensions. The **(10$\bar{1}$0) surface** was prepared from a bulk single-crystal, and as revealed by the AFM in Figure 2, showed a hill and valley structure. The heights from hill peaks to valley lows measure nominally 25 nm; the arrow in Figure 2 indicates a peak of 28.7 nm in height. Research by Mallamaci et. al found the (10$\bar{1}$0) surface to facet into (10$\bar{1}$2) and (10$\bar{1}$1) planes when annealled in the presents of liquid glass.[8] These facet planes are predicted to have an over-all lower surface energy than the original (10$\bar{1}$0) surface, and thus allow for such facetting to occur.[9] Cross-section TEM analysis of the annealed (10$\bar{1}$0) plane also found the surface to facet into the (10$\bar{1}$2) and (10$\bar{1}$1) planes. Both SEM and TEM observations of the (10$\bar{1}$0) surface confirmed the heights and periodicity of the facets to be consistent with those observed by the AFM. The **(11$\bar{2}$0) surface** was prepared in the same manner as the (10$\bar{1}$0) surface and exhibited a terrace and step morphology as seen in Figure 4. The surface terraces were observed to terminate in a saw-tooth fashion with the two edges of the saw tooth corresponding to (0001) and (1$\bar{1}$00) plane traces as identified by previous TEM observation.[7] With the exception of the rounded surface steps, SEM and TEM analysis of the (11$\bar{2}$0) surface was consistent with AFM observations. The AFM did reveal small undulations (~1 nm in height) on the (11$\bar{2}$0) surface terraces.[10]

Proc. Microscopy and Microanalysis 1995, edited by G.W. Bailey, M.H. Ellisman, R.A. Hennigar, and N.J. Zaluzec
Copyright © 1995 MSA. Published by Jones and Begell Publishing, 79 Madison Ave., New York, NY 10016

References:
1. G. Binning, C. F. Quate and C. Gerber, Physical Review Letters, **56** 930-933 (1986).
2. M. D. Antonik and R. J. Lad, Journal of Vacuum Science and Technology, **A 10** [4] 669-673 (1992).
3. H. G. Hansma, J. Vesenka, C. Siegerist, G. Kelderman, H. Morrett, R. L. Sinsheimer, V. Elings, C. Bustamante and P. K. Hansma, Science, **256** 1180-1184 (1992).
4. D. Rugar and P. K. Hansma, Physics Today, **43** 23 (1990).
5. G. Binnig, C. Gerber, E. Stoll, T. R. Albrecht and C. F. Quate, Surface Science, **189/190** 1-6 (1987).
6. T. Thundat, X. Y. Zheng, S. L. Sharp, D. P. Allison, R. J. Warmack, D. C. Joy and T. L. Ferrell, Scanning Microscopy, **6** [4] 903-910 (1992).
7. D. W. Susnitzky and C. B. Carter, Communications of the American Ceramics Society, **69** [9] C-217-220 (1986).
8. M. P. Mallamaci and C. B. Carter, ICEM Proceedings, **2A** 227-228 (1994).
9. P. W. Tasker, "Surfaces of Magnesia and Alumina." Structure and Properties of MgO and Al_2O_3 Ceramics, A. H. Heuer and M. F. Yan, 1984, American Ceramics Society, Columbus, OH.
10. Research is supported by the U. S. DoE under Grant No. DE-FG02-92ER45465. JRH is supported by the Center for Interfacial Engineering, which is a National Science Foundation Engineering Research Center. Both the AFM and TEM are part of this research center. The authors thank Dr. Susie Yang for technical assistance on the AFM.

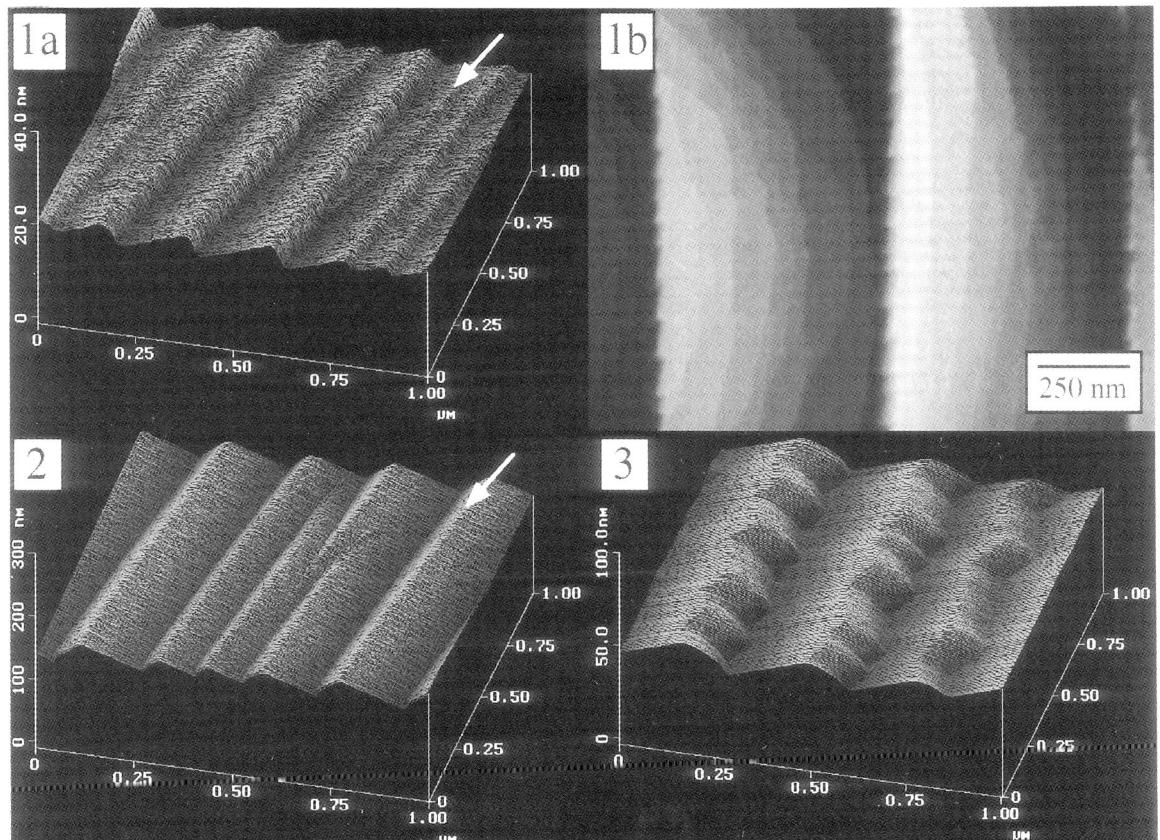

Fig. 1a.   Line-plot AFM image showing raw data of the (0001) surface
Fig. 1b.   BF TEM micrograph of an α-alumina foil as viewed down [0001] pole
Fig. 2.   Line-plot AFM image showing raw data of the (10$\bar{1}$0) surface
Fig. 3.   Line-plot AFM image showing raw data of the (11$\bar{2}$0) surface

AFM STUDY OF THE DYNAMICS OF α-ALUMINA SURFACE FACETING DURING HIGH-TEMPERATURE PROCESSING

Michael W. Bench, Jason R. Heffelfinger, and C. Barry Carter

Department of Chemical Engineering and Materials Science, University of Minnesota
421 Washington Ave. SE, Minneapolis, MN 55455

To gain a better understanding of the surface faceting that occurs in α-alumina during high temperature processing, atomic force microscopy (AFM) studies have been performed to follow the formation and evolution of the facets. AFM was chosen because it allows for analysis of topographical details down to the atomic level with minimal sample preparation.[1] This is in contrast to SEM analysis, which typically requires the application of conductive coatings that can alter the surface between subsequent heat treatments. Similar experiments have been performed in the TEM;[2,3] however, due to thin foil and hole edge effects the results may not be representative of the behavior of bulk surfaces.

The AFM studies were performed on a Digital Instruments Nanoscope III using microfabricated Si_3N_4 cantilevers. All images were recorded in air with a nominal applied force of 10-15 nN. The alumina samples were prepared from pre-polished single crystals with $(000\bar{1})$, $(10\bar{1}0)$, and $(11\bar{2}0)$ nominal surface orientations. Prior to the heat treatments the samples were solvent and acid cleaned in order to remove any contaminants that might affect the surface faceting behavior. During the annealing treatments, performed in air at 1400°C, the samples were protected from any possible contamination in the furnace by placing them inside high-purity alumina crucibles that were themselves packed in high-purity alumina powder.

Two types of experiments were performed in the present study. In one, several samples were cut from a single substrate and a series of anneals were performed for differing times. Two images from an experiment of this type that illustrate the progression of the faceting that occurs on $(10\bar{1}0)$ surfaces are shown in fig. 1. This surface orientation is known to be unstable and typically facets into a hill and valley structure, which is consistent with the 8hr annealed sample shown in fig. 1b. The 2hr annealed sample shown in fig. 1a illustrates that this surface structure nucleates from the nominally flat initial surface through the formation of a small trench and a corresponding hill adjacent to it that then grows in such a manner that its length is extended. The second type of experiment followed the same area of the same sample as the annealing time was increased. This was accomplished by using a diamond stylus to lightly scribe two orthogonal sets of parallel lines spaced 30-50 μm apart forming a grid that could be used to position the AFM tip on the same area after each processing step. A 50μm scan of one of these scribed grids is shown in fig. 2. Imaging in the same area avoids the possibility that differences in images as annealing times progress are not real, but instead are due to other effects such as local variations in initial surface orientation. It also provides the possibility of following the evolution of individual steps and terraces as the annealing progresses. Furthermore, the scribing provides a registry from which the same area of the sample can be imaged using other techniques. In fig. 3 the same area of a (0001) oriented sample is shown after total annealing times of 1, 3, and 7 hrs. Annealing for 1 hr at 1400°C (fig. 3a) has caused the surface to begin to facet and it shows indications of the formation of a banded terrace and step structure. After an additional 2hr anneal (fig. 3b) a series of closely spaced parallel terraces has formed. Increasing the total annealing time to 7 hrs (fig. 3c) has the consequence of bunching the steps and thus coarsening their periodicity. This bunching of steps appears to be occurring through the motion of small steps on the widening terraces. The heights of the steps at each of the annealing times are on the order of 1.5 nm or less, which is considerably smaller than the approximate 4nm height that may be expected at equilibrium.[3] These and other examples demonstrating how the alumina surface evolves during high temperature processing will be described.[4]

References

1. G. Binning, C. Gerber, E. Stoll, T.R. Albrecht and C.F. Quate, Surf. Sci. **189/190**, 1 (1987).
2. K.J. Morrissey and C.B. Carter, Mat. Res. Soc. Symp. Proc. **41**, 137 (1985).
3. D.W. Susnitzky and C.B. Carter, J. Am. Ceram. Soc. **75**, 2463 (1992).
4. This research was supported by the U.S. Department of Energy under Grant No. DE-FG02-92ER45465. The authors also acknowledge the support of the Center for Interfacial Engineering, a National Science Foundation Engineering Research Center. The AFMs used are part of this center.

Proc. Microscopy and Microanalysis 1995, edited by G.W. Bailey, M.H. Ellisman, R.A. Hennigar, and N.J. Zaluzec
Copyright © 1995 MSA. Published by Jones and Begell Publishing, 79 Madison Ave., New York, NY 10016

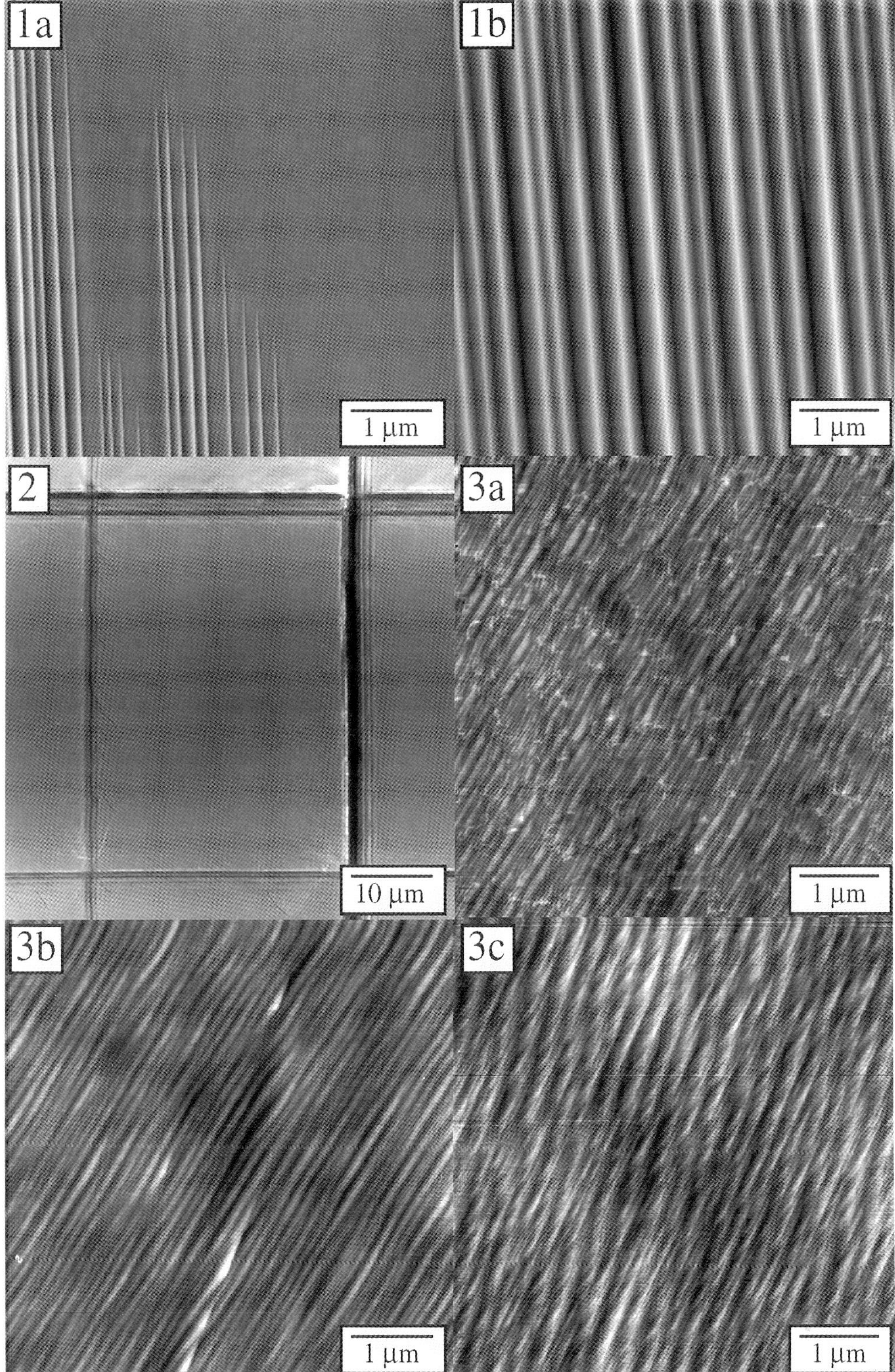

Fig. 1. (10$\bar{1}$0) surfaces imaged (a) after a 2hr anneal and (b) after an 8hr anneal.
Fig. 2. Scribed region of a substrate providing markers to reposition the sample after processing steps.
Fig. 3. Images from the same area of an (000$\bar{1}$) substrate recorded after total annealing times of (a) 1 hr, (b) 3 hr, and (c) 7 hr.

THE ANNEALED (0001) α–ALUMINA SURFACE AND ITS INFLUENCE ON THIN-FILM GROWTH

Michael W. Bench, Paul G. Kotula, and C. Barry Carter

Department of Chemical Engineering and Materials Science, University of Minnesota
421 Washington Ave. SE, Minneapolis, MN 55455

The growth of semiconductors, superconductors, metals, and other insulators has been investigated using alumina substrates in a variety of orientations. The surface state of the alumina (for example surface reconstruction and step nature) can be expected to affect the growth nature and quality of the epilayers. As such, the surface nature has been studied using a number of techniques including low energy electron diffraction (LEED), reflection electron microscopy (REM), transmission electron microscopy (TEM), molecular dynamics computer simulations, and also by theoretical surface energy calculations. In the (0001) orientation, the bulk alumina lattice can be thought of as a layered structure with Al-Al-O stacking. This gives three possible terminations of the bulk alumina lattice, with theoretical surface energy calculations suggesting that termination should occur between the Al layers.[1] Thus, the lattice often has been described as being made up of layers of (Al-O-Al) unit stacking sequences. There is a 180° rotation in the surface symmetry of successive layers and a total of six layers are required to form the alumina unit cell. Considering relaxations at the surface, the LEED studies have revealed that the surface reconstruction is dependent on the annealing parameters (temperature and oxygen pressure).[2] The TEM and REM studies have demonstrated that the surfaces are made up of terraces separated by steps. Reports of the heights of typical steps have ranged from a single unit stacking sequence and small integral multiples of it (i.e. one, two, or three Al-O-Al layers) in one instance[3] to 4-5 nm in others.[4,5]

In the present study, TEM, REM, SEM, and AFM are being used to investigate the nature of the (0001) alumina surface and how it influences the overgrowth of oxide films. To illustrate these effects examples will be presented from the growth of nickel oxide, copper oxide, and titanium oxide films (three systems that we are currently investigating in relation to their growth on and reaction with alumina substrates). The films were grown using pulsed-laser deposition (PLD) onto prethinned TEM specimens, which were prepared from pre-polished single-crystal windows of α–Al_2O_3 (see ref 6 for details). A variety of growth parameters were used in these studies, with values typically being in the following range: 600-900°C substrate temperature, 5-15 mtorr O_2 pressure, and 1-10 Hz laser pulse rate.

Examples of PLD growth using a NiO and TiO_2 targets are shown in the TEM images of Figs. 1 and 2, respectively. The NiO has grown in two twin related rotational variants with $(0001)_{alumina}$ ∥ $(111)_{NiO}$. The twin boundaries are visible in Fig. 1 as randomly curved lines interrupting the moiré fringe contrast present in both twins. Also visible in Fig. 1 is a dark band contrast corresponding to the position of the surface steps on the growth surface of the alumina substrate. It is clear from the twin-boundary and surface-step contrast that the twin-related variants in the NiO grow irrespective of which particular terrace they are on and, indeed, individual grains have grown across the steps. For the growth of NiO on the basal plane of alumina under the present experimental conditions, the surface nature of the alumina seems to have no effect on the nucleation of the respective variants. Fig. 2 illustrates that this is obviously not the case for TiO_2, where the film is distinctly different on adjacent terraces. On one terrace the film is made up rutile grains in three rotational variants. On other terraces the film morphology is considerably different and an additional, as of yet unidentified, phase is present. This result demonstrates that the lattice is not terminating at a similar point in the unit cell on all terraces, which is in contrast to the predictions of the theoretical calculations.

PLD growth of copper oxide also exhibits surface terrace effects on the growth morphology, as is illustrated in Fig. 3. The island nature of the copper oxide enables a more comprehensive investigation the influence of the substrate on the growth, since the islands do not substantially obscure the alumina surface terraces on which they grew. Fig. 3 is a montage following one set of terraces on the growth surface and showing the terrace structure across a single extinction contour on the back surface. By comparing of the number of steps on the back surface between successive intensity minima with the extinction distance for the imaging conditions used, the average step height can be measured. It was found that the step height averaged about 4 nm, which is nearly equal to three times the alumina lattice

Proc. Microscopy and Microanalysis 1995, edited by G.W. Bailey, M.H. Ellisman, R.A. Hennigar, and N.J. Zaluzec

parameter. Contrast comparisons suggest that the step heights between adjacent terraces on the growth surface are also approximately 4 nm high in many instances, although steps less than one unit cell high were also observed. These and other observations will be discussed in relation to the insights they provide into the nature of the (0001) alumina surface.[7]

1. J. Guo, D.E. Ellis and D.J. Lam, Phys. Rev. B **45**, 13647 (1992).
2. T.M. French and G.A. Somorja J. Phys. Chem. **74**, 2489 (1970).
3. G.C. Ndubuisi, J. Liu and J.M. Cowley, Microsc. Res. Tech. **20**, 439 (1992).
4. D.W. Susnitzky and C.B. Carter, J. Am. Ceram. Soc. **75**, 2463 (1992).
5. Y. Kim and T. Hsu, Surf. Sci. **258**, 131 (1991).
6. M.W. Bench, K.B. Sartain, J.R. Heffelfinger and C.B. Carter, Mat. Res. Soc. Proc. **317**, 491 (1994).
7. The authors acknowledge the support of the Center for Interfacial Engineering, which is a National Science Foundation Engineering Research Center. The microscopes used are part of this center.

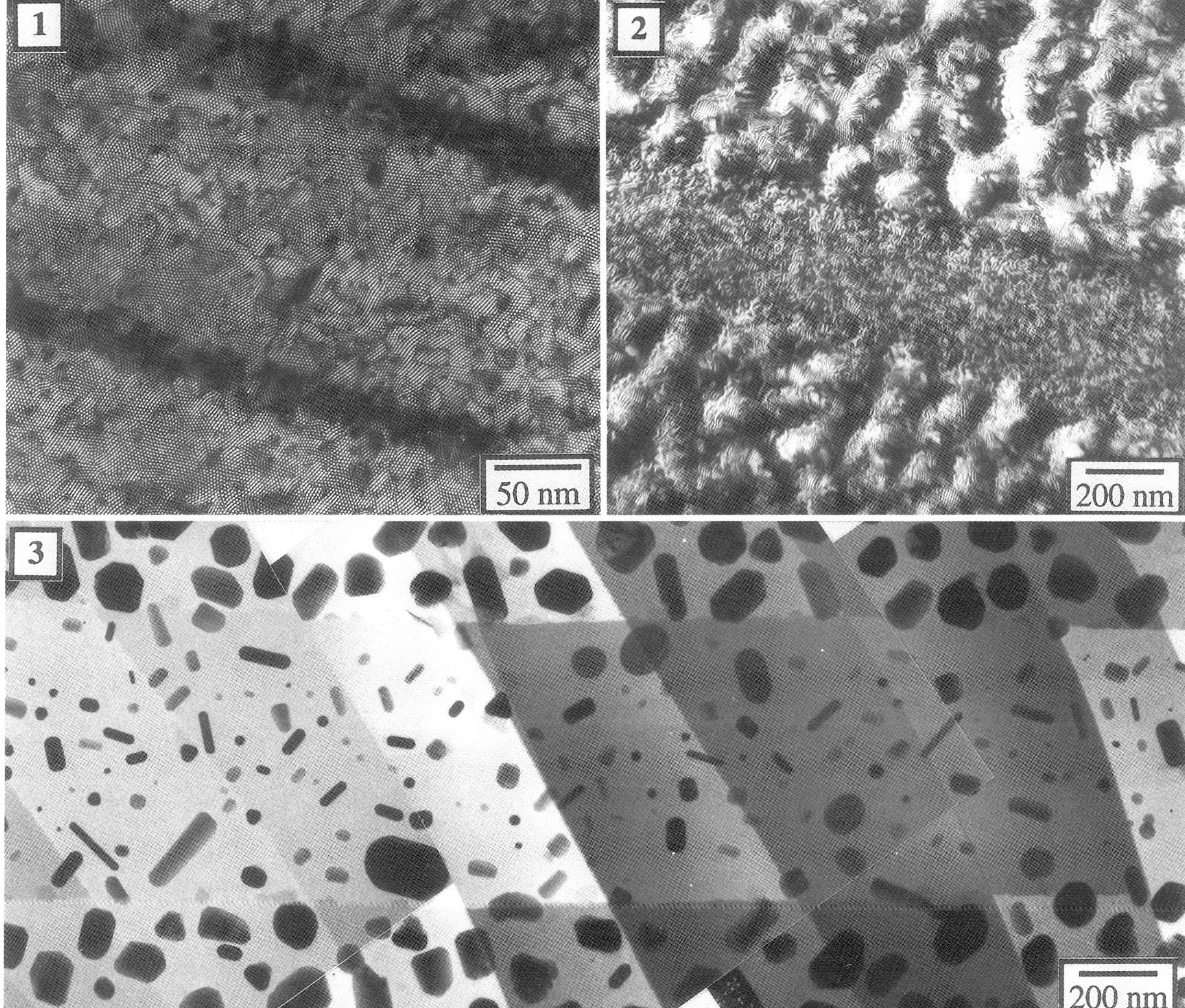

FIG. 1. Bright-field image of NiO film taken at the coincident $[0001]_{alumina}$ and $[111]_{NiO}$ zone axes.
FIG. 2. Dark-field image of titanium oxide film recorded with $\mathbf{g}=(11\bar{2}0)$ in alumina.
FIG. 3. Montage image of Cu_2O islands grown on basal alumina. All images recorded in bright field with the electron beam parallel to the $[000\bar{1}]$ alumina zone axis.

337

CERAMIC THIN-FILM REACTIONS: PHASE BOUNDARIES AND KINETICS

Paul G. Kotula and C. Barry Carter

Department of Chemical Engineering and Materials Science, University of Minnesota, 421 Washington Avenue SE, Minneapolis, MN 55455-0132

Interfacial reactions between epitactic nickel oxide thin films and single-crystal aluminum oxide substrates were investigated as a function of substrate orientation. Two thin-film reaction geometries were employed for this purpose. The first, which was used to study the nucleation of the reaction product, consisted of epitactic nickel oxide thin films on alumina substrates. The second, which was used to determine the reaction kinetics, was similar to the first except that a thin epitactic buffer layer of the reaction product was placed between the nickel oxide film and substrate. In this case, the nucleation was avoided by essentially 'nucleating' the reaction product artificially. As thin films were utilized, and the scale of the salient microstructural features were correspondingly small, electron microscopy techniques were applied to determine the preferred nucleation sites and measure the rate at which the reaction proceeds for the respective geometries.

Both types of thin-film reaction-couple geometry were produced by pulsed-laser deposition (PLD). PLD is a versatile thin-film deposition technique capable of producing high-quality oxide epilayers. The setup for PLD has been described elsewhere.[1] For the nucleation study, 100nm of NiO was deposited while for the kinetics study, 20nm of $NiAl_2O_4$ (spinel reaction product) was deposited followed by 400nm of NiO on four different alumina substrate orientations; (0001), $\{11\bar{2}0\}$, $\{1\bar{1}02\}$, and $\{1\bar{1}00\}$. The reaction couples were characterized with scanning and transmission electron microscopy. The details of the SEM technique used for the measurement of the kinetics are given elsewhere.[2,3]

Figure 1 is an SEM backscattered electron (BSE) image of a cross-section specimen in the nucleation study which has been heated to 1100°C for 8 hours. The reaction layer is quite non-uniform laterally as the nucleation of such reactions has been shown for {111}-NiO films grown on (0001) substrates to occur at the intersection of film $\Sigma=3$ twin-boundaries and the substrate.[4] This type of thin-film reaction couple is not well suited for the measurement of the growth rate of the spinel as this would depend on when and where the spinel nucleated as well as how fast it grows once nucleated. For this reason, the second thin-film reaction geometry has been utilized. Figures 2 and 3 are cross-section FESEM BSE images of the second type of reaction couple, as deposited and heated to 1100°C for 8 hours respectively, for the case where the substrate is (0001). The as-deposited spinel layer is quite smooth while after the heat-treatment, some roughness at the phase boundaries has developed. This roughness is insignificant when compared to that of the sample from the nucleation study shown in Figure 1. It is clear that nucleation must be avoided in order to determine the kinetics of the reaction. Figure 4 is a plot of spinel layer thickness versus time at 1100°C for reaction couples grown on four different alumina substrate orientations. There is clearly a large difference in reaction rate with substrate orientation; the substrate orientation determines the orientation of the overlayers and thereby the structure of the phase boundaries. Figure 4 demonstrates that the structure of the phase boundary can have a dramatic effect on the overall kinetics of the thin-film reaction.[5]

References
1. P. G. Kotula and C. B. Carter, *Proc. 2nd Int. Conf. on Laser Ablation, AIP Conf. Proc.* 288[1993)231-236.
2. P. G. Kotula and C. B. Carter, *These Proceedings.*
3. P. G. Kotula and C. B. Carter, *Scripta Met.* 32[6](1995)863-866.
4. P. G. Kotula and C. B. Carter, *J. Am. Ceram. Soc.* 78[1](1995)
5. The authors acknowledge research support from the Center for Interfacial Engineering (CIE), an NSF Engineering Research Center. PGK acknowledges a graduate student fellowship from IBM and from the Graduate School at the U of MN. The authors also thank Stan Erlandsen for access to the Hitachi S-900 and Chris Frethem for technical assistance.

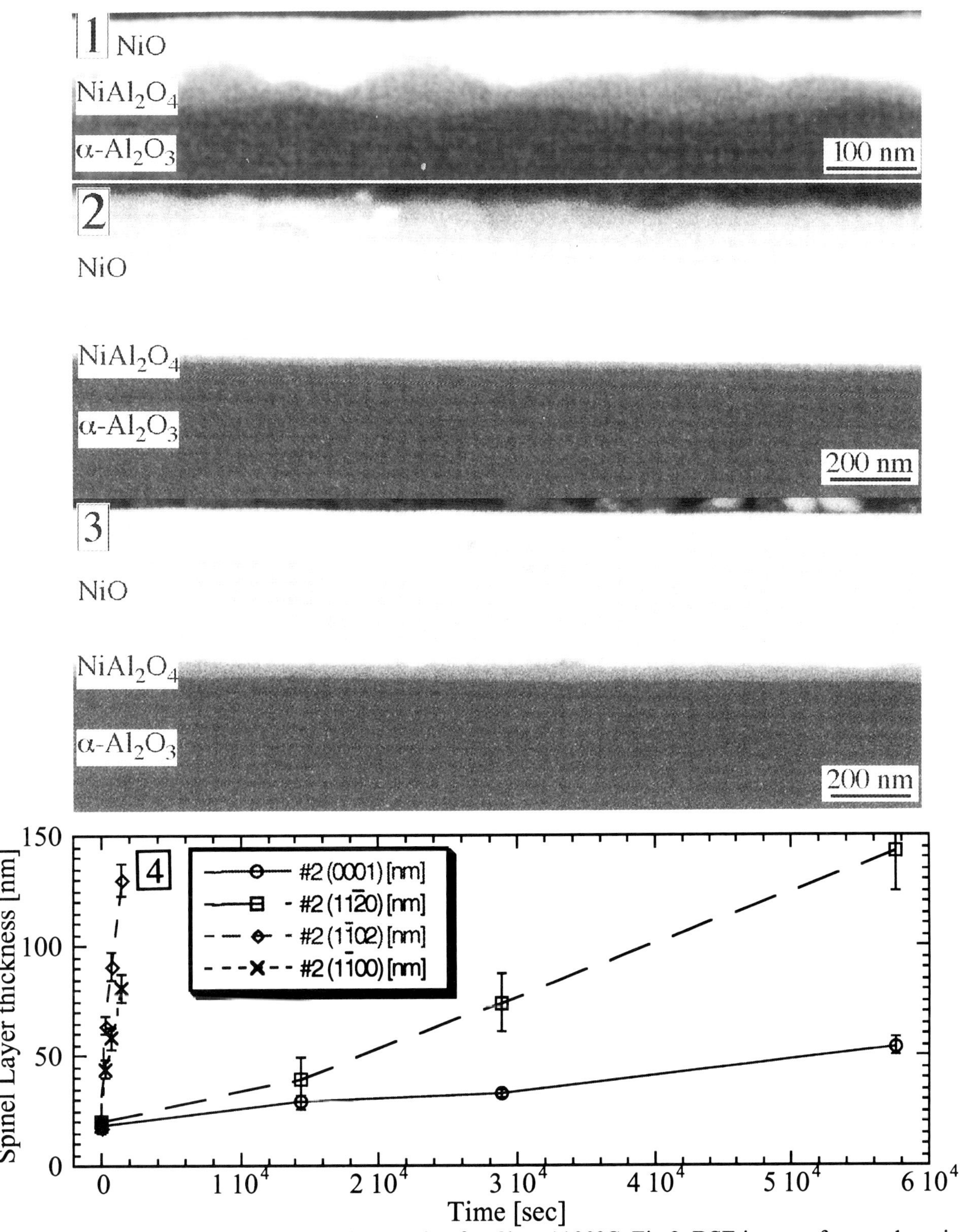

Fig 1. BSE image of a nucleation study sample after 8h at 1100°C. Fig 2. BSE image of an as-deposited specimen from the kinetics study. Fig 3. BSE image of the same sample as in Fig 2 after 8h at 1100°C Fig 4. Plot showing the difference in reaction rate with alumina substrate orientation.

NON-EQUILIBRIUM SILICATE GLASS IN CONTACT WITH FACETED $\{10\bar{1}0\}$ α-Al$_2$O$_3$ SURFACE : AN SEM STUDY

Sundar Ramamurthy, Brian C. Hebert and C. Barry Carter

Department of Chemical Engineering and Materials Science, University of Minnesota, 421 Washington Avenue SE, Minneapolis, MN 55455-0132

Perhaps the most striking feature about α-Al$_2$O$_3$ lies in the anisotropy of its crystal structure. Surface energies, γ, (relaxed and unrelaxed) have been theoretically calculated for different low-index planes in alumina.[1,2] As an example to illustrate the anisotropy, the reported values of γ for two unrelaxed planes ($\{10\bar{1}0\}$ and $\{10\bar{1}2\}$) vary by a factor of about 2 while the factor is only about 15% between $\{110\}$ and $\{111\}$ planes in a fcc metal.[2,3] Among the low-index planes in α-Al$_2$O$_3$, the $\{10\bar{1}0\}$ plane has been reported to have the largest value of γ.[2] Experimental observations of heat-treated $\{10\bar{1}0\}$ surfaces indicate a breakdown of the surface into an array of facets containing inclined steps.[4] This faceting behavior has also been observed when the surface is in contact with calcium aluminosilicate glass at 1600°C.[5] The $\{10\bar{1}0\}$ surface was found to be replaced by two sets of facets parallel to the $\{10\bar{1}1\}$ and $\{10\bar{1}2\}$ planes respectively. In that study the glass was a near-anorthite (CaAl$_2$Si$_2$O$_8$) composition which is in equilibrium with Al$_2$O$_3$ at 1600°C in air.[6]

The present study was aimed at understanding the behavior of crystalline surfaces in contact with a reacting glassy phase. The anorthite-alumina system was chosen since it has been widely studied in relation to liquid-phase-sintering (LPS) of alumina compacts. According to the alumina-anorthite pseudo-binary phase diagram,[7] anorthite dissolves an appreciable amount of alumina at 1700°C.

Al$_2$O$_3$ single-crystals with optically polished surfaces were commercially obtained from *Crystal Systems, Salem, MA*. These samples were acid-cleaned and annealed in air at 1400°C for 8 hours prior to reaction with the silicate glass. The anneal essentially reconstructs the surface into facets that are parallel to low-index planes.[4] The reaction was performed in two different ways. In one setup, glass powder particles were placed on the annealed $\{10\bar{1}0\}$ Al$_2$O$_3$ surface. In the other setup a thin, uniform glass film (approximately 50 nm thick) was deposited onto the alumina surface using Pulsed-Laser Deposition (PLD). The deposition technique is discussed in detail elsewhere.[5] Both these geometries of alumina-glass were heat-treated in air at 1700°C for different times ranging from about 10 minutes up to 20 hours. In order to offer a comparison of the interaction of glass with a mechanically polished surface, identical heat treatments were performed on solvent-cleaned $\{10\bar{1}0\}$ Al$_2$O$_3$ surfaces. This comparison would provide mechanisms of early stages of dissolution in conjunction with the faceting process. The reacted samples were sputter-coated with ~ 1 nm thick Pt and examined using a field-emission scanning electron microscope (FESEM) operating at 5 kV. The microstructures were also characterized using a visible-light microscope (VLM).

Figure 1 is a secondary electron SEM image of anorthite powder particles melted at 1700°C for 10 hours while in contact with an initially polished $\{10\bar{1}0\}$ surface of alumina. The distinctive features in this image are the following. The fine straight lines are associated with inclined steps which have developed uniformly throughout the surface. Almost all the glass particles appear to be flattened along one side. This flattening effect is along $<11\bar{2}0>$, which is the intersection of $\{10\bar{1}1\}$ and $\{10\bar{1}2\}$ facets. In this crystal geometry, the $<11\bar{2}0>$ direction lies in the (0001) plane. The preference for glass to be in contact with basal surface is consistent with earlier work by Simpson and Carter.[8] Figure 2 is an FESEM image showing the surface steps around a melted globule of glass. The surface steps which are usually straight, tend to bend around the flattened side of the glass particle. These curved steps indicate a pronounced interaction of the facets with the molten glass at the reaction temperature. PLD glass films showed a tendency to dewet from the surface and interact in a manner similar to that of powder particles. The fact that all the glass particles show this "flattening" morphology along a specific $<11\bar{2}0>$ direction suggests that the reaction (dissolution) must be asymmetric about $<11\bar{2}0>$ at the anorthite-alumina-air junction line. This asymmetry may be ascribed to the inherent asymmetric faceting of the $\{10\bar{1}0\}$ α-Al$_2$O$_3$ surface.[9]

Proc. Microscopy and Microanalysis 1995, edited by G.W. Bailey, M.H. Ellisman, R.A. Hennigar, and N.J. Zaluzec

Copyright © 1995 MSA. Published by Jones and Begell Publishing, 79 Madison Ave., New York, NY 10016

1. P. Hartman, *J. Cryst. Growth* 49(1980)166-70.
2. P. W. Tasker, *Adv. Ceram.* 10(1984)176-88.
3. C. A. M. Klump and J. T. Klomp, *J. Phys. Colloq.* 46[C4](1985)111-116.
4. D. W. Susnitzky and C. B. Carter, *J. Am. Ceram. Soc.* 75[9](1992)2463-2478.
5. M. P. Mallamaci, Thesis, Cornell U. 1995.
6. T. M. Shaw and P. R. Duncombe, *J. Am. Ceram. Soc.* 74[10](1991)2495-505.
7. Phase Diagrams for Ceramists, Edited by E. M. Levin, C. R. Robbins and H. F. McMurdie. 1964.
8. Y. K. Simpson and C. B. Carter, *J. Am. Ceram. Soc.* 73[8](1990)2391-2398.
9. This research has been suppported by the U.S. Department of Energy under Grant No. DE-FG02-92ER45465. The authors would like to thank Prof. Stan Erlandsen for access to the Hitachi S-900 FESEM and Chris Frethem for technical assistance.

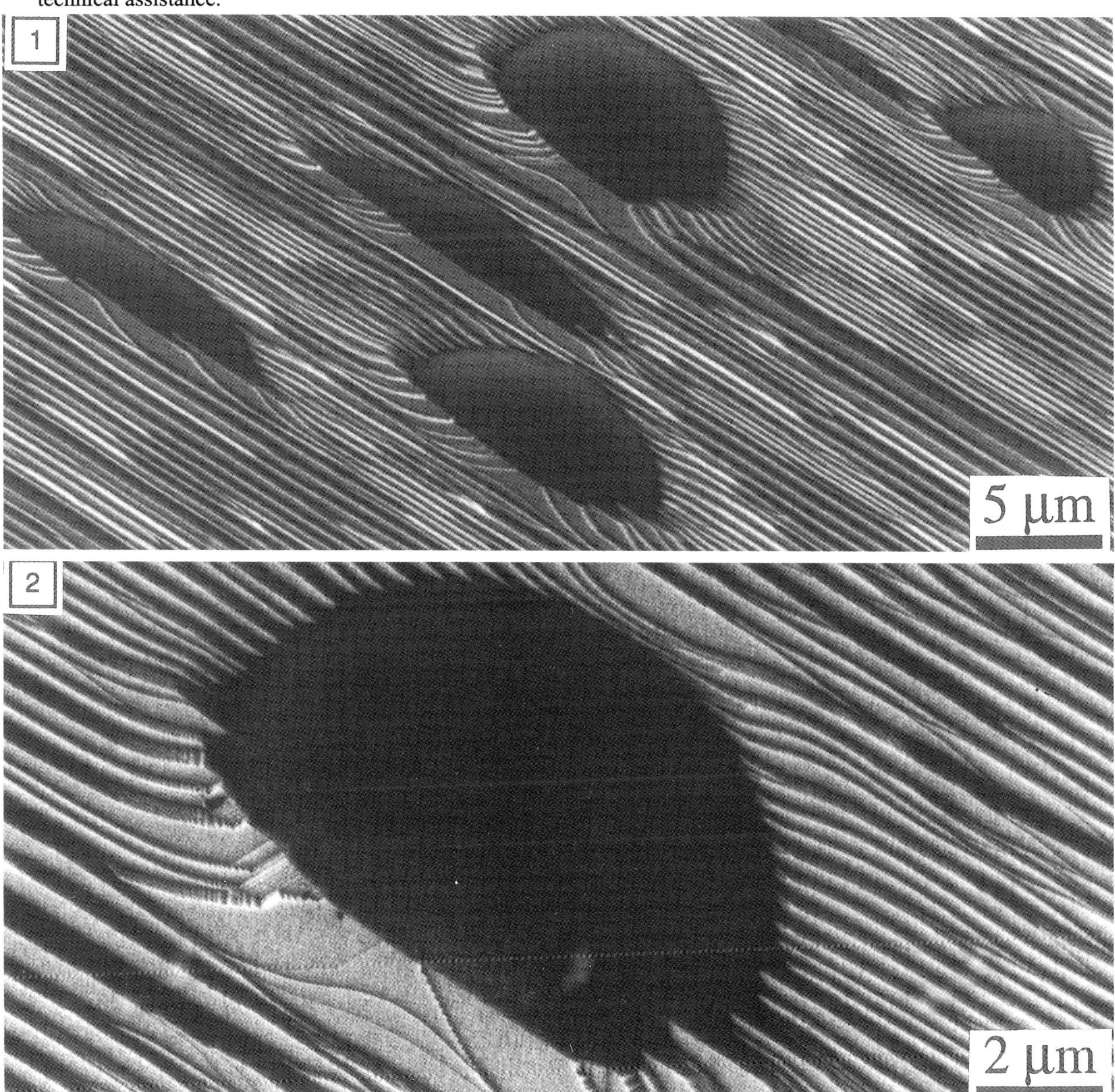

Fig. 1. FESEM image of anorthite powder particles melted at 1700°C for 7 hours. All the glass globules are flattened on side along a specific direction. Fig. 2. FESEM image showing curved surface steps strongly interacting with the flattened side of a glass particle.

OLIVINE-MgO INTERFACES PRODUCED BY CRYSTALLIZATION OF GLASS FILMS ON SINGLE-CRYSTAL MgO SUBSTRATES

Sundar Ramamurthy, Brian C. Hebert and C. Barry Carter

Department of Chemical Engineering and Materials Science, University of Minnesota,
421 Washington Avenue SE, Minneapolis, MN 55455-0132

Glassy silicates are present at grain boundaries in almost all liquid-phase sintered ceramic oxides. In many cases, the amorphous-crystalline interfaces in the sintered microstructure can be modified by inducing crystallization of the glassy phase.[1,2] The intergranular phases in polycrystalline MgO are typically silicates with cations of calcium, magnesium and iron in the silicate network.[3] A systematic approach to study the crystallization behavior of glass-MgO interfaces has been attempted in the present study. Following the work by Mallamaci[4] in which crystallization of a silicate glass in contact with Al_2O_3 was studied, devitrification of glass in contact with single-crystal MgO has been investigated. Transmission electron microscopy (TEM) and scanning electron microscopy (SEM) were used to characterize the resulting microstructures.

A pellet of Mg_2SiO_4 (forsterite) prepared by hot-pressing MgO and SiO_2 powders was used as the target for depositing glass films onto single-crystal MgO substrates by pulsed-laser deposition (PLD). Glass films with the composition of olivine, $(0.9Mg, 0.1Fe)_2SiO_4$, were deposited using a single-crystal target of *San Carlos* olivine. The PLD technique has been described in detail elsewhere.[5] Films ranging between 10 and 100 nm in thickness were deposited at room temperature in an ambient of 15 mTorr oxygen pressure. The laser energy was maintained at 200 mJ/pulse and the pulse-repetition rate was kept constant at 10 Hz. Depositions were made onto bulk as well as TEM-ready substrates. Prior to thin-film deposition, TEM foils of (001) orientation were acid-washed and heat treated at 1350°C for 1 hour leaving a surface characterized with crystallographic steps. Cleaved (100) surfaces of MgO were used as bulk substrates. In addition to (001) substrates, solvent-cleaned and annealed (111)-oriented MgO substrates were also used. After deposition, the samples were heat-treated at 1100°C and 1200°C for varying time intervals. This approach enables examination of the same area of the sample after successive heat treatments.

Forsterite crystallized epitactically on the (111) surface with $[100]_{for} \parallel [111]_{MgO}$. Due to the three-fold symmetry along the [111] direction in MgO, [100] forsterite can nucleate and grow along three different <2$\bar{2}$0> directions producing three equivalent rotational variants of forsterite grains. These three variants are shown in the dark-field image in Figure 1. Figure 2 is a selected-area diffraction pattern corresponding to this image, showing reflections from all the three variants.
Olivine films were found to phase-separate into magnesioferrite spinel ($MgFe_2O_4$) and forsterite. Spinel was found to align with "cube-on-cube" epitaxy on the (001) surface. Figure 3 is a weak-beam dark-field image showing strain contrast due to epitactically aligned spinel precipitates on (001) MgO. Phase separation of spinel has also been found to occur during crystallization of Ca-Mg-Al-Si-O films on MgO in which $MgAl_2O_4$ particles were found in epitactic alignment with the substrate.[6]
SEM studies of the crystallized microstructure show interaction of MgO surface steps with the crystallites. As an illustration, Figure 4 is a secondary-electron image showing curved surface steps strongly interacting with a faceted forsterite particle produced by heat-treating Mg_2SiO_4 powder in contact with annealed (001) MgO surface at 1200°C for 10 hours. The remarkable in-plane epitaxy and the crystallographic symmetry of the substrate surface present interesting homo- and heterophase boundaries in the devitrified microstructure.[7]

1. T. J. Headley and R. E. Loehman, *J. Am. Ceram. Soc.*, 67 [9] 620-625 (1984).
2. C. Liu, S. Komarneni and R. Roy, *J. Am. Ceram. Soc.*, 75 [10] 2665-70 (1992).
3. G. R. Rigby, H. M. Richardson and F. Ball, *Trans. Brit. Ceram. Soc.*, 46 [8] 313-29 (1947).
4. M. P. Mallamaci, Thesis, Cornell U. 1995.
5. P. G. Kotula and C. B. Carter, *Mat. Res. Soc. Symp.*, 285 373 (1993).

Proc. Microscopy and Microanalysis 1995, edited by G.W. Bailey, M.H. Ellisman, R.A. Hennigar, and N.J. Zaluzec
Copyright © 1995 MSA. Published by Jones and Begell Publishing, 79 Madison Ave., New York, NY 10016

6. S. Ramamurthy, M. P. Mallamaci and C. B. Carter, *Mat. Res. Soc. Symp.*, 321 633-638 (1994).
7. This research has been suppported by the U.S. Department of Energy under Grant No. DE-FG02-92ER45465. The authors would like to thank Prof. Stan Erlandsen for access to the Hitachi S-900 FESEM and Chris Frethem for technical assistance. The TEM is a part of the Center for Interfacial Engineering, an NSF Engineering Research Center at the Univeristy of Minnesota.

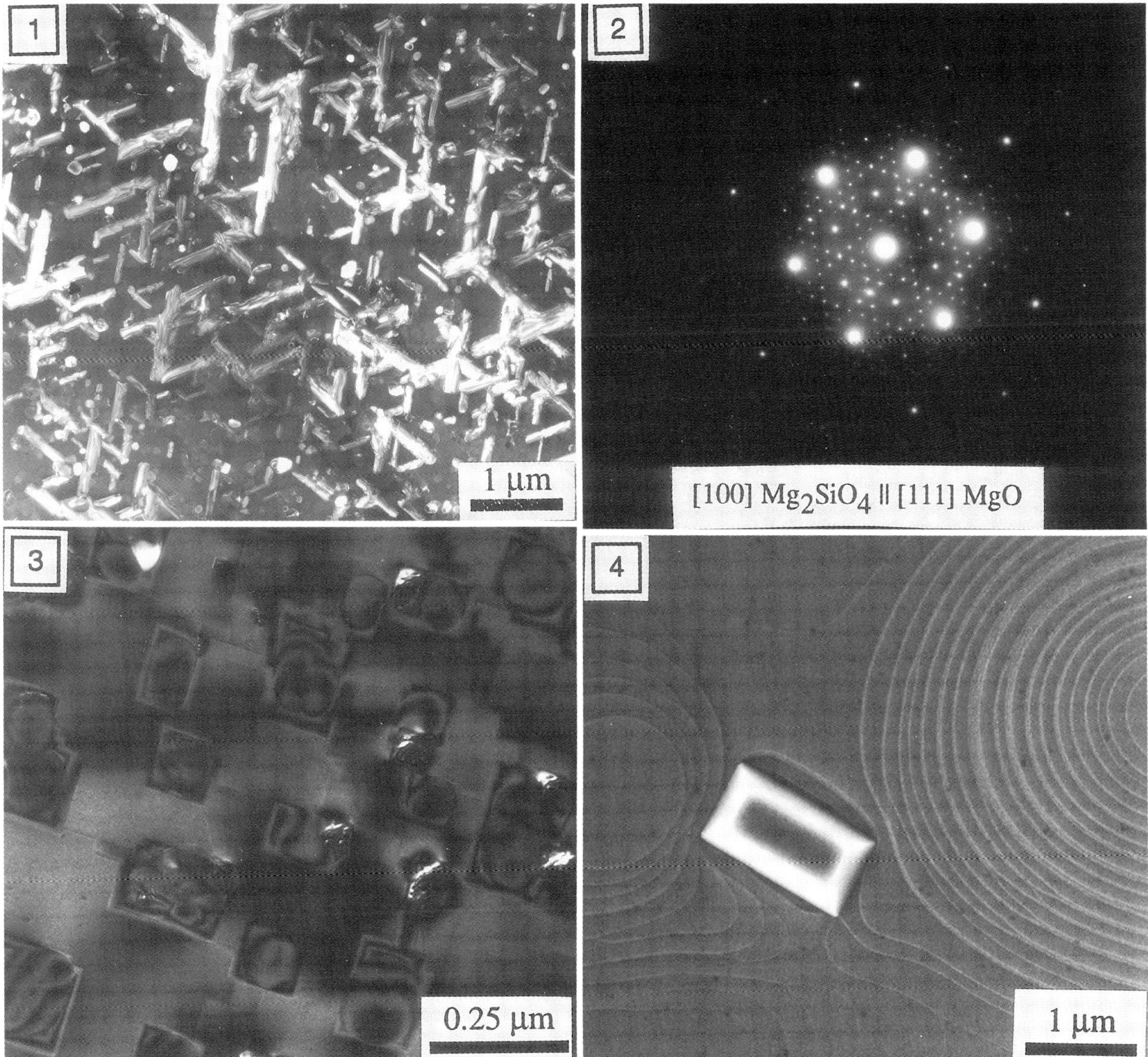

Fig. 1. Dark-field TEM image using $\mathbf{g}$=(040)$_{for}$, a reflection common to all the three variants of forsterite grains.

Fig. 2. Diffraction pattern showing the epitactic alignment of forsterite with (111) MgO. The three variants are rotated by 120° with respect to each other.

Fig. 3. Weak-beam dark-field TEM image ($\mathbf{g}$ = {200}) showing strain contrast around MgFe$_2$O$_4$ spinel precipitates on the (001) MgO surface.

Fig. 4. SEM image showing curved surface steps interacting with a forsterite particle.

FORMATION OF POLYTYPE MICROSTRUCTURE IN PULSED-LASER-IRRADIATED SAPPHIRE SUBSTRATES

Siqi Cao[*], A. J. Pedraza[*], L. F. Allard[**], and D. H. Lowndes[***]

[*]Department of Materials Sci. and Eng., University of Tennessee, Knoxville, TN 37996-2200.
[**]High Temperature Materials Laboratory, Oak Ridge National Laboratory, Oak Ridge, TN 37831.
[***]Solid State Division, Oak Ridge National Laboratory, P.O. Box 2008, Oak Ridge, TN 37831-6056.

Surface modifications of wide-gap materials are produced by pulsed laser irradiation. Under given conditions, these near-surface modifications can promote adhesion enhancement of deposited thin film materials,[1] and activation for electroless deposition.[2] AlN decomposes during laser irradiation leaving a metallic film on the surface. High density dislocations were observed in the surface layer of AlN that was laser melted but not decomposed.[3] The laser melted alumina becomes amorphous at a laser energy density of $\sim$1J/cm^2.[4] In sapphire, γ-alumina is formed when the sample is laser irradiated in Ar/4%H$_2$.[5] Here, we report the formation of a new structure in laser-irradiated sapphire.

Optically polished c-axis sapphire substrates were laser-irradiated in an Ar/4%H$_2$ atmosphere at 4J/cm^2 energy density, using a 308 nm-wavelength laser with a pulse duration of $\sim$40 ns. Sapphire (Al$_2$O$_3$) has a space group $R\bar{3}c$ and can be described as an hcp structure having oxygen and aluminum layers alternately stacking along the c-axis.

During laser irradiation, a surface layer was melted and then solidified as γ-alumina after the laser pulse.[5] The γ-phase has a spinel-type, cubic structure. The γ-phase was epitaxially solidified from the sapphire substrate with an orientation relationship of $(0001)_\alpha//(111)_\gamma$, and $[01\bar{1}0]_\alpha//[1\bar{1}0]_\gamma$ (Fig. 1). Fig. 2 shows a high resolution image with zone axes of $[01\bar{1}0]$ in sapphire. A polytype structure was observed in sapphire just beneath the γ-alumina layer at this orientation. The SAD pattern in Fig. 3 shows extra spots in the c-axis direction of the sapphire $[01\bar{1}0]$ zone axis. Therefore, this polytype structure has a period three times that of the lattice image seen in Fig. 4. We also observed that the polytype structure was distorted because of the rapid cooling after the laser pulse.

Polytypism is a crystallographic property exhibited by a large number of inorganic crystals, such as silicon carbide and zinc sulfide.[6] A periodic formation of stacking faults is commonly used to explain the structural feature of the polytype. In sapphire that has an approximately hexagonal structure, the polytype forms when a periodic stacking disorder is generated. Most of the polytypes are formed during the solidification. In our case, considering that the laser irradiation is a rapid heating and cooling process, nonequilibrium phases may form. The formation of polytype sapphire and γ-alumina during solidification may be related to the change in cooling rate, or may be due to the composition change involved during laser irradiation. Another possibility is that the high thermal stresses induce many stacking faults that form the polytype structure.[7]

References

1. A. J. Pedraza, M. J. DeSilva, R. A. Kumar, and D. H. Lowndes, *J. Appl. Phys*, April, 1995 (in press).
2. M. J. DeSilva, A. J. Pedraza, and D. H. Lowndes, *J. Materials Research*, 9(1994)1019.
3. Siqi Cao, A. J. Pedraza, and L. F. Allard, *J. Materials Research*, 10(1995)54.
4. Siqi Cao, A. J. Pedraza, D. H. Lowndes, and L. F. Allard, in *Structure and Properties of Interfaces in Ceramics*, MRS proceeding, Vol. 357(1995) (in press).
5. Siqi Cao, A. J. Pedraza, D. H. Lowndes, and L. F. Allard, *Appl. Phys. Lett.*, 65(1994)2940.
6. P. R. Buseck, edt., *Minerals and reactions at the atomic scale: TEM*, Mineralogical Society of America, Washington, D.C., 1992.
7. This research has been supported in part by National Science Foundation Grant No. DMR-9116528, by the Division of Materials Science, U. S. Department of Energy (DOE), and by the High

Temperature Materials Laboratory User Program, DOE Office of Transportation Technologies, under contract DE-AC05-84OR21400 with Martin Marietta Energy Systems, Inc.

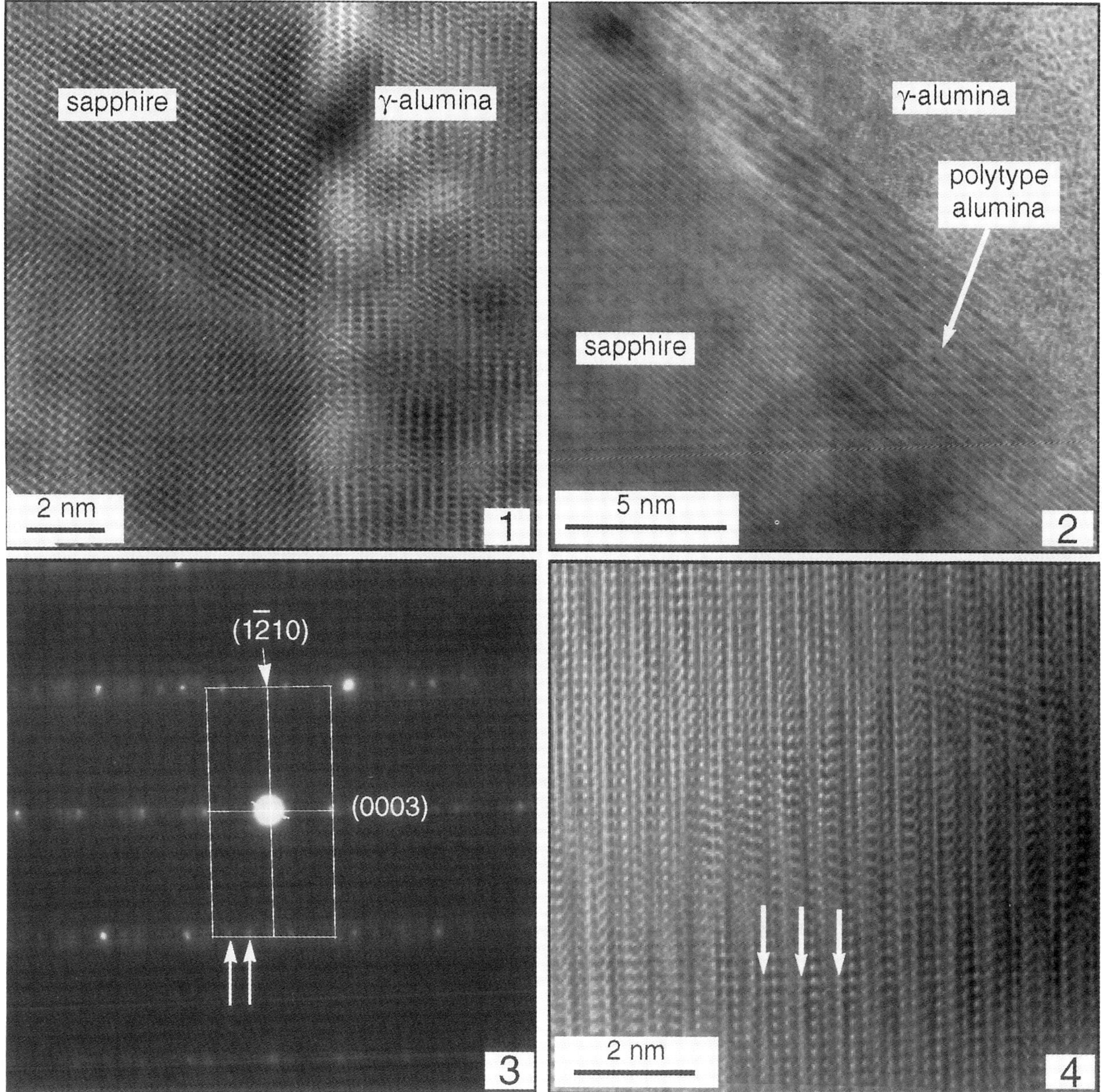

FIG. 1---HRTEM of laser-melted and solidified γ-alumina on a sapphire substrate. No polytype structure contrast was found in sapphire in the [1 2̄ 10] zone axis orientation.

FIG. 2---HRTEM showing polytype structure in sapphire just beneath γ-alumina with [01 1̄ 0] zone axis in sapphire.

FIG. 3---SAD of the [01 1̄ 0] zone axis in sapphire with polytype structure showing extra diffraction spots (arrows) indicating a c-axis periodicity three times that of the original structure.

FIG. 4---HRTEM image showing the polytype structure in laser-irradiated sapphire. Arrows indicate its periodicity.

STEP STRUCTURE AND 5×5 RESTRUCTURE OF ANNEALLED LaAlO3 {100} - A SUBSTRATE FOR ELECTROCERAMIC THIN FILM GROWTH

Z. L. Wang[*] and A.J. Shapiro

Metallurgy Division, National Institute of Standards and Technology, Gaithersburg, MD 20899.
* Currently at: School of Materials Science and Engineering, Georgia Institute of Technology, Atlanta GA 30332-0245.

Nucleation and growth of thin films are strongly affected by the microstructures of the substrate surfaces. Although LaAlO$_3$ {100} surfaces have been widely used as substrates for growing high T$_c$ superconductor and ferroelectric thin films, the surface structure of the material has not been extensively analyzed using electron microscopy techniques. In this paper, reflection electron microscopy (REM) [1] is applied to examine the structures of LaAlO$_3$ {100}. REM experiments were carried out at 300 kV and 120 kV in transmission electron microscopes. The {100} surfaces were mechanically polished and subsequently annealed at 1500 °C for 20 h in air. The LaAlO$_3$ unit cell is the distorted-perovskite structure with lattice constant a = b = c = 0.3788 nm and $\alpha = \beta = \gamma = 90.066°$, in which the La^{+3} ion locates at (000), the Al^{+3} ion at (0.5 0.5 0.5), and the O^{-2} ions at the face-centers {0.5 0.5 0}.

Figure 1 shows a [001] reflection high-energy electron diffraction (RHEED) pattern of the annealed LaAlO$_3$ (100) surface. Two strong superlattice reflections are observed between bulk reflections (10 0 0) and (10 -1 0). The two superlattice reflections are separated by 1/5(0 -2 0) and 1/5(0 2 0), respectively, from the (10 0 0) and (10 -1 0) bulk reflections. On the right-hand side of the (10 -1 0) reflection, four equally-spaced superlattice reflections are seen, corresponding to the formation of the 5×5 reconstruction on the surface (Figure 2). LaAlO$_3$ exhibits {100} twin structures [2], which cause the splitting of the superlattice reflections. Some of the surface terraces are completely covered by the reconstructed surface layer (Figure 3a). Numerous [010] and [001] surface steps were observed on the LaAlO$_3$ (100) surfaces (Figures 3b and 3c). The 5×5 reconstructed regions show darker contrast and the unreconstructed surface areas show bright contrast. The faceted [010] and [001] steps are seen almost everywhere on the surface. The regions which show brighter and darker contrast sometimes share the same terrace, and there is no step between them. No dislocations were found in the flat surface areas, but dislocations were occasionally observed at twin boundaries.

From the structure of LaAlO$_3$, the (100) surface can be terminated with either a La-O layer or an Al-O layer. Thus, there are three possible configurations for forming a [001] or [010] step on the (100) surface. (a) Both the upper and lower terraces of the step are terminated with the La-O layer, and the step height is a multiple of the lattice constant. (b) Both the upper and lower terraces of the step are terminated with the Al-O layer, and the step height is again a multiple of the lattice constant. (c) The upper and lower surface terraces next to the step are terminated with the La-O layer and the Al-O layer, respectively. In configuration (c), the upper and lower terraces adjacent to the step would show different contrast because of the difference in scattering powers between La (atomic number Z = 57) and Al (Z = 13), and the regions which show distinct difference in contrast would be separated by the surface step. However, this was not observed experimentally. The two types of contrast seen in the REM image, as shown in Figure 3b for example, are due to surface reconstruction. If configurations (a) and (b) both occur, a difference in contrast should also be seen due to larger scattering power of La than that of Al, and the regions which show distinct difference in contrast must be separated by surface steps. This result has also not been observed experimentally. Therefore, the termination of the (100) surface must be produced solely by the La-O layer or the Al-O layer, but cannot be the mixture of two.

In conclusion, numerous [010] and [001] surface steps are observed on the {100} surfaces of LaAlO$_3$. The <100> steps have the lowest surface energy. 5×5 reconstruction is formed on annealed LaAlO$_3$ {100}. Most of the {100} surface areas are covered by the reconstructed layer [3].

1. Z.L. Wang, Rep. Prog. Phys. **56** (1993) 997; K. Yagi, Surf. Sci. Rep. **17** (1993) 305.
2. S. Geller and V.B. Bala, Acta Cryst. **9** (1956) 1019.
3. Z.L. Wang and A.J. Shapiro, Surface Science (1995) in press.

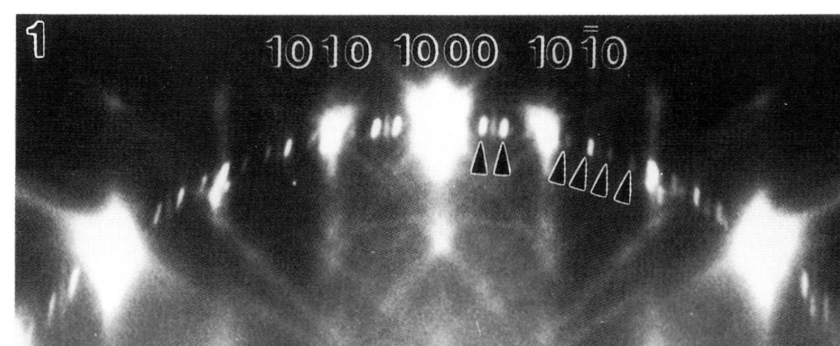

FIG. 1 -- [001] RHEED pattern from annealed LaAlO$_3$ (100) surface showing 5×5 surface reconstruction.

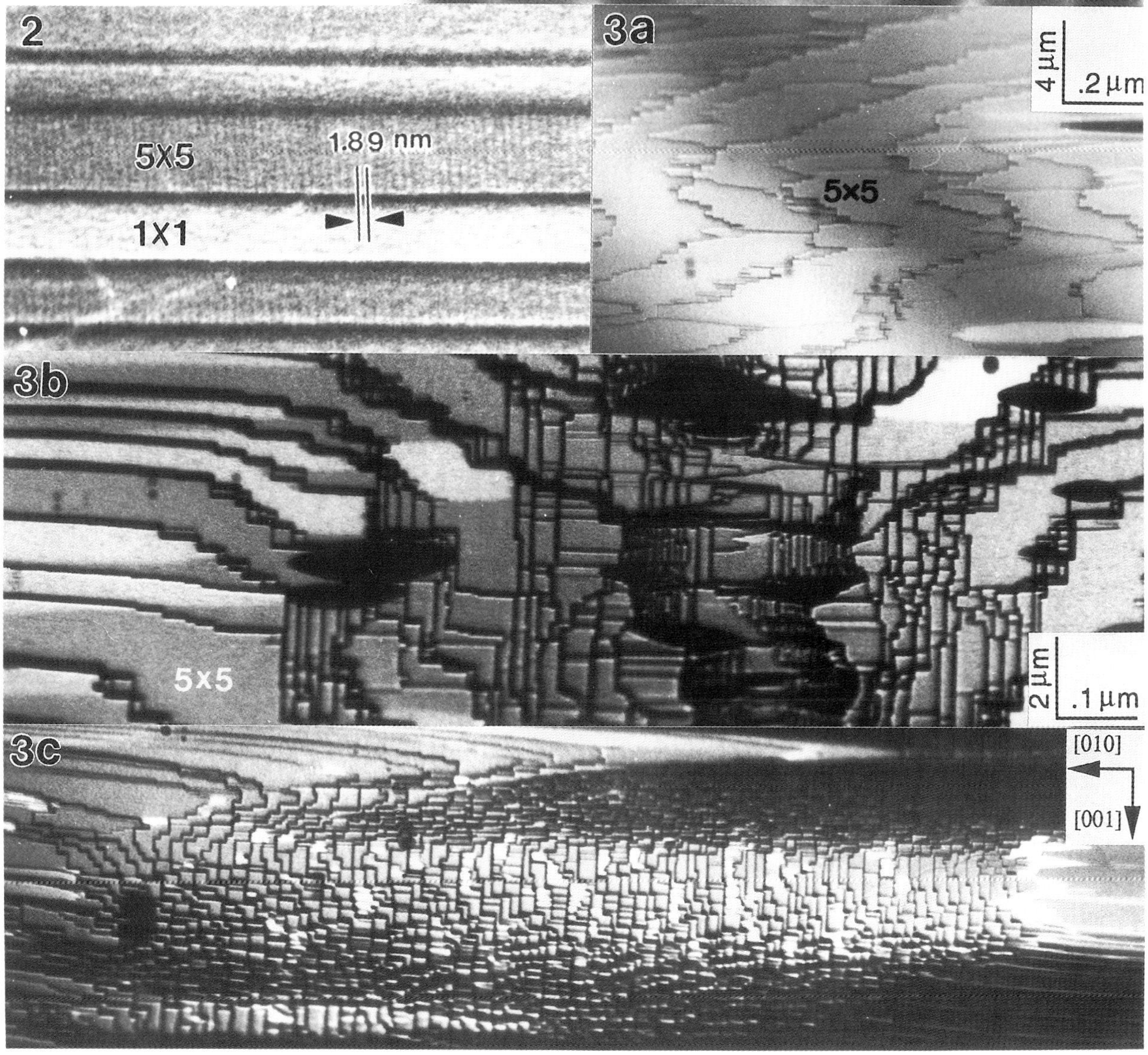

FIG. 2 -- High-resolution REM image of LaAlO$_3$ (100) showing 1.89 nm fringes in correspondence to 5×5 reconstruction, because 5a = 1.89 nm with lattice constant a = 0.378 nm.
FIG. 3 -- REM images of LaAlO$_3$ (100) surface showing [010] and [001] faceted "stair-like" steps. Images were recorded using (10 0 0) specularly reflected beam of azimuth [001].

347

IMPURITIES AND GRAIN BOUNDARIES IN SINTERED TITANIUM DIBORIDE

G. Pettersen,* A. Bardal,** and R.Høier*·**

*Dept. of physics, Univ. of Trondheim-NTH, N-7034 Trondheim, Norway
**SINTEF Applied Physics, N-7034 Trondheim, Norway

In conventional Al-electrolysis the electrodes suffer from degradation when used in the harsh environment of an aluminium electrolysis cell. The carbon cathode, which is placed at the cell bottom, is not well wetted by aluminium, and must be covered by a pad of liquid Al in order to establish a good electrical contact. To avoid short circuiting of the cell the anode-to-cathode distance (ACD) must be kept above 4-6 centimetres. Replacing the carbon cathode with a new, inert cathode, well wetted by liquid aluminium, would allow new cell designs where only a thin film of aluminium covers the cathode during cell operation, thereby allowing the ACD to be lowered. This in turn opens for substantial reductions in the energy consumption of an electrolysis cell,[1] and combined with an inert anode the outlet of CO_2 and fluorides harmful to the environment could be reduced.

The most promising candidate for use as an inert cathode is titaniumdiboride (TiB_2), a ceramic material which exhibits promising properties such as a high electrical conductivity, sufficient mechanical strength, good resistance to chemical attack and low solubility in liquid aluminium, and which is also well wetted by liquid aluminium.[2,3] Research to implement TiB_2 as a cathode has taken place since the 1950s, the major problem being the degradation of the material by cracks and flaws seen after long times of electrolysis. This degradation is believed to be caused by the slow penetration of liquid aluminium into the grain boundaries of the sintered TiB_2. TEM investigations have shown that only some of the grain boundaries are penetrated by the aluminium, others stay clean.[4] Investigating the microstructure of the material, focusing onto the nature of the grain boundaries and utilising the high resolving power of TEM combined with EELS and EDS microanalysis, is hoped to answer the question of what causes the penetration.

In this work commercially available TiB_2-powder from H.C.Stark (grade A) has been hot pressed in an argon atmosphere at temperatures about 1900°C for two hours with a pressure of 50 MPa. Thin films obtained by grinding and subsequent ion beam milling has been investigated in a Philips CM30 TEM/STEM equipped with a PEELS and EDS system for microanalysis at 300 kV.

The sintered material is found to contain different impurity phases and some porosity in addition to the TiB_2-phase. Most of the impurity phases are found at grain boundaries and in triple points, some examples are given here. Figure 1(a) shows a phase identified by PEELS and SAED to be the B_4C-phase. This is found as rather large particles with many stacking faults, and is located in triple points. Along with TiC and BN this phase is believed to have its origin in the carbothermic production of TiB_2 from titanium- and boron-oxides which react with carbon at high temperatures to form TiB_2 and CO. Another impurity commonly encountered in triple points is the amorphous oxide seen covering the rod-like BN phase in figure 1(b). PEELS and EDS shows the main constituents to be oxygen, titanium and boron with a varying, but smaller, content of other elements like iron, chromium and molybdenum. A thin film of oxide, a few nanometres thick, is known to exist on the surface of the TiB_2 powder prior to sintering, in addition some metallic impurities are introduced during the crushing and milling of the very hard TiB_2 material to powder. This may be the origin of the phase in figure 1b. An important question to answer in order to understand the mechanism of the penetration of aluminium into the grain boundaries is whether these boundaries also contains any thin film of an impurity phase which affects the aluminium penetration, or if they are clean. This can only be properly answered by applying lattice imaging. Figure 2 is a HREM micrograph from a grain boundary showing that no impurity phase, neither amorphous nor crystalline, is present here. More grain boundaries must be examined by HREM to know whether this is the case for all boundaries, or only for some of them.

References

1. J.B. Todd, J.Metal 33(1981)42.
2. K. Billehaug and H. Øye, Aluminium 56(1980)642.
3. R.P. Pawlek, Aluminium 66(1990)573.
4. K. Nord-Varhaug, submitted to J.Am.Ceram.Soc.
5. The author gratefully acknowledges the Norwegian Resarch Council for financial support.

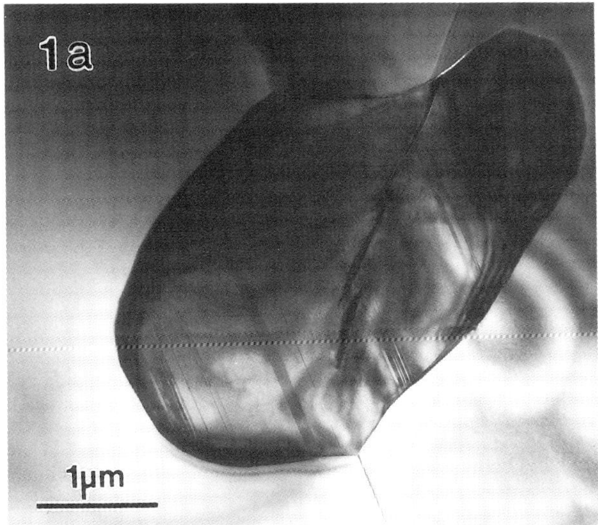
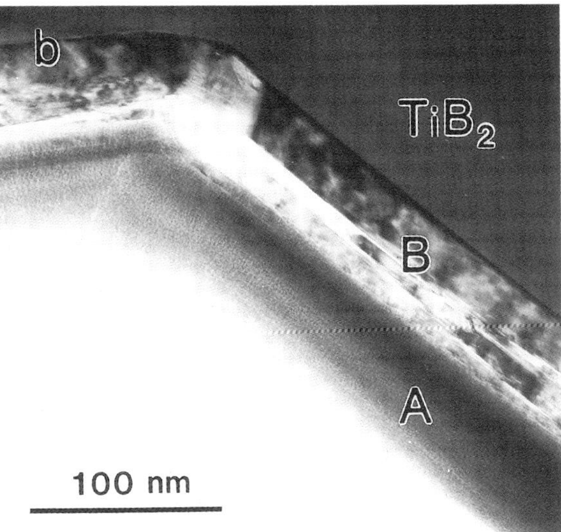

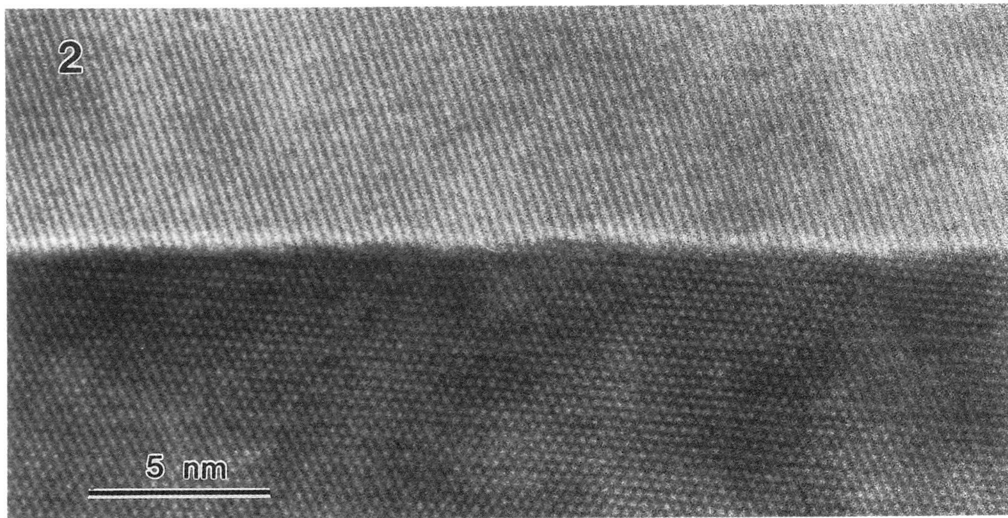

FIG. 1 (a) B_4C-phase located at a triple point in the TiB_2 matrix.
    (b) Amorphous oxide (A) covering the boron nitride phase (B) in a triple point.
FIG. 2 HREM micrograph from a grain boundary in TiB_2. In the lower grain the incident beam direction is parallel to the [0001] axis, in the upper close to the [1212].

TEM OF GRAIN GROWTH AND PHASE TRANSFORMATIONS DURING CREEP OF SCS-6 SILICON CARBIDE FIBERS

L. A. Giannuzzi*, C. A. Lewinsohn**, C. E. Bakis**, and R. E. Tressler**

*Mechanical and Aerospace Engineering, University of Central F orida, Orlando, FL 32816-2450
**Center for Advanced Materials, The Pennsylvania State University, University Park, PA 16802

The SCS-6 SiC fiber is a 142 μm diameter fiber consisting of four distinct regions of βSiC.[1,2] These SiC regions vary in excess carbon content ranging from 10 a/o down to 5 a/o in the SiC1 through SiC3 region. The SiC4 region is stoichiometric. The SiC sub-grains in all regions grow radially outward from the carbon core of the fiber during the chemical vapor deposition processing of these fibers. In general, the sub-grain width changes from 50nm to 250nm while maintaining an aspect ratio of ~10:1 from the SiC1 through the SiC4 regions. In addition, the SiC shows a <110> texture, i.e., the {111} planes lie ±15° along the fiber axes. Previous has shown that the SCS-6 fiber (as well as the SCS-9 and the developmental SCS-50 μm fiber) undergoes primary creep (i.e., the creep rate constantly decreases as a function of time) throughout the lifetime of the creep test.[3] The differences in creep rates between the SCS-6 and the SCS-9 and SCS-50 μm fibers could not be explained by differences in grain sizes among these fibers. It was also determined that the creep behavior was unlikely to be a result of load sharing between the SiC region and the carbon core. As will be shown, significant grain growth, in addition to phase transformations, occur in the SCS-6 fiber during creep testing.

The fibers were prepared in longitudinal cross-section (i.e., the electron beam was perpendicular to the fiber, and therefore, creep axis) for examination by transmission electron microscopy (TEM) using a similar technique previously reported.[1] A detailed description of the specimen preparation may be found elsewhere.[4] The fibers were observed in a Philips EM420T operating at 120kV.

An SCS-6 SiC fiber creep tested for 500h at 1400°C at a stress of 278MPa showed a dramatic change in microstructure as compared with the as-produced fibers, and heat treated fibers at the same time and temperature. FIG. 1a is a bright field (BF) TEM micrograph from the SiC3 region of the crept fiber. The direction of the fiber axis is indicated by the line in the figure. The morphology of the microstructure in FIG. 1a is representative of the SiC1, SiC2, and SiC3 regions of the fiber. A corresponding selected area diffraction pattern (SADP) encompassing the region in FIG. 1a is shown in FIG. 1b. Growth of the individual grains of βSiC occurs primarily along the creep or fiber axis which creates approximately equiaxed grains measuring ~500 nm in width. This represents an observed grain growth compared to the as-produced microstructure by a factor of 2-10 depending on the exact region within the fiber. The βSiC grains retain their textured structure during growth as evident from the SADP in FIG. 1b. Note also the excess carbon which was present in the as-produced fiber has clearly segregated at the βSiC grain boundaries creating the obvious two-phase (C + βSiC) region. The presence of the graphite phase is evident from the polycrystalline rings in the SADP of FIG. 1b. This two-phase region is similar to that observed for a fiber heat-treated at 2000°C for 1 hour.[5] However, the primary difference we observe here is that the entire SiC1-SiC3 region consists of these large βSiC grains and two phase (C + βSiC) region, as opposed to the observation of grain growth and C segregation in only the inner layer of the SiC1 region.

Significant microstructural changes were also observed in the entire SiC4 region of the fiber where exaggerated grain growth occurred during creep. This grain growth is evident in the BF TEM micrograph of FIG. 2a. Note that the grain size in the SiC4 region has increased to several micrometers. In some instances, the grain growth in the SiC4 region represents in increase in grain size by as much as 2 orders of magnitude. The difference in growth rates between the SiC1-SiC3 regions and the SiC4 region suggests that the presence of carbon along the grain boundaries inhibits grain growth. In addition to the exaggerated grain growth observed, portions of the βSiC in the SiC4 region transforms to αSiC. This transformation is evident by examining the SADP of FIG. 2b. which encompasses four individual grains, of which three of these are shown in FIG. 2a. The SADP shows polytypic twinning in both the

Proc. Microscopy and Microanalysis 1995, edited by G.W. Bailey, M.H. Ellisman, R.A. Hennigar, and N.J. Zaluzec.
Copyright © 1995 MSA. Published by Jones and Begell Publishing, 79 Madison Ave., New York, NY 10016

3C βSiC and the 6H αSiC, as well as an orientation relationship (OR) between the βSiC, αSiC, and their twins.   The OR between the four grains may be expressed as: [110]βSiC ‖ [110]βSiC$_{twin}$ ‖ [2$\bar{1}\bar{1}$0]αSiC ‖ [2$\bar{1}\bar{1}$0]αSiC$_{twin}$, ($\bar{1}$11)βSiC ‖ ($\bar{1}$1$\bar{1}$)βSiC$_{twin}$, ($\bar{1}$11)βSiC ‖ (0002)αSiC, ($\bar{1}$1$\bar{1}$)βSiC or ($\bar{1}$11)βSiC$_{twin}$ ‖ (0002)αSiC$_{twin}$. The twinning and α/β polytypic OR is achieved along the close packed {111} βSiC and {0002} αSiC planes by the passage of partial dislocations.   These transformations during creep is consistent with the polytypic phase transformation mechanism involving the motion of partial dislocations accelerated by the presence of stress.[6] Note also that the αSiC grew at the expense of the βSiC (see FIG. 2a).   The above βSiC to αSiC phase transformation during high temperature creep is consistent with the observation of a stress-induced βSiC to αSiC phase transformation followed by exaggerated grain growth.[7,8,9]

References

1. X.J. Ning and P. Pirouz, J. Mater. Res., **6**, [10], 1991, 2234-2248.
2. X.J. Ning, P. Pirouz, and S.C. Farmer, J. Am. Ceram. Soc., **7 6** [8], 1993, 2033-2041.
3. C.A. Lewinsohn, Ph.D. Thesis, The Pennsylvania State University, 1994.
4. L.A. Giannuzzi, et al., to be published, J. Am. Ceram. Soc.
5. X.J. Ning, P. Pirouz, and R.T. Bhatt, Mat. Res. Soc. Symp. Proc. **250** 187-92 (1992).
6. D. Pandey, and P. Krishna, in Silicon Carbide-1973, eds. R.C. Marshall et al., University of South Carolina Press, Columbia, S.C., 1974, 198-205.
7. T.F. Page et al., Proc. Brit. Ceram. Soc., **26**, 1978, 193-208.
8. J.W. Yang et al., Mat. Res. Symp. Proc., **242**, 1992, 531-536.
9. This work was supported by NASA Grant # NAGW-1381.

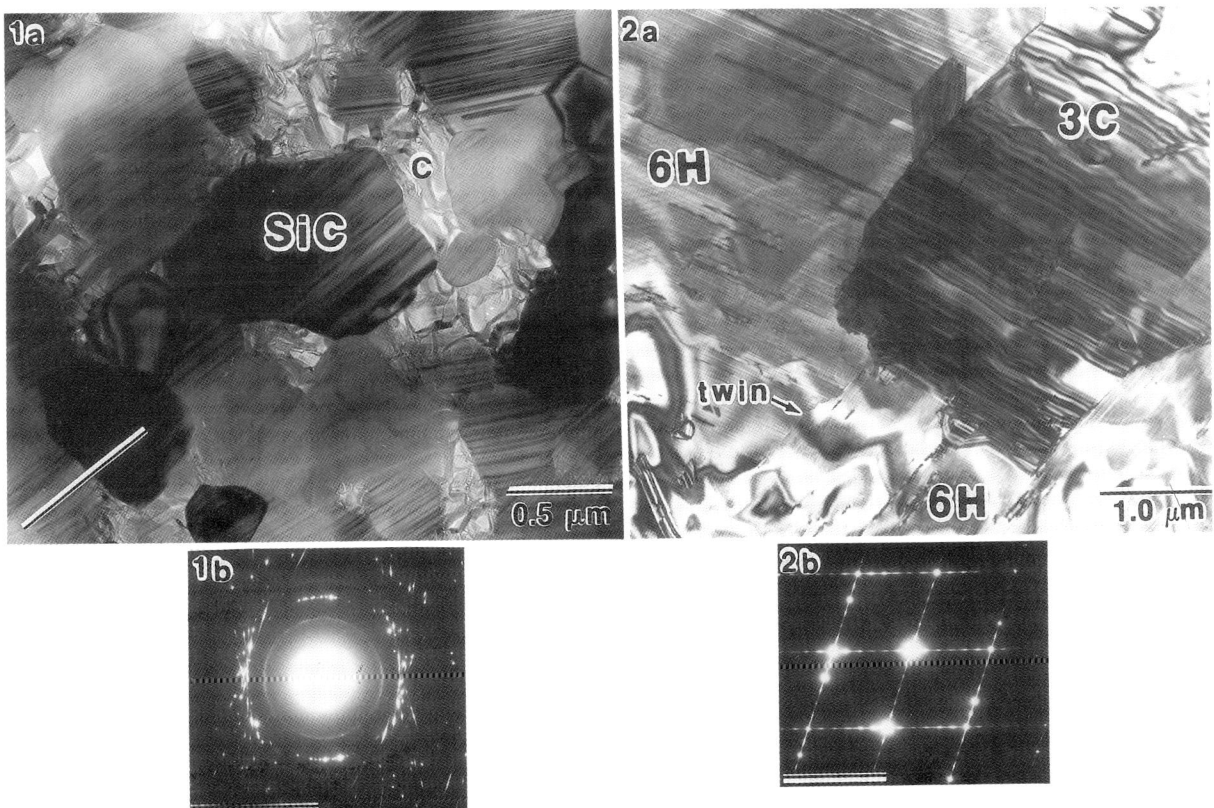

FIG. 1. - (a) BF TEM image and (b) SADP representative of the SiC1-SiC3 regions from a crept SCS-6 fiber. Bar = 5 nm^{-1}.The line in (a) indicates the direction of the fiber axes.
FIG. 2. - (a) BF TEM image and (b) SADP of the SiC4 region of a crept SCS-6 fiber. Bar = 5 nm^{-1}.

CHARACTERIZATION OF MODIFIED CERAMIC FIBER SURFACES USING AFM

C. L. Arnold*, K. L. More**, and E. Lara-Curzio**

*Tennessee Technological University, Cookeville,TN 38505
**High Temperature Materials Laboratory, Oak Ridge National Laboratory, Oak Ridge, TN 37831-6064

Several models have been developed to describe the interfacial behavior of fiber-reinforced ceramic matrix composites.[1] However, these models are limited in that they do not take into account the effect of fiber surface roughness on the fiber sliding resistance. Clearly, the fiber surface roughness will play an important role in extensive fiber/matrix debonding and fiber pull-out behavior which are characteristic of a toughened composite. In this study, a Topometrix TMX2010 Atomic Force Microscope (AFM) was used to quantitatively evaluate the surface topography of a SiC fiber subjected to a series of surface modification treatments. The relationship between interfacial properties determined from fiber push-out tests and the surface roughness will be used to modify existing mathematical models.

In order to systematically study the effect of surface roughness on the interfacial properties of ceramic matrix composites, a "model" single fiber composite having varying degrees of fiber surface roughness was developed. This was accomplished by using a single manufacturer's ceramic fiber which was subjected to a series of treatments in order to alter the surface. SiC fibers prepared by chemical vapor deposition (CVD), designated as Sigma fibers, were chosen for this study for two primary reasons: (1) the surfaces could be altered relatively easily using standard techniques and (2) the fibers have a large diameter, ~120 μm, and thus could be analyzed using AFM and subsequently individually incorporated in a borosilicate pyrex glass matrix for fiber push-out tests to measure the interfacial properties.[2] Individual fibers were subjected to a series of surface treatments consisting primarily of oxidation in air at temperatures ranging from 1200-1400°C for times ranging from 1 to 60 min followed by an HF etch for times ranging from 10 to 30 min. Small lengths of the treated fibers were cleaned using acetone and mounted using a special low melting point wax (Tempfix) on AFM sample mounting discs. Special care was taken to ensure that the mounting wax did not wick up the sides or on top of the fibers. The AFM cantilever, mounted with a Topometrix Super-Tip™ was oriented so that the scanning direction was parallel to the long length of the fiber. Several different areas along the length of the fiber were scanned to verify surface homogeneity.

A typical AFM image of the as-received Sigma fiber is shown in Fig. 1. The AFM data was used·to quantify the surface roughness of the Sigma fibers by taking several line profiles across selected areas of a given image. In order to accurately measure the roughness of a fiber surface, the surface was "levelled" to remove the effects of inherent fiber curvature.[3] For comparative purposes, the primary roughness parameter determined was Ra, the arithmetic mean of the absolute values of the departures of the roughness profile from the mean line. Ra for the as-received Sigma fibers was ~47 nm. After exposing the fiber surface to air oxidation and subsequent HF immersion, different surface morphologies were observed. AFM images for fibers oxidized for 10 min at 1300°C and 1325°C followed by a 10 min HF etch are shown in Fig. 2 and 3, respectively. The Ra values, 70 nm and 121 nm for 1300°C and 1325°C, respectively, increase with increased oxidation temperatures due to thicker oxide formation on the fiber surface. Oxidation for shorter times (<10 min) did not result in significant changes to the surface structure. This is illustrated in Fig. 4 and 5 for a 5 min oxidation at 1300°C and 1325°C, respectively, followed by a 30 min HF etch. The average Ra value for these fibers remained 46 nm, the same as for the untreated Sigma fibers. The formation of large surface features (arrowed in Fig. 4) locally increased the Ra to ~110 nm, but were not very numerous. However, at temperatures greater than 1350°C, short time (≤1 min) oxidation resulted in large Ra values and a more homogeneous distribution of these large surface features. This is illustrated in Fig. 6 and 7 for oxidation at 1350°C and 1400°C for 1 min followed by a 30 min HF etch. The measured Ra values for these fiber surfaces were 89 nm and 132 nm for the 1350°C and 1400°C exposures. Clearly, oxidation effects contribute more to the changes in surface roughness observed than does the HF treatment.

The fibers examined in this study are currently being incorporated individually into borosilicate glass matrices in order to perform fiber push-out tests. Interfacial properties will be determined as a function of varying degrees of fiber surface roughness. These tests will be conducted using the Interfacial Test System designed at ORNL.[2] Work is ongoing in this characterization area.[4]

Proc. Microscopy and Microanalysis 1995, edited by G.W. Bailey, M.H. Ellisman, R.A. Hennigar, and N.J. Zaluzec
Copyright © 1995 MSA. Published by Jones and Begell Publishing, 79 Madison Ave., New York, NY 10016

References

1. C.H. Hsueh, *J. Amer. Cer. Soc.*, 76[12] (1993) 3041.
2. E. Lara-Curzio and M.K. Ferber, *J. Mat. Science,* 29 (1994) 6152.
3. K.L. More, et al., *Proc. Ann. MSA Meeting,* 52 (1994) 1066.

4. Research sponsored by the Office of Industrial Technologies., Improved Energy Prod. Division and Advanced Industrial Concepts Division., and the Assistant Secretary for Energy Efficiency and Renewable Energy, Office of Transportation Technologies., as part of the HTML User Program, both sponsored by the U.S. Department of Energy, under contract DE-AC05-84OR21400 managed by Martin Marietta Energy Systems, Inc.

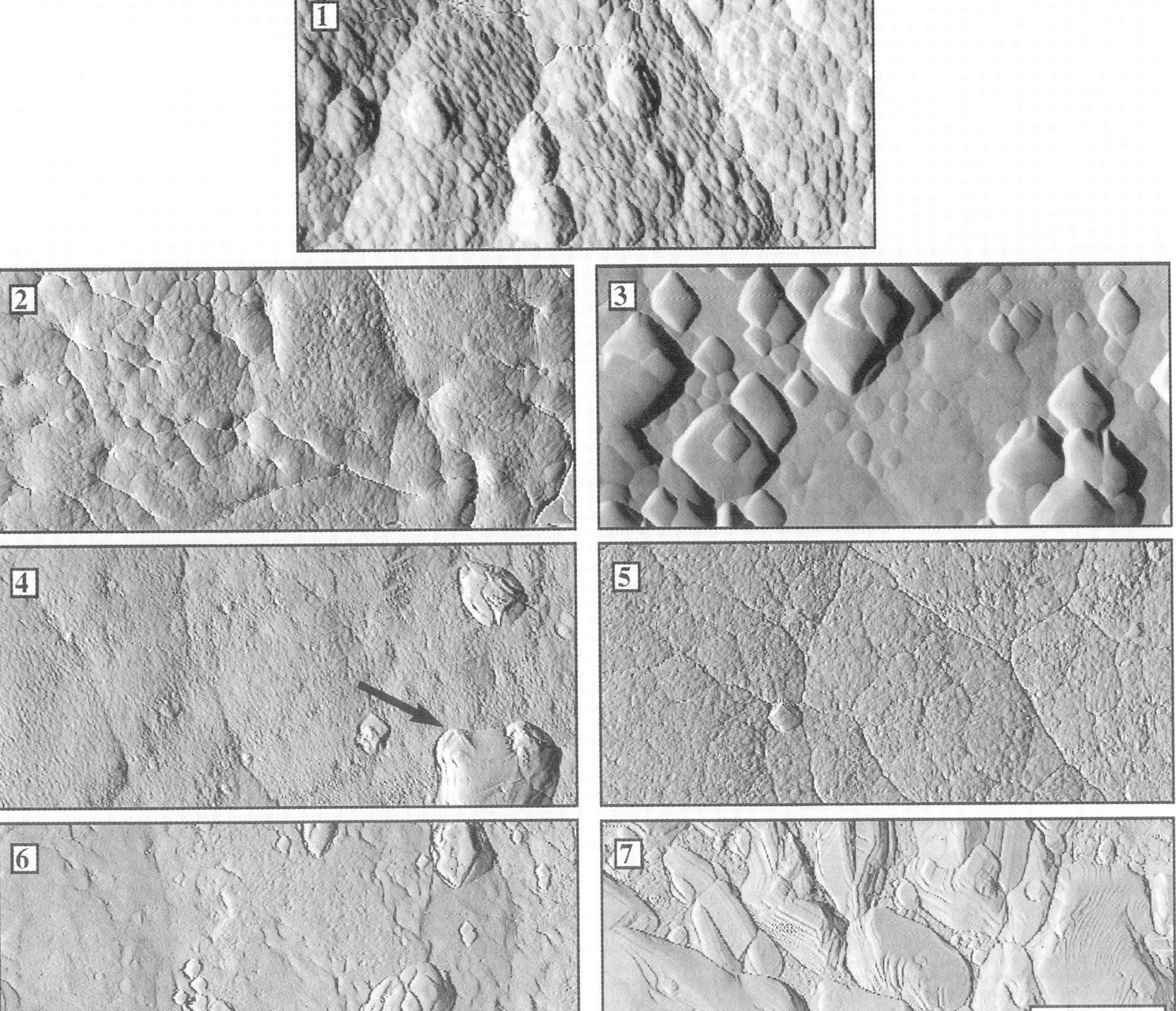

FIG. 1-- AFM image of as-received Sigma Fiber.
FIG. 2-- AFM image of a Sigma Fiber after 10 min oxidation at 1300°C and 10 min HF.
FIG. 3-- AFM image of a Sigma Fiber after 10 min oxidation at 1325°C and 10 min HF.
FIG. 4-- AFM image of a Sigma Fiber after 5 min oxidation at 1300°C and 30 min HF.
FIG. 5-- AFM image of a Sigma Fiber after 5 min oxidation at 1325°C and 30 min HF.
FIG. 6-- AFM image of a Sigma Fiber after 1 min oxidation at 1350°C and 30 min HF.
FIG. 7-- AFM image of a Sigma Fiber after 1 min oxidation at 1400°C and 30 min HF.

A CVI-PROCESSING MARKER STUDY

G. B. Freeman,* W. J. Lackey,** and S. Vaidyaraman**

*Materials Analytical Services, Norcross, GA 30092
** Georgia Institute of Technology, Atlanta GA 30332

Fiber reinforced ceramic composites greatly extend the region of application for ceramic materials because the inclusion of a second phase such as fibers helps to prevent catastrophic failure via the mechanisms of crack branching, deflection, bridging and fiber pull-out.[1] While ceramic composites are excellent candidate materials for high temperature structural applications, many conventional manufacturing processes damage the fibers during fabrication and thereby reduce the strength and toughness from what is theoretically possible. Chemical vapor infiltration (CVI) is an effective and versatile technique for fabrication of ceramic matrix composites which promises to reduce fiber degradation.[2]

While CVI is no longer a truly novel processing technique, it is still difficult to predict the best set of operating conditions for what can be a tedious process. Conventional CVI methods may take up to a week to fully infiltrate a small part and even state-of-the-art forced flow-thermal gradient CVI (FCVI) processing takes the better part of a day.[3] It is, more over, difficult to monitor the evolution of deposition within a, for example, woven preform. One interesting method which has been used is x-ray computed tomography.[4] This, and more commonly used serial sectioning, involve stopping the infiltration process at selected intervals and sequentially examining that sample.

A new process of evaluation has been developed which allows information to be garnered from the finished sample and referenced back to various stages of deposition. In this study, the forced flow-thermal gradient CVI process was used to fabricate a carbon fiber-carbon matrix composite in the shape of a right circular disk. In this process a pressure gradient forces the reagent stream of 50% propylene-50% hydrogen to flow through a preform subjected to a temperature gradient.[5] During the carbon infiltration process, the flow of propylene was interrupted after two, four, six and eight hours to deposit a thin layer of silicon carbide (SiC) by flowing a 9% methyltrichlorosilane-91% hydrogen mixture for four minutes.

The CVI process was continued until the back pressure indicated that the majority of gas passageways had been effectively plugged after about 14 hours. After cooling, the sample was removed from the holder, mounted in epoxy, cut in half, and polished to a relatively smooth finish. This newly exposed cross section ran along the diameter of the disk and presented a rectangular surface, the top of which was held at high temperature during the deposition and the bottom of which was held at the cooler temperature. After carbon coating, the sample was examined in an Hitachi S4100 field emission scanning electron microscope (FESEM) using a GWI channel plate back scattered electron detector with an accelerating voltage of 20 keV.

Proc. Microscopy and Microanalysis 1995, edited by G.W. Bailey, M.H. Ellisman, R.A. Hennigar, and N.J. Zaluzec
Copyright © 1995 MSA. Published by Jones and Begell Publishing, 79 Madison Ave., New York, NY 10016

Figure 1 shows a section of the sample near a large open pore between two tows of fiber and Figure 2 shows a number of carbon fibers in the interior of a tow. By observing the amount of material deposited between layers of SiC, a deposition rate and history can be "backed out" of the data. Results indicate that most of the deposition within a given tow is completed in the first portion of the CVI process and that the remainder of the deposition time is spent in filling large pores. Similarly, using both the thickness of deposition between SiC layers and the thickness of the SiC layers themselves, quantitative data may be generated which yields information about deposition rates as a function of time and position within the sample. These data illuminate the effect of flow patterns and temperature gradients on the overall process and should have similar applicability to a number of other related processing systems.

References

1. R. W. Rice, *Ceram. Eng. Sci. Proc.* 2(1981)661.
2. W. A. Bryant, *J. Mater. Sci.* 12(1977)1285.
3. W. J. Lackey and T. L. Starr, in K. S. Mazdiyasni, Ed., *Fiber Reinforced Ceramics*, Park Ridge, NJ, Noyes Publications (1990)397.
4. J. R. Kinney, T. M. Breunig, T. L. Starr, D. Haupt, M. C. Nichols, S. R. Stock, M. D. Butts and R. A. Saroyan, *Science*, 260(1993)789.
5. S. Vaidyaraman, W. J. Lackey, G. B. Freeman, P. K. Agrawal and M. D. Langman, accepted for publication in *J. Mater. Res.* (1994).

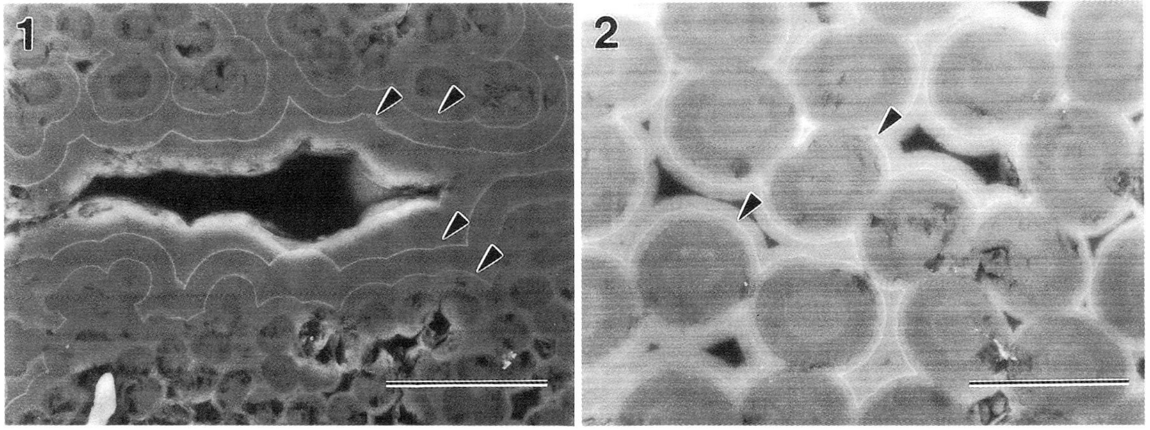

FIG. 1-Backscattered electron (BSE) image of open pore region within carbon-carbon polished cross section. The arrows indicates thin, bright SiC deposition layers. Bar=30 μm.
FIG. 2-BSE image of fibers within tow of carbon-carbon polished cross section. The arrows indicate SiC deposition layers. Bar=10 μm.

SURFACE MORPHOLOGY OF XENON IMPLANTED SINGLE CRYSTAL MAGNESIUM OXIDE OBSERVED BY ATOMIC FORCE MICROSCOPY

Wenbiao Jiang,* M. Grant Norton,* and David B. Poker**

* Department of Mechanical and Materials Engineering, Washington State University, Pullman, WA 99164

** Surface Modification and Characterization Research Facility, Oak Ridge National Laboratory, Oak, Ridge, TN 37831

The implantation of materials with noble gas ions is of particular interest in studying some of the fundamental aspects of the ion implantation process and resultant phenomena because the noble gas ions are immiscible with the matrix and therefore chemical effects are unlikely. Electron diffraction studies have clearly demonstrated that noble gas ions implantated into either metals or ceramics may coalesce to form inclusions.[1,2] These inclusions may be either solid or fluid and their formation often results in swelling of the matrix. This swelling can be readily apparent on the surface of metallic materials. However, in ceramic materials the extent of this swelling, particular at low ion fluences, can be very small and hence only detectable with surface sensitive topographic probes such as atomic force microscopy (AFM).[3]

Polished (001)-oriented MgO single crystals were implanted with xenon ions using an accelerating voltage of 200 keV at a range of fluences. The specimens were then examined with AFM. The AFM results showed that, even at the lowest fluence, there was measurable roughening of the surface. At a fluence of $1 \times 10^{17}/cm^2$ small blisters were observed. These blisters have diameters of $0.5 - 0.8$ μm and extend to heights of up to 80 nm. Figure 1 shows a typical AFM image of such a blister in top view and surface view modes, respectively. The surface of these blisters displayed a characteristic rosette pattern. The arms of the rosette are at angles of 45° to each other producing a symmetrical pattern.

During implantation the xenon ions form small clusters. These clusters then grow by capturing additional xenon ions to form larger inclusions. The growth of the inclusions involves the punching out of dislocation loops into the surrounding matrix material. MgO has the slip system {110}<110>. For (001)-oriented MgO, two of the {110} planes are perpendicular to the surface and may be referred to as $\{110\}_{90}$ planes, four of them are inclined at 45° and referred to as $\{110\}_{45}$ planes. Figure 2 shows a schematic representation of dislocation loops formed on the two types of {110} planes by a xenon inclusion growing during implantation. This process will be occuring on all the {110} planes surrounding the inclusion. Intersection of the dislocation loops with the crystal surface results in the formation of steps and displaced material which reflects the orientation of the slip planes. Thus the eight arms in the rosette pattern oriented in the <100> and <110> directions correspond to the operation of dislocation movement on both $\{110\}_{45}$ and $\{110\}_{90}$ planes, respectively. The appearance of the rosette pattern provides convincing evidence that the xenon inclusions have grown during implantation by a dislocation loop punching mechanism.[4]

References

1. M.G. Norton et al., Phys. Rev. B, 43(1991)9291.
2. M.G. Norton et al., J. Mater. Res., 1(1992)3171.
3. M.G. Norton et al., Scripta Metall. *et* Mater., 31(1994)555.
4. WJ gratefully acknowledges the support of the Materials Science Program through a Graduate Student Assistantship. The work at ORNL was sponsored by the Division of Materials Science, US Department of Energy, under contract DE-AC05-84OR21400 with Martin Marietta Energy Systems, Inc.

Proc. Microscopy and Microanalysis 1995, edited by G.W. Bailey, M.H. Ellisman, R.A. Hennigar, and N.J. Zaluzec
Copyright © 1995 MSA. Published by Jones and Begell Publishing, 79 Madison Ave., New York, NY 10016

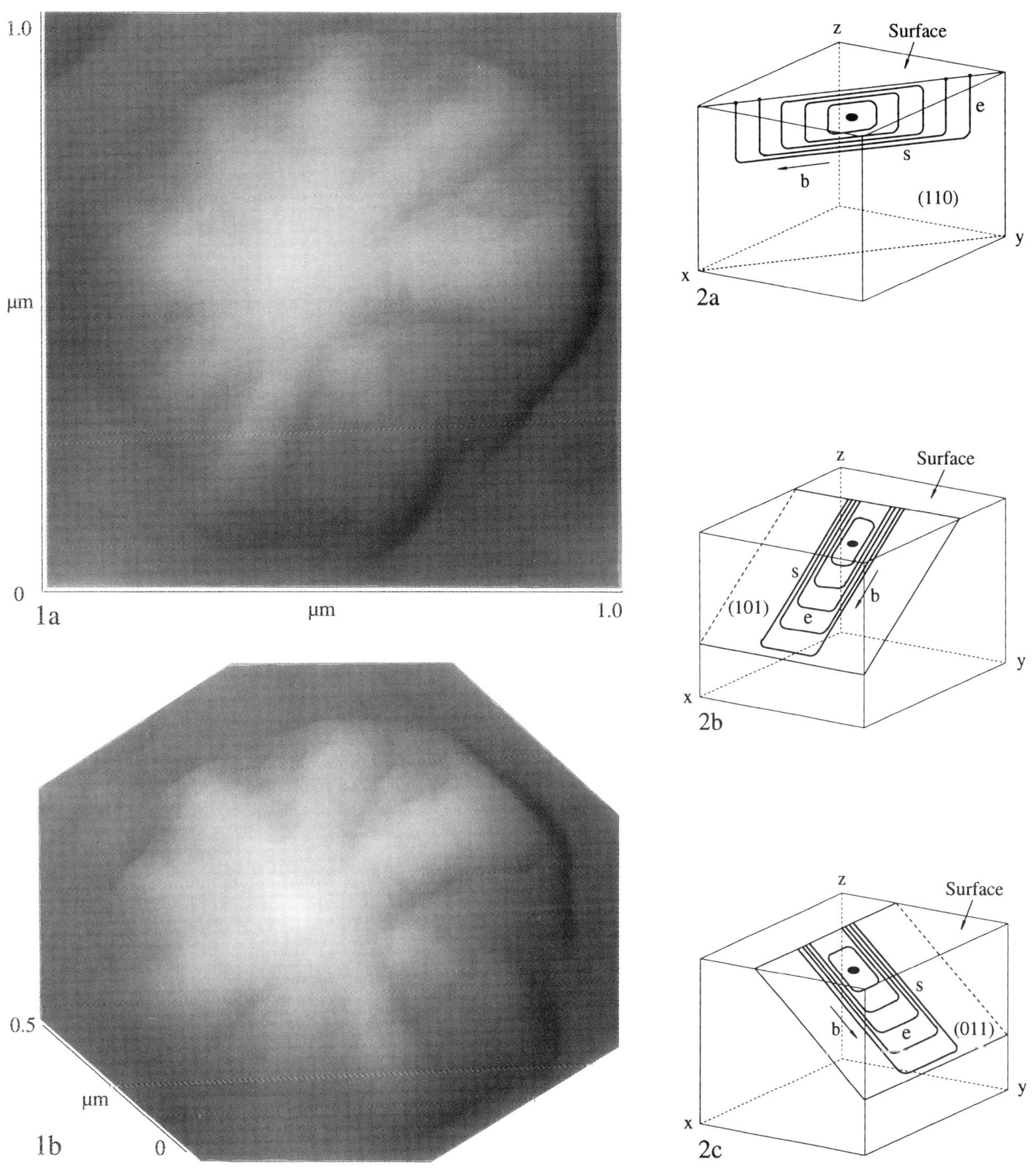

FIG.1—AFM images showing a blister on the surface of MgO implanted with xenon at a fluence of $1.0 \times 10^{17}/cm^2$, (a) top view, (b) surface view with z range of 100 nm/mm.

FIG.2—Schematic showing the arrangment of dislocation loops on three slip planes around the xenon inclusion shown as black spot: b–Burgers vector, e–edge dislocation, s–screw dislocation.

RELATIONSHIP BETWEEN STRUCTURE AND MECHANICAL PROPERTIES OF ION-IRRADIATED MgAl$_2$O$_4$ SPINEL

Ram Devanathan, Ning Yu, Kurt E. Sickafus and Michael Nastasi

Materials Science and Technology Division, Los Alamos National Laboratory, Los Alamos, NM 87545

The relationships between structure and properties of irradiated materials is an important focus of radiation damage studies. A knowledge of these relationships can enable us to tailor the properties of materials to enhance their radiation resistance. Magnesium aluminate spinel is an obvious choice for such a study, because it is known to resist radiation damage when irradiated with fast neutrons or energetic ions.[1-2] Recent work has shown, for the first time, that MgAl$_2$O$_4$ can be amorphized at cryogenic temperatures following ion irradiation.[3] Electron diffraction patterns, obtained after irradiating spinel with 400 keV Xe^{2+} at 100 K to various doses upto 25 dpa, show that the first-order reflections decrease in intensity and become extinct with increasing dose. If the dose is increased further, the material becomes amorphous. We have performed a systematic study of this phase transformation by obtaining the diffraction patterns, and measuring the Young's modulus and nano-hardness of the irradiated layer.

The structure of irradiated spinel was examined using electron diffraction patterns obtained from cross-sectional transmission electron microscopy samples. The mechanical properties were obtained using a state-of-the-art, ultra low-load mechanical properties microprobe (Nano Indenter II). Since the thickness of the irradiated layer in our samples is only about 150 nm, the nano-indentation technique is ideal for probing this layer without any appreciable influence from the substrate. For each dose, three different indentation depths were chosen and ten indentations performed at each depth to obtain good statistics.

Figure 1 shows the Young's modulus of Xe^{2+} irradiated spinel as a function of dose along with the corresponding diffraction patterns. The modulus increases with dose, initially, and reaches a maximum at a dose of about 1.5x10^{15} ions/cm^2. The corresponding diffraction pattern shows that the material has transformed to a metastable state characterized by the disappearance of first-order reflections. We have identified this phase as being made up of a 0.4 nm fcc oxygen lattice with cation interstitials. As the dose is increased further, the material transforms to an amorphous state. At a dose of 1x10^{16} ions/cm^2, a complete halo is seen in the diffraction pattern. The modulus of the material shows a corresponding decrease of more than 30% from the perfect crystal value. The changes in the nano-hardness were found to be qualitatively similar but of a higher magnitude (a decrease of about 60%).

In conclusion, we have performed a comprehensive study of the changes in the structure and mechanical properties of ion-irradiated spinel. Our results show that spinel transforms to a metastable state before undergoing softening associated with amorphization.

References

1. F. W. Clinard Jr., G. F. Hurley, and L. W. Hobbs, *J. Nucl. Mater.* 108-109 (1982) 655.
2. S. J. Zinkle, *J. Am. Ceram. Soc.* 72 (1989) 1343.
3. N. Yu, K. E. Sickafus, and M. Nastasi, *Phil. Mag. Lett..* 70 (1994) 235.
This research is sponsored by the US Department of Energy, Office of Basic Energy Sciences, Division of Materials Sciences.

Proc. Microscopy and Microanalysis 1995, edited by G.W. Bailey, M.H. Ellisman, R.A. Hennigar, and N.J. Zaluzec
Copyright © 1995 MSA. Published by Jones and Begell Publishing, 79 Madison Ave., New York, NY 10016

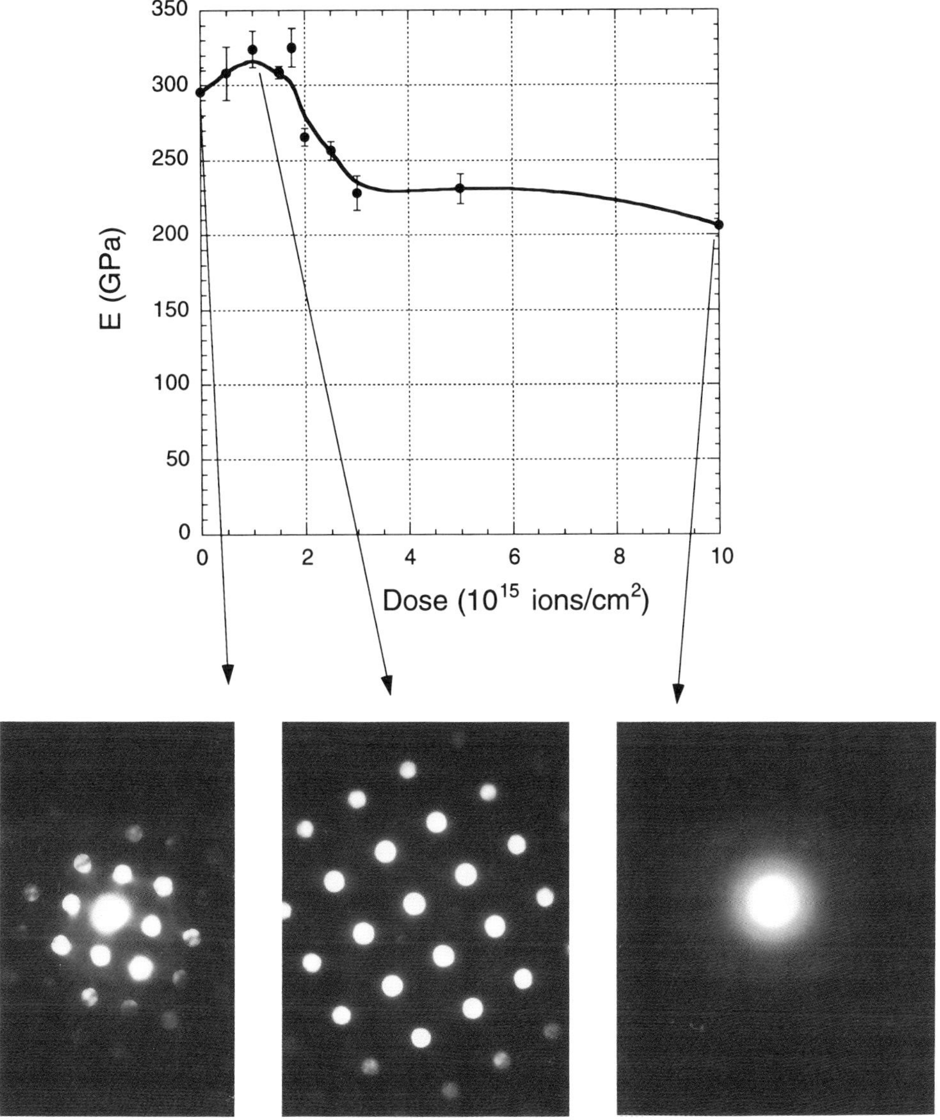

FIG. 1. The changes in the Young's modulus and structure of Xe^{2+} irradiated spinel for various damage doses.

HRTEM STUDY OF ION BEAM IRRADIATION INDUCED AMORPHIZATION IN CERAMIC MATERIALS

L.M. Wang[†], R.C. Ewing[†] and W.J. Weber[*]

[†]Department of Earth and Planetary Sciences, University of New Mexico,
 Albuquerque, NM 87131
[*]Materials Science Department, Pacific Northwest Laboratory, Richland, WA 99352

Radiation damage of nuclear materials (e.g. fast- or fusion-neutron damage in reactor structural components, fission-fragment damage in nuclear fuels and alpha decay damage in nuclear waste forms) has been one of the major challenges faced by the material science community. Ion beam irradiation and implantation experiments have been used extensively in the past few decades not only for simulating these damaging process in materials but also for improving material properties for many technological applications.

As an energetic particle traverses a crystalline target, it loses its energy predominantly through electronic (ionization) and nuclear (elastic collision) interactions with the atoms in the lattice. The target atom which receives sufficient energy from the interactions may get displaced from its lattice site and may further displace other target atoms, thus creating a displacement cascade which is usually a few nanometers in scale. Just at the end of the collision phase, which lasts for only a few tenths of a pico-second, a displacement cascade contains a very dense cluster of point defects and the region may be considered amorphous. However, the type and amount of the radiation damage observable after the irradiation depend strongly on the material's ability to recover during the following relaxation and cooling phases of the cascade. While self-ion irradiation of simple metallic materials often results in void and dislocation loop formation because of the high point defect mobility, heavy ion irradiation of intermetallic compounds, covalently bonded semiconductors and complex ceramic materials with mixed bond types is more likely to induce solid-state amorphization[1]. The amorphization may either occur directly within the displacement cascade (the entire cascade quenching down without much defect annealing) or by the point defect build up (only isolated point defects survive the cooling phase of each cascade), as well as by chemical disordering. In this study, the effects of ion beam irradiation in $Ca_2La_8(SiO_4)_6O_2$, SiC, UO_2 and other ceramic materials are studied with HRTEM and compared at the nanometer scale.

Most HRTEM images taken from 1 to 1.5 MeV Ar^+, Kr^+ and Xe^+ ion irradiated (below 498K) $Ca_2La_8(SiO_4)_6O_2$ shows nanometer scale discrete domains which are believed to be the images of the displacement cascades. Computer processing of the image (Fig. 1) indicates that amorphization occurs directly within the displacement cascade in $Ca_2La_8(SiO_4)_6O_2$. This can be expected because it is easy to understand that such a complex crystal structure (apatite, $P6_3/m$) is hard to recover once it is heavily damaged (all the different atoms have to find the unique lattice sites). The fact that the size of the cascade image increases with the increasing ion mass and decreases with the increasing temperature agree well with the critical amorphization dose—temperature curves determined by *in situ* electron diffraction observations during ion irradiation. Although β-SiC can also be fully amorphized under 1.5 MeV Xe^+ ions at room temperature and the HRTEM images taken below the critical dose contain mottled contrast (Fig. 2), the lack of the discrete amorphous cascade image indicates that amorphization occurs by point defect build up in β-SiC. UO_2 has never been amorphized and the HRTEM image of UO_2 contains no visible damage even after a very high ion dose at near liquid helium temperature (20K) (Fig. 3). The stableness or the high damage recover rate of the structure must be related to its close packing and high symmetry nature and merits further study.

References

1. L.M. Wang and R.C. Ewing, Mat. Res. Soc. Bull. **XVII**, 38 (1992).
2. This work was supported by the Office of Basic Energy Sciences, US Department of Energy under contract DE-FG03-93ER45498 (UNM) and DE-AC06-76RLO 1830 (PNL).

Proc. Microscopy and Microanalysis 1995, edited by G.W. Bailey, M.H. Ellisman, R.A. Hennigar, and N.J. Zaluzec
Copyright © 1995 MSA. Published by Jones and Begell Publishing, 79 Madison Ave., New York, NY 10016

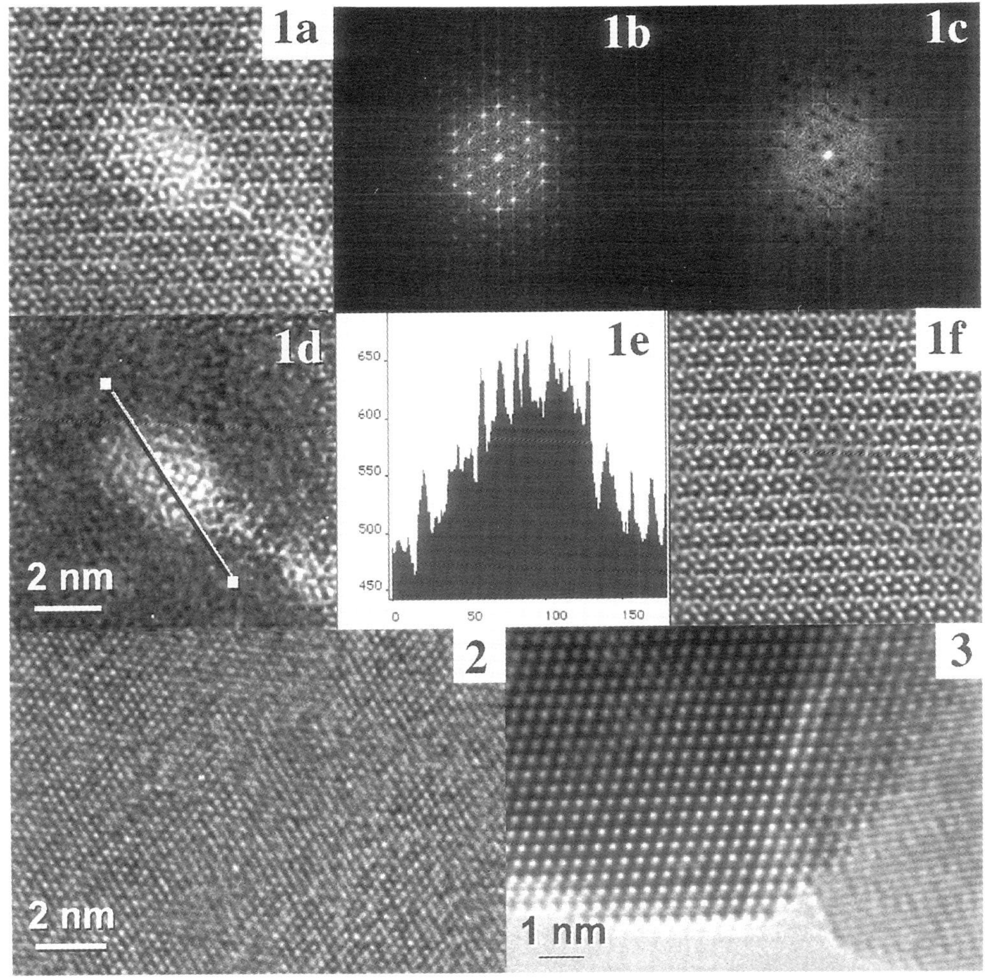

FIG. 1.—(a): Digital HRTEM image of a $Ca_2La_8(SiO_4)_6O_2$ crystal containing a collision cascade damage created by 1 MeV Ar^+ ion irradiation at room temperature; (b): Fourier transform of (a) indicating the image contains a periodic and an aperiodic component; (c): contributions of the periodic component in (b) has been filtered out in Fourier space; (d): inverse Fourier transform of (c) showing the aperiodic (amorphous) component in the original image (a); (e): intensity distribution across the line in (d) showing the cascade damage in the third dimension; (f): the periodic (crystalline) component in the original image (a) obtained by subtracting (d) from (a) pixel by pixel. The discrete nature of the cascade image and the result of the computer processing indicate that amorphization in $Ca_2La_8(SiO_4)_6O_2$ occurs directly with the displacement cascade.

FIG. 2.—HRTEM image of β-SiC irradiated at room temperature with 1.5 MeV Xe^+ to 8.5×10^{13} ions/cm^2. The mottled contrast and lack of discrete amorphous region in this image indicate that amorphization occurs by point defect build up in β-SiC.

FIG. 3.—HRTEM image of UO_2 irradiated at 20K with 1.5 MeV Xe^+ to 1×10^{16} ions/cm^2. No irradiation damage was observed even under such a critical condition (low temperature, high ion dose) indicating that cascade damage can easily be fully recovered in UO_2.

EXPERIMENTAL REALIZATION OF DIAMOND-LIKE CARBONITRIDE

Mehmet Sarikaya,[*] M. Qian,[*] J. M. Gil,[†] F. M. Gil,[†] and M. J. Yacamán[§]

[*]   Materials Science and Engineering, U. of Washington, Seattle, WA 98195, USA
[†]   ETSIT Paseo del Caus s/n, Universidad de Valladolid, España
[§]   Inst. de Física, U. Nacional Autónoma de México, A. P. 20-364, 01000 México, D. F., México

Recently Cohen[1] has predicted, based on an empirical model, that tetrahedral solids of low ionicity with high bulk moduli approaching that of diamond may be candidates as new hard materials. Several groups have attempted to synthesize these covalent C-N solids using vacuum deposition techniques. Although many of the literature results give C-N compounds with amorphous structures, two groups have reported the presence of a few percent of a crystalline phase, the so-called β-C_3N_4, after β-Si_3N_4 (hex.) structure, in an otherwise amorphous matrix.[2,3] It has generally been difficult to synthesize samples with higher crystalline phase content and to confirm these early results. In samples prepared by a chemical precursor route, we report the presence of a crystalline carbonitride phase with zinc blende structure from electron diffraction and spectroscopy investigation. Based on a scaling relationship from Cohen[1], we estimate the bulk modulus of the new crystalline compound (denoted as α-C_4N_4 heretofore) to be comparable to c-BN and diamond from the C - N bond length determined from extended energy loss fine structure analysis, EXELFS, that allows a precise measurement of the first and second nearest neighbor distances between C and N atoms.[4]

Samples were prepared by heating phenyldiamonium sulfate in the presence of N_2 atmosphere and SeO_2 catalyst. The sample temperature was raised at a rate of 10 °C/min. up to 800 °C and then cooled down to ambient.[5] TEM samples were examined with a low temperature holder using a Philips 430T TEM at 200 keV. A parallel detection EELS (GATAN Inc.) was used and spectra were acquired from thicknesses less than 0.3 mean-free-path in the diffraction mode. The net counts in excess of 10^5 per channel were ensured under both C-K and N-K edges.

The samples contained both an amorphous matrix and regions of polycrystalline carbonitride with more than 25% local density (Fig. 1). Both selected area diffraction, SAD, and nanoprobe, ED patterns were obtained. From the peak positions the crystal structure appeared to be a cubic cell with a lattice parameter $a_o = 3.52 \pm 0.15$ Å. The elemental analysis by EEL spectra obtained from the carbonitride crystallites embedded in the amorphous matrix gives an estimated N/C ratio of 1/1 (the value is obtained by subtracting the contribution from the amorphous region). Therefore, the nano-diffraction patterns as well as the stoichiometry from EELS leads to a unit cell best described as a zinc blende structure with C occupying the face centered cubic lattice positions (total of four) and N occupying all the diagonal positions (total of 4). The bonds may not be explained with sp^3 hybridization alone; it is possible, however, that in C_4N_4, electrons may be delocalized with metal-like density of states.[6] EXELFS analysis further confirm the structure, discerned from the ED analysis (Fig. 2). The first shell distance 1.47 Å has agreed with the known C and N covalent radii (0.77 Å, and 0.70 Å, respectively)[6] and based on this, the calculated lattice parameter becomes 3.40 Å, a value differing only by 3.5 % from that obtained by ED. The second shell nn distance was 2.51 Å.

By using the scaling formula by Cohen[1]: $\Omega = (19.71 - 2.20\ \lambda)\ /\ d^{3.5}$ the estimated bulk modulus, Ω, of α-C_4N_4 is 4.1 MBar (for α-C_4N_4 phase, N-C bond length, d is the first nn distance from EXELFS and ionicity, λ, was conservatively assumed to be as 1 (1/2 in β-C_3N_4)[3] including a possible 10% error).
The finding of a cubic carbonitride phase have several significant consequences including its potential applications (if produced in large quantities) as high hardness material as its hardness is estimated to be about that of diamond. It is hoped that our results stimulate further investigations, including their properties and the development of new synthesis techniques for thin film and bulk forms for their possible technological applications.

Proc. Microscopy and Microanalysis 1995, edited by G.W. Bailey, M.H. Ellisman, R.A. Hennigar, and N.J. Zaluzec

1.   Cohen, M. L., *Phys. Rev B* **32**, 7988-7991 (1985).
2.   Niu, C., Lu, Y. Z., and Lieber, C. M., *Science*, **261**, 334-336 (1993).
3.   Yu, K. M., Cohen, et al., *Phys. Rev. B.*, **49**, 5034-5037 (1994).
4.   M. Sarikaya et al., submitted to Nature (1995).
5.   Martin-Gil et al., *Acta Met.*, in press (1995).
6.   Philips, J. C., *Bonds and Bands in Semiconductors*, (Academic Press, Amsterdam, 1973).

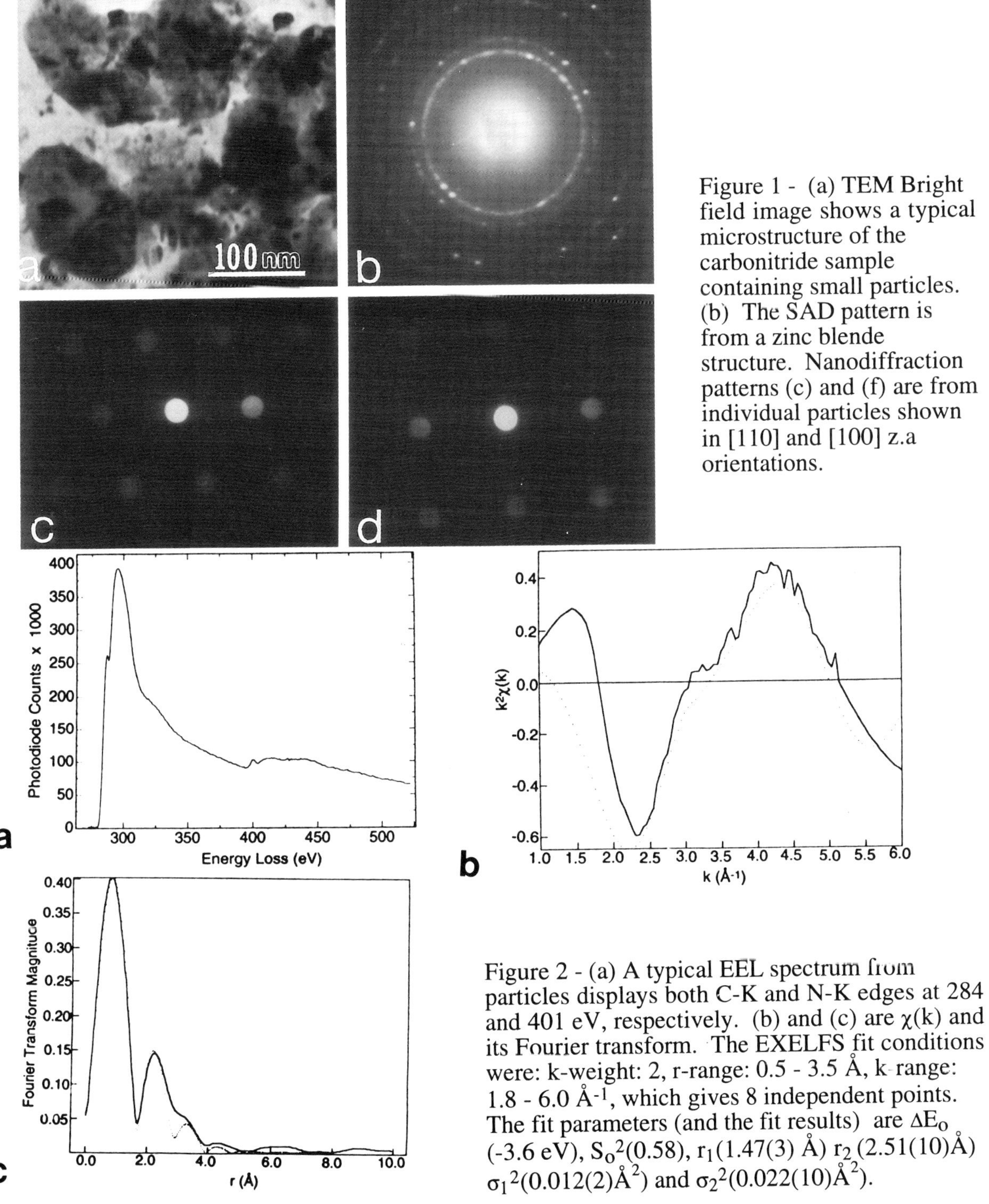

Figure 1 - (a) TEM Bright field image shows a typical microstructure of the carbonitride sample containing small particles. (b) The SAD pattern is from a zinc blende structure. Nanodiffraction patterns (c) and (f) are from individual particles shown in [110] and [100] z.a orientations.

Figure 2 - (a) A typical EEL spectrum from particles displays both C-K and N-K edges at 284 and 401 eV, respectively. (b) and (c) are $\chi(k)$ and its Fourier transform. The EXELFS fit conditions were: k-weight: 2, r-range: 0.5 - 3.5 Å, k-range: 1.8 - 6.0 Å^{-1}, which gives 8 independent points. The fit parameters (and the fit results) are ΔE_o (-3.6 eV), S_o^2(0.58), r_1(1.47(3) Å) r_2(2.51(10)Å) σ_1^2(0.012(2)Å^2) and σ_2^2(0.022(10)Å^2).

STRUCTURE AND SYMMETRY OF CaFeTi$_2$O$_6$ PEROVSKITE

Nan Yao,[1] Alexandra Navrotsky,[1,2,3] and Kurt Leinenweber[4]

[1]Princeton Materials Institute, [2]Department of Geological and Geophysical Sciences, Princeton University, NJ 08540; [3]Center for High Pressure Research, Department of Earth and Space Sciences, SUNY, Stony Brook, NY 11794; [4]Department of Chemistry, Arizona State University, Tempe, AZ 85287

A new calcium iron (II) titanate ordered perovskite (CaFeTi$_2$O$_6$) was recently synthesized from an equimolar mixture of CaTiO$_3$ and FeTiO$_3$ at 12-15 GPa and 1200-1400 $^\circ$C.[1] In the present paper, we discuss the structure and symmetry studies of this new compound CaFeTi$_2$O$_6$ using CBED and HREM techniques. The CaFeTi$_2$O$_6$ powder sample was crushed to small fragments with an agate mortar and pestle under purified methanol. A drop of the resulting suspension was placed on a copper grid coated with holey-carbon film. CBED and HREM studies were performed on a Philips-CM20 ST transmission electron microscope equipped with a double tilt, liquid-nitrogen-cooled specimen holder under moderate vacuum conditions over the range from 10^{-6}-10^{-7} Torr. CBED makes it possible to examine the diffraction symmetry of key orientations of the crystal and therefore determine the point-group symmetry of the crystal. This information, along with the dynamic extinction information on systematic absences, can be used to determine the crystallographic space group uniquely.[2-5]

Figure 1 shows (a) a large-angle and (b) a small-angle CBED pattern at [001] on-zone-axis. These patterns indicate the presence of a fourfold rotation axis along the c-direction with four mirrors coincident with the (100), (010), (110) and (1-10) lattice planes, suggesting a possible diffraction group of 4mm. Figure 2(a) shows a [100] zone-axis CBED pattern collected under the same experimental condition. HOLZ reflections of this pattern show a whole pattern symmetry of 2mm, with (010) and (001) mirror planes, giving a twofold rotation axis in the a-direction. Based on this analysis, the point group can be uniquely determined as 4/mmm.[6] Figure 2(b) reveals that dynamic extinctions (G-M lines) are exactly excited in every alternate reflection (0k0) (k: odd). The presence of G-M lines of A$_2$ type (seen as a radial dark line) indicates that there is a b- or n-glide plane perpendicular to the c-axis and/or a 2$_1$ screw axis parallel to the b- axis. Figure 3(a) is a CBED pattern along the [h0l] (h: 1, l: 1) zone axis. The G-M lines of the A$_2$ type appear in the ZOLZ (010) and (0-10) reflection discs, but not the (-101), (10-1) reflection discs. The CBED pattern at the [hhl] (h: 1, l: 2) zone-axis is also examined (Figure 3(b)). Both G-M lines of A$_2$ and B$_2$ types are seen in alternate reflections (-1-11, 11-1, 33-3, etc.). Combining the above information, the correct space group of the CaFeTi$_2$O$_6$ crystal is uniquely determined to be $P4_2/nmc$ (No. 137),[6] confirming the result from X-ray diffraction study.[1] CBED patterns obtained at 84 K have identical whole pattern symmetry and are consistent with a tetragonal lattice having a=7.51±0.02Å, c= 7.54±0.02Å. Local structure also has been examined by displacing the electron probe across the small crystallite. There are no extra spots or variations in the diffraction pattern, suggesting that the crystal ordering is quite stable over the low temperature range from 300 K to 84 K. High resolution imaging and nanobeam analysis reveal three types of morphology. In addition to the major near-perfect perovskite phase (Figure 4), this material contains iron precipitates and no cavities (Figure 5). The orientation relationship between CaFeTi$_2$O$_6$ and Fe precipitates can be derived as $\{110\}_{CaFeTi2O6}$ // $\{100\}_{Fe}$. Crystalline domains with a high degree of disordering are also observed (Figure 6). The formation of this type of domain may be attributed to local variations of stoichiometry with considerable oxidation. The coexistence of reduced and oxidized regions in the same specimen is noteworthy.[7]

References:

1. K. Leinenweber and J. Parise, *J. Solid State Chem.*, 114 (1995) 277
2. B. F. Buxton, J. A Eades, J. W. Steeds, and G. M. Rackham, *Phil. Trans.*, 281 (1976) 171
3. J. Gjonnes and A. F. Moodie, *Acta Cryst.*, 19 (1965) 65
4. M. Raghavan, J. Y. Koo, and R. Petkovic-Luton, *J. Metals*, 35 (1983) 44
5. M. Tanaka, H. Sekii and T. Nagasawa, *Acta Cryst.*, A50 (1994) 261
6. N. Yao and A. Navrotsky, to be submitted
7. This work was made use of MRSEC Shared Facilities supported by the National Science Foundation under Award Number DMR-940032. This research is supported by the Center for High Pressure Research, an NSF Science and Technology Center.

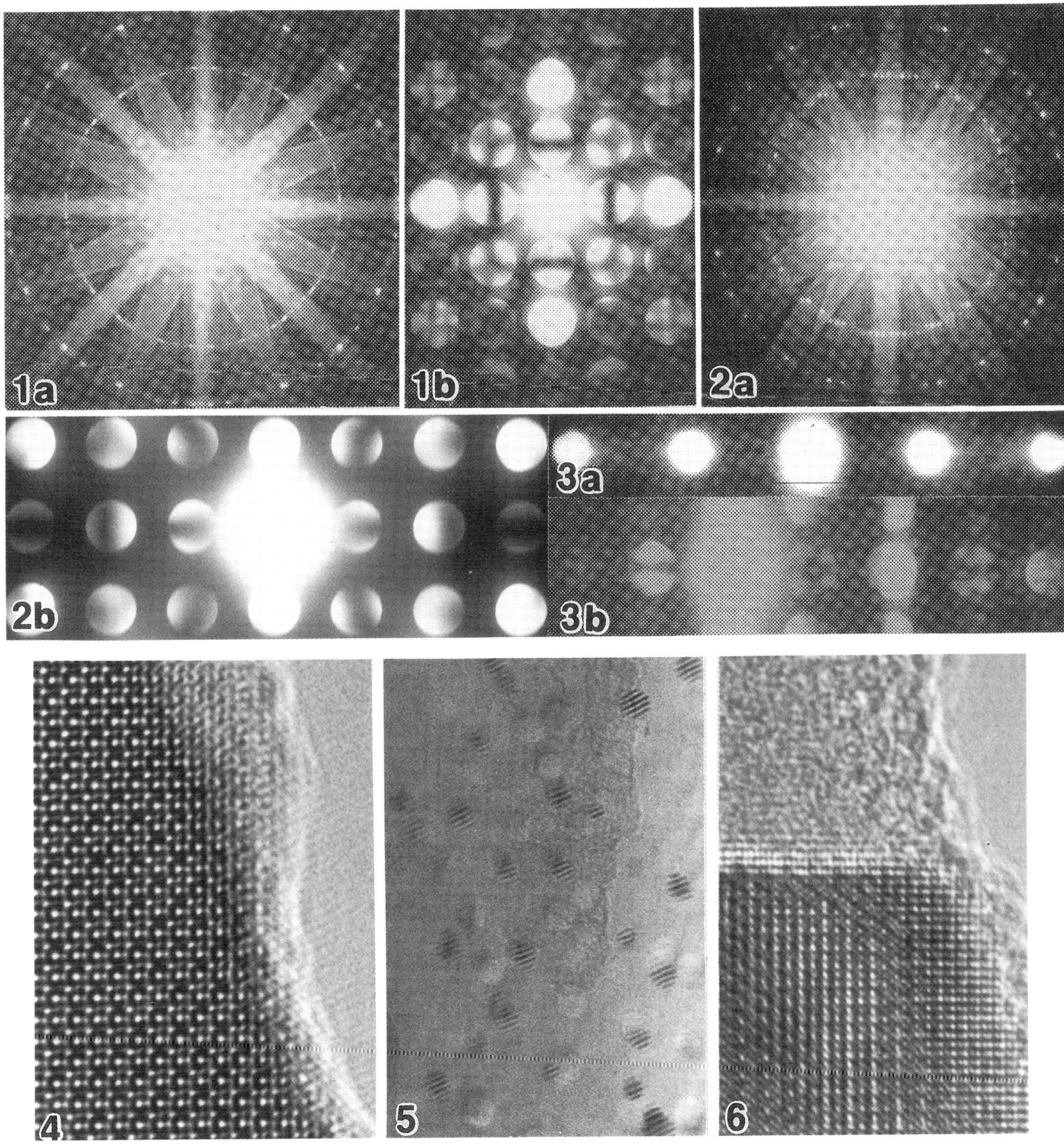

Figure 1. CBED pattern along [001] zone axis, with (a) a small-angle, and (b) a large-angle pattern. Both (a) and (b) showing a symmetry of 4mm.

Figure 2. CBED pattern along [100] zone axis, with (a) a small-angle, and (b) a large-angle pattern. Both (a) and (b) showing a symmetry of 2mm.

Figure 3. CBED pattern showing ZOLZ reflections along (a) [101] and (b) [112] zone axis.

Figure 4. HREM of an ordered $CaFeTi_2O_6$ structure viewed along [001] zone axis at 200 keV.

Figure 5. HREM showing that $CaFeTi_2O_6$ compound contains iron precipitates and no cavities (bar=10nm).

Figure 6. HREM showing a $CaFeTi_2O_6$ crystal consists of both ordered and disordered domain.

LORENTZ MICROSCOPY OBSERVATION OF VORTEX DYNAMICS DUE TO TRANSPORT CURRENT

T.Yoshida,[*] H.Kasai,[*] J.E.Bonevich,[*1] T.Matsuda,[*] A.Tonomura[*]

*Hitachi Advanced Research Laboratory, Hatoyama, Saitama, Japan 350-03

At present, Lorentz microscopy is the only one method to visualize individual vortices in superconductors and observe their motion in real time. Using this technique, we have investigated the fascinating vortex dynamics in superconductors, niobium[2] and BSCCO.[3] And recently we applied it to investigate current driven vortex motion.

The vortices are driven by the Lorentz force due to the transport current, and begin to move. This couses the vortex to dissipate the energy i.e., causes the resistive state in the superconductor. Therefore to clarify the vortex motion is important especially in the relation with vortex pinning.

The sample we examined was niobium thin film, which was single-crystallized by annealing and thinned by chemical etching. It was put in the low temperature specimen stage specially developed for the vortex observation,[1] which was installed in 350 kV FE-TEM based Hitachi H-9000.

We observed current driven vortex motion in the temperature (T) range from 4.5 K to Tc = 9.25 K and in a (externally applied) magnetic field (B) up to 100 Gauss. By changing these parameters, the vortex-pinning (center) and the vortex-vortex interaction could be controlled resulting in various modes of vortices motion. These observations were recorded on video tape. Some of the video frames are reproduced on the next page.

In Lorentz micrographs, the vortices appear as the globules with paired dark and bright hemispheres. The dark curved lines are the bend contours. In each series of images, thin arrows indicate the direction of the current and thick arrows are the direction of vortex motion . The applied magnetic field is parallel to the current in these projected images. The bars in the right bottom corners represent 1μm.

Figures **1a-c** show the result of increasing the current at T= 4.5 K, B= 30 Gauss. Figure **1a** shows the initial configuration of vortices (j= 0) . When the current density exceeded 1.2×10^5 A/cm^2 , the vortices began to move randomly (**1b**). In **1b**, some of vortices are blurred since they are in motion. Finally , the vortex velocity became greater than that can be recorded by the commercial video recorder (30 images per a second). Indeed in **1c** (j= 1.5×10^5 A/cm^2), most of the vortices disappear from the frame. However some globules remain, indicating the existence of the strong pinning centers; Even in the furious vortex motion, the strong pinning centers at least transiently trap the vortices and keep the trace of them on the image.

The existence of such pinning centers was actually confirmed in the photograph series **2a-c** (T= 4.5K, B= 30 Gauss). Although most vortices disappear, still one image marked by the bars can be seen (**2b**, j= 1.5×10^5 A/cm^2). This point is thought to be a crystallographic defect, since a bend contour has a discontinuity in **2c**. This defect will be a dislocation nearly normal to the surface.

We also observed the vortices behavior at higher temperature and in a higher magnetic field (**3a, b**, T = 7.5 K, B = 100 Gauss). Under this condition, vortices formed a fairly ordered lattice and began to flow in the form of lattice at a considerably small current density j = 3×10^3 A/cm^2 (**3a**) . It can be seen that the dislocation in vortex lattice plays an important role in the lattice flow (**3b**, dislocations are indicated by arrowheads); They ran through the vortex lattice to relax the local distortion of the lattice.

From the observations we were able to determine the microscopic critical current at which the vortices depinned and move continuously. We could also calculate the elementary pinning force at a specified pinning center from detailed analysis of the pinning.

References

1. Present Address: National Center for Electron Microscopy, Lawrence Berkeley Laboratory, University of California Berkeley, CA 94720

Proc. Microscopy and Microanalysis 1995, edited by G.W. Bailey, M.H. Ellisman, R.A. Hennigar, and N.J. Zaluzec
Copyright © 1995 MSA. Published by Jones and Begell Publishing, 79 Madison Ave., New York, NY 10016

2. K.Harada, et al. *Nature* 360 (,1992) 51
3. K.Harada, et al. *Phys. Rev. Lett.* 71 (1993) 3371

4. The authors gratefully appreciated the technical assistance of K.Kubota, N.Moriya, and S.Matsunami to develop the low temperature stage.

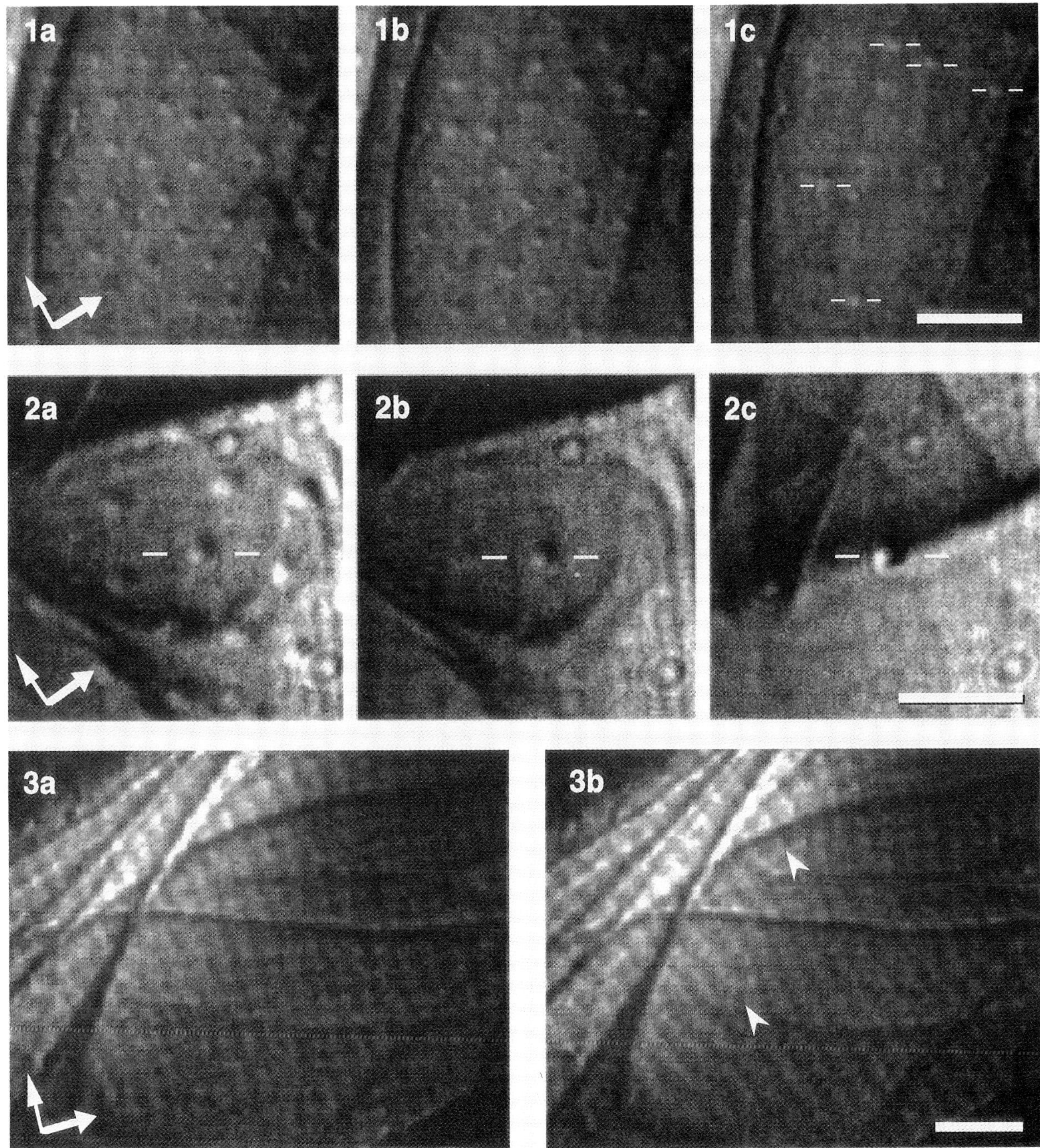

Fig. **1** Vortex motion when current increases at T=4.5 K, B= 30 Gauss. Fig. **2** Elementary vortex pinning at a crystallographic effect. Fig. **3** Flux flow at T=7.5 K, B= 100 Gauss.

A TEM STUDY OF THE INCOMMENSURATE MODULATED STRUCTURE IN $Sr_2CuO_{3+\delta}$

Y.Y. Wang,* H. Zhang,* V.P. Dravid,* L.D. Marks,* P.D. Han,# and D.A. Payne#
* Department of Materials Science and Engineering, Northwestern University, Evanston, IL 60208
# Department of Materials Science and Engineering, University of Illinois at Urbana-Champaign, Urbana, IL 61801

Hiroi et al. have recently reported the occurrence of superconductivity at a transition temperature of ~ 70 K in $Sr_2CuO_{3+\delta}$.[1] They observed a superstructure with lattice constants of $4\sqrt{2}a_p \times 4\sqrt{2}a_p \times c_p$. Neutron diffraction studies on this compound by Shimakawa et al.[2] indicate that the apical oxygen is fully occupied, whereas the oxygen in CuO_2 plane is half occupied. If oxygen does not fully occupy in the conducting CuO_2 planes, it raises doubts about our current understanding of superconductivity in the cuprates, which relies on full oxygen occupancy for the CuO_2 planes.

Recently, Han et al. observed a dependence of T_c on post annealing treatment of the high pressure synthesized $Sr_2CuO_{3+\delta}$.[3] They noticed that heating the sample to 300°C increased the T_c from 70 K to 94 K and that further heating to 450°C caused the sample to lose superconductivity.

In order to investigate the structural and microchemical details of $Sr_2CuO_{3+\delta}$, transmission electron diffraction, high resolution electron microscopy, and electron energy loss spectroscopy are employed.

We noticed an incommensurate modulation with an approximate superstructure of $5\sqrt{2}a_p \times 5\sqrt{2}a_p \times c_p$ The exact wavelength changed with the heat treatment. Figure 1 shows diffraction patterns along $[001]_p$ for specimens with T_c=70 K, 90 K and 0 K. The statistical measurements indicated that the as-made sample has the smallest wavelength, heating the compound to 310°C resulted in an increased wavelength, and overheating the sample at ~ 450°C gave an intermediate wavelength.

It is well known for the hole doped cuprate superconductors that the oxygen K-edge fine structure provides valuable information about the status of the charge carriers. The electron energy loss spectra for different heat treatment specimens contain three excitations as shown in fig.2. For the as-made sample (T_c=70 K) and optimally treated samples (T_c=90 K), the first pre-edge excitation is quite clear and sharp; while it is notably weaker for the over treated sample (T_c=0 K). This is consistent with the superconductor transition temperature for the different heat treated samples and it indicates that the modulated structure is likely to be the superconductor phase in the material.

By approximating the incommensurate modulated structure as a commensurate modulated structure, we proposed a $CuO_{1+\delta}$ network for the CuO plane shown in fig.3 and obtained a favorable comparison between high resolution image and simulated image shown in fig.4. In this model, 50% oxygen atoms are located at the position of Cu-Cu with a bond length of ~4.0 Å and 50% oxygen is missing at the position with the bond length of ~3.6 Å. This atomic structure configuration is consistent with the previous neutron diffraction measurement and calls into questions of the requirement of intact CuO_2 sheets for superconductivity.[4]

The research is supported by the National Science Foundation (DMR 91-20000).

Proc. Microscopy and Microanalysis 1995, edited by G.W. Bailey, M.H. Ellisman, R.A. Hennigar, and N.J. Zaluzec
Copyright © 1995 MSA. Published by Jones and Begell Publishing, 79 Madison Ave., New York, NY 10016

Reference
1. Z. Hiroi et al. *Nature*, **364**, 315 (1993); 2. Y. Shimakawa et al. *Physica C* **228**, 73 (1994); 3. P.D. Han et al. *Physica C* **228**, 129 (1994); 4. Y.Y. Wang et al. and H. Zhang et al. to be published.

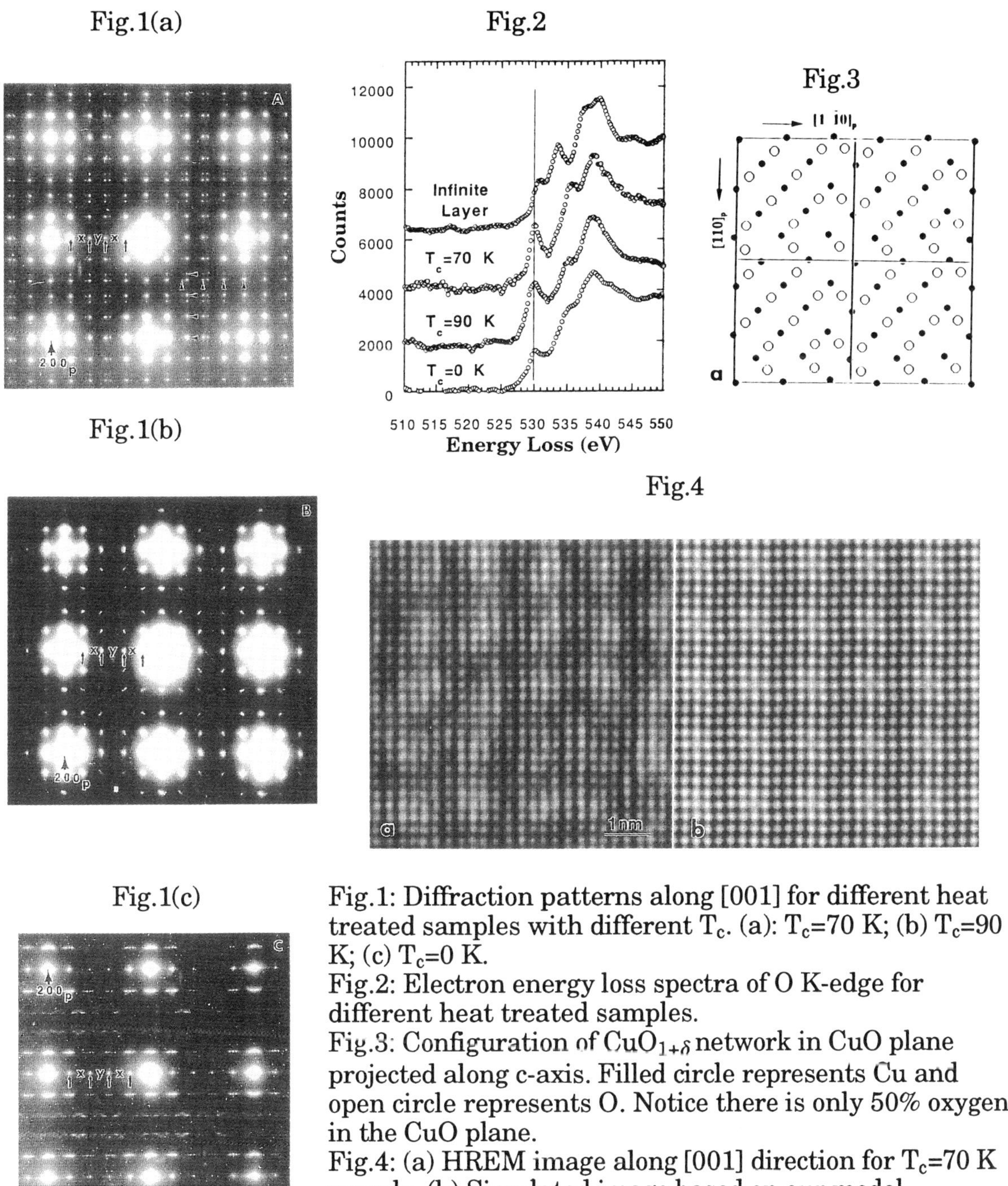

Fig.1: Diffraction patterns along [001] for different heat treated samples with different T$_c$. (a): T$_c$=70 K; (b) T$_c$=90 K; (c) T$_c$=0 K.

Fig.2: Electron energy loss spectra of O K-edge for different heat treated samples.

Fig.3: Configuration of CuO$_{1+\delta}$ network in CuO plane projected along c-axis. Filled circle represents Cu and open circle represents O. Notice there is only 50% oxygen in the CuO plane.

Fig.4: (a) HREM image along [001] direction for T$_c$=70 K sample. (b) Simulated image based on our model.

CRYSTAL STRUCTURE OF Tl-SUBSTITUTED HG-1223 SUPERCONDUCTOR WITH CRITICAL TEMPERATURE OF 138K UNDER ATMOSPHERIC PRESSURE

Y. Feng,* D.C. Larbalestier,* G.F. Sun,[+] Y.Xin,[+] K.W. Wong,[+] M.S. Osofsky,** L.E. Toth,** S. Ikeda,†
H. Fujii,† H. Kumakura† and K. Togano†

*Applied Superconductivity Center, University of Wisconsin, Madison, WI 53706
[+] Midwest Superconductivity, Inc., Lawrence, Kansas 66049
**Materials Science and Technology Division, Naval Research Laboratory, Washington D. C. 20375
†National Research Institute for Metals, 1-2-1 Sengen, Tsukuba, Ibaraki 305, Japan

Since the discovery of the Hg-Ba-Ca-Cu-O superconductors,[1] several members of the system bearing the formula $HgBa_2Ca_{n-1}Cu_nO_{2n+2+x}$ (n=1,2,3) were synthesized. The highest critical temperature T_c of 133.5K under atmospheric pressure was achieved from the Hg-1223 compound (n=3).[2] The Hg-compounds are structurally analogous to the single-layer Tl-based compounds $TlBa_2Ca_{n-1}Cu_nO_{2n+2+x}$.[3] In Hg-1223, oxygen is almost absent in the Hg-O plane at the stoichiometric (x=0) composition. In contrast, the oxygen sites in the Tl-O plane are nearly fully occupied. The difference in oxygen occupancy of the corresponding rock-salt Hg-O and Tl-O planes in Hg-1223 and Tl-2223 is due to the difference in valence of Hg^{2+} and Tl^{3+}.

To create a necessary fractional oxygen occupation in the Hg-O planes, a partial substitution of Tl for Hg was carried out and the T_c of the Tl-substituted sample raised up to 138K.[4] We report here on the crystal structure of the new compound. Synthesis and properties has been published elsewhere.[4] TEM samples were prepared by crushing powder onto a holey carbon foil. TEM work was performed in a JEOL 4000EX at 400KV. Image simulations were calculated using the EMS program.

Electron diffraction patterns were taken from regions where EDS analysis showed the composition to be that of the superconducting phase. From the [100] diffraction pattern (inset to Fig. 1), the structure of this Tl/Hg compound is similar to Hg-1223. The crystal parameters were calculated to be a=0.386 nm and c=1.582 nm, comparable to the X-ray diffraction results (a=0.386 nm and c=1.5794 nm).[4] HRTEM images were taken from the same region with the zone axis of [100], as shown in Fig. 1. The image is from a very thin region (2-3 nm) and corresponds to a defocus value of 25 nm. The first simulations were made using the atom positions reported for Hg-1223.[5,6] These did not provide a satisfactory match between simulation and image. A good match could only be obtained by adjusting the Cu, Ca and Ba positions. Using the central $Cu-O_2$ plane as symmetry plane, the remaining Cu, Ca and Ba atoms were all displaced towards the Hg-O plane, as schematically shown in Fig. 2. A simulated image based on the modified cation positions from the original Hg-1223 structure is inserted in the experimental image. However, these displacements are relative displacements since the c-axis lattice parameter of the Hg/Tl-1223 (1.579 nm) is actually less than that of the Hg-1223 structure (1.592 nm)[4]. This lattice contraction is similar to what occurs in Hg-1223 at pressure above 10 GPa, a treatment which also raises T_c in that compound too.[7,8,9]

REFERENCES
1. S.N. Putilin et al., *Nature* (London) **362**, 226 (1993)
2. O. Schilling et al., *Nature* (London) **363**, 56 (1993)
3. S.S. Parkin, el al., *Phys. Rev. Lett.* **60** 2539 (1988)
4. G. F. Sun et al., Preprint
5. M. Cantoni et al., *Physica C,* **215**, 11 (1993)
6. O. Chmaissem et al., *Physica C,* **217**, 265 (1993)
7. C.W. Chu et al., *Nature* (London) **365**, 323 (1993)
8. M. Nunez-Regueiro et al., *Science* **262**, 97 (1993)
9. Work supported by ARPA and EPRI at UW, by Midwest Superconductor Inc. at MSI, by NRL, ONR and ARPA at NRL.

Proc. Microscopy and Microanalysis 1995, edited by G.W. Bailey, M.H. Ellisman, R.A. Hennigar, and N.J. Zaluzec
Copyright © 1995 MSA. Published by Jones and Begell Publishing, 79 Madison Ave., New York, NY 10016

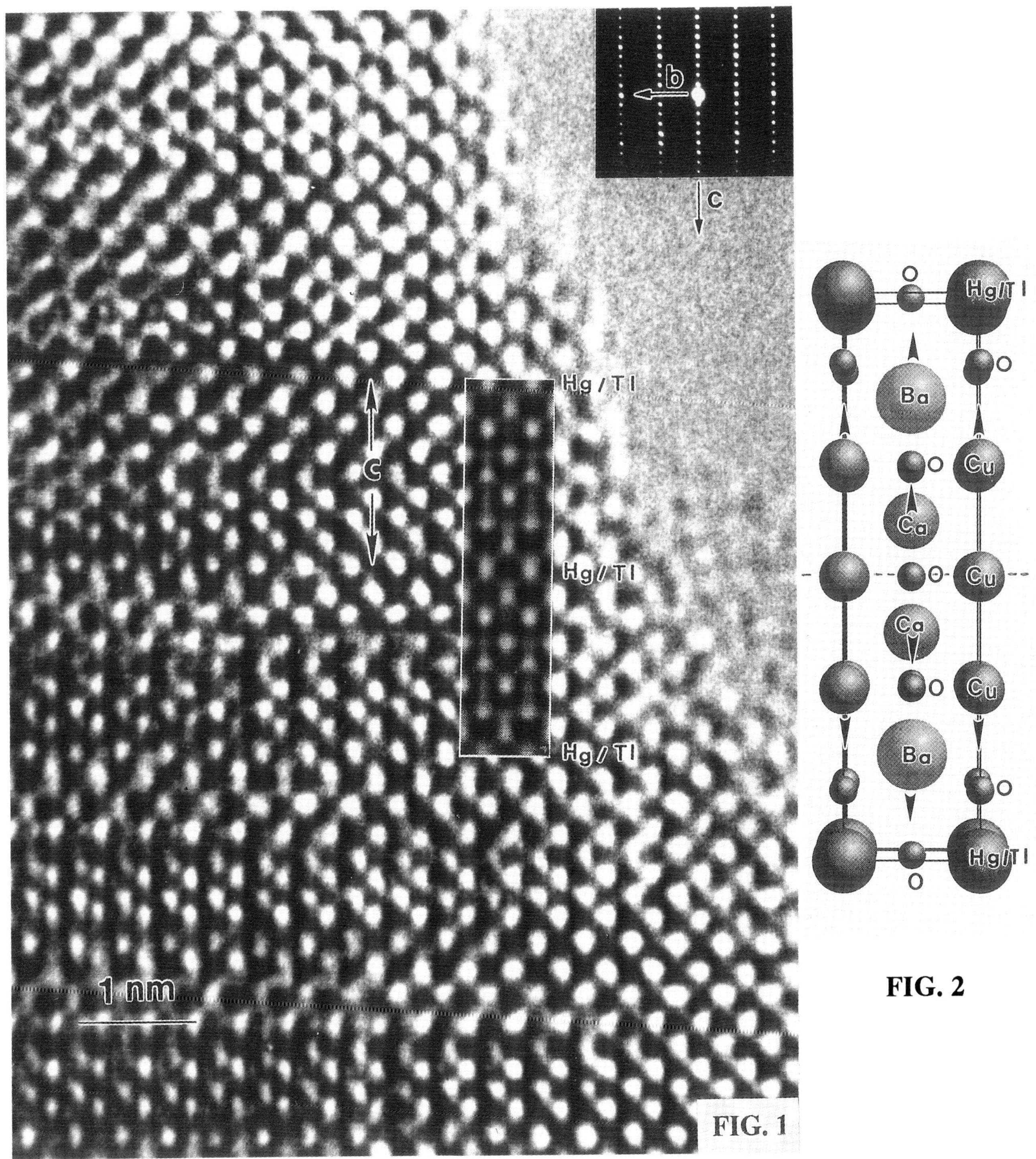

FIG. 1.--[100] HRTEM image of Tl-substituted Hg-1223 superconductor. Inset is the best simulation based on the cation positions of the Hg-1223 structure with modified site positions shown in Fig. 2.

FIG. 2.--A schematical view of the relatively displaced atomic coordinates.

STRUCTURAL ANALYSIS OF Tl/Hg-BASED SUPERCONDUCTOR WITH NOMINAL COMPOSITION Tl$_2$Hg$_1$ Ba$_4$Ca$_4$Cu$_6$O$_x$ HAVING T$_c$ = 120K

Y. Feng,* D.C. Larbalestier,* W.L. Lechter,** M.S. Osofsky,** E.F. Skelton,** S.B. Qadri** and L.E. Toth**
*Applied Superconductivity Center, University of Wisconsin, Madison, WI 53706
**Materials Science and Technology Division, Naval Research Laboratory, Washington D. C. 20375

Recently, Hg-based superconductors HgBa$_2$Ca$_{n-1}$Cu$_n$O$_{2n+2+x}$ (n=1,2,3) were demonstrated to have remarkably high transition temperatures (T$_c$). [1-3] The Hg-compounds are structurally analogous to the previously discovered Tl-based compounds TlBa$_2$Ca$_{n-1}$Cu$_n$O$_{2n+2+x}$.[4] Oxygen site occupation in the rock-salt Hg-O layers and Tl-O layers differ, due to the difference in the valence states of Tl (III) and Hg (II). In the Hg-compounds the Hg-O layer is always single while in the Tl-compounds both single or double Tl-O layers are possible. Up to 4 Cu-O$_2$ layers may be found in both families. Martin et al. [5] investigated the mixed Hg-Tl system and reported a new structure akin to a single cell Hg-1201/Tl-2201 intergrowth, having a T$_c$ about half that of either individual compound. We synthesized mixed Hg-Tl compounds corresponding to the higher order intergrowth formula Tl$_2$Hg$_1$ Ba$_4$Ca$_4$Cu$_6$O$_x$ (Hg-1223/Tl-2223) with the hope of producing new intergrowth structures with higher T$_c$. In this report, we only describe our structural characterization on one of the Tl-Hg compounds with nominal composition Tl$_2$Hg$_1$ Ba$_4$Ca$_4$Cu$_6$O$_x$. A complete report on the synthesis and properties of these compounds will be published elsewhere.[6]

TEM/EDS compositional analysis confirmed that the Hg was incorporated into the superconducting phase (Fig. 1). However, the Tl:Hg ratio was about 3:1 to 4:1, somewhat distant from the nominal composition ratio (2:1). Electron diffraction patterns were taken from the same region where EDS analysis was performed. From the [100] diffraction pattern (Fig. 2), the structure of this Tl/Hg compound is similar to the double-layer Tl-2223 structure. The crystal parameters were calculated based on the diffraction pattern to be a=0.386nm and c=3.636nm, which are comparable to the X-ray diffraction results (a=0.386 nm, c=3.586 nm). After confirming by EDS analysis and electron diffraction that the region was representative of the superconducting phase, HRTEM images were taken along the [100] zone axis. Fig. 3 is a typical [100] HRTEM image (improved by image processing). The image was from a very thin region (2-3 nm) and corresponds to a defocus value of 40 nm. Computer simulations based on different structure models were performed using EMS software with varying thickness and defocus. Only one simulated image matched perfectly with the experimental image. This was based on the Tl-double layer (Tl-2223) structure model. There is a one-to-one correspondence between the white-dot configuration of the HRTEM image and the cation configuration of the Tl-2223 structure model, as shown by the insert in the experimental image. In addition, grains with Tl(Hg)-2234 composition and structure were also identified by EDS, electron diffraction, HRTEM and computer simulations.

From the above analysis, coupled with our earlier X-ray results [6] and the measured physical properties [6] of the sample, we do not have conclusive evidence for the formation of the higher order intergrowth structure of Hg-1223/Tl-2223. The major superconducting phase, which exhibits a T$_c$ of 120K, has a structure similar to the double-layer Tl-2223. However, the compositional analysis clearly indicates that the superconducting phase contains both Tl and Hg. Based on the great similarity between Tl and Hg, it is reasonable to believe that the Hg partially substitutes for Tl in the double Tl-O layer of the Tl-2223 structure. In fact, several groups have reported a T$_c$ enhancement up to 138K under atmospheric pressure in the Hg-1223 systems by partial Tl substitution for Hg.[7,8,9]

REFERENCES

1. S.N. Putilin, E.V. Antipov, O. Chmaissem and M. Marezio. *Nature* (London) **362,** 226 (1993)
2. O. Schilling, M. Cantoni, J.D. Guo and H.R. Ott, *Nature* (London) **363,** 56 (1993)
3. R.L. Meng et al., *Physica C,* **216,** 21 (1993)
4. S.S. Parkin, el al., *Phys. Rev. Lett.* **60** 2539 (1988)

Proc. Microscopy and Microanalysis 1995, edited by G.W. Bailey, M.H. Ellisman, R.A. Hennigar, and N.J. Zaluzec
Copyright © 1995 MSA. Published by Jones and Begell Publishing, 79 Madison Ave., New York, NY 10016

5. C. Martin et al., *Physica C,* **212** 274 (1994)

6. W.L. Lechter, L.E. Toth, M.S. Osofsky, E.F. Skelton, A.R. Drews, C.C. Kim, B. Das, S.B. Qadri, A.W. Webb and R.J. Soulen, Jr. *Physica C,* to appear 1995

7. G. F. Sun, K.W. Wong, B.R. Xu, Y. Xin and D.F. Lu. Preprint

8. N.H. Hur, N.H. Kim, K.W. Lee, Y.K. Park and J.C. Park. *Mater. Res. Bulletin* **29,** 959 (1994)

9. Work supported by ARPA and EPRI at UW and NRL, ONR and ARPA at NRL.

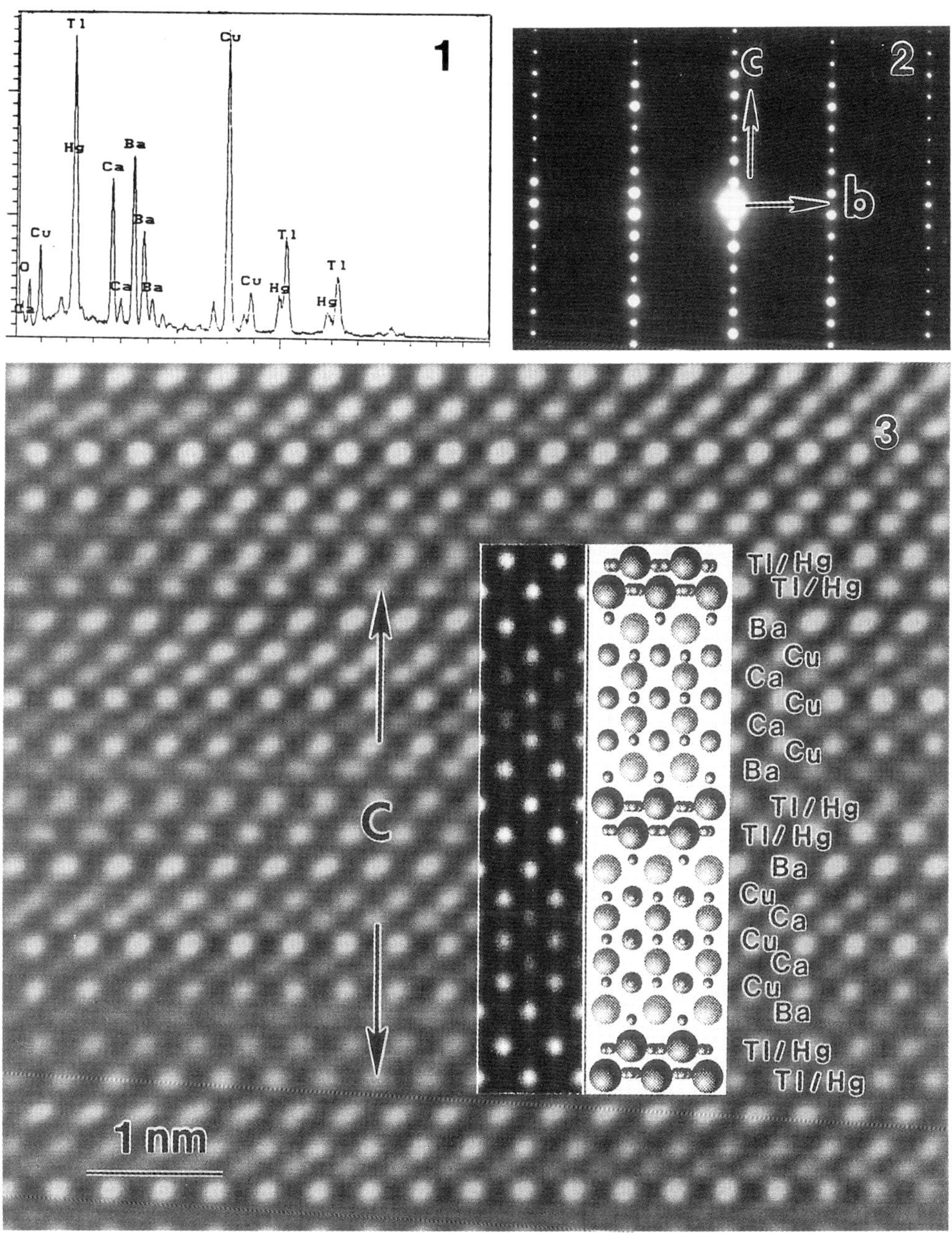

FIG. 1.--TEM/EDS spectra showing that Hg is incorporated into the superconducting phase.

FIG. 2.--[100] diffraction pattern of the Tl/Hg - compound showing similarity to that of the Tl-2223 structure.

FIG. 3.--HRTEM image taken from the same region along [100] zone axis. The simulated image (left inset) corresponds to the cation configuration of the Tl-2223 structure model (right inset).

IDENTIFICATION OF THE INTERMEDIATE PHASE $Bi_4Sr_4Ca_3Cu_5O_{18+\delta}$ DURING Bi/2212 AND Bi/2223 CONVERSION

Weimin Bian,[1] Yimei Zhu, Y.L. Wang, D. Gillette, and M. Suenaga,
Materials Science Division, Brookhaven National Laboratory, Upton, NY 11973

Among all the high-temperature superconductors, Bi/2223 is the most promising material as tapes and wires for large-scale and high-current applications. However, synthesizing Bi/2223 as a single phase is extremely difficult because of its relatively poor thermal stability and the narrow ranges of stoichiometry and temperature for its formation. Recently, to understand the formation mechanism and establish the optimum heat treatment conditions for the growth of Bi/2223, we have systematically studied phase transformation of the Bi-based superconductors. In this article, we report our finding of the intermediate phase of $Bi_4Sr_4Ca_3Cu_5O_{18+\delta}$ (Bi/4435).

Bi/2212 tapes with a Bi/2223 composition, produced by the standard "oxide powder in tube" process, were used in the study. TEM cross-section specimens were prepared from the bulk samples that were quenched or furnace-cooled from different stages during the Bi/2212 and Bi/2223 conversion.

Figure 1 shows a lattice image of the Bi/4435 phase with an alternating narrowly and widely spaced lattice (each black line corresponds to the position of the BiO layer). By comparing the spacing of the pure Bi/2212 and Bi/2223 lattices (also Fig.1), we found that the widely spaced lattice corresponds to a half unit-cell of the Bi/2223 phase, while the narrowly spaced one to a half unit-cell of the Bi/2212 phase. To our knowledge, a discrete Bi/4435 phase has never been reported previously, although planar faults, such as extrinsic stacking-faults (insertion of a CuO/Ca layer) in a Bi/2212 matrix, or intrinsic stacking-faults (absence of a CuO/Ca layer) in a Bi/2223 matrix, have been occasionally observed using high-resolution electron microscopy (HREM). Such faults were considered only as local structural defects.

The extinction rule of the diffraction patterns of the Bi/4435 phase was very different from that of Bi/2212 and Bi/2223 phases. The most noticeable difference was that for Bi/4435 all (00l) reflections were visible, while for Bi/2212 and Bi/2223 they were visible only when l is even. In the (010)* zone (Fig.2(a)), where there was no limiting condition for (h0l) reflections, the absence of the (h00) reflections with h=2n+1 suggests that there was a 2_1 screw rotation along the a-axis. In the (100)* zone (Fig.2(b)), the appearance of the (0kl) reflections including (0k0) for k=2n implied that, in the direction perpendicular to the a-axis, there was a glide operation along the b-axis. The absence of the (hk0) reflections for h+k=2n+1 (Figs.2(a)-(c)) suggested the existence of a n-glide in a direction perpendicular to the c-axis (more precisely, a glide operation with component of $\mathbf{a}/2 + \mathbf{b}/2$), rather than a b-glide ((hk0) allowed with k=2n)) for B-centered Bi/2212 and Bi/2223. The three-dimensional reciprocal lattice of the Bi/4435 phase is sketched in Fig.3. Combining with convergent-beam electron diffraction, we concluded that the space group of the Bi/4435 phase is Pbmn with $a{\approx}b$=5.4A, and c=33.9Å. A calculated crystal structure of the Bi/4435 phase is shown in the inset of Fig.4 agreeing well with the experimental HREM image.[2]

References
1. On leave from Materials Testing Center, Northeastern University, Shenyang, China.
2. The research was supported by the US Department of Energy, Division of Materials, Office of Basic Energy Science, under Contract No.DE-AC02-76CH00016.

Proc. Microscopy and Microanalysis 1995, edited by G.W. Bailey, M.H. Ellisman, R.A. Hennigar, and N.J. Zaluzec
Copyright © 1995 MSA. Published by Jones and Begell Publishing, 79 Madison Ave., New York, NY 10016

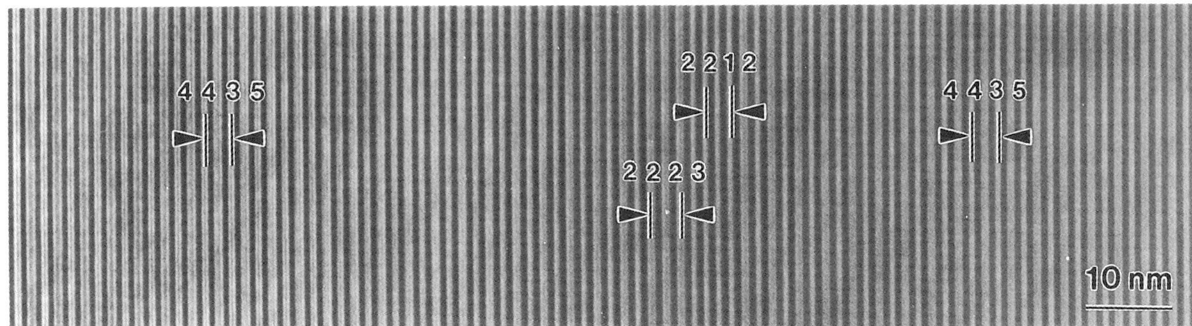

Fig.1 Lattice image of the Bi/4435 phase showing alternating narrow and wide spacing. The narrowly spaced lattice corresponds to the half unit-cell of the Bi/2212 phase, while the widely spaced one corresponds to the half unit-cell of the Bi/2223 phase. The equally spaced pure Bi/2212 and Bi/2223 lattices are also shown.

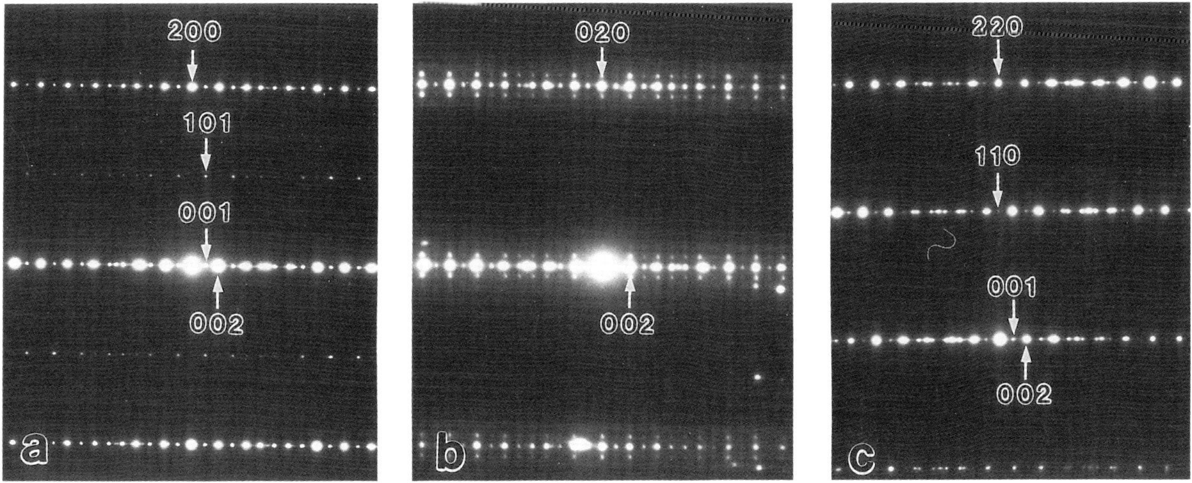

Fig.2 Selected area diffraction of the Bi/4435 phase in the (010)* zone (a), the (100)* zone (b), and (110)* zone (c).

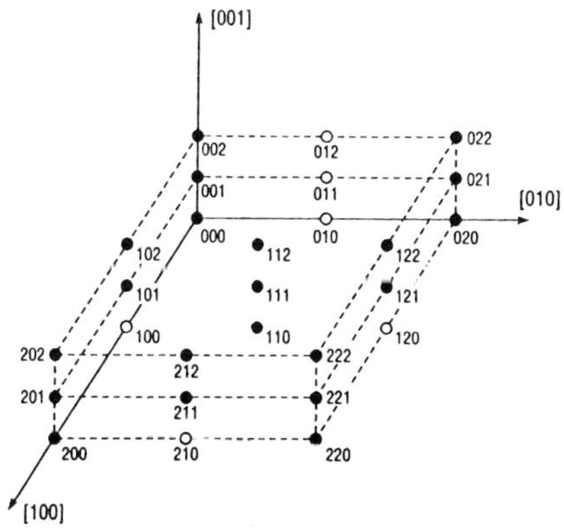

Fig.3 Reciprocal lattice of the Bi/4435 phase showing only the fundamental reflections. The solid dots represent the allowable Bragg reflections, and the open circles represent the forbidden reflections.

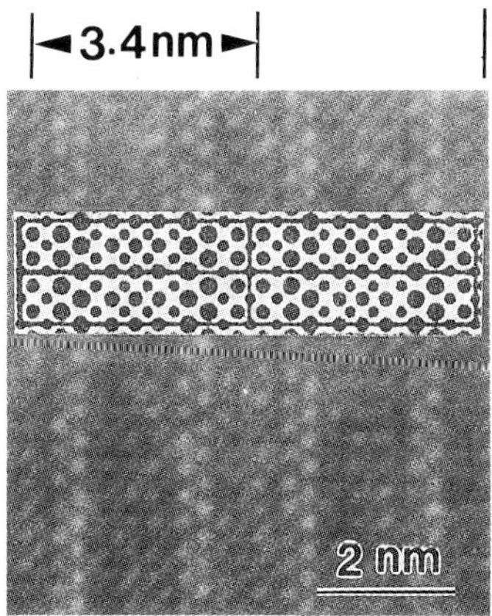

Fig.4 High-resolution image of the Bi/4435 phase in the (110) projection. The inset is the calculated crystal structure (space group: Pbmn) showing good agreement with the observed image.

SUPERCONDUCTIVITY AND OBSERVATION OF ORDERED STRUCTURES IN DEINTERCALATED Li_xNbO_2

M. Fendorf, M.A. Rzeznik[†], and A.M. Stacy[†]

National Center for Electron Microscopy, Lawrence Berkeley Laboratory, 1 Cyclotron Rd., Berkeley CA 94720
[†]Department of Chemistry, University of California, Berkeley, CA 94720

The prevalence of layered structures among the copper oxide superconductors has led us to search for new oxide superconductors with anisotropic structures to evaluate the relationship between structural anisotropy and superconductivity. Because of its layered structure,[1] we identified the dichalcogenide $LiNbO_2$ as a promising candidate for further study. The Nb oxidation state can be readily altered by deintercalation of Li, and we have reported magnetic measurements showing that the resulting Li_xNbO_2 (x<1) is a superconductor with transition temperature T_c=5.5K.[2] Interestingly it is the *Meisner fraction* rather than T_c, which changes as x is varied in this compound. This suggests that as a result of deintercalation, the material separates into two distinct phases, one superconducting and one non-superconducting, although powder x-ray diffraction shows no clear evidence of differences between $LiNbO_2$ and deintercalated material. This type of behavior is often associated with the transition to the superconducting state as the composition is varied in transition metal oxides.[3]

Since x-ray diffraction was unable to distinguish the two phases revealed by magnetic measurement, in the present study we use transmission electron microscopy (TEM) to examine Li_xNbO_2 materials. Because electrons are much more strongly scattered by a solid than are x-rays, electron diffraction will be better able to identify subtle changes in the crystal structure produced during deintercalation. TEM specimens were prepared by crushing the starting powders, suspending the finer particles in dry hexane, and then dispersing on a copper mesh grid with holey carbon film.[4] Except where otherwise indicated, all TEM work was performed using a room-temperature specimen holder. Diffraction patterns were obtained using an accelerating voltage of 100kV, while high-resolution imaging was carried out at 200kV and 800kV.

For material that was not deintercalated (hence x≈1), selected area diffraction (SAD) patterns taken along the [001] axis show 6-fold symmetry and the expected lattice parameter a_0=2.9Å (Fig. 1). In some areas, this sample yielded SAD patterns with extra reflections (also with 6-fold symmetry) corresponding to an extremely large lattice parameter (a≈41Å), as shown in Fig. 2. In deintercalated materials with x ranging from 0.8 down to 0.65, two additional types of extra reflections were observed. The most prevalent of these corresponds to a tripling of the unit cell along <110> directions (yielding a=$\sqrt{3}a_0$), (Fig. 3) and less frequently, a doubling along <100> (yielding a=$2a_0$) also appeared (Fig. 4). Corresponding images of the "perfect" structure and the $2a_0$ structure are presented in Figs. 5 and 6. Preliminary results with samples cooled to 98K indicate that the 41Å reflections disappear, the tripling along <110> and doubling along <100> remain stable, and no additional reflections appear at this temperature. (NOTE: All diffraction patterns are shown with the same scale.)

The occurrence of the extra reflections in these diffraction patterns can be explained either by the presence of charge density waves in Li_xNbO_2, or by ordering of Li-vacancies after deintercalation. Creation and subsequent ordering of oxygen vacancies can be ruled out because neutron diffraction indicates that deintercalated material is fully stoichiometric in oxygen (i.e., oxygen site occupation does not change upon deintercalation).[5] Furthermore, significant changes in niobium occupation would be detected by x-ray diffraction. Of the remaining two possibilities, Li-vacancy ordering is much more probable, since charge density waves are not expected to be stable in a sample at room temperature.[6] The fact that no additional periodicities seem to develop in a cooled sample tends to support this hypothesis. The large periodicity found in some areas of the $LiNbO_2$ sample can possibly be explained by small deviations from exact stoichiometry (x slightly less than 1), possibly beam-induced, with ordering of the small number of corresponding Li vacancies present in this case.

In summary, we have observed the appearance of extra reflections in [001] diffraction patterns obtained from deintercalated Li_xNbO_2 materials, and attribute this to the ordering of vacancies created as Li is removed from the host structure. The ordering of Li, apparently with two preferred periodicities, may be related to the phase separation revealed by magnetic measurement. Due to its sensitivity to scattering from light elements and ability to provide structural information from very small areas, transmission electron microscopy is proving to be an important tool in the study of the Li_xNbO_2 system.[7]

References

1 G. Meyer and R. Hoppe, *J. Less-Common Metals* **46**, 55 (1976).
2 M.J. Geselbracht, T.J. Richardson, & A.M. Stacy, *Nature* **345**, 324 (1990).
3 See for example: a) P.W. Anderson, *Science* **235**, 1196 (1987).
4 M. Fendorf, M. Powers, and R. Gronsky, *Microscopy Research and Technique* **30**, 167 (1995).
5 M.J. Geselbracht, Ph.D. Thesis, University of California, Berkeley (1991).
6 See for example: G.A. Scholz, R.F. Frindt, and A.E. Curzon, *Phys. Stat. Sol (A)* **72**, 375 (1982).
7 This study was carried out using facilities at Lawrence Berkeley Laboratory, funded by the Director, Office of Energy Research, Office of Basic Energy Sciences, Materials Sciences Division of the U.S. Department of Energy under Contract Number DE-AC03-76SF00098.

Proc. Microscopy and Microanalysis 1995, edited by G.W. Bailey, M.H. Ellisman, R.A. Hennigar, and N.J. Zaluzec
Copyright © 1995 MSA. Published by Jones and Begell Publishing, 79 Madison Ave., New York, NY 10016

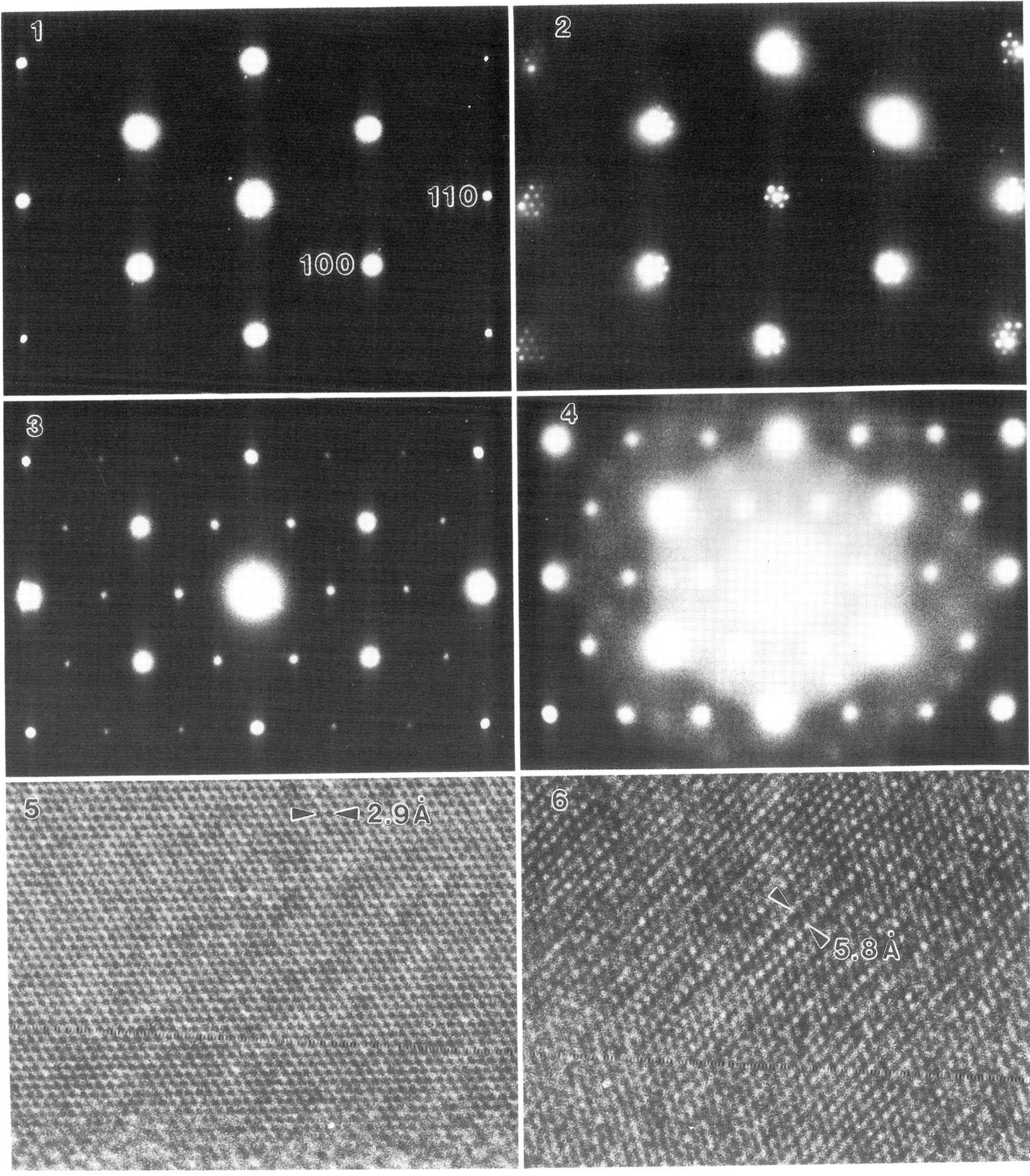

FIG. 1. -- [001] SAD pattern from stoichiometric LiNbO$_2$, showing 6-fold symmetry and $a_o \approx 2.9$Å.

FIG. 2. -- [001] pattern from another LiNbO$_2$ particle, illustrating extra reflections with $d \approx 41$Å.

FIG. 3. -- [001] pattern from deintercalated Li$_x$NbO$_2$ showing tripling along <110> directions; here $x \approx 0.65$.

FIG. 4. -- [001] Li$_{0.65}$NbO$_2$ pattern showing doubling along <100> as well as tripling along <110>.

FIG. 5. -- High-resolution TEM image of "perfect", stoichiometric LiNbO$_2$ structure.

FIG. 6. -- High-resolution image from Li$_{0.8}$NbO$_2$, showing $2a_o$ structure in central area of figure.

THE DETERMINATION OF THE Pr VALENCE STATE IN $(Pr_{1.5}Ce_{0.5})Sr_2Cu_2NbO_{10}$ COMPOUND BY ELECTRON ENERGY LOSS SPECTROSCOPY

S. C. Cheng and V. P. Dravid

Department of Materials Science & Engineering, Science & Technology Center for Superconductivity, Northwestern University, Evaston, IL 60208

The debate on the Pr valence state in PrBCO compounds is well documented.[1] Recently $(Pr_{1.5}Ce_{0.5})Sr_2Cu_2NbO_{10}$(RCeSCNO) system has been synthesized and studied.[2] In this paper, the results of our study to determine the valence of Pr in PrCeSCNO, which has similar property of PrBCO are presented. Typical EELS spectrum of Pr M_{45} edges with Ce M_{45} edges from $(Pr_{1.5}Ce_{0.5})Sr_2Cu_2NbO_{10}$ is shown in Fig. 1. In Fig. 1 the energy position of Ce M_{45} edges were taken at 883.0 eV and 901.0 eV. The Ce M_{45} edges were used to calibrate the energy axis, so that the position of Pr M_{45} edges could be accurately measured. The measured energy positions of Pr M_{45} edges in PrCeSCNO and in $Pr_2Ce_2O_7$ and $Pr_2Ce_2O_8$, which were used to obtain the spectra of trivalent and tetravalent Pr compounds, are also listed in table 1. The spectra taken from these compounds are shown in Fig. 1 (a) and (b). From Table 1 it is found that the energy positions of Pr M_{45} in PrCeSCNO compound match with that of $Pr_2Ce_2O_7$ and have about 1.5 eV difference with that of $Pr_2Ce_2O_8$. The fine structures of Pr M_{45} edges in PrCeSCNO and $Pr_2Ce_2O_7$ are similar: the M_5 edge is symmetric and the M_4 edge is asymmetric with a shoulder at the lower energy side. The fine structure of Pr M_{45} edges in $Pr_2Ce_2O_8$ are different; both M_4 and M_5 peaks are asymmetric with a shoulder at the higher energy side. Therefore, by comparing the fine structure of M_{45} edge and the energy position among these compounds, the valence of Pr in the PrCeSCNO compounds should be +3 rather than +4. However, the Pr M_{45} edges almost overlaps with the Cu L23 edges (see table 1). The contribution of Cu L23 edges to the observed Pr M_{45} edges in the spectrum of $(Pr_{1.5}Ce_{0.5})$SCNO has been evaluated less than 10%. To demonstrate that the contribution from Cu L23 is too low to change the character of observed Pr M_{45} edges, the observed Pr M_{45} edges were decomposed into the pure Pr M_{45} and the Cu L23 peaks. The spectra of PrCeSCNO and EuCeSCNO were recorded at same scattering condition. Both the spectra were normalized by the intensity of the Ce M_5 edge. The resultant spectrum of EuCeSCNO was subtracted from the PrCeSCNO spectrum to yield the pure Pr M_{45} edges, which show the typical trivalent character in Fig. 2. The results imply that the absence of superconductivity in these Pr compounds cannot be attributed to the hole neutralization by Pr cations. The hole density can be studied by the oxygen K absorption pre-edge.[3] Figure 3 shows the O K-edges taken from $(Eu_{1.5}Ce_{0.5})$SCNO, $(Eu_{1.0}Pr_{0.5}Ce_{0.5})$SCNO and $(Pr_{1.5}Ce_{0.5})$SCNO specimens. The shape of the pre-edge for these three compounds are very similar, indicating that the number of holes in these compounds are almost same. The experimental results from the O K pre-edge measurements agree with a trivalent Pr.

References
1. H. B. Radousky, *J. Mater. Res.*, Vol. 7, No. 7, Jul (1992)1917.
2. H. B. Radousky, T. J. Goodwin and R. N. Shelton, *Physica C* 209(1993)155.
3. V. P. Dravid and H. Zhang, *Physica C* 200(1992)349.
4. This research is supported by the NSF 91-20000 through the   Science and Technology Center for Superconductivity.

	Ce M5	Ce M4	Pr M5	Pr M4	Cu L3	Cu L2
• $(Pr_{1.5}Ce_{0.5})Sr_2Cu_2NbO_{10}$	883.0	901.0	929.3	948.5		
• $(Eu_{1.0}Pr_{0.5}Ce_{0.5})Sr_2Cu_2NbO_{10}$	883.0	901.0	929.4	948.2		
• $Pr_2Ce_2O_7$	883.0	901.0	929.2	948.4		
• $Pr_2Ce_2O_8$	883.0	901.0	930.5	950.2		
• $(Eu_{1.5}Ce_{0.5})Sr_2Cu_2NbO_{10}$	883.0	901.0			931.3	951.4

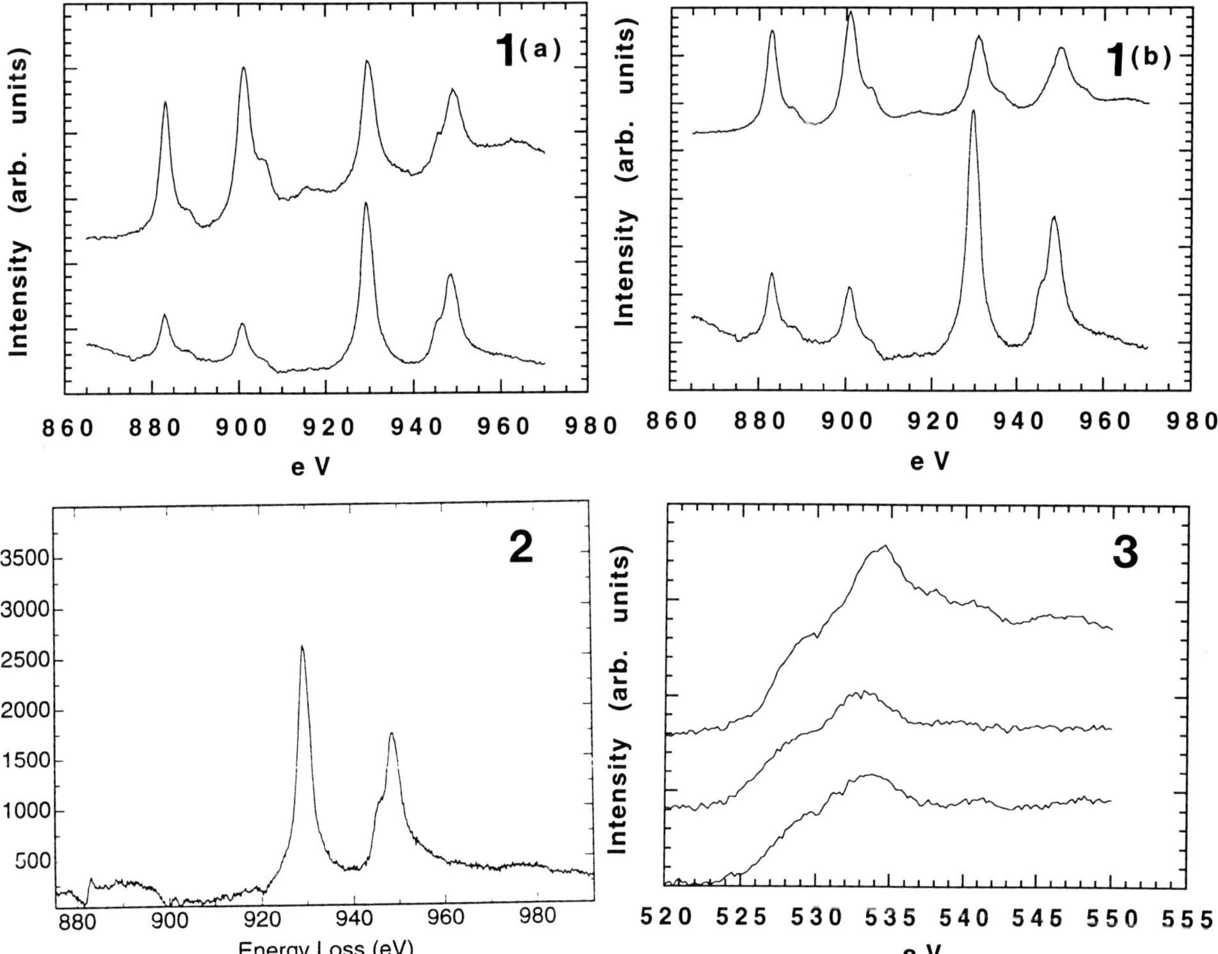

FIG. 1. - (a) and (b) The EELS spectra taken from $(Pr_{1.5}Ce_{0.5})$SCNO displayed on the lower part of the graphs and taken from reference samples of $Pr_2Ce_2O_7$ and $Pr_2Ce_2O_8$ displayed on the top part of the graphs.

FIG. 2. - The resulted Pr M_{45} edge after the removing of Cu L23 edge.

FIG. 3. - The O K-edges taken from $(Eu_{1.5}Ce_{0.5})$SCNO, $(Eu_{1.0}Pr_{0.5}Ce_{0.5})$SCNO and $(Pr_{1.5}Ce_{0.5})$SCNO compounds. They are displayed on the top, middle and bottom of the graph respectively.

SCANNING PROBE MICROSCOPY STUDY OF THE GROWTH OF YBa$_2$Cu$_3$O$_7$ THIN FILMS

M. Grant Norton[*], Rand R. Biggers[**], E.K. Moser[**], T.L. Peterson[**], and I. Maartense[**]

*   Mechanical and Materials Engineering, Washington State University, Pullman WA 99164
**   US Air Force Wright Laboratory, Wright Patterson Air Force Base OH 45433

Previous work in our laboratory has demonstrated the use of scanning probe microscopy (SPM) techniques in the study of microstructure development in YBa$_2$Cu$_3$O$_7$ (YBCO) thin films formed by pulsed-laser deposition (PLD) [e.g., Ref. 1]. Under certain deposition conditions a high density of surface outgrowths was observed. These outgrowths have been identified as a-axis-oriented YBCO grains nucleated at the substrate surface which grow in concert with the c-axis-oriented YBCO grains. This growth mechanism differs from that previously suggested for the formation of mixed a-axis-oriented/c-axis-oriented YBCO films. The microstructure of YBCO thin films deposited on vicinal substrates by PLD is very different from that observed for films grown on oriented substrates (even under equivalent deposition conditions). Films grown on vicinal LaAlO$_3$ substrates consisted of elongated granular features oriented along common directions [2]. The shape of these features is the result of step-flow growth where the ledges on the miscut substrates provide aligned and enhanced nucleation sites. Such features have been observed on films grown using several different deposition conditions: only the width of the features changed.

Figure 1 shows a montage of atomic force microscope (AFM) images of the surface of YBCO films grown using common deposition parameters except for laser pulse frequency and deposition time. The films were grown on (001)-oriented LaAlO$_3$ substrates for appropriate times at different frequencies to ensure that the overall thicknesses of the films were similar (about 360 nm). The AFM images reveal the following features concerning film growth. Firstly, the films form by an island growth mechanism resulting in a microstructure consisting of approximately equiaxed grains. The grain size is larger for films grown at low frequencies and increases, for all films, as a function of film thickness. At the intersection of some of the islands small holes or pits are visible; these may be associated with an anti-nucleation (less energetically favorable) site on the substrate surface. The holes (some up to ~100 nm deep but with relatively shallow 10 - 40° slopes) occur primarily for films grown at low laser pulse frequencies where, it is suggested, the adatoms have enough time between pulses to move to the more energetically favorable sites. Essentially all of these holes can be eliminated by using higher frequencies ($\geq$ 10 Hz) where adatom motion may be significantly reduced by the arrival of the next pulse of material. In addition, the number of holes appears to decrease as the film thickness increases. In Figure 1(a) most of the holes are surrounded by raised grains (the edge of the grain nearest the hole receives the additional plume material repelled by the hole area) which may be part of the process involved in reducing the number of pits by the faster growing grain edges overlapping the holes as the film thickens.

The a-axis-oriented outgrowths also show a significant variation in size, shape, and number as a function of laser pulse frequency. Overall the number and size of the outgrowths increases with increasing frequency. At 20 Hz the number of outgrowths stays relatively constant over a range of thicknesses from ~7 to ~360 nm. At 2 Hz the number of outgrowths decreases by a factor of four as the film thickness increases from ~140 to ~360 nm. Although the outgrowths seem to start out at about the same size (~200 x 300 nm) there are significant size and shape differences after ~360 nm of film growth. At certain frequencies such as 10 and 20 Hz, the outgrowths segregate into basically two sizes and shapes as seen in Figure 1(c). The larger outgrowths have the classic rectangular shape and are oriented at right angles to each other. At 40 Hz, the outgrowths are more square in shape and they show a significant amount of merging as seen in Figure 1(d).

In this presentation, the above features of the growth process, as determined from SPM studies, will be expounded together with the role of other deposition parameters on the microstructural characteristics of the films [3].

References

1. M.G. Norton, et al., Physica C $\underline{233}$, 321 (1994).
2. R.R. Biggers, et al., IEEE Trans. Applied Superconductivity (1995 in press).
3. This research has been supported by the Wright Laboratory, Aeronautical Systems Center, Air Force Materiel Command, USAF, and the Advanced Research Projects Agency (ARPA) under grant number, F33615-94-1-5802. EKM gratefully acknowledges the support of a National Research Council-Wright Laboratory Research Associateship.

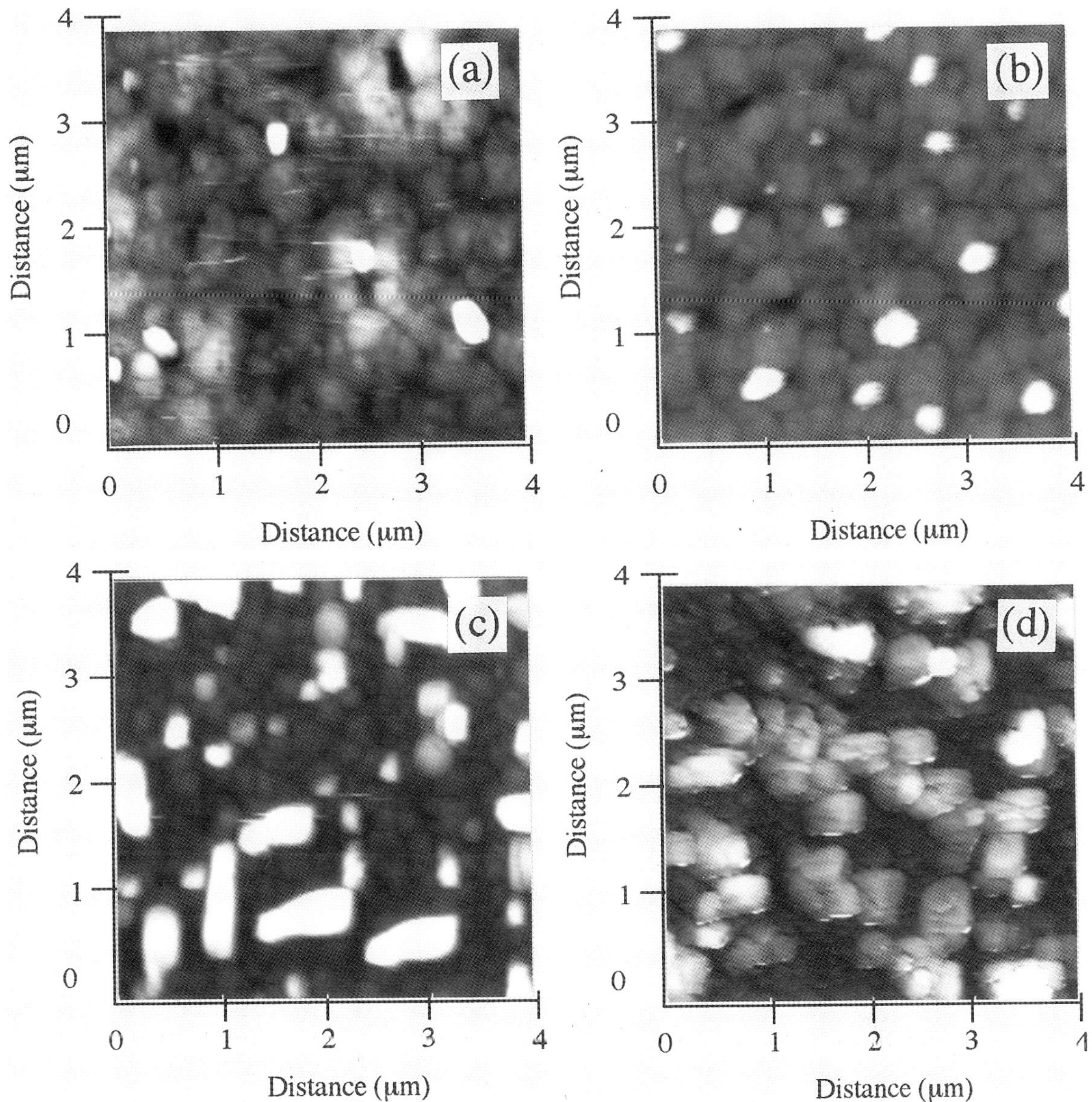

Figure 1. A montage of AFM images of the surface of ~360-nm-thick YBCO films deposited on (001)-oriented LaAlO₃ substrates using the following laser pulse frequencies: (a) 2 Hz; (b) 5 Hz; (c) 10 Hz; and (d) 40 Hz. The z-range is 0 to 50 nm in (a) and 0 to 100 nm in (b) - (d)

THE ROLE OF EPITAXIAL STRAIN IN THE PROPERTIES OF HIGH T_C THIN FILMS

E.J. Williams,*,** A. Cretton,*,** Y. Jaccard,*,*** J.-P. Locquet,* E. Mächler,* P. Martinoli,*** and Ø. Fischer**

*IBM Research Division, Zurich Research Laboratory, CH–8803 Rüschlikon, Switzerland
**D.P.M.C. Université de Genève, CH–1211 Genève, Switzerland
***Institut de Physique, Université de Neuchâtel, CH–2000 Neuchâtel, Switzerland

The difference observed between the critical temperatures measured for bulk material or very thick films (>500 nm for the '214' system) and those measured for thin films (<65 nm) has long been of curiosity. Thin films never attain the high values of T_c achieved in bulk or thick film material. Various mechanisms have been proposed to explain this phenomenon, for example Kosterlitz–Thouless effects, inadequate or unstable oxygenation, and strain itself. Formally the role of strain has remained insubstantiated.

The system on which we have concentrated our efforts is that of $La_{2-x}Sr_xCuO_4$ (LSCO) deposited using MBE, mainly onto (001) $SrTiO_3$ (STO) substrates. These films have been grown at about 700°C at $\sim 3 \times 10^{-6}$ torr oxygen pressure and cooled after growth at $\sim 10°$ min^{-1}. The films have subsequently been examined using several microstructural and electronic probes, including x-ray diffraction, AFM, resistivity, Hall effect and kinetic inductance.[1] The initial observations made by transmission electron microscopy revealed the films to have a unique orientation relation with the substrate such that for the orthorhombic nomenclature (ortho.) (001)LCO‖(001)STO & (110)LCO‖(100)STO. A number of other microstructural features were revealed including a network of interfacial misfit dislocations which were found to be edge type and to have $\mathbf{b}=a/2<110>$ortho.[2] These first observations of misfit dislocations with an $\sim 14 \pm 2$ nm spacing in a specimen with $x=0$ and 30 nm film thickness led us to believe that the films were probably well relaxed since from the room temperature powder diffraction data of undoped film material and substrate, an expected dislocation spacing of 14.5 nm can be calculated. At this stage the only possible influence envisaged of the dislocations on the electronic properties of the thin film was that resultant from the strain fields surrounding the dislocation cores. As can be seen from the 220 ortho. dark field of a cross-sectional specimen of an undoped film shown in Fig. 1 and the disturbance of the contrast of the lattice image of the undoped film shown in Fig. 2, the strain fields penetrate some 2–6 unit cells into the film. However as our studies proceeded, films with varying Sr content $0<x<0.22$ were examined. As the Sr content of LSCO increases, the bulk a and b lattice parameters decrease, the crystal transforms from orthorhombic to tetragonal at $x \approx 0.05$, and the subsequent a parameter continues to decrease (meanwhile, c increases).[3,4] This reduction of the in-plane lattice parameters increases the intrinsic misfit between the LSCO and the STO and so might be expected to reduce the misfit dislocation spacing; for a fully relaxed film of $x=0.2$ the expected dislocation spacing is 11.1 nm. This reduced spacing has, however, not been observed as yet. From Fig. 3, a 220 tet. dark field image from a plan view specimen of an $x=0.16$ film, and from Fig. 4, the 200 tet. dark field of the same film, it can be seen that although the misfit dislocation array is not completely regular, the spacing is around 14 ± 1.5 nm.

A possible explanation for this large misfit dislocation spacing lies with the growth temperature. If the lattice parameters for the substrate and films expected at the growth temperature are compared, it is found that for both undoped and highly doped materials, misfit dislocation spacings of 14.3–14.7 nm are expected. This indicates that the interface structure formed during growth changes very little on cooling, and thus that cooled films are subject to a tensile strain within their plane proportional to the differential thermal expansivities of the materials. This also follows the trends expounded by Speck et al.[5] This in-plane tensile strain implies, via the Poissons ratio, that the c-axis of the films should be less than that observed in the bulk and exactly this has been observed in our x-ray data.[6] This all points to the necessity of considering the operation of the pressure effect in the reductions of T_c of these very thin film materials, even when their thicknesses are greater than the critical thickness.[7]

Proc. Microscopy and Microanalysis 1995, edited by G.W. Bailey, M.H. Ellisman, R.A. Hennigar, and N.J. Zaluzec
Copyright © 1995 MSA. Published by Jones and Begell Publishing, 79 Madison Ave., New York, NY 10016

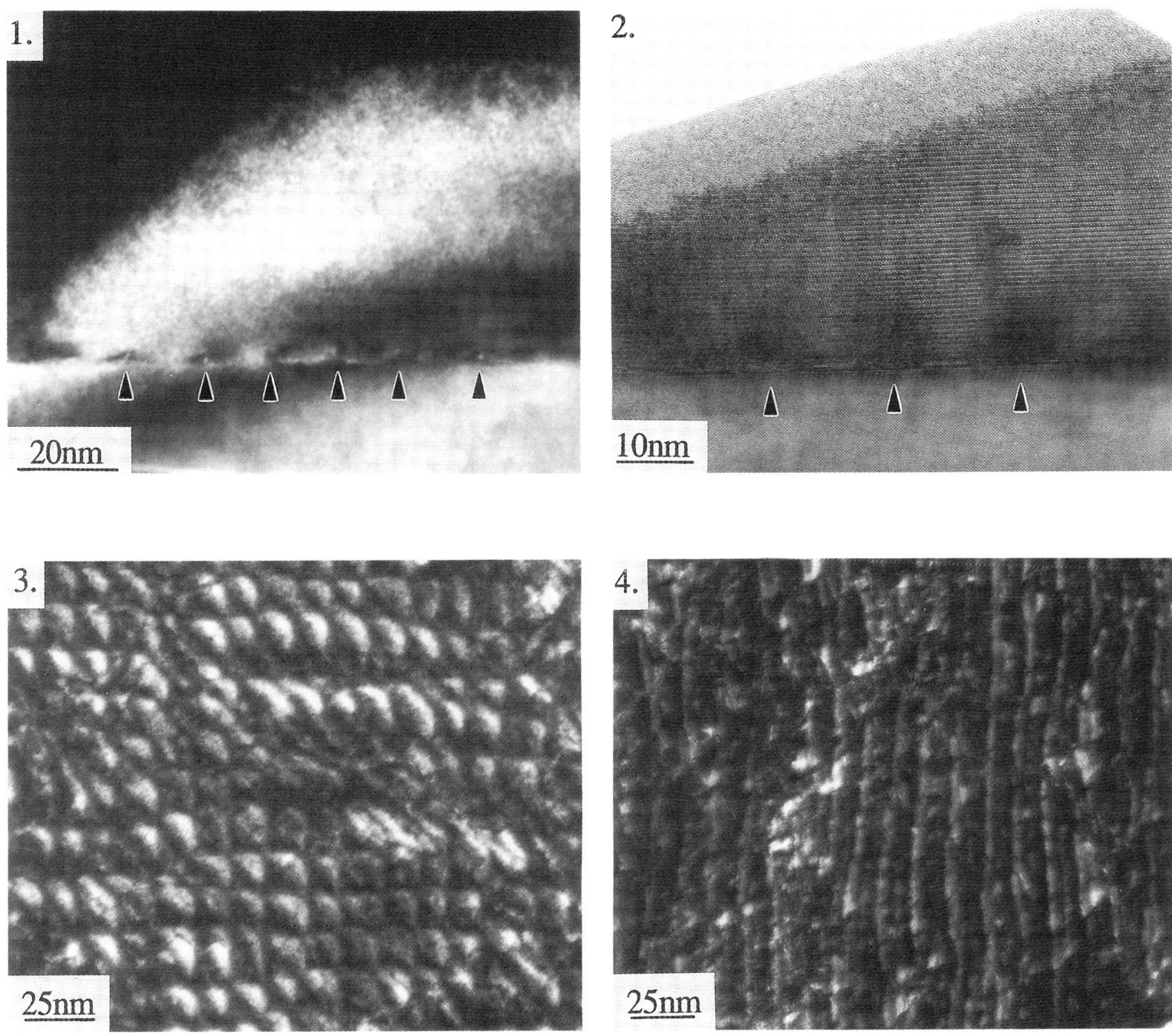

FIG. 1. 220LCO dark field image from a cross section showing strain contrast from misfit dislocations.
FIG. 2. Lattice image from undoped material showing disturbed contrast related to misfit dislocations.
FIG. 3. 220 dark field image from plan view of $x=0.16$ material showing the network of misfit dislocations.
FIG. 4. 200 dark field image from same specimen as in Fig. 3 showing that, although the dislocations are not entirely evenly spaced, their separations are ~14nm.

References

1. Y. Jaccard et al., *SPIE Conf. on Oxide Supercond. Phys. & Nanoeng.* (1994).
2. E.J. Williams et al., *Inst. Phys. Conf. Ser.* 138:7(1993)329.
3. M. Braden *Ph.D. Thesis*, Köln (1992).
4. F. Tresse, *Ph. D. Thesis*, Bordeaux (1990).
5. J.S. Speck et al., *J. Appl. Phys.* 76(1994)447.
6. J.-P. Locquet et al., *Phys. Rev. B.* (submitted).
7. The authors would like to thank the Swiss National Science Foundation for financial support under the NFP30 Program.

NUCLEATION OF PYROXENE AND AMPHIBOLE IN GLASSY VEINS ALONG INTERFACES AND CRACKS IN ULTRAMAFIC INCLUSIONS FROM SAN CARLOS; ARIZONA

R. Wirth

GeoForschungsZentrum Potsdam, Telegrafenberg A 26, D-14473 Potsdam, Germany

Ultramafic inclusions occur in ellipsoidal bombs in basalt flows. The inclusions of the San Carlos ultramafic inclusion locality have been described in detail by Frey & Prinz (1978).[1] They mainly consist of olivine, orthopyroxene (opx), clinopyroxene (cpx) and Cr-spinel. Ultramafic inclusions are important because they are samples of the upper mantle material, and thus they provide direct insight into the upper mantle processes and composition. Glassy veins are fairly abundant along mineral interfaces and cracks. The glass apparently results from incongruent melting of hydrous phases.[1] The glassy veins are between a few nanometer and several microns wide. Veins with a width > 1 µm exhibit crystals of olivine, pyroxene and feldspar. The narrow veins (< 1 µm) only occasionally show crystallization of small crystals (several hundred nanometer). The early stages of crystallization of pyroxene and amphibole and the chemical composition have been investigated by HRTEM and analytical TEM.

Glassy veins along interfaces: small grains (several hundred nanometer) of opx have nucleated at an opx/glass interface. After a short period of growth, cpx has nucleated at opx forming a rim of cpx (~ 500 nm) between opx and glass (**Fig.1**). The nucleation of the pyroxene phases is assumed to occur in the molten state of the glass. Opx has nucleated at the opx/melt interface with the {001}planes parallel oriented. Due to the Ca-content of the melt, cpx, rich in Ca, has nucleated at the opx/melt interface with the orientation relationship (100)opx // (100)cpx (**Fig.2**). The limited growth of cpx is due to rapid quenching of the ultramafic inclusion in the erupting magma. The chemical composition of the glass is inhomogeneous with respect to the elements Mg, Al, Si, Ca, Fe. Quantitative measurements of Mg and Fe show an increase of Fe and a decrease of Mg content, due to an ion exchange with the melt by diffusion, in 500 nm wide seams near the crystal edges. A time estimate for the Mg/Fe exchange in opx (D_0=4.07*10^{-4} (m^2 sec^{-1}),[2] Q=372 (kJ/mol), T= 1000 (°C)) suggests that the interaction of the melt with the minerals could have lasted only for days.

Glass filled cracks in olivine: amphibole crystals (10-20 nm) have nucleated at olivine/glass interfaces growing into the glass (**Fig.3**). Amphibole is oriented with respect to olivine: [001]olivine // [001]amphibole; (020) amphibole is rotated around the common [001]-axis at about 15 ° with respect to the (100) plane of olivine (**Fig.4**). The chemical composition of the glass is inhomogenious but, in the average resembles the composition of the host basalt. In the vicinity of the olivine edges no ion-exchange has been measured. A very short time for interaction of the melt with the silicates or decreasing temperature in the cooling lava flow can explain this observation. It is concluded that olivine suffered brittle deformation (cracks) during the magma eruption and melt of the host basalt filled the cracks .

1. F.A.Frey and M.Prinz, *Earth and Planetary Science Letters*, **38**(1978)129-176.

2. R Freer, *Contrib Mineral Petrol* **76**(1981)444 (Ref.4).

Proc. Microscopy and Microanalysis 1995, edited by G.W. Bailey, M.H. Ellisman, R.A. Hennigar, and N.J. Zaluzec
Copyright © 1995 MSA. Published by Jones and Begell Publishing, 79 Madison Ave., New York, NY 10016

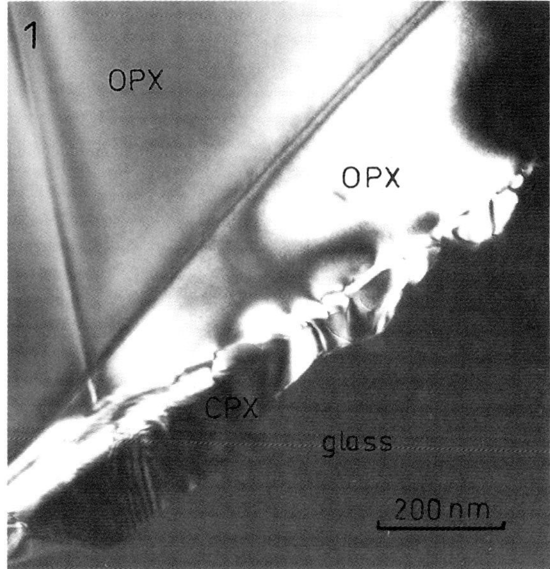

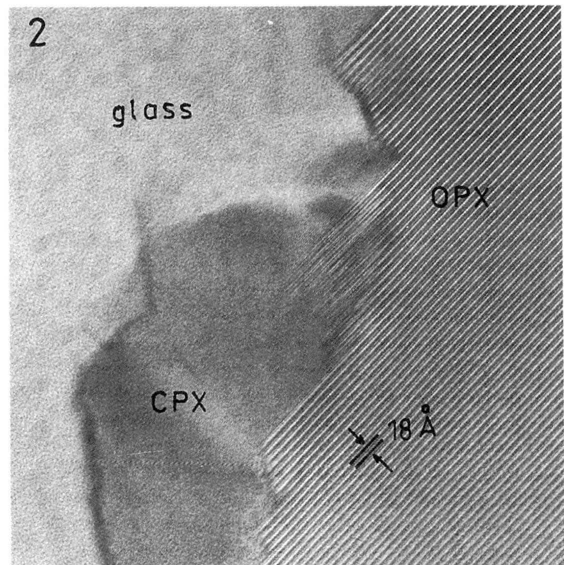

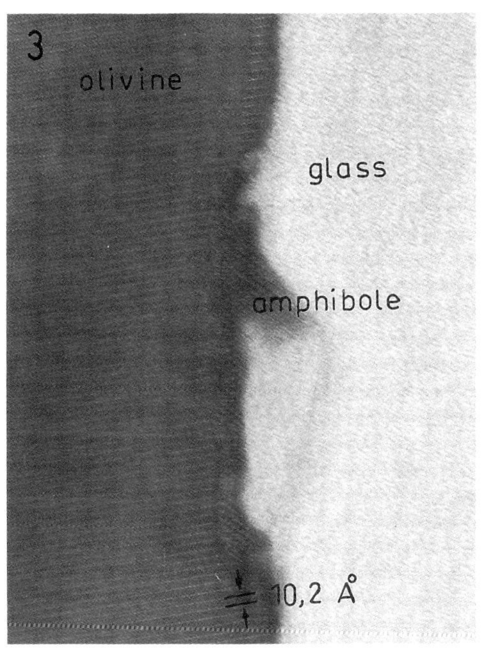

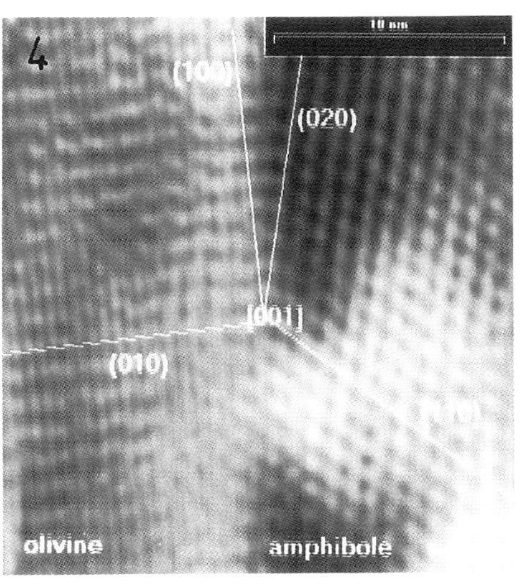

FIG.1.- TEM bright field image of a glassy vein at an opx/olivine interface. Opx has nucleated at the opx/glass interface growing into the glass. The Ca-content of the glass induced the nucleation of cpx at the migrating opx/glass phase boundary.

FIG.2.- HRTEM image of the interface opx/cpx. (100)cpx exhibits half the lattice spacing of (100)opx and is parallel oriented to (100)opx. Numerous steps characterize the interface.

FIG.3.- HRTEM image of amphibole nucleated at an olivine/glass interface.

FIG.4.- Fourier filtered image of amphibole nucleated at olivine. Zone axis in both crystals is [001]. (020) amphibole is rotated about 15° with respect to (100) olivine around the [001] zone axis.

IMPROVING SOIL REMEDIATION THROUGH CHARACTERIZATION

E. C. Buck, N. L. Dietz, and J. K. Bates
Chemical Technology Division, Argonne National Laboratory, 9700 South Cass Avenue, Lemont, IL 60439

Operations at former weapons processing facilities in the U. S. have resulted in a large volume of radionuclide-contaminated soils and residues. In an effort to improve remediation strategies and meet environmental regulations, radionuclide-bearing particles in contaminant soils from Fernald in Ohio and the Rocky Flats Plant (RFP) in Colorado have been characterized by electron microscopy. The object of these studies was to determine the form of the contaminant radionuclide, so that it properties could be established [1]. Physical separation and radiochemical analysis determined that uranium contamination at Fernald was not present exclusively in any one size/density fraction [2]. The uranium-contamination resulted from aqueous and solid product spills, air-borne dust particles, and from the operation of an incinerator on site. At RFP the contamination was from the incineration of Pu-bearing materials. Further analysis by x-ray absorption spectroscopy indicated that the majority of the uranium was in the 6+ oxidation state [3]. Soil samples embedded in a water miscible epoxy were examined by optical microscopy and SEM/ BSE. The radionuclide-rich regions located by SEM/BSE were isolated by trimming the epoxy block and then microtoming with a 45° diamond knife. The microtome was contained in a flow-through hood designed for the preparation of radioactive samples. This method allowed the identical uranium-bearing particle to be tracked from SEM to TEM (Fig. 1). In addition, ultramicrotomy preserved the spatial relationships between particles and provided large electron transparent regions for analysis [4].

The major uranium forms in Fernald site soils were identified as tetragonal calcium uranyl phosphate (meta-autunite), uranium oxide (uraninite), uranium calcium oxide, uranyl silicate (boltwoodite and soddyite), and uranium metaphosphate $[U(PO_3)_4]$. Other types of uranium phases were observed on rare occasions such as uranium silicide, uranium associated with calcium fluoride, and uranium incorporated into silica. The TEM observations indicate that contamination resulted from both dispersion of air-borne particulates as well as precipitation within the soil. The precipitated phases may have formed by interaction of soluble uranium with phosphate in the soil. The source of phosphate was most likely from the use of tributyl phosphate used in uranium purification at Fernald. This resulted in the formation of meta-autunite, ideally $[Ca(UO_2)_2(PO_4)_2.0-2H_2O]$ (Fig. 2). These phosphates control the uranium concentration in groundwater at Fernald, as they have solubilities below those of uranyl silicates and carbonates. In waters where log $([HPO_4^{3-}]/ [HCO_3^-]) > 1 \times 10^{-3}$ uranyl phosphates will dominate, hence immobilization of uranium may be possible by forcing the precipitation of uranium phosphate phases. The local soils around the Fernald plant are actually low in carbonate, however limestone from the building of roads at Fernald, has increased carbonate levels and effectively increased the mobility of uranium in the soils. The refractories UO_2 and $[U(PO_3)_4]$ originated from the incinerator on the site. Electron diffraction identified the uranium oxide phase as uraninite and the phosphite phase as a monoclinic $[U(PO_3)_4]$ (see Fig. 3). Chemical soil washing was found to remove only around 75-80% of the total uranium in the soils, as the chemicals were unable to dissolve the refractory uranium phases. The phases which could not be removed may, however, undergo alteration, resulting in the slow release of uranium into the groundwater. In addition, airborne uranium particulates in the micron-size range formed in the incinerator may have been transported large distances away from the plant. Nevertheless, the presence of phosphate phases suggests that immobilization of uranium could be achieved by forcing the precipitation of uranium phosphates in the soil. Further studies are underway to determine the feasibility of this approach.

Plutonium-bearing incinerator ash from RFP was examined in order to determine the form of Pu in the residues. As in the case of Fernald, incineration resulted in the formation of refractory actinide phases. Most of the Pu was in the form of the reduced non-stoichiometric oxide (PuO_{2-x}), however a proportion of the Pu was found in a titanium silicate phase and in silica glass phases. It is possible that the amorphous Pu-silica phase may have been produced by alpha recoil damage from the Pu decay. PuO_{2-x} is a very insoluble phase possessing the fluorite structure (space group Fm3m) and is isomorphous with uraninite, thoria, and ceria. The cubic unit cell (a = 5.3960 nm) contains four plutonium atoms and eight oxygen atoms. PuO_2 loses a small amount of oxygen at temperatures of 1100°-1200°C in a reducing atmosphere. The glass phase could undergo weathering resulting in the release of Pu, although the size of the oxide particles suggests that colloidal transport may be possible. Studies with Pu loaded borosilicate waste glasses have shown that Pu undissolved in the glass can be released in colloidal form as PuO_{2-x} [5, 6].

[1] E. C. Buck, N. R. Brown, and N. L. Dietz, *Mat. Res. Soc. Symp. Proc.* 333 (1994) 437
[2] S. Y. Lee and J. D. Marsh, Oak Ridge National Laboratory Publication ORNL/ TM-11980 ESD3786 (1992)

Proc. Microscopy and Microanalysis 1995, edited by G.W. Bailey, M.H. Ellisman, R.A. Hennigar, and N.J. Zaluzec

[3] P. M. Bertsch et al., *Environ. Sci. Tech.* 28 (1994) 980
[4] E. C. Buck et al., *Micro. Res. Tech.* 30 (1995)
[5] J. K. Bates et al., 6th Inter. High Level Rad. Waste Mgmt Conf., Las Vegas, NV, April 30-May 5 (1995)
[6] Work was supported by the U.S. DOE, as part of the DOE Technology Development Landfill Focus Area, under contract W-31-109-ENG-38.

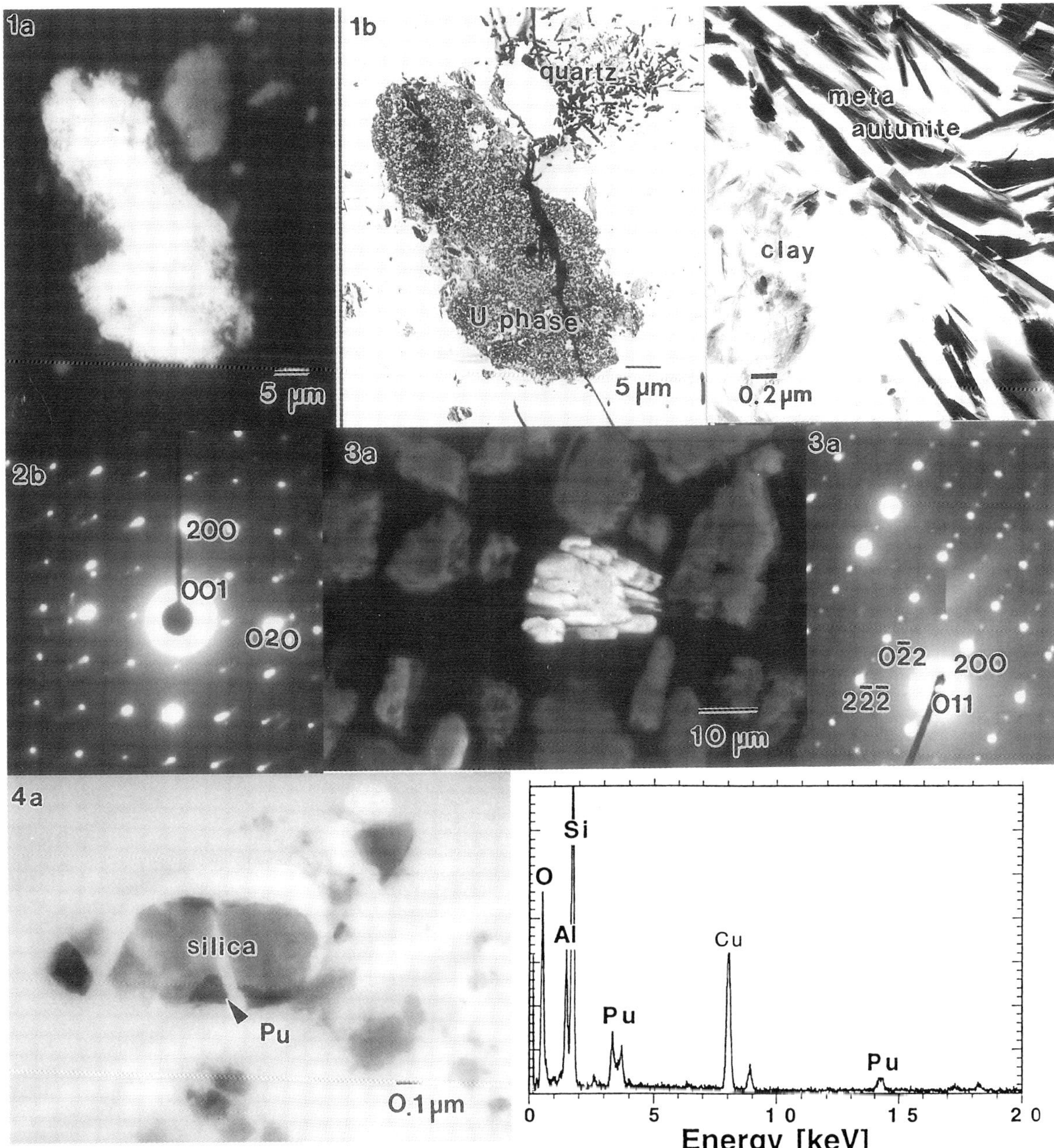

Fig. 1 (a) SEM/BSE image and (b) corresponding microtomed TEM thin section of uranium-bearing phase.
Fig. 2 (a) TEM image of meta-autunite phase and (b) [001] ZAP from phase .
Fig. 3 (a) SEM image of uranium phosphite and (b) SAED taken along [011] from monoclinic variety.
Fig. 4 (a) TEM CCD image of plutonium-bearing silicate phase from RFP and (b) EDS analysis of glassy phase.

Petrofabric study using Electron Channeling Patterns of quartzes and hematites in dual-phase banded iron formations

K. Y. Lee [1] and B. W. Robertson [2]

[1] Department of Geology, University of Nebraska, Lincoln, NE 68588
[2] Center for Materials Research and Analysis, University of Nebraska, Lincoln, NE 68588

Research in deformation of rocks requires the understanding of structure-property relationship. The study of structure-property relationships requires knowledge of the crystallographic texture on a local scale. Electron Channeling in scanning microscope has been used to analyze textures of local features in a microstructure, misorientations between grains and the spatial arrangement of grains of individual crystallographic orientations can be determined. This information is critical for a mechanical understanding of microstructure evolution during the deformation of rocks. Petrofabric analysis using Electron Channeling Patterns can use normal geological thin sections and mounted sections, which allows relatively fast and easy interpretation and allows one to determine the relationship between a microstructural feature and the individual orientations of its grains. The most efficient and satisfactory way of accomplishing this task is by direct electronic detection and indexing of Kikuchi patterns, which requires modification of the microscope, at considerable cost. This study, however, has been undertaken using only an unmodified SEM with electron channeling capability. Thus any structural geologist with access to such an SEM can carry out this kind of texture analysis. The Universal Stage method with a polarized light microscope is limited to a few orientations. The XRD method can only provide bulk crystallographic information. The TEM is generally restricted to very small regions of thin foils.

The purpose of this preliminary study is, by using indexed individual Electron Channeling Patterns of the dual-phase specimens from the Nemo Banded-Iron-Formation (quartz and hematite)(Fig.1 and Fig.2), to compare distribution functions and to deduce their relation about the North-North-West trending foliation in Nemo area, Black Hills, South Dakota.

Recovery and partial recrystallization in the dual-phase specimens were confirmed by dislocation studies using TEM. The slip systems were deduced to be basal-<a> (and perhaps also prismatic-<a>). Even though syntectonic recrystallization has a profound effect on the mechanism of lattice reorientation, the preferred orientations are preserved. The interpretation of the fabrics is controlled by the active slip systems, the type of finite strain, and the kinematic framework. The pole figure data obtained from electron channeling patterns of quartz domains show crossed girdle patterns (type II) (Lister and Dornsiepen, 1982; Fueten,1992) (Fig.3). The patterns indicate the right-handed sense of vorticity along the NNE foliation in the mining area near Nemo. The right-handed sense of vorticity is also recognized by mesoscopic kinematic indicators, such as pressure solution fringes, asymmetrical pinched tails, en echelon arrays of microfaults in the quartzite pebbles in associated Nemo Conglomerate. The sense of vorticity indicates the right-handed motion of blocks (Lister and Hobbs 1980; Schmid and Casey 1986; Law 1990). The pole figure from micaceous (plate-type crystal habit) hematites which were formed syntectonically shows asymmetrical maxima around the Y axis (Fig.4). Even though they are not well resolved as maxima type II, IV or girdle, their asymmetry is attributed to overlapping by a later fabric. This may arise from more equant grains of post-tectonic hematites which grew along the grain boundaries of quartzes and later fractures.

References

1. G. S. Lister and U. F. Dornsiepen, Fabric transitions in the Saxony granulite terrain, J. of Struct. Geol., 4 (1982) 81-92.
2. F. Fueten, Tectonic interpretations of systematic variations in quartz c-axis fabrics across the

Proc. Microscopy and Microanalysis 1995, edited by G.W. Bailey, M.H. Ellisman, R.A. Hennigar, and N.J. Zaluzec
Copyright © 1995 MSA. Published by Jones and Begell Publishing, 79 Madison Ave., New York, NY 10016

Thompson Belt.
3. G. S. Lister and B. E. Hobbs, The simulation of fabric development during plastic deformation and its application to quartzite, J. Struct. Geol., 2 (1980), 355-370.
4. S. M. Schmid and M. Casey, Complete fabric analysis of some commonly observed quartz c-axis patterns. In: Mineral and Rock Deformation: Laboratory Studies—The Paterson Volume (edited by Hobbs, B. E. & Heard, M. C.), Am. Geophys. Un. Geophys. Monogr., 36 (1986), 26-286.
5. R. D. Law, Crystallographic fabrics: a selective review of their applications to research in structural geology. In: Deformation Mechanisms, Rheology and Tectonics (edited by Knipe, R. J. & Rutter, E. H.). Spec. Publ. Geol. Soc. Lond. 54 (1990), 335-352.

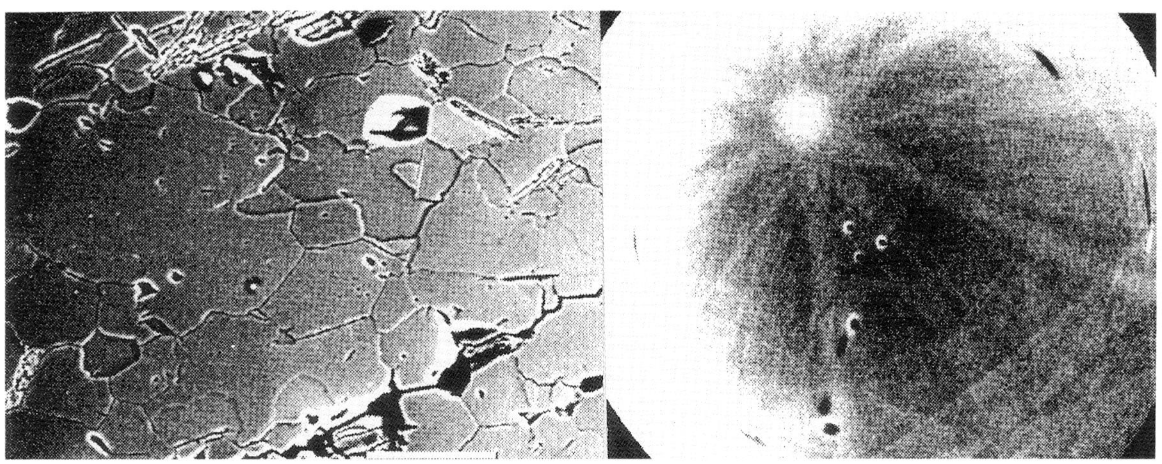

FIG. 1

FIG. 2

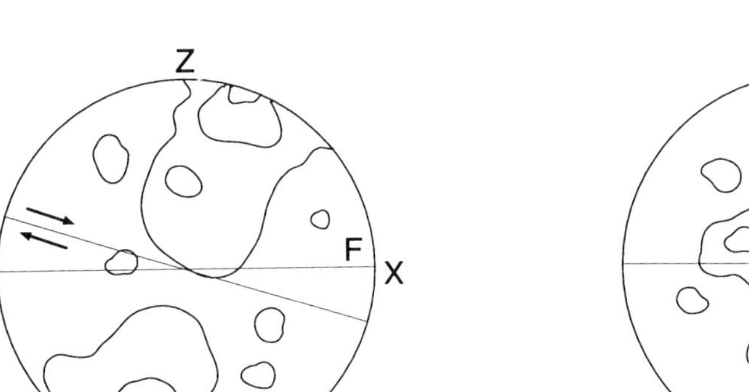

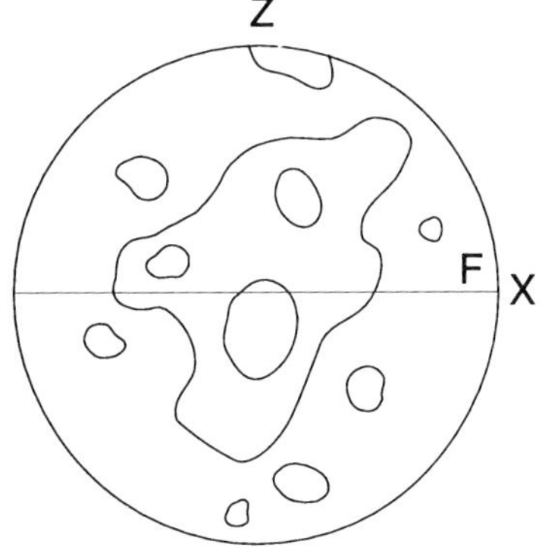

FIG. 3

FIG. 4

FIG. 1 - Backscattered image of the quartz rich domain in the banded iron formation. Bar = 100 μm.
FIG. 2 - Backscattered Electron Channeling Pattern of a quartz (Si_2O) near <0001> zone axis
 (30 KeV, Working Distance = 6 mm)
FIG. 3 - The pole figure of quartz <0001>. F = Foliation, S = Shear plane, Z = Axis normal to foliation, X = Lineation. The relative movement of shear plane is right-handed.
FIG. 4 - The pole figure of hematite (Fe_2O_3) <0001>.

EVIDENCE FOR A RAISED RIM ON PITS IN MICA INDUCED BY keV/NUCLEON IONS

L. Fei* and P. Fraundorf*

*Center for Molecular Electronics and Department of Physics & Astronomy, University of Missouri-St. Louis, St. Louis, MO 63121

Nuclear tracks have been extensively studied and widely used in applications like archeological and geological dating, solar flare iostope determination, extraterrestrial material studies, super heavy elementary particle searches, etc. for more than three decades.[1] Recently, scanning force microscopy (SFM) has proven to be useful for studying the pits induced by ions in the track recording energy range from several keV to several MeV per nucleaon on track recording material (e.g. mica) surfaces.[2-4] SFM can provide topographical information on chemically etched or *unetched* track pits at near atomic resolution in all three dimensions.

We report here using a commercial SFM (NanoScope III, Digital Instruments, CA) in the study of a muscovite mica surfaces irradiated with 150keV Ar^+ ions. The mica was freshly cleaved before the irradiation. The SFM was operated with the amplitude resonance (tapping) mode to lower the interaction between a tip and specimen. We found pits on the mica surface after, but not before the irradiation, like those shown in Fig. 1a. Their number density was orders of magnitude less than the expected argon flux. We expect that pit formation depended on either atom-atom collisions, multiple ionizations, or impurity ions in the beam, and experiments are currently underway to reconcile this discrepancy. All pits showed a raised rim around them (Fig. 1b). Measurements of about 10 pits yielded pit diameters (not including the rim) of around 80-100 Å, and pit depths (refered to the average mean mica surface) of about 3-5 Å. The raised circular rims, which were a surprise to us and which we have not seen reported previously in the literature, were on average about 100 Å in width and 2 Å in height relative to the surrounding surface. Figure 2 is a larger field of view image showing 4 such pits. Judging by the overall curvature of the imaged surface, we concluded that the bottom one third portion of the image has a stronger interaction force between the tip and the specimen than the upper two thirds portion. The fact that the two pits at the bottom portion of the image are larger and deeper in dimensions indicates that pit regions are relatively softer compared with the unbroken mica surface.[2] The size of a pit plus its rim, on the other hand, remain almost the same regardless of the interaction force between tip and specimen. Note: a "ghost" bump associated with all 4 pits near the 8 o'clock position relative to the pit can always be seen in Fig. 2. This is an instrumental double tip artifact.

The structure of a raised rim around a pit with only about monolayer molecule height but extending to 100 Å laterally can not be explained by the ejecta blanket mechanism which of course works for macroscopic craters. Instead, we propose that the structure is due to the surface lattice relaxation because of the breakdown in layer surface tension due to broken bonds in the pit region. Clarification of this possibility is still underway.

References

Proc. Microscopy and Microanalysis 1995, edited by G.W. Bailey, M.H. Ellisman, R.A. Hennigar, and N.J. Zaluzec
Copyright © 1995 MSA. Published by Jones and Begell Publishing, 79 Madison Ave., New York, NY 10016

1. R. L. Fleisher, P. B. Price, and R. M. Walker, *Nuclear Tracks in Solids*, Berkeley:Unversity of California Press (1975).
2. P. Fraundorf and J. T. Tentschert, *Ultramicroscopy*, **37**(1991)125.
3. F. Thibaudau *et al.*, *Phys. Rev. Lett.*, **67**(1991)1582.
4. R. V. Coleman *et al.*, *Surf. Sci.*, **297**(1993)359.
5. This work is supported in part by the U.S. Department of Energy, Grant No. DE-FG02-92CH10499.

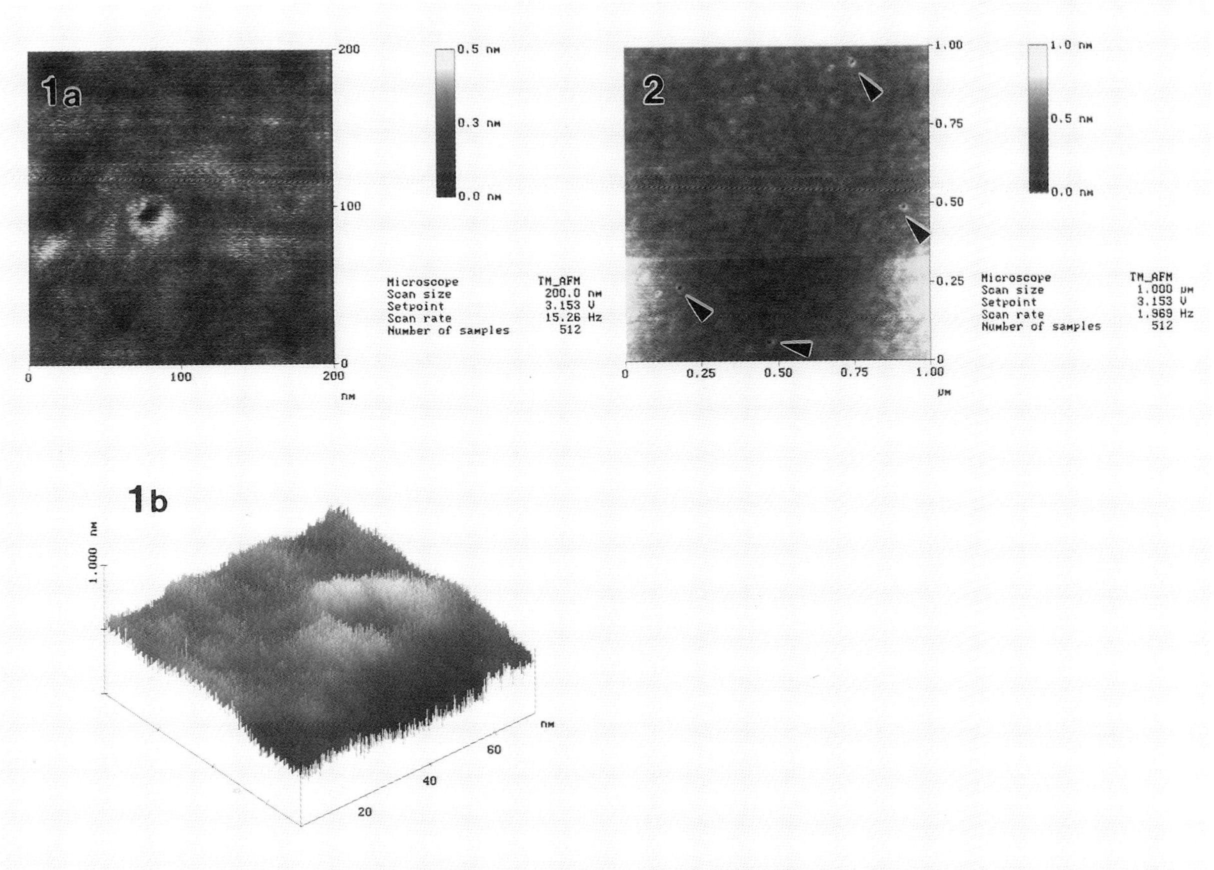

FIG. 1.—(a) Top view of a typical pit on muscovite mica surface induced by a 150 keV Ar$^+$ ion. (b) Surface rendered, yet another such pit showing the raised rim only 2 Å in height.
FIG. 2.—Top view of a 1 × 1 μm field. Arrows point at 4 pits. Two of the pits at the bottom portion of the image have a slightly larger size due to a stronger force interaction between the tip and the specimen than pits on the upper portion of the image. A double tip artifact is seen in the "ghost" bumps associated with each pit.

A Correlation of CL Emissions from Penrith Sandstone with Elemental Distributions obtained by Simultaneous SEM-CL and EDS Analysis

Paul.J.Wright

Oxford Instruments Ltd, Scientific Research Instruments Division, Eynsham, Oxfordshire, OX8 1TL

Introduction

Most industrial and academic geologists are familiar with the beautiful red and orange cathodoluminescence colours produced by carbonate minerals in an optical microscope with a cold cathode electron gun attached. The cement stratigraphies interpreted from colour photographs have been widely used to determine the post depositional processes which have modified sedimentary rock textures.

However to study quartzose materials high electron densities and kV's are necessary to stimulate sufficient emission. A scanning electron microscope with an optical collection system and monochromator provides an adequate tool and gives the advantage of providing secondary and backscattered electron imaging as well as elemental analysis and distribution mapping via standard EDS/WDS facilities.

It has been known that the incorporation of many elements modify the characteristics of the CL emissions from geological materials. They do this by taking up positions between the valence and conduction band thus providing sites to assist in the recombination of electron hole pairs. Potentially CL has much greater sensitivity to the presence or absence of many of these elements than any of the SEM based X-ray techniques but in some instances it is possible to obtain information which correlates CL emission to spatially resolved elemental data.

Experimental

In this study we have examined a quartzose material (Penrith Sandstone). The backscattered electron image in Fig 1 appears to be almost uniform in atomic number contrast implying that this material has similar origins throughout. However the CL image shown in Fig 2 indicates that not only is this material non uniform but that it is composed of grains surrounded by overgrowths of quartz cement. Some of this material has been extruded into grains and has "healed" them. If it is a correct assumption that the CL is generated by minute differences in the chemistry of the cements and grains then it might be possible to see this in an EDS map.

EDS analyses were carried out at positions (a) and (b). These indicated that there was a potential difference in the concentration of Iron (Fe) and Aluminium (Fe) between these two positions.

Mapping of these two elements provided the spatially resolved images in Figs 3 and 4. It is quite evident that in Fig 4 a crack that is not observable in the backscattered image but appears to not luminesce in the CL image has a higher level of aluminium than the surrounding grain and the rim of the grain has also the same element in higher concentration than the surroundings. The increase in the Fe distribution appears to be localised around the edge of the entire original grain below the overgrowth. This would seem to imply that different processes have been responsible for the evolution of the core, rim and overgrowth. Whilst not observable by SEI or BSI, Cathodoluminescence images clearly reveal this texture and can lead to identification of areas that warrant further elemental characterisation.

Proc. Microscopy and Microanalysis 1995, edited by G.W. Bailey, M.H. Ellisman, R.A. Hennigar, and N.J. Zaluzec
Copyright © 1995 MSA. Published by Jones and Begell Publishing, 79 Madison Ave., New York, NY 10016

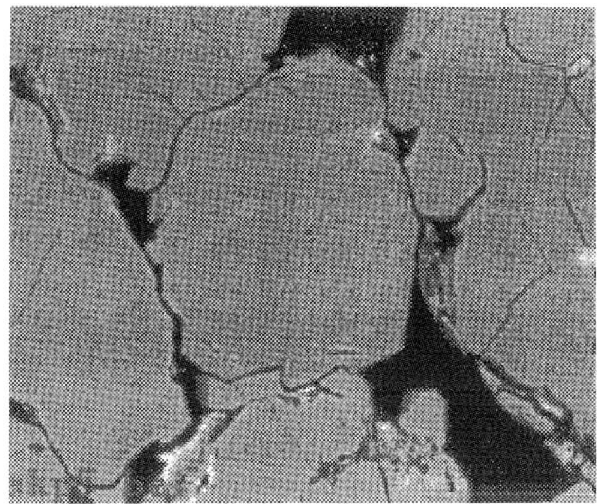

Fig 1; Backscattered Electron Micrograph of a grain from Penrith Sandstone. The uniform contrast level would seem to suggest that the material has a uniform composition

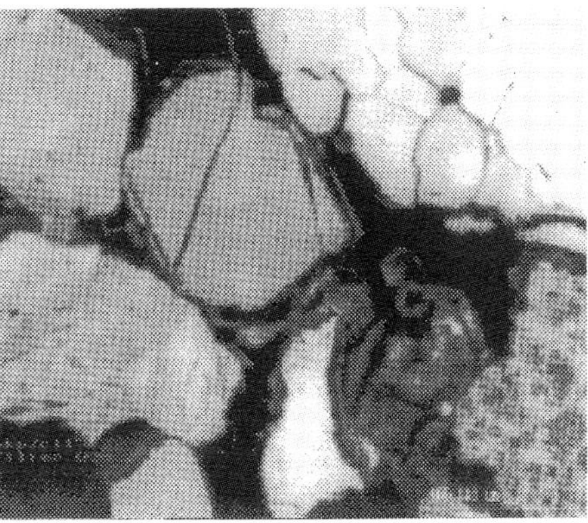

Fig 2; Cathodoluminescence Image of the same area as shown in Fig 1 with a 45° rotation. The differences in contrast are a result of variations in the stoichiometry of the material and it is obvious that there evidence of several different growth processes having occurred with different chemistries.

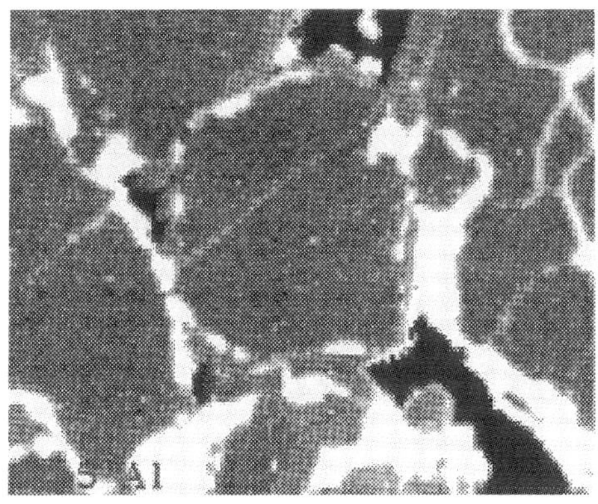

Fig 3; EDS map of the distribution of Aluminium through the same area as shown in Figs 1 and 2. It is obvious that the crack running diagonally across the specimen has been filled at a later date with Aluminium rich (0.5%) cement.

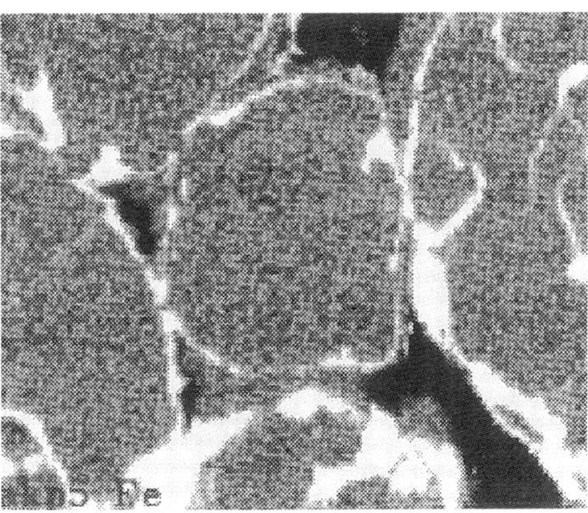

Fig 4; EDS map of the distribution of Aluminium Al through the same area as shown in Figs 1,2 & 3. An Iron rich core appears to surround the central grain and is associated with its environment prior to cementation occurring

Instrumentation

The studies discussed above were carried out on a Jeol 840A SEM equipped with an Link eXL EDS system and an Oxford monoCL cathodoluminescence detector.

Acknowledgements

I would like to thank Anton Kearsley form the Department of Construction and Earth Sciences at Oxford Brookes University, Headington Oxford for these results from this sample.

SURFACE SPECTROSCOPIC STUDIES OF MODEL SUPPORTED-METAL CATALYSTS

D. Wayne Goodman

Department of Chemistry, Texas A&M University, College Station, TX 77843

A new surface science approach to the study of supported-metal catalysts will be described. Thin oxide films (~100 Å) of SiO_2, Al_2O_3, or MgO supported on a refractory metal substrate (e.g., Mo or W) have been prepared by depositing the oxide metal precursor in a background of oxygen (ca. 1×10^{-5} Torr) [1]. The thin-film catalysts facilitate investigation by an array of surface techniques, many of which are precluded when applied to the corresponding bulk oxide. In particular, the oxide films have been characterized by AES, ELS, HREELS, XPS, UPS, ISS, IRAS, and TD spectroscopies and shown to have essentially identical electronic and vibrational properties of the corresponding bulk oxides [1,2]. These studies indicate then that these films can serve as convenient models for oxide catalysts or metal supports. Metal thin films (e.g., Cu, Pd, Ni) have subsequently been deposited onto the oxide films and the properties of the metal/oxide system then studied with the above array of surface techniques [3]. By properly defining the metal thin film thickness, metal particles of varying sizes can be synthesized with dispersions from a few nanometers to tens of nanometers.

These studies have provided access to detailed information with respect to the energetics and morphology of small metal particles supported on oxides. For example, for Pd and Cu supported on SiO_2, the initial copper sublimation energy increases with an increase in the copper coverage, reflecting the change in metal coordination with particle size. Furthermore, the morphology of the Cu and Pd films upon heating could be monitored with TDS and IRAS using CO as a probe. These results show that Cu and Pd form 3-D clusters (the dispersion decreases) upon heating. STM (Fig. 1, for example) and AFM have been utilized and correlated with the IRAS studies to provide detailed information regarding the size and shape of the supported metal particles.

Kinetic studies of CO oxidation on supported Cu [4] and Pd [5] as well as methanation and alkane hydrogenolysis on supported Ni [6] will also be discussed, and the effects of metal particle size and structure addressed.

References

1.   M.-C. Wu, J.S. Corneille, C.A. Estrada, J.-W. He, and D.W. Goodman, *Chem. Phys. Letts.* 472(1991)182; X. Xu and D.W. Goodman, *Appl. Phys. Lett.* 61(7)(1992)774; J.S. Corneille, J.-W. He, and D.W. Goodman, *J. Phys. Chem.* 98(1994)259; P.J. Chen and D.W. Goodman, *Surface Sci.* 312(1994)L767.

Proc. Microscopy and Microanalysis 1995, edited by G.W. Bailey, M.H. Ellisman, R.A. Hennigar, and N.J. Zaluzec
Copyright © 1995 MSA. Published by Jones and Begell Publishing, 79 Madison Ave., New York, NY 10016

2.   M.-C. Wu, C.A. Estrada, D.W. Goodman, *Phys. Rev. Letts.* 67(1991)2910.

3. X. Xu, S.M. Vesecky, and D.W. Goodman, *Science* 258(1992)758; X.Xu and D.W. Goodman, *Appl. Phys. Lett.* 61(1992)1799; X.Xu and D.W. Goodman, *Surface Sci.* 284(1993)103; X. Xu and D.W. Goodman, *J. Phys. Chem.* 97(1993)683.

4. X. Xu and D.W. Goodman, *J. Phys. Chem.* 97(1993)7711.

5. X. Xu and D.W. Goodman, *Catal. Letts.* 24(1994)31; X. Xu, J. Szanyi, Q. Xu and D.W. Goodman, *Catal. Today* 21(1994)57.

6. K. Coulter, X. Xu and D.W. Goodman, *J. Phys. Chem.* 98(1994)1245.

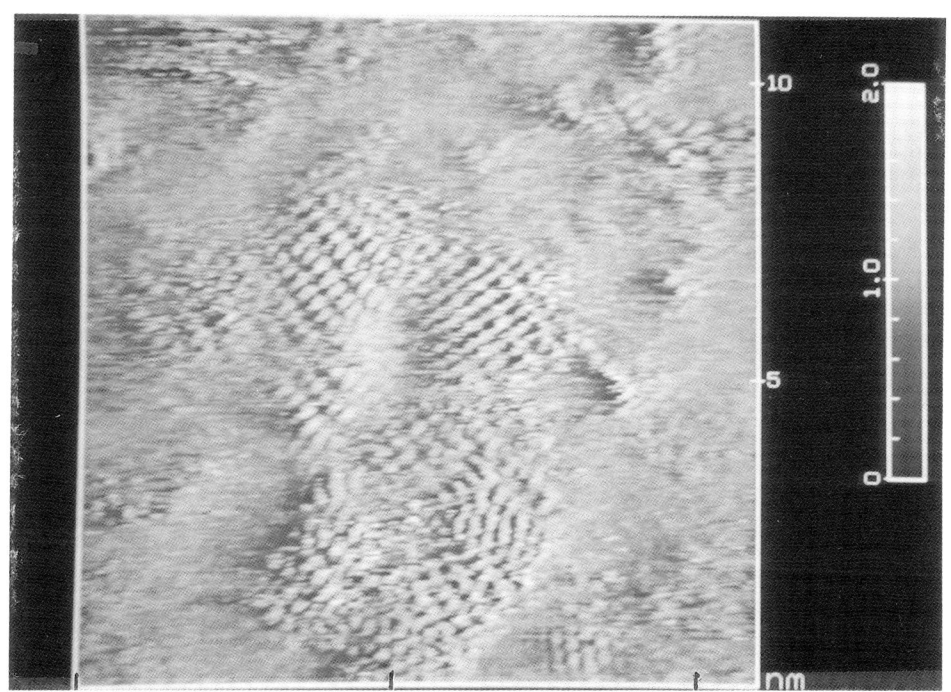

Fig. 1   A STM image (110 X 110Å) of Cu cluster formed by the deposition of 1.2 ML copper on silica(100 Å)/Mo(110).   The Cluster exhibits a Cu-Cu spacing of (4.4 x 2.4) Å.   This image was acquired after the sample was exposed to water and dried in air.

A MICROSCOPIST'S VIEW OF CATALYSIS

W. Targos*

*15777 E. Mustang Dr., Fountain Hills, AZ 85268

Analytical electron microscopy, AEM, has provided valuable information on the structure, composition, location and size of metal clusters and/or crystallites supported on amorphous or crystalline metal oxide materials. Although this information is vital to the understanding of how a catalyst functions, the catalytic reaction is usually more complex. Many variables must be taken into account to evaluate properly a catalyst. Improper selection of process variables can make a good catalyst perform poorly. Consequently, focusing on specific metal site properties may not always be the appropriate direction to take when evaluating a supported metals catalyst problem.

To help the microscopist appreciate the complexity of the problem, the following examples were chosen to illustrate this point. Reforming catalysis involves a complex set of reactions to convert naphtha feed to high octane gasoline. The catalyst is comprised of highly dispersed Pt on a high surface area alumina with Cl. The catalyst is dual functional requiring both metal site activity for dehydrogenation and acidity for isomerization and cyclization. Although metal agglomeration is a primary source of deactivation[1] loss of acidity can also be detrimental.[2] Low Cl levels can results from poor management of Cl during operation, excessive water exposure, or a major loss of surface area.[3] Other factors which can reduce acidic function include excessive metals deposition such as Fe or alkali metals.[4] Use of S is crucial in moderating hydrogenolysis activity of Pt especially during startup.[5] Without the use of S, excessive metal site temperature and selective coke deposition at the site can occur.[6] Addition of dopant metals such as Sn or Re can provide coke management at the metal site.[7] Performance is also related to feedstock composition, particularly those with high paraffin content, hydrogen pressure, water, Cl and S content in the feed, and temperature.[8] Performance changes that may occur when these variables are altered may not readily be observed in metal site changes when comparing good versus poor performing catalysts.

The L zeolite ion exchanged with Pt, PtKL, is used to selectively aromatize paraffins such as hexane to benzene.[9] The Pt clusters are usually 1nm or less in size and are located in the channels of the zeolite.[10] The catalyst is monofunctional which distinguishes it from reforming, however, because it contains a zeolite other factors must be considered. Surface debris and contaminants generated during zeolite synthesis or binding may result in pore blockage and inaccessibility of reactants to the metal sites.[11] Certain amorphous metal oxide contaminants may also alter the regenerability of the catalyst. The low S tolerance of this catalyst has been shown to readily agglomerate Pt thus blocking the pores.[12] Operating variables are also important as they were for reforming, however, the unique aspect of this catalyst system is its sensitivity to loss of performance due to pore blockage.

Hydrotreating and S removal of feedstocks will become increasingly more important as feedstock quality deteriorates.[13] The catalyst commonly used in this process contains sulfided Co and Mo supported on high surface area alumina. There have been studies which established the structure of MoS_2 on alumina and chemical state and functionality of Co.[14,15] Although there is continuing debate as to how the active site desulfurizes molecules, there are several factors that can influence performance. The type of feedstock is particularly important since it can range from low to high metals and molecular weight hydrocarbons.[16]

Proc. Microscopy and Microanalysis 1995, edited by G.W. Bailey, M.H. Ellisman, R.A. Hennigar, and N.J. Zaluzec
Copyright © 1995 MSA. Published by Jones and Begell Publishing, 79 Madison Ave., New York, NY 10016

Depending on the metal content and feedstock composition, activity and stability can be dramatically altered by means of metal and coke deposition.[17] Extent of sulfiding the fresh catalyst can also alter performance.[18] Presence of arsine or phosphine in the feedstock can deactivate the catalyst.[19,20] Excessive coking from heavier feedstocks can block large pores in the alumina thus resulting in loss of activity.[21] Difficulty in desulfurizing feedstocks has been correlated to the structure of S compounds.[22]

Thus just focusing on the metal cluster is insufficient for understanding many catalyst systems. By being aware of the various factors, discussed here, the microscopist should be able to obtain the appropriate information to devise a strategy for evaluating the catalyst.

1.	A.K. Dayte, *Catal. Rev., - Sci. Eng.* 34(1&2)(1992)142.

2.	F.G. Ciapetta and D.N. Wallace, *Catal. Rev.* 5(1972)102.

3.	J. Biswas et.al., *Catal. Rev. - Sci. Eng.* 30(2)(1988)194.

4.	P.G. Menon and T.S.R. Rao-Prasade, *Catal. Rev. - Sci. Eng.* 20(1979)97.

5.	J. Barbier et. al, *Adv. Catal.* 37(1990)298.

6.	J. Barbier and P. Marecot, *J. Catal.* 102(1986)21.

7.	C. Kim and G.A. Somorjai, *J. Catal.* 134(1992)179.

8.	J. Biswas et. al., *Catal. Rev. - Sci. Eng.* 30(2)(1988)192.

9.	W.M.H. Sachtler and Z. Zhang, *Adv. Catal.* 39(1993)129.

10.	E. Mielczarski et. al., *J. Catal.* 134(1992)368.

11.	S. Bhatia et. al., *Catal. Rev. - Sci. Eng.* 31(1989)431.

12.	G.B. McVicker et. al., *J. Catal.* 139(1993)48.

13.	W.J. Eaton et. al., *Cat. Rev. - Sci. Eng.* 33(3&4)(1991)282.

14.	C. Wivel et. al., *J. Catal.* 68(1981)453.4)(1991).

15.	H. Topsoe and B.S. Clausen, *Catal. Rev. - Sci. Eng.* 26(3&4)(1984)395.

16.	W.J. Eaton et. al., *Catal. Rev. - Sci. Eng.* 33(3&4)(1991)308.

17.	Ibid. 295.

18.	J.van Gestel et.al., *J. Catal.* 145(1994)429-436.

19.	S. Bouwers et.al., *J. Catal.* 138(1991)559-568.

20.	R. Merryfied et. al., *ACS Symposium Series* 288(1984)1-14.

21.	T. Ohtsaka, *Catal. Rev. - Sci. Eng.* 16(2)(1972)315-316.

22.	C. Lee and D. Ollis, *J. Catal.* 87(1984)332.

QUANTITATIVE NANO-CHARACTERIZATION OF HETEROGENEOUS CATALYSTS

P.A. Crozier and M. Pan*

Center for Solid State Science, Arizona State University, Tempe, AZ 85287-1704
*Gatan Inc, 6678 Owens Dr., Pleasanton, CA 94588

Heterogeneous catalysts can be of varying complexity ranging from single or double phase systems to complicated mixtures of metals and oxides with additives to help promote chemical reactions, extend the life of the catalysts, prevent poisoning etc. Although catalysis occurs on the surface of most systems, detailed descriptions of the microstructure and chemistry of catalysts can be helpful for developing an understanding of the mechanism by which a catalyst facilitates a reaction. Recent years have seen continued development and improvement of various TEM, STEM and AEM techniques for yielding information on the structure and chemistry of catalysts on the nanometer scale. Here we review some quantitative approaches to catalyst characterization that have resulted from new developments in instrumentation.

HREM has been used to examine structural features of catalysts often by employing profile imaging techniques to study atomic details on the surface. Digital recording techniques employing slow-scan CCD cameras have facilitated the use of low-dose imaging in zeolite structure analysis and electron crystallography. Fig. 1a shows a low-dose image from SSZ-33 zeolite revealing the presence of a stacking fault. The image was recorded with a JEOL 4000 EX equipped with a Gatan 679 slow-scan CCD camera. The change in stacking sequence gives rise to the formation of two different polymorphs in this material. Details of the structural units for each polymorph can be visualized more clearly by performing real-space averaging of the unit cells in each domain. Fig 1b and c show the projected unit cells from polymorph A and B containing structural information on the 2 Å scale. Such images from zeolites are often simply related to the projected potential and can play a key role in determining the structure of new materials (Lobo et al, 1994).

It is often necessary to make a large number of measurements on a catalysts in order to obtain a statistically significant description of microstructural properties. This is the case in heavy metal particle systems where it is desirable to know the particle size distribution. Annular dark field (ADF) or Z-contrast imaging is an established technique for visualizing heavy metal particles on light oxide supports. Fig 2a is an ADF image recorded from a VG HB5 STEM showing Pt particles over a zeolite support. Particle sizes down to 1 nm are seen in this image and the particle size distribution is shown in Fig 2b. For particles on the order of the probe diameter (< 1nm) it is possible to determine the size of the clusters using integrated intensity measurements although signal/noise problems may limit the sensitivity (Treacy and Rice, 1989). Although Z-contrast imaging is a powerful method for revealing heavy elements on a light element substrate it is not useful when the catalysts consist of a number of metals of similar atomic number. Spectroscopic techniques such as energy-filtered imaging can be employed to determine particle size distribution from such materials. Fig 3a is an Cr elemental image from a NiCuCr amination catalyst obtained by use of a Zeiss 912 equipped with an omega filter. The particle size distribution is shown in fig 3b. The technique is limited to a smaller range of sample thickness compared to ADF. However, energy-filtered imaging is a powerful method for sampling large areas at resolutions on the order of a few nanometers (Crozier, 1995).

References

1. R.F. Lobo et al, *J. Phys. Chem.* 98(1994)12040.
2. M. M. J. Treacy and S. B. Rice, *J. Microscopy.* 156(1989)211
3. P. A. Crozier, *Ultramicroscopy* (in press)
4. This work was supported by the HREM Industrial Associates Program of Arizona State University (ASU) and was performed in the Center for HREM at ASU.

Proc. Microscopy and Microanalysis 1995, edited by G.W. Bailey, M.H. Ellisman, R.A. Hennigar, and N.J. Zaluzec
Copyright © 1995 MSA. Published by Jones and Begell Publishing, 79 Madison Ave., New York, NY 10016

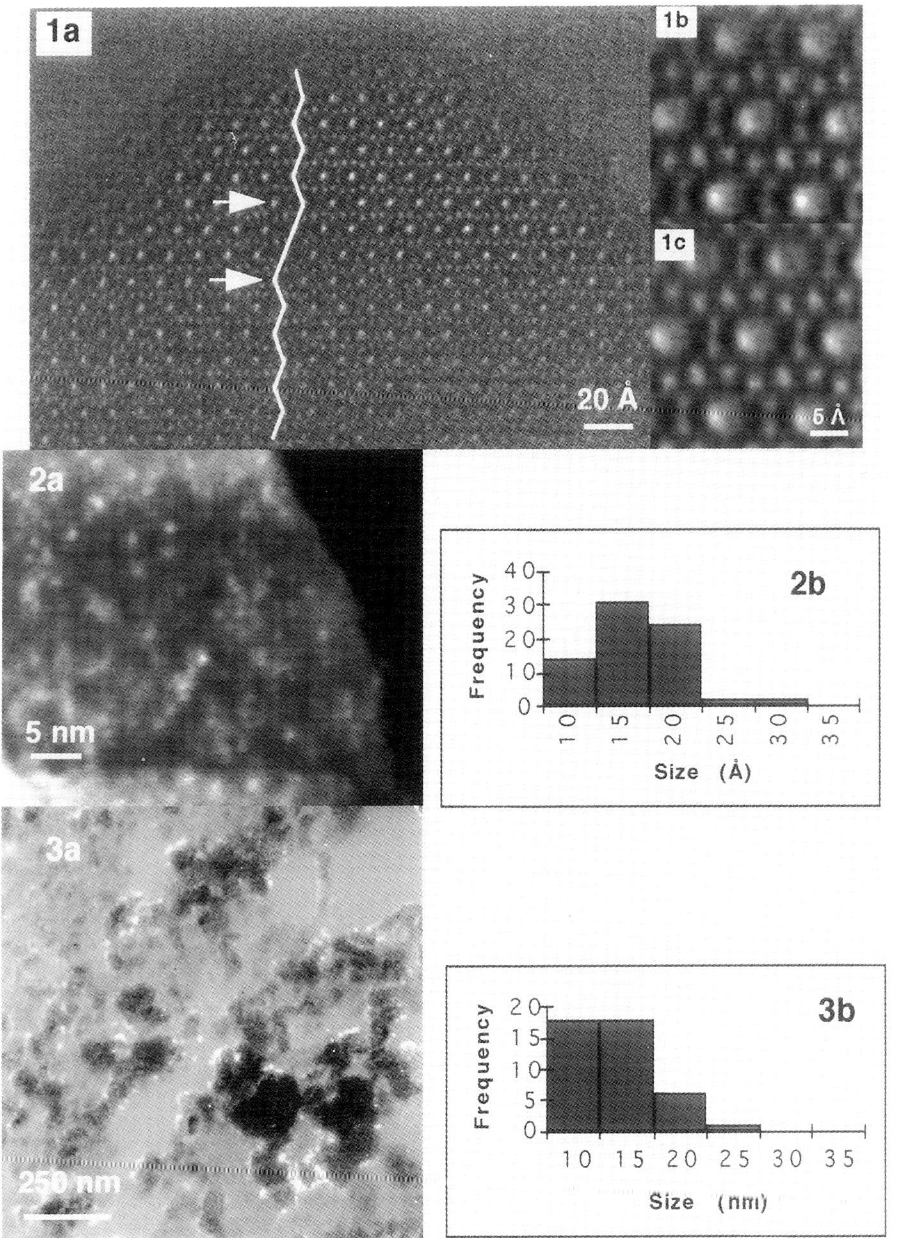

Fig 1:a) Digital HREM image from SSZ-33 as acquired with slow-scan CCD camera. b) and c) Project unit cells determined from a) of two polymorphs in SSZ-33.
Fig 2: a) Annular dark-field image of Pt on zeolite support. b) Distribution of particle sizes determined from image.
Fig 3: a) Elemental map showing Cr particle in NiCuCr catalyst determined by energy-filtered imaging. b) particle size distribution for Cr.

MICROSTRUCTURAL EVOLUTION OF γ-ALUMINA SUPPORTED RHODIUM CATALYSTS

Zara Weng-Sieh*,☆ Ronald Gronsky*,☆ and Alexis T. Bell**,☆

* Department of Materials Science and Mineral Engineering, University of California Berkeley, Berkeley, CA 94720-1760
** Department of Chemical Engineering, University of California Berkeley, Berkeley, CA 94720
☆ Materials Sciences Division, Lawrence Berkeley Laboratory, Berkeley, CA 94720

In an era of increasing environmental awareness, stricter federal and state regulation of pollutant emissions are emerging.[1] A major source of pollution arises from automobiles which inadvertently form gaseous products such as nitric oxide (NO_x), carbon monoxide (CO), and hydrocarbons (HC).[2] Since the early 1980's, these effluents have been converted to safer forms using a three-way catalytic converter that employs a high dispersion of rhodium and platinum particles supported on a large surface area of transitional γ-phase alumina. Unfortunately such a converter is susceptible to decreased performance over time, and this degradation has been attributed to changes in the catalyst microstructure.[2] The nanoscaled nature of the transition metal catalysts and the submicron-scaled size of the transitional alumina necessitates the use of the high spatial resolution analyses made possible by transmission electron microscopy.

A typical high-resolution TEM image of the as-prepared rhodium/alumina catalyst reveals a highly dispersed distribution of fine rhodium particles on a fine-grained γ-alumina substrate (figure 1). The rhodium particles exhibit increased diffraction contrast (due to higher Z), appearing darker than the alumina support. Particle size determination using an image processing method described in reference [3] reveals an average diameter near 9Å. Although at focus conditions deviating from Scherzer, accurate measurement becomes difficult due to strong Fresnel fringe interference [4]. Short duration aging of γ-alumina supported rhodium catalysts at 500 °C in air for 40 hours results in the slight increase in Rh particle size (to 12Å). Upon further aging (500°C for 56 hours), the Rh oxide particle exhibits an ultradispersed raft-like morphology, as shown in figure 2, of approximately one monolayer thickness and 15 Å in width. Aging at 850°C for 25 hours results in the coarsening and phase transformation of the rhodium oxide to its high temperature orthorhombic Rh_2O_3 polymorph (figure 3).

Concomitantly, the transitional γ-alumina support transforms towards its more stable a-phase via the formation of δ- and θ- phases (respectively). One such occurrence of the δ-alumina phase (figure 4) resulted from 56 hours of aging at 500°C, which is approximately 200°C below the homogeneous transformation temperature. This δ-alumina phase is imaged with the incident beam along its [001] direction. Upon aging at 1000°C for 5 hours, the formation of the thermodynamically stable α-Al_2O_3 phase results in the densification of the support (figure 5).

REFERENCES

1. Clean Air Act Amendments of 1990, Detailed Summary of Titles, U. S. EPA, Nov. 30, 1990.
2. K. C. Taylor, Catalysis Reviews— Science & Engineering, **35**, 457 (1993).
3. J. C. Russ, <u>Computer-Assisted Microscopy</u>, Plenum Press, NY, 1990.
4. V. Radmilovic and M. A. O'Keefe, submitted to 53rd MSA conference (1995).
5. Z. Weng-Sieh is thankful for support through a Noyce Foundation Fellowship. The access to the facilities at the National Center for Electron Microscopy is gratefully acknowledged. This work is supported by the Director, Office of Energy Research, Office of Basic Energy Sciences, Materials Sciences Division of the U.S. Department of Energy under contract No. DEAC0376SF00098.

Proc. Microscopy and Microanalysis 1995, edited by G.W. Bailey, M.H. Ellisman, R.A. Hennigar, and N.J. Zaluzec
Copyright © 1995 MSA. Published by Jones and Begell Publishing, 79 Madison Ave., New York, NY 10016

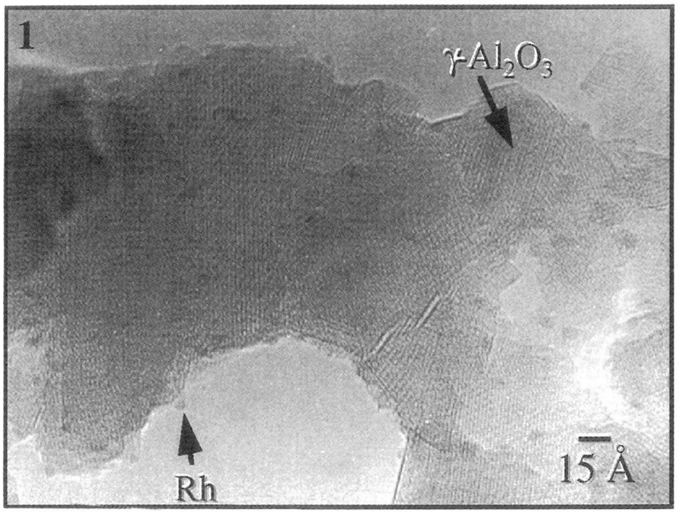

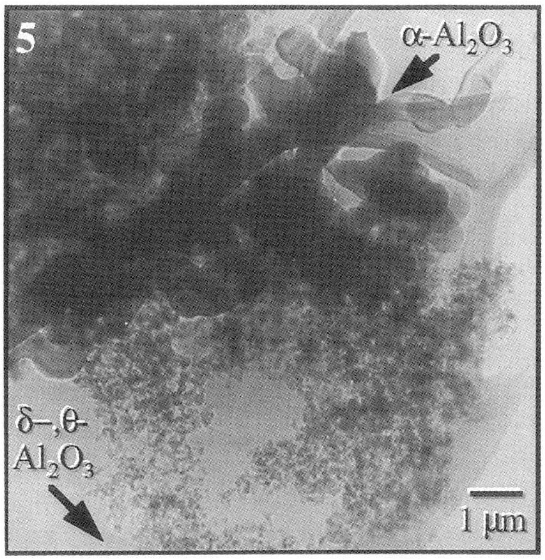

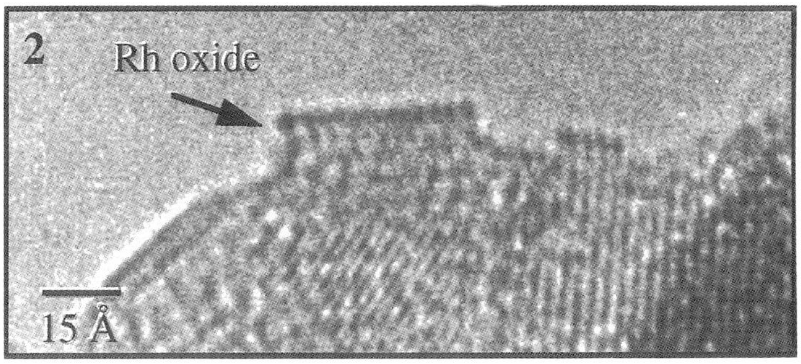

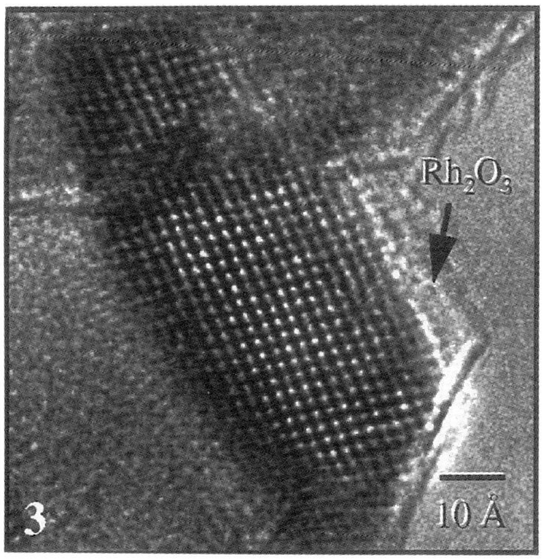

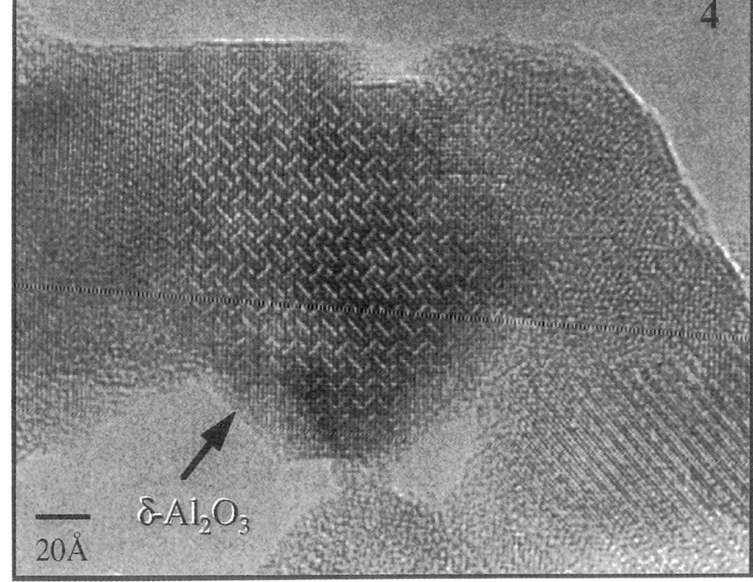

Figure 1- Fresh Rh/γ-alumina catalyst.

Figure 2- Formation of a raft-like
Rh-oxide phase observed
after 56 hours of aging at 500°C.

Figure 3- Coarsening of rhodium
oxide and transformation to its
high temperature polymorph,
orthorhombic Rh_2O_3,
after 25 hours of aging at 850°C.

Figure 4- Grain growth,
sintering and transformation
of the γ-Al_2O_3 support towards
its stable α-Al_2O_3 phase. This
δ-Al_2O_3 grain was observed after
56 hours of aging at 500°C

Figure 5- Densification
of the support upon formation
of α-Al_2O_3 after aging 5 hours
at 1000°C.

ADVANTAGES OF MODERN FIELD EMISSION ELECTRON MICROSCOPES FOR CHARACTERIZATION OF CATALYTIC MATERIALS

L. F. Allard,* S. A. Bradley,** E. Völkl* and S. J. Pennycook*

*Oak Ridge National Laboratory, Oak Ridge, TN 37831-6064
**UOP, Des Plaines, IL 60017-5018

The ability to characterize the structure of a variety of catalytic materials, especially supported metal catalysts, has been dramatically enhanced in recent years by the improved electron microscopy capabilities provided by the new generation of field emission gun transmission (FE-TEM) and dedicated scanning transmission electron microscopes (d-STEM). Although the reciprocity theorem[1] says that a TEM and a dedicated STEM have equivalent, but reciprocal, electron optical geometries so that imaging conditions in one instrument can, in principal, be duplicated in the other, in reality the TEM and d-STEM offer column geometries that are optimized for different kinds of analytical output. For example, the d-STEM permits introduction of an energy dispersive spectrometer, which collects x-rays at a solid angle of 0.2 sr (but at a low take-off angle). A high take-off angle detector in a TEM may give a collection angle of only 0.013 sr. This x-ray collection efficiency coupled with use of annular dark-field techniques[2] makes the imaging and analysis of nanometer-sized metal clusters on supports more effective. However, among several other conveniences, an FE-TEM offers better facility for a technique such as electron holography that is proving useful in the characterization of fine particles in the one to several nanometer range.[3] This paper details the advantages of electron holography techniques in the Hitachi HF-2000 cold FE-TEM for determination of the structure and morphology of nanocrystals, and the use of high-angle annular dark-field imaging in a VG HB-601 dedicated STEM to determine the presence and distribution of sub-nanometer-sized heavy metal clusters on oxide supports.

Electron holography[4] relies on the use of a Möllenstedt biprism placed appropriately in the imaging system so that a reference wave passing one side of the biprism can be bent to interfere with the wave passing through the specimen to form a hologram. The hologram permits the amplitude and phase images to be reconstructed separately, from processing of a Fourier transform of the hologram. For thin objects, the phase component is essentially directly proportional to the thickness t and is related to the mean inner potential V_i of the material by $\Delta\phi = C_E(V) \cdot V_i \cdot t$ where $C_E(V)$ is a constant related to the microscope voltage. Figure 1 shows the phase image reconstructed from a hologram of a nanocrystal of Pd deposited onto the surface of an amorphous silica microsphere (a model combustion catalyst). The particle exhibits a central contrast feature presumed to be a void. Figure 2 shows a phase profile (black line) taken across the phase image as shown, and this matches perfectly with the phase profile (thick grey line) computed using a sphere as the model structure with an internal spherical void present with geometry taken from the original image.[3] This close match is convincing evidence for the presence of a void, and illustrates the power of electron holography to characterize both internal as well as external particle morphologies at the near atomic level.

Many catalyst systems employ low loadings of heavy metals on oxide supports, such as Pt/Al_2O_3, typical of automotive catalysts. Figure 3 shows a bright-field TEM image of such a material taken under optimal conditions , where the Pt loading was 0.75 wt-%. Although several dark spots are seen, no specific assignment can be made to the presence of Pt in those dark spots (which are on the order of 0.8 nm). Even on a FE-TEM precise positioning of the electron beam on an individual spot of such a small size for EDS work is difficult. The lower x-ray collection efficiency can preclude obtaining useful EDS data from particles of this dimension. A BF image of the same material taken in the HB-601 is shown in Fig. 4a, and a similar result is obtained. However, an annular dark field image, which gives mainly amplitude contrast arising from diffuse high-angle Rutherford scattering, of the same area (Fig. 4b) clearly reveals that a distribution of bright spots suggesting a 0.8-1 nm cluster size are likely to be the Pt particles. These clusters can with reasonable facility be located with a point probe (0.22 nm) at high magnifications on the raster image so that EDS data can be taken to confirm the presence of Pt. The high collection efficiency of the d-STEM for x-ray data permits useful EDS results to be obtained from such sub-nanometer particles. However, even though the x-ray geometry appears less favorable,a significant Pt signal can be collected in a 50 s analysis of a particle 1 nm or smaller in this same material, using the HF-2000.[5]

Proc. Microscopy and Microanalysis 1995, edited by G.W. Bailey, M.H. Ellisman, R.A. Hennigar, and N.J. Zaluzec
Copyright © 1995 MSA. Published by Jones and Begell Publishing, 79 Madison Ave., New York, NY 10016

References

1. J. M. Cowley, *Appl. Phys. Lett.* 15(1969)58.
2. S. B. Rice and S. A. Bradley, *Catalysis Today* 21(1994)71.
3. L. F. Allard, E. Völkl, D. Kalakkad and A. K. Datye, *J. Mater. Sci.* 29(1994)5612.
4. H. Lichte, in T. Mulvey and C. J. R. Sheppard, eds., *Adv. in Opt. and Elect. Micro.,* Acad. Press 12(1991)25.
5. Some research sponsored by the Laboratory Directed Research and Development Program of Oak Ridge National Laboratory, managed for the Department of Energy by Martin Marietta Energy Systems, Inc. under contract DE-AC05-84OR21400.

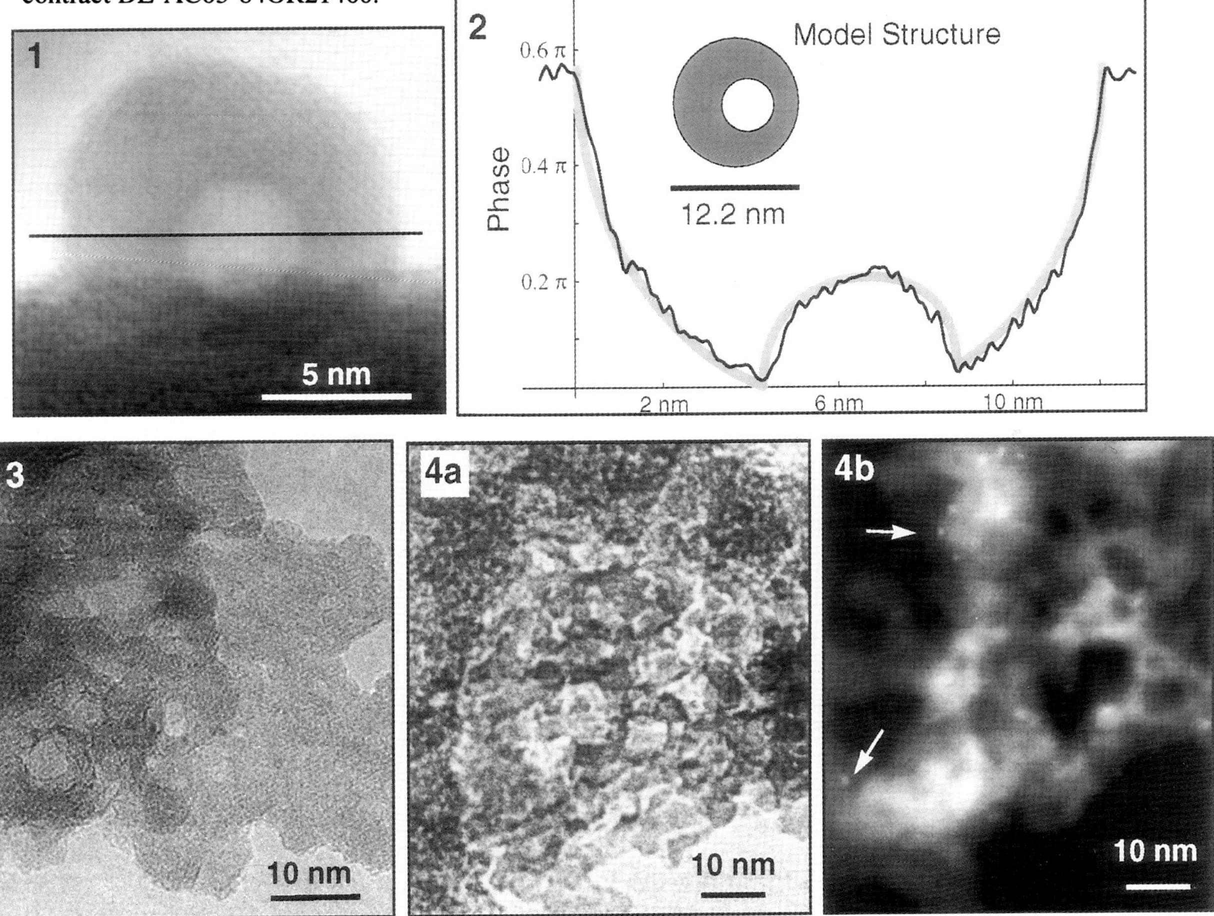

Fig. 1.--Phase image reconstructed from hologram of a nanocrystal of Pd on an amorphous silica microsphere, showing line of trace for phase profile.

Fig. 2.--Experimental phase profile from Pd nanocrystal (solid line) showing close match to phase profile computed as described in text (grey line). This match confirms presence of void in particle interior.

Fig. 3.--Bright-field FE-TEM image of 0.75 wt-% Pt on γ-alumina support. No clear distribution of discrete particles is evident, although some dark spots in the 0.5-0.8 nm size range seen.

Fig. 4a.--Bright-field image from d-STEM, also showing no clear distribution of Pt particles as a result of interference from contrast generated by the underlying support material.

Fig. 4b.--High-angle annular dark-field image from the d-STEM showing clear evidence for a distribution of clusters on the order of 0.8-1 nm in size (e.g. as arrowed).

COMPOSITION-SIZE DISTRIBUTION DIAGRAMS FOR ALLOY CATALYSTS

C. E. Lyman,* R. E. Lakis,*+ and H. G. Stenger, Jr.**

Departments of *Materials Science and Engineering and **Chemical Engineering,
Lehigh University, Bethlehem, PA 18015; +Present address: Laboratory for Research
on the Structure of Matter, University of Pennsylvania, Philadelphia, PA 19104-6202

Catalyst microstructure and catalytic properties can be directly correlated using a composition-size distribution diagram that maps the phases of small alloy particles by composition and size.[1] The application of this diagram given here concerns Pt-Rh/alumina catalysts for reduction of NO in hydrogen at low temperatures. The composition-size diagram provides an understanding of why some alloy catalysts perform worse than pure Pt in the NO reduction reaction, while other preparations exhibit the several times the activity of pure Pt. The composition-size diagram can be used to "fingerprint" active and less active alloy catalysts. Analytical electron microscopy (AEM) has been used to determine the Pt-Rh particle composition-size distribution.

Catalysts were prepared by aqueous impregnation of platinum chloride into gamma-alumina particles with a 4 nm average pore size. After impregnation and air drying, the catalyst was calcined in air at 500°C for 3 hr. Subsequently, rhodium was impregnated from its chloride into the same support material, and the preparation was air-dried. Finally, the catalyst was reduced in hydrogen for 12 hr at 300°C. One catalyst prepared by this route was designated Catalyst 60/40 because in bulk it contained 0.61 wt% Pt and 0.40 wt% Rh, and the relative amounts of Pt and Rh were 60% and 40%. Each catalyst in this series contained about 1 wt% total metal, similar to commercial loadings. A second catalyst was prepared in a similar manner but with a composition designated 95/5. A third catalyst, Catalyst 17/83, was prepared with a higher Rh content. In a fourth preparation, the Rh was impregnated first, followed by calcination and impregnation of Pt (Catalyst 75/25).

The microstructure of each catalyst was examined after several reaction runs. X-ray emission spectrometry was performed in a VG HB-501 STEM at 100 kV with a Link windowless Si(Li) x-ray detector. Figure 1a shows a typical area of Pt-Rh particles. Particles less than 5 nm were analyzed whole (Figure 1b), while particles larger than 5 nm were analyzed in the center (Figure 1c). X-ray counts were collected from each particle until the error in composition from counting statistics was less than 15%. ˙

Figure 2a shows the particle composition-size diagram for Catalyst 60/40 which forms a bimodal composition distribution of Pt-rich and Rh-rich particles. Few particles were observed to have the average 60 wt% Pt bulk composition of the particles. Catalyst 95/5 (Figure 2b) exhibits only Pt-rich alloy particles of high Pt content. Catalyst 17/83 consists primarily of Rh-rich particles (Figure 2c). Catalyst 75/25 (Figure 2d) consisted of Pt-rich alloy particles and a few large Rh-rich particles assumed to be rhodium oxide. The activities of these catalysts for NO reduction in hydrogen are in the following order: 95/5 > 75/25 > pure Pt > 17/83 > pure Rh. Thus, the Pt-rich particle microstructures of Figures 2b and 2d produce higher activity than the other microstructures. The increasing Pt-content with particle size has been related[2] to surface segregation of a partial monolayer of Rh on the particle surface. This supports the hypothesis that the active Pt-Rh alloy particle surface contains both exposed Pt atoms to adsorb hydrogen and exposed Rh atoms to adsorb NO.[3]

References

1. R. E. Lakis et al., Proc. 13th Int. Cong. Elect. Micro., Paris, 1994, 1075.
2. R. E. Lakis et al., Proc. 52nd meeting Micro. Soc. Am., 1994, San Francisco Press, 780.
3. This work was supported by U.S. Dept. of Energy contract DE-FG02-86ER45269.

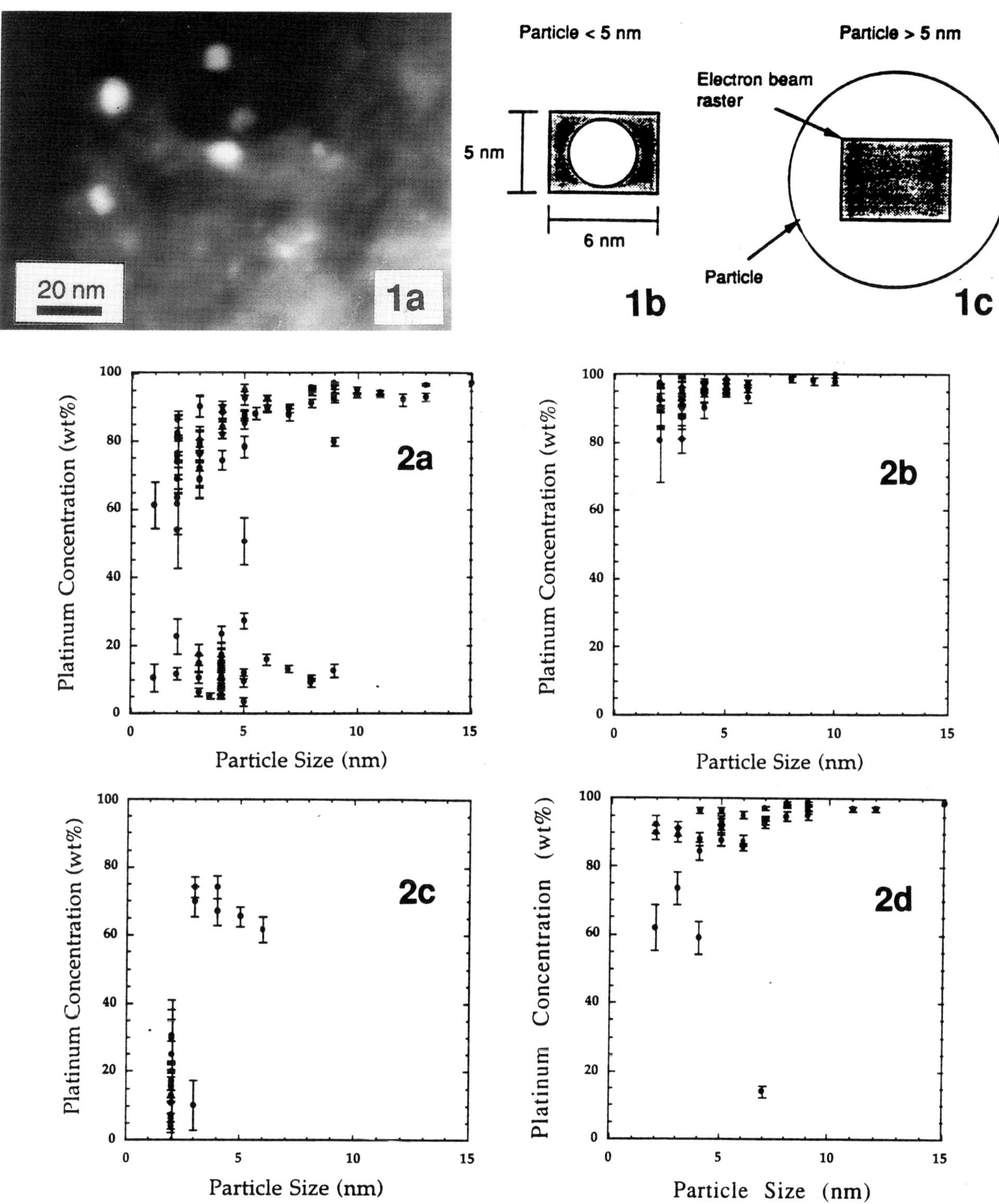

FIG. 1. -- (a) Annular dark-field STEM image of Pt-Rh alloy particles; (b) Effective raster size for whole particles less than 5 nm; (c) Larger particles sampled at the center of the particle.
FIG. 2. -- Composition-size diagrams for (a) 60/40 catalyst showing bimodal phase distribution (activity < pure Pt); (b) 95/5 catalyst (activity > pure Pt); (c) 17/83 catalyst (activity < 60/40 catalyst); (d) 75/25 catalyst (activity > pure Pt).

405

IN OR ON? LOCATION OF NOBLE METAL PARTICLES ON ZEOLITE/BINDER SUPPORT.

G. E. Spinnler,* J. Liu.**

*Shell Development Co., P. O. Box 1380, Houston TX 77251-1380
**Arizona State University, Tempe, AZ 85287

The location of metal particles contained in a zeolite-alumina binder support matrix has been difficult to solve using analytical techniques including electron microscopy. Imaging of metal particles, particularly noble metal particles on relatively light matrices such as zeolites or aluminas, has been easily accomplished using high angle annular darkfield imaging (HAADF). Since transmission imaging provides a projection through the sample, location of the particles in the sample or on the surface is not obvious. Surface-sensitive signals such as secondary electrons (SE) and Auger electrons (AE) are necessary to detect particles on the sample surface. HAADF, SE, and AE imaging have been applied to locate noble metal particles in a zeolite support with an alpha alumina binder.

The samples were analyzed in a UHV HB501S STEM (MIDAS, Microscope for Imaging and Diffraction Analysis of Surfaces) at the Center for High Resolution Electron Microscopy at Arizona State University.[1] The samples were prepared by crushing and dry deposition on a holey carbon grid. The samples were imaged with a variety of STEM signals, HAADF, BFSTEM, SEM, and Auger electrons. All images were recorded under conditions of 100 kV accelerating potential with a -500 V bias applied to the sample. The probe size was approximately 1 nm and the probe current was approximately 0.1 nA. The HAADF signal depends strongly on atomic number, thus the noble metal particles appear as bright areas on a diffuse background. The secondary electron signal on this microscope is surface-sensitive with a resolution similar to the HAADF signal.[2] The Auger electron images were collected using the noble metal's MNN line and have a resolution of approximately 1 nm .[3]

Figure 1a is a HAADF image showing the noble metal particles as bright spots located uniformly across the support. The corresponding SE image (1b) shows a topographically rich surface texture with no indication of metal particles corresponding to the HAADF image. The needle-like features in the SE images are the alpha alumina particles, sitting on top of the zeolite support. Metal particles are visible in the HAADF image of Figure 2a and the SE image of Figure 2b (arrows). The particles observed in the SE images are believed to be located on the surface of the support, in this case the zeolite. The needle-like morphology of the alumina is not evident. Figure 3a is a HAADF image that shows the metal particles uniformly located across the support particle. Figure 3b is an Auger spectral image. Particles corresponding to those in the HAADF image are imaged using an energy selected window around the corresponding MNN Auger peak of the metal particle. Metal particles are imaged in many of the same locations as seen in the HAADF image(arrows). Particles that are not imaged in the SE and the AE images, and do show up on the HAADF images, may be on the underside of the particles, not accessible by the SE and AE images in these imaging modes.

In summary, by using the combination of HAADF, SE and AE images, the location of noble metal particles has been unambiguously determined to be on the outside of the zeolite support and beneath the alumina support. The strength of this approach is in the combination of signals, and the near simultaneous collection of these signals for the same sample without the need of changing instruments.

References

1. G. G. Hembree et al., *Ultramicroscopy* 31 (1989) 111.
2. J. Liu and G. E. Spinnler, *Proc. 49th Annual Meeting of EMSA* (1991) 502.
3. J. Liu et al., *Ultramicroscopy*   52 (1993) 369.
4. This work was conducted at CHREM at ASU supported by NFS-DMR-9115680.

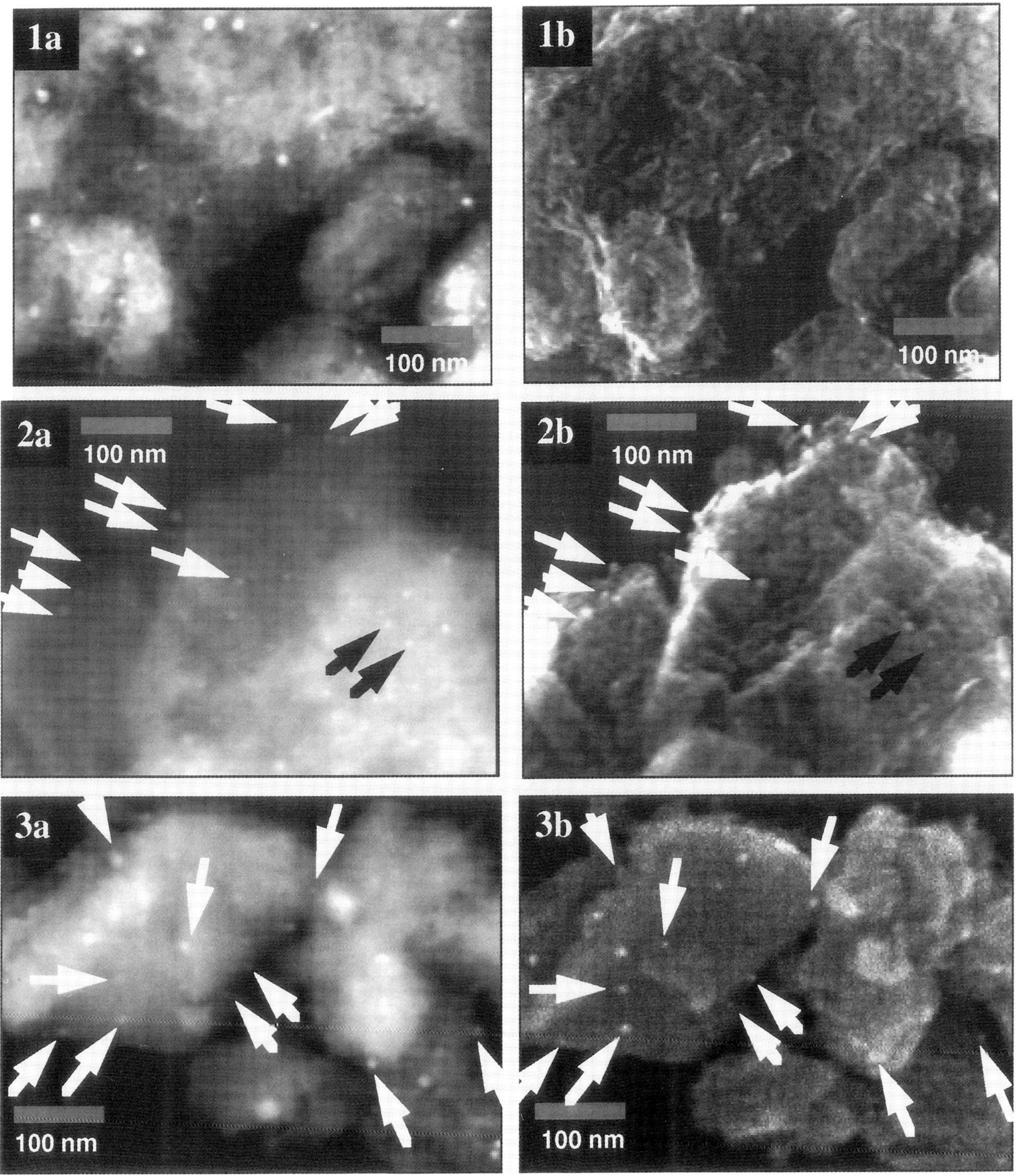

FIG. 1and FIG 2.-(a) HAADF STEM image of noble metal patticles on zeolite/alumina support; (b) SEM image of the area in (a).
FIG. 3.-(a) HAADF STEM image; (b) Metal MNN Auger image of noble metal particles on zeolitesupport.

TRANSMISSION ELECTRON MICROSCOPY STUDIES OF CATALYTICALLY GROWN CARBON NANOSTRUCTURES

R. T. K. Baker and N. M. Rodriguez
Materials Research Laboratory, The Pennsylvania State University,
University Park, PA 16802

In recent years a great deal of research effort has been devoted to the synthesis of carbon structures possessing extensive crystalline order [1]. Methods used for the preparation of such materials center around arc-discharge procedures, however, separation and purification of the desired product has proven rather difficult [2]. Studies carried out in our laboratory have led to the developement of an alternative approach, which involve the participation of a metal catalyst for the production of various carbonaceous structures. Carbon nanofibers are the product of the catalytic decomposition of selected hydrocarbons and carbon monoxide over metal particles [3]. In this process, the hydrocarbon is adsorbed and decomposed on certain faces of the metal, carbon species diffuse through the particle to precipitate at another set of faces to form graphite as illustrated in Figure 1. The structure of the nanofiber is therefore dependent on the geometric orientation and the number of carbon precipitating faces in the metal particle. Changes in the conformation of the nanofibers can be brought about by altering in the pattern of the precipitation step and it is not surprising to find that helixes, coils and ribbons can be formed.

The structural characterization of carbon nanofibers was carried out in a JEOL 2000EXII transmission electron microscope (lattice resolution = 0.14 nm), from which it was possible to determine that the structures consist of graphite platelets aligned in various directions (parallel, perpendicular and at an angle) with respect to the fiber axis. The alignment and perfection of the nanofiber structure has been found to be dependent on a number of factors including, the nature of the catalyst, the composition of the reactant gas and the temperature.

Figure 2 is an electron micrograph of a carbon nanofiber, where the electron dense metal particle responsible for its growth is located within the structure. Graphite platelets have simultaneously precipitated from a pair of faces at both sides of the particle and are aligned at an angle with respect to the fiber axis. By choice of a suitable catalyst it is possible to precipitate the graphite in a direction parallel to the fiber axis and under these conditions the structures that are created are in the form of nanotubes. The role played by the use of high resolution transmission electron microscopy in elucidating the various structural arrangements and concomitant catalyst particle characteristics will be discussed in this presentation.

References

1. S. IIjima, *Nature* 56, (1991) 354.
2. M. Yumura, K. Uchida, H. Niino, S. Ohshima, Y. Kuriki, K. Yase and F. Ikazaki, in *Novel Forms of Carbon II*, C.L. Renschler, D. M. Cox, J. J. Pouch and Y. Achiba Editors, Mat. Res. Soc. Symp. Proc. Vol 349, (1994) 231.
3. N. M. Rodriguez, *J. Mater. Res.* 8 (1993) 3233.

Support from the Department of Energy, Basic Energy Sciences Grant DE-FG02-93ER14358 is gratefully acknowledged.

*Proc. Microscopy and Microanalysis 1995, edited by G.W. Bailey, M.H. Ellisman, R.A. Hennigar, and N.J. Zaluzec
Copyright © 1995 MSA. Published by Jones and Begell Publishing, 79 Madison Ave., New York, NY 10016*

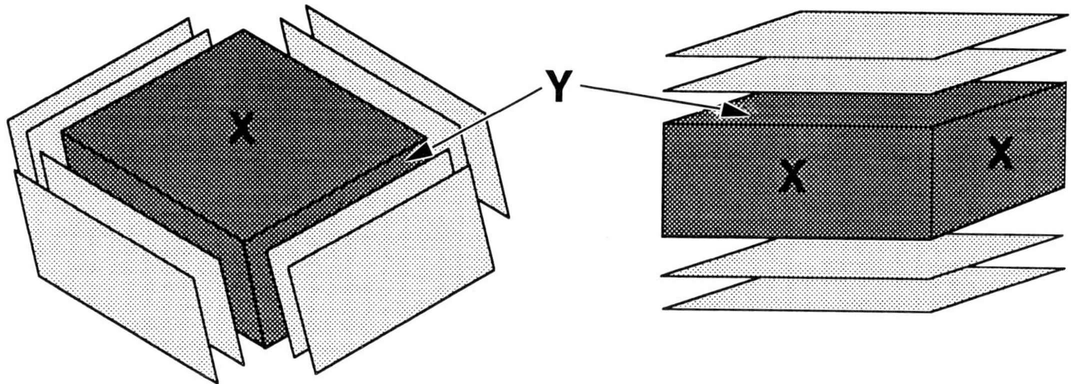

Figure 1. Schematic representation of two metal particles associated with the formation of graphite nanofibers: (X) are faces that decompose hydrocarbons, and (Y) are faces that can precipitate carbon in the form of graphite.

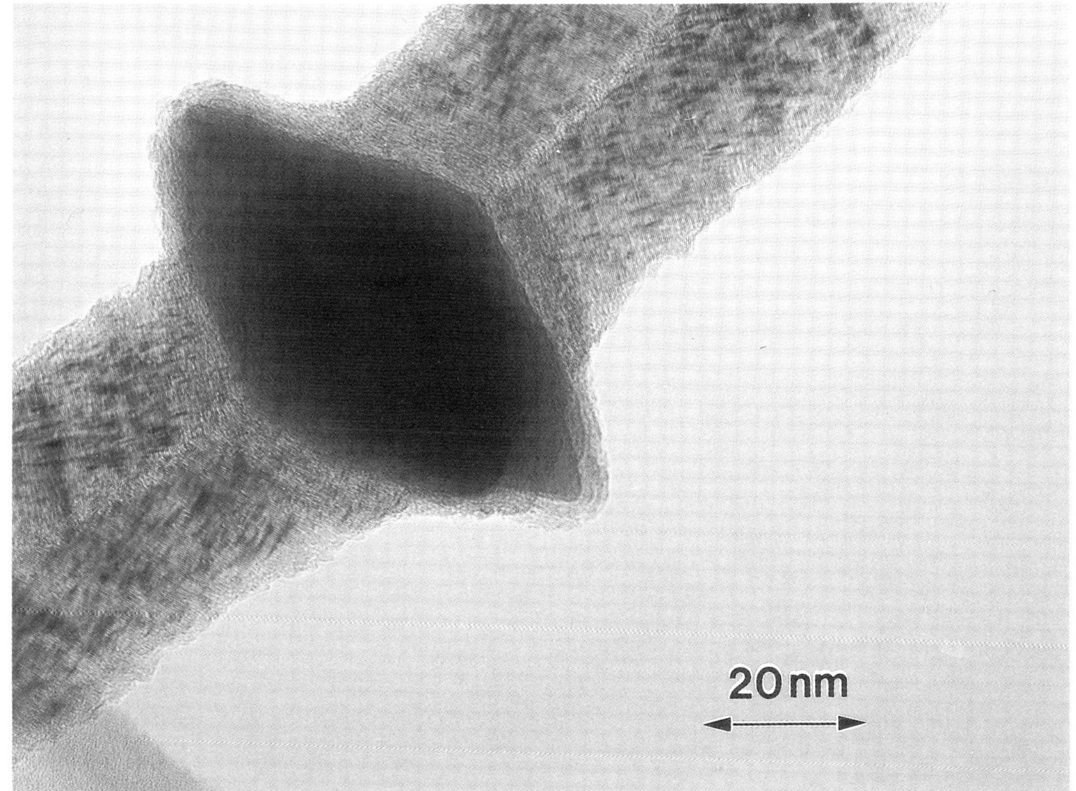

Figure 2. Transmission electron micrograph of a carbon nanofiber that has been grown from the interaction of an ethylene/hydrogen mixture with an iron-based catalyst at 600°C.

TEM PHASE TRANSFORMATION STUDY OF $(VO)_2P_2O_7$ CATALYST

Z. G. Li*, C. C. Torardi*, H. S. Horowitz**

DuPont *Central Research and Development and **Speciality Chemicals, Experimental Station
P. O. Box 80228, Wilmington, DE 19880-0228

$(VO)_2P_2O_7$ (VPO) is known to act as a heterogeneous catalyst in the oxidation of n-butane to maleic anhydride [1]. DuPont has developed a new circulating fluid bed process for this reaction, and first commercial plant using this technology is scheduled to start up in early 1996 [2]. VPO catalyst is typically formed by transforming the vanadyl hydrogen phosphate precursor ($VOHPO_4 \cdot 1/2H_2O$) to the vanadyl pyrophosphate active phase. It was the object of the current study to develop a more fundamental understanding of this phase transformation [3].

Transformation conditions were 500° - 700° C in nitrogen. The hydrated precursors used in this study were in the form of plate-shaped single crystals. The crystallographic convention is given in Table 1. Microstructural and morphological changes associated with the phase transformation were studied at scales ranging from the mm to the nm range using scanning electron microscopy (Hitachi S-4000) and transmission electron microscopy (JEM 2000FX). Figure 1a is an SEM showing the corner of a precursor crystal plate with the c axis indicated. Figure 1b shows the same corner of the resulting psuedomorph after transformation at 500° C. The shape of the crystal has not changed and measurement of crystal dimensions before and after transformation confirmed that the size also was not altered. At a higher magnification (Figure 1c) it is observed that the transformed crystal contains many cracks, most running parallel to the plate surface. Thicknesses of these cracks ranged from less than 0.1 to around 1 micron. Figure 2 shows a TEM image and SAED pattern of a similar $(VO)_2P_2O_7$ pseudomorph with the electron beam directed parallel to the plate surface. Variations in contrast in the TEM image clearly show the microcrystalline nature of this pseudomorph, with microcrystals in slightly different orientations giving rise to the mottled appearance. Some of the brighter areas are due to voids generated during the phase transformation. A high resolution TEM image of the $(VO)_2P_2O_7$ pseudomorph is shown in Figure 3. It illustrates that while the starting single crystals were converted to microcrystalline pseudomorphs, within each constituent micro-crystal the atomic periodicity is still quite high.

Electron microscopy clearly shows the transformation to be topotactic: single crystals are converted to pseudomorphs which are unchanged in size or shape with respect to the starting crystals. Despite these strong similarities, considerable changes in the unit cell dimensions accompany the phase transformation. One of the in-plane (i.e. basal plane) unit cell dimensions increases by 12%, while the unit cell dimension perpendicular to the basal plane (i.e. along the thickness of the plate-shaped crystal) decreases by 32%, leading to a net unit cell volume decrease of 24%. Accommodation of these very significant structural changes while leaving the macroscopic morphology undisturbed is accomplished in several ways. Since expansion of the atomic structure within the basal plane cannot occur throughout the entire length of the crystal, the microcrystals that are formed are slightly buckled, relative to one another, in and out of the basal plane. The large contraction of the atomic structure in the direction perpendicular to the basal plane induces cracks, parallel with the basal plane, and voids within the crystal thereby allowing the macroscopic thickness of the pseudomorph to remain unchanged. The generation of cracks creates additional basal plane surface area, which may be beneficial in terms of catalytic performance.

References

1. E. Bordes, Catal. Today, 1, 572 (1987).
2. W. Stadig, Chemical Processing, Aug., 26 (1992).
3. C.C.Torardi, Z.G.Li, H.S.Horowitz, W.Liang, M.Whangbo, J. Sol. Sta. Chem. (submitted)

Table 1:	a (Å)	b (Å)	c (Å)	V (Å^3)
VOHPO$_4$·1/2H$_2$O	7.416	9.592	5.689	404.7
(VO)$_2$P$_2$O$_7$	16.594 (8.297x2)	9.588	7.735 (3.868x2)	1230.6 (307.65x4)
Δ	+0.88	0.00	-1.82	-97.0
% change	+11.9	0.0	-32.0	-24.0

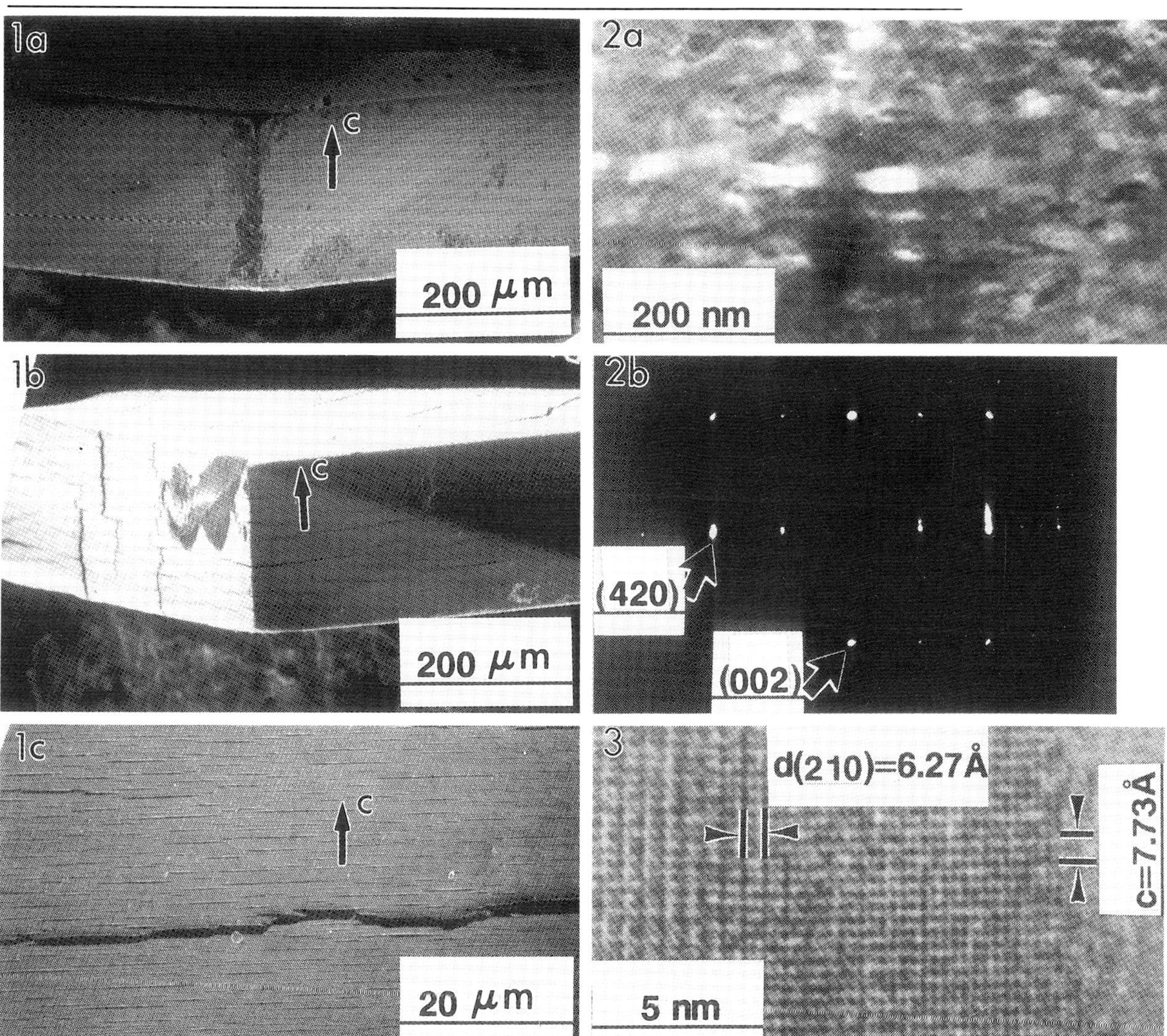

Table 1. Unit cell relationships between VOHPO$_4$·1/2H$_2$O and (VO)$_2$P$_2$O$_7$

Figure 1. -- SEM images of a VPO crystal before (a) and after (b) phase transformation from VOHPO$_4$·1/2H$_2$O. The transformed crystal is shown at higher magnification in (c).

Figure 2. -- TEM image (a) and SAED pattern (b) of a (VO)$_2$P$_2$O$_7$ pseudomorph after transformation from the VOHPO$_4$·1/2H$_2$O single crystal. Electron beam oriented parallel to (001).

Figure 3. -- High resolution TEM of the (VO)$_2$P$_2$O$_7$ pseudomorph with the electron beam parallel with the (001) basal planes.

411

Preparation of Fragile Catalyst Materials for TEM

K.L. More, D.W. Coffey, and T.S. Geer

High Temperature Materials Laboratory, Metals and Ceramics Division, Oak Ridge National Laboratory, Oak Ridge, TN 37831-6064

A novel specimen preparation technique for transmission electron microscopy (TEM) has been developed which allows for the preservation of constituent placement within a variety of diesel and automotive catalyst materials. The standard preparation method for imaging catalyst particles and washcoat constituents has been to use powders scraped from the substrate surface. However, while limited imaging of fine scale structures is possible on clean specimens using this method, all cross-sectional spatial information is lost. Thus, scraped powder specimens cannot be used to directly image surface effects in the TEM or to view large areas of "intact" material in these catalyst systems. Also, for many microscopy investigations such as electron energy loss spectroscopy and high resolution imaging, powders can be too thick. Other preparation techniques have also been used, for example ultramicrotomy and model systems, with some limited success. It is clear that by preparing TEM specimens using this cross-section technique, changes in microstructure to either precious metal particles or washcoat constituents with distance from the exposed surface can be evaluated as a function of aging, engine use, or process modification.

Automotive or diesel catalysts will typically consist of a relatively fragile ceramic substrate, cordierite is used for most of these applications, that is coated with a highly porous "washcoat," which typically consists of a mixture of Al_2O_3 and CeO_2 and holds the precious metal catalyst particles. The substrate and washcoat architecture is shown in Fig. 1(a) and a schematic of the TEM specimen area removed from the catalyst is shown in Fig. 1(b). There are several common problems inherent in the preparation of these types of friable samples: (1) holding the material together during grinding, polishing, dimpling, and ion milling, (2) preserving the surface structure during ion milling, and (3) minimization of artifacts introduced during ion milling. As a result of these problems, ion milling has been discounted as a viable option in the past when compared to other preparation techniques for catalyst materials.[1] However, these problems have been overcome by using an epoxy that can be vacuum impregnated into the highly porous washcoat to hold the structure together and which will remain hard during the arduous sample preparation procedure used for ceramic materials. Dimpling the sample is critical for the proper placement of the hole within a given area of the washcoat. The epoxy will hold the sample together, even when dimpling to a final specimen thickness of 15-20 µm routinely. It is critical to dimple as thin as possible to prevent preferential ion milling of the epoxy, a problem frequently encountered during long milling times, and to minimize damage. Specimens are Ar^+ ion milled at liquid nitrogen temperature, 10° incidence angle, and 5 kV, to reduce the possibility of artifacts during ion milling. If a Precision Ion Polishing System is used, ion milling times are so short that the possibility of artifact introduction are minimized even further.

The value of using these types of specimens can be illustrated using the following examples. The washcoat morphology as well as the catalyst particle distribution in, and association with, various washcoat constituents will play significant roles in the overall performance of the catalyst. Fig. 2(a) and 2(b) are TEM images showing the distribution of catalyst particles in a Pd-based oxidation catalyst for use with natural gas engines. Before engine testing (Fig. 2(a)), the catalyst particles were homogeneously distributed within γ-Al_2O_3 agglomerates and ranged in size from 1-4 nm. After engine testing (Fig. 2(b)), the size of the catalyst particles changed significantly to range from 5-20 nm. Thus, information gained from this specimen included catalyst particle structure identification, size, distribution, and association. The effects of poisoning on the microstructure have also been monitored using these specimens. By protecting the surface during sample preparation, direct imaging of real surfaces after engine use/testing can be performed to identify possible deactivation mechanisms due to poisoning. As shown in Fig. 3, a zinc phosphide glaze having a thickness of ~50 nm was clearly observed on the washcoat surface after testing in a natural gas engine. Also, poisoning effects associated with the catalyst particles themselves, as well as microstructural modifications associated with poisoning, can be characterized. The EDS spectra in Fig. 3 show that the catalyst particles near the surface are poisoned by Cu and Zn, common oil additives. Other deactivation mechanisms, such as sintering, can also be studied using these specimens.
The work presented not only focuses on the development of the specimen preparation technique, but also the type

Proc. Microscopy and Microanalysis 1995, edited by G.W. Bailey, M.H. Ellisman, R.A. Hennigar, and N.J. Zaluzec
Copyright © 1995 MSA. Published by Jones and Begell Publishing, 79 Madison Ave., New York, NY 10016

of important information that can be acquired using these specimens. Work is ongoing to apply these procedures to additional catalyst materials systems.[2]

References
1. S.B. Rice et al., *Mat. Res. Soc. Proc.*, 115 (1988) 15.
2. Research sponsored by (1) the AIM Program and (2) the HTML User Program, U.S. Department of Energy, Assistant Secretary for Energy Efficiency and Renewable Energy, Office of Industrial Technologies, under contract DE-AC05-84OR21400 with Martin Marietta Energy Systems, Inc.

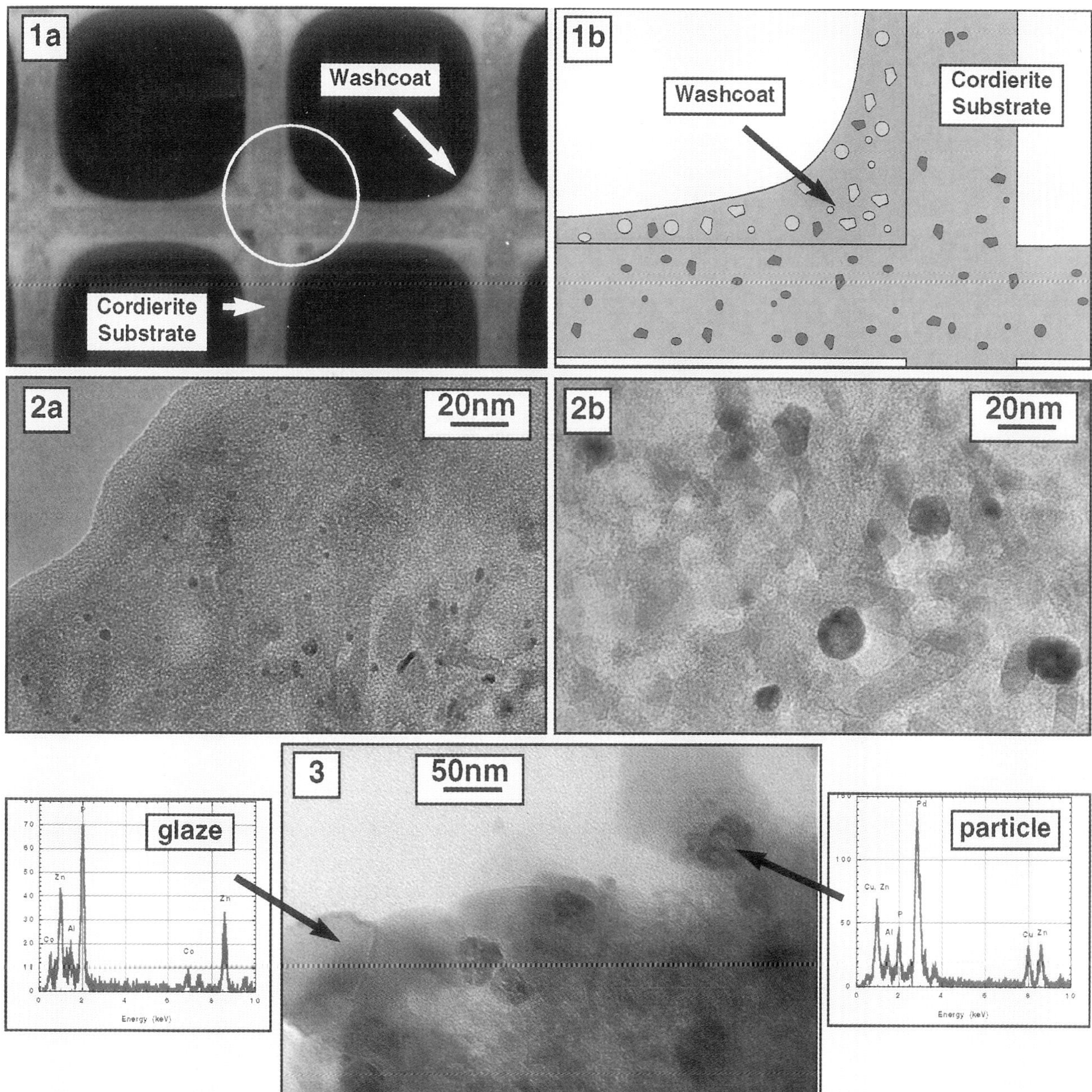

FIG. 1. -- (a) Typical cell structure of automotive catalyst. (b) Schematic of area used for TEM sample.
FIG. 2. -- TEM images of catalyst particles in (a) activated catalyst and (b) engine tested catalyst.
FIG. 3. -- TEM image (and associated EDS spectra) of surface glaze identified after engine testing.

Z-CONTRAST IMAGING OF CATALYSTS IN THE 300 KV STEM

S. J. Pennycook, D. E. Jesson and D. R. Liu*

Solid State Division, Oak Ridge National Laboratory, P.O. Box 2008, Oak Ridge, TN 37831-6030;
*Ford Motor Company, Research Laboratory, 20000 Rotunda Drive, Dearborn, MI 48121.

Z-contrast imaging in the scanning transmission electron microscope has become the accepted technique for imaging sub-nanometer catalyst clusters, utilizing the high angle annular detector introduced by Howie[1]. The lack of coherent phase contrast effects greatly assists the identification of small clusters, especially near the resolution limit of the microscope[2]. The choice of inner detector angle depends on the system being studied. The highest signal to noise ratio is obtained with the smallest inner detector angle, but increasing this angle significantly reduces the contribution of coherently scattered electrons, which is advantageous for crystalline support materials. In this case, small metal clusters may be unambiguosly distinguished from diffracting regions of the support, and their size distributions determined. Fig 1 compares two preparations of 1 wt% Pd on γ-Al$_2$O$_3$, prepared from palladium nitrate solution[3], and aged at 600°C for (a) 6 hours, and (b) 24 hours. Images were taken with a VG Microscopes HB501UX 100 kV STEM, using a probe size of ~3Å. The narrower size distribution resulting from the longer aging time is clearly observed.

Now that cluster sizes are approaching the resolution limit of 100 kV instruments, we might anticipate significant advantages in utilizing the smaller probes available with a 300 kV STEM. Although for size distibution determination we may not require the 1.3Å probe available with this instrument, two potential advantages are immediately apparent. Since the clusters are supported on high surface area support materials, which are naturally three dimensional in form, the microscope depth of focus and the degree of beam broadening in the support are critical considerations. Probe profiles as a function of defocus are shown in Fig. 2. At 300 kV, although the optimum probe profile is much sharper than at 100 kV, the loss of intensity at low defocus values, and the extended tails at high defocus values mean that for a probe size of ~2.5Å the depth of focus is limited to ~800Å in both cases. However, the 300 kV STEM does offer a substantial reduction in beam broadening. Particles on the exit surface of the support will be imaged with a probe that has undergone beam broadening (assuming that no strong channelling condition exists in the support). The thickness dependence of the spatial resolution calculated by the simple formula of Goldstein et. al.[4] is plotted in Fig. 3. Since beam broadening is inversely proportional to accelerating voltage, for a 300 kV probe, 3Å resolution can be expected after passing through 200Å of support, but this reduces to less than 100Å for a 100 kV probe. Evidence of this increased penetration power is seen in the image of Fig. 4a., taken with the VG Microscopes HB603. For comparison, Fig. 4b. shows a bright field phase contrast image, recorded simultaneously, in which no catalyst clusters can be discerned.

In the future it may prove possible to resolve the atomic structure of real catalyst clusters by utilizing the small probe capability of the 300 kV STEM. For sufficient signal to noise ratio it would be necessary to reduce the detection angle to the minimum necessary for incoherent imaging[5], $\theta_i = 1.22 \, \lambda/\Delta R$, or ~12 mrad for a 2Å column spacing. Transverse incoherent imaging would then allow column positions to be determined directly from the image. However, since the image would be formed from coherently scattered electrons, the intensity of each column would be proportional to the square of the number of atoms in the column, up to a thickness of $2\lambda/\theta_i^2$ after which scattering from the top and bottom of the crystal destructively interfere. For a 12 mrad inner detector angle this thickness is 267Å, well above the thickness range of interest. Therefore it should be possible to use maximum entropy image analysis to quantify positional and intensity information, and extract the likely three-dimensional form of the cluster[6].

<u>References</u>

1. A. Howie, J. Microsc. 117 (1979) 11.
2. M. M. J. Treacy and S. B. Rice, J. Microsc. 156 (1989) 211.
3. D. R. Liu et. al. , these proceedings.

Proc. Microscopy and Microanalysis 1995, edited by G.W. Bailey, M.H. Ellisman, R.A. Hennigar, and N.J. Zaluzec
Copyright © 1995 MSA. Published by Jones and Begell Publishing, 79 Madison Ave., New York, NY 10016

4. J. I. Goldstein et. al. SEM/1977, 1, ed O. Johari, IITRI, Chicago, Il.,p315.
5. D. E. Jesson and S. J. Pennycook, Proc Roy Soc Lond A441 (1993) 261.
6. This research was sponsored by the Division of Materials Sciences, U.S. Department of Energy, under contract DE-AC05-84OR21400 with Martin Marietta Energy Systems, Inc.

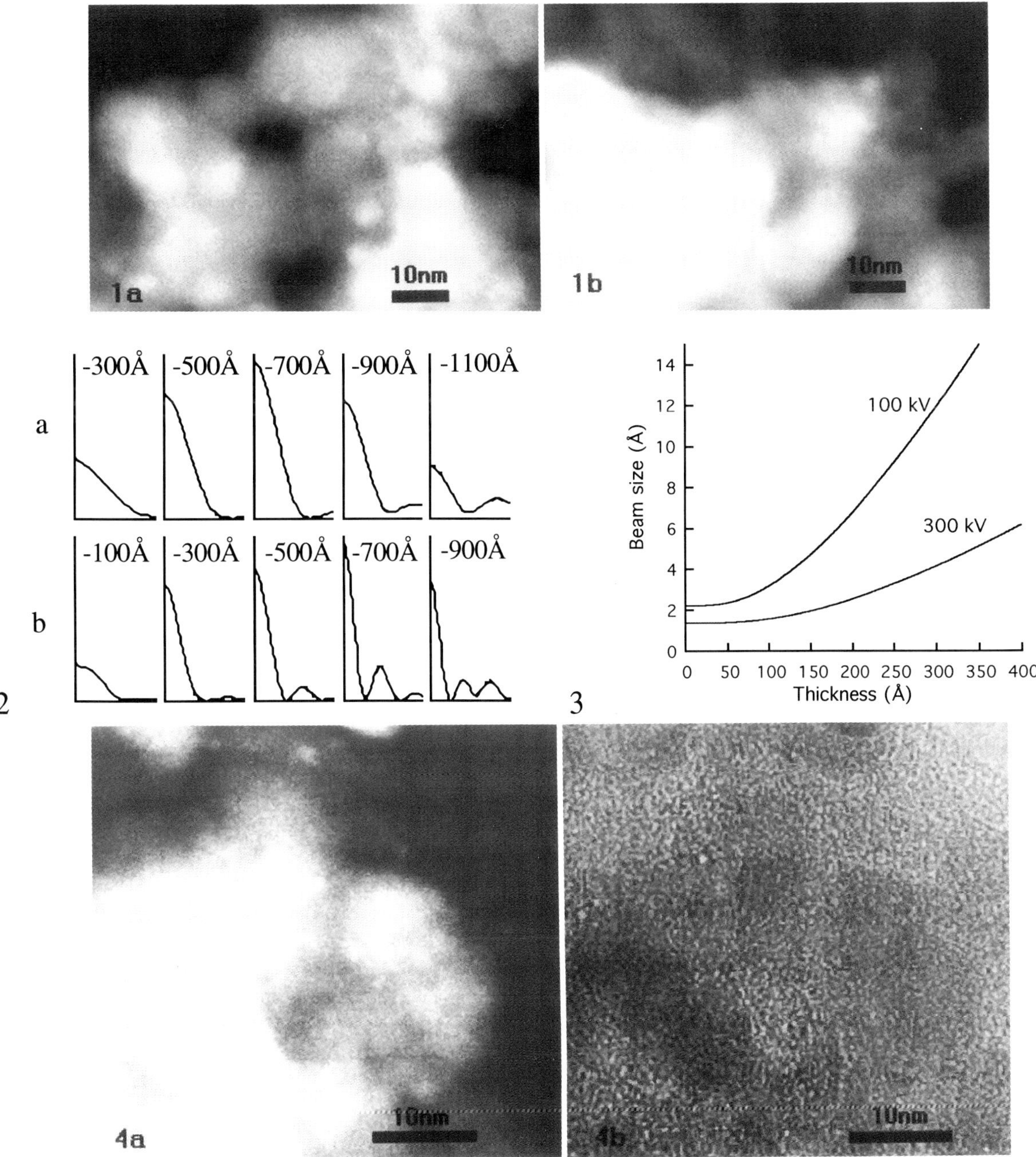

FIG. 1.--100 kV Z-contrast images of Pd clusters on γ-Al$_2$O$_3$, aged at 600°C for (a) 6 hours, and (b) 24 hours. FIG. 2.--Probe profiles as a function of defocus at (a) 100 kV, C$_s$ = 1.3 mm, 10.3 mrad aperture, and (b) 300 kV, C$_s$ = 1 mm, 11.3 mrad aperture. Fig. 3.--Beam broadening in γ-Al$_2$O$_3$. Fig. 4.--(a) 300 kV Z-contrast image of Pd clusters on γ-Al$_2$O$_3$, aged 600°C for 24 hours, (b) Simultaneous bright field image.

ELECTRON HOLOGRAPHY OF CATALYSTS

A. K. Datye,* D. S. Kalakkad,* L. F. Allard** and E. Völkl**

*Center for Microengineered Ceramics, Dept. of Chemical and Nuclear Engineering,
 University of New Mexico, Albuquerque, NM 87131
**High Temperature Materials Laboratory, Oak Ridge National Laboratory, Oak Ridge, TN 37831

The active phase in heterogeneous catalysts consists of nanometer-sized metal or oxide particles dispersed within the tortuous pore structure of a high surface area matrix. Such catalysts are extensively used for controlling emissions from automobile exhausts or in industrial processes such as the refining of crude oil to produce gasoline. The morphology of these nano-particles is of great interest to catalytic chemists since it affects the activity and selectivity for a class of reactions known as structure-sensitive reactions. In this paper, we describe some of the challenges in the study of heterogeneous catalysts, and provide examples of how electron holography can help in extracting details of particle structure and morphology on an atomic scale.

Conventional high-resolution TEM imaging methods permit the image intensity to be recorded, but the phase information in the complex image wave is lost. However, it is the phase information which is sensitive at the atomic scale to changes in specimen thickness and composition, and thus analysis of the phase image can yield important information on morphological details at the nanometer level. Image plane off-axis electron holography using a Möllenstedt biprism in a field emission gun electron microscope can produce holograms of nanoparticles from which the amplitude and phase information can be separately reconstructed.[1] When the specimen is sufficiently thin so that phase changes less than 2π are obtained, a profile across the phase image is essentially a map of specimen thickness.[2] This information can then be directly related to particle morphologies.

The example specimens used here consisted of Pd particles supported on amorphous silica microspheres. Fig. 1a shows a bright field HRTEM image, taken on a Hitachi HF-2000 cold field emission TEM, of one of the Pd particles obtained after H_2 reduction of the calcined catalyst. These particles have been shown by holography on the HF-2000 to contain internal voids, seen as the facetted central contrast feature in this image[2]. The lattice fringes are unaffected by the presence of the internal void and show that the Pd particle is a single crystal with no internal defects. To illustrate that the HRTEM image does not directly provide information on the 3-D shape of the metal particles, we have plotted in Fig. 1b a phase profile along the line marked A-A' in Fig. 1a, which shows only the periodicity in contrast caused by the lattice fringes.

Figure 2a shows the phase image derived from the hologram of this particle, as processed using the software package HoloWorks© now available from Gatan, Inc. HoloWorks allows the choice of an appropriate aperture for the Fourier transformations, to de-emphasize the lattice fringe contrast in favor of the morphological details of the particle. Fig. 2b shows a phase profile derived from the phase image along the line in Fig. 1a. The phase shifts are shown more clearly in Fig. 2c with a model of the metal particle shown in Fig. 2d. The phase shift first changes rapidly over a distance of approximately 0.5 nm which corresponds to the surface roughness on the exposed (111) surface of the particle. After that point, the phase shift changes more gradually corresponding to the gradual change in thickness as we move away from the particle surface, and as the void is encountered there is a reversal in the direction of the phase shift, as expected. The preliminary analysis is of the phase image is consistent with a cubooctahedral particle having a pronounced (111) facet.

References

1. H. Lichte, in T. Mulvey and C. J. R. Sheppard, eds., *Adv. in Opt. and Elect. Micro.*, Acad. Press 12(1991)25.
2. L. F. Allard, E. Völkl, D. Kalakkad and A. K. Datye, *J. Mater. Sci.* 29(1994)5612.

Proc. Microscopy and Microanalysis 1995, edited by G.W. Bailey, M.H. Ellisman, R.A. Hennigar, and N.J. Zaluzec
Copyright © 1995 MSA. Published by Jones and Begell Publishing, 79 Madison Ave., New York, NY 10016

3. Research sponsored in part by the High Temperature Materials Laboratory Faculty Fellowship Program (AKD), DOE Office of Transportation Technologies, and the ORNL Laboratory Directed Research and Development Program, under contract DE-AC05-84OR21400 with Martin Marietta Energy Systems, Inc.

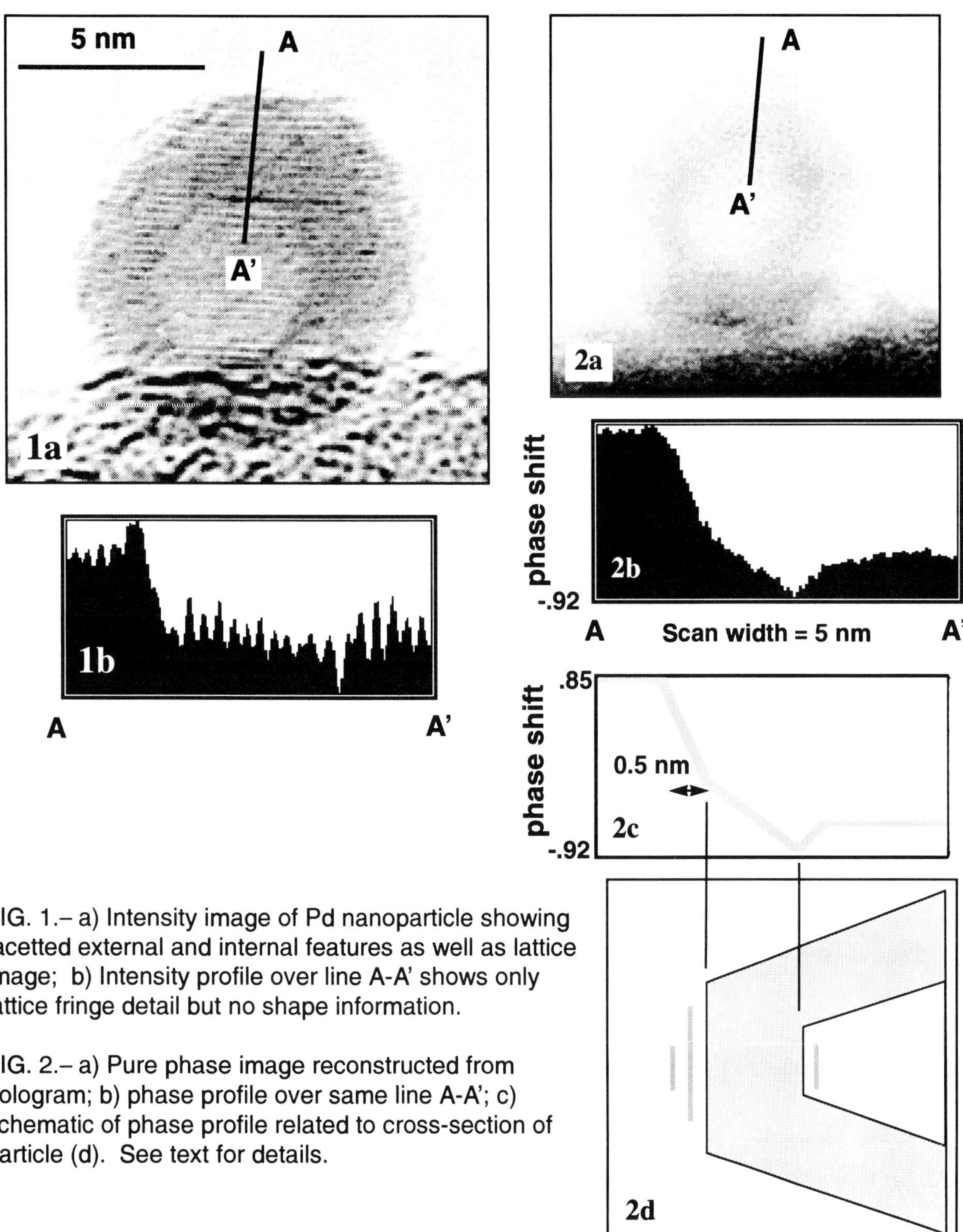

FIG. 1.– a) Intensity image of Pd nanoparticle showing facetted external and internal features as well as lattice image; b) Intensity profile over line A-A' shows only lattice fringe detail but no shape information.

FIG. 2.– a) Pure phase image reconstructed from hologram; b) phase profile over same line A-A'; c) schematic of phase profile related to cross-section of particle (d). See text for details.

STUDIES OF SURFACE AND STRUCTURAL HETEROGENEITY OF CARBON SUPPORTS AND CARBON-SUPPORTED CATALYSTS

J. Liu*, S. M. Colburn*, R. L. Ornberg* and J. R. Ebner**

* Analytical Sciences Center, Monsanto Company, 800 N. Lindbergh Blvd., St. Louis, MO 63167
** Chemical Sciences Center, Monsanto Company, 800 N. Lindbergh Blvd., St. Louis, MO 63167

Activated carbons are generally used as support materials for precious metal catalysts. Properties of carbon that are important to catalyst preparation and application include composition, surface area, microstructure and pore shape and size distribution.[1-3] Macropores (> 50 nm), mesopores (2-50 nm) and micropores (<2 nm) generally coexist in activated carbons. The accessibility of metal particles dispersed in microporous systems is of predominant importance, especially for large molecules that exhibit slow diffusion transport in narrow pores. It is desirable to have metal particles highly dispersed in readily accessible locations. As part of an on-going program of the characterization of carbon-supported catalysts we report some preliminary observations of the microstructure of carbon supports by a variety of electron microscopy techniques

Commercial carbon supports and carbon-supported Pt catalysts were used in this study. High resolution secondary electron (SE) microscopy, low voltage backscattered electron (LVBE) microscopy and high-angle annular dark-field (HAADF) microscopy techniques were employed to extract surface and structural information. Thin sections (~ 60 nm) of the carbon supports and carbon-supported Pt catalysts were prepared through ultramicrotomy for HAADF examination. For SEM observations, activated carbon particles were directly observed in a field-emission SEM (Hitachi-4500) without prior coating. LVBE images were obtained on trimmed block faces of the original sample embedded in an epoxy matrix. All images were digitally acquired.

The surface morphology and the size distribution of carbon particles were examined by high resolution field-emission SEM. Figure 1a shows a typical SE image of the carbon support revealing high rugosity (defined as the ratio of the external surface area to the area of the geometrical envelope of the particle) and mesopores. Flat and smooth carbon surfaces were also observed as shown in figure 1b. The carbon support consists of 'high-surface-area' and 'low-surface-area' regions. The internal structure of the carbon support was examined by HAADF technique. Figure 2a shows a low magnification HAADF image of a carbon-supported Pt catalyst. Areas with bright contrast represent regions with high scattering probability. Figure 2b shows a magnified image of the area framed in figure 2a, clearly revealing the internal structure of the carbon support. It was found that regions of high-density and porous carbon coexist in carbon particles. Both mesopores and micropores exist in porous regions (labeled as region I). In regions of high-density carbon (labeled as region II), however, only ultramicropores were observed. It is noteworthy that regions (labeled as region III) enclosed by thin shells of high-density carbon have pore sizes smaller than those in the porous regions. At higher magnifications HAADF images showed that nano-domains of high-density carbon existed within regions of porous carbon. Small Pt particles located in regions II and III may not be accessible by participating reactants during a catalytic reaction. The different domains existing in carbon supports have also been examined by LVBE imaging technique. Figure 3 shows such an image revealing the morphology and the distribution of domains of high-density/porous carbon in a carbon-supported Pt catalyst.[4]

References

1. A. J. Bird, in : *Catalyst Supports and Supported Catalysts*, London: Butterworths (1987) 107.
2. T. Wigmans, *Carbon* 27 (1989) 13.
3. D. S. Cameron et al., *Catalysis Today* 7 (1990) 113.
4. Part of the electron microscopy work was conducted at the Center for HREM at Arzona State University, supported by NSF-DMR-931432.

Proc. Microscopy and Microanalysis 1995, edited by G.W. Bailey, M.H. Ellisman, R.A. Hennigar, and N.J. Zaluzec
Copyright © 1995 MSA. Published by Jones and Begell Publishing, 79 Madison Ave., New York, NY 10016

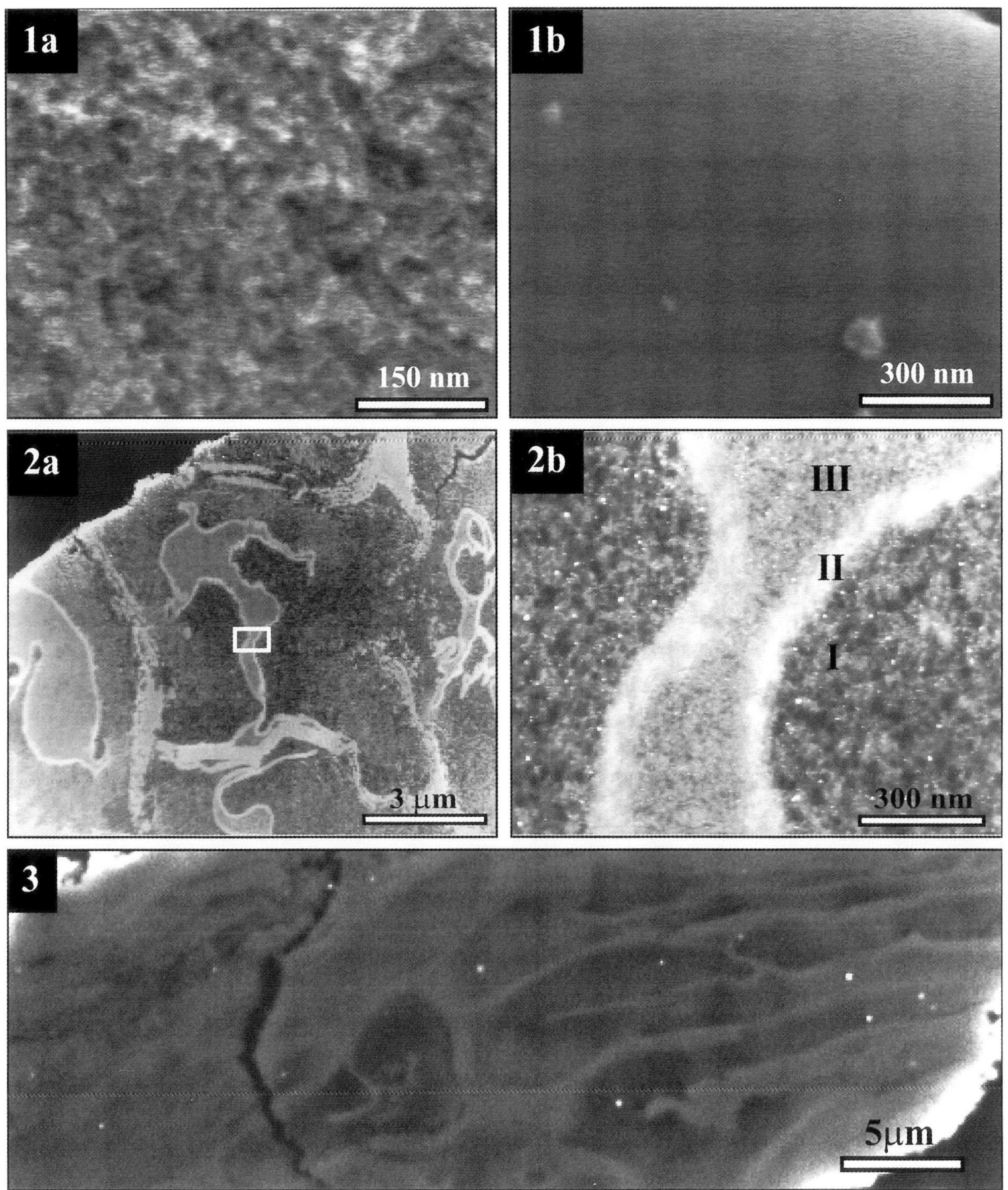

FIG. 1----High resolution SE image of carbon support showing (a) high-surface-area region and (b) low-surface-area region.
FIG. 2----(a) HAADF image of carbon supported Pt catalyst revealing regions of high-density and porous carbon and (b) magnified HAADF image of area framed in (a) showing small metal particles.
FIG. 3----Low voltage backscattered electron image of carbon supported Pt catalyst showing domains of high-density and porous carbon.

ATOM PROBE STUDIES OF CATALYST SURFACES: STRUCTURE AND CHEMISTRY AT ATOMIC LEVEL

G.D.W. Smith[*], A. Cerezo[*] and S. Poulston[**]

*Department of Materials, Oxford University, Parks Road, Oxford OX1 3PH, U.K.
**Department of Chemistry, University of Reading, Reading, Berkshire, U.K.

The imaged (apex) region of a field ion microscope (FIM) specimen is sharply curved and has a radius of less than 100nm. It is thus a reasonably good "model" for one half of a single particle of a metallic catalyst. FIM images show good detail of steps, ledges and kink site atoms (Fig. 1). When combined with a time-of-flight mass spectrometer to form an atom probe (AP),[1] single atom chemical identification becomes possible. The FIM-AP combination has considerable value for the study of heterogeneous catalysts and catalytic reactions, but there are problems due to the high field acting on the specimens during observation, and the need to work in a vacuum environment. The most important applications to date have involved studies of the surface of the catalyst material, and of relatively non-labile adsorbates.[2] However, new developments in AP instrumentation have opened the prospect of seeing catalytic reactions occurring on the atomic scale, and analysing the intermediate reaction products in both spatially- and time-resolved modes with high precision. Two main lines of development have contributed to this exciting prospect. Block and co-workers in Berlin produced the Pulsed Field Desorption Mass Spectrometer (PFDMS).[3] In this instrument, a high electric field is initially applied to a FIM specimen in the presence of a reactive gas mixture. The specimen apex is cleaned by raising the field to a level sufficient to produce field evaporation, and then the field is dropped to zero to allow gas adsorption and reaction to occur. After a preset time interval (typically in the range of microseconds to milliseconds), a voltage pulse is applied to field desorb the reaction products, which are analysed by time-of-flight mass spectrometry. The second innovation is the PoSAP (Position Sensitive Atom Probe), developed at Oxford by Cerezo and Smith.[4] This instrument combines FIM-AP technology with a wide-area, position sensitive detector, so that ions desorbed from a large surface region (up to 20nm diameter) can be mapped and identified with sub-nanometer spatial resolution.

Combination of PFDMS and PoSAP technology should, in principle, allow the study of heterogeneous reactions on the atomic scale. However, one further difficulty had to be overcome. The original PoSAP detector was based on a charge division method, and used a wedge-and-strip anode. This was only capable of positioning one ion at a time, whereas in surface chemical reactions it is common for several molecular fragment ions to be removed during a single desorption pulse. This problem can be partially overcome by the use of a multi-element detector system.[5] A more complete solution is obtainable by the use of a dual detector system of the kind shown in Fig. 2. This has been implemented by the authors.[6] Ion impact on the primary ion detector produces a burst of photons. The light signal is split, with one portion being sent to an 8 x 10 anode array photomultiplier (from which timing signals are obtained), the rest being sent to an image intensified TV camera which contains a 256 x 256 photodiode array (for position measurement). This system has now been shown to operate satisfactorily when multiple ions are desorbed at the same moment. A major advantage of this approach is flexibility of the detector system. All the main components are in air, rather than in the vacuum system, and so the camera and the array detector can, if necessary, be serviced or replaced without difficulty. The most important application of this new instrument is likely to be to the study of the platinum- and palladium-based alloys used in automobile exhaust pollution control applications, extending previous conventional AP studies in this area.[7,8]

References

1. M.K. Miller and G.D.W. Smith, "Atom Probe Microanalysis" Published by M.R.S., Pittsburgh U.S.A. (1989).
2. T.T. Tsong, "Atom Probe Field Ion Microscopy", Published by Cambridge U.P., Cambridge, U.K. (1990)
3. J.H. Block, in "Surface Science of Catalysis: In Situ Probes and Reaction Kinetics", A.C.S. Symposium Series 482 (1992) p.287.
4. A. Cerezo, T.J. Godfrey and G.D.W. Smith, Rev. Sci. Instrum. 59 (1988) 25.
5. D. Blavette et al., Rev. Sci. Instrum., 64 (1993) 2911.
6. A. Cerezo et al., Appl. Surface Science, 76/77 (1994) 374.
7. S. Poulston and G.D.W. Smith, in "Catalysis and Surface Characterisation", ed. T.J. Dienes et al, Published by Roy. Soc. of Chemistry, London (1992) p.228.
8. Work supported by the Paul Instrument Fund of the Royal Society, and by EPSRC under grant number GR/H 38485. A. Cerezo wishes to thank the Royal Society for the award of a Research Fellowship. We also thank T.J. Godfrey and S.J. Sijbrandij for their invaluable assistance with this work.

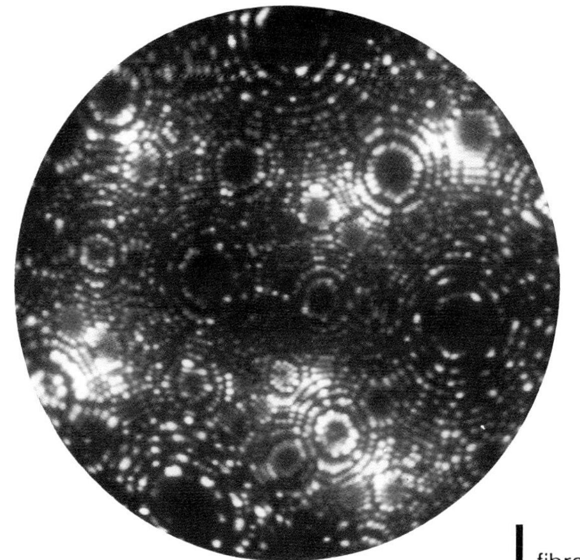

FIG. 1. Neon FIM image of a Pt - 17 at% Rh alloy specimen. (40 K, 8.6 kV).

FIG. 2. Schematic diagram of the design of the new optically coupled position sensitive atom probe. (Courtesy Applied Surface Science)

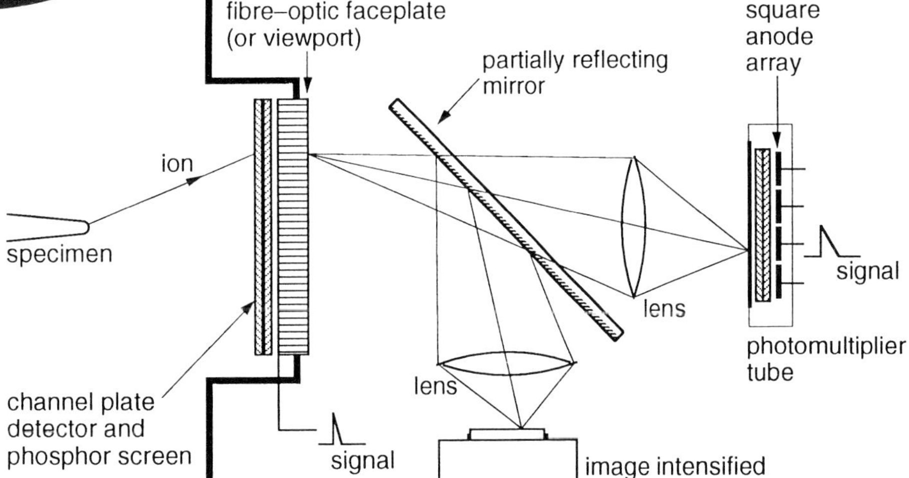

SCANNING TUNNELING MICROSCOPY STUDIES OF REACTIONS AT SURFACES

F.M. Leibsle[*]

[*]Interdisciplinary Research Centre in Surface Science, University of Liverpool,
Liverpool L69 3BX, United Kingdom

Scanning tunneling microscopy (STM) can image clean and adsorbate-covered surfaces with atomic/molecular resolution. This enables us to obtain detailed information on how catalytic reactions proceed on surfaces. Information on such topics as initiation sites for chemical reactions, reaction trends, the development of molecular order, and mesoscopic surface restructuring can be obtained. In addition, variable temperature STM experiments can yield insight into the decomposition of molecules at surfaces.

Studies using STM under ultra-high vaccuum conditions have shown how the oxidation of CO occurs on oxygen-precovered Rh(110) surfaces[1]. Figure 1 shows an STM image of an oxygen-precovered Rh(110) surface displaying areas of oxygen-induced c(2x6) and c(2x8) reconstructions following exposure to 5 Langmuirs of CO. Images like this show that the reaction between the CO and adsorbed oxygen initiates at step edges and defects sites. The CO reaction fronts then proceed in a one-dimensional manner stripping away the oxgyen from two and three atom wide islands.

Below 300 K, methanol reacts with adsorbed oxygen on the (110) face of Cu to create a surface methoxide species[2]. STM images show that methanol reacts with the oxygen at the ends of Cu-O islands removing the oxygen as water and forming an ordered methoxy structure. Images of the ends of the Cu-O islands show that the endmost atoms relax outward (Fig. 2). This relaxation may be the reason that the reaction attacks the ends of the islands as opposed to the sides. Variable temperature STM measurements show that, at temperatures where this methoxide species on Cu(110) should decompose into formaldehyde and hydrogen, the methoxy islands disappear[3]. These islands appear to lose molecules from the island edges while the island centers remains intact (Fig. 3). These images suggests that methoxy molecules diffuse from the islands edges prior to decomposition. Furthermore, images of the breakup of methoxy islands near step edges show the release of Cu atoms demonstrating that methoxy induces a surface reconstruction rather than forming an ordered overlayer.

These same methodologies have been applied to the study of the reaction between formic acid and adsorbed oxygen on the Cu(110) surface to form a surface formate species and yield comparable insights[4]. However, additional findings have come from these studies. Formate adsorption under these conditions appears to cause a major restructuring of step edges giving the edges a "sawtooth" appearance. These "sawteeth" are spaced 76 Å apart and are aligned across step edges. (Fig. 4)

References

1. F. M. Leibsle, P. W. Murray, S. M. Francis, G. Thornton, and M. Bowker, *Nature* 363(1993)706.
2. F. M. Leibsle, S. M. Francis, R. Davis, N. Xiang, S. Haq, and M. Bowker, *Phys. Rev. Lett* 72(1994)2569.
3. F. M. Leibsle, S. M. Francis, S. Haq, and M. Bowker, *Surf. Sci.* 318(1994)46.
4. F. M. Leibsle, S. Haq, and M. Bowker, (in preparation).

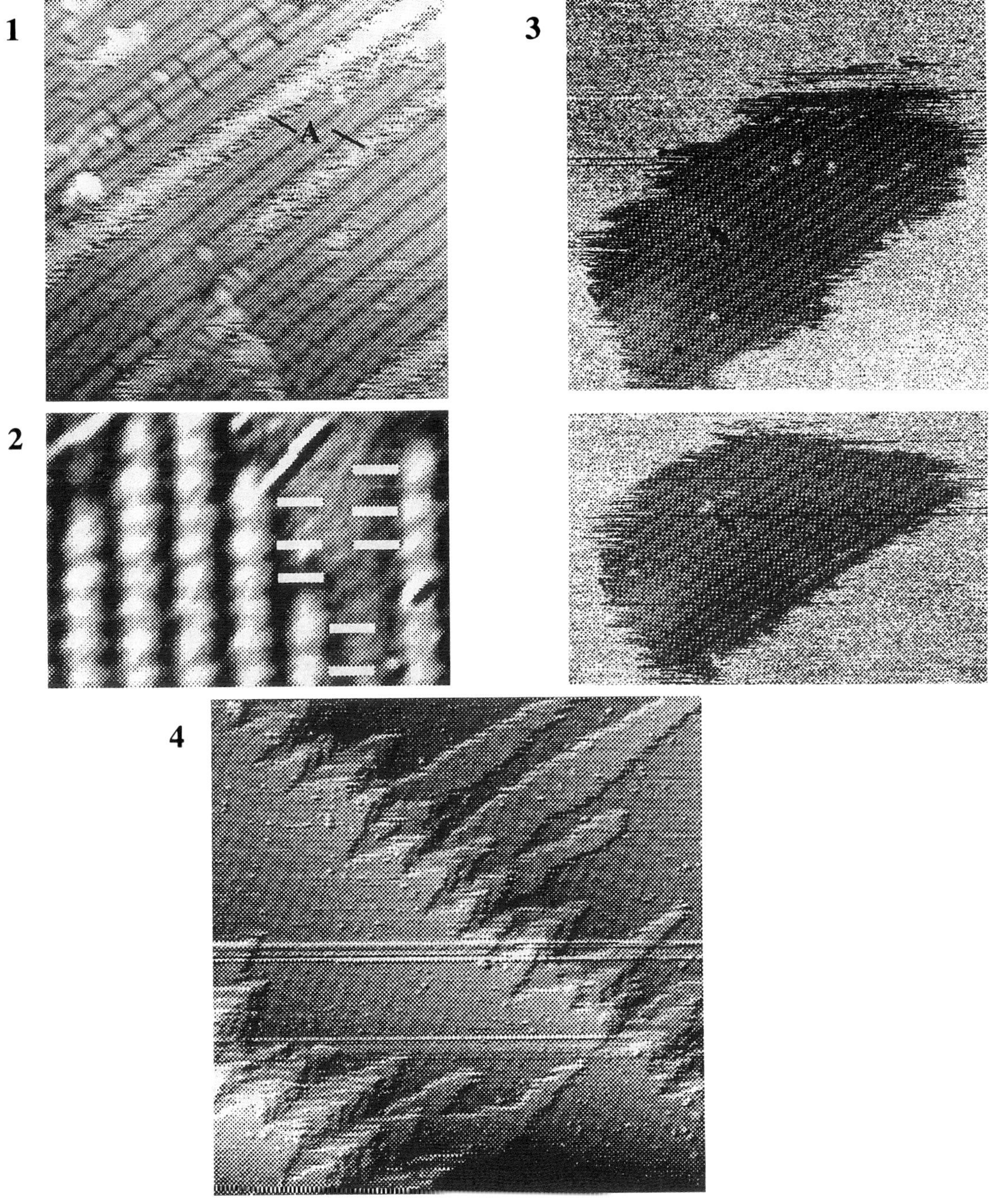

FIG. 1.—200Åx200Å STM image of oxygen covered Rh(110) surface showing propagation of CO reaction fronts (denoted by A).

FIG. 2.—19Åx12.6Å STM image showing Cu-O strings on the Cu(110) surface. Atoms at ends of strings relax outward. This outward relaxation may explain why reactions tend to occur at these sites.

FIG. 3.—400Åx380Å and 400Åx255Å STM images showing the gradual disappearance of a methoxy island from a Cu(110) surface. Methoxy molecules appear to diffuse from the edges of the island.

FIG. 4.—1000Åx1000Å STM image of a Cu(110) surface showing the restructuring of step edges due to formate adsorption. Step edges take on a "sawtooth" appearance.

ANALYSIS OF FLUIDIZED CRACKING CATALYSTS BY ATOMIC FORCE MICROSCOPY

S.A.C. Gould* and M.L. Occelli**

*Keck Science Center, Claremont Colleges, Claremont, CA 91711-5916
**Georgia Tech Research Institute, Atlanta, GA 30332

In the United States, five million barrels of oil per day or one third of all processed crude oil is catalytically converted with fluidized catalysts, a process which requires 500 tons of catalyst daily.[1] Given these immense quantities, a small percentage difference in the efficiency of the oil conversion to liquid fuel can result in a savings of many millions of barrels of oil annually. In this study we have used a contact mode atomic force microscope (AFM) to study to topography of a set of fluidized cracking catalysts (FCC) from the µm level down to the atomic.[2] We selected this technique because it is believed that the cracking in the FCCs occurs mainly on the catalysts top 10-15 µm suggesting a surface which contains numerous pores, something the AFM can well characterize.

We obtained our FCC (GRZ-1) from Davison. The FCC is generated so that it contains and estimated 35% rare earth exchaged zeolite Y, 50% kaolin and 15% binder. Two types of dried FCCs were prepared, the control group and a group contaminated by a solution of vanadyl naphthenate in toluene. After calcining the FCCs at 400° C for 2 hours the microsphere were sprinkled onto a steel disk which had been covered with a thin film of epoxy and placed in the microscope.[3] Figures 1-6 contain some key features our results.

The data from the uncontaminated group (Fig. 1-3) showed a surface that is extremely rough. Upon drying we found the clay mixtures produces large pits and crevices, often on the order of several microns deep (Fig. 1). Upon closer inspection, the surface itself was covered with large oval shaped platelets, 100-200 nm in length and 30-50 nm in width. These platelets were often stacked adjacent to each other or possibly on top of each other. When stacked vertically, the platelets would produce terraces, 200-300 nm in height which could act as possible docking sites for hydrocarbon chains (Fig 2). A 6-9 nm crevice was common in areas where the plates were found adjacent to each other. At the finest detail, we were occasionally able to image the surface with atomic resolution. Fig. 3 shows a square type lattice which can be attributed to the arrangement of oxygen atoms found in AlO_6 layer of kaolin. Other atomic scale images show what appear to be faujusite crystals.

The data from the Vanadium contaminated group (Fig. 4-6) showed what appeared to be islands of and coatings of Vanadium. The valleys and crevices found in the control group data appear to be filled including the channels that were found between platelets. Since Vanadium is known to reduce the catalytic activity of FCCs, its presence within the pores of the FCCs suggests the contamination physically blocks the openings into the microsphere where the catalysis occurs.

In summary, while this preliminary data allowed us to identify the FCCs main components, because of the characteristics of the contact-mode AFM, we are unable to determine actual chemical composition of the catalytic surface.

References

1. Occelli, M.L., in *Keynotes in Energy Related Catalysis*; Kaliaguine, S. Ed.; Elsevier: Amsterdam, 1988, p 101.
2. Nanoscope III, Digitial Instruments, Santa Barbara, CA
3. Occelli, M.L, Gould, S.A.C. and Drake, B., *Proc. Int. Congr. on Advances in FCC*, Amer. Chem Soc. 1994, pg 271.
4. The authors would like to thank Digital Instruments of Santa Barbara, CA (SACG) for their contributions to this study.

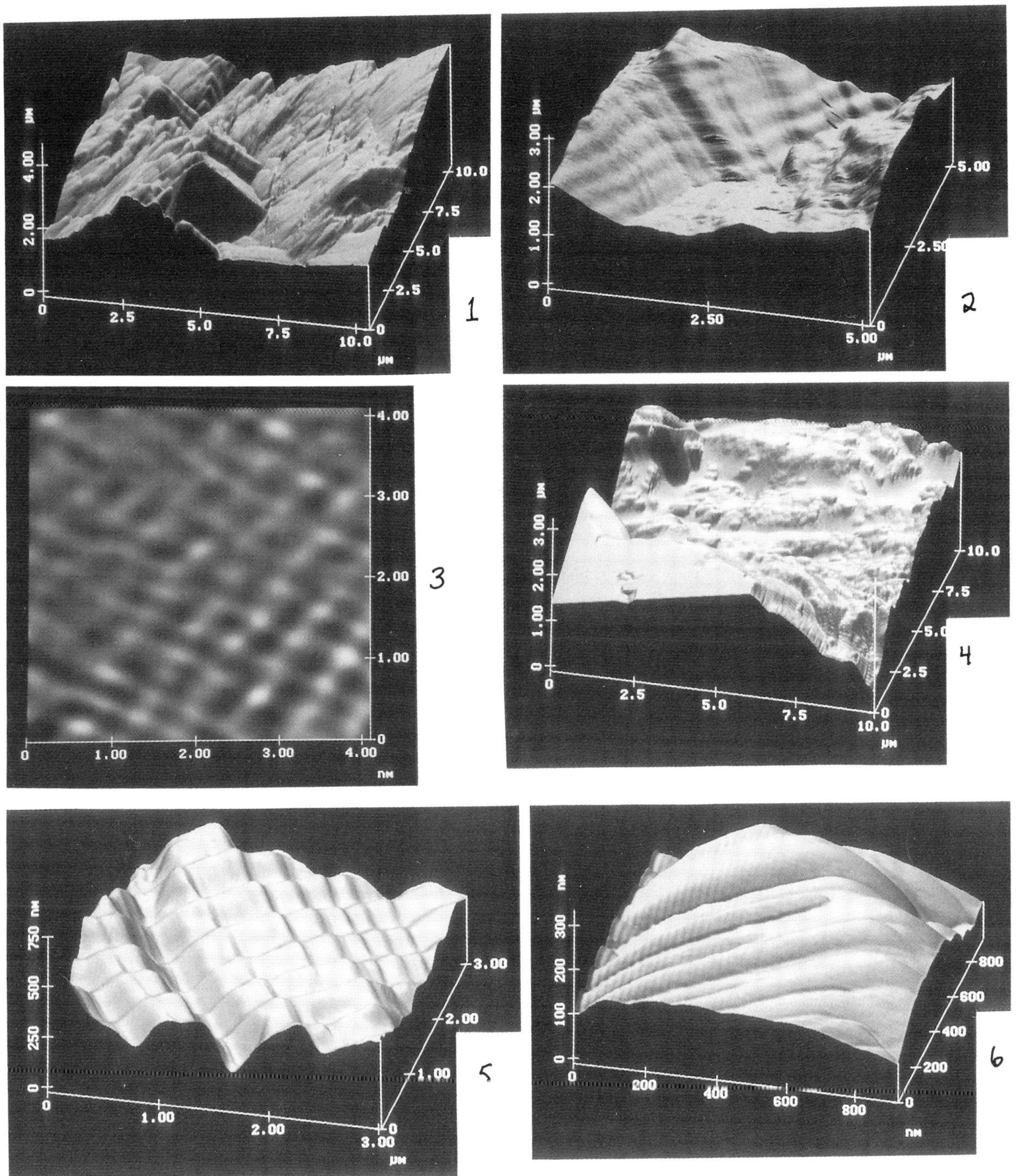

Figure 1. Large scale atomic force microscope (AFM) image of FCC surface. The horizontal size of the image is 10 by 10 μm. The vertical scale is 4 μm.
Figure 2. AFM image of FCC surface: 5 by 5 μm.
Figure 3. High resolution AFM image of FCC surface: 4 x 4 nm.
Figure 4. Large scale AFM image of V-contaminated FCC surface: 10 x 10 μm.
Figure 5. AFM image of V-contaminated FCC surface: 3 x 3 μm.
Figure 6. AFM image of V-contaminated FCC surface: 900 x 900 nm.

SCANNING PROBE MICROSCOPY OF THE ACTIVATION AND SULFUR DEACTIVATION OF Pd MODEL CATALYSTS.

R, Catani[*], K.-H. Lee, and <u>E.E. Wolf</u>

Chemical Engineering Department, University of Notre Dame, Notre Dame, IN, 46556, USA

Scanning tunneling microscopy (STM) and atomic force microscopy (AFM) are used to image the microstructure of surfaces before and after pretreatment and reaction to establish a correlation between activity/ deactivation and morphology. Several surfaces of flat thin Pd films supported on graphite, and single crystals of silica and alumina, show that the morphology of these films is highly sensitive to the method of activating the Pd catalysts. It is shown that the activity and selectivity for the probe reaction, 1,3 butadiene hydrogenation, of a Pd/graphite film catalyst changes with the film microstructure[1]. It is also shown that catalysts prepared via different procedures leading to the same surface structures exhibited the same reactivity, and vice versa, i.e. different microstrcutures lead to different activity, thus demonstrating a correlation between activity and morphology. The structure can be changed by changing the hydrocarbon to hydrogen ratio which causes significant deactivation due to changes in morphology.

AFM studies of a $Pd/SiO_2/Si$ films after oxygen and hydrogen treatment combined with XPS studies, show that the relation between morphology and activity is also a function of surface composition[2]. Figure 1 and 2 show AFM micrographs of a $Pd/SiO_2/Si$ film pretreated at 450°C in H_2 for eight hours and O_2 followed by H_2 each for four hours. In both cases, there is considerable roughening of the surface relative to the unpretreated fresh film. The grain size of the hydrogen pretreated sample is much bigger (~100x125 nm) than those of the sample pretreated in oxygen and hydrogen. The conversion of butadiene at 100°C is about 20% higher for the sample treated in O_2 and H_2 which agrees with its higher dispersion.

Pd thin film deposited on sapphire (Pd/Al2O3) model catalysts were sulfided in thiophene and the surface sulfur content was measured by XPS. It was found that in unreduced films increasing the amount of sulfur added decreased the activity. Kinetic results showed that the activation energy also decreased with sulfur content along with the active area. Turnover frequencies and activation energies for 1,3 butadiene hydrogenation, made at different levels of sulfur uptake are correlated with the changes in geometrical and poisoned metal area. Increasing the pretreatment temperature prior to sulfidation, however, resulted in drastic changes in the surface microstructure and the activity[3]. Fig. 3 shows the microstructure of a sample poisoned after prereduction in hydrogen at 400°C. The amount of sulfur required to completely poison the catalysts (S/Pd=0.2) also decreased in relation to the unreduced catalysts (S/Pd=0.55). The microstructure shows significant sintering of the film that is completely ruptured forming large islands(~200 nm) and fiberlike features (~2000x500 nm).

The studies presented show clearly that the working surface is a function of the pretreatment and the reaction environment. This before extrapolating results from one set of experimental conditions into another it is necessary to ascertain the effect of the reaction environment on the catalysts microstrcuture.

References:
1. K.L. Yeung, & E.E.Wolf, J. Catal., 143 (1993) 409.
2. K.-H. Lee and E.E. Wolf, Catal. Lett., 26 (1994) 297.
3. K.-H. Lee and E.E. Wolf, Proc. Int. Symp. Catal. Deact.

Proc. Microscopy and Microanalysis 1995, edited by G.W. Bailey, M.H. Ellisman, R.A. Hennigar, and N.J. Zaluzec

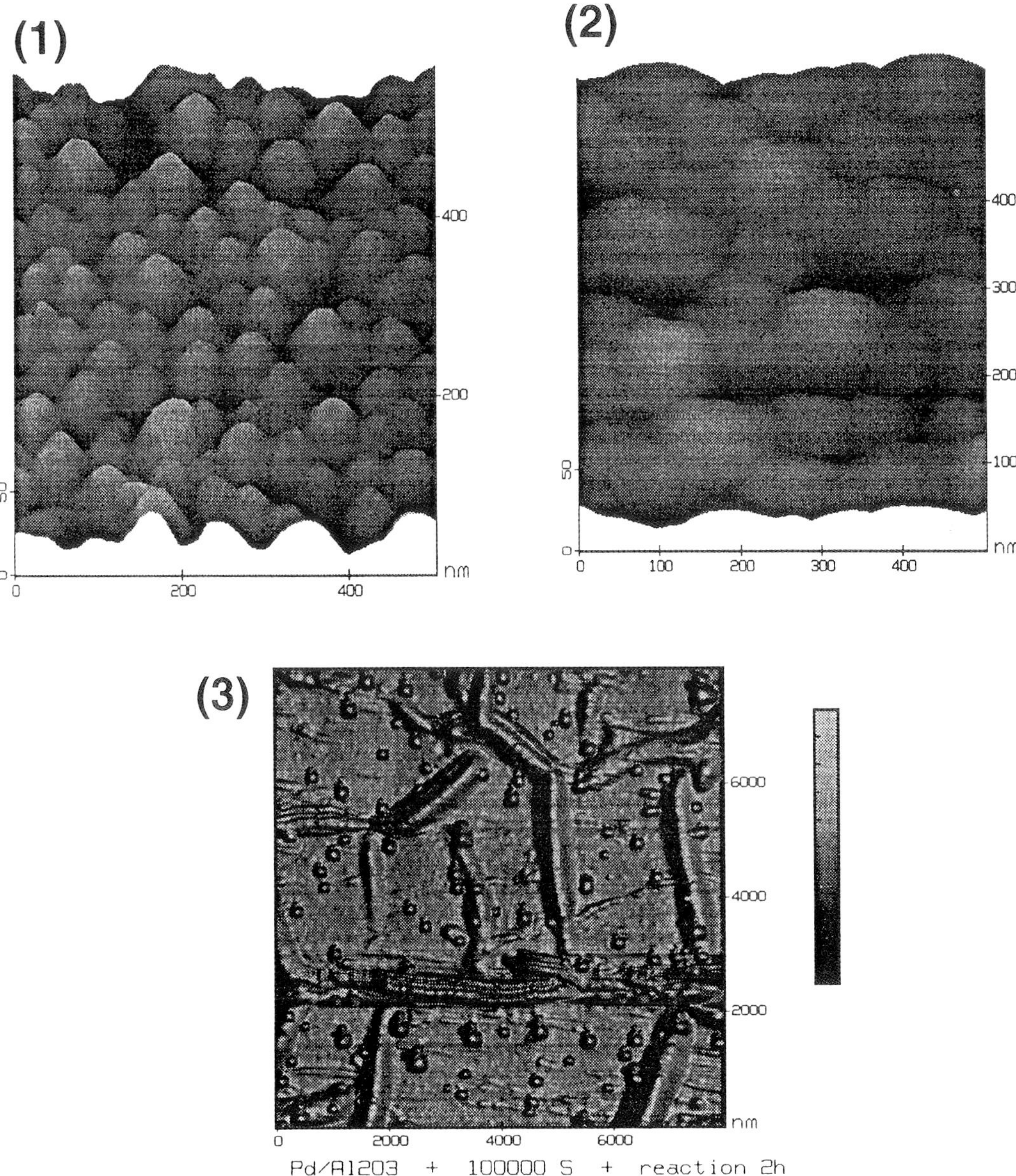

Figures show Pd/SiO$_2$ surface treated in (1).H$_2$ & O$_2$ and (2). H$_2$ at 450°C. The grain particles of the O$_2$+H$_2$ treated sample is much smaller than the H$_2$ treated sample, 40x60 nm^2 vs. 100x125 nm^2. The conversion of 1,3 butadiene hydrogenation for the former case is about 20% higher. (Scale: 500x 500 nm^2)

Figure 3 shows the significant effect of sulfur on the morphology of Pd film. The surface is made of large islands (around 200 nm) and elongated fiber-like features (2000x500 nm^2). (Scale: 8000 x8000 nm^2)

DIVERGENT BEAM MICROANALYSIS STUDIES OF BIMETALLIC PLATINUM CATALYSTS SUPPORTED ON γ–AL$_2$O$_3$

Fryer JR, Huang Z*, Stirling D, and Webb G.
Electron Microscope Centre, University of Glasgow, Glasgow G12 8QQ, Scotland,UK.
* Department of Physics, University of Essex, Wivenhoe Park, Clochester, Essex, UK.

Platinum dispersed on γ-alumina is used as a reforming catalyst to convert linear hydrocarbons to cyclic aromatic products. To improve selectivity and lifetime of the catalyst, other elements are included, and we have studied the distributions of Pt/Re, and Pt/Sn, bimetallic systems on the support both before and after use in octane reforming. Often, one or both of the components are not resolvable by HREM or microanalysis as individual particles because of small size and lack of contrast on the alumina, and divergent beam microanalysis has been used to establish the presence and relationship between the two elements.

In the majority of catalysts the platinum is in the form of small particles, some of which are large enough to be resolvable in the microscope. The ABT002B microscope with Link windowless Pentafet detector, used in this work, was able to obtain a resolvable signal from particles of 2nm diameter upwards. When the beam was concentrated on to such a particle the signal was at a maximum, and as the beam diameter was diverged - at the same total beam intensity and dead time - the signal decreased as shown in Figure 1. If other particles of platinum were present in the exposed area then the curve was altered and prolonged above that of the theoretical 'single particle' curve. Similar curves were given by any other elements present, so that for each element the distribution of each element within the diverged area could be obtained.

Figure 2 shows the distribution for the 0.3w% Pt / 0.3w%Re/γ-Al$_2$O$_3$ catalyst[1], and the platinum experimental curve is above the theoretical one, indicating that some other platinum particles were present evenly spread over the surface, although they could not be resolved in the image, and the Re curve indicates an even dispersion of this element over the alumina surface. There is some indication that the Re is depleted in the vicinity of the centre platinum particle. For catalysts with higher metal loading(3%) there is evidence of sintering of the platinum after reaction, since the platinum curve more closely approximates to the theoretical value, but there is no change in the Re dispersion.

The 1.0w% Pt / 1.0w%Sn catalyst had some pure platinum particles and others which gave the curves shown in Figure 3. Both the experimental platinum and tin curves were the same - after allowing for their relative efficiencies - showing that the elements were in the centre particle in equal concentrations, presumably as PtSn alloy. This was confirmed by microdiffraction.

References

1.Z.Huang, J.R.Fryer, C.Park, D.Stirling, and G.Webb. J.Catal. (1994) 148 478

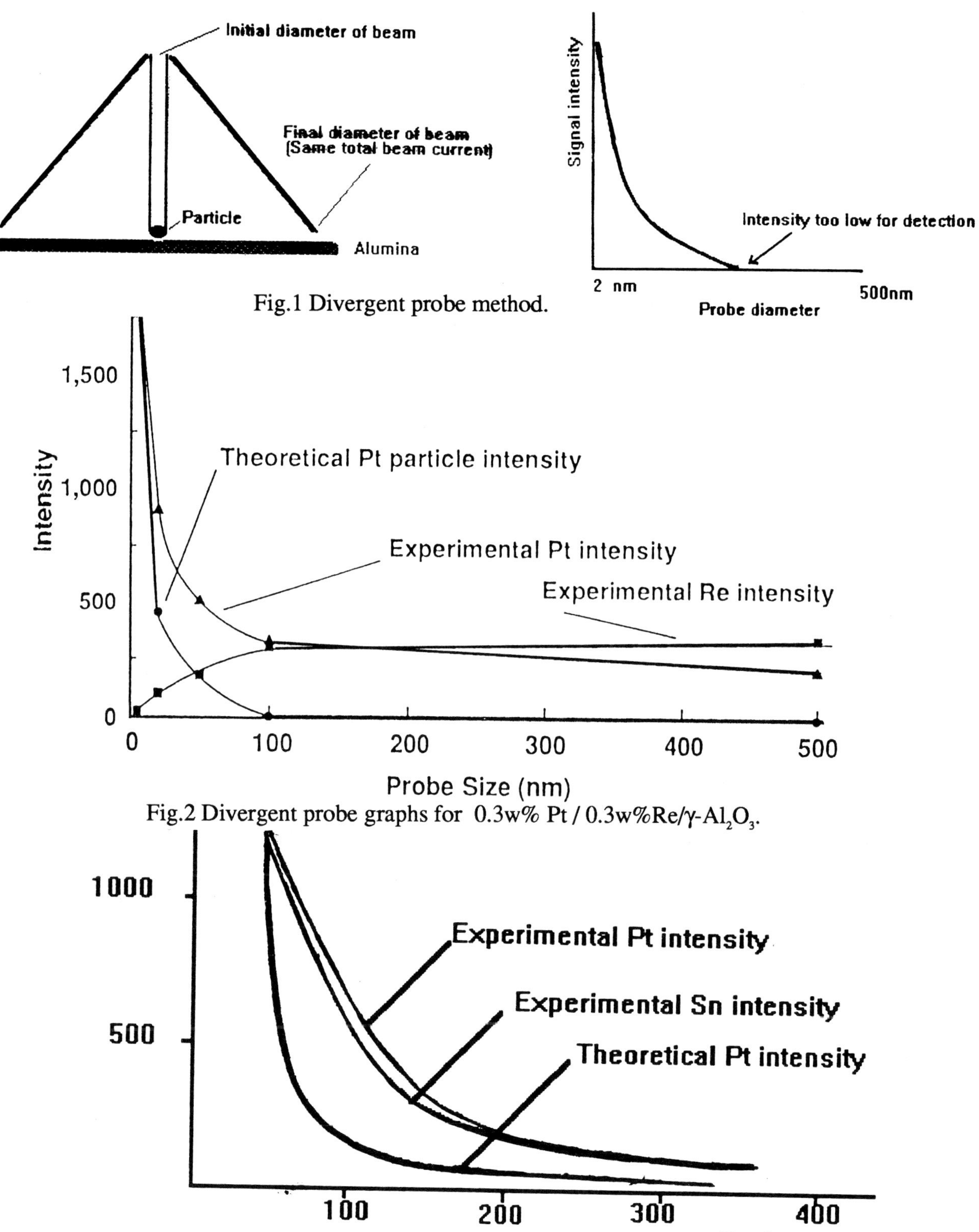

Fig.1 Divergent probe method.

Fig.2 Divergent probe graphs for 0.3w% Pt / 0.3w%Re/γ-Al$_2$O$_3$.

Fig.3 Divergent probe graphs for 1.0w% Pt / 1.0w%Sn/γ-Al$_2$O$_3$.

429

Al$_2$O$_3$/CeO$_2$ Washcoats for Three-Way Automotive Emission Catalysts

E.A. Kenik, K.L. More, W. LaBarge,* and R. Beckmeyer*

Metals and Ceramics Division, Oak Ridge National Laboratory, P.O. Box 2008, Oak Ridge, TN 37831
*General Motors-AC Delco Systems, Flint, MI 48109

Platinum-rhodium based three-way-catalysts (TWC) are the primary catalytic system for control of hydrocarbon, CO, and NO$_x$ automotive emissions. Mixed Al$_2$O$_3$/CeO$_2$ oxides are often dispersed on a cordierite honeycomb monolith as a washcoat and act as a high-surface-area carrier for the heavy metal catalyst clusters. There are both regulatory and marketplace demands for improved performance, lifetime, cost and reliability of these catalytic convertors. The conversion efficiency and lifetime of a convertor is determined by the microstructure of the washcoat/monolith and its evolution during high temperature exposure to the exhaust gas stream. Scanning electron microscopy (SEM), electron microprobe analysis, and analytical electron microscopy (AEM) were utilized to characterize these catalysts. Some AEM was performed on crushed washcoat specimens dispersed on holey carbon films. In addition, an innovative technique was developed for the preparation of transverse section specimens from the monolith,[1] which preserves the spatial correlation of microstructure and elemental distribution with respect to position in the washcoat/cordierite.

Engine dynamometer tests can simulate the ageing a convertor would experience under different driving conditions and mileage. Test 1 simulates 50,000-100,000 miles, depending on the engine/exhaust system; whereas Test 2 simulate exposure to significant over-temperature relative to Test 2. For a new convertor, maximum temperatures are observed near (within ~25 mm) the inlet surface of the monolith. The maximum monolith temperatures typically measured for Tests 1 and 2 are ~1000°C and ~1150°C. Figure 1a shows the general microstructure of the as-prepared washcoat; ceria particles are distributed inhomogeneously as both clumps and individual particles throughout a γ-alumina matrix. Large clumps of the fine-grained (~10 nm) alumina surrounded by a mixture of fine alumina and ceria indicate insufficient blending of the powders. Figures 1b and 1c show the microstructure of the ceria clumps in washcoats as-prepared and after Test 2 ageing, respectively. There is obvious sintering of the ceria phase in the aged washcoat. Sintering was also evident, though less pronounced, after Test 1 ageing. Figure 2a shows a coarse-grained particle (~1 μm) of alumina from the same Test 2 washcoat. Though regions of fine-grained alumina were found in this specimen, a number of large alumina grains were observed that exhibited low dislocation density and fine embedded ceria particles. Electron diffraction indicated that these grains were α-alumina (e.g., Fig. 2b). Similar large α-alumina grains also were observed in the Test 1 aged washcoat, though less frequently. These observations indicate that phase transformation of the alumina matrix and sintering and grain growth of both the alumina and ceria occurred during dynamometer ageing under Tests 1 and 2. These microstructural changes were confirmed by X-ray diffraction on powder samples scraped from the aged washcoats.

The microstructure of the aged washcoats varies with both position along the monolith and depth in the washcoat. Near the inlet surface of the Test 2 aged monolith, the washcoat near the exhaust gas interface exhibited some sintering and grain growth, though both the alumina and ceria had grain sizes <0.1 μm. However, more grain growth was observed in the washcoat close to the cordierite (Fig. 3), with both alumina and ceria exhibiting grain sizes >0.3 μm. In addition, many ceria particles appeared to wet the alumina grain boundaries and some boundaries exhibited a line of bright contrast (arrowed). High spatial resolution X-ray microanalysis with a <2 nm probe (Fig. 4) indicated a marked segregation of cerium (~16 at.%) to such boundaries along with lesser amounts of both silicon (~8 at.%) and titanium (~4 at.%). The slight cerium level (1.4 at.%) indicated in the matrix arises, at least in part, from the excitation of the ceria by both direct (i.e., hole count) and indirect (secondary excitation) routes. The cordierite is an abundant source of silicon. Titanium is an impurity element in both the cordierite and the alumina matrix. The impurity atoms observed at α-alumina boundaries in the washcoat may have diffused long distances. Interphase boundaries can provide fast-diffusion paths for such impurity atoms. In addition, impurity atoms in the matrix may be swept along with the transformation zone and concentrated at the boundary as the alumina transforms from γ → α. The presence of silicon at the grain boundaries of the washcoat may indicate the presence of a glassy, silica phase at the boundaries at elevated temperatures. Such an intergranular phase could promote rapid atom

Proc. Microscopy and Microanalysis 1995, edited by G.W. Bailey, M.H. Ellisman, R.A. Hennigar, and N.J. Zaluzec
Copyright © 1995 MSA. Published by Jones and Begell Publishing, 79 Madison Ave., New York, NY 10016

transport between crystalline phases, which in turn would favor sintering and the $\gamma \rightarrow \alpha$ transformation of the alumina matrix.[2]

References

1. K.L. More, T.S. Geer, and D.W. Coffey, submitted to the 29th Microbeam Analysis Society Meeting.
2. Sponsored by the U.S. DOE, Office of Industrial Technologies, Advanced Industrial Materials Program, under contract DE-AC05-84OR21400 with Martin Marietta Energy Systems, Inc.

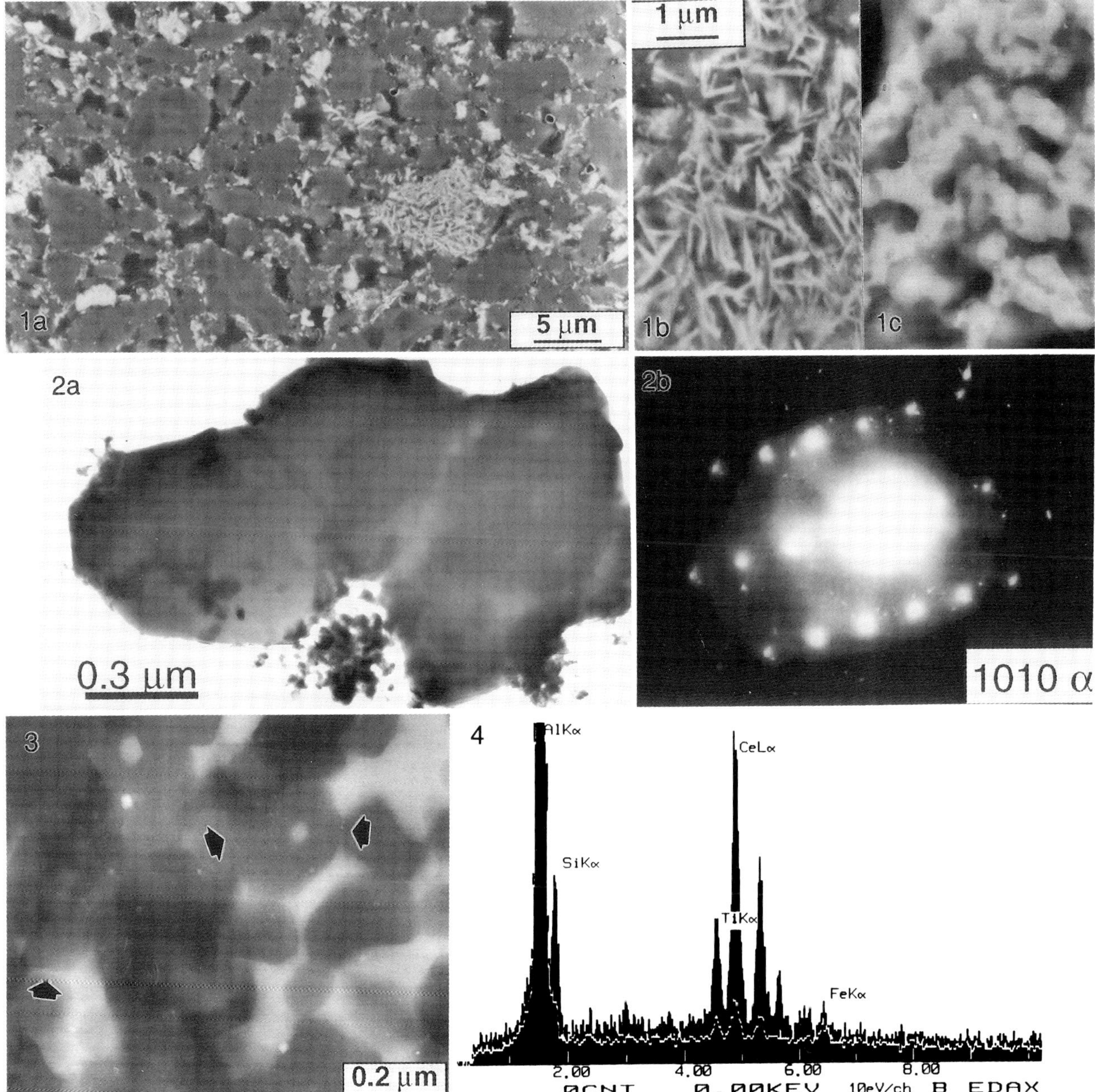

FIG. 1.--SEM micrographs of as-prepared washcoat (a,b) and Test 2 aged washcoat respectively (c).
FIG. 2.--(a) TEM micrograph and (b) convergent beam electron diffraction pattern from large α-alumina particle in Test 2 aged washcoat. FIG. 3.--SEM micrograph of Test 2 aged washcoat near cordierite.
FIG. 4.--EDS spectra for boundary (solid curve) and α-alumina matrix (line curve) of Test 2 aged convertor.

Effect of Oxidation-Reduction Treatments on the Chemisorption Ability and Reactivity of Pd/SiO$_2$.

Dinesh S. Kalakkad and Abhaya K. Datye
Department of Chemical and Nuclear Engineering
University of New Mexico
Albuquerque, NM 87131, USA

Catalytic combustion is known to be the foremost way to oxidize fuels without the simultaneous production of NO$_x$ gases. Supported Pd is an excellent catalyst for low temperature catalytic combustion reactions. However, supported metal catalysts are routinely subjected to oxidation - reduction treatments at high temperatures for the sake of initial activation and regeneration after reaction. It has been shown that supported Pd undergoes drastic morphological changes when subjected to oxidation and reduction at high temperatures. These changes include formation of pits, cavities, horse-shoe shaped morphologies and spreading of the palladium oxide on the support followed by redispersion of the reduced metal[1] . These changes can profoundly affect the catalytic behavior of these catalysts and a thorough understanding of these phenomena is of utmost importance in predicting the behavior of these catalysts under real conditions. However, previous studies reporting these morphological changes have dealt with Pd particles formed by vacuum evaporation onto flat substrates where only plan views of the particles is possible. Since the industrial catalysts are made by chemical impregnation on curved oxide supports, the results obtained previously on such model flat substrates may not correlate well with those seen under practical conditions. We have used model geometry SiO$_2$ (Stöber) spheres to support Pd prepared by chemical impregnation using Pd(acetyl acetonate) as precursor to yield a 5.0 wt.% Pd/SiO$_2$ catalyst. H$_2$ and CO Chemisorption, n-butane hydrogenolysis reactivity and HREM have been used to characterize the catalyst morphology as a function of oxidation and reduction treatments.

Figures 1 and 2 show the chemisorption uptake and reactivity of the catalyst as a function of pretreatment. Reduction at 500°C causes a large drop in both the CO and H$_2$ uptake and the n-butane hydrogenolysis activity of the catalyst. Figure 3 shows a low magnification TEM view of the as-prepared Pd/SiO$_2$ catalyst (oxidized and reduced at 300°C) . Most particles have a central contrast feature suggestive of a second phase which is well evident in the high magnification HREM micrograph shown in Fig. 4. Electron holography[2] shows these central contrast features to be voids in the Pd particles. To isolate the exact conditions under which these voids form, we performed a variety of pretreatments on this catalyst. Figure 5 shows the catalyst after oxidation at 300°C. This figure shows that the catalyst in the oxide form does not contain voids . Figure 6 and 7 show the catalyst particles after reduction at room temperature and reduction at 150°C respectively. The catalyst does not show any voids even after treatment at 150°C suggesting that a higher temperature is necessary to form these voids. Figures 7 and 8 show the catalyst after oxidation at 500°C followed by reduction at 315°C (high uptake and reactivity state) and after reduction at 500°C (low uptake and reactivity state) respectively. The samples do not look very different either from each other or the as-prepared sample except that the catalyst that shows a low reactivity and chemisorption uptake (after reduction at 500°C) shows evidence of an amorphous overlayer. However, such overlayers could either be due to the wetting of the metal by the SiO$_2$ support or due to hydrocarbon contaminants in the system. Nevertheless, the microstructural studies help rule out the possibility of activity variation due to spreading and redispersion of the active metal. Also, since voids are present in both the low activity and the high activity state, these voids do not appear to play a role in activity and chemisorption variations.

[1] J.J. Chen and E. Ruckenstein, *J. Phys. Chem* **85** (1981)1606
[2] L.F. Allard, E. Völkl, D.S. Kalakkad and A.K. Datye, *J. Mater. Sci.* , **29** (1994)5612

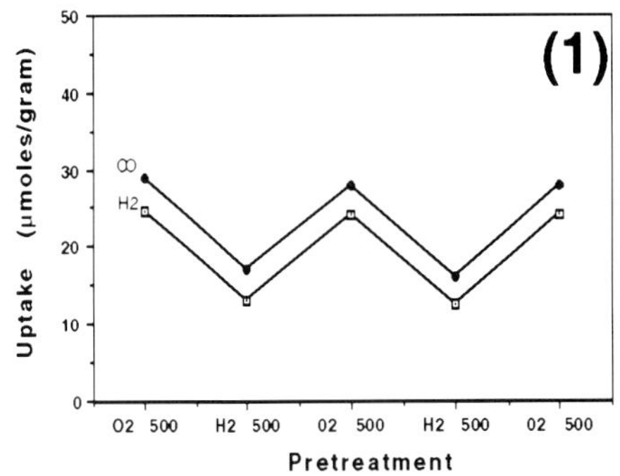

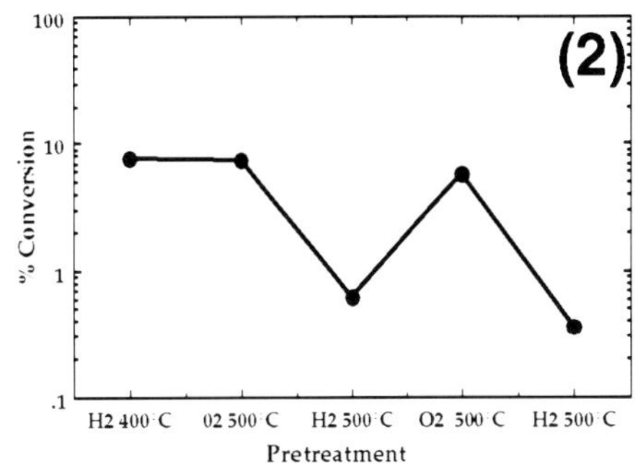

Chemisorption uptake and reactivity as function of pretreatment

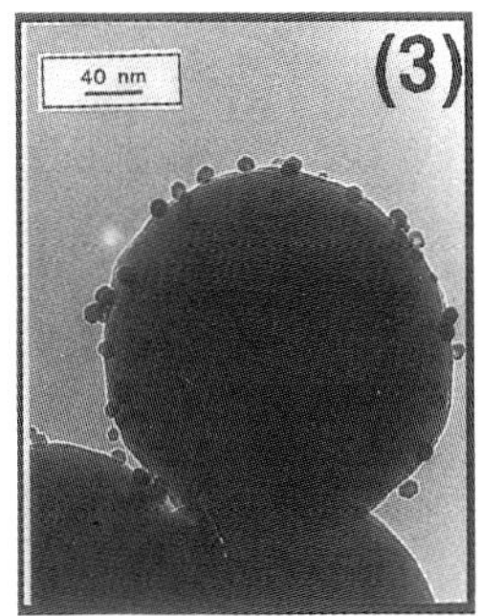

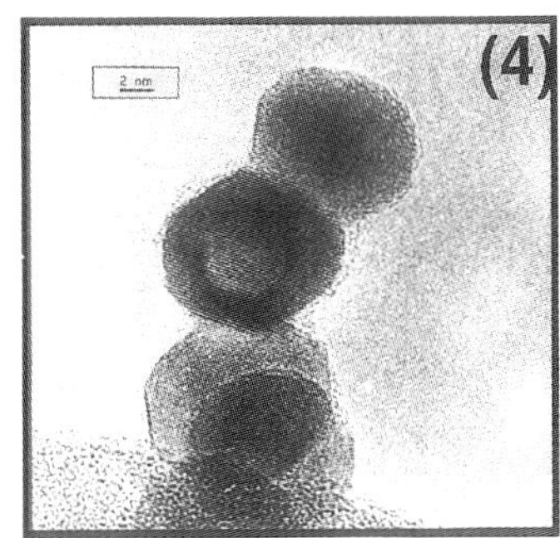

Transmission electron micrographs of the catalyst in the as - prepared state

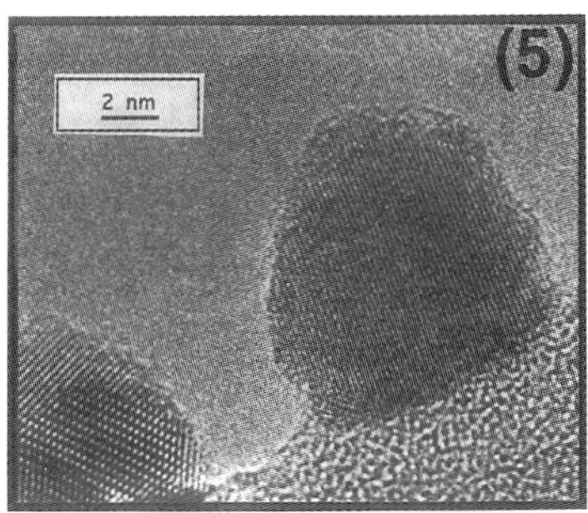

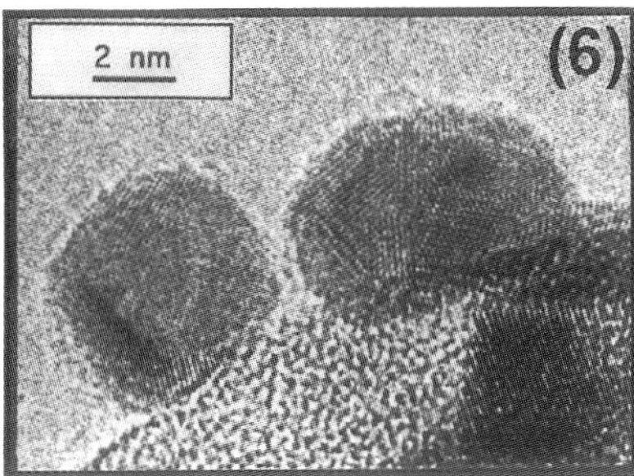

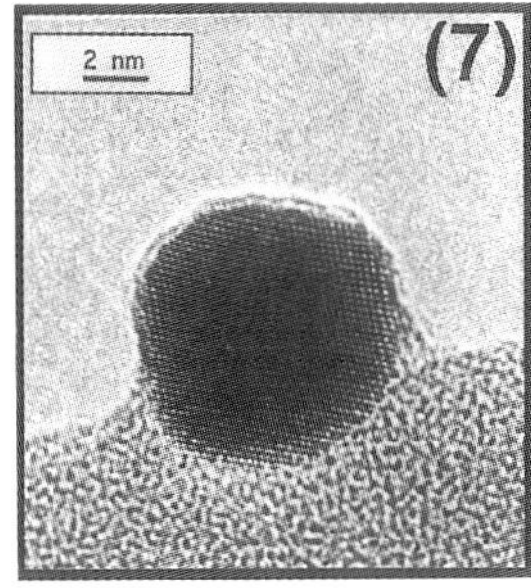

Catalyst after oxidation at 300°C

Catalyst after O2 at 300°C + reduction at room temperature

Catalyst after O2 at 300°C + reduction at 150°C

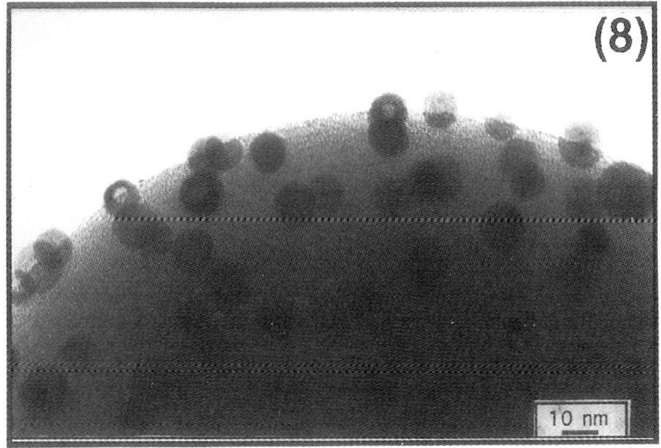

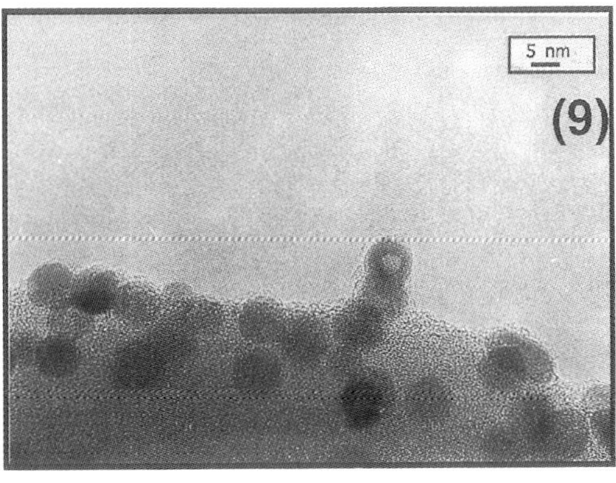

Catalyst in the high uptake and reactivity state

Catalyst in the low uptake and reactivity state

433

Characterization of Carbides in Iron Fischer - Tropsch catalysts

Mehul D. Shroff[1], Dinesh S. Kalakkad[1], Nancy B. Jackson[2], Mark S. Harrington[2], Allen G. Sault[2], and Abhaya K. Datye[1]

[1]: Department of Chemical and Nuclear Engineering and UNM/NSF Center for Microengineered Ceramics, University of New Mexico, Albuquerque, NM 87131.

[2]: Process Research Department, M.S. 0709, Sandia National Laboratories, Albuquerque, NM 87185-0709.

The Fischer - Tropsch synthesis (FTS) for the production of synthetic hydrocarbons from the indirect liquefaction of coal has tremendous potential as an alternative to petroleum - based fuels. The use of iron catalysts is desirable due to their low cost, easy availability, good FTS activity and high water - gas shift activity thus enabling use with low H_2 / CO ratios. However, problems relating to attrition and deactivation need to be addressed. In addition, there has been a controversy in the literature regarding the role of the carbide phase and the identity of the active catalytic phase. The main reason for the existence of this debate has been the use of different characterization techniques. Our results with a commercial, precipitated and spray-dried, Fe_2O_3 - CuO - K_2O Fischer-Tropsch catalyst point to the fact that conventional techniques like X - ray Diffraction (XRD), X - ray Photoelectron Spectroscopy (XPS) and Auger Electron Spectroscopy (AES) which have been traditionally used to characterize these catalysts **(1,2)** are not very successful in detecting the existence of the carbide phase, which generally forms as 20 - 30 nm crystallites on the surface of the micron - sized magnetite crystals. In addition, these carbide nodules are almost invariably covered by carbon films. Both these factors make detection by techniques like XRD and XPS difficult.

High Resolution Transmission Electron Microscopy (HRTEM) provides a powerful tool in resolving the controversies over the nature of the active phase in iron FT catalysts. Our results, based on different activation treatments, elemental analysis and HRTEM, indicate that carbide formation is a prerequisite to obtaining FT activity. The oxide phases do not demonstrate any FT activity. Furthermore, the formation and the morphology of the carbide phase implies the existence of attrition phenomena in the primary particles. This is compounded by the inherent weakness of the primary particle agglomerates. Fig. 1 shows an HRTEM image of the sample activated in H_2 at 543 K for 2 h. followed by exposure to syngas at 523 K for 45 h. The high resolution enlargements of the regions marked (1) an (2) are shown in Fig. 2, where the lattice fringes corresponding to the magnetite and carbide phases are shown in Fig. 2. The existence of large amounts of carbide particles is visible along with the magnetite phase seen immediately after activation. However, XPS analysis of the same sample does not indicate the presence of any carbide (Fig. 3a). This was attributed to the attenuation of the signal due to the presence of the carbon overlayers on the carbide particles, as shown by HRTEM. Only subsequent bombardment of the sample surface with Ar^+ ions to remove the overlayers enabled detection of shoulders in the Fe *2p* and C *1s* regions corresponding to iron carbide and carbidic carbon, respectively (Fig. 3b.)

REFERENCES
1.   C.S. Kuivila, P.C. Stair, and J.B. Butt; *J. Catal.*, **118**, 299, (1989).
2.   C.-S. Huang, L. Xu, and B.H. Davis; *Fuel Sci. and Technol. Intl.*, **11**, 639, (1993).

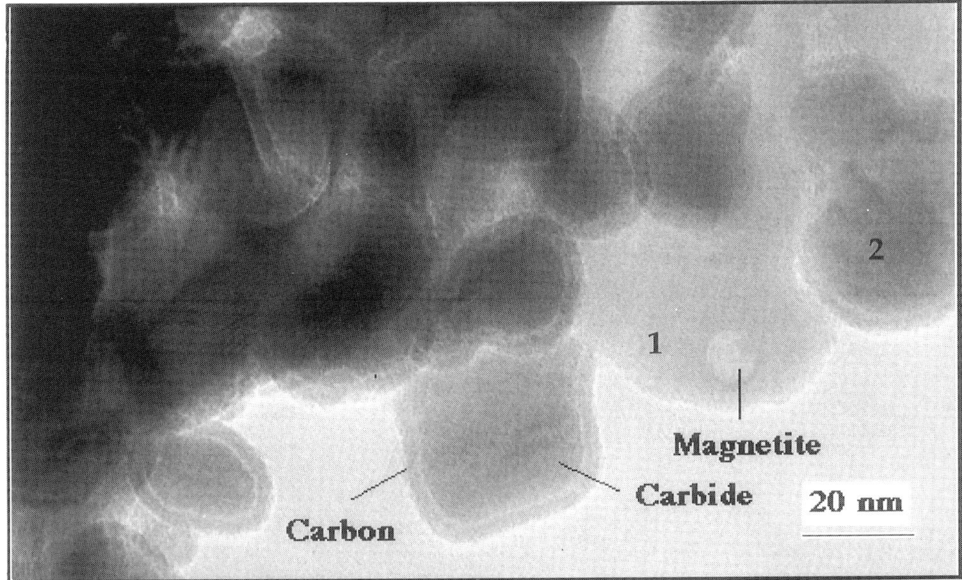

Fig. 1: HRTEM image of sample activated in H$_2$ at 543 K for 2 h. followed by FTS at 523 K for 45 h. showing both the magnetite and carbide phases.

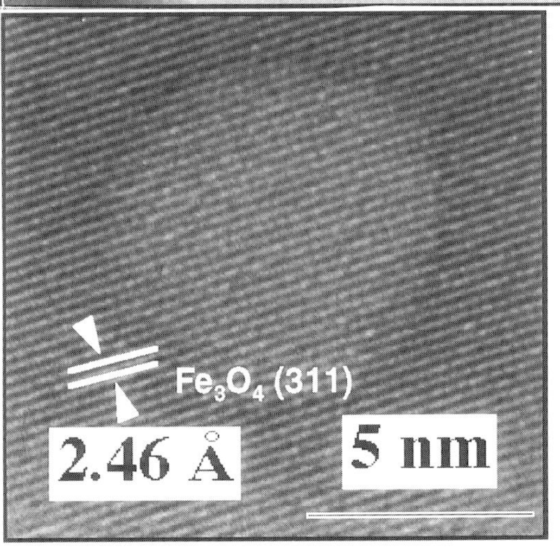

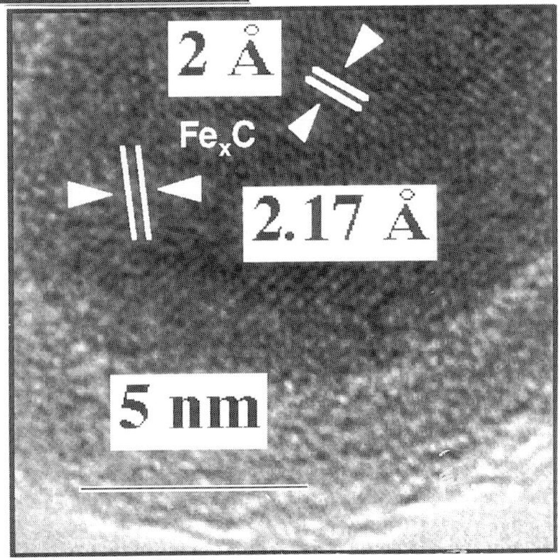

Fig. 2: High resolution enlargements of the regions marked (1) and (2) in Fig. 1, showing lattice fringes corresponding to the magnetite and carbide phases respectively.

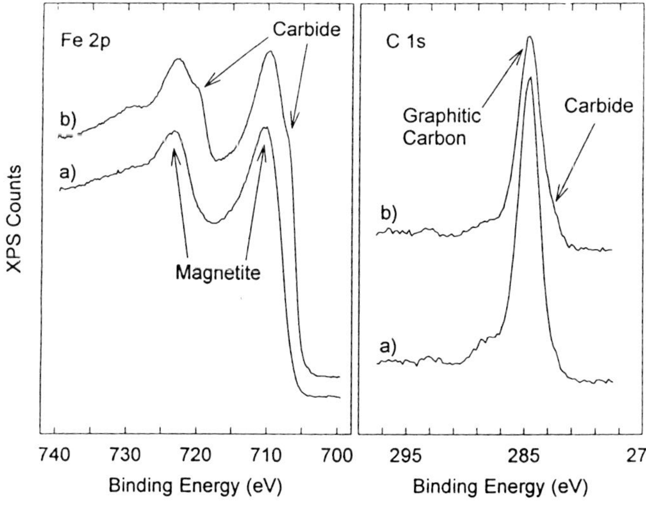

Fig. 3: XPS of the same sample showing the C *1s* and Fe *2p* regions (a) before and (b) after Ar$^+$ bombardment, the latter showing shoulders corresponding to carbidic carbon and iron carbide, respectively.

435

Observation of Grain Superstructure in Thin Aluminum Films by Orientation Imaging Microscopy

Roger Alvis[1], David Dingley, and David Field[2]

[1]Advanced Micro Devices, MS 32
Sunnyvale, California 94088
[2]TexSEM Laboratory
226 West 2230 North #120
Provo, UT 84604 801-344-8990

The correlation of aluminum alloy reliability data to microstructure has long been the goal of those scientists seeking to model electromigration behavior of interconnects. Traditionally, microstructural information has been acquired through x-ray diffraction , and transmission electron microscopy (TEM). However, each of these techniques is capable of delivering only part of the characterization whole. We describe the application of orientation imaging microscopy (OIM) to thin aluminum alloy films and demonstrate its versatility in providing the key microstructural reliability parameters: namely texture and grain size, as well as providing insight to the microstructure of grain boundaries.

OIM was performed on an electromigration test structure (figure 1). The Al-alloy was deposited on titanium and capped with an anti-reflective titanium nitride. Subsequently, the test structure was patterned and capped with a multilayer blanket consisting of silicon nitride (SiN), and SiO2. The structure was annealed after the SOG deposition at 450° C for 90 minutes, seeing no electrical stressing. The die was removed from the package and deprocessed before the OIM was acquired.

The relative orientation between data points is revealed in the OIM from a 10µm x 10µm area of a bond pad. The lines superimposed between the hexagons show the relative orientation between points. A heavy line represents a misorientation greater than 15° and the thin lines show areas of misorientation between 2° and 14°. From these data we measure an approximate grain size of 2.5µm as determined by the large angle grain boundaries. However, upon closer inspection we notice networks of "sub-grain" boundaries that divide the large, high angle grains into smaller grains. An approximate grain size in terms of the average distance between low angle boundaries is 0.5 to 1.0µm--very close to that determined by TEM on a similar blanket Al alloy film (figure 2).

Pole figures were constructed from points on the bond pad with an automated acquisition of approximately 400 points (0.5µm steps) from the pad; furthermore, we were able to acquire orientation data from a 1µm line with a manual acquisition of 65 points (0.5µm steps). The pole figures show both the line and the bond pad to be quite strongly [111] textured (figures 3a-c, 4a-c). A close inspection of the [100] pole figure from the 1µm line reveals there to be two points at the center, indicative of a true [100] orientation. One of these points is from the [100] silicon substrate used to align the sample. The other point is from the line, a small portion (grain) of which exhibits a less desirable [100] orientation.

Thus, in a single experimental set-up, key microstructural information has been acquired which can in turn be related to the reliability of aluminum alloy lines. OIM revealed the [111] texture as well as a "super grain" microstructure of an aluminum film where large 2.5µm grains are composed of clusters of 0.5-1µm smaller sub-grains. Furthermore, pole figures were obtained from an isolated 1µm line which revealed a single grain having an undesirable [100] texture. Therefore, the versatility of OIM is shown for blanket and patterned aluminum thin films--with the potential extention of the method to pre- and post-mortem analyses of electrically stressed lines, with the possible revelations of the microstructural factors in electromigration failures.

Proc. Microscopy and Microanalysis 1995, edited by G.W. Bailey, M.H. Ellisman, R.A. Hennigar, and N.J. Zaluzec

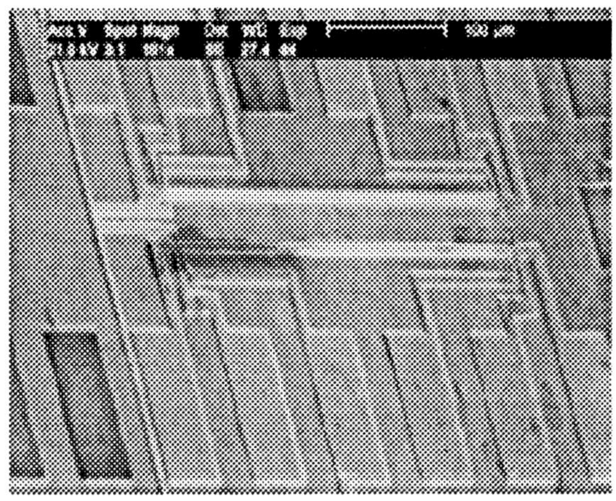

Figure 1: NIST structure providing both "blanket" and patterend aluminum films from which OIM's and pole figures were reconstructed

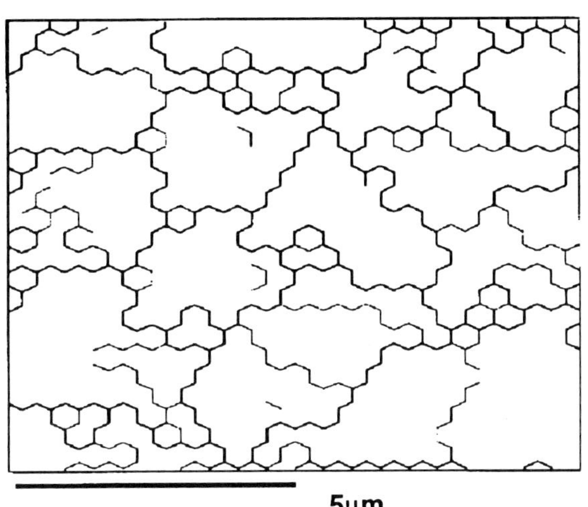

Figure 2: OIM from bond pad showing that ≈2.5μm "super" grains misoriented by >15° (thick lines) are composed of grains with relative misorientations between 2 and 14°.

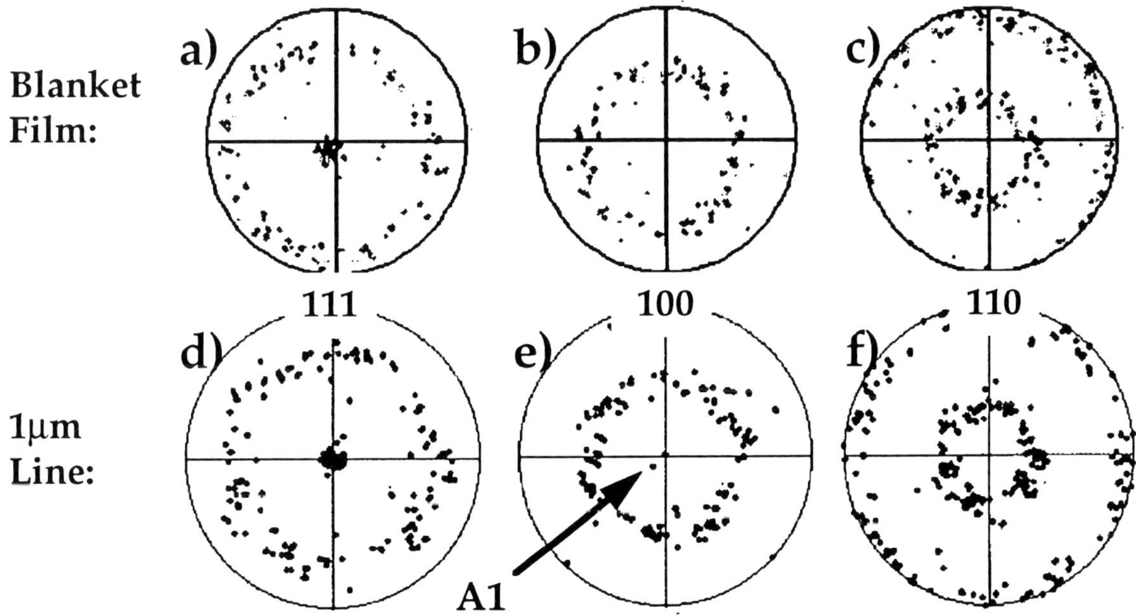

Figure 3: Reconstructed pole figures from indexed EBSPs from blanket (a-c) and patterned(d-f) areas. Both films show strong [111] textures; furthermore, the technique reveals there to be one grain (A1) in the 1μm line which exhibits an almost perfect [100] orientation.

Observation of Silicon Redistribution During TiAl3 Formation in Aluminum Alloy/Titanium Bilayers

Roger Alvis, Paul Besser, John Sanchez, and Ming Tang
Advanced Micro Devices, Sunnyvale, California 94088

Titanium is commonly used as an underlayer to aluminum alloy metal lines in integrated circuit applications, primarily for reliability considerations. The Ti/Al bilayer scheme may exhibit improved electromigration resistance as a combined result of two factors. In the first case, should a void form in Al line, the Ti layer would be capable of carrying a part of the increased current load, effectively shunting the current around the void. The second effect that Ti has is the improvement of the (111) crystallographic texture when compared to Al deposited directly on SiO2 (1). However, the fact that Al and Ti react to form the high resistivity intermetallic TiAl3 at temperatures above 350° C, represents a potentially serious limitation to the practical use of this system. One solution is to alloy the aluminum with silicon, which is known to retard the rate of formation of TiAl3 (2). We propose two mechanisms for the decreased reaction rate involving the diffusion of the Si into the TiAl3 where it acts as a barrier to Al diffusion.

The effect of Si alloying on the rate of TiAl3 formation was studied by preparing films of aluminum-0.5% copper-1% silicon on titanium films in Al/Ti thickness ratios of 40:1, 20:1, and 8:1. These films were prepared by the sputtering of Ti and Al-Cu-Si films onto thermally oxidized silicon wafers at 250° C in a multi-chamber DC magnetron sputtering system at a base pressure of 3e-8 torr. Changes in the sheet resistance were used to calculate the rate of formation of TiAl3 as a function of sequential, 430° C, 90 minute heat cycles.

A kinetic analysis of the AlCuSi/Ti reaction demonstrates the rate of TiAl3 formation to be a function on the ratio of the bilayer thicknesses. Re-plotting these data further reveals a linear relationship between the square of TiAl3 thickness and time, indicative of diffusion limited growth (fig, 1, ref. 3). We thus observe the TiAl3 reaction rate constant to be linearly dependent on the ratio of Ti:AlCuSi thickness (fig. 2). We also see that the addition of 1% Si lowers the reaction rate constant by an order of magnitude, from $\approx$2e-14cm^{2}s^{-1} in Al-Cu to $\approx$3e-15cm^{2}s^{-1} (3).

Microstructurally, we see that there remains unreacted Ti under the intermetallic after 1 thermal cycle. Detailed inspection reveals a thin ($\approx$50Å) amorphous layer at the Ti/TiAl3 interface similar in appearance to the amorphous interfacial layer at the Ti/SiO2 interface (fig 3). An auger electron spectroscopic (AES) depth profile shows a redistribution of silicon into the TiAl3, a constant signal indicative of a uniform concentration throughout the intermetallic layer (fig. 4).

These results suggest two possible mechanisms for the depressed reaction between the Ti and aluminum alloy namely, the formation of a silicon rich layer between the TiAl3 and Ti, and the substitution of Si for Al in TiAl3 (4). While either of these mechanisms could suppress the reaction, it is possible that both operate simultaneously. Substitutional diffusion of Si through the TiAl3 may slow the diffusion of Al to the Ti/TiAl3 reaction front while the presence of the interfacial amorphous diffusion barrier further prevents the aluminum from reaching the titanium. Further, the Ti/TiAl3 interfacial layer may act as a diffusion barrier to aluminum, its presence ensured by the constant source of silicon from the TiAl3 layer. More detailed thermodynamic and analytical studies will be required to confirm these hypotheses.

REFERENCES
1. Tracy, et al, J. Appl. Phys. **76**(5), 2671 (1994)
2. Nahar, et al,Appl. Phys. Let. **50**, 150 (1987)
3. Tardy, et al, Phys. Rev. B **32**(4), 2070 (1985)
4. Shen, et al, MRS Proceedings vol **54**, 103, 1986

Proc. Microscopy and Microanalysis 1995, edited by G.W. Bailey, M.H. Ellisman, R.A. Hennigar, and N.J. Zaluzec
Copyright © 1995 MSA. Published by Jones and Begell Publishing, 79 Madison Ave., New York, NY 10016

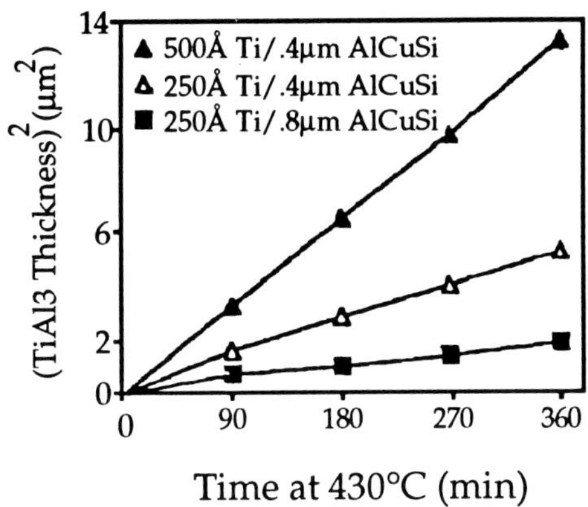

Figure 1: Demonstration of the linear relation between the square of the TiAl3 thickness and the annealing time.

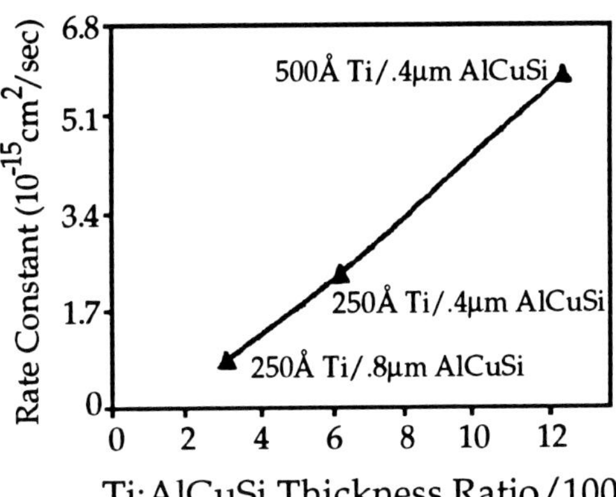

Figure 2: Plot of the rate constant (slopes from fig. 1) showing the linear dependence on the ratio of as-deposited layer thicknesses

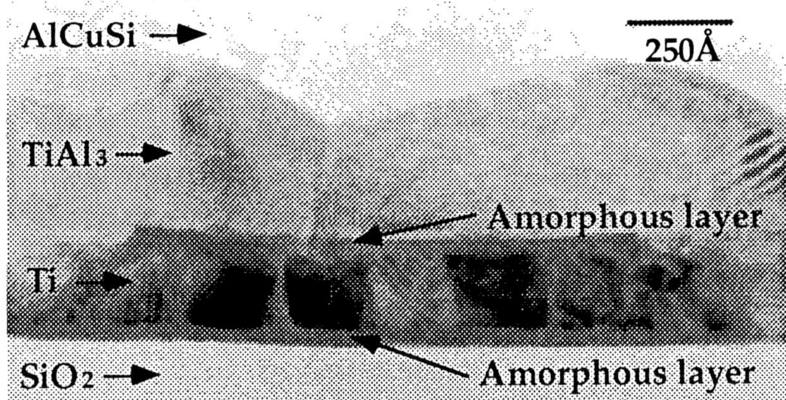

Figure 3: TEM mixcrograph detailing the critical layers after 1 heat cycle. Amorphous layers can clearly be seen between the Ti and TiAl3 as well as between the Ti and SiO2.

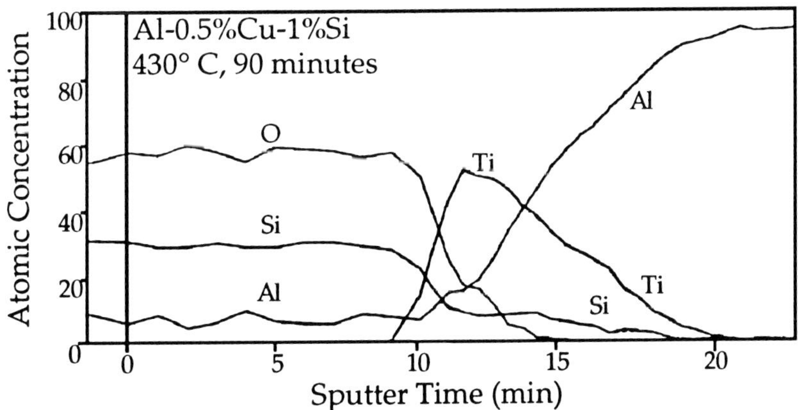

Figure 4: AES Depth profile showing a clear partitioning of Si into the TiAl3 after 1 heat cycle.

ELECTRON MICROSCOPY OF DIAMOND-COATED MOLYBDENUM FIELD EMISSION CATHODES

A. F. Myers, W. B. Choi, M. T. McClure, and J. J. Hren; E. Voelkl,* B. Frost,* and L. F. Allard*

Materials Science and Engineering Department, North Carolina State University, Raleigh, NC 27695
*High Temperature Materials Laboratory, Oak Ridge National Laboratory, Oak Ridge, TN 37831

Molybdenum is a common metal used to fabricate field emission devices.[1,2] However, during extended operation, Mo cathodes exhibit current instabilities, which are thought to arise from adsorption of contaminants on the emitter surface. Coating Mo field emitters with diamond may enhance the current stability, as well as the total emission current.[3] An understanding of the diamond nucleation mechanism is crucial for depositing consistent, uniform coatings on the emitters, and therefore producing reliable devices. To the author's knowledge, little information exists about the mechanism of diamond nucleation on these highly curved Mo surfaces.

In order to study the diamond nucleation mechanism, the morphology of the diamond coating and the structure of the diamond/Mo interface were analyzed by high resolution TEM. Emitters make excellent TEM specimens, since no further sample preparation is required after the diamond deposition; artifacts thus do not become an issue in interpreting the TEM results. Figure 1 shows that the films have a general morphology of clustered nanocrystalline particles. HRTEM lattice images (Fig. 2) and selected area diffraction patterns (Fig. 3) indicate that the coating is polycrystalline diamond, with many crystallites 50-100 Å in size. SAD also showed a Mo_2C interlayer between the diamond and the Mo emitter. The presence of this interlayer suggests that diamond prefers to nucleate on the carbide rather than directly on the Mo surface. In addition to HRTEM, electron holography was employed to analyze the diamond morphology. Scanning electron microscopy cannot be used to determine the 3-dimensional crystal morphology since the diamond crystallites are too small. Thickness profiles from the holography phase reconstructions show highly faceted diamond crystals, as well as unfaceted crystals. These holography results will be discussed in our presentation.

Due to the similarity of several diamond and graphite interplanar spacings, HRTEM lattice imaging alone is not sufficient to identify the carbon coating as diamond. EELS was thus carried out to verify that the films were indeed diamond. A typical EELS spectrum from the coating is given in Fig. 4, which shows distinct peaks that are characteristic of the diamond K-level excitation.[4] As the diamond/Mo interface was approached, weak diamond-like carbon spectral features appeared. The coatings were thus found to be almost entirely polycrystalline diamond, with no amorphous carbon or graphite interspersed among the diamond crystals.

The Mo emitters used in this study were made by electrochemically etching <110>-textured Mo wire in a KOH solution, to a tip radius of less than 100 nm. Diamond was then nucleated on the Mo needles by two different techniques: microwave plasma chemical vapor deposition (MPCVD) using bias-enhanced nucleation[5]; and embedding the Mo with diamond nuclei by immersing the emitter in an ultrasonic bath of diamond powder and alcohol. After nucleation, further diamond growth was carried out in both cases by MPCVD. Microscopy results from the two nucleation methods will be presented and compared.[6]

References

1. I. Brodie and P. R. Schwoebel, Proc. IEEE 82(1994)1006.
2. See recent proceedings of the International Vacuum Microelectronics Conference.
3. J. Liu, et al., Appl. Phys. Lett. 65(1994)2842.
4. R. F. Egerton and M. J. Whelan, Phil. Mag. 30(1974)739.
5. B. R. Stoner, et al., J. Mater. Res. 7(1992)257.
6. The electron holography work sponsored by the HTML User Program, DOE Office of Transportation Technologies, under contract DE-AC05-84OR21400 with Martin Marietta Energy Systems, Inc.

Proc. Microscopy and Microanalysis 1995, edited by G.W. Bailey, M.H. Ellisman, R.A. Hennigar, and N.J. Zaluzec
Copyright © 1995 MSA. Published by Jones and Begell Publishing, 79 Madison Ave., New York, NY 10016

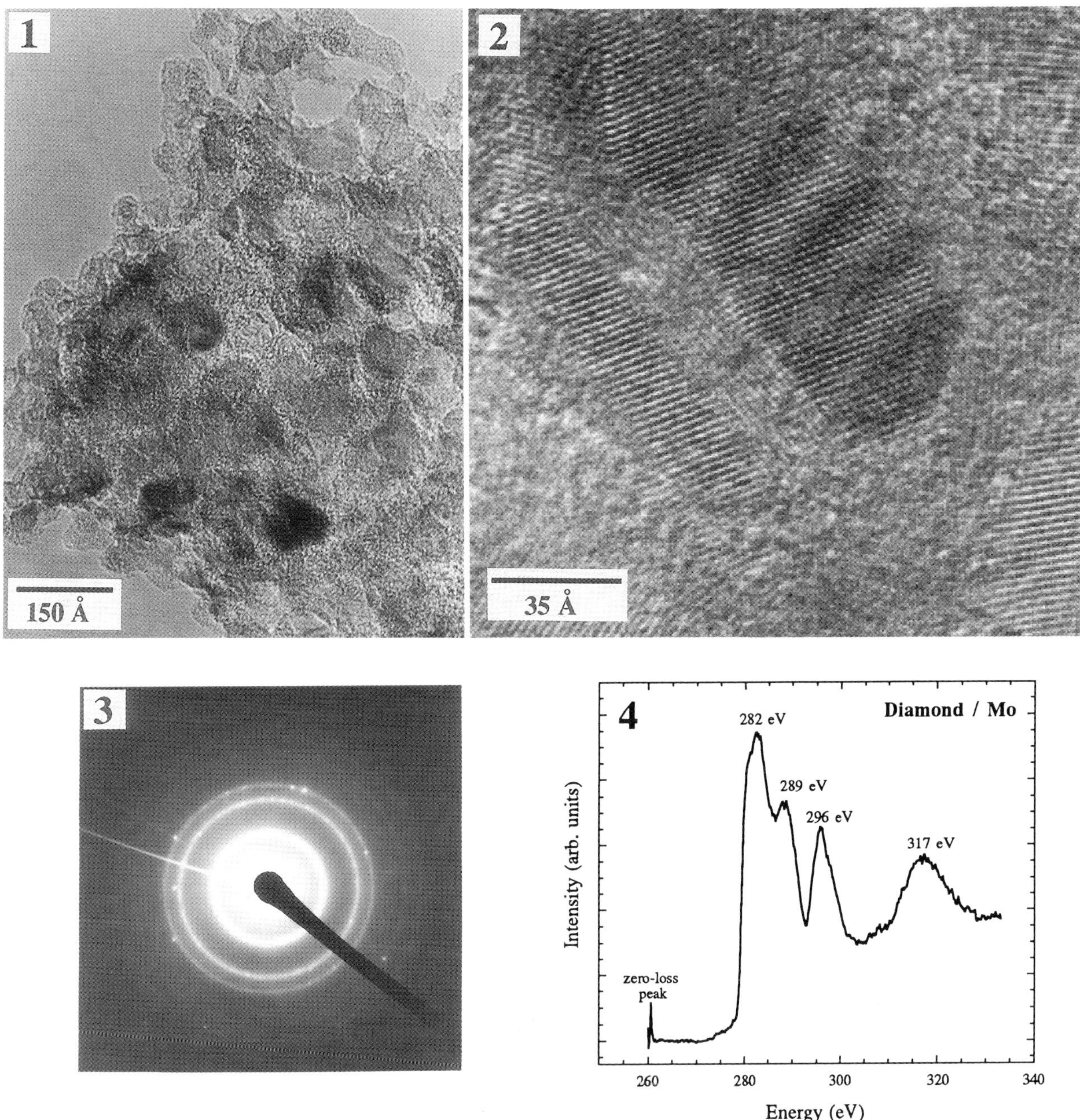

FIG. 1 --- TEM image of diamond coating grown by MPCVD; note polycrystalline nature of film.
FIG. 2 --- High resolution TEM image of film shown in Fig. 1. Diamond {111} planes are visible in several crystals.
FIG. 3 --- Selected area diffraction pattern from coating, indicating presence of polycrystalline diamond.
FIG. 4 --- Typical EELS spectrum from diamond coating on Mo emitter. Neither graphite nor amorphous carbon peaks were observed.

STEM MEASUREMENT OF Ge CONCENTRATION PROFILES IN Si/GeSi/Si HETEROSTRUCTURES WITH 3 Å-RESOLUTION

Tan-Chen Lee, N. David Theodore[+], John Silcox[*]

Department of Materials Science and Engineering, Cornell University, Ithaca, NY 14853
[+]Department of Chem, Bio and Materials Engineering, Arizona State University, Tempe, AZ 85287
[*]School of Applied and Engineering Physics, Cornell University, Ithaca, NY 14853

GeSi alloys are of great interest because of potential applications to high-performance semiconductor devices. For example, the fastest heterojunction bipolar transistor (in Si-based materials) in 1990 was fabricated by using a strained GeSi layer.[1] To accomplish this, it is necessary to keep the Ge concentration and the thickness of the GeSi films below a "critical thickness" to avoid relaxation of the GeSi alloys. It is therefore important to control the Ge concentration and the thickness of GeSi films. When the thickness of the GeSi film is under 10 nm, common characterization techniques such as RBS (Rutherford Backscattering Spectrometry), Auger analysis and SIMS (Secondary Ion Mass Spectrometry) reach resolution limits. The accurate measurement of Ge profiles in GeSi thin films becomes a difficult but important issue. In this paper, we measure Ge concentration profiles by different analytical techniques obtained from STEM and compare them with results from SIMS and Auger analysis.

Measurements of Ge concentration profiles were carried out in a VG HB501A STEM (Scanning Transmission Electron Microscope) with a field emission gun. The STEM is equipped with both Parallel Electron Energy Loss Spectrometry (PEELS) and Serial Electron Energy Loss Spectrometry (SEELS). It is operated with a probe size smaller than 3 Å to ensure a 3 Å spatial resolution throughout the experiment. Fig. 1 presents results from Auger analysis and SIMS versus specified Ge concentration profiles. These spectra are resolution limited. Fig. 2 shows STEM Bright Field (BF) and Annular Dark Field (ADF) images of low magnification. ADF images are formed by collecting electrons scattered to high angles; these reflect atomic number differences.[2] In Fig. 2(b), the ADF image implies a possible Ge-rich region in the interface of GeSi/Si substrate. The high resolution ADF image in Fig. 3 indicates that the Ge-rich region is only two unit cells wide. Energy selected images with an energy width of 5 eV were taken to confirm the existence of the Ge-rich region. In Fig. 4, the intensity ratio from the Ge-L edge (selecting an energy loss of 1258.5 eV) to the Si-L edge (selecting an energy loss of 103.5 eV) are plotted versus ADF intensity. The results from energy selected images and ADF are consistent with each other; both show an unexpected Ge-rich region at the GeSi/Si interface. Similar results were also obtained from PEELS spectra. Ge-L and Si-K edges were used to measure the Ge concentration. The Ge intensity was integrated from 1205 to 1305 eV and the Si intensity was integrated from 1831 to 1845 eV. A GeSi sample with known concentration was used as a concentration standard. The concentration profile from PEELS is consistent with results from energy selected images and ADF.

STEM with 3Å resolution is shown to be a powerful tool for obtaining compositional information on a sub-nanometer scale. STEM analysis for spatial distribution is very accurate compared with other techniques which need to assume film thickness, density, stopping power, etc.[3]

References

1. G. L. Patton, J. H. Comfort, B. S. Meyerson, E. F. Crabbe, G. J. E. DeFresart, J. M. C. Stork, J. Y.-C. Sun, D. L. Harame, Burghartz, Electron Dev. Lett. 11(1990)171.
2. S. J. Pennycook, Ultramicroscopy, 30(1989)58.
3. This work is supported by SRC under Grant No. 95-SC-069. The UHV STEM was aquired through the NSF (DMR-8314255) and is operated by the Cornell MSC. The assistance and improvement of the STEM by Earl Kirkland and Mick Thomas are acknowledged. The sample is provided by Cliff Stein and Curtis Burt. Auger and SIMS data are courtesy of Joy Watanabe and Mike Kottke.

Proc. Microscopy and Microanalysis 1995, edited by G.W. Bailey, M.H. Ellisman, R.A. Hennigar, and N.J. Zaluzec
Copyright © 1995 MSA. Published by Jones and Begell Publishing, 79 Madison Ave., New York, NY 10016

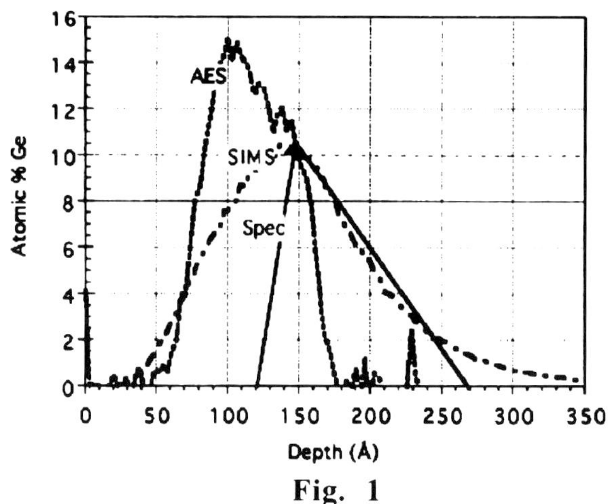

Fig. 1

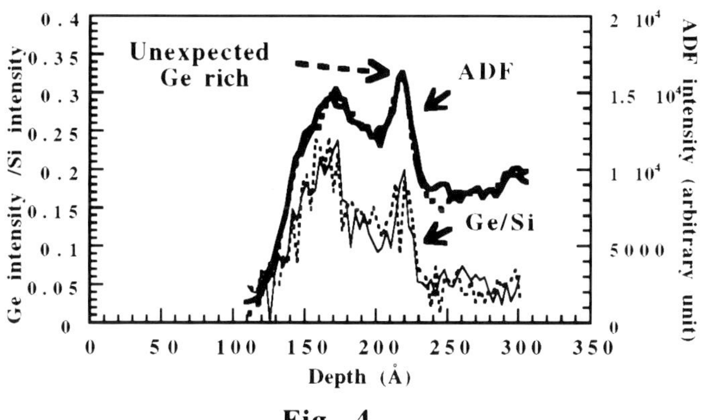

Fig. 4

Fig.1. Comparison of Ge concentration profiles from Auger analysis, SIMS and the specified one.

Fig.2 (a) STEM BF, (b) STEM ADF. An unexpected Ge rich region is observed in (b).

Fig. 3. (a) STEM BF, (b) STEM ADF. The Ge rich region is two unit cells wide.

Fig. 4. Both ADF and energy selected images show the unexpected Ge rich region.

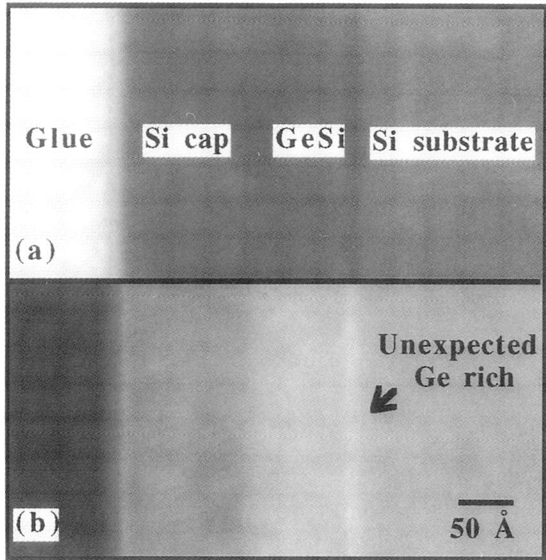

Fig. 2

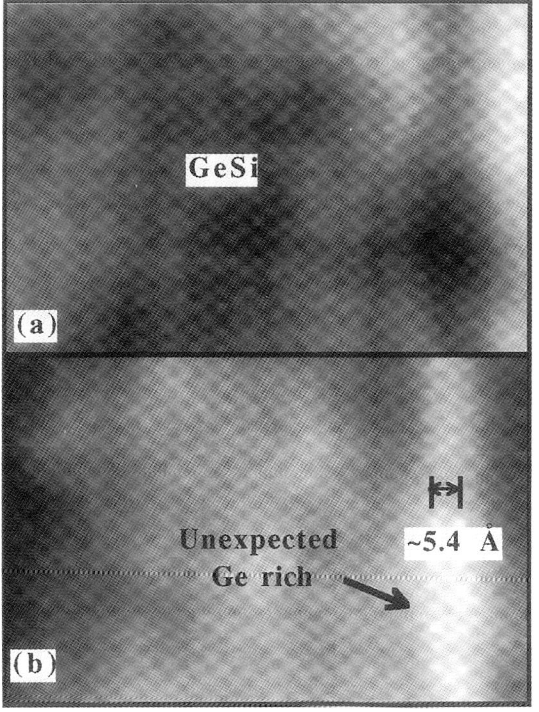

Fig. 3

STRAIN MEASUREMENT IN $In_{0.2}Ga_{0.8}As$/GaAs QUANTUM WIRES BY CONVERGENT BEAM ELECTRON DIFFRACTION

S. Hillyard,* Y.-P., Chen,** J.D. Reed,** W.J. Schaff,** L.F. Eastman** and J. Silcox*

* School of Applied and Engineering Physics, Cornell University, Ithaca, NY 14853
** School of Electrical Engineering, Cornell University, Ithaca, NY 14853

The positions of high-order Laue zone (HOLZ) lines in the zero order disc of convergent beam electron diffraction (CBED) patterns are extremely sensitive to local lattice parameters.[1] With proper care, these can be measured to a level of one part in 10^4 in nanometer sized areas.[2,3] Recent upgrades to the Cornell UHV STEM have made energy filtered CBED possible with a slow scan CCD, and this technique has been applied to the measurement of strain in $In_{0.2}Ga_{0.8}As$ wires.

Semiconductor quantum wire structures have attracted much interest for potential device applications. For example, semiconductor lasers with quantum wires should exhibit an improvement in performance over quantum well counterparts.[4] Strained quantum wires are expected to have even better performance.[5] However, not much is known about the true behavior of strain in actual structures, a parameter critical to their performance.

Figure 1 shows a typical quantum wire. It was constructed by growing a quantum well structure using molecular-beam epitaxy (MBE), dry etching back to form the wire patterning, and then regrowth over the structure by MBE.[6] This forms a defect free structure and puts the wire structure (lattice mismatch 1.4%) under a compressive stress in all three directions. Figure 2 shows a higher magnification picture of a quantum wire with lines marking the location of two CBED linescans, the results of which are plotted in figure 4. Figure 3 shows a typical CBED pattern taken from a sample thickness of approximately 100 nm, a three angstrom probe at a beam energy of 100 kV, and a tilt of about 70 mrad off the (110) axis.

The results in figure 4 are consistent with previous lattice constant measurements of this system performed via high resolution TEM imaging.[7] Specifically, the extension of compressive strain 200 Å outside of the quantum wire sidewall and the lowering of the lattice constant with respect to the quantum well value up to 100 Å inside the wire sidewall is noted. In addition, the added accuracy of this technique has allowed the observation and measurement of a compressive strain in the GaAs on both sides of the wire. No change was detected in the (100) and (010) lattice parameters, indicating that the wire is lattice matched with the substrate in those directions.[8]

[1] P.M. Jones et al., Proc. Roy. Soc. Lond. A 354 (1977) 197.
[2] J.M. Zuo, Ultramicroscopy 41 (1992) 211.
[3] W.T. Pike and L.M. Brown, J. Crystal Growth 111 (1991) 925.
[4] Y. Arakawa et al., Appl. Phys. Lett. 45 (1984) 950.
[5] T. Yamauchi et al., Surf. Sci. 267, (1992) 291
[6] Y.-P. Chen et al., J. Crystal. Growth 134 (1993) 1757.
[7] Y.-P. Chen et al., Appl. Phys. Lett. 65 (1994) 2204.
[8] This work was supported by DoE (Grant #DE-FG02-87ER45322). The UHV-STEM was acquired through NSF (Grant #DMR-8314255) and is operated by the Cornell material Science Center (NSF Grant #DMR-912654).

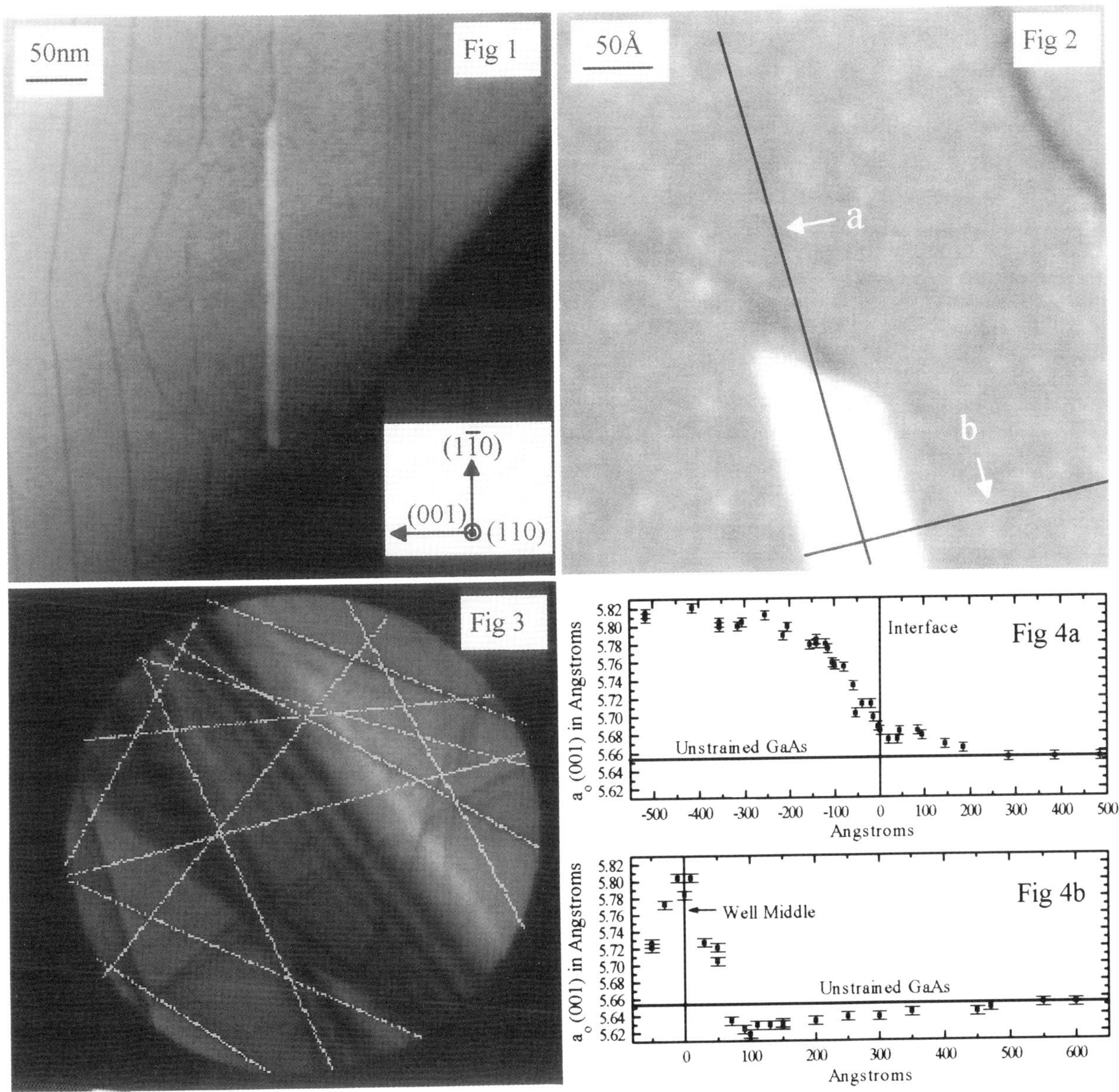

Figure 1. Low magnification annular dark field image showing $In_{0.2}Ga_{0.8}As$ quantum wire (white line). The darker lines are AlAs marker layers showing the pattern of regrowth.

Figure 2. Higher magnification annular dark field image of a different $In_{0.2}Ga_{0.8}As$ quantum wire. The two lines mark the positions of the linescans plotted in figures 4a & b.

Figure 3. Typical CBED pattern from which the the strain was deduced. Some of the HOLZ lines have been marked in white.

Figure 4. Linescans corresponding to figure 2 plotting the (001) lattice constant.

CORRELATION OF DEFECT DENSITIES IN SEMICONDUCTORS MEASURED BY TEM AND OPTICAL PROFILING INSTRUMENTS .

L.Mulestagno[*,**], J.C. Holzer[**], D.E.Hill[**] and P.Fraundorf[*].
* Physics Dept. and Center for Molecular Electronics , University of Missouri - St. Louis, St. Louis , MO 63121.
** MEMC Electronic Materials, Inc., PO Box 8, St. Peters, MO 63376 .

The measurement of oxygen precipitate defect densities in Czochralski (CZ) Silicon wafers after standard heat cycles , or at strategic points in the device manufacturing process is often used as a tool to understand the oxygen precipitation behavior of the wafers on which electronic devices will be manufactured and to measure variability in the precipitation . The relevance of these measurements arises from the fact that controlled oxygen precipitation in the bulk of the wafer is used as a means to 'getter' unwanted impurities from the device layer of the wafer, and control of oxygen precipitation can be an important factor in device yields.

Typically the density of oxygen precipitate in a given wafer and the depth profile is determined by physically counting defects on cleaved slices after preferential etch to reveal the defects . This is a fairly time-consuming and a destructive test. However in the last few years a number of companies have developed computerized, non-destructive IR laser scattering instruments which rely on the fact that silicon is transparent to IR. , and a precipitate (having a different refractive index from silicon), will scatter some of the laser beam which can be detected .[1] Use of IR instruments is becoming more and more common in the field due to their non-destructive nature and the rapid evaluation of material .

Attempts have recently been made to correlate the defect densities determined by these instruments with TEM .[2] While the results from the two techniques usually show similar density patterns (variations) where these exist, it is often noted that the TEM results are higher, by a factor of 2 or more, or even by orders of magnitude in some cases . In the cases were the discrepancy is statistically relevant, the typical explanation has been to attribute these differences to lower sensitivity of the IR instrument to smaller defects , or to say that the IR instrument becomes 'saturated' above a certain density ($\sim 2 \times 10^{10}$ defects/cm^3) . However theory and previous experiments suggest that these IR scattering instruments should be able to detect oxygen precipitates as small as 20-40 nm , so except in some particular case, this explanation does not seem to hold .

A more logical explanation seems to be that the real limitation of optical profiling instruments lies in the volume covered at any time by the focused laser beam (fig 1). At the saturation point ($\sim 2 \times 10^{10}$ defects/cm^3) the specimen has on average 1 or more defects per beam volume . So it seems that the lack of sensitivity at this point arises from the fact that the instrument cannot resolve individual defects anymore . This would also explain why relatively good agreement is obtained between TEM and IR instruments at densities of 10^8 - 5×10^9 defects/cm^3, but the TEM reports a much higher density above

*Proc. Microscopy and Microanalysis 1995, edited by G.W. Bailey, M.H. Ellisman, R.A. Hennigar, and N.J. Zaluzec.
Copyright © 1995 MSA. Published by Jones and Begell Publishing, 79 Madison Ave., New York, NY 10016*

this threshold, where it is observed (in TEM) that n of the defects are within d microns of each other , where d is the diameter of the profiling laser beam . Moreover, since observations in TEM are covering a thin foil, of thickness t, whereas the optical profiler is covering a 3 dimensional volume, there is further clustering of defects which is not visible in the TEM observations, but is leading to a density deficiency in the comparison between the 2 techniques (fig 2). If the beam volume is taken as a cylinder of radius r and height h, the clustering observed in the TEM is only the part occurring in the flat cylinder of volume $\pi r^2 t$, but the actual clustering is occurring over a volume $\pi r^2 h$.

In such comparisons where sufficient data is available to make it statistically significant, the defect density observed in TEM will be larger than that in the IR scattering instrument because (h/t)*n of the defects observed in TEM are not resolvable by the IR scatterer, where n is the fraction of defects within r of other defects in the TEM foil . This correction can be large enough to account for 1 or more orders of magnitude and can account for the discrepancies observed in a comparison between TEM and IR scattering data.

References :

1. J.S. Batchelder and M.A.Taubenblatt, Appl. Phys.Lett. 55,(3), p215 -7,1989.
2. H.Siriwardane et al , ECS proceedings volm. 94-33,1994, p266-276.

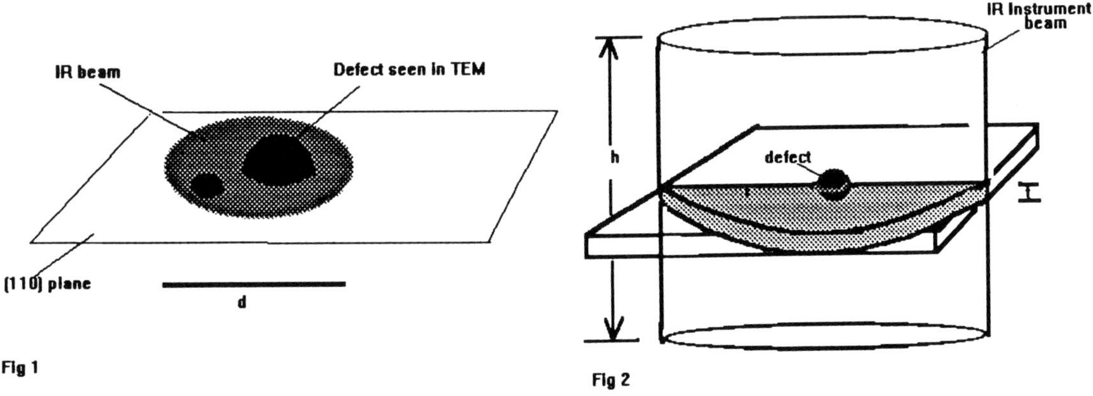

Fig.1 . optical profiling instrument's beam is d μm in diameter, and therefore the instrument would be unable to distinguish 2 defects less than d μm apart.
Fig 2 . Any defect which lies in cylinder shown around another defect would be indistinguishable to an optical profiling instrument of beam of those dimensions . The TEM observer only sees the clustering occurring in the cylinder of height t. Typically r ~1 μm and h ~ 10 μm.

DEFECT FORMATION IN OXYGEN-MULTIPLY IMPLANTED SILICON-ON-INSULATOR MATERIAL

J.C. Park, J.D. Lee*, S.J. Krause*.

Analytical Technology Dept., LG Semicon, Co., Ltd, Cheongju, Korea 360480.
* Chemical, Bio and Materials Eng. Dept., Arizona State University, Tempe, AZ 85287.

High dose oxygen implantation (SIMOX) has been a successful fabrication technology of silicon-on-insulator (SOI) material for CMOS circuits with reduced power consumption and higher operating speed. However, high density ($\sim 10^8$ cm^{-2}) of the through-thickness defects (TTD) in the top Si layer of SIMOX is one of the most serious problems[1]. Hill et al. reported multiple implant/anneal method to remarkably reduce defect densities to $<10^4$ cm^{-2}[2]. In the multiple implant/anneal material, however, $\sim 10^6$ cm^{-2} of the final dominant defects, including stacking fault pyramids (SFP) and the precipitate-dislocation complexes (PDC), still remained after high temperature annealing [3]. In this work, the microstructures and formation mechanism of the final defects were studied by various TEM techniques.

Silicon (100) wafers were sequentially triple implanted to doses of $6/6/6 \times 10^{17}$ at 200kev and 620°C. After each implantation the wafers were held at 1000°C for 2 hours and annealed at 1325°C for 4 hours in argon ambient plus 5% oxygen. Cross-section (XTEM) and plan-view (PTEM) transmission electron microscopy specimens were examined by using a weak beam dark field (WBDF) and high resolution electron microscopy (HREM) techniques in JEM 2000FX and Topcon 002B operating at 200kev.

The precipitate-dislocation complexes (PDC) are the defects formed after incomplete high temperature annealing. The defects consists of narrow stacking faults pinned to residual precipitates at one end, and to the buried oxide at the other hand. It was observed that at intermediate temperature, an almost orthogonal grid of precipitates elongated in <110> directions was formed from multiply faulted defects which were semi-circular plane defects (Figure 1). Further annealing at >1250°C produces dislocations connecting the residual precipitates, resulting in formation of the PDCs at the exact height corresponding to the top of the MFDs in the as-implanted samples (Figure 2). Thus, it is suggested that the origin of PDCs is MFDs in as-implanted sample and they form by the interaction between precipitates and MFDs during thermal annealing.

Stacking fault pyramids (SFP) consist of a pyramid of stacking faults on four {111}planes with stair-rod dislocations along the intersection of the faults. SFPs are formed when the length of the narrow stacking faults(NSF) pinned to a residual precipitate were <530 Å [4]. The length of NSFs was determined by the location of the residual precipitates which varied with the implantation conditions. After each additional implant the tops of the MFDs were closer to the buried oxide. As a result, the residual precipitates are also closer to the buried oxide and the length of the NSFs decreases. After the final implant/anneal, the NSF ribbons are so short that it is energetically favorable for the transformation to SFPs to occur. It is suggested that the transformation of PDCs to SFPs occurs by the dissociation reaction of Shockley partials to form stair-rod dislocations at each edges of a pyramid.

The formation mechanism of various types of SFPs was qualitatively modeled. Four NSF ribbons, eight Shockley partial dislocations on each {111} pyramidal planes of a residual precipitates react at each edge and produce stair-rod dislocations. The excess energy produced during the reaction is consumed in expanding the additional stacking fault area of the pyramid. Thus, a perfect SFP can form by the dislocation reactions. If the formation reaction of stair-rod dislocations from Shockley partial dislocations is incomplete, at any edge of a pyramid, various shapes of imperfect SFPs can form. Four imperfect SFPs were found: I -, Y -, L -, and T - shape SFPs (Figure 3). It is suggested that the shape of SFPs depends on the type of the stair-rod dislocations at edges of the pyramid which is determined by the initial configuration of various Burgers vectors of Shockley partial dislocations.

Proc. Microscopy and Microanalysis 1995, edited by G.W. Bailey, M.H. Ellisman, R.A. Hennigar, and N.J. Zaluzec
Copyright © 1995 MSA. Published by Jones and Begell Publishing, 79 Madison Ave., New York, NY 10016

References

1. A. Auberton-Herve, *Solid State Technology*, 7(1994)89.
2. D. Hill, P. Fraudorf, and G. Fraudorf, *J. Appl. Phys.* 21(1988)4933.
3. J.C. Park, J.D. Lee, D. Venables, S.J. Krause and P. Roitman, *Mater. Res. Soc. Proc.* 275(1993)153.
4. J.D. Lee, J.C. Park, D. Venables, S.J. Krause and P. Roitman, *Appl. Phys. Lett.*, 63 (1993)3330.

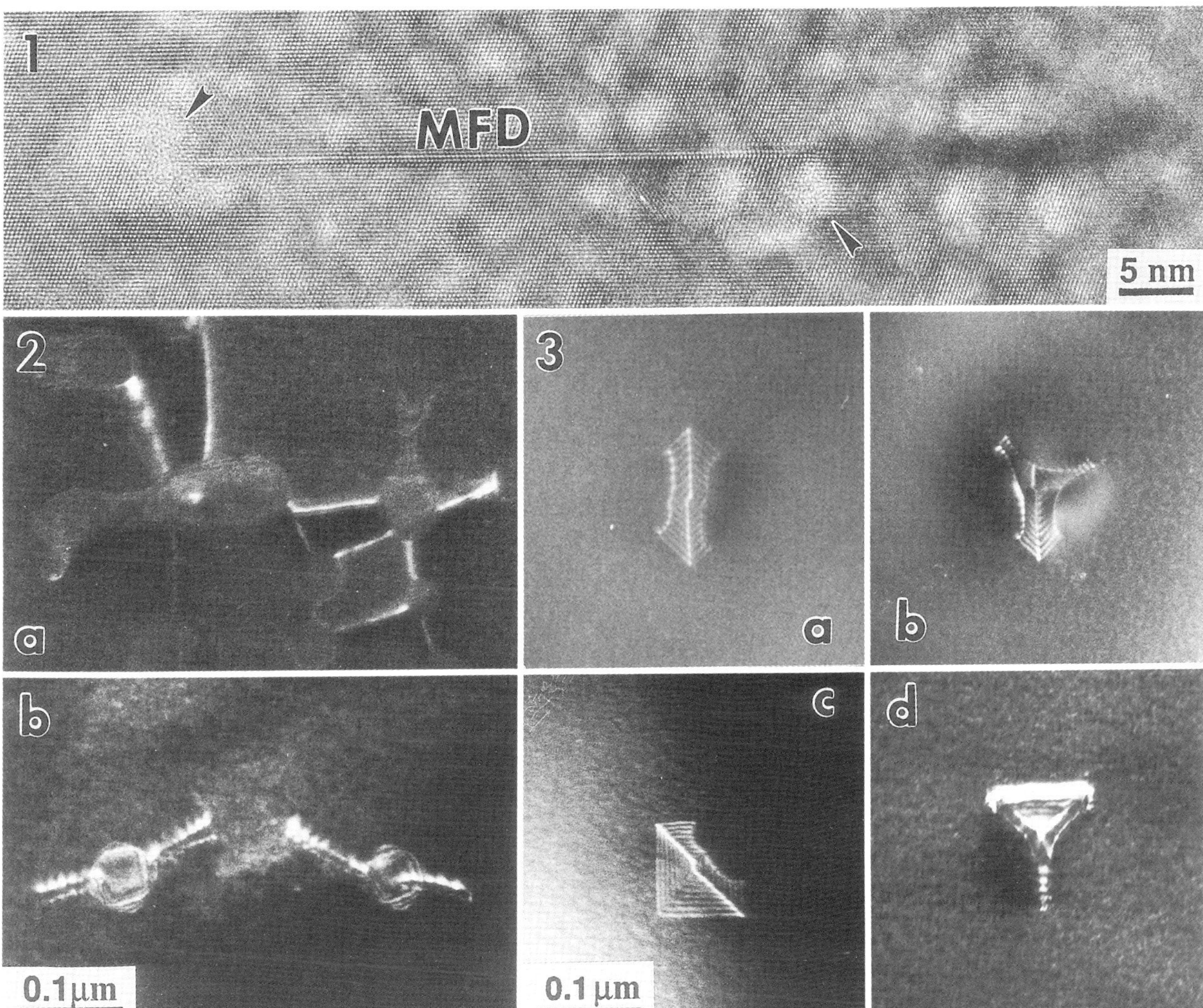

Figure 1.- HREM micrograph of MFDs in SIMOX final-annealed at 1025°C. Notice that oxide precipitates form preferentially at the top of MFD and dissolves at the tail of MFD.

Figure 2.- Plan-view TEM micrographs of the samples third-annealed at (a) 1250°C and (b) 1325°C.

Figure 3.- Plan-view TEM micrographs of various types of imperfect SFPs.
 (a) I- shape SFP   (b) Y- shape SFP
 (c) L- shape SFP   (d) T- shape SFP

MICROSTRUCTURE OF THIN FILM La$_{0.5}$Sr$_{0.5}$CoO$_3$ ELECTRODE FOR MICROELECTRONICS APPLICATIONS

Z. L. Wang, Jiming Zhang*

School of Materials Science and Engineering, Georgia Institute of Technology, Atlanta GA 30332-0245.
*Advanced Technology Materials, Inc., 7 Commerce Drive, Danbury, CT 06810.

Increasing circuit densities in dynamic random access memories (DRAMs) is one of the major interests in microelectronics research. In a ferroelectric capacitor memory, conducting electrodes must make intimate and ohmic contact to both sides of the ferroelectric thin film. The bottom electrode serves as the substrate during growth and the top electrode is deposited on top of the ferroelectric film. Single crystalline La$_{0.5}$Sr$_{0.5}$CoO$_3$ (LSCO) thin films, which are conductive and low electrical resistivity, have recently grown on both sides of Pb-Zr-Ti-O (PZT) ferroelectric film, which has shown a great potential for making fatigue-free small thickness storage capacitors [1]. The present paper reports the microstructure studies of the LSCO thin films grown on MgO(001) and LaAlO$_3$(100) (LAO) substrates by the metalorganic chemical vapor deposition (MOCVD) technique [2].

Cross-section specimens were made to examine the structure of the film and its relationship with the substrate. Select-area electron diffraction (SAD) and high-resolution lattice image show that the LSCO film has an epitaxial relationship with the MgO/LAO substrate, (001)$_{MgO/LAO}$ // (001)$_{LSCO}$ and [100]$_{MgO/LAO}$ // [100]$_{LSCO}$. A remarkable phenomenon seen in Fig. 1 is the appearance of {001}-type reflections of LSCO, which should be forbidden if the isotropic perovskite-type structure remain. TEM image shows the presence of <001> anisotropic structure in correspondence to these "forbidden" reflections (Fig. 2). The (001) and (010) reflections are generated from the regions exhibiting [001] and [010] anisotropic structures, respectively. In comparison to the isotropic perovskite-type structure (Fig. 3a) [3,4], the newly observed structure is denoted as new-LSCO (or n-LSCO), the atomic model of which is shown in Fig. 3b. The model is built by rearranging the Sr and La sites so that the Sr atoms located at z = 0 plane are entirely replaced by La and the La atoms located at z = c/2 plane by Sr, and the Co atom sites remain unchanged. This tetragonal-like structure is proposed with the consideration of the lattice exchange between La and Sr and the preservation of chemical composition of La:Sr:Co:O = 1:1:2:6. The symmetry of the unit cell does not allows (010) and (100) reflections but (001).

Atom structure can be observed directly in HREM image if the specimen is thin. Figure 4 shows a [100] HREM image of the n-LSCO structure. The optic diffracgram shows the appearance of (001) reflection (as indicated by an arrowhead), corresponding to a periodicity of c (= 7.73 Å) along c-axis. For an isotropic perovskite-type structure (Fig. 3a), however, the projected unit cell should exhibit a periodicity of c/2 while viewing along [100]. A simulated image based on the model shown in Fig. 2b is inset, showing reasonable agreement with the observed contrast. The La atom shows the strongest intensity, Sr atom is weaker and Co atom is the weakest. Numerous stacking faults with displacement vectors of 1/4[011] have been observed. The (010) interplanar distance in LSCO is constrained due to larger mismatch between LSCO and MgO, resulting in sharp diffuse scattering streaks along [010] in the [100] diffraction pattern (Fig. 1). This lattice constrain effect, however, is greatly reduced in LSCO grown on LAO due to smaller lattice mismatch. Both LSCO/MgO and LSCO/LAO interfaces show coherent growth, but the LSCO/MgO interface exhibits higher dislocation density

Another new structure that is occasionally seen in LSCO/LAO exhibits a c-axis oriented anisotropic structure with a periodicity of 3/2c (Fig. 5). The corresponding diffraction pattern shows (0 0 2/3) and (0 0 4/3) reflections. A tetragonal structural model corresponding to cation composition of La:Sr:Co = 1:2:3 is shown in Fig. 6. The simulated [100] image of the model shows reasonable agreement with the observed one (Fig. 5). This structure, however, has been observed only in the film grown on LAO. The n-LSCO structure shown in Fig. 3b is a popular structure in the films grown on either MgO or LAO.

1. R. Ramesh et al., Appl. Phys. Lett. **63** (1993) 3952; R. Dat et al., Appl. Phys. Lett. **63** (1994) 2673.
2. J. Zhang et al. in MRS Proc. Vol. **300** (1993) 249.
3. F.S. Galasso, Structure and Properties of Inorganic Solids, Pergamon Press (1970) section 7.4.
4. H.L. Yakel, Acta Cryst. **8** (1955) 394.

Proc. Microscopy and Microanalysis 1995, edited by G.W. Bailey, M.H. Ellisman, R.A. Hennigar, and N.J. Zaluzec.
Copyright © 1995 MSA. Published by Jones and Begell Publishing, 79 Madison Ave., New York, NY 10016

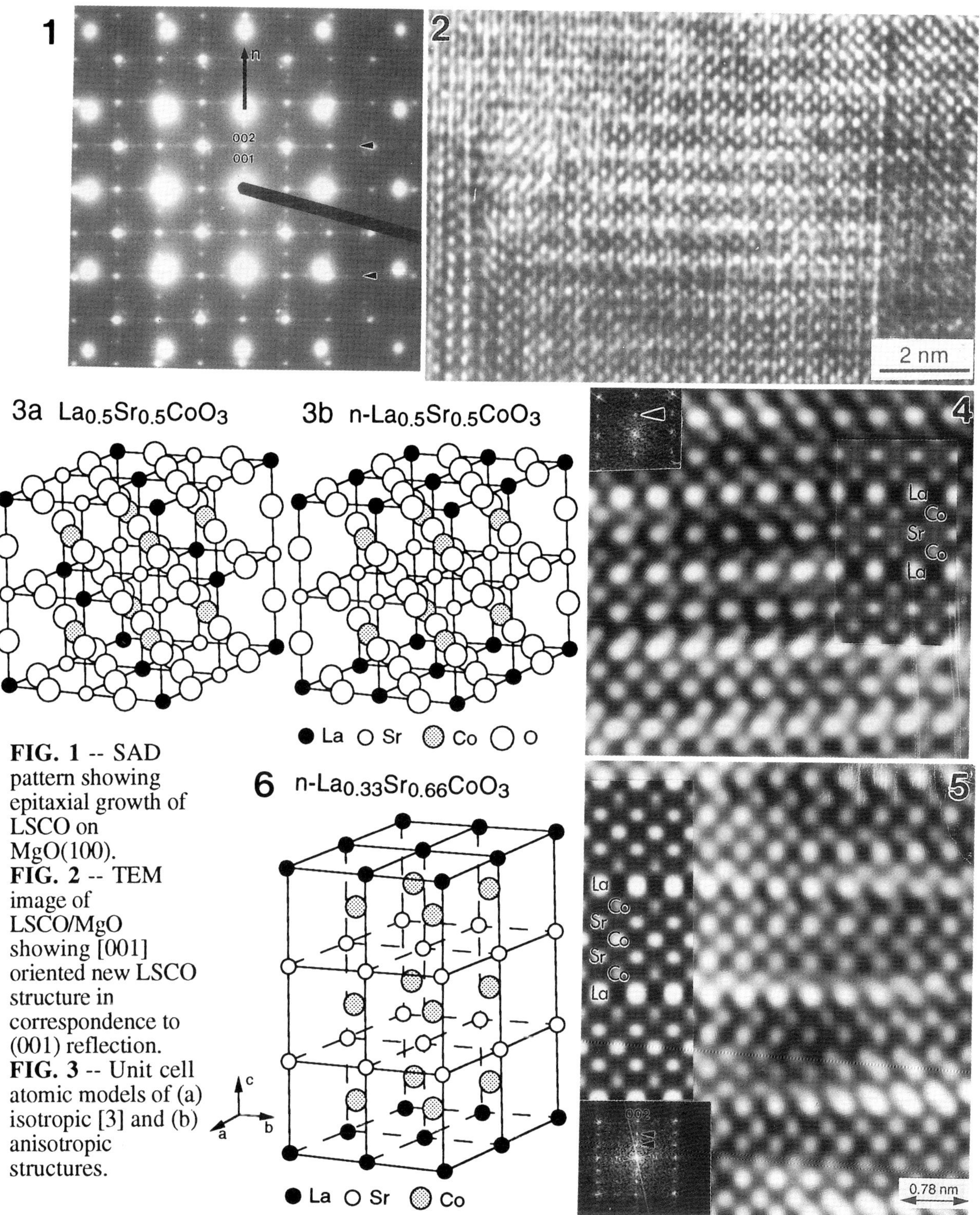

FIG. 1 -- SAD pattern showing epitaxial growth of LSCO on MgO(100).
FIG. 2 -- TEM image of LSCO/MgO showing [001] oriented new LSCO structure in correspondence to (001) reflection.
FIG. 3 -- Unit cell atomic models of (a) isotropic [3] and (b) anisotropic structures.

FIG. 4 -- Comparison of observed and simulated [100] HREM images of n-La$_{0.5}$Sr$_{0.5}$CoO$_3$.
FIG. 5 -- Comparison of observed and simulated [100] HREM images of n-La$_{0.33}$Sr$_{0.67}$CoO$_3$.
FIG. 6 -- Cation atomic model of n-La$_{0.33}$Sr$_{0.67}$CoO$_3$, c/a = 1.62.

451

EVOLUTION OF THE MICROSTRUCTURE OF Cu(Ti)/SiO₂ LAYERS ANNEALED IN NH₃

J. Liu*, N. D. Theodore#, D. Adams#, S. Russell#, T. L. Alford# and J. W. Mayer*

* Center for Solid State Science, Arizona State University, Tempe, AZ 85287
\# Department of Chemical, Bio and Materials Engineering, Arizona State University, Tempe, AZ 85287

Copper-based metallization has recently attracted extensive research because of its potential application in ultra-large-scale integration (ULSI) of semiconductor devices. The feasibility of copper metallization is, however, limited due to its thermal stability issues.[1,2] In order to utilize copper in metallization systems diffusion barriers such as titanium nitride and other refractory materials, have been employed to enhance the thermal stability of copper.[3] Titanium nitride layers can be formed by annealing Cu(Ti) alloy film evaporated on thermally grown SiO_2 substrates in an ammonia ambient. We report here the microstructural evolution of Cu(Ti)/SiO_2 layers during annealing in NH_3 flowing ambient.

The Cu(Ti) films used in this experiment were prepared by electron beam evaporation onto thermally grown SiO_2 substrates. The nominal composition of the Cu(Ti) alloy was $Cu_{73}Ti_{27}$. Thermal treatments were conducted in NH_3 flowing ambient for 30 minutes at temperatures ranging from 450°C to 650°C. Cross-section TEM specimens were prepared by the standard procedure. Figure 1 shows a schematic diagram illustrating the cross-sectional view of the as-prepared $Cu_{73}Ti_{27}$/SiO_2/Si heterolayer structure. High-angle annular dark-field (HAADF) imaging and nanometer resolution energy dispersive X-ray spectroscopy (EDS) were performed on a dedicated scanning transmission electron microscope, VG HB-501, to extract structural and compositional information about the annealed Cu(Ti)/SiO_2 layers.

Figure 2 shows a cross-section HAADF image of a $Cu_{73}Ti_{27}$/SiO_2/Si sample annealed in NH_3 at 550°C for 30 minutes. Interfacial reactions at the $Cu_{73}Ti_{27}$/SiO_2 interface were clearly demonstrated. In order to understand the evolution of the microstructure of $Cu_{73}Ti_{27}$/SiO_2 layers during annealing, high spatial resolution elemental maps of samples annealed at different temperatures were obtained. Figure 3 shows intensity line scans, extracted from digitally stored X-ray elemental maps, of oxygen, silicon, titanium and copper in a sample annealed at 450°C for 30 minutes. There was no clear evidence of reaction between the Cu(Ti) alloy and the underlying SiO_2 layer although slight interdiffusion of elements between the two layers might have occurred. There was a significant formation of titanium oxide and possibly some titanium nitride at the top surface. Based on the variations of Ti and Cu concentration across the layers it was concluded that Ti segregated out of the Cu(Ti) alloy to form a titanium oxide/nitride layer. After 550°C anneal in NH_3 for 30 minutes there was clear evidence that Cu(Ti) alloy had reacted with the underlying SiO_2 (see figure 2) This interfacial reaction was more pronounced after 650°C anneal in NH_3 for 30 minutes as shown in figure 4 which are elemental maps of silicon, oxygen, titanium and copper. The formation of interfacial phases is clearly shown. Quantitative study of the compositional variations of each element across the layers was realized by extracting intensity line scans (figure 5) from the digitally stored elemental maps. The titanium profile clearly follows that of the oxygen in the reaction regions. This suggests that titanium oxide might have formed in the interfacial regions. Copper was also involved in the reaction process, possibly with silicon. The reaction mechanism(s) and reaction products of Cu(Ti)/SiO_2 layers annealed in ammonia will be discussed.[4]

References

1. J. Li, Y. Shacham-Diamand and J. W. Mayer, *Mat. Sci. Rep.* 9 (1992) 1.
2. J. Li and J. W. Mayer, *J. Appl. Phys. Lett.* 6 (1990) 2519.
3. J. Li and J. W. Mayer, *Mat. Res. Soc. Bull.* XVIII, No. 6 (1993) 52.
4. The electron microscopy work was conducted at the Center for HREM at Arizona State University, supported by NSF-DMR-931432.

Proc. Microscopy and Microanalysis 1995, edited by G.W. Bailey, M.H. Ellisman, R.A. Hennigar, and N.J. Zaluzec

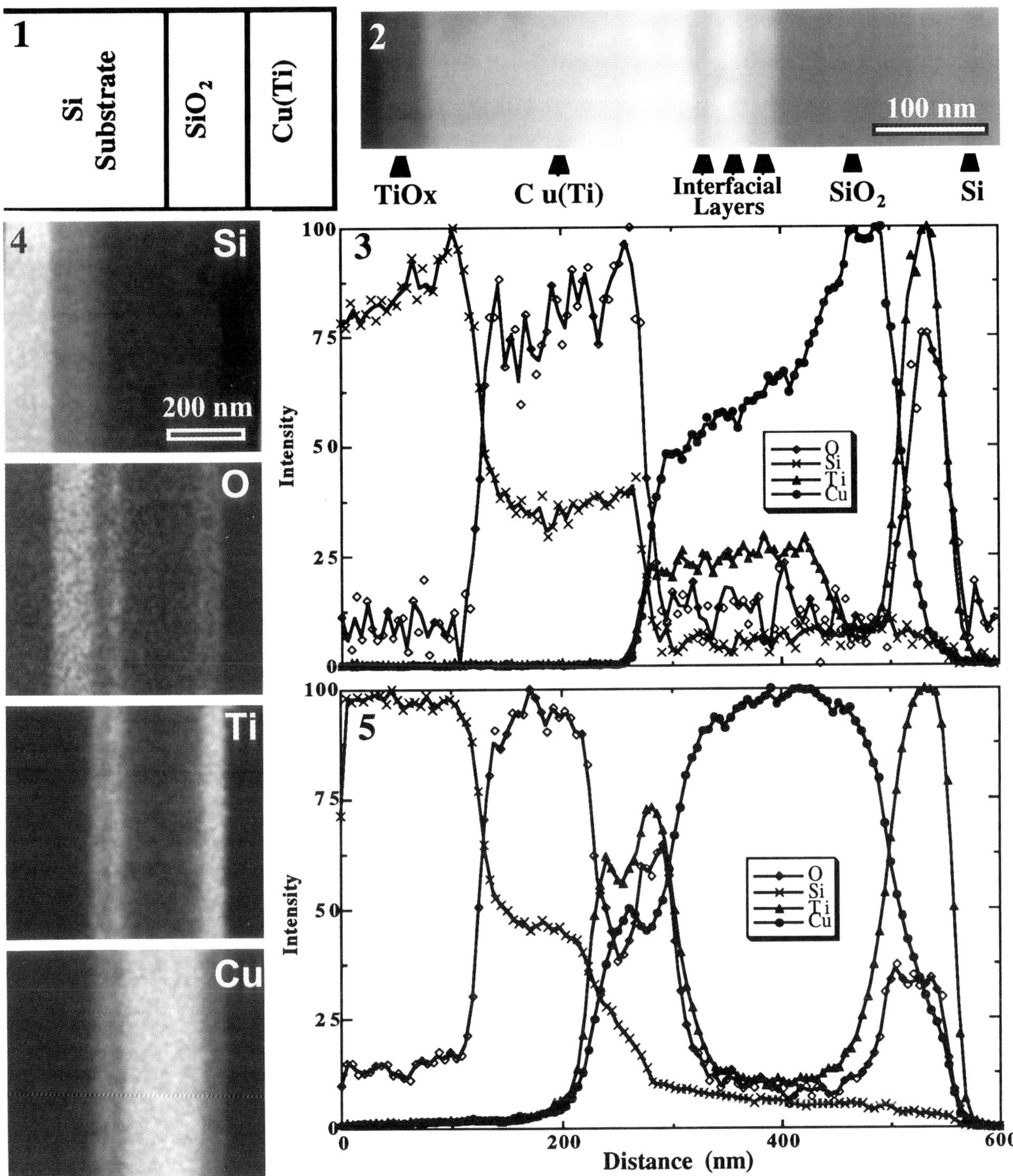

Fig. 1----Schematic diagram illustrating as-prepared Cu(Ti)/SiO$_2$/Si heterolayer structure.

Fig. 2----HAADF image of Cu(Ti)/SiO$_2$/Si sample annealed in NH$_3$ at 550°C for 30 minutes, revealing interfacial reactions at Cu(Ti)/SiO$_2$ interface and formation of TiO$_x$ at free surface.

Fig. 3----EDS intensity line scans of Cu(Ti)/SiO$_2$/Si sample annealed in NH$_3$ at 450°C for 30 minutes.

Fig. 4----Elemental maps of Cu(Ti)/SiO$_2$/Si sample annealed in NH$_3$ at 650°C for 30 minutes.

Fig. 5----EDS intensity line scans of Cu(Ti)/SiO$_2$/Si sample annealed in NH$_3$ at 650°C for 30 minutes.

MICROSTRUCTURAL INVESTIGATION OF SURFACE ROUGHNESS IN MBE-GROWN AlAs

R. Rajesh[*], R. Droopad[**], C. H. Kuo[**], R. W. Carpenter[*] and G. N. Maracas[**]

[*]Center for Solid State Science and Science and Engineering of Materials Program, Arizona State University, Tempe, AZ 85287-1704
[**]Center for Solid State Electronics Research, Arizona State University, AZ 85287-6206

Knowledge of material pseudodielectric functions at MBE growth temperatures is essential for achieving in-situ, real time growth control. This allows us to accurately monitor and control thicknesses of the layers during growth. Undesired effusion cell temperature fluctuations during growth can thus be compensated for in real-time by spectroscopic ellipsometry[1]. The accuracy in determining pseudodielectric functions is increased if one does not require applying a structure model to correct for the presence of an unknown surface layer such as a native oxide[2]. Performing these measurements in an MBE reactor on as-grown material gives us this advantage. Thus, a simple three phase model (vacuum/thin film/substrate) can be used to obtain thin film data without uncertainties arising from a surface oxide layer of unknown composition and temperature dependence.

In this study, we obtain the pseudodielectric functions of MBE-grown AlAs from growth temperature (650°C) to room temperature (30°C). The profile of the wavelength-dependent function from the ellipsometry data indicated a rough surface after growth of 0.5 µm of AlAs at a substrate temperature of 600°C, which is typical for MBE-growth of GaAs. The surface roughness was an indication that low growth temperatures resulted in reduced surface diffusion of the incoming Al atoms. To obtain a smooth surface, the AlAs layers were subjected to post-growth anneals at 650°C and 700°C for 15 minutes each. The profiles of the dielectric function obtained for the layers annealed at 650°C and 700°C show an improvement over the as-grown one. The best profile occurred for the layer annealed at 700°C for 30 minutes, indicating smoothing of the originally rough surface. Annealing for longer times produced little improvement.

RHEED patterns obtained from the as-grown AlAs layer were spotty, confirming the surface roughness, and showed evidence of epitaxial facets along the {311} planes. The roughness and the facets disappeared upon annealing at 700°C for 30 minutes at which time the RHEED pattern became streaky and sharp.

The surface morphology of the as-grown and annealed AlAs was studied directly by HRTEM. Figure 1 shows a low magnification (01$\bar{1}$) cross-sectional view of the as-grown AlAs sample. A pseudo-periodic waviness on the surface of the AlAs layer is clearly seen. Figure 2 shows a low magnification cross-sectional view of the sample annealed at 700°C for 30 minutes. A much smoother and planar AlAs surface is observed. A high-resolution image of the GaAs/AlAs interface is shown in figure 3. Distinct facets along the crystallographic (311) as well as the (511) planes are evident on the surface of the AlAs. The HREM image of the annealed GaAs/AlAs interface shown in figure 4 exhibits good AlAs surface quality and none of the facets observed in the as-grown interface. The relative surface free energies of AlAs for the (311) and (511) planes were determined using the Reverse Wulff method which yields, $\gamma(511)/\gamma(311) \sim 0.9$, where γ is the surface free energy of the low-energy orientation[3]. This confirms the greater propensity to form facets along the (511) planes. In conclusion, AlAs layers grown along the [100] direction at low temperatures (~600°C) need to be annealed at 700°C for 30 minutes to improve the surface smoothness.

References

1. G. N. Maracas et al., J. Appl. Phys., **44** (1995) 77.
2. C. H. Kuo et al., J. Vac. Sci. Technol., **B2** (1995) 13.

Proc. Microscopy and Microanalysis 1995, edited by G.W. Bailey, M.H. Ellisman, R.A. Hennigar, and N.J. Zaluzec
Copyright © 1995 MSA. Published by Jones and Begell Publishing, 79 Madison Ave., New York, NY 10016

3. C. Herring, Structure and Properties of Solid Surfaces (University Press, Chicago, 1953), p. 5-72.
This research was supported by a grant from ARPA, No. N00012-92-J-1931. The microscopy was
conducted at NSF/ASU HREM facility under grant NSF-DMR-9115680.

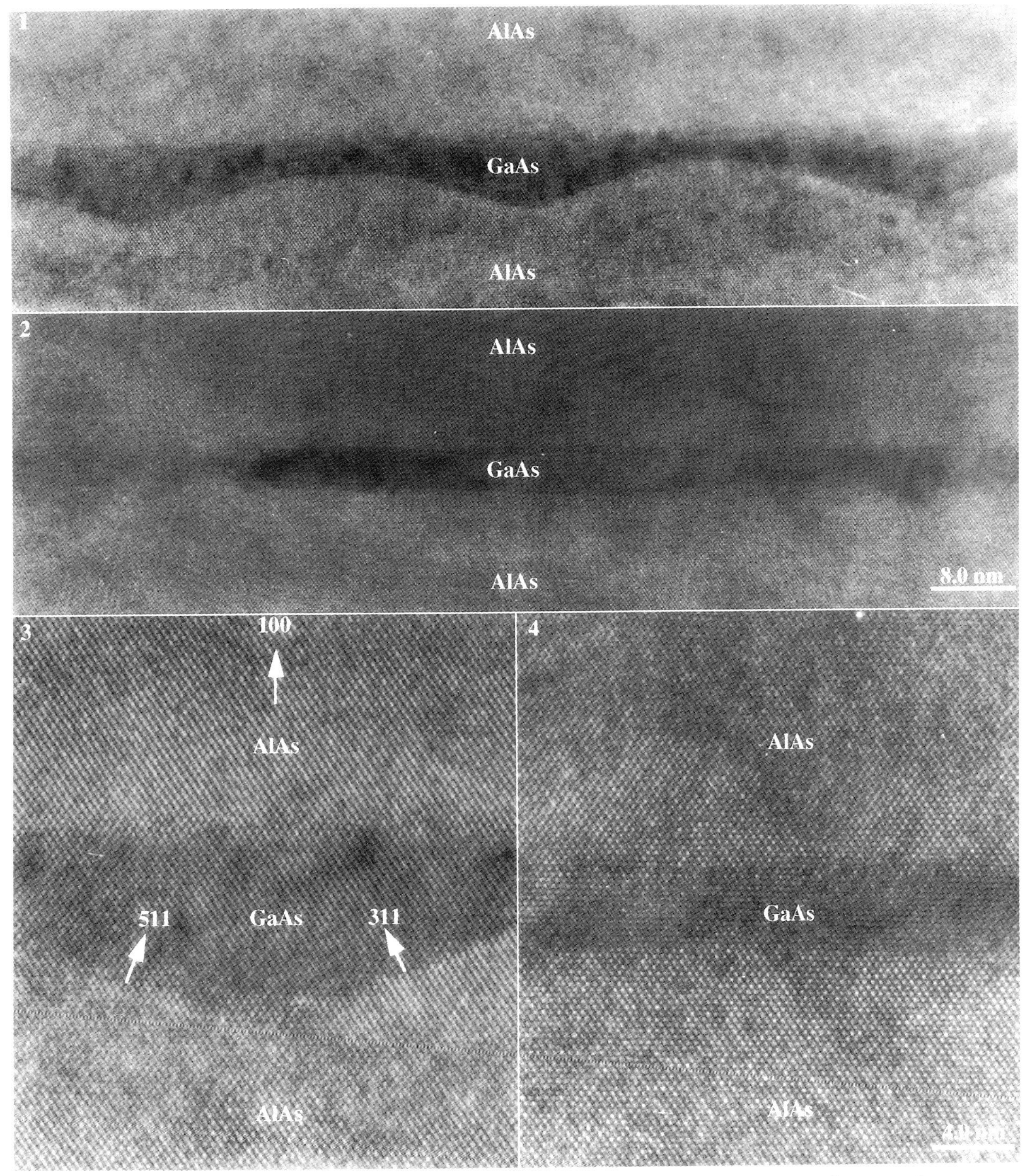

Fig. 1. TEM cross-sectional image of as-grown AlAs showing pseudo-periodic surface roughness.
Fig. 2. TEM cross-sectional image of annealed AlAs showing smooth planar surface.
Fig. 3. HREM image of as-grown GaAs/AlAs showing facets along (311) and (511) on AlAs surface.
Fig. 4. HREM image of annealed GaAs/AlAs showing improved AlAs surface smoothness.

TEM AND AFM STUDIES OF STRUCTURAL DEFECTS IN α-GaN FILMS

W. Qian,* M. Skowronski,* M. De Graef,* G. Rohrer,* K. Doverspike,** L. B. Rowland,**+
and D. K. Gaskill**

*Department of Materials Science & Engineering, Carnegie Mellon University, Pittsburgh, PA 15213
**Laboratory for Advanced Material Synthesis, Naval Research Laboratory, Washington, DC 20375
+now at Westinghouse Science & Technology Center, Pittsburgh, PA 15235

Interest in wide band-gap GaN and related group-III nitrides has grown recently due to their potential applications for short wavelength optoelectronic devices[1]. One of the main obstacles for their application is the high density of defects formed during heteroepitaxy. Though important progress has been made by using a low temperature AlN buffer layer[2], recently reported films still contain a defect density (mainly threading dislocations) on the order of 10^{10} cm^{-2}. This paper reports our current studies of structural defects in α-GaN thin films grown on a-plane $(11\bar{2}0)$ and c-plane (0001) sapphire substrates by organometallic vapor phase epitaxy (OMVPE). Atomic force microscopy (AFM), conventional and high resolution transmission electron microscopy (TEM) were used to investigate the nature and origin of these defects.

Single crystal α-GaN films were grown using a low temperature AlN or GaN buffer layer in an inductively-heated, water cooled, vertical OMVPE reactor operated at 57 torr[3]. On both a-plane and c-plane sapphire substrates, the wurtzite structure GaN and AlN films grew epitaxially along their c-axis. Structural analyses near the buffer layer interfaces and in the GaN overlayer were carried out using cross-sectional and plan-view TEM (JEOL 4000EX, 400 kV; Philips EM420, 120 kV) and AFM (Park Scientific Instruments). Fig. 1 shows the plan-view bright field micrograph taken near the top surface of a 2.8 µm thick GaN overlayer. The predominant defects which penetrate the GaN films are pure edge threading dislocations having Burgers vectors of the $1/3<11\bar{2}0>$ type, lying along the c-axis. Low angle grain boundaries, formed during coalescence of islands at the initial stages of GaN growth[4], are the main source of dislocations. Films with fewer grain boundaries were observed to have dislocation densities as low as 1×10^9 cm^{-2}. TEM and AFM reveal that the initial growth of GaN overlayer was three dimensional. Before coalescence, the island sizes range from 50 to 500 nm, with in-plane misorientation of less than 3°. Low angle grain boundaries consisting of dislocation arrays were thus formed when the neighboring islands coalesced. Fig. 2 delineates the grain structure in a cross-sectional view. Due to the nature of these threading dislocations, there was no appreciable annihilation with further increase of the GaN layer thickness above about 0.5 µm. It is, therefore, critical to optimize the buffer layer deposition conditions to minimize or avoid formation of the grain-like structure.

In addition to dislocations, long hollow pipes of 7 to 100 nm in diameter were observed in the GaN overlayers[5]. They penetrate along the c-axis, with densities between 10^5 to 10^7 pipes/cm^2. The pipes are generally hexagonal, bounded by the six close packed $\{1\bar{1}00\}$ prism planes (Fig.3-4). Most of them exhibit a funnel-like shape with its wider crater at the GaN crystal free surface. A Burgers circuit drawn around the pipe shows no net atomic displacement on the (0001) plane, while two beam analysis shows that the image contrast of the pipe is consistent with that of a screw dislocation. AFM observation reveals steps a few atomic layers high associated with the majority of the pipes. It is thus suggested that these are open-core screw dislocations[6] formed under thermodynamic equilibrium, similar to micropipes (a few microns in diameter) observed in α-SiC crystals (which have been linked to electrical breakdown in pn junction diodes[7]). Since the hollow pipes in GaN films are of nanometer dimension and suitable for TEM observation, their study could also lead to a better understanding of the nucleation mechanism of the micropipes in SiC crystals.

References

1. S. Strite, et al. Thin Solid Films 231 (1993) 197.
2. I. Akasaki, et al. J. Crystal Growth 98 (1989) 209.
3. L.B. Rowland, et al. Inst. Phys. Conf. Ser. 137 (1993) 429.
4. W. Qian, et al, Appl. Phys. Letts. 66 (1995) 1252.
5. W. Qian, et al, J. Crystal Growth (1995) in press.
6. F.C. Frank, Acta Cryst. 4 (1951) 497.
7. J.A. Powell, et al. Inst. Phys. Conf. Ser. 137 (1994) 161.
8. Supported in part by AFOSR Grant No. F29620.94.0392.

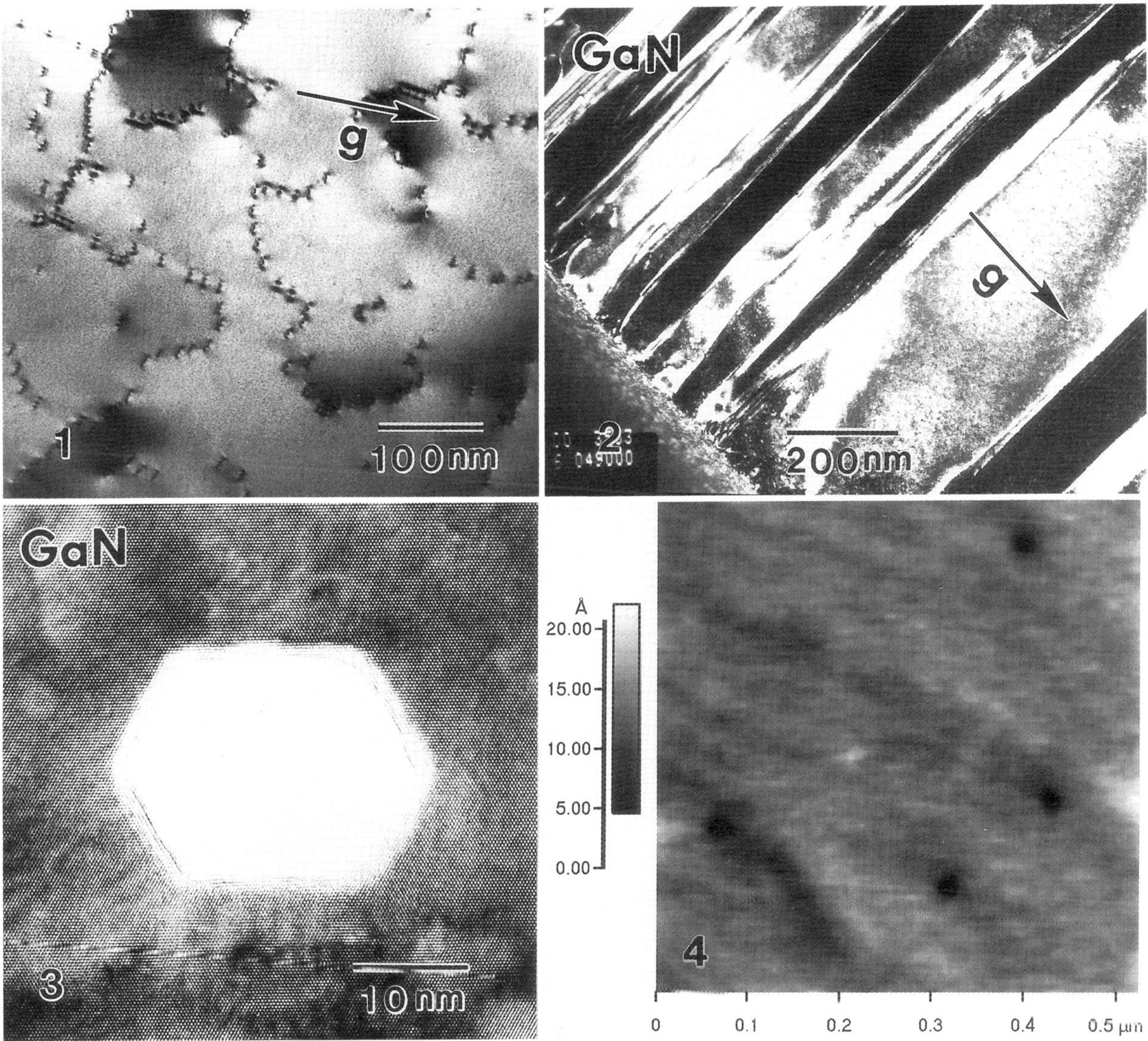

FIG.1 Plan-view bright field micrograph ($\mathbf{g}=[11\bar{2}0]$) taken near the top surface of a 2.8 µm GaN film.

FIG.2 Cross-sectional weak beam image ($\mathbf{g}=[11\bar{2}0]$) of the same sample reveals the grain structure.
FIG.3 HRTEM image of a hexagonal hollow pipe obtained with the [0001] zone axis.
FIG.4 AFM topography of a 2.8 µm α-GaN film showing the nanopipes.

HIGH ENERGY *As* ION IMPLANTATION TO *GaAs*

J. Jasinski*,a, Z. Liliental-Weber*, M. Kaminska**, J. Washburn*, C. Jagadish***

*Lawrence Berkeley Laboratory, University of California, Berkeley, CA 94720, USA
**Institute of Experimental Physics, Warsaw University, 00-681 Warsaw, Poland
***Departament of Electronic Materials Engineering, Research School of Physical Sciences and Engineering, Australian National University, ACT 0200 Canberra, Austaralia

Over the last few years there have been many studies of GaAs layers grown in the temperature range from 180°C to 300°C called LT-GaAs. These studies were motivated by the potential applications of that material in microwave and fast optoelectronic devices. Annealed LT-GaAs layers have interesting electrical properties including high resistivity and short photocarrier lifetime connected with the nonstoichiometry of that material [1]. However, it is still not clear whether arsenic antisite point defects or arsenic precipitates formed during annealing cause these interesting properties of LT GaAs.

A second method of creating highly nonstoichiometric GaAs beside low temperature MBE growth is implantation with a high dose of arsenic. This method first applied two years ago [2,3] showed that 200 keV arsenic ion implantation also leads to semi-insulating properties and that successive post-implantation annealing results in arsenic precipitate formation, similar to that observed in LT GaAs. Recently, more elaborate studies concerning photocarrier lifetime, resistivity, mobility and some structural characteristics were performed on GaAs implanted with high energy arsenic ions [4-6]. It is important to clearly determine the influence of excess arsenic and implantation induced crystalline damage on properties of these materials. This paper reports the first results of TEM investigations on GaAs implanted with high energy arsenic ions.

Cross-sectional samples were prepared from semi-insulating (100) GaAs wafers implanted with 2MeV arsenic at a dose of $1 \times 10^{15} cm^{-2}$ using the Australian National University Tandem accelerator. As-implanted and annealed for 20 min at 300°C, 600°C and 700°C materials were investigated. TEM studies of as-implanted sample showed the existence of a thick amorphous layer containing small crystalline inclusions, buried under a thin fully crystalline layer with a small concentration of defects. The implanted layer was terminated by a region of high density of "end-of-range" defects, as observed earlier in GaAs implanted with 200 keV ions [2,3]. After annealing at 300°C the thin crystalline layer located close to the surface had expanded and the thick amorphous layer almost completely recrystallized forming a large number of microtwins. The structure indicated two fronts of recrystallization of the amorphous material - from the monocrystalline top and bottom. In samples annealed at 600°C and 700°C further changes in structure took place. Microtwins became larger and the size of defect free recrystallized grains increased. The thermal annealing also caused a gradual decrease of the density of planar defects and "end-of-range" defects.The TEM results of ion implanted GaAs are in agreement with the expectation that a high energy implantation produces a highly defective, often amorphous buried layer which can be recrystallized by thermal annealing.

X-ray rocking curve studies showed two peaks: one from the substrate and the second from the implanted material, indicating an expansion of the lattice parameter [7]. In TEM studies the precipitation of As was not observed indicating that the excess As was probably present as point defects or as a component of planar defects.

1. F.W. Smith et al., IEEE Electron. Device Lett, 9, 77 (1988)
2. A. Claverie et al., Appl. Phys. Lett. 62, 1271 (1993)
3. Z. Liliental-Weber et al., Ultramicroscopy 52 N3-4, 570-574(1993)
4. A. Kurpiewski et al., Proc. 8th Conference on Semi-Insulating III-V Materials, Warsaw 1994
5. A. Korotkus et al., Appl. Phys. Lett (submitted)
6. M. Kaminska et al., Proc. 8th Conference on Semi-Insulating III-V Materials, Warsaw 1994
7. H.H. Tan et al., 9th Int. Conf. on Ion Beam Mod. of Mat., Canbera, 1995
(a) on a leave from Institute of Experimental Physics, Warsaw University

Proc. Microscopy and Microanalysis 1995, edited by G.W. Bailey, M.H. Ellisman, R.A. Hennigar, and N.J. Zaluzec
Copyright © 1995 MSA. Published by Jones and Begell Publishing, 79 Madison Ave., New York, NY 10016

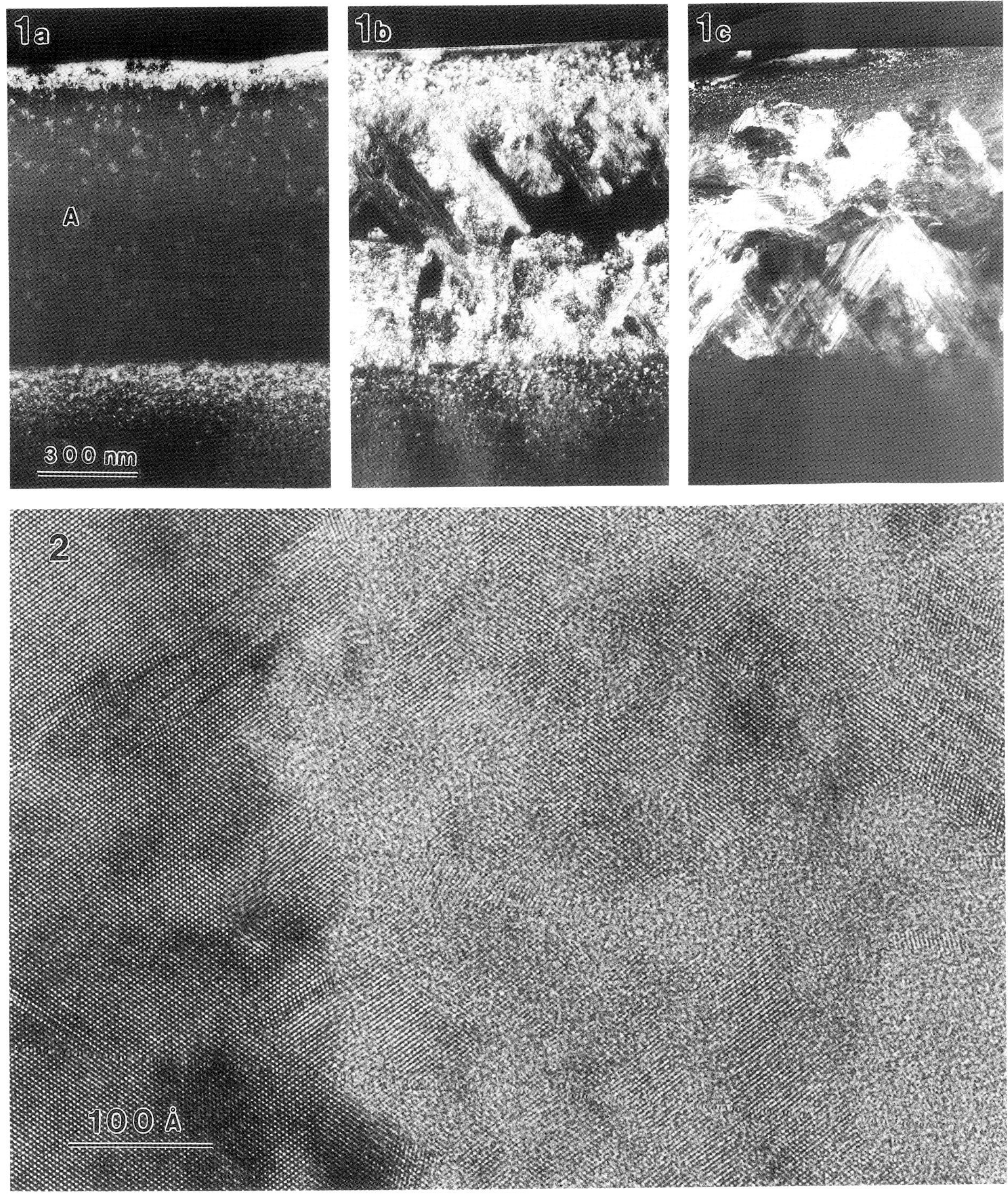

FIG.1.- Dark field images of cross-sectional samples prepared from as-implanted (a) and annealed at 300°C (b) and 700°C (c) 2MeV As implanted GaAs. Note buried amorphous layer "A" with some crystalline grains in as-implanted sample (a). Recrystallization of this layer after annealing lead to the formation of high density of microtwins (b,c).

FIG.2. - High resolution image of interface between top crystalline layer (left part of the micrographs) and amorphous layer in as-implanted GaAs.

TEM STUDY OF BONDED SILICON WAFERS.

L.Mulestagno*, R. Craven** , and P.Fraundorf*.
* Physics Dept. and Center for Molecular Electronics , University of Missouri - St. Louis,
St. Louis , MO 63121.
** SiBond , PO Box 8, St. Peters, MO 63376 .

The promise of higher speed, radiation hard, high temperature capabilities, and increased
device density has made SOI (silicon - on - insulator) an attractive proposition since its
conception several years ago . Until recently the methods for its manufacture had been
plagued by low yields, and poor quality . Remarkable advances have been made lately
resulting in increasing quantities of SOI wafers being sold commercially.[1]

The two main techniques for making SOI are Separation by Implantation of Oxygen
(SIMOX) and direct wafer bonding (Bonded wafers). Little analytical work has yet been
performed on the latter by the microscopy community at large , so we studied 2
commercial quality wafers with the intent of looking for particulate defects which might
arise during annealing, and studying the oxide layer and oxide-SOI interface quality .

Plane view samples were made from the SOI taking advantage of its chemistry, by
dimpling and milling from the back-side, and then dipping in HF which attacks the oxide,
and leaves large areas of silicon film suitable for TEM untouched. The SOI film thickness
was 0.2 and 0.4 μm in the samples studied, and therefore electron transparent without
further treatment. Cross-sectional specimens were prepared in a standard fashion, using a
low-angle ion mill manufactured by Gatan, which resulted in 100 - 300 μm of thin area in
the interface region.

No concrete evidence of particulate defects was present in any of 6 planar specimens
prepared from the 2 samples, putting an upper limit of any precipitates at 10^7 cm^{-3} .
Various observations were made of cross-sectional specimens to investigate the quality of
the oxide layer and the interface (fig 1 - 3) . Fig1 shows an image taken at low
magnification, and it can be seen, that the oxide and the SOI show a great deal of
uniformity with no noticeable change in thickness. Hardly any variation in thickness can be
observed at intermediate magnification (fig 2) , were we started to observe subtle oxide
roughness at the SOI interface and the base wafer interface. At high resolution, the
roughness was measured to be on the order of 2 - 4 nm (fig 3) . The observations were
performed over several areas of cross-sectional specimens with repeatable results.

In summary, we were unable to find any evidence of precipitate formation in bonded
wafers, and a low magnification to high resolution study of cross-sectional specimens,
showed no visible signs of oxide or SOI thickness variations, and subtle roughness of the
oxide in the interface region on the order of 2-4 nm.

Proc. Microscopy and Microanalysis 1995, edited by G.W. Bailey, M.H. Ellisman, R.A. Hennigar, and N.J. Zaluzec
Copyright © 1995 MSA. Published by Jones and Begell Publishing, 79 Madison Ave., New York, NY 10016

References :

1. G.W.Cullen et al , Proceedings of the 6th Interntl symp. on SOI technology and
devices, The Electrochemical Soc. , 1994 .

Acknowledgments :
I would like to thank J.C. Holzer for the many interesting discussions we had over the past
few months.

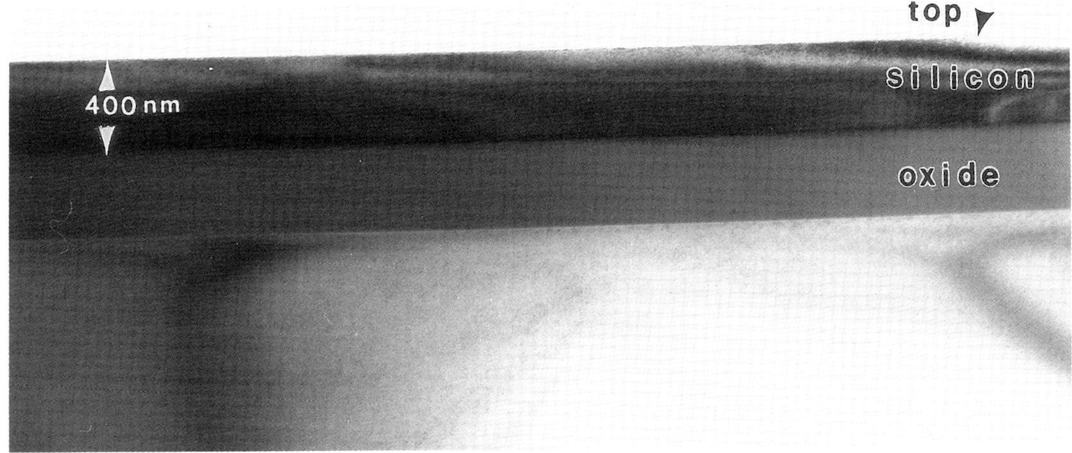

Fig 1

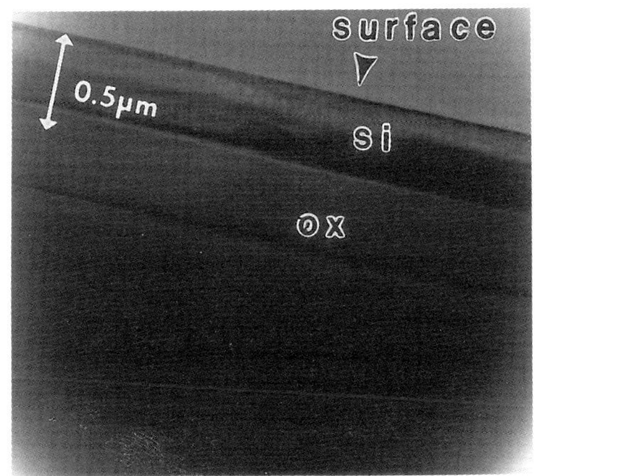

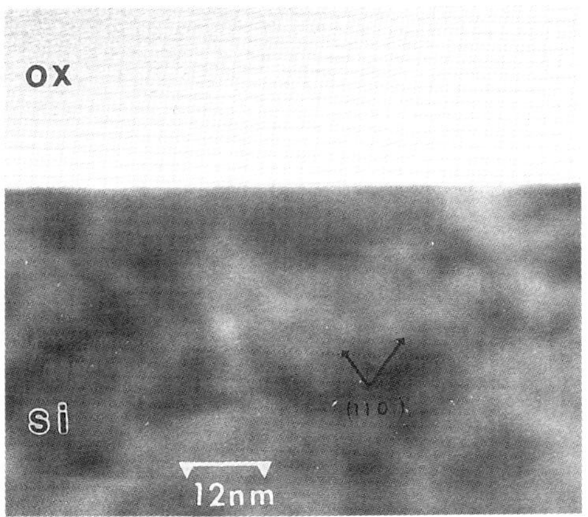

Fig2   Fig 3

Fig 1 . low magnification (10.5K) XTEM view of on of the samples
Fig 2 . higher magnification (43K) in cross - section of the same specimen.
Fig 3 . High resolution image(240K) showing fine (2- 4 nm) roughness at the SOI -
oxide interface.

Cathodoluminescence Studies of Materials for use in Optical Display Devices

P.J.Wright

Oxford Instruments Scientific Research Instruments Division, Oxford, U.K.

Cathodoluminescence studies have been carried out for many years for the assessment of semiconductor materials such as Gallium Arsenide, Gallium Phosphide and Silicon. These investigations have relied on utilising the focused electron beam of an SEM to excite luminescence. This is subsequently directed through a monochromator to allow spectroscopy and also the acquisition of high resolution images showing the spatial distribution of the optically active centres ($\geq 0.5\mu m$).

The recent growth of research in the production and characterisation of optically active materials and devices such as 3/5 or 2/6 LED's for the use in visible display products has fuelled the use of CL to determine the optical properties on a scale that is superior to that available by PL. In addition characterisation of the optical properties as a function of electron beam characteristics, dosage and at temperatures between 4.2K and 300K are possible.

Results from GaN epilayers will be shown that indicate that two separate phases (alpha and beta) can be identified by the position of the Bound Exiton luminescence and that this luminescence is distributed randomly through the sample. The sites responsible for this luminescence are less than $0.5\mu m$ in size.

In addition, in many other materials it will be seen that sites of extraneous luminescence can be correlated to variations in the distribution of elemental impurities which may be present intentionally or unintentionally. These have been identified by simultaneous EDS analysis in the SEM.

Instrumentation
The studies discussed above were carried out on a Jeol 840A SEM equipped with an Link eXL EDS system, an Oxford monoCL cathodoluminescence detector and liquid Helium stage.

Acknowledgements
We would like to thank Prof.T.Foxon and Dr.T.Cheng from University of Nottingham for supplying the GaN epilayer samples to Dr.V.Higgs of Kings College, London who performed and reported the CL experiments.

Proc. Microscopy and Microanalysis 1995, edited by G.W. Bailey, M.H. Ellisman, R.A. Hennigar, and N.J. Zaluzec
Copyright © 1995 MSA. Published by Jones and Begell Publishing, 79 Madison Ave., New York, NY 10016

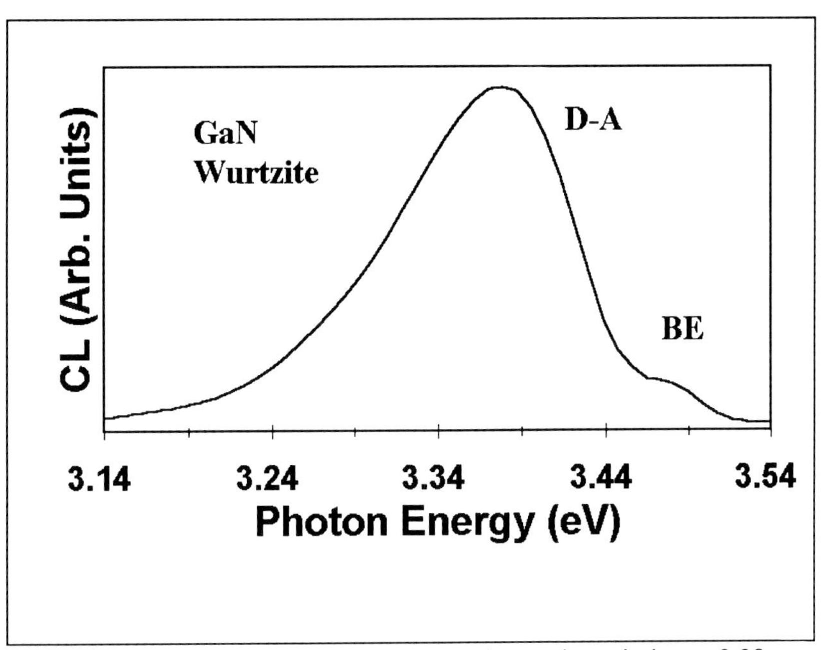

Fig 1; CL spectrum from alpha GaN epilayer showing main emission at 3.38ev and B-E emission at 3.46ev.

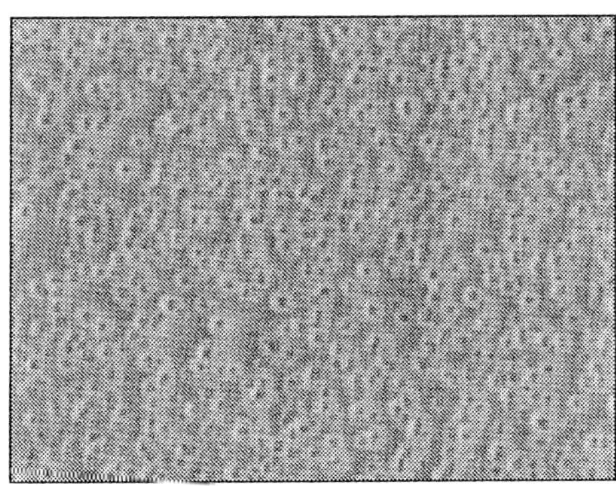

Fig 2; Secondary Electron Image of GaN epilayer at 5000x and 15kV

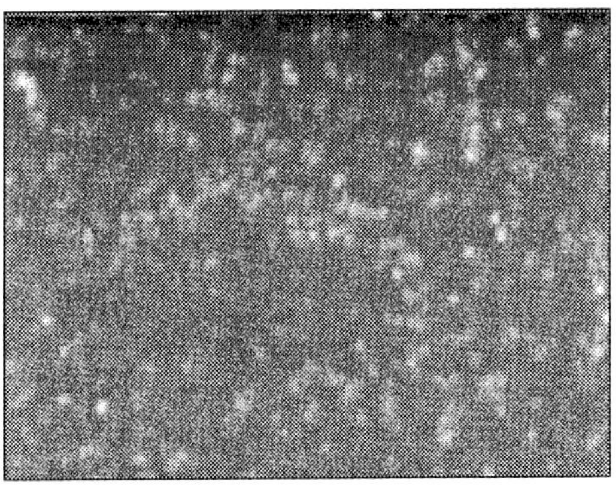

Fig 3: CL image of emissions from GaN epilayer at 3.46ev under same conditions as Fig 2

TEM STUDY OF CoSi$_2$ FORMATION VIA ANNEALING OF Co-Ti BILAYERS ON Si

N. David Theodore, Andre Vantomme*, and Peter Crozier**

Arizona State University, Chemical Bio and Materials Engineering, Tempe, AZ 85287.
*California Institute of Technology, Applied Physics & Electrical Engineering, Pasadena, CA 91125.
**Arizona State University, Center for Solid State Science, Tempe, AZ 85287.

Contact is typically made to source/drain regions of metal-oxide-semiconductor field-effect transistors (MOSFETs) by use of TiSi$_2$ or CoSi$_2$ layers followed by Al(Cu) metal lines. A silicide layer is used to reduce contact resistance. TiSi$_2$ or CoSi$_2$ are chosen for the contact layer because these silicides have low resistivities (~12-15 $\mu\Omega$-cm for TiSi$_2$ in the C54 phase[1], and ~10-15 $\mu\Omega$-cm for CoSi$_2$).[2] CoSi$_2$ has other desirable properties, such as being thermally stable up to >1000°C for surface layers and >1100°C for buried layers[3], and having a small lattice mismatch with silicon, -1.2% at room temperature.[4] During CoSi$_2$ growth, Co is the diffusing species. Electrode shorts and voids which can arise if Si is the diffusing species are therefore avoided. However, problems can arise due to silicide-Si interface roughness (leading to nonuniformity in film resistance) and thermal instability of the resistance upon further high temperature annealing. These problems can be avoided if the CoSi$_2$ can be grown epitaxially on silicon. There is interest therefore in obtaining epitaxial, or near-epitaxial CoSi$_2$ on silicon. Possible applications include contacts and also metal-base transistors. Various techniques have been used to attempt epitaxial growth of CoSi$_2$ on Si, the simplest of which involves reaction of Co/Ti bilayers with silicon.[5] In the present study, we use TEM to investigate the microstructural behavior of Co/ Ti/ Si layers annealed under various conditions. Differences in behavior are noted, and the mechanism of microstructural evolution is discussed.

Silicon (100) wafers were sputter-coated with 8.5nm Ti and 23nm Co, without breaking vacuum, in a sputter system with a base pressure of 5x10^{-8} Torr. Pieces of the wafers were then annealed at 675°C for 10 min or 80 min in vacuum (<5x10^{-7} Torr), or at 900°C for 30 min in vacuum, N$_2$, or N$_2$+5%H$_2$. The samples were analyzed by plan-view and cross-section TEM (XTEM) to study the structure, crystallinity and interface quality of the multilayers. Electron diffraction and high-resolution energy-dispersive spectrometry (EDS) using a VG HB501 STEM were also used to analyze the phases formed.

Figure 1 presents an XTEM micrograph obtained from Co/Ti/Si(100) annealed at 900°C in N$_2$ for 30 minutes. The multilayer has transformed to a CoSi$_2$/ Si structure with Ti-oxide on top. Co and Ti have interchanged places, with Co moving down to react with the underlying Si substrate, and Ti moving up and reacting with oxygen from the ambient. The CoSi$_2$ appears largely single-crystalline and epitaxial. The driving force for Co/Ti layer inversion appears to be reaction of Co with substrate-Si and reaction of Ti with oxygen from the ambient. Figures 2a-b present XTEM micrographs of Co/Ti/Si(100) annealed at 900°C in N$_2$+H$_2$ for 30 min. In this case, occasional polycrystalline grains are present, and the Co-silicide/Si-substrate interface is rough. Electron diffraction indicates that some of the polycrystalline grains are CoSi and/or Co$_2$Si. Once again, a Ti-oxide has formed at the top of the structure. The amount of oxide (measured by RBS) is less when H$_2$ is present in the ambient. The driving forces for Co/Ti inversion are similar to those in the case of the N$_2$-ambient. Figures 3a-b present XTEM micrographs obtained from Co/Ti/Si(100) annealed at 900°C in vacuum for 30 min. In this case, the CoSi$_2$/Si interface is very rough. TEM analysis indicates that the heterolayer contains a mixture of large (~400nm) and small (~30nm) grains identified as CoSi$_2$ and Co$_{0.25}$Ti$_{0.75}$Si$_2$ respectively. The ternary silicide grains tend to be near the surface of the structure. The Co/Ti layer inversion occurs even in vacuum; the inversion is however not complete. The driving force for inversion appears to be reaction of Co with the substrate-Si. Figures 4a-b present XTEM micrographs obtained from Co/Ti/Si(100) annealed at 675°C in vacuum for (a) 10 min, and (b) 80 min. The micrographs show a top layer containing CoSi, Co$_2$Si, and unreacted Co. An intermediate Co$_{0.25}$Ti$_{0.75}$Si$_2$ is present below this top layer. A bottom layer is present, consisting of CoSi + CoSi$_2$ in the case of the 10 min anneal (Fig. 4a) and entirely of CoSi$_2$ in the case of the 80 min anneal (Fig. 4b). In both cases, the intermediate Co$_{0.25}$Ti$_{0.75}$Si$_2$ layer appears to retard diffusion of Co from the top-layer down to the substrate-Si. This limits the supply of Co, and enhances epitaxy of the CoSi$_2$ that does form by reaction with the substrate. Ti can also act to reduce native oxide on Si, and thus enhance epitaxy. In Figs. 4a-b, some Si has diffused up from the substrate to react with Co in the top layer to form CoSi and Co$_2$Si. CoSi present in the bottom layer (Fig. 4a) has reacted with Si from the substrate to form CoSi$_2$ in Fig. 4b. In summary, the driving force for Co/Ti layer inversion (even in vacuum) appears to be reaction of Co with substrate-Si. The Ti layer enhances CoSi$_2$ single-crystallinity and epitaxy by reducing the supply of Co (through the resulting ternary silicide) and possibly by reducing Si native-oxide.

References
1. C.A. Pico, M. Lagally, in *Selected Topics in Electronic Materials* (Mat. Res.Soc., Pittsburgh, 1988) p.19.
2. S.P. Murarka, Silicides for VLSI Applications (Academic Press, Orlando, 1983).
3. M.F. Wu et al, Nucl. Instr. Meth. B 45, 658 (1990).
4. M.-A. Nicolet et al, VLSI Electronics Microstructure Science, Vol. 6 (Academic Press, NY, 1983).
5. M.L.A. Dass et al, Appl. Phys. Lett. 58, 1308 (1991).

Proc. Microscopy and Microanalysis 1995, edited by G.W. Bailey, M.H. Ellisman, R.A. Hennigar, and N.J. Zaluzec
Copyright © 1995 MSA. Published by Jones and Begell Publishing, 79 Madison Ave., New York, NY 10016

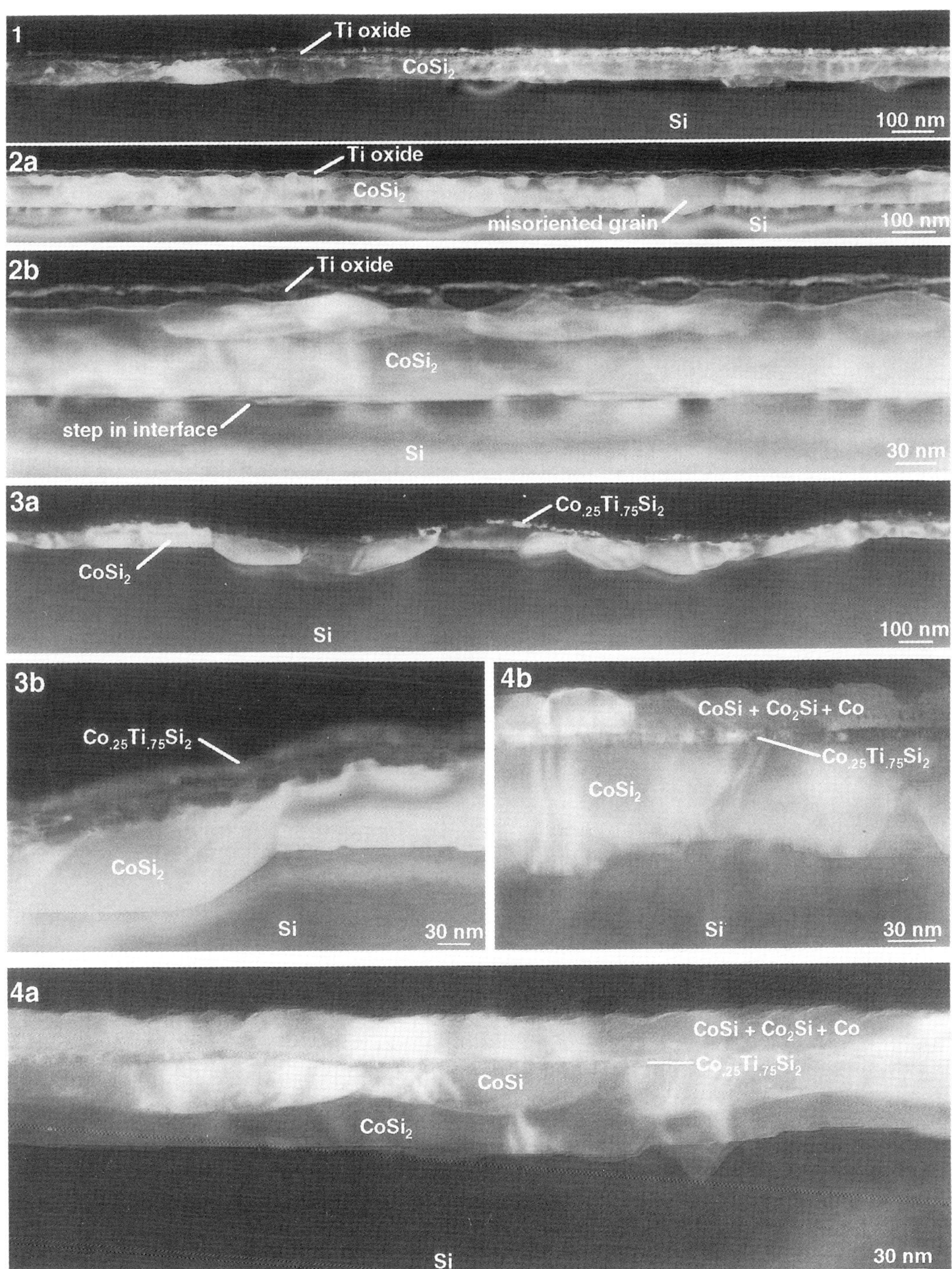

Fig. 1 XTEM micrograph of Co/Ti/Si(100) annealed at 900°C in N_2 for 30 min.
Fig. 2 XTEM micrographs of Co/Ti/Si(100) annealed at 900°C in $N_2 + H_2$ for 30 min.
Fig. 3 XTEM micrographs of Co/Ti/Si(100) annealed at 900°C in vacuum for 30 min.
Fig. 4a XTEM micrograph of Co/Ti/Si(100) annealed at 675°C in vacuum for 10 min.
Fig. 4b XTEM micrograph of Co/Ti/Si(100) annealed at 675°C in vacuum for 80 min.

DETECTION OF A POINT DEFECT IN A SILICON SINGLE CRYSTAL BY HIGH-RESOLUTION TRANSMISSION ELECTRON MICROSCOPY

M.Awaji

Department of Materials, University of Oxford, Parks Road, Oxford, OX1 3PH, U.K.

It is necessary to improve the resolution, brightness and signal-to-noise ratio(s/n) for the detection and identification of point defects in crystals. In order to observe point defects, multi-beam dark-field imaging is one of the useful methods. Though this method can improve resolution and brightness compared with dark-field imaging by diffuse scattering, the problem of s/n still exists. In order to improve the exposure time due to the low intensity of the dark-field image and the low resolution, we discuss in this paper the bright-field high-resolution image and the corresponding subtracted image with reference to a changing noise level, and examine the possibility for in-situ observation, identification and detection of the movement of a point defect produced in the early stage of damage process by high energy electron bombardment.

The high-resolution image contrast of a silicon single crystal in the $[\bar{1}10]$ orientation containing a triple divacancy cluster is calculated using the Cowley-Moodie dynamical theory and for a changing gaussian noise level[1]. This divacancy model was deduced from experimental results obtained by electron spin resonance[2]. The calculation condition was for the 1MeV Berkeley ARM operated at 800KeV. The displacement of atoms around the point defect was deduced by extrapolating the distortion around a stacking fault tetrahedron following the procedure used by Saldin and Whelan[3]. The contrast change and displacement of the images of the point defect are displayed using a subtraction technique[4] between the computed image of the point defect and that from the perfect crystal.

Figure 1 shows the high-resolution image of a Si$[\bar{1}10]$ specimen doped with boron (10^{15}cm^{-3}) after 400KeV electron beam irradiation, and the arrows 1-2, 2-3 and 4-5 show the dumbbell-type, the other arrows 6 and 7 show the single-type contrast anomalies. The basic explanation of Fig.2 is discussed in a previous paper[5]. Suppose the linear size of a resolution element is 0.1mm, the necessary number of electrons[6] for Figs.2c,e and g are 2-6x10^6/mm^2. The calculated images a,c and e in Figs.3-6 correspond to Figs.2c,e and g but contains 1,2,3 and 4% gaussian noise. In each of Figs.3-6, the images p show the perfect crystal lattice image containing the noise, and the images b,d and f show difference images obtained by subtraction of images a,c and e from image p. In Figs.3-6, even if the noise level is 1% of the intensity difference(i.e. the maximum to minimum intensity in the perfect crystal lattice image without noise), it is rather difficult to observe the contrast change due to a defect in a high-resolution image, but the subtracted image predicts that if the noise level is less than 3%, it should be possible to detect the position of the defect. A more accurate detection of a point defect with low contrast depends on the noise elimination technique for high-resolution images, and the value of 3% or less may be a necessary experimental condition for detecting a point defect using this technique.

1. J.M.Gibson and M.L.McDonald,*Mat.Res.Soc.Symp.Proc.*82(1987)109.

2. J.W.Corbett and G.D.Watkins,*Phys.Rev.Lett.*7(1961)314.

3. D.K.Saldin and M.J.Whelan,*Philos.Trans.R.Soc.London* Ser.A292(1979)513.

4. H.Hashimoto et al.,*Ultramicroscopy* 39(1991)171.

5. M.Awaji and H.Hashimoto,*ICEM-13 Proc.*2A(1994)113.

6. W.O.Saxton,*Computer Techniques for Image Processing in Electron Microscopy*, London:Academic Press(1978)30.

Proc. Microscopy and Microanalysis 1995, edited by G.W. Bailey, M.H. Ellisman, R.A. Hennigar, and N.J. Zaluzec
Copyright © 1995 MSA. Published by Jones and Begell Publishing, 79 Madison Ave., New York, NY 10016

7. M.A. is grateful to the JSPS for the award of a Postdoctoral Fellowship for research abroad (1993-5), and to Prof.M.J.Whelan for advice and for reading the manuscript.

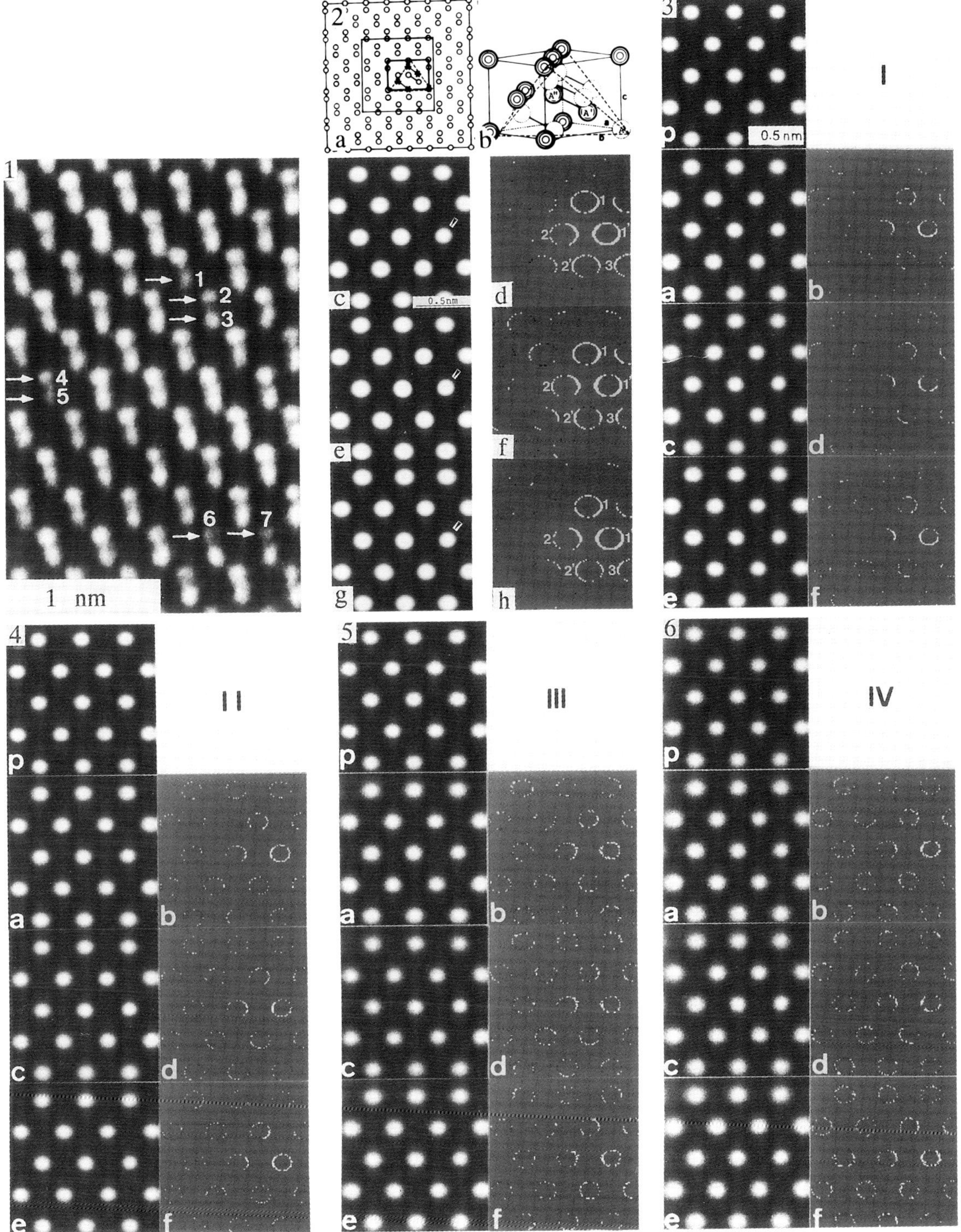

FIG.1.-Two kinds of contrast anomaly due to lattice defects produced by 400KeV electron irradiation.
FIG.2.-The calculated image contrast of a triple divacancy cluster without noise.
FIGS.3-6.-The same image contrast as for FIG.2. but with a gaussian noise ranging from 1% to 4%.

TEM STUDY OF PHOSPHORUS-IMPLANTED PSEUDOMORPHIC $Si_{0.88}Ge_{0.12}$ ON Si(100)

N. David Theodore, Donald Y.C Lie*, J. H. Song*, and Peter Crozier**

Arizona State University, Chemical Bio and Materials Engineering, Tempe, AZ 85287.
*California Institute of Technology, Applied Physics & Electrical Engineering, Pasadena, CA 91125.
**Arizona State University, Center for Solid State Science, Tempe, AZ 85287.

SiGe is being extensively investigated for use in heterojunction bipolar-transistors (HBT) and high-speed integrated circuits.[1] The material offers adjustable bandgaps, improved carrier mobilities over Si homostructures, and compatibility with Si-based integrated-circuit manufacturing. SiGe HBT performance can be improved by increasing the base-doping or by widening the base link-region by ion implantation. A problem that arises however is that implantation can enhance strain-relaxation of SiGe/Si.[2] Furthermore, once misfit or threading dislocations result, the defects can give rise to recombination-generation in depletion regions of semiconductor devices. It is of relevance therefore to study the damage and anneal behavior of implanted SiGe layers. The present study investigates the microstructural behavior of phosphorus implanted pseudomorphic metastable $Si_{0.88}Ge_{0.12}$ films on silicon, exposed to various anneals.

Metastable pseudomorphic $Si_{0.88}Ge_{0.12}$ films were grown ~265 nm thick on a silicon wafer by molecular-beam epitaxy. Pieces of this wafer were then implanted at room temperature with 100 keV phosphorus ions to a dose of $1.5x10^{15}$ cm^{-2}. The energy of the implant was chosen to amorphize only the top ~190 nm of the SiGe layer, while leaving the rest of the film single crystalline. Implanted pieces, together with non-implanted pieces, were then annealed for 30 minutes each in vacuum at temperatures ranging from 400°C to 800°C. Cross-section TEM samples were made by conventional dimpling followed by argon ion-milling. Damage and strain present in the SiGe/Si samples were monitored using TEM and RBS (Rutherford back-scattering spectrometry) ion-channeling.

Figures 1a-h present cross-section TEM (XTEM) micrographs obtained from $Si_{0.88}Ge_{0.12}$/ Si films. Figure 1a is as-grown, Fig. 1b is as-implanted with $1.5x10^{15}$ cm^{-2} phosphorus, and Figs. 1c-h are implanted and then annealed for 30 minutes each at (c) 400°C, (d) 500°C, (e) 550°C, (f) 600°C, (g) 700°C, and (h) 800°C. The as-grown sample (Fig. 1a) is free of defects within the limits of XTEM analysis (~10^5 cm^{-2}). The as-implanted sample is amorphized to a depth of 190±9 nm, well within the thickness of the SiGe layer. The SiGe-Si interface appears unaffected by the implantation. The 400°C annealed sample (Fig. 1c) shows negligible solid-phase regrowth of the amorphized SiGe. Some smoothing of the amorphous-crystalline interface has occurred. The 500°C annealed sample (Fig 1d) shows partial epitaxial regrowth of the amorphized SiGe. Regrowth has occurred using the lower crystalline SiGe as a seed for epitaxy. The amorphous SiGe thickness has decreased to ~85±25 nm. There is implant damage present near the original amorphous-crystalline interface, due to coalescence of excess interstitials. The thickness of the regrown SiGe indicates a regrowth velocity of ~3.6±0.8 nm/min. This regrowth velocity is ~10-20 times faster than regrowth in undoped SiGe. A similar effect has been seen in silicon, where regrowth of P implant-amorphized Si was ~6 times faster than regrowth of undoped Si.[3] The 550°C annealed sample (Fig. 1e) shows complete epitaxial regrowth of the amorphized SiGe. The recrystallized SiGe has ~10^{10}-10^{11} cm^{-2} dislocations, but no defects are observed in the bottom portions of the SiGe which were not amorphized by the implant. There are no signs of misfit dislocations at the SiGe-Si interface; the bottom SiGe is therefore fully strained, within the limits of XTEM analysis. As the regrowth front moves upward, threading dislocations extend from the ends of dislocation half-loops situated near the amorphous-crystalline interface, producing hair-pin dislocations. This is more evident in the 600°C annealed sample (Fig. 1f). Once again, there are no signs of misfit dislocations at the SiGe-Si interface, indicating that the bottom SiGe is fully strained, within the limits of XTEM analysis. Figure 1g presents a micrograph from implanted SiGe annealed at 700°C. For the first time, we now see segments of misfit dislocations at the bottom SiGe-Si interface. The micrograph shows the threading arms of a dislocation half-loop that extend from the implanted top SiGe down to the bottom SiGe-Si interface. Between these threading arms lies a segment of misfit dislocation. It appears therefore that at 700°C dislocation half-loops first form at the bottom of the regrown SiGe layer during annealing. The half-loops expand downwards upon {111} planes until parts of the loops intersect the SiGe-Si interface. After 800°C annealing a high density of threading dislocations is present throughout the entire SiGe layer. The defects are present in the top implanted SiGe and also in the non-implanted bottom SiGe. It appears then that the process of strain-relaxation (by downward glide of dislocation half-loops) that we first observed at 700°C has occurred quite extensively during the 800°C anneal. In summary, the microstructure of phosphorus implanted and annealed pseudomorphic SiGe/ Si has been studied. The microstructure will be correlated with RBS and x-ray results.

References

1. J.D. Cressler et al, IEEE Electron Dev. Lett. 12(4), 166 (1991).
2. R. Hull et al, Appl. Phys. Lett. 56, 2445 (1990).
3. L. Csepregi et al, J. Apl. Phys. 48, 4234 (1977).

Proc. Microscopy and Microanalysis 1995, edited by G.W. Bailey, M.H. Ellisman, R.A. Hennigar, and N.J. Zaluzec
Copyright © 1995 MSA. Published by Jones and Begell Publishing, 79 Madison Ave., New York, NY 10016

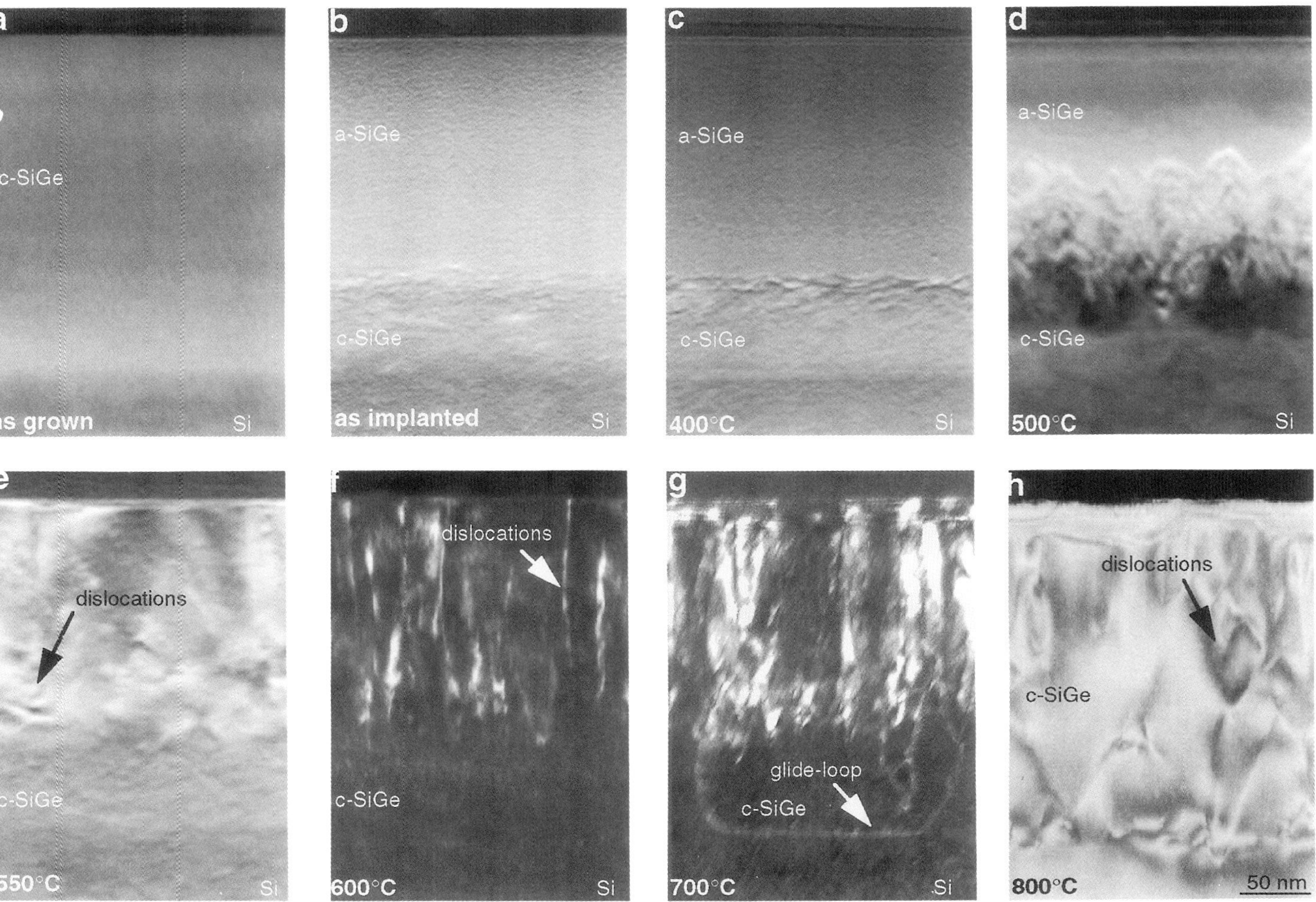

Fig. 1 Cross-section TEM micrographs obtained from $Si_{0.88}Ge_{0.12}$/Si films (a) as-grown, (b) as-implanted with 1.5×10^{15} cm^{-2} phosphorus, (c-h) implanted and then annealed for 30 minutes each at (c) 400°C, (d) 500°C, (e) 550°C, (f) 600°C, (g) 700°C, and (h) 800°C.

A PRECURSOR TO THE BREAKDOWN OF HOMOGENEOUS GROWTH IN LOW TEMPERATURE MBE GaAs DETECTED BY ELLIPSOMETRY

S. D. Walck,* K. G. Eyink,* T. W. Haas,*, and B. G. Streetman**

*Materials Directorate, Wright Laboratory, Wright-Patterson AFB, OH 45433-7750
**Microelectronics Research Center, University of Texas, Austin, TX 78712

GaAs grown by MBE at low temperature (LT) in the range of 150-250°C forms a highly resistive film after it has been annealed. The high resistance of annealed LT-GaAs has been use to provide a semi-insulating layer lattice-matched to GaAs. For this reason, LT layers must have reproducible properties and must be of high crystalline quality for subsequent epitaxial layers. The as-grown LT-GaAs typically has 1-2% excess As which is responsible for an increase in the lattice constant of about 0.1%. The thickness of the LT-GaAs layer is limited by the strain energy in the film which leads to the formation of pyramidal defects which is known as the critical thickness, h_c.[1] *In-situ* single wavelength ellipsometry has been used to precisely set a reference temperature for a given As beam equivalent pressure (BEP) which can be used to grow reproducible LT-GaAs layers.[2] In the past, XTEM has been used to determine h_c, however, *in-situ* ellipsometry can also be used to monitor the film's optical properties during growth and can detect the different regions from breakdown of homogeneous growth through the film's strain relaxation by the generation of the pyramidal defects to the formation of polycrystalline GaAs, if the growth is allowed to continue that far.[2] An XTEM micrograph of such a film with the growth regions defined is shown in Fig. 1.

In this study, a comparison was made for h_c determined by ellipsometry and XTEM for LT-GaAs layers grown by varying the signature temperature, T_{sig} and As_s BEP. The samples were grown in a VG V80H MkII MBE system with As_2 overpressures in the range $0.2\text{-}1.5\times10^{-5}$ Torr and a growth rate of 1.2 μm/hr. The ellipsometer was used to monitor the changes that occurred during the growth of the LT-GaAs. A single wavelength (6328Å) rotating ellipsometer with strain-free windows was used at an angle of incidence of ~70°, near the Brewster angle. Intensity data were taken at a rotational frequency of the analyzer of ~4 Hz with 72 measurements per cycle and averaged over 10 rotations to improve the signal-to-noise ratio. Ellipsometry data (ψ,Δ) were extracted from the intensity data and recorded. The values, ψ and Δ are defined by the relationship, $r_p/r_s = \tan\psi\exp(i\Delta)$, where $r_p(r_s)$ is the complex amplitude reflectivity for light polarized parallel (perpendicular) to the plane of incidence. Figure 2 is an example of the ψ and Δ data that show departure from the calculated homogeneous growth model. The XTEM samples were prepared by standard techniques and examined in a JEOL 2000FX operated at 200 kV and imaged using **g**=400 (parallel to the growth direction) BF and DF conditions. The value of h_c was determined from the GaAs/LT-GaAs interface to the first appearance of the pyramidal defects. The results for the LT-GaAs layers grown are summarized in Table 1.

Originally, the XTEM work was done to calibrate the ellipsometry measurements. The overall thicknesses of the layers measured by XTEM were used to determine a correction factor for the thickness determined by the ellipsometer. When this was done, the critical thicknesses measured by ellipsometry were consistently much less than those measured by XTEM for all samples. Typically, the difference in the adjusted thicknesses was well over 1500 Å. At a loss to explain the discrepancy, the XTEM micrographs were carefully examined. On a few XTEM micrographs in the vicinity where the ellipsometry data showed a departure from homogenous growth, a very faint contrast feature was seen running parallel to the interface. Figure 3 is an example of a sample clearly showing this feature. For this sample, the growth was stopped just after the ellipsometer detected this departure but prior to the nucleation of the pyramidal defects. This strongly suggests that changes are occurring in the film prior to the nucleation of the pyramidal defects that are detectable by ellipsometry. Furthermore, *in-situ* ellipsometry can be used to insure continued epitaxy in LT-GaAs.

References
1. Z. Liliental-Weber et al., *Applied Physics Letters*, 58(1991)2153.
2. K. G. Eyink et al., *Journal of Electronic Materials*, 22(1993)1387.

Proc. Microscopy and Microanalysis 1995, edited by G.W. Bailey, M.H. Ellisman, R.A. Hennigar, and N.J. Zaluzec
Copyright © 1995 MSA. Published by Jones and Begell Publishing, 79 Madison Ave., New York, NY 10016

Table 1. - Comparison of Critical Thicknesses determined by XTEM and Ellipsometry

$h_{tot.}$ XTEM (Å)	$h_{tot.}$ Ellips. (Å)	$h_{c,}$ XTEM (Å)	$h_{c,}$ Ellips. (Å)	Ratio of $h_{tot., XTEM}$ to $h_{tot., Ellips.}$	Corrected $h_{c, Ellips.}$ (Å)	Ratio of $h_{c., XTEM}$ to $h_{c., Ellips.}$
6992	6311	4627	2441	1.11	2705	1.71
3290	2499	7436	4003	1.32	5269	1.41
7198	6301	*	*	1.14	*	*
6761	5477	4756	2745	1.23	3388	1.40
9152	7543	3985	1841	1.21	2234	1.78
2162	1903	3504	2450	1.14	2784	1.26
2900	2578	**	2341	1.12	2634	**

* film did not relax   ** film relaxation was not observed in the

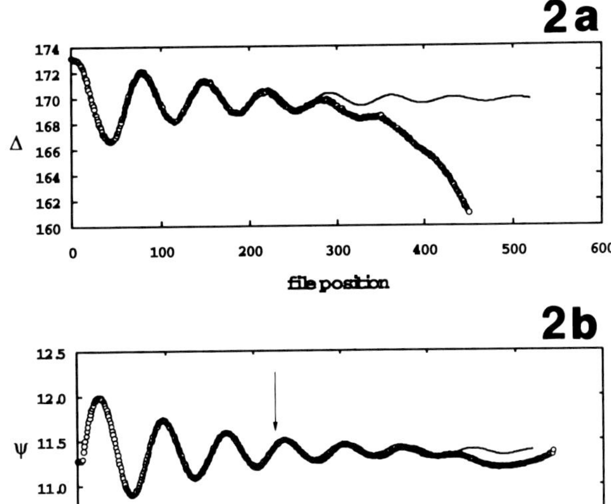

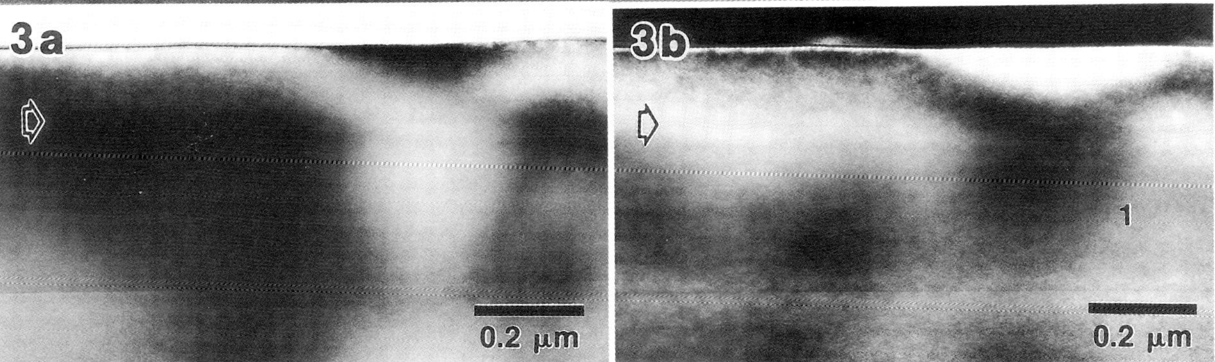

FIG. 1 - XTEM micrograph of LT-GaAs using DF($\mathbf{g}$=400) and hollow cone dark field ($\mathbf{g}$=111).
FIG. 2 - Ellipsometric data for growth of LT-GaAs. (a) $\Delta(°)$ vs file position (thickness), (b) $\psi(°)$ vs file position.
FIG. 3 - XTEM micrograph of LT-GaAs showing a contrast feature at a thicknesses corresponding to ellipsometric h_c.

TEM INVESTIGATION OF THE YSZ-ON-INSULATOR STRUCTURE

D.S. Zhou[1], Q.X. Jia[2], X.D. Wu[2], and T.E. Mitchell[1]

[1]Center for Materials Science and [2]Superconducting Technology Center, Los Alamos National Laboratory, Los Alamos, NM 87545

Silicon-on-insulator (SOI) structures have been widely used in the semiconductor industry as substrates to grow various devices. They are usually prepared by a process called separation by implantation of oxygen (SIMOX)[1,2]. As an alternative, single crystal yttria-stabilized-zirconia (YSZ) can be grown on single crystal silicon substrates, followed by heat treatment in oxygen to form an intermediate SiO_2 layer[3]. This YSZ-on-silicon structure has the potential to be used as a new SOI material in semiconductor device fabrication.

YSZ has a cubic (fluorite type) structure with a lattice parameter of $0.514 \sim 0.523$ nm (as compared to 0.543 nm for silicon) depending on its Y_2O_3 content. The good lattice match between YSZ and silicon makes epitaxial growth possible. In this experiment, the pulsed laser deposition technique has been employed to grow YSZ thin films on single crystal p-Si (100) wafers[3]. Post-deposition heat treatment was carried out at 850°C for 1.5 h in dry oxygen at 1 atmosphere. The structure was first examined by Rutherford backscattering spectrometry and x-ray diffraction, and then by cross-sectional transmission electron microscopy (XTEM) techniques.

Figure 1 is an XTEM image of the $YSZ/SiO_2/Si$ structure. The thicknesses of the YSZ and SiO_2 layers were measured to be 50 nm and 110 nm, respectively. The relatively thick SiO_2 layer was successfully produced by the rapid diffusion of oxygen in YSZ. Stacking faults in silicon are observed (figure 1) lying on {111} planes; those are probably produced by partial dislocation motion in response to stresses generated due to the formation of SiO_2. Electron diffraction analysis shows that the YSZ layer still maintains its single crystallinity after the formation of the amorphous SiO_2 layer, and hence the original cube-on-cube orientation relationship is maintained between the YSZ and silicon substrate. The amorphous SiO_2 region near the YSZ/SiO_2 interface sometimes exhibits a darker contrast, suggesting possible zirconium diffusion into the amorphous SiO_2 layer. However, the amorphous nature of the SiO_2 layer has not been changed.

High-resolution TEM micrographs from the YSZ, SiO_2, and silicon substrate, taken along the [110] direction of the YSZ and silicon, are shown in figures 2(a), 2(b), and 2(c), respectively. The SiO_2 region is indeed amorphous, while the lattice images of the YSZ and silicon substrate have similar configurations, confirming the epitaxial relationship between them. The YSZ image is less clear compared with that of the silicon substrate, indicating the presence of defects which are also apparent in the low magnification micrograph of figure 1. The defects appear to be small clusters produced during the deposition process[4].

References

1. O.W. Holland, et al., *Appl. Phys. Lett.*, **63**(1993), 896.
2. D.S. Zhou, et al., *Appl. Phys. Lett.*, **63**(1993), 3580.
3. Q.X. Jia, et al., unpublished.
4. This research is sponsored by the U.S. DOE-OBES.

Proc. Microscopy and Microanalysis 1995, edited by G.W. Bailey, M.H. Ellisman, R.A. Hennigar, and N.J. Zaluzec
Copyright © 1995 MSA. Published by Jones and Begell Publishing, 79 Madison Ave., New York, NY 10016

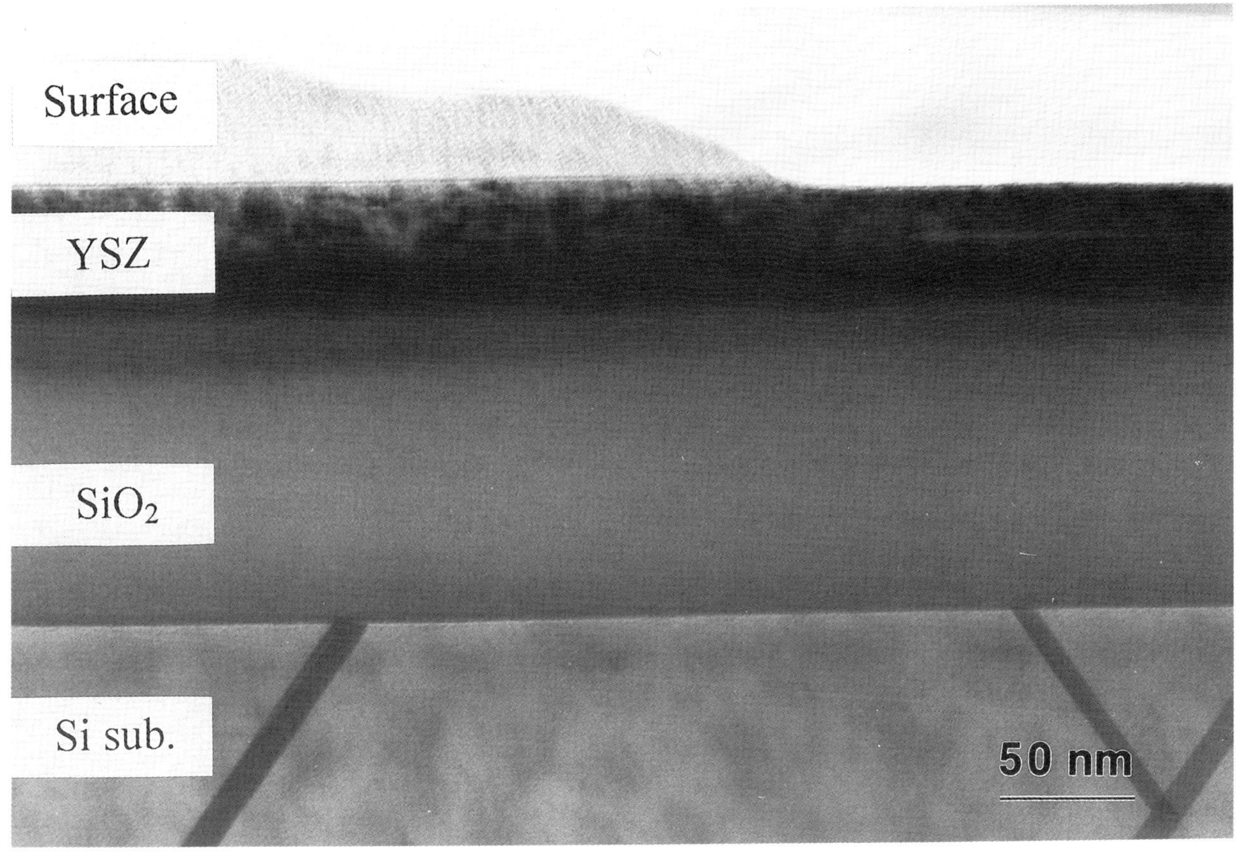

FIG.1. XTEM image of the YSZ/SiO₂/Si substrate structure grown epitaxially
by pulse laser deposition and heat treatment.

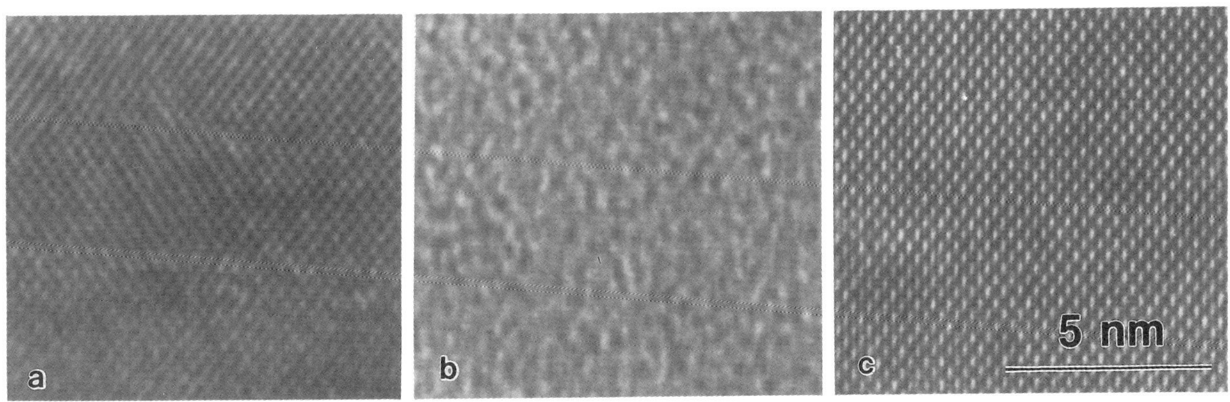

FIG.2. HREM images of the (a) YSZ, (b) amorphous SiO₂, and (c) Si substrate,
taken along the [110] direction of the YSZ and Si substrate.

TEM CHARACTERIZATION OF SUBSURFACE DAMAGE IN SILICON CARBIDE SUBSTRATES

W. Qian,* M. Skowronski,* G. Augustine,** R.C. Glass,** H. McD. Hobgood,** and R.H. Hopkins**

*Department of Materials Science & Engineering, Carnegie Mellon University, Pittsburgh, PA 15213
**Westinghouse Science & Technology Center, Pittsburgh, PA 15235

Because of its wide band gap and physical stability, silicon carbide is an important semiconductor material for high power, high temperature solid state devices[1]. In such applications, any surface damage introduced during wafer fabrication is detrimental to optimum device fabrication and operation. A typical wafer fabrication procedure consists of mechanical lapping and polishing, followed by chemical etching for removal of any residual mechanical damage. However, SiC is a refractory material with strong atomic bonding and high mechanical hardness, chemical etching of its surface is often difficult[2]. Mechanical polishing is thus the crucial step during wafer preparation. Though a number of techniques are currently in use for assessment of polishing induced surface damage, such as grazing incidence x-ray diffraction[3] and Rutherford backscattering[4], none of them is effective for studying the hidden damage beneath the polished surface. In the present work cross-sectional transmission electron microscopy (XTEM) was used to observe directly the nature and extent of subsurface damage in 6H-SiC substrates after different stages of mechanical polishing, and to elucidate the possible mechanism for material removal by the abrasive treatments.

For this study bulk single crystals of 6H-SiC with diameters up to 2-inches were grown in the <0001>-orientation by a modified sublimation technique[5]. Wafers were sliced perpendicular to the [0001] growth direction, and their (0001) Si-face were lapped and polished with consecutively smaller size diamond-based abrasive slurries (denoted by decreasing size as A,B,C and D) at various vertical loads until the final polished surface was attained. XTEM specimens with large thin area which are free of spurious defects have been prepared using our improved procedure. Figures 1- 4 are the typical bright field micrographs ($\mathbf{g}$ =[$11\bar{2}0$]) taken near the [$1\bar{1}00$] orientation from the transverse sections of SiC (0001) Si-face after each polishing stage (Philips EM420-TEM, 120 keV). We observe that both the nature and extent of subsurface damage in 6H-SiC substrates are critically dependent on the abrasive slurry size and the vertical load during polishing. Under severe abrasive treatments of the wafer (lapped with A size (15 µm) diamond abrasives with heavy vertical load), the surface exhibits irregular grooves (~ 1 µm in depth) and heavy subsurface damage (~ 2 µm in depth). The main subsurface damage consists of irregular cracks and dislocations lying on the basal plane due to the excessive shear stress during abrasive action. Most of the cracks appear to be initiated at the surface and propagated deep into the crystal. The observed mechanical damage is similar to that on silicon surface after polishing, where it was characterized as consisting of a heavily damaged layer near the surface and a deeper layer of plastic deformation[2]. However, when polished with reduced size abrasive slurries at reduced vertical loads, the characteristics of SiC surface damage is different (Fig. 2-3). Under these reduced abrasive actions, the surface roughness is much smaller. No dislocations have been observed inside the crystal even though the polishing was carried out on the (0001) slip plane. This may be attributed to the large critical shear stress for plastic deformation in a SiC crystal. It is commonly believed that for most materials the permanent damage caused by mechanical polishing has a depth on the order of the polishing particle sizes[6]. However, we found that both the polish-induced surface roughness and the extent of subsurface damage are lower in 6H-SiC than in Si and GaAs under similar mechanical treatments. The damaged layer in polished 6H-SiC substrate surface extends to less than about 0.1 times the polishing particle size after final polish. Under optimized polishing conditions (Fig.4), a final 6H-SiC wafer surface can be produced which is virtually free of polish induced defects, apart from an approximately 8 nm thick uniform strained layer beneath the polished surface.

References

1. M.E. Lin, et al., *Appl. Phys. Lett.* 62 (1993) 702 .
2. J.W. Faust, in J.R. O'Connor and J. Smiltens, Eds., *Silicon Carbide, New York: Pergamon Press* (1960) 403.
3. V.S. Wang and R.J. Matyi, *J. Electronic Materials* 21 (1992) 23.
4. K.E. Puttick, et al.,*Semicond. Sci. Technol.* 7 (1992) 255.
5. H.M. Hobgood, et al., *J. Crystal Growth* 137 (1994) 181.
6. J.F. Young and R.S. Shane, *Materials and Processes,* Marcel Dekker, Inc., New York (1985).
7. Supported by the US Air Force Wright Laboratory under Contract Nos. F33615-92-C-5912 and F33615-92-C-5963.

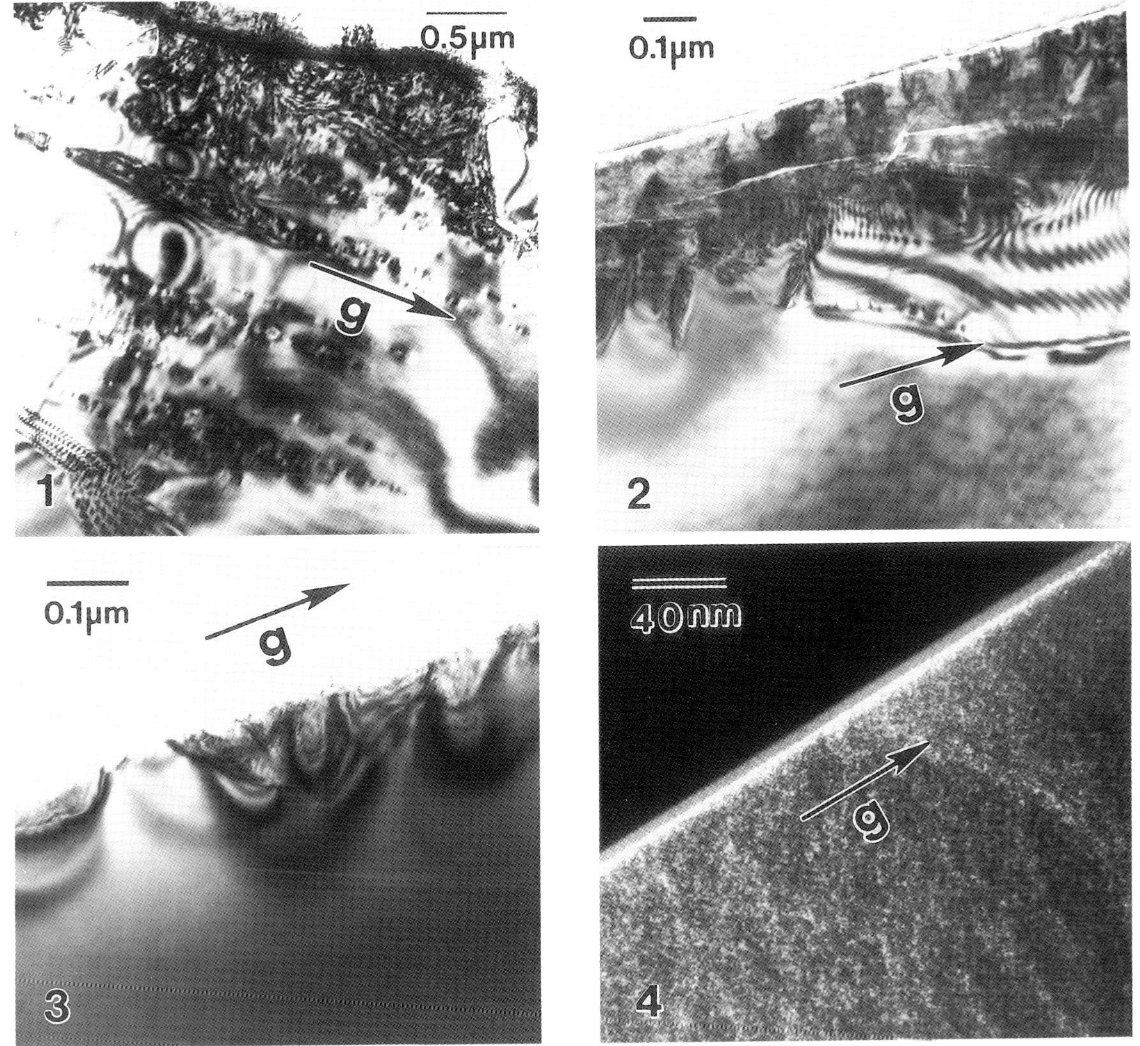

FIG.1. XTEM bright field image of the 6H-SiC(0001) wafer surface lapped with A-size diamond abrasive. The main subsurface damage consists of micro-cracks and dislocations.
FIG.2. Same as figure 1, but the surface was lapped with the smaller B-size abrasive at a lighter vertical load. Cracks are the main defects.
FIG.3. Same as figure 2, but the surface was polished with C-size diamond abrasives.
FIG.4. Weak beam XTEM image of the same surface after final mechanical polishing (with D size abrasives). The surface is flat, but with a 8 nm uniform strained layer.

MICROPATTERN ANALYSIS OF ZnIn$_2$S$_4$ USING ATOMIC FORCE MICROSCOPY AND TRANSMISSION ELECTRON MICROSCOPY

L. Martínez[*], J. M. Briceño-Valero[*], S. A. López-Rivera[*], K. Moore[**] and J. T. Thorthon[***]

[*]Applied Physic Group, Department of Physics, Faculty of Science. Universidad de los Andes. Mérida-Venezuela.
[**]Central Electron Microscopy Research Facility. University of Iowa. Iowa City. IA. 52242. USA.
[***]Digital Instruments. 520 E. Montecito St. Santa Barbara, CA. 93103. USA.

Semiconductor ZnIn$_2$S$_4$ is the only member of the II-III$_2$-IV$_4$ family with layer structure. The crystal structure of this semiconductor reported by Lappe *et. al.*[1] is based upon closed packing of sulfur atoms with octahedral and tetrahedral indium atoms and tetrahedral zinc atoms. Previous studies on this material with high resolution transmission electron microscopy and Ramman spectroscopy demonstrate the existence of challenging problems to be resolved related to its intrinsic nature.[2] There is great interest in this material for possible non-linear optical applications.[3] In the present study Atomic Force Microscopy (AFM) and Transmission Electron Microscopy (TEM) have been utilized to reveal crystallographic aspects of this material.

Yellowish, plate-shaped, single crystals of ZnIn$_2$S$_4$ were prepared with the chemical transport method using iodine as a transporting agent. Specimens suitable for electron microscope analysis in the growth direction were prepared by peeling off layers with a piece of tape. For AFM analysis, freshly cleaved surfaces were used. Figure 1 is a selected area diffraction pattern (SAD) crystal base plane of a hexagonal crystallographic formation. The figure demonstrates the superposition of two identical patterns which correspond to two layers of the material. Fig. 2 is a Fast Fourier Transform picture obtained with a Digital Instruments Nanoscope-III. Fig. 3 shows a cross sectional analysis of image in Fig. 2. Notice that points 1-4 in fig. 3 are the respective points indicated in fig. 2. From this figure, an hexagonal base plane structure with a lattice parameter of 0.38 nm is identified which agrees with the value obtained from electron diffraction and x-rays patterns reported in the literature[1].

References:

[1] F. Lappe, A. Niggli, R. Nitsche, and J. G. White, *Zeitschi fur Kristallographic,* **117**, 146 (1962).
[2] J. M. Briceño-Valero, S. A. López-Rivera, L. Martínez, M. Ogura and R. D. Acosta. *Proceedings of III Workshop on Electron Microscopy, IVIC-Altos de Pipe-Caracas-Venezuela,* page 86 (1988).
[3] S. I. Radautsan, V. F. Zhitar, and V. Y. Railyan, *Soviet Phys., Semic.,* **9**, No. **12**, 1476(1976).

Proc. Microscopy and Microanalysis 1995, edited by G.W. Bailey, M.H. Ellisman, R.A. Hennigar, and N.J. Zaluzec

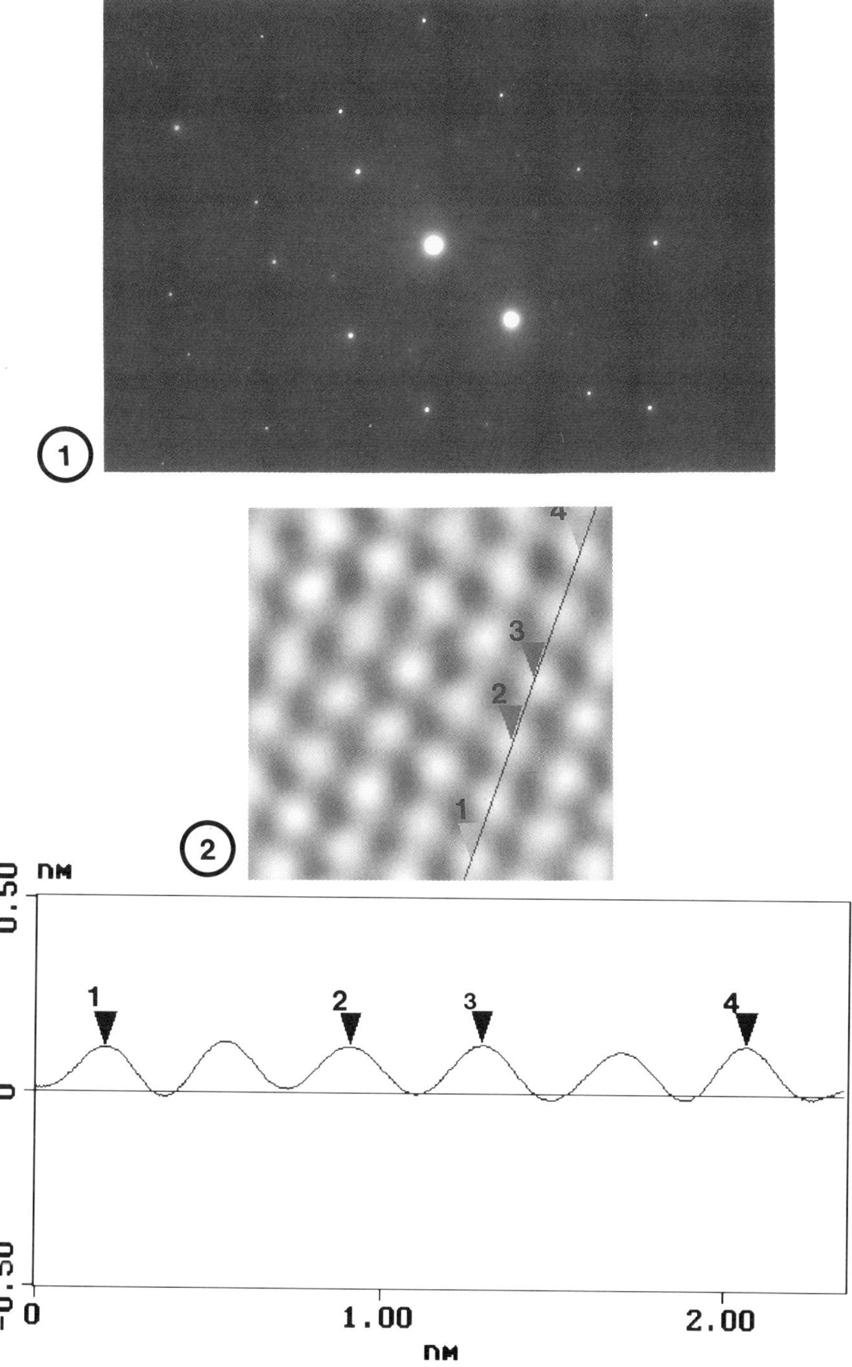

FIG. 1 Selected Area Diffraction Pattern of $ZnIn_2S_4$ single crystal

FiG. 2 Fast Fourier Transform Image obtained from an AFM micrograph.

FIG.3 Cross sectional analysis of image in Fig. 2.

INTERPRETATION OF HREM IMAGES OF M-BARIUM FERRITE TAKEN UNDER DYNAMICAL
SCATTERING CONDITIONS.

Dongchul Kwon, Lew Rabenberg
Center for Materials Science and Engineering, The University of Texas at Austin

In HREM analyses of extended structural defects in complex compounds, it is desirable to be able
to image atomic columns composed of light elements as well as those composed of heavy
elements. For the thinnest possible specimens, HREM image contrast tends to be dominated by
the heavy elements, however, it is possible to increase the contribution of the light elements to
the image by imaging specimens of somewhat greater thicknesses. This paper presents results
from a HREM study of extended defects in M-barium ferrite in which this approach has been
successfully applied.

Hexagonal ferrites, of which M-barium ferrite, $BaFe_{12}O_{19}$, is an example, form a versatile class of
compunds built up from varied stackings of simpler structural units denoted[1] R, S , and T. M-
barium ferrite takes the magnetoplumbite structure - a simple alternation of R blocks and S
blocks. HREM techniques have been widely used to charcterize defects in stacking sequences,
intergrowths, and polytypes in these materials.[2]

TEM specimens of M-barium ferrite were prepared by (a) sintering together powders of barium
carbonate $(BaCO_3)$ and ferric oxide $(\alpha - Fe_2O_3)$, (b) slicing the resulting ceramic body, (c)
mechanical grinding, and (d) ion beam milling. HREM observations were performed using a Jeol
2010 TEM equipped with an UHR pole piece $(C_s = 0.5$ mm) and a LaB_6 emitter operating at 200
kV. Images were recorded using a 1024 X1024 pixel chilled CCD digital camera; numerical
simulation was carried out using the EMS software[3] with nonlinear imaging conditions.

Figure 1 shows (a) a schematic sketch of the atomic structure, (b) the crystal potential of the M

compound projected along the [1$\bar{1}$0] zone axis, and (c) a calculated HREM image taken at
Scherzer defocus with t = 1.2 nm. The strongest crystal potential corresponds to the Ba atom
columns.

Figure 2 shows an experimental HREM image in the [1$\bar{1}$0] zone. Comparison with the calculated
image (1c) (different magnifications) shows that the dark contrast corresponds to the atom
columns and the white contrast corresponds to the interstitial positions. According to the weak
phase object approximation, heavier atoms (Ba) image dark, while lighter atoms (Fe, O) image
bright. As the thickness increases and multiple scattering becomes important, the contribution of
the light atoms to the image increases, and at some thickness, the light atoms, as well as the
heavy atoms can be visualized with good contrast. Thus, the dark spots between small white
spots correspond to Ba atoms, and the other dark contrast regions correspond to Fe and O
atomic columns. These observations enable the R and S blocks to be identified as indicated in
the Figure. Calculations also show that the Ba position begins to image bright at a thickness of 4
nm; this thickness at which contrast reverses represents a limit of applicability of this analysis.[4]

Figure 3 shows an area of the specimen in which defects in the RS stacking sequence are
observed. $(RSS)_2$ (W) intergrowths and the RSSSSR block are associated with the
accommodation of compositional variations during growth.[5]

References

1. P. B. Braun, Philips Research Reports 12(1957)491
2. Y. Hirotsu et al., Trans. Japan Inst. Metals 24(1983)461
3. P. A. Stadelmann, Ultramicroscopy 21(1987)131
4. F. H. Li and H. Hashimoto, Acta. Cryst. B40(1984)454
5. J. Van Landuyt et al., J. Solid State Chemistry 9(1974)103

Proc. Microscopy and Microanalysis 1995, edited by G.W. Bailey, M.H. Ellisman, R.A. Hennigar, and N.J. Zaluzec
Copyright © 1995 MSA. Published by Jones and Begell Publishing, 79 Madison Ave., New York, NY 10016

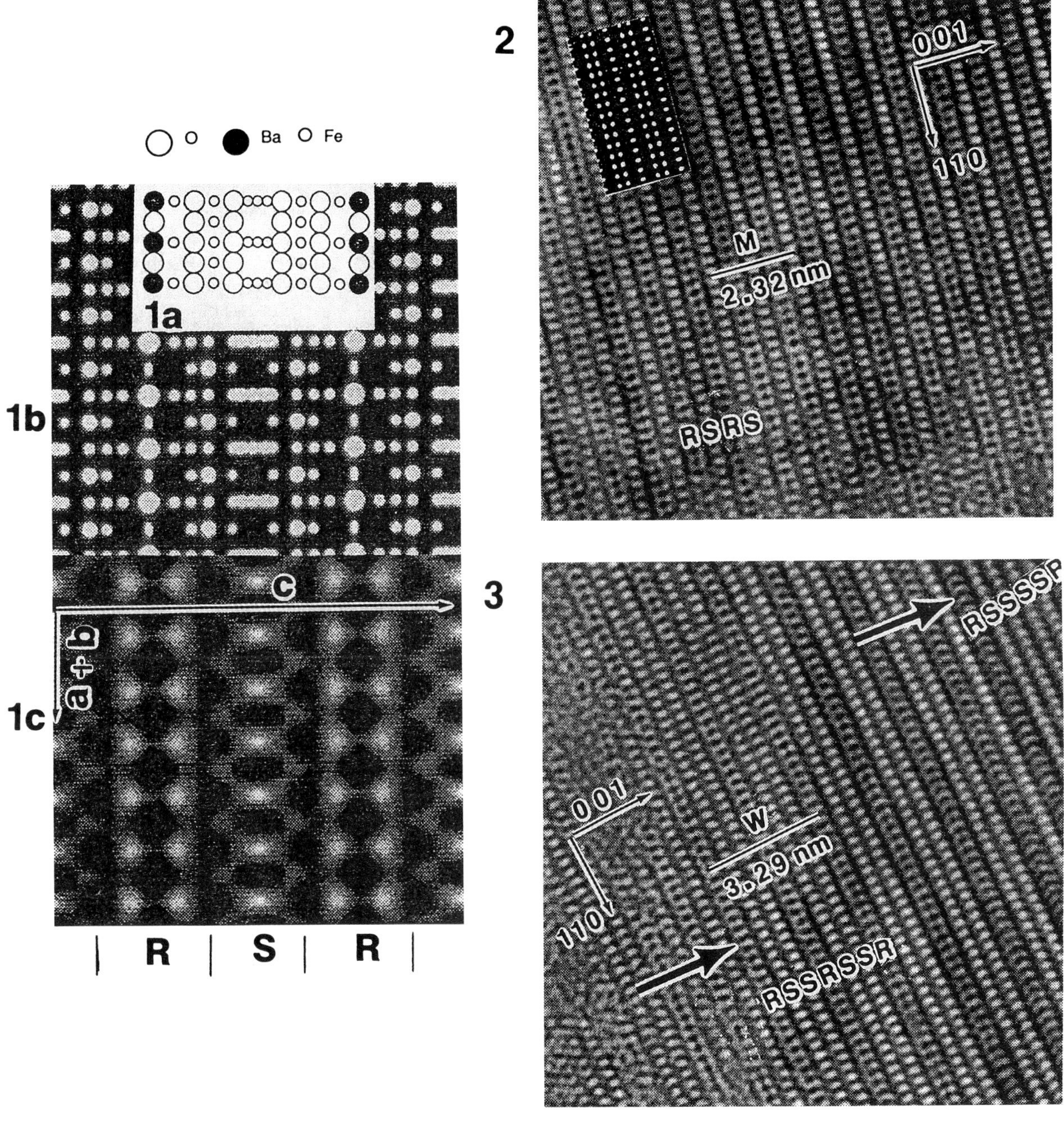

FIG. 1. (a) Atomic structure, (b) projected crystal potential in the $[1\bar{1}0]$ zone, and (c) calculated HREM image at Scherzer defocus with t= 1.2 nm of the M compound.

FIG. 2. Experimental HREM image in the $[1\bar{1}0]$ zone showing the R and S building units in the M compound.

FIG. 3. Intergrowth of W with M compound and a polytype of W compound.

Compositional Separation in CoCrTa Thin Film as Observed Using a Transmission Electron Microscope with an Imaging Filter

K. Kimoto[*], Y. Yahisa[**], T. Hirano[*], K. Usami[*] and S. Narishige[**]

[*]Hitachi Research Laboratory, Hitachi Ltd., Hitachi, Ibaraki 319-12, Japan
[**]Data Storage & Retrieval Systems Division, Hitachi Ltd., Odawara, Kanagawa 256, Japan

Cobalt chromium alloys are current longitudinal high-density recording media. Compositional features in the cobalt chromium thin films directly affect the magnetic and recording properties. Elemental analyses of the films have been performed by several methods,[1,2] however, topological nature and crystallographic information are still ill-defined because of limitations in these methods.

In the present study, elemental mapping of cobalt and chromium in a CoCrTa thin film has been quantitatively performed using a transmission electron microscope (TEM) with an imaging filter.

A chromium underlayer (75nm thick) and a CoCrTa magnetic layer (27nm thick) were deposited on a Ni-P plated Al-Mg alloy substrate using direct current magnetron sputtering at a substrate temperature of 270°C. Target contents for sputtering of the CoCrTa film are shown in Table I. Energy-filtered images and electron energy loss spectra (EELS) were acquired using a TEM (Hitachi, HF-2000) equipped with an imaging filter (Gatan, imaging filter model 678).[3] Because of the high brightness of a field emission gun, this system allows a small collection semiangle (4.4 mrad) and a narrow energy window (20-30 eV) for imaging. Therefore, chromatic broadening caused by the aberrations of the objective lens is reduced, resulting in the high spatial resolution.[4,5] The observations were performed with a liquid-nitrogen cooling holder (-179°C) to reduce specimen heating and radiation damage.

Figure 1 shows the zero-loss image of the CoCrTa film. The film has a polycrystal structure, consisting of grains about 20 nm in diameter. Mean contents in this CoCrTa film were measured by energy dispersive x-ray spectroscopy (EDX) using a 200 nm-diameter probe (Table I). Figure 2 shows the EELS of the CoCrTa film for the whole area observed in Fig. 1. The $L_{2,3}$ edges of Cr and Co are observed at energies of 575 and 779 eV, respectively. The $L_{2,3}$ core-loss images of Cr and Co were obtained by subtracting the pre-edge images from the post-edge images (FIG. 3a and 3b).

Figure 4 shows the atomic ratio map obtained. The quantification of the map is the same as for EELS.[6] The left scale on the color bar is the atomic ratio Cr/Co and the right scale is the Cr content, assuming tantalum homogeneity. Compositional inhomogeneity can be quantitatively observed, consisting of Co-enriched areas and a Cr-enriched phase around those areas.

References
1. Y. Maeda and M. Asahi: J. Appl. Phys. **61** (1987) 1972.
2. K. Hono, Y. Maeda, S. S. Babu and T. Sakurai: J. Appl. Phys. **76** (1994) 1.
3. A. J. Gubbens and O. L. Krivanek: Ultramicroscopy **51** (1993) 146.
4. K. Kimoto et al.: Jpn. J. Appl. Phys. **33** (1994) L1642.
5. K. Kimoto et al.: Journal of Electron Microscopy. **44** (1995) (in press)
6. R. F. Egerton: *Electron Energy-Loss Spectroscopy in the Electron Microscopy.* 1st ed.(Plenum Press, New York, 1986) , p.316

Proc. Microscopy and Microanalysis 1995, edited by G.W. Bailey, M.H. Ellisman, R.A. Hennigar, and N.J. Zaluzec
Copyright © 1995 MSA. Published by Jones and Begell Publishing, 79 Madison Ave., New York, NY 10016

Table I. Co, Cr and Ta contents (at%) in target and deposited CoCrTa film.

	Cobalt	Chromium	Tantalum
Target	78.0	16.0	6.0
Deposited CoCrTa film (EDX measurements)	78.2	17.0	4.8

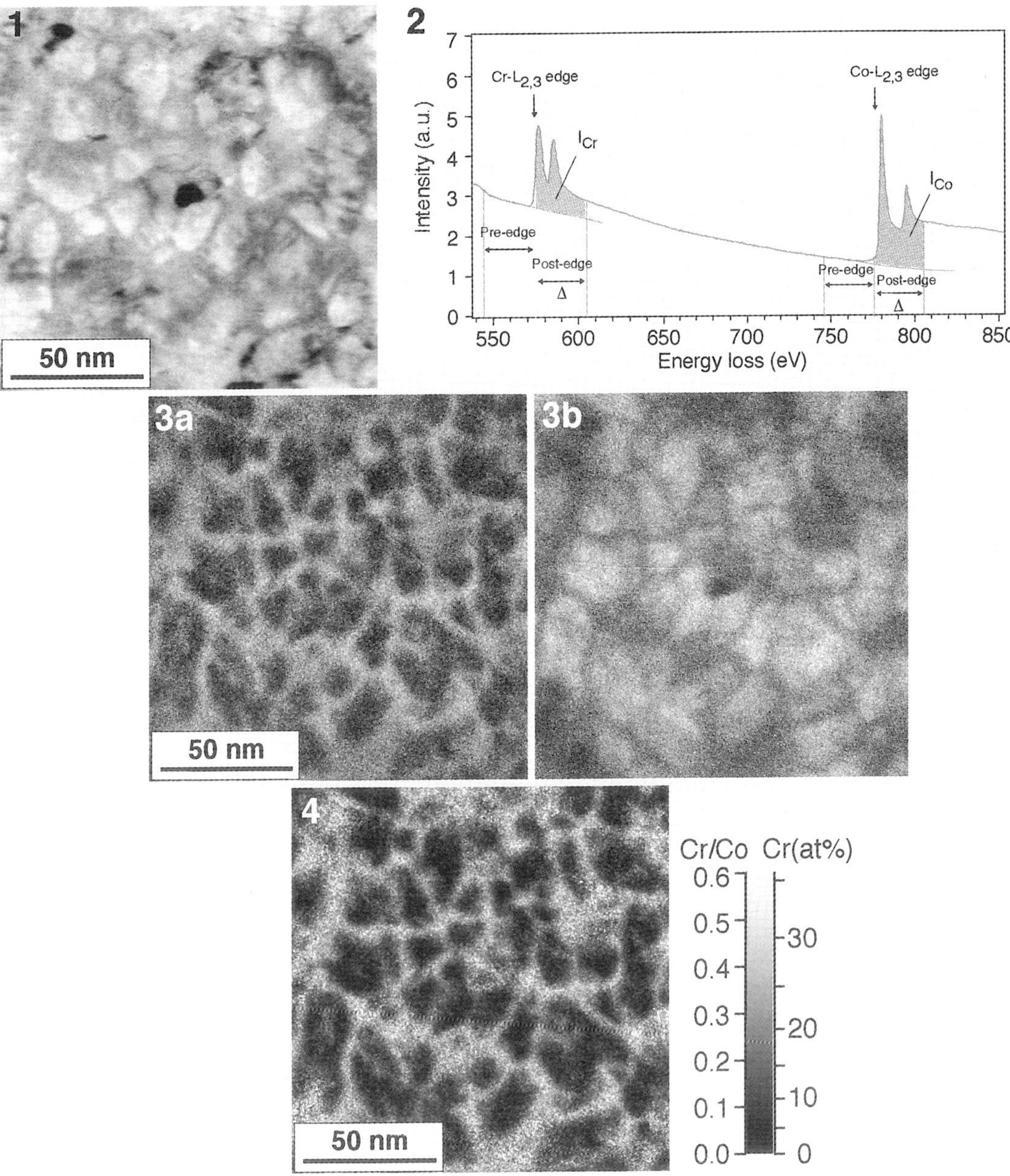

FIG. 1. Zero-loss image of CoCrTa film deposited at substrate temperature of 270°C.

FIG. 2. Energy loss spectrum of the CoCrTa film in the core-loss region.

FIG. 3. The L_{2,3} core-loss images of (a) chromium and (b) cobalt obtained by subtracting the pre-edge images from the post-edge images.

FIG. 4. Atomic ratio (Cr/Co) map of the CoCrTa film.

LORENTZ OBSERVATION OF MAGNETIC DOMAINS IN TERFENOL-D

J. Dooley, N.T. Nuhfer, and M. De Graef

Department of Materials Science & Engineering, Carnegie Mellon University, Pittsburgh, PA 15213

Terfenol ($Tb_{.27}Dy_{.73}Fe_{1.95}$) has the largest known room-temperature magnetostriction coefficient, making it well suited for low frequency, high power sonar applications. Terfenol takes the $MgCu_2$ cubic Laves phase structure, with lattice parameter $a = 0.732$ nm and space group **Fd3m**.[1] Magnetic domain configurations in this material are generally interpreted as 71° and 109° walls, along with 180° walls, with the magnetization along the $\langle 111 \rangle$ directions. At room temperature the cubic unit cell experiences an extremely small rhombohedral distortion along the magnetically soft $\langle 111 \rangle$ directions (e.g., the cubic [100] tilts by 1.13 mrad toward a new orientation). The growth direction, designated [$\bar{2}11$], is taken as a reference throughout this discussion.

The material, prepared by a free-standing float-zone (FSZ) technique,[2] was provided by Etrema Products Inc. in the form of [$\bar{2}11$] oriented rods with a diameter of 3 mm. Since the growth direction is perpendicular to [111] and the growth front is dendritic, growth twins are easily formed by a 180° rotation on the (111) plane. The [$\bar{1}10$] direction was determined by Laue diffraction and [$\bar{1}10$] oriented disks were cut from the rod. The samples were mechanically polished down to 200 mm, double dimpled, ion milled (5kV Ar^+ ions) to perforation, and immediately transferred into the TEM to prevent extensive oxidation. Lorentz microscopy was used to image the domain structure present near twin boundaries. A JEOL 4000EX HRTEM fitted with a Gatan Imaging Filter (GIF) was used to obtain energy-filtered magnetic domain images. The procedure followed to obtain Lorentz images at 400 kV is outlined in another contribution in this volume[3]. Energy filtered magnetic domain images (in both Fresnel and Foucault imaging modes) can be routinely obtained with magnifications of more than ×500,000.

James and Kinderlehrer[4] have proposed a mathematical model for the magnetostriction and have worked out the fine-scale magnetic domain configurations near a twin boundary. There are four magnetic orientations derived from the four $\langle 111 \rangle$ directions, and a rotational growth twin gives rise to an additional four variants. James and Kinderlehrer assume two sets of lamellar domain sequences, one on either side of the boundary. The coherence of the domain boundary is related to the strains associated with any set of magnetization variants. While several combinations are exactly compatible, it can be shown that some of the configurations are only compatible in the fine-phase limit and hence require a finite transition layer. Though not conclusive, our preliminary findings on [$\bar{2}11$] oriented samples[5] can be consistently described through this model.

An energy filtered Fresnel (underfocus) image of the domain configuration near a set of three parallel twin boundaries is shown in Figure 1a, along with a digitized diffraction pattern from the area (1b). The twin plane is (111); the domain wall traces in the lower section of the figure are parallel to the [001] direction. Figure 1c shows an energy filtered Foucault image of the same region; the drawing in Figure 1d indicates the major features in the domain structure. The thin twin band, formed by the planes 2 and 3 in Figure 1d is approximately 8 nm across. The zig-zag boundary in the lower half of the figure corresponds to a 180° domain wall, with average orientation along the (100) plane. The traces of the 180° domain walls are approximately parallel to ($\bar{1}14$) and (112), and the precise orientation may be related to the domain width. The observed domain configuration agrees well with a configuration predicted by James and Kinderlehrer's model and has been observed previously by optical microscopy (differential interference contrast[1]) at a more macroscopic scale. The varying domain widths in the lower half of the figure may indicate that this domain configuration corresponds to an exactly compatible twin configuration. Of the twelve twin-magnetic domain configurations derived by the theory, only two are compatible with the observed configuration. Work is currently underway to document other configurations, in particular the ones that require a thin transition layer.

Proc. Microscopy and Microanalysis 1995, edited by G.W. Bailey, M.H. Ellisman, R.A. Hennigar, and N.J. Zaluzec
Copyright © 1995 MSA. Published by Jones and Begell Publishing, 79 Madison Ave., New York, NY 10016

References

1. M. Al-Jiboory and D.G. Lord, *IEEE Trans. Magn.*, 26(1990)2583
2. O.D. McMasters, J.D. Verhoeven, and E.D. Gibson, *J. Magn. Magn. Mat.*, 54(1986)849
3. N.T. Nuhfer, J. Dooley, and M. De Graef, this proceedings (1995)
4. R.D. James and D. Kinderlehrer, *Phil. Mag. B*, 68(1993)237
5. J. Dooley and M. De Graef, *Mat. Res. Soc. Sym. Proc.*, in press (1995)

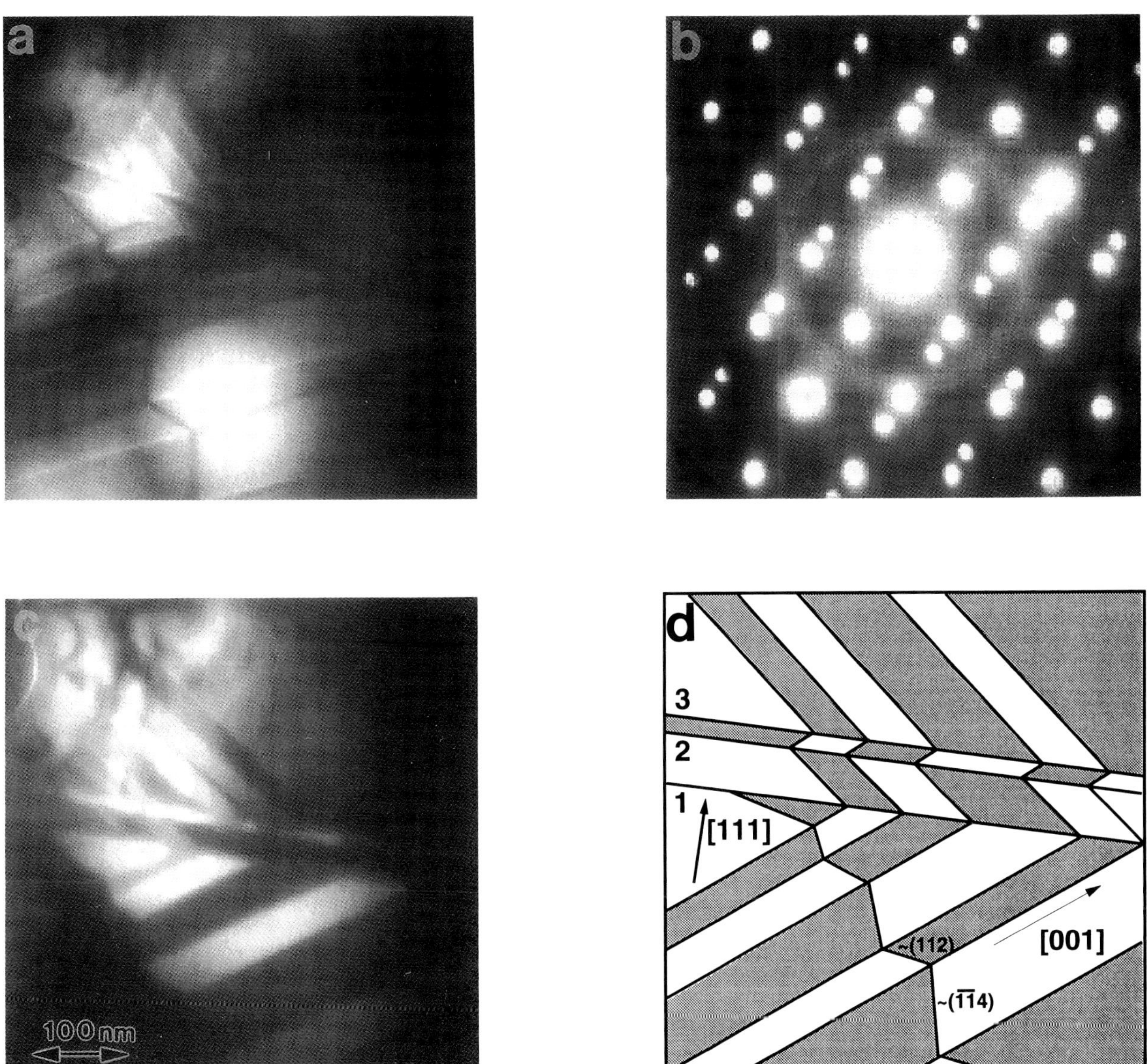

Figure 1: a) Underfocus energy-filtered Fresnel image of a region containing 3 growth twins; b) [$\bar{1}$10] digitized diffraction pattern taken with the conditions described in a companion paper[3]; c) energy-filtered Foucault image of the same region, and d) schematic representation of the various magnetic domains and twin boundaries.

MICROSTRUCTURE OF SPIN-VALVE MR SANDWICHES

L. Tang, M. Xiao*, D.E. Laughlin, and M. H. Kryder*

Data Storage Systems Center, Department of Materials Science and Engineering
*Department of Electrical and Computer Engineering
Carnegie Mellon University, Pittsburgh, PA 15213

Giant magnetoresistance (GMR) effects in magnetic multilayers with spin-valve structures are under intensive investigation.[1] The GMR effects in spin-valve structures originate from the change in the orientation of magnetization in the successive ferromagnetic layers. Of the various types of spin-valve multilayered structures reported, spin-valve sandwiches, in which one of the two ferromagnetic layers separated by a nonferromagnetic metal layer is constrained through exchange coupling to an adjacent antiferromagnetic layer, are most promising for applications in read heads for high density magnetic recording. This is due to their large MR and high sensitivity in low magnetic fields.[2] Study of the correlation between magnetic/magnetotransport properties and the microstructure of spin-valve sandwiches is crucial for a better understanding of the mechanism of the spin-valve effects and for future MR heads design. Here, we present the results of transmission electron microscopy (TEM) studies of the microstructure of a $Ni_{81}Fe_{19}(47Å)\backslash Cu(18Å)\backslash Ni_{81}Fe_{19}(53Å)\backslash FeMn(186Å)$ spin-valve sandwich. The two ferromagnetic $Ni_{81}Fe_{19}$ layers are decoupled by the 18 Å thick Cu layer and the 53 Å thick $Ni_{81}Fe_{19}$ layer is exchange coupled with the antiferromagnetic FeMn layer. This spin-valve sandwich was sputter deposited onto a $Si(100)\backslash Si_3N_4(1000Å)$ substrate using a RF diode sputtering system.

Figure 1a is the bright field plan-view TEM image of the sandwich at 0° tilt. The grain size of the film estimated from this micrograph is about 100 Å. The Moiré fringes in Fig. 1a are due to the grain overlapping of the different layers. Diffraction patterns of the plan-view sandwich at 0°, 30°, and 60° tilt around an axis **OT** in the film plane are shown in Figs. 1b, 1c, and 1d. Two fcc phases, which are believed to be NiFe/Cu/NiFe and FeMn, with slightly different lattice parameters can be clearly identified.[3] Figs. 1 b-d also show that part of the grains are (111) textured with an angular distribution of about 10° and part of the grains are randomly oriented in both the NiFe/Cu/NiFe and FeMn layers. This observation agrees with the diffraction pattern taken from a cross-section sample of the sandwich (Fig. 2b) which shows that the overlapped (111) diffraction arcs of the NiFe/Cu/NiFe/FeMn layers are parallel to that of the Si (200) diffraction spot. Figure 2a is the high resolution cross-section TEM image of the sandwich. A grain with $[1\bar{1}0]$ zone axis (labeled **A** in the image) shows the $(11\bar{1})$ lattice planes are continuous through the NiFe/Cu/NiFe layers. This indicates that the Cu layer is strain matched with the NiFe layers although Cu has a larger bulk lattice parameter (a= 3.61 Å) than that of NiFe (a= 3.55 Å). This is because the thickness of the Cu layer, 18Å, is below the critical thickness where misfit dislocations can be generated. The interface between the NiFe and FeMn layers is not perfectly smooth. The role of the NiFe/FeMn interface roughness is being studied through the comparison of samples with different ΔR/R and different exchange fields.

References

1. B. Dieny, J. Magn. Magn. Mater. 136(1994) 335.
2. C. Tsang et al., IEEE Magn. Trans. Magn., Mag-30(1994) 3801.
3. C. Hwang and T. A. Nguyen, Mat. Res. Soc. Symp., 232(1991)211.
4. This work is supported by the National Science Foundation under Grand ECD-8907068.

Proc. Microscopy and Microanalysis 1995, edited by G.W. Bailey, M.H. Ellisman, R.A. Hennigar, and N.J. Zaluzec
Copyright © 1995 MSA. Published by Jones and Begell Publishing, 79 Madison Ave., New York, NY 10016

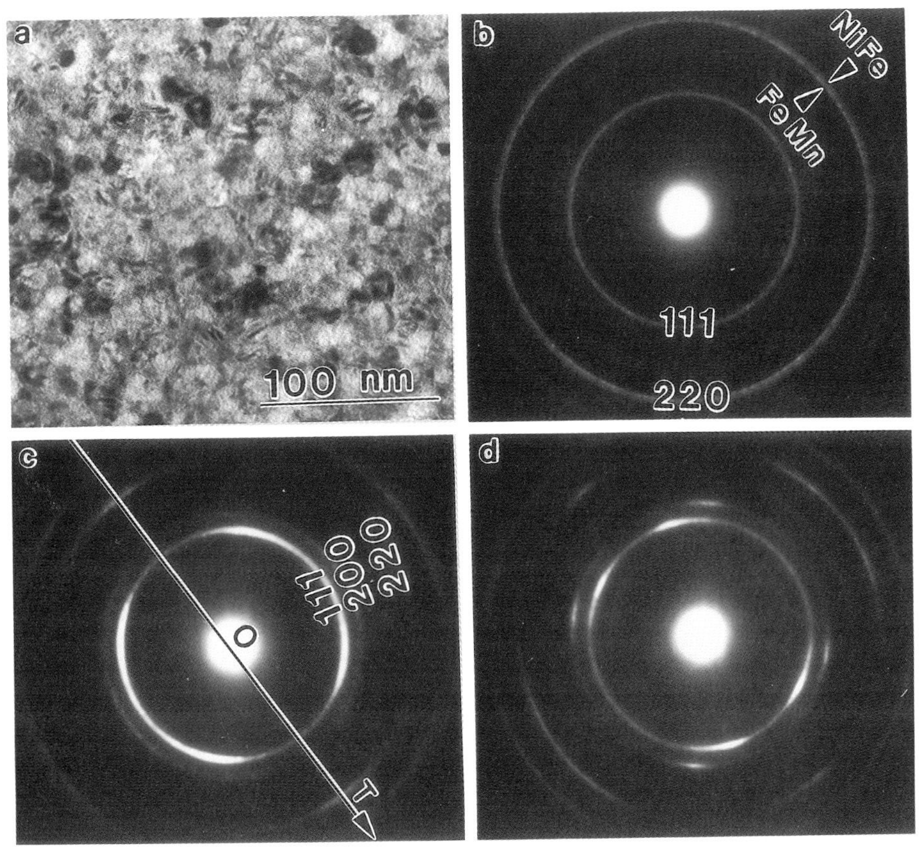

Fig. 1.   Plan -view bright field image (a) and diffraction patterns at (b) 0°, (c) 30°, (d) 60° tilt of the spin-valve sandwich.

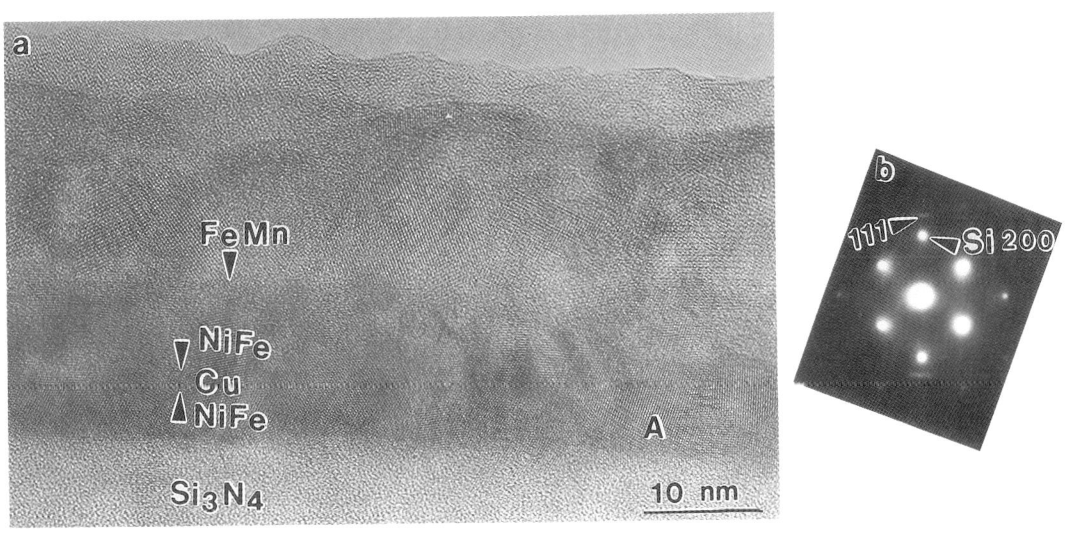

Fig. 2.   Cross-section high resolution image (a) and diffraction pattern (b) of the spin-valve sandwich.

MICROSTRUCTURE AND MICROANALYSIS OF NEODYMIUM IRON BORON MAGNETS SINTERED WITH ALUMINIUM.

M. Chandramouli and G. Thomas

Department of Materials Science and Mineral Engineering, University of California at Berkeley, Berkeley CA 94720 and Lawrence Berkeley Laboratory, 1 Cyclotron Road, Berkeley CA. 94720.

The magnetic properties of NdFeB magnets sintered with 15 vol.% Al at 850 °C and 1000 °C for 1 hour are dependent on microstructure. Complementary use of SEM and TEM techniques are crucial to elucidate the microstructural differences responsible for the properties. The hysterisis loops shown in Fig. 1 indicate that the higher sintering temperature sample has a larger dip near zero applied field and lower coercivity. This behavior is consistent with a larger volume fraction of soft magnetic phases.

Microstructural analysis performed using a JEOL 35 CF SEM does not show significant differences between the samples. As revealed in Fig. 2, a representative backscattered image of the 850 °C sample, the microstructure consists of dark pockets with a gray matrix region with small, bright contrast pockets. Quantitative microanalysis confirmed that the gray matrix phase had a composition consistent with $Nd_2Fe_{14}B$, the ferromagnetic phase in conventionally prepared Nd-Fe-B permanent magnets[1]. The small bright pockets were found to have a high Nd content while the large dark pockets were rich in Al. It was found that the pockets in the 1000 °C sample had a lower overall Al content (Nd-5at.%, Fe-77 at.%, Al-18at.%) than the 850 °C sample Nd-5at.%, Fe-69at.%, Al-26at.%). Due to the small grain size (~ 0.5 µm) it is difficult to determine the composition of the individual grains in the reacted pocket regions using SEM microanalysis.

Transmission electron microscopy investigations were performed using a Philips 400 AEM equipped with a KEVEX quantum window EDS detector. Specific attention was paid to identifying the intermetallic phases present in the reacted pockets since they are thought to be responsible for the magnetic properties. These analyses indicate that for both the 850 °C and 1000 °C, there are two distinct intermetallic phases, labeled as "a" and "b" in Fig. 3, present in the reacted pockets. Convergent beam electron diffraction experiments indicated that the "a" phase was cubic (DO_3) Fe_3Al while the "b" phase has a nearly cubic (a = c = 12.9 Å) tetragonal structure. A more precise identification of the tetragonal "b" phase has not been determined.

The representative EDS spectra shown in Fig. 4a-b show that "a" phase has a higher Fe content than the "b" phase. Quantitative microanalysis with absorption corrections was performed. The two beam CBED method of Kelley et. al. was used to estimate the thickness[2]. The average composition of the tetragonal "b" phase was the same in both samples: Nd-10at.%, Fe-53-at.%, Al-37 at.%. For the Fe_3Al "a" phase, the results indicate that the average Al content is around 30 at.% in the 850 °C material versus 28 at.% in the 1000 °C sample. For Al contents below 30 at.%, Fe_3Al is ferromagnetic and at room temperature. The composition results indicate that there is a higher fraction of Fe_3Al grains which are soft magnetic in the 1000 °C sample than in the 850 °C sample. This explains the poorer magnetic properties of the 1000 °C sample and illustrates the usefulness of quantitative electron microscopy in determining structure property relations in magnetic systems.

References

1. K. D. Durst and H. Kronmüller, Jour. Mag. Mag. Mat., 68, pp. 63 - 75, 1987
2. P. M. Kelley, A. Josten, R. G. Blake and J. G. Napier, Phys. Sta. Sol. (a) 31, pp, 771 - 780, 1975
3. This work was supported by the Director, Office of Energy Research, Office of Basic Energy Sciences, Materials Sciences Division of the U.S. Department of Energy under contract No. DE-AC03-76SF00098.

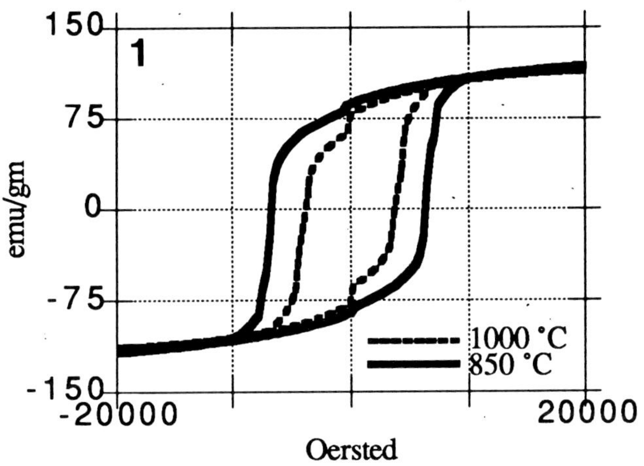
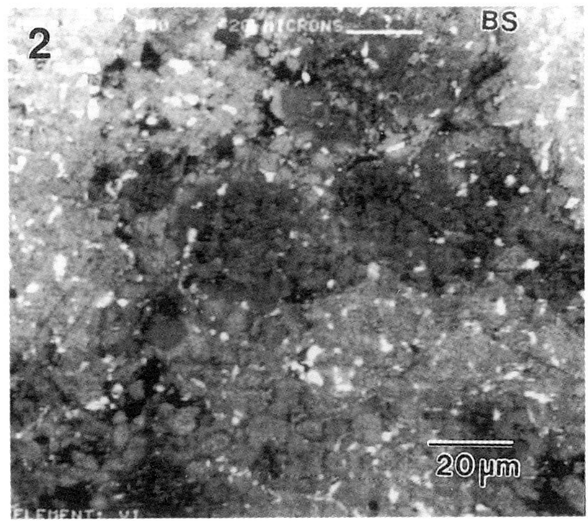

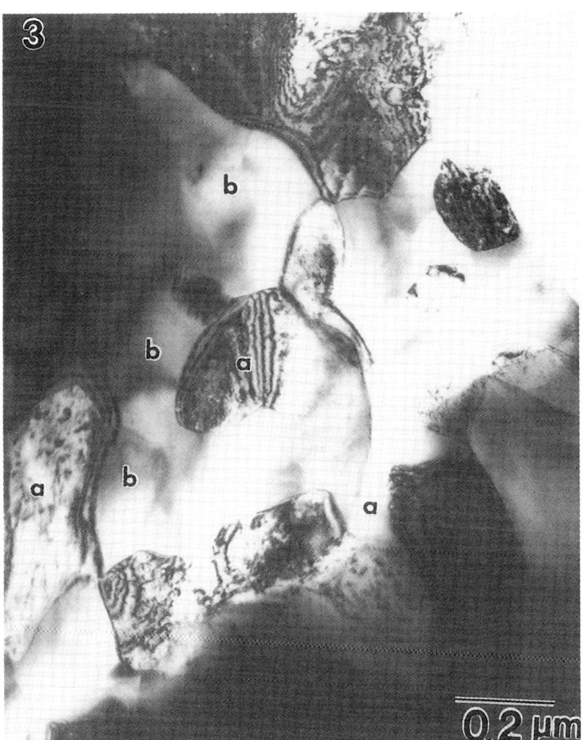

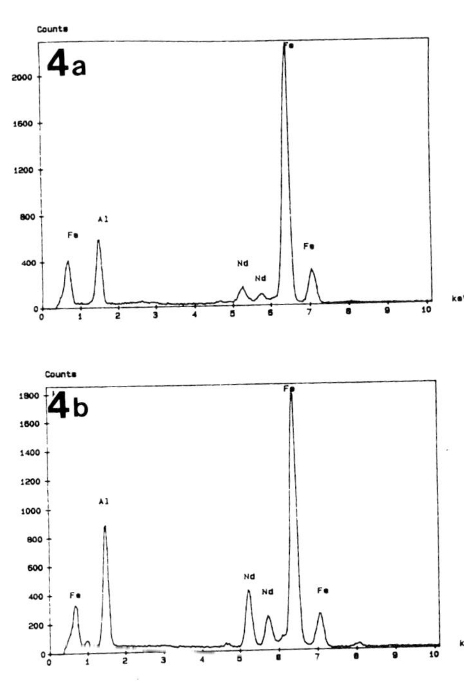

Fig. 1: Hysteresis loops of NdFeB magnets sintered with 15 vol.% Al at 850 °C and 1000 °C for 1 hour
Fig. 2: Backscattered SEM images of 850 °C magnet.
Fig. 3: Representative bright field image of reacted pocket region in 850 °C material.
Fig. 4: Representative EDS spectra of phases labeled as "a" and "b" in Fig. 3.

HREM STUDY OF EPITAXY IN Co-Sm // Cr THIN FILMS

Y. Liu[*,**], B. W. Robertson[*,**] and D. J. Sellmyer[*,***]

[*] Center for Materials Research and Analysis, [**] Department of Mechanical Engineering, [***] Behlen Laboratory of Physics, University of Nebraska, Lincoln, NE 68588.

In developing Co-based magnetic thin films for high density recording, it is well known that the Cr underlayer plays a critical role in generating large in-plane coercivity. Recently it has been shown that Co-Sm films with a nominal composition from Co_4Sm to Co_7Sm_2 exhibit high coercivity and in-plane anisotropy [1,2]. Transmission electron microscopy studies of Co-Sm films have shown that the microstructure of a Co-Sm thin film is composed of the amorphous matrix and crystallites with a grain size of about 5 nm [3]. The crystallites in the as-deposited Co-Sm films have a close-packed structure [4]. Each crystallite has a particular stacking mode which consists of local random stacking, occasionally a few unit cells of two layer stacking AB, three layer stacking ABC, and four layer stacking ABAC. However, the epitaxy relation between the Co-Sm crystallites and the Cr, if any, remains a mystery. In this paper we report our discovery of a new epitaxial relation between the Co-Sm crystallites and the Cr underlayer.

The Co-Sm target used for deposition has a nominal composition of Co_7Sm_2. The film for TEM study was prepared by dc magnetron sputtering under an argon pressure of 5 mTorr at room temperature. The high resolution electron microscopy (HREM) and diffraction work was performed using a JEOL 2010 transmission electron microscope operating at 200 kV.

Figure 1 shows the cross sectional HREM image of a Co-Sm crystallite grown on Cr. The Cr grain is aligned on the [111] zone axis. The Co-Sm film grown on this Cr grain is on the [11$\bar{2}$0] zone axis. The epitaxy is Cr ($\bar{1}2\bar{1}$) // Co-Sm ($\bar{1}$100) and Cr [111] // Co-Sm [11$\bar{2}$0]. The interface between the Cr underlayer and the Co-Sm film is indicated by letter I. The crystal structures of the Cr underlayer (c) and the Co-Sm crystallite (a) are well revealed by the optical diffraction patterns, which are attached to the corresponding regions. The optical diffraction patterns clearly show that the Co-Sm has the close-packed structure and the Cr the BCC structure. The HREM image was obtained by Fourier filtering the original TEM negative. The filtered optical diffraction pattern with the mask (b) is also attached. The stacking sequence in the Co-Sm is ACAC on the left side and then switched to ABAB as indicated by the arrows. Simulated images based on the above stacking sequence for the Co-Sm crystallite and the BCC structure for the Cr underlayer show excellent match to the HREM image.

The authors thank Dr. Z.S. Shan, Todd Voiles and Xueli Zhao for thin film deposition and TEM specimen preparation. This work is made possible by ARPA/NSIC under Grant MDA 972-93-1-0009 and was performed at the CMRA Central Facility for Electron Microscopy .

1.   E. M. T. Velu & D. N. Lambeth, IEEE Trans. Magn., Mag-28, 3249 (1992).
2.   D.J. Sellmyer, Z.S.Shan, Yi Liu, S.H.Liou, S. Malhotra and B.W. Robertson. Acta Metallurgica, in press.
3.   Y. Liu, B. Robertson, Z. S. Shan, S. Malhotra, M. J. Yu, S. K. Renukunta, S. H. Liou, and
      D. J. Sellmyer, IEEE Transactions on Magnetics, Vol. 6. 4035 (1994).
4.   Y. Liu, D. J. Sellmyer, B. Robertson, Z. S. Shan, and S. H. Liou, submitted to IEEE Transactions on Magnetics.

Figure 1 HREM image of a cross-sectional Co-Sm film. The Cr grain is aligned on [111] zone axis while the Co-Sm crystallite is aligned on [112̄0] zone axis. The epitaxy is Cr $(\bar{1}2\bar{1})$ // Co-Sm $(\bar{1}100)$ and Cr [111] // [112̄0]. The interface is indicated by letter I.

INVESTIGATION OF THERMOCHROMIC POLYDIACETYLENES BY ATOMIC FORCE MICROSCOPY

V.Shivshankar,* Daniel J. Sandman,* Jayant Kumar ,[#] Sukant K. Tripathy* and Changmo Sung[+]

Departments of Chemistry *, Physics [#] and Chemical and Nuclear Engineering,[+] Center for Advanced Materials, University of Massachusetts, Lowell, MA 01854

In recent years, Polydiacetylenes (PDAs) have attracted a great deal of attention owing to their extremely large third order non-linear optical susceptibilities, which are believed to result from their molecular arrangement.[1] It is also known that a specific sub-class of these PDAs, namely, the urethane - substituted PDAs such as ETCD (Bis ethyl urethane of 5,7 dodecadiyn 1,12 diol) and IPUDO (Bis isopropyl urethane of 5,7 dodecadiyn 1,12 diol) exhibit thermochromism.[2] This is the underlying principle behind the potential use of such PDAs in optical devices such as holographic elements, spatial light modulators etc. The surface properties of these PDA materials are critical to their performance as optical devices. There are several indications from Raman Spectroscopy that PDA crystals have surface features distinct from the bulk.[3] Also PDA - IPUDO has been identified as a material with a variable surface.[4] We have investigated the free standing single crystals of the monomeric and polymeric forms of PDA - ETCD and IPUDO by Atomic Force Microscopy (AFM). Large scale as well as molecular resolved images were obtained.

AFM imaging was carried out with a commercial Scanning Probe Microscope "Autoprobe CP" (Park Scientific Instruments, CA) under ambient conditions. Constant force (Topographic) imaging mode was used for large scale scans and Constant Height (Error) mode was used for obtaining molecular images. Si_3N_4 cantilevers (Park Scientific Instruments, CA) with a force constant of 0.06 N/m were used. The molecular scale images were processed using the 2-D Fast Fourier Transform (FFT) technique available in the Autoprobe Image Processing software. The large scale scans (Fig. 1 and 2) show a fibrillar structure of the polymeric crystal while the molecular images (Fig. 3 and 4) reveal a backbone repeat of 0.53 nm and a side chain repeat of 1.8 nm. The images of poly IPUDO is quite similar to those of Poly ETCD which is expected due to their similar structures and packing arrangement. These numbers also correspond to their crystallographic data obtained from x-ray diffraction techniques. The interpretation of these images is expected to help us understand the material surface for potential device fabrication. AFM imaging was carried out on poly ETCD and poly IPUDO crystals by heating them locally to various temperatures below and above the thermochromic phase transition to understand whether the thermochromic phenomenon occurs due to surface reconstruction.[5]

References

1. H.S. Nalwa, *Advanced Materials*, 5 (1993) 341.
2. R.R. Chance et al., *J. Chem. Phys.*, 67 (1977) 3616.
3. S.H. Hankin and D.J. Sandman, *Macromolecules*, 24 (1991) 4983.
4. S.H.Hankin and D.J. Sandman in *Electrical, Optical and Magnetic Properties of Organic Solid State Materials, Mat. Res. Soc. Symp. Proc.*, vol. 328 (1994) 101.
5. The authors would like to acknowledge the United States Air Force Office of Scientific Research for partial support of this research.

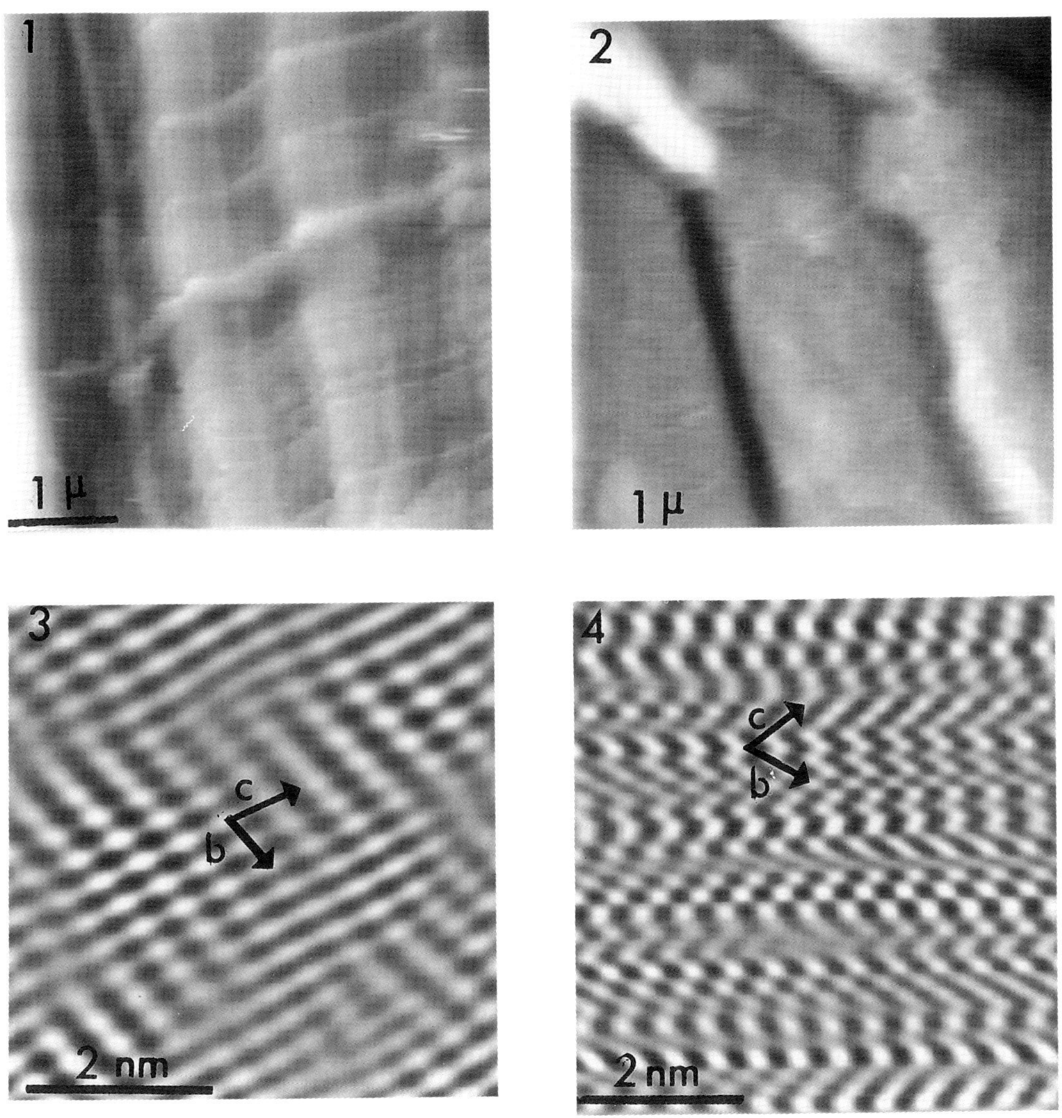

FIG.1 - AFM micrograph (Image size : 5μm x 5 μm) of the surface of poly ETCD single crystal showing a fibrillar morphology.

FIG.2 - AFM micrograph (Image size : 5μm x 5μm) of the surface of poly IPUDO single crystal showing a fibrillar morphology, similar to that of poly ETCD.

FIG.3 - AFM nanograph (Image size : 6nm x 6nm) of poly ETCD single crystal revealing its lattice structure. This image was Fourier filtered to get rid of low frequency noises.

FIG.4 - Fourier filtered AFM nanograph (Image size : 6nm x 6nm) of poly IPUDO single crystal showing molecular resolution.

DETERMINATION OF THE DEGREE OF RADIAL ORDER IN KEVLAR FIBERS USING X-RAY LINEAR DICHROISM MICROSCOPY

A. P. Smith*, H. Ade*, B. Hsiao**, S. Subramoney**

* Dept. of Physics, North Carolina State University, Raleigh, NC 27695-8202
** Experimental Station, DuPont, Wilmington, DE 19800-0302

The Scanning Transmission X-ray Microscope (STXM) located at beamline X1A at the National Synchrotron Light Source at Brookhaven National Laboratory can be used to acquires high resolution, chemically and orientationally sensitive images with 50 nm spatial resolution, as well as point spectra from 0.1 μm^2 areas.[1] Chemical (valence) sensitivity can be achieved via the X-ray Absorption Near Edge Structure (XANES) at the carbon K edge.[2] In addition, linear dichroism microscopy exploits the dependence of x-ray absorption resonances on the bond orientation relative to the linearly polarized x rays. Previous x-ray linear dichroism microscopy work has demonstrated the ability to determine the orientation of specific chemical groups within polymeric systems at high spatial resolution.[3] Here, we have extended this technique to quantitatively determine the degree of radial order in various grades of poly(p-phenylene terephthalamide) (Kevlar™) fibers.

The orientational dependence of the absorption cross-section yields "butterfly" patterns when thin sections of certain grades of Kevlar fibers cut at 45° with respect to the fiber axis are imaged at specific photon energies. This butterfly pattern is due to an average orientational order of specific functional groups in a radially symmetric manner about the fiber axis. Micrographs of three fiber grades imaged at 285.5 eV are shown in figure 1. This photon energy is most strongly absorbed by aromatic groups. The observed butterfly pattern is strongest for Kevlar 149 (fig 1A), intermediate for Kevlar 49 (fig 1B) and weakest for Kevlar 29 (fig 1C). These intensity variations in the micrographs clearly demonstrate the qualitative difference between the fiber grades and the relative degree of radial ordering observed mirrors the crystallinity of the fibers along the fiber axis (Kevlar 149 is largest and Kevlar 29 is smallest)[4].

In order to quantify the degree of radial order in the Kevlar fibers, XANES absorption spectra were acquired at several polar angles. Spectra from vertical locations (dash) and horizontal locations (solid) of both Kevlar 149 and Kevlar 49 are plotted in figure 2 to illustrate the dependence on angular position and fiber grade. By least squares fitting the spectra, the peak intensities of the spectra can be extracted. The "degree of radial order" utilized for the comparisons made here is the difference of the spectral intensities divided over their sum. Utilizing this measure and by analyzing the average orientation of the carbonyl groups in two fibers each, we have found the following preliminary findings: The average radial order in Kevlar 149 varies from 0.23 to 0.27, while in Kevlar 49 it ranges from 0.13 to 0.24.

References

1. H. Ade *et al.*, Science 258, (1992) 972.
2. J. Stöhr, **NEXAFS Spectroscopy** (Springer, Berlin, 1992).
3. H. Ade and B. Hsiao Science 262, (1993) 1427.
4. H. H. Yang, **Aromatic High Strength Fibers** (Wiley-Interscience Pub., New York, 1989).

Proc. Microscopy and Microanalysis 1995, edited by G.W. Bailey, M.H. Ellisman, R.A. Hennigar, and N.J. Zaluzec
Copyright © 1995 MSA. Published by Jones and Begell Publishing, 79 Madison Ave., New York, NY 10016

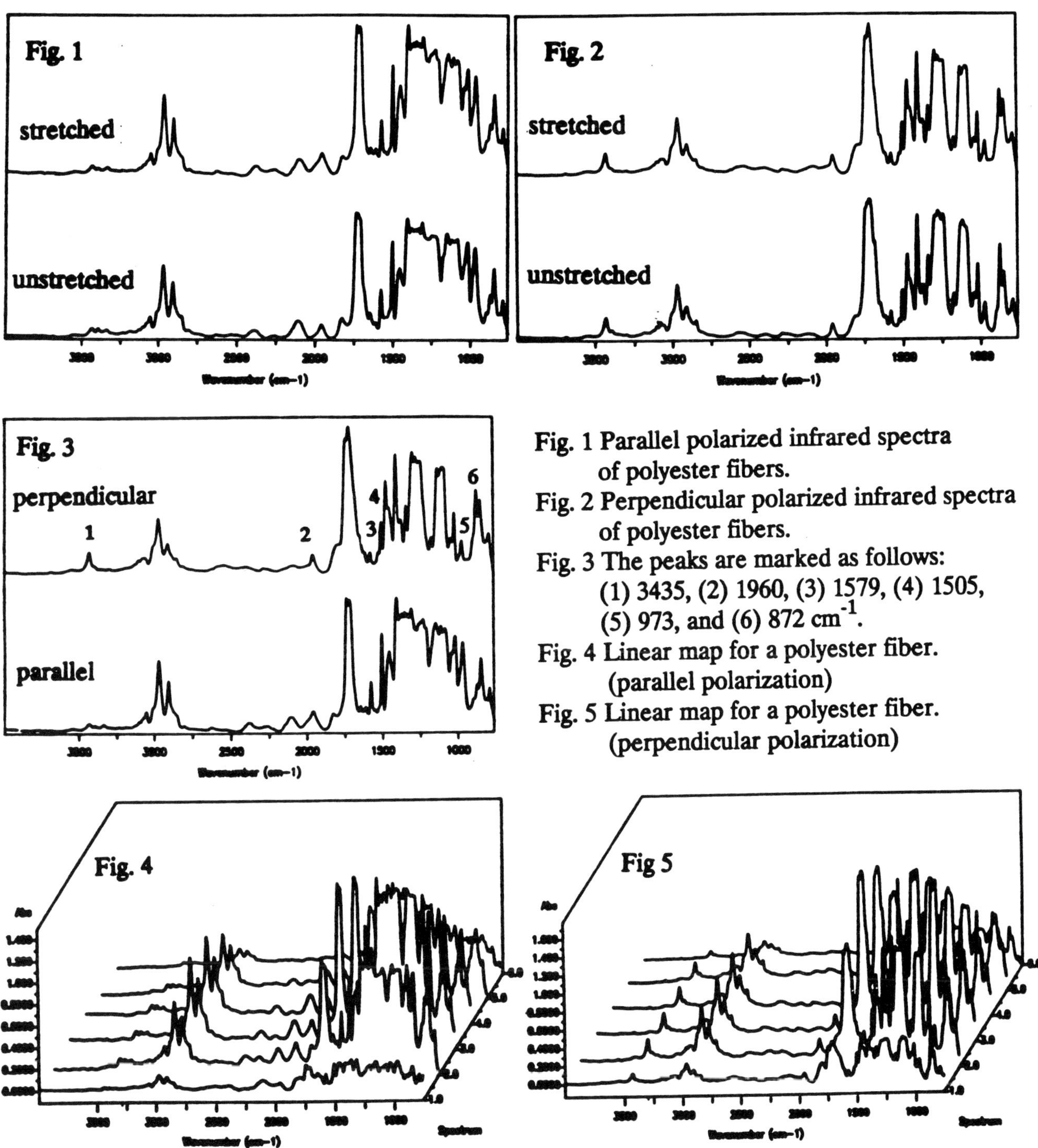

Fig. 1 Parallel polarized infrared spectra
of polyester fibers.
Fig. 2 Perpendicular polarized infrared spectra
of polyester fibers.
Fig. 3 The peaks are marked as follows:
(1) 3435, (2) 1960, (3) 1579, (4) 1505,
(5) 973, and (6) 872 cm^{-1}.
Fig. 4 Linear map for a polyester fiber.
(parallel polarization)
Fig. 5 Linear map for a polyester fiber.
(perpendicular polarization)

References
1. S. P. Church and N. Khan, Polym. Bulle. 30(1991)559.
2. C. W. Myers and S. L. Cooper, Appl. Spectrosc. 48(1994)72.
3. M. W. Tungol et al., in Practical Guide to Infrared Microscpectroscopy (H. J. Humecki, ed.), New York: Marcel Dekker, Inc. (1995)245.

MICROBEAM POLARIZATION INFRARED SPECTROSCOPY: SINGLE FIBER DISCRIMINATION

Liling Cho and David L. Wetzel

Kansas State University, Microbeam Molecular Spectroscopy Laboratory, Shellenberger Hall, Manhattan, KS 66502

FT-IR microspectroscopy of single fiber can produce useful information to distinguish among synthetic fibers. Certain fibers can be readily discriminated on the basis of spectra obtained by this technique. Polarization infrared spectroscopy has been used to measure macromolecular properties such as precipitate crystallinity, polymer film draw ratio, and liquid crystal phase. Measurement of IR dichroism requires light polarized both parallel and perpendicular to a fixed reference direction of the sample. Dichroism may be used to compare and discriminate between different polyester fibers and possibly to establish new groups within a subclass. Four fundamental bands (872, 973, 1505 and 1579 cm^{-1}) and two overtone bands (1960 and 3435 cm^{-1}) were chosen because they were relatively free from overlapping bands, and they were weak enough not to be over-absorbing in the spectra of thick, unflattened fibers.[1,2,3] During industrial production, the process of drawing can impart many useful properties to a synthetic fiber. When this occurs, molecular chains may become oriented in the long direction of the fiber. The infrared dichroism of a particular vibration of known origin has been used to obtain valuable information about molecular orientation. The stretching properties of synthetic polymers are considerably enhanced by molecular orientation, therefore, it is desirable to determine the relationship between the molecular orientation and stretching properties.

In the present study spectra were collected in transmission mode using an IRμs^{TM} infrared microspectrometer. A single fiber stretching device was fabricated for use on the microscope stage. The spectra obtained from a polyester fiber stretched typically 130% of its relaxed dimension showed the transition from unstretched to stretched configuration in spectroscopic terms (Figs. 1 and 2). By comparing the dichroic ratio between the unstretched and stretched fiber, there are significant differences at 973, 1579 and 3430 cm^{-1}. Fig. 3 illustrates the difference between the parallel and perpendicular polarized spectra in those six bands. As the polarizer is rotated from parallel to perpendicular position the spectrum changes. The absorbance values of peaks at 1960, 1579, 1505, and 973 cm^{-1} were decreased, while the absorbance of the peaks at 3435 and 872 cm^{-1} were increased. This technique was applied to a number of fibers. Linear mapping experiments were carried out to determine the orientation within the fibers. Figs. 4 and 5 illustrate the type of data obtained. All six of the spectra in Figs. 4 and 5 are from 6 μm steps of the programmable microscope stage. These six spectra respresent a transition from the edge of the fiber (spectrum 1) across the center to the other edge (spectrum 6) of the fiber.

Proc. Microscopy and Microanalysis 1995, edited by G.W. Bailey, M.H. Ellisman, R.A. Hennigar, and N.J. Zaluzec

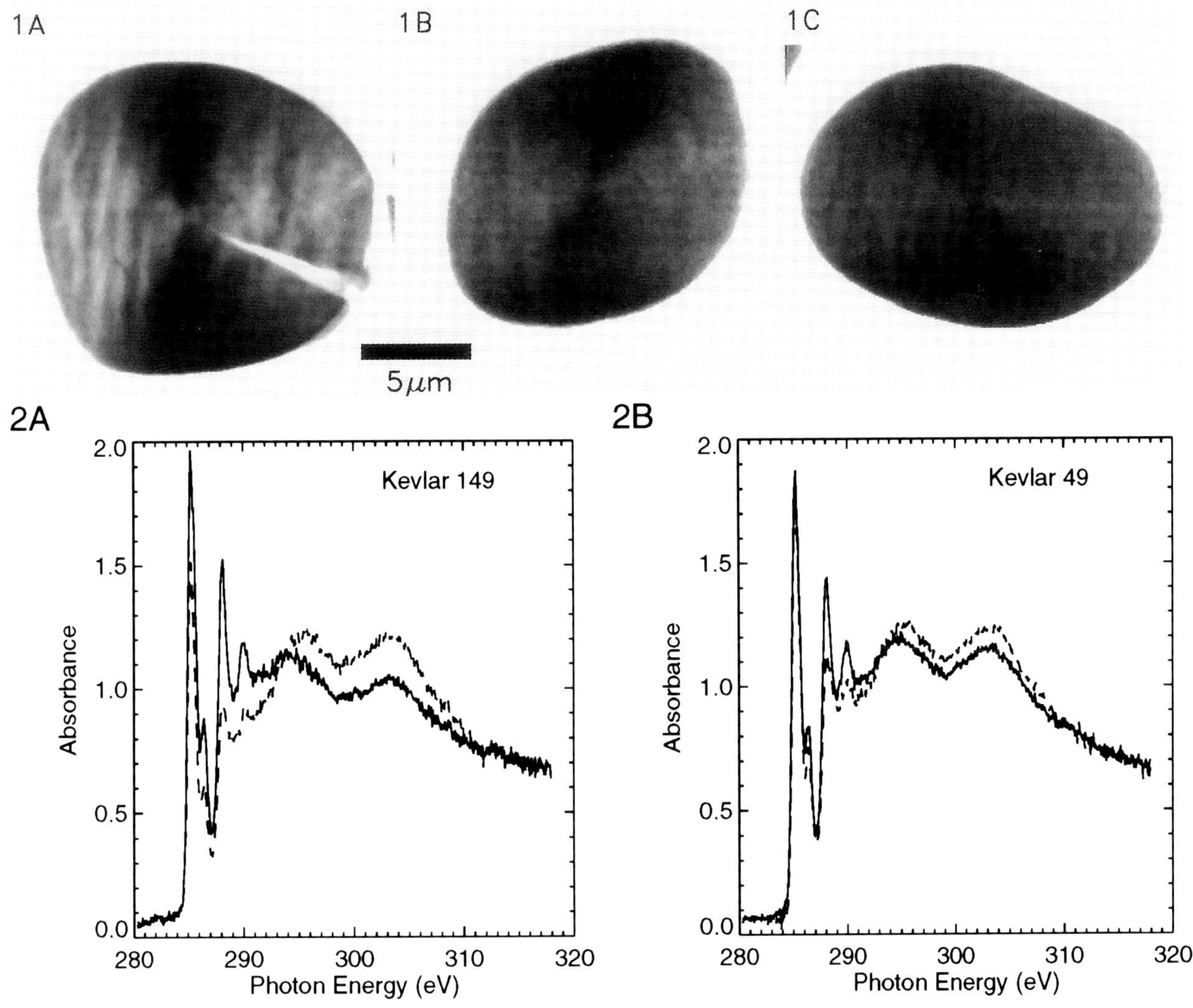

Fig. 1. -- Micrographs of 200 nm thick sections (cut at 45° relative to fiber axis) of (A) Kevlar 149, (B) Kevlar 49, and (C) Kevlar 29, imaged at a photon energy of 285.5 eV. This energy is characteristic of the aromatic groups of the fiber polymer and the butterfly patterns observed in all three grades of Kevlar fibers are due to the radial symmetry and orientational order of the fibers.

Fig. 2. -- Spectra obtained from vertical (dash) and horizontal (solid) locations of Kevlar 149 (A) and Kevlar 49 (B) fibers. The difference in the peak heights is reflecting the degree of radial order in the fibers. Notice that the peak height differences for the two angular locations are much smaller in Kevlar 49 than in Kevlar 149. By determining these peak height differences quantitative information about the degree of radial order can be obtained.

IODOPROPIONIC ACID STAINING OF WATERBORNE EPOXY FILMS

O. L. Shaffer,* F.H. Walker,** and M. S. El-Aasser***

*Emulsion Polymers Institute, Lehigh University, Bethlehem, PA 18015
**Air Products and Chemicals, Inc., Allentown, PA 18195
***Emulsion Polymers Institute, Department of Chemical Engineering, Lehigh University, Bethlehem PA, 18015

The examination of polymer material by transmission electron microscopy (TEM) often relies on staining in order to improve contrast or preferentially stain a polymer phase. Stains such as osmium tetroxide[1,2] for unsaturated polymers and ruthenium tetroxide[3] for unsaturated as well as certain saturated polymers have been used quite extensively. A new stain, 3-iodopropionic acid, has been found to be very useful to stain preferentially the amine rich domains in waterborne epoxy films.[3]

An epoxy dispersion (epoxy equivalent weight, EEW=625) and a polyamine adduct curing agent (AHEW=163) at a 1.5:1 stoichiometry was prepared. The volatile organic content (VOC) was 2.6 lb./gal., and the solvent blend was 90.6% propoxy ethanol and 9.4% Aromatic 150. The particles size of the starting dispersion was obtained by TEM and disk centrifuge and found to be approximately 500 nm. Films (approximately 1.2 mil dry film thickness[DFT]) were cast on a partially fluorinated polyethylene film, 0.5, 2, 4, and 8 hours after hand mixing and cured for two or more weeks at 25° C and 50% relative humidity. After removal from the polyethylene substrate, the film was cryoultramicrotomed. The sections were placed on formvar-copper TEM grids. The sections were then stained by floating the grids, section side down, on a drop of 2% aqueous solution of 3-iodopropionic acid for 10 minutes, washed with distilled-deionized water, and dried before examination by a Philips 400 TEM. The carboxylic acid containing stain should react with the amine groups of the curing agent resulting in darker phases. Solvent-based amine/epoxy films, and the epoxy dispersion without the curing agent were also examined.

An unstained section of an epoxy dispersion cast 30 minutes after mixing (Fig. 1A) shows no morphology. After staining for 10 minutes with 3-iodopropionic acid the section now shows dark spherical domains with a light continuous phase (Fig. 1B). There appears to be a gradient from the bottom of the film to the top. The dark phases are larger at the bottom and gradually decrease in size approaching the top surface. Various staining times from 10 seconds to 20 minutes all gave similar results. The sections from film that was cast two hours (Fig. 1C) and four (Fig. 1D) and eight hours after mixing show similar morphologies. The sections consist of deformed spherical shapes with dark boundary edges which surround a lighter phase. These domains are probably the original epoxy dispersion particles, as they fall in the 500nm diameter range, which contain less amine while the stained dark edges are the amine rich phases. At the shorter time after mixing, amine and epoxy which are inherently incompatible phase separate into the observed stained amine domains surrounded by an epoxy continuous phase. As time after mixing increases the amine reacts with the surface of the epoxy particles which compatibilizes the surface with addtional amine left in the continuous phase.

This staining technique makes it possible to observe the amine rich regions in an epoxy film and study the influence of various parameters such as pot life, rate and temperature of curing, and formulation, and compare the morphology of the films with their physical properties.

References

1. K.Kato, *Polymer Letters*, 4(1966)35.
2. M.S. El-Aasser, J.W. Vanderhoff, S.C. Misra, and J.A. Manson, *J. of Coatings Tech.*, 49(1977)71.
3. J.S. Trent, J.I. Scheinbeim, and P.R. Couchman, *Macromolecules*, 16(1983)589.

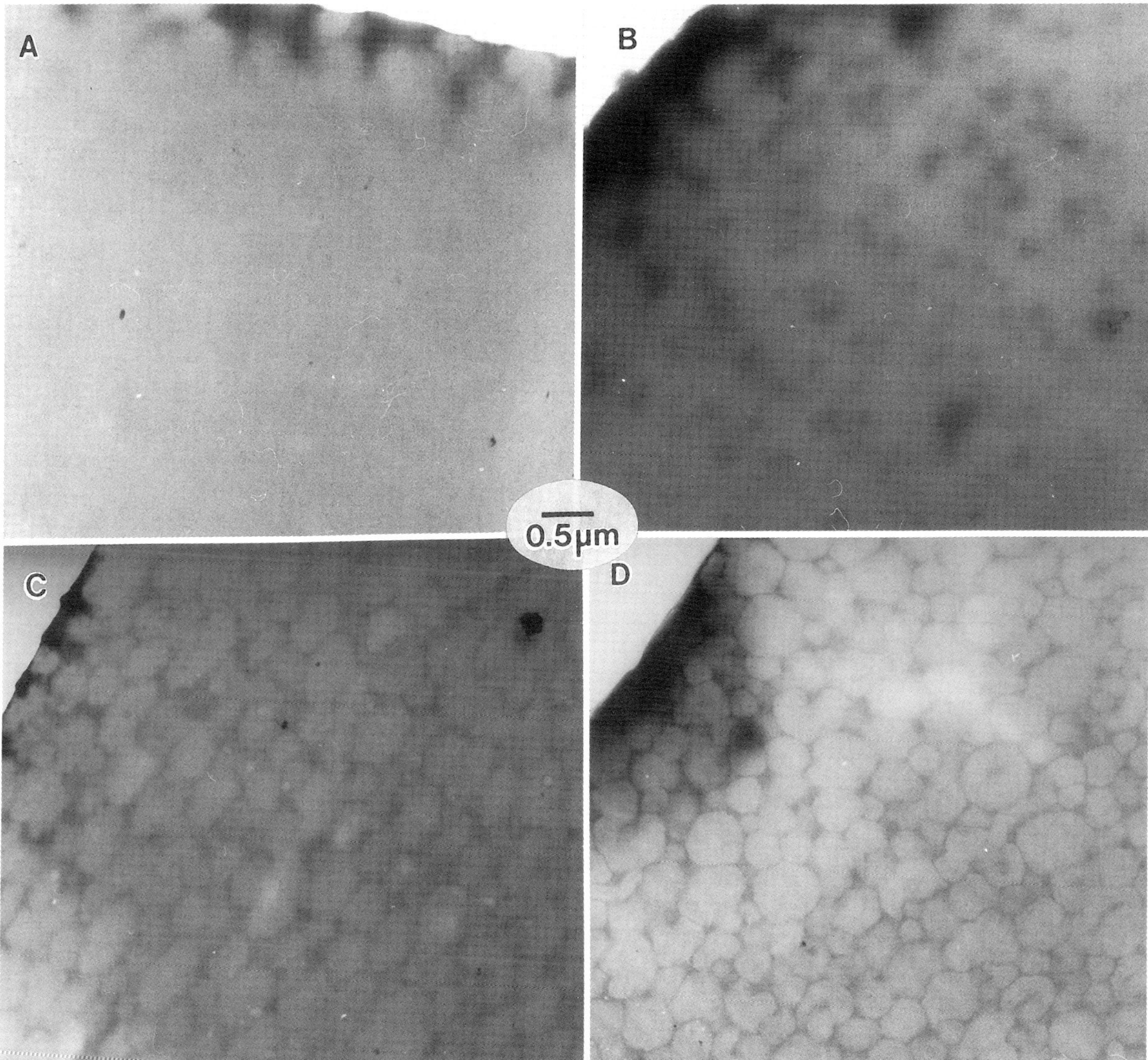

Fig. 1- A cryoultramicrotomed section of a cured epoxy film : (A) cast 30 minutes after mixing; (B) stained with 3-iodopropionic acid (cast 30 minutes after mixing); (C) stained with 3-iodopropionic acid (cast 2 hours after mixing); (D) stained with 3-iodopropionic acid (cast 4 hours after mixing). The morphology is similar in films cast 2 hours (Fig. C), 4 hours (Fig. D) and 8 hours after mixing.

ABSORPTION AND SCATTERING OF ULTRAVIOLET RADIATION BY
TITANIA PIGMENT IN BIS-PHENOL-A POLYCARBONATE

Gudrun A. Hutchins and James E. Pickett

GE Corporate Research and Development, P.O. Box 8, Schenectady, NY 12301.

The surface of bis-phenol-A polycarbonate oxidizes after prolonged exposure to sunlight.
During the laboratory testing of stabilizers and coatings, the effect of sunlight is typically
simulated by a xenon arc lamp, with maximum polycarbonate sensitivity between 300 and
340 nm.[1] Titania pigment is a strong absorber of ultraviolet radiation and in its presence the
affected surface layer will be thinner. The titania pigment utilized in polymer blends
typically ranges from 100 nm to 350 nm in diameter.

Microtomed polycarbonate thin sections stain very well with ruthenium tetroxide vapor and
the degree of staining is a sensitive indicator of the degree of oxidation or chemical
reaction. Ruthenium tetroxide oxidizes the polymer surfaces of the thin section which
results in the localized precipitation of ruthenium dioxide.[2] In addition, the heavily oxidized
surface polymer does not microtome evenly, resulting in a very fine craze pattern that is
accentuated by the RuO4 stain.

With these techniques it was possible to demonstrate the absorption of ultraviolet radiation
by titania particles by imaging the abrupt end of the heavily oxidized layer at the positions
of individual pigment particles (Figure 1). Deeper within the specimens, at lower xenon arc
exposures, the effect of coherent scattering by individual titania particles was also observed
(Figure 2). This unanticipated diffraction effect was confirmed at lower pigment
concentrations and with sections cut parallel to the exposed surface (Figures 3 and 4). It is a
consequence of the similar size of the titania particles and the primary wave length causing
polycarbonate oxidation.

The scattering of arc lamp illumination by titania produces an oxidized layer of non-
uniform thickness due to the relatively small solid angle of the source. Similar scattering
patterns would not be anticipated from cumulative exposure to sunlight because of varying
sun angles during each day and with changing seasons. Thus the arc lamp source produces
a testing artifact that may influence mechanical properties and appearance.

References

1. A. L. Andrady, N. D. Searle, and L. F. E. Crewdson, *Polymer Degradation and
Stability*, **35**, 235 (1992).
2. J. S. Trent, J. I. Scheinbeim, and P. R. Couchman, *Macromolecules*, **16**, 589 (1983).

Proc. Microscopy and Microanalysis 1995, edited by G.W. Bailey, M.H. Ellisman, R.A. Hennigar, and N.J. Zaluzec
Copyright © 1995 MSA. Published by Jones and Begell Publishing, 79 Madison Ave., New York, NY 10016

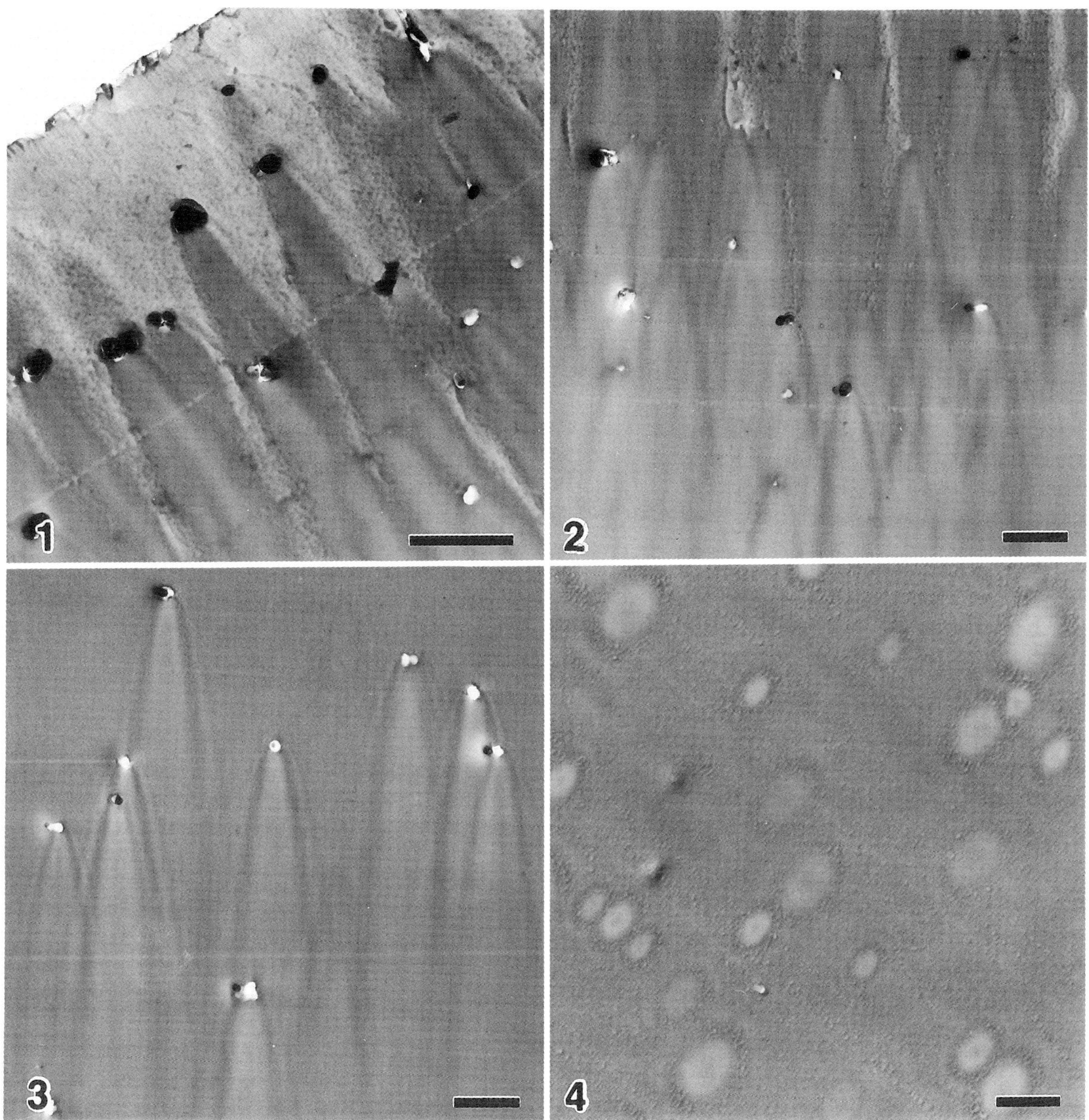

Fig. 1. Polycarbonate with 2 wt % TiO$_2$, exposed to 1200 KJ of xenon arc lamp illumination. The sample is sectioned perpendicular to upper surface which is indicated by an arrow.

Fig. 2. Same specimen as Fig. 1., diffraction of ultraviolet radiation by TiO$_2$ particles located a short distance below the surface. Some of the TiO$_2$ particles have pulled out during sectioning.

Fig. 3. Polycarbonate with 0.2% TiO$_2$, exposed to 826 KJ of xenon arc lamp illumination. Note that several maxima and minima can be observed in the diffraction cones.

Fig. 4. Same sample as Fig. 3., but sectioned parallel to surface to show circular sections of diffraction cones.

All markers equal 1 μm.

MICROSTRUCTURES OF ULTRA HIGH MOLECULAR WEIGHT POLYETHYLENE BY SEM

H. C. Wang,* C. Sung,* and J. Hamilton**

*Center for Advanced Materials, University of Massachusetts at Lowell, Lowell, MA 01854
**Johnson and Johnson Professional, Raynham, MA 02767-0350

Ultra high molecular weight polyethylene (UHMWPE) is widely used in many orthopedic applications because of its good mechanical properties and excellent biocompatability.[1] Mechanical properties are related to its ultra-high molecular weight, but the presence of defects in morphology will cause a decrease in these properties. The aim of this study was to characterize the microstructure and microchemistry of so-called "fusion defects", observed by optical transmission microscopy (Fig. 1), using SEM along with EDXS analysis. The fractured surface of UHMWPE was also studied.

Defects similar to those found in commercially prepared UHMWPE were detected in the hot-pressed and extruded samples prepared at U-Mass Lowell using GUR 412. They consist of holes or cavities and sometimes appear to be circular in shape and are composed of variously sized small holes (Fig. 2). The size of the defects is around 100 µm in diameter which is similar to the particle size of the raw powders. These circular microvoids are expected to be responsible for the deterioration of mechanical properties. Microvoids can be observed by optical microscopy and are granular in shape with light contrast boundaries. SEM micrographs showed in Figures 2 and 3 reveal that fusion defects basically consist of microvoids along the particle boundaries of the sample. The EDXS analysis of these microvoids showed Ca and Cl segregated on the surface of voids (Fig. 4). It was suggested that these microvoids result from the presence foreign impurities which prevent consolidation of the polymer chains. Another possibility could be due to chain scission as a result of processing.

The fractured surface of UHMWPE samples subjected to tensile stress was investigated by SEM/EDXS since fractography provides a permanent record of many key aspects of the fracture process. The primary characteristics in glass and ceramics such as the mirror, mist, hackle, and crack-branching pattern were observed in the UHMWPE samples (Fig. 5), as well.[2] Of note was the presence of fibril and regular undulations in mist region was shown in Figure 6. The deformation of samples during tensile testing is caused by the fracture initiated crazes, i.e., formation at points of high stress concentration due to impurity particles or a network of voids. These impurities, determined by EDXS, associated with a flaw are Ca and Cl. Therefore, it is important to learn how to control these impurities to improve mechanical properties of UHMWPE. The neck of the fracture surface tends to vary depending on tensile properties of the sample. For instance, the neck of the fracture appeared on the center or on the edge depending on the position of initiating crazes. Quantitative analysis of fracture toughness as a function of processing conditions in UHMWPE samples, is being carried out from measurements of the fracture parameters such as size of flaw, mirror and mist.[3]

References

1. H. -G. Willert et al., Eds., *Ultra-High Molecular Weight Polyethylene as Biomaterial in Orthopedic Surgery*, Hogrefe & Huber Publishers (1986)50.
2. V. D. Fréchette and J. R. Varner, Eds., *Fractography of Glasses & Ceramics*, the American Ceramic Society, Inc (1986).
3. Acknowledgment is made to the Johnson and Johnson Professional for matching funds. The authors would like to acknowledge the contributions of A. Crugnola, T. Vasilos, N. Schott and C. Shah to this work.

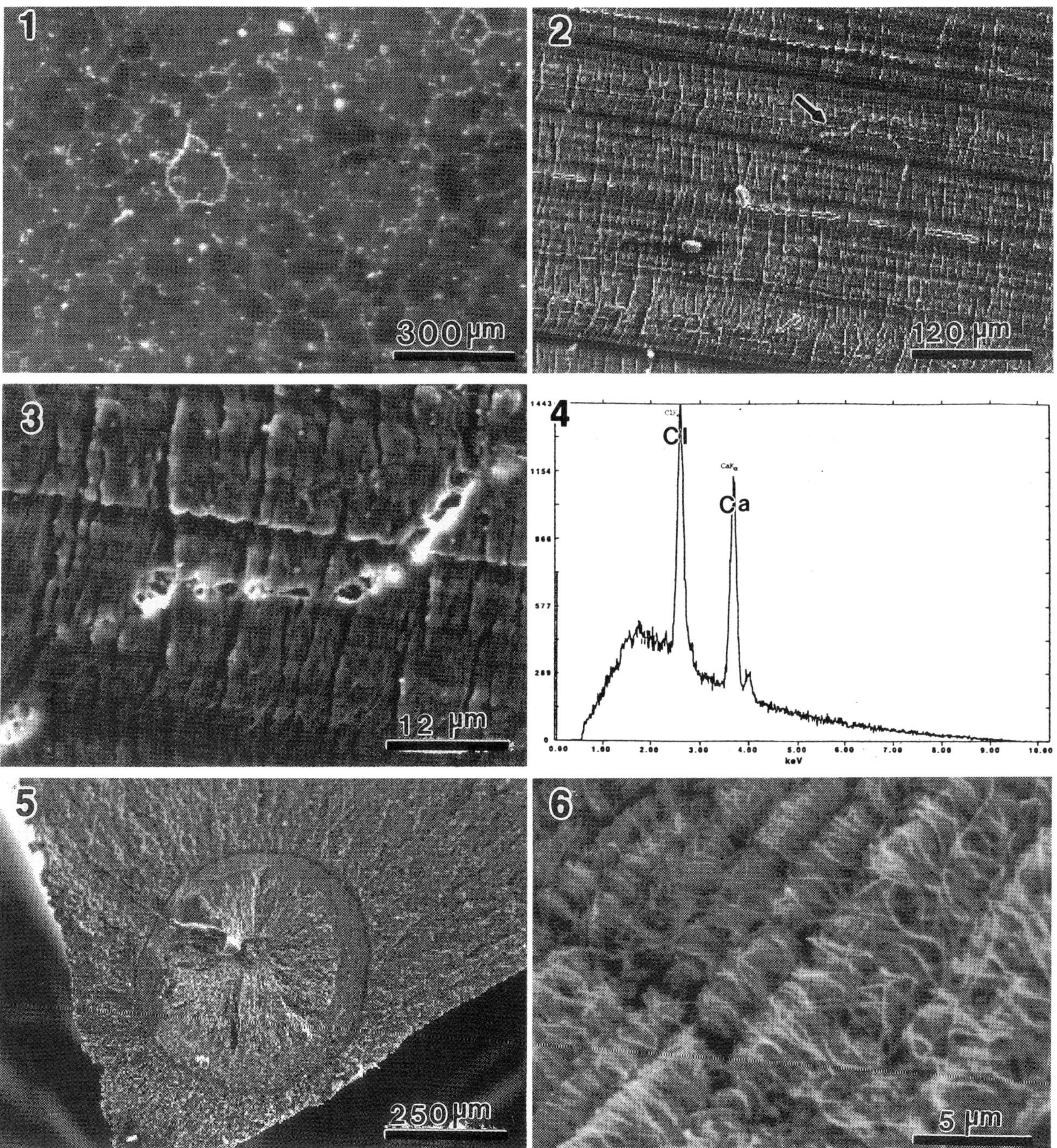

FIG. 1 Optical transmission microscopy image of fusion defects.
FIG. 2 SEM image of fusion defects consisting of microvoids.
FIG. 3 An array of microvoids at high magnification.
FIG. 4 Energy dispersive x-ray spectrum obtained from microvoids showing Ca and Cl.
FIG. 5 Fracture surface of UHMWPE at low magnification.
FIG. 6 Mist region in fracture surface of UHMWPE.

MICROSTRUCTURE OF LASER-IRRADIATED, CONDUCTING KAPTON POLYIMIDE

D. L. Callahan[1], Z. Ball[2], H. M. Phillips[3], and R. Sauerbrey[4]

[1]Dept. Mech. Engg. and Materials Science, Rice University, Houston, TX 77251-1892
[2]Dept. Electrical and Comp. Engineering, Rice University, Houston, TX 77251-1892
[3]Friedrich-Schiller-Universitat, Max-Wien-Platz 1, D-07743 Jena, Germany
[4]Zhengzhou University, 75 Daxuelu, Zhengzhou, Henan 450052, China

Ultraviolet laser-irradiation can be used to induce an insulator-to-conductor phase transition on the surface of Kapton polyimide.[1] Such structures have potential applications as resistors or conductors for VLSI applications as well as general utility electrodes. Although the percolative nature of the phase transformation has been well-established,[2] there has been little definitive work on the mechanism or extent of transformation. In particular, there has been considerable debate about whether or not the transition is primarily photothermal in nature, as we propose, or photochemical. In this study, cross-sectional optical microscopy and transmission electron microscopy are utilized to characterize the nature of microstructural changes associated with the laser-induced pyrolysis of polyimide.

Laser-modified polyimide samples initially 12 µm thick were prepared in cross-section by standard ultramicrotomy. Resulting contraction in parallel to the film surface has led to distortions in apparent magnification. The scale bars shown are calibrated for the direction normal to the film surface only.

Figure 1 of a high conductivity surface film (54 mJ/cm2, 600 shots, 5 Hz) shows two predominant features: a highly porous film on the order of 1 micron thick and laser-induced roughness on a comparable scale. Films treated at a lower repetition rate (1 Hz) show similar roughness but lack both the thin porous layer and high electrical conductivity. Two conclusions may be drawn from this. First, any changes from 1 Hz to 5 Hz are only expected to result from thermal coupling, not photochemical processes which occur at a much faster time scale. Second, the porous film observed is the conducting layer.

The effects of atmosphere on the laser-process supports also supports the pyrolytic / thermal model. Figure 3 a shows a film made with 600 shots in air and Fig 4 a film made with 1000 shots in N_2. Despite the larger number of shots, the film made in N_2 shows considerably less ablation than the film shot in air. This is consistent with the expected gassifaction of a high temperature polymer surface in contact with air. These and associated observations will be discussed in terms of laser-induced pyrolysis of the Kapton.

1. M. Schumann et al., Appl. Phys. Lett 58 (1991) 428.
2. Z. Ball et al., Phys. Rev. Letters 73 (1994) 2099.
3. This material is based upon work supported by the National Science Foundation with Army Research Office cofunding under NSF Grant ECS 9118475.

Proc. Microscopy and Microanalysis 1995, edited by G.W. Bailey, M.H. Ellisman, R.A. Hennigar, and N.J. Zaluzec
Copyright © 1995 MSA. Published by Jones and Begell Publishing, 79 Madison Ave., New York, NY 10016

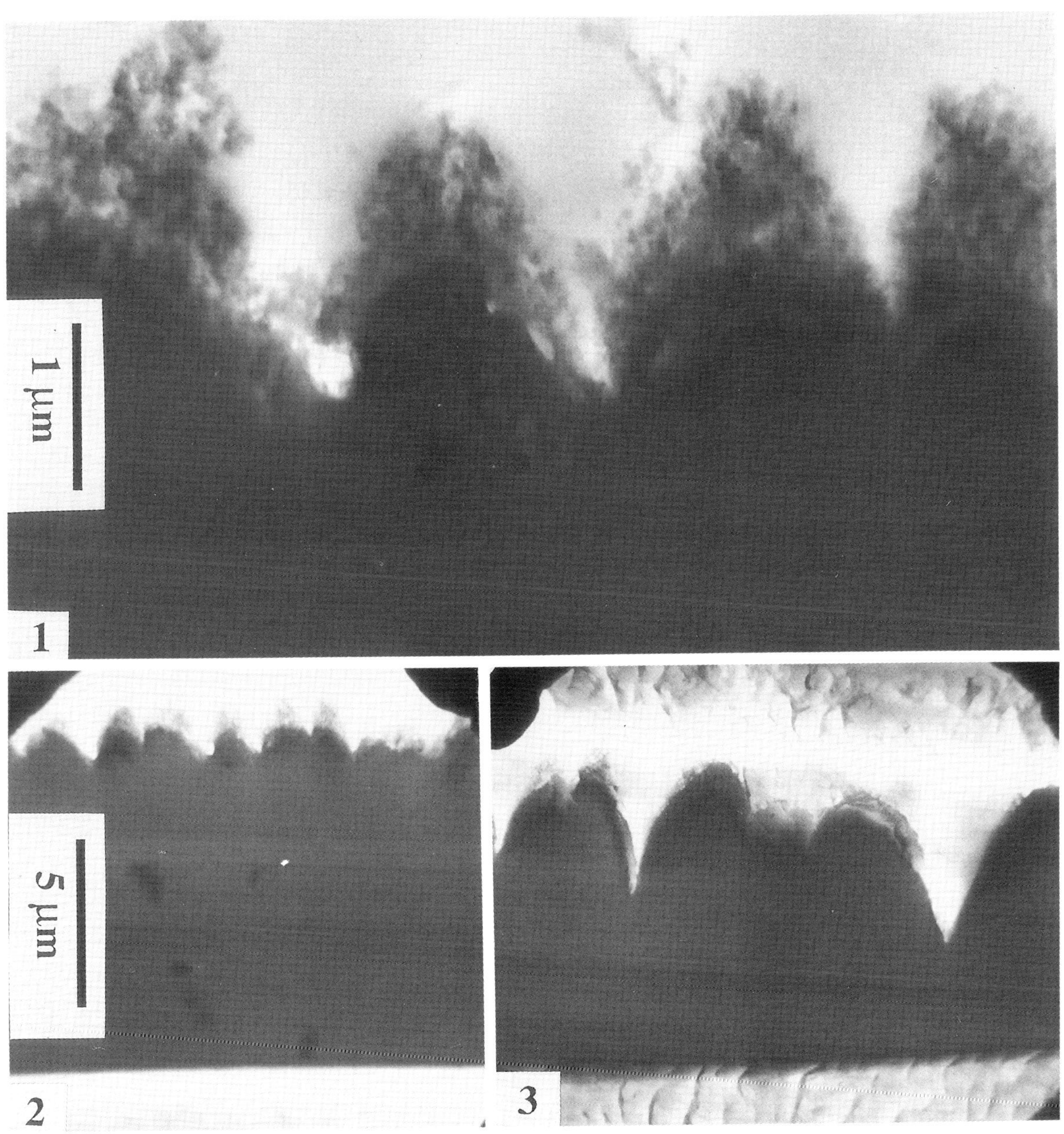

FIG. 1. -- Bright field TEM micrograph of laser-treated, conducting polyimide
FIG. 2. -- Bright field TEM micrograph of polyimide laser-irradiated in N2 gas.
FIG. 3 -- Bright field TEM micrograph of polyimide laser-irradiated in air.

Multitechnique Characterization of Polysulfone Gas Permeation Membranes

M. A. Behrens*, Y. H. R. Jois*, and P. L. Dahlstrom*

*Texaco Research Center, P.O. Box 509, Beacon, NY 12508

Integrally thin skinned asymmetric polysulfone membranes developed for gas separation applications were prepared by casting a dope solution on a glass plate. A dry/wet phase inversion process with a forced convective evaporation step is then used to form a skin[1]. The membrane is immersed in water before undergoing a solvent exchange and drying process. The thin skin of these membranes allows gas to selectively diffuse through the membrane when the surface is without defects. The gas transport properties (selectivity and flux) are measured using standard techniques and are reported on Table II[2].

Scanning electron microscopy is an important method of examining the cross-sections and surfaces of these membranes to determine the surface integrity[3], skin thickness, and porosity of the supporting substructure. The ideal integrally thin skin thickness is 1000 Å or less. We further characterized these membranes using Atomic Force Microscopy to image the membrane surface and X-ray Photoelectron Spectroscopy to obtain chemical information about the polysulfone surface. Polysulfone ultrafiltration membranes for other purposes have been examined by AFM previously[4,5], but not gas permeability membranes. Surfaces are placed on sample stubs with double-sided tape for SEM and XPS, and steel discs for AFM, while cross-sections for SEM are placed in a clip to hold them upright. The membranes are sputter-coated with a thin layer of gold to reduce charging.

Four polysulfone membranes prepared by slightly different methods were examined by these three techniques. Differences in the supporting substructure porosity and in the skin thickness (Table I) could be seen. Sample #2 had nested macropores in its substructure, similar to sample #16. Sample #7 was thinner and had a porous substructure with no macropores, while #1 had a thick substructure which was compressed at the bottom. This non-porous area may interfere with the gas flow through the membrane. In addition, the skin surface varied from sample to sample, depending on preparation treatment. This was evidenced by characterizing the surface with AFM and making roughness measurements. One of the samples (#16) had an extremely rough surface, which could be seen with optical microscopy. Wavy lines seen on the surface of two of the other samples are a result of the forced convection evaporative preparation step. Sample #2 had the least amount of surface structure and was relatively smooth as seen by AFM. XPS of sample #16 revealed the presence of more sulfur and less oxygen on the surface as compared to #2.

We have found that the use of these techniques assists the researcher in improving membrane performance in later preparations.

References:
1. S. C. Pesek and W. J. Koros, *J. Membrane Sci.*, **81** (1993) 71-88, Elsevier Sci. Publishers.
2. Y. H. R. Jois, "Polysulfone Based Asymmetric Membranes for Gas Separation Applications," at APME '95 Meeting in June, 1995.
3. I. Pinnau and W. Koros, *J. Applied Polymer Science*, **Vol. 43**, 1491-1502 (1991), John Wiley & Sons, Inc.
4. A. K. Fritzsche, et. al., *J. Applied Polymer Sci.*, **Vol. 46**, 167-178 (1992), John Wiley & Sons, Inc.
5. A. K. Fritzsche, et. al., *J. Applied Polymer Sci.*, **Vol. 45**, 1945-1956 (1992), John Wiley & Sons, Inc.

Proc. Microscopy and Microanalysis 1995, edited by G.W. Bailey, M.H. Ellisman, R.A. Hennigar, and N.J. Zaluzec

Table I
SEM/AFM RESULTS

Sam. Ref. No.	Membrane Thickness	Thickness of Selective Layer
#1	226 µm	2.0 µm
#2	199 µm	3.2 µm
#7	80 µm	1.2 µm
#16	118 µm	0.61 - 1.4 µm

Table II
GAS TRANSPORT PROPERTIES

Sam. Ref. No.	Flux (P/l) for CH_4	Flux (P/l) for CO_2	Selectivity ($\alpha_{CO2/CH4}$)
#1	0.037	1.184	32
#2	0.53	14.68	27.7
#16	1.117	17.362	15.54

$P/l = V/[t*A*p]$ Reference values: Selectivity = $\alpha_{CO2/CH4} = P_{CO2}/P_{CH4}$ = 22 - 28

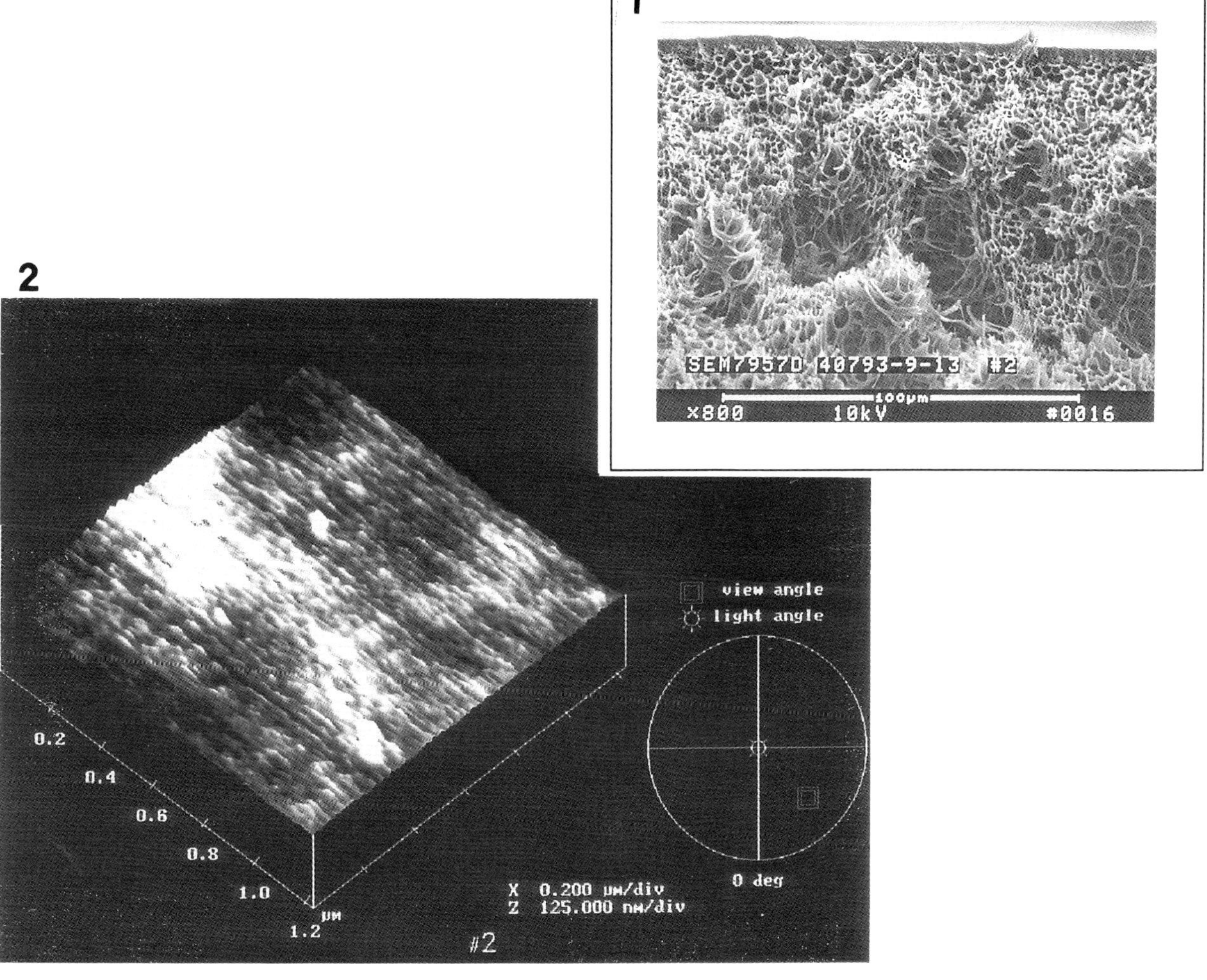

Fig. 1: SEM micrograph of sample #2 cross-section.
Fig. 2: AFM print of sample #2 surface.

PARALLEL ELECTRON ENERGY LOSS SPECTROSCOPY (PEELS) OF POLYMERS IN A TEM: ENERGY CALIBRATION OF THE CARBON-K-EDGE

W. Heckmann

Kunststofflaboratorium der BASF AG, D-67056 Ludwigshafen, Germany

Electron energy loss spectroscopy (EELS) measures the loss in energy of the primary electrons caused by inelastic scatter at the electron shell. The energy loss can be detected with magnetic spectrometers operating in parallel (PEELS) and with imaging magnetic/electrostatic Castaing or magnetic omega filters. To detect EEL-spectra of radiation sensitive polymers a PEEL-spectrometer is useful, because it allows a very fast detection of the spectrum. Dynamic measurements that show the changes in the PEEL spectra during irradiation, yield additional information.

An important part of recording the spectrum at high energy resolution is the energy calibration. In parallel detection, the spectrum is displayed on a diode array with a finite number of diodes (1,024). So at high energy resolution (0.05 eV/ch) it is in general only possible to observe a narrow section of about 50 eV of the total spectrum. At higher energy losses that do not allow simultaneous measurement of the elastic peak, it is only possible to give a very approximate value in terms of eV units as the primary radiation energy can drift during the measurement. It is therefore impossible to compare spectra with one another as the absolute energy scale is not known. But, as the chemical shifts of interest have an order of magnitude ranging from fractions of an eV to several eV, it is essential to know the absolute energy scale.

Since with organic substances, we are mainly interested in the carbon K-edge, a reference edge in the energy range of 320 to 360 eV would be useful. The Ca-L2,3 edge lies within this energy range[1]. It also has the advantage that it consists of two so-called white lines (2p-3d transitions in the transition metals) whose energy can readily be determined.

We have therefore selected calcium chloride as a reference substance. Calcium chloride is soluble in water and can be applied to the specimen drop by drop as a highly diluted solution. When it dries, it forms fine crystals which only slightly contaminate limited areas. An example is shown in Fig. 1a which shows a TEM micrograph of an ultrathin section of polyether ketone contaminated with calcium chloride. The associated PEEL spectrum is shown in Fig. 1b.

With this calibration technique it could be shown, that the CH-resonance (291 eV) in the C-K-absorption edge of polyoxymethylene (POM) is shifted to a 3 eV higher energy (Fig. 2) compared to the corresponding resonance (288 eV) in the polyethylene (PE)[2] spectrum. POM differs from PE in that there is an oxygen atom after each methylene group. This electronegative oxygen atom is responsible for the energy shift. The second maximum in the POM-spectrum (294 eV) is attributed to σ-σ^*- transitions in the C-O-bonds. The absolute energies in all shown spectra are calibrated to the π-π^*-transition in polystyrene (285 eV)[3] or polyether ketone relative to the L 2,3-white lines of calcium chloride.

Proc. Microscopy and Microanalysis 1995, edited by G.W. Bailey, M.H. Ellisman, R.A. Hennigar, and N.J. Zaluzec
Copyright © 1995 MSA. Published by Jones and Begell Publishing, 79 Madison Ave., New York, NY 10016

The acquisition time to detect the shown spectra was 1 sec and the intensity was adjusted so that a signal of 15×10^3 counts/sec was collected into the zero-loss peak. In Fig. 3 a series of POM spectra recorded in a 1sec intervals exhibiting increasing radiation damage is shown. The edge broadens within a few seconds and two additional resonances arise at about 288 and 285 eV. Measuring the same series at -160°C, only the 288 eV shoulder is detected and the relatively sharp resonance at 291 eV is better conserved. The changes of the spectra are not yet completely understood.

References

1. J.M. Martin, *Microsc. Microanal. Microstruct.* 1(1990)93
2. J. Stöhr et al., *Phys. Rev. B* 36,5(1987)2976
3. J.J. Ritsko and R.W. Bigelow, *J. Chem. Phys.* 69,9(1978)4162

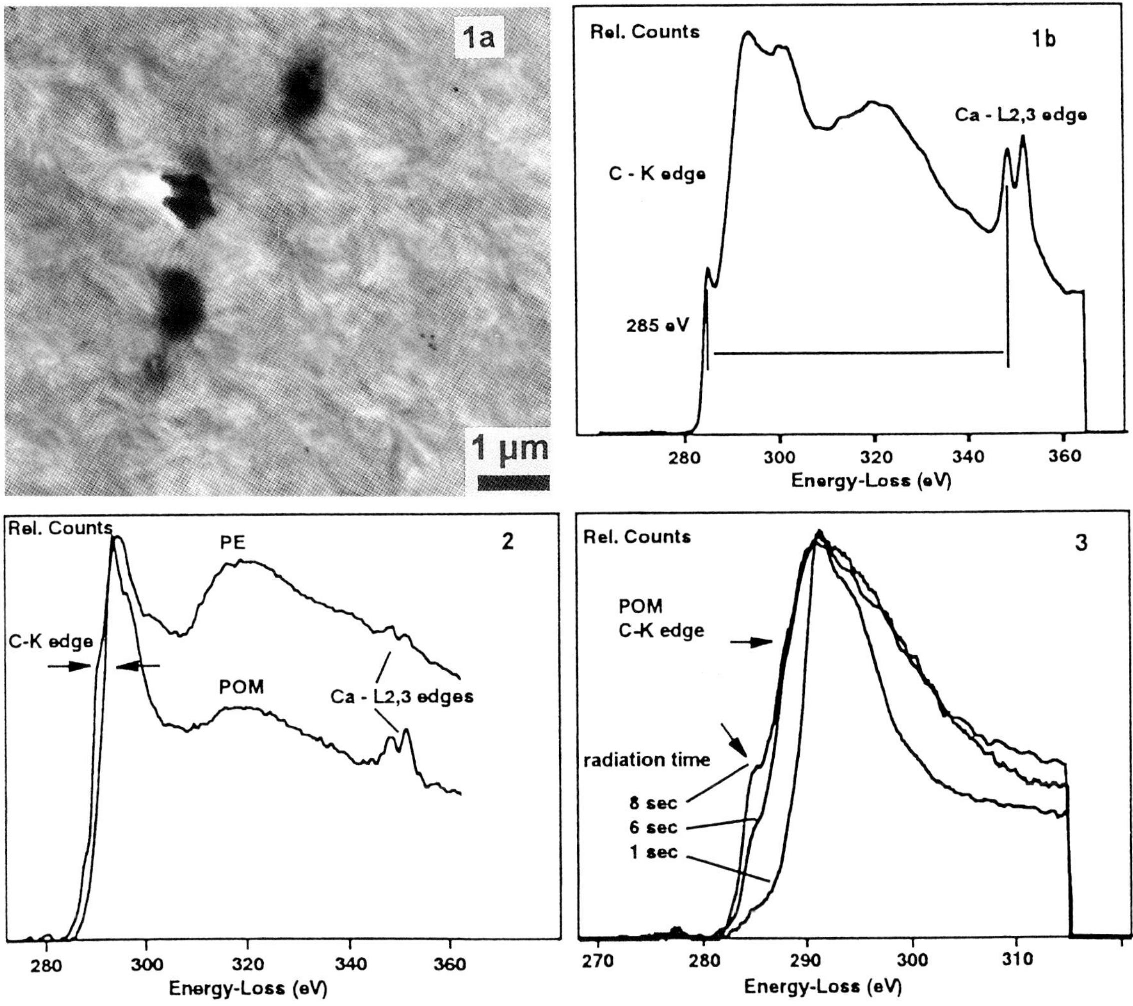

Fig. 1.- (a) TEM image of a calcium chloride contaminated polyether ketone ultrathin section (b) corresponding PEEL-spectrum. Fig.2. - PEEL-spectra of calcium chloride contaminated POM- and PE-sections. Fig.3. - Series of POM-PEEL-spectra with increasing radiation time.

PREPARATION OF SUCCESSIVE THIN SECTIONS OF FINE OXIDE PARTICLES BY AN ULTRAMICROTOME

G-S. Park[*], D. Shindo,[*] Y. Waseda,[*] J-M. Yang,[*] T. Oikawa,[**] and M. Kersker[***]

[*] Insititute for Advanced Materials Processing, Tohoku University, Katahira, Sendai 980-77, Japan
[**] JEOL Ltd., 1-2 Musashino 3-chome, Akishima, Tokyo 196, Japan
[***] JEOL U.S.A. Inc., 11 Dearborn Road, Peabody, MA 01960

A large number of monodispersed fine oxide particles, which are uniform in shape, size and composition, have been developed as ideal advanced materials. Recently, by the newly developed gel-sol method,[1-2] various shapes of micron-sized monodispersed hematite particles (α-Fe_2O_3) were produced, such as peanut-type, ellipsoidal and pseudocubic shapes. To understand the growth mechanism of these particles, it is necessary to investigate their internal structures. In this study, in order to analyze three dimensional internal structure of these particles, we have prepared their successive thin sections with an ultramicrotome and carried out HREM study on these successive thin sections.

Monodispersed peanut-type and pseudocubic particles were prepared by the standard procedures of the "gel-sol" method.[1-2] For carrying out ultramicrotomy, these particles were embedded with acrylic resin (methyl methacrylate: n-butyl methacrylate=3:7 or 4:6, benzoyl peroxide 1.5-1.7%) in a gelatin capsule. After trimming with a glass knife, successive thin sections of the particles were obtained by a diamond knife with the optimum slicing condition. The slicing speed and thickness were set to be 0.6 mm/sec and 70 nm, respectively.[3] Obtained successive thin sections were put on a single oval slot grid without removing the resin for easily identifying the positions of each section. Electron microscopic observation was carried out with a JEM-ARM 1250 and a JEM-2000EX II transmission electron microscope at accelerating voltages 1250kV and 200kV, respectively.

Figures 1 (a) and (b) show successive thin sections of the resin containing peanut-type and pseudocubic particles, respectively. The small dark dots in each section correspond to the respective sections of the particles. Figure 2(a) shows enlarged electron microscope images(A−F) obtained from successive thin sections of a peanut-type particle in Fig. 1(a). As indicated by an arrow, the sections clearly show the successive cross-section of a particle with the peanut shape. The positions of these successive sections in a particle are schematically indicated in Fig.2(b). Figure 3(a) also shows enlarged electron microscope images (A−E) obtained from successive thin sections of a pseudocubic particle in Fig. 1(b). The thickness is taken to be about 70nm assuming that the particle is uniformly sliced with the set thickness of an ultramicrotome. It is found from the shapes of these sections that the sections make an angle of about 37 degrees with the surface of the pseudocubic particle as schematically indicated in Fig.3(b). Moreover, as the case of the peanut-type particles,[4] HREM images of the successive sections of the pseudocubic particles could be obtained by a high-voltage electron microscope as shown in Fig.4. Thus it can be concluded that the preparation of successive thin sections with an ultramicrotome and their observation by a high-voltage electron microscope are very useful to make clear three dimensional atomic arrangement of these fine oxide particles and their growth mechanism.[5]

References

1. T. Sugimoto et al., Colloids Surf. A **70** (1993) 167.
2. T. Sugimoto et al., J. Colloid and Interface Sci. **158** (1993) 420.
3. G-S. Park et al., J. Electron Microsc. **43** (1994) 208.
4. D. Shindo et al., J. Colloid and Interface Sci. **168** (1994) 478.
5. The authors gratefully acknowledge the technical assistance of Mr. S. Aita, JEOL.

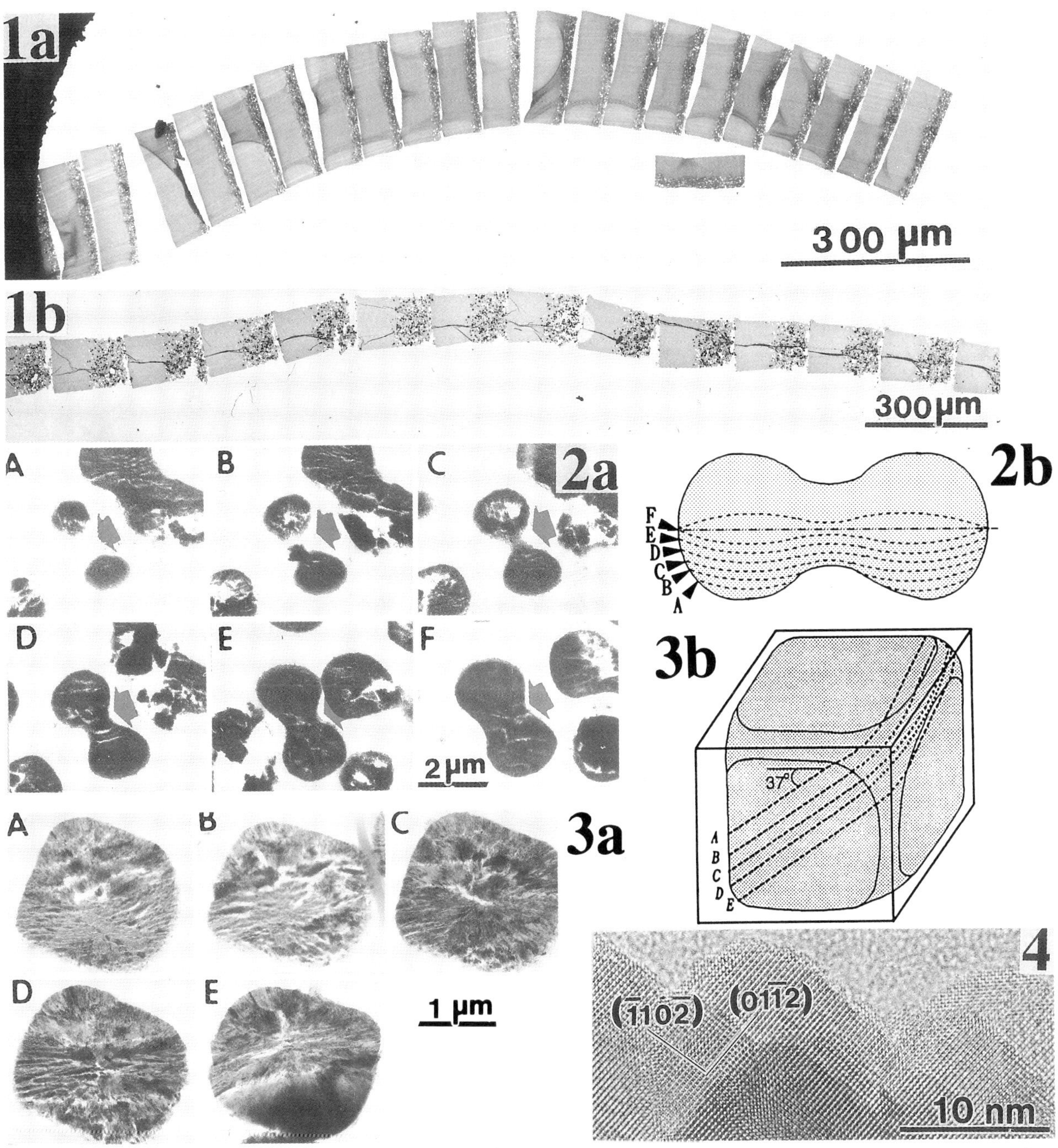

FIG. 1.—Successive thin sections of the resin containing peanut-type (a) and pseudocubic particles (b) obtained by an ultramicrotome.

FIG. 2.—(a) Enlarged electron microscope images obtained from successive thin sections of a peanut-type particle in FIG. 1(a); (b) Schematic illustration showing the positions of the successive sections of (a) in a peanut-type particle.

FIG. 3.—(a) Enlarged electron microscope images obtained from successive thin sections of a pseudocubic particle in Fig. 1(b); (b) Schematic illustration showing the positions of the successive sections of (a) in a pseudocubic particle.

FIG. 4.—High-resolution electron microscope image around the surface of a pseudocubic particle.

509

THIN FOIL PREPARATION FOR AEM STUDY OF SMALL METAL PARTICLES IN STONY METEORITES

Cheol-Woong Yang*, David B. Williams** and Joseph I. Goldstein*

 * Dept. of Mechanical Engineering, University of Massachusetts, Amherst, MA 01003
** Dept. of Materials Science and Engineering, Lehigh University, Bethlehem, PA 18015

The phase transformation processes and phase equilibria of the Fe-Ni and Fe-Ni-S system at low temperature can be better understood by characterizing the microstructure and chemistry of metallic particles in stony meteorites using analytical electron microscopy (AEM) techniques. However stony meteorites have a composite nature consisting of metal particles embedded in an inherently brittle silicate matrix, which causes a differential thinning rate. The thin foil preparation of the metal particles for AEM has proved to be very difficult, and we report on the first successful specimen preparation.

Materials of this type present particular difficulties in the preparation of thin foil specimens because the area of interest is limited to specific regions within the metal particles. The metal particles are usually less than 500 µm in size and the silicate matrix is inherently brittle. Moreover, the interfaces between metal particles and silicates are weak because of corrosion, depending on cooling, etc. In order to overcome a differential thinning rate problem, various ion beam thinning techniques which had been used to thin specimens having similar problems were applied, which included non-rotational ion beam thinning[1], shielded ion beam thinning[2], and low angle ion beam thinning[3]. However, none of them was acceptable for stony meteorites. Since the area of interest in this study is not the silicate matrix but the metal particles, we separated the metal particles from the silicate matrix. A supporting medium which is able to hold the metal particles throughout the thinning process is necessary to make a 3 mm disc which fits in most TEM/AEM instruments. Several embedding materials, such as G-1 epoxy (Gatan), 2 ton clear epoxy (Devcon), Torr Seal (Varian) and Epo-Tek H20E silver glue (Epoxy Tech.) were tested. The different embedding materials have different properties. Epo-Tek H20E silver glue was chosen because it allowed the specimen to be thinned by an electro-chemical polishing technique.

The difficulties in thin foil preparation of metal particles of the stony meteorite were overcome by developing a technique involving electrochemical polishing prior to ion beam thinning. A schematic illustration of the thin foil preparation method is shown in Fig. 1. The extracted metal particles were placed in the center of a shallow brass mold (made of a brass tube; 3 mm outer diameter, 1 ~ 2 mm height) which was stuck on a glass plate. The mold was then filled with Epo-Tek H20E silver glue and cured. After curing, these discs were fixed to a specimen mount (a thick glass plate) using mounting wax called Crystal-Bond (Gatan; melting point ≈ 130˚C). Both sides of the specimen were ground down to a residual thickness of 100 ~ 150 µm. Fig. 2A is a light optical micrograph of a polished AEM sample which shows the metal particles surrounded by the Epo-Tek H20E silver glue. The bright contrast in embedding material is from the Ag particles in the Epo-Tek H20E silver glue. With the dimple grinder (either a Gatan model 656 dimple grinder or a VCR Group DIMPLER® model D500i) the discs were further thinned to a residual thickness of about 30 ~ 50 µm. It is practical to grind a hemispherical depression into the specimen disc to produce a minimum thickness in the center while a thicker border for good mechanical strength and specimen handling is retained. Because the specimen is electrically conducting, the metal particles can be thinned preferentially by electro-chemical polishing prior to ion beam thinning. This technique offered the possibility of overcoming the differential thinning rate problem. Although the ion beam thinning sputtered the embedding material faster, it was possible to preserve the embedding material which held the metal particles during ion beam thinning because the metal particles had been preferentially thinned during electro-chemical polishing prior to ion beam thinning. Taking advantage of the characteristics of the Epo-Tek H20E silver glue, e.g., a high bonding strength and a electro-conductivity, successful thin foil specimens were obtained by electro-chemical polishing or a combination of electro-chemical polishing and ion beam thinning. The electro-chemical polishing was carried out using a Fischione twin jet electropolishing unit with 2% perchloric acid ($HClO_4$) in ethanol at 100V and -20˚C. After jet polishing, the specimen was examined under a light optical microscope comparing to the image taken before thinning to see if the perforation was in the exact area of interest. If it was, the specimen was examined using a TEM. If not, further ion beam thinning was performed until an area of interest was thin enough. Fig. 2B shows the St. Séverin stony meteorite sample prepared by this technique. The arrowhead

Proc. Microscopy and Microanalysis 1995, edited by G.W. Bailey, M.H. Ellisman, R.A. Hennigar, and N.J. Zaluzec.
Copyright © 1995 MSA. Published by Jones and Begell Publishing, 79 Madison Ave., New York, NY 10016

indicates the area of interest which is electron transparent. Specimens prepared by this technique have permitted the direct observation and characterization of the metallic phases in stony meteorites.

References

1. J. Y. Yao and G. L. Dunlop, *J. Elec. Micro. Tech.* 19 (1991) 90
2. P. Chang, M. D. Coviello and A. F. Scott, in *Specimen Preparation for Transmission Electron Microscopy of Materials*, Mat. Res. Soc. Symp. Proc., Vol. 115 MRS: (1988) 93
3. R. Alani, *Proc. 10th Anniversary of Chinese Electron Microscopy Society*, Beijing, China (1990)

Fig. 1

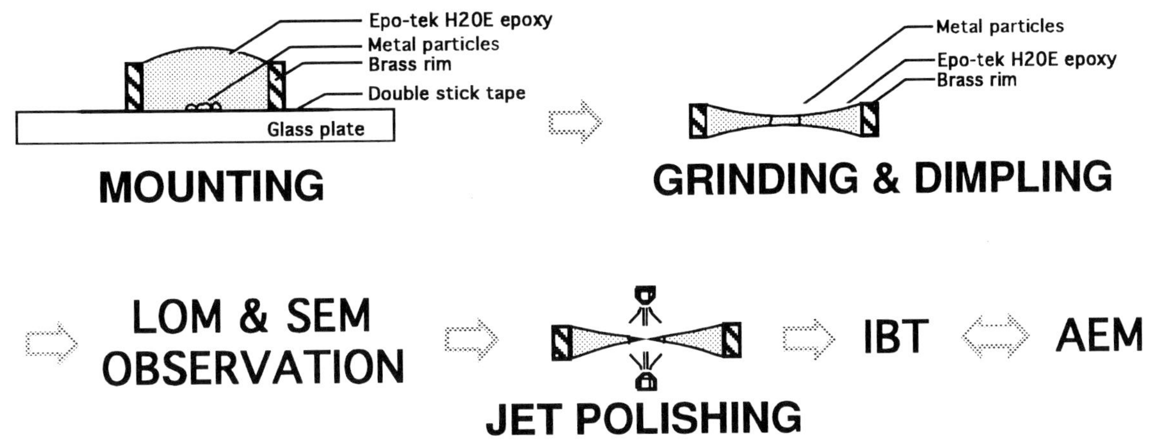

Fig. 2

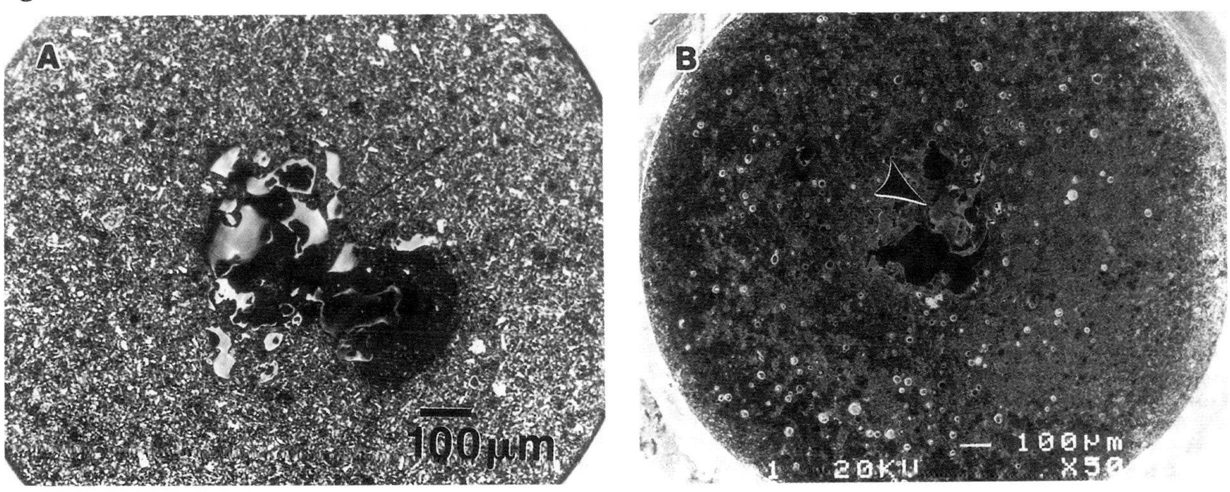

Fig. 1 Schematic illustration of the thin foil specimen preparation technique. Metal particles are extracted from the bulk chondrite sample and then mounted into the electron-conducting epoxy (silver glue). After curing, a 3 mm disc is pre-thinned (grinding and dimpling) and jet polished to produce electron transparent regions in the metal particle. If necessary, further ion beam thinning is performed.

Fig. 2 A) Light optical micrograph of a TEM specimen before dimple grinding (briefly etched with 2% Nital). B) SEM image of a TEM specimen after jet polishing and ion beam thinning.

QUANTITATIVE DEFECT STUDIES IN SEMICONDUCTOR TEM SAMPLES PREPARED BY LOW ANGLE ION MILLING.

L.Mulestagno[*,**] J.C. Holzer[**], and P.Fraundorf[*].
* Physics Dept. and Center for Molecular Electronics , University of Missouri - St. Louis, St. Louis , MO 63121.
** MEMC Electronic Materials Inc, PO Box 8, St. Peters, MO 63376 .

Due to the wealth of information, both analytical and structural that can be obtained from it TEM always has been a favorite tool for the analysis of process-induced defects in semiconductor wafers . The only major disadvantage has always been , that the volume under study in the TEM is relatively small, making it difficult to locate low density defects, and sample preparation is a somewhat lengthy procedure . This problem has been somewhat alleviated by the availability of efficient low angle milling .

Using a PIPS® variable angle ion -mill, manufactured by Gatan, we have been consistently obtaining planar specimens with a high quality thin area in excess of 5×10^4 μm^2 in about half an hour (milling time), which has made it possible to locate defects at lower densities, or, for defects of relatively high density, obtain information which is statistically more significant (table 1) . The variable angle features (0 to 10°) , and the ability of the new model to do double side milling has also eliminated much of the handicap of the previous model , which had a fixed angle of 4°, and was equipped to do single-side milling only .[1]

The specimens were prepared in a standard fashion (mechanical pre-thinning followed by dimpling), but the dimpling was usually terminated at around 10 μm thickness at which point the samples were polished with a felt wheel and 0.05μm alumina or Syton®. This was done not only to improve the quality of surface and reduce the damage prior to ion-milling, but also to flatten out the groove created by the dimpling . If this step is omitted, even though the ion beam is about 350μm at the specimen surface (assuming uniform thinning in this area), the thin area obtained would be significantly smaller due to a local thickness gradient below the dimple (fig 1) . The difference in thickness between the bottom of the dimple and the edge of the beam could be several microns, in which case, the center would become electron transparent before the edge of the beam . Polishing with the soft felt wheel makes this gradient significantly smaller, and thus reduces this problem .

This problem will be even more pronounced as smaller angles are used , where the potential for a ten-fold increase in thin area is a reality (2-3° milling , fig 2), and could increase the usable volume in the specimen by another order of magnitude . We are presently trying to optimize this process, weighing in the increased thin area against the increase in time required for milling .

References:
1. Reza Alani and Peter R.Swann, MRS Proceedings volm 254, p 43 - 63.

Proc. Microscopy and Microanalysis 1995, edited by G.W. Bailey, M.H. Ellisman, R.A. Hennigar, and N.J. Zaluzec
Copyright © 1995 MSA. Published by Jones and Begell Publishing, 79 Madison Ave., New York, NY 10016

density/cm^3	defects present in 1 x 10^4 μm^3	defects present in 1x 10^5 μm^3
1 x 10^{10}	100	1000
1 x 10^9	10	100
1 x 10^8	1	10
1 x 10^7	0.1	1
1 x 10^6	-	0.1

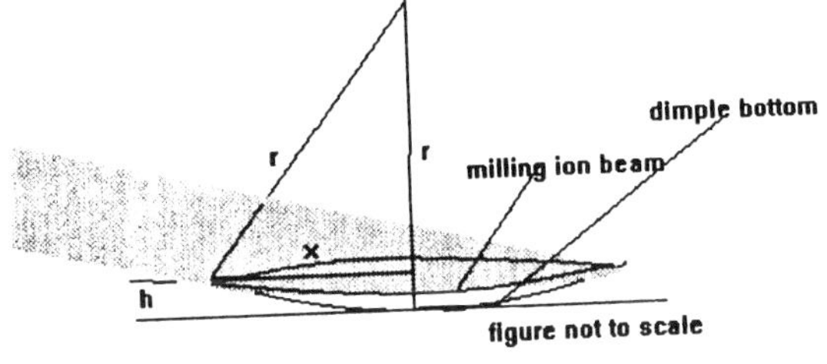

Fig 1

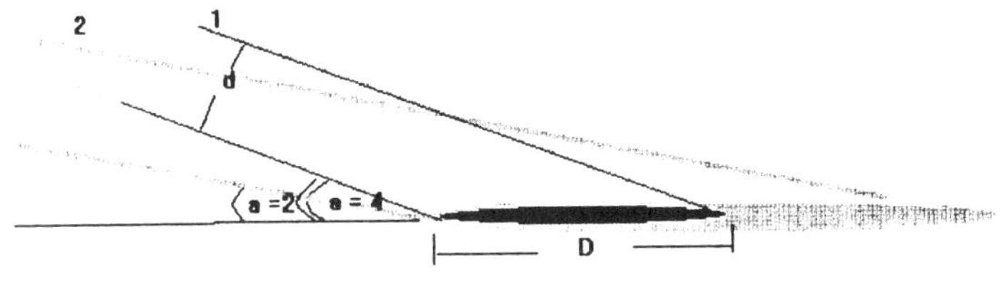

Fig 2

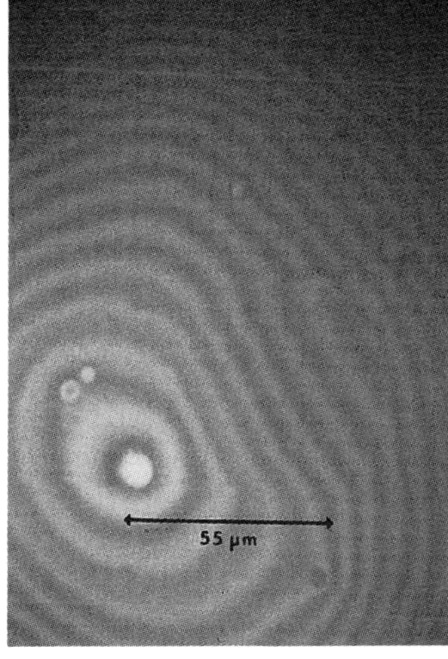

Fig3

fig 1. for a dimpling wheel of radius 7.5mm, and a milling beam of ~300µm, h, the height difference can be as much as 6µm depending on how well it is centered on the dimple.
fig 2. for a beam diameter d of 25µm , the beam at the sample will be ~350µm for 4° milling, and ~ 700 µm for 2°. This would make the gradient at an unsmoothened dimple above 8µm.
fig3. dark-field optical microscope view of thin area in a sample milled at 4-5°.

PREPARATION OF CROSS SECTION SAMPLES OF FERRIMAGNETIC IRON GARNET THIN FILMS FOR ELECTRON MICROSCOPY STRUCTURAL CHARACTERIZATION

B.M. Simion*, R. Ramesh**, and G. Thomas*

* Materials Science and Mineral Engineering Dept., University of California at Berkeley,
 and Lawrence Berkeley Laboratory, Berkeley, CA 94720
**Materials Science and Nuclear Engineering Dept., University of Maryland,
 and The Center for Superconductivity Research, College Park, MD 20742-2115

Ferrimagnetic iron garnet thin film multilayered heterostructures were deposited by pulsed laser deposition on paramagnetic [111]-oriented single crystalline GdGa-garnet (GGG) substrates. Bismuth and rare-earth substituted iron garnet thin films are being presently considered as good potential candidates for the next generation of thermo-magneto-optical (TMO) permanent memories[1], due to their improved magneto-optical (MO) response to the green laser radiation. As reported previously[2], improved magnetic and MO properties were shown by multilayered $Y_3Fe_5O_{12}/Bi_3Fe_5O_{12}$ (YIG/BIG) and $Y_3Fe_5O_{12}/Eu_1Bi_2Fe_5O_{12}$ (YIG/EBIG) heterostructures when compared to their single layer counterparts. The present study tries to elucidate the relationship between the films microstructure and their improved properties. Along with x-ray structural characterization, scanning electron microscopy for topographical characterization and analytical transmission electron microscopy for chemical and microstructural analysis are being used.

Due to the brittleness of both films and substrates, conventional cross section sample preparation becomes extremely difficult. Samples under roughly 20 μm thickness break under the lowest load of the dimpler. Although in principle it would be possible to dimple the specimen down to above 20 μm and then ion mill it until the desired thin area is obtained, because the material is an insulating oxide, the sample charges up in the ion milling machine and starts repelling the ion beam from its surface. An alternative sample preparation is a gentle wedge mechanical polish, obtained by mounting a thinly cut (100 - 200 μm) slice of a typical cross section on a tripod polisher. Figure 1, shows the polishing process and the different mounts the specimen goes through during preparation. Although the wedge polishing technique is not a new one and several commercial enterprises offer materials and instrumentation dedicated to this procedure, the several polishing steps one has to go through in order to achieve optimal TEM thickness conditions vary from one material to another. Hence, while films deposited on Si wafers may get to the desired thickness using as last step the 1 μm diamond film, garnet films still suffer catastrophical cracking during polishing on the 1 μm film. The roughness of the polishing film used during the last step is critical in obtaining a good specimen and in the case of the garnet films this should be lowered to 0.1 μm. Figure 2 shows a SEM backscattered image of a TEM sample for which the last polishing step was done on the 1 μm diamond film. Note that the film appears to have columnar growth. However, the "phi" x-ray scan shown in fig.3, indicates that the film is single crystalline. Figure 4 shows a region of the same film, obtained this time from a much thicker specimen. No columnar growth is observed in this image. The conclusion is that the groves noted in fig. 2 are in reality cracks introduced in the film region during polishing. Such a high dependency of results on the sample preparation technique underlines the importance of this first step in the success of the whole project, as well as indicating the continuous correlation that must be made among the results of all the characterization techniques used.

This work is supported by the Director, Office of Basic Energy Sciences, Division of Materials Science of the   United States Department of Energy under Contract No. DE-AC03--76F00098. Work at Bellcore/UMD is supported by Office of Naval Research.

References

1. P.C. Dorsey, S.E. Bushnell, R.G. Seed, C. Vittoria - *Journal of Applied Physics*, **74** (2), pp.1242-1246, 15 July 1993
2. B.M. Simion, R. Ramesh, V.G. Keramidas, R.L. Pfeffer, G. Thomas, and E. Marinero, *J. Appl. Phys.*, **76**-10, 1994, pp.6287

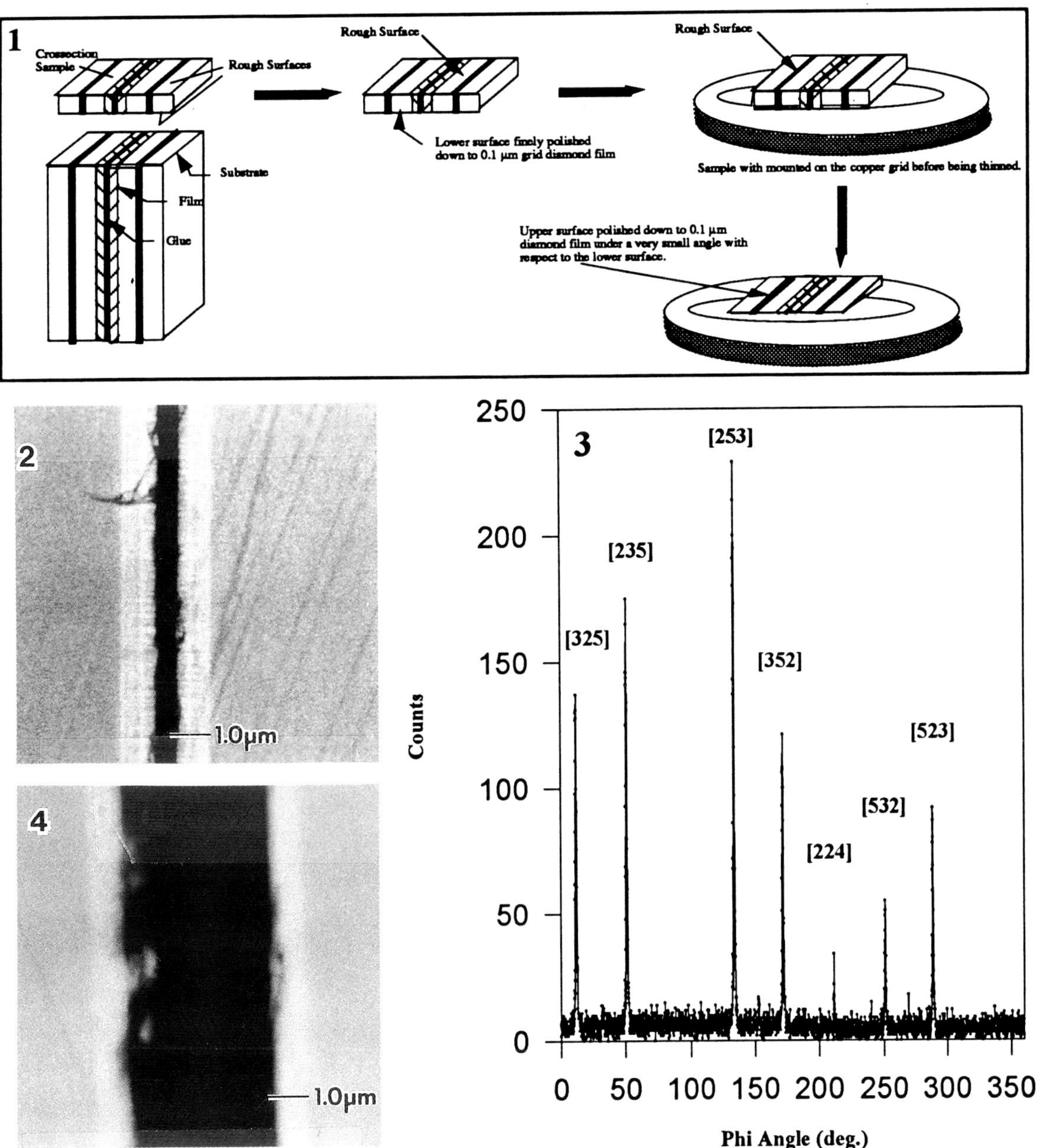

Fig. 1 - Sequence of steps in wedge polishing TEM sample preparation of a thin film cross section specimen. The different layers of the cross section "sandwich" are glued together with an electron beam resistant adhesive. The same adhesive is used to attach the specimen to the Cu TEM grid.

Fig.2 - SEM backscattered image of a TEM cross section specimen, showing what appears to be columnar growth of the garnet film.

Fig.3 - In-plane x-ray Phi scan, showing the single crystalline character of the film.

Fig.4 - SEM backscattered image of a thick (100 - 200 mm) cross section of the same film shown in fig.2. Note that no columnar formations can be observed.

ADVANCED TEM SAMPLE PREPARATION TECHNIQUES FOR SUBMICRON Si DEVICES

P. B. Basham and H. L. Tsai

Texas Instruments, Materials Science Laboratory
P.O. Box 655936, M/S 147, Dallas, TX 75265

The use of transmission electron microscopy (TEM) to support process development of advanced microelectronic devices is often challenged by a large amount of samples submitted from wafer fabrication areas and specific-spot analysis. Improving the TEM sample preparation techniques for a fast turnaround time is critical in order to provide a timely support for customers and improve the utilization of TEM. For the specific-area sample preparation, a technique which can be easily prepared with the least amount of effort is preferred. For these reasons, we have developed several techniques which have greatly facilitated the TEM sample preparation.

For specific-area analysis, the use of a copper grid with a small hole is found to be very useful. With this small-hole grid technique, TEM sample preparation can be proceeded by well-established conventional methods. The sample is first polished to the area of interest, which is then carefully positioned inside the hole. This polished side is placed against the grid by epoxy . Fig.1 is an optical image of a TEM cross-section after dimpling to light transmission. Ion-milling to electron transmission is conducted from the dimpled side and the sample thickness is monitored by the color change under optical microscope. With this technique, a whole thin area inside the hole can be produced for TEM examination. The hole size in the copper grid shown in Fig. 1 is 300 µm. This technique is extemely simple and easy to handle during the process.

Figure 2 shows an optical micrograph of a completed TEM cross section. This TEM cross section is formed by stacking 23 pieces together . We have been able to stack as many as 32 pieces in one TEM cross section. To prepare such a sample, all pieces with suitable sizes are placed on a flat puck and polished from the back side to 10-20 µm During the polishing, a specially designed tool is used to planarize the puck for easy control of obtaining about the same thickness of these pieces. These polished pieces are then carefully removed from the puck and stacked in a particular order so that different samples can be distinguished during the TEM examination. This technique allows more than 10 different production wafers to be prepared at the same time and has significantly reduced the turnaround time.

For the analysis of crystal defects in electrically active substrate areas, it is preferred to prepare planar TEM samples. We have found that this is especially useful to correlate failure spots with dislocations. Figure 3 shows a TEM image with a failure site marked in between two trench holes drilled by focused ion beam. A dislocation is detected (marked by an arrow) under the gate. This dislocation connects source to drain and is responsible for junction leakage, which results in electrical failure. The dislocation was also observed using cross-sectional analysis.

Proc. Microscopy and Microanalysis 1995, edited by G.W. Bailey, M.H. Ellisman, R.A. Hennigar, and N.J. Zaluzec

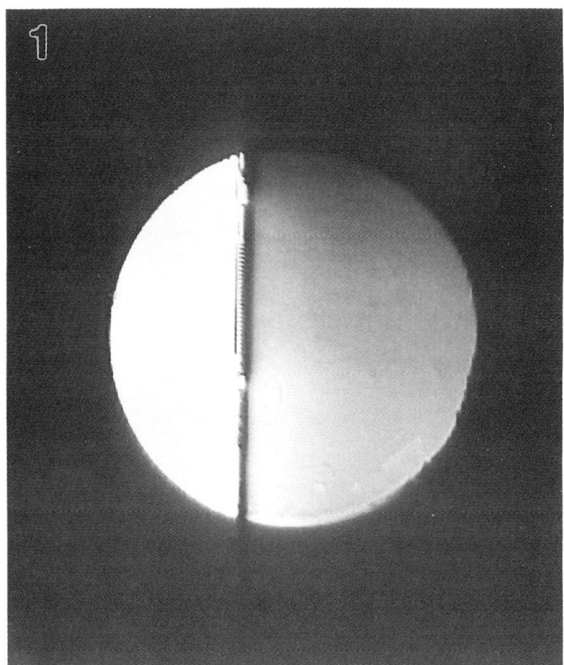 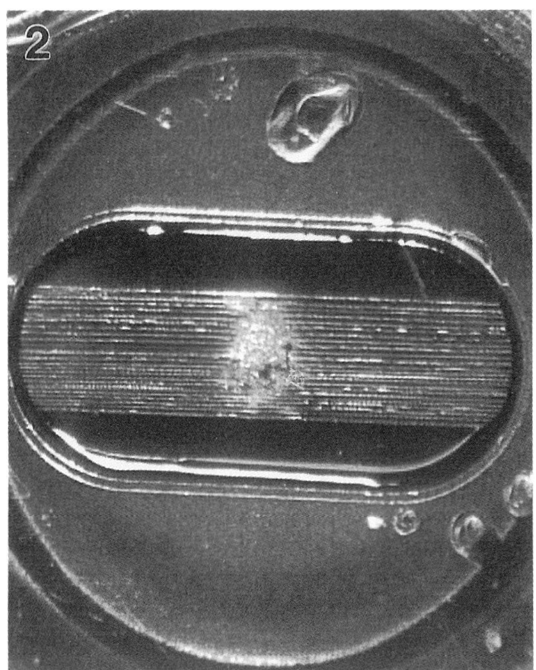

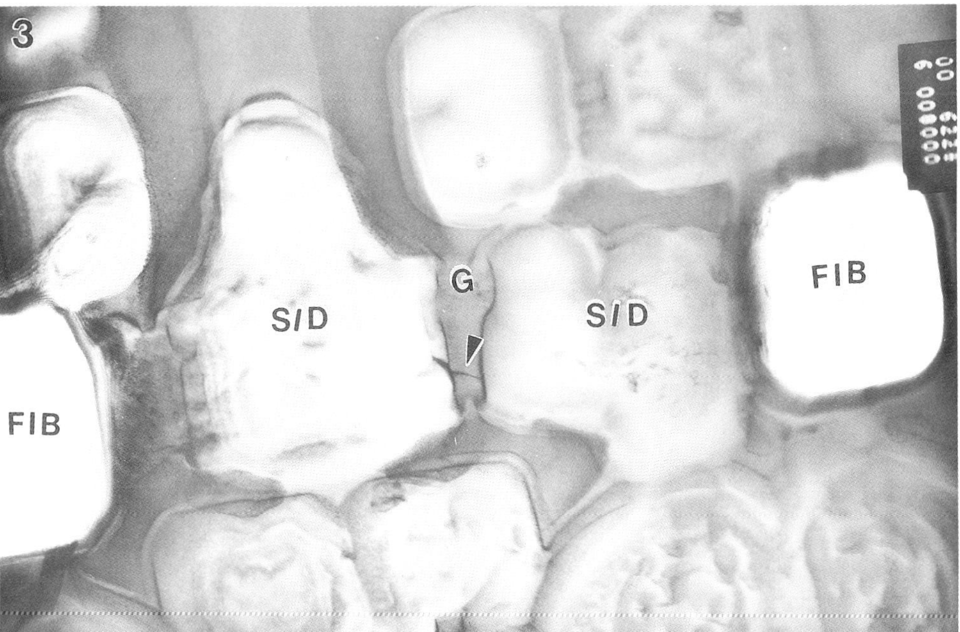

FIG. 1. Optical image of TEM cross-section by small-hole grid technique for specific-area sample preparation. TEM sample for Si device is dimpled to light transmission.

FIG. 2. Optical image showing multi-layer TEM cross-section containing 23 pieces.

FIG. 3. Planar-view TEM image of Si device illustrating correlation between dislocation and specific failure site marked in between two markers(FIB). G: transistor gate, S/D::source or drain

Preparation of TEM samples by FIB

John F. Walker*, J C Reiner** and C Solenthaler**

*FEI Europe Ltd., Cottenham, Cambs. England
**ETH, Zurich, Switzerland.

The high spatial resolution available from TEM can be used with great advantage in the field of microelectronics to identify problems associated with the continually shrinking geometries of integrated circuit technology. In many cases the location of the problem can be the most problematic element of sample preparation. Focused ion beams (FIB) have previously been used to prepare TEM specimens[1-3], but not including using the ion beam imaging capabilities to locate a buried feature of interest. Here we describe how a defect has been located using the ability of a FIB to both mill a section and to search for a defect whose precise location is unknown. The defect is known from electrical leakage measurements to be a break in the gate oxide of a field effect transistor. The gate is a square of polycrystalline silicon, approximately $1\mu m \times 1\mu m$, on a silicon dioxide barrier which is about 17nm thick. The break in the oxide can occur anywhere within that square and is expected to be less than 100nm in diameter. Since no indication of the exact location of the break is visible on the surface, finding the defect using conventional grinding and broad beam polishing techniques would present a serious, if not impossible, problem. The procedure followed involved mounting a small piece of the semiconductor on a TEM grid which has a vee cut out of one side to allow access for the ion beam. This piece should placed such that the gate is positioned near the centre of the grid and the thickness of the piece should be of the order of 100nm or less. One micron of platinum was deposited on the surface of the sample to planarise the surface and to protect the transistor from direct beam damage. The platinum was deposited by ion beam dissociation of a platinum containing organo-metallic gas only in the area immediately adjacent to the gate region. Trenches were milled on either side of the gate using a high current, 7nA, beam. Subsequently, thin slices were milled out of the remaining section in 50nm steps and imaged with a low, 3pA, beam current and short beam dwell time. Despite the low ion dose, necessitated by the need to reduce ion beam damage to an absolute minimum, image quality was sufficient to see the defect as a break in the oxide layer. Subsequent thinning of the remaining section from the other side left a sub 100nm membrane containing the defect and this structure was transferred to the TEM. The TEM image clearly shows the structure of the gate region of the transistor which includes the defect. Diffraction analysis shows that the substrate is single crystal silicon, the gate is polycrystalline silicon and the oxide barrier amorphous silicon dioxide.

References
1. R. Hull, F.A. Stevie, D. Banck, Appl. Phys. Lett. **66** (3) 1 (1994)
2. R. Hull, D. Banck, F.A. Stevie, L.A. Koszi, S.N.G. Chu, Appl. Phys. Lett. **62** (26) 3408 (1993)
3. R. Young, E.C.G. Kirk, D.A. Williams, H. Ahmed, Microelectron. Eng. 11, 409 (1990)

Proc. Microscopy and Microanalysis 1995, edited by G.W. Bailey, M.H. Ellisman, R.A. Hennigar, and N.J. Zaluzec
Copyright © 1995 MSA. Published by Jones and Begell Publishing, 79 Madison Ave., New York, NY 10016

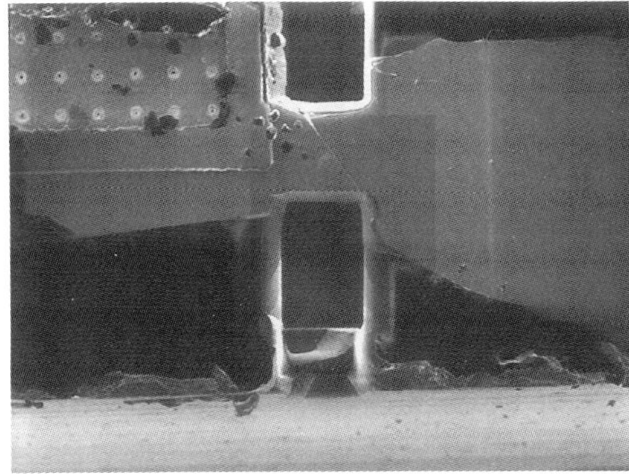

Figure 1. Initial trenches milled on either side of the gate structure

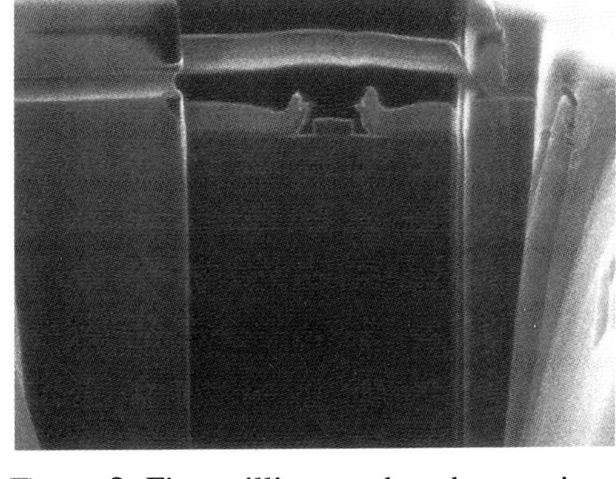

Figure 2. Fine milling reaches the transistor structure. Source, gate and drain visible.

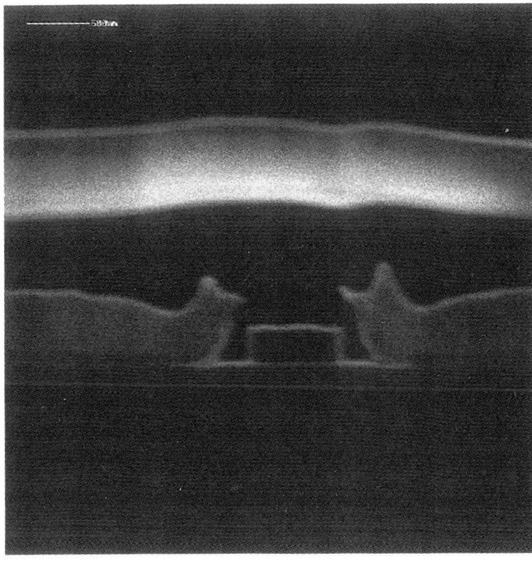

Figure 3. Further milling into the structure. The gate oxide is intact.

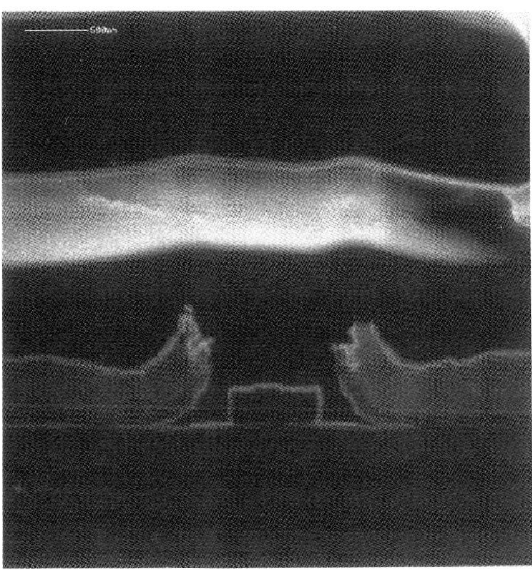

Figure 4. Milling stops as the oxide barrier, visible as a white line between the gate and substrate, is broken.

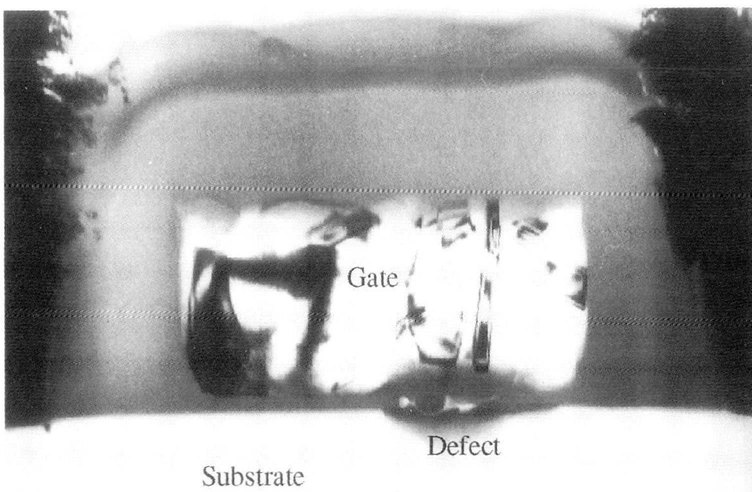

Figure 5. TEM image of the structure described in the text.

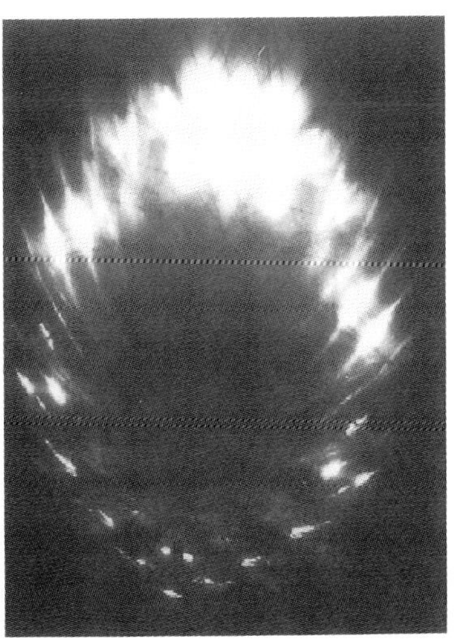

Figure 6. Diffraction pattern from the substrate area.

HYDROGEN-INDUCED PHASE TRANSFORMATIONS IN α_2+B2 Ti-25Al-10Nb-3V-1Mo TITANIUM ALUMINIDE

X. Pierron,* M. De Graef,** T.M. Pollock,** and A.W. Thompson***

* Thayer School of Engineering, Dartmouth College, Hanover NH 03755
** Department of Materials Science and Engineering, Carnegie Mellon University, Pittsburgh, PA 15213
*** Lawrence Berkeley Laboratories, Berkeley CA 94720

Titanium aluminide alloys containing the DO_{19} α_2 (Ti_3Al) phase, such as Ti-25Al-10Nb-3V-1Mo (at.%), are known to be sensitive to hydrogen.[1,2,3] Such alloys could provide an alternative for conventional Ti alloys in a hydrogen-rich environment, and an understanding of the hydrogen embrittlement mechanisms acting in these materials is needed. Hydrides are known to form in conventional Ti alloys. Hence, a systematic study was undertaken to investigate the possibility of hydrogen-induced second phases in the Ti-25Al-10Nb-3V-1Mo alloy. We report here that internal hydrogen results in the precipitation of an orthorhombic O_H phase in the α_2 grains. The morphology and the orientations of the orthorhombic variants can be successfully described using group theory and the phenomenological theory of martensitic transformations.

The as-received material consisted of bars extruded in the $(\alpha_2 + \beta)$ phase field at 1037°C, then oil quenched. A solutionizing treatment at 1085°C, followed by free cooling at 0.1°C/s to room temperature resulted in a microstructure containing primary α_2 grains and α_2 secondary laths in a B2 matrix. Hydrogen charging was performed at 700°C using the gas equilibrium method in a Sieverts apparatus for 48 hours. For hydrogen equilibrium pressures close to 685, 75, and 15 torr, we obtained hydrogen contents of about 2800, 900 and 300 wppm respectively. After charging, TEM thin foils were prepared using a hydrogen-free solution of CaCl in methanol at -48°C/35V.

For hydrogen contents close to 300 wppm, no changes in the microstructure were observed. At 1000 wppm, a hydrogen-induced orthorhombic phase, O_H, precipitates as thin plates in the α_2 grains. The O_H structure can be described as an orthorhombic distortion of the DO_{19} structure, with the **Cmcm** symmetry. X-ray diffraction results show that the distortion consists of an elongation along one of the $\langle 11\bar{2}0 \rangle$ directions, and a contraction along the perpendicular $\langle 10\bar{1}0 \rangle$ direction, with no changes along the [0001] direction. The distortion increases with increasing hydrogen content and is fully reversible. For hydrogen contents higher than 2000 wppm, the α_2 grains are entirely replaced by a complex O_H domain network, Fig. 1. As predicted by group theory, three O_H variants (indicated in Figure 2 by the symbols I, II, and III) were found and their orientations are related to one another by the Variant Generating Group elements associated with the **P6$_3$/mmc** to **Cmcm** transformation. The stress-free interfaces between two neighboring O_H variants were calculated using a technique similar to the phenomenological theory of martensitic transformations and were found to be the {110} and {130} type planes, in agreement with TEM observations. The O_H domains arrange themselves to contact along these stress-free interfaces, Figs. 2 and 3. One can show that such domain configurations introduce almost no internal stresses, as large lattice mismatches across the interfaces are avoided.[4] These results indicate that strain energy minimization is the controlling factor in the formation of the O_H domain configurations.[5]

References

1. X. Pierron et al. in N.R. Moody, A.W. Thompsons Eds., *Hydrogen Effects in Materials*, in press (1995)
2. K. S. Chan, *Metallurgical Transactions A*, 23(1992)183
3. W. Y. Chu, *Metallurgical Transactions A*, 23(1993)1299
4. X. Pierron, Ph.D. Thesis, Carnegie Mellon University (1995)
5. Support for this work was provided by the Alcoa Center for Engineering Materials at CMU and by NASA Ames Grant NAG2-850.

Proc. Microscopy and Microanalysis 1995, edited by G.W. Bailey, M.H. Ellisman, R.A. Hennigar, and N.J. Zaluzec
Copyright © 1995 MSA. Published by Jones and Begell Publishing, 79 Madison Ave., New York, NY 10016

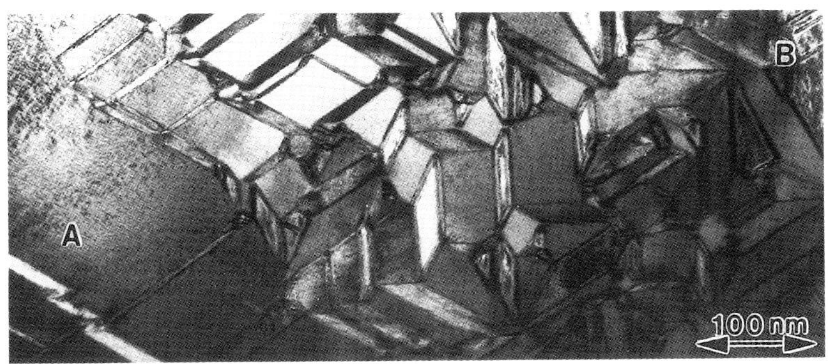

Figure 1: Micrograph showing α_2 grain (1000 wppm hydrogen). Area A contains $(\alpha_2 + O_H)$ mixture. Area B consists of only O_H domains, connected along stress-free $\{110\}$ and $\{130\}$ interfaces.

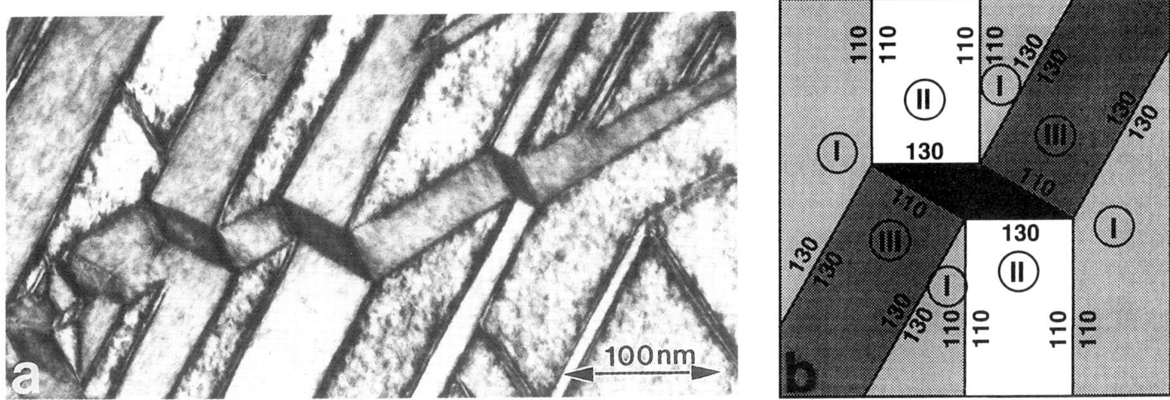

Figure 2: a) TEM micrograph showing O_H domain configuration. b) Schematic showing $\{110\}$ and $\{130\}$ interfaces, and variant types. This configuration does not induce any internal stresses.

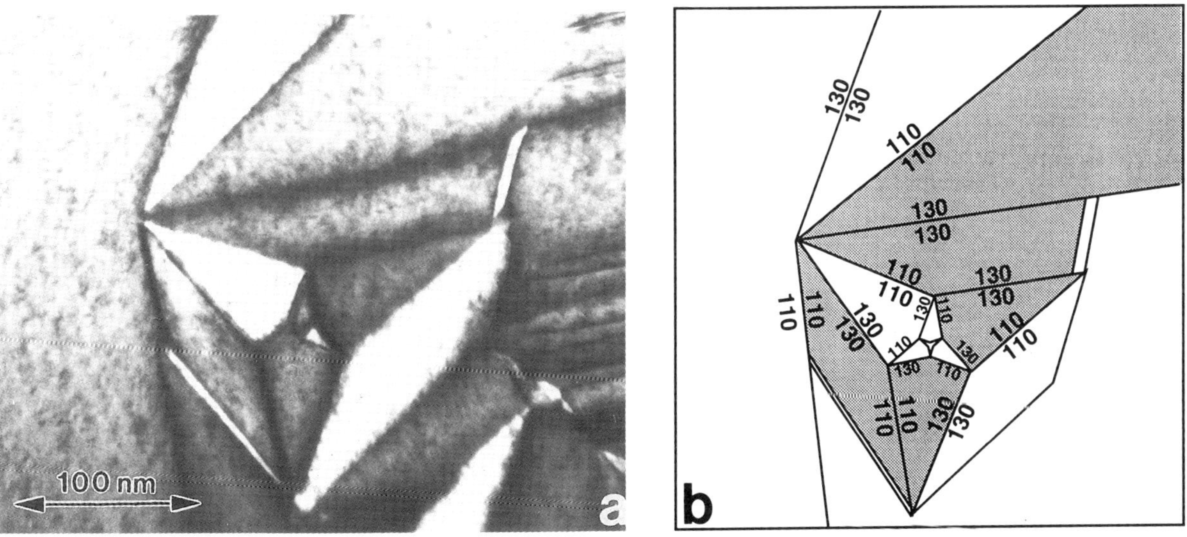

Figure 3: TEM micrograph showing O_H domain configuration. Again, interfaces are only on $\{110\}$ and $\{130\}$ planes. It can be shown that this configuration induces small internal stresses in center of star, and along branches.

DIRECT OBSERVATION OF MECHANICAL TWINNING IN PURE IRIDIUM

C. L. Chen*, S. G. Song*, T. Mitchell*, G. T. Gray III*, and T. T. Tsong†

* Materials Science and Technology Division, Los Alamos National Laboratory, NM 87545
† Physics Institute, Academia Sinica, Taiwan 11529

Mechanical twinning in crystalline solids is understood according to a recent study [1] to occur via two distinct mechanisms; i.e. a step-wise (athermal) and a dislocation pole (thermal) mechanism [1,2]. In f.c.c. structures, the two twinning mechanisms are both operational. Twinning by a dislocation pole mechanism normally takes place at elevated temperatures since thermal activation is needed in structures requiring a high passing stress. When twinning occurs by the step-wise mechanism, the selection of {111} planes as twin boundaries to form a coherent interface between the twin and the matrix lattice is limited. In the past twenty years, Field Ion Microscopy (FIM) has successfully been used to analyze many microstructural defects. However, there has been no detailed analysis of mechanical twin morphology using FIM techniques. We present here the FIM observations of deformation twins in pure Ir and compare the results with predictions of twin thickness and boundary structures according to the two twinning mechanisms.

The sample used for this experiment was selected from 99.998% pure cold extruded iridium wire. Figure 1 shows grains elongated along the wire direction which were developed during extrusion. No subsequent annealing was performed; the microstructure of the specimen remains as-extruded. Mechanical twins have been found in the Ir tip as seen in Fig. 2. The twin exhibits two clearly-defined parallel boundaries. The crystallographic orientation of the twin is apparently different from that of the matrix. It is known that the twinning plane and the twinning shear direction in f.c.c. structures are the {111} plane and <112> direction, respectively. Based on this information on twinning, the major poles of the matrix and twin were calculated and plotted in Fig. 3. The calculated zone axes agree well with the corresponding poles shown in the FIM image. The twin boundary is straight and parallel to {111} planes of both the matrix and twin. The twin terminates at a high angle grain boundary, marked X and Y, at one end. Besides the grain boundary and the twin seen Fig. 2, a Frank partial dislocation, having Burgers vector $\frac{1}{3}$ <111>, is also seen in the matrix, marked E, where a {111} plane terminates in the crystal. Close scrutiny of the structure of the twin boundaries shows that there are no dislocations present on the boundaries since a layer to layer correspondence exists between the matrix and the twin over the length of the boundaries. The thickness of the twins remains constant. One can measure the twin thickness from high quality FIM images by counting the atomic row number along a specific direction on a selected plane because the row separation is known for such a plane. Thus, the twin in Fig. 2 was found to be 21 close packed atomic layers thick along the [1$\bar{1}\bar{1}$] direction.

Figure 4 shows the superposition of the atom arrangements of the matrix and twin on the {011} common plane. Perfectly-matched planes between the two crystals are seen every three {111} planes. According to the step-wise nucleation and growth model [1], only these lattice match planes can be utilized as twin boundaries. Thus, the number of {111} atom planes through the twin thickness has to be ($3n$) (where n is an integer) in order to make use of the lattice match plane as the coherent twin boundary to reduce boundary energy since partial dislocations are not involved in the process. However, when a twin is formed by a dislocation pole mechanism, there are no restrictions on the selection of {111} planes as twin boundaries since atoms are switched from the matrix to twin positions in a layer by layer manner through partial dislocation movement on the twin/matrix boundaries [2-4]. Accordingly, a distinction can be made regarding the formation mechanism for the twins observed in Ir based on the lattice match plane criterion. The above calculation of the twin thickness has shown that the twin seen in Fig. 2 consists of 21 {111} planes which satisfies the ($3n$) layer requirement for coherent twin boundaries. The nucleation and growth of twins occur, according to the step-wise mechanism, through a coordinated movement of all atoms between the two matched planes. During nucleation of a twin, the initial thickness of a twin embryo has to meet the ($3n$) requirement to reduce the interfacial energy.

References:

1. S. G. Song and G. T. Gray III, *Acta Metall. Mater.* in press (1994).
2. T. H. Blewitt, R. R. Coltman, and J. K. Redman, *J. Appl. Phys.,* **28** (1957), 615.
3. K. Yasutake, S. Shimizu, and H. Kawabe, *J. Appl. Phys.,* **61** (1987), 947.
4. S. G. Song and G. T. Gray III, *Phil. Mag. A*: in press (1994).

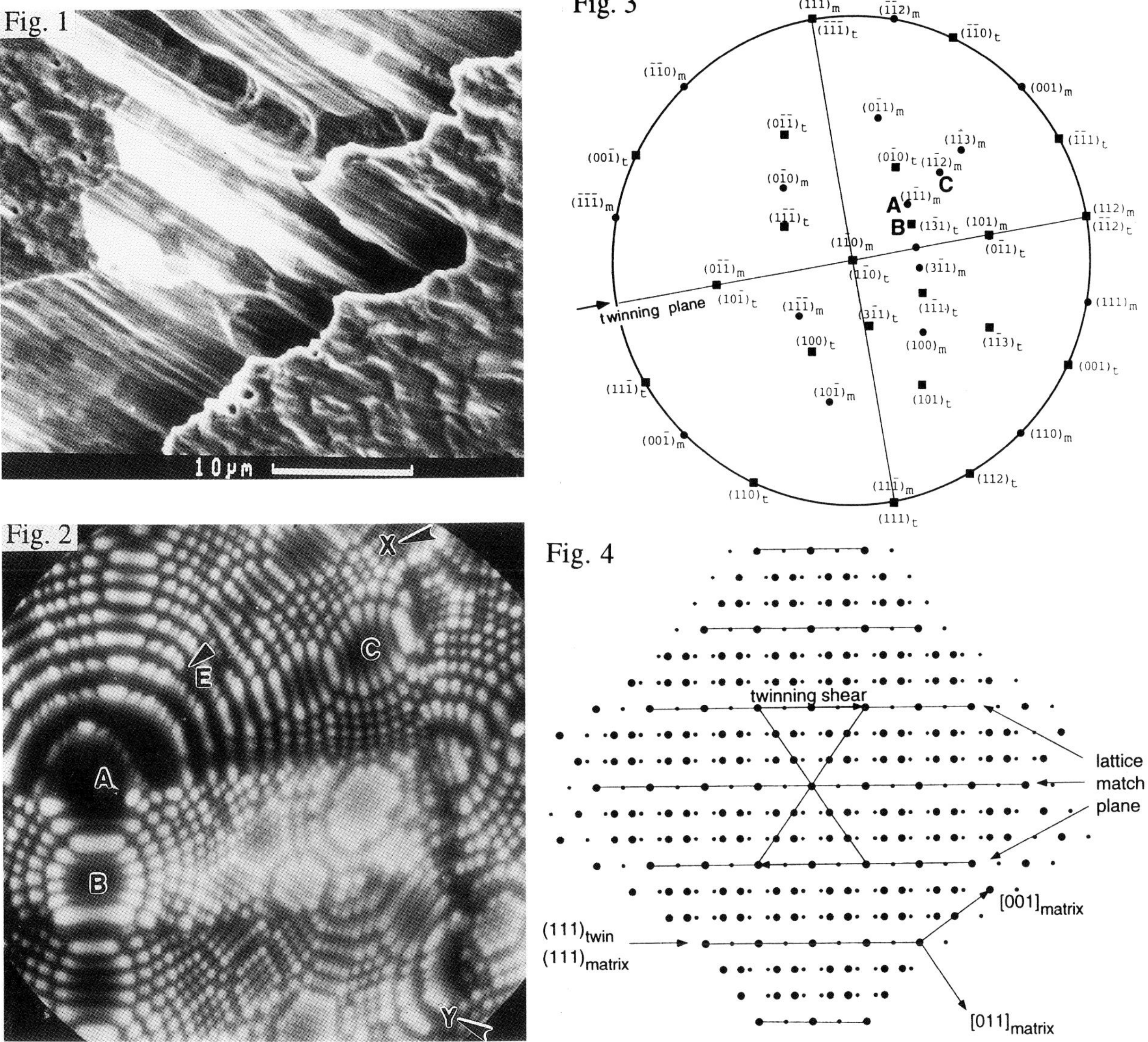

Fig. 1 SEM image showing the etched microstructure of an Ir wire in the as-extruded condititon.
Fig. 2 FIM photograph showing deformation twin (marked T), a partial dislocation (marked E) and a high angle grain boundary (marked X-Y) in pure Ir.
Fig. 3 Stereographic projection of the matrix and twin corresponding to Fig. 2.
Fig. 4 Overlap of the atom positions of the matrix and twin lattices on the {011} common plane. Dark dots - atom positions in the matrix lattice, gray dots - atom positions in the twin; large dots - atoms in the paper plane, small dots - one layer below the paper plane.

HIGH-RESOLUTION ELECTRON MICROSCOPY STUDY OF SUBMICROMETER-GRAINED Al-Mg SOLID SOLUTION ALLOYS

Z. Horita[1], D. J. Smith[2], M. Furukawa[3], M. Nemoto[1], R. Z. Valiev[4] and T. G. Langdon[5]

(1) Department of Materials Science and Engineering, Faculty of Engineering 36, Kyushu University, Fukuoka 812, Japan.
(2) Center for Solid State Science and Department of Physics and Astronomy, Arizona State University, Tempe, AZ 85287, USA.
(3) Department of Technology, Fukuoka University of Education, Munakata 811-41, Japan.
(4) Institute for Metals Superplasticity Problems, Russian Academy of Sciences, Ufa 450001,Russia.
(5) Departments of Materials Science and Mechanical Engineering, University of Southern California, Los Angeles, CA 90089-1453, USA.

It is possible to produce metallic materials with submicrometer-grained (SMG) structures by imposing an intense plastic strain under quasi-hydrostatic pressure.[1] Studies using conventional transmission electron microscopy (CTEM) showed that many grain boundaries in the SMG structures appeared diffuse in nature with poorly defined transition zones between individual grains[2,3]. The implication of the CTEM observations is that the grain boundaries of the SMG structures are in a high energy state, having non-equilibrium character. It is anticipated that high-resolution electron microscopy (HREM) will serve to reveal a precise nature of the grain boundary structure in SMG materials. A recent study on nanocrystalline Ni and Ni_3Al showed lattice distortion and dilatations in the vicinity of the grain boundaries[4]. In this study, HREM observations are undertaken to examine the atomic structure of grain boundaries in an SMG Al-based Al-Mg alloy.

An Al-3%Mg solid solution alloy was subjected to torsion straining[5] to produce an equiaxed grain structure with an average grain size of ~0.09 μm. Thin specimens for transmission electron microscopy were prepared with a twin-jet polisher in a solution of $10\%HClO_4$, $20\%C_3H_8O_3$, and $70\%C_2H_5OH$. Atomic structure observations were conducted using a JEM-4000EX high-resolution electron microscope operating at 400 kV. Lattice images were taken at the optimum defocus condition. Since the grain sizes were much less than 1 μm, individual grains were not tilted during observation but sufficient grains oriented close to <110> could be found for lattice imaging.

Figure 1 shows a typical image of a grain in the SMG structure together with a selected area electron diffraction pattern. All boundaries of this grain are curved or wavy along their length. At the regions indicated by A and B, the grain boundary is corrugated. Lattice images of such regions are shown in Figs 2 (a) and (b), respectively. At region A, the grain boundary exhibits a periodic, stepwise arrangement of facets parallel to (100) and each facet consists of four or five layers of (111). There is also significant distortion or bending of lattice fringes near the grain boundary in the two-dimensional lattice image. Some lattice fringes are missing at the points marked T, suggesting the possible presence of dislocations. At region B, the grain boundary has a zigzag configuration but there appears to be no periodicity in the formation of facets and setps. These observations clearly indicate that the grain boundaries are in high-energy and non-equilibrium state. The facet densities measured in regions A and B are ~10^9 and ~ $2x10^8$ m^{-1}, respectively. It should be noted that the grain boundary in Fig.2(a) migrated towards the one-dimensional grain during observation. At the same time, there is an appreciable decrease in the number of facets and steps on the grain boundary and also in the distortion or bending of the lattice fringes near the grain boundary. It is concluded that the grain boundary migration, while induced by the high-energy electron irradiation, took place to reduce the high grain boundary energy.[6]

Proc. Microscopy and Microanalysis 1995, edited by G.W. Bailey, M.H. Ellisman, R.A. Hennigar, and N.J. Zaluzec
Copyright © 1995 MSA. Published by Jones and Begell Publishing, 79 Madison Ave., New York, NY 10016

References

1.N.A. Smirnova, V.I. Levit, V.I. Pilyugin, R.I. Kuznetsov, L.S. Davydova and V.A. Sazonova, *Fiz. Metall. Metalloved.* 61(1986)1170.
2.R.Z. Valiev, A.V. Korznikov and R.R. Mulyukov, *Mater. Sci. Eng.* A168(1993)141.
3.J. Wang, Z. Horita, M. Furukawa, M. Nemoto, R.Z. Valiev, N.K. Tsenev and T.G. Langdon, *J. Mater. Res.* 8(1993)2810.
4.R.Z. Valiev, *Metal Phys. Metallogr.* 9(1994)1340.
5.R.Z. Valiev, F.Chmelik, F.Bordeaux, G.Kapelski and B.Baudelet, *Scripta Metall.Mater.* 27(1982)855.
6.This work was supported by the Light Metal Educational Foundation of Japan, Monbusho International Scientific Research Program and the National Science Foundation (NSF) of the United States under Grants No. INT-9404693 and DMR-9115443. The Center for High Resolution Electron Microscopy at Arizona State University is supported by NSF Grant No. DMR-9314326.

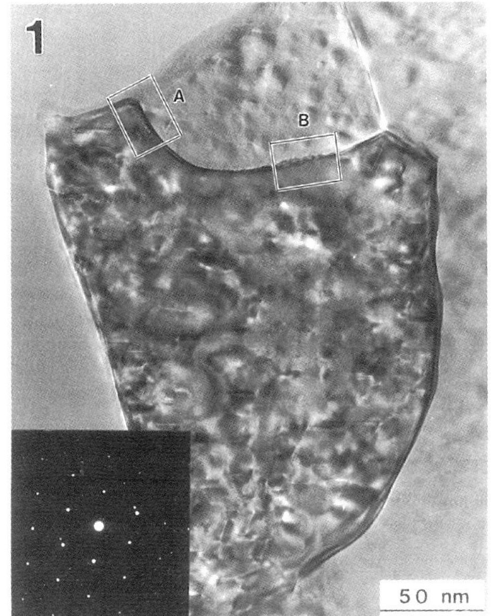

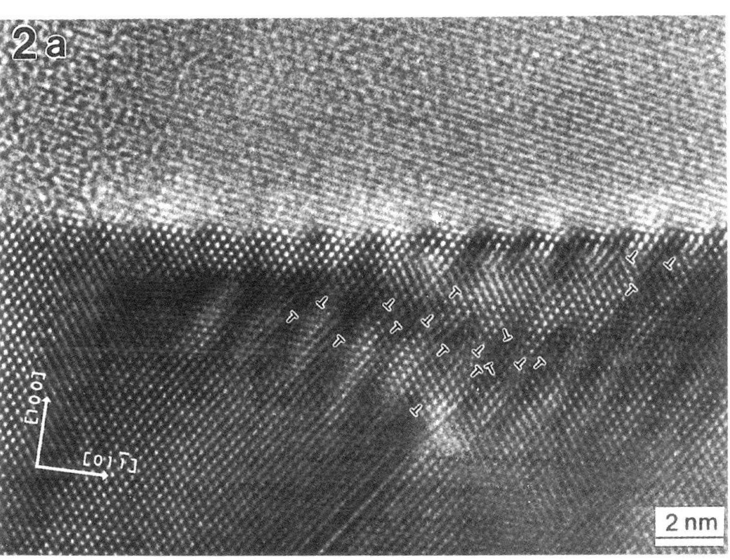

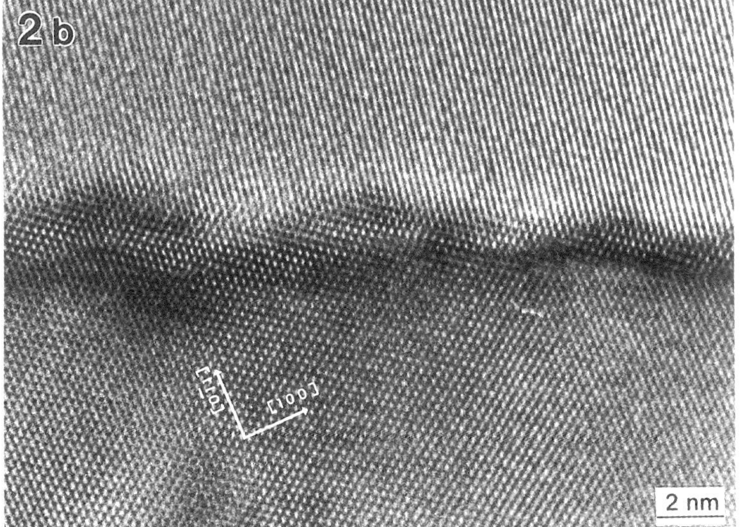

FIG.1.-Electron micrograph of typical grain in Al-3%Mg alloy. SAED pattern (inset) corresponds to [110] grain orientation.
FIG.2-(a) Enlargement of region A shown in FIG.1. Note boundary facetting and terminating lattice fringes labelled T. (b) Enlargement of region B shown in FIG.1.

IDENTIFICATION OF NON-MARTENSITIC TRANSFORMATION PRODUCTS (NMTP) IN A CARBURIZED 8620 STEEL

William T. Donlon and William E. Dowling

Ford Research Laboratories, Dearborn, MI 48121-2053

Additions of Cr and Mn to steel are intended to increase hardenability (the ability to form martensite on cooling from austenite) and thereby improve the mechanical properties. However, conventional gas carburization of alloy steels containing Cr and Mn as well as Si will produce oxides of these elements (internal oxidation) near the surface (due to the oxygen potential of the carburizing atmosphere).[1] Additionally, the oxidation of these elements locally reduces the hardenability of the alloy, creating the potential to produce non-martensitic transformation products (NMTP) near the surface. These oxides and the associated NMTP have been shown to decrease the fatigue resistance of carburized steels if they are not removed by subsequent machining operations.[2,3] The objective of this study is to characterize the oxides, identify the NMTP and assess the degree of local alloy depletion in a carburized 8620 steel (0.21 C, 0.93 Mn, 0.023 S, 0.12 Si, 0.49 Cr, 0.38 Ni and 0.17 Mo).

Flat samples and notched bars were carburized for 4 h at 1550°F (843°C) at a carbon potential of 0.85%. This yielded an effective case depth of ≈ 0.35 mm. The material was quenched to 450°F (232°C) in a salt bath for 4 min and subsequently air cooled to ambient to produce a martensitic structure. Figure 1 is a backscattered electron image (BEI) of the NMTP in cross section and shows its depth to be ≈ 10 μm. Samples of the NMTP were prepared for TEM examination by electropolishing from the back side using a solution of 62cc perchloric acid, 700cc ethanol, 100cc glycerin and 137cc of distilled water at 30V and 0.12A. Figure 2 shows the typical microstructure of this near surface region contains; (a) fine lamellar carbides, (b) oxides along prior austenite grains, and (c) larger (≈ 1 μm) oxides located in the interior as well as along prior austenite grain boundaries. Electron diffraction (ED) identifies the lamellae (FIG. 3) as Fe_3C, cementite (θ) displaying the expected orientation relationship $[001]_\theta | \, | [112]_{\alpha\text{-Fe}}$. Thus the NMTP is identified as pearlite with an interlamellar spacing of ≈ 100 nm. The precipitates denoted as (b) in Figure 2 are identified as amorphous silica (by ED and energy dispersive x ray (EDX) spectroscopy). The larger oxides, which sometimes exhibit a cuboidal morphology (FIG. 4a) are identified as $Mn_3Cr_3O_8$ (a=.845 nm Fd3m) by ED (FIG. 4b&c), however EDX analysis shows the Mn/Cr ratio to be 2/3. EDX analysis of the pearlite regions show that the levels of Cr, Mn and Si have been reduced to approximately 30% that of the bulk 8620 levels. Thus, the formation of the NMTP is due to the oxidation of Mn, Cr and Si from the austenite during carburization. This change in the near surface (10μm) composition of the steel increases the kinetics of the austenite to pearlite transformation, so that pearlite is formed in the alloy depleted zone following carburization, while the bulk remains austenite. The austenite transforms to martensite during subsequent cooling.

References

1. R. Chatterjee-Fisher, Met. Trans. A, **9A**, (1978) 1553.
2. T. Naito, H. Ueda and M. Kikuchi, Met. Trans. A, **15A**, (1984) 1431.
3. K. Namiki and K. Isokawa, Trans. ISIJ, **26** (1986) 642.

Proc. Microscopy and Microanalysis 1995, edited by G.W. Bailey, M.H. Ellisman, R.A. Hennigar, and N.J. Zaluzec

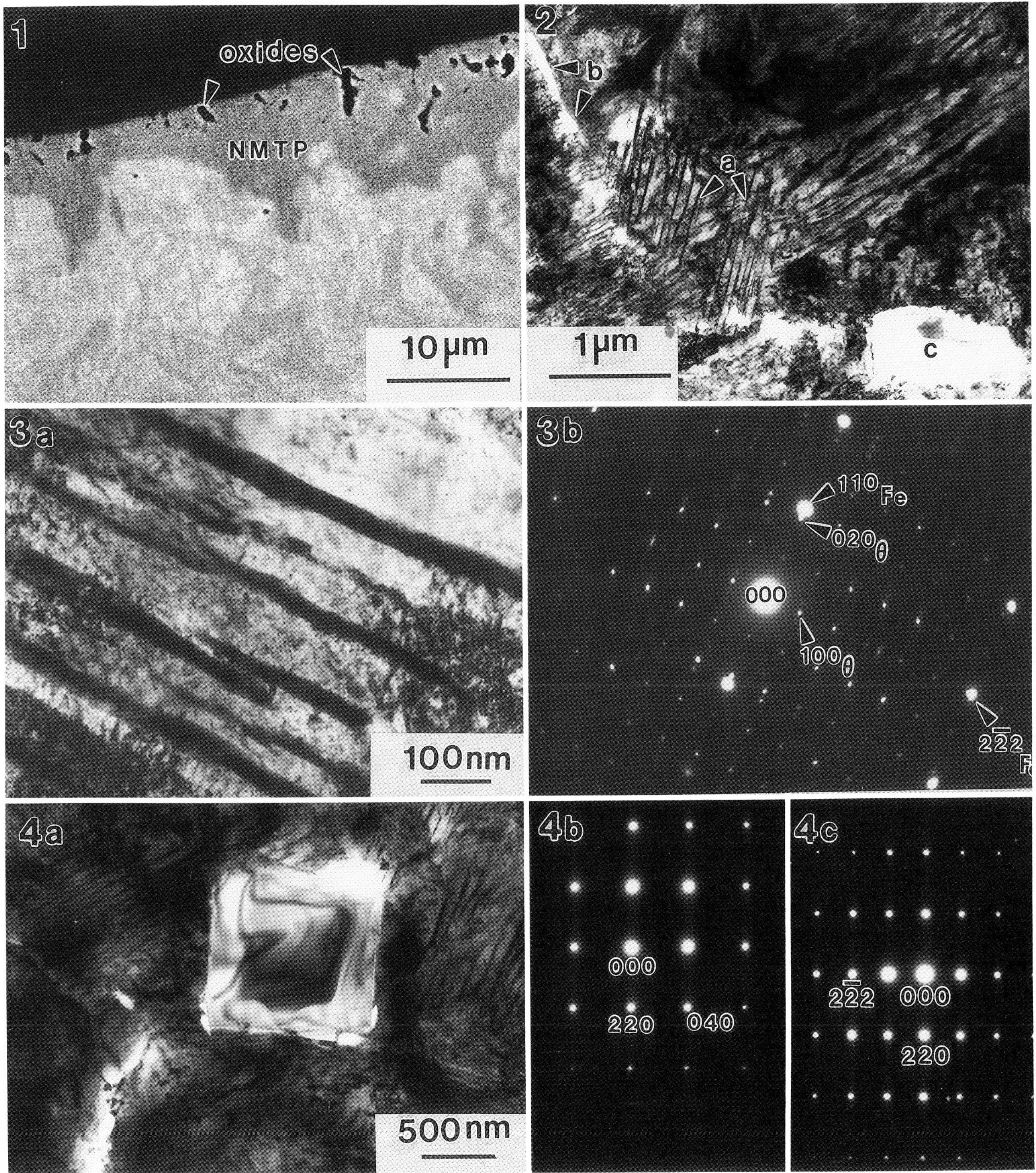

FIG. 1. Cross section of the surface of carburized 8620 steel showing surface oxides and NMTP.
FIG. 2. Low magnification TEM image of NMTP showing; (a) lamellar carbides, (b) oxides on prior austenite grain boundaries, and (c) larger oxides.
FIG. 3. (a) TEM image of lamellar precipitates; (b) diffraction pattern identifying lamellae as cementite with $[001]_\theta$ parallel to $[\bar{1}12]_{\alpha Fe}$ and $(010)_\theta$ parallel to $(110)_{\alpha Fe}$.
FIG. 4. (a) TEM image of cuboidal Mn-Cr oxide; (b) & (c) [001] and $[\bar{1}12]$ diffraction patterns identifying Mn-Cr oxide as Fd3M with a=0.845nm.

THE NUCLEATION OF RECRYSTALLIZATION AT SECOND PHASE PARTICLES IN 7000 SERIES DEFORMED ALUMINUM ALLOY

B. Hidalgo-Prada, J.J. Guilarte and S.Y. Paredes

Instituto de Invest. en Biomedicina y Ciencias Aplicadas
Universidad de Oriente, Cumaná, Edo. Sucre, Venezuela

An investigation was carried out to determine the effect of a thermomechanical process (TMP) in the precipitation of dispersed second phase particles, and to assess the recrystallization behavior of a deformed ALCANVEN-7475 commercial aluminum alloy. The TMP included solution anneal treatment at Ts = 482 °C for 3 hours, overaging at temperatures (To) in the range from 160 - 400 °C for 1 to 30 hours, cold and warm deformation up to 90% reduction of thickness (RA) and recrystallization at Tr = 482 °C for 15 to 60 minutes [1, 2].

A Hitachi H-600 TEM operated at 100 KV was utilized to examine the microstructural evolution of the alloy leading to the nucleation of recrystallization as a consecuence of the previously described four steps TMP.

It was determined that the nucleation of recrystallized grains is promoted by the presence of second phase particles with critical sizes between 0.5 and 1.0 μm . The cold and warm rolling steps induced zones of high deformation around the larger second phase particles. This substructure, shown in Figures 1a and 1b, consisted of elongated cells, typical of Al alloys with a high percentage of mechanical work [3].

The results of the microstructural analysis also show evidence of a close relationship between the density and distribution of precipitates and the size of the recrystallized grain. While the critical size particles act as effective promoters for the nucleation new grains [Figures 2a and 2b], the dispersoids of smaller size (less than 0.5 μm) become barriers that prevent grain boundary migration of the recrystallized grains [Figure 3]. The process results in a stabilized grain size small enough for mechanical behavior of enhanced ductility of the high strength 7475 Al alloy.

References

1. J.A. Wert et al., Met. Trans., 12A(1981)1267.
2. J.F. Humphreys, in R.W. Cahn, P. Hansen, and E.J. Kramer, Eds., Materials Science and Technology, VCH Publisher Inc. 15(1991)373.
3. H.S. Yang et al., J. of Mat. Science 27(1992)2515.

Proc. Microscopy and Microanalysis 1995, edited by G.W. Bailey, M.H. Ellisman, R.A. Hennigar, and N.J. Zaluzec
Copyright © 1995 MSA. Published by Jones and Begell Publishing, 79 Madison Ave., New York, NY 10016

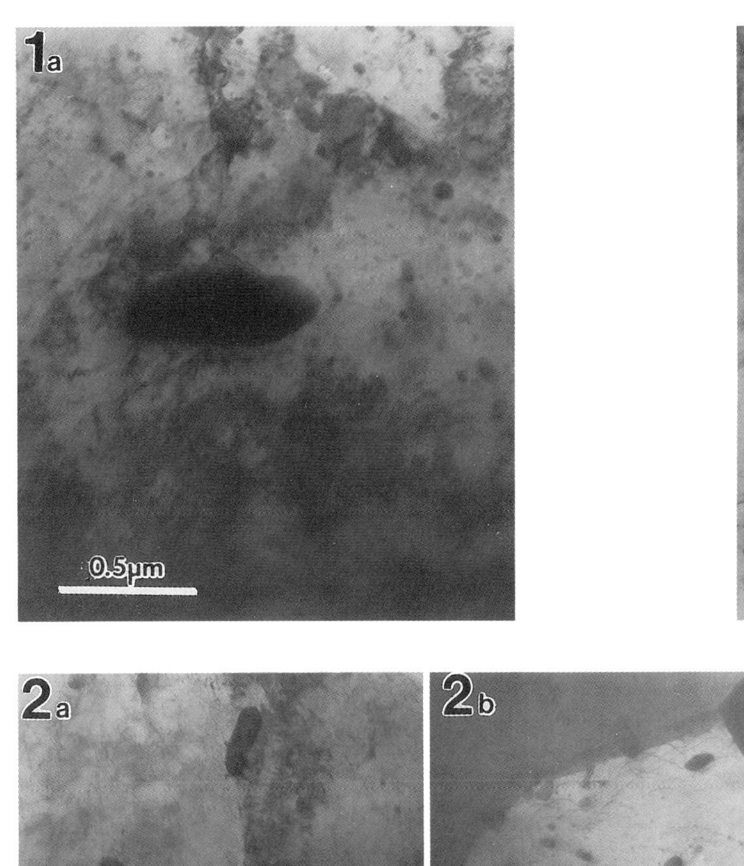

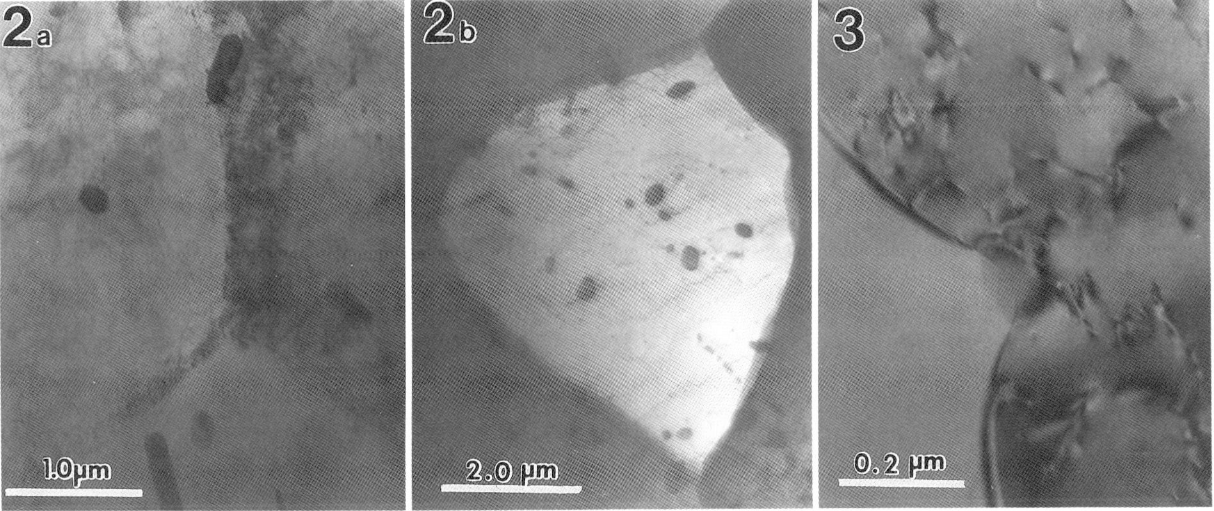

Fig. 1. Deformed area around critical size particles. Formation of cell substructure. Ts = 482 °C for 3 hours; overaging To = 400 °C for 8 hours; Cold rolling and 90% RA (Fig. 1a); Hot rolling at Tr = 220 °C and 90% RA (Fig. 1b).

Fig. 2. Solution treatment, overaging and rolling conditions as above. Recritallization-Solution treatment at Tr = 482 °C for 30 minutes. After cold rolling (Fig. 2a) and hot rolling (Fig. 1b).

Fig. 3. Small dispersoids pinning the grain boundary of the recrystallized grains.

STRUCTURAL DETERMINATION OF L-PHASE WITH AuCu(I) ORDERED STRUCTURE OBSERVED IN A NEW MARAGING STEEL

Ping Liu

Department of Physical Metallurgy, Research and Development Centre
AB Sandvik Steel S-811 81 Sandviken Sweden

A new maraging stainless steel has recently been developed at Sandvik and is termed Sandvik 1RK91 (12Cr-9Ni-4Mo-2Cu-1Ti-0.3Al-0.2Si (wt%)). The anomalous tempering response after aging at 475°C, combined with high strength, ductility, corrosion resistance and a high resistance to overaging were attributed to the formation of a quasicrystalline phase (1). In addition to this an ordered phase designated L-phase precipitated during prolonged aging. The structure type of this phase was determined by electron diffraction combined with EDX analysis. The material was annealed in the temperature range 1050-1150°C with subsequent air cooling and then aged at 475°C for up to 100 hours. Thin foils and carbon replica were studied using a Jeol 2000 FX analytical transmission electron microscope (TEM/STEM) equipped with a Link AN 10000 EDX system operating at 200 kV.

The plate-like L-phase precipitated after prolonged aging at 475°C(100 h) as shown in Fig.1(a). A typical EDX spectrum from L-phase is shown in Fig.1(b). From a number of chemical analysis by EDX L-phase can be expressed as $Ni_{48}Fe_9Cr_5Mo_{10}Ti_{19}Al_9$ or $Ni_{48}M_{52}$, where M=Ti, Al, Mo, Fe, and Cr. A series of tilting experiments of the L-phase crystal gave rise to a number of convergent beam electron diffraction patterns(CBED), four of which are shown in Figs.2(a) to 2(d). From the strong diffraction spots an fcc structure can be recognised and the patterns therefore identified as the <001>, <100>, <111>, and <110> zone axes respectively. However, the weaker spots reduce the symmetries for the <100>(Fig.2(b)) and <111> (Fig.2(c)) patterns from four-fold and six-fold respectively to two-fold while the four-fold symmetry in the <001> zone remains. This reveals that the L-phase has a primary tetragonal structure with point group of 4/mmm. This structure is therefore not any of the ordered structures with primary cubic lattice. Furthermore, in the pattern of <001> (Fig.2(a)) the reflections of {110} appear while in the <100> zone axis the {001} diffraction is present (Fig.2(b)), from which the diffraction condition of h+k= 2n can be deduced. The measurement of diffraction {200} and {002} revealed that the c-axis is longer than the a-axis. The lattice parameters were then calculated as a=3.52 Å and c=3.63 Å. This structure is therefore, isomorphous to FeNi, which has an ordered fcc phase with $L1_0$ structure type AuCu(I); space group P4/mmm and lattice parameters of a=3.579 Å and c=3.579 (2). Fe and Ni atoms occupy alternate (002) planes in FeNi. Assuming that Ni and M occupy alternate (002) plane in L-phase the intensities of electron reflections can be calculated and results are shown in Table1 together with the experimental ones. It can be seen that the calculated intensities match the experimental ones fairly well. Hence the L-phase derived from an fcc(A1) to $L1_0$ (AuCu(I)) can be schematically described as in Fig.3.

REFERENCES

1.   P. Liu, A. Hultin Stigenberg and J.-O. Nilsson , Scripta. Metall. et Mater 31(1994)249-254 .

2.   Pearson's Handbook of Crystallographic Data for Intermetallic Phases, P.Villars and L.D.Calvert, (1985) ASM, Vol.1, pp.416-417; Vol.3, p.2203.

3.   The continuous support of Ms A.Hultin Stigenberg and Dr J.-O. Nilsson are gratefully acknowledged as well as technical assistance by Mr R. Ekrud .

Table.1 The calculated (Ical.) and experimental intensities (Iexp.) for the electron diffraction.

hkl	001	110	111	002	200	112	202	220	003	221	113	222	203	004	400	401
d(Å)	3.63	2.49	205	1.82	1.76	1.47	1.26	1.24	1.21	1.18	1.09	1.03	1.00	.091	0.88	0.86
I exp	0.9	1.4	100	31	30	0.2	13	9	0.9	0.8	32	5	0.2	6	15	0.2
I cal.	0.2	0.2	100	16	30	0.1	17	8	0.1	0.1	11	9	0.1	2	4	0.2

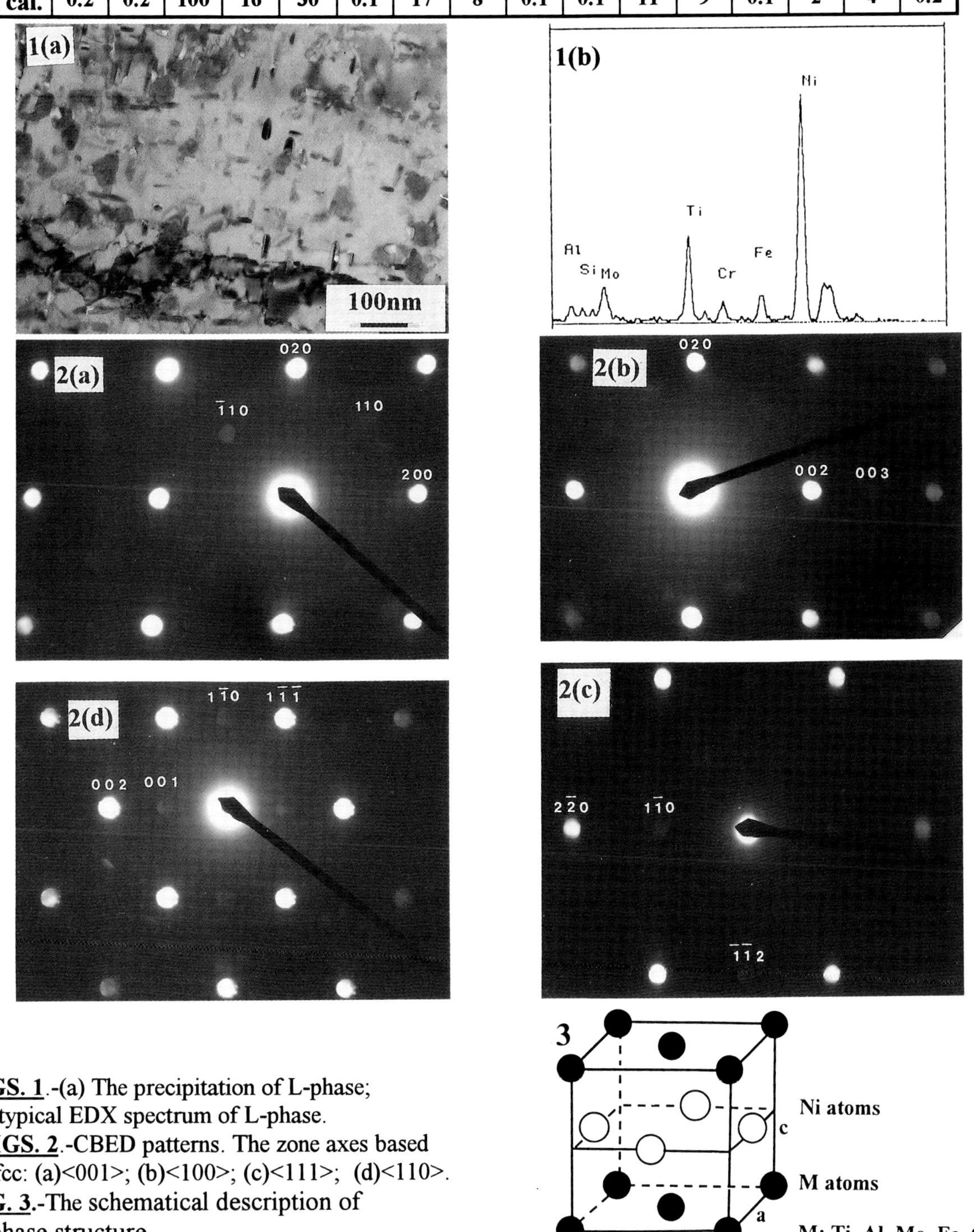

FIGS. 1.-(a) The precipitation of L-phase;
(b) typical EDX spectrum of L-phase.
  FIGS. 2.-CBED patterns. The zone axes based
on fcc: (a)<001>; (b)<100>; (c)<111>; (d)<110>.
FIG. 3.-The schematical description of
L-phase structure.

PRECIPITATION OF NiHfSi PHASE IN NiAl SINGLE CRYSTALS CONTAINING Hf

A. Garg, R. D. Noebe, and R. Darolia*

NASA Lewis Research Center, Cleveland, OH 44135
*General Electric Aircraft Engines, Cincinnati, OH 45215

Small additions of Hf to NiAl produce a significant increase in the high-temperature strength of single crystals[1]. Hf has a very limited solubility in NiAl and in the presence of Si, results in a high density of G-phase ($Ni_{16}Hf_6Si_7$) cuboidal precipitates and some G-platelets in a NiAl matrix[2]. These precipitates have a F.C.C structure and nucleate on $\{100\}_{NiAl}$ planes with almost perfect coherency and a cube-on-cube orientation-relationship (O.R.). However, G-phase is metastable and after prolonged aging at high temperature dissolves at the expense of a more stable Heusler (β'-Ni_2AlHf) phase. In addition to these two phases, a third phase was shown to be present in a NiAl-0.3at. % Hf alloy, but was not previously identified (Fig. 4 of ref. 2). In this work, we report the morphology, crystal-structure, O.R., and stability of this unknown phase, which were determined using conventional and analytical transmission electron microscopy (TEM).

Single crystals of NiAl containing 0.5at. % Hf were grown by a Bridgman technique. Chemical analysis indicated that these crystals also contained Si, which was not an intentional alloying addition but was picked up from the shell mold during directional solidification. These ingots were homogenized for 50h at 1590K and then furnace-cooled at an approximate rate of 10K/min in Ar. Samples for TEM were prepared by electropolishing[1,2] and microstructural characterization and crystallographic analysis were conducted in a Philips 400T TEM operating at 120kV.

A bright-field image close to a $<001>_{NiAl}$ zone-axis is shown in Fig. 1. The microstructure consisted of two types of precipitates which were distinctly different in size, morphology, and density. The high density of fine cuboidal precipitates (size ~ 10-50nm) with $\{001\}_{NiAl}$ habit-plane (H.P.) were identified as the G-phase precipitates. The second type of precipitates had a noticeably lower density but their size was ~10 times larger than that of the G-phase. EDXS showed these precipitates to be rich in Hf and Ni and to contain some Si. In order to identify the crystal structure of these larger precipitates, several low-index microdiffraction (MBED) patterns were recorded from individual precipitates. These patterns could be indexed unambiguously as arising from the orthorhombic NiHfSi phase with unit-cell parameters $\mathbf{a} = 0.64$nm, $\mathbf{b} = 0.39$nm and $\mathbf{c} = 0.72$nm and having the space group $Pnma$[3]. Trace analysis indicated that these precipitates have a $\{111\}_{NiAl}$ H.P. When viewed along a $<110>_{NiAl}$ zone-axis, two sets of edge-on NiHfSi precipitate plates on the $\{111\}_{NiAl}$ planes parallel to the beam and two other sets inclined 35° w.r.t. the beam were clearly visible (Fig. 2). In a $<111>_{NiAl}$ zone-axis, the NiHfSi precipitate plates had a rectangular morphology with sides parallel to $<110>_{NiAl}$ and $<112>_{NiAl}$ directions. MBED from these plates showed superimposed precipitate and matrix spots which suggested an O.R. given by $(100)_{NiHfSi} \parallel (111)_{NiAl}$, $[010]_{NiHfSi} \parallel [10\bar{1}]_{NiAl}$, $[001]_{NiHfSi} \parallel [12\bar{1}]_{NiAl}$ (Fig. 3). Since there are three equivalent $<110>$ directions in a $\{111\}_{NiAl}$ plane, there are 3 variants of NiHfSi precipitate in each $\{111\}_{NiAl}$ plane (as shown in Fig. 3) for a total of 12 variants possible in the NiAl matrix. Common symmetry elements (2/m) and symmetry considerations of the precipitate and NiAl lattices also predict a rectangular morphology and a total of 12 variants for the NiHfSi phase. In each variant, preferential growth was along the $[001]_{NiHfSi} \parallel <112>_{NiAl}$ direction, along which the misfit was least (0.6%) and the corresponding planes $(001)_{NiHfSi}$ and the $\{112\}_{NiAl}$ were exactly parallel. Aging experiments indicated that NiHfSi phase is not a stable phase in this particular alloy and on aging at 1300K for 10h, it begins to dissolve in favor of the β' phase (Fig. 4).

References

1. I.E. Locci, R. Dickerson, R.R. Bowman, J.D. Whittenberger, M.V. Nathal, and R. Darolia, *MRS Symposium Proceedings* 288(1993)685.

Proc. Microscopy and Microanalysis 1995, edited by G.W. Bailey, M.H. Ellisman, R.A. Hennigar, and N.J. Zaluzec
Copyright © 1995 MSA. Published by Jones and Begell Publishing, 79 Madison Ave., New York, NY 10016

2. I.E. Locci, R.D. Noebe, R.R. Bowman, R.V. Miner, M.V. Nathal, and R. Darolia, *MRS Symposium Proceedings* 213(1991)1013.
3. P. Villarsm and L.D. Calvert , *Pearson's Handbook of Crystallographic Data for Intermetallic Phases* (1991)3895.

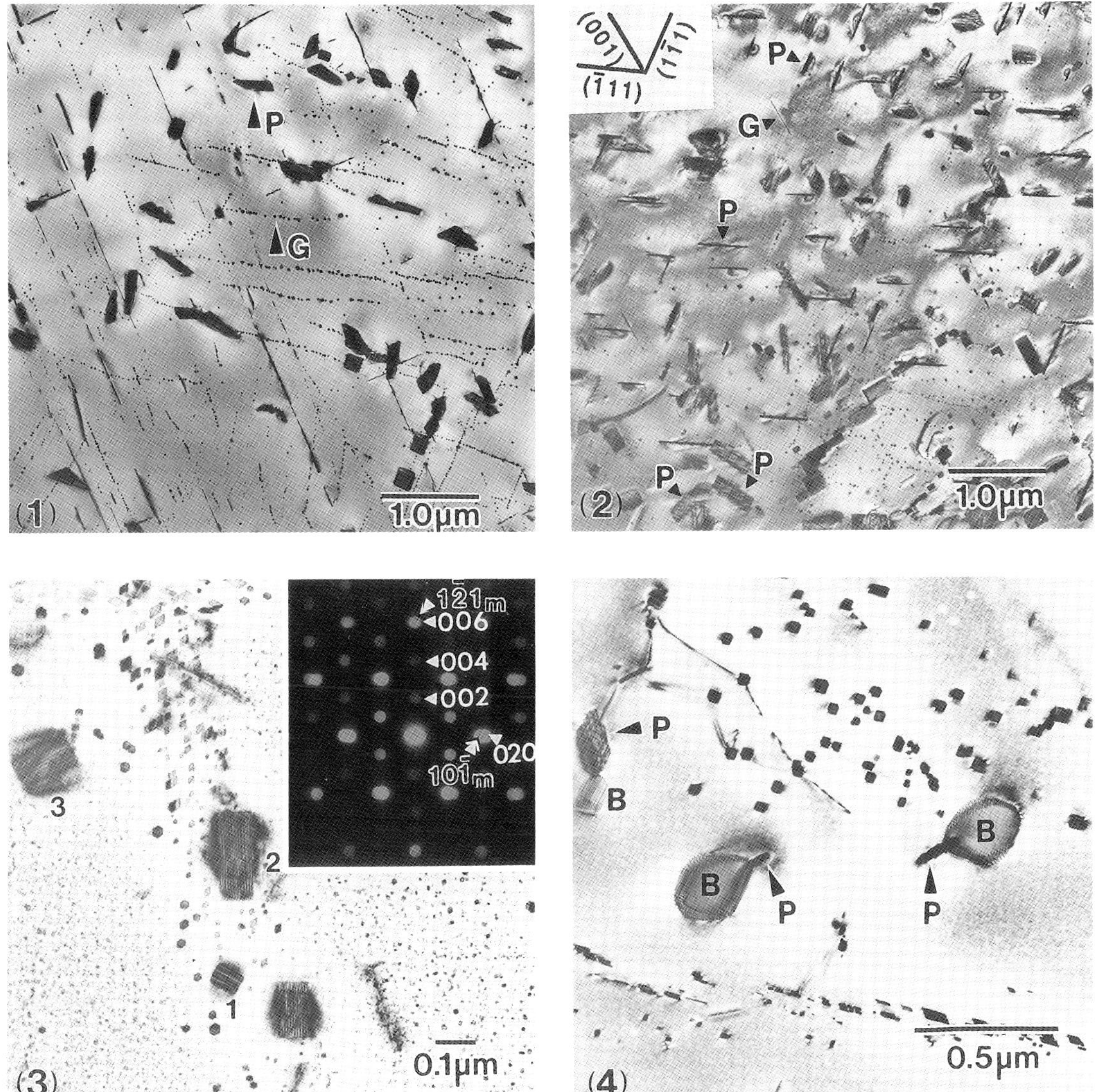

Fig. 1: A bright-field TEM image close to a <001>$_{NiAl}$ zone-axis, showing precipitation of fine Ni$_{16}$Hf$_6$Si$_7$ (marked G) and relatively coarse NiHfSi (marked P) precipitates in a homogenized and furnace-cooled sample.

Fig. 2: A bright-field TEM image close to the [110]$_{NiAl}$ zone-axis showing formation of rectangular plates of NiHfSi phase formed on {111}$_{NiAl}$ planes.

Fig. 3: Transmission electron micrograph showing three variants of NiHfSi phase on a {111}$_{NiAl}$ plane. MBED along the [100]$_{NiHfSi}$ ‖ [111]$_{NiAl}$ direction from plate 2 shows the O.R. between the NiHfSi phase and the NiAl matrix.

Fig. 4: Transmission electron micrograph showing the nucleation and growth of β'-precipitates (marked B) on the original NiHfSi plates in a sample aged at 1300K for 10h. Zone-axis ~ <110>$_{NiAl}$.

STRUCTURAL EVOLUTION OF THE AGING MARTENSITE IN DUPLEX STEEL

Hong-Yuan Chu*, Fu-Rong Chen[+], and Tai-Bor Wu*

* Department of Materials Science and Engineering. National Tsing Hua University, Hsinchu, Taiwan.
[+]Materials Science Center. National Tsing Hua University, Hsinchu, Taiwan.

In the past, the structure and crystallography of needle-like martensite in the as-quenched Fe-Al-Mn-C alloy were studied (1-4). The structure of the needle-like martensite was identified to be a structure between $18R(5\bar{1})_3$ and $18R(4\bar{2})_3$. The orientation relationship between the BCC matrix and martensite phase was determined to be $[\bar{1}1\bar{1}]_{BCC}$ // $[1\bar{1}0]_m$ and $(110)_{BCC}$ // $(\bar{1}\bar{1}8)_m$ (4). The shape strain, habit plane and other crystallography of martensite were also determined (4). Comparison with the as-quenched Fe-Mn-Al-C martensite, there is only little information about the aging behavior of the martensite phase. The purpose of this work is to study the structural evolution of the martensite phase after aged at high temperature.

The wrought Fe-29.4-Mn-8.03Al-0.3C alloy was heated at $1300°C$ for 1 hour in argon atmosphere, and followed by a quenched into water. Aging process were performed at $400°C$, $500°C$ and $600°C$, respectively, for 1 hour in salt bath, and the specimens were then quenched into water. TEM specimens were examined with a JOEL-200CX STEM.

The diffraction patterns of as-quenched and after aged are shown in Fig.1~4. During aging, the structure of the needle-like martensite changes from 18R to FCC structure. The dislocation mechanism is proposed to explain the structure transition. The structure of as-quenched martensite is close to $18R(4\bar{2})_3$. As the martensite is aged at higher temperature, the $18R(4\bar{2})_3$ will be transformed to $18R(5\bar{1})_3$ and the FCC structures by introducing pairs of negative partial dislocations with the Burgers vector equal to negative direction of the lattice invariant shear, as shown in Fig. 5. The negative partial dislocations are responsible of relaxation of shape strain and the condition of undistorted habit plane. The negative partial dislocation will cause the vanishment of stacking fault layer as shown in Fig. 6, and the vanishment of stacking layer results in the structure transition from 18R to FCC. After aged at 400 °C and 500°C for 1 h, the orientation relationship between BCC matrix and martensite phase remains unchanged, which is $[\bar{1}1\bar{1}]_{BCC}$ // $[1\bar{1}0]_m$ and $(110)_{BCC}$ // $(\bar{1}\bar{1}8)_m$. The crystallography of BCC and martensite remains the same as that of as-quenched martensite. After martensite aged at $600°C$ for 1 h, 18R martensite is transformed to FCC, the orientation relationship between BCC matrix and FCC structure is determined to be $[\bar{1}1\bar{1}]_{BCC}$ // $[\bar{1}01]_{FCC}$ and $(\bar{1}12)_{BCC}$ 2° away $(202)_{FCC}$.

References

1. W. S. Yang, T. B. Wu, and C. M. Wan, Scripta Metall., 24, 895 (1990).
2. K. H. Hwang, C. M. Wan and J. G. Byrne, Mater. Sci. Eng., A132, 895 (1991).
3. K. H. Hwang, W. S. Yang, T. B. Wu, and C. M. Wan and J. G. Byrne, Acta Metall., 39 825 (1991).
4. W. B. Lee, F. R. Chen, S. K. Chen and C. M. Wan, Accepted by Acta Metall. (1994).

Proc. Microscopy and Microanalysis 1995, edited by G.W. Bailey, M.H. Ellisman, R.A. Hennigar, and N.J. Zaluzec
Copyright © 1995 MSA. Published by Jones and Begell Publishing, 79 Madison Ave., New York, NY 10016

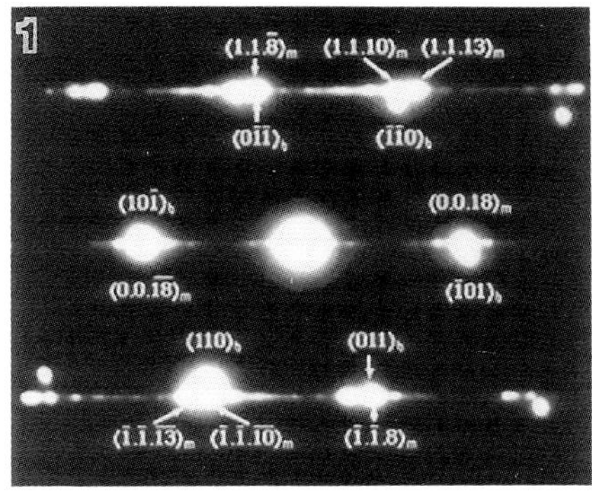
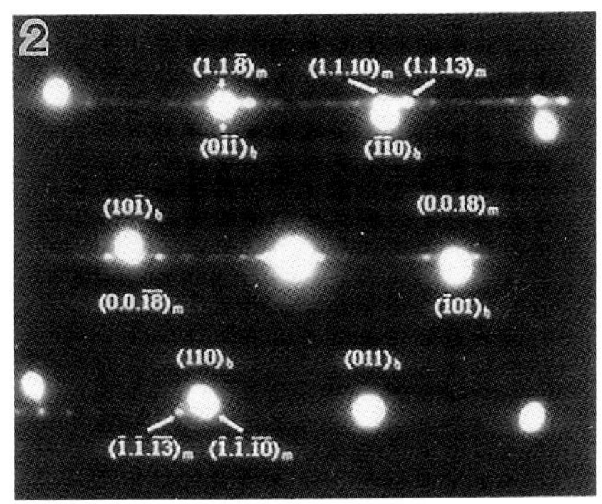
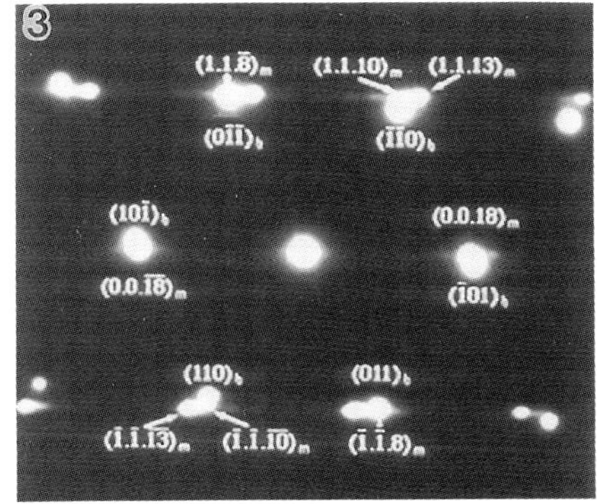
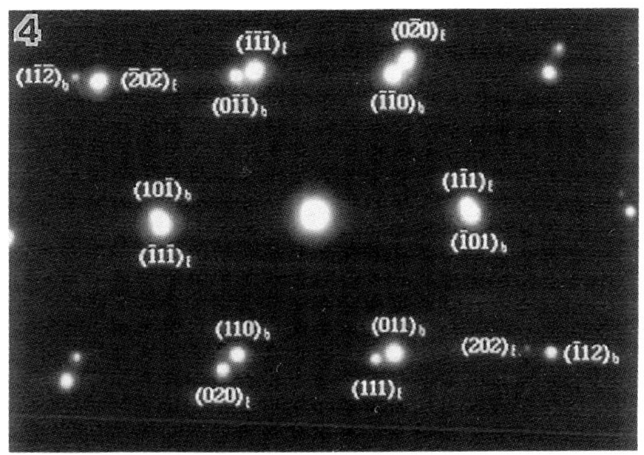
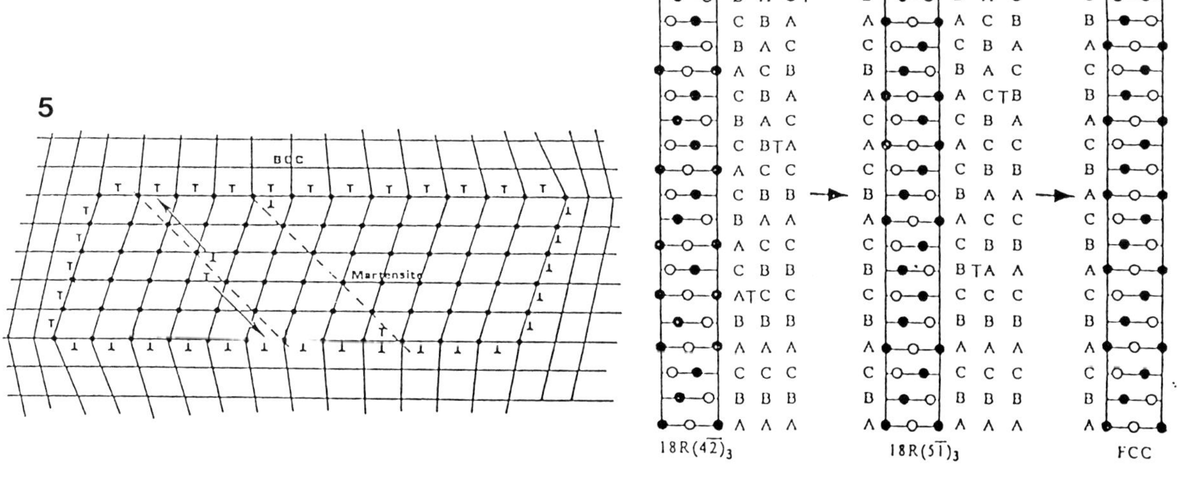

Fig. 1 ~ 4. Diffraction patterns of 1. as-quenched, 2. 400°C. 3. 500°C. 4. 600°C

Fig. 5 (a) The structure change from $18R(4\bar{2})_3 \rightarrow 18R(5\bar{1})_3 \rightarrow$ FCC by the insert of dislocations.

Fig. 6 Illustration of structure transformation by the dislocation mechanism.

A Study of the Dislocation Core Structure in High Nitrogen Austenitic Stainless Steel By Weak Beam Electron Microscopy

M. Zhou,[1] T.F. Kelly,[1] and J.E. Flinn[2]

1. Department of Materials Science and Engineering, University of Wisconsin-Madison, 1500 Johnson Dr., Madison, WI 53706
2. Idaho National Engineering Lab, Idaho Fall, ID 83415

The large strengthening effect of nitrogen interstitials and its strong influence on the dislocation distribution of Fe-Cr-Ni austenite have been rationalized by Owen et al.[1] using a short-range-order-based model. However, the model cannot explain the high population of screw dislocations resulted from a small amount of plastic deformation. Grujicic[2] suggested that nitrogen addition to Fe-Cr-Ni austenitic stainless steels causes spreading of the screw dislocation core into two or more non-parallel planes which results in low dislocation mobility. Higher stress and higher temperature are thus required to overcome the energy barrier for the movement of dislocations. The core structure of dislocations in nitrogen-containing stainless steels is of fundamental importance for understanding the strengthening mechanism of this material. The weak beam dark field imaging technique allows the direct observation of partial dislocations and stacking faults bonded by the partials. When the optimum imaging condition is satisfied, relatively intense, narrow images of less than 2 nm wide can be obtained. Because the intensity peak of a dislocation imaged in this way is closely related to the local strain field, the interpretation of the contrast produced by the dislocation is also relatively straightforward.

The specimens used in this study were prepared from the consolidated powders of rapid solidification processed type 304 stainless steels. The nitrogen concentration in the specimens ranges from 0.032wt% to 0.223wt%. Room temperature tensile tests were performed on these specimens and the tests were interrupted at 1%, 3%, 5%, and 10% strains to achieve various degree of plastic deformation. It is assumed that the dislocations to be discussed here have primary slip systems of {111}/<110> and the dislocation dissociation interaction is of the type a/2<110>=a/6<112>+a/6<112>. For every dislocation analyzed by the weak beam, all three <220> $\mathbf{g}$ vectors corresponding to an arbitrary {111} beam direction were used. The diffraction condition for the imaging was set to give s_g>0.2nm^{-1} using the Kikuchi pattern. Calculated images by Cockayne[3] established that this value of s_g is large enough to obtain partial separation from the weak beam images. The dislocations shown in Fig.1 were taken from a low nitrogen sample. A combination of $\mathbf{g.b}$ and trace analysis show that the Burger's vector of the total dislocation is [1$\bar{1}$0] and the line direction also lies closely to [1$\bar{1}$0]. When $\mathbf{g.b}$=2 is satisfied, both Shockly partials are in contrast as can be seen in Fig.1a. When the other two <220> vectors were used, which corresponds to $\mathbf{g.b}$=1, one partial dislocation is visible. The visibility of the stacking fault depends on the value of $\mathbf{g.R}$, where $\mathbf{R}$ is the displacement vector of the fault. Since $\mathbf{g.R}$ is an integer in this case, the stacking fault is invisible. As nitrogen content increases, the contrast produced by dislocations and stacking faults appears to be more complicated and less straightforward to interpret. Fig.2 gives an example of a dissociated dislocation in the high nitrogen specimen. Only one partial dislocation is visible despite the fact that $\mathbf{g.b}$=2 is satisfied. This leads us to believe that some dissociation mechanism other than a/2<110>=a/6<112>+a/6<112> might have occurred. By referring to the Thompson tetrahedra, the contrast obtained in Fig.2 is consistent with the dislocation dissociated into both the basal {111} plane and one of the side {111} planes. This is also in good agreement with the dissociation reaction of the type a/2<110>=a/6<112>+a/6<112>+a/6<110> proposed by Grujicic.

1. W.S.Owen, Proc. Third Int. Conf. On High Nitrogen Steels, HNS-93, Kiev, 1993
2. M.Grujicic, Materials Science and Engineering, A183, 1994, pp223
3. D.J.H.Cockayne, Diffraction and Imaging Techniques in Material Science, vol.1, North-Holland, 1978,pp153
4. Support for this work from the U.S. Bureau of Mines is gratefully acknowledged.

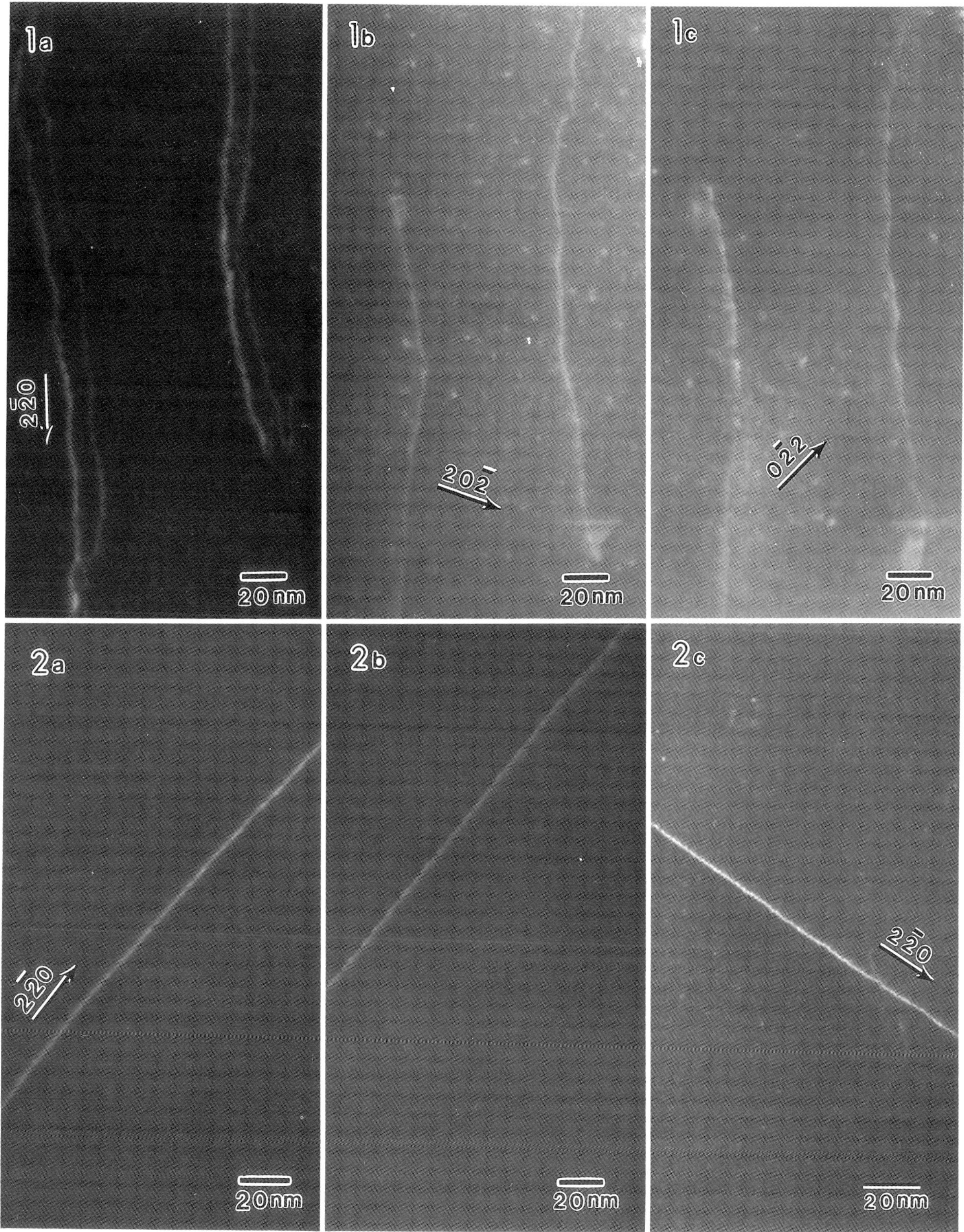

Fig.1. WBDF image of dislocation for 0.032wt% nitrogen, **B**=[111], **g**=<220>
Fig.2. WBDF image of dislocation for 0.223wt% nitrogen, **B**=[111], **g**=<220>

EELS CHARACTERIZATION OF "SMUT" LAYER FILMS ON STEEL SURFACE PREPARED FOR CHROME PLATING BY ANODIC ETCHING.

E. Y. Wang,* J. T. Cherian,** A. Madsen,* and R. M. Fisher*

*Department of Materials Science and Engineering, University of Washington, Seattle, WA 98195
**Boeing Material Technology, Boeing Commercial Airplane Group, Seattle, WA 98124

Many steel parts are electro-plated with chromium to protect them against corrosion and to improve their wear-resistance. Good adhesion of the chrome plate to the steel surface, which is essential for long term durability of the part, is extremely dependent on surface preparation prior to plating. Recently, McDonnell Douglas developed a new pre-treatment method for chrome plating in which the steel is anodically etched in a sulfuric acid and hydrofluoric acid solution.[1] On carbon steel surfaces, this anodic pre-treatment produces a dark, loosely adhering material that is commonly called the "smut" layer. On stainless steels and nickel alloys, the surface is only darkened by the anodic pre-treatment and little residue is produced. Anodic pre-treatment prior to hard chrome plating results in much better adherence to both carbon and alloy steels.

We have characterized the anodic pre-treated steel surface and the resulting "smut" layer using various techniques including electron spectroscopy for chemical analysis (ESCA) on bulk samples and transmission electron microscopy (TEM) and electron energy-loss spectroscopy (EELS) on stripped films. Samples of the surface films on steels pre-treated for plating by anodic etching were prepared using 2 -10 % bromine-methanol solutions which will etch the steel away without attacking the surface oxide film. The pieces of film that floated off were transferred to rinse in clean methanol and caught on copper grids. Carbon coating and freeze-drying helped to prevent the films from breaking due to surface tension effects during drying but sufficient areas for analysis could be located if these methods were not used. Thermal oxide films were also prepared from the same steel using the same stripping technique for comparison purposes.

TEM observations of films stripped from anodically treated 4340 steel samples showed that they were covered with small carbide particles that were released from the steel by the anodic etch and adhered loosely as a smut layer on the surface. TEM of anodic and thermal oxide films stripped from stainless steel showed the presence of many fewer and much smaller carbide particles. EELS spectra from areas well away from particles showed that the major elements in both anodic (Fig. 1a) and thermal films (Fig. 1b & 1c).are Cr, Fe, and O with faint traces of O and Ni. The film thicknesses are about 100Å-200Å.[2] Similar results were found for Inconel alloy samples (Fig. 2a, b & c). Quantitative interpretation of the EELS spectra revealed that the oxygen/metal ratio in the anodic "smut" layers in both stainless steel and Inconel is about 2.1 as in CrO_2 in compareison to 1.5 in Cr_2O_3 thermal oxide films. EELS line-shape analysis of the oxygen and chromium peaks also support the above identification of the anodic and thermal oxide films.[3] ESCA analysis of the surface of anodically-treated 4340 M steel showed similar high oxygen values.

The EELS data discussed above indicates that the anodically-produced surface film has "excess" oxygens which are available to form a very strong bond with positively-charged chromium ions present in the plating bath as polymeric chains.[4] This analysis indicates that the anodic film bonds firmly to both the steel substrate and the electroplated chromium resulting in very good adherence.

References

1. V. J. Vinson, Boeing Specification, BAC 5709, 6 January, 1992.
2. R. F. Egerton, *Electron Energy-Loss Spectroscopy*, New York: Plenum (1986).
3. Manoubi T. et al, Microsc.Microanal. Microstuct. 1, 23-39, 1990.
4. J. P. Hoare, *Plating and Surface Finishing* (Sept. 1989)46.

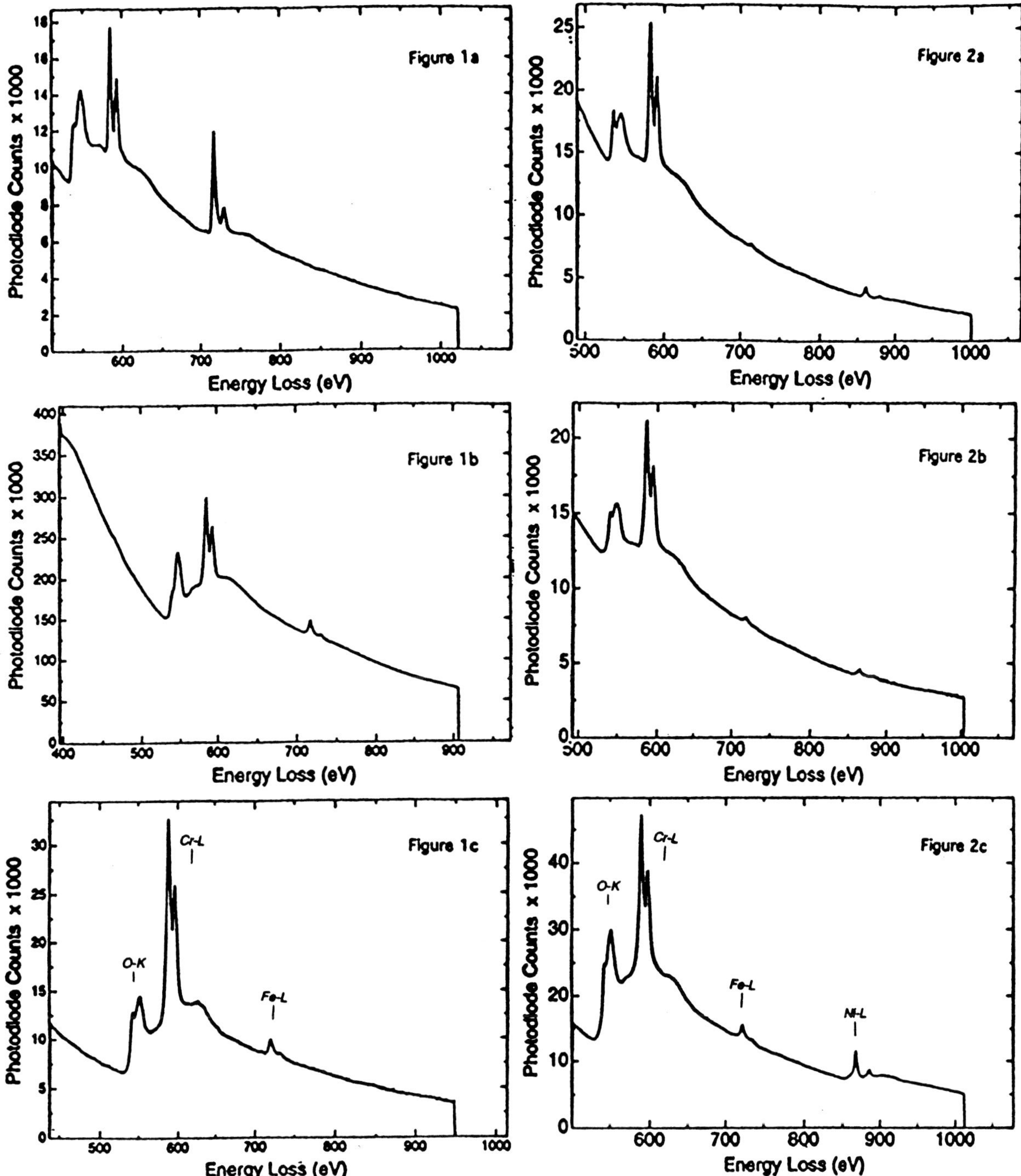

Figure 1 EELS Spectra of Stainless Steel

Fig. 1a, "Smut" layer from anodic etching
    b, Thermal oxide film (200c 15 minutes)
    c, Thermal oxide film (200c 2hours)

Figure 2 EELS Spectra of Inconel Alloy

Fig. 2a, "Smut" layer from anodic etching
    b, Thermal oxide film (200c 41 hours)
    c, Thermal oxide film (300c 25 hours)

TEMPERATURE RAMP DEPENDENT REACTION KINETICS OF INTERNAL TIN PROCESSED Nb₃Sn WIRES

R. L. Sabatini, Yimei Zhu, and M. Suenaga

Materials Science Division, Department of Applied Science, Brookhaven National Laboratory, Upton, NY 11973-5000 USA

Multifilamentary Nb_3Sn wires will be a critical component of a magnetic fusion reactor, which requires a high magnetic field for confinement of the plasma. The international effort to demonstrate the feasibility of the reactor design (International Toroidal Engineering Reactor, ITER) has very stringent requirements for the hysteresis loss and the critical current density J_c ($\leq$ 600 mJ/cm^3 and $\geq$700 A/mm^2, respectively) of the Nb_3Sn wires which are to be used for the central toroidal field coil magnet. Although short wire specimens can meet these requirements, consistently fabricating long lengths of wire meeting these specifications is not trivial. A part of this inconsistency is related to the effects of the temperature ramp rate on these critical properties.[1] Thus, to investigate the observed ramp rate dependencies a metallurgical examination was performed. Samples of wires were subjected to two controlled temperature ramp cycles. Wire cross-sections were examined from samples removed at various stages of the two ramp cycles. Optical and scanning electron microscopy were used to study the reaction kinetics among Cu, Sn, and Nb. The compositional analyses of each of the phases present were carried out using Wavelength Dispersive Spectroscopy (WDS).

The results show a slow ramp rate produces a higher J_c but results in a high hysteresis loss; while a fast ramp rate yields a lower hysteresis loss but produces a poorer J_c. The slow temperature rise causes an outward movement of the inner filament rings. This reduces the interfilamentary spacing and enhances the coalescence of the filaments, resulting in a larger effective filament size. Hysteresis loss is proportional to the filament size. This outward motion of the filaments is associated with the movement of the ε (25 at.% Sn-Cu) / α-bronze (9.1 at.% Sn-Cu) phase interface boundary (Fig. 1). When this interface moves slowly (as in the case during a slow ramp (6° C/h) or while holding at low temperatures (e.g. 375°C)), the filaments are dragged along outward with the advancing interface due to the conversion of the η phase (45 at.% Sn-Cu) to the ε phase. On the other hand, the motion of the filaments is minimal if the temperature is raised faster (e.g. 60°C/h). This results in a lower hysteresis loss due to the maintaining of a larger interfilamentary spacing and less coalescence compared to the slow ramp rate. A further study is needed to understand how the filaments are moved with the advancing ε phase front.

On the other hand, J_c is higher for the wires with a very slow ramp rate (6°C/h) compared to those with a fast rate (60°C/h). This is thought to be associated with the difference in the uniformity of the Sn content in the bronze at ~600°C , where the reaction forming Nb_3Sn becomes significant. The Sn content was found to be more uniform in wires where a slow ramp rate was used (Fig. 2). For example, even the outer locations of the wires (e.g. positions; K, L, M, and N) the Sn content of the matrix is nearly the maximum for the α-bronze phase for the 6°C/h rate. On the other hand, the matrix is essentially pure Cu at the equivalent positions for 60°/h rate sample.

Proc. Microscopy and Microanalysis 1995, edited by G.W. Bailey, M.H. Ellisman, R.A. Hennigar, and N.J. Zaluzec
Copyright © 1995 MSA. Published by Jones and Begell Publishing, 79 Madison Ave., New York, NY 10016

References

1. E. Gregory, et al., in press Proc. of 18th Symp. on Fusion Tech, Karlsrhue, Germany, 1994.
2. We acknowledge E. Gregory, E. and T. Pae for providing the wires for this study and for constant helpful discussions. This work was performed under the auspices of the U.S. Department of Energy under Contract No. DE-AC02-76CH00016

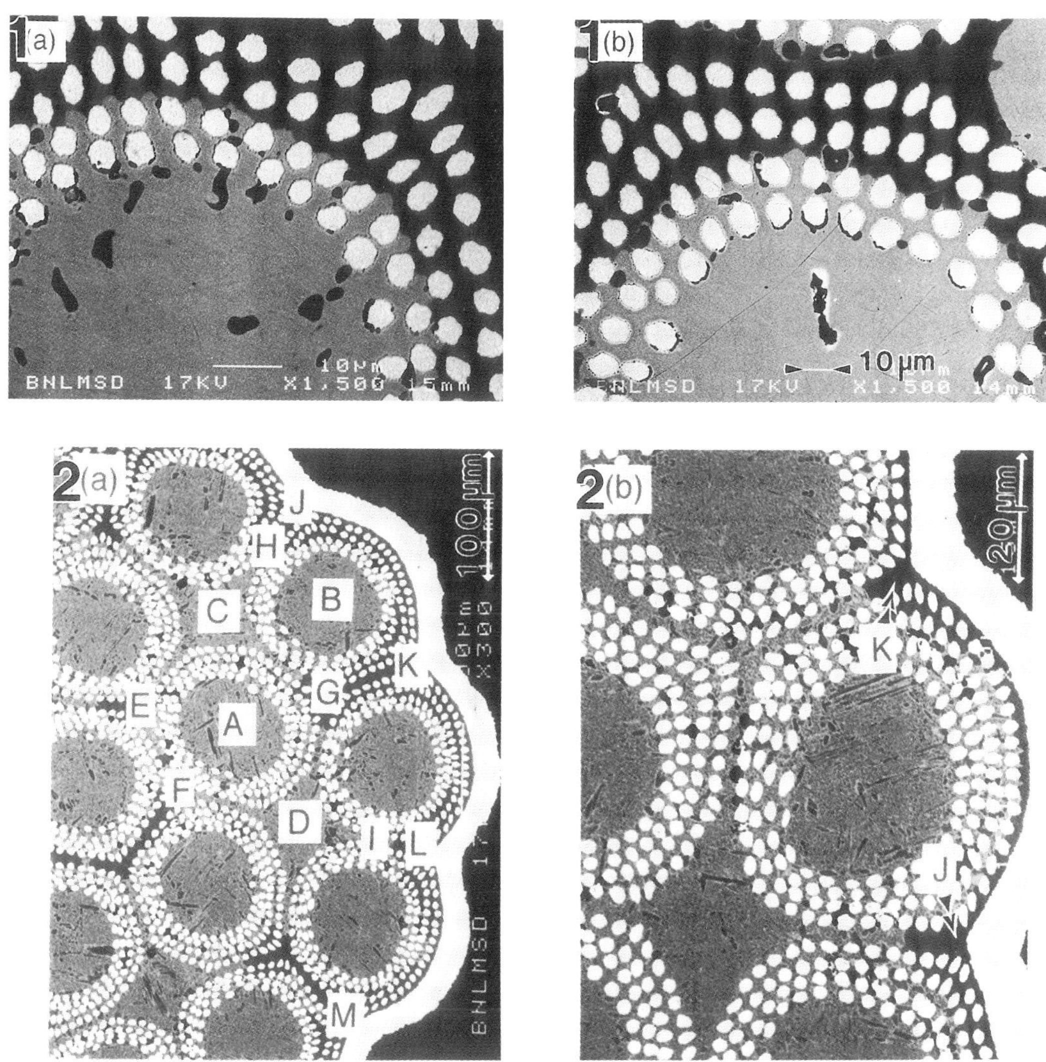

FIG. 1--Shows the difference in the inner filament·motion at 475°C depends on the temperature ramp rate a) 60°C/h and b) 6°C/h. Bar = 10μm.

FIG. 2--(a) Cross section of a wire which was heated to 600°C, the areas marked indicate where the compositional analyses were made. Bar = 100 μm. (b) Backscattered electron image of a wire with 60°C/h ramp rate illustrating non-uniformity in the Sn content. Bar = 20 μm.

ENERGY-FILTERED IMAGING OF PRECIPITATES IN FeCuN ALLOYS

P.M. Rice, R.E. Stoller, and J. Bentley

Metals and Ceramics Division, Oak Ridge National Laboratory, P.O. Box 2008,
Oak Ridge, TN 37831-6376

The microstructures of a series of model ferritic alloys are being characterized by transmission electron microscopy (TEM) techniques. The alloys have varying concentrations of Cu (0 to 0.9 wt%), N (5 to 120 appm), and other solutes to allow a systematic study of the effects of various elements on embrittlement. The first step in this process is the characterization of the alloys in their as-received state: solution treated at 775°C for 17 h, quenched in a salt bath at 450°C for 180 s, and subsequently air cooled.

The imaging portion of this initial study is complete; however, several questions about the chemical content of many of the precipitates remain unanswered. The heat treatment described above resulted in extensive precipitation, especially in the high N alloys. It is difficult to distinguish between α'' nitride platelets and the ε-carbide platelets which formed in many of the alloys. High-angle tilting is difficult with these ferromagnetic specimens, and the $\mathbf{g} \cdot \mathbf{R} = 0$ invisibility criterion previously used for distinguishing the α'' nitrides[1] has given ambiguous results. It is necessary to use analytical methods to determine the chemical content of the platelets. To this end, energy-filtered imaging is being used to identify these and other precipitates in these alloys. A Gatan Imaging Filter (GIF), installed on a Philips CM30 operated at 300 kV, was used. Figure 1(a) shows a conventional TEM image of a 0.17 wt% carbon alloy with $\mathbf{g} =$ 011 and beam direction near $[100]_\alpha$; a large edge-on (010) platelet runs diagonally across the image. Figure 1(b) shows the C elemental map of the same platelet acquired with a 30eV window centered at 298 eV and with the background calculated from two pre-edge, 30eV windows centered at 217 eV and 258 eV. The image clearly shows that the platelet contains C, identifying it as an ε-carbide, as well as the fact that the platelet apparently formed on or near a spherical carbon-containing precipitate.

To study the affect of Cu precipitation on the hardening of the alloys, a 0.9 wt% Cu alloy was annealed at 550°C for times up to 15 h. The Cu precipitates that form under-go a phase change from bcc to a faulted fcc (9R) crystal structure when they are about 6 nm in diameter.[2] These fcc precipitates have been well characterized by conventional TEM techniques for these alloys. However, the smaller bcc Cu precipitates are difficult to image in these strongly oxidation-prone specimens. Energy-filtered imaging is being evaluated as a possible technique to image and measure the size distribution of the small Cu precipitates. Figure 2 shows examples of energy-filtered images from an annealed .9 wt% Cu alloy. Figure 2(a) shows the Fe jump-ratio image calculated by dividing an image acquired with a 30eV post-edge window centered at 725 eV by an image acquired with a 30eV pre-edge window centered at 685 eV. Figure 2(b) shows the Cu jump-ratio image for the same region with images acquired at 948 eV and 908 eV. Black arrows mark the same precipitates in (a) and (b) clearly showing that the Cu precipitates appear as a lack of signal in the Fe image. These preliminary results are very promising, and the search for smaller Cu precipitates is in progress.[3]

References

1. K.F.Hale and D. McLean, *J. of Iron and Steel Institute*, (1963) 337.
2. P. J. Othens et. al., *Phil. Mag. Lett.* 64(1991)383.
3. This research is sponsored by the Division of Materials Sciences, U.S. Dept. of Energy and the Office of Nuclear Regulatory Research, U.S. Nuclear Regulatory Commission under inter-agency agreement DOE 1886-8109-8L with the U.S. Dept. of Energy under contract DE-AC05-840R21400 with Martin Marietta Energy Systems Inc. and by an appointment of P.M.Rice to the ORNL Postdoctoral Research Associates Program administered jointly by ORNL and Oak Ridge Institute for Science and Education.

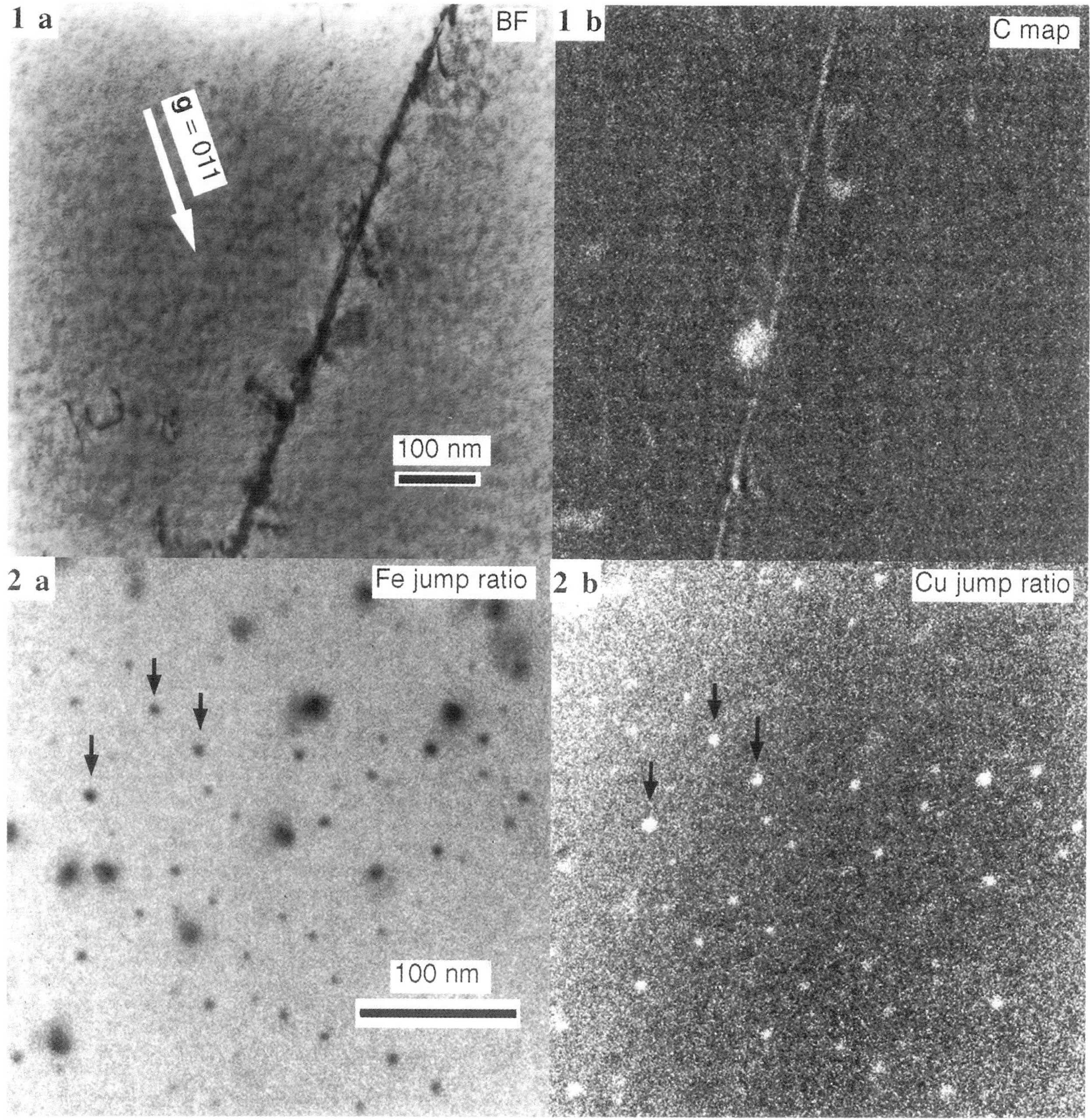

Fig. 1.- Edge-on platelet which precipitated in a high-C (0.17 wt%) alloy.
(a) Conventional TEM image with $\mathbf{g} = 011$ (white arrow) and beam direction near $[100]_\alpha$.
(b) Carbon elemental map calculated from energy-filtered images recorded with 30eV pre-edge windows centered at 217 eV and 258 eV, and a 30eV post-edge window centered at 298 eV.

Fig. 2.- Fcc Cu precipitates in an annealed high-Cu (0.9 wt%) alloy. The black arrows point the same precipitates in both (a) and (b).
(a) Fe jump-ratio image calculated by dividing an energy-filtered image acquired with a 30eV post-edge window centered at 725 eV by a pre-edge image centered at 685 eV.
(b) Cu jump-ratio image of the same region calculated by ratioing images acquired with 30eV post-edge and 30eV pre-edge windows centered at 948 eV and 908 eV, respectively.

MICROSTRUCTURE OF ROUGHENED SURFACE IN METAL ROLLING

W. M. Lee*, M. Meshii**, W. R. D. Wilson*, W. A. Chiou**, and C. S. Lin**

* Department of Mechanical Engineering, Northwestern University, Evanston, IL 60208
* * Department of Materials Science and Engineering, Northwestern University, Evanston, IL 60208

Flat rolled aluminum products have a $25 billion market worldwide annually. The topography of the rolled surface is a key measure of surface quality. A dull surface finish can be generated by non-homogeneous surface deformation when the sheet is rolled under conditions of hydrodynamic lubrication. This is undesirable for some applications where high reflectivity is required, though it improves paintability. Wilson and Schmid have investigated the influence of lubrication conditions on roughening [1]. Although the texture evolution during the hot and cold rolling of aluminum has been studied [2], the basic mechanisms of the roughening process are not fully understood. The purpose of the present research is to explore the microstructural changes associated with the evolution of surface roughness and to use this information to develop mathematical models which can accurstely describe the surface roughening evolution process.

A TEM study of cross-sectional microstructure evolution of aluminum alloy Al 5052-0 material during cold rolling in the presents of a thick lubricant film has been conducted. Commercial aluminum alloy AL 5052-0 as-received was used in the experiments. The material was heat treated (annealed) after rolling in the mill. In the laboratory, the aluminum alloy was rolled with 30 % reduction using a highly viscous mineral oil as lubricant. The oil drop test method [3] was employed to determine the lubricant film thickness which was from 3 to 10 times the surface roughness. The microstructure of both the as-received and rolled specimens were studied by using TEM. The method of preparing the TEM specimens was adapted from that of Lin and Meshii [4].

The montage of TEM micrographs shown in Fig.1 reveals the cross-sectional microstructure of as-received aluminum alloy. The typical grain size is approximately 10 μm, and the grains are equiaxed in shape. No specific grain orientation was observed in the body of the material although some preferred orientation appeared near the surface of the sample. The lack of subgrain structure and the rather even (tangled) dislocation distribution is a typical phenomenon from static recrystallization. A fair amount of small rectangular precipitates (Mg) were also found scattered throughout the sample.

Another montage of TEM micrographs, shown in Fig. 2, illustrates the cross-sectional microstructure of the aluminum alloy after 30% reduction. The averaged grain size is about 1μm, one tenth that of the original material, and the grains are aligned in the rolling direction. Sub-grains of near equiaxed shape can be found inside many large elongated grains suggesting the occurrence of dynamic recovery during the rolling process. The relatively high density of low angle subgrain boundaries near the surface compared with the well-defined high angle subgrain boundaries in the body of the material, is an indication of inhomogeneous deformation during rolling.

REFERENCES

1. Wilson, W.R.D., and Schmid, S. Engineered Surfaces, PED-Vol. 62 ASME. 91-100, (1992)
2. Ren, B. and Morris, J. G., Metal. and Mat. Trans., Vol. 26A, No.1, (1995)
3. Azushima, A., Bulletin of JSME., Sep., (1978)
4. Lin, C.S. and Meshii, M. 35th MWSP Conf. Proc., ISS-AIME, Vol.XXXI, 99-105, (1993)
5. The research was supported by Comalco, Inc. and Kaiser Aluminum and Chemical, Inc.

Proc. Microscopy and Microanalysis 1995, edited by G.W. Bailey, M.H. Ellisman, R.A. Hennigar, and N.J. Zaluzec
Copyright © 1995 MSA. Published by Jones and Begell Publishing, 79 Madison Ave., New York, NY 10016

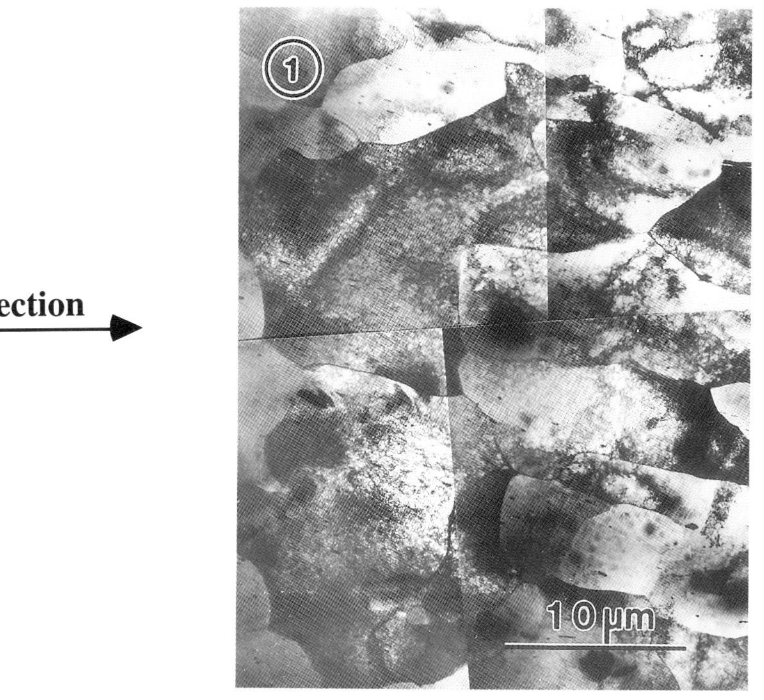

Figue 1. Cross-sectional TEM image of commercial aluminum alloy AL 5052-0.
Figue 2. Cross-sectional TEM image of commercial aluminum ally AL 5052-0 after rolling with 30% reduction and using viscous mineral oil Thuban 250 as lubricant.

MICROANALYSIS AND CHARACTERIZATION OF PROTECTIVE COATINGS USED IN METAL-MATRIX COMPOSITES

Jason R. Heffelfinger, Robert R. Kieschke* and C. Barry Carter

Department of Chemical Engineering and Materials Science, University of Minnesota, 421 Washington Ave. S.E., Minneapolis, MN 55455

* Metal-Matrix Composites Program, 3M Company, 3M Center Bldg. 60-1N-01, St. Paul, MN, 55144-1000

Transmission electron microscopy (TEM), scanning electron microscopy (SEM) and parallel-aquistion electron-energy loss spectrometry (PEELS) have been used to investigate various coating schemes used in alumina-fiber reinforced titanium alloys. The high-temperature properties of metal-matrix composites (MMC's) depend greatly on the interface between the ceramic fiber and metal matrix.[1,2] The chemical and mechanical properties of this interface may be tailored by coating the alumina fibers with various materials.[3] In this work, emphasis has been placed on the chemical aspect of the coatings and their role in inhibiting the reaction between the alumina fiber and the titanium matrix. The brittle nature of the reaction products, Ti_3Al and $TiAl$, weakens the composite and thus leads to its ultimate failure.[4]

The composites are made of α-alumina fibers (Nextel 610 with a mean diameter of ~12 mm) which are first coated with thin protective layers such as Y_2O_3 and Nb followed by a coating of β-titanium alloy (TIMETAL 21S Ti -15Mo -3Al -2Nb -0.2Si.) Several thousand fibers are then placed together inside a glass tube and HIP-ped (Hot Isostatic Press) at temperatures of ~900°C and times of ~2 hours to attain a dense composite. From these composites TEM samples were prepared by conventional dimpling and ion milling techniques with special care taken to ensure that the fibers remained normal to the surface of the sample.

The coating's role in the composite is to act as a diffusion barrier between the fiber and matrix. Coating materials must be selected which are thermodynamically stable in contact with the fiber and/or the matrix. Figure 1 shows such a coating scheme using Y_2O_3 (stable in contact with Ti) and Nb (stable in contact with alumina). With a total thickness of ~250nm, the coating seen in Figure 1 has provided adequate protection against the interdiffusion and reaction of the fiber and matrix. EELS has been used to analysis this type of interface and provide chemical profiles of Ti and Al across the coating. Information on the diffusion of the reacting species can help to predict the life time of the composites. Not all regions of the coating are successful, as seen by the reaction in Figure 2. The formation of Ti-Al intermetallics is inevitable in areas where the coating has been damaged or is too thin to provide protection. The microstructure and chemical nature of the reaction product is best seen by the TEM, Figure 3. As identified by selected-area diffraction and energy-dispersive spectrometry, the reaction product was Ti_3Al which is consistent with previous research.[3] In order to study the reaction between alumina and titanium matrix more closely, a model system in which a Ti 21S alloy is sandwiched between two alumina single-crystals was developed. The sandwich was exposed to similar conditions to those experienced by the composites during processing. Figure 4 shows the interface between the alumina single-crystal and Ti 21S alloy which has been exposed to a temperature of 900°C for 2 hours. The brighter regions seen in the Ti 21S alloy represent Mo rich regions (Mo stabilizes the β phase of Ti), and the darker areas are the less dense α-Ti. Small protrusions of a Ti-Al intermetallic were observed near the alumina interface. A larger destabilized α-Ti region was also observed near the alumina interface and is believed to be the result of oxygen diffusing into the Ti 21S matrix. This model system is now being used as an aid in determining new protective coatings and their effectiveness.[5]

Proc. Microscopy and Microanalysis 1995, edited by G.W. Bailey, M.H. Ellisman, R.A. Hennigar, and N.J. Zaluzec

References:
1. Y. Lu, S. L. Sass, Q. Bai, D. L. Kohlstedt and W. W. Gerberich, *Acta Metallurgical Materials*, **43** [1] 31-41 (1995).
2. J. A. DeKock, Y. A. Chang, M.-X. Zhang and O. Y. Chen, in *Proc. MRS Symposium*, 1990.
3. Y. A. Chang, R. Kieschke, J. DeKock and M. X. Zhang, Interfacial Stabilities of High-Temperature Composite Materials, 1994.
4. R. E. Tressler, T. L. Moore and R. L. Crane, *Journal of Material Science*, **8** 151 (1973).
5. The authors acknowledge the research support from the Center for Interfacial Engineering, which is a National Science Foundation Engineering Research Center. The TEM is part of this research center. The authors thank Prof. Stan Erlandsen for access to the Hitachi S-900 SEM and Chris Frethem for technical assistants.

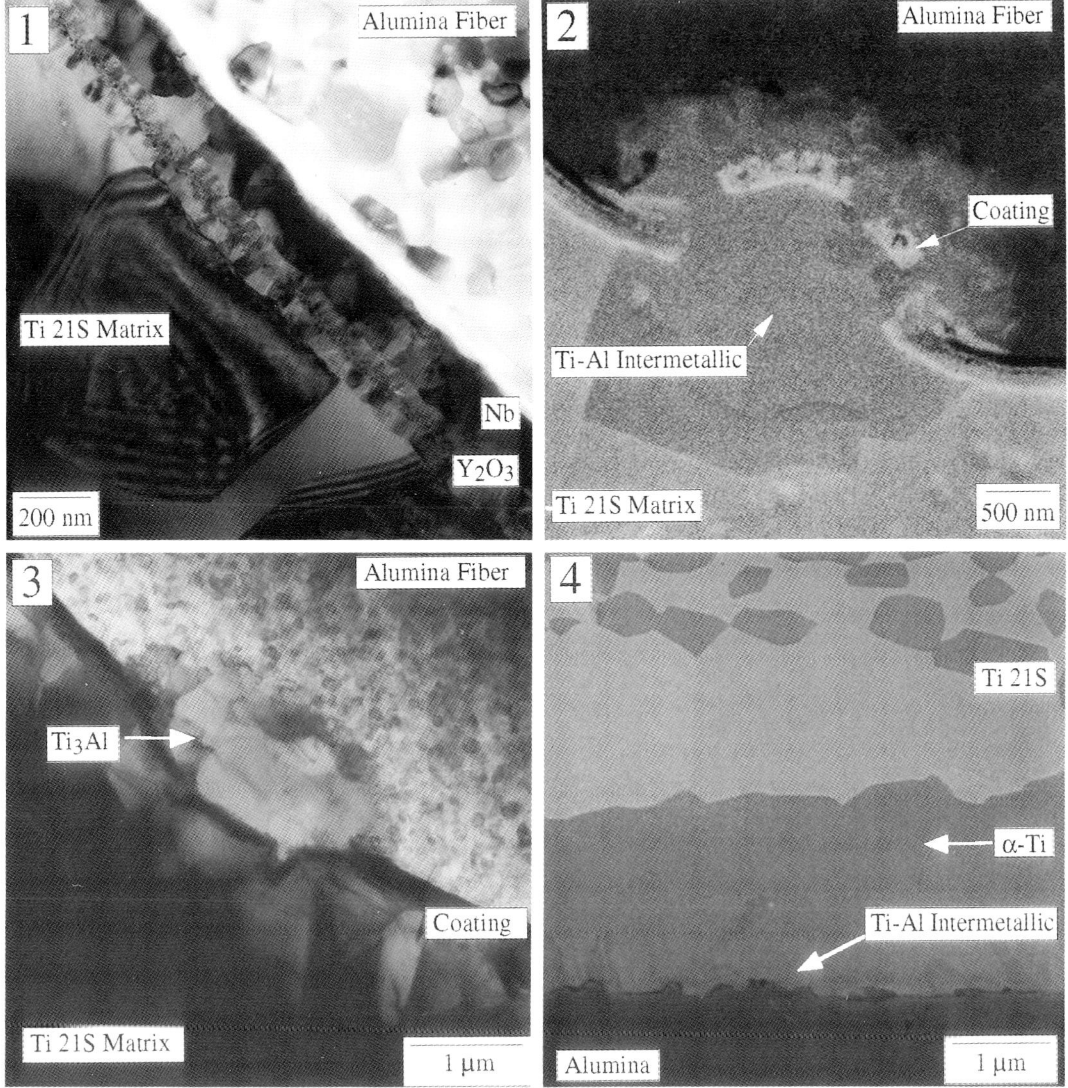

Fig. 1. Bright-field TEM image showing the coating between the Ti 21S matrix and alumina fiber.
Fig. 2. Back-scattered SEM image showing the darker contrast of the intermetallic phase.
Fig. 3. Bright-field TEM image showing the microstructure of the reaction product Ti3Al.
Fig. 4. Back-scattered SEM image showing the alumina/Ti 21S interface.

Characterization of Corrosion Scales in Water Distribution Pipes Using Scanning Electron Microscopy (SEM) and Energy Dispersive X-ray Spectroscopy (EDS)

J. L. Shull,* W. M. Kriven,* and A. S. McAnally**

*Department of Materials Science and Engineering, University of Illinois at Urbana-Champaign, Urbana, IL 61801
**Department of Civil Engineering, University of South Carolina, Columbia, SC 29208

A series of water treatment programs involving different combinations of chemical additives was evaluated with the aim of reducing the levels of lead and copper in the drinking water of several communities in South Carolina.[1,2] Test coupons from 7 pilot systems and 3 municipal distribution systems involved in the study were analyzed using SEM and EDS in order to determine the structure and composition of the corrosion scales and other deposits which formed on the interior pipe surfaces.

Sample coupons consisted of short segments of copper pipe, each of which was coated on one end with 50-50 wt.% lead-tin solder. Small (3 mm x 6 mm) pieces cut from both the soldered and unsoldered ends of the coupons were used as specimens for SEM surface analysis. Cross-sectional specimens were prepared by embedding small pieces in an acrylic resin and using an ultramicrotome to obtain high quality surfaces.

Analysis of the coupon surfaces produced results that were consistent with the corresponding water quality analyses.[1] Coupons from systems in which high lead concentrations were detected in the water displayed significant depletion of lead from the solder. These coupons also contained large numbers of cerussite ($PbCO_3$) and/or hydrocerussite ($PbCO_3 \cdot 2H_2O$) crystals on their inside surfaces (Fig. 1). Coupons from systems in which zinc orthophosphate was used in the water treatment showed very little loss of lead from the solder (Fig 2). Corrosion of lead from the solder was very evident in the cross-sections. An example is shown in Fig. 3. Here the bright (lead-rich) phase was noticeably absent in a portion of the solder layer. The x-ray spectra in Fig. 4a and 4b compare the relative amounts of lead and tin in a typical large area of uncorroded solder versus the corroded region indicated by the arrow in Fig. 3.

In several of the pilot systems which showed promising lead and copper control, thick films were found on the coupon surfaces. These films were mainly non-crystalline and had compositions similar to the treatment additives. These films, however, did not protect the copper and solder surfaces from a physical point of view. In all cases the films contained interconnected porosity which could not have prevented water from reaching the coupon surfaces (Fig. 5). The protective qualities of these films may have been more chemical than physical. The presence of zinc phosphate precipitates at the coupon surfaces required that the adjacent water be saturated with zinc and phosphate ions. This condition would have made the corrosion of lead and copper electrochemically unfavorable.

References

1. C.A. Chen et al., *Journal of Environmental Science and Health*, A29(1994)1587.
2. A.S. McAnally et al., *Proc. Ann. Conf. American Water Works Association*, (1993)973.
3. The authors gratefully acknowledge the use of facilities at the Center for Electron Microscopy and the Center for Microanalysis of Materials, UIUC. This work was funded by the U.S. Environmental Protection Agency.

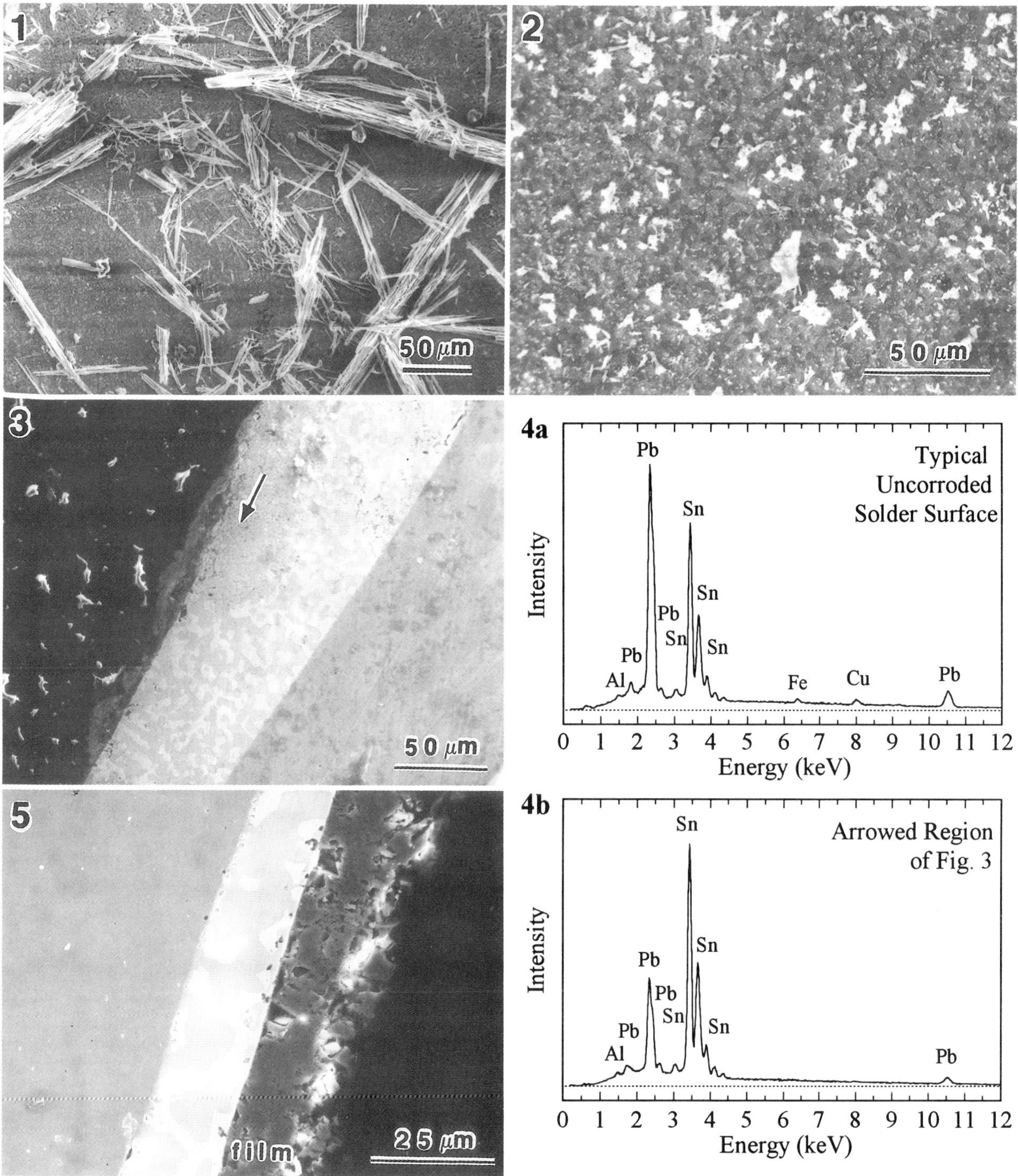

FIG. 1 Cerussite (PbCO$_3$) crystals on the soldered surface of a test coupon.
FIG. 2 Zinc phosphate particles on a relatively unaltered solder surface. Note eutectic microstructure.
FIG. 3 Cross-section showing a region of lead depletion in the solder layer (arrowed).
FIG. 4 X-ray spectra taken at 20kV of (a) uncorroded solder and (b) corroded region arrowed in Fig. 3.
FIG. 5 Cross-section showing interconnected porosity in the deposited film.

AEM ANALYSIS OF STRESS CORROSION CRACKS

N. Lewis, D. J. Perry, M. L. Bunch

Knolls Atomic Power Laboratory, Martin Marietta, Schenectady, New York 12301-1072

Cross-sectional analytical electron microscopy (AEM) was used to study stress corrosion cracking (SCC) cracks in pressurized water in an attempt to understand the mechanism of cracking. AEM is particularly well suited to study the crack tip, the crack sides and the corrosion product inside the crack. These analyses provide clues about the stress (i.e., plasticity), the environment (corrosion product) and material (precipitates/grain boundary composition) which are the key factors controlling SCC. In this investigation SCC cracks were filled with epoxy, dimpled and ion-milled in argon until electron transparent. A cross-section of an SCC crack in Alloy 600 generated in deaerated water is shown in Figure 1. The crack is intergranular and generally propagates down grain boundaries without carbides as shown in Figure 1. The crack seen in Figure 2 is an exception and propagated down a grain boundary containing Cr_7C_3 carbides. Although dislocations are clearly observed in both figures, no deformation is observed uniquely associated with the cracks and no voids have been observed ahead of the cracks. Interestingly, analysis of the oxides inside the crack shows differences. Sometimes a chromium rich oxide containing Ni is observed and other times NiO is observed without Cr as seen in Figure 3. This result is in contrast to the oxides formed on the outside surfaces of the specimens. On the specimen surface, the oxide adjacent to the Alloy 600 base metal is rich in Cr and Ni while the outer layers exhibit NiO, Ni metal and Ni-Fe_2O_4 spinels. The implication of this result is that the crack environment is different than the bulk chemistry environment.

In a separate analysis, a stainless steel compact tension (CT) specimen was prepared. Cracks in this material were generated in a simulated crevice environment with dissolved solids. The stress corrosion cracks in this specimen were all transgranular with oxides completely filling the crack tip region as seen in Figures 4 and 5. One interesting feature of the cracks is that they do not intersect matrix carbides, rather they appear to go between them. Furthermore, dislocations are observed in the specimen matrix but there is no evidence of deformation uniquely associated with the crack. Additionally, no voids, microcracks or evidence of slip are observed ahead of the crack or along the crack length. The observations cast doubt on some mechanisms involving hydrogen such as enhanced plasticity, microcracking ahead of the crack or hydride formation under these experimental conditions. The best candidates appear to be mechanisms involving oxide rupture and dissolution or possibly some type of oxidation and decohesion.

Over 100 cracked test specimens have been examined using this technique. A number of cross-sections for each experimental condition should be prepared to eliminate the variability which does occur from test specimen to test specimen to avoid misinterpretation. Nevertheless, the AEM results are exciting and this technique can provide unique insight into the cracking phenomenon.

Proc. Microscopy and Microanalysis 1995, edited by G.W. Bailey, M.H. Ellisman, R.A. Hennigar, and N.J. Zaluzec

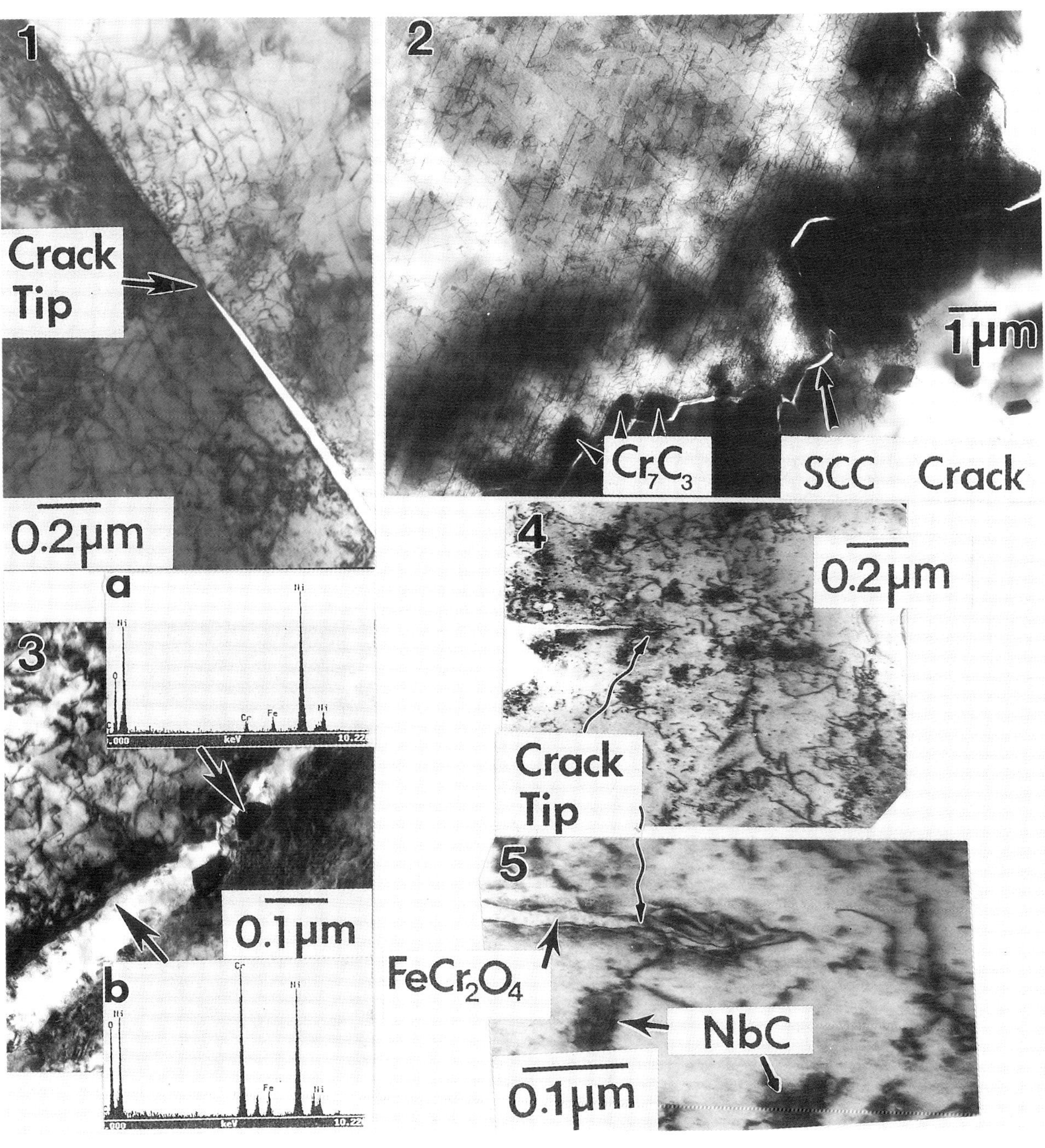

Figure 1.   AEM cross-sectional image showing SCC crack-tip in Alloy 600.

Figure 2.   SCC crack along grain boundary containing Carbides in Alloy 600.

Figure 3.   SCC crack and corrosion product away from the crack-tip in Alloy 600.

Figure 4.   Cross-sectional AEM image showing transgranular SCC crack-tip in 347SS.

Figure 5.   Higher magnification image of Figure 4.

ELECTRON MICROSCOPY OF MECHANICAL PLATED TI AND TIN COATING ON STEEL

K. Kashihara, W. A. Chiou, C. S. Lin, and M. Meshii
Department of Materials Science and Engineering, Northwestern University, Evanston, IL 60208

Mechanical alloying, originally developed as a means to produce oxide dispersion strengthened alloys[1], has since been adopted for producing a wide variety of materials such as solid solution alloys, metal matrix composites, intermetallics, nanocrystalline and amorphous materials[2]. It has also demonstrated as an efficient method for alloying gaseous elements by solid-gas reaction[3]. Interest in using metal and metalloid nitrides as coating materials has been increased considerably in the past few years because of their special properties including extreme hardness, high temperature stability, high corrosion and wear resistance, and excellent conductivity. A new coating technique through the marriage of mechanical plating and nitriding has recently been developed in this laboratory. This paper presents an electron microscopic study of mechanically plated Ti and TiN coating.

Commercially produced (Johnson Matthey, Inc.) fine-grained (325 mesh, or 44 μm) Ti (99%) powders were used as the starting coating material. Interstitial-free steel sheet (provided by Inland Steel) were cut to a size of 1.5 cm x 1.5 cm as the substrate. The Ti powder and steel balls were mixed with the pre-anchored steel sheets and ball milled in a modified stainless steel vial using SPEX 8000 Mixer/Mill machine. The entire process including Ti powder cleaning and mechanical milling/plating were carried out in the nitrogen atmosphere with constant nitrogen gas flow of 5 psi. To monitor the progress of mechanical plating, samples were visually examined at every 6 hrs. of milling. Selected samples (after plating of 6, 12, 24, and 48 hrs.) were also examined by X-ray diffraction (XRD). Sample used for electron microscopy study were taken after mechanically plating for 24 and 48 hrs. To examine the microstructure of the mechanically plated coating layer, cross-sectional specimens for both SEM and TEM study were prepared following the technique developed by Lin et al.[4]

Low magnification cross-sectional SEM image shows the general view of a specimen after 12hrs. of mechanical plating. Two distinct surface layers, although not uniformly, were observed to cover the steel surface. Results of EDS analyses on SEM specimens indicate that steel surface was initially coated by Ti powders. But with increased plating time, TiN crystals formed on the outer layers of coating and mixed with original Ti powder. The transformation of elemental Ti powder into TiN compound is demonstrated in Fig. 2. The first visible change in the XRD pattern indicating the presence of the TiN phase was noted after 12 hrs. of plating, and intensified to become well defined peaks after 48 hrs. of milling. The broadening of TiN peaks indicates the nanostructural character of the compound. Cross-sectional TEM image (Fig. 3) also depicts very fine grained particles on the surface of coating after 48 hrs milling. Selected area electron diffraction patterns also showed that those very fine grained particles were TiN (Fig. 4). HREM image (Fig. 5) and EDS analyses (Fig. 6) illustrate that these fine grained particles were composed mainly of nanocrystalline TiN and Ti ranging from 2 to 10 nm in size. In addition, a small amount of an amorphous compound was also observed.

It has been known that chemisorption of nitrogen onto rough metal surface is much higher than on a perfect smooth surface[5]. The mechanical plating process generates a large number of small particles with atomically clean but rough surfaces. These surfaces were exposed to molecular nitrogen and absorb it. During milling of pure metal powders, a large negative heat of mixing provides the driving force for solid-state reactions, nanocrystalline formation, and amorphization. More research on the mechanism of nitriding is in progress.

1. J. S. Benjamin, Sci. Am., 234, 40 (1976)
2. M. Arzt and L. Schultz, eds., New Materials by Mechanical. Alloying Techniques: Verlag (1989)
3. A. Calka, Appl. Phys. Lett., 59, 1568 (1991); Also Y. Ogino et. al., Scripta Metall., 28, 967 (1993)
4. C. S. Lin et al., Proc. of TMS Symp. on the Physical Metallurgy of Zinc Coated Steels, 31 (1994)
5. R. Liu and G. Ehrlich, Surface.Sci., 119, 207 (1982)
6. This research made use of MRL Central Facilities supported by the NSF, at the Materials Research Center of Northwestern University, under award No. DMR-9120521.

Proc. Microscopy and Microanalysis 1995, edited by G.W. Bailey, M.H. Ellisman, R.A. Hennigar, and N.J. Zaluzec

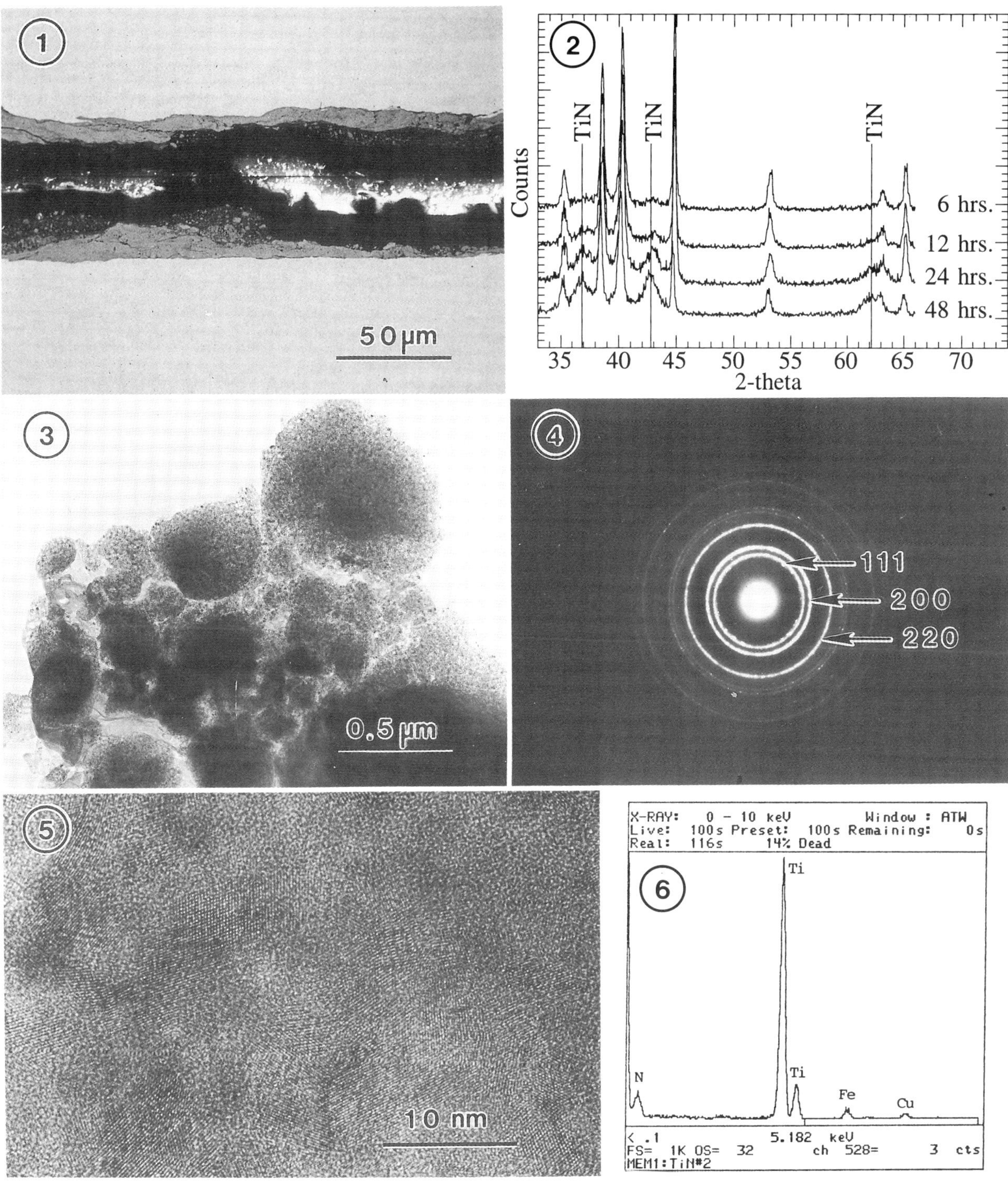

Fig. 1 SEM low magnification cross-sectional view showing the coating layer after 24 hrs. plating.
Fig. 2 XRD patterns of Ti powders milled/plated for different period of time.
Fig. 3 TEM micrograph revealing very fine grained particles on the surface after 48 hrs. plating.
Fig. 4 SAD pattern from the central area of Fig. 3 depicting matched TiN diffraction peaks (rings).
Fig. 5 HREM image of coating surface illustrating nanocrystalline grains and amorphous materials.
Fig. 6 EDS pattern of a nanocrystalline particle.

HRTEM AND AEM STUDY IN (Au, TiN) THIN FILM ON (100) SILICON

J.S. Bow*, Y.C. Hung**, M.J. Kim*, R.W. Carpenter*, W.M. Kim[#], S.K. Lee[#], S.G. Kim[#]

*Center for Solid State Science, Arizona State University, Tempe, AZ 85287-1704
**Materials Research Laboratory, Industrial Technology Research Institute, Taiwan
[#]Materials Design Laboratory, Korea Institute of Science and Technology, Cheongryang, Seoul, Korea

The cross-sectional microstructure of a (Au, TiN) thin film deposited on a (100) Si substrate without further heat treatment was studied by CTEM, HRTEM, and AEM. HTREM was performed in a Topcon 002B microscope with interpretable resolution limit of 0.18 nm, and high spatial AEM was done in a Philips 400ST field emission gun microscope at 100 kV using a Gatan 666 parallel-detection electron energy loss spectrometer. Cross-section specimens of the interface were prepared by traditional polishing and ion milling. Temperatures used in the whole process of TEM specimen preparation were below 100°C and a liquid-nitrogen-cooled cold stage was used in ion milling to prevent interaction between Au and Si[1].

Ti was co-deposited with Au by sputtering in a nitrogen atmosphere to increase the hardness of the thin Au contact film. Fig.1a shows the microstructure of the (Au, Ti, N)/Si contact. The features of this microstructure are very similar to the those of pure gold thin films on Si substrates annealed below the Au-Si eutectic temperature (363°C)[2]. The thin, white layer was considered to be the original Au/Si interface by Chang et al. The large islands of dark contrast below the interface in the Si indicate that they are an Au-rich phase. Some of these islands have well defined habit planes with the Si substrate, as shown in Fig. 1a. The film is polycrystalline with grain size of about 15 nm, measured from HREM images, and Au-rich from EDX spectra. SADP and CDF images (Fig.1b, c, and d) show that the most of Au grains have $\{111\}$ planes parallel to the substrate Si$\{200\}$ planes, and some Au grains with $\{111\}$ planes parallel to the Si$\{111\}$ planes.

The HRTEM image in Fig. 2 shows that the white line in Fig.1a is an amorphous phase of about 3nm thick. Oxygen is detected from this amorphous interface layer by EELS. Fig. 3 shows a set of EELS and EDX spectra collected simultaneously from the point marked "*" in Fig. 1a. The signals of Ti, N, and Si(Fig. 3b) result from the neighbor (Au, Ti, N) film and Si substrate covered by the probe (about 3 nm). The high intensity of titanium indicates that Ti is rich close to the (Au, Ti, N)/Si interface. This is consistent with intensity distribution in the CDF image in Fig.1c. The depletion of Au in the film close to the interface indicates diffusion of Au into silicon substrate to form the large Au-rich islands. Chang et al.[2] reported this phase to be Au$_3$Si according to the work of Magee and Peng[3]. The mechanism of Au transport through this amorphous layer into the Si substrate will be discussed.

References

1. M. J. Kim and R. W. Carpenter, Ultramicroscopy **21**(1993) 327.
2. P. H. Chang, G. Berman, and C. C. Shen, J. Appl. Phys., 63 (1988) 1473.
3. T. J. Magee and J. Peng, Phys. Stat. Sol., A49(1978) 313.
4. This research was partly supported by USDOE grant DE-FG03-94ER45510 and partly supported by the KIST (WST 8588). The microscopy was performed at NSF/ASU HREM facility under grant NSF-DMR-9115680.

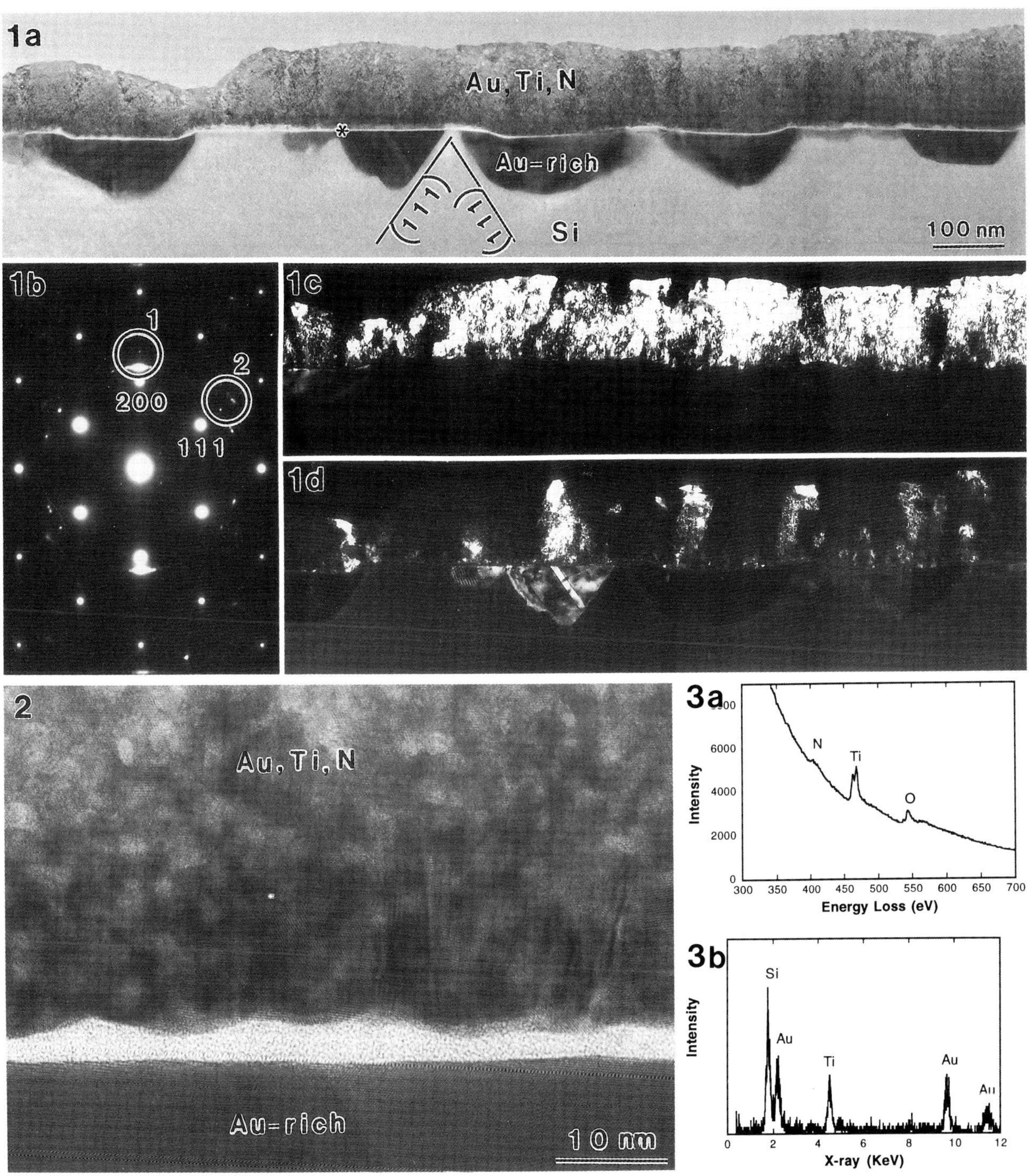

FIG. 1 Cross-sectional TEM micrographs showing the structure of as-deposited (Au, Ti, N)/Si. (a) BF image, (b) SADP, (c) and (d) CDF images formed by spots encirled and marked 1 and 2 respectively in (b).

FIG. 2 HREM image of interface.

FIG. 3 Spectral from the (Au, Ti, N)/Si interface, (a) EELS, (b)EDS.

MICROSTRUCTURE OF PLASMA-DEPOSITED IRON-NITRIDE FILMS

H. Siriwardane,[*] C. Polly,[•] A. Sago,[¶] P. Fraundorf,[*] O.A. Pringle, [•] J. W. Newkirk,[¶] and W.J. James[↑↑]

[*] *Center for Molecular Electronics, and Department of Physics, University of Missouri-St. Louis, St.Louis, MO 63121 (U.S.A.)*

Departments of Physics,[•] Metallurgical Engineering,[¶] and Chemistry,[↑↑] and The Graduate Center for Materials Research, University of Missouri-Rolla, Rolla, MO 65401 (U.S.A.)

Iron nitride films are potential candidates for information storage media.[1] In order to use the films as a perpendicular magnetic recording media, it is necessary to know about the micro-, crystal, and magnetic structures, physical properties, and processing of these thin films.[2] Here we describe the micro- structure of plasma-deposited iron nitride films.

The microstructure of films deposited on glass and carbon support films on copper grids were investigated by X-ray diffraction (XRD), transmission electron microscopy (TEM), scanning electron microscopy (SEM), and Auger electron Spectroscopy (AES). According to XRD, single phase Fe_3N was successfully produced on glass substrates at temperatures 200, 275, 400, & 500 °C, and flow ratios between 0.4 to 15.3 of $Fe(CO)_5$, and NH_3. Electron diffraction peak positions of films were compared with Fe_3C, Fe_7C_3, Fe_2N, Fe_3N γ-Fe_4N, Fe_5N_2 and Fe_3O_4 crystal structure d-spacings. The presence of a carbide satellite peak in AES data and by matching all d-spacings in the JCPDS Fe_3C card file with every experimental diffraction pattern, enabled us to confirm the presence of Fe_3C in these films at all deposition temperatures. However the EDX data collected on individual grains show both carbon and nitrogen, Fig. 1, indicating the possibility of carbon sharing the sites in the parent iron nitride matrix. This suggests XRD shown Fe_3N films on glass to contain either of Fe_3C as a secondary phase (Fig. 2), or the carbide phase seen in AES and in electron diffraction data could be explained as carbon occupying lattice sites for nitrogen in the iron nitride parent matrix. Evidence of a graphitic phase was not found in films on glass, indicating the graphite we see in films deposited on carbon films to be a substrate effect.

Due to much shorter deposition times, grain sizes on carbon support films (15 minutes) were smaller compared to TEM grain size measurements on glass substrates (4 hours). In the early stage of growth process, it is expected the grains to be smaller and increase their size with progression of deposition, which we see in films on glass substrates.[3] The effect of substrate temperature on grain size is shown in Fig. 3. The crystallite sizes in the TEM were in the range (20-100 nm). The bumps on the surface seen in SEM were in the range (150-900 nm) in size, Fig. 4. This may result because a smaller number of larger and larger crystals remain exposed, as the film increase in thickness.

Cross-sectional images of a plasma-deposited Fe_3N film indicate its surface to be rougher than the glass-film interface. In Figs. 5a & 5b, the angle between the largest crystal dimension and the substrate normal indicated that 63% of the 65 crystals align with the surface-normal, while only 29% are perpendicular to the surface-normal. Having the large dimension of most crystals aligned with the surface-normal is consistent with the growth method described above, and suggests that these crystals may have magnetic moments in the surface normal direction.

References:

[1] George J-S Gau, J. of Metals, 3 (1990), 35-38.

[2] J.L. Li, T.J. O'Keefe, and W. J. James, Materials Science and Engineering, B15 (1992) 203-208.

[3] H. Siriwardane, W. James, O. A. Pringle, and J.W Newkirk, Materials Research Society Symposium, Proceedings, 313 (1993), 425-430.

[4] Authors wish to thank Dr. Daniel Schwatz at McDonnell Douglas Research Laboratory in St. Louis, for his useful discussions and help getting EDX data in this study. Support for this research was provided by the U. of Missouri research board, and by the U.S. Department of Energy, Grant # DE-FG02-92CH10499.

Proc. Microscopy and Microanalysis 1995, edited by G.W. Bailey, M.H. Ellisman, R.A. Hennigar, and N.J. Zaluzec

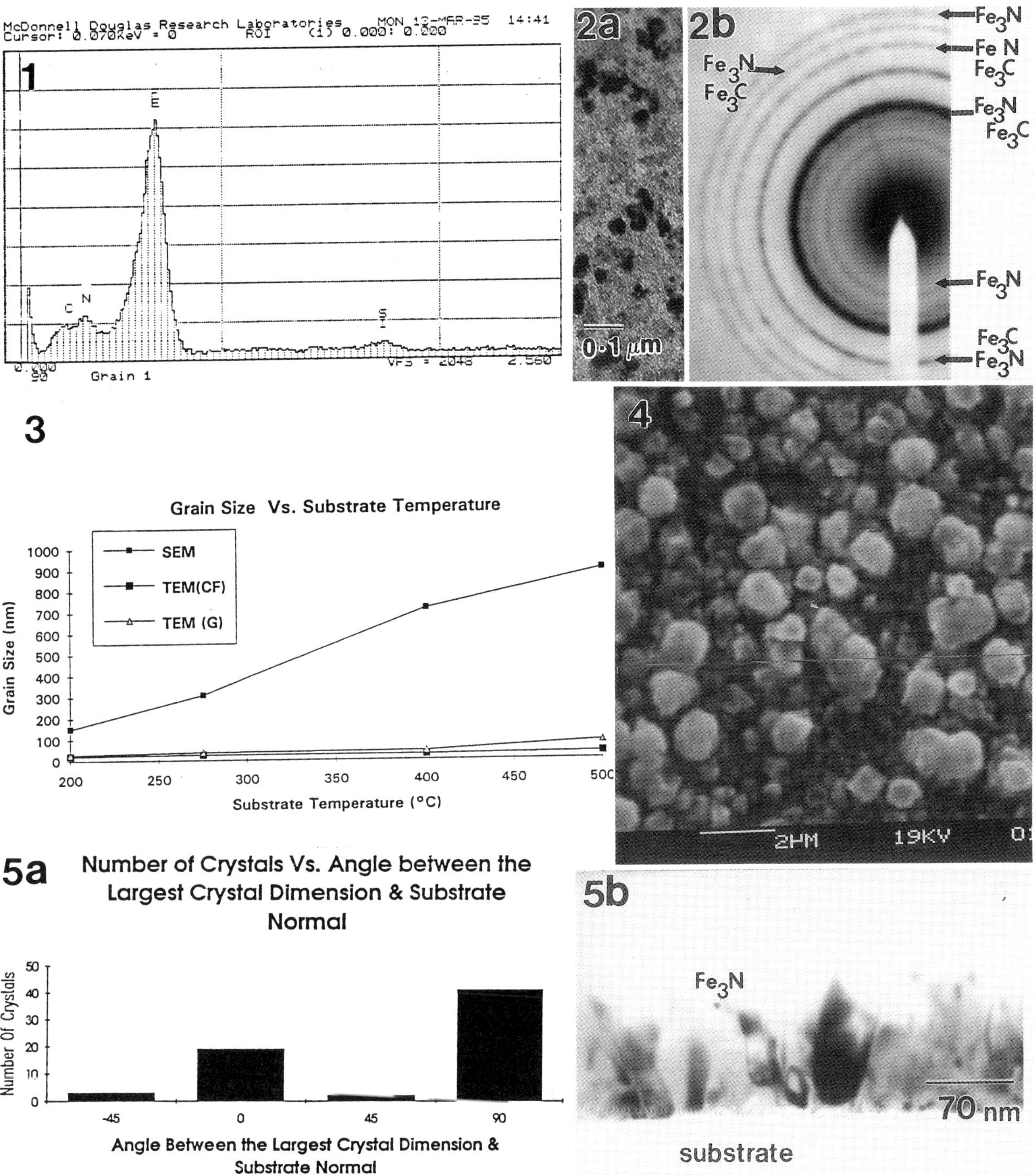

Fig. 1. EDX spectra from an approximate 100 nm size crystal.

Fig. 2. (a) Bright field TEM image (b) its SAD pattern showing the presence of Fe_3N and Fe_3C phases.

Fig. 3. Grain Size Vs. Substrate Temperature plot for plasma-deposited iron nitride films.

Fig. 4. SEM surface image of a plasma-deposited Fe_3N film.

Fig. 5. (a) Crystal Orientation Vs. Number of Crystals (b) Cross-sectional TEM image of a Fe_3N film; indicating most crystals in the film to be oriented $\perp$ to the substrate surface.

557

CHARACTERIZATION OF PULSED LASER DEPOSITED ZnO-WS$_2$ THIN SOLID FILM LUBRICANTS

S. D. Walck, *J. S. Zabinski,* N.T. McDevitt,** and J. E. Bultman***

*Materials Directorate, Wright Laboratory, Wright-Patterson AFB, OH 45433-7750
**RAMSPEC Research, Dayton, OH 45431
***Research Institute, University of Dayton, Dayton, OH 45469-6533

ZnO-WS$_2$ is a potential high temperature solid film lubricant for aerospace applications that exhibits adaptive lubricant behavior. An adaptive lubricant undergoes phase and/or chemical changes in response to thermal, environmental, and tribological conditions; with the resulting phase or wear debris also being lubricious. Pulsed laser deposited (PLD) ZnO-WS$_2$ thin films deposited at room temperature (RT) and wear-tested at room temperature have been shown to have coefficients of friction of 0.04 or less which are comparable to WS$_2$ films, but have much longer wear lives. In the as-deposited state, PLD ZnO-WS$_2$ films are amorphous, but when wear-tested, the phases WS$_2$, WO$_3$, and ZnWO$_4$ are produced.[1] Of these, WS$_2$ is a lubricant phase at low temperatures ($\leq$ ~450°C) while ZnWO$_4$ is a lubricant phase above about 600°C. The purpose of this work was to characterize the microstructural and chemical changes that occur when the RT-PLD ZnO-WS$_2$ films are heated in air.

The RT-PLD ZnO-WS$_2$ films were deposited in a system having a base pressure of 9×10^{-7} Pa with a typical pressure during deposition of 6×10^{-5} Pa. The laser was a Lambda Physik LPX 110i excimer laser operating with a Ne-KrF mixture producing a wavelength of 248 nm. The beam was pulsed at a repetition rate of 10 Hz and a power level of 100 mJ per pulse. The target was a cold pressed and polished target of ZnO and WS$_2$ powders mixed in a molar ratio of 1:1. A calibrated off-axis quartx crystal oscillator (XTC) was used to monitor film thickness during deposition. The films were deposited on as-received, polished, polycrystalline silicon carbide substrates which were subsequently found to have about 1% porosity. Previous experience with solid film lubricants suggested depositing the film through a 200 mesh mask for improved adherence during preparation of XTEM specimens. The samples were heat treated in air at 400, 500, 600, 700 and 800°C for ½ hr. The samples were then examined by Raman spectroscopy and x-ray photoelectron spectroscopy (XPS). Because of the tungsten content in the film, tripod polishing followed by low angle ion milling proved to be the best way of preparing the XTEM samples.

Figure 1 shows an XTEM micrograph (Fig. 1a) of the as-deposited RT-PLD ZnO-WS$_2$ films with the associated SAD pattern (Fig. 1b). The film is amorphous with an unexpected compositional modulation. The spherical particles are inherent in the PLD process and are due to liquid droplets that are expelled from the target, cool on their way to the substrate, and are incorporated in the growing film. The reason of the compositional modulation is uncertain. Figure 2 shows the numerical derivative of XTC data averaged over 10 intervals and shows a corresponding periodic modulation in deposition rate. Since the laser power was constant throughout the deposition, it is speculated that some dynamic process is occurring at the target that in turn, affects the deposition rate. The self-layering microstructure seen in this system is of potential use in engineering layered tribological thin films for specific applications encountering thermal cycling.

Figure 3 shows the Raman spectra of the heat treated samples. The as-deposited sample yields the same spectral response as the 400°C sample. No active Raman phases were seen until 500°C where a small WO$_3$ peak is observed. At 600°C, well defined peaks of both WO$_3$ and ZnWO$_4$ are present. At 700°C, the amount of WO$_3$ has increased until at 800°C it is the dominate phase. Interestingly, ZnO and WS$_2$ phases were not seen in any of the heat treated samples.

References
1. J. S. Zabinski et al., submitted to *Thin Solid Films*, March 1995.

Proc. Microscopy and Microanalysis 1995, edited by G.W. Bailey, M.H. Ellisman, R.A. Hennigar, and N.J. Zaluzec
Copyright © 1995 MSA. Published by Jones and Begell Publishing, 79 Madison Ave., New York, NY 10016

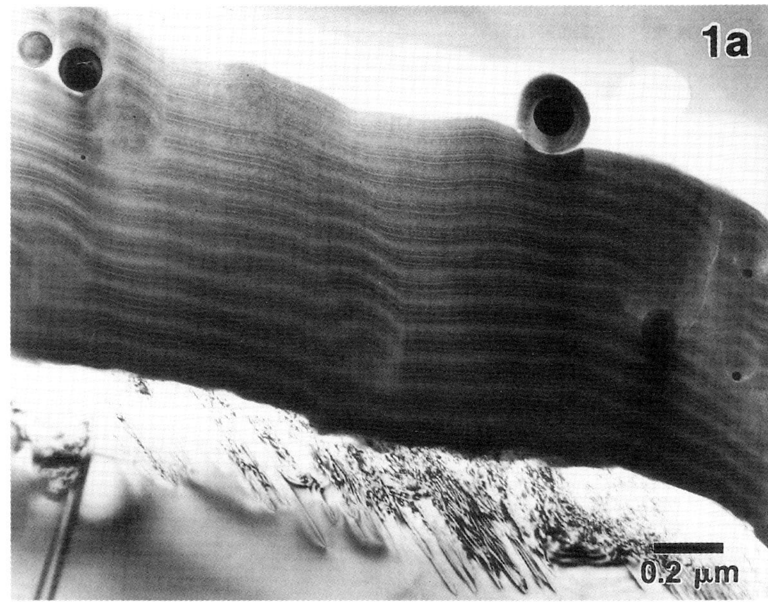
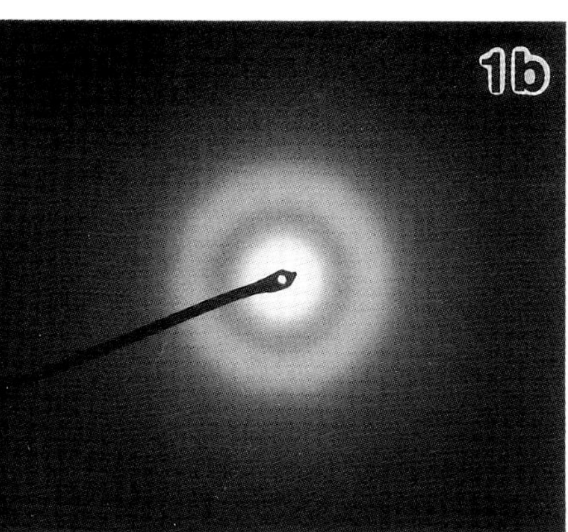
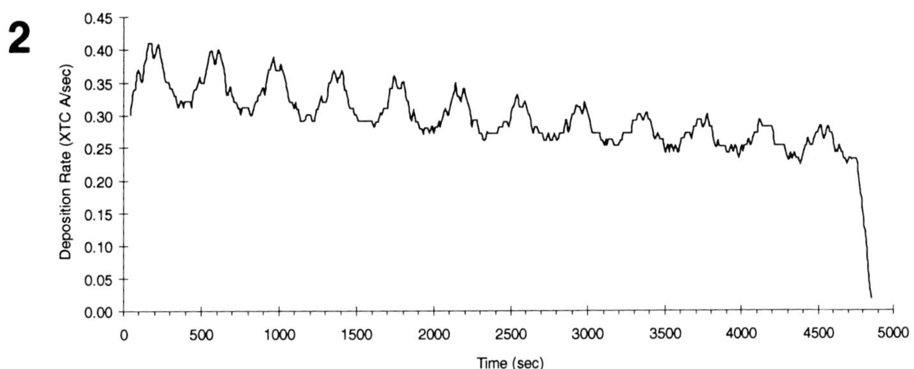
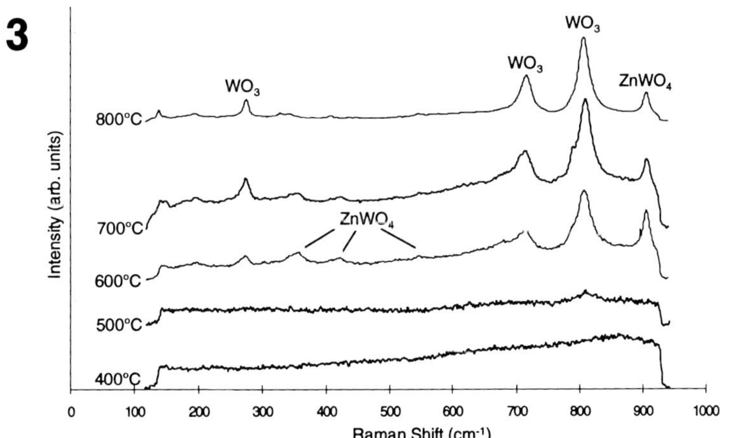

FIG. 1- (a) XTEM micrograph of RT-PLD ZnO-WS$_2$ on SiC; (b) SAD of film.
FIG. 2- Numerical time derivative of quartz crystal thickness monitor output.
FIG. 3- Raman spectra of heat treated RT-PLD ZnO-WS$_2$ films on SiC.

TEM EXAMINATION OF WEAR DEBRIS FROM A SELF-LUBRICATING METAL-MATRIX COMPOSITE PREPARED BY ULTRAMICROTOMY

S. V. Prasad,* S. D. Walck*, and P. F. Lloyd**

*Materials Directorate, Wright Laboratory, Wright-Patterson AFB, Ohio 45433-7750
**UES, Inc., Dayton, OH 45432

Self-lubricating materials are usually multi-phase composites comprising of a solid lubricant dispersed in a matrix of a metal or a polymer. During sliding contact, chemical reactions among the constituent phases of the composite, the counterface and the gases in the surrounding environment are inevitable. The wear debris is a product of such chemical (or tribochemical) reactions. Determination of the chemistry and the crystal structure of the wear debris is therefore crucial to the understanding of the wear mechanisms. In self-lubricating composites, the quantity of the debris generated is often too small to perform x-ray diffraction.

It has been previously demonstrated that self-lubricating aluminum metal-matrix composites can be synthesized by dispersing silicon carbide and tungsten disulfide particles in commercial aluminum alloy matrices.[1] A typical microstructure of an Al-$0.10SiC$-$0.05WS_2$ MMC is shown in Fig. 1. The friction and wear test was performed, in a ball-on-disk configuration, on a polished disk of the aluminum MMC against a 440C steel ball. The test environment was air with a relative humidity of 65%, and the test was run for duration of two million cycles. The coefficient of friction was 0.1.[1] The wear scar was first analyzed in the electron microprobe. Figure 2 shows a typical BSE micrograph of the wear track. Microanalysis was performed on the different phases seen in the wear scar, the sulfur to tungsten (S/W) atomic ratios of the phases were computed. None of the phases on the scar had an S/W ratio close to the stoichiometric value of 2.0, confirming the fact tungsten disulfide had undergone chemical changes during sliding contact.

The wear debris was examined in a JEOL 2000 FX microscope, operated at 200 kV. Initially, a replicating technique was used to prepare the TEM samples. A small piece of an acetate tape was first softened and pressed on the debris. After the acetate tape was hardened, it was stripped from the surface, coated with carbon and placed on a TEM grid. When the acetate was dissolved with acetone, the debris was left behind in the carbon film. The major drawback of this technique was that most of the debris was too thick for TEM. This is a problem for both imaging and microanalysis. To circumvent this problem, an ultramicrotoming technique was adopted to prepare the TEM samples. The wear debris was brushed into a BEEM capsule and encapulated with Spurr's epoxy resin. The samples were ultramicrotomed with a Reichert Ultracut ultramicrotome using a 55° Diatome diamond knife. The sections were collected on 400 mesh, thin bar TEM grids and coated with a thin layer of carbon in order for the resin to be stable under the electron beam.

The debris comprised of both crystalline and amorphous phases. The crystalline phases were of extremely small in crystallite size, and some of these crystallites were interdispersed with the amorphous material. No crystalline WS_2 were detected in sufficient quantity to confirm the improvement in tribological behavior due to slippage of WS_2 platelets. One single crystal phase with an extremely large d-spacing (8.7 Å) was detected in several debris particles. A typical debris particle is shown in Fig. 3. The faint concentric rings in Fig. 3c are due to diaspore. Diaspore is an aluminum hydroxide, $AlO(OH)$, and has an orthorhombic crystal structure. The major conclusion of this study has been that tungsten disulfide is not directly responsible for the improved friction behavior, but there are several new phases formed as a result of tribochemical reactions. TEM is the most appropriate technique for analyzing the extremely small tribochemical phases in self-lubricating composite materials.

References

1. S. V. Prasad and K. R. Mecklenburg, *Lubrication Engineering*, 50(1994) 511.

Proc. Microscopy and Microanalysis 1995, edited by G.W. Bailey, M.H. Ellisman, R.A. Hennigar, and N.J. Zaluzec
Copyright © 1995 MSA. Published by Jones and Begell Publishing, 79 Madison Ave., New York, NY 10016

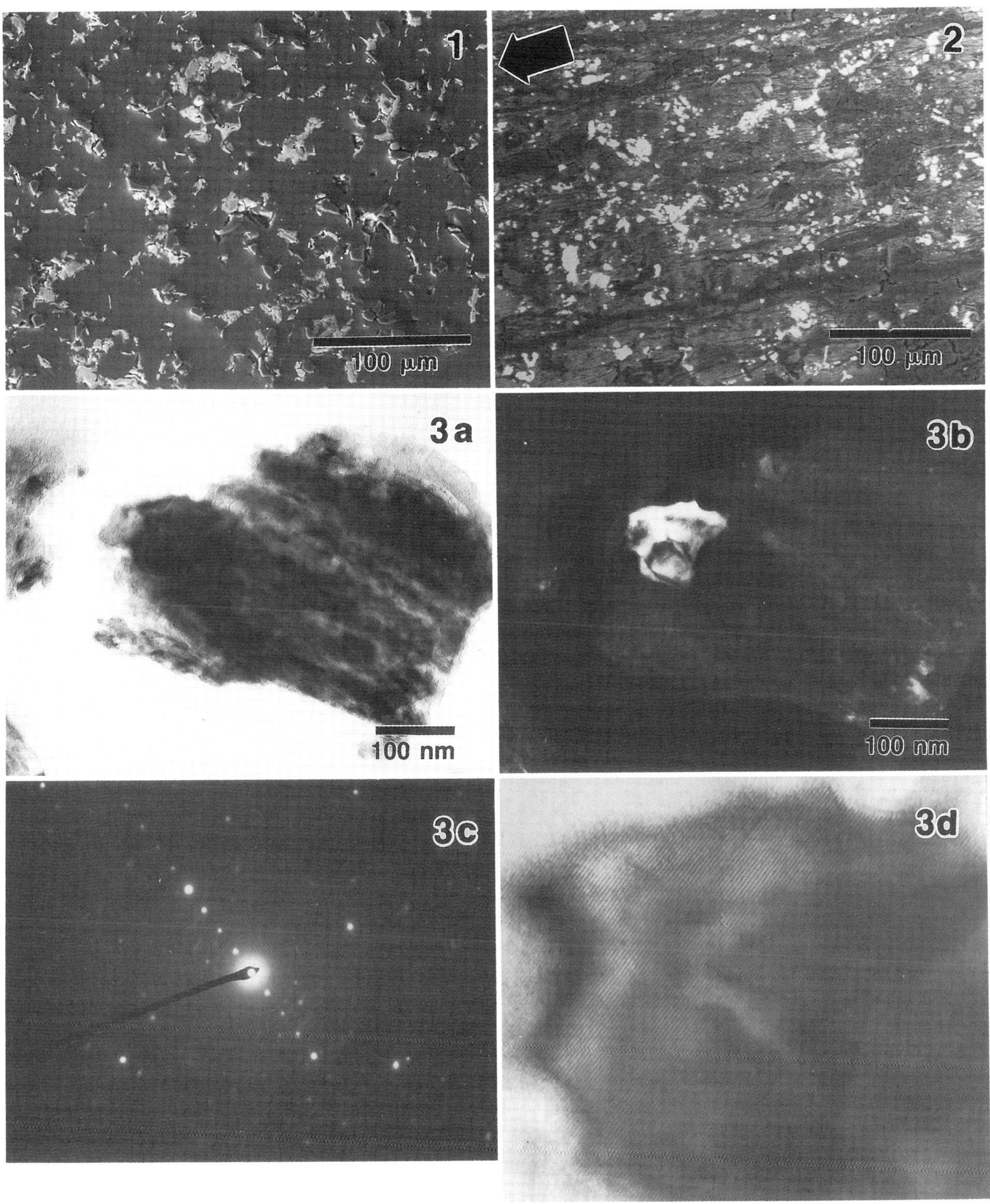

FIG. 1 - SEM micrograph of an Al-0.10SiC-0.05WS$_2$ MMC.
FIG. 2 - BSE SEM micrograph of the wear track. Arrow indicates the direction of sliding.
FIG. 3 - (a) BF TEM micrograph of wear debris particle; (b) DF; (c) SAD; (d) lattice image, d=8.7 Å.

THE FIRST TEST ON A MAGNETICALLY FOCUSED MIRROR CORRECTOR

A.V. Crewe, F. Tsai, P. Korda, and S. Ruan

Enrico Fermi Institute and the Department of Physics, University of Chicago, 5640 S. Ellis Ave., Chicago, IL 60637

In some recent publications we have discussed the possibility of correcting both spherical and chromatic aberrations using a magnetically focused mirror.[1,2,3,4] If such a scheme could be made to work it would represent the simplest possible type of aberration corrector and for this reason we decided to test the principles of operation using a simple experimental system (Fig. 1). Electrons are obtained from a TFE source operating at 3.6kv which is immersed in a region of uniform magnetic field which refocuses them at a point 5cm away. At this position we placed a simple specimen consisting of a copper grid with a holey plastic film. Electrons which pass through a hole in the specimen enter a region where the magnetic field is supplemented by a uniform electrostatic field which can act as a mirror. By a proper adjustment of the strength of this field the electrons can be refocused on the reverse side of the specimen where the total aberration should be zero. We have provided two scan coils, one external to the vacuum system which scans the whole beam and one internal to the vacuum system which is located in the region of the electrostatic mirror. These two scan coils can be driven independently or synchronously.

We have recently completed a series of tests on this system in order to explore the first order focusing properties. Our experiments indicate that the electron optical system behaves almost exactly as predicted, the electrons focus on the front side of the specimen at the proper setting of the magnetic field and can be refocused on the back side of the specimen at the calculated value of the electrostatic field. We have operated this system in many different modes but we are currently operating with a 750μm diameter aperture located approximately midway between the source and the specimen so that the convergence angle of the electron beam at the specimen is approximately 27mr. Image contrast is provided by the specimen current, which is approximately 5nA or the current on a copper washer in the aperture plane. The various contributions to the size of the probe at the first focus are

Diffraction $0.61\lambda/\alpha = 0.45$nm

Source size $= 15$nm

Spherical aberration $0.25C_s\alpha^3 = 147$nm

Chromatic aberration $C_c\,\alpha\Delta v/v = 42$nm

Space charge[5] $= 2.1$nm

We have also performed some stochastic calculations on the shape of the intensity distribution to be expected in the region of the best focus. These calculations indicate that we should expect a distribution with a sharp peak in the forward direction. We have not observed this peak, presumably due to the fact that the use of specimen current as our contrast mechanism does not permit such fine detail.

We are just beginning studies on third order imaging properties and the results to date are encouraging but not conclusive.

References

1. A.V. Crewe, *J. Microscopy* 172(1993)97.
2. A.V. Crewe, *Ultramicroscopy* 41(1992)269.
3. A.V. Crewe, *Ultramicroscopy* 41(1992)279.
4. A.V. Crewe, *Optik* 94(1993)180.
5. J.A. Venables and G. Cox, *Ultramicroscopy* 21(1987)33.

Proc. Microscopy and Microanalysis 1995, edited by G.W. Bailey, M.H. Ellisman, R.A. Hennigar, and N.J. Zaluzec
Copyright © 1995 MSA. Published by Jones and Begell Publishing, 79 Madison Ave., New York, NY 10016

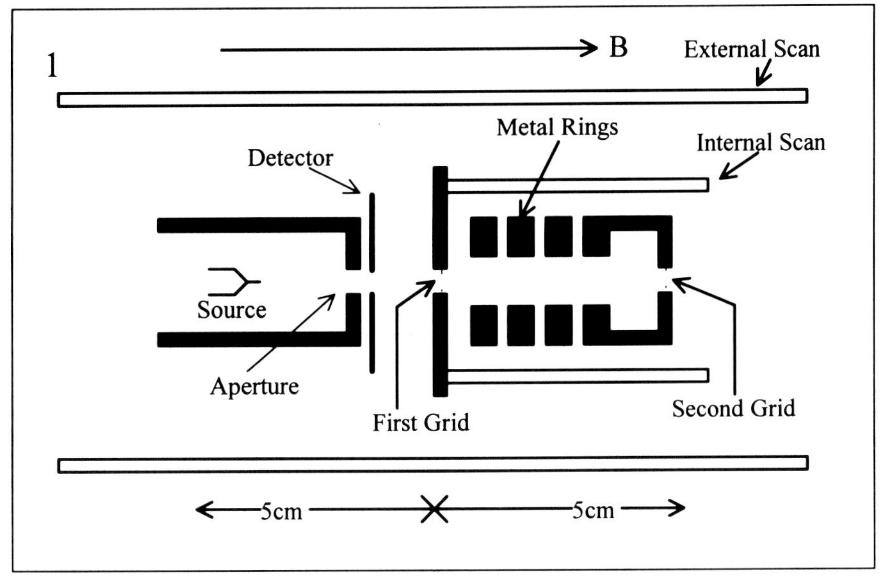

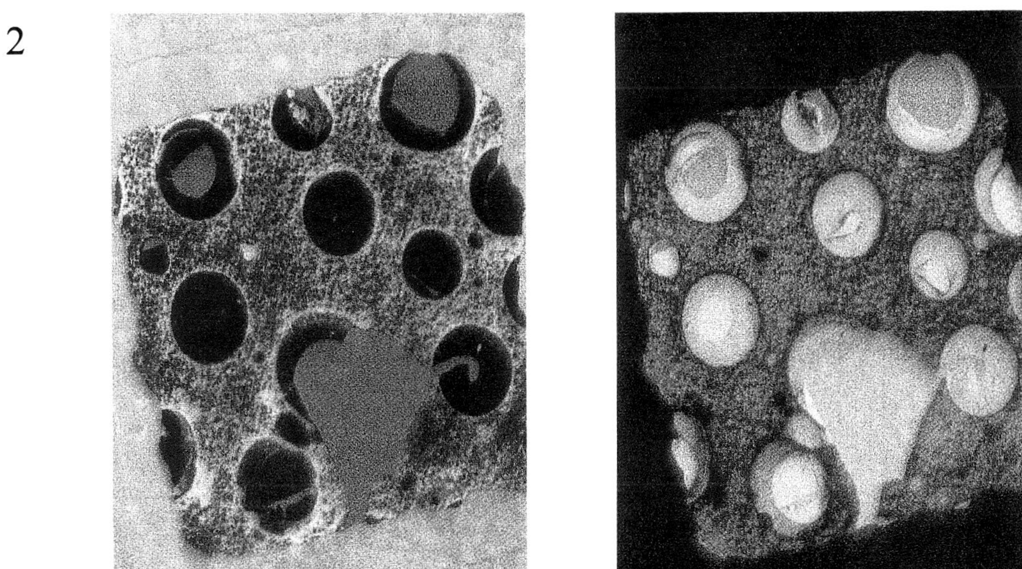

FIG. 1.- Schematic diagram of configuration used for first tests of mirror system. Metal rings are connected to potential divider to produce uniform retarding field.

FIG. 2.- The left image is of front side of specimen using specimen current for the signal. The image on right is of back side of specimen using mirror and aperture plane detector. Approximately 40μm full scale.

FRESNEL EFFECT IN HIGH RESOLUTION TEM IMAGING OF SMALL PARTICLES

Velimir Radmilovic◇ and Michael A. O'Keefe☆

◇University of Belgrade, Dept. of Physical Metallurgy, Karnegijeva 4, 11001 Belgrade, Yugoslavia
☆National Center for Electron Microscopy, University of California, LBL B72, Berkeley, CA 94720

It is established that analysis of Fresnel contrast visible at the edges of nanoparticles in the high resolution electron microscope provides a valuable tool for determining the size of the particles with near atomic spacing accuracy.

Methods based on the Fresnel effect are enjoying increased use in the study of flat precipitates, multi-layers, and grain boundaries.[1-4] In profile imaging studies of solid surfaces, Fresnel effects have been observed by Marks[5], and by O'Keefe et al.[6] who showed that image structure within Fresnel fringes from MgO cubes can produce "ghost" atoms far outside the specimen surface. However, consequences of the Fresnel effect appear not to have been included in the study of small particles at atomic resolution. In this work, we have used HRTEM image simulation to explore the changes in images of a nanoparticle under various imaging conditions, in particular to relate the model particle size and its apparent size as derived from the HRTEM image.

We used a particle consisting of 309 atoms of platinum and ruthenium, arranged to form a four-shell cube-octahedron. The cube-octahedron was placed in a supercell of a 40Å lattice parameter (fig.1), as in the work by Malm and O'Keefe,[7] to ensure that particle images were not affected by electrons scattering from the unavoidable adjacent particles generated by the periodic boundary conditions imposed by the computation.[8] Computations, using microscope parameters corresponding to those of the JEOL ARM-1000 at Berkeley over the defocus range of 1000Å (overfocus) to -1000Å (underfocus) at an electron energy of 800keV, were carried out with the MacTempas package.[9]

Examples of images simulated from the model described above are shown in figure 2. Digital measurements of the spacings of the first Fresnel fringe were used to produce an estimated size for the particle. Such measurements produce figures for the particle size that are larger than the true size, but are very close to it near Scherzer defocus (fig.3a). For this particle, the minimum size measured from the image was ~18.5Å, which is essentially the same as the 18.47Å distance between the 001 facets of the particle. It is interesting to note that the minimal difference is obtained at a defocus approximately -400Å below the Gaussian image plane, i.e. close to the Scherzer value for the JEOL ARM-1000 of $\sqrt{(C_S\lambda)} = -448$Å. This falls within the results obtained by Ross and Stobbs[2-4], in which Fresnel fringes exhibited minimal widths at zero defocus (to within ±2100Å). Contrast of the fringes (measured over 50 pixels, equivalent to 11Å), shows a similar behavior; minimum contrast is obtained close to Scherzer defocus (fig.3b).

Finally, it is important to note that at nearly all defocus settings above and below the Scherzer defocus, the spot intensity distribution does not correspond to the positions of the atoms in the particle, especially near a thick edge; such behavior can be observed experimentally in small particles of MoS_2.[10] We believe that this effect is due to a strong Fresnel influence on the image, since the deviations between atoms and dots are more pronounced in regions close to the thick edges of small particles (fig.2d), i.e. in regions where the potential drop is sharper. This non-uniform "relaxation" may even change the aspect ratio of the particle (e.g. it appears "taller" in fig.2d when compared with the Scherzer image of fig.2c).[11]

1. J.N. Ness, W.M. Stobbs and T.F. Page, *Phil.Mag.A* **54** (1986) 679-703.
2. F.M. Ross and W.M. Stobbs, *Phil.Mag.A* **60** (1991) 1-36.
3. F.M. Ross and W.M. Stobbs, *Phil.Mag.A* **60** (1991) 37-72.
4. F.M. Ross and W.M. Stobbs, *Ultramicroscopy* **36** (1991) 331-354.
5. L.D. Marks, *Surface Science* **139** (1984) 281-298.
6. M.A. O'Keefe, J.C.H. Spence, J.L. Hutchison & W.G. Waddington, *43rd Ann. Proc. EMSA*, Louisville, Kentucky (1985) 64-65.
7. Jan-Olle Malm and Michael A. O'Keefe, *SCANDEM-93*, Lund, Sweden (1993) 131-132; and *Ultramicroscopy (submitted)*.

Proc. Microscopy and Microanalysis 1995, edited by G.W. Bailey, M.H. Ellisman, R.A. Hennigar, and N.J. Zaluzec
Copyright © 1995 MSA. Published by Jones and Begell Publishing, 79 Madison Ave., New York, NY 10016

8. M.A. O'Keefe and S. Iijima, 9th Internat. Cong. on Electron Microscopy, Toronto (1978) 282-283.
9. R. Kilaas, Proc. 45th Ann. Proc. EMSA, Baltimore, MD (1987) 66-69.
10. M. José-Yacaman, Journal of Nano-Structured Materials (submitted) (1994).
11. Work supported by the Director, Office of Energy Research, Office of Basic Energy Sciences, Material Sciences Division of the U.S. Department of Energy, under contract No. DE-AC03-76SF00098.

Figure 1. View of the model particle inside its supercell. Particle "size" is defined as the distance between the top and bottom 001 facets (18.47Å). Changes in crystal potential (projected in the [001] direction) are abrupt at the thick edges of the particle (top & bottom), but less so at its tapering edges (left & right).

Figure 2. HRTEM simulations of the model particle viewed in [001] for four values of defocus (marked). Fresnel effects produce "ghost" atoms outside the particle (dots show atom positions of the model). Fringes are more intense at the thick (top & bottom) edges of the particle, and less at its thin tapering (left & right) edges. At Scherzer defocus (c), interplanar spacing is almost regular, but with further underfocus (d) the image shows an apparent "relaxation" of the outer planes of atoms, especially at the top and bottom surfaces where potential change is steep and Fresnel contrast strong.

Figure 3. Variation, as a function of defocus, of (a) particle size as measured from the HRTEM image, and (b) Fresnel fringe intensity. Plots show a minimum in the neighborhood of Scherzer defocus. Size is measured "across flats", i.e. between the top and bottom facets, where the potential change is most abrupt.

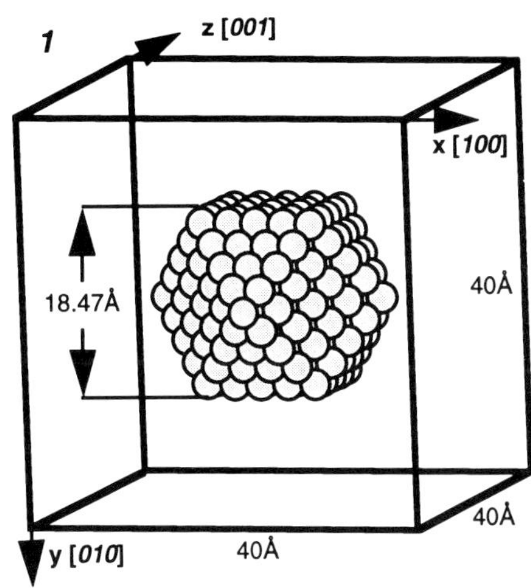

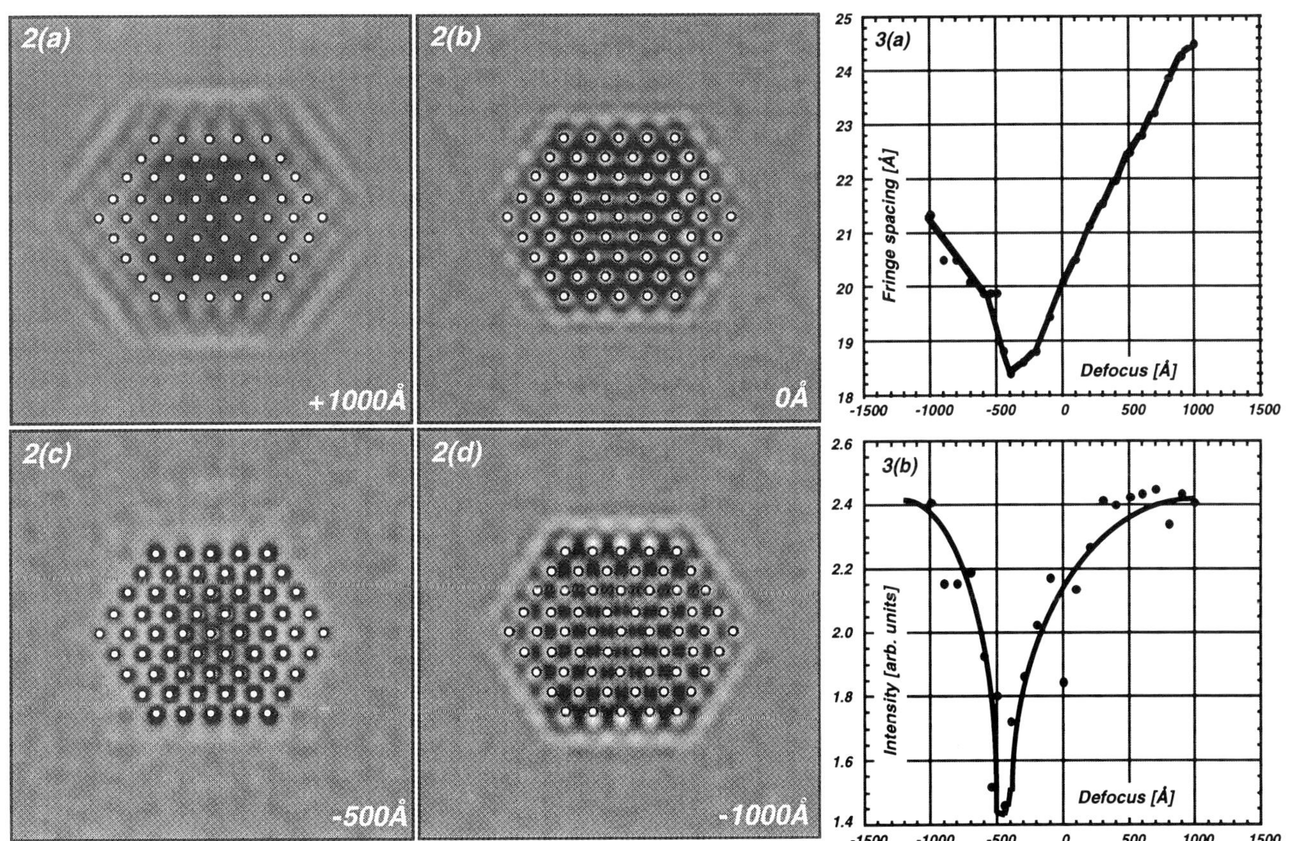

QUANTITATIVE MEASUREMENT OF EXPERIMENTAL PARAMETERS FOR HIGH RESOLUTION ELECTRON MICROSCOPY IMAGES IN FOURIER SPACE

T. Zheng* and J. Murray Gibson*

*Seitz Material Research Laboratory, University of Illinois at Urbana-Champaign, Urbana, IL 61801

Due to the multiple scattering and interference intrinsic in high resolution transmission electron microscopy (HRTEM) images, interpretation of HRTEM images is difficult. In order to interpret contrast details in a HRTEM image, simulated images for various proposed structure have to be obtained until good agreement is reached. The usual multislice method for image simulation is extremely time-consuming and has to be repeated for each structure under consideration. Also, the experimental parameters, for example objective lens defocus and sample thickness etc., are related to image details in HRTEM images. Knowledge of these experimental parameters is essential for the image interpretation and is important for quantitative interface analysis, for example chemical mapping by Ourmazd and colleagues.[1] Thust and Urban[2] have reviewed the conventional method for measuring sample thickness and objective lens defocus value and proposed a method for sample thickness and defocus determination. But their method requires synchronous sampling of experimental images under consideration, which is not always possible for on-line digitized images in order to retain the resolution. And experimental and simulated images have to be perfectly aligned. Also in their matching only sample thickness and defocus value have been considered, without inclusion of other parameters such as beam tilt, crystal tilt and Cs value for the specific experimental setting. In all the current quantitative matching methods of experimental HRTEM images, error assessment has never been concerned.

In this paper a Fourier space refinement method which allows the determination of all the experimental parameters with error evaluation is proposed. This method also avoids the unrealistic simplifications of synchronous sampling and the alignment of experimental and simulated images. In order to show the reliability of extracting information from an experimental image, the method is tested on a computer generated sinusoid function with random noise incorporated, as shown in figured 1a. If under continuous and infinite Fourier transformation, amplitudes and phases of four delta peaks should be 0.25 and 0 or π respectively in Fourier space. Figure 1b, the diffractogram of this image, the extension of the spots is displayed and the insert at the bottom left shows the amplitude and phase of the indicated spot. In order to get amplitudes and phases of a nonsynchronously sampled image like this, a non-linear curve fitting algorithm using the Levenberg-Marquardt[3] method was developed. A small area around the spot, usually 12pixel X 12pixel, was selected and the intensity values of these pixels were fitted to a two dimensional sinc function. From this fitting the amplitude and phase were obtained, as shown at the top left corner in fig. 1b. This method has been tested on simulated images with very strong noise and displays reliable results. Fourier transformation has been used as a noise filtering method previously. In our method, this advantage of Fourier transformation can be easily used. Our noise analysis shows that inclusion of a constant term in the objective function can give us satisfactory results about noise without loosing information of the signal. A defocus series of experimental images is acquired and the amplitudes and phases of all the frequencies in each image are obtained using this method.

These experimental data are read into an optimization program for the determination of the experimental parameters. This optimization program is developed based on Stadelmann's[4] EMS electron image simulation package. Theoretical image intensities, I^t_g, are calculated and compared with experimental data, amplitudes a^e_g and phases ϕ^e_g. The optimization algorithm tries to find the minimum of the function (r_0 is the relevant shift between the experimental and the theoretical images):

Proc. Microscopy and Microanalysis 1995, edited by G.W. Bailey, M.H. Ellisman, R.A. Hennigar, and N.J. Zaluzec
Copyright © 1995 MSA. Published by Jones and Begell Publishing, 79 Madison Ave., New York, NY 10016

$$R=\sum_{\mathbf{g}} (Re(I_{\mathbf{g}}^{t})-a_{\mathbf{g}}^{e}\cos(\phi_{\mathbf{g}}^{e}+2\pi\mathbf{g}.\mathbf{r}_{0}))^{2}+(Im(I_{\mathbf{g}}^{t})-a_{\mathbf{g}}^{e}\sin(\phi_{\mathbf{g}}^{e}+2\pi\mathbf{g}.\mathbf{r}_{0}))^{2}$$

with 14 variables; λ, defocus spread, beam divergence, Cs, absorption constant, crystal tilts, beam tilts, specimen thickness, objective lens defocus, image shifts etc. The quasi-Newton method[5] is used for the optimization. The results of the optimization algorithm using quasi-Newton method will be tested on both simulated and experimental data and will be presented in the talk. The outcome of this research will show the validity of current image calculation theory and can greatly improve the reliability of HRTEM image analysis. The application of our method on experimental images will be extensively discussed in the demonstration[6].

References

1. A. Ourmazd, F.H. Baumann, M. Bode and Y. Kim, Ultramicroscopy 34 (1990) 237
2. A. Thust and K. Urban, Ultramicroscopy 45 (1992) 23
3. William H. Press, Brian P. Flannery, Saul A. Teukolsky and William T. Vetterling, *Numerical Recipes in C* (Cambridge university Press, 1988)
4. P. A. Stadelmann, Ultramicroscopy 21 (1987) 131
5. David Kahaner, Cleve Moler and Stephen Nash, *Numerical Methods and Software* (Prentice Hall, 1989)
6. This work is supported by DOE under grant No. DEFG02-91ER45439/10.

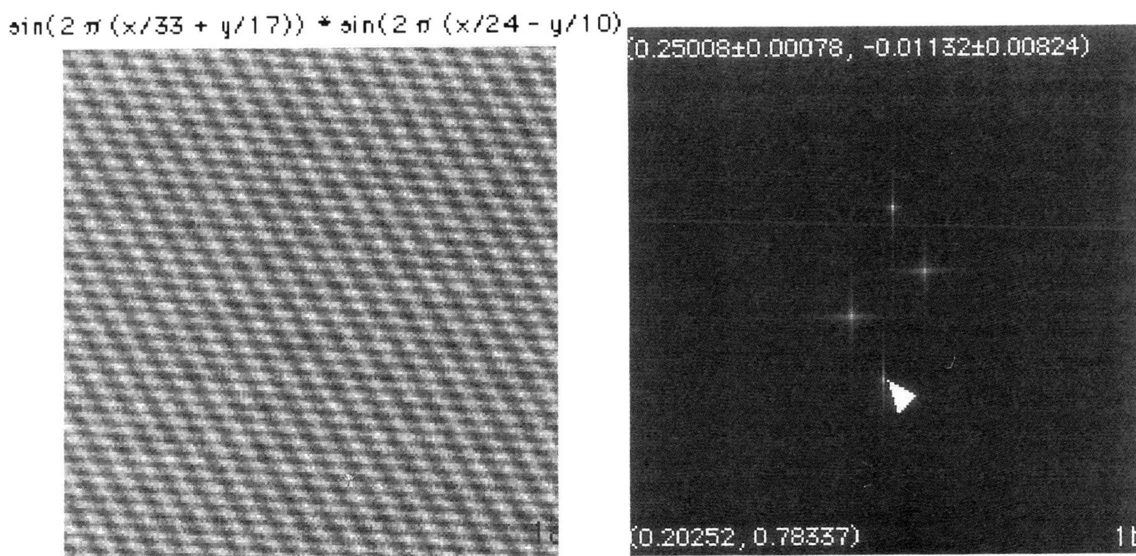

FIG. 1--(a) A computer generated image using function displayed on the top of the image. For this sinusoid function, the continuous and infinite Fourier transformation should give four delta peak with amplitudes 0.25 and phases either 0 or π. (b) Digital Fourier transformation of 1a. The left corner insert shows the amplitude and phase value which a commercial software package, Gatan DigitalMicrograph, produced for the indicated spot. The top insert displays our analysis result as (amplitude, phase), which are statistically close to the values from the continuous and infinite Fourier transformation.

HREM STRUCTURE ANALYSIS OF Σ13 SYMMETRICAL BOUNDARIES IN STRONTIUM TITANATE BICRYSTALS

J.E. Bonevich, S.Q. Xiao and U. Dahmen

National Center for Electron Microscopy, Lawrence Berkeley Laboratory, Berkeley, CA 94720

Strontium titanate (STO) has found a myriad of technological applications, most notably as a substrate material for superconductors such as $YBa_2Cu_3O_{7-\delta}$ (YBCO). One example is the growth of thin film YBCO devices whereby grain boundaries introduced in the films are used to create SQUID devices.[1] The nature of the interfacial structure plays an important role in determining the behavior of these materials, e.g., their electrical properties.[2] We have thus endeavored to investigate the atomic structure of grain boundaries in STO and to explore the structure implications on the properties of YBCO films.

The Σ13 STO bicrystals with two symmetrically inclined boundary planes of (510) and (320) were produced by Shinkosha Ltd.(Japan); the bicrystals were rotated by 22.6°±0.5° about the common [001] zone axis. Cross-sectional TEM specimens were prepared in the usual manner and examined in the NCEM's Atomic Resolution Microscope operating at 800 kV. The HREM images were subsequently digitized for template matching (cross-correlation) analysis.[3]

The Σ13 (320) boundary (Figure 1a) consists of long terraces with relatively few steps. While the rotation of the (001) planes in this bicrystal was expected to be 22.6°, the measured value was ~22.1°. Closer examination of the boundary revealed a rigid body shift of the crystals along the boundary plane of ~0.11 nm as measured by the extension of the crystal directions in the bulk up to the boundary. This shift corresponds to a DSC dislocation vector of $1/13[2\bar{3}0]$ leading to the creation of a mirror glide symmetrical boundary . To further elucidate the local structure, template matching of image motifs was investigated, see Figure 1b. One such motif (inset) had a periodicity of 1.4 nm as expected for the (320) boundary and displays a pattern consistent with a mirror glide symmetry. The boundary regions with best fit to the periodic template were found in ~10 nm long segments. However, the segments themselves were spaced aperiodically, shifted along the boundary by integral multiples of the DSC vector. This observation is consistent with the presence of DSC dislocations in the boundary.

The Σ13 (510) boundary, although macroscopically flat, had a measured rotation of ~21.7° giving rise to 50 nm terraces separated by steps (Figure 2a). Although the boundary appeared slightly asymmetrical about the (510) plane, a rigid body shift could not be detected. Structural units with 1 nm periodicity were observed, the expected characteristic spacing for the (510) boundary plane. This structure can be understood by considering the effects of the angular deviation from perfect coincidence on the periodic structure units. A misorientation of 0.9° in one crystal can be accommodated by the introduction of a series of DSC steps, of height 0.15 nm, perpendicular to the boundary plane. Pairs of steps of opposite sign maintain an overall boundary flatness while shifting the periodicity of the characteristic structure unit along the boundary. Template matching of the boundary (Figure 2b) reveals that the periodic image motif (2b inset) is indeed shifted by an integral multiple of the DSC lattice vector.

In summary, we have made detailed HREM observations of two Σ13 [001] tilt boundaries in strontium titanate, grown in the (320) and (510) symmetrical orientations, respectively.[4] A rigid body translation has been found only for the (320) boundary which was shown to exhibit mirror glide symmetry. Both interfaces contained steps and pattern shifts consistent with the presence of DSC lattice dislocations. Cross correlation with a structural unit pattern template was applied as a useful technique to analyze the degree of similarity and precise location of characteristic motifs with the HREM images.

References
1. J. Clarke, in H. Weinstock and R.W. Ralston, Eds., *The New Superconducting Electronics*, Amsterdam: Kluwer Academic Publishers (1993)123.

Proc. Microscopy and Microanalysis 1995, edited by G.W. Bailey, M.H. Ellisman, R.A. Hennigar, and N.J. Zaluzec
Copyright © 1995 MSA. Published by Jones and Begell Publishing, 79 Madison Ave., New York, NY 10016

2. D. Dimos et al., *Phys. Rev. B* 41(1990)4038.
3. S. Paciornik, R. Kilaas, J.H. Turner and U. Dahmen, submitted to *Ultramicroscopy*.
4. The NCEM is supported by the Director, Office of Basic Energy Science, Materials Science Division of the Department of Energy under Grant No. DE-AC03-76SF00098.

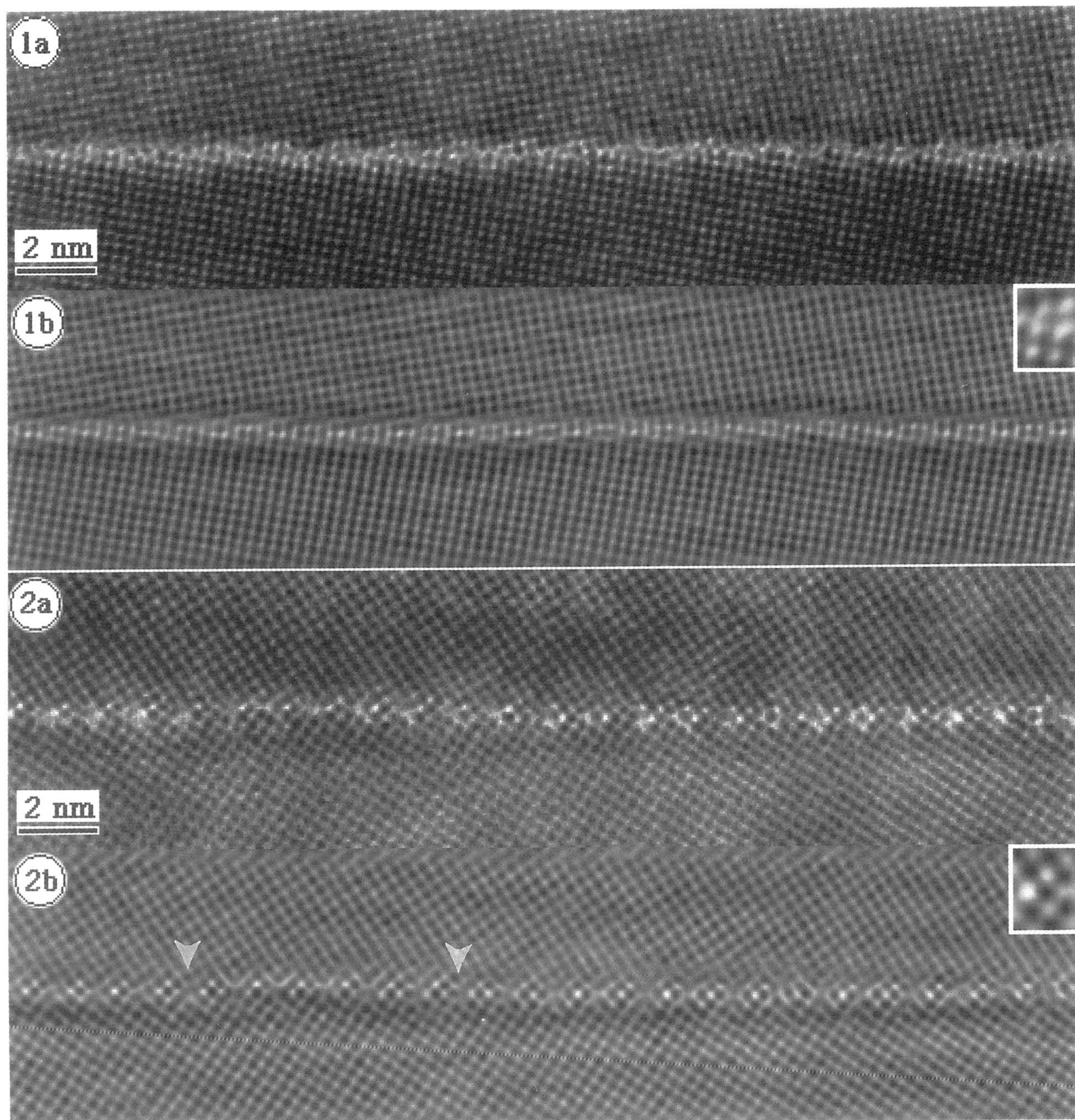

FIG. 1.--HREM image of STO Σ13-(320) boundary (a) . Cross-correlation (b) of image with template (inset) shows 1.4 nm periodicity and reveals the mirror-glide symmetry of boundary.
FIG. 2.--STO Σ13-(510) boundary (a) . Template matching (b) reveals a motif with 1.0 nm periodicity and the presence of steps (arrowed) in the boundary. Rigid body shifts could not be detected.

REFINEMENT OF RIGID SHIFT COMPONENT NORMAL TO A Σ5 GRAIN BOUNDARY IN RUTILE BY QUANTITATIVE HREM

S. Paciornik[1,2], D. Michel[3] and U. Dahmen[1,3]

[1]National Center for Electron Microscopy, University of California, LBL B72, Berkeley CA 94720,
[2]DCMM, PUC-Rio, P.O BOX 38008, Rio de Janeiro, RJ 22452, Brazil
[3]CECM-CNRS, 15, rue Georges Urbain, F94407 Vitry sur Seine Cedex, France

In contrast to the macroscopic parameters of a grain boundary, its microscopic parameters are difficult to determine with sufficient accuracy to make critical comparisons between models and experimental observations. Whereas the axis and angle of misorientation and the orientation of the boundary plane are easily measured from diffraction patterns or high resolution images, the rigid body shift and the localized atomic relaxation are far more complicated to determine and are more sensitive to image or sample artifacts. Yet it is only these microscopic parameters that differentiate the behavior of the same crystallographic interface in different materials. While the angle of misorientation and the plane of inclination are fixed macroscopically, the rigid body shift and localized atomic relaxations depend directly on the type of bonding that characterizes a given material.

Several methods for the measurement of rigid shifts at interfaces have been given recently [1,2]. Their accuracy is limited by instrumental parameters such as beam tilt, sample tilt, residual two-and three-fold astigmatism, specimen noise or twist components. Because atomic relaxations often extend over several atomic planes near the interface, the measurement of the rigid shift is best performed by measuring the relative position of each grain at a certain distance from the interface and extrapolating to the boundary plane [2]. Disadvantages of this procedure are that elastic distortions due to dislocations or steps in the interface can contribute to the measured shift, and that it combines the rigid shift immediately at the boundary with the more extended relaxations in the vicinity of the boundary.

In the present work we describe an alternative method that optimizes the fit between simulated and experimental images with respect to the magnitude of the rigid shift as a variable. The technique compares experimentally observed with modeled images using the normalized cross correlation coefficient as a measure for the goodness of fit. To avoid complications from elastic distortions, misorientations or extended relaxations, this optimization is applied only to a small segment of the image depicting a structural unit characteristic for the interface.

A section of the experimental image used in this comparison is shown in figure 1. As shown previously [3], a global 2° deviation of the whole boundary from the exact Σ5 misorientation and a 5° deviation from the exact (210) symmetrical orientation are accommodated by DSC lattice dislocations of 1/5[210] at steps of height 1/2[210], spaced at about 50Å along the boundary.

Image simulations were performed for the precise Σ5 misorientation and the exact symmetry position of the interface using experimentally determined imaging conditions of the Berkeley ARM operated at 800kV, a defocus of -690Å, and a foil thickness of 20Å. The in-plane component of the rigid shift was determined previously by the observation that the interface possessed glide mirror symmetry. Subject to this constraint, a structural model of the interface was built and subsequently relaxed to optimize spacings between ions. In order to avoid boundary effects during relaxation or image simulations, a unit cell size of about 40Å perpendicular to the interface was employed. The image/model comparison was then performed for a small region of the interface containing a characteristic structural unit pattern. This region is framed white in the figure. The size of the unit cell used in the model simulations was several times larger.

Proc. Microscopy and Microanalysis 1995, edited by G.W. Bailey, M.H. Ellisman, R.A. Hennigar, and N.J. Zaluzec
Copyright © 1995 MSA. Published by Jones and Begell Publishing, 79 Madison Ave., New York, NY 10016

Because the region of interest was extremely localized it was possible to vary the rigid shift through a uniaxial expansion of the entire model unit cell normal to the interface. Figure 2 shows the resulting plot of the correlation coefficient as a function of rigid shift perpendicular to the boundary. A clear peak is observed for an expansion of 0.44Å (a contraction with respect to the Ti sublattice [3]), corresponding to a local change in atomic volume. As will be shown in detail elsewhere, this change is almost precisely that required to accommodate the deduced O deficiency which changes the local stoichiometry from TiO_2 towards Ti_2O_3. The procedure presented here can thus provide highly accurate m+easurements of rigid shifts at an interface even in situations where global deviations in misorientation and inclination destroy the perfect symmetry and periodicity of a symmetrical grain boundary.

References
1.G.J. Wood, W.M. Stobbs and D.J. Smith, Phil. Mag. A50, 375 (1984)
2.K. Merkle, Scripta Met. 23, 1487 (1989), Ultramicroscopy 40, 281 (1992), M.I. Buckett and K.L. Merkle, Proc. MSA 51, 982 (1993)
3.U. Dahmen, S. Paciornik, I.G. Solorzano and J.B. Vandersande, Interface Science 2, 127 (1994)
4.This work is supported by the Director, Office of Energy Research, Office of Basic Energy Sciences, Materials Sciences Division of the U.S. Department of Energy under Contract No. DE-ACO3-76SFOOO98.

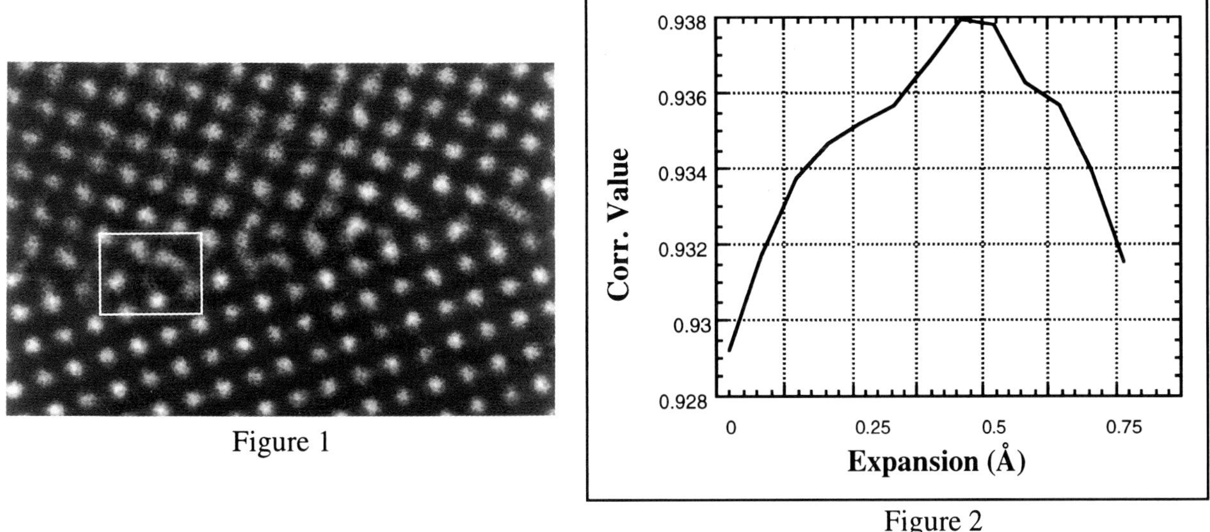

Figure 1

Figure 2

Figure 1: Section of experimental image (Berkeley ARM, 800kV, -690Å defocus) showing two flat segments of Σ5 interface separated by a step containing DSC lattice dislocation.

Figure 2:Normalized cross correlation coefficient between structural unit patterns in the experimental image (fig.1) and in model simulations, plotted as a function of rigid shift normal to the interface (given here by the size of model unit cell in the direction normal to the interface). A clear peak appears at a local expansion from the rutile lattice of 0.44Å.

THE MEASUREMENT OF THE PERIOD AND STRAIN OF SEMICONDUCTOR SUPERLATTICE SYSTEMS USING HREM

M.D. Robertson,* J.M. Corbett,* J.E. Currie,* and J.B. Webb**

*Guelph-Waterloo Program for Graduate Work in Physics, University of Waterloo, Waterloo, Ontario, Canada N2L 3G1
**Institute for Microstructural Sciences, National Research Council of Canada, Montreal Road, Bldg. M-50, Ottawa, Ontario, Canada K1A 0R6

The quantification of local strains in semiconductor epilayer systems is of technological importance since strain can affect the physical, electrical and optical properties of the material. For the measurement of strains in epitaxial structures on the nanometer scale, high resolution electron microscopy (HREM) images recorded using slow scan charged-coupled device cameras and analyzed using Fourier filtering techniques can provide reliable, quantitative results that are independent of image contrast.[1,2]

Fig. 1a is a HREM image of about 3 periods from a 20 period $In_{.82}Al_{.18}Sb$ / InSb coherently strained superlattice in [110] zone axis projection. The superlattice was grown by magnetron sputter epitaxy and the growth direction is along [001].[3] The image was recorded at 200 kV using a Philips CM20 TEM. Displayed is a 128x697 pixel subset of the original 256x1024 image. This image was Fourier transformed to obtain the digital diffractogram and all reflections apart from the [002] were filtered out using elliptical masks of approximately the first Brillouin zone in size. The inverse Fourier transform, displaying the [002] lattice fringes, is presented in Fig. 1b. Any reflection from the diffractogram can be filtered in this manner depending on the component of strain that is to be measured. Reflections with their components of the g-vector parallel / perpendicular to the interface normal will yield strain information along these respective directions. Since the superlattice under study is coherently strained, only strain information parallel to the interface normal is required. The pixel spacing between two consecutive fringes, Z_i, and the average fringe spacing $\bar{Z}$, can be accurately determined by using polynomial fits to the lattice fringe intensity data. Polynomial fits were found to be more accurate than sinusoidal fits since more than one Fourier frequency is always present in the elliptical masks. Fig. 1c is a plot of the cumulative sum (CUSUM) of the deviations in lattice fringe spacing from the average fringe spacing, $\sum (Z_i - \bar{Z})$, versus the pixel position and Fig. 1d is a plot of the raw lattice fringe spacings versus pixel position. The vertices of the CUSUM plot indicate regions where the spacings of the lattice fringes have changed and in strained epitaxial systems, this marks the position of the interface between the two materials. Regions of positive slope in the CUSUM plot indicate the InSb layers, while regions of negative slope indicate the lower lattice parameter $In_{.82}Al_{.18}Sb$ layers. A comparison of Figs. 1c and 1d demonstrates the superior ability of the CUSUM technique to detect small changes in strain even in the presence of significant noise arising from the lattice fringe spacing measurements. Using the fringe spacing of the InSb layers as an internal calibration standard, the average period of the superlattice was measured to be 11.4(0.6) nm. This compares well to the average period determined by x ray methods of 11.2 nm for all 20 periods. Also, from the slopes of the CUSUM plot,[2] and correcting for surface relaxation effects,[4] the average lattice misfit was found to be -0.010(0.001). X ray results for the entire 20 periods gave a lattice misfit of -0.0093(0.0005), a value within the 1 standard deviation error limits.[5]

References

1. R. Bierwolf et al., *Ultramicroscopy*, 49(1993)273.
2. M.D. Robertson et al., *Ultramicroscopy*, in press.
3. J.B. Webb et al., *J. Cryst. Growth*, 137(1994)405.
4. M.M.J. Treacy and J.M. Gibson, *J. Vac. Sci. Tech. B*, 4(1986)1458.
5. M.D.R and J.M.C. would like to acknowledge the support of this research by the Natural Sciences and Engineering Research Council of Canada. Also, M.D.R. thanks the National Research Council of Canada for the generous support provided by the visiting research graduate program.

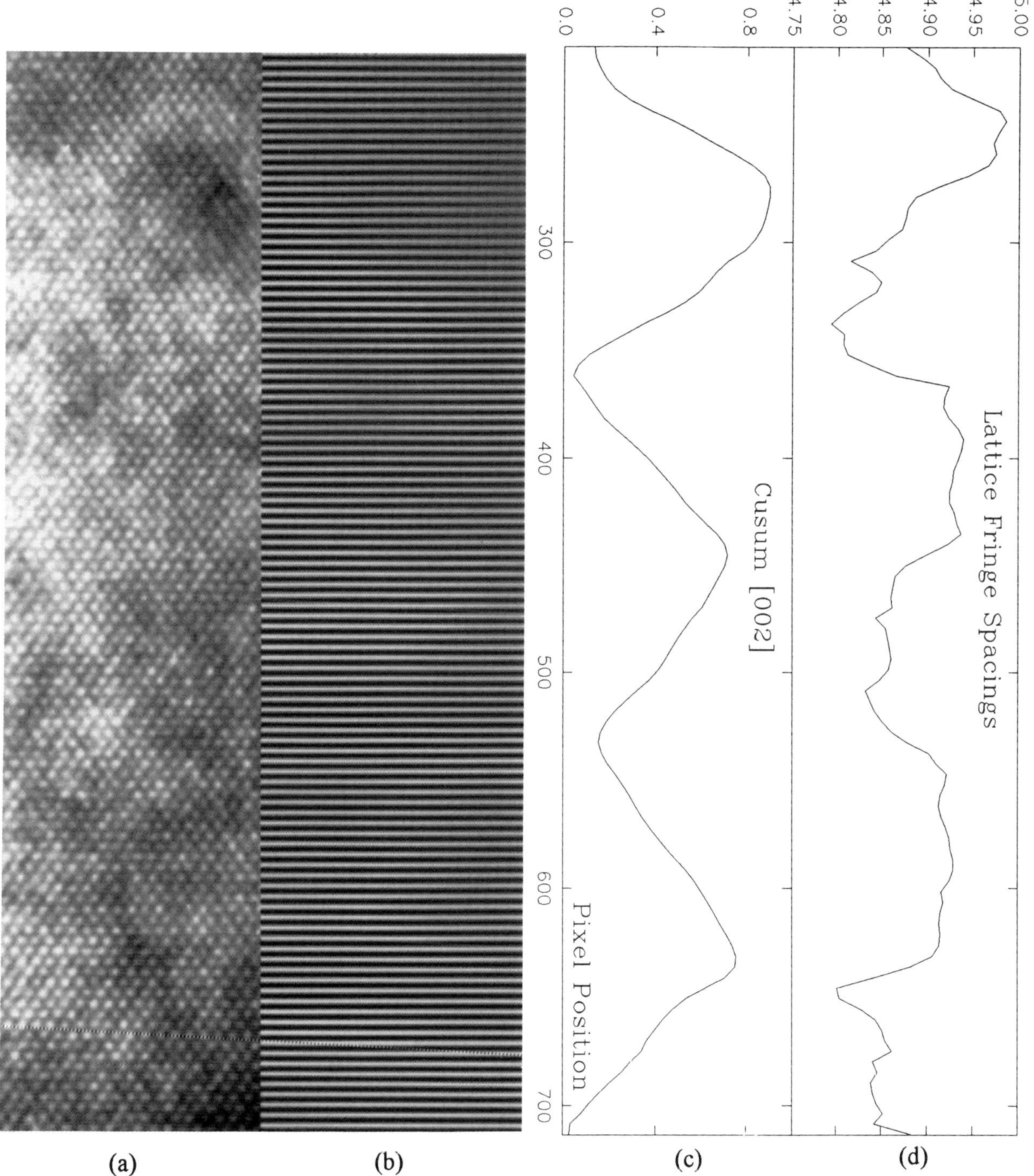

Fig. 1. (a) 128x697 pixel HREM image of three periods of a 20 period coherently strained $In_{.82}Al_{.18}Sb$ / InSb superlattice. (b) [002] Fourier filtered lattice fringe image. (c) CUSUM [002] lattice fringe spacing versus pixel position plot. (d) [002] lattice fringe spacing vs pixel position plot.

DETERMINATION OF THE SI(111)-($\sqrt{3}$X$\sqrt{3}$) AU SURFACE STRUCTURE USING UHV TRANSMISSION ELECTRON MICROSCOPY

R. Plass and L. D. Marks

Department of Materials Science and Engineering, Northwestern University, Evanston,IL 60208

For the past few decades the Si(111)-($\sqrt{3}$x$\sqrt{3}$) Au surface (denoted hereafter as $\sqrt{3}$-Au) has received considerable attention in the literature as one of the prototypical Si(111)-($\sqrt{3}$x$\sqrt{3}$) Metal systems. The $\sqrt{3}$-Au system appears for gold coverages from roughly 0.5 ML to 1.0 ML after annealing of the substrate over 280°C. Several authors have noted the small surface domains of this structure and several possible diffraction spot features associated with different coverages and annealing temperatures.[1,2,3] The two dominant models which have been proposed for this surface are the missing top layer twisted trimer (MTLTT) and the conjugate honeycomb chained trimer (CHCT) models.[4,5]

The $\sqrt{3}$-Au surface was prepared on bulk Si(111) TEM samples thinned by mechanical polishing and HNO_3/HF solution chemical etching. Under UHV conditions the samples were cleaned by cycles of Ar^+ ion milling and electron beam annealing until well defined Si(111)-(7x7) diffraction patterns were visible. On these surfaces roughly 0.7 ML of gold was evaporated followed by light electron beam annealing to produce the $\sqrt{3}$-Au surface. Through exposure electron diffraction patterns from three different sample locations and crystal tilts were collected and the negatives quantified using a cross correlation analysis technique. Theoretical beam intensities were calculated for various proposed models of the structure using double precision dynamical multislice methods. Parameters for each model were adjusted to obtain a best χ^2 fit to the experimental data.

Shown in Figure 1a is the structure with the best χ^2 fit (2.76) to all three diffraction data sets fitted simultaneously. This structure is of the MTLTT type with the key parameters being a gold trimer rotation of 1.9°, a Au-Au interatomic spacing of 2.71 Å, a second layer silicon trimer rotation of 3.4° and a Si-Si spacing of 2.85 Å. One of the interesting features of this structure is that when Au^+ scattering factors are used a better fit, $\chi^2 = 2.76$, is obtained than when Au scattering factors are used, 3.92, in good agreement with the results of Dobrodey.[6] If the three data sets are fitted individually we find a significant drop in the χ^2 values, typically by 0.5, with the key difference in model parameters being the increase in gold trimer rotation to 3.57° for one data set, while the gold trimer rotation drops to essentially zero for the other two data sets. This difference correlates to the degree of sharpness of the diffraction spots and the general surface domain morphology shown in Figure 2 with more gold trimer rotation corresponding to sharper diffraction spots. Additional details concerning the nature of the surface domain walls will also be presented.[7]

References
1. J. Nogami et al., *Phys. Rev. Let.* 65(1990)1611.
2. S. Takahashi et al., *Surf. Sci.* 242(1991)73.
3. S. Ino, *Jpn. J. Appl. Phys.* 16(1977)891.
4. M. Chester and T. Gustafsson, *Surf. Sci.* 256(1991)135.
5. Y. G. Ding et al., *Surf. Sci.* 275(1992)L691.
6. N. V. Dobrodey et al. *Surf. Rev. and Let.* 1(1994)273.
7. This work was supported by the AFOSR under Grant #F49620-92-J-0250.

Proc. Microscopy and Microanalysis 1995, edited by G.W. Bailey, M.H. Ellisman, R.A. Hennigar, and N.J. Zaluzec
Copyright © 1995 MSA. Published by Jones and Begell Publishing, 79 Madison Ave., New York, NY 10016

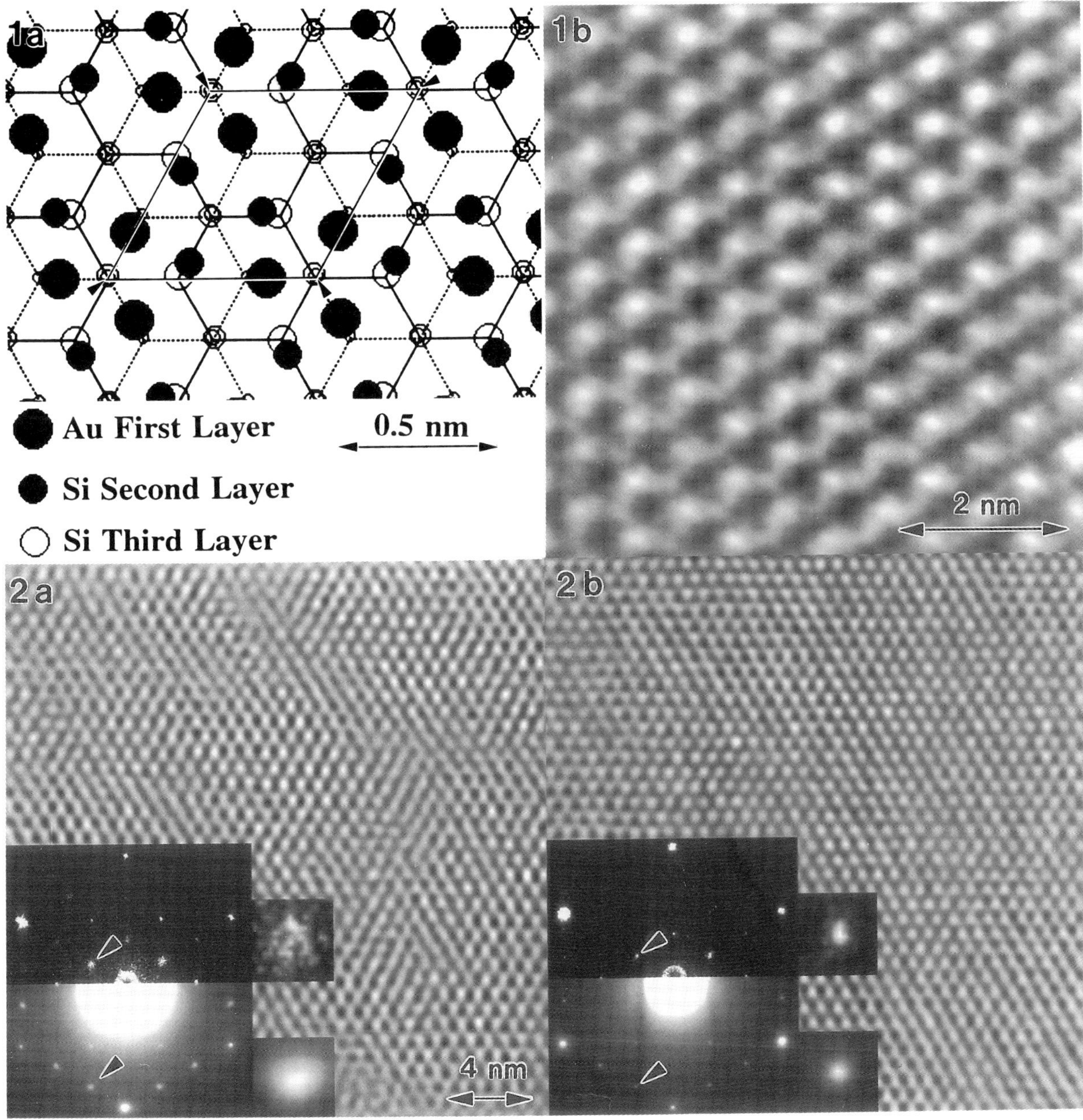

FIG. 1.-(a) Top view schematic diagram of the average Si(111)-($\sqrt{3} \times \sqrt{3}$) Au atomic structure.
(b) (111) zone axis high resolution image of the Si(111)-($\sqrt{3} \times \sqrt{3}$) Au surface in which the gold trimers appear as dark triangles showing the local three fold symmetry of the structure.
FIG. 2.-Strongly low pass filtered HREM (111) zone axis images showing two regions of the $\sqrt{3}$-Au surface. (a) is from a region with diffuse diffraction spots as shown by its corresponding diffraction pattern in the top half inset. (b) is from a region with sharp diffraction spots. The bottom halves of the insets are from corresponding experimental diffraction patterns while to the right of each main inset are 8X magnifications of the arrowed spots.

PRACTICAL ASPECTS ABOUT HOLLOW-CONE IMAGING

T. Geipel and W. Mader

Institut für Anorganische Chemie, Universität Bonn, 53117 Bonn, Germany

Hollow-cone imaging (HCI) as a possibility to improve the resolution of a TEM has already been proposed in the late 40ties [1] and besides others, there have been extensive hollow-cone experiments 10 years back using a low resolution TEM with a non-tilt specimen holder [2]. In a recent paper [3] the optimum imaging parameters for HCI were determined leading to an improvement of the resolution by a factor of two. However, there are contrast limitations and experimental problems for HCI which were only partly considered in Ref. [3] and which will be discussed in this paper for a modern electron microscope. Preliminary experiments were performed which are not shown in the abstract.

In Fig. 1 $\int ctf(u)du$ is plotted versus defocus Δf for different cone radii Θ_c and a fixed aperture radius $\Theta_o = 1/\delta = 5\ nm^{-1}$ (ctf is the phase contrast transfer function (PCTF) for HCI [3] and $\delta = 0.2\ nm$ is the resolution of a CM30 supertwin microscope). Our optimum HCI defocus is defined that $\int ctf(u)du$ has a global minimum (maximum) for negative (positive) phase transfer. In Fig. 2 the corresponding PCTF for the optimum defocus is plotted for different Θ_c. In the axial PCTF ($\Theta_c = 0$) the optimum defocus corresponds to the Scherzer defocus. It can be seen that the resolution in HCI is $\delta_{HCI} = 1/(\Theta_o + \Theta_c)$. The maximum value of Θ_c is Θ_o, in this case the improvement of the resolution is a factor two. However, Fig. 2 also shows a problem of HCI for large Θ_c: with increasing Θ_c the contrast decreases. If the global minimum (maximum) of $\int ctf(u)du$ is below ≈ 2.5 the HCI contrast is low. This limits the maximum cone radius to $\Theta_c \leq 1.5\ nm^{-1}$. Fig. 3 shows HCI simulations (using the EMS software [4]) for $\langle 011 \rangle$ projected Ge in a CM30 microscope illustrating the increasing resolution and the decreasing contrast for increasing cone radius. It can be shown that in addition to the contrast problems the experimental alignment of the TEM for larger cone radii becomes more difficult compared to smaller radii.

HCI requires a very good adjustment of axial coma and good circular alignment of the cone. If all the limitations of HCI are considered, an improvement of the resolution of about $20 - 30\%$ (compared to the axial Scherzer resolution) is achievable. For a CM30 microscope with supertwin lens this should lead to a resolution of $\delta = 0.16\ nm$.

First attempts of a combination of HCI and reconstruction techniques for retrieving the complex phase information of exit-wave functions have been presented recently [5].

References
1. O. Scherzer, Journal of Applied Physics (1948) 20, 20,
2. W. Kunath, F. Zemlin, and K. Weiss, Ultramicroscopy (1985) 16, 123,
3. C. Dinges, H. Kohl, and H. Rose, Ultramicroscopy (1994) 55, 91,
4. P.A. Stadelmann, Ultramicroscopy (1987) 21, 131,
5. Y. Takai et al., Proceedings of 13th ICEM (1994) vol 1,153.

Proc. Microscopy and Microanalysis 1995, edited by G.W. Bailey, M.H. Ellisman, R.A. Hennigar, and N.J. Zaluzec
Copyright © 1995 MSA. Published by Jones and Begell Publishing, 79 Madison Ave., New York, NY 10016

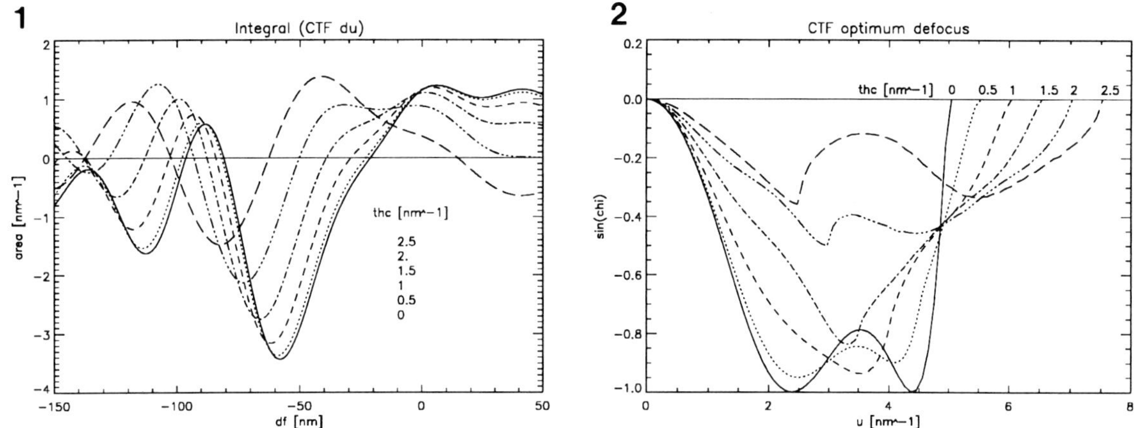

FIG. 1 - $\int_0^\infty ctf(u)du$ versus Δf for different HCI radii $\Theta_c = 0...2.5\ nm^{-1}[0.77]$ and $\Theta_o = 5\ nm^{-1}[1.55]$. The minima (maxima) of the curves give the optimum defoci for negative (positive) phase transfer. For better comparison with Refs. [2,3] $\Theta_{c,o}$ and Δf are given in so-called reduced units also.

FIG. 2 - ctf(u) for different cone radii $\Theta_c = 0 ... 2.5\ nm^{-1}\ [0.77]$, $\Theta_o - 5\ nm^{-1}\ [1.55]$ and optimum defocus each
$$(\Delta f = -58\ [1.19], -59\ [1.21], -62\ [1.28], -66\ [1.36], -73\ [1.5], -82\ [1.69]\ nm).$$

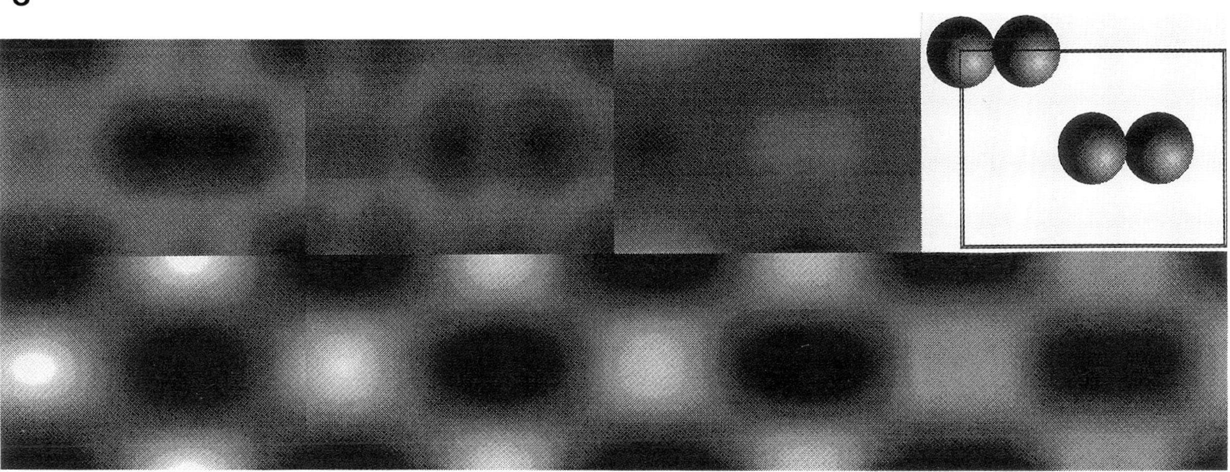

FIG. 3 - Simulated HRTEM images of Ge along $\langle 011 \rangle$, specimen thickness $H = 1.6nm$, $\Theta_o = 5\ nm^{-1}[1.55]$, $\Theta_c = 0[a] - 3\ nm^{-1}[0.93][g]$ and $\sqrt{2}\sigma(\Delta f) = 10\ nm$, $\sqrt{2}\sigma(\alpha) = 1\ mrad$ for fixed contrast and optimum defocus each. It can be seen that with increasing Θ_c the resolution increases, making the characteristic dumbbells visible in [f] and not in [e] but on the other hand the contrast decreases. Image [g] corresponds to reversed contrast, i.e. in this case white spots correspond to atoms. The image in the upper right corner shows the projected unit cell of Ge.

STRUCTURE DETERMINATION OF THE INCOMMENSURATE INTERCALATION COMPOUND Hg_xTiS_2 USING TRANSMISSION ELECTRON MICROSCOPY

M. Sidorov
Center for Solid State Science; Arizona State University; Tempe, AZ 85287-1704

Although metal intercalated transition-metal dichalcogenides have been generally viewed as ionic compounds with guests occupying discrete lattice sites, the Hg_xTiS_2 intercalates ($1.24 \geq x \geq 0.00$) have been recently discovered to have vastly different character, including essentially-neutral covalently-bound guests, superstoichiometric Hg uptake ($x>1.00$), complete thermal reversibility of the intercalation process and unusually complex structures.[1-3] A combination of HREM and electron diffraction was used to determine the structure of the intercalates formed during the *in situ* deintercalation process. Digital Fourier image processing provided HREM image analysis and contrast enhancement.

Hg_xTiS_2 has a remarkable in-plane guest-layer structure, which is composition independent. It results from the presence of partially incommensurate monoclinic Hg and TiS_2 sublattices, with common *a* and *c* axes and incommensurate *b* axes (a= 5.92 Å; b_{TiS_2} = 3.41 Å; b_{Hg}= 2.76 Å; c= 8.70 Å, and β= 96.5°).[2] Figure 1a shows a typical *c*-axis SAD pattern from a thin crystal indicating the presence of two subsets of reflections incommensurate along b^* direction. In the corresponding HREM images this gives rise to the 14.5Å structure modulation along *b* (Fig. 1b). Fourier Transforms of these images show the same geometry as the SAD patterns presenting a superposition of two sets of Fourier peaks produced by Hg and TiS_2 sublattices (Fig. 1c). The in-plane structure of the Hg and TiS_2 sublattices can be directly resolved by Fourier deconvolution of the HREM image into constituent periodical components (Fig. 2). The 3D structure is solved by imaging the commensurate *a-c* plane (Fig. 3). The insets in Figures 3 and 1b show the results of multislice image simulations calculated for Hg_xTiS_2 proving the validity of the proposed model (Fig. 3 and 4).

The Hg guest layers/islands in Hg_xTiS_2 are comprised of 1D Hg chains that are incommensurate with the host layers along chain direction (*b*) (Fig. 4). As can be seen in Figure 3, the chains reside in sulfur channels of rectangular cross-section. Thicker crystals are observed to contain three equivalent orientational variants of the above Hg domains rotated by 120° with respect to each other. Starting with this model, refinement of XPD data revealed fully-intercalated $Hg_{1.24}TiS_2$ has a similar four-dimensional structure consisting of interpenetrating 3D TiS_2 and Hg sublattices.[3]

References:

1. E. Ong, et al., *Chem. Mater.*, 4 (1992)14.
2. M. Sidorov, et al., *Chem. Mater.*, (1995), in press.
3. P. Ganal, et al., *Chem. Mater.*, (1995), in press.
4. I would like to thank my PhD committee co-chairs, M. McKelvy and W. Glaunsinger, for their guidance and many discussions. We wish to acknowledge the National Science Foundation for support through grant DMR 91-06792 and acknowledgment is made to the donors of the Petroleum Research Fund administered by the American Chemical Society, for partial support of this research.

Proc. Microscopy and Microanalysis 1995, edited by G.W. Bailey, M.H. Ellisman, R.A. Hennigar, and N.J. Zaluzec
Copyright © 1995 MSA. Published by Jones and Begell Publishing, 79 Madison Ave., New York, NY 10016

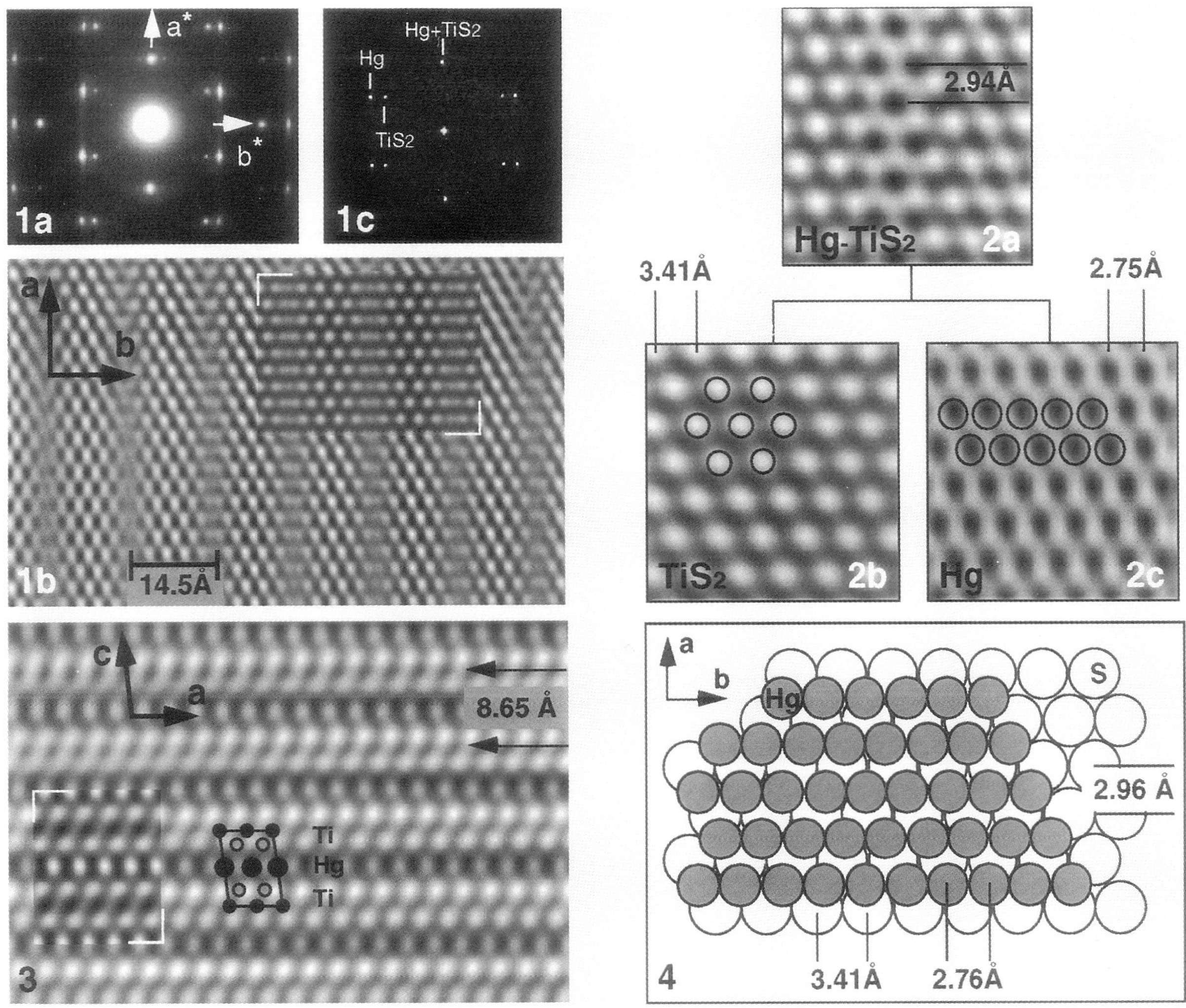

FIG. 1. (a) Typical SAD pattern for thin Hg_xTiS_2 crystal along c showing reflections from Hg and TiS_2 sublattices. Note diffuse streaks along a^* accompanying Hg reflections. (b) HREM image of partially commensurate a-b plane. Note ~14.5Å long-range modulation along incommensurate b axis indicating near coincidence of two incommensurate sublattice unit cells. Inset: calculated image. (c) Fourier transform of 1b.

FIG. 2 Results of Fourier image deconvolution: (a) highly magnified HREM image of Hg_xTiS_2, (b,c) Fourier components corresponding to the TiS_2 and Hg sublattices extracted from (a). Note hexagonal symmetry and chain-like structure of TiS_2 and Hg sublattices respectively.

FIG. 3 HREM image of commensurate a-c plane, as viewed along Hg chains. Unit cell is superimposed onto the image. Inset: calculated image.

FIG. 4 In-plane model of partially incommensurate Hg_xTiS_2 structure. Note that 2.96 Å= 2.94 / sinβ.

SURFACE CHEMICAL SENSITIVITY COMBINED WITH HREM

E. Landree*, C. Collazo-Davila*, D. Grozea*, R. Plass*, G. Jayaram*, L.D. Marks*, P.C. Stair**

*Department of Material Science and Engineering, Northwestern Univ., Evanston Il 60208
**Department of Chemistry, Northwestern University, Evanston Il 60208

Although high resolution electron microscopy (HREM) has proved to be an invaluable tool in the study of surfaces, one critical drawback is its inability to obtain specific surface chemical state information. The ideal investigation would combine the imaging capabilities of HREM with the surface chemical information obtainable from X-ray Photon Spectroscopy (XPS) and Auger Electron Spectroscopy (AES) without exposing the sample surface to possible contaminants. We have taken a step in this regard with the addition of a new Specimen Preparation Evaluation and Analysis for Research (SPEAR) side chamber to the existing Hitachi UHV-9000 microscope[1], designed by Superior Vacuum Technology (Figure 1). Currently we are able to combine all of the imaging tools of HREM with surface chemical sensitivity, in an attempt to achieve the ideal surface analysis instrument.

Samples are introduced and retrieved from the system through a load-lock mechanism. Pumped by a 210 Liter/sec Balzers' Turbomolecular pump, this introduction chamber is capable of going from atmospheric pressure to 10^{-10} torr in four to eight hours. This procedure eliminates the lengthy bake-out procedure required by our previous surface side chamber when exchanging samples. Once UHV is achieved, this load-lock is then opened to the transfer chamber. The transfer chamber consists of a storage module, capable of holding up to eight samples, and a central transfer module used for shuttling samples among the various chambers. From here, the sample may be transferred to either the Molecular Beam Epitaxy (MBE) chamber, the analytical chamber or to the microscope. The MBE chamber has four effusion cells and is currently configured for the thin film deposition of GaAs. It also contains a resistive heating stage in order to facilitate epitaxial thin film growth during deposition. The analytical chamber, as its name suggests, consists of various analytical tools such as XPS, AES as well as an ion gun and a secondary electron detector for purposes of surface cleaning and Scanning electron microscopy (SEM). The analytical stage is also designed to resistively heat or cool the sample, allowing us to acquire surface information and vary the sample temperature simultaneously. The transfer chamber is pumped by a Perkin Elmer 400 Liter/sec Ion pump, while the MBE and the analytical chambers are each pumped by a Perkin Elmer 220 Liter/sec Ion pump, rating the entire system at 10^{-11} torr.

XPS spectra are acquired with a Perkin-Elmer Dual anode X-ray source and Spherical Capacitor Analyzer with multichannel detector. The X-ray source is capable of producing either Mg or Al Kα x-rays, and the analyzer has a maximum energy resolution of 0.1 eV and a minimum scanning area of 70 μm X 70 μm. A FEI Thermal Field Emitter or "Schottky" Emitter (SE), is used in combination with the spherical capacitor analyzer to acquire AES spectra. The SE source emits up to 15 kV electrons and has a minimum probe size of 0.05 μm. The SE source is also used with a Channeltron secondary electron detector to produce SEM images. This combination of the SE source, secondary electron detector and spherical capacitor analyzer allows us to obtain AES with precise spatial resolution. Similarly, a PHI Electronics Duo-plasmatron Ion Gun has been installed, with a variable gas source to produce either Oxygen, Argon or Xenon ions. It has a maximum extraction voltage of 10 kV and a minimum probe size of <5 μm with a maximum current density rated at 20 mA/cm^2 for a 50 μm beam diameter. During sputtering the secondary electrons emitted from the sample are imaged using the secondary electron detector to obtain very precise control over the sputtered region.

SPEAR has already demonstrated it usefulness in the characterization of the sample surface during the surface cleaning procedure. A silicon (111) sample, etched using a 90% HNO$_3$-10% HF solution and scanned with XPS, indicated the presence of carbon, oxygen and fluorine. This sample was then etched for five minutes with 1 kV oxygen ions at 60 degrees to the surface normal. Figure 2a and Figure 2b illustrate the XPS spectrum acquired from the silicon surface before and after the ion sputter. Trace contaminants are still present along with a noticeably high oxygen peak. Figure 2c is an SEM image of the surface following the oxygen etch.

---

[1] L.D. Marks et. al., *Ultramicroscopy* 37 (1991) 90-102.

The problems of combining HREM and surface chemical state sensitivity without risk of contaminating the surface of the specimen has made this type of procedure difficult at best. With the addition of SPEAR, we are now able to combine HREM and PEELS along with XPS, AES, SEM and Ion Beam analysis while maintaining UHV conditions[2].

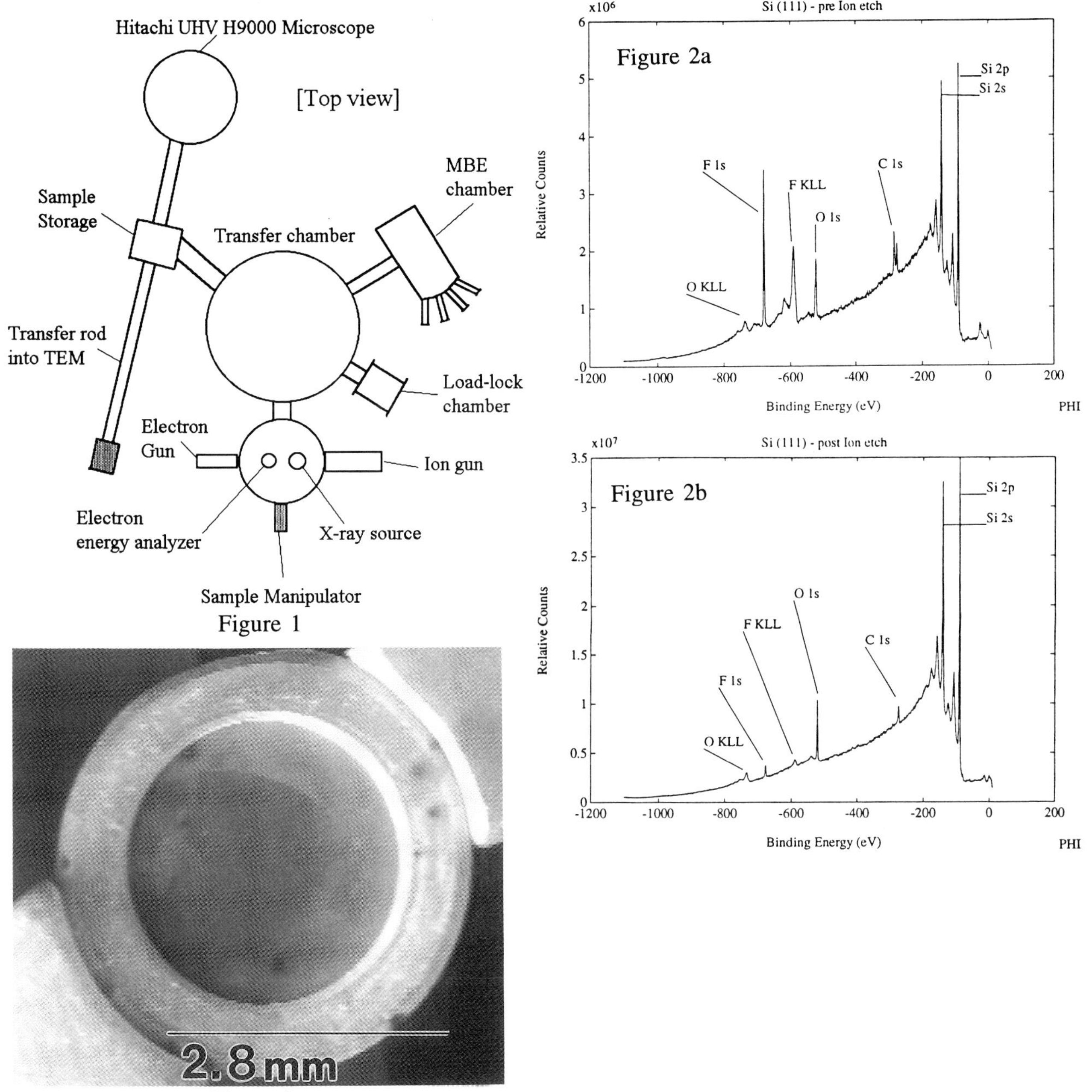

Fig. 1 - Schematic of SPEAR surface side chamber.
Fig. 2 - (a). XPS spectrum acquired from Si (111) prior to oxygen ion sputtering; (b). XPS spectrum acquired from same specimen following 5 minute oxygen sputter at 1 kV; (c). SEM image of sample surface following ion sputter.

---

[2] This work supported by Air Force Office of Scientific Research on grant # F49620-94-1-0164.

Modification of A Transmission Electron Microscope for Ultra-High-Vacuum Reflection-Electron-Microscope (UHV-REM) Applications [1]

Tung HSU, Min-Yi SHIH, A. V. Latyshev [2],

Materials Science Center, National Tsing-hua University, Hsin-chu 30043, TAIWAN, R.O.C.

A JEOL JEM-100C electron microscope was modified by adding a cryogenic UHV specimen holder for studying clean crystal surfaces with the reflection high energy electron diffraction (RHEED) and REM techniques [3]. The Si(111) (1x1) and (7x7) phase transitions have been successfully observed (Fig. 1). Further modification is in progress for better resolution and other functions. Fig. 2.a shows the unmodified specimen holder and the objective lens of the microscope. The cryogenic holder based on the Novosibirsk design [4] is shown in Fig. 2.b. Liquid nitrogen is continuously pumped through the shell of the holder for achieving UHV inside. The tilt/rotation controls and the current for heating of the specimen are fed through the holder. In this modification, the specimen was not placed at the normal position of the lens and therefore is not at the best position for imaging and diffraction.

A new holder is shown in Fig. 2.c. This holder is inserted into the pole piece to place the specimen at the normal position. A gear assembly permits rotate and tilt of the specimen and wires for DC current provide heating. A reversible liquid/gas pumping circuit is designed for pumping the LN_2 during operation. The pumping direction is reversed for pumping dry air or N_2 into the specimen holder for returning it to the ambient temperature before opening the chamber.

This instrument will be used to in situ investigation of surface atomic structure during growth, sublimation, and other processes.

REFERENCE

1. Supported by National Science Council 84-2112-M-007-047, R.O.C. Use of the JEOL JEM-100C microscope is provided by Dr. Jian-yih Wang of Chung Shan Institute of Science and Technology.
2. Permanent address: Institute of Semiconductor Physics, Russian Academy of Science, Siberian Branch, 630090 Novosibirsk, RUSSIA.
3. A special issue on current research on REM, Microscopy Research and Techniques, 20/4 (1992), guest ed. T. Hsu.
4. A.V. Latyshev, A.B. Krasilnikov and A.L. Aseev, Appl. Surface Sci., 60/61 (1992) 397-404.

Proc. Microscopy and Microanalysis 1995, edited by G.W. Bailey, M.H. Ellisman, R.A. Hennigar, and N.J. Zaluzec
Copyright © 1995 MSA. Published by Jones and Begell Publishing, 79 Madison Ave., New York, NY 10016

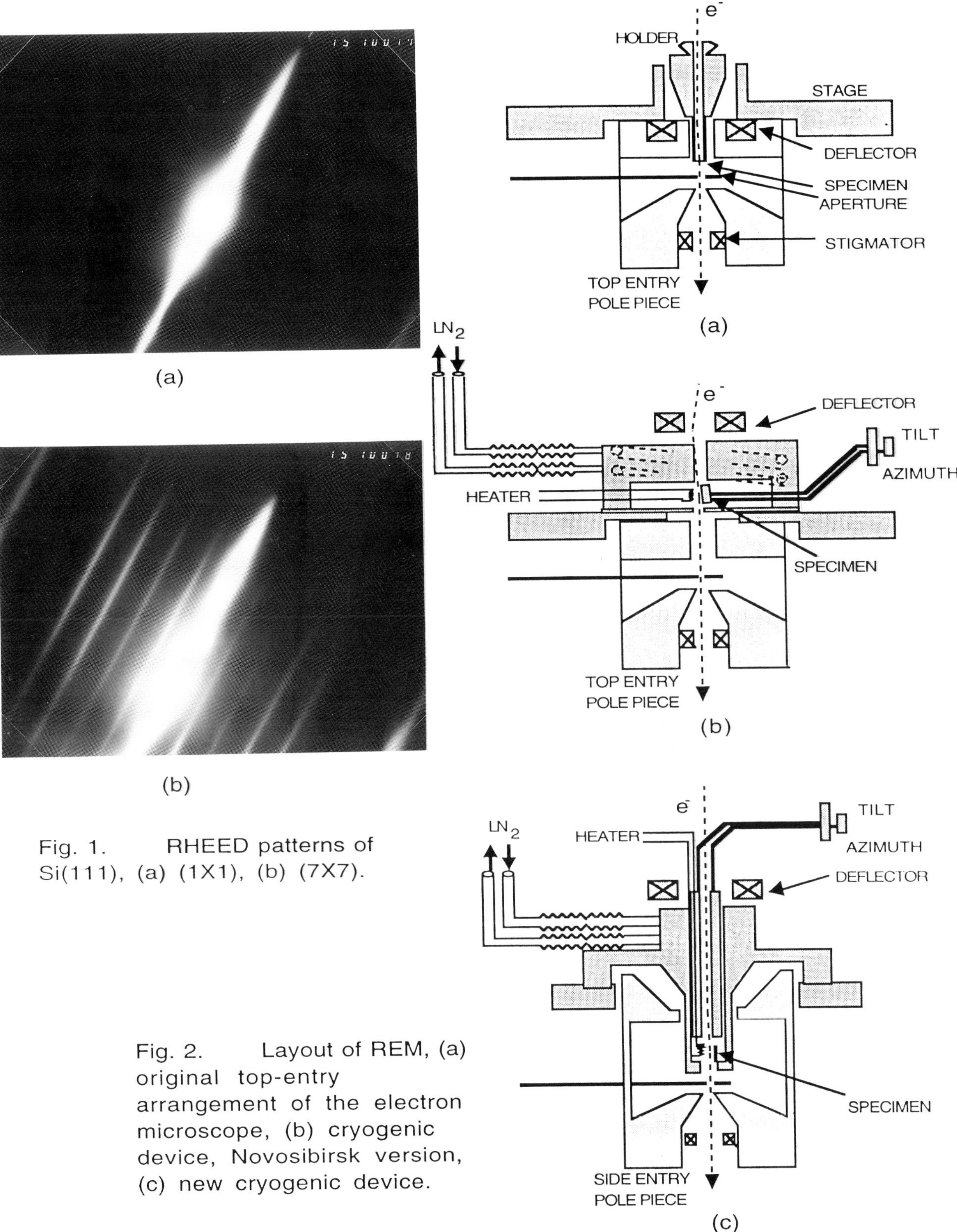

(a)

(b)

Fig. 1.   RHEED patterns of Si(111), (a) (1X1), (b) (7X7).

Fig. 2.   Layout of REM, (a) original top-entry arrangement of the electron microscope, (b) cryogenic device, Novosibirsk version, (c) new cryogenic device.

THE CM120 BIOTWIN: A HIGH-CONTRAST OBJECTIVE LENS, OPTIMUM CRYO-VACUUM AND IMPROVED EDX PERFORMANCE

Uwe Lücken, Michael Felsmann, Wim M. Busing and Frank de Jong.

Philips Electron Optics Applications Laboratory, Building AAE, 5600 MD Eindhoven, The Netherlands

A new microscope for the study of life science specimen has been developed. Special attention has been given to the problems of unstained samples, cryo-specimens and x-ray analysis at low concentrations.

Contrast

A new objective lens with a Cs of 6.2 mm and a focal length of 5.9 mm for high-contrast imaging has been developed. The contrast of a TWIN lens (f = 2.8 mm, Cs = 2 mm) and the BioTWIN are compared at the level of mean and SD of slow scan CCD images. Figure 1a shows 500 +/- 150 and Fig. 1b only 500 +/- 40 counts/pixel. The contrast-forming mechanism for amplitude contrast is dependent on the wavelength, the objective aperture and the focal length. For similar image conditions (same voltage, same objective aperture) the BioTWIN shows more than double the contrast of the TWIN lens. For phase-contrast specimens (like thin frozen-hydrated films) the contrast at Scherzer focus is approximately proportional to the $\sqrt{}$ Cs. The contrast transfer function of the BioTWIN shows a 50% increase compared to the TWIN for frequencies below 0.5nm^{-1} (i.e. spacings above 2 nm). A complete theory of the BIO-contrast in the BioTWIN will be presented.

Cryo-vacuum

Imaging and analysis of biological specimens at liquid-nitrogen temperature is nowadays a routine method. Cryo protection describes the effect that one can image or analyse a specimen under cryo conditions with a higher electron dose than at room temperature because of the reduced rate of diffusion. The damage due to the interaction with the electron beam, like breaking chemical bonds, still takes place, but the effect, which can be the loss of high-resolution frequencies in crystalline specimens is less obvious, because the damaged molecules do not move as fast as at room temperature.

Long-term imaging and analysis does require an excellent vacuum and low rates of contamination. To meet these requirements the CM120 BioTWIN is equiped with the following features: standard, fixed sandwich cryo-blades reduce the open solid angle of warm surfaces towards the specimen; the Compustage has no sliding O-ring and therefore the goniometer is no longer a source of microleakages and drift is minimised; the 200 µm pumping aperture separates the camera vacuum from the clean column vacuum. These features provide excellent cryo-vacuum under liquid-nitrogen conditions. The quality of this cryo-vacuum can be measured from the rate of contamination on a cold surface like a holey carbon film (Fig. 2a,b). After 3 hours there is no significant ice-growth visible and the rate of contamination is garanteed at less than 10 nm/hour. To verify the contrast at Scherzer focus and the low rate of contamination a frozen hydrated image of TMV is shown in Fig. 3.; the 2.3nm fringes are clearly visible.

X-ray analysis

With a large gap of 20 mm between the pole pieces the BioTWIN lens offers new possibilities for X-ray detection. The EDX detector can be brought much closer to the specimen resulting in a high solid angle of 0.4 sterad for X-ray collection. This makes spot analysis and X-ray maps by at least a factor of three faster compared with standard systems (see Fig. 4a,b). When working with similar analysis times, much lower beam currents can be used in order to protect the specimen. The sandwich cryo-blades are made of Be and together with a collimator and a special low-background holder they ensure a clean X-ray spectrum. The minicondenser of the BioTWIN lens allows fast switching between high-contrast imaging and the analytical spot mode with spots down to 5 nm.

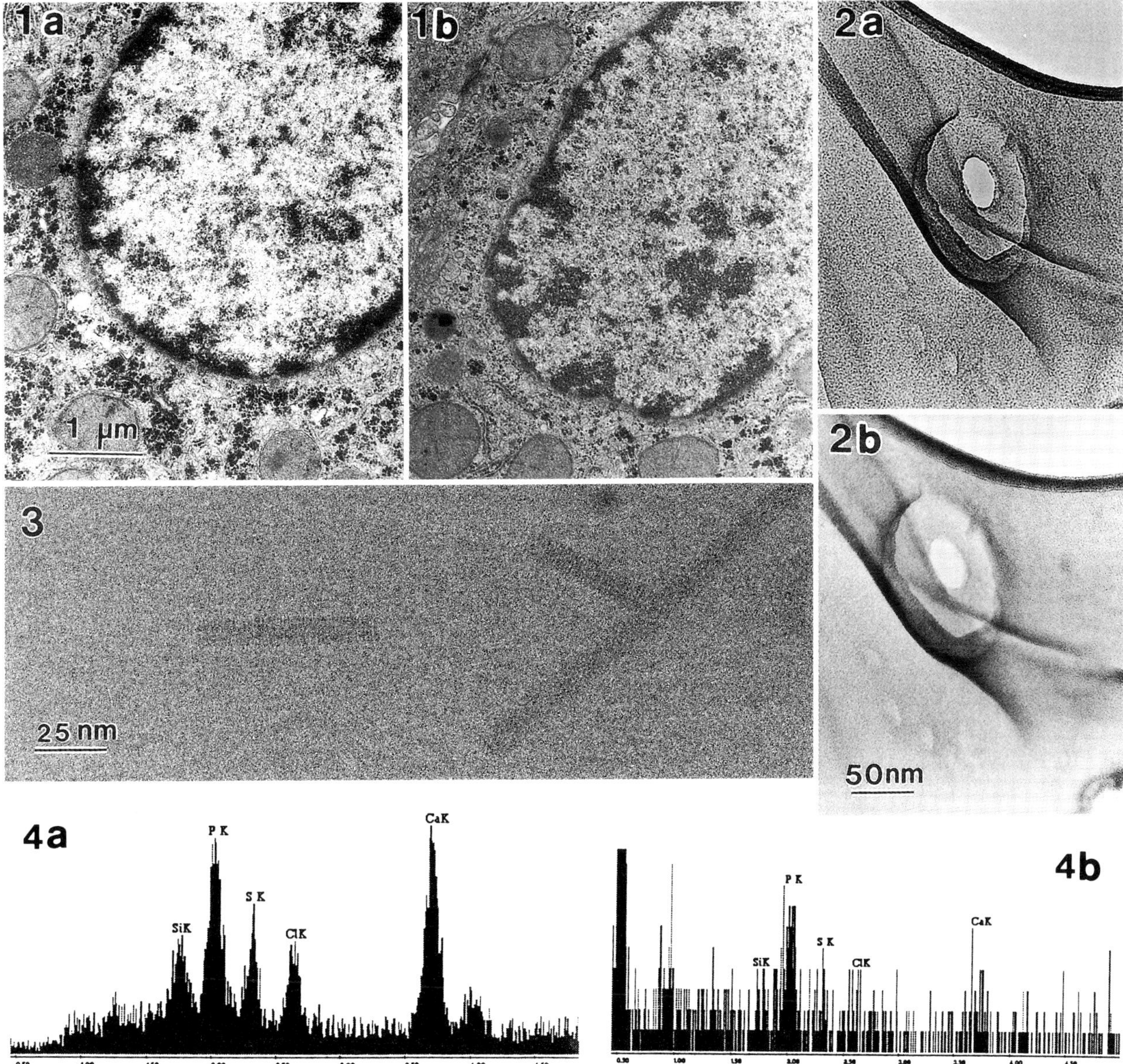

FIG. 1. Contrast comparison of stained liver sections at 120 kV between (a) BIOTWIN and (b) TWIN.
FIG. 2. Contamination experiment, holey carbon film at -186°C under 45° tilt, (a) at t = 0, (b) after 3 hours.
FIG. 3. Frozen hydrated image of tobacco mosaic virus (TMV) at Scherzer focus and 52kx primary magnification.
FIG. 4.EDX spectra of spleen taken under the same conditions (a) with BioTWIN detector and (b) with standard S-UTW detector. Conditions: beam current 1.2 nA, collection time: 20s, high tension 100kV.Specimen courtesy of Dr Jeremy Rees, Sanofi Winthrop.

THE INFORMATION LIMIT OF A 200KV FE-TEM

T. Kaneyama, M. Kawasaki, T. Tomita, T. Honda and M. Kersker

JEOL Ltd., Musashino 3-1-2, Akishima, Tokyo, 196 Japan
JEOL(USA) Inc., 11 Dearborn Road, Peabody, MA 01960

Introduction

The Point resolution of a transmission electron microscope is normally defined by the reciprocal of the spatial frequency of the first zero in the phase contrast transfer function at the Scherzer defocus condition.   When a field emission gun (FEG) is used as the electron source, the information limit, that point at which the contrast beyond the first zero goes to zero contrast, becomes equally important. We have investigated the primary microscope parameters that affect the information limit.

Experimental

A 200kV FE-TEM (JEM-2010F) equipped with a ZrO/W shottkey emitter and Gatan Parallel EELS (PEELS) was used for the experiments.   The aberration coefficients of the objective lens are $Cs = 1mm$ and $Cc = 1.4mm$.   The specimen used is an evaporated amorphous Ge thin film with small gold islands.

Results

The resolution performance of the microscope depends not only on the performance of the objective lens, the high voltage stability, stability of the lens and deflector power supplies, operating parameters of the FEG, and the overall mechanical stability of the microscopes.   Improvements of power supplies were not much fruitful because they have sufficient performance for high-resolution work operating at 200kV.   The extraction voltage (V_{ex}) of the FEG was found to be the most effective factor.

Figures 1a and 1b are amorphous film images taken at 260nm defocus at $V_{ex} = 3.1kV$ and $V_{ex} = 2.3kV$.   Optical diffractograms are inset.   The difference in image contrast beyond $4nm^{-1}$ spatial frequency between the two images can clearly be seen.   The information limits were measured to be about 0.165nm for 1a and about 0.15nm for 1b.

We tried to measure the energy spread dependence of V_{ex} using PEELS and succeeded in catching the δE (FWHM) change as a function of V_{ex} as shown in Figure 2.   No other operating conditions of the microscope and the FEG including the specimen position, etc., were changed so the change in energy spread as the extraction voltage is changed.   Theoretical calculations of the phase contrast transfer function support this supposition (Figure 3).   Though it is conceivable that the Boersch effect is the origin of the energy spread difference with the V_{ex}, the change is too large to ascribe the observed effect only to the Boersch effect.   We are investigating other possible reasons of this energy spread.

Proc. Microscopy and Microanalysis 1995, edited by G.W. Bailey, M.H. Ellisman, R.A. Hennigar, and N.J. Zaluzec
Copyright © 1995 MSA. Published by Jones and Begell Publishing, 79 Madison Ave., New York, NY 10016

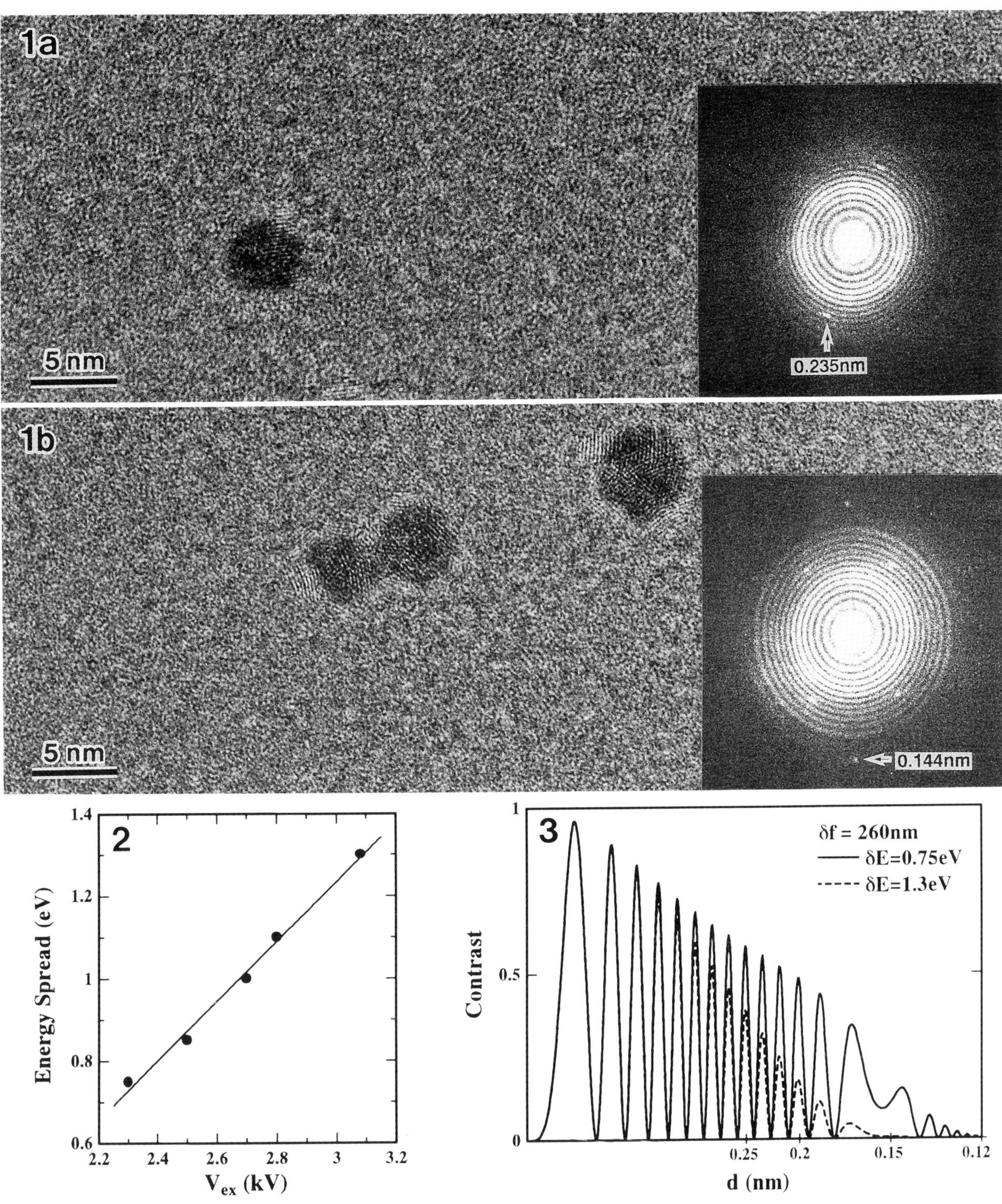

FIG. 1. —Micrograph of amorphous Ge thin film taken at about 260nm defocus and its optical diffractogram.   (a) V_{ex} = 3.1kV, (b) V_{ex} = 2.3kV.

FIG. 2. —Relation between V_{ex} and energy spread (FWHM).

FIG. 3. —Theoretical calculations of phase contrast transfer function.

INFORMATION LIMIT OF 300kV FETEM

I.Ishikawa,* K.Kaneyama,* T.Tomita,* T.Honda* and M.Kersker**

*JEOL, Ltd., Musashino 3-1-2, Akishima, Tokyo 196, Japan.
**JEOL U.S.A Inc., 11 Dearborn Road, Peabody, MA 01960, U.S.A.

The use of an FEG on a high resolution microscope will increases the spacial and temporal coherence of the microscope due to the improvement in coherence of the electron source. Additionally, the improved energy spread will extend the envelope of the phase contrast transfer well beyond the first zero point at Sherzer focus. Between the Sherzer point and the ultimate point of zero contrast (the information limit) there is useful information for the analysis of high-resolution images. We have already obtained a resolution of 0.194nm and an information limit of 0.14nm by using the 200kV FETEM (JEM-2010F).[1] In the present experiment, we attempted to obtain a higher resolution and information limit by using the 300kV FE TEM (JEM-3000F).[2]

The electron gun is configured with a ZrO/W Shottky emitter which has a small energy spread with a stable emission.[3] The objective lens has a spherical and chromatic aberration of 0.6mm and 1.3mm respectively. The specimen used was gold islands on an amorphous Ge thin film.

The contrasts as a function of spacial frequency is calculated by the amount of defocus, energy width, illumination angle and objective lens aberration coefficient as shown in Figures 1 and 3. At high frequency, other factors, e.g. mechanical vibration, stability of the lens and deflector currents and specimen contamination, have serious effects on the contrast. These negative factors were dealt with by
1. isolating the column from through the high voltage cable by assembling the high-voltage tank and column on the same frame,
2. by improving circuit filtration for the lenses and deflectors,
3. and by baking out the column several times prior to conducting the experiment to reduce contamination.

After accounting for the factors mentioned above, a theoretical resolution of 0.17nm (figure 2) and an information limit of 0.11nm were obtained experimentally at 0.9eV energy spread and a 0.1mrad illumination half angle. Shown in Figure 5 shows a lattice image of a $\langle 110 \rangle$ Si single crystal showing full doublet splitting.

References

1) Ultramicroscopy 54(1994)132-144 T.Honda, T.Tomita, T.Kaneyama and Y.Ishida.
2) EMAG93 Liverpool(1993) 535-538 T.Tomita, T.Honda and J.Critchell
3) J.Vac.Sci.Technol. 12,6(1975) 1228 L.W.Swanson

Proc. Microscopy and Microanalysis 1995, edited by G.W. Bailey, M.H. Ellisman, R.A. Hennigar, and N.J. Zaluzec
Copyright © 1995 MSA. Published by Jones and Begell Publishing, 79 Madison Ave., New York, NY 10016

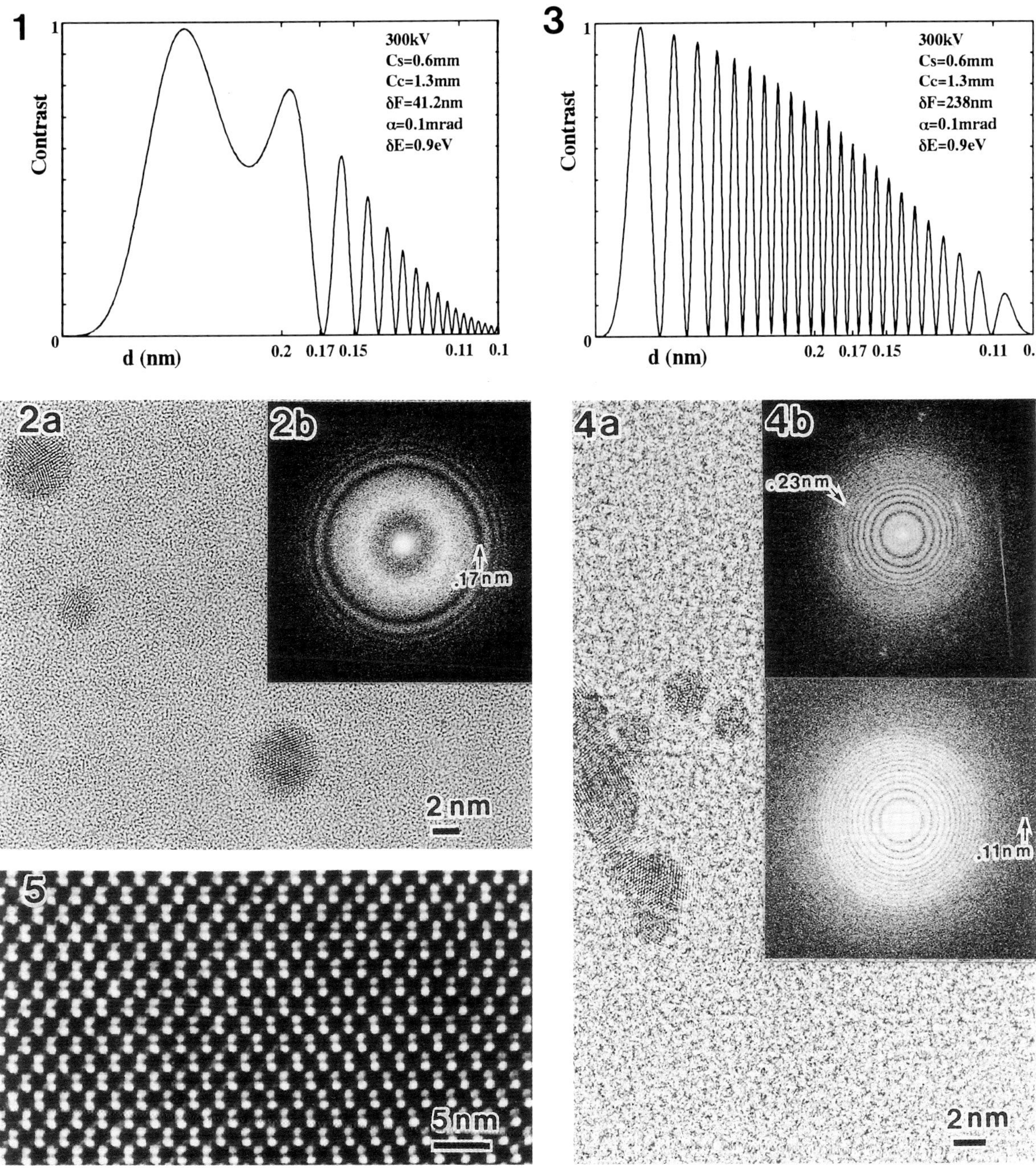

FIG. 1.—Phase contrast transfer function calculated under the condition of the Sherzer focus.

FIG. 2.—(a) TEM image of an amorphous Ge thin film under the condition near the Sherzer focus; (b) ODM pattern of the TEM image (a).

FIG. 3.—Phase contrast transfer function calculated under the condition of 248nm under focus.

FIG. 4.—(a) TEM image of an amorphous Ge thin film under the condition of 248nm under focus; (b) ODM pattern of the TEM image (a).

FIG. 5.—Lattice image of a single crystal Si viewed along the ⟨110⟩ orientation.

1 NM RESOLUTION X-RAY ELEMENTAL MAPPING USING A 300-KV FE-TEM

H. Murakoshi, T. Tsuchiya and H. Kakibayashi

Central Research Laboratory, Hitachi, Ltd., Kokubunji, Tokyo 185, Japan

The characteristics of advanced semiconductor devices are affected by contamination due to impurity atoms with sizes at the nanometer level, so it is very important to evaluate such contamination with nanometer-level resolution. In order to cope with this task, we have developed an x-ray elemental mapping system, attached to a 300-kV FE-TEM[1], with an information limit of 0.11 nm.

Both small spot size and sufficient probe current are necessary for high-resolution and high-sensitivity x-ray elemental mapping. The elemental mapping is performed with the scanning transmission electron microscope(STEM) mode. For high-resolution STEM imaging, the objective lens works with high excitation as a condenser-objective single lens[2], where the aberrations of the other illuminating lenses are demagnified to negligible level, allowing the extremely small spot size to be obtained. The probe current, however, decreases in proportion to the square of the demagnification of the illuminating system. In this study, the optimum illuminating condition for elemental mapping is examined.

Figure 1 shows spot size as calculated from the root-mean-square of source size, spherical aberration, chromatic aberration and diffraction error. M_t indicates the magnification of the illuminating system. For $M_t = 0.016$, calculated spot size reaches a minimum of 0.27 nm at a probe current of 0.1 nA. With increase in probe current, spot size drastically increases due to the influence of the spherical aberration. For $M_t = 0.029$, spot size reaches 0.37 nm at a probe current of 0.3 nA and remains in the less than 0.5 nm range for probe currents between 0.07 nA and 0.7 nA. The latter condition is suitable for high-resolution and high-sensitivity elemental mapping.

Figure 2 shows the TEM micrograph of a InP/InGaAs superlattice grown by organometalic vapor phase epitaxy. The widths of the InP layers are measured via their (200) spacing of 0.29 nm. The measured widths are 9.7, 4.4, 2.4, 1.2 and 0.6 nm, respectively. The characteristic x ray is detected by a side take off EDX system with an x-ray collection solid angle of 0.11 sr. The x-ray elemental map of a 200 nm × 200 nm area was acquired with an acquisition time of 400 sec. Figure 3 shows the x-ray elemental map obtained with the P $K\alpha$ line. Bright lines correspond to the InP layers. The layers with 1.2 nm to 9.7 nm widths are clearly visible and the 0.6 nm layer is indistinctly observed. Figure 4 shows the x-ray elemental map obtained from the sum of all x-ray energies between 1.0 and 1.4 keV including the Ga $L\alpha$ and the As $L\alpha$ lines. InP layers can be recognized as dark lines with almost the same resolution as Fig. 3.
Thus, x-ray elemental mapping with c. 1 nm resolution is attained, by optimizing the illuminating conditions of the 300 kV FE-TEM.

References
1. H. Murakoshi, M. Ichihashi and H. Kakibayashi, *Proc. Ann EMSA Meeting* 50(1992)936.
2. L. Reimer, *Transmission Electron Microscopy*, Berlin: Springer-Verlag (1993)120.

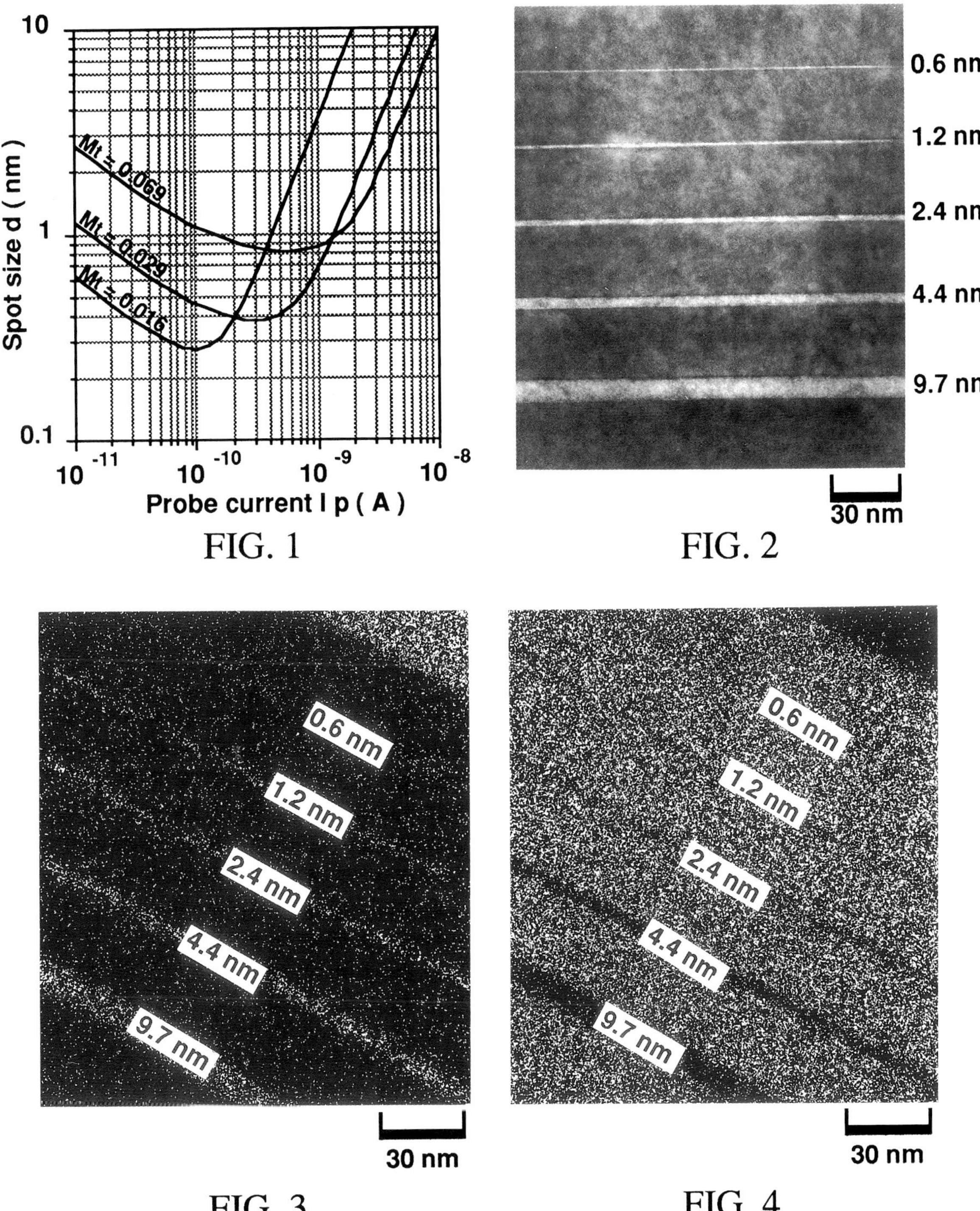

FIG. 1--Calculated spot sizes .
FIG. 2--TEM micrograph of InP/InGaAs multilayers.
FIG. 3--X-ray elemental map of InP/InGaAs obtained with the P $K\alpha$ line.
FIG. 4--X-ray elemental map of InP/InGaAs obtained with the Ga $L\alpha$ and As $L\alpha$ lines.

Position-Tagged Spectrometry: A New Approach For EDS Spectrum Imaging

Richard B. Mott, Charles G. Waldman, Rosa Batcheler, and John J. Friel

Princeton Gamma-Tech, Princeton, NJ 08540

Graphical display of the spatial distribution of elements in a specimen has been recognized as a powerful technique since the earliest days of electron-beam x-ray microanalysis. With recent advances in computing power and mass storage, it has become practical to save complete spectra at each pixel rather than simple window counts, providing great flexibility for analytical post-processing. The term "spectrum imaging" has been coined to describe such a data structure.[1,2]

Hunt and Williams give a concise summary of the advantages of spectrum imaging, including the ability to map an element which was not known to be present when the data were acquired, and the ability to integrate spectra from image regions of arbitrary shape to improve statistical precision.[1] Quantitative analysis may also be applied at each pixel to produce a concentration map, which is more informative than the raw or perhaps background-subtracted window counts usually displayed.[3] Sensitive discrimination of phases is possible using various fitting techniques on the pixel spectra, matching them against each other or against standards.[1]

Until now, the drawback has been that data acquisition and reduction times are very long. Therefore, in order to avoid wasting that time, the microscopist must do a great deal of analytical work before acquiring the spectrum image. In a sense, the spectrum image is taken to document what is already known, rather than being a useful tool for exploring the specimen.

New hardware has been developed which combines digital beam control with digital x-ray pulse processing to tag each photon energy measurement with the X and Y position of the beam at the time the photon was emitted from the sample, while the beam is scanned at full speed. As a result, a frame-averaged high-resolution digital image, multiple spectra from different phases, and multiple element maps can all be acquired and displayed simultaneously in real time while a spectrum image is streamed to disk for more detailed post-processing. On-the-fly compression of the spectrum image reduces the disk space required, which could otherwise reach hundreds of megabytes at high spatial resolution.

The real-time element maps initially resemble digital versions of old-fashioned dot maps. During the first seconds of acquisition, only a few speckles appear; as more photons come in, the contrast of each map is adjusted automatically so that the pixel with the highest photon count for each element is scaled to the maximum display intensity. Over time, a full gray-scale image emerges for each element. Pseudo-coloring and contrast may also be manually adjusted independently for each x-ray image during acquisition.

The spatial resolutions of the digital electron image, the displayed element images, and the stored spectrum map are all independent; however, the position tags carry the full scan resolution of the digital image. This raises an intriguing possibility for future work. Image deblurring algorithms are well known, given the point-spread function which causes the blurring.[4] Electron interaction volumes have been extensively studied, and constitute the "point spread function" from the point of impact of the beam. It might be possible (if computationally painful, given spatially varying composition) to use image-restoration techniques to sharpen x-ray images. For example, this might improve detectability for particles which are small compared to the excitation volume.

Proc. Microscopy and Microanalysis 1995, edited by G.W. Bailey, M.H. Ellisman, R.A. Hennigar, and N.J. Zaluzec
Copyright © 1995 MSA. Published by Jones and Begell Publishing, 79 Madison Ave., New York, NY 10016

References

1. J. A. Hunt and D. B. Williams, *Proc. Ann. MSA Meeting* 50(1992)1200.
2. C. Jeanguillaume and C. Colliex, Ultramicroscopy **28**(1989), 252.
3. J. I. Goldstein et. al., *Scanning Electron Microscopy and X-Ray Microanalysis*, N.Y.:Plenum(1992)525.
4. K. R. Castleman, *Digital Image Processing*, Englewood Cliffs, N.J.: Prentice-Hall (1979)277.

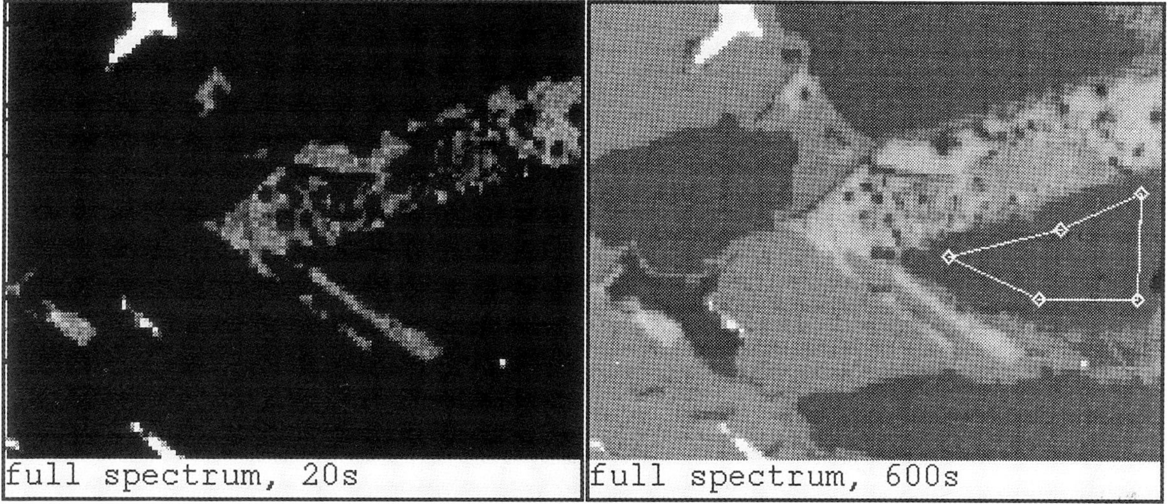

Fig. 1. Two stages of dynamic x-ray map display of the full spectrum. At left is the image after 20 seconds, at right after approximately 10 minutes. Displayed spatial resolution is 100 x 128 pixels.

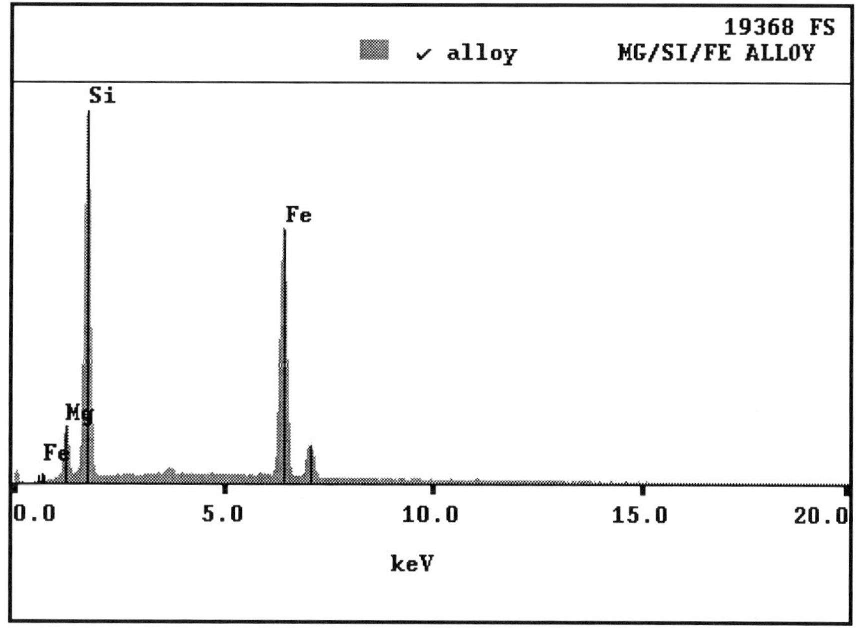

Fig. 2. Integrated spectrum from the Fe/Si-rich region outlined in the right-hand image of Figure 1.

DEVELOPMENT OF ELECTRON HOLOGRAPHY SYSTEM IN FIELD EMISSION ELECTRON MICROSCOPE

M.Takeguchi, T.Tomita,T.Honda and M.Kersker*

JEOL Ltd., Musashino 3-1-2, Akishima, Tokyo, 196 Japan
*JEOL(USA) Inc., 11 Dearborn Road, Peabody, MA 01960

Introduction

Electron holography is a powerful and practical method to know the phase of electron wave quantitatively and enables us to obtain various kinds of physical information such as magnetic and electric field distribution inside/outside of the specimen, thickness of the crystal etc.. Since this method requires an FE-TEM equipped with an electron biprism and some electron hologram reconstruction methodology, the general EM users is typically not yet familiar with electron holography practice. The recent introduction of commercial type FE-TEM, however, brings some users the opportunities of practical electron holography applications. We have recently developed an electron holography system for our FE-TEMs, the JEM-2010F and the JEM-3000F. The components, specifications and some application data are presented below:

Components and specifications

Our electron holography system consists of electron biprism, power supply and hologram reconstruction processor(PC + software). The electron biprism is rotatable and the beam-splitting electrode(a Pt wire of 0.6 μ m in diameter) can be cleaned by direct heating. The electron biprism is of the cartridge type for easy maintenance, thus removing the necessity of wire exchange. Hologram is digitized by an image/film scanner or a CCD-camera and the 1024×1024(maximum) image is numerically reconstructed at 256 gray-levels by using the Holography Image Processing Software (HIPS) which runs on MS-Windows in an IBM PC/AT compatible computer. There are three kinds of modes of operation: ①High magnification mode for the atomic revel resolution, ②Medium magnification mode for the observation of fine particle etc. and ③low magnification mode for the observation of wide field of view such as magnetic or electric field distribution. These modes are fully interchangeable through the lens optics of microscope eliminating the need to change the biprism position. The fringe spacing/interference width is ①0.05～0.2/10～50, ②0.2～1.0/20～150 and ③10～50/1000～10000 nm respectively.

Application data

Fig.1 (a) shows an electron hologram and (b) reconstructed phase image of MgO particles in the medium magnification mode. Fig.2 shows a double-exposure electron hologram of Fe_3O_4 particles on a carbon microgrid in low magnification mode, revealing magnetic field distribution around some Fe_3O_4 particles. Figs 3 (a) is electron hologram and (b) reconstructed phase image of one particle shown by the arrow in Fig.2. The phase of the reconstructed image is magnified by 4 times.

Proc. Microscopy and Microanalysis 1995, edited by G.W. Bailey, M.H. Ellisman, R.A. Hennigar, and N.J. Zaluzec

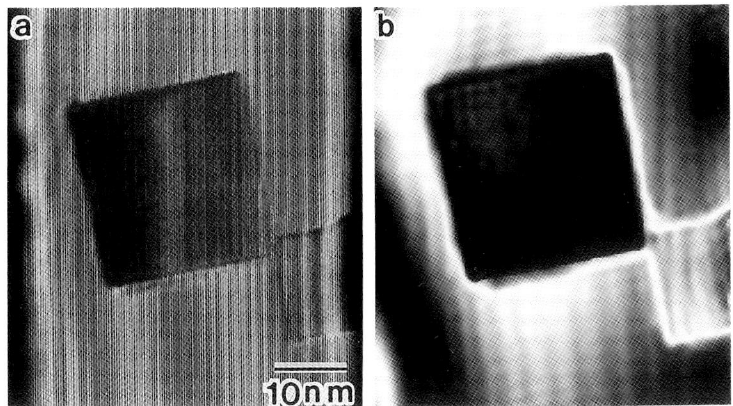

FIG.1　(a)Electron hologram and (b)reconstructed phase image of MgO smoke particles, observed in medium magnification mode.

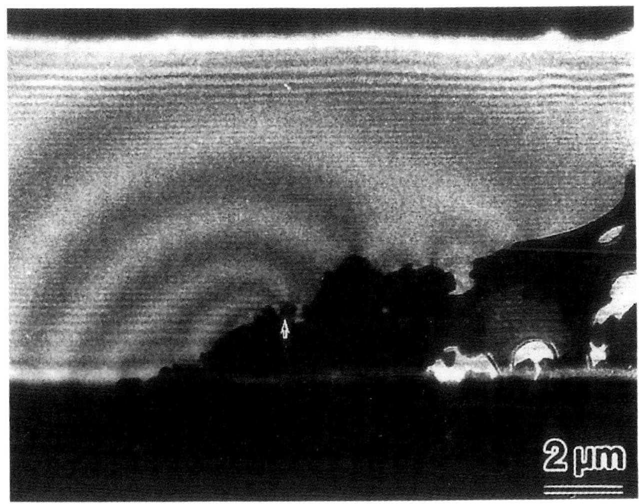

FIG.2　Double-exposure electron hologram of Fe$_3$O$_4$ particles on a carbon microgrid, showing magnetic field distributed over 9 μ m around some particles.

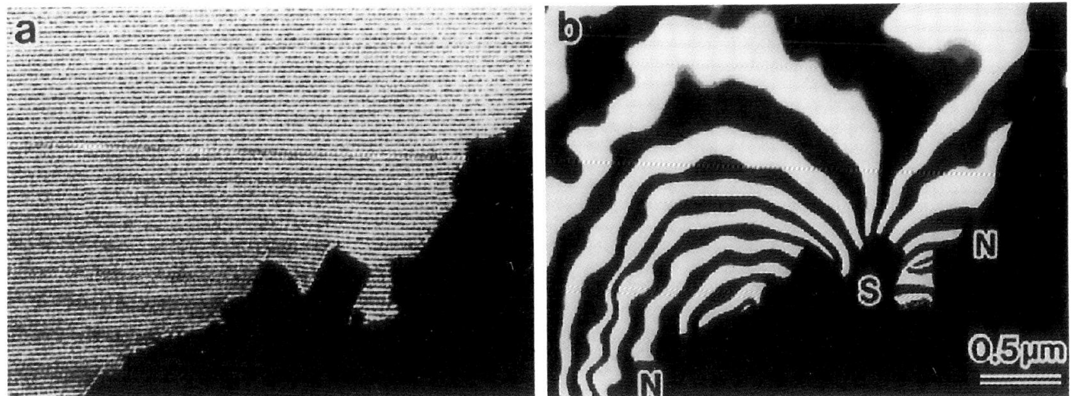

FIG.3　(a)Electron hologram and (b)reconstructed phase image of one particle shown by arrow in Fig.2. The phase of (b) is magnified by 4 times.

STATE OF THE DEVELOPMENT OF MULTIPOLE CORRECTORS FOR A PROBE-FORMING SYSTEM AND A HIGH RESOLUTION 200 kV TEM

M. Haider* and J. Zach*

*European Molecular Biology Laboratory, Postfach 102209, D-69012 Heidelberg, FRG

The development of modern high resolution electron microscopes has shown the emergence of new modern microscopes which are easy to operate and, more importantly, with which one can attain high resolution on a routine basis. However, the physical limit set by the inherent aberrations of electron lenses could not be overcome by simply optimising the geometry of the pole pieces. The only direct way to get rid of the aberrations of the round lenses is to use a corrector with which the resolution limiting aberrations can be compensated.

For the observation of uncoated biological surfaces we started a project to develop a Low Voltage Scanning Electron Microscope (LVSEM). The main problem with using low energies is to design electron optics with which one can achieve high resolution in the range of 1 nm at energies of 1 keV and below[1]. Hence, the two main axial aberrations of such a probe forming system: the chromatic and the spherical aberration C_C and C_3, respectively, have to be considered as the limiting factors of the probe diameter. Therefore, the compensation of these two aberrations is the best choice as one does not want to run into other limitations if, for example, the geometry of the objective lens is scaled down in order to obtain small aberration coefficients.

The multipole-corrector[2] consists of four combined electric and magnetic twelve pole-elements with which the astigmatic ray path can be adjusted. With these multipole-elements all required fields of symmetry and orientation can be generated (e.g. from dipole up to duodecipole fields). The combined quadrupole fields allow the correction of the chromatic aberrations and the spherical aberration C_3 can be corrected by exciting octopole fields within the multipole-element.

This type of multipole corrector can be used for any type of probe forming system such as SEMs as well as STEMs[3]. The primary energy can go up to around 100 kV but it would be difficult to go up to 300 keV with the dimensions of the existing corrector. The resolution we obtained so far with the present set up of the LVSEM was d < 2.0nm at 1 keV (see Fig.1). The theoretical limit is 1 nm at 1 keV which we expect to approach at the end of this year.

One way to improve the resolving power of a **TEM is to correct the spherical aberration** as proposed by Rose 1990[4]. We are following his hexapole-corrector concept. The possibility of correcting aberrations of the third order with long hexapoles was already mentioned in the late '70's by Beck[5] and later by Crewe[6] and Rose[7] for probe forming systems. However, in our case, two hexapole fields are used to generate a negative round axial aberration of the third order with which the spherical aberration of an imaging device can be compensated.

The developmental project of the spherically corrected TEM is divided into two stages: first, the feasibility study and second, the setting up of a high resolution 200 kV TEM

Proc. Microscopy and Microanalysis 1995, edited by G.W. Bailey, M.H. Ellisman, R.A. Hennigar, and N.J. Zaluzec
Copyright © 1995 MSA. Published by Jones and Begell Publishing, 79 Madison Ave., New York, NY 10016

with a point resolution of d ≤ 0.1 nm. The first stage was successfully finished by the demonstration of correcting C3 of a Philips CM 20 super twin (see Fig.2). The second phase with the incorporation of the corrector into a commercial C3 limited 200 kV TEM has just been started.

References :

1 J. Zach, Optik (1989) **83**, 30
2 S. Lanio & M. Haider, Optik (1989) **83** Suppl. 4, 54
3 J. Zach & M. Haider, Optik (1995) **98**, 112
4 H. Rose, Optik (1990) **85**, 19
5 V. D. Beck, Optik (1979) **53**, 241
6 A. V. Crewe, Optik (1982) **60**, 271
7 H. Rose, Nucl. Instr. a. Methods (1981) **187**, 187
8 Part of this work (correction of the spherical aberration in a TEM) is supported by a grant from the Volkswagen Stiftung, Hannover, FRG.

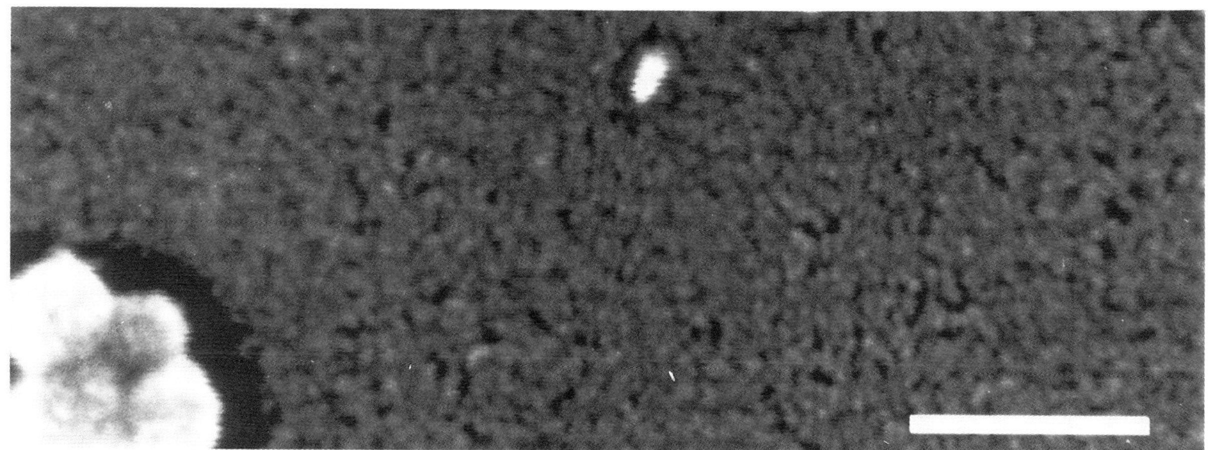

Fig. 1 -Resolution test at 1 keV: The test sample consists of 30 and 60 nm latex spheres and 8 nm colloidal gold, slightly sputter-coated with gold. The finest details in the grainy gold-film show a size of below 2 nm. The bar shows 100 nm.

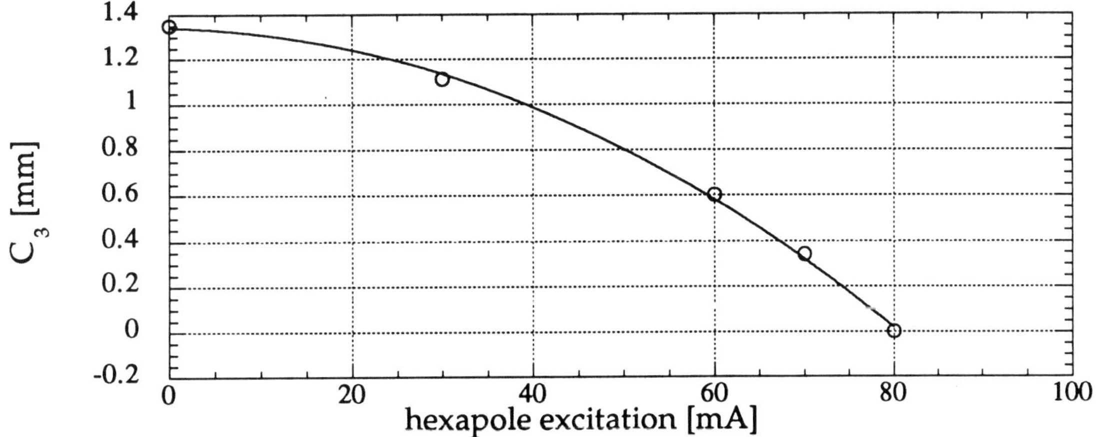

Fig. 2 - Spherical aberration constant Cs of a 200 kV TEM measured for various excitations of the hexapole field. This spherical aberration decreases with the second power of the hexapole excitation and as can be seen Cs ≈ 0 at 80 m A hexapole current.

INDIRECT MEASUREMENTS IN MICRO-SPACE WITH THE ELECTRON MICROSCOPE

Sterling P. Newberry

CBI Labs., 11 S. Westcott Road, Schenectady NY 12306, USA

At the 1958 meeting of our society, then known as EMSA, the author introduced the concept of micro-space and suggested its use to provide adequate information storage space and the use of electron microscope techniques to provide storage and retrieval access. At this current meeting of MSA, he wishes to suggest an additional use of the power of the electron microscope.

The author has been contemplating this new use for some time and would have suggested it in the EMSA fiftieth year commemorative volume, but for page limitations. There is compelling reason to put forth this suggestion today because problems have arisen in the "Standard Model" of particle physics and funds are being greatly reduced just as we need higher energy machines to resolve these problems. Therefore, any techniques which complement or augment what we can accomplish during this austerity period with the machines at hand is worth exploring.

We are all aware that a problem which cannot be answered experimentally is often resolvable by indirect measurements, even though the items of interest may be too small, too far away, or of too short duration to be observed directly. Historically, insights from these indirect measurements have often been available long before the technology for direct observation becomes available. Now the basic particles and organization of the nucleus are not only too small for us, and often of brief duration, but the force (the glue) which holds the protons and neutrons together has not yet been detected.

The crux of this suggestion is that we create an "atomic reactor" in or near the field of view of an electron microscope and measure observables such as electron wave front distortions and atom displacements to infer the behavior and properties of the disintegration products. One may reasonably ask how can we put an atomic reactor in a microscope or who wants to place a specimen chamber in a beam line? Actually these have already been done occasionally as for example, placing an alpha emitter in a field ion microscope, but the results have not been spectacular. No nucleus has been split or fused nor have traces of strange particles been observed. Placing samples in a reactor and post mortem examination of the damage has also been done but is again of marginal usefulness.

What can be done, without a radical change of the microscope or moving it from its location, is to induce particle/anti-particle annihilations in a controlled fashion. For example, one can literally carry a "bottle" of anti-protons from a large particle accelerator site to the microscope laboratory.[1,2] These anti-protons have been used in hospitals for tumor imaging. By raising the intensity level from image formation to tissue destruction levels, cancerous tissue has been destroyed without harming normal tissue a few mm away. Much closer control should be possible, thus making the use of anti-particles our torch for a small reactor in the specimen chamber. The reason this is possible is that annihilation only takes place very near the end of the anti-particle's range. Apparently the particle and anti-particle must be stationary with respect to one another for annihilation to occur. If one starts with a monoenergetic beam, then the end of range is very tightly packed and can be adjusted very precisely by interposing thin sheets of homogenous material. An anti-proton can destroy a proton or a neutron. These two are the major mass components of an atom. The products of disintegration from annihilation

Proc. Microscopy and Microanalysis 1995, edited by G.W. Bailey, M.H. Ellisman, R.A. Hennigar, and N.J. Zaluzec
Copyright © 1995 MSA. Published by Jones and Begell Publishing, 79 Madison Ave., New York, NY 10016

events are generally, charged and neutral pions, nuclear gamma rays, and nuclear fragments.

Even if one grants that disruptions of atoms occur from annihilation events, why try to measure these events with a microscope? Why not use a Wilson cloud chamber or directional detectors at the beam line? The first reason is that some of the particles have a very short half life in free space so that even if they are created in abundance at a target, the number of survivors are almost nil at laboratory distances. In the microscope we may be able to enjoy a flux of particles per second rather than particles per day. Still at this higher flux we must save only useful image frames by coincidence triggering of accumulating frames to keep from being swamped by noise from quiet periods. Another reason is that the very few measurements we have gained in laboratory sized experiments, indicate that the behavior of some particles depends upon their interaction with near by particles. What we really want to know is therefore, how they interact while still close to the other debris of the disturbed nucleus and to neighboring atoms. For example, if two protons are detected from a single event, do their paths project back to the same origin in space or is there evidence that one or both have curved paths initially? As another example, when two neutral particles are close together is there evidence of a strong attractive force?

Can we be sure that any useful information will be gained by trying to make indirect measurement in the microscope field of events too small and brief of themselves to be resolved; of course not. Do we know enough about the behavior of the fundamental particles to predict that nothing will be gained by trying, the answer again must be negative. By publishing this suggestion, it is hoped that a dialogue can be started which may lead to some promising experiments at one of our national facilities.

1. B. W. Augenstein, Letters to the editor, *Physics Today* 46(1993)9.
2. B. Augenstein, B. Bonner, F. Mills and M. Nieto, eds., *Anti-proton Science and Technology* World Scientific Publishers (1988)640.

THE PERFORMANCE OF ANALYTICAL ELECTRON MICROSCOPES AND EDX SYSTEMS, MEASURED WITH A NICKEL OXIDE TEST SPECIMEN

Kaikee Wong and R.F. Egerton

Department of Physics, University of Alberta, Edmonton, Canada T6G 2J1

A test specimen consisting of a 50nm film of NiO on a 200-mesh molybdenum grid can be used to evaluate the diffraction, EELS and EDX capabilities of an analytical TEM.[1] Several laboratories have utilised this specimen and their results have been compared.[2] One measurement is the Mo-Kα/Ni-Kα ratio, which characterizes the amount of stray radiation within a TEM column; values are typically in the range 3 to 10 for conventional TEM's and as high as 40 for instruments in which beryllium sheet is used to shield the upper and lower polepieces; see Fig. 1. Another criterion is the peak/background ratio of the Ni-Kα peak, the background being measured within a 10eV window beneath the peak.[3] This ratio reflects electronic noise as well as column radiation but correlates well with the Ni-Kα/Mo-Kα ratio (see Fig. 1), indicating that column radiation can substantially affect the EDX background.

The variations in performance visible in Fig. 1 are partly due to differences in TEM accelerating voltage. Fig. 2 shows that higher peak/background is possible at higher voltage, as would expected if the background were entirely bremsstrahlung radiation.[4] But the fact that many of the P/B values are substantially below this limit again suggests that column radiation makes an important contribution.

The Mo-Kα/Mo-Lα ratio gives an indication of the character of the column radiation. We find that, as the diameter of the second-condenser aperture is reduced, this ratio decreases from a high value (typical of x-radiation in the column) to a low value (typical of electron impact); see Fig.3. With the beam passing through the NiO film, the Mo-Kα/Mo-Lα ratio is somewhat lower than with the beam passing through an empty grid square (see Fig. 3), suggesting irradiation of the Mo grid by electrons which have been scattered in the film and backscattered by the lower objective polepiece. For the case of incident electrons, we find that the K/L ratio depends on the electron energy; see Fig. 4.

The Ni-Kα/Ni-Lα ratio R(K/L) can be used to monitor hydrocarbon or ice buildup on the EDX detector or its protective window, since both of these materials preferentially absorb the low-energy Ni-Lα photons. The Ni-Kα/O-K ratio R(Ni/O) is more affected by a carbon layer than by ice, since O-K photons (observable with a light-element detector) have energies below the O-K absorption edge. Solving two simultaneous equations provides estimates of the ice and hydrocarbon thickness t (in nm):

$$t(\text{ice}) \approx 1850 \ln[R(K/L)/R_0(K/L)] - 532 \ln[R(Ni/O)/R_0(Ni/O)]$$

$$t(\text{carbon}) \approx 485 \ln[R(Ni/O)/R_0(Ni/O)] - 87 \ln[R(K/L)/R_0(K/L)]$$

Here, R_0 denotes ratios measured for a clean detector (soon after installation or after removing adsorbed layers by thermal treatment). These equations assume that the layer thicknesses are insufficient to appreciably absorb the higher-energy Ni-Kα x-rays, that the hydrocarbon layer is predominantly carbon and that the linear absorption coefficients (in cm⁻¹) for Ni-Lα photons are 5696 (in ice) and 6240 (in carbon), while those relevant to O-K photons are 1022 (for ice) and 21720 (for carbon).

Proc. Microscopy and Microanalysis 1995, edited by G.W. Bailey, M.H. Ellisman, R.A. Hennigar, and N.J. Zaluzec

1. R.F. Egerton and S.C. Cheng, *Ultramicroscopy* 55(1994)43.

2. J.C. Bennett and R.F. Egerton, *J. Microsc. Soc. Amer.*, in press.

3. C.E. Fiori, C.R. Swyt and J.R. Ellis, in K. Heinrich, Ed., *Microbeam Analysis - 1982*, San Francisco: San Francisco Press (1982)57.

4. J.N. Chapman, C.C. Gray, B.W. Robertson and W.A.P. Nicholson, *X-ray Spectrometry* 12(1983)153

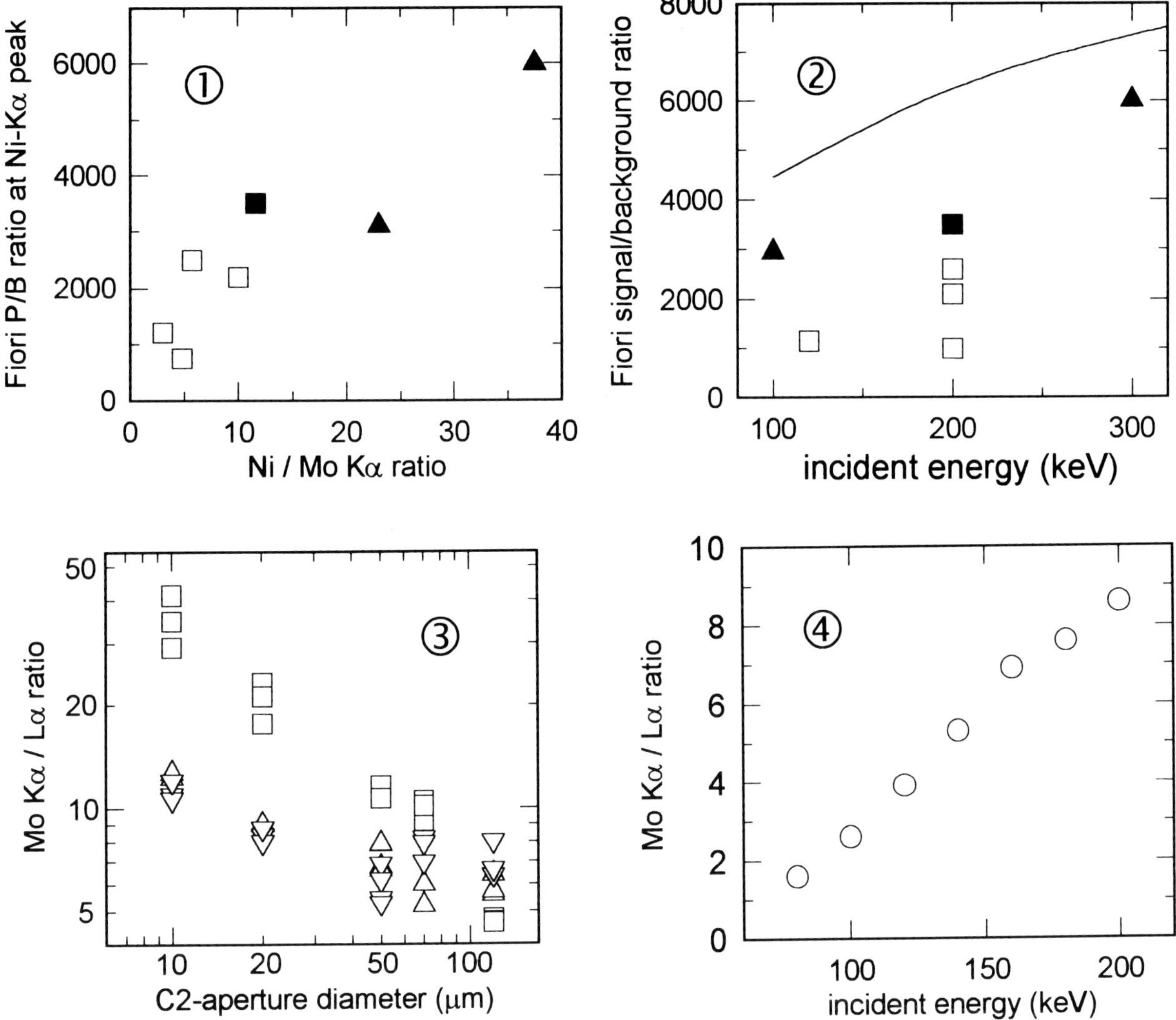

FIG. 1. -- Performance parameters measured in various microscopes. Squares represent conventional TEM's, triangles denote dedicated STEM's. Open squares are from thermionic-source instruments, whereas solid data points are from microscopes which use a field-emission electron source.

FIG. 2. -- Fiori peak/background ratio as a function of incident-electron energy. The shapes of the data points identify the type of instrument, as in Fig.1. The curve was calculated using a modified Bethe-Heitler formula for the bremsstrahlung background.[4]

FIG. 3. -- Mo K/L ratio measured for 200keV electrons passing though the centre of a grid square with no NiO film (squares), film on the exit surface (triangles) and the entrance surface (inverted triangles).

FIG. 4. -- Mo K/L ratio for electrons focussed onto a grid bar, as a function of the incident energy.

Multiple Scattering Calculations at the Oxygen K-edge for SiO_2.

DJ Wallis[*1], R Brydson[**] and PH Gaskell[*]

[*]University of Cambridge, Cavendish Laboratory, Madingley Road, Cambridge, CB3 OHE, U.K.
[**]University of Surrey, Dept. Materials Science & Engineering, Guildford, Surrey, GU2 5XH, U.K.

Electron Energy Loss Spectroscopy (EELS) in the STEM is potentially a very powerful technique for studying the structure of materials. The Near Edge Structure (NES) in core loss edges carries information about triplet, and higher order correlation functions and under certain conditions is sensitive to long range atomic correlations. However, since the NES contains strong contributions due to multiple elastic scattering, a simple interpretation in terms of atomic structure is not possible.[2] It is therefore necessary to perform some form of modeling to obtain structural information. One approach to modeling NES is Multiple Scattering (MS) theory.[3]

MS calculations have been performed to model the O K-edge ELNES for the α-quartz phase of SiO_2.[4] The calculations use a real-space atomic cluster and model the atomic potentials using the Muffin Tin approximation.[3] Effects of the core hole produced upon excitation were accounted for using the $(Z+1)^*$ approximation for the excited atom.[5] Energy scales of the calculated data have also been compressed to correct for the energy dependence of the exchange correlation between the photoelectron and the valence electrons which is not accounted for in the calculation.[6] The compression factor used was 0.9 and is in agreement with that used by other authors.[6-7] Figure 1 show a full MS calculation for a 1nm diameter cluster of atoms, along with an O K-edge for α-quartz. It can be seen that the calculation reproduces all the features seen in the experimental edge, except at the edge onset. Incorrect estimation of the intensity of features near the edge onset is due to the failure of the Muffin Tin approximation.

Having shown that the MS calculations reproduce the experimental NES it is now possible to determine the structural origins of particular features by performing calculations for reduced clusters with restricted scattering paths. Figure 2 (a) shows calculations for a single Si-O^*-Si unit (the asterisk indicates the excited atom) with only single scattering events considered, and also a full MS calculation. It can be seen that only a single peak b is reproduced by single scattering and that its intensity is only a little enhanced in the full MS calculation. Peak b therefore seems to be a so-called "scattering shape resonance" - mainly due to single scattering by first neighbour Si atoms. In figure 2 (b) calculations for two corner-sharing tetrahedra (i.e. O_3Si-O^*-SiO_3) are presented. The calculations shown are for a single scattering event from the 2nd shell (containing oxygens) only; single scattering in both shells 1 and 2; and a full MS calculation for both shells. In the curve for single scattering from oxygens in shell 2 only, a peak labeled a is seen; peak b being present only when single scattering in both shells 1 and 2 is considered. This suggests that peak a has a large single scattering contribution from the six surrounding second neighbour oxygen atoms. The results of the full MS calculation show that peak a is sharpened by MS within this cluster. Addition features seen in the experimental edge are not reproduced in the calculations for these small clusters and therefore their origins must lie in medium range (i.e. 0.3-0.5nm) atomic correlations.

In conclusion, using MS theory the O K-edge NES for α-quartz has been successfully modeled allowing the assignment of spectral features to their atomic origins. The calculations also demonstrate that information about the atomic environment of oxygen in SiO_2 beyond second neighbours is available from EELS spectra.[8]

References
1 Contact address ORNL, Solid State Division, P.O. Box 2008, Oak Ridge, TN 37831 - 6031.

Proc. Microscopy and Microanalysis 1995, edited by G.W. Bailey, M.H. Ellisman, R.A. Hennigar, and N.J. Zaluzec

2 R.F. Egerton, Electron Energy Loss Spectroscopy in the Electron Microscope (1986).
3 D.C. Koningsberger, X-Ray Absorption: Principles, Applications, Techniques of EXAFS, SEXAFS and XANES, (1987).
4 D.J. Wallis et. al. To be published.
5 R. Brydson, J. Chem. Soc. Chem. Commun. (1989)1010.
6 I. Davoli, Phys. Rev. B 33(1986)2979.
7 H. Kurata, and C. Colliex, Phys. Rev. B 48(1993)2102.
8 This research was sponsored by Pilkington Plc and the Science and Engineering Research Council.

FIG. 1.

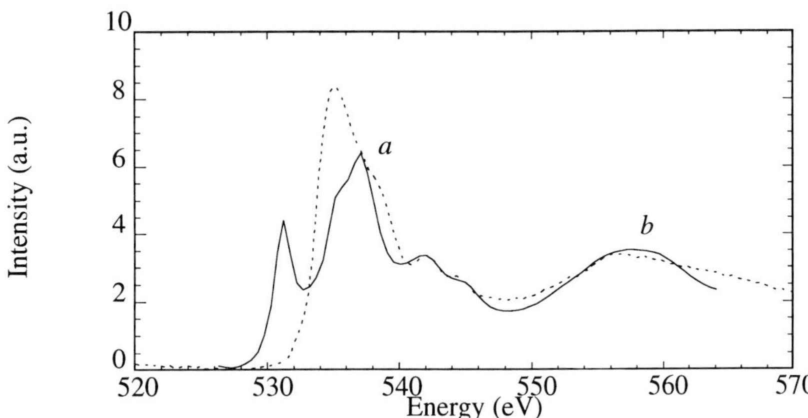

FIG. 2.

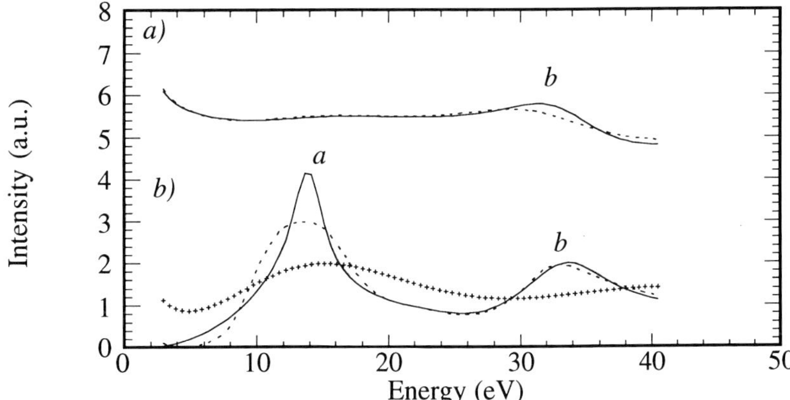

FIG. 1. Full multiple scattering calculation for a 1nm cluster of atoms modeling the structure of α-quartz (solid line) and an experimental O K-edge after removal of inelastic multiple scattering and a background of the form AE^{-r}.

FIG. 2. (a) Multiple scattering calculation for a Si-O*-Si unit; single scattering (dotted line); full scattering (solid line). (b) Multiple scattering calculation for an O_3Si-O*-SiO_3 unit; single scattering from oxygen shell only (crosses); single scattering from both shells (dotted line); full multiple scattering calculation (solid line).

TRIAL PRODUCTION OF γ-TYPE ENERGY FILTERING TEM

Y. Taniguchi,* E. Nakazawa** and S. Taya*

*Instrument Division, Hitachi Ltd., 882 Ichige, Hitachinaka, Ibaraki 312, JAPAN
**Techno Research Laboratory, Hitachi Instruments Engineering Co. Ltd., 882 Ichige, Hitachinaka, Ibaraki 312 JAPAN

Imaging energy filters[1] can add new information to electron microscopic images with respect to energy-axis, so-called electron spectroscopic imaging (ESI). Recently, many good results have been reported using this imaging technique.[2] ESI also allows high-contrast observation of unstained biological samples, becoming a trend of the field of morphology. We manufactured a new type of energy filter as a trial production. This energy filter consists of two magnets, and we call γ-filter since the trajectory of electrons shows 'γ'-shape inside the filter. We evaluated the new energy-filter TEM with the γ-filter.

Figure 1 shows schematic view of the electron optics of the γ-type energy filter. For the determination of the electron-optics of the γ-type energy filter, we used the TRIO[3] (Third oRder Ion Optics) program which has been developed for the design of high resolution mass spectrometers. The TRIO takes the extended fringing fields (EFF) into consideration. EFF makes it difficult to design magnetic energy filters with magnetic sector fields. We optimized the trajectory of electrons and the geometry of the magnets by eliminating or minimizing the second-order aberrations both in the achromatic image plane and in the energy dispersive plane. Moreover, we kept in mind to design as small as possible in size. The energy filter including the yoke is 110 mm high and 225 mm long, which is the smallest in the in-column type energy filters proposed so far. The energy filters of such concept have the advantage of easy handling and maintenance for daily use.

We installed the γ-type energy filter in a conventional TEM (CTEM), and observed electron spectroscopic images of some test samples at 100 keV. The filtered images were recorded on imaging plates. As the first experiment, we applied energy filtering TEM to unstained biological samples. This kind of sample shows weak contrast in CTEM images because most part of the sample consists of low-Z elements such as carbon, etc. and is useless for morphologic observation. It is well known that ESI with great contrast of unstained biological samples is obtained with energy-loss electrons just before the K-edge of carbon (284 eV). Figure 2 shows the energy-filtered images of immature cranial bones. These samples were isolated from 1 day postnatal rats under deep anaesthesia and fixed with 2 % paraformaldehyde and 2 % glutaraldehyde in 0.1 M phosphate buffer (pH 7.4). Fixed tissues were dehydrated with an ascending series of ethanol, and then embedded in LRWhite or Polybed 812 resin. Thin sections of 70 nm thick was observed without any staining. Figure 2a shows the zero-loss image obtained by removing the plasmon-loss electrons and core-loss electrons and figure 2b was obtained by setting the energy-selecting slit on the energy dispersive plane at an energy range of 250 eV ~ 270 eV. Figure 2b shows negative contrast since the electrons with primary energy don't contribute to the image formation, and high contrast yielding the possibility of morphological information. In these preliminary experiments, the fundamental performances of the γ-filter as an energy-filtering TEM was successfully confirmed.

1. H. Rose, *Ultramicroscopy* 28(1989)184.
2. K. Kimoto, et al., Jpn. J. Appl. Phys. 33(1994)L1642.
3. T. Matsuo, et al., *Mass Spectroscopy* 24(1976)19.
4. The authors gratefully acknowledge the offer of the unstained biological samples and evaluation of the results of Prof. Dr. J. Usukura of Nagoya Univ. One of the authors (Y. T.) would like to express his thanks to Prof. Dr. H. Rose of Technische Hochschule Darmstadt for the useful advice about alignment procedure of energy filters.

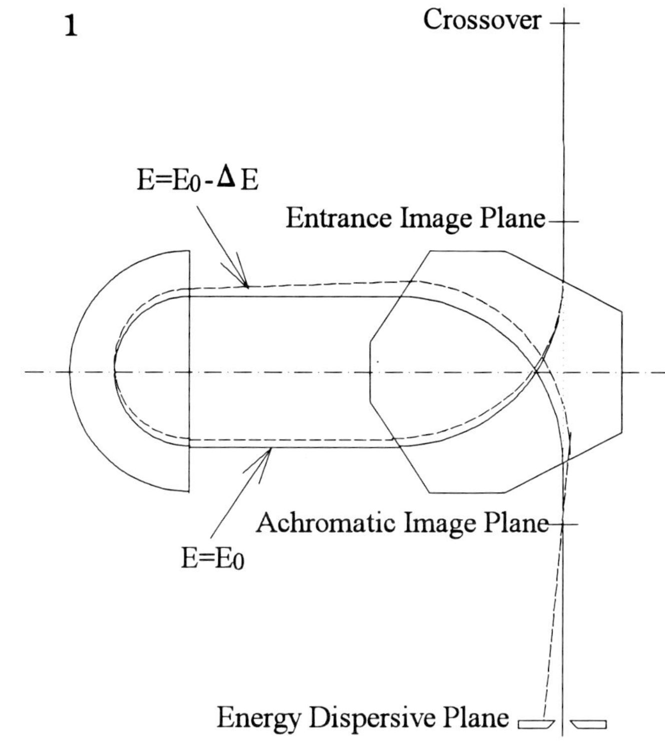

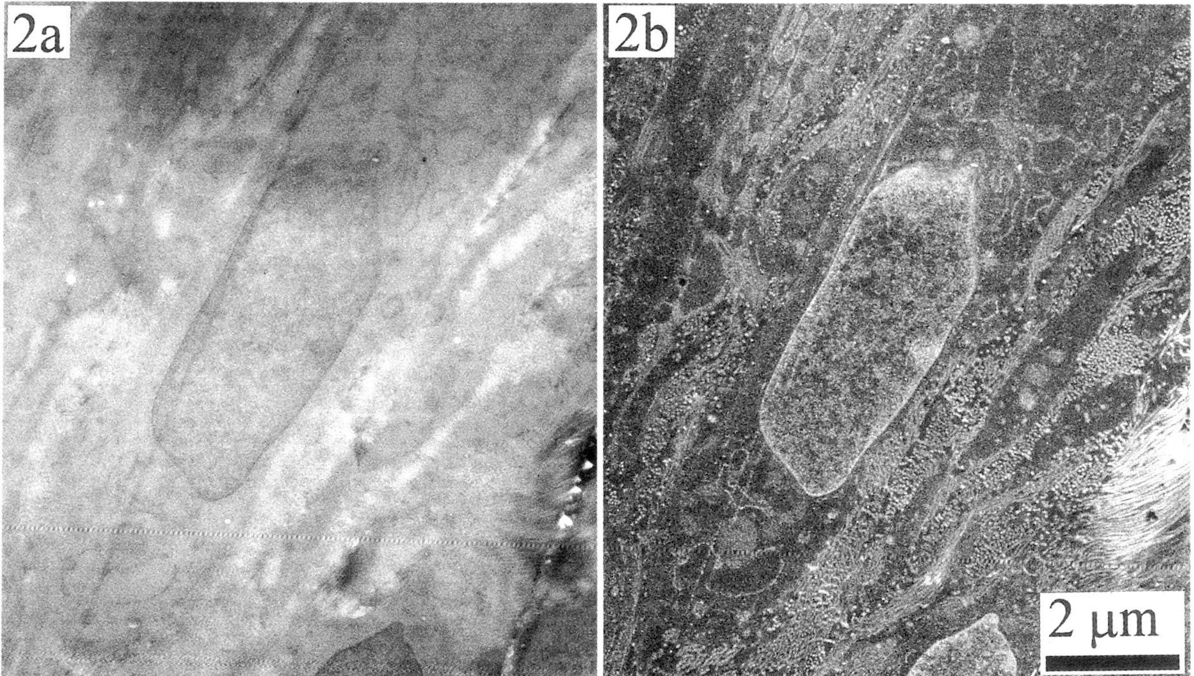

FIG. 1. — Schematic view of the electron optics of the γ-filter. The solid line indicates the trajectory of electrons of energy E_0; the broken line that of energy E_0-ΔE.
FIG. 2. — Energy filtered images of immature cranial bones isolated from 1 day postnatal rats obtained by the γ-type energy filtering TEM. (a) Zero-loss image and (b) 250 eV ~ 270 eV loss image.

DIFFERENTIAL PHASE CONTRAST IN TEM FOR MAGNETIC MICRO STRUCTURE OBSERVATION

P.Kruit*, A.H.Buist*, M.R.McCartney** and M.R.Scheinfein***

*Delft University of Technology, Department of Applied Physics, Delft, NL 2628 CJ, The Netherlands
**Center for Solid State Science, Arizona State University, Tempe, AZ 85287-9004
***Arizona State University, Department of Physics and Astronomy, Tempe, AZ 85287

Phase contrast in TEM can only give information in a limited band of spatial frequencies. The differential phase contrast mode in STEM[1,2] does not have this limitation, since one simply measures the angle over which the electron beam is deflected by the specimen. This is useful, for example, for determining the magnetic field distribution in magnetic thin films[3]. Recently, a form of electron holography was developed, also in STEM, to obtain differential phase contrast images[4]. In this method, the illuminating beam is split by a biprism in such a way that there are two mutually coherent electron source images close to the specimen. On a far away CCD detector, two shadow images are formed which interfere to give a fringe pattern. It is this fringe pattern that contains the differential phase information. Here, we show that differential phase contrast can also be obtained in a TEM.

In the TEM, two coherent illuminating beams are created by a beam splitter in the condenser system. These beams can be treated as two plane waves, creating a fringe pattern in the plane of observation[5],as shown in figure 1. The observation plane is defocussed by a distance D_z with respect to the specimen plane. The classical beam paths P_1 and P_2 contributing to the formation of the fringe pattern have gone through different points on the specimen. It is the phase difference between these two points which determines the position of the fringe in the observation plane. In a classical approximation, the field inside the specimen deflects the electron beam, resulting locally in a different position of the fringe. In principle, for small values of defocus, the line pattern could also be created by scanning a focussed beam or imaging a lattice. For large values of the defocus, one of the paths can pass through a hole in the specimen. In this case, one measures the phase difference with respect to vacuum, in an almost identical manner as in regular off-axis holography. Because of that large defocus, the resolution in that mode is intrinsically poor[6], although it is sometimes possible to reconstruct the in-focus image[7,8].

To test the principle, we inserted a biprism in the condenser aperture of a Philips EM400, equipped with a field emission source. The fringes were imaged on a cobalt film, which was deposited on carbon. In order not to disturb the magnetic fields in the Co, the objective lens was switched off and the microscope was operated in low magnification mode (1000X). The images were recorded on a slow scan CCD camera. Figure 2 is a traditional Lorentz microscopy image showing contrast at the domain boundaries. The fringes appear in this same image when the biprism is switched on. We were not able to fill the whole image with high contrast fringes. However, we could easily shift the coherent fringe band to different positions. In figure 3, eight of these bands are pasted together to form one image. In the diffractogram of this image, a sideband was selected and used for the reconstruction of the phase image. Since the phase varied over about 6π it was necessary to follow a phase unwrapping procedure, which was sometimes difficult at the domain boundaries. The result is given in figure 4. The bright and dark areas in this picture must be interpreted as domains with the magnetic field parallel, and anti-parallel to the fringes in figure 3. In conclusion, we have demonstrated that differential phase contrast in TEM is possible.

Proc. Microscopy and Microanalysis 1995, edited by G.W. Bailey, M.H. Ellisman, R.A. Hennigar, and N.J. Zaluzec.
Copyright © 1995 MSA. Published by Jones and Begell Publishing, 79 Madison Ave., New York, NY 10016

1. H.Rose, *Optik* 39(1974)416.
2. N.H.Dekkers and H.de Lang, *Optik* 41(1974)452.
3. J.N.Chapman, *J.Phys.D.* 17(1984)623
4. M.Mankos, M.R.Scheinfein and J.M.Cowley, *J.Appl.Phys.* 75(1994)7418
5. P.Kruit and A.H.Buist, *ICEM 13-PARIS* (1994)335
6. G.Matteucci, G.F.Missirolli and G.Pozzi, *Ultramicroscopy* 6(1981)109
7. Q.Ru, J.Osakabe, J.Endo and A.Tonomura, *Ultramicroscopy* 53(1994)1
8. M.Mankos, *PhD-thesis,* Arizona State University (1994)

9. This work was supported by the Dutch Technology Foundation (Stichting voor Technische Wetenschappen) and utilized facilities at CHREM of CSSS at ASU.

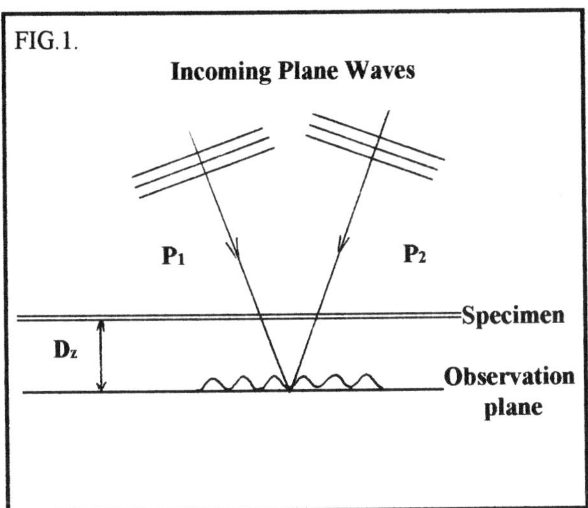

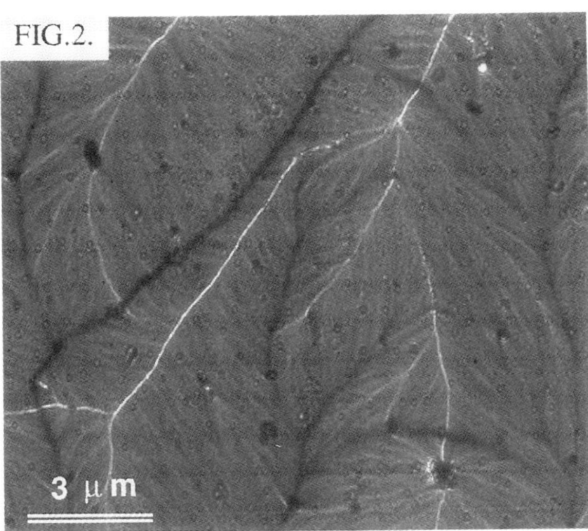

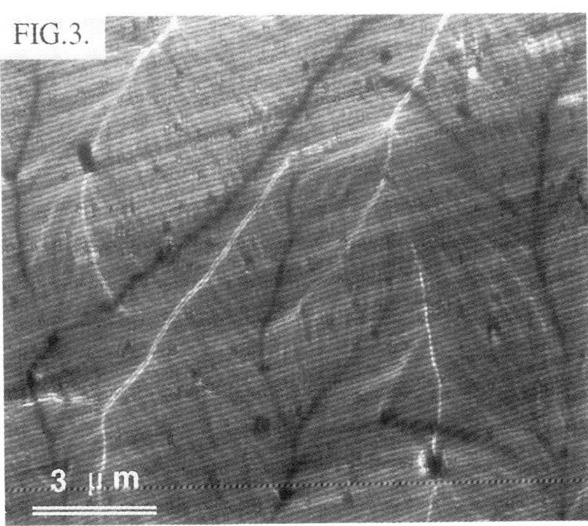

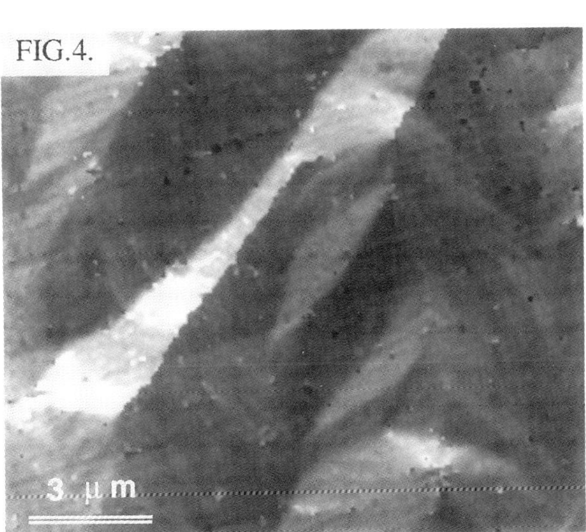

FIG.1.--Principle of differential phase contrast in TEM
FIG.2.--Slightly defocussed TEM image of Co on C film, magnetic contrast in Lorentz microscopy mode.
FIG.3.--Same as fig.2, with standing wave illumination, notice fringe displacement in magnetic domains.
FIG.4.--Differential phase contrast image obtained by processing fig.3

RECONSTRUCTION OF LOW-VOLTAGE POINT-PROJECTION HOLOGRAMS OF MASK-LIKE OBJECTS

X. Zhang, P. Kruit*, W. Qian and J. C. H. Spence

Department of Physics and Astronomy, Arizona State University, Tempe, AZ 85287-1504
*Department of Applied Physics, Delft Univeristy, Lorentzweg 1, Delft 2628CJ, The Netherlands

Point-projection shadow images (PPM) are obtained by placing a nanotip field emitter at a potential of a few hundred volts close to a grounded sample, which acts as the anode.[1] It can be shown[2] that the shadow image is identical to an conventional out-of-focus TEM image with a defocus equal to the tip-to-sample distance z_1, and is an in-line hologram. Recent work[3] shows that most objects used for PPM may be treated as opaque masks containing holes, since the inelastic mean free path (IMF) for 100 eV electrons is 6-8Å in carbon.[4] The question arises if the mask shape can be reconstructed from the shadow image, i.e. if the image can be brought back into focus. For masks, Fresnel edge fringes constitute most of the hologram (apart from interference between different holes). Fresnel fringes and shadow images were first observed by Boersch[5], prior to Gabor's demonstration that, for small opaque objects, they could be removed by holographic reconstruction or "focus restoration".[6] Shadow images at low energy were first observed in 1939.[7]

Conventional high-voltage in-line electron holography[8] uses the transmitted beam as the reference wave. At low voltage for masks, the reference wave is the beam transmitted through the holes. If the mask transmission function is $q(x) = 1 - p(x) = 1$ within holes and $= 0$ within opaque regions, then the wavefunction at defocus z_1 is $\psi_h(x) = q(x)*t_{z1}(x)$ for unity magnification, with $t_{z1} = exp(-i\pi x^2 / \lambda z_1)$ the Fresnel propagator. The recorded hologram intensity is[9]

$$I(x) = \psi_h(x)\psi_h^*(x) = 1 - p(x)*t_{z1}(x) - p^*(x)*t_{z1}^*(x) + |p(x)*t_{z1}(x)|^2.$$

Thereconstruction is achieved by back-progating the hologram by distance z_1. One obtains:

$$1 - p(x) - p^*(x)*t_{2z1}^*(x) + |p(x)*t_{z1}(x)|^2 *t_{z1}^*(x) = 1 - p(x) - p(x)*t_{-2z1}(x) + |p(x)*t_{z1}(x)|^2 *t_{-z1}(x).$$

The first two terms give a perfect reconstruction. The third term gives the twin image, out of focus by $2z_1$, the fourth term is a second-order term, a reconstruction of the complementery object. Reconstruction is possible if the twin image and second-order term are much smaller than p(x), which they are under the Fraunhoffer condition[10] (Fresnel number $N = \pi d^2 / \lambda z_1 \ll 1$ for a compact object of width d) when smearing by t_{z1} reduces their amplitude. The empty area (and coherence width) must also be much larger than d so that the hologram intensity is everywhere close to unity. Both conditions are satisfied if there are no real shadows in the shadow image. Figures 1a and b show a simulated shadow image and the reconstruction for two square masks. Profiles in figures c and d show sharper edges for smaller N. Figure g shows an experimental PPM image of a carbon film obtained at 90 volts on a channel plate 14 cm away from the sample. CCD camera exposure times were 0.1 seconds with about 1 nA beam current. A reconstruction with defocus z_1 of 1850 nm is shown in figure h with profiles shown in figure e and f. Figure i shows a forward-simulated hologram using mask shapes retrieved from the reconstruction in figure h, for comparison with the original image. Integration over the source includes partial coherence effects and improves agreement. The profiles in e and f show that the reconstruction is sharper along the thin fiber than the boundaries of extended opaque areas, as predicted above. Resolution in the reconstruction is governed by the electron source size (which determines the highest order Fresnel fringe). In order to avoid the restriction to small objects (N<<1), we are exploring the use of recursive calculation between two defocii, and the imposition of the a-priori constraint that q=1 or 0 in image processing. The use of lower voltage[11] (where IMF increases) may increase transmission, but reconstruction is needed to test this, since defocus produces delocalised images, and Young's fringes between pinholes may otherwise give a false impression of transmission.[9,12] Work is also in progress on an instrument designed to operate in the reflection mode (see Spence et al, these proceedings).

Proc. Microscopy and Microanalysis 1995, edited by G.W. Bailey, M.H. Ellisman, R.A. Hennigar, and N.J. Zaluzec

References

1. H. W. Fink et al. Phys. Rev. Lett. 67 (1991) 1543. Spence et al., Ultramicroscopy 52 (1993) 473.
2. J. C. H. Spence, Optik 92 (1992) 57; Spence et al J. Vac Sci. A12, (1994) 542.
3. G. M. Shedd, J. Vac. Sci. Technol. A12 (1994) 2595.
4. C. Martin et al., J. Electr. Spectr. Rel. Phenom. 42 (1987) 171.
5. H. Boersch, Phys. Zeit. 44 (1943) 202 and Z. f. Techn. Physik. 12 (1939) 346.
6. D. Gabor, Proc. Roy. Soc. A197 (1949) 454.
7. G. A. Morton and E. G. Ramberg, Phys. Rev. 56 (1939) 705.
8. J. Munch, Optik 43 (1975) 79; J. Lin and J. M. Cowley, Ultramic. 19 (1986) 31; M.Haine and T. Mulvey, J. Opt. Soc. Am. 42 (1952) 763.
9. J. Spence et al., Knoxville Workshop on Electron Holography. Ed. A. Tonomura. Elsevier. 1995.
10. B.J.Thomson et al., Applied Optics 6 (1967) 519.
11. R. Morin and A. Gargani, Phys. Rev. B48 (1993) 6643.
12. This research was supported by NSF award DMR 9116362.

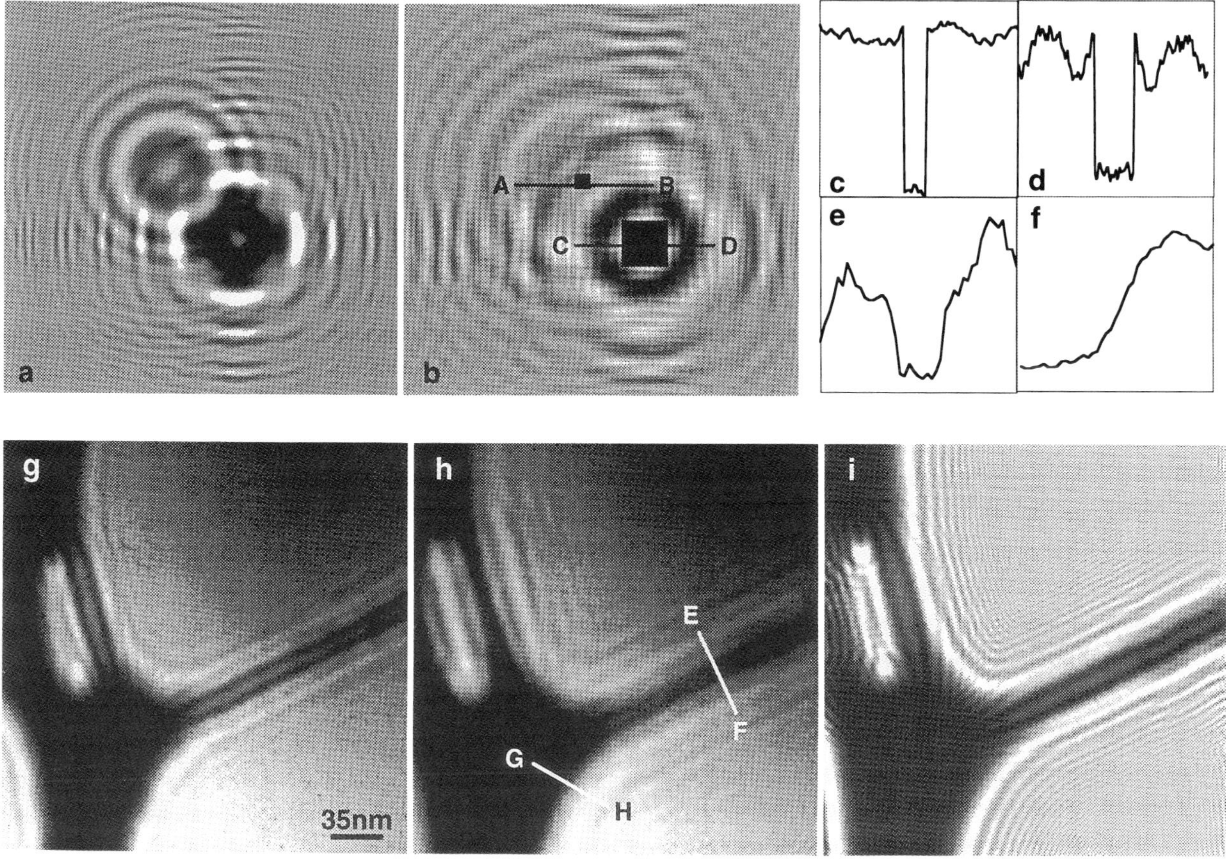

Figure 1. (a) Simulated hologram of opaque squares at 100 volts, and (b) its reconstruction (real part of wavefunction). Defocus z_1= 60nm. Square widths 1nm and 3nm (Fresnel number 0.43 and 3.86). (c) Line profiles along AB and (d) along CD in figure b. (g) Experimental hologram of a holey carbon film obtained at 90V and (h) its reconstruction (shown as amplitude) with defocus 1850 nm; line profiles along EF and GH are shown in figures e and f. (i) Simulated hologram using object shapes retrieved from the reconstruction (h), defocus 1850 nm.

LOW-ENERGY POINT-REFLECTION ELECTRON MICROSCOPY

J. C. H. Spence, X. Zhang, J. M. Zuo and U. Weierstall
Department of Physics and Astronomy, Arizona State University, Tempe, AZ 85287-1504
E. Munro and J. Rouse
Munro's Electron Beam Software Ltd., 14 Cornwall Gardens, London SW7 4AN, England

The limited penetration of the low-voltage point-projection microscope (PPM) may be avoided by using the reflection geometry to image clean surfaces in ultra-high vacuum. Figure 1 shows the geometry we are using for experimental point-reflection (PRM) imaging. A nanotip field-emitter at about 100 - 1000 volts is placed above a grounded atomically flat crystalline substrate, which acts as a mirror and anode. Since most of the potential is dropped very close to the tip, trajectories are reasonably straight if the sample is in the far-field of the tip. A resolution of 10 nm is sought initially. The specular divergent RHEED beam then defines a virtual source S' below the surface, resulting in an equivalent arrangement to PPM (or defocused CBED). Shadow images of surface asperities are then expected on the distant detector, out of focus by the tip-to-sample distance. These images can be interpreted as in-line electron holograms and so reconstructed (see X. Zhang et al, these proceedings). Optical analog experiments confirm the absence of foreshortening when the detector is parallel to the surface. We have calculated the specular ray paths using a ray-tracing program as shown in figure 2, where the tip, the grounded plate, equipotentials, emitted and specular rays are shown. The incident angles range from 90 to 35 degrees for rays striking the plate. The failure of virtual rays projected back to meet at a point S' defines the image aberrations. Our experiments are performed at a much smaller glancing angle (in order to increase intensity), resulting in less curvature of trajectories due to the tip field after reflection. The experimental full emission angle is also smaller (14 degrees)[1] producing smaller image aberrations. Since most of the potential drop occurs within a few hundred nm of the tip, the potential variation at the sample is small, and the surface orientation has little influence on the ray trajectories.

Since the contrast of surface steps of a few atoms high in REM images is known to be dominated by phase contrast[2], we have computed the PRM image of surface steps by adding a phase-grating step as a perturbation to Bloch-wave multiple-scattering RHEED computations with spherical-wave illumination. Figure 3 shows the case of a down step. Neglecting diffuse scattering at the step, reflections from the upper and lower surfaces generate two virtual sources, s' and s", with longitudinal separation $d = 2h \sin\theta_s$ and transverse separation $a = 2h \cos\theta_s$, each contributing half the wavefront at the step. Since h is only a few angstroms, the transverse separation a can be ignored for emitters of nanometer or greater dimensions. (The image resolution is fixed by this source size). Equivalently the scattering can be modeled as a spherical wave from one point source illuminating a phase grating distance z_1 away with transmission function $q(x) = e^{i\phi}$ for $x < 0$ and $q(x) = 1$ elsewhere for a step at $x = 0$ with $\phi = 2\pi d / \lambda$. Similar result can be obtained for an up step, which shows reversed phase contrast. Coherent CBED theory[3] then applies. We assume no overlap of diffracted orders (illumination angle less than Bragg angle). If the detected amplitude is $\psi_D(u)$ for the specular disk, a defocus z_1 will introduce an additional spherical wave phase factor $exp(i2\pi\lambda z_1 u^2)$. To incorporate the effect of the step which now becomes visible, we first Fourier transform this amplitude to obtain the wavefield in real space lying on the crystal surface which contributes to the central disk, then multiply by the step transmission function, and finally transform back to the detector to give the shadow image diffraction pattern. The final amplitude is

$$\Psi(u) = FT\left\{FT^{-1}\left[\psi_D(u) exp(i2\pi\lambda z_1 u^2)\right] \cdot q(x)\right\} = \left[\psi_D(u) exp(i2\pi\lambda z_1 u^2)\right] * Q(u).$$

Here the RHEED reflectivities $\psi_D(u)$, the complex reflection coefficients from the surface for different incident directions, were found using dynamical two-dimensional Bloch waves computations.[4] Figure 4 shows a simulated PRM shadow image of a step with the height of one lattice plane spacing 5.65 Å on a GaAs(100) surface with the beam near azimuth [001], using an electron energy of 500 volts and a defocus z_1 of 1000 nm. The image corresponds to a beam divergence of 7 degrees in the x and y

Proc. Microscopy and Microanalysis 1995, edited by G.W. Bailey, M.H. Ellisman, R.A. Hennigar, and N.J. Zaluzec
Copyright © 1995 MSA. Published by Jones and Begell Publishing, 79 Madison Ave., New York, NY 10016

dimension and the incident angle at the step is 9 degrees, so the phase shift $\phi=1.43$. We see that the Fresnel fringe contrast of the step (as also seen in REM) is modulated by the reflection rocking curve. The relative size of the fringes and the rocking curve features will depend on magnification z_2/z_1. Optical analog experiments for phase-grating steps show similar Fresnel fringes. Simulations have also been completed for a superconducting vortex, showing good phase contrast. The design of the instrument is given elsewhere[5] - a crucial feature is the stick-slip eucentric tip goniometer. NSF award DMR-9116362.

1. M. R. Scheinfein et al., J. Appl. Phys. 73 (1993) 2057.
2. J. M. Cowley and L.-M. Peng, Ultramicroscopy 16 (1985) 59.
3. J. C. H. Spence and J. M. Zuo, Electron Microdiffraction (Plenum Press, New York, 1992).
4. L.-M. Peng et al., Proc. R. Soc. Lond. (1995) in press.
5. J. C. H. Spence, 13th Int. Congr. Electr. Micros. Paris. (1994) 329.

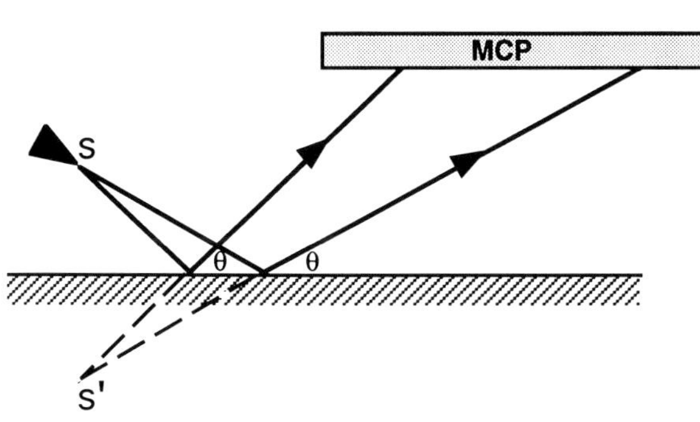

Figure 1. Principle of point-reflection electron microscope, showing shadow image of surface detail projected from virtual source S'.

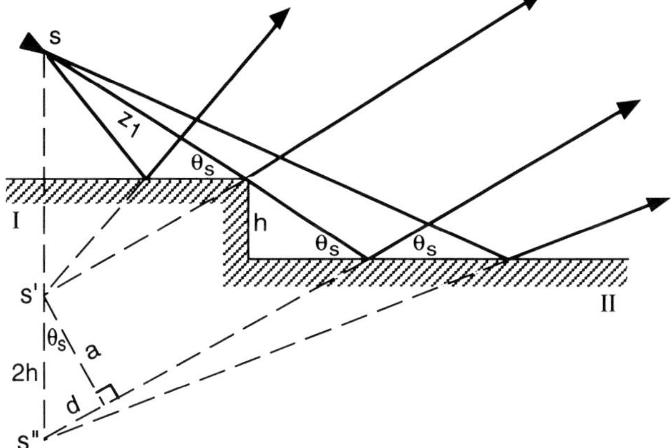

Figure 2. Path differences for phase shift at step illuminated by a spherical wave. h is the step height and θ_s is the incident angle at the step.

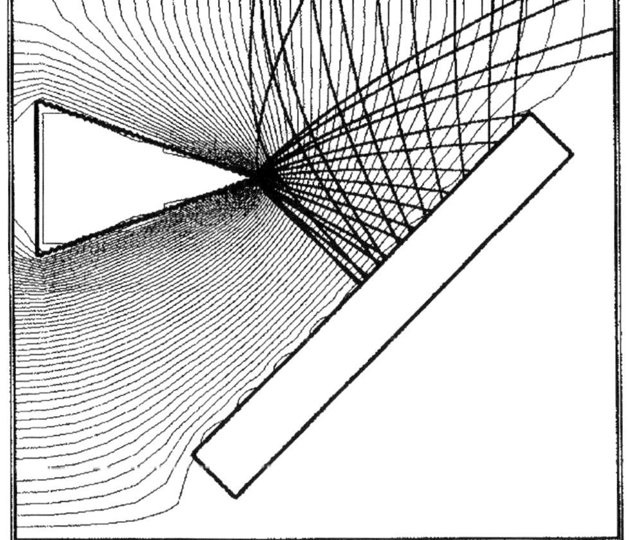

Figure 3. Equipotential lines and electron rays for a field-emission tip above a grounded mirror. Tip voltage 100, tip-to-mirror distance 1000 nm.

Figure 4. Spherical-wave RHEED computation for a surface step showing Fresnel fringes across rocking curve modulation. The step runs across the middle.

X-RAY GABOR HOLOGRAPHY

Steve Lindaas[‡], Chris Jacobsen[†], Alex Kalinovsky[†] and Malcolm Howells[‡]

† Physics Department, State Univ. of New York at Stony Brook, Stony Brook, NY 11794-3800
‡ Advanced Light Source, Lawrence Berkeley Laboratory, Berkeley, CA 94720

Soft x-ray microscopy offers an approach to transmission imaging of wet, micron-thick biological objects at a resolution superior to that of optical microscopes and with less specimen preparation/manipulation than electron microscopes[1]. Gabor holography has unique characteristics which make it particularly well suited for certain investigations: it requires no prefocussing, it is compatible with flash x-ray sources, and it is able to use the whole footprint of multimode sources. Our method serves to refine this technique in anticipation of the development of suitable flash sources (such as x-ray lasers) and to develop cryo capabilities with which to reduce specimen damage. Our primary emphasis has been on biological imaging so we use x-rays in the water window (between the Oxygen-K and Carbon-K absorption edges) with which we record holograms in vacuum or in air.

The hologram is recorded on a high resolution recording medium; our work employs the photoresist poly(methylmethacrylate) (PMMA). Following resist "development" (solvent etching), a surface relief pattern is produced which an atomic force microscope is aptly suited to image[2,3]. The work reported here involves a custom atomic force microscope with a highly linear ($\sim 0.05\%$) scanning stage we have developed for hologram readout[3]. Following readout using this atomic force microscope we reconstruct numerically[4] on a computer. By reconstructing numerically, we obtain both the magnitude *and* phase of the reconstructed image. Reconstructed holograms of test objects show sub 50 nm resolution (we believe this to be the highest resolution obtained by an x-ray holographic technique to date). Figure 1 shows a subimage from a reconstructed hologram of a critical point dried NIL cell. An image from a visible light microscope (100x N.A.= 0.9) of the same region is shown for comparison. Small subcellular features as well as a tendril are observed in the x-ray image that are not resolved in the visible light micrograph. We are continuing to refine x-ray Gabor holography for the examination of biological specimens.

1. J. Kirz, C. Jacobsen, and M. Howells. Soft x-ray microscopy. *Quarterly Reviews of Biophysics*, 0, 1995. Also available as Lawrence Berkeley Laboratory report LBL-36371.
2. C. Jacobsen, M. Howells, J. Kirz, and S. Rothman. X-ray holographic microscopy using photoresists. *Journal of the Optical Society of America*, A 7:1847–1861, 1990.
3. S. Lindaas, C. J. Jacobsen, M. R. Howells, and K. Frank. Development of a linear scanning-force microscope for x-ray Gabor hologram readout. In C. Jacobsen and J. Trebes, editors, *Soft X-ray Microscopy*, volume 1741, pages 213–222, Bellingham, Washington, 1992. Society of Photo-Optical Instrumentation Engineers (SPIE).
4. J. W. Goodman and R. W. Lawrence. Digital image formation from electronically detected holograms. *Applied Physics Letters*, 11:77–79, 1967.

5. *This research was supported by the NSF under Grant No. DIR-9006893 (S.L.,C.J.,A.K.), Presidential Faculty Fellow Award RCD 92-53618 (C.J.), DOE Alexander Hollaender distinguished postdoctoral fellowship (S.L.), and the Office of Health and Environmental Research of the DOE under Subcontract No. 431-3378A of Grant No. DE-FG02-89ER60858 (M.H.).*

Reconstructed X-ray Gabor Hologram

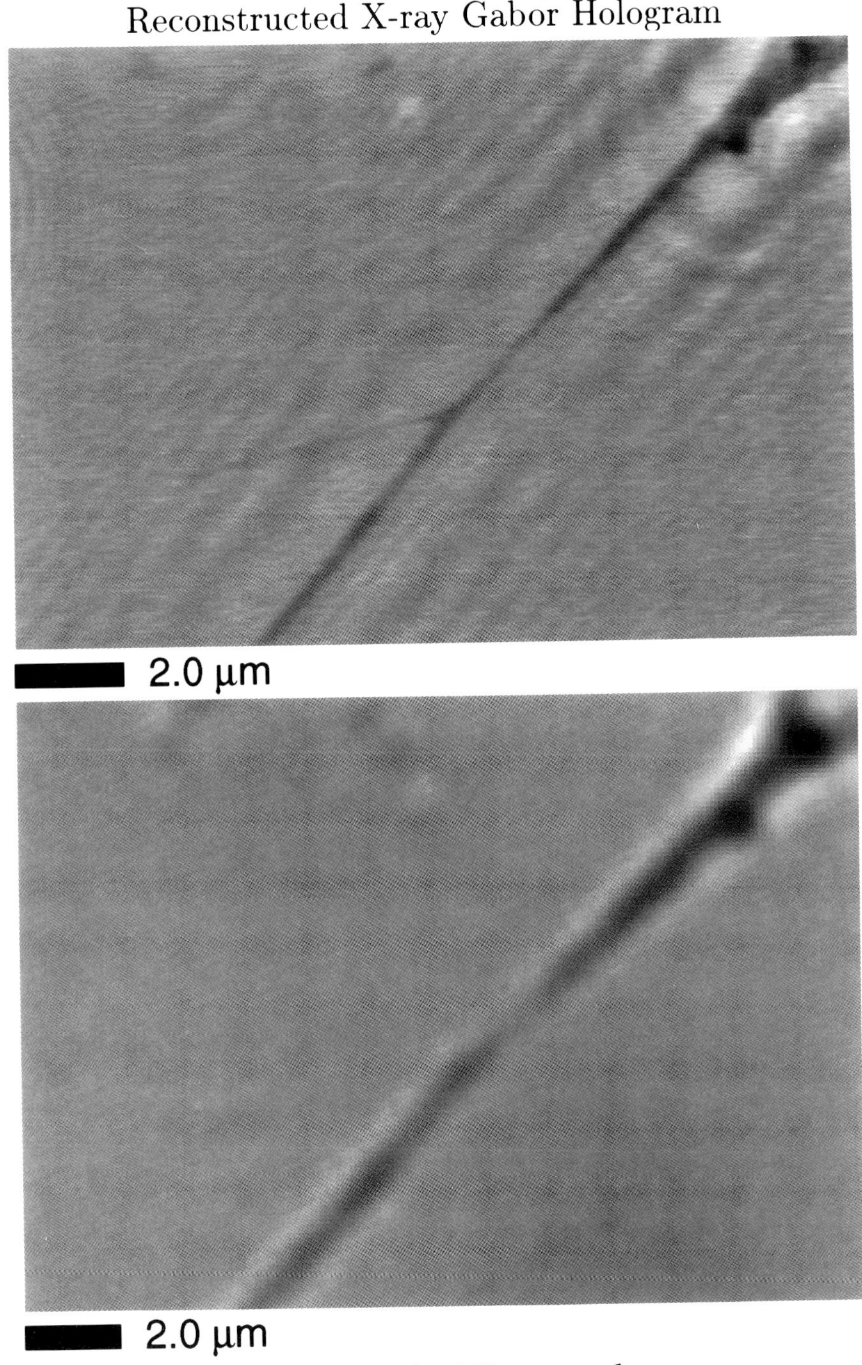

Visible Light Micrograph

Figure 1: Images of a critical point dried NIL cell. The top picture is from a reconstructed x-ray hologram recorded using $\lambda = 1.9\,\mathrm{nm}$. The bottom picture is from a visible light microscope (100x N.A.= 0.9) of the same region and is shown for comparison.

THE INFLUENCE OF THE OBJECTIVE LENS CURRENT IN LOW MAGNIFICATION ELECTRON HOLOGRAPHY

B.G.Frost,* D.C.Joy,*,** E.Völkl,** and L.F.Allard**

*EM Facility, University of Tennessee, Knoxville, TN 37996
**Oak Ridge National Laboratory, Oak Ridge, TN 37831-6064

In order to align an electron microscope for low magnification holography we usually completely switch off the objective lens and image the sample by the first intermediate lens. In addition, to achieve a highly coherent electron beam we highly excite the condensor lens resulting in a divergent illumination of the sample and the intermediate lens. Now negatively biasing the fiber of a Möllenstedt type biprism placed between the first an second intermediate lenses of our Hitachi HF-2000 field emission electron microscope creates two virtual sources below the back focal plane of the first intermediate lens. These two sources are necessary to form off-axis holograms. Slightly exciting the objective lens and still imaging the sample by the first intermediate lens results in two major changes in our holograms.

First: Due to an electron beam less divergent or even convergent illuminating the first intermediate lens when exciting the objective lens (compare Fig.1 to Fig.2) the angle β at which object wave and reference wave are superimposed decreases. Therefore the fringe spacing s of the hologram fringes connected to β by $s = \lambda/\beta$ increases. Without changing the fiber voltage U_f of the biprism the interference width Δ given by $\Delta = c \cdot U_f$ remains constant, the number of fringes decreases. Increasing the fiber voltage to obtain an identical fringe spacing as without excitation of the objective lens increases the interference width. The upper limit of this width which increases from $8\mu m$ at $0.0\,A$ objective lens current to $15\mu m$ at $0.37\,A$ is now given by the coherence of our electron source rather than by instabilities of the microscope or the resolution limit of the electron detector. A large interference width is especially important for investigating long range electric fields for the reference wave is less influenced by the field.

Second: The measured phase distribution of electric fields strongly depends on the excitation of the objective lens. Increasing the objective lens current decreases the measured phase shift. For example, an increase from $0.00\,A$ (Fig.3) to $0.25\,A$ (Fig.4) decreases the phase shift of about 50%, while the magnification of the sample increases about 3%. Even the smallest possible change in the lens current of our microscope ($10\,mA$) changes the phase shift about 5%. When taking the holograms of Fig.3 and Fig.4 no parameters were changed except thc objective lens current. As sample we used a p-n junction in silicon. The phase lines represent equipotential lines of the electric potential projected onto the image plane. The electric field lines are perpendicular to the phase lines. It is remarkable that increasing the objective lens current decreases only the phase shift caused by the electric field inside and outside the sample. The phase shift due to thickness variations inside the sample stays constant. This means, that we have to be very careful when evaluating electric field parameters using phase distributions reconstructed from holograms aquired with the objective lens excited.

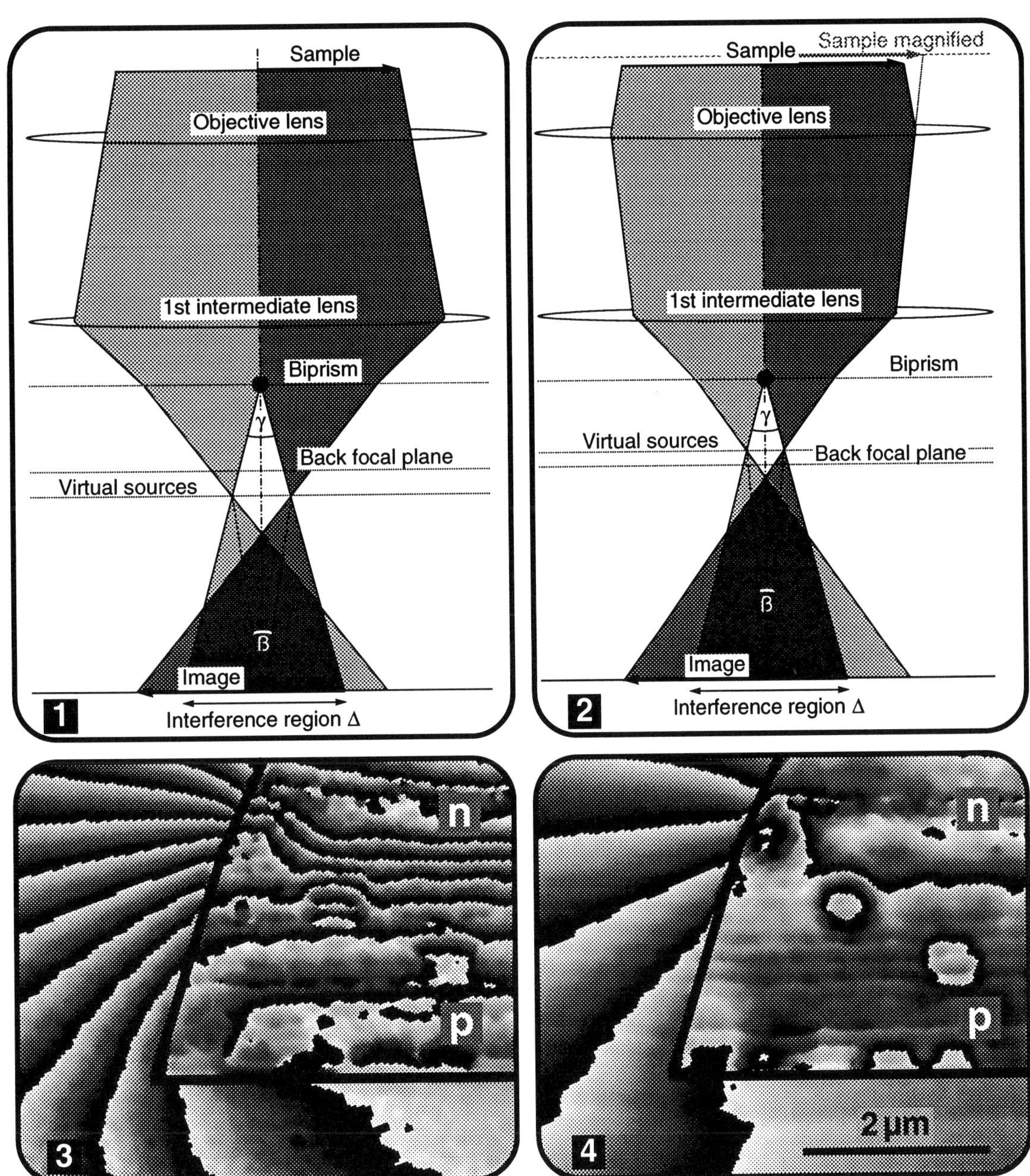

Fig.1: Ray path for taking holograms with objective lens switched off.

Fig.2: Ray path with objective lens slightly excited. Angle β decreases resulting in increasing fringe spacing. Interference width stays constant.

Fig.3: Hologram of p-n junction aquired following Fig.1.

Fig.4: Hologram of identical p-n junction and identical lens conditions as in Fig.3 but objective lens current of 0.25A. Phase shift caused by electric field decreases about 50%, phase shift due to thickness variation stays nearly constant.

This work was performed at the Oak Ridge National Laboratory, sponsored by the Directed R&D Program of Oak Ridge Natioanl Laboratory, managed for the DOE by Martin Marietta Energy Systems, Inc. under contract DE-AC05-84OR21400 and supported by an appointment to the Oak Ridge National Laboratory Postdoctoral Research Program administered by the Oak Ridge Institute for Science and Education.

QUANTITATIVE ASPECTS OF ELECTRON DIFFRACTION USING ELECTRON HOLOGRAPHY

E. Völkl, L.F. Allard, B. Frost and T.A. Nolan

High Temperature Materials Laboratory, Oak Ridge National Laboratory, Oak Ridge, TN 37831-6064

Off-axis electron holography has the well known ability to preserve the complex image wave within the final, recorded image.[1] This final image described by $I(x,y) = I(r)$ contains contributions from the image intensity of the elastically scattered electrons $I^{el}(r) = |A(r) \exp(i\Phi(r))|^2$, the contributions from the inelastically scattered electrons $I^{inel}(r)$, and the complex image wave $\Psi = A(r) \exp(i\Phi(r))$ as:

$$(1) \qquad I(r) = I^{el}(r) + I^{inel}(r) + \mu A(r) \cos(2\pi \Delta k\, r + \Phi(r))$$

where the constant μ describes the contrast of the interference fringes which are related to the spatial coherence of the electron beam, and Δk is the resulting vector of the difference of the wavefront vectors of the two overlaping beams. Using a software package like HoloWorks,[2] the complex image wave Ψ can be extracted. A consecutive absolute square of the Fourier transformed Ψ yields the power spectrum $S(u,v)$:

$$(2) \qquad S(u,v) = | FT\{ A(r) \exp(i\Phi(r)) \} |^2$$

If we ignore the chromatic envelope function as well as the spatial coherence envelope function and retain only isoplanatic aberrations described by $\chi(u,v)$, then equation (2) can be re-written as:

$$(3) \qquad S(u,v) = | FT\{ a(r) \exp(i\varphi(r)) \} \exp(i\chi(u,v)) |^2 = | FT\{ a(r) \exp(i\varphi(r)) \} |^2$$

$a(r) \exp(i\varphi(r))$ describes the complex object function leaving the object. $S(u,v)$ is therefore, cum grano salis, the power spectrum of the complex object function, i.e. the image that is found in the diffraction plane of the electron microscope and which is relevant for electron diffraction. Looking at electron diffraction from the point of view of electron holography offers several advantages: 1) Intensities of the diffraction maxima can be evaluated with an accuracy limited by the signal / noise ratio in the complex image wave Ψ. 2) The dynamic range of these intensities is not limited by the dynamics of the camera that records conventional electron diffraction patterns. 3) The holographic approach to electron diffraction offers, in addition to the determination of the position and intensity of the diffraction maxima, access to the amplitude and the phase of the diffraction maxima. 4) The holographic approach offers the ability to obtain diffraction patterns of very small object areas.[3,4]

This ability to record an electron diffraction pattern from areas of 20 nm by 20 nm down to the area of a unit cell is shown for CeO_2 in Figures 1-2. The disadvantage of the holographic approach to electron diffraction should also be mentioned: the spatial resolution is limited, as large spatial frequencies beyond the information limit are cut-off by the envelop functions of the electron microscope.[5]

References

1. G. Möllenstedt and H. Düker, Zeitschrift für Physik, 145(1956), 377-397
2. E. Voelkl and L.F. Allard, (1995) Journal of Microscopy, in press
3. E. Voelkl, PhD thesis, University of Tübingen, 1991
4. H. Lichte, E. Voelkl and K. Scheerschmidt, Ultramicroscopy 47 (1992) 231--240
5. Research sponsored in part by the Laboratory Directed R&D Program of Oak Ridge National Laboratory, managed for the DOE by Martin Marietta Energy Systems, Inc. under contract DE-AC05-84OR21400 and supported by an appointment (E.V.) to the Oak Ridge National Laboratory Postdoctoral Research Program administered by the Oak Ridge Institute for Science and Education.

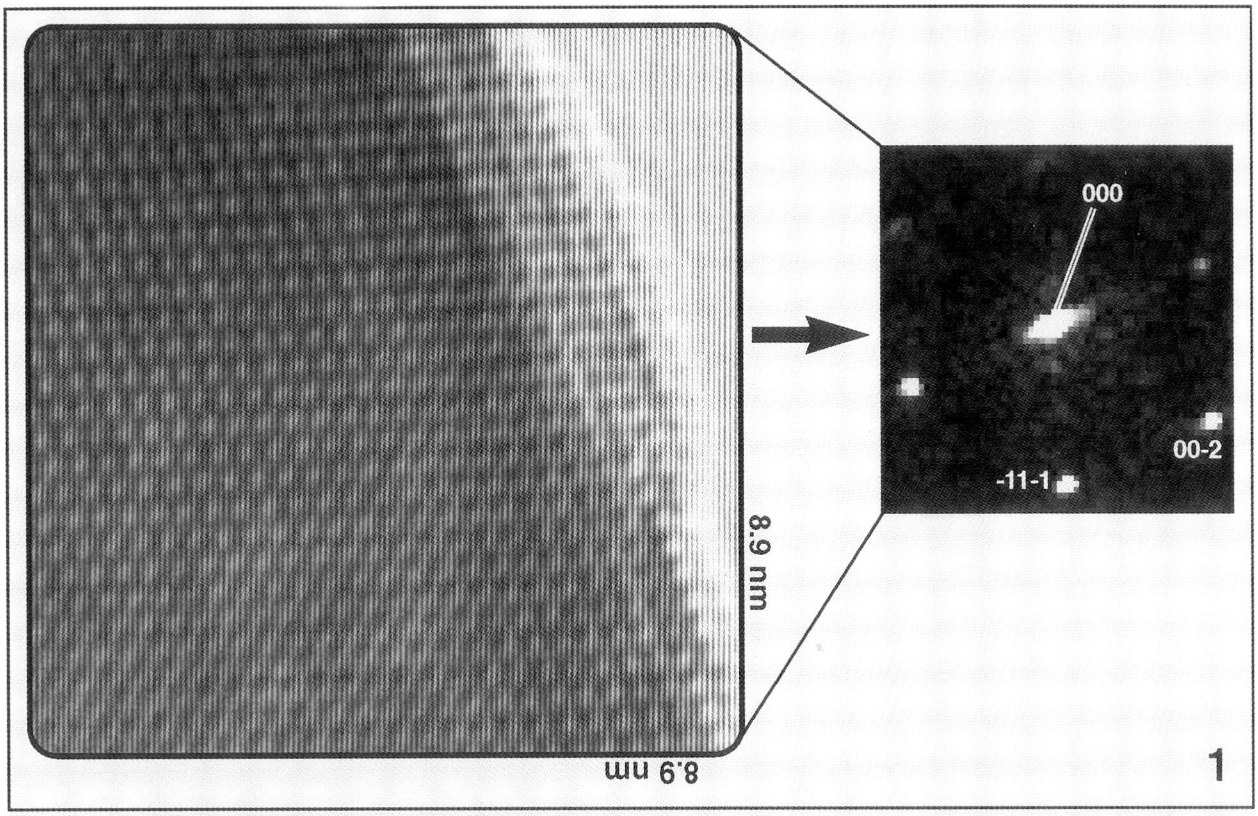

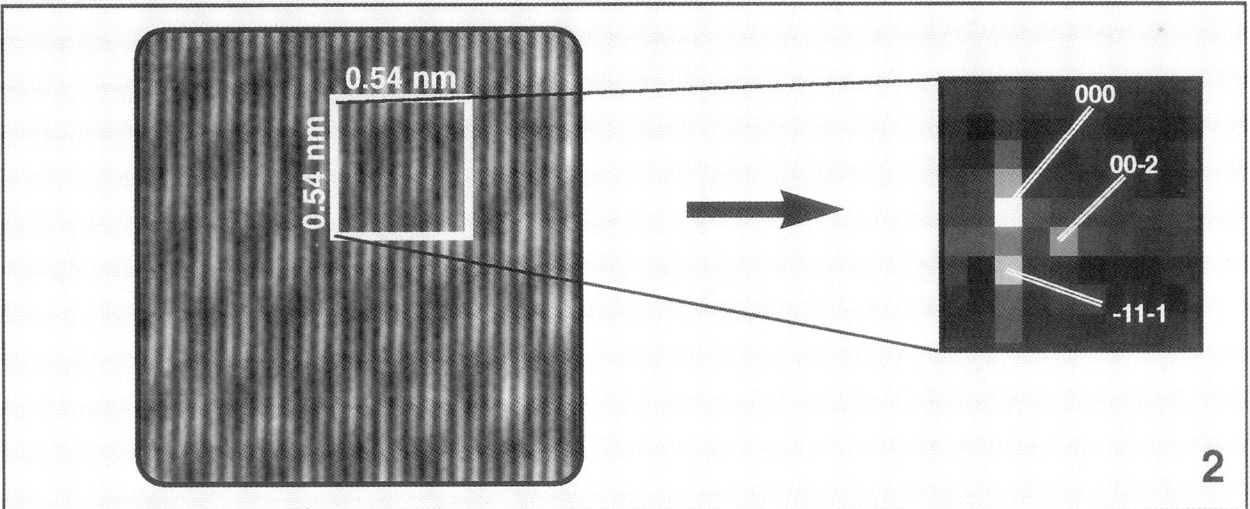

FIG. 1. – Left: 8.9 nm by 8.9 nm sub-area of electron hologram of CeO_2 in <110> direction. Right: Electron diffraction pattern as computed from subarea. All diffraction maxima can be investigated in position, intensity-, amplitude- and phase- values. Ratio of intensities of center peak over background is $\sim 10^5$. Ratio of intensities of center peak over diffraction maxima is $\sim 10^2$.

FIG. 2. – Left: 1.8 nm by 1.8 nm sub-area of same electron hologram. Right: Electron diffraction pattern computed from unit cell outlined. Evaluation of diffraction maxima requires complex Fourier processing. Hologram was recorded at 100kV, on Philips 420 TEM in Tübingen, Germany.

RECONSTRUCTION OF ELECTRON HOLOGRAMS RECORDED IN A NON-FEG, NON-BIPRISM TEM

Z.L. Wang

School of Materials Science and Engineering, Georgia Institute of Technology, Atlanta GA 30332-0245.

An experimental technique for performing electron holography using a non-FEG, non-biprism transmission electron microscope (TEM) has been introduced by Ru et al. [1]. A double stacked specimens, one being a single crystal foil and the other the specimen, are loaded in the normal specimen position in TEM (fig. 1). The single crystal, which is placed onto the specimen, is responsible to produce two beams that are equivalent to two virtual coherent sources illuminating the specimen beneath, thus, permitting electron holography of the specimen. In this paper, the imaging theory of this technique is described. Procedures are introduced for digitally reconstructing the holograms.

Theory of hologram formation If the incident wave from the electron gun onto the crystal film is expressed as a plane wave $\Psi_0 = \exp(2\pi i \mathbf{K}\cdot\mathbf{r})$, where $\mathbf{K} = 1/\lambda$ is the electron wave number and λ the electron wavelength, the wave which exits the crystal film is given by Eq. (1) under the two-beam diffracting conditions,

$$\Psi_1 = \alpha \exp (2\pi i \mathbf{K}\cdot\mathbf{r}) + \beta \exp [2\pi i(\mathbf{K} + \mathbf{g})\cdot\mathbf{r})], \tag{1}$$

where α and β are wave coefficients determined by the diffracting condition set up for the crystal film. The transmission function of the object is denoted by $p(\mathbf{b})$, with $\mathbf{b} = (x, y)$. For a phase object, the wave at the exit face of the object is $\Psi_2 = \Psi_1 p(\mathbf{b}) = \exp (2\pi i \mathbf{K}\cdot\mathbf{r}) [\alpha + \beta \exp (2\pi i \mathbf{g}\cdot\mathbf{b})] p(\mathbf{b})$. We introduce Φ, which is related to Ψ_2 by $\Psi_2 = \Phi(\mathbf{b}) \exp (2\pi i \mathbf{K}\cdot\mathbf{r})$, thus $\Phi(\mathbf{b}) = [\alpha + \beta \exp (2\pi i \mathbf{g}\cdot\mathbf{b})] p(\mathbf{b})$. From Abbe's imaging theory, the wave function at the image plane (i.e., viewing screen plane) is

$$\Phi(\mathbf{b}) = \{[\alpha + \beta \exp (2\pi i \mathbf{g}\cdot\mathbf{b})] p(\mathbf{b})\} \otimes t(\mathbf{b}), \tag{2}$$

where $\otimes$ denotes a convolution calculation, $t(\mathbf{b})$ is the inverse Fourier transform ($\mathbf{FT^{-1}}$) of the optical transfer function $T(\mathbf{u}) = \exp [\pi i \Delta f \lambda u^2 + 0.5\pi i C_s \lambda^3 u^4]$, where $\mathbf{u}$ is a vector in reciprocal space, and C_s is the spherical aberration coefficient of the objective lens. Finally, the intensity distribution in the recorded hologram is $I(\mathbf{b}) = |\Phi(\mathbf{b})|^2 = | \alpha p(\mathbf{b}) \otimes t(\mathbf{b}) + \beta [\exp (2\pi i \mathbf{g}\cdot\mathbf{b}) p(\mathbf{b})] \otimes t(\mathbf{b}) |^2$. $\tag{3}$

We now consider the following calculation with the use of the Fourier transforms $T(\mathbf{u})$ and $P(\mathbf{u})$ of functions $t(\mathbf{b})$ and $p(\mathbf{b})$, respectively.

$[\exp (2\pi i \mathbf{g}\cdot\mathbf{b}) p(\mathbf{b})] \otimes t(\mathbf{b}) = \int d\mathbf{b}' \exp [2\pi i \mathbf{g}\cdot(\mathbf{b}-\mathbf{b}')] p(\mathbf{b}-\mathbf{b}') t(\mathbf{b}')$

$\qquad = \int d\mathbf{b}' \exp [2\pi i \mathbf{g}\cdot(\mathbf{b}-\mathbf{b}')] \int d\mathbf{u} \exp [2\pi i \mathbf{u}\cdot(\mathbf{b}-\mathbf{b}')] P(\mathbf{u}) \int d\mathbf{u}' \exp [2\pi i \mathbf{u}'\cdot\mathbf{b}'] T(\mathbf{u}')$

$\qquad = \exp (2\pi i \mathbf{g}\cdot\mathbf{b}) \int d\mathbf{u} \exp (2\pi i \mathbf{u}\cdot\mathbf{b}) P(\mathbf{u}) T(\mathbf{u}+\mathbf{g}) = \exp (2\pi i \mathbf{g}\cdot\mathbf{b}) [p(\mathbf{b}) \otimes s(\mathbf{b},\mathbf{g})], \tag{4a}$

where $\qquad\qquad\qquad\qquad s(\mathbf{b},\mathbf{g}) \equiv \int d\mathbf{u} \exp (2\pi i \mathbf{u}\cdot\mathbf{b}) T(\mathbf{u}+\mathbf{g}). \tag{4b}$

Using Eqs. (4a), Eq. (3) is written as

$I(\mathbf{b}) = | \alpha p(\mathbf{b}) \otimes t(\mathbf{b}) |^2 + |\beta [p(\mathbf{b}) \otimes s(\mathbf{b},\mathbf{g})] |^2 + \alpha\beta^* [p(\mathbf{b}) \otimes t(\mathbf{b})] [p^*(\mathbf{b}) \otimes s^*(\mathbf{b},\mathbf{g})] \exp (- 2\pi i \mathbf{g}\cdot\mathbf{b})$

$\qquad + \alpha^*\beta [p^*(\mathbf{b}) \otimes t^*(\mathbf{b})] [p(\mathbf{b}) \otimes s(\mathbf{b},\mathbf{g})] \exp (2\pi i \mathbf{g}\cdot\mathbf{b}). \tag{5}$

In Eq. (5), the first and the second terms are the images of the object formed by the 0 and $\mathbf{g}$ beams, respectively. The third and fourth terms are the interference of the 0 and $\mathbf{g}$ beams, resulting in the fringes observed in the hologram (figure 2). The fringes act as the carrier fringes as in off-axis holography.

Digital reconstruction of the hologram Take a Fourier transform of Eq. (5),

$\mathbf{FT}\{I(\mathbf{b}\} = \mathbf{FT}\{ | \alpha p(\mathbf{b}) \otimes t(\mathbf{b}) |^2 + |\beta [p(\mathbf{b}) \otimes s(\mathbf{b},\mathbf{g})] |^2\} + \alpha\beta^*[P(\mathbf{u}) T(\mathbf{u})] \otimes [P^*(-\mathbf{u}) S^*(-\mathbf{u},\mathbf{g})] \otimes \delta(\mathbf{u}+\mathbf{g})$

$\qquad + \alpha^*\beta [P^*(-\mathbf{u}) T^*(-\mathbf{u})] \otimes [P(\mathbf{u}) S(\mathbf{u},\mathbf{g})] \otimes \delta(\mathbf{u}-\mathbf{g}), \tag{6}$

where $S(\mathbf{u},\mathbf{g})$ is the Fourier transform of $s(\mathbf{b},\mathbf{g})$. In Eq. (6), the first term is the result of auto-correlation that is centered at $\mathbf{u} = 0$. The second and third terms are the side bands centered at $\mathbf{u} = -\mathbf{g}$ and $\mathbf{u} = \mathbf{g}$, respectively, which carry the phase information of the object. If a small aperture, which passes only the information carried by the right-hand side band centered at $\mathbf{u} = \mathbf{g}$, is introduced, after shifting the origin of reciprocal space to $\mathbf{u} = \mathbf{g}$, the filtered hologram in reciprocal space is

$$F(\mathbf{u}) \equiv \alpha^*\beta [P^*(-\mathbf{u}) T^*(-\mathbf{u})] \otimes [P(\mathbf{u}) S(\mathbf{u},\mathbf{g})]. \tag{7}$$

If the object transmission function is in a form $p(\mathbf{b}) \equiv 1 + q(\mathbf{b})$, Eq. (7) can be rewritten as,

$$F(\mathbf{u}) = \alpha^*\beta \{P^*(-\mathbf{u}) T^*(-\mathbf{u}) T(\mathbf{g}) + Q(\mathbf{u}) S(\mathbf{u},\mathbf{g}) + [Q^*(-\mathbf{u}) T^*(-\mathbf{u})] \otimes [Q(\mathbf{u}) S(\mathbf{u},\mathbf{g})]\}. \tag{8}$$

In comparison to the first and the second terms, the third term in Eq. (8) is a second order effect for weak scattering object ($q \ll 1$), thus, it is dropped. Based on the experimental set up shown in figure 1, since Δf is

Proc. Microscopy and Microanalysis 1995, edited by G.W. Bailey, M.H. Ellisman, R.A. Hennigar, and N.J. Zaluzec
Copyright © 1995 MSA. Published by Jones and Begell Publishing, 79 Madison Ave., New York, NY 10016

the distance between the crystal film and the object, the condition of $\pi\Delta f\lambda u^2 \gg 0.5\pi C_s\lambda^3 u^4$ holds if u is small (also u $\ll$ g). Therefore, the spherical aberration effect is ignored, and $T(u+g)/[T^*(-u)T(g)] \approx \exp[2\pi i\Delta f\lambda u\cdot g]$. Equation (8) is finally approximated as

$$F(u) \approx \alpha^*\beta\, T^*(-u)\, T(g) \{P^*(-u) + Q(u) \exp[2\pi i\Delta f\lambda u\cdot g]\}. \tag{9}$$

To see the physical meaning of Eq. (9), F(u) is transformed back to real space

$$\mathbf{FT}^{-1}[F(u)] = \alpha^*\beta\, T(g) \{p^*(b) \otimes t^*(b) + [q(b) \otimes t^*(b)] \otimes \delta(b-\Delta f\lambda g)\}. \tag{10}$$

The first term is the object image centered at the origin in real space, which is produced by the central transmitted 0 beam. The second term is the object image centered at $\mathbf{R} = \Delta f\lambda g = \Delta fg/K$, which is produced by the g beam (figure 2). If an aperture is applied to select the image formed by the 0 beam only, we have

$$F_o(b) = \alpha^*\beta\, T(g)\, p^*(b) \otimes t^*(b). \tag{11}$$

The object transmission function p can be obtained from Eq. (11),

$$p(b) = [\alpha^*\beta\, T(g)]^{-1}\, F_o^*(b) \otimes t(b), \tag{12}$$

where $t(b) \approx \exp(\pi i b^2/\Delta f\lambda)/[i\Delta f\lambda]$. The constant factor $[\alpha^*\beta T(g)]^{-1}$ does not affect the spatial variation of the reconstructed phase image. Therefore, the final reconstructed object function p(b) is independent of the diffracting condition of the crystal as long as the two beam condition is met.

As a summary, the procedures for reconstructing the transmission function of the object wave are: First, take a Fourier transform of the recorded hologram; Second, apply an aperture in reciprocal space which selects only the band centered at $u = g$. Then, shift the origin of the reciprocal space to $u = g$; Third, take an inverse Fourier transform of the selected band. Then, select the image formed by the 0 beam only, and take a complex conjugate of the image; Finally, the convolution of the image with the t(b) function is the transmission function of the object. Experimental examples will be shown to illustrate this process.

Reference

1. Q. Ru, N. Osakabe, J. Endo and A. Tonomura, Ultramicroscopy **53** (1994) 1.

FIG. 1 -- Schematic ray-diagram showing formation of two-beam hologram in non-FEG, non-biprism TEM. Crystal film, located in focal plane, is responsible for producing two beams (0, g), which are virtual sources illuminating specimen at E-F. Interference fringes between 0 and g beams are effectively "carry-fringes" of hologram. In image recording plane, two images of specimen are formed, one from 0 beam "source" and other from g beam. Distance between two images is determined by Δf, distance between crystal film and specimen. Specimen (or object) is out-of focus for Δf in recorded image plane.

FIG. 2 -- Low-magnification two-beam hologram of MgO cube. Inset is two-beam diffracting condition of Si crystal film. Lattice fringes of Si are visible at high magnification. Electron energy 300 keV. From R, Δf is calculated as 29 µm.

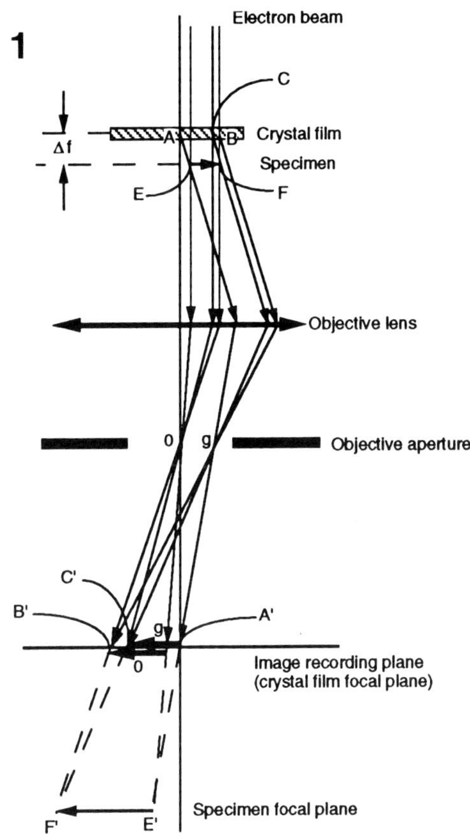

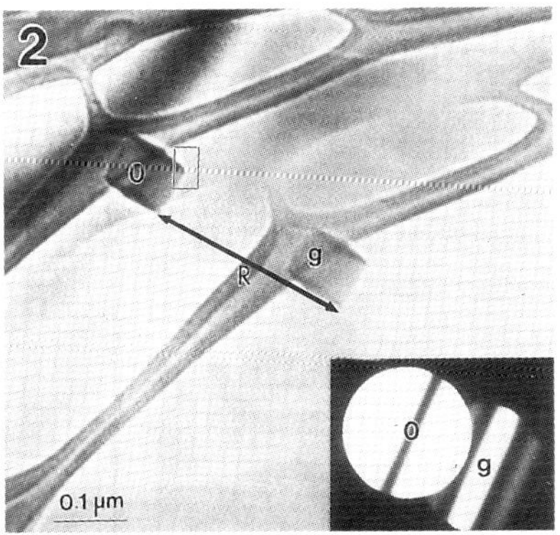

DIFFRACTED BEAM INTERFEROMETRY FOR ELECTRON MICROSCOPY AND MATERIALS RESEARCH

Rodney A. Herring

604 Aldershot Rd., Baltimore MD 21229 USA

Diffracted beam interferometry, DBI, is a practical method of interferometry which removes the strict requirement for a bright, coherent source and extends the availability of holography to a wider community of electron microscopists [1,2,3]. The only alteration to the microscope is the installation of an electron biprism into a post-objective lens position. DBI is very flexible as the interferogram can be produced from the conditions of the infocus diffraction (Fraunhofer plane), through the Fresnel (de)focus region to the infocus specimen plane by varying the strength of the post-objective lenses, Fig. 1. As well the convergence of the interfering beams can vary from highly convergent (Fig. 1) to pseudo-parallel (Fig. 2). Although the phase in the interferogram due to beam tilt and the contrast transfer function of the prefield objective lens cancel out when the interfering beams are exactly overlaid [2], the phase due to the spatial frequency of the biprism, K_B, the contrast transfer function, $u_{L,R}$, of the postfield objective lens for each beam, L, R, and the inherent material phase, $F_{L,R}$, remain. The intensity of the beam in the interferogram is then given by

$$I = A_L^2 + A_R^2 + 2|A_L||A_R|\cos\begin{bmatrix} 4\pi K_B r_B \\ -u_L(\mathbf{k} + \mathbf{g}_L) + u_R(\mathbf{k} - \mathbf{g}_R) \\ +(F_L + F_R) \end{bmatrix} \tag{1}$$

where the beams amplitudes are given by $A_{L,R}$, r_B is the distance from the optic axis to the biprism, $\mathbf{g}_{L,R}$ are the diffraction vectors and $\mathbf{k}$ is the reciprocal lattice vector(s) of the beams. The contrast transfer functions of the postfield objective lens for the two beams interfering in equ. 1 are given by

$$\mp u_{L,R}(\mathbf{k} \pm \mathbf{g}_{L,R}) = \mp\frac{\pi}{2}C_S\lambda^3(\mathbf{k} \pm \mathbf{g}_{L,R})^4 \mp \pi\lambda Z(\mathbf{k} \pm \mathbf{g}_{L,R})^2 \tag{2}$$

where Z is the (de)focus distance from the specimen to the biprism (Z$_{OB}$-Z$_O$ in Fig.2) and C_S is the spherical aberration of the post-objective lens system. From equ. 1) and equ. 2) it can be seen that for the special imaging condition of $\mathbf{g}_L = \mathbf{g}_R$, where the beams are either equal and opposite in sign or the main beam is tilted to -1/2 the diffracted beam's vector, $-u_L + u_R$ reduces to zero. At this condition for convergent beams, where $\mathbf{k}$ = a range of spatial frequencies, $-u_L + u_R$ reduces to zero at a point located on the optic axis, whereas for pseudo-parallel beams, where $\mathbf{k}$ = one spatial frequency, $-u_L + u_R$ reduces to zero over the entire interferogram. It has long been a goal of microscopists to correct for the microscope's spherical aberration, C_S. The DBI method does not correct C_S but can compensate for it at the biprism's spatial frequency. In essence, the biprism is acting as a positive lens in the negative-lens system of the microscope. It has also been a goal of material scientists to accurately measure the inherent phase of a crystal. For the special case where $-u_L + u_R$ reduces to zero over the entire interferogram and where the biprism is centered on the optic axis, i.e., $K_B r_B = 0$, which are both easily met experimentally and independent of Z, the phase of the interferogram is only due to the inherent material phase, $F_{L,R}$. Thus a material's most basic structure, i.e., its unit cell, can be accurately reconstructed from a series of interferograms which are produced by the main beam, having

Proc. Microscopy and Microanalysis 1995, edited by G.W. Bailey, M.H. Ellisman, R.A. Hennigar, and N.J. Zaluzec
Copyright © 1995 MSA. Published by Jones and Begell Publishing, 79 Madison Ave., New York, NY 10016

F_L or $F_R = 0$, interfering sequentially with a number of low-index diffracted beams.

The support given by the Tonomura Electron Wavefront Project, ERATO, JRDC, its members, and Professor Giulio Pozzi are gratefully acknowledged.

1. R.A. Herring, G. Pozzi, T. Tanji, and A. Tonomura, Ultramicroscopy 50(1993)94.
2. R.A. Herring, G. Pozzi, T. Tanji, and A. Tonomura, Ultramicroscopy, in press.
3. R.A. Herring, G. Pozzi, T. Tanji, and A. Tonomura, ICEM 13-Paris (1994) 321.

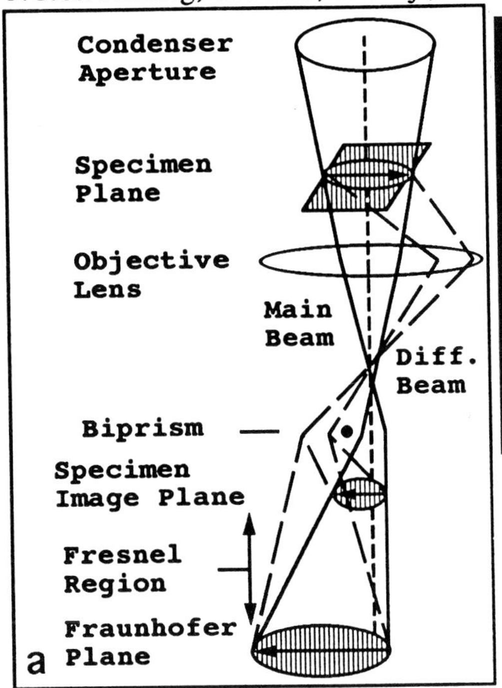

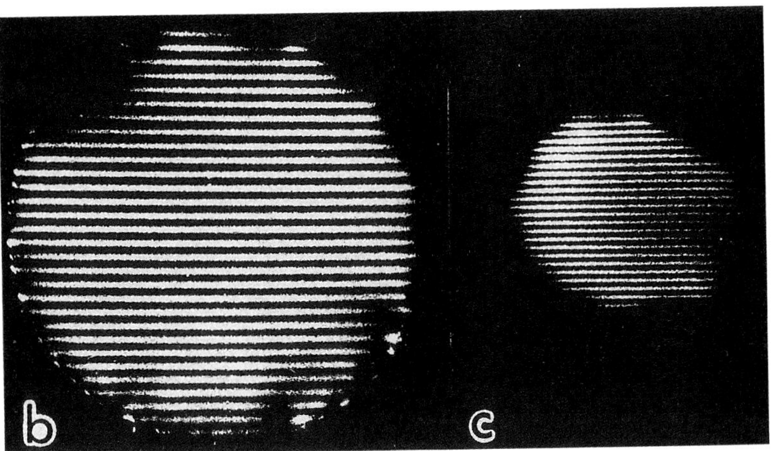

FIG. 1. The DBI method using convergent beams and showing in a) an electron ray diagram, and in b) and c) interferograms taken at the Fraunhofer plane and the specimen infocus image plane, respectively.

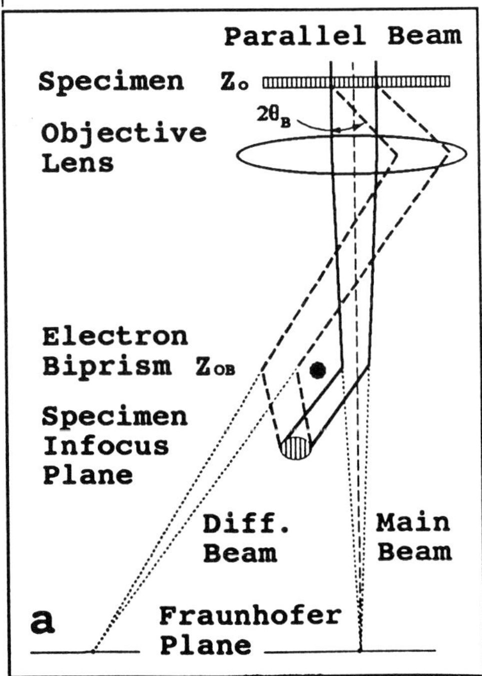

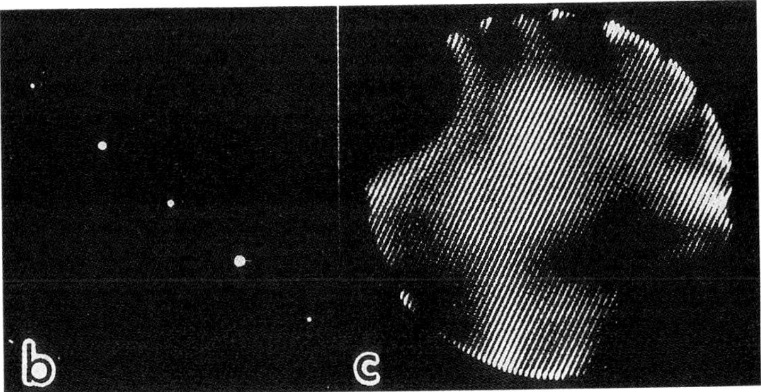

FIG. 2. The DBI method using parallel beams at the specimen, showing in a) an electron ray diagram, and in b) and c) interferograms taken at the Fraunhofer plane and a Fresnel plane, respectively.

AN EXPERIMENTAL MEASUREMENT ON THE MECHANISM OF CHARGE BALANCE IN THE ENVIRONMENTAL SCANNING ELECTRON MICROSCOPE

D. W. Phifer, Jr.* and D. C. Joy**

*Technical Division, Martin Marietta Energy Systems, Inc., Oak Ridge K-25 Site, Oak Ridge, TN 37831-7272
**EM Facility University of Tennessee, Knoxville, TN 37996-0810 and Oak Ridge National Laboratory, Oak Ridge, TN 37831-6064

Increasingly, electron microscopes are being used to look at non-conducting samples. These samples are prone to buildup of surface charge that must be drained away to prevent distortion of the image (due to the large potential buildup) and damage to the sample surface (due to dielectric breakdown). In conventional high vacuum scanning electron microscopes, the microscopist control beam energy to produce an acceptable yield of electrons while minimizing sample charging on non-conducting samples. This charge balance is achieved by reducing incident beam voltage to produce detectable interactions without excessive charge buildup on the sample surface. Electron yields at the low voltages required are sufficient for imaging when using a field emission gun, but it is not, in general, possible to perform microanalysis because the x ray lines of interest cannot be excited. The environmental scanning electron microscope (ESEM) has demonstrated the ability to overcome this problem by placing the sample in a low vacuum environment.[1] Non-conducting samples can readily be examined at high beam energies in a weak vacuum (a few Torr of gas) where the gas molecules present can influence the charge buildup.[2]

The purpose of this paper is to examine the mechanism which allows charge balance (charge neutralization) to take place in the ESEM and to determine the optimum operating conditions; that is to determine the best compromise between gas pressure and charge balance in order to obtain optimum spatial resolution (i.e. minimize beam broadening due to gas interaction).

The incident beam current of the ElectroScan ESEM 2020 was measured utilizing a Keithley pico-ammeter connected to a Faraday cup. Variables included beam energy (10-30 KeV) and pressures ("Hi-Vacuum" setting to 10 Torr). The variation in the sample current as a function of the beam energy and the pressure was monitored to determine optimum microanalysis conditions.

Figure 1 illustrates a typical experimental result. As gas pressure increases the negative specimen current rises, passing through the charge balance point to become positively charged, and essentially levels off. At low N_2 pressure (<0.7 Torr) there is a constant negative specimen current. As the gas pressure is increased (>0.7 to ~3 Torr) there is a rapid rise in specimen current which dissipates some at higher pressure (>3 Torr). To eliminate charging in the low pressure region (<0.7 Torr) would require lowering the incident beam energy similar to conventional high vacuum microscopes. As the chamber pressure is increased, the specimen current increased by a factor proportional to the pressure to about the 3.5 power and changed from negative to positive. This is due to low energy ions from the gas striking the specimen and producing additional secondary electron emissions. Increasing the pressure further causes specimen current to saturate. At very high pressures the current starts to fall slowly while maintaining a relatively high positive charge on the sample.

It can be seen that optimal conditions for observing non-conducting specimens are the lowest pressure at which charge balance can be achieved. This pressure is also optimum for microanalysis because

Proc. Microscopy and Microanalysis 1995, edited by G.W. Bailey, M.H. Ellisman, R.A. Hennigar, and N.J. Zaluzec
Copyright © 1995 MSA. Published by Jones and Begell Publishing, 79 Madison Ave., New York, NY 10016

beam broadening rises as a function of gas pressure. Therefore achieving charge balance at the lowest pressure will produce the highest spatial resolution and optimize signal information. Going to higher pressures poses potential problems for examining samples due to the introduction of a positive charge on the sample. The positively charged sample will absorb more secondary electrons, greatly reducing secondary electron yield even at high beam energies. This positive charge can also introduce sample damage if reactive gasses are used for pressure equilibration. For example if water vapor is used, the water molecule can dissociate forming negatively charged hydroxyl radicals which will be drawn to the positively charged sample surface causing significant surface etching damage.

By knowing the variation of the charge balance point for various materials, optimum conditions can be established for various beam energies and used to extrapolate operating conditions for optimum spatial resolution with non-conductive materials. This technique should also help prevent unnecessary surface damage due to reactive molecule attraction to excessive positive charge buildup on the sample.

References

1. G. D. Danilatos, Foundations of Environmental Scanning Electron Microscopy, in:"Advances in Electronics and Electron Physics", Vol. 71, p. 109, Academic Press, Boston.
2. R. B. Bolon et. al. The Environmental SEM: A New Way to Look at Insulators, in: Microbeam Analysis, San Francisco: San Francisco Press, Inc. (1989),449.
3. D.C. Joy wishes to express his gratitude to D.P. Hoffmann for the use of laboratory facilities.
4. This submitted manuscript is an account of work by Martin Marietta Energy Systems, Inc. for the U. S. Department of Energy, under Contract DE-AC05-84OR21400. Accordingly, the U.S. Government retains a nonexclusive, royalty-free license to publish or reproduce the published form of this contribution, or allow others to do so, for U. S. Government purposes.

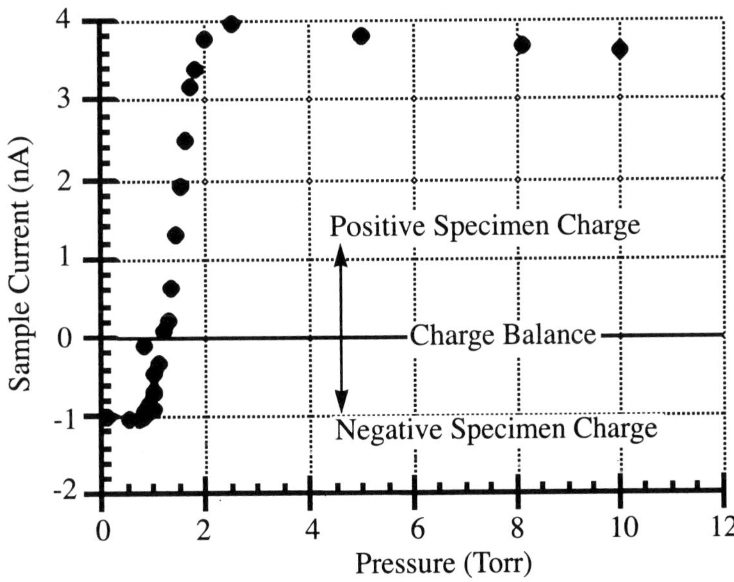

FIG. 1. Representative plot of specimen current vs. N_2 gas pressure at 20KeV in the Environmental Scanning Electron Microscope.

SCANNING ELECTRON MICROSCOPY IN VERY HIGH ELECTRIC FIELDS NEAR FIELD ION/EMISSION SPECIMENS

David J. Larson[*,**], Patrick P. Camus[**,***] and Thomas F. Kelly[*,**,***]

[*] Materials Science Program, [**] Applied Superconductivity Center and [***] Department of Materials Science and Engineering, University of Wisconsin, Madison WI 53706

An atom probe field ion microscope (APFIM) has been constructed inside a NORAN Instruments Automated Digital Electron Microscope (ADEM).[1] The ADEM is a scanning electron microscope (SEM) with a field emission source and a very large vacuum chamber. The APFIM has positive and negative high voltage capability and uses a microchannel-plate/phosphor screen assembly as an imaging and single-ion detector. The APFIM specimen can be cooled by a cryogenic refrigerator. The motivation for this study was the need to deliver an electron beam to the apex of an APFIM specimen while a high field is applied. The beam will be used to thermally pulse the field evaporation rate.[2] The expected field-induced image shift and distortion has been studied previously in a transmission EM with a liquid metal field emission source as a specimen.[3]

Fig. 1 shows the interior of the instrument. Computer simulations were done for electron trajectories with negative and positive voltages applied to the emitter based on a simple paraboloidal electric field model described previously.[4] Fig. 2 is a schematic of the geometry used for the trajectory calculations which is modeled after the layout of Fig. 1. The electron beam originates approximately 10 cm above the specimen apex. The electron beam source and the secondary electron detector shown in Figs. 1 and 2 are normally at and very near ground potential, respectively. These potentials are not included in the paraboloidal model.

A spatial shift in the image of the emitter is expected with a change in emitter potential[3,4]; the image moves from its base toward its apex for a positive voltage and in the opposite direction for a negative voltage. Fig. 3 shows the predicted trajectories of electrons in the y-z plane for positive and negative specimen voltages. The predicted displacements in the z direction are compared with experimentally determined values for a 12 nm radius specimen in Fig. 4 using a 10 keV beam energy and a positive emitter voltage. The shift in slope between the theory and the 12 nm radius tip is likely due to the simple model neglecting the effects of the chamber and other nearby ground planes on the electric field distribution. The effect of the electric field on deflection decreases significantly with distance from the emitter apex. In the x-y plane, the field produced by the emitter acts as a lens creating focusing effects on the features below. This produces a highly astigmatic, position-dependent electrostatic lens element. Fig. 5 shows the FIM image (7.2 kV and 7×10^{-5} torr He), from which the radius of curvature was determined by the ring counting technique[5], before the data from Fig. 4 were obtained.

Fig 6. shows the rather bizarre effects of the electric field on imaging the emitter (shown by arrow) and a 5 mm square aluminum mesh positioned in the chamber bottom above the vacuum pumping apparatus. The emitter voltage is approximately 1 kV. The image of the mesh is drastically altered as the electron trajectories pass the high field region near the emitter, travel toward the mesh and are backscattered back to the electron detector. However, the image of the emitter itself is not distorted a large amount.

The conclusions of this work are as follows. The high electric fields (1-10 V/nm) near field ion/emission specimens have very large effects on a SEM image. Electrons are deflected astigmatically by the fields in both the pre- and post- emitter areas and to greater degree in close proximity to the apex. Electrons can be made to impact the emitter at a desired position by translating the specimen in the direction of its long axis.[6]

1. T. F. Kelly et al., Proc. 50th Ann. EMSA Mtg., (San Francisco Press, Inc., 1992) p. 1616.
2. P. P. Camus et al., Appl. Surf. Sci., **67** (1993) 467 .
3. K. Hata et al., Appl. Surf. Sci., **76/77** (1994) 36 .
4. D. J. Larson et al., Appl. Surf. Sci., **67** (1993) 473 .
5. E. W. Müller and T. T. Tsong, *Field Ion Microscopy Principles and Applications*, (Elsevier, Amsterdam, 1969) p. 187.
6. This work is sponsored by the National Science Foundation under grant #DMR-8911332.

Proc. Microscopy and Microanalysis 1995, edited by G.W. Bailey, M.H. Ellisman, R.A. Hennigar, and N.J. Zaluzec
Copyright © 1995 MSA. Published by Jones and Begell Publishing, 79 Madison Ave., New York, NY 10016

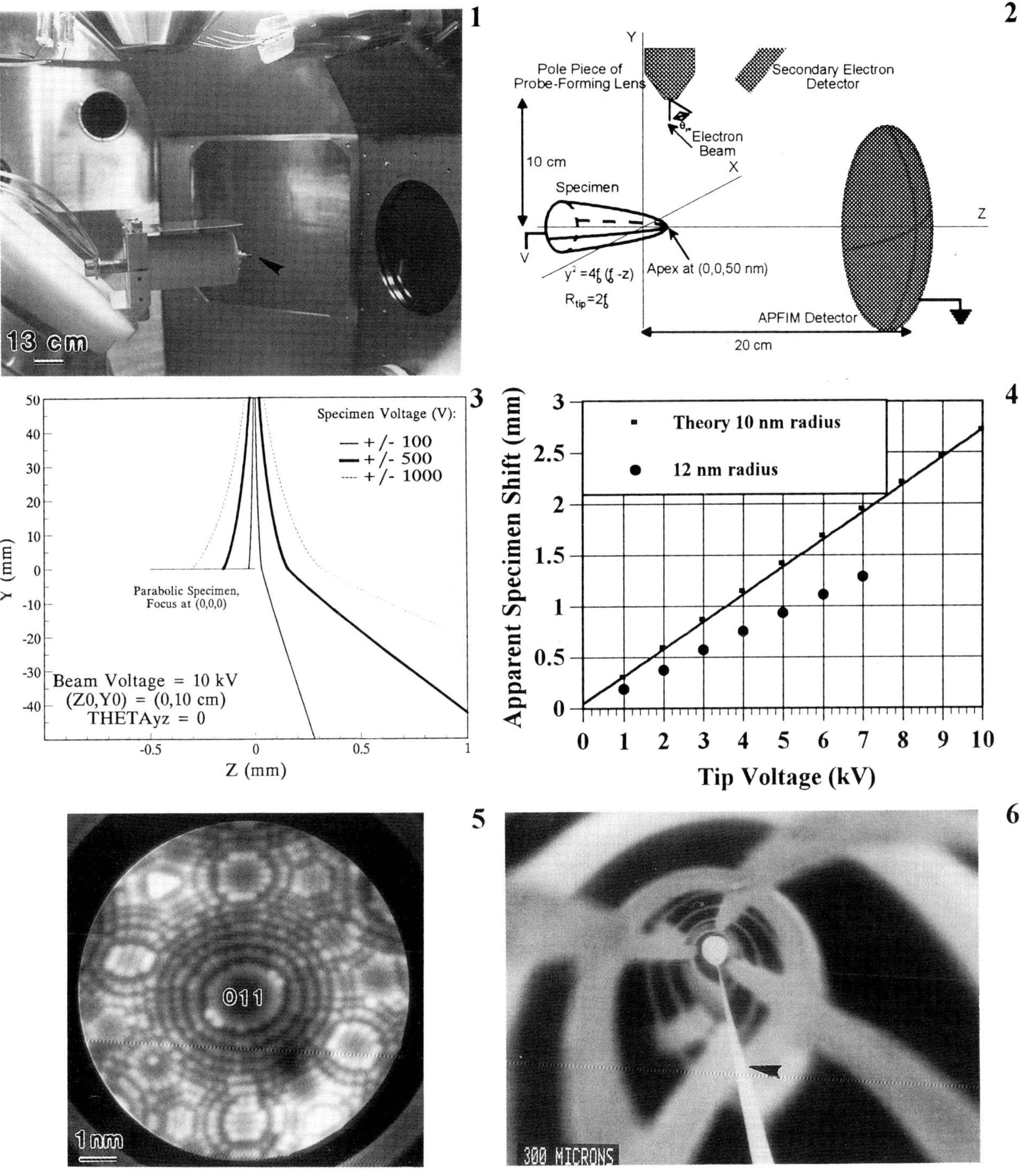

FIG 1   The interior of the ADEM as modified for use as an APFIM (arrow shows specimen position).

FIG 2   Paraboloidal specimen and detector geometry used for electron trajectory calculations.

FIG 3   Computer simulation of electron trajectories near the emitter.

FIG 4   Spatial shift of the SEM image in the emitter apex direction for various positive emitter voltages.

FIG 5   FIM image of (011) oriented tungsten emitter from which radius of curvature was determined.

FIG 6   Electric-field-altered SEM image of an emitter (arrow) and a 5 mm square aluminum mesh approximately 100 mm below .

625

STUDIES OF DENTAL SAMPLES BY THE LOW-LOSS ELECTRON METHOD IN THE FIELD EMISSION SCANNING ELECTRON MICROSCOPE (SEM).

Oliver C. Wells**, Stephen A. Rishton* and Tim Bromage.***

*IBM Research Division, Yorktown Heights, NY 10598.
**Research Staff Member Emeritus.
***Hunter College, City University of New York.

A low-loss electron (LLE) detector[1-3] has been installed with a redesigned specimen stage in a Hitachi S800 field-emission SEM (Fig. 1). This has been evaluated by comparing secondary electron (SE) and LLE images of abrasive striations known as "microwear" on an uncoated human incisor mounted at 45° to the beam (Fig. 2).

Microwear features typically appear as long linear scratches or discrete pits in the range 0.2 μm to 2.0 μm in width. These reflect both the specific contacts that one tooth makes against another when chewing (jaw movements and tooth wearing patterns) and the food eaten (seed, plant fiber, fruit, etc.). The density of scratches per unit area has been widely used as a measure of the dietary proclivities of an organism. Thus grazers have been distinguished from browsers on the basis of the greater density of scratches caused by opaline phytoliths in grasses that are harder than tooth enamel.

Fig. 2 shows comparison SE and LLE images of a human incisor surface (buccal) showing abraded Tomes Process Pits (remnant pits left from ameloblasts once occupying them), and possibly toothbrushing stria (very fine stria) and gross stria resulting from mechanical damage. It is clear to us that the conventional SE image as shown in Fig. 2(a) underrepresents the number and character of microwear features. As compared with this, the LLE image in Fig. 2(b) provides better topographic contrast and increases our confidence to a very high level, that those microwear features we see are a representation of their true density on the tooth surface. A similar conclusion with reference to these two imaging methods was reached when examining uncoated botanical samples also.[2]

In the LLE images, small scratches cannot be seen if they are aligned towards the detector. It may then be necessary to rotate the sample in the plane of the surface to be sure of seeing scratches in all possible directions. The beam energy $E_0 = 2.0$ keV was adjusted to minimise charge effects in the SE image but even so, some charge effects are still present. The LLE image is not affected by this degree of charging and clearly shows topographic features that cannot be seen in the SE image.

The specimen stage shown in Fig. 1 is mounted on an adaptor plate (H in Fig. 1(b)) that can be changed to fit other microscopes (such as the Hitachi S570, for example). A low-profile x-y stage is located at the lowest level on the replacement specimen stage. The LLE detector occupies the space above this compact mechanism. The specimen can be rotated about a vertical axis to face either towards the LLE or towards the existing SE detector. The existing air lock mechanism can still be used (with an adaptor) to change specimens without breaking the vacuum. The LLE detector used in this work is soon to be replaced by a high voltage unit to extend this study to the imaging of defects in single crystals.[4]

Proc. Microscopy and Microanalysis 1995, edited by G.W. Bailey, M.H. Ellisman, R.A. Hennigar, and N.J. Zaluzec
Copyright © 1995 MSA. Published by Jones and Begell Publishing, 79 Madison Ave., New York, NY 10016

1. O.C. Wells, *Appl. Phys. Lett.* 19(1971)232.
2. O.C. Wells and P.C. Cheng, *Proc. Ann. MSA Meeting* 50(1992)1608.
3. O.C. Wells and S.A. Rishton, *Proc. Ann. MSA Meeting* 52(1994)1022.
4. P. Morin et al, *J. Microsc. Spectrosc. Electron.* 6(1981)257.

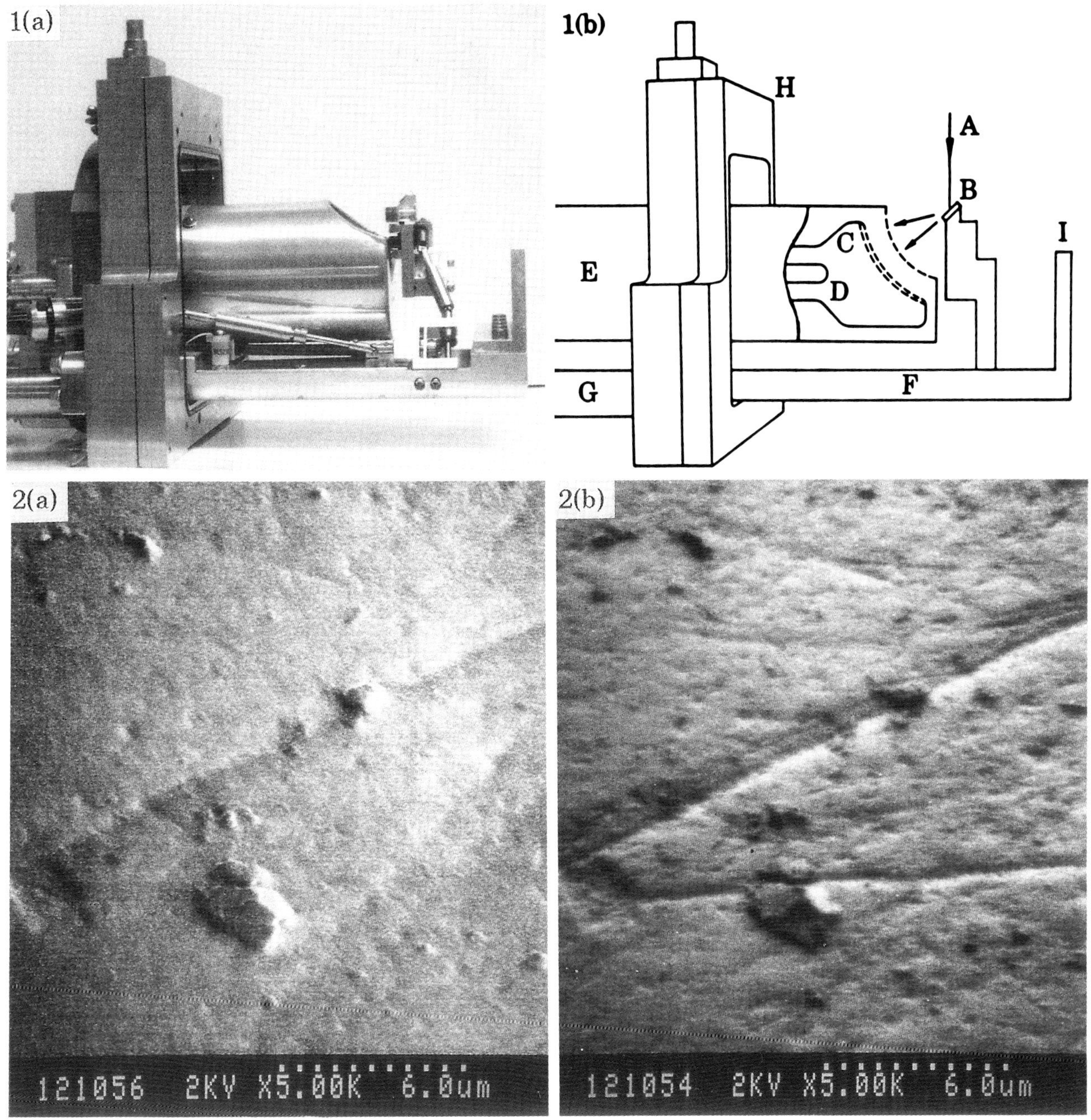

FIG. 1. (a) Specimen stage with LLE detector. (b) A=electron beam. B=specimen. C=filter grids. D=scintillator. E=photomultiplier assembly. F=low profile x-y stage. G=micrometer(s). H=adaptor plate. I=plate for anti-vibration clamp.

FIG. 2. Uncoated human incisor examined in a Hitachi S800 FE SEM. E_0=2.0 keV. Tilt angle = 45°. (a) SE image. (b) LLE image with 300 eV energy window.

THE NCEM PUBLIC DOMAIN SOFTWARE LIBRARY OF EXTENSIONS TO DIGITAL MICROGRAPH

Roar Kilaas & Sidnei Paciornik*

National Center for Electron Microscopy, Lawrence Berkeley Laboratory, Berkeley CA 94720, USA
*on leave from Departamento de Ciência dos Materiais e Metalurgia, Pontifícia Universidade Católica do Rio de Janeiro, P.O. Box 38008, Rio de Janeiro, R.J. 22452 Brazil

The National Center for Electron Microscopy (NCEM) develops software for use in image processing and image interpretation as they are needed in ongoing research at the center. Some routines are esoteric and of little use to others, while some are more general and could be considered to be of interest to a large number of scientists. This represents the first announcement of the NCEM's dedication to making software developed at the center available to the general microscopy community.

An unfortunate by-product of software development is the often necessary connection of the software to a hardware platform and/or another software platform. Most of the image processing carried out at the NCEM is done under Digital Micrograph[1] running on the Macintosh, either through the built in functionality of the program or through extensions that are written at the center. Some of the functionality that we added were routines that were available under the SEMPER[2] image processing software which was retired not because of limitations of the program, but because of the death of the hardware on which it was running. The routines that are described below are mostly written in "C" and run as custom functions under DM. Some are partially implemented as scripts.

The first four routines below are often used together. Lattice Fit requires the presence of a list of peaks, Lattice Annotation requires the lattice obtained through Lattice Fit, and Displacement List requires the presence of a fitted lattice and a set of peaks.

<u>Peak Finding:</u> This is a general routine for finding extrema (maximum or minimum) in an image[3]. The normal controls for setting threshold values are all included as are options for setting minimum distance between peaks and a radius for refinement of peak positions through a center of intensity calculation. The routine first finds peaks and plateaus in the image and proceeds to reduce plateaus to single peaks before refining the peak list. Options include saving the peak list as a standard text file, a SEMPER peak list file or a MacTempas[4] structure file (containing atom positions corresponding to peak positions).

<u>Lattice Fit:</u>   Starting with a list of peaks as obtained above, this routine takes as input initial estimates for the lattice parameters and an estimate for an origin of the lattice, and proceeds to fit a perfect lattice that minimizes the summed square difference between peak positions and lattice positions. The routine can be run iteratively in order to further refine the lattice.

<u>Lattice Annotation:</u> This is a small utility routine that annotates the image with a grid of lines corresponding to the perfect lattice found above.

<u>Displacement List:</u>   Using the data for the perfect lattice and peak-positions, this routine will create a file containing perfect lattice positions, actual peak positions, horizontal, vertical and actual displacements and the numbers n and m corresponding to which lattice point the peaks correspond to.

Proc. Microscopy and Microanalysis 1995, edited by G.W. Bailey, M.H. Ellisman, R.A. Hennigar, and N.J. Zaluzec
Copyright © 1995 MSA. Published by Jones and Begell Publishing, 79 Madison Ave., New York, NY 10016

This file is suitable for input to a spreadsheet or plotting program which then can sort or manipulate the data and plot displacements in various ways.

<u>Template Matching:</u> This routine takes as input two images or an image with a selection. The smaller image or the selection will be used as a motif (template) and the routine produces a new image containing the normalized cross-correlation coefficients between the template and the underlying image. An examples of the use of this routine is the search for repeating structural units along an interface[5].

<u>Unit Averaging:</u>   This script uses the location of the templates identified by Template Matching and located by Peak Finding to obtain an average motif with improved signal-to-noise ratio.

<u>Radial Average:</u>   Produces a 1-D plot of the radial average of an image (usually a recorded diffraction pattern). Values of exact zeroes are not counted in the average and thus various filters can be applied to the image to exclude regions from being considered.

<u>Radial Average Noise Reduction Filter:</u>   This is a routine that is useful for reducing noise in images containing both crystalline structures and amorphous material. The routine calculates the average radial intensity distribution and uses the obtained data to manipulate the Fourier transform of the image by either thresholding or vector subtraction prior to an inverse Fourier transform. The standard deviation of the radial average can be used to determine threshold levels. The algorithm has been proven particularly useful for thin crystalline regions with amorphous surface layers and crystalline particles on amorphous substrates.

<u>Custom Function Help:</u>   All the functions listed above are detailed through an on-line help mechanism. The help itself is made available as an installed custom function and explains the functionality and use of each particular routine.

The routines described above are all available through the new NCEM World Wide Web home page which will be on line by or before mid-summer 1995. In addition to the routines mentioned here, there will be other programs made available as they are prepared for distribution. The exact address for the NCEM home page will be made available by the time of the MSA meeting.

References

1.   Trademark of Gatan, Inc, 6678 Owens Drive, Pleasanton, CA 94588, USA.

2.   W.O. Saxton, T.J. Pitt and M. Horner, Ultramicroscopy 4 (1979), 343-350

3.   R. Kilaas, S. Paciornik, A. J. Schwartz and L. E. Tanner, Journal of Computer Assisted Microscopy, Volume 6, Number 3, 1994

4   Trademark of Total Resolution, 20 Florida Ave., Berkeley, CA 94707, USA.

5.   S. Paciornik, R. Kilaas, J. Turner and U. Dahmen, Ultramicroscopy (submitted).

6.   This work is supported by the Director, Office of Energy Research, Office of Basic Energy Sciences, Materials Sciences Division of the U.S.Department of Energy under Contract No. DE-AC03-76SF00098.

IMPROVED 3D RECONSTRUCTION USING STEREO COMPUTER GRAPHICS AND MULTIPLE TILT ELECTRON MICROSCOPE IMAGES

Lee D. Peachey,* Lou Fodor,* John C. Haselgrove,** Stanley M. Dunn,*** and Junqing Huang***

*Dept. of Biology, University of Pennsylvania, Philadelphia, PA 19104-6018
**Childrens Hospital of Philadelphia, Philadelphia, PA 19104-6018
***Dept. of Biomedical Engineering, Rutgers University, Piscataway, NJ 08855

Stereo pairs of electron microscope images provide valuable visual impressions of the three-dimensional nature of specimens, including biological objects. Beyond this one seeks quantitatively accurate models and measurements of the three dimensional positions and sizes of structures in the specimen. In our laboratory, we have sought to combine high resolution video cameras[1] with high performance computer graphics systems[2] to improve both the ease of building 3D reconstructions and the accuracy of 3D measurements, by using multiple tilt images of the same specimen tilted over a wider range of angles than can be viewed stereoscopically. Ultimately we also wish to automate the reconstruction and measurement process, and have initiated work in that direction.

Figure 1 is a stereo pair of 400 kV images from a 1 micrometer thick transverse section of frog skeletal muscle stained with the Golgi stain[3]. This stain selectively increases the density of the transverse tubular network in these muscle cells, and it is this network that we reconstruct in this example. Figure 2 shows the digitized form of these images, as initially displayed stereoscopically on the screen of the Silicon Graphics Indigo computer. The stereoscopic display is a special feature of this computer, not employed in the system we described earlier,[2] and is accomplished by rapidly alternating between the left-eye and right-eye images on the display screen. The viewer wears glasses that alternately blocks the right and the left eye, in synchrony with the display, resulting in a stereoscopic view of the image pair. As shown here, the images often are displayed as negatives for greater clarity of the reconstruction as it is built.

First we perform an automatic alignment of the stereo axis of the images from fiducial marks.[4] Then the operator creates a reconstruction by moving a 3D cursor within the 3D image on the display, using a mouse. When the cursor appears to be coincident in three dimensions with a junction (node) in the network, the operator clicks on the mouse, and a node is recorded and displayed by the computer at that 3D location (small squares in Figure 2). Moving the cursor in 3D along a tubule (link) in the network leaves a multisegmented vector (lines in Figure 2) from the first node to the next, where another node (square) is deposited. Multiple links can be connected to a single node, as occurs in the biological network being reconstructed. This process is continued until the entire network has been traced in 3D (Figure 2). Errors can be corrected by "capturing" nodes and links with the mouse and cursor, and editing them as needed. When the tracing and editing process has been completed, the images themselves are removed from the display, so that only the reconstructed nodes and links are seen. The reconstruction can be rotated and displayed from other viewing directions for better visualization of the data and for editing the reconstruction in conjunction with other stereoscopic pairs of images of the same specimen taken at different tilt angles. Several kinds of 3D measurements can be made from the data, such as relative 3D positions of nodes, distances between nodes, etc.

Our latest development is the automation of the above process, beginning with segmentation of the network by adaptive thresholding, skeletonization of the network, intersection detection, and, finally, determination of correspondence of nodes in the two images by geometric hashing (not illustrated here).[5] Comparison of the speed and accuracy of the two approaches (manual and automatic) when applied to the same images is in progress.

References

1. E. Horn et al., Proc. Ann EMSA Meeting 50 (1992) 960-961
2. L.D. Peachey et al., Proc. Ann EMSA Meeting 50 (1992) 1056-1057
3. N. Fujimaki et al., Bioimages 1 (1993) 167-174
4. J. Haselgrove et al. Proc. Ann MSA Meeting 52 (1994) 502-503
5. J. Huang et al., J. Computer-Assisted Microscopy 6 (1994) 85-102
6. This work was supported by Grant No. RR-02483 from the National Institutes of Health, U.S.A.

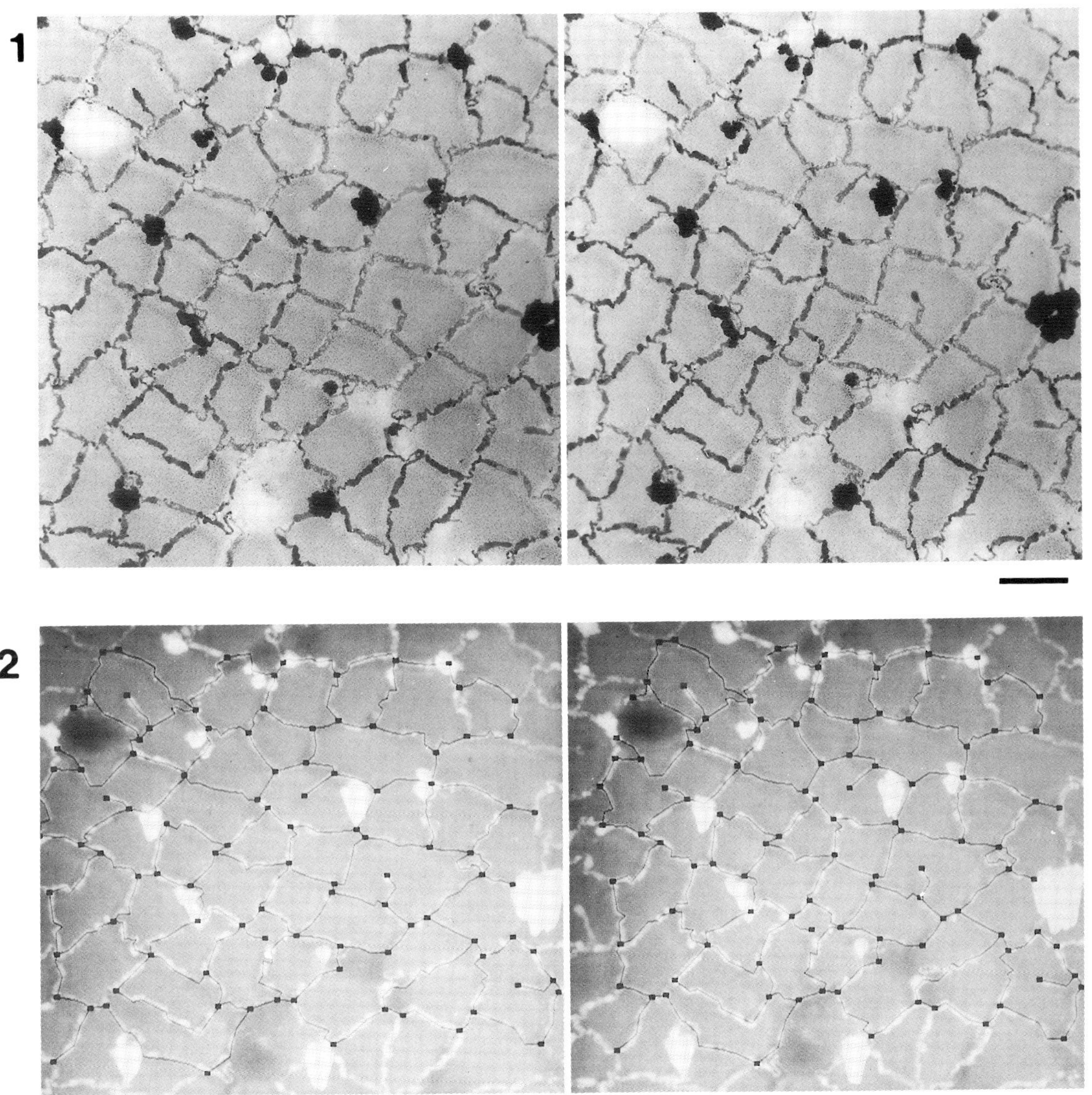

FIG. 1-Intermediate voltage electron micrographs (400 kV) of a transverse section of a Golgi stained frog leg skeletal muscle, taken at tilt angles of 8 and 16 degrees. The images can be viewed stereoscopically with crossed eyes, or with a prismatic stereo viewer. The darkly stained network is the transverse tubular system, which surrounds the less densly stained myofibrils. The larger dark objects are contamination from the stain. Line indicates 1 micrometer.

FIG. 2.-Same images digitized at a resolution of 1024 by 1024 pixels and displayed as negatives on the computer display. Some blooming was allowed in the contamination particles in order to obtain good contrast in the network. Part of the transverse tubular system network has been reconstructed using the computer graphic system, as described in the text. The reconstruction is shown superimposed on the original micrographs, as it would appear to the person making the reconstruction. Black squares represet nodes in the network, and thin, black lines are links between nodes, representing the tubules of the network.

THE DETECTION OF Xe GAS BUBBLES IN MgO HIGH-RESOLUTION LATTICE IMAGES

Stuart McKernan and M. Grant Norton[†]

Center for Interfacial Engineering Microscopy Facility, University of Minnesota, 100 Union St. S. E., Minneapolis, Minnesota 55455
[†] Mechanical and Materials Engineering Department, Washington State University, Pullman, WA 99164

Room temperature implantation of noble gases into metal and ceramic targets results in the formation of gas bubbles within the crystalline matrix. The pressure within these bubbles is sufficient to stabilize the gas in a crystalline form which, in general has an epitactic relation with the matrix[1-3]. The size of the bubbles increases with increasing noble gas fluence. As the size of the bubbles increases, the pressure within the bubble decreases. There is therefore a limiting size for the bubble above which the trapped noble gas exists in the fluid phase rather than as a crystalline inclusion. Because of the large size of the noble gas atoms they do not diffuse through the material at an appreciable rate after the implantation event, and thus probably rely on the presence of structural defects in the matrix to coalesce into bubbles. The ability to detect and measure small bubbles in a crystalline matrix is therefore an important step in characterizing the implanted material and understanding the formation mechanism of the bubbles.

Pre-thinned and annealed, single crystal MgO TEM specimens were implanted with monoenergetic 200keV Xe[+] ions. These samples were then examined, without further specimen preparation, in a JEOL 4000EX operating at 400kV. The negatives were then digitized using a Kodak Megaplus 4.2 CCD camera, and the images were analyzed using NIH Image and Gatan's Digital Micrograph.

A typical high-resolution electron micrograph is shown in figure 1. The MgO {200} lattice fringes are clearly visibly throughout the whole image, but several localized areas with a different periodicity are also visible. The Fourier transform (figure 2) of this region shows the same features as is seen in the electron diffraction patterns obtained on the microscope. Strong, sharp diffraction spots corresponding to the MgO matrix are evident at N. A weaker diffuse ring is visible peaked at the radius denoted by X. Within this ring the intensity is strongly peaked in the directions parallel to the MgO axes. Additional weak, diffuse intensity is observed between these peaks and the origin. Similar patterns have been observed by Donnelly et. al [4] where they performed image processing on an optical bench to remove the matrix periodicities to reveal the structure of the trapped solid noble gas.

The digitized images were processed in a variety of ways to extract as much of the Xe signal as possible and suppressing the MgO signal and the noise. Figure 3 shows the result of removing the MgO frequencies as well as the extremely high and low frequencies. The effect of different processing schemes will be discussed. Well-defined regions of high local contrast are visible (some have been marked B in figure 3). These correspond to the trapped Xe bubbles in the MgO film. Enlargements of two of these regions and their diffractograms are shown in figure 4. Having detected the individual Xe bubbles in the lattice images, the diffractogram from each bubble can be used to calculate the lattice parameter of the Xe. This can then be correlated with the bubble size to show the variation of pressure with particle size. The contrast from the bubbles in the processed images is very well defined, and indicates that the lateral extent of the Xe inclusions is ~20-25 nm. The diffractograms from the individual inclusions shows a much more well-defined Xe {100} diffraction spot than the diffractogram from the whole image. This can be interpreted if the specimen consists of a few isolated, crystalline Xe inclusions, providing the intensity in the Xe {100} spots, separated by a MgO matrix containing small, randomly oriented Xe clusters, which contribute to the diffuse ring of intensity. The visibility of these very small bubbles is in agreement with simulation studies of spinel particles in NiO [5].[6]

References

1. M. G. Norton, E. L. Fleischer, W. Hertl, C. B. Carter, J. W. Mayer, and E. Johnson, Phys. Rev. B **43**, 9291 (1991)

Proc. Microscopy and Microanalysis 1995, edited by G.W. Bailey, M.H. Ellisman, R.A. Hennigar, and N.J. Zaluzec.
Copyright © 1995 MSA. Published by Jones and Begell Publishing, 79 Madison Ave., New York, NY 10016

2. E. L. Fleischer, M. G. Norton, M. A. Zaleski, W. Hertl, C. B. Carter, and J. W. Mayer, J. Mater. Res. **6**, 1905 (1991)

3. G. A. Hismeh, L. Cartz, F. Desage, C. Templier, J. C. Desoyer, and R. C. Birtcher, J. Mater. Res. **9**, 3095 (1994)

4. S.E. Donnelly, C.J. Rossouw, and I.J. Wilson, Radn. Eff. **97**, 265 (1986)

5. D.R. Rasmussen, S.R. Summerfelt, S. McKernan, and Carter, C.B., Submitted to J.Microsc.

6. This research is supported, in part, by the Microscopy Facility at the University of Minnesota Center for Interfacial Engineering, an NSF Engineering Research Center.

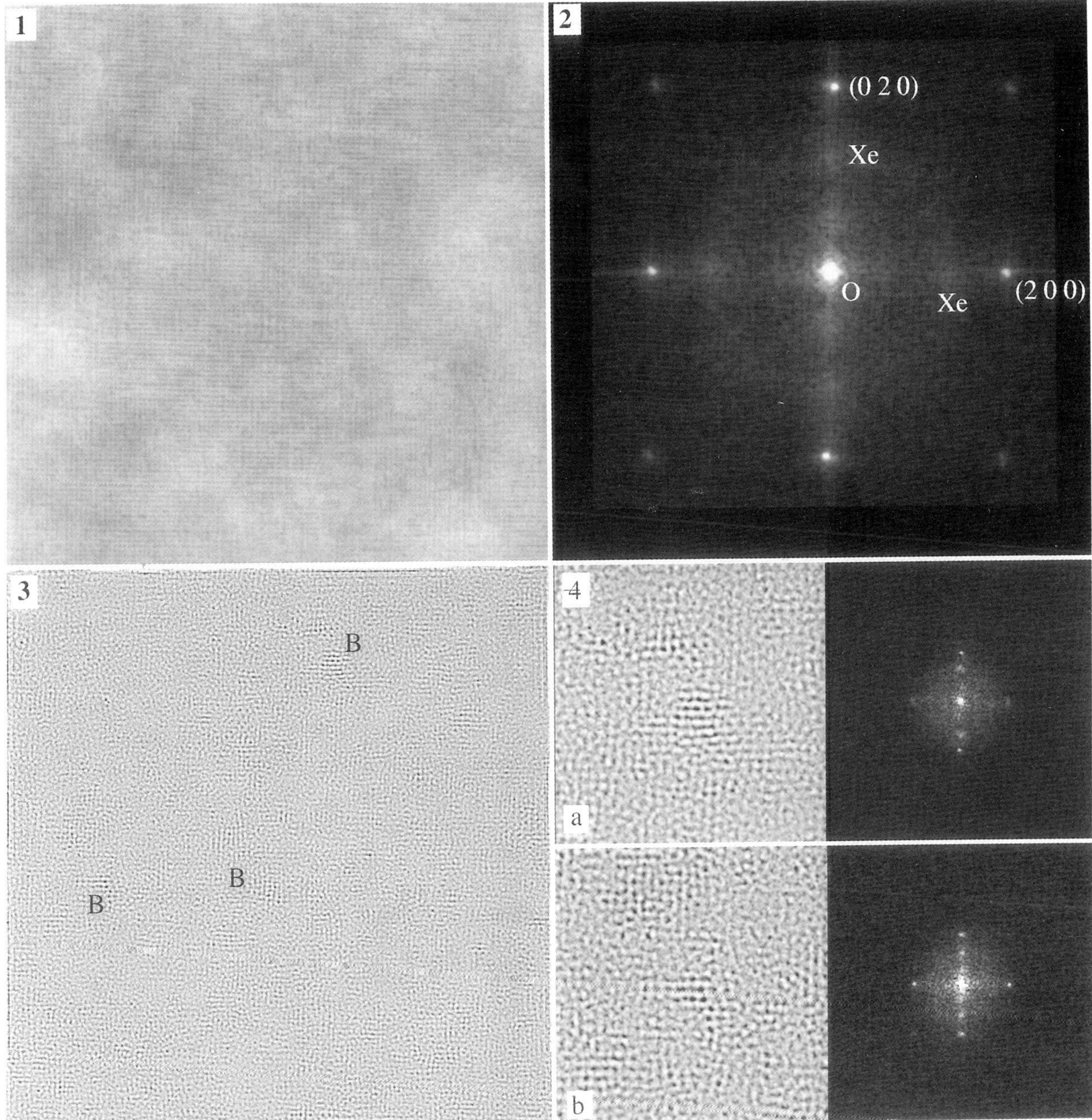

FIG. 1 High-resolution lattice image of Xe irradiated MgO at the (001) zone axis
FIG. 2 Diffractogram of Fig. 1 with indexed MgO spots and diffuse Xe diffraction spots (Xe)
FIG. 3 Image-processed version of Fig 1 with MgO periodicities suppressed.
FIG. 4 Enlarged images of individual gas bubbles and their diffractograms.

AN INTEGRATED DIGITAL IMAGE ACQUISITION AND ARCHIVING SYSTEM FOR DIAGNOSTICS AND EDUCATION IN PATHOLOGY

F. He, E. Schubert, W. Gross, M. Mancini, T. Mahoney, F. Wimberley, R. Roskies, M. Levine and M. Becich. Division of Informatics and Facility for Imaging and Research Microscopy, Department of Pathology, University of Pittsburgh, School of Medicine, Pittsburgh, Pennsylvania 15261

Background Advances in digital imaging technology and instruments have profoundly enhanced the way of decision-making in diagnostic pathology. The image that was typically recorded on film and stored in different divisions of the department in the past can now be stored via a centralized image file server. Since pathology images are derived from gross surgical specimens, light, immunoflourescence, electron microscopy and clinical laboratory data in graphical forms, it is essential to have an integrated system to manage and distribute the images. The images along with attached text information should be rapidly stored in a fashion of pathologist's perspective, and be viewed on their desktop computers with an absolutely minimal impact on existing work flow. In addition, the images originated from individual patients should be easily reorganized or grouped for the purposes of medical training and on-line conferences. Having identified these basic requisitions, we have recently implemented such a system in our department[1,2].

System Configuration The configuration of the system is depicted Figure 1. It consists of a wide area network (WAN) which manage imaging traffic for optimal speed and multiple image acquisition and viewing stations. Ethernet connection is adopted for peripheral stations to communicate with the site WAN. For on-line storage, a central image server is currently configured with 32 gigabytes (GB) redundant array of inexpensive disk (RAID) storage device. It also involves a 50 GB 8mm tape backup device and a 20GB optical disc jukebox for "near-line" storage. Pittsburgh Supercomputing Center (PSC) provides us additional "near-line" storage space. All of this ensures substantial storage room and optimal high performance for archiving. The total images captured per year is projected at 100,000 and requires 100 GB of new storage space per year. The capture station are equipped with either high resolution CCD or digital cameras and a minimal 486-PC with 16 megabytes (MB) RAM is furnished for both capture and viewing stations. The front end software (Fig. 2) that manages the images is the key to the abilities and usefulness for archiving and retrieval. It was developed by Applied Machine Vision and The Image Source. The Windows-based software captures image simply by several mouse button clickings and saves image in BMP or TIFF format. It also has the ability to import individual or batched images in different format. Retrieval of images from a viewing station takes less than 10 seconds for an 1 MB image. Everything needed to annotate the image is included in the system. Sound, text, boxes and line drawing can be easily added to an image in a non-destructive fashion. It also can quickly index image information from individual patient's folder via query search.

Main Advantages 1.) *Improving patient care*: the availability of gross images at time of microscopic signout, verifying recurrence of malignancy from archived images, monitoring bone marrow engraftment and immunosuppressive intervention following bone marrow/solid organ transplantation on repeat biopsies, and the ability to seek instantaneous "consultation" with any pathologist on the network; 2.) *Enhancing the teaching education environment*: the building of a digital surgical pathology atlas, improving the availability of images for conference support and sharing cases across the network with colleagues or trainees, on-site and off-site, even worldwide within seconds; 3.) *Enhancing research*: case study compilation, metastudies, and availability of digitized images for quantitative analysis and 4) *Economic considerations*: the considerable cost savings in photographic media for conferences, improved quality assurance by porting control stains across the network and a multiplicity of other advantages that enhance image and information management in pathology.

References

1. M. Becich et al., *Sem. Diagnostic Pathol.* 11(1994)237-244.
2. E. Schubert et al., *Sem. Diagnostic Pathol.* 11(1994)263-273.

Proc. Microscopy and Microanalysis 1995, edited by G.W. Bailey, M.H. Ellisman, R.A. Hennigar, and N.J. Zaluzec
Copyright © 1995 MSA. Published by Jones and Begell Publishing, 79 Madison Ave., New York, NY 10016

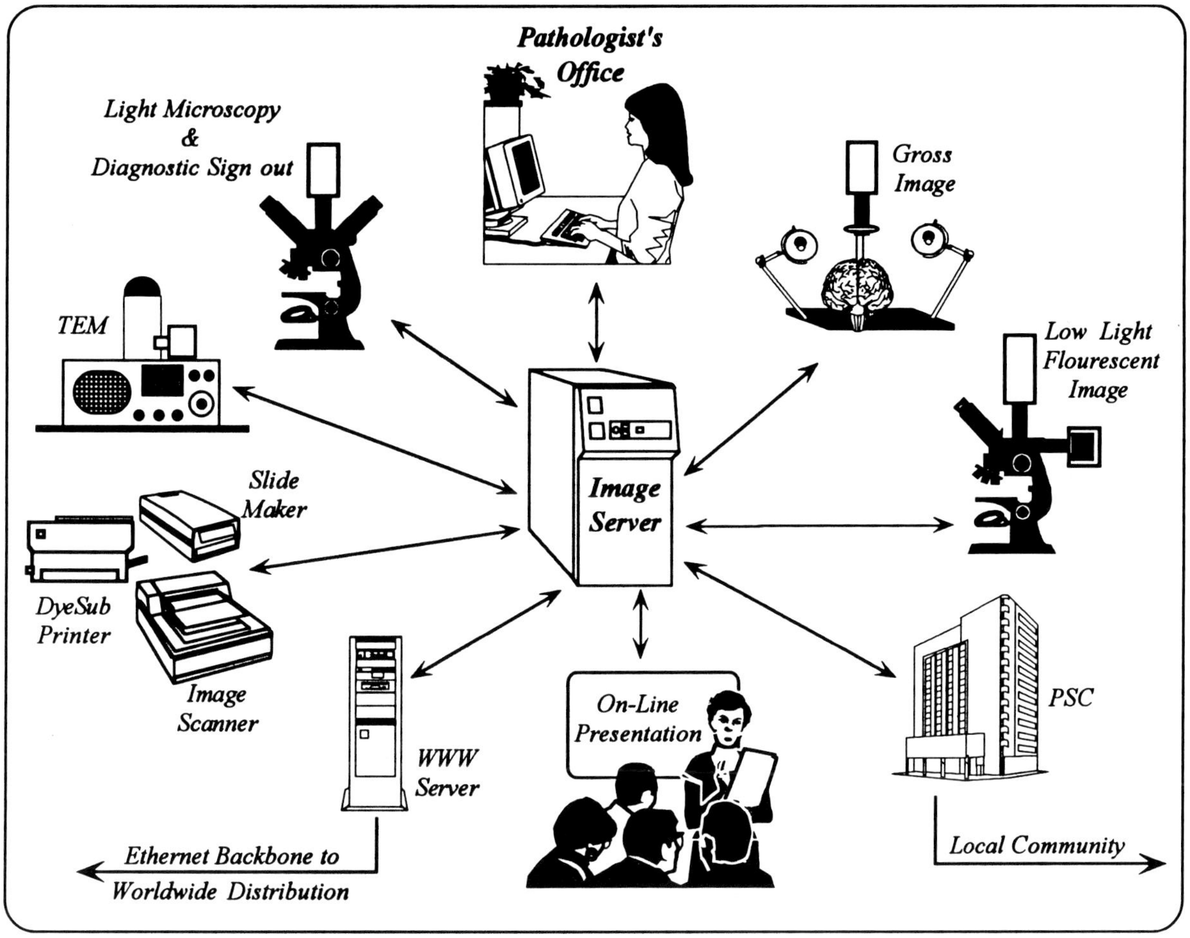

Figure 1. Schematic diagram of system configuration.

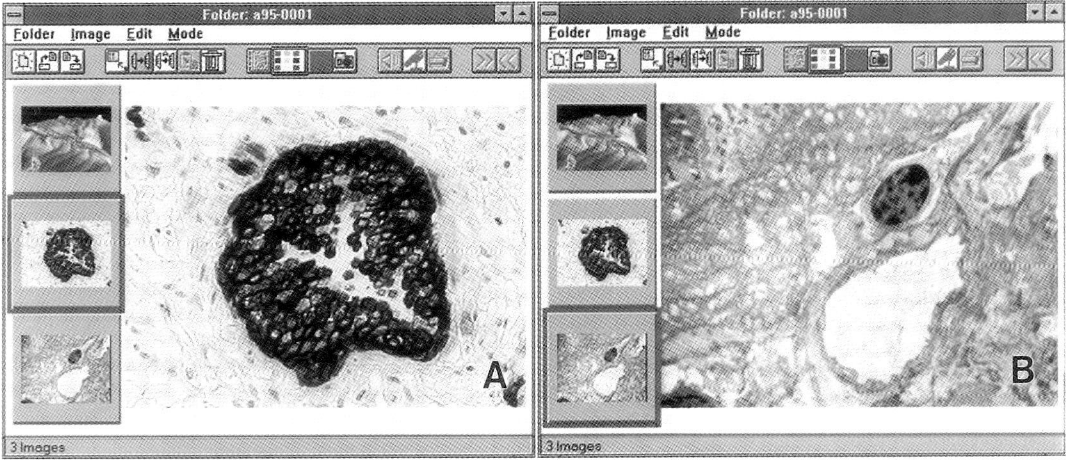

Figure 2. Desktop of window based front end. A: Reduced thumbnail images (left) and a full size image (LM) in viewing mode. B: Reduced thumbnail images (left) and a full size image (TEM) in viewing mode.

EXPERIMENTAL STUDY ON THE RESTORATION FROM A TILT-AZIMUTH SERIES FOR IMPROVEMENT IN RESOLUTION

Norio Baba, Norihiko Ichise and Syunya Watanabe

Department of Electrical Engineering, Kogakuin University, 1-24-2 Nishishinjuku, Shinjuku-ku, Tokyo 163-91, Japan

The tilted beam illumination method is used to improve the resolution comparing with the axial illumination mode. Using this advantage, a restoration method of several tilted beam images covering the full azimuthal range was proposed by Saxton[1], and experimentally examined[2].   To make this technique more reliable it seems that some practical problems still remain. In this report the restoration was attempted and the problems were considered. In our study, four problems were pointed out for the experiment of the restoration. (1) Accurate beam tilt adjustment to fit the incident beam to the coma-free axis for the symmetrical beam tilting over the full azimuthal range. (2) Accurate measurements of the optical parameters which are necessary to design the restoration filter. Even if the spherical aberration coefficient Cs is known with accuracy and the axial astigmatism is sufficiently compensated, at least the defocus value must be measured. (3) Accurate alignment of the tilt-azimuth series images. (4) Restraint of excessive enhancement of noise due to the inverse restoration filter. For the item (1), the diffractogram-based method was used[3]. To further precise adjustment the nature of the 3-fold axial astigmatism was used[4]. By detecting the 3-fold symmetry of the astigmatism it was judged that the beam tilt adjustment was achieved at worst less than 0.5mrad. For the (2), at present the diffractogram-based method was also used. However, because its method is not specimen independent and in general restricted within a range of the defocus and the axial astigmatism, the alternative method is examined in which the phase spectra between tilted beam images are measured and from comparisons of them with a mathematical model the TEM parameters are measured and controlled[5]. For the (3), though the cross-correlation function (XCF) is generally used, in this case it may not be a powerful tool because the illumination condition changes for each image.   Therefore a new alignment method was practiced[6] in which a phase spectrum between two images recorded under different azimuths in the tilted illumination was measured and the image drift was corrected based on the criterion deduced from the theoretical nature of the phase spectrum, e.g., using the symmetry of the phase distribution. The phase spectrum is the phase part of the cross-spectrum which is defined as the Fourier transform of the XCF. For the (4), in general the S/N ratio of HRTEM images are low, and when the inverse restoration filter is simply applied the noise possessing high frequency components are excessively enhanced. Therefore, in this study the restoration filter was applied after the noise removal which was performed using the phase spectrum. The tilted beam images were recorded serially, and Fourier components keeping the phase constant through the series were chosen as signal.

An experiment is shown in the next figures using a HRTEM H-8100 at 200kV and a slow scan CCD camera system (Gatan model 694). In this case an alloy thin film made with the ion beam sputtering was used, and the tilting angle was 10mrad. A resolution improvement of about 43% relative to the axial mode is demonstrated.

Reference

1. W.O.Saxton, *Scanning Microscopy* **Suppl. 2** (1988) 213

2. A.I.Kirkland et al., Proc. 13th ICEM **1** (1994) 463

3. O.L.Krivanek and G.Y.Fan, *Scanning Microscopy* **Suppl. 6** (1992) 105

4. K.Ishizuka, *Ultramicroscopy* **55** (1994) 407

5. A.J Koster et al., *Ultramicroscopy* **27** (1989) 251

6. N.Ichise et al., *Microscopy Research and Technique* **30** (1995) 350

Proc. Microscopy and Microanalysis 1995, edited by G.W. Bailey, M.H. Ellisman, R.A. Hennigar, and N.J. Zaluzec
Copyright © 1995 MSA. Published by Jones and Begell Publishing, 79 Madison Ave., New York, NY 10016

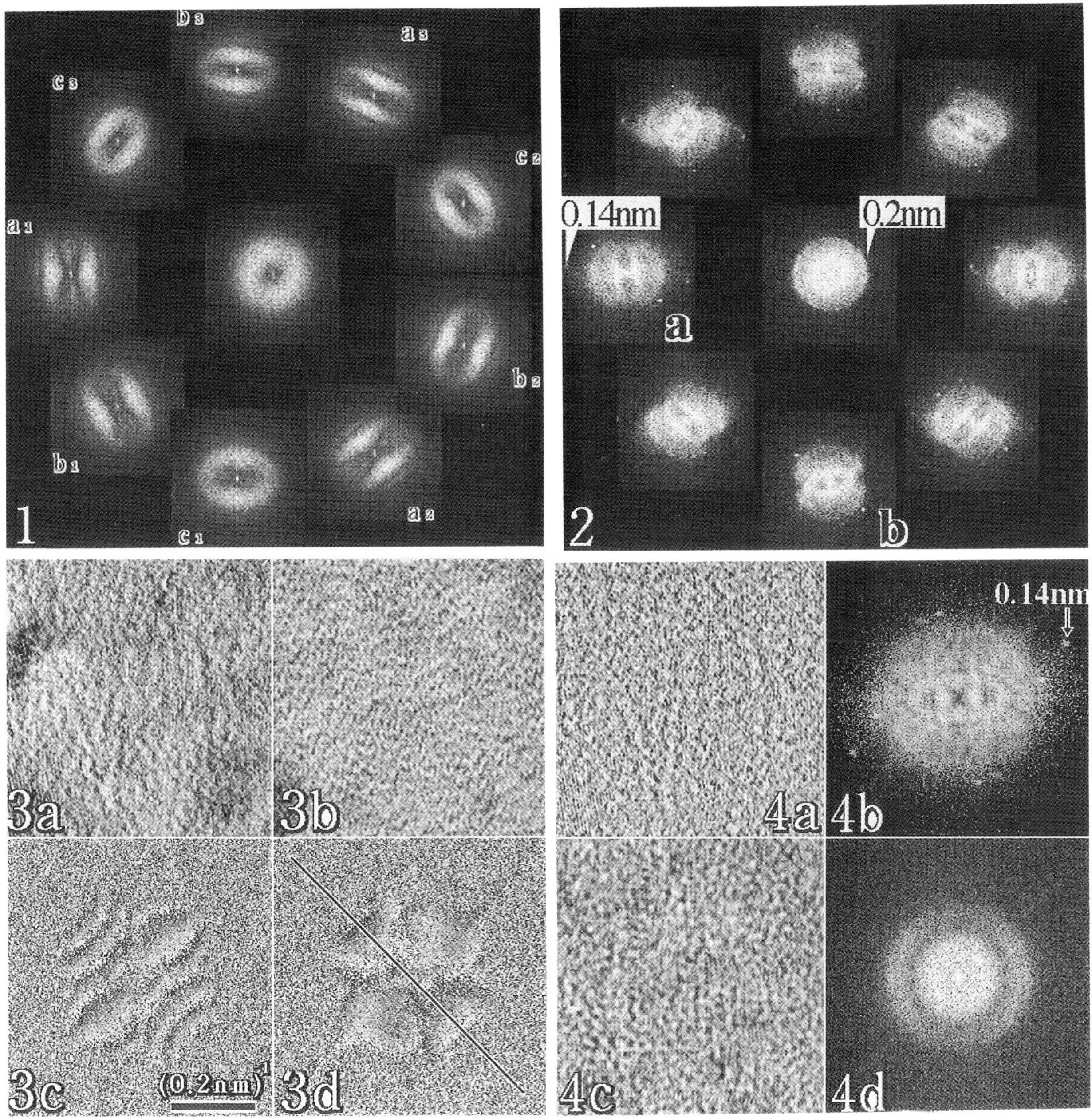

Fig.1 - Coma-free alignment with 3-fold astigmatism (6mrad tilting).

Fig.2 - A tilt-azimuth series for recording images to be restored (10mrad tilting).

Fig.3 - Alignment of tilted beam images (a) and (b) whose illumination condition are indicated in Fig.2 by the symbols a and b; (c): shows a phase spectrum before the alignment and (d): after the alignment using the line symmetry of the phase distribution.

Fig.4 - A final restored image (a) and the corresponding diffractogram (b) which are compared with (c) and (d) at axial mode.

PROBLEMS IN INTERPRETING SUB-Å-RESOLUTION IMAGES STUDIED BY SIMULATED EVOLUTION BASED DIGITAL IMAGE MATCHING

G. Möbus, G. Dehm, G. Gutekunst and M. Rühle

Max-Planck-Institut für Metallforschung, Inst. für Werkstoffwissenschaft, 70174 Stuttgart, Germany

1. Introduction: Experimentally determined[1,2] CTF-data caused predictions[1] that atomic resolution microscopy (JEM ARM1250) of sapphire should become possible with single atomic columns resolved (Fig. 1). The smallest projected bond-length (ca. 0.85 Å) lies beyond the point-resolution of 1.05 Å but within the information limit of this instrument (around 0.85 Å). After extensive trials such images could not be verified experimentally. Instead, images like Fig. 3 b typically appear, containing still high frequencies but different from predictions.

2. Ideal and realistic image simulations: Reasons for this deviations have been examined following [3,4,5] with special respect to sub-Å-resolution, including specimen tilt, beam tilt, achromatic anisotropic damping (MTF, vibration, drift), non-ideal columns and amorphous contamination. The stepwise EMS[6]-contrast generation from the crystal single slice over the specimen exit plane into the image plane is compared in FIG. 1 (ideal alignment) and FIG. 2 (realistic simulation). Predominant factors lie in specimen misalignment or other disturbancies (e.g. strain fields) during multislice followed by trade-off effects during fitting of imaging parameters and achromatic damping. The single slice potential is of less influence in the present case. Iterative digital image matching [5] has been used to match experimental templates (FIG 3). As optimization algorithm to find the EMS- parameters for a pattern of perfect sapphire *simulated evolution* by Schwefel [7] has been applied which is capable to find global optima in highly multi-variant search spaces.

3. Atom location by intensity matching: As a consequence, single atom location *beyond* the Rayleigh-limit is studied testing full gray-level matching of a simulated image series with one oxygen atom shifted in 0.1Å steps. The capability of finding the correct atom position is tested as a function of Poisson-noise level and damping envelope. Two examples are printed in FIG. 4 showing that a Poisson-width of $\sigma = 2$ only leads to a mis-location of 0.1 Å and a doubling of the confidence level. This makes it likely that resolution-limit information can be successfully extracted by intensity matching at low noise level. The capability of locating many atoms at an interface under poisson-noise environment is studied applying *simulated evolution* with coordinates as variables instead of imaging parameters. The examples under study are metal-sapphire interfaces[8].

1. G. Möbus, F. Phillipp and M. Rühle, *Proc. 52nd MSA*, 1994,748.

2. F. Phillipp, R. Höschen, M. Osaki, G. Möbus, and M. Rühle, *Ultramicroscopy*, **56**,1994,1.

3. M.J. Hytch, W.M. Stobbs, *Ultramicroscopy*, **53**,1994,191.

4. M.A. O'Keefe and V. Radmilovic, *Proc. 51st MSA*,1993,980.

5. G. Möbus and M. Rühle, *Ultramicroscopy*, **56**, 1994,54.

6. P. Stadelmann, *Ultramicroscopy* **21**,1987,131.

7. H.-P. Schwefel, Numerical Optimization of computer models, J. Wiley, New York, 1981.

8. Support from BMFT under NTS O230-0 is gratefully acknowledged.

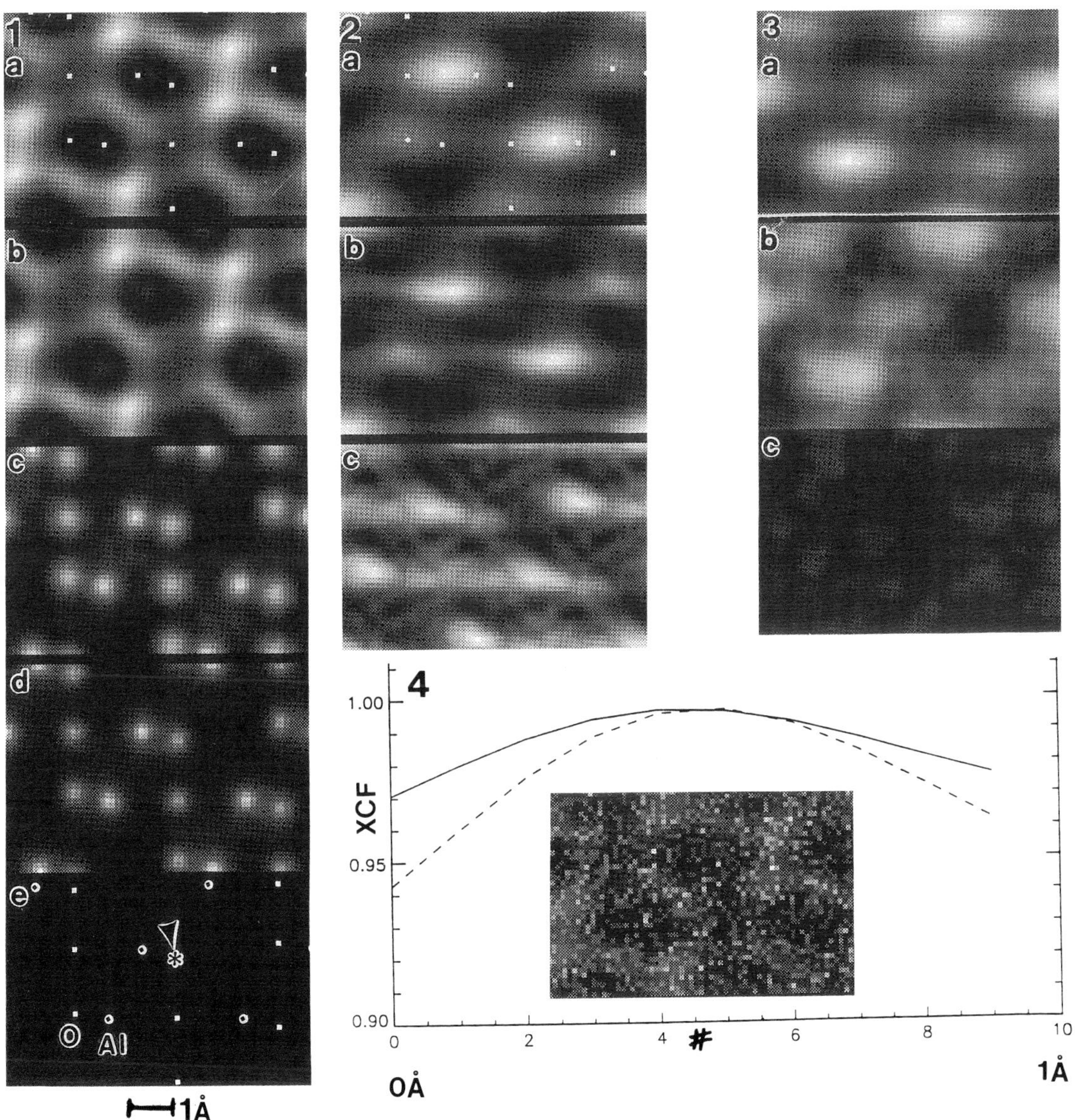

FIG. 1. - Ideal ARM1250 EMS-Simulation for sapphire $< 2\bar{1}\bar{1}0 >$ - (a) image with atom coord., (b) undamped image, (c) exit wave (imag. part), (d) single slice wave (imag. part), (e) structure.

FIG. 2 - Realistic ARM1250 simulation including misalignment and achrom. damping, (a-c to compare to Fig.1 - d,e unchanged).

FIG. 3. - Simulation, Experiment and Difference image matched by simulated evolution (cross correlation factor of $XCF = 0.84$).

FIG. 4. - Atom location out of Poisson noise image (inset with Poisson-width of $\sigma = 2$). XCF as a function of oxygen-position (marked by $\star$ in Fig. 1e) for $\sigma = 0$ (dashed) and $\sigma = 2$.

THE EFFECT OF TEMPERATURE ON HRTEM IMAGE CONTRAST

T. Geipel*, W. Mader*, and P. Pirouz**

* Institut für Anorganische Chemie, Universität Bonn, 53117 Bonn, Germany
** Dept. of Mat. Sci. and Eng., Case Western Reserve University, Cleveland OH 44106, USA

Temperature affects both elastic and inelastic scattering of electrons in a crystal. The Debye-Waller factor, B, describes the influence of temperature on the elastic scattering of electrons, whereas the imaginary part of the (complex) atomic form factor, $f_c = f_r + if_i$, describes the influence of temperature on the inelastic scattering of electrons (i.e. absorption). In HRTEM simulations, two possible ways to include absorption are: (i) an approximate method in which absorption is described by a phenomenological constant, μ, i.e. $f_i = \mu f_r$, with the real part of the atomic form factor, f_r, obtained from Hartree-Fock calculations, (ii) a more accurate method in which the absorptive components, f_i, of the atomic form factor are explicitly calculated [1]. In this contribution, the inclusion of both the Debye-Waller factor and absorption on HRTEM images of a $\langle 011 \rangle$-oriented $GaAs$ crystal are presented (using the EMS software [2]).

Fig. 1 shows the the amplitudes and phases of the dominant 111 beams as a function of the specimen thickness, t, for the cases when $\mu = 0$ (i.e. no absorption, solid line) and $\mu = 0.1$ (with absorption, dashed line). The lower part of this figure shows a plot of the defocus of minimum contrast versus crystal thickness [3,4]. The regions between the defoci of minimum contrast correspond to conditions at which the phase transfer is negative (positive) and the black (white) spots correspond to atomic positions. It can be seen that the absorption leads to change in the phase of the Bragg beams close to the extinction distance, ξ, and effectively to a change of the defoci of minimum contrast. Fig. 2 is similar to Fig. 1 except that the calculated absorptive form factors from [1] have been used; the solid lines correspond to the case when $B = 0$, and the dashed lines to the case when $B = 0.009 \ nm^2$ (which is an upper estimate for $T = 295 \ K$ [5]). According to these figures, when the Debye-Waller factor is taken into account (with $B = 0.009 \ nm^2$), the extinction distance changes by almost 50% compared to the case when it is not ($B = 0$). On the other hand, these figures show that the inclusion of absorption in the atomic form factors changes the extinction distances by a few percents. The effect of absorption on the phase of the Bragg beams, and subsequently on the HRTEM contrast, can be explained by the Bloch wave theory; this is not considered here. Fig. 3 ($t = 4 \ nm$) and Fig. 4 ($t = 13.6 \ nm$) show the effect of absorption and the Debye-Waller factor on simulated images of $\langle 011 \rangle$-oriented $GaAs$. In Fig. 3, corresponding to a very thin crystal, the inclusion of B and f_i hardly make a difference in the image contrast. On the other hand, in Fig. 4, for a relatively shick crystal, both B and f_i affect the image contrast and the images for $B = 0$ and $B = 0.009 \ nm^2$ are very different. It may be concluded that for $t < \xi$, the effect of temperature on HRTEM image contrast is negligible, but for $t < \xi$, both the Debye-Waller factor and absorption can change the contrast and the effect becomes more important with increasing crystal thickness.

References

1. D.M. Bird and Q.A. King, Acta Cryst. A (1990) 46, 202,
2. P.A. Stadelmann, Ultramicroscopy (1987) 21,131,
3. R.W. Glaisher et al., Ultramicroscopy (1989) 27, 131,
4. T. Geipel and M. Hohenstein, Ultramicroscopy (1992) 45, 167,
5. T. Matsushita and J. Hajashi, Phys.Stat.Sol. A (1977), 41, 139.

Proc. Microscopy and Microanalysis 1995, edited by G.W. Bailey, M.H. Ellisman, R.A. Hennigar, and N.J. Zaluzec
Copyright © 1995 MSA. Published by Jones and Begell Publishing, 79 Madison Ave., New York, NY 10016

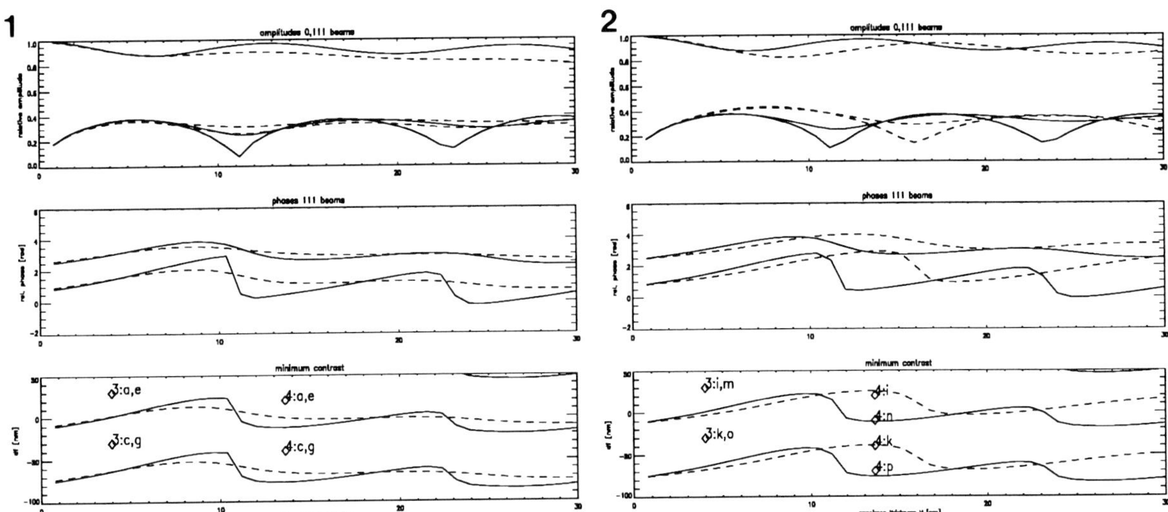

FIG. 1 - Upper two plots: Amplitudes and phases of 0 beam and 111 Bragg beams in $GaAs$ along $\langle 011 \rangle$. Lower plot: Defoci of minimum image contrast for the 111 beams. The solid lines correspond to $\mu = 0$, the dashed lines correspond to $\mu = 0.1$. The Debye-Waller factor is $B_j = 0$. The numbers and letters indicate the defoci of maximum contrast of the simulated images in Figs. 3 and 4.

FIG. 2 - Amplitudes, phases and defoci of minimum image contrast as in Fig. 1 using the calculated absorptive form factors of Ref. [1]. The solid lines correspond to $B_j = 0$, the dashed lines correspond to $B_j = 0.009 \ nm^2$.

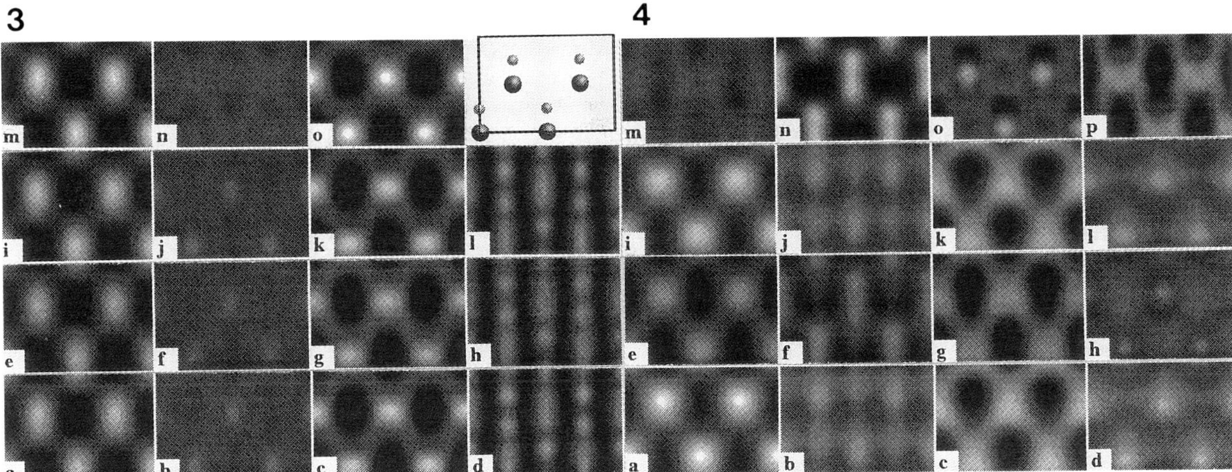

FIG. 3 - Simulated HRTEM images of $GaAs$ along $\langle 011 \rangle$ with specimen thickness $t = 4 \ nm$ for different defoci, Debye-Waller factors and absorptive form factors. The rows correspond to the same Debye-Waller factor and absorptive form factor each, the columns correspond to the same defocus each. Column [a,e,i,m]: $\Delta f = +30 \ nm$; [b,f,j,n]: $\Delta f = 0$; [c,g,k,o]: $\Delta f = -30 \ nm$; [d,h,l]: $\Delta = -60 \ nm$. Row [a-d]: $B_j = 0, \mu = 0$; [e-f]: $B_j = 0, \mu = 0.1$; [i-l]: $B_j = 0, \mu = [1]$; [m-o]: $B_j = 0.009 \ nm^2, \mu = [1]$. The corresponding unit cell of $GaAs$ along $\langle 011 \rangle$ is also shown. The large circles correspond to Ga, the small circles correspond to As.

FIG. 4 - Simulated HRTEM images as in Fig. 3 for $t = 13.6 \ nm$. Column [a,e,i,m]: $\Delta f = +20 \ nm$; [b,f,j,n]: $\Delta f = -10 \ nm$; [c,g,k,o]: $\Delta f = -40 \ nm$; [d,h,l,p]: $\Delta f = -70 \ nm$.

DIGITAL DIFFERENTIAL HYSTERESIS IMAGE PROCESSING DISPLAYS WHAT THE MICROSCOPE ACQUIRES BUT THE EYE CAN'T SEE

Klaus-Ruediger Peters

Molecular Imaging Laboratory, Biomolecular Structure Analysis Center, University of Connecticut Health Center, Farmington, CT 06030-2017

Differential hysteresis processing is a new image processing technology that provides a tool for the display of image data information at any level of differential contrast resolution. This includes the maximum contrast resolution of the acquisition system which may be 1,000-times higher than that of the visual system (16 bit versus 6 bit). All microscopes acquire high precision contrasts at a level of <0.01-25% of the acquisition range in 16-bit - 8-bit data, but these contrasts are mostly invisible or only partially visible even in conventionally enhanced images.[1] The processing principle of the differential hysteresis tool is based on hysteresis properties of intensity variations within an image.

Differential hysteresis image processing moves a cursor of selected intensity range (hysteresis range) along lines through the image data reading each successive pixel intensity. The midpoint of the cursor provides the output data. If the intensity value of the following pixel falls outside of the actual cursor endpoint values, then the cursor follows the data either with its top or with its bottom, but if the pixels' intensity value falls within the cursor range, then the cursor maintains its intensity value. Each line is read in both directions and the output values are averaged producing the hysteresis line.[2] Hysteresis lines are generated in many directions by radial processing and averaged generating the hysteresis image.[3] Using 100-200 or more different radial reading directions produces artifact-free images of the difference between two hysteresis images, each processes with a different hysteresis range (differential hysteresis range = DHR). The differential hysteresis image displays data contrasts of minimum and maximum range (DHR) as coherent, discrete and additive differential hysteresis patterns. The patterns were very similar to image contrast components generated by the different contrast mechanisms. Consecutively extracted hysteresis patterns can be summed and then reproduce the raw image data set without producing any artificial structures or structural distortions. On this account, any hysteresis pattern can be added to the raw data at amounts at which the corresponding contrast components become visible in the enriched image. This procedure would be comparable to a hypothetical amplification or suppression in the microscope of individual signal components by summing or changing corresponding detector signals.

The 12-bit atomic force microscope (AFM) image data (IR = 1-1110) of a bit hole (pit) in a compact disk was displayed as linearly scaled 8-bit image (Fig. 1a) but revealed only the large image contrasts of the image feature (hole). All high precision structural detail information was found within a DHR of only 4% IR and was not visible. Two minute isometric AFM contrasts, which were part of the raw data, imaged globular surface components as well as an elevation at the rim. Enrichment of the image with the corresponding differential hysteresis patterns (selected spacial detail pattern DHR 35-36, and selected intensity detail pattern DHR 65-67) displayed the detail information at the highest contrast resolution level while maintaining the overall (12-bit) image contrast (lowest contrast resolution) (Fig. 1b).

The differential hysteresis processing provides on-line data assessment, background elimination, noise management, maximizing of data acquisition and detection efficiency, and most importantly enrichment of the high precision contrasts for improvement of the image resolution of a microscope.[4]

References
1. K.-R Peters, *Proc. Ann. MSA Meeting* 52(1994)416.
2. R.S. Ledley et al., in J.T. Tippett, Ed., *Optical and Electro-Optical Information Processing*, Cambridge:MIT Press (1965)591.
3. E. Oho et al., Scanning (1995) accepted.
4. The digital AFM image data set was kindly provided by TopoMetrix Corp., Santa Clara, CA.

Proc. Microscopy and Microanalysis 1995, edited by G.W. Bailey, M.H. Ellisman, R.A. Hennigar, and N.J. Zaluzec
Copyright © 1995 MSA. Published by Jones and Begell Publishing, 79 Madison Ave., New York, NY 10016

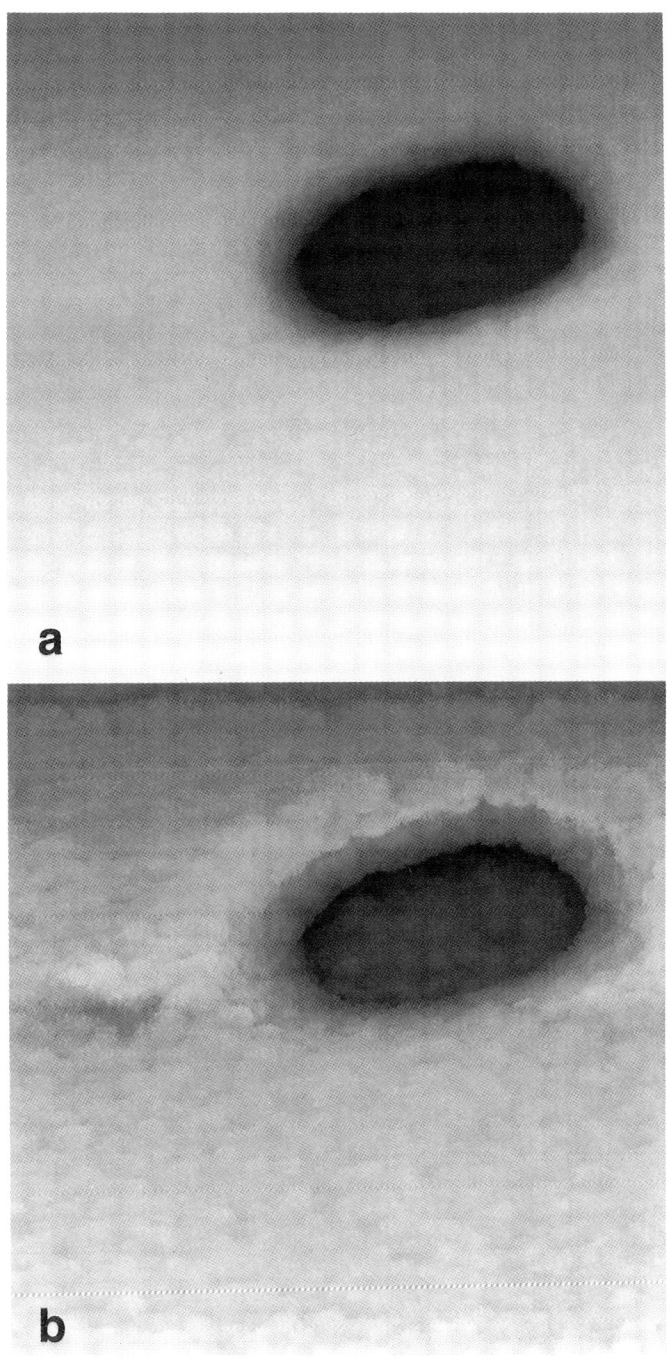

FIG. 1.--(a) AFM image of bit hole on compact disk acquired at 500x500x12-bit (IR 1-1111.), scaled to 8-bit. (b) Contrast enriched image revealing the small topographic details captured in the image data. Two isometric contrast components were found and extracted from the raw data by differential hysteresis processing and then enriched in the raw data by interactive addition to levels at which they became visible: 9% selected spacial hysteresis details of DHR = 39-41, 21% selected intensity hysteresis details of DHR = 55-57 + 70% raw data. Further enhancement is possible but would lose the high 12-bit contrast.

SIZE DISTRIBUTION STANDARDS MEASURED BY DIGITAL SEM IMAGING

James F. Kelly,

National Institute of Standards and Technology, Gaithersburg, MD 20899

The National Institute of Standards and Technology (NIST) certifies several size and size distribution Standard Reference Materials (SRMs) using spherical particles ranging in diameter from submicron to several millimeters. Glass beads are used as SRMs for a variety of laboratory applications. Their broad size distribution and spherical shape make them a model material for calibration of instruments for powder measurements.

The glass bead standard (SRM1017b) covering the diameter range from 50μm to 350μm is currently being certified using SEM backscatter electron imaging. The digital images are acquired using an AMRAY SEM connected to a Apple Power Macintosh via a 4pi[1] digital interface. The public domain computer program, NIH Image[2], is used in the image acquisition and size analysis. Figure 1 shows the spherical nature of the glass beads. Measurement of the circular area of each particle permits the calculation of the particle volume and diameter. The system magnification in terms of pixels per micrometer using an optical stage micrometer calibrated at NIST by laser interferometry.

Sieving is used to divide the starting sample into eight size fractions. An optimal image magnification can then be selected for each size fraction. The selection of number of pixels per image is determined by the required precision in measured area. Figure 2. shows the reduction in variability in the measured areas of twelve particles as the pixels per field is increased from 256x256 to 2048x2048. The results are referenced to measurements at 4096x4096 pixels.

Several hundred beads are analyzed for each size fraction and the total cumulative volume distribution is obtained by a weighted average. Figure 3 shows the computed volume distribution for the analysis of five test samples. The measured variation results from both differences between samples and measurement errors. These two factors are combined to give a total uncertainty in the certified values. The use of digital imaging simplifies the measurement process while improving the precision of the results.

1. 4pi Analysis, Inc. Durham, NC 27707.
2. NIH Image program (written by Wayne Rasband at the U.S. NIH) is available by anonymous
 FTP from zippy@nimh.nih.gov

Proc. Microscopy and Microanalysis 1995, edited by G.W. Bailey, M.H. Ellisman, R.A. Hennigar, and N.J. Zaluzec
Copyright © 1995 MSA. Published by Jones and Begell Publishing, 79 Madison Ave., New York, NY 10016

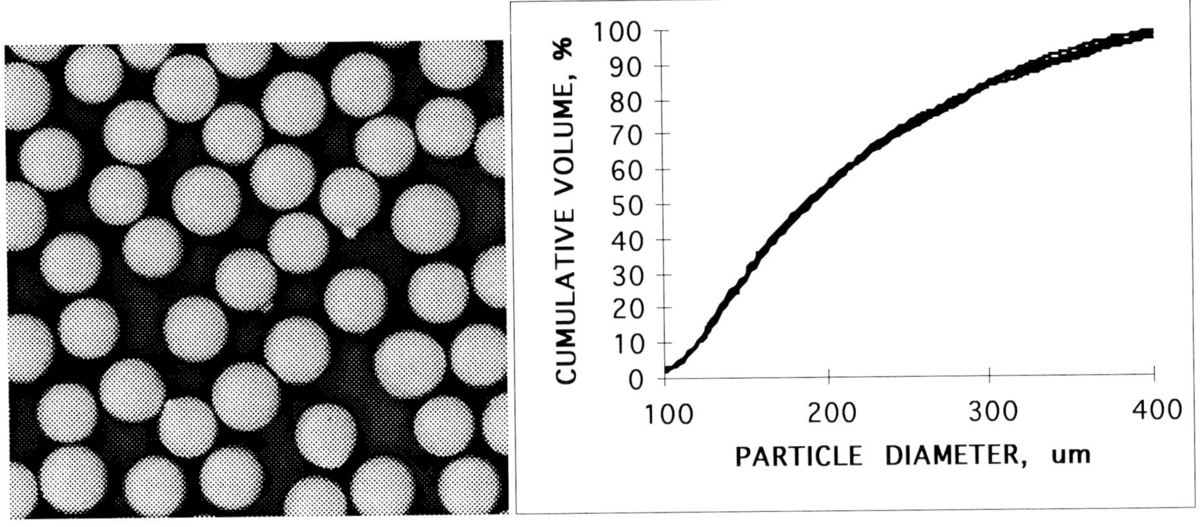

Figure 1. Backscattered electron image of glass beads.

Figure 3. Size distribution for 5 samples of SRM1017b

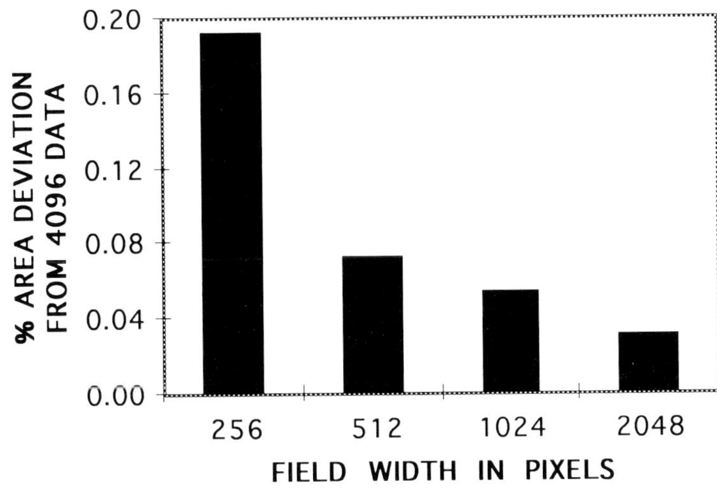

Figure 2. Deviation of measured areas for 12 particles analyzed at 5 pixel densities.

MODELING AND SIMULATION OF OCTAHEDRAL Pb INCLUSIONS IN Al

S.Q. Xiao[*], S. Paciornik[#], R. Kilaas[*], E . Johnson[&] and U. Dahmen[*]

* National Center for Electron Microscopy, University of California, LBL B72, Berkeley CA 94720
# DCMM, PUC-Rio, P.O BOX 38008, Rio de Janeiro, RJ 22452, Brazil
& Niels Bohr Institute, Ørsted Laboratory, Copenhagen, Denmark

Pb inclusions in Al have been extensively studied for their unusual melting/solidification behavior.[1,2] Pb inclusions have a cube on cube parallel orientation relationship with the Al matrix and assume cuboctahedral shapes faceted on {111} and {100}. Al and Pb are both fcc structures but with very different lattice parameters: $a_{Al} = 0.405$ nm, $a_{Pb} = 0.495$ nm. Thus 5 Al spacings match approximately 4 Pb spacings giving rise to a moiré pattern visible in HREM images.

High resolution electron micrographs in the <110> zone axis orientation were recorded on the Berkeley ARM at an accelerating voltage of 800 kV. In this orientation the cuboctahedra project as truncated parallelograms as shown in Fig. 1. Although the four (111) interfaces revealed in Fig. 1 are imaged edge-on, the Al lattice overlaps the Pb lattice above and below, because the other four (111) interfaces are inclined. Therefore, even though the $(111)_{Al}$ lattice is clearly resolved, the determination of inclusion size is not straightforward because the contrast depends on defocus (Δf), particle size (s), depth of the inclusion in the matrix (z) and total sample thickness (t). To study the influence of these parameters modeling and HREM imaging simulation of 3D octahedral Pb inclusions in Al have been performed. This is a first approximation to a cuboctahedron .

A variety of supercells was built, containing inclusions of 2,3,4 and 5 moiré spacings in size, each in a matrix with 5 total thicknesses from ~0.5 to ~1.5 Al extinction distances ($\xi_{Al} = 24$ nm at 800 kV)[3], and at three different depths for each size and each thickness. The largest supercells included ~70000 atoms. Each supercell was composed of a number of thin slabs with plane normal parallel to <110>. Using CrystalKit, each slab was built to contain an AB stacking of Al atoms (2.86Å thick), the smallest repeat distance along <110> in Al. While the Al lattice repeats in each slab, the Pb lattice does not, due to the lattice mismatch between the two lattices and this was considered in creating subsequent slabs. Finally, the supercell was built by stacking all the slabs together thus creating 4 edge-on and 4 inclined interfaces. Atomic relaxation at the interfaces was not considered in these models. In order to test the visibility of steps at the interfaces, supercells containing an atomic ledge in one of the {111} interfaces were built. Using NCEMSS running on a DEC Alpha station, image simulations were carried out for 9 defoci ranging from -30 nm to -110 nm.

Of the more than six hundred images thus simulated, Fig. 2 shows a subset of images of an inclusion, four moiré spacings in size, at three different depths in a matrix of two different thicknesses , at two different defoci. The best visibility of the inclusions occurs when t ~ ξ_{Al} for which the visibility of the Al lattice is at a minimum. The octahedral shape is generally preserved but for certain conditions the inclusions can appear as cuboctahedra. The visibility is also z-dependent, with a stronger dependence for larger t. A method for quantitative analysis of the size and visibility of the inclusions is described in a separate contribution to these proceedings.[4,5]

References

1. K.L. Moore, D.L. Zhang and B.Cantor, *Acta metall. mater.* 38(1990)1327.
2. E. Johnson et al., *MRS Proc.* 157(1990)247.
3. M.A. O'Keefe, U. Dahmen , C.J.D. Hetherington, *MRS Proc.* 159(1989)453.
4. S. Paciornik et al., *MSA Proc.* (1995).
5. This work is supported by the Director, Office of Energy Research, Office of Basic Energy Sciences, Materials Sciences Division of the U.S. Department of Energy under Contract No. DE-ACO3-76SFOOO98.

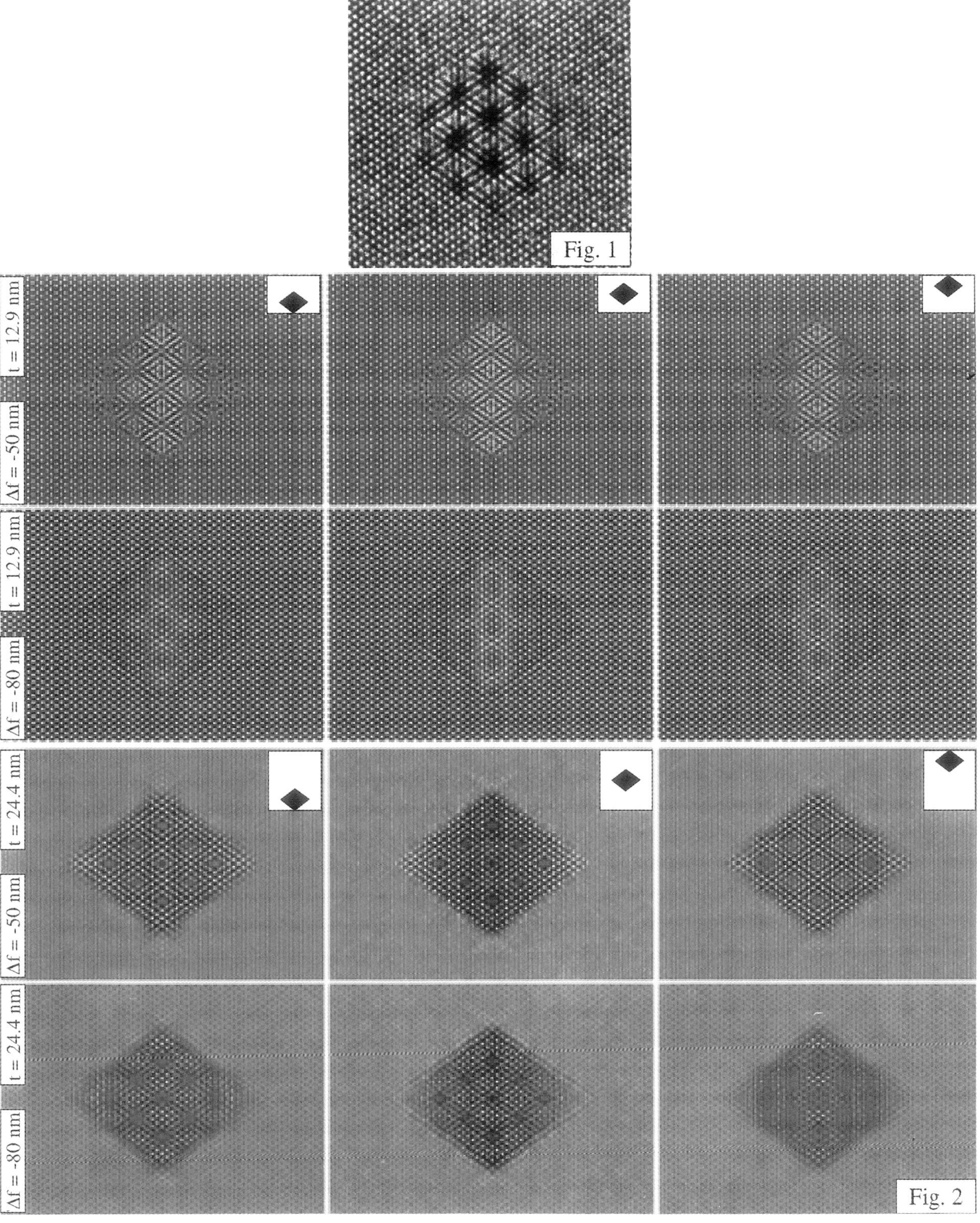

Figure 1 - HREM image of an octahedral Pb inclusion in Al viewed along a <110> at Δf~-60nm.
Figure 2 - Set of simulated images of an octahedral Pb inclusion in Al of two different thicknesses viewed along <110> at two different defoci and three different inclusion depths.

AUTOMATED MEASUREMENT OF SIZE AND VISIBILITY OF SMALL INCLUSIONS IN HREM IMAGES

S. Paciornik[#], S.Q. Xiao[*], S. Hinderberger[*], E. Johnson[&] and U. Dahmen[*]

\# DCMM, PUC-Rio, P.O BOX 38008, Rio de Janeiro, RJ 22452, Brazil
* National Center for Electron Microscopy, University of California, LBL B72, Berkeley CA 94720
& Niels Bohr Institute, Ørsted Laboratory, Copenhagen, Denmark

This work describes an automatic image analysis method for measuring the visibility and apparent size of inclusions in a matrix, imaged by HREM. The method is based on intensity changes between the matrix and inclusion regions of the image. It provides a quantitative way of describing changes that are generally interpreted only in a qualitative sense. It also allows for measurements in situations of very low visibility where visual estimates are nearly impossible. The method measures the mean squared intensity difference between lines of an image defined as

$$R = \text{mean}[(I_{nl}(x) - I_{nl}(x_0))^2]$$

$I_{nl}(x)$ is a line, selected by the user, which is scanned across the image. $I_{nl}(x_0)$ is the first line of the scan, taken in a region away from the inclusion. For each position of $I_{nl}(x)$, $I_{nl}(x_0)$ is subtracted pixel by pixel, and the average of all squared intensity differences, is obtained (R).

As the line is scanned across the image, R will oscillate with a period given by the lattice spacing along the scan direction. When the line enters the inclusion, the baseline value of R will change, indicating the position of the apparent interface between inclusion and matrix. When the line leaves the inclusion, a reverse change occurs indicating the other end of the inclusion. The visibility is given by the magnitude of the change while the size is given by the difference between the onset and offset positions of the change. The program allows the user to choose the length and inclination of $I_{nl}(x)$ and the direction of scan. The inclination should be chosen according to the particular shape of the inclusion. The scan direction defines the direction in which the inclusion dimension is measured (which needs not be, in general, perpendicular to $I_{nl}(x)$). Orientation dependence of the inclusion size can be measured by choosing different scan directions.

To allow the interpretation of measurements on experimental images, it is necessary to first apply the method to simulated images because the image contrast of an inclusion depends on various parameters such as inclusion depth in the matrix, total matrix thickness and defocus.[1] Image simulations of a Pb inclusion 16 $d_{111}{}^{Pb}$ = 20 $d_{111}{}^{Al}$ in size were carried out, as exemplified in Figures 1(a), (b) and (c). Size measurements were performed on this set of images using a line parallel to one of the {111}, scanned along the other {111} as indicated in Figure 1(a). The measurement traces are shown in Figures 1(d), (e) and (f). The oscillation period corresponds to the Al d_{111} spacing. The onset and offset of change in the baseline of the oscillation gives the apparent size of the inclusion, as shown by arrows in the figures. The measured apparent size is equal to the modeled size in these cases, even for very low inclusion visibility as in Figure 1(e) and (f). In certain conditions, the measured size differs from the modeled size by up to 2 d-spacings. Further measurements of simulated and experimental images will be presented.[2]

1   S.Q. Xiao et al., *MSA Proc.* (1995).
2.   This work is supported by the Director, Office of Energy Research, Office of Basic Energy Sciences, Materials Sciences Division of the U.S. Department of Energy under Contract No. DE-ACO3-76SFOOO98.

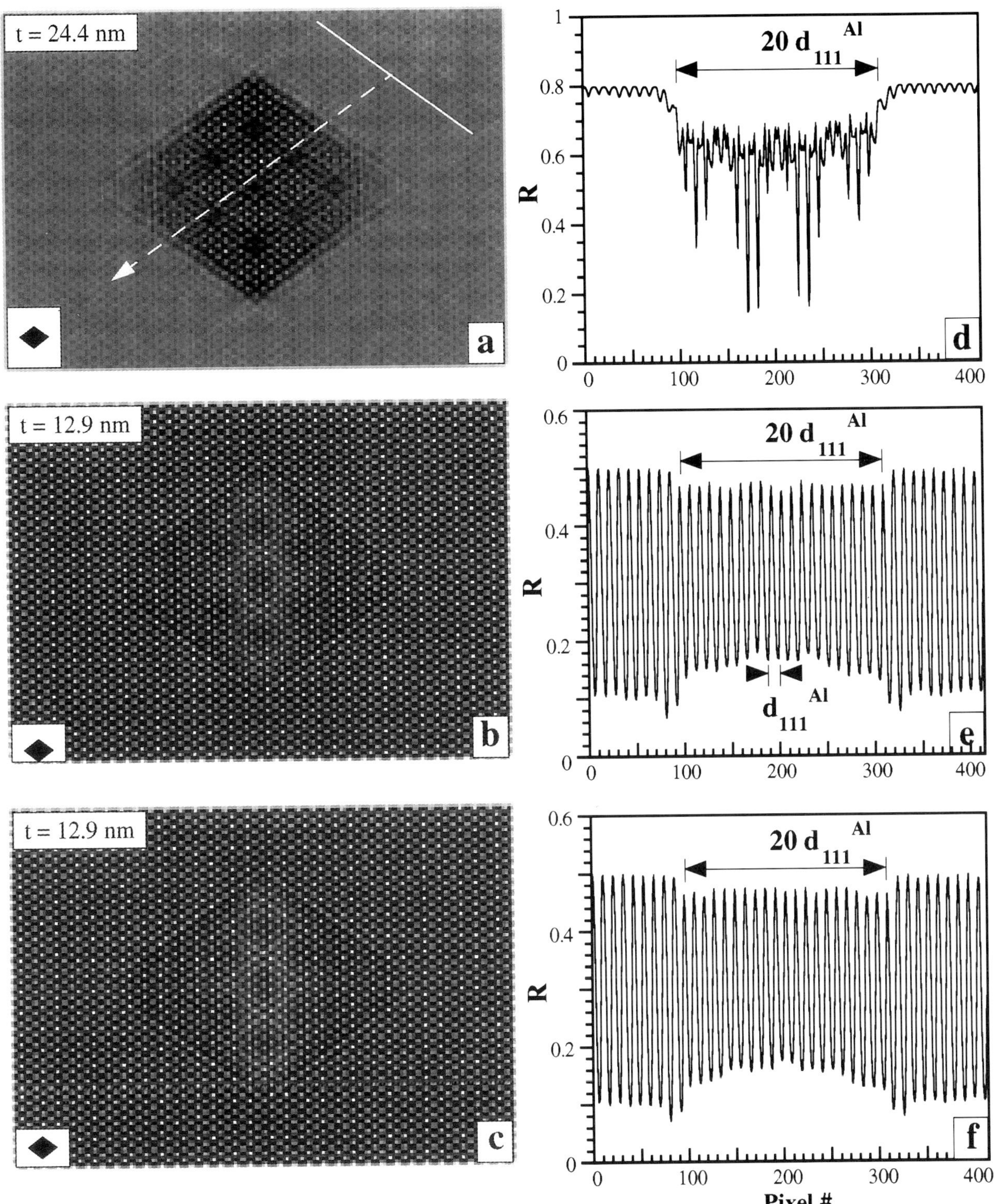

Figure 1 - (a), (b), (c) Simulated images of a Pb inclusion 20 d_{111}^{Al} in size in an Al matrix with two different thicknesses (t) and inclusion depths as shown in the insets, at a defocus of -80 nm. The measurement line and scan direction are indicated in Figure 1a. (d), (e), (f) Measurement traces corresponding to (a), (b) and (c) respectively. As shown in the figures the measured size is equal to the modeled size.

ALGORITHMS FOR AUTOMATED MONTAGE SYNTHESIS OF IMAGES FROM LASER-SCANNING CONFOCAL MICROSCOPES

D. E. Becker

Rensselaer Polytechnic Institute, Troy, New York 12180-3590

An efficient, robust, and widely-applicable technique is presented for computational synthesis of high-resolution, wide-area images of a specimen from a series of overlapping partial views. This technique can also be used to combine the results of various forms of image analysis, such as segmentation, automated cell counting, deblurring, and neuron tracing, to generate representations that are equivalent to processing the large wide-area image, rather than the individual partial views. This can be a first step towards quantitation of the higher-level tissue architecture. The computational approach overcomes mechanical limitations, such as hysterisis and backlash, of microscope stages. It also automates a procedure that is currently done manually. One application is the high-resolution visualization and/or quantitation of large batches of specimens that are much wider than the field of view of the microscope.[1]

The automated montage synthesis begins by computing a concise set of landmark points for each partial view. The type of landmarks used can vary greatly depending on the images of interest. In many cases, image analysis performed on each data set can provide useful landmarks. Even when no such "natural" landmarks are available, image processing can often provide useful landmarks.[2] The next step in the processing aligns partial views by searching for possible correspondences between landmarks in one set and landmarks in the other using a point matching algorithm.[3] This algorithm is robust enough to allow some missing and/or spurious landmarks. It hypothesizes correspondences between the landmarks, and evaluates the spatial transformations implied by them. Each set of hypothesized correspondences is evaluated by computing the transformation and examining the result of aligning the landmark points in one partial view with the points in the other. The optimal transformation is determined as the one that produces the largest number of coincident points. The point matching algorithm is known to be combinatorial, and computationally intensive. In this work, the computational requirements were greatly reduced by pruning the search using image overlap constraints, and by using pose clustering.[3] The two partial views are then merged, and duplicate landmarks are removed. As this occurs, the results of image analysis on each partial view are also combined. Further partial views are combined similarly, until a montage consisting of all the partial views merged together is obtained along with the combined image analysis results. This result is equivalent to analyzing a hypothetical single large image of the specimen.

Fig. 1 is a computationally synthesized montage of two confocal microscope images of overlapping regions of 50 μm thick section of a rat hippocampus stained with Acriflavin. A 3-D cell population analysis system was used to compute the centroids of the nuclei, to be used as landmarks.[4] To indicate the overlap, one image in the montage is displayed in shades of red, while the other is displayed in shades of green. The overlap region is shown in yellow. Fig. 2 illustrates the result of montaging the results of image analysis performed individually on the two images shown in Fig. 1. Each colored spot in Fig. 2 is the 2-D projection of a nucleus identified by an automated segmentation and cell counting system.[4] Not shown here is the fact that tabular results generated by the image analysis over partial views can also be merged once the point correspondences are known. Fig. 3 is a montage of seven 2-D projections of confocal microscope images similar to those in Fig. 1. The knife scratch in the third image was deliberately placed as an identifying mark for visual confirmation.[5]

References
1. D.E. Becker et al., *Proc. Ann. MSA Meeting*, 52(1994)224.
2. W. E. L. Grimson, *Object Recognition by Computer*, Cambridge:MIT Press (1990).
3. Stockman, G. and Esteva, J.C., *Pattern Recognition Letters*, 3:279-286, 1985.
4. Roysam et al., *J. Microscopy,* Vol. 173, Pt. 2, February 1994.
5. Acknowledgments: H. Ancin, B. Roysam, J. Turner, D. H. Szarowski, NSF, Procter & Gamble

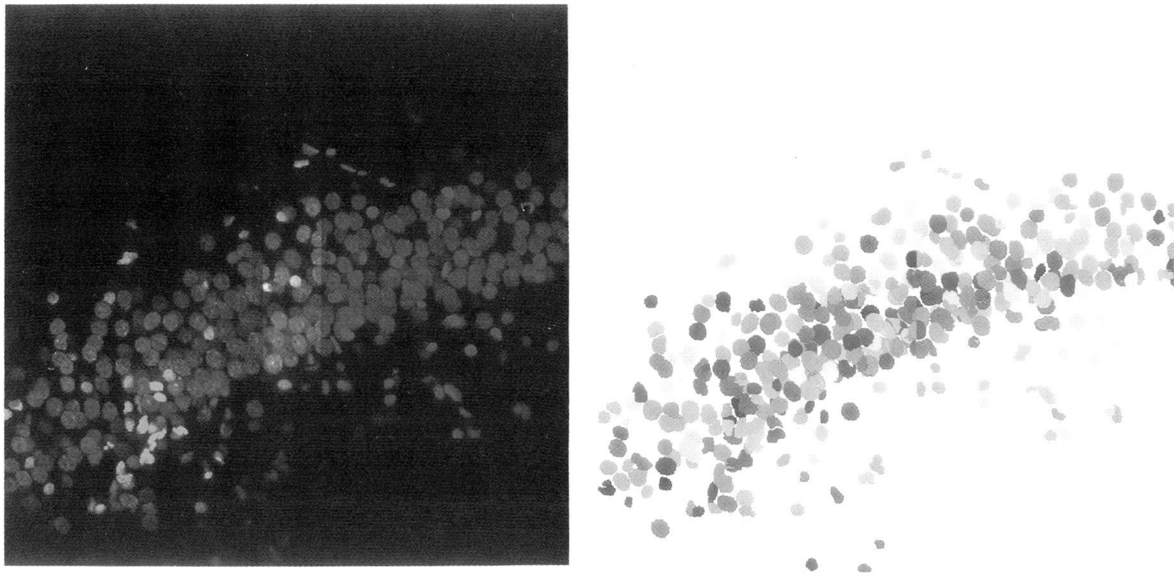

Figure 1   Figure 2

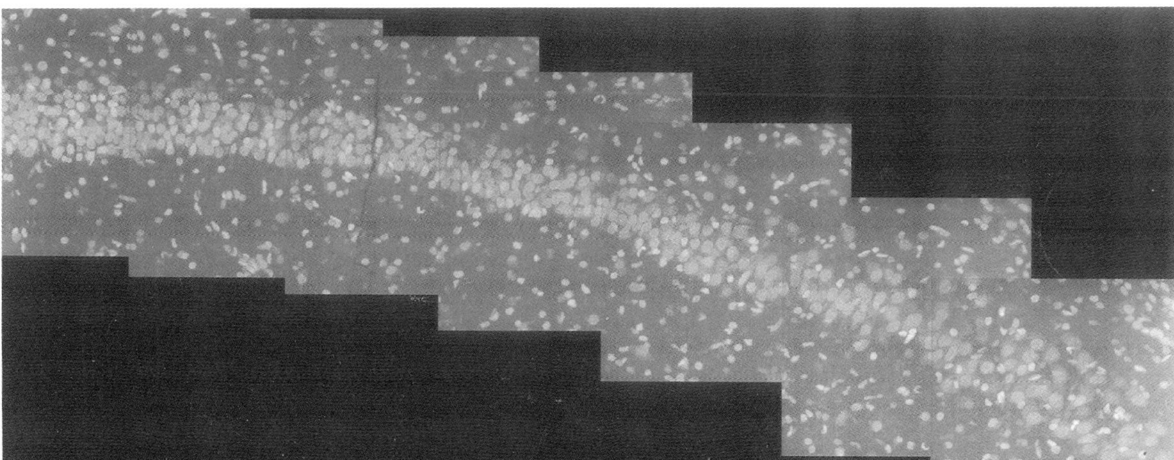

Figure 3

Fig. 1: Montage of two confocal microscope images of overlapping regions of 50 μm thick section of rat hippocampus stained with Acriflavin. One image displayed in shades of red, the other in shades of green. Overlap region is shown in yellow.

Fig. 2: Result of montaging the results of image analysis performed individually on the two images shown in Fig. 1. Each colored spot is the 2-D projection of a nucleus identified by an automated segmentation and cell counting system.

Fig. 3: Montage of seven 2-D projections of confocal microscope images of a 50 μm thick section of a rat hippocampus stained with Acriflavin. The knife scratch in the third image was deliberately placed as an identifying mark for visual confirmation.

ESTIMATION AND ANALYSIS OF 3D CHROMOSOME MOTION

W.F. Marshall, D.A. Agard, and J.W. Sedat

Dept. Biochemistry and Biophysics, University of California, San Francisco, CA 94143

The mechanical properties of chromosomes in vivo are at present poorly characterized, yet an understanding of such properties is likely to be important for understanding both the behavior of chromosomes during mitosis[1] as well as the establishment and maintenance of specific chromosome configurations in the nucleus[2]. In order to better investigate the mechanics of chromosomes inside living cells we have developed algorithms for quantitative analysis of chromosome motion. This allows us to directly determine chromosome mobility, and to visualize the trajectories of specific regions, such as telomeres or centromeres. Previously developed tools for analysis of 3D structure can then be extended to 4 dimensions, allowing quantitative description of chromosome behavior. The motion vector field itself can also be analyzed for specific patterns of motion.

Three dimensional motion estimation is difficult in fluorescence microscopic images, due to limited spatial resolution, particularly in the axial direction, which result in a significant aperture problem because each part of a chromosome looks identical to any other part of any other chromosome, such that intensity patterns are not useful in motion estimation. To circumvent these difficulties, we have developed a model-based approach[3,4] in which the chromosomes are represented by a graph, and motion estimation is reformulated as graph-matching. To carry out the matching, we locate all chromosome arms as simple paths within the graph, and then find the best match between arms at time t and arms at t+1 such that the sum of the mean-squared displacements between matching arms is minimized. This minimization is accomplished using simulated annealing. This method works well on actual images. Figure 1 depicts the graph models of chromosomes from two successive time-points in a time-lapse three-dimensional dataset. The chromosome images were obtained by injecting Cy-5 labeled histones into Drosophila embryos, and collecting a series of three-dimensional datasets using a wide-field optical sectioning microscope. The chromosomes were then modeled using an interactive modeling program to generate the graph structure. Fig 2 depicts the result of the motion estimation algorithm. Only motion vectors for a single chromosome are displayed. When all such vectors were compared with the original images, correct performance of the algorithm was verified.

Previously, several scalar measures of local 3D chromosome conformation were developed[5]. By computing the rms change in each of these measures for a single point on a chromosome over its trajectory, we can determine which aspects of chromosome organization are maintained, and thus deduce chromosome mechanical properties. The rms change in position yields mobility (Fig 3a). Flexibility is estimated by finding the rms change in curvature or unpacking ratio (Fig 3b). The rms change in intrachromosomal distance (Fig 3d) can be used to detect interactions between two points on the same chromosome. Finally, it is possible to directly analyze the pattern of motion. In order to detect possible sliding motion of telomeres on the inner surface of the nuclear envelope, we decompose velocities into components normal and tangential to the surface of the nucleus. The fraction of radial motion is defined as the rms normal component divided by the sum of the rms normal and rms tangential components, and is expected to be low for points constrained to slide on a spherical shell. Results of this computation for all points on one chromosome are shown in figure 3c.

1. R.V. Skibbens, V.P. Skeen, and E.D. Salmon, *J. Cell Biol.*, 122(1993)859-875.
2. D.E. Comings, *Human Genetics*, 53(1980)131-143.
3. W.F. Marshall, D.A. Agard, and J.W. Sedat, *SPIE proc.*, 2184(1994)149-158.
4. W.F. Marshall, D.A. Agard, and J.W. Sedat, *SPIE proc.*, 2412(1995) in press.
5. D. Mathog, *J. Microscopy*, 137(1985)253-273.

Proc. Microscopy and Microanalysis 1995, edited by G.W. Bailey, M.H. Ellisman, R.A. Hennigar, and N.J. Zaluzec

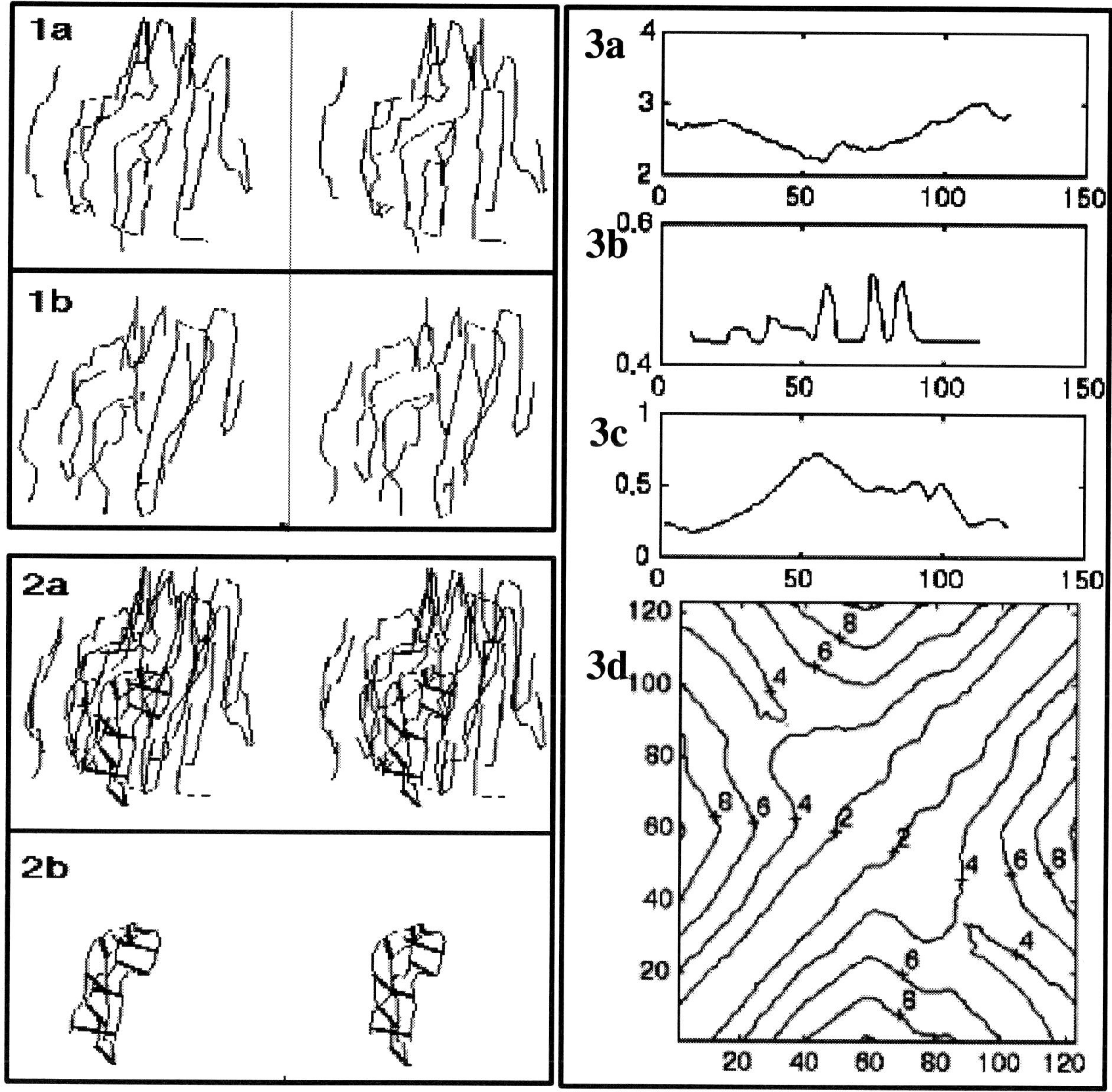

FIG 1. Wireframe model of chromosomes at two timepoints. (a) stereo pair of chromosomes at first timepoint; (b) stereo pair of next time point 25 sec later.

FIG 2. Result of motion estimation. (a) displacement vectors for selected points (heavy lines) superimposed on wireframe models of two successive timepoints (light lines). (b) displacement vectors and wireframe model of single chromosome.

FIG 3. Chromosome mechanical properties derived from motion estimate. All quantitites plotted versus position on a selected chromosome in arbitrary units of arc length. (a) rms displacement; (b) rms change in unpacking ratio; (c) fraction of radial motion; (d) rms change in intrachromosomal distance. See text for definitions.

653

Development of a Multi-User Teaching SEM

L. S. Chumbley,*[†] M. K. Meyer,* K. Fredrickson,** and F. C. Laabs[†]

*Materials Science Department, Iowa State University, 3053 Gilman, Ames, IA 50011
**Now at MTS Systems, 14000 Technology Drive, Eden Prairie, MN
[†]Ames Laboratory, 214 Wilhelm, Ames, IA 50011

The goal of this project is to develop a next-generation laboratory for the teaching of scanning electron microscopy (SEM). Although SEM is one of the most heavily used analytical techniques in the materials field, it has been extremely difficult to teach because of the single-user nature of SEM instruments up to now. Hence, 20 students in a 3-hour weekly SEM laboratory with one machine can hope to get only 10 minutes of hands-on experience per week. To overcome this problem, we have developed a teaching laboratory that makes use of a local area network to provide a large number of students with direct, interactive access to the microscope. Figure 1 is a photograph of the laboratory in operation. This laboratory was field tested Fall semester, 1994 in three separate classes and the students appear enthusiastic about the new method.

The laboratory is based on the JEOL 6100 SEM owned by the Materials Science and Engineering Department. While dated, the microscope does allow a limited number of operational parameters to be controlled through means of an RS-232 serial port by a computer. The computer platform chosen, an Apple Macintosh Quadra with a 500 MB hard drive, 40 MB RAM, and a frame grabber board, was selected due to its ease of use, student familiarity, and the availability of the shareware program "Image", developed by researchers at NIST, for image analysis. This computer controls the EDS spectrometer attached to the microscope using equipment purchased from Dapple, Inc. An interface program named "Scope" has been written that allows window-based control of the operating parameters made available by JEOL, e.g. magnification, focus, probe current (Fig. 2). The Quadra is connected via ethernet to a series of computer-based remote workstations for student use, as well as to the internet, allowing essentially world-wide access. Each remote work-station consists of a TV monitor carrying the digital image from the SEM, a joystick that allows the student to move the sample within the SEM, a thermal printer for obtaining images, and a Macintosh computer to control the microscope (Fig. 3). The TV image is accessed through use of an RS-170 port provided by JEOL. The network program "Timbuktu"® is used to allow students at each station to control the server, thus gaining access to all controls, functions, and features of SEM and energy dispersive x-ray spectroscopy (EDS) available to the server. The individual stations also have frame grabber boards, allowing images to be acquired and directly inserted into documents. Finally, a large display imaging system is available for the instructor's use that can display either the microscope image or the computer screen. This allows the instructor to point out relevant features or make comments to the class without having to visit each individual station.

The great advantage of the system is that it has been designed to be flexible enough to allow a multitude of extensions. Images from the microscope can be acquired by the server and viewed from any computer on campus possessing the necessary software. Present plans include linking the city high school vie internet to the SEM laboratory, allowing students in classes there to see images and change operating conditions. The concept of a multi-user SEM with remote access should have applications, not just in education, but in research and industrial settings as well.[1]

References

1. The authors gratefully acknowledge the assistance of Chad Miller and Jeff Jensen in setting up the computer network. Funding for this project is provided by the National Science Foundation as part of their Instrumentation and Laboratory Improvement Program. The authors wish to thank J.D. Verhoeven, S.W. Martin, and M.A. Akinc for their assistance in preparation of the original proposal. The microscope was purchased on a grant from the Office of Naval Research, M.A. Akinc being the principal investigator.

Fig. 1

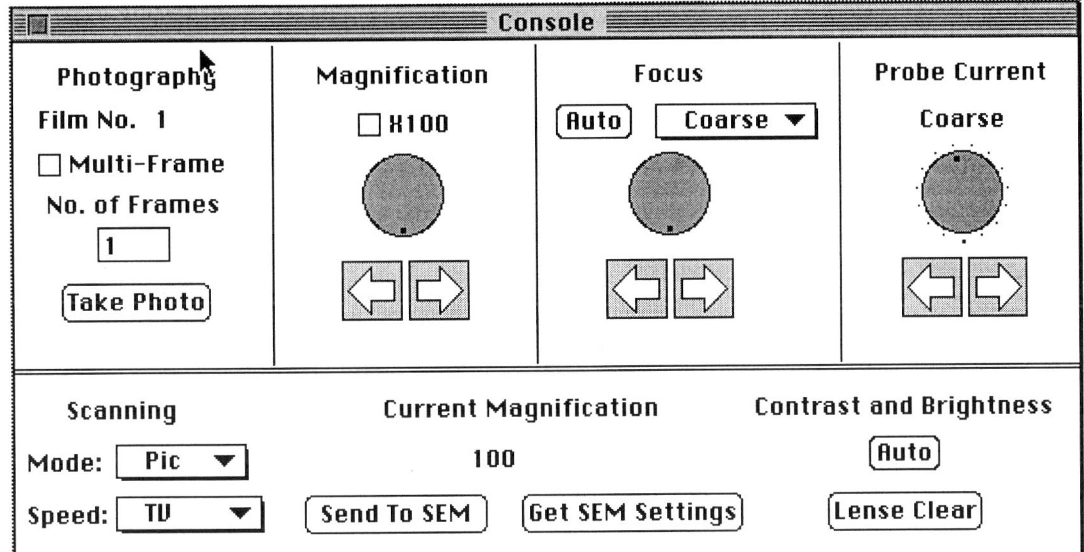

Fig. 2

Fig. 3

FIG. 1.-- Photograph of students in the multi-user SEM teaching laboratory.
FIG. 2.-- Main screen from program "Scope". Microscope parameters are changed by simply clicking on the appropriate arrows.
FIG. 3.-- Photograph showing a remote teaching station.

EXPERIMENTAL STUDY ON THE MEASUREMENT OF TEM PARAMETERS WITH THE PHASE SPECTRUM

Norihiko Ichise,* Syunya Watanabe,* Mitsuo Ogasawara** and Norio Baba*

*Department of Electrical Engineering, Kogakuin University, 1-24-2 Nishishinjuku, Shinjuku-ku, Tokyo 163-91, Japan.
**Instrument Division, Hitachi Ltd., 882, Ichige, Nakaminato, Ibaraki 313, Japan

For HRTEM works, it seems to be indispensable to measure and control TEM parameters : defocus, astigmatism and beam tilt misalignment. The autotuning method with phase spectra was proposed by Koster[1], which has theoretical advantages to the previous methods as described in his paper (e.g., the method is specimen independent). However, it seems that experimental study using the phase spectrum itself is insufficient. In our study, it was found that to make the method practical use some additional modifications are necessary.

One among modifications was the correction of the image drift during recording tilted images. The image drift is a serious problem for the method because it is based on measuring the beam tilt induced image displacement[2]. The phase spectrum is so sensitive as not to neglect one pixel drift. The image drift also disturbs recoding images in on-line. If the drift is considered to be parallel movement, the phase spectrum, $P(\mathbf{k})$, for image drift $\mathbf{d}$ is simply given by

$$P(\mathbf{k}) = \pi \; \mathbf{d} \cdot \mathbf{k} \; / \; k_{max} \qquad (1)$$

where, vector $\mathbf{k}$ and k_{max} represent the spatial frequency and its maximum, respectively. Because the phase is numerically calculated within only the range of 2π, discontinuous lines appear and the pattern is seen like a band (see Fig.1) that is useful to visually estimate the magnitude and direction of the drift. Therefore, for a case of the simple image drift some compensation algorithms may be devised because a simple criterion is found, i.e. when the alignment is completed the phase differences are totally zero. For the alignment between tilted images, other criteria are employed such as some symmetricalness of the phase distribution or a minimum of the total phase change (because the image drift causes an additional phase change) in which, of course, some restriction conditions are necessary. As a result of these corrections of the image drift the TEM parameters could be accurately measured(Fig.2, 3).

In the original work[1], the strategy to estimate the TEM parameters was to compare an experimental phase spectrum with an approximated mathematical model based on the DSB (double side band) imaging[4]. In our study, to rise the sensitivity of the method to the TEM parameters, the tilting angle k_0 is increased and the rigorous mathematical model composed of the DSB and the SSB(single side band) imaging was used. Unfortunately, at present this modification requires much time for the estimation, but there may be some feasibilies to be solved by a proper numerical calculation technique and advance of computer power.

Typical examples of the TEM parameter measurement with the phase spectrum are demonstrated for an amorphous specimen (Fig.2) and a crystalline specimen (Fig.3). (The images are recorded using a Gatan 694 SSC camera mounted on a Hitachi H-8100.) For the former case the estimated defocus value was comparable to that by means of the diffractogram-based method[5]. The diffractogram-based method was restricted within a region of defocus, but the present method worked at the current focus position. In the latter case the diffractogram-based method was unsuitable.

References

1. A.J.Koster et al., *Ultramicroscopy* **27** (1989) 251.
2. A.F. de Jong, Proc. 50th Ann MSA Meeting (1992) 136
3. N.Ichise et al., *Microscopy Research and Technique* **30** (1995) 350
4. A.J.Koster and W.J.de Ruijter, *Ultramicroscopy* **40** (1992) 89
5. O.L.Krivanek and G.Y.Fan, *Scanning Microscopy* **Suppl. 6** (1992) 105

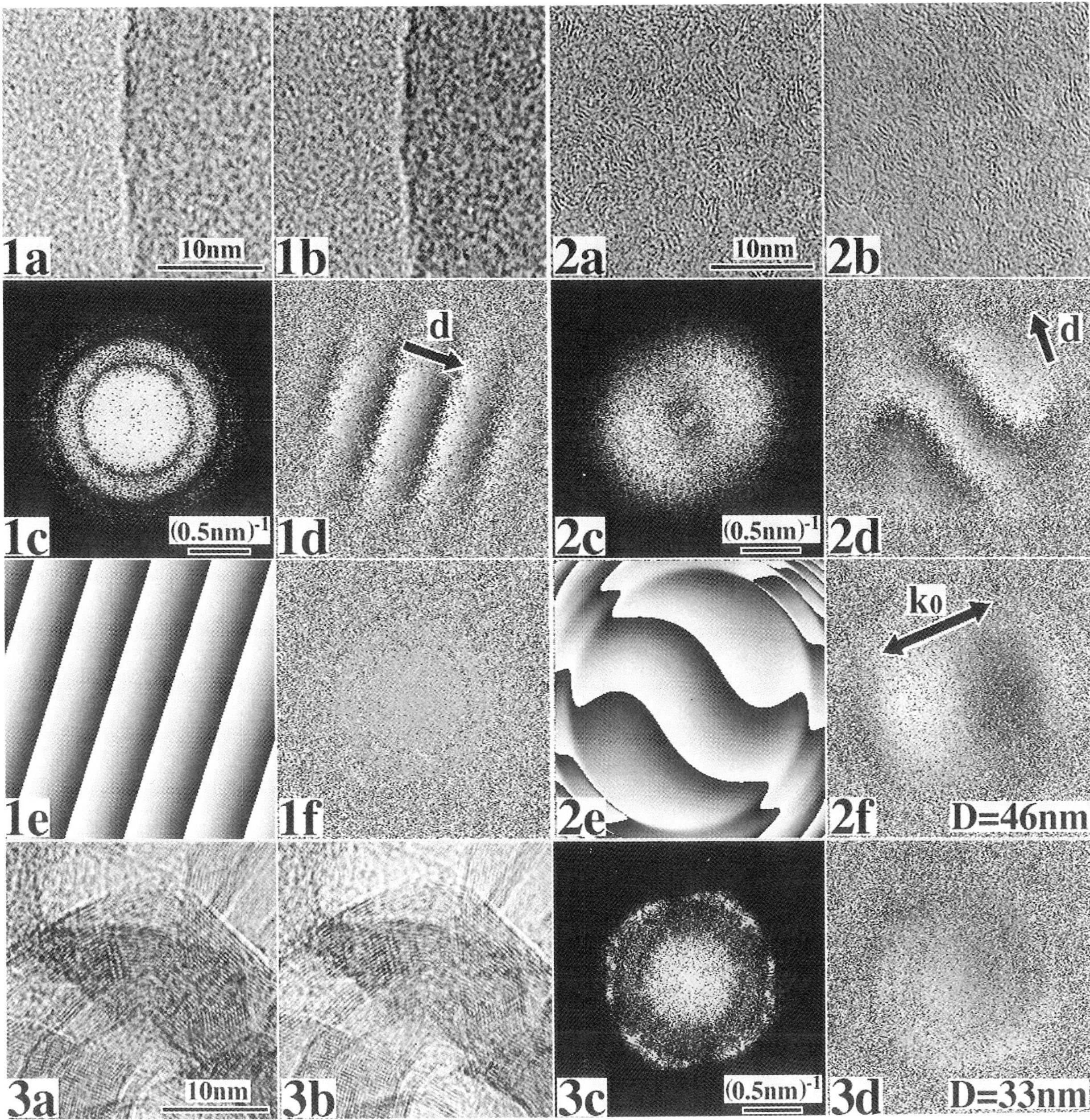

Fig.1 - Correction of image drift with the phase spectrum; (a) and (b) are an image series, (c) and (d) modulus and phase part of the cross spectrum between (a) and (b), respectively, (e) a fitting result of the phase spectrum (d) to equation (1), and (f) the phase spectrum after the alignment.

Fig.2 - A measurement of defocus with the phase spectrum; (a) and (b) are tilted beam images of amorphous carbon thin film (2.3mrad tilting oppositely each other), (c) and (d) the cross spectrum between (a) and (b) without alignment, (e) a demonstration by a simulation to explain the phase spectrum pattern (d) disturbed by image drift, and (f) a phase spectrum between aligned images which was used to fit with the mathematical model.

Fig.3 - A measurement with the phase spectrum for a crystalline specimen; (a) and (b) are tilted beam images, (c) and (d) modulus and phase part of the cross spectrum after the image alignment, and the measured defocus value is indicated in the figure.

THE DEVELOPMENT OF AN ELECTRONIC MICROSCOPY JOURNAL ON THE WORLD WIDE WEB.

John F. Mansfield

University of Michigan Electron Microbeam Analysis Laboratory
413 Space Research
2455 Hayward
Ann Arbor
MI 48109-2143
http://www.engin.umich.edu/~jfmjfm/jfmjfm.html

Developments in digital camera technology and digital data acquisition have allowed many of us to record most, if not all of our data entirely electronically. We are able to present this data at conferences and meetings direct from the desktop of a small personal computer. Why should we not think of publishing the data in the same manner? Indeed the assembly of the traditional scientific manuscript; printing out the text according to the journals strict format and printing and annotating the micrographs may really be thought of as a backward step. Frequently images, spectra and diffraction patterns appear much better on the screen of the computer than when they are printed out on paper. While this may be simply a limitation of developments in the printing technology lagging behind those in the display technology, why wait for the printing technology to catch up? We should publish electronically.

Since it's inception in 1989, The World Wide Web (typically shortened to the WWW or just The Web) has proved to be a tremendously popular system for the publication and dissemination of an incredibly broad spectrum of information[1]. Initially started by Tim Berners-Lee, when working at CERN, the European Particle Physics Laboratory, to share physics data. The Web now contains everything from electronic art museums to features on wrestling[2]. The secret to the it's success is the elegant, yet relatively simple, way that presentations can be prepared with text, pictures, sounds and video clips. All data transmission on The Web is sent via Hypertext Transport Protocol (HTTP) and the documents displayed are formatted in HyperText Markup Language (HTML), which is a subset of Standard Generalized Markup Language (SGML)[3,4]. The Web is constructed on a client/server model. The servers are the computers on the Internet that make available hypertext documents and the clients are those machines running browser software to view documents.

There are a number of issues to be addressed before it is possible to publish on the Web as effectively as publishing in the traditional manner. These issues largely relate to authentication and commerce. There is, as yet, no way of safely and effectively performing commercial transactions on the Web and there is no common authentication scheme. Most journals have a subscription charge. If the journal is based on a Web Server, the server software has no way to determine if the client attempting to connect has paid for a subscription. In addition, library subscriptions to journals are often many times the price of individual subscriptions. However, if a library has paid for a subscription, how does the server software determine that the user attempting to connect is connecting via a workstation in the subscribing library? Ideally anyone should be able to access any article within the journal provided they are willing to pay a small fee per page. This can be likened to paying to photocopy a traditional article so that you can peruse it later and add your own annotations or notes to it. This also raises the question of copyright, how is copyright defined for Web publications and how is it to be enforced? The authentication question also works in the other direction. How does the user know if the journal to which he/she is connecting is in fact an official publication sanctioned by the stated organization? Fortunately, although, at the present time, there are no definitive answers to these questions, they are the subject of a great deal of thought and discussion and it is likely that these issues are going to solved within the next year.

In anticipation of the solution of the aforementioned issues, a sample journal is under development by The Electron Microbeam Analysis Laboratory at The University of Michigan. The figure below shows a sample page from that experimental electronic journal. A preliminary issue may be viewed by connecting to the

Proc. Microscopy and Microanalysis 1995, edited by G.W. Bailey, M.H. Ellisman, R.A. Hennigar, and N.J. Zaluzec
Copyright © 1995 MSA. Published by Jones and Begell Publishing, 79 Madison Ave., New York, NY 10016

following URL: http://www.engin.umich.edu/~jfmjfm/journal/.   The use of the Netscape web browser is recommended. All members of the microscopy community are encouraged to examine this issue and send the editor comments via the electronic form that is included in the journal.

References
1. Tim Beners-Lee "http://info.cern.ch/hypertext/WWW/TheProject.html".
2. Tim Beners-Lee "http://info.cern.ch/hypertext/WWW/People.html#BernersLee"
3. Tim Beners-Lee "http://info.cern.ch/hypertext/WWW/Protocols/Overview.html"
4. Tim Beners-Lee "http://info.cern.ch/hypertext/WWW/MarkUp/MarkUp.html"

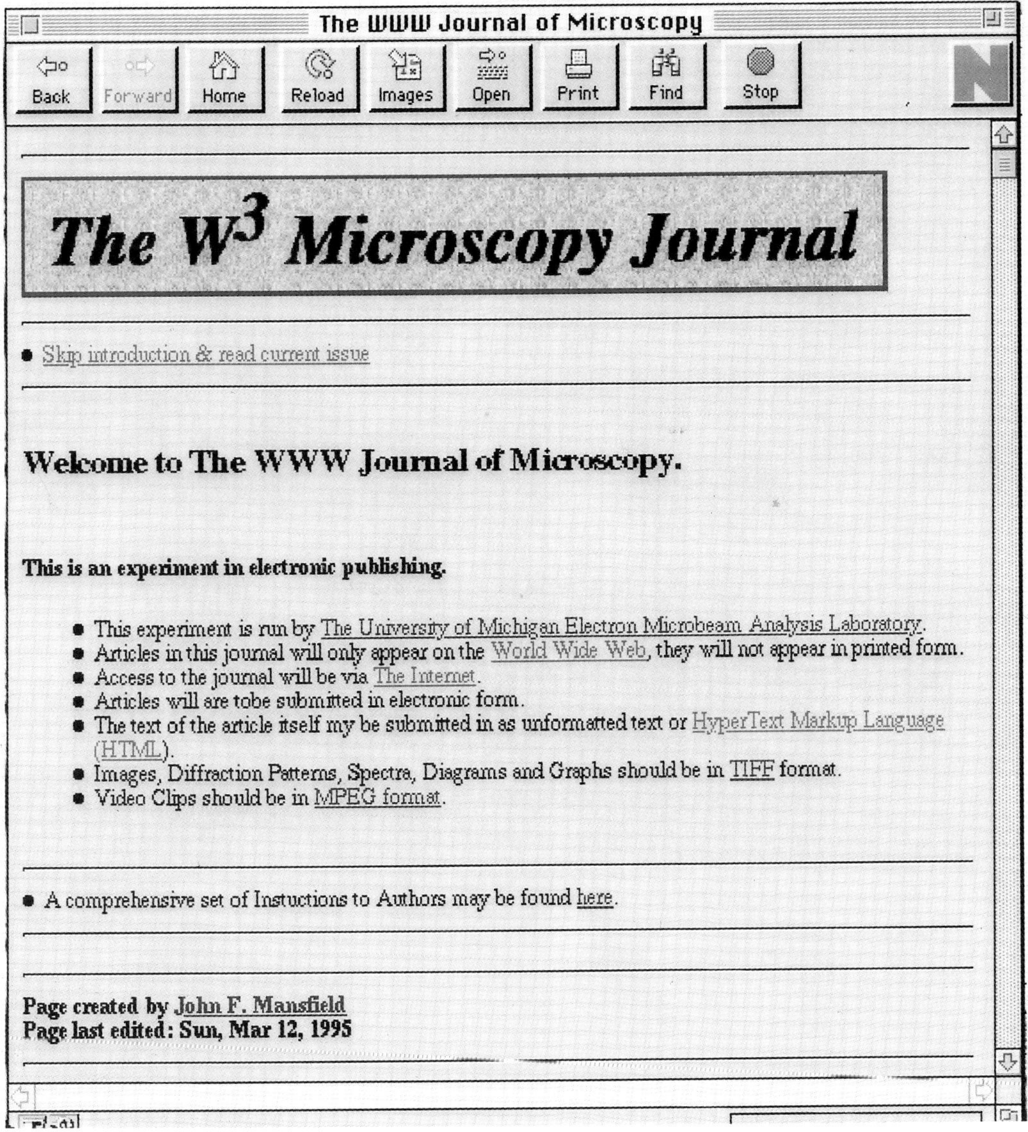

Figure 1. Example World Wide Web Microscopy Journal page. This is simply an introductory page outlining how articles are to be presented. Underlined phrases are links to other documents or other parts of the current document.

A SECURITY CIRCUIT FOR THE IMAGE INTENSIFIER ON THE ELECTRON MICROSCOPE [1]

Min-Yi SHIH and Tung HSU [2]

Material Science Center, National Tsing-hua University, Hsin-chu 30043, TAIWAN, R.O.C.

Image intensifiers are widely used on the electron microscopes, mostly for observation and recording of weak images. Usually, the image intensifier is not used in diffraction mode, for the strong direct beam may damage the YAG crystal. However, neither the microscope nor the image intensifier is equipped with any protection mechanism, leaving the safety of the intensifier as the responsibility of the operator.

While modifying a JEOL JEM-100C microscope for UHV-REM, we designed a security circuit to blank the electron beam in order not to damage the image intensifier accidentally. The required logic, shown in Fig. 1, indicates the electron beam blanked (1) or unblanked (0) under all possible combinations of three parameters. The requirement is that the beam must be blanked (1) while the view screen is fully raised (1), film not advanced (0), and the SAD button is depressed (1). Blanking is accomplished with a 5V signal applied to a relay and short the deflection coil [2]. The circuit is shown in Fig. 2.

This security circuit does not alter the normal operation of the microscope. Viewing the SAD pattern on the screen or recording it on the film is not affected. The circuit only makes it impossible to accidentally shoot the direct beam in SAD mode down through the camera without advanced film, which serves no useful purposes but may damage the image intensifier.

A switch is added (not shown in Fig. 2.) to enable blanking under any condition. Consequently, the circuit serves as an extra shutter for recording. Beam sensitive specimens can be translated into the position with the beam blanked. Then unblank the beam for a short period of time, recording the image before any beam induced effects become noticeable [3]. This circuit is also applicable to other models of electron microscopes, as long as the corresponding leads are located.

REFERENCE

1. Supported by National Science Council 84-2112-M-007-047, R.O.C. Use of the JEOL JEM-100C microscope is provided by Dr. Jian-yih Wang of Chung Shan Institute of Science and Technology.
2. Tung Hsu, in Proceedings of the 52nd Annual Meetings of Microscopy Society of America, 1994, New Orleans, pp. 808-809.
3. Private communication, F. S. Liao, San Kwang Instruments Co., Ltd., Taipei, Taiwan.

Proc. Microscopy and Microanalysis 1995, edited by G.W. Bailey, M.H. Ellisman, R.A. Hennigar, and N.J. Zaluzec
Copyright © 1995 MSA. Published by Jones and Begell Publishing, 79 Madison Ave., New York, NY 10016

SAD (D)	Film advance (F)	Screen (S)	Beam blanking (D*F*S)
0	0	0	0
0	0	1	0
0	1	0	0
0	1	1	0
1	0	0	0
1	0	1	1
1	1	0	0
1	1	1	0

Fig. 1.   The required logic indicates that the electron beam is blanked (1) while the SAD button is pressed (1), the view screen is fully raised (1) and the film is not advanced (0). All signals are +5V for (1) and 0V for (0).

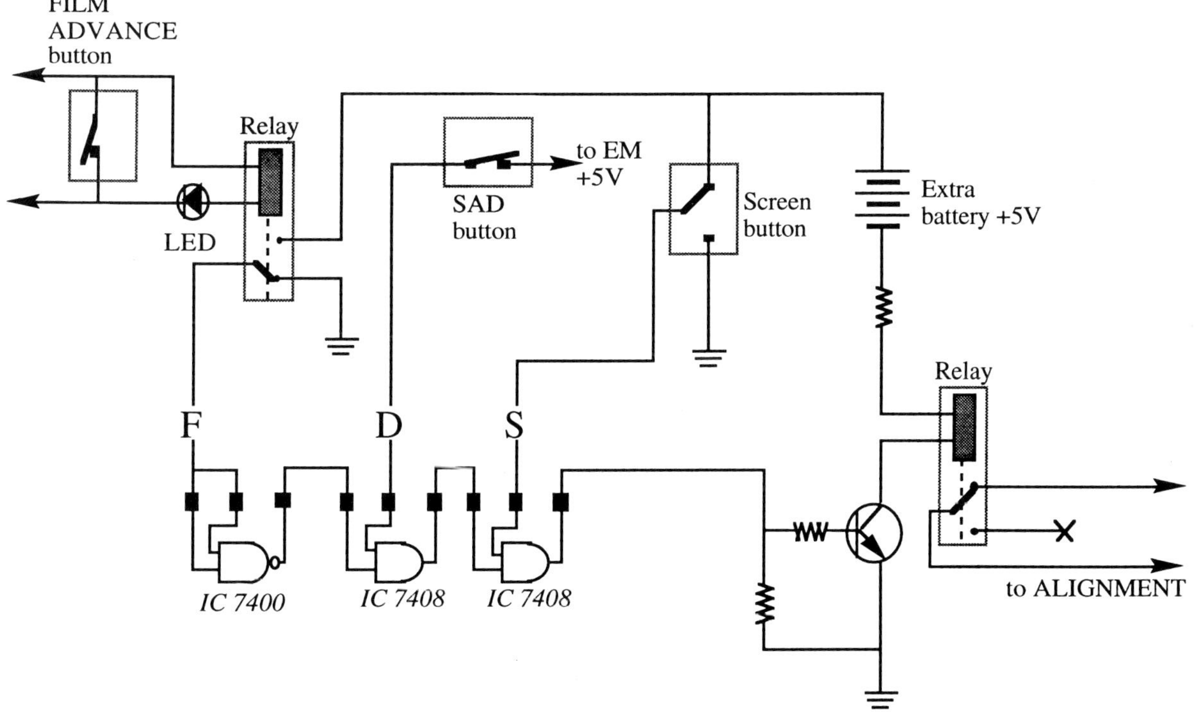

Fig. 2. The layout of security circuit.

EVALUATION METHOD FOR IMAGING PLATE RESOLUTION BY MEANS OF PHASE CONTRAST TRANSFER FUNCTION

T. Oikawa,* H. Kosugi,** F. Hosokawa* D. Shindo** and M. Kersker***

*JEOL Ltd., 1–2 Musashino 3–Chome, Akishima, Tokyo 196, Japan
**Institute for Advanced Materials Processing, Tohoku University, Katahira, Sendai 980–77, Japan
***JEOL U.S.A. Inc., 11 Dearborn Road, Peabody, MA 01960

Evaluation of the resolution of the Imaging Plate (IP) has been attempted by some methods.[1-3] An evaluation method for IP resolution, which is not influenced by hard X-rays at higher accelerating voltages, was proposed previously by the present authors.[3] This method, however, requires truoblesome experimental preperations partly because specially synthesized hematite was used as a specimen, and partly because a special shape of the specimen was used as a standard image. In this paper, a convenient evaluation method which is not infuenced by the specimen shape and image direction, is newly proposed. In this method, phase contrast images of thin amorphous film are used.

Several diffraction rings are obtained by the Fourier transformation of a phase contrast image of thin amorhous film, taken at a large under focus. The rings show the spatial–frequency spectrum corresponding to the phase contrast transfer function (PCTF). The envelope function is obtained by connecting the peak intensities of the rings. The evelope function is offten used for evaluation of the instrument, because the function shows the performance of the electron microscope (EM). In this experiment, firstly, a phase contrast image was recorded on the IP at a sufficiently high magnification at which the information–limit was not limited by the IP resolution. The envelope function of the PCTF from the image was defined as a standard information–function. Secondly, another phase contrast image was recorded at low magnification at which the information–limit was limited by the IP resolution. Here, the observation conditions were kept the same as in the high–magnification observation, except for the magnification change. By comparing the envelope function obtained from this image and the standard function obtained from the high–magnification image, the "modulation transfer function (MTF)" of the IP used was obtained.

In the experiment, both 50 μm-pixel IP (PIXsysTEM) and a newly-developed 25 μm-pixel IP[4] (PIXsysTEM II) were used. The accelerating voltage was varied from 100 kV to 1250 kV.

Figure 1 shows a comparison of spatial–frequency spectra obtained from the images recorded on the 25 μm-pixel IP. The spectrum obtained from the high–maginification image (a) is not limited by the IP resolution. However, in the spectrum from the low–magnification image (b), high spatial–frequency is limited by the IP resolution.
Figure 2 (a) and (b) show the intensity distributions of the power spectra and envelope functions corresponding to Fig. 1 (a) and (b), respectively.

1. N. Mori et al., J. Electron Microsc. 39 (1990) 433.
2. S. Isoda et al., Ultramicroscopy 41 (1992) 99.
3. T. Oikawa, D. Shindo and K. Fukushima, Proc. 13th ICEM (Paris) Vol. 1 (1994) 221.
4. N. Ogura et al., Proc. 13th ICEM (Paris) Vol. 1 (1994) 219.

 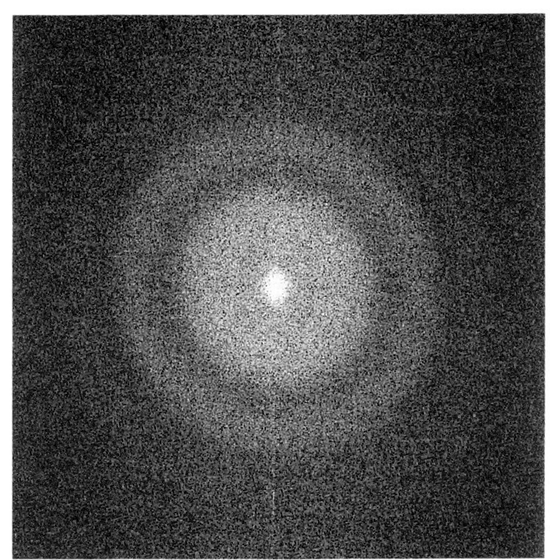

(a) Magnification: × 800,000 (b) Magnification: × 250,000

Fig.1 Comparison of spatial-frequency spectra obtaied from images recorded at different maginification. Specimen, thin amorphous Ge film; accelerating voltage, 200 kV; defocus, + 120 nm. (images were taken with JEM-2010F/PIXsysTEM II)

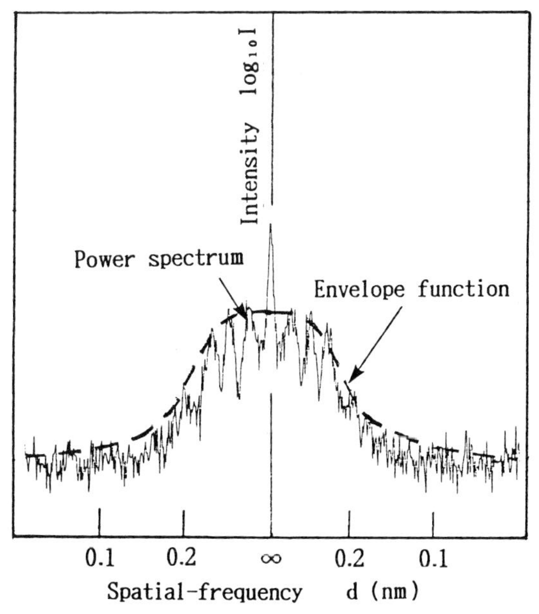

(a) Magnification: × 800,000 (b) Magnification: × 250,000

Fig.2 Intesity distributions of power spectra and envelope functions corresponding to Fig. 1.

EXPLORING LOCAL PROPERTIES OF BIOLOGICAL SURFACES WITH THE ATOMIC FORCE MICROSCOPE

Jan H. Hoh

Department of Physiology, School of Medicine, Johns Hopkins University, 725 N. Wolfe Street, Baltimore, MD 21205

The atomic force microscope (AFM) is developing into a very general tool for exploring biological surfaces. At present there are four broad areas of applications for the AFM: imaging surfaces, manipulating surfaces, measuring intermolecular forces and determining micromechanical properties. Because the dimensions of the apex of the tip used are typically on the order of 10's of nanometers, local measurements of properties are possible on a much smaller length scale than previously. Several reviews that provide further details on AFM instrumentation and biological applications have been published.[1,2,3,4,5,6,7,8] These reviews also provide specific references for the range of applications that will be reviewed in this tutorial.

AFM imaging modes can be divided into two groups based on the detection method, the DC method detects the absolute deflections of a cantilever while the AC method detects the RMS amplitude of an oscillating cantilever. Conventional contact AFM, sometimes called repulsive mode, is based on DC detection of the cantilever movement as the tip moves across a surface, usually in direct contact with the surface. The deflection of the cantilever is used to control the height of the sample, and the force applied to the sample, through a feedback loop. A topograph of a surface can then be generated by plotting the z position of the sample required to maintain a constant deflection of the cantilever (constant force) at each point in a field. In the AC imaging modes the cantilever is oscillated at or near its resonant frequency, typically 100-400 kHz. Interactions of the tip with the surface cause a change in the amplitude or phase of the resonance, which can be used to control the height of the sample and produce a topograph. The interactions that perturb the sample can either be attractive for non-contact mode, or repulsive for tapping (or intermittent contact) mode. The recent development of AC type imaging in liquids is similar to AC methods in air or vacuum, but the nature of the interaction between the tip and sample in liquids is more complex. Using these modes of operation biological structures such as cells, chromatin, lipid bilayers, ion channels, solution proteins, and nucleic acids have been imaged. The lateral resolution varies greatly, atomic resolution (<0.3 nm) can be obtained on a hard flat sample, but on softer samples, such as living cells, the lateral resolution is >10 nm. The z resolution is also sample dependent, but the instrumental limit is about 0.01 nm for conventional AFMs.

There are two features of the AFM that make it particularly useful for imaging biological samples. First, imaging can be performed in almost any liquid, from organic solvents to physiological buffers, and in a range of temperatures from below 100 °K to above 400 °K. It is therefore possible to image biological samples under highly physiological conditions. Second the signal to noise ratio of the AFM is extremely high, which provides for excellent image contrast. This allows the direct visualization of individual molecules without the use of signal averaging methods. Together, these features have allowed the direct visualization of processes and dynamics on a molecular level.

The AFM is also a tool for moving and manipulating material on a nanometer length scale. By using the tip in a purely mechanical fashion, several types of biological structures have been modified. For example, membranes can be manipulated by pushing the lipids to the side, or scraping material off the surface. Nucleic acids, ranging from whole chromosomes to single plasmids, can be cut with the tip. This is a novel method for generating specific sized fragments of nucleic acids since it appears to be independent of sequence. In addition, protein molecules can be positioned on a surface. As work on modifying tips proceeds, we are likely to soon see bio-active manipulation using proteases, lipases and other novel probes attached to the tip for surface modification.

Proc. Microscopy and Microanalysis 1995, edited by G.W. Bailey, M.H. Ellisman, R.A. Hennigar, and N.J. Zaluzec.

Measuring intermolecular forces with the AFM is another exciting and rapidly developing field. Although glass needles and other springs have been used to measure forces between molecules, the small tip-sample contact area and lateral scanning ability of the AFM give it unique capabilities. There are several ways in which forces between the AFM tip and a sample can be monitored. Holding the tip above the surface of a sample results in deflections of the cantilever corresponding to the force between the two surfaces. For small deflections the cantilever behaves as a simple spring and the force (F) on the tip can be calculated from Hooke's Law (F=k•Δd, where k is the spring constant for the cantilever and Δd is the deflection of the cantilever). This allows plots of force as a function of sample position (force curve), or as a function of separation distance between tip and sample, to be collected. Cantilevers with spring constants of 0.01 N/m are widely available, and with a detection limit for cantilever deflection of 0.01 nm, forces of 10^{-13} N can be measured (in DC). Although this is well below the thermal noise limit, measurements below kT can be achieved by time averaging the cantilever position. The AFM force curves are sensitive to van der Waals, hydration, electrostatic and other forces. By collecting 2D arrays of force curves a volume of force near a biological structure can be described. More specific interactions can be studied by attaching molecules of interest to the tip. For example, by attaching biotin to an AFM tip and collecting force curves relative to immobilized streptavidin, the rupture force, which is related to the depth of the potential energy well, between streptavidin and biotin has been determined. Forces between the tip and sample also result in changes in friction as the tip is scanned across a surface. The friction is typically detected by the torqueing of the cantilever. Using this approach chemically modified tips have been used to identify chemically distinct regions of surfaces, including Langmuir Blodgett films. However, because of difficulties in calibrating the frictional forces, and the geometry of the interaction, it is not yet possible to quantify intermolecular forces through friction measurements.

The tip of the AFM can also be used as a very small indentor to determine the local elastic properties of biological materials. There are two approaches that are being used for this application. In so called force modulation the sample is pulsed against the tip, typically at frequencies of many kHz and amplitudes of several nanometers. The amplitude and phase of the response of the cantilever are monitored and used to evaluate the local visco-elastic properties. In general the amplitude of the response is sensitive to elasticity while the changes in phase are sensitive to viscosity. However, both viscosity and elasticity can have frequency dependent components, so complete separation of these to parameters is difficult. It is also possible to produce indentation curves at specific points on a surface, by measuring cantilever deflections as a function of the depth of indentation. These curves are similar to the force curves described above. By attaching micron sized spheres to the tip, giving it a controlled geometry, it is possible to make reasonably quantitative measurements of local visco-elastic properties. Again, it is possible to collect 2D arrays of these curves, which will allow for high resolution quantitative visco-elastic mapping of biological surfaces, including living cells.

References

1. P.K. Hansma, et al., *Science* 242(1988)209.
2. A. Engel, *Ann. Rev. Biophys. Biophys. Chem.* 20(1991)79.
3. H.-J. Butt, ct al., *Ultramicrosc.* 46(1992)375.
4. J.H. Hoh and P.K. Hansma, *Trends Cell Biol.* 2(1992)208.
5. M. Radmacher, et al., *Science* 257(1992)1900.
6. J. Yang et al., *J. Microsc.* 171(1993)183.
7. H.G. Hansma and J.H. Hoh, *Ann. Rev. Biophys. Biomol. Struct.* 23(1994)115.
8. R. Lal and S.A. John, *Am. J. Physiol.* 166(1994)C1.

HISTOCHEMICAL METHODS FOR LOCALIZING AND CHARACTERIZING CELL AND TISSUE GLYCOCONJUGATES

B.A. Schulte

Medical University of South Carolina, Charleston, SC 29425

Glycoproteins, glycolipids and proteoglycans, collectively termed glycoconjugates, account for a large percentage of all tissue constituents.[1] The carbohydrate moieties associated with these macromolecules have long been known or suspected to play specific biologic roles such as protection of the peptide backbone from recognition by antibodies and proteases, regulation of proper folding, assembly and exocytosis of secretory glycoproteins and lubrication of mucosal surfaces.[1,2] Structural studies of complex carbohydrates have gained momentum over the past two decades with the realization that they participate in a much wider range of dynamic biological processes than was previously envisioned.[3] Expressed predominantly at the external surface of plasma membranes and in the extracellular matrix glycoconjugates possessing unique saccharide sequences serve as recognition sites (ligands) for endogeneous or exogeneous carbohydrate binding proteins (lectins).[4,5] Lectin-ligand interactions are involved in cell-cell and cell-substrate recognition events which occur for instance, during development and morphogenesis, wound healing and activation of immune responses.[3,6] These interactions also play a critical role in pathologic processes such as inflammation, tumor metastasis, and recognition and binding of various pathogens including viruses, bacteria and parasites. The discovery of much of this new knowledge has been driven, in part, by advances in available methodology for localizing and characterizing glycoconjugates *in situ* and this is the subject of the present tutorial.

The earliest methods for demonstrating glycoconjugates in tissue sections were based on the chemical properties of the different functional groups present in their carbohydrate moities. These methods have been reviewed extensively and although still useful are mentioned here primarily to provide an historical perspective.[7-9] Substances rich in hexoses with periodate labile *vicinal diols* can be selectively stained with the periodic acid-Schiff (PAS) procedure. Complex carbohydrates containing an abundance of acidic groups are recognized by their selective affinity for basic dyes such as alcian blue and high iron diamine. Tissue constituents rich in sialic acids can be differentiated from those heavily esterified with sulfates through combined application of the high iron diamine-alcian blue sequence, control of the pH of the cationic dye solutions or pretreatment of sections with sialidase to eliminate basophilia attributable to sialic acids. Employed in various combinations, these techniques provide useful generalized information about the relative abundance of neutral vs acidic and different types of acidic glycoconjugates in tissues. Variations of all of these light microscopic procedures have been developed and applied successfully at the ultrastructural level.[8,9]

The discovery and characterization of a large group of carbohydrate-binding proteins now known as lectins has provided a powerful new tool to the carbohydrate histochemist. Easily isolated and purified from plants and less so from animals the lectins, now numbered in the hundreds, bind specifically to a wide range of terminal and internal saccharide sequences in glycoconjugates.[10-12] Labelled directly or indirectly with fluorochromes, enzymes, ferritin, colloidal gold, or other visible or electron dense markers, numerous lectins have been employed to define the tissue, cell, or organelle specific distribution of glycoproteins and to analyze changes in glycosylation patterns in both normal and pathologic conditions. Used in conjunction with chemical pretreatments such as acidification, methylation or saponification or following digestion of tissues with highly specific endo- and exoglycosidases, the lectins have provided valuable new knowledge concerning the precise cellular and subcellular distribution of specific saccharide sequences in glycoconjugates. Such information is not obtainable biochemically and provides a basis for formulating hypotheses about the functional roles specific glycoconjugates play in these various sites. The use of lectins as histochemical probes has been the subject of several comprehensive reviews.[13-17]

The more recent availability of a large selection of polyclonal and monoclonal antibodies with well characterized affinities for specific saccharide structures now provides another important resource for carbohydrate histochemists. Use of these antibodies, many of which are directed against carbohydrate determinants in blood group and histocompatibility antigens and cell adhesion molecules, is rapidly expanding information about the precise biological function of diverse cell and tissue glycoconjugates.

As with any histochemical procedure, the analysis of glycoconugates *in situ* is subject to certain limitations. Optimizing tissue fixation and processing conditions for retention of histochemical reactivity usually results in some degree of sacrifice in morphological preservation. Failure to optimize tissue processing and staining conditions and to utilize appropriate control procedures can produce false-negative and false-positive results. Methods for performing many of these procedures along with possible pitfalls and examples of their application in biomedical research will be presented.

References

1. J. Montreuil et al., in M.F. Chaplin and K.F. Kennedy, Eds., *Carbohydrate Analysis. A Practical Approach*, Washington, DC:Oxford (1986)143.
2. N.R. Thotakura and D.L. Blithe, *Glycobiology* 5(1995)3.
3. A. Varki, *Glycobiology* 3(1993)97.
4. S.H. Barondes, *Science* 223(1984)1259.
5. F.L. Harrison, *J. Cell Sci.* 100(1991)9.
6. S.E. Zalik, *Anat. Embryol.* 183(1991)521.
7. S.S. Spicer et al., in B.M. Wagner and D.E. Smith, Eds., *The Connective Tissue*, Baltimore:Williams & Wilkins (1967)251.
8. G.N. Thomopoulos et al., *J. Elec. Microsc. Tech.* 5(1987)17.
9. J.E. Scott, *Collagen Res. Rel.* 5(1985)541.
10. I.J. Goldstein and R.D. Poretz, in I.E. Liener, N. Sharon, and I.J. Goldstein, Eds., *The Lectins*, Orlando:Academic Press (1986)33.
11 H. Lis and N. Sharon, *Ann. Rev. Biochem.* 55(1986)35.
12. N. Agrawal et al., *J. Biol. Chem.* 268(1993)14932.
13. S.S. Spicer et al., in B.M. Wagner, R. Fleischmajer, and N. Kaufman, Eds., *Connective Tissue Diseases*, Baltimore:Williams & Wilkins (1983)163.
14. J. Roth, *J. Microsc.* 143(1986)125.
15. I. Damjanov, *Lab Invest.* 57(1987)5.
16. S.S. Spicer and B.A. Schulte, *J. Histochem. Cytochem.* 40(1992)1.
17. A. Danguy and H.J. Gabius, in H.J. Gabius and S. Gabius, Eds., *Lectins and Glycobiology*, Berlin:Springer Verlag (1993)241.

THE ENERGY FILTERING TEM (EFTEM) IN MODERN BIOLOGICAL TRANSMISSION ELECTRON MICROSCOPY

W. Probst and V.E. Bayer

Carl Zeiss, Electron Optics Division, P.O. Box 1380, D-73446 Oberkochen, Germany

Modern biological electron microscopy can no longer be a static tool merely describing morphology. In addition to ultrastructural information, insights into the molecular and chemical composition of a sample are needed so that new findings stemming from molecular biological and biochemical analyses can be given meaning in an ultrastructural context. Biological electron microscopy will be an essential tool for future discoveries involving the ultrastructural localization of molecules and chemical elements, and it will provide a means to identify the ultrastructural basis for a variety of reaction mechanisms. Many messenger compounds are currently known which can produce dynamic changes of either a subtle or dramatic nature at the ultrastructural level, but only the most basic of these can be examined using a conventional transmission electron microscope (CTEM). CTEMs provide limited information because they perform conventional imaging and do not employ all the signals available for analysis.

Unlike a CTEM, an EFTEM permits the selection of a defined energy (wavelength) of electrons which are then used for imaging. The ability to select a given wavelength of electrons is analogous to the ability to select a particular color from the spectrum of visible light in light microscopy and provides similar gains in information about the composition of a specimen. In addition, the use of monochromatic electrons for imaging eliminates the problems associated with chromatic aberration of the objective lens in a TEM. Separation of electrons of different energies is achieved by sector magnets which operate using a similar principle to that of a prism in light optics.[1] At present, there are two methods for integrating a filter into a TEM. A post-column filter utilizes more than 15 additional electron optical elements, necessitated by its location, and is employed as an accessory mounted to the bottom of different CTEMs[2], where it can provide energy-filtered data under dedicated use conditions. This dedicated use is required since the mode of operation of the microscope under energy-filtering conditions is substantially different from that employed during conventional imaging.[3] Also, detection systems such as TV cameras and negatives cannot be used for collecting data with a post-column filter. The second design, that of an in-column filter, has been widely used for more than a decade and consists of a spectrometer (either a magnetic prism or four sector magnets) integrated into the basic electron optical column of the microscope.[1,4,5] Because the spectrometer is integrated into the optics of the column, operation is the same for conventional imaging and energy-filtering except that one additional aperture must be inserted to select which electrons will be used for imaging.

Because all modes of operation are conjugate to the microscope's fluorescent viewing screen, any detection system which is normally employed can be utilized to obtain energy-filtered data.

By making optimum use of a number of analytical signals, EFTEMs can be used in a wide range of applications.[6-9] Some of the techniques which can be employed using EFTEMs are discussed below:

Selective visualization of elastically scattered electrons: By separating the elastically scattered electrons from the inelastically scattered ones, the highest quality imaging, both in terms of contrast and resolution, can be obtained. Using this technology allows for:

- Greater freedom from difficulties encountered as a result of variation in staining and section thickness, yielding more uniform images and making possible improved quantitative image analysis.

   - Imaging of thick sections (>2 microns) which cannot be done using CTEMs. This ability allows for the fast collection of 3-dimensional (3-D) information and permits more reliable screening for

Proc. Microscopy and Microanalysis 1995, edited by G.W. Bailey, M.H. Ellisman, R.A. Hennigar, and N.J. Zaluzec
Copyright © 1995 MSA. Published by Jones and Begell Publishing, 79 Madison Ave., New York, NY 10016

immunolabelled compounds or ultrastructural features of rare occurrence.
- Collection of electron diffraction data from such samples as frozen-hydrated protein crystals. The diffraction images are easily quantifiable and can be used to obtain structural data which would otherwise only be accessible through the use of x-ray diffraction.
- High contrast imaging of frozen-hydrated specimens (such as liposomes, biological molecules and tissue) with high resolution. Cryo microscopy preserves specimens as close to their native conformations as is possible. Cryo is one of the most important tools for investigating the 3-D shape of biological molecules in order to elucidate their mechanism of interaction with drugs , etc.

Selective visualization of inelastically scattered electrons: The ability to image electrons selected from different regions of the energy-loss spectrum allows the use of techniques such as inelastic darkfield imaging, contrast tuning, element-specific contrast enhancement, elemental imaging, and electron energy-loss spectroscopy (EELS). These techniques allow for:
- High contrast imaging of unstained specimens. This is essential for immunocytochemical localization of membrane-bound proteins and for cytochemical localization of enzymes, RNA ,and drugs or their derivatives, where the use of heavy metal counterstains could affect accurate interpretation.
- Imaging using elemental contrast, employed in the detection of multiply-labelled samples (where each antibody utilizes a different marker element), or in examining histochemical reactions.
- Elemental imaging which yields highly spatially resolved images of the various elements contained within the specimen under investigation. Such analysis can be used to localize both endogenous elements of interest (such as Ca, P, N, O) and marker elements (such as La, Cer, B).
- EELS for the confirmation of the identity of imaged elements and for molecular discrimination.

Element-specific x-ray emission: Used for elemental analysis in the following ways:
- Fast determination of elemental composition using fingerprint-type analysis of x-ray peaks.
- EDX mapping resulting in element imaging , which is particularly useful for heavy elements.

Secondary electrons: Used in STEM mode for topographical analysis and imaging of 3-D samples.

Backscattered electrons: Used in STEM mode for localizing immunogold in cell cultures and tissues.

Filtered bright/darkfield STEM: Provides images with higher contrast and better signal/noise ratios.

One EFTEM, the Zeiss EM 912 Omega, provides an excellent example of the new direction in biological electron microscopy. This EFTEM is unique in that it is the only one which also provides Koehler illumination. Because of its fully computerized optics, this EFTEM provides completely parallel illumination across the entire viewing area, with specimen irradiation limited to only the area actually being viewed. Parallel illumination results in images with greater crispness and brilliance, and by limiting specimen irradiation to only the area being viewed, radiation-induced damage of areas not yet examined is eliminated. Because of its digital control, all of the filtering techniques as outlined above can be employed with greater ease and less alignment requirements than are necessary with most of today's CTEMs. The ability to merge such sophisticated technology into microscopes which are easy to use means that the advantages of EFTEMs are more accessible now than ever before.

1.   L. Reimer, *Advances in Electronics and Electron Physics*, 81 (1991) 43.
2.   O.L. Krivanek et al., *Microsc. Microanal. Microstruct.*, 2 (1991) 315.
3.   A.J. Gubbens and O.L.Krivanek, *Ultramicroscopy*, 51 (1993) 146.
4.   S. Lanio, *Optik*, 73 (1989) 99.
5.   W. Probst et al., *Inst. Phys. Conf. Ser.*, 130 (1992) 295.
6.   G. Wilson, Ed., *Journal of Microscopy*, Blackwell Scientific, 174:3 (1994), various.
7.   G. Wilson, Ed., *Journal of Microscopy*, Blackwell Scientific, 166:3 (1992), various.
8.   E. Zeitler, Ed., *Ultramicroscopy*, 32:1 (1990), various.
9.   R.D. Leapman and J.A. Hunt, *Microscopy: The Key Research Tool*, 22:1 (1992) 39.

A HitchHiker's Guide to Microscopy & Microanalysis Using Telecommunications, Email and the Internet - Part II

Nestor J. Zaluzec

**Materials Science Division, Kinetics & Irradiation Effects Group
Argonne National Laboratory, Argonne, Illinois 60439 , USA**

In this tutorial, we will update, review and demonstrate the means of finding and accessing resources of electronic information which are maintained by the Microscopy Society of America and related national/international organizations. During the course of this live demonstration over the Internet, we illustrate the use of Electronic Mail, Newsgroups, File Transfer Protocols, Gophers, and Information Servers using the World Wide Web protocols. Although the MSA electronic bulletin board system (BBS) has be recently discontinued, MSA has recently decided expanded it's role in other telecommunications areas. The procedures needed by individuals or organizations to access this wealth of information will be presented in detail. The following list of topics summarizes some of the areas which will be covered.

Electronic Mail: (EMail) is textual communications which is sent over networks from one individual to another. Each user is assigned a unique address by the computer system administrator. The message composed of normal ASCII (text) characters which is sent from one computer to another. The mail message is composed of header lines which identify the sender, and receiver , routing information and a body of text. BINARY files can not be sent by simple Email without first converting them into ASCII characters, which the receipent must then convert back into binary format to be usable. There are encoding and decoding programs available which perform these operations routinely.

MailServers: An organized electronic mail system which connects a group of individuals by simple electronic mail. Messages sent to a central location are forwarded to all members of the group. All users of mailing lists subscribe to a particuliar topical group and automatically receive messages from the server.

File Transfer Protocol (FTP) is a universally adopted method for the exchange of text and binary data files between computersystems over the Internet. FTP access can be restricted to an individual user account, or it can be made ANONYMOUS. An anonymous site allows any person to access (usually downloading only) files on a remote computer, without actually having an account on that system. It is a rapid and efficient way used world wide to share information.

Proc. Microscopy and Microanalysis 1995, edited by G.W. Bailey, M.H. Ellisman, R.A. Hennigar, and N.J. Zaluzec
Copyright © 1995 MSA. Published by Jones and Begell Publishing, 79 Madison Ave., New York, NY 10016

Gopher: A search and find system which is based upon a simple menu structure allowing you to access textual information at a specific site, by searching for documents identified by key words.

World Wide Web (WWW) : A graphical user interfaced based , hypertext information service which is platform independant, but allows the user to retrieve text and images on a variety of computer systems. Links are setup by the varioius system administrators to allow the user to connect to associated information.

Wide Area Information Server (WAIS) is a search and retrieve listing of articles in specially configured databases.

This work was supported in part by US. DoE under contract BES-MS W-31-109-Eng-38 and the Microscopy Society of America.

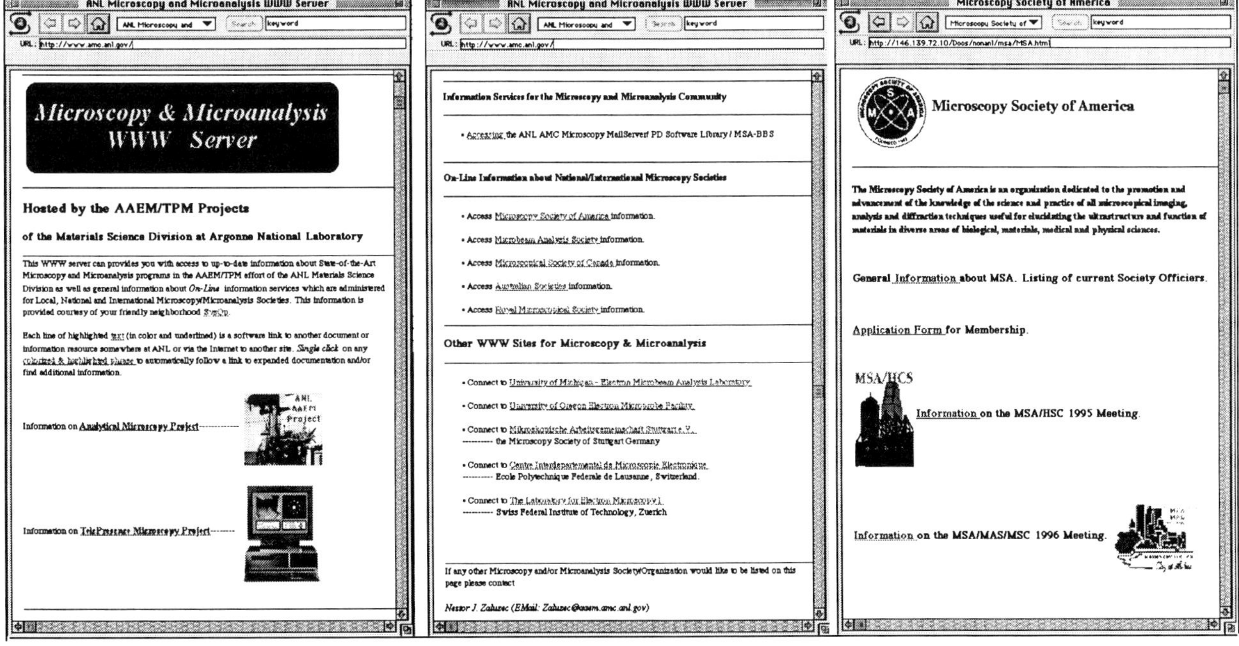

Figure 1.): World Wide Web (WWW) Pages, illustrating the Graphical User Interface (GUI), access to sites dealing with Microscopy and Information Pages supported by the Microscopy Society of America.

A FEW WORDS ON BITS & BYTES: A TUTORIAL ON IMAGE AND SPECTRAL PROCESSING FOR THE NOVICE

John F. Mansfield

University of Michigan Electron Microbeam Analysis Laboratory
413 Space Research
2455 Hayward
Ann Arbor
MI 48109-2143
http://www.engin.umich.edu/~jfmjfm/jfmjfm.html

We are now in a position, in the world of microscopy and microanalysis, to record all of our data digitally and often, the first time that the data is printed out is when it is submitted for publication, although there are distinct possibilities that we shall soon be publishing electronically too[1]. Digital image acquisition schemes are largely the realm of the microscope manufacturers, however, there are also a multitude of "homemade" acquisition systems in microscope laboratories around the world. Such a wide variety of systems leads to a wide variety of data storage formats. For example, a any one time there have been at least 6 manufacturers of X-ray Energy Dispersive Spectroscopy (XEDS) systems, each with their own proprietary data format, which, frequently, was not explained to the user. The philosophy was: "Our system is all you will ever need, so you don't need to know how the data is stored". Fortunately, largely due to the efforts of a Standards Group organized by The Microscopy Society of America, there is a standard format for spectral data and a corresponding image data format is under development. Meanwhile, however, there is a large quantity of data that continues to be acquired from systems that have proprietary formats and there are a correspondingly large number of people wishing to analyze it. In addition, not everyone is a computer "hacker" and many people are still perplexed and apprehensive when it comes to computer use. This tutorial will gently introduce the novice to digital data acquisition and processing, how it is stored, how it can be transferred from one computer to another and how it may be analyzed with a wide variety of inexpensive and often free software.

For a wide range of image processing and measurement tasks, the *Macintosh* software application *NIH-Image,* developed by Wayne Rasband at the National Institutes of Health in Bethesda MD, is indispensible[2]. However, many users and vendors of so-called "high-end" or "advanced" image acquisition systems will note that NIH-Image is an 8-bit program. This means that the intensity value of each pixel can vary only from 0 to 255. Also, these intensities may only be integer values. Clearly this is a limitation for microscopists, particularly those who wish to analyze electron diffraction patterns, where the dynamic ranger may be $\sim 10^5$, or work with compositional maps, where each pixel represents a true chemical composition. If you need to manipulate images containing "real data" there is an application, again written for the *Apple Macintosh,* that is capable of handling such data, it is called *ImagNSpect*[3]. ImagNSpect fully supports the TIFF6 implementation that allows for floating point data and it will also read integer TIFF files that are more than 8 bits deep. The application also contains extensive image acquisition and manipulation tools and can perform Fast Fourier Transforms on any size image. Both Image and ImagNSpect have their limitations when it is necessary to prepare data for publication and it is wise to invest in the commercial application Photoshop for these "desktop publishing" tasks[4].

References
1. J.F. Mansfield, "The Development Of An Electronic Microscopy Journal On The World Wide Web", this proceedings.
2. *NIH-Image*, Wayne Rasband, NIH, Bethesda MD. Public domain software, copies (application, source and sample images) are available by anonymous ftp from zippy.nimh.nih.gov in the directory /pub/nih-image.
3. *ImagNSpect*, David Kopf, Pittsboro, North Carolina & Scott Davilla, 4-Pi Analysis, Durham, North Carolina. The current version is available by anonymous ftp from zippy.nimh.nih.gov in the directory /pub/nih-image/contrib.

Proc. Microscopy and Microanalysis 1995, edited by G.W. Bailey, M.H. Ellisman, R.A. Hennigar, and N.J. Zaluzec
Copyright © 1995 MSA. Published by Jones and Begell Publishing, 79 Madison Ave., New York, NY 10016

4. *Photoshop*, Adobe Systems Incorporated, 1585 Charleston Road, Mountain View CA 94039-7900. Current version is 3.0.1.

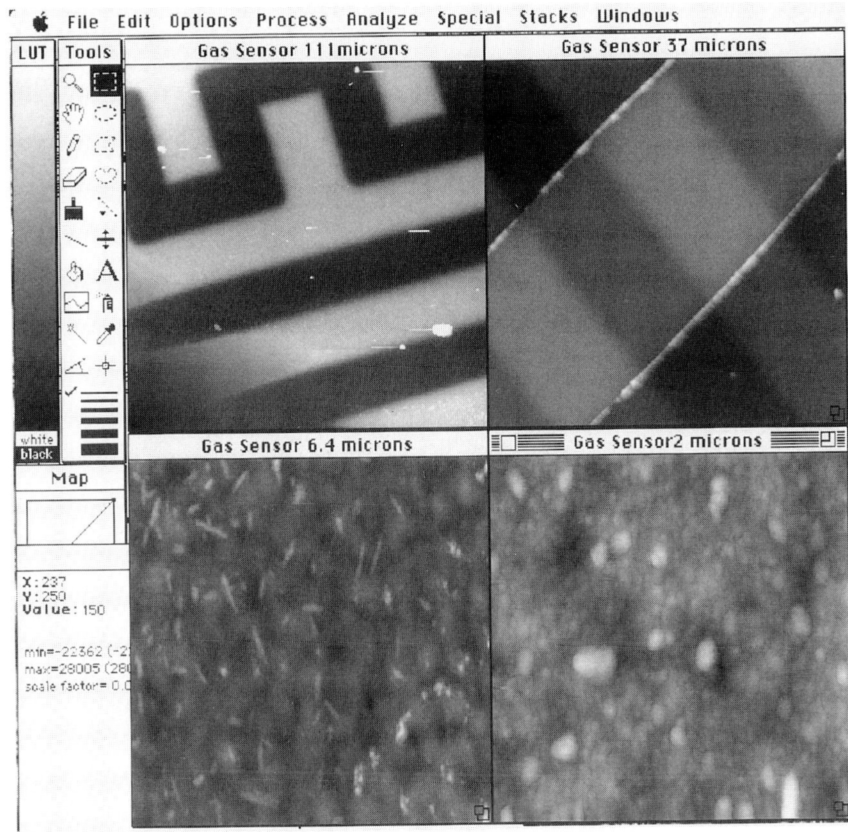

Figure 1. Atomic Force Microscope images imported into NIH-Image. Original Data is 16-bit, but Image is capable of reading the data and displaying the images compressed into 8-bits.

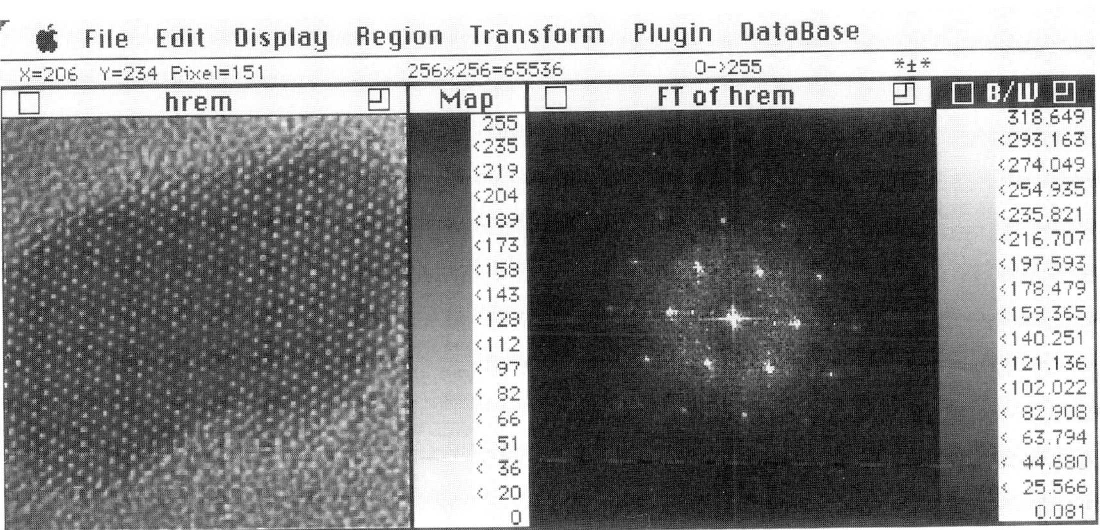

Figure 2. A high resoution TEM image of a Pt/Sn particle with its accompanying FFT as displayed by ImagNSpect.

ELECTRON BACKSCATTER DIFFRACTION A NEW TOOL FOR MATERIALS CHARACTERISATION

D.J.Dingley*

TexSEM Laboratories Inc. Provo Utah 84604
*On leave from University of Bristol, Bristol BS81TL United Kingdom.

Electron backscatter diffraction is an experimental technique that provides the investigator with a tool to determine crystallographic data from sub-micron areas of a polycrystalline sample. The first relevant experiments in elctron backscatter diffraction, were conducted by Venables and co-workers[1] followed later by Dingley and co-workers[2] who developed methods of on-line orientation measurement and later phase identification procedures and methods for internal strain measurement.

A specimen is mounted in a scanning electron microscope within which are mounted a phosphor screen positioned vertically and close to the specimen and a low light video camera to view it.   The specimen is tilted steeply from the horizontal and oriented to face the phosphor screen. It is normal to enhance the contrast observed in the diffraction pattern using digital background subtraction procedures. A standard arrangement is shown in figure 1.

Electron diffraction occurs within the sample as a natural consequence of the scattering of the electron beam beneath the sample surface to form a divergent source of radiation. Electrons are diffracted into cones of enhanced radiation emanating from each set of crystal planes for which the Bragg condition $\lambda = 2d\sin\theta$ can be satisfied. There are two cones for each set of planes corresponding to diffraction of electrons incident from above and from below respectively. Those cones directed out of the upper surface of the specimen that fall across the phosphor screen form the observed diffraction pattern. An example pattern is shown in figure 2.

A typical sequence of procedures to obtain diffraction patterns would be first to view the sample in secondary electron mode, to select an area of interest and then to position the electron beam as a stationary probe over that area. The diffraction pattern is observed without further microscope adjustment. Automated methods have been developed to index the diffraction patterns[3] and this has lead to mapping of   crystallographic variations over the sample surface. The maps, called Orientation Image Micgrographs, OIMs, can be drawn to show crystallographic phase changes, gradual and abrupt changes in crystal orientation, grain boundary character, and residual strain. Examples of these features are depicted in the OIM shown in figure 2.

References

1. J.A.Venables and C.J.Harland, Philos. Mag. 27(1973)1193.
2. D.J.Dingley, M Longdon, J Wienbren and J Alderman, Scanning Elect. Mic.1,2 (1987)451
3. B L Adams, S.I.Wright and K Kunze, Metall. Trans.,24A(1993)819.

Proc. Microscopy and Microanalysis 1995, edited by G.W. Bailey, M.H. Ellisman, R.A. Hennigar, and N.J. Zaluzec
Copyright © 1995 MSA. Published by Jones and Begell Publishing, 79 Madison Ave., New York, NY 10016

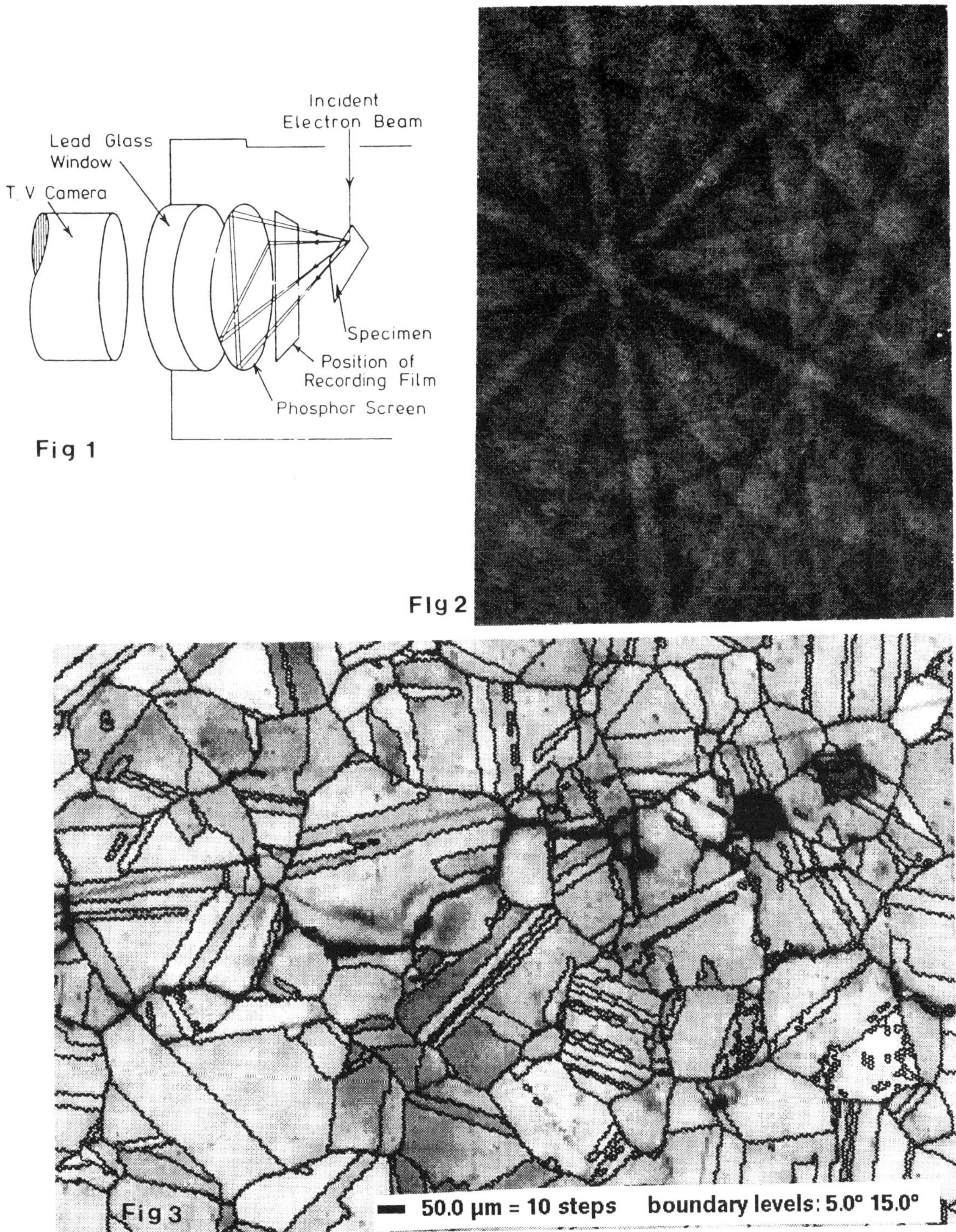

FIG.1. Schamtic, illustrating arrangement for observing electron backscatter patterns.
FIG.2. Electron backscatter diffraction pattern from nickel.
FIG.3. Orientation image micrograph from nickel shaded to show residual strain, with grain boundaries of different thickness to distinguish low angle from high angle cases.

ATOM-PROBE ANALYSIS OF THE SOLID-LIQUID INTERFACE

J.A. Panitz*

*Department of Physics and Astronomy and Department of Anatomy, School of Medicine.
The University of New Mexico. Albuquerque, NM 87131.

The first few atomic layers of a solid can form a barrier between its interior and an often hostile environment. Although adsorption at the vacuum-solid interface has been studied in great detail, little is known about adsorption at the liquid-solid interface. Adsorption at a liquid-solid interface is of intrinsic interest, and is of technological importance because it provides a way to coat a surface with monolayer or multilayer structures. A pinhole free monolayer (with a reasonable dielectric constant) could lead to the development of nanoscale capacitors with unique characteristics and lithographic resists that surpass the resolution of their conventional counterparts. Chemically selective adsorption is of particular interest because it can be used to passivate a surface from external modification or change the wear and the lubrication properties of a surface to reflect new and useful properties. Immunochemical adsorption could be used to fabricate novel molecular electronic devices or to construct small, "smart", unobtrusive sensors with the potential to detect a wide variety of preselected species at the molecular level.[1-3] These might include a particular carcinogen in the environment, a specific type of explosive, a chemical agent, a virus, or even a tumor in the human body.

The liquid-solid interface has been difficult to study because the majority of surface techniques that provide an analytic capability operate most effectively in high vacuum. The desire to characterize structure and morphology on a molecular scale further limit the possibilities to Scanned-Probe and Field-Ion Microscopes. Although Scanned Probe Microscopes can be configured to operate within a liquid environment, Field Ion Microscopy has the advantage of providing surface analysis by Atom-Probe spectroscopy.[4] The Imaging Atom-Probe (Fig. 1) provides an additional capability: a surface can be probed for constituent species whose individual distributions can be mapped laterally, and as a function of depth.[5] Although Atom-Probe analysis is destructive; the lateral and depth resolution of the technique is impressive, usually exceeding 0.5 nm.

The transfer of a sample from a liquid environment to the ultrahigh vacuum conditions of the atom-probe must be accomplished without contamination and without allowing the sample to dry. Surface tension forces encountered during the drying process are inhomogeneous, and morphologically destructive. For example; an adsorbate, 10 nm in size, will experience a pressure, $P \sim 2 \times 10^7$ Nm^{-2} (~ 1 ton/in^2) upon crossing a water-air interface. Organic monolayers, macromolecular adsorbates, and the distribution of macromolecular assemblies at a surface can irreversibly deform under such conditions. Artifacts can be minimized by critical-point drying techniques, but these techniques require exposure to harsh environments (acetone and liquid CO_2 are typical) if reasonable conditions of pressure and temperature are maintained.[6] Alternately, the dying problem can be eliminated by embedding the monolayer within a thin layer of amorphous ice formed from its native liquid environment. The amorphous state is unique in that freezing occurs so rapidly that ice crystal formation and solute partitioning in the frozen solid are absent. Species within the liquid (and adsorbates at a surface in contact with the liquid) are preserved on a microsecond time scale as the freezing front advances through the liquid. This approach has been used successfully to visualize frozen, hydrated specimens by cryoelectron microscopy.[7]

A similar approach has been developed for Atom-Probe analysis.[8] After the interface is frozen, it is cryogenically transferred (in an inert atmosphere) into the Atom-Probe. The interface is exposed for analysis only when the ambient pressure is reduced to a point where the time to adsorb gas-phase contaminants is much longer than the time to collect data. Under these conditions, an electric field can be used to isothermally "sublime" the layer of amorphous ice that encases the interface. Mass spectra obtained during this process can provide chemical analysis of the original liquid environment of the interface (Fig. 2a). Maps of constituent species can also be produced with a lateral resolution of several nanometers (Fig. 2b). As the encasing layer of ice is ionized and removed, the interface is exposed for analysis. Quantitative spectra reflecting this process are produced by recording the ion species as the electric field is ramped linearly with time (Fig 3a).[9] Vitrified benzene, desorbed from the surface, produces a fiducial marker of known field strength. The TEM is used to assess the resulting change in surface morphology (Fig 3b-3c).

Proc. Microscopy and Microanalysis 1995, edited by G.W. Bailey, M.H. Ellisman, R.A. Hennigar, and N.J. Zaluzec
Copyright © 1995 MSA. Published by Jones and Begell Publishing, 79 Madison Ave., New York, NY 10016

References

1. E.L. Carter, J. Vac. Sci. Technol. (4), (1983)959.
2. S. Middelhoek and A.C.Hoogerwerf, in *Transducers '85*: IEEE #85CH21279. New Jersey:IEEE(1985)2.
3. J.A. Panitz, J. Physique. 45(1985)285.
4. E.W. Müller, J.A. Panitz and S.B. McLane, Rev. Sci. Instrum. 39(1968)83.
5. J.A. Panitz. in: R.L. Park and M.G. Lagally Eds., *Methods of Experimental Physics*, New York:Academic Press (1985)414.
6. A.A. Bartlett and H.P. Burstyn, Scanning Electron Microsc. 75(1975)305.
7. N. Unwin. in: A.P. Somlyo, Ed., *Recent Advances in Electron and Light Optical Imaging in Biology and Medicine*, New York: NYAS (1986).
8. J.A. Panitz and A. Stintz, Surface Sci. 246(1991)163.
9. A. Stintz and J.A. Panitz, J. Appl. Phys. 72(1992)741.
10. This research was supported by the United States Department of Energy (Office of Basic Energy Sciences) under Grant DE-FG04-88ER45348.

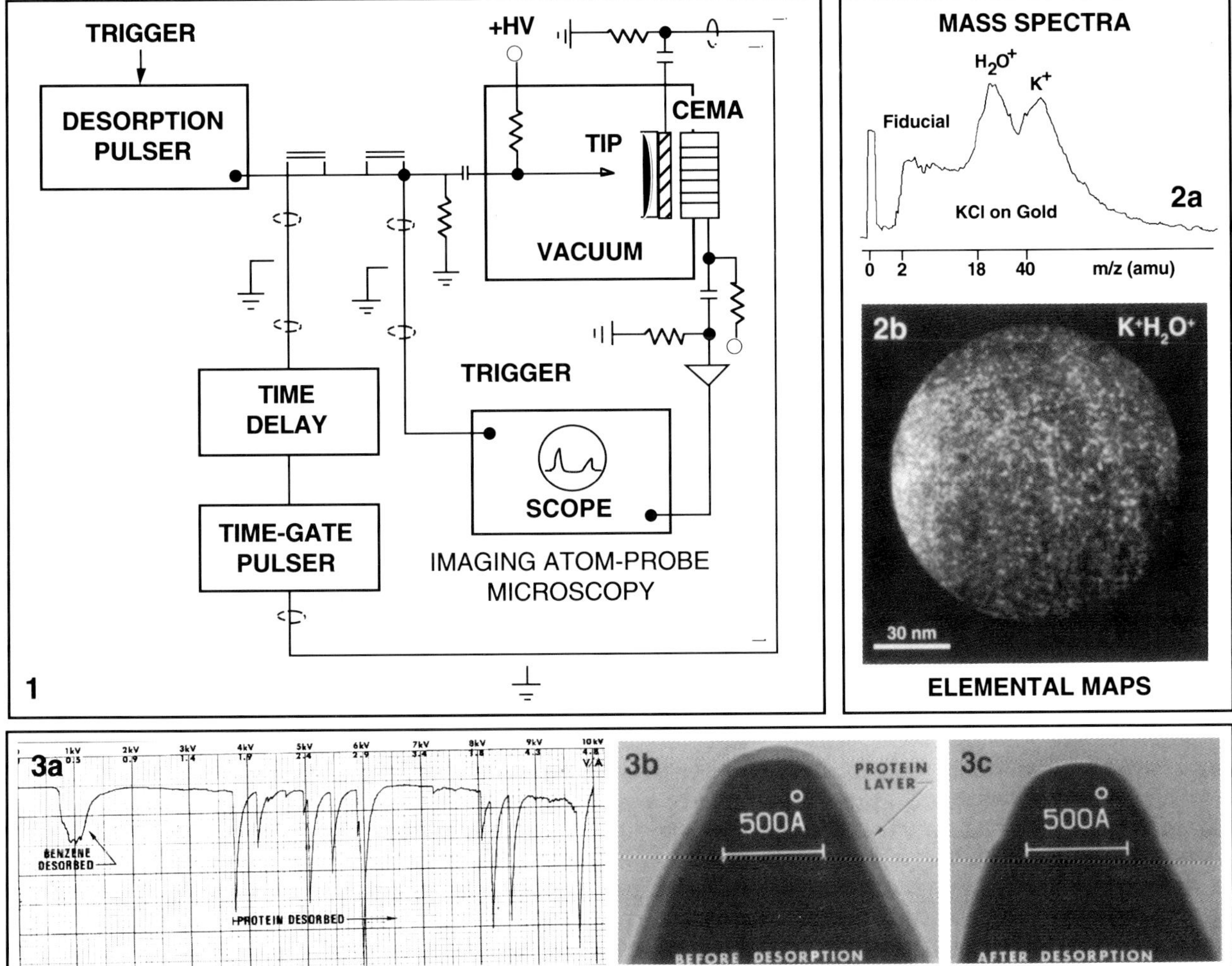

Fig.1—Schematic diagram of the Imaging Atom-Probe mass spectrometer.
Fig.2—(a) A mass spectrum of a vitrified KCL solution. From reference 8.
Fig.2—(b) An elemental map of a vitrified KCL solution. From A. Stintz and J.A. Panitz, JVST. A9(1991)1365.
Fig.3—(a) Isothermal ramped field desorption spectrum of BSA on tungsten. (b-c) TEM micrographs of surface.

IMAGE PROCESSING AND ANALYSIS FUNDAMENTALS FOR MICROSCOPY

Bruce D. Newell

Analytical Technology Division, Eastman Kodak Company, Rochester, NY, 14652-3712

Advances in computers and related digital hardware, coupled with sophisticated software techniques have resulted in microscopy migrating from its historical roots as a subjective, qualitative science towards a more robust position as a truly quantitative technique. Granted, we will probably never totally remove the microscopist from the process of image interpretation (at least those at this conference hope not) but we will certainly continue to progress from describing our image data in qualitative terms (e.g., many/few, large/small, equiaxis/elongated, ordered/random, rough/smooth) to quantitative measurements of number, size, shape, location, texture, and so on.

To move along the path toward quantitative image interpretations requires an understanding of image processing and analysis (IP/A) fundamentals to insure that the data obtained is of the required accuracy and precision. A generalized model of the critical steps in the image processing and analysis chain is given in Figure 1. This tutorial will examine the fundamental issues in each step that impact the quality of the final result and provide a broad overview of techniques that may be applicable.

Image Generation

In the early days of IP/A technology development the attitude was that digital processing could overcome poor analog imaging, i.e., one could obtain meaningful results from poor images. Perhaps microscopists didn't fall as far into this trap because our experience had taught us that optimum image quality was essential to our work and while sophisticated techniques have been developed to extract data from out of focus, low contrast, or otherwise distorted images, we have in general learned that optimization of the imaging process provides a major advantage in simplifying the subsequent processing and analysis. Thus, one must consider the information required from the sample and select/optimize the image system that produces the highest possible isolation of the required features. This may involve the use of new sample preparation methods and/or imaging schemes involving multiple detectors.

Image Capture

To the first approximation, the selection of an image capture device is dictated by the mode of imaging. Video camera technology (tube and CCD) is generally the choice for light microscopy and some TEM applications while analog-to-digital converters are most appropriate for SEM applications. Photographic film (with subsequent scanning to obtain the digital image) still provides advantages in some applications. Evaluation of criteria such as image intensity range, image resolution, capture speed, and the measurement resolution required are a few of the parameters that need to be evaluated when selecting the mode of image capture.

Image Segmentation

Isolating the features of interest from all other features and background information is perhaps the most difficult step in the chain. Basic techniques relying on grey level thresholding are sufficient in only those cases were relatively high contrast differences can be obtained between the features of interest and other image features/background. More common is the situation where low and/or varying feature contrast, and varying background intensity (e.g., resulting from illumination non-uniformities and/or sample thickness variations) call for more elaborate image processing procedures. In these cases both linear (convolution)

Proc. Microscopy and Microanalysis 1995, edited by G.W. Bailey, M.H. Ellisman, R.A. Hennigar, and N.J. Zaluzec
Copyright © 1995 MSA. Published by Jones and Begell Publishing, 79 Madison Ave., New York, NY 10016

and non-linear (mathematical morphology) methods may be utilized.

Feature segmentation may also be achieved through image information other than pixel intensity. For example, segmentation based on texture, shape, and height are other approaches.

Feature Measurement/Classification

The algorithms required to obtain geometric measures of the segmented features are readily available however, these procedures must be utilized with an understanding of their inherent limitations which in large part result from the fact that we are working with a digitized image. Which measurement(s) best characterize the features is a second consideration. Frequently a single parameter is insufficient to describe the feature and multi-parameter, joint probabilistic methods come into play. These can be further enhanced by statistically based feature classifiers or modeling.

Data Analysis and Presentation

Perhaps the goal of any quantitative analysis is to produce a single number that fully and uniquely describes the sample. Obviously this is most often not possible and those suffering from "mononumerosis" can easily mislead themselves. But how do we deal with the large amount of data that is available to us from our image processing and analysis procedures? What are the advantages of non-parametric versus parametric statistics? Can data modeling provide useful advantages?

Automation

Once we have considered all the above factors and have established an accurate and precise image processing and analysis procedure a final consideration might be automation. What additional issues must be considered if we wish to reduce the microscopist's involvement in the process? For example, effective, robust automation of the entire process may require additional controls at each step of the procedure, as well as a re-design of imaging, illumination, and/or sample holding devices.

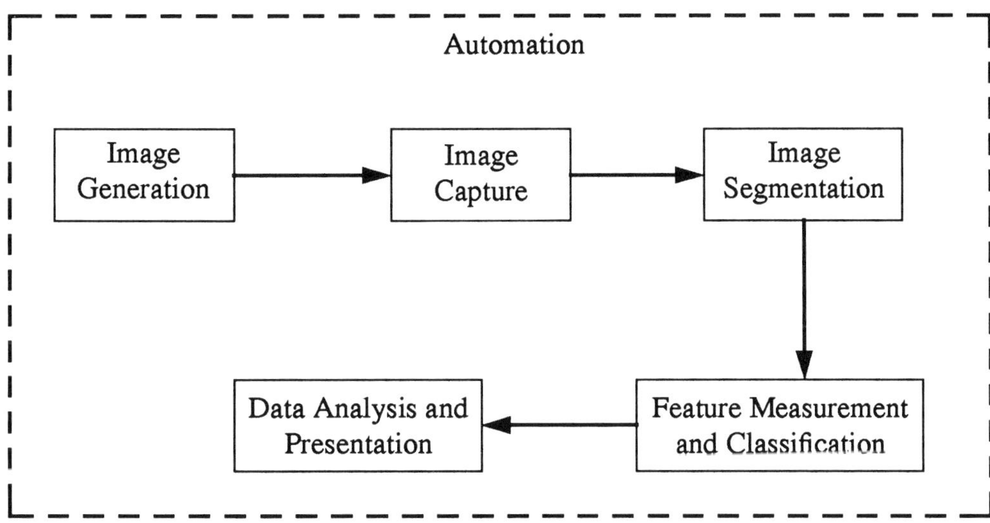

FIGURE 1.- The Image Processing and Analysis Chain

SPECIMEN PREPARATION: THE ACHILLES HEEL OF MICROSCOPY

B. Kestel

Materials Science Division, Argonne National Laboratory, Argonne, IL 60439

As the variety of microscopy techniques to reveal new information increases, the difficulty of preparing suitable specimens for examination has also grown. Some microscopy requires very uniform thickness, smooth surfaces, and minimal preparation induced damage or contamination. The variety of materials studied, their different physical properties, and the number of techniques available makes this process more troublesome than some project planners realize. New employees of an organization are often given this task, but their inexperience may compound the problem. A worker's success in this process may be increased by experience with tedious hobbies, teamwork events, or the need to record pertinent details of processes.

A starting point for optical microscopy, mechanical dimpling before ion milling, and the making of thin slices or wedges is mechanical polishing. The relatively new tripod polishing technique is based on controlled mechanical polishing using very fine abrasives. Examples of modified equipment and simple fixtures which make surface polishing easier, faster, and of high quality with reduced labor will be shown. Use of various lubricants during intermediate and final polishing to improve results will be addressed. Fixtures that support specimens properly, are lightweight for final polishing, and move in a truly random fashion to prevent scratches or directional polishing, have been made. The advantages and labor savings possible with the use of vibratory polishing on abrasive slurry covered cloths and some possible pitfalls of the process will be presented.

The transmission electron microscopy (TEM) specimen must be thin enough for electrons to penetrate in vacuum at 100 kV accelerating voltage. Both surfaces must be very smooth and clean to prevent extraneous scattering of the image producing electrons. Published information in this field is beginning to increase.[1-3] A common technique for thinning metallic materials is jet electropolishing. It can produce excellent surfaces, good thin area with automatic polishing termination, and minimal induced damage which is important in defect or amorphization studies. Recent process improvements include a non-acid electrolyte, BK-2, which thins a wide variety of materials, including some superconductors. A specimen's surface may be rapidly prepared for optical inspection of grain size, precipitate distribution, etc. by briefly electropolishing it, then reducing the cell voltage briefly to produce a scratch free etched surface. A larger jet matched to the specimen size and a vented holder are all that is required.

A rolled nickel foil thinned with this safer method rather than in a perchloric acid/alcohol bath is shown (Fig. 1). The electrolyte viscosity, chemical content, and temperature can be selected for improved retention of precipitates or uniform polishing of two-phase material such as the depleted uranium-silicon specimen shown (Fig. 2). A few ideas for improving the exchange of information among workers will also be mentioned.[4]

Proc. Microscopy and Microanalysis 1995, edited by G.W. Bailey, M.H. Ellisman, R.A. Hennigar, and N.J. Zaluzec
Copyright © 1995 MSA. Published by Jones and Begell Publishing, 79 Madison Ave., New York, NY 10016

References

1. Materials Research Society Symposium Proceedings, Vols. 115, 199, 254.
2. K. C. Thompson-Russell, J. W. Edington, Monograph 5, Electron Microscope Specimen Preparation Techniques in Materials Science (1977).
3. B. J. Kestel, Polishing Methods for Metallic and Ceramic Transmission Electron Microscopy Specimens, Argonne National Laboratory Report, ANL 80-120, Rev. 1.
4. Work supported by the U. S. Department of Energy, BES-Materials Sciences, under Contract W-31-109-Eng-38.

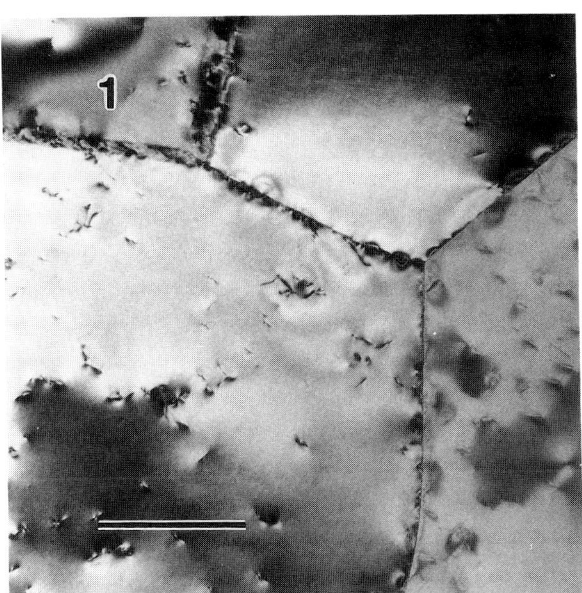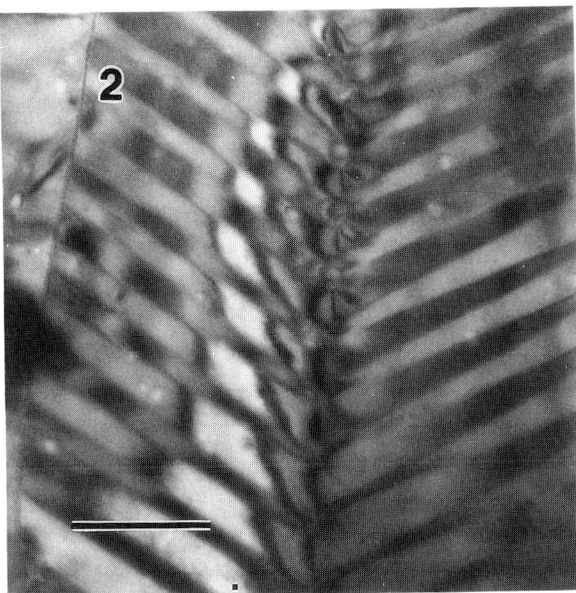

FIG. 1.–TEM image of rolled nickel jet electropolished with non-acid BK-2 electrolyte shows well preserved grain boundaries and precipitates. Bar = 0.5 μm.

FIG.2.–Electron micrograph of annealed 96 w/o depleted uranium, 4 w/o silicon foil jet electropolished in hydrochloric acid, methanol, and 2-butoxy ethanol. Bar = 0.3 μm.

CORRELATIVE MICROSCOPY IN DISTRIBUTED (CLIENT/SERVER) COMPUTING ENVIRONMENTS: TOOLS FOR CATALYZING THE RAPID DISSEMINATION OF INFORMATION ABOUT THE CELL BIOLOGY OF THE HUMAN PROSTATE

A. Hakam, J.T. Gau, M.L. Grove, B.A. Evans, M. Shuman, S.C. Strom, R. Bahnson*, P.K. Sethi, K.C. Song, C.J. DiGiorgio, F. He, C.S. Hatton, F.C. Wimberly[+], R.Z. Roskies[+], L. Deckenbaugh[#], W. Gross, E. Schubert and M.J. Becich.

The Divisions of Anatomic Pathology, Cellular and Molecular Pathology, Pathology Informatics and Facility for Imaging and Research Microscopy, Department of Pathology, Pittsburgh Supercomputing Center[+], Informatics Services Division[#], and Division of Urological Surgery, University of Pittsburgh School of Medicine and Carnegie Mellon University, Pittsburgh, PA 15261*.

Prostate adenocarcinoma is the most common malignant tumor of men in the United States and is the third leading cause of death in men. Despite attempts at early detection, there will be 244,000 new cases and 44,000 deaths from the disease in the United States in 1995. Therapeutic progress against this disease is hindered by an incomplete understanding of prostate epithelial cell biology, the availability of human tissues for in vitro experimentation, slow dissemination of information between prostate cancer research teams and the increasing pressure to "stretch" research dollars at the same time staff reductions are occurring.

To meet these challenges , we have used the correlative microscopy (CM) and client/server (C/S) computing to increase productivity while decreasing costs. Critical elements of our program are as follows:

1) Establishing the Western Pennsylvania Genitourinary (GU) Tissue Bank which includes >100 prostates from patients with prostate adenocarcinoma as well as >20 normal prostates from transplant organ donors.[1]

2) Expansion of cells from normal and neoplastic prostate epithelium and/or stroma available for research by producing primary prostate cultures for in vitro experiments.[2]

3) Installation of a client/server architecture throughout the medical center and research facilities to facilitate rapid transmission of images to clinical and basic science research teams.[3]

4) Developing a centralized microscopy facility (the Facility for Imaging and Research Microscopy) with the ability to produce and distribute digital images.[4] Coupled with the availability of prostates from the GU tissue bank it is possible to rapidly and without negatives or print film produce gross and light microscopic images which can quickly and efficiently be shared with collaborators. This includes cost intensive specialized correlative microscopic techniques such as low light fluorescence, in situ hybridization, autoradiography, and light and electron microscopic immunolocalizations.

5) Encouraging aggressive use of the internet including developing department wide expertise in e-mail, gopher servers and the World Wide Web (WWW).[5] These tools allow us to easily share experimental protocols, manuscripts and digital images with collaborators in other institutions via the internet.

The use of the human prostate as our experimental model has allowed us to gain insight into the normal cell biology of the prostate and gain a foothold on understanding several important aspects of epithelial growth control. With correlative microscopic techniques we will demonstrate via our client/server technology and the WWW how we have established that:

1) Prostate columnar epithelial cells are derived from basal cells and maintain the ability to retrodifferentiate in culture.[6]

2) Normal prostate cells secrete prostate specific antigen (PSA, a serine protease) in a regulated fashion while neoplastic prostate epithelial cells secrete in a constitutive fashion.[7]

3) Normal prostate cells overexpress c-Met (a proto-oncogene) not only in metaplastic areas but at the invasion front of the tumor as well as in areas of lymphatic and perineural invasion indicating a dynamic regulation of this receptor during normal and neoplastic prostate growth and repair.[8]

Cost effective, efficient distribution of many image types is now possible using CM & C/S via the WWW.

Supported by the Pathology Education and Research Foundation and the CaP (CAncer of the Prostate) Cure Foundation.

References

1. P.K. Sethi et al., *Lab. Invest.* 70(1994)83.
2. P.K. Sethi et al., *Lab. Invest.* 70(1994)82.
3. M.J. Becich et al., *Sem Diagnostic Pathol.* 11(1994)237-244.
4 E. Schubert et al., *Sem Diagnostic Pathol.* 11(1994)263-273.
5 C.J. DiGiorgio et al., *Sem Diagnostic Pathol.* 11(1994)294-304.
6. J.T. Gau et al., Proc. *Micros. Soc. of Amer.* 1994(1994)86-87.
7 J.T. Gau et al., AACR Special Conference on Basic and Clinical Aspects of Prostate Cancer (1994).
8 K.C. Song et al., *Lab. Invest.* 70(1994)83.

1a) Facility for Imaging and Research Microscopy ethernetted Quantitative Image Analysis and Morphometry unit is outfitted with a high-grade research light/fluorescence/darkfield microscope, a digital imaging workstation, flatbed scanner, slide maker, dye sublimation printer and laser printer.

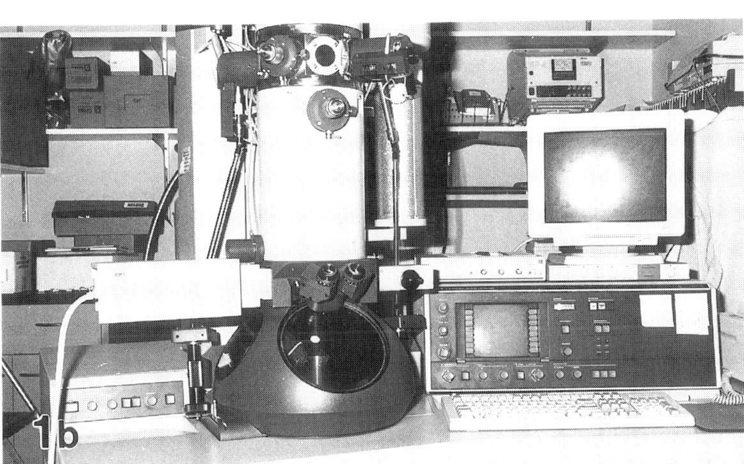

1b) Phillips CM-12 transmission electron microscope with an ethernetted digital imaging workstation.

1c) "Filmless" digital image of normal prostate epithelium showing columnar cells with abundant secretion granules in the apical cytoplasm and inconspicuous basal cells. (H&E, 1000X)

1d) "Filmless" digital image of normal prostate epithelium stained with anti-PSA ABC immunoperoxidase showing abundant immunoreactivity in the apical cytoplasm. (1000X)

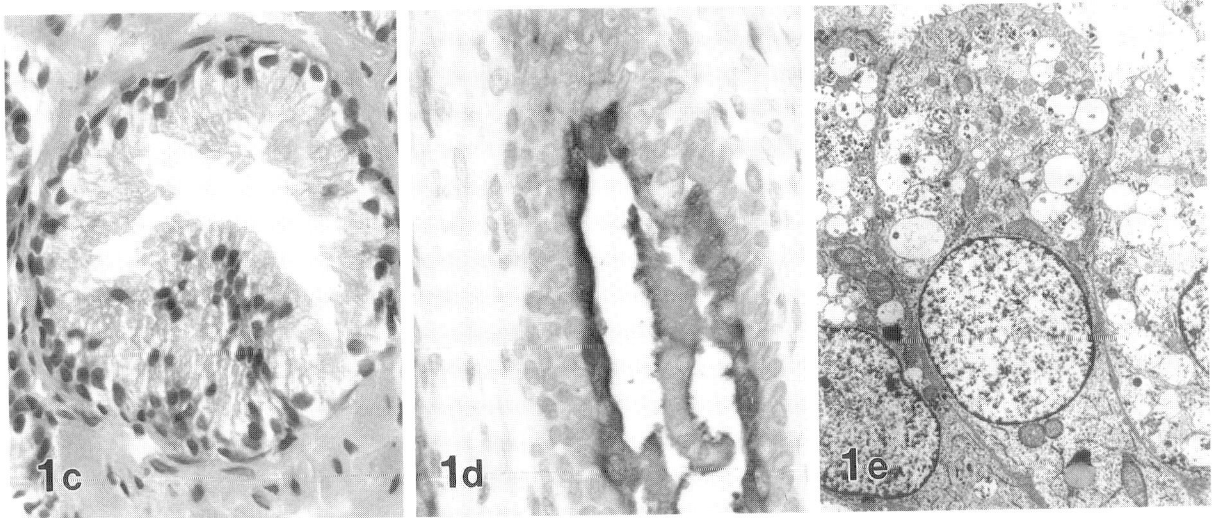

1e) "Filmless" digital image of a transmission electron micrograph of a prostate columnar epithelial cell showing abundant secretion granules within the apical cytoplasm. This correlates well with PSA immunostains. (uranyl acetate and lead citrate, 2500X)

USING CORRELATIVE MICROSCOPY (LM, EM, and SFM) IN BIOLOGICAL PROBLEM SOLVING

R.M. Albrecht* and B. Wetzel**

*Department of Animal Health and Biomedical Sciences, University of Wisconsin, 1655 Linden Dr., Madison, WI 53706
**Office of Review Activities, National Institutes of Health, Bethesda, MD 20892

Conventional morphologic approaches generally place a heavy reliance on standardized preparative methodology and often are narrowly designed to provide supporting documentation for preexisting lines of evidence or for a particular interpretation thereof. Visual images constitute a particularly convincing form of data and several well chosen micrographs can be the difference between studies which are deemed publishable and those that are not.

The potential, however, for imaging methodology to go beyond a "supporting" role and allow investigators to ask and answer questions not approachable by other methodologies is considerable. This has always been the case but relatively recent developments have strongly reinforced this potential. The application of several imaging based technologies to parallel specimens or, preferably, to the exact same specimen in series is the basis for correlative microscopy. Appropriate tracking and imaging of living cells and the subsequent correlative observation of the same cells following immobilization by various physical or chemical based preparative procedures can provide a wealth of information, which bears not only on the biological question at hand but on the likelihood and/or extent of preparative "artifact" as well.[1]

Correlative microscopy has derived considerable support and benefit from the development of high sensitivity, high resolution video cameras, computer driven image enhancement/analysis /reconstruction, rectified interference based light microscopy, and confocal light microscopy. These developments provide a variety of methodologies to extract structural and physiological information from living and variously prepared cells, tissues, and whole organisms. More recently the emergence of two photon microscopy greatly improves the capability for imaging fluorescent labels in living cells while near field scanning optical microscopy , NSOM, holds the promise of photon based imaging at EM levels of resolution, on living material. Newer field emission high resolution scanning electron microscopes, HR-SEM, environmental SEM, ESEM; and microprocessor controlled TEM, IVEM, and HVEM have substantially increased the information available regarding surface and internal ultra structure, at molecular and sub molecular levels of resolution. Similarly scanning force microscopy, SFM, can provide information pertaining to surface detail in dry or hydrated specimens at submolecular levels of spatial resolution in addition to providing other force based measurements.[2,3]

Improved labeling technology has also played a key role in furthering the utility of correlative microscopy. Labeling methodologies using heavy metal colloids, such as gold, coupled to antibodies, to ligands, or to active fragments of antibodies or ligands provide markers that can be visualized in various forms of light microscopy , including, but not limited to: bright field; dark field; phase; interference based, (DIC/AIC); interference reflectance; and confocal. These labels can be clearly identified in SEM, HR-SEM, ESEM, TEM, IVEM, and HVEM due to their round regular shape, increased coefficient of secondary and backscattered electron emission, as well as their increased density and associated electron scattering capabilities. Their shape and rigidity also permits identification of all but the smaller colloidal particles, less than 10 nm, in SFM. The smallest particles intercalate themselves within the conjugated molecular species and thus are not readily apparent on the "surface" of the molecule. Identification in NSOM, particularly transmitted NSOM, should also be possible although additional studies on labeled biological material are required to confirm limits of detectability and size.
Thus, for example, an investigator wishing to follow ligand receptor binding and associated ligand receptor movement , as well as related changes in cell structure, internal free ions, or pH can use small 10-20 nm gold labels coupled to ligand. Binding and subsequent movement of groups of labels or

individual labels on or in living cells can be followed by video enhanced interference based LM. Individual gold particles cannot be resolved but can be detected and their movement tracked by observation of their inflated diffraction images. Cellular structures are resolved in the living cell via interference based imaging or by other light microscopies with suitable modulation transfer functions. Simultaneous or near simultaneous low level florescent imaging of internal free calcium ions or pH is possible with dual illumination and imaging pathways coupled to high resolution ISIT or CCD type cameras. Computer control of illumination and image acquisition permits direct comparisons of label position and movement, cellular structure, and fluorescent signals. At present, localization of labels at high spatial resolution and relative to cell surface and internal ultrastructure requires preparative procedures which more or less attempt to maintain the structure as close to the living state as possible yet stabilize and provide contrast such that imaging via SEM, TEM, and AFM are possible.[2,3] Preparation, however, can involve an extremely wide range of possibilities often, but not necessarily encompassing some sort of physical (cold temperature) or chemical "fixation", followed variously by dehydration designed to minimize surface tension and facilitate subsequent removal of the liquid phase or its replacement with solidifiable polymers. Certain additional labeling, cytological and/or cytochemical procedures can be pursued at various points in the preparation process. Newer SFM technology and possibly NSOM procedures may provide high resolution information regarding surfaces, in the case of the former; and the surface and/or internal information in the latter in fully hydrated or even living specimens and hence both SFM and NSOM may soon become important links in the correlative process.

During or following LM examination, SFM, with a suitably configured tip can be used to provide the higher, molecular level resolution on hydrated or dried cells, and can be used to identify individual labels, tracked by LM in living cells. Under appropriate circumstances SFM can provide considerable surface detail and label localization often in the absence of extensive sample preparation. Low voltage HR-SEM, while requiring more extensive preparation of the sample, can obviate the need for coating and provide correlative images of the same cell or individual labels viewed first by LM and SFM. HR-SEM provides a clear, high resolution image of the surface structure; labels can be identified by their distinctive spherical shape at low, 1.0-1.5 kv, accelerating voltage. Increasing the accelerating voltage reduces surface detail but permits unambiguous identification of the gold label via imaging of increased secondary and/or backscattered electrons. Additional information regarding internal structure can be collected at the higher accelerating voltages if specimens are appropriately stained.[3] Higher resolution EDX can provide compositional information. Continued correlative examination of the same cell and labels by TEM at standard, intermediate, or high (1 MeV) accelerating voltages can clearly identify the position of the label relative to internal ultrastructure, for example cytoskeletal elements or cell organelles.

Thus in this single, limited example the use of relatively straightforward correlative microscopy not only permits identification of positively labeled cells but provides considerable additional information regarding initial sites of binding, movement of ligand receptors or antigen-antibody complexes, associated changes in cell structure and cell adhesion, changes in intracellular free calcium and pH associated with the movement. Rates and direction of ligand-receptor complex movement can be determined. The ultimate destination of the complex can be correlated relative to surface and internal structure and ultrastructure identified variously by selective stains or labels.

There are a variety of correlative strategies that can be pursued. Some involve relatively simple combinations of standard procedures, while others employ more extensive use of stains and labeling procedures coupled with multiple imaging modes. Although care must be exercised so that conditions necessary for any one procedure do not interfere with the success of other correlative steps; nearly any combination of methodologies can be designed to be compatible via a correlative approach.

1 B. Wetzel and R.M. Albrecht, *Scanning Microscopy Sup.* 3(1989)1.
2.R.M. Albrecht et al., in *Immunocytochemistry: A practical Approach.* J. Beesley(ed). Oxford Press.
3.R.M. Albrecht et al., *Proc . Ann MSA Meeting* 52(1994)1016.

ANALYTICAL SOLUTIONS FOR COMPLEX PROBLEMS
USING MULTIPLE DIAGNOSTIC TECHNIQUES

L. A. Heimbrook

AT&T Bell Laboratories, Reading, PA 19604

The ability to apply multiple diagnostic techniques to complex material, biological, and device problems with the goal of obtaining analytical solutions is the daily objective of the typical analytical laboratory. This paper will describe the use of both microscopy and surface analysis diagnostic tools to evaluate routine and highly complex material and device problems often found in the semiconductor industry. The characterization requirements for silicon and III-V materials and devices cover a wide range of technology research and development programs. These programs involve the characterization of starting materials, doping and implant technologies, thin film technology, particle and contamination issues and final device inspection. Continued advances in ultra-shallow silicon devices and multi-quantum-well (MQW) lasers rely on the accurate introduction of doping elements, the ability to deposit high quality materials of specified layer thicknesses and on advances in fabrication and characterization tools.[1-4]

The complex analysis problems of evaluating ultra shallow junctions in sub-micron silicon devices and the measurement of grating depths and duty cycles in semiconductor lasers are the characterization challenges which will be addressed in this paper. Figures 1 and 2 show cross sectional views of a multi-quantum-well (MQW) laser and a metal oxide semiconductor field effect transistor (MOSFET) respectively. The key diagnostic tools currently used in evaluating ultra-shallow junctions and semiconductor laser gratings are: Scanning Electron Microscopy (SEM), Transmission Electron Microscopy (TEM) with chemical staining techniques, Atomic Force Microscopy (AFM), Focused Ion Beam (FIB) sample preparation, Secondary Ion Mass Spectrometry (SIMS), Auger Electron Microscopy (AES) and a variety of other surface and electrical characterization methods. Examples will be presented of the current use of these diagnostic tools to provide data on the actual delineation of a junction, grating or active area, on the quality of the ion implantation process, and on layer thickness and interface quality of the structures shown in Figures 1 and 2. The high resolving power of SEM and TEM have often made them the characterization tools of choice to examine chemically-etched ultra-shallow junctions of silicon devices at high magnification. The accuracy of these microscopy methods for identifying p-n junctions will be compared to the combined analysis of SIMS, electrical measurements and theoretical simulations. The use of the non-destructive characterization method of AFM in the analysis of wet-etched gratings in multi-quantum-well lasers will also be examined in relations to cross-sectioned SEM and TEM analysis.. The correlation of microscopy and surface analysis data as well as the gaps in characterization capability versus industry requirements will be summarized.

References

1. S. M. Sze, editor, VLSI Technology, McGraw Hill, New York, 1983.
2. S. M. Sze, Semiconductor Device Physics and Technology, Wiley and Sons, New York, 1985.

Proc. Microscopy and Microanalysis 1995, edited by G.W. Bailey, M.H. Ellisman, R.A. Hennigar, and N.J. Zaluzec

3. M. Geva, SIMS Analysis of InP and Related Materials, in Sodium Phosphide and Related Materials, A. Katz, Ed., Artech House, Boston, 1991.

4. T. T. Sheng and R. B. Marcus, Transmission Microscopy of Silicon: VLSI Circuits and Structures, John Wiley and Sons, New York, 1983.

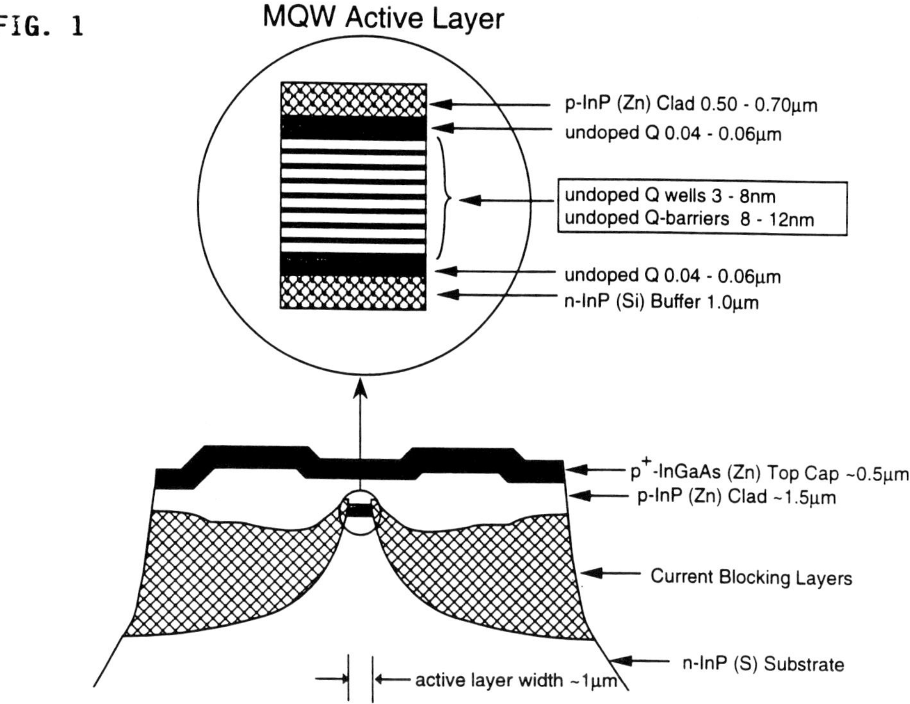

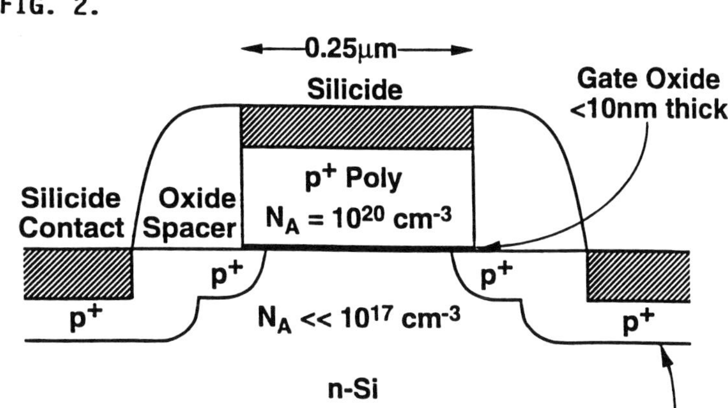

FIG. 1.—Cross sectional view of a Multi-Quantum Well (MQW) Laser.
FIG. 2.—Cross sectional view of a Metal-Oxide-Semiconductor Field Effect Transistor (MOSFET).

CORRELATIVE MICROSCOPY TECHNIQUES FOR MATERIAL SCIENCE

J.S.Hudson

Electron Microscope Facility, 120 Kinard Lab , Clemson University,
Clemson, SC. 29634

The microscopy center at Clemson University recently invested funds to provide a computer network system that incorporates all of its microscopes. The facility connects SEM, TEM, STM/AFM, Auger Microprobe and the light microscope to Sun workstations equipped with chemical analysis and imaging programs. Images from the network system microscopes can be sent to any of the workstations. I should like to review a few applications of correlative microscopy techniques related to material science; this is a technology that allows the acquisition of multiple data from a given sample. Often a given technique can be augmented by the use of complimentary microscopy technique. Since electron microscopy is subject to interpretation, correlative microscopy methods will prove to be useful in reaching conclusions regarding the image micrographs. Additionally, more than one type of information may be necessary for a given material and this can be found with the different systems of microscopy. In my presentation I will discuss instrumentation and methods by demonstrating advantages and disadvantages of applications as they apply to materials such as polymers, ceramics, microstructures and textiles. The success of correlative microscopy is determined by the user's knowledge of the instrumentation and the sample preparation.

The use of light microscopy for the visualization of large surface areas taking into consideration various staining techniques, can be combined with scanning and transmission electron microscopy. The backscattered images along with the secondary images from the scanning microscope will demonstrate their usefulness as intermediate studies between conclusions observed with the light microscope and the transmission microscope. The scanning tunneling atomic force microscope may be used to analyze materials at the atomic level. Recently the STM/AFM has proven itself to be a useful instrument in materials research. With it, information regarding microstructures of films, studies of grain boundaries and surface structures of ceramics is being investigated. For example, characterization of the surface roughness of ceramic fibers, a factor critical to the overall mechanical behavior of the material may be accomplished with STM/AFM.

The combined resources of the Auger and SEM may be used to generate chemical maps for analysis. Correlative microscopy allows preselection of specific areas of interest for chemical microanalysis. In some studies, the correlation between morphological and elemental information is necessary. Discrete second-phase particles may be evaluated by volume fraction, number per area and standard measurements of area, perimeter and average diameter. However, problems arise in the evaluation process when analyzing clusters which appear in multiple manifestations. A new and powerful approach to this problem is Dirichlet Tessellation, which is simply a mosaic of which we are able to evaluate particles by their shape. Correlative microscopy of materials may require the resolution of problems by employing the applications of three or more forms of microscopy. The advantage of corroborative data is significant.

THE MULTIANALYTIC APPROACH; OR YES, YOU CAN GET THERE FROM HERE!

PROBLEM SOLVING IN SEMICONDUCTOR MANUFACTURING

M. M. Ecker

AT&T Microelectronics, Reading, PA 19604

Manufacturing, development and failure mode analysis problems, in the semiconductor industry, present unique challenges and opportunities, not only to utilize several analytical techniques, but to process and deprocess samples into an appropriate format for analysis. In today's' cleanrooms a great deal of development is integrated into the manufacturing line.   This, combined with the ever decreasing size and increasing density of components on a single integrated circuit, creates the need for accurate as well as precise measurements. The method of measure must be appropriate to the task, compatible with the manufacturing environment, and useful in a process control system.

Current complimentary bipolar technologies, utilizing shallow junction implants, require additional as well as less robust processing in order to maintain the depth and latitude of the junction. Additional processing includes the formation of a contact window spacer. The spacer is formed, following implant and thermal anneal, by growing and subsequently removing a thermal oxide layer. In the manufacturing process, oxide removal is verified by optical measure using visible light. Results on the "in-line" development product identified an oxide film in the "P" type (BF2 implant) window. The "N" type windows were clean.

The film was measured by several methods including visible, UV and thermal wave. All identified a film of approximately 40A thickness.   In shallow junction technology a 40A film is considered continuous and is fatal to the device performance. More robust processing to remove the film did not preserve the shallow junction and was fatal to device performance. This was shown by SEM imaging of polished cross sections of completed devices.

Here, we present data from several methodologies to determine the composition and structure of the oxide film in attempting to understand the failure to etch.

Samples with the thermal oxide, prior to etch, were analyzed for composition of the film. Auger spectroscopy and SIMS provided surface and depth data. The AES analysis included surface scans, depth profiling and band gap techniques. The SIMS analysis was done at a slow erosion rate of 1.6A/sec. using a 2KV Cs+ beam. The data obtained identified the film as SiO2, with no evidence of contamination. The thicknesses measured were in agreement with our optical methods.

Structure analysis of the film by TEM included both planar and cross section preparations.   The preparations were chemically thinned to prevent any damage to the oxide. TEM data identified an amorphous, uniform film and confirmed our thickness measurements.

Proc. Microscopy and Microanalysis 1995, edited by G.W. Bailey, M.H. Ellisman, R.A. Hennigar, and N.J. Zaluzec
Copyright © 1995 MSA. Published by Jones and Begell Publishing, 79 Madison Ave., New York, NY 10016

This film should have responded to the manufacturing process etch. Was the film really there or were our optical methods misleading us? Samples were prepared from the etched product to investigate the surface morphology.

AES and SIMS analyses were repeated on the etched samples. The data from both methods indicted that there was no oxide present. Investigations of the surface morphology were made by AFM, HRSEM and TEM. The SEM images showed little contrast and a slight erosion of the surface. No distinct surface structure was found. The AFM data identified surface features randomly dispersed on the "P" type surface. The individual features were of the order of 20nm elevation.

TEM planar sections confirmed damage at the BF2 implant surface. The surface exhibited occlusions which were determined to be richer in B and contained low concentrations of C.

The result of all analytical methods indicated that the optical method is compromised by the surface irregularity of the BF2 implant regions. Electrical data from completed product supported this conclusion. The optical method is valid for the "as deposited" oxide as well as for the etched oxide in the "N" regions, however, the surface characteristics of the "P" regions create false readings.

The next challenge is the implementation of an accurate, non-destructive method of measure, compatible with the manufacturing environment.

The author acknowledges the technical expertise and assistance of E. Lane, H. S. Luftman, S. Nakahara, J. L. Shearer and C. H. Tzinis.

SCANNING ELECTRON MICROSCOPY AND X-RAY ANALYSIS OF MICROPARTICLES AND EARLY HYDRATION EFFECTS

Yun Lu[*] and David C. Joy[+]

[*]Albert Einstein College of Medicine, Bronx, New York.

[+]Electron Microscope Facility, The University of Tennessee, Knoxville.

High resolution scanning electron microscopy (SEM) and energy dispersive x-ray analysis (EDXA) were performed to investigate microparticles in blended cements and their hydration products containing sodium-rich chemical wastes. The physical appearance of powder particles and the morphological development at different hydration stages were characterized by using high resolution SEM Hitachi S-900 and by SEM S-800 with a EDX spectrometer. Microparticles were dispersed on the sample holder and glued by 1% palomino solution. Hydrated bulk samples were dehydrated by acetone and mounted on the holder by silver paste. Both fracture surfaces and flat cutting sections of hydrating samples were prepared and examined. Some specimens were coated with an 3 nm thick Au-Pd or Cr layer to provide good conducting surfaces. For high resolution SEM S-900 observations the accelerating voltage of electrons was 1-2 KeV to protect the electron charging. Microchemical analyses were carried out by S800/EDS equipped with a LINK detector of take-off angle = 40°. Other settings selected were: a window width of 10 channels, where each channel is 10 eV wide; detector resolution: 158 eV; probe current: 0.5 nA; counting time: 100 secs. The quantitative microanalysis was performed by using Desktop Spectrum Analyzer (DTSA) software acquisiting with a Macintosh II computer.

SEM/EDXA characterization showed heterogeneous morphologies of microparticles and resulting gels. The microparticles in fly ashes are mainly spherical. By contrast, other starting materials generally are composed of various irregular shaped particles. The spheres in fly ash, with a large fraction of amorphous and crystalline phases of mullite, quartz, hematite, etc., exhibit a broad range of sizes and many of them have particularly complex internal structures. The morphology of these particles may be classified as: (1) Solid spheres with various sizes; (2) Cenospheres and plerospheres; An example of the microspheres encapsulated in a parent hollow sphere is showed in Fig.(1). (3) Coated spheres; Almost all of fly ash spheres are coated with thin layers or tiny particles [1,2]. (4) Needle-like crystals and irregular particles. The behavior of fly ash in blended cements is attributed to its fine particle dispersing effect on the matrix and its high pozzolanic activity.

Both physical appearances and chemical changes in hydration samples are pronounced due to the interaction between the waste solution and cementitious products. During early stage of hydration, the fly ash behaves as a precipitation nucleus for cementitious gels. Reactions from other source particles appear to be much slower start. Fig.(2) shows that at 28 days of hydration, the surfaces of fly ash spheres were all covered by hydrating products. Higher magnification demonstrates that, as shown in Fig.(3), those tiny crystals are tree-like and the SEM/EDXA characterization demonstrated these fine tree-like crystals are Na-Al-Si-rich. In later hydration samples large areas of Na-rich crystal aggregates and layers were observed at the fracture surfaces. High concentration of Na was found frequently associated with Al-Si rich regions, implying that the interaction between Na-rich solution and the Al-Si fly ash microparticles played an important role in improving the retention of the blended system for Na-rich wastes. Various early hydration gels, particle aggregates, and reaction shells around the relict particles have been observed. As shown in Fig.(4), large white calcium carbonate (calcite $CaCO_3$) crystals with irregular shapes embed in the matrix. No Na-rich crystals deposited on $CaCO_3$ particle surfaces were observed. The surfaces of $CaCO_3$ particles remain smooth and have loosely bonding interfaces with their neighbors. The surrounding hydration gels, which are Si-Ca rich, generally have honeycomb or irregular flake structure and contain high trace elements from the chemical wastes. The SEM/EDXA results are in agreement with our previously TEM observations and electron diffraction patterns[3], which indicated that $CaCO_3$ is the predominant crystalline phase and the fly ash Al-Si glasses make main contributions to the nucleation of early hydration products.

REFERENCES

(1)C.M. Dobson et al, J. Mater. Sci., 1988, 23, 4108-14.
(2)W. Sinclair & G. W. Groves, J. Mater. Sci. Lett., 1986,5,101-102.
(3) Y. Lu and D. C. Joy, Proc. 52th annual Meeting of MSA, 1994,658-659.

Fig.(1): Fly ash microspheres encapsulated in a parent hollow sphere. Diameters of plerospheres vary from 5 µm to 0.1 µm. The wall thickness of the parent sphere (indicating by arrays) is about 1 µm. Bar = 800 nm.
Fig.(2) Al-Si spheres coated by tiny Na-Al-Si hydration crystals, supposing that the Na-rich waste was stabilized by reacting with Al-Si glasses, and trapped within the cementitious structure. Bar = 3.8 µm.
Fig.(3) The length of main branches of tree-like crystals from 1 µm to 1.5 µm, and the thickness from 50 nm to 150 nm. Dimensions of the second branches: lengths from 200 nm to 600 nm, and thickness around 20~50 nm. Bar = 600 nm.
Fig.(4) White calcium carbonate crystals with smooth surfaces embed in honeycomb hydration gel matrix. Bar = 1.6 µm.

Proc. Microscopy and Microanalysis 1995, edited by G.W. Bailey, M.H. Ellisman, R.A. Hennigar, and N.J. Zaluzec.
Copyright © 1995 MSA. Published by Jones and Begell Publishing, 79 Madison Ave., New York, NY 10016

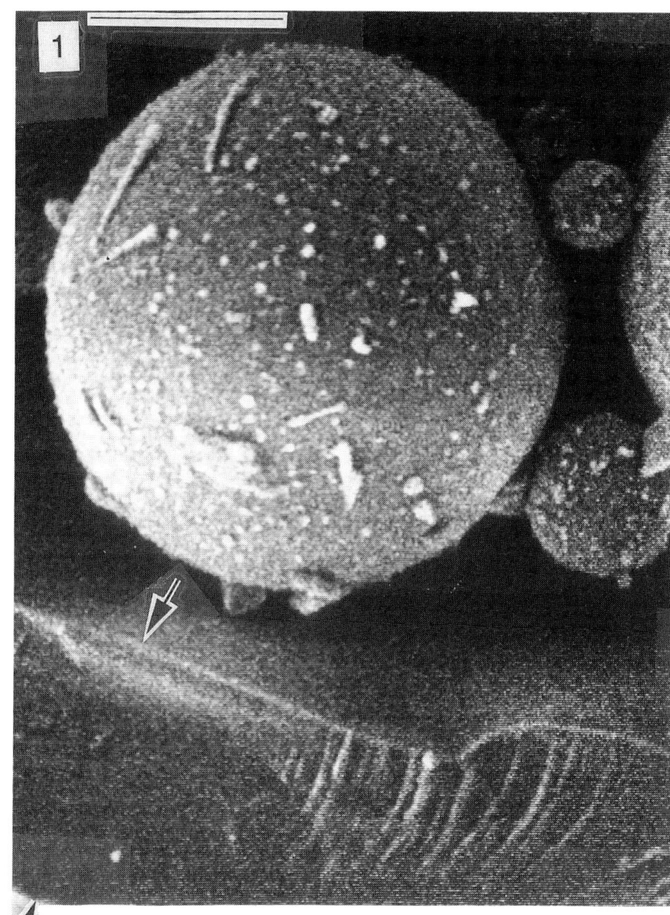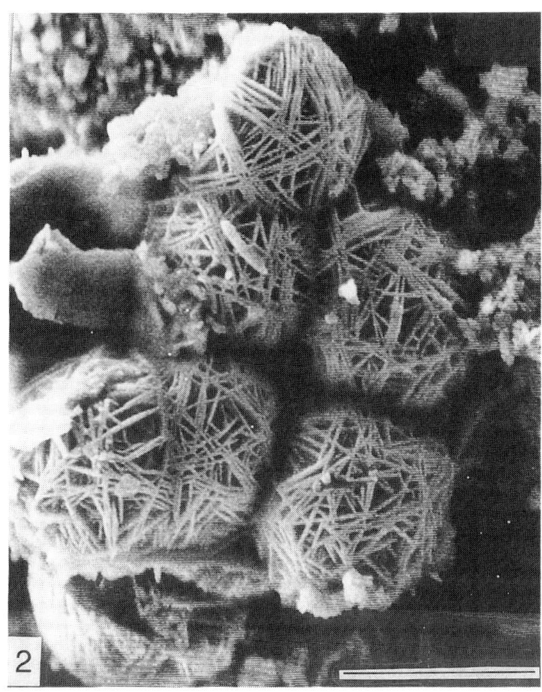

COMPARATIVE INVESTIGATION OF A Pd/Al_2O_3 CATALYST SYSTEM WITH TEM AND STEM

D. R. Liu[*], S.J. Pennycook[**], M. H. Yao[*] and T. E. Hoost[*]

[*]Ford Motor Company, Research Laboratory, 20000 Rotunda Drive, Dearborn, MI 48121;
[**]Oak Ridge National Laboratory, Solid State Division, Oak Ridge, TN 37831.

Pd-only catalysts are attractive for automotive applications because of the low cost and wide availability of Pd compared with that of Pt and Rh. Analytical electron microscopy may be used to investigate the dispersion, which is a very important parameter in catalyst characterization, of the Pd particles on γ-Al_2O_3 supports. It was of interest to compare how two different techniques, TEM and STEM, might be used for the Pd-particle size determination in a model catalyst. The model system of $1wt\%Pd/\gamma$-Al_2O_3 used in the present study was prepared by the incipient wetness technique from a palladium nitrate solution. It was then dried at 120°C and calcined at 400°C, followed by steam aging at 600°C for 24 hours. The TEM used was a JEOL 2000FX operated at 200 kV whereas the STEM used was a VG Microscopes HB-501UX operated in the Z-contrast mode at 100 kV.

It was not difficult to see Pd particles in the size range of 5 to 40 nm with TEM. Shown in Fig. 1 is such a Pd-particle, where one does not see a very clear Pd particle boundary with the support, which was true in most cases for this model catalyst. About 100 such Pd-particles were measured. The particle size distribution (PSD) of the measured particles is shown in Fig. 2. The average size (simple averaging) is 19 nm. The area-weighted average and volume-weighted average sizes are calculated to be 26 nm and 29 nm, respectively, using TEM.[1] The size (the volume-weighted average size) as determined with XRD (x-ray diffraction) was 32 nm, comparable with the TEM-determined volume weighted average. However, due to the fact that palladium has a relatively small atomic number which makes Pd less contrasty against alumina background in a TEM micrograph, it would be difficult to see the presence of Pd particles in an alumina substrate if they are around 1 nm or less. For example, it was not clear whether some small dots in the micrograph in Fig. 3 were actually Pd particles.

The same catalyst was examined using the HB-501UX STEM. A micrograph is shown in Fig. 4, where the Pd particles down to about 0.4 nm are clearly seen. The PSD of the measured particles is shown in Fig. 5. The simple average size is 1.3 nm whereas the area-weighted average and the volume-weighted average sizes are 1.6 nm and 1.8 nm, respectively. However, without prior knowledge or careful search, the large particles seen with TEM might have been missed with a STEM. Thus, with the two different techniques, TEM and STEM, we were able to see that there were two groups of Pd particles of very different sizes in this model catalyst. One group had an average size of 19 nm whereas the other showed an average size of 1.3 nm.

It is easy to see large Pd particles but not the small Pd particles with TEM because the weak phase and diffraction contrasts of the small Pd particles may change and even disappear as the focusing is changed. In a STEM equipped with a cold field emission gun, the probe current is in general low, which leads to a low single/noise ratio in a micrograph compared with the ratio in a TEM micrograph. Since the probe current in STEM is independent of magnification, low magnification images also have poor statistics. Thus, small Pd particles with bright contrast and round shape on the edge of the Al_2O_3 can be easily identified against the substrate with gray contrast (Fig. 4) whereas a large Pd particle might be missed because such particles are rare and low magnification must be used. Therefore, it may be necessary to use both TEM and STEM to investigate a Pd/Al_2O_3 catalyst if the PSD of the Pd particles in the catalyst are suspected to be so broad that some are very small and suitable for STEM study whereas others are large and better suited for TEM study.[2]

References

1.   M. H. Yao et al., *Proc. Ann. EMSA Meeting*, San Francisco Press, 53(1994)776.
2.   Part of this research was supported by the Division of Materials Sciences, US Department of Energy, under contract DE-AC05-84OR21400 with Martin Marietta Energy Systems, Inc.

Proc. Microscopy and Microanalysis 1995, edited by G.W. Bailey, M.H. Ellisman, R.A. Hennigar, and N.J. Zaluzec
Copyright © 1995 MSA. Published by Jones and Begell Publishing, 79 Madison Ave., New York, NY 10016

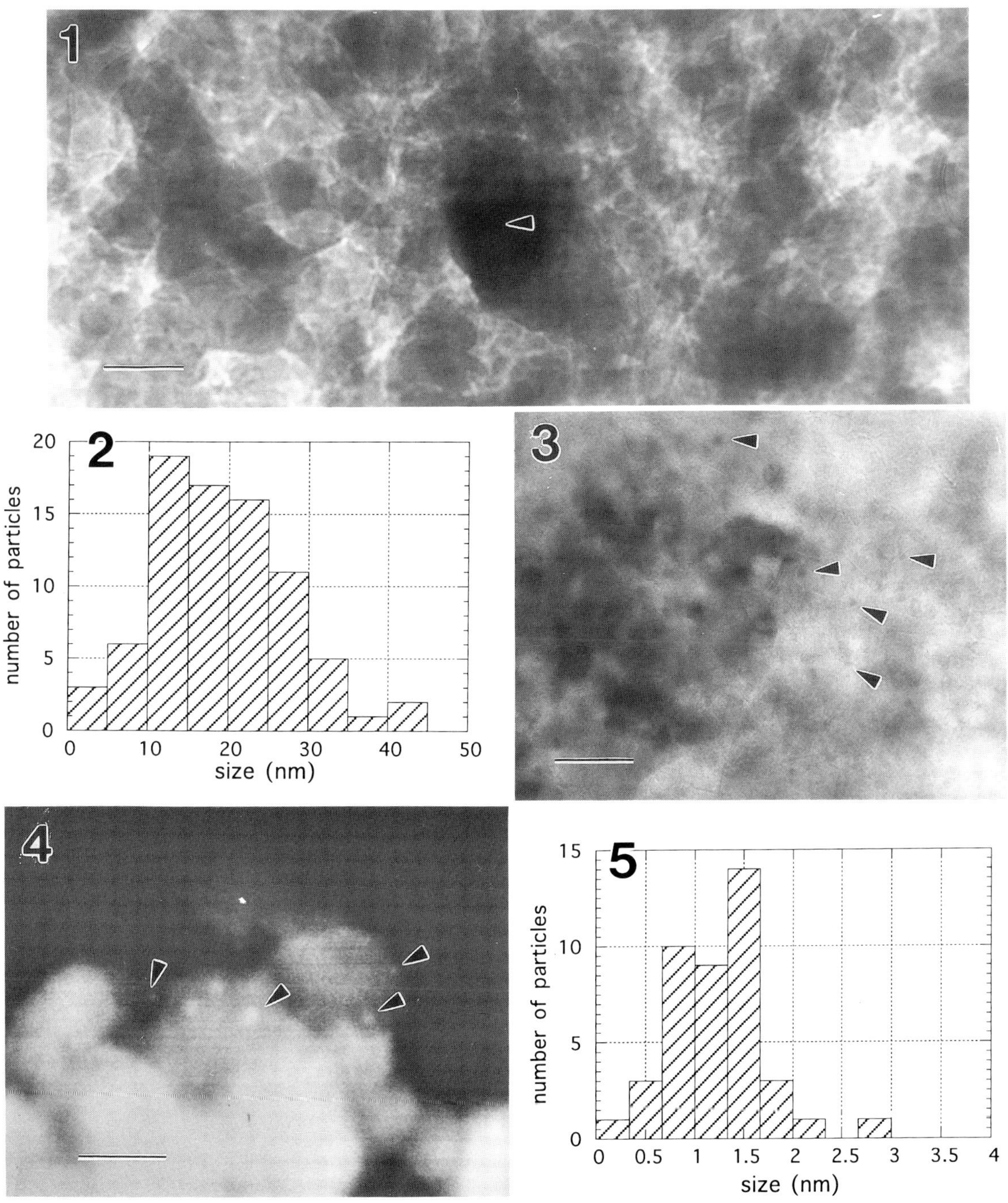

FIG.1.--TEM micrograph showing large Pd particle, marked by arrow. Bar=20 nm.
FIG.2.--PSD from TEM counting.
FIG.3.--TEM Micrograph showing small 'dots', with some being marked by arrows. Bar=20 nm.
FIG.4.--STEM micrograph showing small Pd particles in Z-contrast. Bar=10 nm.
FIG.5.--PSD from STEM counting.

OBSERVATIONS OF UNUSUAL STACKING FAULT CONFIGURATIONS USING COMPLIMENTARY STRONG AND WEAK BEAM IMAGES

H. S. Kim* and S. S. Sheinin**

* Department of Physics, Kyungsung University, Pusan 608-736, Korea
**Department of Physics, University of Alberta, Edmonton, AB, Canada T6G 2J1

In this paper, the authors present both strong and weak beam images of stacking faults in FCC cobalt which indicate the presence of an unusual stacking fault configuration. Experimental images have been obtained at 100 kV from a specimen in a [123] orientation with (1-1-1) stacking fault planes. Fig. 1 shows the (-1-11) strong beam dark field image. It can be seen that the contrast in faults A, B and D are very similar, while the fault at C is different. Using conventional techniques for identifying the nature of stacking faults in FCC materials[1] A, B and D have been identified as single or intrinsic faults with a displacement vector R=1/3[-111]. On the other hand, the fault at C can be identified as a double or extrinsic fault. At this point, the interpretation of the faults in the image in Fig. 1 appears to be completely straightforward and in keeping with results obtained from a conventional analysis. The next image of interest is shown in Fig. 2 which shows the same four faults as in Fig. 1 obtained under weak beam diffraction conditions. The operating reflection in this case is (11-1) and the deviation from the Bragg condition is given by the deviation parameter $S_g =$ 0.03 A^{-1}. In this image, faults B and D exhibit very similar contrast, as would be expected from the analysis of Fig. 1 which indicated that both faults are intrinsic. C on the other hand is quit different in contrast, again as would have been expected from the analysis based on Fig. 1 which indicated that this fault is extrinsic. The fault at A however exhibits very weak contrast and is virtually invisible. Clearly, therefore, the weak beam image in Fig. 2 indicates that the fault at A is different from those at B and D whereas the results obtained from an analysis of the strong beam image in Fig. 1 indicates that these three faults are all the same. The question which arises, of course, is what is the nature of the stacking fault at A in Figs 1 and 2.

In order to answer this question, theoretical calculations of the image contrast obtained from various stacking fault configurations have been carried out (details of the methods involved in these calculations will be presented elsewhere).[2] The theoretical images obtained were then compared to the contrast exhibited by the experimental images of the fault at A in Figs. 1 and 2. The stacking fault configuration which gave agreement between theory and experiment had a stacking sequence ABCABIAICIBIABC... (I indicates a slip plane) where the configuration represented by the fault plane consists of a quadruple fault or four adjacent intrinsic faults (equivalent to a microtwin). The result obtained for this fault configuration under the weak beam diffraction conditions used to obtain the experimental image in Fig. 2 is shown in Fig. 3. It can clearly be seen that image contrast is very weak, in agreement with experiment. In addition, theoretical image contrast obtained for this fault configuration under strong beam diffraction conditions showed that image contrast was very similar to the fault at A in Fig. 1 (this image is not shown).

In conclusion, it can be said that analysis of the nature of stacking faults based on either strong beam or weak beam images alone can be in serious error. A strong beam image indicates an intrinsic fault at A in Fig. 1 while in the weak beam image in Fig. 2, the presence of this fault would be difficult to detect. If both strong and weak beam images are obtained and appropriate theoretical calculations are carried out, the nature of the fault can be identified as being four adjacent intrinsic faults.

References

1.P. B. Hirsch, A. Howie, R. B. Nicholson, D. W. Pashley and M. J. Whelan, Electron Microscopy of Thin Crystals, London: Butterworths, 1965 (233)
2.H. S. Kim and S. S. Sheinin, to be published
3.The authors are grateful to the Foundation of Kyungsung University and the Natural Sciences and Engineering Research Council of Canada for support.

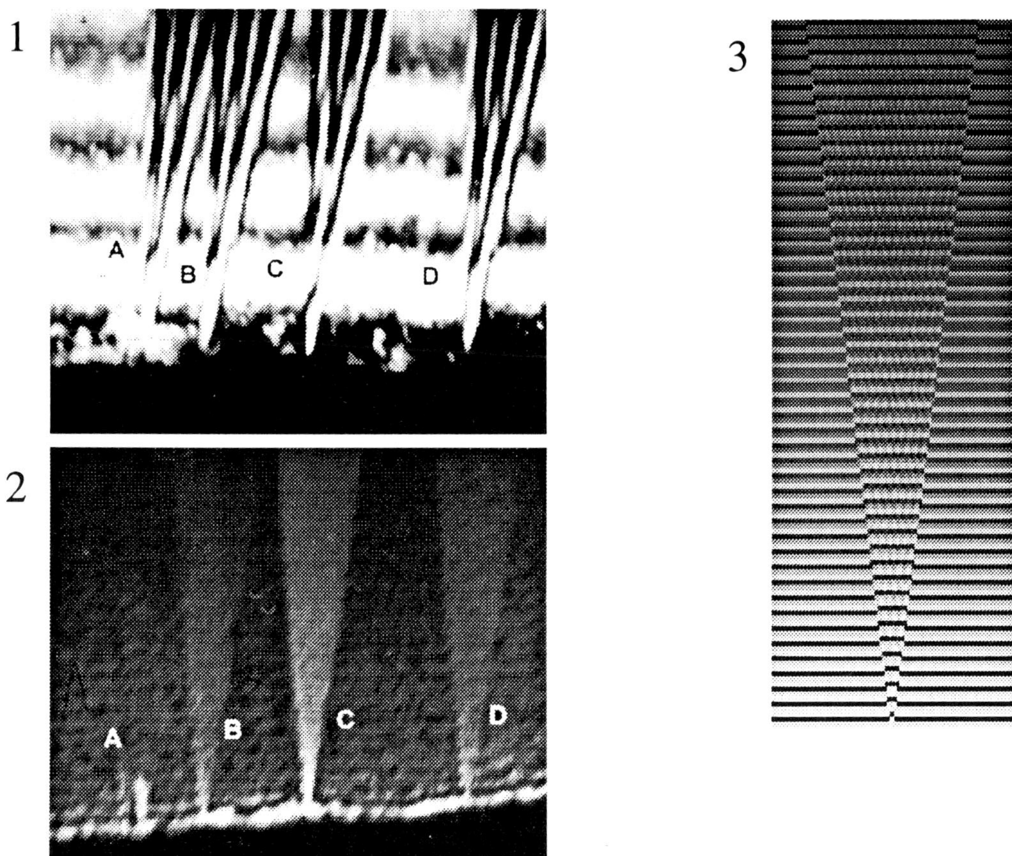

FIG. 1.-Experimental strong beam images of stacking faults in FCC cobtalt. A, B, and D exhibit the same contrast while C is different.
Fig. 2.- Experimental weak beam image of the same faults as in Fig. 1. Note that the contrast exhibited by A is very weak and quite different from B and D.
Fig. 3.- Theoretical weak beam image of a quadruple fault configuration showing contrast which is in good agreement with the experimental image at A in Fig. 2.

Structural Studies of Immunodeficiency Viruses

Beverly E. Maleeff[1], Timothy K. Hart[1], James A. Hoxie[2] and Peter J. Bugelski[1]

[1]Department of Toxicology-US, SmithKline Beecham Pharmaceuticals, King of Prussia, PA, and
[2]Department of Medicine, University of Pennsylvania School of Medicine, Philadelphia, PA

Immunodeficiency viruses (IV) are retroviruses belonging to the subfamily Lentivirinae and include human (HIV) and simian (SIV) immunodeficiency viruses.[1] In this study, we focus on the structure of proteins in SIV isolated from macaques (SIV_{mac}). IV form by budding through the surface of an infected cell, where viral proteins and RNA are enveloped by the host cell bilayer membrane. Immature virions consist of this outer membrane studded with surface glycoprotein projections and a thick sub-envelope protein plaque composed of unprocessed $Pr55^{gag}$ protein.[2] Viral maturation is marked by cleavage of this protein by a viral protease into the structural proteins p17, p24 and p15. Mature virus particles are icosahedral, approximately 120 nm in diameter, consisting of membrane surface glycoprotein projections, a thin submembrane p17 protein layer, a centrally located core capsid composed of p24 protein subunits surrounding the viral ribonucleic acid, and nucleocapsid proteins p7 and p6, derived from processed p15. One step in the development of anti-IV treatments is to understand the organization of the proteins in order to determine specific sites to target chemotherapeutic agents.

Hut78 cells chronically infected with SIV_{mac} were used as the source of virus for this study. Cell cultures were fixed with 2.5% glutaraldehyde in 0.1M phosphate buffer. For embedded preparations, cells were postfixed with osmium tetroxide, *en bloc* stained with tannic acid and uranyl acetate, and processed into epoxy resin. Thin sections (50-60 nm) were cut and stained with lead citrate. For negative staining, cell-free supernatants were treated with saponin to permeabilize the viral membrane and ammonium molybdate to negatively stain the viral particles. Specimens were examined with a JEOL 1200EX operating at 80 kV.

Transmission electron microscopic (TEM) analysis of immature SIV_{mac} in cross-section (Figure 1) reveals an amorphous, electron-dense protein plaque underlying the viral membrane. In mature SIV_{mac} (Figure 2a), longitudinal sections reveal an electron-dense trapezoidal core capsid that is centered in the virion. In cross-section (Figure 2b), the core appears round and may be oriented centrally or eccentrically in the virion. There is no evidence of a substructure in either orientation. Examination of negatively-stained mature SIV_{mac} (Figure 3) reveals a capsid similar to that seen in sectioned material with the same absence of structural detail.

We have been able to visualize a macrostructure in the viral plaques and cores using image analysis of TEM images of virions. Images were digitized from photographic negatives using an image processing workstation running OPTIMAS™ software (Optimas Corp., Edmonds, WA).[2] Masks were created of the plaques of immature virions or of the protein core of mature virions. Two-dimensional fast Fourier transform analysis (2D-FFT)[3] (Figure 4) was used to filter out information that concealed fine structural detail in images of individual SIV_{mac} particles. The resultant power spectrum of each mask was analyzed for periodic components present in the original image. Periodic structures were enhanced by editing the power spectra to remove non-periodic components and returned to the spatial domain using an inverse 2D-FFT. Analysis of immature SIV_{mac} revealed no internal plaque substructure, but there appeared to be anchors between the surface membrane glycoproteins and periodic structures on the plaque (Figure 5). In contrast, analysis of viral cores in SIV_{mac}, either in section (Figure 6) or negatively stained, revealed a periodicity consistent with the capsids being composed of six subunits.

TEM of SIV_{mac} virions, either by thin section or negative staining, provides an image of the viral structure, but the fine features remain obscure. The use of image processing, such as 2D-FFT, allows for the correlation of fine structure as observed in traditional TEM images with fine detail containing periodic information relevant to a substructure in the virus. The information thus obtained from these samples may be helpful in the development of a new vaccine or chemotherapeutic agent.[4]

<u>References</u>
1. Nermut, MV, Microscopy and Analysis 7:7-9, 1994.
2. Bugelski, PJ, et al. AIDS Res. Human Retroviruses 11(1): 55-64, 1995.
3. Russ, JC. The Image Processing Handbook, pp. 165-224. CRC Press, Boca Raton, FL, 1992.
4. This work was supported by SmithKline Beecham Pharmaceuticals.

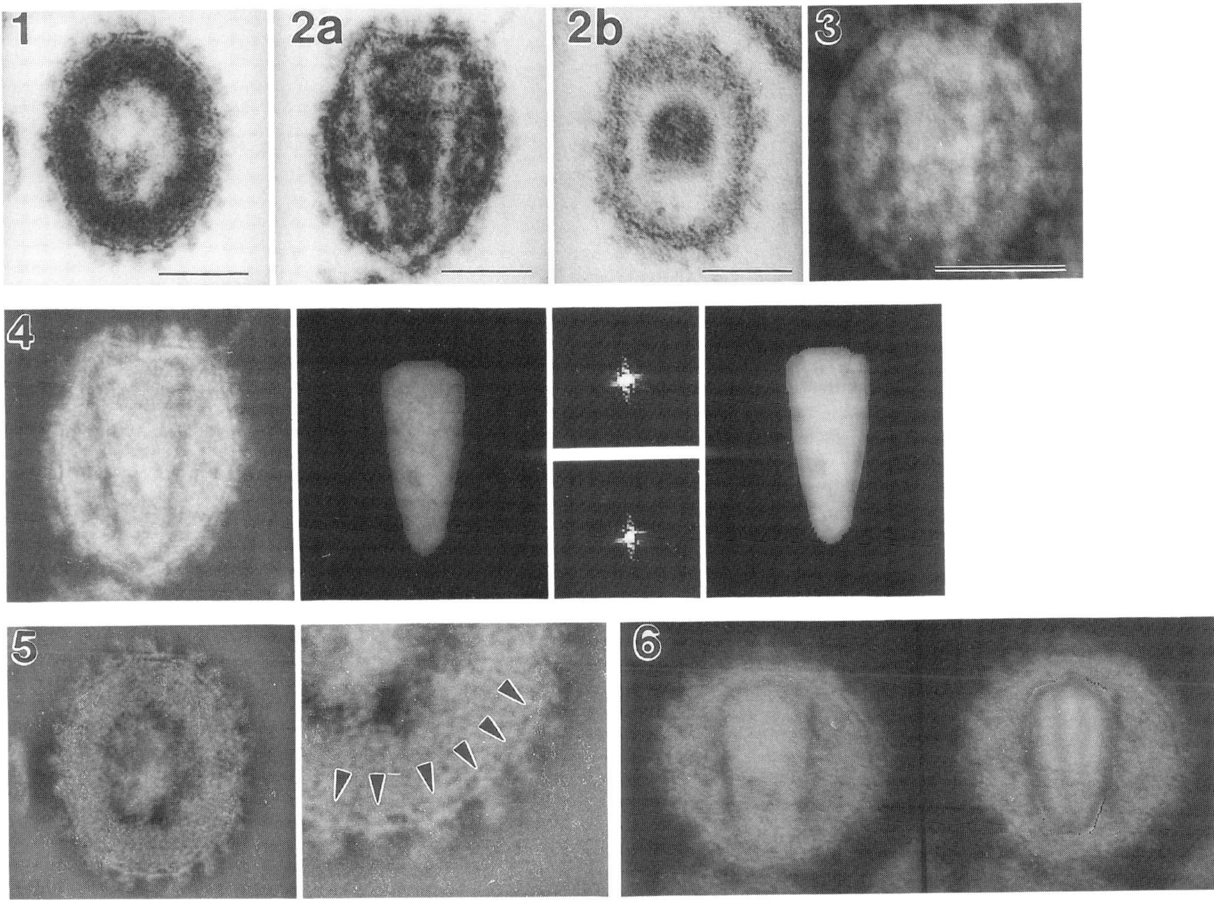

Figure 1   Transmission electron micrograph of an immature SIV$_{mac}$ particle. Note the protein plaque underlying the membrane. Bar = 50 nm.

Figures 2a, 2b Transmission electron micrographs of mature SIV$_{mac}$ particles in longitudinal and cross-section. Note the trapezoidal core capsid in longitudinal section and the centrally located core in cross-section. Bar = 50 nm.

Figure 3   Transmission electron micrograph of a negatively stained mature SIV$_{mac}$ particle. Bar = 50 nm.

Figure 4   The 2D-FFT process, as applied in this study. A TEM negative is digitized, a mask of the core created, a forward 2D-FFT applied, the resultant power spectrum edited, an inverse 2D-FFT applied, and the edited image is displayed in the spatial domain.

Figure 5   Transformed image of immature SIV$_{mac}$ revealing anchors between the protein plaque and the surface glycoprotein projections (arrows).

Figure 6   Transformed image of mature SIV$_{mac}$ revealing substructures in the protein core capsid.

MICROSCOPY AND MICROANALYSIS OF CALCIFICATION IN THE DEVELOPING AND ADULT BARNACLE

Crystal M. Brimer,^* Edward Roberts,* Daniel Hockett,+* Craig C. Freudenrich,+* and Ann LeFurgey+*#

^University of North Carolina, Chapel Hill, NC 27510; *Department of Cell Biology, Box 3709, Duke Univ. Medical Center, Durham, NC 27710; +School of the Environment, Duke University, Durham, NC 27706; #VA Medical Center, Durham, NC 27705

The barnacle begins its development in its first larval stage, the naupilus, and matures to the larval cyprid stage. Permanent attachment of the barnacle occurs between the cyprid and juvenile stages. Soon after, calcification takes place and the shells of the barnacle grow upward. The application of secondary electron imaging (SEI) and electron probe microanalysis (EPXMA) has defined the time course of calcification and shell growth in the barnacle.[1] Freeze-dried cryosections of the larval cyprid stage can also be employed for localization of calcium at the cellular level with scanning transmission electron microscopy (STEM)/EPXMA; however these sections are unstained, thereby offering very little contrast and making the recognition of ultrastuctural compartments difficult. In an attempt to identify these structures, conventional fixation staining and transmission electron microscopy (TEM) will be used. The light microscopic study of the cyprid has been extensive,[2] but very little ultrastructural research has been conducted. The goal of this study is twofold: to use light microscopy (LM) and TEM as guides in identifying structures of the STEM/EPXMA cryosections; and to follow the growth of the barnacle by using STEM/EPXMA to detect calcium tracers in the shell.

After fixation in 2% buffered glutaraldehyde and embedding in epoxy, the cyprid larvae were cut into semi-thick (350 nm) and thin (70nm) sections. To observe the three-dimensional structure of cells and subcellular regions, a computer reconstruction was created from one hundred and eight serial semi-thick longitudinal sections that were collected and digitized (Fig.1). An additional computer reconstruction using the cross-section of the cyprid is currently in progress. Characteristic body regions including the oil cells, appendages, and compound eye were identified in the light microscopic and TEM images. An enlarged montage of a longitudinal cyprid section was assembled from the TEM images (Fig.2). From this heterogeneous cell population, cell substructures have been identified including nuclei, mitochondria, muscle fibers, etc.

EPXMA is useful to further understand biomineralization and growth in the cyprid. EPXMA cannot distinguish between isotopes of calcium; therefore, several calcium substitutes have been tested as tracers of calcium.[3] Manganese and strontium have been successfully incorporated into the barnacle shell and detected by EPXMA, proving to be useful calcium tracers (Fig.3). We are exploring the possibilities of using barium as an additional tool in tracing calcium movement, but previous studies have shown that Ba at micromolar concentrations induces structural deformation.[4]

We are still working to identify the ultrastructure of the cyprid larvae and to understand the methods of biomineralization in the barnacle. The recognition of cellular organelles in the TEM image allows us to refer to the correlating location on the unstained cryosection and grasp the relative distribution levels of calcium given by the EPXMA image. The inclusion of metals into the shell depicts the flow of calcium as the barnacle undergoes calcification as a means of growth. This combined approach employing TEM, STEM and EPXMA as well as using SEI and EPXMA to detect calcium tracers will allow for better comprehension of the mechanics of biomineralization and growth of cyprids and adult barnacles.

Proc. Microscopy and Microanalysis 1995, edited by G.W. Bailey, M.H. Ellisman, R.A. Hennigar, and N.J. Zaluzec

References
1.   N. Wallace et al., *Proc. Ann MSA Meeting* 52(1994)178.
2.   L.J. Walley, *Phil. Trans. Roy. Soc. London* 256(1968)237.
3.   C.C. Freudenrich et al., *Proc. Ann MSA Meeting* 52 (1994)180.
4.   S. Gallager, et al., *Proc. of the 5th Intern. Symp. on Biomineralization* (1990)7.
5.   This work supported by the National Science Foundation (NSF BIR9106607) and the Office of Naval Research (N00014-92-J-1211)
The authors extend special thanks to Dr. Peter Ingram for his assistance.

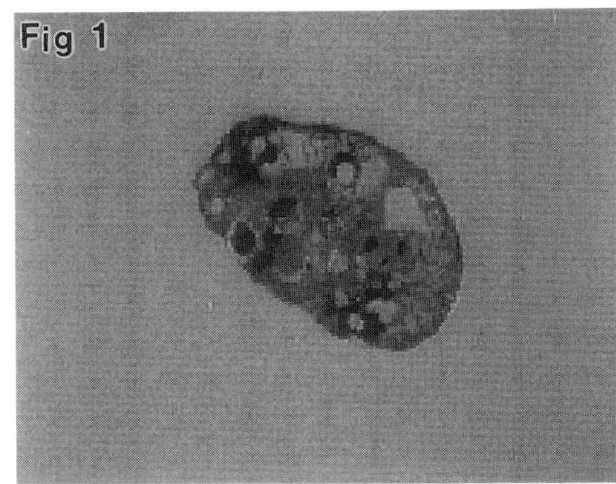

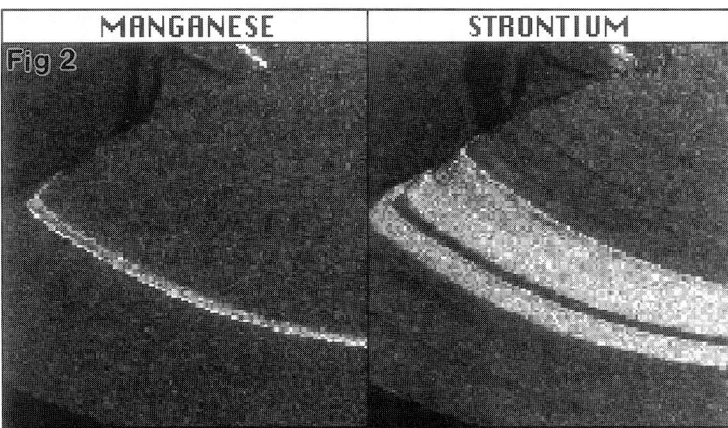

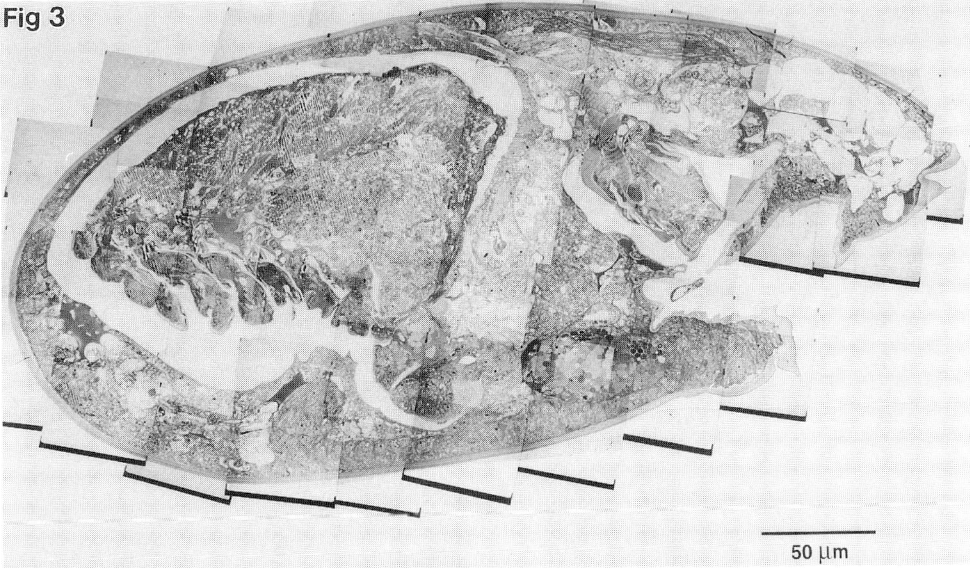

Figure 1- A digitized image taken from 3-D computer reconstruction of the cyprid.
Figure 2- An enlarged montage of a longitudinal cyprid section assembled from TEM images.
Figure 3- EPXMA map of a barnacle shell pulsed with Mn and Sr. The bands show detection of Mn and Sr respectively.

CORRELATIVE IMMUNOLOCALIZATION WITH HIGH RESOLUTION LIGHT MICROSCOPY AND ELECTRON MICROSCOPY USING LONDON RESIN GOLD EMBEDDED TISSUE

M. L. Grove*, B. A. Evans*, D. N. Misra**, J. Zhao***, D. H. Alpers***, and M. J. Becich*

*Univ. of Pittsburgh School of Medicine, Dept. of Cellular and Molecular Pathology, Pittsburgh, PA 15261
**Children's Hospital of Pittsburgh, Univ. of Pittsburgh School of Medicine, Dept. of Pathology, Pittsburgh, PA 15213
***Washington University Medical School, Gastroenterology Division, St. Louis MO 63110

Immunoelectron microscopy specimens are often embedded in hydrophilic resins, like London Resin Gold(LRG), which permit antibody staining without etching (or deplasticizing) the sections.[4] This characteristic of hydrophilic resins also allows for immunohistochemistry at the light level on semi thin sections (0.5μm - 1μm).[5,6] The ability to do immunohistochemistry at both the light and electron microscopic level on the same tissue block allows focused ultrastructural study. Immunohistochemistry on semi-thin sections displays cellular localization of macromolecules, permitting more specificity in the selection of areas for studying intracellular localization ultrastructurally. We have developed a method for immunoperoxidase staining of LRG embedded tissues, utilizing anti-human polyclonal antibodies directed against intrinsic factor (Fig. 1). Intrinsic factor (IF), a cobalamin binding protein, is known to be produced in the stomach, pancreas and salivary glands of most mammals. We are interested in distribution of IF in gastric epithelium, small intestine (ileum) and supporting tissues in both gastrointestinal tract sites. Previously, we have described the cellular localization of IF in human and rat tissues.[1,2] We are using immunoperoxidase staining to study changes in the distribution of IF in rat stomach treated with H_2 blockers (omeprazole, cimetidine, or ranitidine) and it's subsequent fate in ileal adsorption.

High Resolution Light Microscopic Immunostaining (Modified from protocol developed by B. A. Evans[3]):
1) The tissue is processed for IEM, embedded in LrGold resin and UV polymerized at -20° C. 0.5 μm sections are cut and mounted on plus coated slides.
2) Trace sections with a PAP pen, hydrate overnight with dH_2O.
3) Wash in PBS for 5 min followed with a dH_2O rinse for 5 min.
4) Treat slides with protease (Type XXIV: Sigma; St. Louis, MO) for 3 min at 25° C.
5) Treat slides for 30 min at 25° C with 0.5% H_2O_2 in MetOH.
6) Rinse in dH_2O for 5 min, wash in PBS for 5 min and then wipe around sections.
7) Apply Protein Blocking Agent (Blue Block: Shandon Lipshaw; Pittsburgh, PA) for 1 hr at 25° C in a humid chamber.
8) Drain blocking agent and place primary antibody (rabbit α-human IF, 1:50: Dr. Alpers; St. Louis, MO) on sections, incubate for 16-24 hrs at 25° C in a humid chamber.
9) Drain antibody, wash in 2 changes of PBS (5 min each) and wipe around sections.
10) Place secondary antibody (biotinylated goat α-rabbit IgG, 1:200:Vector; Burlingame, CA) on sections and incubate for 2 hrs at 25° C in a humid chamber.
11) Drain antibody, wash in 2 changes of PBS (5 min each) and wipe around sections.
12) Place ABC Elite (Avidin Biotin Complex: Vector; Burlingame, CA) solution drop wise onto sections and incubate for 30 min at 25° C in a humid chamber.
13) Drain slides, wash in 2 changes of PBS (5 min each) and wipe around sections.
14) Treat with DAB(3, 3'-diaminobenzidine: Vector; Burlingame, CA) for 10 min + at 25° C until sections turn brown, then rinse in dH_2O for at least 1 min.
15) Counterstain for 30 min with Hematoxylin (Instant Hematoxylin: Shandon Lipshaw; Pittsburgh, PA) which has been filtered and rinse in 3 changes of dH_2O.
16) Dehydrate through a graded series of ethanols and 3 xylene's. Coverslip with Accumount (Baxter).

Proc. Microscopy and Microanalysis 1995, edited by G.W. Bailey, M.H. Ellisman, R.A. Hennigar, and N.J. Zaluzec
Copyright © 1995 MSA. Published by Jones and Begell Publishing, 79 Madison Ave., New York, NY 10016

References

1. D.H. Alpers et al., *Endocrinology*, in press.
2. T. A. Howard et al., *Proc. of Mic. Soc. Amer.*, (1993)32.
3. B. A. Evans et al., *Proc. of Mic. Soc. Amer.*, (1994)230.
4. J. Roth et al., *J. Histochemistry and Cytochemistry* 26(1978)1074.
5. R. W. Horobin, *J. Microscopy*, 131(1983)173.
6. J. H. Beckstead, *J. Histochemistry and Cytochemistry* 33(1985)954.

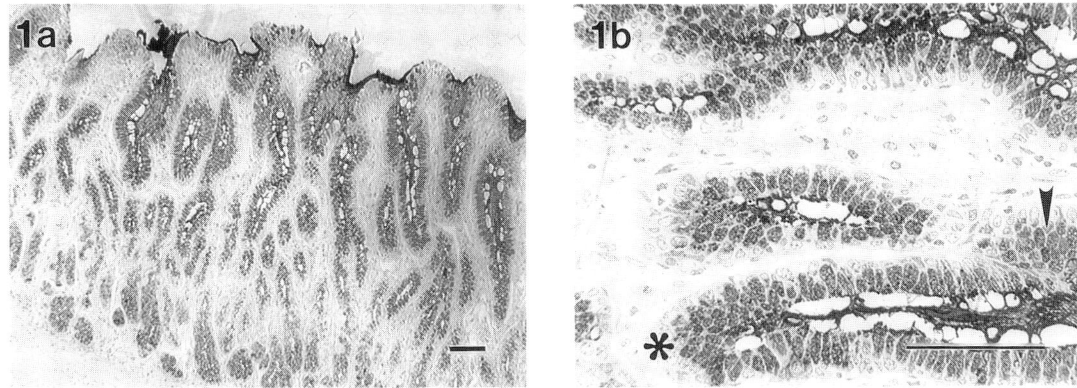

Fig. 1. LM images of normal human stomach showing immunoperoxidase staining for IF. (a) Gastric epithelium. (b) Chief cells (*) show a granular staining of the secretory vesicles and parietal cells (arrowhead) display disperse cytoplasmic staining. Bar = 100 μm.

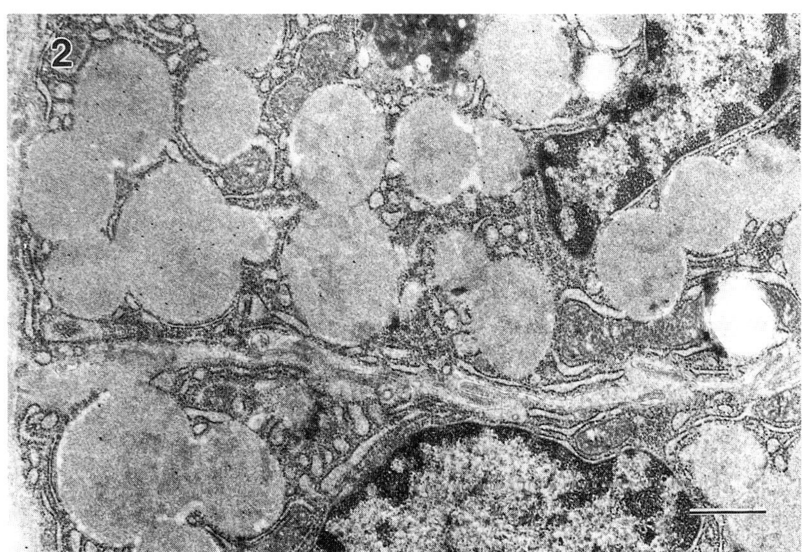

Fig. 2. TEM image displaying immunogold staining for IF in secretory vesicles.   Bar = 1 μm.

ACQUISITION AND ANALYSIS OF FLUORESCENCE RATIO IMAGES USING A MULTISPECTRAL MICROANALYTICAL IMAGING PROGRAM

Craig C. Freudenrich,*+ Robert T. Boyle,+ David A. Kopf,+ Peter Ingram,** and Ann LeFurgey*+¶

*School of the Environment, Duke University, Durham, NC 27706; +Division of Physiology, Department of Cell Biology, Box 3709, Duke University Medical Center, Durham, NC 27710; **Research Triangle Institute, Research Triangle Park, NC 27709; ¶Veterans Affairs Medical Center, Durham, NC 27705

Any study regarding the influence of physiological, pathological, or pharmacological/toxicological factors on cellular ion homeostasis requires knowledge of multiple parameters including cytoplasmic ion concentrations, ion fluxes, and subcellular elemental distribution so that an accurate assessment of the state of ion homeostasis can be obtained. These multiple parameters are often obtained by using different types of microscopy and microanalysis. For example, cytoplasmic ion concentrations are measured by fluorescence digital imaging microscopy using various ion indicators,[1] while cellular and subcellular element contents are determined by electron probe X-ray microanalysis (EPXMA) digital imaging.[2] For correlating information from various microscopies, it would be advantageous to have one software package and image format that operates on the same computer platform. Here, we describe the modification of an existing software (ImagNSpect) developed for EPXMA digital imaging[2,3,4] for use in analyzing fluorescence digital images.

Fluorescence images (490 and 440 nm excitation, 540 nm emission) of BCECF-loaded cultured rat hepatocytes (Fig. 1a, b) were obtained on an inverted fluorescence microscope (Carl Zeiss, Inc., Thornwood, NY) equipped with a rotating excitation filter wheel, a CCD camera (Photometrics, Ltd., Tucson, AZ), a controller box, and a Macintosh Quadra 800 computer (Apple Computer Corp., Cupertino, CA). A plug-in module was written into ImagNSpect that, at time intervals specified by the operator, would rotate the filter wheel, open and close the camera shutter, acquire the dark current and fluorescence images, subtract the dark current image from the fluorescence images, calculate a ratio image for display, and write the raw data to disk. At the end of the experiment, the cells were exposed to solutions containing 115 mM KCl, 10 mM Hepes (pH 6.49-8.04), 10 µM nigericin to calibrate the BCECF fluorescence.

For data analysis, ratio images were generated *manually* from the dark current-corrected images (Fig. 1a, b) by subtracting the background (non-cell) intensities before calculating the ratio. These ratio images (Fig. 1c) were then analyzed by three different methods. In Method 1, a histogram of the non-zero pixels was generated and fitted to a Gaussian distribution within the program to determine the mean ratio of the cell population for that image (Fig. 1d). In Method 2, individual cells (72 out of ~250 total) within the fluorescent and ratio images were *manually* outlined with the mouse function, the individual cell ratios were recorded, the distribution was plotted (Fig. 1e), and the mean ratio was calculated. In Method 3, a software module was developed for the program that would, for a time-series of images, *automatically* subtract the backgrounds from each dark current-corrected fluorescence image set, calculate the ratio image, calculate a weighted average of the pixels in a user-defined region of interest, and display the average ratio. Furthermore, intracellular pH values were calculated from the calibration images by each method and compared.

The ratios obtained by methods 1, 2, and 3, were 9.625, 9.48, and 9.457, respectively. The intracellular pH values obtained by methods 1, 2, and 3 were 7.975, 7.953, and 7.97, respectively. There were no significant differences between any of the methods with respect to ratios or pH values. The three methods of data analysis have different advantages and disadvantages. Method 1 has the advantage of obtaining a single number average for the entire cell population, but requires some manual manipulation in the data analysis; the manual manipulation can be time-consuming when

Proc. Microscopy and Microanalysis 1995, edited by G.W. Bailey, M.H. Ellisman, R.A. Hennigar, and N.J. Zaluzec

several image sets are present in the time-series. Method 2 has the advantage of detecting the variation between individual cells under various experimental manipulations, but is the most time-consuming because it requires manual data analysis almost exclusively; this is especially true for images containing many cells and time-series containing many image sets. Method 3 has the advantage of obtaining a single number for the entire cell population rapidly for any given time-series, but does not provide information about individual cells. Regardless of which method of data analysis is chosen, all of them can be conducted with the same software program used for acquisition of both fluorescence and EPXMA images; with its modifications, ImagNSpect becomes a versatile tool for the microscopy laboratory.

1.   D.L. Taylor et al., *American Scientist* 80 (1992) 322.
2.   A. LeFurgey et al., *J. Microscopy* 165 (1992) 191.
3.   P. Ingram et al., *Microbeam Analysis* 3 (1994) 111.
4.   P. Ingram et al., *ICEM* 13 (1994) 815.
5.   This work was supported by the Office of Naval Research (#N00014-93-F-0067, #N00014-94-1-0818), NIH S10RR06692, and NSF DIR9106607.

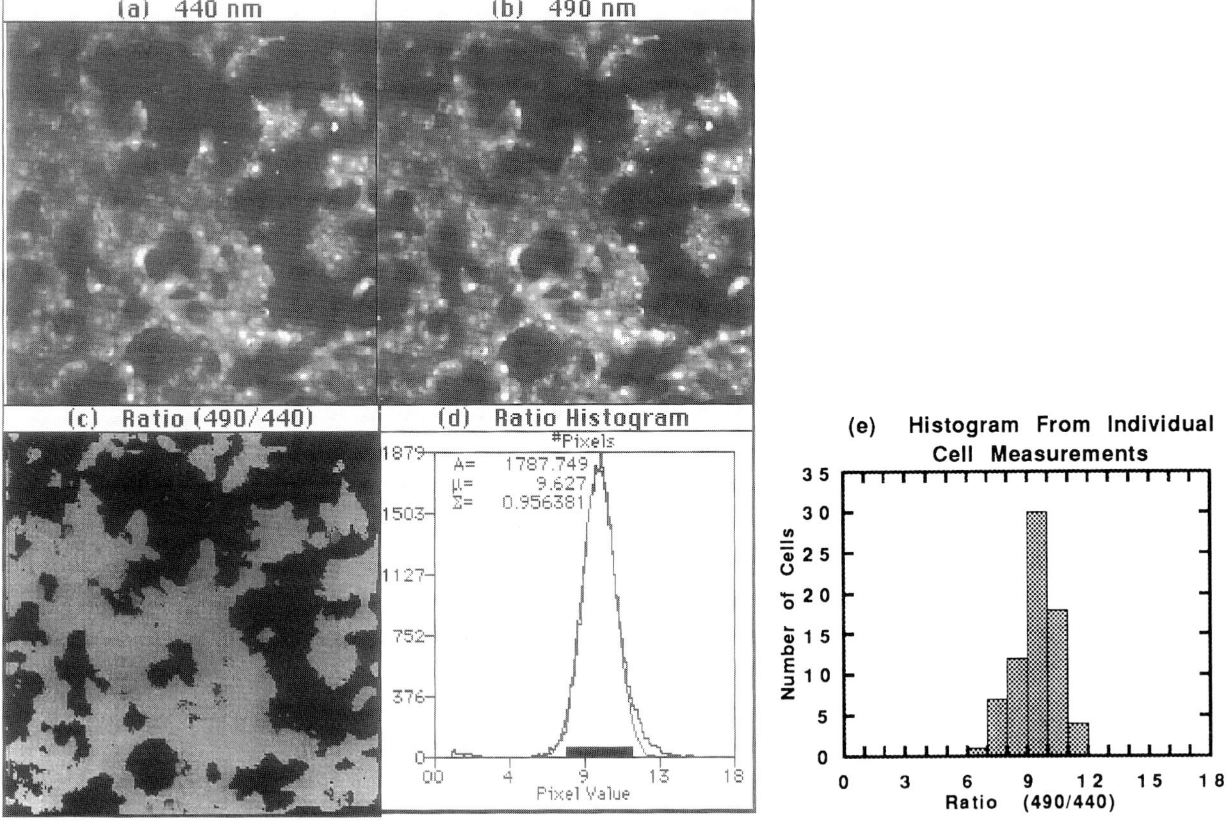

FIG. 1 - Fluorescence images (a,b) of BCECF-loaded rat hepatocytes along with the subsequent ratio image (c), pixel histogram from the ratio image (d), and a histogram of ratios obtained from individual cells within the image (e).

THREE MUSCLE ISOFORMS OF α–ACTININ FORM UNIPOLAR ACTIN BUNDLES

K. A. Taylor and D. W. Taylor

Department of Cell Biology, Duke University Medical Center, Durham, NC 27710

α-Actinin is an F-actin binding protein that is ubiquitous in eucaryotic cells. It is the smallest member of the spectrin superfamily and consists of two, identical polypeptide chains of 95-103 kDa. α-Actinin is an elongated molecule of 34 nm length and consists of a string of peptide domains with different functions. The N-terminal domain of ~255 residues has actin binding activity, the central core consists of four 122 residues triple-helical repeats and the C-terminal domain contains two E-F hand Ca^{2+} binding motifs[1].

The physiological importance α-actinin lies in its subcellular location and its interactions with other proteins. The F-actin crosslinking activity of α-actinin is its best known function. In striated muscle α-actinin is found in the Z-disk. In smooth muscle, α-actinin is found in both cytoplasmic dense bodies, which are analogs of the Z-disk, and adhesion plaques, which are cell membrane anchoring sites of actin filaments. In dense bodies and Z-disks, oppositely oriented, unipolar bundles of actin filaments overlap so it has been commonly assumed that α-actinin is functioning as a bipolar crosslinker but the unipolar orientation of actin filaments at the cell membrane implies a unipolar bundling activity. A unipolar crosslinking function for α-actinin is consistent with both Z-disk and adhesian plaque locations but has not been demonstrated in vitro with all muscle isoforms.

We have recently developed a method[2] for forming 2-D F-actin bundles (Fig. 1) to test the crosslinking orientation of 3 muscle isoforms of α-actinin, i.e. chicken gizzard (smooth), rabbit erector spinae (skeletal) and rabbit cardiac. Evidence for crosslinking orientation comes from 3 image features. (a) Diffraction patterns of 2-D bundles reveal sampling of the same spacing on both equatorial and non-equatorial layer lines. If the unit cell contained two oppositely oriented actin filaments, the spacing on the equator would be twice the spacing on the outer layer lines (Fig. 2). If the actin filaments were oriented randomly, the outer layer lines would be unsampled. The uniform spacing indicates one actin filament in the unit cell. (b) Actin filaments inserted in the 2-D bundle do not disrupt the crosslinking (Fig. 3). If the crosslinking were between antiparallel actin filaments and the crosslinking constitutively bipolar, insertion of an actin filament would have to disrupt the crosslinking to one of the neighboring filaments. (c) Single actin filaments coiled into spiral figures by α-actinin crosslinking can only occur if α-actinin has a unipolar bundling activity (Fig. 4). These data have been obtained for both smooth and striated muscle isoforms demonstrating in contrast to earlier conclusions that there is apparently no difference in the ability of α-actinin to form unipolar bundles.[2]

These data taken together indicate at the least that α-actinin forms unipolar actin bundles although they do not exclude a bipolar bundling activity. We have as yet seen no evidence of bipolar bundling with this assay. In addition several reports suggest that after treatments to remove tropomyosin, α-actinin can bind to the I-band of striated muscle and presumably crosslink actin filaments.[3,4] However, the 35-40 nm interfilament spacing in the bundles is much longer than the ~24 nm spacing between similarly oriented actin filaments in the Z-disk.[5] The only spacing between similarly oriented actin filaments in the Z-disk that appears to be compatible with the in vitro bundles is the spacing across the unit cell diagonal, i.e. ~34 nm. These observations would suggest a more complicated crosslinking pattern in the Z-disk than is evident from the in vitro studies reported here.[6]

1. A. Blanchard, et al., *J. Muscle Res. and Cell Motil.* 10(1989)280.
2. K. A. Taylor and D. W. Taylor, *Biophysical J.* 67(1994)1976.
3. J. W. Sanger et al., *J. Cell Biol.* 98(1984)825.
4. M. H. Stromer et al., *J. Cell Biol.* 40(1969)167.
5. Goldstein et al., *Electron Microsc. Rev.* 3(1990)227.
6. Supported by NIH grant AR42872

Proc. Microscopy and Microanalysis 1995, edited by G.W. Bailey, M.H. Ellisman, R.A. Hennigar, and N.J. Zaluzec
Copyright © 1995 MSA. Published by Jones and Begell Publishing, 79 Madison Ave., New York, NY 10016

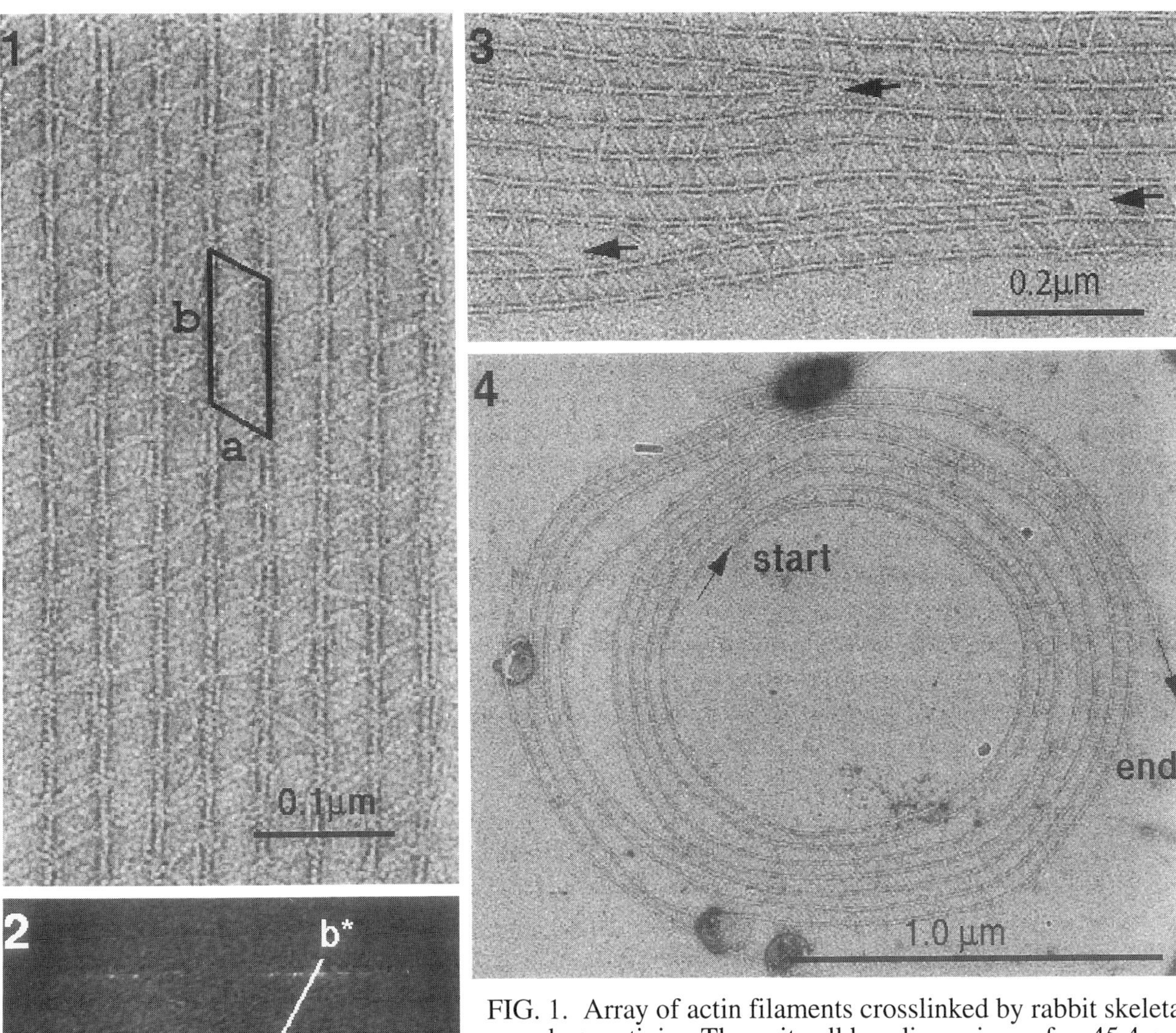

FIG. 1. Array of actin filaments crosslinked by rabbit skeletal muscle α-actinin. The unit cell has dimensions of a=45.4 nm, b=112.1 nm, γ=120.4° and contains only one actin filament. The interfilament spacing is 39.1 nm. The unit cell is defined by the more regularly arranged actin filaments than by the more disordered crosslinkers. Typically there are 2 α-actinin molecules per crosslink most of which are angled at ~60° to the filament axis, but occasionally α-actinins are angled oppositely.

FIG. 2. Computed diffraction pattern from the region shown in Fig. 1. Sampling on all layer lines indicates that the actin filaments in the array are oriented in the same direction. The helical structure of the actin filaments is 41 subunits in 19 turns of the 5.9 nm genetic helix.

FIG. 3. Another crosslinked array of actin filaments with insertion defects denoted by arrows. Note that insertion does not disrupt the crosslinking after the filament spacing has adjusted to the optimal amount. There is also no alteration in the angled direction of the crosslinkers.

FIG. 4. Spiral figure of one actin filament crosslinked with rabbit skeletal muscle α-actinin. Since the two segments of crosslinked filament are always parallel to one another, the crosslinking orientation must be unipolar.

LIGHT AND ELECTRON MICROSCOPIC LOCALIZATION OF INDUCIBLE NITRIC OXIDE SYNTHASE IN THE HEART OF RATS WITH MYOCARDITIS

T. Nishikawa,* S. Ishiyama,* K. Takeda,* T. Shimojo,* M. Hiroe,** and T. Kasajima*

*Department of Pathology, Tokyo Women's Medical College, Tokyo, 162 Japan
**Department of Internal Medicine, Tokyo Medical and Dental University, Tokyo, 113 Japan

Nitric oxide synthase (NOS) is an enzyme involved in the synthesis of nitric oxide (NO). The inducible form of NOS (iNOS) is induced in the presence of cytokines and produces massive amounts of NO which may be harmful to living tissue.[1,2] In acute myocarditis, a serious disease which can be fatal, extensive myocardial cell damage associated with massive inflammatory cell infiltration is usually observed and large amounts of cytokines are released from inflammatory cells.[3] However, the mechanism of the heart tissue injury is still unclear. In this study, we investigated the role of iNOS in the myocardial tissue injury in acute myocarditis.

The heart tissue specimens were taken from Lewis rats with experimentally induced autoimmune myocarditis.[4] An immunohistochemical study was performed using iNOS antibody raised in New Zealand white rabbits immunized with synthetic peptides (17 amino acids) corresponding to the C-terminal of macrophage iNOS. The specificity of the antibody was confirmed by Western blot analysis. Immunoelectron microscopic studies were carried out by the gold-labeled IgG method using ultrathin sections of LR white resin-embedded material. The distribution of iNOS mRNA in the myocardial tissue was analysed by in situ hybridization using an iNOS cDNA probe cloned from murine macrophages. The probe was labeled with digoxigenin.

Hematoxylin-eosin stained histologic sections of the heart from rats with experimental myocarditis showed severe myocyte necrosis with massive inflammatory cell infiltration composed of lymphocytes, neutrophils and macrophages (Fig. 1). Immunohistochemical staining with anti- iNOS antibody revealed positive reactivity in macrophages, neutrophils, vascular endothelial and smooth muscle cells. Myocardial cells adjacent to inflammatory cells were also positive (Fig. 2). Negative immunoreactivity was observed in uninflamed areas. Control staining, including the immunoabsorption test, showed negative reactivity. In the ultrathin sections, gold particles corresponding to iNOS were detected in the cytoplasm of myocytes (Fig.3) as well as macrophages, vascular endothelial and smooth muscle cells. An in situ hybridization study revealed positive reactivity for iNOS mRNA in the myocardial cells, vascular endothelial cells, vascular smooth muscle cells and macrophages in the inflammatory lesion (Fig.4). The myocardium which was not involved in the inflammation in the same heart revealed negative reactivity. The negative control using the sense probe also showed no reactivity.

These results indicate that iNOS is expressed in myocytes, macrophages, vascular endothelial cells and vascular smooth muscle cells in inflammatory lesion of the heart with myocarditis and suggest that the myocardial damage may be related to iNOS expression and subsequent NO overproduction in myocarditis.

References

1. M. Feelisch et al., *Nature* 368(1994)62
2. R. J. Rivers et al., *Am. J. Physiol.* 258(1990)H606.
3. A. Matsumori et al., *J. Mol. Cell Cardiol.* 24(1992)S122.
4. M. Kodama et al., *Clin. Immunol. Immunopathol.* 57(1990)250.
5. This study was supported by Grant-in-Aid for Scientific Research from the Ministry of Education, Science and Culture, Japan (No. 06670746), by a grant from the Japan Private School Promotion Foundation and by a grant from the Japan Research Promotion Society for Cardiovascular Disease.

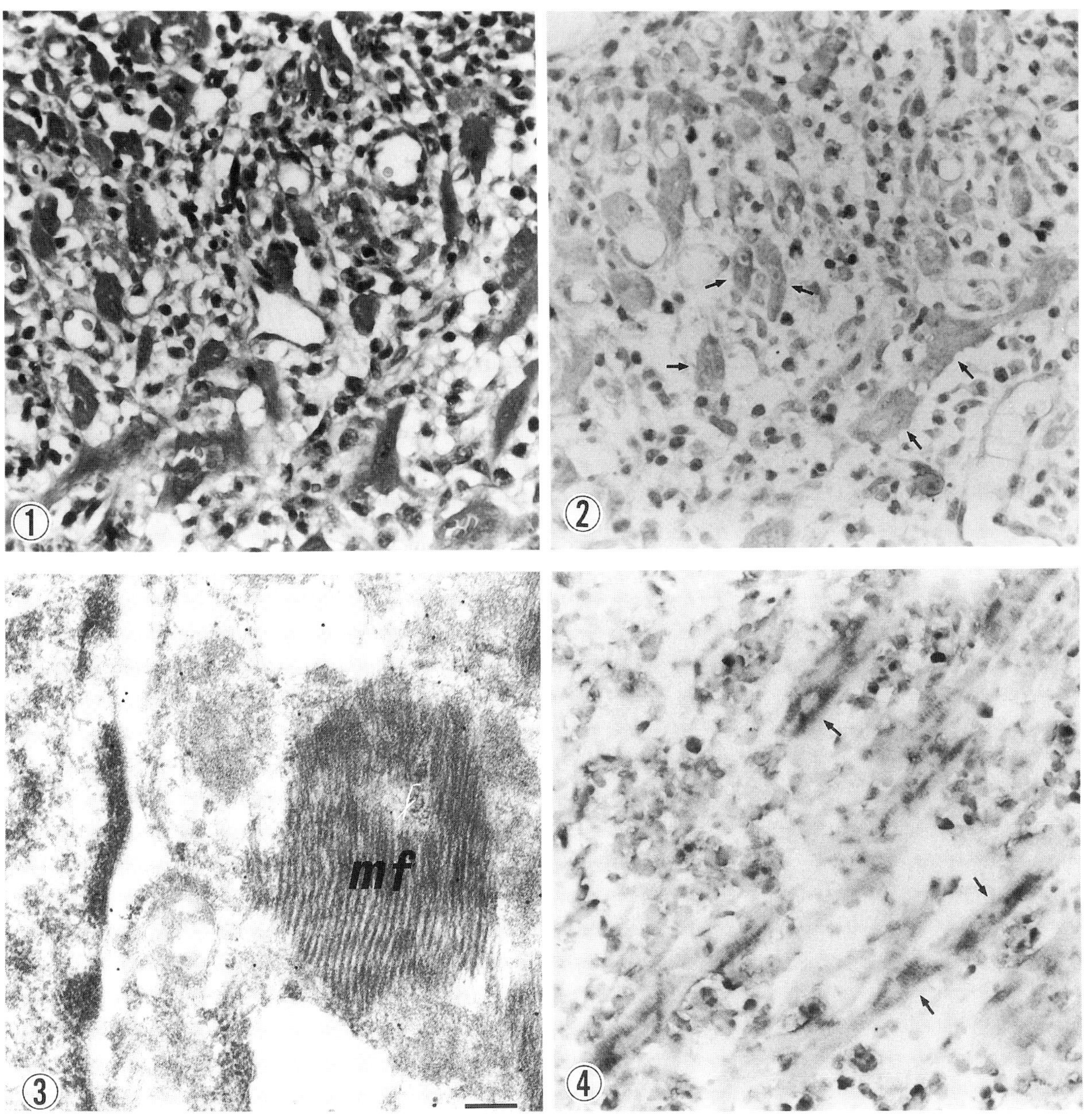

FIG. 1 - Histopathologic findings of the heart tissue from rats with experimental autoimmune myocarditis. Extensive myocardial necrosis with massive inflammatory cell infiltraion composed of lymphocytes, neutrophils and macrophages is observed. H & E stain. Original magnification x200.
FIG. 2 - Immunohistochemical staining with anti-iNOS antibody. Positive immunoreactivity is demonstrated in macrophages, neutrophils and vascular endothelial cells. Myocytes are also positive (arrows). Original magnification x200.
FIG. 3 - Immunoelectron microscopy of a cardiac myocyte in the affected area. Gold particles corresponding to iNOS are observed in the cytoplasm of myocyte. mf=myofibrils, Bar=0.2 μm.
FIG. 4 - In situ hybridization of iNOS mRNA in the heart tissue of myocarditis rat. Expression of iNOS mRNA is demonstrated in myocytes (arrows) and macrophages. Original magnification x200.

STRUCTURAL AND CHEMICAL ANALYSIS BY TRANSMISSION AND SCANNING ELECTRON MICROSCOPY OF PARTICLES FOUND IN EYES OF CATARACT PATIENTS

D. Barron, U.S. Army Center for Health Promotion and Preventive Medicine, Aberdeen Proving Ground, Maryland 21010

There has been a problem associated with removal of cataracts noted by a number of ophthalmologists, in particular, from Dr. David George at the Madigan Army Medical Center in the state of Washington (ref. 1-3). Particles were found lodged in the eyes of patients after the cataract operation (see fig.1). There was serious concern about the origin of the particles. Were they from the ultrasonic surgical head, the surgical blade, or were they from a long time accumulation of the metals found in the body? In an effort to find answers, Dr. George sent samples of the cataract tissue filtered through a Whatman filter to the U.S. Army Center for Health Promotion and Preventive Medicine (CHPPM) for analysis by transmission electron microscopy (TEM), scanning electron microscopy (STEM), x-ray elemental spectral analysis (EDAX), and selected area electron diffraction (SAED). The results were compared with the composition of sample particles from the ultrasonic head and the surgical blade. These particles were filed off the head and blade and secured to a formvar film stretched over a TEM grid. Each set of particles was then analyzed by TEM and a x-ray elemental spectrum was generated.

Techniques were also developed to isolate the tiny 0.1 to 5 micron particles found in the eye for preparation for analysis by TEM, SEM, EDAX, and SAED. In order to prepare the samples, surgical tweezers were used to pluck the particles from the cataract tissue under a Wild stereo microscope equipped with Halogen fiber optic light. The particles were then deposited on a formvar film stretched over a TEM copper grid coated with a carbon film using a Denton high vacuum carbon coater. Other methods were developed to analyze the particles by EDAX, SAED, and SEM. Preliminary results indicate a number of particles emanating from the ultrasonic surgical head display a strong peak of titanium (see fig. 2). Other particles from the eye contain several elements including Fe, Cr, Ni, Ca, etc. (see figures 3 & 4) which indicates that the source of some of the particles is the stainless steel surgical blade.

References

1. L. B. Sabbagh, "Phaco Fallout: Beware of Metallic Shrapnel"., Ophthalmology Times. 18,1993,69
2. C. H. Mathey et al., "Polishing methods for the lens capsule: Histology and scanning electron microscopy", Journal of Cataract Refraction Surgery. 20(1994)64
3. George, D. P., Ltr. dated 31 Jan.'95, MAMC, Tacoma, Washington

Proc. Microscopy and Microanalysis 1995, edited by G.W. Bailey, M.H. Ellisman, R.A. Hennigar, and N.J. Zaluzec
Copyright © 1995 MSA. Published by Jones and Begell Publishing, 79 Madison Ave., New York, NY 10016

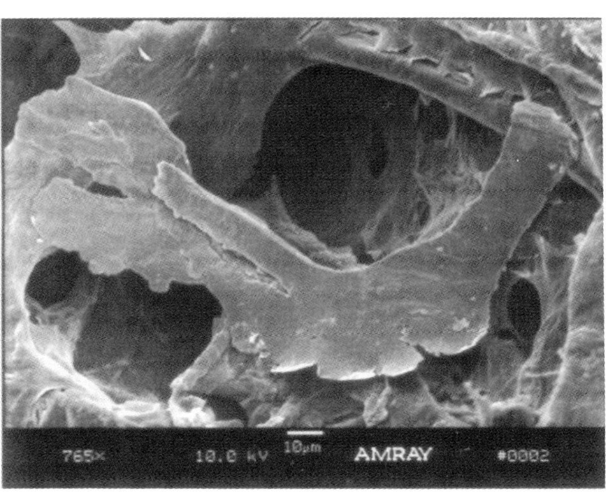

Fig. 1

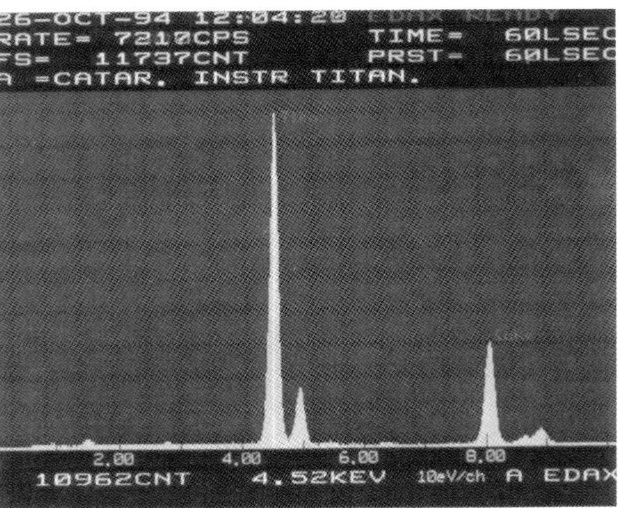

Fig. 2

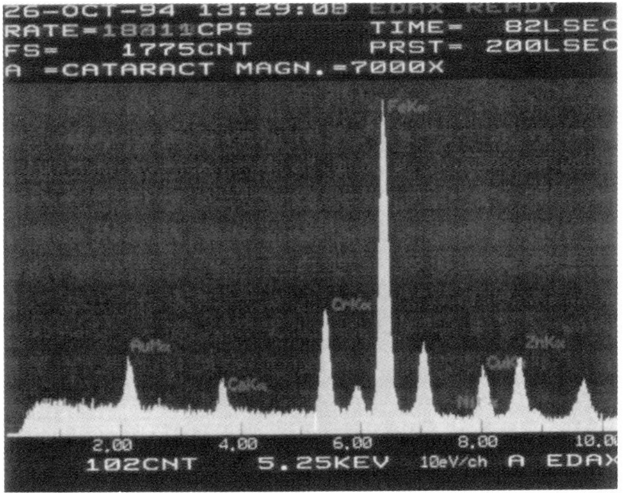

Fig. 3

Elemental composition of typical particles taken from the eye and from instruments (weight per cent)

Element	Part. eye (T8679) SEM	Stainless blade (T8819) TEM	Particle #1 (T8820) SEM	Particle #2 (T8820) SEM	Particle #3 (T8820) SEM	Blank Specimen Plate SEM
Al	----	0.781	-----	6.15	4.04	----
Si	----	4.743	11.047	25.06	----	----
P	----	0.791	-----	-----	----	----
Mo	----	11.114	-----	-----	----	----
Cl	----	16.480	-----	3.62	----	----
K	----	11.104	-----	10.2	----	----
Ca	----	3.299	-----	-----	41.02	----
Ti	----	0.333	40.131	15.8	----	----
Cr	19.125	1.292	-----	-----	----	----
Fe	66.598	4.092	-----	5.8	10.236	----
Cu	2.497	*41.211	27.340	24.2	25.2	46.7
Zn	2.913	4.760	8.436	-----	18.95	35.1
Ni	----	----	6.289	-----	9.45	17.8
Na	----	----	-----	-----	----	----
Zr	----	----	-----	-----	----	----
S	----	----	6.757	8.26	----	----
Mg	----	----	-----	-----	23.129	----
Au	4.056	----	-----	-----	----	----
Mn	0.796	----	-----	-----	----	----

Fig. 4

FIG. 1.- Scanning electron microphotograph of typical particle found in the eye after cataract operation at 710X magnification. [by permission of Dr. John Petrali, APG]

FIG. 2.- EDAX x-ray elemental spectrum of typical particle taken from eye of cataract patient showing strong peak of titanium, which is characteristic of the ultrasonic surgical head.

FIG. 3.- EDAX x-ray elemental spectrum of another typical particle taken from eye of cataract patient showing peaks of Fe, Cr, Zn, Ni, Au & Ca which characterize stainless steel.

FIG. 4.- Table summarizing elemental composition of particles from cataract patient's eye and instruments used in the operation.

CRYO-ELECTRON AND OPTICAL MICROSCOPY STUDY OF LIPID-ENCAPSULATED MICROTUBULES

M. E. Bisher*, D. K. Fygenson[†], F. Booy[#], A. Libchaber[†]* and M. M. J. Treacy*

*NEC Research Institute, Inc., 4 Independence Way, Princeton, NJ 08540-6685
[†]The Rockefeller University, 1230 York Avenue, New York, NY 10021-6399
[#]National Institutes of Health, NIAMS, LSB, Building 6, Bethesda, MD 20892-2755

The dimeric protein *tubulin* is found in every eukaryotic cell. In the presence of GTP and Mg^{++} cations, tubulin polymerizes into long hollow cylinders known as *microtubules*, that are 24 nm in diameter and can grow as long as 10–100 µm in length. Microtubules play an important role in living cells: they act as guides for internal molecular transport (most notably the separation of genetic material during anaphase in cell division), they are a major component of the cytoskeleton, and are important structural constituents of cilia and flagellae.[1] Locomotion, morphogenesis and reproduction, are fundamental cellular processes that rely on the polymerization of microtubules and on their ability to re-organize – a feature termed *dynamic instability*.[2]

Tubulin, purified from fresh cow brain, exhibits this dynamic instability in the laboratory in the *absence* of microtubule-associated protein. Length fluctuations can be observed by differential interference contrast (DIC) optical microscopy. The instability persists when tubulin is encapsulated in lipid vesicles (liposomes), formed by adding the phospholipids DOPC and DOPS, and applying the freeze-thaw technique to the mixture.[3-5] Fig. 1 shows a DIC optical micrograph of a ~5 µm lipid vesicle that is distorted, or impaled, by a 12 µm rod which is presumably a microtubule, or a bundle of microtubules. Frequently, such microtubules are observed to buckle because of reaction forces from the membrane.

These experiments raise the following questions; (1) Can the growing microtubule exert sufficient pressure to rupture the lipid membrane? (2) It is known that polymerization occurs only at the microtubule ends, and encapsulation is expected to slow the transport of tubulin dimers to the ends. How does the microtubule continue to polymerize after it has begun pressing against the membrane?

Specimens were prepared for TEM by placing a droplet of the polymerizing tubulin/liposome mixture onto a holey carbon film, blotting and rapidly immersing into liquid-nitrogen-cooled ethane using a Reichert-Jung KF80 cryo-immersion unit. Frozen specimens were transferred to a Gatan liquid nitrogen holder. Specimens were examined in a Hitachi H9000 NAR at 100 kV and 300 kV, and a Phillips 400 at 120 kV. It was found that vesicles that are thin enough for electron transmission (< 1 µm) rarely contained microtubules, even though microtubules were abundant exterior to the vesicles. Fig. 2 shows two rare exceptions to this rule. In each case, one end of the microtubule has clearly interacted with the bi-layer membrane, which has distorted or even ruptured (Fig 2a). Significantly, the microtubule end has not penetrated far before rupture, < 10 nm. Membrane rupturing would explain why some microtubules continue to polymerize. However, there is the remote possibility that the rupture is an artifact of the specimen preparation induced by differential rates of contraction of the components on cooling.

References
1. See for example, J. D. Watson, N. H. Hopkins, J. W. Roberts, J. A. Steitz and A. M. Weiner, *Molecular Biology of the Gene* 4th edition (Benjamin/Cummings; Reading, Massachusetts) (1987).

Proc. Microscopy and Microanalysis 1995, edited by G.W. Bailey, M.H. Ellisman, R.A. Hennigar, and N.J. Zaluzec
Copyright © 1995 MSA. Published by Jones and Begell Publishing, 79 Madison Ave., New York, NY 10016

2. T. Mitchison and M. Kirschner, *Nature* 312 (1984) 232; 312 (1984) 237.
3. H. Hotani and M. Miyamoto, *Advances in Biophysics*, 26 (1990) 135.
4. D. K. Fygenson *et al.*, to be published (1995).
5. DOPC = 1,2-Dioleoyl-sn-Glycero-3-Phosphocholine
   DOPS = 1,2-Dioleoyl-sn-Glycero-3-Phospho-L-Serine.
The authors wish to thank Alasdair Steven for his support and encouragement.

Fig. 1—Differential interference contrast micrograph of a microtubule (or microtubule bundle) inside a 5 μm phospholipid vesicle. The microtubule polymerized within the vesicle, and continued to grow after the microtubule ends had made contact with the phospholipid bi-layer membrane.

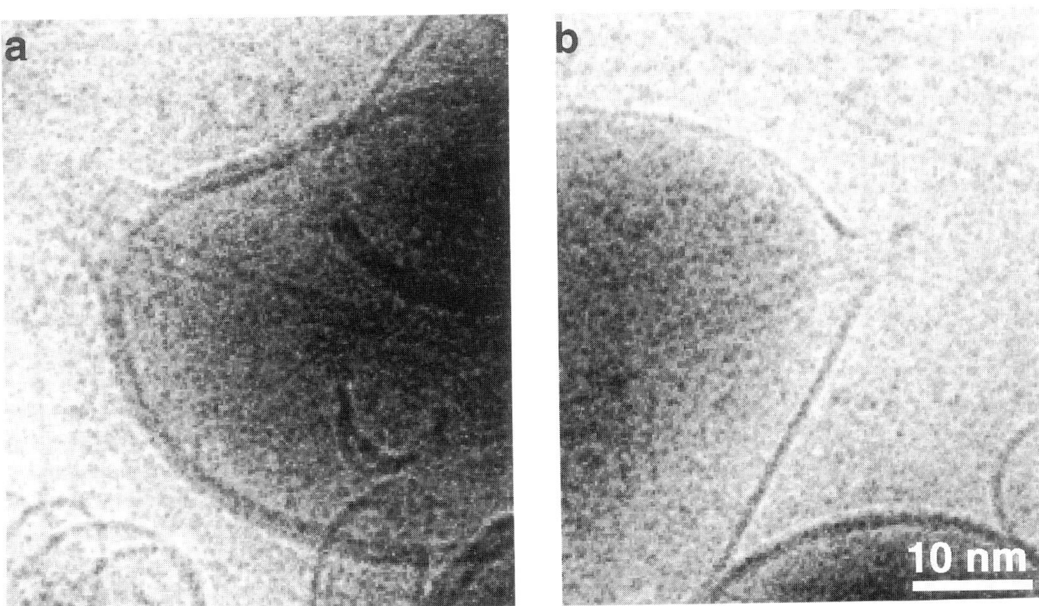

Fig. 2. Cryo-electron micrographs showing encapsulated microtubules protruding through bi-layer vesicle membranes. In *a*), the phospholipid vesicle membrane is bi-lamellar. The outer bi-layer membrane is clearly ruptured, whereas the inner membrane appears (in projection) to be undistorted. In *b*) the membrane appears to be intact up to the rim of the microtubule, and may be forming an open loop at the microtubule tip. These microtubules appear to have direct access to the tubulin solution outside the membrane. In such instances, polymerization would be not be mitigated by any membrane diffusion barrier.

THE MOLECULAR STRUCTURE OF THIOL-MONOLAYERS ON GOLD IMAGED WITH AN ATOMIC FORCE MICROSCOPE

M. Jaschke*, H. Wolf**, H Schönherr**, H. Ringsdorf**, E. Bamberg*, H.-J. Butt*

* Max-Plank-Institut für Biophysik, Kennedyallee 70, 60596 Frankfurt, Germany
** Institut für Organische Chemie, J. Gutenberg Universität, 55099 Mainz, Germany

Thiols spontaneously form monolayers on gold.[1-4] An important question is which factors determine the structure of the thiols. The two extreme positions are: (a) the structure of the underlying gold determines the packing of the thiol molecules; (b) the geometry of the thiol molecules and the interaction between thiols determine their two-dimensional structure.

To help answer this question two series of thiol molecules on gold (111) were imaged by atomic force microscopy. In order to reduce the force between tip and sample, and to prevent distortion of the thiol monolayer, all images were taken in ethanol at forces between 0.5 nN and 3 nN.

The first series consisted of several azobenzene-thiols:

i. $HS\text{-}(CH_2)_6\text{-}O\text{-}(C_6H_4)\text{-}N{=}N\text{-}(C_6H_5)$

ii. $HS\text{-}(CH_2)_{11}\text{-}O\text{-}(C_6H_4)\text{-}N{=}N\text{-}(C_6H_5)$

iii. $S\text{-}(CH_2)_{11}\text{-}O\text{-}(C_6H_4)\text{-}N{=}N\text{-}(C_6H_5)$
$$|$$
$S\text{-}(CH_2)_{11}\text{-}O\text{-}(C_6H_4)\text{-}N{=}N\text{-}(C_6H_5)$

Although these molecules possess a terminal alkyl chain, their lattice is different from the well known hexagonal ($\sqrt{3} \times \sqrt{3}$) R 30° lattice (with respect to the underlying gold (111) lattice) of alkane-thiols. Instead, we found two stable rectangular lattices which could both be described by lattice constants of 0.61 nm and 0.79 nm, and an angle of 89°. These structures were also found on polycrystalline gold. Also, varying the length of the alkyl chain (molecules (i) and (ii)) or using a disulfide instead of a thiol (molecules (ii) and (iii)) did not change the lattice constants. Thus we conclude that the lattice for azobenzenethiols is end-group dominated.[5]

In a second series we compared the structures of disulfides of hydrocarbons and disulfides of fluorocarbons:

iv. $S\text{-}(CH_2)_2\text{-}O\text{-}CO\text{-}(CH_2)_{16}\text{-}CH_3$
$$|$$
$S\text{-}(CH_2)_2\text{-}O\text{-}CO\text{-}(CH_2)_{16}\text{-}CH_3$

v. $S\text{-}(CH_2)_2\text{-}O\text{-}CO\text{-}(CH_2)_2\text{-}(CF_2)_7\text{-}CF_3$
$$|$$
$S\text{-}(CH_2)_2\text{-}O\text{-}CO\text{-}(CH_2)_2\text{-}(CF_2)_7\text{-}CF_3$

vi. $S\text{-}(CH_2)_2\text{-}O\text{-}CO\text{-}(CH_2)_{10}\text{-}CH_3$
$$|$$
$S\text{-}(CH_2)_2\text{-}O\text{-}CO\text{-}(CF_2)_8\text{-}CF_3$

Proc. Microscopy and Microanalysis 1995, edited by G.W. Bailey, M.H. Ellisman, R.A. Hennigar, and N.J. Zaluzec
Copyright © 1995 MSA. Published by Jones and Begell Publishing, 79 Madison Ave., New York, NY 10016

The hexagonal monolayer lattice of monolayer molecules (iv) is slightly larger (lattice constant of about 0.52 nm) than the lattice of alkane-thiols (lattice constant of about 0.50 nm). This might be due to the ester group in the alkyl chain. The fluorocarbons (v) also show a hexagonal lattice. However, the lattice constant in this case is close to 0.58 nm which is in accordance with the results of Liu et al.[6] who imaged $CF_3(CF_2)n(CH_2)_2SH$ on gold. In this case the lattice structure is determined by the fluorocarbon chains and not distorted by the ester group. The mixed disulfide (vi) was very difficult to image. This might be due to the mismatch of the hydrocarbon chain or to the difference in chain length.

References

1. L. Strong and G.M. Whitesides, *Langmuir* 4(1988)546.
2. C.E.D. Chisdey et al., *J. Chem. Phys.* 91(1989)4421
3. N. Camillone et al., *J. Chem. Phys.* 94(1991)8493
4. C.A. Alves et al., *J. Am. Chem. Soc.* 114(1992)1222
5. H. Wolf et al., *J. Phys. Chem.* (1995), in press.
6. G.-Y. Liu et al., *J. Chem. Phys.* 101(1994)4301

NEW TOOLS FOR ATOMIC FORCE MICROSCOPY

C. B. Prater

Digital Instruments, Inc., 520 E. Montecito St., Santa Barbara CA 93103

The Atomic Force Microscope (AFM) has been widely used in the physics, chemistry, and materials science communities, and is becoming more common in life sciences research. To better serve the biological community, new instruments have been developed recently that combine AFM and various forms of optical microscopy including EPI-fluorescence, DIC, and confocal microscopy. In addition, new techniques like fluid TappingMode™ have been developed to allow gentle, non-destructive imaging of biological samples, including live specimens in physiological conditions. Other new techniques can provide information about sample elasticity or molecular adhesion along with nanometer-scale topography measurements.

Combined AFM/Optical Microscopy

Until recently, most AFMs scanned the sample under a stationary probe using a small piezoelectric scanner. This arrangement placed serious limits on the size and type of sample that could be used as a sample substrate. Now instruments have been developed that scan the AFM probe over a fixed sample that then allows imaging of larger, more convenient sample substrates, including cover slips, slides, and even petri dishes. These transparent substrates also now allow the AFM to be combined with inverted optical microscopes (Fig. 1), commonly used in biological research. This combination allows the AFM probe to be easily placed over a region of interest on a sample, and allows direct comparison of 3-D AFM images and optical measurements (Fig. 2).

Fluid TappingMode

Previously, most AFM measurements have been made in so-called "contact mode" where the AFM probe is in constant contact with the sample surface. This constant contact often damaged soft or adhesive samples due to frictional forces between the probe and the sample surface.

Fluid TappingMode is a technique that can reduce or eliminate damage to soft samples imaged in liquid. TappingMode, which is also used in ambient air, sets the AFM probe into oscillation so that the AFM tip only intermittently contacts the sample surface. The intermittent contact eliminates frictional forces that can distort image data and damage the sample in contact mode. Because the damaging frictional forces are eliminated, soft biological specimens—even live cells—can be imaged with this technique (Fig 3).

Adhesion and Elasticity

Since the AFM fundamentally measures forces between the AFM probe and a sample surface, it can provide both topographic structure and information about nano-mechanical properties of samples. For example, Fig. 4 shows a comparison of a the surface topography (left) and elasticity (right) of a triblock copolymer. The images were made using a technique called interleaved scanning where the AFM alternates line by line between a TappingMode measurement of topography and a force modulation image of elasticity. The AFM can also measure, and even map surface adhesion. By measuring the phase change of the oscillating cantilever in TappingMode, the AFM becomes sensitive to viscous forces from the sample, including adhesion. Fig. 5 shows a TappingMode phase image of a patterned self-assembled monolayer on gold. While the entire gold surface is covered by an organic monolayer, the 10 μm squares have a different chemical termination, resulting in larger adhesive forces.

The ability to map topography, elasticity and adhesion, and combining AFM with optical microscopy provide additional tools for a variety of fields in the biosciences, including cellular structure and dynamics, and binding of biomolecules.

Proc. Microscopy and Microanalysis 1995, edited by G.W. Bailey, M.H. Ellisman, R.A. Hennigar, and N.J. Zaluzec
Copyright © 1995 MSA. Published by Jones and Begell Publishing, 79 Madison Ave., New York, NY 10016

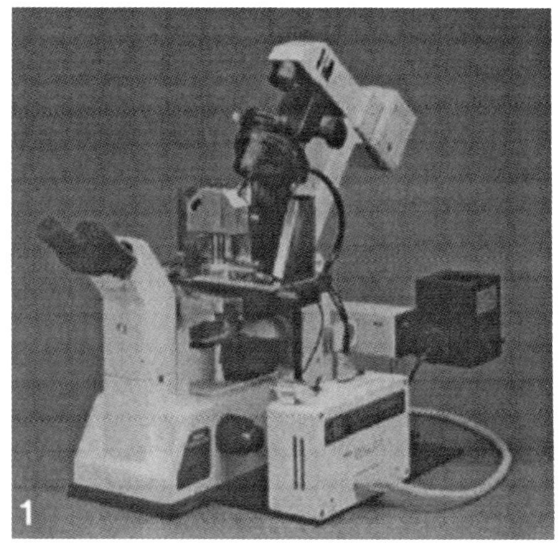

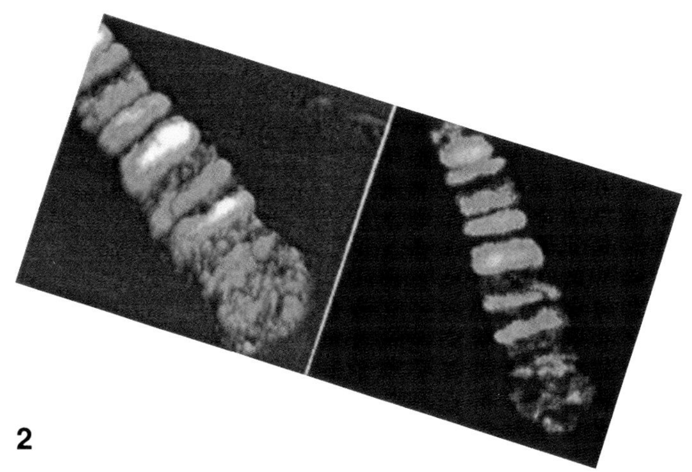

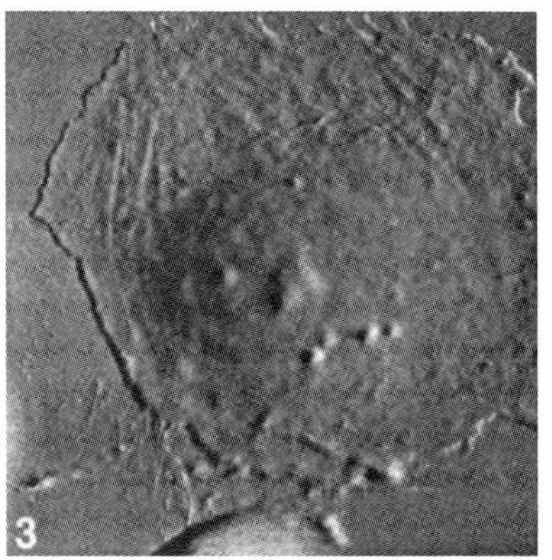

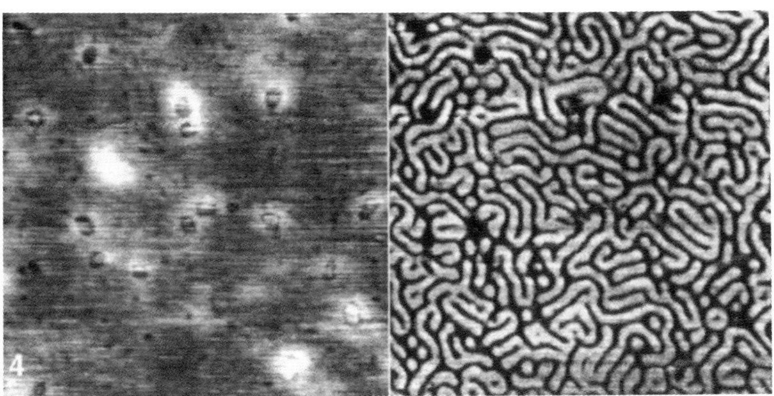

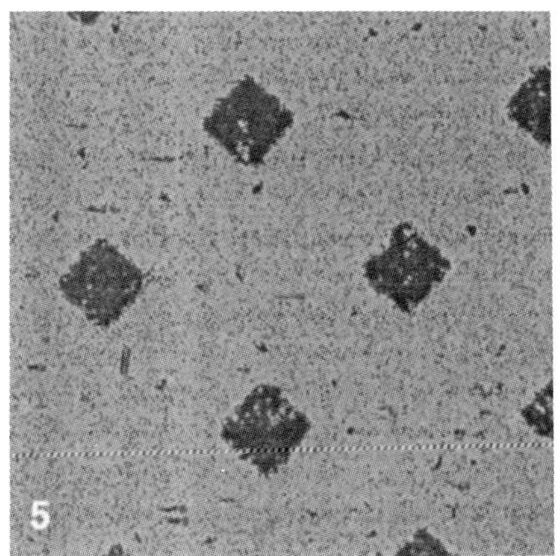

Figure Captions

FIG. 1 - The BioScope Atomic Force Microscope (AFM), combining AFM and an inverted optical microscope.

FIG. 2 - Comparison of AFM image (left) and a confocal optical microscope image (right) of a polytene chromosome. Images courtesy of J. Vesenka, C. Mosher, S. Schaus, and E. Henderson, Iowa State University and BioForce Labs.

FIG. 3 - Live cell imaged in fluid TappingMode. Sample has not been stained or fixed in any way. Image courtesy Iowa State University and BioForce Labs (above).

FIG. 4. - Interleaved AFM measurements of topography (left) and elasticity (right) of a triblock copolymer. Sample courtesy of David Denley, Shell Oil.

FIG. 5 - TappingMode/Phase image of patterned self assembled monolayer on gold. Sample courtesy of Aleksandr Noy and Charles Lieber, Harvard University.

717

Measuring Forces Between Biological Macromolecules with the Atomic Force Microscope: Characterization and Applications

Gil U Lee†, Linda Chrisey* and Richard J. Colton†

†Naval Research Laboratory, Chemistry Division, Code 6177, Washington, DC 20375-5342.
*Naval Research Laboratory, Center for Molecular Science and Engineering.

Structure and function in biological macromolecular systems such as proteins and polynucleotides are based on intermolecular interactions that are short ranged and chemically specific. Our knowledge of these molecular interactions results from indirect physical and thermodynamic measurements such as x-ray crystallography, light scattering and nuclear magnetic resonance spectroscopy. Direct measurement of molecular interaction forces requires that the state of a system be monitored with near atomic resolution while an independent force is applied to the system of 10^{-12} to 10^{-9} Newton magnitude. The atomic force microscope (AFM) has recently been applied to the study of single molecular interactions. [1,2] The microfabricated cantilever of the AFM, a force transducer of small yet variable stiffness and high resonance frequency, produces a transducer of 10^{-15} N/Hz$^{1/2}$ force sensitivities and 0.01 nm position accuracy.

This presentation describes the AFM measurement of the molecular interaction forces in the model ligand-receptor system streptavidin-biotin and between complementary strands of DNA. In these experiments, the ligand and receptor molecules were immobilized on either a spherical probe or flat surface and the forces of interaction were measured as the surfaces were brought into contact and separated (Figure 1). The probability of observing single molecular interactions was maximized by controlling the area of contact through the diameter of the probe and applied load. Streptavidin and biotin were immobilized on opposing surfaces using biotinylated bovine serum albumin (Figure 2). The observed streptavidin-biotin rupture forces fell in distributions that were dependent on the rate at which the forces were applied. The distribution of rupture forces is believed to result from variations in the orientation of the molecules on the surface while the rate dependence of the rupture forces appears to be characteristic of the intermolecular interaction.

Interaction forces between complementary strands of DNA were measured by covalently immobilizing oligonucleotides to the silica surfaces. The resulting DNA-DNA interaction force is significantly larger then the streptavidin-biotin force making it necessary to covalently anchor the oligonucleotides. Adhesive forces measured between complementary (ACTG)$_5$ and (CAGT)$_5$ strands (Figure 3) fell into three distinct distributions centered at 1.5, 1.1 and 0.8 nN, which may be correlated with the rupture of the intermolecular interaction between a single pair of molecules involving 20, 16 and 12 base pairs, respectively. When a third long DNA molecule was coupled between complementary surfaces both intra- and intermolecular forces were observed. The

Proc. Microscopy and Microanalysis 1995, edited by G.W. Bailey, M.H. Ellisman, R.A. Hennigar, and N.J. Zaluzec
Copyright © 1995 MSA. Published by Jones and Begell Publishing, 79 Madison Ave., New York, NY 10016

intramolecular interaction results from the molecule's elasticity and manifests itself as a long-range cohesive force.

In conclusion, these measurements promise to produce new insight into the fundamental forces responsible for structure and function in molecular biology and have lead us to develop a novel sensor based on force amplification.

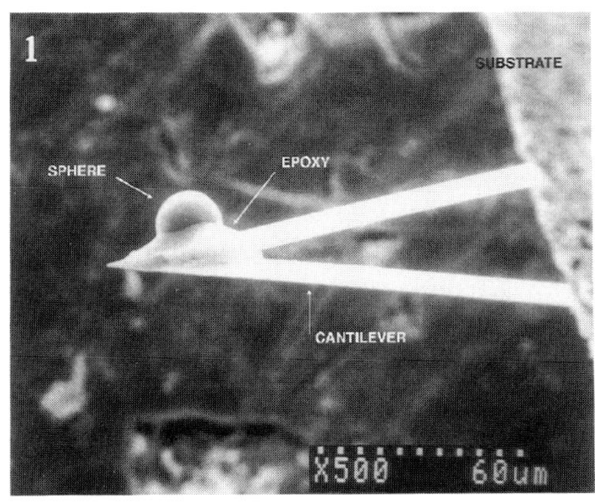

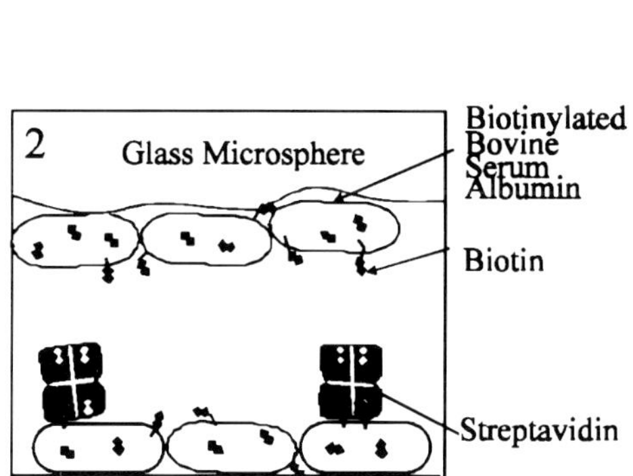

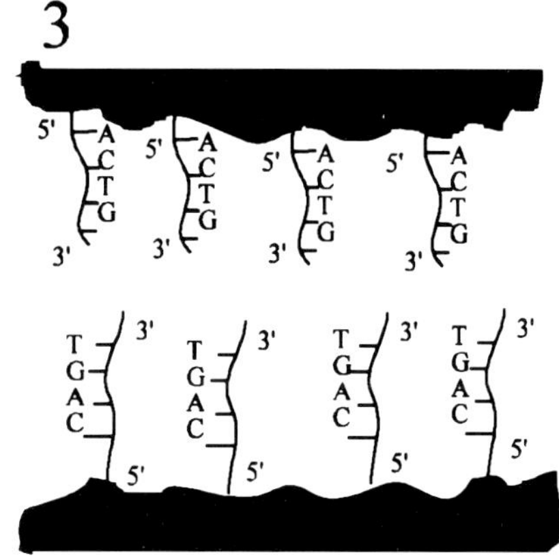

Fig. 1. - Microfabricated cantilever with spherical probe of controlled chemistry and diameter.

Fig. 2. - Schematic of surface chemistry used to measure streptavidin-biotin intermolecular forces.

Fig. 3. - Schematic of surface chemistry used to measure the intermolecular forces between single strands of DNA.

References

1 . G.U Lee, D.A. Kidwell, and R.J. Colton *Langmuir* **1994** 10, 354.
2 . G.U Lee, L.A. Chrisey and R.J. Colton *Science* **1994**, 266, 771.

Funded by the Office of Naval Research and the work of GUL supported by the American Society for Engineering Education

MAPPING VISCOELASTICITY, (SPECIFIC) ADHESION AND NEAR-FIELD FLUORESCENCE OF
CELLULAR STRUCTURES WITH SCANNING PROBE MICROSCOPY

B.G. de Grooth, N.F. van Hulst, J. Greve, C.E.H. Berger, M.H.P. Moers, and K.O.van der Werf

Department of Applied Physics,
University of Twente, P.O. Box 217, 7500 AE Enschede, The Netherlands

One of the major features of scanning probe microscopy is the ability to produce high-resolution images in air
and in liquid. This makes these microscopes potentially very useful for the study of biological materials. Indeed the
number of reports that use these microscopes, especially the atomic force microscope (AFM), has increased
exponentially in the past decade [1]. However, in order to become a routine apparatus that is able to image live
processes at a scale of a few nanometers, the instrumentation of the AFM has to be improved. First of all, under
normal operation, the movement of the tip across the surface of fragile biological structures such as cell membranes,
often results in movement or even destruction of the object. Secondly, the lack of specificity of the AFM makes it
difficult to identify the observed structures. Other problems include the low imaging speed and problems associated
with the tip-sample convolution. Here we will report on attempts to improve the first two points: sample destruction
and specificity.

It has become clear recently that for operation in air the so-called tapping mode AFM is much more gentle for
the sample than the standard operation mode [2]. In tapping mode AFM the tip is forced to oscillate and taps the
surface of the sample only during a small fraction of the oscillation period. We have recently shown that tapping
mode AFM can be operated in liquid. This allows us to study cellular processes such as cell growth and
phagocytosis [3]. The succes of the tapping mode is due to at least two properties: although the maximum force
excerted by the tip on the sample is comparable for tapping and normal operation ('contact mode'), in contrast to
normal opperation during tapping the tip does almost not move laterally while in contact. Therefore, the work done
by friction forces is much smaller with tapping mode. This lateral work is probably the main cause of sample
movement and destruction. The second advantage of tapping mode is a consequence of the viscoelastic properties of
the cellular membrane. When tapped at high frequency, the stiffness of the membrane increases and the cell can be
imaged with low deformation (figure 1).

By measuring the force necessary to pull the tip from the surface at each pixel, it is possible to obtain adhesion
images simultaneously with topographical images [4]. In figure 2 we show the height image and the adhesion image
of a DMPE Langmuir-Blodgett monolayer [5]. By direct comparison with surface plasmon microscopic images we
could conclude that the regions with a high adhesion correspond to the liquid phase of the monolayer and regions
with low adhesion correspond to the crystalline, solid phase. In the AFM adhesion images we can observe
theoretically predicted nanodomains of solid phase in the liquid phase. The ratio of the adhesion forces for both
domains agreed quantitatively with theory; where in the solid domain the tip will interact with the CH_3 groups at the
ends of the hydrocarbonchains (see sketch in figure 2), while in the fluid domain the interaction will be with the
CH_2 groups along the hydrocarbonchains. Experiments are in progress to provide the tip with specific molecules
such as antibodies in order to obtain very specific images in combination with the height images.

A different method to obtain specific information is to combine a scanning probe microscope with a scanning
near-field optical microscope, NSOM [6]. The instrument is based on the illumination of the sample through a sub-
wavelength sized aperture at the end of a tapered, aluminum coated, oscillating fiber. By keeping the amplitude of
this oscillating fiber constant, topographic information (shear force) is obtained. The fluorescence signal from the
exposed volume, which is not limited by diffraction, is detected with conventional optics. We have used this to
obtain a better localisation of *in situ* hybridisation probes on human metaphase chromosomes. The images (figure 3)
clearly reveal extensive substructures of the telomer probe p1-79.

1. H.G. Hansma and J. Hoh, *Ann. Rev. Biophys. Biomol. Struct.*, (1994).
2. Q. Zhong, D. Inniss, K. Kjoller and V.B. Elings, *Surf. Sci. Lett.* 290 (1993).
3. C.A.J. Putman, K.O. van der Werf, B.G. de Grooth, N.F. van Hulst, and J. Greve, *Biophys. J.* 67 (1994) 1749.
4. K.O. van der Werf, C.A.J. Putman, B.G. de Grooth, and J. Greve, *Appl. Phys. Lett.* 65 (1994) 1195.
5. C.E.H. Berger, K.O. van der Werf, B.G. de Grooth, and J. Greve, submitted.
6. E. Betzig and J.K. Trautman, Science 257, (1992) 189.

Proc. Microscopy and Microanalysis 1995, edited by G.W. Bailey, M.H. Ellisman, R.A. Hennigar, and N.J. Zaluzec
Copyright © 1995 MSA. Published by Jones and Begell Publishing, 79 Madison Ave., New York, NY 10016

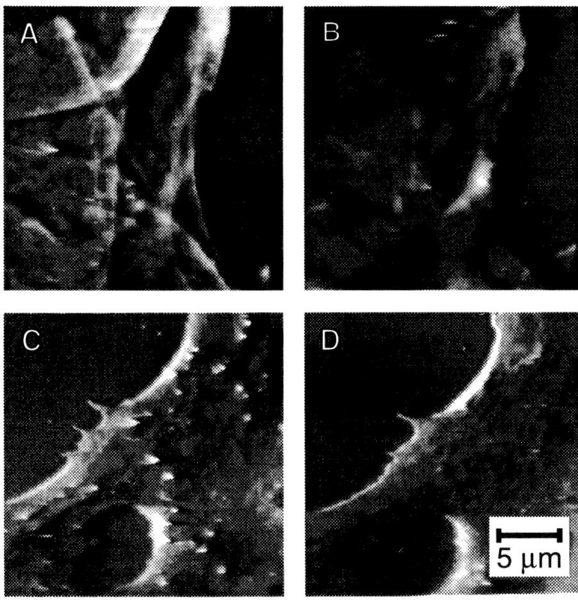

Figure 1

The same monkey kidney cell is imaged in contact mode AFM (a) and tapping mode AFM (b). It is clear from the figure that in contact mode the cell membrane was pushed to the substrate considerably so that the stiff cytoskeleton became apparent. In contrast, the tapping image shows much less sample indentation without loss of resolution. The frequency dependence of the viscoelastic properties of different structures can be used to change contrast (see figure 1c and d). By changing the tapping frequency from 14 kHz to 80 kHz, certain structures can become visible.

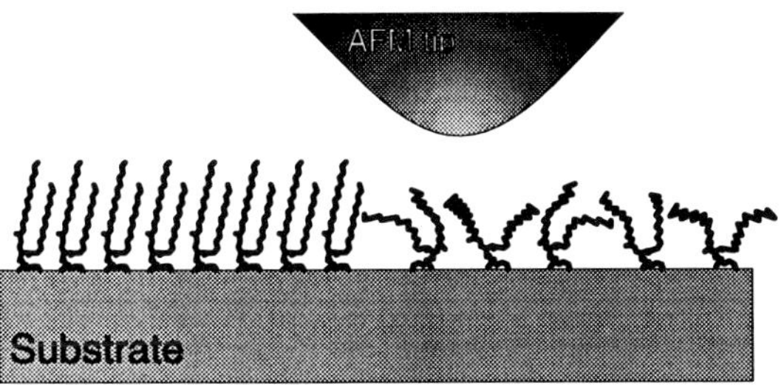

Figure 2

The sketch above shows the AFM tip interacting with DMPE molecules in the solid (left) and liquid (right) domains of the monolayer, resulting in the theoretically expected contrast in Fig. 2 (a) and (b). Simultaneously measured topography images are shown as well (on the right; height difference: 0.5 nm).

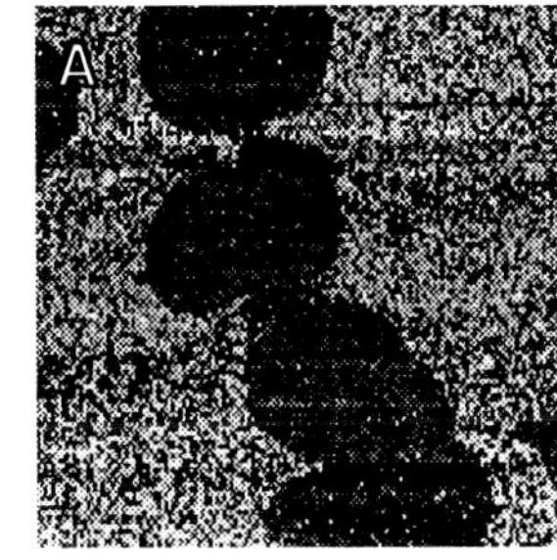

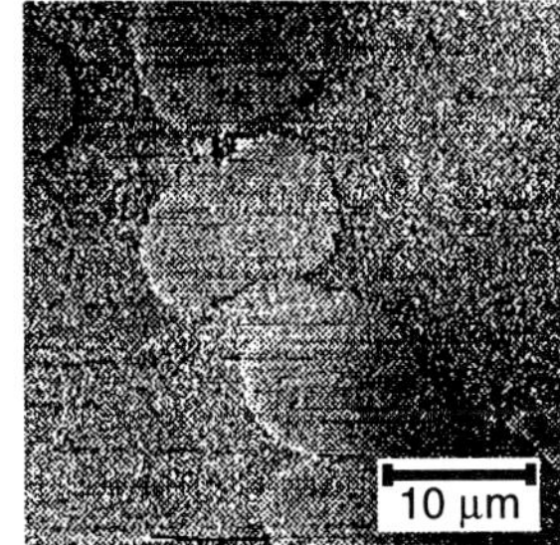

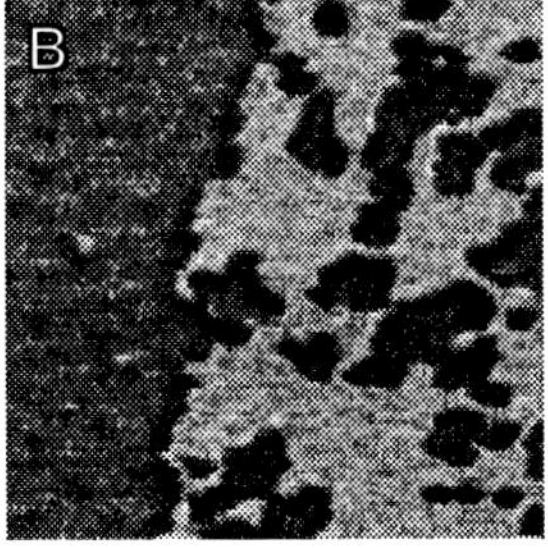

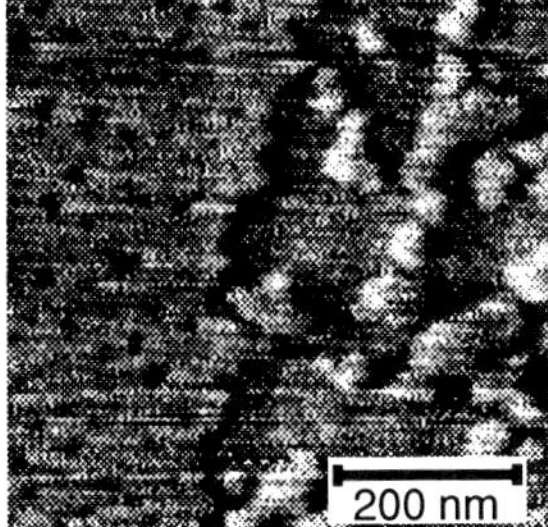

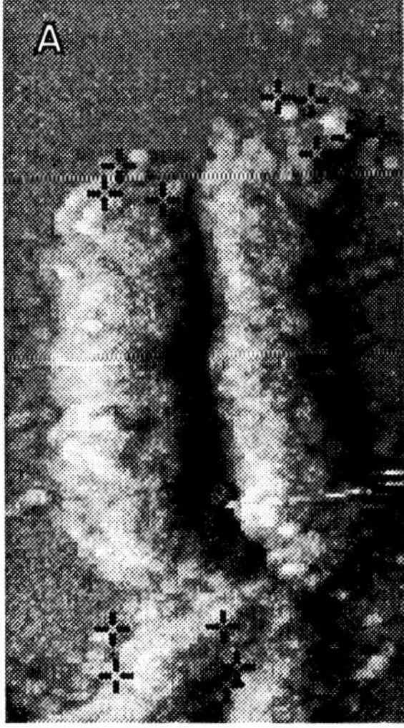

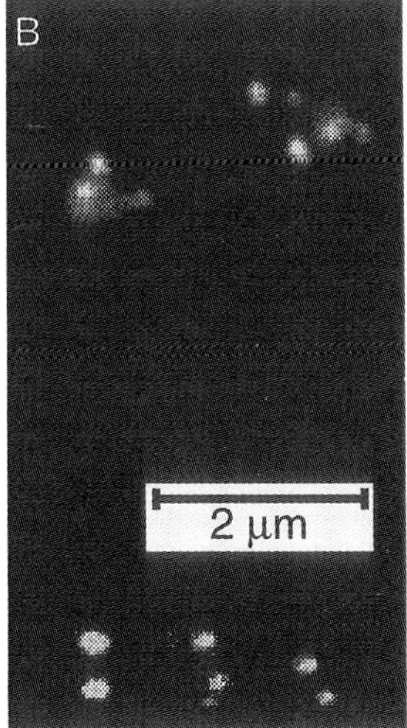

Figure 3

Fluorescence in situ hybridization (FISH) on a human metaphase chromosome. (a) shear force image. (b) corresponding near-field fluorescence image (excitation at 521nm; fluorescence > 550 nm) showing extensive substructure in the CY-3 labeled telomeres and isolated probes in the centromeric region.

STRUCTURE OF ATP SYNTHASE BY SFM AND SINGLE-PARTICLE IMAGE ANALYSIS

David Keller*, Seema Singh*, Paola Turina[†], Roderick Capaldi[†], and Carlos Bustamante[†]

*Department of Chemistry, University of New Mexico, Albuquerque, NM 87131
[†]Institute of Molecular Biology, University of Oregon, Eugene, OR 97403

F_1F_0 ATP synthases are the proteins responsible for the synthesis of ATP in oxidative phosphorylation, and are present in some form in all aerobic organisms, both prokaryotic and eukaryotic. They use the energy stored in a transmembrane proton gradient (which is generated by other members of the oxidative phosphorylation pathway) to synthesize ATP from ADP and P_i, or, working in reverse, to pump protons across the membrane using the energy of ATP hydrolysis. The full protein has two sectors, F_1 and F_0. F_1 is normally bound to F_0 (which is membrane integrated), but is water soluble when dissociated. The F_1 sector contains the sites which bind ADP and catalyze its conversion to ATP. The Fo sector contains a channel which allows protons to to cross the membrane, dissipating the transmembrane chemical potential. By an unknown mechanism this translocation of protons through Fo is coupled to the hydrolysis or synthesis of ATP in F_1, so that the energy released in hydrolysis of ATP can drive the motion of protons against an electrochemical potential, or the energy of translocating protons can be used to form high energy ADP-P_i bonds (1-3).

The F_1 sector has been analyzed by both electron microscopy and X-ray diffraction, and its structure is much better established than F_0. It has the same basic subunit structure in all known organisms, including bacteria, chloroplasts, and mitochondria (4). It is composed of five polypeptide subunits called α, β, γ, δ, and ϵ. In the full F_1 protein there are three copies of α, three of β, and one each of γ, δ, and ϵ. The three α and three β subunits combine to form a hexagon in which α's and β's alternate around the vertices. The locations of γ, δ, and ϵ are less certain, but seem (by chemical modification, antibody binding, and TEM studies) to be arranged "beneath" the α-β hexagon (that is, on the side that binds to F_0, near the membrane and thus beneath the α-β unit when viewed from above) with γ centrally located, δ off-center near γ, one α and one β, and ϵ located near one of the β's. There are three catalytic sites on the three β's, and there are also ATP binding sites of unknown function on the α's. The γ subunit plays a role in binding to F_0 and also in the catalytic functions of F_1 ($\alpha_3\beta_3\gamma$ is the minumum catalytically functional unit (5). Subunit δ is thought to form part of the stalk that connects F_1 and F_0, and is likely to be involved in coupling proton translocation with ATP catalysis.

The F_0 sector is more variable from one organism to another. In E.coli it is composed of three subunits a, b, and c, with stoichiometry $a_1b_2c_{10\pm1}$ (6). The F_0's of other organisms have additional subunits and different stoichiometries. F_0 has been studied in chemical modification experiments and by TEM, but the subunit arrangement is still unknown.

The subunit arrangement of F_0 is directly tied to key questions regarding the mechanism of ATP synthase function. For example, both the a and c subunits show H^+ conduction in reconstituted vesicles, but the proton channel in F_0 seems to be the result of interactions between the two subunits (7). The position of the a subunit relative to the c subunits would help confirm these results and to build better models of H^+ conduction. The mechanism of *coupling* between proton translocation and ATP synthesis is thought to involve a conformational change in Fo that is transmitted through the stalk to F_1 (7). Several models have been proposed for this conformational change, and all depend strongly on the subunit arrangement and involve movements of the F_0 subunits. Clearly, more

Proc. Microscopy and Microanalysis 1995, edited by G.W. Bailey, M.H. Ellisman, R.A. Hennigar, and N.J. Zaluzec
Copyright © 1995 MSA. Published by Jones and Begell Publishing, 79 Madison Ave., New York, NY 10016

information on F_O will help complete the structural picture of ATP synthase, and allow better mechanistic models to be built.

For SFM imaging the F_1 sector (separated from the Fo sector) containing the a, b, c and d subunits was deposited on cleaved mica by brief exposure to an F_1 solution. The sample was then rinsed, placed in the SFM's liquid cell in buffer, and imaged in fluid tapping mode (Si_3N_4 cantilevers, nominal force constant 0.4 N/m, 30-35 kHz). The images show that the F_1 preferentially binds to the mica surface with the minor subunits (c, d) underneath the a-b hexamer. In the best images most molecules (60-70%) appear as round or hexagonal features with a small dimple in their centers. The apparent molecular width (broadened by tip effects) is 30-40 nm. The apparent heights are more variable, from 1 to 7 nm. The known height of the F_1 sector is about 10 nm, so these apparent heights are low, perhaps due to compression of the molecules or, more likely, the effects of changes in surface properties between the mica substrate and the protein on image "contrast" mechanisms. The dimple that appears in the center of many molecules has an apparent depth of about 5 Å, in qualitative agreement with the depth of the hollow at the top of the a-b hexamer as measured from the X-ray structure.

The F_O sector was prepared for imaging by reconstituting the full F_1F_O protein into egg PC vesicles. The F_1 sector was then removed, and the resulting F_O membranes were deposited by exposing the mica to the vesicle suspension for several minutes. During exposure the vesicles burst and spread on the substrate. The vesicles were prepared with the F_O subunits initially oriented "inward", so that after adsorption to the substrate the F_O sectors are oriented upward. SFM imaging was carried out under conditions similar to those used for F_1. The preliminary images show F_O sectors interspersed with a few unstripped complete F_1F_O proteins. The apparent width and height of the F_O sectors are 25-35 nm and approximately 1 nm, respectively. A significant fraction (roughly 5%) of the F_O sectors show a characteristic recurring structure: a ring broken by three distinct large masses surrounding a central hollow. If this structure is supported by further data and image analysis, it represents the first direct, real-space information on the subunit arrangement of the F_O sector.

The initial images of both F_1 and F_O are of sufficient quality that single-particle image classification and averaging (using SPIDER software) is being attempted. In this procedure a large number of small images, each containing only one protein particle, are cut from larger images containing many particles. The single-particle images are then classified by statistical correspondence analysis, and the particles belonging to each correspondence class are aligned and averaged. The procedure will be tested using the data from the F_1 sector, using the X-ray structure and previous electron microscope data for comparison. If successful the method will then be applied to the determination of the subunit arrangement of the F_O sector.

1. A. E. Senior Annu. Rev. Biophys. Biophys. Chem. 19, 7 (1990).
2. J. Hoppe and W. Sebald Biochim. Biophys. Acta 768, 1 (1984).
3. E. Schneider and K. Altendorf Microbiol. Rev. 51, 477 (1987).
4. E. J. Boekema and B. Bottcher Biochim. Biophys. Acta 1098, 131 (1992).
5. D. J. Klionski, W. S. A. Brusilow, and R. D. Simoni J. Bacteriol. 160, 1055 (1984).
6. D. L. Foster and R. H. Fillingame J. Biol. Chem. 257, 2009 (1982).
7. E. P. Gogol, E. Johnston, R. Aggeler, and R. A. Capaldi Proc. Natl. Acad. Sci. U. S. A. 87, 9585 (1990).

MEASURING INTERMOLECULAR FORCES USING THERMAL NOISE IN AN ATOMIC FORCE MICROSCOPE

J.P. Cleveland, T. Schaeffer, P.K. Hansma

Physics Department, University of California, Santa Barbara, CA 93106

Understanding intermolecular forces in liquids at distances below a few nanometers is extremely important in many fields because forces on this length scale determine properties such as chemical bonding, wetting, and specific molecular recognition. However, it is difficult to measure forces over these length scales and perhaps even more difficult to understand them theoretically. The problem stems from the fact that it is hard to make simplifying approximations on these length scales. For example, continuum models usually break down because the finite size of the solvent molecules (typically several angstroms) can not be ignored. Molecular dynamics must often be used to study these systems, limiting the accessible time scales to, at most, many nanoseconds with current computing power.

Atomic force microscopy (AFM)[1] is a relatively new technique that is proving useful in the measurement of forces at these small length scales. Probably the oldest and most direct way to measure a force is through the deflection of a spring, and in many ways, the AFM is simply a much smaller version of more familiar spring based instruments that measure force such as a weight scale. As one might expect, the forces between individual atoms and molecules are small and an instrument capable of measuring them must be sensitive. A quick estimate of the forces involved comes from dividing a typical bond energy by a bond length. Using one electron volt (about 100 kJ per mole, typical of a weak covalent bond) as a bond energy and 1 Angstrom for a bond length yields about one nanoNewton. Biological interesting receptor-ligand bonds can be much weaker, with corresponding forces of tens to a few hundred picoNewtons.[2,3] For comparison, one picoNewton is about the weight of a red blood cell.

In AFM, the deflection of the spring (a small cantilever, typically 100 µm long) is usually measured using a laser beam reflected off the back of the cantilever. Typical deflection sensitivities range from 0.1 to 1 Å. At first glance it would appear that by choosing a weak enough spring arbitrarily small forces could be measured. For instance a cantilever with a spring constant 0.1 N/m will deflect one Ångstrom under a 10 pN force. However, there is a constraint on spring constant; if the attractive force gradients being measured exceed the spring constant of the cantilever, a mechanical instability develops and the cantilever will jump to the surface. A simple estimate of the force gradients involved in a single molecular bond comes from dividing the typical force by the length scale of the bond. Using our numbers from above gives a typical force gradient of about 10 N/m for a 1 eV bond with a 1 Å range. Using a spring of the same stiffness, forces from 0.1 to 1 nN could be resolved. But since the expected forces were about 1 nN our force resolution is similar to the total magnitude of force. The situation is even a bit worse than this. Since we must hold the molecules or atoms on some surfaces there can be additional attractive background forces that we are not interested in measuring, but which force us to use stiffer cantilevers, making our force resolution worse. If we want to measure weaker bonds, the gradients involved get smaller so we can use weaker springs, but the forces also get smaller, so the same problem arises

If a spring weak compared to the force gradients at least one important number can still be extracted from the force measurement: the force required to pull it from the surface (the adhesive force). This is somewhat akin to pulling two pieces of velcro apart using a long rubber band. You get little information about how far the velcro moves before it breaks (perhaps a millimeter or less), but because the rubber band stretches so much (many centimeters) you can accurately measure the force at which it breaks. This technique has been the one used thus far in most sensitive force measurement.[2-4]

To measure the entire intermolecular potential, ideally a spring stiffer than the gradients involved should be used. From the numbers given above the detection sensitivity must also be increased to make good measurements. We have improved the low frequency (0.1 Hz to 1 kHz) sensitivity by using a laser diode coupled to a single mode optical fiber to eliminate angular noise in the laser beam; however, as the AFM matures significant improvements in sensitivity are hard won. There is a way of using weaker springs and still measuring the entire potential. If a spring not too much weaker than the force gradients (say a factor of ten or less) is used then the mechanical instability can become small enough that thermal noise can actually cause the cantilever to jump back off the sample. Statistical anaylsis of this thermal noise signal can yield information about the potential the tip is moving in. This method has the advantage that the deflections remain measurable, but data can still be collected in a region "mechanically unstable".

We have used this technique to measure oscillatory hydration forces between a cleavage plane of a barite ($BaSO_4$) crystal and a silicon oxide AFM tip in water. These forces arise from the layering of water and/ or hydrated ions at the surfaces.[5] (Fig. 1). The cantilever used had a spring constant of about 1 nN/nm. Figure 2 shows a "gray scale" force curve. This is simply the deflection of the cantilever measured as the sample is moved toward the cantilever. The flat region on the right hand side of the graph is the region where the tip is farthest from the crystal and the 45° line on the left hand side is where the tip is in "contact" with the hard sample (a one nm movement of the sample results in a one nm deflection of the cantilever). The region in between these two is the most interesting. Here an attractive force is seen, but there are also tip-sample separations spaced about 2 Å apart where the cantilever tip is more likely to be found. These correspond to layering of water and/or hydrated ions at the crystal surface.

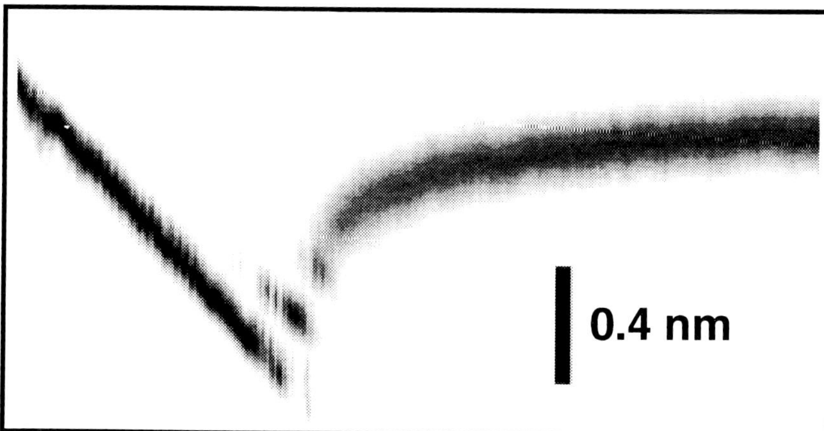

Figure 1. A cantilever tip, held within a nanometer of a crystal surface in water, experiences an oscillatory potential near the surface due to layering of the water molecules and/or hydrated ions.

References

1.   G. Binnig et al., *Phys. Rev. Lett.* 56 (1986) 930.
2.   E.-L. Florin et al., *Science* 264 (1994) 415.
3.   G.U. Lee et al., *Langmuir* 10 (1994) 354.
4.   J.H. Hoh et al., *J. Am. Chem. Soc.* 114 (1992) 4917.
5.   J.N. Israelachvili, *Intermolecular and surface forces* (Academic Press, New York, ed. 2nd, 1992) chapter 13
6. This work is supported by a grant from the Materials Research Division of the National Science Foundation under grant #NSF-DMR-9221781. We thank Digital Instruments for generous equipment donations.

Figure 2. A plot of cantilever deflection (vertical) vs. sample position (horizontal). Since the cantilever was thermally hopping between several stable minima, the cantilever deflection is not well defined. Instead the probability of the cantilever being at a particular deflection is plotted in grayscale. This probability was determined by digitizing the cantilever deflection (1.4 million points at 140 kHz sampling rate) and then histogramming every 3500 points.

MUSSEL ADHESIVE PROTEIN AS A NOVEL SUBSTRATE FOR ATOMIC FORCE MICROSCOPY

Neill P. D'Costa and Jan H. Hoh

Department of Physiology, School of Medicine, Johns Hopkins University, 725 N. Wolfe Street, Baltimore, MD 21205

Atomic Force Microscopy (AFM) requires immobilization of samples to a suitable substrate. Immobilized samples must withstand forces, particularly lateral forces, exerted by the tip during imaging.[1,2] Air-drying is the simplest way of immobilizing molecules onto a substrate. However, this often compromises the integrity of structural features because of forces at the air-liquid interface. In solution the most common method is simple adsorption. While very convenient, this method works only for a limited number of samples. To overcome this problem procedures involving covalent chemical cross-linking have been developed.[3] We have developed an alternative AFM substrate based on the mussel adhesive protein (MAP) from *Mytilus edulis*. This substrate is simple to prepare and can immobilize a wide range of samples, including gold particles, silica spheres, proteins, and nucleic acids.

MAP is a polyphenolic marine adhesive by which the mussel attaches itself to different surfaces.[4] Studies of purified MAP indicate that it possesses a unique chemistry.[5] Primary sequence data reveal that MAP is predominantly made up of the amino acids tyrosine, lysine, proline, serine, threonine, 3- and 4-hydroxyproline and dihydroxyphenylalanine. Thus, it presents a variety of polar side-chain moieties to the external environment that mediate a tight non-specific binding. It has been shown to be biocompatible with a number of cell types and is widely used in tissue culture.[6]

To produce MAP substrates, MAP was adsorbed onto the mica at a concentration of 3 µg/cm^2 of surface area. The required amounts of MAP (Collaborative Research, Bedford, MA) in acetic acid were diluted with 100µl of purified water (MilliQ-UV, >18 MOhm). This solution was pipetted onto the freshly cleaved mica disks (~1 cm^2), following which, it was neutralized with 200 µl 0.5 M sodium bicarbonate for 20 - 30 minutes. MAP substrates were washed by repeatedly by pipetting several 100 µl volumes of purified water onto the surface. It is important that the surface of the newly adsorbed MAP layer be kept wet at all times to reduce contamination. A series of test structures, including 14 nm colloidal gold particles, 150 nm silica spheres, ferritin and plasmid DNA were attached to the MAP film by adsorption for 10-20 minutes. Samples were imaged with a Nanoscope IIIa atomic force microscope (Digital Instruments Inc, Santa Barbara, CA) in constant force mode or tapping mode, in aqueous solutions. Adsorbed layers of MAP were characterized for thickness and roughness, as well. as, for the ability to immobilize the test structures. For the thickness measurements, the force applied to the surface was increased to produce an imaging square in the MAP layer. The height of this layer could then be obtained by section analysis using the Nanoscope software. Roughness data were also obtained using the Nanoscope software. Data were pooled from five different experiments.

MAP films made at 3 µg/cm^2 produced a ~4nm thick protein layer with an RMS roughness of 1.1 nm. The thickness of the MAP layer increased with concentration, although 3 µg/cm^2 was sufficient to form a complete film. At lower concentrations the MAP film was incomplete. At higher concentrations a thin film of material formed at the liquid-air interface, indicating that some of the MAP was not adsorbing to the surface. The MAP preparations did contain some contaminating structures of unknown origin. The data shown are for MAP adsorbed to mica, but similar results were obtained on silicon oxide surfaces. As shown in Fig. 1a-c, the MAP substrate could immobilize 14 nm gold particles, ferritin and plasmid DNA for conventional contact mode (constant force) AFM in solution. Silica spheres (150 nm) were not stable during contact imaging, but could be imaged by tapping in solution Fig 1d. These results demonstrate that MAP substrates provide a simple, yet robust, method of immobilizing samples for AFM. We also note that because samples immobilized to a surface of MAP are stable to interactions with the AFM tip, then coating the AFM tip with MAP should provide a simple method for tip modification.

Proc. Microscopy and Microanalysis 1995, edited by G.W. Bailey, M.H. Ellisman, R.A. Hennigar, and N.J. Zaluzec
Copyright © 1995 MSA. Published by Jones and Begell Publishing, 79 Madison Ave., New York, NY 10016

References

1. J.H. Hoh and P.K. Hansma, *Trends Cell Biol.* 2(1992)208.
2. H.G. Hansma and J.H. Hoh, *Ann. Rev. Biophys. Biomol. Struct.* 23(1994)115.
3. S. Karrasch et al, *Biophysical Journal* 65(1993)2437.
4. R. Laursen , in S.T. Case, Ed., *Results and problems in cell differentiation*, Berlin:Springer-Verlag (1992)55.
5. M. P. Olivieri, R. E. Baier and R. E. Loomis, *Biomaterials* 13 no. 14(1992)1000.
6. M. P. Olivieri et al, *J. Adhesion Sci. Technol.* 4 (1990),197

1 a   **1 b**

1 c   **1 d**

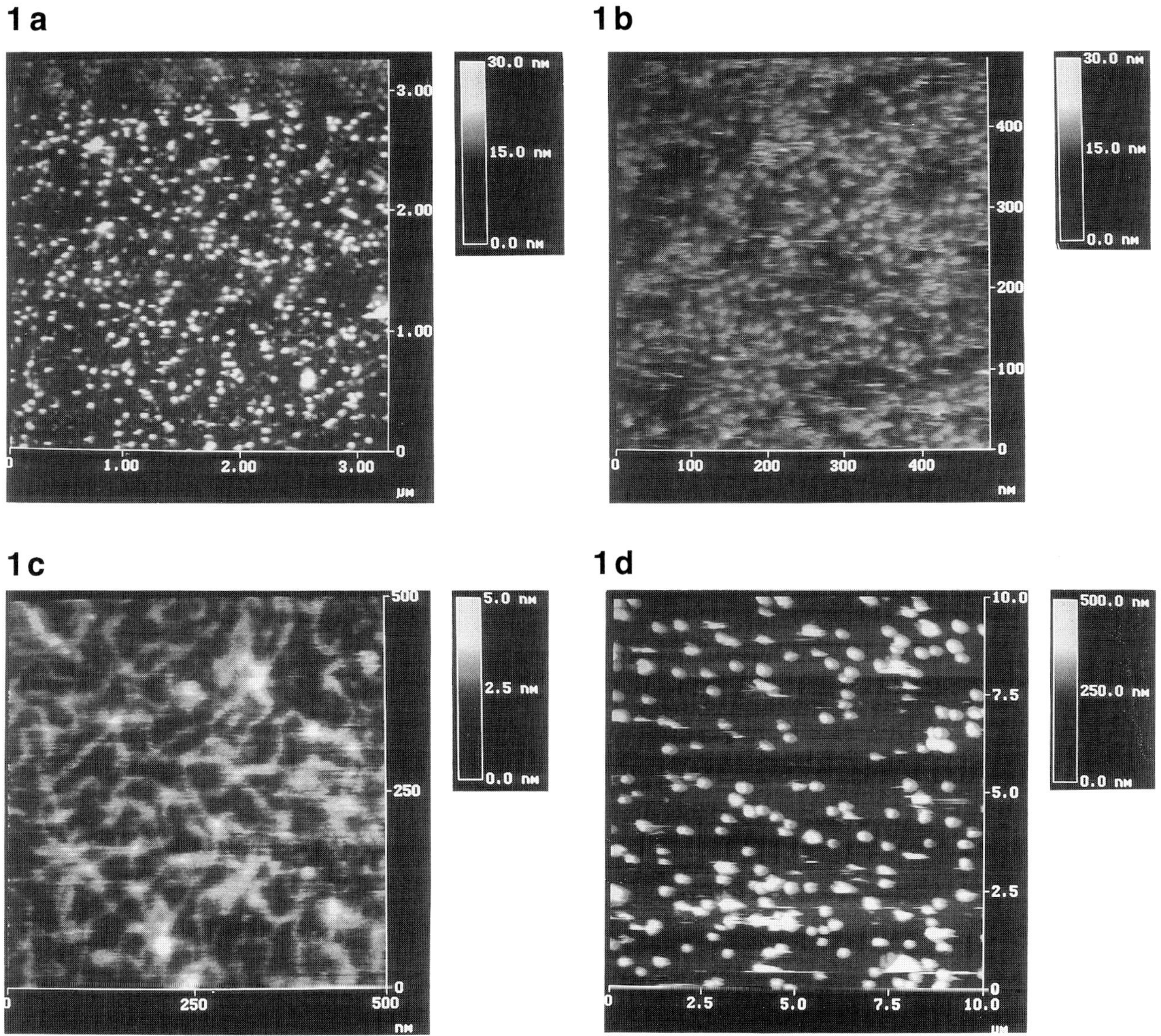

Figure 1. Samples immobilized on MAP (3 µg/cm^2 on mica) substrates and imaged by AFM in solution. a) Colloidal gold particles (14 nm) imaged in contact mode. b) Ferritin (horse spleen) imaged in contact mode. c) Network of plasmid DNA imaged in contact mode. **d**) silica spheres (150 nm) imaged in tapping mode.

DEVELOPMENT OF A SCANNING TUNNELLING ATOM PROBE.

J. C.H.Spence, U. Weierstall and W. Lo[1]

Department of Physics and Astronomy, Arizona State University,
Tempe, AZ 85287. USA.
[1]Applied Physics, Cornell University, Ithaca, N. Y. 14853.

An STM has been constructed which will allow atomic clusters of interest to be transferred into a time-of-flight spectrometer for species identification. Atoms will first be transferred onto the tip, using a small voltage pulse. The sample is then removed, and these atoms ejected into a time-of-flight (TOF) analyser for mass identification (see[1] for an excellent review of TOF methods). Figure 1 shows the experimental arrangement. The STM is based on a miniature Burleigh inchworm, to which a piezo tube is directly attached. This is used in bending mode to provide scan motion of the stage plate in two lateral dimensions, and fine motion in z. The inchworm accounts for the coarse approach. The sample holder sits on three balls, (to one of which the tunnel current amplifier is connected) and may be removed using a wobble-stick. Electrical insulation for 15 kV at the tip is provided, which does not scan. The tip is not cooled. A 75mm chevron dual channel plate is positioned L = 200 mm above the tip. This is used both as a time-of-flight detector, and to form field-ion images and electron field emission images of the tip. A D.C. voltage (up to 10 kV) is applied to the tip for TOF analysis, to which a small additional ten nanosecond wide pulse is applied to stimulate field evaporation of adatoms at the tip. 50 Ohm termination is provided below the inchworm. We use a Blumlian pulse generator and mercury reed switch[2] driven by a repetitive pulsing circuit. Similar instruments have been described for different purposes[3,4]. The output from the phosphor screen is led via a capacitor to a 2 ns sampling digital oscilloscope, whose trace is triggered by the tip pulse. For a total effective potential V_{eff} at the tip, the mass to charge ratio m/n is obtained in the usual way from the equation

$$n \, e \, V_{eff} = (1/2) \, m \, v^2 = (1/2) \, m \, (L/t)^2$$

where t is the flight time (typically about a microsecond for heavy elements and sharp tips). We take L = 200mm on the assumption that the ions are accelerated to their final velocity within a few microns of the tip. At present the STM and TOF capabilities of the system are being evaluated, and atom transfer from a sample followed by TOF analysis has yet to be attempted. It remains to be seen whether adatoms collected from the sample will remain at the tip apex within the collection angle of the detector. Figure 2 demonstrates the excellent performance of this small STM on the 2X1 cleavage surface of (111) Si. Figure 3 shows the TOF spectrum obtained from an undecorated tungsten tip. The transit times suggests that these are W^{4+} and W^{3+}, and the ratio of times allows determination of V_{eff}. Current developments include the provision of a tip exchange system to allow indirect tip heating and an upper stage for FIM and FEM examination of the tip with larger angular view. By limiting the coarse lateral motion to one dimension we hope to be able to find the same region after removing the sample for analysis and returning the sample holder to the STM. This method of microanalysis is destructive[5].

References.
1. M. Miller and D. A. Smith,Atom Probe Microanalysis (Materials Research Society) 1989
2. A. R. Waugh, J. Phys. E14 615 (1981).
3. T.Sakurai et al Progr. Surf. Sci. Vol 33 (1990) 3.
4. J. A. Panitz, Rev. Sci. Instr. 44 1034 (1973).
5. Supported by NSF award DMR91-16362 .

Proc. Microscopy and Microanalysis 1995, edited by G.W. Bailey, M.H. Ellisman, R.A. Hennigar, and N.J. Zaluzec
Copyright © 1995 MSA. Published by Jones and Begell Publishing, 79 Madison Ave., New York, NY 10016

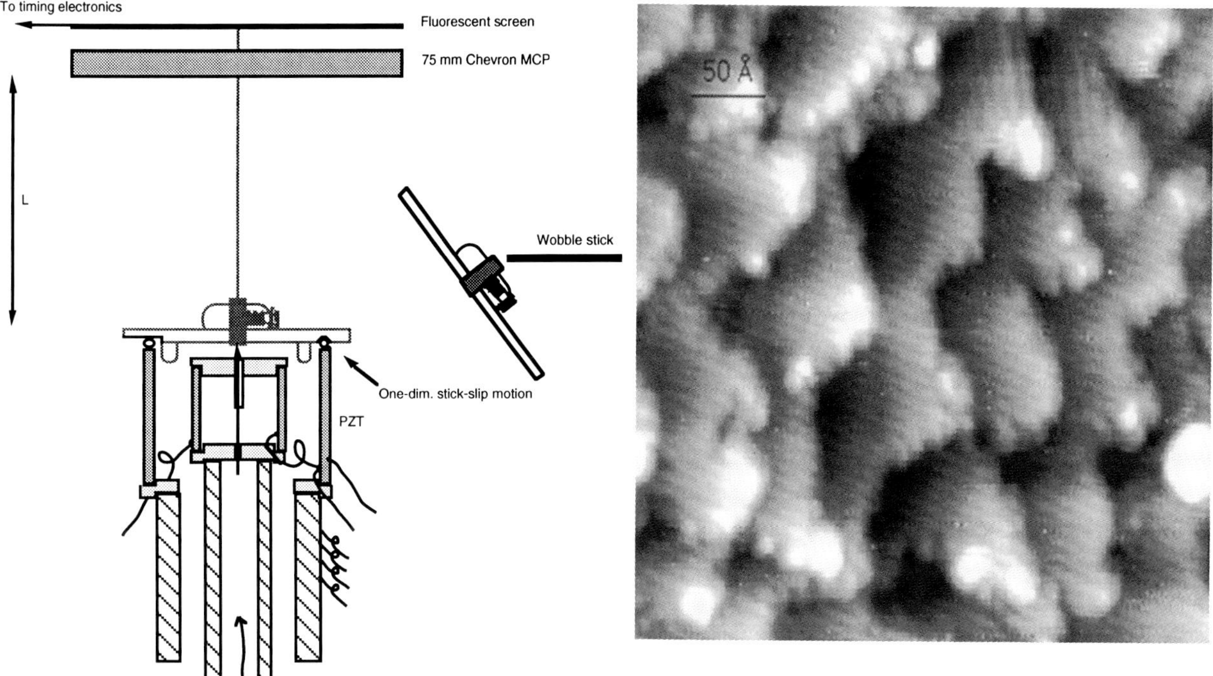

Figure 1. STM with time-of-flight spectrometer. The hatched region is an inchworm. Sample is removed (after transfering atoms to tip) for TOF.

Figure 2. STM image of Si(111) 2X1 cleavage surface, showing dense cleavage steps.

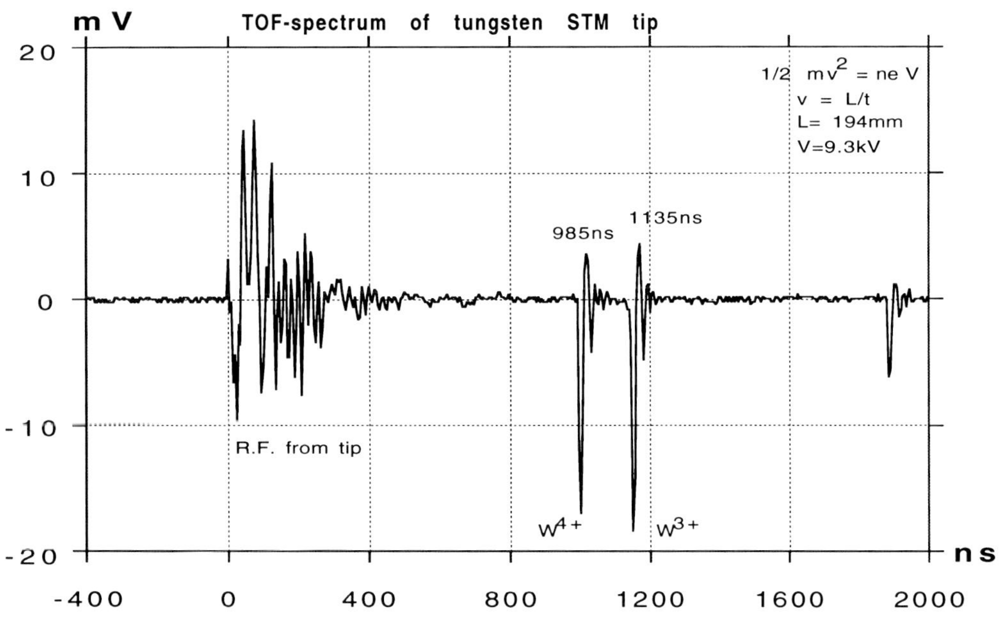

Figure 3. TOF spectrum from bare tungsten tip at about 9.3 kV. Two W ions are identified by the ratio of their transit times. Noise at left is direct RF pulse from tip which preceeds ions.

CAPACITANCE IMAGING
WITH A SCANNING FORCE MICROSCOPE

R. ANDERSON,* ¶ H. SIRIWARDANE,* P. FRAUNDORF,* T. STIVERS¶

*Department of Physics, and Center for Molecular Electronics, University of Missouri - St. Louis, St. Louis, Mo. 63121 (USA)
¶ ITT Technical Institute Earth City, Mo. 63045 (USA)

Scanning force microscopes show topography, but alone cannot provide insight into the electrical properties of either a specimen or particular features of a specimen. In this project we will look at a specimen that consists of conductive spheres in a base of insulating material (FIG. 1a). A theoretical capacitance image corresponding to this particular arrangement of conductive spheres is shown in Figure 1b. Figure 2 is an x-derivative image of Figure 1b, displayed with a logarithmic gray scale to show more of the image structure. The mechanism that generates the raw data image is an electronic "black box" (capacitance differentiator) which produces a voltage proportional to the change in capacitance measured from one point on the specimen to the next. We will talk more about the workings of the differentiator below. The voltage from the differentiator, modeled in the derivative image of Figure 2, is recorded to provide a map of the specimen's electrical properties.

The system described above exploits capacitance to characterize the electrical coupling of the conductive features on the specimen. In the model specimen, for example, it can tell us which of the spheres are completely penetrating the base material, and which are "afloat" in the substrate. The major mechanism for contrast is capacitance because of its ability to show variation when DC resistance measurements do not provide differences large enough to obtain a signal to noise ratio that is useful.[1-2] The image in Figure 3 is a Scanning Force Image (topography) taken from a specimen of polyanaline spheres in a non-conducting base material. Figure 4 shows a Capacitance Image that corresponds to the same area.

The basic function of the capacitance differentiator is to create an electrical   potential difference when a change in capacitance is sensed. The specimen is placed in a circuit that adjusts other parameters in this circuit to resonate at some frequency. As the probe traverses the specimen, the change in capacitance causes the circuit to be "knocked" off of this resonance point. The circuit then responds by producing a potential difference to regain the previous resonant frequency. It is this potential difference that is recorded as an auxiliary image. This specimen is an integral part of a continuous electronic feedback circuit.

References

1. J.A. Slinkman et al., *Lateral dopant profiling in MOS structures on a 100 nm scale using scanning capacitance microscopy*, International Electron Devices Meeting 1990. Technical Digest (Cat. No. 90ch2865-4) , p. 986, 73-6.

Proc. Microscopy and Microanalysis 1995, edited by G.W. Bailey, M.H. Ellisman, R.A. Hennigar, and N.J. Zaluzec
Copyright © 1995 MSA. Published by Jones and Begell Publishing, 79 Madison Ave., New York, NY 10016

2. Abraham, D. W. et al., *Lateral dopant profiling in semiconductors by forcing microscopy using capacitive detection,* Journal of Vacuum Science & Technology B, vol. 9, no. 2, pt.2, p. 703-6.

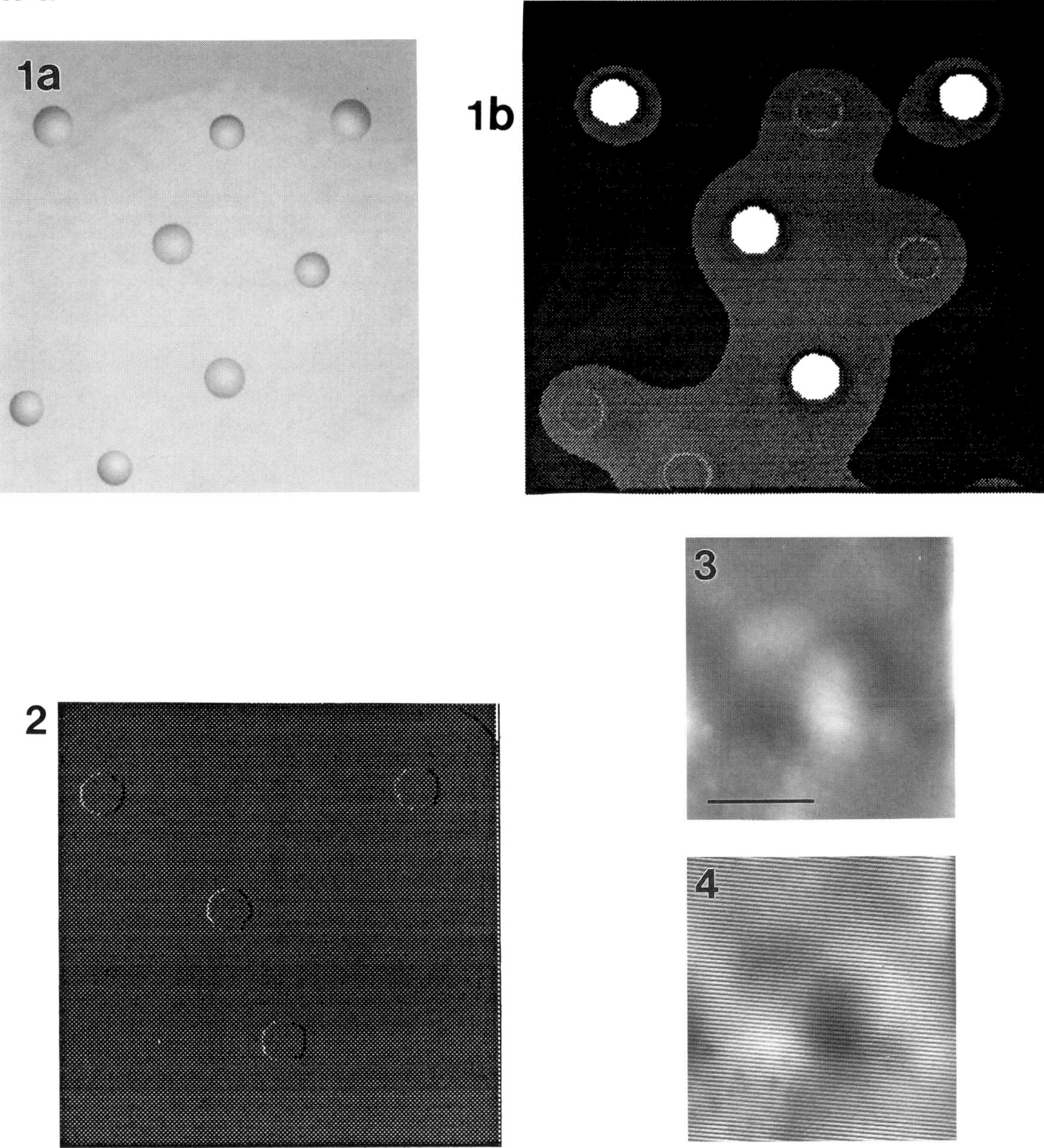

FIG. 1.— (a) Topography image of conducting Au conducting spheres in an insulating substrate; (b) Capacitance image of the theoretical image of FIG. 1a.

FIG. 2.— X-derivative of the image in FIG. 1b.

FIG. 3.— Height image from scanning force microscope. Bar = 2 μm.

FIG. 4.— Capacitance image taken on the scanning force microscope simultaneously with the image in FIG. 4.

THREE DIMENSIONAL IMAGING OF BIOLOGICAL MACROMOLECULES

A.J. Koster, J. Walz, D. Typke, M. Nitsch and W. Baumeister

Max-Planck-Institute for Biochemistry, Am Klopferspitz, D-82152 Martinsried, Germany

3D imaging of both cellular structures as well as molecular assemblies of biological molecules has become an increasingly useful tool to study structure-function relationships of biological systems[1,2]. In this paper instrumental and methodological developments are discussed towards automated 3D imaging, which will be illustrated by examples of structures studied in Martinsried. To image individual structures with dimensions in the range of 10-500 nm with a resolution of 1-5 nm, electron tomography is the only technique available. The strategy of choice depends on size and shape of the structure to be reconstructed. Single-tilt axis tomography is suitable for the reconstruction of unique structures (for example, irregularly shaped viruses or cellular structures). Random conical-tilt data collection, as well as angular reconstitution techniques, can be used to reconstruct the average structure of many copies of a particle, such as those present in suspension of one kind of protein. To reconstruct a unique structure with single-tilt axis tomography the tilt range and tilt increments are chosen to meet the resolution desired within the constraint of the allowable electron doses (Table 1). In principle, when 61 views are required over a tilt range of $\pm60°$ to reconstruct a structure with a diameter of 60 nm, a resolution of 3 nm can be obtained. In spite of its great potential, single-tilt axis tomography is used by a relatively small number of laboratories compared to those applying conical tilt techniques.

In our laboratory we realized an automated single-tilt axis tomography setup capable of collecting low dose tilt series under cryo conditions[3]. The images are recorded digitally, and lateral specimen displacements together with focus changes are automatically measured and compensated for. Data are collected using a Photometrics slow scan CCD camera (1024^2 pixels of 19 μm) attached to a Philips CM200 FEG microscope both controlled by a TVIPS image processing computer. Recently our cryo-holder was modified to enlarge the tilt range over $\pm70.°$ The total dose required to collect a tilt series of 61 views of ice embedded structures can be as low as 15 e/Å^2, as was demonstrated by data collected on ice-embedded DPPC vesicles (approximately 300 nm in diameter). Practical limitations in collecting a tilt series of ice-embedded structures are strong beam-induced movements, and, for tilted specimens, charging of the ice at high specimen tilts. Currently we are evaluating the possibility to overcome these problems by covering the ice with a thin conducting layer (e.g. carbon or a metal), before inserting the cryo-holder with the sample into the microscope.

Recently, we have extended the capabilities of the system with the possibility to automatically collect random conical tilt series with the CCD camera. At high tilt of the specimen (say 60°) an image is recorded, followed by a corresponding image at 0° tilt. For optimal signal to noise ratio of the reconstruction it is important to obtain projections of a large number of the single particles scattered over the grid. Also, one needs to relate the high tilt view with that of the untilted specimen. Unfortunately, the CCD camera has a considerably smaller field of view compared to photographic film. Therefore, to collect a sufficiently large number of projected particles we record the high tilt data by taking a CCD spot scan series[4] along the tilt axis of the holder to record images with equal defocus. The corresponding views at 0° tilt are acquired by taking per CCD spot three partly overlapping images perpendicular to the tilt axis, to match the whole area covered with the high tilt view. Fig. 1 shows a tilt pair (60° and 0°) of *phosphoenolpyruvate synthase* (2 MDa, negatively stained) from *Staphylothermus marinus* (in collaboration with G. Harauz and C. Cicicopol[5]). Our present work is aimed at reconstructing single molecules embedded in ice, such as the 26 S proteasome[6] of *Dictyostelium* which has a length of 45 nm, using both the automated single-tilt axis tomography and random conical tilt data collection strategies.

Proc. Microscopy and Microanalysis 1995, edited by G.W. Bailey, M.H. Ellisman, R.A. Hennigar, and N.J. Zaluzec

Cellular (300-500 nm)	Molecular (10-100 nm)
plastic embedded	frozen hydrated
dose: <500 e/Å^2 (shrinkage)	dose: <20 e/Å^2 (damage)
high tension: 200-400 kV	high tension: 120-200 kV, FEG
image size: 1-2 µm (montage)	300-500 nm (one image)
tilt range: ±75°	tilt range: ±75°
obtainable resolution: 5-10 nm	obtainable resolution: 2-5 nm
#views: 50-150	#views: 20-50

TABLE 1. - Typical data collection specifications and possibly obtainable resolution after 3D reconstruction using automated single-tilt axis tomography data collection of cellular and molecular structures.

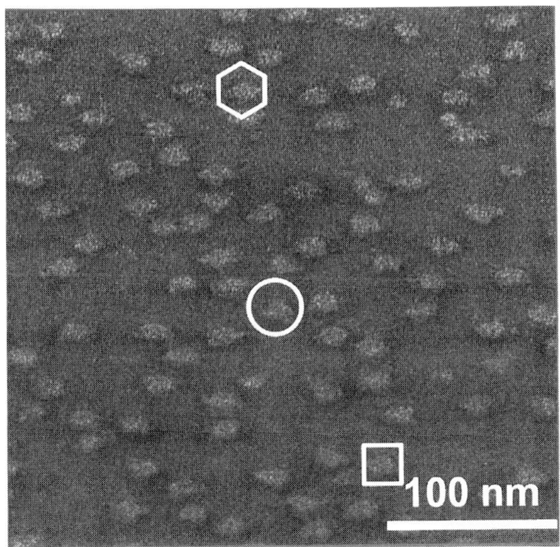

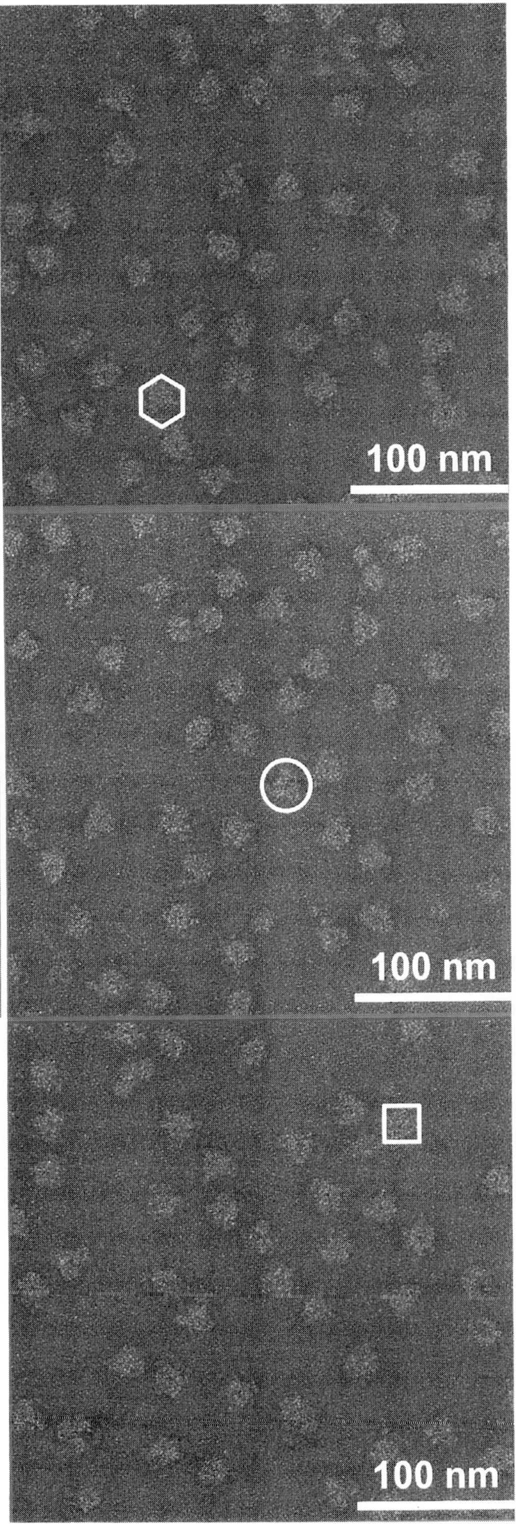

FIG. 1 - Tilt pair of the automated random conical tilt data collection. On the left the tilted view is shown (60°) on the right the three corresponding untilted views.

References

1. K. Dierksen et al., *Ultramicroscopy* 40(1992)71.
2. A.J. Koster et al., *Ultramicroscopy* 46(1992)207.
3. K. Dierksen et al. *Ultramicroscopy* 49(1993)109.
4. D. Typke et al., *Ultramicroscopy* 46(1992)157.
5. C. Cicicopol et al., *FEBS Lett.* 356(1994)345.
6. A. Lupas et al., *Enzyme & Protein* 47(1993)252.
7. This work is supported by the Deutsche Forschungsgemeinschaft, project Ty 2/4-1.

3-D RECONSTRUCTION OF PARACRYSTALLINE BIOLOGICAL SPECIMENS BY TOMOGRAPHY

K. A. Taylor and H. Winkler

Department of Cell Biology, Duke University Medical Center, Durham, NC 27710

The calculation of 3-D images of paracrystalline objects is often done using methods appropriate to 2-D crystals.[1] However, this method leads to a loss of information about the structural variations within the paracrystal. Structural variations are often referred to as noise but may represent important features. In our own work with 3-D imaging of sectioned insect flight muscle (IFM) structural variations represent the adjustment in crossbridge form imposed by the filament lattice.[2] In order to complement our averaged 3-D images obtained by 2-D crystal methods, we have experimented with performing essentially the same procedures using all the data, both sampled and continuous, to generate the 3-D image. While the method embodies no new principles, the combination of protocols appears to be unique.

The procedure uses the following steps: (1) A uniaxis tilt series ranging to ±75° is recorded. The increment between tilt angles decreases as tilt angle increases and ranges from ~8° at low angles to ~2° at high tilts.[3] (2) A reciprocal lattice is fit to the sampled diffraction in the transform of a selected region from each image. This region is identified in all the images of the tilt series using naturally occurring fiducials. (3) The images are aligned using a cross correlation procedure that sets the amplitudes of each Fourier coefficient to unity.[4] This makes the origin search insensitive to the pseudorepeat from the sampled part of the transform. (4) Merging is done starting with the lowest tilts. As part of the merging process, the sampling in each transform is matched to the sampling of the preceeding tilts by a 4 parameter minimization involving 2 stretching, 1 skew and 1 rotation mode (Fig. 1,2). The sampling is matched by bilinear interpolation of the images. The interpolation distorts the images requiring that a low pass filter be applied to the cross-correlation function to remove spurious high frequencies produced by the distortion. Scale factors are calculated from the diffraction intensities. Reference transforms are calculated using the previously merged data. (5) After merging and scaling, the 3-D transform is calculated by sinc function interpolation of the "lattice lines" (Fig. 3).[5] (6) The 3-D image is obtained by inverse 3-D Fourier transformation.

The 3-D images obtained by this method revealed that the sections, which are 25-30 nm thick are not flat on the EM grid (Fig. 5). Though not surprising, this phenomenon would reduce the resolution in the 3rd dimension in an averaged 3-D image unless it was corrected. We therefore developed a 3-D unbending procedure which both flattened the reconstruction and at the same time straightened the filaments (Fig. 6). The effect of this 3-D unbending process is to sharpen the transform in the X-Y plane as it removes the wrinkles (Fig. 4).

Tomography of 2-D crystals by this or other methods could obviate the need to make specimens flat once it is realized that the tomogram contains the information on specimen flatness thereby facilitating computational correction. Alternatively, correlation averaging methods could be used to average 3-D motifs after 3-D alignment. This tomography method is not restricted to paracrystalline objects and we have been able to obtain 3-D images of entire half-sarcomeres, including the A-band of the muscle as well as the Z-disk and the M-line, both structures that are completely different from the A-band and thus do not contribute to the sampled diffraction. Obviously, the method does require an initial guess for the dimensions of the area in each image and that is facilitated by the presence of sampled diffraction.[6]

1. L. A. Amos et al., *Prog. Biophys. molec. Biol.* 39(1982)183.
2. K. A. Taylor et al., *J. Cell Biol.* 109(1989)1085.
3. W. O.Saxton et al., *Ultramicroscopy* 13(1984)57.
4. P. J. Shaw et al., *Biophys. J.* 55(1989)101.
5. R. A. Crowther et al., *Proc. Roy. Soc. Lond. A.* 317(1970)319.
6. Supported by NIH grant GM 30598

Proc. Microscopy and Microanalysis 1995, edited by G.W. Bailey, M.H. Ellisman, R.A. Hennigar, and N.J. Zaluzec

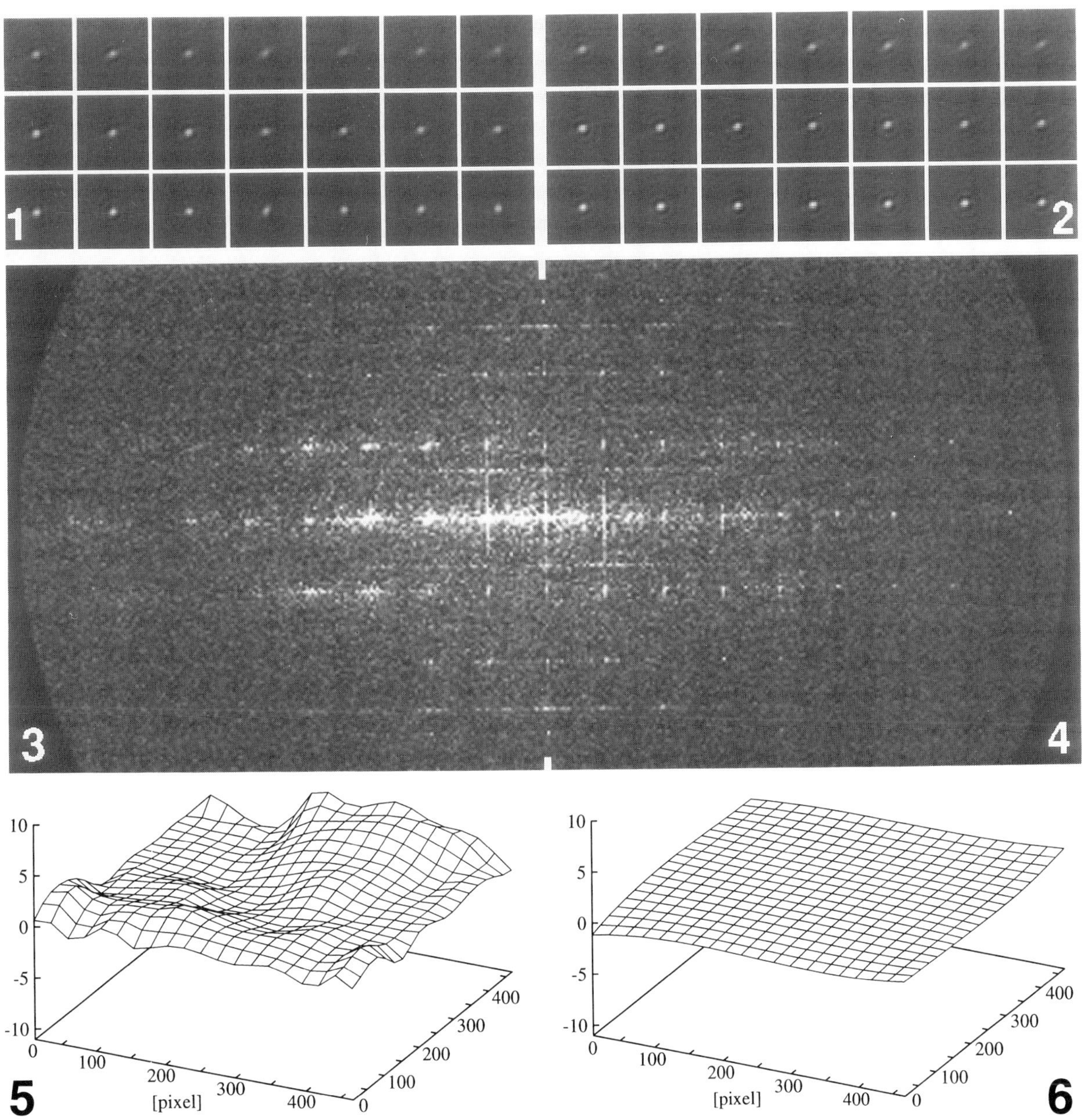

Figure 1. Montage of correlation peaks from the initial merging prior to dimension minimization. The peaks have been orthogonalized and normalized. The zero tilt is at the lower left hand corner. Tilt angle increases from left to right and up.

Figure 2. Montage of correlation peaks after dimension minimization. Note that the subsidiary rings are sharper and the peak height is generally higher after this process. The faint streak through the peak is in the direction perpendicular to the tilt axis.

Figure 3. Central section from 3-D transform after sinc function interpolation. Note that the higher order peaks are diffuse.

Figure 4. Central section from the 3-D transform after 3-D unbending. Note how this process sharpens up the higher order peaks that were diffuse in the original data.

Figure 5. 3-D plot of correlation peaks from the 3-D reconstruction before 3-D unbending. The vertical scale has been stretched to amplify the wrinkling.

Figure 6. 3-D plot of correlation peaks from the 3-D reconstruction after 3-D unbending.

MAXIMUM ENTROPY EM TOMOGRAPHY: DEMONSTRATION OF IMPROVED RESOLUTION WITH ADENOVIRUS.

Ulf Skoglund[*], Lars-Göran Öfverstedt[*], Roger Burnett[**] and Gérard Bricogne[***]

[*]Dept. of Cell and Molecular Biology, Karolinska Institute, S-17177 Stockholm, Sweden
[**]The Wistar Institute, 3601 Spruce Street, Philadelphia, PA 19104, U.S.A.
[***]MRC Laboratory of Molecular Biology, Hills Road, Cambridge CB2 2QH, England; and LURE, Bâtiment 209D, 91405 Orsay, France.

Electron microscope tomography (EMT)

EMT is a method for three-dimensional (3-D) reconstruction of single objects from electron microscope pictures in a tilt series[1]. The EMT method is general and can be applied to any transparent object, and is not restricted to symmetrical or regularly arranged objects, nor to objects with preferred orientations on a support grid. In its present shape, the EMT method allows reproducible 3-D reconstructions of molecular objects with a resolution in the range of 5 nm[2]. Currently, the EMT method covers the intermediate resolution range where there is no other physical technique available to analyze single molecular complexes.

New prospects

The recent availability of techniques for high quality automatic data collection from vitrified specimens at low dose[3] promises an improvement of the reproducible resolution as well as considerable expansion of the number of users of the EMT technique. A serious drawback of EMT, however, has been the notoriously low quality of single-object 3-D reconstructions when compared to symmetrized reconstructions. An inherently high noise level, 'the missing cone problem'[4] and a non-quantitative aspect have rendered the interpretation of single-particle reconstructions difficult and slow. Here we report the advent of an objective and quantitative procedure that significantly improves the quality and resolution of 3-D reconstruction as well as the fit to their projection data.

The COMET procedure

We have developed a new numerical algorithm to improve the fidelity of 3-D tomographic reconstructions made from electron micrographs while at the same time filtering much of the noise present in the raw data. The new technique is called COnstrained Maximum Entropy Tomography (COMET). The essence of the method[5,6] is that it will produce the most featureless reconstruction that fits the raw projection data within their observational accuracy. This objective is achieved by maximising the entropy of the final 3-D image relative to a non-informative prior under the constraint that all its calculated projections fit the raw 2-D data with a reduced chi-squared of 1.0, i.e. that the sum of the inverse-variance weighted squared differences be equal to the number of independent pixel observations. The constraint on chi-squared ensures that the reconstructed density is consistent - but not overconsistent -with its observed projections; while entropy maximisation ensures that the density modulations are as conservative as possible, giving the smoothest achievable density by eliminating detail which is not significantly above noise and also flattening artefacts arising from systematic errors. In particular, the COMET procedure will minimise the detrimental effects of the 'missing data' from non-recorded regions on 3-D reconstructions (treating them as absent, rather than measured as zero) and of errors due to bad statistics in the measured data.

The algorithm used for constrained entropy maximisation was first devised for crystallographic phase refinement against reciprocal space data[5]. It was adapted by replacing the Fourier transformation (which relates model to data in the crystallographic setting) by line projections in real space. The COMET program package can handle projections at any angle. Thus the program is independent of specific data sampling strategies like single-axis tilt series, conical tilt series, random tilt series or tilts from symmetric samples where the projection angles are calculated later (as is the case for the adenovirus, our test object). General symmetrization and averaging schemes are handled by a user

Proc. Microscopy and Microanalysis 1995, edited by G.W. Bailey, M.H. Ellisman, R.A. Hennigar, and N.J. Zaluzec

supplied symmetrization routine. Currently a full icosahedral symmetrization within an envelope has been developed and was used in the adenovirus teststructure. The symmetrization is used both for the densities and the derivatives calculated to improve the density.

The COMET programs also compensate for the different contrast transfer effects on the different views of a tilt series, using an analytical contrast transfer function (CTF) currently specified through input parameters like defocus. At a later stage in the development of the COMET programs, the CTF could be supplied also as an experimental curve or as parameters refined from a parametrized model.

The input data to the programs consist of two parts. The first is an initial estimate of the reconstruction (i.e. the 'regular' 3-D reconstruction), and the second is the actual projection data (the EM images). These "observations" are the values from the scanned micrographs and it is important to have a good statistical model for the variances of the data. For diffraction data in crystallography, the variance is often equal to the observed intensity (Poisson-distributed counting statistics). As a first approximation, the electron micrograph variances have been estimated as proportional to the square root of the observed (scanned) density. This approximation is rather good since the scanned data is a form of contrast and the variance of a contrast is nearly proportional to its square root[7]. For CCD data the variance would be modelled as proportional to the intensity.

The adenovirus reconstruction test.

As hinted above, we used the adenovirus as our test object. It has been analysed as an unstained and vitrified specimen with cryo-EM[8,9] and the structure of its major coat protein, the hexon, has been determined with X-ray crystallographic techniques[10] to 2.9 Å resolution. Thus a COMET application on the cryo-EM adenovirus structure can be objectively compared with both the cryo-EM and the crystallographic description of the hexon structure. In our tests we used the data from digitized original micrographs[8] without modifications, the low noise adenovirus structure defined at approx. 35 Å resolution[8] and an analytically defined contrast transfer function damped to zero at 25 Å resolution in accordance with the published work[9]. We have calculated a 3-D reconstruction of the adenovirus by the COMET procedure, and (sofar) compared with one of the four icosahedrally independent hexons in the virus coat with a suitably aligned model density for the hexon, calculated from the hexon coordinates in the Brookhaven data bank and lowpass filtered to 25 Å resolution. This test shows that the hexons reconstructed by COMET are in significantly better agreement with the 25Å resolution crystallographic map than those in the original reconstruction.

The generality and potential power of the COMET procedure represent a substantial advance in the field of image reconstruction because it makes possible quantitative, low noise 3-D reconstructions of low contrast specimens (such as unstained, vitrified non-symmetric *in situ* specimens) with better resolution than was previously achievable.

References

1. U. Skoglund et al., *Nature* 319(1986)560.
2. L.-G. Öfverstedt et al., *Cell* 79(1994)629.
3. K. Dierksen et al., *Biophysical Journal* 68(1995)1.
4. M. Radermacher *J. Electron Microsc. Tech.* 9(1988)359.
5. G. Bricogne *Acta Crystallographica* A40(1984)410.
6. G. Bricogne, in B. Buck and V.A. Macaulay, Eds., *Maximum Entropy in Action*, Oxford, Clarendon Press (1991)187.
7. H. Lichte, in A. Ríos et al., Eds., *Electron Microscopy 92, Proceedings of EUREM 92*, Granada, Secretariado de Publicaciones de la Universidad de Granada 1(1992)637.
8. P.L. Stewart et al., *Cell* 67(1991)145.
9. P.L. Stewart et al., *EMBO J.* 12:7(1993)2589.
10. F.K. Athappilly et al. *J. Mol. Biol.* 242(1994)430.

TOWARDS EM TOMOGRAPHY OF LARGE CELLULAR VOLUMES

A. Delaney,[*][#] Y. Bresler,[#] A.S. Belmont[*]

*Department of Cell and Structural Biology, University of Illinois, Urbana-Champaign
#Department of Electrical and Computer Engineering, University of Illinois, Urbana-Champaign

Three-dimensional reconstructions of large cellular subvolumes would be quite valuable in addressing a number of basic questions related to cell ultrastructure. More specifically, we are interested in obtaining reconstructions with several nm resolution over projected areas of 5-50 um diameter and 1-10 um depth. Recently, EM tomography has been combined with serial sectioning as a means of accomplishing this goal.[1] In practice, however, a number of technical problems must still be solved to realize the full potential of this approach, and most applications of EM tomography to date have been limited to high resolution reconstructions over much smaller volumes or lower resolution reconstructions over larger volumes.

A major limitation is the limited depth of focus of the electron microscope. Although this depth of focus usually exceeds the section thickness, at very high tilt the range of underfocus through overfocus in the image will approach the diameter of the reconstructed area. Spot-scanning using dynamic focusing provides one method for obtaining in focus high tilt projections, but application of this approach is limited by the changes in image magnification and rotation caused by focusing over a range of 10-30 um. As an alternative approach, we have developed an algorithm to deblur high tilt projections which merges several images at different focus to produce one in focus, high tilt projection, assuming a spatially varying defocus blur, knowledge of the PSF, and the absolute defocus images of the micrographs.[2] An iterative method was used to determine these absolute defocus images, starting from knowledge of the relative defocus values for each micrograph in the focus series. Numerical results using simulated and real data demonstrate the effectiveness of this procedure. This method is currently being used to deblur high-tilt micrographs up to 80 degrees tilt, prior to their use in EM tomography.

We are also interested in developing new tomography algorithms, not only to improve the quality of the reconstruction, but also to allow reconstructions from significantly fewer projections, particularly from sparse images produced by immunostaining of particular cell components. We have developed a new tomographic reconstruction algorithm designed for objects with sharp boundaries but smooth internal density variations. This algorithm uses a regularized, least squares minimization approach with cost functionals for smoothness, edge sharpness, density constraints, and consistency with projection data. Numerical simulations suggest greatly improved reconstructions as compared to either a standard R-weighted back projection or a projection on convex sets reconstruction algorithm, particularly for the cases of missing angular data and/or sparse angular sampling. We demonstrate using simulated data the improvements attained with increased data and the use of this edge preserving, regularized, least squares approach. We also demonstrate with real data, improved reconstruction results with reduction of effects produced by the missing wedge of angular data. Encouragingly, in these real reconstructions we appear to be recovering sharp, horizontal edges corresponding to the edges of the actual physical section. This reduction of z-axis blurring should greatly increase the accuracy of alignment of tomographic reconstructions from adjacent physical sections.

Finally, we discuss additional problems including alignment of reconstructions of adjacent sections and the problem of projection alignment.

References

*Proc. Microscopy and Microanalysis 1995, edited by G.W. Bailey, M.H. Ellisman, R.A. Hennigar, and N.J. Zaluzec
Copyright © 1995 MSA. Published by Jones and Begell Publishing, 79 Madison Ave., New York, NY 10016*

1. G.E. Soto et al., NeuroImage(1994) 1: 230-243.
2. A. Delaney and A.S. Belmont, Ultramicroscopy(1994) 56: 319-335.

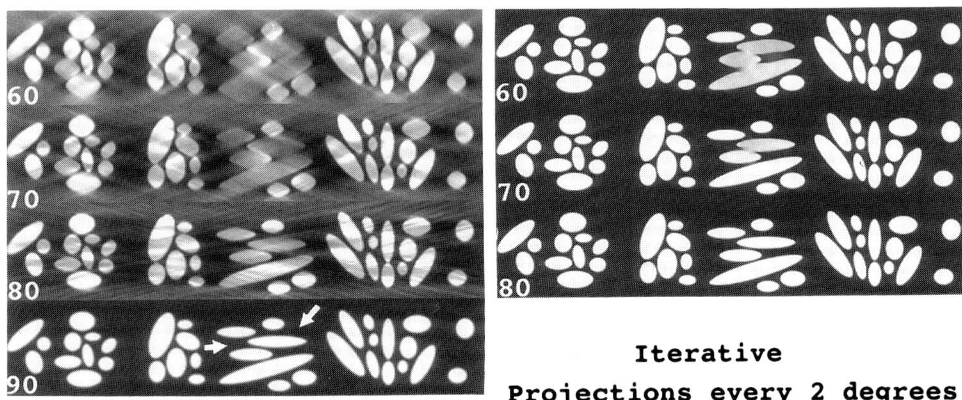

Fig. 1- Comparison using simulated data of standard filtered back projection reconstruction and iterative, edge preserving algorithm. Reconstructions based on varying size of the missing wedge of tilt angles are shown. Note the recovery of horizontal edges with the iterative approach (middle, elongated ellipses). The iterative algorithm allows separation of closely packed features. Also note the improved image segmentation and background densities.

TOMOGRAPHIC RECONSTRUCTION OF MITOCHONDRIA

C.A. Mannella, M. Marko, K. Buttle, P. Penczek, A. Leith, J. Frank

*The Biological Microscopy and Image Reconstruction Resource, Wadsworth Center, Empire State Plaza, Box 509, Albany, NY 12201-0509
**Department of Biomedical Sciences, School of Public Health, Albany, NY 12201-0509

During oxidative phosphorylation, free energy of substrate oxidation is used to generate electrochemical gradients across the mitochondrial inner membrane. Full understanding of this process requires knowing the pathways for internal diffusion of ions and metabolites inside the organelle. This, in turn, requires detailed information about the organization of the membranes that compartmentalize the mitochondrion and the distribution of transport proteins (proton pumps, ion channels, metabolite carriers) on these membranes.

We have undertaken the study of the compartmentation of rat-liver mitochondria (conventionally fixed and plastic-embedded) using electron microscopic tomography[1]. Reconstructions of isolated mitochondria have been computed using projection images collected on the Albany high-voltage electron microscope from sections (0.5-1-μm thick) tilted around one axis over $+/-$ 70° at 2° increments. A problem encountered with this approach is the directional loss of resolution due to the "missing wedge" of information in Fourier space in the direction parallel to the tilt axis. To overcome this problem, we have developed an alignment algorithm that allows projections collected over two or more tilting directions to be used in the same modified back-projection calculation. As a result, Fourier space can be sampled more evenly and directional resolution losses are reduced. This is essential for study of the organization of mitochondria, which contain many highly convoluted and nested membrane compartments. As seen in Figs. 1-3, the definition of several membrane cross-sections in slices normal to the single tilt axis greatly improves when data from a second tilt series (axis approximately perpendicular to the first) is included in the reconstruction.

After computing the reconstructions, the membranes have to be segmented so that they can be studied in detail. This process is described in an abstract by Marko, et al., in these proceedings. After segmentation, volume and surface area measurements can be made, and individual inner-membrane invaginations, crista, can be studied in detail. In addition, the proximity between the inner and outer membrane can be mapped. This is done by projecting onto the outer membrane a density value proportional to the separation between the inner and outer membranes. The outer membrane is then "unrolled" and projected onto a flat surface. By adjusting the contrast threshold of this map, different ranges of membrane closeness can be displayed. Fig. 4 is a "contact map", i.e., white loci are those at which the the center-to-center distance between outer and inner membranes is equal to or less than 2 nm (less than one-half the thickness of a phospholipid bilayer). These results are providing important new insights into the organization of the cristae and of the physical interactions between the outer and inner mitochondrial membranes.[3]

References

1. Mannella et al., *Microsc. Res. Tech.* 27(1994)278
2. P. Penczek et al., *Ultramicroscopy* (submitted)
3. The Biological Microscopy and Image Reconstruction Resource is supported by NIH PHS grant R01219 (P.I. Dr. C. Rieder). This work was also supported by NSF BIR 921 9043 (P.I. Dr. J. Frank), NSF MCB 921 9353 (P.I. Dr. C. Mannella), and NIH GM29169 (P.I. Dr. J. Frank).

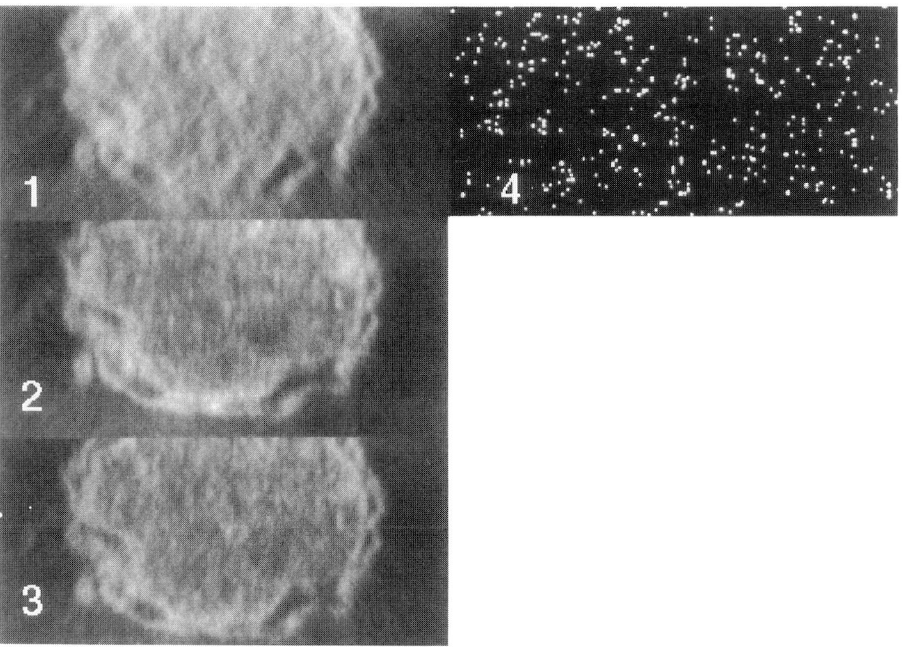

FIGS. 1-3—Equivalent 6-nm-thick slices through three different tomographic reconstructions of the same mitochondrion. (1) Single-tilt reconstruction, tilt axis normal to page. (2) Single-tilt reconstruction, tilt axis horizontal. (3) Double-tilt reconstruction about both axes. Note severe loss of continuity of membranes in (1) and general reduction in artifacts in (3) vs (2). Image width 0.9 μm.

FIG. 4—"Contact map" for an unrolled segment of the outer membrane from the double-tilt reconstruction. Each point represents a location at which the the center-to-center distance between outer and inner membranes is equal to or less than 2 nm (less than one-half the thickness of a phospholipid bilayer). Image width 1.5 μm.

PRACTICAL ELECTRON TOMOGRAPHY

C.L. Woodcock
Department of Biology, University of Massachusetts, Amherst. MA 01003

Despite the potential of the technique, electron tomography has yet to be widely used by biologists. This is in part related to the rather daunting list of equipment and expertise that are required. Thanks to continuing advances in theory and instrumentation, tomography is now more feasible for the non-specialist. One barrier that has essentially disappeared is the expense of computational resources. In view of this progress, it is time to give more attention to practical issues that need to be considered when embarking on a tomographic project. The following recommendations and comments are derived from experience gained during two long-term collaborative projects[1,2].

The specimen

Tomographic reconstruction results in a three dimensional description of an individual EM specimen, most commonly a section, and is therefore applicable to problems in which ultrastructural details within the thickness of the specimen are obscured in single micrographs. Information that can be recovered using tomography includes the 3D shape of particles, and the arrangement and dispostion of overlapping fibrous and membranous structures. It is a mistake to assume that tomography will compensate for poor contrast in the different structures that must be discriminated. Although contrast enhancement can be used on the final reconstruction, it is advisable to invest time in maximizing the contrast and definition of the original specimen, rather than in trying later to find an image processing solution. For sections, this may involve exploring alternatives to standard uranyl-lead staining. The early literature on staining includes a number of formulations with great promise[3]. Exploratory tests must also be carried out to detemine the most reliable method of stabilizing the specimen and adding gold particles as fiduciary markers[2]. The final specimen grids should be unbent, and contain flat, untorn sections. The selection of nominal section thickness depends on the nature of the question, and must be matched to the KV available, bearing in mind that at high tilt, the electron beam traverses about 3 times the specimen thickness. Although in principle, HVEM is ideal for handling thicker specimens with minimal beam damage, in practise, the IVEM range from 150KV to 300KV is often more suitable given the inherent low contrast of HVEM instruments. Also, HVEMs tend to be of older design, making it difficult to collect adequate tilt series.

Data collection

Before starting to collect data, it is important to consider the resolution required by the project, since this is directly related to the angular separation between the tilt images. . Resolution is limited by the angular separation of the tilt data, the maximum tilt angle, and the voxel size. Of these, the latter is trivial, and can be selected at the time of image acquisition to be non-limiting. The angular separation of the tilt data sets a formal limit to resolution (d) in the plane (xy) of the specimen:

$$d = \pi D/N \quad (1)^4$$

where D is specimen 'diameter', and N the number of equally spaced tilt images between -90° and +90°. As discussed elsewhere[5], the appropriate value of D depends on the nature of the specimen. In the 'worst case' situation of a section completely filled with specimen detail, D is the maximum distance traversed by the beam at maximum tilt, while for a spherical particle that is not overlapped with other specimen elements throughout the tilt range, D is effectively the particle diameter. The lack of a complete angular range further limits the resolution in the 'z' direction, approximately twofold for a +/-60° data set[6].

Simple modeling experiments in which projections are made from computer generated images, and then 'reconstructed' by weighted back projection suggest that there are conditions where valid information is present beyond the limit set by equation (1). As pointed out[4], this is to be expected for images that are not densely packed with information. It is also possible to use modeling to examine the effects of different high tilt restrictions. As expected, reducing the angular range leads to circular elements becoming elongated in the 'z' direction as the edges become ill-defined at the top and bottom. Another effect of tilt restriction is the severe degradation of linear features in a wedge-shaped region defined by the missing angular range. To what extent the results from reconstructions of rather simple images should be heeded when selecting resolution levels for real reconstructions is not clear. The modeling data do underscore the need for tomographic data sets of specimens with known structure. Sections containing a mixture of spherical and rod-like viruses could produce very informative results.

Most specimens lose mass during electron beam exposure, the damage leading to substantial shrinkage of sections in the 'z' direction. Data sets collected while the sample is shrinking yield invalid reconstructions, a conclusion verified by modeling experiments. At present, two strategies can be used to minimize the effect: pre-irradiate the specimen until most of the shrinkage has taken place, or reduce the shrinkage using low-dose microscopy of cooled specimens. The latter approach must be regarded as still in the development stage, and the stability of cryo-holders at high tilt remains a problem. If the pre-shrinking method is used, what is the effect on the specimen. Experiments using Lowicryl K11M sections containing a dispersion of colloidal gold beads or osmium ammine stained nuclei suggest that beam damage does not result in a simple uniform contraction in 'z'. Rather, stable components such as stain aggregates tend to stay in place, as the embedding medium collapses around them.

Automated collection of tilt series images is becoming a reality in several laboratories[7,8] , and promises to make tomography simpler, more accurate, and less damaging to the specimen. Without this advance, data collection is tedious and time consuming, and one must expect to reject a high proportion of data sets that cannot be aligned, or show obvious specimen changes between the start and finish. Nevertheless, even at worst, the time spent on data collection will be small compared to the time spent on interpretation! If possible, data should be collected beyond the +/-60° limits of standard goniometers through the use of a high tilt holder.

Alignment and reconstruction

A number of algorithms for alignment and reconstruction are in active use, and have been demonstrated to yield faithful 3D information[9]. It would be a valuable service to the EM community if the different reconstruction strategies could be compared using identical input data sets.

Interpretation

Interpreting the resulting cube of numbers (voxels) representing the 3D density distribution of the specimen involves consideration of the resolution in the x,y, and z directions, and ways to turn the numerical data into images that best represent the salient features of the volume. Strategies have to be developed on a case by case basis. Often, it is not possible to obtain useful information from shaded surface representations, where a voxel threshold that distinguishes specimen from background, and more sophisticated tools such as volume rendering where levels of opacity and contrast can be used to view the entire volume. The final challenge for the tomographer is to find a way to impart the interpretation in the printed page medium that is convincing to reviewer and reader. Perhaps this problem will be alleviated in the future by the further development and use of depositories for volume data that can be accessed over the network (e.g. http//indy.cnb.uam.es), and form an essential part of a publication.

References

1. C.L. Woodcock, B.F. McEwen, and J. Frank, *J. Cell Sci.*99(1991)107.
2. R.A. Horowitz et al,. *J. Cell Biol.* 125(1994)1.
3. R.A. Horowitz and C.L. Woodcock, *J. Histochem. Cytochem.* 40(1992)123.
4 R.A. Crowther, D.J. DeRosier, and A. Klug. *Proc. Roy. Soc. London A 317(1970)319.*
5. C.L. Woodcock, in J Frank, Ed, *Electron Tomography* New York:Plenum Press (1992)313.
6. M. Radermacher, in J Frank, Ed, *Electron Tomography* New York:Plenum Press (1992)91.
7. A.J. Koster et al., *MSA Bulletin* 23(1993)176.
8. D. Typke et al., in B. Jouffrey and C. Colliex, Eds, *Proc. 13th Intl. Congress EM* Vol 1 Les Ulis:Les Editions de Physique (1993)523.
9. J. Frank, Ed. *Electron Tomography* New York:Plenum Press (1992)
10. Supported in part by NIH GM 43786

SAME-CELL CORRELATIVE VIDEO LIGHT MICROSCOPY/ELECTRON MICROSCOPY TOMOGRAPHY: AN APPROACH TO UNDERSTANDING KINETOCHORE BEHAVIOR DURING MITOSIS

B.F. McEwen, G. Osorio, R. Cole, and C.L. Rieder

Division of Molecular Medicine, Wadsworth Center, Albany, NY 12201-0509

The goal of structural biology is to determine structure, structural relationships, and changes in these parameters in an effort to comprehend mechanism of function. In this context same-cell correlative LM/EM provides a powerful approach for elucidating the mechanisms responsible for the behavior of cell components. In this method the event of interest is followed *in vivo* by video-LM, and the cell then fixed at a critical time during the observational period for a subsequent 3D EM analysis. In this manner the history of a particular event or response can be correlated with the 3D ultrastructure underlying the event.

Biologists have long sought to elucidate the mechanism(s) that generate, control and coordinate the poleward and away-from-pole motion of sister kinetochores on each chromosome during mitosis.[1-3] These "congression" movements require the association of microtubules with each kinetochore (kMTs), and ultimately align the chromosome on the spindle equator. Motion in each direction is distinct because the slow growing (-) ends of kMTs end near the spindle pole while the rapidly growing (+) ends terminate in the kinetochore plate. Thus the (-) end directed movement towards a spindle pole requires that the kMTs shorten (depolymerize) while the (+) end directed motion away from the pole requires that the kMTs elongate (polymerize). Evidence indicates that the kinetochore is the primary site for both force production and addition/subtraction of kMT subunits.[1-3]

It has been proposed that the direction of chromosome motion during congression is mediated by the number of MTs attached to its sister kinetochores.[4] Another hypothesis predicts that kMTs grow through the kinetochore when it moves away from the pole but terminate on its surface during motion towards the pole.[1] Evaluating these hypotheses requires 3D EM, but cannot be readily or reliably approached using thin serial sections: individual MTs cannot be traced with certainty between serial thin sections when cut parallel to the spindle long axis and, although this problem is minimized by cross sectioning spindles,[5] it is extremely difficult to identify kinetochores followed *in vivo* in these views. To overcome these problems we are developing same-cell correlative LM/3D EM methods using EM tomography of thick sections. In this approach a congressing chromosome is followed by video DIC LM prior to and during fixation (see 6). The cell is then post-fixed in OsO_4, flat embedded and serial thick sectioned using correlative LM/EM techniques.[6] Next, tomographic 3D reconstructions of kinetochores followed *in vivo* are computed and analyzed as previously described.[7] Preliminary experiments reveal that, for a given chromosome, both sister kinetochores are completely contained within a single 0.50 to 0.75 μm thick section 40-50% of the time (Fig. 1). Moreover, under these conditions all of the MTs within the vicinity of each kinetochore can be reliably identified and tracked. Similarly studies with thinner (0.25 μm) sections reveal that, for an individual kinetochore, some MTs terminate on the interior surface of the plate while others terminate on its distal surface (Fig. 2). Currently we are computing tomographic 3-D reconstructions of sister kinetochores whose motion has been recorded by video LM up to the point of fixation.

1. T.J. Mitchison, *Ann. Rev. Cell Biol.*, 4(1988)527.
2. C.L. Rieder and E.D. Salmon, *J. Cell Biol.*, 124(1994)223.
3. W.C. Earnshaw and A.F. Pluta, *BioEssays,* 16(1994)639.
4. A.A. Hyman and T.J. Mitchison, *Cold Spring Harbor Symp. Quant. Biol.*, 56(1991)745.

Proc. Microscopy and Microanalysis 1995, edited by G.W. Bailey, M.H. Ellisman, R.A. Hennigar, and N.J. Zaluzec
Copyright © 1995 MSA. Published by Jones and Begell Publishing, 79 Madison Ave., New York, NY 10016

5. K.L McDonald et al., *J. Cell Biol.*, 118(1992)369.
6. C.L. Rieder and S.P. Alexander, *J. Cell Biol.*, 110(1990)81.
7. B.F. McEwen et al. *J. Cell Biol.*, 120(1993)301.
8. Supported by the following grants: NSF MCB-9420772 to BFM, NIH R01 40198 to CLR, and NIH RR01219 that partly supports the Wadsworth Center's Biological Microscopy and Image Reconstruction facility as a National Biotechnological Resource.

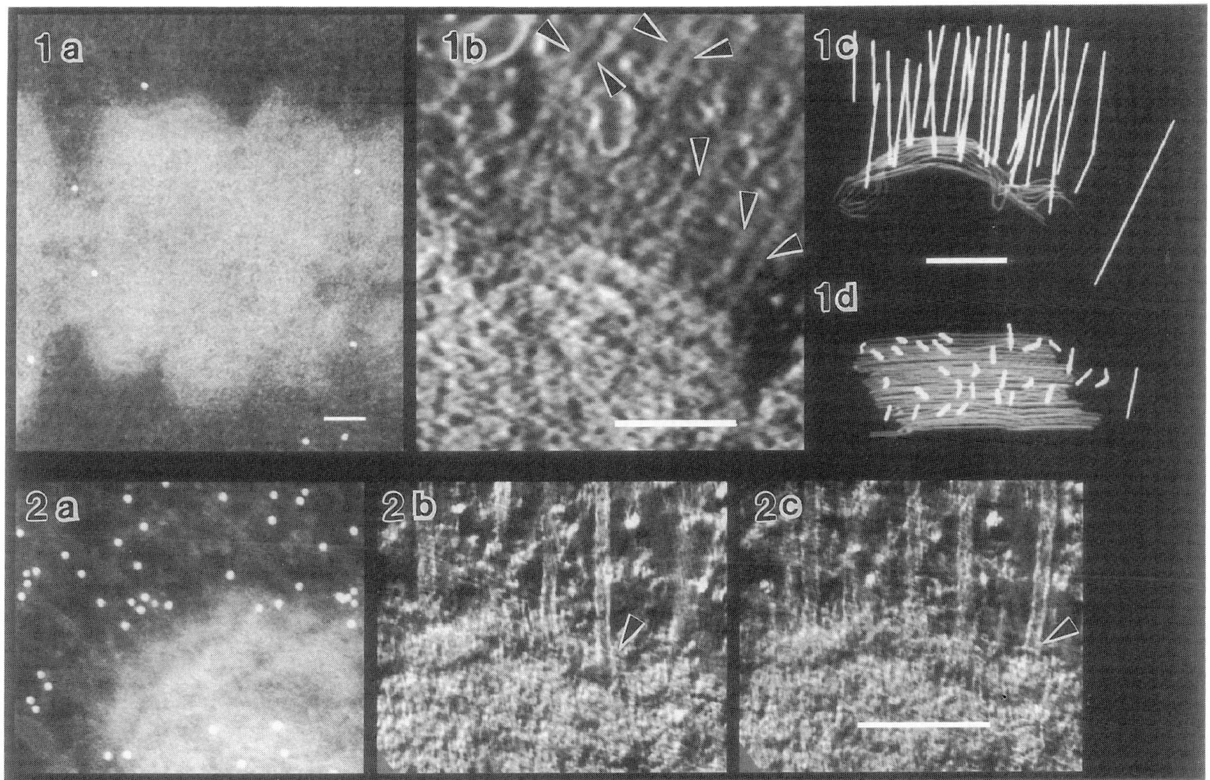

FIG. 1—Tomographic reconstruction of kinetochores from metaphase PtK cells: thicker (~0.50 μm) sections. (a) Untilted view from the input data set. (b) Single 3.2 nm thick slice from 3D reconstruction of the upper kinetochore in (a) with MTs indicated by arrowheads. (c) Tracing of the outer plate with each kMT represented by a single line. (d) Illustration in (c) rotated by 70°. From this view it is clear that 30 MTs were scored. One of those passes outside the outer plate and four more terminate distal to its surface. Hence 25 kMTs were detected on this kinetochore. Similarly 23 kMTs were detected on the sister kinetochore.

FIG. 2—Tomographic reconstruction of kinetochores from metaphase PtK cells: thinner (~0.10 μm) sections. (a) Untilted view from the input data set. (b) Single 2.5 nm thick slice from the reconstruction illustrating a kMT terminating at the interior surface of the outer plate (arrowhead). (c) A different 2.5 nm thick slice illustrating a kMT that terminates at the distal surface of the outer plate (arrowhead).

All bars = 0.25 μm.

HISTOCHEMICAL ASPECTS OF NUCLEIC ACID DETECTION

B. A. Hamkalo* and Elizabeth R. Unger**

*University of California at Irvine, Irvine, California
**Emory University, Atlanta, Georgia

This symposium brings together several approaches for the detection of specific nucleic acid sequences that have potential applications at the histochemical level.

Trask et al. report on the use of fluorescence in situ hybridization (FISH) techniques to study the arrangement of DNA sequences in normal and disease-related chromosomes. The sites of specific DNA sequences can be fluorescently tagged. Different sequences can be labeled with different fluorochromes so that their arrangement can be studied using fluorescence microscopy. The distances between points on the same or different chromosomes can be determined in a large number of interphase nuclei or metaphase chromosomes. A variety of probe types, ranging from single-copy sequences to highly repeated sequences can be employed.

Hamkalo and co-workers have used non-radioactive methods at the EM level for the detection of nucleic acid sequences by in situ hybridization. Analysis of metaphase chromosomes by electron microscopy allows for high resolution mapping of chromosomes. A variety of labelling procedures have been employed to illustrate the utility of high resolution nucleic acid sequence mapping in these preparations.

Huang and Spector have used in situ hybridization methods to detect specific RNAs and immunocytochemical methods for the detection of splicing factors in order to explore the functional organization of RNAs in the mammalian cell nucleus. These studies indicate that the spatial association of RNA transcripts with splicing factors is intron-dependent.

Unger et al. note that surgical pathology tissue archives are a valuable resource for investigators to analyze molecular markers in well characterized patient material. Comparisons between in situ hybridization ad polymerase chain reaction for detection of human papillomaviruses in archival tissues have been conducted.

Macechko et al. have applied various in situ hybridization methods for the detection of Giardia trophozoites in archival pathology specimens of human

Proc. Microscopy and Microanalysis 1995, edited by G.W. Bailey, M.H. Ellisman, R.A. Hennigar, and N.J. Zaluzec
Copyright © 1995 MSA. Published by Jones and Begell Publishing, 79 Madison Ave., New York, NY 10016

small intestine. The use of strain specific DNA probes permit the determination of whether or not multiple strains of Giardia can infect individual patients.

Bagasra describes application of in situ polymerase chain reaction (PCR) methods for detection of defined nucleic acid sequences. The PCR method for amplification of defined gene sequences has proved a valuable tool for baic as well as clinical research.

The spatial organization of specific nucleic acid sequences within cells and tissues represents a powerful approach for the analysis of numerous questions in basic and clinical sciences.

USE OF FLUORESCENCE IN SITU HYBRIDIZATION TO STUDY THE ARRANGEMENT OF DNA SEQUENCES IN NORMAL AND DISEASE-RELATED CHROMOSOMES

Barbara J. F. Trask, Hillary Massa, Cynthia Friedman, Richard Esposito, Ger van den Engh, and Hiroki Yokota

Department of Molecular Biotechnology, University of Washington, Seattle

The sites of specific DNA sequences can be fluorescently tagged by fluorescence in situ hybridization (FISH). Different sequences can be labeled with different fluorochromes so that their arrangement can be studied using epifluorescence microscopy. The distances between points on the same or different chromosomes can be determined easily in a large number of interphase nuclei or metaphase chromosomes. A variety of probe types, ranging from single-copy sequences to highly repeated sequences can be employed. Our work has focussed on the analysis of hybridization patterns in two dimensions using conventional fluorescence microscopy.

We have used FISH to study various aspects of genome organization that are difficult to study using other techniques. Examples of these applications will be presented.

1) DNA sequence location

FISH is now the method of choice for determining the chromosomal location of DNA sequences. DNA sequences can be positioned in the genome with <1:1000 accuracy (to a 3-Mbp region within a 3000-Mbp genome). Through FISH, the cytogenetic, physical and genetic maps of chromosomes can be linked.

2) Determination of clone integrity.

A prevalent problem in genome analysis is the formation of chimeric clones during library construction. This cloning artifact is especially troublesome if the clone is used to find genes suspected to reside in a particular region of the chromosome. FISH is the most rapid technique to detect grossly chimeric clones. The clones produce signals on ≥ 2 chromosomal locations. Some chimerism is not detected using FISH if one of the chromosomal locations is represented by only a very small portion of the clone.

3) Polymorphism analysis.

Although most fragments of the genome hybridize only to a single locations in the haploid genome (after suppression of distributed repeated sequences such as Alu), some DNA sequences are found on more than one chromosomes. FISH useful adjunct to molecular analyses for studying these regions. For example, alpha-satellite sequences label different, but specific subsets of chromosomes at their centromeres. We have studied a 40-kbp sequence that is present near the ends of several chromosomes. This sequence is found on additional chromosomes in some individuals.

Proc. Microscopy and Microanalysis 1995, edited by G.W. Bailey, M.H. Ellisman, R.A. Hennigar, and N.J. Zaluzec

4) Organization of chromosomes within nucleus.

FISH can also be used to study the folding and packaging of chromatin within the interphase nucleus. Our approach is to measure the distance in interphase between DNA sequences, whose separation on the linear DNA molecule is known. The observed linear relationship between mean-square interphase distance and genomic separation indicates a high degree of randomness in chromosomal folding at scales >100 kbp. We have also studied the arrangement of chromosomes with respect to each other.

5) Detection and characterization of chromosomal rearrangements.

Chromosomal abnormalities can be readily detected using FISH and suitable DNA-sequence probes. For example, deletions are obvious from the absence of a FISH-signal on one of the two homologs. Translocations result in a redistribution of the FISH-signals from DNA sequences that flank or cross translocation breakpoints in interphase or metaphase chromosomes. We have applied FISH in several positional-cloning projects in an effort to identify the genes involved in human disease. The FISH results aid in the construction of long-range maps of clones and the breakpoints of different chromosomal rearrangements.

This work was supported by NIH grant R01 HG00256, DOE grant FG06-9361553, and DOE grant FG06-93ER61662.

THE FUNCTIONAL ORGANIZATION OF RNAs IN THE MAMMALIAN CELL NUCLEUS

S. Huang, and D.L. Spector

Cold Spring Harbor Laboratory, Cold Spring Harbor, New York 11724.

Several models have been proposed for the functional organization of RNA within the eukaryotic nucleus and for the relationship of this organization to the distribution of pre-mRNA splicing factors. One model suggests that RNAs which must be spliced are capable of recruiting splicing factors to the sites of transcription from storage and/or reassembly sites[1]. In order to further evaluate this model we have transiently transfected HeLa cells with constructs which express RNA transcripts containing introns, lacking introns, or containing an intron with a deletion at the 3' splice site. The expression of RNAs was detected by in situ hybridization and their association with splicing factors was evaluated by immunostaining using specific antibodies (Y12, SC35) in the same cells. We have found that the majority of the RNA transcripts produced from constructs which express intron-containing genes such as β-globin, tropomyosin, and HIV tat are associated with splicing factors. In contrast, RNAs lacking introns, such as β-galactosidase, and adenovirus VAI, are not associated with splicing factors in the nucleus. Furthermore, the majority of HIV tat RNA, which contains a deletion in its intron at the 3' splice site, showed little association with splicing factors. These observations suggest that the spatial association of RNA transcripts with splicing factors is intron-dependent. Such an association is functionally significant as transcripts containing mutated introns, which are incapable of being spliced, are not associated with splicing factors. In addition, we have also found that splicing factors associate with nascent RNA transcripts at the sites of transcription suggesting that splicing factors are actively recruited to splice newly synthesized RNA. Since the transcription of intronless RNA did not colocalize with splicing factors, the transcription of intron containing RNA may represent a signal to trigger the movement of splicing factors towards the sites of transcription. We propose that the organization of RNA and splicing factors in the cell nucleus reflects the transcriptional activity of the cell and the localization patterns of splicing factors will change to reflect changes in transcriptional activity.

Reference:

1. Jiménez-García, L. F. and D. L. Spector, *Cell.* 73(1993)47.

Proc. Microscopy and Microanalysis 1995, edited by G.W. Bailey, M.H. Ellisman, R.A. Hennigar, and N.J. Zaluzec
Copyright © 1995 MSA. Published by Jones and Begell Publishing, 79 Madison Ave., New York, NY 10016

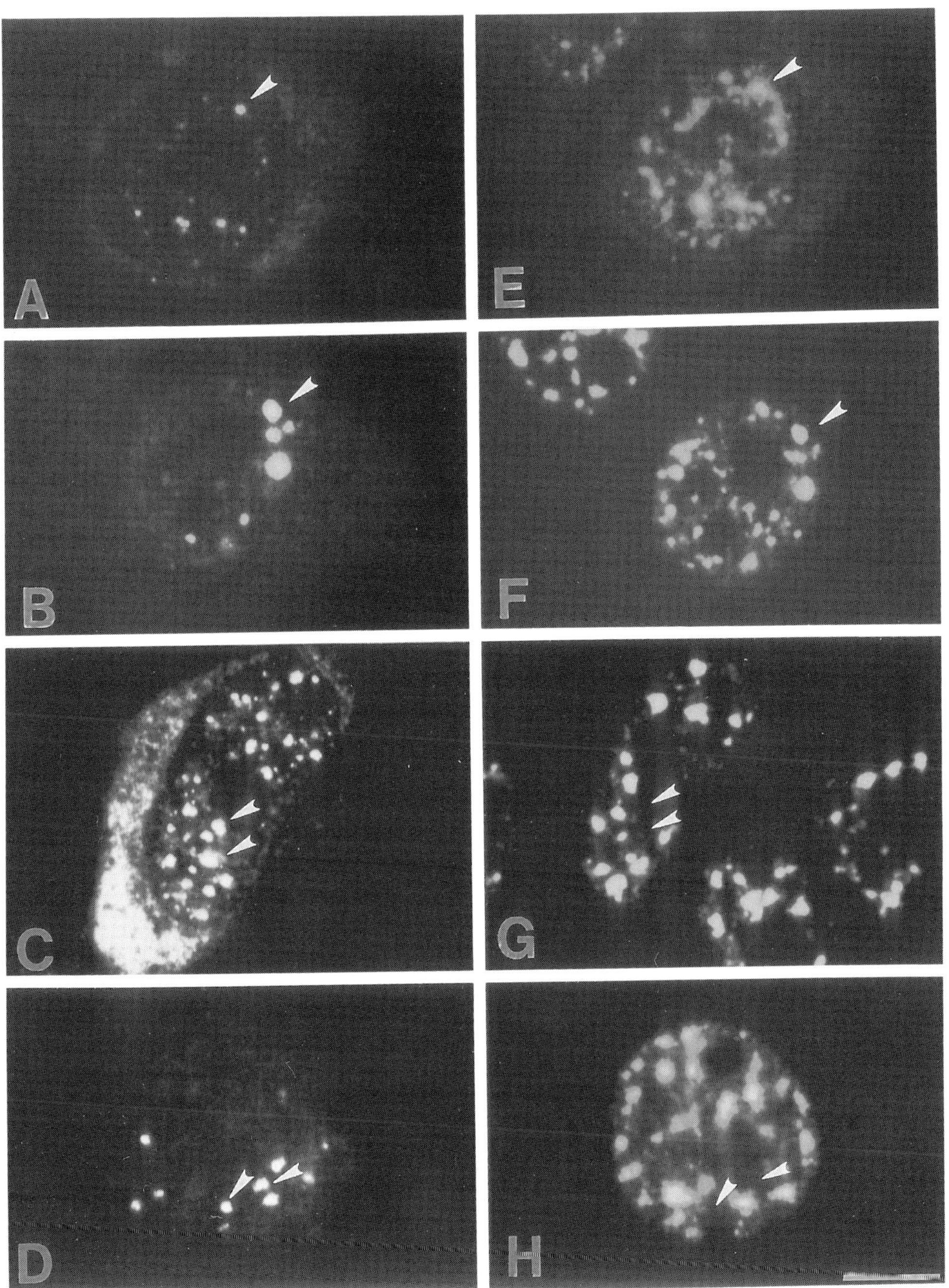

Figure 1. A close association is observed between the localization of transiently expressed RNA transcripts containing introns and the localization of splicing factors. Such association is not observed when RNA transcripts without introns or with a trucated intron are expressed. HeLa cells were transfected with expression vectors which encode for RNA transcripts with introns such as CG-Tat (A and E) and β-globin (C and F) or vectors which encode for transcripts without introns, β-glactosidase (C and G), or with a truncated intron, CMV-Tat (D and H). The localization of these RNAs were examined at 7-8 hours after transfection by in situ hybridzation with biotinylated probes (A-D). The localization of splicing factors in the same cells was detected by immunostaining with monoclonal antibody specifically recognizing SC35 (E-H). Arrowheads indicate the corresponding localization of RNA in the panel for SC35 staining. The bar represents 10 μm.

HIGH RESOLUTION NUCLEIC ACID SEQUENCE MAPPING VIA *IN SITU* HYBRIDIZATION AT THE ELECTRON MICROSCOPE LEVEL

B.A. Hamkalo,[*], S. Narayanswami,[**] and A.P. Kausch[***]

[*] Department of Molecular Biology & Biochemistry, University of California, Irvine, Irvine, CA 92717
[**] The Jackson Laboratory, Bar Harbor, ME 04609
[***] DeKalb Plant Genetics, Mystic, CT 06355

The availability of nonradioactive methods to label nucleic acids an the resultant rapid and greater sensitivity of detection has catapulted the technique of in situ hybridization to become the method of choice to locate of specific DNA and RNA sequences on chromosomes and in whole cells in cytological preparations in many areas of biology. It is being applied to problems of fundamental interest to basic cell and molecular biologists such as the organization of the interphase nucleus in the context of putative functional domains[1]; it is making major contributions to genome mapping efforts[2]; and it is being applied to the analysis of clinical specimens.[3] Although fluorescence detection of nucleic acid hybrids is routinely used, certain questions require greater resolution. For example, very closely linked sequences may not be separable using fluorescence; the precise location of sequences with respect to chromosome structures may be below the resolution of light microscopy(LM); and the relative positions of sequences on very small chromosomes may not be feasible.

In order to exploit the increased resolution afforded by electron microscopy we have combined specimen preparation using the Miller spreading technique with in situ hybridization and immunogold labelling of hybrid sites[4] to define the arrangement of a variety of DNA sequences in metaphase chromosomes from several sources. Highly repeated sequences such as those restricted to centromeric heterochromatin are specifically and efficiently detected with hundreds of gold particles at the appropriate site (Fig. 1); intermediately-repetitive sequences also are readily detected. Figure 2a illustrates the detection of a cluster of tRNA genes in a Xenopus chromosome and illustrates the mirror image pattern of labelling often observed in sister chromatids which is characteristic of higher order chromosome organization. The simultaneous hybridization of two alternately-labelled probes and two different sized gold particles permits double-labelling as illustrated in Figure 2b which shows a small chromosome derived from human chromosome 1 after labelling of two centromere-linked sequences which could not be discriminated in the LM. Finally, we are investigating the utility of various types of magnetic particles to both label in situ hybrids and, ultimately, isolate individual chromosomes that are refractile to traditional isolation methods. Figure 3 shows a group of mouse chromosomes after hybridization with biotin-substituted major centromeric satellite DNA and labelling with beads (ca 30 nm diameter) coated with streptavidin. Labelling is both specific and as intense as colloidal gold.

Proc. Microscopy and Microanalysis 1995, edited by G.W. Bailey, M.H. Ellisman, R.A. Hennigar, and N.J. Zaluzec
Copyright © 1995 MSA. Published by Jones and Begell Publishing, 79 Madison Ave., New York, NY 10016

References

1. S. Huang et al., *J. Cell Biol.*126(1994)877.
2. B. Trask et al., *Genomics* 15(1993)133.
3. C.T. Thompson et al., *Am. J. Path.* 144(1994)237.
4. S. Narayanswami et al., *Meth. Cell. Biol.* 35(1991)110.
5. Research supported by NIH GM23241.

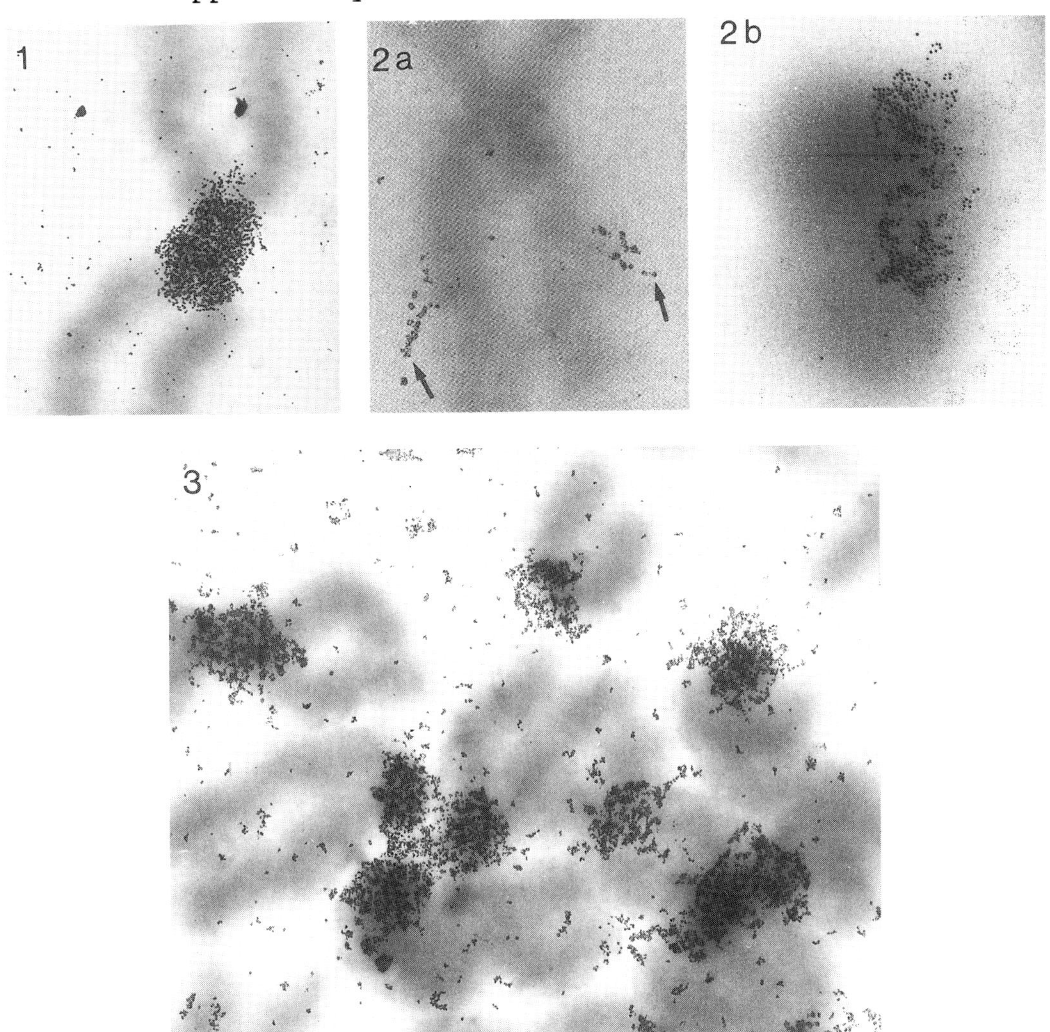

FIG. 1-Mouse metaphase chromosome hybridized with biotin-satellite DNA and detected with anti-biotin and secondary antibody-20nm gold.
FIG. 2-(a) <u>Xenopus</u> <u>laevis</u> metaphase chromosome hybridized with bio-tRNA, detected as in Fig. 1. Arrows point to mirror image pattern; (b) human chromosome 1 centromere fragment hybridized to bio-alpha satellite (5nm gold) and digoxigenin-satellite III (20nm gold).
FIG. 3-Mouse metaphase chromosomes hybridized as in Fig. 1 but detected with strepavidin directly coupled to 30 nm ferrofluid.

APPLICATION ON IN SITU PCR METHODS IN HISTOCHEMISTRY:

Omar Bagasra, M.D., Ph.D.

Center for Human Retrovirology & Molecular Therapeutics Section Chief, Molecular Diagnostics Department of Medicine, Thomas Jefferson University, Philadelphia, PA 19102

Since the publication of the first report regarding the in situ amplification of HIV-1 *gag* gene in a HIV-1 infected cell line in 1990 (1), there has been an explosion of research in the area of in situ PCR. There are over 200 publications describing various forms of in situ gene, identifying various infectious agents, tumor marker genes and other genetic elements of interest, in peer reviewed journals (reviewed in 2-15). The polymerase chain reaction (PCR) method for amplification of defined gene sequences has proved a valuable tool not only for basic researchers but also for clinical scientists. Using even a minute amount of DNA or RNA and choosing a thermostable enzyme from a large variety of sources, one can enlarge the amount of the gene of interest, which can be analyzed and/or sequenced. Thus genes or portions of gene sequences present only in a small sample of cells or small fraction of mixed cellular populations can be examined. However, one of the major drawbacks of standard PCR technique is that the procedure does not allow the association of amplified signals of a specific gene segment with the histological cell type(s). For example, it would be advantageous to determine what types of cells in the peripheral blood circulation carry HIV-1 provirus at various stages of HIV-1 infection and what percent of HIV-1-infected cells actually are expressing viral RNA. Similar approaches have been used to detect the presence of other gene sequences in tissue materials and pathological specimens.

The ability to identify individual cells, expressing or carrying specific genes of interest in a tissue section, under the microscope, provides a great advantage in determining various aspects of normal, as opposed to pathological, conditions. For example, this technique could be used in determination of tumor burden, before and after chemotherapy, in lymphomas or leukemias, where specific aberrant gene translocations are associated with certain types of malignancy. In case of HIV-1 infection or other viral infections, one can determine the effects of therapy or putative anti-viral vaccination by evaluating the number of cells still infected with viral agent, post-vaccination. Similarly, one can potentially determine the pre-neoplastic lesions by examining p53 mutations associated with certain tumors or oncogenes or other aberrant gene sequences which are known to be associated with certain types of tumors. In the area of diagnostic pathology determination of origin of metastatic tumors are a perplexing problem. By utilizing the proper primers for genes which are expressed by certain histological cell types, one can potentially determine the origin of metastatic tumors by performing reverse transcriptase-initiated in situ PCR.

Our laboratories are using in situ PCR (ISPCR) techniques since 1988 and we have developed a simple, sensitive ISPCR which has proved reproducible in multiple double-blinded studies. One can use this method for amplification of both DNA or RNA gene sequences. By use of multiple labeled probes, one can detect various signals in a single cell. In addition, under special circumstances, one can perform immunohistochemistry, RNA and DNA amplification at a single cell level (the so-called "Triple-Labeling").

To date, we have successfully amplified and detected HIV-1, SIV, HPV, HBV, CMV, EBV, HHV-6, HSV, LGV, p53 and its mutations, mRNA for surfactant Protein A , estrogen receptors, inducible nitrous oxide synthesis (iNOS)- gene sequences associated with multiple sclerosis, by DNA and/or RNA (RT-ISPCR), in various tissues, including: PBMCs, lymph nodes, spleen, brain, skin, breast, lungs, cytological specimens, tumors, cultured cells and numerous others formalin-fixed, paraffin-embedded tissues. In this article, we have provided a detailed account of the ISPCR procedure which can be used for routine research investigations. In the subsequent article, we will provide a detailed procedure for special applications of ISPCR. For example, its use in cytogenetics, to localize a single gene in the chromosomal bands, its use in dual and triple labeling of cells, where more than one signals can be detected at the single cell level, its use in combination with EM, immunohistochemistry and in other special situations [1-15].

Proc. Microscopy and Microanalysis 1995, edited by G.W. Bailey, M.H. Ellisman, R.A. Hennigar, and N.J. Zaluzec
Copyright © 1995 MSA. Published by Jones and Begell Publishing, 79 Madison Ave., New York, NY 10016

References

1.   Bagasra, O., Polymerase Chain Reaction In Situ. Amplifications (March 1990), Editorial note, 20-21
2.   Bagasra, O., Hauptman, S.P., Lischner, H.W., Sachs, M. & Pomerantz, R.J. 1992 Detection of HIV-1 Provirus in Mononuclear Cells by In Situ PCR. New England Journal of Medicine 326 :1385-1391
3.   Bagasra, O., Seshamma, T. & Pomerantz, R.J. 1993. Polymerase Chain Reaction In Situ: Intracellular Amplification and Detection of HIV-1 Proviral DNA and Other Specific Genes. Journal of Immunological Methods 158:131-145
4.   Bagasra O and RJ Pomerantz. 1993. HIV-1 provirus is demonstrated in peripheral blood monocytes in vivo: A study utilizing an in situ PCR. AIDS Research and Human Retroviruses. 9:69-76.
5.   Bagasra, O.,T. Seshamma, J. Oakes, R.J. Pomerantz. 1993. Frequency of Cells Positive for HIV-1 Sequences Assessed by In Situ Polymerase Chain Reaction AIDS 7:82-86
6.   Bagasra, O.,T. Seshamma, J. Oakes, R.J. Pomerantz. 1993. High Percentages of CD4-positive Lymphocytes Harbor the HIV-1 Provirus in the Blood of Certain Infected Individuals. AIDS 7 :1419-1425.
7.   Bagasra O, M.N. Qureshi, B. Joshi, I. Hewlett, C.E. Barr, D. Henrad. High Prevalence of HIV DNA and RNA and Localization of HIV-Proviral DNA in Oral Mucosal Epithelial Cells in Saliva From HIV (+) Subjects in 1994 Annual Meeting United States and Canadian Academy of Pathology #742.
8.   Bagasra, O.,T. Seshamma, R.J. Pomerantz. 1993. In Situ PCR: A Powerful New Methodology" in In Situ Hybridization and Neurology, Oxford Univ. Press, New York, pp. 143-156
9.   Bagasra, O., R.J. Pomerantz,. 1994 In Situ PCR: Applications in the Pathogenesis of Diseases" Cell Vision 1:13-16
10.   Bagasra, O.,H. Farzadegan, T. Seshamma, J. Oakes, A. Saah, R.J. Pomerantz. 1994. Human Immunodeficiency Virus Type 1 Infection of Sperm In Vivo. (In Press) AIDS
11.   Bagasra O, Pomerantz RJ. 1994. In situ polymerase chain reaction and HIV-1. Clinics of North America. p351-66. Edit R.J. Pomerantz. W.B. Saunders Publishers, Philadelphia.
12.   Bagasra O, R.J. Pomerantz. 1993. Detection of HIV-1 in the brain tissue of individuals who died from AIDS. PCR in Neuroscience, Edited by Gobinda Sarkar. Academic Press, Orlando, FL (In Press).
13.   Bagasra O, Seshamma T, Pestanar JP and Pomerantz R. 1994. Detection of HIV-1 Gene Sequences In the Brain tissues by In Situ Polymerase Chain Reaction. Technical Advances in AIDS Research in the Nervous System. Editors Eugene Majors & JA Levy, Plenum Press, (In Press).
14.   Bagasra O, T. Seshamma. 1994. In Situ PCR: Applications in the pathogenesis of diseases- A practical manual. Edited by J. Gu. Eaton Publication Press, MA. In Press.
15.   Bagasra O, Seshamma T, Hansen J and Pomerantz R. 1994. In situ PCR: A manual. In "Current Protocols in Molecular Biology" John Wiley & Sons, New York. (in press).

ANALYSIS OF DNA IN ARCHIVAL TISSUES: COMPARISON OF IN SITU HYBRIDIZATION AND POLYMERASE CHAIN REACTION FOR DETECTION OF HUMAN PAPILLOMAVIRUSES

E.R. Unger,* S. D. Vernon,** D.R. Lee,* D. Miller, ** and W.C. Reeves**

*Department of Pathology and Laboratory Medicine, Emory University School of Medicine, Atlanta, GA 30303
**Division of Viral and Rickettsial Disease, Center for Infectious Disease, Centers for Disease Control and Prevention, Atlanta, GA 30333

Surgical pathology tissue archives have been discovered to be a valuable resource for investigators seeking to study molecular markers in well characterized patient material. The potential to analyze patient material stored for several years means that correlation of results of testing with disease outcome can be accomplished rapidly. The use of archival tissue is also important for relatively uncommon diseases which are difficult to obtain in a prospective fashion. The accuracy of molecular tests in formalin-fixed paraffin-embedded tissues can be compromised by variations in fixation and processing as well as by time of storage and storage conditions. To be reliable, molecular assays must be optimized to compensate for variations in tissue preservation. Both in situ hybridization (ISH) and polymerase chain reaction (PCR) assays have been designed to study DNA in archival tissues. The advantage of ISH is that the analysis is placed within a morphological context. This allows exclusion of cases where the marker is present in non-lesional tissue and allows detection of sub-populations of cells within the lesion. ISH is also not hampered by problems of cross-contamination between samples or from the environment. However concerns about the sensitivity of ISH and its relatively tedious nature has tended to favor application of PCR for large scale studies.

We were interested in the molecular epidemiology of human papillomavirus in cervical carcinoma and designed a study of archival tissues using both ISH and PCR to directly compare the results of two independent assays in the same material and to assure accuracy of results. The study material consisted of archival tissues from cervical cancer patients diagnosed and treated at Grady Memorial Hospital from January 1986 through June 1992. All samples had been fixed in formalin and processed into paraffin blocks according to routine surgical pathology practice. Serial sections were cut for both ISH and PCR with histologic evaluation of beginning and ending levels to confirm that both assays were performed on identical lesional tissue. Assays were performed and results recorded independently, so as not to bias interpretation of either assay.

An automated colorimetric ISH assay was used.[1,2] Conditions of digestion with pepsin were optimized for each tissue using biotin labeled placental DNA as a probe for availability of tissue DNA. Tissues which could not produce a satisfactory signal with this endogenous control probe were considered to demonstrate unsatisfactory DNA preservation and were eliminated from analysis. Biotinylated pBR322 DNA was used as the negative control for each tissue. The HPV probes contained the entire genome in plasmid vectors. The ISH assay tested for HPV types 16, 18 and 31/33/35. Formalin-fixed paraffin-embedded cell blocks of CaSki (500 copies HPV 16/cell), HeLa (50 copies of HPV 18/cell) and SiHa cells (1-2 copies HPV 16/cell) were used to monitor the sensitivity of each ISH analysis. Detection was

Proc. Microscopy and Microanalysis 1995, edited by G.W. Bailey, M.H. Ellisman, R.A. Hennigar, and N.J. Zaluzec
Copyright © 1995 MSA. Published by Jones and Begell Publishing, 79 Madison Ave., New York, NY 10016

achieved with an avidin-alkaline-phosphatase conjugate and bromochloroindoyl-phosphate/nitroblue tetrazolium color development.

For PCR analysis, extracts were prepared from two 5µm sections using deparaffinization and proteinase K digestion.[3] Amplification of the HPV L1 consensus region was performed using consensus primers MY11 and MY09 to generate a PCR product of approximately 450 base pairs.[4] Every PCR assay included a positive control of CaSki cellular DNA (HPV 16 DNA), a negative control of human placenta DNA (HPV negative DNA, Sigma Chemical Co), and a negative template control. DNA quantity and integrity was monitored by amplification of a portion of the beta-globin gene (Perkin Elmer Cetus) in replicate tubes. In the event of unsuccessful globin amplification, template concentrations were increased and or Mg^{++} concentration adjusted to optimize amplification. Products from the L1 consensus PCR were typed by using a high-stringency dot blot hybridization assay on nylon membranes with individual HPV type-specific (16, 18, 31, 33, 35, 56) oligonucleotide probes synthesized with digoxigenin at the 3' end.[5] Hybrids were detected by reaction with antibodies to digoxigenin, conjugated to alkaline phosphatase, followed by chemiluminescent detection with Lumiphos (Boehringer Mannheim). L1 consensus PCR products that did not hybridize to any of the type-specific probes were sequenced using cycle sequencing with fluorescent-labeled dideoxynucleotides. Each PCR product was sequenced in both directions, using MY11 and MY09 primers and Taq DyeDeoxy Terminators Cycle Sequencing Kits (Applied Biosystems) according to the manufacturer's instructions. Sequence information was collected with Applied Biosystems 373A DNA Collection and Analysis Software and analyzed using the University of Wisconsin Genetics Computer Group package.

Table 1: Results of ISH and PCR.

NA = Not amplifiable

	PCR +	PCR -	PCR NA
ISH +	61	20	6
ISH -	19	28	8

A total of 142 blocks were analyzed by both methods. As shown in Table 1, the assays showed a concordance of 69.5%, with nearly equal numbers of PCR+/ISH- and ISH+/PCR- instances. In eight of the ISH-/PCR+ cases, the PCR assay detected HPV types not included in the ISH assay. Despite careful optimization 9.8% of the blocks failed to yield amplifiable DNA as evidenced by failure to amplify beta-globin. These results demonstrate the need for careful optimization of conditions when testing archival tissues for DNA targets. Both ISH and PCR required significant adjustment of assay conditions. These results support using these complementary methods to yield the most accurate results.

References
1. E.R. Unger et al., *J Histotechnol* 11(1988)253.
2. E.R. Unger, M.L. Hammer, M.L. Chenggis, *J Histochem Cytochem* 39(1991)145.
3. D. Shibata, *Hum Pathol* 25(1994)561.
4. Y. Ting, M. Manos, in *PCR Protocols*, New York:Academic Press (1990)356.
5. A. Hildesheim et al., *J Infect Dis* 169(1994)235.

MICROSCOPIC DETECTION OF *GIARDIA* TROPHOZOITES IN ARCHIVAL PATHOLOGY SPECIMENS OF HUMAN SMALL INTESTINE: LIGHT MICROSCOPY, LOW VOLTAGE SEM, AND CONFOCAL SCANNING LASER MICROSCOPY USING FLUORESCENT IN SITU HYBRIDIZATION

P.T. Macechko*, T. Mulgrew*, H. van Keulen**, E.L. Jarroll**, D.C. Snover***,and S.L. Erlandsen*

*Department of Cell Biology and Neuroanatomy, ***Department of Laboratory Medicine and Pathology, University of Minnesota Medical School, Minneaplis, MN 55455; **Department of Biology, Cleveland State University, Cleveland, OH 44115

In situ hybridization methods have been widely used for the analysis of chromosomal markers[1], and for the detection and identification of various species of bacteria in environmental samples[2]. In situ hybridization probes to ribosomal RNA have been used as a diagnostic tool for the study of various parasitic protozoa[3], and fluorescent in situ hybridization (FISH) probes have been used to study the intracellular distribution of mRNA[4].

The intestinal protozoan, *Giardia*, colonizes the small intestine of a variety of different animals, including humans, and forms an infective cyst stage which is responsible for host-to-host transmission. Light microscopic detection in environmental or fecal samples has relied on the detection of cysts whose presence may be infrequent. In human giardiasis, confirmation of diagnosis has often required microscopic analysis of intestinal biopsies for detection of trophozoites. Analysis of the molecular sequence of 16s rDNA (small subunit) in *Giardia* isolates has permitted the development of species specific probes for *Giardia*[5]. Also, FISH methods have been developed for detection of *Giardia* cysts[6] and here we report the use of these FISH methods to study archival formalin-fixed, human intestinal biopsies for the detection /identification of *Giardia* trophozoites.

Paraffin-embedded sections of human intestinal biopsies were studied after deparaffinization by phase and brightfield (BF) microscopy (H&E), by FISH using 18-23 bp oligo rDNA probes conjugated at the 5' end with the cyanine dye Cy3 and imaged on a Biorad 600 confocal scanning laser microscope (CSLM), and with a Hitachi S-900 FESEM operated at low voltage.

Using phase and BF microscopy, *Giardia* trophozoites were seen adherent to the microvillous border of epithelial cells on intestinal villi (Figure 1). Clusters of trophozoites were also detected between villi and often appeared to be associated with strands of mucus. In favorable planes of section the binuceate appearence of individual trophozoites could be recognized, and outlines of the ventral disc were also apparent. Using FESEM to examine selected portions of the section containing trophozoites, specific features such as flagella or the ventral adhesive disc could be recognized, but due to random orientation section thickness, none of the trophozoites were observed to be completely intact. FISH analysis using a rDNA specific for *G. lamblia* (human species) revealed high levels of signal within individual trophozoites, but no detectable signal was observed within intestinal epithelial cells (Figure 2). Controls including omission of the probe or use of a *G. muris* specific rDNA probe resulted in no detecable staining within trophozoites.

The successful application of FISH for the detection of *Giardia* in human intestinal biopsies suggests that this approach may be useful in determining whether or not multiple strains of *Giardia* may infect a patient at one time. Homan et al.[7] and Andrews et al.[8] have used ISH and isoenzyme analysis, respectively, to demonstrate that multiple strains of *G. lamblia* exist, but as of yet, the

Proc. Microscopy and Microanalysis 1995, edited by G.W. Bailey, M.H. Ellisman, R.A. Hennigar, and N.J. Zaluzec
Copyright © 1995 MSA. Published by Jones and Begell Publishing, 79 Madison Ave., New York, NY 10016

microscopic confirmation of these multiple strains of *G. lamblia* within a single host has not yet been accomplished. Development of rDNA probes for specific strains may help to address this question, and likewise, the use of *G. lamblia* specific FISH probes may be helpful in answering the question of whether or not human giardiasis is a zoonotic disease.[9]

References

1. J.G. Dauwerse et al., *Hum. Mol. Genetics* 1(1992)593 .
2. M. Wagner et al., *J. Micros.* 176(1994)181.
3. Y. Hayashi et al., *Lab. Invest.* 63(1990)576.
4. J.B. Lawrence and R.H. Singer, *Cell* 45(1986)407.
5. H. van Keulen et al., *FASEB* 7(1993)223.
6. S.L. Erlandsen et al., in R.C.A. Thompson, J.A. Reynoldson and A.J. Lymbery Eds., *Giardia: From Molecules to Disease and Beyond*, Wallingford: CAB International (1994)64.
7. W.L. Homan et al., *Parasitol. Res.* 78(1992)316.
8. R.H. Andrews et al., *Parasitol.* 105(1992)375.
9. This research was supported by the Environmental Protection Agency (EPA 1R821404).

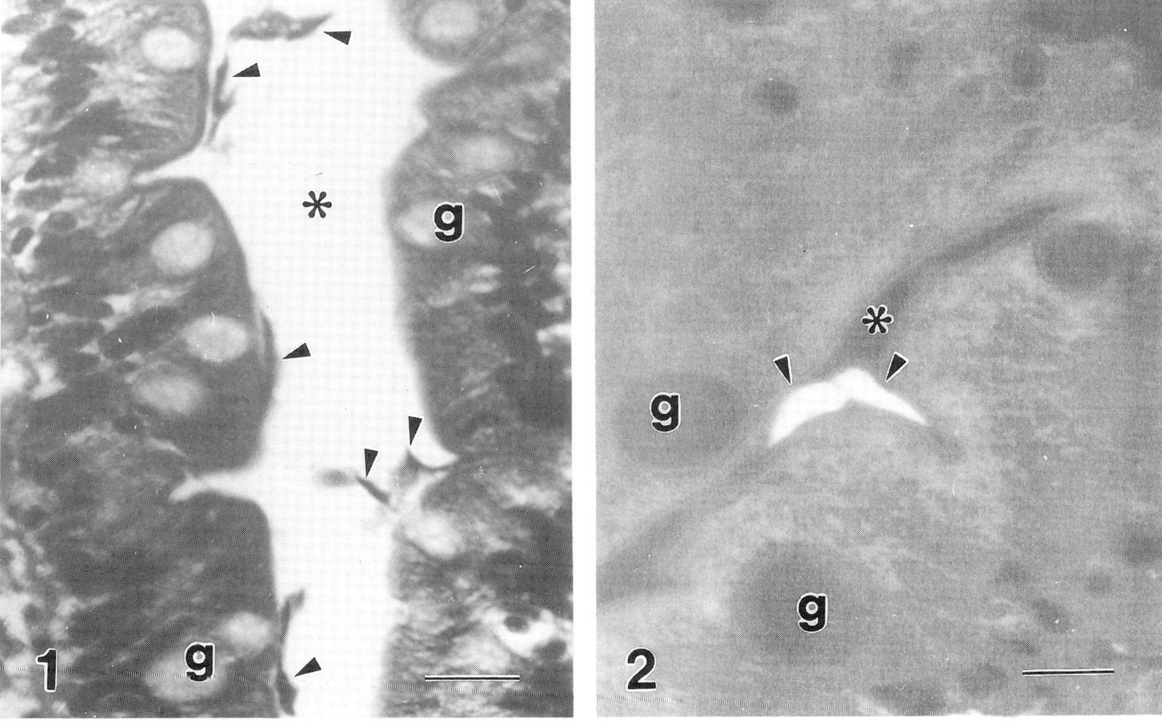

Figure Legend:

1. Phase LM of human duodenum showing presence of *Giardia* trophozoites (arrowheads) between adjacent villi. g, goblet cell; * lumen. Bar, 10 μm.
2. Intense FISH staining for *Giardia* rDNA oligo within trophozoites (arrowheads) in lumen and lack of hybridization signal within intestinal epithelial cells. g, goblet cell; * lumen. Bar, 5 μm.

THE NUCLEAR ENVELOPE AND NUCLEAR PORE COMPLEX AS SEEN BY HIGH RESOLUTION FIELD EMISSION SEM AND HIGH VOLTAGE TEM

Hans Ris

Dept. of Zoology and Integrated Microscopy Resource, University of Wisconsin Madison WI 53706

The nuclear envelope (NE) consists of an inner membrane and an outer membrane continuous with the endoplasmic reticulum.The membranes are fused at the nuclear pores, which connect cytoplasm with nuclear space. Located in these pores is the nuclear pore complex (NPC) which controls the transport of macromolecules across the NE.[2] Attached to the inner surface of the NE is a network of 10 nm lamin filaments, the nuclear lamina.[1] Recent studies using high voltage TEM and field emission "in lens" SEM have greatly advanced our knowledge of NPC structure, particularly the components facing the cytoplasm and nuclear space. The cytoplasmic face consists of a 120 nm ring carrying 8 twisted filaments condensed into 20x40 nm cylinders. On the nuclear side is another 120 nm ring, from which 8 filaments, 5 nm thick and 50 nm long, extend into the nuclear space. Each filament ends in a 10 nm bead. The 8 beads are joined into a 50 nm ring, forming the top of a fishtrap-like structure (Fig.2) . Usually this ring is obscured by distorted pore connecting fibers (Fig.2). The common low salt buffer results in distortion of pore connecting fibers and the lamin network. I have used McGregor's buffer, isotonic to cytoplasm, to isolate *Xenopus* oocyte nuclei. Adding 5% sucrose prevents contraction of the nuclear gel and makes it soluble. Treatment of the isolated nucleus with 0.1% Triton X-100 in McGregor-sucrose buffer containing 1% glutaraldehyde removes the membrane and reveals well preserved fishtraps inserted into the orthogonal lamin network. This method now permits a study of relationships of NPCs and lamins. It also reveals a new nuclear structure : attached to the small ring of the fishtraps we see 50 nm thick fibers which interconnect groups of fishtraps (Fig.1). In whole mounts we can see the branching fibers with a periodic structure. These fibers are especially clear in isolated membranes plunge-frozen in ethane, freeze substituted and embedded in epon. Thick sections were imaged by FESEM after epon extraction.[4] In cross sections of the NE we see side views of fishtraps with attached 50 nm fibers. Unextracted sections imaged by HVEM show that this fiber consists of eight 5 nm fibrils forming the wall of a hollow tube. Folding of these fibrils in the long axis results in a 50 nm periodicity(Fig.3). In summary, on the intranuclear side, groups of NPCs are interconnected by a system of branching hollow cables. Goldberg and Allen[3] described a fibrous network (the NE lattice) connecting tops of adjacent fishtraps in oocytes of a newt. No such structure is present in *Xenopus*. A collaborative study is in progress to see whether these hollow cables play a role in nuclear transport.

References

1. U. Aebi .et al..*Nature* 323 (1986) 560.
2. D.J. Forbes, *Ann.Rev.Cell Biol.* 8 (1992)495.
3. M.W. Goldberg and T.D. Allen, *J. Cell. Biol.* 119(1992)1429.
4. H. Ris and M. Malecki , *J. Struct. Biol.* 111(1993) 148.

This Research was supported by grant DRR-570 from NCRR-NIH to the Integrated Microscopy Resource (IMR) Madison, WI

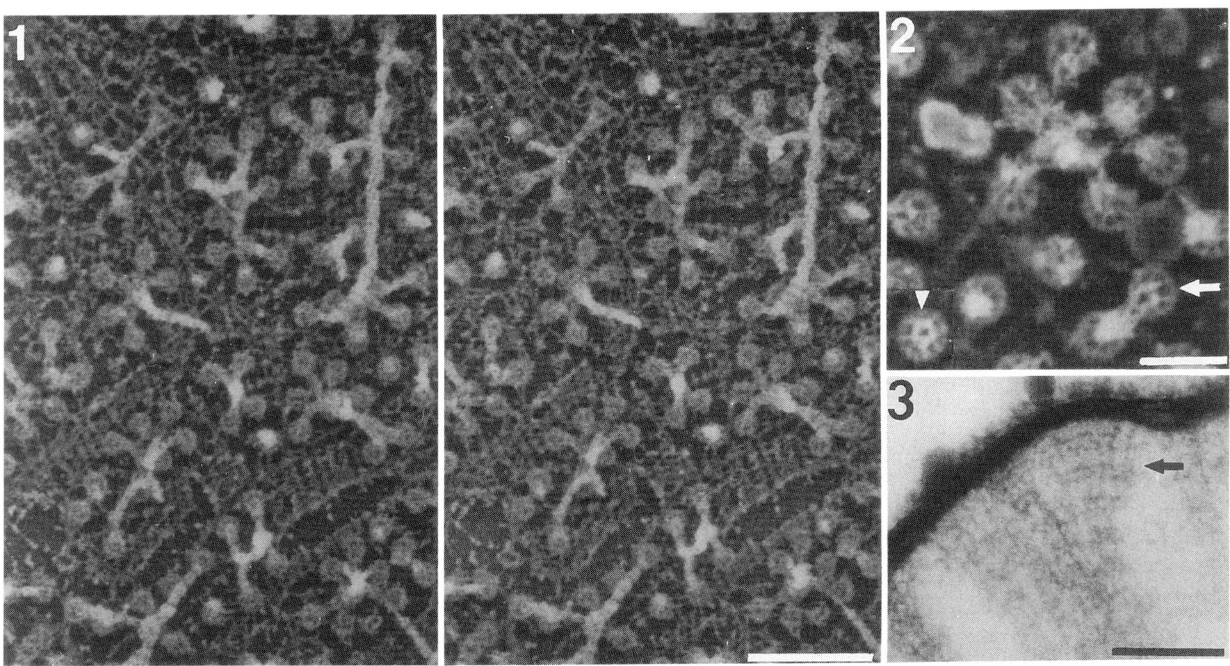

Fig.1.-Stereo-image of intranuclear surface of an isolated *Xenopus* oocyte nuclear envelope.The nucleus was isolated in McGregor buffer containing 5% sucrose, attached to glass and spread open with glass needles. The membrane was then extracted with 0.1% Triton X-100 in isolation buffer containing 1% glutaraldehyde. Both lamin network and fishtraps are now preserved. Branching 50 nm thick fibers with a 50nm periodicity interconnect groups of fishtraps. Hitachi S-900 FESEM, 1.5 kV,   Bar = 500 nm

Fig.2- Intranuclear surface of a *Xenopus* oocyte NE. The nucleus was isolated in McGregor buffer, attached to glass, spread open in low salt buffer(LSB) and fixed in 1% glutaraldehyde in LSB containing 0.2% tannic acid. We see a group of fishtraps connected by distorted, irregular pore connecting fibers . At the arrow, these fibers are torn off, revealing the beads at the ends of the fishtrap filaments. These beads are joined to form the top ring of the fishtrap. At the arrowhead , the 50 nm fiber was ruptured above the fishtrap. The resulting cross section of the 50 nm fiber shows that it is a hollow tube. Hitachi S-900 FESEM, 1.5 kV, bar = 200 nm

Fig.3- 200 nm cross section of *Xenopus* oocyte NE , fixed in 1% glutaraldehyde in LSB buffer, plunge frozen in melting ethane, freeze substituted in 0.5% osmium tetroxide in acetone and embedded in epon. At the arrow we see several overlapping 50 nm fibers. They consist of 6 nm fibrils arranged to produce a 50 nm periodicity in the long axis.
AEI EM -7 HVEM, 1000 kV,   Bar= 200 nm

THE STRUCTURE AND FUNCTION OF THE NUCLEAR LAMINS.

R.D. Goldman, A. Goldman, S. Khuon, M. Montag-Lowy, R. Moir, and T. Spann.

Northwestern University Medical School, Department of Cell and Molecular Biology, Chicago, IL.

The nuclear lamins are the Type V intermediate filament proteins comprising the nuclear lamina. The lamina is located subjacent to the nucleoplasmic face of the nuclear envelope where it interfaces with chromatin. The lamins are major karyoskeletal proteins which are thought to play important roles in the formation and maintenance of nuclear shape and architecture, as well as in the supramolecular organization of chromatin (1). The lamins have long been thought to be stable polymeric constituents of the interphase nuclear matrix, due to their insolubility in solutions containing detergents and high salt concentrations. During mitosis, however, the nuclear lamins depolymerize during nuclear envelope breakdown. Subsequently, the lamins repolymerize around the decondensing chromosomes as the nuclear envelope reassembles at the end of mitosis (2). Although there is a significant amount known about the properties and potential functions of the lamins during mitosis, surprisingly little is known about their properties during interphase. In light of this, we have undertaken experiments which are aimed at determining the properties of the lamins in interphase cells. These experiments involve the microinjection of biotinylated or myc-tagged bacterially expressed human lamins into live cells. The results show that microinjected lamins are assembled into the lamina during interphase. This illustrates that the lamina is a dynamic structure, capable of subunit/polymer exchange. The incorporation process appears to involve discrete lamin-rich foci located deep within the nucleoplasm (3). We now have evidence that these foci may be involved in the post-translational processing of the lamins. Similar nuclear foci are also seen as normal constituents of uninjected cells, and we have been able to link their distribution to different stages of the cell cycle. One of our most intriguing observations is that the DNA replication foci which typify mid-late S phase appear to contain nuclear lamin B, but not lamins A/C, indicating that lamin B may play a role in DNA synthesis (4). Other studies have been initiated which attempt to determine the dynamic properties of the nuclear lamins in transiently transfected cells. In order to gain insights into the biochemical

Proc. Microscopy and Microanalysis 1995, edited by G.W. Bailey, M.H. Ellisman, R.A. Hennigar, and N.J. Zaluzec

mechanisms regulating lamin dynamics and function, we have been using two cell free systems. These involve studies of nuclear envelope breakdown in the surf clam oocyte in an attempt to show the relationship between nuclear lamin disassembly and the congression of chromosomes prior to the formation of the metaphase plate (5). In addition, a hybrid system consisting of Xenopus laevis extracts spiked with bacterially expressed human lamins, is employed to study lamin-lamin interactions during the formation of the lamina. This work has been supported by the NCI.

References:

1. Moir, R., and R.D. Goldman (1993). Lamin dynamics. (ed. Ron Evans and John Newport). Current Opinions in Cell Biology. 5: 408-411.

2. Chou, Y.-H., Opal, P. and Goldman, R.D. (1995). Vimentin phosphorylation in mitotic BHK cells: phosphorylation of the N-terminal, but not the C-terminal domain is essential for IF disassembly. J. Cell Biol. Submitted.

3. Goldman, A., Moir, R.D., Lowy-Montag, M., Stewart, M., and R.D. Goldman (1992). Pathway of incorporation of microinjected lamin A into the nuclear envelope. J. Cell Biol. 119(4): 725-735.

4. Moir, R.D., Lowy-Montag, M. and Goldman, R.D. (1994). Dynamic properties of nuclear lamins: lamin B is preferentially associated with sites of DNA replication during S phase. J. Cell Biol. 125: 1201-1212.

5. Dessev, G., Palazzo, R., Rebhun, L., and R.D. Goldman (1989). Disassembly of the nuclear envelope of spisula oocytes in a cell free system. Develop. Biol. 131: 496-504.

DETECTION AND MAPPING OF INTERACTIONS BETWEEN CHROMATIN AND THE NUCLEAR ENVELOPE IN DROSOPHILA EMBRYOS

W.F. Marshall, A.F. Dernburg, B. Harmon, and J.W. Sedat

Dept. Biochemistry and Biophysics, University of California, San Francisco, CA 94143

Interactions between chromatin and nuclear envelope (NE) have been implicated in chromatin condensation[1], gene regulation[2], nuclear reassembly[3], and organization of chromosomes within the nucleus[4]. To further investigate the physiological role played by such interactions, it will be necessary to determine which loci specifically interact with the nuclear envelope. This will not only facilitate identification of the molecular determinants of this interaction, but will also allow manipulation of the pattern of chromatin-NE interactions to probe possible functions. We have developed a microscopic approach to detect and map chromatin-NE interactions inside intact cells.

Fluorescence in situ hybridization (FISH) is used to localize specific chromosomal regions within the nucleus of Drosophila embryos[5] and anti-lamin immunofluorescence is used to detect the nuclear envelope[6]. Widefield deconvolution microscopy is then used to obtain a three-dimensional image of the sample (Fig. 1). The nuclear surface is represented by a surface-harmonic expansion[7] (Fig 2). A statistical test for association of the FISH spot with the surface is then performed.

Using this approach, we have mapped the localization of 36 probes in cycle 13 embryos, revealing 4 euchromatic regions and 1 heterochromatic region which are associated with the NE. The NE-associated regions are large (approx. 500kb). Experiments are currently underway to further localize the actual NE binding site within this large region. Based on these results, we predict that there are approximately 10 NE contacts per chromosome arm. We have also found two sites which are in fact nonrandomly far from the NE, implying that they are targeted to the nuclear interior.

We do not find any general tendency of scaffold-attachment regions (SARs) to be associated with the NE. Two known SARs[8] (Hsp70 and Adh) are within NE associated regions, but another (ftz) is randomly localized, and the histone region is essentially never in contact with the NE. This contradicts results of other workers which indicate that histone SAR binds lamins in vitro[9] and illustrates the importance of studying biological interactions in intact cells using microscopy. We also do not find a universal association of heterochromatin with the NE. One heterochromatic site (the rDNA locus) is strongly associated with the NE, but the satellite region AACAC is randomly localized, and the dodecasatellite region is in fact nonrandomnly far from the NE. Finally, we note that of two sites mapped so far, which in polytene nuclei were shown to frequently lie near the nuclear surface[10], both are associated with the NE in diploid embryo nuclei, implying that these interactions can be maintained throughout development.

1. Y. Hiraoka et. al, *Nature* 342(1989)293-296.
2. G. Blobel, *Proc. Natl. Acad. Sci. U.S.A.* 82(1985)8527-8529.
3. C. Wiese and K. Wilson, *Curr. Opin. Cell Biol.* 5(1993)387-394.
4. D.E. Comings, *Human Genetics* 53(1980)131-143.
5. Y. Hiraoka et. al., *J. Cell Biol.* 120(1986)591-600.
6. M.R. Paddy et. al., *Cell* 62(1990)89-106.
7. C. Purcell et. al., *IEEE Trans. Biomed. Eng.* 38(1991)303-305.
8. S.M. Gasser and U.K. Laemmli, *Cell* 46(1986)521-530.
9. M.E. Luderus et. al., *Mol. Cell. Biol.* 14(1994)6297-6305.
10. M. Hochstrasser et. al., *J. Cell Biol.* 120(1986)591-600.

Proc. Microscopy and Microanalysis 1995, edited by G.W. Bailey, M.H. Ellisman, R.A. Hennigar, and N.J. Zaluzec

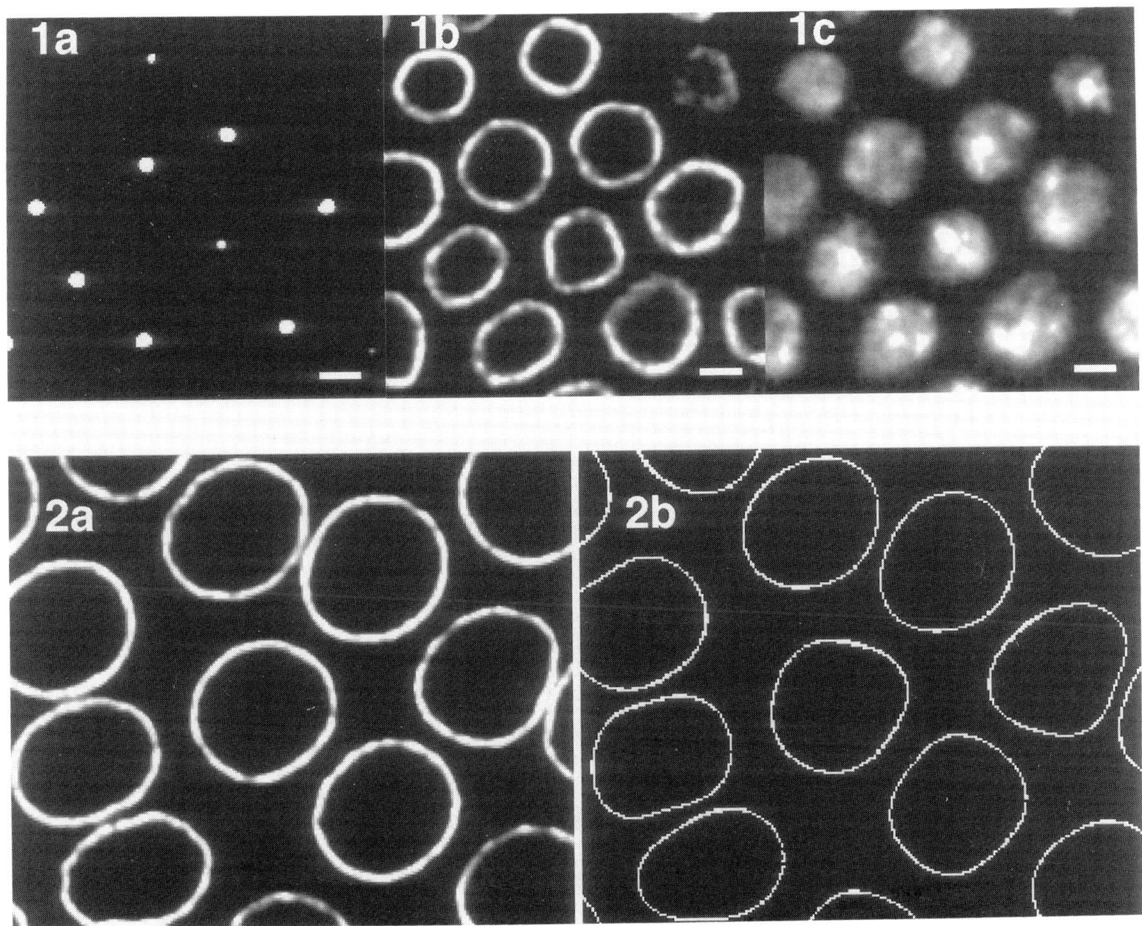

FIG 1. Example of simultaneous FISH and lamin immunofluorescence in cycle 14 embryo. Image depicts a single optical section from a three-dimensional three-wavelength dataset, following deconvolution. (a) Histone locus detected by hybridization with biotinylated probe; (b) Nuclear lamins detected with anti Drosophila lamin monoclonal antibody; (c) DAPI stain. Bar=2.0 μm.

FIG 2. Representation of nuclear surface via surface harmonic expansion. (a) Lamin immunofluorescence signal; (b) Surface generated by least-squares fitting algorithm.

DOMAIN ORGANIZATION OF ALLELE-SPECIFIC DNA REPLICATION WITHIN THE GABA$_A$ RECEPTOR GENE CLUSTER

Janine M. LaSalle and Marc Lalande

Howard Hughes Medical Institute; Genetics Division, Children's Hospital; and Department of Pediatrics, Harvard Medical School, Boston, MA 02115

Parental imprinting is a gamete-specific modification that distinguishes the paternal and maternal chromosomes in higher eukaryotes, resulting in allele-specific changes in chromatin organization, transcription and replication.[1] One example of parental imprinting in humans is revealed by two distinct genetic diseases, Prader-Willi syndrome (PWS) and Angelman syndrome (AS) which both map to chromosome 15q11-13. PWS is caused by the absence of a paternal contribution to 15q11-13, while AS results from the lack of a maternal copy of the region.[2,3] Within this chromosomal subregion lies a cluster of GABA$_A$ receptor β3 and α5 subunit genes (*GABRB3* and *GABRA5*) which are separated by about 100 kb and arranged in opposite transcritional orientations (Figure 1). Allele-specific asynchronous DNA replication has previously been found to be associated with imprinted chromosomal regions.[4,5] In order to further study the association between DNA replication and imprinting, allele-specific replication was assayed by fluorescence in situ hybridization (FISH). Biotin-labeled phage probes detected by FITC hybridized to each chromosome as either a singlet (unreplicated state) or a doublet (replicated state). Cells demonstrating asynchronous replication (one singlet and one doublet) for each probe are shown in Figure 2. The parental origin of each chromosome 15 replication event was simultaneously detected by the size polymorphism of a digoxigenin-labeled chromosome 15 centromeric probe (*D15Z1*) detected with rhodamine. The replication kinetics of each allele was determined by using a flow sorter to fractionate mitogen-stimulated lymphocytes on the basis of cell cycle progression prior to FISH analysis. These kinetic studies reveal a 50-60 kb chromosomal domain extending from the middle of the *GABRB3/A5* intergenic region into the *GABRA5* 5'-UTR which displays maternal replication in early S with paternal replication delayed until the end of S (Figure 3). In contrast, genomic regions on either side of this maternal early replication domain exhibit the opposite pattern with paternal before maternal replication and both alleles replicating in the latter half of S. Uniparental disomy or hemizygous deletion of chromosome 15 results in altered allele-specific replication kinetics compared to normals, suggesting that allele-specific replication within the *GABRB3/A5* region may be regulated by reciprocal imprints on the maternal and paternal chromosomes.[6,7]

References

1. H.V. Crouse, *Genetics* 45(1960)1429.
2. M.G. Butler and C.G. Palmer, *Lancet* 1(1983)1285.
3. J.H.M. Knoll et al., *Am.J.Med.Genet.* 32(1989)285.
4. D. Kitsberg et al., *Nature* 364(1993)459.
5. J.H.M. Knoll et al., *Nature Genet.* 6(1994)41.
6. J.M. LaSalle and M. Lalande, *Nature Genet.* (1995) in press
7. We are grateful to Drs. A. Bottani, S. Langlois, R. Wharton, B. Horsthemke, A. Beaudet, and K. Orstavik for generously providing blood samples or lymphoblast lines from AS and PWS patients; to Drs. D. Sinnett and E.W. Kirkness for their contributions to constructing the phage contig; and to A. Flint, K. Glatt, W. Xie, and P. Bailey for expert technical assistance. M.L. is an assistant investigator of the Howard Hughes Medical Institute. This work was also supported by NIH grant RO1NS30628-01 (ML).

Proc. Microscopy and Microanalysis 1995, edited by G.W. Bailey, M.H. Ellisman, R.A. Hennigar, and N.J. Zaluzec
Copyright © 1995 MSA. Published by Jones and Begell Publishing, 79 Madison Ave., New York, NY 10016

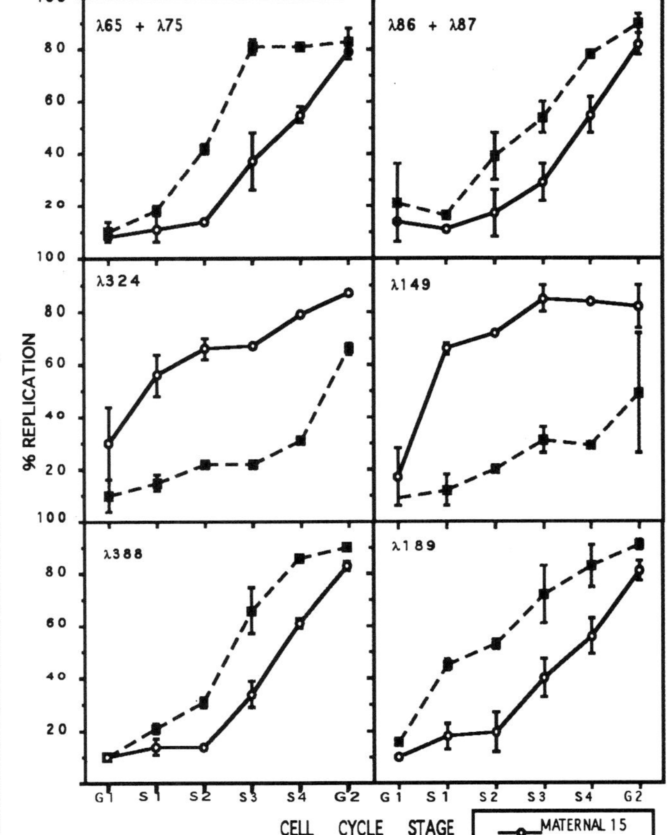

Figure 1. Physical map of the GABA$_A$ receptor subunit β3 and α5 genomic region.
A contiguous physical map of lambda phage (λ) and P1 (Π) clones was constructed to completely span the genomic region between GABRB3 and GABRA5. The intergenic region extending from λ87 on the proximal side to λ388 on the distal side forms a contiguous map of phage clones to be used as probes for high resolution FISH replication analysis in this region. Phage used for FISH analysis of replication in Fig. 2 are highlighted in gray.

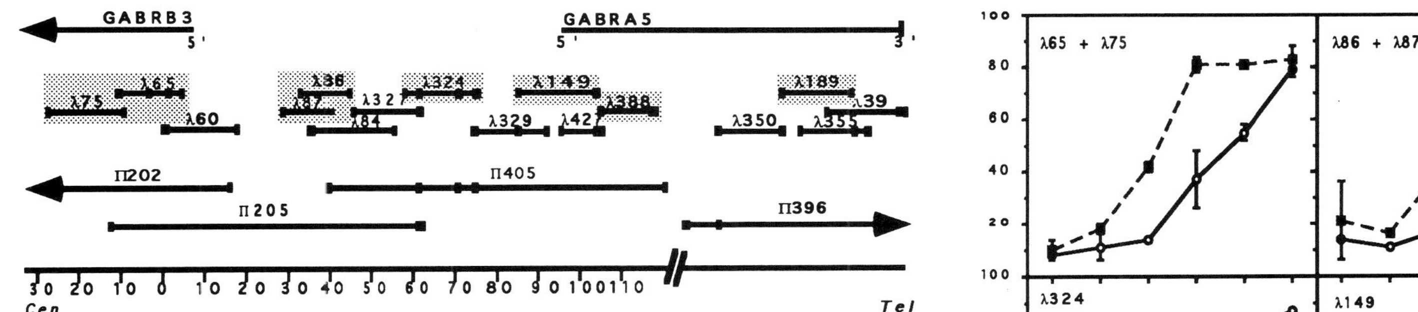

Figure 2. Allele-specific replication patterns in the GABRB3/A5 region.
FISH detection of allele specific replication was performed on interphase nuclei from a normal subject heterozygous for the chromosome 15 satellite size polymorphism detected by digoxigenin-labeled D15Z1 (larger gray signal). The chromosome 15 with the larger (brighter) D15Z1 signal is maternally derived and the shorter (less intense) D15Z1 signal is paternal (data not shown). Biotin-labeled phage DNA derived from the regions shown in Fig. 1 were simultaneously hybridized to the nuclei to detect replication events (small white spots). Phage DNA signals appeared as either a singlet (unreplicated) or doublet (replicated) on each parental chromosome. Representative nuclei demonstrating the predominant parental allele specificity of asynchronous replication (singlet-doublet) are shown for six probes. Probes λ65 + λ75, λ86 + λ87, λ388, and λ189 display earlier replication on the paternal chromosome 15, as the doublet is associated with the shorter D15Z1. In contrast, probes λ324 and λ149, which recognize intergenic genomic sequences, detect earlier replication on the maternal chromosome 15 shown by the proximity of the doublet to the larger D15Z1 signal.

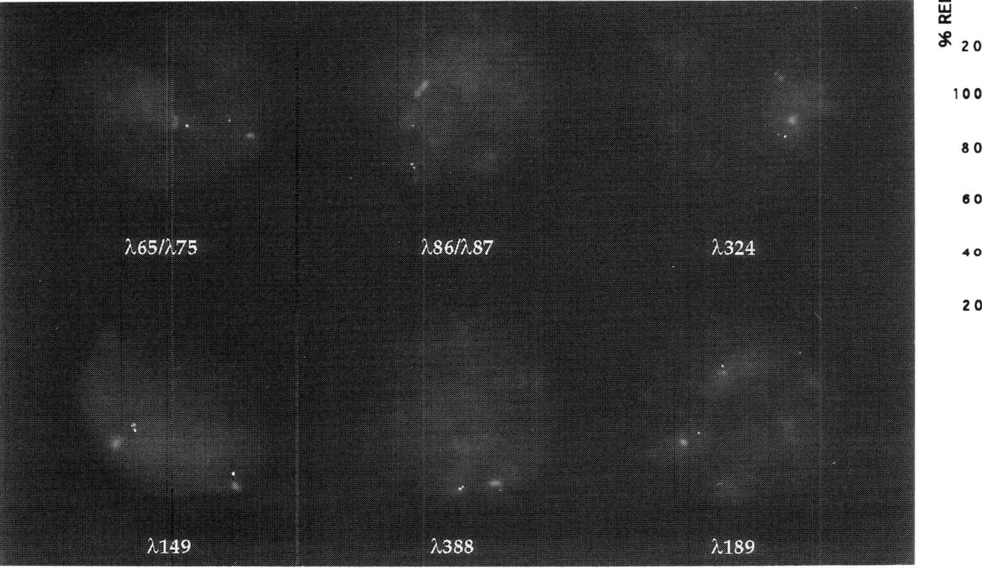

Figure 3. Kinetics of allele-specific replication.
Peripheral blood lymphocytes from NS-1 were fractionated into six stages of cell cycle progression on the basis of DNA content. Allele-specific replication events are represented on a kinetic time line based on FISH analysis of flow sorted fractions. Nuclei from each cell cycle fraction (G1,S1,S2,S3,S4,G2) were hybridized with D15Z1. For each graph, the abscissa (% replication) is the percentage of cells displaying a doublet on either the maternal chromosome 15 (open circles, solid line) or the paternal chromosome 15 (closed squares, dashed line). The ordinate (cell cycle stage) is the fraction of flow sorted cells and is a measure of relative time of progression through S phase. This kinetic analysis reveals that for probes demonstrating earlier replication of the paternal allele (λ65 + λ75, λ86 + λ87, λ388, and λ189), both alleles show a linear progression of replication events in mid to late S phase. In contrast, the intergenic region recognized by λ324 and λ149 replicates in the beginning of S phase on the maternal allele and the end of S on the paternal allele. The control probe λ347 demonstrates that the distal region at D15S46 replicates between S1 and S3 on both maternal and paternal chromosome 15 (data not shown).

ORGANIZATION OF TRANSCRIPTION AND PRE-mRNA SPLICING WITHIN THE MAMMALIAN CELL NUCLEUS.

D.L. Spector, S. Huang, and S. Kaurin

Cold Spring Harbor Laboratory, P.O. Box 100, Cold Spring Harbor, New York 11724

We have been interested in the organization of RNA polymerase II transcription and pre-mRNA splicing within the cell nucleus. Several models have been proposed for the functional organization of RNA within the eukaryotic nucleus and for the relationship of this organization to the distribution of pre-mRNA splicing factors. One model suggests that RNAs which must be spliced are capable of recruiting splicing factors to the sites of transcription from storage and/or reassembly sites[1]. When one examines the organization of splicing factors in the nucleus in comparison to the sites of chromatin it is clear that splicing factors are not localized in coincidence with heterochromatin (Fig. 1). Instead, they are distributed in a speckled pattern which is composed of both perichromatin fibrils and interchromatin granule clusters. The perichromatin fibrils are distributed on the periphery of heterochromatin and on the periphery of interchromatin granule clusters as well as being diffusely distributed throughout the nucleoplasm. These nuclear regions have been previously shown to represent initial sites of incorporation of ^{3}H-uridine[2]. The interchromatin granule clusters do not appear to incorporate ^{3}H-uridine after short pulses, however, they are highly enriched in pre-mRNA splicing factors. Upon inhibition of RNA polymerase II transcription or pre-mRNA splicing, splicing factors accumulate at interchromatin granule clusters and these factors will return to their normal distribution upon reinitiation of transcription and splicing. Studies are currently underway to examine the organization of several RNA polymerase II transcription factors in the context of the above described nuclear components. In the course of our examination of splicing components in nuclei we have identified a series of filaments 7-10 nm in diameter that extend up to 2 μm in length within the nucleoplasm. From 2-20 individual filaments are observed per 80 nm thin-section. These filaments may represent, or be a part of, nuclear architectural elements involved in the 3-dimensional organization of the mammalian nucleus.

References

1. L.F. Jiménez-García, L.F. and D.L. Spector, Cell 73(1993)47.
2. S. Fakan and E. Puvion, Int. Rev. Cytol. 65(1980)255.

Proc. Microscopy and Microanalysis 1995, edited by G.W. Bailey, M.H. Ellisman, R.A. Hennigar, and N.J. Zaluzec
Copyright © 1995 MSA. Published by Jones and Begell Publishing, 79 Madison Ave., New York, NY 10016

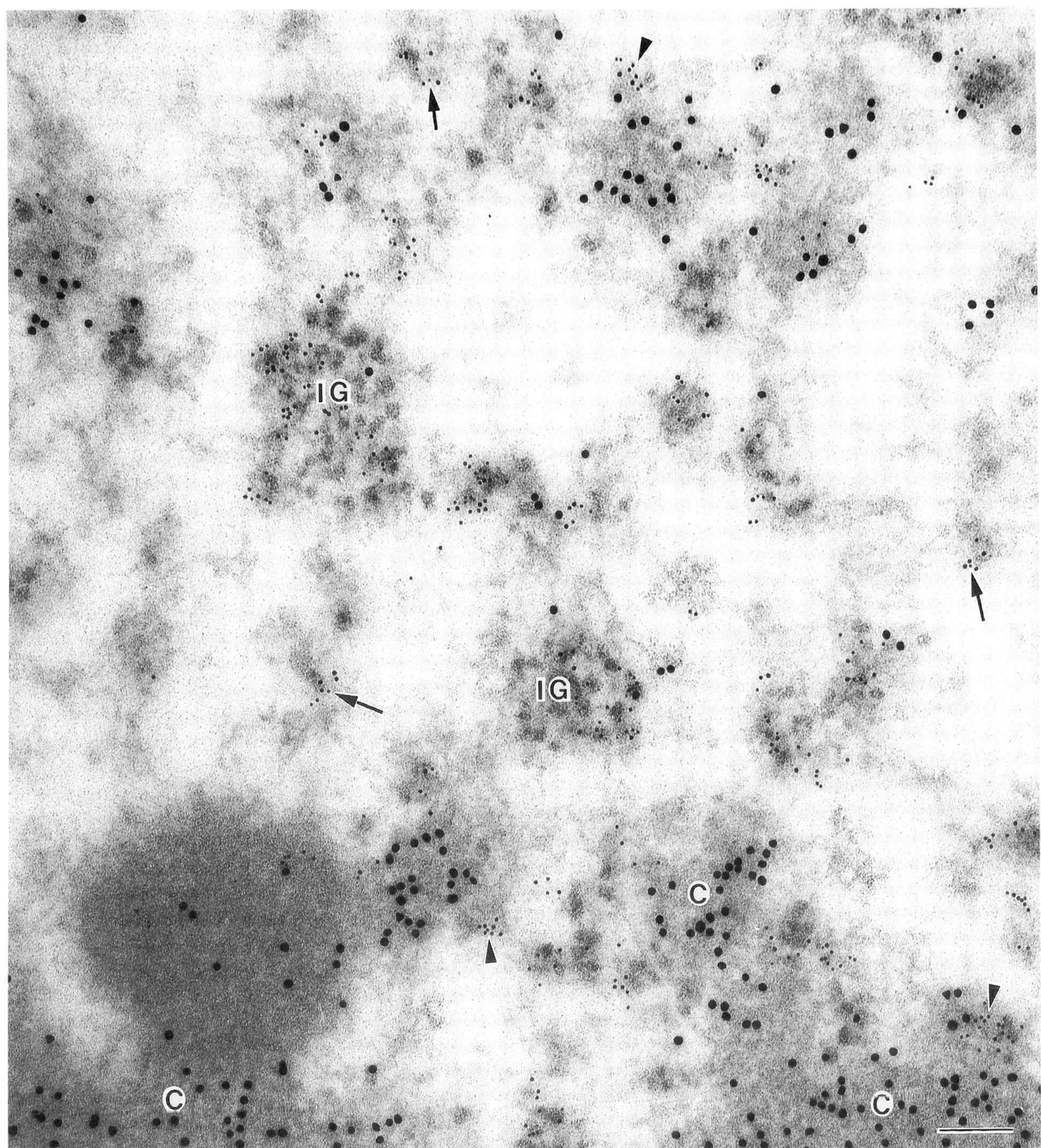

FIG. 1. The localization of chromatin is complementary to the localization of pre-mRNA splicing factors. HeLa cell sections were immunolabeled with antibodies to DNA (10 nm colloidal gold) and the splicing factor SC35 (5 nm colloidal gold). Splicing factors are localized to interchromatin granule clusters (IG) and to perichromatin fibrils (arrowheads) on the edge of condensed chromatin (C). Immunolabeled perichromatin fibrils are also scattered throughout the nucleoplasm (arrows). Bar = 100 nm.

NUCLEOSKELETAL PROTEINS WITH BIFUNCTIONAL ROLES IN THE NUCLEUS AND MITOTIC APPARATUS: A PARADIGM FOR PROTEIN DYNAMICS IN THE CELL CYCLE

B. R. Brinkley,* D. He,* C. Zeng,** B. Scott,* and D. Turner*

*Department of Cell Biology and **Verna and Marrs McLean Department of Biochemistry, Baylor College of Medicine, Houston, TX 77030

The eukaryotic cell nucleus, once believed to contain structureless nucleoplasm surrounding chromatin and the nucleolus is now thought to contain an extensive nucleoskeletal matrix on which chromatin, RNP and a complex array of transcription factors, hormone receptors and other regulatory factors are spatially arranged. A distinct nucleoskeleton has been difficult to identify in the intact nucleus, due in part to masking by a dense array of chromatin fibers. However, if chromatin is extracted by nuclease digestion and high salt, an underlying anastomosing network of 9-13 nm core filaments can be demonstrated.[1] The nucleoskeleton resembles the cytoplasmic intermediate filament complex but is confined entirely within the nucleus where it connects the nuclear lamina with various nuclear organelles, forming an integral lattice of fibers collectively called the nuclear matrix.[2] Although the molecular composition of the nucleoskeleton remains elusive, its integrity apparently requires RNP and a growing list of nuclear matrix proteins.

Although relatively insoluble within the nuclear compartment, the entire nucleoskeletal framework and associated chromatin is efficiently dismantled, packaged, partitioned and reassembled into daughter nuclei during mitosis. The dissolution of the nucleus is accompanied by the assembly of the mitotic apparatus composed of a spindle, chromosomes and many nuclear-derived proteins. We have investigated a unique array of nuclear matrix proteins that depart the nucleus at the onset of mitosis and become structural and functional components of the mitotic apparatus. A paradigm has been developed whereby proteins associated with the nuclear matrix in interphase become targeted to one or more specific domains (Fig. 1) of the mitotic apparatus, where they assume new functional roles in the mitotic process. One such protein is the nuclear mitotic apparatus protein (NuMA) which associates with the nuclear matrix in interphase but becomes targeted to the mitotic spindle poles at the onset of mitosis.[3-4] NuMA, a 200-230 kD protein with a large α-helical domain, is present in multiple isoforms and displays either a diffuse or punctate distribution in the nucleus as shown by indirect immunofluorescence using various antibodies. We have found that at least one isoform appears to interface spliceosomes with the nuclear matrix.[5] At the onset of mitosis, NuMA exits the nucleus and becomes localized at the minus ends of microtubules in the pericentriolar region of the spindle, where it may regulate microtubule assembly/disassembly or function in nuclear reassembly at telophase.

The centromere/kinetochore is a locus that functions to attach chromosomes to the mitotic spindle in mitosis and consists of at least four constitutive proteins, CENP-A, -B, -C and -D. During interphase, prekinetochores are functionally inactive and display programmed structural changes and nonrandom movements while remaining firmly attached to the nuclear matrix or chromosome scaffold. Although CENP-A is extracted by treatment with nucleases and high salts required to reveal the nucleoskeleton, CENP-B and CENP-C remain with the matrix and scaffold and may function in the anchoring and

Proc. Microscopy and Microanalysis 1995, edited by G.W. Bailey, M.H. Ellisman, R.A. Hennigar, and N.J. Zaluzec.
Copyright © 1995 MSA. Published by Jones and Begell Publishing, 79 Madison Ave., New York, NY 10016

positioning of chromosomes in the interphase nucleus. When the nuclear envelope is disrupted, prekinetochores differentiate into trilaminar plates and capture spindle microtubules required for chromosome movement and partitioning on the mitotic spindle. Recent studies in our laboratory and elsewhere have identified cell cycle specific (facultative) proteins, some derived from the nuclear matrix, that associate with the kinetochore during mitosis. The latter may regulate kinetochore activity and function in mitotic checkpoints that control chromosome alignment and the metaphase-anaphase transition.[6]

References

1. D. He et al., J. Cell Biol. 110(1990)569.
2. R. Berezney and D. S. Coffey, Biochem. Biophys. Res. Com. 60(1974)1410.
3. D. He, C. Zeng and B. R. Brinkley, Intl. Rev. Cytol. (Sup.)(1995) in press.
4. B. Lydersen and D. Pettijohn, Cell 22(1980)489.
5. C. Zeng et al., Proc. Natl. Acad. Sci. USA 91(1994)1505.
6. This work is supported in part by NCI/NIH Grant CA 41424.

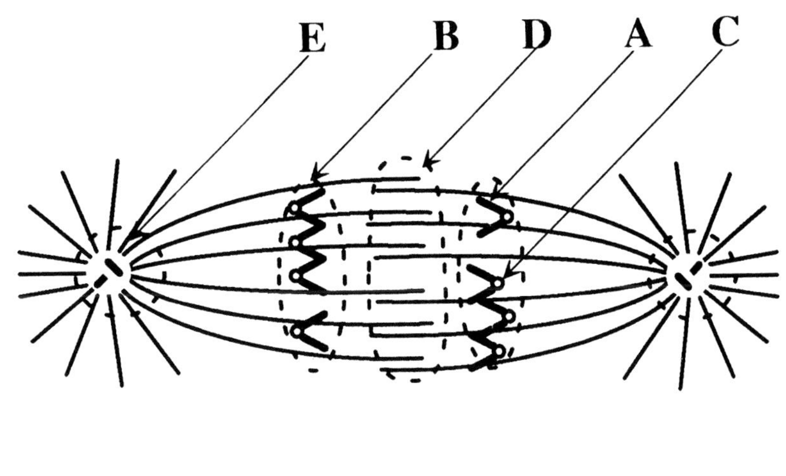

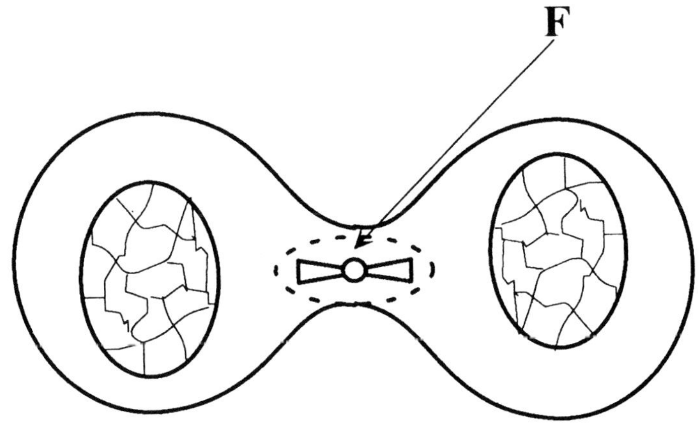

FIG. 1. – Arrows indicated domains within the mitotic apparatus occupied by proteins derived from the nuclear matrix. (A) chromosome scaffold; (B) perichromosome region; (C) centromere; (D) spindle midzone; (E) spindle pole; and (F) midbody.

SUBCELLULAR ORGANIZATION OF snRNPS IN MAMMALIAN CELLS

A. Gregory Matera

Department of Genetics, Case Western Reserve University, School of Medicine, Cleveland OH, 44106-4955

Our laboratory is interested in aspects of eukaryotic gene expression that may be regulated at the level of nuclear architecture and/or localization. Understanding the spatial organization of genes and gene products within mammalian nuclei will provide valuable insight into how the various metabolic processes such as replication, transcription, processing and transport are orchestrated. In order to begin to address these questions, we have chosen to study the cell biology of RNA processing. Small RNA-protein complexes known as ribonucleoproteins (RNPs) are central factors in numerous RNA processing reactions. Much of what is currently known about the subcellular organization of RNPs is based upon the locations of the protein components of RNPs, not the RNA components. Furthermore, since many RNAs share common protein subunits it is essential to develop probes which are specific to the individual RNAs. We utilize highly specific antisense oligonucleotide probes and fluorescence *in situ* hybridization (FISH) to ascertain the organization of RNPs and their associations with particular subdomains within mammalian cells.

The most prominent subnuclear domain is the nucleolus, in which the ribosomal genes from different chromosomes come together to perform the critical task of ribosome synthesis. This complex process involves not only rRNA transcription and processing, but ribonucleoprotein (RNP) assembly and transport. The nucleolus is a dynamic organelle. Its size and organization depend upon ribosome biogenesis. The various steps in ribosome synthesis are thought to be partitioned between nucleolar subregions that are identified by their morphology. Among the small nucleolar RNPs (snoRNPs), U3 is by far the most abundant and the most studied (reviewed in [1]). Most of the small nucleolar ribonucleoproteins (snoRNPs) share a common 34 kD protein called fibrillarin (Fb). These so-called Fb snRNPs include U3, U8, U13, U14, U15, U16, U18, U20, U21, U22 and the list is still growing. Perhaps the most interesting aspect of the above-listed Fb snRNPs is that only U3, U8 and U13 are transcribed from their own promoters; the remaining RNAs are encoded within the introns of other genes. Aqueous fractionation methods reveal that the snoRNPs involved in rRNA processing are localized in the nucleolus; such studies give no information regarding the organization of these particles *within* the nucleolus. Investigation of the distribution of U3, U8 and U13 RNAs within the nucleolus revealed that they are differentially organized [2]. Colocalization studies with anti-pol I antibodies and anti-hUBF/NOR-90 (a pol I transcription factor) suggest that U8 is recruited to discrete subnucleolar regions that surround the fibrillar centers and resemble the dense fibrillar component [2].

Recently we have focussed our attention on two different nuclear organelles: coiled bodies (CBs) and perinucleolar compartments (PNCs). Small RNPs are major components of each of these subnuclear domains. Coiled bodies (CBs) are nuclear organelles whose morphology has been conserved from plants to animals. In rapidly cycling mammalian cells, they are typically located in the nucleoplasm, but are often

Proc. Microscopy and Microanalysis 1995, edited by G.W. Bailey, M.H. Ellisman, R.A. Hennigar, and N.J. Zaluzec
Copyright © 1995 MSA. Published by Jones and Begell Publishing, 79 Madison Ave., New York, NY 10016

found in contact with the nucleolus. The CBs in human cells contain a unique protein, called p80-coilin. We present evidence that, in addition to the spliceosomal RNAs, CBs contain U7 snRNA and associate with specific chromosomal loci in interphase cells [3]. Using an antisense 2'-OMe oligomer complementary to U7 snRNA and FISH, we found that signal was detected throughout the nucleoplasm, excluding nucleoli. Double-labeling with anti-p80 coilin antibodies reveals that, like the other Sm class snRNPs, U7 also intensely stains CBs. Interestingly, we found that CBs often associate with subsets of the histone, U1 and U2 snRNA gene loci in interphase HeLa-ATCC and HEp-2 monolayer cells. Control hybridizations showed that CBs do not associate with 5S rRNA, NGFR, c-myc, pRb or Down syndrome loci. Possible roles for CBs in the metabolism of these various histone and small nuclear RNAs are discussed.

Human Ro RNPs consist of four small RNA molecules of 85–112 nt, each complexed with a 60 kD protein (and perhaps other proteins). The RNA components of human Ro RNPs are called hY RNAs. Although all four Y RNAs were primarily cytoplasmic, oligonucleotides directed against three of the RNAs hybridized to discrete structures near the nucleolar rim. We have termed these structures "peri-nucleolar compartments" (PNCs). Double labeling experiments with appropriate antisera revealed that PNCs are distinct from coiled bodies and fibrillar centers. Co-hybridization with a genomic DNA clone spanning the human Y1 and Y3 genes showed that PNCs are not stably associated with the transcription site for these Y RNAs [4]. 5S rDNA was often located near the nucleolar periphery, however, PNCs were not found to be associated with 5S gene loci. Two additional pol III transcripts, the RNA components of RNase P and RNase MRP, did colocalize within PNCs. Most interestingly, the polypyrimidine tract-binding protein hnRNP I/PTB was also concentrated in this compartment. The observation that PNCs contain several different small RNAs suggests that, like coiled bodies, they may represent cellular compartments in which some aspect of RNA biogenesis or assembly occurs. As all of the RNAs detected in PNCs are at least partially cytoplasmic, we speculate that this novel nuclear subdomain may represent the place at which cytoplasmic RNAs are assembled and sorted after entry into the nucleus or before export to the cytoplasm.

References:

1. Baserga, S.J. and Steitz, J.A. (1993). *The Diverse World of Small Ribonucleoproteins.*, in The RNA World, R. Gesteland and J. F. Atkins, (eds.), Cold Spring Harbor Laboratory Press, Cold Spring Harbor, NY., pp. 359-381.

2. Matera A.G., Tycowski K.T., Steitz J.A. and Ward D.C. (1994). Organization of Small Nucleolar Ribonucleoproteins (snoRNPs) by Fluorescence *In Situ* Hybridization and Immunocytochemistry. Mol. Biol. Cell **5**: 1289-1299.

3. Frey M.R. and Matera A.G. (1995). Coiled Bodies Contain U7 snRNA and Associate with Specific DNA Sequences in Interphase Cells. Proc. Natl. Acad. Sci., U.S.A., (**in press**).

4. Matera A.G., Frey M.R., Margelot K.and Wolin S.L. (1995). A Perinucleolar Compartment That Appears to Contain Several RNA Polymerase III Transcripts and the Polypyrimidine Tract-Binding Protein, hnRNP I. J. Cell Biol., (**in press**).

NUCLEAR LOCALIZATION OF INITIATION FACTOR 2 IN NEURON PRIMARY CULTURES.

F.J. Martinez Alonso**, M.V. Toledo Lobo**, S. Rodriguez Martínez*, F.M. Muñoz Postigo*, J.J. López-Fando Castro*.

* Biochemistry and Molecular Biology Dept. Univ. of Alcala de Henares. Madrid Spain.
** Morphological Sciences and Surgery Dept. Univ. of Alcala de Henares. Madrid Spain.

The dominant mechanism that controls protein synthesis is the phosphorylation/dephosphorylation of initiation and elongation factors, with a translational control function. Each phase of protein synthesis is promoted by some of these factors that transiently interact with ribosomes, mRNAs and aminoacyl-tRNAs. Eukaryotic initiation factor-2 (eIF2, 130 kD) is one of these proteins and it is composed of three subunits: alpha, beta and gamma. eIF2 forms a ternary complex (GTP-eIF2-Met tRNAi) that can then interact with the 40S ribosomal subunit which in turn binds mRNA and the 60S ribosomal subunit to form the 80S initiation complex[1]. The relation between eIF2 and the ribosomes is then a well established aspect of protein synthesis, but there are no previous studies about the distribution of eIF2 within the cell.

Using immunocytochemical techniques, we show the distribution of eIF2 within the cell found in primary cultures of rat embryo brain neurons, in which eIF2 and eIF2-kinases have been identified[2]. Primary culture neuron cells were grown in D15 and N2 mediums for 8 days. Cells were fixed using paraformaldehyde 4%-glutaraldehyde 0,5% and embedded in Unicryl. For post-embedding immnunocytochemical staining, sections were incubated with monoclonal anti-alpha eIF2 (courthesy of the late Dr. Henshaw) and a polyclonal antibody developed in our Department of Biochemistry. After rinsing in TBS and incubating with 10 nm and 15 nm gold-conjugated IgG as secondary antibodies, sections were examined with TEM. Immunoblotts demonstrate the specifity of the immunoreaction. Controls of the immunocytochemical reaction were done by ommiting the primary antibody. Other controls were set up using the supernatant obtained of immunoabsortions of the primary antibody instead of the primary antibody to incubate the sections.

The distribution of immunolabeled eIF2 within the neurons was the same using both monoclonal (Figs.2,3) and polyclonal antibodies (Fig.1). eIF2 immunoreactivity was located in the cell nucleus, cell cytoplasm and mitochondria. Cytoplasmic eIF2 was found in areas with ribosomes. Cell membranes such as nuclear envelope or plasma membrane, can not be observed with the TEM methods used. However, the space between the inner and the outer nuclear membranes appears as a clear halo surrounding the nucleus (Figs.1,2). The immunolabelling in the interphasic cell nucleus was clearly located in the heterochromatin and frecuently found in the border between euchromatin and heterochromatin (Fig.3). Mitoses were not seen in the cells cultured. Thus, we were not able to observe the distribution of immunolabelled eIF2 during the mitosis period. Gold particles were occasionally observed occupying a narrow site of the nuclear halo, at bridges between the nucleus and the cytoplasm, resembling nuclear pores (Fig.2). Since eIF2 is synthesized at the cytoplasm, its presence in the nucleus would involve a cytoplasm/nucleus transport. Some studies have demonstrated that factors involved in protein synthesis control, such as eIF-4E, are located not only in the cytoplasm but in the cell nucleus[3]. These data, along with the presence of eIF2 in the cell nucleus, suggest that nuclear eIF2 might be involved in transcription or another unknown function[4].

References
1.- S. Ochoa, Archs. Biochem. Biophys. 223 (1983) 325.
2.- A. Alcazar et al., Int. J. Devl. Neuroscience 4 (1986) 525.
3.- F. Lejbkowicz, Proc. Natl. Acad. Sci. USA, 89 (1992) 9612.
4.- The authors gratefully acknowledge the collaboration of Dr. González Santander.
Supported by DGICYT PB 91-157 grant.

Proc. Microscopy and Microanalysis 1995, edited by G.W. Bailey, M.H. Ellisman, R.A. Hennigar, and N.J. Zaluzec
Copyright © 1995 MSA. Published by Jones and Begell Publishing, 79 Madison Ave., New York, NY 10016

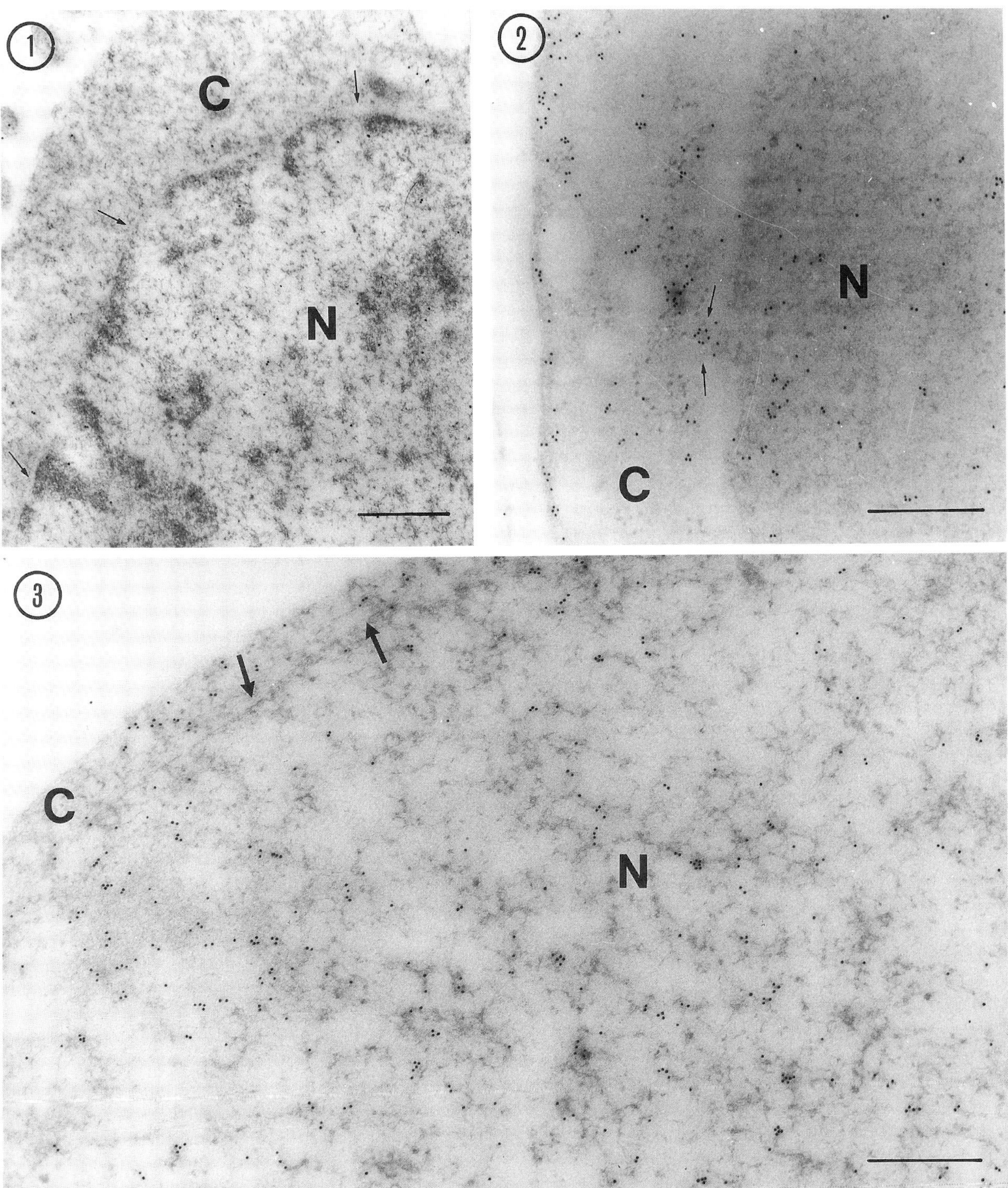

Fig.1.- Rat embryo brain neuron (primary culture). Immunolocalization of eIF2 with a polyclonal antibody. Nuclear envelope (arrows); N: nucleus; C: cytoplasm. Scale bar: 0.5 μm (8,000 X).
Figs.2,3.- Nuclear and cytoplasmic immunolocalization of alpha-eIF2 with a monoclonal antibody. Gold particles occupying a site into the nuclear halo that resembles nuclear pores (small arrows). N: nucleus. C: cytoplasm. Nuclear envelope (big arrows). Scale bar: 0.5 μm (20,000 X).

MEMBRANE MICRODOMAINS: LESSONS FROM MICROSCOPY

J.M. Robinson[*] and J.M Oliver[**]

[*]Dept. Cell Biology, Neurobiology, Ohio State Univ., Columbus, OH 43210
[**]Dept. Pathology, University of New Mexico. Albuquerque, NM 87131

Specialized regions of plasma membranes displaying lateral heterogeneity are the focus of this Symposium. Specialized membrane domains are known for certain cell types such as differentiated epithelial cells where lateral heterogeneity in lipids and proteins exists between the apical and basolateral portions of the plasma membrane. Lateral heterogeneity and the presence of microdomains in membranes that are uniform in appearance have been more difficult to establish. Nonetheless a number of studies have provided evidence for membrane microdomains and indicated a functional importance for these structures.

This symposium will focus on the use of various imaging modalities and related approaches to define membrane microdomains in a number of cell types. The importance of existing as well as emerging imaging technologies for use in the elucidation of membrane microdomains will be highlighted. The organization of membrane microdomains in terms of dimensions and spatial distribution is of considerable interest and will be addressed in this Symposium.

Edidin has employed near-field scanning optical microscopy (NSOM) of fluorescent-labeled cells to visualize membrane patchiness on the nanometer scale. Spatial resolution of 50 nm or less can be achieved by NSOM. Microdomains for lipid and protein have been observed in cells. NSOM examination of model lipid monolayers revealed that a fluorescent-labeled lipid was uniformly distributed within a single-phase monolayer but was patchy in its distribution in a two-phase monolayer.

Erlandsen has examined the distribution of cell adhesion molecules (CAMs) on the surface of leukocytes. Immunocytochemical techniques with colloidal gold probes have been utilized for detection of CAMs. The distribution of gold particles has been recorded using high resolution backscatter imaging with field emission SEM. This approach permits detection and high spatial resolution of small gold probes by atomic number contrast. These studies have shown that CAMs involved in the regulation of extravascular migration of leukocytes occupy different microdomains on the plasma membrane these cells.

Proc. Microscopy and Microanalysis 1995, edited by G.W. Bailey, M.H. Ellisman, R.A. Hennigar, and N.J. Zaluzec
Copyright © 1995 MSA. Published by Jones and Begell Publishing, 79 Madison Ave., New York, NY 10016

Wofsy et al. have also employed colloidal gold probes and SEM to determine the distribution of IgE receptors on basophilic leukemia cells. This approach has been used for quantitative analysis of receptor aggregation in these cells. SEM micrographs of gold-labeled basophilic leukemia cells are complex images that may contain thousands of gold particles. A software system has been developed for the statistical analysis of these complex images in order to investigate the underlying biological problem that relates to receptor aggregation.

Anderson has examined caveolae which are specialized regions of the plasma membrane that are present in many cell types. While caveolae are readily observed by TEM, their function in cells has remained elusive and controversial. Recent evidence shows that these specialized membrane domains are dynamic structures that are enriched in certain protein species (e.g., GPI-linked proteins). Moreover, evidence is emerging that suggests caveolae are involved in a number of signal transduction events.

Webb has employed microscopy and biophysical techniques to investigate IgE-receptor microdomain function. Measurements of the trajectories of individual cell surface molecules as they diffuse spontaneously on the surfaces of living cells give evidence for heterogeneity of cell surface microdomains on all spatial scales down to a few nanometers. The observed anomalous dynamics imply evanescent intermolecular interactions that can restrain diffusion so as to account for the immobile fraction reported by fluorescence photobleaching recovery experiments. Observations of coaggregation of receptor clusters and selected lipid derivatives again imply ubiquitous modulatable intermolecular interactions in the cell membrane.

These studies illustrate the importance of various types of microscopy for the analysis of membrane domains in cells and model membrane systems. Information derived from these and other imaging methodologies is crucial for more complete understanding of membrane domains in cell function.

NEAR-FIELD MICROSCOPY OF MEMBRANE DOMAINS

J. Hwang[*,] E. Betzig[**], and M. Edidin[*]

* Biology Department, The Johns Hopkins University, Baltimore, MD 21218
** ATT Bell Laboratories, Murray Hill, NJ 07974

Results from several different methods for probing the lateral organization of cell surface membranes indicate that these membranes are patchy, divided into domains.[1,2] The data suggest that on average these domains are 0.1-1 μm across and that they persist for 10's to 1000's of seconds.[3] At least some domains in this size range, when labeled by fluorescent proteins or lipids ought to be detectable by conventional, far-field, fluorescence microscopy. However, though some images are consistent with a domain structure for membranes,[4,5] most far-field images of fluorescent cell surfaces lack the detail necessary to define domains.

We have used near-field scanning optical microscopy, NSOM,[6] of fluorescent-labeled cells to visualize membrane patchiness on the nanometer scale. This method yields images with resolutions of 50 nm or less. In our near-field microscope the labeled sample is illuminated by a optical fiber probe, with an aperture of 50-80nm. The probe is scanned over the cell surface at a distance of ˜ 10 nm from the surface. Only surface fluorescence is excited by the scanned probe. Emitted light is collected by a conventional objective. An image of the scanned surface is developed from the 2-dimensional intensity map of the scans. Resolution in this image is approximately equal to the radius of the fiber aperture. Besides the fluorescence image, a shear force image which shows the surface contours is obtained simultaneously. NSOM images of normal human fibroblasts labeled with the fluorescent lipid analog, BODIPY-PC and with a fluorescent antibody to an abundant surface protein, class I MHC molecules, display a pattern of ragged fluorescent patches, consistent in size with the domains predicted by earlier experiments (Figure 1). Protein and lipid labels colocalize in some regions of the surface, but not in others. This patchiness is not an artifact of cell preparation. The fluorescence of BODIPY-PC is uniform when the probe is incorporated into single-phase lipid monolayers, though it is patchy when the probe is incorporated in 2-phase (gel and fluid) monolayers.[7] We also find that a different lipid probe, diI C12, which did not seem to localize to domains, does not show the same patchy distribution in cell membranes as BODIPY-PC . Rather, its intensity follows the folds of the membrane defined in the shear force image.

References

1. M. Edidin, *J. Cell Sci. Supplement* 17(1993)165.
2. M. Edidin, *Trend in Cell Biology* 2(1992)376.
3. M. Edidin, M. Zuniga and M.P. Sheetz, *Proc. Natl. Acad. Sci. USA* 91(1994)3378.
4. E. Yechiel and M. Edidin, *J. Cell Biol.* 105(1987)755.
5. W. Rodgers and M. Glaser, *Biochemistry* 32(1993)12591.
6. E. Betzig and J.K. Trautman, *Science* 257(1992)189.
7. J. Hwang, C. Bohm, L. Tamm, E. Betzig and M. Edidin, *Biophys. J.* 68(1995)A293.

Supported by NIH Grants 5 R37 AI14584 and 5 P01 DK 44375

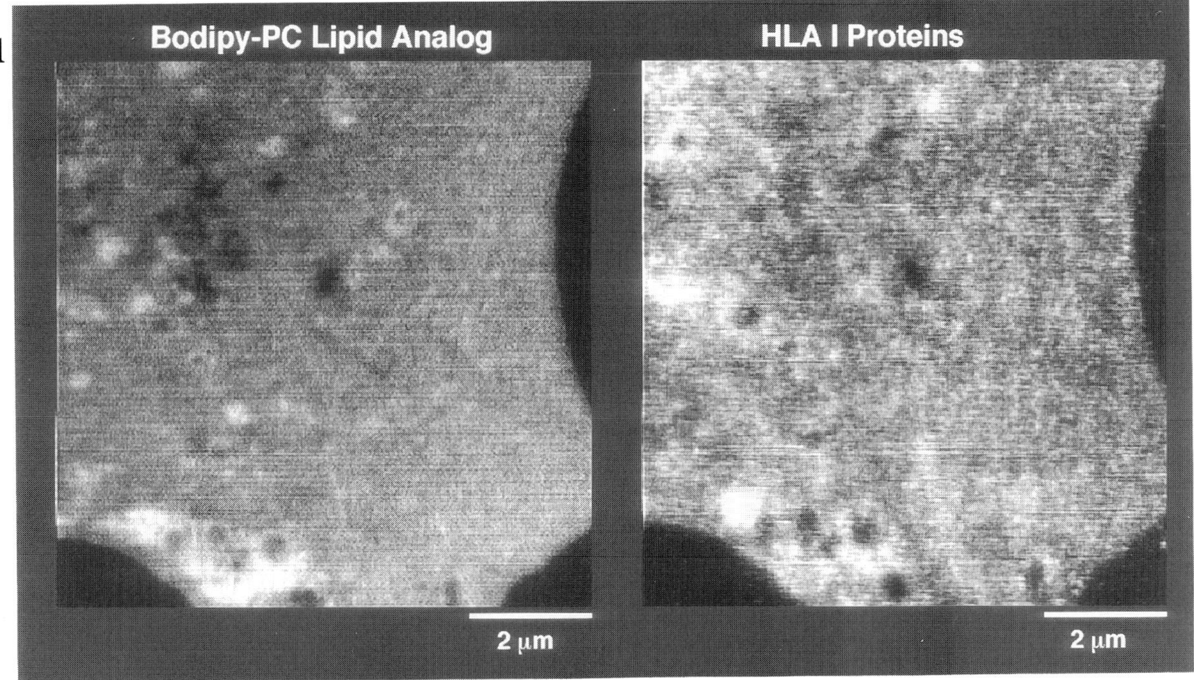

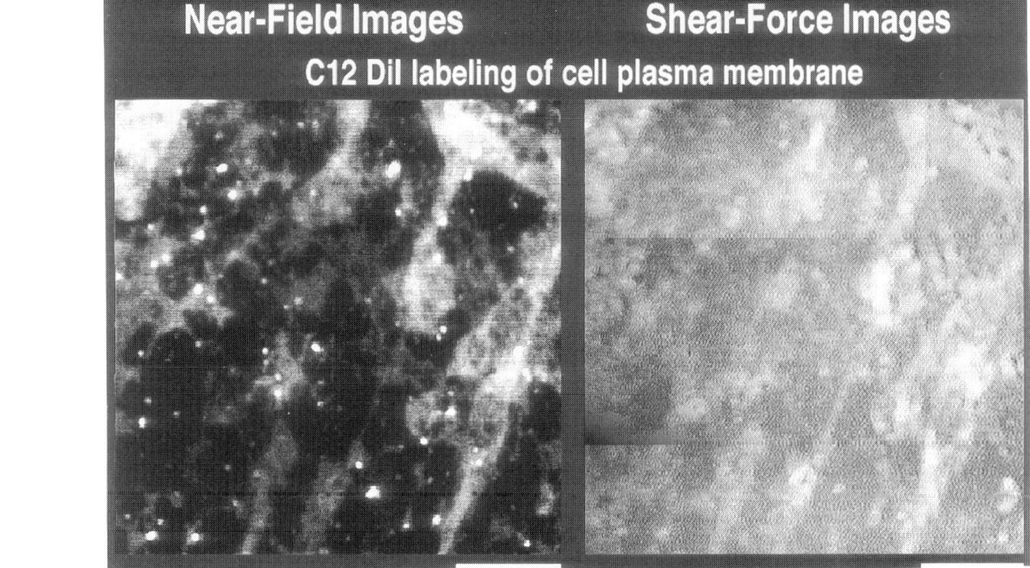

Fig. 1. NSOM of the leading edge of a human skin fibroblast labeled with BODIPY-PC and with rhodamine-conjugated anti-HLA mAb. Though some areas coincide, the coarse patching of the HLA label does not match most of the pattern of the BODIPY-PC fluorescence.

Fig. 2. Comparison of near-field fluorescence and shear force images of a region of fibroblast membrane labeled with the lipid analog diI C12. The intensity of fluorescence follows the contours of the surface seen in the shear force image. The very bright dots are micelles of diI that have precipitated from the labeling solution. Bar = 2µm.

SEGREGATION OF CELL ADHESION MOLECULES INTO MEMBRANE MICRODOMAINS FACILITATES LEUKOCYTE-ENDOTHELIAL INTERACTIONS

S.L. Erlandsen

Department of Cell Biology and Neuroanatomy, University of Minnesota Medical School, Minneaplis, MN 55455

Cells interact with their extracellular environments by means of a variety of cellular adhesion molecules (CAM) and surface ligands. In many instances, CAMs interact in a sequential temporal fashion which suggests that these adhesion molecules may occupy or be polarized to various membrane microdomains on the cell surface. Detection of CAMs can be accomplished by a variety of methods including immunofluorescent microscopy and flow cytometry, and by the use of immunocytochemical markers (i.e. colloidal gold) in electron microscopy. The development of high resolution field emission SEM in the mid 1980's and the Autrata modification of the YAG detector for backscatter electron detection at low voltage has greatly facilitated the recognition of colloidal gold probes for detection of surface CAMs.[1] Low voltage FESEM with Bse imaging provides increased resolution of cell surface topography (~3nm at 3-4 keV) which can be observed in 3-dimensions, and simultaneously permits detection/high spatial resolution of immunogold label by atomic number contrast.

Evidence for an important functional role(s) of CAMs in membrane microdomains exists in the mechanisms regulating the extravascular migration of leukocytes from blood vessels at sites of inflammation. Circulating leukocytes possess microvilli and surface membrane ruffles which interact with the endothelial surface of vessels, particularly in capillaries and postcapillary venules where the reduced flow of blood brings them into close contact with the endothelium.

The first step in the extravascular migration of leukocytes from vessels involves the "rolling", or functional contact of CAMs present on the leukocyte/endothelial surface, leading to momentary tethering of the cells with one another, followed by upregulation of β_2 integrin leading to firm adherence.[2] Three different CAMs or ligands (CD62L, a4 integrin, and high affinity sites for CD62P) have been shown to promote "rolling" activity under sheer forces similar to that in the microvascular environment, and in each case, these CAM/ligands have been shown to occupy specific membrane domains on microvilli. In unactivated neutrophils, a unique polarization of L-selectin (CD62L) molecules occurs on the membrane domain associated with the apical surfaces of microvilli and membrane ruffles[3,4,5], and this selectin mediates the rolling response of many leukocytes on upregulated endothelial cells. In

Proc. Microscopy and Microanalysis 1995, edited by G.W. Bailey, M.H. Ellisman, R.A. Hennigar, and N.J. Zaluzec
Copyright © 1995 MSA. Published by Jones and Begell Publishing, 79 Madison Ave., New York, NY 10016

contrast to L-selectin, β_2 integrins on neutrophils are restricted to the body of the unactivated cell, but become widely spread on the surface after upregulation.[3] Upregulation of endothelial cells by cytokines leads to increased expression of a variety of ligands including E-selectin (CD62E), P-selectin (CD62P), ICAM-1, and V-CAM. The rolling response of neutrophils can be mediated by interactions of L-selectin with carbohydrate ligands on endothelial cells and by simultaneously presenting sialyl-Lewis[x] to E-selectin on endothelial cells.[2] L-selectin independent rolling can occur in leukocytes, such as lymphoblasts and memory lymphocytes. In these cells, the rolling phenomenon appears to be mediated through α_4 integrins which also possess a unique distribution on the membrane domain associated with microvilli[6], and these CAMs are thought to interact with V-CAM on the endothelial surface. In addition to L-selectin and α_4 integrin, the rolling of human leukocytes has been also shown to occur in vitro by interaction of high affinity binding sites (P-selectin glycoprotein-1) on neutrophils for P-selectin. Within minutes after stimulation with thrombin, endothelial cells express P-selectin on the surface by secreting stored forms from Weibel-Palade bodies, and thus may regulate the initial interactions of leukocytes with upregulated endothelium. Again, the distribution of these high affinity sites for P-selectin on the neutrophil is similar to that for L-selectin and α_4 integrin, being associated with the membrane domain on the tips of microvilli.[7]

Multiple forms of CAMs have been shown to be involved in the "rolling" of leukocytes on upregulated endothelial cells, and in each case the CAM has been shown to occupy the membrane microdomain associated with microvilli or ruffles. The common topographical presentation of these different receptors/ligands on the microvillar surface suggests that this spatial distribution may be a critical factor in adhesive interactions to overcome sheer forces generated under physiologial conditions.

References:

1. R. Autrata, EMSA Bulletin 22(1992)54.
2. U. von Andrian et al., Blood 82(1993)182.
3. S.L. Erlandsen et al., J. Histochem. Cytochem. 41(1993)327
4. L.J. Picker et al., Cell 66(1991)921.
5. S. Hasslen et al., Histochem. J. (in press), 1995.
6. C. Berlin et al., Cell 80(1995)1.
7. K.L. Moore et al., J. Cell Biol. 128(1995)661.

QUANTIFYING IgE RECEPTOR AGGREGATION FROM SEM-IMMUNOCYTOLOGY

Carla Wofsy[†*], Margaret L. Sanders[†], Gregory W. Donohoe[‡], Matthew Pujol[‡], and Janet M. Oliver[*]

Departments of [†]Mathematics and Statistics, [‡]Electrical Engineering and Computer Engineering, School of Engineering, and [*]Pathology, School of Medicine, University of New Mexico, Albuquerque, NM 87131

Improved procedures for imaging immunogold markers on scanning electron micrographs, new procedures for extracting gold particle coordinates from the images, and both new and standard statistical methods for analyzing spatial point patterns make it possible to study quantitatively the aggregation of cell surface molecules and their localization in specialized membrane domains. The imaging procedures have been described previously.[1] Here we present the methods we have developed or adapted for image processing and for statistical analysis of the immunogold point patterns. We apply the methods to study the aggregation of the high affinity receptor for IgE ($Fc\epsilon RI$) on rat basophilic leukemia (RBL) cells.

The aggregation of cell surface receptors is a necessary step in most known receptor-mediated cellular processes. For example, activation of the intrinsic kinase activity of growth factor receptors requires receptor aggregation, as does the initiation of a variety of antigen-induced responses in the immune system. In the most common form of allergic reaction (immediate hypersensitivity), the aggregation of $Fc\epsilon$ receptors on basophils and mast cells triggers a sequence of biochemical events, culminating in the release of histamine and other inflammatory agents from intracellular granules. RBL cells, with monoclonal IgE bound to $Fc\epsilon$ receptors, provide a well-defined experimental system for studying immediate hypersensitivity and, more generally, receptor-mediated cell signalling pathways.

We used anti-IgE antibody to cross-link receptor-bound IgE into aggregates on RBL cells, then labeled the complexes with 15 nm protein A – gold particles, and finally used backscattered scanning electron microscopy (SEM) to observe the gold particle distributions that form under different experimental conditions. Qualitative analysis of the resulting images, interpreted in connection with secretion data, suggests that smaller aggregates are involved in cell activation while large scale aggregates are inhibitory.[2] To go further in identifying distinct aggregation states with distinct functions, or in testing models for the underlying aggregation process, we need quantitative tools that make contact between the data and the models.

The main statistical function we have used to provide this link, since it can be estimated from the data and calculated from the models, is known in the statistical literature as Ripley's K

Proc. Microscopy and Microanalysis 1995, edited by G.W. Bailey, M.H. Ellisman, R.A. Hennigar, and N.J. Zaluzec
Copyright © 1995 MSA. Published by Jones and Begell Publishing, 79 Madison Ave., New York, NY 10016

function and is defined by

$$\lambda K(r) = \text{the average number of particles within a distance } r \text{ of a random particle}$$

where λ denotes the overall particle density.[3-5] The difference between λK and its expected value if the point distribution is random has been used in various ways to detect clustering or regularity in a spatial point pattern. We propose a new estimator of mean aggregate size, based on the first local maximum of the difference function. Simulations of a chemical reaction model, with chemical aggregates located according to a Spatial Poisson Process, show that with our procedure we can recover mean aggregate size and a related chemical parameter. This is true even if immunogold labeling is incomplete.

In order to estimate the K function, or to determine distances and angles required for other statistical tests, we need two dimensional coordinates of gold particle locations in the images. A single SEM image may contain thousands of particles. We developed a software system to automatically find x-y coordinates of centers of mass of gold particles in a digitized image. This was implemented on a Sun workstation using the Khoros software environment.

We can visualize the digitized grayscale image as a surface whose height at a point corresponds to the corresponding brightness level in the image: the darkest black has a "height" of zero, and brightest white, 255. The gold particles jut out from the background as bumps whose profile is approximated by a two-dimensional Gaussian function. Finding them is complicated by random noise. The extraction routine consists of a set of filters to isolate the spots from the background and noise, an adaptive threshold to find regions containing the spots, and a peak-finding routine to locate the peak of each particle and produce a list of individual particles.

References
1. R. Stump et al., *Am. J. Anatomy* 185(1989)128.
2. JC. Seagrave et al., *J. Cellular Physiology* 148(1991)139.
3. P. J. Diggle *Statistical Analysis of Spatial Point Patterns*, New York: Academic Press (1983).
4. R. D. Ripley, *Spatial Statistics*, New York: Wiley (1981).
5. R. D. Ripley *Statistical Inference for Spatial Processes*, Cambridge: Cambridge University Press (1988).
6. The authors gratefully acknowledge support from the National Science Foundation and the National Institutes of Health.

COMPARTMENTALIZED CELL SIGNALING IN CAVEOLAE

Richard G.W. Anderson

University of Texas Southwestern Medical Center at Dallas
Dallas, Texas 75235

Caveola undergo an internalization cyclic that begins with the conversion of a flat segment of membrane into a vesicle and ends when that vesicle returns to the cell surface. During this cycle, resident molecules are sequestered in the interior of the vesicle where they perform special functions. Previously we showed that the internalization step is interrupted by activators of PKC such as phorbol-12,13-dibutyrate (PDBu). Recently we found that caveolae contain a resident population of PKCα molecules that phosphorylate a 90-kDa protein during invagination. In unstimulated cells, virtually all of the detectable PKCα in plasma membranes is in caveolae. Treatment of the cells with PDBu or agents that raise the concentration of diacylglycerol (DAG) in the cell displaced PKCα from caveolae, which blocked the phosphorylation of the 90-kDa protein and membrane invagination. Caveolae also contain a protein phosphatase that dephosphorylated the 90-kDa protein once PKCα was gone or inactive. PKCα was also dissociated from caveolae and invagination was inhibited after cells were incubated in the presence of histamine. This effect was blocked by pyrilamine but not cimetidine, suggesting that histamine was acting through H$_1$ receptors. Therefore, the caveolae internalization cycle can be hormonally regulated by displacing PKCα from the membrane. Another molecule we found to be highly enriched in caveolae is sphingomyelin. We have now found that interleukin 1β (IL-1β) stimulation of normal human fibroblasts initiates a lipid messenger cascade in caveolae. Hormone binding first stimulated the formation of diacylglycerol (DAG) in caveolae. This was immediately followed by the conversion of a resident population of sphingomyelin to ceramide. The ceramide produced in caveolae, in turn, blocked PDGF-stimulated DNA synthesis. IL-1β stimulated DAG production at other locations within the same cell, but this DAG was not coupled to ceramide production. These results suggest that caveolae can compartmentalize ceramide production for delivery to specific sites within the cell. The identification of two different signaling pathways in caveolae is strong evidence that this organelle has a major role to play in coordinating both intercellular and intracellular communication.

Proc. Microscopy and Microanalysis 1995, edited by G.W. Bailey, M.H. Ellisman, R.A. Hennigar, and N.J. Zaluzec
Copyright © 1995 MSA. Published by Jones and Begell Publishing, 79 Madison Ave., New York, NY 10016

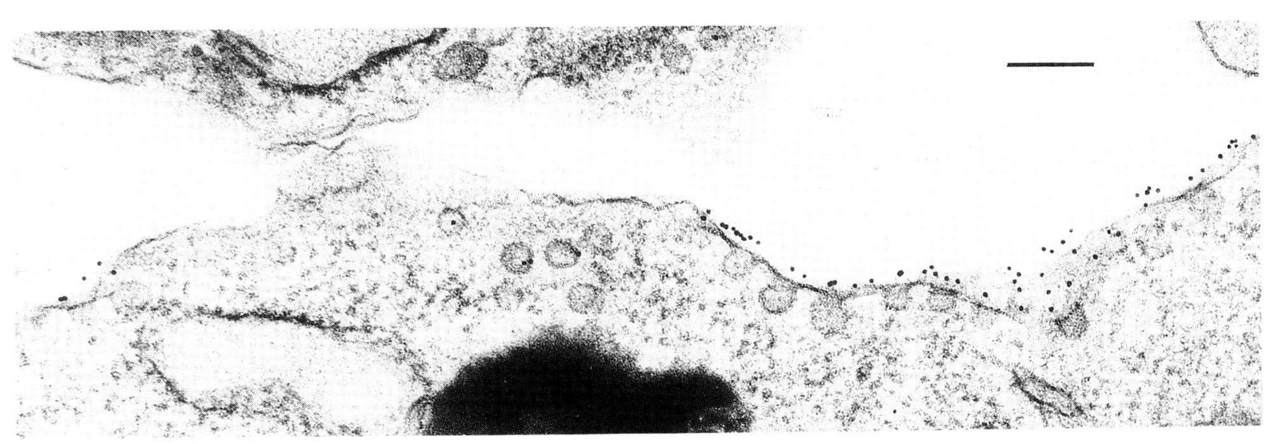

Immunogold labeling of Thy-I antigen in 3T3 cells is associated with caveolae. This protein is one of many GPI anchored membrane proteins that are associated with caveolae. Some of the gold label can be seen within the lumen of a few of the vesicles. Caveolae are abundant in fibroblast cells and tend to be organized into patch at the leading edge of the cell. Bar = 0.2 μm.

WHAT DO NANOMETER MOLECULAR TRAJECTORIES TELL US ABOUT HETEROGENEOUS CELL SURFACE DOMAINES?

Watt W. Webb

Cornell University, School of Applied and Engineering Physics, Ithaca, NY 14853

Plasma membrane heterogeneity is implicit in the existence of specialized cell surface organelles which are necessary for cellular function; coated pits, post and pre-synaptic terminals, microvillae, caveolae, tight junctions, focal contacts and endothelial polarization are examples. The persistence of these discrete molecular aggregates depends on localized restraint of the constituent molecules within specific domains in the cell surface by strong intermolecular bonds and/or anchorage to extended cytoskeleton. The observed plasticity of many of organelles and the dynamical modulation of domains induced by cellular signaling evidence evanescent intermolecular interactions even in conspicuous aggregates. There is also strong evidence that universal restraints on the mobility of cell surface proteins persist virtually everywhere in cell surfaces, not only in the discrete organelles. Diffusion of cell surface proteins is slowed by several orders of magnitude relative to corresponding protein diffusion coefficients in isolated lipid membranes as has been determined by various ensemble average methods of measurement such as fluorescence photobleaching recovery(FPR). Results are usually expressed in terms of a (small) diffusion coefficient $10^{-12}<D<10^{-9}$ cm^2/s and a mobile fraction $0<R<1$ which describes a fraction of fluorescence photobleaching recovery after the longest experimental diffusion times and is interpreted as the fraction of labeled protein that appears capable of diffusion. In contrast with proteins diffusion coefficients of membrane lipids in the cell surface are usually slowed only a little relative to pure mixed lipid membranes thus $D\sim 10^{-8}$ cm^2/s and $R\sim 1$. Although lipid linked proteins diffuse relatively fast, $D\sim 10^{-9}$cm^2/s, there is usually some marker immobilization, $R<1$.

The search for understanding of these ubiquitous restraints on membrane protein diffusion have lead us to develop experimental procedures for tracking simultaneously the trajectories of thousands of individual cell surface receptor molecules with spatial resolutions δR (rms) ~ 30nm and temporal resolution $\delta t\sim 0.3$s with dynamic range of about 1000. Spontaneous random migration and sometimes actively driven motion of the low density lipoprotein receptors (LDL-R) and of immuoglobulin E receptors (IgE-R) have been studied. Most individual molecular trajectories for both receptors appear at first glance to be random walks as expected for conventional diffusion in a uniform medium where the mean square displacement is expected to grow linearly in time, $<(\Delta R)^2> = Dt$. However in the untreated cell surfaces the temporal correlation of most trajectories displayed "anomalous diffusion" in which the mean square displacements increase as a power law in time with the exponent $0< \alpha <1$. Thus $<(\Delta R)^2> = D(1)t^{\alpha} = [D(1)t^{\alpha-1}]\,\Delta t$ defines a time scale dependent diffusion coefficient $[D(1)\,\Delta t^{\alpha-1}]$ where $D(1)$ specifies the effective value at $t=1$ say one second for the units selected. A plot of $\log <(\Delta R)^2>$ versus $\log t$ yields a slope α. The ubiquity of straight line log log plots over 2 or 3 orders of magnitude implies the absence of any characteristic scale of time or space within the experimental range. No sign of discrete barriers could be seen in these experiments on integral membrane proteins in fiberblasts and secretory cells. The LDL receptors studied are mutants molecules that do not bind to coated pits; one mutant has its cytoplasmic tail shortened to 3 amino acids thus excluding direct cytoskeletal coupling; similar trajectories, the second (JD) mutant is decoupled from coated pits by a single aminoacid change and values of α and $D(1)$ were found for these two cases.

Anomalous diffusion with $\alpha < 1$ leads to "long time tails" in diffusion limited processes. This effect can account quantitatively for the limited mobile fraction $R < 1$ customarily used to fit FPR experiments. FPR curves were synthesized from tracking data by constructing the appropriate ensemble average from measured probability distributions of trajectory parameters $D(1)$ and α. Comparison with actual FPR data on the same preparations with data fitted by D and R values show the clear correspondence of $< \alpha >$ and R. Small values of α and R both correspond to long time tails.

Numerous theoretical investigations of the statistical physics of random walks in inhomogenous media explore possible models for protein diffusion on cell surfaces. Only one model of many models studied unequivocally fits the observed power law correlations. This model describes a two dimensional random walk in a field of random potentials wells that vary such that the distribution of dwell times (τ) is sufficiently broad that its average value (τ) does not exist; for example if the probability distribution has the form $\chi(\tau) \approx \tau_0{}^\alpha \tau^{-(1-\alpha)}; (\tau \rightarrow \infty)$. Anomalous power law diffusion is found in two dimensions only if the broad dwell time distributions vary randomly at each site for each visit by a diffusing particle. This implies randomly varying potential well depths of such broad distribution of well energies that the range of dwell times exceeds the dynamic temporal range of the experiment. This "annealed" model in which the potential field varies with time and space contrasts with "quenched" models in which the field of random potential wells consist of fixed traps. One can understand the annealed random potential well model in terms of weak energies of interaction amongst mobile particles.

Is there other relevant evidence for weak interactions amongst cell surface molecules? It has long been known that global changes in diffusability can be induced by immunological stimulation. Recent experiments have shown profound segregation to clusters of cross linked cell surface receptors, including IgE-R by certain fluorescent lipid analogs mimicking saturated phospholipids. These localized lipid-protein aggregates restrict mobility of both lipid and protein. Similar localization has been attributed to evanescent associations with caveolae and with certain membrane permeabilization reagents. Evidence for these evanescent membrane domain associations is to be repeated.

Experimental methods for measurement of molecular trajectories and for detection of evanescent aggregation in immunological responses will be described.

References:
Watt W. Webb, Biophys. J. <u>66</u>, A18(1994).
Richik N. Ghosh and Watt W. Webb, Biophys. J. 55, 498a (1989).
James L. Thomas, David Holowka, Barbara Baird and Watt W. Webb, J. Cell Bio., 125, 795-802 (1994).
Richik N. Ghosh and Watt W. Webb, Biophys. J., 66, 1301-1318 (1994).
Toni J. Feder, En-Yuh Chang, David Holowka and Watt W. Webb, J. of Cell. Physiol., 158:7-16 (1994).
James L. Thomas, Toni J. Feder and Watt W. Webb, Biophys. J. 61, 1402-1412 (1992)
Watt W. Webb and David Gross, Applications of Fluorescence in the Biomedical Sciences, eds. D. Lansing Taylor, Alan S. Waggoner, Robert F. Murphy, Frederick Lanni and R"
David Gross and Watt W. Webb, Spectroscopic Membrane Probes, ed. Leslie M. Loew (CRC Press, Inc., Boca Raton, FL, 1988), Vol. II, Cha"
David W. Tank, W. J. Fredericks, L. S. Barak and W. W. Webb, J. Cell Biol. 101, 148 (1985).
L. S. Barak and W. W. Webb, J. Cell Biol. 95, 846 (1982).
L. S. Barak and W. W. Webb, J. Cell Biol. 90, 595 (1981).

QUANTITATIVE BIOMEDICAL MICROSCOPY: A TECHNOLOGY FOR BOTH BASIC AND CLINICAL SCIENCES

B. Herman,* M.R. Pillai,[+] S. Halabi,[#] A. McKalip,* A. Periasamy,* P. Wodnicki*, J. Gordon,* and X.F. Wang*

*Department of Cell Biology & Anatomy, University of North Carolina, Chapel Hill, NC 27599
[+]Regional Cancer Center, Thiruvananthapuram Kerala, India
[#]Department of Biostatistics and Epidemiology, Tulane University Medical Center, New Orleans, LA 70112

The development of quantitative optical microscopy has led to the ability to obtain accurate, highly sensitive spatial and temporal measurements of cell structure and function as well as the detection of low levels of pathogenic agents. One application of this capability has been in studies examining the role of high risk Human Papillomaviruses (HPV) in cervical cancer. Cervical cancer is the second most prevalent cancer in women worldwide, and epidemiological studies have strongly implicated the sexually transmitted HPV as a causative agent. So far, approximately 70 different genotypes have been identified, and those types associated with genital lesions have been classified as either "low-risk" or "high-risk" based on their association with benign or malignant lesions. The "high-risk" HPVs, such as HPV-16 and 18, have been found in approximately 84% of cervical carcinomas and are associated with cervical dysplasia and cervical intraepithelial neoplasia (CIN). HPV-16 is the most prevalent high-risk genotype and is present in about 50% of all genital lesions.

The method commonly used to detect cervical dysplasia/carcinoma is the PAP smear. However, for various reasons, it is difficult to determine on the basis of a PAP smear whether or not HPV is present. Polymerase chain reaction (PCR) could be used to assess HPV prevalence, but this requires destruction of cellular morphology and does not allow staging of cervical disease if present. To overcome this limitation, **Quantitative Automated Image Cytometry** has been developed. In combination with Fluorescent In Situ Hybridization (FISH) techniques, this microscope instrument is capable of detecting as little as one copy of HPV DNA per cell nucleus and is linearly quantitative up to ~1000 copies of HPV DNA. In addition, cellular morphology is preserved, allowing cytological assessment of the PAP smear. In examination of a large number of cervicovaginal samples, the combination of quantitative automated image cytometry and FISH was found to have higher sensitivity and equivalent specificity in terms of staging cervical disease status compared to PCR (1).

A common problem in all studies employing fluorescent probes is background fluorescence, including: 1) nonspecific autofluorescence from solvents, solutes, cell and tissues, 2) fixative induced fluorescence of cells and tissues, and 3) autofluorescence of the optical components of the microscopic system which effectively serve to decrease the signal to noise ratio of detection. To remove these sources of nonspecific information, **Fluorescence Long Lifetime Imaging Microscopy** has been developed. The principal of this type of microscopy is that the fluorescence emitted from the reagent probe, which is labeled with a fluorescent reporter whose lifetime is on the order of micro- to milliseconds, is measured after background fluorescence (autofluorescence) has completely decayed. The ability to discriminate between probe-specific or protein-specific fluorescence and autofluorescence and other nonspecific fluorescence leads to suppression of background by two orders of magnitude. This approach has successfully been applied to the detection of high risk HPV in cervical cancer cells (2).

The ability of the high-risk HPVs to contribute to malignant progression appears to depend on expression of the E6 and E7 oncogenes. The E7 protein of high-risk HPVs undergoes complex formation with the cellular product of the retinoblastoma (Rb) tumor suppressor gene. E7 binds preferentially to the hypophosphorylated form of Rb, disrupting complex formation between Rb and the transcription factor E2F. Based on data generated using in vitro translation systems, the E6 oncoproteins have been documented to form complexes with the cellular tumor suppressor protein

Proc. Microscopy and Microanalysis 1995, edited by G.W. Bailey, M.H. Ellisman, R.A. Hennigar, and N.J. Zaluzec
Copyright © 1995 MSA. Published by Jones and Begell Publishing, 79 Madison Ave., New York, NY 10016

p53. E6 binding to p53 is thought to lead to degradation of p53 through the ubiquitin-dependent proteolysis system. To test this hypothesis, both quantitative **Confocal Microscopy** and **Fluorescence Resonance Energy Transfer Microscopy** have been employed. Recent findings using immunofluorescence in conjunction with confocal and fluorescence resonance energy transfer microscopy have demonstrated a similar *cytoplasmic* localization for E6 and p53 as well as evidence of a close physical association between the two proteins in HPV-positive cervical carcinoma cell lines, suggesting that E6/p53 complex formation in the cytoplasm prevents p53 from exerting its tumor-suppressing function in the *nucleus* (3).

Inactivation of p53 has been found to lead to loss of responsiveness of cells to radiotherapy and chemotherapy. The mechanisms responsible for this are not understood, but the findings of a number of investigators indicate that non-functional p53 can lead to alterations in the levels of bcl-2 and bax, two regulators of apoptosis (programmed cell death). The presence or absence or p53, HPV 16 E6 and bcl-2 protein has recently been examined in PAP smears from women with varying degrees of cervical disease using **Immunofluorescence Microscopy**. Results form these studies suggest that: 1) there is an inverse association between the presence of p53 and cervical disease; 2) there is an inverse association between the presence of p53 and bcl-2; 3) there is a strong association between the presence of HPV 16 E6 and bcl-2 with high grade disease; 4) women who are positive for HPV 16 E6 have a 3-fold higher risk of having high grade cervical disease; 5) women who are positive for bcl-2 have a 6 fold higher risk of having high grade disease; and 6) when p53, HPV-16 E6, and bcl-2 are collectively compared, women who are negative for p53 but positive for both HPV-16 E6 and bcl-2 have a 12-fold higher risk of having high grade cervical disease. These results suggest that the presence or absence of these 3 proteins may be useful for identifying women at high risk for the development of cervical cancer.

These studies, all of which have been made possible by the use of a variety of different optical microscopic techniques, underscore how basic research can be translated into practical applications providing important information regarding the mechanism of diseases as well as leading to the development of potentially useful diagnostic reagents. The figure below summarizes current findings and illustrates a testable model of how high risk HPV may contribute to cervical disease.

References

1) Siadat-Pajouh, M. et al., *J. Clin. Microbiol* (1995) **submitted**.
2) Periasamy, A. et al., *Micro. & Anal.* (1995) **in press.**
3) Liang, X. H. et al., *Oncogene* 8: (1993) 2645-2652.

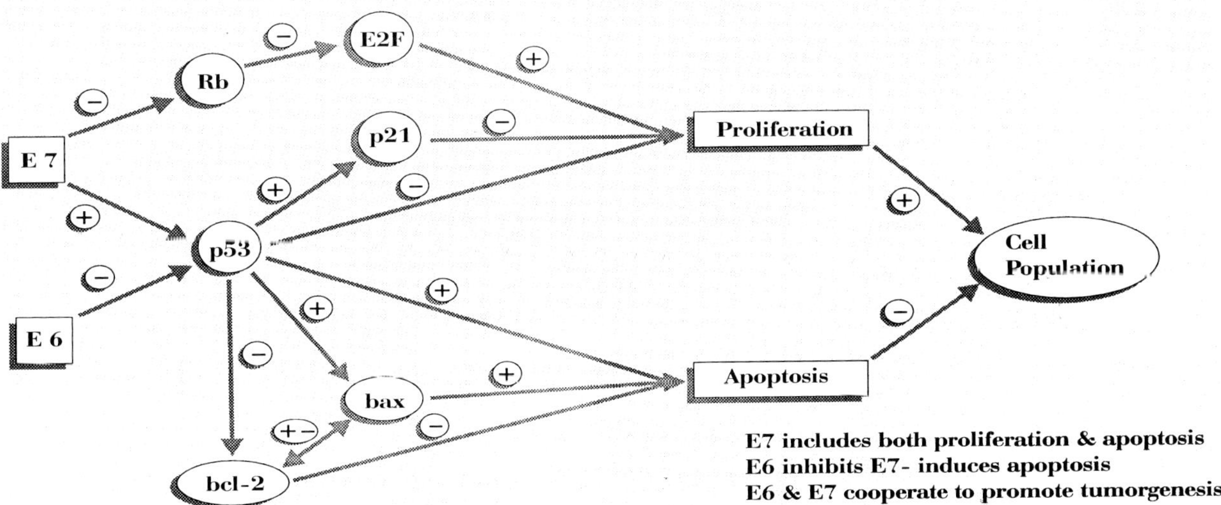

HIGH RESOLUTION LIGHT MICROSCOPY OF LIVING ORGANS *IN SITU*

R. S. McCuskey

Department of Cell Biology and Anatomy, College of Medicine, University of Arizona, Tucson, AZ 85724

Most organs in anesthetized small laboratory animals can be studied *in vivo* by light microscopy of relatively thin (3-5mm), transilluminated areas of the organ. Thicker areas of the organs in these species, as well as thicker organs of larger animals can be examined only by epi-ilumination. However, the resolution obtainable with epi-ilumination usually is inferior to that realized with transillumination. This paper reviews these methods using the liver as an example of the organ of study. A standard compound trinocular microscope is used which is modified for *in vivo* microscopy and is equipped for both transillumination and epi-illumination.[1] After the animal is anesthetized, the liver is gently exteriorized through a subcostal, abdominal incision and positioned over a window of optical grade mica or glass on a specially designed, heated microscope stage having provisions for draining irrigation fluids. The window overlies a long working distance condenser. The liver is covered by a piece of Saran or Mylar film which holds it in position and limits movements induced by respiration, the heart and the intestines. Homeostasis is maintained by constant suffusion of the organ with Ringer's solution which is maintained at body temperature by proportional regulating heaters electronically clamped to rectal temperature. Then, the organ is transilluminated with monochromatic light (400-800 nm) obtained from a prism monochromater equipped with a xenon lamp (XBO150) through the long working distance condenser. The microscopic images of the microvasculature and its surrounding tissue are secured at magnifications of 100-1500X using appropriate water immersion objectives and oculars. The resulting optical images are televised using a silicon or silicon intensified target (SIT) vidicon television camera, or a newvicon camera equipped with an image intensifier and then recorded using a 3/4" video tape recorder or on 16 mm film using a motion picture camera which is synchronized with the video framing rate. Figure 1 diagrams the imaging system.

The use of specific wavelengths of monochromatic light enhance definition of cellular detail through the selective absorption or transmission of these wavelengths by specific tissue and cellular components. When such monochromatic, microscopic images are televised, the contrast between tissue and cellular components can be enhanced further by readjustments of the brightness and contrast controls on the video monitor. Thus, the images of a particular structure(s) can be enhanced or suppressed depending upon the wavelength of light selected and the adjustments of the television system. The system permits differentiation of the microvasculature into arterioles, capillaries or sinusoids, and venules; patterns of blood flow through these vessels; the shape and deformation of individual blood cells; the endothelium of most vessels; and identification of most intra-and extravascular cell types and some of the cytoplasmic and nuclear detail in these cells (e.g., secretory granules, fat droplets, mitochondria, and lysosomes, nucleoli) (Fig. 2). The measured resolution of this system under optimal conditions is 0.3-0.5 µm when using 80-100x water immersion objectives.

Proc. Microscopy and Microanalysis 1995, edited by G.W. Bailey, M.H. Ellisman, R.A. Hennigar, and N.J. Zaluzec
Copyright © 1995 MSA. Published by Jones and Begell Publishing, 79 Madison Ave., New York, NY 10016

The liver also can be epi-illuminated through the objective lens using a Ploempak illuminator containing a variety of selectable excitation and barrier filters. Epi-illumination is most useful for studying the patterns and distribution of fluorescent probes. In the liver, for example, fluorescent probes have been used to study the phagocytic and endocytotic properties of Kupffer cells under a variety of conditions, the transport of material from the sinusoid into parenchymal cells and from parenchymal cells into bile canaliculi, the patterns of flow and entrapment of leukocytes and tumor cells, as well as to measure temporal and spatial alterations in the fluorescence of probes sensitive to reactive oxygen radicals, mitochondrial energization, cell death and autofluorescence of NADH as an indicator of mitochondrial oxygen consumption.[1] The definition obtained using epi-ilumination, however, is never as good as that obtained by transillumination. For intensely fluorescing materials, epi- and trans- illumination may be combined to provide improved definition of the cellular localization subsequently identified by transillumination. Alternatively, weakly fluorescing probes may first be imaged and recorded by epi-illumination and their localization subsequently identified by transillumination. In many cases, the use of intensified (SIT, ISIT or cooled CCD) video cameras coupled with digital image processing and/or filtering techniques are necessary to obtain images of reasonable quality and for extraction of the desired information, especially if this information is to be quantified.

References

1. R.S. McCuskey, in B. Herman and J.J. LeMasters, Eds., *Optical Microscopy: New Technologies and Applications*, Orlando: Academic Press (1993) 355.

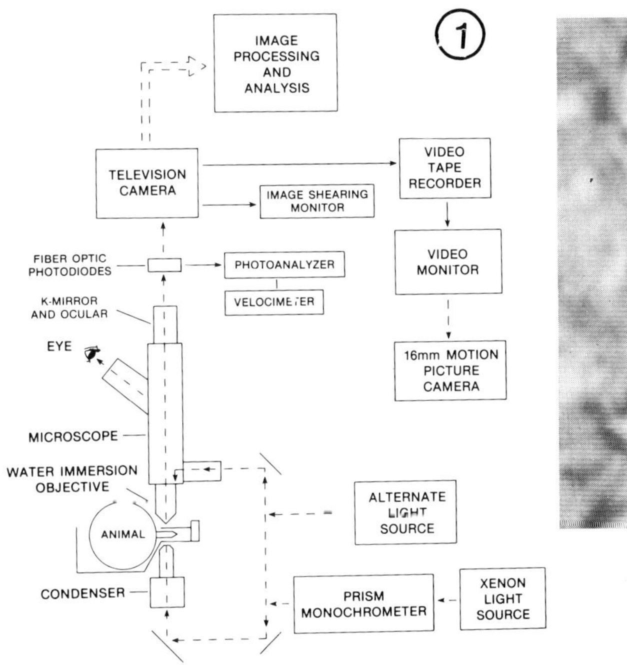

FIG.1. Diagram of the imaging system used for high resolution *in vivo* microscopy of organs.[1]

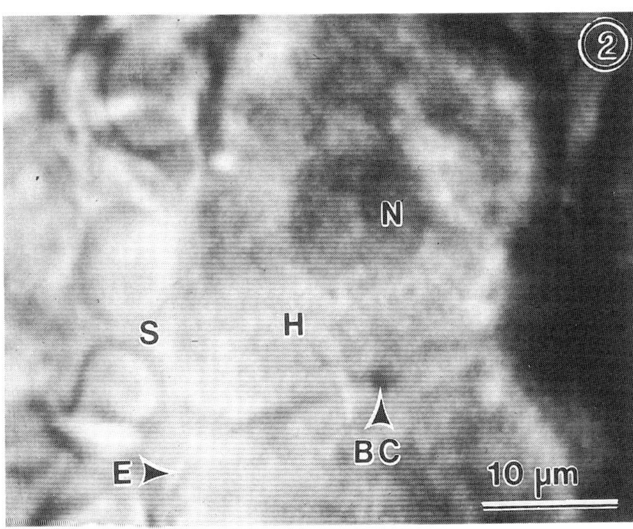

FIG. 2. *In vivo* photomicrograph of the liver. H, hepatocyte; N, nucleus; BC, bile canaliculus; E, endothelium; S, sinusoid containing erythrocytes.

CONFOCAL MICROSCOPY OF SINGLE LIVING CELLS

John J. Lemasters

Department of Cell Biology & Anatomy, University of North Carolina, Chapel Hill, NC 27599-7090

The advent of laser scanning confocal microscopy solves the dilemma of studying thick specimens with optical microscopy by creating optical slices less than 1 μm in thickness. Increasingly, confocal microscopy is an essential analytical tool for studying the structure and physiology of living cells. Because confocal microscopy collects light from only a fraction of the specimen volume, greater illumination is required. Consequently, photodamage and photobleaching are greater considerations, especially for study for living cells where repeated measurements over time are desired. To minimize photodamage, laser intensity should be attenuated by 100-1000 fold, photomultiplier circuits should be operated at highest sensitivity, and stable fluorophores should be used. When these conditions are met, literally hundreds of high resolution confocal images can be obtained from single cells loaded with parameter sensitive fluorophores (1).

The number of parameter-specific fluorophores useful for observing single living cells by confocal microscopy is large and increasing. By labeling with calcein and collecting serial images, the volume, shape and surface topography of single living cells are reconstructed with results rivaling scanning electron micrographs (Fig. 1) (2,3). Calcein can also indicate changes in mitochondrial membrane permeability associated with oxidative stress (4). The Nernstian distribution of cationic fluorophores like rhodamine 123 and tetramethylrhodamine methylester marks the distribution of electrical potential within single cells, identifying both mitochondrial and plasmalemmal membrane potentials (2). Ion-indicating fluorophores like Fluo-3, Indo-1 and SNARF-1 determine the intracellular distribution of Ca^{2+} and pH (Fig. 2) (2). Dual-labeling techniques permit cytosolic, mitochondrial and lysosomal parameters to be distinguished individually. Use of line-scanning techniques permits millisecond temporal resolution. In this way, the time course of Ca^{2+} transients in both cytosolic and mitochondrial compartments can be followed after electrical stimulation of muscle cells (5). Many other parameters (e.g., oxygen radicals, protein and non-protein thiols, membrane lipid fluidity) may be followed similarly. Application of confocal microscopy promises to provide unique insights into the physiology of single cells and organelles. Its overall impact on physiology may rival that of single cell electrical recording.

References

1. Lemasters, J.J., E. Chacon, G. Zahrebelski, J.M. Reece and A.-L. Nieminen (1993) In *Optical Microscopy: Emerging Methods and Applications*, B. Herman and J.J. Lemasters, Eds., Academic Press, New York, pp. 339-354.
2. Chacon, E., J.M. Reecc, A.-L. Nieminen, G. Zahrebelski, B. Herman, and J.J. Lemasters (1994) *Biophys. J.* **66**, 942-952.
3. Zahrebelski, G., A.-L. Nieminen, K. Al-Ghoul, T. Qian, B. Herman and J.J. Lemasters (1995) *Hepatology,* in press.
4. Nieminen, A.-L., A.K. Saylor, S.A. Tesfai, B. Herman and J.J. Lemasters (1995) *Biochem. J.* **307**, 99-106.
5. Lemasters, J.J., E. Chacon, H. Ohata, I.S. Harper, A.-L Nieminen, S.A. Tesfai and B. Herman (1995) *Methods in Enzymology,* in press.

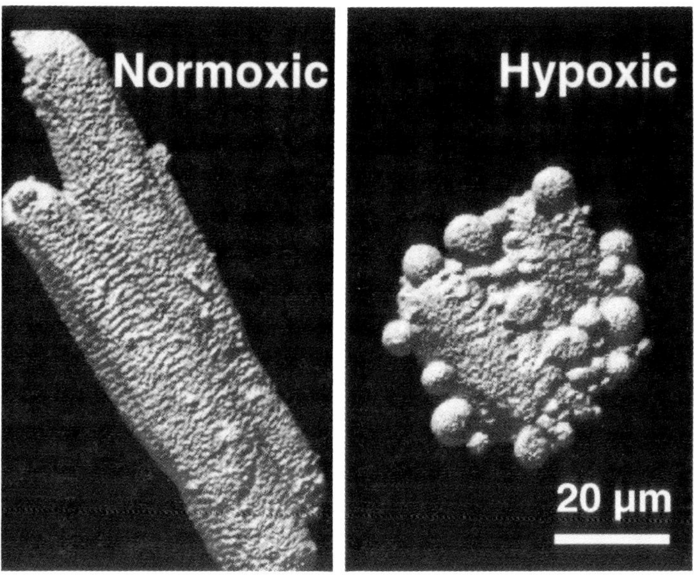

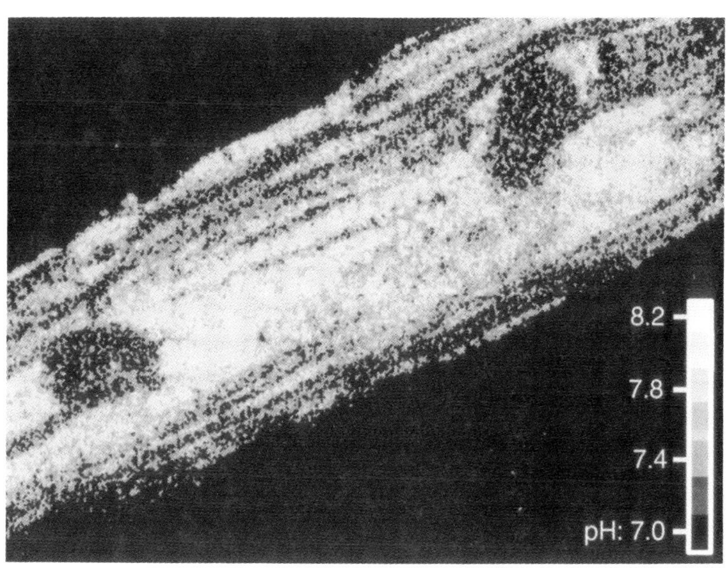

Fig. 1 (top). Reconstruction of a calcein-labeled myocyte - A cultured adult rabbit cardiac myocyte was loaded with calcein and imaged before and 40 min after 'chemical hypoxia' with 2.5 mM NaCN and 20 mM 2-deoxyglucose. A confocal image series was collected in 1.2 µm increments. Three-dimensional distribution of calcein fluorescence was reconstructed using VoxelView software. Software-generated shading enhances perception of depth and surface detail.

Fig. 2 (bottom). Intracellular pH in a cultured cardiac myocyte - A cultured adult rabbit cardiac myocyte was loaded with SNARF-1. Using an in situ calibration, the confocal image is displayed to represent distribution of pH. Mitochondria are 0.8-1 pH unit more alkaline than the cytosol.

CRITICAL ISSUES IN THE APPLICATION OF IMAGE CYTOMETRY TO THE EARLY DETECTION OF LUNG AND CERVICAL CANCER

D.M. Garner*, C.E. MacAulay**, and B. Palcic**

*Xillix Technologies Corp., Richmond, B.C., Canada V6V 2V4
**Cancer Imaging Dept., Medical Physics Division, British Columbia Cancer Agency, Vancouver, B.C., Canada, V5Y 3Y3

Screening for cervical cancer ("Pap screening") is probably the largest single application of absorbance microscopy in medicine, with 50-60 million samples screened per year in the United States alone[1]. Pap screening also represents one of the great success stories in cancer management having reduced the morbidity and mortality due to cancer of the uterine cervix by 50 - 80% in regions where mass Pap screening is carried out[2-4]. Despite this success, the subjective manual examination of stained cellular specimens with a microscope in current Pap screening practice results in significant over treatment. We estimate that for every 30 women currently treated for early cervical cancer, only one would actually develop invasive cancer in the absence of the Pap screening programme.

One key question then is: "Could image cytometry, the objective measurement of cells, improve the current Pap screening programmes by reducing the number of unnecessary medical interventions, without subjecting the patients to undo risk?"

Lung cancer, on the other hand, which is the greatest source of cancer death striking 170,000 Americans in 1995, and killing 150,000 of them[5], has no screening programme for its early detection. Although the average five year survival for all lung cancer patients is only 13%, for localized disease it is 48%[5] and could be even higher if detected in its very early, pre-invasive state. A major study of sputum in the 1970's[6] indicated that it was not effective as a screening method (the study of asymptomatic patients). However, other studies[7] have indicated that sputum could detect early lung cancer with high sensitivity and specificity if all phases of the analysis were refined, including sample collection and preservation, slide preparation, staining, and microscopic examination.

Another key question then is: "Could image cytometry help make the testing of sputum samples effective as a screening method for lung cancer?"

The first steps in answering these questions are in the form of retrospective studies by image cytometry of historical materials. We have postulated that, to answer these questions, image cytometry must be:

Quantitative for DNA: The standard Pap stain (hematoxylin) is **not** quantitative for DNA[8]. New stains are required to improve the current screening programme.

High Resolution: Both high photometric and spatial resolutions are required to measure the subtle changes in nuclear texture that appear to characterize malignancy associated changes (MACs)[9]. Early evidence suggests that MACs are not only of diagnostic utility but may also predict prognosis as well. Figure 1 shows an ROC (receiver operator characteristic) curve for lung sputum based only on MACs.

Proc. Microscopy and Microanalysis 1995, edited by G.W. Bailey, M.H. Ellisman, R.A. Hennigar, and N.J. Zaluzec
Copyright © 1995 MSA. Published by Jones and Begell Publishing, 79 Madison Ave., New York, NY 10016

Automated: A statistically significant sample of up to several thousand cell nuclei per patient must be measured. This is completely impractical by interactive means. Furthermore, automation is necessary to achieve the repeatability in focus, segmentation and unbiased sample selection that are implied by high resolution and MACs measurements.

The establishment of objective slide grading criteria based on image cytometry is the initial step in both screening questions posed above. Preliminary data from retrospective studies will be presented and some comprehensive retrospective studies currently being planned will be described.

1. National Cancer Institute. *National Health Interview Survey on Cancer Control.* Bethesda MD, National Center for Health Statistics, 1989.
2. B. Stenkvist *et al.*, *J.A.M.A.* 252(1984)1423.
3. G.H. Anderson *et al.*, *Br. Med. J.*, 296(1988)975.
4. A.J. Kirby *et al.*, *Lancet* 339(1992)828.
5. L.A.G. Reis *et al.* (eds.), *SEER Cancer Statistics Review, 1973-1991: Tables and Graphs,* National Cancer Institute. NIH Pub No 9402789, Bethesda, MD, (1994).
6. B.J. Flehinger *et al.*, *Am. Rev. Respir. Dis.*, 130(1984)555.
7. A. Böcking *et al.*, *ACTA Cytologica*, 36(1992)37.
8. A. Tezcan *et al.*, *Cytometry*, accepted (1995).
9. H.E. Neiburgs *et al.*, *Lab Invest*, 11(1962)80.

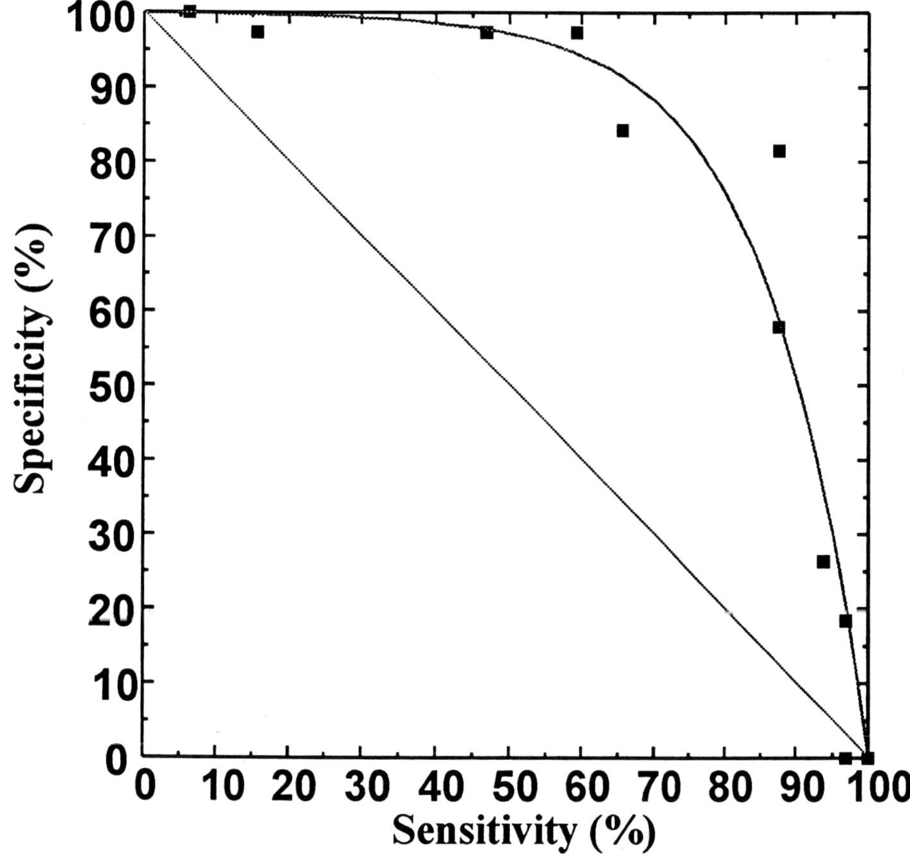

Fig. 1 ROC curve for sputum testing for lung cancer using MACs only, measured on normal cells.

Identification of Antigenic Epitotes on Hepatitis E virus by Immune Electron Microscopy

C.D. Humphrey, K.A. McCaustland, P. Paltel, Yu. E. Khudyakov, and H.A. Fields

Division of Viral and Rickettsial Diseases, CDC, Atlanta, Ga 30333

Outbreaks of enterically transmitted hepatitis involving many thousands of cases have occurred in southern and central Asia, North Africa, and Latin American countries.[1,2] Hepatitis E virus (HEV), a major cause of the disease is a 27-32nm virus that has features resembling those of "small round structured viruses" (e.g. Norwalk agent). Usually, for identification by electron microscopy, immune electron microscopy (IEM) is required to trap or aggregate the virus.[3]

Analysis of the HEV genome has identified 3 open reading frames (ORF) within the positive single-stranded RNA 7.5kb molecule.[4] ORF1 appears to code for nonstructural proteins, while ORF2 and ORF3 are considered to code for structural capsid-associated proteins. Recently, the antigenic nature of the ORF2 and 3 coded putative structural proteins was shown in various immunoassays by the application of synthetic peptides and recombinant proteins.[5] In this study, we tested whether antisera (guinea pig) to immunoreactive synthetic peptides encoded by ORF2 and ORF3 would identify HEV by IEM.

Traditional methods of IEM were used to trap HEV onto formvar-carbon grids from stool suspensions and to label virus with either biotinylated antisera to appropriate peptides or HEV antisera.[3,6] Biotinylated IgG (guinea pig, diluted 1:50 or 1:500) to peptides of HEV representing ORF2 (GP23.2) and ORF3 (GP6.2) were used as primary labels. Biotinylated IgG to hepatitis E (human + control) or biotinylated non-immune IgG (guinea pig, - control) were used as controls. 5nm Streptavidin gold was used as the secondary label. Specimens were stained with 2% phosphotungstic acid and viewed in a Philips 201 electron microscope.

HEV labeling results are shown in figure 1 and table 1. No labeling or only sparse nonspecific labeling occurred when biotinylated IgG was omitted (Fig. 1a) or when incubated with nonimmune biotinylated IgG (Fig. 1b). HEV particles readily labeled with human IgG against HEV (Fig. 1c,d) and also when incubated with biotinylated IgG against structural peptides (Fig. 1e-h). Immune IgG and IgG against structural peptides to HEV labeled significantly over the negative control treatments.

These observations provide evidence that HEV ORF2 and ORF3 have important antigenic roles in the disease and that antisera to peptides of these regions may be used in immunoassays to identify the virus and to associate antigenic epitotes with the virus.

Proc. Microscopy and Microanalysis 1995, edited by G.W. Bailey, M.H. Ellisman, R.A. Hennigar, and N.J. Zaluzec
Copyright © 1995 MSA. Published by Jones and Begell Publishing, 79 Madison Ave., New York, NY 10016

REFERENCES:

1. H.A. Fields et al., *J. Clin. Immun.*, 16(1993)215.
2. D.W. Bradley, *Rev. Med. Virol.*, 2(1992)19.
3. C.D. Humphrey et al., *J. Virol. Meth.*, 29(1990)177.
4. A.W. Tam et al., *Virology*, 191(1991)120.
5. Yu.E. Khudyakov et al., *Virology*, 194(1993)89.
6. K.G. Murti, in A.D. Hyatt, and B.T. Eaton, eds., *Immuno-Gold Electron Microscopy in Virus Diagnosis and Research*, Boca Raton, CRC Press. (1993)289.

Table 1: **IMMUNOLABELING OF HEPATITIS E STRUCTURAL PEPTIDES**

Dil	Human IgG	P[1]	IgG 23.2	P	IgG 6.2	P	NGP[2] IgG
1:50	141/255	<.001	82/256	<.001	73/264	<.001	16/258
1:500	81/256	<.001	24/256	<.003	15/256	<.200	7/258

Proportion of captured HEV labeled with biotinylated IgGs against HEV and peptides of HEV structural proteins compared with negative control. Comparisons of immunolabeling by the different IgG preparations were made by counting HEV particles that had one or more gold particles within 10nm of the virus particle.
1. P values were determined by Chi square statistics.
2. NGP = Normal guinea pig.

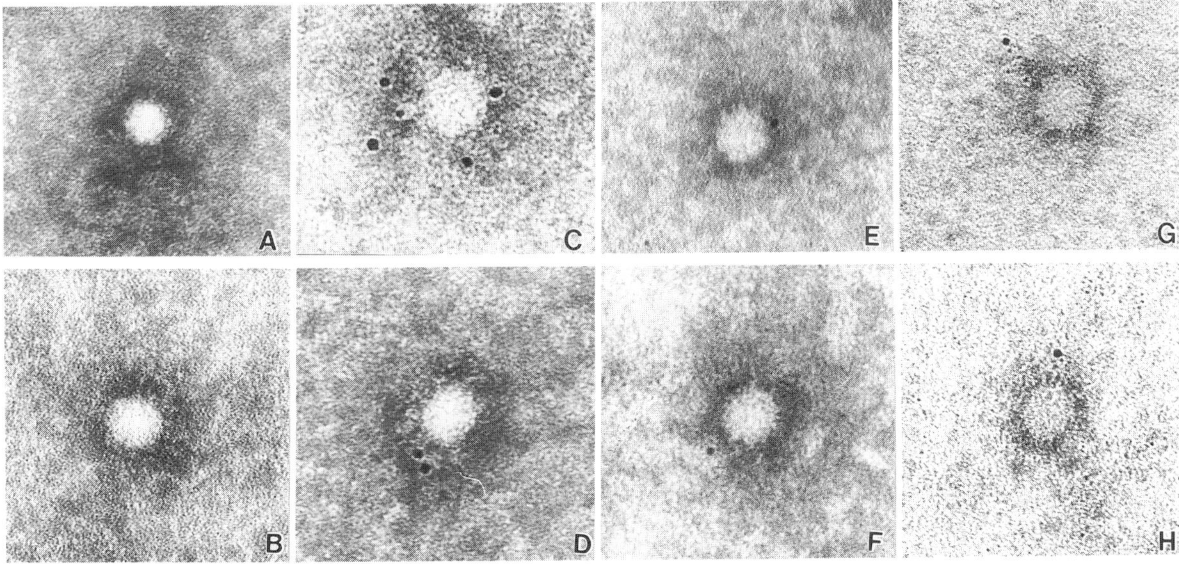

Fig.1--Electron micrographs of solid phase immune trapped HEV incubated with either non-immune biotinylated IgG (guinea pig), immune biotinylated IgG (human sera) against HEV, or biotinylated IgGs (guinea pig sera) against structural peptides to HEV, and labeled with diluted 5nm gold conjugated streptavidin (Alphaquest 1:5). A, no IgG; B, 1:50 diluted nonimmune guinea pig IgG; C, 1:50 diluted anti-HEV IgG; D, 1:500 diluted anti-HEV IgG; E, 1:50 diluted anti-peptide GP23.2 IgG; F, 1:500 diluted anti-peptide GP23.2 IgG; G, 1:50 diluted anti-peptide GP6.2 IgG; H, 1:500 diluted anti-peptide GP6.2 IgG.

STRUCTURAL ANALYSIS OF ENDOTHELIAL CELL MEDIATED PROLONGATION OF t-PA DRIVEN LYSIS OF AN "AGED CLOT"

W.G. Jerome,[*] S. Handt,[**] and R.R. Hantgan[*]

[*]Bowman Gray School of Medicine, Winston-Salem NC and [**]Technical University of Aachen, Germany.

In an effort to reestablish blood flow in blocked coronary arteries, heart attack victims are now routinely treated with thrombolytic (clot busting) drugs. However, thrombolytic therapy fails to restore sufficient flow in approximately 20% of cases. The reasons for these failures are undoubtedly multifactorial and difficult to study *in vivo*. We have developed an "in vitro" model blood vessel which allows controlled study of factors influencing thrombolysis. In the studies reported here, we investigated the 3-D distribution of HUVEC secreted PAI-1 within the clot using immuno-gold cytochemistry and intermediate voltage electron microscopy (IVEM).

Human umbilical vein endothelial cells (HUVEC) were grown on the inside surface of fibronectin-coated glass tubes. The cells are stimulated for 4 hours with 1000 U/ml of TNF-α. Clot formation was initiated by addition of 2.3 μM fibrinogen, 100 nM plasminogen, and then 2 nM thrombin in HEPES buffered Gey's salts + 0.5% human serum albumin. After aging the clot for up to 4 hours (to mimic the time delay between heart attack and treatment), 0.5 nM rt-PA (thrombolytic agent) was perfused through the clot for 10 min. at 5 ml/hr. Formation and dissolution of the clot were monitored by laser light scattering (Figure 1).

Stimulated HUVEC prolonged lysis 3.6 fold. PAI-1 is secreted by HUVEC and is a potent inhibitor of t-PA. A genetically constructed PAI-1 resistant t-PA showed no delay in lysis in the presence of HUVEC, suggesting secretion of PAI-1 is the mechanism by which HUVEC prolong lysis. Immuno-gold IVEM demonstrated PAI-1 was specifically bound to individual strands of fibrin throughout the clot (Figure 2). Immunocytochemical controls verified that less than 1% of staining was non-specific.

To quantify PAI-1 distribution we used Tuned Aperture Computed Tomography (TACT), our computer program which converts IVEM 3-D volumes into planar slices for analysis and quantitation (Figure 1). Although PAI-1 permeated the entire clot, quantitation of labeling revealed that PAI-1 was 2.6 times more concentrated near the periphery in the vicinity of endothelial cells than in more central areas away from the endothelium. This distribution fit that predicted for time dependent diffusion. This suggests that the time dependent increase in lytic resistance is, at least in part, the result of increased diffusion of PAI-1 into the center of the clot.

Proc. Microscopy and Microanalysis 1995, edited by G.W. Bailey, M.H. Ellisman, R.A. Hennigar, and N.J. Zaluzec

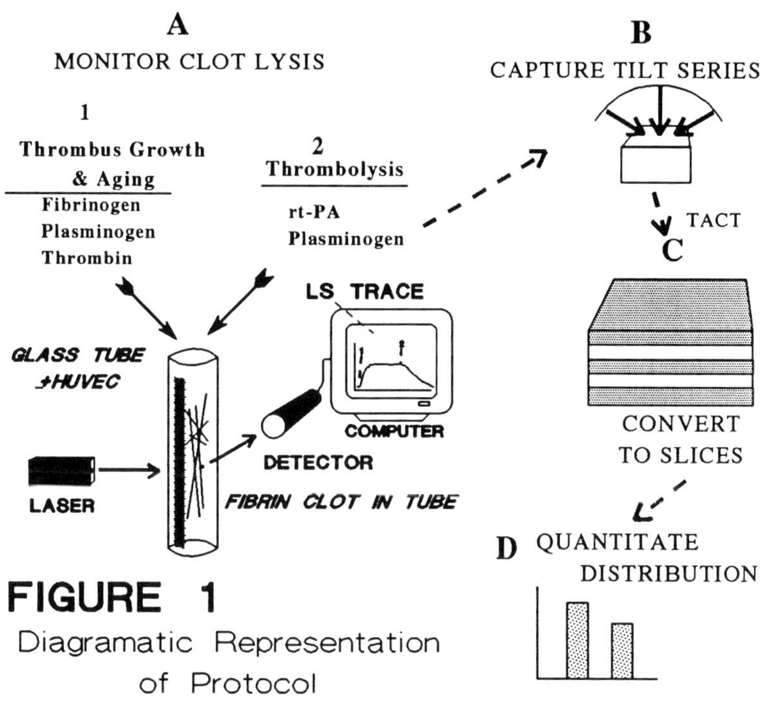

FIGURE 1

Diagramatic Representation
of Protocol

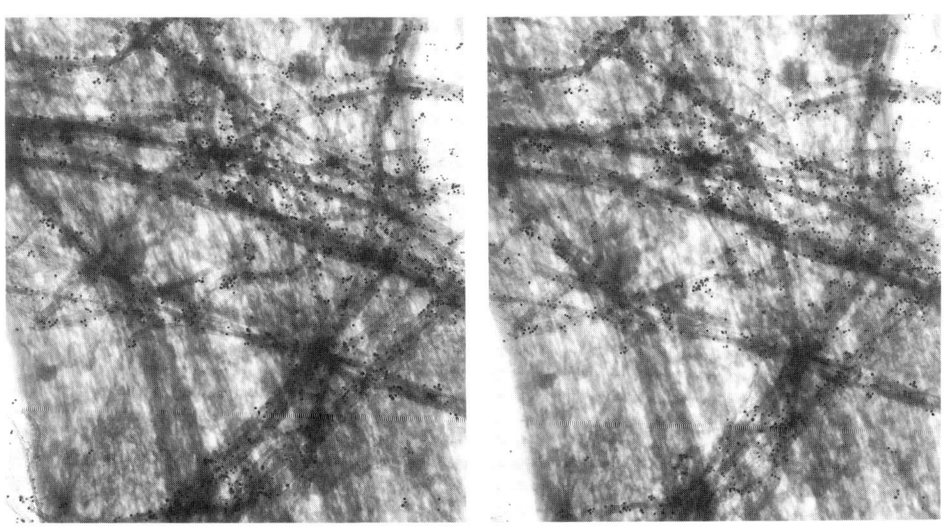

Figure 2. Stereopair of fibrin clot lying on top of endothelial cell. Immunogold stained PAI-1 is specifically bound to fibrin fibers radiating up from surface of endothelial cell. Magnification= 25,000; gold particle diameter= 15nm.

SURGICAL REPAIR OF LONG BONE DEFECTS OR TRAUMA BY GUIDED
TISSUE ENGINEERING WITH COMPOSITE IMPLANTS OF POROUS
BIOACTIVE GLASS-CERAMIC (A-W·GC) PARTICLES BOUND BY A
PLASTER/CaO/P_2O_5 CEMENT CONTAINED BY A RESORBABLE VICRYL™
MESH

J.S. Hanker*,** L.C. Hanker* and B.L. Giammara***

*Dental Research Center and **Dept. of Biomedical Engineering, The University
of North Carolina, Chapel Hill, NC 27599-7455
***Center for Clinical Investigation, Medical College of Ohio, Toledo, OH 43699

One of the biggest problems associated with the repair of trauma in long
bones is their requirement to withstand greater mechanical stress than any other
bones of the body (1,2). After the success we achieved with composite
hydroxylapatite (HA)/plaster(PP) in craniofacial bone repair in approximately 200
animals (cats and rats) and 300 humans in our laboratory, clinics and surgeries, it
was felt that an area which could perhaps benefit from our experience in bone
repair might be orthopedic surgery. When first informed of our desire to apply our
biomaterials interest and experience to long bone trauma patients, others expected
little success with these biomaterials that were so successful in craniofacial
surgery.

We felt, however, that success might be achieved for long bone repair with
plaster of Paris ($CaSO_4 \cdot \frac{1}{2}H_2O$) and/or calcium phosphate. The feeling of my
laboratory received some support from Larry Hench's 1988 article (3). In this
article he pointed out that bioactive ceramics such as hydroxylapatite (HA)
possibly could bond to bone. But even if HA couldn't, Kokubo's 1990 article (4)
pointed out that the Ca, P layer, formed on the surface of his high strength
bioactive glass-ceramic A-W, bonds with living bone in a short period (5). This
glass-ceramic A-W contains crystalline apatite and wollastonite in a MgO-CaO-
SiO_2-P_2O_5 glassy matrix. The glass ceramic A-W·GC maintains high mechanical
strength for an extended period of time, even under load-bearing conditions; it also
forms a strong chemical bond with bone tissue in a short period of time.

The metal alloys employed for the internal reinforcement of long bones are
relatively strong, ductile and inert; but they do not provide opportunities for
chemical bonding to bone unless the surfaces of the alloys are roughened. When
this is done, the textured surfaces can then be adhered to by compounds such as
calcium hydroxide or hydroxylapatite to promote the adherence of bone. The
biologic calcification sites can then readily be seen by light or electron microscopy
(Figs. 1, 2) with our silver methenamine variation of the von Kossa stain (6).

Proc. Microscopy and Microanalysis 1995, edited by G.W. Bailey, M.H. Ellisman, R.A. Hennigar, and N.J. Zaluzec
Copyright © 1995 MSA. Published by Jones and Begell Publishing, 79 Madison Ave., New York, NY 10016

Fig. 1

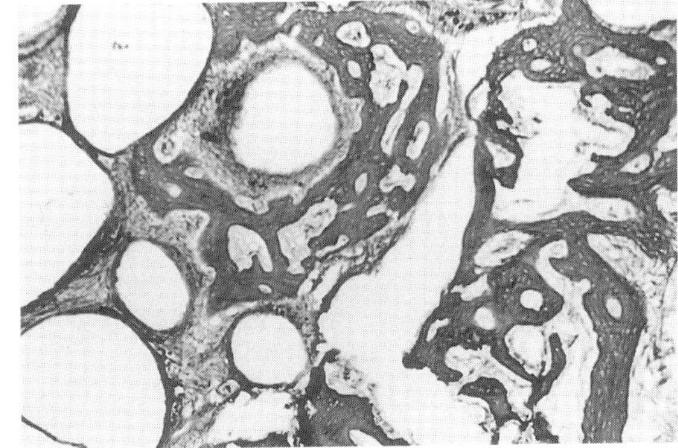

Fig. 2

Fig. 1. Undecalcified section of femur of a 20-day-old mouse that had been implanted in this area for 15 days with a composite of porous bioactive glass-ceramic (A- W·GC) particles bound by a plastered CaO/P_2O_5 cement contained by a Vicryl™ mesh. This section was stained by our 2 min. variation of the 30 min. von Kossa silver nitrate stain for biologic calcification which requires silver methenamine solution for only 1 min. at 2450 MHz [Proc. 49th Ann. Meet. EMSA, 300-301 (1991)].

Fig. 2. A much lower magnification of an undecalcified section of the femur of a 50-day-old mouse that had been similarly implanted for 45 days. This section was also stained by our 1 min. silver methenamine stain [Proc. 49th Ann. Meet. EMSA, 300-301 (1991)].

REFERENCES

1.   G. Daculsi et al. J. Biomed. Mater. Res. <u>24</u>, 379-396 (1990).

2.   E.E. Johnson et al. Clin. Orthop. Rel. Res. <u>277</u>, 229-237 (1992).

3.   L.L. Hench, Ann. N.Y. Acad. Sci. <u>523</u>, 54-71 (1988).

4.   T. Kokubo et al. J. Biomed. Mater. Res. <u>24</u>, 331-343 (1990).

5.   T. Nakamura et al. J. Biomed. Mater. Res. <u>19</u>, 685-698 (1985).

6.   J. Hanker and B. Giammara, Proc. 49th Ann. Meet. Electr. Microsc. Soc. Amer. 300 (1991).

7.   Mr. Donald A. Randolph, Manager-Business Development, United States Gypsum Company, Chicago, IL since 1984 has worked together with us and supported this work.

GROWTH CONE DYNAMICS OF EMBRYONIC GRASSHOPPER PIONEER NEURONS <u>IN SITU</u>.

Tim P. O'Connor

Department of Anatomy,
University of British Columbia,
2177 Wesbrook Mall,
Vancouver, B. C.
Canada
V6T 1Z3

During development of the nervous system, neurons extend axons over relatively long distances to contact their targets. A variety of molecules in the extracellular environment are instrumental in guiding a neuronal process. The motile tip of the process, the growth cone, senses and transduces this guidance information, resulting in a local reorganization and consolidation of the cytoskeleton. Although much work has been dedicated to isolating the molecules that guide a neuronal growth cone, relatively little is known about the dynamic processes that occur when a growth cone turns in response to guidance information. Recently, a number of biological systems have been developed that enable time lapse imaging of growth cones as they extend axons <u>in situ</u>. One of these systems is the embryonic grasshopper limb fillet.

In the grasshopper embryo, a pair of sibling neurons, named the Ti1 pioneers, are the first neurons to extend axons toward the central nervous system (CNS). Born at the distal tip of the limb bud, they extend axons proximally along a stereotyped route to the CNS. As the growth cones migrate they contact a complex environment that includes various kinds of epithelial cells, basal lamina, and pre-axonogenesis neurons called guidepost cells. This pathway in the embryonic limb bud can be rendered accessible for manipulation using an opened epithelial fillet preparation. This preparation exposes the Ti1 cell bodies, but preserves the guidance information present in the limb. In addition, the preparation allows the growth cones to be imaged as they migrate [1].

Previous work using this system has shown that finger-like microspikes (filopodia), that extend from the growth cone, play an instrumental role in steering the extending axon. Observations of Ti1 growth cone dynamics as they extend <u>in situ,</u> have indicated that a single filopodial contact with a high affinity substrate (guidepost cell) is sufficient to reorient the entire growth cone [1]. In addition, filopodia have been shown to be necessary for correct growth cone steering. Filopodia, which are tubular actin-based structures, can be eliminated with addition of cytochalasins to the fillet preparation. In the absence of filopodia (after cytochalasin addition) Ti1 pioneer growth cones do not steer correctly [2].

One model of growth cone steering indicates that differential adhesion of filopodia with the surrounding substrate may direct growth cone steering. This would suggest that filopodia extending in the direction a growth cone is turning may have relatively stronger adhesive contacts

Proc. Microscopy and Microanalysis 1995, edited by G.W. Bailey, M.H. Ellisman, R.A. Hennigar, and N.J. Zaluzec

with the surrounding environment than filopodia extending in other directions. The Ti1 pioneer pathway is an ideal model system to test this hypothesis. Pioneer growth cones make stereotyped turns at precise locations in the limb allowing for tests of filopodia adhesion while observing a growth cone as it changes its direction of growth (turns).

To image growth cones as they extend in their _in situ_ environment, Ti1 cell bodies were labeled with the fluorescent lipophillic dyes DiI or DiO (Molecular Probes). Cells were contact-labeled with fine crystals of dye previously dried on the tip of a micropipette. All labeling and injections were done using a micromanipulator (Narishige) mounted on an inverted Nikon compound microscope (Diaphot). The dyes readily diffused throughout the membrane. The majority of images were taken with a cooled charge-coupled device (CCD) camera system (Photometrics, Tucson, AZ) at the Univ. Of California, Berkeley. Images were relayed using conventional optics onto a 1320 x 1024 pixel Kodak chip (KAF-1400), digitized, and stored using conventional computer hardware. The chip, light path shutter and stage focal position were controlled by Perceptics BioVision imaging software (Knoxville, TN).

The strength of adhesive contacts between filopodia and the surrounding environment were tested by bathing pioneer growth cones in cytochalasin D (0.01 - 10.0 ug/ml). Some pioneer neurons were injected with rhodamine labeled phalloidin (Molecular Probes, 2.5U/ul; phalloidin preferentially labels actin filaments) to observe the disassembly of the actin cytoskeleton. Despite a rapid loss of the actin cytoskeleton in all of the filopodia, individual filopodia showed varying time courses of retraction to the growth cone that ranged from seconds to hours. Using these time courses as indices of adhesive strength with the surrounding environment, pioneer growth cone filopodia were analyzed for a correlation between filopodia adhesion and growth cone steering.

References

1. O'Connor TP, Duerr JS and Bentley D (1990) J. Neurosci. 10, 3935-3946.

2. Bentley D and Toroian-Raymond A (1986) Nature 323: 712-715.

USE OF CALCIUM IMAGING TECHNOLOGIES TO DISSECT MECHANISMS UNDERLYING NEURONAL GROWTH CONE RESPONSES TO ENVIRONMENTAL CUES

C.V. Williams, S.B. Kater

Department of Anatomy and Neurobiology, Colorado State University, Fort Collins, CO 80523

Since calcium is a key second messenger in both the developmental formation and adult function of the nervous system, the ability to rapidly image changes in this molecule has added greatly to our understanding of how development of the nervous system is regulated. The nervous system is comprised of billions of neurons and glial cells that establish characteristic patterns of connections during development. Neurons extend processes that often must grow long distances to establish appropriate synaptic connections. Neurons perform a pathfinding behavior largely via the highly dynamic behavior of the neuronal growth cone at the distal tip of elongating processes. The motile behavior characteristic of growth cones allows the growth cone to survey the local environment, read local cues and respond to those cues with a change in behavior. A variety of cues are now known to direct growth cones (e.g. electrical activity, depolarization, growth factors, mechanical factors, neurotransmitters, substrate factors). This collection of factors includes both growth promoting and growth inhibitory influences. Some environmental cues induce subtle changes in outgrowth (i.e. navigational changes such as turning, branching, stalling). Other cues cause more dramatic effects (i.e. growth cone collapse). We now know that the effects of many of these cues are mediated by calcium signals within the growth cone.[1]

A major question in neurobiology is how the growth cone "reads" environmental signals and how these signals are transduced to behavioral responses. Growth cone filopodia, the finger-like protrusions emanating from the leading edge of the growth cone, possess many of the necessary cellular systems to respond to environmental cues.[2] Filopodia serve as antennae that sample the environment surrounding the growth cone. Signals received at individual filopodia can be transmitted back to the growth cone, providing guidance information to the growth cone. We have been dissecting the intracellular mechanisms underlying filopodial-induced growth cone behaviors. We have found that both filopodia and growth cones can independently utilize the intracellular molecule, calcium, in convergent or divergent second messenger signalling cascades.

Fluorescent dyes, such as fura-2, have made possible the measurement and subcellular localization of free intracellular calcium in living growth cones and filopodia.[3,4] Membrane-permeable fura-2 AM is taken up by cultured cells, hydrolysed and trapped internally. Dual wavelength excitation of the dye provides a reliable means for obtaining relative levels of intracellular calcium.[4] Changes in fluorescence within the cell reflect changes in intracellular calcium levels (nM to mM range). Using these techniques, calcium levels in living neurons can be measured and simultaneously correlated with changes in motility and navigational behaviors.

We have utilized calcium imaging technologies to investigate the role of intracellular calcium in growth cone guidance. We have found that a local change in intracellular calcium in growth cone filopodia can result in very different growth cone behaviors. Quantitative comparisons of calcium

Proc. Microscopy and Microanalysis 1995, edited by G.W. Bailey, M.H. Ellisman, R.A. Hennigar, and N.J. Zaluzec

rises in filopodia and in growth cones encountering either inhibitory or growth promoting molecules has revealed clear differences. Filopodial contact with inhibitory molecules (e.g. NI-35) can produce a large calcium rise in the filopodia followed by a massive rise in intracellular calcium in the growth cone (order of magnitude).[5] This is followed by growth cone collapse. Such induced calcium rises are global in nature, occurring throughout the entire growth cone and often spreading into the neurite. In contrast, filopodial contact with beads coated with laminin, a growth promoting cue, leads to a nearly undetectable calcium rise in the filopodia and much more subtle changes in calcium (2-fold rise) in the growth cone.[6] This is followed by accelerated growth cone advance. Intracellular calcium rises induced by laminin are much more restricted and localized to discrete subregions of the growth cone itself. As yet another example, stimulation of the neurite shaft behind the growth cone causes more intermediate alterations in intracellular calcium (5-fold rise), followed by the formation of a new neurite.[7] These findings suggest that the location and magnitude of the calcium rise provide the major difference by which the same second messenger can result in dramatic alterations in neuronal behavior. The illustration below summarizes these findings.

References

1. S.B. Kater and L.R. Mills, *J Neurosci* 11 (1991) 891-899.
2. R.W. Davenport et al., *Nature* 361 (1993) 721-724.
3. R.Y. Tsien, *TINS* 11 (1988) 419-424.
4. D. Grynkiewicz et al., *J Biol Chem* 260 (1985) 3440-3450.
5. C.E. Bandtlow et al., *Science* 259 (1993) 80-83.
6. T.B. Kuhn et al., *Neuron* (1995) submitted.
7. C.V. Williams et al., *Journal of Neurobiology* (1995) in press.
8. This work was supported by NIH Grants NS24683 and NS28323.

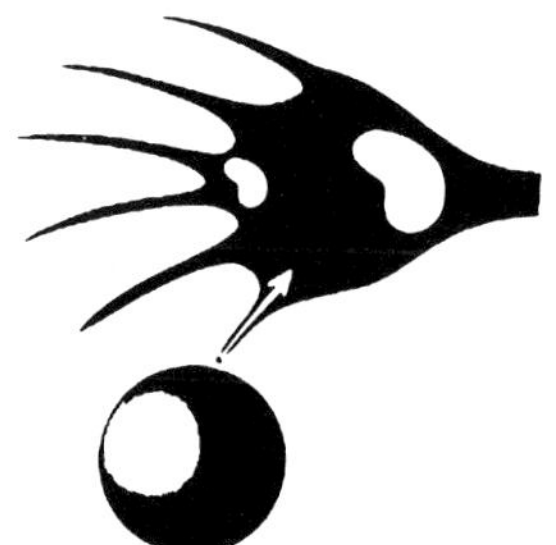

Different growth cone behaviors can be regulated by a single intracellular second messenger molecule. *Left* Filopodial signals may not affect the growth cone. *Middle* Filopodial signals could lead to local calcium events in the growth cone and in local growth cone behaviors. *Right* Filopodial signals could lead to global calcium and behavioral responses in growth cones.

DYNAMICS OF DENDRITE DEVELOPMENT VISUALIZED BY TIME-LAPSE CONFOCAL IMAGING IN BRAIN SLICES

M.E. Dailey and S.J. Smith

Department of Molecular & Cellular Physiology, Stanford University Medical School, Stanford, CA 94305

In the mammalian CNS, dendritic neuronal branches typically are studded with numerous short (<3μm), lateral protrusions called "spines". Such spines are the primary sites of excitatory synaptic input, and changes in spine morphology are thought to play important roles in plasticity of synaptic function in both the developing and adult animal.[1] However, dynamic changes in spine number and structure are not easily determined by electron microscopy, and the small size of spines has made them difficult to study by conventional light microscopy. Recent advances in vital fluorescent staining[2] and high resolution confocal imaging in tissue slices[3] now afford the possibility of assessing changes in morphology of individual spines on single dendrite branches over time.[4]

To investigate the dynamics and plasticity of dendritic structure during development, vital fluorescent staining and time-lapse confocal imaging methods were applied to preparations of live brain slices from developing rat.[3] Tissue slices were prepared[5,6] from hippocampus of neonatal rat (postnatal day 2–7) and cultured for variable periods of time (hours to weeks). Such *in vitro* slices support growth and development of axonal and dendritic arbors as well as formation of synaptic connections.[5,6] To label pyramidal neuron dendrites, a solution (0.5% in dimethylformamide) of a fluorescent lipid, $DiIC_{18}(3)$ (Molecular Probes), was injected directly into slices, or slices were pierced with glass micropipets coated with dye crystals (Fig. 1). Within a few hours, the surface membranes of numerous pyramidal cells were labeled, revealing elaborate dendritic arbors (Fig. 2). Confocal imaging permitted high resolution views of dendritic microstructure up to 100 μm into tissue. Many dendrites were covered with lateral filopodia and spine-like protrusions of variable shapes and sizes. To follow changes in the number and structure of individual spines over time, live slices were maintained on the microscope stage for up to 24 h.[3,7] Images were captured at 5min intervals. Time-lapse sequences indicated that, at early times in culture [i.e., 1 day *in vitro* (d.i.v.)], many lateral dendritic protrusions had a filopodial form (>2 μm) and were fleeting (limetime: ~10 min), whereas by two weeks in culture dendrites were studded with short (<2 μm), relatively-stable (>22 h) spine-like protuberances. During an intermediate stage (~7 d.i.v.), individual spine-like protrusions were found to undergo very dynamic changes in shape while persisting for many hours (Figs. 3 & 4). These data reveal a transition from a highly dynamic to a more stable dendritic structure coincident with dendritic differentiation and synaptogenesis. Such observations also underscore the feasibility of examining plasticity in dendritic structure, at high spatial and temporal resolution, within live brain slice preparations.[8]

References

1. K. Harris and S.B. Kater, *Annu. Rev. Neurosci.* 17(1994)341.
2. M. Honig and R.I. Hume, *J. Cell Biol.* 103(1986)171.
3. S.J. Smith et al., in *XXIII Symposia Medica Hoechst, Biology of Memory* (L. Squire & E. Lindenlaub, eds.) Stuttgart: Schattauer (1990)49.
4. Hosokawa et al., *NeuroReport* 3(1992)477.
5. B.H. Gahwiler, *Neuroscience* 11(1984)751.
6. Dailey et al., *J. Neurosci.* 14(1994)1060.
7. M. Terasaki and M.E. Dailey, in *Handbook of Biological Confocal Microscopy, 2nd Edition* (J.B. Pawley, ed.) New York: Plenum (1995).
8. Supported by NIH grants NS09027 (M.E.D) & NS28587 (S.J.S), & NIMH grant MH48108 (Silvio Conte).

Proc. Microscopy and Microanalysis 1995, edited by G.W. Bailey, M.H. Ellisman, R.A. Hennigar, and N.J. Zaluzec
Copyright © 1995 MSA. Published by Jones and Begell Publishing, 79 Madison Ave., New York, NY 10016

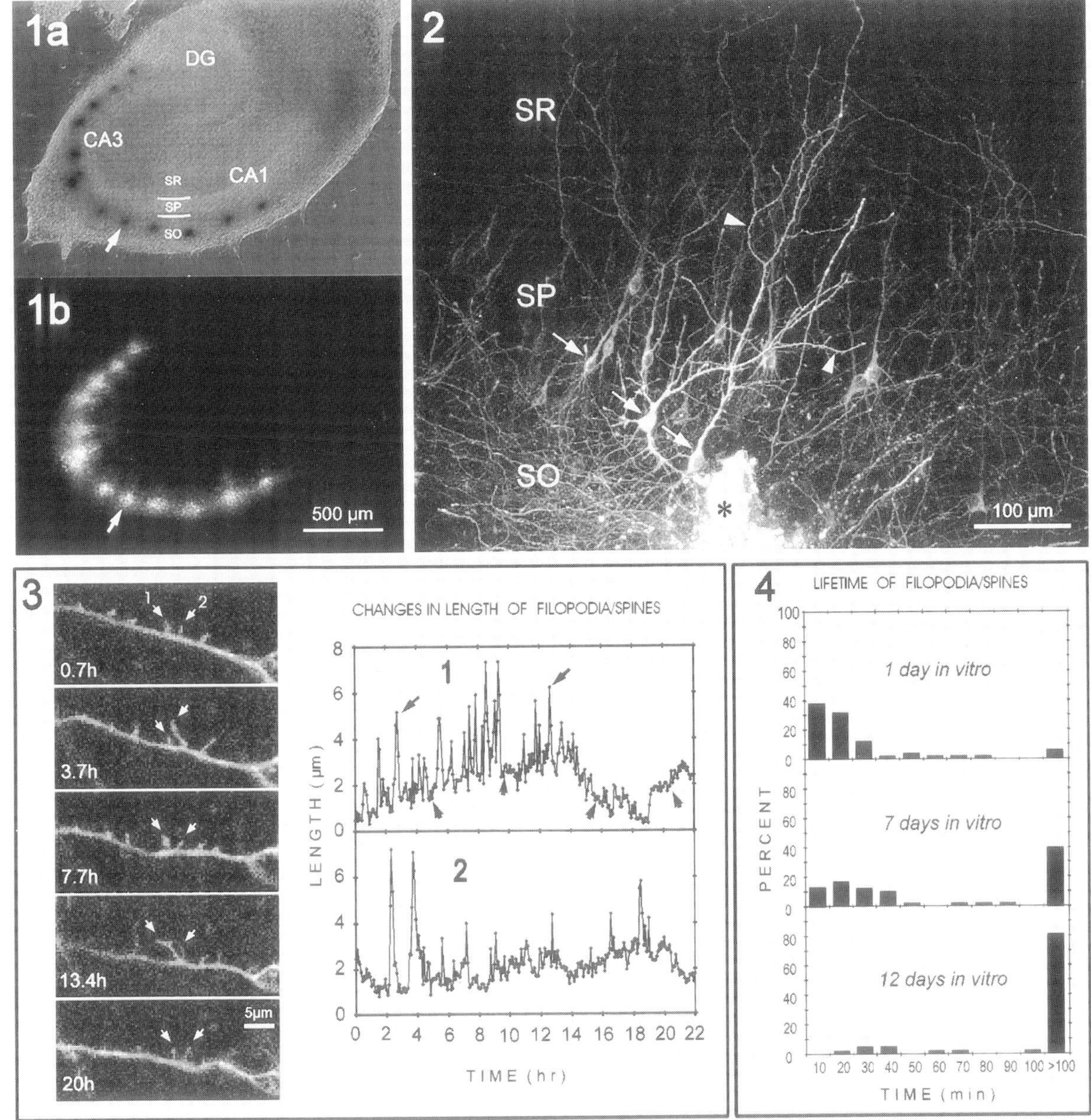

FIG. 1—Fluorescence labeling of live tissue slices for confocal observation. Corresponding brightfield (a) and fluorescence (b) images show sites of dye injection (arrow). DG, *dentate gyrus*. SR, *stratum radiatum*. SP, *stratum pyramidale*. SO, *stratum oriens*. Slice from a P3 rat, 1 day *in vitro* (d.i.v.).

FIG. 2.—Higher magnification view of DiI-labeled pyramidal cell bodies (arrows) and apical dendrites (arrowheads) near site of labeling (*) in area CA1. Slice from a P5 rat, 5 d.i.v.

FIG. 3.—Dynamic changes in length and shape of lateral filopodia/spines (arrows) on a differentiating pyramidal cell dendrite (slice from P5 rat, 7 d.i.v.). Images were selected from a 22h sequence. Lengths of the two filopodia/spines indicated (1,2) are plotted. Note transient length excursions (arrows) from relatively stable, persisting bases (arrowheads).

FIG. 4.—Lifetimes of filopodia/spines on pyramidal cells in acute (1 d.i.v.) and cultured (7 & 12 d.i.v.) tissue slices. The fraction of long-lived (>100 min) filopodia/spines increased substantially over time *in vitro*, coincident with synaptogenesis. The long-lived filopodia/spines may be sites of synaptic contact.

TIME-LAPSE CONFOCAL CALCIUM IMAGING IN AN INTACT UNROLLED HIPPOCAMPUS PREPARATION

A. H. Cornell-Bell, R. M. Villalba, R.H. Selinfreund, B. D. Stein, J.L. Cornell and L.A. Riblet

Lion Imaging Inc. Westbrook, CT 06498

The excitatory neurotransmitter glutamate (100 uM) induces intracellular calcium transients in cultured hippocampal astrocytes that can be imaged using the calcium indicator (Fluo3AM) and time-lapse microscopy [1]. In response to glutamate (Fig. 1A), cultured astrocytes exhibit distinct patterns of intracellular Ca^{2+} oscillations and long-distance intercellular waves. Two distinct types of intercellular Ca^{2+} waves are attributed to excitation by different agonists of the glutamate receptor subtypes. A long-distance regenerative intercellular wave is induced by the ionotropic glutamate receptor, kainate (Fig. 1B). This is a true wave lasting for 50-125 sec with a constant velocity of 10-20 um/sec. This wave requires extracellular Ca^{2+} and Na^+ and is driven by the Na^+/Ca^{2+} exchanger [2]. A fast Ca^{2+} wave which travels at speeds from 10 to 200 um/sec is dependent upon the metabotropic glutamate receptor, is inducible by t-ACPD and is dependent upon cytoplasmic release of Ca^{2+} regulated by IP3 (Fig. 1C). This wave is not dependent upon extracellular Ca^{2+} and is stopped by MCPG, a specific inhibitor of IP3-mediated intracellular Ca^{2+} release. This Ca^{2+} transient has a variable velocity, slowing down as it crosses cell membranes and speeding up as it traverses the cytoplasm.

The physiologic relevance of Ca^{2+} waves still remains elusive due to the need to study these responses in the intact brain. These signals most likely affect neurons and astrocytes and travel in three dimensions *in situ*. Astrocytes have been studied in the roller tube- slice preparation [3] and in the acute *in vitro* slice, but tissue damage still remains a major detriment in these studies. The intact unrolled hippocampus allowed us to study astrocyte Ca^{2+} responses in a preparation that is relatively free of damage. Neonatal rat hippocampus was removed from the cortex, the presubiculum was separated from the dentate gyrus and the hippocampus was "unrolled" without slicing to preserve the morphology. Following Fluo3AM staining (40 uM with Pluronic in ACSF) the unrolled hippocampus was imaged using a BioRad MRC600 CSLM. Cells within the 300-400 um thick preparation responded to glutamate addition with intracellular Ca^{2+} oscillations. Ca^{2+} transients traveled along cell processes and between cell bodies within the same layer. Initial intercellular Ca^{2+} waves start in a region of hippocampus and propagate rapidly across this tissue (Fig 2). Fast Ca^{2+} waves spread radially from an initiation site involving 25-50 cells and then retract back [2]. The intact unrolled hippocampus is ideal for deciphering the physiologic role of these Ca^{2+} transients.

1. Cornell-Bell, AH, Finkbeiner, S., Cooper, MS., and Smith, S.J (1990) Science, 247: 470-473.
2. Cornell-Bell, A.H. and Finkbeiner, S.M. (1991) Cell Calcium 12: 185-204.
3. Dani, J.W., Chernjavsky, A. and Smith, S.J (1992) Neuron 8: 429-440.

Proc. Microscopy and Microanalysis 1995, edited by G.W. Bailey, M.H. Ellisman, R.A. Hennigar, and N.J. Zaluzec
Copyright © 1995 MSA. Published by Jones and Begell Publishing, 79 Madison Ave., New York, NY 10016

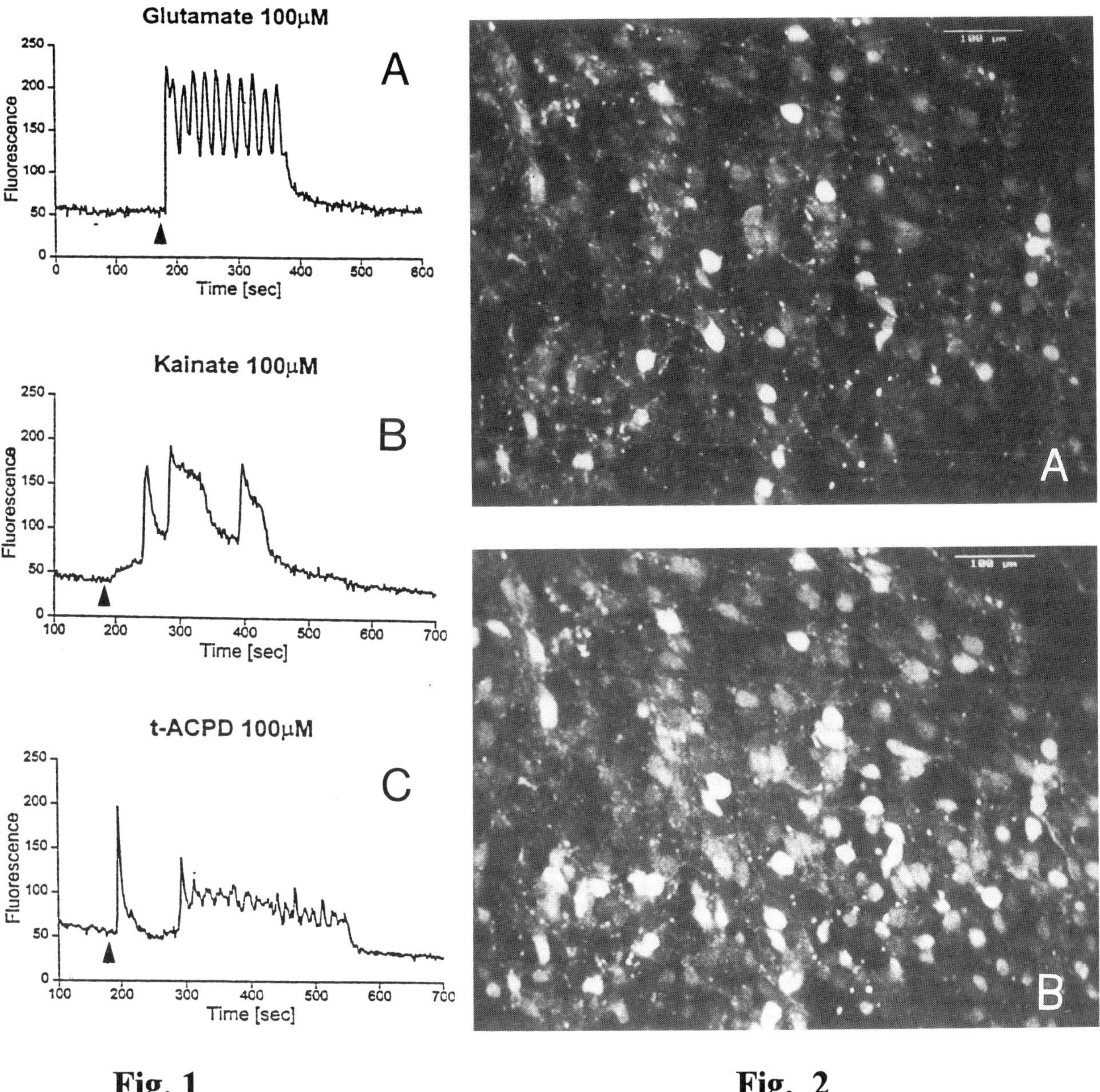

Fig. 1   **Fig. 2**

Fig. 1. A. Astrocytes exhibit intracellular Ca^{2+} oscillations when superfused with glutamate. The glutamate agonist kainate (Fig. 1B) induces an iontropic intercellular Ca^{2+} wave that last for 50-100 sec. The metabotropic agonist, t-ACPD (Fig. 1C) induces a fast intercellular Ca^{2+} wave and very rapid intracellular oscillations.

Fig 2. The intact unrolled hippocampus is loaded with the Ca^{2+} dye, Fluo3AM and imaged on the CSLM. Using time-lapse programs this tissue is imaged before agonists addition (Fig2A) and the 20 sec following addition of agonists (Fig. 2B).

OPTICAL IMAGING OF BRAIN SLICE PREPARATIONS USING VOLTAGE SENSTIVE DYES

D.A. Nelson and L.C. Katz

Department of Neurobiology, Duke University Medical Center, Durham, NC 27710

Fast responding, biological membrane-soluble voltage sensitive dyes are probes which make it possible to directly monitor the electrical activity of neuronal tissues with both high temporal and high spatial resolution[1]. High temporal resolution is desirable because of the millisecond time scale on which neuronal events occur. High spatial resolution is desirable to better reveal the complex interactions between different regions of neural circuits.

Taking full advantage of the promise of voltage sensitive dyes can be both difficult and expensive. Tissues stained with these dyes undergo very small (often less than one part in a thousand) relative changes in fluorescence in response to changes in transmembrane potential. The signal-to-noise ratio is similarly small for these responses. Finally, the responses are very fast. Most major components of the signal rise to their maximum and then disappear within 15 milliseconds of stimulation (figure 1).

Because standard video cameras are inadequate for recording the very small and very fast optical signals transduced by voltage sensitive dyes, specialized detection equipment is required. This entails a high cost per pixel, often limiting the practial number of detector elements to a few hundred or even less. Additionally, digitizing and processing voltage sensitive dye signals from multiple detector elements can require considerable computing power. In answer to these problems, we have developed a technique for optically recording from living brain slices which uses a small (10x10), relatively inexpensive detector array; relies on off-the-shelf electronics for the rest of the apparatus; and yet allows recordings of slice activity which consist of thousands of pixels.

Typically, optical recordings are aquired over the course of several trials which are then averaged to increase the signal-to-noise ratio[1, 2]. We also aquire over several trials, but between each trial, a motorized, computer controlled x-y stage is used to change the region of the slice from which our 10x10 detector array samples activity . We then montage recordings taken from different places in the slice into a single large data array, which now contains the overall pattern of activity evoked by the stimulus. We typically average together five separate passes over the whole recording area. Each pass is offset by 1-5 pixel locations to help eliminate montaging artifacts.

Thus, a typical recording requires from 100 to 200 trials to be aquired, and may contain from 2000 to 4000 pixels. Total aquisition time for a recording ranges from 10 to 20 minutes. When we record through a 25x objective lens with our equipment, each pixel represents light sampled from an area of the slice 45μm x 45μm in dimension. This allows us to rountinely image slice activity extending over regions of more than 5 square millimeters.

Software is an integral part of aquiring and analyzing optical recording data. We have developed a National Instruments LabView based application for controlling all aspects of data aquisition, including stimulus control, automated montaging, and pre-processing of data. This application also allows on-line monitoring of data aquisition.

Once the data have been aquired, they must be analyzed. Optical recording data have often been treated as collections of individual voltage traces[2, 3]. Processing, analyzing and interpreting optical recordings is greatly facilitated by treating them instead as sequential images. This allows the use of flexible image analysis software packages such as Spyglass Transform and NIH Image not merely to filter the data, but also to extract latency and maximum amplitude information, quickly examine large data sets, and rapidly generate frame-by-frame comparisons between recordings.

Finally, we have also used computer graphics to combine optical recording with anatomical techniques. For some experiments, we determined which laminae correspond to which elements of an

Proc. Microscopy and Microanalysis 1995, edited by G.W. Bailey, M.H. Ellisman, R.A. Hennigar, and N.J. Zaluzec
Copyright © 1995 MSA. Published by Jones and Begell Publishing, 79 Madison Ave., New York, NY 10016

optical response map by superimposing optical recording data on a videomicrograph of the Nissl stained slice (figure 1). In another set of experiments, we have used localized biocytin injections to determine the pattern of axonal radiation from an electrical stimulus site, then compared this with the optical recording results to show that the two are well correlated in spatial extent. Combining techniques in this way promises to open up new avenues of investigation utilizing voltage sensitive dye-based optical recording.[4]

1.   A. Grinvald et al., *Physiol Rev* 68(1988)1285.
2.   A. Grinvald et al., *J Physiol* 333(1982)269.
3.   A. Grinvald et al., *J Neurosci* 14(1994) 2545.
4.   This research was supported by NIH grant EY07690 (L.C.K.) and a Howard Hughes Predoctoral Fellowship (D.A.N.)

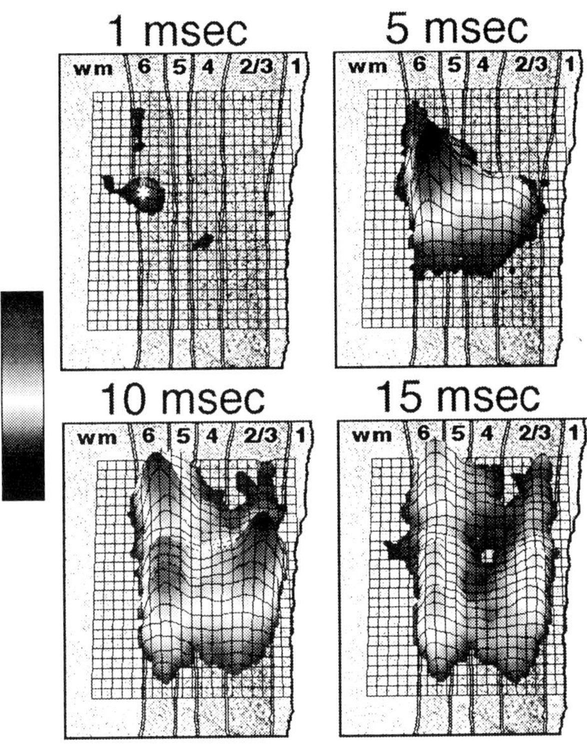

FIG. 1—Time Course of the Pattern of Activation Resulting from Electrical Stimulation Near the Layer 6/White Matter Boundary. Four frames from optical recordings made of slices of mature ferret visual cortex.Time in milliseconds after stimulus delivery is indicated along the top. Surface plots of thresholded, filtered and spatially interpolated optical recording data have been superimposed on a Nissl counterstained section of the slice from which the recording data was taken. Cytoarchitectonically defined layers have been labeled and are outlined in light grey. The stimulus site is marked by an asterisks in the first frame. Each square of the surface plot is 90 μm on a side. The color scale bars shows linear relative amplitude.

PREDICTING SPATIAL ACTIVITY PATTERNS IN A NEURAL MAP FROM A PROBABILISTIC ATLAS OF 3D RECONSTRUCTED NEURONS

G. Jacobs and F. Theunissen

Department of Molecular and Cell Biology, University of California, Berkeley, CA. 94720

In order to understand how the algorithms underlying neural computation are implemented within any neural system, it is necessary to understand details of the anatomy, physiology and global organization of the neurons from which the system is constructed. Information is represented in neural systems by patterns of activity that vary in both their spatial extent and in the time domain. One of the great challenges to microscopists is to devise methods for imaging these patterns of activity and to correlate them with the underlying neuroanatomy and physiology. We have addressed this problem by using a combination of three dimensional reconstruction techniques, quantitative analysis and computer visualization techniques to build a probabilistic atlas of a neural map in an insect sensory system. The principal goal of this study was to derive a quantitative representation of the map, based on a uniform sample of afferents that was of sufficient size to allow statistically meaningful analyses of the relationships between structure and function.

In order to achieve this goal, two key morphological and functional characteristics of large sample of receptor neurons was measured: 1) the optimal stimulus direction for each receptor, and 2) the spatial distribution of each afferent's synaptic sites within the central nervous system. The optimal stimulus direction was directly observable for each receptor. The anatomical characteristics of the neurons were measured quantitatively in three dimensions using a computer-assisted light-microscopic anatomical reconstruction system. The reconstructions of all neurons were aligned and scaled to a common standard set of dimensions, according to a highly reproducible set of intrinsic fiducial marks. The database therefore preserves accurate information about spatial relationships *between* the neurons within the ensemble. Algorithms were then applied to integrate electrophysiological data about the stimulus/response characteristics of the reconstructed neurons into the database. The algorithms essentially map a physiological function onto a "field" representing the continuous distribution of synaptic terminals throughout the neural structure. In particular, quantitative and testable predictions were made about ensemble response patterns within the map.

Individual sensory afferents were stained with cobalt, and silver-intensified according to the method described in Jacobs and Nevin (1991). The three dimensional reconstruction techniques used to digitize individual stained afferents were developed in our laboratory and have been described in detail elsewhere (Jacobs and Nevin, 1991). Although afferents were reconstructed in their entirety, only the terminal varicosities were used for our analysis of the neural map. The 3-D spatial coordinates and diameter of each varicosity were measured with 1 micron accuracy.

A general methodology was developed to quantify the anatomy of the terminal varicosities or *output sites* of identified sensory neurons. The measures are based on the expected density of membrane surface area of the output sites in the 3D space of the terminal ganglion. The terminal arborization of each identified type of sensory afferent was represented as a 3D continuous probability distribution. The distribution was calculated from 5 samples of each identified afferent. A series of measures

Proc. Microscopy and Microanalysis 1995, edited by G.W. Bailey, M.H. Ellisman, R.A. Hennigar, and N.J. Zaluzec
Copyright © 1995 MSA. Published by Jones and Begell Publishing, 79 Madison Ave., New York, NY 10016

including center of mass and amount of anatomical overlap between different afferent types were calculated using this method. In addition, the functional property of any small volume of neural substrate can be calculated from the overlapping density functions, as determined by the co-localization of output sites from different afferents.

There are three major results of this study. First, we found that there was very low inter-animal variance in several structural features of the cells, including the position of the center of mass of their synaptic varicosities and the detailed shape of their termination arborization within the map. Cells with similar directional sensitivities arborized near each other within the map, and their terminal arborizations showed significant anatomical overlap. The amount of anatomical overlap decreased as the difference in directional tuning increased. The terminal arbors of the afferents were very broad with respect to the dimensions of the entire map which effectively spreads their functional contribution over a large area.

Secondly, there was a clear global organization pattern of afferents within the central nervous system: they were organized in a spiral shape, with stimulus direction mapped continuously around the spiral. Each afferent had a specific location on the spiral, and shared space with neighboring arborizations, which were "adjacent" in terms of both their physical location around the spiral *and* in their directional tuning characteristics. The entire map is composed of two hemi-maps, each organized identically and each restricted to one side of the ganglion. The two hemi-maps are mirror images of one another, and the representation of direction around the spiral rotates in the mirrored directions in the two maps. The observed overlap of synaptic arborizations from adjacent afferents, in combination with the extreme breadth of the receptors' directional tuning curves, allows for a continuous and nearly uniform "interpolated" representation of stimulus direction progressing around the spiral.

Thirdly, based on the global reconstruction of the entire afferent map, it was possible to predict the steady-state spatial patterns of activity within the *ensemble* of afferents which would be expected to result from several different stimuli. Each different stimulus activated a broad region of the map, but each region was specific to the particular stimulus. There are two major aspects of system structure that are responsible for the apparent size of these activation patterns: the size of the terminal arbors of the sensory afferents and the breadth of their tuning curves. These curves are essentially sinusoidal, with a spatial period (i.e., "receptive field") spanning the entire 360 degree range of directions in the horizontal plane. An air current stimulus presented to the cricket will change the activity levels of all of the mechanosensory afferents, by either increasing or decreasing their firing rates. This will, in turn, cause a change in the global pattern of activation of all terminal varicosities within the neural map. This specific change in the global activation pattern results in a corresponding unique spatial pattern of activity within the map.

Jacobs GA and Nevin R (1991) Anatomical relationships between sensory afferent arborizations in the cricket cercal system. Anat Rec 231: 563-572.

Troyer TW, Levin JE and Jacobs GA (1994) Construction and analysis of a data base representing a neural map. Microscopy Research and Technique : 29:329-343.

CHANGES IN RAT PITUITARY GONADOTROPES AFTER IMMUNIZATION WITH PEPTIDES DERIVED FROM THE RECEPTOR FOR GONADOTROPIN-RELEASING HORMONE

P. Petrusz, G. Grossman and C. Weaver

Department of Cell Biology & Anatomy, University of North Carolina, Chapel Hill, NC 27599

Gonadotropin-releasing hormone (GnRH) is a neuropeptide which regulates hormone release from pituitary gonadotropes. The recently cloned pituitary GnRH receptor consists of a linear chain of amino acids in which seven transmembrane segments separate 4 extracellular and 3 intracellular (cytoplasmic) sequences. The rat receptor[1] is approximately 90% homologous with that of the human.[2]

Observations made in several autoimmune diseases and experimental models suggest that anti-receptor antibodies can interfere with receptor function. The present studies were designed to test this hypothesis in female rats (n=9x3) after immunization with synthetic peptides whose overlapping sequences replicated all putative extracellular domains of the cloned human GnRH receptor (Table 1). The peptides (A-H) were conjugated to keyhole limpet hemocyanin and the conjugates were injected i.m. at monthly intervals for four months. A control group (K) received adjuvant only. Anti-peptide antibodies in the sera of immunized rats were detected by ELISA. Recognition of the "native" GnRH receptor was tested in immunocytochemistry on normal pituitary and brain tissue.

Daily vaginal smears showed that all rats continued to cycle throughout the study except those in Group D, which showed continuous proestrus-like smears for the last two months. Body weights and weights of pituitaries, ovaries, uteri, and adrenals did not differ from controls except for animal D1 in which the pituitary was significantly larger. Serum FSH, LH, estradiol, and progesterone levels (RIA) were within normal range in all rats. The only consistent change in all immunized groups was the presence of altered gonadotropes in the anterior pituitary. Gonadotropes were identified by immunocytochemistry with an anti-human chorionic gonadotropin serum. In the immunized groups, some gonadotropes were enlarged and appeared to store excess amounts of hormone. Instead of the normally visible individual secretory granules, immunostaining revealed large irregular spaces filled with dense immunoreactive material (Fig. 1). Furthermore, instead of the normal, oval or rounded shape, some of these cells were greatly enlarged, assumed irregular or bizarre shapes, and were filled with large vacuoles containing unstained material (Fig. 2). The largest of these cells measured as much as 44 μm in diameter. They were reminiscent of castration cells observed after long-term gonadectomy. Because of their large size and large vacuoles, they could be recognized even in routine hematoxylin-eosin-stained sections.

Proc. Microscopy and Microanalysis 1995, edited by G.W. Bailey, M.H. Ellisman, R.A. Hennigar, and N.J. Zaluzec
Copyright © 1995 MSA. Published by Jones and Begell Publishing, 79 Madison Ave., New York, NY 10016

TABLE 1. Primary structure of synthetic peptides used for immunization. These peptides replicate the proposed sequence of all extracellular domains of the human GnRH receptor.[1] The last column indicates % homology of the peptides with corresponding portions of the rat GnRH receptor.[2]

CODE	AA NO.	SEQUENCE	HOM.(RAT)
A	1-18	M A N S A S P E Q N Q N H C A I	82%
B	9-27	Q N Q N H C S A I N N S I P L M Q G	89%
C	18-36	N N S I P L M Q G N L P T L T L S G	89%
D	101-115	D G M W N I T V Q W Y A G E L L C K	94%
E	179-196	R M I H L A D S S G Q T K V F S Q C	67%
F	191-206	K V F S Q C V T T H C S F S Q W W	88%
G	197-212	V T T H C S F S Q W W H Q A F Y	88%
H	292-306	F D P E M L N R L S D P V N H	87%

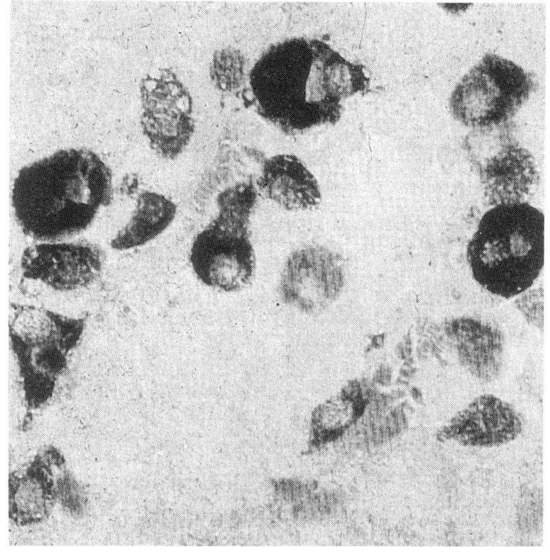

Fig. 1

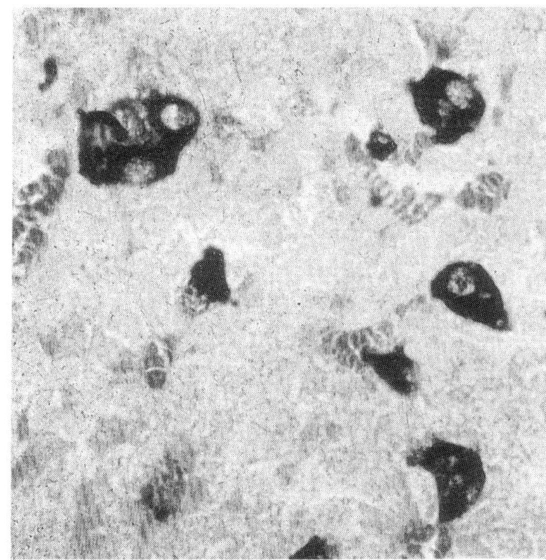

Fig. 2

The mechanism of these changes in pituitary gonadotropes remains to be clarified. The results obtained to date indicate that immunization with GnRH receptor peptides may interfere with pituitary cytology (all peptides) and with estrous cycles (peptide D) without other deleterious effects. Thus, this approach has a potential as a new method of immunocontraception.

References

1. K.A. Eidne et al., Mol. Cell. Endocrinol. 90(1992)125.
2. S.S. Kakar et al., Biochem. Biophys. Res. Comm. 189(1992)289.

815

LOCALIZATION OF C-FOS AND C-JUN PROTO-ONCOGENE PROTEINS IN RAT PLEURAL MESOTHELIAL CELLS BY IMMUNOFLUORESCENCE AND CONFOCAL SCANNING LASER MICROSCOPY

K.A. BéruBé, D.J. Taatjes, A. Hoag, and B.T. Mossman

Department of Pathology, University of Vermont College of Medicine, Burlington, VT 05405

The proto-oncogenes c-*fos* and c-*jun* are members of a multigene family known as immediate early genes (IEG) that are transiently expressed in response to a wide variety of environmental stimuli. Homodimeric complexes of *jun* gene-family products and heterodimeric complexes of the *jun* and *fos* gene-family products form the transcription factor AP-1, which can bind to the promoter region of genes involved in cell proliferation.

Here we report on the expression and localization of the nuclear proteins c-*fos* and c-*jun* in rat pleural mesothelial (RPM) cells. Previous work in our lab has shown that in RPM cells crocidolite asbestos is a potent and persistent inducer of c-*fos* and c-*jun* mRNA and AP-1 DNA-binding activity.[1] Furthermore, asbestos may act as a mitogen in carcinogenesis by persistently activating the early response gene pathway.[1] Although c-*fos* and c-*jun* message levels are increased in RPM cells after exposure to asbestos, it is unclear whether their proteins localize exclusively in proliferating RPM cells. In studies here, we used antibodies to c-*fos* (Fig. 1) and c-*jun* (not shown) to localize these proto-oncogenes in *in vitro* cell cultures of RPM cells at near confluency. Our goal was to determine whether c-*fos* and c-*jun* were localized in cells incorporating the cell proliferation marker 5-bromo-2'-deoxyuridine (BrdU), an indicator of DNA synthesis. Immunodetection was performed via fluorescence microscopy on RPM cells grown on glass coverslips, pulsed with BrdU prior to immunostaining, and stained first for c-*fos/jun* proteins followed by BrdU staining. The *fos* proteins were predominately localized within the nuclei of cells (Fig. 2), whereas the *jun* proteins were also observed in a peri-nuclear location (Fig. 3). Cells showing immunoreactivity for c-*fos*, c-*jun* and BrdU were quantified by random sampling of microscope fields. Assessment of 1000 cells revealed that both proto-oncogenes were expressed in all proliferating RPM cells, as determined by their co-localization with BrdU (Fig. 4). Confocal scanning laser microscopy was used to further corroborate that co-localization of these IEG within BrdU-labeled RPM cells occurred (Fig. 4). We are presently investigating the *in vivo* localization of both proto-oncogenes in mesothelial cells in rat lungs after exposure to asbestos using confocal microscopy.

References

1. N. Heintz et al., *Proc. Natl. Acad. Sci.* 90(1993)3299.
2. The authors gratefully acknowledge the excellent technical assistance of Jennifer Torino.
3. This work is supported by NIEHS grant #ESO6499 and NHLBI grant #HL39469.

Proc. Microscopy and Microanalysis 1995, edited by G.W. Bailey, M.H. Ellisman, R.A. Hennigar, and N.J. Zaluzec
Copyright © 1995 MSA. Published by Jones and Begell Publishing, 79 Madison Ave., New York, NY 10016

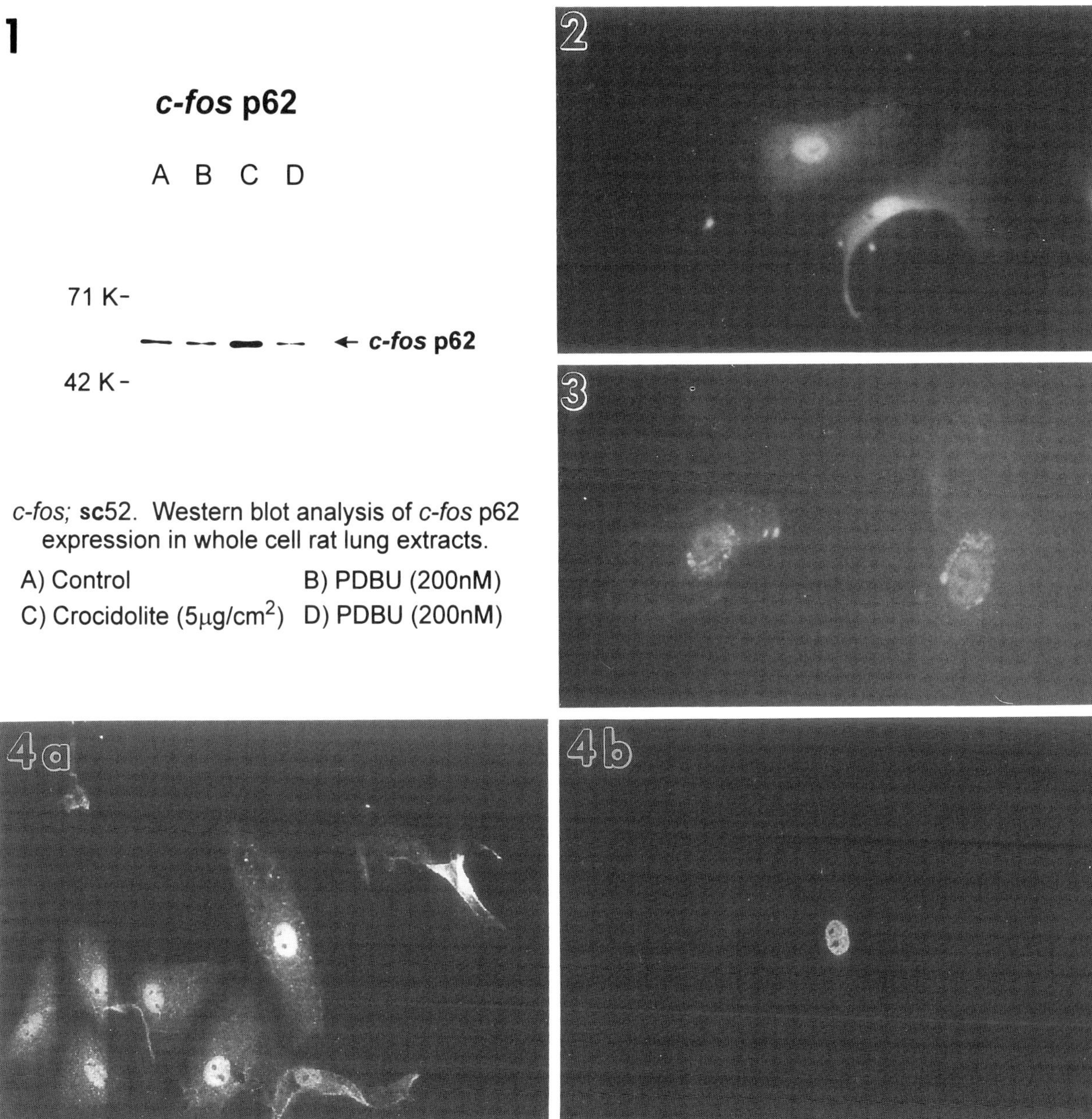

FIG. 1.Western blot analysis of c-*fos* p62 protein expression in whole cell rat lung extracts. Phorbol dibutyate (PDBU) was used as positive control.

FIG. 2.–Immunofluorescence localization of c-*fos* using rabbit polyclonal antibody followed by goat anti-rabbit lissamine-rhodamine conjugate. Original magnification 400X.

FIG. 3.–Immunofluorescence localization of c-*jun* using rabbit polyclonal antibody followed by goat anti-rabbit lissamine-rhodamine conjugate. Original magnification 400X.

FIG. 4.–Co-localization of c-*fos*/BrdU in RPM cells demonstrated by confocal laser scanning microscopy. (a) Same field of view imaged with 568nm laser line (c-*fos*); (b) imaged with 488nm laser line (BrdU). Note several cells positive for c-*fos* are negative for BrdU.

THE PROTEASE PHENOTYPE AND CRYSTALLINE NATURE OF THE GRANULES OF INTRAEPITHELIAL MAST CELLS IN *TRICHINELLA SPIRALIS*-INFECTED MICE

D.S. Friend,* N. Ghildyal,** M.F. Gurish,** K.F. Austen,** and R.L. Stevens**
Departments of *Pathology and **Medicine, Harvard Medical School, and Department of Rheumatology and Immunology, Brigham and Women's Hospital, Boston, MA 02115

Trichinella spiralis induces a profound mastocytosis[1] and eosinophilia in the small intestine of the infected mouse. Mouse mast cells (MC) store in their granules various combinations of at least five chymotryptic chymases [designated mouse MC protease (mMCP) 1 to 5], two tryptic proteases designated mMCP-6 and mMCP-7 and an exopeptidase, carboxypeptidase A (mMC-CPA). Using anti-peptide, protease -specific antibodies to these MC granule proteases[2-6], immunohistochemistry was done to determine the distribution, number and protease phenotype of the MCs in the small intestine and spleen 10 to >60 days after *Trichinella* infection of BALB/c and C3H mice. TEM was performed to evaluate the granule morphology of the MCs between intestinal epithelial cells and in the lamina propria (mucosal MCs) and in the submucosa, muscle and serosa of the intestine (submucosal MCs).

As noted in the table below, the number of submucosal MCs remained constant throughout the study. In contrast, on day 14, the number of MCs in the mucosa increased ~25 fold. Increased numbers of MCs were observed between epithelial cells in the mucosal crypts, in the lamina propria and to a lesser extent, between epithelial cells of the intestinal villi. As assessed immunohistochemically, all of the intraepithelial MCs and 85% of the MCs in the lamina propria express predominantly mMCP-1 and mMCP-2. The other 15% in the lamina propria resemble MCs found in connective tissue in that they express predominantly mMCP-5 through mMCP-7 and mMC-CPA. Considering analysis of only the small number of MCs in the normal mouse intestinal mucosa, the MCs in the intestine of the infected mice were phenotypically similar to those in uninfected mice. Thus, helminth infection induces a dramatic proliferation of MCs in the mucosa and in the intensity of the immunoreactivity but no change in the spectrum of granule proteases expressed.

The MCs in the intestinal lamina propria and in the epithelium have fewer and larger granules than those MCs in the connective tissue. The MCs in the lamina propria and among the mucosal epithelial cells have granules that contain mMCP-1 and mMCP-2. Despite their similar protease content, the granules in MCs in the lamina propria are mainly spherical while those in the MCs between epithelial cells are primarily angular. We observe that granule crystallization occurs after MCs traverse the epithelial basal lamina. As assessed by TEM, the crystals in the angular granules have parallel lattice spacing of 8.3 ± 1.5 nm and 5.4 ± 1.2 nm. Tilting the sections by 20° on the goniometer stage reveals that the same crystalline unit contains both periodicities intersecting at a 78° angle. Thus, mMCP-1 and mMCP-2 reside in a single type of crystal in intraepithelial MC granules. This is the first report of two mouse MC serine proteases packaged in a granule as a crystal.

References

1. H. Alizadeh and D. Wakelin, *Int. J. Parasitol.* 12(1982)65.
2. N. Ghildyal and R.L. Stevens, unpublished data.
3. N. Ghildyal et al. *J. Immunol.* 151(1993)3206.
4. H.P. McNeil et al. *J. Immunol.* 149(91992)2466.
5. N. Ghildyal et al. *J. Immunol.* 153(1994)2624.
6. M.F. Gurish et al. *J. Allergy Clin. Immunol.* 89(1992)309.

Table

		MCs /mm^2	
Day	n	Mucosa	Submucosa
0	6	4±3	22±10
10	4	64±34	30±5
14	7	100±16	25±8
20	2	98±9	31±2
30	2	22±6	22±4
>60	2	6±2	24±3

Proc. Microscopy and Microanalysis 1995, edited by G.W. Bailey, M.H. Ellisman, R.A. Hennigar, and N.J. Zaluzec
Copyright © 1995 MSA. Published by Jones and Begell Publishing, 79 Madison Ave., New York, NY 10016

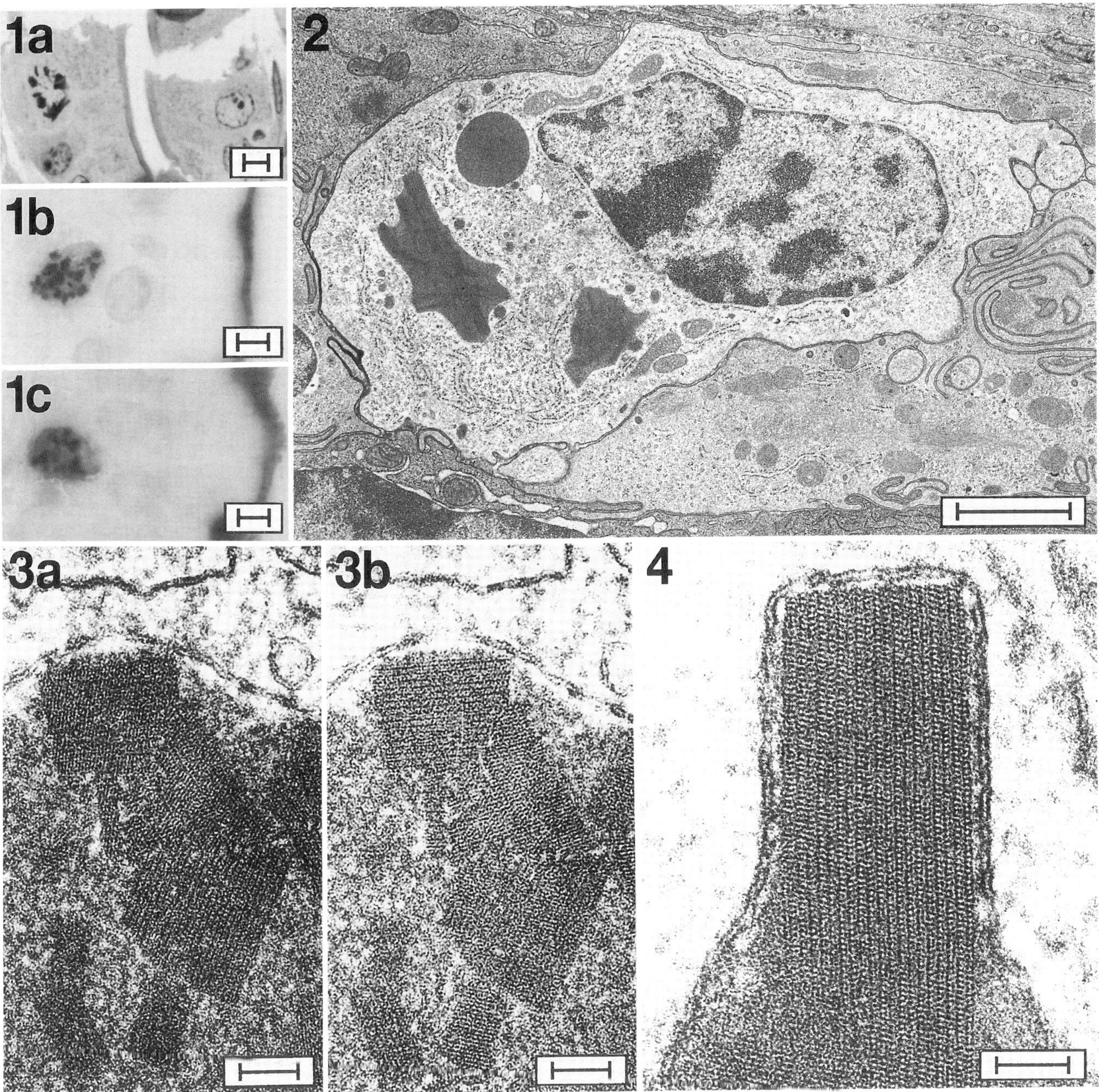

FIG. 1—Toluidine blue-stained semithin epoxy section (**a**). Anti-mMCP-1 Ig (**b**) and anti-mMCP-2 Ig (**c**) immunoalkaline phosphatase-reacted JB4 sections. Bar = 10 μm.

FIG. 2—Electronmicrograph showing profiles of crystalline granules in an intraepithelial mucosal mast cell. Bar = 2 μm.

FIG. 3—Higher magnification of a portion of a crystalline granule at goniometer tilts of minus and plus 10°. The 5.4 μm spacing, vertical (**a**) and the 8.3 μm spacing, horizontal (**b**) appear in the same crystal. Bar = 0.2 μm.

FIG. 4—Higher magnification view of another crystal showing the 78° angle at lattice intercepts. Bar = 0.1 μm.

ASSOCIATION OF γ-GLUTAMYL TRANSPEPTIDASE (γ-GTP) POSITIVE CELLS WITH TUBE-FORMING STRUCTURES IN RAT BRAIN ENDOTHELIAL CELL CULTURES: A BRIGHT FIELD AND LASER CONFOCAL MICROSCOPY STUDY

L.E. De Bault,[*] B.L. Wang,[***] P. Grammas,[*] D.P. Ringer,[**] and J. Kong[**]

[*] Department of Pathology, University of Oklahoma Health Sciences Center, Oklahoma City, OK 73190
[**] Noble Center for Biomedical Research, Oklahoma Medical Research Foundation, Oklahoma City, OK 73104
[***] Bligh Cancer Research Laboratory, Fung University Health Science Center/Chicago Medical School, North Chicago, IL 60064

Gamma-glutamyl transpeptidase (γ-GTP), an enzyme purported to regulate amino acid transport, is expressed at high levels in blood-brain-barrier (BBB) capillary endothelium[1,2], but is a property that is variably expressed in cell culture. The ability of endothelial cells to form capillary-like tubes under different culture conditions has been reported by several investigators[3,4]. In this study, a rat cerebral resistance vessel endothelial cell strain (RV-150 ECT)[5], that spontaneously forms cellular aggregates and connecting tubes, has been shown to contain a subpopulation of γ-GTP positive cells[6], was evaluated for the spatial relationship between γ-GTP and its mRNA, and the tube-forming structures in culture.

Cells were grown on coverslips and fixed in acetone for enzymecytochemistry or in 3% paraformaldehyde and 0.5% glutaraldehyde in PBS pH 7.4 for immunocytochemistry and *in situ* transcription (IST), respectively. The enzyme substrate/acceptor were γ-glutamyl-4-methoxy-2-naphthyl-amine and glycylglycine. The primary antibody was affinity purified goat anti-γ-GTP IgG. The primers for the IST were 25-nt oligonucleotides complementary to γ-GTP mRNA. After the primers were hybridized to the mRNA, the mRNA was transcribed *in situ* in a mixture of dNTPs containing DIG-labeled dUTP and reverse transcriptase. The DIG was bridged with sheep anti-DIG, and the sheep Ig detected with 10nm gold labeled rabbit anti-sheep IgG followed by silver enhancement, immunogold silver (IGS) staining.

The spontaneously formed tube structures consist of a complex of multiple cell layered cords that connect larger spherical cellular aggregates, all resting on the apical surface of a contact inhibited monolayer. Cells containing γ-GTP enzyme activity (Fig. 1), its protein (Fig. 2), and its mRNA (Fig. 3) were associated with the cellular aggregates and tubes but not with the monolayer cells. In cultures stained for γ-GTP mRNA by IST-IGS and examined by confocal microscopy (Figs. 4-18), showed groupings of cells that surround an apparent lumen in the connecting tubes. These results showed that while endothelial cells in monolayer lost γ-GTP expression, a subpopulation of cells associated with aggregates and cords express the enzyme even in long term culture. These data suggest that the transcriptional regulation of BBB properties in cerebral endothelial cells may vary with phenotype.

References

1. Z. Albert, M. Orlowski, Z. Rzucidlo and J. Orlowska. *Acta Histochem. Jena* 25(1966)312.
2. L.E. De Bault and A. Mitro. *Histochem. J.* 26(1994)447.
3. D.H. Robinson, Y.H. Kang, S.H. Deschner and T.B. Nielsen. *In Vitro Cell. & Dev. Biol.* 26(1990)169.
4. I.E. Goetz, J. Warren, C. Estrada, E. Roberts and D.N. Krause. *In Vitro Cell. & Dev. Biol.* 21(1985)172.
5. C.A. Diglio, W. Liu, P. Grammas, F. Gracemelli and J. Wiener. *Tissue & Cell* 25(1993)833.
6. B.L. Wang, P. Grammas and L.E. De Bault. *J. Cell. Physiol.* 156(1993)531.

(Supported in part by NIH grants NS-18775 to L.E.D. and NS-30457 to P.G.)

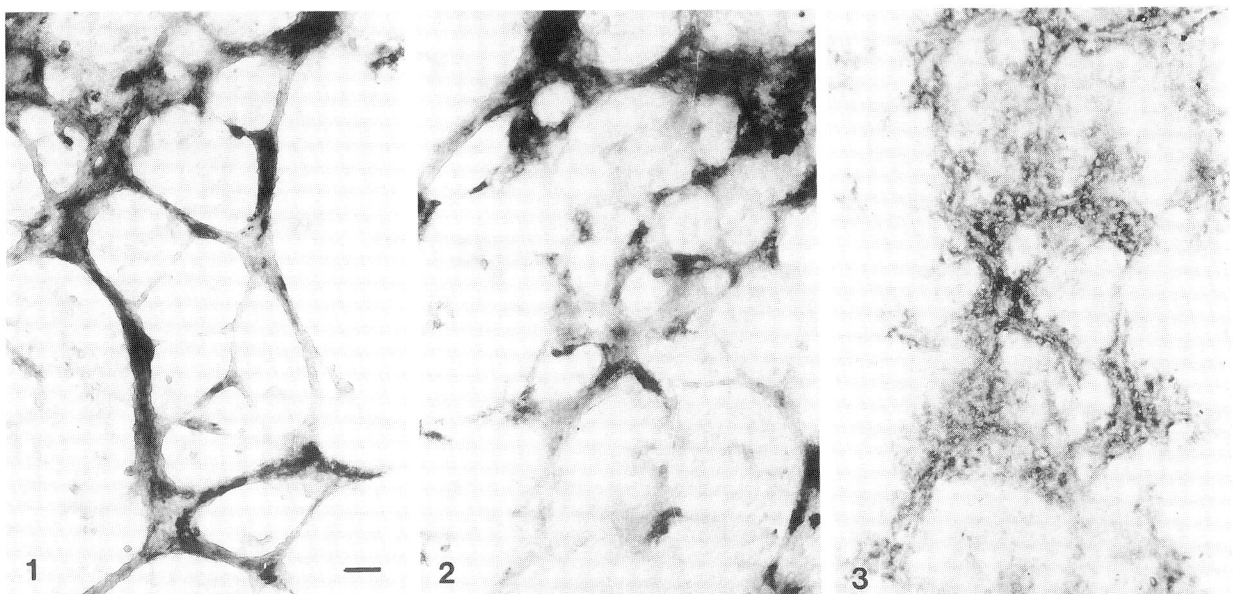

Localization of γ-GTP, its Protein, and its mRNA by bright field microscopy tube-forming endothelial cell cultures of: FIG.1. - Enzymecytochemistry. FIG.2. - Immunocytochemistry. FIG.3. - *In Situ* Transcription Immunogold Silver Staining (IST-IGS). Bar= 10μm.

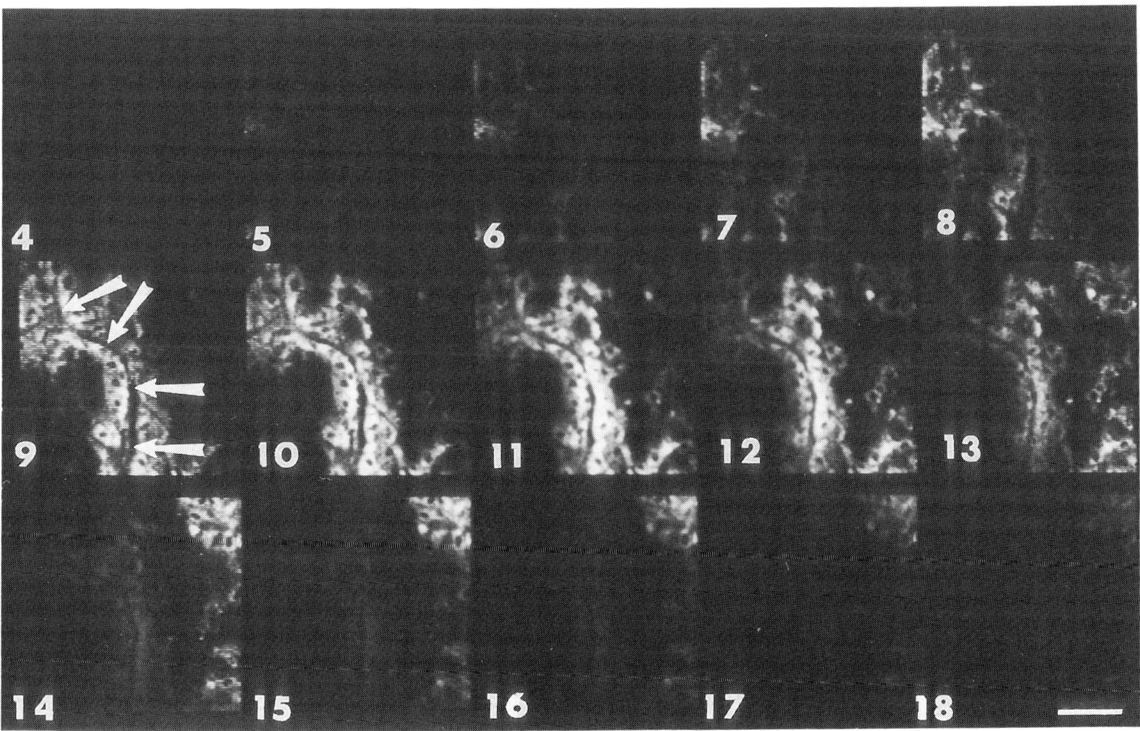

FIGS.4-18. - Laser confocal microscopy of γ-GTP mRNA IST-IGS staining in endothelial cells in a branched tube-forming structures in culture, taken from a series of confocal images scanned at 1μ intervals on the z-axis. Arrows indicate location of lumen. Bar=25μm

IMMUNOCYTOCHEMICAL DOUBLE STAINING WITH THE ELF METHOD

D. G. Baskin,[1,2] C. Vathanaprida,[1] and W. L. Stahl[1,3]

V.A. Medical Center, Seattle, WA 98108
Departments of Medicine[1], Biological Structure[2], and Physiology & Biophysics[3],
University of Washington, Seattle, WA 98195

The reagent 2-(5'-chloro-2'-phosphoryloxyphenyl)-6-chloro-4-[^{3}H]-quinazolinone has recently become commercially available as the ELF (enzyme-labeled-fluorescence) phosphatase substrate. The ELF method is basically an alkaline phosphatase enzyme immunostaining procedure in which the reaction product is fluorescent instead of chromogenic. The ELF substrate produces a photostable, bright yellow-green fluorescence (500-580 nm) with low background.[1] The use of the ELF substrate as a green label may have advantages over FITC in double-labeling immunocytochemical staining procedures that employ Cy3, which frequently shows residual orange-red fluorescence when viewed with conventional FITC filter sets.

To evaluate the suitability of ELF for dual immunofluorescence with Cy3-labeled second antisera, we used paraffin sections of rat pancreas fixed in 4% paraformaldehyde. Primary antibodies were an anti-insulin monoclonal antibody (1:50) (Biogenex) and anti-glucagon polyclonal antibody (1:1K)[2], in Tris-HCL, 1% BSA, 0.1% TX-100, pH 7.4. The insulin antibody was applied overnight at 6 °C, followed by goat anti-mouse IgG-Cy3 (Jackson Immunochemical Research) for 60 min. at room temperature (RT). The glucagon antibody was then applied for 1 hr. at RT, followed by reagents for the ELF-AP Immunohistochemistry Kit (Molecular Probes, Eugene, OR). This procedures uses biotinylated goat anti-rabbit IgG for 30 min. (RT), followed by streptavidin (1:250) for 30 min. (RT), and biotinylated-alkaline phosphatase conjugate (1:250) for 30 min., RT, and the ELF-AP substrate reagent (1:20) for 8 min. Sections were mounted in the mounting media supplied with the kit.

The results were analyzed with a Zeiss Axioplan fluorescence microscope and recorded with a Hamamatsu #C4880 dual mode cooled CCD camera and MCID image analysis system (Imaging Research, St. Catherines, ONT) (Fig. 1). ELF fluorescence was visualized with UV illumination, using a Hoechst long-pass filter set: peak excitation at 360 nm, a 400 nm barrier filter, and emission filter >535 nm (green). The bright green ELF fluorescence detected insulin-like immunoreactivity in beta cells of the pancreatic islets. Background fluorescence was very low. Glucagon immunoreactivity was localized by red-orange Cy3 fluorescence to islet alpha cells, using filters for green excitation (535 peak excitation), a 565 nm dichroic filter, and 575 nm barrier filter (orange). There was no Cy3 fluorescence visible with the ELF filter set, and ELF fluorescence was not visible with the Cy3 filter set. Successful double staining was achieved when the ELF-AP labeling was done after the Cy3-labeled second antibody. When the ELF labeling was done first, no ELF fluorescence resulted, although Cy3 fluorescence was present. Preliminary experiments indicate that the ELF reaction product is extracted in aqueous media.

Proc. Microscopy and Microanalysis 1995, edited by G.W. Bailey, M.H. Ellisman, R.A. Hennigar, and N.J. Zaluzec
Copyright © 1995 MSA. Published by Jones and Begell Publishing, 79 Madison Ave., New York, NY 10016

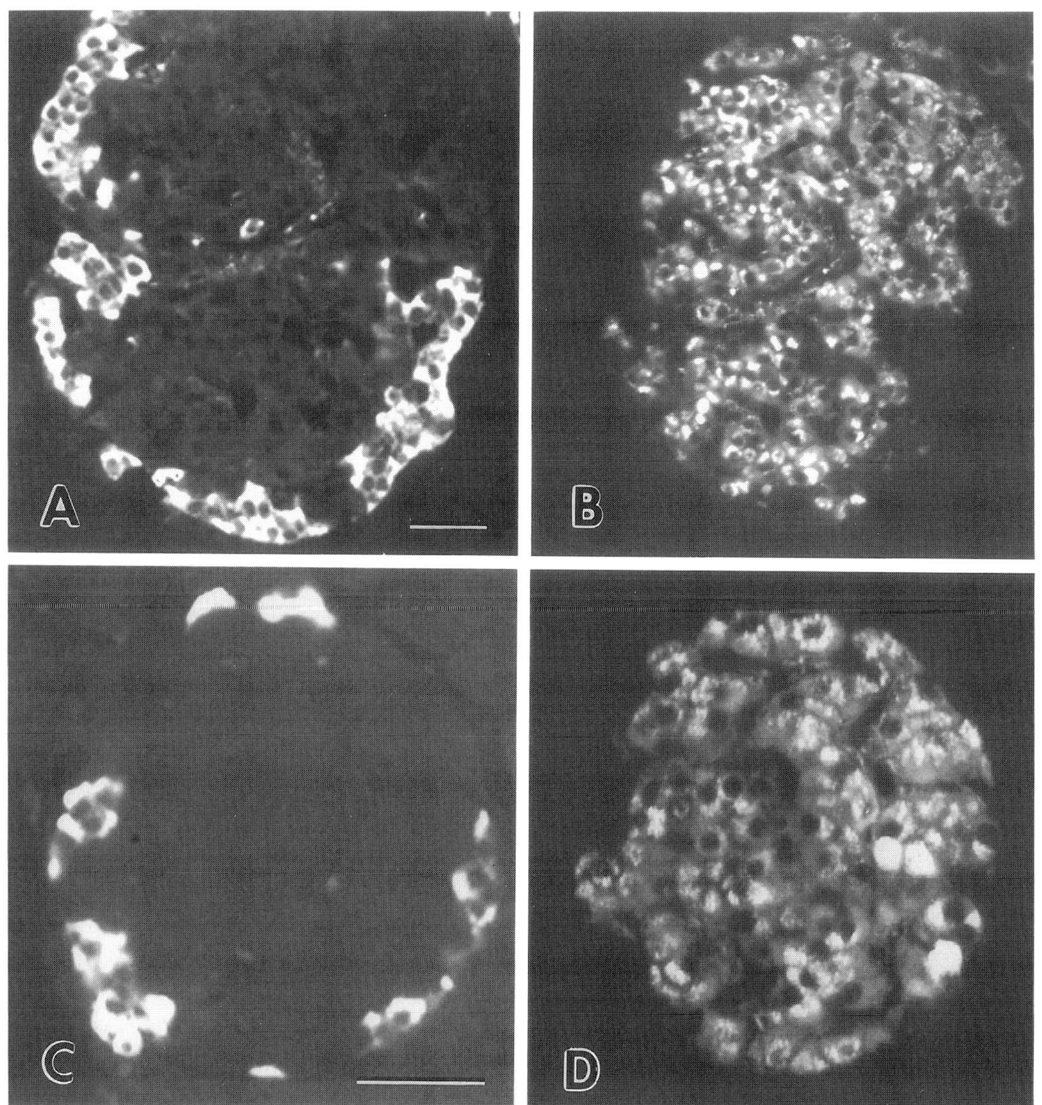

FIG. 1 Rat islets immunostained with antibodies to glucagon (A,C) and insulin (B,D), recorded with Hamamatsu CCD camera. Cells with glucagon-like immunoreactivity were detected with ELF-AP fluorescence. Cells with insulin-like immunoreactivity were detected with Cy3 fluorescence. A & B and C & D are same islets, respectively. Images with ELF-AP and Cy3 were made sequentially by changing fluorescent filters on microscope. Bars = 50 μm

Cy3 fluorescence was visibly undetectable when viewed with the ELF filter set. The results suggest that the ELF substrate is an alternative to FITC for double immunostaining with Cy3. The bright fluorescent signal produced by ELF staining indicates that this technique may be suitable for detecting antigens present in low concentrations.[3]

References

1. K.D. Larison et al., *J. Histochem. Cytochem.* 43:77-83, 1995
2. D.G. Baskin et al., *Anat. Rec.* 208:567-578, 1984
3. Supported by the Department of Veterans Affairs and NIH grant DK17047.

IMMUNOELECTRON MICROSCOPY OF THE CONNECTIVE TISSUE MATRIX FOLLOWING CRYOTECHNICAL PROCESSING

D.R. Keene* K. McDonald** and C.C. Ridgway*

*Shriners Hospital for Crippled Children, Portland, Oregon 97201
**University of California at Berkeley, Berkeley, California

Our understanding of the microarchitecture of connective tissues is based primarily on micrographs taken of tissues stabilized by immersion fixation. Considerable extraction of tissue components, particularly small matrix molecules, occurs even during fixation in glutaraldehyde. An alternative to immersion fixation is the technique of cryostabilization whereby tissue components are instantaneously immobilized and surrounded by non-crystaline (amorphous) or microcrystaline ice. The ice is replaced at low temperature (-100 to -85°C) by acetone containing fixative, and the tissue finally embedded in epoxy. A comparison of micrographs taken from tissue embedded following immersion fixation and those taken following cryotechnical processing reveals striking differences in the structure of cells, basement membranes, and matrix (Fig. 1a and 2a).[1,2,3,4]

Our recent objective has been to utilize sections from samples which have been stabilized by high pressure freezing for immunocytochemical labeling experiments with the expectation that such studies would result in a more precise localization of matrix components. The results to date suggest that many of the difficulties normally experienced in section surface labeling technique are still a problem: Antigens may be affected by the chemistry of freeze-substitution and choice of embedding media. Among the successes that we have had are with a monoclonal antibody to elastin and a polyclonal antibody to type VII collagen. As compared to previous experiments localizing these antigens on the surface of immersion fixed, lowicryl K4M embedded tissue, the architecture of the labeled area appears considerably different following cryostabilization. Anchoring fibrils, to which type VII monoclonal antibodies localize [5], are not visible in tissues prepared by cryotechnique, yet antibody applied to the surface of sections cut from such preparations bind to a similar zone relative to the epithelium (compare figs. 1a and 1c). Likewise, although the architecture of elastin bundles is considerably different in samples prepared by immersion fixation as compared to those prepared by cryotechnical procedures, the pattern of elastin localization on the surface of sections cut from these preparations is remarkably similar (compare figs. 2a and 2c).

References:

1. R. Dahl and L.A. Staehelin, J. Electron Microsc. Tech. 13(1989)165.
2. E.B. Hunziker and R.K. Shenk, J. Cell Biol. 98(1984)277.
3. E.B. Hunziker et al., J. Cell Biol. 98(1984)267.
4. D.R. Keene and K. McDonald, J. Histochem. Cytochem. 113(1993)41.
5. D.R. Keene et al., J. Cell Biol. 104(1987)611.
6. The authors gratefully acknowledge the expert assistance of Mary K. Morphew and Professor L.A. Staehelin (University of Colorado, Boulder, CO) for the use of his high-pressure freezer.

Proc. Microscopy and Microanalysis 1995, edited by G.W. Bailey, M.H. Ellisman, R.A. Hennigar, and N.J. Zaluzec
Copyright © 1995 MSA. Published by Jones and Begell Publishing, 79 Madison Ave., New York, NY 10016

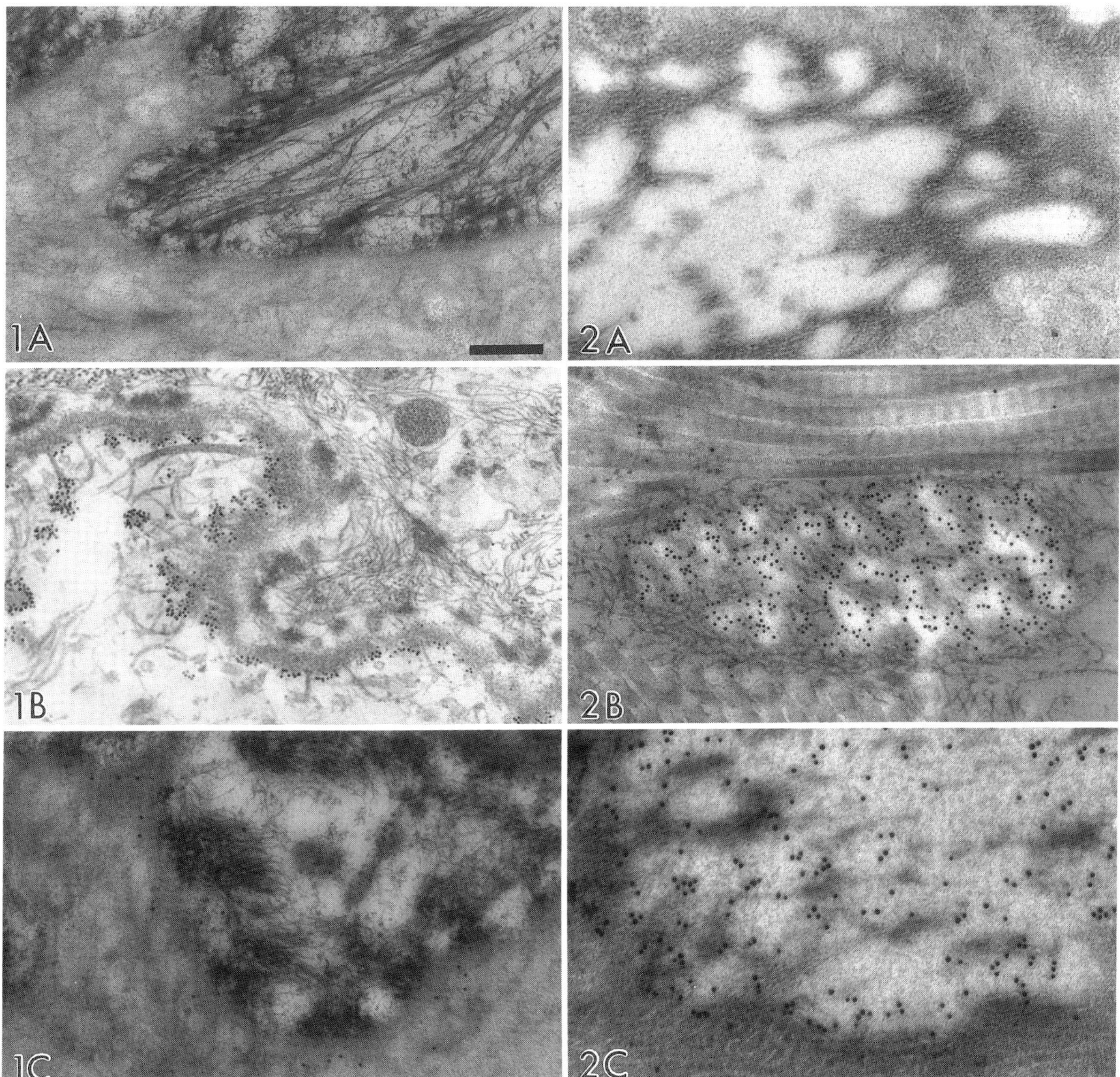

FIG 1A. Following high pressure freezing(HPF), freeze substitution(FS) and embedding in lowicryl HM-20, the DEJ of human skin lacks many of the ultrastructural features seen following conventional fixation, including the lamina densa, lamina lucida, and anchoring fibrils. Instead, the zone beneath the epithelial cell is uniformly dense.

Fig 1B. Section surface immunocytochemistry performed using an anti-type VII antibody on the surface of sections cut from HM-20 embedded tissue following cryostabilization reveals a pattern of binding which is in the dermis, immediately adjacent to the basal portion of the epithelial cell.

FIG 1C. Enbloc immunolocalization of type VII collagen in human skin followed by conventional fixation and embedding reveals specific labeling of the basement membrane region.

FIG 2A. Following HPF, FS, and embedding in epon 812, elastin bundles in skin appear much larger than those in conventional preparations, and are completely surrounded by amorphous material.

FIG 2B. Section surface immunocytochemistry of HPF, FS skin embedded in Lowicryl HM-20 reveals dense gold labeling specific to elastin bundles following exposure to anti-elastin antibody.

FIG 2C. Following aqueous fixation and embedding in Lowicryl K4M, elastin bundles are severely shrunken in size and surrounded by apparently open space. Elastin antibody localizes to the interior of these bundles.

All Figures magnified x 45,000; bar = 250 μm

EXHAUSTIVE PHOTON REASSIGNMENT™ - A Method for Performing High-Resolution, Quantitative 3D Fluorescence Microscopy

Authors : Ramkumar K. Moorthy and David C. Hitrys.

Scanalytics, 40 Linnell Circle, Billerica, MA 01821.

Fluorescence microscopy is an increasingly popular technique in the field of light microscopy. However, when one uses an objective lens of high magnification and numerical aperture, only a very shallow plane within the specimen is in optimal focus. Out-of-focus fluorescent objects in planes above and below this plane of focus contribute light haze to this plane, compromising the visual quality of the specimen image. This haze also renders it impossible to accurately quantify light intensities of fluorescent objects present in this plane.

Confocal microscopy is one technique that allows researchers to obtain a highly resolved, high-contrast image of a specimen plane by excluding or rejecting light emanating from out-of-focus planes. Most biologists, however, can ill-afford to discard valuable fluorescence energy, because its efficient use would allow for optimal imaging conditions, and because it does contain useful information even if emanating from out-of-focus structures. There are also other drawbacks associated with confocal technology. The laser used as the excitation source in a confocal microscope is not really suitable for live cell studies, and also limits the range of fluorochrome labels that may be used. Moreover, the very efficiency of confocal microscopes in rejecting out-of-focus light means that it is often difficult to avoid severe photobleaching of specimens, to visualize fine or faintly-labeled structures, or to acquire images using short exposure times.

Dr. Fredric S. Fay and members of the Biomedical Imaging Group at the University of Massachusetts Medical School in Worcester, MA have spent the last fifteen years developing and perfecting a digital imaging system that helps scientists overcome these problems. However, obtaining blur-free images using this process was computationally intensive and therefore not easily implemented. Advances in computer technology have now made it possible to perform this technique using standard desktop PCs. Scanalytics (a division of CSPI) is the exclusive worldwide licensee of this patented technology which allows researchers to obtain high-resolution, quantitatively accurate three-dimensional images of fluorescent specimens.

The CELLscan™ system from Scanalytics employs this patented procedure which is called Exhaustive Photon Reassignment or EPR™. This is a method in which the distortion and depth-of-field properties of an individual wide-field microscope are characterized and mathematically reversed, yielding quantifiable, high-resolution images of the true object being visualized. Images are usually obtained using a standard (wide-field) epifluorescence microscope and high dynamic range cooled CCD camera, resulting in short exposure times and minimized photobleaching. Use of a standard Hg or Xe lamp as the excitation source results in minimal stress to live cells and allows the use of any available fluorochrome. In addition, EPR can be used to enhance the resolution of confocal microscopy images that were acquired under suboptimally confocal conditions (as in cases where increased sensitivity to low signal or speed of acquisition are important). Under these conditions, some haze from neighbouring planes is accepted and then reassigned to its proper points of origin using EPR, as described below.

The optical system (wide-field or pseudo-confocal) is characterized by acquiring an empirical Point Spread Function (PSF) using a sub-resolution-sized fluorescent bead. This knowledge about how the image of a point source appears when the source is in focus or at various distances out of focus is utilized to *reassign* out-of-focus photons in the 3D image stack back to their points of origin within the specimen image. Since all available out-of-focus photons are reassigned rather than removed as is common with subtractive methods or filters, and since this procedure is performed utilizing a true, i.e. empirical PSF, the result of the process is a three-dimensional image in which out-of-focus light haze is reassigned to its proper locations; one in which fluorescent objects may be accurately quantitated (except in case of pseudo-confocal images) and one where fine or faint structures are clearly visualized. Research performed utilizing this approach has been regularly published in international journals and the technology is thus well-characterized, and widely known and accepted.

Scientists worldwide are now finding that EPR can let them perform experiments that were impossible until recently due to the limitations of technology. Improvements in Cooled CCD cameras and computers are now matched by the development of new fluorophores and staining/visualization techniques that can take advantage of these improvements (Green Fluorescent Protein or GFP being a good example). Imaging live cells over time under gentle conditions and with high resolution is thus an exciting frontier that is being explored by more and more researchers. This presentation seeks to familiarize scientists with the EPR approach and discuss ways it may apply to their present and future research. Examples of different applications that utilize the EPR technology will be illustrated and described in some detail.

Proc. Microscopy and Microanalysis 1995, edited by G.W. Bailey, M.H. Ellisman, R.A. Hennigar, and N.J. Zaluzec
Copyright © 1995 MSA. Published by Jones and Begell Publishing, 79 Madison Ave., New York, NY 10016

References:

1. W. Carrington et al. *Three-dimensional Imaging on Confocal and Wide-field Microscopes*. <u>Handbook of Biological Confocal Microscopy</u>, Chapter 14.
2. K. C. Carter et al. *A Three -dimensional View of Precursor Messenger RNA Metabolism within the Mammalian Nucleus*. <u>Science</u>. Vol. 259, February 26, 1993, pp. 1225-1368.
3. E. D. Moore et al. *Coupling of the Na^+/Ca^{2+} Exchanger, Na^+/K^+ Pump and Sarcoplasmic Reticulum in Smooth Muscle*. <u>Nature</u>. Volume 365, No. 6447, October 14, 1993, pp 657-660.
4. L. M. Loew et al. *Imaging in Five Dimensions: Time-Dependendent Membrane Potentials in Individual Mitochondria*. <u>Biophysical Journal</u>. Volume 65, December, 1993, pp. 2396-2407.
5. S. Corvera at al. *Disruption of PDGF Receptor Trafficking by Mutation of its PI-3 Kinase Binding Sites*. <u>Science</u>. Volume 263, February 4, 1994, pp.684-687.

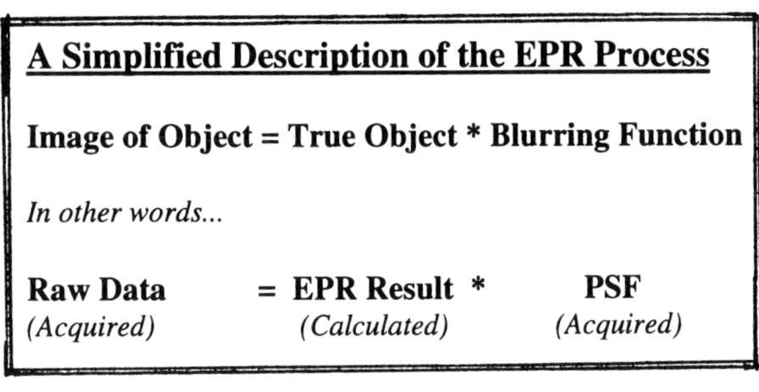

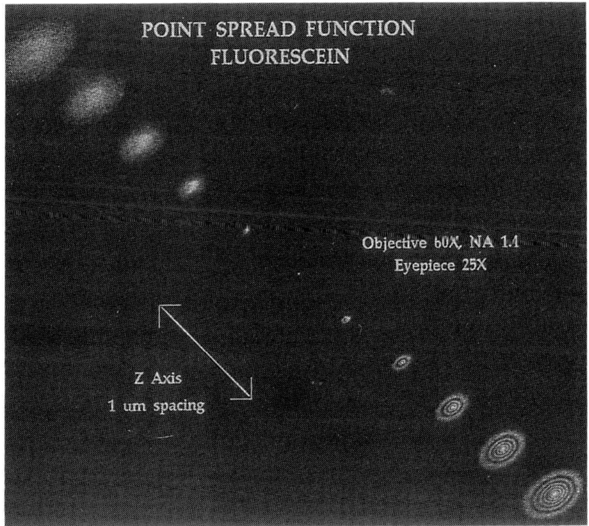

Fig 1. Representation of Fluorescein PSF

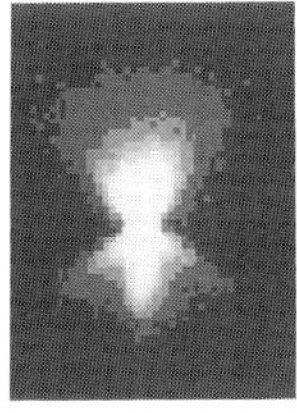

Fig 2. X/Z View of Point-Spread Function (PSF)

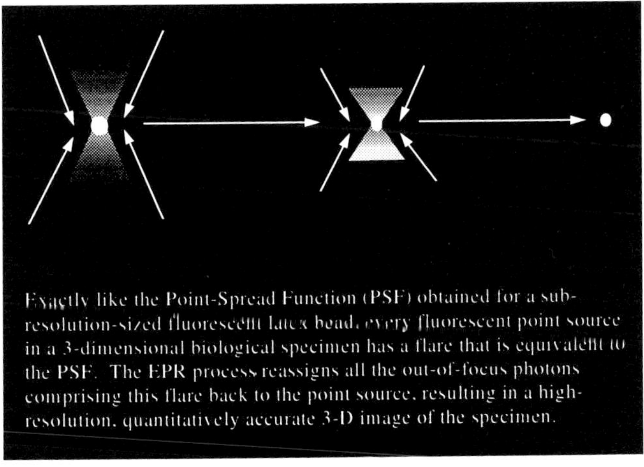

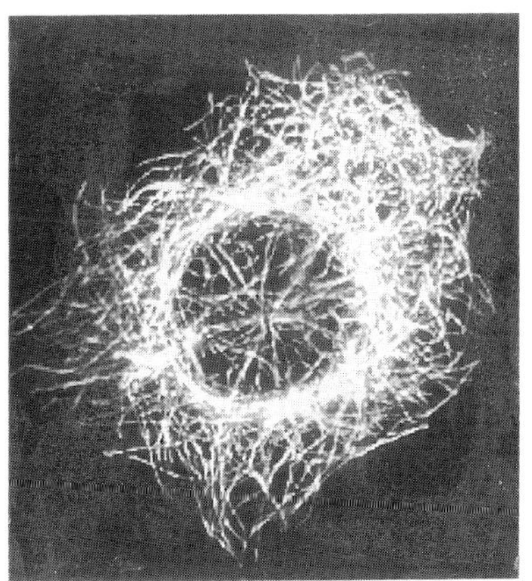

Fig. 3. Example of EPR-restored specimen :
3-D Volume View (Maximum Intensity Projection view) of Normal Rat Kidney (NRK) Epithelial Cell displaying Rhodamine-labeled Tubulin. Image courtesy of D. Bowman of the Biomedical Imaging Center at U. Mass. Medical School, Worcester, MA.

DEVELOPMENT OF OPTIMAL PROTOCOLS FOR ANTIGEN RETRIEVAL IMMUNOHISTOCHEMISTRY BASED ON THE EFFECTS OF VARIATION IN TEMPERATURE AND PH: USE OF A 'TEST BATTERY'

S.-R. Shi,* R.J. Cote,* L. Young,* C. Yang,* C. Chen,* G.D. Grossfeld,** D.A. Ginsberg,** F.L. Hall,*** C.R. Taylor*

*Departments of Pathology, **Urology, University of Southern California School of Medicine, Los Angeles, CA 90033. ***Division of Orthopaedic Surgery, Childrens Hospital of Los Angeles, CA 90054.

Following development of the microwave antigen retrieval (AR) technique in 1991 (1) for enhancement of immunohistochemical (IHC) staining, there have been numerous articles published worldwide describing the use of AR-IHC on archival tissue sections (2). The AR technique provides a method for retrieval of formalin-masked antigens which otherwise yield negative or very weakly positive IHC staining on archival paraffin sections. AR appears to be superior to other unmasking procedures, and serves to reduce detection thresholds of IHC for a wide range of antigen antibody systems (3). One important result is that AR increases the accuracy of diagnosis when using IHC for pathology (4). However, certain critical issues remain concerning the optimal protocol, the choice of AR solution, pH and the precise heating conditions. These aspects of AR-IHC require further study in order to improve standardization of AR-IHC, as emphasized by Taylor (5). Based on our recent study of AR-IHC under the influence of pH (6), we designed a new study specifically to examine the effect of different microwave heating conditions. A modern laboratory microwave (MW) oven (H2550, Energy Beam Sciences, Agawam, MA), was employed, the device has a screen which records the exact temperature, and allows programming for specific time periods. All test tissues were routinely processed paraffin sections. The pH value of Tris-HCl or citrate buffer solutions was measured carefully by using an electronic pH meter (Orion). Under different controlled heating conditions, we demonstrated that: (1) Optimal result of AR-IHC = T (temperature) x t (time of heating); (2) Actual temperature achieved is the critical factor; there were no differences between microwave, conventional and autoclave heating methods, if the heating conditions, T x t, were the same; (3) The pH value of the AR solution also is an important factor as demonstrated previously (6). In order to find an optimal protocol of AR for every antibody of interest, we designed a 'test battery', combining 3 pH values: pH 1, 6 and 10 using Tris-HCl buffer as the AR solution, with 3 microwave heating conditions: 100, 90 and 80°C. A total of 9 slides were used for this test (Table 1). In every instance a frozen section was also stained as the gold standard. Table 1 summarizes the results of 14 antibodies tested by using this battery. One striking finding is that the pH 6 is not the best condition for most antibodies tested in spite of the fact that PH 6 citrate buffer is widely used. Also tested the lower temperature (80°C) was of value for only one of the antibodies tested (cell surface antigen TGFβ receptor). This study also raises another interesting issue, namely that the immunostaining results may be dramatically altered by different AR protocols. This test battery provides a rapid approach to find an optimal AR protocol for a new antibody, and may contribute to the optimization of the AR-IHC in analysis, (Fig. 1). It is recommended to use negative control, and frozen sections as the gold standard for both intensity and pattern of stainings, we must note that the possibility of diffusion of antigen results in non-fixed tissues.

References

1. S-R. Shi et al., *J. Histochem. Cytochem.* 39(1991)741.
2. S-R. Shi et al., *Cell Vision* 2(1995)2.
3. P.H. McKee et al., *Histopathology* 23(1993)377.
4. A.M. Gown et al., *Appl. Immunohistochem.* 1(1993)256.
5. C.R. Taylor, *Hum. Pathol.* 25(1994)2.
6. S-R. Shi et al., *J. Histochem. Cytochem.* 43(1995)193.

Table 1. 'Test Battery' used for 14 antibodies

Temperature	pH 1	pH 6	pH 10
100°C	AE1,AE3,ER,MIB1, Thrombospondin		RNP,AE1,ER,MIB1,Ki67, PAb-cyclin A
90°C	AE1,MIB1		MT1,P105,HMB45,p53-1801, DO7
80°C		PAb-TGFß receptor	

Note. H2550 MW oven (EBS, Agawam, MA) was used for heating control, Tris-HCl buffer was used for all test. The heating time was: 100°C and 90°C for 10 min (when using 100°C, two cycles as 5 min x 2 were used), 80°C for 5 min. All antibodies are monoclonal antibodies except those noticed as PAb (polyclonal antibody). RNP = Ribonucleoprotein Ab-1 (Clone 58-15.6).

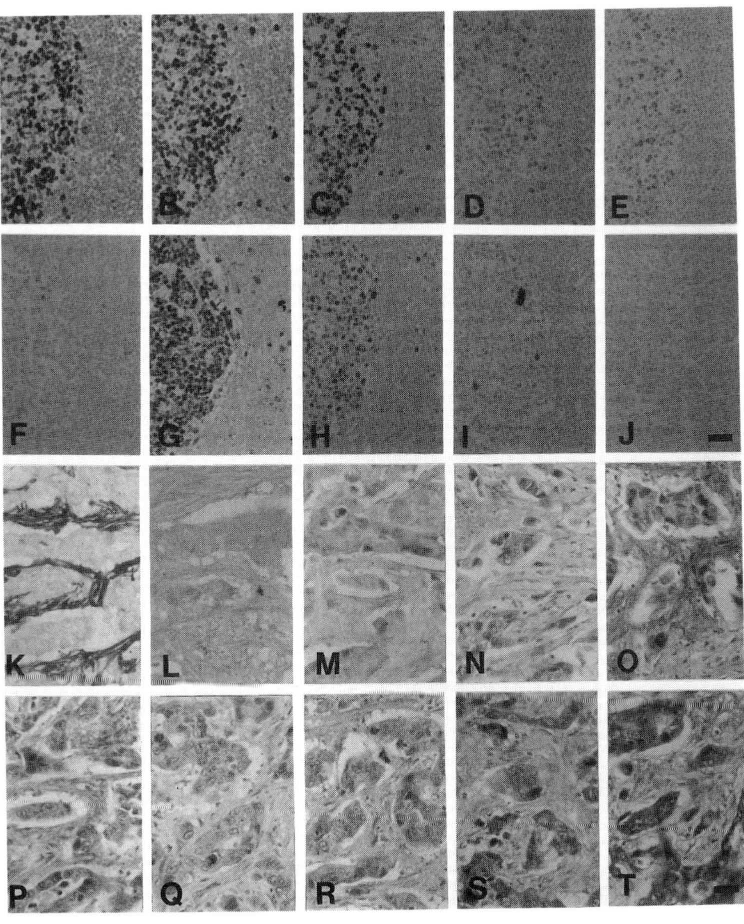

Fig. 1. Comparison of intensity of AR-IHC by using the 'test battery' for monoclonal antibodies to MIB1 of tonsil (A-J) and thrombospondin of bladder carcinoma (K-T). The AR protocols used are arranged in the following order: pH 1, 100°C (A,K), pH 1, 90°C (B,L), pH 1, 80°C (C,M), pH 6, 100°C (D,N), pH 6, 90°C (E,O), pH 6, 80°C (F,P), pH 10, 100°C (G,Q), pH 10, 90°C (H,R), pH 10, 80°C (I,S), None of AR pre-treatment (J,T). For MIB1, the strongest intensity can be found in Fig. 1A, B and G, however, pH 1 showed weak background of false nuclear staining, and hematoxylin was not well stained, as comparing to pH 10 + 100°C (G). The strongest extracellular labelling of thrombospondin was found in Fig. 1 K, at lower temperature (L,M), the intensity was decreased progressively (K>L>M). pH 6 (N,O,P) showed some cytoplasmic positive staining for the cancer cells in addition to a much weaker extracellular staining. pH 10 (Q,R,S) and none of AR pre-treatment (T) also showed cytoplasmic staining.

FIELD EMISSION SCANNING ELECTRON MICROSCOPY AND BACK-SCATTERED ELECTRON IMAGING OF MICROWAVE SILVER-STAINED BACTERIAL LIPOPOLYSACCHARIDE ON CLINICAL SLIDES OF SEPTIC PATIENTS

B. L. Giammara,* R. C. Bone,* D. H. Eubanks,* and J. S. Hanker**
*Center for Clinical Investigation, Medical College of Ohio, Toledo, OH 43699
**Biomedical Engineering Dept. and Dental Research Center, The University of North Carolina, Chapel Hill, NC 27599

Several stains have emanated from the classic work of Hanker et al.(1) which provide highly visable and electron opaque end products. These allow back-scattered electron imaging (BEI) and visualization of polymerized 3, 3'-diaminobenzidine (DAB) or p-phenylenediamine/pyrocatechol (PPD/PC) on endogenous enzyme or nonenzyme hemoproteins, or visualization of exogenous horseradish peroxidase (HRP) transported in tissue or attached to cells, or immunoperoxidase localized to tissue antigen or receptor sites(2,3). Additionally, the Giammara-Hanker PATS reaction (Pat.Appl. No. 08/225,058), originally used to show biomacromolecules or structures such as glycogen, basement membranes, reticular fibers or lipopolysaccharide, appears to be a positive stain for Gram (-) bacteria, including spirochetes, as well as for neutrophils and activated macrophages (4,5). The PATS reaction also demonstrates sites of calcification in tissues and certain calcium compounds by both light and electron microscopy(6). Likewise, an improved FETS reaction uses silver as an electron opaque Feulgen-type reaction for DNA and demonstrates DNA in bacteria and *Candida albicans* (7) as well as other cellular nuclei.

In this field emission study, 21 blood (and other) smears were obtained from patients with sepsis syndrome along with 11 separate bacterial specimens characterized by cell culture. Specimens were prepared on standard 3 x 1 in. glass microscope slides or on coverslips coated with poly-L-lysine to aid adherence and fixed with 4% formaldehyde-1% glutaraldehyde in a 0.1M phosphate buffer. Slides were processed in plastic Columbia jars through the standard series of buffer, periodic acid, and thiocarbohydrazide. The silver methenamine was applied as a droplet to each slide placed horizontally on a Silastic elevated stand. Staining was accomplished in a household microwave oven (GE JEM 23L) using 30 sec. at 10% power level to a target temperature of 50 degrees. Specimens were rinsed, dehydated through increasing concentrations of ethanol to 100%, placed in hexamethyldisilazane (HMDS) for 5 min as a substitute for critical point drying (8) and then allowed to dry in a dessicator overnight. Slides were carbon coated, gold coated or left uncoated. They were examined by commercially available Field Emission systems having BEI attachments and electron micrographs were obtained as analog or digital images.

This step-wise field emission scanning electron microscopy staining procedure accelerated by microwave irradiation, provides a useful rapid method that allows correlative light microscopy (LM), scanning electron microscopy (SEM) and BEI to be performed on whole cells or sectioned specimens carried on coverslips or glass slides. As BEI provides elemental Z Contrast imaging, when used in the reversed polarity mode, the images are consistently and instantly informative, and provide ultrastructural information that can be correlated with other microscopies, including LM and even with images provided by the more difficult and time-consuming techniques of transmission electron microscopy (TEM). Field Emission SEM with BEI offers a new generation of high resolution, high magnification opportunities to examine biological surface cell adhesion molecules (9) and also now, directly relevant clinical specimens on glass microscope slides.

Proc. Microscopy and Microanalysis 1995, edited by G.W. Bailey, M.H. Ellisman, R.A. Hennigar, and N.J. Zaluzec
Copyright © 1995 MSA. Published by Jones and Begell Publishing, 79 Madison Ave., New York, NY 10016

REFERENCES
1. J. Hanker et al., Science 146 (1964) 1039-1043.
2. J. Hanker et al., 43rd Electron Micros. Soc. Am. (1985) 704-705.
3. B. Giammara et al., Scanning 95 (1995) In Press.
4. B. Giammara et al., 47th Electron Micros. Soc. Am. (1989) 1082-1083.
5. J. S. Hanker and B. L. Giammara, XIIth Intl. Cong. Elec. Micros. (1990) 764-765.
6. J. S. Hanker and B. L. Giammara, 49th Elec. Micros. Soc. Am. (1991) 300-301.
7. B. Giammara et al., XIIth Intl.Cong. Elec. Micros. (1990) 762-763.
8. B. Giammara et al., 45th Electron Micros. Soc. Am. (1987) 878-879.
9. S. Erlandsen et al., Scanning 95 (1995) In Press.

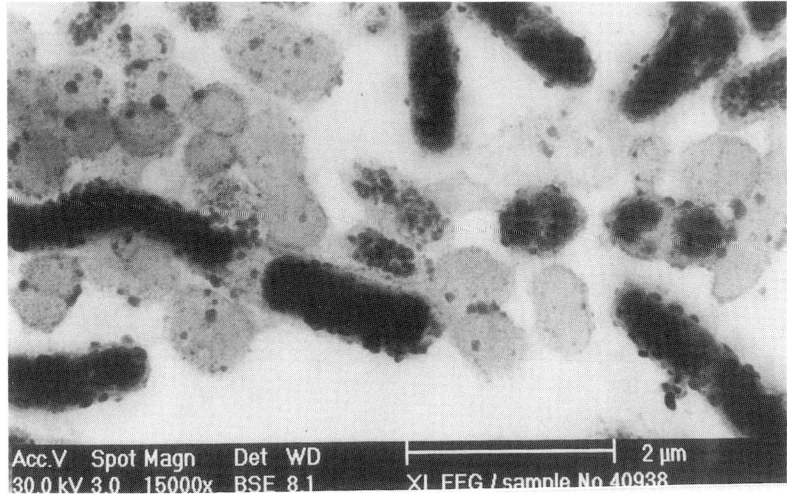

FIG. 1. Microwave applied PATS reaction of a gingival smear from a patient with acute myeloid leukemia shown by reverse polarity back-scattered electron imaging. Note the prominently stained mixed population of cocci and rods. 15,000 X Magnification.

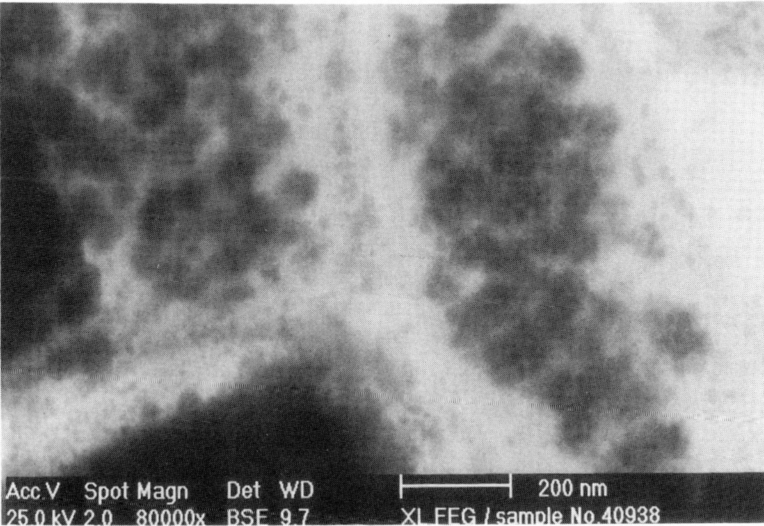

FIG. 2. At higher magnification, on the same slide, the back-scattered electron image shows good definition of bacterial DNA separated by an electron lucent coating of two touching bacteria. 80,000 X Magnification.

DIFFERENCE FOURIER MAPS REVEAL STRUCTURAL CHANGES FOR INTERMEDIATES IN THE BACTERIORHODOPSIN PHOTOCYCLE

R. M. Glaeser,* B.-G. Han,* F. M. Hendrickson, and J. Vonck

Life Sciences Division, Lawrence Berkeley Lab, University of California, Berkeley, CA 94720
*Department of Molecular and Cell Biology, University of California, Berkeley, CA 94720

Bacteriorhodopsin (bR) is a light-driven proton pump that exists in the cell membrane of halobacteria. Proton pumping across the membrane is achieved through a photochemical cycle that involves several discrete structural intermediates. These intermediates are characterized by discrete changes in the visible and the resonance Raman spectra of the chromophore (a retinal group linked to the protein by a Schiff base) and in the infrared spectrum of the protein itself. Protein structural changes are thought to change access to the protonated Schiff base via inward and outward facing aqueous channels, respectively, during different stages of the photocycle, thus allowing for vectorial proton movement across the cell membrane. A schematic model of the functional changes in structure that occur throughout the photocycle is given by the cartoon shown in Figure 1.

We have recently begun electron diffraction studies on the intermediates in the bR photocycle, with the goal of using difference Fourier maps to see directly the structural changes that correspond to the previously observed spectroscopic changes. These structural changes, in combination with the atomic resolution model of the resting state of the protein (Henderson et al., 1990) should give a clear picture of the molecular mechanism involved in proton pumping by this protein. We have used a modification (Perkins et al., 1993) of the glucose-embedded specimen technique because of the relatively high quality of flat (i.e. planar) specimens that can be prepared in this way (Han et al., 1994). The majority of work completed so far has been with the M-state intermediate, which is characterized by 13-cis, isomerized retinal; a deprotonated Schiff base; and distinct changes in the peptide backbone for some of the amino acid residues. The structural changes that are shown in a two-dimensional projection of the difference map are much greater than the noise level in the map, as is shown in Figure 2. Encouraged by this favorable signal-to-noise ratio, we have collected a preliminary three-dimensional (3-D) data set for the M-state. As a first step, electron diffraction patterns were collected at only relatively small tilt angles, up to 30 degrees. The data, collected at 400 kV and with a liquid nitrogen-cooled specimen stage, has an R_{merge} of only 12 per cent, and thus is of comparable quality to data used in x-ray crystallography. The preliminary 3-D difference Fourier map shows that there are small structural changes that are distributed throughout the protein, as is summarized on the graph shown in Figure 3. The resolution in the direction perpendicular to the membrane is limited, of course, by the 30 degree tilt angles used. As our next step, therefore, a second, independent (3-D) data set is being collected at a tilt angle of 45 degrees. The second data set will allow us to determine the statistical reliability of the difference peaks seen in the preliminary map, and the improved resolution will facilitate the interpretation in terms of the atomic resolution model of the structure. Conditions for trapping the L-state intermediate and the N-state intermediate have also been worked out, so that it will be possible to obtain difference Fourier maps for these structures, once it has been established from the work on the M-state intermediate that the signal-to-noise ratio in the 3-D maps is sufficient to permit an interpretation at atomic resolution.

References

Han, B.-G., Wolf, S. G., Vonck, J. & Glaeser, R. M., *Ultramicroscopy* 55(1994)1.
Henderson, R., Baldwin, J. M., Ceska, T. A., Zemlin, F., Beckman, E. & Downing, K. H., *J. Mol. Biol.* 213(1990)899.
Perkins, G., Burkard, F., Liu, E. & Glaeser, R. M., *J. Microscopy* 169(1993)61.

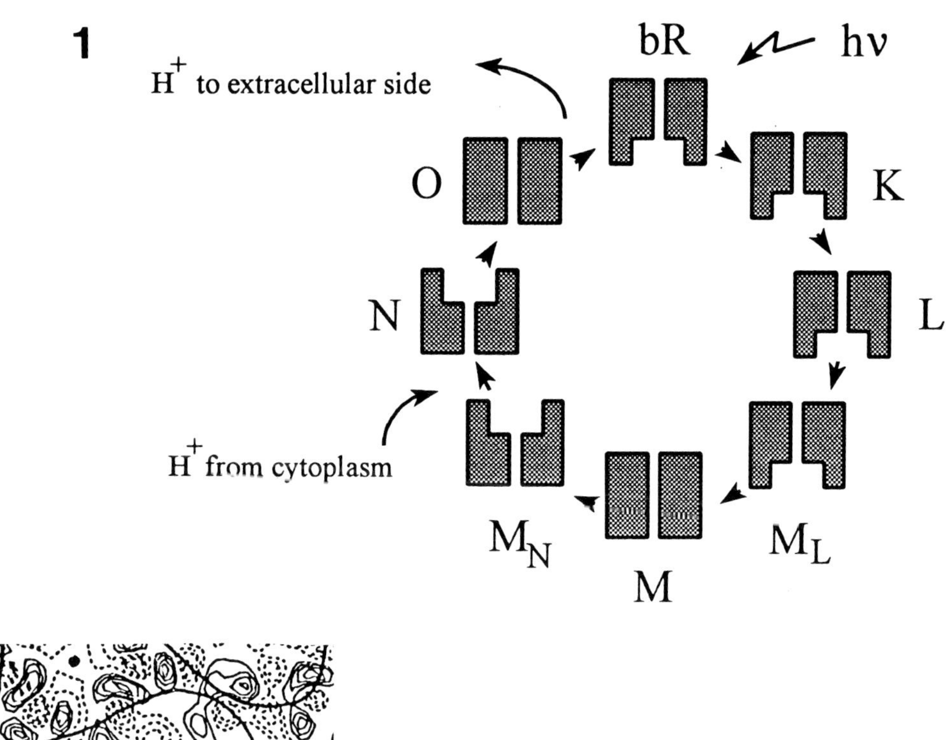

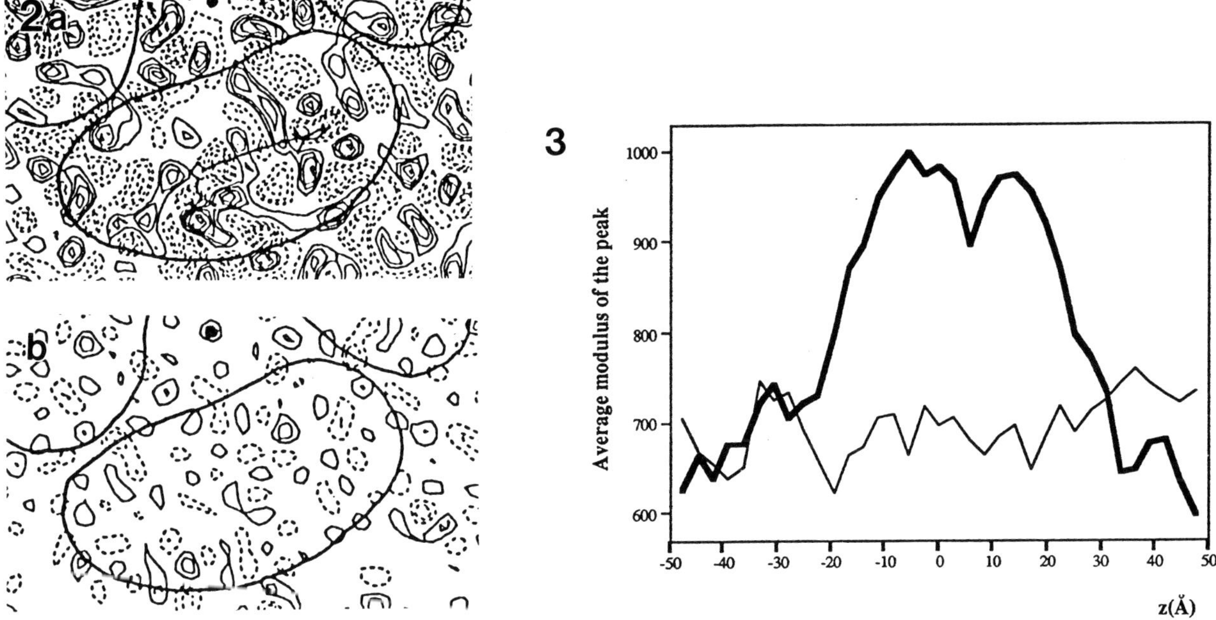

FIG. 1 - The bacteriorhodopsin photocycle.

FIG. 2 - (a) M-bR difference map at 3.8 Å resolution. The outline of the monomer is shown, as well as the position of the retinal. Positive differences are shown as solid lines and negative differences as broken lines. (b) Noise map calculated from two independent M data sets.

FIG. 3 - Distribution of differences in a 3-D M-bR difference map. The horizontal axis shows the height along z from extracellular to cytoplasmic side. The protein extends from -20 Å to 35 Å. The vertical axis shows the average of the absolute values of the peak height in each vertical plane. Thick line: M-bR amplitudes, phases from Henderson et al. (1990); thin line: random amplitudes, same phases.

CHARACTERISTIC FEATURE ON STRUCTURE ANALYZED BY HIGH-RESOLUTION ELECTRON CRYSTALLOGRAPHY

Y. Fujiyoshi,* K. Mitsuoka,* T. Hirai,* K. Murata,* A. Miyazawa,** and Y. Kimura**

*International Institute for Advanced Research, Matsushita Electric Industrial, 3-4, Hikaridai, Seika, Kyoto, 619-02, JAPAN
**Protein Engineering Research Institute, 6-2-3, Furuedai, Suita, Osaka, 565, JAPAN

The structure of bacteriorhodopsin (bR), which was already analyzed by Henderson et al.[1], is studied by our new electron cryo-microscope equipped with Field Emission Gun (FEG) and Slow Scan CCD camera (SSCCD), because our system together with ice embedding technique enable us to solve the structure of bR at various pH conditions between pH 4.0 and 10.0. Ionization of amino acid is naturally closely related to the translocation of proton and then the function of the proton pump of bR. Therefore, observation of translocation of proton in bR is very important, if possible. Both ice embedding and high resolution techniques are essential to achieve this intention. Therefore, we intended to develop an electron cryo-microscope fit to these techniques and recently we had succeeded it.

We collected whole sets of diffraction patterns for bR up to 70 degree tilt at pH 5.5 by using SSCCD, and merged these data of 300 diffraction patterns. The three-dimensional data sets were calculated as shown in Fig. 1 a. where the resolution is 2.8Å. We also collected whole sets of bR including 60 degree tilt at pH 4.0. We analyzed these 200 patterns and merged them. The merged data sets were shown in Fig. 1 b at the resolution of 3.2Å.

Recently, we took many images of bR at pH 5.5 including highly tilted images to 60 degree. A typical example of nice image is shown in Fig. 2 which is Fourier transform of an image of a two-dimensional crystal of bR tilted at 45 degree. Numerical grades (IQ value) for each spot are shown as the size of the symbols in this figure. These Fourier patterns suggest that we can get phase information for structure analysis of bR close to 3.0Å.

The calculated density map of Fig. 3 is indicating a typical example of a side chain contrast assigned to arginine 175 of bR, and on this map the arginine can be particularly well defined because of the strong contrast. On the other hand, some side chains of glutamic acid aspartic acid give very weak or negative contrasts in our map. We can observe these characteristic feature on the structure of plant light-harvesting complex (LHC-II) solved by Kühlbrandt et al.[2]. In LHC-II the arginine side chains extend from a position near the centre of membrane towards the stroma side so that the positively charged side chains are close to the negatively charged glutamic acid side chains from the opposite helix. The two arginine side chains give extremely high contrast, whereas the two glutamic acids give no contrast. Atomic scattering factors for electron are extremely different from those for X-ray as shown in Fig.4 a. Though positively charged guanidinium group of arginine cause higher contrast, the negatively charged carbonyl oxygens give very weak or negative contrast on the density map. Actually in simulated map shown in Fig. 4 b, negatively charged oxygen atom forms negative contrast, on the other hand neutral one gives dark contrast.

These results suggest that electron crystallography could reveal the location and motion of the proton(s) and the pumping mechanism of bR through characteristic atomic scattering factors for charged atoms. The feature is one of strong point of electron crystallography to X-ray crystallography.

References

1. R. Henderson et al., *J. Mol. Biol.* 213 (1990) 899.
2. W. Kühlbrandt et al., *Nature* 367 (1994) 614.

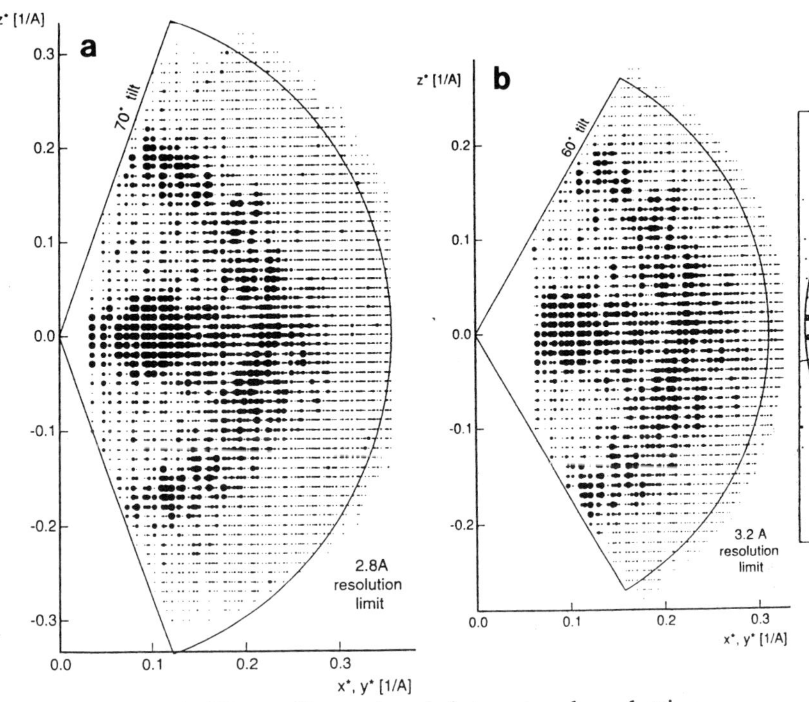

FIG. 1. Three-dimensional data sets where lattice line were sampled at 1/100Å and rotationally projected about z*-axis normal to the membrane plane. (a) bR at pH 5.5 and (b) at pH4.0.

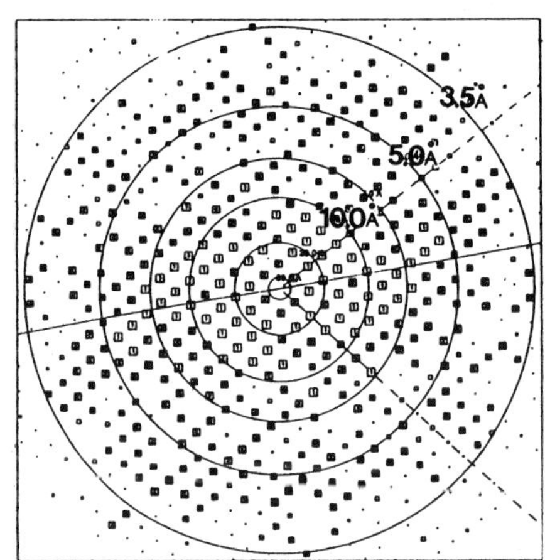

FIG. 2. Fourier transform of an image of bacteriorhodopsin at 45°tilt. The size of the squares is proportional to the quality of the phase information. Tilt axis is indicated by solid line.

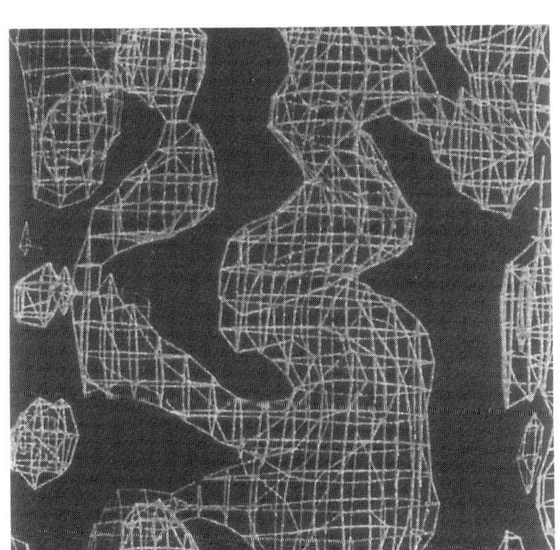

FIG. 3. A part of calculated density map of bR.

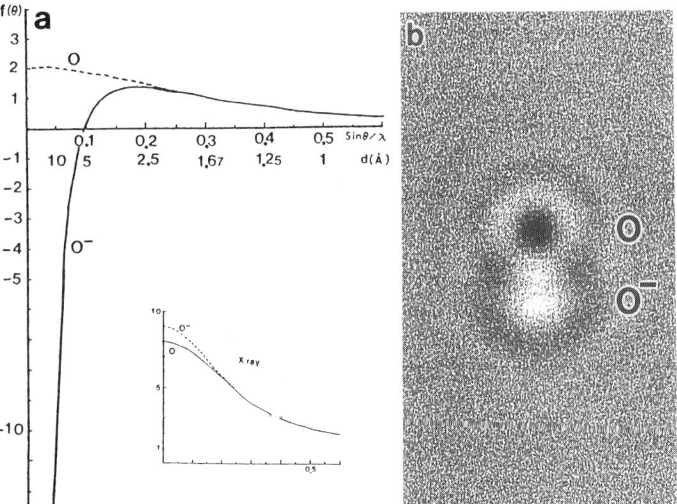

FIG. 4. Atomic scattering factor of neutral and ionized oxygen atoms (a) and computer simulation of 2 oxygen atoms(b). Ionized oxygen gives completely different contrast, negative contrast.

STRUCTURE DESIGN OF OSMO-REGULATED H₂O-CHANNEL INFERRED FROM HIGH RESOLUTION PROJECTION MAP OBTAINED BY ELECTRON CRYSTALLOGRAPHY

Bing K. Jap and Huilin Li

Life Sciences Division, Lawrence Berkeley Laboratory, University of California, Berkeley, CA 94720

The aquaporins (AQP) are a class of water channel proteins belonging to the major intrinsic protein (MIP) superfamily.[1] This class of proteins includes the osmo-regulated H_2O-channel protein AQP-CHIP, which was first identified in erythrocyte membranes and later found to be present in a variety of tissues from organs such as the kidney, gall bladder, lung, intestine and eyes. These channels are believed to be osmotically regulated and to transport water across a number of endothelial and epithelial cell layers during fluid secretion and absorption. AQP-CHIP is extremely specific for the transport of water; small solutes such as urea and ions have been shown to be impermeable to the channels.

We have chromatographically purified AQP-CHIP from bovine red blood cell membranes to homogeneity with the use of octylglucoside as the detergent. We have also reconstituted the purified protein with dimyristoyl phosphatidylcholine, yielding crystalline ordered arrays. To increase the percentage of highly coherent crystalline patches, the reconstituted membrane patches were further treated with phospholipase A_2 to remove excess lipids.

The crystalline patches embedded in glucose yielded electron diffraction patterns with reflections extending to a resolution of about 3.0Å. The crystals belong to the space group $P42_12$ with the unit cell dimensions of a = b = 96.4 ± 0.2 Å and γ = 90°, determined by the use of ice diffraction as the internal standard. A typical electron diffraction pattern shows strong intensities to a resolution of about 10Å, beyond which the intensities are rather weak. The particularly strong reflections within a zone, centered at about 10Å, would suggest the presence of α-helical bundles having a helix-axis to helix-axis spacing of about 10Å. The intensities from sixteen diffraction patterns were merged with an overall R(merge) of 25% to a resolution of 3.5Å and used in the reconstruction of the high resolution map.

High resolution images of the crystalline patches embedded in glucose were recorded, digitized and processed. Processing includes unbending distortions of the crystalline lattice and applying corrections for the effect of image defocus and astigmatism. A typical image gives phase information extending to 3.4Å resolution as shown in Figure 1. Phases obtained from the images and amplitudes derived from the diffraction patterns were combined to calculate the projection map at 3.5Å resolution, as shown in Figure 2. The map shows that AQP-CHIP exists as a tetramer in reconstituted membranes, confirming the conclusion drawn from previous biochemical and biophysical studies as well as low resolution electron microscopy. The map shows that each monomer is clearly defined and has a trapezoid-like envelope. Within the trapezoid, there is also a substructure that may play a crucial role in the channel structure itself. The trapezoid-like envelope is primarily composed of high density regions that can be interpreted as projections of α-helices along their axes.

References

1. P. Agre et al., *Am .J. Physiol.* 265(1993)F463.
2. R. Henderson et al., *Ultramicroscopy* 19(1986)147.
3. This work was supported by NIH grant #GM51487 and by the Office of Health and Environmental Research, U.S. Department of Energy, under contract DE-AC03-76SF00098.

Proc. Microscopy and Microanalysis 1995, edited by G.W. Bailey, M.H. Ellisman, R.A. Hennigar, and N.J. Zaluzec
Copyright © 1995 MSA. Published by Jones and Begell Publishing, 79 Madison Ave., New York, NY 10016

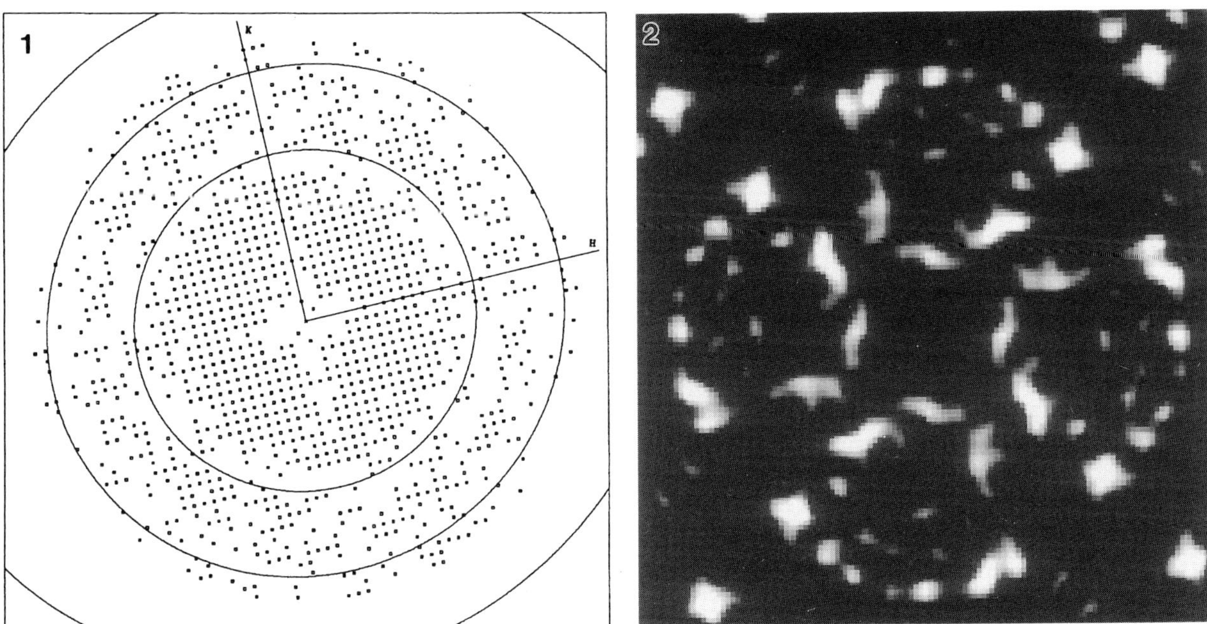

FIGURE 1. Structure factors obtained from image after correction of distortions. Strength of reflections is categorized into 9 ranges with IQ = 1 to be the strongest.[2] In this plot, reflections with IQ $\geq$ 8 are not shown. Nearly circular rings represent zero values of contrast transfer function. Edge of plots corresponds to resolution of 3.1Å.

FIGURE 2. Gray-scale plot of projection map of AQP-CHIP at 3.5Å resolution (protein shown in white). Map shows that osmo-regulated water channel exists as tetramer. Each monomer has a well defined, trapezoid-like envelope formed by high density regions, many of which have peak-to-peak separation of about 10Å between their nearest high density regions, and weaker density peak near center of trapezoid. Many of high density regions on trapezoid-like envelope can be interpreted as projection of transmembrane α-helices.

3D STRUCTURE OF SINGLE MACROMOLECULES AT 15 Å RESOLUTION BY CRYO-MICROSCOPY AND ANGULAR RECONSTITUTION

Prakash Dube, Holger Stark, Elena V. Orlova, Michael Schatz, Erich Beckmann, Friedrich Zemlin, and Marin van Heel,

Fritz-Haber-Institut der Max-Planck-Gesellschaft, D-14195, Berlin, Germany.

Electron cryo-microscopy of individual non-crystallized macromolecules ("single particles") is a very rapid technique for probing the three-dimensional ("3D") structure of biological macromolecules. By exploiting the different orientations of the macromolecules in the embedding medium, one may extract 3D information from the data without ever collecting tilt series in the microscope. The angular reconstitution approach [1], designed for this purpose, was recently extended with a number of refinements which allow its use as a routine technique for finding 3D structures of macromolecules with arbitrary pointgroup symmetry [2,3], from entirely asymmetric ribosomes to viruses with icosahedral symmetry.

The specimen preparation technique associated with the angular reconstitution approach is simple and fast since crystallization experiments are avoided altogether. The vitreous-ice embedding specimen preparation technique remains one of our favorite specimen preparation techniques. We are, however, currently experimenting with specimens embedded in glucose and ammonium molybdate or other heavy-metal salts. Collecting good micrographs can also be quite straightforward since the images are taken from untilted specimens, while exposing each image area only once. Due to these practical advantages over other techniques of 3D analysis, the angular reconstitution approach is well suited for studying macromolecular systems in different conformational states [2]. Moreover, thanks to the simplicity of data acquisition, collecting the high number of molecular images (~10.000) needed for obtaining high-resolution results is unproblematic.

The approach is illustrated by our analysis of the *Lumbricus terrestris* hemoglobin which is the oxygen-carrying protein in the blood of the common earthworm. This huge oligomer, with a molecular weight of about 3.9 million Dalton [4,5], was shown to have 622 point-group symmetry by X-ray crystallography [6] and by electron microscopy [6,3]. A publication of its 3D structure at 30Å resolution from vitreous-ice embedded specimens is forthcoming [3]. Our new 15Å resolution structure of glucose/heavy-metal embedded hemoglobin (almost identical to the earlier vitreous-ice reconstruction when filtered down to the appropriate resolution) reveals many new details including intricate interconnections between the 1/12th subunits. Careful interpretation and refinement of the new structure may help solve a controversy concerning the stoichiometry of the assembly [4,5]. Our refined electron microscopical structure is currently being used to bootstrap the phases of X-ray diffraction data [6] to high resolution.

For attaining high-resolution results with our approach a number of issues need to be addressed carefully including: **1**. The electron microscopical imaging conditions must be such that the desired spatial frequencies are registered with a sufficiently high contrast in the micrographs. The SOPHIE FEG-equipped helium-cooled microscope fulfills all requirements in this respect [8]. **2**. The particles must be imaged with high contrast in the low spatial-frequency region, to allow for automatic or visual localization of the particles against the background. **3**. The sampling grid must be sufficiently fine to support the desired high frequencies in spite of interpolation artifacts introduced during processing.

Proc. Microscopy and Microanalysis 1995, edited by G.W. Bailey, M.H. Ellisman, R.A. Hennigar, and N.J. Zaluzec

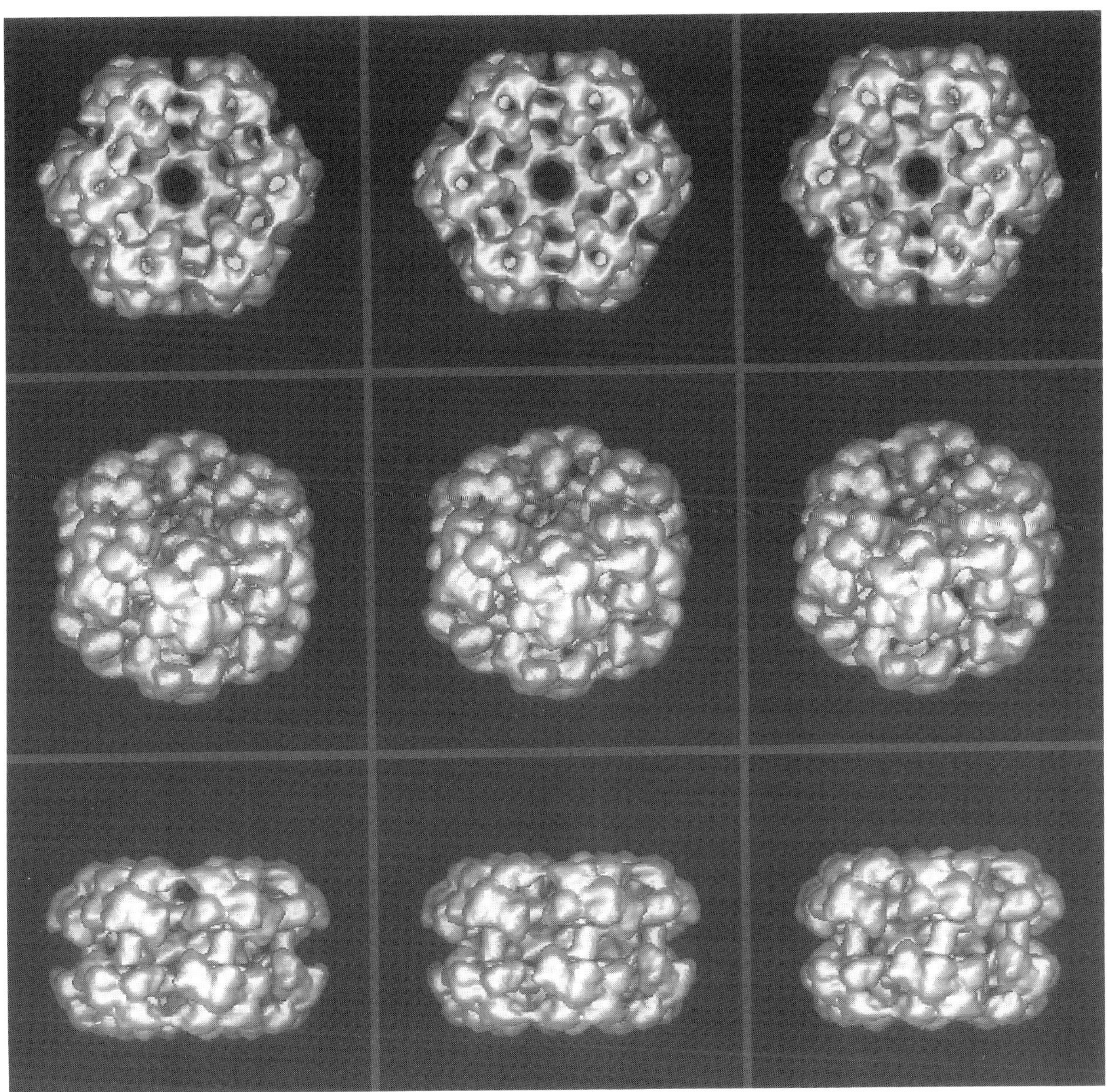

Figure: Stereo views of the 3D Structure of *Lumbricus terrestris* hemoglobin at 15 Å resolution. The 622 pointgroup symmetry structure contains a well-defined central "doughnut" feature, thought to consist of non-heme linker chains. The 1/12th "asymmetric" subunit contains a local 3-fold axis [3] which is confirmed here at higher resolution. This 1/12th, subunit is connected to the neighboring 1/12th subunits and to the central doughnut by a number of masses which may, again, be linkers chains.

[1] M. van Heel, *Ultramicroscopy* 21 (1987) 111-124
[2] I. Serysheva et al., *Nature Struct. Biol.* 2 (1995) 18-14
[3] M. Schatz et al. *J. Struct. Biol.* (1995) in press
[4] D.W. Ownby et al. *J. Biol. Chem.* 268 (1993) 13539-13547
[5] S.N. Vinogradov et al. *Comp. Biochem Physiol.* 98b (1991), 187-194
[6] W.E. Royer, et al. *J. Mol. Biol.* 197 (1987) 149-153
[7] E.J. Boekema and M. van Heel *Biochim. Biophys. Acta* 957(1989) 370-379
[8] H. Stark et al, these proceedings

ASSEMBLY OF VP26 IN *HERPES* SIMPLEX VIRUS-1 CAPSID IMPLICATED BY 3D STRUCTURE OF RECOMBINANT CAPSID

Z. Hong Zhou*, Jing He**, Joanita Jakana*, J. D. Tatman***, Frazer J. Rixon***, and Wah Chiu*,**

*Dept. of Biochemistry, **Program of Structural and Computational Biology and Molecular Biophysics, W.M. Keck Center for Computational Biology, Baylor College of Medicine, Houston, TX 77030; and ***MRC Virology Unit, Institute of Virology, Glasgow G11 5JR, Scotland

Herpes simplex virus-1 (HSV-1) is a ubiquitous virus which is implicated in diseases ranging from self-curing cold sores to life-threatening infections. The 2500 Å diameter herpes virion is composed of a glycoprotein spike containing, lipid envelope, enclosing a protein layer (the tegument) in which is embedded the capsid (which contains the dsDNA genome). The B-, and A- and C-capsids, representing different morphogenetic stages in HSV-1 infected cells, are composed of 7, and 5 structural proteins respectively. The three capsid types are organized in similar T=16 icosahedral shells with 12 pentons, 150 hexons, and 320 connecting triplexes.[1,2] Our previous 3D structure study at 26 Å revealed domain features of all these structural components and suggested probable locations for the outer shell proteins, VP5, VP26, VP19c and VP23.[3] VP5 makes up most of both pentons and hexons. VP26 appeared to bind to the VP5 subunit in hexon[3,4] but not to that in penton[3].

In order to confirm the location of VP26 and to obtain insights into its assembly, we have carried out comparisons of the 3D structure of the HSV-1 B-capsid with a recombinant baculovirus generated capsid lacking VP26.[5] The combined use of 400kV spot-scan electron cryomicroscopy and a projection-based refinement procedure has enabled us to process very close to focus images (Fig. 1) and thus to improve the resolution of the reconstruction beyond 26 Å. The 3D structure of the −VP26 recombinant capsids revealed pentons, hexons and triplexes organized on a surface lattice similar to that of the naturally occurring capsids (Fig. 2). The horn-shaped mass densities observed at the upper distal end of each hexon subunit in the naturally occurring capsid are not seen in recombinant capsids lacking VP26 (Figs. 2, 3). As shown in Fig. 3, comparison of a hexon isolated from the 3D map of the naturally occurring B-capsid and that of the −VP26 recombinant capsid revealed pronounced difference at the upper distal end. The horn-shaped density is completely missing from the −VP26 hexon, confirming the exact location for VP26 which we previously proposed based on comparison of penton and hexon subunits. The structure of VP26 appears to consist of a major and a minor domain which attach to the face and vertex respectively, of the upper domain of VP5 (Fig. 3).

An interesting question is why VP26 does not bind to the penton subunit. It appears that there is no steric hindrance affecting the equivalent position in the VP5 of penton subunits. The difference map illustrating the position of VP26 in the hexon, suggests that there are important interactions between adjacent VP26 molecules as well as between VP5 and VP26. One possible pathway of VP26 assembly which might account for this behavior is the pre-assembly of VP26 into hexamers (Fig. 3c) before attaching to VP5. Attachment would then occur exclusively through binding to the hexons of a preformed −VP26 capsid.[6]

References

1 J.D. Schrag et al., *Cell* 56(1989)651.
2 T.S. Baker et al., *J. Virol.* 64(1990)563.
3 Z.H. Zhou et al., *J. Mol. Biol.* 242(1994)456.
4 F.P. Booy et al., *Proc. Natl. Acad. Sci. USA* 91(1994)5652.
5 J.D. Tatman et al., *J. Gen. Virol.* 75(1994)1101.
6 This project is supported by W.M. Keck Foundation, NLM, NCRR and GM of NIH.

Proc. Microscopy and Microanalysis 1995, edited by G.W. Bailey, M.H. Ellisman, R.A. Hennigar, and N.J. Zaluzec
Copyright © 1995 MSA. Published by Jones and Begell Publishing, 79 Madison Ave., New York, NY 10016

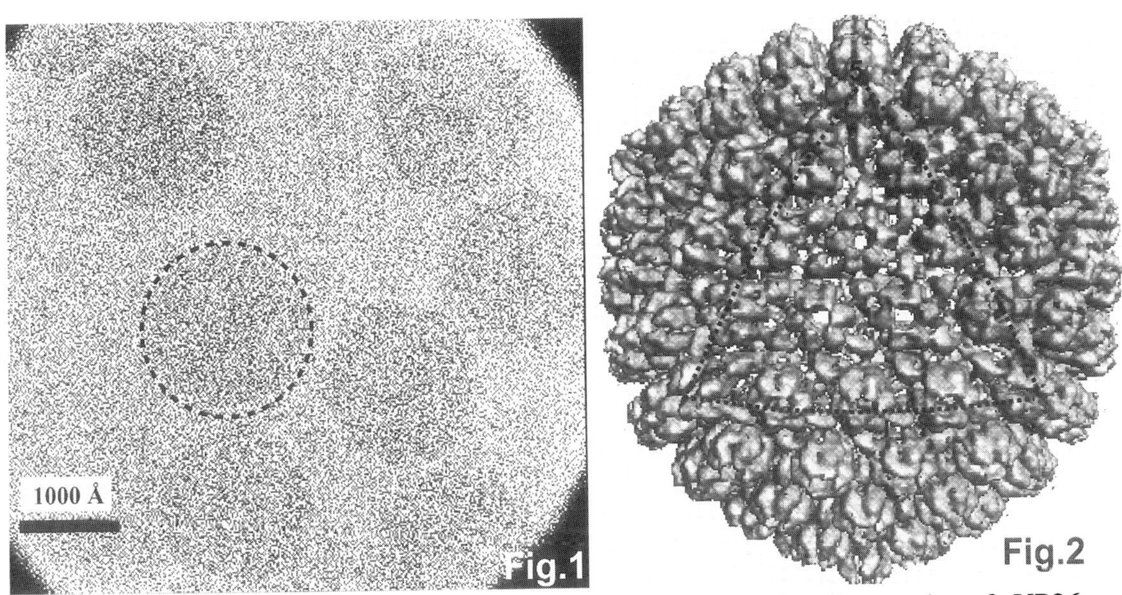

Fig. 1 400 kV spot-scan image of –VP26 capsids embedded in vitreous ice. Preparation of –VP26 capsids using the recombinant baculovirus expression system has been described[5]. A panel of recombinant baculoviruses each expressing one HSV capsid gene were generated. Co-infecting insect cells with baculoviruses expressing 5 of the 6 capsid protein genes results in the formation of capsids lacking VP26. The capsids were embedded in a layer of vitreous ice and imaged in a JEOL4000 electron cryomicroscope. A computer-controlled spot-scan procedure was used to minimize electron beam-induced movement of the specimen. Shown here is the closer to focus (defocus=1.8 μm) image in a focal pair. Indicated by dashed circle is one capsid particle.

Fig. 2 3D structure of –VP26 capsid reconstructed from 290 particles to a nominal resolution of 17 Å. The T=16 shell is composed of 60 asymmetric units ($\frac{1}{3}$ dotted triangle) each of which contains 1 penton subunit, $2\frac{1}{2}$ hexons, and $5\frac{1}{3}$ connecting triplexes.

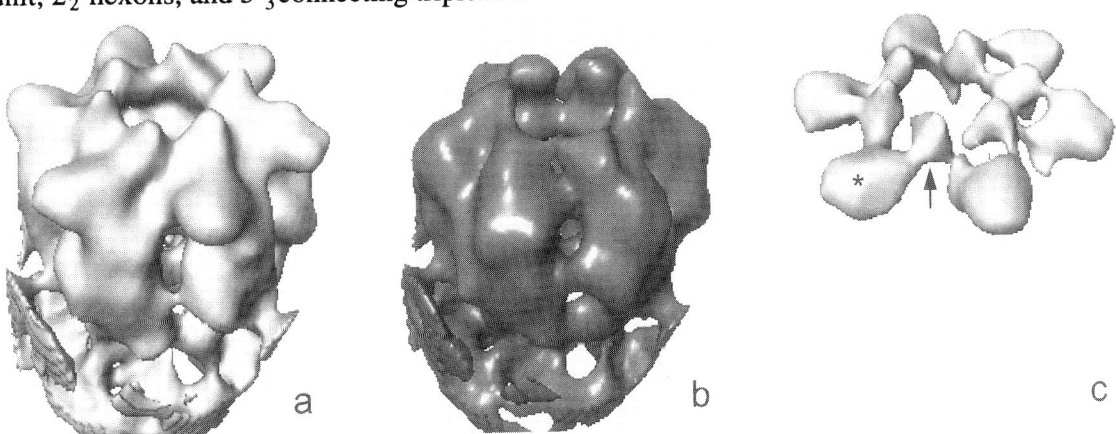

Fig. 3 Hexons computationally isolated from naturally occurring B-capsid (a), –VP26 recombinant capsid (b), and their difference map (c). The horn-shaped density at the distal end exists ONLY in the naturally occurring capsid, confirming the exact location of VP26. Each VP26 appears to consist of a major (*) and a minor (⦿) domain which attach to the face and the vertex respectively, of the upper domain of VP5. The major and minor domains of the adjacent VP26 subunits associate with each other to form an intact gear-like hexamer.

CRYO ELECTRON MICROSCOPY OF SSV1 PHAGE PARTICLES

U. Ziese, D. Typke, R. Hegerl and W. Baumeister

Max-Planck-Institut für Biochemie, Am Klopferspitz 18a, 82152 Martinsried, Germany

SSV1 phages are lemon-shaped particles, normally about 90 nm x 40 nm in size, with short tail fibres attached to one pole, produced by the thermophilic archaeon *Sulfolobus shibate*, isolate B12.[1] They are made of 3 different proteins and DNA (15.5 kbp). Two proteins, together with host lipid, form the envelope, the third protein is associated with the DNA. Interestingly, this virus produces particles of varying size and shape. We have investigated the mass of the virions by STEM mass determination, the inner structure by cryo-electron microscopy, and the shape variability by electron tomography. Automatic electron tomography (AET) has been shown to be a useful technique for collecting 3D structural data of individual biological particles under low dose conditions, in negative stain as well as in frozen-hydrated preparations.[2-5]

Taking electron micrographs of vitrified samples we obtained images revealing some details of the inner structure and, on some particles, a periodic structure of the envelope. (Fig. 1a-b) The inner structure has periodicities of about 2.5 nm, which is in agreement with that found by Lepault et al [6] on vitrified samples of the phages lambda and T4. The period found in the SSV1 envelope is somewhat smaller, 2.2 nm. We are now recording tilt series at a resolution sufficient to visualize the inner structure. The main problem is the high sensitivity of these structures, which fade at doses higher than about 10 e/Å^2. The image intensity of STEM dark field images, which are made using elastically scattered electrons, is directly proportional to the projected mass of the sample.[7] We investigated 155 SSV1 particles by this method using the STEM set-up at the Maurice E. Müller-Institut in Basel. The images were acquired with 200 kV, 200 kX magnification, and 3.3 e/Å^2 dose.The distribution of total masses was calculated by the IMPSYS program package.[8] It shows one well defined peak. (Fig. 2) Therefore we conclude that the varying shape of SSV1 particles is not caused by different masses, but only by their flexibility. The phage particles have an average mass of 22.8 ± 1.1 MDa.

Several tilt series were recorded semi-automatically using the AET set-up in Martinsried, including a CM20FEG electron microscope and a CCD camera. We used the following acquisition parameters: 120 kV accelerating voltage, 70 μm objective aperture, 4 μm defocus, 27 kX EM magnification, 0.72 nm pixel size. Typically a tilt series consists of 31 projections in the range + 60 ° to - 60 ° with 4 ° tilt angle increments. 3D reconstructions were calculated by weighted backprojection using the EM software package [9] on a Silicon Graphics Indigo computer. The envelopes of the particles and the tail structures can clearly be recognized. (Fig. 3) The determination of the particles' shape is impaired by artifacts due to the missing data in the high-tilt angle range. This produces an elongation of the particles in the z-direction. Therefore, the true shapes can not be determined. However, tomography allows us to determine the orientation of the particles. The shape and sizes can be evaluated assuming axial symmetry of the particles. We recently implemented a new goniometer stage (Compu-Stage) into the CM20FEG in Martinsried, together with modifications of the cryoholder, thus enabling tilt angles in the range - 70 ° to + 70 °.

References

1. C. Schleper, K.Kubo, W. Zillig, Proc. Natl. Acad. Sci 89 (1992) 7645.
2. D. Typke, K. Dierksen, W. Baumeister, Proc. 49th Ann. M. of EMSA, San Jose (1991) 544.
3. K. Dierksen, D. Typke, R. Hegerl, A.J. Koster and W. Baumeister, Ultramicro. 40 (1992) 71.
4. K. Dierksen, D. Typke, R. Hegerl and W. Baumeister, Ultramicroscopy 49 (1993) 109.
5. K. Dierksen, D. Typke, R.Hegerl, J. Walz, E. Sackmann and W. Baumeister, Biophys. J., in press
6. J. Lepault, J. Dubochet, W. Baschong and E. Kellenberger, EMBO Jour., 6,5 (1987), 1507.
7. A. Engel, Ultramicroscopy 3 (1978) 273-281.

Proc. Microscopy and Microanalysis 1995, edited by G.W. Bailey, M.H. Ellisman, R.A. Hennigar, and N.J. Zaluzec

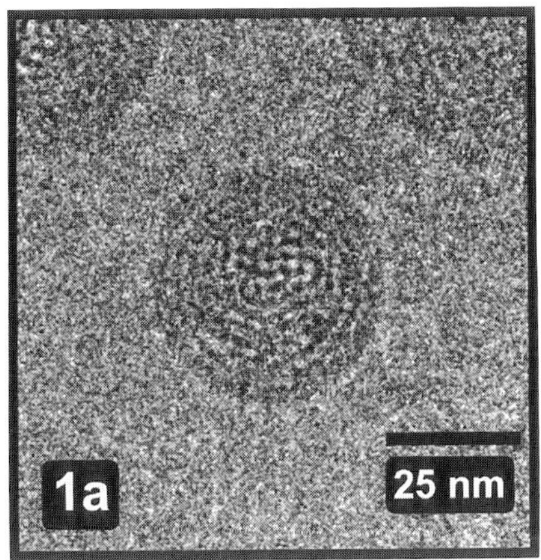
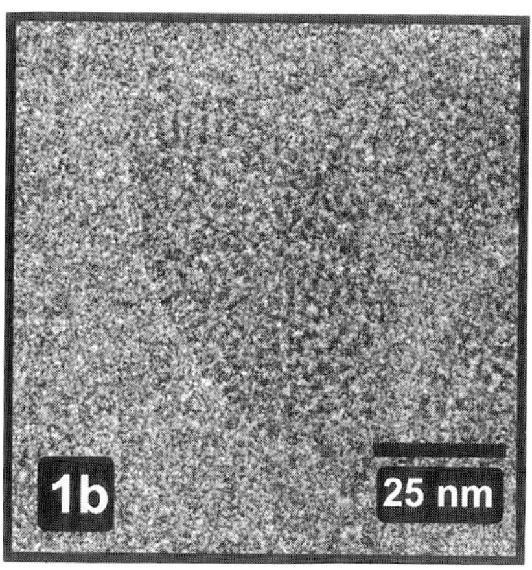

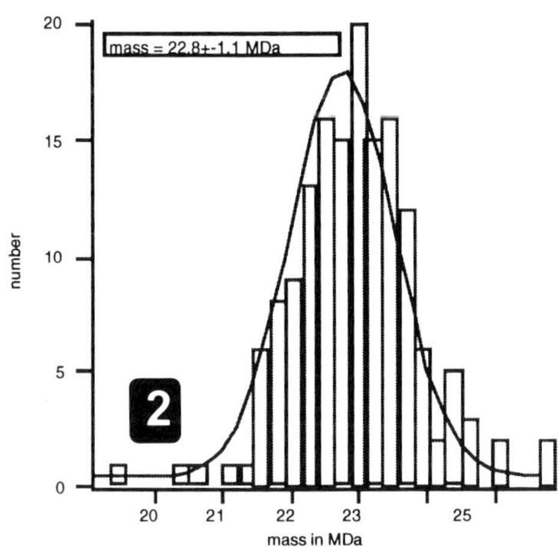
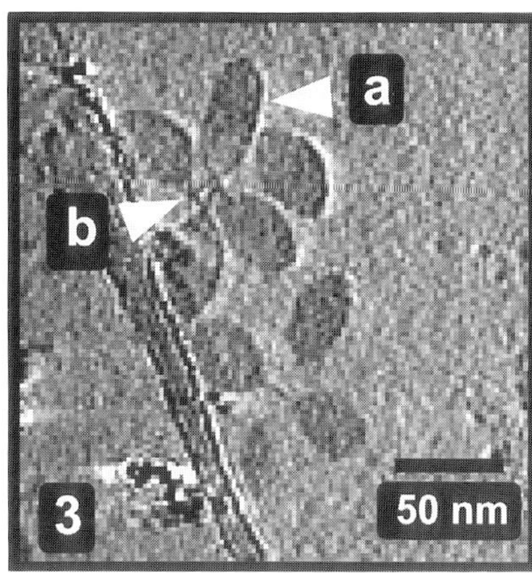

FIG. 1 - Projections of ice-embedded SSV1 particles. (a) Particle in top view. (b) Particle in side view.

FIG. 2 - Mass distribution of SSV1 phage particles. The average mass according to the Gauss-Fit is 22.8 ± 1.1 MDa.

FIG. 3 - Slice through the low-resolution 3D reconstruction of ice-embedded SSV1 particles. (a) Envelope. (b) Tail structures.

8. S. Müller, K. Goldie, R. Bürki, R. Häring and A. Engel, Ultramicroscopy 46 (1992) 317-334
9. R. Hegerl, A. Altbauer, Ultramicroscopy 9 (1982) 109.
10. We are grateful to C. Schleper, who helped to grow the bacterial cells (MPI für Biochemie, Germany), M. Cyrklaff (EMBL Heidelberg,Germany) for his help with cryoelectron microscopy, and to K. Goldie and S.Müller (Maurice E. Müller-Institut, Biocentrum, University of Basel, Switzerland) for their work with the STEM mass determination.
This work was supported by the DFG (SFB 266, A2).

LOW-DOSE ENERGY-FILTERED TEM OF FROZEN HYDRATED VIRUS PARTICLES USING A POST-COLUMN ENERGY FILTER

R. Grimm and W. Baumeister

Max-Planck-Institut für Biochemie, Molekulare Strukturbiologie, 82152 Martinsried, Germany

Zero-loss filtered imaging of frozen hydrated macromolecular structures has become a commonly applied method over the last ten years.[1,2,3] Instruments with in-column filters have been dominant due to their commercial availability (maximum voltage 120 keV), and because they allow the microscope to be used as for standard imaging, with essentially no modification in the procedure for low-dose microscopy. Post-column filters have been used mainly in materials science and for intermediate and high voltage microscopes.

At the MPI in Martinsried we have recently added a Gatan post-column imaging filter[4] (model 678) to an existing CM 12 electron microscope (120 keV, LaB_6 cathode). The lowest magnification of the microscope with the objective lens excited is 600x at the viewing screen (lowered from standard 3000x by free lens control). The imaging filter uses a central portion of about 2 mm diameter of the image on the screen, and adds an additional magnification of 18x. For extreme low-dose conditions, the image intensity should be adjusted so that only this very small area on the viewing screen is illuminated. Behind the energy selective elements, the filter offers a tv-rate CCD camera for real time viewing and a slow-scan CCD camera (SSC) for image acquisition. One computer essentially controls the energy filter, the microscope, and the SSC.

Two procedures for low-dose imaging have been developed at the MPI, both of which work with three different preset microscope states (which do not correspond to the states defined on the CM 12):

"Overview State" - offers an overview at 3000x magnification with a spread beam on the viewing screen. It is used to select an appropriate mesh for observation. The dose on the sample is 0.1-1 electrons/$Å^2$s.
"Search State" - allows energy filtered imaging at a magnification of 12000x. The image can be viewed either on the tv-rate camera (at a dose of 0.3-1 electrons/$Å^2$s) or by taking single frames with the SSC (0.1 electrons/$Å^2$ per exposure). On this image, an area can be selected for higher magnification imaging using the computer mouse.
"Exposure State" - is the state with the desired final magnification for imaging (72000x in the case stated below). One frame is taken at a dose of 1-10 electrons/$Å^2$.
In the "Search" and "Exposure" states the beam is contracted to about 3 mm diameter on the viewing screen, in order to minimize radiation damage to adjacent areas on the specimen.

A first evaluation of the filter's performance is illustrated here using a frozen hydrated virus, *SIRV*.[5] As the filter is mounted below the microscope column, cryo work is conducted as usual. Fig. 1 shows images of the virus without (a) and with (b) energy filtering. The ice thickness was more than 100 nm. The energy slit width was 10 eV centered around the zero-loss peak. The scaling is mean ± 2 standard deviations in both cases. Individual viruses are depicted in fig. 2. The enhancement in contrast is visible to the naked eye and can be seen more clearly in the power spectra in fig. 3. The power spectrum of the unfiltered image shows a weak first order peak (4 nm). The spectrum of the filtered image reveals a strong first order peak and a second order peak that can be made visible with careful contrast scaling.

Future work will include a more quantitative examination of the contrast enhancement by zero-loss energy filtering, investigation of thicker structures, and three-dimensional reconstruction of frozen hydrated particles using random conical tilt or tomographic methods.

Proc. Microscopy and Microanalysis 1995, edited by G.W. Bailey, M.H. Ellisman, R.A. Hennigar, and N.J. Zaluzec

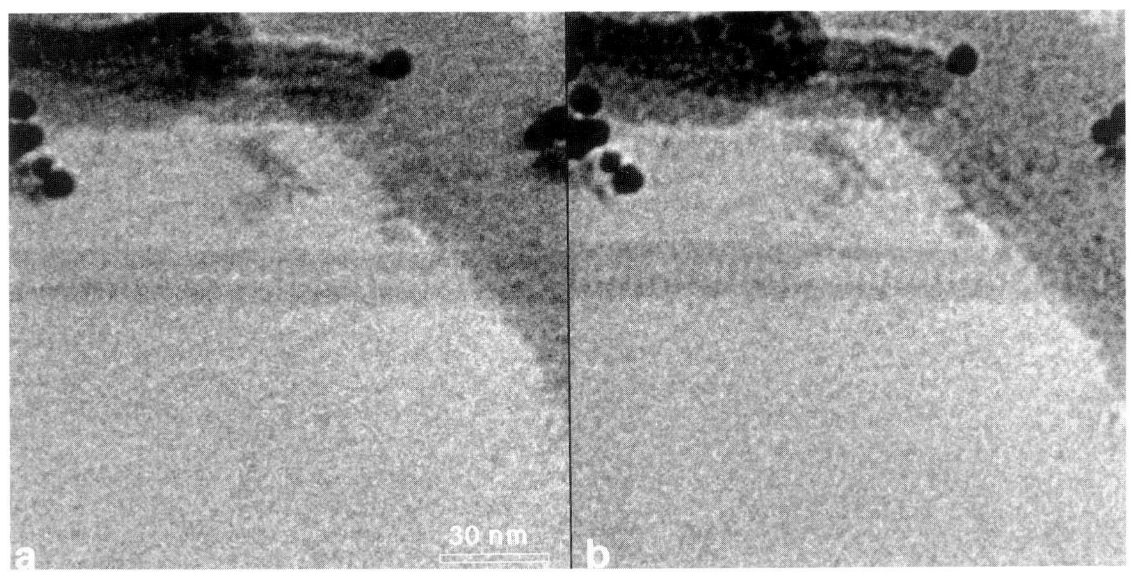

FIG. 1.—Comparison of low-dose EFTEM images, (a) unfiltered, (b) filtered. Ice thickness 200 nm.

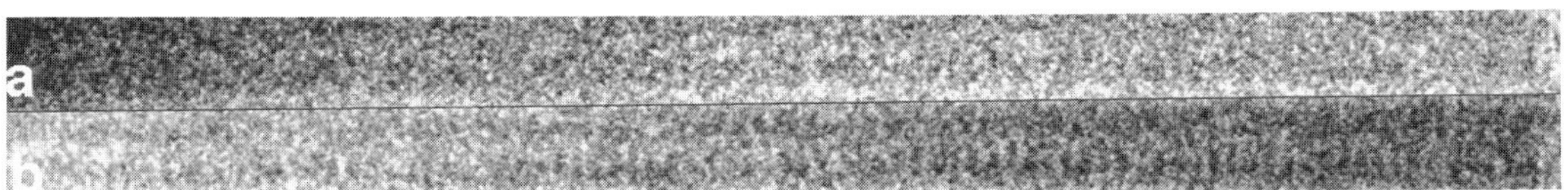

FIG. 2.—Individual virus strains, (a) unfiltered, (b) filtered. Ice thickness 140 nm.

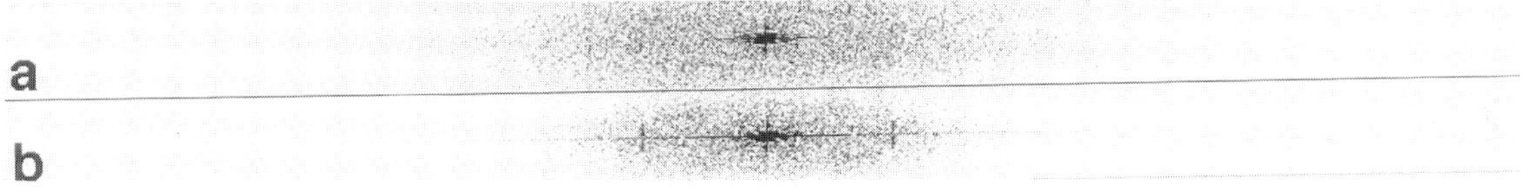

FIG. 3.—Power spectra of fig. 2.

References

1. R.R. Schröder et al., *Journal Of Structural Biology* 105 (1990) 28.
2. J.P. Langmore and M.F. Smith, *Ultramicroscopy* 46 (1992) 349.
3. W.C. de Bruijn et al., *Scanning Microscopy* 7 (1993) 693.
4. Krivanek et al., *Microscopy, Microanalysis, Microstructures* 2 (1991) 315.
5. W. Zillig et al., *Systematic And Applied Microbiology* 16 (1994) 609.
6. The author thanks Prof W. Zillig and D. Prangishvilli for providing the purified virus particles. This work was supported by a scholarship from the Stiftung Stipendien-Fonds des Verbandes der Chemischen Industrie.

FRESNEL INTERFERENCE IN POINT-PROJECTION IMAGING OF PURPLE MEMBRANE

X. Zhang, J. Spence, W. Qian, D. Taylor* and K. Taylor*

Department of Physics and Astronomy, Arizona State University, Tempe, AZ 85287-1504
*Department of Cell Biology, Box 3011, Duke University Medical Center, Durham, NC 27710

Experimental point-projection shadow microscope (PPM) images of uncoated, unstained purple membrane (PM, bacteriorhodopsin, a membrane protein from Halobacterium holobium) were obtained recently using 100 volt electrons.[1] The membrane thickness is about 5 nm and the hexagonal unit cell dimension 6 nm. The images show contrast around the edges of small holes, as shown in figure 1. The interior of the film is opaque. Since the inelastic mean free path for 100V electrons in carbon (about 6 Å)[2] is much less than the sample thickness, the question arises that how much, if any, transmission of elastically scattered electrons occurs. A large inelastic contribution is also expected, attenuated by the reduced detection efficiency of the channel plate at low energies. Quantitative experiments using an energy-loss spectrometer are planned. Recently Shedd[3] has shown that at about 100V contrast in PPM images of thin gold films can be explained as Fresnel interference effects between different pinholes in the film, separated by less than the coherence width. The gold films may thus be treated as opaque masks containing holes. The large defocus in the PPM produces a highly delocalised image - as in high energy electron holography, bright areas do not necessarily indicate locally high electron transmission. By bringing images back into focus and eliminating this spreading of image information, holographic reconstruction (or simulation) may be used to estimate the degree of local transmission. The experimental PM image in figure 1(a) shows an *isolated* hole, however the fringes running normal to the edge (into the film) in a lattice-like pattern seem inconsistent with Fresnel edge fringes, which must follow along the edge. Assuming the PM to act as an opaque mask, we see that the CCD detector was highly saturated inside the hole (bright band) due to the strong direct electron beam, which may have washed out all the Fresnel edge fringes within the hole. Thus we may treat the shadow edge in the image as an approximate geometric projection, i.e., as the "in-focus" edge of the hole. This edge shows considerable roughness. Under this assumption the mask shown in figure 1(b) was retrieved and holograms simulated for different defocii, as shown the one in figure 1(c) for defocus 550 nm. Because of the equivalence between the PPM geometry (with spherical-wave illumination) and the conventional TEM geometry (with plane-wave illumination)[4] (see also Zhang et al, this proceedings), the simulation used plane-wave illumination with the object size adjusted with the variation of defocus (tip-to-sample distance), since this changes the magnification in the PPM. Lattice-like crossed fringes, which are similar to those in the experimental image, are clearly seen in the simulation, obviously due to the roughness along the edge which results in a complex *two dimensional* pattern of fringes quite different from the Fresnel fringes found parallel to a smooth edge. The rough agreement (e.g. figs 1(d), 1(a)) suggests that there may be negligible electron transmission through the film. By comparing the simulations with the original image, especially by judging the size of dots, we may also estimate the experimental tip-to-sample distance (defocus) to be about 550 nm. Adjusting the contrast in the simulated image, in order to eliminate fringes inside the hole region, shows that the idea that shadow edge in the experimental image is a direct geometric projection of boundary of the hole in the film may not be entirely justified. We have shown, however, that edge roughness can produce crossed Fresnel fringes which resemble a lattice image.

By preparing a specially blunt tip, the PPM may be operated at higher voltage (but lower resolution). By this method we find that PM is reasonably transparent above about 1 kV.

Recently we have also imaged DPPC C16 lipid monolayers whose thickness is about 2 nm. Figure 1(f) shows an experimental image of the monolayer across a hole in a supporting carbon film. Although the lipid region appears much more transparent than the surrounding carbon, it will be necessary to reconstruct these images (in-line holograms) in order to determine how much transmission occurs. Experiments aimed at observing increased transmission at very low energies are in progress.

Proc. Microscopy and Microanalysis 1995, edited by G.W. Bailey, M.H. Ellisman, R.A. Hennigar, and N.J. Zaluzec
Copyright © 1995 MSA. Published by Jones and Begell Publishing, 79 Madison Ave., New York, NY 10016

References

1. J. C. H. Spence et al., Ultramicroscopy 55 (1994) 19.
2. C. Martin et al., J. Electr. Spectr. Rel. Phenom. 42 (1987) 171.
3. G. M. Shedd, J. Vac. Sci. Technol. A(12) (1994) 2595.
4. J. C. H. Spence, Optik. 92 (1992) 57.
5. Supported by NSF award DMR-9116362. We thank Y. Fujiyoshi for the PM samples.

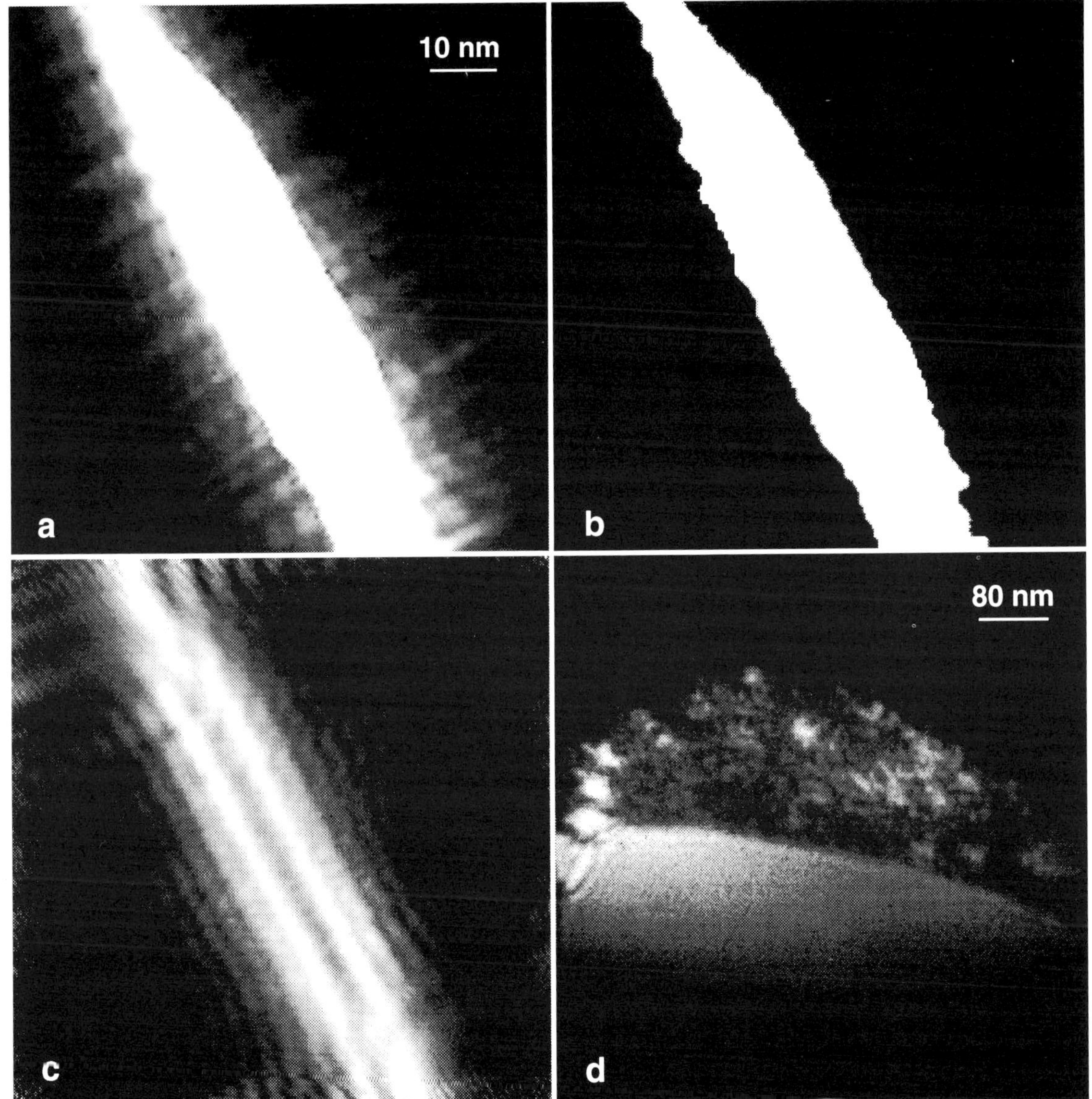

Figure 1. (a) Experimental image of Purple Membrane at 100 volts. (b) The mask shape extracted from (a) used to simulate the hologram in figure (c) (defocus 550 nm). (d) Experimental image of Lipid Monolayer obtained at 160 volts.

OBLIQUE SECTION 3-D RECONSTRUCTION FROM THIN SECTIONED BIOLOGICAL SPECIMENS

K. A. Taylor and H. Winkler

Department of Cell Biology, Duke University Medical Center, Durham, NC 27710

Three dimensional (3-D) image reconstruction of plastic embedded and sectioned tissue has become an important tool for imaging complex biological structures *in situ*. Most 3-D imaging methods combine multiple views of tilted specimens which leads to radiation damage and mass loss. On the other hand oblique section 3-D reconstruction (OSR) can produce a 3-D image from a single view because an oblique section through a 2- or 3-D periodic specimen contains 3-D information which can be folded into a 3-D image[1]. The resolution achievable in an OSR depends on section thickness and is lowest in the direction perpendicular to the section plane. In OSR there is no missing cone or wedge in the 3-D image because of tilt angle restrictions, but finite section thickness limits the vertical resolution. Two OSR methodologies have been developed, one called Crystallographic Serial Section Reconstruction or CSSR[2,3], and another called Super Lattice Reconstruction or SLR[4].

The resolution limitations of OSR can be overcome by several methods. When the section thickness is known, the reconstruction can be deconvoluted to maximizing the resolution. In OSR, the section thickness is as important a parameter as defocus is for 2-D imaging. This importance derives from the fact that section thickness as well as the section orientation defines the point spread function. Fortunately, the section thickness is encoded in the image itself and can be recovered by several methods.[3-6] Deconvolution can be done on individual reconstructions but leads to at best an improvement in the vertical resolution to about half of the section thickness.[1] Further improvements can be obtained by averaging deconvoluted data from sections cut at different angles.[5] This averaging is most easily done using the SLR method which combines image transforms to recover the 3-D transform of the original object. The resolution improvement obtained by averaging data from sections cut oblique to only one axis is somewhat limited (Fig. 5). The highest possible resolution is obtained by combining data from images cut oblique to different unit cell axes.[6] For example, a section cut oblique to the *a b* plane of the unit cell (Fig. 1) is combined with sections cut oblique to the *b-c* plane (Fig. 3) and the *a-c* planes. Their Fourier transforms reveal modulated structure factors from the 3-D transform of the fixed and embedded crystal (Figs. 2,4). When these data are merged and deconvoluted a reconstruction with nearly isotropic resolution is obtained (Fig. 6). Deconvolution to recover the structure factors of the fixed and embedded crystal is facilitated when data from as many section orientations as possible are combined because the measurements for a particular structure factor will not usually be clustered at nodes in the point spread function.

Combination of data from oblique sections cut at different angles using OSR gives a new capability to 3-D imaging of sectioned crystals. Once the section thickness is known, data from sections of different thickness can be readily combined. Thus, minimal dose images can be used for data recording and combined in a manner analogous to 2-D crystal methods. If X-ray diffraction data from the native crystal are available, power spectra comparisons can be made to detect changes after fixation, embedding and staining or the phases can be combined with X-ray amplitudes to produce a 3-D image limited only by the quality of the EM preservation.[7]

1. R. A. Crowther. *Ultramicroscopy* 13(1984)295.
2. R. A. Crowther and P. K. Luther, *Nature* 307(1984)569.
3. K. A. Taylor and R. A. Crowther. *Ultramicroscopy* 38(1991)85.
4. K. A. Taylor and R. A. Crowther. *Ultramicroscopy* 41(1992)153.
5. K. A. Taylor et al., *J. Mol. Biol.* 233(1993)86.
6. H. Winkler and K. A. Taylor. *Ultramicroscopy* 13(1994)295.
7. Supported by NIH grant GM-30598

Proc. Microscopy and Microanalysis 1995, edited by G.W. Bailey, M.H. Ellisman, R.A. Hennigar, and N.J. Zaluzec
Copyright © 1995 MSA. Published by Jones and Begell Publishing, 79 Madison Ave., New York, NY 10016

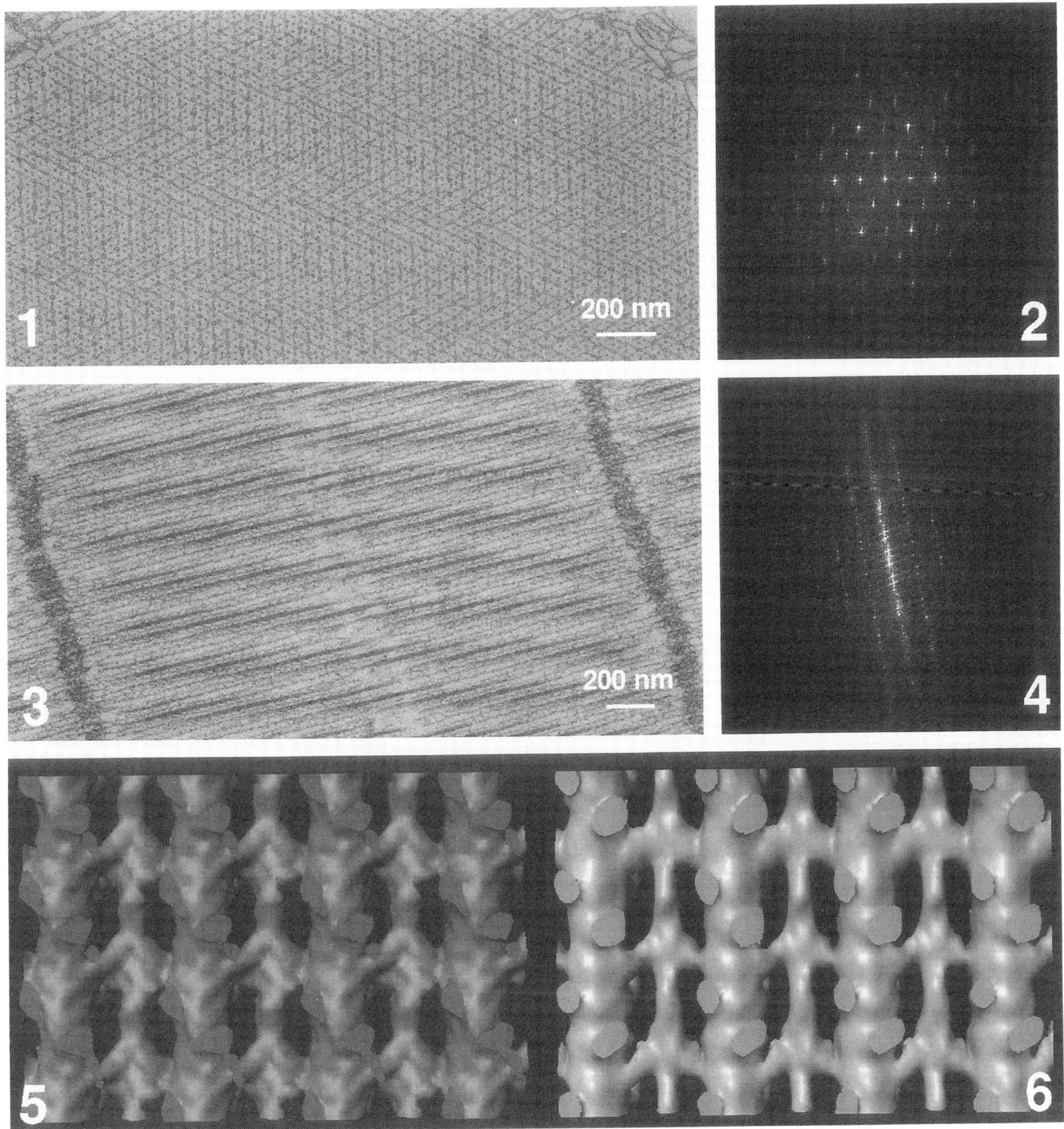

Figure 1. EM of a thin section cut oblique to the *a-b* plane of the IFM filament lattice.

Figure 2. Fourier transform of the area shown in Fig. 1. The superlattice spots are modulated structure factors from upper layer planes of the 3-D crystal.

Figure 3. EM of a thin section cut oblique to the *a-c* plane of the IFM filament lattice.

Figure 4. Fourier transform of the area shown in Fig. 3. The data shown in Fig. 2 has its best resolution along a^* and b^*. This transform has best resolution along a^* and c^*.

Figure 5. 3-D reconstruction of rigor IFM obtained by combining only sections cut oblique to the *a-b* plane of the unit cell.

Figure 6. 3-D image of IFM in the presence AMPPNP obtained by combining sections cut oblique to both the *a-b* planes and the *a-c* planes.

COMPARISON OF 13Å HELICAL AND 7Å CRYSTALLOGRAPHIC PROJECTION MAPS OF FROZEN, HYDRATED ACROSOMAL BUNDLE FROM *LIMULUS* SPERM

Michael F. Schmid*, Joanita Jakana*, Paul Matsudaira** and Wah Chiu*

* Verna and Marrs McLean Department of Biochemistry and The W.M. Keck Center for Computational Biology, Baylor College of Medicine, Houston, TX 77030
** Whitehead Institute for Biomedical Research and Department of Biology, Massachusetts Institute of Technology, Cambridge, MA. 02142

Actin associates with crosslinking proteins to form bundles and networks. These assemblies can provide a scaffold to anchor organelles, a support for the cell membrane or can attach the cell to the extracellular matrix or to other cells. Each crosslinker must have at least two actin binding sites, one for each actin filament.[1] Scruin, in the *Limulus* acrosomal bundle, shows no homology with other actin cross-linking proteins.[2] However, it has two homologous domains also found in the sequence encoded by *kelch*, a gene in *Drosophila* that is important in nutrient transport into the oocyte,[3] and in MIPP, a mouse placental protein.[4]

We have determined a 13Å helical 3-dimensional structure of the actin-scruin complex, which is the basic repeat unit of the acrosomal bundle in *Limulus* sperm[5] and have fitted the F-actin filament of Holmes.[6] Each scruin binds to two adjacent actin molecules along a single F-actin filament helix. In the present investigation, we used our high resolution (7Å) spot scan electron images[7] of ice-embedded acrosomal bundles for further analysis of the actin atomic map and scruin-scruin contacts. Spot-scan images[8] were obtained from the frozen, hydrated acrosomal bundles in a 400 kV electron cryomicroscope.[7] A P1 projection map was calculated with all the reflections IQ≤7 up to 7 Å resolution. The 2_1 phase residual for each image averaged 20°.

Figure 1a is a projection of our previously determined 3D reconstruction[5] which covers one complete actin crossover of 381Å, and used in the cross-correlation with the high resolution projection map. Figure 1b shows three side-by-side unit cells of a 7Å projection map calculated from three merged spot scan images. An outlined box locates the cross-correlation of Figure 1a with this projection. Since we determined the actin orientation in the helical reconstruction, we can bring the Holmes F-actin model into the same orientation (Figure 1c) as that in Figures 1a and 1b. The fitting suggests possible scruin-scruin contacts between adjacent filaments, but until we further analyze this packing in the crystal, we cannot pinpoint the interactions that make up the bundle.

References:

1. Matsudaira, P. *Trends Biochem. Sci.* **16**, 87-92 (1991).
2. Way, M. et al. *Journal of Cell Biology* **128**, 51-60 (1995).
3. Xue, F. & Cooley, L. *Cell* **72**, 681-694 (1993).
4. Chang-Yeh, A. et al. *Nucleic Acids Research* **19**, 3667-3672 (1991).
5. Schmid, M.F. et al. *Journal of Cell Biology* **124**, 341-350 (1994).
6. Holmes, K.C. et al. *Nature* **347**, 44-49 (1990).
7. Schmid, M.F. et al. *Journal of Molecular Biology* **230**, 384-386 (1993).
8. Downing, K.H. & Glaeser, R.M. *Ultramicroscopy* **20**, 269-278 (1986).
9. Supported by the NIH NCRR grants (RR02250), DK35306 and W. M. Keck Foundation.

Captions:

Figure 1a. A projection map calculated directly from the 3-D helical reconstruction.

b. Projection mass density map of three unit cells of acrosomal bundle seen in *h0l* view (*a*= 147 Å and *c*= 762 Å). This map was generated from the merged structure factors including reflections out to 7 Å resolution. The light color represents the protein (high) density and the dark color represents the solvent (low) density. Outlined by a dotted line is the position of the best cross-correlation of the projection map (Figure 1a) with the central unit cell.

c. The F-actin filament model based on the X ray fiber diffraction and the monomer crystal structure[6] which has been brought to coincidence with Figure 1b from previous results.[5]

Scale bar: 127Å (one unit cell translation in this direction).

Proc. Microscopy and Microanalysis 1995, edited by G.W. Bailey, M.H. Ellisman, R.A. Hennigar, and N.J. Zaluzec
Copyright © 1995 MSA. Published by Jones and Begell Publishing, 79 Madison Ave., New York, NY 10016

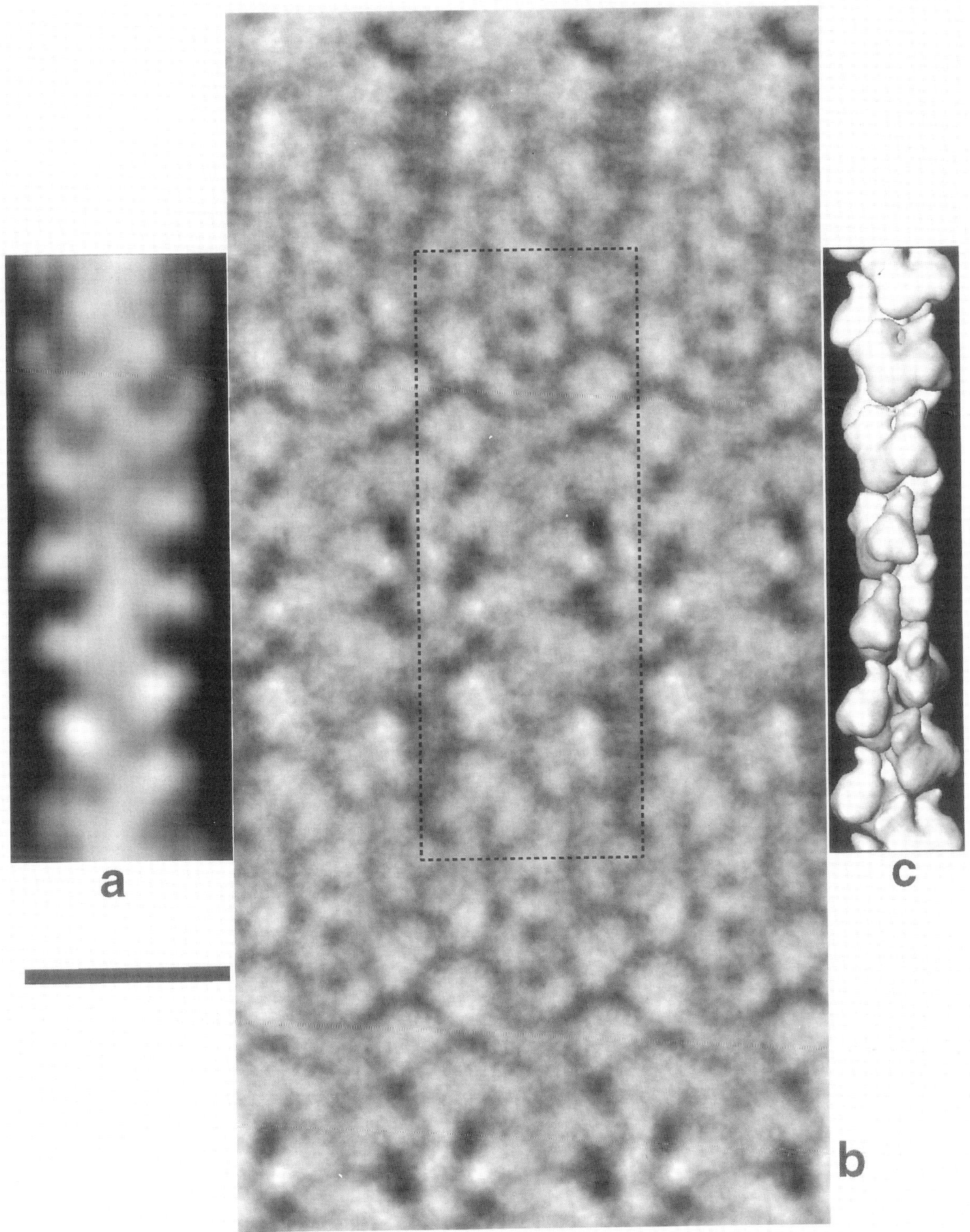

851

TUBULIN STRUCTURE AT INTERMEDIATE RESOLUTION

K. H. Downing, S. G. Wolf and E. Nogales

Life Science Division, Lawrence Berkeley Laboratory, Berkeley, CA 94720

Microtubules are involved in a host of critical cell activities, many of which involve transport of organelles through the cell. Different sets of microtubules appear to form during the cell cycle for different functions. Knowledge of the structure of tubulin will be necessary in order to understand the various functional mechanisms of microtubule assemble, disassembly, and interaction with other molecules, but tubulin has so far resisted crystallization for x-ray diffraction studies. Fortuitously, in the presence of zinc ions, tubulin also forms two-dimensional, crystalline sheets that are ideally suited for study by electron microscopy. We have refined procedures for forming the sheets and preparing them for EM, and have been able to obtain high-resolution structural data that sheds light on the formation and stabilization of microtubules, and even the interaction with a therapeutic drug.

Tubulin sheets had been extensively studied in negative stain, demonstrating that the same protofilament structure was formed in the sheets and microtubules. For high resolution studies, we have found that the sheets embedded in either glucose or tannin diffract to around 3 Å. Projection maps have been computed to 3.8 Å from images of the sheets under a variety of conditions, and all show consistently the same features, although these features have been difficult to interpret in projection. Contrast matching by the embedding medium produces very low contrast for the molecular envelope. By comparison with samples either negatively stained or first rinsed in uranyl acetate and then glucose embedded, we can relate the high-resolution details to the overall molecular shape.

Amplitudes of structure factors that are obtained from the images show a falloff with increasing resolution similar to what has been observed for images of other 2-D crystals, notably purple membrane. When the structure factor amplitudes from the images are replaced by electron diffraction amplitudes in computing the projections, low resolution features are nearly obscured by features at around 4.8 Å that correspond to a set of particularly intense spots in the diffraction patterns. Figure 1 illustrates the appearance of the projections when the either image or diffraction amplitudes are used, and when the image amplitudes are scaled by a temperature factor that nearly matches the falloff of the image amplitudes.

Difference images between the different sets of projection data have proven quite informative. For example, the binding site of the anti-mitotic drug taxol has been identified. Difference images have also been used to identify the region of the C-terminus of the amino acid chain, by locating a segment that is cleaved by subtilisin, and to investigate differences between isotypes of the tubulin monomers. In computing these difference images, we have used only information from the images, rather than including electron diffraction data. This approach is in contrast to the usual approach to difference image calculation in x-ray crystallography, where it is assumed that phases of structure factors are unaffected by the modifications to the proteins and the difference is calculated from the amplitude changes alone. In our case, the amplitudes from the images are not as well defined as they would be from electron diffraction, but the phases are so accurate that the difference maps show a high degree of reliability.

Understanding the structure of the molecule requires building a three-dimensional density map. From a collection of images of specimens tilted up to 60 degrees, we have generated a 3-D map. We initially faced the same problem with tubulin that has often plagued other 2-D crystals, that as the specimen is tilted, diffraction in the direction perpendicular to the tilt axis becomes limited. For this reason, we have truncated the resolution in the 3-D map to 6.5 Å. Up to this resolution we have data that is reasonably isotropic and complete .

Proc. Microscopy and Microanalysis 1995, edited by G.W. Bailey, M.H. Ellisman, R.A. Hennigar, and N.J. Zaluzec
Copyright © 1995 MSA. Published by Jones and Begell Publishing, 79 Madison Ave., New York, NY 10016

At the resolution of our map, a number of features are seen that have the shape expected for alpha helix or beta sheet structures. One cylindrical density, with the axis nearly perpendicular to the plane of the sheet, appears similar in shape to regions of alpha helix in similar orientations that have been seen in previous reconstructions at a similar resolution. There are also several flattened slabs of density, apparently beta sheet, that lie in regions where, in projection, sets of stripes are observed at a spacing around 4.8 Å, characteristic of the inter-strand spacing of beta sheet.

Interpretation of the map at this resolution must, of course, be tentative, but several of the features suggest a possible interpretation. Secondary structure prediction algorithms suggest the presence of only one alpha helical region long enough to fit the cylindrical density mentioned above. This segment of the amino acid chain is near the C-terminus. The positioning of this section in the apparent helix, which is near both the intermonomer and interprotofilament contacts, would be consistent with the fact that variations in the sequence near the C-terminus have a strong effect on polymerization characteristics. Identification of the taxol binding site indicates the likely region of the N-terminus, since there is evidence that taxol binds within the last 31 residues of the sequence. There is also evidence suggesting that GTP also interacts with residues near the N-terminus, indicating one region in which we may expect to see bound nucleotide. While these interpretations may be speculative, they demonstrate that even at a resolution not nearly high enough to resolve amino acid side chains we can already begin to formulate hypotheses about the functional significance of the structure, and these hypotheses may guide further experiments.

Ultimately, confirmation of the structure requires that we extend the resolution to near 3.5 Å. Recent improvements in specimen preparation now allow collection of isotropic data from specimens tilted to at least 45 degrees. Thus we are well poised to produce a density map that can be interpreted in terms of an accurate chain-tracing model of the structure.

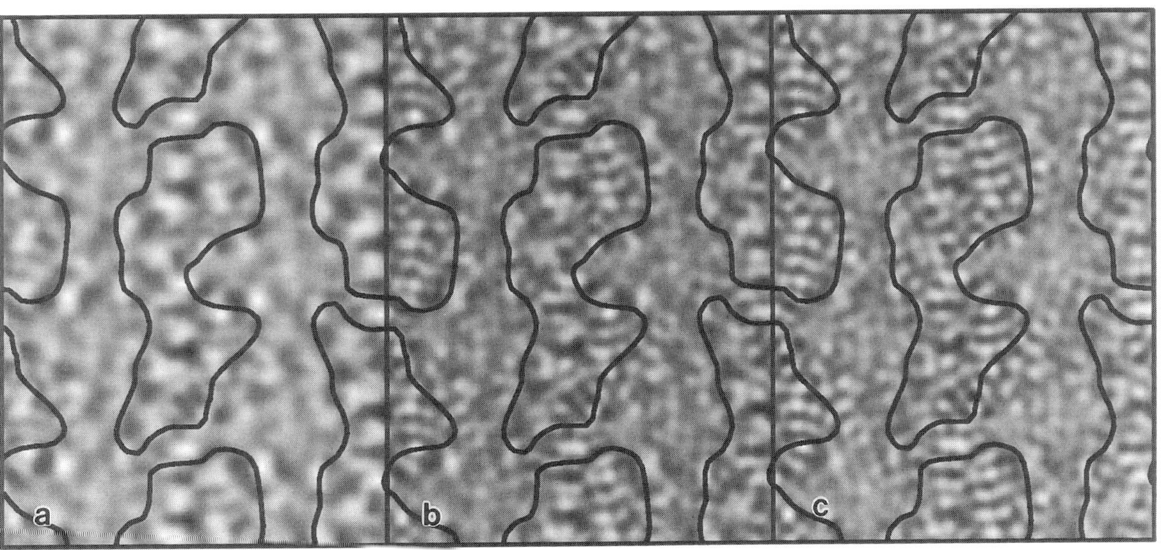

Figure 1. Projection maps of tubulin illustrating effects of scaling Fourier amplitudes. (a) Reconstruction using unscaled image amplitudes. (b) Temperature factor of -120 applied in reconstruction to restore high-resolution features. (c) Reconstruction using amplitudes obtained from electron diffraction. Curves are from contour map of negatively stained tubulin sheet, to indicate molecular outline

DIRECT PHASE EXTENSION IN PROTEIN ELECTRON CRYSTALLOGRAPHY - TESTS WITH DATA FROM *ESCHERICHIA COLI* OMP F PORIN

Douglas L. Dorset

Electron Diffraction Department, Hauptman-Woodward Medical Research Institute, Inc., 73 High Street, Buffalo, NY 14203-1196

Phase determination in the high-resolution electron crystallographic analysis of membrane protein structures has been based most often on the Fourier transforms of low-dose electron micrographs. With increasing resolution, this task becomes increasingly difficult because of inherent paracrystalline lattice distortions, requiring some 'unbending' operation to access information beyond e. g. 6 Å. In addition, the specimen itself will be more prone to radiation damage as the demands for greater resolution are imposed. For this reason, Gilmore, *et al.*[1] have suggested that direct methods, e. g. maximum entropy and likelihood, be employed to extend easily available lower resolution phase information to the resolution of the electron diffraction pattern. More recently, we have carried out phase extensions[2] for bacteriorhodopsin and halorhodopsin, *via* the Sayre equation[3], which is simply a convolution of phased structure factors. Despite pessimistic estimates from x-ray crystallography[4], the extension, e. g. from 15 to 10 Å to a resolution limit of 6 Å, is relatively easy, so that new phases for bacteriorhodopsin have a mean error of 50°, whereas those corresponding to the most significant magnitudes from halorhodopsin have only a 36° error (but 72° for all new reflections). Thus, relatively low-resolution images define phase values for the region of reciprocal space most difficult to access in x-ray protein crystallography, which corresponds to the most intense region of the diffraction pattern. This means that definition of the molecular envelope is defined by electron microscopy while higher resolution details can be obtained from this basis by conventional crystallographic procedures.

The bacterial porins represent a class of protein distinct from the ones already tested - i. e. they are composed of mainly β-sheet rather than α-helix. Experimental (image-based) phases and amplitudes[5] from the *E. coli* Omp F porin were used in trials of phase extensions with the Sayre equation. Starting from a 10 Å basis set, extension to 6 Å resulted in an overall phase error of 53°, with 83° error in the new phases. Extension to 5 Å gives an overall mean error of 66° but 85° for the new phases. The agreement of the derived and 'actual' structures (Figure 1) is sufficient (as it had been for other proteins) to begin phase refinement. The nearness of the high-resolution phase error to the random estimate (90°) may be an over-estimate of actual error, because the experimental phase deviations in this region themselves are rather large (ca. 60°).

Phase extensions to a resolution higher than 5 Å from a lower resolution basis set are somewhat more problematic, as recognized also by x-ray crystallographers[6]. For bacteriorhodopsin, a 'phase-node' was encountered, requiring a multisolution branching point in the phase determination[2]. The actual solution was found by testing the smoothness of the resultant potential maps. For the Omp F porin, extension to the nominal 3.7 Å resolution of the pattern cannot be distinguished from the solution generated by a random phase generator beyond 6 Å (while the distinction can be made at lower resolutions). This problem is both related to the large experimental deviation of image phase values and the low diffraction amplitudes at higher resolution. Hence the actual mean phase error for new phases may be as good as 55°, an observation that cannot be tested until more accurate measurements are made to permit a valid comparison.

Proc. Microscopy and Microanalysis 1995, edited by G.W. Bailey, M.H. Ellisman, R.A. Hennigar, and N.J. Zaluzec
Copyright © 1995 MSA. Published by Jones and Begell Publishing, 79 Madison Ave., New York, NY 10016

References

1. C. J. Gilmore, K. Shankland, and J. R. Fryer, Ultramicroscopy 49 (1993) 132.
2. D. L. Dorset, S. Kopp, J. R. Fryer, and W. F. Tivol, Ultramicroscopy, in press.
3. D. Sayre, Acta Cryst. 5 (1952) 60.
4. M. M. Woolfson and J. X. Yao, Acta Cryst. A44 (1988) 410.
5. H. J. Sass, G. Büldt, E. Beckmann, F. Zemlin, M. van Heel, E. Zeitler, J. P. Rosenbusch, D. L. Dorset, and A. Massalski, J. Mol. Biol. 209 (1989) 171.
6. A. D. Podjarny and A. Yonath, Acta Cryst. A33 (1977) 655.
7. Thanks to Dr. J. Sass for providing the experimental data for this study. Research was supported by a grant from NIGMS (GM-46733) which is gratefully acknowledged.

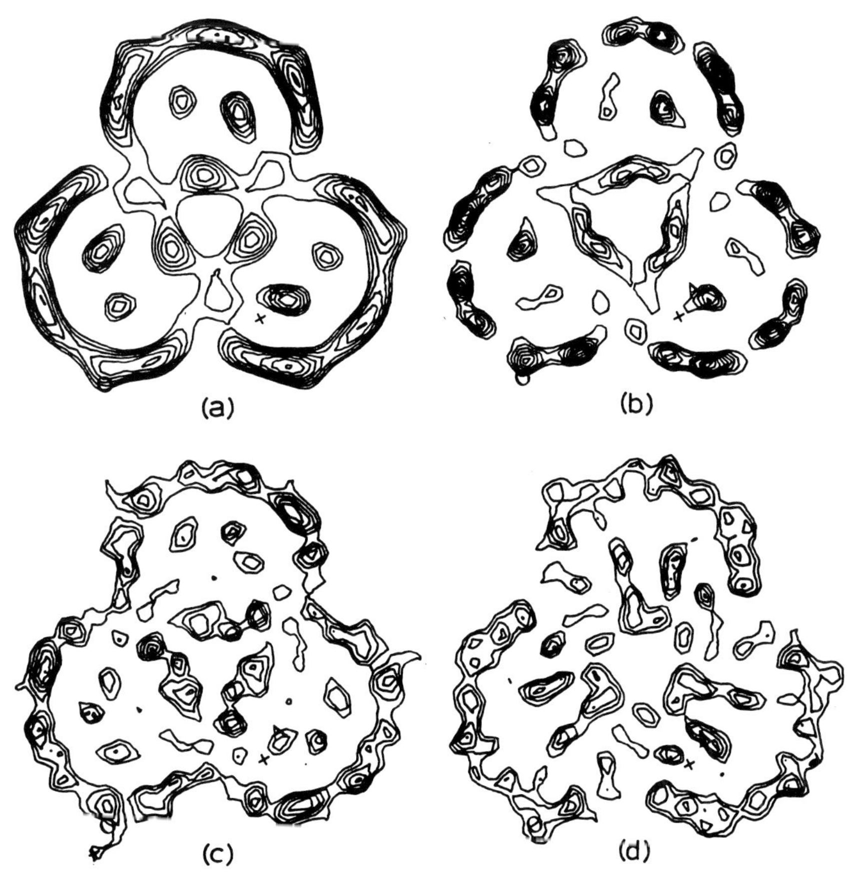

Figure 1. Potential maps for Omp F porin. a. 10 Å resolution (basis set); b. 5 Å resolution (Sayre expansion); c. 5 Å resolution (image phases); d. 5 Å resolution (random phases beyond 10 Å).

MULTIPLE ISOMORPHOUS REPLACEMENT FOR PHASE DETERMINATION IN PROTEIN CRYSTALS

Christoph Burmester, Kenneth C. Holmes, and Rasmus R. Schröder

Department of Biophysics, Max-Planck-Institute for Medical Research, P.O. Box 103820, 69028 Heidelberg, Germany

Electron crystallography of 2D protein crystals can yield models with atomic resolution by taking Fourier amplitudes from electron diffraction and phase information from processed images.[1,2] Imaging at atomic resolution is more difficult than the recording of corresponding high resolution electron diffraction patterns. Therefore attempts have been made to retrieve phase information from diffraction from heavy atom labelled protein crystals[3]. The expected differences between native and labelled crystals are small, therefore a high experimental accuracy is necessary. This is achieved by the use of energy filter TEM and image plates, as dicussed in[4]. Here we present electron diffraction data obtained from frozen hydrated 3D protein crystals with an energy filter microspcope and a specially designed image plate scanner[5]. Data were recorded for the native crystal as well as for two different heavy atom derivatives. Differences between the native and the derivate forms can be detected and are significant.

Electron diffraction patterns from frozen hydrated catalase crystals were recorded on an EFTEM Zeiss 912 Ω (120kV, zero loss mode, energy width $\Delta E=10eV$, electron dose 5 e⁻/A²) using image plates and a quasi confocal scanner readout[5]. Derivatives were obtained by soaking the crystals in 0.05mM $HgCl_2$ for 0.5h, 3h and 7h or in 1.25mM K_2PtCl_4 for 1h. It is known that mercury and platinum can in principle bind to catalase and form derivatives[6].

Crystal thickness was determined by taking low magnification images of the crystals in focus without objective aperture in the zero loss mode. The image is then formed by absorption contrast only. It could be seen that the contrast varies in discrete steps, which could be assigned to the unit cell thickness. From these experiments we found that the crystal thickness varied from 2 to 5 layers giving a total specimen thickness of 50 to 100 nm including supporting carbon film.

Merged data sets were calculated for the native crystal form, for mercury with its different soaking times and for platinum. Each data set was subdivided into two arbitrary groups. Merged data sets were then compared with all other data sets by calculating an R-factor and a difference Patterson map.

The R-factor between subgroups, as well as between soaking times, was significantly lower than between native and any derivative data. This indicates a change of intensities above noise level if heavy atoms are substituted. The same holds for the Patterson maps. They showed low intensity for the difference sets obtained within one data group (e.g. native vs native) and between different soaking times (e.g 0.5h vs 3h for mercury). These maps did not show reproducible features, thus their signal can be interpreted as noise. In contrast, the difference patterson maps between native and any derivative showed strong spots, which were more than 3 standard deviations above mean. Difference Patterson maps for native vs mercury with different soaking times showed reproducible spot positions.

References

1. Henderson R. et al. (1990) *J. Mol. Biol.* **213**, 899.
2. Kühlbrandt W., Wang D. and Fujiyoshi Y. (1994) *Nature (London)*, **367**, 614.
3. T.A. Ceska and R. Henderson, *J. Mol. Biol.* (1990) **213**, 539.
4. Burmester C. et al. (1994) *Proc 52nd MSA meeting*, 102.
5. Burmester C. and Schröder R.R. (1994) *Ultramicroscopy*, **55**, 55.
6. Murthy M.R.N. et al. (1981) *J. Mol. Biol* **152**, 465.

Proc. Microscopy and Microanalysis 1995, edited by G.W. Bailey, M.H. Ellisman, R.A. Hennigar, and N.J. Zaluzec

Table 1--Crystallographic R-factor for comparison of different data groups. Within one group, diffraction patterns with identical parameters (native, derivative [Hg, Pt], soak time [t = 0.5h, 1h, 3h, 7h]) are merged. Each group is divided into two subgroups A and B. In the table average figures for R-factors between datasets differing only in either soaking time, subgroup, or derivative/native status are presented. The table shows that differences between native and derivative data groups are higher than the differences of any other combination. This indicates, that the changes in signal produced by the heavy atom substitution is above noise level. $R = \Sigma\ |I_1 - I_2| / \Sigma\ (I_1 + I_2)$, summing over spot indices h,l. I_1 from first dataset, I_2 from second dataset.

resolution	5 .. 3.5 [Å]	3.5 .. 2.5 [Å]
soak time t_1 vs t_2	8.5%	7.5%
subgroup A vs B	8.9%	9.5%
native vs derivative	14.8%	12.9%

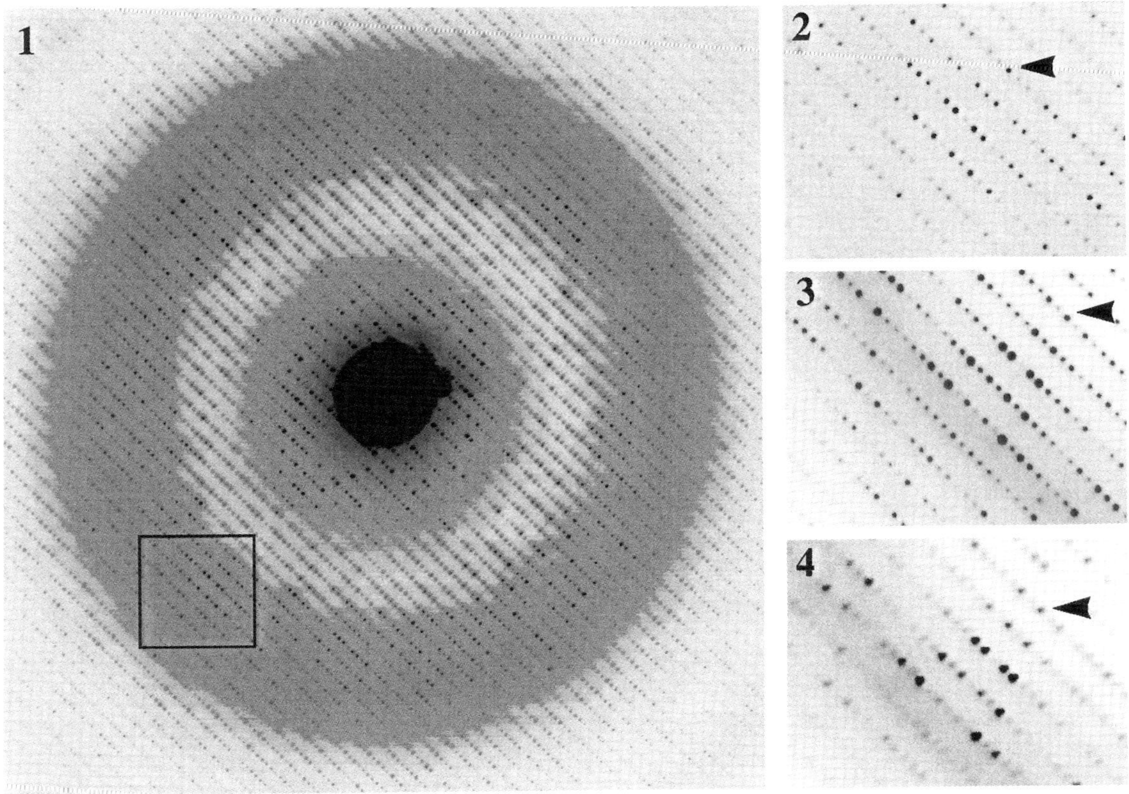

FIG. 1.--Zero-loss filtered (h,0,l) diffraction pattern of a native catalase 3D microcrystal embedded in vitreous ice ($\delta E=0$, energy width $\Delta E=10eV$) recorded on image plate[4,5]. Diffraction extends to about 2Å resolution. Shown is an original data set without background correction.

FIG. 2.--Detail of FIG. 1, as marked by the box in FIG. 1

FIG. 3.--Detail of the same reciprocal space as in FIG. 2, but for a platinum derivative diffraction pattern. Note the difference of the pattern compared with FIG. 2.

FIG. 4.--Detail of the same reciprocal space as in FIG. 2, but for a mercury derivative diffraction pattern. The difference is very similar to that of FIG. 3. This indicates, that platinum and mercury may have bound to the same site in the protein, as suggested by x-ray[6] studies.

FIG. 2, 3, 4: Arrows mark spot (14,0,0).

ALDEHYDE GOLD CLUSTERS FOR MOLECULAR LABELING

James F. Hainfeld and Frederic R. Furuya*

Biology Department, Brookhaven National Lab., Upton NY 11973
*Nanoprobes, Inc., 25 East Loop Rd., Suite 124, Stony Brook, NY 11790-3355.

Glutaraldehyde is a useful tissue and molecular fixing reagents. The aldehyde moiety reacts mainly with primary amino groups to form a Schiff's base, which is reversible but reasonably stable at pH 7; a stable covalent bond may be formed by reduction with, e.g., sodium cyanoborohydride (Fig. 1). The bifunctional glutaraldehyde, (CHO-(CH$_2$)$_3$-CHO), successfully stabilizes protein molecules due to generally plentiful amines on their surface; bovine serum albumin has 60; 59 lysines + 1 α-amino[1]. With some enzymes, catalytic activity after fixing is preserved[2]; with respect to antigens, glutaraldehyde treatment can compromise their recognition by antibodies in some cases. Complicating the chemistry somewhat are the reported side reactions, where glutaraldehyde reacts with other amino acid side chains, cysteine, histidine, and tyrosine[3]. It has also been reported that glutaraldehyde can polymerize in aqueous solution[4]. Newer crosslinkers have been found that are more specific for the amino group, such as the N-hydroxysuccinimide esters, and are commonly preferred for forming conjugates. However, most of these linkers hydrolyze in solution, so that the activity is lost over several hours, whereas the aldehyde group is stable in solution, and may have an advantage of overall efficiency.

In order to explore the potential advantages and unique features of aldehyde linking, gold clusters were prepared with one or more aldehydes on their surface. A phosphine was first synthesized that contained a dihydroxy terminating group, tris [p-2,3-dihydroxypropylcarboxamido) phenyl] phosphine. This was used to form undecagold clusters, which were then oxidized with NaIO$_4$ to produce a polyaldehyde gold cluster, having the formula Au$_{11}$(P(C$_6$H$_4$)CONHCH$_2$CHO)$_3$)$_7$, which was then purified by column chromatography. This showed high reactivity with Schiff's reagent for testing aldehydes, indicating multiple aldehydes per cluster.

The aldehyde-gold was reacted with BSA or Fab' fragments in varying ratios (5 gold clusters to 1 protein (5:1), or 1:1), along with 20 mM NaCNBH$_3$ in 0.1 M HEPES pH 7.5 and incubated overnight. The unreacted aldehyde was blocked with 0.1 M glycine for 1 hr, and the reaction purified on a Superose-12 (Pharmacia) gel exclusion column in PBS to separate protein from unreacted gold, and to separate larger complexes, such as protein dimers or multimers. With a 5:1 mixing ratio, most of the protein was in the monomer peak with gold labeling; with a 1:1 mixing ratio, the monomer protein peak was largest, but smaller peaks corresponding to dimer, trimer, and aggregate protein were seen. In one case, the monomer peak had ~20% gold labeling, and the dimer peak had a calculated labeling of 2.2 gold clusters per BSA.

Scanning transmission electron microscopy (STEM) of the chromatographic peaks showed respectively, protein monomers (Fig. 2), which had one Au$_{11}$ attached, oligomers, or small aggregates of BSA-gold (Fig. 3). A similar preparation of aldehyde-gold using the 1.4 nm Nanogold cluster showed similar results (Fig. 4); Fig. 5 shows high multiple gold labeling of one or a few protein moleucules.

Aldehyde gold clusters therefore provide an interesting method of preparing conjugates which may also be of interest in producing gold clusters with multiple small molecules attached, since for example, the undecagold cluster discussed has 21 aldehyde groups around its surface. A monofunctional aldehyde cluster has also been synthesized, and should eliminate any aggregation or oligomer formation.

References

1. B.F. Erlanger, Meth. Enzym. 70(1980)85.
2. F.A. Quiocho and F.M. Richards, Biochemistry, 5(1966)4062.

Proc. Microscopy and Microanalysis 1995, edited by G.W. Bailey, M.H. Ellisman, R.A. Hennigar, and N.J. Zaluzec
Copyright © 1995 MSA. Published by Jones and Begell Publishing, 79 Madison Ave., New York, NY 10016

3. A.F.S.A. Habeeb and R. Hiramoto, Arch. Biochem. Biophys. 126(1968)16.
4. M.A. Hayat, in Principles and Techniques of Electron Mocroscopy: Biological Applications, v.1, NY:Van Nostrand Reinhold Co. (1970)79.
5. J.C. Saccavini et al., in S.C. Srivastava, Ed., Radiolabeled monoclonal antibodies for imaging and therapy, NY:Plenum Press (1988)239.
6. The authors would like to thank N.I. Feng for biochemical assistance, J. Marecek for synthetic chemistry, and M. Simon and B. Lin for STEM microscopy. This work was partially supported by US Dept of Energy, OHER.

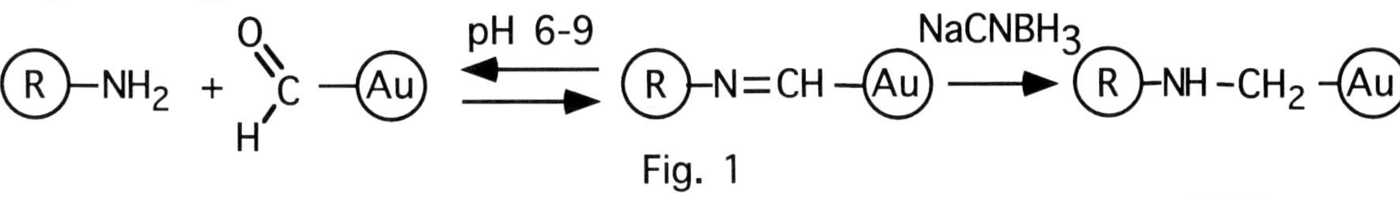

Fig. 1

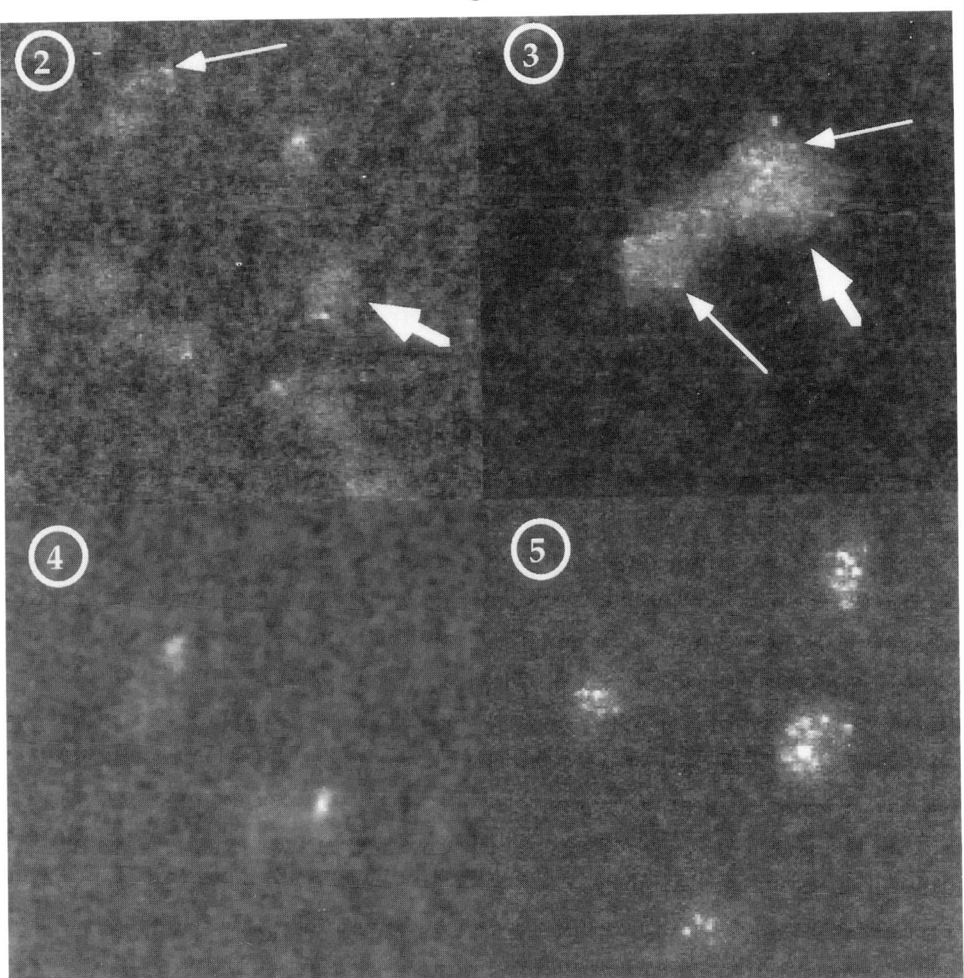

Fig. 1. Reaction scheme of coupling aldehyde gold to proteins; R = protein; Au = gold cluster.
Fig. 2. Darkfield STEM micrograph of unstained Fab' labeled with aldehyde Au_{11}. Thin arrow points to Au_{11} cluster (bright spot); thick arrow to protein (grey mass). Full width, 64 nm.
Fig. 3. STEM micrograph of Fab' aggregate crosslinked with aldehyde Au_{11}. Full width, 64 nm.
Fig. 4. STEM micrograph of unstained BSA labeled with aldehyde $Au_{1.4nm}$. Full width, 64 nm.
Fig. 5. STEM micrograph of unstained BSA labeled with multiple $Au_{1.4nm}$ clusters. Full width, 64 nm.

HIGH MAGNIFICATION IN-LENS FIELD EMISSION SE-I SEM OF CARBON BASED STYRENES, DIAMONDS, AND ENDOTHELIAL CELL MEMBRANES

Robert P. Apkarian,[*] Jonathan A. Tarr,[1] Myron J. Kaufmann[1] and Fredric M. Menger[1]

Integrated SEM & Microanalytical Facility[*], Department of Chemistry[1]. Emory University, Atlanta GA 30322

High resolution topographic imaging of carbon based samples have been problematic primarily due to the low secondary electron-I yield from specimens of low atomic mass. Both cold cathode and Schottky field emission (SFE) in-lens SEMs produce small beam diameters (5-10 Å) and achieve their highest resolution when operated at high (20-30 keV) accelerating voltages. The Topcon in-lens SFE-SEM/STEM has been in service at Emory U. for four years, and has generated quality images of carbon based materials and biological membranes. Enrichment of the SE-I signal from hydrocarbon based specimens was achieved by the application of a 1 nm thick, fine grain Cr film and provided images of 1-2 nm particulate features without decoration or enlargement[1]. Specimen beam interaction and the resultant particle contrast in a high magnification analog image recording was compared for graphitic carbon, diamond, and chromium coated polyvinylstyrene and fenestrated endothelial cell membranes (FECM).

Chemical vapor deposition (CVD) was used to deposit various forms of carbon on Si (111) wafers. The wafers were coated in a quartz flow tube in a temperature controlled furnace at 500-700° C. Small amounts of acetaldehyde were added to a flowing mixture of 10% H_2 in Ar at linear velocities of 30-100 cm/sec after microwave discharge. Depositions were carried out for 4-120 hours. Photochemically induced free radical deposition of styrene produced polyvinylstyrene (PVS). The lumenal aspect of fenestrated adrenocortical endothelial cell surfaces were exposed by cryofracture method. Adrenal fragments and PVS specimens were coated with a 1nm thick, fine grain Cr film by planar magnetron sputtering to enrich the SE-I contrast.[1] Specimens were staged in-lens of the DS-130 SFE-SEM and operated at 25 keV.

When acetaldehyde was added close to the Si wafer over a range of CVD conditions, cubo-polyhedral diamond crystal surfaces were observed (Fig. 1 a&b). Further down stream of the acetaldehyde inlet 1-2 μm spherical graphite islets composed of 1-10 nm particulate features were observed (Fig. 1 c). Cr coated PVS hydrocarbon specimens, photographed at a CRT magnification of 800,000 times, exhibited 2 nm particle contrasts (Fig 2). Bulk samples of Cr coated FECM contained particle contrasts of hydrocarbon ectodomains <10 nm (Fig. 3) which defined the fenestral brim.

The high angular current intensity of the SFE in-lens SEM produced SE-I rich images despite the low atomic mass of uncoated diamond and graphitic carbon.[2] High brightness analog images contained contrasts of 1 nm carbon particles. The mass density of hydrocarbon specimens, PVS and FECM, generated low numbers of large angle inelastically

scattered electrons and therefore required application of an ultrathin
Cr (z=24) film.[3] Resultant images of PVS attained 2 nm particle
contrasts when metal films were used to define the SE production.
Current studies involve digital imaging and processing of SE-I images.

References

1. R.P. Apkarian, Scanning Microsc. 1994 8:2, 289.
2. L.W. Swanson, 1989, FEI Company internal communication.
3. D.C. Joy, 1994, Proc 52nd Ann. Meeting MSA, 1032.
4. This research was supported in part by NIH grant RR-00165 to Yerkes
Research Center and NIH grant GM21457 to F.M. Menger.

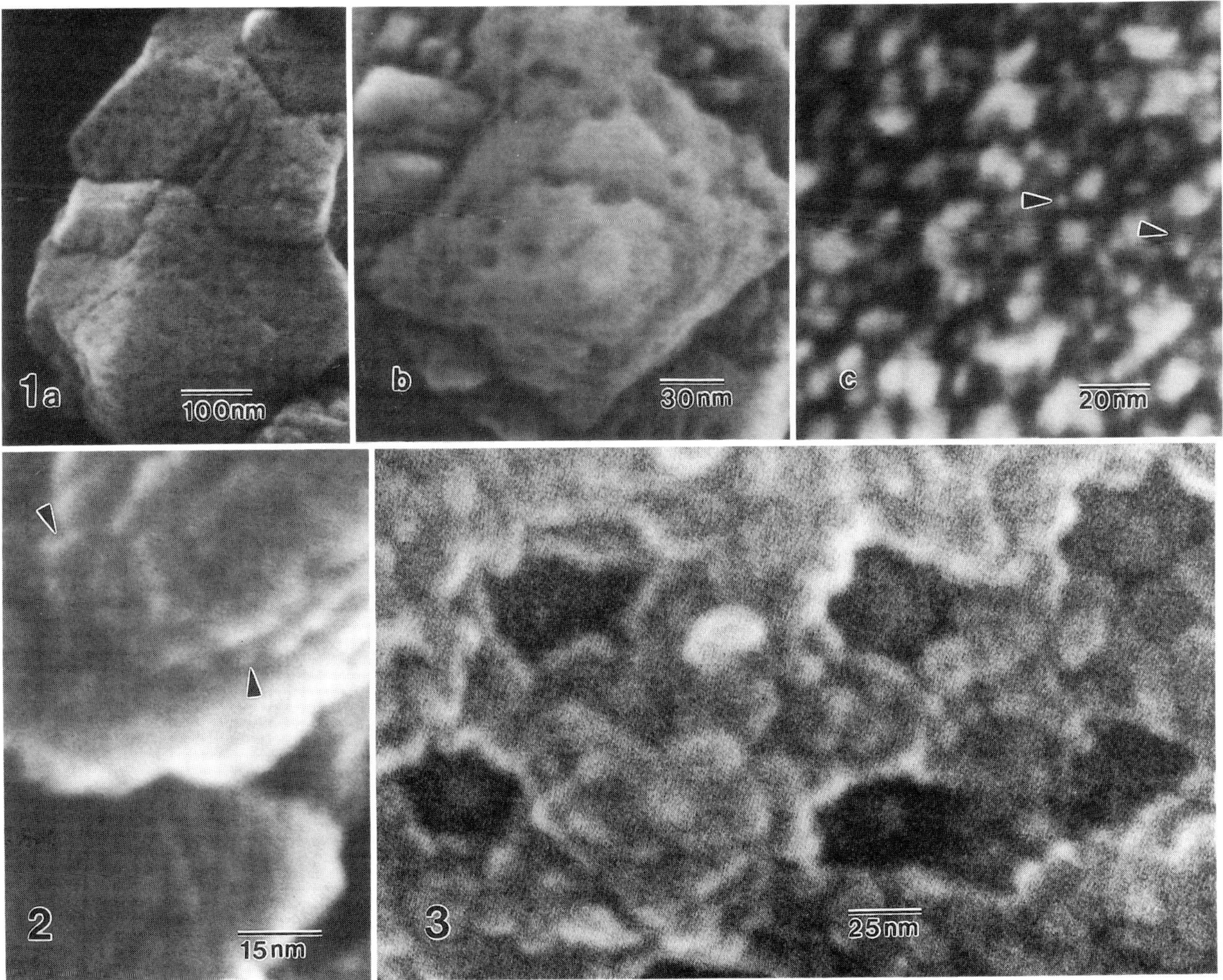

Fig. 1 CVD of (a&b) diamond and (c) graphitic carbon. Note 1-2 nm
(arrows) carbon particles.
Fig. 2 High resolution SE-I enriched image of 1-2 nm (arrows)
PVS features.
Fig. 3 FECM features routinely display SE-I particle contrast <10 nm.

EFFECTS OF THE PSEUDOMONAD PHYTOTOXIN CORONATINE ON
TOMATO LEAF STRUCTURE AND ULTRASTRUCTURE

D.A. Palmer and C.L. Bender
Department of Plant Pathology, Oklahoma State University, Stillwater, OK 74078.

Coronatine is a non-host-specific phytotoxin produced by several members of the *Pseudomonas syringae* group of pathovars. The toxin acts as a virulence factor in *P. syringae* pv. *tomato*, allowing the organism to multiply to a higher population density and develop larger lesions than mutant strains unable to produce the toxin.[1] The most prominent symptom observed in leaf tissue treated with coronatine is an intense spreading chlorosis; this has been attributed to a loss of chlorophylls a and b in tobacco.[2] Coronatine's effects on membrane integrity and cell ultrastructure have not been previously investigated. The present study describes changes in tomato leaves in response to treatment with purified coronatine, infection by a coronatine-producing strain of *P. syringae* pv. *tomato*, and infection by a cor⁻ mutant.

In contrast to H_2O-treated tissue, coronatine-treated tissue showed a diffuse chlorosis extending approximately 5 mm from the inoculation site. Leaf thickness, cell number, and cell dimensions were similar for both healthy and coronatine-treated, chlorotic tissue; however, the epidermal cell walls were consistently thicker in coronatine-treated leaves (Figs. 1a and 1b). Although chloroplasts in coronatine-treated tissue were significantly smaller, often had reduced starch deposits, and stained more intensely than those in healthy tissue, the membrane systems appeared to be intact (Figs. 2 and 3). Coronatine-treated tissue contained cubical and spherical particles in the mesophyll cells (Figs. 1b, 1d, and 2). The cubical particles were large (often up to 6 μm per side), moderately electron-dense, granular, and were apparently surrounded by numerous ribosomes and a thin layer of cytoplasmic ground substance. The spherical particles were visible as electron-dense granular spheres and were present in the vacuoles of mesophyll cells treated with coronatine. These spherical particles were clustered into aggregates, with each sphere up to 1 μm in diameter. On the basis of enzyme assays and structural similarities, the cubical crystals were identified as accumulations of chymopapain inhibitors, and the spherical bodies as accumulations of chymotrypsin inhibitors.

Lesions induced by the coronatine-producing *P. syringae* pv. *tomato* PT23.2 had a sharply defined necrotic center (1 mm diameter) and were surrounded by diffuse chlorotic areas extending approximately 2 mm from the lesion center. Bacterial populations were restricted to the area within the necrotic lesion and to the external surface of the lower leaf cuticle. The chlorotic tissue surrounding the necrotic lesion exhibited the same symptoms observed in coronatine-treated tissue: shrunken chloroplasts, thickened walls, and a massive accumulation of spherical bodies and cubical crystals (Fig. 1d). *P. syringae* pv. *tomato* PT23.21, a mutant of PT23.2 unable to synthesize coronatine, induced necrotic lesions substantially smaller than those observed with PT23.2. Lesions were approximately 0.2-0.5 mm in diameter and were not surrounded by a chlorotic area. Parenchyma and epidermal cells immediately adjacent to the lesion appeared identical to healthy tissue (Fig. 1c).

The findings presented in this study thus indicate that the enhanced virulence of coronatine-producing *P. syringae* pv. *tomato* is not the result of massive cellular destruction in the host.[3]

References

1. C.L. Bender et al., *Physiol. Mol. Plant Pathol.* 30(1987)273.
2. J. Kenyon and J.G. Turner, *Physiol. Mol. Plant Pathol.* 37(1990)463.
3. We thank Ginger R. Baker and the OSU Electron Microscopy Laboratory for supplying valuable technical assistance. This work was supported by grants from the OSU Electron Microscopy Laboratory, Oklahoma Agricultural Experiment Station Hatch Project 2009, and National Science Foundation grants EHR-9108771 and MCB-9316488.

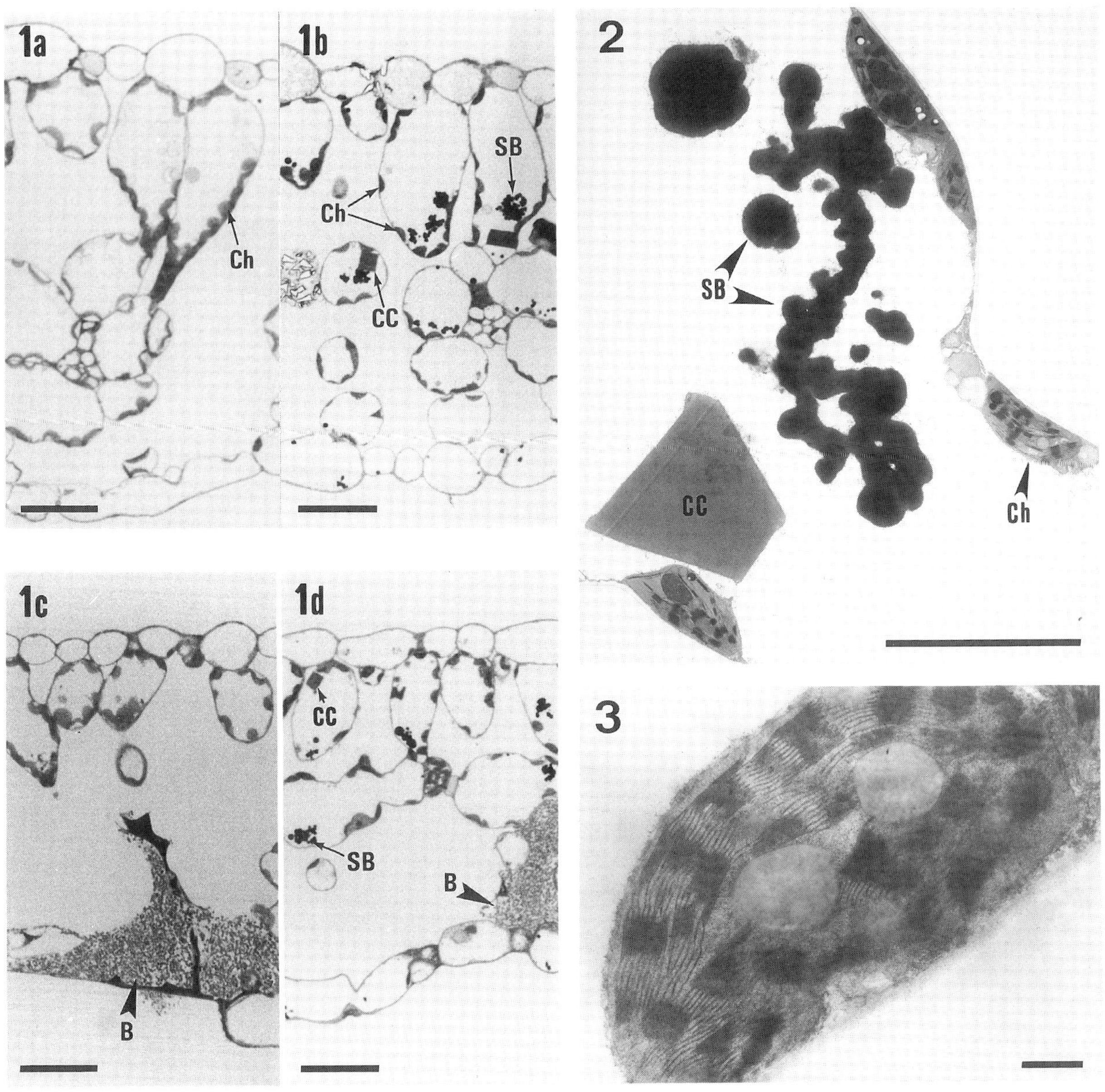

FIG. 1. -- Thick (1 μm) sections of tomato leaf tissue treated with (a) sterile H_2O, (b) 125 ng purified coronatine, (c) cor$^-$ PT23.21, and (d) cor$^+$ PT23.2. CC = cubical crystal, SB = spherical body, Ch = chloroplast, B = bacterial mass. Bars = 25 μm.

FIG. 2. -- TEM image of palisade parenchyma cell in tomato leaf tissue treated with 125 ng coronatine. 120 nm thin section. Bar = 5 μm.

FIG. 3. -- TEM detail of chloroplast in palisade parenchyma cell from coronatine-treated leaf. 120 nm thin section. Bar = 0.5 μm.

EN FACE DUAL FLUORESCENCE STAINING OF FATTY STREAKS ALLOWS
OBSERVATION OF INITIATION AND PROGRESSION OF ATHEROSCLEROTIC LESIONS

Anne M. Klinkner, Crystal R. Waites, Peter J. Bugelski and William D. Kerns

Dept. of Toxicology, SmithKline Beecham Pharmaceuticals, King of Prussia, PA, 19406

A primary effort in the understanding of the progression of atherosclerotic disease has been methods development for visualization of the atherosclerotic plaque.[1,2] We introduce a new method for the qualitative analysis of lipids in atherosclerotic fatty streaks which also retains those lipids for biochemical evaluation. An original aspect of the process is the ability to view an entire fatty streak *en face,* selectively stained for specific lipid classes within the lesion.

New Zealand white rabbits were fed a high cholesterol diet(0.15%-0.3% for 14 wks). The aorta was removed and fixed in Carson's phosphate buffered formaldehyde followed by dual staining in the fluorescent dyes Nile red and filipin. Stock solutions of nile red(0.5mg/ml acetone) and filipin(2.5mg/ml dimethyl formamide) were prepared and kept at -20°C; all subsequent steps were at RT. 0.5cm x 1.0cm pieces of aorta were trimmed and adventitia removed. The pieces were then washed 3x15 min in PBS w/o CaMg, soaked in Nile red(NR)/filipin(Fl) stain(100µl NR stock + 200µl Fl stock in 10 ml PBS for 30 min, washed in PBS 3x30 min, rinsed with distilled water, mounted(Crystal Mount, Biomedia) and coverslipped and viewed by fluorescence microscopy. Nile red staining(459-490nm ex) causes neutral lipids to fluoresce golden-yellow(570-590nm em) and polar lipids to fluoresce orange-red(590-620nm em).[3] Filipin staining(340-380nm ex) causes free cholesterol to fluoresce blue(460-500nm em). The area of plaque attributable to each of the fluorescence colors was quantitated using image analysis to determine relative amounts of esterified cholesterol(neutral lipid), phospholipid (polar lipids) and free cholesterol. Fatty streaks similar in size and location to those fluorescently stained were processed for LM/TEM to investigate cell types and ultrastructure of lipid inclusions in the cells. Cholesterol ester and phospholipid inclusions were also demonstrated by NR/Fl staining in thioglycollate elicited peritoneal macrophages cultured in the presence of oxidized LDL for 72 hrs.

The fatty streak may originate as an area of free cholesterol accumulation, possibly serum derived lipoproteins[4,5], in the intima between smooth muscle cells and beneath an intact endothelium. These areas fluoresce blue with filipin, but do not fluoresce with Nile red and correspond with areas shown by transmission electron microscopy(TEM) to be filled with phospholipid liposomes(Fig. 2). As lesion size increases, macrophage derived foam cells that fluoresce both blue(Fig. 1,dk. gray). and orange-red(Fig.1, lt. gray) appear. Macrophages in these areas have many phospholipid inclusions(Fig. 4). Filipin staining also reveals intimal areas with free cholesterol in the form of cholesterol crystals, seen as clefts by TEM(Fig. 3). In larger lesions, foam cells have accumulated esterified cholesterol and begin to fluoresce golden-yellow(Fig.1, white). These foam cells have many cytoplasmic pools of neutral lipid(Fig. 5). Overlays of the three fluorescent images reveal "growth rings" with a halo of free cholesterol(dk. gray) at the periphery(Fig.1).

This method has enabled us to study *en face* fluorescent staining of lesions of various sizes. As the lesion evolves, characteristic aspects of the atherosclerotic process of lipid accumulation and cellular change may be observed and quantitated.

References

1. H. S. Kruth, Lab. Invest. 50(1983)87.
2. M. E. Rosenfeld et. al., Arteriosclerosis 7(1987)9.
3. S. D. Fowler and P Greenspan, J. Histochem Cytochem 33(1985)833.
4. R. Mora et. al., Atherosclerosis 67(1987)143.
5. F. Chao et. al., Am J. Pathol. 131(1988)73.

Proc. Microscopy and Microanalysis 1995, edited by G.W. Bailey, M.H. Ellisman, R.A. Hennigar, and N.J. Zaluzec

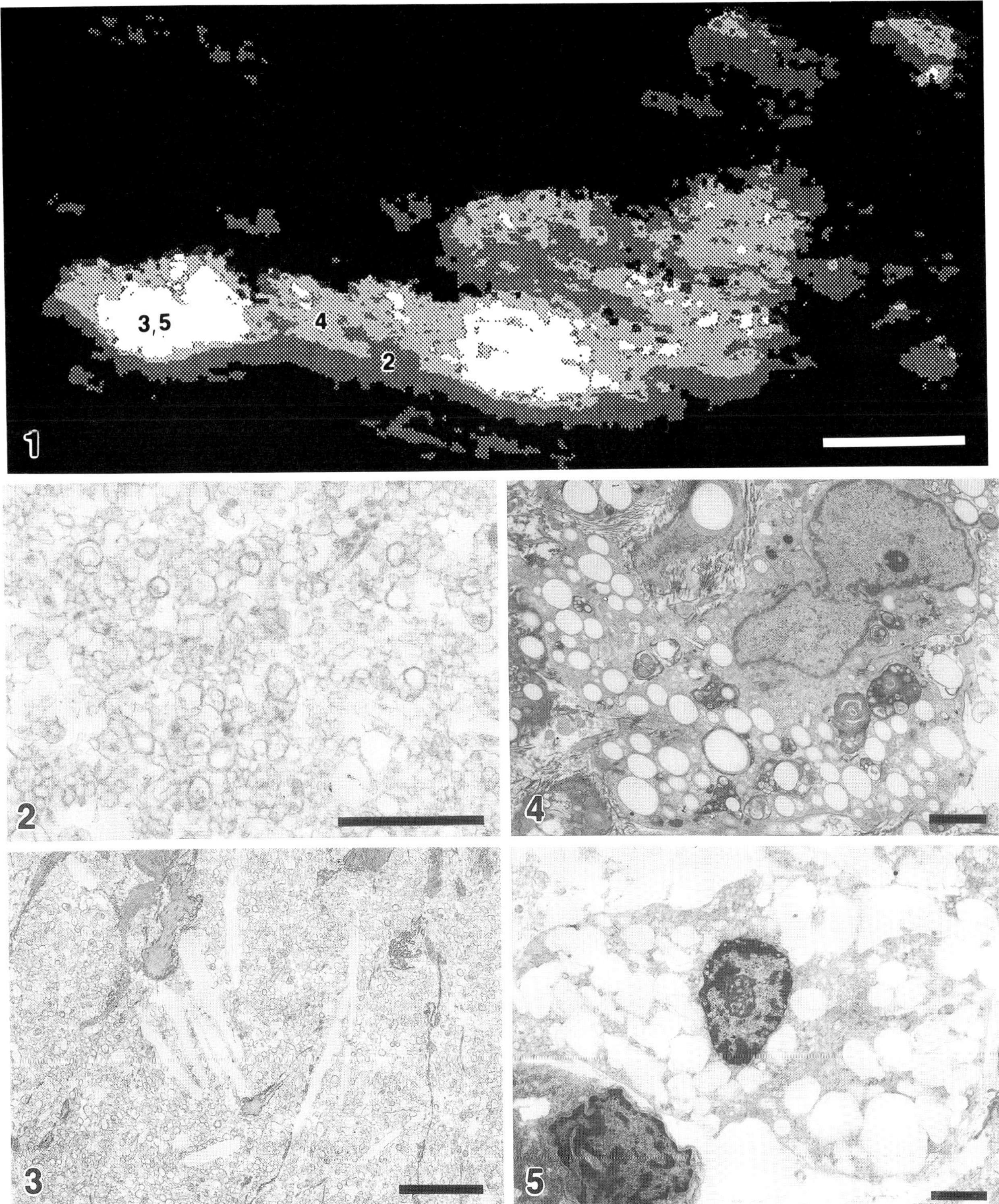

FIG. 1.-Composite dual fluorescence LM of fatty streak. Bar =0.5mm. FIG. 2.-TEM of phospholipid liposomes. Bar =1µm. FIG. 3.-TEM of cholesterol clefts. Bar =2µm. FIG 4.-TEM of macrophage with phospholipid inclusions. Bar =2µm FIG. 5- TEM of macrophage with neutral lipid inclusions. Bar =2µm. Numbers in Fig. 1 correlate with TEM figures.

X-RAY MICROANALYSIS OF MINERALIZED DEPOSITS IN A VARIETY OF PATHOLOGICAL CONDITIONS IN THE BRAIN.

C. A. Ackerley* and L. E. Becker*

*Department of Pathology, The Hospital for Sick Children, 555 University Avenue Toronto, Ontario, Canada M5G 1X8.

Although a small degree of mineralization can be a common occurrence without associated pathological symptoms, certain diseases of the brain do however exhibit distinct increases in mineralization with characteristic distributions [1,2]. In this study, tissues from a number of these disorders were prepared for x-ray microanalysis in several ways. Where possible, material was slam frozen on a liquid nitrogen cooled polished copper block, cryosections prepared and freeze dried in the scanning transmission electron microscope (STEM) using a cold stage prior to analysis by energy dispersive x-ray spectrometry (EDS). In addition, samples were freeze substituted for several days, embedded in LR white and cut on dry knives before analysis. Where only formalin fixed paraffin embedded materials were available, .5 μm sections were cut and mounted on carbon planchets. The specimens were then deparaffinized with xylene and viewed with the backscatter electron detector (BEI) in the scanning electron microscope (SEM) and analyzed by EDS. Tissues examined included mineralized tumor (a choroid plexus papilloma and a craniopharyngioma), degenerative disease, (2 cases of Cockayne's syndrome and 7 cases of Sturge Weber syndrome), biochemical defect (complex 1 mitochondrial cytopathy with neuronal mineralization), a paediatric seizure disorder (non cortical dysplasia with mineralization), 3 cases of viral cerebral inflammation of unknown etiology, and a case of megaloencephaly.

No changes were found in individual cases in the elemental composition of the mineralized areas between the methods used in preparation where material was available for all procedures (Table 1). This was probably due to the degree of mineralization and the insolubility of the salts formed in the mineral. Perhaps of more importance was the similarity of elemental composition in mineralized areas of tissues from different patients affected with the same disorder (Table 1). In some disorders (i.e. Sturge Weber Syndrome) (Fig, 1a, b, c, d), mineralization content was different between cell types. Elemental distribution maps of frozen freeze-dried materials demonstrated the presence of Ca and P in the endothelial cell, while in the neuropile concretions contained Al, S, P and Ca[3]. Further study is required to fully understand the process of mineralization in the brain. The elemental content of mineralization being used as a characteristic feature in some of these disorders has some potential as a diagnostic feature.

References

1. H. Urich: Malformation of the nervous system, perinatal damage and related condition in early life. In, Greenfields Neuropathology, W. Blackwood and J. Corsellis editors, Edward Arnold Publishers Ltd, Edinburgh 1976, pp 361-470.

2. FN Ghadially: Calcification material lipidic debris, matrix vesicle and calcifying secretions. In, Ultrastructural Pathology of the Cell and Matrix. Vol II, 3rd Edition, Butterworth, Toronto 1988, p 1278.

3. JP Provias, CA Ackerley, V Jay and LE Becker: Aluminum in the mineralized foci in the cortex of Sturge Weber Syndrome. Submitted J. Neuro. Res. 1995.

Proc. Microscopy and Microanalysis 1995, edited by G.W. Bailey, M.H. Ellisman, R.A. Hennigar, and N.J. Zaluzec

SUMMARY OF RESULTS

CONDITION	FREEZE DRIED CRYOSECTION	FREEZE SUBSTITUTED		FORMALIN FIXED
Choroid plexus papilloma	Ca, S, P, O	Ca, S, P, O		Ca, S, P, O
Craniopharyngioma	Ca, S, P, O	Ca, S, P, O		Ca, S, P, O
Cockayne's syndrome	Not done	Not done		Ca, Mg, P, K, O
Sturge Weber syndrome	Al, Ca, P, S, O	Al, Ca, P, S, O		Al, Ca, P, S, O
Complex Mitochondrial cytopathy	Mg, Ca, O	Mg, Ca, O		Mg, Ca
Viral cerebral inflammation	Not done	Ca, Mg, P	1 case	
		Ca, Fe, Mg, P, S	2 cases	
Megaloencephaly	Not done	Ca, K, Mg, Na, S		Ca, K, Mg, Na, S

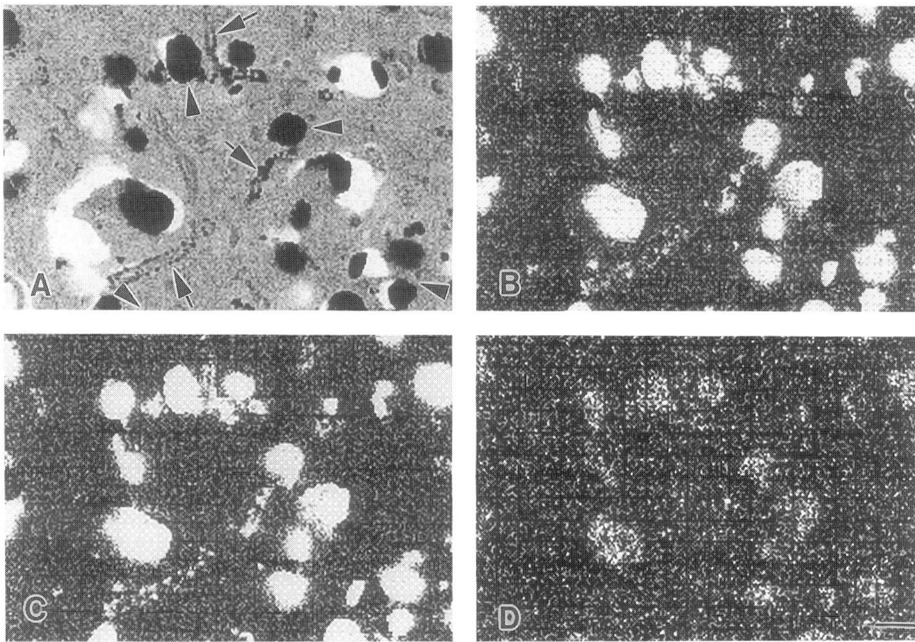

FIG 1:　(a)　Elemental distribution maps of Sturge Weber Syndrome. A STEM image of freeze dried cryosection of mineralized area. Note the numerous concretions (arrowheads) and mineral deposits in the endothelial cells (arrows).

(b)　Ca elemental distribution map.

(c)　P elemental distribution map.

(d)　Al elemental distribution map. Note that it is only present in the concretions.

Bar equals 5 μm

ULTRASTRUCTURAL IDENTIFICATION OF
OPPORTUNISTIC ORGANISMS IN AUTOPSY TISSUE

S. Siew, W. deMendonca-Calaca, J. Penner

Michigan State University, East Lansing, MI 48824
St. Lawrence Hospital, Lansing, MI 48915

Opportunistic infections pose a progressively greater threat to the immunocompromised patient both as a cause of morbidity and mortality. Difficulty may be experienced with the establishment of a definitive diagnosis of such infections during life. We report the ultrastructural identification in autopsy lung material of Pneumocystis carinii of two immunocompromised patients and Herpes nucleocapsids in a third case. These three patients were young adults, two men and one woman. All of them died of the pulmonary complications of the acquired immunodeficiency syndrome (AIDS). One of the men and the woman had a history of sexual promiscuity. The former was a homosexual, whose lover had predeceased him, six months ago. The other young man was an innocent victim of medical progress. As a hemophiliac, he had been on intravenous therapy with anti-hemophilic globulin.

At autopsy, their lungs were increased in size and weight and they were indurated. The light microscopic findings of the lungs of the first two cases were similar. They varied in the degree of reactive fibrosis. The prominent feature was the presence of intra alveolar spongy, eosinophilic masses. Some of these had an infiltration of macrophages and non-granular cells on the periphery; occasional ones were encapsulated by a monolayer of elongated fibroblastic cells; yet, others, were completely surrounded by an ingrowth of fibroblasts. This was associated with an irregular interstitial fibrosis. There was cuboidalization of most of the alveolar epithelium. Some of the alveolar cells were enlarged, binucleate, with an "owl eye" appearance of the nuclei. Hyaline membrane formation was noted in some foci. Transmission electron microscopy showed the presence of intra alveolar masses containing cysts of Pn. carinii. Some of these retained their rounded structure, but, most of them had collapsed into irregular masses (Fig.1) Some of this distortion may be attributable to the therapy that these patients had received. Occasional trophozoites were present. There was an admixture of fibrin within these masses. Fibroblasts and macrophages were noted in the alveoli. Some of the latter were in close proximity to the cysts.

Light microscopy of the lungs of the third patient showed the presence of focal hemorrhages with an extravasation of blood into the alveoli. Hyaline membrane formation was more extensive than in the other two cases. Desquamation of the alveolar epithelium was more marked, with cuboidalization of the remaining cells. Some of these were enlarged, with large vesicular nuclei, which had an owl eye appearance. A very scanty non-granular cell reaction was present.

Transmission electron microscopy showed the presence of cells with large nuclei, which had an electron lucent appearance due to the clumping of the chromatin in the center of the nucleoplasm. This formed an irregular meshwork, in the interstices of which there were hexagonal particles, whose structure was compatible with that of Herpes nucleocapsids. (Fig. 2)

Proc. Microscopy and Microanalysis 1995, edited by G.W. Bailey, M.H. Ellisman, R.A. Hennigar, and N.J. Zaluzec
Copyright © 1995 MSA. Published by Jones and Begell Publishing, 79 Madison Ave., New York, NY 10016

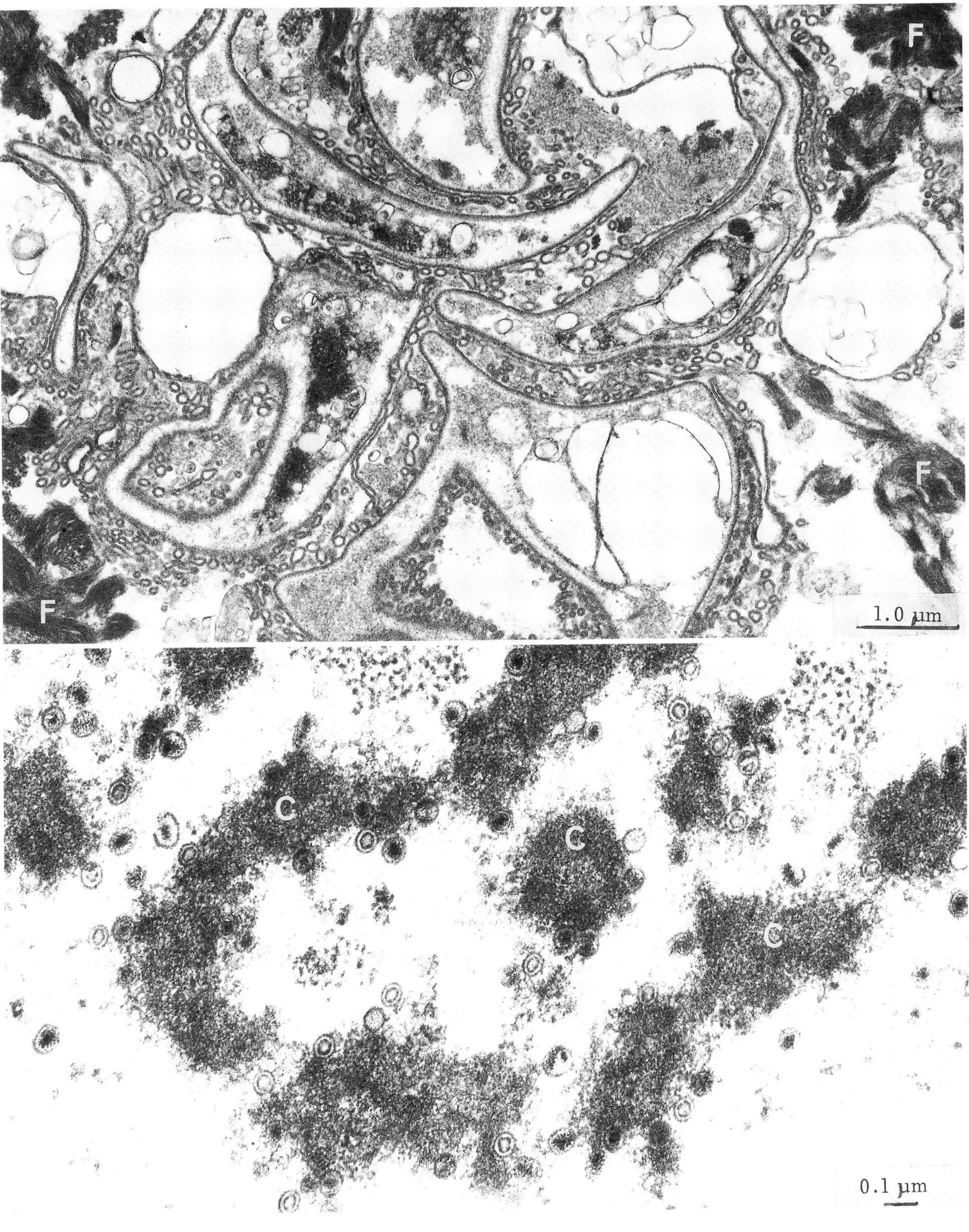

FIG. 1.- Collapsed forms of Pn. Carinii, F. Fibrin.
FIG. 2.- Intranuclear hexagonal Herpes nucleocapsids C-condensed nucleochromatin.

THE EFFECT OF DATES ON THE ULTRASTRUCTURE OF HEPATOCYTES OF MICE INOCULATED WITH _CERASTES CERASTES_ VENOM

A. Al-Sallal*, Z. Amr*, and F. Al-Bagdadi**

*Department of Biology,Faculty of Science, Jordan University of Science and Technology, Irbid, Jordan and **Department of Veterinary Anatomy and Cell Biology, College of Veterinary Medicine, Louisiana State University, Baton Rouge, Louisiana, 70803, U.S.A.

Upon fractionation of _Cerastes Cerastes_ venom it was found to contain L-aminoacidoxidase, phosphodiesterase, hyaluronidase and phospholipase A.[1] The pathological effects of snake venom on the kidney,[2,3] muscle tissues[3], liver,[4] spleen and lymph nodes[5] were investigated. In this study we report the effects of dates on the ultrastructural morphology of hepatocytes of mice envenomized intraperitoneally with _Cerastes Cerastes_ venom.

The venom used in this study was collected from _Cerastes Cerastes_ snake, which were usually found in Wadi Araba, south of Jordan. The extracted venom was lyophilized and kept at -20 degrees centigrade. Three white albino mice were starved for 60 hours and then fed dates. Then after twelve hours the mice were injected intraperitoneally with a lethal dose of _Cerastes Cerastes_. After 42 hours the mice were still alive. Control mice were not starved nor fed with dates,they were also envenomized intraperitoneally with lethal dose of _Cerastes Cerastes_ venom, and the liver of all mice was dissected immediately after death. Liver specimens were minced in a lake of fixative, 1.25% glutaraldehyde and 2% formaldehyde in sodium cacodylate buffer PH 7.4. Fixed for ninety minute in the same fixative and post fixed in 1% osmium tetroxide.Dehydrated in graded series of ethyl alcohol cleared in propylene oxide and embedded in epon. Silver sections were stained with lead citrate and urynal acetate. The sections were examined by Zeiss-10 transmission electron microscope and photographed by black and white kodak film.

The hepatocyte of the control mice showed hyperplastic nuclei contained no heterochromatin. All of the mitochondria were enlarged, irregular and lost their cristae (Fig. 1). Some of the mitochondria have lost part of their outer and inner membranes. Mitochondria seen phagocytizing other mitochondria and some contained membrane bounded round vesicles. The hepatocytes of the mice fed dates and envenomized intraperitoneally showed thick plasmalemma and their cytoplasm contained rounded lipid globules (Fig. 2). Some of the mitochondria were moderately enlarged and a few showed disrupted cristate. The nuclei were rounded and contained a thin layer of marginated heterochromatin.

The effect of snake venom on the liver has been reported causing vacuolation of the hepatocytes and patchy necrosis of the hepatocytes.[7] The conclusion of this preliminary study indicates that dates have interfered with the degenerative effect of the lethal dose of _Cerastes Cerastes_ venom on the fine structure of the hepatocytes. Further investigation is very encouraging and necessary in this respect, especially in the desert countries were snake bite is a common incident every day.

Proc. Microscopy and Microanalysis 1995, edited by G.W. Bailey, M.H. Ellisman, R.A. Hennigar, and N.J. Zaluzec

References

1.   S. Labib et al., Toxicon 17(1979)337.
2.   K. Chugh, et al., Am. J. Trop. Med. Hyg. 24(1975)692.
3.   V. Cattle, et al., Am. J. Path. 37(1977)511.
4.   M. Homma, et al., Br. J. Exp. Path. 52(1971)538.
5.   S. Gitter, et al., Am. J. Trop. Med. Hyg. 11(1962)861.
6.   A. Mohamad et al., Toxicon 18(1980)374.
7.   A. Mohamad et al., Toxicon 16(1978)253.
8.   The authors are very thankful to Mr. Muner Al-Khdaur for the excellent tissue preparation for electron microscopy and Mrs. Virginia Duncan for typing the article.

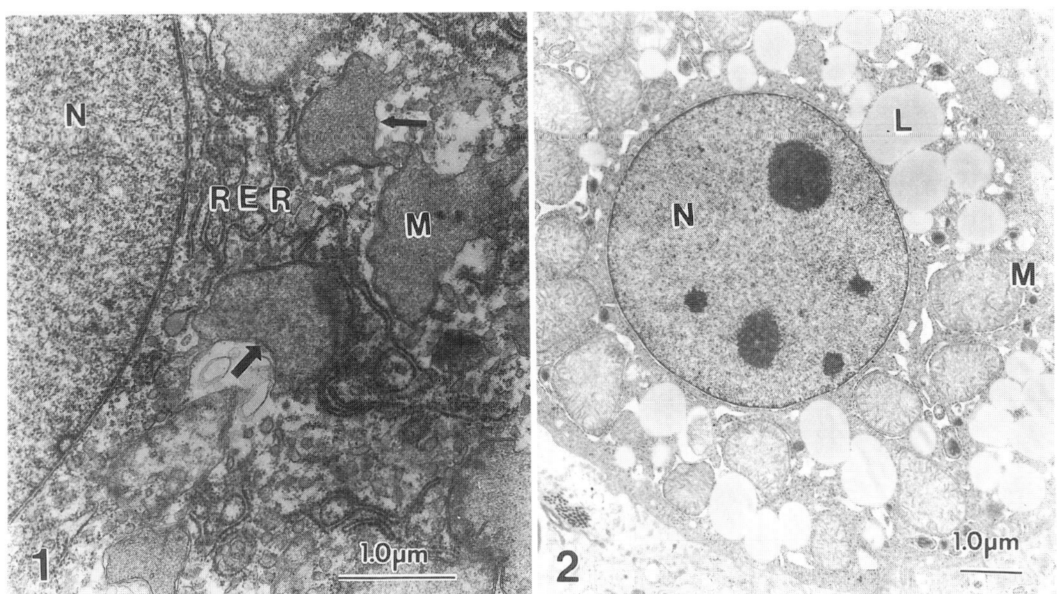

FIG. 1.-Part of a hepatocyte obtained from a mouse envenomized intraperitoneally with *Cerastes Cerastes* venom. The nucleus is hyperplastic (N) has only euchromatin. The mitochondria (M) are hypertrophied, irregular and have no cristae. Part of the mitochondria is ruptured (arrows). Long dark stained profiles of rough endoplasmic reticulum (RER). Stained with lead utrate and urynal acetate.

FIG. 2.-Part of a hepatocyte obtained from a mouse fed dates and envenomized intraperitoneally with *Cerates Cerates* venom. The nucleus is rounded, (N) two nucleoli and thin layer of marginated heterochromatin and the rest is euchromatin. The mitochondria are moderately hypertrophied (M), some have disrupted cristae. Lipid globules (L), are in scattered groups in the cytoplasm. Stained with lead citrate and urynal acetate.

HUMAN T-LYMPHOCYTES (JURKAT CELLS) EXPRESSING TAT-72: LM and TEM.

J. Gilloteaux*‡, J. Koch**‡, J. M. Jamison**‡, and J.L.Summers‡. Departments of Anatomy*, Microbiology & Immunology**, and Urology‡, Northeastern Ohio Universities College of Medicine, P.O. Box 95, Rootstown OH. 44272, U.S.A.

The human immunodeficiency virus (HIV-1) is dependent upon the synthesis of TAT for efficient transcriptional activation and replication[1]. TAT serves to transactivate the HIV-LTR which leads to an increased transcriptional activation of approximately 1000 fold[2].

We have transfected human T-lymphocytes with an expression vector containing the HIV-1 LTR driving the synthesis of TAT-72. A stable cell line was generated by selection with G418.

Cells were centrifuged with a cytospin and stained for nucleic acids with methyl green and pyronin Y[3,4] in order to visualize and to localize DNA and RNA respectively in each preparation examined by light microscopy (LM). In addition, cells were fixed with a 3.5% glutaraldehyde buffered solution (0.1M Na cacodylate) for 5 min at room temperature, 35 min at 4°C, centrifuged in a microfuge (400 rpm, 3 min) to obtain a cellular pellet. Cell pellets were washed in a buffer-sucrose solution and postfixed in a 1.5% aqueous solution of OsO_4. Dehydration and epoxy embedding was followed by sectioning for TEM, following treatment with uranyl and lead salts.

LM examination showed typical T-lymphocytes (Figs. 1 and 2 a-b). In contrast, TAT-72 transfected cells contain an enlarged nucleus with perinuclear vacuoles and cell excrescences rich in ribonucleoproteins. Cell pleomorphism affects cell size, shape, number of nuclei and nuclear shape. Note also the presence of nucleoproteins in the perinuclear space (arrow in Fig 3a).

TEM examination confirms the LM characteristics i.e. cells maintain a more or less ovoid to spherical outline in both control groups and typical morphology of T-lymphocyte. However, TAT-72 transfected cells often display an enlarged euchromatic nuclei. In these cells, mitochondria are often largely vacuolated and perinuclear pockets appear as dilated intermembranous spaces created by the nuclear envelope (arrows in Fig. 3b). These spaces are also often filled with a fuzzy but electron dense material which is in continuity with the nucleoplasm (Fig. 3b). Immunocytochemistry is currently being conducted in addition to molecular biological techniques to identify the content of these perinuclear structures present in TAT-72 cells[5].

References

1. A.L.Dayton, J.G. Sodroski, C.A. Rosen, W.C. Goh, W.A. Haseltine. *Cell* 44 (1986) 941-947.
2. A. Gatignol, A. Kumar, A. Rabson, and K-T. Jeang. *PNAS* 86 (1989)   7828-7832.
3. J. Brachet *J. Microsc. Sci.* 94 (1954) 1-10.
4. J. Chayen, J. *Exptl. Cell Res.* 3 (1952) 652-655.
5. Supported by Summa Health System Foundation, Akron, OH.

Proc. Microscopy and Microanalysis 1995, edited by G.W. Bailey, M.H. Ellisman, R.A. Hennigar, and N.J. Zaluzec
Copyright © 1995 MSA. Published by Jones and Begell Publishing, 79 Madison Ave., New York, NY 10016

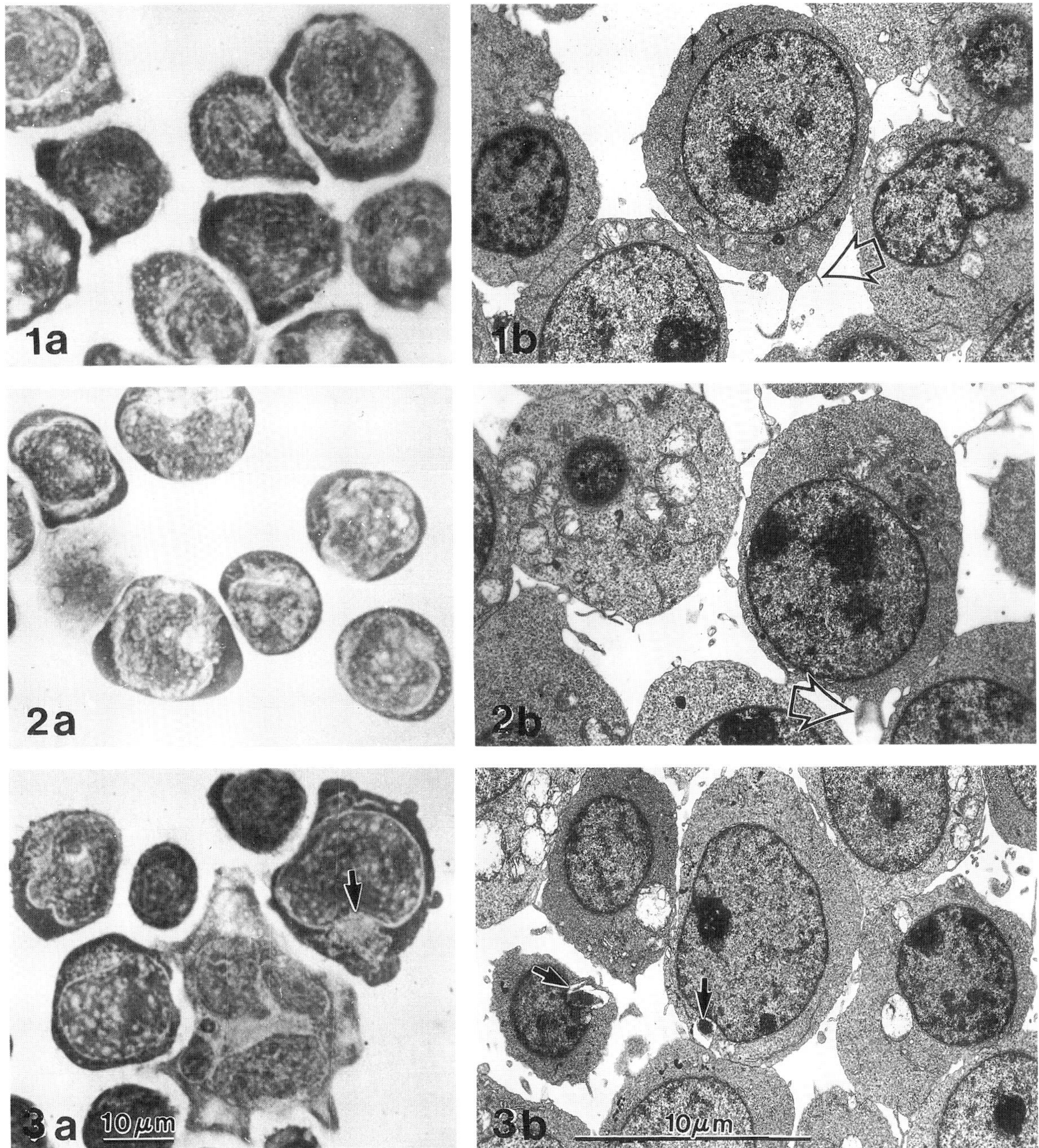

LM (Figs. 1-3 a) and TEM (Figs. 1-3 b) aspects of control and TAT-72 expressing human T-lymphocytes. Scales are 10 μm in columns a and b.
FIGS. 1 and 2.- Wild type (Fig. 1 a-b) and Control (Fig. 2 a-b) human T-lymphocytes. Typical morphological characteristics are shown, including the presence of uropods (open arrows).
FIG. 3 a-b.- Human T-lymphocytes expressing TAT-72 showing euchromatism and perinuclear vacuolated mitochondria. Notice extruded nuclear electron dense material in the perinuclear space (arrows).

FORMATION OF POLYRIBOSOMAL CRYSTALS IN *ENTAMOEBA HISTOLYTICA* BY HYPEROSMOTIC SOLUTION OF NaCl.

* C. Argüello, ** Reyes, J.L.

* Departments of Experimental Parasitology and ** Physiology and Biophysics, CINVESTAV, I.P.N. 07000, Mexico, City.

Entamoeba histolytica is a parasitic protozoan that during its life cycle presents two different biological forms; the trophozoite that is a pleomorphic, motile and phagocytic organism, and the cyst that is small, nonmotile and rounded form covered by an external fibrous coat [1]. Differentiation of trophozoites into the cyst involves a series of complex events including the organization of helical polyribosomes within a crystal known as the chromatoid body [2]. Induction of this differentiation has been achieved experimentally in hypoosmotic medium composed of $Mg\ SO_4$ and bovine serum (100 mOs/kg) in *Entamoeba invadens*, a reptilian parasite[3]. However, *Entamoeba histolytica* is unable to survive when treated on the above conditions. Due to the biological importance of understanding which are the physiological conditions that may elicit encystation of this organism, we cultured trophozoites in TYI-S-33 medium containing NaCl to a final osmolarity of 400, 500, 600 and 800 mOs/kg. Parasites were observed directly through a videosystem and noticed that at 600 mOs the trophozoites stop moving and rounded in about 3 min, similar results were obtained at 800 mOs. An immediate reversion of the effects occurred when the medium was replaced by the normal one. Samples of parasites treated under the above conditions were processed for histological and ultrastuctural studies. Statistical analysis of the number and size of cytoplasmic vacuoles was conducted and the organization of polyribosomes into chromatoid bodies by transmission electron microscopy was sequentially followed. One striking effect of hyperosmotic treatment was the formation of multiple chromatoid bodies all over the cytoplasm , the reduction in size of vacuoles and the polymerization of glycogen (Fig. 1). The inner organization of polyribosomes into crystalline arrangement was detected both in longitudinal and transverse sections of the bodies (figs. 2 and 3). A hexagonal pattern of polyribosomes is clearly seen on figure 3, where the external aspect of each ribosome seems to associate to each other. A localization of RNA in the chromatoid body was visualized by a reaction with osmium amine [4,] revealing that RNA condensed into structures similar to small stars (Fig.4). The mechanism by which ribosomes form the crystals is unknown, however, our experimental conditions permit a reproducible analysis of this event and consider for instance the participation of microfilaments in the assembly process of ribosomes.

1. A. Martínez-Palomo, The Biology of *Entamoeba histolytica*, Chichester: John Wiley & Sons, 1982.
2. R.S. Morgan, Science.162(1968)670.
3. B. Avron et al., Arch. Invest. Med. 17(1986)195.
4. M. Derenzini and F. Farabegoli, J.Histochem. Cytochem. 38(1990)1495.

Proc. Microscopy and Microanalysis 1995, edited by G.W. Bailey, M.H. Ellisman, R.A. Hennigar, and N.J. Zaluzec
Copyright © 1995 MSA. Published by Jones and Begell Publishing, 79 Madison Ave., New York, NY 10016

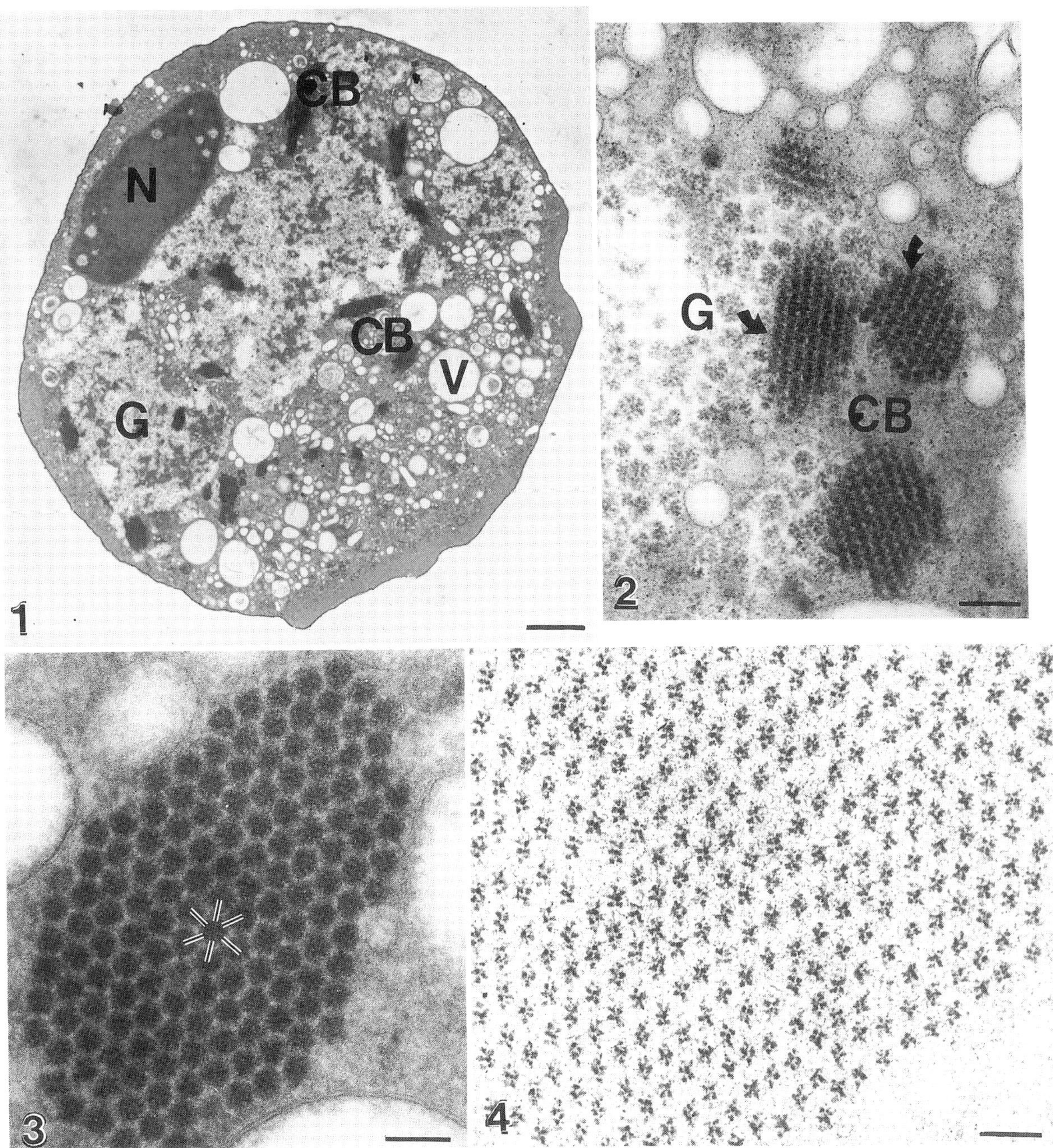

FIG. 1.- *E. histolytica* trophozoite under hyperosmotic conditions forms multiple chromatoid bodies (CB), small vacuoles (V) and polymerization of glycogen granules (G). Nucleus (N). Bar = 2.2 μm
FIG. 2.- Longitudinal section of chromatoid bodies (CB) showing the helical organization of polyribosomes (arrow). Glycogen granules are closely associated to the bodies (G). Bar = 200 nm
FIG. 3.- The hexagonal arrangement of ribosomes is evident in transverse sections of chromatoid bodies. Bar = 100 nm
FIG. 4.- Regular pattern of RNA organization into the chromatoid body revealed by osmium amine reaction. Bar = 100 nm

HOW DOES LYSOZYME AFFECT *ESCHERICHIA COLI:* Questions Arising from Impact Frozen and Freeze-substitutet Bacteria
A. Gabrieli*, P. Wild*, E.M. Schraner*, A. Pellegrini** and R. von Fellenberg**

* Institute of Veterinary Anatomy and
** Institute of Veterinary Physiology, University of Zürich, 8057 Zürich, Switzerland

Since the discovery by Fleming in 1922 that lysozyme has the ability to lyse Gram-positive bacteria but not Gram-negative bacteria the assumption was established that lysozyme per se cannot affect Gram-negative bacteria. One basis of this assumption is that lysozyme is an enzyme with muramidase activity acting on the peptidoglycan. In *E. coli*, the peptoglycan layer is protected by the outer membrane, and, thus lysozyme has no access to it unless the outer membrane is destroyed, e.g. by EDTA or complement[1]. Accidentally, Pellegrini et al.[2] found that lysozyme kills but does not lyse *E. coli*. Electron microscopy of bacteria fixed with aldehydes revealed that affected *E. coli* were often enlarged, the cytoplasm destroyed to various extend, but the cell membranes seem to remain intact (Fig. 1).

Chemical fixation is proved to induce dramatic changes in bacterial structure[3]. Thus we immobilized bacteria after exposure to lysozyme at 37°C for 15 min by impact freezing on a liquid nitrogen cooled copper block. After freeze-substitution in pure acetone and embedding in HM20 or epon, most of the *E. coli* were found to be intact (Fig. 2). Immunolabeling with a polyclonal antibody raised in rabbits against lysozyme showed that labeling was associated with the bacterial wall. Many gold particles were present in the periplasmic space of bacteria incubated with lysozyme in the presence of 0.7M sucrose which resulted in slight (Fig. 3a) or enhanced plasmolysis and deformation of the cell wall (Fig. 3b). A few bacteria were enlarged exhibiting disrupted cell wall and destroyed cytoplasm. Labeling is associated with the cell wall but also with material within the cytoplasm.

These findings demonstrate that lysozyme exerts the bactericidal effect, hat was proved by viability tests, without destroying the bacterial body unless other factors are involved. How and where lysozyme is active is unknown. The cationic property may play a significant role. Lysozyme may perturb membrane function in a selective manner. Lysozyme may even penetrate the outer cell membrane. Penetration is obviously enhanced under high osmolar condiotions. In the periplasmic space, lysozyme could act as muramidase. In this case, however, one has to consider that the bacterial body still remains intact.

1. L.A. Wilson and J.K. Spitznagel, J. Bacteriol. **96** (1968) 1339.
2. A. Pellegrini, U. Thomas, R. von Fellenberg and P. Wild, J. Appl. Bacteriol. **72** (1992) 180.
3. J.A. Hobot, W. Villiger, J. Escaig, M. Maeder, A. Ryter and E. Kellenberger, J. Bacteriol. **162** (1985) 960.

This study is supported by Stiftung für wissenschaftliche Forschung an der Universität Zürich.

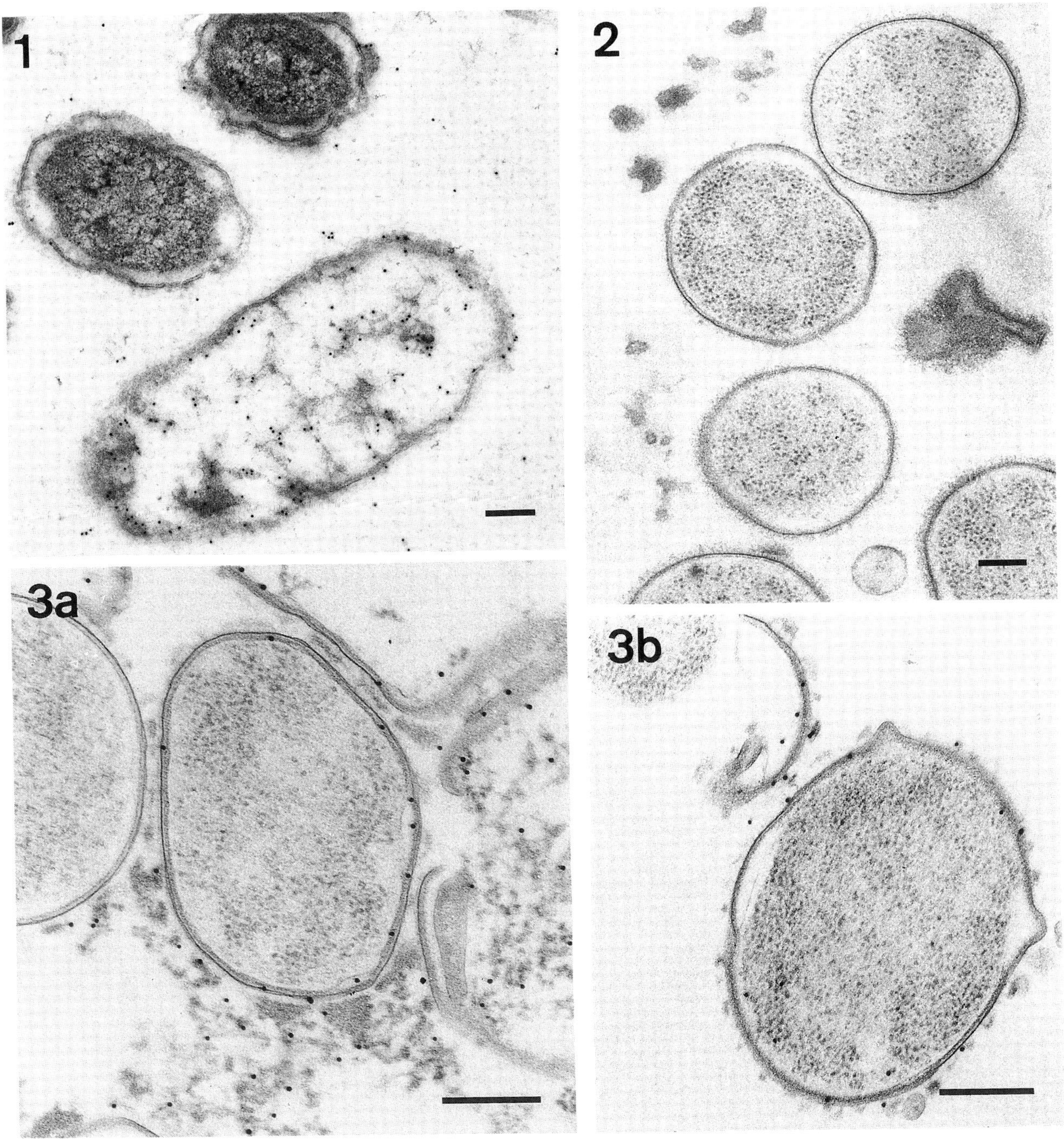

Fig. 1. Intact and swollen *E. coli* after incubation with lysozyme and fixation with aldehydes followed by immunolabeling to localize lysozyme.

Fig. 2. Intact *E. coli* after incubation with lysozyme, impact frozen, freeze-substituted in pure acetone and embedded in epon.

Fig. 3. *E. coli* incubated with lysozyme in the presence of sucrose, impact frozen, freeze-substituted and embedded in epon. Labeling for lysozyme is associated with the cell wall in intact (a) or plasmolysed bacteria (b) and with material within the cytoplasm of destroyed bacteria (b). Bars in Fig. 1-3 = 0.2 µm.

ULTRASTRUCTURAL ANALYSIS OF THE PHYSICAL INTERACTIONS THAT OCCUR BETWEEN *BACILLUS SP.* AND POLYURETHANE DURING BIODEGRADATION

W.N. Norton,* G. Howard,* and R. Blake**

* Biology Department, Southeastern Louisiana University, Hammond, LA 70402
** College of Pharmacy, Xavier University, New Orleans, LA 70746

The scientific community is confronted with a number of significant environmental issues, including the dilemma of how to maintain, safely and effectively, the enormous quantity of organic and inorganic hazardous wastes that are produced in the U.S. alone. Substantial interest in the concept of bioremediation, the use of microorganisms to accelerate the degradation of environmental contaminants, has been generated during the past decade. *Bacillus sp.*, has demonstrated an ability to degrade water dispersible polyurethane, a molecule normally exceedingly difficult for microbial organisms to metabolize.[1] The primary objective of this investigation is to obtain basic ultrastructural information on the physical nature of polyurethane biodegradation conducted by *Bacillus*.

A bacterial suspension of 1×10^9 cells was placed in one liter of a stock solution of polyurethane (3mg polyurethane/L distilled water). A comparable number of cells was maintained in 1 L of growth media and served as a control. At 2, 8, and 24 h subsequent to the initial exposure, samples were obtained from the experimental and control flasks and processed for phase-contrast microscopy, TEM and SEM. Bacteria examined by phase-contrast microscopy at the 2 hour period appeared as small chains of cells. Large aggregates of linked cells were a distinguishing feature of samples observed at the 8 and 24 hour periods (Fig. 1). *Bacillus* observed by SEM at the 2 hour period of exposure were usually observed as chains consisting of two to six cells. The bacteria appeared to be coated with a viscous material or with particles of various dimensions (Fig. 2). Bacteria at the 8 and 24 hour periods were characterized by the formation of large cellular aggregates coated with a viscous material (Fig. 3). A developing transverse septum was a feature of *Bacillus* observed by TEM at the 2 and 8 hour periods of exposure. None, however, were noted among cells at the 24 period. Small amounts of electron-dense amorphous material was located, occasionally, along the outer surface of the cell wall of bacteria collected at the 2 hour period of exposure, but was rarely present on the surface of cells investigated from the 8 or 24 hour periods. Maturing endospores were a prevalent feature of cells examined at 24 hours (Fig. 4), while none were noted at the earlier time periods. The consistent formation of transverse septa in *Bacillus* maintained in the polyurethane solution for either 2 or 8 hours signifies an ability of the bacteria to utilize the synthetic polymer as a nutrient source, at least through 8 hours of exposure. The prevalence, however, of endospores at the 24 hour period indicates the *Bacillus* are incapable of further metabolizing the compound. Inconclusive information was obtained by electron microscopy on the physical appearance of polyurethane. Images obtained by phase-contrast microscopy and scanning electron microscopy indicate polyurethane, at least partially, coalesces to form irregular particles which adhere to the cell wall. A significant number of the cells are also coated with a viscous material indicating polyurethane may be present in two different physical states.

References

1. G.L. Griffin, Pure and Appl. Chem. 52(1980)389.

Proc. Microscopy and Microanalysis 1995, edited by G.W. Bailey, M.H. Ellisman, R.A. Hennigar, and N.J. Zaluzec

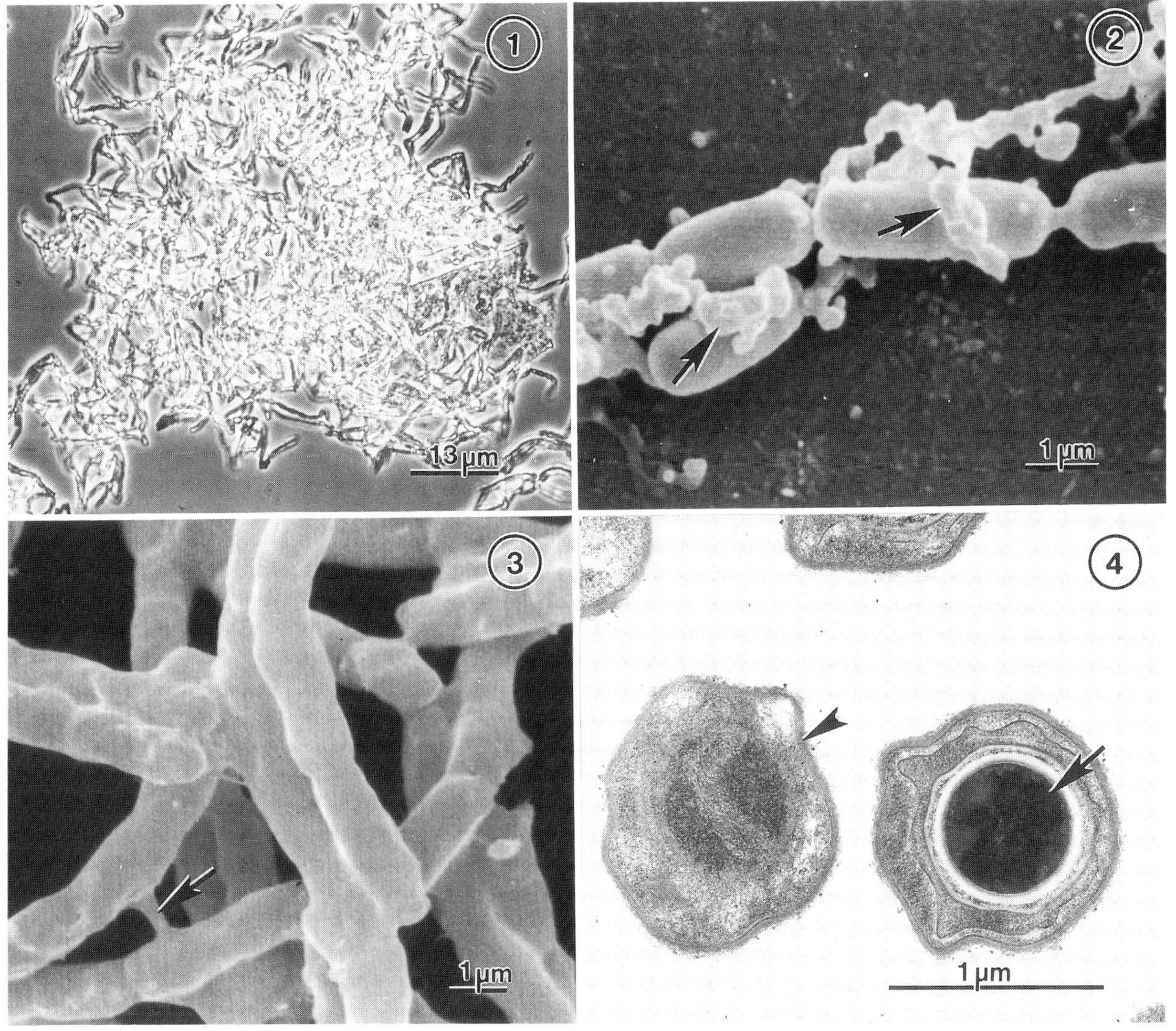

FIG. 1 - Aggregates of *Bacillus* are present in large concentrated masses. Phase-contrast microscopy: Twenty-four hour exposure to polyurethane.

FIG. 2 - Irregularly shaped particles of polyurethane (arrows) lie along the surface of several cells that lack fimbriae. SEM: Two hour exposure to polyurethane.

FIG. 3 - Cells arranged in multiple chains appear coated with viscous material (arrow). SEM: Eight hour exposure to polyurethane.

FIG. 4 - One cell has generated endospore (arrow), while the other (arrowhead) appears in the process of spore formation. TEM: Twenty-four exposure to polyurethane.

ION COMPOSITION OF THE CONTRACTILE VACUOLE OF *ACANTHAMOEBA CASTELLANII* AS DETERMINED BY X-RAY MICROANALYSIS

B.Bowers*, S.L.Shi, S.Q.Sun, and R.D.Leapman.
Biomedical Engineering and Instrumentation Program, NCRR, and *LCB, NHLBI, National Institutes of Health, Bethesda, MD 20892

The contractile vacuole (CV) is an osmoregulatory organelle found in free-living protozoa that functions to pump water out of the cell.[1] In amoebae it comprises at least two distinct systems, a spongiome of tubules and/or vesicles that surround a central collecting vacuole (Fig 1). The filling phase and the discharge of the engorged vacuole appear to be independently regulated.[2] Calmodulin[3], alkaline phosphatase[4] ,and a proton pump[5,6] are known to be specifically associated with the CV complex in amoebae, but the mechanism of water pumping is not known. In order to gain insight into possible ion transport mechanisms, we have examined the ion content of the CV of the small soil amoeba, *Acanthamoeba castellanii*, by analytical electron microscopy.

For analytical electron microscopy amoebae were incubated in growth medium with 5% polyvinyl pyrrolidone (PVP), MW 360,000, for 1.5 hours, then concentrated by centrifugation and rapidly frozen by plunging into liquid nitrogen-cooled ethane. The frozen samples were cryosectioned and cryotransferred at $-180°$ C into a VG-HB501 STEM equipped with a Gatan parallel energy loss spectrometer (EELS) and a Noran energy dispersive X-ray spectrometer (EDXS). EELS was used to determine the mass of the frozen hydrated sections.[7] The sections were then dried in the microscope at $-100°$ C and quantitative dark-field STEM was used to determine the mass of the dried salts in the CV relative to the mass of the support film. The EELS and dark-field measurements were performed at low electron dose. EDXS measurements were then made at higher dose in order to determine the ratios of ions in the dried salts. By combining these data it was possible to estimate the molar concentrations in the hydrated CV.

The results showed that the contractile vacuole contains Na^+ and K^+ in the ratio of about 6:1, very low Ca^{++} and Mg^{++} and insufficient Cl^- to act as counter ion for the anions (Fig.2). Measurements on standards indicated that the observed low Cl^- was not due to mass loss from radiolysis of the alkali halides. Heuser et al.[8] have suggested the compensating counter anion may be bicarbonate by analogy with other transport epithelia. Total osmolality calculated from the sum of the anions and cations was about 21 mosmoles, somewhat lower than the 30-50 mosmoles determined by micropuncture studies[9,10]. This new method of analytical EM offers the possibility of quantitatively determining ion concentrations in dilute cellular compartments for the first time. We plan to use it to explore how changes in the external ionic environment affects the ion ratios in the CV.

1. T.Zeuthen, *Biochim.Biophys.Acta* 1113(1992)229.
2. D.J.Patterson, *Biol.Rev.*55(1980)1.
3. Q. Zhu & M. Clarke, *J.Cell Biol.*118(1992)347.
4. B.Bowers & E.D.Korn, *J.Cell Biol.*59(1973)784.
5. A.K.Fok et al., *J.Cell Sci.*106(1993)1103.
6. K.V. Nolta & T.L.Steck, *J.Biol.Chem.*269(1994)2225.
7. S.Q.Sun et al., *J.Micros.*177(1995)18.
8. J.Heuser et al., *J. Cell Biol.*121(1993)1311.
9. B.Schmidt-Nielsen & C.R.Schrauger, *Science* 139(1963)606.
10. D.H.Riddick, *Am.J.Physiol.*215(1968)736.

Proc. Microscopy and Microanalysis 1995, edited by G.W. Bailey, M.H. Ellisman, R.A. Hennigar, and N.J. Zaluzec
Copyright © 1995 MSA. Published by Jones and Begell Publishing, 79 Madison Ave., New York, NY 10016

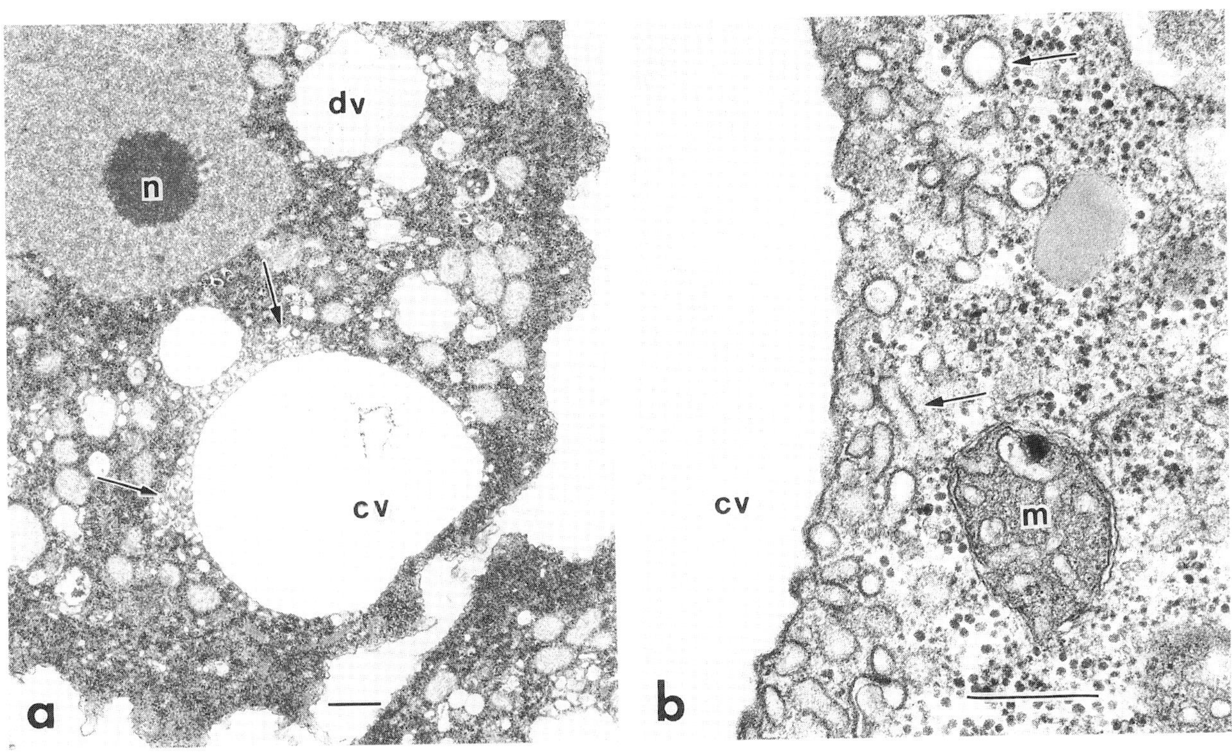

Fig.1.**a**,Thin section of *Acanthamoeba* showing contractile vacuole(cv)and spongiome (arrow).Bar is 1 μm. **b**,Detail of spongiome. dv,digestive vacuole;n,nucleus; m,mitochondrion. Bar is 0.5 μm.

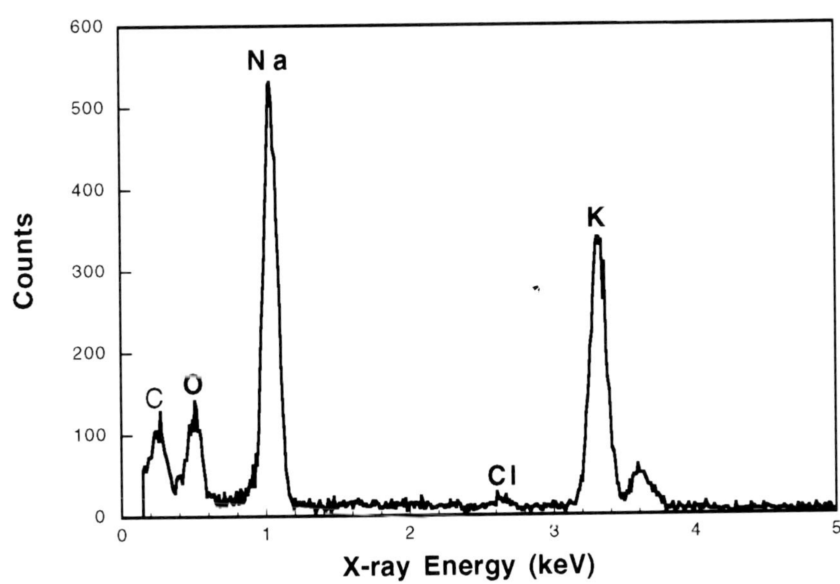

Fig.2-Energy-dispersive X-ray spectrum from dried salts in the *Acanthamoeba* contractile vacuole. Sodium is the major cation; calcium appears as a weak shoulder above the potassium K_β peak.

SEM OF THE TONGUE AND RELATED VOMERONASAL STRUCTURES IN *Coluber constrictor* AND *Uta stansburiana*

J. Froerer* and J. S. Gardner**

* Provo High School, 1125 N. University Ave., Provo, Utah 84604
** Microscopy Laboratory, 128 WIDB, Brigham Young University, Provo, Utah 84602.

The forked tongue in snakes and lizards serves to follow the pheromone trails of both prey and conspecifics.[1] The three phases of the tongue flicking cycle are interrelated with three functional structures involved with the transportation of stimulus particles from the environment to the vomeronasal organ (VNO). (1) The sample phase involves protrusion of the tongue, collection, and transfer of stimulus particles to the anterior processes. (2) A transfer phase in conjunction with the anterior processes serves to facilitate the collection and localization of stimulus particles from the tongue. (3) The delivery phase includes anterior process elevation and the delivery of particles to the VNO.[2] The VNO simultaneously deciphers the chemical concentrations, enabling the animal to follow pheromone trails in relation to the most volatile side, right or left.[3] The morphology of the tongue, anterior processes, and VNO of a previously unevaluated snake and lizard are presented here.

Snake species *Coluber constrictor* and lizard species *Uta stansburiana* were used in this study. Snakes were collected in Northern Utah and lizards were collected in Southern Utah. All specimens were prepared for scanning electron microscopy using standard techniques.

All three functional structures increase surface area and provide a more particle-adherent surface. Evaluation of tongue morphology in *C. constrictor* showed microfacets and micropores which were abundant throughout the anterior portion of the tongue (fig. 1). The microfacets increase surface area and the micropores secrete mucous which collects the stimulus particles.[4] Instead of the bifurcated tongue in snakes, lizards exhibit two ventral pads. Present on these pads in *U. stansburiana* was a latticework of micropores (Fig. 2).

During the transfer of stimulus particles from the tongue, the anterior processes collect and concentrate chemicals. The oblique folds increased the surface area of the anterior processes and the small opening of the sublingual gland may lubricate the surfaces providing a particle-adherent surface.[3] The folds in the snake are smaller and more numerous than the broad folds in the lizard (fig 3&4).

The paired VNO is directly above and complimentary to the bilateral anterior processes. In the *C. constrictor* the anterior processes are approximately 1 mm long, which is similar to the length of the VNO (figs. 3&5). The *U. stansburiana* has the same size correlation as *C. constrictor*; anterior processes, 2mm and the VNO, 2 mm (figs. 4&6). However, the laterally located VNO in the lizard, as well as the size of the ventral pads and the lack of tongue bifurcation disproves Broman's hypothesis of direct tongue insertion into the VNO.[5] This study suggests an alternate possibility of the transfer of stimulus particles to the VNO, which is in agreement with Gillingham and Clark.[2]

References

1. K. Schwenk, *Science* 263(1994)1573.
2. J. C. Gillingham and D. L. Clark, *Can. J. Zool.* 59(1981)1651.
3. M. Halpern, *Biology of the Reptilia* 18(1992)423.
4. S. H. Mao et.al., *J. Morph.* 208(1991)279.
5. J. Broman, *Anat. Heft.* 158(1920)137.

Proc. Microscopy and Microanalysis 1995, edited by G.W. Bailey, M.H. Ellisman, R.A. Hennigar, and N.J. Zaluzec

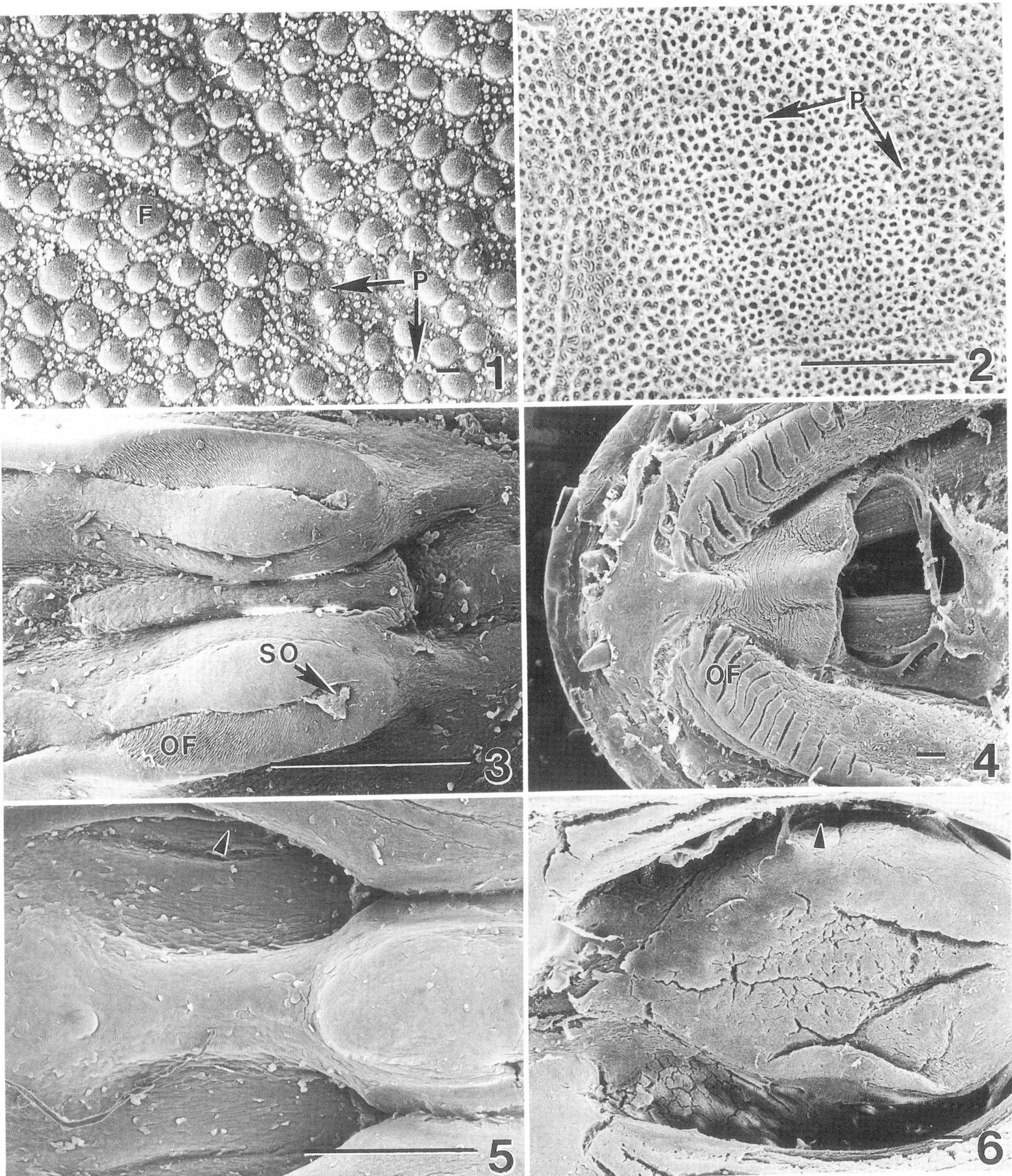

Fig. 1-6. SEM micrographs. Figs. 1,3,5. *C. Constrictor* and Figs. 2,4,6 *U. Stansburiana*. Fig. 1. Microfacets (F) and micropores (P). Bar = 1 μm. Fig. 2. Micropores (P). Bar = 10 μm. Fig. 3. Bilateral anterior processes possessing oblique folds (OF) and a small opening (SO). Bar = 1mm. Fig. 4. Broad-surfaced oblique folds (OF) of the anterior processes. Bar = 100 μm. Fig.5. Paired vomeronasal organ openings (arrowhead). Bar = 1mm. Fig. 6. Paired openings of the vomeronasal organ (arrowhead). Bar = 100 μm.

ULTRA-SPATIALLY RESOLVED SYNCHROTRON POWERED FT-IR MICROSPECTROSCOPY OF INDIVIDUAL CELLS OF BIOLOGICAL MATERIAL

David L. Wetzel*, John A. Reffner**, and Gwyn P. Williams***

*   Kansan State University, Microbeam Molecular Spectroscopy Laboratory,
    Shellenberger Hall, Manhattan, KS 66502.
** Spectra-Tech, Inc., 2 Research Drive, Shelton, CT 06484.
***Brookhaven National Laboratory, National Synchrotron Light Source, Bldg. 725B,
    Upton, NY 11973.

Synchrotron radiation is 100 to 1000 times brighter than a thermal source such as a globar. It is not accompanied with thermal noise and it is highly directional and non-divergent. For these reasons, it is well suited for ultra-spatially resolved FT-IR microspectroscopy. In efforts to attain good spatial resolution in FT-IR microspectroscopy with a thermal source, a considerable fraction of the infrared beam focused onto the specimen is lost when projected remote apertures are used to achieve a small spot size. This is the case because of divergence in the beam from that source. Also the brightness is limited and it is necessary to compromise on the signal-to-noise or to expect a long acquisition time from coadding many scans. A synchrotron powered FT-IR Microspectrometer does not suffer from this effect. Since most of the unaperatured beam's energy makes it through even a 12 x 12 μm aperture, that is a starting place for aperture dimension reduction..

Attenuation of the beam by the specimen and reduction of incidental diffraction sampling of adjacent specimen by double aperaturing may also reduce the energy somewhat. Otherwise, it is practical to work with 6 x 6 μm aperaturing and obtain excellent spectra in less than two minutes. In situ single cell infrared microspectroscopy can be achieved readily rather then painstakingly. Also, the mapping of microscopic specimens by accumulating data from many small spots in a reasonable time allows construction of functional group maps from spectra from discrete well defined subsamples within the tissue.

The figures show spectra of various single cells from plant and mammalian tissue.

The maps of the cross section of a kernel of rye show differences in lipid and protein groups in reference to the germ (to the rear), the outside portion of the seed cross, section and the inner portion containing endorsperm.

Proc. Microscopy and Microanalysis 1995, edited by G.W. Bailey, M.H. Ellisman, R.A. Hennigar, and N.J. Zaluzec
Copyright © 1995 MSA. Published by Jones and Begell Publishing, 79 Madison Ave., New York, NY 10016

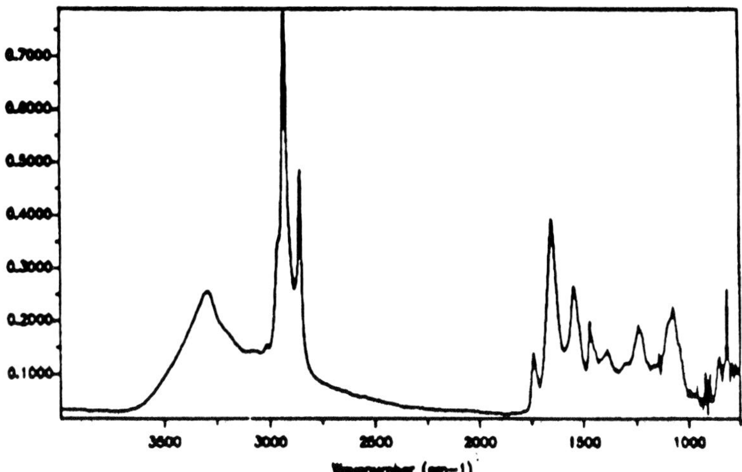

Spectrum of retina tissue, lipid rich.

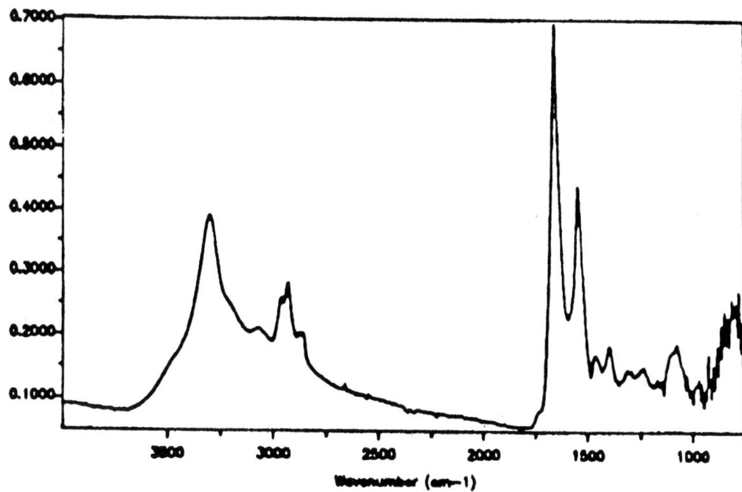

Spectrum of a different spot in the same retina

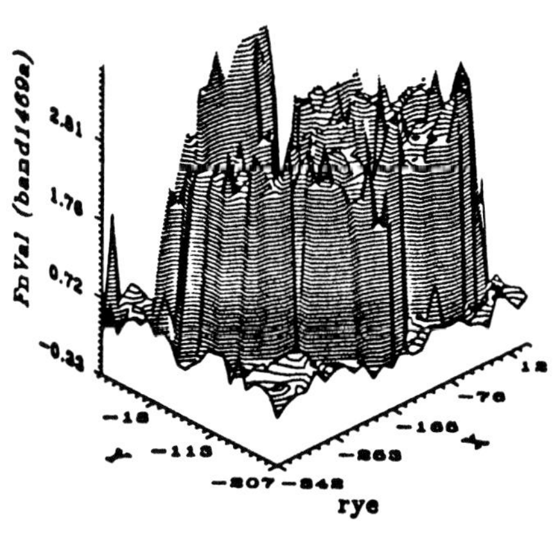

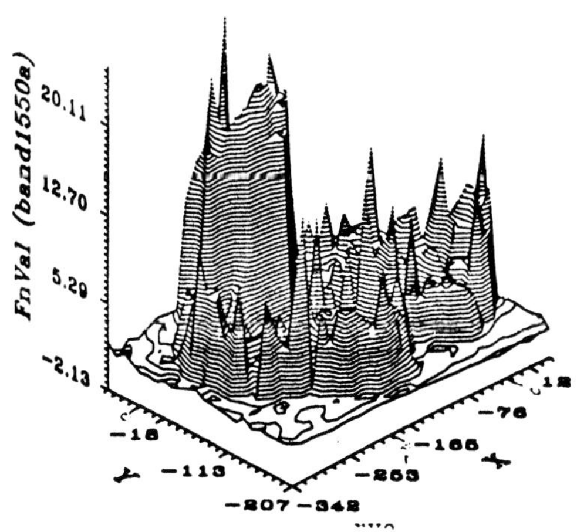

AN ENHANCED CELLULAR ORGANELLE VISUALIZATION PROCEDURE WHEN STAINING FOR PEROXISOMES

J. W. Horn* and B.J. Dovey-Hartman**

*Lilly Research Laboratories, Eli Lilly & Company, 2001 W. Main St., Greenfield, IN 46140
**Schering-Plough Research, Schering-Plough, P.O. Box 32, Lafayette, NJ 07848

Reliable visual identification of peroxisomes is important in developmental, clinical, and investigational research. The current technique employed in most laboratories uses a specific electron dense label for the demonstration of peroxisomes by transmission electron microscopy by applying 3,3'-Diaminobenzidine Tetrahydrochloride (DAB) directly to freshly fixed tissue samples to react with endogenous peroxisomal catalase. After routine processing, ultrastructural examination of tissue sections is conducted either with light staining or without post-staining of grids. While peroxisomes are easily identified using this method, remaining tissue architecture is difficult to visualize due to the opacity of the tissue. Additionally, if grids are post-stained with heavy metal solutions, they must be modified to allow for enough staining to visualize cellular components without compromising the quality of the peroxisome label. We will describe a technique whereby DAB-reacted tissues are stained with a post-fixative solution including potassium ferricyanide that imparts density to cell membranes and cellular components thereby enhancing identification and interpretion of data.

Freshly necropsied or biopsied tissues were fixed in 2% formaldehyde/2.5% glutaraldehyde, reacted with DAB, and rinsed with an appropriate buffer.[1&2] A secondary fixative solution of 1% osmium tetroxide/1.5% potassium ferricyanide is prepared fresh and the tissues are placed in this solution for 1 to 2 hours at room temperature with gentle stirring.[3] The tissues can then be buffer rinsed and routinely processed by dehydration with serially graded ethanols, infiltrated, and embedded in fresh 100% resin. Ultrathin sections are then cut, mounted on appropriate grids and evaluated with a transmission electron microscope without post-staining.

Unstained sections were evaluated and photographed with a Philips 410LS transmission electron microscope. Peroxisomes were densely stained with DAB in hepatocytes of control tissues (Figs. 1 through 4). Additionally, cell membranes and cellular organelles such as endoplasmic reticulum, golgi, and nuclei, were moderately stained (Figs. 2 and 4). Cell types within this tissue were easily distinguishable.

This method resulted in enhanced uniformity of moderately dense-stained cell organelles without the need to post-stain grids with lead citrate and uranyl acetate. The technique using 1% osmium tetroxide/1.5% potassium ferricyanide is a fast and reliable method compared to other methods of lightly stained grids or no staining at all. Eliminating the post-staining step is advantageous due to the need to develop a staining protocol that would reveal satisfactory cellular stain density but not mask DAB-stained peroxisomes. The overall advantage of evaluating tissue samples with this method is that it yields convincing information as to the involvement of cell constituents with the target DAB-stained peroxisomes, the location of DAB-stained peroxisomes within a cell type, and the ability to identify cell types.

References

1. H.D. Fahimi, *J. Histochem. Cytochem.* 27(1979)1265.
2. A.B. Novikoff and S. Goldfischer, *J. Hisotchem. Cytochem.* 17(1969)675.
3. J. Turek, Technical Section, *Fall MSEM Newsletter.* (1984) 22.

Proc. Microscopy and Microanalysis 1995, edited by G.W. Bailey, M.H. Ellisman, R.A. Hennigar, and N.J. Zaluzec

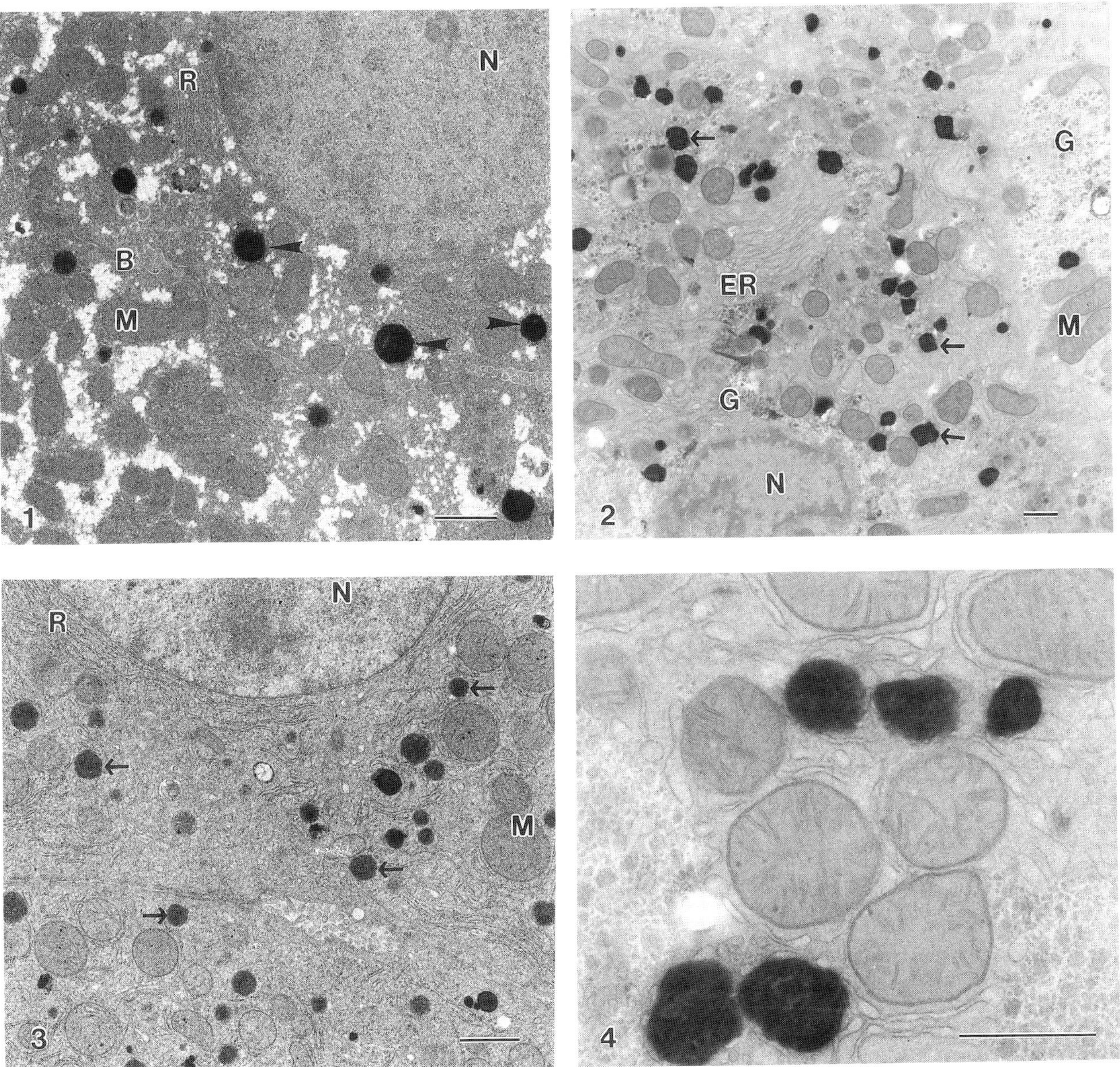

FIG. 1.   Hepatocytes from the liver of a control rat. Note the ultrastructural detail in sections that were not post-stained with heavy metals, DAB-stained peroxisomes (arrowheads) and cellular organelles mitochondria (M), nucleus (N), bile caniliculi (B), and rough endoplasmic reticulum (R) are clearly visible. Bar = 1 μm

FIG. 2.   Mouse hepatocytes postfixed with osmium tetroxide and potassium ferricyanide, reacted with DAB for peroxisomes (arrows) but sections were not post-stained. Peroxisomes are densely stained while endoplasmic reticulum (ER), mitochondria (M), nucleus (N), and glycogen (G) are moderately stained. Bar = 1 μm

FIG. 3   Hepatocytes from the liver of a control rat. Unstained sections reveal intensely labelled endogenous peroxisomal catalase easily seen by the electron dense reaction product of DAB (arrows). Cell types are easily identified including the cellular organelles; mitochondria (M), nucleus (N), and rough endoplasmic reticulum (R). Bar = 1 μm

FIG. 4   Higher magnification reveals mitochondria with well defined cristae, densely stained peroxisomes and well defined membranes. Bar = 1 μm

HIGH RESOLUTION CHEMICAL MAPPING USING TAPPING MODE AFM WITH PHASE CONTRAST

Donald A. Chernoff

Advanced Surface Microscopy, Inc., 6009 Knyghton Road, Indianapolis, IN 46220-4955

Introduction. A challenging problem in materials analysis is imaging the microstructure of composite surfaces. Some examples of composite surfaces are: contamination deposits, discontinuous (i.e. defective) thin films, devices built of composite materials, and cross-sectional specimens of composite materials. Techniques such as XPS and SIMS can provide detailed information about the chemical species present on the surface, but only average information is obtained for areas smaller than a few μm wide. Auger can have sub-micron spatial resolution, but it provides little useful information about organic materials and it is difficult to use on insulating surfaces. In contrast, we have developed a new method for imaging chemical domains with spatial resolution of 10-20 nm. In principle, this method may be used on any surface.

The TappingMode[TM] AFM[1] uses an oscillating cantilever whose motion can be characterized by its amplitude and phase relative to its driver. In its most common mode of operation, this AFM uses the amplitude signal to sense the specimen surface, control the Z piezo and create a height image. This report describes the additional contrast obtained by also capturing the phase signal.

Methods. We used the Digital Instruments NanoScope III Scanning Probe Microscope, fitted with the Phase Extender Module and the Dimension 3000 large sample microscope. We used silicon cantilevers for TappingMode[TM] work, capturing dual images using the height and phase data channels. We used silicon nitride cantilevers for contact mode work, repositioning the probe to the same spot previously examined and capturing triple images of height and lateral force (trace and retrace directions).

Results. Figure 1 shows height and phase images of a bleached wood pulp fiber, a natural composite structure. Because of the relatively rough topography, fine structure was nearly invisible in the height image. However, the phase image had much fine structure. The edge- and slope-sensitivity of the phase signal revealed large areas of closely-spaced microfibrils. We identify these as cellulose, because the microfibrillar nature of cellulose is well-known. In addition, there were small patches of smooth, amorphous material, present as a thin coating on top of the cellulose. We identify these features as residual lignin. Note that the lignin appeared bright in the phase image. We interpret the distinctive phase contrast levels for cellulose and lignin as follows. The probe experiences momentary adhesion when it touches the surface. Greater adhesion increases the damping and retards (decreases) the phase of the oscillating cantilever. The lower phase observed on cellulose means that the Si tip adhered more strongly to the cellulose than to the lignin. This is probably due to the larger number of -OH groups in cellulose as compared with lignin. -OH groups are available on the Si probe for hydrogen bonding to the -OH groups on the fiber. Lateral force images (not shown) demonstrated lower friction on the low adhesion (high phase) regions.

Figure 2 shows height and phase images of a magnetic recording head, a composite device fabricated by joining and polishing glass and ferrite. The height image shows that the surface of both regions was covered with polishing marks and fine debris. The phase image allows us to classify the debris as glass particles, because the phase contrast level of the particles matched that of the glass. Lateral force images (not shown) were uninformative because the contact mode scans swept the particles away.[2]

1. Q. Zhong et al., *Surf. Sci. Lett.*, 290(1993)L688.
2. The author thanks V. Elings for helpful discussions and encouragement.

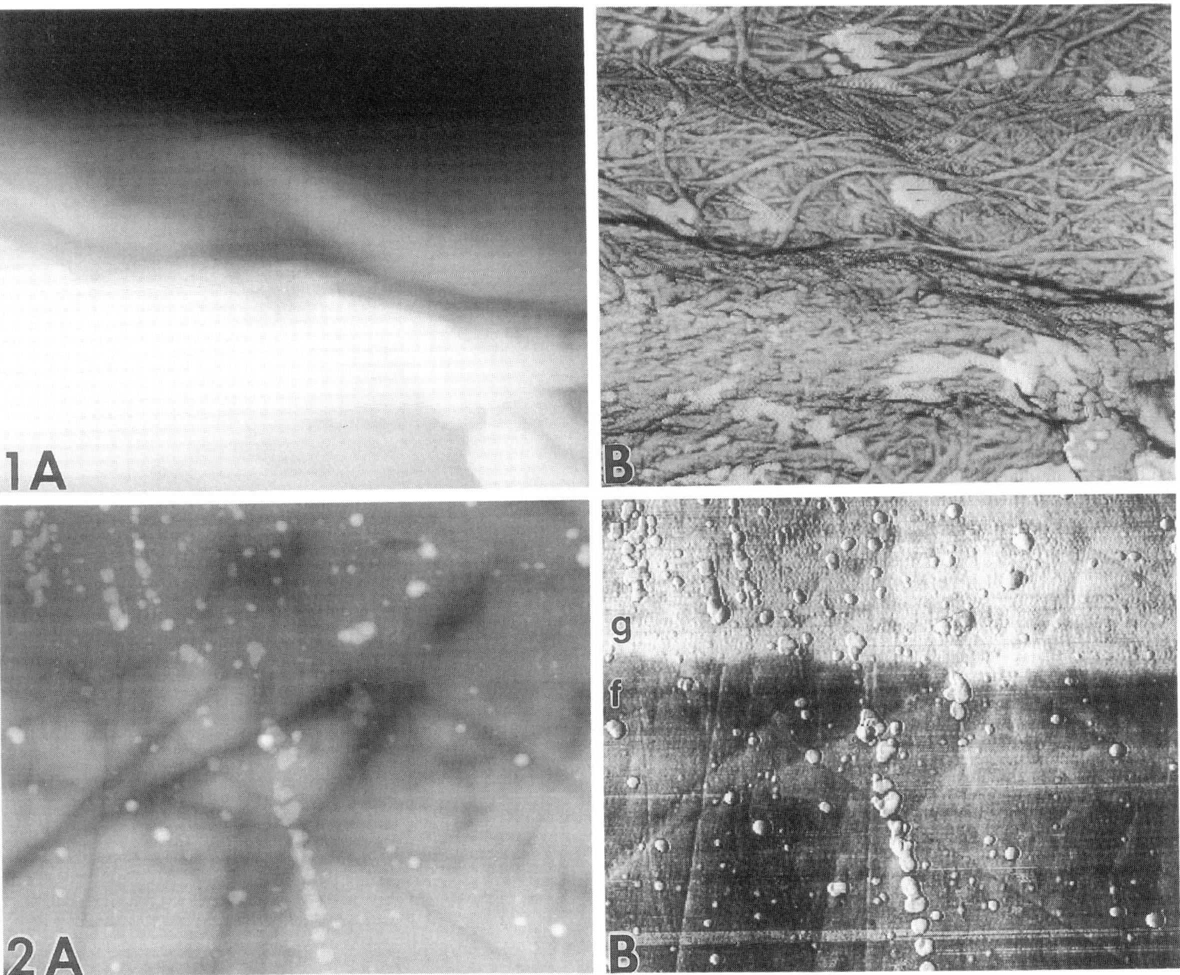

FIG. 1.-- Bleached wood pulp fiber imaged by TappingMode[TM] AFM. A) Height image. B) Phase image. Note the fine microfibrils (cellulose) and the amorphous, bright deposits (lignin). Specimen courtesy of Dr. Doro-Pereira (Aracruz Celulose). Bar = 1μm.

FIG. 2.-- Minicomposite magnetic recording head, showing a ferrite pole (f) and the glass gap (g). A) Height image. B) Phase image. Small particles on the surface have the same contrast level as the glass. Bar = 1 μm.

ATOMIC FORCE MICROSCOPIC INVESTIGATION OF ENDOTHELIAL FENESTRAE

E. Wisse,[*] F. Braet,[*] W. Kalle,[**] B. de Grooth,[***] A. Raap,[**] and H. Tanke[**]

[*]Laboratory for Cell Biology and Histology, Free University of Brussels (VUB), Laarbeeklaan 103, 1090 Brussels-Jette, Belgium; [**]Laboratory for Cytochemistry and Cytometry, State University of Leiden, Wassenaarseweg 72, 2333 AL Leiden, The Netherlands; [***]Department of Applied Physics, Technical University of Twente, P.O. Box 217, 7500 AE Enschede, The Netherlands.

Endothelial cells of rat liver sinusoids (LEC) contain fenestrae, which are clustered in sieve plates. These open fenestrae, with an average diameter in the order of 0.1 μm, control the exchange of fluids, solutes and particles between the blood and the parenchymal cells.[1] The surface of LEC can optimally be imaged by scanning electron microscopy (SEM). SEM can also be used to study dynamic changes in fenestrae by comparing specimens subjected to different experimental conditions.[2] However, the SEM uses fixed, dried and coated specimens. Recently, the atomic force microscope (AFM) was introduced, enabling the analysis of cell surfaces under fluid. We used the AFM for the investigation of fenestrae in cultured LEC, whereas SEM served as a reference.

LEC were isolated by collagenase perfusion and purified by a Percoll gradient and selective adherence[2]. After spreading on collagen, LEC were cultured for 8 h. Cells were fixed in 2% glutaraldehyde (12 h), followed by 1% tannic acid (1 h) and postfixed in 1% osmium. Samples were dehydrated in alcohol, critical point dried in ethanol / CO_2, and sputter coated with gold. We compared imaging of dried-coated (SEM and AFM), dried-uncoated (AFM) and wet-fixed (AFM) specimens in a series of experiments, with the purpose to prepare for depicting living cells in vitro.

AFM images of LEC in culture show structures that correlate well with SEM images. Dried-coated, dried-uncoated, and wet-fixed cells show flat, fenestrated cellular processes. We measured the diameter of fenestrae in dried-coated specimens in SEM (213 ± 57 nm, Fig. 1), dried-coated ones in AFM (211 nm ± 40 nm, Fig. 2), dried-uncoated ones in AFM (197 ± 40 nm, Fig. 3) and wet-fixed in AFM (269 ± 44 nm, Fig. 4). After treatment with ethanol or serotonin, the diameters of fenestrae increased 6% or decreased 15% respectively. The same alterations of fenestrae could be distinguished by measuring AFM images of dried-coated, dried-uncoated and wet-fixed LEC. Comparison of measurements on dried-coated SEM preparations and wet-fixed AFM preparations indicated a shrinkage of about 20% in tannin-treated SEM preparations.

In conclusion, high resolution imaging of LEC by AFM show cellular details, such as fenestrae in sufficient detail to perform measurements, which was hitherto only possible with fixed, dried and coated preparation in SEM.

References

1. E. Wisse et al., *Hepatology* 5(1985)683.
2. F. Braet et al., *Lab. Invest.* 70(1994)944.
3. This research was supported by the Belgian National Fund for Scientific Research, grant no. 3.0053.92. The support of R. De Zanger and C. Derom is gratefully acknowledged.

Proc. Microscopy and Microanalysis 1995, edited by G.W. Bailey, M.H. Ellisman, R.A. Hennigar, and N.J. Zaluzec
Copyright © 1995 MSA. Published by Jones and Begell Publishing, 79 Madison Ave., New York, NY 10016

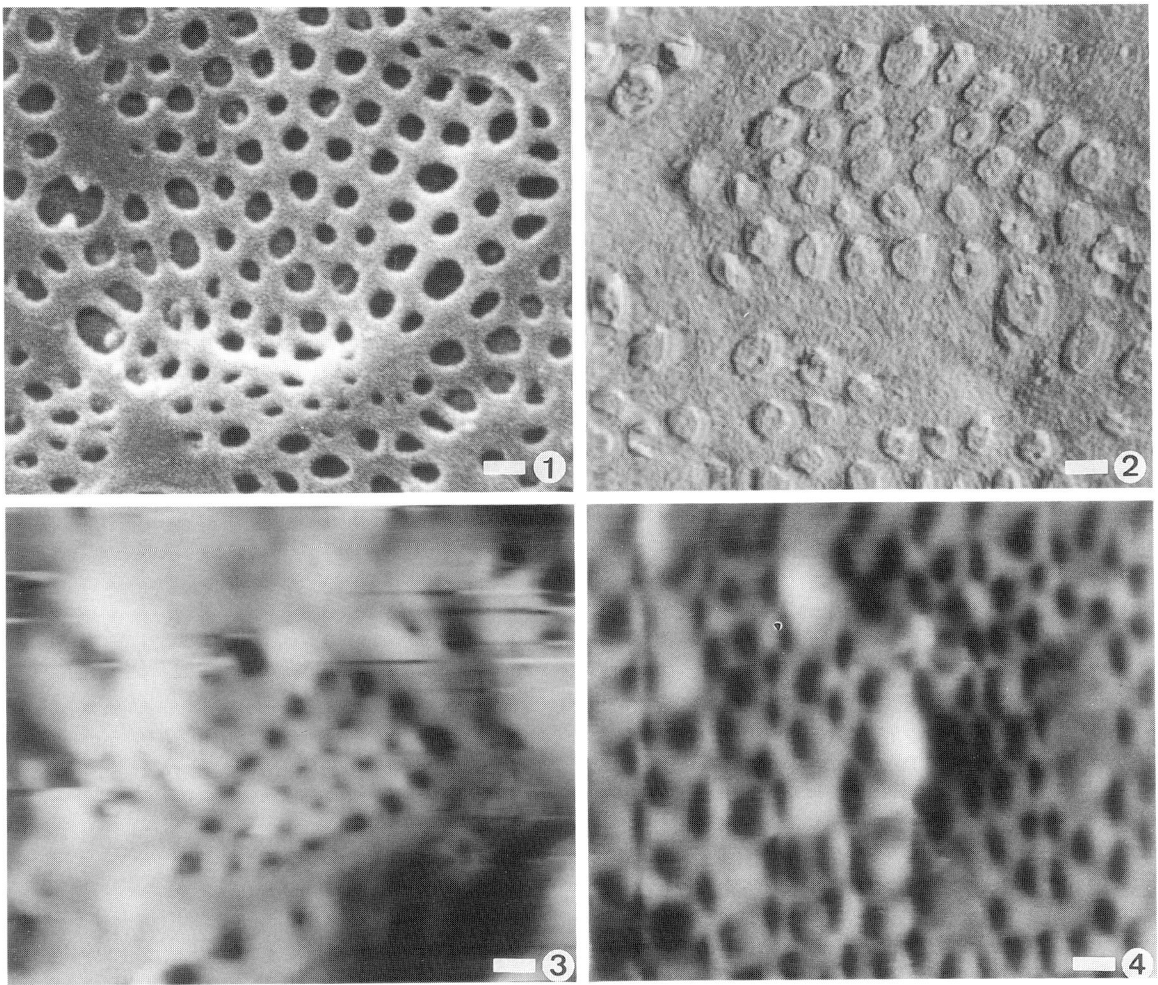

FIG. 1.-SEM micrograph of fenestrae in cultured endothelial cells which were isolated from rat liver. Cells were fixed, dehydrated, critical point dried, sputter coated with gold and photographed in a SEM at 30 KV. Bar = 200 nm.

FIG. 2.-Atomic force microscopic image of the same preparation as in Fig. 1. The granular surface structure suggest the imaging of the sputtered gold layer. The image was scanned in contact mode with a standard tip. This force image was not processed by software. The image shows some distortion in vertical alignment. Bar = 200 nm.

FIG. 3.-Atomic force microscopic image of a preparation, as in Fig. 1, but not sputter-coated with gold. During scanning, the tip apparently collected material from the surface of the cells, causing linear streaks, disturbing the image. This force image differs in quality and character from the one in Fig. 2. Bar = 200 nm.

FIG. 4.-Atomic force microscopic image of fixed cells, which were not treated any further. Cells were scanned in buffer in contact (not tapping) mode. Fenestrae are depicted, but image distortions are clearly present as alternating compressed and extended rows of fenestrae. Bar = 200 nm.

MAPPING TRACTION FORCES GENERATED BY MOTILE CELLS

Tim Oliver,* Olga J. Pletjuushkina,*** Juri M. Vasiliev,*** Micah Dembo** and Ken Jacobson*

*Department of Cell Biology and Anatomy, University of North Carolina at Chapel Hill, Chapel Hill, NC 27599-7090. **Theoretical Biology and Biophysics Group T-10, Mail Stop K710, Los Alamos National Laboratory, Los Alamos, NM 87545. ***Belozersky Institute of Physical and Chemical Biology, Moscow State University, Moscow 119899, Russia and Cancer Research Center, Acadamy of Medical Sciences of Russia, Moscow 109801, Russia.

In a continuing effort to understand how cell-generated traction forces are utilised for locomotion, we have applied our modified silicone rubber traction force assay to rapidly locomoting fish epidermal keratocytes executing turns and shape changes, and negotiating obstacles. The resulting maps show that these cells can redistribute tractions from the "steady-state" pattern (previously observed during unobstructed, gliding locomotion),[1-2] into a variety of transient patterns, with lifetimes of less than 1 minute (Figs. 1-4). The map for a "steady state" locomoting keratocyte shows a maximum traction force density of $\sim 5 \times 10^{-5}$ dynes/μm^2 (data not shown). This value was derived from cell-free experiments in which elastic films were manipulated with a pair of micronneedles. Such experiments, in which all forces were known, showed that both the magnitude and direction of traction forces applied to the film could be closely predicted, and that the Young's modulus of elasticity for the silicone substratum could be calculated.[3] The consequences for understanding the underlying molecular basis for shape change and cell motility from this type of analysis will be discussed.

Traction mapping on silicone substrata has also been used to compare traction forces generated before and after treatment of C3H-10T1/2 fibroblasts with colcemide (Figs. 5-6). Colcemide (0.2 $\mu g/ml$) induced a marked increase in tractions within 20 minutes of application. These experiments corroborate earlier experiments performed on glass substrata, in which colcemide treated fibroblasts shortened their long axes and showed elongated focal adhesions.[4]

References
1. T. Oliver et al., *Proc. Ann. MSA Meeting* 52(1994)162.
2. T. Oliver et al., *Cell Motility and Cytoskeleton* (in press).
3. T. Oliver et al., (manuscript in preparation).
4. O.J. Pletjushkina et al., *Experimental Cell Research* 212 (1994) 201-208.
5. Supported by NIH grants: GM 35325 (KJ) and A121002MD (MD) and FIRCA grant TW00401 (JV).

FIGS. 1-2-Two separate keratocytes with spontaneously adherent tails, each locomoting in direction of large arrow. Note maximum tractions (longest arrows) either side of cell body and at tail tip. Tail subsequently breaks free. Tractions amplified x3 and x5 respectively, for display.
FIG. 3-Keratocyte making a small radius turn anticlockwise about a point marked **O**. Note maximum tractions are located at the tail. Tractions amplified x2 for display.
FIG. 4-Keratocyte moving in direction of large arrow, undergoing a complex sequence of shape changes. This particular map (part of a 10 frame sequence) was preceeded by a more rounded morpholgy (2 minutes earlier), and followed by the re-establishment of "steady state' gliding morphology (within 4 minutes). Tractions amplified x3 for display.
FIG. 5-Control fibroblast shows moderate tractions. Tractions amplified x3 for display.
FIG. 6-Same cell as figure 5, after application of colcemide. Note strong axially aligned tractions compared to control. Tractions amplified x3 for display. All scale bars =10 μm.

Proc. Microscopy and Microanalysis 1995, edited by G.W. Bailey, M.H. Ellisman, R.A. Hennigar, and N.J. Zaluzec
Copyright © 1995 MSA. Published by Jones and Begell Publishing, 79 Madison Ave., New York, NY 10016

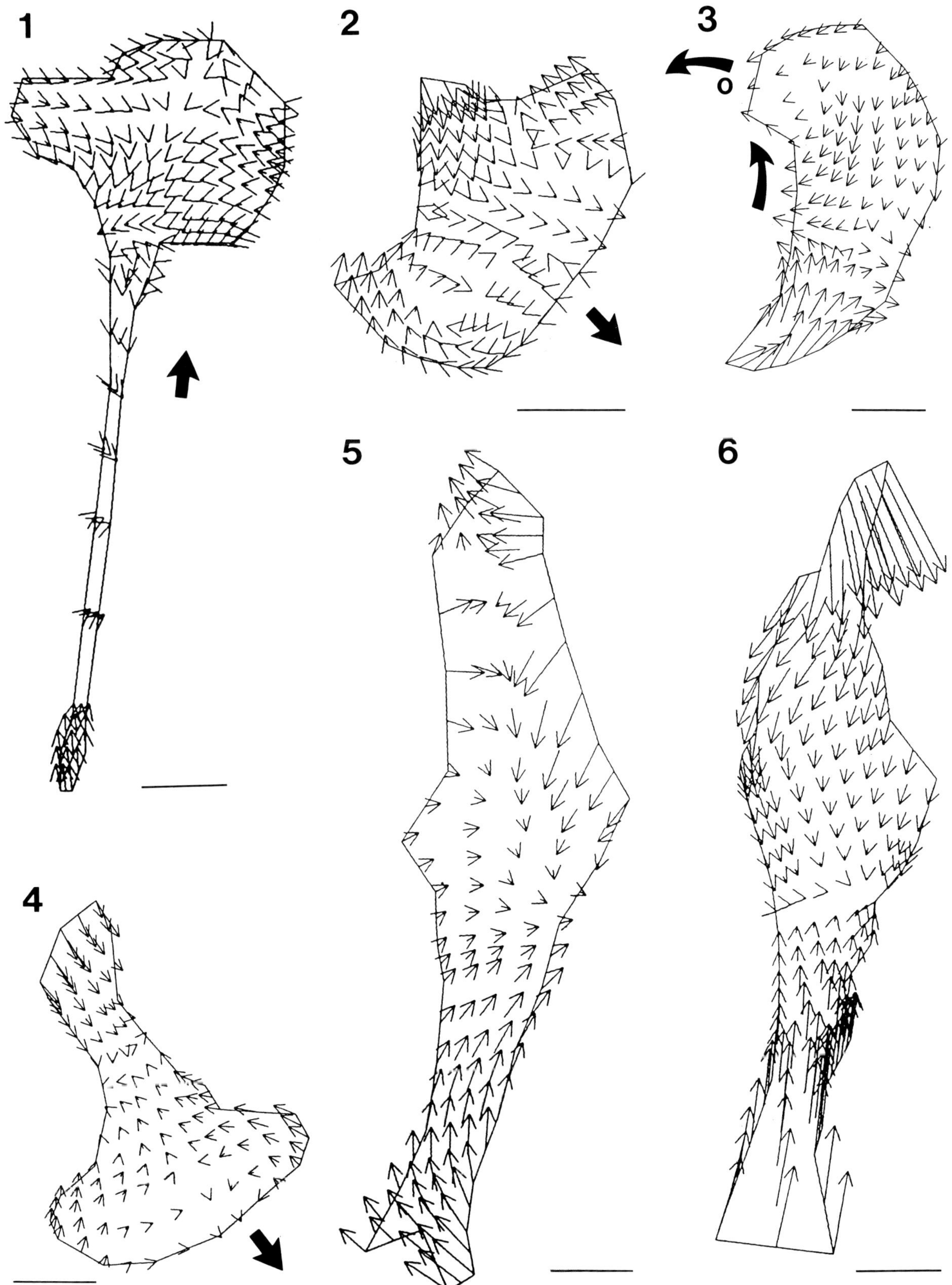

893

HIGH RESOLUTION SEM OF CLOSTRIDIAL ATTACHMENT STRUCTURES

B. Panessa-Warren, G.T.Tortora* and J.B.Warren**

School of Health, Technology and Management, Divisions of Diagnostic and Therapeutic Sciences and Graduate Studies, State University of New York, Stony Brook,N.Y., 11794-8205.
*School of Health, Technology and Management, Dept. of Medical Technology and Head, Clinical Microbiology Laboratory, University Medical Center, State University of New York, Stony Brook, N.Y.11794-8205.
**Division of Instrumentation, Bldg 535B, Brookhaven National Laboratory, Upton, N.Y. 11973.

Clostridium sporogenes forms small (1 μm), dehydrated endospores that are resistant to the effects of solvents, acids, alkali, pressure, UV radiation and most sporocides[1]. In the past it was felt that the exosporial membrane served no function in the infection and colonization process, however our recent work on intact spores has demonstrated that the exosporial membrane seems to play a significant role in the initial attachment of the spore to a nutritive substrate[2,3]. With the improved resolution of the SEM, it has been possible to examine intact spores attached to various substrates and characterize each stage of attachment[3]. This paper examines the exosporial membranes of activated C.sporogenes spores using conventional specimen preparation with imaging using a lanthanum hexaboride gun, and specimens imaged with a field emission SEM. Spores were grown on Trypticase Soy agar with 0.5% glucose (TSA) added and blood CDC (BCDC) agar and attachments and viability monitored by plate counts and light microscopy of unfixed spores (both phase and malachite green/safranin staining).

C.sporogenes endospores were suspended in distilled water and 10 μl aliquots (0.5 McFarland std.) were plated onto TSA agar plates for colony counting and light microscopy, or 10μl of spore suspension was placed onto aluminum SEM stubs that had TSA agar embedded into grooves cut into the aluminum. All samples were incubated in an anaerobic incubator at 35°C for 1, 1.5 or 2 hours. Agar plates were removed from the anaerobic hood and read at 24 and 48 hrs and photographed. Smears were made of the agar surface, stained with malachite green/safranin, photographed and read using an Olympus photomicroscope. Two milliliters of 2.8% glutaraldehyde in 0.1 M cacodylate buffer (pH 7.2) were placed onto the scanning stub and the stub covered with a small glass vial to prevent evaporation of the fixative in the incubator. The spores were fixed for 30 min at 35°C, after which they were removed from the anaerobic hood and incubator, and placed in fresh fixative at 4°C. Spores were subsequently washed in buffer, dehydrated in acetone, dried by the critical point method in liquid CO_2 and immediately coated with gold-palladium (AuPd) using a Polaron EJ100 sputter coater, or tungsten (W) coated in an Edwards Vacuum Evaporator.

The attachment of spores to human cells was also studied using shell vials of MRC5 (Bio-Whittaker) lung fibroblasts. Coverslips with viable fibroblasts were placed in a small volume of media in a sealed petri dish. Ten microliters of spore suspension were added to each fibroblast coated coverslip in the anaerobic hood, and the petri dishes incubated for 1 hr at 35°C. To remove unattached spores and media from the cell surface, preparations were washed in either Dulbecco's phosphate buffered saline with calcium and magnesium added (pH7.2) or HEPES buffer (pH 7.2). The coverslips were then post-fixed in 1% aqueous osmium tetroxide, washed in deionized water, dehydrated in acetone, dried by the critical point method, and the coverslips attached to the stub with silver conducting paint. Samples were either sputter-coated with 8-10nm AuPd, or rotary coated with 3nm W. AuPd coated samples were examined at 20kV in an AMRAY SEM 1000 with a lanthanum

Proc. Microscopy and Microanalysis 1995, edited by G.W. Bailey, M.H. Ellisman, R.A. Hennigar, and N.J. Zaluzec

hexaboride gun. W-coated samples were imaged in an AMRAY 1845 field emission SEM at 5 or 10kV.

Spores were found to attach to the TSA agar-coated aluminum scanning stubs by extensions of the exosporial surface (Fig. 1,2). Spores appeared to prefer to attach to the active borders of the fibroblast cells, rather than their smooth surface of the fibroblast plasmalemma. And these attachments were primarily by delicate exosporial projections from any part of the exosporium. By field emission SEM it was possible to see the intricate surface structure of the activated spores (Fig. 2) and discern the individual exosporial attachment structures that adhered to the protein coated coverslip, agar coated aluminum stub and active periphery of the fibroblast cells.

Because the spore attachment period was limited to an hour when using the fibroblast preparations, there were less examples of the final stage of endospore attachment in which the spore tail region forms a single anchoring structure to the fibroblast surface. More work needs to be done to clearly delineate the nature of all of these attachments at high resolution with as "structureless" a metal coating as possible. The field emission SEM had significant advantages for imaging the very delicate, hairlike attachment structures between the activated bacterial endospore membrane and a nutritive surface. Although some of the endospore attachments could be clearly seen by conventional SEM with AuPd coatings (Fig. 1), an order of magnitude more information was obtainable by field emission SEM of these small endospores, due to the reduction in beam induced problems, and the capability to use a thinner, less structured coating that did not obscure the surface fine structure.

1. C. Philips,Bacteriol. Rev. 16(1952)135.
2. B.Panessa-Warren, G. Tortora and J.Warren,SCANNING 16(1994) 227.
3. B.Panessa-Warren, G. Tortora and J. Warren, Proc. Ann. MSA Meeting 52 (1994) 354.
4. The authors wish to gratefully acknowledge AMRAY,Inc. and the assistance of Mr. Bill Riley.This research was partially supported by the U.S.Department of Energy: Contract No. DE-AC02-76CH00016.

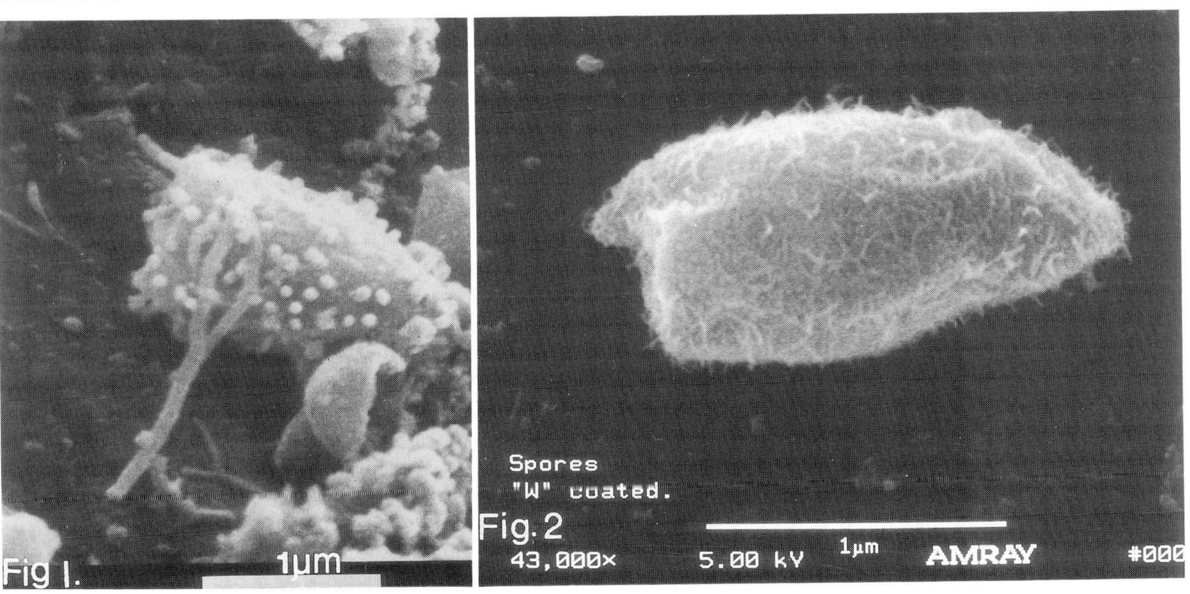

FIG. 1.-LaB$_6$ image of AuPd-coated endospore with attachments to the agar-coated stub.
FIG. 2.-Field emission image of W-coated similarly attached endospore.

MOLECULAR CHAPERONES HSP-70 AND BIP ARE PRESENT IN THE SECRETORY PATHWAY OF EXOCRINE GLANDS

T. Menini and M. Bendayan.

Department of Anatomy, Université de Montréal. Montréal, Québec. Canada.

In the past few years the study of a new group of proteins, the molecular chaperones, has demonstrated their relevant roles in cell biology. These proteins have been extremely well conserved during evolution, with percentages of homology reaching 50% between members of the same family (1).

The heat shock 70 kDa family of proteins belonging to the molecular chaperone group includes at least 4 members which are permanent residents of several cellular compartments. Besides being induced by stress, these proteins fulfil many important functions such as collaborating with the proper folding and assembly of newly synthesized proteins and facilitating their transport and translocation into organelles (2).

Different studies have demonstrated the involvement of these proteins in cell secretion in bacteria and yeast (3). Few data are however available for mammalian cells (4). In this regard, we have set out a study to demonstrate the presence and localization of two members of this family of proteins, the HSP-70 and the BiP, on several exocrine glands. By using a combined biochemical and immunoelectron microscopical approach, the salivary and lacrimal glands as well as the exocrine pancreatic tissue have been studied.

Results from immunoblot analysis have shown that both proteins, the HSP-70 and the BiP, are present in all the glands investigated. By applying the immunogold approach (5), the labeling for the HSP-70 was found mainly over the Golgi apparatus for all of the glands examined (Fig. 1,a). In the exocrine pancreatic cells, however, it displayed a particular concentration in the trans-Golgi-network (Fig. 1,b). BiP was also present in all the tissues studied and except for the exocrine pancreatic cells, the localization was restricted to the endoplasmic reticulum (Fig. 2,a). In pancreatic cells, the labeling for BiP was found not only over the endoplasmic reticulum but also over the Golgi apparatus, secretory granules and even in the acinar lumen (Fig. 2,b).

The results obtained strongly suggest relevant fonctions for these two proteins in the secretory process. HSP-70 could participate in preventing misaggregation of zymogen proteins through the Golgi apparatus or during the concentration process taking place in the trans-cisternae. Moreover, in the trans-Golgi-network it could also be involved in protecting and maintaining some specific signals on the protein, needed for targeting and sorting.

The role of BiP in the endoplasmic reticulum implies the participation in the proper folding of newly synthesized proteins. However, along with previous reports (6) our findings suggest that BiP is transported throughout the secretory pathway in exocrine pancreatic cells. In fact, the differences in location between pancreatic cells and the other exocrine glands are probably due to differences in the nature of the secretory proteins. Our data point out to a cooperative interaction between BiP and pancreatic zymogens.along the exocrine secretory pathway in pancreatic cells. BiP could protect these proteins against a premature activation until their secretion.

The immunocytochemical demonstration of chaperones along the secretory pathway of various acinar cells indicates the important roles played by chaperones in the proper processing of secretory proteins.

REFERENCES
1.   Hendrick, J.P. & Hartl, F.-U. *Ann Rev Biochem* **62**, 349-384 (1993).
2.   Becker, J. & Craig, E.A. *Eur J Biochem* **219**, 11-23 (1994).
3.   Parlati, F., Dominguez, M., *et al. J Biol Chem* **270**, 244-253 (1995).
4.   Vélez-Granell, C.S., Arias, A.E., *et al. J Cell Science* **107**, 539-549 (1994).
5.   Bendayan, M. *Prog. Histochem. Cytochem. "In press"* (1995).
6.   Takemoto, H., Yoshimori, T., *et al. Arch Biochem Biophys* **296**, 129-136 (1992).

Proc. Microscopy and Microanalysis 1995, edited by G.W. Bailey, M.H. Ellisman, R.A. Hennigar, and N.J. Zaluzec

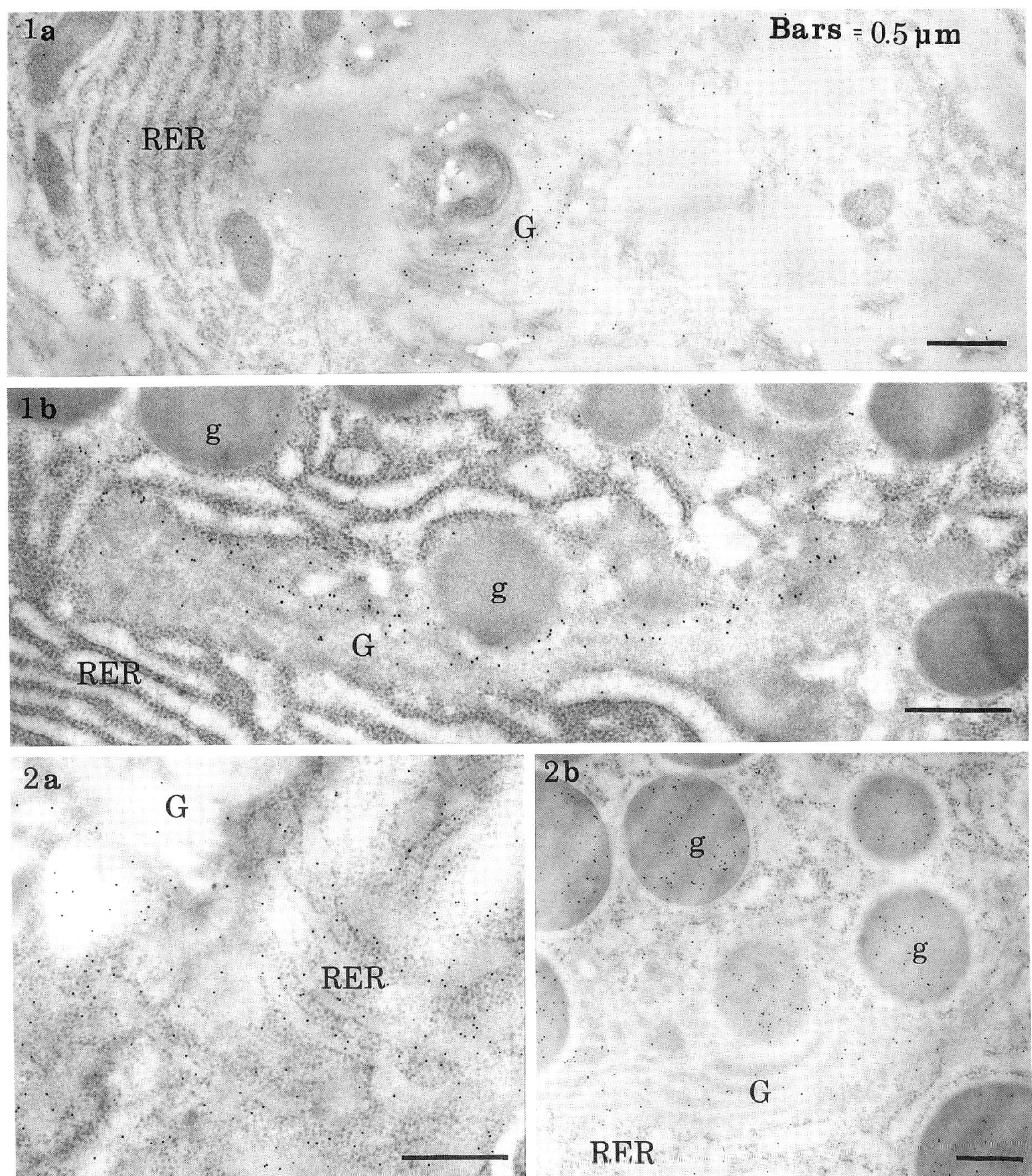

Fig. 1: Immunocytochemical localization of HSP-70 antigenic sites in exocrine glands. The labeling is present mainly over the Golgi apparatus (G) in submaxillary gland (1,a) and particularly over the TGN in pancreatic acinar cells (1,b). Secretory granules (g), endoplasmic reticulum (RER).

Fig. 2: Immunocytochemical localization of BiP antigenic sites in exocrine glands. The labeling is present over the endoplasmic reticulum (RER) in parotid gland (2,a) and extends over the entire secretory pathway in pancreatic acinar cells (2,b). Golgi apparatus (G), secretory granules (g).

COMPLIANCE MAPPING OF CELLS USING FORCE-DISTANCE (F/S) CURVE IMAGING WITH THE ATOMIC FORCE MICROSCOPE (AFM).

S. P. Marchese-Ragona and M. Wörtge

TopoMetrix, 5403 Betsy Ross Drive, Santa Clara, CA. 95054.
(408)-982-9700.

The Atomic Force Microscope is an extremely sensitive force measuring instrument which routinely measures the force exerted on the AFM tip by the sample, with a routine force resolution in the pico Newton range. By plotting the distance that the z-piezo is extended vs. cantilever deflection (force), one can produce what is traditionally known as a force distance (F/S) curve. Analysis of the F/S curve allows one to extract information on micro elasticity and plasticity of cell surfaces as well as the magnitude of any tip-sample binding that may occur. To increase the utility of the F/S curve for cell biological studies an to overcome the lateral distortion experienced by soft material in the scan direction we have developed a technique that images by performing a F/S curve for every pixel of the image. For a 200 x 200 pixel image, 40,000 F/S curves are generated with 25 data points in each F/S curve. The data obtained by this type of imaging can be conveniently analyzed in the following four ways:

1) TOPOGRAPHY. A topographic image is generated by determining the feed back position at each pixel of the image immediately before the F/S curve is produced (Fig. 1.). This method of topographic data acquisition has the advantage of reducing lateral deformation of the cells, even at comparatively high loading forces, compared to constant force imaging which continually exerts a force on the sample and causes lateral shearing of the sample even at minimum force settings.

2) COMPLIANCE MAPPING. A compliance map can be generated for a given z-position of the F/S curve. Since 25 data points per F/S curve were collected, 25 compliance maps, each with a different z-piezo extension can be viewed. Figure 2 shows the compliance map for z-piezo extension of 350 nm beyond the established feedback point. In the compliance map, the plastic substrate that the cells are grown on appears bright, whereas the fibroblasts which are more compliant appear dark. The compliance maps can also be viewed as a function of z-piezo extension by viewing each of the 25 compliance maps in quick succession.

3) HARDNESS PROFILING. Analysis of Hardness profiles for a given z-piezo extension. By drawing a line across a compliance image, a hardness profile can be generated as depicted in figure 3. The hardness profile is obtained from the line in figure 2. This type of analysis allows the relative compliance of any data point within the profile to be compared with any other data point for a given piezo extension.

4) F/S ANALYSIS. Extraction and comparison of any of the 40,000 F/S curves in the image. A comparison of the slope of the F/S curves taken from the substrate and fibroblast (figure 4) allows the relative compliance's to be determined, furthermore the lateral separation between the two curves at a given loading force allows the magnitude of the indentation of the cell to be measured at that location.

This imaging methodology enables topographic and F/S data to be acquired simultaneously. The advantage of this technique is two fold. 1) An exact pixel to pixel registration for the topographic data and the F/S curve is produced. This degree of accuracy simply cannot be achieved by scanning the entire image first, because piezo hysteresis and drift prevents the user from returning to the precise topographic location to perform the F/S curve. 2)Lateral deformation of samples is virtually eliminated, even at comparatively high loading forces (40 nN).

Proc. Microscopy and Microanalysis 1995, edited by G.W. Bailey, M.H. Ellisman, R.A. Hennigar, and N.J. Zaluzec
Copyright © 1995 MSA. Published by Jones and Begell Publishing, 79 Madison Ave., New York, NY 10016

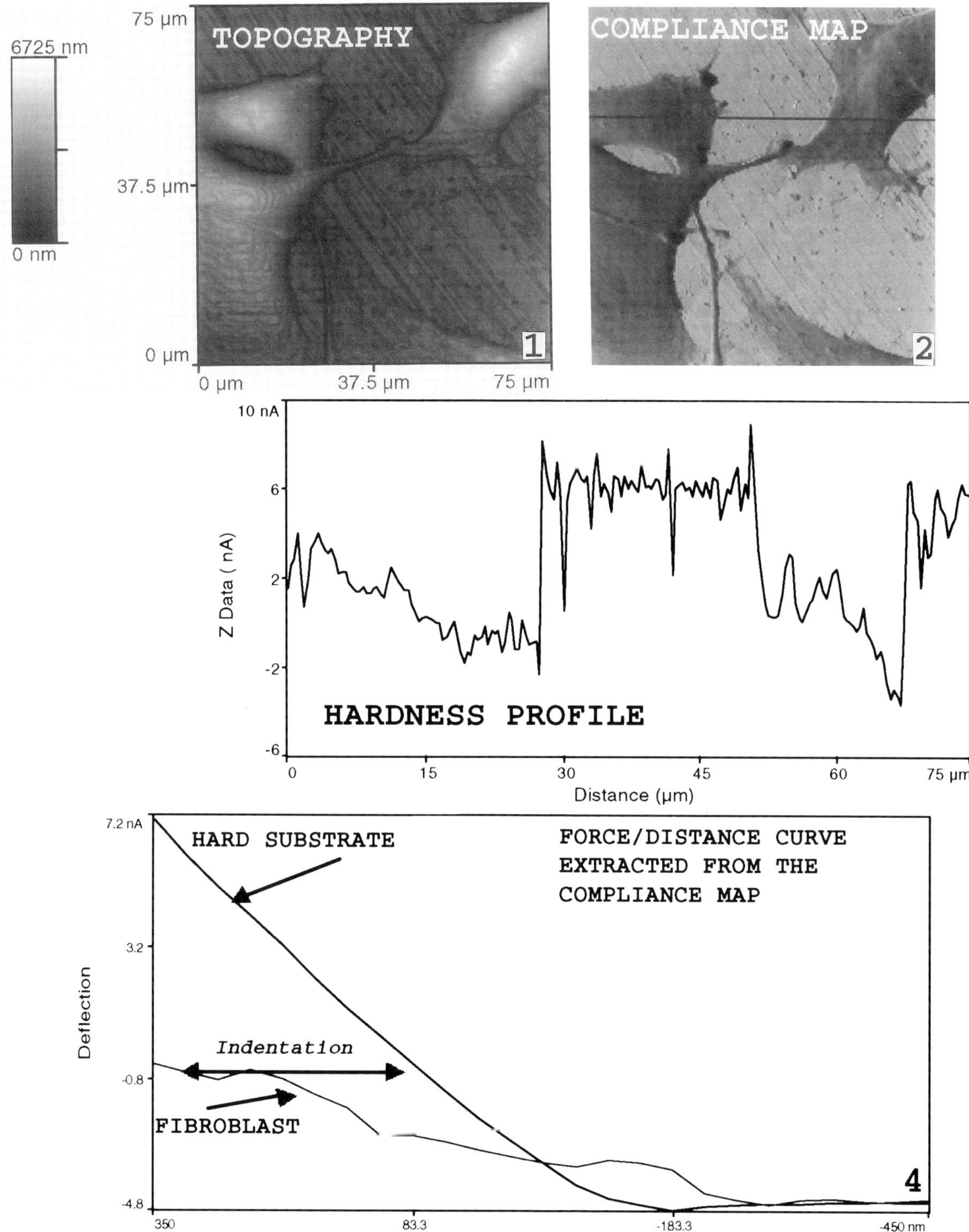

FIG. 1. Topographic image of fibroblast cells imaged in PBS.
FIG. 2. Compliance image of fibroblasts, dark areas are soft, bright areas are hard.
FIG. 3. Hardness profile derived from the line in fig.2.
FIG. 4. F/S curve's taken from fibroblast cells and the hard substrate. The x-axis represents nm of piezo movement from the feedback point.

RECOGNITION-ALIGNMENT AND ADHESION IN MYOBLAST FUSION

R. González Santander*, M.V. Toledo Lobo*, F.J. Martínez Alonso*, G. Martínez Cuadrado*, M. González-Santander Martínez**, M. Monteagudo*.

* Morphological Sciences and Surgery Dept. Univ. of Alcalá de Henares. Madrid, Spain.
** Toxicology and Sanitary Legislation Dept. Univ. of Alcalá de Henares. Madrid, Spain.

Myoblast fusion results from a sequence of different stages, previously demonstrated "in vitro"[1]. After withdrawal from the cell cycle, myoblasts align forming long chains, in a process termed "recognition-alignment". This stage is extracellular Ca^{2+} and N-Cadherin dependent.[2] Alignment is followed by adhesion, defined as the stage prior to membrane fusion when aggregates are resistant to dispersal by EDTA[1]. Adhesion is extracellular Ca^{2+}-independent and N-CAM-dependent.[3] Membrane fusion originates multinucleate myotubes.

We have studied these stages of myoblast fusion at the brachial myotome of chick embryo from 51 to 105 h. of incubation. Samples were obtained by embryo microdissection and included the neural tube, the notochord and the brachial somites. These samples were embedded in araldite by conventional methods. Some samples were embedded in Unicryl, a recently formulated GMA derived resin [4].

The first myoblasts clusters were observed in the ventral-lateral region of the brachial myotome in 22-24 Hamburger and Hamilton stages embryos. Clusters of pre-fusion myoblasts are usually surrounded by "electrondense blast cells" within a basal lamina in process of formation. It was concluded that the alignment stage in vivo is the process that leads single cells to form these clusters. Recognition-alignment is mediated by finger-like projections and adherens junctions (Fig.4), these being the first membrane junctions observed in relation with myoblast fusion. It is proposed that cell projections moving through the intercellular space make adherens junctions with adjacent cells. Then, active forces generated by the cytosqueleton and transmited through the actin microfilaments associated to adherens junctions produce a movement in the direction of these projections. These active forces could be responsible for the process of alignment or cluster formation. In some diamond-knife sectioned adherens junctions the 15-25 nm gap between cell membranes is periodically bridged by electron-dense bars, which may represent the cell adhesion molecules of this junction. These data are consistent with previous studies since N-cadherin (a Ca^{2+} dependent cell adhesion molecule), that is probably the same molecule as A-cadherin (adherens junction-specific cell adhesion molecule) participate in this Ca^{2+}-dependent alignment stage[2], probably through adherens junctions. Once formed, myoblast clusters start the adhesion stage. Initially the plasmalemmas of the adjoining cells of the cluster run parallel and separated by an intercellular space of uniform width (30-45 nm). Later, this space is ocluded by gap junctions. Gap junctions closing the intercellular space with a zipper-like mechanism, and junctions of different length from small gap junctions to long gaps can be observed (Figs.1,2,3). The participation of two types of gap junctions in the adhesion stage is demonstrated: typical gap junctions (Figs.1,2,3), as previously described in myoblast[5] and other tissues, and "gap junctions with electron-dense associated material". The latter possess an unusual fibrilar dense material on both cytoplasmic sides of junctional area. We conclude that adhesion is mainly mediated by these two types of gap junctions, which provide an ionic and metabolic coupling to the myoblast cluster (functional syncytium) before the membrane fusion (anatomical syncytium).

References

1. J.O. Wakelam, Biochem. J. 228 (1985) 1.
2. K.A. Knudsen et al., Exp. Cell Res. 188 (1990) 175.
3. K.A. Knudsen et al., Dev. Biol. 138 (1990) 159.
4. C. Scala et al., J. Histochem Cytochem. 40 (1992) 1799.
5. N. Kalderon et al., J. Cell Biol., 75 (1977) 788.

Proc. Microscopy and Microanalysis 1995, edited by G.W. Bailey, M.H. Ellisman, R.A. Hennigar, and N.J. Zaluzec

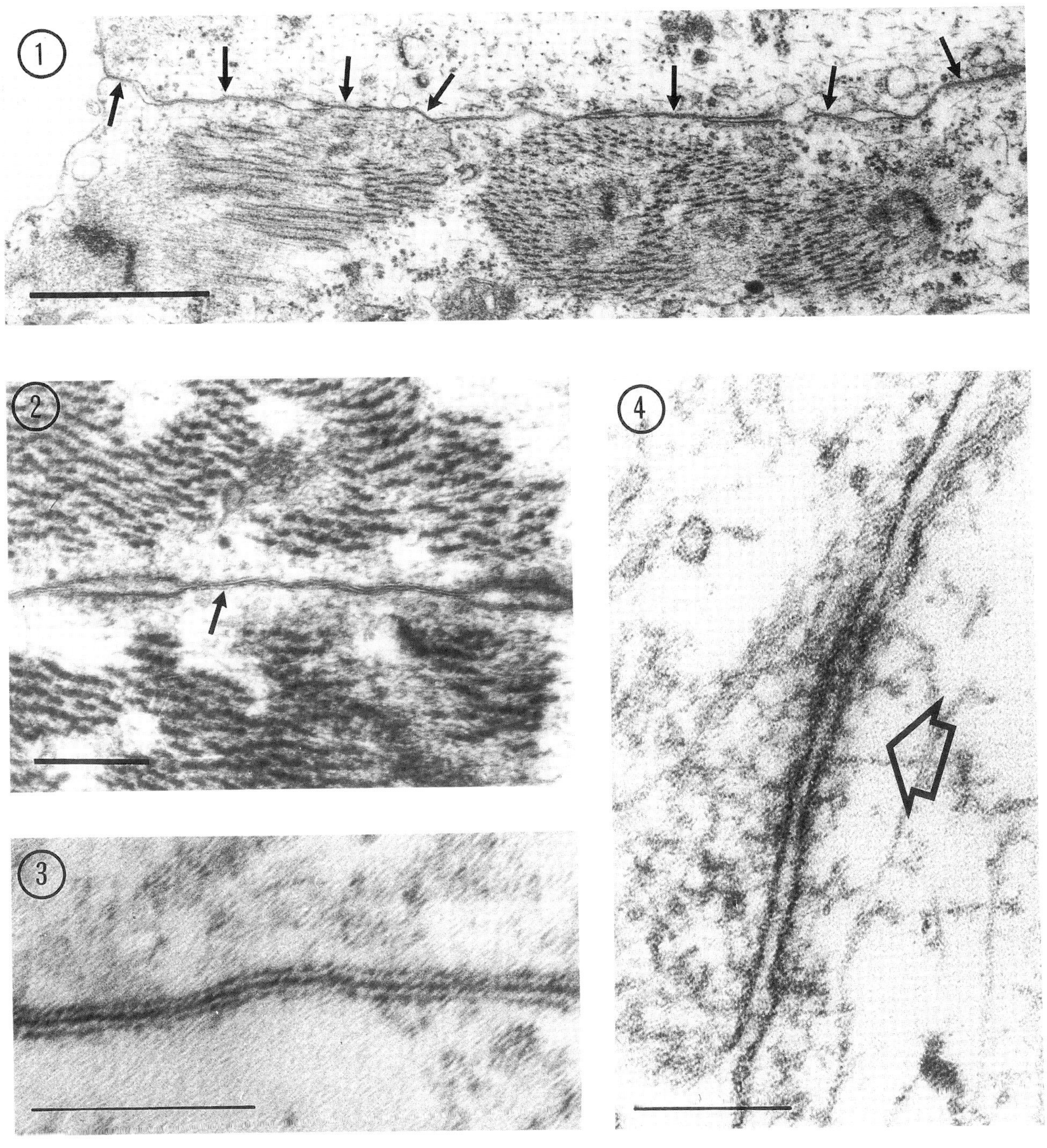

FIGURES.- Myoblasts in the myotome of 24 H.H. stage chicken embryos.
Figs. 1,2,3.- Gap junctions without electrondense associated material (arrows). Fixation: tannic acid-glutaraldehyde. 1: scale bar 1 μm (4,000 X); 2: scale bar 0.25 μm (25,000 X); 3: scale bar 0.125 μm (63,000 X)
Figs. 4.- Adherens junction (arrow). Fixation: glutaraldehyde. Scale bar: 0.125 μm (50,000 X)

MODULATED POLARIZATION MICROSCOPY DISPLAYS THE DYNAMICS OF CYTOSKELETAL STRUCTURES IN LIVING, MOTILE CELLS.

J. R. Kuhn, and M. Poenie

Department of Zoology, The University of Texas at Austin, Austin, TX 78712

Cell shape and movement are controlled by elements of the cytoskeleton including actin filaments and microtubules. Unfortunately, it is difficult to visualize the cytoskeleton in living cells and hence follow its dynamics. Immunofluorescence and ultrastructural studies of fixed cells while providing clear images of the cytoskeleton[1], give only a static picture of this dynamic structure. Microinjection of fluorescently labeled cytoskeletal proteins has proved useful as a way to follow some cytoskeletal events[2], but long term studies are generally limited by the bleaching of fluorophores and presence of unassembled monomers.

Polarization microscopy has the potential for visualizing the cytoskeleton. Although at present, it has mainly been used for visualizing the mitotic spindle. Polarization microscopy is attractive in that it provides a way to selectively image structures such as cytoskeletal filaments that are birefringent. By combining standard polarization microscopy with video enhancement techniques it has been possible to image single filaments. In this case, however, filament intensity depends on the orientation of the polarizer and analyzer with respect to the specimen. Thus, only filaments oriented within a narrow range of angles are clearly seen.[3]

We have extended the potential of polarization microscopy by exploiting the dependence of birefringence on the angular orientation of the illuminating beam. We sought to selectively image birefringent elements based on their change in intensity as the angle of polarization is varied. By altering the angle of polarization at known frequencies, signal processing techniques can be used to isolate signals in the image that change at the same rate. When this procedure was applied to a stream of video images, cytoskeletal components within the cell were clearly evident.

The modulated polarization microscope (MPM) was designed around a Nikon inverted Diaphot microscope. The polarizer and analyzer consisted of two Glan Thompson prisms. Two Faraday effect magneto-optic polarization rotators[4] were place between the polarizer and analyzer; one above and one below the specimen. Current is supplied to the Faraday rotators so that the rotation produced by the first Faraday Rotator is canceled by the second. Thus polarization angle presented to the specimen can be varied while maintaining crossed polars. The polarization angle is modulated at a known frequency (from 0.5 to 4 Hz). The resulting digital images are Fourier filtered and background subtracted on-line to extract signals whose frequency matches the modulation frequency.

The power of MPM can be demonstrated by comparing a conventional polarized light image to MPM images of rat fibroblasts (Fig. 1). Fig. 1a shows a simple polarized light image of a REF-52 fibroblast. Figure 1b shows the change in filament brightness as polarization angle is varied under computer control. Figure 1c shows the fully processed MPM image. The birefringent stress fibers of these cells are seen more clearly in the MPM image. MPM microscopy shows filaments organized at a variety of different angles in the cell, including both circular and linear arrays of actin (Fig. 2a). In addition to cytoskeletal filaments, small vesicles become greatly enhanced within the image due to their edge birefringence. MPM images of living R6-C1 fibroblasts, reveal membrane ruffling and filaments associated with filopodia (Fig. 3). The microtubule organizing center can be detected in P388i cells as a small region of intense perinuclear birefringence (Fig. 4).

Polarization modulation microscopy is a powerful tool that should be a great aid in understanding the dynamics of the cytoskeleton. Thus far, it has been useful in visualizing cytoskeletal changes at the leading edge of migrating cells as well as MTOC rearrangements and vesicular traffic.

REFERENCES

[1] G. Rinnerthaler, et al., *J. Struct. Biol.* **106**, 1-16 (1991).
[2] M. H. Symons, T. J. Mitchison, *J Cell Biol.* **114**, 503-513 (1991).
[3] S. Inoué, *Video Microscopy* (Plenum Press, New York, 1986).
[4] Azzam, *J. Opt. Soc. Am.* **68**, 518-521 (1978).

Proc. Microscopy and Microanalysis 1995, edited by G.W. Bailey, M.H. Ellisman, R.A. Hennigar, and N.J. Zaluzec
Copyright © 1995 MSA. Published by Jones and Begell Publishing, 79 Madison Ave., New York, NY 10016

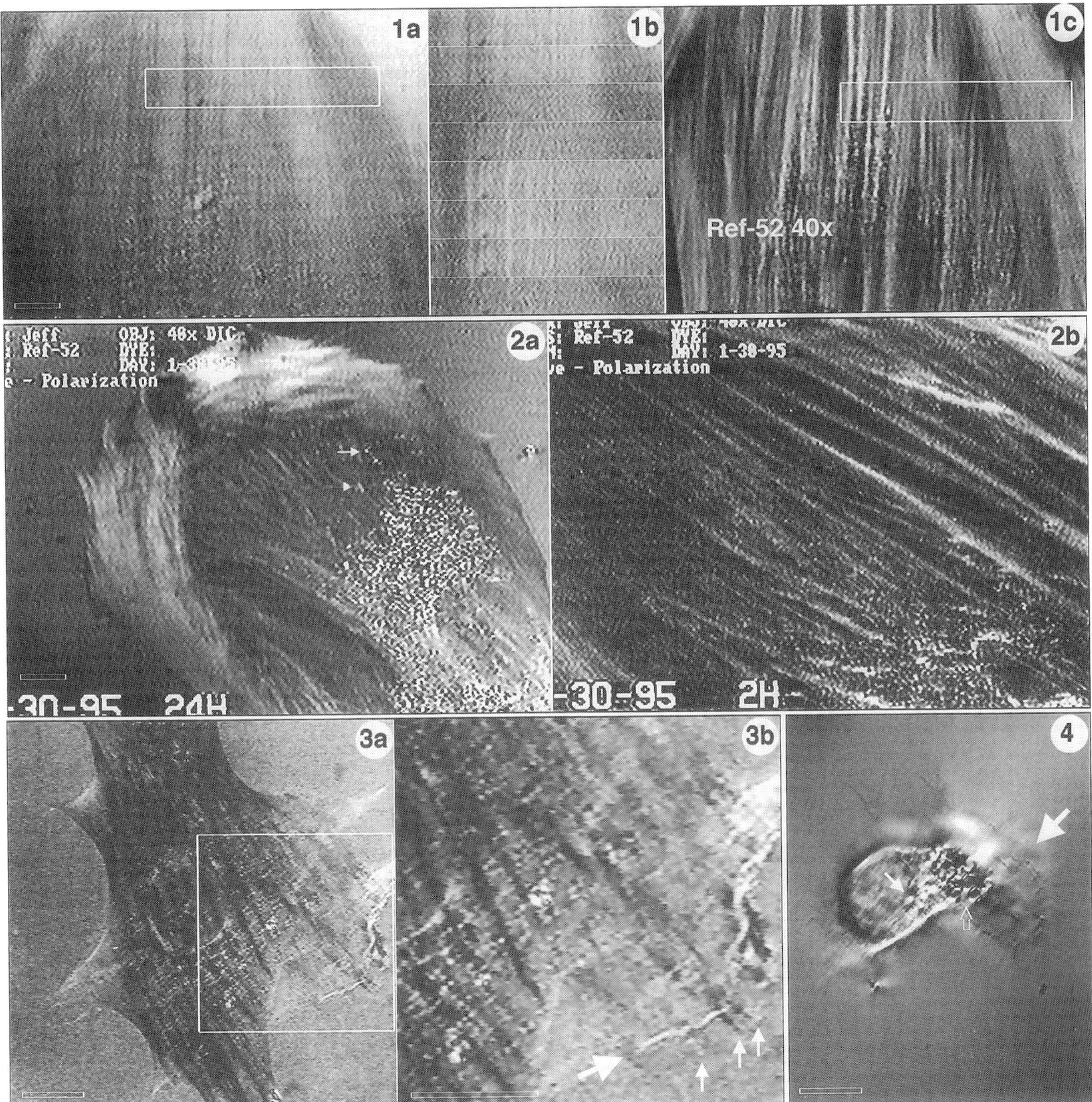

Fig. 1: Ref-52 fibroblasts cells. 40x, 0.65 NA DIC objective. (a) Unprocessed polarization image. (b) Change in filament brightness as the polarization angle is rotated. (c) Fully processed MPM image showing birefringent stress fibers. Scale bar 10 μm.

Fig. 2: Ref-52 fibroblast cell. 40x 0.65 NA DIC objective. (a) Circular and linear actin filaments in spreading fibroblast. Vesicular movement along filaments (arrow) was readily detected in this cell. (b) Perinuclear filaments of fully spread cell. Scale bar 10 μm

Fig. 3: MPM image of R6-C1 Fibroblast. 60x 1.4 NA DIC objective. Boxed area in (a) was digitally magnified 2x in (b). Membrane ruffling (large arrow) and filopodia (small arrow) were observed. Scale bars 10 μm.

Fig. 4: Mouse macrophage-like P388i cell showing leading edge (large arrow), microtubule organizing center (small arrow), and large granules (open arrow). 60x 1.4 NA DIC objective. Scale bar 10 μm.

ACCUMULATION OF SMOOTH TUBULAR MEMBRANE AGGREGATES IN THE VITELLOGENIC OOCYTES OF AN INSECT, *DYSDERCUS CINGULATUS*

László G. Kömüves[†,‡] and Miklós Sass[†]

† Department of General Zoology, Eötvös University, Budapest, Hungary
‡ Microscopy Laboratory, Department of Dermatology, University of California San Francisco, VA Medical Center, San Francisco CA 94121 USA

The accumulation of vitellogenins in the oocytes of insects is a receptor-mediated endocytic process, which leads to the formation of yolk granules.[1,2,3] We have been studying the hormonal regulation of vitellogenin accumulation and yolk granule formation in the red cotton stainer, *Dysdercus cingulatus* (Insecta, Hemiptera). During our investigation we observed the appearance of extensive aggregates of smooth tubular membranes in the vitellogenic oocytes. In this report we describe the ultrastructural features of these new cellular organelles.

In *D. cingulatus* (as in other insects) the onset of vitellogenesis (i.e. yolk accumulation) is marked by the appearance of coated pits between the microvilli (Fig. 1). The coated pits transform into electron dense coated vesicles upon invagination into the cortical ooplasm and their subsequent fusion leads to the formation of yolk granules. At a later stage the fusion and continuous growth of yolk granules gives rise to large yolk platelets. During the early phase of yolk accumulation the cortical ooplasm also contains non-coated vesicles of different size in a large number (Fig. 1). Furthermore, we observed accumulation of unusual, large aggregates of smooth tubular membranes at this stage of oocyte development (Figs. 1, 2, 3). Depending on the section, the smooth tubular membranes might display a wavy, circular, or zig-zag pattern (Figs. 2, 3). The aggregates of smooth tubular membranes do not have an isolating membranes. However, at the periphery of the aggregates the smooth tubular membranes frequently connected to non-coated vesicles and event to yolk granules (Figs. 2, 3). The aggregates of smooth tubular membranes were no longer present in fully-grown oocytes. Although the ultrastructural features of vitellogenesis and yolk formation are described in great details in a wide variety of insects,[2,3] this is the first observation of the smooth tubular membrane aggregates. The role of these membranes is not clear yet. Nevertheless, their connection to non-coated vesicles and yolk granules suggests that they might serve as precursors and/or membrane reserve for non-coated vesicle membranes during early vitellogenesis, when membrane traffic occurs at a high rate.

References

1. A.S. Raikhel and T.S. Dhadialla, *Ann. Rev. Entomol.* 37(1992) 217.
2. W.H. Telfer et al., in R.C. King and H. Akai, Eds., *Insect Ultrastructure*, New York: Plenum 1(1982)149.
3. E. Huebner, in R.C. King and H. Akai, Eds., *Insect Ultrastructure*, New York: Plenum 2(1984)3.

FIG. 1.— Cortical area of a vitellogenic oocyte of *D. cingulatus*. The extracellular space between the oocyte (O) and the surrounding follicular epithelium (F) is filled by microvilli. Coated pits (arrows) and electron-dense coated vesicles (open arrows) indicate the endocytic uptake of vitellogenins from the hemolymph. In the deeper region of the oocytes yolk granules and a large aggregate of smooth tubular membranes are present. Bar = 1 μm.
FIG. 2.— An aggregate of smooth tubular membranes (arranged in a wavy pattern in this section) in a vitellogenic oocyte of *D. cingulatus*. At the periphery of the aggregate some tubules are connected to non-coated vesicles (open arrows). Bar = 1 μm.
FIG. 3.— Aggregates of smooth tubular membranes in a vitellogenic oocyte of *D. cingulatus*. Some of the tubules are connected to yolk granules and lamellar bodies (open arrows). Different cross-sectional areas display a zig-zag (arrows) or circular (white arrows) pattern. Bar = 1 μm.

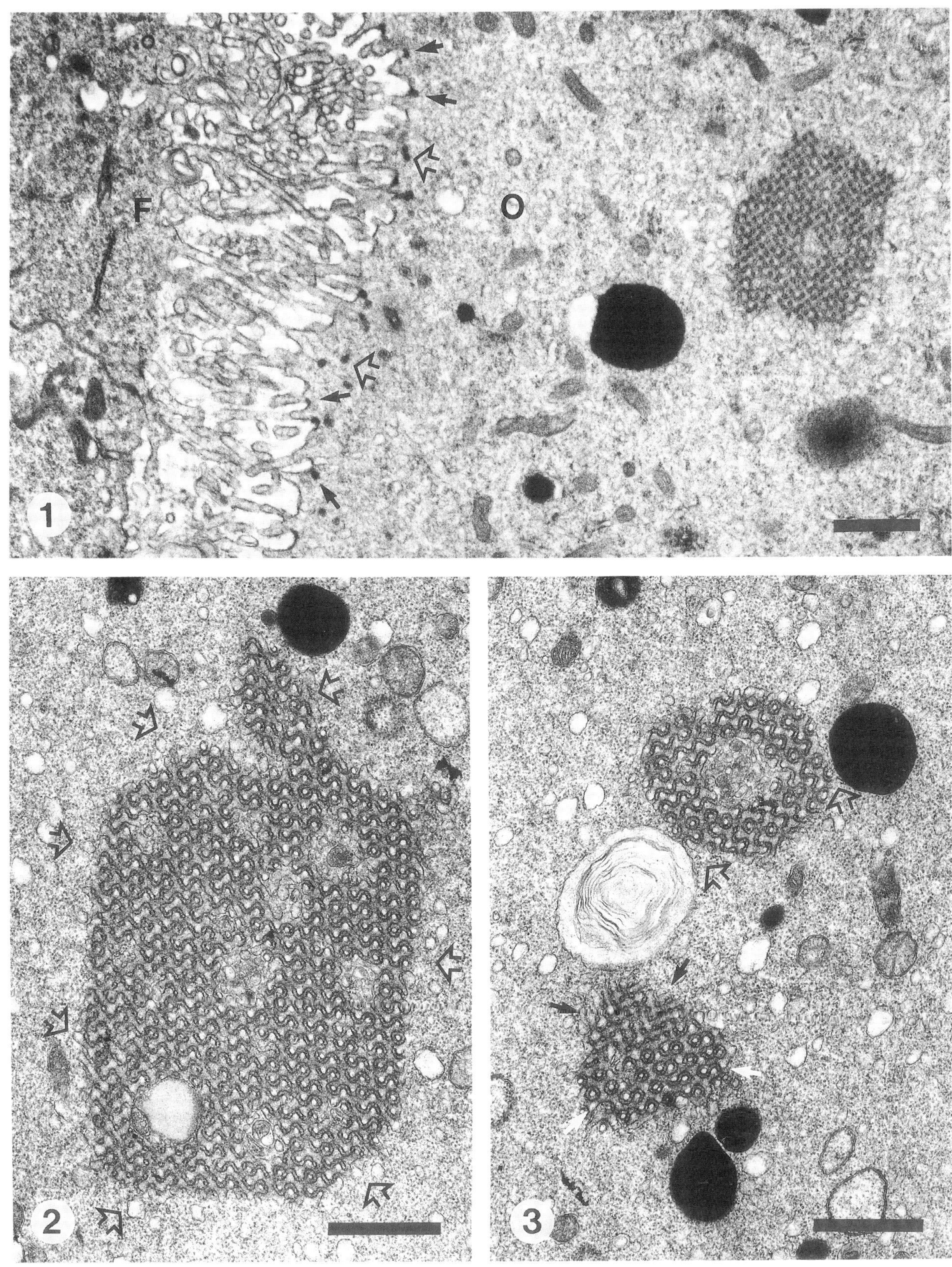

ULTRASTRUCTURE OF THE BASAL CELLS OF THE UTERINE GLANDS OF MARES
INFUSED WITH POVIDONE IODINE SOLUTION

F. Al-Bagdadi,* G. Richardson,** B. Eilts,*** D. McCoy,*** W. Braun,**** L. Archbald,*****
L. Olsen,****** D. Thompson,******* and C. Titkemeyer*

*Department of Veterinary Anatomy and Cell Biology, ***Department of Veterinary Clinical
Sciences, *******Department of Animal Science, Louisiana State University, Baton Rouge, LA
70803, **Department of Health Management, Atlantic Veterinary College, University of Prince
Edward Island, Charlottetown, P.E.I., Canada, C1A4P3 ****Department of Veterinary Clinical
Science, College of Veterinary Medicine, Columbus, MO 65211, *****Veterinary Teaching
Hospital, College of Veterinary Medicine, Gainsville, FL 32610 ******Foxland Farm, P.O. Box
117, Keen, KY 40339

Gentamicin and povidone iodine infusion solutions have been used separately, to treat endometritis
in mares.[1,2] The effect of povidone iodine on the histology of the endometrium has been reported.[2]
This study describes the morphology of the fine structure of the endometrial epithelium and its
response to povidone iodine infusion with the emphasis on the basal cell of the uterine gland.

Ten light horse mares of breeding age were selected and divided into two groups. Five mares were
used as controls and five mares were infused with 1% povidone iodine solution. Estrous of the
mares was controlled with two doses of prostaglandin, 11 days apart, so that all the mares were
in the first 3 days of estrus at the beginning of the experimental period (day 0). Uterine biopsies
were obtained on days 0, 3, 5, 7, 10, 15, 20, and 30 of the experimental period. The biopsies
were obtained from the endometrium at the junction of the uterine body and the uterine horn using
alternate sides for subsequent biopsies. A small piece of each biopsy was minced into 1-2 mm
sizes in a lake of 1.25% glutaraldehyde and 2% formaldehyde in 0.1M phosphate buffer pH 7.4.
Then the tissue was fixed in the same fixative and post fixed in 1% osmium tetroxide, dehydrated
in alcohol and embedded in Epon. Silver sections were stained with lead citrate and urynal acetate
and examined with a Zeiss-10 electron microscope.

The control mares had basal cells between the cells of the uterine glands (Fig. 1). These cells
contained several darkly stained mitochondria, and were underlined by the basement membrane of
the uterine gland. The nuclei were irregular, indented and had mainly heterochromatin. Mast cells
were present outside the border of the uterine glands (Fig. 2) They were separated from the basal
cells by the uterine gland basement membrane. The mast cells contained granules at different stages
of maturation. The basal cells of the uterine glands in the mares infused with povidone were
located beneath the basement membrane of the uterine gland (Fig. 3). The cytoplasm had large
vacuoles, several mitochondria and almost no cytoplasmic organelles. The nucleus was pear shaped,
and contained mainly euchromatin.

The mast cells seemed to be migrating towards the uterine gland [3], and were morphologically
different than the basal cell. The fine structural morphology of the basal cells indicate that they
do not resemble mast cells as has been suggested. [4] This preliminary study indicates that mast
cells, leukocytes and basal cells are three different types of cells. The basal cell could be a resting

or degenerating cell or a cell having the potential to develop into a glandular cell. The repeated infusion of povidone iodine caused degenerative damages to the basal cells of the uterine glands of the mares examined in this study.

References

1. B.E. Eilts et al., Therio. 29(1988)1253.
2. L.M. Olsen et al., Therio. 37(1992)1311.
3. F.K. Al-Bagdadi et al., Micros. Soc. Am. 44(1986)338.
4. J.S. Perry et al., J. Anat. 134(1982)339.

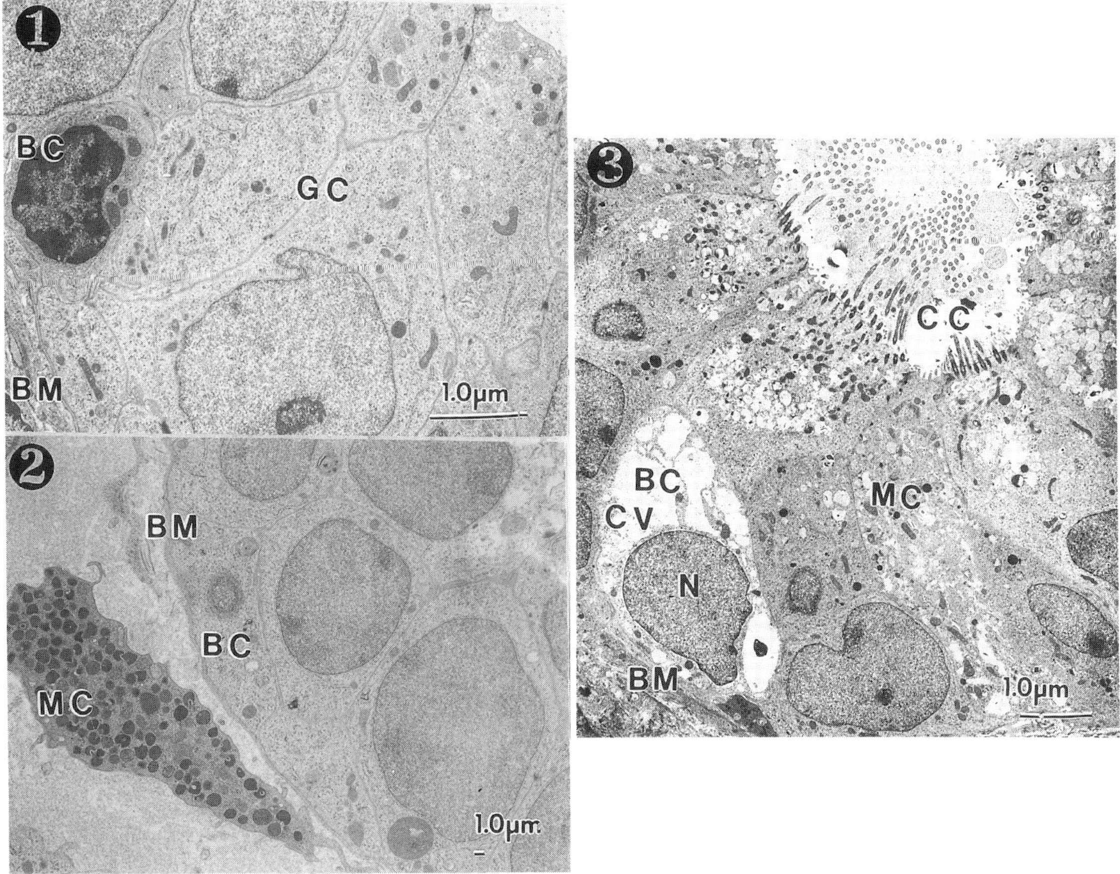

FIG. 1, 2 and 3 are electron micrographs of the uterine glands of the uterus of mares. FIG.1-shows part of a uterine gland of a control mare, is surrounded by a basement membrane (BM). The basal cell (BC) is between the uterine gland cells (GC). FIG. 2 shows part of a uterine gland of a control mare, is surrounded by a basement membrane (BM). The basal cell (BC) is underlincd by the basement membrane which separates it from a periglandular mast cell (MC). FIG. 3-shows part of a uterine gland of a mare infused with povidone iodine. The section was cut near the orifice of the uterine gland. The basement membrane is irregular (BM) and most of the cells are ciliated (CC) and some cells have lost their microvilli (MC). The basal cell (BC) is degenerating with abundant cytoplasmic vacuoles (CV). The nucleus (N) is pear shaped and contains mainly euchromatin.

MYOFIBRILLOGENESIS AND MYOFIBER PATTERN IN NORMAL AND ANGIOTENSIN INDUCED CARDIAC INVERSION IN DEVELOPING RAT HEART

RL Price, L Fu, DG Simpson, TK Borg and L Terracio

Dept of Dev Bio & Anat, USC School of Med, Columbia, SC 29209

Myofibrillogenesis, myofiber pattern formation, and heart looping respond to physical, hormonal and teratogenic agents. In cell culture, myofibrillogenesis and myofiber pattern formation depend on coordinated expression of cytoskeletal and myofibrillar components and signals from extracellular matrix (ECM). However, these experiments have been conducted with isolated cardiac myocytes in which phenotype of myocytes is stellate rather than tubular as found *in situ*.

Whole embryo culture (WEC) allows *in situ* interactions to occur during controlled application of perturbatory agents. Application of 10 μg/ml angiotensin to WEC results in 10% cardiac inversion (controls - 0% inversions). In this condition, hearts loop right, rather than left as in normal development. We are using confocal scanning laser microscopy (CSLM), transmission electron microscopy (TEM), and image reconstruction techniques to examine myofibrillogenesis and myofiber pattern formation in embryos allowed to develop *in utero* and WEC.

Rat hearts 9.5 to 11.5 days *in utero* development, and hearts 9.5 days *in utero* and 48 H WEC were examined by CSLM and TEM. Culture techniques have been published previously.[1] Embryos (CSLM) were fixed in 4% paraformaldehyde, embedded in 13% acrylamide, and vibratome sectioned (200 μm). Sections were stained with rhodamine phalloidin and optically sectioned (0.5 to 2 μm) with a BioRad MRC 1000 CSLM. Z-series were reconstructed with BioRad COMOS software and Vital Image's Voxel View Silicon Graphics. Embryos (TEM) were fixed in 2.5% glutaraldehyde, 1% OSO_4, ETOH dehydrated, and embedded in Poly/Bed 812. Sections (0.25 μm) were viewed at 200kV on a JEOL 200CX TEM.

TEM of 10 day *in utero* heart shows a gradation of myofibrillar development (1A-B). Actin is present in outflow tract, but organization of myofibrils and Z-bands is poor. Myofibrils in the ventricle of the same heart are formed with modest Z-bands and intercalated disks. CSLM indicates myofibrillogenesis is not affected by cardiac inversion. Reconstruction of 11.5 day myocytes (2) shows myofibrils end on membranes in stress fiber like structures as in cultured myocytes. In cells with few myofibrils, they follow contours of cell membranes. As myofibrils mature, cells change from round to spindle shaped. Ultrastructural analysis of myofibrillogenesis in cultured hearts is ongoing, but cardiac inversion can result in a functional heart indicating normal myofibrillar development. Stereos (3A-B) of the same heart shown in Fig. 2 show myofibrils of ventricular walls are arrayed in a circumferential pattern around the heart lumen and peripheral networks of myofibrils radiate toward the pericardium. Myofibril organization is greatest where heart looping is ongoing (3B). Morphometric studies are ongoing to determine if angiotensin treatment alters seqential development of myofibrils in left versus right sides of the heart which may affect looping pattern. Adrenergic agents, which increase hemodynamic load during development, have been implicated in cardiac inversion defects.[2] Angiotensin treatment, which also increases hemodynamic load, provides an additional model for studying serious congenital heart defects associated with hormonal imbalance during development.

1. New, DAT. 1978. J Reprod Fert. 35:135-138.
2. Flynn, TJ et al. 1993. Teratol 48:161-168.

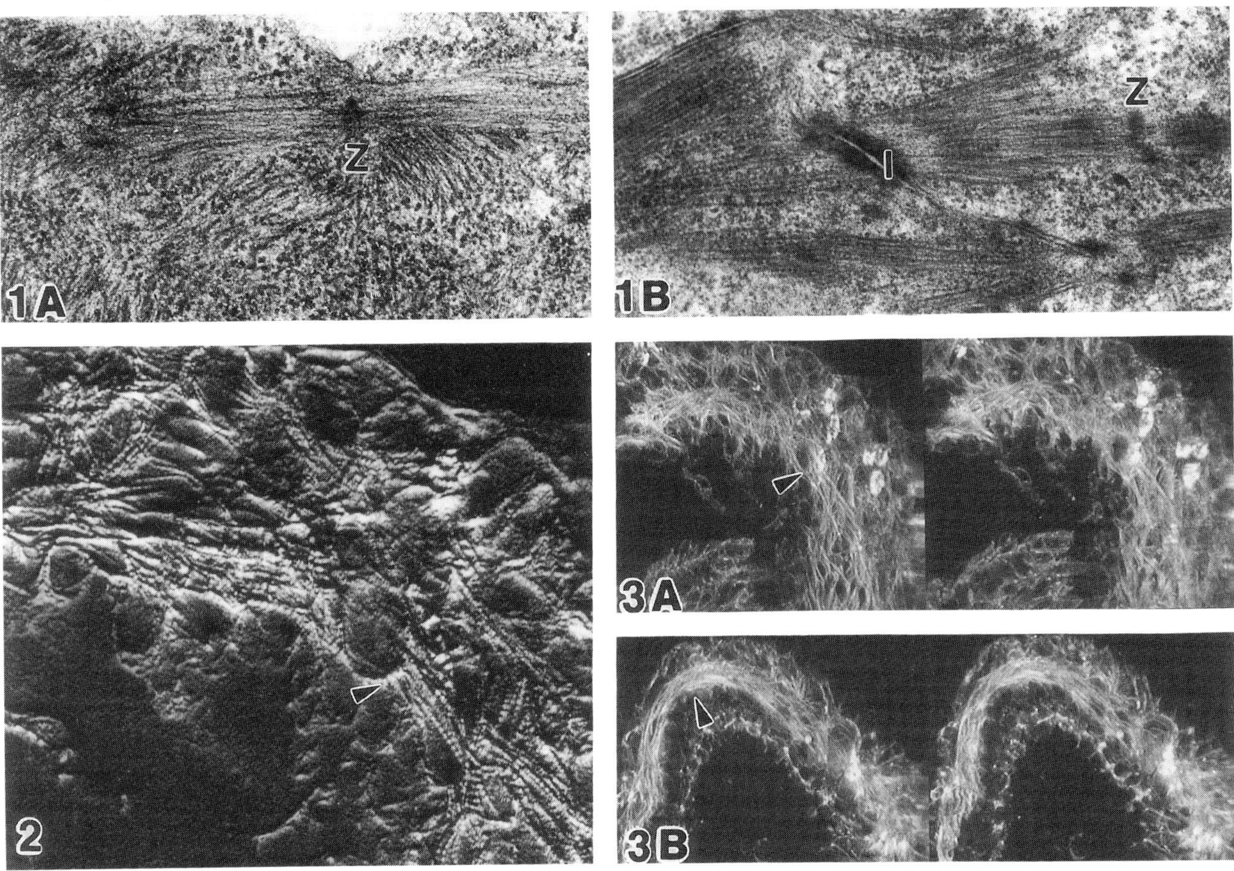

FIG 1: TEM of 10 day outflow tract (A) and ventricle (B). Actin fibrils in outflow tract are poorly organized with immature Z-bands (Z). In ventricle (B) of the same heart, myofibrils and Z bands are advanced and developing intercalated disks (I) are found.

FIG 2. Reconstruction of area shown in 3A (arrows show same cell). Myofibrils end in SFLS (arrow) similar to cultured myocytes. Early myofibrils follow contours of the sarcolemma and phenotype of myocytes changes from round to spindle as myofibrils develop.

FIG 3: CSLM stereo images of ventricle (A) and loop near atrial ventricular junction (B). Myofibrils in ventricle arrange circumferentially around heart lumen and in outer regions radiate in a network toward pericardium. In areas of looping (B), myofibrils are arranged circumferentially in the middle of the heart wall (arrow).

CALCITONIN EFFECTS ON MYOBLASTS. AN ULTRASTRUCTURAL STUDY.

R. González Santander*, M.V. Toledo Lobo*, F.J. Martínez Alonso*, G. Martínez Cuadrado*, M. González-Santander Martínez**, M. Monteagudo*.

* Morphological Sciences and Surgery Dept. Univ. of Alcalá de Henares. Madrid, Spain.
** Toxicology and Sanitary Legislation Dept. Univ. of Alcalá de Henares. Madrid, Spain.

Several studies have provided detailed descriptions about the effects of calcitonin (CT) on different cell types such as osteoclasts, multinucleate giant cells and trophoblastic cells[1]. CT main role is associated with the maintenance of peripheral Ca^{2+} homeostasis and it may also modulate free intracellular Ca^{2+}. Myoblasts are particularly sensitive to changes in extra and intracellular Ca^{2+} concentration, so we have studied CT effects on myotome myoblasts differentiation.

Chick embryos from 56-59 h. of incubation (Hamburger and Hamilton stages 16-17) were injected with 0,25 IU of CT dissolved in 125 ul of 0,9 % sterile saline solution (CALSYNAR-50, from Rorer Labs.- synthetic lyophilized salmon CT made by Armour Pharmaceutical Co., Kankakee, Illinois, USA). CT was injected into the ammniotic cavity through a window in the egg shell. Afterwards, the window was closed and the eggs returned into the incubator, until embryos had incubated for 103 h., to reach H.H. stage 24. Some embryos were only injected with saline solution (controls). Embryo samples were embedded in Araldite by conventional methods. This experimental model does not interfere with endogenous embryonic CT since ultimobranchial gland secretory activity (CT secretion) starts at around the 11th day of incubation[1].

The embryos showed no macroscopic malformations. However, when observed by TEM, myoblasts from CT treated embryos showed significative differences compared with myoblasts from control embryos. CT treated myoblasts showed a Golgi complex (Figs.1,2) with multiple dictyosomes and vesicles being more developed than Golgi complex from control cells. CT treated myoblasts had a larger number of electrondense secretory granules than controls. These granules have previously been related with natriuretic peptides[2], since myotome shows ANP immunoreactivity[2], and myoblast granules being ultrastructurally similar to the natriuretic peptide-containing granules found in other tissues. Vesicular evaginations of myoblast plasma membranes were only found in CT treated samples (Figs.3,4). These evaginations had a homogeneous slightly electrondense content, and some of them were observed detached from the cell surface and liberated into the extracellular space. Recently, using light and confocal microscopy, it has been proposed that an endogenous 14 kD lectin (L14) is secreted by myoblasts at the time of fusion by means of an unusual mechanism. This mechanism involves a concentration of the lectin in vesicular evaginations of plasma membrane, which pinch-off releasing L14 into the extracellular space[3-5]. The observed evaginations in CT treated pre-fusion myoblasts may represent the secretion of L14, which had not been previously observed by TEM methods.

Myoblasts from CT treated embryos showed a higher microtubule concentration than controls (an increase in the number of microtubules, although these microtubules are normal in size and morphology). Previous LM studies have shown that CT induces a rearrangement of osteoclast microtubules[6], but we have not observed a similar effect on myoblasts using TEM.

The mechanism by which CT induces the effects described remains unknown, but CT may act directly on myoblasts since myoblasts have receptors to molecules structurally homologous to CT, such as CGRP or amylin[7].

References

1. J.P. Barlet, in J.A. Parsons Ed., Endocrinology of Calcium Metabolism. Raven Press, NY (1982).
2. R. González Santander et al., Acta Anat., 150 (1994) 87.

Proc. Microscopy and Microanalysis 1995, edited by G.W. Bailey, M.H. Ellisman, R.A. Hennigar, and N.J. Zaluzec
Copyright © 1995 MSA. Published by Jones and Begell Publishing, 79 Madison Ave., New York, NY 10016

3. D.N. Cooper and S.H. Barondes, J. Cell Biol., 110 (1990) 1681.
4. D.N. Cooper et al., J. Cell Biol., 115 (1991) 1437.
5. F.L. Harrison and T.J. Wilson, J. Cell Sci., 101 (1992) 635.
6. M.H. Zheng et al., Histochem J., 23 (1991) 180.
7. G. Zhu et al., Biochem. Biophys. Res. Commun., 177 (1991) 771.

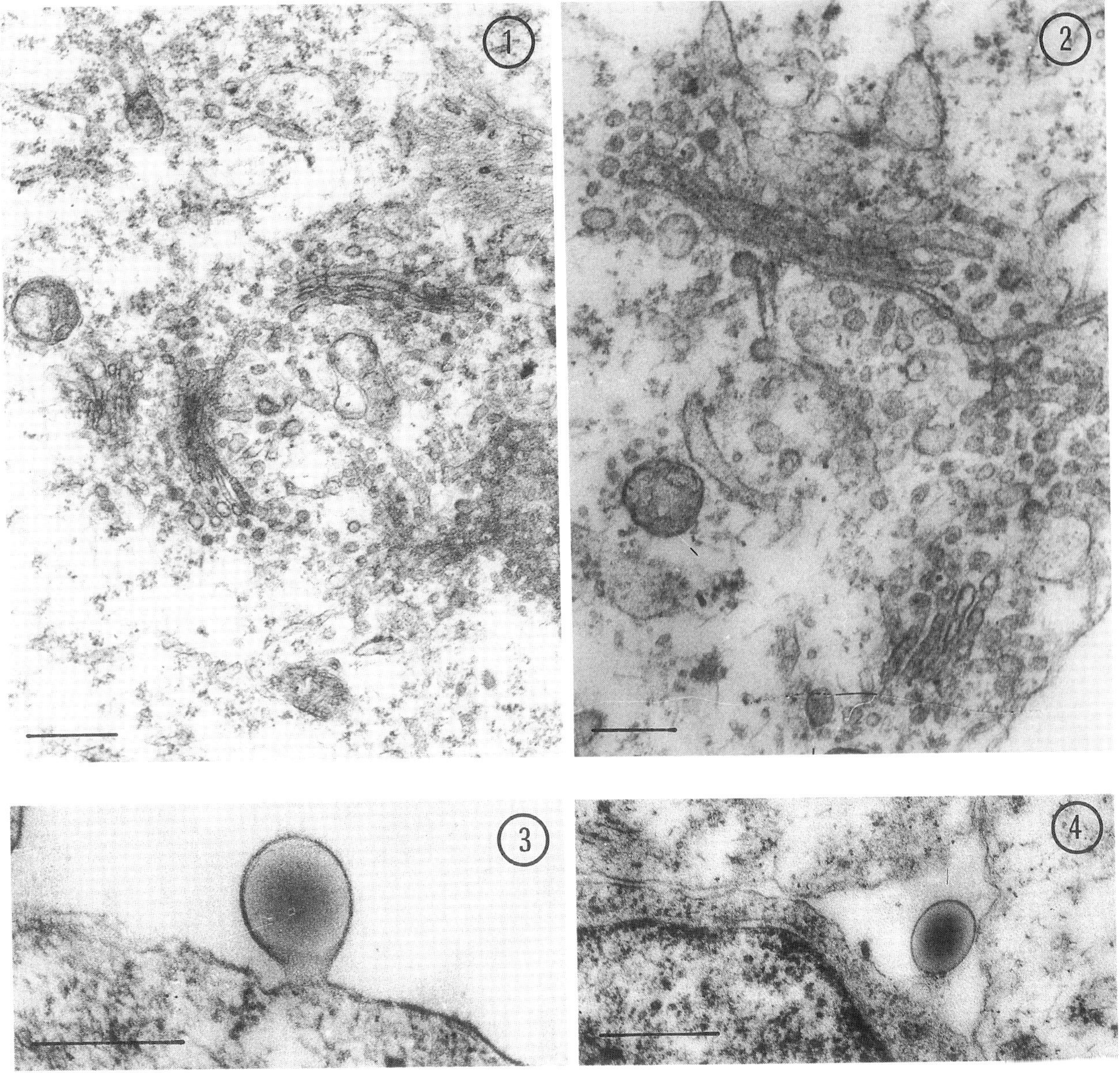

FIGURES.- Myoblasts in the myotome of 24 H.H. stage chicken embryos.
Figs. 1,2.- Highly developed Golgi complex. CT treated embryos. Fixation: glutaraldehyde.
1: scale bar 0.5 μm (8.000 X); 2: scale bar 0.25 μm (28,000 X).
Figs. 3,4.- Vesicular evaginations of plasma membrane (arrows). 3: scale bar 0.25 μm (31,500 X);
4: scale bar 0.5 μm (10,000 X).

CYTOCHEMICAL STUDY OF GLYCOGEN IN MYOBLASTS OF THE CHICK EMBRYO MYOTOME

R. González Santander*, M.V. Toledo Lobo*, F.J. Martínez Alonso*, G. Martínez Cuadrado*, M. González-Santander Martínez**, M. Monteagudo*.

* Morphological Sciences and Surgery Dept. Univ. of Alcalá de Henares. Madrid, Spain.
** Toxicology and Sanitary Legislation Dept. Univ. of Alcalá de Henares. Madrid, Spain.

Potassium ferrocyanide-reduced osmium has been used as secondary fixative to study the presence of glycogen in different tissues[1-3]. However, this method has not been previously applied to skeletal muscle development research. This study reports the use of this method to analyze the structure and cytoplasmic distribution of glycogen particles in the myoblasts of the brachial myotome of chicken embryos at 51-105 h. of incubation (Hamburger and Hamilton stages 16 to 25)[4].

In those sections post-fixed with potassium-ferrocyanide reduced osmium it is observed that glycogen particles stain in contrast to the rest of the structures of the cell (Figs.1,2). In those sections processed for T.E.M., following conventional methods, glycogen particles are occasionally observed in less contrast and number that those with potassium-ferrocyanide reduced osmium. An explanation for this differences might be in the effects of potassium-ferrocyanide reduced osmium fixation preventing the loss of glycogen during dehydration and staining with uranyl acetate. We conclude that this cytochemical method may be specific for glycogen detection.

Glycogen particles are electrondense, slightly irregular in shape and with a diameter between 15-45 nm. These particles are observed either randomly dispersed and distributed along the cytoplasm or in areas where they accumulate. Glycogen is also present within cytoplasmic vacuoles surrounded with a membrane, sometimes as part of multivesicular bodies (Fig.2). Accumulation of glycogen particles is only observed in "pre-fusion myoblasts" (myoblast clusters that will fuse originating primary myotubes). Pre-fusion myoblasts, containing accumulation of glycogen (Fig.2), are preferentially found in the ventral-lateral region within the myotome (H.H. stages 23-24). Glycogen particles are however, in smaller numbers and randomly dispersed through the cytoplasm in those less-differenciated myoblasts, preferentially in ventral-medial and dorsal regions within the myotome (Fig.1). The presence of pre-fusion myoblasts showing accumulation of glycogen starts at the ventral-lateral region of the somite, progressing in a ventral-dorsal and lateral-medial wave. These results lead us to suggest the existence of a ventral-dorsal and lateral-medial gradient of differentiation within the myotome besides the well documented craneal-caudal gradient.

The significance of that increase in glycogen content within pre-fusion myoblasts is unknown. However, several studies show an increase in the activity of those enzymes involved in glycogen metabolism (such us glycogen-phosphorylase and beta-enolase) as well as others (myokinase, creatin-phosphokinase).[5-7] These data, along with the presence of glycogen particles in pre-fusion myoblasts, suggest an increase in glycogen metabolism and probably an increase in the myoblast metabolism as a whole during these stages of early muscle development and the process of myoblast fusion.

References
1. N. Shepard and N. Mitchell, J. Electron. Microsc. Tech. 11 (1989) 83.
2. M. Z. Le, Chung-hua Ping Li Hsueh Tsa Chih. 19 (1990) 291.
3. M. P. Goheen et al., Biotech. Histochem. 67 (1992) 140.
4. V. Hamburger and H. Hamilton, J.Morphol. 88(1951)49 reprinted in Dev.Dynam. 195(1992)231.
5. M. Oguni et al., Acta Histochem. Cytochem. 24 (1991) 47.
6. M. Oguni, Acta Histochem. Cytochem. 24 (1991) 389.
7. A. Shainberg et al., Dev. Biol. 25 (1971) 1.

Proc. Microscopy and Microanalysis 1995, edited by G.W. Bailey, M.H. Ellisman, R.A. Hennigar, and N.J. Zaluzec
Copyright © 1995 MSA. Published by Jones and Begell Publishing, 79 Madison Ave., New York, NY 10016

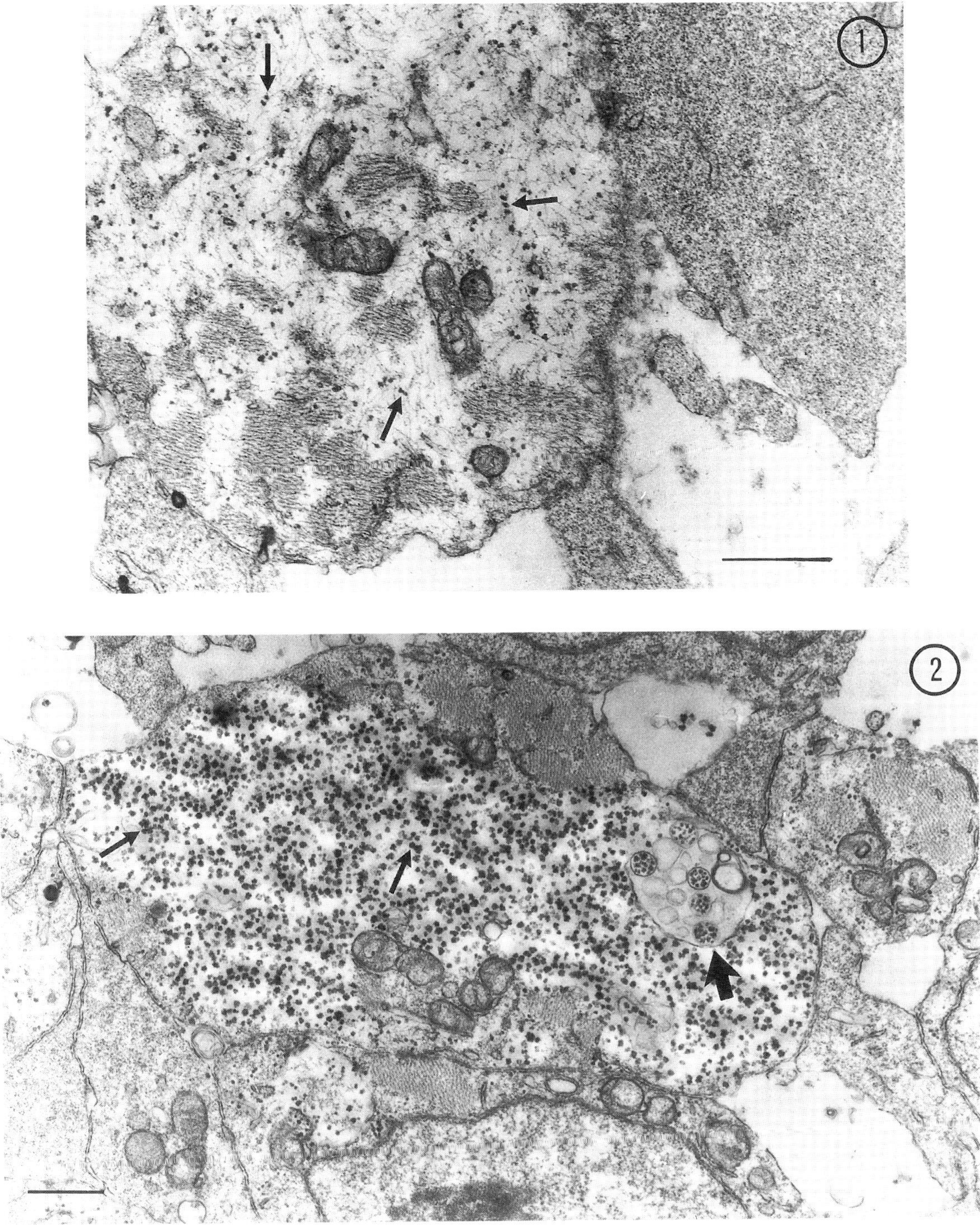

FIGURES: Myoblasts in the myotome of 24 H.H. stage chicken embryos.
Fig. 1.- Glycogen granules randomly dispersed throughout myoblast cytoplasm (arrows). Scale bar: 1 μm (6,300 X).
Fig. 2.- Pre-fusion myoblast. Glycogen accumulation (small arrow) and glycogen-containing vacuoles (big arrow). Scale bar: 1 μm (5,000 X)

ULTRASTRUCTURAL ANALYSIS OF SATELLITE CELLS IN SKELETAL MUSCLE OF OBESE AND LEAN LA/N-fa^{cp} RATS.

D. W. Domaschko, R. E. Edelmann, E. B. Kahle

The biochemical phenomena associated with obesity have been studied extensively. Hollands and Cawthorn[2] have reported that the back postural muscle in the ob/ob obese mouse model are responsible for nearly as much fatty acid production as the liver. The hindlimb musculature has been shown to produce higher levels of fatty acids in venous return blood when compared to the arterial blood[1]. These studies have shown the ability of obese muscle to serve as major sites of fatty acid synthesis in the obese animals. This study has sought to identify any ultrastructural variations in red skeletal muscle from the hindlimb and back postural muscles.
These studies have revealed aberrations in the satellite cells between the obese and lean phenotypes as well as the presence of lipid droplets in the capillaries of the obese animals' psoas major muscle.

Samples were taken from the bellies of the psoas major representing the back postural muscles and the gastrocnemeius and soleus muscles representing the hindlimb muscle. The satellite cells from obese skeletal muscle showed morphological evidence of higher metabolic activity than those from lean skeletal muscle. Satellite cells from obese skeletal muscle exhibited a retention of cytoplasmic volume and organelle development resemblinng the embryonic state more so than the mature muscle state (fig 1). Lean muscle satellite cells resembled the more classic apperance of reduced cytoplasm and poorly developed or absent organelles[3,4] (fig 2). The obese satellite cells exhibiting the more embryonic ultrastructure may be caused by the activity of the obesity gene. Furthermore, the capillaries of the psoas major muscle contained lipid droplets in the obese animals (fig 3) and not the hindlimb muscles (fig 4). The lean animals showed no lipid droplets in any of the muscles surveyed in this study. Based on these obsevations it is possible that these obese satellite cells show that fatty acid synthesis in muscle tissue results in the export of those fatty acids and not its storage in the muscle tissues. The absence of any appreciable fat stores found in the muscle of these animals support this explination.

References

1. S. Harapanahalli, *Mechanisms for lipogenesis in Running Muscles of the Obese LA/N-cp Rat hindlimb and Total Lipid Exchange in These Tissues.* Masters Thesis, Marshall University, August 1993.
2. M. A. Hollands and M. A. Cawthorn, *Important Sites of Lipogenesis in the Mouse Other than White Adipose Tissue,* Biochem. J. 196(1981)645.
3. A. Mauro, *Satellite Cell of Skeletal Muscle fibers,* J. Biophys. Biochem. Cytol. 9(1961)493.
4. E. Schultz, K. M. McCormick, *Skeletal Muscle Satellite Cells,* Rev. Physiol. Biochem. Pharm. 123(1994)213.

Proc. Microscopy and Microanalysis 1995, edited by G.W. Bailey, M.H. Ellisman, R.A. Hennigar, and N.J. Zaluzec
Copyright © 1995 MSA. Published by Jones and Begell Publishing, 79 Madison Ave., New York, NY 10016

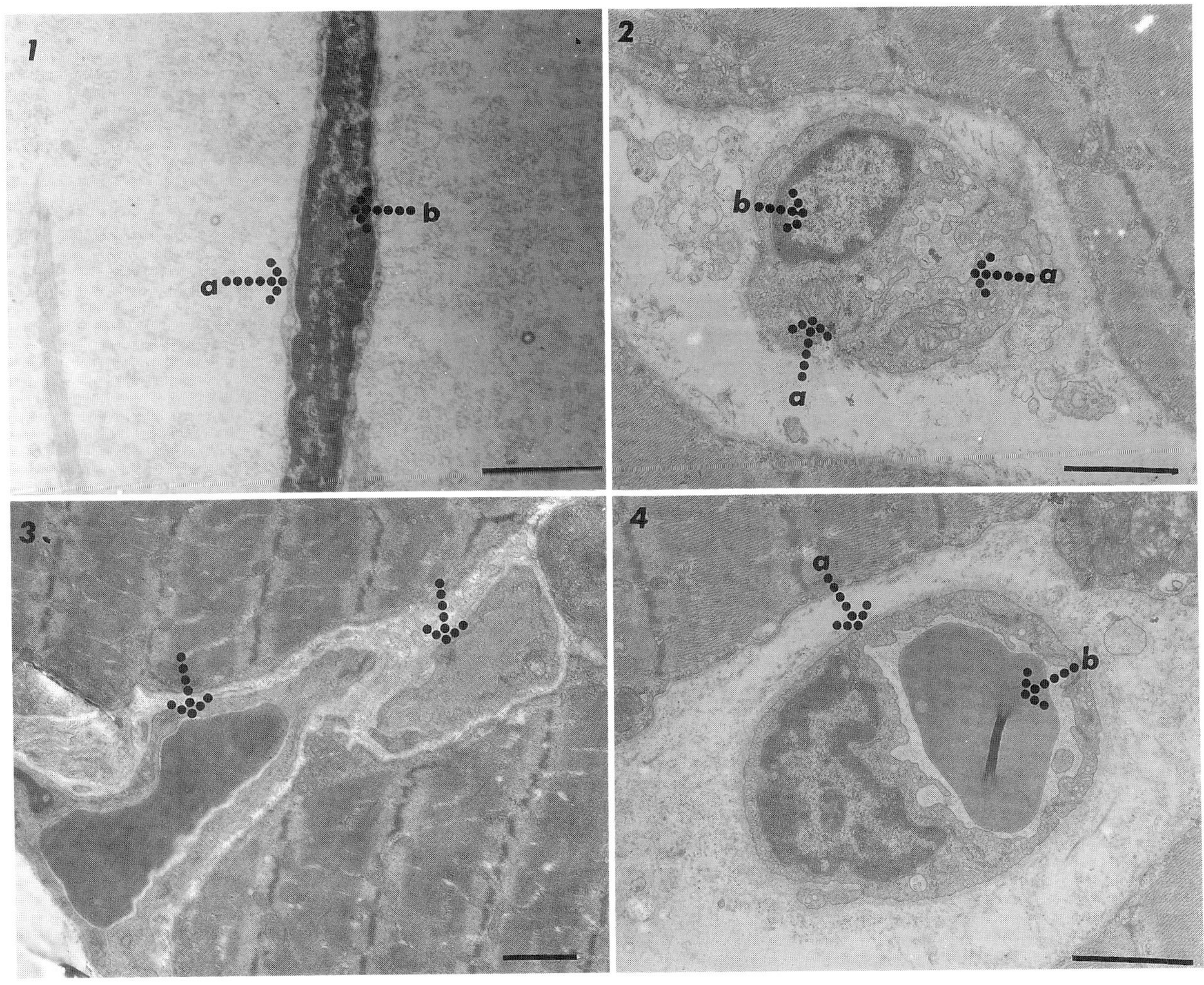

FIG. 1.-(a) Reduced cytoplasm, organelles greatly reduced; (b) Condensed chromatin. Bar = 1µm.

FIG. 2.-(a) Well developed organelles; (b) Less condensed chromatin. Bar = 1µm.

FIG. 3.- Capillaries with no lipid droplets. Bar = 1µm.

FIG. 4.- (a)Capillary; (b) Lipid droplet. Bar = 1µm.

A METHOD FOR ENCYSTATION OF **Entamoeba invadens**

M. Morales-Vallarta*, L. Villarreal-Treviño*, E. Ramírez-Bon**, S. Said-Fernández***, B. Mata-Cárde-
nas***.
* Departamento de Biología Celular, Facultad de Ciencias Biológicas, Universidad Autónoma de Nuevo
León. Monterrey, Nuevo León, México. Ap. Postal 2790 y 119 S/N.
** Departamento de Patología, Facultad de Medicina, Universidad Autónoma de Nuevo León, Monterrey, N.L.
México.
*** Centro de Investigación Biomédica del Noreste, I.M.S.S., México.

Some axenic encystation media for **Entamoeba invadens** have been described for different authors:
Rengpien and Bailey medium (AEM (1), Vázquezdelara-Cisneros and Arroyo-Begovich (low glucose medium)
(2) and Das et al. medium (3). The first medium must aging for one month for reach maturity, moreover
its efficiency is also variable, on the other hand, the three media require cells at middle log-phase
for good results. In this paper we report not a medium but a method for encystation of **Entamoeba
invadens.**

Trophozoites of **Entamoeba invadens** IP-1 strain were incubated with 10^3 cells/ml in 16 X 150 mm
screw capped tubes, each containing 10 ml of culture medium (TP-S-1 (4) or TYI-S-33 (5) or PEHPS (6))
without vitamines (except PEHPS medium), and gassed with CO_2 (0.2 1/min/2 min) through a pasteur pi—
pette with a cotton filter and incubated to 25°C. After 4 days the culture medium is eliminated and
sustituted with 10 ml of the same medium but without glucose, newly gassed as above mentioned and incu
bated to 25°C. After 4-5 days, formation of cystic structures reach a maximum efficiency of 70-80%.

The cystic structures of this manner obtained were characterized and: a) They were resistant to
0.25% triton X-100 for 20 min. (figs. 1 and 2), b) They produced an intense fluorescence with ultravio-
let-light (figs. 2 and 3), c) They presented a wrinkled surfase when they were observed at scanning –
electron microscope, equally as the cysts obtained with AEM medium, d) The transmission electron – –
microscopy showed the presence of a thick wall, multiple nuclei and big chromatoid bodies (fig. 5),
e) When they were incubated in growth medium to 25°C after washed with 0.25% triton X-100 they produced
abundant trophozoites after 3-5 days and a new population of trophozoites was obtained. This – –
caracterization showed that the cystic structures obtained are cysts and that they are mature and via-
bles.

The methodology here described have some adventages over encystation cultures media before mentioned because: a) There are not drastic changes of osmolarity, b) The encystation method can begin with
cells near stacionary-phase, and are not necessary middle log-phase cells, c) It is not necessary a maturation time.

REFERENCES

1. Rengpien, S., Bailey, G.B. J. Parasitol. 61:(1975) 824-830.
2. Vázquezdelara-Cisneros, L.G., Arroyo Begovich, A.J., Parasitol. 70:(1984) 620-633.
3. Das, S.R., Rastogi, A.K., Sagar, P., Singh, M.P. Indian J. Exp. 18:(1980) 333-336.
4. Diamond, L.S., J. Parasitol. 54:(1968) 1047-1056.
5. Diamond, L.S., Harlow, D.R., Cunnick, C. Trans. of the Royal Soc. of Trop. Med. and Hyg.
   72(1978) 431-432.
6. Said-Fernández, S., Vargas-Villarreal, J., Castro-Garza, J., Mata-Cárdenas, B., Navarro-Marmo-
   lejo, L., Lozano-Garza, G., Martínez-Rodríguez, H., Trans. of the Royal Soc. of Trop.Med. and
   Hyg. 82(1988) 249-253.

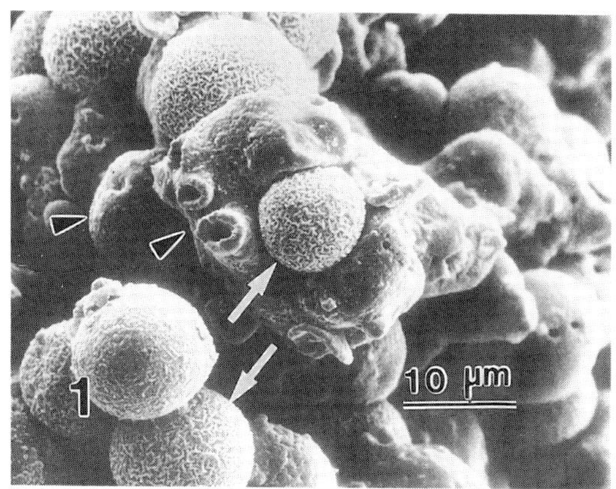
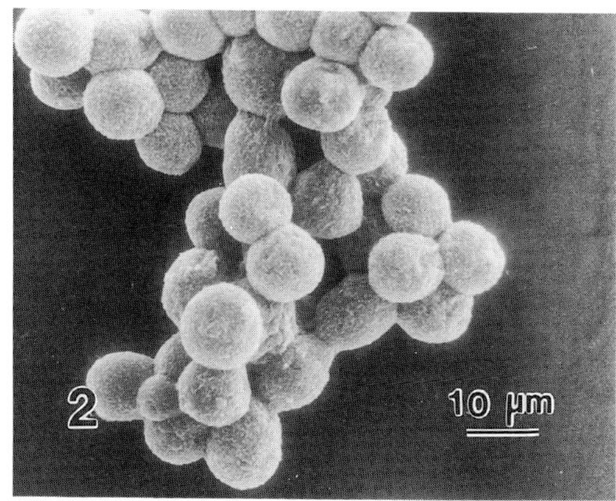

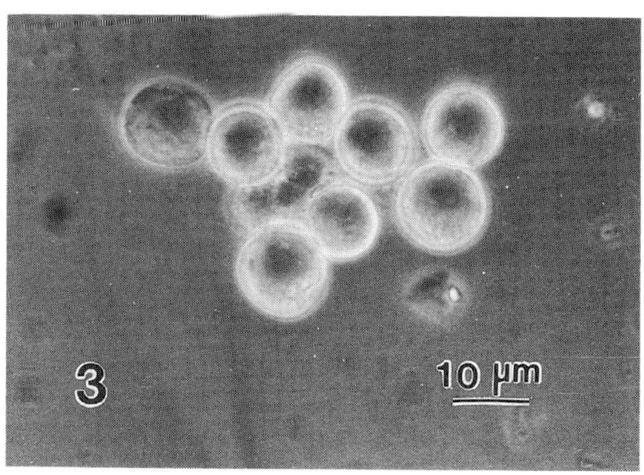
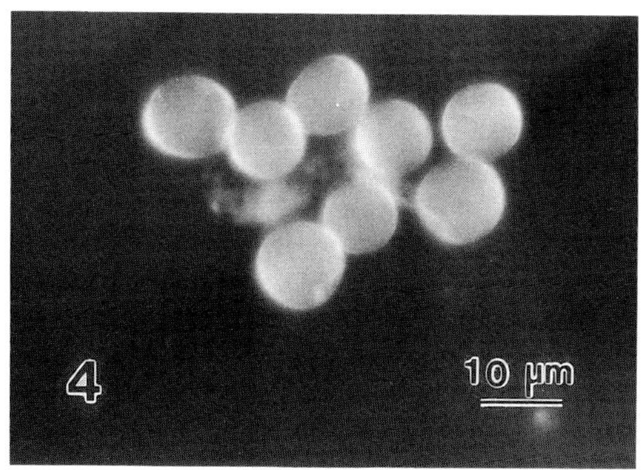

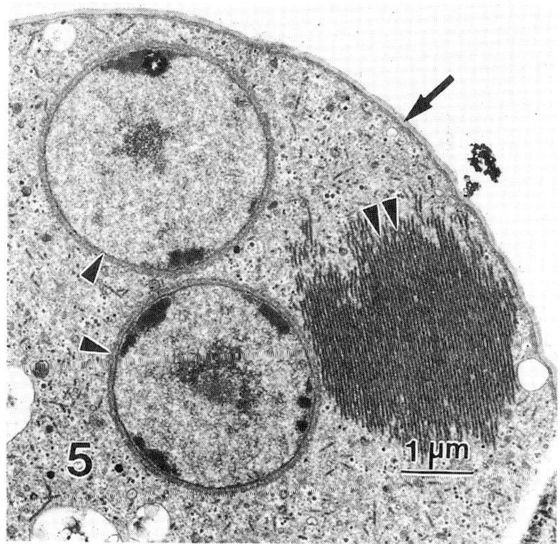

Figs. 1 and 2. Figure 1 and 2 show respectivelly pellets of cysts of **E. invadens** before and after treatment with 0.25% triton X-100. Figure 2 dont show trophozoites. Cysts (arrows), trophozoites (head arrow).

Figs. 3 and 4. Figure 3 show an aggregate of cysts of **E. invadens** observed by phase-contrast microscopy, and figure 4 show the same aggregate stained with calcoflour white M2R observed by fluorescence microscopy.

Fig. 5 Ultraestructure of a cyst of **E. invadens** obtained by the method here described. Cell wall (arrow), Nuclei (head arrows), chromatoid bodies (double head arrow).

ULTRASTRUCTURAL OBSERVATIONS ON HUMAN SECONDARY YOLK SAC.

D.A. Beckman*, YP. Wang**, G. Yang**, M. Zeng***, YP. Wang***, KZ. Xiao***, J. Holshek* and R.L. Brent*.

*Department of Medical Cell Biology, Nemours Research Programs, Wilmington, DE 19803
**Electron Microscope Laboratory, West China Univ. of Medical Sciences, Chengdu, Sichuan 610041, P.R., China.
***National Center for Birth Defects Monitoring, West China Univ. of Medical Sciences, Chengdu, Sichuan 610041, P.R., China

The human secondary yolk sac persists for a short but critical period of development during early organogenesis. Although it is established that the yolk sac is the site of origin of several cell lines, a locus of hematopoiesis and secretes several proteins, the yolk sac may have other as yet less well defined functions. Here we focus on ultrastructural evidence of the role of the human yolk sac as an organ of exchange with the extraembryonic coelom and the yolk-sac cavity. Yolk sacs of 6 to 10 weeks developmental age were fixed and processed for light microscopy, TEM and SEM by standard procedures. Numerous microvilli, coated and uncoated vesicles, and lysosomes in the mesothelial layer together suggest a high capacity for pinocytosis of coelomic fluid and lysosomal digestion of internalized substrate. It is not known whether this process is a significant nutritional source for the yolk sac and embryo as it is in the rodent.[1,2] Both SEM and TEM observations provide evidence of cilia on dispersed cells in the endodermal layer lining the yolk sac cavity. To our knowledge, there is only one other brief report of cilia on the human yolk sac.[3] TEM observations suggest that there is a network in the endodermal cell layer comprised of microvilli-lined intercellular spaces that open to the yolk-sac cavity and intracellular tubules sparsely lined with short microvilli that open to the intercellular spaces. The intracellular tubules are 2-4 um in diameter and, when present in endodermal cells lining the yolk sac cavity, extend from the apical to the basolateral cell membrane thus bypassing tight junctions between cells. Because serial sectioning was not commonly performed in previous studies of human yolk sac, it is not surprising that there are no other definitive reports of these intracellular tubular structures. Intracellular tubules were also observed in the columns of endodermal cells near the mesothelial cell layer. We speculate that the intercellular and intracellular canalicular-like system facilitates exchange between the endodermal cells and the yolk-sac cavity.[4]

References

1. D.A. Beckman et al., Placenta 11(1990)109.
2. D.A. Beckman et al., Placenta 12(1991)37.
3. J. Pereda et al., Arch. Histol. Cytol., 57(1994)107.
4. The support of Harry Bock Charities is gratefully acknowledged.

Proc. Microscopy and Microanalysis 1995, edited by G.W. Bailey, M.H. Ellisman, R.A. Hennigar, and N.J. Zaluzec
Copyright © 1995 MSA. Published by Jones and Begell Publishing, 79 Madison Ave., New York, NY 10016

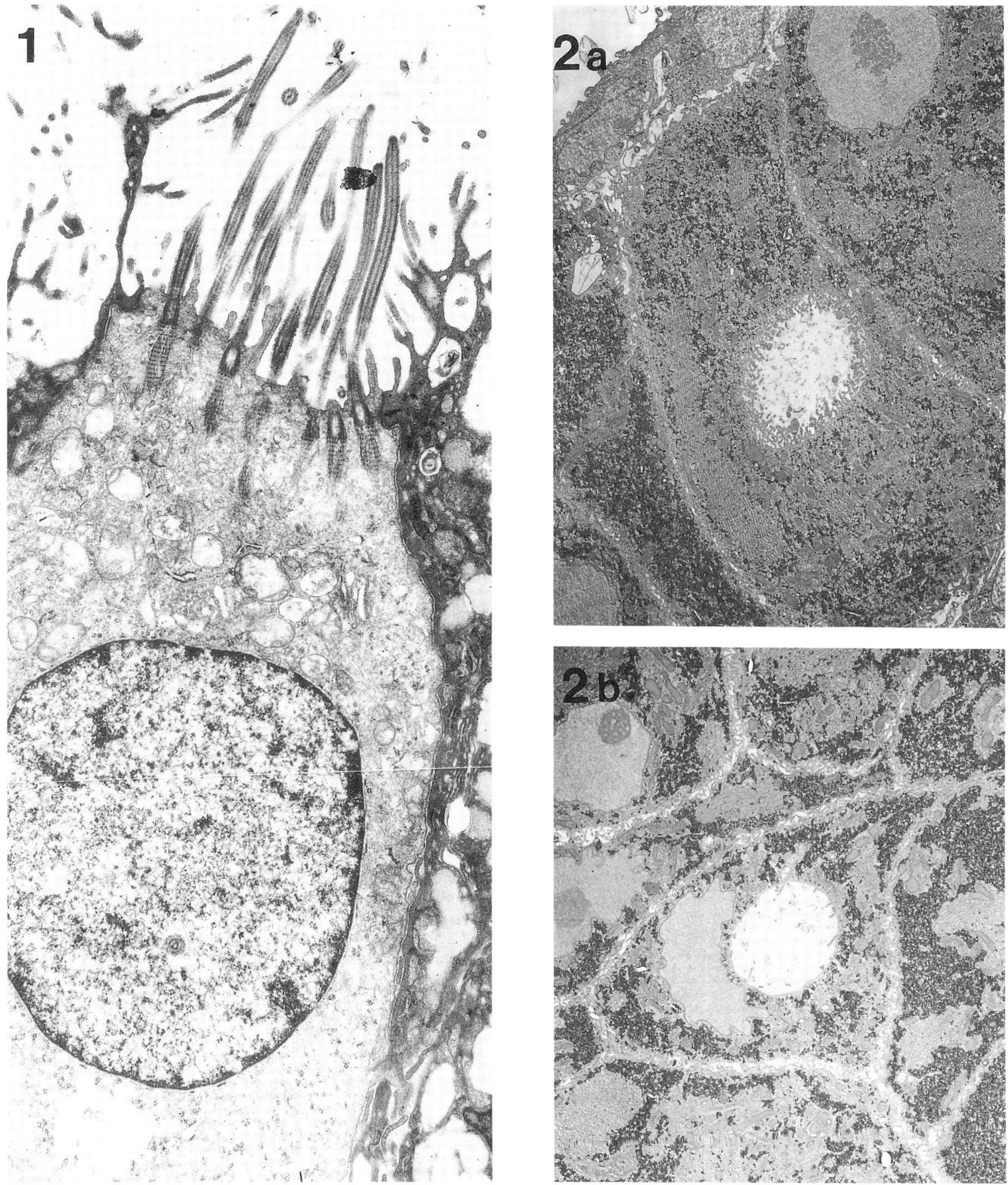

FIG. 1. -- Cilia on cell in endodermal layer lining yolk sac cavity, original magnification x 8,000.
FIG. 2. (a & b) -- Intracellular tubules, 2-4 um in diameter and sparsely lined with short microvilli, were observed in some endodermal cells, original magnification x 2,000.

METALLOTHIONEIN EXPRESSION DURING WOOL FOLLICLE DEVELOPMENT

N.Nishimura*, H.Nishimura**, C.Tohyama***, G.R.Cam* and D.L.Adelson*

*Animal Production, CSIRO, Prospect, NSW 2149, Australia
**Human Science, Aichi Mizuho University, Hiradobashi 470-03, Japan
***Environmental Health Sciences Div., NIES, Onogawa, Tsukuba 305 JAPAN

This study investigated metallothionein(MT) expression in proliferating or differentiating cells of developing wool follicles. Dividing cells in foetal sheep skin were identified by incorporated BrdU, which was injected into the jugular vein of a ewe before sacrifice. MT gene expression was assayed using a digoxigenin-labelled sheep MT-1b riboprobe as described previously[1]. Immunohistochemical staining for MT protein and BrdU was carried out using rabbit-anti rat MT-1 serum and anti-BrdU, as described elsewhere[2,3]. Foetal sheep skin was obtained from 60 to 130 days of gestation, fixed in HistoChoice or Bouin's and embedded in paraffin.

At 71 days of gestation, MT-1b mRNA was detected in the epidermis(EP), primarily in the spinous cells (Fig. 1a). At 90 days, MT-1b mRNA was concentrated at the tip of the follicle rudiment(F), but not in the presumptive dermal papilla(DP) (Fig. 1b). At 112 days, a symmetric distribution of MT-1b mRNA was observed in the follicle bulb matrix (Fig. 1c). However, very little MT-1b mRNA was detected in the matrix cell layers adjacent to DP. At 130 days, strong MT-1b gene expression was localized to the cytoplasm of cells in the follicle bulb(B) and keratogenous zone(K)(Fig. 1d).

MT protein was localized to the epidermis at 71 days (Fig. 2a), co-localizing with BrdU in the some cells. The MT immunohistochemical signal intensified and became concentrated at the tip of the follicle rudiment by 81 days (Fig. 2b). At 95 days (Fig. 2c), MT protein was localized to the presumptive outer root sheath(ORS) and follicle rudiment lumen, but not to the presumptive dermal papilla cells. At 105 days (Fig. 2d), MT protein was found in the ORS, keratogenous zone and follicle bulb matrix, but did not co-localize with the BrdU-proliferative zone surrounding the DP (Fig. 2d). The staining pattern for MT in the bulb and keratogenous zone was symmetrical in distribution. After 120 days of gestation, MT distribution was found to be asymmetric in the cytoplasm and some nuclei of cells in the follicle bulb(Fig. 2d). Further, MT protein was localized to ORS and the basal cell layer of sebaceous glands and sweat glands, but not to the DP (Fig. 2e). MT protein was not observed in the epidermis or keratogenous zone after this stage, even though MT-1b mRNA was present in the follicle bulb and keratogenous zone (Fig. 1d). Sense probe and pre-immune serum controls did not detect MT-1b mRNA or protein (Fig. 3a and b).

These findings suggest 1) that MT-1b may be important in cell differentiation in the follicle bulb, 2) that translation of MT-1b mRNA is inhibited during cell differentiation and migration.

References
1. C. Tohyama et al.,*Histochemistry* 101(1994)341.
2. H. Nishimura et al.,*J Histochem Cytochem* 37(1989)715.
3. D.L. Adelson et al., *Aust J Agric Res* 43(1992)843.

Fig. 1 *a-d*. Expression of MT mRNA at several stages of wool follicle development. *a*. 71 days, *b*. 90 days, *c*. 112 days, *d*. 130 days. (*EP* epidermis, *F* follicle, *DP* dermal papilla, *B* follicle bulb, *K* keratogenous zone, *ORS* outer root sheath). *a-d* x250.

Fig. 2 *a-e*. Localization of MT protein at several stages of wool follicle development. *a*. 71 days, *b*. 81 days, *c*. 95 days, *d*. 105 days, *e*. 120 days. (*EP* epidermis, *F* follicle, *DP* dermal papilla, *B* follicle bulb, *K* keratogenous zone, *ORS* outer root sheath). *a-c* x250; *d,e* x125.

Fig. 3 *a*. Detection of MT-1b mRNA expression using a sense riboprobe (90 days). x250.
b. Pre-immune rabbit serum control and BrdU double staining (95 days). x125.

Proc. Microscopy and Microanalysis 1995, edited by G.W. Bailey, M.H. Ellisman, R.A. Hennigar, and N.J. Zaluzec
Copyright © 1995 MSA. Published by Jones and Begell Publishing, 79 Madison Ave., New York, NY 10016

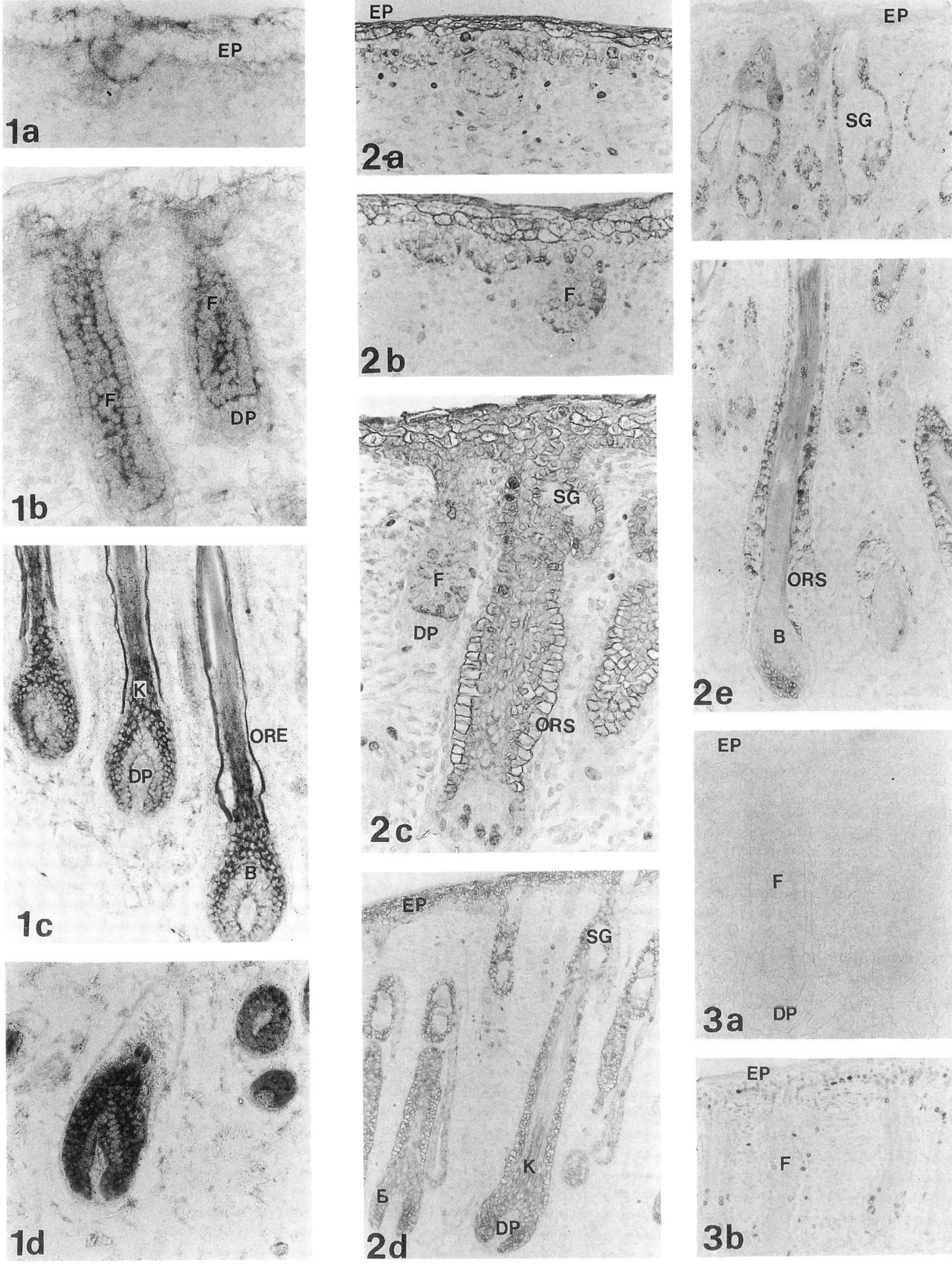

EP
1a
F
F
DP
1b
K
DP
ORE
B
1c
1d
EP
2a
F
2b
SG
F
DP
ORS
2c
EP
SG
K
Б
DP
2d
EP
SG
ORS
B
2e
EP
F
DP
3a
EP
F
3b

SCANNING ELECTRON MICROSCOPY OF SHELL-FREE CYSTS OF THE BRINE SHRIMP *ARTEMIA FRANCISCANA* (ANOSTRACA) IN THE PROCESS OF HATCHING

James R. Rosowski,* Mark A. Gouthro,* Denton Belk,** and Kit W. Lee*

*School of Biological Sciences, University of Nebraska-Lincoln, Lincoln, NE 68588-0118
**Our Lady of the Lake University of San Antonio, San Antonio, Texas 78207-4689

Air-dried, shell-free cysts of brine shrimp occasionally are used in aquaculture as a food substitute for live nauplii hatched from shelled cysts. Imbibed embryos of shell-free cysts are surrounded by an embryonic cuticle 1 (EC1)[1] composed of an outer and inner cuticular membrane with a fibrous layer in between.[2] Such imbibed embryos, when never air dried but used immediately or dehydrated and stored in a hypertonic salt solution, retain high viability. Although the hatching stages of shelled cysts have been documented with scanning electron microscopy (SEM),[3] no SEM study has examined the hatching of nauplii from cysts without shells, which was the purpose of the present study.

Great Salt Lake shelled cysts (Sanders Brine Shrimp Company, Ogden, Utah, U.S.A.) were treated with a mixture of bleach and sodium hydroxide to dissolve their shells.[4] After washing in tap water, the shell-free cysts were placed in a Petri dish on a filter paper that had been saturated with a salt solution consisting of 28 g NaCl plus 6 g of $NaHCO_3$ / liter, for 15 hr at 28° C in a lighted incubator. The nauplii hatch percentage at 15 hr was 89% (SD of 3.5%) and was similar to that of shelled cysts. The 15 hr-old unhatched embryos and nauplii were fixed for 48 hr in 3% glutaraldehyde in 3.4% saltwater, rinsed throughly in tapwater, and dehydrated in 20%, 50%, 75%, 95%, and 100% ethanol (3X). The specimens were air dried, or critically-point dried in liquid CO_2 and coated with 400 Å gold/palladium in a Denton Desk II sputter coater for SEM examination.

The EC1 polygonal plates, barely visible in fully hydrated shell-free cysts (Fig. 1), are clearly evident in air-dried, shell-free cysts (Fig. 2). Spotte and Anderson[5] speculated that the polygonal septa forming these plates perhaps mechanically reinforce the EC1 enhancing its function of protecting the embryo. Imbibed, shell-free cysts on damp filter paper hatched upward creating a jagged edged, 180° rupture arc in the EC1 along the suture of the polygonal plates (Figs. 3, 4) just as they do when shelled. As the nauplii emerge in their EC2 membrane (Fig. 3), they tend to tip over and lie flat on the filter paper near the EC1 membrane. During hatching, the inner cuticular membrane peals off the interior surface of the EC1 polygonal plates and the entire EC1 and part of the EC2 are left (Fig. 4) as the nauplius escapes (Fig. 5). This study shows that (1) the absence of the shell (tertiary envelope) does not effect the nauplii hatch percentage, (2) the spherical shape of the EC1 is not determined by the shell for it retains its shape without the shell (as in *A. franciscana*, San Francisco biotype[5]), during and after hatching, and (3) nauplii hatching stages and membrane movements from shell-free cysts are similar to those with shells.[3]

References

1. D. Belk, *J. Crust. Biol.* 7(1987)691.
2. J.E. Morris and B.A. Afzelius, *J. Ultrastruct. Res.* 20(1967)244.
3 R. Wheeler et al., *Aquaculture* 18(1979)59.
4. Aquafauna BioMarine, "Decapsulating Artemia Cysts" brochure, P.O. Box 5, Hawthorne, CA 90250.
5. S. Spotte and G. Anderson, *J. Crust. Biol.* 8(1988)221.

Proc. Microscopy and Microanalysis 1995, edited by G.W. Bailey, M.H. Ellisman, R.A. Hennigar, and N.J. Zaluzec
Copyright © 1995 MSA. Published by Jones and Begell Publishing, 79 Madison Ave., New York, NY 10016

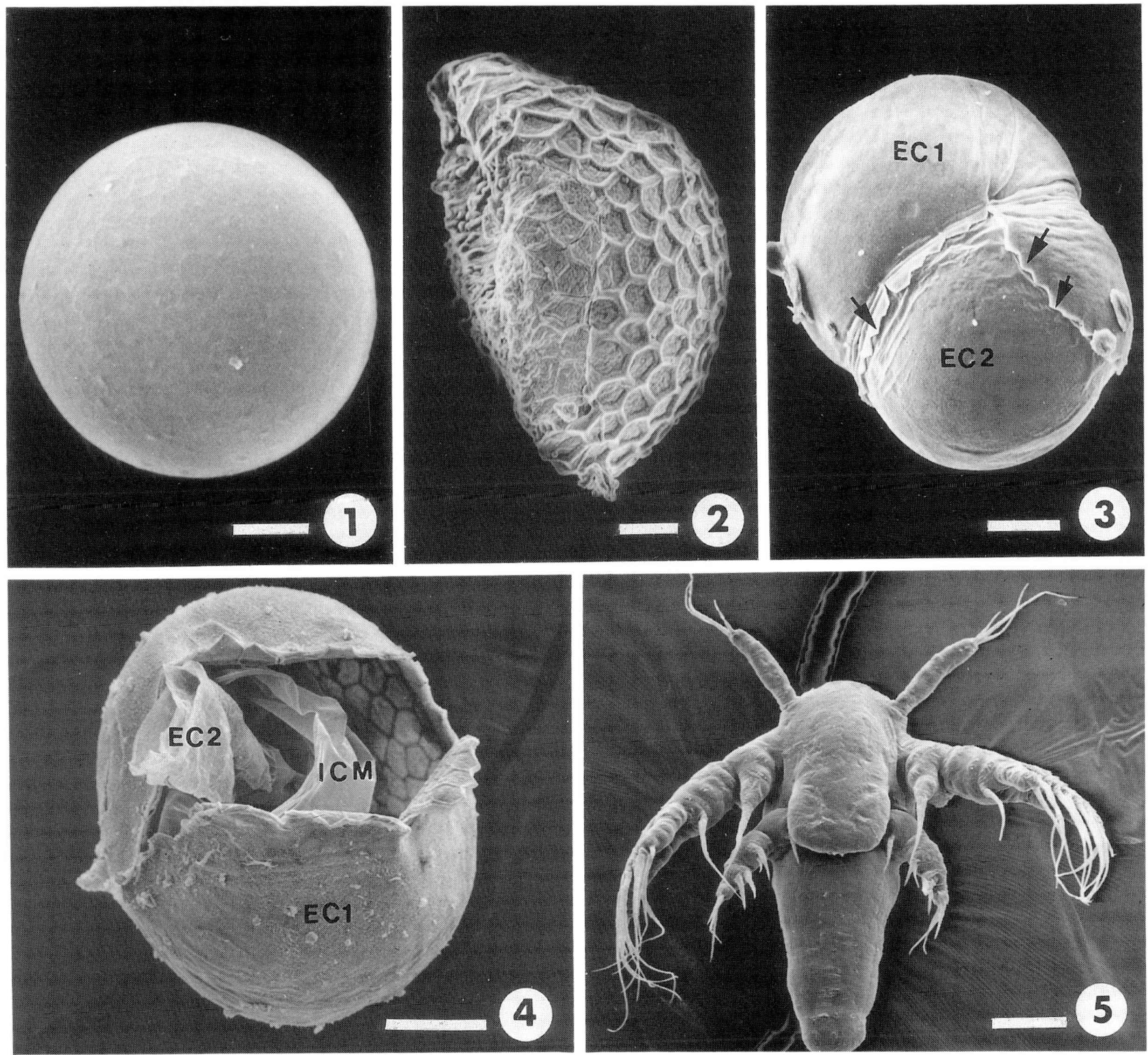

FIG. 1.--Shell-free cysts with EC1 exterior; polygonal plates barely visible. CPD. Bar = 50 μm.
FIG. 2.--Shell-free cysts with EC1 exterior; polygonal plates evident. Air-dried. Bar = 50 μm.
FIG. 3.--Wrinkled embryonic cuticles (EC1, EC2). Nauplius in hatch process; the EC1 has a jagged edge (arrows) from separation of its polygonal plates. The head of the nauplius is beneath the EC2 hatching mcmbrane. CPD. Bar = 50 μm.
FIG. 4.--The nauplius has left the protective EC1 membrane; the ICM has pulled away from the polygonal plates of the EC1. The inner surface of the polygonal plates is visible to the right of the ICM. CPD. Bar = 50 μm.
FIG. 5.--Brine shrimp nauplius hatched from a shell-free cyst. CPD. Bar. = 250 μm.
Abbreviations: EC1 = embryonic cuticle 1; ICM (inner cuticular membrane of the EC1); EC2 = embryonic cuticle 2; CPD = critical-point dried.

DNA LEVELS IN SOMATIC AND GERM LINE CELLS OF A FRESHWATER CRUSTACEAN, *Mesocyclops edax*.

E. M. Rasch * and G.A. Wyngaard**

*Dept. Anatomy & Cell Biology, East Tenn. State Univ., Johnson City, TN 37614
**Dept. of Biology, James Madison Univ., Harrisonburg, VA 22807

Chromatin diminution--the fragmentation and elimination of chromosome regions--provides an unusual opportunity to study genomic reorganization during development.[1,5] Some species of copepods regularly excise and discard large amounts of nuclear DNA from presumptive somatic cell lines during early cleavage stages [1-5]. To study this phenomenon in *M. edax* we determined DNA-Feulgen levels for more than 5,600 individual nuclei from squash preparations of 30 female and 25 male adults collected from lakes in Nova Scotia, Virginia and Florida. Fixation in 3:1 methanol/acetic acid was followed by squashing individual specimens in 45% acetic acid, freezing each slide in liquid N_2 and thawing in absolute ethanol before air drying. Each series of slides was stained with the Feulgen reaction for DNA and measured with a Vickers M86 scanning and integrating microdensitometer at 560 nm, using chicken RBC nuclei as an internal reference standard of 2.5 pg DNA per cell. This allowed us to ask several questions: are there differences in genome size (1) among specimens from different collecting localities, (2) between females and males at any single locality, and (3) between cells of germ line and somatic cell lineages?

We find no significant differences in diploid (2C) genome sizes among Nova Scotia, Virginia and Florida females (Table 1, Table 2). Nor are there differences between somatic cell genomes of Nova Scotia and Virginia males. Calculated haploid (1C) genome size for females (1.43 $\pm$ 0.015 pg DNA, n = 30) does not differ significantly from the haploid (1C) genome size estimated for males (1.48 $\pm$ 0.025 pg DNA, n = 25). If DNA values for putative sperm (3.07 $\pm$ 0.151, n = 16 are compared with 1C levels calculated from values of male somatic nuclei (1.48 $\pm$ 0.012, n = 25), we find the DNA content of sperm to be about twice the expected level, suggesting retention of about 1.6 pg DNA per male germ cell during later stages of spermatogenesis. Studies are now in progress to determine DNA levels in female germ line cells.

Haploid (1C) genome size estimates for *M. edax* (1.4-1.5 pg DNA or roughly 1.34-1.35 x 10^9 bp) are similar to those reported for *Cyclops furcifer* (1.4 pg)[1], but differ somewhat from 1C values for *Cyclops strenuus* (0.9 pg DNA) and *Cyclops divulsus* (1.8 pg DNA)[1].

References

1. S. Beermann, Chromosoma 24(1977)297.
2. J. Robins and I.A. McLaren, Canad. J. Genet. Cytol. 24(1982)529.
3. H. Stich, Exp. Cell Res. 26(1962)136.
4. B. Terpilowska, Zool. Poloniae 21(1971)163.
5. G.A. Wyngaard and C.C. Chinnappa, in Developmental Biology of Freshwater Invertebrates, A.R. Liss, Inc. NY, (1982) 485.

Proc. Microscopy and Microanalysis 1995, edited by G.W. Bailey, M.H. Ellisman, R.A. Hennigar, and N.J. Zaluzec
Copyright © 1995 MSA. Published by Jones and Begell Publishing, 79 Madison Ave., New York, NY 10016

Table 1. Source, geographical location of *Mesocyclops edax* populations, amount of DNA per cell, estimated genome size expressed in pg DNA and approximate number of base pairs DNA per nucleus for somatic and germ cells. N denotes number of adult specimens examined.

Nuclear Type	N	Geographical Population	2C DNA per Cell $\pm$ S.E. (pg = 1 x 10^{-12} g)	Base Pairs (1 x 10^{9})
Somatic				
Female	8	Nova Scotia	2.87 $\pm$ 0.026	2.62
Female	17	Virginia	3.00 $\pm$ 0.039	2.74
Female	5	Florida	2.92 $\pm$ 0.075	2.66
All Females	30	NS,VA,FL	2.95 $\pm$ 0.027	2.69
Somatic				
Male	3	Nova Scotia	2.95 $\pm$ 0.029	2.69
Male	22	Virginia	2.96 $\pm$ 0.029	2.70
All Males	25	NS, VA	2.96 $\pm$ 0.026	2.70
Sperm				
Sperm	1	Nova Scotia	2.96 $\pm$ 0.034	2.70
Sperm	16	Virginia	3.07 $\pm$ 0.151	2.80

Table 2. Analysis of Variance of Nuclear DNA Contents of Female Somatic Cells of Nova Scotia, Virginia and Florida Populations of *Mesocyclops edax*.

Source of Variation	df	SS	MS	Fp
Population	2	0.09465	0.04732	2.240.1263
Error	27	0.57140	0.02116	
Among slides within population				
Total	29	0.66606		

WHAT MICROSCOPIC STUDIES OF CONFRONTING CISTERNAE AND
ANNULATE LAMELLAE REVEAL ABOUT THEIR SIMILARITIES AND
DIFFERENCES

John R. Palisano* and Karen S. Renzaglia**

*Department of Biology, University of the South, Sewanee, TN 37383-1000
**Department of Biological Sciences, East Tennessee State University, Johnson City, TN
37614

Knowledge of the origin and function of confronting cisternae (CC) and annulate lamellae
(AL) has been limited because they are so infrequently observed. CC were first described
by Porter in 1955 and have since been observed in a variety of rapidly proliferating tumor
cells and selected fetal tissues.[1,2,3] AL were first reported in 1952 by McCullough in sea
urchin eggs.[4] Although AL have since been observed in a variety of cells, they have been
predominately found in oocytes, spermatocytes, and tumor cells.[5] A microscopic study
was initiated to gain a better understanding of the origin and role of these two membranous
organelles that are frequently found in rapidly dividing cells that lose the ability to express
these organelles once their rapid proliferative stage ceases.[6,7] Because it has proven
difficult to interpret the flow of membrane in the mitotic cell by electron microscopy alone,
data obtained by electron microscopy was correlated with fluorescence microscopy
wherever possible.

CC were studied in HeLa cells that were obtained from American Type Culture Collection
and cultured in minimum essential medium containing 10% calf serum in a mixture of 5%
CO_2 in air. Cells were harvested by mitotic shake-off 3 or 4 days after passage while they
were in exponential growth phase. The mitotic cells were fixed, dehydrated, and
embedded for routine electron microscopy. Axenic gametophytes of <u>Lycopodium
obscurum</u> were the source of AL. Small pieces of tissue were fixed in 6% glutaraldehyde
containing 0.05% caffeine buffered in 50 mM PIPES, pH 7.2 for 2h at room temperature.
The remainder of the fixation followed routine procedure of dehydration through a graded
ethanol series and infiltration by gradually increasing the resin to propylene oxide over the
course of several days. Movement of nuclear envelope in HeLa cells was followed by
fluorescence microscopy. Cells were grown on glass coverslips for 3 or 4 days at which
time they were rinsed in Delbecco's phosphate buffered saline, fixed in 3.7%
paraformaldehyde buffered in 50 mM PIPES containing 5 mM EGTA and
2 mM $MgSO_4$. The cells were permeabilized with Triton X-100 and incubated with a
monoclonal antibody to lamin B in a humidified chamber for 30 min. After this incubation,
the cells were washed to remove any excess monoclonal antibody to lamin B from them
and incubated for 30 minutes in the secondary antibody which was goat anti-mouse
conjugated to fluorescein. The coverslip was then mounted on a glass slide with
Fluoromount and viewed with an Olympus IMT-2 microscope fitted for fluorescent
microscopy.

Electron microscopy observations indicate that both CC and AL consist of stacked cisternae
that are, with few exceptions, found in close association with the nuclear envelope.
Although they both are observed in rapidly dividing cells, the CC are almost always found
during mitosis of the cell cycle, never contain pores, and are always studded with
ribosomes. AL have only been described during interphase, always have pores in their
cisternae, and are never studded with ribosomes. Because of the close proximity of CC
and AL to the nucleus, investigators have assumed that both organelles arise from the
nuclear envelope (NE). Preliminary observations using the monoclonal antibody to lamin

*Proc. Microscopy and Microanalysis 1995, edited by G.W. Bailey, M.H. Ellisman, R.A. Hennigar, and N.J. Zaluzec
Copyright © 1995 MSA. Published by Jones and Begell Publishing, 79 Madison Ave., New York, NY 10016*

B indicate that the CC in HeLa cells probably arise from the NE since the antibody to lamin B appears to label the CC; however, this must be comfirmed by immunoelectron microscopy. Although Saunders has reported that this same monoclonal antibody recognizes lamin B in the NE of higher plants,[8] all attempts to detect either NE or AL in <u>Lycopodium</u> with anti-lamin B antibody have been unsuccessful. The inability to label AL in <u>Lycopodium</u> may be the result of AL having lost the lamin B during an earlier stage of development, not having originated from NE, or the inability of this monoclonal to recognize <u>Lycopodium</u> lamin B.[9]

References

1. K.R. Porter, *Symposium on Fine Structure of Cells*, New York: Interscience Publishers, Inc. (1955) 236.
2. J.F.M. Numes et al., *J. Submicrosc. Cytol.* 18(1986)441.
3. J.R. Palisano, *Cell Biology International Reports* 14(1990)1025.
4. D. McCullough, *J. Exp. Zool.* 119(1952)47.
5. R.G. Kessel, *International Review Cytology* 133(1992)43.
6. J.R. Palisano, *Cell Biology International Reports* 17(1993)653.
7. K.S. Renzaglia et al., *Can. J. Bot.* (In press.)
8. M.J. Saunders, Personal Communication.
9. This research was supported in part by NSF Grant DEB-9207646 to KSR and an ROA supplement to this grant enabling JRP to work at ETSU during the summer of 1993. JRP was the recipient of a faculty development grant from the University of the South.

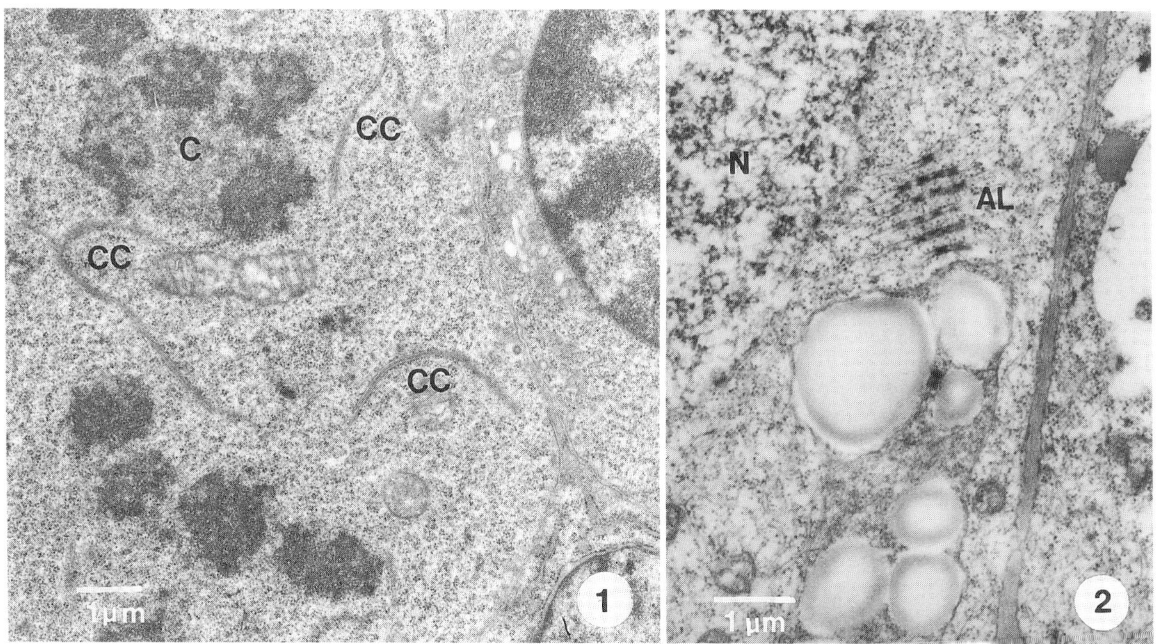

FIG. 1.--Electron micrograph of HeLa cell in prophase. Confronting cisternae (CC) are in close association to the chromosomes (C). Bar = 1μm

FIG. 2.--Section through <u>Lycopodium</u> cell. The annulate lamellae (AL) are closely associated with the nucleus (N). Bar = 1μm

NUCLEAR PORES IN APOPTOSIS AND NECROSIS INDUCED BY CISPLATIN INTOXICATION

V.J.A. Montpetit, N. Mikhael, R. Goel, D.J. Stewart, J. M. Molepo and B. T. Luck

Department of Pathology & Laboratory Medicine, University of Ottawa, Ottawa, Canada, K1H 8M5

Nuclear pore complexes (NPC) are an extensively studied cell component involved in the trafficking of macromolecules between the nucleus and cytoplasm[1]. The distribution of NPC is reported to be altered during apoptosis in five different apoptotic models[2]. Cisplatin (DDP) is an important anticancer drug that induces apoptosis in renal epithelial cells[3]. Falcieri's findings prompted us to study the distribution and structural alterations of NPC during DDP intoxication in both necrotic and apoptotic cells. Six BDF_1 mice were injected intraperitoneally at a dose of 10 or 20 mg/Kg and were sacrificed after 2 or 4 days. The kidneys were removed and processed for conventional electron microscopy. An examination of random sections of the renal cortex from different mice showed that there was significant apoptosis although the predominant process of cell death was necrosis. The characteristics of apoptosis were readily identified by the chromatin margination adjacent to the inner nuclear membrane, clearing of the granules in the nucleoplasm as well as the relatively intact cytoplasm. Earlier changes appeared to be preceeded by nucleolar segregation. NPC were easily identified in the compact chromatin margination due to basket shaped areas of low density along the inner part of the nuclear envelope. NPC were also seen in heterochromatin (HC) free zones (Figs. 1, 2,). The perinuclear cisterns occasionally revealed focal mild dilation in sharp contrast with necrotic cells where dilation of the cisterns was an early and prominent finding (Figs. 3, 4). Cellular necrosis was identified by the early swelling of the cytoplasm and mitochondria, increased primary lysosomes and vesiculation. These degenerative features appeared to be followed by the shrinkage of the cell including nuclei, hyperdensity of the cytoplasm and clumping of the chromatin. The perinuclear cisterns were noticeably dilated with prominent NPC due to the evagination of the low density areas on the nucleoplasmic side and invagination of the cytoplasmic ring (Figs. 3, 4). In profile the NPC appeared to be randomly distributed in both apoptosis and necrosis. This distribution of NPC during apoptosis is at variance with Falcieri's findings of pore-free dense chromatin areas and clustered pores in the diffuse chromatin zones[2]. However, it is possible that at a later stage there would be some disassembly of pores adjacent to HC due to impaired nucleocytoplasmic exchange.

References

1. L. Gerace, *Curr. Opin. Cell Biol.* 4(1992)637.
2. E. Falcieri et al., *Histochem. J.* 26(1994)754.
3. V.J.A. Montpetit et al., *Proc. Mic. Soc. Can.* (1994)142.
4. Supported by National Cancer Institute of Canada with funds from the Canadian Cancer Society.

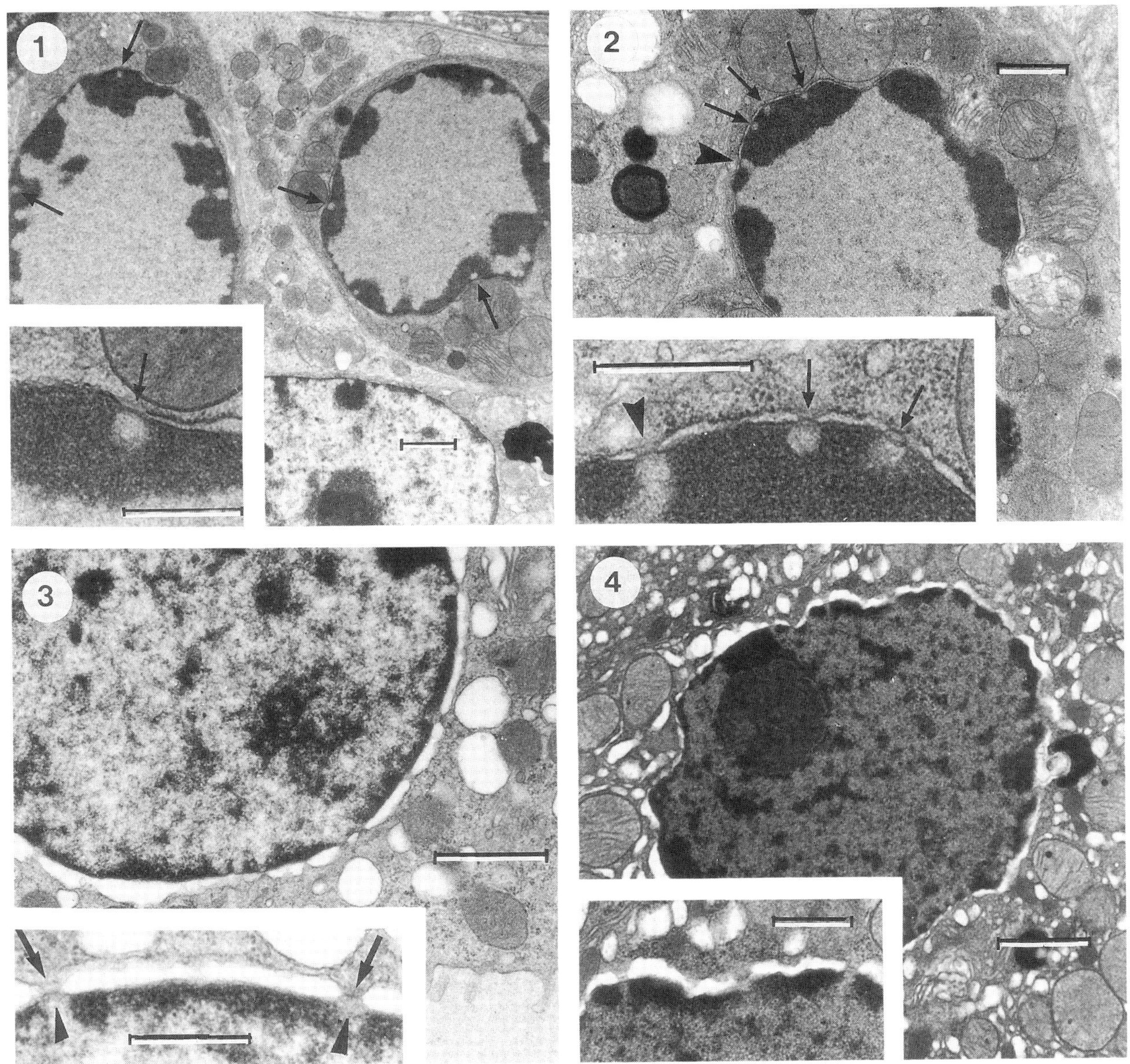

Figs. 1, 2, 3. 20 mg/Kg DDP 4 days.
Fig. 1. Two apoptotic cells with margination of HC and clearing of the nucleoplasm. Note the NPC adjacent to the inner nuclear membrane enclosed by HC (arrows). Inset. Higher magnification of a NPC.
Fig. 2. Apoptotic cell with NPC embedded in HC (arrows) and NPC in HC free zone (arrowhead). Inset. Higher magnification of NPC.
Fig. 3. Cell in early necrosis with dilation of the perinuclear cistern (PC) and stretched NPC. Inset. Higher magnification of two NPC that traverses the dilated PC due to invagination of the cytoplasmic ring (arrow) and evagination of the nucleoplasmic cage (arrowhead).
Fig. 4. 10 mg/Kg DDP 2 days. More advanced stage of necrosis showing a shrunken necrotic cell with compact nucleolus and clumped chromatin. Note the shrunken and hyperdense mitochondria. Inset. Higher magnification of NPC. Bars 1 μm figs. 1-4; bars 0.5 μm insets.

BISMUTH HEMATOXYLIN STAINING OF NUCLEIC ACIDS

A.A. Smith

Barry University School of Podiatric Medicine, Miami
Shores, FL 33161

Hematoxylin is readily oxidized to the blue pigment
hematein. Hematein can complex with a wide variety of
metal salts or mordants to form mordant dyes. These
dyes stain various components of animal tissues. The
component stained depends on the metal in the mordant
dye[1,2].

Alum hematoxylin uses trivalent aluminum ion as a
mordant. At extreme dilution, requiring long staining
times, alum hematoxylin is a specific stain for nucleic
acids.[3] As usually used, alum hematoxylin stains
nucleoproteins, staining them nearly as well after the
removal of the nucleic acids as before.[4,5]

Bismuth shows a high affinity for nucleic acids.[6,7]
Although all common bismuth salts are insoluble in
water,[8] sodium bismuthate reacts with an aqueous
solution of hematoxylin to produce hematein and
trivalent bismuth ions. The bismuth ions complex with
the hematein to form a water-soluble "bismuth
hematoxylin." HEPES (N-[2-hydroxyethyl]piperazine-
N'-2-ethanesulfonic acid), 8 mg/ml of solution, greatly
increases the amount of bismuth hematoxylin formed.
Glycerol, 1/8 v/v, improves the stability of the
complex.

A weakly alkaline solution of bismuth hematoxylin is a
specific stain for nucleic acids. At pH 8, DNA is much
better stained than RNA. At pH 10, both nucleic acids
are well stained. Staining is abolished by prior
removal of the nucleic acids with 4% trichloroacetic

acid at 90° C.

Proc. Microscopy and Microanalysis 1995, edited by G.W. Bailey, M.H. Ellisman, R.A. Hennigar, and N.J. Zaluzec
Copyright © 1995 MSA. Published by Jones and Begell Publishing, 79 Madison Ave., New York, NY 10016

References

1. A.B. Lee, et al, *The Microtomist's Vade Mecum* , *10th ed.*, Philadelphia: Blakiston (1937) 161.

2. J.A. Kiernan, *Histological and Histochemical Methods, Theory & Practice, 2nd ed.* Oxford: Pergamon (1990) 95.

3. C. Bettinger and H.W. Zimmermann, *Histochemistry* 96 (1991) 215

4. R.D. Lillie, et al, *Histochemistry* 49 (1976) 23.

5. R.W. Horobin, *Understanding Histochemistry*, London: Ellis Horwood Ltd. (1988) 81.

6. P. Albersheim and U Killias, *J. Cell Biol.* 17 (1963) 93.

7. M.A. Hayat, *Principles and Techniques of Electron Microscopy, Biological Applications, vol. 1*, New York: Van Nostrand Reinhold (1970) 301.

8. J.A. Dean, *Lange's Handbook of Chemistry, 13th ed.*, New York: McGraw-Hill (1985) 4-29.

ADVANCES IN IMAGING SIMS STUDIES OF STAINED AND BrdU-LABELLED HUMAN METAPHASE CHROMOSOMES

R. Levi-Setti,* J. M. Chabala,* R. Espinosa,** and M. M. Le Beau**

*The Enrico Fermi Institute and Department of Physics,   **Section of Hematology/Oncology, Department of Medicine, The University of Chicago, Chicago, IL 60637

We have shown previously that isotope-labelled nucleotides in human metaphase chromosomes can be detected and mapped by imaging secondary ion mass spectrometry (SIMS), using the University of Chicago high resolution scanning ion microprobe (UC SIM). [1-2] These early studies, conducted with BrdU- and ^{14}C-thymidine-labelled chromosomes via detection of the Br$^-$ and ^{28}CN$^-$ (^{14}C^{14}N$^-$) label-carrying signals, provided some evidence for the condensation of the label into banding patterns along the chromatids (SIMS bands) reminiscent of the well known Q- and G-bands obtained by conventional staining methods for optical microscopy. The potential of this technique has been greatly enhanced by the recent upgrade [3] of the UC SIM, now coupled to a high performance magnetic sector mass spectrometer in lieu of the previous RF quadrupole mass filter. The high transmission of the new spectrometer improves the SIMS analytical sensitivity of the microprobe better than a hundredfold, overcoming most of the previous imaging limitations resulting from low count statistics.

In the present study, we have repeated the SIMS analysis of BrdU-labelled human metaphase chromosomes, prepared with the previously adopted protocol [2], with the goals, now within reach, to construct a karyotype for the BrdU density distribution revealed by SIMS, and to assess the feasibility of sister chromatid exchange (SCE) and fragile sites studies. We have also investigated the chemical effects of Giemsa and trypsin-Giemsa stains on BrdU-labelled chromosomes. Previous problems associated with wetting the gold substrates on which the chromosomes are deposited have been overcome by the use of gold-coated glass coverslips, treated with a surfactant (Triton-X, 0.5 %).

Fig.1 (a) shows, in inverted contrast, a ^{81}Br$^-$ SIMS map of a portion of a complete metaphase cell in second division in the presence of BrdU, the latter identified by the label intensity ratio between sister chromatids, close to 2:1. The two chromosomes in the lower right hand corner exhibit clear examples of SCE (indicated by arrows). This image originates from the partial sputter erosion of the sample. In effects, due to the low current of our Ga$^+$ probe (here ~6 pA), it is feasible to explore the in-depth sample composition by sequential mappings (SIMS tomography). Fig. 1 (b) is a CN$^-$ map for the same area of Fig. 1 (b). As previously shown[2], this abundantly emitted ion species maps the overall distribution of proteins and DNA, with rather uniform intensity along the chromatids in the tightly packed human chromosomes. The comparison between Figs. 1 (a) and 1 (b) indicates that the protein distribution extends somewhat beyond the chromatin revealed by the BrdU label, possibly also due to cytoplasmic residues surrounding the chromosomes.

Fig. 2 (a) is a ^{81}Br$^-$ SIMS map of BrdU-labelled chromosomes stained with Giemsa, shown in inverted contrast, and Fig. 2 (b) a CN$^-$ map for the same sample area. The added Giemsa stain, a mixture of methylene blue, eosin Y and azure B, affects the original BrdU label in several ways, not yet entirely understood. Possibly through the Br content of eosin Y ($C_{20}H_6Br_4Na_2O_5$), the differentiation of BrdU label intensity between sister chromatids is no longer visible. At the same time, the banding patterns become more discernible. Striking is the reduction (by a factor ~2) in the CN$^-$ intensity within the chromosomes relative to that prior to Giemsa stain, accompanied by the appearance of intense diffuse CN$^-$ emission from the local substrate (> ten times that present prior to Giemsa staining), as established from a large number of observations similar to the content of Fig. 2 (b). The above suggests that the Giemsa stain induces a profound chemical alteration of the chromosomal structure, not entirely surprising since the dyes in Giemsa can intercalate into DNA. Even more dramatic are the chemical changes observed in BrdU-labelled chromosomes treated with trypsin-Giemsa. In this case, both the Br and CN chromatid maps (not shown here) appear surrounded by much brighter narrow halos, such as to take the appearance of absorption rather than emission maps. This suggests

Proc. Microscopy and Microanalysis 1995, edited by G.W. Bailey, M.H. Ellisman, R.A. Hennigar, and N.J. Zaluzec
Copyright © 1995 MSA. Published by Jones and Begell Publishing, 79 Madison Ave., New York, NY 10016

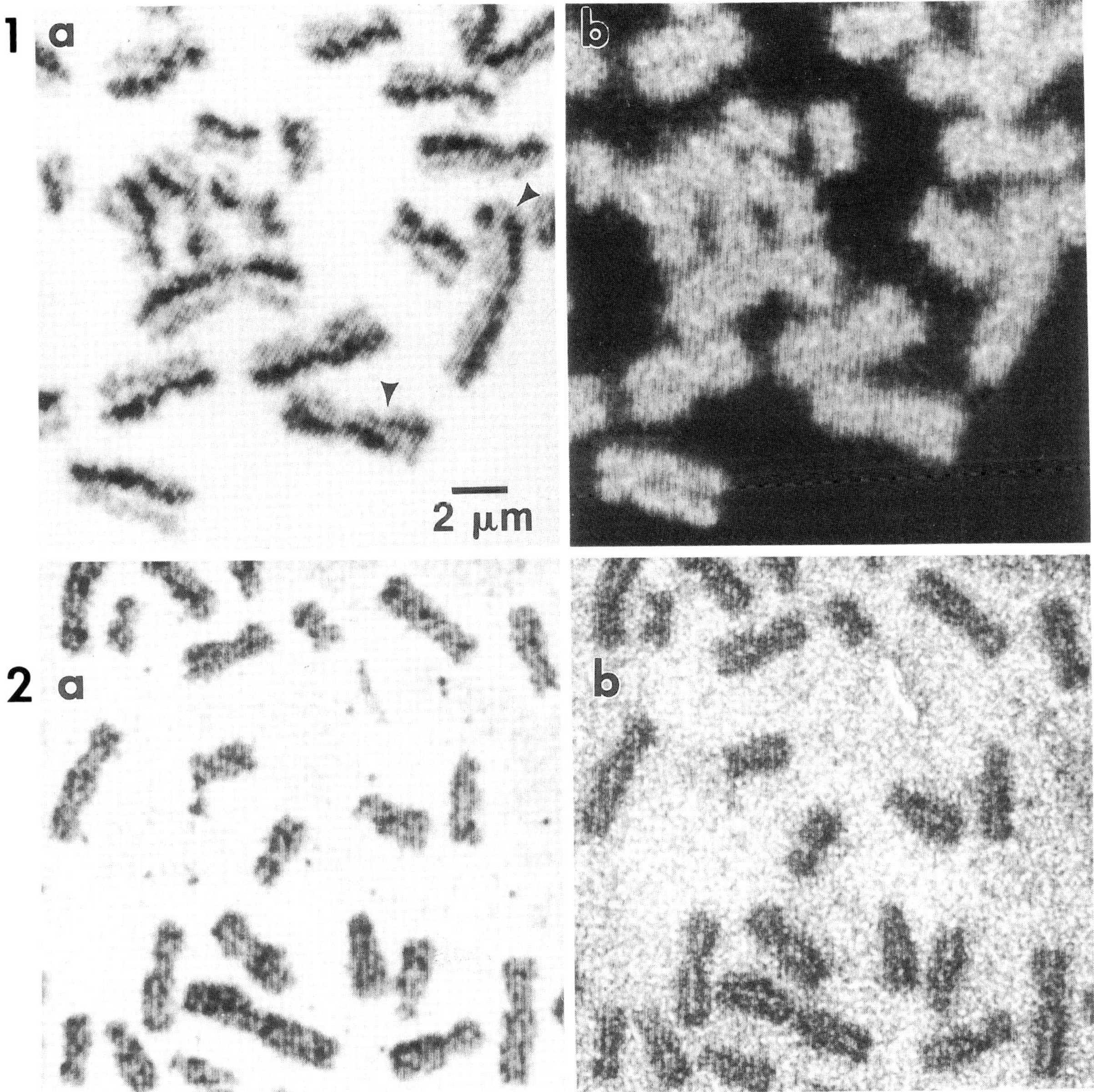

FIG. 1. (a) ^{81}Br⁻ SIMS map of BrdU-labelled human metaphase chromosomes. Inverted contrast.
(b) CN⁻ SIMS map of the same sample area shown in (a).
FIG 2. (a) ^{81}Br⁻ SIMS map of BrdU-labelled, Giemsa-stained chromosomes. Inverted contrast.
(b) CN⁻ SIMS map of the same sample area shown in (a).

extraction of the BrdU label and proteins from the chromosomal interior to form a halo onto the substrate. While elucidation of the staining mechanisms from our SIMS data is still elusive, we have established that our SIMS methodology has advanced to the level of being potentially useful in actual studies of chromosome structure.

1. R. Levi-Setti, et al. *SIMS VIII*, A. Benninghoven et al., eds. Chicester: Wiley (1992) 669.
2. R. Levi-Setti and. M. M. LeBeau, *Biol. Cell* 74 (1992) 51.
3. R. Levi-Setti, et al. *SIMS IX*, A. Benninghoven et al., eds. Chicester: Wiley (1994) 233.
4. This report is bases on work supported by NSF Grant BIR-9317959.

RABBIT EXTRASINOVIAL REGENERATING TENDON. AN EXPERIMENTAL STUDY.

R. González Santander*, M.A. Plasencia Arriba*, A. Lopez Alonso*, M.V. Toledo Lobo*, G. Martínez Cuadrado*, F.J. Martínez Alonso*, M. González-Santander Martínez**, M. Monteagudo*.

* Morphological Sciences and Surgery Dept. Univ. of Alcalá de Henares. Madrid, Spain.
** Toxicology and Sanitary Legislation Dept. Univ. of Alcalá de Henares. Madrid, Spain.

The roles of the extrasynovial tendon cells (Achilles tendon) throughout the repairing process following injury are controversial. This study was performed to investigate the ultrastructural features of activated tenocytes and collagen synthesis experienced while healing after partial tenotomy and repair.

Adult rabbit Achilles tendons were partially tenotomized. Before the suture of the paratenon, the borders of the surgical dissection were treated with different substances copolimerized with hydrogel. Hydrogel non-treated, tenotomized tendons were considered as controls. The rest of the samples were treated with hydrogel alone or hydrogel with E-vitamin. Rabbit legs were immobilized for 3, 10, 20 and 30 days. Samples from tenotomized and normal tendons were embedded in parafin and in Spurr and Araldite resins by conventional methods.

With LM, a more pronounced healing process was observed in tendons treated with hydrogels if compared with controls, but this comparative study needs further investigation. During the process of repair, different stages can be identified: inflammation, revascularization and angiogenesis, cell proliferation, collagen synthesis and fibrillogenesis, and alignment/organization of collagen fibers into bundles oriented in the longitudinal axis of the tendon. The repair of the lacerated tendon progresses from areas close to the paratenon towards the center of the tenotomy. Mitosis were infrequently observed except in hydrogel-E vitamin treated samples, in which mitosis were found in the reparing tissue near the paratenon, but an increase on cellularity was appreciated in all the samples. In all the samples angiogenesis process was similar to that observed in other tissues[1].

With TEM, activated tenocytes showed an increase in its rough endoplasmic reticulum (RER) as main feature, compared with non-activated tenocytes from normal tendon. This increased RER is located at the cell periphery along the longitudinal axis of the cell. Collagen fibres with different lenghts and an evident cross striation were observed into tenocyte plasma membrane invaginations. Besides, activated tenocytes showing intracytoplasmic collagen fibril-bearing vacuoles were observed (Fig.). These collagen fibrils located in the fibroblast cytoplasm, in vacuoles and/or cell invaginations, have been previously reported under certain conditions that accelerate collagen turn-over (such as certain pathological conditions or drug accelerators of collagen synthesis, among other situations)[2,3]. No such structures were observed in normal, non tenotomyzed samples.

Transmission and scanning electron microscopy revealed large amounts of immature and mature collagen and other fibers in the intercellular space from tendons during the healing process.

References

1. L. Diaz-Flores et al., Histol. Hitopathol., 9 (1994) 807.
2. C.S. Enwemeka, Tissue Cell, 23 (1991) 173.
3. C.S. Enwemeka, Tissue Cell, 24 (1992) 511.

Proc. Microscopy and Microanalysis 1995, edited by G.W. Bailey, M.H. Ellisman, R.A. Hennigar, and N.J. Zaluzec
Copyright © 1995 MSA. Published by Jones and Begell Publishing, 79 Madison Ave., New York, NY 10016

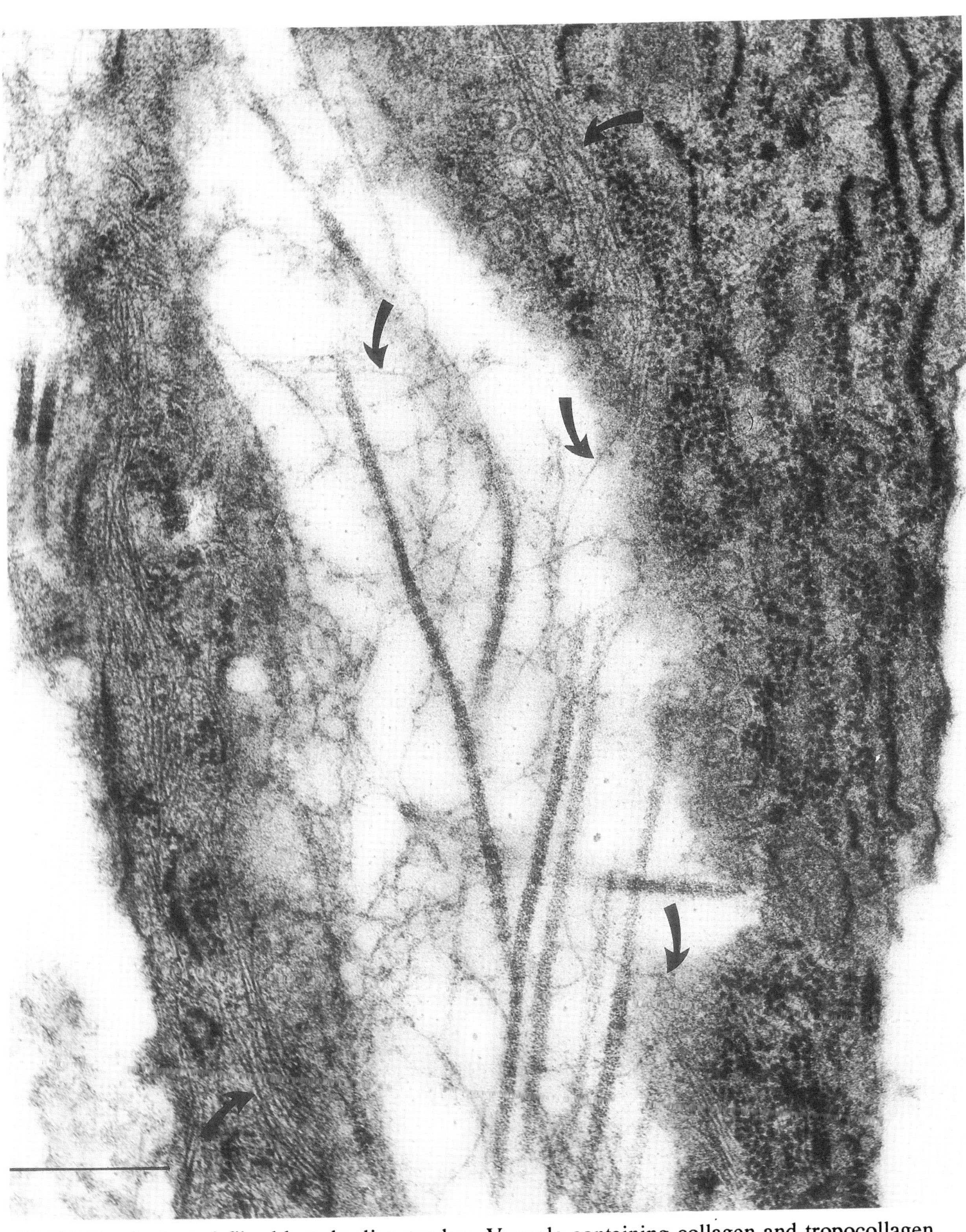

FIGURE.- Activated fibroblast, healing tendon. Vacuole containing collagen and tropocollagen. Fibers (11 nm diameter) in the cytoplasm of the cell (arrows).Scale bar: 0,5 µm (12.000 X).

DIFFERENT CARBOHYDRATE SUBSTRATES AFFECT CELL WALL ADHESION IN *CANDIDA ALBICANS*

Sara L. Browning*, Richard F. E. Crang*, Aaron D. Burns* and Charles R. Vossbrinck**

*Department of Plant Biology and **Office of Agricultural Entomology,
University of Illinois, Urbana, IL 61801

Candida albicans is an imperfect dimorphic yeast which is an opportunistic human pathogen that can cause diseases ranging from superficial to life-threatening systemic infections. Its ability to infect, however, depends upon its ability to adhere to host cells. A previous study has shown that adhesion of certain strains of *C. albicans* is made possible through a major cell wall antigen of approximately 49 kDa [1]. This protein is also found to be associated with a carbohydrate (believed to be mannose), resulting in a glycoprotein. Growing the yeast in different carbon substrates has demonstrated differences in host cell adhesion [2,3].

C. albicans (strain SC-5314) from the Bristol Myers Squibb Culture Collection (Wallingford, CT) was grown separately in media of several carbon substrates: sucrose, fructose, glucose, and mannose, at 37°C for 24 hr. Each group of cells were washed in phosphate-buffered saline and suspended in freshly isolated human buccal cells for 1 hr. During that time, counts were made every 10 min to determine an average number of adhered yeast cells/buccal cell based on counts of 100 buccal cells. Representative scanning electron micrographs were taken (Fig. 1).

Yeast cells from the treatments that showed both the greatest and the least ability for adhesion were chosen for cell wall protein isolation and immunogold labeling. The yeast cells were grown as before, but in 500 ml volumes of media for sufficient growth. It was found that sucrose-grown yeast cells were more than two times adherent to buccal cells than mannose-grown yeast (*e.g.* at 30 min. growth time there was 15.0% yeast adhesion from cells grown in sucrose media *vs.* 6.5 % from mannose). After washing, the cells were lysed in 1N sodium chloride at 60°C for 40 min, and the supernatant was collected and dialyzed to yield a cell wall extract. The concentrations of crude cell wall extracts obtained were 65.2 mg/ml from growth in mannose, 53.8 mg/ml from the growth in sucrose, and the absorbency of the isolated purified proteins (205 nm) were 0.9 from mannose and 1.1 from sucrose. The cell wall extracts were run along with standards on separate wide-combed 10% acrylamide gels and, in each run, the band corresponding to 49 kDa was cut and electroeluted. Comparisons of yeast cell wall proteins were made after growth in the two selected sugar media (Fig. 2). The isolated protein from each growth condition was concentrated by ultrafiltration and used to produce polyclonal antibodies from mice. Subsequent immunogold-labeled samples were prepared to compare relative amounts and localization of the 49 kDa major cell wall antigen from the two growth conditions of the yeast as measured against controls (*e.g.* Figs. 3,4).

References:

1. Jenq, W., C.L. Chen, C.C. Chang and R.F.E. Crang, Archiv. Microbiol. **162** (1994) 33.
2. McCourtie, J. and L.J. Douglas, J. Gen. Microbiol. **131** (1985) 495.
3. Tosh, F.D. and L.J. Douglas, Infect. and Immun. **60** (1992) 4734.
4. The authors gratefully acknowledge the support of the MICROSCOPY SOCIETY OF AMERICA through granting an undergraduate scholarship to Sara Browning for 1994-95.

Proc. Microscopy and Microanalysis 1995, edited by G.W. Bailey, M.H. Ellisman, R.A. Hennigar, and N.J. Zaluzec

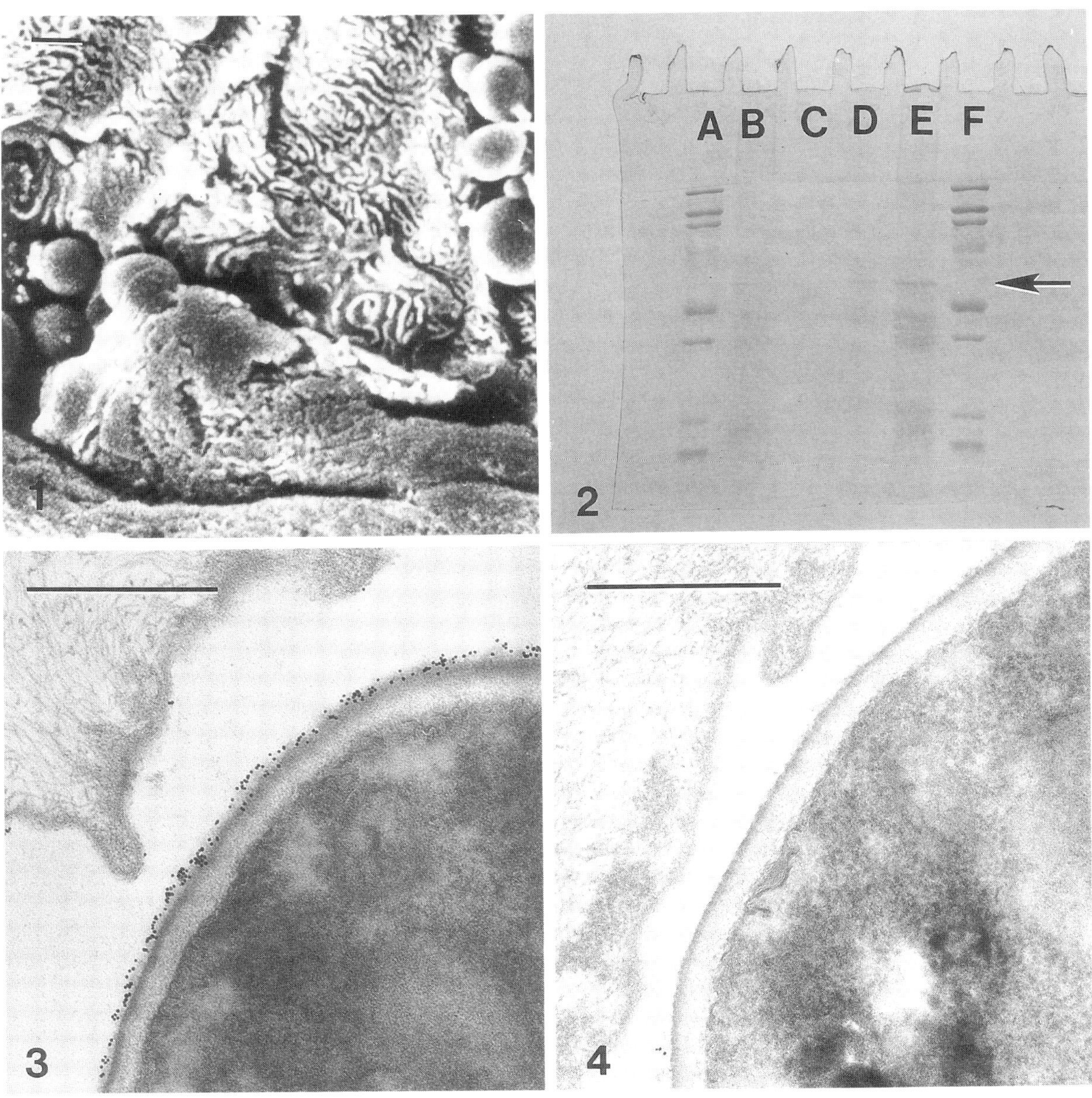

Fig. 1. Scanning electron micrograph representing adherent *C. albicans* cells on the surface of a human buccal cell after 30 min. contact in phosphate-buffered saline solution. Bar represents 1 μm. **Fig. 2.** Comparative 10% SDS/PAGE of cell wall extract. Lanes A and F represent protein weight standards. Lane B represents crude cell wall extract from growth in sucrose medium, lane C is the isolated protein from growth in sucrose. Lane D represents the isolated protein from growth in mannose medium, and lane E represents the crude cell wall extract from growth of the yeast in mannose. Arrow represents site approx. 49 kDa molecular weight. **Fig. 3.** Immunogold labeling revealing antigenic site on surface of *C. albicans* cell wall. Adjacent human buccal cell is evident. Bar represents 0.5 μm. **Fig. 4.** Same as Fig. 3, but representing control situation in which yeast cell sections were reacted with unimmunized antiserum prior to immunogold labeling. Bar represents 0.5 μm.

SUITABILITY OF ZINC-BASED BIOCERAMIC FOR REPAIR OF BONE DEFECTS

C. Taylor,* B. Mehling,* A. Longo,** D. Snead,* D. Reynolds,*
P. Bajpai** and P. Barre.*

*Wright State University School of Medicine, Dayton, Ohio 45329
**University of Dayton, Dayton, Ohio 45409

The advantages of a synthetic bone-graft substitute include reduction
of risks due to immunological rejection and transmission of diseases,
such as AIDS.[1] A satisfactory synthetic material must be non-toxic,
allow adequate re-vascularization, and be capable of bearing weight
and tolerating strain. Our experiments using a resorpable, zinc-
calcium-phosphorous-ceramic oxide (ZCAP)[2] in composite with malic
acid, vitamin E, and gentamicin sulfate, indicate that this zinc-
based bioceramic implant fulfills the requirements of an osteocon-
ductive filler between two fractured ends of rat femurs.

Assessment of the suitability of ZCAP involved the following techniques:
1) Radiological examination showing the position of the implant,
bridging of the defect with bone, and dissolution of the bioceramic.
2) Histology of serial sections indicating resorption of the ZCAP (a
few granules remain after 10 weeks) and replacement by endochondral
ossification.
3) Scanning electron microscopy revealing that the intact ZCAP first
becomes surrounded by fibrous tissue and then infiltrated by cell
processes.
4) Energy dispersive x-ray analysis of zinc confirming its disappearance during the resorption of the implant.

ZCAP bioceramic is suitable for repair of fractured femurs in rats
when optimal pore size (500-600 um), specific particle size (45-63
um), and temporary immobilization pins are used. Infusion of the
composite with testosterone or bone morphogenic protein[3] may increase
the rate of bone growth.

References

1. M.B. Habal, Biomater. Med. Devices Artif. Organs 7(1979)229.
2. J.R. Cooper et al., Fourth World Biomaterials Congress (1992)544.
3. S.Stevenson et al., JBJS 76-A(1994)1676-87.

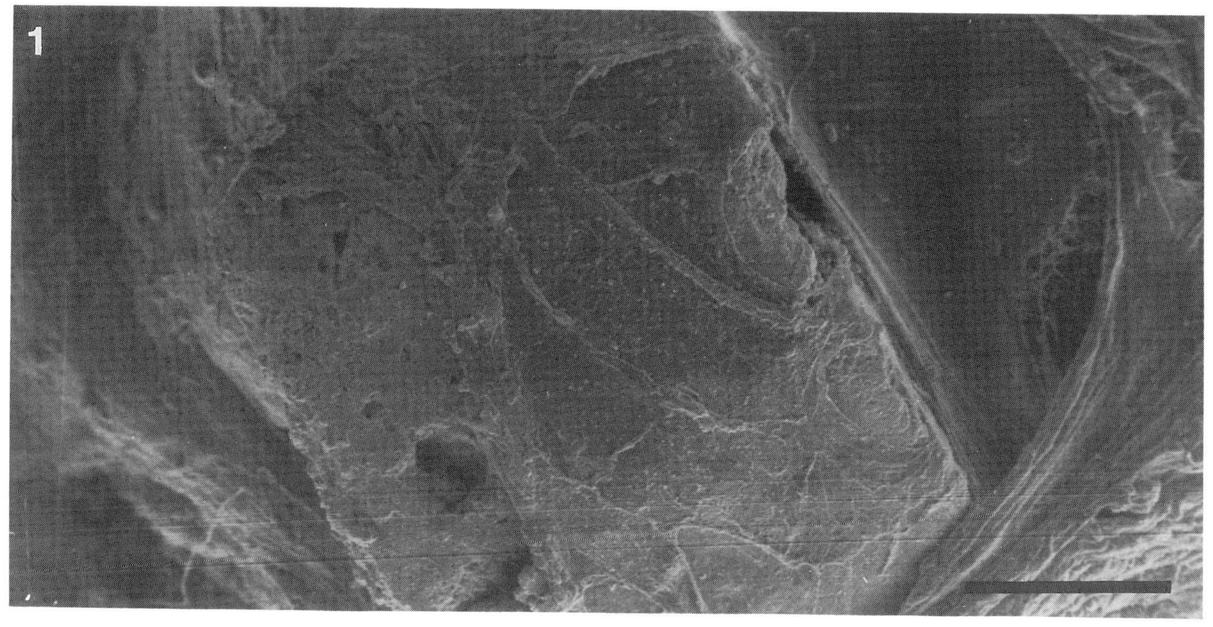

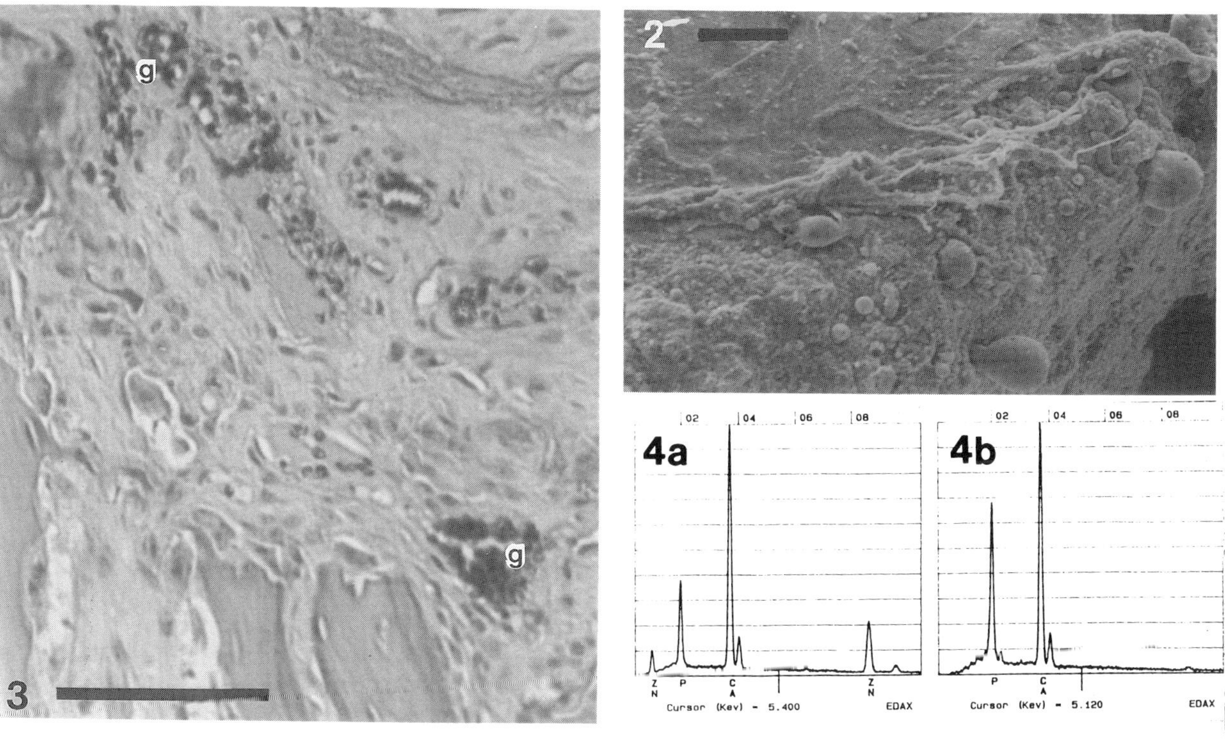

FIG.1--SEM of ZCAP implant surrounded by fibrous tissue. Bar=1000 um.
FIG.2--SEM of ZCAP implant infiltrated by cell processes. Bar=100 um.
FIG.3--Histology of growing bone and remaining ZCAP granules at 10 weeks post-implantation. Bar=100 um.
FIG.4--Semiquantitative electron x-ray analysis of ZCAP ą)before and b)10 weeks after implantation.

THE GEOMETRY OF THE EJSR Z-RETE IN AVIAN CARDIAC MUSCLE

J.R. Sommer, T. High, and I. Taylor

Departments of Pathology, Duke University and V A Medical Centers, Durham, N.C., 27710

The junctional SRs (JSR) in skeletal and cardiac muscle in general, including extended junctional SR (EJSR) in avian hearts, are morphologic and functional homologues[1,2,3] as shown by biochemical and morphometric evidence, including the recent demonstration that EJSR contains ryanodine receptors and binds [^{3}H]-Ryanodine.[4] Bird hearts have two aberrant membrane features: 1. absence of transverse tu-bules (TT) and, 2. extension of JSRs (thus: E-JSR) from peripheral couplings into the interior of the myocytes forming EJSR Z-retes that both surround and pervade Z-discs.[1,2,3] Absence of TT calls for an alternative mechanism for global muscle activation in bird hearts for which EJSR may provide the anatomical substratum, a possibility supported by both experimental evidence,[5,6] and recent theoretical considerations.[7] Propagated calcium-induced calcium release (CICR) has been difficult to reconcile with the graded response of cardiac muscle to stimulation.[2,cf.5] Propagation of an action potential along the Z-retes can be excluded.[3] In the absence of TT, saltatory CICR as an alternative to propagated cardiac activation, is made plausible by the existence of EJSR which, (a) is demonstrably calcium-sensitive, (b) is well within < than 0.5µm of the regulatory proteins of cardiac myocytes and, (c) carries rows of proven, vicinal calcium release channels. JSRs extending from couplings as EJSRs, and the presence of only one type of ryanodine receptor in birds,[4] intimates identical mechanisms for the initial CICR at peripheral couplings (one step) and its saltatory propagation (two step) through EJSRs along Z-discs for global cardiac activation. Timely propagation is facilitated in avian cardiac myocytes by their small diameters (finch: ≈7µm, mouse: ≈15µm). **Methods**: to measure, in longitudinal and transverse EM sections, distances between (a) vicinal JSRs in mouse and, (b) vicinal EJSRs in avian cardiac muscle. **Findings:** There are virtually contiunuous 3D-nets (Z-retes[1,3]) of EJSR through/around Z-discs across an entire cell, most often in connected parallel rows, with free SR segments of from <<0.1 µm to ≈ 0.3 µm (which is also ≈ width of the convoluted Z-retes' meshes) between JSR/EJSR. JSRs of couplings in mouse hearts do not extend through/around Z-discs; thus, inter-JSR distances in transverse direction are determined by the contractile material spread between interior couplings at TT (width between 0.5 and >1µm). Longitudinally, JSRs and EJSRs are separated by sarcomere lengths. **Conclusions:** A hypothetical propagated, quasi saltatory CICR for global activation of avian hearts could be supported by the geometry of avian EJSR Z-retes. Corbular SR,[3] present mainly in cardiac cells specialized for conduction, not contraction, seems atavistic by comparison, both anatomically and functionally.

References:

1. Jewett, P. et al. *J.Cell Biol.* 49(1971)50
2. Anderson, P.A.W. et al. *J. Mol. Cell. Cardiol.* 8(1976)123
3. Sommer, J.R. et al., *Acta Physiol. Scand.* 142, Suppl.599(1991)5
4. Junker, J. et al. *J. Biol. Chem.* 269(1994)1627
5. Fabiato, A. *J. Gen. Physiol.* 85(1985)189
6. Allbritton, N.L., Meyer, T. & Stryer, L., *Science* 258(1992)1812
7. Stern, M.D. *Biophys. J.* 63(1992)497
8. Supported by NIH grant HL12486 and the VA Research Service

Proc. Microscopy and Microanalysis 1995, edited by G.W. Bailey, M.H. Ellisman, R.A. Hennigar, and N.J. Zaluzec
Copyright © 1995 MSA. Published by Jones and Begell Publishing, 79 Madison Ave., New York, NY 10016

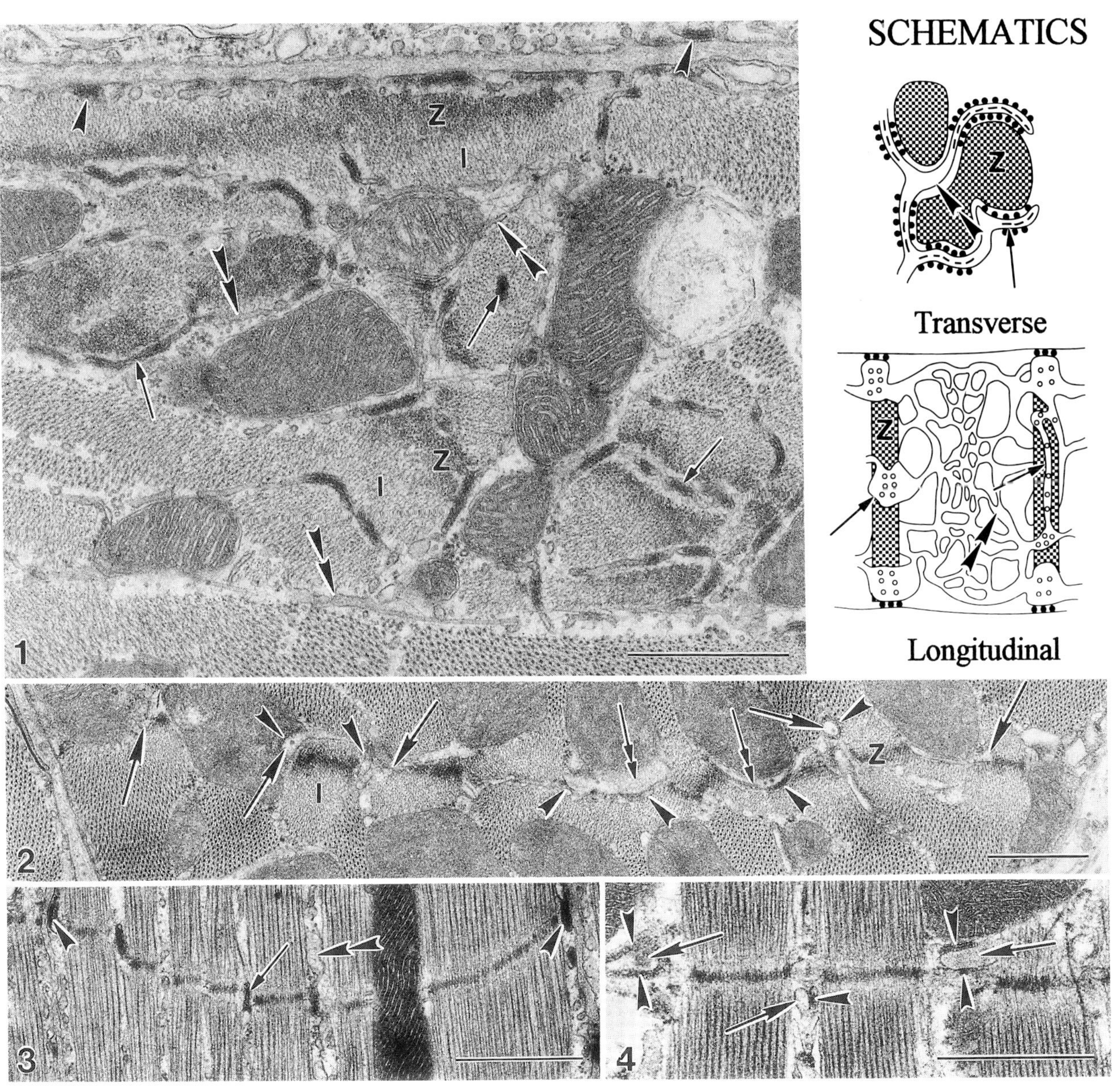

FIG. 1 - Finch cardiac muscle. Transverse section through Z-disc (Z) and adjacent I bands (I). EJSR-rete (thin arrows). Couplings (arrowheads). Free SR (double arrowheads). EJSR both surrounds and penetrates Z-disc. Bar: 1.0μm. Inset: Schematic geometry of EJSR-retes. Labels same in all figures.

FIG. 2 - Mouse cardiac muscle. Transverse section through Z-disc and adjacent I bands. Distances between transverse tubules (T<large arrows>; T cut longitudinally<double arrow>) with JSR at couplings are much farther apart than individual segments of the continuous EJSR-retes in finch hearts. Both T and EJSR run a tortuous course, often hidden outside the plane of sectioning. Bar: 1.0μm

FIG. 3 - Finch cardiac muscle. Longitudinal section. EJSR-retes in mainly tangential distribution around Z-disc. Interfibrillary distances comparable to those found in mouse cardiac muscle. Bar: 1.0μm

FIG. 4 - Mouse cardiac muscle. Longitudinal section. T in interfibrillary spaces bearing interior couplings (arrowheads) whose JSRs do not extend around/through Z-discs. Bar: 1.0μm

SUBSTRUCTURAL FEATURES OF LLC-PK$_1$ PORCINE KIDNEY CELLS GROWN ON
ANOCELL FILTER SUPPORTS

A.J.Mia*, S.Johnson*, J. Franklin*, J.F.Johnson*, K.Puttaparthi*, L.X.Oakford** and T.Yorio**

*Jarvis Christian College, Hawkins, Texas 75765
**UNTHSC at Fort Worth, Fort Worth, Texas 76107.

LLC-PK$_1$ cells, derived from porcine kidney cortex, have been used for a variety of studies including sodium-coupled hexose transport, vasopressin/calcitonin sensitive adenylate cyclase (1) and vasopressin V$_1$ and V$_2$ receptor-mediated endocytosis involving clathrin coated pits and vesicles (2,3,4,5). Recently, we reported on the morphological profile of these cells using the techniques of scanning electron microscopy (SEM). SEM studies of confluent LLC-PK$_1$ cells exhibited surface morphology similar to intact tissue. These cells also responded to ADH (antidiuretic hormone, vasopressin) or MZ (mezerein), a non-phorbol activator of protein kinase C (PKC) with a propagation of numerous microvilli over the apical plasma membrane typical of ADH responsive tissue.   The current study evaluates the cyto-morphological features of cultured LLC-PK$_1$ cells following hormone addition.

LLC- PK$_1$ cells were grown on Anocell filter supports using DMEM media in a 5% CO$_2$/air environment in an incubator at 37° C. Upon confluency, cultured cells were washed in buffer and then treated with 100 mU/ml ADH or MZ10^{-6}M for 15 min.   Control and stimulated tissues were fixed in 2% glutaraldehyde in PIPES with a postfixation in 1% osmium tetroxide for 1 hr for either SEM or TEM.

SEM observations of confluent LLC-PK$_1$ cells on Anocell filters are presented in Fig. 1. The presence of scattered microridges over the apical plasma membrane, with no well defined cellular junctions between cells are depicted (Fig. 1). Ultrathin sections across cellular junctions revealed cellular tight junctions (Fig. 2, t) and diffused desmosomes. The presence of typical clathrin coating (arrows) and the formation of clathrin coated pits (Fig. 2, cp) from the plasma membrane were observed. Presence of clathrin coated pits and vesicles was reported in the dissociated LLC-PK$_1$ cells, involving ADH receptor-mediated endocytosis (2,3,4). Microtubules were often cited in close proximity of the coated pits and vesicles (Figs. 2,3,4, m) suggesting their possible role in the translocation of clathrin coated vesicles. Coated pits (cp) were seen to form not only from the apical plasma membranes (Figs. 2,4) but also from the basolateral (Fig. 3), basal plasma membranes (Fig. 5). Occasionally, a nucleolus may appear in close contact with the inner nuclear membrane (Fig. 6, arrow) as seen in ultrathin sections. The significance of their relationship is not known. These cells appear to be useful models for studying the processes involved in the formation of clathrin coated pits and vesicles particularly in response to ADH.

References
1. J. S. Handler. J. Exp. Biol. (1983) 106:55-69.
2. W. Lutz et al., j. Biol. Chem. (1990) 87: 6507-6511.
3. W. Lutz et al., J. Biol. Chem. (1992) 267:1109-1115.
4. B. Hocher et al., Biochem. Biophys. Res. Comm. (1992) 3:1376-1383.
5. A. J. Mia et al. FASEB J. (1995), in press.
6. Supported by grants: S14 GM48114 and DAAL03-92-G-0075.

Proc. Microscopy and Microanalysis 1995, edited by G.W. Bailey, M.H. Ellisman, R.A. Hennigar, and N.J. Zaluzec
Copyright © 1995 MSA. Published by Jones and Begell Publishing, 79 Madison Ave., New York, NY 10016

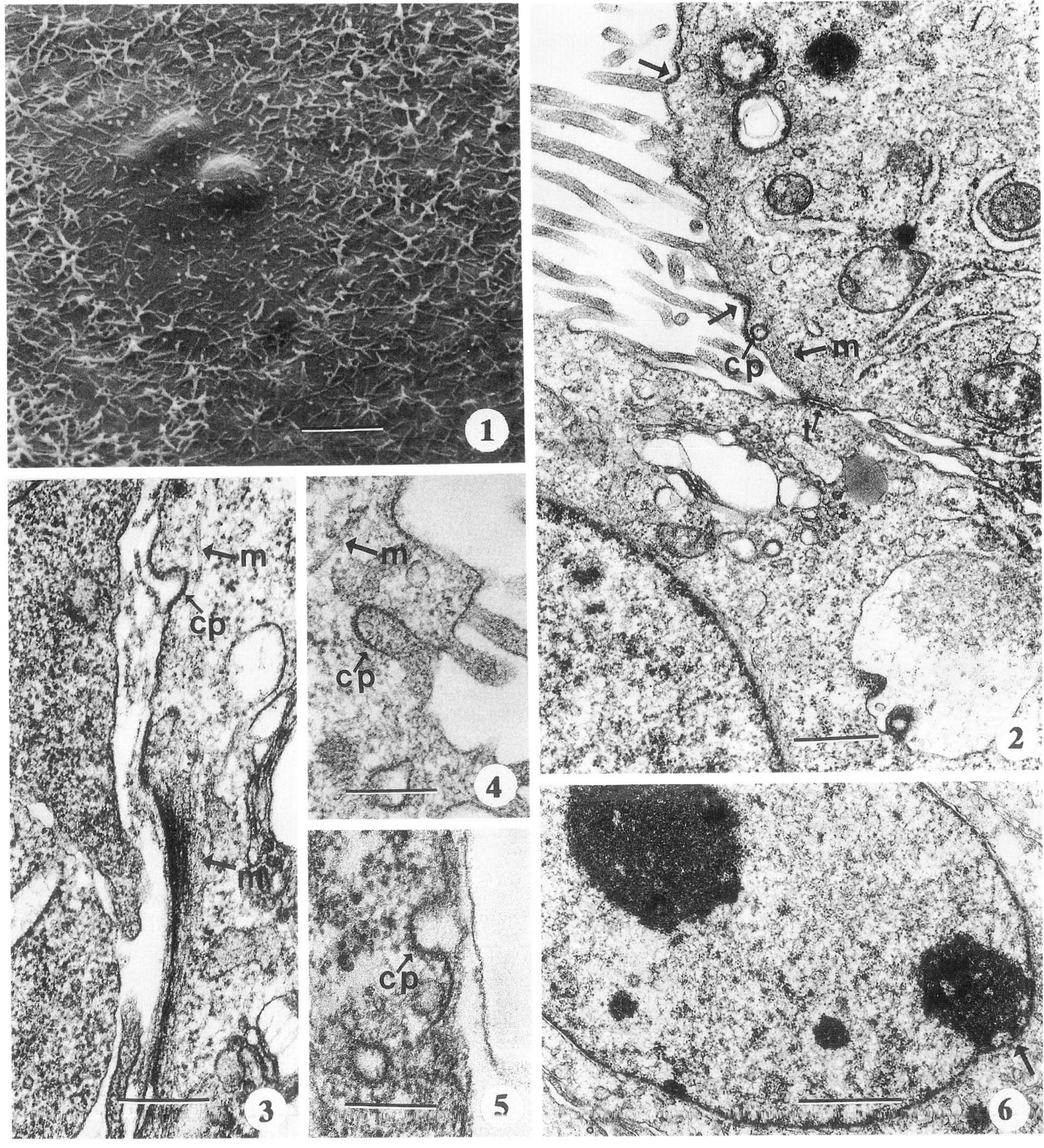

See text for detailed description of the figures.

Fig.1.SEM of several LLC-PK$_1$ cells showing microridges, nucleoli and no distinct cell junctions. Bar= 5μm.

Fig.2.TEM section shows clathrin coated apical membrane and the cytoplasmic structures. Bar= 0.5μm.

Fig.3.TEM shows a clathrin-coated pit on basolateral membrane and golgi bodies. Bar= 0.5μm.

Fig.4.TEM section shows apical membrane with a coated pit, a coated vesicle and a microtubule.
Bar= 250nm.

Fig.5. TEM section shows basal plasma membrane with a coated pit and a coated vesicle. Bar= 250nm.

Fig.6. TEM through the nucleus showing a nucleolus attached to the inner muclear membrane. Bar= 2μm.

MICROSTRUCTURAL MORPHOLOGY OF SPHINGOLIPID DISPERSIONS AS INVESTIGATED BY FREEZE-FRACTURE ELECTRON MICROSCOPY

Vitthal S. Kulkarni, Wayne H. Anderson, and Rhoderick E. Brown

The Hormel Institute, University of Minnesota, 801 16th Avenue NE, Austin, MN 55912

The biological significance of the sphingomyelins (SM) and monoglycosylated sphingolipids like galactosylceramides (GalCer) are well documented[1,2]. Our recent investigation showed tubular bilayers[3] in the aqueous dispersions of N-nervonoyl GalCer [N-(24:1$^{\Delta 15,cis}$) GalCer] (a major fatty acyl moiety of natural GalCer). To determine the influence of lipid head groups on the resulting mesophasic morphology, we investigated microstructural self-assemblies of N-nervonoyl-SM [N-(24:1$^{\Delta 15,cis}$) SM; the second most abundant sphingomyelin in mammalian cell membranes], 1-palmitoyl-2-nervonoyl phosphatidylcholine [PNPC] (the lipid species with the same acyl chain configuration as in N-(24:1) GalCer) and also compared it with egg-SM by freeze-fracture EM.

Procedures for synthesizing and purifying N-(24:1) GalCer, N-(24:1) SM, and PNPC have been reported[4]. Egg-SM was purchased from Avanti Polar Lipids, Alabaster AL. All lipids were >99% pure as checked by thin layer chromatography. Lipid dispersions were prepared by hydrating dry lipid with phosphate buffer (pH 6.6) at 80-90°C (3-5 min), vigorously vortexing (1 min) and repeating this procedure for three times prior to three freeze-thaw cycles. The samples were rapidly cryofixed, from the gel state (determined by microcalorimetry[4]), in liquid propane cooled by liquid nitrogen and fractured in Balzer 300 apparatus at -120°C followed by coating with platinum-carbon at 45° angle and by carbon. The replicas were washed in chloroform/methanol (2/1; v/v) and observed in JEOL 100-S transmission electron microscope.

Fig. 1A shows that N-(24:1) GalCer formed tubular bilayers with average diameter of 30 nm (±5) and the length varied from 5-10μm. The N-(24:1) SM formed multilamellar spherical vesicles with a peculiar surface morphology that resembled multi-faceted, zig-zagging, ridge-like structures (Fig.1B). Egg-SM which has different N-fatty acyl chains (80% 16:0, 5% 18:0; 3.9% 24:1) formed spherical vesicles with rippled surface morphology (Fig.1C). Previous studies by Hui et al. have also shown rippled vesicle surface morphology for bovine brain SM[5]. In contrast PNPC showed spherical vesicles with smooth surface morphology as seen in Fig.1D.

This study shows that a change in lipid head group from galactose (as in GalCer) to phosphatidylcholine (as in SM and PC) dramatically affects the morphology of the microstructures. Comparison between N-(24:1) SM and egg-SM also indicates that the molecular packing which is strongly influenced by the fatty acyl chains significantly alters the surface morphology of the lipid dispersions. Efforts aimed at determining the role that intermolecular hydrogen bonding plays in bilayer tubule formation are underway.

[Supported by USPHS grant GM 45928 and the Hormel Foundation]

References:

1. Smaby et al. Biochemistry 33, 9135-9142 (1994)
2. Brown et al. Biophys. J. 68(4), in press (1995)
3. Kulkarni V. S. et al. Microscopy and Analysis January 1995, pp. 13-15
4. Kulkarni V. S. et al. Biophys. J. in preparation 1995
5. Hui S-W et al. Biochim. Biophys. Acta 601, 271-281 (1980)

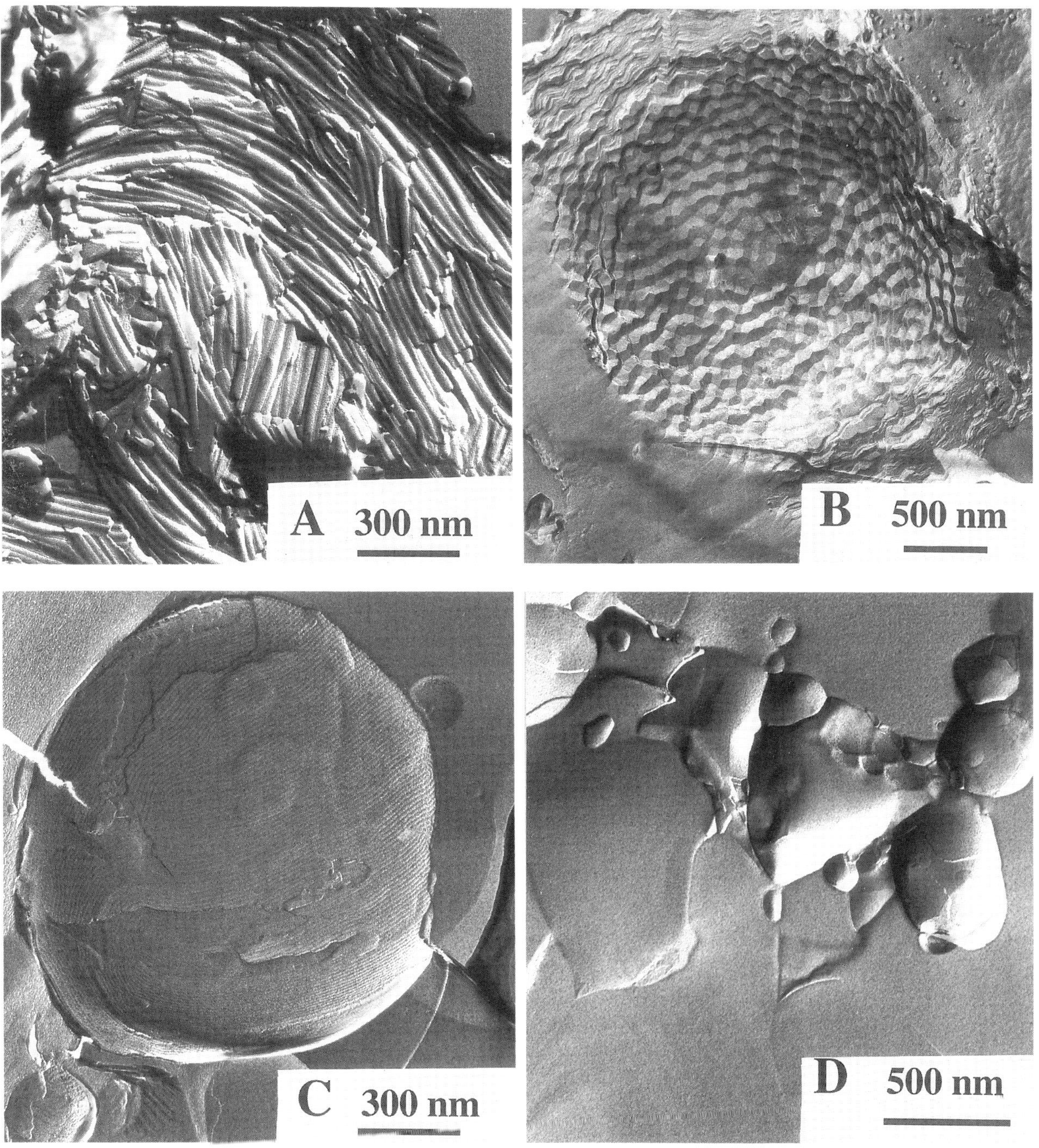

Fig.1 Freeze-fracture electron micrographs showing: A: tubular bilayers in N-(24:1) GalCer, cryofixed from 20°C (phase transition temperature T_m=59.3°C); B: spherical vesicles with zig-zag ridges in N-(24:1) SM, cryofixed from 10°C (T_m=22.3°C); C: spherical vesicles with ripples in egg-SM, cryofixed from 20°C (T_m=39°C); D: smooth spherical vesicles in PNPC, quenched from 10°C (T_m=27.7°C).

IMMUNOCYTOCHEMICAL STUDIES ON THE RENAL FATE OF CIRCULATING ADVANCED GLYCATED PRODUCTS

A. Gugliucci and M. Bendayan.

Department of Anatomy, University of Montreal, Montreal, Quebec, Canada, H3C 3J7.

A growing body of evidence points to a causal relationship between advanced glycation end products (AGE) accumulation in tissues and the development of chronic diabetic complications (1). Particularly impressive are data showing a dramatic increase in the levels of circulating AGE peptides in diabetic end stage renal disease that correlates with the severity of the nephropathy (2). We have recently shown that AGE-BSA probes tagged with colloidal gold display great reactivity towards glomerular structures (3). In the same way, a recent study has shown that chronic injection of AGE-BSA but not BSA produces the classical picture of diabetic glomerulopathy in otherwise euglycemic animals (4). It is apparent from these and other studies that small metabolic fragments of AGE proteins are implicated in these phenomena. It is certain that the kidneys play an important role in the clearance of circulating AGE products from the bloodstream. However, so far little is known about the fate of AGE proteins and AGE peptides in renal tissue. In order to get further insight on this issue we undertook the present sudy. AGE-bovine serum albumin (AGE-BSA) and AGE peptides were prepared and injected to otherwise normal rats. Using colloidal gold post embedding immunoelectron microscopy an anti-AGE antibody generated in our laboratory, we followed, in a short term kinetic study, the cellular and subcellular localisation of circulating AGE products throughout the nephron. AGE peptides or AGE-BSA (10 mg/animal) were injected i.v to Sprague Dawley rats. After 15, 30 or 45 minutes of circulation the animals were killed by terminal anaesthesia, small pieces of kidney were excised and fixed by immersion in 1 % glutaraldehyde solution for 2 h at 4°C. The tissue fragments were then embedded in Lowicryl K4M at -20°C. Protein A gold immunocytochemistry was performed on thin sections of renal tissues (5).

The anti-AGE antibodies employed were prepared by immunization of rabbits with AGE-BSA and were affinity purified. They were characterized by dot blotting and double diffusion on agarose. They reacted with AGE-BSA, AGE-Hb and AGE-IgG but not against BSA, Hb or IgG. Reaction with AGE-BSA could be abolished by preincubation with AGE-Hb or AGE IgG and viceversa. The distribution of AGE-BSA over the glomerular basement membrane is shown in figure 1a. When injected to normal rats most of AGE-BSA was found in the lumen of capillary vessels and distributed along the endothelial side of the glomerular basement membrane. Some AGE-BSA is filtered and reabsorbed by the proximal tubules (Fig 1b). Presence on mesangial matrix was also apparent. A short term follow-up of AGE peptides localisation at the kidney level has shown that they are easily filtered and that they are actively reabsorbed by the proximal convoluted tubule. At 15 min little labeling is found in the glomerulus. Instead, the labeling is present in the urinary space and microvilli (Fig 1c). Early endosomes display intense labeling as well. At 45 min, late endosomes and lysosomes add to the pattern of labeling (Fig 1d). The distal tubules are devoid of labeling for any of the intervals studied.These observations point out to a participation of the endo-lysosomial apparatus of the proximal convoluted tubule to the disposal of AGE peptides.

In conclusion, our data confirm the clearance role of the kidney as regards to AGE products, give ultrastructural evidence for the mesangial accumulation of AGE-BSA and demonstrate for the first time that filtered AGE peptides are readily reabsorbed by proximal convoluted tubular cells.

References

1. Bucala, R. & Cerami, A. *Adv Pharmacol* **23**, 1-19 (1992).
2. Makita, Z., Bucala, R., *et al. Lancet* **343**, 1519-1522 (1994).
3. Gugliucci, A. & Bendayan, M. *J Histochem Cytochem* (1995) in press.
4. Vlassara, H., Striker, L.J., *et al. Proc Natl Acad Sci USA* **91**, 11704-11708 (1994).
5. Bendayan, M. *J Elec Micros Tec* **1**, 243-270. (1984).

Proc. Microscopy and Microanalysis 1995, edited by G.W. Bailey, M.H. Ellisman, R.A. Hennigar, and N.J. Zaluzec

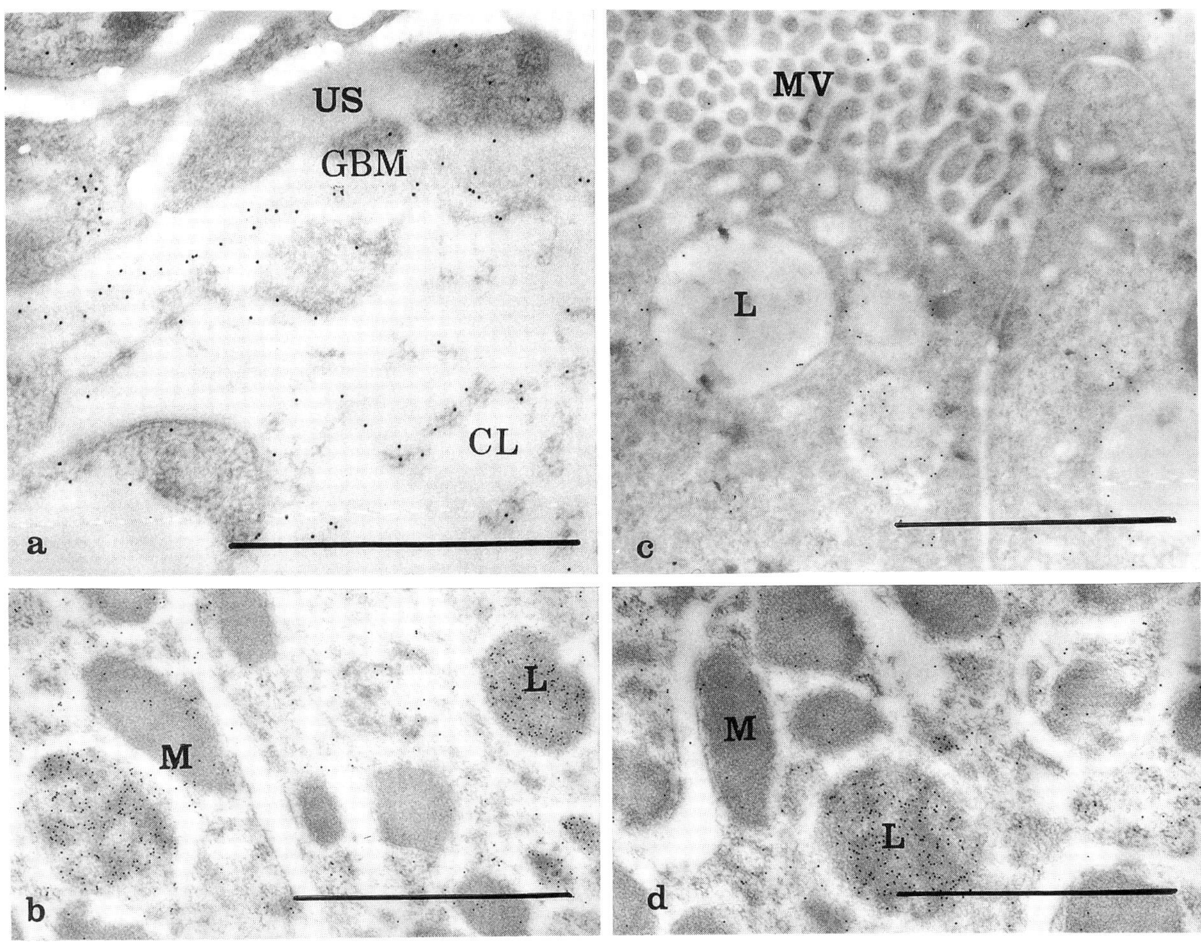

Figure 1. Detection of circulating injected AGE products in renal tissue by protein A-gold immunocytochemistry.

Renal tissue sections from a normal rat, injected respectively with AGE-BSA (**a** and **b**) or AGE peptides (**c** and **d**).
The probes were injected in the inferior vena cava and the animals were sacrifized after different intervals up to 45 min. After 45 min AGE-BSA was found in the lumen of capillary vessels and distributed along the endothelial side of the glomerular basement membrane (a). Some AGE-BSA is filtered and reabsorbed by the proximal tubules (b). At 15 min AGE peptides were found in the urinary space and microvilli, but mainly in early endosomes and lysosomes which display intense labeling.(c). At 45 min, late endosomes and more lysosomes add to the pattern of labeling (d). CL, capillary lumen; GBM, glomerular basement membrane; US, urinary space; L, lysosome; MV, microvilli; M, mitochondria. Bars=1μm.

RENAL HANDLING OF NATIVE AND GLYCATED ALBUMINS BY THE NORMAL MOUSE

I. Londoño and M. Bendayan

Dept. d'anatomie, Université de Montréal, Montréal, Québec, Canada H3C 3J7

One of the major pathophysiological sequela of long-term hyperglycemia is the increased glycation of proteins. The glycation reaction consists in the covalent attachment of glucose molecules to amino groups in proteins forming a stable product, the Amadori adduct. Albeit further transformations can occur,[1] Amadori adducts constitute the predominant form of circulating glycated proteins in normal and diabetic subjects, and seem to be directly involved in the pathogenesis of diabetic nephropathy.[2]

In an attempt to determine the role of glycated proteins in glomerular hyperfiltration and proteinuria characteristics of diabetic nephropathy, we have evaluated the renal handling of glycated albumins by the normal mouse, at two levels: the glomerular filtration and the tubular epithelial reabsorption. Bovine serum albumin (BSA) was glycated *in vitro* within conditions allowing for the formation of Amadori products.[3] Both native and glycated albumins were tagged to haptens and intravenously injected in mice. Upon 10 and 30 min of circulation, the renal cortex was fixed and processed for immunocytochemical labeling which combined anti-hapten antibodies and protein A-gold complexes.[4]

Labeling of native BSA over the glomerular basement membrane (GBM) (Fig. 1a) showed a preferential distribution towards the endothelial side, similar to the one observed for endogenous albumin.[5] The morphometric evaluation of the labeling confirmed this result, showing a peak of labeling in the endothelial side of the GBM (Fig. 1b). In the case of glycated albumin, labeling was found throughout the GBM, gold particles being also present in the urinary space (Fig. 2a). The labeling distribution (Fig. 2b) revealed a shift towards the epithelial side, suggesting a further penetration in the GBM. When native and glycated albumin were co-injected, a shift in the labeling distribution was also observed for native BSA, indicating that glycated BSA, when present, traverses the GBM towards the urinary space and induces the passage of the native protein.

At the level of the proximal tubule epithelium, both native and glycated albumin were found in the endocytic compartments (Fig. 3). However, quantitative determinations revealed differences in labeling densities ratios for native and glycated albumin when comparing assorted compartments. Higher proportions of glycated BSA were found in the urinary space, confirming results from the GBM labeling distributions. Conversely, higher proportions of native BSA were detected in the endocytic compartments of the epithelial cells. These results suggest an increased passage of glycated BSA through the GBM with a preferential absorption of the native one by the proximal tubules.

In conclusion, the present study suggests that, upon glycation of serum proteins, (a) the properties of the normal mouse glomerular wall are altered allowing for the passage of native and glycated proteins, and that native serum albumin appears to be reabsorbed preferentially to the glycated one. These results highlight the role of early glycation products in the onset of renal dysfunction during diabetes.

References
1. Monnier VM et al. *Diabetes* 41 (Suppl.2)(1992)36
2. Ziyadeh FN and Cohen MP. *Mol. Cell. Biochem.* 125(1993)19
3. Murtiashaw MH and Winterhalter KH. *Diabetologia* 29(1986)366
4. Ghitescu L and Bendayan M. *Microsc. Res. Techn.* 22(1992)392
5. Londoño I and Bendayan M. *Am. J. Physiol. (1995) (in press)*
6. The authors gratefully acknowledge the assistance of M. Jean Godbout, M. Gaétan Mayer and M. Jean Leveillé. This work has been supported by a grant from the MRC of Canada

Proc. Microscopy and Microanalysis 1995, edited by G.W. Bailey, M.H. Ellisman, R.A. Hennigar, and N.J. Zaluzec
Copyright © 1995 MSA. Published by Jones and Begell Publishing, 79 Madison Ave., New York, NY 10016

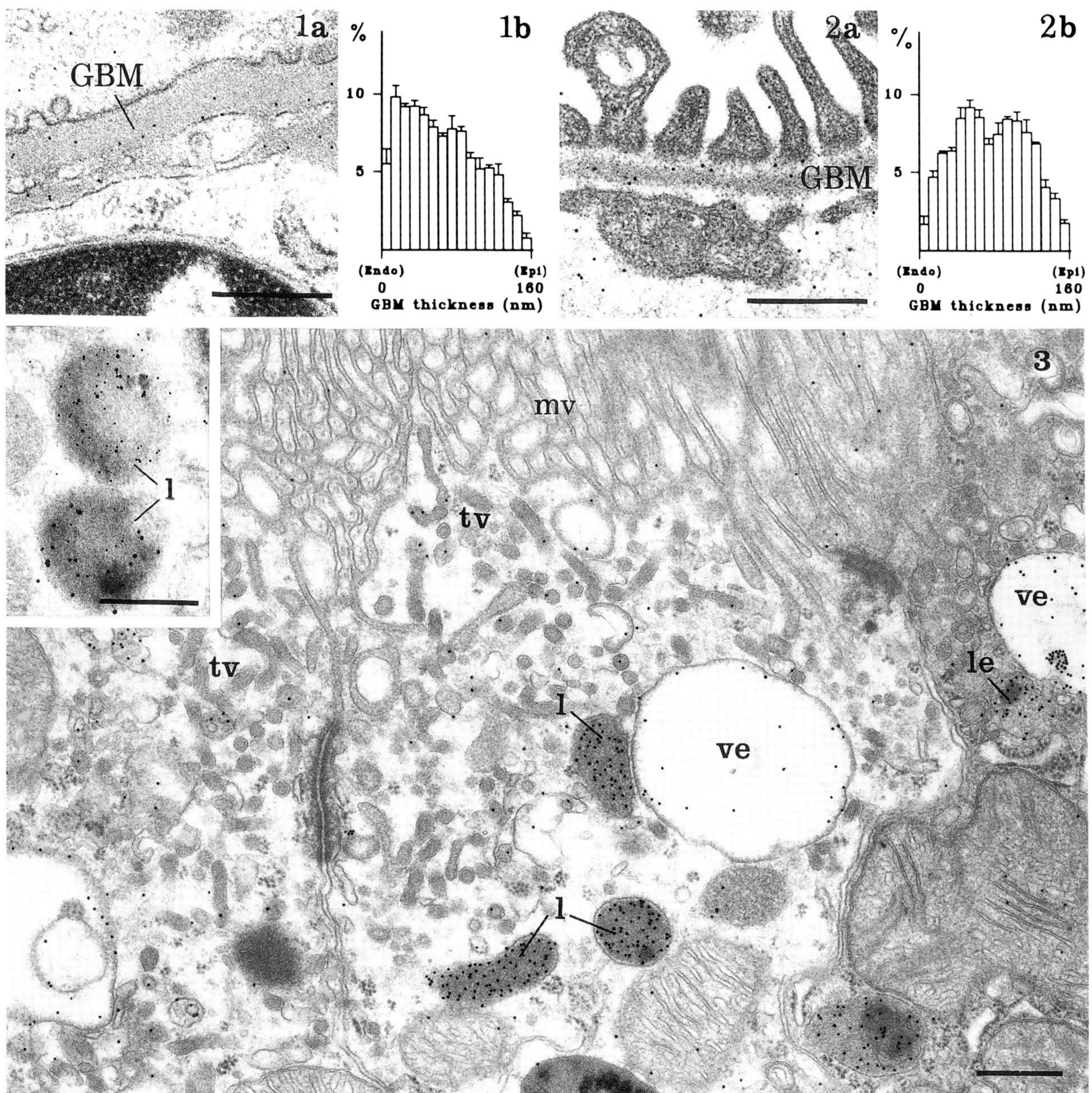

FIG. 1.- Localization of native BSA in the glomerular basement membrane (GBM). (a) Gold particles are present within the capillary lumen and over the GBM, mostly in the endothelial side. This is confirmed by the morphometrical evaluation (b). %, percent of labeling.

FIG. 2.- Localization of glycated BSA in the GBM. (a) Gold particles are located throughout the GBM. Compared with native BSA, a shift in the labeling distribution towards the epithelial side is observed for glycated BSA(b).

FIG. 3.- Localization of native BSA in the proximal tubule epithelial cell. Gold particles are distributed over different endocytic compartments: tubulo-vesicular endosomes (tv); vacuolar endosomes (ve); late endosomes (le) and lysosomes (l). mv, microvilli. Inset: double labeling for native (15 nm particles) and glycated BSA (10 nm particles); both albumins co-localized in the same endocytic compartments. Bars = 0.5 μm

ISOFORM PMCA3 OF THE PLASMA MEMBRANE CALCIUM PUMP (ATPASE) IS
LOCALIZED IN THE INNER PLEXIFORM LAYER OF RAT RETINA

M. C. Antonelli,* T. J. Eakin** and W.L. Stahl** ***

*Instituto de Quimica y Fisicoquimica Biologicas (UBA-CONICET), Universidad de Buenos Aires,
Buenos Aires, Argentina
**Veterans Affairs Medical Center, Seattle, WA 98108
***Departments of Medicine(Neurology) and Physiology and Biophysics, University of Washington
School of Medicine, Seattle, WA 98195

Regulation of cytosolic Ca^{2+} is of critical importance for maintaining normal cellular function in the
nervous system. Control of calcium homeostasis involves the Ca^{2+}-ATPase of plasma membrane
(PMCA)[1] and other transporters. The presence of these regulatory molecules may not be a fixed
characteristic of cells in a general sense. Rather, each class of cells may utilize a unique combination
of transporters to maintain homeostatic control. In addition, unique distributions of transport
molecules and their isoforms may exist in individual parts of cells imparting unique transport
signatures to specific cell types. PMCA is a major regulator of calcium transport under normal
physiological conditions. In order to understand localization of PMCA isoforms in specific cells we
have localized specific PMCA mRNAs by in situ hybridization[1] and translated PMCA protein by
immunocytochemistry.[2,3]   The present work is focused on localization of PMCA3, an isoform
predominantly known to exist in brain and skeletal muscle, but not previously investigated in retina.
We previously showed that mRNA encoding rat plasma membrane Ca^{2+}-ATPase isoform PMCA3
was localized in the granule cell layer of the cerebellum. PMCA3-like immunoreactive sites were
found in the granule cell and molecular layers of rat cerebellum and the pattern of staining suggests
that immunoreactive sites are associated with granule cell processes.

As an initial step for identifying the role of PMCA transporters in retina we have used site-directed
antibodies to the PMCA3 isoform of the Ca^{2+}-ATPase. The laminar organization of the mammalian
retina provides an especially suitable model for studying the cellular localization and of PMCA3 (Fig.
1a). As shown in Figure 1b prominent immunoreactivity was found in the inner plexiform layer, a
region containing synaptic contacts between bipolar, amacrine and ganglion cells. Immunoreactivity
was not associated with lamina containing photoreceptor cells or ganglion cell bodies. In a control
tissue section (Fig. 1c) little immunoreactivity was found except for non-specific binding in the
photoreceptor outer segments. In the inner plexiform layer at least two strata contained PMCA3-like
immunoreactive sites were found (Fig. 1b). Other studies[4] suggest that these strata contain γ-
aminobutyric acid-positive processes of amacrine cells and we hypothesize that amacrine presynaptic
processes are the major sites containing PMCA3 protein. Further immunocytochemical work with
selected cell markers and in situ hybridization studies to localize PMCA3 mRNA will be needed for
verification. This new finding of PMCA3 immunoreactivity in the inner plexiform layer, probably in
amacrine cell processes, may provide a useful approach for study of the function of this isoform of
the plasma membrane calcium ATPase in retina[5].

References

1. W. L. Stahl, T. J. Eakin, J.W.M. Owens, Jr. et al., *Mol. Brain Res.* 16(1992)223.
2. W. L. Stahl, T. P. Keeton, and T. J. Eakin, *Neuroscience Lett.* 178(1994)267.
3. T. J. Eakin, M. C. Antonelli, E. L. Malchiodi et al., *Brain Res.*, in press.
4. A. H. Milam, A.M. de Leeuw, V.P. Gaur et al., *J. Comp. Neurol.* 296(1990)123.
5. This research was supported in part by the Department of Veterans Affairs(WLS), and by grants
from the NIH (NS 20482 to WLS), the National Science Foundation(WLS), and CONICET (MCA).

Proc. Microscopy and Microanalysis 1995, edited by G.W. Bailey, M.H. Ellisman, R.A. Hennigar, and N.J. Zaluzec
Copyright © 1995 MSA. Published by Jones and Begell Publishing, 79 Madison Ave., New York, NY 10016

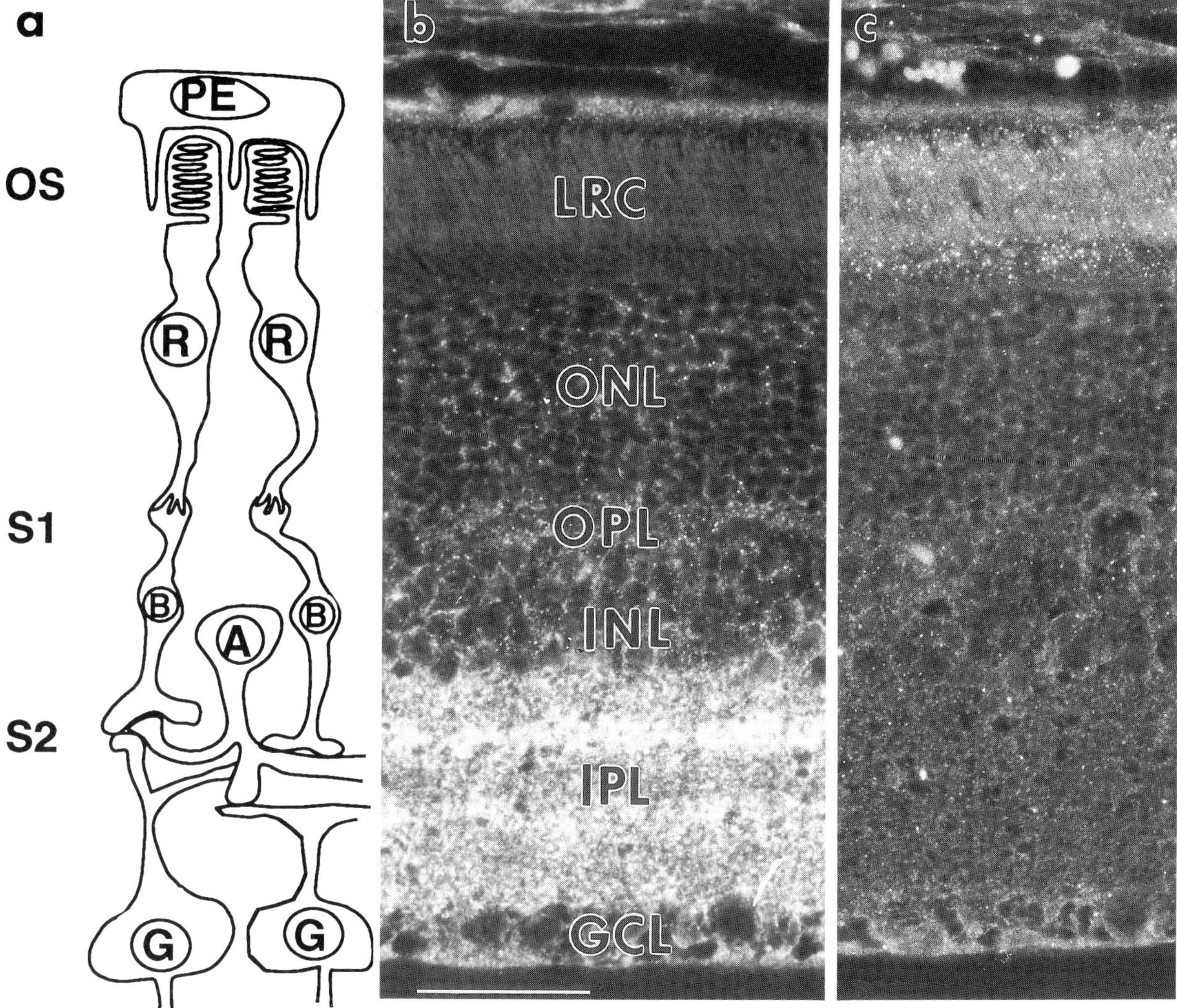

FIG. 1. (a) Drawing of retina showing relative positions of pigment epithelium, photoreceptor, bipolar, amacrine and ganglion cells. Horizontal and Müller cells are not shown. (b) Micrograph showing immunofluorescent localization of PMCA3 in the inner plexiform layer of rat retina. Paraffin tissue sections were incubated with affinity purified rabbit anti-PMCA3 antibodies[3] (anti-PMCA3$_{1142-1159}$ corresponds to a C-terminal 18 amino acid sequence of PMCA3; primary antibody was diluted 1:10,000); these were incubated with 1:100 dilution of goat anti-rabbit IgG(H+L) conjugated with Cy3 (Jackson ImmunoResearch, West Grove, PA); (c) Micrograph showing control immunoreactivity with 1-10,000 dilution of non-immune rabbit serum. Bar = 50 μm.

Abbreviations: A, nucleus of amacrine cell; B, nucleus of bipolar cell; G, nucleus of ganglion cell; GCL, ganglion cell layer; INL, inner nuclear layer; IPL, inner plexiform layer; LRC, layer of rods and cones; ONL, outer nuclear layer; OPL, outer plexiform layer; OS, outer segment of photoreceptor cells; PE, nucleus of pigment epithelium; R, nucleus of rod photoreceptor; S1, synapse between rod and bipolar cells; S2, synapses between bipolar, amacrine and ganglion cells.

CONTINUED EXPRESSION OF NEURONAL TRAITS IN CHROMAFFIN CELLS OF THE ADULT BOVINE ADRENAL GLAND

V. Kriho, H.-Y. Yang, N. Lieska, N. Kriho and G. D. Pappas

University of Illinois at Chicago, Dept. of Anatomy and Cell Biology, Chicago, Illinois 60612

Intermediate filaments (IF) are a major constituent of the cytoskeleton of most eukaryotic cells. While there are five major classes of IF, amino acid sequence and secondary structure homologies suggest that all IFs evolved from a single precursor gene early in the history of life. Characteristically, IF are 10 nm filaments whose distribution, in most cases, is tissue-type specific.[1] For example, the IFs of muscle cells contain the muscle-specific IF protein desmin, and those of epithelial cells contain subunits of the keratin family. Neurofilaments (NF) are a class of IFs expressed in most neuronal cells and are composed of three polypeptide subunits with apparent molecular weights of 68,000 (NF-L), 150,000 (NF-M) and 200,000 (NF-H).[2] Interestingly, NFs are present in chromaffin cells cultured from bovine adrenal medulla. . Chromaffin cells are derived from the neural crest, and the presence of NFs confirms their relatedness to neurons.[3] However, the presence of NFs in _vivo_ has been the subject of debate.[4,5,6] There has been the suggestion that chromaffin cells in _vivo_ transiently express NF during development, but that once the cells acquire the ability to synthesize epinephrine, the neurofilament-specific proteins disappear.[4] It has also been suggested that this inhibition of neuronal traits is reversible since adult chromaffin cells in culture both synthesize epinephrine and re-express NFs. However, these in _vivo_ studies were done on avian and rat adrenal gland and these results were compared to studies done on bovine chromaffin cells in culture. The present study maintains that interspecies comparisons can be difficult and perhaps misleading, since it is known that there are marked differences in NF protein sequences even in closely related species. For a more reliable comparison to bovine chromaffin cells in culture, the presence of NF-L in adult bovine adrenal gland has been investigated.

Small pieces of bovine adrenal gland were excised and either frozen in liquid freon for indirect immunofluorescent localization of NF-L in cryostat sections or fixed in 1% glutaraldehyde, 4% paraformaldehyde in 0.1M phosphate buffer, pH 7.4, for immunogold labelling of NF-L in LR White sections.

Immunofluorescence microscopy clearly showed the presence of NF-L immunoreactivity in all chromaffin cells in the adult bovine adrenal medulla (Fig. 1). There was no NF-L immunoreactivity associated with the adrenal cortex. Post-embedding immunogold labeling of adult bovine adrenal gland with the same polyclonal NF-L antibody, confirmed the presence of NFs in adult bovine chromaffin cells in _vivo_ (Fig. 2). NF-L labeled gold particles were prominently localized along NFs.

While it is possible that the epinephrine-synthesizing chromaffin cells of the adult avian and rat adrenal gland do, indeed, lack NF subunits, results must be interpreted with caution.[5] A false-negative result may be due to low levels of NF antigens, the method of fixation and embedding, as well as the functional state of the cell at the time of fixation. Therefore, while our immunohistochemistry results strongly suggest that adult chromaffin cells retain neuronal

Proc. Microscopy and Microanalysis 1995, edited by G.W. Bailey, M.H. Ellisman, R.A. Hennigar, and N.J. Zaluzec
Copyright © 1995 MSA. Published by Jones and Begell Publishing, 79 Madison Ave., New York, NY 10016

characteristics, our data are not conclusive since antibodies can react with epitopes shared by more than one class of IF's. However, the observations are important and merit further investigation, especially regarding the formulation of an hypothesis concerning the structure and function of NF proteins in chromaffin cells.

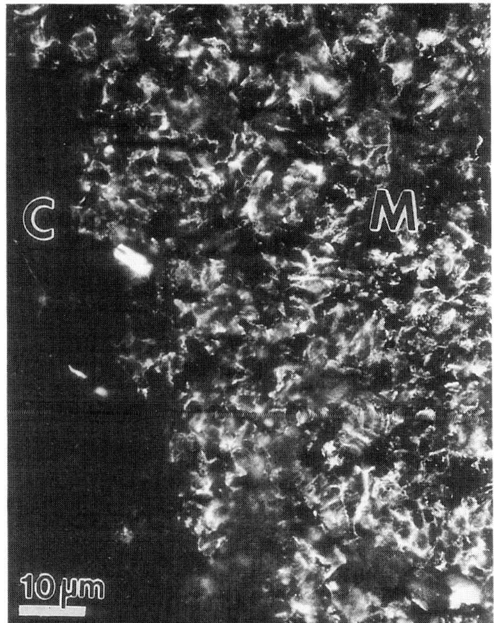

FIG.1.--Immunofluorescent localization of NF-L in a cryosection from adult bovine adrenal gland. NF-L immunoreactivity is present in the chromaffin cells of the medulla (M), but not in cells of the adrenal cortex (C).

FIG.2.--Postembedding immunogold localization of NF-L on intermediate filaments present in chromaffin cells of the adrenal medulla.

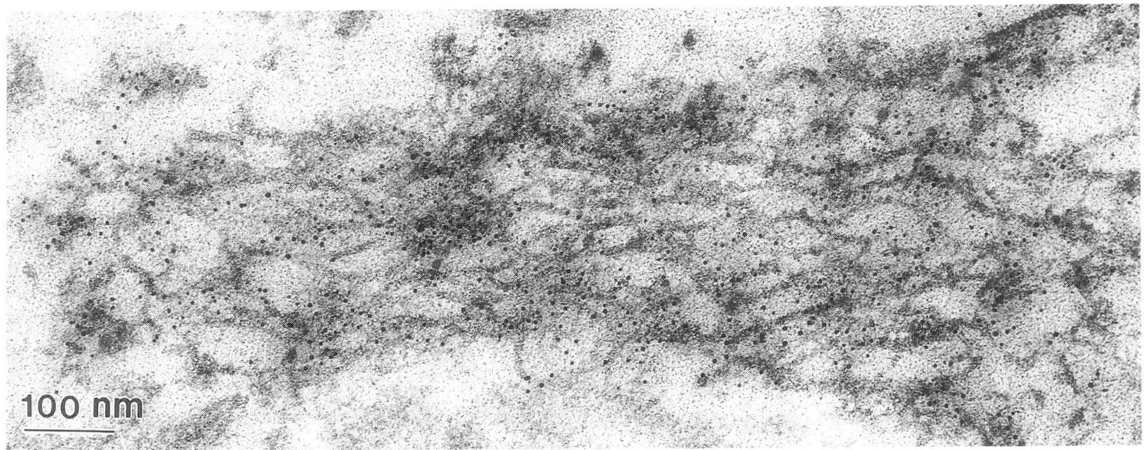

References

1. E. Lazarides, Nature (Lond.) 283(1980)249.
2. G. Shaw, Neurofilament proteins, The Neuronal Cytoskeleton, Wiley-Liss (1991)185.
3. M. F. Bader et al., J. Neurosci. 43(1984)1180.
4. M. E. Ehrlich et al., Devel. Biol. 163(1994)480.
5. J. Q. Trojanowski et al., J. Neurosci. 6(1986)650.
6. K. S. Vogel et al., Devel. Biol. 139(1990)1.
7. This study was supported by NIH Grant NS26395.

ULTRASTRUCTURAL ANALYSIS REDEFINES LIGHT MICROSCOPIC FINDINGS RELATIVE TO MORPHOLOGICAL RELATIONSHIPS AMONG CELLS OF THE CEREBRAL VASCULATURE IN THE RAT

Ruth V. W. Dimlich

Departments of Emergency Medicine and Cell Biology, Neurobiology and Anatomy, College of Medicine, University of Cincinnati, Cincinnati, OH 45267-0769

A previous light microscopic study determined that ~70% of all connective tissue mast cells (CTMCs) in the thalamus of the rat were adjacent to periodic acid-Schiff staining cells[1] which also were ED2 positive therefore identified as a type of macrophage.[2] Using transmission electron microscopy, the goal of the present study was to examine that association at the ultrastructural level and to determine the relationship of each of these cells to other components of the cerebrovasculature.

Brains of anesthetized adult male Sprague-Dawley rats were fixed by intracardiac perfusion with paraformaldehyde and glutaraldehyde [2%:2.5% (phosphate buffer 0.1 M, pH 7.40 or 2%:2.5% for 5 min followed by 1%:1.25% (phosphate buffer 80 mM, pH 7.4) for the remainder of the fixation]. Brains were sliced into ~1 mm sections and samples of thalamus were postfixed in OsO_4 (1%, 0.1 M phosphate buffer), dehydrated, and flat embedded in LX112. Semi-thin (1 µm) and thin (60-90 nm) sections were cut and stained with toluidine blue and uranyl acetate/lead citrate respectively. Grids were viewed on a JOEL 100CXII electron microscope at 80kV.

Electron microscopy enabled us to determine the location of CTMCs and macrophages relative to other components of the central nervous system. Both cell types were in the adventitial compartment of the cerebrovasculature (Fig. 1). To distinguish this macrophage from others in the brain, it will be referred to as a resident adventitial macrophage (RAM) because of its presence in non-pathologic tissue and the adventitia of cerebral blood vessels. Because CTMCs and RAMs were together in their own connective tissue compartment, i.e. the adventitia, they could be easily differentiated from pericytes and/or smooth muscle cells in the media and from endothelial cells in the intima. CTMCs and RAMs also were separated by a distinct basal lamina from the brain parenchyma, thus could be differentiated from neurons, astrocytes, oligodendroglia, microglia and other brain parenchymal macrophages, even when each of those components was immediately adjacent to the outer layer of the blood vessel wall. Thus, based on location alone, CTMCs and RAMs could be distinguished especially from pericytes of the vessel wall as well as microglia and other brain parenchymal macrophages with which the RAM in particular has been confused.

Specific ultrastructural characteristics of each type of cell allowed us to differentiate the RAM from the CTMC, another cell for which the RAM has been mistaken partly because each cell posses 'granules'. The granules of the CTMC are generally of the same size and density whereas 'granules' in the RAM that actually are lysosomes and vacuoles, are generally of various sizes and densities (Fig. 1). Not identified at the light microscopic level but quite evident using the electron microscope, was a third cellular component to this adventitial layer, the fibroblast (Fig. 1). Nuclei and mitochondria were similar in these three connective tissue cells; however, the CTMC and RAM could be distinguished from the fibroblast based on their 'granules' which were numerous in both cell types but almost lacking in the fibroblast. Fibroblasts also possessed dilated RER not observed in the RAM. This unique characteristic allows one to differentiate accurately between fibroblasts and RAMs even when only a small process of either cell type is visible (Fig. 1).

The ability to identify even small pieces of these three types of cells in the adventitia enabled us to initiate analysis of the morphological relationships among them. As expected from the light microscopic data, CTMCs and RAMs were found adjacent to one another in numerous microscopic images and as stated above, frequently there was an image of a fibroblast also within this adventitial

Proc. Microscopy and Microanalysis 1995, edited by G.W. Bailey, M.H. Ellisman, R.A. Hennigar, and N.J. Zaluzec
Copyright © 1995 MSA. Published by Jones and Begell Publishing, 79 Madison Ave., New York, NY 10016

compartment (Fig. 1). Interestingly, often processes of fibroblasts were interposed between the RAM and the CTMC (Fig. 1). This may suggest that although the majority of CTMCs are near RAMs and there appears to be some contiguity between those two cells, the more relevant interactions may be between CTCMs and fibroblasts as well as perhaps fibroblasts and RAMs.

References

1. R.V.W. Dimlich, *Soc. Neurosci. Abst.* 20(1994)612.
2. R.V.W. Dimlich, *J. Histochem. Cytochem.* 38(1990)1040.

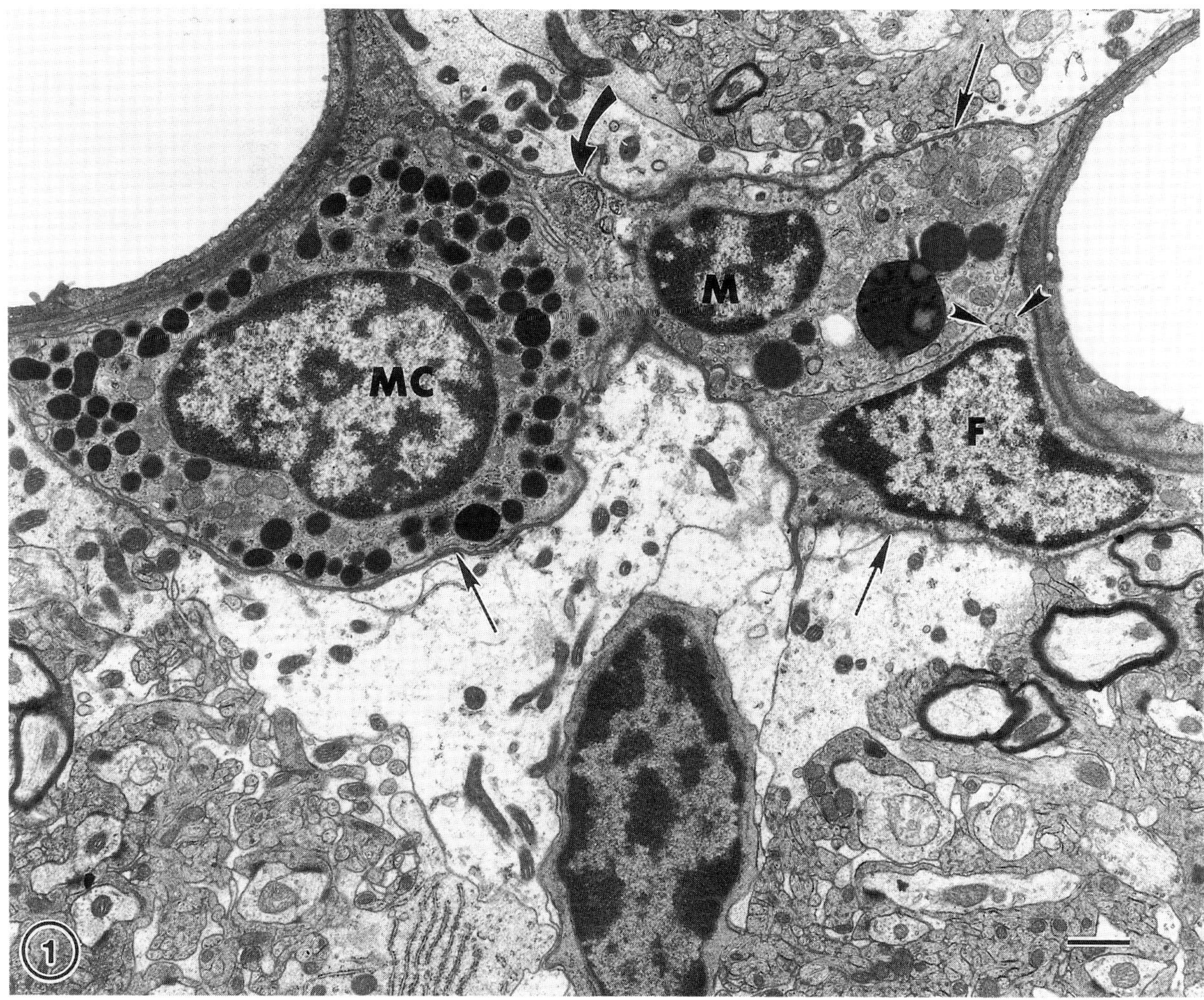

FIG. 1.—Resident adventitial macrophage (M), connective tissue mast cell (MC) and fibroblast (F) enclosed within the same basal lamina (straight arrows) and in direct contact with one another. Nuclei and mitochondria are similar in all three cells. The CTMC can be identified by small finger-like processes at its surface and by granules that are relatively similar in size and density. By contrast, the RAM has a relatively smooth surface and contains a variety of lysosomes and vacuoles that vary in size and density. Although the fibroblast also has a smooth surface and can contain a few lysosomes, it contains dilated RER (arrowheads) not found in the RAM. In this micrograph, the presence of dilated RER (curved arrow) enabled us to identify a portion of a fibroblast that is interposed between the CTMC and the RAM. Bar = 1 μm.

ULTRASTRUCTURAL IDENTIFICATION OF THREE TYPES OF SENSORY SETAE ON COPEPOD ANTENNAE

T. M. Weatherby* and P.H. Lenz**
*Biological Electron Microscope Facility and **Békésy Laboratory of Neurobiology, Pacific Biomedical Research Center, University of Hawai'i, Honolulu, HI 96822

Crustaceans, as well as other arthropods, are covered with sensory setae and hairs, including mechano- and chemosensory sensillae with a ciliary origin. Calanoid copepods are small planktonic crustaceans forming a major link in marine food webs. In conjunction with behavioral and physiological studies of the antennae of calanoids, we undertook the ultrastructural characterization of sensory setae on the antennae of *Pleuromamma xiphias.*

Distal mechanoreceptive setae exhibit exceptional behavioral and physiological performance characteristics: high sensitivity (<10 nm displacements), fast reaction times (<1 msec latency) and phase locking to high frequencies (1-2 kHz).[1] Unusual structural features of the mechanoreceptors are likely to be related to their physiological sensitivity. These features include a large number (up to 3000) of microtubules in each sensory cell dendrite, arising from or anchored to electron dense rods associated with the ciliary basal body microtubule doublets. The microtubules are arranged in a regular array, with bridges between and within rows. These bundles of microtubules extend far into each mechanoreceptive seta and terminate in a staggered fashion along the dendritic membrane, contacting a large membrane surface area and providing a large potential site of mechanotransduction. In contrast, in insects the microtubules are more randomly arrayed and terminate in a tubular body at the base of the seta. Proximal to the setae, within the antennal shaft, the ciliary portions of the one or two dendrites per seta are enclosed within a rigid tube shown by immunoelectron microscopy to be composed of actin fibers. This scolopale cell is firmly anchored to the antennal cuticle by bands of microtubules and to other accessory cells by desmosomes, providing a stiff encasement for the sensory cell dendrites. These setae, as seen in SEM, tend to be long, sometimes whip-like, and have a thick and rigid cuticle.

In contrast, the cylindrical chemoreceptive setae each contain numerous dendrites of small diameter, arising from several sensory neurons. These dendrites also contain microtubules arising from ciliary basal bodies, but in far fewer numbers than in the mechanoreceptors. The distal dendrites may branch once to several times. There is no scolopale cell and the sensory neuron cell bodies lie closer to the setae than do those of the mechanoreceptors. These setae readily take up azocarmine dye.[2] The cuticle is thin and apparently permeable to the external medium.

Spiniform setae located along the antennal shaft possess one or two mechanoreceptive dendrites enclosed by a scolopale tube, and two or more chemoreceptive dendrites outside of the tube. The setae readily stain with azocarmine and methylene blue; the latter travels into the neuronal cell bodies located near the bases of the setae.[2] These setae have the thicker cuticle of the mechanoreceptors, but each possesses an apical pore which appears to give access to the external medium. Their characteristics indicate contact chemo + mechanoreception, suggesting a gustatory function.

The distribution of these three types of receptors along the antennae reflects their putative functions. Sensitive mechanoreceptors occur at the distal tip, optimally placed for detection of vibrations generated by predators. The chemosensory and bimodal setae occur in highest densities in the area of high water velocities generated by feeding currents. The pure chemosensors along most of the length of the antenna seem to be involved in distance chemoreception (olfaction). The higher concentration of the bimodal sensors proximally suggests they are involved in contact chemoreception (tasting).

1. Lenz, P.H. and J. Yen, *Bull. Mar. Sci.* 53(1993)170.
2. Lenz, P.H. *et al., Mar. Freshw. Behav. Physiol.* (in press).
3. Supported by NSF OCE 89-18019 & UH Proj. Dev. Grant to D. Hartline, and NIH RCMI RR03061

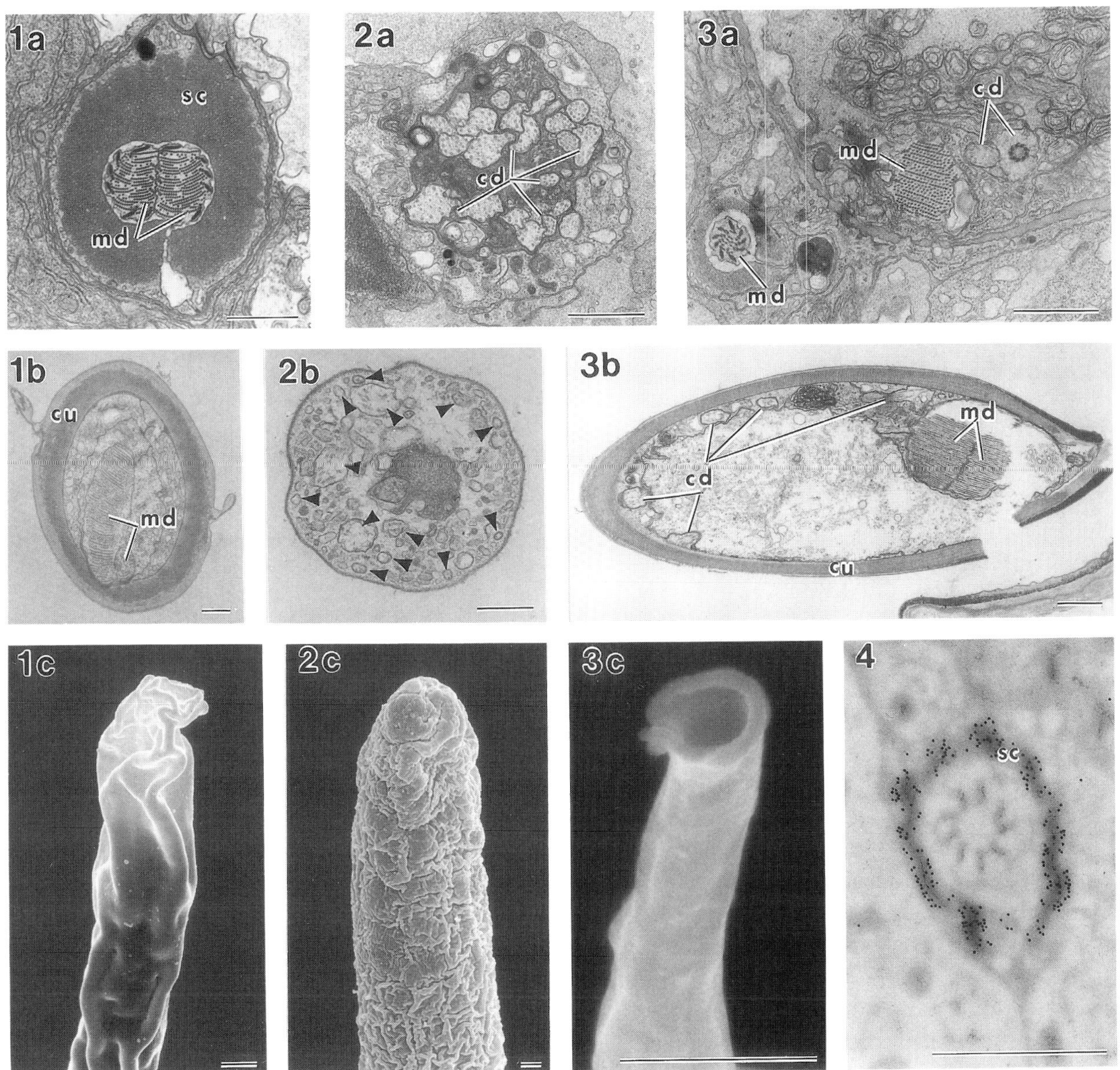

FIG. 1. Mechanoreceptive sensillae. (a) TEM of cross-section through microtubular array in two dendrites within antennal shaft, just distal of ciliary basal body. Dense-cored microtubules appear to have arisen in rows from each of nine microtubule doublets. Scolopale tube encloses dendrites throughout their course within antennal shaft; (b) Cross-section through microtubular array of seta; (c) SEM of tip of seta, showing no cuticular specializations.

FIG. 2. Chemoreceptive sensillae. (a) TEM of ciliary portions of dendrites, within antennal shaft, just distal to basal bodies. At this level there are 24 dendrites arising from two or three sensory cells. (b) Cross-section through seta containing dozens of dendritic endings (arrowheads), some still exhibiting 9+0 arrangement of microtubules. (c) SEM of tip of seta, showing thinner and more convoluted cuticle.

FIG. 3. Bimodal (chemo- + mechanoreceptive) sensillae. (a) TEM of microtubular array of mechano-sensor plus two dendrites of chemosensor in antenna near base of seta. The single dendrite of another mechanosensor lies nearby. (b) Section of seta containing both types of dendrites. (c) SEM of apical pore on tip of seta.

FIG. 4. TEM of a single mechanoreceptive dendrite. The scolopale tube labels clearly with anti-actin. cu: cuticle, sc: scolopale, md: mechanoreceptive dendrite, cd: chemoreceptive dendrites   Bars = 1.0 μm

GLUCURONYL 3-SULFATE OCCURS EXCLUSIVELY ON DISTAL UNMYELINATED TERMINALS OF SELECTED SENSORY NEURONS IN THE PERIPHERAL NERVOUS SYSTEM

S.S. Spicer and B.A. Schulte

Medical University of South Carolina, Charleston, SC 29425

Generation of monoclonal antibodies (MAbs) against tissue antigens has yielded several (VC1.1, HNK-1, L2, 4F4 and anti-leu 7) which recognize the unique sugar epitope, glucuronyl 3-sulfate (Glc A3-SO$_4$).[1] In the central nervous system, these MAbs have demonstrated Glc A3-SO$_4$ at the surface of neurons in the cerebral cortex, the cerebellum, the retina and other widespread regions of the brain.[2-4]

Here we describe the distribution of Glc A3-SO$_4$ in the peripheral nervous system as determined by immunostaining with a MAb (VC1.1) developed against antigen in the cat visual cortex.[2] Outside the central nervous system, immunoreactivity was observed only in peripheral terminals of selected sensory nerves conducting transduction signals for touch, hearing, balance and taste. On the glassy membrane of the sinus hair in murine nasal skin, just deep to the ringwurt, VC1.1 delineated an intensely stained, plaque-like area (Fig. 1). This previously unrecognized structure of the nasal vibrissae presumably serves as a tactile end organ and to our knowledge is not demonstrable by means other than its selective immunopositivity with VC1.1 and its appearance as a densely fibrillar area in H&E stained sections. By electron microscopy, the vibrissa plaque consisted of collagenous stroma infiltrated by unmyelinated nerves that were extensively invested by processes of glia-like cells (Fig. 2).

Immunostaining with VC1.1 revealed another heretofore undescribed, corpuscle-like structure at the base of hair follicles in gerbil skin. This structure was recognized only by means of the linear VC1.1 immunostaining presumably reflecting antibody binding to cell surfaces and by the dense collection of nuclei of which it appears to be largely composed (Figs. 3 and 4). The nature of this structure remains unexplained, since ultrastructural examination failed to disclose nerve processes in the corpuscle. In the gerbil and mouse inner ear, VC1.1 bound selectively to unmyelinated afferent radial fibers underlying cochlear inner hair cells, to outer spiral bundle fibers directed toward cochlear outer hair cells and to nerves innervating hair cells of vestibular maculae and cristae. In the mouse tongue, VC1.1 stained selectively linear entities consistent with nerves innervating neurosensory epithelial cells in the taste buds of the foliate papilla.

Nerves transmitting signals for touch, hearing, balance and taste thus expressed Glc A3-SO$_4$ characteristically on their surface. Although the significance of this unique sugar on sensory terminals remains enigmatic, the strongly anionic nature of the epitope suggests an influence on the channels that mediate cation transport essential to nerve impulse conduction.

References

1. D.K.H. Chou et al., *J. Biol. Chem.* 261(1986)11717.
2. Y. Arimatsu et al., *J. Neurosci.* 7(1987)1250.
3. Kosaka et al., *Exp. Brain Res.* 82(1990)566.
4. Yamamoto et al., *Neurosci. Res.* 5(1988)273.

Proc. Microscopy and Microanalysis 1995, edited by G.W. Bailey, M.H. Ellisman, R.A. Hennigar, and N.J. Zaluzec
Copyright © 1995 MSA. Published by Jones and Begell Publishing, 79 Madison Ave., New York, NY 10016

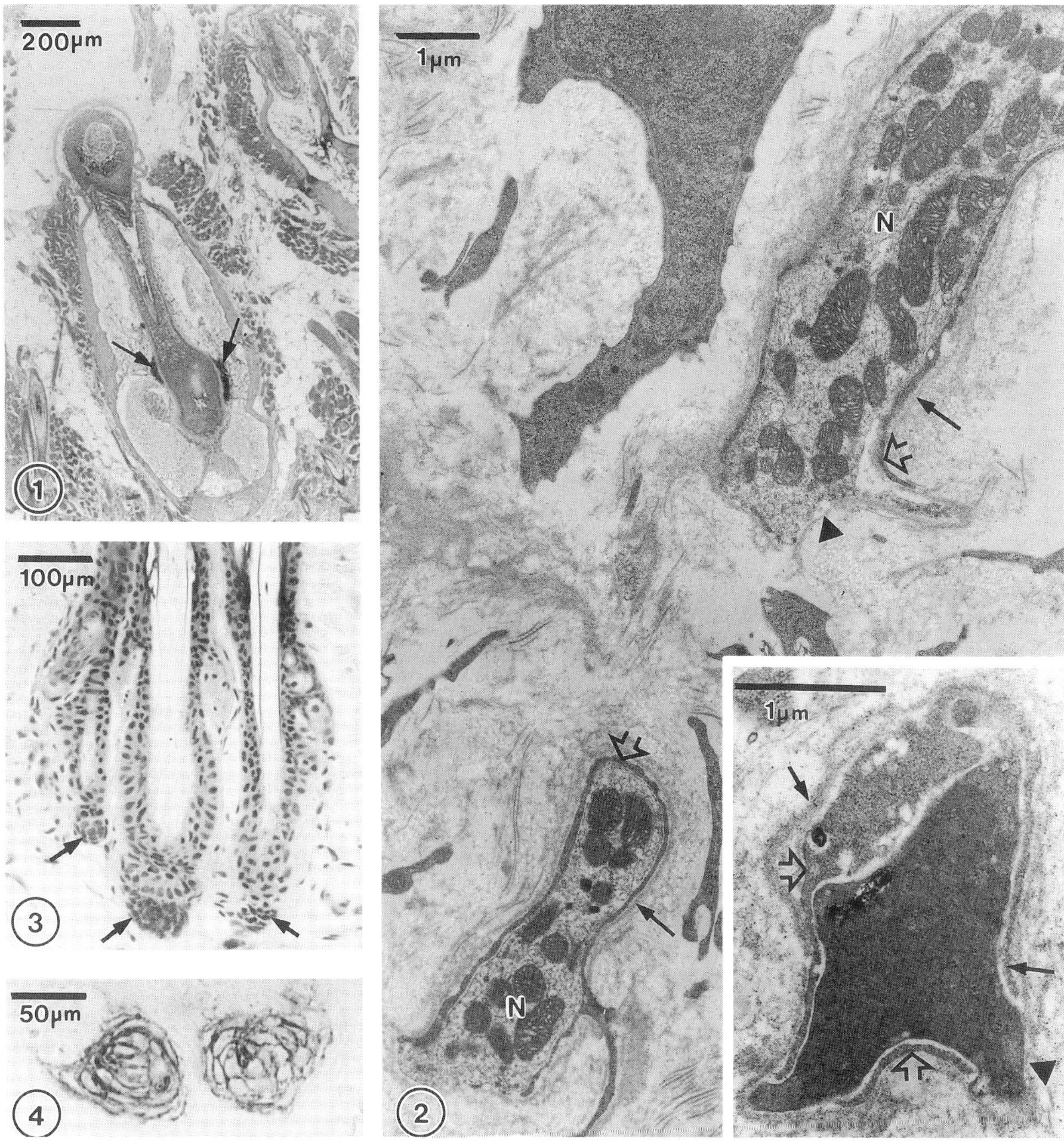

FIG. 1. Two plaque-like profiles (arrows) on the vibrissa's glassy membrane at junction of blood and cavernous sinuses and below the ringwurt stain for Glc A3-SO$_4$. VC1.1 stain, H&E counterstain.
FIG. 2. Processes of glia-like cells (open arrows) surrounded by basement membrane (closed arrows) and filaments closely invest nerves (N) in a plaque on a murine vibrissae (cf Fig. 1) except at branch points (arrowheads).
FIG. 3. Dense corpuscles (arrows) composed of small cells with dark staining nuclei are visible below hair follicles in gerbil skin. H&E stain.
FIG. 4. Corpuscles (cf Fig. 3) below hair follicles show an immunopositive network. VC1.1 stain.

MORPHOLOGIC CHANGES IN THE ANTEROVENTRAL COCHLEAR NUCLEUS OF
ACOUSTICALLY-DEPRIVED GERBIL

Shang-Ming Yu and Shian-Jyue Du
Institute of Anatomy, National Yang-Ming University, Taipei,
Taiwan, Republic of China

Microcystic lesions are reported as a neurodegenerative disorder in
the cochlear nucleus of the Mongolian gerbil[1]. These lesions
resemble the spongioform degeneration[2,3]. Such lesions are large
holes, cavities or vacuoles and are also known as the microcysts.
The microcysts develop in the neuropil or appear in the neuronal
perikarya, degenerated axons and dendrites. The number and size of
the microcysts is greatly reduced either by the elevation of
auditory thresholds or by the conductive hearing loss induced by
ligation of the external auditory canal[4]. Furthermore, absence of
the microcysts is found in the old deaf animals[5]. This study was
to investigate the morphologic changes in the anteroventral
cochlear nucleus (AVCN) of the acoustically-deprived gerbil
following the sequence of time courses after ligation of the
external auditory canal during postnatal development.

The mongolian gerbil, *Meriones unguiculatus*, had been acoustically
deprived on the right side by a surgical ligation of the external
auditory canal at postnatal day 12-14. At one month after
ligation, the microcyst contained the myelin sheath and the
membranous structure in the contralateral AVCN of the acoustically-
deprived gerbils (Fig. 1a). The myelin sheath protruded into the
neuropil or the adjacent axon (Fig. 1b). At three months after
ligation, the myelin sheath detached from the axolemma of the
retracted axon in the contralateral AVCN (Fig. 2a). A degenerative
capillary contained the membranous structure and surrounded by
several multivesicular bodies (Fig. 2b). At six months after
ligation, a degenerative capillary contained the membranous
structure in the contralateral AVCN (Fig. 3a). The lumen of
another degenerative capillary was divided by a thin layer of the
cytoplasm and a few cytoplasm protruded into the lumen (Fig. 3b).
At nine months after ligation, the detachment of the myelin sheath
from the axolemma were severe and the disruption of the myelin
sheath were found. The small and fused vacuoles were present in
the contralateral AVCN (Fig. 4a). A degenerative capillary with
cytoplasmic protrusion were juxtaposed by the microcyst (Fig. 4b).
These microcysts are presumably derived from the vacuoles of
retracted axons, or from the detachment and disruption of the
myelin sheath, or from the degenerative capillary.

References

1. K.D. Statler et al., *Hear. Res.* 50(1990)275.
2. M.D. McGinn and B.T. Faddis, *Hear. Res.* 31(1987)235.
3. E.-M. Ostapoff and D.K. Morest, *Hear. Res.* 37(1989)141.
4. B.T. Faddis and M.D. McGinn, *Exp. Neurol.* 120(1993)160.
5. A. Czibulka and I.R. Schwartz, *Hear. Res.* 52(1991)43.

Proc. Microscopy and Microanalysis 1995, edited by G.W. Bailey, M.H. Ellisman, R.A. Hennigar, and N.J. Zaluzec

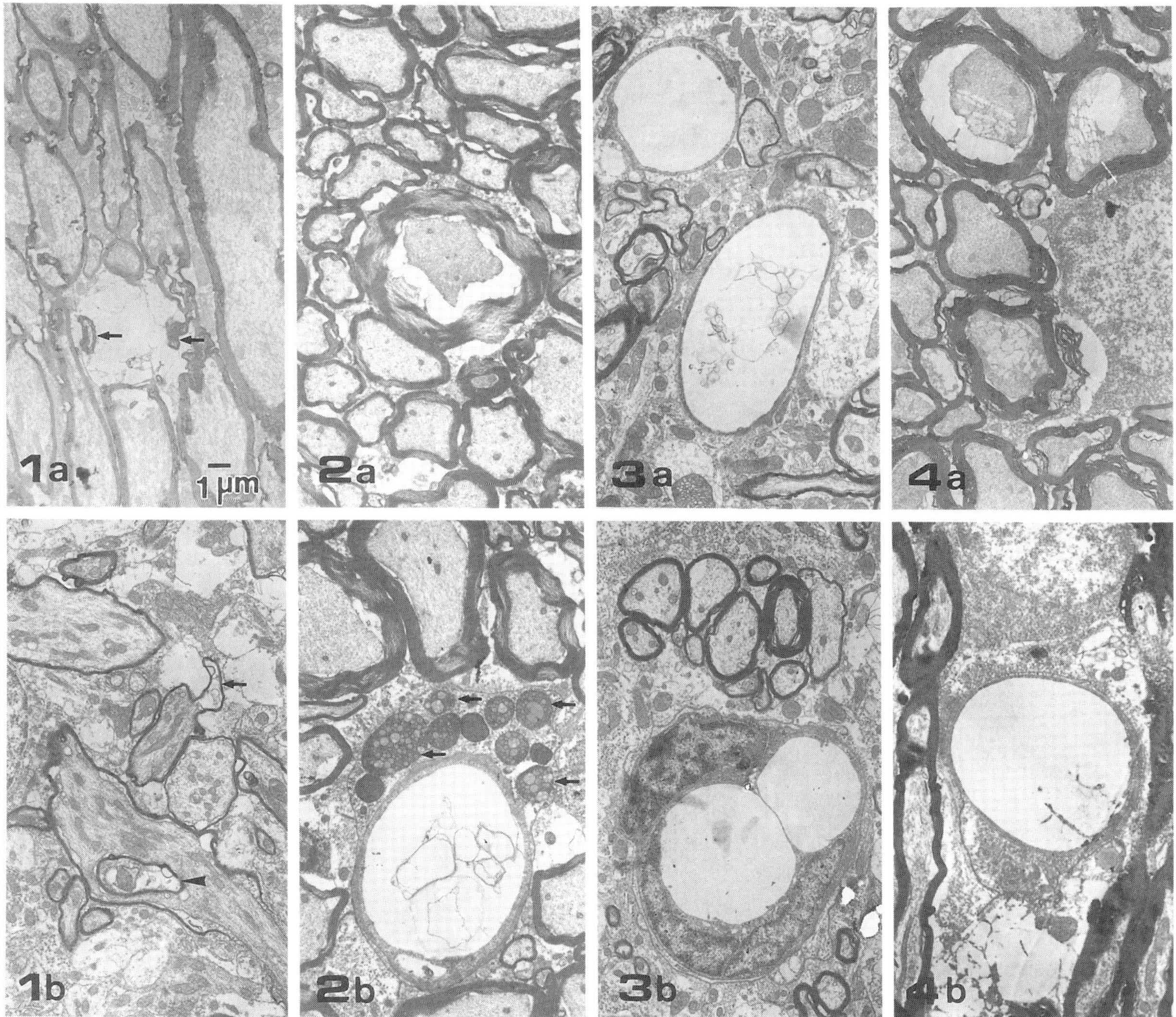

FIG. 1 - Left AVCN at one month after right ligation. (a) The microcyst was found in the neuropil and contained the myelin sheath (arrows). (b) The myelin sheath (arrow) protruded into the neuropil or into the myelin sheath (arrowhead). Bar=1μm.

FIG. 2 - Left AVCN at three months after right ligation. (a) The myelin sheath detached from the axolemma of the retracted axon. (b) A degenerative capillary contained the membranous structure and was surrounded by several multivesicular bodies (arrows).

FIG. 3 - Left AVCN at six months after right ligation. (a) A degenerative capillary contained the membranous structure. (b) The lumen of another degenerative capillary was divided by a thin layer of the cytoplasm and a few cytoplasm protruded into the lumen.

FIG. 4 - Left AVCN at nine months after ligation. (a) Several vacuoles were present in the axon. (b) A degenerative capillary with cytoplasmic protrusion were juxtaposed by the microcyst.

DNA NICK END LABELING: A RELIABLE MARKER FOR APOPTOSIS?

Anne F. Bushnell, Sarah Webster, and Lynn S. Perlmutter

Institute for Dementia Research, Bayer Corporation, West Haven, CT 06516

Apoptosis, or programmed cell death, is an important mechanism in development and in diverse disease states[1]. The morphological characteristics of apoptosis were first identified using the electron microscope. Since then, DNA laddering on agarose gels was found to correlate well with apoptotic cell death in cultured cells of dissimilar origins[2]. Recently numerous DNA nick end labeling methods have been developed in an attempt to visualize, at the light microscopic level, the apoptotic cells responsible for DNA laddering[3].

The present studies were designed to compare various tissue processing techniques and staining methods to assess the occurrence of apoptosis in post mortem tissue from Alzheimer's diseased (AD) and control human brains by DNA nick end labeling methods. Three tissue preparation methods and two commercial DNA nick end labeling kits were evaluated: the Apoptag kit from Oncor and the Biotin-21 dUTP 3' end labeling kit from Clontech. The detection methods of the two kits differed in that the Oncor kit used digoxigenin dUTP and anti-digoxigenin-peroxidase and the Clontech used biotinylated dUTP and avidin-peroxidase. Both used 3-3' diaminobenzidine (DAB) for final color development.

Extensive modification of the original methods, both those included with the Apoptag kit and those utilized by Gavrieli, et al.[3], were necessary. These included omitting the protease digestion step and, in the case of the Apoptag kit, diluting the reaction mixture. Paraffin, cryostat and vibratome sections were reacted and examined. Both positive (DNase treated) and negative (minus terminal deoxynucleotidyl transferase [TdT]) controls were included with every run. Samples were examined with the light microscope and selected examples were processed for electron microscopy.

In all cases the Oncor Apoptag kit was superior. It gave consistent results on all types of sections. The only disadvantage was that not enough reaction buffer was included to run the appropriate controls so reagents had to be mixed among kits. The final protocol used mixed aldehyde fixed autopsy material, which was vibratome sectioned at 50μ. They were reacted with equilibration buffer for 10 minutes, then half strength TdT enzyme mixture for one hour at 37°C. The reaction was stopped in supplied stop/wash buffer, rinsed, reacted with digoxigenin-peroxidase, rinsed, and the color developed in nickel intensified DAB.

The number of positive cells was markedly different when comparing AD cases with controls and various brain regions within AD cases. Unfortunately these results did not correlate with either the presence of amyloid plaques, neurofibrillary tangles, or the pattern of neurodegeneration known to occur in these regions in AD. Upon electron microscopic examination, many stained nuclei had normal morphology. Several exhibited chromatin clumping, but not margination. Classic apoptotic morphology was not seen. Coates[4] noted that DNA nick end labeling is not always a reliable marker of apoptosis. This study amply confirms his cautionary remarks especially in regard to the study of neurodegeneration in Alzheimer's disease.

Human brain samples	Biotin-21-dUTP	Apoptag
Cryostat sections	variable labeling, poor morphology	variable labeling, poor morphology
Paraffin sections	little labeling, high background	good labeling, poor morphology
Vibratome sections	little labeling	good labeling, good morphology

References

1. M.R. Alison, C.E. Sarraf, *J. Roy. Coll. Phys. Lond.*, 26 (1992) 25-35.
2. R. Gold, et al., *J. Histochem. Cytochem.*, 41 (1993) 1023-1030.
3. Y. Gavrieli, Y. Sherman, and S.A. Ben-Sasson, *J Cell Biol.*, 119 (1992) 493-501.
4. P.J. Coates. *J. Histotechnology*, 17 (1994) 261-267.

Proc. Microscopy and Microanalysis 1995, edited by G.W. Bailey, M.H. Ellisman, R.A. Hennigar, and N.J. Zaluzec
Copyright © 1995 MSA. Published by Jones and Begell Publishing, 79 Madison Ave., New York, NY 10016

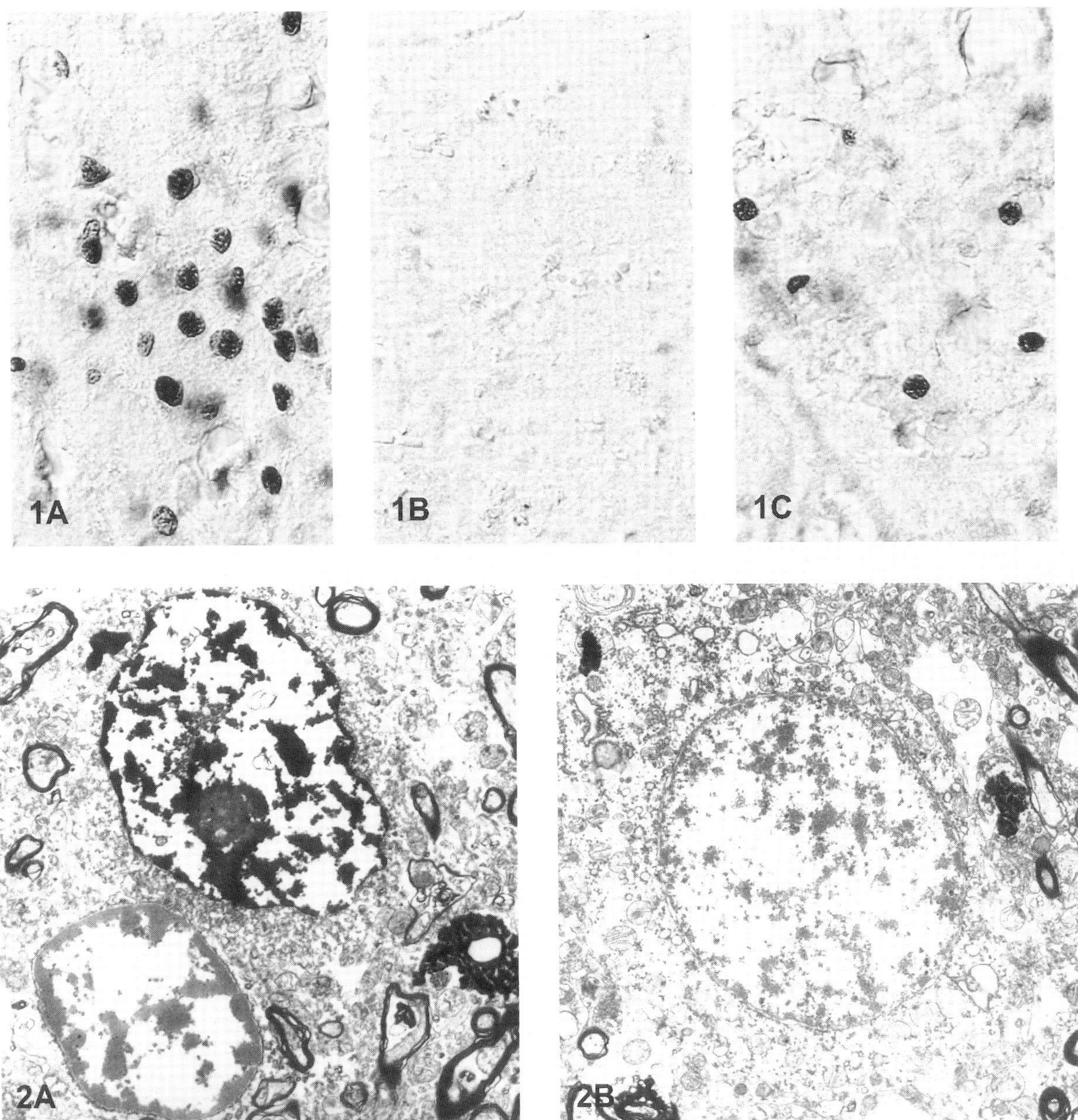

Figure 1. Apoptag staining in human Alzheimer's disease. A.) DNase reacted positive control. Note that most nuclei are labeled. B.) -TdT enzyme. Note total lack of staining. C.) Staining in a severe case of AD. Note that many nuclei stain although this section is from 1^O visual cortex where there is reportedly little cell loss.

Figure 2. Electron micrographs from AD and non-AD cases stained with Apoptag pre-embed. A.) Section from the same case shown in Fig. 1. Note heavily stained nucleus and intact nucleolus. Although the chromatin is clumped it is not marginated. B.) Section from a non-AD case also labeled for nicked DNA. Note the absence of staining. Ultrastructure is compromised due to post mortem changes.

MICROANATOMICAL CHANGES OF THE CEREBRAL CAPILLARY NETWORK IN AGED RATS: INFLUENCE OF TREATMENT WITH A DIHYDROPYRIDINE-TYPE CALCIUM ANTAGONIST

D. Zaccheo* and F. Amenta**

* Istituto di Anatomia Umana, Università di Genova, Genoa, Italy
**Sezione di Anatomia Umana, Istituto di Farmacologia, Università di Camerino, Camerino, Italy

A variety of age-dependent changes affect cerebral vasculature. Morphological changes of the brain microvasculature characterized by microvascular fibrosis , vascular convolutions , thickening or membranous changes of the capillary basal membrane and gliofibrillary proliferation have been reported with aging[1]. These changes may be related with age-dependent impairment in behavioral performance[1]. Increasing evidence suggests that Ca^{+2} antagonists of the dihydropyridine family, which are used primarily in the therapy of cardiovascular disorders may be useful for treating neurologic diseases including cerebral ischemia, age-related neurodegenerative disorders, senile dementia and epilepsy[2]. This in view of the capability of these compounds to decrease Ca^{+2} overload which can result in increased cell death[2] .

Recent studies have shown that the dihydropyridine Ca^{2+} antagonists delayed the expression of age-related microvascular changes in the rat forebrain and sciatic nerve without reducing systolic blood pressure. The present study was designed to assess the influence of long term treatment with the dihydropyridine derivative darodipine[3] [diethyl 4-2,1,3- (benzoxadiazal-4-yl)-1,4-dihydro-2,6-dimethyl-pyridine-3,5-dicarboxylate, PY 108-068] on age-related microvascular changes occurring in the rat cerebral cortex and hippocampus. The brain capillary network was investigated using alkaline phosphatase histochemistry associated with image analysis.

Male Wistar rats of 12 months (adult reference group) and of 27 months (old rats) were used. Old rats consisted of one group of control animals and of another one of rats receiving an oral daily dose of 5 mg/kg of darodipine from the 21st month of life. Animals were sacrificed by decapitation under ether anaesthesia and frozen sections of the frontal cortex, hippocampus and occipital cortex were processed for alkaline phosphatase histochemistry[4]. Volumes of the cerebral areas and of the hippocampus, the number, diameter, length and intercapillary distance of alkaline phosphatase-reactive capillaries were evaluated by image analysis[5].

Body and brain weight and systolic arterial pressure values were increased significantly in old in comparison with adult rats. Treatment with darodipine did not affect body and brain weight or systolic arterial pressure values of old rats.

Volumes of the frontal and occipital cortex, of the Ammon's horn or of the dentate gyrus were not significantly different in the three animal groups investigated. Sections processed for the histochemical detection of alkaline phosphatase developed a black-brown staining in the endothelium of cerebral capillaries.

The number of alkaline phosphatase-reactive capillary profiles and the intercapillary distance were significantly decreased and increased respectively in old in comparison with adult rats. The diameter of alkaline phosphatase-reactive capillaries was slightly increased in the frontal cortex and in the Ammon's horn and unchanged in the occipital cortex or in the dentate gyrus of old rats . The average length of capillaries was reduced in the frontal cortex and in the Ammon's horn and unchanged in the occipital cortex and in the dentate gyrus of old rats. Treatment with darodipine significantly increased the number and the length of alkaline phosphatase-reactive capillary profiles and decreased the intercapillary

Proc. Microscopy and Microanalysis 1995, edited by G.W. Bailey, M.H. Ellisman, R.A. Hennigar, and N.J. Zaluzec
Copyright © 1995 MSA. Published by Jones and Begell Publishing, 79 Madison Ave., New York, NY 10016

distance in the different cerebral areas investigated. The diameter of alkaline phosphatase-reactive capillaries was decreased in the frontal cortex and in the Ammon's horn, but not in the occipital cortex or in the dentate gyrus of darodipine-treated old rats in comparison with age-matched controls.

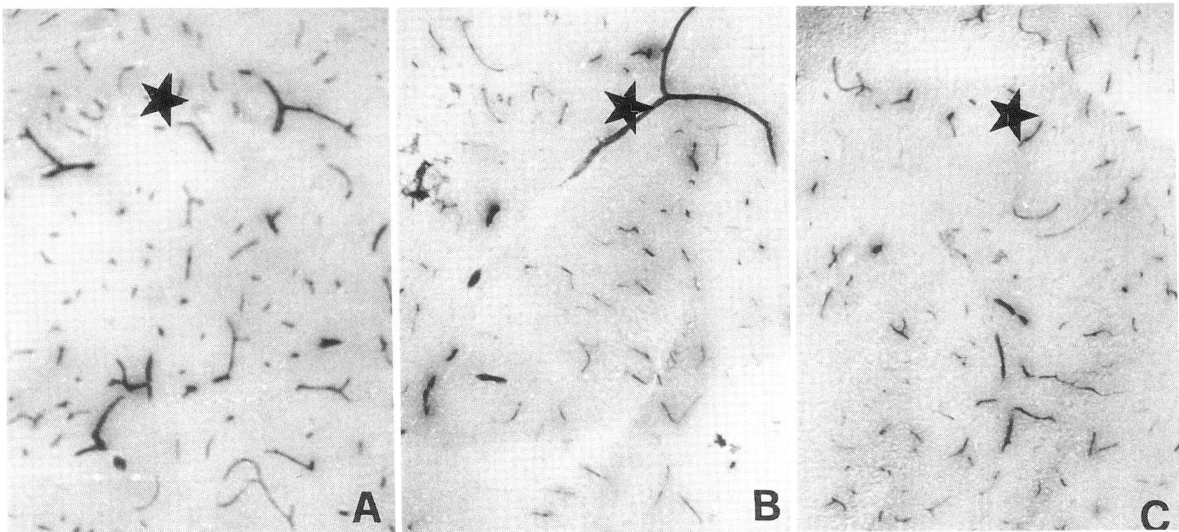

FIG, 1 - Alkaline phosphatase-reactive capillaries in the Ammon's horn (★) of adult (A), old (B) and darodipine-treated old rats (C).The density of alkaline phosphatase-reactive capillaries is higher in the Ammon's horn of adult or darodipine-treated old rats in comparison with control old rats. (x 140)

The restoration of some microvascular parameters impaired in aging by treatment with darodipine suggests that the compound, similarly as reported for nimodipine, may counter to some extent the expression of cerebrovascular alterations occurring in aging brain. The observation that the occipital cortex and the hippocampus were the areas more sensitive to treatment with darodipine is probably related with the higher density of dihydropyridine-type Ca^{2+} channels in these cerebral regions than in the frontal cortex .

References

1. J.Cervos-Navarro et al., *Mech. Ageing Dev., 39* (1987) 223-231.
2. A.Scriabine et al., *FASEB J. 3,*(1989) 1799-1806.
3 R.P. Hof, *Gen. Pharmacol., 16* (1985) 1-6.
4. R.G. Bannister and F.C.A. Romanul, *J. Neurol. Neurosurg. Psychiat.,26* (1963)333-340.
5. O.Hunziker et al., *Brain Res., 65* (1974) 1-11.

INSULIN-LIKE GROWTH FACTOR I AND DEMYELINATION: CYTOCHEMICAL EVIDENCE FOR FREE RADICAL DERIVED OXIDANT MEDIATION

E. Ann Ellis, Maria B. Grant, and Robert N. Mames

Division of Endocrinology, Department of Medicine, College of Medicine, University of Florida, Gainesville, FL 32610

The peptide hormone, insulin-like growth factor I (IGF-I) is a growth factor which has been shown in elevated concentrations in the serum and vitreous of diabetic patients with proliferative diabetic retinopathy.[1] Studies with an animal model of IGF-I induced proliferative retinopathy[2] showed demyelination in the optic nerves and along the medullary rays of rabbit eyes which received intravitreal injections of IGF-I, basic fibroblast growth factor (bFGF), or a combination of the two growth factors. Biochemical studies have demonstrated a role for oxidative injury in complications of diabetes including lipid peroxidation and demyelination in peripheral neuropathy.[3] The cerium NADH-oxidase technique[4] has been used to demonstrate free radical derived oxidant production in several pathological conditions including demyelination in experimental allergic encephalomyelitis (EAE).[5] We investigated the role of free radical derived oxidants in growth factor-induced demyelination.

Pigmented rabbits were injected intravitreally with 500 μg of IGF-I or bFGF and sacrificed at 4, 8, 12 and 18 days. Eyes were enucleated, fixed and processed for NADH-oxidase localization. Extensive demyelination throughout the optic nerve head and and along the medullary ray coincided with areas of cerium perhydroxide reaction product (Figs. 1, 2). There was an influx of activated inflammatory cells and macrophages into the vitreous which resulted in extensive localization of cerium perhydroxide within the cytoplasm, on the plasmalemma, and in the vitreous (Figs. 3, 4).

In vitro studies have shown that IGF-I activates macrophages and neutrophils to secrete superoxide (O_2^{-}) via the respiratory burst.[6,7] Detoxification of O_2^{-} produces the free radical derived oxidant, hydrogen peroxide (H_2O_2) which is localized by the cerium NADH-oxidase cytochemical technique. These *in vivo* studies provide evidence for growth factor mediated oxidative injury as a potential mechanism for demyelination in optic and peripheral nerves of diabetic patients.

References

1. M. B. Grant et al., *Diabetes* 35(1986)416.
2. M. B. Grant et al., *Diabetologia* 36(1993)282.
3. P. A. Low and K. K. Nickander, *Diabetes* 40(1991)873.
4. R. T. Briggs et al., *J. Cell Biol.* 67(1975)566.
5. J. Guy et al., *Arch. Ophthalmol.* 108(1990)1614.
6. C. K. Edwards et al., *Science* 239(1988)769.
7. Y-K. Fu et al., *J. Immunol.* 146(1991)1602.
8. The authors acknowledge support from NEI EY07739-0181 and American Heart Grant 92GIA/862.

Proc. Microscopy and Microanalysis 1995, edited by G.W. Bailey, M.H. Ellisman, R.A. Hennigar, and N.J. Zaluzec
Copyright © 1995 MSA. Published by Jones and Begell Publishing, 79 Madison Ave., New York, NY 10016

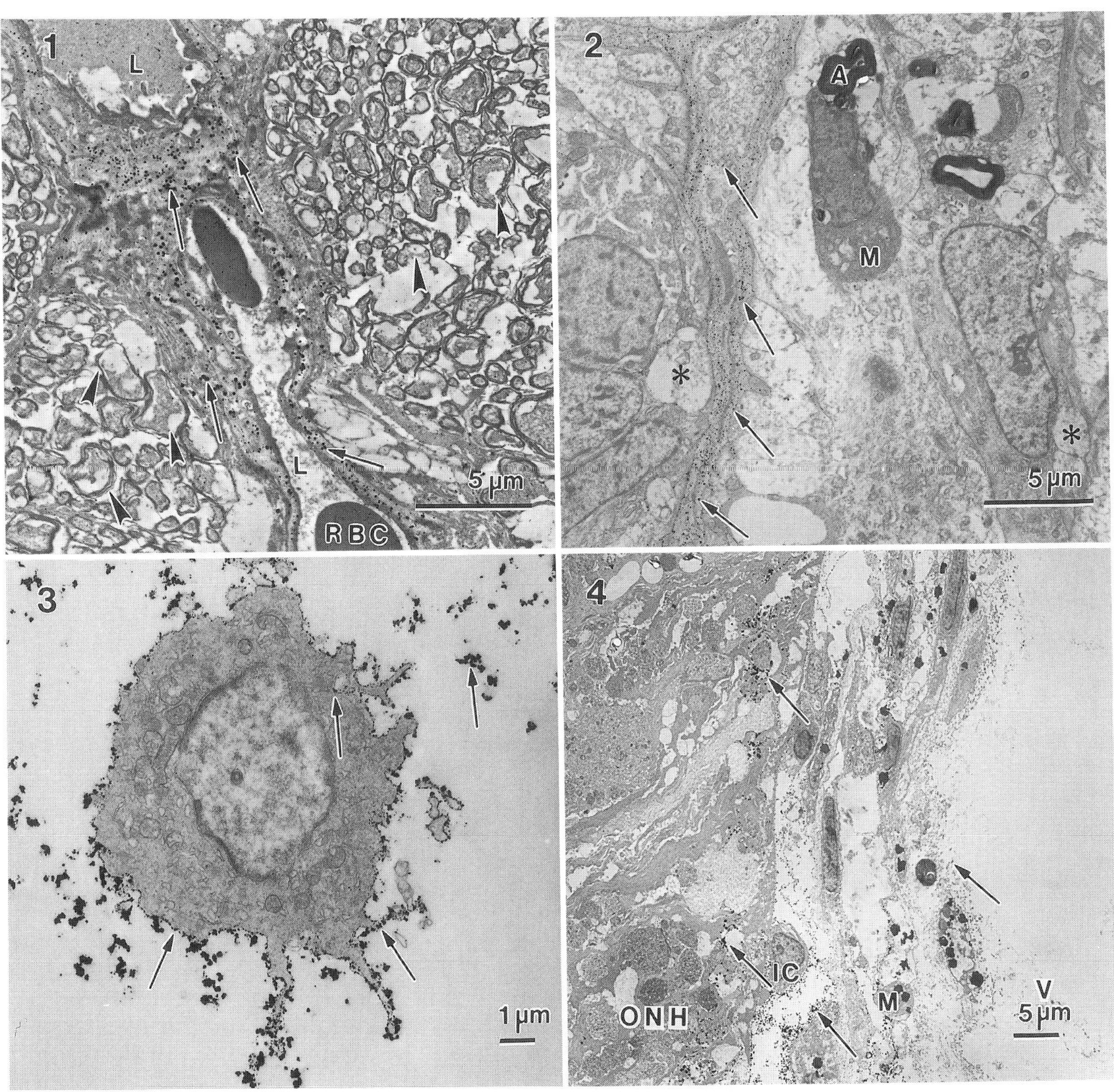

FIG. 1.-Demyelination (arrowheads) and perivascular localization of cerium perhydroxide (arrows) in the optic nerve head of a rabbit eye 4 days post injection with 500 μg IGF-I. Red blood cell (RBC); vessel lumen (L).

FIG. 2.-Localization of cerium perhydroxide (arrows) in the optic nerve head of a rabbit eye 18 days post injection with 500 μg bFGF. There are completely demyelinated axons (*) and gliosis. Note the macrophage (M) adjacent to an axon (A) with damaged myelin.

FIG. 3.-Extensive cerium perhydroxide localization (arrows) around macrophages (M) and inflammatory cells (IC) in the vitreous (V) adjacent to the optic nerve head (ONH) of a rabbit eye 18 days post injection with 500 μg IGF-I. The optic nerve head is completely demyelinated.

FIG. 4.-Cerium perhydroxide localization (arrows) in the cytoplasm and on the plasmalemma of an inflammatory cell in the optic nerve head of a rabbit eye 18 days post injection with 500 μg IGF-I.

FT-IR MICROSPECTROSCOPIC ANALYSIS OF DISEASED WHITE MATTER BRAIN TISSUES

Steven M. LeVine* and David L. Wetzel**

* University of Kansas Medical Center, Detpartment of Physiology, 3901 Rainbow Blvd., Kansas City, KS 66160
**Kansasn State University, Microbeam Molecular Spectroscopy Laboratory, Shellenberger Hall, Manhattan, KS 66506.

In situ FT-IR microspectroscopy has allowed spatially resolved interrogation of different parts of brain tissue. In previous work[1], the spectrrscopic features of normal barin tissue were characterized. The white matter, gray matter and basal ganglia were mapped from appropriate peak area measurements from spectra obtained in a grid pattern. Bands prevalent in white matter were mostly associated with the lipid. These included 2927 and 1469 cm^{-1} due to CH_2 as well as carbonyl at 1740 cm^{-1}. Also 1235 and 1085 cm^{-1} due to phospholipid and galactocerebroside, respectively (Figs 1and2). Localized chemical changes in the white matter as a result of white matter diseases have been studied. This involved the documentation of localized chemical evidence of demyelination in shiverer mice in which the spectra of white matter lacked the marked contrast between it and gray matter exhibited in the white matter of normal mice (Fig. 3).

The twitcher mouse, a model of Krabbe's desease, was also studied. The purpose in this case was to look for a localized build-up of psychosine in the white matter caused by deficiencies in the enzyme responsible for its breakdown under normal conditions. Spatial resolution enabled finding a shift of select absorption bands at points within the white matter (Fig. 3). Spectral subtraction (Fig. 4) of normal white matter from the psychosine bearing tissue produced the spectrum consistent with that compound. The shift toward the psychosine frequency was also observable from examination of the second derivative spectra of both twitcher and litter mate normal mice.

Extravasated blood produced in white matter by injection of blood from the tail of a rat into the brain white matter of the same rat was analyzed. Spectra were obtained at varying distances from the resulting lesion (Fig. 5). The lipids that are a part of the myelin sheath were deminished more the closer to the lesion. These observations were consistent with attack of the double bonds by reactive oxygen species (ROS) traceable to the blood. Additionally, the amide I and II bands were enhanced at the location of the blood deposition. The differences were more pronounced in animals 2 days after injection than in those examined 6 days after.

Tissue fixed in formalin and mounted in parafin of human brain tissue with HIV infection was examined and shown to differ significantly from nomal human brain tissue treated in the same way. A map of lesion site is showed in Fig. 6.

Proc. Microscopy and Microanalysis 1995, edited by G.W. Bailey, M.H. Ellisman, R.A. Hennigar, and N.J. Zaluzec
Copyright © 1995 MSA. Published by Jones and Begell Publishing, 79 Madison Ave., New York, NY 10016

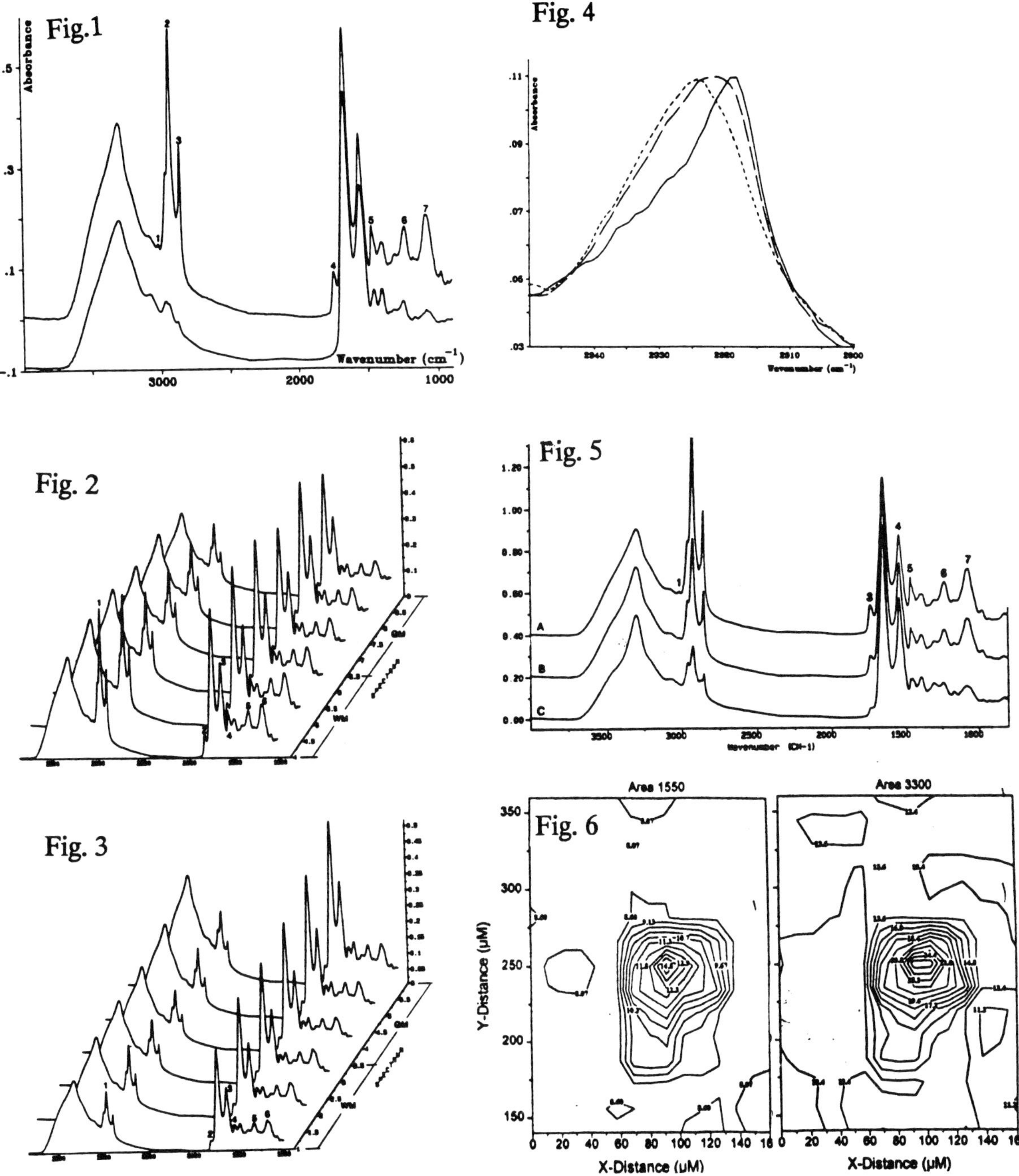

Fig.1 Spectra from white matter (top) and gray matter (bottom) from a normal tissue. The peaks are marked as follows: (1) 3015, (2) 2926, (3) 2853, (4) 1740, (5) 1469, (6)1235 cm-1 .

Fig. 2 WM firat two spectra, GM last three spectra for normal tissue.

Fig. 3 WM first two spectra, GM last three spectra for shiverer tissue.

Fig. 4 Normal, twitcher and differenence at 2919.6 cm^{-1} correspond to psychosine.

Fig. 5 Spectra of white matter away from(A), near (B) and (C) in the extravastes blood lesion.

Fig. 6 Functional group maps at the site of lesioned in diseased tissue.

Reference
1. D. L. Wetzel and S. M. LeVine, Spectroscopy, 8(1993)40.

ALTERATIONS IN THE NUMBER OF MOTONEURONS CONTAINING IMMUNOREACTIVE CALCITONIN GENE-RELATED PEPTIDE (CGRP) AND CHOLINE ACETYLTRANSFERASE (ChAT) IN THE CERVICAL SPINAL CORD OF THE WOBBLER MOUSE DURING THE DEVELOPMENT OF THE MOTONEURON DISEASE.

L. Vacca-Galloway, Y.Q. Zhang, P. Bose and S.H. Zhang

Department of Anatomy, University of Hong Kong, Hong Kong

The Wobbler mouse (wr)[1] has been studied as a model for inherited human motoneuron diseases (MNDs). Using behavioral tests for forelimb power, walking, climbing, and the "clasp-like reflex" response[2], the progress of the MND can be categorized into early (Stage 1, age 21 days) and late (Stage 4, age 3 months) stages[3]. Age-and sex-matched normal phenotype littermates (NFR/wr) were used as controls (Stage 0), as well as mice from two related wild-type mouse strains: NFR/N and a $C_{57}Bl/6N$. Using behavioral tests, we also detected pre-symptomatic Wobblers at postnatal ages 7 and 14 days. The mice were anesthetized and perfusion-fixed for immunocytochemical (ICC) of CGRP and ChAT in the spinal cord (C_3 to C_5).

Using computerized morphomety (Vidas, Zeiss), the numbers of IR-CGRP labelled motoneurons were significantly lower in 14 day old Wobbler specimens compared with the controls (Fig. 1). The same trend was observed at 21 days (Stage 1) and 3 months (Stage 4). The IR-CGRP-containing motoneurons in the Wobbler specimens declined progressively with age. In contrast, the number of IR-CGRP labelled motoneurons counted in the normal phenotype (NFR/wr) specimens was greater by age 21 days compared with younger NFR/wr specimens, and lower by age 3 months. The wild-type NFR/N specimens also exhibited significantly lower numbers of IR-CGRP-containing motoneurons at age 3 months compared with age 21 days.

Comparing Wobbler and control specimens at postnatal days 7 and 14, there were no significant differences in the number of IR-ChAT labelled motoneurons (Fig. 1). By 21 days, IR-ChAT labelled motoneurons were significantly greater in Wobbler (Stage 1) specimens compared with the controls. By age 3 months, the number of IR-ChAT motoneurons was significantly lower in the Wobbler (Stage 4) specimens, but did not decrease in the controls, with advancing age.

The increase in the number of ChAT labelled neurons in Wobbler by the time symptoms are expressed suggests an over-production of motoneurons occurs, and coincides with previous biochemical data showing an increase in ChAT activity[5], followed by severe motoneuron activity losses as the disease advances. The decline in CGRP labelled motoneurons suggests a selective loss occurs around postnatal day 14[6].

References

1.   D.S. Falconer, Mouse News Letter 15 (1956) 23.
2.   D.J. Lange, P.F. Good and W.G. Bradley, J. Neurol. Sci. 78 (1983) 253.
3.   L.L. Vacca-Galloway and C.C. Steinberger, J. Neurosci. Res. 16 (1986) 657.
4.   Y.Q. Zhang and L.L. Vacca-Galloway, Brain Res. 578 (1992) 169.
5.   K.K.L. Yung, F. Tang, R. Fielding, Y.H. Du and L.L. Vacca-Galloway, Neurosciences 50(1992)209.
6.   This work was supported by RGC Grant No. 338/031/0010 and Croucher Founder Grant No. 360/031/0814

Proc. Microscopy and Microanalysis 1995, edited by G.W. Bailey, M.H. Ellisman, R.A. Hennigar, and N.J. Zaluzec
Copyright © 1995 MSA. Published by Jones and Begell Publishing, 79 Madison Ave., New York, NY 10016

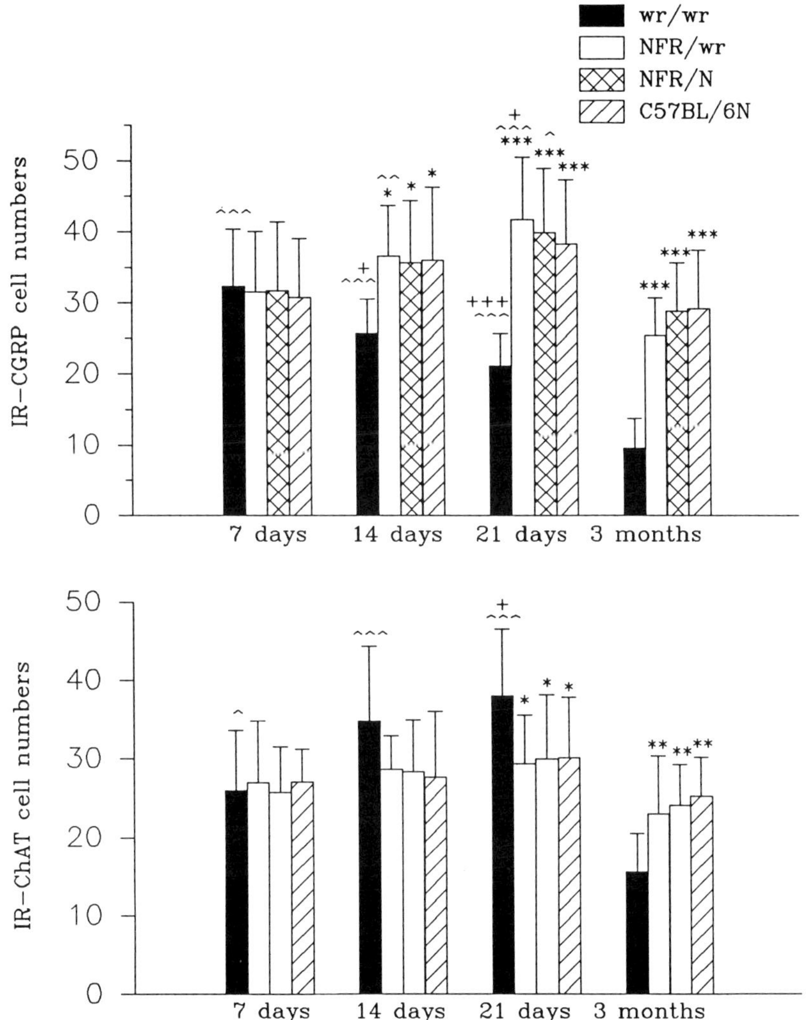

* Differs from the Wobbler (wr/wr) specimens in the same age group, *P<0.05, **P<0.01, ***P<0.001.
^ Differs from 3-month-old specimens from the same type mouse, ^P<0.05, ^^P<0.01, ^^^P<0.001.
+ Differs from 7-day-old specimens from the same type mouse, +P<0.05, ++P<0.01, +++P<0.001.

Fig. 1.—Cell numbers (mean+standard deviation) counted for motoneurons containing immunoreactive (IR) calcitonin gene-related peptide (CGRP) and choline acetyl-transferase (ChAT) in cervical spinal cord of Wobbler (wr/wr) mice and normal phenotype (NFR/wr) littermates as well as pair-matched wild-type (NFR/N; C57BL/6N) mice at postnatal ages 7 days, 14 days, 21 days (Stage 1) and 3 months (Stage 4).

MEASUREMENT OF SPONTANEOUS NEURONAL MEMBRANE ACTIVITY BY TIME LAPSE CONFOCAL MICROSCOPY.

R. H. Selinfreund and A. H. Cornell-Bell
Lion Imaging Inc. Westbrook, CT. 06498

Cellular electrophysiological properties are normally monitored by standard patch clamp techniques [1]. The combination of membrane potential dyes with time-lapse laser confocal microscopy provides a more direct, least destructive rapid method for monitoring changes in neuronal electrical activity. Using membrane potential dyes we found that spontaneous action potential firing can be detected using time-lapse confocal microscopy. Initially, patch clamp recording techniques were used to verify spontaneous electrical activity in GH4\C1 pituitary cells. It was found that serum depleted cells had reduced spontaneous electrical activity. Brief exposure to the serum derived growth factor, IGF-1, reconstituted electrical activity. We have examined the possibility of developing a rapid fluorescent assay to measure neuronal activity using membrane potential dyes. This neuronal regeneration assay has been adapted to run on a confocal microscope. Quantitative fluorescence is then used to measure a compounds ability to regenerate neuronal firing.

The membrane potential dye di-8-ANEPPS was selected for these experiments. Di-8-ANEPPS is internalized slowly, has a high signal to noise ratio (40:1), has a linear fluorescent response to change in voltage [2]. The linear response is in the 100mV range which corresponds to the membrane voltage change during neuronal action potential firing. Therefore, di-8-ANEPPS can be used to measure the resting membrane potential and changes in neuronal electrical activity.   Changes in Fluorescence is detected using a Zeiss and a Bio-Rad MRC 600 confocal microscope. Excitation at 488nm. Subsequent experiments using time lapse confocal microscopy demonstrate that the mV change in electrical activity could be directly correlated to the change in fluorescence of the membrane potential dye. Using this fluorescent based imaging process we can monitor the electrical coupling and spontaneous activity in a large field of cells. This technology is directed towards resolving the complex interactions of growth factors in the central nervous system.

References:

1. O. P. Hamill et al., *Pflug. Arch.* 391 (1981) 85.
2. S. Rohr and B.M. Salzberg , *Biophysical Journal* 67 (1994) 1301

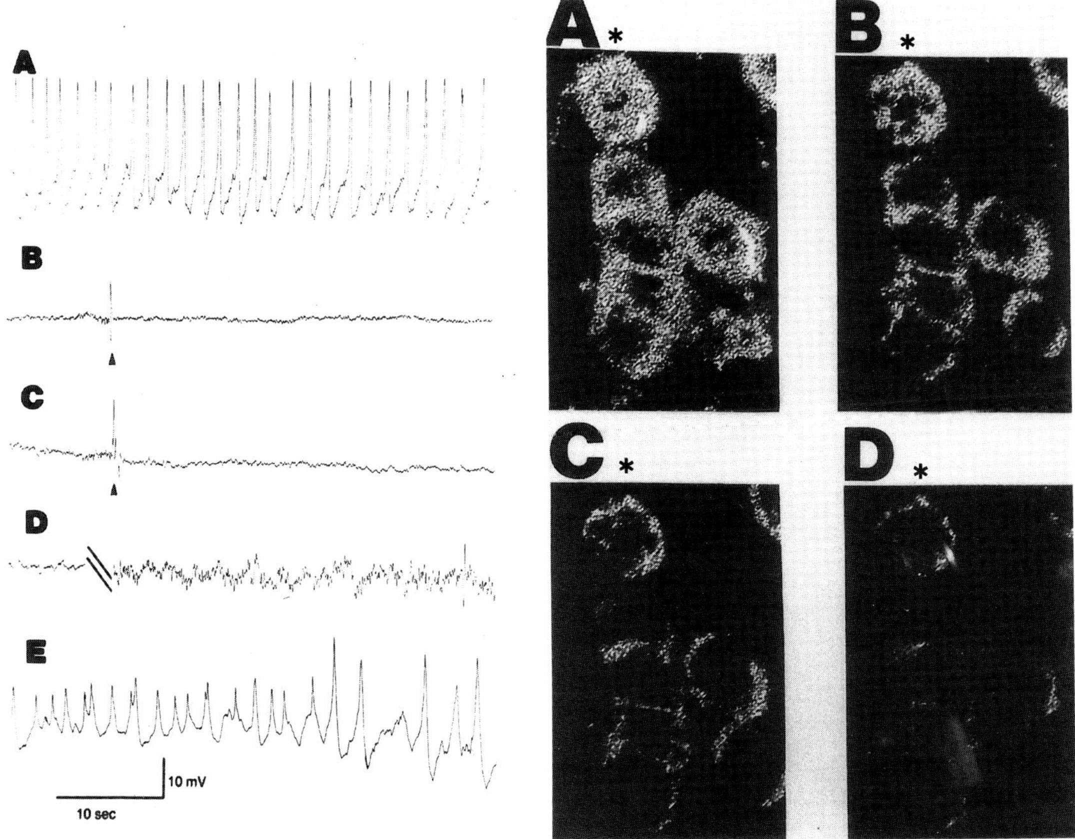

Fig 1.
Electrical activity was initially monitored using the whole-cell patch recording technique in the current clamp mode. **A**. GH4/C1 cells fire spontaneous action potentials. **B**. Serum depleted neurons have low spontaneous electrical activity. **C**. Low concentration of the growth factor IGF-1 (0.2ng/ml) fail to regenerate activity. **D**. IGF-1 at 20ng/ml rapidly increases activity. **E**. 20ng/ml IGF-1 Calibrations: vertical, 10mV; horizontal, 10 seconds.

Figure 2A Neurons made electrically quiescent; (Fig 2B) 20 seconds after drug addition; (Fig. 2C) 4.5 minutes after drug addition; (Fig 2D) 14.5 minutes after drug addition. Cell identified by an asterisk is followed through the time series illustrating dramatic dye loss as the cell depolarizes.

AUTOMATED CELL COUNTING OF ASTROCYTES ON PATTERNED SUBSTRATES CONTAINING ALIPHATIC AND CHARGED PROPERTIES

W. Shain*, H. Ancin**, H.C. Craighead***, M. Isaacson***, L. Kam**, B. Roysam**, D. Szarowski*, J.N. Turner*, S.W. Turner***

*Wadsworth Center, NYS Dept. of Health and School of Public Health, The University at Albany, Albany, NY 12201
**Rennselear Polytechnic Institute, Troy, NY 12180
*** National Nanofabrication Facility, Applied and Engineering Physics, Cornell University, Ithaca, NY 14853

Neural protheses have potential to restore nervous system functions lost by trauma or disease. Nanofabrication extends this approach to implants for stimulating and recording from single or small groups of neurons in the spinal cord and brain; however, tissue compatibility is a major limitation to their practical application. We are using a cell culture method for quantitatively measuring cell attachment to surfaces designed for nanofabricated neural prostheses.

Silicon wafer test surfaces composed of 50-μm bars separated by aliphatic regions were fabricated using methods similar to a procedure described by Kleinfeld et al[1]. Test surfaces contained either a single or double positive charge/residue. Cyanine dyes (diIC$_{18}$(3)) stained the background and cell membranes (Fig 1); however, identification of individual cells at higher densities was difficult (Fig 2). Nuclear staining with acriflavine allowed discrimination of individual cells and permitted automated counting of nuclei using 3-D data sets from the confocal microscope[2,3] (Fig 3). For cell attachment assays, LRM55 astroglial cells[4] and astrocytes in primary cell culture were plated at increasing cell densities on test substrates, incubated for 24 hr, fixed, stained, mounted on coverslips, and imaged with a 10x objective. At the highest cell densities$\leq$5% of the cells attached to the hydrophobic background. LRM55 astroglial cells attached to the surface with saturating kinetics, while astrocytes in primary culture attached in a biphasic fashion (Fig 4). At cell plating densities below 1.5x10^5 cells/35 mm culture dish, these cells demonstrated saturating kinetics; however, above these plating densities attachment began to increase. Stereo-pairs (Fig 3) showed that cells attached to each other producing 'mounds'. Thus, at lower plating densities our results describe attachment to the test surfaces, and at higher densities attachment to other cells. When astrocytes from different brain regions were assayed for cell attachment, differences were observed when using lower cell densities (Fig 5). These results indicate (i) that 3-D data sets obtained by confocal imaging can be used in an automated assay for counting nuclei in a cell attachment assay and (ii) that astrocytes from different brain regions may have different attachment specificities.

References

1. D. Kleinfeld et al., J. Neurosci. 8(1988)4098.
2. B. Roysam et al., *J. Microsc.* 173(1994)115.
3. J.N. Turner et al., *Microsc. Res. Tech.* 29(1994)269.
4. W. Shain et al.,Glia 5 (1992)223.
5. Work partially supported by NSF DIR 9108492 and NIH RR01219.

Proc. Microscopy and Microanalysis 1995, edited by G.W. Bailey, M.H. Ellisman, R.A. Hennigar, and N.J. Zaluzec
Copyright © 1995 MSA. Published by Jones and Begell Publishing, 79 Madison Ave., New York, NY 10016

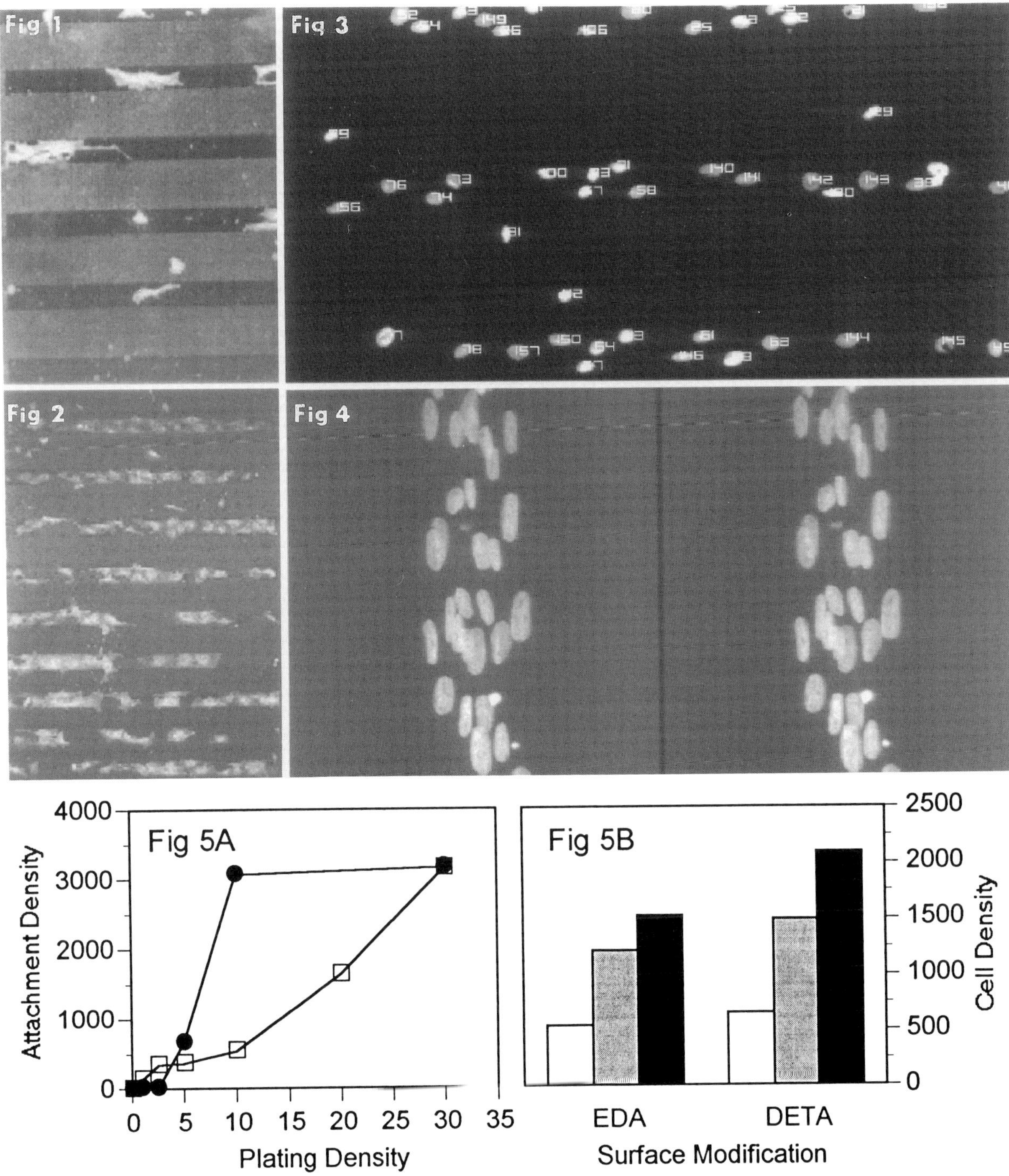

Fig 1. A test surface stained with diIC$_{18}$(3) demonstrating areas of the charged surfaces.
Fig 2. Difficulty in discerning individual cortical astrocytes stained with diIC$_{18}$(3).
Fig 3. Acriflavine-stained cortical astrocytes following cell counting.
Fig 4. Stereo-pair of acriflavine-stained cortical astrocytes illustrating cell-to-cell attachment.
Fig 5. (A) Cell attachment of LRM55 cells (◊) and primary cultures of cortical astrocytes (♦).
     (B) Cell attachment of the cortical (□), cerebellar (▨), and hippocampal (■) astrocytes.

FUNGAL INFECTION OF SEED--A SOURCE OF COTTON TEXTILE IMPERFECTIONS

W. R. Goynes, B. F. Ingber, and D. P. Thibodeaux

USDA, ARS, Southern Regional Research Center, New Orleans, LA

Cotton seed develop within a thick-walled boll that is usually divided into three or more compartments called locules. The seed and fibers that grow within a locule are called a lock. An unopen boll is shown in figure 1. The cotton fiber is a single cell and develops from the epidermis of the seed. The cell wall, or primary wall, of the fiber, a complex mixture of cellulose, protein, waxes, pectins, and other plant related materials, elongates for approximately 17-25 days. Completion of this elongation is overlapped by the beginning of secondary wall synthesis which deposits successive layers of cellulose inside the primary wall.[1] This main body of the fiber is composed almost entirely of cellulose. Production of a commercially useful fiber depends on the completion of this secondary wall development since it provides both fiber strength and dyeability. If this growth process is inhibited at any stage, less mature fibers with thin secondary walls are produced, and if it is interrupted before secondary wall production begins, undeveloped fibers are produced that can cause problems during processing into textiles.

Imperfections on surfaces of textiles appear as light or dark specks (neps), and can render fabrics commercially unusable. Dark specks are often composed of nonfiber plant materials, and can be removed by chemical treatments. The source of white specks on fabrics dyed dark colors has recently been shown to be undeveloped fibers.[2] Because of the absence of the cellulosic secondary wall in these fibers, they fail to dye in cellulose dyes normally used in cotton textiles. These undeveloped fibers remain in compact bundles rather than separating, or fluffing, on boll opening as normal fibers do, and can be carried through fabric production processes and remain attached to yarn surfaces. An example of a "white speck" fiber mass on a fabric surface is shown in figure 2. Figure 3 shows the bundle at higher magnification and reveals the flat, undeveloped, compressed fibers that form it.

One cause of undeveloped fibers is thought to be plant growing conditions that cause the abortion of a seed in early stages of growth. These may be environmental causes, or inherent plant qualities. An additional source of undeveloped fibers has now been shown to be fungal invasion of the seed within a boll locule. Early attack on the seed by fungus can cause undeveloped seed similar to those caused by abortion of the seed due to environmental conditions, and similarly undeveloped fibers as well. A sliced cotton seed showing an incompletely developed cotyledon, and fungal hyphae (H) is shown in figure 4. Some fungal-infected seed are so undeveloped that they have no cotyledon at all, and often have shrunken, twisted seedcoats. Typical fibers from these seed, as shown in figure 5 at low magnification, and at higher magnification in figure 6, appear in exactly the same structural conformations as do those undeveloped fibers from early aborted seed. Fungus that attacks cotton has been a problem in both seed and fiber quality for many years. With this evidence that fungus infection of cottonseed causes undeveloped fibers that can become white speck defects on dyed fabrics, control of fungal attack must be considered in attempts to eliminate white speck sources.

1. M. L. Rollins, *The Cotton Fiber,* in <u>The American Cotton Handbook,</u> Dame S. Hamby, ed., Interscience Publishers, New York,(1965)45.
2. W. R. Goynes et al., *Proc. Ann MSA Meeting,*51(1993)352.

Proc. Microscopy and Microanalysis 1995, edited by G.W. Bailey, M.H. Ellisman, R.A. Hennigar, and N.J. Zaluzec

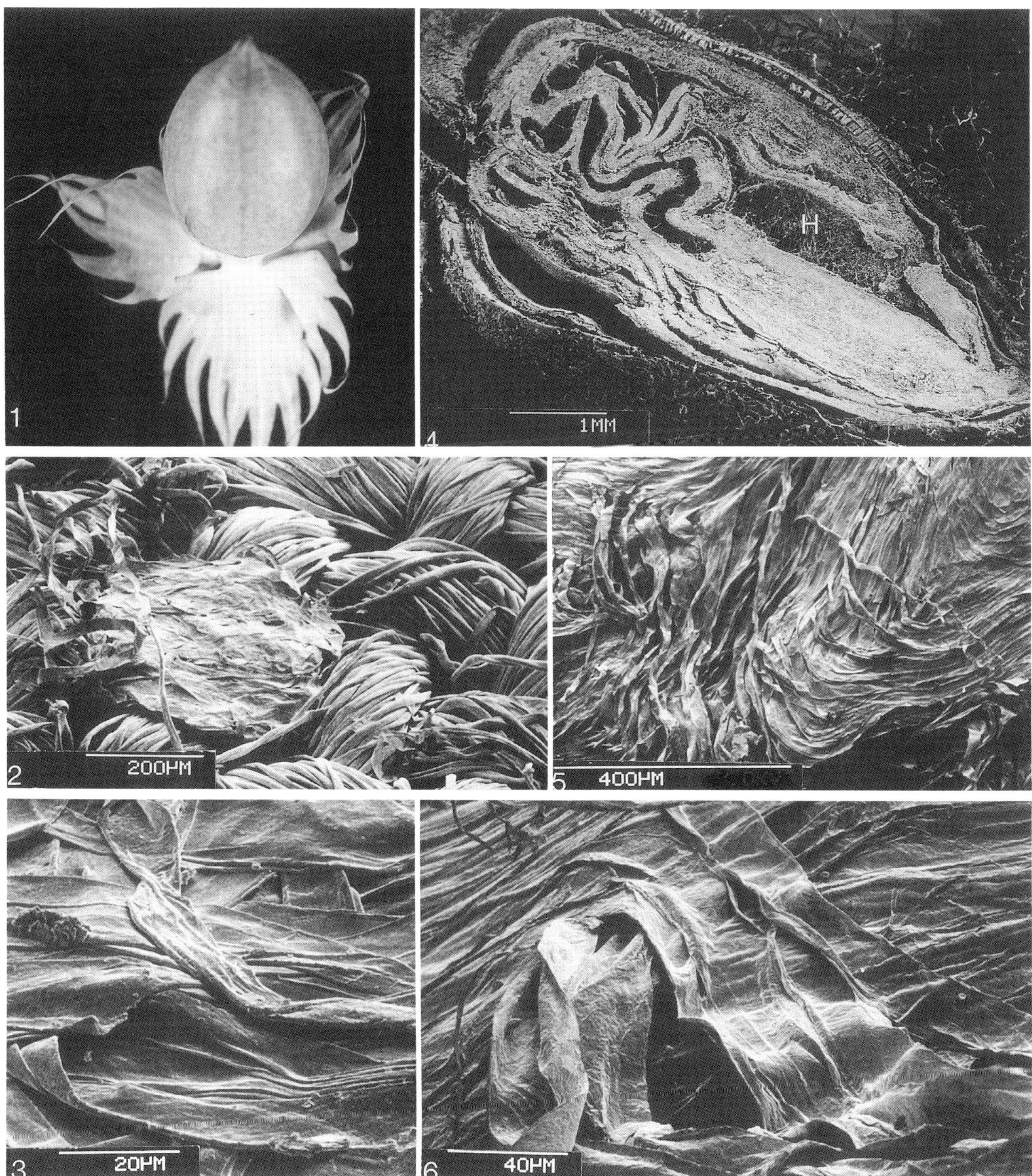

FIG. 1. An unopened cotton boll
FIG. 2. White speck mass on a fabric surface
FIG. 3. White speck bundle at higher magnification
FIG. 4. Sliced, fungal infected cotton seed
FIG. 5. Undeveloped fiber mass from fungal infected seed
FIG. 6. Undeveloped fiber mass as higher magnification

SEM STUDY OF THE MORPHOLOGICAL EFFECTS OF KINETIN AND ZEATIN ON THE GROWTH AND DEVELOPMENT OF THE APICAL MERISTEM IN WHEAT

G. M. Hutchins* and J. S. Gardner**

* Provo High School, 1125 N. University Ave., Provo, UT, 84604
** Microscopy Lab, 128 WIDB, Brigham Young University, Provo, UT, 84602

Cytokinins are plant hormones that play a large and incompletely understood role in the life-cycle of plants.[2] The goal of this study was to determine what roles cytokinins play in the morphological development of wheat. To achieve any real success in altering the development and growth of wheat, the cytokinins must be applied directly to the apical meristem, or spike of the plant. It is in this region that the plant cells are actively undergoing mitosis. Kinetin and Zeatin were the two cytokinins chosen for this experiment. Kinetin is an artificial hormone that was originally extracted from old or heated DNA. Kinetin is easily made from the reaction of adenine and furfuryl alcohol.[3] Zeatin is a naturally occurring hormone found in corn, wheat, and many other plants.[2]

Chinese Spring Wheat (*Triticum aestivum* L.) was used for this experiment. Prior to planting, the seeds were germinated in a moist environment for 72 hours. The wheat plants were grown in a growth chamber at 20° C with a 16 hour photoperiod until the formation of the third leaf (25 - 29 days). The full development of the third leaf (perpendicular to the stalk) signals the transformation of the apical meristem from the vegetative to the reproductive stage.[1] At the third leaf stage 5 - 15 ng of cytokinin were applied in solution directly to the apical meristem. Kinetin was injected in a solution of ETOH and distilled water. Zeatin was injected in a solution of NaOH and distilled water. The control solutions contained only ETOH or NaOH in distilled water. Ten to thirteen days following the injections the apical meristems were removed and placed in 2% glutaraldehyde buffered at pH 7.3 with sodium cacodylate. The samples were then critical point dried and gold plated using standard SEM techniques. Measurements were obtained with a Link Analytical ExL system.

The application of kinetin created no substantial effect on the growth or development of the apical meristem. There was no correlation between the amount of kinetin applied and the height of the apical meristem. Zeatin, when applied in 5 (fig. 2) and 10 ng (fig. 3) amounts caused an increase in the height and number of spikelets of the apical meristem when compared with the control (figs. 1,5,6). At 15 ng per plant Zeatin proved detrimental and caused a substantial decrease in height, and number of spikelets (fig 4,5,6).

A lack of kinetin receptors in the developing wheat spike is a possible explanation for the ineffectiveness of kinetin to initiate a change in the growth or development. Cytokinins are normally present in only very small quantities. Fifteen nanograms of zeatin was outside the normal range and thus the stunted growth that can be seen in figure 4 occurred.

Kinetin, or a similar naturally occurring hormone, seems to play no part in the early growth or development of the apical meristem of Chinese Spring Wheat. Zeatin appears to influences the height and number of spikelets of the apical meristem (fig 5, 6).

References:

1. J. S. Gardner et al., *Amer. J. Bot.* 72(1985)548.
2. M. Kaminek et al., *Physiology and Biochemistry of Cytokinins in Plants.* SPB Academic Published, Czechoslovakia, (1990)7.
3. T. C. Moore, *Biochemistry and Physiology of Plant Hormones.* Springer-Verlag, New York: (1989)160.

Proc. Microscopy and Microanalysis 1995, edited by G.W. Bailey, M.H. Ellisman, R.A. Hennigar, and N.J. Zaluzec

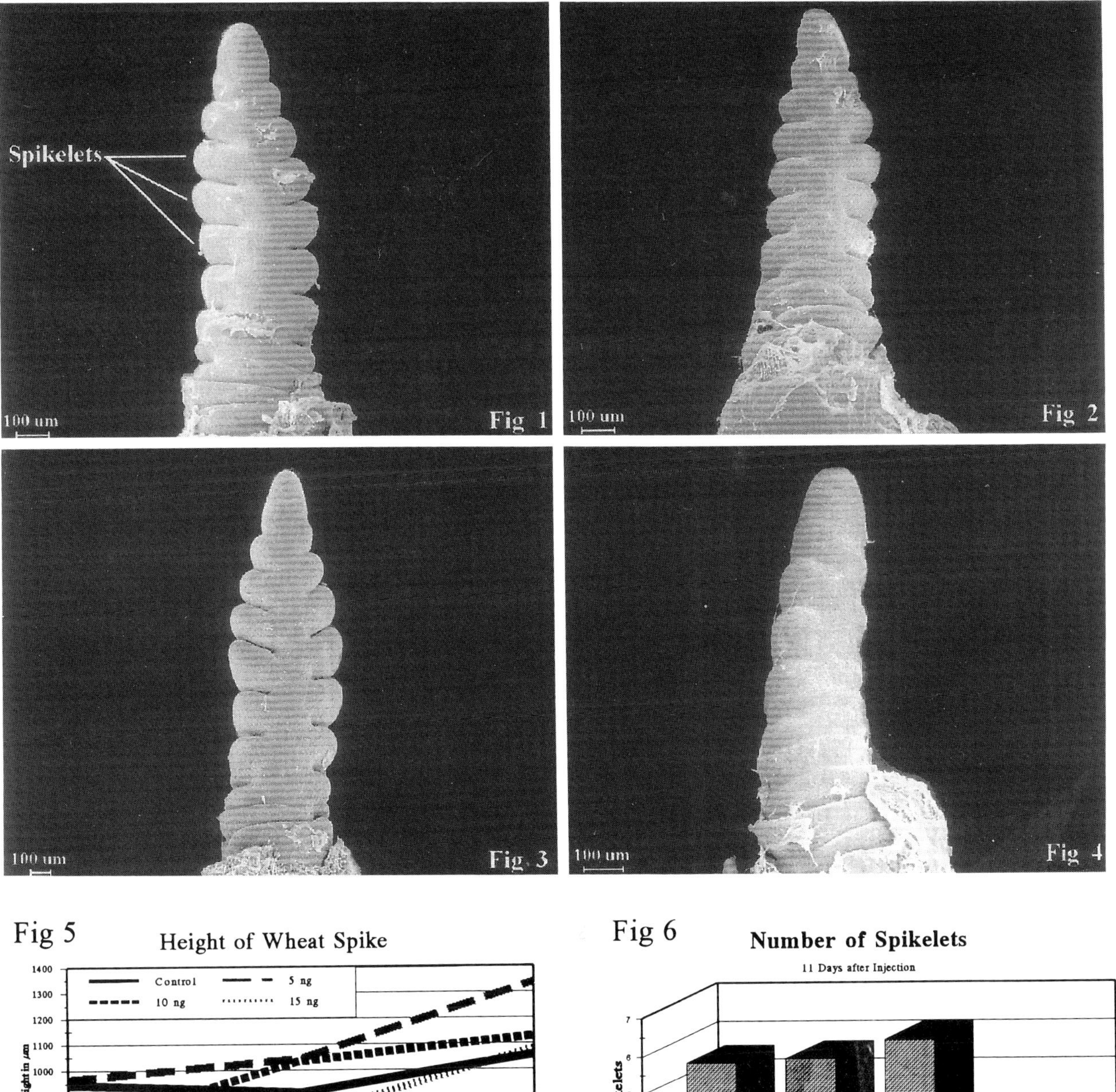

Figures 1-4: SEM Images of the Apical Meristem of Wheat
Figure 1: Control wheat spike at x95
Figure 2: Wheat spike injected with 5 ng Zeatin at x95
Figure 3: Wheat spike injected with 10 ng Zeatin at x80
Figure 4: Wheat spike injected with 15 ng Zeatin at x140
Figure 5: Plot of the height of the wheat spike when injected with zeatin
Figure 6: Plot of the number of spikelets when injected with zeatin

Evaluation of the Emerging Coleorhiza and Radicle of *Bromus tectorum* Using Scanning Electron and Light Microscopy

Maren Christensen*, John S. Gardner**, and Phil S. Allen*

* Dept. of Agronomy and Horticulture, 275 WIDB, Brigham Young University, Provo, UT 84602
** Microscopy Laboratory, 128 WIDB, Brigham Young University, Provo, UT 84602

The coleorhiza is a nonvascular sheath that encloses the embryonic radicle in Gramineae, and is generally the first tissue to emerge during germination. Coleorhiza cells develop extensions similar to root hairs, which are difficult to study quantitatively *in situ* because they are extremely sensitive to desiccation and breakage. Protective, absorptive and adhesive roles have been postulated for this tissue, but have received limited attention.[1] This study used scanning electron microscopy (SEM) and light microscopy (LM) to characterize the coleorhiza and radicle during germination of *Bromus tectorum,* a species that often emerges from the soil surface in semiarid habitats.

Caryopses (seeds) were imbibed at various water potentials over solutions of polyethylene glycol (PEG) or in soil until coleorhizae had emerged.[2] For SEM, seeds imbibed over PEG were frozen in liquid freon and then freeze dried.[1] Seeds imbibed in soil were frozen in liquid nitrogen and then freeze dried. For LM, seeds imbibed over PEG were subjected to freeze substitution in dry acetone with 1% OsO_4 at -80°C for 19 days. Seeds were then embedded in Spurr's resin, sectioned 1-2μm thick, and stained with Toluidine Blue and Azure II.

SEM allowed observation of several features of the emerging coleorhiza and radicle. Variation in coleorhiza development as a function of water availability was quantified, as were features of coleorhiza hairs. Hairs were often the first part of the coleorhiza to emerge from the seed. On expanded coleorhizae, cells distal to the seed showed the most hair growth (Fig. 1,2). Hair proliferation was more extensive in soil than over PEG solutions. Mucilage was observed on the surface of the coleorhiza (Fig. 1,2), surrounding the root cap of emerging radicles (Fig.2), and at the interface between hairs and soil particles (Fig. 3,4).

In cross sections of the radicle and coleorhiza viewed with a light microscope, emerged coleorhiza cells were larger in diameter than the radicle cells they surrounded (Fig. 5). This was apparently due to large vacuoles contained within coleorhiza and coleorhiza hairs. However, cells of the coleorhiza and radicle were similar in diameter prior to emergence from the seed (Fig.6). Even at this early stage, the radicle appeared to be surrounded by mucilage (Fig. 5).

These observations provide additional support for the adhesive role of coleorhiza hairs in facilitating germination of *Bromus tectorum*. In addition, they indicate an important role of mucilage as an adhesive agent as well as a lubricant.

References

1. S.B. Debaene-Gill et al., *Amer. J. Bot.* 81(1994)739.
2. S.P. Hardegree and W.E. Emerich, *J. Exp. Bot.* 43(1992)233.

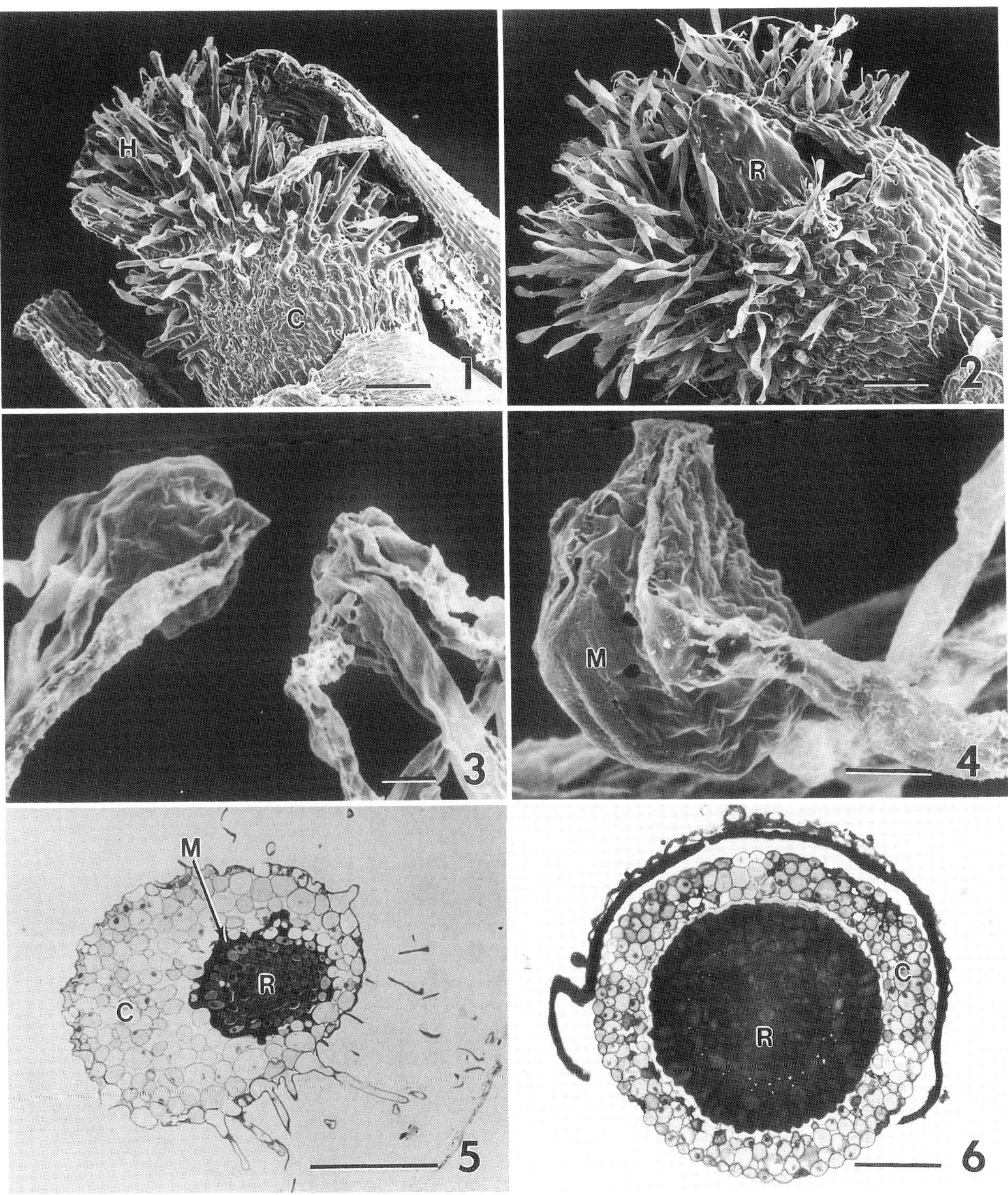

Figs. 1-4 SEM images. Figs. 5-6 LM images. Fig. 1 Hairs (H) are longest near the tip of the emerging *Bromus tectorum* coleorhiza (C). Bar = 100µm Fig. 2 Mucilage covers the emerging radicle (R). Bar = 100µm Fig. 3 Coleorhiza hairs adhere to soil particles. Bar = 10µm Fig. 4 Mucilage (M) is abundant at the soil-coleorhiza hair interface. Bar = 10µm Fig. 5 Mucilage surrounds the radicle in an emergent coleorhiza. Bar = 100µm Fig. 6 Coleorhiza cells surround the radicle within the seed. Bar = 100µm

ULTRASTRUCTURAL CHANGES IN ROOTS OF PEPPERS RESISTANT AND SUSCEPTIBLE TO *PHYTOPHTHORA CAPSICI.*

H. Ilarslan,* A.S.Ustun and R. Yilmazer**

*Departments of Agronomy and Genetics,Iowa State University,Ames,IA 50011,USA
**Ankara University,Department of Biology,06100 Ankara,TURKEY

The infection by *Phytophthora capsici* Leonian causes foliar blight and crown and root rot of pepper plants.[1] The ultrastructural examination of resistant and susceptible host-pathogen interactions was conducted in the pepper cultivars Ince Sivri-35, PM217, and PM702=CM 334 following inoculation with *Phytophthora capsici*. Responses were characterized and compared with healthy non-inoculated controls at 2, 4 and 6 days after inoculation. Numerous ultrastructural studies have been made of the interaction of various host plants with *Phytophthora* spp.[2-3] No ultrastructural studies comparing the resistant and susceptible reactions of pepper cultivars to *P. capsici.* have been reported. It is important to examine the ultrastructural changes in inoculated and infected tissue of resistant and susceptible pepper cultivars to *P. capsici*. Information reported here characterizes the processes of pathogen containment in resistant interactions and compares these with the processes occuring in susceptible interactions.

After 2 days in susceptible interactions, the pathogen grew intercellularly in roots, whereas in resistant interactions only a few intercellular hyphae were observed penetrating the host cells and forming haustoria. The consistent features of ultrastructural changes in susceptible interactions at 2, 4 and 6 days after inoculation were degeneration of cell organelles and dissolution of host cell walls (Figs. 1,2 and 3). The host cell walls were poorly stained and swollen, possibly due to alteration of fungal macerating enzymes. Infected parenchyma cells of susceptible cultivars were completely macerated whereas lignified walls of the xylem vessels were not severely disrupted, even though the vessels were filled with fungal hyphae. In the resistant cultivars PM 217, at 2 and 4 days after inoculation, the organelles appeared healthy, but at 6 days after inoculation organelles had degenerated (Fig. 4). In another resistant cultivar, PM 702 (=CM 334), at 2, 4 and 6 days after inoculation fungal hyphae grew usually intercellularly, and mitochondria and chloroplasts, but not microbodies, were degenerated and papilla had formed (Fig. 5). In general, reactions of the resistant cultivar to *P. capsici* were intercellular, rather than intracellular fungal growth (therefore host cell walls were not severely damaged), papilla (callose) formation, resistance of the cell wall to dissolution and microbodies with normal appearing inclusions. These observations could be either the cause or the result of the resistance of the cultivars to the fungus. The undistorted microbodies may be a requisite for normal peroxidase activity. In order to understand the mechanisms behind the resistance it may be necessary to investigate the peroxidase activity and isozymes. Several investigations indicate that peroxidase may affect resistance through cell wall lignification.[5] In resistant genotypes no fungal hyphae are observed in the xylem lumens of the root rots and in our opinion this can be considered as an indication that the vascular conduction is not inhibited and consequently this can also be a factor in the resistance of these plants againtst the disease.

References

1. T.H. Barksdale, *Plant Dis.* 68(1984)506.
2. B.K. Hwang et al., *J.Phytopat.*127(1989)305.
3. D.R. Jones et al., *Phytopathology* 64(1974)1084.
4. R. Hammerschmidt et al., *Physiol.Plant Pathol.* 20 (1982)73.
5. This research was supported by "The Scientific and Technical Research Council of Turkey (TUBITAK),"Project Nos. TBAG-1171.

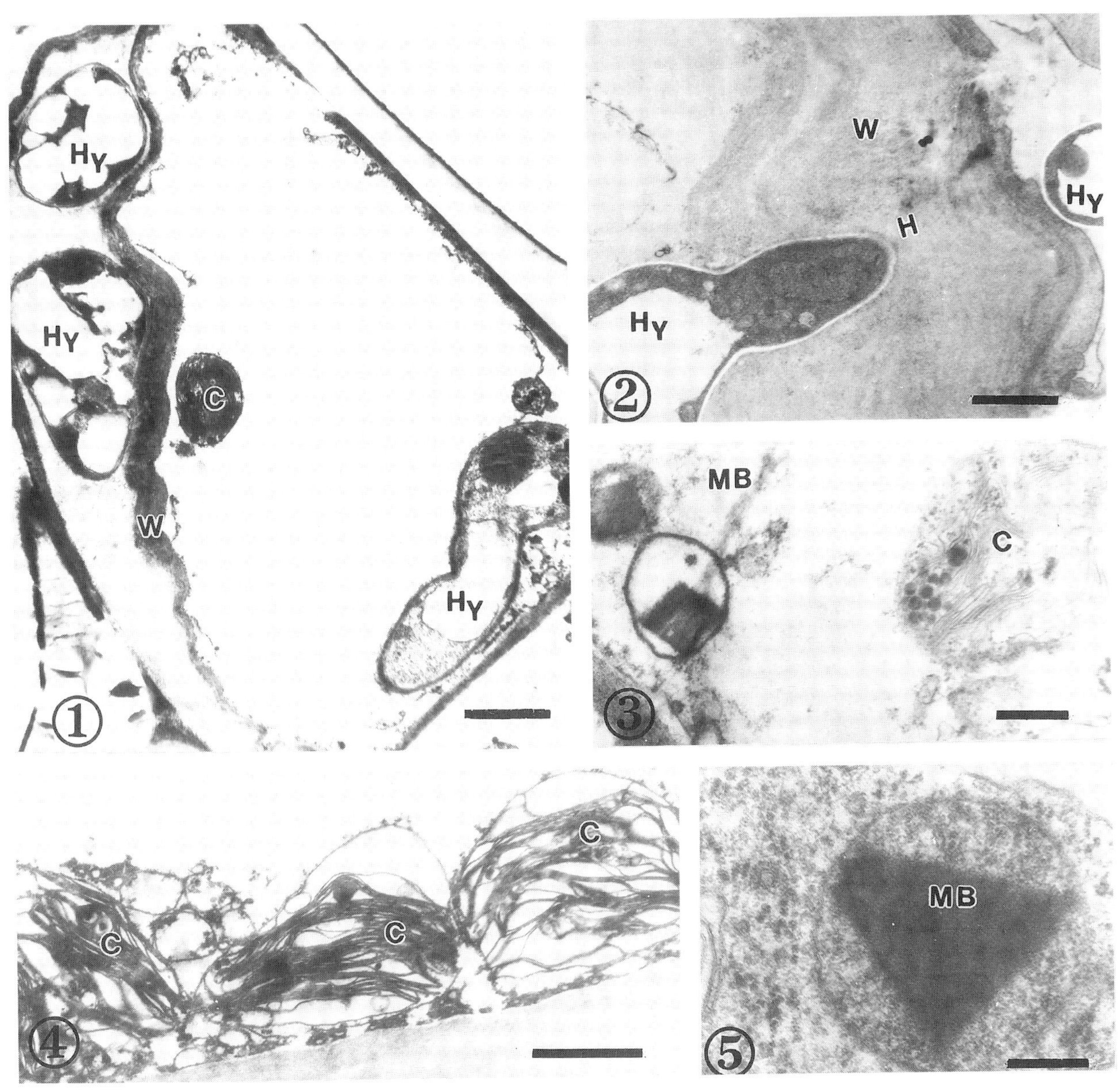

Electron micrographs of resistance reactions of susceptible Ince-Sivri 35 (1,2 and3) and resistant PM 702 cultivars (4 and 5) two days after inoculation.

FIG. 1.-The fungal hyphae(Hy) which have grown intercellularly, within the host middle lamella adjacent to the cell wall (W) , are seen in a phloem cell.Bar= 2 μm.

FIG. 2.- Hyphae (Hy) penetrating the parenchyma wall (W). Note the swollen and dissolved host walls (W) and middle lamella. Bar= 2μm.

FIG. 3. -Degenerated cytoplasmic organelles [microbody (MB), chloroplast (C)] are seen in the resistant PM 702 cultivar. Bar=.5μm.

FIG. 4.-Electron micrograph of resistant reaction 4 days after inoculation. Note the swollen, sac-like chloroplast (C) containing disorganized lamellae. Bar= 1.5μm.

FIG.5.-Normal microbody (MB) with crystalline inclusion. Bar= 0.2 μm.

AN SEM STUDY OF RAPHIDE CRYSTAL INITIALS IN THE LEAVES OF *VITIS* (GRAPE)

H. J. Arnott,* M. A. Webb** and L. E. Lopez*

*Department of Biology and Center for Electron Microscopy, The University of Texas at Arlington, Arlington, TX 76019
**Department of Botany and Plant Pathology, Purdue University, West Lafayette, IN 47907

Many papers have been published on the structure of calcium oxalate crystals in plants,[1-3] however, few deal with the early development of crystals.[4] Large numbers of idioblastic calcium oxalate crystal cells are found in the leaves of *Vitis mustangensis, V. labrusca and V. vulpina*.[3] A crystal idioblast, or raphide cell, will produce 150-300 needle-like calcium oxalate crystals within a central vacuole. Each raphide crystal is autonomous, having been produced in a separate membrane-defined crystal chamber; the idioblast's crystal complement is collectively embedded in a water soluble glycoprotein matrix which fills the vacuole.[1-3] The crystals are twins, each having a pointed and a bidentate end (Fig 1); when mature they are about 0.5-1.2 μm in diameter and 30-70 μm in length. Crystal bundles, i.e., crystals and their matrix, can be isolated from leaves using 100% ETOH. If the bundles are treated with H_2O the matrix surrounding the crystals rapidly disperses.[3]

Crystal bundles are prepared for viewing by scanning electron microscopy (SEM), by placing bundles in 100% ETOH on a glass slide and allowing them to dry. H_2O is pipetted on to the specimen, which is allowed to dry and then is sputter coated with gold palladium. When the crystal bundles are examined, mature raphide crystals are seen as well as 5-40 very small raphide initials and several raphides of intermediate length (Fig.1). Raphide initials less than 1μm in length have been observed, however, most are 2-3 μm in length and about 0.5 μm in diameter (Figs.1-3). The average raphide initial is torpedo-shaped, with a blunt-pointed and a bidentate end (Figs. 2-3). The bidentate end of raphide initials is a replica of that seen in mature crystals. The bidentate end has a characteristic reentrant angle indicating that each initial, like the mature crystals, is a twin. The twin character of the initial, intermediate and mature crystals has been confirmed using polarization microscopy. Separate extinction points can be demonstrated for each half of very small initials (data not shown). Crystals twins of intermediate length apparently represent stages in the raphide growth. Compared with young initials, intermediate crystals initials have a more extended, sharper, pointed end than the small initials (Figs. 1, 4-5). In contrast to views of previous workers,[4] the change in shape of the pointed end, exhibited by initials, intermediate and mature crystals, indicates that some growth must take place at that end. Less commonly, small initials have two pointed ends (Fig. 1) and sometimes holes in or grooves on their surfaces.

References

1. H. J. Arnott and F. J. E. Pautard, in H. Schraer, Ed., *Biological Calcification.* Appleton-Century-Crofts, New York. (1970) 375.
2. H. J. Arnott. in G.H. Nancollas, Ed., *Biological Mineralization and Demineralization* (1982) Springer-Verlag, Berlin.
3. M .A. Webb, et al., The Plant Journal (1995) *in press.*
4. A. Cody and H. T. Horner, Bot. Gaz. 144 (1983) 318.
5. We acknowlege the continuing support of The Graduate School of The University of Texas at Arlington.

Proc. Microscopy and Microanalysis 1995, edited by G.W. Bailey, M.H. Ellisman, R.A. Hennigar, and N.J. Zaluzec

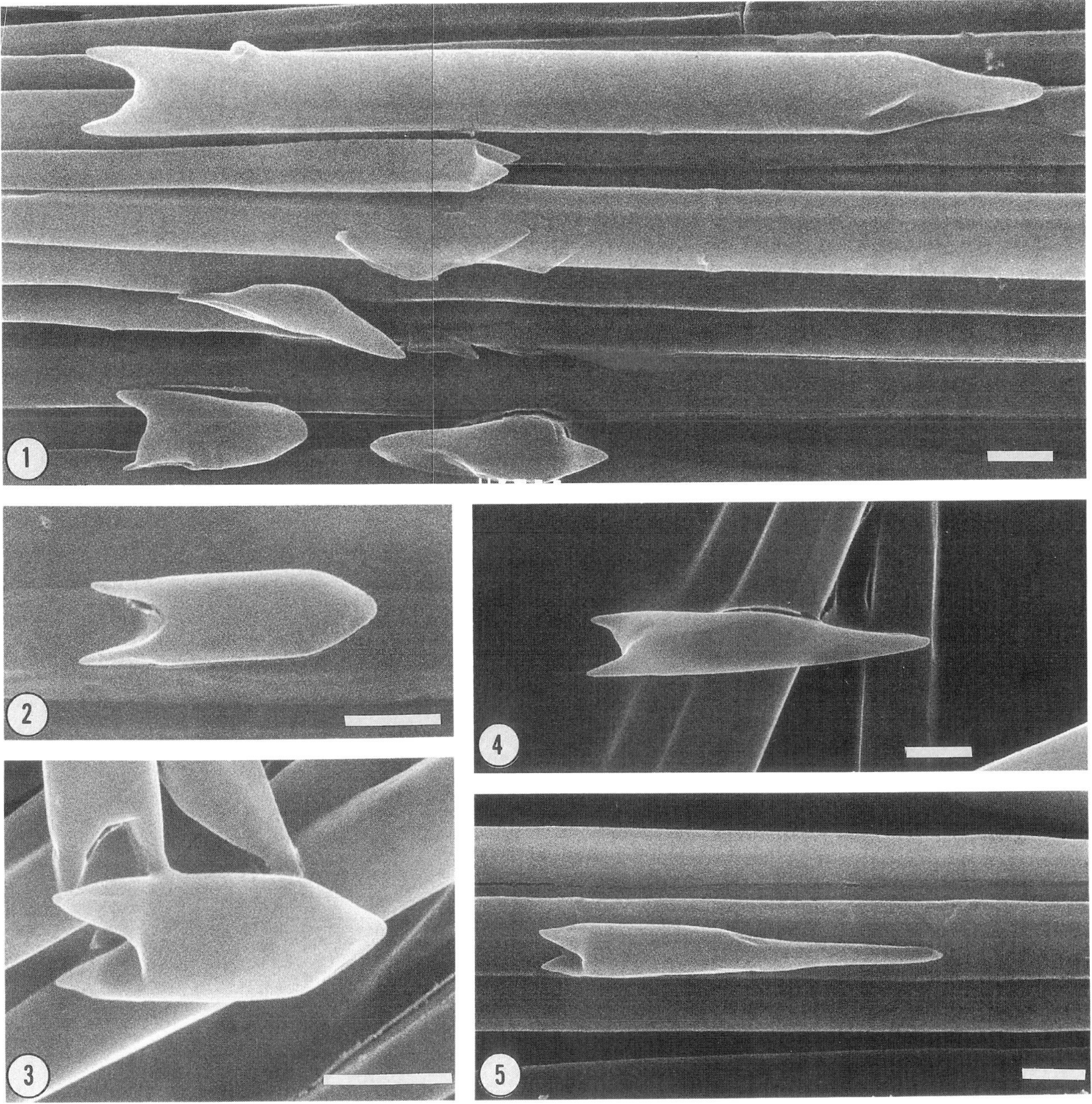

FIGS. 1.-5.--SEM images of raphide crystals isolated from *Vitis mustangensis*. Bars = 1 μm.
FIG. 1.--Portion of a bundle of raphide crystals. Mature crystals extend to left and right beyond field of view. Intermediate crystal and three raphide crystal initals can be seen on surface of mature crystals.
FIGS. 2.-3.--Raphide initials. Note bidentate and blunt-pointed ends characteristically found as a part of raphide initials.
FIG. 4.--Developing raphide initial with more sharply pointed end. Note unchanged and characteristic bidentate end. Portions of mature crystals in background.
FIG. 5.--Developing raphide initial in which significant growth of pointed end has occurred; bidendate end appears unchanged. Initial lies on surface of several mature crystals.

ELECTRON MICROSCOPIC AND MICROSPECTROPHOTOMETRIC ANALYSIS OF BIOLOGICAL DELIGNIFICATION OF PLANTS

D. E. Akin,* L. L. Rigsby,* W. H. Morrison III*, A. Sethuraman,** and K.-E. L. Eriksson**

*Russell Research Center, Agricultural Research Service-USDA, Athens, GA 30604
** Department of Biochemistry and Molecular Biology, University of Georgia, Athens, GA 30602

Aromatic constituents such as lignin bind to carbohydrates within plant cell walls and thus render the plant carbohydrates less utilizable as food and energy.[1] Chemical methods used to upgrade the quality of plant biomass are costly, expensive, and unsafe. White rot fungi, which are the only known microorganisms that, to any extent, can remove lignin from plant cell walls, offer a biological solution to upgrading plant quality.[2] Microscopic analyses provide information on the site of delignification that is strategically important in improving use of plant biomass.

Stems of grasses and a legume were treated with the white rot fungi *Ceriporiopsis subvermispora* and *Cyathus stercoreus* for 6 weeks. Treated residues were analyzed for structural modifications using scanning and transmission electron microscopy and for aromatic constituents using ultraviolet (UV) absorption microspectrophotometry and gas chromatography. These modifications were related to improved utilization of cell walls by rumen microorganisms.

UV absorption microspectrophotometry, in conjunction with gas-chromatography of alkali-treated plants, indicated that ester-linked ferulic and *p*-coumaric acids were particularly susceptible to removal by both fungal species. These ester-linked phenolic acids appeared to comprise most of the aromatic constituents in parenchyma cells of bermudagrass stems. Electron microscopy showed that the delignified parenchyma cells, which were not removed already by the fungi, were totally degraded later by rumen microorganisms (Fig. 1). In wheat stem parenchyma, a similar phenomenon occurred in that ester-linked ferulic acid was removed (Fig.2) with a subsequent increase in biodegradation. In more highly lignified cell types, such as grass sclerenchyma, phenolic acid esters were removed from the walls, but the more polymerized aromatics were often still present to some extent. The result was an improved biodegradation of these cell types but not a complete removal (Fig. 1). Stems of alfalfa, which is a legume and has a different type of lignin than grasses, showed little effect by fungi using UV absorption analysis (Fig. 3) and by SEM.

Combining UV absorption microspectroscopy and electron microscopy with gas chromatography provided information on the sites and types of aromatics most susceptible to biological delignification. These microscopic techniques allow structural analysis of strategic barriers to improved utilization and could provide rapid analytical screening of fungal impact on plants.

References

1. Akin, D.E. *Agron. J.* 81:17-25 (1989).
2. Eriksson, K.-E.L. et al. *Microbial and Enzymatic Degradation of Wood and Wood Components.* New York: Springer-Verlag (1990).

Proc. Microscopy and Microanalysis 1995, edited by G.W. Bailey, M.H. Ellisman, R.A. Hennigar, and N.J. Zaluzec
Copyright © 1995 MSA. Published by Jones and Begell Publishing, 79 Madison Ave., New York, NY 10016

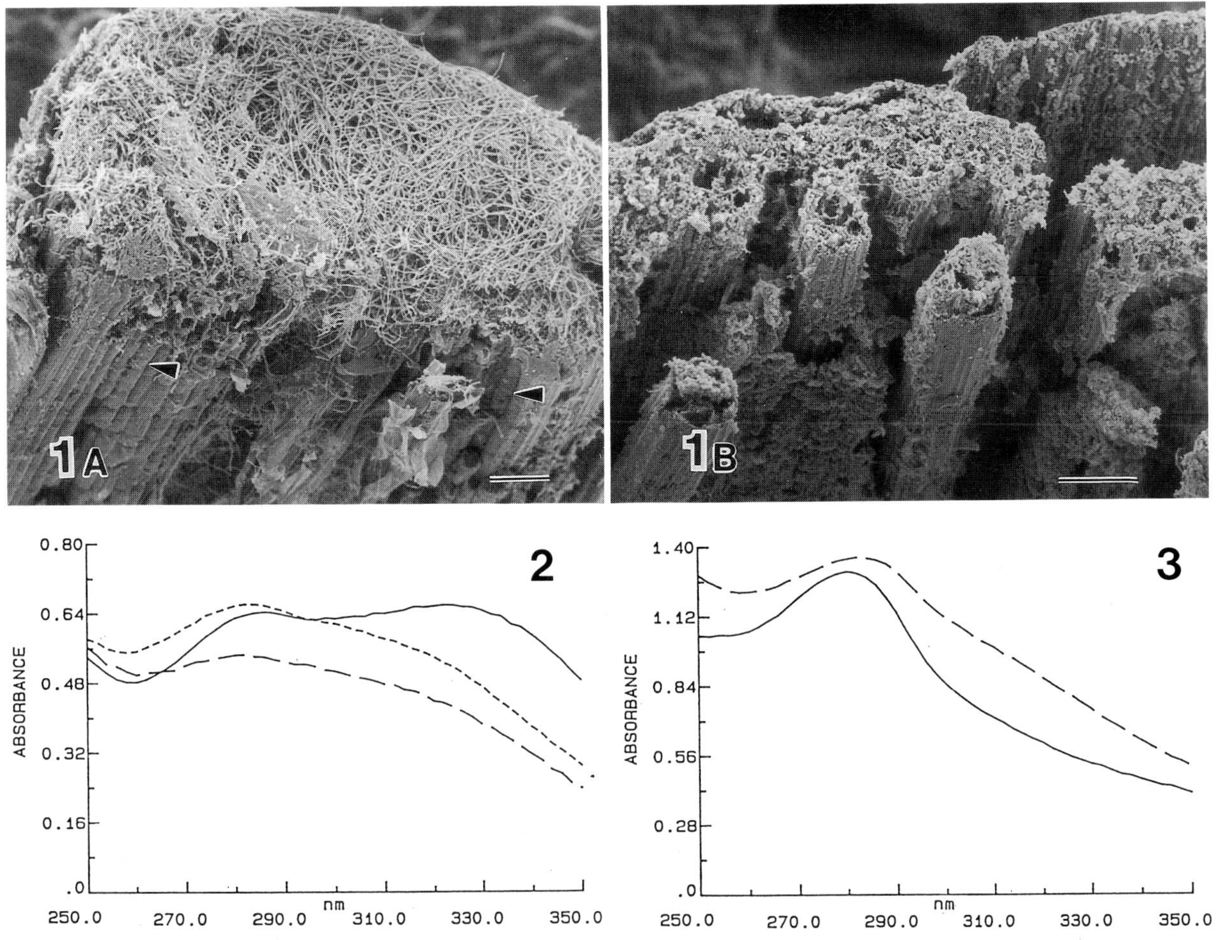

FIG. 1.—SEM of bermudagrass stem. (A) Incubated for 6 weeks with the white rot fungus *Ceriporiopsis subvermispora* showing colonization of end of stem but with little loss of plant biomass. Arrows indicate some loss of parenchyma walls. (B) Treated as in (A) and then incubated 48 h with rumen microorganisms showing loss of all parenchyma and residues of sclerenchyma and vascular bundles. Bar = 200 μm.

FIG. 2.—UV absorption spectra of parenchyma cell walls of wheat stem untreated (——), treated 6 weeks with *C. subvermispora* CZ-3 (– – –), or *C. subvermispora* 90031-sp (- - -). Loss of UV absorption near 320 nm indicates removal of ester-linked phenolic acids.

FIG. 3.—UV absorption spectra of interfascicular cells in alfalfa stem showing similar spectra in untreated (——) and C.subvermispora CZ-3-treated (– – –) walls.

ENDOPHYTIC RELATIONSHIP BETWEEN AN ENTOMOPATHOGENIC FUNGUS *BEAUVERIA BASSIANA* AND CORN (*ZEA MAYS*)

Bruce L. Wagner * and Leslie C. Lewis**

*Bessey Microscopy Facility, Iowa State University, Ames, IA 50011
**USDA-ARS, Corn Insects Research Unit, Iowa State University, Ames, IA 50011

Fungi are known to colonize plants, usually as phytopathogens or symbionts. This is the first report of the entomopathogenic fungus, *Beauveria bassiana* (Balsamo) Vuillemin (Deuteromycotina: Hyphomycetes) penetrating and colonizing a plant in a manner similar to when it invades a typical insect host *Ostrinia nubilalis* Hübner (Lepidoptera: Pyralidae). Electron microscopy was used to document this fungus penetrating the leaves of corn (*Zea mays*) and establishing an endophytic relationship within the plant.

Several studies have shown *B. bassiana* to adhere to insect cuticle and penetrate with or without forming appressorial penetration structures.[1] Investigations have concluded that the young, germinating hyphae may produce sequentially a proteinase followed by chitinase to dissolve the cuticular envelope of target insects.[2] Within three days after inoculation, viable *B. bassiana* conidia germinated and formed vegetative mycelia which grew randomly over the corn leaf surface (Fig. 1). Often a germ tube is formed from a conidium and elongates only a short distance before terminating its growth and penetrating the leaf surface (Figs. 2 and 3). Ultrastructural characteristics of penetration by fungal hyphae are similar in both insect and plant cells.[3] This study indicates that *B. bassiana* is well adapted to penetrate leaf surfaces. Upon penetration, hyphae quickly elongate and spread throughout the leaf apoplast. Long hyphal structures were observed to follow the apoplast in any direction from the point of penetration. In some cases, the hyphae were observed within zylem elements (Fig. 4). Mortality studies indicated that *B. bassiana* retained its virulence against *O. nubilalis* after colonization of the corn plant.

Beauveria bassiana applied to whorl-stage corn by foliar application or injection colonized the plants, moved within the plant and persisted to provide season-long suppression of the European corn borer.[4] This association then provides internal protection against invading insects. Understanding this unique relationship will be invaluable in further development and utilization of such fungi to manage insect pests of agricultural crops.[5]

References

1. M. J. Bidochka and G. G. Khachatourians, *J. Invertebr. Pathol.* 58(1991)106.
2. R. J. Smith et. al., *J. Invertebr. Pathol.* 38(1981)335.
3. R. L. Nicholson, in D. W. Roberts and J. R. Aist, Eds., Infection Processes in Fungi, New York: The Rockefeller Foundation.(1984)74.
4. L. A. Bing and L. C. Lewis, *Biocontrol Sci. Technol.* 2(1992)39; L. C. Lewis and B. A. Bing, *Can. Entomol.* 123(1991)387.
5. The authors wish to acknowledge the Bessey Microscopy Facility of Iowa State University where the microscopic aspects of this study were performed.

Proc. Microscopy and Microanalysis 1995, edited by G.W. Bailey, M.H. Ellisman, R.A. Hennigar, and N.J. Zaluzec
Copyright © 1995 MSA. Published by Jones and Begell Publishing, 79 Madison Ave., New York, NY 10016

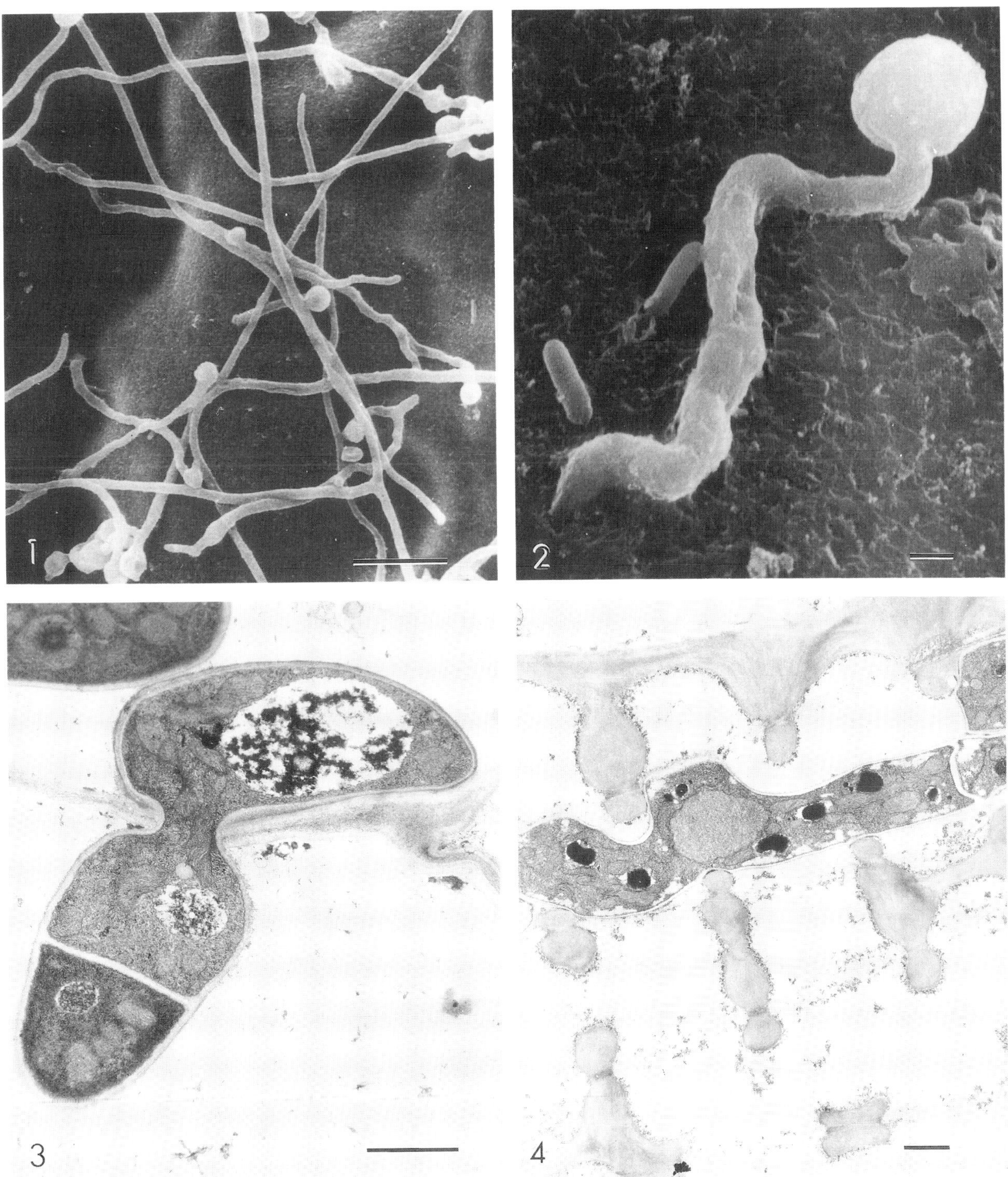

FIG. 1. - *Beauveria bassiana* conidia growing on corn leaf surface. Bar = 10µm.

FIG. 2. - Germ tube growing from conidia and terminating at penetration site. Bar = 1µm.

FIG. 3. - Hyphal penetration of epidermal cell wall. Bar = 1µm.

FIG. 4. - Fungal hyphae in zylem elements. Bar = 1µm.

WHEAT NECROSIS INDUCED BY *PYRENOPHORA TRITICI-REPENTIS* TOXIN

Thomas Freeman*, Jack Rasmussen*, Leonard Francl*, and Steve Meinhardt**

*Plant Pathology and **Biochemistry Departments, North Dakota State University, Fargo, ND 58105

Tan spot is a serious foliar disease of wheat caused by *Pyrenophora tritici-repentis*. This fungal pathogen produces a proteinaceous phytotoxin that induces necrosis in susceptible but not resistant cultivars.[1] Larez et al. reported generalized membrane breakdown and cell death in advance of fungal spread, an effect presumably due to excreted fungal products, including the necrosis toxin.[2] The direct ultrastructural effects of necrosis toxin isolated from *P. tritici-repentis* have not been previously examined.

Crude extract from a culture of isolate 86-124 (1) was thawed and centrifuged for 20 min at 35000g. The supernatant was filtered through a 0.45 μm Whatman nylon filter and placed in dialysis tubing with a 3,500 molecular weight cut off. The crude extract was dialyzed overnight against 10 volumes of 20 mM sodium acetate buffer, pH 4.8. The dialyzed extract was loaded onto a fresh 10 ml S-sepharose Fast Flow column (Pharmacia) previously equilibrated with 20 mM sodium acetate. After loading, the column was washed with 4 column volumes of buffer at a flow rate of 1 ml per min. The sample was eluted with a linear gradient from 0 to 300 mM NaCl over 4 column volumes. The protein elution was monitored spectrographically at 280 nm. Samples from 80 to 150 mM NaCl were combined and concentrated approximately 10-fold by vacuum distillation at room temperature. The concentrated sample was chromatographed on a HiLoad 16/60 Superdex-30 gel filtration column (Pharmacia) at a flow rate of 1 ml per minute. A peak centered near 14,000 Da was collected. This preparation, approximately 80% pure based on an analysis of relative peak size on an HPLC, was used in the following experiments.

The hard red spring wheats Erik (resistant) and ND 495 (susceptible) were grown in soilless media for two weeks. The second emerged leaf was infused with about 100μl of toxin diluted 50-fold with distilled water. Leaf tissue was harvested 0, 4, 12 and 24 hr after infiltration and prepared for electron microscopy using standard techniques.

Ultrastructural effects of the toxin were almost immediately evident in the susceptible wheat variety. Some of the earliest ultrastructural changes were expressed in the chloroplasts and included a swelling of the granal membranes and an evagination of the outer limiting membrane (Fig. 1). Within 12 hr after infiltration large groups of cells showed significant ultrastructural changes. The stroma portion of chloroplast were significantly enlarged, granal membranes were dilated, and limiting chloroplast membranes were frequently lost (Fig. 2). Twenty-four hours after infiltration there were general areas of leaf necrosis in the susceptible plants. In these areas all cellular organelles were severely damaged (Fig. 3). Resistant plants showed a similar pattern of necrosis, however, ultrastructural damage was limited to individual cells or small groups of cells and general necrosis did not occur. Even 12 to 24 hr after infiltration, most cells of the resistant variety appeared unaltered ultrastructurally (Fig. 4).

References

1. Lamari, L., and Bernier, C. C. 1991. Genetics of tan necrosis and extensive chlorosis in tan spot of wheat caused by *Pyrenophora tritici-repentis*. Phytopathology 81:1092-1095.
2. Larez, C. R., Hosford, R. M., Jr., and Freeman, T. P. 1986. Infection of wheat and oats by *Pyrenophora tritici-repentis* and initial characterization of resistance. Phytopathology 76:931-938.

Proc. Microscopy and Microanalysis 1995, edited by G.W. Bailey, M.H. Ellisman, R.A. Hennigar, and N.J. Zaluzec
Copyright © 1995 MSA. Published by Jones and Begell Publishing, 79 Madison Ave., New York, NY 10016

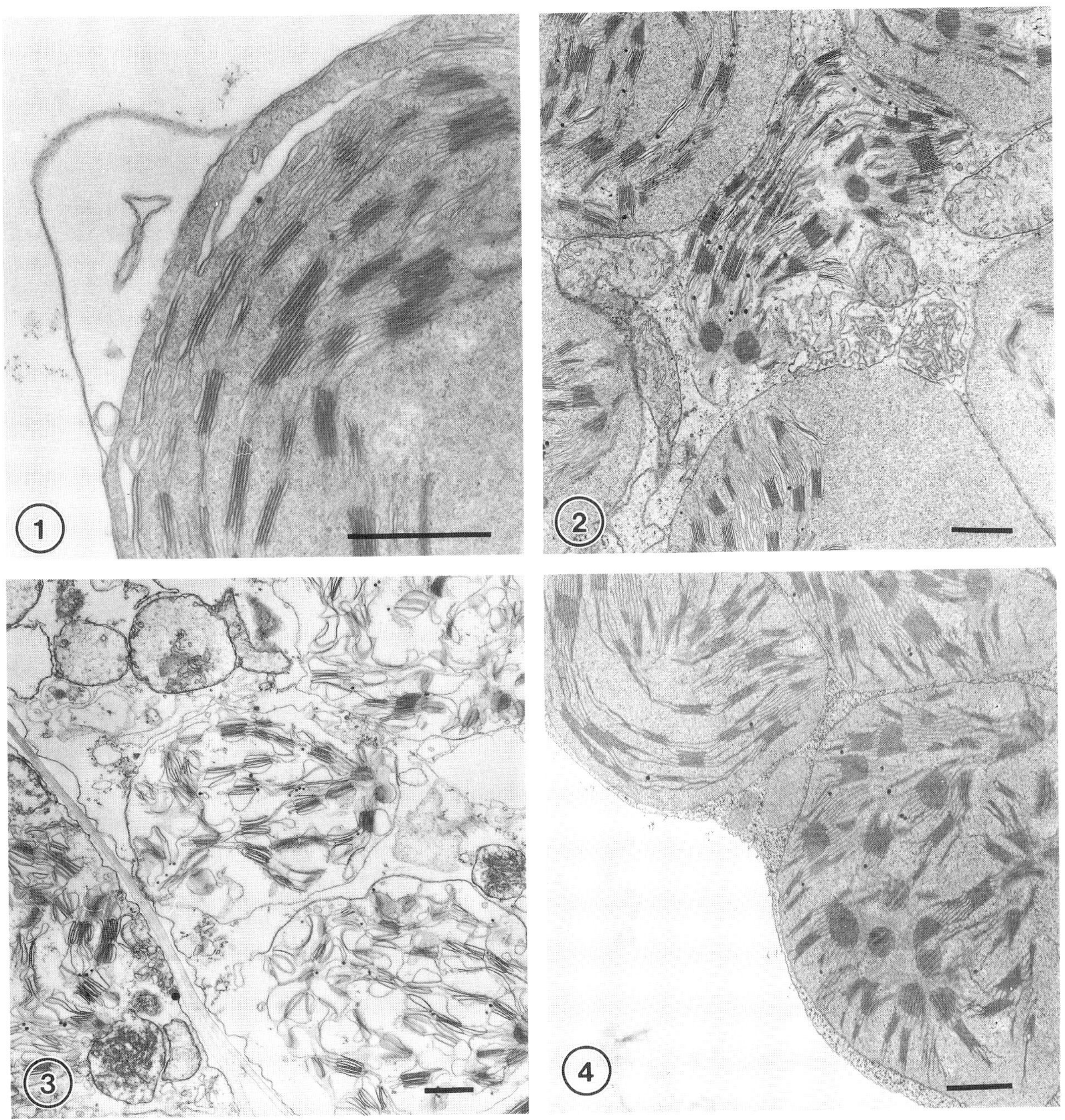

FIG. 1-3 - Toxin induced damage in the susceptible wheat variety ND 495.
FIG. 1 - Chloroplast with swollen grana and evaginated outer membrane, 4 hr after infiltration.
FIG. 2 - Toxin induced ultrastructural damage to cells 12 hr after infiltration.
FIG. 3 - Necrotic cells 24 hr after infiltration with toxin.
FIG. 4 - Undamaged cells of the resistant variety (Erik) 12 hr after infiltration.

Bars = 1 μm.

EFFECTS OF THAXTOMIN A ON THE FINE STRUCTURE OF POTATO TUBERS

B. D. Stein*, R. R. King**, and R. Hammerschmidt***

*Lion Imaging, Inc., 1921 Boston Post Road, Westbrook, CT 06498
**Agriculture Canada, P.O Box 20280, Fredericton, NB, E3B 4Z7, Canada
***Botany and Plant Pathology, Michigan State University, East Lansing, MI 48824

Streptomyces scabies[2] and *S. acidiscabies*[3] cause common scab in potato plants. At the start of infection a brown lesion develops ahead of the actual site of infection on tubers. This observation led to the isolation of the toxin thaxtomin A from infected potato tubers.[4] Application of the toxin to tubers replicates many of the symptoms seen in infected tubers. Potato mini-tubers were cultured by the methods of C. H. Lawrence (personal communication) where nodal segments of stolons from sprouted tubers were placed in White's media with 8% sucrose and 0.4% agar.[1] Thaxtomin A, $20ul$ at $10^{-5}M$, was applied to the surface of the mini-tubers. Samples, 24 hours later, were fixed in 2.7% glutaraldehyde in 0.1M Na cacodylate buffer, pH 7.0, and postfixed in 1% OsO_4. Part of the samples were dehydrated by an ethanol series, critical point dried, placed on aluminum stubs, gold coated, viewed and photographed by SEM. The remainder of the samples were dehydrated in a graded acetone series, infiltrated in epoxy resin (as formulated by John Heckman, 5.0g VCD, 7.4g Quetol 651, 26g NSA, and 0.4g DMAE). Silver-gold sections were viewed and photographed for TEM. For light microscopy $0.5um$ semi-thick sections were stained with 0.1% toluidine blue, 1.0% sodium borate in dH_2O. Lesions produced by thaxtomin A are brown.[4] The cells within the lesion are sunken (Figures 1 and 2). The cytoplasm of the cells is disrupted and the cells are filled with a flocculent material that appears to be associated with the tonoplast and plasmalemma during the earliest events observable (Figure 3). The site of injury to the cells is therefore, possibly, associated with these membranes. Alternately, weakening of cell walls by interaction with the toxin may have caused the cell collapse since the walls of affected cells are deformed and more electron-lucent than normal cells (Figure 3). This effect, though, may be a result of the breakdown of the cell. Due to the nature of standard microscopic techniques it is difficult to assess what the initial events are after thaxtomin A application. For this reason it is hoped that confocal microscopy used in conjunction with ion sensitive dyes will allow the actual site of action by thaxtomin A to be determined.

References

1. W. G. Barker. Science 118:384-385.
2. D. H. Lambert and R. Loria. 1989. International J. of Systematic Bacteriology. 39: 387-392.
3. D. H. Lambert and R. Loria. 1989. International J. of Systematic Bacteriology. 39: 393-396.
4. C. H. Lawrence, M. C. Clark, and R. R. King. 1990. Phytopathology 80:606-608.

Proc. Microscopy and Microanalysis 1995, edited by G.W. Bailey, M.H. Ellisman, R.A. Hennigar, and N.J. Zaluzec
Copyright © 1995 MSA. Published by Jones and Begell Publishing, 79 Madison Ave., New York, NY 10016

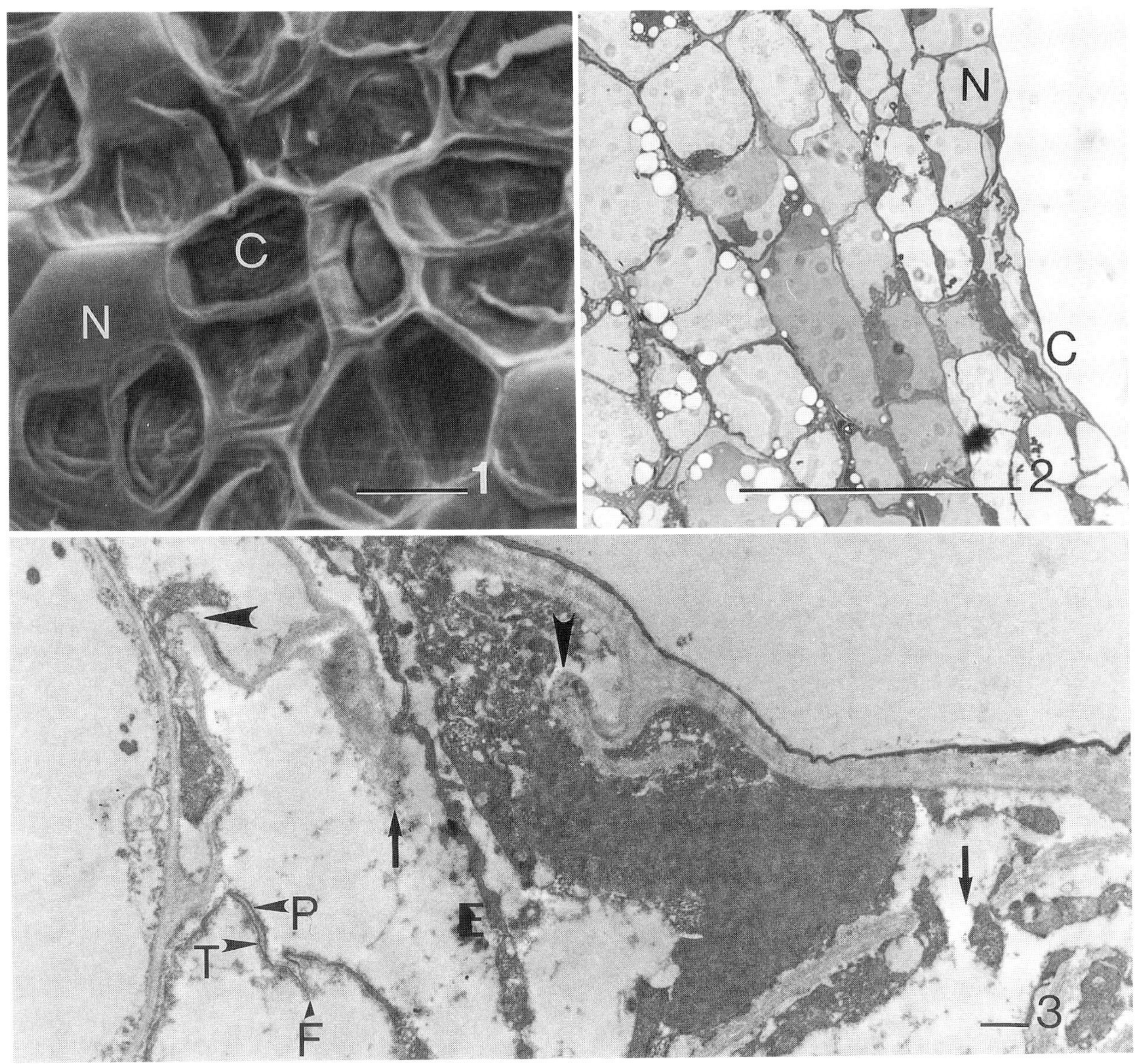

Figure 1. SEM micrograph of potato mini-tuber treated with thaxtomin A. Some of the epidermal cells are collapsed (C) though others are still normal (N) in appearance. Bar = 25*u*m.

Figure 2. LM in cross section where thaxtomin A application has resulted in collapsed (C) and normal (N) epidermal cells in treated mini-tuber, similar to what appears in Figure 1. Bar = 100*u*m.

Figure 3. TEM of epidermal cell (E) which has been affected by thaxtomin A in a manner similar to what is seen in Figure 2. Cellular structure is disrupted by thaxtomin A with much of the cell being filled with a flocculent material (F) apparently associated with the tonoplast (T) and plasmalemma (P). The cell wall is broken in places (arrows) and is deformed (arrowheads). Bar = 1*u*m.

STEREOLOGY OF LIVER LESIONS INDUCED BY FEEDING PCB 153 [2,2',4,4',5,5'-HEXACHLOROBIPHENYL] TO THE RAT

J. Peng, A. Singh, W. Ireland, and I. Chu*

*Atlantic Veterinary College, Charlottetown, PE, C1A 4P3, and *Environment Contaminants Section, Health Canada, Ottawa, ON, K1A 0L2*

Polychlorinated biphenyls (PCBs) are persistent environmental contaminants.[1] A potential health hazard exists for humans and animals because of the PCBs long half-life and capacity to bioaccumulate in the aquatic and terrestrial food chains.[2] The liver is known to be the most significantly affected organ following experimental exposure to PCBs.[2] We have reported the ultrastructure of liver of rats orally administered PCB 153 for 13 weeks.[3] In the present work stereology was applied to estimate measurements of the ultrastructural lesions in the liver.

Twelve Sprague-Dawley rats of both genders were distributed into four groups comprising three animals of each gender. Animals in treated groups received diets containing PCB 153 (dissolved in 4% corn oil) at 50,000 ppb concentration; animals in control groups received diets containing corn oil only. The animals were euthanized 13 weeks after the onset of the experiment. Liver specimens were harvested from the animals, fixed in buffered 2% glutaraldehyde and post-fixed in 2% osmium tetroxide. Thick and thin sections of the specimens were prepared by routine procedures for transmission electron microscopy (TEM). Thick sections of the specimens, which contained a terminal hepatic venule, were selected and subjected to measurements by "Bioquant IV" image analysis system. Optical microscope images of the field were projected on a monitor by a video camera. Hepatocytes and nuclei of zone 3 were sampled with a point grid. A volume weighted mean volume of the hepatocytes or nuclei was estimated.[4] Six electron micrographs of cells in zone 3 were taken at 5,000 diameters. A 1-cm² grid was superimposed on the micrographs to estimate the volume density of different organelles in the cytoplasm. Finally, the quantitative data were expressed as absolute dimensions per hepatocyte.

PCB treatment resulted in an increase in average hepatocyte volume in the female from 5780 to 9240 μm^3 and in the male from 5480 to 9210 μm^3 (a result that was presented at VII International Congress of Toxicology). The cytoplasm played an overwhelming part in the volume increase (Table 1). The average volume of SER per cell increased threefold in treated animals of both genders (Fig. 1); increase of mean RER volume per cell was noted in female treated animals only (Fig. 2). Significant increase of mean peroxisome volume was detected in treated animals of both genders (Fig. 3). Average normal mitochondria volume slightly decreased in male treated animals (Fig.4), for the abnormal mitochondria which showed the longitudinally oriented cristae, the mean volume increased significantly in female treated animals (Fig. 5). The current quantitative data supported our results that were obtained by qualitative TEM.[3]

References
1. L. Kasza et al., *J. Toxicol. Environ. Hlth.* 1(1976)689.
2. S. Safe, *Toxicology* 21(1990)51.
3. K. MacLellan et al., *Histol Histopathol.* 9(1994)461.
4. H. J. G. Gundersen and E. B. Jensen, *J. Microsc.* 138(1985)127.

Proc. Microscopy and Microanalysis 1995, edited by G.W. Bailey, M.H. Ellisman, R.A. Hennigar, and N.J. Zaluzec.

Table 1

Volume	Control		Treatment	
(μm^3)	Female	Male	Female	Male
Hepatocyte	5779	5482	9243	9213
(standard error)	234	124	370	386
Nuclei	297	293	357	357
(standard error)	12	13	20	17
Cytoplasm	5482	5190	8886	8856
(standard error)	57	96	148	155

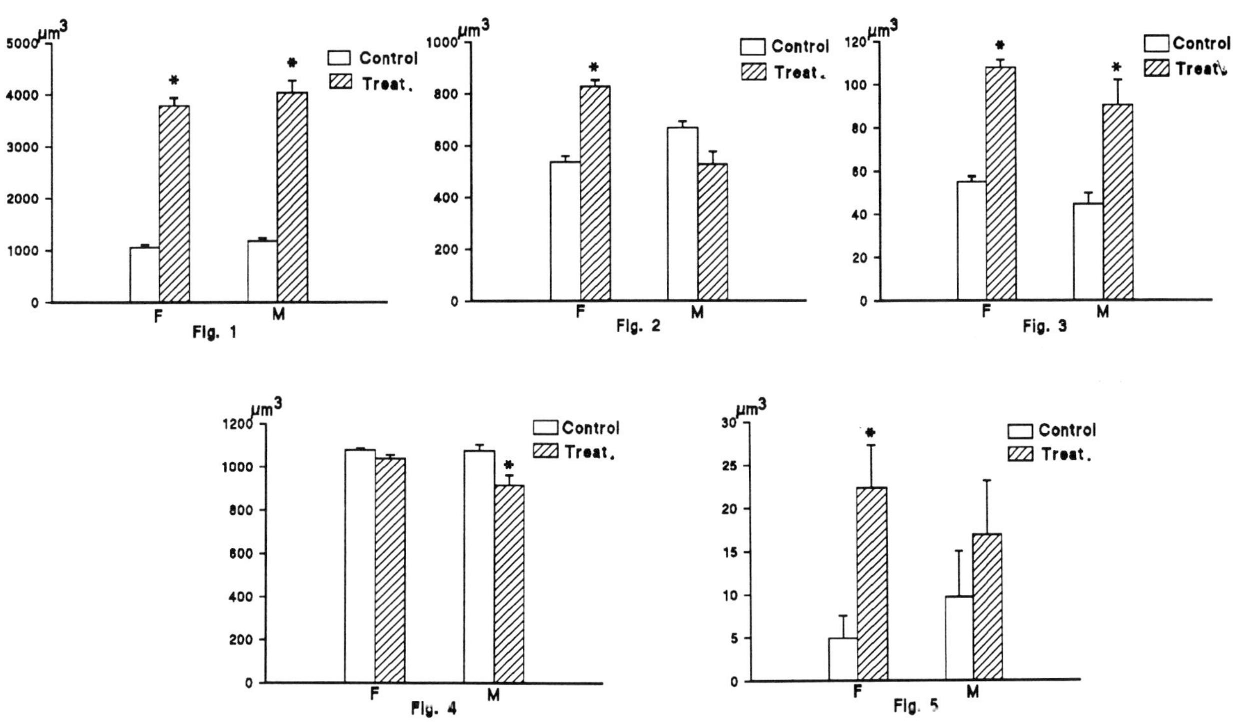

*Significant at $p < 0.05$
F: Female; M: Male

Fig. 1. Mean volume of smooth endoplasmic reticulum per hepatocyte.
Fig. 2. Mean volume of rough endoplasmic reticulum per hepatocyte.
Fig. 3. Mean volume of peroxisome per hepatocyte.
Fig. 4. Mean volume of normal mitochondria per hepatocyte.
Fig. 5. Mean volume of abnormal mitochondria per hepatocyte.

AMELANOTIC MALIGNANT MELANOMA: RECOGNIZING ABERRANT PREMELANOSOMES

Karen K. Bovard & Chhanda Bewtra

Pathology Department, Creighton University, Omaha, NE, 68131

The key to diagnosing cases of melanoma, a tumor of melanocytes, is to recognize cytoplasmic pigment called melanin. In most cases this can be done at the light microscopy level; however, in the rarer amelanotic (non-pigmented) melanomas, melanin is not present in detectable amounts. Thus it is necessary to do an ultrastructural investigation to detect melanosomes, the specific organelles of the melanocyte that synthesize melanin. Other non-melanoma cells can also possess melanin by internalization of melanin granules by phagocytosis or passive transfer (e.g. macrophages, keratinocytes) but they lack the premelanosome organelle, the precursor to the fully developed melanosome. In premelanosomes, the internal structures are not obscured by melanin. Thus it is the diagnostic feature of choice for the pathologist. In malignant melanoma, aberrant forms of premelanosomes and melanosomes are more typical than in normal melanocytes.[1] These aberrant forms may not be well recognized. Amelanotic melanomas give us a good opportunity to study these structures. In studying premelanosomes,the investigator must be aware of variations of size, shape, appearance in different planes of sections, and knowledge of the four stages of normal melanosome development (melanogenesis).[2] The Stage I premelanosome is a relatively structure-free vesicle which elongates and enlarges into an oval structure that develops a characteristic periodicity and is known as a Stage II premelanosome. It is the Stage II premelanosome that is considered the classic diagnostic feature of melanoma. In the Stage III premelanosome, melanin is deposited on the internal structure. Completion of the melanin deposition process produces a mature Stage IV melanosome in which melanin completely obliterates the internal structure.[3]

The amelanotic melanoma tumor cells are mainly spherically shaped with no desmosomal or other junctional structures. There are many well-developed golgi apparatuses with numerous membrane-bound primordial vesicles, often with no internal structure corresponding to Stage I premelanosomes. Many pleomorphic forms of premelanosomes were found including the classic ellipsoidal shape with periodic transverse striations (Fig. 1 B), spherical forms with the same periodic internal structure (Fig. 2 A), and ellipsoidal shapes with a zig-zag internal pattern (Fig. 2 B). Other forms included internal concentric coils of filaments (Fig. 2 C-F), vesicles (Fig. 2 G,H), and many other forms not easily categorized. In amelanotic malignant melanomas, most premelanosomes are Stage I and II premelanosomes whereas pigmented Stage III and IV melanosomes are essentially absent.

In this study we show some of the aberrant forms of premelansomes as these structures may not be commonly recognized by the pathologist. Identification of these structures may aid in diagnosis of poorly differentiated, anaplastic tumors.

1. A.M. Dvorak & R.S. Monahan-Earley, *Diagnostic Ultrastructural Pathology I*, 1992.
2. F.N. Ghadially, *Ultrastructural Pathology of the Cell and Matrix*, 1982.
3. F.N. Ghadially, *Diagnostic Electron Microscopy of Tumours*, 1980.

Proc. Microscopy and Microanalysis 1995, edited by G.W. Bailey, M.H. Ellisman, R.A. Hennigar, and N.J. Zaluzec

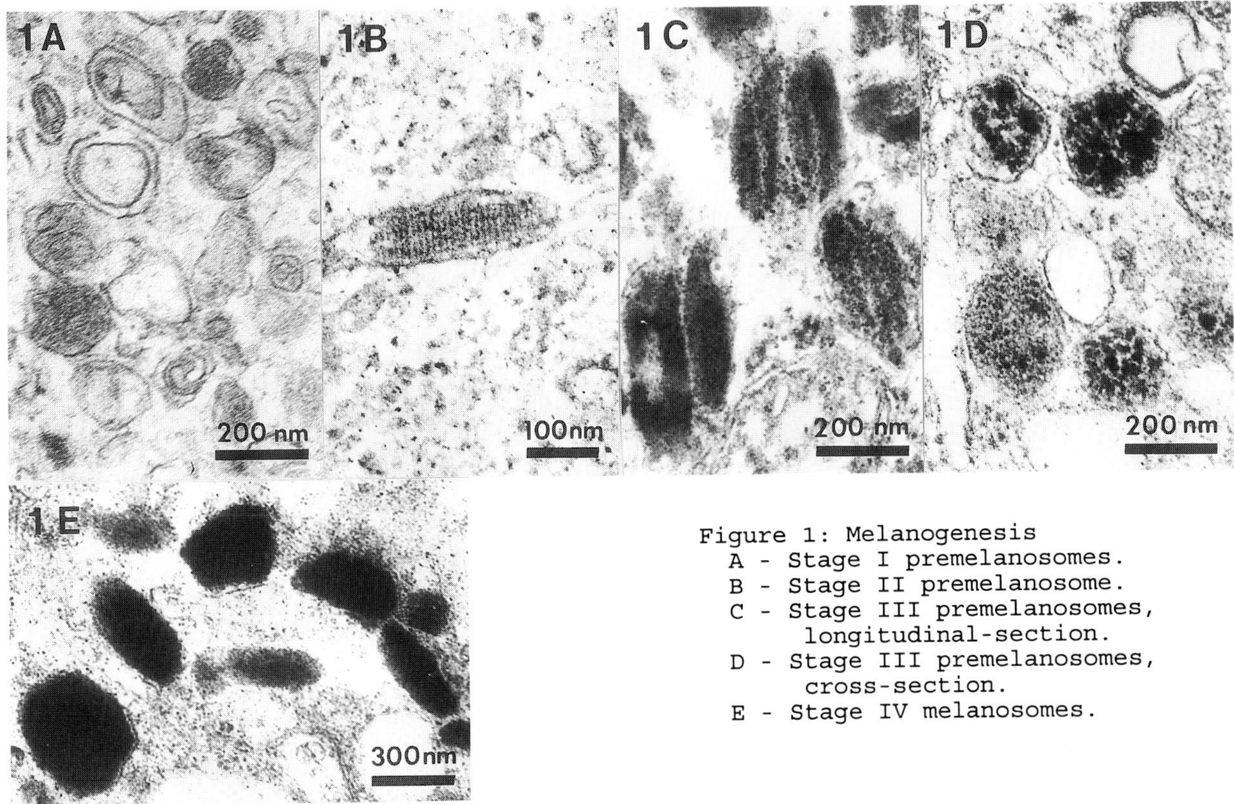

Figure 1: Melanogenesis
  A - Stage I premelanosomes.
  B - Stage II premelanosome.
  C - Stage III premelanosomes,
        longitudinal-section.
  D - Stage III premelanosomes,
        cross-section.
  E - Stage IV melanosomes.

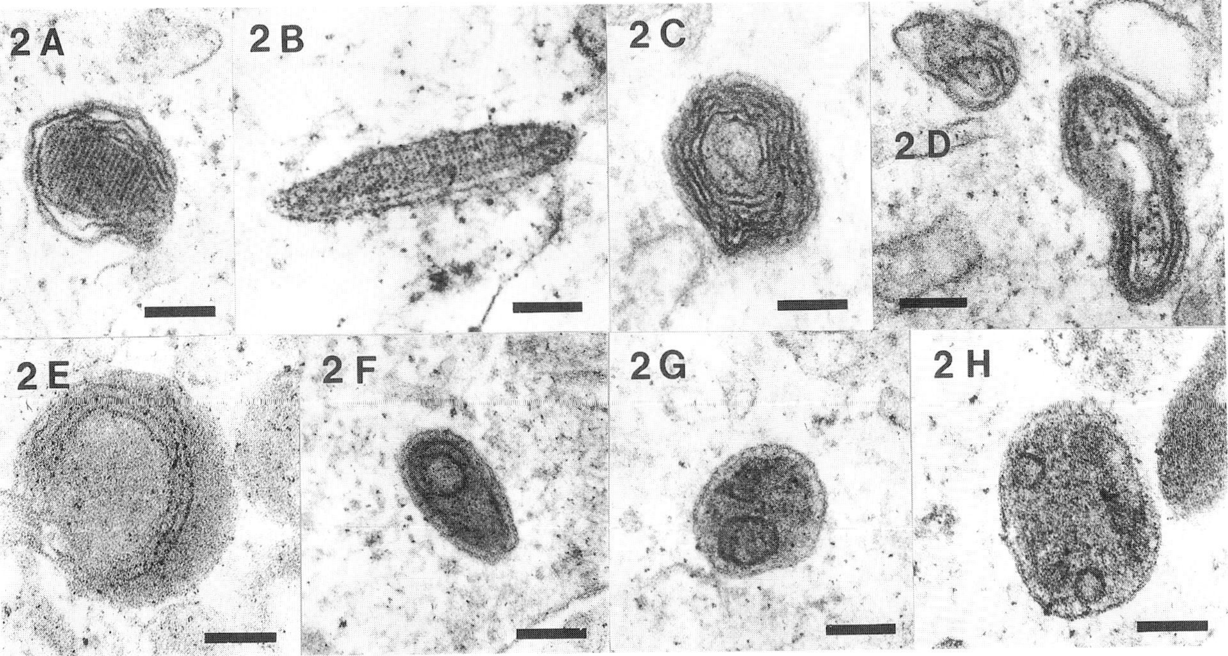

Figure 2, A - H: Pleomorphic premelanosomes of amelanotic malignant melanoma
Scale bar = 100 nm.

IMMUNOGOLD LOCALIZATION OF HMB-45 ANTIGEN IN PULMONARY LYMPHANGIOMYOMATOSIS

J.M. Minda, M.S., E. Dessy[*], M.D., G. G. Pietra, M.D.

Division of Anatomic Pathology, Department of Pathology and Laboratory Medicine , University of Pennsylvania Medical School, Philadelphia, Pa.
[*] current address: Instituto di Medicina Legale e di Anatomia Patologica, Universita di Cagliary, Italy.

Pulmonary lymphangiomyomatosis (PLAM) is a rare disease occurring exclusively in women of reproductive age. It involves the lungs, lymph nodes and lymphatic ducts. In the lungs, it is characterized by the proliferation of smooth muscle cells around lymphatics in the bronchovascular bundles, lobular septa and pleura.[1,2] The nature of smooth muscle proliferation in PLAM is still unclear. Recently, reactivity of the smooth muscle cells for HMB-45, a melanoma-related antigen has been reported by immunohistochemistry.[3-6] The purpose of this study was the ultrastructural localization of HMB-45 immunoreactivity in these cells using gold-labeled antibodies.

Lung tissue from three cases of PLAM, referred to our Institution for lung transplantation, was embedded in either Poly/Bed 812 post-fixed in 1% osmium tetroxide, or in LR White, without osmication. For the immunogold technique, thin sections were placed on Nickel grids and incubated with affinity purified, monoclonal anti-melanoma antibody HMB-45 (1:1) (Enzo Diag. Co) overnight at 4^0C. After extensive washing with PBS, grids were treated with Goat-anti-mouse-IgG-Gold (5nm) (1:10) (Amersham Life Sci) for 1 hour, at room temperature. Grids were washed with PBS, distilled water, air dried and stained and stained with 1% aqueous uranyl acetate and 0.5% lead citrate. Controls were incubated with a similar protocol omitting the primary antibody. Grids were examined with a transmission electron microscope Hitachi H-600 (Nissei Sangyo).

Electron microscopy revealed bundles of smooth muscle cells arranged in short fascicles around lymphatics in the bronchovascular bundles and pleura (Fig 1). The proliferating smooth muscle cells ranged in appearance from typical spindle-shaped cells to plump myoblast-like cells. The spindle cells were characterized by abundant myofilaments, dense bodies, abundant RER and large amount of glycogen. The plump cells had centrally placed nuclei with nucleoli and only few cytoplasmic organelles and actin filaments (Fig.2). Both cell types contained round to ovoid membrane bound vesicles with osmiophilic concentric internal filaments or fine crystal-like lamellar structures, resembling stage I or II melanosomes and rare stage III or IV(Fig. 3).[5] Immunoelectron microscopy with anti-melanoma HMB-45, in our cases, revealed clusters of gold particles within SER vesicles and stage I or II melanosomes with incomplete melanin deposition (Figs. 4). The presence of melanocytic organelles in vascular-associated smooth muscle cells with HMB-45 has been reported also in tuberous sclerosis involving the lung and kidney.[5,6] Our findings suggest that HMB-45 positivity in the smooth muscle cells of PLAM, may be a marker of a genetically controlled hamartomatous proliferation.[3]

References

1. N.B. Eisera, E. Venegoni, R. Colombi, *Journal of Royal Society of Medicine* , 81 (1988)607.
2. T. Urban, F. Kuttenn, A. Gompel, et al., *Chest,* 102 (1992)472.
3. J.K.C. Chan, W.Y.W. Tsang, M.Y. Pau, et al., *Histopathology,* 22(1993)445.
4. A.M. Gown, A.M. Vogel, M. Hoak, et al., *American Journal of Pathology,* 123 (1986) 195.
5. F. Bonetti, P.L. Chiodera, M. Pea, et al., *American Journal of Surgical Pathology,* 17(1993)1092.
6. E. Kaiserling, S. Krober, J. Xiao, et al - *Histopathology,* 25(1994) 41.
7. M.J. Gaffey, S.T. Mills, R.J. Zarbo, et al., *American Journal of Surgical Pathology,* 15(1991) 644.

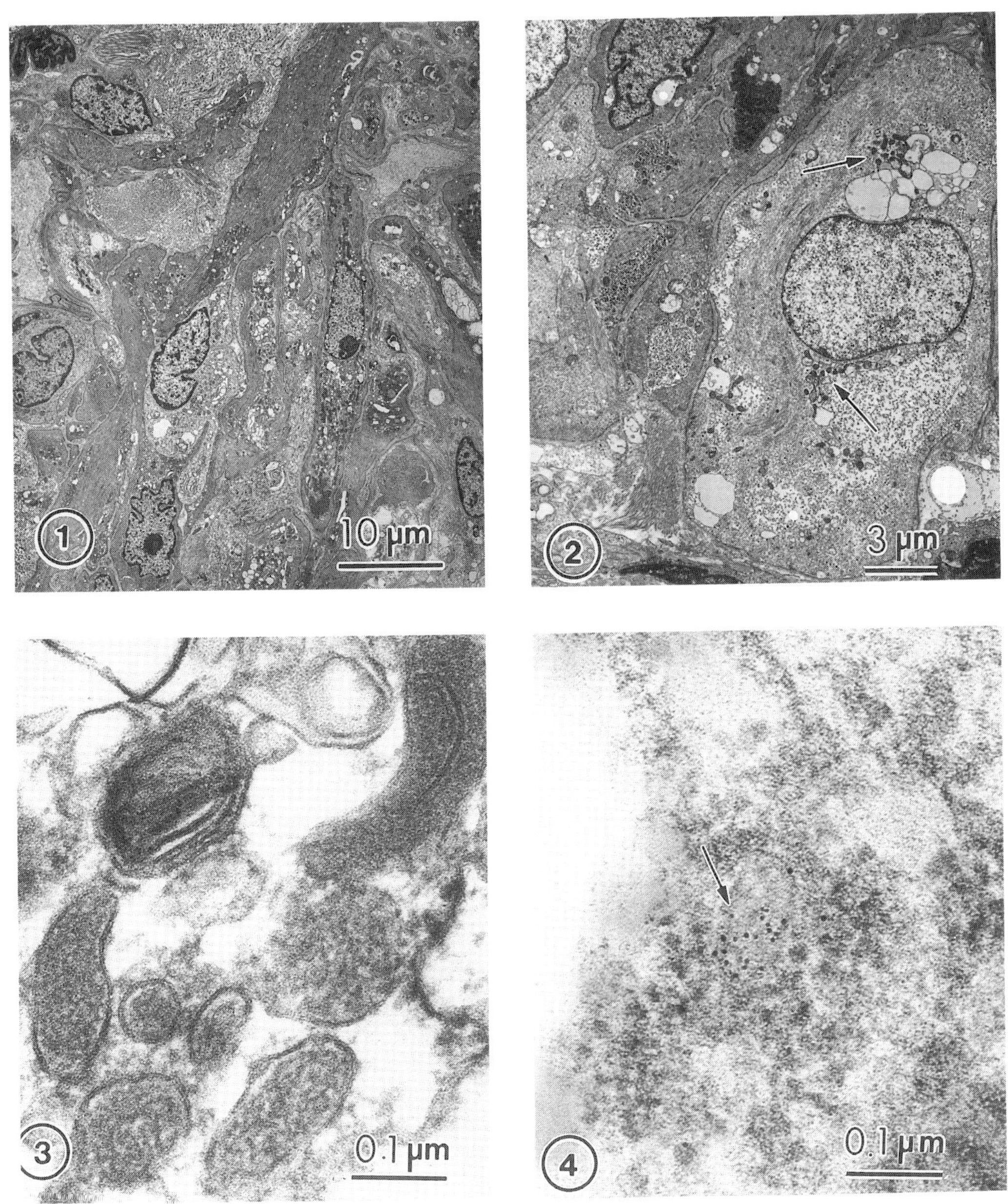

Fig.1 - Large number of spindle-shaped smooth muscle cells, in the vicinity of a blood vessel, in PLAM. Methanolic 20% Uranyl acetate and aqueous 0.5% Lead citrate stain. Magnification = 1,500

Fig.2 - Ultrastructural appearance of a myoblast-like cell, revealing vacuolization, mitochondria, melanosomes (arrow) and actin filaments. Magnification = 3,500

Fig.3 - Stage II melanosomes in the perinuclear region of a smooth muscle cell. Magnification = 120,000.

Fig.4 - Stage II melanosome labelled with monoclonal HMB-45 and GaM-IgG-Gold (arrow). LR White embedded unosmicated tissue. Aqueous 1% Uranyl acetate and 0.5% Lead citrate stain. Magnification = 120,000.

CYTOPLASMIC INCLUSION BODIES IN FAMILIAL DILATED CARDIOMYOPATHY

W.T. Gunning*, A.F.Gohara*, T.E. Walsh*, and E.P. Calomeni*

Departments of Pathology* and Medicine**, Medical College of Ohio, Toledo, OH, 43614

A thirty-six year old male with severe congestive heart failure was given a cardiac transplant. The findings of the evaluation of his native heart is the basis for this presentation. Our findings included the identification of unusual cytoplasmic inclusion bodies found throughout the heart. These inclusions, found most often in a perinuclear location, were eosinophilic and crystalloid at the light microscopic level (Figure 1.). Special histologic stains were not supportive of a diagnosis of nemaline rod disease as a trichrome stain failed to differentiate inclusions from surrounding cytoplasm, yet could be accomplished with PAS staining (Figure 2.) [1]. Furthermore, the ultrastructural morphology of the inclusions did not show any similarity to z-band material, the electron microscopic characteristic of inclusion bodies in nemaline rod myopathy. The shape of these bodies was usually rectangular or elliptical (Figure 3.) and size was found to be quite variable, some measured in excess of $15\mu \times 4\mu$. Other characteristics included an amorphous appearance with occasional electron densities within the inclusions (Figure 3), an apparent single limiting membrane (Figure 3 and 4), and for some of the inclusions, there was a suggestion of secondary lysosomal association (Figure 4). The latter observation may be evidence that the idiopathic cardiomyopathy this patient had suffered for 14 years may have been a complication of a potential metabolic storage disease process.

A variety of metabolic and inherited diseases are definitively known to cause cardiomyopathies. Some of these include carnitine deficiency, Fabry's disease, glycogen storage disease, hemochromatosis, mucolipidoses and mucopolylipidoses, oxalosis, and Wilson's disease [2]. The case which is presented here does not correlate with any metabolic or inherited storage disease with which we are aware; furthermore, we have yet to find any report in the literature to date which correlates with our observations. The inherited nature of this patient's idiopathic cardiomyopathy, is elucidated in his family history. His father at age 62 and his sister at age 31 were diagnosed with idiopathic cardiomyopathy. His sister's cardiomyopathy was found to be reversible as she was found to have thyroid dysfunction. Two cousins are known to have dilated cardiomyopathy and both his grandparents and great-grandparents have a history of congestive heart failure and myocardial infarction. There is no history of coronary artery disease. Other interesting (and possibly important) family histories include two siblings with protein C deficiency and three aunts (two maternal and one paternal) with hypothyroidism. We are actively investigating the etiology of these unusual inclusion bodies, but to date we are only able to describe our observations.

References
1. Jones, J.G. and Factor, S.M. (1985) *Vircows Arch (Pathol Anat)* 408:307-312.
2. Dustin, P. (1992) In: *Diagnostic Ultrastructure of Non-Neoplastic Diseases.*
   Eds.:Papadimitriou, J.M., Henderson, D.W., and Spannolo, D.V. Curchill Livingstone Publ.,
   London pp. 110-132.

Proc. Microscopy and Microanalysis 1995, edited by G.W. Bailey, M.H. Ellisman, R.A. Hennigar, and N.J. Zaluzec
Copyright © 1995 MSA. Published by Jones and Begell Publishing, 79 Madison Ave., New York, NY 10016

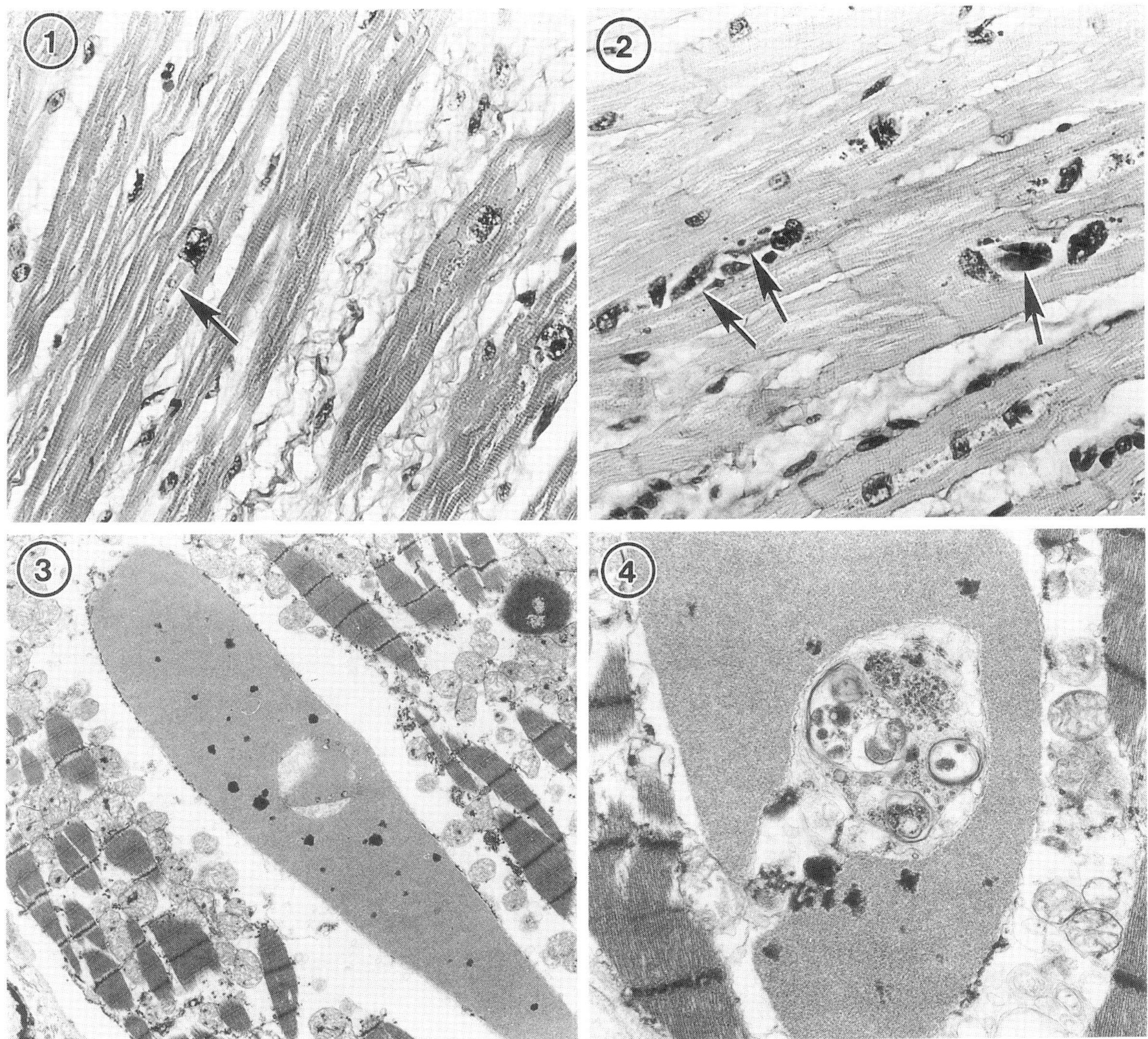

Figure 1.   H&E stained section of ventricular wall demonstrating a cytoplasmic inclusion body (arrow) in degenerative and hypertrophic cardiomyocytes. (160X)

Figure 2.   PAS stained section of ventricular wall with many cytoplasmic inclusions (arrows). (160X)

Figure 3.   Ultrastructural morphology of an inclusion body shows an amorphous, membrane limited, non-crystalline material. (1,450X)

Figure 4.   Some of the inclusion bodies appear to be associated with secondary lysosomes. (45,500X)

ULTRASTRUCTURE OF EXPERIMENTAL CRUSH INJURY IN RATS

Raymond Coleman*, Ludmilla Mazor*, Michael Silbermann*, Irit Rubinstein** and Ori S.Better**

*Division of Morphological Sciences & **Rebecca Chutick Crush Syndrome Center, Bruce Rappaport Faculty of Medicine, Technion-Israel Institute of Technology, Haifa 31096, Israel

Crush Syndrome is a potentially life-threatening condition that develops in humans trapped under debris of collapsed building such as occur in mass catastrophes such as earthquakes, volcanic eruptions and following explosions such as those of bombings or missile attacks. Traumatic pressure damage on limbs can result in myopathy and rhabdomyolysis. The consequences of muscle breakdown lead to major physiological disturbances and may result in acute renal failure if the condition is not recognised and dealt with as an emergency. Whereas there is now considerable progress in treating the condition, much less is known about the basic biology of the muscle pathology and the potential for repair of the traumatized muscle. In the present study we have developed a rat model in which controlled external pressure is used to mimic crush damage in humans, and we have followed the early stages of muscle damage and repair using histological and ultrastructural techniques.

A novel apparatus is used to apply controlled pressure (4.25 atmospheres) for 2 hours on the hindlimbs of adult Sprague-Dawley rats. The anterior tibialis muscle was examined at various time intervals : immediately after the trauma, 4 and 7 days after treatment and up to several weeks after treatment. The muscle was processed by standard means for histology and TEM. The anterior tibialis muscle after 4 and 7 days post-injury showed a swelling of muscle fibers, which had marked signs of edema. The normal sarcomere arrangement was severely disrupted (Fig.1) and this was accompanied by marked myofilament loss. The mitochondria showed swelling and indications of damage. In addition most fibers showed the development of many large lipid inclusions. These were easily seen by light microscopy. With TEM it was shown that most of these lipid inclusions appeared to be secondary lysosomes and were most commonly distributed close to the periphery of the myofibers (Fig.2.). The nuclei of the myofibers appeared enlarged with extremely enlarged nucleoli. In many cases the nuclei had abandoned their peripheral location and migrated to the center of the myofibers.At high magnifications, it was possible to find the early stages of myofilament loss from the central portions of myofibers (Fig.3.). In the connective tissue surrounding the damaged myofibers, there were many macrophages, which appeared to be active in clearing up the debris of damaged myofibers, and these macrophages frequently were swollen with many lipid inclusions. Muscle damage also caused changes in the endothelial cells of the capillaries of the endomysial connective tissue. These capillaries appeared to remain intact, though they were considerably enlarged and demonstrated post-traumatic vasodilation. By 4-days post-trauma, there appeared to be a proliferation and stimulation of satellite cells in the connective tissue. Other signs of early attempts of post-injury regeneration,included the development of myotubes similar to those found in embryonic muscle. These myotubes were characterized by their strings of centralized nuclei.

The authors gratefully acknowledge the support for this project provided by the Chief Scientist's Office, Israel Ministry of Health (Technion 184-197), The Fund for Promotion of Research at the Technion (R.C.) and the Center for Absorption in Science, Ministry of Immigrant Absorption, State of Israel (L.M.) and the Chutick Crush Syndrome Center (I.R., O.S.B.).

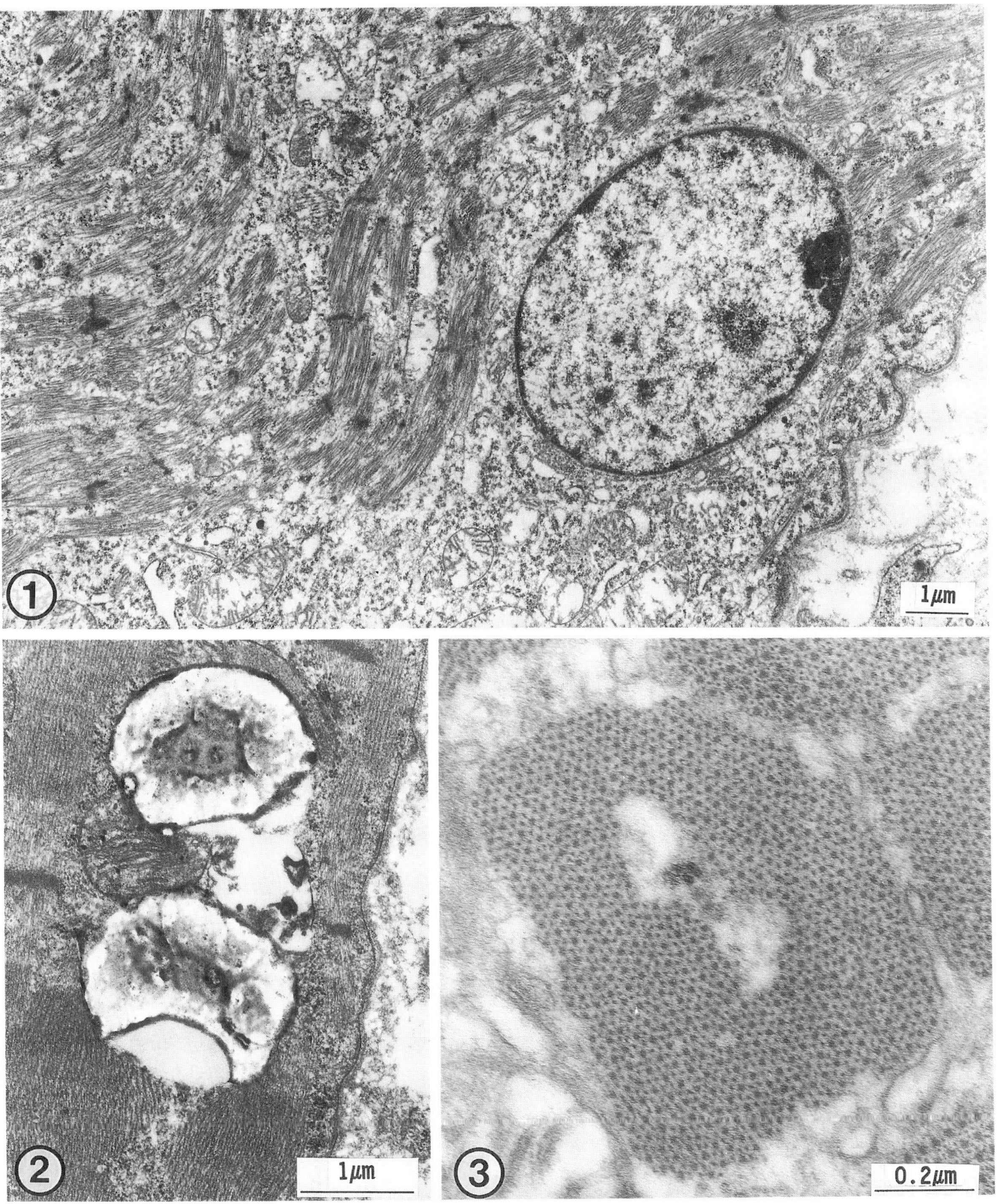

FIGS 1-3. TEM micrographs of crush damage to rat anterior tibialis muscle four days following controlled external pressure of 4.25 atmospheres for two hours.
FIG.1.- Typical myocyte with myopathy including sarcomere destruction and loss of myofilaments
FIG.2.- Development of large lysosome-like lipid-rich bodies near periphery of myocyte.
FIG.3.- High-power view of transverse section of myofiber showing early stages of myofilament breakdown and loss in the central region.

ULTRASTRUCTURAL CHARACTERIZATION OF AN *IN VITRO* MODEL OF HUMAN EPIDERMIS: ITS POTENTIAL FOR SULFUR MUSTARD STUDY

J.P. Petrali, P.T. Lopolito, T.A. Hamilton, S.B. Oglesby-Megee and L.S. Rhoads

Comparative Pathology and Drug Assessment, USAMRICD, Aberdeen Proving Ground, Maryland 21010-5425

Morphological correlates of sulfur mustard toxicity continue to be studied in *in vivo* models such as the haired and hairless guinea pig and the domestic weanling pig. These animal models provide useful evidence for the mechanisms of sulfur mustard-induced lesions, however the correlation with human skin is not exact and the availability of a continuing supply of any animal model is problematic. *In vitro* studies have largely been performed with cultured monotypic human cells such keratinocytes and lymphocytes. Although monotypic cellular models have added data to the mechanisms of the observed cytopathology, the acclimatized nature of media-nourished cells may present a different phenotype for study than *in vivo*. To bridge the information gap between monolayers of cells and *in vivo* studies, *in vitro* skin equivalents have been proposed as suitable models. We have used several commercially available human skin equivalents in the past[1], however the lack of basement membrane components, the absence or paucity of hemidesmosomes, general problems with replication and inconsistences of availability, question the usefulness of commercial preparations at this time. We are now evaluating morphological presentations of a human epidermal model developed in-house by Rhoads[2] and present here ultrastructural impressions of that model.

Millipore Milli-Cm inserts were precoated with 50 ul of Type I collagen gelled by exposure to ammonium hydroxide, dehydrated in 70% ethanol and cross-linked by incubation in 2.5% glutaraldehyde in PBS. Normal human epidermal keratinocytes, suspended in keratinocyte growth medium supplemented with or without 10% fetal bovine serum and 1.5 mM calcium chloride, were dispensed onto the coated cell culture inserts at a density of 1 x 10^5 per ml. Inserts were maintained at 37 °C with daily changes of medium in an atmosphere of 5% CO_2. At five days, the media was removed from the apical surface creating an air-liquid interface. Specimens selected for study were fixed at 4 °C in 1.6% formaldehyde and 2.5% glutaraldehyde in cacodylate buffer for 24 h, post fixed in buffered osmium tetroxide, dehydrated in graded ethanols and embedded in epoxy resin. Thin sections differentiated with uranyl acetate and lead citrate were examined with a JOEL 1200EX.

The model presented a replicate display of epidermal structural components found *in vivo*. There was evidence of stratification, differentiation, keratohyaline formation and keratinization. Typical epidermal cell types were recognized to each stratum: stratum germinativum (basal cells), stratum spinosum (spinosal cells), stratum granulosum (keratohyalin-bearing squames), stratum corneum (keratinized squames). Also well defined were desmosomes, tonofilaments, human-like keratohyaline granules and junctional complexes. Of special interest was the presentation of morphologically complete hemidesmosomes at the basolateral surfaces of basal cells (*Fig. 1*).

Proc. Microscopy and Microanalysis 1995, edited by G.W. Bailey, M.H. Ellisman, R.A. Hennigar, and N.J. Zaluzec
Copyright © 1995 MSA. Published by Jones and Begell Publishing, 79 Madison Ave., New York, NY 10016

References

1.   J.P. Petrali, S.B. Oglesby and T.A. Justus, J. Toxicol. Cut. & Ocular Toxicol.(1991) 10, 315-324.

2.   L.S. Rhoads, J. Toxicol. Cut.& Ocular Toxicol. (1994) 13, 231-248.

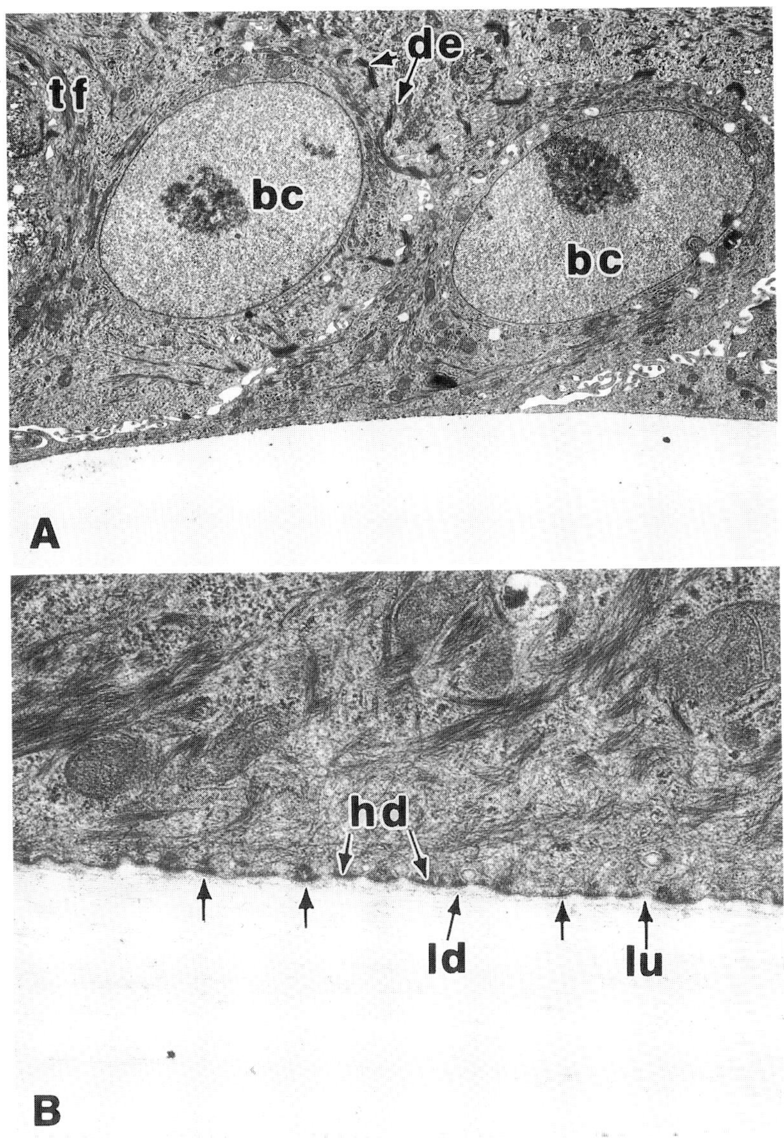

Fig 1. *A.* Stratum germinativum with basal cells (bc); tonofilaments (tf), desmosomes (de). *B.* Basement membrane zone with hemidesmosomes (hd) and anchoring filaments (arrows) in association with a rarefied lamina lucida (lu) and lamina densa (ld). Magnifications 4000X, 26,000X.

THE USE OF *GRIFFONIA SIMPLICIFOLIA* LECTIN, GSA-I-B-4, TO DETERMINE THE EXTENT OF NEOVASCULARITY IN MURINE NEOPLASMS

C. D. Bucana, D. Reynolds, K. Dunner, Jr., R. Radinsky and I. J. Fidler

UTMDACC, Dept. Cell Biology, Houston, Texas 77030

Angiogenesis plays a critical role in many physiological and pathological processes and is measured by a vascularity index derived from quantification of newly formed blood vessels. Although the identification of human endothelial cells can be accomplished by binding of antibodies against Factor VIII or binding of *Ulex europeaus* I agglutinin (UEA-I), in murine tissues these markers produce poor results. Since the lectin (GSA-I-B4) from the seeds of *Griffonia simplicifolia* which binds to -D-galactose, has been shown to bind preferentially to murine endothelial and some epithelial cells,' we determined if binding of this lectin will identify mouse endothelial cells within and around tumors growing in different organs of the mouse such as lung, kidney, bladder, and subcutaneous space.

Tumors were excised, fixed in 10% formalin and processed for histology. The original procedure was modified to include a microwave antigen retrieval procedure which appears to enhance the signal better than trypsin treatment of the deparaffinized sections (Fig.1 and 2). The sections were then incubated with biotinylated GSA-I-B4 (Vector Labs, Burlingame,CA) in PBS supplemented with 2mM $CaCl_2$ and 2mM $MgCl_2$. Binding of the lectin was visualized by further incubation with peroxidase labeled streptavidin followed by diaminobenzidine (DAB). A positive reaction was indicated by a brown precipitate. Sections were counterstained with Mayer's hematoxylin. Control sections incubated with peroxidase-labeled streptavidin plus DAB showed no brown precipitate.

Lectin binding was observed in endothelial cells of large pulmonary vessels (Fig. 2a), capillary endothelium, alveolar macrophages, intratumoral macrophages in lung tumors (Fig. 3),epidermal keratinocytes, follicular region, dermal blood vessels (Fig. 4), glomerular capillaries, renal blood vessels, proximal tubule brush borders of the kidney (Fig. 5), and bladder transitional urothelium (Fig. 6). In addition, positive reactions were also observed in mononuclear cells and neutrophils inside blood vessels (Fig. 7) and in intratumoral blood vessels (Fig. 3-6) of tumors regardless of where they were grown. None of the tumors tested bound the lectin. Epithelial cells of the distal bronchioles did not bind the lectin but occasional macrophages seen between epithelial cells were strongly positive (Fig. 2a). Lectin binding to selected epithelial cells does not appear to interfere with binding to vascular endothelium, however lectin binding to monocytes and alveolar macrophages could lead to an overestimation of the vascularity index when these cells infiltrate the tumor. Therefore, application of GSA-I-B-4 lectin in the determination of vascularity index in growing tumors should be used in conjunction with monocyte/macrophage detection.

References:

1. L. Laitinen, Histochemical J. 18(1987)225.

Proc. Microscopy and Microanalysis 1995, edited by G.W. Bailey, M.H. Ellisman, R.A. Hennigar, and N.J. Zaluzec
Copyright © 1995 MSA. Published by Jones and Begell Publishing, 79 Madison Ave., New York, NY 10016

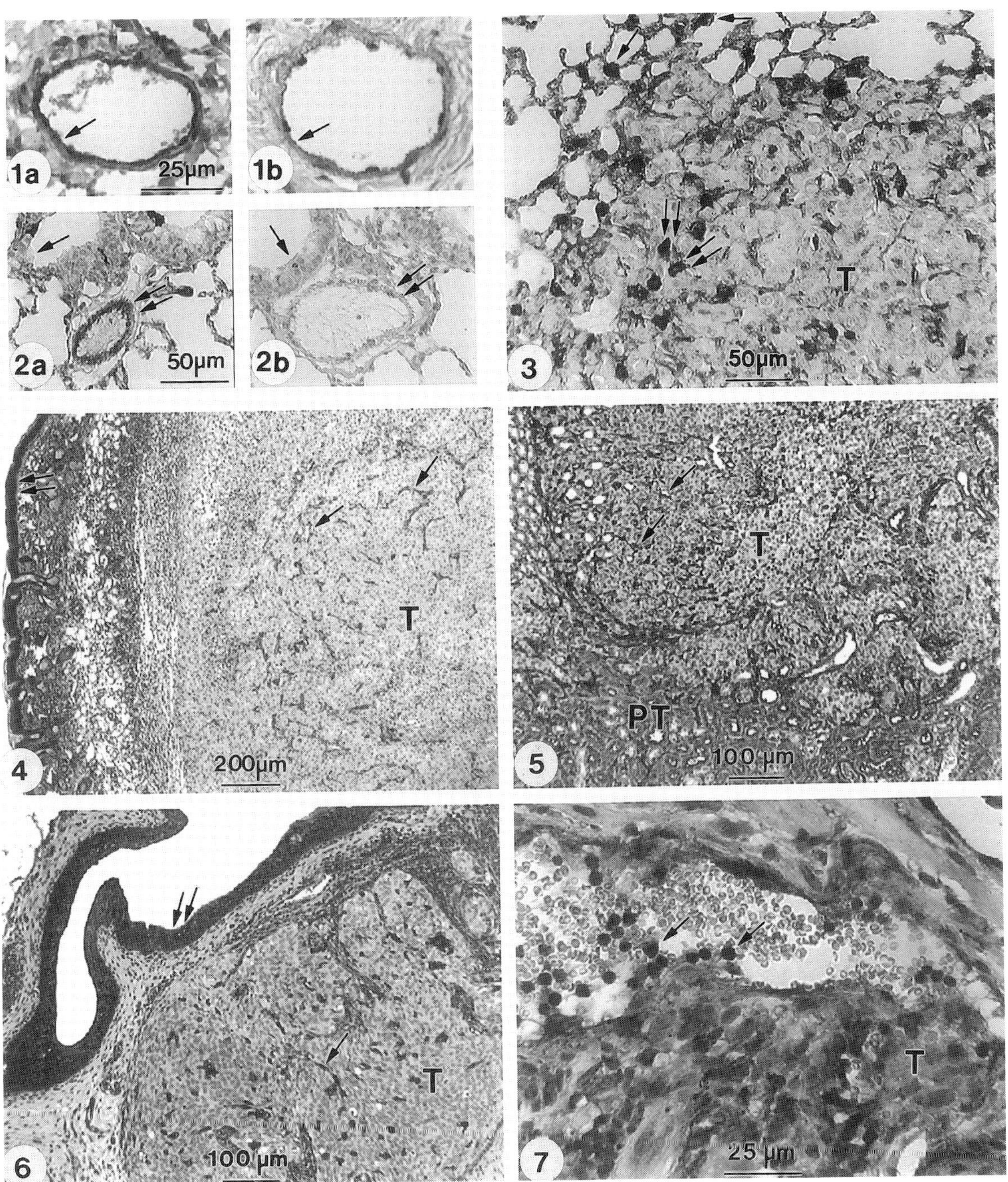

FIG. 1. Lectin binding to dermal blood vessel (↑) after microwave (1a) or trypsin (1b) treatment. **FIG. 2.** Staining of pulmonary blood vessel (↑) and bronchioles (↑↑) in the presence(2a) or absence (2b) of lectin. **FIG. 3.** Lectin binding to alveolar macrophages (↑) and macrophages (↑↑) infiltrating a tumor (T). **FIG. 4.** Lectin binding to epidermis and blood vessels (↑↑) in a subcutaneous tumor (T). **FIG. 5.** Lectin binding to proximal tubules (PT) and blood vessels (↑) in a renal tumor (T). **FIG. 6.** Lectin binding to transitional epithelium (↑↑) and blood vessels (↑) of bladder tumor (T). **FIG. 7.** Lectin binding to mononuclear cells (↑) found inside a blood vessel in a bladder tumor (T).

SEQUENTIAL DETECTION OF mRNA FOR THE CHEMOKINE IL-8 AND MACROPHAGES (F4/80) IN HUMAN MELANOMA

C.D. Bucana, R. Sanchez, R. Singh and I.J. Fidler
UTMDACC, Department of Cell Biology, Houston, Texas 77030

The purpose of this study was to demonstrate by ISH the presence of IL-8 mRNA, and by immunohistochemistry (IHC) the presence of the chemokine IL-8 and the distribution of infiltrating macrophages in subcutaneous melanomas in the same tumor. IL-8 is a multifunctional cytokine produced by melanoma cells , activated macrophages and monocytes and it has been shown to be a growth and angiogenic factor for tumor cells.   More recently it was shown that constitutive expression of IL-8 correlated directly with metastatic potential of human melanoma cells in nude mice[1]. IL-8 content of a solid tumor as determined by Western blot analysis does not take into account the contribution of macrophages.   Previous studies showed that murine tumors contain many infiltrating cells interspersed among tumor cells whereas human tumors growing in nude mice exhibit macrophages at the periphery or between tumor islands[2]. In this study we demonstrate the expression of IL-8 and the distribution of macrophages by immunoperoxidase assay and IL-8 mRNA by ISH.

Human melanoma, A375, growing subcutaneously in nude mice were excised, embedded in OCT mounting medium, snap frozen in liquid nitrogen and stored in a -80°C freezer. Cryosections were cut and fixed for either ISH or IHC. Expression of IL-8 was determined using a rabbit antibody against human IL-8 and macrophages were detected using a rat anti-F4/80 antibody by immunoperoxidase assay.   IL-8 mRNA was detected by a colorimetric ISH assay[3]. Because optimal fixation for F4/80 is not favorable for ISH, several fixation protocols were tested and we found that a fixative containing 0.0125 % glutaraldehyde and 1 % paraformaldehyde allowed sequential staining and detection of macrophages and IL-8 mRNA.

Macrophages were seen around the tumor and in the stroma between tumor islands but rarely seen between tumor cells (Fig. 1). IL-8 protein expression was observed in A375 tumor cells as well as in macrophages (Fig. 2). IL-8 mRNA was also observed in both tumor cells and in macrophages (Fig. 3). Sequential labeling of cells show the distribution of macrophages relative to the tumor cells that are positive for IL-8 mRNA (Fig. 4-6). The section was not counterstained.

The role of macrophages in tumor growth, migration and vascularization is not clear but macrophage expression of certain cytokines in concert with cytokines produced by tumor cells and the distribution of macrophages in the tumor may facilitate the metastatic process.

References:

1. R.K.Singh, et al., Cancer Res. 54(1994)3242.
2. C.D.Bucana, et al., Am. J. Pathol. 141(1992)1225.
3. C.D. Bucana, et al., J. Histochem. Cytochem. 41(1993)499.

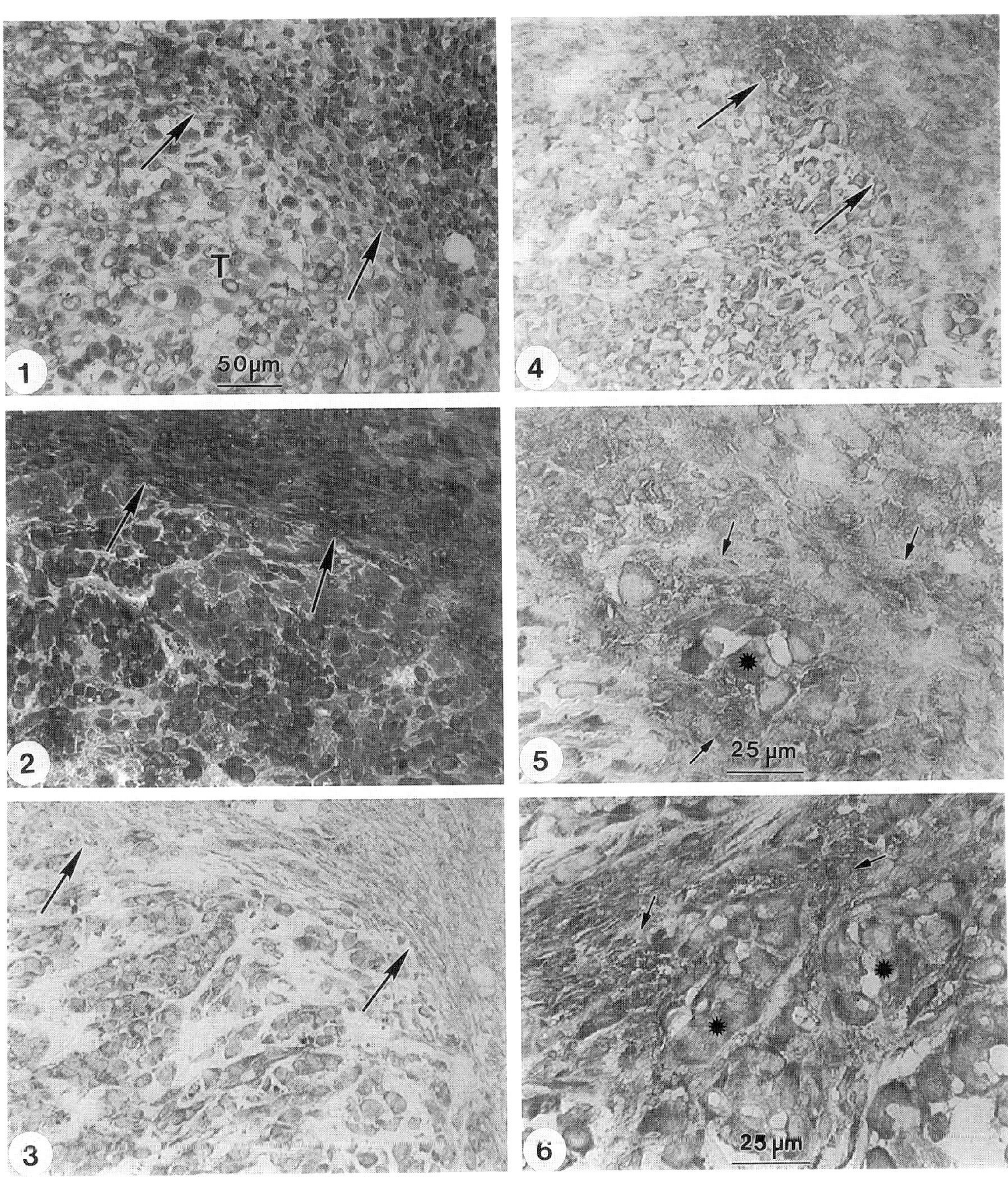

FIG. 1   F4/80⁺ cells (arrows) at the periphery of the tumor (T).
FIG. 2   Anti-IL-8 reactivity of tumor cells and infiltrating macrophages (arrows).
FIG. 3.   IL-8 mRNA seen in both tumor cells and macrophages (arrows).
FIG. 4.   Macrophages at the margin of tumor island.
FIG. 5-6.   Tumor cell colonies (*) surrounded by macrophages (arrows).

EFFECTS OF NUCLEAR POLYHEDROSIS VIRUS FORMULATIONS ON *LYMANTRIA DISPAR* LARVAL PERITROPHIC MEMBRANE

K. S. Shields and J. D. Podgwaite

USDA Forest Service, Northeastern Forest Experiment Station, Northeastern Center for Forest Health Research, Hamden, CT 06514

The peritrophic membrane is a continuous tube that encases the food in the midgut and hindgut of Lepidopteran larvae. It presumably protects midgut cells from abrasion by food particles, and perhaps acts as a barrier to infection. The gypsy moth, *Lymantria dispar*, is susceptible to a nuclear polyhedrosis virus (LdMNPV), which is transmitted *per os*. Once ingested, viral occlusion bodies (OB) dissolve in the alkaline midgut, liberating virions that ultimately invade susceptible cells. Recent studies have shown that selected stilbenedisulfonic acid derivatives administered in combination with LdMNPV significantly enhance viral potency, but the precise mode of action is unknown.[1] We tested the effect of the stilbene derivative, Blankophor BBH (Burlington Chem. Co., Burlington, NC), on potency of LdMNPV in gypsy moth larvae, and determined the effects of ingested BBH/LdMNPV combinations on the surface structure of the peritrophic membrane.

Newly molted second instar larvae were fed high wheat germ diet overlaid with either water, LdMNPV (10^2-10^6 OBs/ml), BBH (0.5%), or LdMNPV + BBH. Larvae were dissected at intervals ranging from 30 minutes to 8 days post-ingestion of inoculum. Intact peritrophic membranes were removed, fixed, dehydrated, and critical-point dried prior to examination by scanning electron microscopy (SEM). Remaining larvae (50/dilution) were reared for 21 days.

The addition of BBH to LdMNPV resulted in decreased kill times and enhanced potency of LdMNPV at all dose levels, representing about a 50-fold difference in potency.

Peritrophic membranes from untreated larvae (Fig. 1) appeared slightly wrinkled but had no other obvious surface perturbations. LdMNPV treatment had little, if any observable effect on peritrophic membrane (Fig. 2). BBH treatment resulted in changes in surface structure that were apparent 72 h post-ingestion (Fig. 3); the peritrophic membrane appeared thickened and fissured, and at later times appeared to be made up of multiple lamellae. Treatment with BBH + LdMNPV resulted in surface anomalies that were first detectable 2 h post-ingestion. By 72 h post-ingestion (Fig. 4), the peritrophic membrane had greatly deteriorated, and bits of food could be seen within a disorganized membranous matrix. At the time of death, only a few strands of membrane remained around the food bolus.

Our data suggest that BBH acts synergistically with LdMNPV to rapidly degrade and ultimately disintegrate gypsy moth larval peritrophic membrane. We speculate that the damaged peritrophic membrane permits large numbers of virions to pass into the ectoperitrophic space and ultimately to invade susceptible cells and replicate, thus accounting for the observed increases in viral potency.

References

1. M. Shapiro and J. L. Robertson. *J. Econ. Entomol.*, 85(1992)1120.

Proc. Microscopy and Microanalysis 1995, edited by G.W. Bailey, M.H. Ellisman, R.A. Hennigar, and N.J. Zaluzec

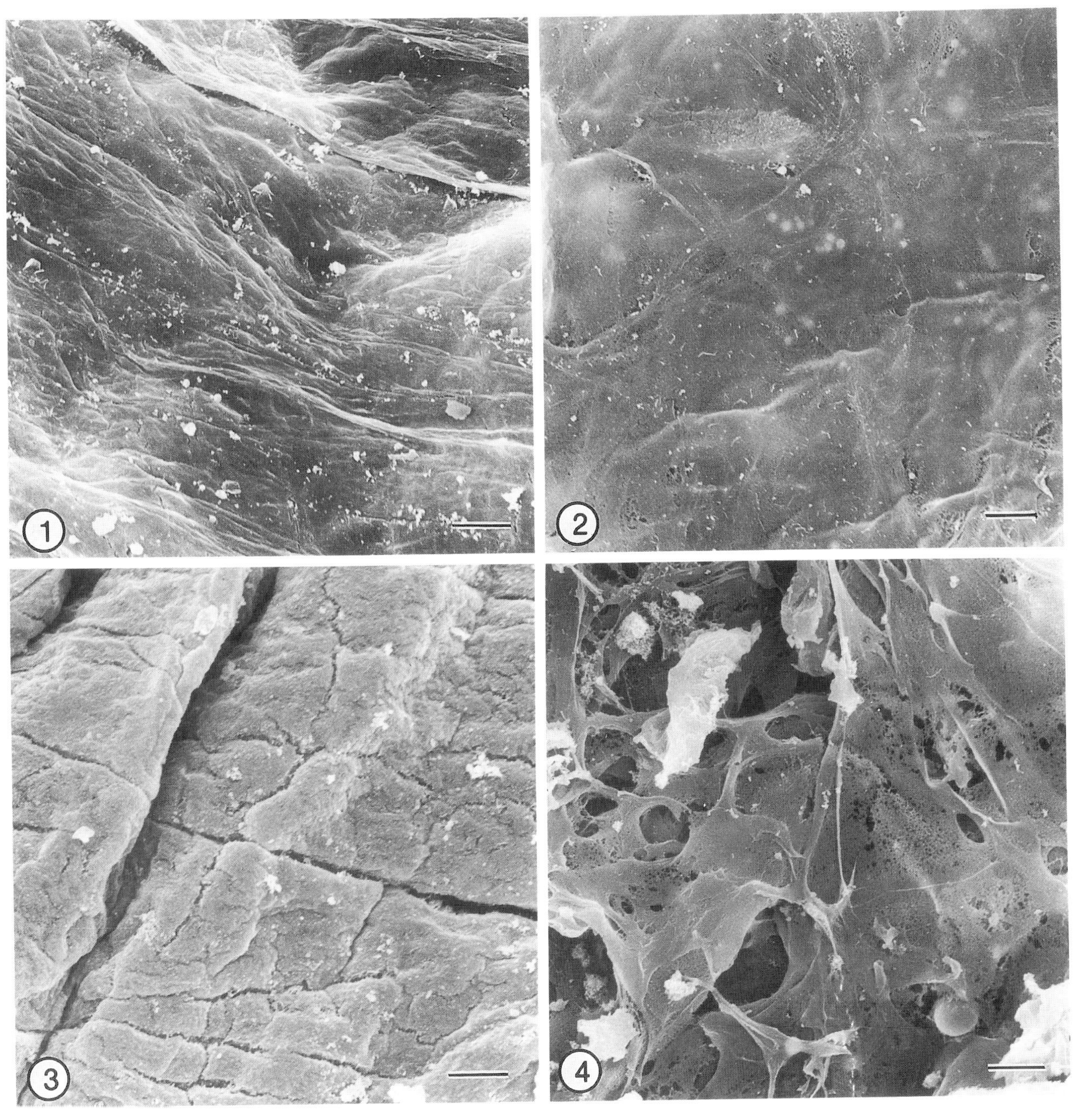

FIGS. 1-4 — SEM images of epithelial surface of peritrophic membranes 72 h post-ingestion of inoculum. Bars = 10 μm.

FIG. 1. — Control larva
FIG. 2. — LdMNPV-treated larva
FIG. 3. — BBH-treated larva
FIG. 4. — LdMNPV + BBH-treated larva

A SEM STUDY OF ATRAZINE EFFECTS ON RED SHINER (<u>CYPRINELLA LUTRENSIS</u>)

Ibrahim A. Messaad,[*] Edward J. Peters,[*] Douglas G. Rogers[**] and Kit W. Lee[***]

[*]Dept. of Forestry, Fisheries and Wildlife Science, University of Nebraska--Lincoln, NE 68583-0814.
[**]Dept. of Veterinary and Biomedical Sciences, University of Nebraska--Lincoln, NE 68583-0907.
[***]School of Biological Sciences, University of Nebraska--Lincoln, NE 68588-0118.

Atrazine is widely used in agriculture to control weeds.[1,2] However, little is known about the effects of long-term exposure in fish. Normal fish gill morphology and ultrastructure have been studied using scanning electron microscopy (SEM).[3] Pollutants including pesticides can cause lesions in gills, which ultimately affect osmoregulation and oxygen consumption.

Specimens of red shiner (<u>Cyprinella</u> <u>lutrensis</u>) were collected from the Platte River, Nebraska, by seining and treated with sodium chloride and malachite green to reduce infections. Fish were randomly distributed to aerated glass aquaria. Biological sponge filters were used to keep aquaria clean during the 14 day bioassay. Fish were maintained at a 22°C on a 12:12 photoperiod, and fed a commercial fish food once daily. Atrazine concentrations of 0, 10, 100, and 1000 μg L^{-1} were used. For scanning electron microscopy, fish were fixed in 3% glutaraldehyde in 0.1 M phosphate buffer. Then, gills were removed and post-fixed in 1% OsO_4. After dehydration in a graded ethanol series, samples were critical point dried with CO_2, sputter-coated with gold-palladium, and observed with Cambridge S-90 stereoscan at 15 kV.

The gill is composed of three parts: arch, filament, and lamellae. The lamellae are more important in respiratory function than the filament. However, both filament and the lamellae are characterized by the presence of many microridge patterns (Fig. 1). Gills from atrazine-stressed fish revealed ultrastructure alterations such as curling and disorganization of lamellae (Fig. 2), clubbing, swelling of filaments and lamellae in association with invagination of epithelial cells (Fig. 3) , fusion of lamellae (Fig. 4), goblet cell proliferation, and dilation of the connective tissue sinus at low and medium concentrations (10 and 100 μg L^{-1}). Loss of microridge patterns (Fig. 5), excessive mucus secretion and complete fusion of lamellae (Fig. 6), formation of a rugose epithelial surface, and mortality occurred at high atrazine concentrations (1000 μg L^{-1}).

The present study showed that the gills of <u>C.</u> <u>lutrensis</u> have lesions similar to those seen in other fish species that are exposed to toxicants other than atrazine, which have been reported in response to adverse physiological and biochemical changes.[4] These alterations in gill morphology may have affected osmoregulation and oxygen exchange capacity leading to anoxia and death.

References

1. M. C. Tortorelli et al., Environ. Contam. Toxicol. (1990)523.
2. F. DeNoyelles et al., Ecol. 63(1982)1285.
3. G. J. Hughes, J. Zoo. (London), 187 (1979)43.
4. D. E. Hinton and D. J. Lauren, Amer. Fish. Soci. Symp. 8(1990)51.

Proc. Microscopy and Microanalysis 1995, edited by G.W. Bailey, M.H. Ellisman, R.A. Hennigar, and N.J. Zaluzec

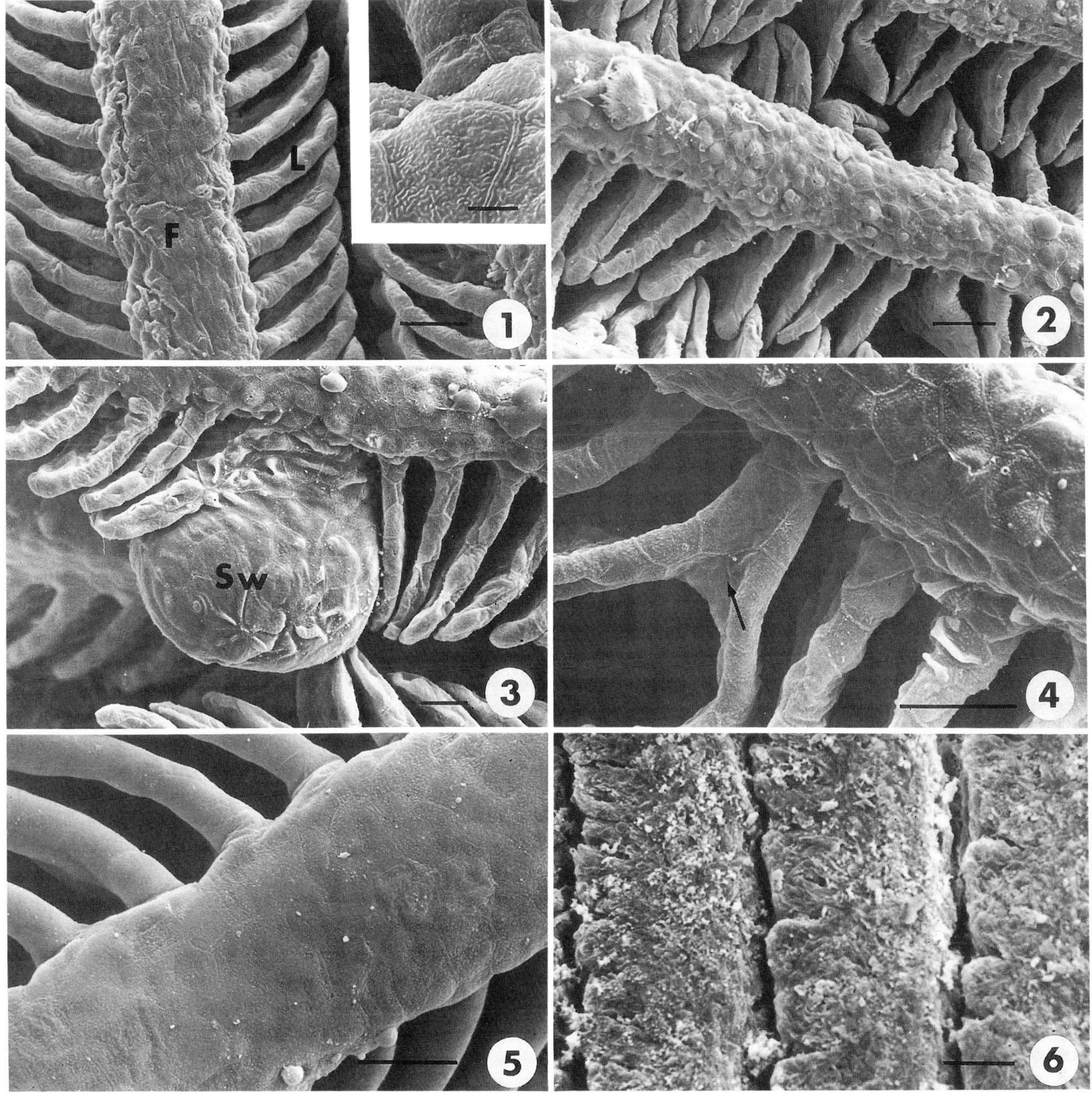

FIG. 1.--Normal gill filament of <u>C. lutrensis</u> showing filament (F) and lamella (L). Bar = 20 μm. Inset: microridge patterns. Bar = 5 μm.

FIG. 2.--Gill filament: curling and disorganization of lamellae at 10 μg L^{-1} atrazine. Bar = 20 μm.

FIG. 3.--Gill filament: swelling of lamellae (Sw) at 10 μg L^{-1} atrazine. Bar = 20 μm.

FIG. 4.--Gill filament: fusion of lamellae (arrow) at 100 μg L^{-1} atrazine. Bar = 20 μm.

FIG. 5.--Gill filament: complete loss of microridge patterns at 1000 μg L^{-1} atrazine. Bar = 20 μm.

FIG. 6.--Gill filaments: complete fusion of lamellae, mucus , and a rugose epithelial surface at 1000 μg L^{-1} atrazine. Bar = 50 μm.

HUMAN T-LYMPHOCYTES (JURKAT CELLS) EXPRESSING HIV-1 TAT-72: SEM.

J. Koch*‡, J. Gilloteaux**‡, J. M. Jamison*‡, and J.L.Summers‡. Departments of Microbiology & Immunology*, Anatomy**, and Urology‡, Northeastern Ohio Universities College of Medicine, P.O. Box 95, Rootstown OH. 44272, U.S.A.

The human immunodeficiency virus (HIV-1) is dependent upon the synthesis of TAT for efficient transcriptional activation and replication[1]. Following translocation to the nucleus TAT interacts with a cis-acting TAT-responsive element located between nucleotides +1 to +42 within the HIV-1 long terminal repeat[2]. This region contains an RNA stem loop structure TAR, which interacts with TAT and cellular proteins to increase transcription 1000 fold[3].

We have transfected human T-lymphocytes with an expression vector containing the HIV-1 LTR driving the synthesis of TAT-72. A stable cell line was generated by selection with G418. Cells were grown on round glass coverslips coated with a solution of fibronectin overnight (at 37°C). In addition, cells on coverslips were fixed with a 3.2% glutaraldehyde buffered solution (0.1M Na cacodylate) for 5 min at room temperature, then 15 min at 4°C, then washed gently in a buffer-sucrose solution. Postfixation was performed with an aqueous mixture containing 1.5% OsO_4+ 0.5% RuO_4 for 20 min at 18°C. The coverslips were then washed in buffer before dehydration with graded ethanol and critical point drying. Samples were then coated with gold and viewed in a Jeol 35C SEM.

SEM observations of Wild type ("W": Fig 1a-b) or Control ("C", Fig. 2 a-b): cells transfected with the vector backbone and selected with G418) Jurkat cells have the typical features of T-lymphocytes, i.e. cells appear spherical, and show some typical microvilli. However, when examined at SEM higher magnification, one can detect small to long microvilli closely apposed on the cell surface of the "C" cells compared to the "W" cells. Both cell types display a prominent uropod. In contrast, the TAT-72 cells are a heterogenous cell population (from 12 to more than 100 µm in diam.) and display pleomorphism (Fig. 3a arrows). Their surface is usually devoid of long microvilli, and they develop wing-like excrescences often resembling mouse ears which make contact with other cells (Fig.3a-b) [4].

References

1. A.L.Dayton, J.G. Sodroski, C.A. Rosen, W.C. Goh, W.A. Haseltine. *Cell* 44 (1986) 941-947.
2. J.A. Garcia, D. Harrich, E. Soultanakis, F. Wu, R. Mitsuyasu, R.B. Gaynor. *EMBO J.* 8 (1989) 765-778.
3. A. Gatignol, A. Kumar, A. Rabson, and K-T. Jeang. *PNAS* 86 (1989) 7828-7832.
4. Supported by Summa Health System Foundation, Akron, OH.

Proc. Microscopy and Microanalysis 1995, edited by G.W. Bailey, M.H. Ellisman, R.A. Hennigar, and N.J. Zaluzec
Copyright © 1995 MSA. Published by Jones and Begell Publishing, 79 Madison Ave., New York, NY 10016

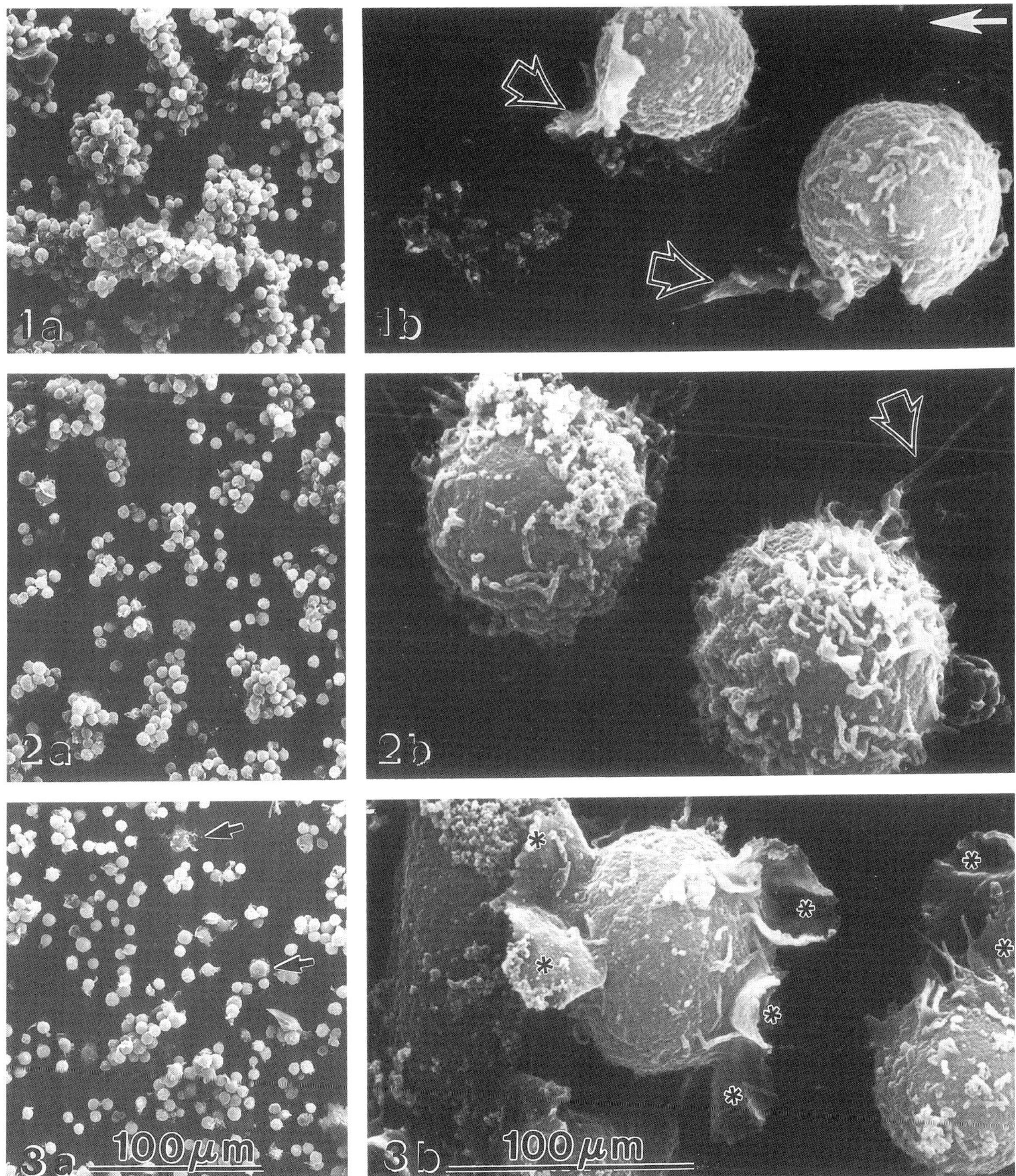

SEM aspects of control and human T-lymphocytes expressing TAT-72. Scales are 100 µm in columns a and b.

FIGS. 1 and 2.- Wild type (Fig. 1a-b) and Control (Fig. 2a-b) human T-lymphocytes. Typical morphological characteristics are shown, including the presence of uropods (open arrows). In Fig. 1b, white arrow indicates the origin of electron beam.

FIG. 3. a-b - Human T-lymphocytes expressing TAT-72 showing pleomorphic size and peculiar cellular appendages (*).

ULTRASTRUCTURAL CHANGES IN THE MOUSE PLACENTA AFTER ADMINISRATION OF TRICHOSANTHIN

V.E.C. Ooi*, C.F. Lau**, and H.W. Yeung***

Departments of *Biology and ***Biochemistry, The Chinese University of Hong Kong;
**Institute of Molecular Biology, University of Hong Kong, Hong Kong.

The root tuber of <u>Trichosanthes</u> <u>kirlowii</u> Maxim (Family curcurbitaceae) has long been used in Chinese traditional medicine as an abortifacient and in therapy for choriocarcinoma.[1] The active principle of the root tuber extract was recently purified and characterized. It is a basic protein of 24 KDa and is named trichosanthin.[2] Investigations on trichosanthin action showed that it affected directly on the placenta, selectively causing necrosis of the syncytiotrophoblasts of placental villi.[3] The present study attempts to determine the initial action of trichosanthin and follow through from the earliest sign of histopathological and ultrastructural changes of the mouse placentae after administration of trichosanthin.

Mature ICR female mice were used. On the tenth day of gestation, a single dose of trichosanthin was injected intraperitoneally. The dosage used was 0.1 mg of trichosanthin in physiological saline per 25 g body weight of the mouse. At various prescribed intervals after injection, the mice were sacrificed by cervical dislocation. The bicornate uterus of the mouse was then cut opern to count the number and conditions of the fetuses. The placentae together with fetuses were quickly removed and processed for obserbvation under the transmission electron microscope (JOEL 1200 EX2).

The earliest sign of cell injury could be traced in the basal zone of the placenta as early as 4 hours after administration of trichosanthin. Alterations in cells of the decidua basalis facing the maternal blood spaces was detected before the labyrinth showed any observable change. Some mitochondria of these cells appeared swollen and the cristae were broken. Electron-dense granules were seen discharged into the intercellular spaces which were filled with fibrillar materials. Damage to the trophoblastic cells in the labyrinth was found in the specimens which had been treated with trichosanthin for 8 hours or longer. The changes included cytoplasmic budding, vacuolation and leucocyte invasion in the decidua basalis. However, intensive damage occurred only in the middle layer of the labyrinthine syncytiotrophoblast whereas the outer and inner layers remained relatively intact (Fig. 1). In the trophoblastic cells of the middle layer, cytoplasm was shrunk but mitochondria and RER were swollen. Blood flow in the fetal capillaries appeared to be blocked as exemplified by the accummulation of nucleated erythrocytes of the fetal blood. The trophoblastic cells of the middle layer were seen sloughed off into the meternal blood space but the inner syncytiotrophoblasts were still well preserved. The sloughed-off cytoplasmic bud contained contracted RER and fragmented mitochondria (Fig. 2). The placentae from the specimens sacrificed 20-24 hours after treatment showed intensified and widespread degeneration and disorganization in the middle trophoblast layer. The electron-dense granules discharged from the degenerated cells were accummulated in the maternal blood spcaes alongside with other debris of detached cytoplasm and organelles. Autophagic vacuoles were found among the dissolved cytoplasm of the middle trophoblastic cells. Their shape, size and number of layers of myelination were not uniform (Fig. 3).

Proc. Microscopy and Microanalysis 1995, edited by G.W. Bailey, M.H. Ellisman, R.A. Hennigar, and N.J. Zaluzec

References

1. T.B. Ng, W.Y. Chan and H.W. Yeung, *Gen. Pharmac.* 11(1992)00-00.
2. J.M. Maraganore, M. Joseph and M.C. Bailey, *J. Biol. Chem.* 262(1987)11628.
3. Wang, et al., *Acta Zool. Sinica*, 22(1976)156.

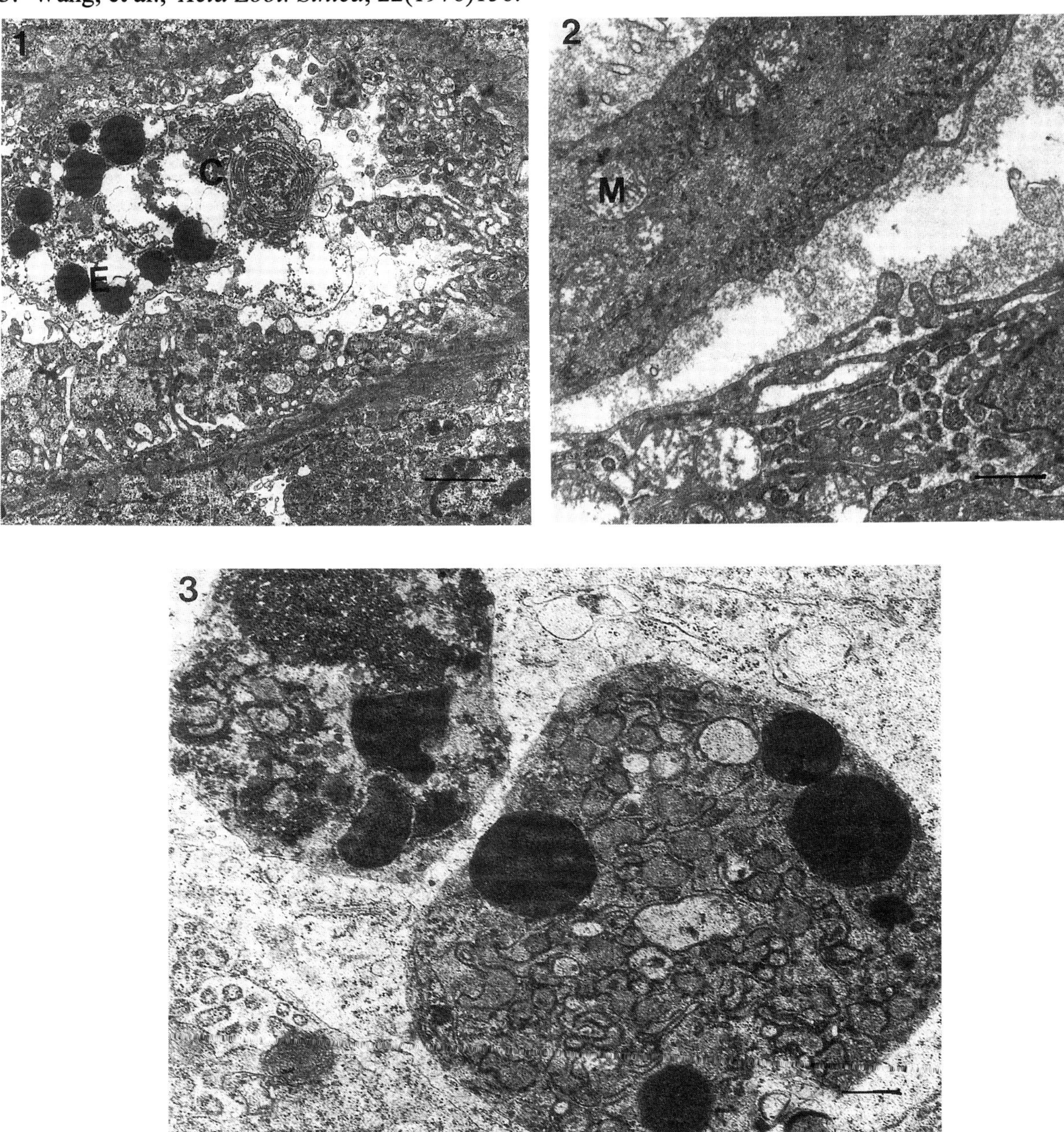

Fig. 1 - Damage in middle labyrinthine syncytiotrophoblast after 8 hours' treatment with trichosanthin. Cytosegresome (C) and electron dense granules (E) are found in the space. Bar=2 μm.
Fig. 2 - Sloughed-off syncytiotrophoblast bud containing swollen mitochondria (M) and contracted RER (arrow). Bar=2 μm.
Fig. 3 - Two autophagic vacuoles in the cytoplasm of degenerating trophoblast. Bar=1 μm

3-D IMAGING OF RESPIRABLE SIZE ASBESTOS FIBERS IN RAT AND HUMAN LUNG TISSUE.

R.A. Rogers, [*] J.M. Antonini, [*] J. Lai, [*] B.A. Ekstein, [*] E.H. Oldmixon, [**] B.S. Ducatman, [***] and J.D. Brain[*]

[*] Physiology Program, Harvard School of Public Health, Boston, MA 02115
[**] Dept. Med., Memorial Hospital of R.I., Pawtucket, R.I. 02860
[***] National Institute of Occupational Safety and Health, Morgantown, W.V. 26505

Pulmonary exposure to natural and artificial fibers can result in serious health impacts to many segments of the occupational work force in the U.S. and worldwide. The toxicological response of a fiber on the respiratory tract depends upon the site at which the fiber makes initial contact, through which it migrates, where it deposits, and the route through which it is cleared. Persistence of fibers in lung tissue influences the severity of lung disease[1]. The pulmonary deposition and clearance of inhaled fibers is a major determinant in understanding the health outcome of environmental and occupational exposures to these materials. Current microscopic methods of fiber analysis[2] result in disruption of fiber orientation relative to lung structures and may influence our understanding of fiber toxicity within lungs.

The purpose of this study was to examine the *in situ* pulmonary fate of respirable fibers in lungs using simultaneous nonspecific fluorescent tissue staining and reflected light confocal microscopy. Serial optical sections were recorded using a Sarastro 2000 confocal laser scanning microscope (Molecular Dynamics, Inc.) optimized for reflected light imaging. Three-dimensional images were rendered from serial optical sections of lung using ImageSpace (Molecular Dynamics, Inc.) and Voxel View Ultra software (Vital Images, Inc.) to examine the nature and distribution of fibers in lung tissue.

Lung samples were prepared for microscopic analysis from 1) adult Sprague-Dawley rats following intratracheal instillation with amosite asbestos (1 mg/0.15 ml/100 gm body wt), and 2) asbestos-containing human lung specimens[5]. Lung pieces ($10mm^2$ x 5mm thick) were fixed in 2% glutaraldehyde, fluorescent stained (Lucifer Yellow CH; 1 mg/ml), and embedded in Spurr's epoxy[3]. Images were recorded using polarized light (<510 nm) to examine fibers. Fluorescent emission spectra >510 nm was used to image cellular morphology and lung structure. An optical sectioning/tissue removal procedure[4] was applied to collect enough serial optical sections to record fiber distribution among airways spanning hundreds of microns in length. Digital images were superimposed and then reconstructed in 3-D to show surface and interstitial fiber positions in relation to tissue. Fiber number, length, diameter, and location were readily obtained while preserving anatomic relationships and fiber geometry in situ.

References
1. J.M.G. Davis et al., Br.J. Exp. Pathol. 67(1986)415.
2. D.B. Warheit et al., Environ. Res. 54(1991)183.
3. R.A. Rogers, et al., Mol. Biol. Cell 3:s,185a, 1992.
4. H. Brismar, et al., ISandT/SPIE Proceedings, Vol.2412 (Feb 8, 1995) In Press.
5. Supported by NIEHS-ES00002, NIOSH-109979, and NIH HL26863. Human lung specimens were obtained from the Pathology files, Division of Respiratory Disease Studies, NIOSH.

Proc. Microscopy and Microanalysis 1995, edited by G.W. Bailey, M.H. Ellisman, R.A. Hennigar, and N.J. Zaluzec
Copyright © 1995 MSA. Published by Jones and Begell Publishing, 79 Madison Ave., New York, NY 10016

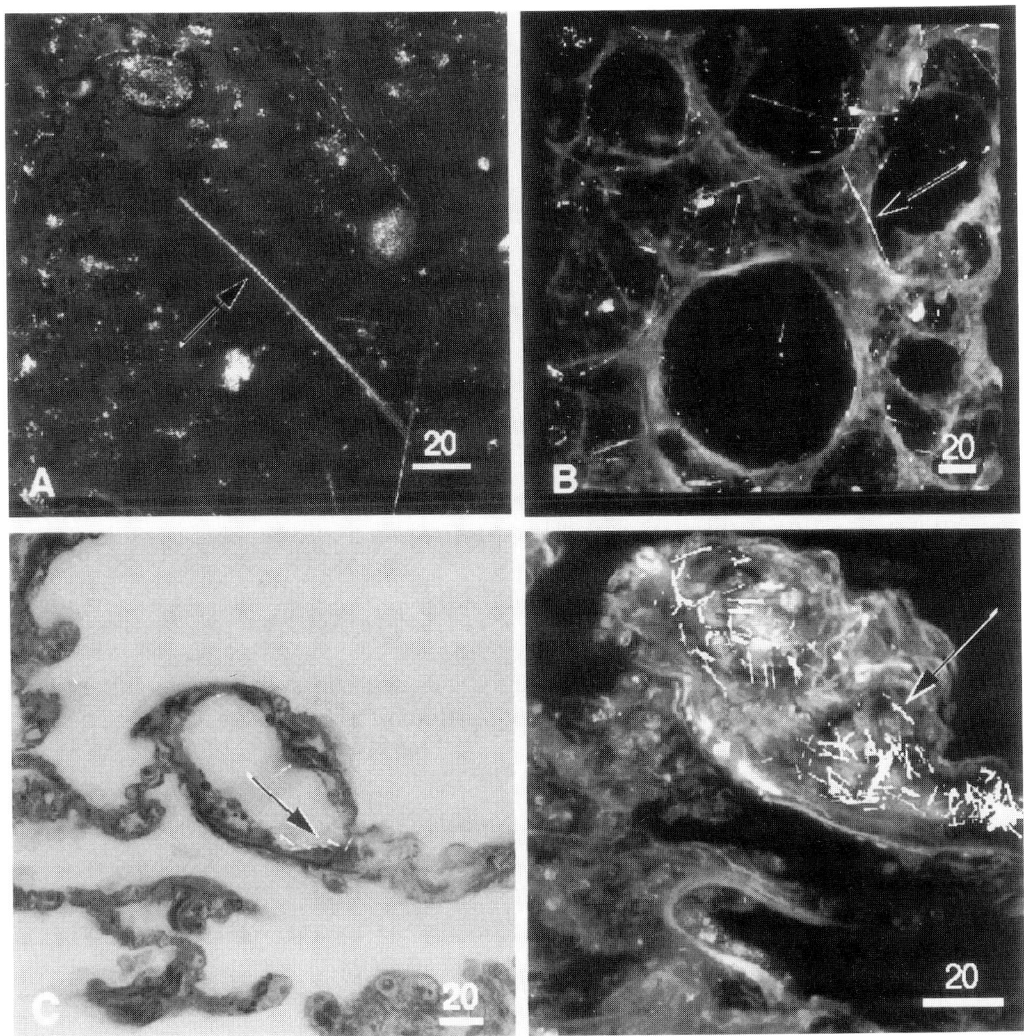

Figure 1. Confocal micrograph of asbestos fibers recorded using reflected light mode. **(A)** Distinctive long narrow characteristics of amosite fiber morphology are shown (arrow). Bar is 20 μm. **(B)** In situ amosite fibers (arrow) are present among airways in this three-dimensional projection of rat lung. Fibers appearing in the center of airways extend from regions outside the boundaries of the reconstruction. The appearance of this reconstruction was adjusted to highlight the connective tissue architecture (grey line structures) to help identify fibers penetrating airway walls (arrow). The reconstruction was made from 240 optical sections collected at 1 μm intervals. Rats were intratracheally dosed with 1 mg/100 g body weight of 70% amosite for one hour then killed. Serial optical sections were collected, and 3-D projections made. Bar is 20 μm. **(C)** Confocal micrograph of a human lung revealing asbestos fibers. Individual fibers are clearly seen penetrating the alveolar wall (arrow). Bar is 20 μm. **(D)** Three-dimensional reconstruction of human lung containing asbestos fibers (arrow) in alveoli. Bar is 20 μm.

THE REMOVAL AND MORPHOLOGY OF INTACT BIOFILMS FROM IMPLANTED VENOUS ACCESS DEVICES

M.A.Gregory and G.P.Hadley*

EM Unit, University of Durban-Westville, P.Bag X54001, Durban 4000. *Department of Paediatric Surgery, University of Natal, PO Box 17039, Congella 4013, South Africa

The insertion of implanted venous access systems for children undergoing prolonged courses of chemotherapy has become a common procedure in pediatric surgical oncology. While not permanently implanted, the devices are expected to remain functional until cure of the primary disease is assured. Despite careful patient selection and standardised insertion and access techniques, some devices fail. The most commonly encountered problems are colonisation of the device with bacteria and catheter occlusion. Both of these difficulties relate to the development of a biofilm within the port and catheter. The morphology and evolution of biofilms in indwelling vascular catheters is the subject of ongoing investigation. To date, however, such investigations have been confined to the examination of fragments of biofilm scraped or sonicated from sections of catheter[1]. This report describes a novel method for the extraction of intact biofilms from indwelling catheters.

15 children with Wilm's tumour and who had received venous implants were studied. Catheters were removed because of infection (n=6) or electively at the end of chemotherapy. Catheters had been in place from 11-300 days. At explantation, 1cm. sections were harvested from the proximal, mid and distal portions of each catheter and immediately immersed in Karnovsky's fixative. Following dehydration, the cores were removed by gentle agitation in propylene oxide prior to subsequent conventional preparation for LM, TEM and SEM.

Biofilms from non-infected patients were thin, especially distally and composed of fibrin, normal and necrotic platelets, white cells, erythrocytes and viable and necrotic cocci (Fig.1). Occasional colonies of encapsulated, perhaps dormant cocci were detected near the biofilm/catheter interface (Fig.2). Cocci were present in and/or on all films studied but were more often observed by SEM (Fig.3) and TEM (Fig.4) on or near the luminal surface of the thick-walled biofilms from infected children.

The novel method of core removal enabled wall thickness and the spatial relationships of cells, structures and bacteria on/within intact biofilms to be determined. Observations of intact biofilms may further our knowledge regarding their evolution in access devices and help in the prevention and management of biofilm mediated bacteremia.

References

1. I.Raad et al., J. Inf. Dis. 168(1993)400.

Proc. Microscopy and Microanalysis 1995, edited by G.W. Bailey, M.H. Ellisman, R.A. Hennigar, and N.J. Zaluzec
Copyright © 1995 MSA. Published by Jones and Begell Publishing, 79 Madison Ave., New York, NY 10016

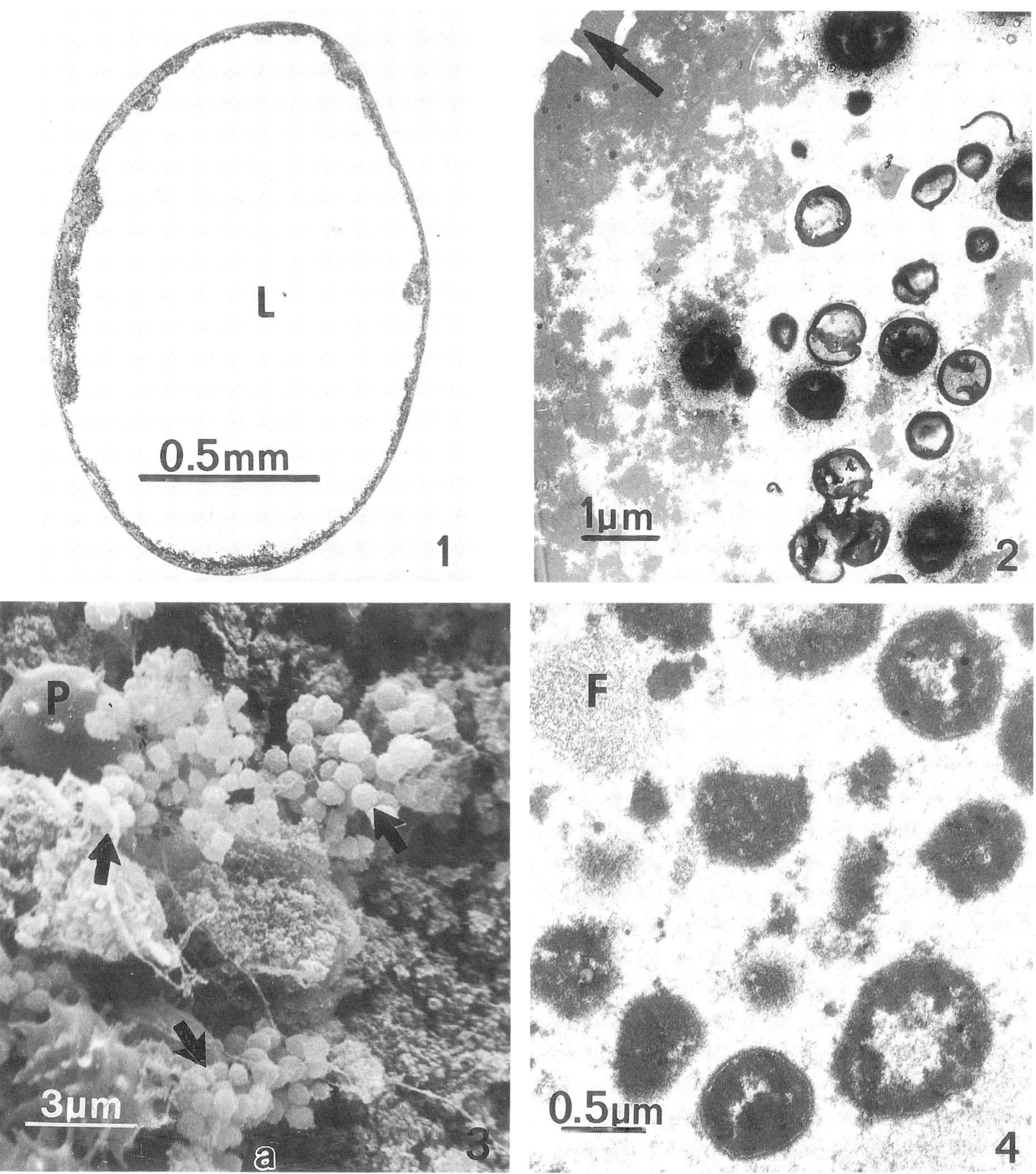

Fig.1.- LM: Transverse/obliquely sectioned biofilm from the distal portion of a catheter from a non-infected patient. L = Lumen.
Fig.2.- TEM: Encapsulated cocci from the deeper regions of a biofilm from a non-infected patient. Outer surface of biofilm (arrowed).
Fig.3.- SEM: Colonies of cocci on the surface of a biofilm from a child with bacteremia (arrowed). P = platelet.
Fig.4.- TEM: Healthy and necrotic cocci near the surface of a biofilm from a patient with bacteremia. F = fibrin.

ADHERENCE OF ENTEROHEMORRHAGIC *ESCHERICHIA COLI* IN A BEEF SYSTEM

T. S. Schwach* and E. A. Zottola*

*Department of Food Science and Nutrition, University of Minnesota, St. Paul, MN 55108

Microbiologists have traditionally studied pure cultures growing in laboratory media. Food microbiologists, interested in interactions between the food system, pathogen, and host have tried to extrapolate information in the same manner. However, clinical isolates readily lose virulence characteristics when repeatedly subcultured into laboratory media[1]. Expression of many virulence factors, such as fimbriae, extracellular polymers and outer membrane adhesins are affected by growth environment; an environment which encompasses not only nutrient availability and growth temperature, but also physical surroundings, moisture, oxygen and pH levels, and the presence of competitive flora. The ability of an organism to adhere to a surface is a major virulence factor[2].

Enterohemorrhagic *Escherichia coli* (EHEC) are one of five groups of enteropathic *E. coli* which cause intestinal disease world-wide. The most well known of this group, serotype O157:H7, has been implicated in major food-borne illness outbreaks primarily due to the consumption of undercooked hamburgers[3]. They are found in the intestinal tracts of cattle, feces, hanging carcasses, and frozen ground beef. *E. coli* O157:H7 adheres to large intestinal epithelium with characteristic lesions and produces verocytoxins which damage internal organs, such as the kidney.

The objective of this study was to monitor growth and adherence of six EHEC strains and one laboratory *E. coli* (strain K-12) in ground beef and beef media using microscopy and standard enumeration methods. Ground beef patties (25 g) were inoculated with 10^3-10^4 organisms/g and incubated at 9.5°C for 12 d. Patties were placed on sterile nitrocellulose or polycarbonate filters in sterile petri dishes. Beef Extract Broth (BEB) was also inoculated and incubated at 9.5°C. Two g from each patty and 2 ml from each BEB tube were removed each day for enumeration and microscopy studies. Beef and filter samples were fixed in a glutaraldehyde-paraformaldehyde-ruthenium red fixative for 48 h at 4°C, post-fixed in 1% OsO_4 and dehydrated in a grade ethanol series. Samples for LVSEM were critical point dried, gold coated in an ion-beam sputterer and viewed in a Hitachi S-900 FE LVSEM at 3 keV. Samples for TEM were embedded in Epon and viewed in a Philips CM12 at 80 keV. BEB samples (0.5 ml) were placed into sterile Ependorf tubes to which 0.5 ml of a warm 2% Nobel Agar solution was added. When the agar cooled, 0.5 ml fixative was added. After 24 h at 4°C, tubes were centrifuged (8,000 rpm for 1 min) and all subsequent rinses and fixative changes were done with this semi-solid pellet, centrifuging between each change.

Throughout the study, numbers of EHEC organisms in both the ground beef and BEB systems remained within one log of inoculation levels (Figure 1). In the ground beef system, gram-positive cocci and bacilli were noted after 1 d, a 1-2 log increase in other gram-negative flora was seen after 4 d, and yeasts after 7 d (Figure 2). A biofilm layer was seen on filter surfaces after 1 d as evidenced by fat globules, muscle fibers, and entrapped bacteria (Figure 3). Initial results indicate that EHEC and other gram negative flora are intimately associated with beef muscle fiber and fat globules while gram-positive cocci and bacilli appear to be less "covered" by biofilm polymers (Figure 4). Macroscopically, EHEC do not produce noticeable extracellular polymer in either ground beef or BEB, although fibrils are evident by both TEM (Figure 5) and LVSEM.

Currently, adherence studies using the same system as described above, are being conducted to HT-29 cultured intestinal cells from inoculated ground beef incubated at 9.5°C and inoculated BEB at 9.5 and 35°C. A Fluorescent Actin Stain[4] is used to screen for intestinal lesions and adherence levels are quantified both microscopically and using enumeration techniques[5].

Proc. Microscopy and Microanalysis 1995, edited by G.W. Bailey, M.H. Ellisman, R.A. Hennigar, and N.J. Zaluzec

1. Klemm, P. *Rev. Inf. Dis.*, 7(1985)321.
2. Smith, H. *Can. J. Microbiol.* 38(1992)747.
3. Doyle, M. P. *Int. J. Food Microbiol.* 9(1991)105.
4. Knutton, S. et al., *Infect. Immun.* 57(1989)1290.
5. Sherman, P. M. and R. Soni, *J. Med. Microbiol.* 26(1988)11.

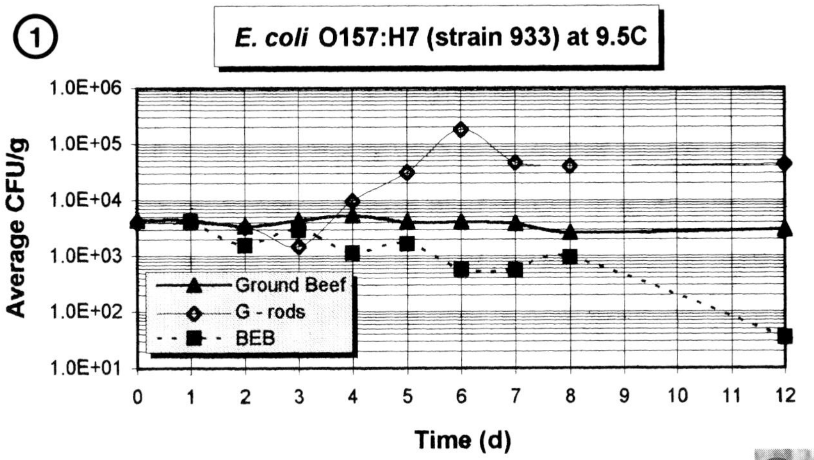

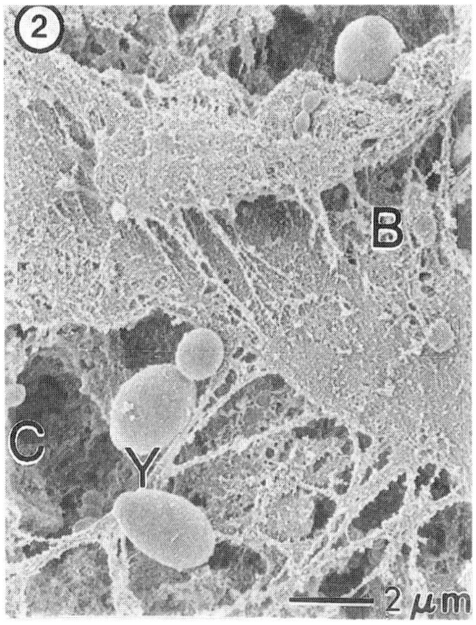

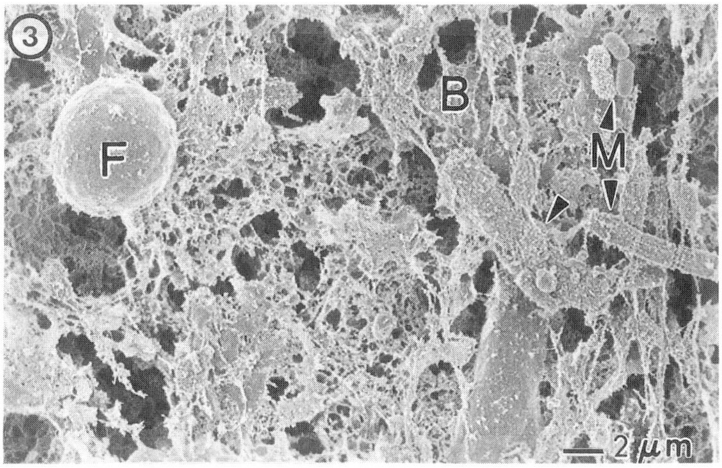

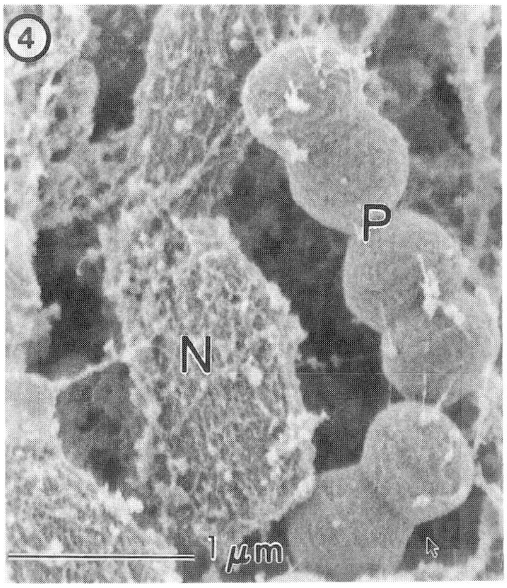

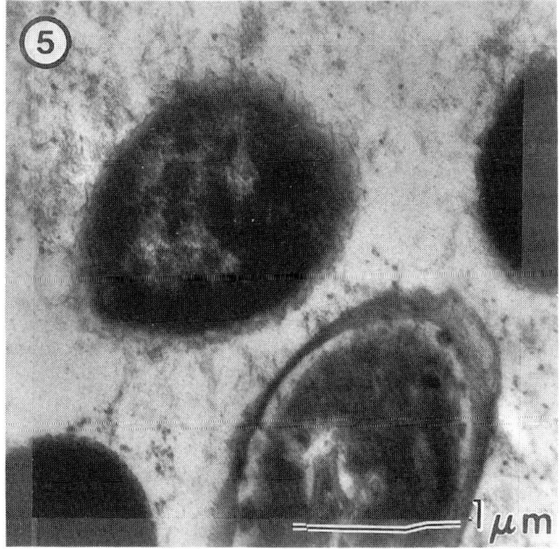

Fig. 1.- *E. coli* O157:H7 growing in ground beef and BEB at 9.5°C.

Fig. 2.- LVSEM image of nitrocellulose filter surface after 7 d contact with beef patty showing yeast (Y), gram-positive cocci (C), and gram-negative bacilli (B).

Fig. 3.- LVSEM image of biofilm layer on nitrocellulose filter showing fat globule (F), beef fiber (B), and bacteria (M).

Fig. 4.- High magnification LVSEM image of gram-negative bacilli (N) and gram-positive cocci (P) surface morphology.

Fig. 5.- TEM image of *E. coli* O157:H7 (strain 933) from BEB at 9.5°C processed using the Nobel Agar technique.

ULTRASTRUCTURAL PROCESS OF INTESTINAL WOUND HEALING

Rick L. Vaughn, Shailendra K. Saxena and John G. Sharp

Department of Cell Biology and Anatomy, University of Nebraska Medical Center
Omaha, NE 68198-6395

We have developed an intestinal wound model that includes surgical construction of an ileo-cecal patch to study the complex process of intestinal wound healing .[1] This allows approximation of ileal mucosa to the cecal serosa and facilitates regeneration of ileal mucosa onto the serosal surface of the cecum. The regeneration of ileal mucosa can then be evaluated at different times. The wound model also allows us to determine the rate of intestinal regeneration for a known size of intestinal wound and can be compared in different situations (e.g. with and without EGF[2,3] and Peyer's patches[4]).

At the light microscopic level it appeared that epithelial cells involved in regeneration of ileal mucosa originated from the enlarged crypts adjacent to the intestinal wound and migrated in an orderly fashion onto the serosal surface of the cecum.[1] The migrating epithelial cells later formed crypts and villi by the process of invagination and evagination respectively. There were also signs of proliferation of smooth muscles underneath the migratory epithelial cells.[1]

In the present study, we evaluated the process of intestinal wound healing using SEM and TEM techniques in rats on the 5th, 7th, and 14th day after construction of an ileo-cecal patch. SEM pictures on the 7th and 14th day showed a well-defined process of epithelial cell migration from the adjacent normal mucosa on the serosal surface of the cecum (Fig. 1). The leading edge of squamous type epithelial cells was clearly identified. Fibroblast-type cells were also seen with processes joining epithelial cells to the underlying connective tissue mass. SEM pictures also demonstrated the presence of granulation tissue associated with migrating epithelial cells (Fig. 2). Semithin plastic sections on day five revealed enlarged crypts next to intestinal wound and villi bending over onto the intestinal wound. The epithelial cells involved in regeneration not only appeared to arise from the adjacent crypts but also from the villi that have bent over onto the wounded area. Semithin sections also showed that the outer musculature of the cecum may be a potential source of muscle cells involved in regeneration underneath the migratory epithelial cells. TEM pictures on day five demonstrated increased vascularization at the junction between the regenerating ileal mucosa and cecal serosa. There was also evidence that blood cells were migrating out into the wound area. There was an abundance of lymphocytes, neutrophils, apoptotic and necrotic cells, and cellular debris on the serosal surface of the cecum. This was presumably the granulation tissue observed in SEM. Proliferation of smooth muscle along the serosal surface was also observed in the TEM.

The present study has confirmed the migration of epithelial cells in the regeneration of ileal mucosa. However, it suggests an additional source of epithelial cells from adjacent villi as well as the movement of smooth muscle cells and inflammatory cells from the serosal surface of the cecum. Future challenges will be to characterize various cells involved in intestinal regeneration and define the factors that they produce which appears to regulate the process of intestinal wound healing.

Proc. Microscopy and Microanalysis 1995, edited by G.W. Bailey, M.H. Ellisman, R.A. Hennigar, and N.J. Zaluzec

References:

1. S.K. Saxena et al.,in *Medical Intelligence Unit*, R.G. Landes company, (1993) 1-118.
2. S.K. Saxena et al., *Surgery*, (1992) 111: 318-325.
3. J.S. Thompson et al., *Surgery* (1989) 106: 45-51.
4. S.K. Saxena et al., *The Wound Healing Society*, (1995). Abstract.

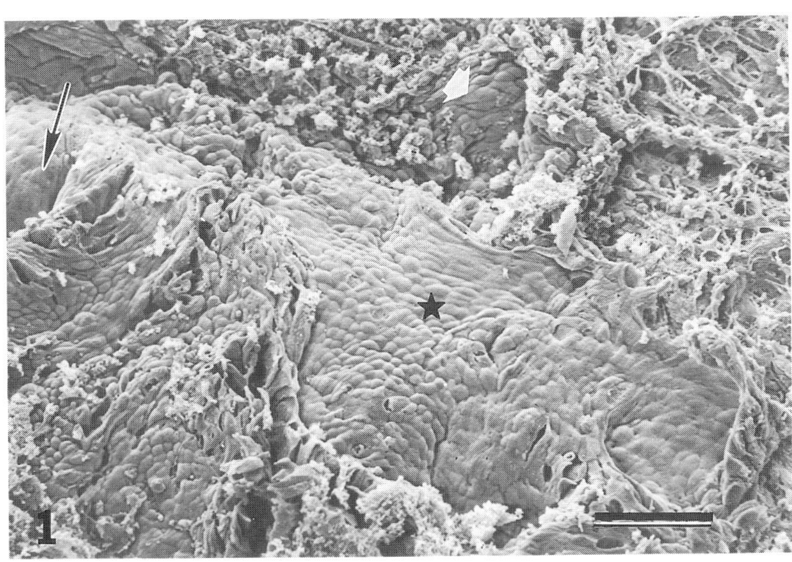

Fig. 1 - Regenerating ileal mucosa (star) adjacent to suture line (long arrow). Additional patches of regenerating mucosa can be identified underneath granulation tissue (short arrow). Bar = 50μm

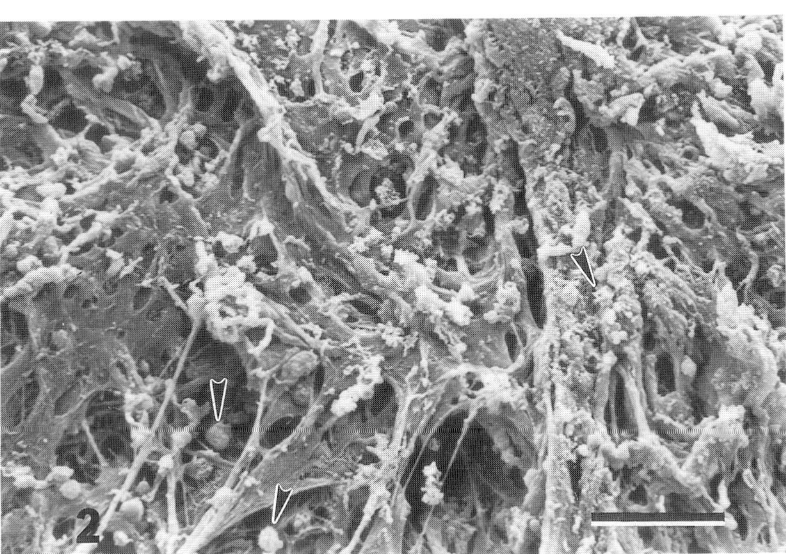

Fig. 2 - Area of granulation tissue showing fibrous matrix with an abundance of inflammatory cells (arrow head). Bar = 20 μm

Microscopic Study of the Plerocercoid Larva of *Otobothrium* (Trypanorhyncha)

N. Abdou*, R. A. Heckmann *, J. S. Gardner**, and A. A. Ashour***

* Department of Zoology, 109 WIDB, Brigham Young University, Provo, Utah 84602
** Microscopy Lab, 128 WIDB, Brigham Young University, Provo, Utah 84602
*** Department of Zoology, Faculty of Science, Ain Shams University, Egypt

This study represents the first scanning electron microscope (SEM) evaluation of cestode larvae infecting Red Sea fish. General structures of the larvae and different types of microtriches on the integument are described.

Plerocercoids of Trypanorhyncha were collected from the mesenteries and liver of *Cephalopholis oligostricta*. The capsule of each larva was opened, the plerocercoid was removed and fixed in 10% formalin. Following buffer (pH. 7.3) wash, they were fixed in OsO_4 12 hr, washed in buffer, immersed in 2% tannic acid 8 hr, washed in buffer, fixed again in OsO_4 2 hr, buffer washed, dehydrated, critical point dried, and sputter coated with gold.

The plerocercoids were identified as belonging to the genus *Otobothrium*. Each organism was divided into a scolex and body. The scolex consisted of two large bothria, each bothrium was nearly round in outline and possessed two circular invaginations on its posterior margin. On the apex of the scolex, four extended tentacles covered with hooks were observed (Fig 1).

Microtriches covered the surface of the scolex and the body of the larvae. Each of these structures consisted of a narrow proximal region and a wide terminal region with finger-like extensions. The number of these extensions varied according to the particular region of the plerocercoid. On the majority of the adhesive surface of the bothrium, microtriches had four to five extensions, whereas microtriches near and on the border of the bothrium had three extensions (Figs 2 & 3). In the invagination, the microtriches had two short extensions (Fig. 2). On the outside side of the invagination and on the opposite side of the adhesive surface, all of the observed microtriches possessed three extensions (Fig 4). Tentacles were covered with large hooks in distinct rows. Each hook was approximately 15 µm long and curved (Fig. 5).

Previous studies of microtrich-like structures have shown that there are polymorphic forms covering the surface of the adult and larvae of cestoda.[1,2,3] Whittaker, 1985, described the adult worm of *Parachristianella* (Trypanorhyncha) using SEM.[4] He observed that the entire surface of the scolex and neck were covered with various types of tegumental structures. He found microtriches with a variety of shapes including elongate filamentous, short cylindrical, and flabelliform with 6-8 digitiform extensions. In 1991, Ashour studied different types of microtriches of *Otobothrium* obtained from the fish *Liza tade* in Sri-Lanka.[5] He described cylindrical microtriches, which end in two short extensions. These were also observed in the present study. He reported tridigitate structures covering the adhesive surface of the *Otobothrium sp.* which he studied. The specimens of *Otobothrium sp.* in the present investigation possessed microtriches on the adhesive surface with three, four, or five extensions.

References

1. K. Andersen, *Int. J. of Parasit.* 5(1975)293.
2. J. Berger and D. F. Mettrick, *Trans. Amer. Microsc. Soc.* 90(1971)393.
3. R. C. A. Thompson et al., *Int. J. for Parasit.* 13(1982)579.
4. F. H. Whittaker, *J. Parasit.* 7(1985)376.
5. A. A. Ashour, *J. Of Egypt. Society of Parasit.* 21(1991)277.

Proc. Microscopy and Microanalysis 1995, edited by G.W. Bailey, M.H. Ellisman, R.A. Hennigar, and N.J. Zaluzec
Copyright © 1995 MSA. Published by Jones and Begell Publishing, 79 Madison Ave., New York, NY 10016

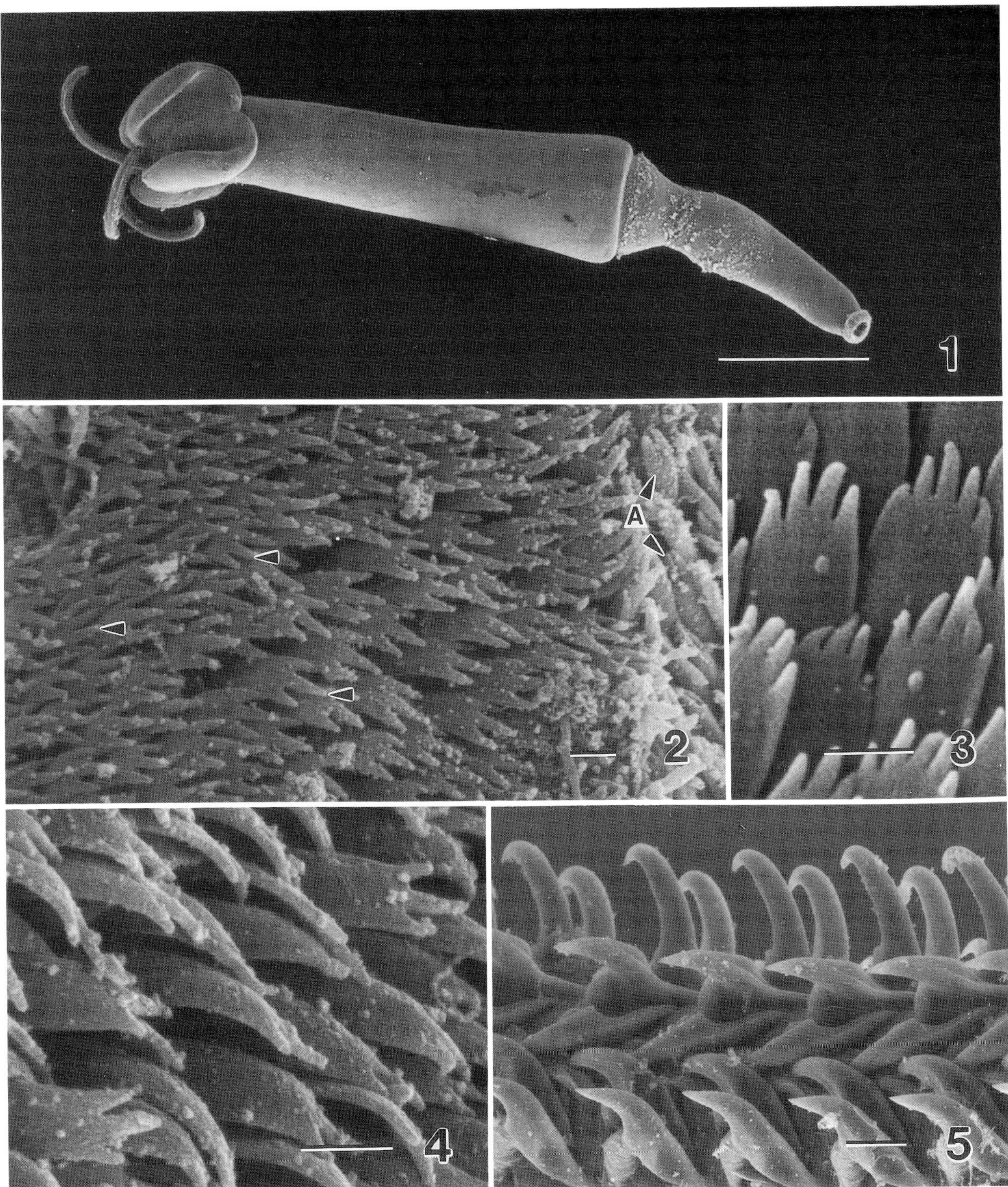

Figs. 1-5 SEM images of *Otobothrium*. larva. Fig. 1 Whole larva. Bar = 1mm. Figs. 2-4 Bar = 1μm. Fig 2. Microtriches on bothrium invagination showing 2 extensions (A) and 3, 4, and 5 extensions in the center of the adhesive surface (arrows). Fig. 3. Enlarged microtriches on adhesive surface showing 4 and 5 extensions. Fig. 4. Microtriches on outer edge of bothrium showing 3 extensions. Fig. 5. Portion of tentacle exhibiting several rows of hooks. Bar = 10 μm.

Microscopic Study of *Procamallanus elatensis*, a Parasite in Siganids, Red Sea Fish in Egypt.

N. Abdou *, R. A. Heckmann*, J. S. Gardner**, and A. A. Ashour***

* Department of Zoology, WIDB 109, Brigham Young University, Provo, Utah 84602
** Microscopy Lab, Brigham Young University, Provo, Utah 84602
*** Department of Zoology, Faculty of Science, Ain Shams University, Cairo, Egypt

Camallanidae is a common nematode family found in a variety of Red Sea fish. Procamallanids were found to be specific to *Siganus luridus* and *S. rivulatus* which are considered to be commercially important fish. *Procamallanus elatensis* collected from Siganids in the northern Gulf of Elat were first described by Fusco and Overstreet using light microscopy.[1] This report includes the first SEM description of Camallanid nematodes.

The parasites were washed in 7% saline and saved in 7% formalin until brought to Brigham Young University.They were washed with sodium cacodylate buffer (pH. 7.2), fixed in OsO_4 for 12 hr, washed in cacodylate buffer, immersed into 2% tannic acid for 8 hr, washed in buffer, fixed again in OsO_4 for 2 hr, washed in buffer, dehydrated with critical point drying and coated with gold.

Adult worms were long, slender, and reddish in color. At the anterior end the mouth was round and surrounded by the cephalic plate. The mouth was also surrounded by two rings of cephalic papillae (Fig 1). The inner ring consisted of six small papillae and the outer ring consisted of four large papillae. A small cephalic pore was located between each small papilla and the mouth opening. Two large amphids were observed adjacent to and on opposite sides of the peribuccal collar (Fig 2).

The posterior end of the male curved ventrally. Eight pairs of caudal papillae were seen on each side of the ventral surface. Two asymmetrical lateral alae extended on both sides of the posterior end. Two spicules were observed on other specimens, but they were retracted inside this worm (Fig 3). The poseterior end of the female worm was a pointed tail. The anus was curved slit opening (Fig 4). Figure 5 is a transverse section through the female worm showing the uterus full of eggs and a portion of intestine situated near the center of body (Fig 5). Eggs are oval to round in shape (Fig 6).

Procamallanus elatensis possesses similarities to *D. grundmanni* described by Chitwood in the number and arrangement of cephalic papillae.[3] It is also similar to *Gongulonema aegypti* found in Egyptian rodents as reported by Ashour and Lewis.[2] They reported the presence of cephalic papillae in two rings, however the number of papillae differed from the present study. The number of caudal papillae was the same as described by Fusco and Overstreet.[1] The presence of alae and spicules was shown in *G. aegypti* using scanning electron microscopy.[2] However, 9-12 pairs of caudal papillae were reported in their investigation and only eight pairs were observed on *Procamallanus elatensis* .

References:

1. A. C. Fusce and R. M. Oversteet ,*J . Nat. Hist.* 13 (1979)35.
2. A. A . Ashour and J. W. Lewis, *Syst. parasit.* 8(1986)199.
3. M .B. Chitwood, *Proc. Helmin. Soc. of Wash..* 30(1963) 70.

Proc. Microscopy and Microanalysis 1995, edited by G.W. Bailey, M.H. Ellisman, R.A. Hennigar, and N.J. Zaluzec

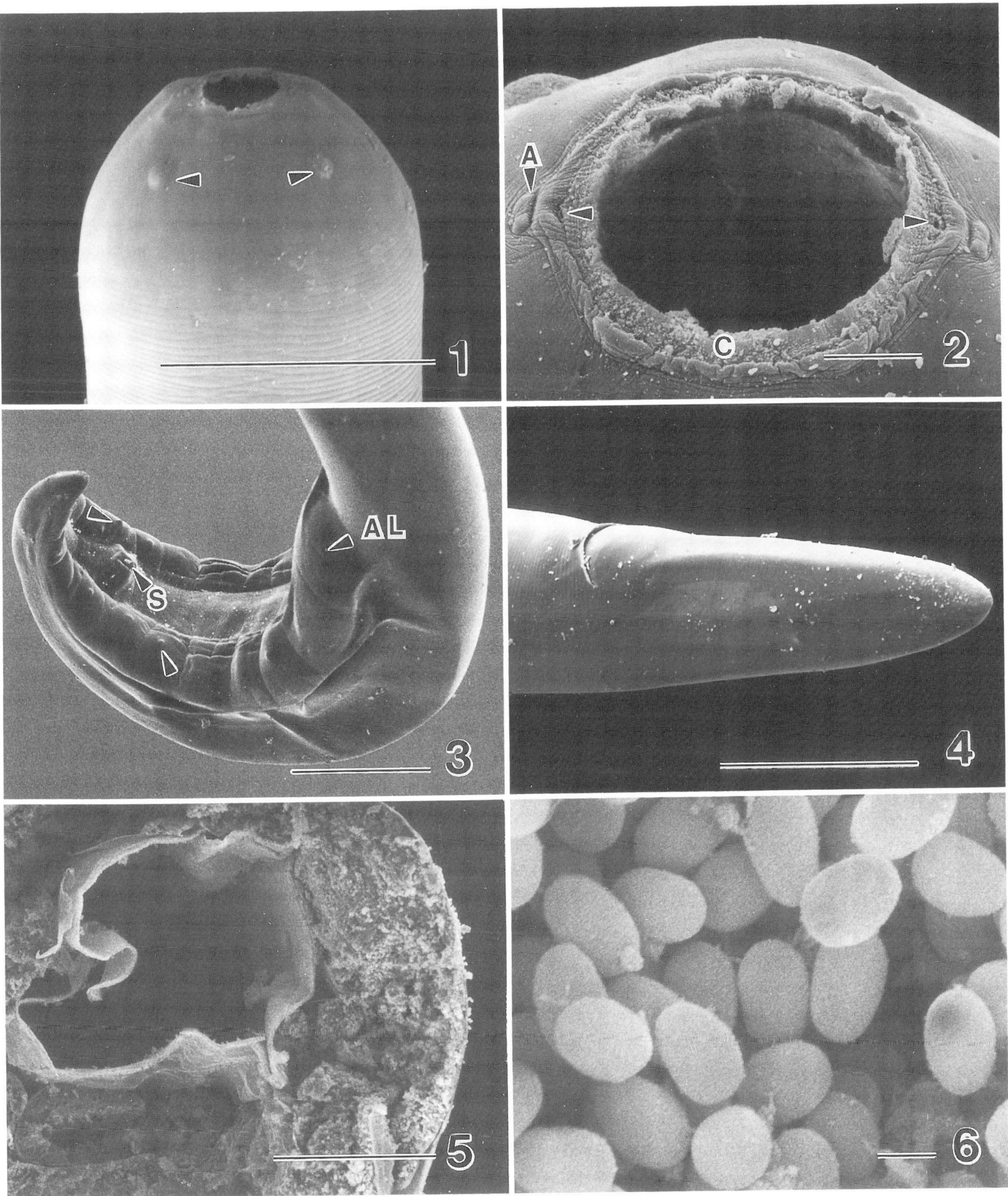

Figs. 1-6. SEM images of *P. elatensis* . Figs. 1 and 3-5. Bar = 100 μm. Fig. 1. Anterior end showing the cephalic plate and papillae (arrows). Fig. 2. Peribuccal collar (C), amphids (A), and cephalic pores (arrows). Bar = 10 μm. Fig. 3. Posterior end of male showing caudal papillae (arrows), alae (AL), and spicule (S). Fig. 4. Posterior end of female showing curved anal slit. Fig. 5. Transverse section of female showing uterus and intestine. Fig. 6. Eggs. Bar = 1 μm.

BACULOVIRUS *ACMNPV* AS WELL AS RNA- AND PROTEIN-SYNTHESIS INHIBITORS INDUCE APOPTOSIS IN A CONTINUOUS MIDGUT CELL LINE, FPMI-CF-203, OF *CHORISTONUERA FUMIFERANA* (LEPIDOPTERA:TORTRICIDAE).

A.J. Brownwright, S.R. Palli, G.F. Caputo, and S.S. Sohi,

Forest Pest Management Institute, Sault Ste. Marie, Ontario, Canada, P6A 5M7.

Apoptosis is an active cellular self-destruction regulated by expression or repression of certain genes[1]. Apoptosis can be caused by a variety of both external and internal stimuli. Whether these different stimuli that can cause apoptosis converge into a final pathway that leads to self-destruction is not known. This paper compares apoptosis caused by a baculovirus, an RNA-synthesis inhibitor and a protein-synthesis inhibitor.

Inoculation of IPLB-SF-21 (SF-21) and FPMI-CF-203 (CF-203, Fig.1,) cells with *Autographa californica* multicapsid nuclear polyhedrosis virus (*Ac*MNPV) and *Choristoneura fumiferana* multicapsid nuclear polyhedrosis virus (*Cf*MNPV), respectively, results in successful infection as visualized by formation of occlusion bodies (OBs). However, inoculation of CF-203 cells with *Ac*MNPV, or SF-21 cells with *Cf*MNPV, is unsuccessful and no OBs are seen. Inoculation of CF-203 cells with *Ac*MNPV results in premature lysis of cells beginning at 12 hr post-inoculation (pi) and most of the cells are lysed by 48 hr pi. These cells exhibit characteristics typical of apoptosis including formation of apoptotic bodies and fragmentation of cytoplasm and nuclei (Fig. 2). Agarose gel electrophoresis of DNA from the infected cells showed a characteristic DNA ladder beginning at 12 hr pi. Thus, AcMNPV triggers but does not block the apoptosis in CF-203 cells. This is similar to the effect observed when SF-21 cells were infected with AcMNPV p35⁻ mutants[2].

CF-203 cells incubated with an RNA-synthesis inhibitor, actinomycin D (1 ug/ml) showed the signs of DNA fragmentation in 6 hr post-treatment (pt) and the maximum fragmentation was seen by 12 hr pt. These cells incubated with a protein-synthesis inhibitor, anisomycin (5 ug/ml) showed considerable DNA fragmentation in 6hr pt and maximum was seen at 24 hr pt. Figures 3 and 4 show the light micrographs of CF-203 cells incubated with actinomycin D (Fig.3) or anisomycin (Fig.4) for 12 hr. Numerous cells that are at various stages of apoptosis were observed in both treatments. Unlike SF-21 cells, where RNA-synthesis inhibitors but not protein-synthesis inhibitors induce apoptosis[3], in CF-203 cells both RNA and protein synthesis inhibitors induce apoptosis.

The results obtained so far show that the way CF-203 cells self-destruct in response to a virus stimulus is different from their response to the stimulus provided by inhibitors of either RNA- or protein-synthesis. Cellular response due to the stimulus of inhibitors of either RNA- or protein-synthesis is more typical of apoptosis showing condensation of cytoplasm, convolution of the cellular and nuclear outlines, segregation of chromatin and finally formation of membrane-bound apoptotic bodies. The cells exposed to virus stimulus show condensation of cytoplasm followed by fragmentation of the cytoplasm. Moreover, membrane-bound apoptotic bodies containing intact cytoplasmic organelles including mitochondria are seen throughout cytoplasm and nucleus.

References:
1. L.D.Tomei and F.O.Cope, Ed. *The Molecular basis for apoptosis*. Curr. Cumm. Cell Mol. Biol. Cold Spring Harbor Laboratory Press, (1994) 430.
2. R.J.Clem, M.Fechheimer and L.K.Miller, *Science* 254 (1991) 1388.
3. R.J.Clem and L.K.Miller *Mol. & Cell. Biol.* 14 (1994) 5212.

Proc. Microscopy and Microanalysis 1995, edited by G.W. Bailey, M.H. Ellisman, R.A. Hennigar, and N.J. Zaluzec
Copyright © 1995 MSA. Published by Jones and Begell Publishing, 79 Madison Ave., New York, NY 10016

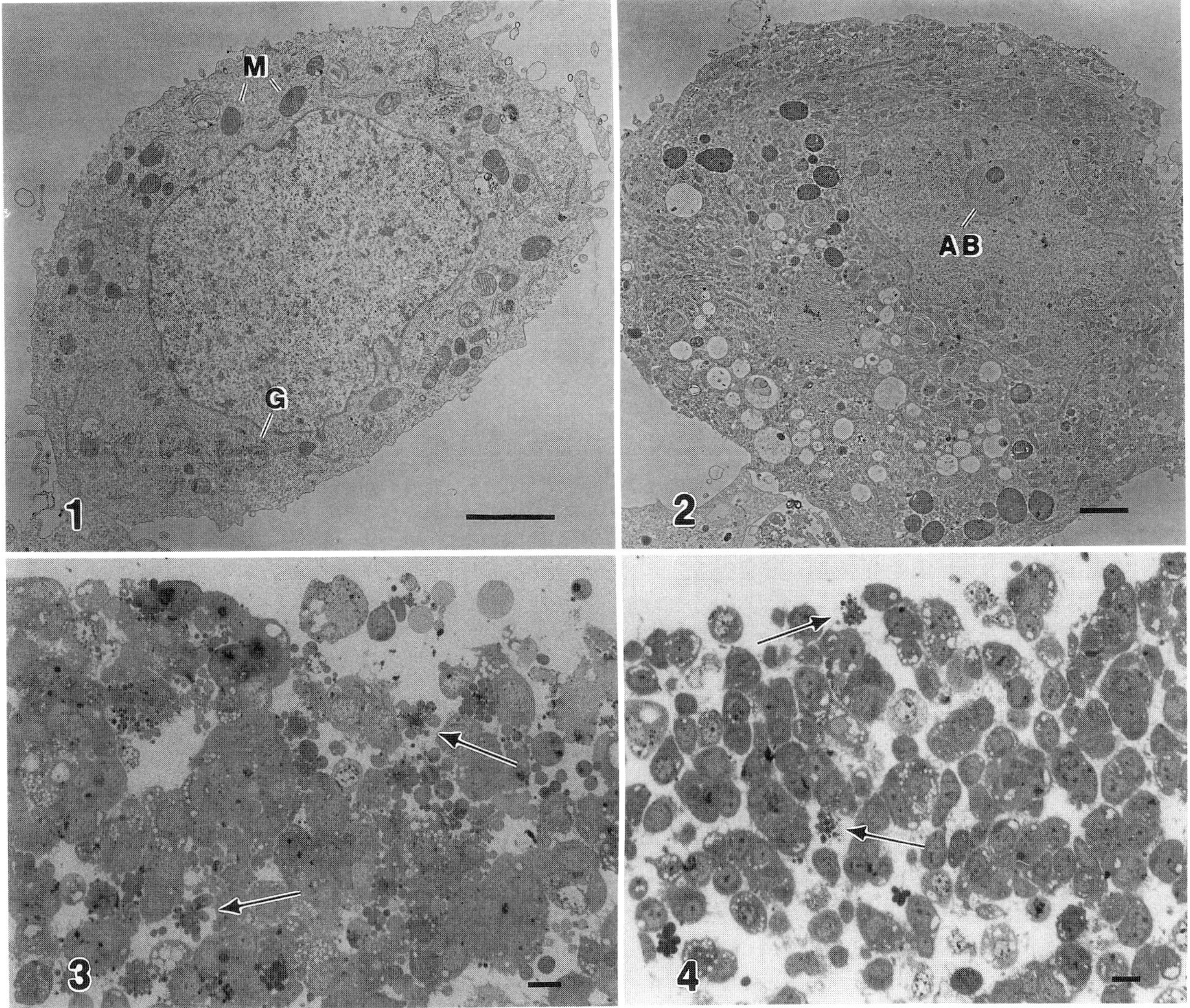

Fig. 1 TEM of CF-203 cells. The cells were grown in SF-900 medium.   Bar= 2 um.
Note the healthy cytoplasmic organelles including mitochondria (M) and active
Golgi complex (GC).

Fig. 2 TEM of CF-203 cells infected with AcMNPV (72 hr pi). Bar = 2 um.   Note the
presence of apoptotic bodies (AB) in the nucleus containing intact cytoplasmic
organelles such as endoplasmic reticulum and condensed chromatin.

Fig. 3 LM of actinomycin treated CF-203 cells (1ug/ml, 12 hr pt). Bar = 20 um.
Arrows point to the cells in final stages of apoptosis.

Fig. 4 LM of anisomycin treated CF-203 cells (5 ug/ml, 12 hr pt). Bar = 20 um.
Arrows point to the cells in final stages of apoptosis.

ULTRASTRUCTURE OF HUMAN HERPESVIRUS 7

C. S. Goldsmith, D. Burns, P. E. Pellett, S. R. Zaki, and J. B. Black

Division of Viral and Rickettsial Diseases, Centers for Disease Control and Prevention, Atlanta, GA 30333

Human herpesvirus 7 (HHV-7) was originally isolated in 1989 from the peripheral blood lymphocytes of a healthy individual.[1] Most children between 2 and 5 years of age develop antibodies against HHV-7,[2] and the virus has recently been implicated as the causative agent in secondary episodes of the childhood disease roseola infantum.[3] Approximately 90% of the general population is seropositive,[2] and virus can be isolated from the saliva of up to 75% of healthy adults.[4,5] This study describes the ultrastructural growth properties of an HHV-7 isolate obtained from a saliva specimen.

Cord blood lymphocytes (CBLs) were inoculated with cell-free virus and harvested at days 1, 3, 4, 5, and 7. The growth curve for HHV-7 was determined by examining the ultrastructure of the cells for signs of viral production and by monitoring the cells with an anti-complement immunofluorescence assay for the presence of viral proteins.

The nuclei of HHV-7-infected CBLs typically had viral nucleocapsids scattered throughout, although the capsids would occasionally converge into organized structures (Fig. 1). Some nuclei had enlarged rarefied areas, a feature often seen in herpesvirus-infected cells. Once the capsids moved into the cytoplasm, they acquired a dense tegument coat, which contained a very ragged outer edge. Cytoplasmic particles would then obtain the viral envelope by budding into the Golgi (Fig. 2). A distinct electron-lucent space was observed between the tegument and the envelope. Virus particles presumably are released by an induced cell lysis, and extracellular particles were seen to accumulate in large masses (Figs. 3,4).

The process of viral replication and assembly was determined by sequential harvesting of infected culture cells over a 7-day period (see Table 1). No cells showed signs of virus assembly on day 1 postinoculation, and only 2% showed signs of infection on day 3. By day 4, 21% of cells gave signs of replicating virus particles; the majority of these cells had viral nucleocapsids and cytoplasmic particles, and extracellular particles were also beginning to appear. Viral replication continued and by day 7, 89% of culture cells showed signs of virus production.

HHV-7 replicates and assembles virus particles in a manner similar to that seen for HHV-6,[6] the causative agent of most cases of roseola infantum.[7] However, HHV-7 can be somewhat differentiated when grown in CBLs in that mature virus particles maintain a distinct space between the viral envelope and the ragged outer edge of the viral tegument.

References

1. N. Frenkel, et al., *Proc. Natl Acad. Sci. USA* 87(1989)748.
2. L.S. Wyatt, et al., *J. Virol.* 65(1991)6260.
3. K. Tanaka, et al., *J. Pediatrics* 125(1994)1.
4. L.S. Wyatt and N. Frenkel, *J. Virol.* 66(1992)3206.
5. J.B. Black, et al., *Virus Res.* 29(1993)91.
6. S. Nii, et al., in C. Lopez, et al., Eds., *Immunobiology and Prophylaxis of Human Herpesvirus Infections*, New York:Plenum Press (1990)19.
7. K. Yamanishi, et al., *Lancet* i(1988)1065.

Proc. Microscopy and Microanalysis 1995, edited by G.W. Bailey, M.H. Ellisman, R.A. Hennigar, and N.J. Zaluzec

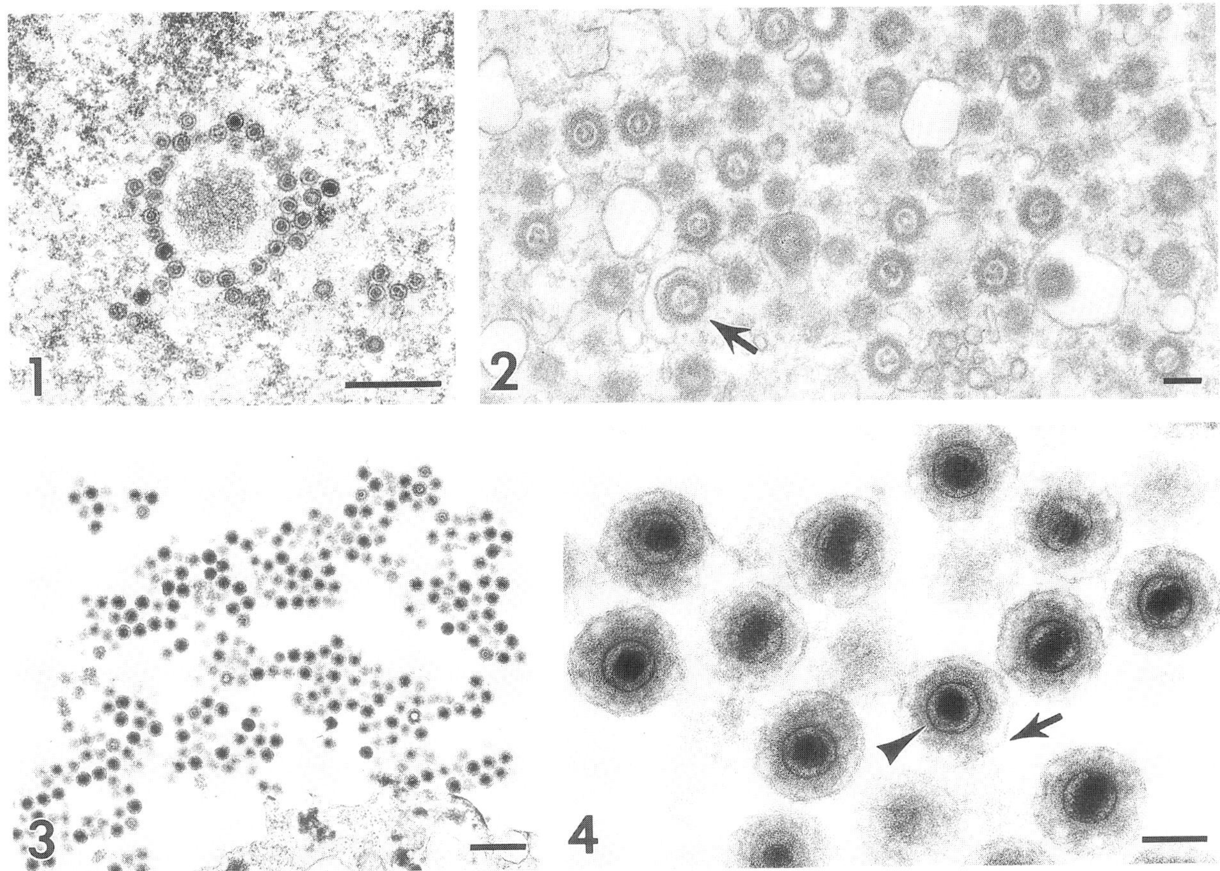

FIG. 1.-Human herpesvirus 7 nucleocapsids arranged in nucleus of infected cord blood lymphocyte. Bar=0.5um. FIG. 2.-Cytoplasmic particles showing virus budding into Golgi (arrow). Bar=100nm. FIG. 3.-Accumulation of extracellular HHV-7 virus particles. Bar=0.5um. FIG. 4.-Higher magnification of boxed area of figure 3. Whole virus particles consist of nucleocapsid (arrowhead) surrounded by electron-dense tegument contained within the viral envelope (arrow). Bar=100nm.

Table 1: Growth Curve of HHV-7 (% cells)

	Nuc	Nuc Cyto	Nuc Cyto Extra	Nuc Extra	Cyto	Cyto Extra	Extra	Total Infected	Total Uninfected
Day 1									100
Day 3		1		1				2	98
Day 4	5	7	1	1	7			21	79
Day 5	20	15	3	1	18	2	1	60	40
Day 7	9	49	13	4	9	4	2	89	11

Virus structures observed in: Nuc, nucleus; Cyto, cytoplasm; Extra, extracellular space.

PROLIFERATING CELLS IN THE VERTEBRATE SMALL INTESTINE: AN IMMUNOHISTOCHEMICAL STUDY

László G. Kömüves

Morphology Laboratory, Department of Dermatology, University of California San Francisco and Veterans Administration Medical Center, San Francisco CA 94121

In the small intestinal mucosa of healthy adult mammals proliferating cell are confined to the crypts of Lieberkühn.[1] Earlier radioautographic studies identified proliferative cells in the small intestine of several non-mammalian vertebrates. However, it is still not clear whether cell renewal is confined to proliferative compartment within the small intestinal mucosa in non-mammalian vertebrates. In the present study proliferative cells were identified using an immunological marker of cell proliferation, the proliferating cell nuclear antigen (PCNA)[2] in the small intestine of several non-mammalian vertebrate species, including birds (zebrafinch, *Poephila guttata*), reptiles (green anole, *Anolis carolinensis*), amphibia (axolotl, *Ambystoma mexicanum*), and fishes (goldfish, *Carassius auratus*).

Segments of the small intestine were fixed in 4% formaldehyde in 0.86 M phosphate buffer, pH=7.2 and embedded in paraffin. Deparaffinized and rehydrated sections were microwaved in citrate buffer.[3] The immunohistochemical detection method used in this study based on the capillary action principle, as developed by Brigati.[4] The ABC method was used to detect the binding of the anti-PCNA antibody, and the peroxidase activity was developed with DAB chromogen. All of the reagents and buffers were supplied as a kit by Bio-Tek Solutions (Santa Barbara, CA). Immunolabeling was done with an automated workstation, TechMate500 (developed and manufactured by Bio-Tek Solutions, Santa Barbara, CA, available from Curtin Matheson Scientific). The following set of controls were used to ensure the specificity of immunolabeling: omission of the first antibody, use of a murine isotype-matched antibody in lieu of the anti-PCNA antibody, omission of the biotinylated second antibody, omission of both first and second antibodies. No endogenous peroxidase activity was observed in the tissue samples and all of the controls resulted in no staining. The slides were counterstained with hematoxylin. Photographs were taken with a Nikon Microphot FX photomicroscope.

Immunohistochemical detection of PCNA, a technique widely used on mammalian tissues, proved to be useful to identify proliferating cells in the small intestine of several non-mammalian species, examined in this study. In the mucosal epithelium in the small intestine of *P. guttata* PCNA expressing cells were confined to the crypt area, and no proliferating enterocytes were detected in the villus epithelium (Fig. 1). In reptiles, amphibia and fishes there are no villi present in the small intestine. The mucosal surface, however, is increased by numerous longitudinal folds. In *A. carolinensis* cells expressing PCNA occurred in the basal regions of the folds (Fig. 2). However, some epithelial cells along the folds also expressed PCNA. In the small intestine of *A. mexicanum* PCNA expressing cells were found scattered along the mucosal epithelium (Fig. 3). Furthermore, the subepithelial buds showed strong staining for PCNA. Albeit PCNA-expressing cells were more abundant in the basal areas of mucosal folds in the small intestine of *C. auratus*, these cells were also found closer to the lumenal surface (Fig. 4). Apart from the mucosal epithelium, other cells also showed PCNA staining in the small intestine of these species. Proliferating cells in the lamina propria were present in large number in *P. guttata* and *C. auratus* (Figs. 1, 4). Furthermore, intraepithelial lymphocytes also stained for PCNA (Figs. 1, 2, 3, 4).

References

1. C. P. Leblond, *Protoplasma* 160(1991)5.
2. P. A. Hall and A. L. Woods, *Cell Tissue Kinet.* 23(1990)505.
3. S.-R. Shi et al., *J. Histochem. Cytochem.* 41(1993)1599.
4. J. A. Reed et al., *BioTechniques* 13(1992)434.

Proc. Microscopy and Microanalysis 1995, edited by G.W. Bailey, M.H. Ellisman, R.A. Hennigar, and N.J. Zaluzec

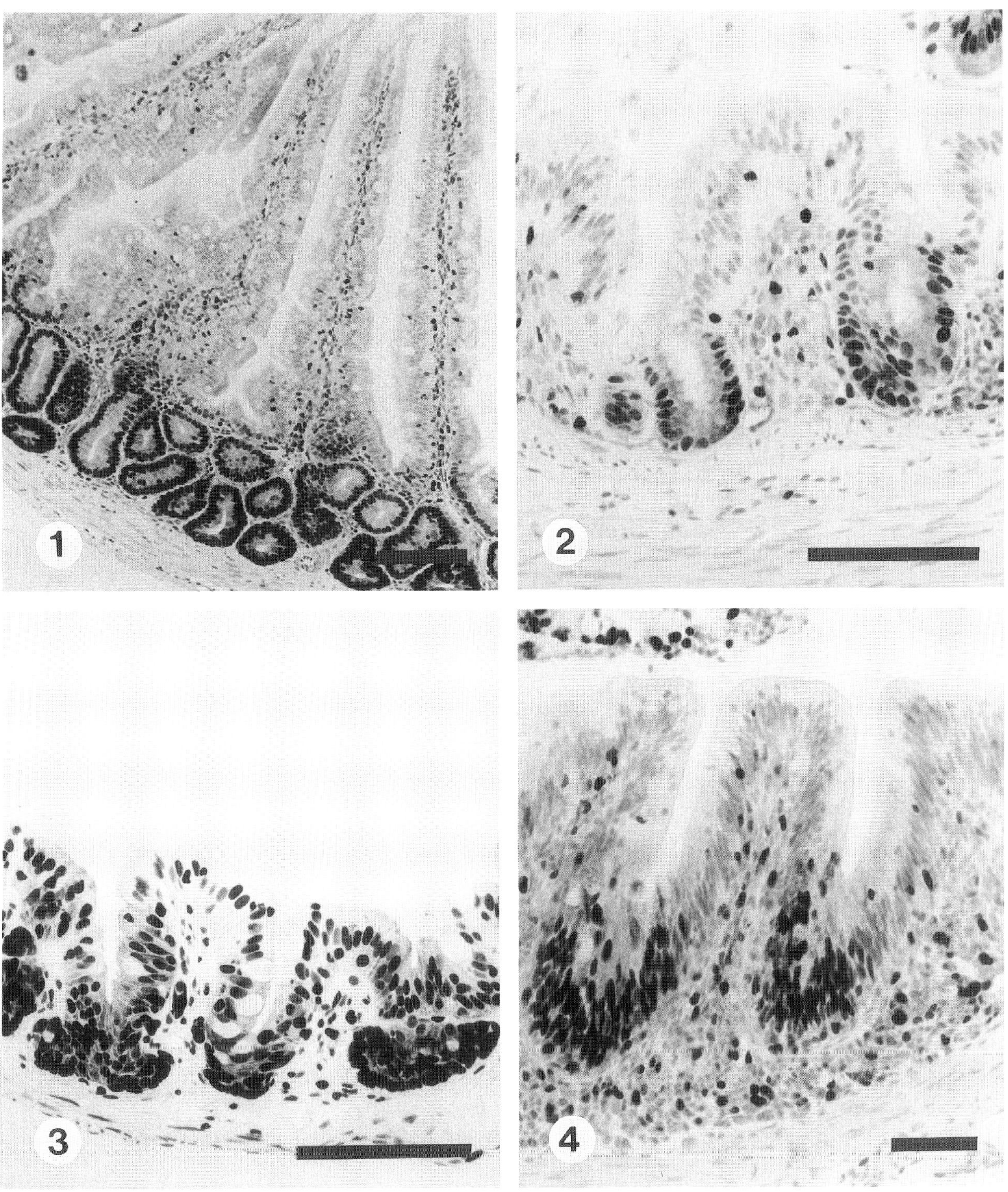

FIG. 1.— Localization of PCNA in the small intestine of zebrafinch, *P.guttata* . Bar= 100 μm.
FIG. 2.— Localization of PCNA in the small intestine of green anole, *A. carolinensis*. Bar= 100 μm.
FIG. 3.— Localization of PCNA in the small intestine of axolotl, *A. mexicanum.* Bar= 100 μm.
FIG. 4.— Localization of PCNA in the small intestine of goldfish, *C. auratus.* Bar= 200 μm.

IMMUNOHISTOCHEMICAL LOCALIZATION OF PROLACTIN RECEPTOR IN HUMAN
DIGESTIVE TISSUES

T. García-Caballero*, R. Gallego*, M. Fraga**, E. Pintos**, A. Beiras*

* Department of Morphological Sciences, Faculty of Medicine, 15705 Santiago de
Compostela, Spain.
** Department and Service of Pathology, Faculty of Medicine and General Hospital of Galicia,
15705 Santiago de Compostela, Spain.

Prolactin (PRL) is primarily recognized for its lactogenic effect. However, it is known that this polypeptide
hormone exerts a great variety of biological functions acting on reproduction, osmoregulation, growth,
metabolism, immunomodulation and even on behavior.[1] The actions of PRL initiate with hormone binding to a
specific cell surface receptor that belongs to the cytokine/GH/PRL receptor family or the hematopoietin
receptor family.[2] The PRLR is widely distributed in numerous tissues.[1,4-5] The aim of the present work was to
investigate by immunohistochemistry the cellular distribution of PRLR in the human gastrointestinal tract and
associated glands (liver and pancreas).

Samples of normal esophagus, stomach, small and large intestine, liver and pancreas were obtained from
surgical pieces or recent autopsies. These samples were immersion-fixed in 10% buffered formalin for 24 hr,
dehydrated and embedded in paraffin routinely. B6.2 anti-PRLR and avidin-biotin-peroxidase complex (ABC)
procedure were employed. The B6.2 mouse monoclonal antibody (prepared against a membrane enriched
fraction of human metastatic breast cancer), was generously provided by Dr. Barbara K. Vonderhaar[6] and
used at a dilution of 1:250, for 1 hr. In the gastrointestinal tract, immunoreactivity was seen in the mucosa,
mucous glands, muscularis layers and nervous plexuses. The immunostaining pattern found in the mucosa of
the different segments was as follows. *Esophagus.* An intense PRLR immunoreactivity was observed in the
superficial layers of the stratified squamous epithelium, and no immunostaining was found in the basal and
suprabasal layers (Fig. 1). *Stomach.* Strong immunoreactivity was found in parietal cells of the fundic glands.
No immunoreactivity was observed in the surface epithelium nor in the mucous neck cells, but a weak staining
was shown in the chief cells (Fig. 2). *Small intestine.* Absortive cells showed a moderate cytoplasmic
immunostaining, with an intense focal immunoreactivity in the supranuclear area (Golgi area). Luminal
membrane of cells was also stained. In globet cells and columnar epithelial cells of the intestinal glands
(crypts of Lieberkühn) only focal immunostaining of the Golgi area was observed (Fig. 3). Paneth cells
showed an intense and diffuse cytoplasmic immunostaining. *Large intestine.* Immunoreactivity was found in
the surface epithelium and in the superficial half of the crypts of Lieberkühn (Fig. 4). *Liver.* A diffuse and
moderate immunoreactivity was observed in the hepatic parenchymal cells. Epithelial cells of bile ducts were
also positive as well as sinusoidal cells (Fig. 5). In some samples a nuclear immunoreactivity was detected in
hepatocytes. *Pancreas.* Both exocrine and endocrine pancreas were positive. The intensity of
immunoreactivity was higher in the islet of Langerhans than in the acini (Fig. 6). By serial consecutive
sections immunostained for PRLR, insulin and glucagon, we could prove that the pattern of PRLR
immunostaining in the islet of Langerhans was similar to the pattern obtained for insulin and was very different
to that observed for glucagon. The results obtained are in agreement with the previously reported in the rabbit
gastrointestinal tract.[7] The major difference was found in the immunostaining of fundic glands, i. e., the more
intense immunostained cells in our study were parietal, whereas in the Lobie' s study were chief cells.

1. P.A. Kelly et al., *Endocr. Rev.* 12:235-251.
2. D. Cosman, *Cytokine.* 51(993)95.
3. M. Nagano and P.A. Kelly, *J. Biol. Chem.* 269 (1994)13337.
4. A. Ouhtit et al., *Endocrinology.* 133(1993)135.
5. M. Freemark et al. *J. Endocrinol.* 144(1995)285.
6. R. Banerjee et al., *Int J Cancer* ; 55(1993)712.
7. Lobie et al., *J. Endocrinol.* 139(1993)371.

Supported by the Xunta de Galicia (grant No. 20811B94).

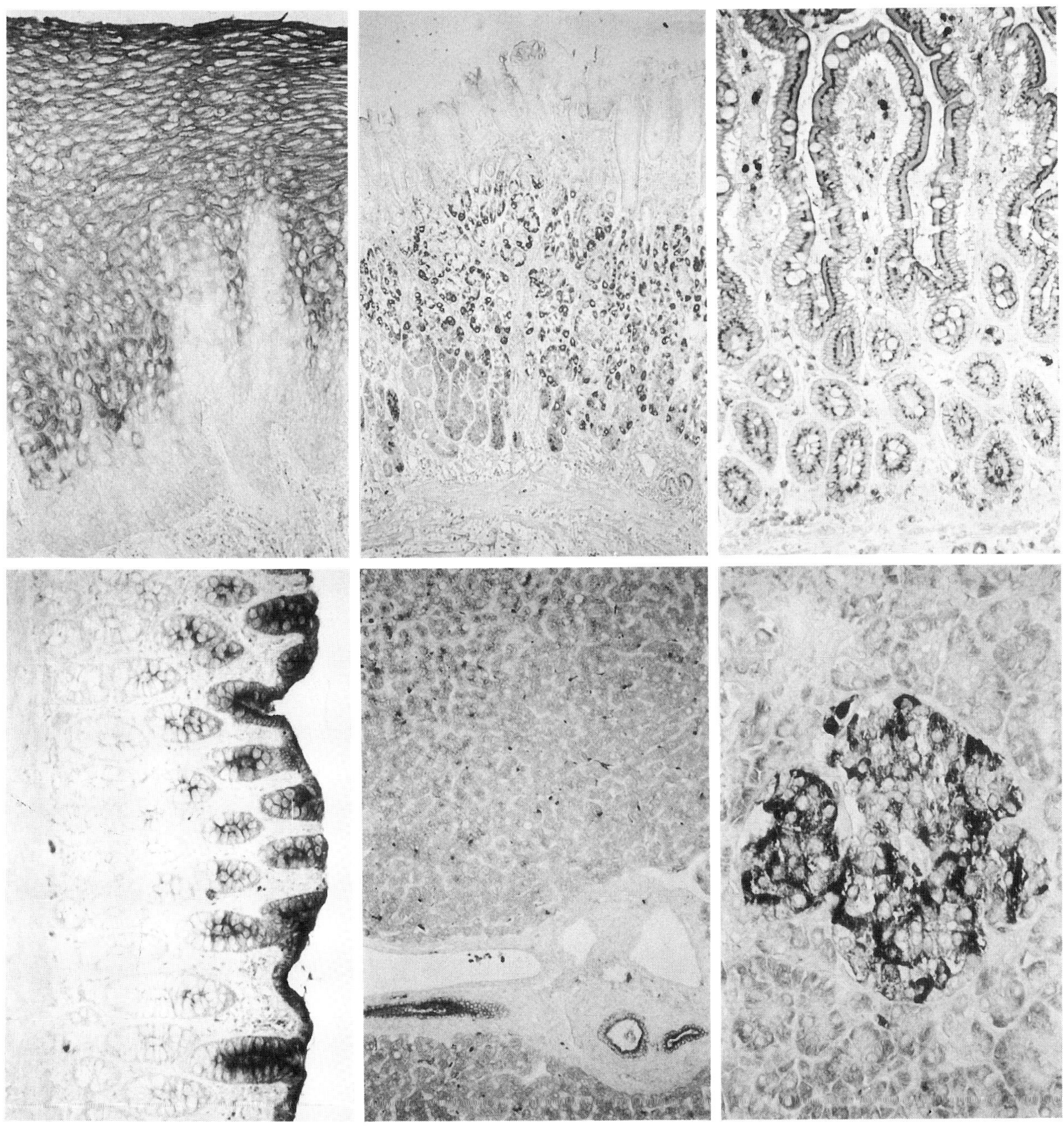

FIG. 1.- *Esophagus*. PRLR immunostaining in the superficial epithelial layers. x130
FIG. 2.- *Stomach*. Immunoreactivity for PRLR mainly localized in the parietal cells. x60
FIG. 3.- *Small intestine*. PRLR positivity localized in epithelial cells of villi and glands. x130
FIG. 4.- *Large intestine*. PRLR expression in the superficial half of Lieberkühn's crypts. x100
FIG. 5.- *Liver*. Note PRLR immunostaining in hepatocytes, sinusoidal cells and bile ducts. x60
FIG. 6.- *Pancreas*. Imunoreactivity for PRLR is prominent in islets and faint in acini. x250

PH DEPENDENT GASTRIN SECRETION MECHANISM. AN IMMUNOCYTOCHEMICAL STUDY.

J.A. Galán Torres*, A. LLorca Gallego*, R. González Santander**, M.V. Toledo Lobo**, F.J. Martínez Alonso**, G. Martínez Cuadrado**.

* Veterinary Military Center. Microbiology and Clinical Analysis Service. Madrid, Spain.
** Morphological Sciences and Surgery Dept. Univ. of Alcalá de Henares. Madrid, Spain.

Gastrin secretion is induced by vagal stimuli, local distension, peptides or aminoacid solutions, and Ca^{2+} concentration both in stomach lumen and blood [1,2]. However, the cellular and molecular mechanisms of gastrin-release are unknown. We have tested the effects of different pH fixative solutions on gastrin distribution by the immunolabeling of antral gastrin cells (G cells).

Samples from different portions of adult dog digestive tract and related glands (salivary glands, liver and pancreas) were embedded in paraffin by conventional methods. Another samples from the antral mucosa of adult dogs and rats were fixed using mixtures of glutaraldehyde and paraformaldehyde at different pH (from 4 to 7,6 pH), and embedded in araldite and Unicryl. For post-embedding immnunocytochemical staining purposes, sections were incubated with pre - diluted policlonal rabbit anti-human gastrin (Zimed Labs.Inc., San Francisco, USA), rinsed in TBS and incubated with a 15 nm gold-conjugated IgG as secondary antibody. These sections were stained with uranil acetate and examined with TEM. Paraffin sections were incubated with the same primary antibody and the immunoreaction was revealed either by estreptavidin-peroxidase and silver enhancement methods.

With LM, a possitive immunoreaction was only found in the piloric antrum and the first cm of duodenum. No possitive immunoreaction was found in other areas of the digestive tract. With TEM, G cells showed different types of granules (150-400 nm): electron-dense, electron-lucent and intermediate forms with pale cores. It was demonstrated that after fixation at low pH (acid solutions) most of the granules are electron-dense and are immunopositive to gastrin (Figs.1,2), while fixation at neutral pH (pH around 7) results in most of the gastrin granules with their content degranulated into the cytoplasm, and the inmunopositive reaction found in the cytoplasm (Figs.3,4,5). Neither in the present study nor in previous ones[3] we have been able to find evidence of gastrin secretion by the fusion of the granule membrane with the plasma membrane. The positive immunoreactivity of the cytoplasm at neutral pH leads us to suggest the hypothesis that gastrin is not secreted by exocytosis with full fusion. We propose that pro-gastrin and gastrin are accumulated in the granules and given a neutral pH the granules degranulated its gastrin content into the cytoplasm. This cytoplasmic gastrin is then released into the extracellular space. This gastrin-release may be mediated either by its diffusion or by a plasmalemmal device (mediatophore). We consider that the storage of gastrin into specific granules is required not for the exocytosis, which does not occur, but for other functions such as the post-translational processing of gastrin, as suggested by recent studies[4]. The pH shift of the granule from acid to pH around 7 may activate the enzime required for the conversion of pro-gastrin to biologically active gastrin. If the fixation of antral glands with solutions at different pH represents an experimental model of the real antral gastrin-release mechanism, the hypothesis proposed might explain some aspects of the feed-back inhibition of gastrin release by gastric acids, and this would be consistent with the hypergastrinemia reported after treatment with acid inhibiting drugs[2].

References

1. J.H. Walsh, Ann. Rev. Physiol., 50 (1988) 41.
2. J.H. Walsh, Gastroenterology, 104 (1993) 653.
3. J. Rahier et al., Gastroenterology, 92 (1987) 1146.
4. C.J. Dickson et al., Am. J. Physiol., 264 (1993) G 553.

Proc. Microscopy and Microanalysis 1995, edited by G.W. Bailey, M.H. Ellisman, R.A. Hennigar, and N.J. Zaluzec
Copyright © 1995 MSA. Published by Jones and Begell Publishing, 79 Madison Ave., New York, NY 10016

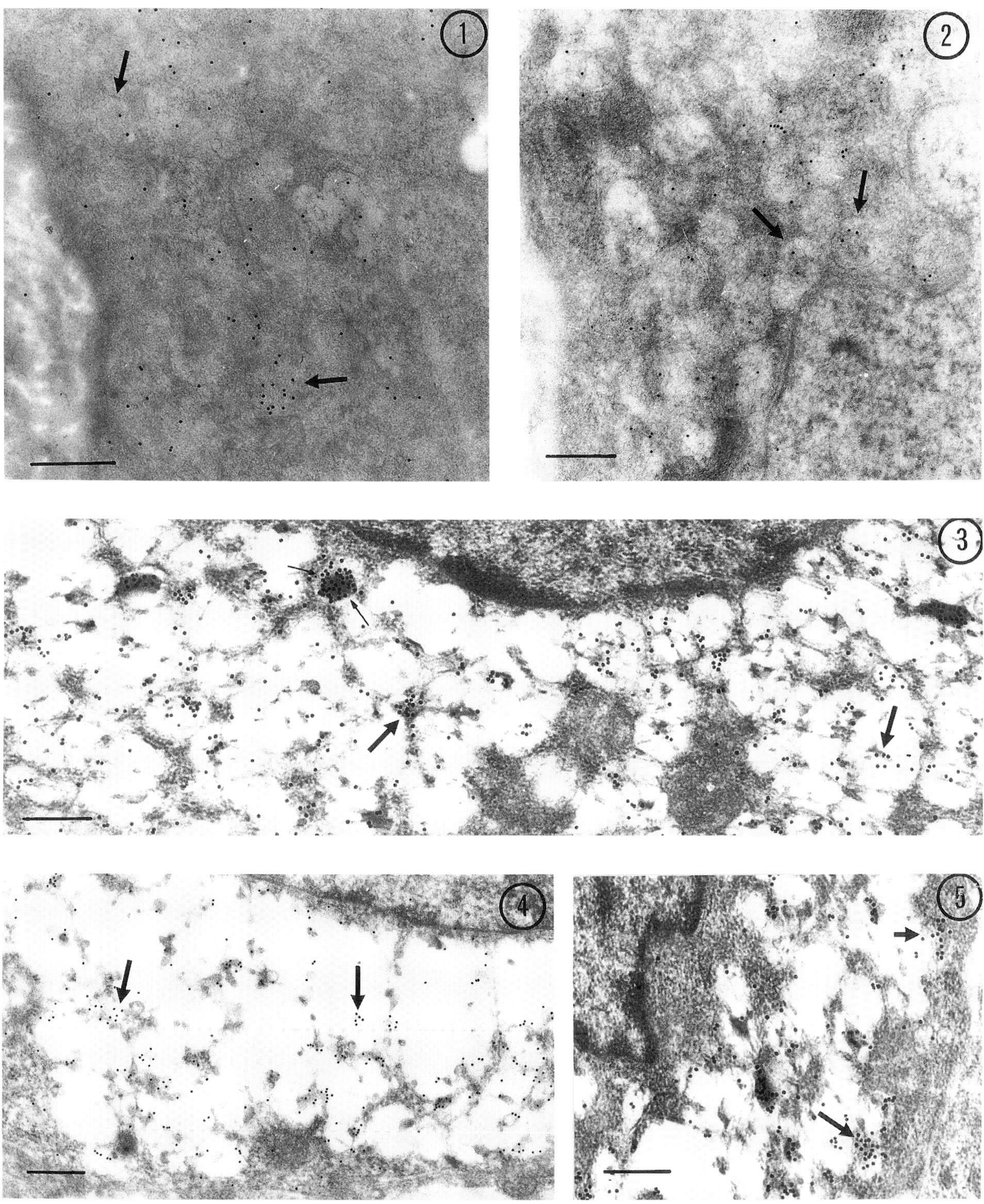

Figs.1,2.- Undegranulated G cells. Fixation at acid pH. Arrows: granules. Scale bar: 0.5 μm.
Figs. 3,4,5.- Degranulated G cells. Fixation at neutral pH. Small arrows: remanent granules; Big arrows: immunoreactivity to gastrin into cytoplasm. Scale bar: 0.5 μm.

IMMUNOELECTRON MICROSCOPIC STUDIES OF THE MUSCLE PROTEIN TITIN

Theresa A. Fassel and Marion L. Greaser

Muscle Biology Laboratory, University of Wisconsin-Madison, Madison, WI 53706

Titin, a skeletal and cardiac muscle protein, is almost 3 million daltons in mass and about 1 μm in length. It extends approximately parallel to the long axis of the sarcomere from the Z line to the M line. A series of monoclonal anti-titin antibodies and a polyclonal anti-titin antibody were used to localize several titin epitope positions. A pre-embedding approach was used in which primary and secondary antibody incubations preceded chemical fixation, ethanol dehydration and LR White (room temperature infiltration, heat polymerization) or LR Gold (low temperature infiltration, ultraviolet light polymerization) processing for transmission electron microscopy. A post-embedding step provided a third layer of colloidal gold label. Monoclonal antibody incubations were followed by treatment with sheep anti-mouse IgG and finally with donkey anti-sheep IgG conjugated to 15 nm colloidal gold. Polyclonal antibody incubations were followed by biotinylated anti-chicken and streptavidin-gold (15 nm).

Rabbit psoas bundles were tied to glass capillaries in relaxing solution and incubated in a skinning solution for 6 hours at 4°C. Bundles were soaked in primary antibody (0.7 mg/ml in skinning solution) overnight at 4°C followed by washes to remove unbound antibody. Secondary antibody (0.3 mg/ml) incubations were also overnight at 4°C. Samples were washed and fixed in buffered 4% paraformaldehyde for 1 hour at room temperature or overnight at 4°C. The procedure to this point is similar to that of Furst et al. (1988, 1989). The tissue was embedded and thin sections (on nickel grids) were blocked (1% BSA in rigor buffer) and incubated with the colloidal gold label overnight. Sections were washed and stained in 25% ethanolic uranyl acetate and Reynolds lead citrate.

Primary and secondary antibodies provide visualization of lines of increased electron density perpendicular to the long axis of the sarcomere. Multiple lines in the I band and A band are seen in Fig. 1 for the polyclonal 534 chicken anti-titin antibody. A single line is seen near the A-I junction for 891 mouse anti-titin (Fig. 3). Multiple lines occur in the A band for mouse anti-titin H4 (Fig. 5). The third layer of colloidal gold improves visualization against the background density of the sarcomere. Colloidal gold particles appear to overlay the lines due to the primary and secondary antibodies for 534 (Fig. 2), 891 (Fig. 4) and H10 (Fig. 6) and are associated with Z and M lines for 03G12 mouse anti-titin (Fig. 7). Figure 8 shows a control sarcomere which lacks any antibody lines.

References

1. Furst, D.O., Osborn, M., Nave, R., Weber, K. *Journal of Cell Biology* 106(1988)1563.
2. Furst, D.O., Nave, R., Osborn, M., Weber, K. *Journal of Cell Science* 94(1989)119.
3. Supported by USDA (91-37206-6743) and NIH (HL47053).

Proc. Microscopy and Microanalysis 1995, edited by G.W. Bailey, M.H. Ellisman, R.A. Hennigar, and N.J. Zaluzec
Copyright © 1995 MSA. Published by Jones and Begell Publishing, 79 Madison Ave., New York, NY 10016

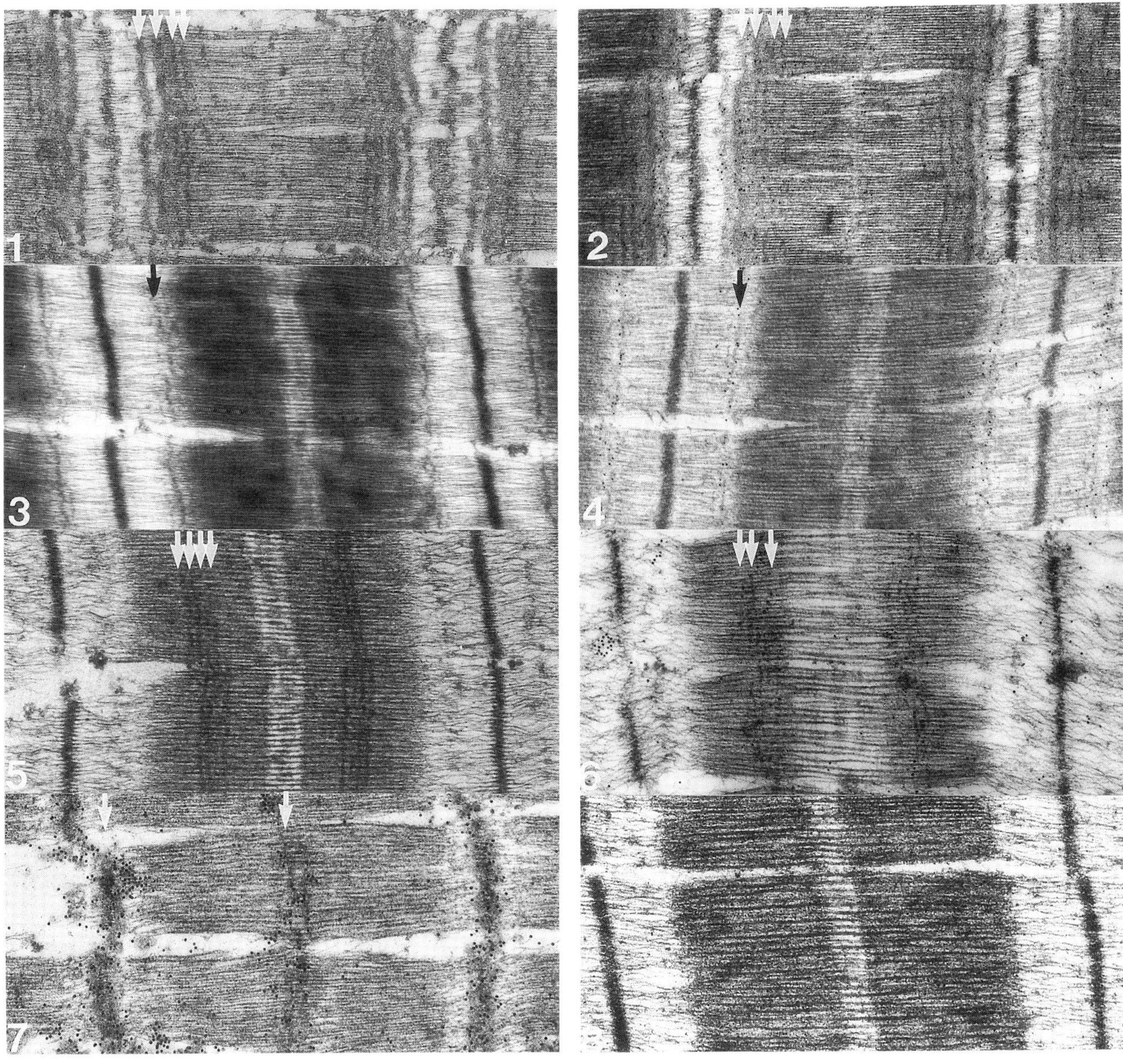

Fig. 1. and Fig. 2. Polyclonal 534 chicken anti-titin labels multiple epitope sites in the I and A bands shown without (Fig. 1) and with colloidal gold label (Fig. 2).
Fig. 3. and Fig. 4.-Monoclonal 891 anti-titin labels single line near A-I junction shown without (Fig. 3) and with colloidal gold label (Fig. 4).
Fig. 5. and Fig. 6.-Monoclonal H4 anti-titin label multiple lines in A band shown without (Fig. 5) and monoclonal H10 with colloidal gold label (Fig. 6).
Fig. 7.-Monoclonal 03G12 labels Z and M line areas shown with colloidal gold layer.
Fig. 8.-Control sarcomere shows absence of dense antibody lines perpendicular to the long axis of the sarcomere.

COMPARISON OF LOCALIZATION OF METALLOTHIONEIN IN TISSUES OF METALLOTHIONEIN - DEFICIENT MICE

H.Nishimura[1], N.Nishimura[2], D.L.Adelson[1], A.E.Michaelska[3] , K.H.A.Choo[3] and C.Tohyama[4]

[1]Animal Production, CSIRO, Prospect, NSW 2149, Australia,   [2]Human Science, Aichi Mizuho University, Hiradobashi 470-03, Japan,   [3]Murdoch Inst., Royal Child, Hosp., Parkville, 3052 Australia,   [4]Environmental Health Sciences Div., NIES, Onogawa, Tsukuba 305 JAPAN

Metallothionein (MT), a cysteine-rich heavy metal binding protein, has several isoforms designated from I to IV. Its major isoforms, I and II, can be induced by heavy metals like cadmium (Cd) and, are present in various organs of man and animals. Rodent testes are a critical organ to Cd and it is still a controversial matter whether MT exists in the testis although it is clear that MT is not induced by Cd in this tissue. MT-IV mRNA was found to localize within tongue squamous epithelium. Whether MT-III is present mainly glial cells or neurons has become a debatable topic. In the present study, we have utilized MT-I and II gene targeted mice and compared MT localization in various tissues from both MT-deficient mice and C57Black/6J mice (C57BL) which were used as an MT-positive control. For MT immunostaining, we have used rabbit antiserum against rat MT-I known to cross-react with mammalian MT-I and II and human MT-III. Immunohistochemical staining was conducted by the method described in the previous paper[2] with a slight modification after the tissues were fixed in HistoChoice and embedded in paraffin. To both strains of mice we have administered $CdCl_2$ (1.0 mg Cd/kg b.w., s.c.) for 2 consecutive days and sacrificed mice under pentobarbital anesthesia 72 hrs later. In the liver, kidney and pancreas, MT and cadmium levels were measured by RIA and inductively-coupled plasma emission spectrometry methods, respectively. Cadmium administration resulted in a remarkable increase in MT levels in the liver, kidney and pancreas in C57BL mice, but not in the MT deficient mice, as determined by MT radioimmunoassay. However, in MT deficient mice that these tissues accumulated substantial amounts of Cd, between 50% and 80% of the concentrations found in C57BL mice. MT was found in the cytoplasm and some nuclei of hepatic cells of C57BL , and MT staining intensity was increased by Cd treatment. In the distal tubular epithelial cells of untreated C57BL mice we found cytoplasmic MT staining, but Cd treatment resulted in a marked increase in MT staining both in the epithelial cells of the proximal tubules and of the descending loop of Henle. The pancreas of untreated C57BL mice showed a diffuse cytoplasmic MT staining in acinar cells with a different staining intensity, but not in the cells of pancreatic islets (Fig. 1a). However, Cd treatment increased MT staining intensity in the acinar cells. In the intact C57BL brain, MT was found to localize in ependymal cells, pia matter, glial cells and in the molecular layer of the cerebellum (Fig. 1b). Cadmium injections resulted in the appearance of intense MT staining in capillary endothelial cells. In untreated C57BL mice the basal cells and cells of the suprabasal cell layer of the epithelium of the tongue exhibited strong immunostaining. MT was also found in the muscle halfway between root and tip of tougue (Fig. 1c). MT was found in the spermatogenic cells and Sertoli cells in the seminiferous tubules of untreated C57BL mice (Fig. 1d).

In contrast, almost no MT immunostaining was detected in these organs from MT-deficient mice (Fig. 2a-d). However, immunostaining for MT showed that MT is present in the muscle of the tongue from untreated and Cd-treated mice (C57BL; MT-deficient mice), and in macrophages from Cd-treated testis, suggesting that these tissues may express unknown MT isoforms. In addition, the lack of MT immunostaining in tongue epithelium and brain from MT deficient mice suggests that MT-III and IV expression may be down regulated in MT-deficient mice. This hypothesis needs to be tested using MT isoform specific antibodies or in situ hybridization. Histological lesions caused

Proc. Microscopy and Microanalysis 1995, edited by G.W. Bailey, M.H. Ellisman, R.A. Hennigar, and N.J. Zaluzec
Copyright © 1995 MSA. Published by Jones and Begell Publishing, 79 Madison Ave., New York, NY 10016

by Cd were observed in the testes of MT deficient mice. Cadmium treatment destroyed seminiferous tubular structure and caused necrosis of spermatogenic cells in MT deficient mice (Fig. 3b) with a striking contrast to no testicular damage of Cd-treated C57BL mice (Fig. 3a). The present study indicates that, in the testis, MT isoforms, I and/or II could play a key role in detoxification against cadmium toxicity.

References
1. A.E. Michaelska and K.H.A. Choo *Proc. Natl. Acad.Sci* 90(1993)8088
2. H. Nishimura et al., *J Histochem Cytochem* 37(1989)715.

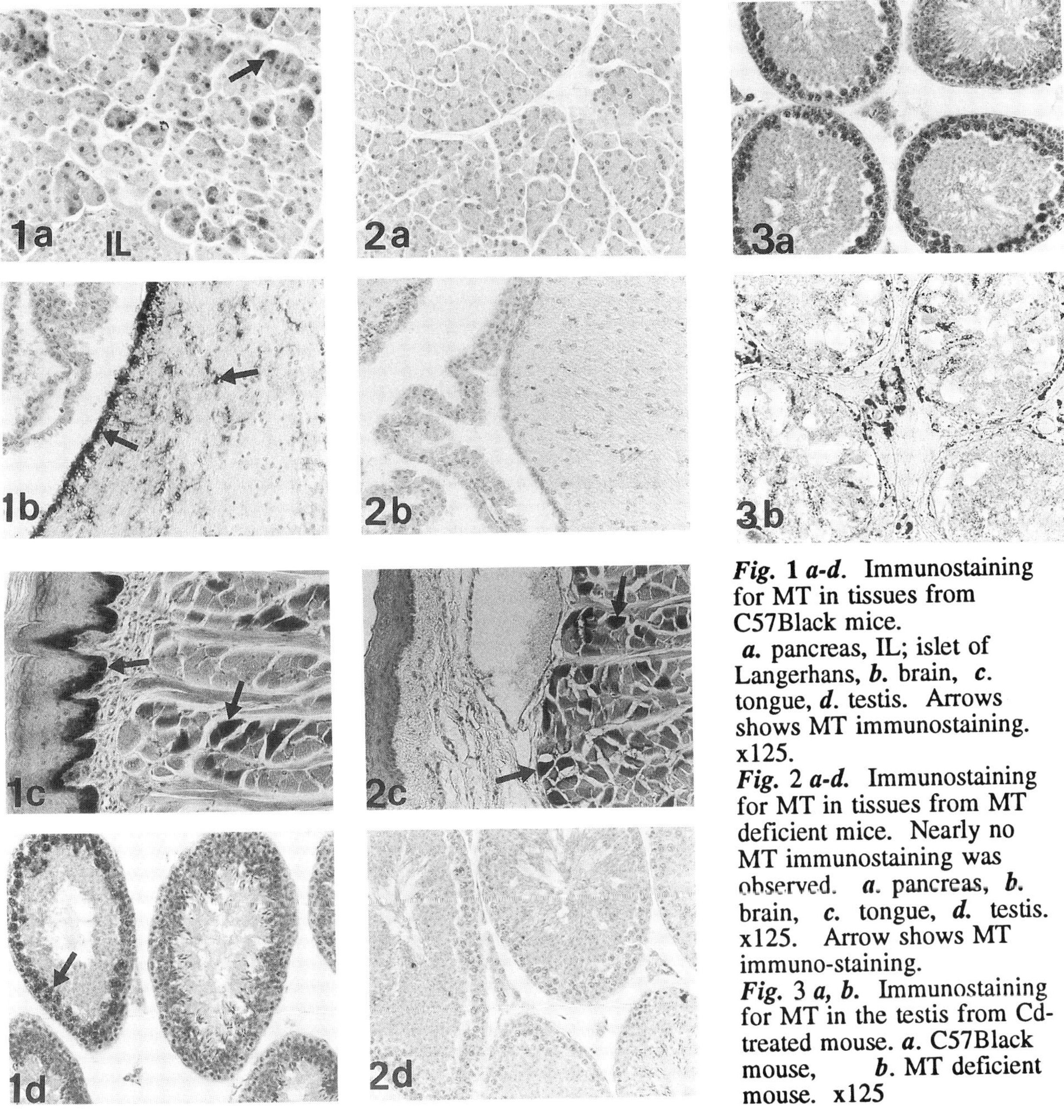

Fig. 1 *a-d.* Immunostaining for MT in tissues from C57Black mice. *a.* pancreas, IL; islet of Langerhans, *b.* brain, *c.* tongue, *d.* testis. Arrows shows MT immunostaining. x125.

Fig. 2 *a-d.* Immunostaining for MT in tissues from MT deficient mice. Nearly no MT immunostaining was observed. *a.* pancreas, *b.* brain, *c.* tongue, *d.* testis. x125. Arrow shows MT immuno-staining.

Fig. 3 *a, b.* Immunostaining for MT in the testis from Cd-treated mouse. *a.* C57Black mouse, *b.* MT deficient mouse. x125

GLUCOSE METABOLISM IN CANCER CELLS: IMMUNOGOLD LOCALIZATION OF HEXOKINASE TO THE OUTER MITOCHONDRIAL MEMBRANE

Krishan K. Arora, Glenn L. Decker, and Peter L. Pedersen
Department of Biological Chemistry, Johns Hopkins University,
School of Medicine, 725 North Wolfe Street, Baltimore, MD 21205

Hexokinase (ATP: D-hexose 6-phophotransferase EC 2.7.1.1) is the first enzyme of the glycolytic pathway which commits glucose to catabolism by catalyzing the phosphorylation of glucose with ATP. Previous studies have shown that hexokinase activity is markedly elevated in rapidly growing tumor cells exhibiting high glucose catabolic rates. A large fraction (50-80%) of this enzyme activity is bound to the mitochondrial fraction (1,2) where it has preferred access to ATP (3). In contrast, the hexokinase activity of normal tissues is quite low, with one exception being brain which is a glucose-utilizing tissue (4). Biochemical evidence involving rigorous subfractionation studies have revealed striking differences between the subcellular distribution of hexokinase in normal and tumor cells [See review by Arora et al (4)].

In the present report, we have utilized immunogold labeling techniques to evaluate the subcellular localization of hexokinase in highly glycolytic AS-30D hepatoma cells and in the tissue of its origin, i.e., rat liver. The primary IgG fraction used for this purpose was raised against tumor hexokinase and was previously shown to recognize a major band near 100 kDa on a western blot, which is consistent with the molecular weight of this enzyme (5). Liver and tumor mitochondria were prepared (2), and fixed for 60 min in 1.0% paraformaldehyde and 0.1% glutaraldehyde in 0.1M sodium cacodylate (adjusted to pH 7.4 with dilute HCl). The fixed specimens were then dehydrated through a graded series of dimethylformamide (50-100%), and embedded in Lowicryl K4M, and prepared for immunolabeling as described earlier (6). Subsequently, thin sections of these specimens were incubated for 60 min with 40 μg/ml of the primary antibodies (rabbit antihexokinase IgG, or normal rabbit IgG). Phosphate buffered saline containing 0.5% Tween 20 and 0.1% Triton X 100 was used for diluting the antibodies and for washing the grids between antibody applications. A 1:100 dilution of goat antirabbit IgG coupled to colloidal gold was applied to the grids for 60 min to localize bound primary antibody (6). After washing the grids in the buffered detergent mixture, the sections were poststained for 10 min in 2.0% uranyl acetate. Micrographic data was recorded with a Phillips 410 transmission electron microscope operated at 60kV.

Figure 1 illustrates representative micrographs from this study. Our results show that, based on the relative lack of antibody-gold particles bound to sections treated with preimmune serum (Panel A), antibodies raised against hexokinase bound significantly to the surfaces of tumor mitochondria (Panel B). Similar results were obtained with intact hepatoma cells (not shown). Further, as shown in Panel C, in sections prepared from tumor mitochondria treated with glucose-6-P, which is known to release the bound enzyme (1,2), antihexokinase antibodies failed to bind above apparent background levels under these conditions. Rat liver mitochondria, also used as a control, showed relatively low levels of bound gold particles (Panel D).

Thus, in agreement with earlier studies from this laboratory (4), the results of the current study indicate that binding of antihexokinase antibodies to tumor mitochondria is largely dependent on the level of hexokinase present in the outer mitochondrial membranes.

References

1. Bustamante, E., Morris, H.P., and Pedersen, P.L. (1981) J. Biol. Chem. 256, 8699-8704.
2. Parry, D.M., and Pedersen, P.L. (1983) J. Biol. Chem. 258, 10904-10912.
3. Arora, K.K., and Pedersen, P.L. (1988) J. Biol. Chem. 263, 17422-17428.
4. Arora, K.K, Parry, D.M., and Pedersen, P.L. (1992) J. Bioenerg. Biomemb. 24, 47-53.
5. Arora, K.K., Fanciulli, M., and Pedersen, P.L. (1990) J. Biol. Chem. 265, 6481-6488.
6. Decker, G.L., Valdizan, M.C., Wessel, G.M., and Lennarz, W.J. (1988) Develop. Biol. 129, 339-349.

Proc. Microscopy and Microanalysis 1995, edited by G.W. Bailey, M.H. Ellisman, R.A. Hennigar, and N.J. Zaluzec
Copyright © 1995 MSA. Published by Jones and Begell Publishing, 79 Madison Ave., New York, NY 10016

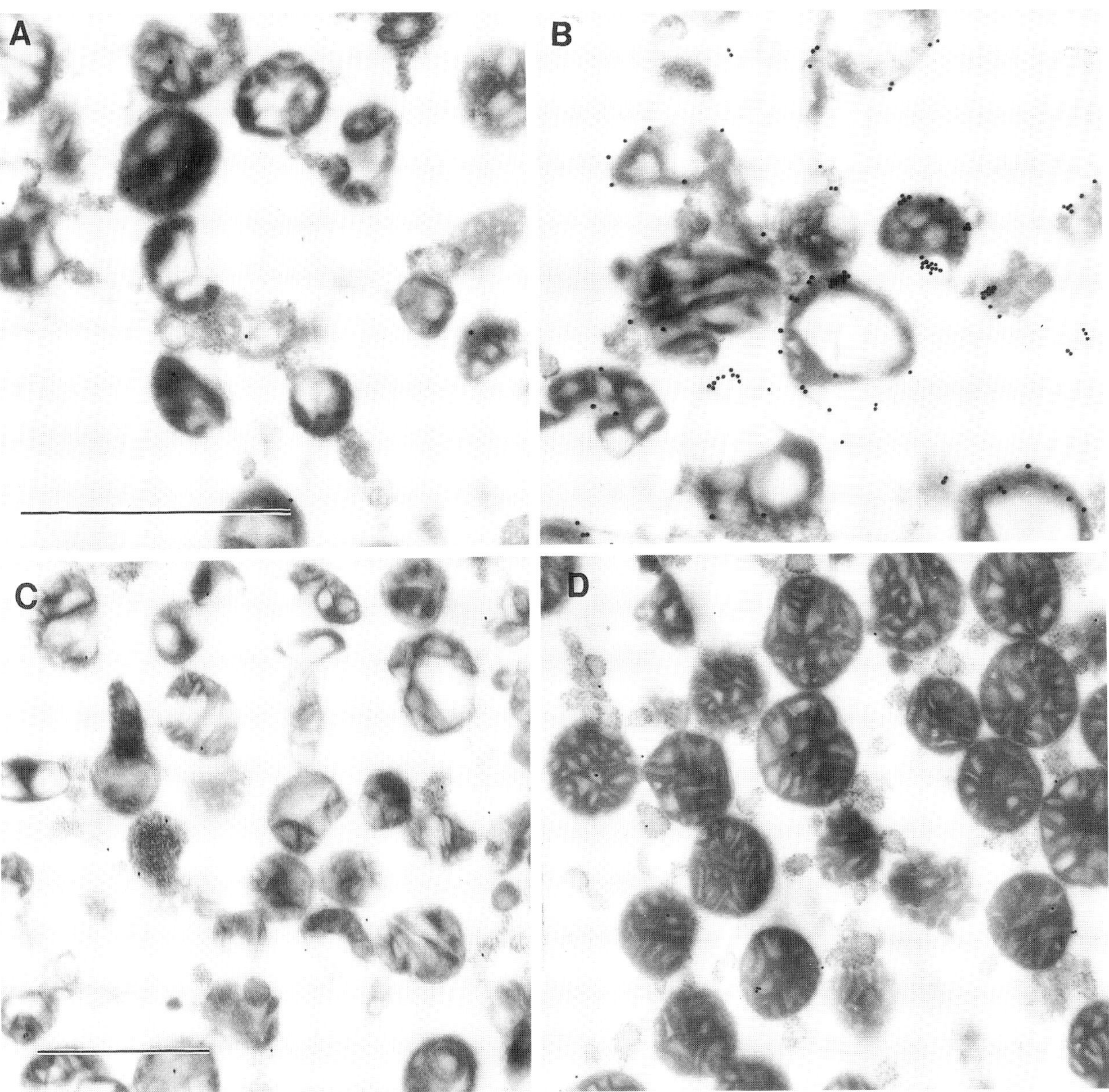

Fig. 1 Immunogold labeling of mitochondria from AS-30D hepatoma cells treated with <u>A</u>. preimmune serum or <u>B</u>. an antibody against tumor hexokinase. <u>C</u>. Immunogold labeling of AS-30D hepatoma cell mitochondria which had been prior treated with glucose-6-phosphate to release bound hexokinase before treatment with the hexokinase antibody. D. Immunogold labeling of rat liver mitochondria that had been treated with the tumor hexokinase antibody. The bar in A and C = 1 μm. A and B are the same magnification and C and D are the same magnification.

IMMUNOLOCALIZATION OF LACTASE IN THE CRYPT AND VILLUS ENTEROCYTES OF THE NEONATAL PIG INTESTINAL EPITHELIUM

Julian P. Heath and László G. Kömüves*

Microscopy Laboratory, USDA/ARS Children's Nutrition Research Center, Department of Pediatrics, Baylor College of Medicine, Houston, Texas 77030, and *Dermatology Service and Research Unit, VA Medical Center, San Francisco, CA 94121-1598

The crypt is the proliferative compartment of the intestinal epithelium where pluripotent stem cells continuously divide giving rise to committed daughter cells that follow a differentiation pathway as they migrate out of the crypt and along the villus axis. Certain cell antigens and markers are crypt-cell specific. Others, such as brush border enzymes, are thought to be expressed only on fully differentiated enterocytes, a theory supported by *in situ* hybridization analyses that reveal an absence of enzyme mRNAs in the crypt. But recent studies using immunocytochemical detection methods indicated that the brush border enzymes sucrase-isomaltase, aminopeptidase N and dipeptidylpeptidase IV are in fact expressed by both crypt and villus enterocytes.[1] We are using immunogold electron microscopy to examine the effects of diet on enzyme expression in the newborn pig intestine.

Tissues were collected from the jejunum of newborn and 9 day-old piglets. Cryosections (100 nm thick) were incubated in monoclonal anti-pig lactase antibody (PBB.3.7 from Dr. A. Quaroni, Cornell Univ) followed by 17 nm gold-conjugated anti-mouse immunoglobulin G.

Figures 1 and 2 contrast the morphology of villus and crypt enterocytes respectively. Fully differentiated enterocytes with a mature brush border are found at the crypt-villus junction and along the length of the villi. In the crypt, undifferentiated enterocytes are characterized by a smaller number of short microvilli and numerous apical secretory vesicles. There is a high density of lactase in the mature brush border of villus enterocytes (Figure 3). The average length of the microvilli on these cells was 1.5 um; the mean gold particle density was about 28 per micrometer of microvillus. Lactase was found on the microvilli and in the apical compartment of all crypt cells (except goblet, endocrine and tuft cells) (Figure 4). Gold particle densities were around 18 per unit length of microvillus, which is more than half the density on mature microvilli. The early expression of lactase (and other brush border enzymes) on undifferentiated enterocytes that are still proliferative may be in preparation for the rapid growth of the intestine in the first days of life. Interestingly, maturation of the brush border appears to be concomitant with insertion of the enzyme into the microvilli.[2]

1. Hansen G, et al. (1994) J. Submicros. Cytol. Path. **26**:453-460
2. Supported by USDA/ARS Cooperative Agreement 50-6250-1-003
Figure 1. Mature villus enterocyte in jejunum of 9 day old pig intestine
Figure 2. Undifferentiated enterocytes at the base of a crypt, with apical vesicles
Figure 3. Immunogold localization of lactase on microvilli of a mature villus enterocyte
Figure 4. Immunogold detection of lactase on crypt cell microvilli and in apical vesicles

Proc. Microscopy and Microanalysis 1995, edited by G.W. Bailey, M.H. Ellisman, R.A. Hennigar, and N.J. Zaluzec
Copyright © 1995 MSA. Published by Jones and Begell Publishing, 79 Madison Ave., New York, NY 10016

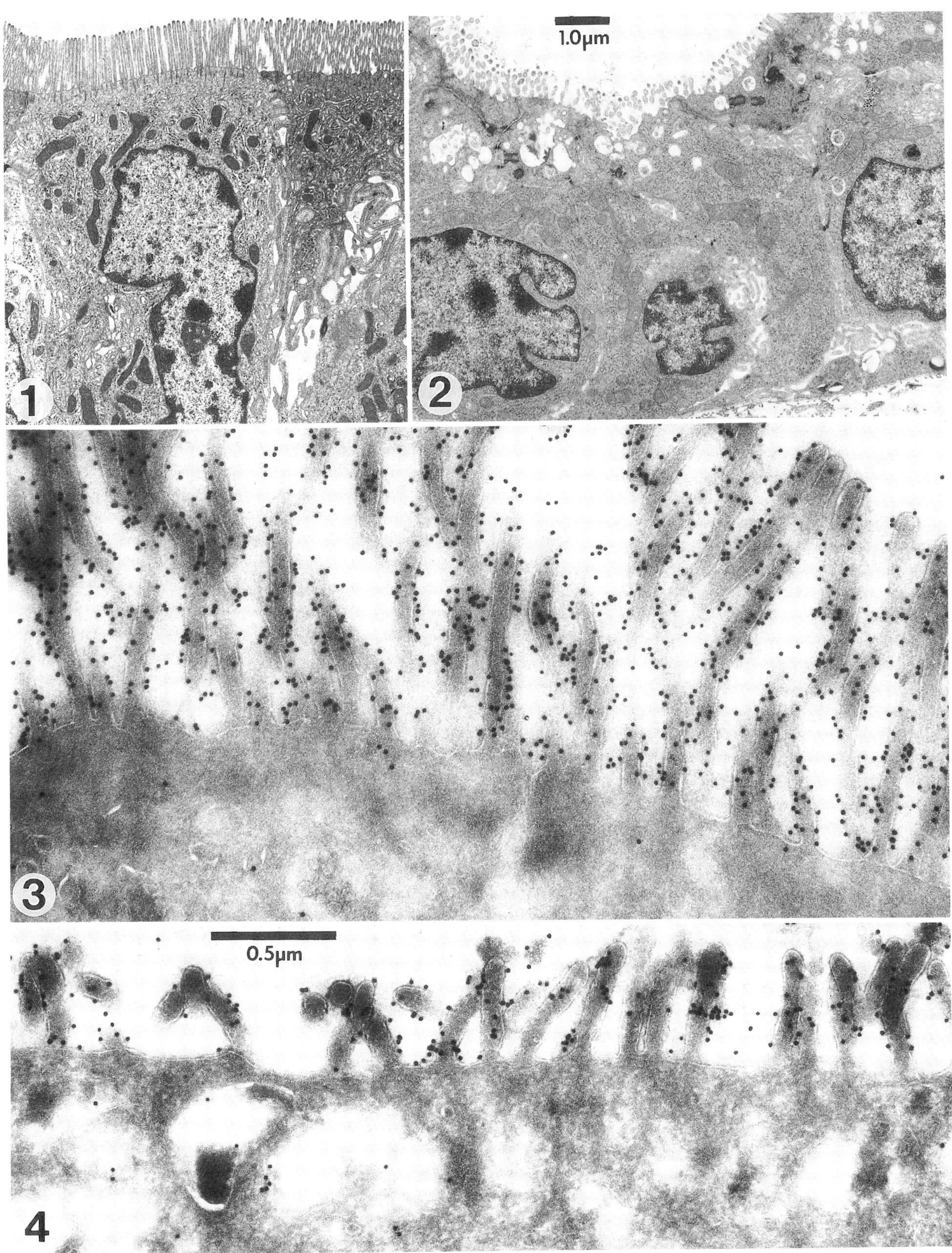
1.0µm
0.5µm
1
2
3
4

IMMUNOGOLD LOCALIZATION OF β_1 INTEGRIN IN MATURE AVIAN SMOOTH MUSCLE

M. Stromer

Muscle Biology Group, Iowa State University, Ames, IA 50011

Integrins are $\alpha\beta$ heterodimeric transmembrane glycoproteins.[1,2] The extracellular aminoterminal portion of the dimer contains the ligand-binding region and provides a linkage to the extracellular matrix and to other cells.[1,2] The relatively smaller cytoplasmic carboxyterminal region interacts with the cytoskeleton and specifically interacts with α-actinin[3] and with talin.[4] Integrins are, thus, thought to be involved in signal transduction at the surface of many cell types.[2] Smooth muscle cells contain $\alpha_1\beta_1$ integrin which has been immunolocalized in fibroblast focal contacts.[5] Because α-actinin and talin are components of attachment plaques which are sometimes referred to as plasma membrane-bound dense bodies in smooth muscle, we used immunogold labeling and transmission electron microscopy (TEM) to determine the location of β_1 integrin at the smooth muscle cell surface.

Strips of smooth muscle from mature chicken gizzards were isometrically clamped and were fixed in 2% paraformaldehyde in modified Krebs Ringer that contained 2mM Ca^{2+} and 2mM Mg^{2+} for 2½ hrs. at 2°C. Fixed strips were cut into 0.5-0.8mm cubes that were immersed in 0.15 M glycine for 1 hr., dehydrated in a graded methanol series by the progressive lowering of temperature method, embedded in Lowicryl K4M, and polymerized with UV light at -20°C for 7 days.[6] Thin sections on nickel grids were labeled with an antibody to β_1 integrin.[5] Protein A-15nm colloidal gold complexes identified the antibody localization. Attachment plaques are heavily labeled both on the plaque and on the extracellular side of the plaque (Fig. 1). Labelling is also prominent where collagen fibrils are concentrated near the cell (Fig. 1). Attachment plaques near the ends of cells are extensively labeled and trace the contour of the cell profile (Fig. 2). The density of the label on the plaque and on the extracellular side of the plaques in Fig. 3 is greater than in the interplaque regions. No labeling is observed at other intracellular sites which indicates that the reaction is specific.

Our studies indicate that β_1 integrin is localized in adult smooth muscle in attachment plaques but is also located where collagen fibrils approach the cell and at lower density in interplaque regions.

References

1. R.O. Hynes, *Cell* 69(1992)11.
2. F.W. Luscinskas and J. Lawler, *FASEB J.* 8(1994)929.
3. C.A. Otey et al., *J. Biol. Chem.* 268(1993)21193.
4. A. Horwitz et al., *Nature* 320(1986)531.
5. T. Kelly et al., *J. Biol. Chem.* 262(1987)17189.
6. M. Bendayan, *J. Electron Microsc. Tech.* 1(1984)243.
7. The β_1 antibodies were generously provided by K. Burridge. The excellent technical assistance of M.S. Mayes is gratefully acknowledged. This work was supported in part by a grant from the AHA-Iowa Affiliate. Journal Paper No. J-16258 of the Iowa Agriculture and Home Economics Experiment Station, Project 2921.

Proc. Microscopy and Microanalysis 1995, edited by G.W. Bailey, M.H. Ellisman, R.A. Hennigar, and N.J. Zaluzec
Copyright © 1995 MSA. Published by Jones and Begell Publishing, 79 Madison Ave., New York, NY 10016

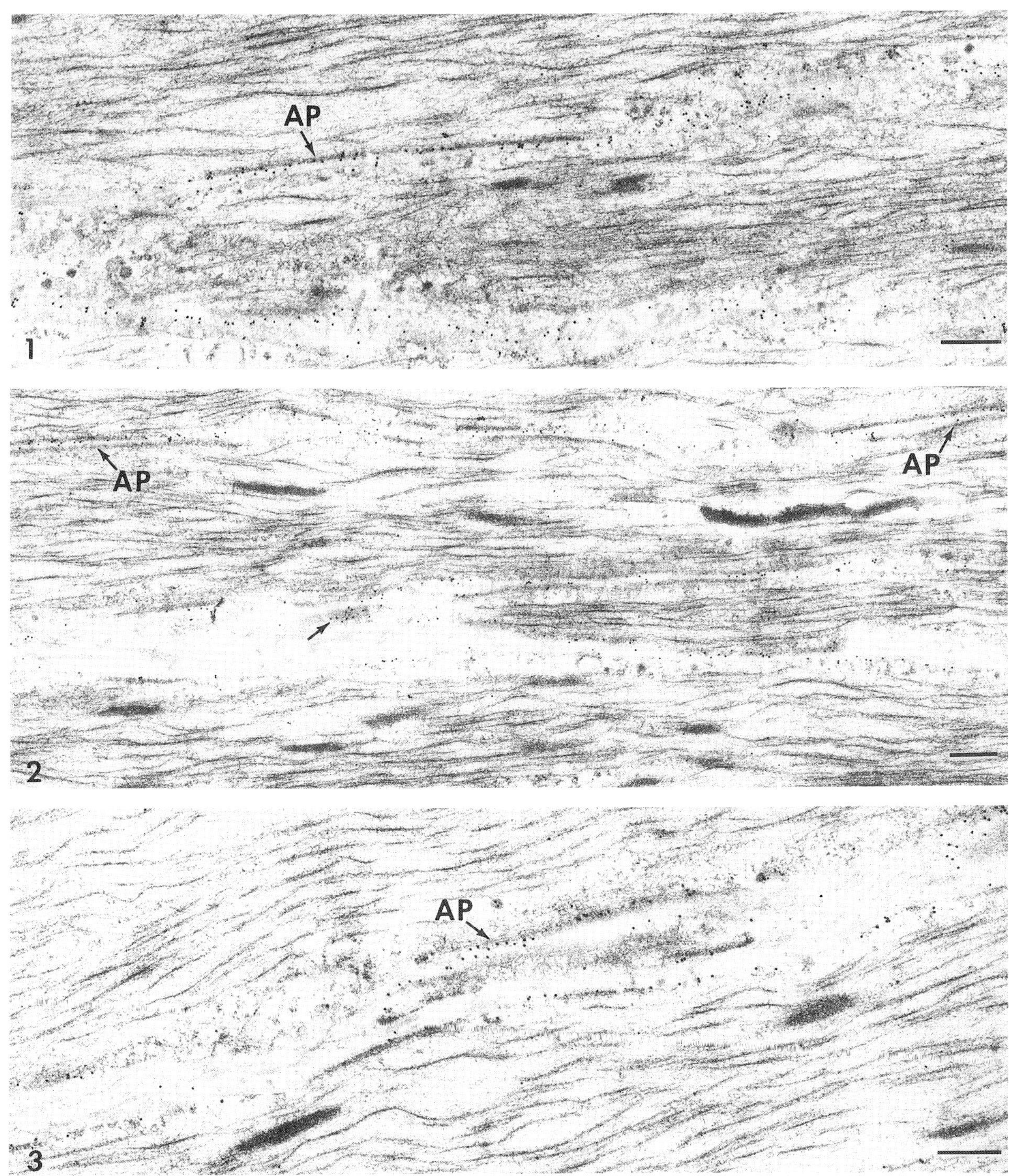

FIG. 1. Attachment plaques (AP) are heavily labeled with anti-β_1 integrin. The label is also associated with regions where collagen fibrils (C) approach the cell surface. Bar = 0.5 μm.

FIG. 2. Although all attachment plaques (AP) are labeled with anti-β_1 integrin, plaques near the end of the cell are often prominently labeled (arrow). Bar = 0.5 μm.

FIG. 3. The label associated with attachment plaques (AP) is both on the plaque per se and in the extracellular space next to the plaque. Interplaque regions are less extensively labeled. Bar = 0.5 μm.

TRIS-HCL BUFFER PH 9.5 MIXED WITH 5% UREA USED FOR ANTIGEN RETRIEVAL IMMUNOHISTOCHEMISTRY

S-R. Shi, R.J. Cote, L. Young, S.A. Imam, C.R. Taylor

Department of Pathology, University of Southern California School of Medicine, Los Angeles, CA 90033

The antigen retrieval (AR) method plays an important role in the high temperature heating AR immunohistochemistry (IHC). It has been demonstrated that the temperature, heating time and pH value of the AR solution are critical factors in determining the effectiveness of AR-IHC (1,2). For example, the intensity of AR-IHC staining of certain antibodies varies greatly according to the pH value of AR solution (2). Although the original discussed metal solution gave good results of AR-IHC (3,4), the potential for toxicity caused concern. A variety of other AR solutions have been tested as possible replacements for the metal solution, in the hope of obtaining better AR-IHC staining results without toxic side effects. Reports include the use of citrate buffer pH 6, urea, and a commercial 'TUF' solution. Currently, citrate buffer pH 6 is the most commonly used AR solution. We have posed the question as to whether this is the optimal AR solution for most antibodies used in surgical pathology. Following our recent studies (2,5), we have recommended the use of Tris-HCl buffer at higher pH (pH 8 - 10) in order to avoid false negative immunostaining results, that may occur with certain solutions at low, and mid range pH values (2,5). In an attempt to find an optimal AR solution with broad utility, we conducted a three-step research based on the AR-IHC method for routine formalin-fixed, paraffin-embedded tissue sections. As a first step, we conducted a screening study comparing Tris-HC1 buffer as the AR solutions at a variety of pH values, ranging from pH 1 to 10. results with citrate buffer pH 6 and with 0.1 M Tris-HCl buffer pH 9.5 mixed with the following chemicals: 5% urea, 2 M guanidine hydrochloride, 1% zinc sulfate, 4% aluminum chloride, saturated 5,5-dimethyl-1,3-cyclohexanedione (dimedone). Each of these chemicals in water solution alone was also tested. The monoclonal antibody MIB1 was used as a basis for comparison because of well recognized dependence on effective AR for good results. By comparison of the intensity of AR-IHC for MIB1, both Tris-HC1 pH 9.5 and mixed with 5% urea showed the strongest intensity of nuclear staining. As a second step the intensity of the AR-IHC using Tris-HCl buffer pH 8 to 10 was carried out in order to determine whether a pH value of 9.5 was indeed optimal; the results demonstrated the strongest intensity and best morphology at pH 9.5. This finding was then confirmed using several other antibodies including ER, Keratin, and vimentin. As a third step, a total of 34 antibodies were studied comparing 0.1 M Tris-HC1 buffer, with mixed Tris-HCl and 5% urea solution as the AR solution for immunostaining on archival tissue sections. Citrate buffer solution (0.01 M, pH 6.0) was used as the gold standard in a side-by-side comparison. In order to confirm the immunostaining intensity more accurately, all antibodies tested were based on two different tissue sections, and the test was repeated 3 times at higher dilutions of the primary antibody in order to confirm the existence of significant differences between the three AR solutions. The final results of this comparison are summarized in Table 1. For 12 antibodies the mixed AR solution produced the strongest intensity; for another 7 antibodies the Tris-HCl/urea was superior to the citrate buffer and as good as Tris-HCl alone for another 12 antibodies the intensity of staining was similar for all three AR solutions. In the remaining three cases Tris-HCl/urea was still superior to citrate, but was marginally inferior to Tris-HCl, pH 9.5 alone. We concluded that the mixed Tris-HCl and urea solution may be recommended as an AR solution for many antibodies employed commonly in routine surgical pathology, yielding results that were superior to citrate buffer pH 6.0 (currently the most widely used AR solutions) for 65% (22/34) antibodies that were tested, and equivalent results in the remainder immunostaining intensity for more antibodies tested.

References

1. P. Evers and H.B.M. Uylings, *J. Histochem. Cytochem.* 42(1994)1555.
2. S-R. Shi et al., *J. Histochem. Cytochem.* 43(1995)193.
3. C.R. Taylor et al., *Hum. Pathol.* 25(1994)263.
4. A.J.H. Suurmeijer and M.E. Boon, *Eur. J. Morphol.* 31(1993)144.
5. C.R. Taylor et al., *Hum. Pathol.* 25(1994)1107.

Table 1. Comparison of intensity of AR-IHC by using 3 AR solutions

Antibody	Dilution	Citrate	Tris-HCl	Tris-HCl + 5% urea
Keratin cocktail[1]	1:100	++	++	+++
Bcl-2	1:2000	+	+	++
Vimentin	1:1000	+	+	++
PSA	1:60000	++	++	+++
CD31	1:100	+	+	++
PCNA	1:100	++	++	+++
CD43	1:100	+	+	++
Cam5.2	original	++	+++	++++
PAb-S-100	1:100	+	+	++
Synaptophysin	1:500	++	+++	++++
TdT	1:10	+	++	+++
PAb-CD3	1:400	++	+++	++++
MIB1	1:100	+++	++++	++++
Ki-1	1:30	++	+++	+++
CD68	1:5000	-	+	+
PR	1:100	++	++++	++++
UCHL-1	1:10	+	++	++
L26	1:10	++	++++	++++
LN-1	1:40	++	+++	+++
LN-2	1:20	±	++	+
A n d r o g e n receptor	1:10	++	++++	+++
p53 (DO7)	1:800	++	++++	+++
NF	1:100	+	+	+
PAb-kappa	1:100	+	+	+
PAb-lambda	1:100	+	+	+
PAb-AFP	1:10000	+	+	+
Actin	1:100	+	+	+
Her-2/neu	1:4000	+	+	+
EMA	1:10	++	++	++
HMB-45	1:100	+	+	+
Lcu 7	1:10	+	+	+
Leu M-1	1:10	++	++	++
CD34	1:500	++	++	++
Collagen IV	Original	++	++	++

Note. Most antibodies were purchased from BioTek (Santa Barabara, CA) which are pre-diluted antibodies. As most antibodies tested were diluted into a much lower concentration than regular concentration, the intensity of AR-IHC listed above was weaker. Citrate = citrate buffer pH 6.0, Tris-HCl = 0.1 M Tris-HCl buffer pH 9.5. PAb = polyclonal antibody. (1) Keratin cocktail = AE1 + AE3 + Cam 5.2 + 35BH11.

IMMUNOGOLD LABELING OF THE BACTERIAL SURFACE COMPONENTS OF SHIGELLA FLEXNERI 2a

D.H. Chen, G.H. Wang, H.Q. Zhang and H. Li

Institute of Basic Medical Sciences, P.O. Box 130(3), Beijing 100850, P.R. China

Some bacterial outer membrane components are efficient protective antigens against Shigella spp. Electron microscopic immunogold labeling provides a possibility for investigating the surface exposure of antigens (1). The aim of the present study was to localise surface antigens of Shigella flexneri 2a by using an electron microscopic immunogold labeling technique.

Seven monoclonal antibodies (MAbs: 2A3, 3A6, 2E6, 1G8, 1A1 and 4F1) used in this study were produced as described previously (2). Pure cultures of Shigella flaxneri 2a were grown in broth, harvested by centrifugation, prefixed and then immunolabelled by MAbs and anti–mouse IgG conjugated to colloidal gold particles (GAMG15 and GAMG5) respectively. Ultrathin sections of labelled bacteial cells were cut and viewed in a transmission electron microscope (Philips EM 400T).

Electron microscopic examination of the sectioned organisms showed that gold labeling was confined to the outer surface of the bacterial cell wall. Binding of the MAb on whole bacterial cell indicated the antigens recognized by MAbs 3A6, 2E6, 4F1 and 2A3 were exposed on the outer surface of cell membrane (Fig 1, a–c), and those by MAbs 1A1 and 1G8 were not. However, the labeling densities varied largely in each cell. It was probabely caused by the growth conditions such as growth phase and nutrient limitations which exerted an influence on the bacterial surface structure. Comparing with the biological activities of these MAbs, the exposure of the antigen determints on the bacterial surface had a direct bearing on their blocking activities of the contact haemolysis and the passive protecting capacities in mice (2). The results suggested that the quantitative immunoelctron microscopic analysis allow a better understanding of the accessibility and numbers of copies to the bacterial protecting antigenic determinents.

References

1. M. Paques et al., Infect. Immun. 57(1989)582.
2. H. Li and J.M. Zhou, Chinese J. Epid. 11(1989)148 (In Chinese).

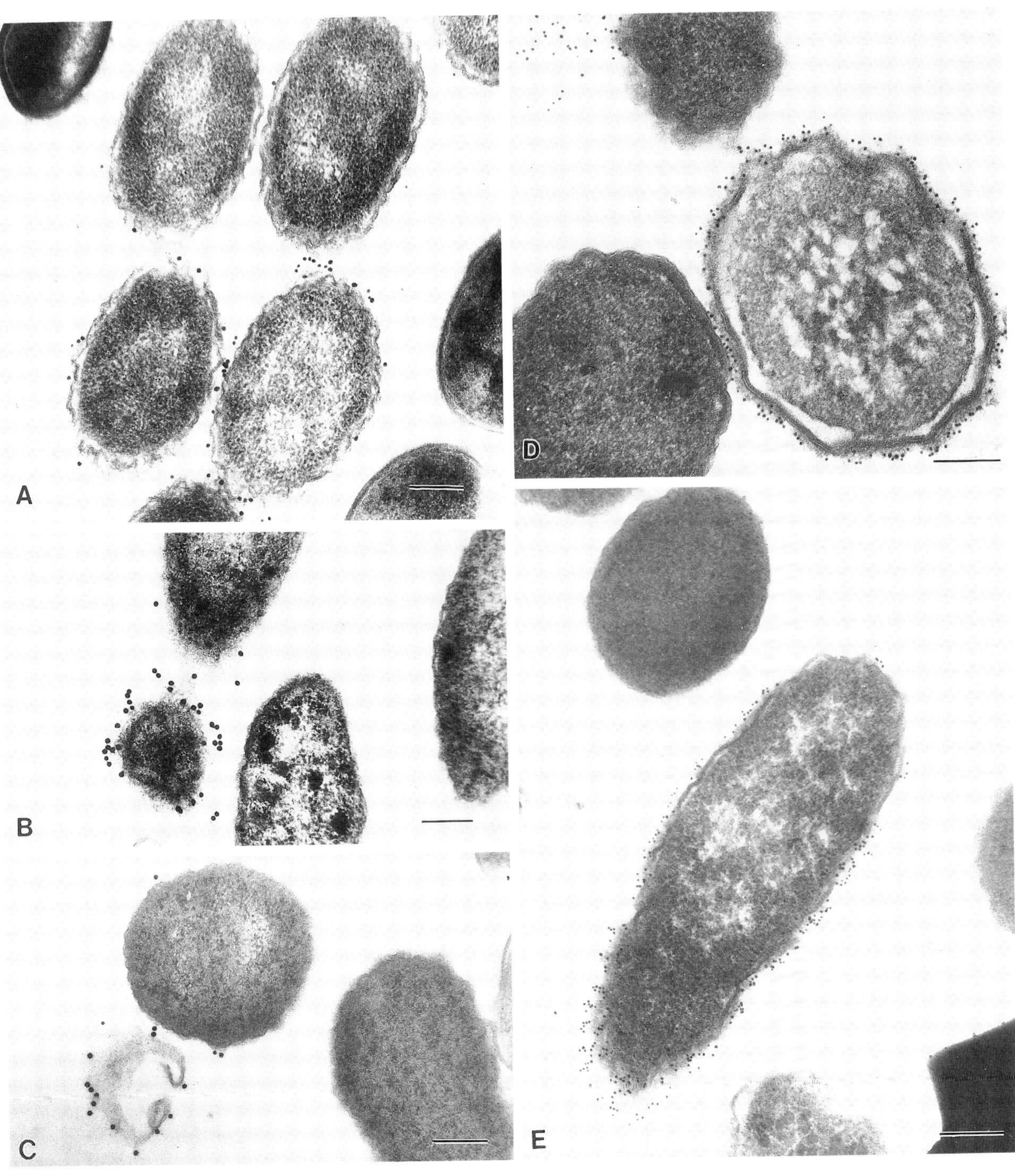

FIG. 1. **A–E** Immunogold labelling of bacterial surface antigens of Shigella flexneri 2a cells after incubation with MAb 3A6 followed by incubation with GAMG15 (A,B,C) and with MAb 2E6 followed by GAMG5 (D,E). The variation in labeling density per cells is striking. Bar = 0.2μm.

SEQUENTIAL STAINING AND DE-STAINING OF A SINGLE SECTION BY A PANEL OF FLUORESCENTLY-TAGGED LECTINS

C. Michael Stanley and Thomas E. Phillips
Division of Biological Sciences and Molecular Cytology Core Facility, University of Missouri, Columbia, MO 65211

Numerous previous studies have used fluorescently tagged lectins to demonstrate the mucin glycoprotein secretory product of intestinal goblet cells are heterogeneous in respect to their lectin binding sites[1]. Lectins stain the mucin secretory product in partially overlapping subpopulations of goblet cells. There are several difficulties in using a panel of fluorescently tagged lectins to determine the overlapping sites of binding. First, steric hindrances resulting from the binding of the first lectin compete with the subsequent binding of other lectins to the same or adjacent sites. Secondly, the number of fluorescent tags that can be practically differentiated by routinely configured microscopes is generally limited to two or three. As an alternative approach, we investigated whether 8 M urea could remove previous bound lectins without damaging the carbohydrate binding sites. Microwave treatment with 8 M urea has previously been used to recover antibody binding sites masked by fixation and embedding[2].

Small pieces of distal colon were removed from anesthetized rats and fixed for 2 hours at room temperature in 2% freshly-depolymerized paraformaldehyde in 70 mM NaCl, 30 mM HEPES, 2 mM $CaCl_2$, pH 7.4. Following rinsing, the tissues were dehydrated with a series of ethanol and embedded in JB-4 methacrylate resin. Semi-thin (0.9 µm) sections were mounted on aminopropyltriethoxysilane coated glass slides. A 100 µl droplet of a fluorescein-tagged lectin was placed on top of the sections for 2 hours in a humidified chamber at room temperature. The following lectins (and their binding specificities) were applied, examined, and removed in the following order: fluorescein-*Dolichos bifluoros* (α-GalNac); rhodamine-soybean agglutinin (α and ß GalNac); rhodamine-*Bandeiraea simplicifolia* I (α–Gal and α–GalNac); rhodamine-*Phaseolus vulgaris* L (Gal with an internal Man residue); rhodamine-*Lens culinaris* (α-Man> αGlu>αGlcNAc) and fluorescein-wheat germ agglutinin (GlcNAc and sialic acid). All lectins except wheat germ agglutinin were used at 10 µg/ml in 70 mM NaCl, 30 mM HEPES, 2 mM $CaCl_2$, pH 7.4; wheat germ agglutinin was used at 5 µg/ml.

Following rinsing, sections were examined using a BioRad MRC-600 confocal fluorescent microscope equipped with a krypton-argon laser. Digital images of several fields were collected from within each section at a standard setting for the pinhole aperture, gain, and neutral density filter (1% transmission) in the excitation path. To de-stain the sections, the slides were immersed in 8 M urea in a plastic Coplin jar and heated at full power for 90 seconds in a GE microwave (Model JE48A; 575 watts). The solution was then agitated by hand for 30 seconds and re-microwaved for an additional 90 seconds. Following rinsing, the original microscopic fields of the sections were re-examined in both the FITC and rhodamine channels to confirm the complete removal of label. By increasing the laser excitation to 100% transmission and manipulating the black level, one could increase the background fluorescence sufficiently to identify the appropriate microscopic field. Even when the background staining was artificially increased in this manner, the absence of staining of the mucin secretory product was clear. The results clearly demonstrate that staining by fluorescently tagged lectins can be removed by microwaving in 8 M urea.

We are currently investigating whether this sequential staining and de-staining strategy will work with a panel of mouse monoclonal antibodies.

References:

1. Boland CR, Montgomery CK, and Kim YS (1986) Proc. Nat. Acad. Sci. (USA) 79:2051-2055.
2. Shi SR, Chaiwum B, Young L, et al., (1993) J. Histochem. Cytochem. 41(11)1599-1604.

Proc. Microscopy and Microanalysis 1995, edited by G.W. Bailey, M.H. Ellisman, R.A. Hennigar, and N.J. Zaluzec
Copyright © 1995 MSA. Published by Jones and Begell Publishing, 79 Madison Ave., New York, NY 10016

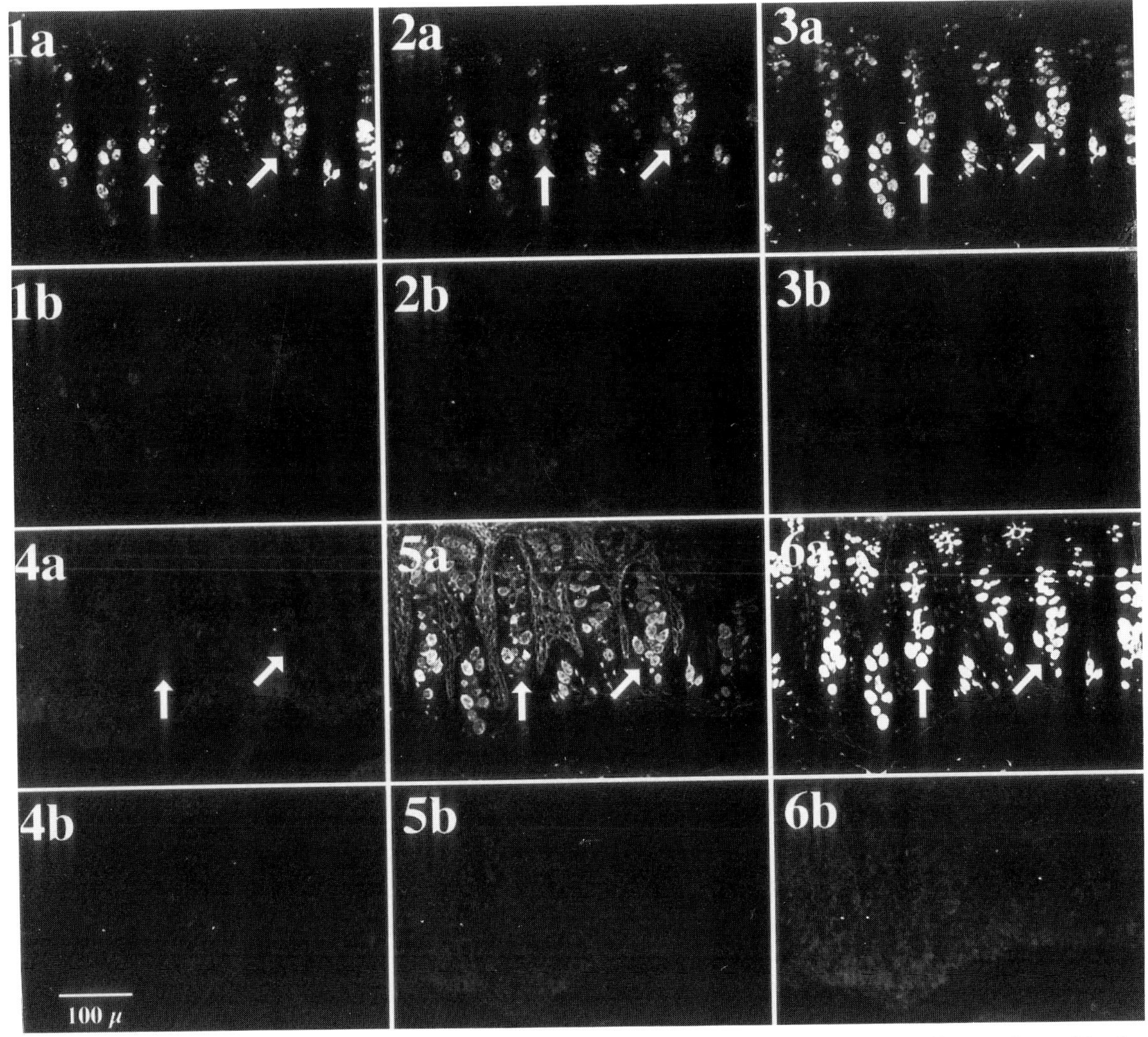

All images represent the same field of view of a single section of rat descending colon. Each pair of images (a + b) show the staining pattern by a fluorescently tagged lectin (a) followed by the same field after de-staining with 8 M urea (b). The micrographs depict approximately 7 crypts facing downward towards the colonic lumen (near the base of the arrows) along the lower margin of the image. The arrows point to identical clusters of cells in each micrograph.

Figure 1: Fluorescein-*Dolichos bifluros* agglutinin (DBA) stained primarily the mucin secretory product stored in more mature goblet cells closer to the lumen.

Figure 2: Rhodamine-soybean agglutinin (SBA) stains, with slightly less intensity, a subset of goblet cells similar to that stained by DBA.

Figure 3: Rhodamine-*Bandeiraea simplicifolia* I (BSL-I) had a more intense staining pattern than either DBA and SBA which allowed recognition of a greater number cells is the mid-crypt zone.

Figure 4: Rhodamine-*Phaseolus vulgaris* L (PHA-L), which has a binding specificity requiring an internal mannose residue common in N-linked glycoproteins, fails to stain the O-linked mucin glycoproteins. This lectin illustrates that the repetitive urea de-staining step does not artificially create non-specific binding sites.

Figure 5: Rhodamine-*Lens culinaris* (LCA) preferentially labeled the secretory product of more mature goblet cells but resulted in higher background staining of the lamina propria.

Figure 6: Fluorescein-wheat germ agglutinin (WGA) intensely labeled all goblet cells.

ULTRASTRUCTURAL DISTRIBUTION OF STEROID SULFATASE ACTIVITY AND mRNA IN THE HUMAN KIDNEY

M. Nishikawa,* T. Nakazawa,* N. Kanda,* E. Aikawa,* and H. Toma**

*Department of Anatomy and Developmental Biology, and **Department of Urology, Tokyo Women's Medical College, Tokyo, 162 Japan

Steroid sulfatase (STS) is a microsomal enzyme catalyzing hydrolysis of 3β–hydroxysteroid sulfatases, including dehydroepiandrosterone sulfate, cholesterol and estrogen precursors.[1] STS exists widely in various tissues and is involved in the metabolism of steroid hormones which play a critical role in cell proliferation and may be related to cancer induction in several organ systems.[2,3] In the kidney, the proximal and distal portions of the nephrons have been suggested as possible sites of synthesis or as target tissues for steroid hormones.[4,5] However, the presence and the role of STS in the human kidney have not been fully investigated. In this study, we examined the distribution of STS activity and mRNA in human renal tissue by enzyme histochemistry and in situ hybridization at the light and electron microscopic level.

Renal tissue specimens were taken from normal areas in the surgically resected kidney from patients with renal cancer. STS activity was analyzed by enzyme histochemistry using 4-methylumbelliferyl sulfate. STS-mRNA distribution was detected by in situ hybridization using an STS cDNA probe cloned from human placenta. The probe was labeled with digoxigenin, and hybridization was detected by an antidigoxigenin antibody conjugated with 5 nm gold particles. STS activity was clearly demonstrated in the proximal tubules (Fig. 1a) and glomerular epithelial cells. The reaction products were located at the endoplasmic reticulum (Fig. 1b). STS mRNA was predominantly distributed in the cortex-medullary junction of the kidney (Fig. 2). It was not detected in the mesangial or endothelial cells of the glomeruli. RNase-treated control tissue showed negative reactivity. The ultrastructural localization of STS mRNA was mainly adjacent to the rough endoplasmic reticulum between the mitochondria in the proximal tubules (Fig. 3).

These results indicate that STS is synthesized in the proximal tubules and glomerular epithelial cells of the human kidney, and suggest that STS may be related to cell proliferation and the metabolism of steroid hormones in these areas of renal tissue.

References

1. P.H. Yen et al., *Cell* 55(1988)1123.
2. G. Wilding et al., *Cancer Res.* 48(1998)802.
3. P.S. Rennie et al., *Cancer Res.* 48(1988)6309.
4. H.F. DeLuca, in J. DuMont, and J. Nunez, Eds., *Hormones and Cell Regulation,* Amsterdam: Elsevier. vol 2(1978)249.
5. W.E. Stumph et al., *Proc. Natl. Acad. Sci. USA* 77(1980)1149.
6. This study was supported by Grant-in-Aid for Scientific Research from the Ministry of Education, Science and Culture, Japan (No.05670031), by a grant from the Japan Private School Promotion Foundation and by Yayoi Yoshioka Award.

Proc. Microscopy and Microanalysis 1995, edited by G.W. Bailey, M.H. Ellisman, R.A. Hennigar, and N.J. Zaluzec
Copyright © 1995 MSA. Published by Jones and Begell Publishing, 79 Madison Ave., New York, NY 10016

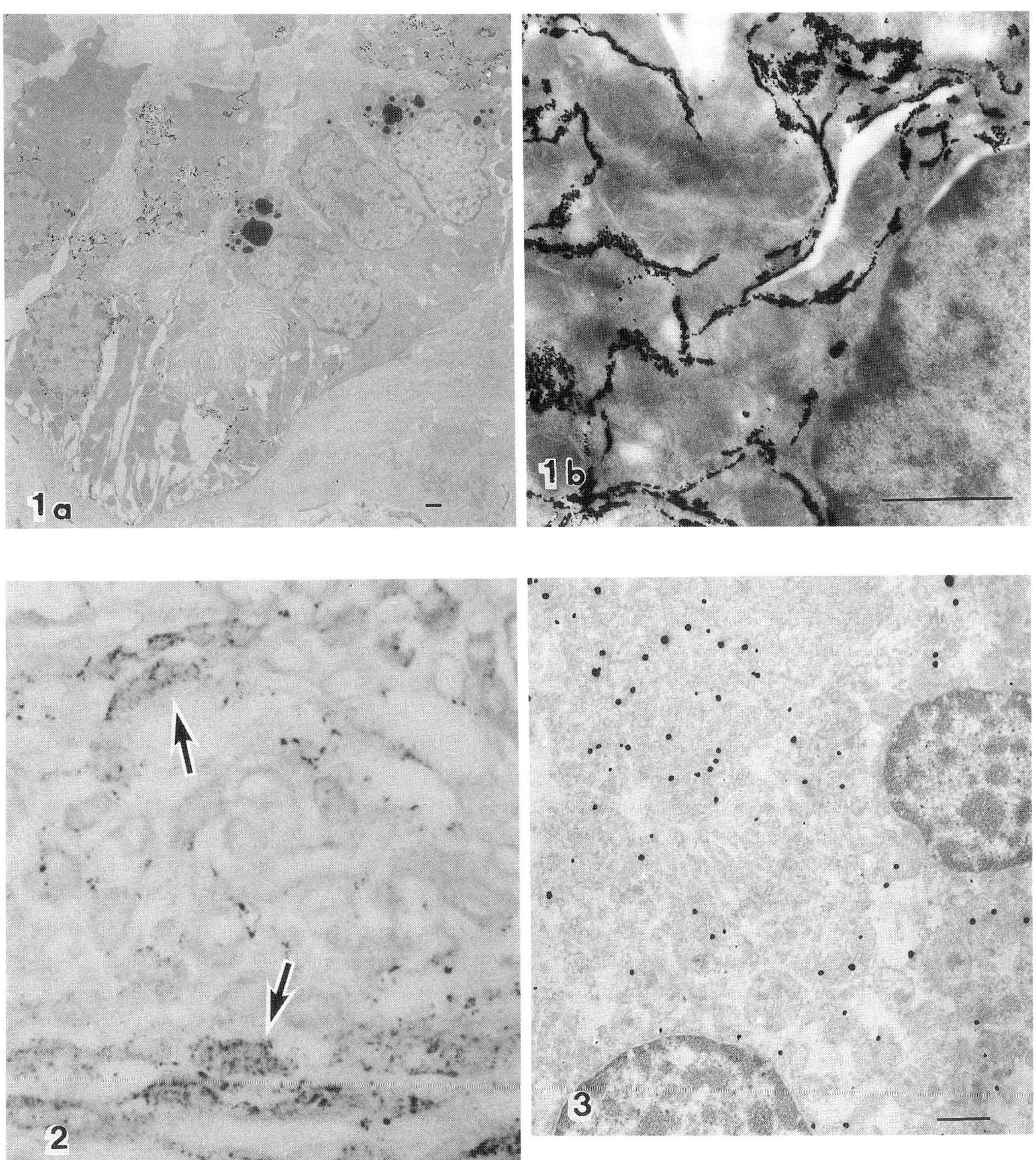

FIG. 1 - Enzyme histochemistry for STS activity. (a)The reaction products are demonstrated in the proximal tubules; (b)these products are located at the endoplasmic reticulum. Bar=1μm.
FIG. 2 - In situ hybridization. STS mRNA is predominantly distributed in the cortex-medullary junction. Arrows indicate the positive signal of STS mRNA in the proximal tubules. x20
FIG. 3 - Ultrastructural localization of STS mRNA in the proximal tubule. Gold particles adjacent to the rough endoplasmic reticulum between the mitochondia are seen. Bar=1μm.

CYTOCHEMICAL LOCALIZATION OF ATPase ACTIVITY IN THE LEAVES OF A SEAGRASS AND A SUBMERGED HALOPHYTE

A.D. Barnabas and M. Coopoosamy

Department of Botany, University of Durban-Westville, Private Bag X54001, Durban 4000, South Africa

Flowering plants that grow submerged in seawater are known as seagrasses, while those that grow submerged in water of generally lower and varying salinity as in estuaries, are known as submerged halophytes.[1] Knowledge of salt tolerance mechanisms in both groups of plants is important to our understanding of their biology. Recently, high ATPase activity was reported to be associated with the copiously-invaginated plasma membrane of leaf epidermal cells of a seagrass (*Zostera marina*), suggesting that ATPase played a role in salt regulation.[2] In the present study, the seagrass *Halophila ovalis* and the submerged halophyte *Ruppia maritima* growing at salinities of 35°/oo and 25°/oo respectively, were used. The subcellular distribution of ATPase in leaf epidermal cells of both submerged aquatics was accomplished using a lead precipitation technique specific for ATPase activity.[3] Lead deposits in the cells were an indication of sites of ATPase activity. Transmission electron microscopy (TEM) and energy dispersive X-ray (EDX) microanalysis using a Joel 6100 SEM with a Noran Voyager 2100 EDX microanalyser, were used to study lead distribution. Ultrathin sections for the TEM were examined unstained. EDX microanalysis was performed on thick sections of resin-embedded material mounted on glass slides.

Leaf blade epidermal cells resembled transfer[4] cells in that the plasma membrane was invaginated along the outer tangential wall in *H. ovalis* (Figs. 1,3) and along the radial walls in *R. maritima* (Figs. 2,4). In *H. ovalis* a narrow electron-dense layer was closely associated with the plasma membrane, especially where the membrane was invaginated, suggesting a high level of ATPase activity (Figs. 1,3, arrows). X-ray mapping and spot EDX microanalysis showed that this electron-dense layer was composed of lead (Figs. 5-7). In *R. maritima*, ATPase activity was more widespread. Heavy deposits of lead occurred throughout the cytoplasm, within vacuoles, and in mitochondria (Figs. 2,4). Based on the results obtained, it would appear that salt control is better developed in seagrasses, plasma membrane associated ATPase presumably playing a crucial role in salt regulation. Salt tolerance in *R. maritima* seems to be dependent on the activity of ATPase located in various places in the cells. This might represent a less efficient mechanism of salt control. Hence seagrasses would be better adapted to survive in the sea than other submerged halophytes.[5]

References

1.   C. Den Hartog, *Helgol. Wiss. Meeresunters* 15(1967) 648.
2.   J.Y. Pak et al., *Planta* (1995) In press.
3.   J.L. Hall et al., *Protoplasma* 104(1980) 193.
4.   B.E.S. Gunning and J.S. Pate, *Protoplasma* 68(1969) 107.
5.   The authors gratefully acknowledge the financial support of the University of Durban-Westville and Foundation for Research and Development, South Africa.

Proc. Microscopy and Microanalysis 1995, edited by G.W. Bailey, M.H. Ellisman, R.A. Hennigar, and N.J. Zaluzec
Copyright © 1995 MSA. Published by Jones and Begell Publishing, 79 Madison Ave., New York, NY 10016

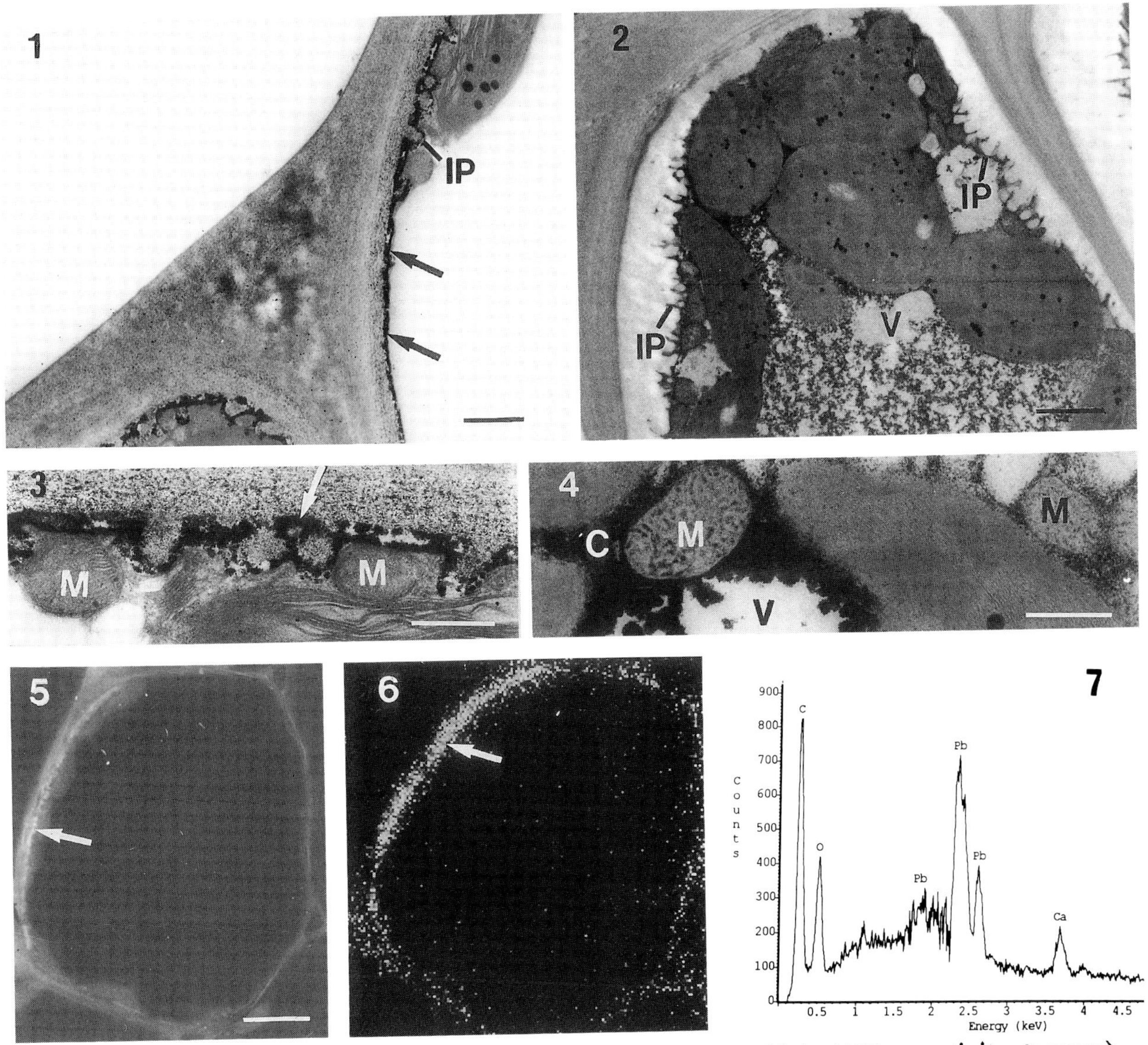

FIG. 1.-- TEM image of epidermal cells of *H. ovalis* showing high ATPase activity (arrows) associated with plasma membrane. IP, invaginated plasma membrane. Bar = 1μm.

FIG. 2.-- TEM image of epidermal cell of *R. maritima* showing ATPase activity (electron-dense material) in cytoplasm and within vacuole (V). IP, invaginated plasma membrane. Bar = 2μm.

FIG. 3.-- TEM image of portion of outer epidermal wall of *H. ovalis* showing ATPase activity (arrow) associated with invaginated plasma membrane. M, mitochondria. Bar = 0.5μm.

FIG. 4.-- TEM image adjacent to radial wall of *R. maritima* showing ATPase activity (electron-dense deposits) in vacuole (V), mitochondria (M) and cytoplasm (C). Bar = 0.5μm.

FIG. 5.-- SEM image of resin section of *H. ovalis* indicating narrow bright layer (arrow) adjacent to outer wall. Bar = 2μm.

FIG. 6.-- X-ray map of Fig. 5 showing distribution of lead. Highest lead concentration (arrow) corresponds to bright layer.

FIG. 7.-- X-ray spectrum (minus background) from spot analysis of bright layer of Fig. 5 showing high concentrations of lead.

USE OF DIFFERENT HUMAN MYELOID LEUKEMIA CELL LINES TO DETERMINE THE SPECIFICITY OF 3,3'-DIAMINOBENZIDINE CYTOCHEMISTRY FOR THE ULTRASTRUCTURAL LOCALIZATION OF MYELOPEROXIDASE AND CATALASE

J. E. Boyd[1], C. S. Gilbert[2], I. B. Pinnix[3], C. A. Ballinger[3], J. M. Kinkade, Jr.[3], R. T. Parmley[2]

[1]The University of Texas Health Science Center, Dept. of Pediatrics, San Antonio, TX 78284
[2]Carolinas Medical Center, Dept. of Pediatric Research, Charlotte, NC 28203
[3]Emory University, School of Medicine, Dept. of Biochemistry, Atlanta, GA 30322

Myeloid cells are known to contain myeloperoxidase (MPO) and catalase.[1,5] This study has used MPO and catalase replete and deficient myeloid cell lines to clarify the localization of these components using 3,3'-diaminobenzidine (DAB) ultrastructural cytochemistry. Conditions of DAB incubation can be modified to preferentially stain catalase (alkaline at pH 9.7) or MPO (neutral at pH 7.0-7.6), but cross-reactivity persists, preventing the discrimination between catalase and peroxidase.[6] Biochemical assays demonstrated both MPO[5] and catalase[1] in HL60 cells; similar amounts of catalase but no MPO activity in the A7 cell line; increased amounts of catalase but no MPO activity in the HP50 and HP100 cell lines; and neither MPO nor catalase in the KG1 cell line. Neutral DAB stained MPO (pH 7.4; [DAB] 5 or 20 mg/10 mL 0.05 M Tris; 30 min or 120 min; 24° or 37°C; 0.01% H_2O_2)[2,4,7] in HL60 (Fig. 1), but not in A7. Alkaline DAB intensely stained catalase (pH 9.7; 20 mg/10 mL; 120 min; 37°C; 0.01% or 0.03%)[3,7] in A7 (Fig. 3). MPO cross reactivity with alkaline DAB obscured catalase localization in HL60 (Fig. 1,2), although alkaline DAB without H_2O_2 stained catalase moderately, but only weakly stained MPO in HL60 (Fig. 3 inset). Incubation of viable cells with neutral DAB[7] stained HL60 containing MPO but failed to stain A7(catalase only). Catalase containing organelles were larger in HP50 and HP100 (Fig. 4). Neither catalase nor MPO could be stained in KG1, consistent with biochemical results. Preincubation of cells with 0.2M 3-amino-1,2,4-triazole (AT; pH 9.0; 3 hrs; 37°C) followed by alkaline DAB-AT[2.5mM] incubation[3,7] resulted in inhibition of catalase in A7, but did not completely inhibit alkaline DAB staining of presumed MPO in HL60. Comparisons in this study better define the localization of catalase and MPO in myeloid cells and characterize cell lines which may be useful in future MPO and catalase gene transfer experiments.

References

1. C.A. Ballinger et al., *Blood* 83(1994)2654.
2. R.C. Graham and M.J. Karnovsky, *J Histochem Cytochem* 14(1966)291.
3. E.T. Nishimura et al., *Lab Invest* 34(1976)60.
4. A.B. Novikoff and S. Goldfischer, *J Histochem Cytochem* 17(1969)675
5. S.O. Pember et al., *Arch Biochem Biophys* 221(1983)391.
6. F. Roels et al., *Histochemistry* 41(1975)281.
7. D.M. Zellmer and W.A. Shannon, *Histochem J* 15(1983)211.

8. The authors are grateful to the following individuals who made cell lines available: HP Koeffler (University of California, Los Angeles), AK Sullivan (McGill University, Montreal), M Yamada (Yokohama City University), R Kawaguchi (SRL Laboratories, Tokyo). We also thank Mei-Whey Wu for expert management of the tissue culture facility and for providing cells for this study.

Proc. Microscopy and Microanalysis 1995, edited by G.W. Bailey, M.H. Ellisman, R.A. Hennigar, and N.J. Zaluzec
Copyright © 1995 MSA. Published by Jones and Begell Publishing, 79 Madison Ave., New York, NY 10016

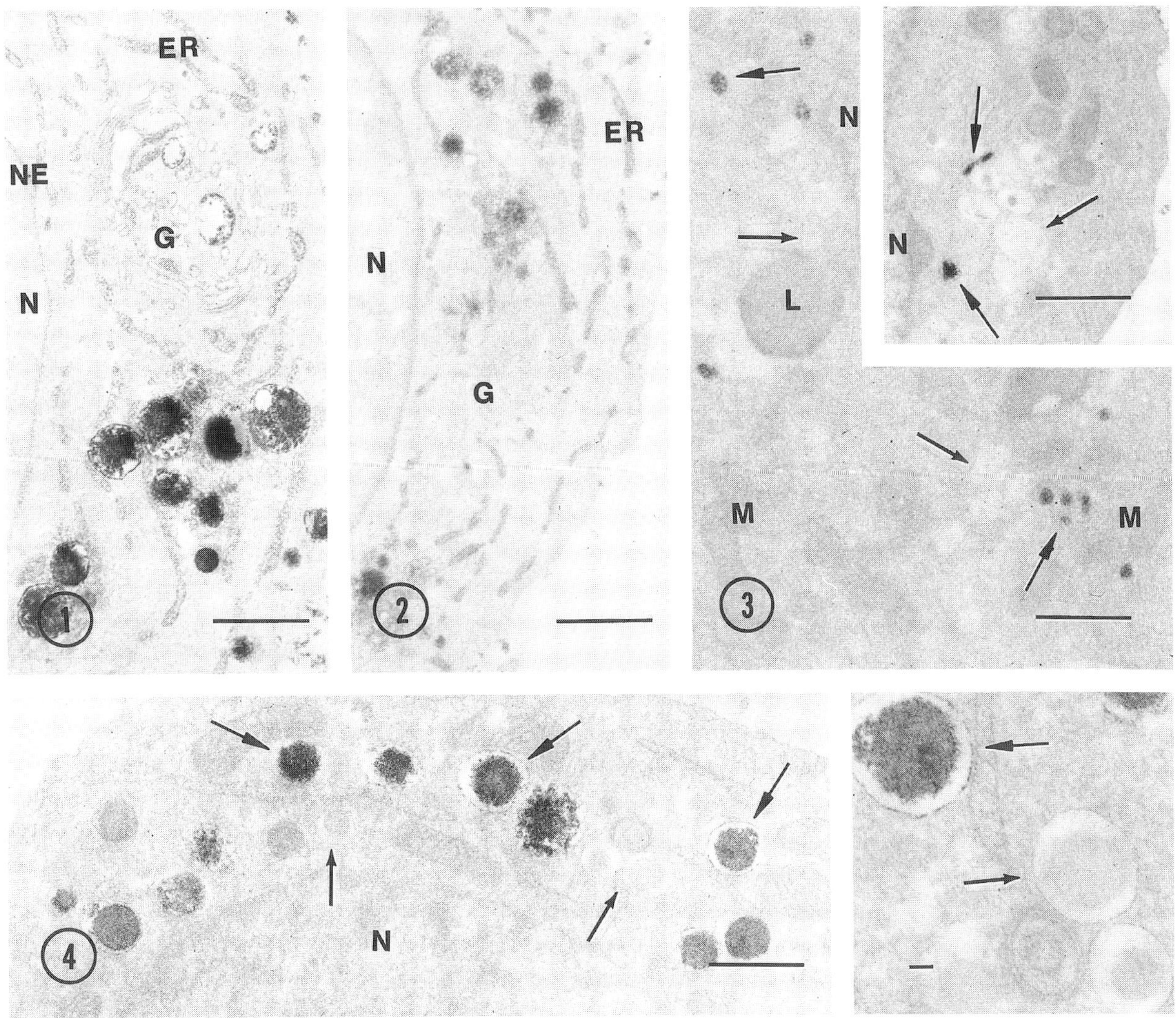

FIG. 1.--Neutral DAB 20mg/10mL(30min; 37°C), stains MPO in both large and small granules of HL60. Endoplasmic reticulum (ER) and nuclear envelope (NE) stain moderately. Golgi lamellae (G) stain weakly. Extracellular debris on the plasmalemma is intrinsically dense. Nucleus (N). Bar = 1 μm
FIG. 2.--Alkaline DAB 20mg/10mL(120min; 37°C) stains MPO and catalase in ER and granules. The similar ER staining observed at neutral pH (cf Fig 1) suggests that catalase conditions are not specific and stain MPO as well. Golgi (G) lacks staining. Nucleus (N). Bar = 1 μm
FIG. 3.--Alkaline DAB resulted in staining of small peroxisomes (large arrows) in A7 (MPO deficient). Large granules (small arrows) do not stain. Lipid droplets (L), Nucleus (N), mitochondria (M). Inset: Alkaline DAB-staining (no H_2O_2) persists in small granules of A7 (1arge arrows). Bar = 1 μm
FIG. 4.--Several large granules stain for catalase in this MPO deficient HP50 cell (1arge arrow). Reactive granules are larger than in A7(cf Fig. 3) consistent with catalase gene amplification in HP50. Inset: enlarges granules. Some granules do not stain (small arrow). Nucleus (N). Bar = 1 μm
Inset Bar = 0.1 μm

CYTOCHEMICAL LOCALIZATION OF CALCIUM IN MYOBLASTS OF THE CHICK EMBRYO MYOTOME.

R. González Santander*, M.V. Toledo Lobo*, F.J. Martínez Alonso*, G. Martínez Cuadrado*, M. González-Santander Martínez**, M. Monteagudo*.

* Morphological Sciences and Surgery Dept. Univ. of Alcalá de Henares. Madrid, Spain.
** Toxicology and Sanitary Legislation Dept. Univ. of Alcalá de Henares. Madrid, Spain.

Muscle fibers are derived from multinucleated myotubes which are themselves formed during embryonic development by the fusion of mononucleated myoblasts. Myoblast fusion results from a sequence of different and highly orchestrated stages demonstrated previously in vitro: recognition-alignment, adhesion and membrane fusion.[1] Like many other fusion systems, myoblast fusion is Ca^{2+} -dependent.[1] The role of Ca^{2+} is multiple since it is needed for muscle cell differentiation, for the alignment stage and it has also been demonstrated that Ca^{2+} influx precedes fusion increasing free intracellular Ca^{2+}.[2-3] It has been proposed that this increase in free intracellular Ca^{2+} may activate an enzimatic cascade which leads to membrane fusion.

The present study, using the K-pyroantimonate method, describes Ca^{2+} localization and storage in myoblasts before fusion for the first time, since this method had not been applied to skeletal muscle development studies before. Chick embryos from 51 to 108 h. of incubation (Hamburger and Hamilton stages 16 to 25) were used. Samples were fixed in 2.5% glutaraldehyde-2% K-pyroantimonate, postfixed in 1% OsO_4 - 2% K-pyroantimonate, dehydrated and embedded in araldite or Unicryl. This method is sensitive to Ca^{2+} at 10^{-6} M levels, since K-pyroantimonate forms precipitates with Ca^{2+}. Controls of the reaction were done treating sections with 10 mM EDTA. In all the sections treated with a Ca^{2+} chelator, EDTA, an evident destaining of the antimoniate precipitates was documented. Some samples were embedded in araldite by conventional TEM methods.

Using conventional TEM methods, some cisternae closely aposed and parallel aligned to the cell membrane were observed in myoblasts (Fig.1). The gap between the cisternae and the cell membrane was approximately 15 nm wide and bridged periodically by electron-dense structures (junctional feet). The cisternae were in continuity with endoplasmic reticulum. These structures were identified as "peripheral couplings", previously described by other authors mainly in relation with the triad formation.[4-5] It was not observed any structure similar to T-tubules, but the immature sarcomeres showed sarcoplasmic reticulum membranes.

Samples treated with K-pyroantimonate showed electrondense precipitates both in the basal membrane of the myotome and in the cytoplasm of the myoblasts. Precipitates were observed in the basal lamina that separates the myotome myoblasts from the adjacent mesenchyme. The significance of Ca^{2+} localization in the myotome basal lamina is unknown. However, it has been demonstrated the presence of "sialic acid calcium binding sites" along the basal lamina of the muscle fibres from adult muscle and from Duchenne and Becker muscular dystrophies[6]. Some precipitates periodically distributed at the cell periphery close to the cell membrane were observed in myoblasts before fusion (Fig.2). These round-oval shaped precipitates showed a regular contour. Electrondense lines were frequently observed starting from these precipitates (one line for each of the precipitates) and oriented towards the center of the cell (Fig.2). These results suggest that those precipitates observed in the myoblast may represent the Ca^{2+} storaged in the preripheral couplings. However, we have not been able to observe the pool of Ca^{2+} at the myoblast surface suggested by Przybylski and cols., using Ca^{2+} antagonist lanthanum[7], although this could be due to the different methods used. Our results and the data reported by other authors led us to suggest that the Ca^{2+} storaged at the peripheral couplings of myoblasts is mobilized before fusion, and this massive Ca^{2+} entry is the "Ca^{2+} influx" that iniciates the enzimatic cascade leading to membrane fusion and other methabolic events during early muscle development, as nitric oxide synthase activation which has been reported in relation with myoblast fusion[8].

References

1. J.O. Wakelam, Biochem. J. 228 (1985) 1.
2. J.D. David et al., Dev. Biol. 82 (1981) 297.
3. R.J. Przybylski et al., In Vitro Cell Dev. Biol. 25 (1989) 830.
4. B.E. Flucher, Dev. Biol. 154 (1992) 245.
5. C. Franzini-Armstrong and A.O. Jorgensen, Annu. Rev. Physiol. 56 (1994) 509.
6. M. Moggio et al., Muscle Nerve 15 (1992) 325.
7. R.J. Przybylski et al., Cell Calcium 15 (1994) 132.
8. K.H. Lee, J. Biol. Chem. 269 (1994) 14371.

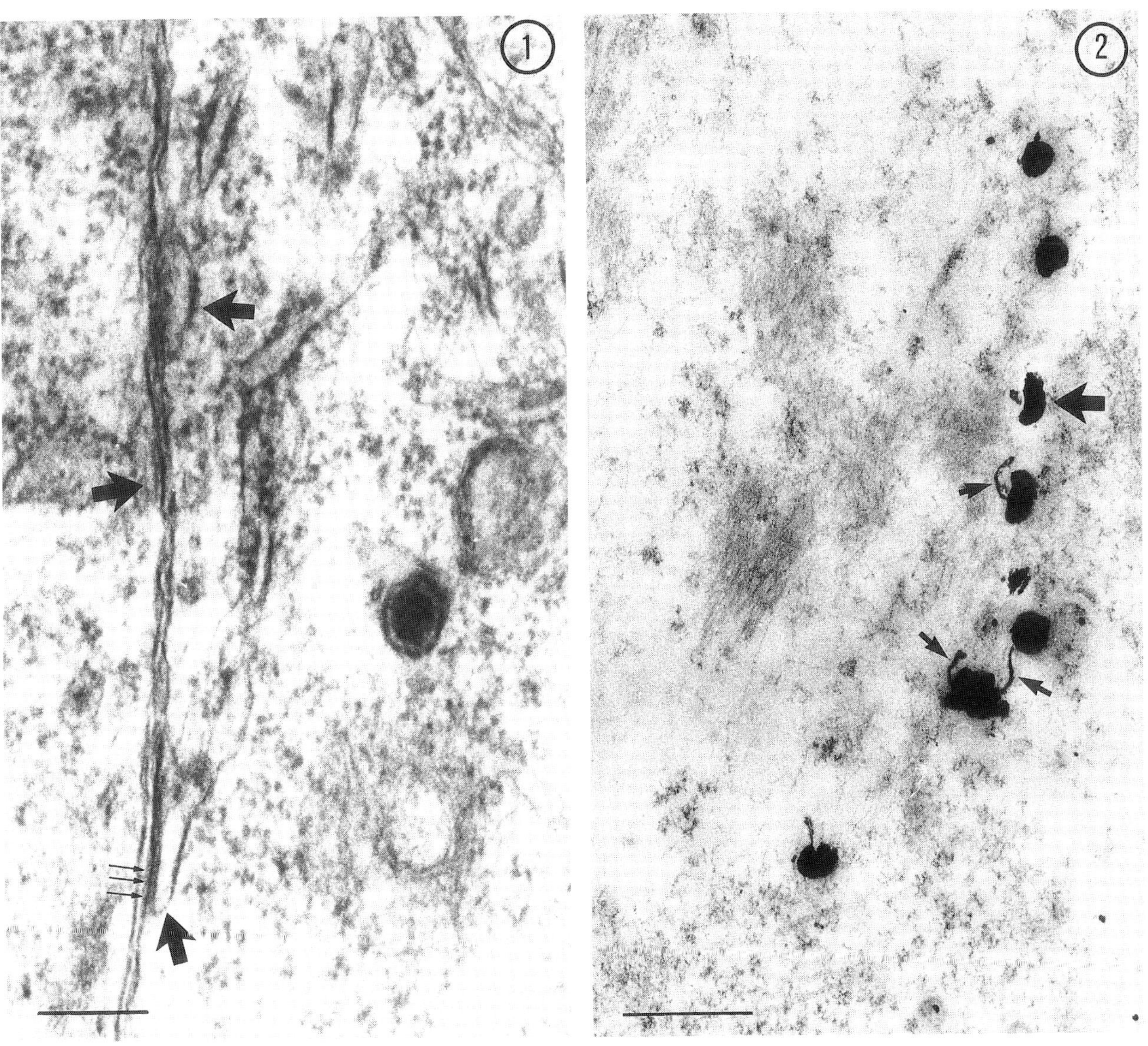

Fig. 1.- Myoblasts in the myotome of 18 H.H. stage chicken embryos. Peripheral couplings (big arrow) and junctional feet (small arrows). Fixation: glutaraldehyde. Scale bar: 0.25 μm. (12,500 X)
Fig. 2.- Myoblasts in the myotome of 18 H.H. stage chicken embryos. Ca^{2+} distribution in myoblasts. Electrondense precipitates related to plasma membrane (big arrows). Electrondense lines (small arrows). Fixation: K-pyroantimoniate. Scale bar: 0.5 μm (8,000 X).

A NEW MICROSCOPY METHOD TO ASSESS MASS TRANSFER DURING FREEZING IN RAT LIVER TISSUE SLICES.

D. Raha, P. Pazhayannur, and J. C. Bischof·

University of Minnesota. Department of Mechanical Engineering. Minneapolis, MN 55455

In cryofixation of biological samples it is typically desirable to minimize or avoid ice crystals in the sample by one of several rapid cooling techniques. Using these techniques freezing rates of 1,000 -10,000 °C/s are commonly achieved at a cooling boundary. At these rates, cellular dehydration is typically absent, and intracellular ice crystals are so small as to be considered negligible within an "ice-free" zone typically ranging 5-25 μm into the sample from the cooling boundary [1]. However, during medical applications of freezing including cryopreservation as well as cryosurgery, the rate of freezing is orders of magnitude less than the idealized freezing procedures used in cryofixation [2,3]. During slower freezing at rates between 1 - 100 °C/min, significant water transport between the cells and the extracellular/vascular space of a tissue can occur. This dehydration affects how and where the ice will form in the tissue which in turn can influence the post-freeze viability and therefore the overall success of a medical freezing treatment. The purpose of this work is to obtain a more quantitative understanding and prediction of the biophysics of water transport in liver tissue during freezing.

Experiments have shown that significant mass transfer occurs at slower rates of freezing in liver (1-22 °C/min) [2,3]. However, there is a lack of dynamic experimental data available to model this phenomenon. In this work we use two separate techniques to extract quantitative data on water transport in liver tissue of male Sprague Dawley (175-250 g) rats during freezing. The first method incorporates a two-step cooling protocol where the first step involves controlled rate freezing (5 °C/min) on a directional solidification device described elsewhere [2,3], followed by copper block slam freezing at intermediate high subzero temperatures (-4, -6, -8, -10, -20 °C) where dehydration is dynamically occurring. Samples were freeze substituted, embedded, sectioned, and stained by standard protocols [1] prior to digitized photomicrography of the samples. Typical results of the tissue morphology at different intermediate temperatures is shown in Figure 1. Using NIH Image^{TM} image analysis package (NIH, Bethesda, MD), the darkly stained cellular space in the samples was quantified and graphed as dark circles (•) in Figure 2 (Error bars are the standard deviation of the data).

Mass transfer from cellular to extracellular/vascular space in a two-compartment Krogh cylinder model is assumed to adequately predict mass transfer within the liver [4,5]. Non-linear regression analysis of this model to the experimental data for freezing at 5 °C/min is shown as the solid line (—) in Figure 2. The permeability of the membrane, L_p , is given as

$$L_p = L_{pg} \; exp \; [- E_{Lp}/R(1/T_R - 1/T)]$$

Where L_{pg} is the permeability of the membrane at the reference temperature (273.15 K) T_R , E_{Lp} is the apparent activation energy for the process, and R is the gas constant [4]. Our regression analysis predicts $L_{pg} = 1.2$ μm/min-atm and $E_{Lp} = 80$ kcal/mole, with chi-squared variance of < 0.0085. These values for the biophysical parameters are within the range of reported values for isolated cells of rat liver [6,7].

A second technique using a DSC (differential scanning calorimeter) (Perkin Elmer, DSC 7) was used to monitor the amount of heat released/weight of rat liver tissue as a function of temperature. The tissue was allowed to equilibrate at different temperatures (-4, -6, -8, -10 and -20 °C) in the presence of extracellular ice which was seeded by an ice nucleating strain of *Pseudomonas Syringae* bacteria (ATCC, Rockville, MD). The heat released upon further cooling at 5 °C/min was measured and plotted as shown in Figure 3 for typical experiments. Notice how the amount of heat released per mg of tissue decreases as the temperature decreases. No further heat release is detected after equilibration below -10 °C. By integrating the area under each heat release curve, the amount of total energy released is obtained, ΔE. The nonfrozen cellular volume corresponding to this energy release can then be calculated as V = ΔE/ρ L, where ρ is the density of the water, and L is the heat of fusion of water. The volumes calculated from this technique are shown in Figure 2 as open circles (o). This method yields data which lies along the equilibrium dehydration curve (---) in Figure 2 which represents the cell volume as a function of temperature under infinitely slow cooling conditions. This curve can be calculated by equating the osmotic pressure within the unfrozen cell to the osmotic pressure of the rejected saline in the unfrozen fraction of the extracellular space at any subzero temperature. Notice that both the experimental data as well as the modeled curves for the equilibrium case are less than the corresponding values for the constant cooling curve at 5 °C/min.

Future plans include using these parameters in the quantitative prediction of biophysical response throughout a tissue exposed to boundary cooling during applications such as cryosurgery, cryopreservation as well as cryofixation.

Proc. Microscopy and Microanalysis 1995, edited by G.W. Bailey, M.H. Ellisman, R.A. Hennigar, and N.J. Zaluzec

References:
1. Echlin, P. *Low-Temperature Microscopy and Analysis.* Plenum, N.Y. (1992).
2. Rubinsky, B. et al., *Cryobiology* 27, 85-97 (1990).
3. Bischof, J.C. et al., *Cryobiology* 30, 482-492 (1993).
4. Rubinsky, B. and Pegg, D.E., *Proc. R. Soc. Lond.* B234, 343-358 (1988).
5. Bischof, J.C. and Rubinsky, B. *ASME J.H.T.* 115, 1029-1035 (1993).
6. Toner, M. et al., *AICHE J.* 38, 1512-1522 (1992).
7. Yarmush et al. , *Ann. N.Y. Acad. Sci.* 665:238-252 (1992).
8. This work was supported by NSF/CTS-9410004.

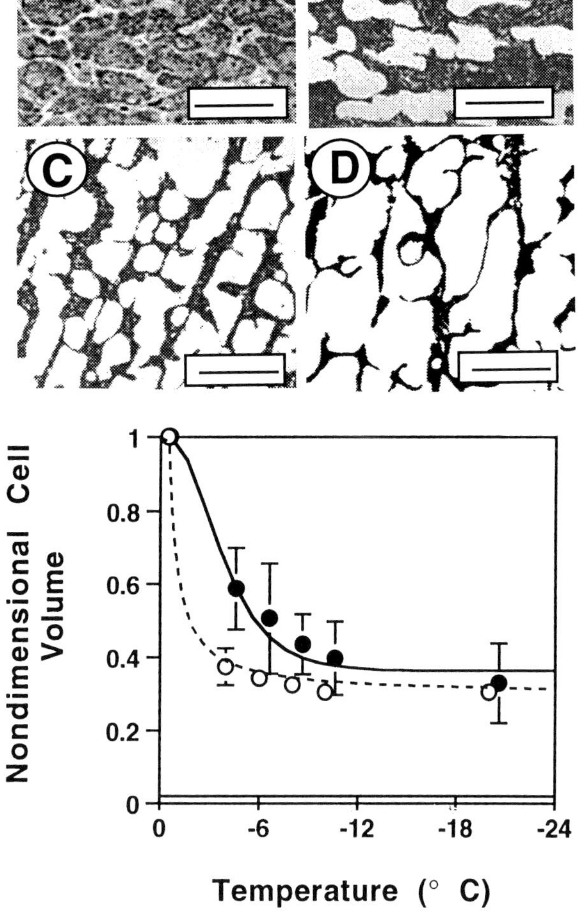

Figure 1. Two-step freezing response in rat liver tissue . The first step involved cooling the tissue at 5 °C/min to A.) 0°C (no freezing), B.) -4 , C.) -8 °C, and D.) -20°C followed by a second slam freezing step. Scale bar 50 μm.

Figure 2. Temperature vs. volumetric response of cells in rat liver tissue (a) during freezing at 5 °C/min with experimental data (•), and Krogh cylinder model (—), and (b) during equilibrium freezing with experimental data (o) and equilibrium freezing model (----).

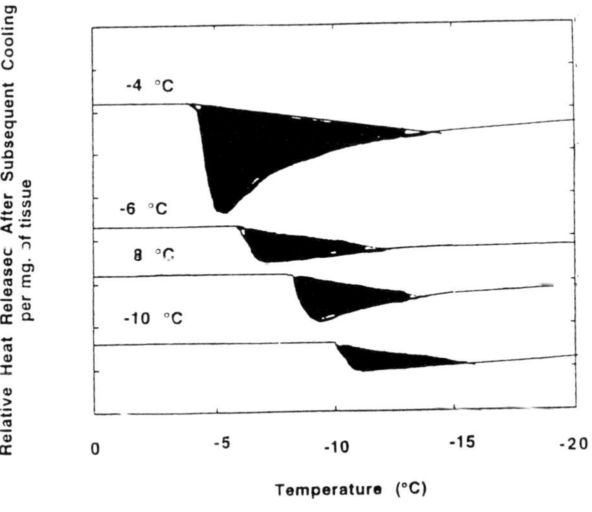

Figure 3. DSC thermograms of heat release/mg of rat liver tissue cooled at 5 °C/min from A.) -4, B.) -6, C.) -8 and D.) -10 °C after equilibration with extracellular ice.

LOW TEMPERATURE SEM COMPARISONS OF ADJACENT FREEZE-FRACTURED AND FREEZE-ETCHED AREAS ON FROZEN, HYDRATED SAMPLES

William P. Wergin*, Eric F. Erbe* and Robert W. Yaklich**

*Electron Microscopy Lab & **Soybean and Alfalfa Research Lab; USDA-ARS; Beltsville, MD 20705

Most biological samples contain 70-95% water, consequently cryofixation and freeze-fracturing result in relatively smooth surfaces that exhibit few structural details. Freeze-etching, a technique that solved this problem, was initially developed for TEM observations of virus particles by Steere nearly 40 years ago[1]. The technique, which sublimes water-ice from the surface of a fractured sample, produces surface topography that corresponds to the structural components on the freeze-etched face. This technique was further enhanced by recovering the complementary halves of a fractured sample, etching one of the surfaces and then comparing the complementary replicas from the freeze-fractured and freeze-etched faces[2]. Recently, similar techniques were used on frozen, hydrated samples to examine complementary halves of freeze-fractured, freeze-etched specimens by low temperature SEM[3]. Imaging complementary images of frozen, hydrated specimens in the SEM was faster than imaging complementary replicas in the TEM, however the procedure required specialized holders and was technically demanding.

To simplify comparisons of freeze-fracture, freeze-etch images, samples were frozen, fractured and etched in the prechamber of an Oxford CT 1500 HF Cryotrans system that was attached to a Hitachi S-4100 FESEM. Next, a scalpel blade was used to refracture one-half of the sample. Then the entire sample was coated and inserted onto the cold stage of the microscope.

Observations of adjacent freeze-fractured and freeze-etched areas provide direct comparison of these two techniques in the same field of view. Comparisons of the two images in cheddar cheese (**Fig. 1**) and pizza dough (**Fig. 2**) quickly identify the relative location and the extent of hydration in the sample and easily distinguish the free water from other particulate components such as lipid droplets and protein bodies. Comparisons of images obtained from a cherry tree twig (**Figs. 3 & 4**) indicate that under winter conditions free water is generally confined to thin layers around the plasma membrane, tonoplast and other cellular organelles. The vacuoles appear to be cryoprotected by natural sugars.

These results demonstrate that this procedure can be used to distinguish air from water as well as to locate the position of free water in a sample. In addition the procedure can be used to determine the effects of cryoprotectants and to ascertain the affects of preparation variables such as various cryogens and the ideal times and temperatures that are required for optimum etching.

References:

1. R.L. Steere, J Biophs. Biochem. Cytol. **3**(1957)45.
2. R.L. Steere and E. F. Erbe, J Microsc. **117**(1979)211.
3. W.P. Wergin and E. F. Erbe, Scanning **14**(1992)17.

Figure 1. Cheddar cheese. Bar = 5μm. (All Figs. have been mounted with the freeze-fractured image on the upper left and the freeze-etched portion on the lower right.)

Figure 2. Commercial pizza dough. Bar = 15μm.

Figure 3. Cherry twig (winter collection). Bar = 15μm.

Figure 4. Single cell from cherry twig. Bar = 5μm.

Proc. Microscopy and Microanalysis 1995, edited by G.W. Bailey, M.H. Ellisman, R.A. Hennigar, and N.J. Zaluzec

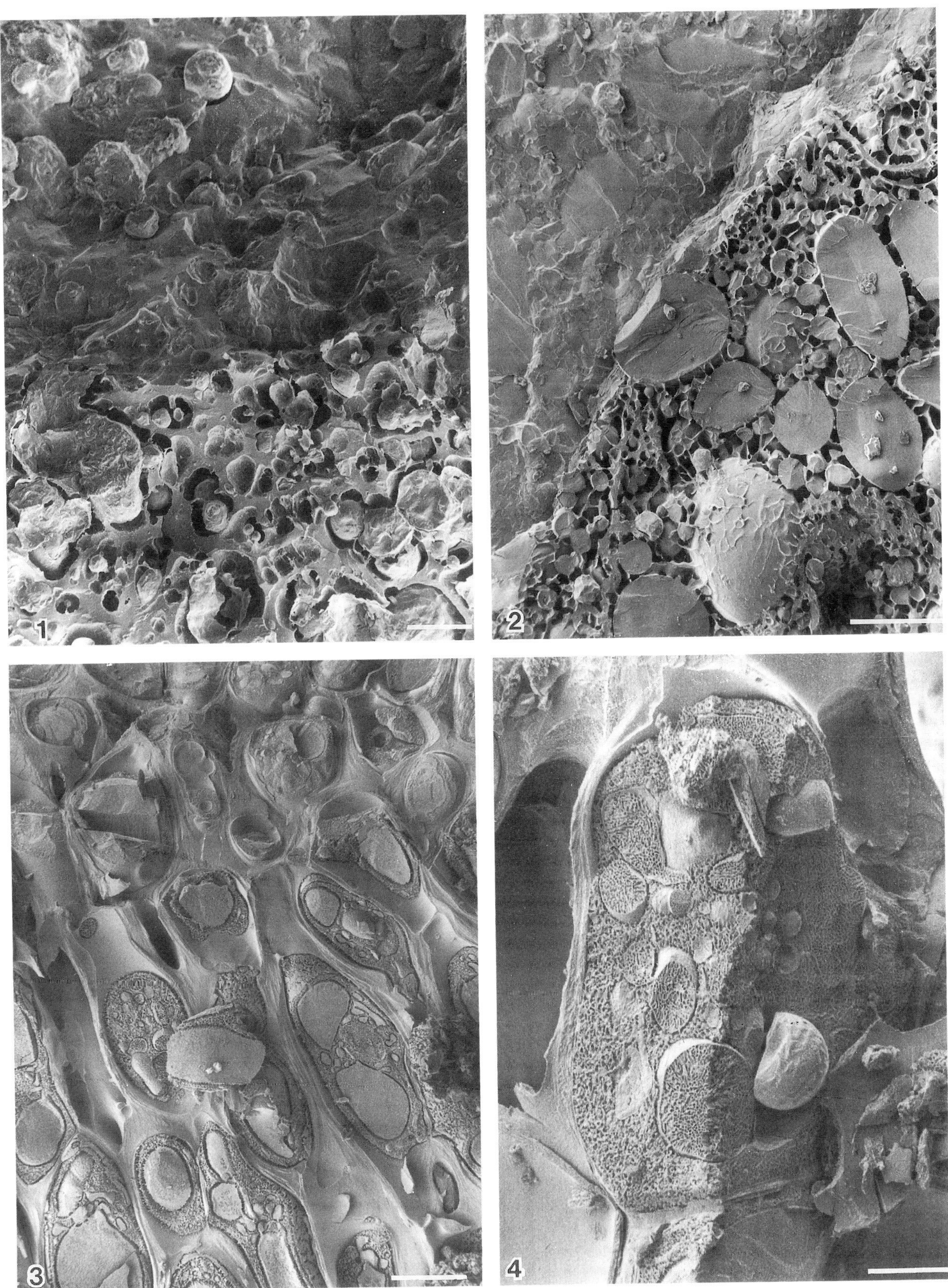

AN IMPROVED CRYOFIXATION TECHNIQUE FOR OPTIMAL ORIENTATION OF *Caenorhabditis elegans* FOR CRYOSECTIONING AND EPXMA

Edward Roberts,* Nancy Wallace,* Jonathan Freedman,** and Ann LeFurgey**,***

*Division of Physiology, Department of Cell Biology, Duke University Medical Center, Durham, NC 27710; **School of the Environment, Duke University, Durham, NC 27706; and ***Department of Veterans Affairs Medical Center, Durham, NC 27705

INTRODUCTION: Plunge freezing into liquid cryogens has been frequently employed with success for cryofixation of biological specimens $<1mm^3$ in size; however, positioning of cells or organisms with a precise orientation for subsequent cryosectioning is difficult. In these studies, a metal mirror cryofixation device, e.g. cryogun, has been tested to determine its applicability for cryofixation of small organisms such as nematodes or larval invertebrates with specific positioning. The objective of the experiments was to determine the optimum technique for rapid cryofixation and cryosectioning of the intestinal tract and lumen of the nematode *Caenorhabditis elegans*, prior to electron probe x-ray microanalysis (EPXMA) of subcellular elemental distribution.

METHODS: Freezing and sectioning quality obtained using plunge freezing into -190°C liquid propane (Figure 1a)[2] was compared with that obtained with -196°C metal mirror fixation using the cryogun (PS 1000, Delaware Diamond Knives, Wilmington, DE)(Figure 1b)[1]. *C. elegans* were grown in culture to the adult stage, and individual nematodes were placed on either (1) a wooden specimen stub compatible with the plunge freezing device (Figure 2a) or (2) a cryogun specimen mount (Figure 2b). In both instances the specimen support surface was coated with a small mound of 20% gelatin containing carbon black.[3] For cryosectioning, the plunge frozen specimens were inserted into the microtome stub sample holder and cryosectioned without any possibility of further orientation. However, the cryogun specimen, after alignment in the parallel jaws of the microtome specimen holder, allowed precise orientation of the organism for transverse, as well as anterior/posterior positioning. In addition the gelatin disk containing the gun frozen specimen was trimmed to achieve a block face of optimal dimensions (width, ~25-50 μm). Frozen, hydrated cryosections on carbon-coated nickel grids were then transferred to a vacuum (10^{-3} Torr), freeze dried, and coated with carbon. All cryosections were examined in a conventional transmission electron microscope (100CX-II, JEOL) at an operating voltage of 80 kV and magnifications of 100 to 10,000 X. Selected sections were then examined by STEM and EPXMA as previously described.[1,2]

RESULTS: The quality of freezing demonstrated by the metal mirror device, in these experiments, appears superior to the plunge freezing method especially in those regions of the specimen where first contact with the copper anvil occurred (Figure 3b). Because of the physical support provided by the frozen gelatin, ribbons of sections were consistently obtained from cryogun-frozen specimens; the ribbons remained intact during transfer from the knife edge to the grid, and adhered to the carbon substrate during freeze drying. Only fragments of sections were obtained with the plunge frozen specimens, with limited areas of good cryofixation (Figure 3a). EPXMA images were obtained from both types of specimens with appropriate maintenance of cell K:Na ratio (Figure 4). The structural definition of organelles and cell types was superior in the metal mirror frozen specimens, and the extent of the sections available for mapping of complete organism regions was much larger.

IN SUMMARY, this improved cryofixation technique for *C. elegans*, utilizing a gelatin specimen support along with a metal mirror device, provides several advantages over the standard plunge freezing method. These include specimen orientation and control of block face size and shape, thereby aiding the sectioning process as well as providing optimal sections for electron probe x-ray mapping.

REFERENCES:

1. Freudenrich et al., J. Struct. Biol., 112 (1994) 173.
2. LeFurgey et al., Molec. Biochem. Parasitology, 40 (1990) 77.
3. Nassar et al., SEM, I (1986) 309.
4. Funded by ONR N00014-93-F-0067 and -94-1-0818, NIH 1S1ORR06692, and NSF BIR 9106607.

Proc. Microscopy and Microanalysis 1995, edited by G.W. Bailey, M.H. Ellisman, R.A. Hennigar, and N.J. Zaluzec
Copyright © 1995 MSA. Published by Jones and Begell Publishing, 79 Madison Ave., New York, NY 10016

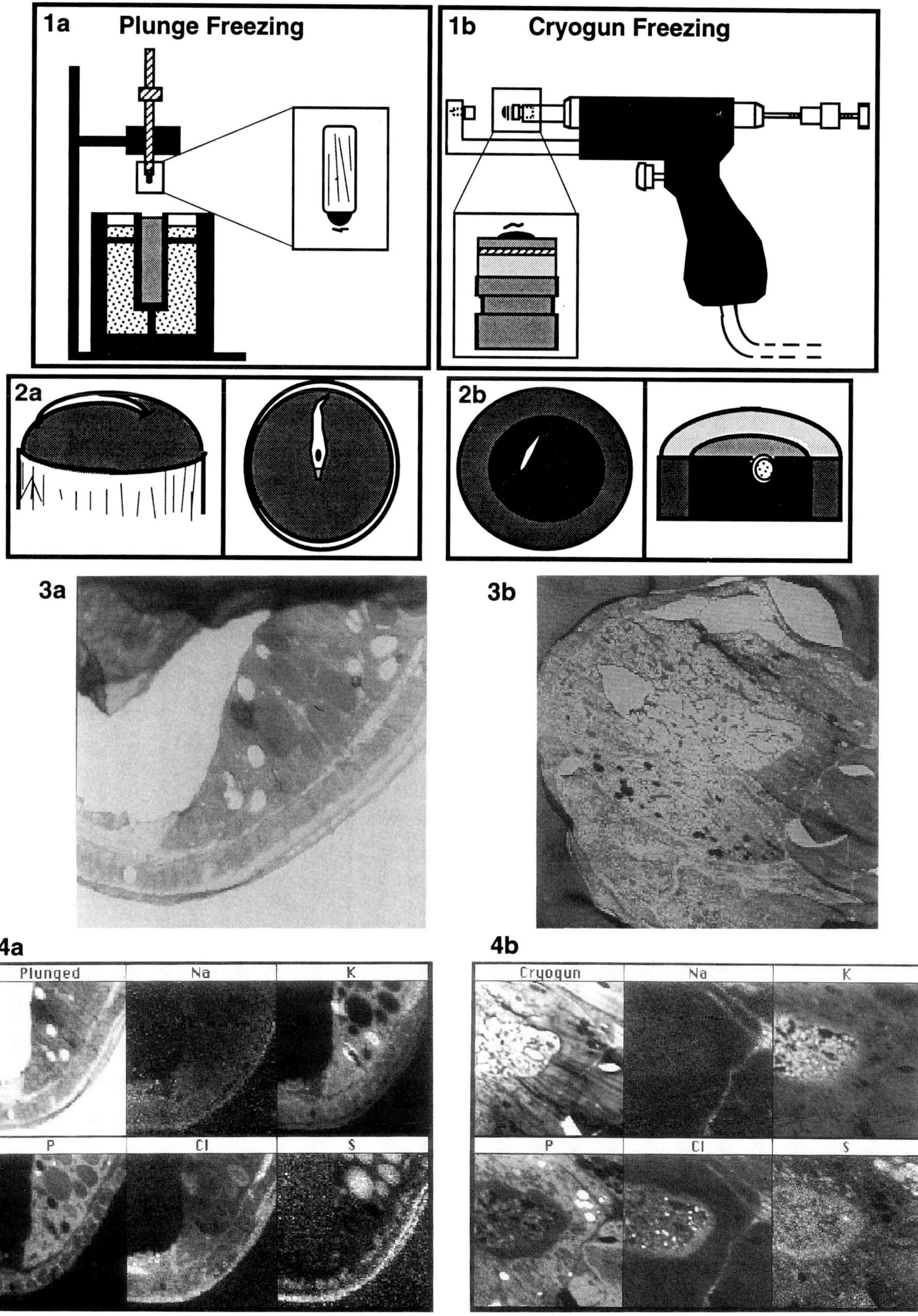

1a Plunge Freezing
1b Cryogun Freezing
2a
2b
3a
3b
4a
Plunged Na K P Cl S
4b
Cryogun Na K P Cl S

ADVANTAGES OF LOW TEMPERATURE SCANNING ELECTRON MICROSCOPY FOR BACTERIAL STUDIES

Stéphane Roy *†, Isabelle Babic*, Alley E. Watada* and William P. Wergin†

*Horticultural Crops Quality Lab and †Electron Microscopy Lab, USDA-ARS, Beltsville, MD 20705.

The application of transmission electron microscopy (TEM) has greatly increased our understanding of structure-function relationships in bacteriology[1]. However, to achieve further advancements investigators are seeking preparation procedures that would avoid the artifacts associated with conventional chemical fixation, dehydration and critical point drying or embedding[2]. In our laboratory a field emission scanning electron microscope (SEM) was recently equipped with a cold stage. This combination of techniques, referred to as low temperature (LT) SEM, allowed us to examine frozen, fully hydrated biological specimens[3]. The present investigation images bacteria that were cryofixed for LTSEM observations and then freeze-substituted for TEM observations. In addition an attempt was made to culture the samples that had been cryofixed and observed with LTSEM so that taxonomic identification and further microscopic observations could be made.

Bacteria used in this study were isolated from spinach leaves (variety New Jersey). LTSEM observations of cryofixed samples were performed on a Hitachi S-4100 field emission SEM equipped with an Oxford CT-1500HF Cryotrans System. After observation in the SEM, frozen hydrated samples were either freeze-substituted with 2% (w/v) osmium tetroxide in acetone at -90°C or transferred to sterile culture plates as summarized in **Fig. 1**. Our results indicated that cryofixation and LTSEM observation: i) preserved the surface glycocalyx of the bacteria (**Fig. 2**); ii) permitted recovery of the sample for freeze-substitution and TEM observation (**Fig. 3**) and; iii) did not appear to affect the viability of the bacteria, which could be cultured after LTSEM observation (**Fig. 4**).

In conclusion, bacteria in infected tissues that were observed with the LTSEM remained viable. They could be subsequently processed for TEM observations to observe the fine structural details of thinly sectioned material and could also be cultured for further studies. These results indicate that under our conditions LTSEM is non destructive technique that has advantages for in bacteriological investigations.

1. J. W. Costerton, Ann. Rev. Microbiol. **33**(1979)459.
2. S.P. Collins et al., Microscopy Res. Tech. **25**(1993)398.
3. W.P. Wergin and E.F. Erbe, Scanning Microsc. **5**(1991)927.

Figure 1. Summary diagram indicating the possibilities for processing and observing bacteria.
Figure 2. LTSEM image of frozen-hydrated palisade parenchyma cells (C) in a spinach leaf. The bacteria cells are embedded in the glycocalyx (arrows). Bar = 30μm.
Figure 3. Thin section of parenchyma cell and bacteria (B) that were freeze-substituted after cryofixation and LTSEM examination. The fine structure of the plant cell and the bacteria appear well preserved. Pl = Chloroplast; CW = Cell Wall; ER = Endoplasmic Reticulum. Bar = 0.37μm.
Figure 4. LTSEM of a freeze-fractured bacterial colony that was established by isolating and culturing bacteria from infected tissue that had been previously observed in the LTSEM. Bar = 16.7μm.

Proc. Microscopy and Microanalysis 1995, edited by G.W. Bailey, M.H. Ellisman, R.A. Hennigar, and N.J. Zaluzec

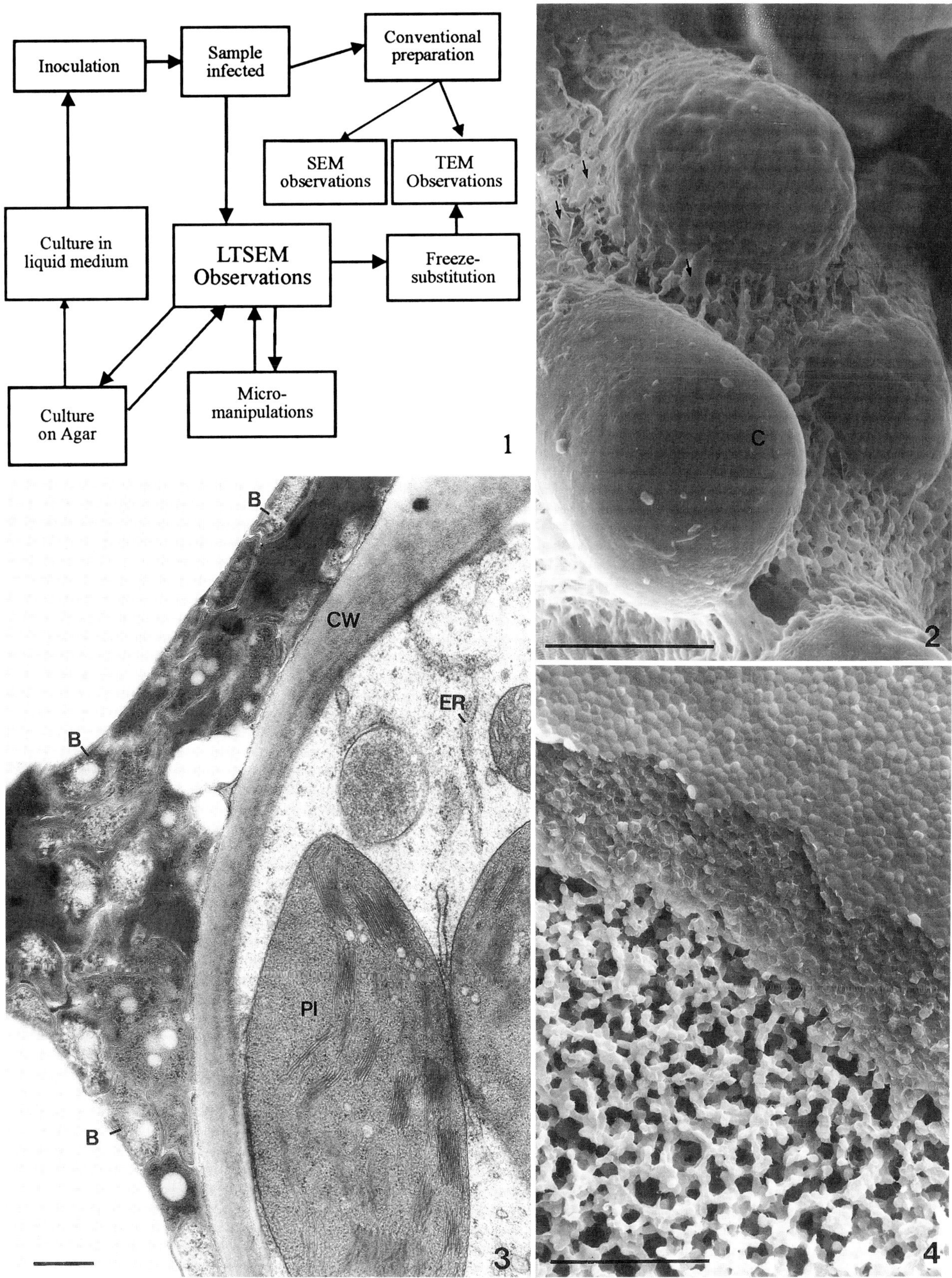

Inoculation
Sample infected
Conventional preparation
SEM observations
TEM Observations
Culture in liquid medium
LTSEM Observations
Freeze-substitution
Culture on Agar
Micro-manipulations
1
C
2
B
CW
ER
B
Pl
B
3
4

IMPROVEMENT IN LYSINE EFFECT ON ALDEHYDE-RUTHENIUM RED PRESERVATION AND STAINING OF THE STAPHYLOCOCCAL GLYCOCALYX

Theresa A. Fassel[1], Paul E. Mozdziak[1,2], James R. Sanger[3], and Charles E. Edmiston[4]

[1]Muscle Biology Laboratory, and [2]Department of Anatomy, University of Wisconsin-Madison, Madison, WI 53706, [3]Department of Plastic and Reconstructive Surgery, and [4]Department of Surgery, Medical College of Wisconsin, Milwaukee, WI 53226

The cationic reagent ruthenium red (RR) has been used to enhance ultrastructural preservation and staining of the bacterial glycocalyx. Lysine was needed in prefixation to preserve the polysaccharide glycocalyx material for the gram-positive, coagulase-negative staphylococci species. The inclusion of lysine in the prefixative of glutaraldehyde (GA)-RR enhanced observation of elaborate and extensive glycocalyx material[1,2]. However, prefixation in 75 mM lysine in GA-RR is restricted to 20 minutes because gelling or solidification frequently occurs at longer intervals resulting in loss of sample. To improve the utility of this approach, the effect of the Karnovsky ratio of 2% paraformaldehyde to 2.5% GA was investigated.

Cells of gram-positive, coagulase-negative staphylococci species, *Staphylococcus epidermidis* RP62, a polysaccharide producer, and *Staphylococcus epidermidis* M187-SN3, a polysaccharide negative mutant, were recovered from frozen storage and grown on blood agar plates. After 24 hours, cells were transferred to trypticase soy broth for 18 hours, and incubated at 35°C. Half of the cells were prefixed in 75 mM L-lysine, 0.075% RR, 2.5% GA in 0.1 M sodium cacodylate pH 7.2 for 20 minutes. The remaining cells were prefixed with 2% paraformaldehyde added to 75 mM L-lysine, 0.075% RR, 2.5% GA in 0.1 M sodium cacodylate buffer (20 minutes). Some samples were maintained in Karnovsky-RR-lysine solution overnight. Fixation was in 0.075% RR, 2.5% GA or Karnovsky-RR in 0.1 M sodium cacodylate pH 7.2 for 2 hours, followed by buffer washes. Control fixation excluded RR or lysine. For all samples, postfixation was in 1% OsO_4 in cacodylate buffer for 2 hours. Samples were washed in buffer, dehydrated in a graded ethanol series and infiltrated in LR White resin. Thin sections were post-stained in 25% ethanolic 2% uranyl acetate and Reynolds' lead citrate.

Polysaccharide glycocalyx material was not seen by traditional GA fixation for *Staphylococcus epidermidis* RP62 (FIG. 1), nor by aldehyde-RR-lysine for *Staphylococcus epidermidis* M187-SN3 (FIG. 2). Elaborate glycocalyx material was observed for *Staphylococcus epidermidis* RP62 (FIG. 3) by GA-RR-lysine processing. By Karnovsky-RR-lysine (FIG. 4) and Karnovsky-RR-lysine acetate salt (FIG. 5), extensive glycocalyx material was also observed. The latter samples when fixed in the presence of lysine overnight showed extensive glycocalyx material (FIG. 6), were almost always recovered and rarely gelled. Thus, inclusion of paraformaldehyde at the Karnovsky ratio with GA, RR and lysine improves the utility of this method.

References

1. Fassel, T.A. et al., 1992 *Cells and Materials* 2: 37-48.
2. Fassel, T.A. et al., 1993 *Cells and Materials* 3: 327-336.

Proc. Microscopy and Microanalysis 1995, edited by G.W. Bailey, M.H. Ellisman, R.A. Hennigar, and N.J. Zaluzec

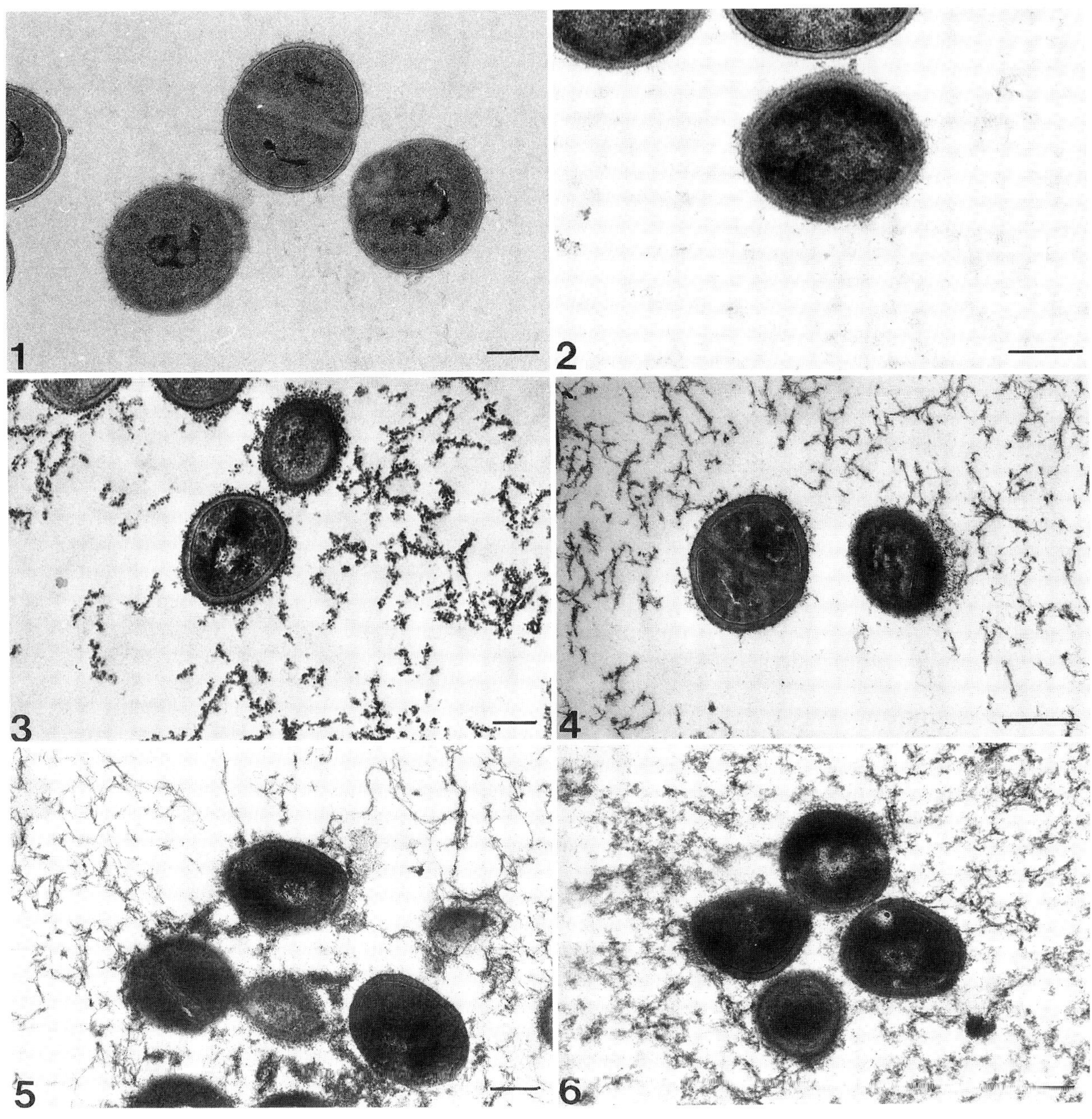

FIGS. 1. and 2.- *Staphylococcus epidermidis* RP62 by GA fixation (FIG. 1) and *Staphylococcus epidermidis* M187-SN3 by aldehyde-RR-lysine (FIG. 2) do not show preservation of glycocalyx material. FIGS. 3. and 4.- *Staphylococcus epidermidis* RP62 reveals elaborate glycocalyx by GA-RR-lysine (FIG. 3) and Karnovsky-RR-lysine (FIG. 4). FIGS. 5. and 6.- For *Staphylococcus epidermidis* RP62, elaborate glycocalyx is also preserved by Karnovsky-RR-lysine acetate salt at 20 minutes (FIG. 5) and overnight (FIG. 6). Magnification bars are 0.25 μm .

MICROVASCULATURE AND STRUCTURE OF HEMAL LYMPH NODES IN THE WALL OF THE RABBIT BLADDER: A VASCULAR CORROSION CASTING AND EM STUDY

F.E. Hossler[*], M.I. McKamey[*], and F.C. Monson[**]

[*]Department of Anatomy and Cell Biology, J.H. Quillen College of Medicine, East Tennessee State University, Johnson City, TN 37614
[**]Division of Urology, University of Pennsylvania School of Medicine, Philadelphia, PA 19104

A comprehensive study of the microvasculature of the normal rabbit bladder[1], revealed unusual "capillary glomeruli" along the lateral walls. Here they are characterized as hemal lymph nodes using light microscopy, SEM, TEM, ink injection, and vascular casting.

Bladders were perfused via a cannula placed in the abdominal aorta with either 2% glutaraldehyde in 0.1M cacodylate buffer (pH 7.4) for fixation, 10% India ink in 0.9% saline and 0.1M phosphate (pH 7.4) for vessel tracing, or resin (Mercox:methylmethacrylate: catalyst, 4:1:0.3; Ladd Research Industries) for vascular corrosion casting. Infusion pressure was 100mm Hg. Fixed tissue was sectioned from epon-araldyte resin, and stained with toluidine blue for light microscopy, and lead and uranium for TEM. Ink injected tissue was photographed directly from saline-filled bladders illuminated from below. Resin-filled tissue was macerated in 5% KOH and distilled water. Casts were critical point dried, sputter coated with gold-palladium, and examined by routine SEM at 10 KV.

Hemal lymph nodes were located along the lateral surfaces of the basal half of the bladder in close association with the main vesicular vessels (Figs. 1 & 2). Nodes were rarely seen in the apical half of the bladder. They received blood from tertiary branches of the main vesicular vessels, although direct vascular connections could not always be traced. Numbers ranged from about 2 to 12 nodes for each side of the bladder although all were likely not identified. Nodes measured about 1 to 3 mm in length, consisted of 1 to 5 lobes (best seen with vascular casts; Figs. 3 & 4), and exhibited prominent subcapsular sinuses (Fig.5). Nodes contained blood and lymphatic vessels (Fig.5). To the best of our knowledge hemal lymph nodes have not previously been described in the bladder wall using vascular corrosion casting. These nodes may serve to monitor and combat urothelial infections, especially common in the female bladder (e.g. interstitial cystitis[2]). Studies of the immune properties of the nodal lymphocytes is under way.

References

1. F.E. Hossler and F.C. Monson, Proc. 51st Ann. Mtg. of Micros. Soc. Amer. p.238, 1993.
2. F.E. Hossler, L. Sun, and F.C. Monson, Interstit. Cystitis. Sympos., NIH, Bethesda, MD, p.95, 1995.
3. This research is supported by NIH: 1-RO1-DK-47563-1.

Proc. Microscopy and Microanalysis 1995, edited by G.W. Bailey, M.H. Ellisman, R.A. Hennigar, and N.J. Zaluzec
Copyright © 1995 MSA. Published by Jones and Begell Publishing, 79 Madison Ave., New York, NY 10016

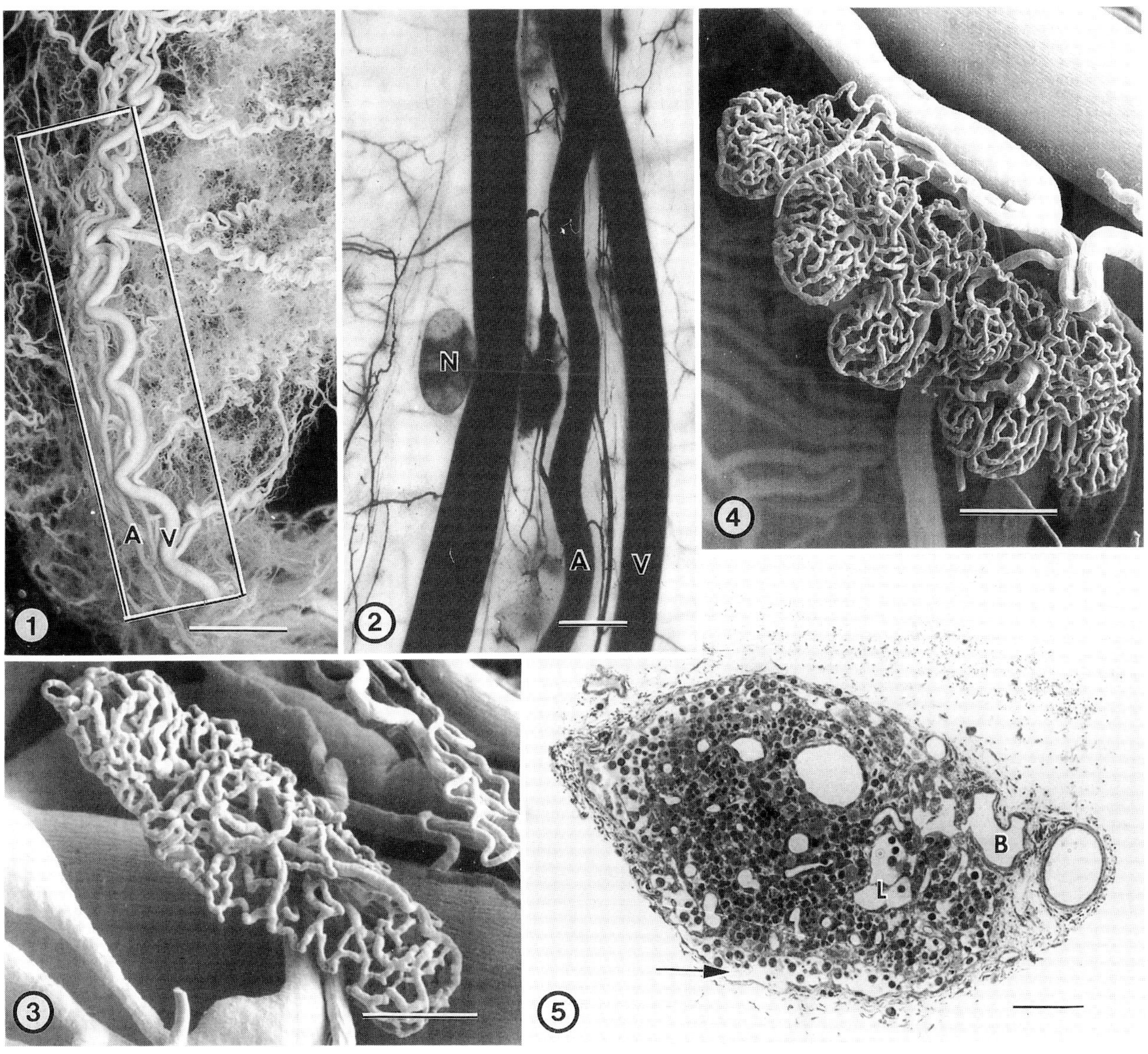

FIG.1.- Vascular cast of lateral bladder wall. Box defines general region where hemal nodes are found. A and V, vesicular artery and vein. Bar = 500μm.

FIG.2.- Hemal node (N) in ink injected bladder. A and V, vesicular artery and vein. Bar = 150μm.

Fig. 3. - Vascular corrosion cast of hemal node with 1 lobe. Bar = 100μm.

FIG. 4. - Vascular corrosion cast of hemal node with 5 lobes. Bar = 200μm.

Fig. 5.- Hemal node. Arrow, subcapsular sinus; B, blood vessel; L, lymph vessel. Bar = 20μm.

METHODS FOR PREPARATION OF CRIBELLATE SPIDER SILK FIBRILS FOR TRANSMISSION ELECTRON MICROSCOPY

B. Cutler.

Electron Microscopy Laboratory, and Department of Entomology, University of Kansas, Lawrence, KS 66045-2106

Spider silk has been examined by transmission electron microscopy (TEM) for over 50 years.[1,2] A later, more detailed study produced higher resolution images and included a discussion of fixation with osmium tetroxide (OsO_4) vapor.[3] The fixation resulted in only minor effects compared to unfixed specimens. Cribellate spiders produce a nonviscous adhesive silk that has as a significant component very fine fibrils produced by a structure called a cribellum.[4] An opportunity to study cribellate silk in a spider family not previously investigated also led to a study of different fixation methods. Immature specimens of <u>Titanoeca nigrella</u> (Chamberlin) (Araneae, Titanoecidae) were collected under rocks at the Cimarron National Grasslands, Morton Co., Kansas. Spiders were kept in the laboratory and spun normal appearing webs. Copper and gold (for OsO_4 treatments) 200 m grids were dragged through the webs. Some grids with fibrils were left in the native state, or were exposed to vapors of 37% formaldehyde and/or 2% aqueous OsO_4 for 3 hours at 20°C. Some of these grids were sputter coated with gold/palladium (60/40) so that 20 nm were deposited. A thickness monitor was not available so thickness was estimated by the voltage setting and duration of coating. All combinations of treatments were used, but formaldehyde exposure always preceded OsO_4 exposure which preceded sputter coating. Grids were examined with a JEOL 1200 EXII TEM. An accelerating voltage of 60kv and 100 um objective aperture were used for imaging. Whether native or fixed by various means, all fibrils appeared similar, confirming and extending previous observations.[3] Fibrils were nodular, corresponding to the expected structure of fibrils produced by advanced cribellate spiders (Fig.1). The exceptions were when specimens were sputter coated regardless of previous treatment. Beam damage in the form of electron dense areas appeared, and increased in number and size with increasing duration of beam exposure (Fig. 2). An extreme case is shown (Fig. 3) which was caused by exposure to an 80kv beam and no objective aperture. To summarize the results of this study, it is not necessary to fix spider silk specimens for TEM examination if accelerating beam voltages are held to moderate levels and appropriate objective apertures are used.

1. T. F. Anderson and A. G. Richards, Scientific Monthly 55 (1942) 187.
2. R. Lehmensick and E. Kullman, in <u>Electron Microscopy</u>: Proc. Stockholm Conference, Sept. 1956. New York: Academic Press (1957) 123.
3. V. Friedrich, Jr. and R. M. Langer, Am. Zool. 9 (1969) 91.
4. W. Eberhard and F. Pereira, J. Arachnol. 21 (1993) 161.

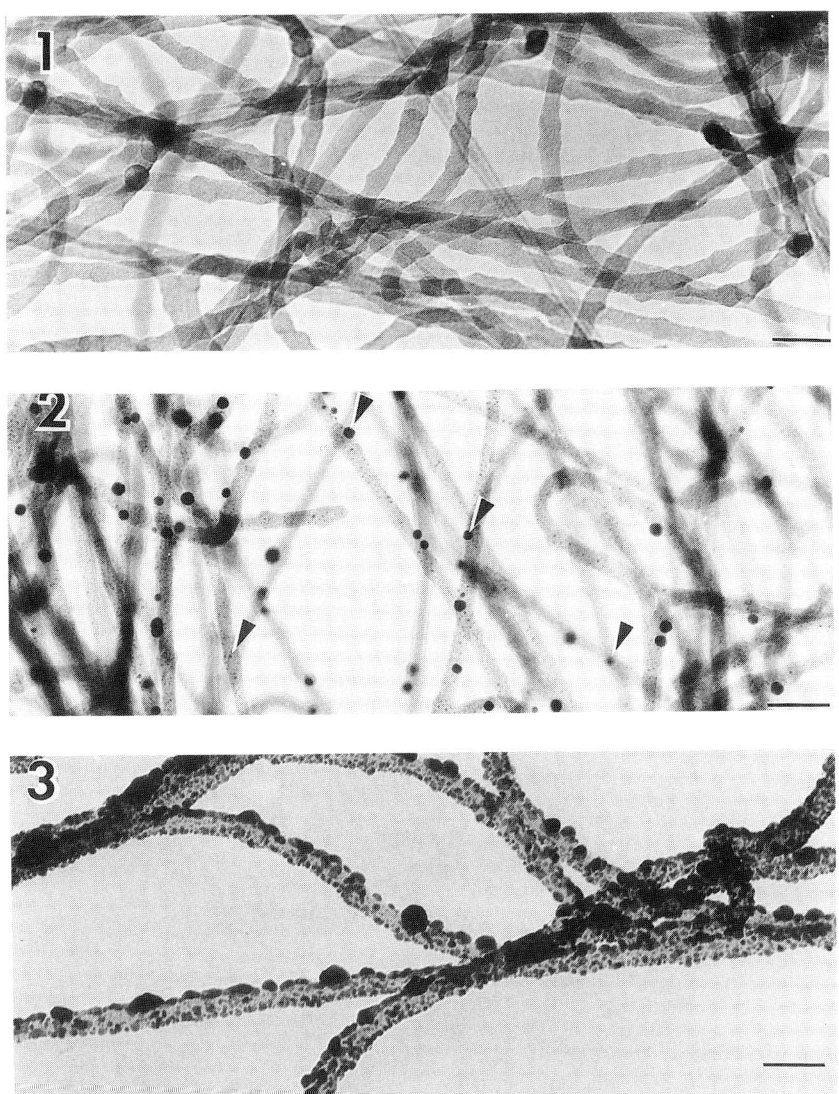

Fig. 1. Specimen of formaldehyde vapor exposed fibrils, 60kv accelerating voltage, 100 μm objective aperture, exposed to beam for 5 m. Bar = 50 nm.

Fig. 2. Specimen of formaldehyde vapor exposed fibrils coated with gold/palladium. Note large and small areas of beam induced damage (arrowheads). All conditions and bar as in Fig. 1.

Fig. 3. Specimen of native fibrils, 80kv accelerating voltage, no objective aperture inserted, exposed to beam for 5 m. Note extensive beam induced damage and swelling. Bar = 50 nm.

BENZYLDIMETHYLAMINE (BDMA) AND DIMETHYLAMINOETHANOL (DMAE) AS THE CATALYSTS OF CHOICE FOR EPOXY RESIN EMBEDDING MEDIA

J.A. Mascorro and G.S. Kirby

Department of Anatomy, Tulane University School of Medicine, New Orleans, LA 70112

INTRODUCTION: By long held tradition, many epoxy resin embedding media are catalyzed with 2,4,6-Tri(dimethylaminomethyl)phenol (DMP-30). Embedding media polymerized in this manner are widely-used and produce acceptable results,[1] nevertheless they show high viscosity, a rapid loss in fluidity, and ultimately a shortened useful pot-life necessary for complete specimen infiltration. Perhaps using less viscous catalysts, such as Benzyldimethylamine (BDMA) or Dimethylaminoethanol (DMAE), would yield embedding media with lowered viscosity and prolonged fluidity, both very important characteristics that would ensure complete infiltration and strengthening of tissues subsequent to microtomy and ultrastructural study.

MATERIALS & METHODS: The amine catalysts BDMA, DMAE, and DMP-30 were tested for *Average Flow Time (AFT)* and *Volume Flow Rate (VFR)* by recording the time necessary for 9 ml of each component to flow from a vertically oriented viscosimeter. *Viscosity (cp)*, which is a relative measure of fluidity and a real correlate of *AFT/VFR*, then was calculated with the Hagen-Poiseville relation for laminar flow as applied to fluids. Embed 812 and LX112 media were prepared and catalyzed either with BDMA, DMAE, or DMP-30 and their *AFT, VFR,* and *cp* similarly determined @ 5, 30, and 60 minute intervals after initial mixing.

RESULTS & SUMMARY: BDMA (0.84 *cp*, Table I) and DMAE (0.96 *cp*, Table I) demonstrated very favorable *AFT* and *VFR* characteristics and a *cp* many times lower than DMP-30 (25.0 *cp*, Table I). Emed 812 in combination with Nonenyl Succinic Anhydride (NSA) and Nadic Methyl Anhydride (NMA) was less viscous and most fluid at all time trials when the medium was catalyzed with BDMA (22.5 *cp* @ 60 min, Table II). But, when BDMA was replaced by DMP-30 as the catalyzer, *cp* increased by as much as 25% at 5, 30, and 60 minute time trials (*cp*, Table II). DMAE produced viscosity characteristics intermediate between BDMA and DMP-30. For example, @ 60 minutes the viscous character of LX112/NSA/NMA increased from 60.7 *cp* (with BDMA, Table III), to 123.4 *cp* (with DMAE, Table III) to a higher 160.7 *cp* (with DMP-30, Table III). Any individual component in an embedding medium will have a single effect on final viscosity.[2] Dodecynyl Succinic Anhydride (DDSA), as well as the resins Polybed 812, Scipoxy 812, and Eponate 12 are known to impart high viscosity but can be replaced by lower-viscosity alternatives.[2] Similarly, DMP-30 now can be easily replaced either by BDMA or DMAE in order to produce embedding media that would remain fluid for longer periods. Even if utilizing lower viscosity alternatives, it seems clear from present results that, once combined, any combination of ingredients will cause a two- and three-fold viscosity increase within 30 to 60 minutes with a subsequent loss in fluidity and impaired ability to penetrate tissues.

REFERENCES:
1. J.H. Luft, *J Biophys. Biochem. Cytol.* 9(1961)409.
2. Mascorro, J.A. and G.S. Kirby, *XII ICEM Proceedings,* 3(1990)138.

Proc. Microscopy and Microanalysis 1995, edited by G.W. Bailey, M.H. Ellisman, R.A. Hennigar, and N.J. Zaluzec
Copyright © 1995 MSA. Published by Jones and Begell Publishing, 79 Madison Ave., New York, NY 10016

TABLE I - CHARACTERISTICS OF AMINE CATALYSTS

	Average Flow Time[A]	Volume Flow Rate[B]	Viscosity (centipoise)[C]
BDMA	4.57 sec/9ml	1.97 ml/sec	0.84 cp
DMAE	5.10 sec/9ml	1.76 ml/sec	0.96 cp
DMP-30	2:11:96 min/9ml	0.0682 ml/sec	25.0 cp

TABLE II - VISCOSITY/HARDENING CHARACTERISTICS
EMBED 812/NSA/NMA CATALYZED WITH BDMA OR DMP-30

Time Trial	Average Flow Time	Volume Flow Rate	Viscosity (centipoise)
@ 5 min	51.16 sec/9ml (BDMA)	0.1759 ml/sec (BDMA)	10.8 cp (BDMA)
	57.53 sec/9ml (DMP-30)	0.1564 ml/sec (DMP-30)	14.5 cp (DMP-30)
@ 30 min	1:24:92 min/9ml (BDMA)	0.1052 ml/sec (BDMA)	18.0 cp (BDMA)
	1:36:02 min/9ml (DMP-30)	0.0937 ml/sec (DMP-30)	24.2 cp (DMP-30)
@ 60 min	1:46:45 min/9ml (BDMA)	0.0837 ml/sec (BDMA)	22.5 cp (BDMA)
	1:58:98 min/9ml (DMP-30)	0.0756 ml/sec (DMP-30)	30.0 cp (DMP-30)

TABLE III - VISCOSITY/HARDENING CHARACTERISTICS
LX112/NSA/NMA CATALYZED WITH BDMA, DMAE, OR DMP-30

Time Trial	Average Flow Time	Volume Flow Rate	Viscosity (centipoise)
@ 5 min	1:19:85 min/9ml (BDMA)	0.1127 ml/sec (BDMA)	19.1 cp (BDMA)
	1:30:08 min/9ml (DMAE)	0.0999 ml/sec (DMAE)	28.5 cp (DMAE)
	1:46:47 min/9ml (DMP-30)	0.0845 ml/sec (DMP-30)	37.1 cp (DMP-30)
@ 30 min	2:42:53 min/9ml (BDMA)	0.0553 ml/sec (BDMA)	38.9 cp (BDMA)
	3:58:23 min/9ml (DMAE)	0.0378 ml/sec (DMAE)	75.4 cp (DMAE)
	5:36:97 min/9ml (DMP-30)	0.0267 ml/sec (DMP-30)	117.4cp (DMP-30)
@ 60 min	4:13:83 min/9ml (BDMA)	0.0355 ml/sec (BDMA)	60.7 cp (BDMA)
	6:29:53 min/9ml (DMAE)	0.231 ml/sec (DMAE)	123.4 cp (DMAE)
	7:41:09 min/9ml (DMP-30)	0.0195 ml/sec (DMP-30)	160.7 cp (DMP-30)

[A] Time necessary for 9 ml to flow from a vertically oriented viscosimeter.
[B] Amount of the individual component flowing per time.
[C] A relative measure of fluidity (how fast or slow a fluid moves because of molecular adhesion)

A COMPARISON OF THE THREE-DIMENSIONAL STRUCTURES OF Z BANDS IN UNSTIMULATED AND RIGOR SKELETAL MUSCLE

J.P. Schroeter[*], R.J. Edwards[**], and M.A. Goldstein[*]

[*]Department of Medicine, Baylor College of Medicine, Houston, TX
[**]Department of Cell Biology, Duke University, Durham, N.C.

Previous studies (reviewed in ref. 1) lead to the expectation that Z bands from unstimulated skeletal muscle exhibit the unactivated small square form of the Z band lattice. Rigor Z bands, on the other hand are expected to exhibit the basket-weave form of the Z band lattice associated with activation in skeletal muscle. This Z band structural transition has been investigated by three-dimensional reconstruction of Z bands and nearby I bands in unstimulated and rigor rat soleus muscle. The reconstructions were calculated using the tomographic method of weighted back-projection on a series of electron micrographs of longitudinal thin sections of muscle.

Examination of Z band cross-sections of the reconstructions reveals that the unstimulated muscle does indeed exhibit the expected small square lattice form. Furthermore, Z band cross sections of the rigor reconstructions reveal the expected basket-weave lattice form. The lattice dimensions were 20 +- 1 nm for the small square lattice and 27 +- 4 nm for the basket-weave lattice, consistent with the results from electron micrographs of cross-sections.[1-3] Longitudinal projections of the reconstructions showed a longitudinal repeat of 32 +- 2 nm for the small square lattice and 38 +- 4 nm for the basket weave. Thus, the reconstructed regions show an increase in both the longitudinal and cross-sectional dimensions of the lattice, as has been predicted on the basis of a simple "overlap" model of the Z band cross-connecting filaments.[4]

Shaded solid renderings (Figure 1) of parts of the Z band reconstructions reveal substantial re-arrangement of the cross-connecting Z-filaments in the small-square to basket-weave transition. The density due to the connecting structures has decreased, while the longitudinal spacing between cross-connections has increased. This suggests that the ss and bw Z bands are related through a more complex re-arrangement than is predicted by the "overlap" models presented by us and other authors.[1,4,5]

1. M.A. Goldstein, J.P. Schroeter, and R.L. Sass, Electron Microsc. Rev. 3(1990)227-248
2. R.J. Edwards, M.A. Goldstein, J.P. Schroeter, and R.L. Sass, J. Ultrastruct. Mol. Struct. Res. 102(1989)59-65
3. M.A. Goldstein, L.H. Michael, J.P. Schroeter, and R.L. Sass, J. Musc. Res. Cell Motil. 7(1986)527-536
4. M. Yamaguchi, M. Izumimoto, R.M. Robson, and M.H. Stromer, J. Molec. Biol., 184(1985)621-644
5. M.A. Goldstein, J.P. Schroeter, and R.L. Sass, J. Musc. Res. Cell Motil. 3(1982)333-348

Proc. Microscopy and Microanalysis 1995, edited by G.W. Bailey, M.H. Ellisman, R.A. Hennigar, and N.J. Zaluzec
Copyright © 1995 MSA. Published by Jones and Begell Publishing, 79 Madison Ave., New York, NY 10016

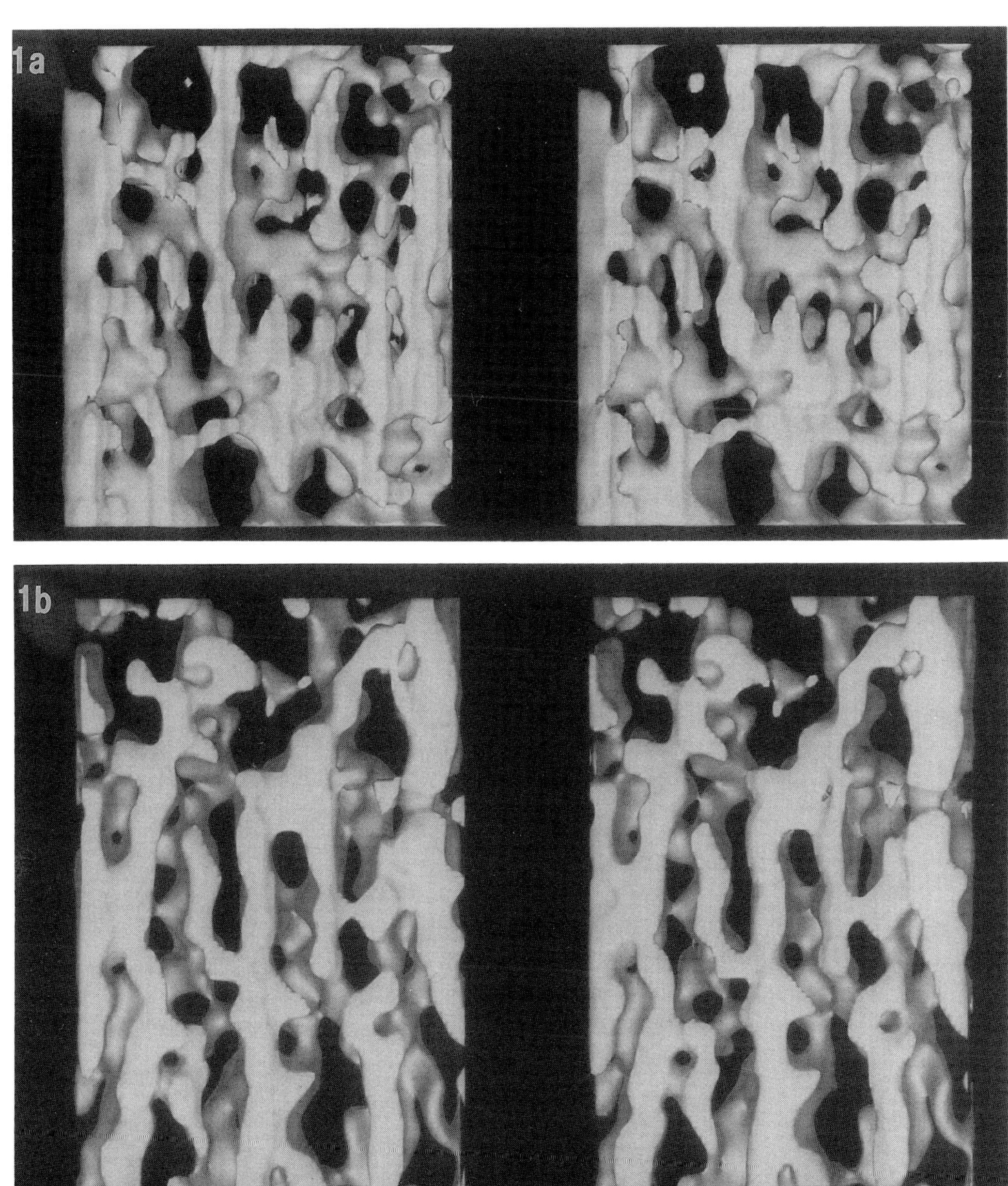

FIG 1. - (a) Stereo shaded solid rendering of a 15 nm thick portion of one of the Z band reconstructions from unstimulated rat soleus muscle. Thin filaments enter the Z band from the top and bottom of the figure and are joined by a dense net of cross-connecting Z filaments; (b) Similar rendering of a Z band from rigor skeletal muscle. Horizontal cross-connecting Z filaments are now mostly located at the Z band edges. Bar = 10 nm.

CONTOUR-BASED SEGMENTATION AND VISUALIZATION OF ELECTRON TOMOGRAPHIC VOLUMES

M. Marko, A. Leith, K. Buttle, and Y. Li

Biological Microscopy and Image Reconstruction Resource, Wadsworth Center, P.O. Box 509, Albany, NY 12201-0509

Electron tomography[1] is increasingly employed to obtain highly-detailed 3-D structural information from complex biological specimens, when such information cannot be obtained otherwise. Thick sections (ca. 0.25-3 μm), which require an intermediate- or high-voltage EM, are often used so that an appreciable portion of the structure of interest can be contained within the section. Because of this relatively large amount of material, volumes made by tomographic reconstruction are often filled with a tremendous amount of detail, not all of which is of interest. Often, in the absence of special stains, the structures of interest have low contrast compared to surrounding or adjoining structures. Typically, all the structures are embedded in the "ground cytoplasm", which has nonuniformly-dispersed small structural elements that have the same density and spatial-frequency range as the structures of interest. In addition, the density of a given structure may vary within an image due to non-uniform staining. In these situations, the segmentation of the volume to isolate the interesting portions of the structure presents a formidable problem.

Typically, structures of interest such as organelles or vesicles are bounded by membranes. The structure is recognized primarily by the pattern of its membranes. In the case of filamentous structures such as cytoskeletal elements, the structures are recognized because they form a linear pattern. In other cases, a structure will be recognized as a region having a certain texture, whose boundary is not sharp. These patterns are quite apparent to the trained eye, yet if looked at closely they are often interrupted or indistinct in typical electron micrographs. Computer algorithms for segmenting and identifying structures under these conditions have to be very sophisticated, and there are often cases in which computerized segmentation fails. Furthermore, in the case of electron tomography, these procedures should work in 3-D. At best, computer segmentation is only partially automatic since some interaction will be needed to label extracted objects so that they can be combined in a flexible way for display or quantitation.

In many cases the most effective way to segment structures from tomographic volumes is by interactive, visual tracing. We use a general-purpose 3-D tracing and reconstruction system, Sterecon.[2,3] This system was developed in our laboratory, originally for 3-D contouring of thick serial sections from stereo pairs, and has recently been redesigned to run on a Silicon Graphics workstation. Sometimes we simply treat the slices of a tomographic volume as serial sections, and trace contours outlining objects on each. Note that the "thickness" of these tomographic slices is typically much less than can be cut by an ultramicrotome. However, Sterecon has the capability of viewing the tomographic volume stereoscopically, either as a whole or as sub-volumes, while tracing contours in 3-D. This is preferable to the serial 2-D approach in that the 3-D continuity of extended or branching structures is easer to maintain. The stereoscopically viewed sub-volumes are overlapping, so there is no discontinuity in tracing. Contours are identified as they are traced, so it is easy to group them into separate objects for display and measurement.

Once all the contours are traced, they are displayed, using Sterecon, as stacked contours, which may be filled (Figs. 1,2). For more sophisticated visualization, the contours can be ported to other software such as SPIDER.[4] The contours are stacked and converted into filled binary volumes. The binary volumes are each the same size as the whole tomographic volume, but each usually contains

Proc. Microscopy and Microanalysis 1995, edited by G.W. Bailey, M.H. Ellisman, R.A. Hennigar, and N.J. Zaluzec
Copyright © 1995 MSA. Published by Jones and Begell Publishing, 79 Madison Ave., New York, NY 10016

only one segmented structure. These volumes can be used as 3-D masks to cut out defined portions of the tomographic volume for subsequent volume rendering, using a variety of available software. More often, we convert the volume to a surface model, using tiling between contours on sucessive levels, or the "marching cubes" algorithm. If desired, the binary volumes are low-pass filtered before surface extraction to give the objects a smoother appearance. A surface description file is made for each object. Subsequently, we can create a scene with any combination of surface objects. We can alter the color, transparency, and material properties of each object independently, to obtain the most meaningfull display.

References

1. J. Frank, *Electron Tomography*, New York:Plenum (1992).
2. M. Marko, et al., in A. Kriete, Ed., *Visualization in Biomedical Microscopies*, Weinheim/New York:VCH (1992)45.
3. M. Marko, et al., *J. Electron Microsc. Tech.* 9(1988)395.
4. J. Frank, et al., *Ultramicroscopy* 6(1981)343.
5. C. Mannella, et al., in these Proceedings.
The Biological Microscopy and Image Reconstruction Resource is supported by NIH PHS grant R01219 (P.I. Dr. C. Rieder). This work was also supported by NSF BIR 921 9043 (P.I. Dr. J. Frank), and NSF MCB 921 9353 (P.I. Dr. C. Mannella).

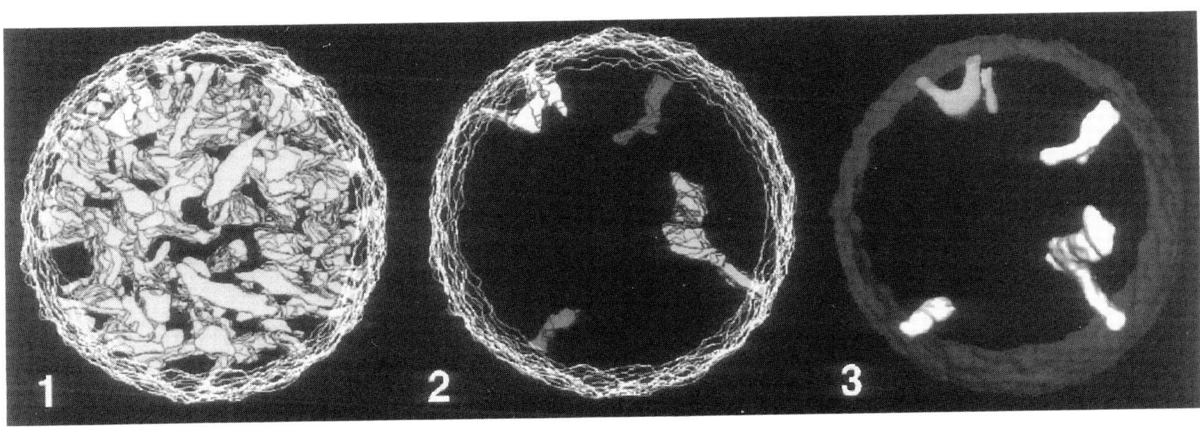

FIG. 1.—Filled-contour display showing cristae in tomographic reconstruction of 0.6um-thick section of rat-liver mitochondrion,[5] based on contours traced visually. Diameter is approximately 1 μm.
FIG. 2.—Selected cristae from Fig. 1. Each crista has unique identifying number, making selection easy.
FIG. 3.—Surface-rendering of cristae and outer membrane (inner membrane not shown) using SGI Explorer software.

INTENSITY ATTENUATION OF THREE-DIMENSIONAL, CONFOCAL FLUORESCENCE IMAGES OF THICK TISSUE

S. Kayali*, H. Ancin*, B. Roysam*, W. Shain**, D.H. Szarowski**, J.N. Turner**

*Depts. of Biomedical Engineering and ECSE, Rensselaer Polytechnic Institute, Troy, NY 12180
**Wadsworth Center, NYS Dept. of Health, School of Public Health, The University at Albany, Albany NY 12201-0509

The confocal scanning laser microscope (CSLM) provides three-dimensional (3-D) images from fixed tissues. These images are obtained by stacking consecutive confocal planes through the depth of a specimen. We have developed software for quantitative analysis of these data sets and have applied these methods to cell counting[1,2]. A major issue in this analysis is the relative intensity of signal measured from like objects through the depth of the specimen[3].

The rat hippocampus was chosen to test the attenuation of the fluorescence signal allowing us to make observations in areas of higher density—pyramidal cell layer—and lower density —extra-pyramidal regions. Paraformaldehyde fixed specimens were stained with Feulgen-Schiff's acriflavine. This stain was selected for its fluorescent properties and its high DNA specificity[4]. Specimens were mounted in media with different glycerol concentrations. Significant attenuation was observed using a 50:50 mixture of glycerol and buffer, while 100% glycerol provided significantly less attenuation. All data presented here were collected using sections mounted in 100% glycerol. 3-D images were collected from 50-, 75-, and 100-μm thick sections with a 40x oil objective, having an (x,y) resolution of 2 pixels/μm and a distance of 1 μm between optical sections. Fig. 1 is a projection of a 100-μm section of the rat hippocampus illustrating pyramidal and extra-pyramidal areas. Individual nuclei segmented by the software are indicated by a unique number (Fig. 1). The variability in the average intensity of nuclei in both pyramidal and extra-pyramidal regions are similar (Fig. 2). Signal attenuation is also similar in both areas (Fig. 3). Thus nuclear density appears not to influence the fluorescence signal. To determine if attenuation was due to photobleaching or tissue absorption, the same area was imaged again from the opposite z-direction by flipping the specimen on the stage. The locations of the nuclei in the z-dimension of the first image ($\Diamond$) were normalized to the total number of optical sections and paired with the corresponding nuclear image in the second image ($\blacklozenge$) (Fig. 4). The average intensity of the nuclei are similarly attenuated in both images. If the slope of the regression lines (Fig. 4) represents tissue absorption and the displacement represents photobleaching, these two parameters can be corrected in the software and thus provide accurate quantitative measurement through out the specimens.

References

1. B. Roysam et al., *J. Microsc.* 173(1994)115.
2. J.N. Turner et al., *Microsc. Res. Tech.* 29(1994)269.
3. J.P. Rigaut et al., *Visualization in Biomedical Microscopies*, New york:VCH (1992)205.
4. W. Lin et al., *Proc. 51st Ann. Mtg. Microsc. Soc. Am.* (1993)268.
5. Work partially supported by NSF DIR 9108492 and NIH RR01219.

Proc. Microscopy and Microanalysis 1995, edited by G.W. Bailey, M.H. Ellisman, R.A. Hennigar, and N.J. Zaluzec
Copyright © 1995 MSA. Published by Jones and Begell Publishing, 79 Madison Ave., New York, NY 10016

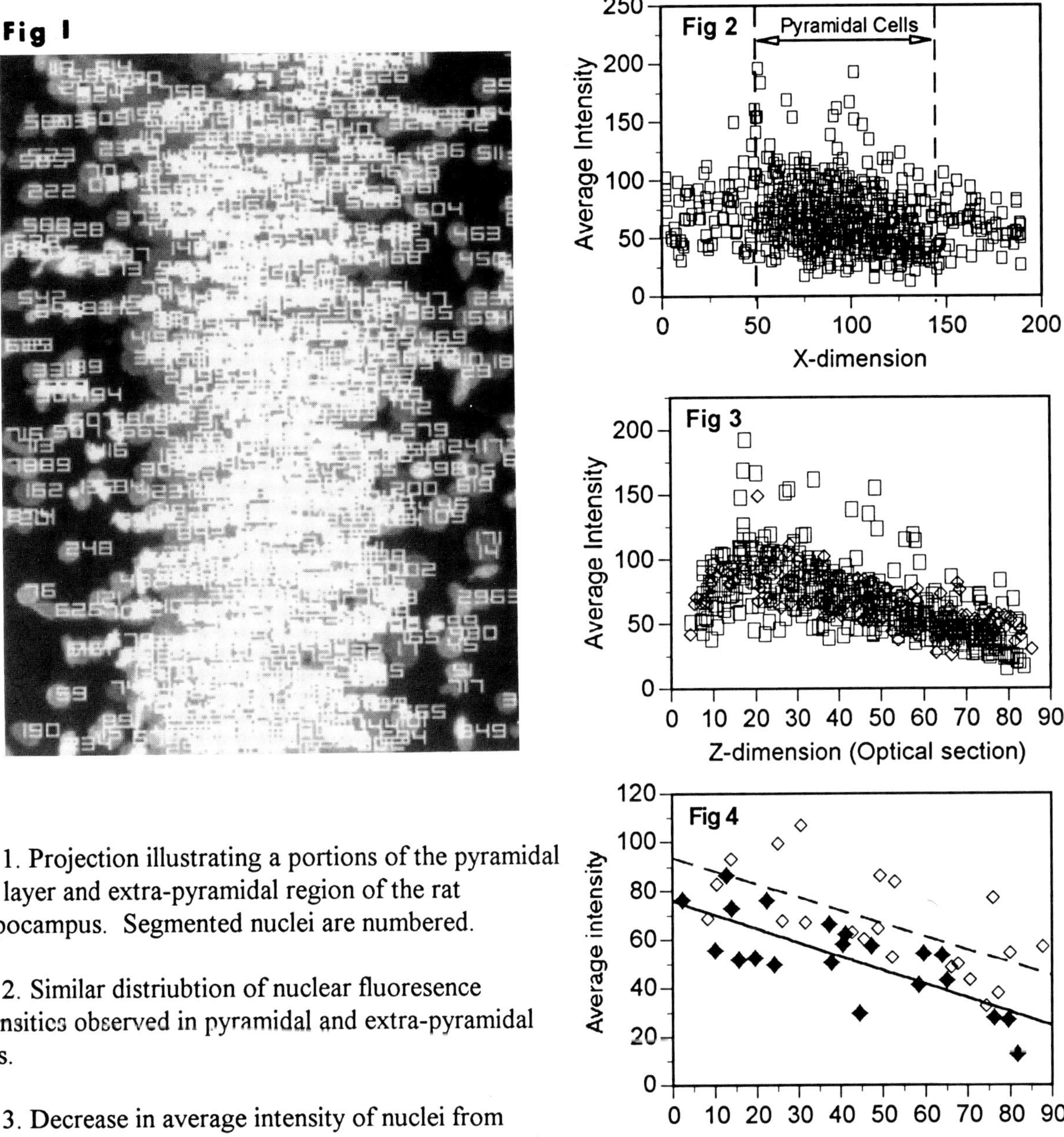

Fig 1. Projection illustrating a portions of the pyramidal cell layer and extra-pyramidal region of the rat hippocampus. Segmented nuclei are numbered.

Fig 2. Similar distriubtion of nuclear fluoresence intensities observed in pyramidal and extra-pyramidal cells.

Fig 3. Decrease in average intensity of nuclei from pyramidal cell layer (□) and extra-pyramidal region (◊) through the specimen depth (z-dimension)

Fig 4. Average intensities of nuclei identified in mages collected first from the top (◊) and then the bottom (◆) of the tissue section projected in Fig 1.

CRYOELECTRON MICROSCOPY AND IMAGE RECONSTRUCTION OF SPHERICAL VIRUSES WITH SPOT SCAN AND FEG TECHNOLOGIES

N. H. Olson,[*] U. Lücken,[†] S. B. Walker,[*] M. T. Otten,[†] and T. S. Baker[*]

[*]Department of Biological Sciences, Purdue University, West Lafayette, IN 47907 USA
[†]Philips Electron Optics, Applications Laboratory, P. O. Box 218, 5600 MD Eindhoven, The Netherlands

The field emission gun electron microscope (FEG) is a tool that has the potential to achieve near atomic resolution information of biological macromolecules[1,2]. The FEG provides a beam with higher spatial and temporal coherence and a better phase contrast transfer function than do microscopes with either tungsten or LaB_6 filaments. The FEG is also ideal for spot scan imaging applications because it can produce a small, coherent and very bright spot. In spot scan mode the specimen is exposed to an array of non-overlapping spots rather than a flood beam. This significantly reduces beam-induced specimen drift[3,4].

Frozen-hydrated samples of cowpea chlorotic mottle (CCMV, Fig. 1A) and cowpea severe mosaic virus (CPSMV, Fig. 1B) were examined on a Philips CM12 transmission electron microscope equipped with a standard LaB_6 gun and on a Philips CM200 equipped with a field emission gun, respectively. The CM12 was operated at 120kV and was externally controlled by means of a spot scan imaging program which produced a series of 250 nm diameter spots on Kodak SO-163 sheet film. The CM200 FEG was operated at 200kV and the micrographs were exposed with flood beam illumination. The CM12 and the CM200 FEG were at the Philips Electronics application labs in Mahwah, NJ and in Eindhoven, The Netherlands, respectively. Samples of CCMV and cowpea mosaic virus (CPMV), a virus in the same family with significant sequence homology to CPSMV, were examined in a Philips EM420 for comparison. The EM420 was equipped with a tungsten filament and was operated at 80kV with flood beam illumination. Selected micrographs were digitized, particle images were selected, and three-dimensional reconstructions of each sample were calculated as previously described[5].

CCMV, a member of the Bromoviridae family, has a capsid with 180, identical, protein subunits arranged in a T=3 lattice. The subunits are organized into pentameric capsomers on each of the 12 vertices and into hexameric capsomers on the icosahedral threefold axes[5] (Fig. 1C, E). CPSMV and CPMV, both of the Comoviridae family, consist of 60 copies each of a large and a small protein subunit organized into a pseudo T=3 icosahedral lattice[6,7] (Fig. 1D, F). Preliminary examination of reconstructions from images recorded with either FEG illumination (Fig. 1D) or with spot scan imaging (Fig. 1C) indicates an improvement over that of similar reconstructions computed from "conventional" images (Fig. 1E, F). Fine details on the surface of the virions indicate that features may be faithfully represented to a higher resolution in both the CM12 and CM200 FEG data.

Although many factors may have contributed to the apparently improved resolution in the FEG and spot scan data, the preliminary results are encouraging. The possible advantages of combining spot scan imaging and field emission technology will be tested on a Philips CM200 FEG microscope. In addition, a CCD camera will be used to acquire images directly from the instrument. This should provide a more effective way to collect the much larger data sets that will be required to achieve higher resolution[8]. The EM data can then be compared with the available, atomic resolution structures of CCMV[5] and CPMV[7].

References

1. Z. H. Zhou and W. Chiu, *Ultramicroscopy* 49(1993)407.
2. F. Zemlin, *Micron* 25(1994)223.
3. K. H. Downing, *Ultramicroscopy* 24(1988)387.
4. Z. H. Zhou et al, *J. Mol. Biol.* 242(1994)456.
5. J. A. Speir et al., *Structure* 3(1995)63.
6. X. Chen and G. Bruening, *Virology* 187(1992)682.
7. Z. Chen et al., *Sem. Virol.* 1(1990)453.

8. J. Brink and W. Chiu, *J. Struct. Biol.* 113(1994)23.
9. T. S. Baker et al., *Proc. EMSA* 50(1992)454.
10. The authors thank J. E. Johnson (Purdue University) for virus samples and the Philips application laboratories in Mahwah, NJ and Eindhoven, The Netherlands for assistance. This work was supported by NIH and NSF grants to TSB and a grant from the Lucille P. Markey Foundation.

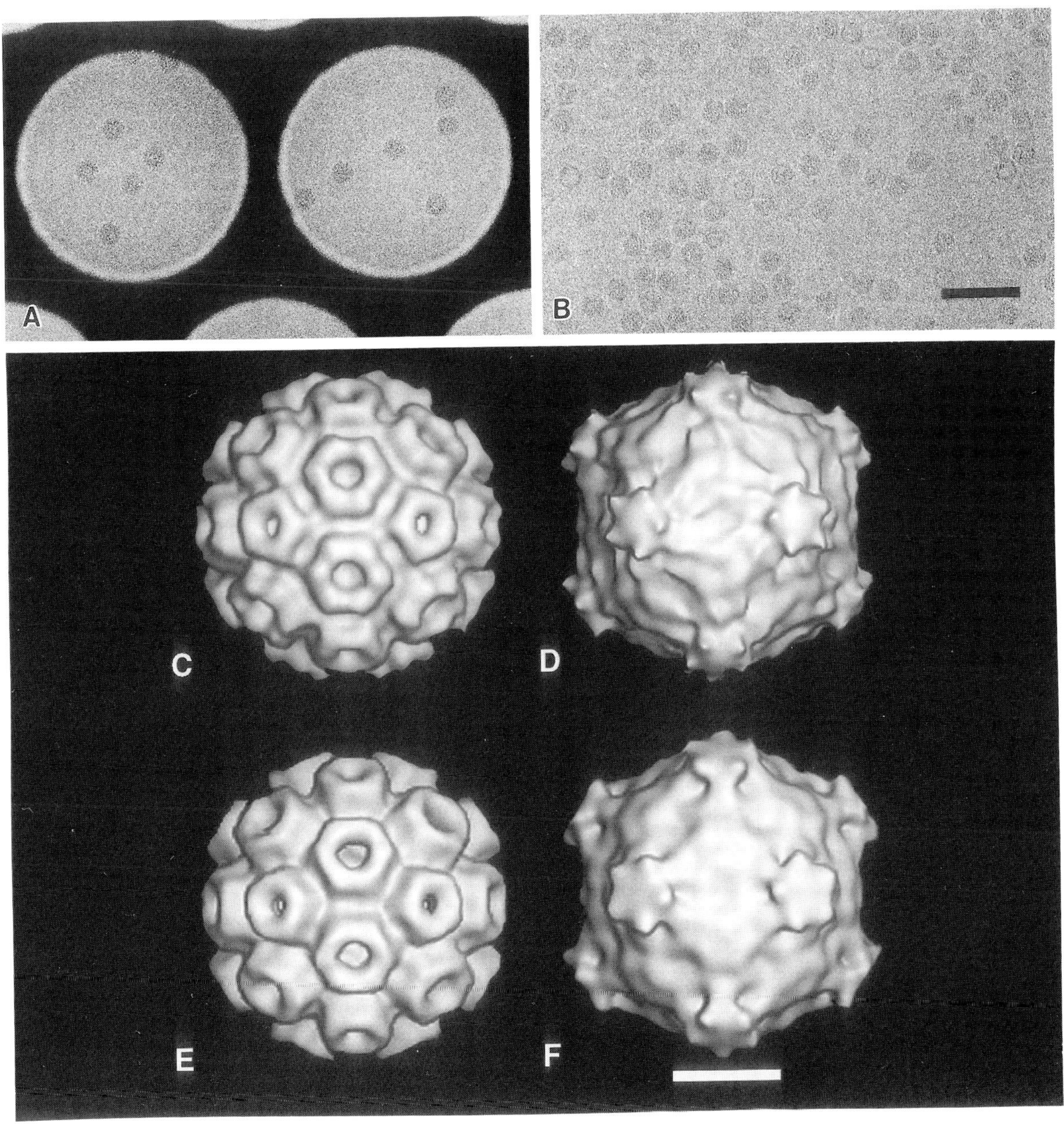

Figure 1. (A) Frozen-hydrated CCMV in two spots from a spot scan micrograph recorded with a Philips CM12 electron microscope operated at 120kV. (B) Frozen-hydrated CPSMV recorded with a Philips CM200 FEG electron microscope operated at 200kV. (C - F) Three dimensional reconstructions of (C, E) CCMV, (D) CPSMV and (F) CPMV[9]. The micrographs for reconstructions in (E, F) were recorded with a Philips EM420 electron microscope operated at 80kV. Bar = 100 nm (A, B), 10 nm (C - F).